Proceedings of the
XIV INTERNATIONAL GRASSLAND CONGRESS

Proceedings of the
XIV INTERNATIONAL GRASSLAND CONGRESS

Held at Lexington, Kentucky, U.S.A.
June 15–24, 1981

Edited by J. Allan Smith and Virgil W. Hays

Routledge
Taylor & Francis Group

LONDON AND NEW YORK

First published 1983 by Westview Press, Inc.

Published 2019 by Routledge
52 Vanderbilt Avenue, New York, NY 10017
2 Park Square, Milton Park, Abingdon, Oxon OX14 4RN

Routledge is an imprint of the Taylor & Francis Group, an informa business

Library of Congress Catalog Card Number 50-37750

ISBN 13: 978-0-367-28439-8 (hbk)

Supporting Organizations

The XIV International Grassland Congress would not have been possible without the support of many organizations, both private and public. We especially wish to thank those listed who have financially or otherwise supported the Congress in amounts of $1,000 or more as benefactors, sponsors, and contributors.

Many other organizations and individuals have financially supported the Congress in amounts of less than $1,000. Their support is appreciated as well.

Benefactors

Deere and Company
Kentucky State Government
U.S. Department of Agriculture
  Science and Education Administration

Sponsors

Hesston Corporation
International Harvester
Sperry New Holland
U.S. Department of Interior
  Bureau of Land Management
U.S. Department of State
  Agency for International Development

Contributors

A.O. Smith Harvestore Products
Agway Foundation
Brown-Forman Distillers Corp.

Chevron Chemical Company
Dairymen, Inc.
Farm Credit Banks of Louisville
Gehl Company
International Minerals and Chemical Corporation
Kentucky Beef Cattle Association
Kentucky Farm Bureau Federation
Kentucky Farm Bureau Mutual Insurance Company
Kentucky Forage and Grassland Council
Louisville Seed Company and CIBA-GEIGY Corporation
North American Plant Breeders
Pioneer Hi-Bred International, Inc.
Ramsey Seed Company
U.S. Department of Agriculture
  Economics and Statistics Service
U.S. Department of Agriculture
  Forest Service
U.S. Department of Agriculture
  Soil Conservation Service
Western Seedsmen's Association
Winn-Dixie Louisville, Inc.
Winrock International Livestock Research and Training Center

Continuing Committee, 1977–1980

Region 1/North America, except region 2 (Canada, United States of America):

Dr. W.R. Childers, Canada

Region 2/Central America (Mexico, all Caribbean countries, all Central American countries southwards to include Panama):

Dr. Juan J. Paretas, Cuba

Region 3/South America (all countries south of Panama):

Dr. Augusto Gallardo Z., Venezuela

Region 4/Southeast Asia (Bangladesh, Burma, Cambodia, India, Indonesia, Laos, Malaysia, Northern Himalayan countries, Philippines, Sri Lanka, Thailand, Vietnam):

Vacant

Region 5/Australia, New Zealand:

Dr. L.E. Humphreys, Australia

Region 6/East Asia (China, Japan, North Korea, South Korea):

Dr. Yoshisuke Maki, Japan

Region 7/Middle East (Afghanistan, Iran, Jordan, Pakistan, Saudi Arabia, Syria, Turkey, Egypt, Yemen):

Prof. Fahrettin Tosun, Turkey

Region 8/Mediterranean and Near East (Albania, Algeria, Bulgaria, Greece, Israel, Italy, Libya, Morocco, Portugal, Spain, Yugoslavia):

Prof. Antonio Corleto, Italy

Region 9/Europe, excluding regions 8 and 10 (Austria, Belgium, Czechoslovakia, Denmark, Federal Republic of Germany, Finland, France, German Democratic Republic, Iceland, Ireland, Netherlands, Norway, Sweden, Switzerland, Great Britain):

Dr. R.J. Wilkins, Great Britain

Region 10/Northern Eurasia (Hungary, Mongolia, Poland, Soviet Union):

Dr. Vladilen G. Iglovikov, Soviet Union

Region 11/Africa, excluding regions 7 and 8:

Prof. V.A. Oyenuga, Nigeria

Immediately preceding host country:

Prof. Dr.Sc. Eberhard Wojahn, German Democratic Republic

CONTENTS

PLENARY PAPERS

INVITED PAPERS

SECTION PAPERS

SECTION I:
PLANT INTRODUCTION, EVALUATION, AND BREEDING

SECTION II: SEED PRODUCTION

SECTION IV: THE NITROGEN CYCLE

SECTION V: MULTIPLE USE OF GRASSLAND

SECTION VI: PHYSIOLOGICAL PROCESSES

SECTION VII: GRASSLAND ECOLOGY

SECTION IX:
MANAGEMENT OF GRAZED AND CONSERVED FORAGES

SECTION X:
MECHANIZATION AND TREATMENT OF FORAGES

SECTION XIII:
TRANSFER OF GRASSLAND RESEARCH FINDINGS

OPENING CEREMONY

The XIV International Grassland Congress opened 15 June 1981 in Concert Hall of the Center for the Arts at the University of Kentucky in Lexington, Kentucky, U.S.A., with Dr. Robert F Barnes, Congress President, presiding. Mr. Maurice E. Heath, retired agronomist, Purdue University, gave the invocational prayer and dedication. Mr. Garland M. Bastin, Executive Director of the XIV Congress, read a proclamation by the Honorable John Y. Brown, Governor of Kentucky, designating June 1981 as *Forage and Grassland Month in Kentucky.*

Dr. Walter R. Childers, Chairman of the Continuing Committee, extended a cordial welcome to all in attendance, and after introducing the members of the Continuing Committee and the Congress participants by a roll call of nations and areas of the world, he briefly described the activities of the Continuing Committee since the previous Congress.

Statements of welcome, explanation, and appreciation next were made by the Congress cohosts, the American Forage and Grassland Council, represented by its president, Dr. William C. Templeton, Jr., and the University of Kentucky, represented by the Dean and Director of the College of Agriculture, Dr. Charles E. Barnhart. Their comments follow.

WELCOMING COMMENTS BY DR. TEMPLETON

On behalf of the American Forage and Grassland Council (AFGC), it is a distinct honor and pleasure to welcome each of you to the XIV International Grassland Congress. The Council has been well represented at several previous congresses and has assumed leadership in obtaining financial assistance for a number of participants from the U.S.A. However, this is the first opportunity for the Council to serve in the role of cohost for the Congress, and we are pleased to do so.

Planning for the XIV Congress started in early 1978 when a small group of dedicated U.S. grasslanders, primarily members of AFGC, met at Beltsville, Maryland, to consider possible U.S. hosting of the Congress. At that time, a statement was formulated recommending to AFGC's Board of Directors that the Council assume responsibility for seeking a cohost for the Congress and, if that effort were successful, for extending an invitation to the Continuing Committee to hold the XIV Congress in the U.S.A. The AFGC Board accepted the challenge and shortly thereafter received a very strong invitation from administrators of the University of Kentucky and the Commonwealth of Kentucky to hold the Congress on the Campus of the University here at Lex-

ington. After acceptance by the AFGC Board and approval by appropriate U.S. authorities, an invitation was extended to hold the Congress on the University of Kentucky Campus in June 1981, to be cohosted by AFGC and the University of Kentucky. As you now know, that invitation was accepted.

On behalf of AFGC's Board of Directors and members, I express sincere appreciation to the Commonwealth of Kentucky and the University of Kentucky for the cooperation and support extended in behalf of the Congress. Special appreciation is extended to the many members of the faculty of the College of Agriculture who have cooperated in the pre-Congress planning and organizing to assure that the Congress is successful and that your time in the Bluegrass Region is enjoyable.

I wish to acknowledge, also, the assistance of the many firms, organizations, and individuals who are supporting the Congress. The private business sector is providing outstanding financial assistance and leadership, for which we are indeed grateful.

As we are all so well aware, forages and grasslands are tremendously important in world agriculture. Owing primarily to the complexities of producing and utilizing plants for animal feed, often under marginal conditions for growth, progress has been slower than in some other areas of agriculture. Improved communication and increased interaction among grassland scientists, educators, producers, and many other groups and organizations are required to exploit more fully the world's grassland resources in animal production.

In recognition of this need, AFGC has made an effort to contact all parallel organizations known to us, inviting them to name a representative to meet with us during the Congress. We hope that this effort will serve as a stimulus to the initiation of closer working relations between the individual organizations of national and regional scope within the countries and areas of the world represented at this Congress. We are delighted with the responses and extend a special welcome to each of the representatives.

AFGC wishes for each of you a pleasant and profitable visit in Kentucky and continuing success for the Congress.

William C. Templeton, Jr.
President, AFGC

WELCOMING STATEMENT BY DR. BARNHART

As Dean and Director of the University of Kentucky College of Agriculture, I am very pleased to extend my official and personal greetings. We take great pride in the fact that Kentucky was chosen as the location for the XIV International Grassland Congress and are pleased to serve as cohost of this prestigious event.

Agriculture is Kentucky's largest industry. Kentucky's history is based on a strong and productive agriculture, and its future is tied closely to developing further this tremendous resource. Kentucky has traditionally been recognized as a leading grassland state. Almost 65% of the agricultural land is occupied by pastures, with additional acreages used for cropping of hays and silages. Forage production utilizes approximately 9-1/2 million acres (or about 3.5 million hectares) annually in Kentucky in support of rapidly expanding livestock numbers. In addition, high-quality burley tobacco is the leading cash crop, and corn and soybeans produced in no-till double-cropping systems have shown considerable growth. In total, Kentucky's farm income now exceeds $2 billion annually. During the past decade, Kentucky's agricultural production has grown at a rate almost twice that of the U.S. average and has the potential to continue this growth during the 1980s and into the next century.

The University of Kentucky has been an essential part of Kentucky's agricultural progress in partnership with a strong farm people and a dynamic industry. As the land-grant institution, our University has a unique mandate by state and federal legislation to serve people through programs of research and education. Since 1865 the University of Kentucky has continued to advance higher education throughout the state in the basic sciences as well as the applied sciences. Since 1885 the College of Agriculture has served as the state's agricultural research agency through the programs of the Kentucky Agricultural Experiment Station. These research programs have generated knowledge and developed advanced technology that support the growth and the continually increasing productivity of this progressive industry. Since 1910 the College of Agriculture has provided a close link with the farmers in the state through the Cooperative Extension Service. These unique educational programs in agriculture, youth, home economics, and community development bring the University in close contact with increasing numbers of Kentuckians daily. The records clearly indicate that these programs have had tremendous influence in advancing Kentucky agriculture.

I recognize that each of you might speak of similar happenings and opportunities in your country's or state's agriculture, and we look forward to sharing such experiences as this Congress proceeds. However, we must recognize that such developments and accomplishments do not just happen. They result from commitments, hard work, and investments of time and resources. I am pleased that world needs for food and fiber are receiving increased attention and that we in research and education are recognizing new challenges and responsibilities. We should be excited with the potentials and very much aware of the problems and opportunities that can only be approached through scientific investigations and learning pursuits.

Your presence indicates a professional commitment to scientific excellence. The tremendous values of research

and education are obvious. We invite you to share your experiences and ideas relating to grasslands and to enjoy the hospitality that is traditional for the people of this state. For this Congress, whose theme is "to strengthen the Forage-Livestock Systems of the World," it is indeed appropriate to welcome you to Kentucky and to a University that is committed to developing the full potential of its agricultural resources.

Charles E. Barnhart
Dean and Director
University of Kentucky
College of Agriculture

Following the remarks by Drs. Templeton and Barnhart, Dr. Gerald E. Carlson, National Research Program Leader for forage, pasture, and range on the National Program Staff of the USDA-SEA-AR, read a letter of greetings from the U.S. Secretary of Agriculture, John R. Block, during his comments.

GREETINGS FROM THE U.S. DEPARTMENT OF AGRICULTURE

President Barnes, colleagues, and friends of grassland agriculture, it is my privilege to be here today as a representative of the U.S. Department of Agriculture and to bring to you a message from the Secretary of Agriculture. In a letter dated 9 June 1981, addressed to the President of the XIV International Grassland Congress, Secretary Block had this to say:

> Thank you for the opportunity to participate in this important International Congress. I regret that I can't be there personally to welcome the delegates.
>
> Grasslands have been and will continue to be a vital part of a total food production system not only in the United States but throughout the world. Grasslands are the major source of feed for the livestock industry, which provides us with wholesome meat, milk, wool, and other animal products.
>
> The demand for food will become greater as the world's population increases; all of our resources will be challenged to meet these expanding needs and to maintain and enhance the quality of life.
>
> Therefore, I commend you, scientists, educators, ranchers, farmers, and industry representatives from around the world, for your efforts to meet these challenges facing us. Your willingness to commit your ideas and valuable time and resources speaks well of your commitment to grassland agriculture.
>
> I wish the XIV International Grassland Congress well and know that it will be a success.

We appreciate those words of welcome and thank the Secretary for his support.

The Department of Agriculture has welcomed and accepted this opportunity to support the Congress. Its employees have devoted many hours to Congress committees. Some of the Department's employees will be speaking at technical sessions, presenting new ideas and technologies. Also, the Department has programs that help make new technology available to farmers, ranchers, and consumers; programs that provide economic incentives; and programs that support research that explores the frontiers of science. It is the intent of the Department of Agriculture that these programs work together to solve people's problems.

Today, through the advances of modern technology — electronics and transportation — the world is becoming smaller. Thus, we share common problems and common challenges, such as the efficient use of energy, water, and land resources and the effective transfer of technology.

Agriculture is the foundation of society. Grassland agriculture has been and will always be of major importance to the well-being of the people of the world. We must continue to work closely with one another — scientist to scientist, institution to institution, and country to country — to meet the challenges before us. I speak for myself, but I know that my colleagues in the Department of Agriculture and in other governmental agencies will continue to work to meet this challenge. We look forward to working with you.

Gerald B. Carlson
USDA-SEA-AR

GREETINGS FROM
THE AGRICULTURAL INDUSTRY

To bring greetings and a message to the Congress from the agricultural industry, Dr. Samuel Kincheloe, Manager of Agronomic Services at International Minerals and Chemical Corporation, addressed the Congress.

Forages — An Industry Viewpoint

"When tillage begins, other arts follow. The farmers, therefore, are the founders of human civilization." That statement, made over 100 years ago by U.S. diplomat Daniel Webster, summarizes the role agriculture has played in the development of man . . . and of the U.S. The country's first families were farmers. Today farmers are a minority, yet the most important industry in the U.S. is still farming. Agriculture is the shining light in U.S. productivity.

The success of the farming industry is due largely to political and agricultural systems founded on cooperation: a team effort, with the players being the public, private industry, academic institutions, and the government, cooperating to help U.S. farmers build and operate their own farms for the production of food and fiber, feed, and now fuel, cooperating to adjust to changes in the market, to conserve soil and water, to use energy wisely, and, most importantly, to allow U.S. farmers to make capital gains and honest profits.

Profit is part of the reason U.S. agriculture, throughout its history, has demonstrated the ability to meet world demands for farm products. Efficiency of production of food and fiber is tied to the profit motive. Farm-management systems are complicated. And the profit motive must be strong enough for those who farm to be willing to work within those management systems. It is this motivation, this energy, and this cooperative spirit that have made the U.S. a leading world food supplier.

This country's private industry and consuming public recognize the importance of forage and grasslands as a tremendous resource that requires proper management. For people, and for the livestock that are so much a part of agriculture, the grasses and legumes represent the two most important families of food and feed-producing plants. That is why we supply time and money to forages: because we in industry, like farmers, have a profit motive.

Good forage productivity — in fact, the production of all farm goods, in all countries — depends on three things: natural resources, human energy, and tools. Of these three factors, the only one that can be increased without limit is tools. The productivity of tools, or the efficiency of the human energy applied within their use, has always been highest in a competitive society — a society in which the economic decisions are made by millions of progress-seeking individuals, rather than by a few people, regardless of how well-meaning, unselfish, sincere, and intelligent those few may be.

Historically, the U.S.'s capacity to produce farm products has stayed well ahead of the growth in domestic and foreign demand. At our nation's agriculture-outlook conference held in Washington last November, it was stated that domestic and foreign demand for U.S. farm products could grow as much as 3% per year during this new decade. It was also stated that it is within our capacity to expand output at least 3% per year through acreage and productivity gains.

This agricultural abundance will be achieved by a combination of education, research, capital, technology, and labor. But hunger will not vanish from the earth until that combination has been sufficiently developed throughout the world.

U.S. agriculture may justly take pride in the excellence it has achieved and the contribution it has made to mankind's welfare, culture, and dignity. But the job is not finished. World population continues to rise; at the same time, good farmland continues to be taken out of production. To meet the demands for more food, we must rely heavily on the development and application of new technologies.

That is what this meeting is all about: to form new cooperative efforts, to exchange our knowledge with others having common interests, and to form international teams for the improvement of forages — their development, application, production, distribution, and utilization. In short, to transfer the words that will be said here today so that the world's peoples may be fed tomorrow.

It is my hope that your participation in this Congress will aid in meeting this challenge, that it will be both informative and enjoyable to you.

Thank you.

Sam Kincheloe
Manager of Agronomic Services,
International Minerals
and Chemical Corporation

President Barnes then introduced Dr. J.E. Baylor, Chairman of the Governing Board, who made the welcoming comments that follow.

WELCOMING COMMENTS BY DR. BAYLOR

Thank you, President Barnes. On behalf of the Board of Governors I would like to add my word of welcome to all of you attending this XIV International Grassland Congress. We sincerely hope your stay in the U.S. and in Kentucky will be a most pleasant one.

Early activities relative to planning and preparing for this Congress have been referred to by Dr. Templeton. After the acceptance of the invitation from the AFGC and the University of Kentucky, organizational planning for the Congress proceeded immediately. Following a meeting of representatives of the cohosts, the official organizing committee, later called the "Board of Governors," was formed, and officers were elected. Dr. John E. Baylor was asked to serve as Chairman, with Dr. C. Oran Little acting as Vice Chairman, and Dr. A.J. Hiatt as Secretary. Mrs. Elizabeth Thompson, Executive Secretary of the American Forage and Grassland Council, was asked to serve as the Congress Treasurer. Mr. Garland Bastin came aboard in March 1980 as Executive Director.

The Governing Board and committees have been actively planning for this Congress for over two years, and we look forward to meeting many of you personally over the next ten days.

John E. Baylor
Chairman of the Governing Board

In his presidential address, the text of which follows, Dr. Robert F Barnes emphasized the importance of grassland science and its complexity. The potential for increased grassland productivity in the world was discussed, as were the major constraints facing the world today. The importance of viable local and national grassland organizations, which contribute to the strengthening of international relationships and the exchange of information, was stressed.

PRESIDENTIAL ADDRESS:
GRASSLAND AGRICULTURE – SERVING MANKIND

I sincerely thank our previous speakers for their expressions of welcome and their statements concerning grassland agriculture and its significant role in serving mankind. Grassland agriculture may be described as the art and science of cultivating forage crops, pasture, and rangelands for food and fiber production. Grassland systems are dependent upon grasses, legumes, and some woody sources of forage, as well as upon managers for proper land use and increased animal profitability.

It is important to understand the terminology associated with grasslands. Efforts have been made in the past to document it, and I commend those efforts, for I feel that there is a continuing need for clarifying terms and their use. Although there may be some discrepancies, I urge that we orient our thinking toward the following definitions:

- *Grasslands:* Moore (1970) used the term to denote all plant communities on which animals are fed, annually sown crops excepted.
- *Forages:* Henzell (1982) uses the term broadly to apply to all plant materials eaten by herbivores, including those that are grazed (pastures) and those that are cut before being fed (hay and fodder). Crop residues such as straw and the foliage of trees and shrubs also fall within this broad definition.
- *Forage crops:* this two-word term, which has a much narrower meaning, refers to any crop of vegetative plants or plant parts harvested before being fed to animals. Thus, forage crops include hay, dehydrated forage, haylage, silage, greenchop, as well as soilage, fodder, and certain byproducts, including crop residues.
- *Pastures:* Primarily refers to plant communities predominantly of introduced species, whether sown or volunteer, on which animals are grazed (Moore 1970). A more restrictive definition is "fenced area of domesticated forages, usually improved, on which animals are grazed."
- *Rangeland:* This is a term of American origin. It means land on which the native vegetation is predominantly grasses, grasslike plants, forbs, or shrubs suitable for grazing or browsing use; land not dominated by trees.
- *Range:* This term is difficult to define precisely since it has evolved into a collective word with broad definitions, such as "the region throughout which a plant or animal naturally lives." Range, in this context, encompasses all rangelands and forest range, that is, those forest lands that support an understory of herbaceous or shrubby vegetation providing native forage for grazing and browsing animals.

Thus, grassland agriculture, in the broad sense, constitutes the largest land-use practice in the world, covering more than half the total land surface of the earth. Grasslands also remain one of the largest undeveloped resources for increased agricultural productivity in the world today.

The basic natural resources associated with the production of forages include land, climate, water, and energy. Sound husbandry of these natural resources will be required if increased grassland productivity is to be attained while maintaining a quality environment. As Dr. Gerald Thomas will emphasize in his plenary paper, increasing population, changing attitudes of people, and increased levels of affluence are having a decided influence upon the development and use of the earth's resources.

During this Congress, many speakers will identify a multitude of problems concerning the development, production, and use of the grassland resources of the world. Among the major constraints facing the world today are:

1. Shortage of fossil fuel energy.
2. Scarcity of water and deteriorating water quality.
3. Soil losses.
4. Insufficient knowledge and technology reserve.
5. Failure to apply existing technology.
6. Increasing competitive uses for resources.

I suggest that we look upon these constraints not as problems but rather as opportunities. I trust that each of you will strive to define these opportunities clearly and to establish the research, extension, and educational programs needed to effect major improvements in our grassland resources. Increasing pressures for goods and services to meet the needs of society require that these resources be given full attention. Our grasslands must be improved and maintained in an ecologically and economically sound manner in order to meet national and international needs for food, fiber, environmental quality, wildlife, and outdoor recreation.

An array of scientific disciplines is required to tap the tremendous potential that exists for increasing agricultural productivity through judicious use of grassland resources. Moreover, a sound national grassland philosophy is required by any nation before an efficient grassland agricultural program can be developed. We all have an opportunity and a responsibility, whether we are scientists, technicians, administrators, farmers, ranchers, or consumers, to influence our nation's grassland philosophy and, in turn, the establishment of a sound agricultural policy that allows the effective development and use of those grassland resources. The importance of establishing strong local and national grassland organizations as a means of providing leadership for such efforts cannot be overemphasized. I have experienced the importance and impact that the American Forage and Grassland Council and the Society for Range Management have had in the U.S.A. Representatives from many such forage, grassland, and rangeland organizations from throughout the world are present here today. Many of you have seen the effects of your organizations in your own countries. We salute you and commend your efforts.

I would now like to address briefly an issue that developed at the Final Business Meeting of the XI International Grassland Congress, held in Australia in 1970. It was noted that the arid and semiarid areas of the world's land masses were "receiving increasing pressure to produce forage for livestock and wildlife, water for downstream needs, and services for man's enjoyment. Research efforts into problems of arid and semiarid lands are rapidly increasing, and a worldwide need exists to communicate the results of this research and a practical management." It was recommended that "future Grassland Congresses contain contributed papers, discussions, and plenary sessions concerning this important

area of the world's grasslands."

Parenthetically, I would like to note that a conscientious effort has been made to develop a program for the XIV Congress that will encompass the needs of the full continuum of the arid, semiarid, subhumid, and humid areas of the world, as well as of the temperate, subtropical, and tropical regions. It remains for you to determine and for history to document whether this goal is achieved. In 1978 the First International Rangeland Congress was convened in Denver, Colorado, U.S.A., due, at least in part, to the failure of the International Grassland Congress to encompass the full complexity and diversity of the grassland agricultural systems, particularly arid and semiarid rangelands.

A Committee for the Continuation of the International Rangeland Congress (IRC) has been actively involved in identifying a host for the Second IRC. A report concerning the status of these activities will be made during the business meeting of this XIV Congress.

I personally support the concept of two congresses, provided their programs are complementary and their meetings are held in alternating years. Also, it is highly desirable that a close liaison be maintained between the two continuing committees. I will be serving on both committees for the next three to four years and thus hope to be able to aid in that continuity. However, I strongly recommend that the two committees specifically provide for a formal liaison on a continuing basis.

I would also like to speak to the question of the founding of an international grassland organization. A resolution was passed by the XII International Grassland Congress, which met in Moscow, USSR, in 1973, recommending that the Continuing Committee study the question of the advisability of founding an international grassland organization and to report the results to the XIII International Grassland Congress. At that Congress, in Leipzig, GDR, in 1977, the Resolution was addressed superficially at the final business meeting, and it was concluded that "there were many considerations and aspects which were not in favor of setting up such an organization for the time being. The Continuing Committee, however, recommended that grassland organizations should be established at national levels."

I personally have a dream that I would like to share with you. I envision the establishment of a coordinating body for the Grassland Congress and the Rangeland Congress. Perhaps it might best be called the International Grazing Lands Organization or the International Forage Pasture and Range Organization.

I recognize the complexities, difficulties, and obstacles to be overcome in the establishment of such an organization. My wish and my prayer are that there are enough like-minded individuals gathered here today who may cause it to happen. It may not come to pass for another decade—but if it is to succeed, it must be started at the earliest possible date. I look forward to hearing your reaction to such a proposal. For it is only as we work together for good that we can truly serve mankind.

I am confident that the interchange of experience and knowledge of those attending this Congress will result in tremendous benefits to this and future generations. I thank you for your attention and now declare the XIV International Grassland Congress duly open for business.

Robert F Barnes
Associate Regional Administrator
USDA-SEA-AR, Southern Region
New Orleans, Louisiana

LITERATURE CITED

Henzell, E.F. 1982. Contribution of forages to worldwide food production: now and in the future. Proc. XIV Int. Grassl. Cong. (on press).

Moore, R.M. (ed.) 1970. Australian grasslands. Austral. Natl. Univ. Press, Canberra.

Thomas, G.W. 1982. Resource allocation for animal grassland systems. Proc. XIV Int. Grassl. Cong. (on press).

OPENING BUSINESS MEETING
15 June 1981

The initial business meeting was called to order by Dr. W.R. Childers, Chairman of the Continuing Committee, immediately following the Opening Ceremony. Dr. Childers announced that the rules of procedure as outlined in the International Grassland Congress Constitution published in the Proceedings of the XIII International Grassland Congress would be followed. Since no amendments to the Constitution had been received for presentation to the Congress, he announced that there would be no designation of voting delegates for individual countries. For adopting resolutions and recommendations, full members of the Congress were entitled to one vote by the showing of hands.

Members then stood in silence in remembrance of grassland scientists who had died since the previous Congress.

Dr. Childers went on to announce that an invitation to hold the XV International Grassland Congress in Japan in 1985 had been duly received and accepted by the Continuing Committee. The XV Congress will be held at the Kyoto International Conference Hall in Kyoto, Japan, in the autumn of 1985. The theme "Evolutional Application of Grassland Sciences" was selected, with recognition that the future, and even the survival, of mankind depends largely on the progress of grassland science and technology associated with food production, the human environment, and energy-saving problems throughout the world. It was noted that any inquiries or other correspondence concerning the XV IGC should be addressed to:

The Japanese Society of Grassland Science
c/o National Grassland Research Institute
768 Nishinasuno, Tochigi 329-27, Japan

Congress participants were encouraged to attend the Closing Ceremony to hear the invited presentation, "Grassland Development in Japan and other Asian Countries," by Dr. I. Nikki, President of the Japanese Society of Grassland Science.

A Nominations Committee was duly appointed and accepted by the Congress for selecting candidates to fill vacancies on the Continuing Committee. Dr. Josef J. Noesberger (Switzerland) was designated Chairman. Other members were Geraldo De Rocha (Brazil), Eberhard Wojahn (GDR), Syutaro K. Kawabata (Japan), and Carl B. Hoveland (U.S.A.). The following were appointed and accepted as members of a Resolutions Committee: T. Alberda (Netherlands), Chairman; W. Stern (Australia); V.G. Iglovikov (USSR); and Y. Maki (Japan). The chairmen of the two committees were charged with presenting reports at the final business meeting.

Following the premiere showing of a short documentary film entitled "From Hopeful Greenstuff Woven," the Initial Business Meeting adjourned. The film, recently completed by the USDA Agricultural Research Service, tells the story of the application of science to grassland agriculture by a team of scientists at Tifton, Georgia, U.S.A.

FINAL BUSINESS MEETING
24 June 1981

The final business meeting was called to order by the Congress President, Dr. R.F Barnes. It was noted that Dr. R.H. Hart (U.S.A.) was designated parliamentarian. The following resolution was presented by Dr. T. Alberda (Netherlands), Chairman of the Resolutions Committee, and accepted by the Congress:

Resolution: The participants of the XIV International Grassland Congress express their sincere thanks to the government of the United States of America, to the State of Kentucky, to the American Forage and Grassland Council, to the Governing Board of the XIV International Grassland Congress, and to all other persons or organizations in this country who enabled the Congress to be held at this time in Kentucky, and in this way assisted in making it a success, as in reality it has been.

Both in the plenary sessions and in the separate paper sections, the papers and discussions were of high standards and served to promote all aspects of grassland science and contacts among participating scientists.

Thanks also to the organizers of this Congress, and of the pre-, mid-, and post-Congress tours, and to all those involved in making it a success. It is impossible to personally thank everyone and it is suggested that thanks be given through honoring the President of the XIV International Grassland Congress, Dr. Robert F Barnes, and hence to all those involved.

Respectfully,
Resolutions Committee:

T. Alberda, The Netherlands, *Chairman*
W. Stern, Australia
Y. Maki, Japan
V.G. Iglovikov, USSR

A further resolution for consideration by members of the Congress was submitted and read by Dr. R.W. Brougham (New Zealand):

Resolution: That the Continuing Committee of the International Grassland Congress consider the desirability of reconstituting the International Grassland Congress to consist of a *Central Governing Body* of similar constitution to the Continuing Committee but having responsibility for the formation and coordination of a number of *Chapters* representing and taking responsibility for smaller international meetings embracing the different *climatic* and *topographic* regions of the world.

It was moved and seconded that the above resolution be accepted. Considerable discussion followed, during which a number of suggestions were made for consideration by the Continuing Committee in taking action upon the resolution. Although the resolution was not modified, Dr. Brougham indicated that he embraced in general the discussion. The resolution passed upon a showing of hands. Notes taken during the discussion period will be forwarded by the President to the Chairman of the Continuing Committee for its use.

Dr. Childers, Chairman of the Continuing Committee, announced the new members of the Continuing Committee as selected from the nominations provided by the Nominations Committee. The new members included Dr. Robert F Barnes (U.S.A.), Region 1; Dr. R. Martinez (Mexico), Region 2; Dr. I.M. Nitis (Indonesia), Region 4; Dr. Y. Maki (Japan, reelected), Region 6; and Mr. A. Hentgen (France), Region 9. All new members except Dr. Maki were to serve two terms. Dr. Childers announced that Dr. John E. Baylor will also serve on the Continuing Committee as ex-officio member for one term, representing the XIV Congress Host Committee, as designee appointed by the Congress President.

The new Continuing Committee thus becomes:

Continuing Committee, 1981–1984

Region 1/North America (Canada, United States of America):

Dr. Robert F Barnes, Associate Regional Administrator, Southern Region, USDA-SEA-AR, P.O. Box 53326, New Orleans, La. 70153, U.S.A.

Region 2/Central America (Mexico, all Caribbean countries, all Central American countries southward to include Panama):

Dr. R.A. Martinez, Technical Sub-Director, Centro de Investigaciones Agricoles del Norte Central, Apartado Postal Numero 18, Calera, Zacatecas, Mexico

Region 3/South America (all countries south of Panama):

Dr. Augusto Gallardo Z., Estacion Experimental "El Cuji," Apartado 592, Barquisimeto, Venezuela

Region 4/Southeast Asia (Bangladesh, Burma, Cambodia, India, Indonesia, Laos, Malaysia, Northern Himalayan countries, Philippines, Sri Lanka, Thailand, Vietnam):

Dr. I.M. Nitis, Dept. F.K.H.P., Udayana University, Jl. Jendral Sudirman, Denpasar, Bali, Indonesia

Region 5/Australia, New Zealand:

Dr. L.R. Humphreys, Department of Agriculture, University of Queensland, St. Lucia, 4067, Australia

Region 6/East Asia (China, Japan, North Korea, South Korea):

Dr. Yoshisuke Maki, Hokkaido National Agricultural Experiment Station, Ministry of Agriculture and Forestry, Jitsujigaoka, Sapporo, 061-01, Japan

Region 7/Middle East (Afghanistan, Egypt, Iran, Iraq, Jordan, Pakistan, Saudi Arabia, Syria, Turkey, Yemen):

Prof. Dr. Fahrettin Tosun, 19 Mysis Universitesi, Zirast Facültesi, Lise Cad. No. 6, Samsun, Turkey

Region 8/Mediterranean and Near East (Albania, Algeria, Bulgaria, Greece, Israel, Italy, Libya, Morocco, Portugal, Spain, Yugoslavia):

Antonio Corletto, Associate Professor of Agronomy, Institute di Agronomia e Coltivazoni Erbacea, Universita degli Studi, 70126 Bari, Italy

Region 9/Europe, excluding regions 8 and 10 (Austria, Belgium, Czechoslovakia, Denmark, Federal Republic of Germany, Finland, France, German Democratic Republic, Iceland, Ireland, Netherlands, Norway, Sweden, Switzerland, United Kingdom):

Dr. A. Hentgen, Dept. A.F.P.F., INRA, Route de St. Cyr, Versailles, France

Region 10/Northern Eurasia (Hungary, Mongolia, Poland, Rumania, USSR):

Dr. Vladilan G. Iglovikov, Deputy Director of Scientific Work, All-union Williams Fodder Research Institute, 141740 Lugovaya Moscow region, USSR

Region 11/Africa, excluding regions 7 and 8:

Prof. V.A. Oyenuga, Department of Animal Science, University of Ibadan, Ibadan, Nigeria

Immediately preceding host country:

Dr. John E. Baylor, Agronomy Department, 106 Agr. Admin. Bldg., Pennsylvania State University, University Park, Pa. 16802, U.S.A.

DR. ROBARDS'S ANNOUNCEMENT: THE SECOND INTERNATIONAL RANGELAND CONGRESS

We regret that Dr. Victor R. Squires, Dean of the Faculty of Natural Resources, Roseworthy, South Australia, could not be with us to participate in the Congress and to bring this announcement. Dr. Squires was elected as the Federal President of the Australian Rangeland Society in May 1980 and has been active in Australia's effort to host the Second International Rangeland Congress.

Our basis for bringing an announcement on the status of the proposed scheduling for the Second International Rangeland Congress is a letter of invitation from Dr. Squires to the Chairman of the committee for continuation of rangeland congresses, Dr. Henri Le Houerou. The Australian Rangeland Society has extended an invitation to stage the Second International Rangeland Congress in Australia in May 1984. The chosen venue noted in the letter is Adelaide, the capital city of South Australia, which is a centrally located city in relation to rangelands. A number of pre- and post-season tours are being planned.

Geoff Robards
Australia

THE EXECUTIVE DIRECTOR'S REPORT

The XIV International Grassland Congress has been in development for over two-and-one-half years. Much time, effort, and planning have been devoted by the Governing Board, which includes representatives from eight states. Eight national committees and twelve local committees were formed to develop the specific plans and activities for the Congress.

Initially, seven pre- and post-Congress tours were offered for the participants. Due to the economic situation and resulting insufficient numbers of participants indicating interest, two of the pre-Congress and two post-Congress tours were cancelled. Two pre-Congress tours and one post-Congress tour were conducted, with a total of 185 people taking part.

The seven mid-Congress tours were conducted as planned and the tours to the University of Kentucky Agricultural Experiment Station's research farms were well received. Attendance figures were as follows:

Saturday	553
Sunday	400
Tour guides	56
Experiment Station tour	365
TOTAL	1,374

The "special activities" were also well attended: Wednesday night 1,050 came to the cookout and viewed the exhibits, Thursday night 1,200 participated in the Horse Park outing, and Friday noon 1,040 attended the beef cookout and visited the exhibits again.

The records show that attendance at the Congress as a whole may be categorized as follows:

Full members	718
Associate members	135
Students	86
Day members	72
Forage Producers' Forum	113
Tours only	2
TOTAL	1,126

International full members	394
U.S.A. full members	324
Countries and areas represented	59
States (U.S.A.) represented	47

The foregoing figures are exclusive of University of Kentucky personnel, children accompanying their parents, and persons staffing exhibits. Over 400 Kentucky residents have been identified as assisting in some way with the Congress.

From the outset the figure of 1,500 had been projected as the number of people who would be involved in the XIV International Grassland Congress. That number was exceeded.

A special thanks to all who were involved in making the XIV Congress a success.

Garland M. Bastin
Executive Director

CLOSING CEREMONY

The closing ceremony, immediately following the final business meeting, was called to order by Dr. J.E. Baylor, Chairman of the Governing Board and Vice President of the Congress.

Dr. C.O. Little, Vice Chairman of the Governing Board and Vice President of the Congress, introduced Dr. I. Nikki, President of the Japanese Society of Grassland Science, host for the XV International Grassland Congress to be held at the Kyoto International Conference Hall in Kyoto, Japan, in the autumn of 1985. Dr. Nikki's paper, "Grassland Development in Japan and other Asian Countries," is included in the Proceedings.

Dr. Gordon C. Marten, Chairman of the Program Committee, then commented on the highlights of the Congress program.

DR. MARTEN'S REPORT ON CONGRESS PROGRAM HIGHLIGHTS

First, I would like to express my gratitude to the nine other members of the Program Committee of the XIV International Grassland Congress for their inputs into this summary. With their assistance, I have summarized just a few of the technical program highlights of the XIV International Grassland Congress. The plenary papers were among the best we have heard at any conference anywhere in the world, and the excellent discussion of them was very gratifying. The plenary papers and a summary of the discussions that followed are fully documented in the Congress *Proceedings*.

The tours and the Producers' Forum clearly showed us that today's successful farmer is an articulate, intellectually astute, and technically competent business person who is a better gambler than most of us who took part in International Grassland Congress night at the Red Mile Racetrack.

The Sectional Sessions and lead-off Plenary Sessions included a breadth and depth of forage and grassland topics rarely achieved at any conference. I will identify but a few typical highlight items by section:

I. We were informed of the need for expanded germ plasm bases and told to search for and use ecologically unadapted germ plasm to allow large advances in the breeding of adapted cultivars of forage species and hybrids. Evidence was given that grasses and legumes are being identified and developed to improve pest resistance, forage quality, and production by use of new breeding procedures. An exciting potential surfaced concerning use of interspecific hybridization and apomixis in forage-breeding programs. We were reminded that protected natural grassland can provide a valuable genetic resource. Several times, team research among scientists of diverse disciplines was acclaimed to enable future breakthroughs in forage plant breeding, physiology, management, and utilization.

II. The potential for seed production of forage species was revealed to be only partially realized in marketable commercial seed. The problems may be overcome somewhat by management, but more likely genetic improvements will be required for substantial progress.

III. Scientists reported that increasing the levels of chloride or phosphate in solution culture markedly depressed the concentrations of nitrate in orchardgrass. This information could aid in the prevention of nitrate toxicity in grazing animals. Rhodesgrass provided high yields and excellent nitrogen recovery in a municipal effluent irrigation system. The hazard of nitrate pollution in such a system was low when the grass was grown in a fine-textured soil, but high in sand dune soil. Differences in absorption of soil phosphorus among plant species were traced to differences in their abilities to exploit the same available pool of phosphorus. The available pool increased with time after wetting the soil.

IV. We were informed of the importance of studies of nitrogen cycling in grasslands and of the newly prominent status of the N-fixing legume in forage-livestock systems throughout the world. The old "nitrogen fertilizer on grass" emphasis of the international conferences of the 1970s has been quickly supplanted by the new legume emphasis for the 1980s and beyond. Major advances were revealed in understanding the magnitude of the N inputs to and N losses from grassland systems. Rates of N input were reported to vary markedly with species of forage legume and with methods of managing grass-legume mixtures. Losses, by volatilization of ammonia from urine, urea fertilizers, or animal slurries, were reported to often exceed 70% of the N applied. These findings have crucial implications for the productivity of grasslands and for the maintenance of environmental quality.

V. We were reminded of the responsibility we have as grassland scientists to maintain valuable natural grasslands as we consider their multiple-use potential. Two grass species were revealed to be superior for reclamation of saltpans and waterlogged soils in a semi-arid environment. The wise use of government funds was reported for the revegetation of 52,000 hectares in a western state of the U.S.A. Assessment of drainage potential, species adaptation, and topsoil depth was shown to be crucial in revegetation of diverse sites throughout the world.

VI. Major advances were reported in understanding some of the physiological limitations to productivity of pasture and forage grasses. Growth rate of grasses appears to be sync-limited in that rates of leaf-area expansion and elongation were much more closely linked to seedling vigor and yield of vegetative regrowth than was carbon exchange of leaves. A major problem is maximizing productivity while maintaining a photosynthetically efficient canopy of young, highly palatable grass leaves. Again, the importance of collaborative physiology-breeding-management research was revealed.

VII. The growth regulator mefluidide was reported to be effective in suppressing tall fescue to allow establishment of perennial legumes and in inhibiting fescue seedhead formation, leading to improved cattle weight gains and increased animal productivity per hectare. Improved management schemes were provided for a large number of grass and legume species in natural grasslands, mountain meadows, and pastures. Properly managed alfalfa was reported to persist for 34 years on certain rangeland sites. In furtherance of the development of grazing-land site potential throughout the world, scientists were requested to develop specific information on introduction and adaptation of legumes and on fertilizer needs in future research endeavors. Scientists were advised to measure every grassland management practice in terms of its use of water, especially in arid and semiarid regions.

VIII. Several speakers outlined the need to give more attention to the physical properties of forages as they influence intake and feed quality generally. Characteristics such as bulk volume, particle size, and rate of particle-size breakdown during digestion were recognized as important quality features. Considerable progress was reported in the development of near-infrared reflectance spectroscopy as a new technique for assessment of both in-vitro and in-vivo herbage quality. This new technology must be researched in depth before it can be routinely applied to practical programs such as new hay-grading standards that were proposed. The potential impact of near-infrared reflectance spectroscopy of forage-quality evaluation is substantial. Presentations during a special session on forage utilization of the Southern Pasture and Forage Crop Improvement Conference (mid-Congress) revealed that chemical estimates of forage quality have vastly different predictive value depending on genotypes, environments, and maturation ranges. Local regressions must be developed when these procedures are used. New evidence suggests that the rate of digestion both in vitro and in vivo is slower in bloat-safe legumes than in bloat-causing legumes. Intensive efforts are in progress to breed bloat-safe alfalfa by selection for slower rates of cell rupture and by searches for the presence of condensed tannins.

IX. Legumes received major attention throughout forage-management sessions. Particularly emphasized was the establishment of temperate, subtropical, and tropical legumes in all types of seedbeds and intact swards. Topics relating to the seasonal productivity of swards, especially for supply of feed during periods of shortage, stimulated much discussion, as did the influence of management practices on herbage quality. Comprehensive papers concerned the dynamics of herbage

growth and its utilization by livestock in a vast array of systems. Forage managers were charged to be certain that plant leaves are eaten by the grazing animal before the leaves have an opportunity to senesce. We received a unique view of the grasslands of China, not available at previous grassland congresses.

X. The introduction, treatment, improvement, and utilization of crop residues and other low-quality forages in livestock feeding systems were featured. Chemical treatment of these forages with ammonia was proposed as a means of quality improvement and of reducing the need for protein supplementation in cattle-feeding systems. Research has revealed that field drying of hay can be improved by microwave treatment, plastic "brush" conditioners, and chemical treatment with methyl esters and/or potassium carbonate.

XI and XII. Great interest again centered on the use of legumes in animal-production systems, especially in tropical areas. Concern exists about how to establish, maintain, and evaluate legumes. Utilization systems that maximize intake by animals received considerable attention. We were given a very objective and scholarly view of the comparative strengths and weaknesses of temperate and tropical forages and suggestions of mechanisms that should help scientists and practitioners to use each type of forage so as to realize its greatest potential. For example, tropical grasses provide a high yield of maintenance feed for grazing animals.

XIII and XIV. We received a realistic reassessment of the role of modeling in grassland research and improvement. Modeling is not a panacea; rather, it is one of numerous tools that gives us insights for future research and practical application of research. Socio-economic aspects of grassland systems and policies must be considered during extension of research results to achieve society's needs. Presentations revealed that creative combination of research technology or sets of practices into extension programs for innovative farm managers has resulted in more livestock products per unit area from pasture and range than is indicated by research alone. Again, the use of legumes as improved feed was a general theme from several countries. The use of leucaena in Australia is an example. Conservation of high-quality forage as silage generally gave the greatest economic gain in appropriate temperate environments. New cropping systems were proposed to reduce water use in low-rainfall regions, to increase plant cover of soils, to reduce fire hazards, and to increase feed supplies for both domestic and wild animals. We received suggestions about efficient ways to utilize scarce fossil fuels and a warning that we must adjust our research, extension, and farming practice programs to shortages of these fuels and to higher energy costs. The suggestion was made that we are more in need of improved extension input than additional research to preserve our natural grasslands.

Finally, we were reminded that ruminant animals are now competing with humans for grain and protein supplements and that the human population is growing faster than the food-supplying ruminant animal population in the world. The potential for forages to replace grains in ruminant feeding systems must be realized. Scientists must join the planners and policy makers and be more assertive in making research information more generally available.

Gordon C. Marten
Program Chairman

RESPONSES OF DELEGATES
TO CONGRESS ACTIVITIES

Following several general comments by Dr. W.R. Childers, Chairman, Drs. B.R. Watkin of New Zealand and C.E. Wright of Australia spoke briefly about their impressions of the Congress.

Comments by Dr. Watkin

Thank you for this honor and opportunity to comment on a few of the highlights and impressions I gained during the Congress sessions.

First and foremost, I believe the papers and presentations have been, in general, of a high standard, and I give full marks to all contributors, it being no mean effort to stand up in front of such a highly qualified and critical audience as this and present a paper plus discussion. They deserve our full commendation. I firmly believe that they have done much to justify our continuing support and efforts for the future success of the International Grassland Congress.

In expressing my genuine praise and appreciation to the Organizing Committee for its tremendous effort, I would also like to offer one or two helpful suggestions for future gatherings. As stated, we all value immensely the contributions of the speakers, but I'm sure we also recognize the tremendous importance of the informal postsessional discussions that those same speakers initiate. I would suggest that the Congress Committee give further thought to facilitating and thereby exploiting this real opportunity for further elaborating, understanding, disseminating, and expanding these topics through subsequent informal discussions — and I stress the informal rather than the formal.

In terms of the topics covered, it was very pleasing to me, as a New Zealander, to see the resurgence of interest and effort in forage legumes, nitrogen fixation, and nitrogen transfer around the world — no doubt accelerated by the sharply rising energy costs of the N fertilizer approach. Along with this, of course, was the significant emphasis on better and cheaper ways of introducing legumes into established swards. In fact, approximately 40 papers dwelt on aspects of pasture establishment and renovation for better legume performance alone.

It was also very pleasing to see many well-thought-out and well-designed experiments incorporating detailed assessment of meaningful parameters being conducted on tropical pastures in many developing countries. Such work must be so much more demanding than the temperate equivalent and is therefore that much more commendable.

I would also like to congratulate the Program Organizing Committee for including the short, sharp, yet telling program of the Forage Producers' Forum of AFGC. This was rightly placed, in my opinion, in the very center of the Congress activities; it highlighted to me the ultimate objective and purpose of such a technical Congress — to serve and improve our respective grassland farming industries.

I'm no agricultural economist and must admit some intolerance towards such folk at times, but they or at least their measurement parameters have a vital role to play in research application in these days of significant and growing economic dimensions. It was encouraging to see this coming through in a few field experiments reported.

There were so many individual papers worthy of note, but it is just not possible to do more than mention or tilt at a few. Some examples of things that seemed particularly noteworthy are:

- The tremendous potential for animal products from grass-legume pastures in Brazil, especially when extended to include the many other countries reporting similar promise.
- The tremendous contribution that grassland products can make towards feeding the starving millions of the world, if the planners and politicians meet their responsibilities.
- The useful and critical evaluation of grazing techniques being undertaken by some of our American colleagues.
- The delightfully simple yet effective technique used to assess one of the many difficult parameters in grazing studies — i.e., the use of a 10-cent ruler from Woolworth's.
- The surprising lack of papers on pasture pest problems and management, which may be a good or a bad omen.
- The interesting possibilities in less well-known tropical grasses, with exciting flexibility in terms of temperature tolerance, and so on.

Finally, Prof. Wright and I both are closely involved with the education of our most valuable resource — our agricultural students. We believe that the impact, the expertise, and the experiences of the world's top grassland scientists in such a gathering as this could and should have much greater influence and value to our talented graduates than at present. We would like to recommend, therefore, that an appropriate Congress committee look at possible ways and means of exploiting the tremendous educational potential of such a Congress through some form of support scheme to enable at least a representative section of top young graduate students in grassland science from around the world to attend future International Grassland Congresses.

In conclusion might I again congratulate the speakers for making such an excellent Congress possible and the many Congress committees for their months of hard work on our behalf.

B.R. Watkin
New Zealand

Comments by Dr. Wright

Ladies and gentlemen, fellow scientists,

It is an honor, a privilege, and a pleasure for me to come to the microphone to convey on your behalf thanks to the organizers of this Congress.

There have been many good things at this Congress, which has been an undoubted success. I wish to thank especially the Program Committee for the scientific content of the Congress.

Each of you will have, according to your interests, your own personal favorites from among the plenary papers, but taken together I'm sure you will agree that these papers provided a stimulating commentary on matters of current importance in grassland science and formed a comprehensive background to the offered papers of the Congress.

Regarding the sessions, many of you will join me in deprecating the absence of those who, having had a paper accepted, did not arrive to present it and left gaps in the program. We commiserate with the organizers where this occurred without notification and share their disappointment and frustration.

For the most part I attended the breeding sessions, though sometimes the physiology sessions drew me away. Breeders have been particularly well served in this Congress. There has been a remarkable cross-fertilization of ideas. Not only have many of us picked up schemes or techniques of potential value for inclusion in our own breeding programs but several fundamental issues have been raised that may well have far-reaching implications in the ensuing decade on breeding aims and objectives. In particular, the possible potential of inducing or introducing apomixis into normally out-breeding species will be examined by breeders of temperate forages following reports of its successful exploitation by breeders of tropical grasses. The progress made with tropical legumes has helped to increase the reviving momentum in legume research and breeding, especially of white clover, for temperate regions.

Quite a number of papers, as well as the Forage Producers' Forum (Was it not revealing to hear how succinctly the farmers described their work?), emphasized the need to ensure seed production capability as an essential criterion in breeding programs, although the possible use of vegetative material rather than seed as a means of producing pastures is likely to be more widely assessed in countries outside the U.S.A.

The test of the success of a conference such as this, however, must rest not only in disseminating and acquiring knowledge, but also in meeting people and making friends internationally. I have had a lot of good times at the Congress, engendered by the hospitality and goodwill of our U.S.A. hosts and the camaraderie of colleagues new and old. To use the local idiom, I have visited with many people with much enjoyment and benefit.

All the tours and ladies' events added to the social atmosphere and were much appreciated. On your behalf I thank everyone involved in the Congress, including all who contributed to our well-being and enjoyment.

We look forward to our reunion at the XV Congress in Japan.

C.E. Wright
Australia

INTRODUCTION OF NEW CHAIRMAN OF CONTINUING COMMITTEE

Dr. W.R. Childers, retiring Chairman of the Continuing Committee, introduced Dr. L.R. Humphreys of Queensland, Australia, newly elected Chairman of the Committee. Dr. Humphreys, in his acceptance remarks, challenged the Congress in regard to its future opportunities and responsibilities.

Statement by Dr. Humphreys

I am honored by this appointment, which I accept as recognition of the attention paid to grassland science in Australia and New Zealand. I add my congratulations to the U.S.A. for an admirable conference with excellent standards of science and hospitality.

Now we have to work to produce good papers for Japan and especially to find funding so that younger scientists can also experience the educational opportunities of an International Grassland Congress. We have to ensure that countries active in grassland science that were not represented at this Congress will be able to come to Japan.

Invitations for hosting the XVI International Congress will need to be submitted to me by September 1984. I should like to correspond with any interested members before that time, to ensure that the Continuing Committee can be properly informed about the physical facilities available for the Congress, the capacity of a national organization to run it, the scientific interest in the regional work, the probable funds available to meet the objectives of the Congress, and the support of the national government for facilitating the entry of delegates from all over the world.

I hope that in the intervening period before the next International Grassland Congress the work of the national grassland associations will be strengthened, especially in Third World countries that have a shorter history of grassland research than have Europe and North America. Regional and specialist meetings, especially as promoted by FAO and other agencies, will build on many of the new insights and techniques discussed at this Congress. I hope we may see an increasing rapport between forage and range scientists. Range scientists and forage scientists sometimes emphasize different things, but they cover exactly the same scientific fields.

The distinct identity of grassland science needs to be maintained; it survives as a holistic discipline in an age of increasing specialization. The farmers of the world look for innovations that work; such innovations rarely enter farm practice as the outputs of single-discipline research. While the immense volume of new science presented at this Congress is making real contributions to farming progress, to increased food supply, and to the conservation of the natural resources upon which community welfare depends, we must all be freshly aware of the partial character of our knowledge of grassland ecosystems. This Congress will strengthen ongoing research programs, especially through the education of the younger scientists present and those of us still receptive to change.

We look for long-term industry and government support for grassland research and education and for grassland development programs to help harness the immense biological potential of the world's grasslands. The realization of these hopes depends first of all upon innovative grassland scientists and responsive farmers.

L.R. Humphreys
Chairman of the
Continuing Committee

SPECIAL RECOGNITIONS

As a final formal part of the Closing Ceremony, Dr. W.C. Templeton, Jr., President of the American Forage and Grassland Council, cohost for the XIV Congress, expressed the thanks of the AFGC and the University of Kentucky to the Congress Governing Board and the office staff.

Comments by Dr. Templeton, Jr.

Mr. Chairman, President Barnes, ladies and gentlemen,

I wish to echo the sentiments of several previous speakers. The XIV International Grassland Congress has provided an excellent opportunity for sharing information, exchanging views, and becoming better acquainted. It is my sincere hope that the enthusiasm, contacts, and friendships developed at the XIV Congress may be maintained and renewed at the XV Congress in Japan in 1985.

On behalf of the University of Kentucky and the American Forage and Grassland Council, it is a pleasure to express appreciation to the Governing Board and office staff of the XIV International Grassland Congress. Their dedicated efforts have, in large measure, helped to insure the success of the Congress, and I am sure I speak for all Congress members in saying "thank you" to each of them. I have the honor, at this time, of presenting certificates that officially recognize their contributions and express our appreciation for them. The wood bases for the certificates were hand-crafted by Warren C. Thompson from native black walnut lumber produced in Lincoln County, Kentucky. Mr. Thompson fashioned them in memory of the late Russell F. Cornelius, long-time Agricultural Extension Agent in that County and a renowned leader in grassland agriculture in the Commonwealth of Kentucky.

William C. Templeton, Jr.
President, AFGC

The certificates we will now present read as follows: "The American Forage and Grassland Council and the University of Kentucky gratefully recognize (*name*) for outstanding service as a member of the Governing Board [or office staff] in planning and conducting the XIV International Grassland Congress, June 15–24, 1981."

Following several concluding remarks by President Barnes and a warm invitation to Japan in 1985 from the Japanese delegation, the Congress was officially adjourned at 11:30 A.M., Wednesday, 24 June 1981.

TOURS AND EXHIBITS

PRE- AND POST-CONGRESS TOURS

Two pre-Congress tours and one post-Congress tour were included as part of the XIV Congress.

One pre-Congress tour, eight days in length, into northeastern U.S.A. and Canada included visits to the states of Virginia, Pennsylvania, New York, Ohio, West Virginia, and Kentucky and to the Province of Ontario, Canada. The second pre-Congress tour, six days in length, featured tropical agriculture and included stops in the states of Florida and Georgia.

An eight-day post-Congress tour to the Pacific Coast featured legume and grass-seed production and processing in the states of Oregon and California.

MID-CONGRESS TOURS

Seven one-day field trips to visit commercial forage-livestock-production systems and surface-mine reclamation sites in Kentucky were conducted simultaneously on Saturday, 20 June, and again on Sunday, 21 June. In addition, bus tours were conducted most afternoons during the Congress to the Kentucky Agricultural Experiment Station research farms to provide an overview of forage-livestock research in progress.

MID-CONGRESS FORAGE PRODUCERS' FORUM

A special mid-Congress Forage Producers' Forum was held midway through the Congress featuring six U.S. forage producers who spoke to Congress participants concerning their production, management, and utilization programs. The forum provided a unique opportunity for an exchange of views among farmers and scientists, extension specialists, agribusiness personnel, technicians, and others.

Papers presented by the six producers have been published in a separate *Proceedings* by the American Forage and Grassland Council.

SPOUSES PROGRAM

A number of special activities were included specifically for the approximately 180 spouses and children attending the Congress, including a special homemakers' reception, organized tours to places of scenic and historical interest, and various recreational activities. Spouses also attended the opening and closing ceremonies and various social functions of the Congress.

SPECIAL ACTIVITIES

A variety of special activities were held during the Congress for all participants, including an informal reception the evening prior to the opening sessions, several sponsored food functions, an outing at Kentucky Horse Park, and the closing banquet held the night before the closing business meeting and ceremony. Special entertainment was provided most evenings during the Congress.

EXHIBITS

A wide range of educational and commercial exhibits were on display during the Congress, including an up-to-date display of books and journals directly concerned with grassland agriculture, exhibits prepared by state forage and grassland councils affiliated with the American Forage and Grassland Council, and displays of many of the newest commercial products for establishing, harvesting, handling, preserving, and evaluating forages and grasslands.

COMMITTEES FOR THE XIV CONGRESS

The Congress hosts and the Board of Governors are indebted to the many individuals throughout Kentucky and the United States who have given unselfishly of their time and talent to serve on one or more of the many committees that have helped to make the Congress a success.

PROGRAM COMMITTEE

G.C. Marten, Chairman, USDA-SEA-AR, Agronomy & Plant Genetics Dept., University of Minnesota

W.W. Hanna, USDA-SEA-AR, Georgia Coastal Plain Experiment Station

David Grunes, USDA-SEA-AR, Cornell University

G.H. Heichel, USDA-SEA-AR, Agronomy & Plant Genetics, University of Minnesota

K.L. Von Bargen, Agricultural Engineering Dept., University of Nebraska

A.G. Matches, USDA-SEA-AR, Agronomy Dept., University of Missouri

V.L. Lechtenberg, Agronomy Dept., Purdue University

J.E. Moore, Animal Science Dept., University of Florida

C.H. Herbel, USDA-SEA-AR, Las Cruces, NM

D.A. Rohweder, Agronomy Dept., University of Wisconsin

BUDGET AND FINANCE COMMITTEE

W.F. Wedin, Chairman, Dept. of Agronomy, Iowa State University

E.C. Charron, Technical Dir., Food, Beverage & Tobacco Div., The Chase Manhattan Bank, N.A.

Harry Gamage, Product Manager, GRASLAN-Elanco Range Products

J.F. Harrington, President, IRI Research Institute, Inc.

J.R. Porter, Director, Crop Management-Agway, Inc.

N.B. Raun, Director, Production Programs, Winrock International

C.N. Cooley, President, The Dodson Mfg. Company, Inc.

J.H. Hamilton, Manager, Technical Development, Advanced Harvesting Systems, International Harvester Company

L.S. Murphy, Potash-Phosphate Institute

T.H. Ryan, Director, Forages & Turf Seeds, North American Plant Breeders

PUBLICATIONS COMMITTEE

V.W. Hays, Chairman, Department of Animal Sciences, University of Kentucky

J. Allan Smith, Editor, Emeritus Professor, University of Kentucky

1. Plant Introduction, Evaluation, and Breeding for Production, Quality, and Utilization

A.W. Hovin, Department of Agronomy & Plant Genetics, University of Minnesota

R.R. Hill, U.S. Regional Pasture Research Lab., University Park, Pennsylvania

James Elgin, Field Crops Lab., USDA-SEA-AR-BARC

Reed Barker, Agronomy Department, USDA-SEA-AR, Northern Great Plains Research Lab.

K.H. Asay, Crops Res. Lab., USDA-SEA-AR, Utah St. University

2. Seed Production and Dissemination

D.O. Chilcote, Crop Science Department, Oregon St. University

R.R. Smith, Agronomy Department, University of Wisconsin

3. Soil Fertility and Plant Nutrition

R.D. Munson, Potash-Phosphate Institute

C.L. Rhykerd, Agronomy Department, Purdue University

4. The Nitrogen Cycle in Grasslands

L.K. Porter, USDA-SEA-AR, Fort Collins, CO

Joseph Legg, Rogers, AR

5. Multiple Use of Grassland Resources

G.A. Jung, U.S. Regional Pasture Res. Lab.

C.T. Dougherty, Department of Agronomy, University of Kentucky

6. Physiological Processes in Plants

C.J. Nelson, Department of Agronomy, University of Missouri

N.J. Chatterton, Plant Physiologist, USDA-SEA-AR, Utah St. University

7. Grassland Ecology

David Cummins, Agronomy Department, Georgia Agricultural Exp. Station

J.C. Burns, Department of Crop Science, North Carolina St. University

J.B. Washko, Em. Prof. Agronomy, The Pennsylvania St. University

8. New Evaluation Techniques for Pastures, Natural Grassland, and Conserved Forage

R.H. Hart, USDA-SEA/FR.HPGRS, Cheyenne, WY

S.W. Coleman, Research Scientist, USDA-SEA-AR, El Reno, OK

9. Management of Grazed and Conserved Forages for Improved Yield, Quality, and Persistence

R.W. Van Keuren, Department of Agronomy, Ohio Agr. Res. & Dev. Center

T.H. Taylor, Agronomy Department, University of Kentucky

M.B. Tesar, Department of Crop & Soil Sciences, Michigan St. University

10. Mechanization and Treatment During Harvest, Processing, and Storage for Improved Forage Quality and Yield

D.R. Waldo, Ruminant Nutrition Lab, Beltsville, MD

T.J. Klopfenstein, Animal Science Department, University of Nebraska

11. Utilization of Pasture, Natural Grassland, and Conserved Forage in Animal Production

R.L. Reid, Animal and Veterinary Sciences, West Virginia University

L.L. Wilson, Department of Animal Science, The Pennsylvania St. University

12. Grassland Agriculture in Tropical Regions

A.E. Kretschmer, Jr., Agricultural Research Center, Fort Pierce, FL

N.B. Raun, Winrock International Livestock Res. & Training Center

13. Transfer of Grassland Research Findings to Farm and Ranch Practice

W.J. Moline, Department of Agronomy, University of Arkansas

J.B. Washko, Em. Prof. Agronomy, The Pennsylvania St. University

14. Socio-Economic Aspects of Grassland Systems and Policies

David Petritz, Department of Ag. Economics, Purdue University

E.M. Smith, Ag. Engineering Department, University of Kentucky

TOURS COMMITTEE

C.S. Hoveland, Chairman, Agronomy & Soils Dept., Auburn University

H.E. White, Agronomy Dept., VPI & SU

N.P. Martin, Agronomy/Plant Genetics, University of Minnesota

R.L. Haaland, Agronomy & Soils Dept., Auburn University

D.W. Jones, Dept. of Agronomy, University of Florida

J.N. Pratt, Soil & Crop Sciences Center, Texas A & M University

Harold Youngberg, Ext. Certif. Specialist, Oregon State University

G.D. Lacefield, Agronomy Dept., University of Kentucky

COMMUNICATIONS AND PUBLIC RELATIONS COMMITTEE

W.C. Thompson, Chairman, North American Plant Breeders

W.E. Barksdale, Barksdale Agri-Communications

M.A. Balas, Publications Supervisor, Sperry New Holland

E.H. Row, Associate Editor, Hoard's Dairyman

D.M. Springer, Assistant Director for Ag. Information, University of Kentucky

Vernon Bourdette, Regional Information Officer, USDA

Elizabeth Thompson, Exec. Secretary, American Forage & Grassland Council

COMMERCIAL EXHIBITS COMMITTEE

A.M. Best, Chairman, Director, Engineering Research, Sperry New Holland

F.C. Bergamini, Regional Sales Manager, Northrup King Co.

Lindsay Brown, Marketing Manager, Fertilizer Group, Int. Mineral & Chemical Corp.

R.J. Buker, V.P. & General Manager, FFR Cooperative

R.D. Child, Winrock International

C.N. Cooley, President, The Dodson Mfg. Co., Inc.

Tony Rutz, District Marketing Specialist, Chevron Chemical Co.

EDUCATIONAL EXHIBITS COMMITTEE

J.L. Ragland, Chairman, Assistant to the Dean for Green Thumb, College of Agriculture, University of Kentucky

LOCAL ARRANGEMENTS COMMITTEE

C.O. Little, Co-Chairman; Associate Dean for Research

A.J. Hiatt, Co-Chairman; Chairman, Department of Agronomy

P.P. Appel, Coordinator; Assistant to the Dean

Registration: C.T. Dougherty, H.F. Massey, Co-Chairmen; P.G. Woolfolk, J.L. Sims, T.P. Pirone, R.H. Dutt

Housing: B.C. Pass, D.E. Peaslee, Co-Chairmen; R.C. Buckner, J.D. Fox, H.G. Love, P. Burrus, W.H. McCollum

Meeting Rooms & Visual Aids: J.C. Robertson, N.L. Taylor, Co-Chairmen; K.V. Yeargan, W. Springer, J.A. Boling, J.B. Williams, W.E. Wise

Transportation: J.R. Calvert, D.G. Ely, Co-Chairmen; R.L. Blevins, L.G. Wells, G.D. Pendergrass

Information & Emergencies: K.L. Wells, R.E. Tucker, Co-Chairmen; J.G. Rodriguez, T.H. Taylor, K.E. Pigg, J.C. Redman

Entertainment: G.W. Thomas, F.E. Justus, Co-Chairmen; L.M. Busch, D.L. Davis, N. Gay

Special Events: B.F. Parker, L.P. Bush, Co-Chairmen; R.W. Hemken, F.W. Knapp, J.E. Leggett, S.H. Jackson, D.E. LaBore, P.D. Warner

Publicity: D.M. Springer, I.E. Massie, Co-Chairmen; D.C. Wolf, G.W. Quinn, D.B. Witham

Tours: G.D. Lacefield, J.T. Johns, Co-Chairmen; E.M. Smith, G.T. Lane, J.K. Evans, N.W. Bradley, G.M. Chappell, R.E. Sigafus, J.P. Baker, C.M. Christensen, S.H. Phillips

Displays & Exhibits: J.N. Walker, D.M. TeKrony, Co-Chairmen; A.J. Powell, W.L. Crist, O.J. Loewer, J.T. Davis, R.M. Jacques

Spouses & Children: C.M. Coughenour, J.D. Kemp, Co-Chairmen; D.A. Tichenor, L.C. White, G. Heersche, D.M. Richey

The local arrangements committees are composed of staff members of the College of Agriculture, University of Kentucky.

PLENARY PAPERS

Forage Quality: Assessing the Plant-Animal Complex

D. J. MINSON

Senior Principal Research Scientist, Division of
Tropical Crops and Pastures, CSIRO, Brisbane, Australia

Summary

At the VI International Grassland Congress held at Pennsylvania State College in 1952, the late Dr. C.P. McMeekan reviewed the many problems associated with the plant-animal complex. Great stress was placed on the need to measure forage intake accurately as a prerequisite to an understanding of the plant-animal complex.

Progress in the development of more accurate methods of estimating the forage intake of grazing animals is reviewed, including the introduction of different methods of administering chromic oxide, the esophageal fistula for obtaining samples of selectively grazed material, the in-vitro technique for estimating the digestibility of the forage eaten, and the use of grazing behavior and bite-size data. Many different methods of measuring forage intake are available, but none of these has a high degree of accuracy.

The absence of accurate methods of measuring forage intake has led to use of other approaches to study the plant-animal complex. The existence of a critical bite size below which intake will be depressed is considered. It is shown that the critical bite size will be achieved only when the yields of *desired* forage exceed about 1,000 kg/ha. Grazing animals prefer green leaf, and in many cases rejection of the stem and dead material is so great that only green leaf should be considered when determining if sufficient forage is available for the grazing animal to achieve the critical bite size. In mixed pasture, animals will tend to prefer certain species, and the yield of these species will determine whether maximum voluntary intake of forage will be achieved. The possibilities of using fertilizers to change this preference rating in favor of a plant of higher nutritive value is discussed, together with problems associated with the low intake of the stem fraction.

It is concluded that major advances have been made in our knowledge of the plant-animal complex and that most of these advances have been achieved without using accurate methods of measuring the forage intake by animals. This progress was achieved by bringing together the skills of many different disciplines, and there is a continuing need for individual scientists who have trained in both the plant and the animal sciences.

KEY WORDS: intake, grazing behavior, chromic oxide, feces, fertilizer, odor, leaves, stem, preference.

INTRODUCTION

At the VI International Congress held at Pennsylvania State College in 1952, the late Dr. C.P. McMeekan described himself as neither a grassland man nor an animal man but a "hybrid, standing midway between the keepers of the grass swards, the agrostologists, and the keepers of the animal, the veterinarians and animal husbandmen." Dr. McMeekan (1952) listed many problems associated with the seasonal and variable nature of pasture production and their interaction with the grazing animal. Many of these problems have their origin at the pasture-animal interface, and Dr. McMeekan stressed the need for accurate techniques for measuring pasture intake if these problems were to be solved. Other aspects of the pasture-animal interface were also discussed at the Congress: the need for methods of measuring the quality of selectively grazed pasture, the value of studies of grazing behavior—grazing time, speed of grazing, and rumination time—and differences in forage palatability or animal preference.

Studies with cut forage fed to penned animals have provided information on many factors limiting forage intake.

These same factors will also limit the forage intake of grazing animals, but there are additional factors unique to the pasture-animal interface. This paper will review the progress made in the last 29 years in developing better methods of measuring forage intake and in our understanding of forage intake as it relates to grazed pasture. I have adopted this limited approach in order to demonstrate the progress that has been made on the plant-animal interface despite the known weaknesses of the various methods used. It is hoped that this review will lead to a better definition of problems associated with the plant-animal complex and of how these might be solved in the next decade.

MEASURING PASTURE INTAKE

In 1952 the usual method of estimating daily intake of grazing animals (I) was from the quantity of feces produced (F) and digestibility of the pasture selectively grazed (D) (Reid 1952):

$$I = F \times \frac{100}{100 - D} \qquad (1)$$

Fecal output is usually determined by dosing twice daily with known quantities of the indigestible tracer chromic oxide (Cr_2O_3) in ready-filled 1- and 10-g Cr_2O_3 capsules (Raymond and Minson 1955), with average annual sales of 30,000 and 60,000, respectively, or with Cr_2O_3 paper, which gives a more even distribution of Cr_2O_3 in the feces (Corbett et al. 1958).

The other main change in measuring forage intake has been in the method of estimating the digestibility of the selectively grazed pasture. In 1952 digestibility was estimated from either the nitrogen or the chromogen content of the feces (Reid 1952). The development in the early 1960s of a reliable technique for fistulation of the esophagus (Van Dyne and Torell 1964) made it possible to collect samples of the grazed forage that could be used to estimate digestibility by the in-vitro technique (Tilley et al. 1960). This technique requires calibration of the results with samples of known in-vivo digestibility usually determined with sheep. The different digestive efficiency of cattle is usually ignored (Playne 1970, Ternouth et al. 1979).

At the VI International Grassland Congress, Hancock (1952) said that forage intake could be estimated from a knowledge of grazing time, rate of biting, and bite size. Grazing time and rate of biting were determined visually, but there was no method of measuring bite size. During the past 10 years equipment has been developed for measuring both number of grazing bites (N) and bite size (S), and this has enabled forage intake (I) to be measured using the equation $I = N \times S$ (Chacon et al. 1976).

Of the various methods now available for measuring the intake of grazing animals, none has achieved the accuracy desired by McMeekan. Despite their limitations, the techniques have helped develop many of our current concepts even if they have failed to provide accurate numerical data.

FACTORS CONTROLLING FORAGE INTAKE

In a pen-feeding study with cut forage it was shown that voluntary intake is positively correlated with the digestibility of the energy (Blaxter 1960). Since then many studies have confirmed that this principle also applies to pastures grazed by cattle and sheep (Hodgson 1977). A positive relationship of this type is to be expected, since with increasing fiber content of the grass there is a decline in digestibility (Van Soest 1965) and a proportional increase in the quantity of indigestible residues that have a longer transit time in the rumen (Blaxter et al. 1961). The physical form of the fiber is also important, as is illustrated by the higher intake of animals for forages that have been ground and pelleted (Wainman and Blaxter 1972).

Only a healthy animal will consume the maximum quantity of forage. The animal requires an adequate supply of all essential nutrients, including protein; this principle applies to both the penned and the grazing animal. The intake of both penned and grazing animals will also be depressed if the forage contains any toxic compounds.

At the pasture-animal interface there are additional factors that can limit the intake of a forage. Both Hancock (1952) and Hubbard (1952) drew attention at the Congress to the limit on the time that animals will spend grazing. This limit, plus the relative constancy of rate of biting (Hancock 1952), means that intake will be less than maximum once pasture density falls below the level required to allow cattle to have a bite size exceeding about 300 mg of organic matter (OM) (Stobbs 1973).

An ideal grazing pasture may be defined as one on which bite size exceeds the critical level for that species of animal. As these ideal grazing conditions rarely occur in practice, there is considerable interest in the extent to which the intake of pasture by grazing animals is reduced when the pasture falls short of this ideal.

Five different situations will be considered.

1. Low pasture yields. On young pastures with $>$ 1,000 kg dry matter (DM)/ha, animals generally have no difficulty in satisfying their appetites, taking in large quantities of pasture with each bite. Under these conditions grazing is not restricted by the soil surface, and cattle and sheep can achieve maximum intake of dry matter, provided the yield of dry matter is between 1,000 and 1,500 kg/ha and the area grazed is not restricted (Woodward 1936, Johnstone-Wallace and Kennedy 1944, Waite et al. 1950, Arnold and Dudzinski 1966).

When pasture yields fall below about 1,000 kg/ha, bite size falls below the critical 300 mg OM. With a *Setaria anceps* pasture yielding 650 kg/ha, bite size was only 130 mg OM compared with 300 mg OM for an adjacent *Setaria* sward yielding 3,120 kg/ha (Stobbs 1973).

2. Contrasting leaf and stem. In most tropical pastures and mature temperate pastures there are large differences between leaf and stem fractions. The differences in chemical composition are well recognized, but there are also differences in physical composition that increase the energy required to sever the stem compared with that to sever the leaf (Hendricksen and Minson 1980).

Where there are large quality differences between leaves and stems in a pasture, animals prefer to eat the leaf. Selective grazing is usually regarded as a nutritional advantage, but recently the disadvantages have been recognized (Chacon and Stobbs 1976, Hendricksen and Minson 1980). Once the grazing animal is accustomed to eat the leaf fraction, it appears to continue to search out the leaf even when very little is present, and this can lead to very low feed intake. This phenomenon is clearly demonstrated in grazing studies with spring-grown *Setaria anceps* cv. Kazungula (Chacon and Stobbs 1976) and the annual legume *Lablab purpureus* (Hendricksen and Minson 1980). In these studies it was reported that when cattle entered a new plot with leaf yields ranging from 1,700 to 2,400 kg/DM/ha, daily intake was 6.1 to 11.5 kg/OM/day; but daily intake was only 2.8 kg/OM/day after 12–15 days, when leaf yields were 300–400 kg/ha. Although intake declined, the quantity of stem on offer remained almost constant because little was eaten ($<$ 1.5 kg/OM/day). Associated with the decline in daily intake was a decrease in the size of bite from 300–400 mg to 90–120 mg/bite. Grazing time and rate of biting also changed, but only to a small extent.

3. *Green vs. dead material*. Most pastures contain both green and dead material. With temperate pastures the live-weight gain of sheep is more closely related to the quantity of green dry matter/unit area than to the total forage present (Willoughby 1958, Arnold and Dudzinski 1966). The apparent failure of dead material to contribute significantly to animal production has also been shown with cattle grazing the tropical grasses *Sorghum almum* (Yates et al. 1964) and *Cenchrus ciliaris* ('t Mannetje 1974). Studies with cattle fistulated at the esophagus showed that cattle eat very little dead material provided some green leaf is available ('t Mannetje 1974, Chacon and Stobbs 1976, Hendricksen and Minson 1980).

Although the difference in preference by animals between dead and green pasture is widely recognized, the cause of this difference has not yet been determined. A difference in odor seems to be the most likely explanation for animals' disliking dead forage, but there are no direct experimental data to support this hypothesis.

4. *Grass vs. legume*. In pastures containing more than one plant species the situation is further complicated. In some cases a species will be completely rejected, and the effective yield for grazing purposes will then depend on the yield of the preferred plant. The more usual situation is for all species to be eaten to some extent.

In the tropics many legumes are virtually rejected in the spring but readily eaten later in the year. With pastures containing siratro (*Macroptilium atropurpureum*) cattle selected a diet containing 2%–10% legume in the spring but selected 62%–73% of it during the autumn (Stobbs 1977). Cafeteria studies using frozen forage showed cattle preferred autumn-grown siratro to summer-grown siratro (Stobbs 1977). This indicates that some factor(s), probably odor, associated with the siratro was influencing preference. It is also possible that the preference rating of the companion grass changed throughout the season, and this change could result in very large seasonal differences in the legume:grass ratio in the diet selected.

The difference in preference between one grass and one legume is a simple example. The situation is much more complex in rangelands containing many plant species grazed by more than one animal species (Van Dyne and Heady 1965).

Can animal preferences be changed in favor of plants of higher nutritive value? Van Soest (1965) has shown that stall-fed animals will eat more legume than grass of similar digestibility, so any method of increasing the intake of legume will probably increase animal production. Recently it has been shown that the proportion of legume in a diet may be increased by a factor of five if superphosphate is applied to grass-legume pastures (McLean et al. 1981). The legume content of the pasture on offer was doubled, but this increase was poorly correlated with preference for the legume (r = 0.13). Which of the elements in superphosphate affect preference and the precise mechanisms are not known.

5. *Restricted area*. When pastures are rotationally or strip-grazed the quantity of pasture initially available usually exceeds 1,000 kg/DM/ha, so pasture density will not restrict intake. With restricted grazing systems only a limited quantity of pasture is available for grazing compared with set stocking where each day the animal will eat only 1% or 2% of the pasture on offer. If the area allotted is too small, then all the pasture is removed down to a level determined by the dental configuration, and intake will be less than the animal's capacity.

In intensive grazing systems the area allotted is a compromise; sufficient pasture is offered without "excessive waste." Forcing the animal to eat a high proportion of the pasture leads to reduced intake (Raymond et al. 1956, Tayler and Rudman 1965, Greenhalgh 1966, 1970), live-weight gain (Blaser et al. 1960, Tayler and Rudman 1965), and milk production (Greenhalgh 1966, 1970).

All studies have shown a decline in pasture intake and/or animal production with increasing utilization of the available pasture, but there is no agreement on the cause of this decline. Voluntary intake is often positively correlated with digestibility (Blaxter 1960), so the decline could be caused by a lower digestibility of the lower part of the plants. This suggestion is supported by fecal index estimates of digestibility (Raymond et al. 1956, Blaser et al. 1960). However, in other experiments where digestibility has been determined by an in-vivo or in-vitro technique, there was little difference in OM digestibility between the feed selected from leniently and intensively grazed pastures (Tayler and Rudman 1965; Greenhalgh 1966, 1970; Nicol et al. 1976). Another suggestion is that soil contamination reduces the intake of the lower parts of the pasture (Nicol et al. 1976), an idea supported by the well-recognized dislike by stock of pasture contaminated with volcanic ash. To test whether pasture availability restricted the intake of bottom grazers, a comparison was made by Nicol et al. (1976) between cattle-grazed or pen-fed pasture harvested to stubble heights of 8 and 20 cm. The depression in live-weight gain with reduced stubble height was similar in both cases. This finding indicated that provided no attempt was made to graze the pasture down to less than 8 cm, pasture availability did not restrict intake or performance.

The most likely reason for the lower intake and animal production as pastures are grazed down is the change in the physical composition of the pasture eaten. At low stocking pressure the animals have ample opportunity to select leaf, but as grazing intensity is increased there is less opportunity for selecting leaf, and the proportion of stem in the diet increases (Chacon and Stobbs 1976, Stobbs 1978, Hendricksen and Minson 1980). Previously it was thought that the voluntary intake of leaf and stem were similar provided both fractions had the same digestibility. However, studies with sheep and cattle fed separated leaf and stem fractions of grasses and a legume have clearly demonstrated that leaf is eaten in greater quantity (Laredo and Minson 1973, 1975; Poppi et al. 1981; Hendricksen et al. 1981). This difference appeared to be related to the longer time the stem fraction was retained in the rumen. However, the difference in voluntary intake between leaf and stem fractions is not constant. With the fine stems of ryegrass the difference in intake was only 20% (Laredo and Minson 1975), but it was 87% with the thick stems of *Setaria splendida* (Laredo and Minson 1973).

THE FUTURE OF FORAGE-QUALITY RESEARCH

This review of 29 years of research demonstrates the large number of factors that influence pasture intake. A review of pasture digestibility or the efficiency with which digested nutrients are utilized would reveal a similar degree of complexity.

The problems of measuring pasture intake and understanding the factors controlling it are obviously far more difficult than was appreciated in 1952. I see no chance of our ever measuring pasture intake in the field with the same accuracy as can be achieved in the pen. Confronted with this situation, should we continue to use our limited research resources to search for an accurate technique of measuring pasture intake in the field? Do we need to measure pasture intake in the field with a high degree of accuracy to solve most of our pasture problems, since forage intake and level of animal production are closely related (Blaxter 1960)?

Our future aims should be to exploit the recent findings on forage intake and to determine the mechanisms responsible for these effects. I believe there are three areas of research that should prove very rewarding.

1. Australian work has shown that the preference of animals for legumes in a mixed pasture can be increased by applying superphosphate and that this leads to improved growth rates of cattle. There is a need to determine whether this effect also applies to pastures grown on fertile soils and to temperate as well as tropical grass–legume pastures. Superphosphate contains phosphorus, sulfur, and calcium; with the greater use of high-analysis fertilizers there is a need to know which element(s) are causing the changes in preference rating. There is also a need to study other fertilizers to determine whether they also may be used to change the animals' preference for species in mixed pastures.

Changing the preference of animals for species in a mixed pasture can have beneficial effects on animal production if the change is in favor of a species with a higher nutritional value. However, this change could have adverse effects, increased grazing pressure reducing survival of the legume in the pasture.

2. Animals prefer the leaf fraction of pastures; this preference has a beneficial effect on animal production, provided sufficient of the desired leaf is present to satisfy the animals' appetites. However, when the quantity of leaf is low the animals prefer to restrict intake as opposed to increasing the quantity of stem in the diet. These observations were made with tropical species, and there is a need to repeat the work with temperate species, which generally have finer stems. Fourteen-day grazing periods were used in these studies, and there is a need to determine the relevance of this conclusion to strip-grazed and set-stocked grazing systems.

Leaf is eaten in larger quantities than stem of the same digestibility because it is retained for a shorter time in the rumen, but how much stem can be included in the diet without an adverse effect on intake and production? Pasture yields are usually expressed in terms of total dry matter, but if stem is so poorly utilized it may be more valuable to quote separate yields for leaf and stem fractions.

3. The preference of animals for pasture species or plant parts is probably controlled by physical effects and odors. The relative importance of these factors needs determining, and the odors that make plants attractive to animals need to be positively identified. A knowledge of what odors make animals prefer pasture plants will aid in the development of improved grazing systems and possibly the use of special chemical sprays to improve animal preference.

To advance our knowledge of the plant-animal complex the scientist in this area must be more than a plant man, more than an animal man; he needs to be, to use McMeekan's term, a hybrid. He must be very versatile, with a very broad knowledge of the developments in many fields of science, and must be able to call on the help of the specialist physicist, chemist, and veterinarian if the problems of the plant-animal complex are to be solved.

LITERATURE CITED

Arnold, G.W., and M.L. Dudzinski. 1966. The behavioural response controlling the food intake of grazing sheep. Proc. X Int. Grassl. Cong., 367–370.

Blaser, R.W., R.C. Hammes, H.T. Bryant, W.A. Hardison, J.P. Fontenot, and R.W. Engel. 1960. The effect of selective grazing on animal output. Proc. VIII Int. Grassl. Cong., 601–606.

Blaxter, K.L. 1960. The utilization of the energy of grassland products. Proc. VIII Int. Grassl. Cong., 479–484.

Blaxter, K.L., E.W. Wainman, and R.S. Wilson. 1961. The regulation of food intake by sheep. Anim. Prod. 3:51–61.

Chacon, E., and T.H. Stobbs. 1976. Influence of progressive defoliation of a grass sward on the eating behaviour of cattle. Austral. J. Agric. Res. 27:709–727.

Chacon, E., T.H. Stobbs, and R.L. Sandland. 1976. Estimation of herbage consumption by grazing cattle using measurements of eating behaviour. J. Brit. Grassl. Soc. 31:81–87.

Corbett, J.L., J.F.D. Greenhalgh, I. McDonald, and E. Florence. 1958. Excretion of chromium sesquioxide administered as a component of paper to sheep. Brit. J. Nutr. 14:289–299.

Greenhalgh, J.F.D. 1966. Studies of herbage consumption and milk production in grazing dairy cows. Proc. X. Int. Grassl. Cong., 351–355.

Greenhalgh, J.F.D. 1970. The effects of grazing intensity on herbage production and consumption and on milk production in strip-grazed dairy cows. Proc. XI Int. Grassl. Cong., 856–860.

Hancock, J. 1952. Grazing behaviour of identical twins in relation to pasture type, intake and production of dairy cattle. Proc. VI Int. Grassl. Cong., 1399–1407.

Hendricksen, R.E., and D.J. Minson. 1980. The feed intake and grazing behaviour of cattle grazing a crop of *Lablab purpureus* cv. *Rongai*. J. Agric. Sci. 95:547–554.

Hendricksen, R.E., D.P. Poppi, and D.J. Minson. 1981. The voluntary intake, digestibility and retention time by cattle and sheep of leaf and stem fractions of a tropical legume (*Lablab purpureus*). Austral. J. Agric. Res. 32:389–398.

Hodgson, J. 1977. Factors limiting herbage intake by the grazing animal. Proc. Int. Mtg. Anim. Prod. Temp. Grassl., 70–75.

Hubbard, W.A. 1952. Following the animal and eye-estimation method of measuring the forage consumed by grazing animals. Proc. VI Int. Grassl. Cong., 1343–1347.

Johnstone-Wallace, D.B., and K. Kennedy. 1944. Grazing management and behaviour and grazing habits of cattle. J. Agric. Sci. 34:190-197.

Laredo, M.A., and D.J. Minson. 1973. The voluntary intake, digestibility and retention time by sheep of leaf and stem fractions of five grasses. Austral. J. Agric. Res. 24:875-888.

Laredo, M.A., and D.J. Minson. 1975. The voluntary intake and digestibility by sheep of leaf and stem fractions of *Lolium perenne*. J. Brit. Grassl. Soc. 30:73-77.

McLean, R.W., W.H. Winter, J.J. Mott, and D.A. Little. 1981. The influence of superphosphate on the legume content of the diet selected by cattle grazing *Stylosanthes*-native grass pastures. J. Agric. Sci. 96:247-249.

McMeekan, C.P. 1952. Interdependence of grassland and livestock in agricultural production. Proc. VI Int. Grassl. Cong., 149-161.

Nicol, A.M., D.G. Clarke, J. Munro, and M.C. Smith. 1976. The influence of stubble height on digestibility, intake and liveweight gain of beef steers. Proc. N. Z. Soc. Anim. Prod. 36:81-86.

Playne, M.J. 1970. Differences in the nutritional value of three cuts of buffel grass for sheep and cattle. Proc. Austral. Soc. Anim. Prod. 8:511-516.

Poppi, D.P., D.J. Minson, and J.H. Ternouth. 1981. Studies of cattle and sheep eating leaf and stem fractions of grasses. I. The voluntary intake, digestibility and retention time in the reticulo-rumen. Austral. J. Agric. Res. 32:99-108.

Raymond, W.F., and D.J. Minson. 1955. The use of chromic oxide for estimating the faecal production of grazing animals. J. Brit. Grassl. Soc. 10:282-296.

Raymond, W.F., D.J. Minson, and C.E. Harris. 1956. The effect of management on herbage consumption and selective grazing. Proc. VII Int. Grassl. Cong. 123-132.

Reid, J.T. 1952. Indicator methods, their potentialities and limitations. Proc. VI Int. Grassl. Cong., 1334-1339.

Stobbs, T.H. 1973. The effect of plant structure on the intake of tropical pastures. I. Variation in the bite size of grazing cattle. Austral. J. Agric. Res. 24:809-819.

Stobbs, T.H. 1977. Seasonal changes in the preference by cattle for *Macroptilium atropurpureum* cv. Siratro. Trop. Grassl. 11:87-91.

Stobbs, T.H. 1978. Milk production, milk composition, rate of milking and grazing behaviour of dairy cows grazing two tropical grasses pastures under a leader and follower system. Austral. J. Expt. Agric. Anim. Husb. 18:5-11.

Tayler, J.C., and J.E. Rudman. 1965. Height and method of cutting or grazing in relation to herbage consumption and live-weight gain. Proc. IX Int. Grassl. Cong., 1639-1644.

Ternouth, J.H., D.P. Poppi, and D.J. Minson. 1979. The voluntary food intake, ruminal retention time and digestibility of two tropical grasses fed to cattle and sheep. Proc. Nutr. Soc. Austral. 4:152.

Tilley, J.M.A., R.E. Deriaz, and R.A. Terry. 1960. The in vitro measurement of herbage digestibility and assessment of nutritive value. Proc. VIII Int. Grassl. Cong., 533-537.

't Mannetje, L. 1974. Relations between pasture attributes and liveweight gains on a subtropical pasture. Proc. XII Int. Grassl. Cong., 3:299-304.

Van Dyne, G.M., and H.F. Heady. 1965. Botanical composition of sheep and cattle diets on a mature annual range. Hilgardia 36:465-492.

Van Dyne, G.M., and D.T. Torell. 1964. Development and use of the esophageal fistula: A review. J. Range Mangt. 17:7-19.

Van Soest, P.J. 1965. Symposium on factors influencing the voluntary intake of herbage by ruminants: Voluntary intake in relation to chemical composition and digestibility. J. Anim. Sci. 24:834-843.

Wainman, F.W., and K.L. Blaxter. 1972. The effect of grinding and pelleting on the nutritive value of poor quality roughages for sheep. J. Agric. Sci. 79:435-445.

Waite, R., W. Holmes, Jean I. Campbell, and D.L. Ferguson. 1950. Studies in grazing management. II. The amount and chemical composition of herbage eaten by dairy cattle under close-folding and rotational methods of grazing. J. Agric. Sci. 40:392-402.

Willoughby, W.M. 1958. A relationship between pasture availability and animal production. Proc. Austral. Soc. Anim. Prod. 2:42-45.

Woodward, T.E. 1936. Quantities of grass that dairy cows will graze. J. Dairy Sci. 19:347-357.

Yates, J.J., L.A. Edye, J.G. Davies, and K.P. Haydock. 1964. Animal production from a *Sorghum almum* pasture in southeast Queensland. Austral. J. Expt. Agric. Anim. Husb. 4:325-335.

DISCUSSION

Remarks of Discussion Leader, J.E. Moore

As Dr. Minson indicated, a very important factor affecting the performance of grazing animals is their intake of pasture. It is very difficult, however, to estimate the intake of pasture by individual animals. The conventional approach is either to measure or to estimate fecal output and digestibility of dry matter and then to calculate intake. The late Dr. Stobbs, Dr. Minson, and their colleagues developed an alternative approach involving estimates of both number of bites/day and bite size. Their contribution is much more than just another method to estimate intake, however, because by separating number of bites from size of bite it may be possible to identify some of the reasons for differences in intake among pastures.

Fig. 1 illustrates the plant-animal interface as it pertains to factors determining pasture intake by grazing animals. The plant component is expressed in terms of the quantity and character of forage available at time of grazing. Quantity may be expressed either on a per-head or per-hectare basis; although these quantities are not the same, they will be combined for general discussion. Character includes all those factors affecting quality potential, including both chemical composition (nutrients, cell-wall, anti-quality factors, odors, etc.) and structural characteristics (microanatomy, leaf:stem ratio, grass:legume ratio, living:dead ratio, bulk density, and growth habit or the manner in which the forage is presented to the animal).

The animal component (Fig. 1) is expressed in terms of those physiological mechanisms that act as limits or controls on daily dry-matter intake. In the "bites" mechanism, there are upper limits to both the number of bites and the size of bite. The "distention" mechanism assumes some upper limit on rumen fill. When that limit is reached, retention time (t_R) determines intake. Retention time depends on rates of digestion and passage. The "metabolic" mechanism assumes some upper limit on digestible energy intake (DEI). When that limit is reached, intake is determined by the digestible energy (DE) concentration of the diet (i.e., digestibility).

Fig. 1. Conceptual illustration of the plant-animal interface in grazed pastures: (A) intake limited by distention or metabolic mechanisms; (B) intake limited by bite number and/or size; and (C) combined limits, plus the effect of intake on quantity and character of available forage (t_R = retention time; DEI = digestible energy intake; DE = digestible energy content of forage).

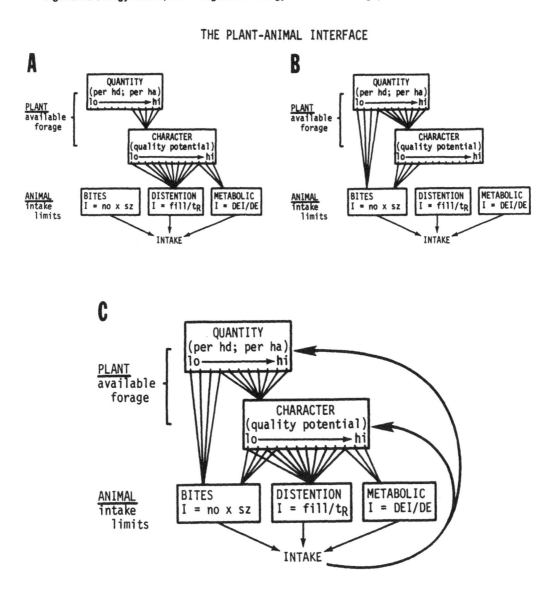

THE PLANT-ANIMAL INTERFACE

Fig. 1A describes the conventional theories, which suggest that when the quantity of forage available is high enough for ad-libitum intake, the character of the forage determines intake through the distention mechanism or, for very high-quality forages, the metabolic mechanism. Fig. 1B describes intake limitation operating through the bites mechanism. When quantity of forage is low, forage character may have little effect on intake. At higher quantities, even ad-libitum levels, the "availability of desired forage" or forage "accessibility" may affect intake through limits on number and/or size of bites. Fig. 1C combines the many possible intake-limiting factors. In addition, Fig. 1C illustrates a "feedback" in terms of the effect of intake (the act of consumption) on the quantity and character of forage available in the following time period.

Although Fig. 1 illustrates an oversimplified view of a very complex system, it may help to visualize the answer to a question attributed to the late Dr. H.L. Lucas, Jr. (M.E. Riewe, Proc. Southern Pasture Forage Crop Imp. Conf., 1980). Dr. Lucas asked: "How, since the components of a forage and animal system interact so strongly, can knowledge of the behavior of the individual components be integrated to provide proper input-output relationships for the total system?" His answer: "By study, not just of the components themselves, but also of the mechanisms whereby the components are coupled with each other, that is, the interfacing of sward and animal in grazing systems." Dr. Minson and his colleagues have made important contributions to our understanding of the plant-animal interface.

Questions and Responses

1. *Question:* What is the present status of 7% CP (crude protein) in relation to pasture intake? (U.S.A.) *Response:* The minimum CP level with respect to forage intake will depend on the type of animal in question. With dairy cows this level may be 12%, but with beef and sheep fed mature feeds, 7% is a good working figure to use. Twenty-five percent of all the tropical forages that have been analyzed contain less than 7% CP, and this in itself may limit intake by the grazing animal, indicating the widespread occurrence of protein deficiency, even at the 7% level.

2. *Question:* Does high availability of pasture reduce intake? (Australia) *Response:* No, not unless there is an accompanying decline in pasture quality. However, the leaf-to-stem ratio may be an important consideration with differences in the amount of pasture on offer. One may speculate that high yields of stems could reduce intake. However, there is no evidence in the literature to support this hypothesis, and there is a need to design experiments to evaluate the effects of stem yield on leaf consumption.

3. *Question:* Is there much progress with the development of models to predict intake? (U.S.A.) *Response:* Unfortunately, there remain many unidentified and/or poorly understood factors that influence forage consumption. Consequently, predictions made by models will be of very little value.

4. *Question:* While cutting techniques may not be suitable for intake measurements under Australian conditions, are not the intensive grazing systems of Europe more appropriately evaluated by cutting techniques? (Netherlands) *Response:* Cutting techniques are suitable for intake measurements with strip-grazing and short-term management practices. In fact, under these conditions cutting techniques are as good as some other methods of measuring intake. However, when estimating intake by the cutting technique, the effects of plant decay must be eliminated. The effects of plant decay on subsequent intake are a disadvantage of cutting techniques. However, cutting techniques are unsuitable for measuring intake under grazing systems where only a small portion of the pasture is used, and any estimates of intake are going to be severely affected by both pasture growth and decay.

5. *Question:* Since grazing animals select leaves above stems, what characteristics of the leaf are of importance to the grazing animal, and what should we be measuring? (UK) *Response:* Ideally we should measure all those factors that affect intake, including all the elements, protein, digestibility, etc. Unfortunately, there is no single analysis that plant breeders are always looking for.

6. *Question:* What is the relative importance of grass with a high leaf content compared to the legume component in the sward? (Colombia) *Response:* Legumes are more important. Apart from high protein and other nutrient levels, intakes of legumes are also higher than those of grasses. Ruminal digestion of legumes is faster than that of grasses, and passage rates increase with a concomitant increase in intake. As a result, there is usually a positive relationship between the proportion of legumes in the sward and animal performance.

7. *Question:* Is bite size an alternative for estimating pasture intake compared with the more conventional methods of measurement? (UK) *Response:* Both bite size and number of bites are required to measure intake. However, the choice of method will depend on the nature of the problem. The object is to identify those factors that limit intake, and we must therefore select those techniques that best fit the objectives. The more recent approaches to intake measurements, including the use of rare earth markers, are far from perfect. In fact, many of the older techniques might be superior.

Use of Genetic Resources for Improvement of Forage Species

J.R. HARLAN
Professor of Plant Genetics, University of Illinois,
Urbana, Ill., U.S.A.

Summary

The use of genetic resources of forage species poses some special problems because of the number of species involved, the inadequacy of our knowledge about them, the fact that they are often used in combination and should be evaluated in terms of livestock product, and so on. Collections should be as comprehensive as possible at the outset of improvement programs; otherwise, there may be much use of inferior material and much waste of time and effort.

Most of the forage species in use come from (1) Europe (excluding Mediterranean zones), (2) Mediterranean Basin and Near East, (3) African savannas (mostly tropical grasses), and (4) tropical America (mostly tropical legumes). They tend to be members of grazing subclimax or plants with weedy or colonizer habits.

Utilization strategies depend on stage of development of research programs: (a) useful species are introduced unintentionally; (b) a range of materials is assembled, tested, and selected entries used; (c) a breeding program is developed using extensive hybridization to identify elite parents; (d) a full-scale team is developed with pathologists, entomologists, biochemists, animal scientists, soil scientists, and so on. Such teams are unlikely to develop in the Third World without international cooperation.

International cooperation should be encouraged, and the logical coordinator at the present time is the International Board for Plant Genetic Resources.

KEY WORDS: germ plasm, collections, centers of origin, plant breeding, biosystematics, international cooperation.

INTRODUCTION

The improvement of most forage species poses some special problems in comparison with the improvement of other crops. Research on forage species is relatively new, and we lack an appropriate inventory of information and experience for efficient use of genetic resources. A forage scientist must usually deal with several species simultaneously, thereby diluting his efforts and effectiveness. Many of the species are wild or only partially domesticated, and we do not know the ecological, geographic, or even taxonomic limits to the available gene pools. Seed production of many species is poor and erratic, and establishment ecology is poorly understood. Collections of forage plants are often difficult and expensive to maintain, and genetic erosion of collected material is common. Many collections are small and inadequate. Evaluation procedures are difficult, slow, and expensive. The ultimate goal is an increase in amount or dependability of livestock yield so that forage evaluation is indirect. Forage species must often be evaluated as mixtures of two or more species. Plant-breeding techniques tend to be difficult and tedious with a slow generation turnover.

All of these complaints, and many more, are familiar to every forage-crop worker. There is no need to catalogue them here, except that they pose limitations to the use of genetic resources. There is a more positive side to forage improvement. Experience in Europe, New Zealand, Australia, the U.S.A., Canada, and elsewhere has indicated that the discipline of forage-crop improvement can mature with time and effort. A body of theory and information can be built up that makes our work more efficient. We have been able to learn a good deal about the ryegrasses (*Lolium* spp.) and white clover (*Trifolium repens*). We have generated a considerable inventory of information about alfalfa (*Medicago sativa*), orchardgrass (*Dactylis glomerata*), tall fescue (*Festuca arundinacea*), smooth bromegrass (*Bromus inermis*), and bermudagrass (*Cynodon dactylon*). Some of the annual forages such as oats (*Avena sativa*), rye (*Secale cereale*), sorghum (*Sorghum bicolor*), and pearl millet (*Pennisetum americanum*) are fairly well defined by now. We have even learned a good deal about native North American species since my introduction to them some 40 years ago.

But information and theory concerning tropical forages are only beginning to emerge. We are still exploring for species and even genera that might be useful. We may not even have identified the most useful species for some regions. It is difficult to collect if we do not know what to collect. It will take time and a great deal more effort than has been allocated to date for a body of information and theory about tropical forages to mature. We have a great deal to learn and, perhaps, not much time to learn it in. Even so, some principles and strategies may be proposed at this stage.

COMPREHENSIVE COLLECTIONS

Perhaps the most obvious principle is that *you cannot use what you do not have.* Obvious as it is, program after program is restricted in effectiveness for lack of adequate or appropriate genetic resources. We still have not done the work of germ plasm assembly that we should have done long ago. It is fruitless to engage in forage improvement programs with inadequate germ plasm collections. I know, because I've tried it. I have spent as much as 10 years making modest progress only to acquire new accessions several times better than my best productions. I have also worked with some species for years before I found that other species were more appropriate. It pays to assemble a full range of variation of the species or species complex at the start of the program. The first step of any program to use genetic resources is to assemble the material.

It is critically important for the plant breeder to get into the field and observe variation where the species is native and to sample the whole range of its distribution and variation. Much of it might not be adapted to his area and be of no use to him, but some of it might be useful elsewhere. In any case, he needs to know the extent of variation available. What applies to the plant breeder applies as well to the plant pathologist, the entomologist, the nematologist, and so on. It is too much to expect of an entomologist that he should cope with outbreaks of insects when he does not know what predators or parasites might control them. The pathologist must be aware of the natural defenses deployed by a species in order to be effective in dealing with epidemics in artificially established forages. To stay home and work with a narrow endemic range of germ plasm will, in the long run, be self-defeating. One must explore the outer ranges of usable materials in order to define the optimum range.

For example, temperate, continental climates provide an eternal problem of adaptation. Southern sources introduce vigor, productivity, and more fall growth at the expense of winter hardiness. Northern sources are hardy, tend to become dormant in the fall, and are less productive. The plant breeder is constantly attempting to fine-tune his productions to achieve an optimum compromise.

But, in order to do the tuning, an adequate sample of germ plasm must be available, ranging from material too unproductive to be used directly to material too cold-tender to be used as such. The range of variation should exceed adaptation to the site of the program.

For example, Coastal bermudagrass is surely one of the most important contributions in the history of U.S. forage-crop improvement. Glenn Burton selected it as a chance nursery hybrid between a local naturalized, rather worthless "wire-grass" type and an introduction from South Africa (Burton 1947). The South African accession was later lost, but it can be reconstructed with some accuracy by extrapolation. It must have been coarse and robust with widely spaced rather erect stems with cauline leaves forming an open turf. Rhizomes were few, long, and straight. The leaves had a bluish cast, and the plant was less cold-tolerant than Coastal. This description fits in every respect the botanical variety *Cynodon dactylon* var. elegans Rendle (Harlan and deWet 1969). Neither parent was of any use to the southeastern U.S. — the one lacked productivity and the other lacked cold hardiness — but the combination was outstanding (Harlan 1970).

Later, Burton produced Coastcross-1 with Coastal as one parent and an accession of *C. nlemfuensis* var. Robustus from Kenya as the other (Clayton and Harlan 1970, Harlan 1970). The latter has neither rhizomes nor any frost tolerance at all, but it does have more tender growth and forage of higher digestibility. Coastcross-1 has considerably higher DM digestibility than Coastal (Burton et al. 1967). The point is that if we confine our collections to materials that are adapted and perform well, we are putting the breeding program in a straitjacket.

Another distinguished forage breeder and cytogeneticist, Douglas Dewey, has had remarkable success orchestrating species among the Triticinae. One of the most

useful parents in his program is *Agropyron repens*, a species that is likely to arouse more passion than admiration. But, in suitable combinations, the creeping habit can be contained or even eliminated. The great vigor, aggressiveness, and excellent forage quality can be exploited (Dewey 1980). Dewey has collected extensively in Eurasia and North America and has assembled large collections representing most of the perennial species in the tribe. This is of fundamental importance. If you don't have it, you can't use it. Accessions that have little or no merit in themselves can make outstanding parents.

It is necessary to try many, many combinations before patterns become apparent. Eventually, after a species complex has been studied enough, it may become clear which combinations are best for specified conditions and which combinations are hopeless. To reach this state of enlightenment, however, a lot of travel, a lot of field work, a lot of collecting, and a great deal of crossing and evaluation must be done. It is impossible to gain much understanding without adequate field work and adequate collections of materials.

THEORY AND PRINCIPLES

Because of the special problems with forage species, because we must deal with so many species, because we know so little about many of them, because our support is so minimal for the tasks to be done, the most urgent need, it seems to me, is to develop a body of theory and information that would help us to focus on the most likely regions of the world for collection and on the most likely taxa for useful materials. The world is a big place, and the grass and legume families are vast. Our resources are always limited; we cannot afford to approach the utilization of genetic resources in randomized fashion. Are there geographic "centers" where a search for useful germ plasm would be more rewarding than in other parts of the world? Are there groups of grasses and legumes more likely to provide useful materials than others? We need to address these quesions seriously (Harlan 1981).

We are not altogether without experience in these matters, even in the tropics. Our experience may not yet be broad enough, and we may find many exceptions to the general principles, but as of now the most important forage resources have come from (1) Europe (excepting zone of Mediterranean climate), (2) the Mediterranean Basin and Near East, (3) African savannas for tropical grasses, and (4) tropical America for tropical legumes (Table 1). We need to look at these areas in a little detail to see if we can understand why these regions have contributed so much more than other regions.

Europe (Except Zone of Mediterranean Climate)

The most widely grown forages from this region are the annual and perennial ryegrasses, white and red clover, alfalfa, sweetclover (*Melilotus* spp.), orchardgrass, tall fescue, timothy (*Phleum pratense*), rye, oats, lupine (*Lupinus* spp.), birdsfoot trefoil (*Lotus corniculatus*), sainfoin (*Onobrychis*), and so on. These are, perhaps, the most

Table 1. Geographic 'centers' of forage species.

(a) EUROPE (except zone of Mediterranean climate)

Lolium, Festuca, Bromus, Dactylis, Agropyron, Cynodon, Phalaris, Phleum, Trifolium, Medicago, Melilotus, Lotus, Onobrychis, Vicia, Lupinus

(b) MEDITERRANEAN BASIN AND NEAR EAST (winter rainfall)

Lolium, Festuca, Bromus, Dactylis, Agropyron, Elymus, Cynodon, Phalaris, Trifolium, Medicago, Melilotus, Onobrychis, Vicia

(c) AFRICAN SAVANNAS

Cynodon, Setaria, Cenchrus, Chloris, Eragrostis, Brachiaria, Digitaria, Pennisetum, Sorghum, Urochloa, Hyparrhenia, Panicum, Vigna, Lablab, Lotononis, Clitoria, Glycine, Trifolium

(d) TROPICAL AMERICA

Paspalum, Tripsacum, Axonopus, Eriochloa, Stylosanthes, Macroptilium, Aeschynomene, Desmodium, Centrosema, Phaseolus, Arachis, Desmanthus, Leucaena

familiar of all forages to grasslanders. They have been studied longer and have been characterized in more detail than other forage crops. Some, or perhaps most, of these are probably naturalized and not really native to Europe, having spread out of the Near East with early Neolithic farmers as they cleared the forests for their crops and livestock. But that was some 6,000 to 7,000 years ago; there has been ample time to evolve races adapted to temperate summer rainfall conditions, and these races have been widely introduced to regions of similar climate.

In these regions—e.g., North America, New Zealand—the forests were cleared for farming and especially for pasturage at a very late date, and there had been no opportunity for coevolution of native forages and livestock. We do not know if the 6,000- to 7,000-year history of European agriculture and stock rearing was long enough to evolve superior forages or if the species were better to begin with, having evolved earlier under heavy grazing in western Asia. At any rate, Europe has served as a center of origin for these crops even if some of the species originated elsewhere.

Mediterranean Basin and Near East (Winter Rainfall Zone)

Here we encounter the winter annual species of *Trifolium, Medicago, Secale,* and *Phalaris,* as well as *Vicia* and *Avena.* The various burmedics, snailmedics, berseem (*Trifolium alexandrinum*), Persian clover (*T. resupinatum*), crimson clover (*T. incarnatum*), subterranean clover (*T. subterraneum*), lupines, and vetches (*Vicia* spp.) are important contributions from the region. Among the perennials are alfalfa, sainfoin, white clover, wheatgrasses (*Agropyron* spp.), fescues (*Festuca* spp.), bromegrasses (*Bromus* spp.), and bermudagrass. These plants, with their weedy companions, have become established around the world wherever temperate winter rain–summer drought climates are found. The region is clearly a center of origin for this class of materials.

One must ask why these plants are more productive, more persistent under grazing, and of better quality than their counterparts in California, Chile, Australia, or South Africa. One might suggest that these forages are the end product of a long history of coevolution with grazing ruminants and that the intensity of interaction was much greater over a longer period of time than in other regions of similar climate. Australian species were never subjected to grazing pressure by ruminants at all, and the pressures on the floras of California, Chile, and the southern tip of Africa were probably minimal. The great abundance of annual legumes may be associated with the prevalence of limestone and calcareous soils. The high percentage of annuals in the flora suggests that the winter rainfall pattern may have been more extensive and of longer duration in this region than in the others.

African Savannas

For tropical forages, the African savannas have contributed guineagrass (*Panicum maximum*), elephantgrass (*Pennisetum purpureum*), pearl millet, kikuyugrass (*P.*

clandestinum), pangolagrass (*Digitaria decumbens*), other digitgrasses, rhodesgrass (*Chloris gayana*), bermudagrass, stargrasses (*Cynodon* spp.), buffelgrass (*Cenchrus ciliaris*), molassesgrass (*Melinis minutiflora*), lovegrasses (*Eragrostis* spp.), sorghum, jaraguagrass (*Hyparrhenia rufa*), and species of *Setaria, Brachiaria,* and *Urochloa.* The list is impressive, and some of these are widely used around the world. Some are widespread through the savannas of Africa and some are more restricted; indeed, pangolagrass is narrowly endemic. East and South Africa have contributed relatively more than the sub-Saharan Sahel and Guinea savannas. Africa has some legumes that have not been used much up to now but might well be exploited more in the future. These include: *Lotononis bainesii, Lablab purpureum, Glycine wightii, Trifolium semipilosum,* and *Vigna unguiculata.* The latter might be especially useful because of its high tolerance to aluminum and other toxic metals (Cock and Howeler 1978).

Diffuse though they are, the African savannas contain a wealth of useful tropical forages and can qualify as a "center," or more precisely as a "noncenter" as I have used the term elsewhere (Harlan 1971). Each species has its own range of distribution, of course; some are confined to southern Africa, some to the East African highlands, some to dry savannas, and others to the wetter and derived savannas. But why are African grasses so much more aggressive, persistent, and productive than savanna grasses elsewhere? Again, one could suggest that these are end products of a long coevolution and intense interaction with grazing ruminants. This also took place at higher levels of soil fertility than in the llanos of South America or the savannas of tropical Australia.

Tropical America

The contributions of this region are dominated by legumes. Here we find the stylos (*Stylosanthes* spp.), centros (*Centrosema* spp.), siratro (*Macroptilium atropurpureum*), the desmodiums (*Desmodium* spp.), and species of *Calopogonium, Desmanthus, Aeschynomene, Leucaena, Arachis,* and *Phaseolus.* No doubt there are many more since we have only just begun to sort through them. Many are abundant and weedy and tend to increase with disturbance of the habitat, at least up to a point. Some are widespread and others restricted in distribution. Turkey is the only other place in the world where I have seen such concentrations of herbaceous legumes as I have seen in parts of South America.

What are the reasons for such an abundance of potential forage legumes? This time, grazing pressure would seem to have nothing to do with it. The controlling selection pressure is more likely to be the soils of low fertility and often high toxicity. Plants that could fix nitrogen symbiotically with microorganisms could at least survive if not thrive. It is, perhaps, not by accident that grasses with symbionts that fix nitrogen were first identified in Brazil (Döbereiner et al. 1972).

Tropical, and more especially subtropical, South America has contributed grasses that are used, e.g, bahiagrass (*Paspalum notatum*), dallisgrass (*P. dilatatum*), rescuegrass (*Bromus catharticus*), and the tropical guatemalagrass (*Tripsacum andersonii*). Probably there are more that have not been tested, but Central and South America appear to qualify as another noncenter of forage plants with tropical adaptation (Harlan 1981).

Exceptions

Plant geography suggests to us, then, two centers and two noncenters toward which we should direct our attention. There is much more exploration and collecting to be done in all of them, but the tropical noncenters have been explored the least. On the other hand, we must keep in mind that important materials have come to us from other parts of the world. The steppes of Central Asia have yielded wheatgrasses and races of alfalfa of great value. India has contributed alyce clover (*Alysicarpus*), and the *Pueraria* spp. are native to East Asia and Oceania. The North American grasslands have provided a number of species of value to the region, although they seem to have little use outside of North America. Perhaps we have had enough experience to know more or less where to look for materials to fill a particular niche.

Ecology and Taxonomy

Knowing where to look is only a part of our problem; we must also know what to look for. In general, species and races of a grassland climax flora are marginally useful as cultivated forages. They are difficult to establish and tend to decrease under use. Grasses and legumes of a grazing subclimax or plants with weedy and colonizer tendencies are more satisfactory. They are more easily established, are more persistent under use, and are usually better seed producers. No doubt there are exceptions to this guideline as there are to our geographic sources, but the principle may make our collecting and germ plasm assembly more efficient. By and large we are looking for plants that have evolved to fill a niche under grazing by ruminants, plants that persist when repeatedly defoliated by grazing animals.

Those of you who have collected forage plants in Asia and parts of Africa are all too familiar with a landscape in which the native grasses and legumes have been grazed down to the dust and the only seed stalks or pods available for collecting are found in the protection of thorn bushes. These few seeds are the ones most worth collecting, even though it means scratches on hands and arms. Plants available but not grazed under these conditions are antipastorals and of no value. They are unpalatable and perhaps toxic. Plants that survive heavy use are the ones we are looking for.

Taxonomically, our most used forages are grasses and legumes, although species of other families may also be useful. Chenopodiaceae, in particular, might be explored more than they have been. Some of the most useful genera are shown in Table 1. Genera such as *Panicum, Festuca, Bromus, Setaria, Pennisetum, Eragrostis, Digitaria,* and *Paspalum* are very large and contain a motley assortment of species. Within the same genus one may find a few valuable forages, many that have little or no exploitable value, and some that are antipastoral or even

noxious. But, if we find a useful species, it is important not only to assemble a collection of it but also to collect those species that can be crossed with it. The biosystematics of the species complex to which a forage belongs must be worked out in order to exploit the germ plasm efficiently.

UTILIZATION

The simplest and most naive use of genetic resources takes place when species insinuate themselves into a region without conscious human effort. Not only wild oats but a substantial sample of the Mediterranean herbaceous flora were introduced to California without trying. Subterranean clover came to Australia uninvited and had spread widely and developed considerable variation before its value was discovered. Although we have legends about who introduced bermudagrass and johnsongrass into the U.S.A., they are quite capable of arriving without deliberate action and, I am convinced, have done so many times. Whether or not Kentucky bluegrass (*Poa pratensis*) is native to the U.S., it has insinuated itself over large regions where it was never native. Guineagrass, despite its shy seeding habits, has found its way to the wetter tropics all over the world, often accompanied by elephantgrass and wild races of *Sorghum bicolor*. A weedy leucaena has taken over parts of Hawaii, and in fact these islands have become a dumping ground for a remarkable array of grasses and some legumes from all continents of the world.

The next step in complexity—one that is often as naive —is to deliberately introduce a range of materials, evaluate them, and use those that appear to be the best. This is almost a mandatory first step for a region without a formal forage program or where one has just begun. This is the only way to get a feel for a species that has not been previously evaluated. Sometimes it is a feeble step, and poor materials may be used that would be difficult to replace with better sorts. Early use of leucaena concentrated on low-producing weedy types high in mimosine. Stylo pastures were devastated by disease in Colombia. Siratro was abused by rust in Australia, rhodesgrass was taken out of Texas by scale insects. King Ranch bluestem (*Bothriochloa ischaemum* var. songarica), after looking like a world-beater in south Texas, was roughly handled by rust. Lines that look good in a space-planted nursery often give a poor performance under pasture or range conditions.

We become more efficient in use of genetic resources when we enter practical breeding programs. Hybridization is the key that unlocks doors and reveals mysteries. The art and science of plant breeding consists, to a large extent, in finding parents whose offspring perform better than they do. Parents that resemble each other very closely are seldom satisfactory for the purpose; those that are too divergent may produce lethals, weak or sterile offspring. The problem is to seek the optimum level of divergence and the best combinations. The solution requires a great deal of experimental crossing and evaluation, and each case is unique. Sometimes it is appropriate to work at the species level; in other cases it is impossible to work at the species level. In any case, it is necessary to explore the limits of genetic reach in order to define how far is too far. With interspecific crosses it is important not to give up too soon. Early generation materials may look terrible and later generations provide cultivars that perform very well.

In Dewey's program in Utah, over 250 interspecific hybrid combinations have been made, some of which give an excellent performance (Dewey 1980). This appears to be an effective level of operation in the Triticinae. Useful interspecific crosses have been made in *Cynodon, Pennisetum, Bothriochloa, Lolium, Digitaria, Poa, Festuca, Paspalum,* and many others. Pangolagrass seems to be a natural interspecific hybrid, and guatemalagrass (*Tripsacum andersonii*) appears to be a natural hybrid involving *Zea.* Many of these hybrids can be reproduced asexually either vegetatively or by seed apomixis, and sterility is not a problem. When seed production is required for propagation, colchicine-induced polyploidy may result in adequate fertility. Where wide crosses are useful, it is imperative to work out the biosystematic relationships among species.

In other groups it may be that interspecific hybridization has no place at all. Each forage is a case of its own, but experience with enough crosses will eventually establish a pattern. Certain materials will be identified as elite either by performance or as parents. But if the collection does not include the elite, patterns may not be detectable. Breeding success depends a good deal on the quality of the genetic inventory.

We begin to approach full utilization of genetic resources when we have large, comprehensive collections available and the programs include not only plant breeders but cytogeneticists, plant pathologists, entomologists, nematologists, microbiologists, biochemists, soil scientists, agronomists, livestock specialists, and so on. Teams of this dimension are only occasionally assembled in developed countries, and are financially out of reach in most developing countries. International cooperation could make utilization of genetic resources much more efficient in the Third World and could be of considerable benefit to the developed countries as well.

INTERNATIONAL COOPERATION

The assembly of germ plasm and the maintenance of genetic collections would be made easier by international exchange. Once collecting teams have brought materials together and increased them, it seems inefficient not to share them with others who might find them useful. Our tasks of evaluating collections would be more efficient if we shared information generated in the process. If collections were shared there would be less loss and genetic erosion within existing collections. Even in the field work of collecting, much duplication of effort could be avoided by cooperation. I recall a situation some 10 years ago in Ethiopia. I was collecting sorghum and millets; Howard Gentry was collecting peas; G.A. Wiebe was collecting barley; Ladizinsky and Zohary of Israel were collecting oats; there was a team from USSR and another from

Japan. We were all chasing each other around the same roads (there weren't many roads). The Ethiopians thought they were being invaded, and they had reason to believe we were up to no good. A little cooperation and coordination would have been sensible. There are, in fact, so many compelling reasons for cooperation that one wonders why there has not been more of it. The International Board for Plant Genetic Resources could play a vital role in international coordination.

LITERATURE CITED

Burton, G.W. 1947. Breeding Bermuda grass for the southeastern United States. Am. Soc. Agron. J. 39:551–569.

Burton, G.W., R.H. Hart, and R.S. Lowrey. 1967. Improving forage quality in Bermudagrass by breeding. Crop Sci. 7:329–332.

Clayton, W.D., and J.R. Harlan. 1970. The genus *Cynodon* L.C. Rich in tropical Africa. Kew Bul. 24:185–189.

Cock, J.H., and R.H. Howeler. 1978. The ability of cassava to grow on poor soils. Am. Soc. Agron. Spec. Pub. Madison, Wis.

Dewey, D.R. 1980. Quackgrass—a species with unusual plant-breeding potential. Agron. Abs. 1980:53.

Döbereiner, J.M., M. Day, and P.J. Dart. 1972. Nitrogenase activity and oxygen sensitivity of the *Paspalum notatum –Azotobacter paspali* association. J. Gen. Microbiol. 71:103.

Harlan, J.R. 1970. *Cynodon* species for grazing and hay. Herb. Abs. 40(3):233–238.

Harlan, J.R. 1971. Agricultural origins: centers and noncenters. Science 174:468–474.

Harlan, J.R. 1981. The scope for collection and improvement of forage plants. *In* J.G. McIvor and R.A. Bray (eds.), Genetic resources of forage plants. Townsville.

Harlan, J.R., and J.M.J. deWet. 1969. Sources of variation in *Cynodon dactylon* (L.). Pers. Crop Sci. 9:774–778.

DISCUSSION

Responses to Questions

1. There is no apparent advantage to evaluating plant species for meat production by using coevolving animals. It makes little difference what species of animal graze the plants.

2. Breeding programs for controlled pastures and rangelands are very different. Breeding plants for rangelands is very difficult, and a program needs to evaluate each situation. Intensive breeding programs are probably restricted to cultivated pastures.

3. It would be helpful if more work were done in studying insect predators, diseases, and sources of resistance in the native habitats.

4. A meeting was held in Townsville in 1979 to address the problems of germ plasm–collection duplication and the need to make collection records, data, and materials available. Also emphasized was the need to identify desirable information and the value of distributing information about collected material. An international effort will take money and dedication to be successful.

5. It is difficult to get improved germ plasm back into its area of origin because of a lack of a long-standing commitment to forage programs in many countries.

Potential Productivity of Temperate and Tropical Grassland Systems

G.O. MOTT
Professor of Agronomy, University of Florida,
Gainesville, Fla., U.S.A.

Summary

Grasses of temperate regions of the world belong primarily to the subfamily *Festucoideae*, whereas tropical regions are populated by tribes of the *Panicoideae* subfamily. Temperate grasses have a C_3 biochemical pathway of photosynthesis in contrast to tropical grasses, which have a C_4 photosynthetic mechanism. Both tropical and temperate legumes are C_3 plants. Compatible mixtures of legumes and grasses have long been successful in many temperate regions, but combining C_4 grasses and C_3 legumes as mixtures in the tropics has been only moderately successful.

Maximum net primary productions of temperate and tropical grasslands are in the order of 25 and 80 tons/ha/yr, respectively. Thirty tons/ha/yr is considered an exceptionally high yield in the wet-dry tropics, and no more than 10 to 15 tons/ha/yr can be expected at 40° N latitude.

Rates of conversion of net primary production to animal yield in temperate and tropical grasslands are quite different. Although net primary production in tropical grasslands may be several times that of the most productive temperate grasslands, a much smaller proportion of the production may be recovered by the grazing animal. Animal yields/ha/yr on high-producing cultivated temperate and tropical pastures seldom exceed 1,000 kg. Failure of tropical pastures to give higher yields of meat and milk commensurate with their greater net primary production can be partially accounted for by much greater losses in the grazing process. Other factors are lower daily intake of energy above maintenance, which results in a much larger proportion of the energy consumed/ha being expended for maintenance than for performance.

It is concluded that more nearly maximum yields of animal product for any particular environmental circumstance will be attainable when management schemes are devised to match feed supply with demand and to recover maximum net primary production mutually beneficial to both pasture and animal. Other strategies include improving the quality of consumed forage by genetic manipulation, developing legume and grass cultivars that will coexist in mixtures, conserving forages, and supplying the mineral requirements of the grazing animals.

KEY WORDS: production, animal yield, nutritive value, voluntary intake, grasses, legumes.

INTRODUCTION

It is generally accepted that tropical grassland systems have lower animal-output potentials than temperate grassland systems. This paper examines a few components of these systems in an attempt to identify dominant factors that serve as constraints to their production potential.

Taxonomists have recognized wide morphological differences in the grass family by subdividing the family into two subfamilies—*Festucoideae* and *Panicoideae* (Hitchcock 1950). Within *Festucoideae* are to be found cultivated forages of temperate grasslands. Almost without exception grasses used in cultivated pastures in the tropics are from the *Panicoideae* subfamily (Dirven 1977). It now is well documented that these two subfamilies dominate the temperate and tropical grasslands.

Natural grasslands in temperate and tropical zones are highly diversified with respect to their botanical composition. This is in marked contrast to cultivated pastures, which frequently are monospecific or, at best, very simple mixtures. Single species, cultivars, or simple mixtures can be maintained only by man's intervention, since environmental pressures, including the effects of grazing by domestic animals, eventually result in a widely diversified association of species. In each of the temperate and tropical grasslands from 20 to 25 species of grasses are currently in use in cultivated pastures (Bula et al. 1977, Crowder 1977).

CHARACTERISTICS OF TROPICAL AND TEMPERATE GRASSES

Some tropical grasses have coarse stems and broad leaves, such as guineagrass (*Panicum maximum*) and napiergrass (*Pennisetum purpureum*); others, e.g., bermudagrasses (*Cynodon* spp.) and the digitgrasses (*Digitaria* spp.), are fine-stemmed and have much smaller leaves. The tiller density of temperate species is greater than that of the tropical grasses. For example, ryegrass (*Lolium*

This is a contribution of the Agronomy Department of the University of Florida, Gainesville.

perenne) has from 8,000 to 13,000 tillers/m^2 as compared with Pangola digitgrass (*Digitaria decumbens*), which has from 4,460 to 6,720 tillers/m^2 (Dirven 1977). The leaf:stem ratio is higher for temperate grasses. Leafiness and tiller density are important characteristics of grasses as they relate to their nutritive value, bulk density, and prehension rate under grazing and to rate of voluntary intake.

Of great importance in productivity of grasslands are the biochemical pathway of photosynthesis and the differences between grasses of *Festucoideae* (C$_3$, temperate grasses) and those of *Panicoideae* (C$_4$, tropical grasses) (Whiteman 1980). The rate of CO$_2$ fixation in isolated leaves of C$_4$ grasses is about twice that of C$_3$ grasses. Leaves of C$_3$ grasses are light-saturated at about one-third to one-half the light intensity of C$_4$ grasses. Present evidence suggests that some C$_4$ grasses are not saturated even in full sunlight. Ryegrass (*Lolium* spp.) appears to be light-saturated at about 30,000 lux; however, even light intensities greater than 60,000 lux did not saturate dallisgrass (*Paspalum dilatatum*) (Cooper and Tainton 1968). Photorespiration in C$_4$ plants is near zero during photosynthesis in the light, whereas C$_3$ plants have respiration values of 7–15 mg CO$_2$/dm^2/hour. The CO$_2$ compensation point for C$_4$ grasses is near zero in the light compared with that of C$_3$ plants of about 40 ppm (Ludlow and Wilson 1972).

CHARACTERISTICS OF TROPICAL AND TEMPERATE LEGUMES

All pasture legumes, temperate and tropical, are C$_3$ plants and show similar responses to light. However, optimum temperatures for tropical legumes are in the low 30°C range, whereas the range for temperate legumes is from the low to middle 20°C. In contrast the tropical grasses have optimum temperatures for growth of near 40°C (Whiteman 1980). The physiological differences between tropical grasses and legumes have important implications for legume-grass associations. Since their optima for light, temperature, and moisture differ, it is much more difficult to select compatible grasses and legumes in the tropics than among the temperate species where the responses to environmental factors are similar. The viny growth habit of several genera of tropical legumes (*Calopogonium*, *Centrosema*, some species of *Desmodium*, *Neonotonia*, *Macroptilium*, and *Pueraria*) may confer an advantage over C$_4$ tropical grasses in that they are able to climb to the top of the canopy. Devising defoliation strategies that will maintain the regrowth potential of viny legumes is very important.

AERIAL BIOMASS PRODUCTION

Maximum net primary production (Milner and Hughes 1970) of forages in temperate and tropical grasslands is in the order of 25 and 80 tons/ha/yr, respectively. These production levels have been attained under the most favorable conditions within the limits set by light energy and temperature regimes and under circumstances in which soil nutrients and water availability were not limiting. However, in the wet-dry tropics both available moisture and temperature may become limiting, thus reducing the maximum production to about 30 tons/ha/yr. In most temperate regions forages have less than 12 months to grow and frequently are limited to less than 6 months. For example, at about lat 40° N, maximum yields of dry matter appear to be in the order of 10 to 15 tons/ha/yr (Rhykerd et al. 1967, Greenhalgh 1975). In arid regions where moisture supply is minimal, annual primary production may be as low as 0.5 tons/ha.

IMPROVED CULTIVARS

New cultivars result from introduction of ecotypes from areas where a species is indigenous and from breeding programs designed to generate better-adapted and more productive genotypes. Crowder (1977) lists 25 tropical grasses and 14 tropical legumes that are at present the species most often used in cultivated pastures. Of these, 16 grasses and 12 legumes have been included in selection and breeding programs, and the number is increasing. The number of temperate grasses and legumes used in cultivated pastures is about the same as for tropicals, and most have been subjected to selection pressure in plant-improvement schemes. It should be noted that most tropical grasses now in use in cultivated pastures had their origin in Africa, whereas the majority of tropical legumes originated in tropical America. Compared with tropical species, temperate grasses and legumes have a much longer history of study and exploitation; in addition, the range of adaptation and production of many species greatly exceeds their areas of origin.

CONVERSION OF AERIAL BIOMASS TO ANIMAL YIELD

"Net primary production" is the term frequently ascribed to aerial biomass production of pasture. For the producer, net primary production is less important than forage yield—the amount of forage that can safely be recovered (consumed) by the grazing animal (Lucas 1962).

Forage Yield

In the best interest of the pasture as well as of the animal, not all net primary production can be harvested. However, the optimum amount of production that must remain in the pasture at the end of a grazing period may not be the same for the pasture and the animal. In natural grasslands and particularly in those where moisture stress frequently occurs, the optimum stocking rate for the pasture is probably less than the optimum stocking rate for the animals. Conversely, on highly productive cultivated pastures in both tropical and temperate regions, the optimum stocking rate for the animal is probably less than that for the pasture.

Grazing Pressure

Optimum grazing pressure is contingent upon the amount of aerial biomass required to provide animals with enough acceptable forage for ad-libitum intake. In both temperate and tropical cultivated pastures of high quality, animals are able to satisfy their appetites when pasture availability ranges between 1,200 and 1,600 kg/ha; this is equivalent to 4–6 kg/100 kg live weight/day (Vohnout and Jimenez 1975, Peterson et al. 1965, Mott 1981). The latter range of availability is about twice the consumption rate of the bovine on pasture. These forage allowances are slightly lower than those reported by Marsh (1977); his data indicate that values greater than 7.5 kg/100 kg live weight/day may be necessary for maximum intake on ryegrass–white clover pastures.

Intake of Digestible Energy

Cattle on temperate pastures usually give higher performance in live-weight gain or milk production than do cattle grazing only on tropical pastures (Minson 1980). This difference is due to the lower intake of digestible energy on tropical pastures. Although some *Panicoids* have as high a rate of digestible energy intake as some *Festucoids*, as a group tropical forages provide a lower rate of intake of digestible energy to grazing animals than do temperate species (Moore and Mott 1973, Minson 1980). A major step in solving the problem of low animal performance on tropical pastures will be attained when the intake of digestible energy is increased to a level of 2 to 3 times the maintenance requirement.

Rate of voluntary intake. Since gross energy values of tropical and temperate forage plants are similar, they obviously cannot explain differences in voluntary intake and animal production. Using sheep as a reference animal and expressing intake on a metabolic weight (W) basis (metabolic weight is the 0.75 power of body weight in kilograms), Moore and Mott (1973) reported a range of 30 to 94 g DM/day/W for temperate grasses (Table 1)

with high frequencies of consumption of 80 g/day/W or above. Tropical grasses had a range from 28 to 98 g/day/W, but very few exceeded 70 g/day/W. The data suggest that tropical grasses have consumption rates of 10 to 15 units lower than those of temperate grasses. This finding agrees with data reported by Minson (1980) for six tropical grasses (28 samples). Intakes were found to range from 47 to 68 g/day/W with a mean value of 56. Comparable evaluations of six temperate species (23 samples) revealed intakes ranging from 58 to 85 g/day/W with a mean value of 71. This difference of 15 g/day/W is very significant with respect to animal performance.

• Fiber content: Detailed studies during the past 20 years strongly suggest that fiber content may be largely responsible for quality differences between temperate and tropical grasses. The higher fiber content of tropical grasses reduces intake, lowers the digestible dry-matter percentage, and increases the time that ingesta remain in the reticulo-rumen when compared with temperate grasses (Minson 1980). A frequency distribution comparing crude-fiber contents (Table 2) of forage from Latin America (mostly tropical species) with forage from the U.S. and Canada (mostly temperate species) indicates a higher percentage of forages above 30% crude fiber among tropicals than among temperates.

• Sward structure: Since tiller density of temperate grasses is greater than that of many cultivated tropical grasses, bulk density of forage presented to grazing animals may be greater for temperate grasses, facilitating their prehension rate and increasing their bite size. Vertical distributions of tropical pasture swards are loosely packed compared with those of temperate swards, and bulk densities (kg/ha/cm) vary from 14 to 200 for tropical pastures, compared with a range of 160 to 410 kg/ha/cm for temperate swards. High sward bulk density, low stem

Table 1. Ranges in voluntary intake by sheep and nutrient digestibility of temperate and tropical grasses.†

Grasses	Intake g/day/W	Digestibility %
Temperate		
Lolium species	48–89	53–88
Festuca arundinacea	58–91	52–83
Dactylis glomerata	30–94	45–83
Bromus inermis	62–64	53–68
Phleum pratense	52–92	48–77
Tropical		
Paspalum spp.	41–62	38–65
Cynodon dactylon	46–88	45–71
Digitaria decumbens	40–73	42–72
Cenchrus ciliaris	29–45	35–70
Pennisetum purpureum	61–66	60–65
Brachiaria spp.	49–98	57–71
Pennisetum clandestinum	28–83	45–74
Panicum maximum	40–81	45–66
Setaria sphacelata	29–66	41–68

†Adapted from J. E. Moore and G. O. Mott (1973).

Table 2. Crude-fiber and crude-protein comparisons between Latin American (LA) and United States-Canadian (US-Can) forages (DM basis), showing frequency distributions of means (%).†

Crude fiber		
	No. of entries	
% in	LA	US-Can
Dry matter	1450	1652
	Frequency %	
0–20	10	20
21–30	36	39
31–40	47	37
40	7	4

Crude protein		
	No. of entries	
% in	LA	US-Can
Dry matter	1993	1729
	Frequency %	
0–7	25	25
8–10	24	21
11–15	22	22
15	29	32

†Summarized from McDowell et al. 1977.

content and a high leaf:height ratio greatly influence bite size and prehension rate (Stobbs 1975). It is quite understandable that intake of digestible energy and animal performance may be greater for temperate than for tropical grasslands.

• Protein: Forage intake also may be reduced if forage is deficient in protein. The frequency distribution for crude protein appears to be very similar for Latin American and U.S.-Canadian forages (Table 2). However, crude-protein contents of temperate and tropical grasses studies by Minson (1980) indicated about 22% of all observations were below 6% protein for tropical species, but only 6% of all observations of temperate species fell below this level. Dry-matter intake and animal-performance increase show a linear relationship with crude-protein content up to some maximum level at which protein is no longer limiting. The protein level of diets may be increased by including a legume in the pasture. Protein contents of legumes, both temperate and tropical, have a range of 15% to 20%, and in some cases the value is even higher. A small percentage of legumes in swards has the effect of supplying nitrogen to the associated grass and also increasing levels of protein in the animals' diet.

• Minerals: Very few data are available documenting the primary effects of minerals—deficiencies and excesses—upon the rate of voluntary intake. The indirect evidence indicating their influence upon animal performance is considerable. If we use mineral requirements of the bovine as indicated by the National Research Council of the National Academy of Sciences as a basis for analysis, it is clear that deficiencies may be very frequent. A recent survey by Fick et al. (1978) of mineral concentrations in Latin American forages (Table 3) strongly indicates that the incidence of deficiencies in phosphorus, sodium, zinc, cobalt, and copper may be high. In addition, the frequency of calcium and magnesium deficiencies may be greater than 30%. Deficiency of any one or more of these elements may result in a reduction in rate of intake and a decrease in efficiency of forage utilization. The most widespread mineral deficiency in the Latin American tropics is that of phosphorus. Calcium deficiency is much less frequent. Copper, cobalt, and zinc

deficiencies are common, but much more research is needed to delineate these areas.

Digestibility. Tropical forages are usually somewhat lower in digestibility than temperate forages. Most of this difference may be attributed to differences in the rate and extent of digestion of the cell wall. Minson (1980) states that in young temperate pastures with a low fiber content, only 20% of the daily intake of dry matter may appear in the feces (80% dry-matter digestibility), whereas with mature tropical pastures as much as 60% of the intake may be lost in the feces (40% dry-matter digestibility). Digestibility of temperate forages ranges from about 45% to 88% (Table 1), with the highest frequencies between 55% and 80%; with regard to tropical forage the range extended from 35% to 74%, with the highest frequencies occurring between 45% and 65% dry-matter digestibility. The average difference in dry-matter digestibility is in the order of 10% to 15%.

The two most important factors determining both digestibility and the intake of digestible dry matter in temperate and tropical grasses and legumes are the genetic composition of the species and their stage of growth. Each of these factors is to a degree under the control of man.

ANIMAL PRODUCTION

Potential productivity of temperate and tropical grasslands ultimately is measured in terms of live-weight gain (meat), milk, and wool production. Since these products/ha are functions of yield/animal and the number of animals that the pasture will support, we may write:

$$\text{Yield/animal} \times \text{animals/ha} = \text{animal yield/ha} \quad (1)$$

Yield/animal is determined primarily by intake of digestible energy. The number of animals that the pasture will support is a function of net primary production, the portion of that production that is accessible and acceptable to the grazing animal, and the consumption/animal.

It follows that:

$$\text{Consumption/animal} \times \text{animals/ha} = \text{forage yield/ha} \quad (2)$$

Table 3. Mineral concentrations of Latin American Forages, including tropical and temperate grasses and legumes.†

Mineral	No. of entries	Deficient concentrations††	Percent of entries Deficient	Percent of entries Adequate
Calcium	1123	< 0.30%	31	69
Potassium	198	< 0.80%	15	85
Sodium	146	< 0.10%	60	40
Phosphorus	1129	< 0.30%	86	14
Magnesium	290	< 0.20%	35	65
Cobalt	140	< .10ppm	43	57
Copper	236	< 10ppm	47	53
Iron	256	< 100ppm	24	76
Manganese	293	< 40ppm	21	79
Zinc	177	< 50ppm	75	25

†Adapted from Fick et al. 1978.

††Concentrations expressed on dry-matter basis. Approximate requirements are those published by National Research Council of the National Academy of Sciences, Washington, D.C. 1976, for the average bovine.

Forage yield/ha then represents that portion of the net primary production that is recovered from the pasture (Mott 1981).

As to yield of animal product, meat and milk are secondary products of grassland systems, yet they are the products that are marketed. The quantity of energy utilized for maintenance is the processing cost for converting forage into animal product. If the quality of forage is sufficient only for maintenance, then yield/animal will be zero and animal yield/ha will also be zero. When translated into animal yield, the proportion of digestible energy expended for maintenance usually is greater in tropical forages than in temperate species. If forage digestibility and rate of intake have lower values in tropical than in temperate forages, intake of digestible energy will obviously be lower. In increasing animal yield, the critical point is to increase the intake of digestible energy to as high a level as possible above maintenance requirements so that energy intake for performance will be maximized (Table 4). A 300-kg growing-finishing steer with a daily live-weight increase of 1.0 kg/day divides its intake of net energy about 56% for maintenance and 44% for gain. The same size steer with a gain of only 0.25 kg/day utilizes only 16% for gain in live-weight, and 84% is allocated to maintenance. To produce the same amount of animal yield would require 4 steers at 0.25 kg/day or one steer at 1.0 kg/day, but the net energy expended for maintenance would be nearly 4 times as great for the 4 steers.

A similar calculation for the lactating cow indicates that only 16% of the net energy intake is utilized for milk yield by a 400-kg cow producing milk at 2 kg/day. The remainder (84%) is expended for maintenance. At a milk production level of 10 kg/day, net energy is divided equally for maintenance and milk. Growing-finishing steers gaining about 1 kg/day or lactating cows producing about 10 kg/day of milk represent about the maximum yield for the highest quality of tropical pastures.

STRATEGIES FOR INCREASING YIELD / ANIMAL

Matching Feed Demand with Supply

In grazed pastures, intake may be restricted by availability of acceptable forage. In temperate pastures and many tropical pastures, availability of forage must be at least double maximum consumption rates. This level of availability will assure that intake will not be limited by quantity of feed and that yield of animal product will be at a maximum, given the constraints imposed by forage quality. It should be noted that under circumstances of ad-libitum forage availability, quality of pasture (feed) determines intake and animal yield.

Improving Quality of Consumed Forage

Quality of available forage can be improved by breeding and selecting grasses and legumes of high digestibility and rates of intake. Development of in-vitro simulated digestion procedures has provided an additional tool for estimating forage quality in the genetic improvement of new cultivars (Barnes and Marten 1979).

Legumes have long represented an important component of temperate pastures; they contribute substantially to the relatively high quality of temperate grassland systems. Legumes benefit the sward through added symbiotic nitrogen and also by providing a high protein component to the animal's diet. In tropical, semiarid, and arid grasslands, legumes either are absent or represent a very small component of available feed. Exploration for legumes adapted to these grasslands has only begun. That some species will be successful is very encouraging, since the protein in a grass diet is frequently below the critical level of about 7%. The addition of as little as 10% legume in the diet may lead to large increases in intake and animal performance (Minson 1980).

Forage quality of both temperate and tropical species is also greatly influenced by the physiological age of the forage. Systems of grazing management that provide the grazing animal with maximum leaf of high digestibility usually will result in higher rates of intake and animal performance.

Table 4. Percentage of net energy requirements for maintenance and performance at several levels of animal yield.†

Growing-finishing steer (300 kg liveweight)		
Daily gain, g/day	Net energy Maintenance %	Gain %
0	100	0
250	84	16
500	72	28
750	63	37
1000	56	44
1500	46	54
Lactating Dairy Cow (400 kg liveweight, 3.5% fat)		
Milk yield, 3.5% fat kg/day	Net energy Maintenance %	Milk %
0	100	0
2	84	16
5	67	33
10	51	49
15	41	59
20	34	66
25	29	71

†Adapted from Nutrient Requirements of Beef Cattle No. 4, Fifth Revised Edition 1976, and Nutrient Requirements of Dairy Cattle No. 3, Fifth Revised Edition 1978. National Research Council, National Academy of Sciences.

Increasing Intake of Digestible Energy

Selection of species and cultivars that have a high leaf:stem ratio and high bulk-density swards facilitate the gathering of herbage by the grazing animal and thus increase the intake of digestible energy.

Conserving forage as hay or silage during periods when there is surplus of forage may prevent overaging and a reduction in nutritive value. The discovery of new methods for accelerating the loss of water in harvested forages, and new methods of processing, deserve intensive investigation.

Concentrate feeds offer about the only means of increasing the daily gains of growing-finishing animals to 1.5 kg/day or more and milk yield of lactating cows to 25 kg/day. In the tropics such energy sources as waste bananas (Vohnout and Jimenez 1975), molasses, roots of cassava, and yams need further investigation as possibilities for increasing the intake of digestible energy.

STRATEGIES FOR INCREASING ANIMAL YIELD / HA

Since animal yield/ha is a function of yield/animal and carrying capacity of the pasture (Eq. 1), in addition to increasing yield/animal, factors that increase forage yield/ha (Eq. 2) may also provide an opportunity for increasing animal yield/ha. Two general strategies for increasing carrying capacity of pasture swards are (1) increasing the net primary production and (2) increasing the percentage of net primary production that is recovered by grazing animals.

Net Primary Production

Net primary production may be increased by selecting species and cultivars with greater production potential for a given set of environmental circumstances. Detecting nutrient deficiencies and timely application of nutrients in appropriate quantities are compelling provisions for their efficient utilization (Brougham 1977). Net primary production also may be increased by the introduction of a legume into swards to supply much-needed nitrogen to the system.

Utilization

In many grassland ecosystems less than one-half the net primary production is utilized each year. This is especially true in tropical grasslands. Unless forage-management schemes are developed to maintain a uniform feed supply, periods of surplus and of scarcity are certain to occur in cycles each year. Conservation of forages, feeding of concentrates, or preservation of corn as silage are some strategies for attaining efficient use of seasonally produced pasture in temperate grasslands. Comparable strategies in many tropical grassland areas have not yet been developed but are badly needed.

LITERATURE CITED

Barnes, R.F, and G.C. Marten. 1979. Recent developments in predicting forage quality. J. Anim. Sci. 48:1554–1560.

Brougham, R.W. 1977. Maximizing animal production from temperate grassland. In Proc. Int. Mtg. on Anim. Prod. from Temp. Grassl., Dublin, 140–146.

Bula, R.J., V.L. Lechtenberg, and D.A. Holt. 1977. Potential of temperate zone cultivated forages and pastures for ruminant animal production. Winrock Rpt. 7–28. Winrock Int. Livest. Res. and Train. Cntr. Morrilton, Ark.

Cooper, J.P., and N.M. Tainton. 1968. Light and temperature requirements for the growth of tropical and temperate grasses. Herb. Abs. 38:167–176.

Crowder, L.Y. 1977. Potential of tropical zone cultivated forages for ruminant animal production. Winrock Rpt., 49–78. Winrock Int. Livest. Res. and Train. Cntr. Morrilton, Ark.

Dirven, J.G.P. 1977. Beef and milk production from cultivated tropical pastures. A comparison with temperate pastures. Stikstof 20:2–14.

Fick, K.R., L.R. McDowell, and R.H. Houser. 1978. Current status of mineral research in Latin America. In J.H. Conrad and L.R. McDowell (eds.) Proc. Latin Am. Symp. on Min. Nutr. Res. with Graz. Rumin., 149–162. Univ. of Fla., IFAS, Gainesville.

Greenhalgh, J.F.D. 1975. Factors limiting animal production from grazed pasture. J. Brit. Grassl. Soc. 30:153–160.

Hitchcock, A.S. 1950. Manual of the grasses of the United States, 2d ed., revised by A. Chase. U.S. Dept. Agric. Misc. Pub. 200.

Lucas, H.L. 1962. Determination of forage yield and quality from animal responses. In Range research methods. U.S. Dept. Agric. Misc. Pub. 940, 43–54.

Ludlow, M.M., and G.L. Wilson. 1972. Photosynthesis of tropical pasture plants. IV. Basis and consequences of differences between grasses and legumes. Austral. J. Biol. Sci. 25:1133.

McDowell, L.R., J.H. Conrad, J. Thomas, L.E. Harris, and K.R. Fick. 1977. Nutritional composition of Latin American forages. Trop. Anim. Prod. 2:273–279.

Marsh, R. 1977. How much pasture should be offered to beef cattle? Proc. Ruakura Fmrs. Conf., Hamilton, N.Z., 53–55.

Milner, C., and R.E. Hughes. 1970. Methods for the measurement of the primary production of grassland. IBP Handb. 6, Blackwell Sci. Pubs., Oxford.

Minson, D.J. 1980. Nutritional differences between tropical and temperate pastures. In F.H.W. Morley (ed.) Grazing animals, 143–157. Elsevier Sci. Pub. Co., Amsterdam.

Moore, J.E., and G.O. Mott. 1973. Structural inhibitors of quality in tropical grasses. In A.G. Matches (ed.) Anti-quality components of forages. CSSA Spec. Pub. 4, 53–98. Crop Sci. Soc. Am.

Mott, G.O. 1981. Measuring forage quantity and quality in grazing trials. Proc. 37th South. Past. Forage Crop Impr. Conf., 3–9. U.S. Sci. and Educ. Admin., New Orleans, La.

National Academy of Sciences. 1976. Nutrient requirements of beef cattle, No. 4 (5th rev. ed.). Subcommittee on Beef Cattle Nutr. Washington, D.C.

National Academy of Sciences. 1978. Nutrient requirements of dairy cattle, No. 3 (5th rev. ed.). Subcommittee on Dairy Cattle Nutr. Washington, D.C.

Peterson, R.G., H.L. Lucas, and G.O. Mott. 1965. Relationship between rate of stocking and per animal and per acre performance on pasture. Agron. J. 57:27–30.

Rhykerd, C.L., C.H. Noller, J.E. Dillon, J.B. Ragland, B.W. Crowl, G.C. Naderman, and D.L. Hill. 1967. Managing alfalfa-grass mixture for yield and protein. Purdue Univ. Agri. Expt. Sta. Res. Bul. 839, 7.

Stobbs, T.H. 1975. Factors limiting the nutritional value of grazed tropical pastures for beef and milk production. Trop. Grassl. 9:141–150.

Vohnout, K., and C. Jimenez. 1975. Supplemental by-product feeds in pasture livestock feeding systems in the tropics. Am. Soc. Agron. Spec. Pub. 24, 71–82.

Whiteman, P.C. 1980. Tropical pasture science. Ch. 1, 5–33. Oxford Univ. Press.

DISCUSSION

Remarks of Discussion Leader, A.G. Matches

Thank you, Dr. Mott, for a most informative address on the potential productivity of temperate and tropical grasslands. Dr. Mott, you have clearly shown us that temperate and tropical grasses are quite different in many respects. Particularly significant are the vast differences in tiller production, leafiness, stem size, and leaf:stem ratios — characteristics that, as pointed out by Dr. Minson yesterday morning, may be closely associated with level of herbage consumption by livestock.

Dr. Mott also dealt with physiological characteristics of these C_3 and C_4 plants. Noteworthy is the difference in carbon pathways. The temperate grasses possess photo-respiration and have nearly half the CO_2 fixation rate of the tropical species. Dr. Mott showed that the level of primary production is 2 to 4 times greater from the tropical grasses. However, the conversion of this production into animal yield is generally less, because tropical grasses have a much greater maintenance cost for animal gain and milk production.

Strategies were also offered for increasing animal yield/ha. I believe an examination of the Congress program will reveal that many of Dr. Mott's strategies are related in part to topics being presented in more than half of the technical sessions in Section IX. This is indeed a very timely topic.

Comments, Questions, and Responses

1. *Question:* Temperate pasture species have evolved over several hundred years, but man's involvement in the evolution of tropical pastures has been over a comparatively short time. Do you consider this will affect future development? (Nigeria) *Response:* I am very optimistic that both the yield and quality of temperate and tropical forages will be improved. The temperate grasses and legumes have been subjected to man's intervention over a much longer period than the tropicals, but there is no doubt that there is still considerable latitude for improvement. Selection pressure by man has only begun with many tropical species. Genetic improvement must be accompanied by improved systems of defoliation management similar to those that have evolved for temperate pastures.

2. *Question:* Digestible protein levels in tropical and temperate species are near equal, but there are variations in crude fiber. There is an inverse relationship between fiber and protein — could this be explained? (Nigeria) *Response:* The information that I presented was for total crude fiber and crude protein and not digestible CP and CF. A negative relationship between CP and CF occurs because each is strongly related to stage of maturity or age. Crude fiber increases with age, and CP decreases.

Increasing temperature also is a factor in increasing CF and decreasing CP. Selection of ecotypes or cultivars that with age retain a relatively low crude-fiber and high protein content should be sought.

3. *Question:* With respect to mineral requirements, do C_4 plants have lower critical levels for N and P than C_3 plants, and can they survive higher exchangeable Al levels? (U.S.A.) *Response:* If the answer is restricted to those grasses used in cultivated pastures, it would probably be safe to say that most C_4 grasses have lower critical levels for N and P and greater tolerance of high Al levels. There are, however, wide differences within C_4 and C_3 grasses, and my own experience would suggest that some of the indigenous C_4 grasses of the South American savannas have much greater tolerance for low N and P and high Al levels than tolerances of the African introduced grasses. For example, guineagrass will not survive the seedling stage on some newly cultivated savanna that has supported a natural grass cover for thousands of years.

4. *Question:* The discussions about temperate and tropical species are valid when grass grows all year, but there are many regions where this does not occur and conservation in the form of hay, silage, or dehydrated grass is necessary. There is a lack of knowledge of the value of conservation in the wet-dry tropics where grass is unable to grow in the dry season. (Federal Republic of Germany) *Response:* The dry season of the wet-dry tropics imposes no greater constraint upon growth of pastures than does cold temperature in temperate regions. Conservation is practiced to some extent in the tropics but only under the most intensive type of operation. High temperatures and humidity make it much more difficult to conserve hay in the tropics during the wet season. Silage and dehydration are possible, but the economic return is usually not favorable.

5. *Question:* Are legumes more important in tropical regions than in temperate regions, as your data suggest? *Response:* I did not mean to imply that legumes were more important in the tropics than in the temperate regions but that a good legume-grass mixture is much more difficult to attain and maintain in the tropics than in temperate climates.

6. *Question:* Could you comment on mixing temperate and tropical species in a system? (U.S.A.) *Response:* In subtropical regions this is definitely a possibility and should be exploited by extending the region of adaptation by selecting appropriate cultivars within both the temperate and tropical species. Under intensive management, planting temperate and tropical grasses in separate pastures may be more successful, but ryegrass overseeded on bahiagrass pastures for winter–cool-season pasture in peninsular Florida is a very successful practice. In the high-altitude tropics, mixtures of ryegrass and kikuyugrass are common.

7. *Comment:* Protein values for tropical and temperate plants have been shown to be similar. There are physiological reasons associated with C_4 and C_3 metabolism that give lower protein levels in the tropical grasses. (Australia)

8. *Question:* There is a wide range of adaptability within temperate legumes and grasses. Is there a similar range in the tropical species? (New Zealand) *Response:* Yes. As an example, *Panicum virgatum* is found from the Canadian-U.S. border to the Amazon. There are a number of these wide-ranging species that need intensive investigation. The *Cynodon* species range from 40° N to near the Equator. The potential for selecting ecotypes and the development of new cultivars by breeding for specific

niches of the environment appear very promising.

9. *Question:* The future for tropical plants appears bright, but is there a similar future for temperate plants? *Response:* Ask the New Zealanders. It was not many years ago that 16,000 kg DM/ha was considered the absolute maximum. Today 20,000 kg DM/ha is quite common, and experimentally nearly 30,000 kg DM/ha/yr has been attained. There is still room to advance.

Contribution of Forages to Worldwide Food Production: Now and in the Future

E.F. HENZELL
Chief, Division of Tropical Crops and Pastures,
CSIRO, St. Lucia, Queensland, Australia

Summary

Herbivorous livestock provided during the 1970s an estimated minimum 7%-17% of the total dietary protein in Asia and Africa compared with about 30% in Latin America and just over 30% in the developed countries. Cattle accounted for approximately 70% of the world's domestic animals (calculated in units weighted for size), buffaloes for 10%, and sheep and goats for 11%. About 69% of these animal units were in the developing countries, but productivity was higher in the developed countries. Forages, defined broadly to include all edible plant materials except grains and concentrates, provided more than 90% of the feed energy consumed by herbivorous livestock.

Forages can make an important contribution to future world food supplies by providing greater amounts of meats and dairy products at acceptable prices. They can also help to increase supplies of plant foods for people by (1) replacing grains and other potential human foods now fed to herbivores, (2) allowing food crops to be grown on the arable land now used for forage production (through increased yield of forages on nonarable land), (3) improving the fertility of arable land, and (4) allowing economies in the use of fuel and nitrogenous fertilizer.

Improvement of forage productivity is likely to be limited by the availability of labor and capital and by the decreasing quality of grazing land. Population growth alone will substantially increase the demand for foods of animal origin by the year 2000. Demand will also be increased by any growth in purchasing power, especially among the people of the developing countries. The failure of domestic herbivore numbers to keep pace with the human population during the 1970s indicates that the world's forage scientists face a major challenge.

KEY WORDS: protein foods, herbivorous livestock, pastures, fodders, feedgrains, economic development.

INTRODUCTION

The chief role of forages in world food production is to feed the domestic herbivores that provide meat and dairy products. It is the herbivores with the ability to digest plant fiber, especially the ruminants, that are of particular interest to grassland scientists.

The term forages is used broadly in this paper to mean all the plant materials that are eaten by herbivores, including those that are grazed (pastures) and those that are cut before being fed (fodders). Crop residues such as straw and the foliage of trees and shrubs fall within this broad definition, but it is convenient to treat cereal grains and other potential human foods as a separate category.

THE PRESENT CONTRIBUTION OF FORAGES

Food from Animal Products

FAO (1979a) statistics show that animal products make a larger contribution to dietary energy in the developed countries than in the developing ones, the range being from 6% in south and southeast Asia to 41% in Australia and New Zealand. For animal protein, which is especially important in human diets, the corresponding range is 15% to 74%.

These figures are for food from all types of animals, poultry and pigs as well as herbivores. Assuming that all the milk was obtained from herbivorous animals, and using the data of Jasiorowski (1976) and FAO (1977) to calculate their contribution to meat consumption, herbivores provided an average 55%–60% of the animal protein consumed in the developed countries during the early 1970s. In the developing countries, the proportion is estimated to have ranged from less than 40% in the People's Republic of China (China) to more than 80% in the Near East (as defined in FAO 1979a).

In combination with the statistics for consumption of animal protein from all sources, these estimates indicate that herbivorous livestock provided during the 1970s a minimum 7%–17% of total dietary protein in the developing countries of Africa and Asia, compared with an average of about 30% in Latin America and just over 30% in the developed world.

These averages conceal a great deal of variation – e.g., between the people of Argentina and Uruguay and others in Latin America, between the pastoral and nonpastoral peoples of Africa and the Near East, and between the Hindu people of India, who do not eat beef, and other Asians, who do. Generally, as personal incomes rise above a minimum of approximately $250–$300/yr (1980 U.S. $), animal products, along with fats, sugar, fruits, and vegetables, play an increasingly important role in people's diets (Scrimshaw and Taylor 1980).

The more affluent consumers are usually found in the cities. In Indonesia, the daily per-capita consumption of meat in the cities during 1975 was around 27 g and growing at 8% a year, compared with 5 g and 4% in the rural areas (FAO 1978). The respective annual per-capita incomes were then about $650 and $50.

Herbivorous Livestock

In order to compare the numbers and productivities of different kinds of livestock in different countries and regions, the statistics have been converted to animal units using the weighting factors listed in the footnote to Table 1. In 1978, cattle accounted for approximately 70% of the world's domestic animal units (AU); buffaloes for 10%; sheep and goats combined for 11%; horses, mules, and asses together for 8%; and camels for 1%. The developing countries had about 69% of the world's AU and India alone 16%. Useful herbivorous animals omitted from these calculations included 40–50 × 10⁶ deer, 5 × 10⁶ of the cameloid species found in South America, and several hundred million wild ruminants (McDowell 1977, Fitzhugh et al. 1978).

Between 1969–1971 and 1978 the total number of AU in the world increased only half as fast as the human population did (8.1% vs. 16.3%). Livestock numbers increased in each major region, with the greatest relative increase in Latin America (16.8%). Numbers decreased by 2.2% in Italy; they also decreased during recent years in the intensive, irrigated rice-growing areas of southeast Asia (FAO 1978, Crotty 1980).

The productivity of herbivorous livestock was higher in the developed countries than in the developing – 4.8 times as high for cattle and buffalo meat, 1.6 for sheep and goat meat, and 4.6 for milk production/cow (Table 1).

Those relative productivities take no account of the fact that herbivorous livestock also supply fiber and power for farming and transport. The sheep in Australia and New Zealand produce more than a third of the world's wool. Data published about a decade ago showed that over 80% of the world's cultivated land was tilled by man with the help of draft animals. If one disregards human energy, which was and is also of major importance, farm animals accounted for 75% or more of the power employed in agriculture in the developing countries (McDowell 1977). China has over 30 million buffaloes used almost exclusively as work animals (Cockrill 1980).

Forages and Feedstuffs

A study by the Winrock Foundation (Fitzhugh et al. 1978) estimated that forages provide more than 90% of the feed energy consumed by the herbivorous livestock of the world. Even in the developed countries of the temperate zone, forages supply some 75% of the feed consumed by beef cattle and 60% of that consumed by dairy animals (Bula et al. 1977). Australia and New Zealand have the highest ratio of permanent pasture land/livestock unit (Table 2), whereas Europe has only 0.67 ha/AU (0.44 in Eastern Europe) and the Far East the very low ratio of 0.11 ha/unit (0.06 in India).

Unimproved native and naturalized grasslands are the main source of forage in Africa, the Near East, and the subtropical and tropical areas of Latin America and Australia. In the temperate zone, forage from permanent grassland is supplemented by pasture and fodder crops grown on arable land, including alfalfa, which occupied 33 × 10⁶ ha in the 1960s (Bolton et al. 1972). Southern Australia and New Zealand have about 30 × 10⁶ ha of legume-based pastures, and nearly all their meat and dairy products are produced from forages (Table 2).

In the countries of south and southeast Asia, with their small proportion of pasture land, the main forages are crop residues (such as rice straw), fodder grown on arable land, green feed cut from roadsides and wastelands, and leaves of trees. It has been calculated that in India in

Table 1. Herbivorous livestock and their productivities.

Region	Herbivorous livestock 10⁶ a.u.†	Beef & buffalo meat kg/a.u./yr	Sheep & goat meat kg/a.u./yr	Milk kg/hd/yr
Developing countries	953.3	20	40	666
Developed countries	424.6	96	65	3085

Source: FAO 1979a. †Animal unit: camels 1.1, buffaloes 1.0, horses and mules 1.0, asses 0.8, cattle 0.8, sheep and goats 0.1.

Table 2.   Relative areas of permanent pasture land and quantities of grain fed to livestock.

Region	Permanent pasture land ha/a.u.	Grain fed to livestock 10^6 t
Developing		
Africa	4.39	2.8
Near East	2.41	12.3
Far East	0.11	4.4
China	1.80	32.1
Latin America	2.02	30.8
Developed		
Europe	0.67	159.3
USSR	3.35	104.4
Canada & USA	2.31	139.5
Australia & New Zealand	9.19	2.7

Source: FAO 1979a, 1979b.

1970, 4% of the feed energy for ruminants came from pasture and wasteland, 55% from fodders grown on arable land, and 36% from crop residues (Fitzhugh et al. 1978).

The grain fed to livestock (principally maize and barley) accounted for more than 40% of world production during recent years. Over 80% of it was used in developed countries (Table 2). FAO (1979b) estimated that 20% of the grain used by livestock in the developing countries during 1972–1974 was fed to dairy cattle and 10% to beef cattle. In the developed countries, dairy and beef production each accounted for about half of the 48% that was fed to livestock. Nonruminants would have consumed most of the remainder. Other potential human foods used for cattle feed included cassava chips, by-products of cereal milling, oilcakes, and meals (FAO 1979b).

One of the important interactions between forage plants and human nutrition is through the energy obtained for cooking from woody forage residues and dung. In 1981, about 2×10^9 of the world's people were thought to be eating food cooked with commercial energy, 1.5×10^9 principally with woodfuels, and 1×10^9 with agricultural residues, including dung (Hughart 1979).

Another valuable interaction between forages and food production is through their effects on the fertility of land used for crops. Livestock influence the fertility of land because they excrete most of the plant nutrients that they ingest. Thus, the grazing of pastures and feeding of crop residues in situ help to maintain soil fertility, and there can be a net gain of nutrients by soils that receive the dung and urine of animals fed on plant material brought from elsewhere.

THE FUTURE CONTRIBUTION OF FORAGES

The important contribution that forages make to world food production, through the production of meat and dairy products that add flavor and protein of high biological value to people's diets, clearly must be main-

tained and if possible increased to help feed an increasing population with a strong desire for a higher standard of living. While this paper deals only with the future of forages, there is great scope also for animal research to improve breeds and to control pests and diseases, e.g., trypanosomiasis, which limits livestock production in Africa between lat 15°N and 21°S.

Increased forage productivity can assist in two ways: directly, by providing people with greater amounts of meat and dairy products at acceptable prices, and indirectly, by permitting larger quantities of plant foods to be grown and consumed. Although the direct benefits are obvious, the potential indirect benefits require further explanation.

Indirect Effects on Food Production

There are at least four ways in which improved yield and quality of forages can increase the supply of plant foods to needy people:

1. By substituting forages for the grains and concentrates that are now fed to herbivorous animals. Reference has been made already to the fact that substantial quantities of potential human foods are fed to cattle. In theory all those fed to beef cattle, and some fed to dairy cattle, could be replaced by high-quality forages if they were available.

2. By increasing the yield of forages from land that is not suitable for food crops, so releasing the arable land on which forages are now grown. Hodgson (1977) estimated that 15% to 20% of the world's cropland is used to grow forages. Even in India with its 661×10^6 people (in 1978) forages are grown on arable land; in the U.S.A. they occupy as much as 34% of the cropland (Fitzhugh et al. 1978).

3. By improving the fertility of arable land. In the Mediterranean climatic zone of southern Australia, cropping practices involving rotation with annual legume-based pastures have enabled wheat yields to be maintained without nitrogenous fertilizers (Clarke and Russell 1977). Each year, North Africa and the Near East have approximately 40×10^6 ha under fallow in the arable areas, on which the same annual legumes might be grown (Oram 1975). Some progress has been made in research on pasture-legume-based farming systems for the semiarid tropics (Shelton and Humphreys 1975, McCown et al. 1980).

4. By allowing economies in the use of commercial energy for food production and preparation. Forages save fossil fuels by substituting biological nitrogen fixation for industrial synthesis, by supporting the large herbivores used for draft (a pair of draft animals and a man can do about 13 times as much as a man alone—Norman 1978), and by serving through their residues as fuels for cooking.

Exporting animal products to pay for importing food grains is another possibility. Unfortunately, most of the countries with surplus food grains are also either major exporters of animal products (Argentina, Australia) or strong protectors of their own livestock industries (EEC).

Against that background, the main factors likely in the future to influence the supply of forages and the demand for meat and dairy products can be discussed.

Future Supply of Forages

Economic, social, and technological factors are likely to exert a strong influence on the future supply of forages. Land, labor, and capital are the principal economic factors.

In the foreseeable future, the quality rather than the quantity of land is likely to be the more important constraint on forage production. It has been calculated that there are some 300×10^6 ha of improvable permanent grazing land with humid temperate climates (Bula et al. 1977) and a similar-sized area of improvable tropical savannas in Latin America (Cochrane 1979). Even in the most densely populated parts of Asia, there is usually some land on which forages could be grown.

In contrast, the quality of land remaining available for forage production must decline as herbivorous livestock industries are restricted progressively to regions with less-fertile soils, steeper slopes, and more difficult climates. Unless scientists can find economical methods for improving the productivity of the lower-quality land that will be available for forage production, the prices of meat and milk could rise rapidly in the future. It will not be easy, especially in semiarid and arid areas.

Replacement of the existing vegetation and use of fertilizers are the chief means of improving forage productivity. The first has a substantial capital or labor cost, depending on whether seed production, land preparation, and sowing are mechanized or not, and the second a significant capital cost. Even the wealthiest developed countries may have difficulty in finding the necessary capital in the future. For instance, in the UK it is feasible to more than double the stocking rate, but the capital cost would exceed £1,000/ha (Wilkins et al. 1981). Although labor-intensive systems are appropriate for the developing countries, the costs of seed (initially at least) and fertilizer are probably inescapable, and it may be difficult to obtain that capital in competition with other urgent developmental priorities.

The importance of social factors is epitomized in the quotation "Society, in adapting cattle to its needs, has made them money in Africa and gods in India" (Crotty 1980). Baker (1976) argues that the reluctance of African cattle-owners to sell their stock for slaughter is not irrational if one understands the social and economic conditions under which they operate. Their communal use of grazing land, however, is certainly a complicating factor for forage improvement.

Two general points can be made about the role of technology in improving forage production. First, the major gaps in forage science and technology are in the subtropics and tropics, which have large herbivore populations (Table 1) and tremendous potential for increased beef production from forages (Stobbs 1976). Recently, encouraging progress has been made in Latin America (CIAT 1980), though much less has been achieved in Africa and Asia. Grassland scientists usually find that the broad-scale, mechanized pasture and fodder crop technologies of the temperate northern hemisphere countries, with their emphasis on nitrogenous fertilizers, and of Australia and New Zealand, with their emphasis on herbaceous legumes, are not directly relevant to the forage problems of Asia.

An appropriate technology for intensive Asian farming systems will probably require research on improving forage production from trees and shrubs (which also provide firewood—e.g., *Leucaena leucocephala*) and from pasture among fruit and coconut trees (Asia has about 6 $\times 10^6$ ha under coconuts—Plucknett 1979) or along paths and roads, and on breeding and selection of crop varieties whose residues have a higher feeding value. None of these areas of research is well developed, a fact that is surprising in view of the important role of large bovines in the cropping systems of monsoonal Asia.

The second point concerns the strong likelihood of a continuing high rate of return from forage research in the developed countries of the temperate zone (Bula et al. 1977). Thus, while the output of ruminant products might be increased two-and-a-half times in the UK (Wilkins et al. 1981), such increases are unlikely to be achieved without further research.

Future Demand for Meat and Dairy Foods

Population growth alone will substantially increase the demand for foods of animal origin. The world's population, which was about 4×10^9 in 1975, is projected by the United Nations to increase to between 5.8 and 6.6 $\times 10^9$ in 2000; most of the increase is expected to occur in the developing countries.

The demand for meat and dairy products is also influenced by their cost and by the consumers' purchasing power. Consumption of meat generally increases with rising income, even among quite affluent people. The same is true for whole milk at low income levels, but its consumption tends to decline sharply with increasing affluence (Fitzhugh et al. 1978, CIAT 1980). The monotonous cereal- or root crop–based diets of the poor in many developing countries are thus a matter of economic necessity rather than of choice (Scrimshaw and Taylor 1980). Tracey (1975) concluded that there is a potential demand for more animal protein among some 85% of the world population.

Economic development, which will be influenced strongly by the cost of commercial energy, affects not only individual purchasing power but also the availability of foreign exchange to pay for imported foods. Currently there are many barriers to such trade, and only 9% of the world's bovine meat production and 11% of its sheep and goat meat production were traded in 1978 (Anon. 1979, FAO 1979a).

Cultural attitudes and religious beliefs usually do not have a major influence on the demand for animal products. Even in India milk supplies about 9% of dietary protein (Fitzhugh et al. 1978), and there are no significant religious taboos anywhere on sheep or goat meat. Another example of a human factor limiting the consumption of foods of animal origin is the lactose intolerance found widely among adults (Simoons 1978). The "sleeping" issue is the possible hazard to human health of diets high in saturated animal fats and cholesterol, on which the evidence is currently inconclusive (Wade 1980).

Fitzhugh et al. (1978) of the Winrock Foundation have estimated the demand for ruminant products in the year 2000. They project increases over 1972 figures of 77% for meat and 72% for milk production. Increases of 174% and 289% are projected for beef and milk, respectively, from cattle in the developing regions. The failure of domestic herbivore numbers to keep pace with the human population during recent years, and increasing pressure on the world's resources generally, indicate that it will be difficult to attain those projections. Their increased feed requirement constitutes a formidable challenge to the world's forage scientists.

LITERATURE CITED

Anon. 1979. Statistical review of live-stock and meat industries. Compiled by Austral. Meat and Live-stock Corp. for year ended 30 June 1979.

Baker, P.R. 1976. The social importance of cattle in Africa and the influence of social attitudes on beef production. *In* A.J. Smith (ed.) Beef cattle production in developing countries, 360–374. Cent. Trop. Vet. Med. Edinburgh.

Bolton, J.L., B.P. Goplen, and H. Baenziger. 1972. World distribution and historical developments. *In* C.H. Hanson (ed.) Alfalfa science and technology. Agron. Monog. 15:1–34. Am. Soc. Agron., Madison, Wis.

Bula, R.J., V.L. Lechtenberg, and D.A. Holt. 1977. Potential of temperate zone cultivated forages. *In* Potential of the world's forages for ruminant animal production. Winrock Rpt., 7–28.

CIAT. 1980. Trop. Pastures Prog. 1979 Ann. Rpt. Cali, Colombia.

Clarke, A.L., and J.S. Russell. 1977. Crop sequential practices. *In* J.S. Russell and E.L. Greacen (eds.) Soil factors affecting crop production in a semi-arid environment, 279–300. Qld. Univ. Press. St. Lucia.

Cochrane, T.T. 1979. An ongoing appraisal of the savanna ecosystems of tropical America for beef cattle production. *In* P.A. Sanchez and L.E. Tergas (eds.) Pasture production in acid soils of the tropics, 1–12. CIAT.

Cockrill, W.R. 1980. The ascendant water buffalo—key domestic animal. World Anim. Rev. 33:2–13.

Crotty, R. 1980. Cattle, economics and development. CAB: Farnham Royal.

FAO. 1977. The fourth world food survey. FAO Statis. Ser. 11, FAO Food and Nutr. Ser. 10.

FAO. 1978. Report of the Indonesian livestock sector survey. Vol. 1. FAO/World Bank Coop. Prog.

FAO. 1979a. 1978 production yearbook. Vol. 32.

FAO. 1979b. Utilization of grains in the livestock sector: trends, factors and development issues. Committee on Commodity Problems, Intergovernmental Group on Meat. Rome, 7–11 May 1979.

Fitzhugh, H.A., H.J. Hodgson, O.J. Scoville, Thanh D. Nguyen, and T.C. Byerly. 1978. The role of ruminants in support of man. Winrock Rpt.

Hodgson, H.J. 1977. Food from plant products—forage. Presented at Symposium on the Complementary Role of Plant and Animal Products in the US Food System. Natl. Acad. Sci. Washington. 29–30 November 1977.

Hughart, D. 1979. Prospects for traditional and nonconventional energy sources in developing countries. World Bank Staff Working Paper 346.

Jasiorowski, H.A. 1976. The developing world as a source of beef for world markets. *In* A.J. Smith (ed.) Beef cattle production in developing countries, 2–28. Cent. Trop. Vet. Med. Edinburgh.

McCown, R.L., R.K. Jones, and D.C.I. Peake. 1980. A ley farming system for the semi-arid tropics. Proc. Austral. Agron. Conf., Lawes, 1980, 188.

McDowell, R.E. 1977. Ruminant products: more than meat and milk. Winrock Rpt.

Norman, M.J.T. 1978. Energy inputs and outputs of subsistence cropping systems in the tropics. Agro-Ecosystems 4:355–366.

Oram, P.A. 1975. Livestock production and integration with crops in developing countries. *In* R.L. Reid (ed.) Proc. III World Conf. Anim. Prod., 309–330.

Plucknett, D.L. 1979. Managing pastures and cattle under coconuts. Westview Trop. Agric. Ser. 2. Westview Press, Boulder, Colo.

Scrimshaw, N.S., and L. Taylor. 1980. Food Sci. Am. 243:74–84.

Shelton, H.M., and L.R. Humphreys. 1975. Undersowing rice (*Oryza sativa*) with *Stylosanthes guyanensis*. I. Plant density. Expt. Agric. 11:89–95.

Simoons, F.J. 1978. The geographic hypothesis and lactose malabsorption: a weighing of the evidence. Am. J. Dig. Dis. 23:963–980.

Stobbs, T.H. 1976. Beef production from sown and planted pastures in the tropics. *In* A.J. Smith (ed.) Beef cattle production in developing countries, 164–183. Cent. Trop. Vet. Med. Edinburgh.

Tracey, M.V. 1975. The future of animal production from the point of view of food technology. *In* R.L. Reid (ed.) Proc. III World Conf. Anim. Prod., 458–463.

Wade, N. 1980. Food Board's fat report hits fire. Science 209:248–250.

Wilkins, R.J., J.E. Newton, P.J. James, and J.M. Walsingham. 1981. Possibilities for change in the output from grassland: an overall view. *In* J.L. Jollans (ed.) Grassland in the British economy, 375–389. CAS Paper 10, Reading. Cent. Agric. Strategy.

DISCUSSION

Remarks of Discussion Leader, G.C. Marten
(presented by G.E. Carlson)

Dr. Henzell comprehensively considered the numerous avenues whereby forages have contributed and will contribute to worldwide food production. He alerted us to the disturbing recent failure of domestic herbivore populations to keep pace with human populations; but he also pointed to the potential for forages to replace grains and other human foods in the diets of domestic herbivores.

As we interpret the true meaning of the relative ruminant animal and human population trends, I suggest that we closely examine statistics concerning the changing ratio of *animal populations* to *productivity/animal* over time. This investigation should reveal aspects of the biologic as well as economic efficiency of evolving forage-livestock systems. Frequently, as animal populations have decreased, productivity/animal has increased. A crucial concern to grassland scientists is the forage:grain ratio used to gain seeming "system efficiency." If vast amounts of concentrate feeds utilizable by nonruminants, and easily marketed, are needed to gain increases in productivity/ruminant animal, the efficiency of the system may not be biological even though it is economical at the moment.

For example, in the U.S.A. between 1950 and 1973, the number of dairy cattle decreased by 47%, while the milk production/cow more than doubled, resulting in a slight increase in total milk produced. However, increasing human population during this 23-year period reduced the available milk/person from 352 kg to 261 kg. Also, the increased milk produced/cow was accomplished by sizable increases in the proportion of concentrate feed grains in the typical ration. Similar trends have occurred in other ruminant feeding systems, but predictions of return to "grass-fed beef" and even pastures for high-producing dairy cows have often been heard in the U.S.A. in recent years.

Reference to recent lists of research priorities in the U.S.A., including those compiled by dairy-forage, beef-range, meat animal–crop residue, and animal-feed committees, gives considerable evidence that the importance of grain as a feed base is expected to decline and that its place, especially for ruminant animals, is expected to be taken by forages or alternative feeds.

Comments, Questions, and Responses

1. *Question:* What role do you see for fertilizer in raising total forage yield in light of economic limitations and resource scarcity? *Response:* Goals are probably unobtainable without the use of fertilizer. Whether or not it is used will depend largely on fertilizer price relative to that of food, and fertilizer price may not go up faster than other costs.

2. *Question:* What are the relative advantages of dairy versus beef cattle under irrigation in the Mediterranean regions? (Italy) *Response:* Irrigation is most likely to be first used for food crops rather than for any type of forage production, and economic factors will be the determinant of its use. A pertinent factor is that in tropical areas where the dry season is absolutely dry, herbage quality is better than where there is more moisture, which may lower quality. There are areas in the Mediterranean where there is this absolute drought.

3. *Comment:* Although microprocessors are common, there is a need in many developing countries for "micro-harvesters" or other micro-machinery. There is a distinct need for some machinery to serve the need of individual farmers who need to chop feed for animals: e.g., preparation of elephantgrass. (Kenya) *Response:* There are indeed many people in Asia who would be relieved of a burden with use of machinery for cutting forage. However, they should also be made aware of the benefits to be derived from allowing animals to harvest material for themselves.

4. *Question:* Why, after such a large effort in stimulating production in the tropics, is there so little animal production there? (FAO) *Response:* There are few places where good research and development on applicability of techniques have been carried out. One important reason is market uncertainty; e.g., northern Australia sells beef cattle to affluent countries, and the economic status of those countries limits exports. Many other regions, however, could develop animal production for their internal benefit.

5. *Question:* Would you comment on problems related to red-meat production by developing countries, particularly on communal pastures? There seems to be a need for a policy on communal pastures. *Response:* Cattle herdsmen are behaving rationally. If there is security and profit in feeding more cattle at a better level, then they will respond. In the more difficult climates, there is uncertainty; the people there are matching production to long-term survival.

6. *Comment:* The place of forages in relation to crops is important, considering aspects of soil fertility and soil structure. We should be careful in following developed-country equipment programs, because some developing countries are better off without mechanization. It is better to use people in a sensible way than to create unemployment by over mechanization. (U.S.A.) *Response:* Yes, technology must be appropriate.

7. *Comment:* Much of the development work in temperate legumes is in alfalfa. It seems that it is important to provide crops for cutting, and legumes with the habit of alfalfa would be most appropriate. (Netherlands)

8. *Comment:* Liberia is interested in grassland productivity, but apparently has not been interested much in grassland research. There is now a challenge after attending this Congress to slant research programs and to plan for the spread of technology there. The possibility of developing regional research programs may be explored. (Liberia)

9. *Question:* It was stated that the relative production of sheep and goats is superior to that of cattle. Is there any unique physiological property that could make beef cattle better producers? (U.S.A.) *Response:* It is probable that sheep and goats are better performers because they are not used for work such as ploughing or hauling carts.

Practical Livestock-Forage Systems: Model to Manager

R.W. BROUGHAM

Grasslands Division, Department of Scientific and Industrial Research,
Palmerston North, New Zealand

Summary

The developments in practical livestock-forage systems research during this century are outlined, and present-day approaches are discussed. Some examples of production levels presently being obtained from research stations, top farmers, and average farmers in different regions of the world are presented. These indicate that production levels being obtained from livestock-forage systems research remain appreciably higher than levels being obtained from the best farmers. Comments are made on the many different research techniques used to demonstrate these levels in both developed and developing countries. The value of this type of research to the farm adviser and farmer is also noted.

Some results of beef-oriented systems research in New Zealand are then presented and used to outline future needs of livestock-forage systems research. The need for more hard, factual data on all aspects of livestock-forage systems for all regions is emphasized.

KEY WORDS: livestock-forage systems, research developments and methods, production levels, research needs in developed and developing countries, research/extension, research/farmer, beef systems research.

INTRODUCTION

The scientific investigation of aspects of practical livestock-forage systems is a 20th-century development. Before then the basis of present-day forage research was being laid by some of the early ecologists and the more perceptive farmers. As early as 1598 the basic idea of livestock being moved at regular intervals from one small enclosure to another was advocated in Scotland by Archibald Napier (Smith 1956). Smith records that between 1598 and the beginning of the 20th century only very occasional references to controlled systems of livestock farming occur in the literature.

In the less agriculturally developed regions of the world, experience coupled with good observation also enabled shepherds and managers to develop systems of forage management that produced good feed for livestock during the growing season yet maintained plant communities in a stable but productive state. In some regions patterns of management were associated with conservation systems, and management tools such as burning were usefully employed.

DEVELOPMENTS IN LIVESTOCK-FORAGE SYSTEMS RESEARCH

Since the turn of the century prominent ecologists in many countries have turned their attention to forage-livestock systems (e.g., Cockayne 1919, Sampson 1923, Stapledon and Hanley 1927, Boerger 1935, Weaver and Clements 1938, Trumble 1949, Levy 1951). From their observations and inductive reasoning, especially on aspects such as the stability, composition, and productivity of pastures under different livestock systems, tremendous progress was made, and highly productive livestock-forage systems were established.

Development of statistical methods from the 1920s (e.g., Fisher 1925, Snedecor 1937) gave research workers the methodology to test and perhaps demonstrate the validity of derived hypotheses. In agricultural science this methodology led to livestock-forage systems and compartmental research (e.g., Jones 1933, Johnstone-Wallace 1937) that resulted in the very large increase in livestock-forage systems research that occurred after World War II in most agricultural countries (e.g., McMeekan 1956, 1960; Willoughby 1959; Blaser et al. 1960; Kennedy et al. 1960; Riewe 1961; McMeekan and Walshe 1963; Browne et al. 1970; Lowe 1970; Hutton 1973; Suckling 1975; Eadie 1976). In Australia, for example, during the 1960s there were more than 150 stocking-rate experiments in progress. Significantly, the factorial design predominated, and some useful information, especially on a regional basis, was obtained.

The author is indebted to Dr. R.J. Clements, CSIRO, Brisbane, Australia; Dr. C.S. Hoveland and Dr. D. Ball, Auburn University, Auburn, Ala., U.S.A.; and Dr. A.G. Matches and Dr. V. Jacobs, University of Missouri, Columbia, Mo., U.S.A. for making data available. He is also grateful to his colleagues in the Grassland Division for assistance in the preparation of this paper.

Interpretation of whole systems/farmlet-style research was aided by the large number of scientists researching various aspects or components of livestock-forage systems (e.g., Wheeler 1962, Rae et al. 1963, Campbell 1964, Conway 1968, Holmes 1968). In some locations individuals or teams of researchers were simultaneously involved in whole systems and compartmental research.

The next development in livestock-forage systems research occurred during the 1960s. Simulation modeling was vigorously promoted by some as a means to formalize concepts or hypotheses of the complex livestock-forage systems. As a result there was a proliferation of computer-based livestock-forage models (e.g., Spedding 1970, Van Dyne 1970, Christain et al. 1974, Noy-Meir 1976). These models often illustrated the paucity of quantitative data on critical aspects of systems. However, any ideas that the models would shortcut the route to improved systems by removing the need for detailed, time-consuming, tedious, and expensive field and laboratory research were rapidly dispelled. In fact during the 1970s much has been written about the overreach that occurred around modeling (Innis et al. 1980, Seligman and Arnold 1980).

Although it is difficult to quantify the impact of research developments of the last 50 years on farm production levels, much of the research has led to increases in productivity (Swain et al. 1970), particularly in countries or states very dependent on agricultural production for economic survival.

These developments have also led us to the present time where many are now looking for leads to the next areas of progress in livestock-forage systems. The 1970s have been a period of consolidation and soul-searching by livestock-forage systems researchers in developed countries. Much of this has occurred because in the 50 years or so preceding 1970 major increases in production levels were regularly obtained at individual farm, regional, and national levels. The 1970s, on the other hand, have seen much smaller or static rates of progress.

Similarly, there is also much debate on the relative merits of different approaches in grassland research and extension and hence farming progress. Some of this debate stems from pessimism, some from economic considerations, some from arrogance, and some from genuine interest in "fast track" progress. There are many other reasons.

RESEARCH, TOP FARMER, AND AVERAGE FARM PRODUCTION LEVELS FROM DIFFERENT REGIONS OF THE WORLD

In many regions of the developed world there remain large differences between research production levels, top farmer production levels, and average farm production levels. The following examples illustrate this:

Australia (Table 1): In the South Burnett Region (coastal Queensland), top research levels were obtained from research stations where either siratro-based pastures or fine-stem stylo (*Stylosanthes* sp.) pastures were used. The best farmers have some 20% to 40% of their properties in legume(siratro)-based pastures.

U.S.A. (Table 1): In the Tennessee Valley of northern Alabama, the large difference between research and farmer production levels is attributed to the low usage by farmers of cocksfoot (*Dactylis* sp.) and Ladino clover (*Trifolium* sp.). In a study in Missouri, in which no legumes were used, the pastures were tall fescue (*Festuca* sp.) fertilized with 224 kg N/ha/yr.

Eire (Table 2): Significantly, the factorial design associated with regression analyses was the dominant experimental approach used to determine research optima. The impact of this research on Irish and European agriculture was very large indeed.

Table 1. Beef production levels of average, top-farmer, and best research.

	Average farm	Best farm	Highest research
South Burnett Region, coastal Queensland, Australia[1]			
Carrying capacity (cattle/ha)	0.11	0.16	0.44
Annual live-weight gain (kg/ha)	27.5	60	165
Tennessee Valley, northern Alabama, U.S.A.[2]			
Beef gain (kg/ha/year)	275	385	585
Average daily gain (kg)	0.3	0.55	0.8
Missouri, U.S.A.[3]			
Calf gain (kg/ha)	101	178	280

[1]Data supplied by R. J. Clements, C.S.I.R.O., Brisbane, Australia.

[2]Data supplied by C. S. Hoveland and D. Ball, Auburn Univ., Ala. Grazing season: 26 Sept.–19 Dec. and 29 March–30 Aug. Forage: average farm, mostly tall fescue (*Festuca*) plus 110 kg N/ha; best farm, cocksfoot (*Dactylis*) plus 110 kg N/ha; highest research, cocksfoot (*Dactylis* sp.) plus ladino clover (*Trifolium* sp.) plus no nitrogen.

[3]Data supplied by A. G. Matches and V. Jacobs, Univ. Missouri. Cow-calf operation; fall-dropped calves; year-long grazing.

Table 2. Dairy production levels of average, top-farmer, and best research.

	Average farm	Best farm	Highest research
Eire (Cunningham 1979)			
Lactation			
Milk (kg)	3057	4500	6000
Butterfat (kg)	108	162	240
Cows/ha	1.1	2.0	3.1
New Zealand (Brougham 1973)			
Milk production (kg/ha)	250–300	550	650[1]

[1]Hutton 1973.

New Zealand (Table 2): The levels shown are for year-round outdoor grazing systems based on *Lolium-Trifolium* pastures.

The data presented in Tables 1 and 2, when associated with many similar sets of data that could have been produced for other regions, confirm that there are large gaps between average and best farmers' production levels, and in most regions a substantial gap between best farm and research production. In some regions the reasons for this have been identified in simple terms; in others the reasons are more complex.

To the theme of this paper, however, it is significant that the research levels of production were demonstrated using a range of experimental techniques. The techniques ranged over put-and-take methods, "compartmental" research approaches, simple factorial studies of stocking rates, etc., associated with linear regression interpretations, and "whole systems" approaches, some of which were of an evolutionary nature. All approaches have contributed significantly to progress in research, in extension, and hence in developing more productive livestock-forage systems of farming in the different regions of the world.

Also significantly, formal simulation modeling has not yet been of much value in establishing production levels such as those presented. The complexity of a livestock-forage system does not readily lend itself to simulation modeling. Most systems are far too complex. As the multitude of factors influencing the system vary from day to day, they are affected by social and economic conditions, as well as by biological elements; the mode of operation of the factors controlling the system and the interactions among them are still not well understood in many areas (Brougham 1970, 1981; Wolfe 1979).

Views similar to these have also been expressed recently by groups strongly involved in simulation modeling. Innis et al. (1980), discussing a number of ecosystem-level models, concluded that the availability of data adequate for construction of a useful ecosystem model is rare. They also cautioned regarding the diminishing returns from continued modeling efforts and argued that biologically realistic models for resource management are still a thing of the future. Seligman and Arnold (1980), although more optimistic about the developments and role of simulation modeling, have expressed somewhat similar views. Such views are more cautious than those expressed a decade or so ago when simulation modeling and the euphoria that went with it were being strongly advocated.

This initial overplay and subsequent cooling has been very similar to that associated with the introduction and development of statistics and biometrics into agricultural research approximately 40 years earlier. It has also been very costly in resource terms. However, present-day perspectives are much healthier, and simulation modeling (hypothesis generation), like statistics, is more likely to be used primarily as a research and teaching aid.

For emerging or developing countries, research needs are markedly different. Of importance is the need to research at levels and in ways that have direct relevance to current farming practices (McMeekan 1966). There is little value, for example, in researching livestock-forage systems by any of the conventional ways on a large scale in regions where subsistence agriculture is practiced. Yet examples of these approaches are evident in many developing regions, approaches that are often the legacy of visiting "experts." Further, there is little value in attempting to use a modeling approach to livestock-forage systems research in developing countries. Yet this approach is being advocated and used in South American countries in environments in which simple management techniques (such as subdividing farms into just a few paddocks) on pasture and animal production have not yet been attempted, let alone researched to discover their value and effects.

There are few shortcuts in research. In developing countries the best progress will be made if research and farming are developed in parallel with social and economic developments. Then components of the total system (biological, social, and economic) will evolve in a much more stable and controllable manner.

VALUE TO THE EXTENSION WORKER AND FARMER

The value of studies and results such as those shown in Tables 1 and 2 to the farm adviser and farmer is highly variable. It is common experience that studies like these are usually of relevance only to the best farmers (Morley 1979, McKenzie 1980). Yet the long-term impact is frequently much greater because of the "spin-off" effect of top-farmer practice to others. In spite of this, unless potentials are determined, goals cannot be achieved, and in livestock-forage systems there is no better way than the approach taken by the researchers quoted above. In parallel is the need to continue producing data on all facets of livestock-forage systems so that it is available in an easily retrievable form for the practitioners' uses (McMeekan 1966). This is the way we've made progress in the past, and it is unlikely to change in the immediate future.

It can be a slow process, as is illustrated in Fig. 1 for New Zealand dairy production. Each point on this graph

Fig. 1. Milk production levels (kg butterfat/ha) of "top" commercial farmers with time.

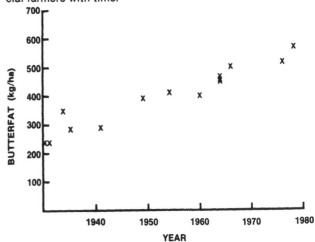

is top-farmer production level for the year shown. If this relationship is correct, today's average New Zealand dairy farmer is producing at about the level of the top farmer 50 years ago. Is this proportion the same for other regions of the world? The data presented in Tables 1 and 2 suggest it is. In some countries it is probably much worse. This is the challenge to the extension worker and most probably to planners and governments if we are to take world food production seriously.

The techniques needed to meet these challenges in different regions of the world will differ. Two examples are given.

Developing or Emerging Countries

In emerging countries, as in research, there are no shortcuts in extension except perhaps through the use of modern aids. Extension must evolve with the needs of the country. This has been clearly shown by a recent survey (Leagans 1979) of 2,300 small farmers in India, Kenya, Pakistan, Thailand, Togo, Trinidad, and Turkey, and of low-income farmers in the U.S. The principal findings and implications of this survey indicated that the adoption of modern techniques depends on the farmer's being convinced (a) that such techniques are technically sound, economically feasible, physically possible, and politically and socially acceptable, and (b) that a favorable environment is created when farmers receive technical assistance and production-related services (e.g., credit, markets, and education). Provided these factors were recognized, the survey indicated that previous tendencies *to start programs with the technology instead of with the purpose* or *to provide a solution before adequately identifying the technical and nontechnical elements of the problem* would be eliminated. Similar views have been previously .stressed by McMeekan (1966).

Developed Countries

Techniques for dissemination of information in developed countries are reaching high levels of sophistication, with the computer-based retrieval system being the ultimate. Application of this type of technology will demand extremely competent extension workers with a wide range of skills and a breadth of knowledge. However, the impact of such developments will be slow because most farmers will not be capable of seeking out pertinent information. This has recently been indicated by an assessment undertaken by Morley (1979) of the potential of grazing systems in the western district of Victoria, Australia (3.5 million ha). It indicated that the potential livestock carried could be increased by 70% without major difficulties or excessive investment. However, he concluded that realization of this potential would be dependent mainly on the educational resources available to the farmers and the extent to which these resources are used by farmers and their families.

SYSTEMS-ORIENTED BEEF RESEARCH IN NEW ZEALAND

In the remainder of this paper some systems-oriented beef research undertaken at Grasslands Division, New Zealand, this last decade will be discussed in the context of the theme of this paper—i.e., experimental approach, sample results, extension of results, implications for economics of farming, and further emphasis.

The research has assessed a range of pasture and animal parameters that either enhance or act against pasture and animal production. The approach taken has been to use small farmlets and in different years to compare replicated and, more recently, duplicated systems (Brougham et al. 1975). Importantly, management procedures are not fixed from year to year, although in any year the same management procedures are applied to the different treatments being compared. The management meets the needs of, and changes in, climatic, animal, pasture, and economic considerations. Using this approach, the aim is to maximize animal production/ha, and to assess differences due to treatment.

One set of data is shown in Table 3. Two treatments were compared: (1) old permanent pasture (*Lolium, Agrostis,* and *Trifolium* species), and (2) 4- to 5-year-old pasture with Pitau white clover (*Trifolium repens*) and Pawera red clover (*Trifolium pratense*) as the clover base. These two legumes are recent-bred varieties of Grasslands Division.

Table 3. Net hot carcass meat yields (kg/ha) from old and new pastures.

Year	Old pasture	New pasture	Difference (%)
1977/78	908	1052	146 (16)
1978/79	844	938	94 (11)

At high levels of meat production/ha some very significant production increases have been obtained, attributable mainly to the use of the improved cultivars.

Investigations such as these can be the center point for widespread dissemination of research results. In our environment many methods of extension are used and the impact on farming is high. Of particular interest is the discussion group of farmers who have met regularly over 10–12 years to discuss and study these investigations and evaluate their use in farming practices. The results obtained also have high value to the economics of dairy-beef farming in the region, in that they frequently are used as a reference point.

The sequence described above, from research to economic assessment, is no different from that employed by other groups in the world. Significantly, when assessed in detail, studies of this type give very real evidence of the controlling and limiting factors in livestock-forage farming. They also highlight gaps in knowledge and hence areas for further study.

Some of the management factors likely to be involved in future developments of livestock-forage systems that improve on those outlined in Table 3 include:

1. The use of improved cultivars and new species, especially at high levels of production at which their genetic potential is most likely to be expressed.
2. The development of improved systems of management that bring together combinations of species that will, for example, utilize more incoming radiation (de Wit 1960, Harris 1968) or more efficiently use water.
3. Improvements in the feeding value of pasture, especially by increasing the contribution of legumes in pastures.
4. The need to understand better and evaluate more thoroughly the many factors associated with utilization of pasture (Hodgson 1977) in livestock-forage systems at different animal pressures.
5. A requirement to better understand aspects of nitrogen fixation, nitrogen and phosphorus cycling, and other aspects of soil fertility and plant nutrition in pastoral agriculture. Of special interest is the role of the grazing animal in nutrient cycling and soil fertility buildup (Ball 1979).

The impact of results such as these on methods of utilization of grass and their interaction with grazing, flock or herd management, and fertilizer usage could be very large indeed.

There are many other facets of the soil-plant-animal complex of pastoral farming that need more intensive research in the New Zealand environment. Control of pasture pests and diseases will require approaches that integrate the different methods of control into systems. Similarly, as better feeding and breeding of livestock occur there will be a need to evaluate the significance of these in livestock-forage systems. Similarly, developments around synchronization of mating and induced lactation for better matching of feed supply and mineral supplementation for better stock performance and improved stock health will all need integration into livestock systems. These and many others, including interactions, will have their influence on future production levels and will require many different research approaches to elucidate them and their significance in forage-livestock systems.

CONCLUSIONS

Importantly, for all regions of the world we need more facts. To obtain them, research of excellence is needed. It doesn't really matter what approach is used as long as the research objectives are realistic and meaningful and are based on the "Null Hypothesis" principle, the research is enthusiastically and thoroughly executed, and the results of the research are put in an easily retrievable form readily available to all. For livestock-forage systems research the researchers who will succeed are those who get outside, get muck on their boots, and are not too bedeviled by the biometricians and the modelers. The best researchers (the generalists) will be those who also associate with the extension workers and the farmers and, if possible, work in teams.

If these conditions continue to be fulfilled the extension workers and the generalists will be satisfied. They will put the bits into their practical models for the managers. The simulators will also be satisfied, as they will have more hard data to feed into their models. Enthusiastic and dedicated approaches to research that produces meaningful results are also the surest way of maintaining the support of administrators and governments.

LITERATURE CITED

Ball, P.R. 1979. Nitrogen relationships in grazed and cut grass-clover systems. Ph.D. Thesis, Massey Univ., N.Z.

Blaser, R.E., R.C. Hammes, H.T. Bryant, W.A. Hardison, J.P. Fontenot, and R.W. Engel. 1960. The effect of selective grazing on animal output. Proc. VIII Int. Grassl. Cong., 601–606.

Boerger, A. 1935. Grassland improvement in Uruguay. Herb. Rev. 3:47–52.

Brougham, R.W. 1970. Agricultural research and farming practice. Proc. XI Int. Grassl. Cong., A120–A126.

Brougham, R.W. 1973. Pasture management and production. Proc. Ruakura Fmrs. Conf. N.Z., 169–184.

Brougham, R.W. 1981. Pasture management and animal production. Proc. N.Z. Grassl. Assoc. 42:191–193.

Brougham, R.W., D.C. Causley, and L.E. Madgwick. 1975. Pasture management systems and animal production. Proc. Ruakura Fmrs. Conf. N.Z., 65–69.

Browne, D., M.J. Walshe, and D. Conniffe. 1970. Irish research on problems in animal production experiments comparing legumes with fertilizers as the source of pasture nitrogen. Proc. XI Int. Grassl. Cong., A101–A107.

Campbell, A.G. 1964. Grazed pasture parameters. Dead herbage, net gain and utilization of pasture. Proc. N.Z. Anim. Prod. Soc. 24:17–28.

Christain, K.R., J.S. Armstrong, J.L. Davidson, J.R. Donnelly, and M. Freer. 1974. A model for decision making in grazing management. Proc. XII Int. Grassl. Cong., 3:106–110.

Cockayne, L. 1919. New Zealand plants and their story. Govt. Print., Wellington, N.Z.

Conway, A. 1968. Grazing management in relation to beef production. IV. Effect of seasonal variation in the stocking rate of beef cattle on animal production and on sward composition. Irish J. Agric. 7:93–104.

Cunningham, E.P. 1979. New technology in animal production. Agric. Record, Eire. 7–10 Sept.

de Wit, C.T. 1960. On competition. Versl. Landbouwk. Onderz. No. 66 8:1–82.

Eadie, J. 1976. Animal production systems from hill country in the United Kingdom. Proc. Int. Hill-lands Symp., Morgantown, W.Va., 687–691.

Fisher, R.A. 1925. Statistical methods for research workers. Oliver and Boyd Ltd. Edinburgh.

Harris, W. 1968. Pasture seeds mixtures, competition, and productivity. Proc. N.Z. Grassl. Assoc. 30:143–153.

Hodgson, J. 1977. Factors limiting herbage intake by the grazing animal. Proc. Int. Mtg. on Anim. Prod. Temp. Grassl. Dublin, 70–75.

Holmes, W. 1968. The use of nitrogen in the management of pasture for cattle. Herb. Abs. 38:265–277.

Hutton, J.B. 1973. Developments in nutrition and management and their relation to the future of the New Zealand dairy industry. Proc. Ruakura Fmrs. Conf. N.Z., 220–232.

Innis, G.S., I. Noy-Meir, M. Gordon, and G.M. Van Dyne. 1980. Total-system simulation models. *In* Grasslands systems analysis and man, 759–797. Cambridge Univ. Press.

Johnstone-Wallace, D.B. 1937. The influence of white clover on the seasonal production and chemical composition of pasture herbage, and upon soil temperature, soil moisture, and erosion control. Proc. IV Int. Grassl. Cong., 188–196.

Jones, M.G. 1933. Grassland management and its influence on the sward. Emp. J. Expt. Agric., 43–58.

Kennedy, W.K., J.T. Ried, M.J. Anderson, J.C. Wilcox, and D.G. Davenport. 1960. Influence of system of grazing on animal and plant performance. Proc. VIII Int. Grassl. Cong., 640–644.

Leagans, J.P. 1979. Adaption of modern agricultural technology by small farm operators: an interdisciplinary model for researchers and strategy builders. Cornell Univ., Ithaca, N.Y.

Levy, E.B. 1951. Grasslands of New Zealand. Govt. Print., Wellington, N.Z.

Lowe, J. 1970. Comparative efficiency of pastures and crops for animal production. Proc. XI Int. Grassl. Cong., A88–A94.

McKenzie, S.A. 1980. Changing pattern of advisory work. Proc. N.Z. Grassl. Assoc. 42:191–193.

McMeekan, C.P. 1956. Grazing management and animal production. Proc. VII Int. Grassl. Cong., 146–156.

McMeekan, C.P. 1960. Grazing management. Proc. VIII Int. Grassl. Cong., 21–26.

McMeekan, C.P. 1966. International financing of pasture production. Proc. X Int. Grassl. Cong., 45–49.

McMeekan, C.P., and M.J. Walshe. 1963. The interrelationships of grazing method and stocking rate in the efficiency of pasture utilization by dairy cattle. J. Agric. Sci. 61:147–166.

Morley, F.W. 1979. Agricultural systems and advances in technology conference. Anim. Res. Inst., Werribee, Victoria, Austral.

Noy-Meir, I. 1976. Rotational grazing in a continuously growing pasture: a simple model. Agric. Systems 1(2):87–112.

Rae, A.L., R.W. Brougham, A.C. Glenday, and G.W. Butler. 1963. Pasture type in relation to live-weight gain, carcase composition, iodine nutrition and some rumen characteristics of sheep. I. Live-weight growth of sheep. J. Agric. Sci. 61:187–190.

Riewe, M.E. 1961. Use of relationships of stocking rate to gain of cattle in an experimental design for grazing trials. Agron. J. 53:309.

Sampson, A.W. 1923. Range and pasture management. John Wiley & Sons, New York.

Seligman, N.G., and G.W. Arnold. 1980. Simulation of intensively managed grazing systems. *In* Grasslands system analysis and man, 853–880. Cambridge Univ. Press.

Smith, J.H. 1956. Some early advocates of rotational grazing. J. Brit. Grassl. Soc. 11:199–202.

Snedecor, G.W. 1937. Statistical methods. Iowa State Univ. Press, Ames, Iowa.

Spedding, C.R.W. 1970. The relative complexity of grassland systems. Proc. XI Int. Grassl. Cong., A126–A131.

Stapledon, R.G., and J.A. Hanley. 1927. Grassland—its management and improvement. Clarendon Press, Oxford.

Suckling, F.E.T. 1975. Pasture management trials on unploughable hill country in Te Awa. III. Results for 1959–69. N.Z. J. Agric. Res. 3:351–436.

Swain, F.G., P.T. Mears, F.H. Drawe, G.J. Mortagh, J.G. Bird, R.L. Colman, and G.M. Yabsley. 1970. Commercial evaluation of a new farming system. Proc. XI Int. Grassl. Cong., 925–929.

Trumble, H.C. 1949. The ecological relations of pastures in South Australia. J. Brit. Grassl. Soc. 4:133–159.

Van Dyne, G.M. 1970. A systems approach to grasslands. Proc. XI Int. Grassl. Cong., A131–A143.

Weaver, J.E., and F.E. Clements. 1938. Plant ecology. McGraw-Hill Book Co., New York.

Wheeler, J.L. 1962. Experimentation in grazing management. Herb. Abs. 32:1–7.

Willoughby, W.M. 1959. Limitations to animal production imposed by seasonal fluctuations in pasture and by management procedures. Austral. J. Agric. Res. 10:248–268.

Wolfe, E.C. 1979. Opportunities for improving the integration of crops and livestock. J. Austral. Int. Agric. Sci. 45:2.

DISCUSSION

Remarks of Discussion Leader, K. Von Bargen

As Dr. Brougham has indicated, our aim as researchers, educators, or suppliers to producers is to develop and disseminate information concerning plants, animals, production practices, and equipment. A livestock-forage producer then evaluates these data and makes decisions to achieve productivity or profit goals.

A systems approach is necessary to develop good solutions to complex problems, especially those that involve several disciplines. The Livestock-Forage Grazing System certainly qualifies. The ultimate goal of the systems approach is a mathematical model that will accurately predict performance.

Livestock-forage producers who have been successful in reaching the higher levels of productivity given by Dr. Brougham have done so because, I believe, they are good intuitive modelers and optimizers. Likewise, I would suggest that highly successful scientists, teachers, and extension workers are also good intuitive modelers and optimizers. This ability to intuitively perceive a system is aided immensely by a working experience with actual production systems.

System models are slow in developing. For example, linear programming, a relatively simple model when compared to a forage-livestock system model, required about 25 years to become widely used to formulate rations. Application of linear programming was also dependent upon producers' having access to computers.

One of the greatest benefits of a system model is an improved understanding of a system. Another is a means of improved communication between disciplines. Dr. J.E. Moore, in the discussion following the first plenary paper by Dr. D.J. Minson, developed a graphical model of the plant-animal complex. After specifically identifying each element of the system in this way, one can determine what data are available or what areas need to be researched.

As an example of improved communications, consider an element of a system to be reducing crop residue. Specifying a hammer mill screen size does not describe the process adequately for an engineer trying to design an energy-efficient machine. Hammer speed is of equal or perhaps more importance for larger screen openings. In other situations collecting and reporting more complete data on operating conditions will aid in the development of models and improvements in systems.

President Barnes stressed a need for standardized terminology. System models can be an aid in developing this terminology because all disciplines must communicate and agree upon the same information. Systems modelers would also benefit from the use of standardized terminology.

Your attention is directed to the papers on mathematical ecology if you would like to learn more about the state of model development and application. A simulation model, the Kentucky Beef Model, should be brought to your attention. Crop growth, livestock growth, and livestock reproduction are simulated, and energy as well as economic measures is determined. This model and an illustrative example are reported in recent conference proceedings of the American Forage and Grassland Council.

With the rapid increase in availability of microcomputers and access to computers through network systems, producers will want to utilize models for livestock-forage system management. Dr. Brougham was asked, "How can individuals be educated or prepared to meet the need for competent extension workers with many skills and a wide breadth of knowledge needed to apply computer-based model technology?"

Comment, Questions, and Responses

1. *Comment:* The people who are going to do effective modeling are those who get outside, know the forage-livestock systems they're researching, have the experience, and understand what it's all about. These are the sorts of people, I believe, who are going to make progress and lead this "revolution" that you've mentioned. I don't believe it's going to be a revolution. The progress already made, especially between, say, 1900 and 1950, before we became too involved with biometrics and, more recently, with modeling, was tremendous. It was achieved mainly by those very good researchers (the ecologists, agronomists, and livestock specialists) associated with perceptive extension workers and farmers, who knew the environment in which they were researching, saw what was required, and achieved. So perhaps there's a conflict in our approaches to the use of simulation modeling at the beginning of this discussion.

2. *Question:* I would like for you to comment on how we are going to resolve the dilemma alluded to by Dr. Von Bargen—the producers in developed countries are looking to the microcomputers and computer programs to tell them when to plant, when to harvest, what fertilizer to apply, and how to compile livestock feed rations—and they hope that as a result they may be able to do better in terms of grassland management. Yet we, even those of us who have had some experience in looking at the place of modeling, have trouble quantifying these matters. I think we are going to be into a hectic time trying to fight off the computers and their use in grassland management. Do you have comments on this? (U.S.A.) *Response:* I agree. Modeling has its place, although as I see it there is nothing new in modeling. Dr. Von Bargen referred to this. There were many people in research before the computer was developed who were very good modelers. I've worked with a number of them. They were good because they had perception and developed concepts that they researched, yet some of them were hopeless with calcula-

tors. The computer doesn't do much more than that, except it allows one to do the calculating so much more quickly. The computer could also allow one to cover more factors. But before factors are drawn together into concepts or models, a data base is needed. If the data base is insufficient, many will be misled by simulation. It concerns me that too many people are coming into research who are going straight to the computer before they know what they are modeling and what factors make up the system, and before they have even participated in making measurements. As an example, we have a team of researchers at Palmerston North, some of whom are involved in modeling. We also have a large data base, yet the models that our fellows tackle are models only of components of livestock-forage systems and not of the whole system.

I would also like to emphasize that I talked about livestock-forage systems in my paper. I didn't talk about compiling livestock rations or growing a crop, and I didn't talk about some of the things that Dr. Von Bargen referred to—e.g., indoor feeding and the complications of feed rations. The computer is a great aid for these types of assessments. Importantly, the subject I was asked to talk on was livestock-forage systems, and that is what I did. The comments in the paper about simulation should be assessed in that context.

3. *Question:* There is one thing that generates heat and that is talking about models. I agree with Dr. Brougham's comments. There is a large deficit of hard information and a need to get outside and get more. I can also see the role of modeling, but it's one thing to model the growth of a greenhouse crop of, say, lettuce, and another thing to model productivity in an uncertain environment in the field, at least under our conditions.

I would like to refer to one of your slides, where you were talking about maximizing production. In research we usually attempt to establish the biological potential in productivity, whereas the farmer usually optimizes his return/man/ha at a much lower level. I wonder if you have information on that or would care to comment. Of course, this is where the computer can be useful. But the farmer can usually work out permutations in his own mind at the level at which he operates. We need a range of options, and this is particularly important in the developing countries. There is no point in doing highly sophisticated research in many developing countries. A range of options is needed in research to give practical information to the man in the field. (Australia) *Response:* A couple of comments I think are relevant, and I did make reference to this in the paper. In many of the developed regions of the world the pressure is not on agriculture. There are, for example, some states in the U.S.A. where the pressure is not on agriculture. Kentucky is an example. Production could be doubled (some say trebled) if the incentives and national need were there.

When research becomes important is when the pressure is on the land. In some countries in the developed as well as in the underdeveloped world, that extra bit of production means much to economic survival, or even to human survival. That is where research becomes important.

Relating to the New Zealand scene for a moment: Our specialist dairy farmers are right up with the research workers, pushing them pretty hard. The research workers have to work hard to remain ahead. This is ideal, as the good farmer and the research worker then associate to best advantage.

The same applies in some of the developing countries, where production and sometimes human survival are at stake. I don't want to be too hard, but I do believe that much research carried out in some of the developing countries has been aimed too high. It should be aimed at levels where it will help the social and economic structure of those countries. I would love to hear some expression from people in the audience from some of those countries.

3. *Question:* Ray, you took an example of a special situation where you could see the potential for increasing production/ha by perhaps 2- to 3-fold and suggested that the problems associated with achieving increases in production in such situations are physical and that excessive capital investment would probably not be required. I would like to take an example from the UK, where, in our thinking, the actual requirements for improvement in grassland production are rather greater than perhaps you have indicated. In the UK the most obvious route for increasing milk production/ha is to put on more nitrogen fertilizer, then more stock. Yet a 10% increase in stocking rate would require a capital input of another U.S. $200/ha. We could, on average, at least double our stocking rate, thus requiring $2,000/ha. We have 7 million ha of grassland in the country. That would require some $14,000 million of extra capital in the agricultural industry if it were our goal to maintain grassland area and increase dairy production up to the potential. (England) *Response:* A couple of things would easily alter this. You don't have great food needs, and you have energy at this stage. But if your energy needs change, the pressures for increasing production using other methods could change. I've observed this in the U.S.A. In Texas, some of that research is really mission-oriented. It has to be, because they are running out of water in some sections, and the cost of production, especially in the feedlot systems, is very high. This has really put some enthusiasm into research. The same thing could happen in other countries, including the UK. In some ways I hope it doesn't, because nitrogen-fertilized pasture is predictable agriculture. Relying on clover can be much more unpredictable, especially at the start of the spring growing season. One other point in relation to your comments: In your environment, or similar environments, you still have to set goals, taking account of all factors, then set about

researching them with appropriate techniques. The data obtained are hard data that can and will be manipulated by economists and simulators alike. But it's the hard data that are important.

4. *Question:* I come from a country where the pressure is on agriculture for survival. Ray Brougham referred in his paper to the transmission of research results to farmers and to extension people. I would like to hear him comment a bit more on this. How could this be done more effectively? I think in his own country it is being done very effectively, but in many countries there is much information kicking around that is not being applied on farms. I would also like to comment in regard to the underdeveloped world. Often research people move in from outside countries and advocate programs that have no application whatsoever to the problems of those countries. I believe there is need for guidelines in setting out priorities in terms of the problems of those developing countries. (Ireland) *Response:* Thanks, Dr. Conway. Again, I think there is agreement here. As to how we do it in New Zealand, I don't think we do it much differently from others, except that we have a 70% dependence on agricultural income for export income. That's where we are: 70% of our export income from agriculture. So it is very, very important that we apply the knowledge we have to our agriculture. That makes a big difference. Some of the techniques we use are not much different from those used in other countries, except perhaps the discussion-group concept. In this situation groups of 8 or 10 farmers with similar interests, who have the ability to get on, associate for discussion to mutual advantage. It is peer criticism and it works very effectively. The good extension workers are the catalysts, and I certainly enjoy participating in those groups. That is one way we do it in New Zealand. But there are many ways, and the computer will certainly help by making data retrieval easier to achieve.

I'm a little disappointed that people from one or two of the developing countries have not participated in the discussions, but there it is.

5. *Question:* I am involved in computer simulation and could argue the point of the usefulness of models. Instead, I will ask if you had a choice between a modeler and a biological scientist, which would you choose? (Australia) *Response:* Obviously, the choice is dependent upon the needs and the balance of the staff. In New Zealand, positions are filled mainly with biologically oriented individuals. However, those with added background in modeling and computers would be encouraged to use those skills. But all would be required to participate in the collection of hard research data.

Nitrogen Relationships in Grassland Ecosystems

A. LAZENBY

Grassland Research Institute, Hurley,
Berkshire, England, UK

Summary

Some recent findings affecting nitrogen (N) cycling in grassland ecosystems are discussed, and an indication is given of major problems requiring research and development. The importance of modern techniques is stressed to enable further quantification of the pathways and processes of the nutrient in grassland; such data are needed both to improve understanding and to enable more efficient use of N in practice.

A brief description of the N in soil organic matter is followed by an account of symbiotic N fixation by legumes, including estimated levels and rates of fixation and factors affecting the process. Data are presented on N transfer from legumes to soil and availability for grass growth.

Studies on nutrient uptake indicate that grass can attain its potential growth rate with very low N concentrations in solution, provided they are kept constant; this suggests a possible role for slow-release fertilizers. Selection for more efficient conversion by grasses of N to dry matter (DM) also appears to be a possibility. DM response to N fertilizer and its recovery in harvested herbage are considered.

The importance of pathways of N loss are evaluated, the most significant being volatilization of ammonia, both from excreta deposited directly onto the pasture and from slurry. Losses from denitrification and leaching are generally less than from volatilization, though they can be significant under certain conditions.

Legumes and fertilizers are compared as sources of N in grassland ecosystems.

KEY WORDS: nitrogen, grassland ecosystems, fixation, fertilizer, recovery, transfer, losses, animal utilization.

INTRODUCTION

Much has been written on nitrogen (N) relationships in grassland, and there are comprehensive reviews of our knowledge of the element in temperate (e.g., Whitehead 1970) and tropical (e.g., CSIRO 1962) pasture ecosystems. However, information is accumulating rapidly, and in this review an attempt is made to highlight the more important recent developments in the field and to indictate major areas where our understanding is incomplete or where accurate information is lacking and hence where further research is needed.

The main pathways of N in grassland are well known (Woodmansee 1979), as are the processes involved. The N necessary for plant growth is derived mainly from soil organic matter, symbiotic fixation, or fertilizer. Grass is normally grazed, the ruminant retaining a relatively small proportion of the element contained in the ingested herbage; the rest is excreted, some becoming available for uptake by the plant almost immediately, and some being returned to the soil organic matter, together with that from decaying plant parts. Some nitrogen is lost from the system in gaseous form or by leaching.

The author is pleased to express his thanks to his colleagues, particularly Dr. L.H.P. Jones and Mrs. K.M. Down, for their assistance in preparing this paper and to the late Mr. D.W. Cowling, Dr. J.C. Ryden, and Dr. D.C. Whitehead for helpful discussions.

Quantification of the N pathways and processes is less complete. Whereas there is a fund of information in some areas, such as N recovery in herbage, and a reasonable amount on others, such as leaching losses, less is known of the incorporation into, and release from, soil organic matter, and little reliable data exist on gaseous losses. These differences in the level of our knowledge reflect both the difficulty of measuring the various parts of the N cycle and the availability of suitable techniques (Fig. 1). Some components (e.g., N recovery in herbage or removal in animal products) have been easy to measure for a number of years; in contrast, techniques have only recently been developed to quantify volatilization and denitrification, while field measurement still presents problems for processes such as N fixation.

Techniques such as lysimetry (Burford 1977), acetylene reduction (Ham 1977) and inhibition (Ryden et al. 1979), and the use of stable isotopes (^{15}N and ^{14}N) (Bremner 1977) will play a key role in future research developments. Their increasing use is essential to collect the accurate information needed both to improve our understanding of the N cycle and to provide bases for more efficient use of the element in practice.

SOIL N

Most soil N is contained in organic matter, which may vary in amount from less than 0.01% in infertile sand (Henzell and Ross 1973) to more than 30% in peat soils

Fig. 1. Quantification of the N cycle (after J.C. Ryden, personal communication).

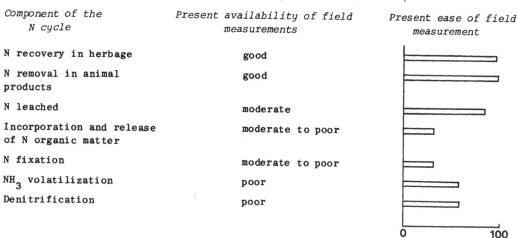

Component of the N cycle	Present availability of field measurements	Present ease of field measurement
N recovery in herbage	good	
N removal in animal products	good	
N leached	moderate	
Incorporation and release of N organic matter	moderate to poor	
N fixation	moderate to poor	
NH$_3$ volatilization	poor	
Denitrification	poor	

(Bramao and Riquier 1968). In the early stages of grassland development, the usual objective is to build up soil fertility in order to increase the supply of N for grass growth. In contrast, well-developed grassland soils may contain 3% to 6% organic matter (Allison 1973)—i.e., from 4 to more than 20 t N/ha in the root zone (Henzell and Ross 1973), some being more readily available to the plant than the rest.

There is a good general understanding of the processes of mineralization (microbial release of inorganic from organic forms of N) and immobilization (synthesis of organic from inorganic N) and of the effect on the balance between them of soil factors such as the C:N ratio, pH, aeration, temperature, and water. However, much work remains to be done on the complex relationships between soil organic matter and the supply of inorganic N to plants, and a major research effort is needed to improve our understanding of the processes involved. Further, if the rate of turnover of the large reserves of N under some of our permanent pastures could be increased, making more N available for growth without either losing it from the system or destroying the soil structure, it would be a tremendous breakthrough. This objective must be a considerable challenge to research and development (R&D).

SYMBIOTIC N FIXATION

Many estimates have been made of the annual levels of N fixed by legumes, varying from 0 (Henzell 1970) to nearly 900 kg/ha (Younge et al. 1964). Nutman (1976), consolidating data from a number of experiments, concluded that the mean annual fixation levels for four categories of temperate and tropical legumes fell into a not dissimilar range of 150–200 kg N/ha; yet there was a considerable variation in individual estimates, from some 30 to almost 600 kg N/ha/yr, which reflected local conditions.

Many of these estimates of fixation levels are based on the N content of harvested herbage and thus at first sight might appear to be low. However, it cannot be assumed that all the nitrogen in legumes has been fixed symbiotically by the plant. For example, early estimates of annual fixation levels in New Zealand made in the 1950s (Sears et al. 1965) were as much as 650 kg N/ha; they were con-

siderably higher than those of Hoglund et al. (1979), whose data, collected 15 to 20 years later from a number of sites, ranged from 1 to 400 kg N/ha. The buildup in soil fertility under many New Zealand pastures over the last 2 decades appears to have coincided with a fall in the level of N fixation. Such a result seems logical; legumes, like any other plants, will take up NO$_3$ and NH$_4$ ions if they are available.

Symbiotic fixation is a complex process. High rates of fixation require the presence of a rhizobium strain, both compatible with the host plant and sufficiently competitive with less efficient N-fixing strains to penetrate the legume roots in quantity. The considerable variation in the efficiency with which different genotypes of legumes and rhizobia fix N provides an opportunity for the plant breeder to select for increased fixation; however, rapid improvement of existing bacteria-host associations is unlikely.

In the longer term there is the possibility of using genetic engineering to develop a symbiotic relationship in grasses similar to that in legumes. The first step toward this objective—the introduction of the "nif" (N-fixing) gene cluster from *Klebsiella pneumoniae* into non-N-fixing bacteria—has already been taken (Postgate 1979). However, considerable technical problems remain to be overcome in achieving such an objective, which would rank as a major breakthrough in R&D.

As N fixation involves both host plant and bacterium, any factor interfering with the growth of either is likely to affect the process adversely. The effects on fixation of two major factors—soil temperature and defoliation—are illustrated in Fig. 2. Fixation occurred almost entirely between March and October when soil temperatures were above 6°C, achieving daily rates of 3.5 and 2.7 kg N/ha, respectively, prior to cutting the herbage in late May and August. A marked drop in fixation rates occurred immediately after defoliation.

The effects on both nodulation and N fixation have been summarized for soil temperature and pH (Gibson 1976), water (Sprent 1976), and nutrients (Sprent 1979). It is significant that NO$_3$ and NH$_4$ ions reduce nodulation. The effects of defoliation appear to be variable (Masterton and Murphy 1976, Gibson 1977).

Fig. 2. Seasonality of N fixation and soil temperature in white clover-ryegrass cut twice for silage. Dumdrum, Northern Ireland, 1972 (from Halliday and Pate 1976).

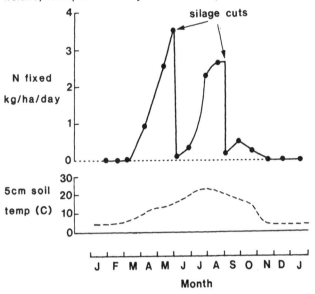

NITROGEN TRANSFER

Vigorously growing legumes provide little N for other plants. Excluding the grazing animal—the major pathway for N cycling in grassland—transfer occurs mainly as a result of the death and decay of plant parts (Vallis 1978).

Simpson's (1976) data (Table 1) demonstrate the considerable differences that may exist between legumes in the proportion of fixed-N transferred to the soil and to a companion grass. These differences are relatable to the growth characteristics of the plants. Annual legumes typically show a low rate of transfer prior to senescence; rapid release of N characterizes decomposition of the roots and leaves. In a Mediterranean climate, white clover releases little N until it succumbs to summer heat and moisture stress, the rate of such transfer being accelerated by defoliation. The deep-rooting system of lucerne appears to be associated with reduced tissue death and

thus to the rate of N transfer, which, in contrast to white clover, is further reduced by frequent cutting. Despite the considerably lower level of N fixation by the annual subterranean clover, the amount transferred to grass was almost twice that of the perennial white clover and three times that of lucerne (Table 1).

The complex and relatively little-researched field of N transfer needs further study. The use of isotopes of N should make possible better understanding of the processes involved and more accurate quantification both of differences between plants in their N-transfer levels and of the amount of N available for subsequent grass growth.

NITROGEN UPTAKE BY GRASS

Nitrogen is taken up by grass as NO_3 and NH_4. Because NO_3 is the more mobile ion it is likely to be the more readily available, at least in fertile soils. However, in short-term studies, Clarkson and Warner (1979) showed a more rapid uptake of NH_4 than of NO_3 by Italian ryegrass; further, at temperatures below 14°C there was a greater fall in uptake of the NO_3. More recent work with perennial ryegrass (L.H.P. Jones, personal communication) has shown that while NO_3 uptake is negligible below 8°C, NH_4 uptake can continue at temperatures as low as 3°C. Early spring growth of grass may thus be a consequence of the plant's ability to take up NH_4 at low soil temperatures.

The practice of applying fertilizer results in considerable fluctuation in concentration of N in the soil solution. While it may be at least 1,000 ppm N immediately following an application of 100 kg N/ha, rapid uptake by grass (daily rates of up to 10 kg/ha being recorded by Clement et al. [1978]) results in very low concentrations after 2 or 3 weeks. However, in flowing solution culture studies with *constant* N concentrations, a 1,000-fold difference in such N concentration (i.e., over the range of 0.2 to 200 ppm) produced less than a 10% difference in the yield of perennial ryegrass (Fig. 3).

The ability of a plant to achieve maximum growth rate with a very low concentration of N, provided the concen-

Table 1. Cumulative effects of three legumes on nitrogen balance over a 3-year period (from Simpson 1976).

	Subterranean clover	White clover	Lucerne
	kg N/ha		
Legume N Yield	437	851	1162
Increase in soil N at 0–10 cm (relative to nil legume controls)	159	483	322
N transferred to grass	142	88	48
	738	1422	1532
	percent		
% transferred to grass & soil	41	40	24
% transferred to grass	20	6	3

Fig. 3. Dry weight of shoots and roots of perennial ryegrass grown in flowing nutrient solution at different concentrations of NO_3 (from Clement et al. 1978).

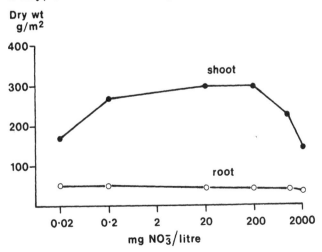

tration is kept constant (Clement et al. 1978), taken with the recent findings that there are two pools of NO_3 in plants (Martinoia et al. 1981), one ("stored N") apparently less available for plant metabolism than the other ("current uptake") (Clement et al. 1979), provides two good reasons for exploring the possibility of developing a fertilizer that produces a slow, constant supply of N. Such an innovation would both reduce the risk of pollution through leaching and denitrification and make possible better use of N by plants.

Differences exist both between and within grass species in their ability to convert N into DM; C_4 plants, with their lower N content (Jensen and Brown 1980), are, in this sense, more efficient than C_3 species. Exploitation of differences in genotypes of perennial ryegrass (Lazenby and Rogers 1965) provides a worthwhile objective for the plant breeder. Efficient conversion of N into DM should result in high yields of herbage with a low N content, though taking this concept to extremes has obvious dangers for animal production.

DM YIELDS AND N RECOVERY

The level of available soil N is indicated by the DM yields from · unfertilized (N_0) grass swards, recorded results ranging from less than 0.25 t (Morrison et al. 1980) to more than 30 t/ha/yr (Vicente-Chandler et al. 1959). The attainment of a particular yield level using fertilizer is dependent on predicting the amount of soil N likely to be available during the growing season, which, in turn, requires a suitable method. Recent developments in this area are encouraging; the use of one such technique (Whitehead 1981) accounted for more than 60% of the differences in N uptake and DM yields of perennial ryegrass grown on N_0 plots at 18 sites in England and Wales (Whitehead et al. 1981).

Irrespective of the level of soil N, the DM response to fertilizer N is more or less linear up to about 300 kg/ha in the UK (Morrison et al. 1980) and the Netherlands (Prins et al. 1980) and some 400 kg/ha in the humid tropics (Vicente-Chandler et al. 1959). The average response to N by temperate pastures — whether cut or grazed — is 22-25 kg DM/kg N (Richards 1977, Morrison et al. 1980), but in the humid tropics it can be as high as 70 kg (Vicente-Chandler et al. 1959). It is also possible to manipulate seasonal distribution of growth, to some extent, without affecting annual yield, by the timing and pattern of N application. Thus, applying a high proportion of N in spring will increase early growth at the expense of summer and autumn growth; such spring growth could be exploited for conservation. Application of a higher proportion of N in summer will produce a more even distribution of the herbage available for grazing (Morrison et al. 1980).

Even assuming conditions favoring N recovery, it is unreasonable to expect much more than 70% of the applied N to be contained in herbage that is cut and harvested in the year of application. Long-term field experiments, using isotopes of N, are needed to discover the fate of the remainder of the applied N in a grassland ecosystem, including the proportion ultimately recovered in cut herbage.

The above assumptions make no allowance for N losses, which undoubtedly occur. Further, as most grass is grazed, studies must include the effects of the grazing animal, even though this adds to the complexity and uncertainty of the ecosystem and places greater emphasis on the pathways and extent of N losses.

LOSSES OF N

In most grazing situations, volatilization of N as ammonia is the major pathway of loss. While there are few data, the proportion of N lost from urine appears to vary between 5% (Ball 1979) and 80% (Watson and Lapins 1969), depending on prevailing environmental conditions. Clearly, more studies are needed.

Vast quantities of N are voided in animal excreta every year. In the UK alone, Cooke's (1979) estimate is 840,000 t — i.e., considerably more than the total N applied as fertilizer to grassland. About 500,000 t of this N is voided directly onto grassland, and much of the rest is contained in slurry. U.S.A. data (Porter 1975) revealed a loss of N from pastures receiving heavy dressings of farmyard manure over a 15-month period amounting to 65% of the ammoniacal N and 17% of the total N applied. Data from U.S.A. and Western Europe (Cooke 1979) suggest that about half the N in animal excreta is lost by volatilization, though the proportion can vary from 0% to almost 100% (Porter 1975).

Applying animal manure, particularly slurry, to grassland is not without problems for both the plant and the animal (Schechtner et al. 1980). Nevertheless, slurry is a valuable, cheap source of N that should not be wasted. While there is evidence that acidification reduces volatilization as ammonia (Tunney and Molloy 1978), more studies are needed, especially on slurry. Their objectives should be not only to minimize N losses and increase the efficiency of animal excreta as a fertilizer but also to determine any wider environmental effects of high levels of ammonia volatilization.

DENITRIFICATION

Very few data exist on denitrification (the microbial reduction of nitrate to nitrous oxide or dinitrogen) in grassland; they indicate that the process is a less important source of loss than volatilization, though it can sometimes be significant. Daily losses of 0.3-0.7 kg N/ha, occasionally rising to 1 kg, were recorded by J.C. Ryden (personal communication) in southern England, especially in wet and warm conditions, with approximately 70% of total denitrification occurring between mid-May and mid-July. In temperate conditions it is estimated that denitrification may account for 5% to 8% of fertilizer N applied at rates of 250 kg/ha/yr.

Table 2. Nitrogen balance sheet under intensive milk production.

	kg N/ha	
	(i) *Without losses*	(ii) *With losses*
N available for plant growth:		
Fertilizer	250	
Loss by leaching (4%)	—	
Loss by denitrification (7%)	—	223
Mineralization (1.5%)	150	150
	400	373
Herbage N (80% in shoots)	320	298
N intake:		
Herbage (75% utilization)	240	224
Concentrate	60	60
	300	284
N retention by animal (25% of intake)	75	71
Gaseous losses of excretal N (50%)		106
Balance added to soil N:		

	Additions	retention	gaseous and leaching losses		
				235	
(i)	310	75			
(ii)	310	71	133		106

LEACHING

There is conflicting evidence of the importance of leaching as a source of N loss, with estimates varying from nil (Alberda 1972, van Steenbergen 1977) to 23% (Sluijmans 1978). In a UK study over a period of 8 years (Garwood et al. 1980), the average annual loss of the 250 kg N/ha applied to cut plots was about 8 kg, i.e., 4%. Sensible management will normally reduce such losses but cannot overcome the effects of freak conditions such as prevailed in Britain in 1976, when a severe summer drought followed by heavy rains resulted in the leaching of more than half of the 250 and 750 kg N/ha applied to experimental plots (Garwood and Tyson 1977).

The effects of gaseous and leaching losses can be illustrated in a system of intensive milk production on an all-grass pasture. Assuming an annual fertilizer rate of 250 kg N/ha plus 150 kg N from mineralization (1.5% of a soil organic pool of 10 t N/ha), a grass yield of about 11 t DM/ha should be produced with about 80% of the nitrogen in the above-ground herbage. Such a yield, plus concentrate supplementation of 30 kg N/cow, should allow a stocking rate of two 600-kg dairy cows, each yielding 5,500 kg of milk. Allowing for 25% N retention, over 200 kg N would be excreted by the animals.

The resultant annual N-balance sheets, with and without losses, are presented in Table 2. Assuming a 50% loss of excretal N by volatilization, 7% from denitrification, and 4% from leaching (this figure, because it was taken from cut-plot data, may be low), the soil N would be enriched by over 100 kg N/ha following the addition of 310 kg N/ha (250 kg from fertilizer, 60 kg from concentrates). If all losses could be eliminated, 240 kg N/ha would be added to the soil. Clearly, the greater the reduction of such losses, the more rapid the buildup of soil N reserves.

LEGUME N AND FERTILIZER N

Grassland systems based respectively on legume N and fertilizer N each have their particular uses, advantages, and disadvantages (Table 3).

There is little doubt that legumes will continue to form the basis of pasture improvement and production in many areas of the world. Indeed, it is possible to predict with some confidence their increased use following introductions, adaptive breeding, and agronomic and animal production studies, not only in tropical and subtropical areas of South America and Australia but also in the more temperate parts of Australia and New Zealand.

N-fertilized grass pastures, with their more predictable, higher yields, are likely to remain the basis of some inten-

Table 3. Advantages and disadvantages of grassland systems based on legumes and on fertilizers.

Legumes	Fertilizer N
Present use	
Increase fertility under extensive systems Where returns on animal products are low or unreliable	Achieve high reliable herbage yields for intensive systems
Advantages	

Cheap source of N
High quality
	Pastures—
	  more predictable yields
	  easier to manage

Disadvantages	
Less predictable yield Difficult to manage Pests and diseases Animal disorders	High inputs of costly support energy

sive livestock systems in parts of Western Europe and southeast U.S.A. for some time at least. Research programs designed to improve the efficiency of N fertilizer use will still be required although, in the long term, the future of high N use on grassland must be questionable unless alternative energy sources are found for fertilizer manufacture.

The advantages of legumes as a cheap source of N and their superior quality, coupled with the relatively recent release in Western Europe of a longer-petioled, larger-leafed white clover variety (Blanca) that is better able to resist shading and defoliation than varieties previously available, have led to a resurgence of interest in legumes in the UK. A major research effort is underway designed to overcome some of the problems preventing their wider usage. Introducing white clover into permanent pasture, and developing a management system enabling it to make a sustained contribution to grassland yields, are two practical R&D objectives. They are especially challenging because of the plant's susceptibility to intensive grazing and to competition from the more vigorous grass (Curll and Wilkins 1982).

It is doubtful that the basis for superior quality of legumes is as yet fully understood. Higher animal intake levels are certainly attributable in part to the better use by the ruminant of the energy content of legumes. However, the high protein content of legumes and lower rate of protein breakdown in the rumen, which combine to produce a greater flow of nonammonia N into the small intestine than a grass diet (Beever et al. 1980), seem associated on occasion with the superior growth rates of young cattle. Major experimental programs are in progress to reveal what happens to forage N in the ruminant; they include an examination of (1) losses in the early stages of digestion, (2) microbial protein production in the rumen, (3) the protection of protein from breakdown, and (4) the measurement of protein supply from rumen to ileum. They should make possible both a better understanding of ruminant digestion and an improvement in the efficiency with which forage N is utilized as a feed.

EPILOGUE

No reference has been made in this review to the role of below-ground invertebrates in the N cycle, which can be significant, or to associative N fixation, which could have real potential for tropical and subtropical pastures. The possible pollution effects of slurry, silage, leached nitrate, and volatilized ammonia have only been touched upon, and the analysis of the role of forage N in the ruminant diet is far from comprehensive. However, I have attempted to provide an indication of the magnitude and complexity of some of the challenges facing the research worker in the field of N in the grassland ecosystem. The solution of even some of these problems would increase understanding of the N cycle and, by making possible "finer tuning" of management practices, must surely lead to a better use of N in grassland farming.

LITERATURE CITED

Alberda, T. 1972. Nitrogen fertilization of grassland and the quality of surface water. Stikstof 15:45-51.

Allison, F.E. 1973. Soil organic matter and its role in crop production. Elsevier, Amsterdam.

Ball, R.P. 1979. Nitrogen relationships in grazed and cut grass-clover systems. Ph.D. Thesis, Massey Univ.

Beever, D.E., M.J. Ulyatt, D.J. Thomson, S.B. Cammell, A.R. Austin, and M.J. Spooner. 1980. Nutrient supply from fresh grass and clover fed to housed cattle. Proc. Nutr. Soc. 39:66a.

Bramao, D.L., and J. Riquier. 1968. Characteristics of the organic matter in the major soils of the world and its importance to soil fertility. *In* V.H. Fernandez (ed.) Organic matter and soil fertility, 45-53. North-Holland, Amsterdam.

Bremner, J.M. 1977. Use of N-tracer techniques for research on nitrogen fixation. *In* A. Ayanaba and P.J. Dart (eds.) Biological nitrogen fixation in farming systems in the tropics, 335-352. J. Wiley & Sons, Chichester.

Burford, J.R. 1977. Determination of losses of nitrogen from soils in the humid tropics by lysimeter studies. *In* A. Ayanaba and P.J. Dart (eds.) Biological nitrogen fixation in farming systems in the tropics, 353-363. J. Wiley & Sons, Chichester.

Clarkson, D.T., and A.J. Warner. 1979. Relationships between root temperature and the transport of ammonium and nitrate ions by Italian and perennial ryegrass (*Lolium multiflorum* and *Lolium perenne*). Pl. Physiol. 64:557-561.

Clement, C.R., M.J. Hopper, and L.H.P. Jones. 1978. The uptake of nitrate by *Lolium perenne* from flowing nutrient solution. J. Expt. Bot. 29:453-464.

Clement, C.R., L.H.P. Jones, and M.J. Hopper. 1979. Uptake of nitrogen from flowing solution culture: effect of terminated and intermittent supplies. *In* E.J. Hewitt and C.V. Cutting (eds.) Nitrogen assimilation of plants, 123-133. Academic Press, London.

Cooke, G.W. 1979. Losses of ammonia from agricultural systems. Royal Society study group on the nitrogen cycle. Roy. Soc. London.

CSIRO. 1962. A review of nitrogen in the tropics with particular reference to pastures. Bul. 46. Commonw. Bur. Past. and Field Crops. Commonwealth Agric. Bur. Farnham Royal, Bucks.

Curll, M.L., and R.J. Wilkins. 1982. The effect of treading and the return of excreta on a perennial ryegrass/white clover sward defoliated by continuously grazed sheep. Proc. XIV Int. Grassl. Cong. (in press).

Garwood, E.A., J. Salette, and G. Lemaire. 1980. The influence of water supply to grass on the response to fertilizer nitrogen and nitrogen recovery. *In* W.H. Prins and G.H. Arnold (eds.) The role of nitrogen in intensive grassland production, 59-65. Pudoc, Wageningen.

Garwood, E.A., and K.C. Tyson. 1977. High loss of nitrogen in drainage from soil under grass following a prolonged period of low rainfall. J. Agric. Sci. 89:767-768.

Gibson, A.H. 1976. Recovery and compensation by nodulated legumes to environmental stress. *In* P.S. Nutman (ed.) Symbiotic nitrogen fixation in plants. IBP. 7:385-405. Cambridge Univ. Press.

Gibson, A.H. 1977. The influence of the environment and managerial practice on the legume-*Rhizobium* symbiosis. *In* R.W.F. Hardy and A.H. Gibson (eds.) A treatise on nitrogen fixation. Section 4. Agronomy and ecology, 393-450. J. Wiley & Sons, New York.

Halliday, J., and P.S. Pate. 1976. The acetylene reduction assay as a means of studying nitrogen fixation in white clover under sward and laboratory conditions. J. Brit. Grassl. Soc. 31:29–35.

Ham, G.E. 1977. The acetylene-ethylene assay and other measures of nitrogen fixation in field experiments. *In* A. Ayanaba and P.J. Dart (eds.) Biological nitrogen fixation in farming systems in the tropics, 325–334. J. Wiley & Sons, Chichester.

Henzell, E.F. 1970. Problems in comparing the nitrogen economies of legume-based and nitrogen fertilized pasture systems. Proc. XI Int. Grassl. Cong., A112–A120.

Henzell, E.F., and P.J. Ross. 1973. The nitrogen cycle of pasture ecosystems. *In* G.W. Butler and R.W. Bailey (eds.) Chemistry and biochemistry of herbage. 2:227–246. Academic Press, London.

Hoglund, J.H., J.R. Crush, J.L. Brock, and R.P. Ball. 1979. Nitrogen fixation in pastures. XII Gen. Discussion. N.Z. J. Expt. Agric. 7:45–51.

Jensen, J.P., and R.H. Brown. 1980. The efficiency of nitrogen use by C_3 and C_4 grasses. Agron. Abst. 85. Am. Soc. Agron. Madison, Wis.

Lazenby, A., and H.H. Rogers. 1965. Selection criteria in grass breeding. V. Performance of *Lolium perenne* genotypes grown at different nitrogen levels and spacings. J. Agric. Sci. 65:79–89.

Martinoia, E., U. Heck, and A. Wiemken. 1981. Vacuoles as storage compartments for nitrate in barley leaves. Nature 289:292–293.

Masterton, C.L., and P.M. Murphy. 1976. Application of the acetylene reduction technique to the study of nitrogen fixation by white clover in the field. *In* P.S. Nutman (ed.) Symbiotic nitrogen fixation in plants. IBP 7:299–316. Cambridge Univ. Press.

Morrison, J., M.V. Jackson, and P.E. Sparrow. 1980. The response of perennial ryegrass to fertilizer nitrogen in relation to climate and soil. Grassl. Res. Inst. Hurley, Tech. Rpt. 27.

Nutman, P.S. 1976. IBP field experiments on nitrogen fixation by nodulated legumes. *In* P.S. Nutman (ed.) Symbiotic nitrogen fixation in plants. IBP 7:211–237. Cambridge Univ. Press.

Porter, K.S. 1975. Nitrogen and phosphorus—food production, waste and the environment. Ann Arbor Sci., Mich.

Postgate, J. 1979. The genetics of nitrogen-fixing bacteria. Royal Society study group on the nitrogen cycle. Roy. Soc., London.

Prins, W.H., P.F.J. van Burg, and H. Wieling. 1980. The seasonal response of grassland to nitrogen at different intensities of nitrogen fertilization, with special reference to methods of response measurements. *In* W.H. Prins and G.H. Arnold (eds.) The role of nitrogen in intensive grassland production, 35–49. Pudoc, Wageningen.

Richards, I.R. 1977. Influence of soil and sward characteristics on the response to nitrogen. Proc. Int. Mtg. Anim. Prod. from Temp. Grassl., Dublin, 45–49.

Ryden, J.C., L.J. Lund, J. Letey, and D.D. Focht. 1979. Direct measurement of denitrification loss from soils. II. Development and application of field methods. Soil Sci. Soc. Am. J. 42:110–118.

Schechtner, G., H. Tunney, G.H. Arnold, and J.A. Kenning. 1980. Positive and negative effects of cattle manure on grassland with special reference to high rates of application. *In* W.H. Prins and G.H. Arnold (eds.) The role of nitrogen in intensive grassland production, 77–93. Pudoc, Wageningen.

Sears, P.D., V.C. Goodall, R.H. Jackman, and G.S. Robinson. 1965. Pasture growth and soil fertility. VIII. The influence of grasses, white clover, fertilizers and the return of herbage clippings on pasture production of an improverished soil. N.Z. J. Agric. Res. 8:270–288.

Simpson, J.R. 1976. Transfer of nitrogen from three pastures under periodic defoliation in a field environment. Austral. J. Expt. Agric. Anim. Husb. 16:863–870.

Sluijmans, C.M.J. 1978. De mest-en gieverspreiding op landbouwgrond in de E.G. Commissie Europese Gemeenschappen. Informatie over landbouw. 47:154.

Sprent, J.I. 1976. Nitrogen fixation by legumes subjected to water and light stresses. *In* P.S. Nutman (ed.) Symbiotic nitrogen fixation in plants. IBP 7:405–420. Cambridge Univ. Press.

Sprent, J.I. 1979. The biology of nitrogen-fixing organisms. McGraw-Hill, Maidenhead.

Tunney, H., and S. Molloy. 1978. Nitrogen loss from cattle slurry. Soils Res. Rpt. 49. Agric. Inst., Dublin.

Vallis, R.J. 1978. Nitrogen relationships in grass/legume mixtures. *In* R.J. Wilson (ed.) Plant relations in pastures, 190–201. CSIRO, Melbourne.

van Steenbergen, T. 1977. Influence of type of soil and year on the effect of nitrogen fertilization on the yield of grassland. Stikstof 20:29–35.

Vicente-Chandler, J., S. Silva, and J. Figarella. 1959. The effect of nitrogen fertilization and frequency of cutting on the yield and composition of three tropical grasses. Agron. J. 51:202–206.

Watson, E.R., and P. Lapins. 1969. Losses of nitrogen from urine on soils from southwestern Australia. Austral. J. Expt. Agric. Anim. Husb. 9:85–91.

Whitehead, D.C. 1970. The role of nitrogen in grassland productivity. Commonw. Bur. Past. and Field Crops, Bul. 48. Commonwealth Agric. Bur., Farnham Royal, Bucks.

Whitehead, D.C. 1981. An improved chemical extraction method for predicting the supply of available soil nitrogen. J. Sci. Food Agric. 32:359–365.

Whitehead, D.C., R.J. Barnes, and J. Morrison. 1981. An investigation of analytical procedures for predicting soil nitrogen supply to grass in the field. J. Sci. Food Agric. 32:211–218.

Woodmansee, R.G. 1979. Factors influencing input and output of nitrogen in grasslands. *In* N. French (ed.) Perspectives in grassland ecology, 117–134. Springer Verlag, New York.

Younge, O.R., D.L. Plucknett, and P.P. Rotar. 1964. Culture and yield performance of *Desmodium intortum* and *D. canum* in Hawaii. Hawaii Agric. Expt. Bul. 59.

DISCUSSION

Remarks of Discussion Leader, D.L. Grunes

Dr. Lazenby has presented a very nice overview of N relationships in grasslands. The study of nutrient cycling through soil-plant-animal systems is important worldwide and is being considered by many groups in a number of countries. Methods of decreasing N losses are needed. Slow-release N fertilizers and nitrification inhibitors may be useful.

Some rather fascinating questions still remain concerning the reasons for differential transfer of N from different legumes to companion and succeeding graminaceous forages. Also, the comparative effectiveness of N fertilizers and legumes will require much research before it is fully understood. Much more research with legumes is needed.

One very interesting slide was that showing similar growth of ryegrass plants with a 1,000-fold range of N concentration in the nutrient culture solution—that is, from 0.2 to 200 ppm. Now I understand that the total N concentration in these plants was similarly unaffected over this range. It certainly points out that much-needed, important information can be obtained using flowing solution cultures. The equipment developed at the Grassland Research Institute in Hurley, England, is especially suitable, since the ion concentrations are automatically monitored and controlled. Solution temperatures are also controlled, permitting ion-uptake studies over a 3°–25°C temperature range. Development and use of such facilities require specialized technical staff and equipment.

Research concerning the differential effects of temperature and N sources on the uptake of N by plants is important and will require much additional effort. The present availability of both ^{15}N and enriched ^{14}N stable isotopes opens up many new areas for potential research. This will allow for separate labeling of both the ammonium and the nitrate N.

Another topic of concern is the potential for N pollution of groundwater, streams, and lakes due to leaching or runoff from agricultural lands. Appreciable N losses from soil to groundwater and streams occur when land that has been in grassland is tilled and planted to succeeding crops. There is also potential pollution from the application of N, manures, or slurries to pastures.

Finally, research is needed on soil tests to estimate N in the soil that will become available to plants during the growing season; on methods of determining N fixation; on rapid and accurate methods of determining N loss; and on the very difficult task of measuring nitrate and ammonium concentrations, as well as pH, at the soil-root interface in the field and in growth chamber and greenhouse studies.

Questions and Reponses

1. *Question:* Is it possible to have antibiosis of *Rhizobium,* with a resultant loss of N fixation? Is such research in progress? (U.S.A.) *Response:* The answer to your first question is that it is possible to have antibiotic effects in *Rhizobium* bacteria. I know of no research that presents positive evidence of an antibiotic effect. Competition between different *Rhizobium* strains does affect the effectiveness of infection and N fixation by the host plant. Current research developments in fingerprinting methods help in selecting the more competitive strains.

2. *Question:* Is there a move to discourage research on the use of legumes to fix N? (Liberia) *Response:* There was no intent to discourage research on biological N fixation through legumes. Legumes are tremendously important

in both developing and developed countries, and research should be continued on the use of legumes to fix N.

3. *Question:* We found that following a 120 kg/ha N application (ammonium nitrate) to June grass, the soil nitrate concentration dropped to zero in three weeks. During the same period, the concentration of ammonium dropped to the level in unfertilized soil. Nevertheless, the nitrogen uptake of the fertilized grass continued to be much higher than that of the unfertilized grass. We believe this was caused by rapid incorporation of N into microbial bodies, followed by remineralization. What is your opinion? (France) *Response:* Your analysis seems very reasonable and your observation illustrates the need to research the transformations of fertilizer N into and out of soil organic matter, as well as the role played by soil microbes and soil invertebrae in these processes.

4. *Question:* There is a need for continued research directed toward better methods for determining the amount and timing of nitrogenous fertilization. (New Zealand) *Response:* We do need better predictive tools to achieve the best management of nitrogenous fertilizers, including, if possible, procedures that the farmer can use with confidence to make his own decisions.

5. *Question:* We have obtained growth with legumes comparable to that obtained with 250 kg N/ha from fertilizer. However, we have difficulty in consistently repeating these findings. Consequently, we depend on fertilizer N. (Netherlands) *Response:* The objective of the Hurley program is to develop systems based on white clover providing enough N for pastures for many years. There are dangers in taking results from cut plots and applying them to grazed conditions. There are real difficulties in maintaining white clover in high-intensity grazing systems.

6. *Question:* Is the estimated loss of some 4% of fertilizer N by leaching derived from lysimeter studies where the forage was cut and removed or from grazed swards? (New Zealand) *Response:* These data were obtained from lysimeters where the forage was cut and removed. This result is the average of 7 years of data. That average does include one year with very abnormal rainfall distribution.

7. *Comment:* Information obtained from lysimeters where the forage was cut and removed ignores the effect of concentrating excess dietary N into urine patches by grazing animals. Research in New Zealand indicates that approximately 25% of urine N may be leached from a grazed pasture annually (up to 75% from irrigated pastures). About 75% of nitrate in groundwater in intensively farmed areas results from leaching of urine patches. (New Zealand) *Response:* Certainly one must be careful in extrapolating from one set of conditions to another.

Multiple Use of Grassland Resources (Grasslands to Provide Natural Resource Conservation and a Quality Environment for Mankind)

P.J. EDWARDS

Head, Pasture Research, Natal Region, Department of Agriculture and Fisheries,
Pietermaritzburg, Republic of South Africa

Summary

The grasslands of the world (and communities closely akin to them, such as shrub and savanna types) provide not only a very effective mantle for protection of the soil, but also the main source of feed for herbivores. They also cater to many other needs of mankind. Increasing population results in competition between types of grassland use for specific sites. Sound planning is necessary to resolve this conflict in the way most beneficial to mankind.

In order to plan logically it is necessary to make an inventory of the resources, to consider the various uses to which grasslands can be put, and to match the resources to the uses with the least disruption of the grassland. The inventory of resources should identify the main characteristics of the climate, soil, physiography, and vegetation at each grassland site. The uses to which natural grasslands can be put are divided on the basis of the extent to which they are disrupted. Those uses that maintain the natural grassland are grazing land, habitats for wild animals, recreation areas, genetic pool reserves, water source areas, and several other direct uses. The group of uses that result in drastic alteration include improved grasslands for grazing and sap extraction, cropping and afforestation, and urban development.

Management of natural grasslands is a vast subject but one that is critical for the maintenance of this steadily deteriorating vegetation. It is very difficult to restore degenerated natural grassland to its original composition and productivity. It is suggested that the most useful strategies available to the manager of natural grassland are fencing, control of stocking rates, fire, stocking density and rotation of grazing, and long periods of deferment from grazing.

The maintenance of existing grassland that is in good condition should take precedence over reclamation. The reclamation of areas where soil loss is not excessive should be catered for before attention is given to severely degraded areas.

Three groups are identified in the planning of a grassland. These are grassland scientists, planners, and policy makers. It is the responsibility of the grassland scientist to acquire the necessary data and to present them in an acceptable form to the other groups. He must also ensure that his voice is heard when his data have been ignored in the planning process.

KEY WORDS: grassland inventory, grassland uses, management, reclamation, planning.

INTRODUCTION

A large proportion of the vegetated landmass of the world is covered by grassland or plant communities closely akin to it (i.e., shrub and savanna that have a considerable grass component). These grasslands provide the main source of feed for herbivores, which produce most of man's supply of red meat and milk and much fiber. They also assist in providing for other less obvious aspects of man's well-being such as his aesthetic enjoyments, his relaxation, his water, and his physical and mental health. The grasslands of the world provide a most effective mantle, protecting the soil of the earth and slowing the rate of runoff of surplus water. Grasslands also provide sites for the so-called development of the nation (i.e., housing, industry, mines, and transport). Because of the variety of uses to which grasslands can be put, and because the types of usage are often mutually exclusive, conflict arises, particularly as population pressures increase.

It would seem to be in keeping with the objectives of this session to consider ways and means of ensuring the optimum allocation of grassland resources in order to achieve a quality environment for mankind. In achieving this requirement, available grassland resources should first be identified and described. Then we must consider the possible uses to which these resources can be put, as well as their site and management requirements. Next, the ways and means of reclaiming and making more productive those areas that have been degraded through abuse must be discussed. Finally, the use of grassland resources should be planned. It is also appropriate to consider the role, or perhaps the responsibility, of the grassland scientist in ensuring that allocations are made in a planned manner on the basis of sound information and not on an ad hoc haphazard basis.

This paper is largely concerned with the allocation, maintenance, improvement, and reclamation of grasslands.

RESOURCE INVENTORY

If we wish to provide for the conservation and optimum utilization of our natural grasslands to the ultimate benefit of mankind, the first step is to examine, describe, and classify our various grassland sites. There is a vast diversity of grassland sites, each of which is capable of supporting its own variety of plant communities and has its own optimum use. However, the identification and description of sites not only are vital for deciding on the best use of the site, but also provide the grassland ecologist with much of the information needed for its management.

A feature common to all systems of site assessment for grasslands is a description in varying detail of the climate, soil, physiography, and vegetation. Edwards et al. (1980) suggested that this should include precipitation (season and amount), temperature (extremes), soil climate (moisture), degree of leaching, erosion and flood hazards, effective depth, salinity, mechanical limitations, slope, rockiness (degree and size), landscape position, stock water sites, proportion of bushy, nonedible, or poisonous plants, uniformity of vegetation and its palatability at maturity, shade, and shelter.

USES OF NATURAL GRASSLAND

Grassland can be put to a number of uses. It is appropriate at this stage to look at these uses in relation to their degree of intensification and resulting disturbance.

Uses That Maintain the Natural Grassland

Into this category fall many pastoral and recreational activities in which management plays a vital role in determining stability and production.

1. *Grazing for domestic animals.* This use probably occupies the greatest proportion of existing grassland. Veld or rangeland occupies much of the land that is unsuitable for cultivation due to environmental limitations of soil, climate, and topography. In developing countries it may still occupy potential cropping land where it will, in due course, be replaced. This method of utilization is of considerable economic importance, particularly in the more arid zones.

2. *Grazing for wild animals.* In the past all grassland was used by wild animals. However, with the advance of time this type of utilization has been confined to the less accessible, drier, and rougher sites. In some instances, however, the tide has turned and game ranching has become more popular and occasionally as economical as the ranching of domestic animals. Thus, some ranching land has been allocated to wild animals, or these animals have been integrated with ranching.

3. *Habitats for wild animals.* Grasslands provide habitats for wild animals that do not graze (carnivores, birds, and insects). This type of land use is often combined with other uses, such as recreation and wilderness areas. The sites available for this use are generally those discarded for other uses.

4. *Recreation and wilderness areas.* There is a growing need for areas where people can "get away from it all."

The demand for nature and wildlife parks and wilderness areas seems to grow in proportion to the increase in the urban population and its affluence. Such "parks" may vary from small areas with urban facilities to large wilderness areas. Thus, a large variety of sites is required to cater to the various classes of recreation as well as to provide the desired vegetation and scenic variation.

5. *The direct sale of vegetation.* In some areas the flowers in grassland are plucked and sold commerically (the flowering component usually is not graminaceous). The plants of grasslands may be used for thatching dwellings and in medicines, herbal remedies, and even perfumes. Sites suited to the particular commercial component are required, although there is a tendency for such wild sites to be replaced by monospecific commercial plantings. A second category of this use is the harvesting of grasslands for fodder, for the extraction of nutrients for direct nonruminant and human consumption, and possibly for fuel. Due to the low yields of natural grassland and the requirements of an even surface for mechanical harvesting, herbage for these uses is usually supplied by higher-yielding cultivated crops. It is interesting to note that the heat energy contained in South African grass species was in the region of 17,500 kJ/kg (Trollope 1980), compared with 20,000 kJ/kg estimated as a general figure for forest and bush fires (Brown and Davis 1973, Luke and McArthur 1978). But because grasslands for fuel require annual harvesting of low yields (1/2 to 3 ton/ha), their potential as a source of fuel is unlike that of forest.

6. *Sources of genetic material.* Grasslands represent a vast reservoir of untapped genetic potential. Apart from the plants, a variety of microorganisms, insects, and animals exists in specific grassland habitats, and some of this pool of genetic material is liable to disappear should the vegetation on these sites be destroyed. This material could be used in the breeding of new fodder, fiber, medicinal, perfume, or fuel types. It is certainly desirable that examples of this material be retained. Consideration must be given to the location, size, and distribution of such reservoirs. From management, security, and access considerations a single large block encompassing a variety of sites is most practical. However, natural disasters or man-made intrusions could destroy the whole complex. Therefore more numerous moderate-size blocks are favored as a compromise between security and ease of management. Replications should be available to safeguard against disasters, and access to these areas should be reserved to those who wish to study the flora in detail.

7. *Soil and water conservation.* Natural grasslands, when in good condition, form a very effective cover to protect the soil from erosion and to encourage the infiltration of rain water. Thus, although grasslands are seldom used solely to conserve soil or water, this conservation is usually a secondary objective in any form of grassland use.

Uses That Radically Alter the Natural Grassland

Into this category fall agricultural uses, industrial uses, mining, and urban development. The consequences of these uses usually are irreversible changes in, or destruction of, the grassland. These developments are particularly relevant to this paper because they are continually making inroads into natural grassland. While such progress cannot be halted, it can often be planned and directed into the least disruptive (to grasslands) localities. It is essential that the allocation of grassland for these uses be carried out on a planned basis.

1. *Improved grassland for grazing.* Planted pastures that are established following the cultivation, fertilization, and seeding of land considerably outyield natural grasslands. Consequently their area expands, as does the demand for pastoral products. The accessible, more productive, and less erodable sites are usually selected for this development, although, as Edwards (1978) indicates, appropriate methods of veld reinforcement and replacement that also radically alter the native sward are available for most natural grassland sites.

2. *Direct production of food from grasslands for nonruminant and human nutrition.* This category refers to the mechanical and chemical processing of grassland herbage for direct consumption by pigs, poultry, and humans. This subject has been under investigation for some years. According to Pirie (1975), leaf protein was isolated by Rouelle in 1873 and was suggested as a food by Ereky in 1925.

It is interesting to note that in many aspects the nutrient content of milk and of lush clover pasture are similar. From Table 1 it is apparent that the moisture, protein, and "other carbohydrate" contents of the two substances are similar. Clover has less fat and more fiber and minerals than milk, and it is possible that the pasturage may be deficient in certain amino acids essential for human nutrition (Eggum 1970). However, if the fiber is removed, we are left with a low-fat milk substitute. Most interest has centered around use of this product as a concentrate for pigs and poultry, but it can also be used successfully in reinforcing the protein content of human food (Pirie 1975). The fractionation process is simple, with a low energy requirement. Because this process requires harvested material relatively high in protein and low in fiber, herbage suitable for reasonable extract yields would most likely come from planted pastures on high-quality sites (i.e., replaced natural grasslands).

3. *Cropping* (food, fiber, perfume, medicinal, and fuel crops). Because of mechanization of planting, manage-

ment, and harvesting of these crops, and because of the crops' high production potential, the most accessible, productive, level, and stable sites are used.

4. *Afforestation.* In some nations having inadequate supplies of timber from forested areas, grassland is replaced by forests. Sites used for this purpose are often, although not always, steep and rocky and are situated in high-rainfall areas.

5. *Urban development.* Natural grasslands must often give way to urban, industrial, and mining development. The sites for this development are usually determined by the proximity and the requirements of the existing infrastructure and population.

Mines must be placed where the minerals are located. Nevertheless, certain options are open that can reduce the impact of this development on natural grassland. Where possible, new factories and urban development should be placed on poor-quality sites such as eroded areas. When planning road and rail networks, the agricultural potential of various alternative routes should at least be considered, not completely ignored as is usually done today.

MANAGEMENT OF NATURAL GRASSLANDS

Prevention of further deterioration (and ultimately desertification) of our existing grasslands is, to my mind, a far higher priority than the reclamation of severely denuded areas. Of the four natural resources involved in rangeland—soil, climate, topography, and vegetation—it is the last that is most easily changed by modifications or misuse. It is disturbing to realize that, in the natural grassland with which I am familiar, it is extremely difficult to restore to more advanced successional stages, within 30–40 years, grasslands that have deteriorated to secondary stages of succession. Edwards et al. (1979) have estimated that a large portion of the eastern seaboard of South Africa is susceptible to colonization by the unpalatable *Aristida junciformis*. Added to this is the fact that such change can be initiated by poor grazing management in a relatively short period of 7 years (Tainton et al. 1978). We have also recently recorded a 65% increase in secondary grasses after only 2 years of overgrazing. A similar situation was noted in the southern Great Plains of the U.S.A., where some 30–40 years after abandonment, formerly cultivated fields had not recovered to the climax composition of true prairie (Savage and Runyon 1937, Booth 1941). The almost irreversible replacement of productive natural grassland by poorly producing, more monospecific secondary grassland is rife in southern Africa and a similar situation probably exists in many other natural grasslands. This change from productive to unproductive grassland, although not as spectacular as the final degradation from pioneer plants to bare earth, nevertheless represents a grave loss to grassland productivity and stability. Thus, it is essential that we understand the management requirements and tolerance limits of vegetation on each site of natural grassland that occurs under our jurisdiction.

The management of natural grassland is a vast subject; it is possible to deal here with only a few basic principles

Table 1. Comparison of the nutrient content of milk and of a lush clover pasture (Morrison, 1949).

Constituent	Milk		Lush clover	
	%	%	%	%
Water		87		84
Dry matter		13		16
Composition of dry matter:				
Protein	27		27	
Fat	29		2	
Fiber	0		14	
Other carbohydrates	38		44	
Minerals	6		13	
TOTAL	100	100	100	100

that we consider to be essential to management. We have found a modification of the Quantitative Climax Method of Dyksterhuis (1952) to be very useful for assessing the condition and trend of the vegetation on each of the units (Foran et al. 1978, Tainton et al. 1980). This method allows for three increaser groups and one decreaser group of species. These are:

Increaser I: Those species that increase in abundance when grazing or fire is infrequent.

Increaser II: Those species that increase in abundance when veld is overutilized.

Increaser III: Those species that increase in abundance when grazing is selective.

Decreasers: Those species typically found in veld that is in good condition (in this case the fire grazing subclimax) and that decrease in abundance as condition of the veld deteriorates with reference to production from grazing animals.

Decreaser dominance is considered to be optimum for animal production, while large amounts of Increaser Is may be desirable for certain types of wildlife and sightseeing. Excesses of Increaser IIs and IIIs are difficult to reduce. The proposed use of grassland will determine the optimum condition for each site. Objectives must be set and management designed to achieve these.

Because of the heterogeneity of our vegetation (due to site and palatability variations) over short distances it is necessary either to identify and separate (by fences) the more palatable and thus more vulnerable areas for individual management, *or* to manage the whole area for the survival of the palatable patches, *or* to sacrifice the most palatable areas to degradation (Edwards 1972). With domestic herbivores fencing should be used, but with game very light stocking may provide the solution. *Correctly placed fences are one of the most effective ways of preventing desertification.*

The intensity of utilization of natural vegetation is critical in its development or retrogression. Thus at each site it is essential to know the effects of stocking rate on the vegetation and on the production of the grazing animal. Our experience tends to indicate that a stocking rate that maintains optimum veld condition for domestic livestock (i.e., grazing capacity) is close to the one that gives maximum gain/animal. Fire can play an important role in maintaining natural grasslands in a particular stage of development, and consequently its effect and interaction with grazing for the various sites must be understood.

For the grazing of domestic animals we believe that the withdrawal of grassland from grazing for reasonably long periods to allow for specific physiological rests for seeding, translocation, and seedling establishment is necessary. Because of mixed composition and palatability of the sward on each site we find that a high stocking density (animals/unit area at a specific time) for short periods, with variable (depending on the vegetation) regrowth periods between the grazings, is advantageous on our natural grasslands. This allows more uniform utilization

of the mixed sward and a rationing of fodder. We thus advocate rotational grazing within a system of rotational rests.

The vegetation that is eventually present on each site depends on the environmental potential of the site, the vegetation originally present, the management applied, and the response of the vegetation to this management. To achieve success in our grassland management it is necessary to identify and separate sites of varying potential. The vegetation present must be identified in terms of its probable reaction to a variety of management practices, vegetation objectives must be defined, and the appropriate management must be applied to achieve these objectives.

REVEGETATION

I hope it is apparent from the foregoing that I consider prevention of further deterioration of grassland to be the first step in containing desertification. However, there are already large areas that have been denuded to the extent that they will not recover without a helping hand. Where possible, these areas should be utilized for those purposes that destroy the natural grassland, e.g., roads, industry, dams, and urban development. Even on individual farms such possibilities exist. A gully may form a useful site for a dam or bunker silo even though other, slightly better sites that are covered by good grassland do exist. Farm roads can be sited over denuded patches in preference to good grazing sites, and where other factors such as soil are equal, new cropping land should be ploughed on secondary grassland rather than on good grassland.

The reclamation of denuded areas may require simple measures such as overseeding or fertilization. On the other hand, it may require massive inputs of capital and energy in the form of mechnical works. The areas that require small inputs should be reclaimed and secured first, for it is these areas that usually show the greatest benefit/unit cost and that are the most vulnerable to further rapid deterioration. Would not the thousands of dollars spent on reclaming one ha of open-cast mine yield greater benefit in reclaiming a hundred ha of land that has been overgrazed and requires $50/ha for stabilization and reclamation? Because of public awareness the relatively small areas of roadsides and mine dumps are far better taken care of in terms of revegetation than are vast areas of denuded grassland that are not directly in the public eye. There is room to reassess our priorities.

The revegetation of denuded areas is a well-researched subject, and many books, starting with those of Ayres (1936) and Bennett (1939), have been published on this subject. Many techniques have been developed, and new materials are constantly being introduced in this field. Of these we will no doubt hear during this session. At the same time new types of revegetation problems are being created by advances of mining, industry, and transport. However, the success of revegetation must, as is the case with land use, depend on identification of the characteristics of each site, the restrictions these conditions impose on plant growth, and the resources available to remove the restrictions. Where the soil is relatively undisturbed, revegetation is usually not a difficult process. However, where soil has been eroded or cut away and removed, revegetation poses considerable problems.

The methods used in revegetation must be aimed at overcoming the shortcomings of the site for good plant growth. They are usually associated with lack of suitable plant material (seed), lack of ability of plants or their seed to remain attached to the site, lack of fertility, and lack of a sustained water supply. The lack of supplies of seed of species suitable for revegetation appears to be a problem in many subtropical and tropical countries.

It is suggested that the order of priorities should be, first, to retain existing good-quality grasslands; second, to revegetate those areas where topsoil is still available; and third, to reclaim those areas where topsoil is absent.

PLANNING THE USE OF NATURAL GRASSLAND

"Planning is essentially the allocation and balancing of resource potentials with resource demands within social, political, biological, physical, economic and legislative constraints" (Barry and Egging 1977, as quoted by Lyon 1977). To achieve planned grassland, use objectives must be established, the grassland resource must be defined, and an assessment must be made of the possibility of meeting the objectives with the available resources. If the resources are inadequate, the possibility of their modification and its cost must be considered. This assessment must include the long-term effects of the modifications, the cost of maintenance, and the cost of reclamation. In the light of this information the objective must be reassessed.

It is possible to distinguish at least three groups of people who should contribute to the planning of use of grassland. There are the scientists, the planners, and the policy makers. It is the job of the grassland scientist to provide an inventory of available resources and to describe their current condition, their potential, and their possible uses. He must indicate what modifications to the resource are possible and the consequences and cost of such modifications. He must also propose management systems to be applied during the use of the resource. The planner should have available information supplied by the grassland scientist and also information from a variety of other sources, such as demographers and economists, from whom he can derive the requirements of the nation within the curent political constraints. The planner must match the resources with the requirements and assess the need to modify either the requirements or the resources. The policy maker must keep the populace satisfied in the short term and, if he is wise, will attempt also to satisfy the long-term requirements of future generations. He may have to choose between short-term expediency and long-term wisdom. The grassland scientist, by providing information on the stability, elasticity, and reclamation possibilities of each site, can simplify the tasks of the planner and the policy maker.

It is doubtful that the scientist has always played his rightful role in the planning and decision-making process in land use. There rests upon the scientist an important responsibility in land use. First, he must ensure that the type of information he gathers is that required by the planner. Second, he must ensure that his information is collated and made available to those who will use it in a form that is useful to them. Third, he must make his voice heard through his scientific societies if proposals for land use do not match the environmental tolerances he has determined.

CONCLUSION

Planning of multiple uses of grassland resources is in the hands of the grassland scientist, the planner, and the policy maker. The role of the grassland scientist is to make an inventory of the grassland resources that are available. He must consider the possible uses of grassland and match these with the least disruption to the grassland sites he has identified. He must understand the response of the vegetation on each grassland site to management variables and must be in a position to recommend those management practices necessary to achieve a particular desired grassland situation. The grassland scientist must also consider priorities in reclamation and make his proposals on this and other subjects known to the planners and the policy makers.

LITERATURE CITED

Ayres, Q.C. 1936. Soil erosion and its control. McGraw-Hill Book Co., New York.

Bennett, H.H. 1939. Soil conservation. McGraw-Hill Book Co., New York.

Booth, W.E. 1941. Revegetation of abandoned fields in Kansas and Oklahoma. Am. J. Bot. 28:415-422.

Brown, A.A., and K.P. Davis. 1973. Forest fire: control and use. McGraw-Hill Book Co., New York.

Dyksterhuis, E.J. 1952. Determining the condition and trend of range (natural pasture). Proc. VI Int. Grassl. Cong. 2:1322-1327.

Edwards, P.J. 1972. A system of veld classification and management planning. So. Afr. Dept. Agric. Tech. Serv. Tech. Comm. Rpt. 102, 12.

Edwards, P.J. 1978. Methods of veld reinforcement, their action and adaptability to various sites. Proc. Grassl. Soc. So. Afr. 13:71-74.

Edwards, P.J., R.I. Jones, and N.M. Tainton. 1979. *Aristida junciformis* (Trin. et Rupr.): a weed of the veld. Weeds. Third Natl. Conf. A.A. Balkema, Cape Town, 25-32.

Edwards, P.J., D.M. Scotney, P.E. Bartholomew, and N.M. Tainton. 1980. Environmental modifications for improved grassland production. Proc. Grassl. Soc. So. Afr. 15:19-30.

Eggum, B.O. 1970. The protein quality of cassava leaves. Brit. J. Nutr. 24:761-768.

Foran, B.D., N.M. Tainton, and P. de V. Booysen. 1978. The development of a method for assessing veld condition in three grassveld types in Natal. Proc. Grassl. Soc. So. Afr. 13:27-33.

Luke, R.H., and A.G. McArthur. 1978. Bush fires in Australia. Austral. Govt. Pub. Serv., Canberra.

Lyon, W.N. 1977. The information about fire needed by the land management planner. Proc. Fire Working Group, Soc.

Am. Foresters. U.S. Dept. Agric. Forest Serv., New Mex. Gen. Tech. Rpt. I.N.T. 49:3-5.

Morrison, F.B. 1949. Feeds and feeding. Morrison, New York.

Pirie, N.W. 1975. Leaf protein: a beneficiary of tribulation. Nature 253:239-241.

Savage, D.A., and H.E. Runyon. 1937. Natural revegetation of abandoned farmlands in the central and southern Great Plains. Proc. IV Int. Grassl. Cong., 178-182.

Tainton, N.M., P.J. Edwards, and M.T. Mentis. 1980. A revised method for assessing veld condition. Proc. Grassl. Soc. So. Afr. 15:37-42.

Tainton, N.M., B.D. Foran, and P. de V. Booysen. 1978. The veld condition score: an evaluation in situations of known management. Proc. Grassl. Soc. So. Afr. 13:35-40.

Trollope, W.S.W. 1980. Characteristics of fire behaviour. *In* Ecological effects of fire in South Africa. Natal Univ. Press.

DISCUSSION

1. *Question:* The rangelands and grasslands in many areas of the Old World (e.g., Near East and North African regions) are devastated by overgrazing and uncontrolled cropping. This frequently reflects the rapid rise in human and livestock populations. There is insufficient control of grazing on rangelands and grassland that are mainly common lands. Both governments and international research centers seem to be largely dodging the problems because of social and economic difficulties. The tragedy is that, if nothing is done to stop the abuse of common lands, more and more productive agricultural land will be destroyed (e.g., tropical sands are covering good croplands in the Sudan). Dr. Edwards, will you comment on the problems of grazing the common lands? (Australia) *Response:* It appears that no one is responsible for common lands. The problem is social restraint. We should not try to introduce difficult and complex systems for improvement of good grazing lands. The two most important management factors that can be best used are controlled stocking rate and introduction of improved species.

2. *Question:* You have described three classes of increaser species. Some of us are not familiar with your nomenclature. Please expand on your definition of the Class I increaser species. (U.S.A.) *Response:* Fire has decreased Class I increasers. Class I cannot be burned without damage. Class I does not tolerate heavy grazing and is basically unpalatable to livestock.

3. *Question:* In your presentation you did not suggest the introduction of clover or legumes into grasslands. Was this in your paper and you did not have time to cover it, or would you comment on why it was not covered? (Canada) *Response:* I think we should improve species, but I don't want to see legumes introduced into a site that is not adapted to that legume species.

4. *Question:* Why don't you want to see the grassland go to a good pasture instead of leaving it as a climax species? (U.S.A.) *Response:* I agree with you on sites that are adapted to improved species, but many problems resulting from multiple use of grassland resources result from the high requirements of the management dictated by maintaining a good pasture. Introducing a good pasture with no management skills to maintain it results in something worse than what the climax species originally provided.

5. *Question:* In our current requests for research support, our legislators often remind us of and ask about cost:benefit ratios. Would you comment on the cost:benefit ratio of the multiple-use programs you suggested? (U.S.A.) *Response:* We do not have cost:benefit ratio information. Generally, the procedure followed is setting priorities and then making decisions on saving an area for ecological purpose or allotting it for industry development. We must decide that we need to preserve the resource on the basis of keeping that resource rather than on the economic value of that resource.

6. *Question:* You referred to three types of grassland in your presentation. Do you know what proportion of these types of grassland are in the world? (U.S.A.) *Response:* We don't have figures that relate to the treatment or description of grassland as I have discussed them because people are not trained well enough to assess their definition or use.

Physiological and Morphological Advances for Forage Improvement

J.P. COOPER
Director, Welsh Plant Breeding Station,
University College of Wales, Aberystwyth, UK

Summary

Forage production basically involves the conversion of environmental inputs to provide digestible energy and other nutrients for ruminant livestock. The plant breeder needs to consider the seasonal inputs or constraints in his particular environment, the efficiency of the crop in responding to these factors, those features of the plant that contribute to this efficiency, and the degree to which those features can be modified by selection and breeding.

The primary climatic variable determining crop production is the seasonal input of light energy, but in most grassland environments growth or even survival can be limited by temperature or water stress, while shortage of soil nutrients, particularly nitrogen, is usually important.

Photosynthetic production will be influenced by the photosynthetic rate of the individual leaves, by their arrangement in relation to light perception, and by the extent of respiratory losses. In forage crops, the harvestable yield, including regrowth, is also influenced by the partition of assimilates between shoot and root development and between flowering and continued vegetative growth. Response to seasonal limitations of temperature or water stress often involves two conflicting requirements, the ability to continue active growth under moderate stress and/or survival under more extreme conditions. Furthermore, the physical and chemical constitution of the harvested material is important in determining digestibility and intake, and hence animal production.

Considerable genetic variation is available within forage species for many physiological or morphological features. The plant breeder has to consider which features are most important in determining seasonal production and how far reliable selection criteria can be developed that compare favorably with the standard methods of field assessment. Such rapid screening techniques can be of particular value where production is limited by one major but unpredictable climatic factor, such as low temperature or water shortage, or where rapid laboratory techniques can be developed for the assessment of nutritional components.

KEY WORDS: assimilate partitioning, leaf arrangement, nutritive quality, photosynthesis, plant breeding, respiration, water stress, temperature.

INTRODUCTION

Forage production basically involves the conversion of environmental resources, such as light energy, CO_2, water, and soil nutrients, to provide digestible energy and other nutrients for farm livestock. The energy in the feed is derived ultimately from crop photosynthesis, which provides both the structural material of the crop and the energy required to synthesize it. In attempting to improve this conversion the plant breeder must consider the seasonal inputs or constraints in his particular environment, the efficiency of the crop in responding to these factors, those physiological or morphological features of the plant that contribute to this efficiency, and how far those features can be improved by selection and breeding (Evans 1975, Cooper 1981).

The primary climatic variable determining crop photosynthesis and production is the distribution of light energy, which differs greatly with climatic region, both in total input and in seasonal amplitude. In many grassland environments, however, the use of this energy for photosynthesis and growth is severely limited by seasonal temperature or water stress, while shortage of soil nutrients, particularly nitrogen, is usually an important limitation.

CROP PHOTOSYNTHESIS

The efficiency of a forage crop in converting the seasonal inputs of light energy and CO_2 into total dry matter will be influenced by the photosynthetic rate of the individual leaves, by their arrangement in the canopy, and by the extent of respiratory losses.

Individual Leaf Photosynthesis

The photosynthetic response of the individual leaf to light intensity is well documented (Cooper 1976, Ludlow 1978). At low light intensities the limitations are largely photochemical and some 15% of the incoming light energy can be fixed, but as light intensity increases, CO_2 supply becomes limiting and eventually light saturation and the maximum photosynthetic rate are reached. As light saturation is approached the percentage of incoming light energy fixed is steadily reduced and may fall to values of 2%–3%.

In forage crops, a major distinction in the photosynthetic response to light energy lies between the forage legumes and most temperate grasses that possess the standard C_3 photosynthetic pathway, and many subtropical grasses in which the rather more efficient C_4 pathway is present (Ludlow 1978). In C_3 species light saturation is reached at fairly low intensities (some half of full sunlight), with a maximum photosynthetic rate of 50–100 ng CO_2/cm^2 (of ground surface)/s corresponding to some 2%–3% fixation of light energy. This comparatively low value results at least in part from marked photorespiration in the light, which may take up 30%–40% of the carbon initially fixed.

In C_4 species, on the other hand, the initial carboxylating enzyme PEPcarboxylase has a greater affinity for CO_2 and there is little or no loss by photorespiration. As a result, light saturation is rarely reached even in bright sunlight, and a much higher maximum photosynthetic rate of up to 200 ng CO_2/cm^2/s can be attained, corresponding to some 3%–4% fixation of light energy.

Temperate and tropical species also show marked differences in their temperature response for both photosynthesis and growth. Most temperate grasses and legumes show an optimum of about 20-25°C and photosynthesis and growth are greatly reduced at 5-10°C, while the optimum for many subtropical and tropical species can be as high as 30–35°C, with a marked reduction below about 15°C (McWilliam 1978).

Even so, appreciable variation has been detected within many grass and legume species both in the maximum photosynthetic rate (Wallace et al. 1972, Wilson 1981) and in the temperature response of photosynthesis

(McWilliam 1978). In few cases, however, has it been possible to correlate these differences in photosynthetic rate with differences in dry-matter production of the whole plant or crop, possibly because of other compensatory changes, such as an inverse correlation between photosynthetic rate/unit area and leaf size.

In view of the large photorespiratory losses in C_3 plants, the possible value of selecting for lower rates of photorespiration has been suggested (Zelitch 1976). Marked differences in photorespiration have in fact been detected in a number of species, including perennial ryegrass (Wilson 1972), but again no regular correlation has been obtained with dry-matter production or indeed with individual leaf photosynthesis itself.

Leaf Arrangement and Light Interception

The photosynthetic rate of the whole crop will also be greatly influenced by the way in which the leaves are arranged in relation to the distribution of light down the canopy (Saeki 1975, Ludlow 1978). In species with flat prostrate leaves, as with many legumes, most of the incoming light is absorbed in the top few layers of the crop, while in the grasses with longer, more erect leaves the light can penetrate to a greater depth. The greater the leaf area over which the light can be spread, the more efficiently light will be converted. The crop canopy can therefore attain a much greater potential photosynthetic rate than the individual leaf, and light saturation is rarely reached even in bright sunlight. In temperate C_3 species, photosynthetic rates of 200–300 ng CO_2/cm²/s have been recorded, while even higher values are recorded for tropical C_4 grasses in high-insolation environments. Taking account of respiratory losses, these would correspond to crop growth rates of 20–30 g/m²/day and some 40–50 g/m²/day, respectively (Cooper 1976, Ludlow 1978).

Appreciable variation has been reported between and within forage species for such canopy characteristics as leaf size, leaf and stem angle, and leaf rigidity, which affect the structure of the canopy and hence its maximum photosynthetic rate (Wallace et al. 1972, Ludlow 1978), and in some cases, as in perennial ryegrass, appear to be related to differences in crop growth rate (Rhodes 1975, Rhodes and Mee 1981).

Respiratory Losses

Although much of the carbohydrate initially produced by photosynthesis is incorporated into the structure of the crop, an appreciable part is utilized to provide energy for the synthesis of new material and maintenance of the existing structure. Recent studies on the carbon balance of perennial ryegrass, for instance, indicate that some 50% of the carbon and energy initially fixed is lost by respiration (Robson 1973).

It is important, however, to distinguish between "synthetic" respiration that is coupled to active growth processes and "maintenance" respiration concerned with the maintenance of the existing structure. While it may be difficult to reduce synthetic respiration without also

reducing the growth of the crop, there is some evidence that maintenance respiration can be reduced, with a consequent increase in dry-matter production. In perennial ryegrass, for instance, selection for reduced dark respiration of mature leaves has resulted in increases of 10%–15% in annual dry-matter production (Wilson 1975b, 1981).

HARVESTABLE YIELD

In practice, however, the farmer is interested not so much in total dry-matter production as in *harvestable yield*, usually over a series of harvests, and in nutritive quality for farm livestock.

Partition of Assimilates

The optimum partition of assimilates will depend on the expected duration of the crop. For a short-term crop that is to be utilized at a single harvest, the maximum partition to harvestable shoot, rather than to root and shoot bases, is desirable. For a crop that is expected to continue production over a series of harvests, under either grazing or cutting, a balanced partition between harvested material, shoot bases, and roots is important in order to provide sufficient residual photosynthetic area and/or energy reserves for active regrowth. The relative importance of carbohydrate reserves and current photosynthesis will depend on the frequency and level of defoliation (Ryle and Powell 1975, Harris 1978).

The partition of dry matter is also markedly influenced by the timing and extent of reproductive development. In the grasses, the extension of flowering stems is associated with an increased harvestable yield above the level of cutting or grazing. On the other hand, the development of an inflorescence on any shoot puts an end to further leaf production and may also inhibit the development of axillary tiller buds, leading to a subsequent depression on yield (Harris 1978).

Similarly in the legumes, the balance between harvested yield and remainder of the plant is important. In white clover, for instance, the relative partition of dry matter to the stolon or to the growth of new leaves and petioles has a marked effect on varietal persistence from year to year (Rhodes and Harris 1979).

Varietal differences in the seasonal partition of assimilates are often related to the effects of past climatic or agronomic selection. In many temperate grasses, for instance, populations from the Mediterranean region grow actively at moderately low temperatures through the winter and flower comparatively early in the spring before the summer drought begins. Conversely, populations from northern or central Europe, which are exposed to severe winters, show a degree of winter dormancy; active leaf growth begins later in the spring and flowering is comparatively late (Cooper 1964, McWilliam 1978).

Superimposed on these climatic responses are the selective effects of different farming systems. In the ryegrasses, for instance, intensive grazing tends to select plants that are prostrate and highly tillering and that flower late or sparsely, whereas management for hay and even more so for seed production will favor early-flowering types with a higher proportion of reproductive shoots.

Nutritive Value

In forage crops, which are grown to provide feed for farm livestock, the nutritive quality of the harvestable yield is also important, in particular digestibility and those as yet ill-defined characteristics that determine voluntary intake (Cooper 1973, Cooper and Breese 1980). In most temperate grasses, vegetative material and young developing inflorescences are highly digestible, but as the inflorescence matures and lignification proceeds, digestibility decreases steadily.

Forage species can, however, differ in their digestibility even at the same stage of development. Most subtropical grasses are regularly less digestible than temperate species, while among the temperate species cocksfoot is usually less digestible than ryegrass. Genetic variation in digestibility has, however, been reported within several forage species, including cocksfoot, bromegrass, and lucerne (Cooper 1973).

Voluntary intake is influenced greatly by the digestibility of the feed, but even at the same level of digestibility, forage species and varieties can differ significantly in intake (Walters 1974). These differences may be based on factors affecting the rate of breakdown in the rumen, such as the proportion of soluble to structural material in the feed, which can be modified by selection, or the rate of passage out of the rumen, such as particle size and shape during breakdown, which varies greatly between legumes and grasses. Furthermore, under grazing conditions, intake may be affected by differences in the structure of the canopy, which influences the presentation of herbage to the animal (Cooper and Breese 1980).

RESPONSE TO CLIMATIC STRESS

Most grasslands are exposed to major seasonal limitations of temperature or water stress that limit the achievement of their potential. The effects of water stress are usually more serious than those of low temperature, since low temperatures are often associated with periods of reduced solar insolation and hence lower potential production. In coping with these climatic stresses, the crop is often presented with two conflicting requirements: (1) the ability to continue active growth under moderate stress, and (2) survival under more extreme conditions, often involving a degree of dormancy.

Temperature Stress

The response of forage species or ecotypes to temperature is usually closely related to their climatic origin (McWilliam 1978). In the temperate grasses, collections from north or central Europe show extreme cold-hardiness, usually associated with a degree of winter dormancy, while Mediterranean or southern forms can continue leaf growth actively at moderately low temperatures (c.5°C) but are more cold-susceptible.

Similar variation in low temperature responses for active growth, related to ecological origin, has been reported for such subtropical grasses as *Setaria* and *Paspalum,* though their critical temperatures are naturally rather higher (McWilliam 1978).

Water Stress

Water stress results from high evaporative demand and/or shortage of water within the root range of the crop. Continued active growth can thus be achieved by either reducing transpiration losses or increasing the availability of water. C_4 grasses show an increased CO_2 uptake relative to water-vapor loss and hence a greater water-use efficiency than C_3 species. Even within the same photosynthetic group, however, forage species and varieties differ in their response to water stress both in leaf extension growth and photosynthesis (Turner and Begg 1978), while in the C_3 perennial ryegrass, selection for either fewer or smaller stomata results in more efficient water use and higher dry-matter production during moderate water stress (Wilson 1975a, 1981).

Variation in root range in relation to soil water supply is less well documented, though forage species and varieties may vary in both root extension and depth and in the response of these features to soil water deficits (Turner and Begg 1978).

In environments with more severe water stress, however, as in Mediterranean climates, continued growth is not possible and other strategies are required. One frequent option is the development of an annual habit, as in many Mediterranean grasses and legumes such as subterranean clover, which germinate with the autumn rains, grow actively during the winter and early spring, and flower and produce seed before the summer drought begins. In these species, the lifecycles of locally adapted populations are usually closely linked to the possible growing season as determined by water supply (Cooper 1964).

An alternative strategy is summer dormancy, as developed by many perennial species in Mediterranean or subtropical environments. In *Phalaris tuberosa,* for instance, the length of the active growing season and the time of flowering, before dormancy sets in, is closely related to the onset of the period of drought (McWilliam 1978).

RESPONSE TO NITROGEN

In most grasslands, a shortage of soil nutrients, especially of nitrogen, is a major limitation. The nitrogen supply for the crop is derived from organic matter in the soil, from fertilizer nitrogen, or from biological N fixation, including the *Rhizobium*-legume association, or to a lesser extent the symbiotic bacteria associated with the roots of certain subtropical grasses.

Grass species and varieties differ in their ability to respond to inorganic nitrogen. In general, response is less in tropical than in temperate species, and less in native grasses adapted to infertile soils (Andrews and Johansen 1978). Considerable variation in response to nitrogen (and other nutrients, including phosphorus) has also been reported among edaphic ecotypes in both temperate and tropical forage species.

The level of nitrogen fixation by the *Rhizobium*-legume association differs with species and environment, a recent New Zealand survey indicating a range from under 100 to over 300 kg N/ha, depending on locality and soil type (Hoglund et al. 1979). A problem in many environments, however, is lack of predictability among seasons and locations, both for the fixation of nitrogen by the *Rhizobium*-legume association and for the growth of the legume itself.

In addition to the specificity of certain rhizobia on particular legume species, considerable variation in both infectivity and effectiveness has been reported among different strains of *Rhizobium* on the same host species (Date and Brockwell 1978), while in white clover, Mytton (1975) found that over 30% of the variation in nitrogen fixation was due to specific interaction between host genotype and *Rhizobium* strain. Such variation clearly offers scope for the breeding and selection of improved clover-*Rhizobium* associations, but in many soils the problem remains of establishing such improved associations in the face of competition from the indigenous *Rhizobium*.

The nitrogen-fixing bacteria, such as *Azotobacter* spp. and *Spirillium* spp., associated with the roots of certain tropical grasses fix rather less nitrogen than does the *Rhizobium*-legume symbiosis, but here also there is evidence of specific host-strain interactions (Dobereiner and Day 1976).

UTILIZATION IN FORAGE IMPROVEMENT

Considerable genetic variation thus exists within forage species for many of the physiological and morphological characteristics that influence their potential for providing digestible energy and other nutrients for farm livestock.

In attempting to make use of this variation the plant breeder has to consider which of these features are most important in improving production and quality of the crop in the field (Wallace et al. 1972, Cooper 1981).

Their relative importance can be assessed (1) by examining the physiological or morphological basis of known species and varietal differences in field performance, (2) by selecting for high and low expression of a particular characteristic and examining the effect of such selection on performance in the field—an approach that should also reveal any undesirable correlated responses and may provide material for early incorporation into a practical breeding program, and (3) by using recent models of crop photosynthesis and crop growth to predict the likely result of changing particular physiological components (Monteith 1977).

The value of a particular physiological or morphological feature as a selection criterion, however, also depends on the development of rapid and reliable screening techniques that are capable of handling large numbers of individuals, preferably at an early stage in the life cycle, and

that offer some practical advantage over the more traditional field assessment.

Such techniques can be of particular value where production is limited by one major environmental factor, such as winter cold or summer drought, the effects of which may be rather unpredictable from year to year, and where response can be assessed in the field only at one time of the year. Rapid screening techniques for cold-hardiness, for instance, have been developed by using controlled environments, the results of which correlate well with field performance. Similar techniques can also be used to screen for leaf growth over a range of light and temperature regimes (McWilliam 1978). Similarly, it is often possible to screen in sand or water culture for response to major or minor nutrients, though the interpretation of the results in terms of field performance needs caution. Rather less is known of the possibilities of rapid screening for response to water stress, though in perennial ryegrass selection for smaller and fewer stomata has resulted in improved water-use efficiency and dry-matter production (Wilson 1975a, 1981).

The development of rapid laboratory screening is also of value in the assessment of nutritive quality of forage, particularly digestibility. Standard digestibility assays based on long-term feeding trials using sheep or cattle are clearly not practicable for the large number of small samples that need to be screened in a breeding program. Various laboratory techniques have therefore been developed, including the assay of various fiber fractions, digestion with rumen liquor, and more recently the use of cellulase-pepsin digestion (Cooper and Breese 1980). A major problem, however, remains the development of suitable laboratory techniques for assessing those characteristics that contribute to voluntary intake, particularly under grazing.

The potential benefits of screening for the various components of crop photosynthesis, however, are rather more speculative. Although appreciable genetic variation has been detected in maximum photosynthetic rate and in photorespiration, no regular relationship between these characteristics and crop photosynthesis or crop growth rate has yet been obtained. Screening for respiratory losses, on the other hand, particularly the "maintenance" respiration of the mature leaf, shows more promise (Wilson 1975b, 1981). Similarly, certain of the plant characteristics that affect light interception by the canopy and hence crop growth rate can be readily selected in the young seedling stage (Rhodes and Mee 1981).

CONCLUSIONS

It is becoming possible to identify some of the limiting steps in the efficient conversion of available environmental inputs by forage crops to feed of high nutritive value for farm livestock and to specify some of the physiological and morphological features of the plant that determine this efficiency. The importance of some of these characteristics in limiting or determining production is clear, and rapid and reliable early screening techniques have already been developed. For others, such as many components of crop photosynthesis and those features determining the partition of assimilates, neither their relative importance nor the practicability of using them as selection criteria is as yet completely clear.

Much of our information on the physiological and morphological features determining seasonal yield and quality of forages has, however, been based on studies on temperate grasses, growing without major limitations of water and soil nutrients. It is encouraging that comparable information is now rapidly accumulating for the tropical and subtropical grasses, for the forage legumes (including both temperate and tropical species), and also for environments with greater limitations of temperature, water stress, or low nutrient status.

LITERATURE CITED

Andrew, C.S., and C. Johansen. 1978. Differences between pasture species in their requirements for nitrogen and phosphorus. *In* J.R. Wilson (ed.) Plant relations in pastures, 111-127. CSIRO, Melbourne, Austral.

Cooper, J.P. 1964. Species and population differences in climatic response. *In* L.T. Evans (ed.) Environmental control of plant growth, 381-403. Academic Press, London.

Cooper, J.P. 1973. Genetic variation in herbage constituents. *In* G.W. Butler and R.W. Bailey (eds.) Chemistry and biochemistry of herbage. 2:127. Academic Press, London.

Cooper, J.P. 1976. Photosynthetic efficiency of the whole plant. *In* A.N. Duckham, J.G.W. Jones, and E.H. Roberts (eds.) Food production and consumption, 107-126. North-Holland Pub. Co., Amsterdam.

Cooper, J.P. 1981. Physiological constraints to varietal improvement. Roy. Soc. London, Phil. Trans. B292:431-440.

Cooper, J.P., and E.L. Breese. 1980. Breeding for nutritive quality. Proc. Nutr. Soc. 39:281-286.

Date, R.A., and J. Brockwell. 1978. Rhizobium strain competition and host interaction for nodulation. *In* J.R. Wilson (ed.) Plant relations in pastures, 202-216. CSIRO, Melbourne, Austral.

Dobereiner, J., and J.M. Day. 1976. Associative symbioses in tropical grasses: characterization of micro-organisms and dinitrogen fixing sites. Proc. I, Int. Symp. Nitrogen Fixation. 2:518-538. Washington State Univ. Press, Pullman.

Evans, L.T. 1975. The physiological basis of crop yield. *In* L.T. Evans (ed.) Crop physiology: some case histories, 327-355. Cambridge Univ. Press.

Harris, W. 1978. Defoliation as a determinant of the growth, persistence and composition of pasture. *In* J.R. Wilson (ed.) Plant relations in pastures, 67-85. CSIRO, Melbourne, Austral.

Hoglund, J.H., J.R. Crush, J.L. Brock, R. Ball, and R.A. Carran. 1979. Nitrogen fixation in pasture. XII. General discussion. N.Z. J. Expt. Agric. 7:45-51.

Ludlow, M.M. 1978. Light relations of pasture plants. *In* J.R. Wilson (ed.) Plant relations in pastures, 35-49. CSIRO, Melbourne, Austral.

McWilliam, J.R. 1978. Response of pasture plants to temperature. *In* J.R. Wilson (ed.) Plant relations in pastures, 17-34. CSIRO, Melbourne, Austral.

Monteith, J.L. 1977. Climate and the efficiency of crop production in Britain. Roy. Soc. London, Phil. Trans. B281:277-294.

Mytton, L.R. 1975. Plant genotype × Rhizobium strain interactions in white clover. Ann. Appl. Biol. 80:103-107.

Rhodes, I. 1975. The relationships between productivity and some components of canopy structure in ryegrass (*Lolium* spp). IV. J. Agric. Sci. 84:345-351.

Rhodes, I., and W. Harris. 1979. The nature and basis of differences in sward composition and yield in ryegrass-white clover mixtures. *In* A.H. Charles and R.J. Haggar (eds.) Changes in sward composition and productivity. Occ. Symp. Brit. Grassl. Soc. 10:55-60.

Rhodes, I., and S.S. Mee. 1981. Changes in dry matter yield associated with selection for canopy characteristics in ryegrass. Grass and Forage Sci. 35:35-39.

Robson, M. 1973. The growth and development of simulated swards of perennial ryegrass. II. Carbon assimilation and respiration in a seedling sward. Ann. Bot. 37:501-518.

Ryle, G.J.A., and C.E. Powell. 1975. Defoliation and regrowth in the graminaceous plant: the role of current assimilate. Ann. Bot. 39:73-78.

Saeki, T. 1975. Distribution of radiant energy and CO_2 in terrestrial communities. *In* J.P. Cooper (ed.) Photosynthesis and productivity in different environments, 297-322. Cambridge Univ. Press.

Turner, N.C., and J.E. Begg. 1978. Responses of pasture plants to water deficits. *In* J.R. Wilson (ed.) Plant relations in pastures, 50-66. CSIRO, Melbourne, Austral.

Wallace, D.H., J.L. Ozbun, and H.M. Munger. 1972. Physiological genetics of crop yield. Adv. Agron. 24:97-146.

Walters, R.J.K. 1974. Variation between grass species and varieties in voluntary intake. Proc. V Gen. Mtg. Europ. Grassl. Soc. 1973, 184-192.

Wilson, D. 1972. Variation in photorespiration in *Lolium*. J. Expt. Bot. 23:517-524.

Wilson, D. 1975a. Leaf growth, stomatal diffusion resistances and photosynthesis during droughting of *Lolium perenne* populations selected for contrasting stomatal length and frequency. Ann. Appl. Biol. 79:67-82.

Wilson, D. 1975b. Variation in leaf respiration in relation to growth and photosynthesis of *Lolium*. Ann. Appl. Biol. 80:323-328.

Wilson, D. 1981. Breeding for morphological and physiological traits, *In* K.J. Frey (ed.) Plant Breeding Symposium II, 233-290. Iowa State Univ. Press, Ames, Iowa.

Zelitch, I. 1976. Biochemical and genetic control of photorespiration. *In* R.H. Burris and C.C. Black (eds.) CO_2 metabolism and plant productivity, 343-358. Univ. Park Press, Baltimore.

DISCUSSION

Remarks of Discussion Leader, G.H. Heichel

I should like to enlarge upon the subject of nitrogen, which Prof. Cooper highlighted as having a major role in the productivity of grasslands.

Replacing fertilizer N sources with N derived from pasture and forage legumes is receiving increasing interest in North American agriculture. The potential for savings of fossil energy is immense, and likewise the potential for reducing the costs of production.

Several research teams in North America have embarked upon high-risk research programs to combine plant breeding and genetics with crop physiology, biochemistry, and microbiology in attempts to *breed* forage legumes for improved N fixation on a crop-community basis. These research teams are located in Minnesota (collaboration of the USDA and the University of Minnesota), in Mississippi (collaboration of the USDA and Mississippi State University), at the University of California, and at the Canadian Department of Agriculture in Ontario.

These programs are showing that conventional plant-breeding procedures can be used to select for N-fixation capability in greenhouses and controlled environments. As Prof. Cooper pointed out, the crux of the situation is making these selection schemes relevant to the field environment. In Minnesota, our USDA group is finding that the important link between greenhouse performance and field performance can be made. Greenhouse selection has improved N fixation of experimental genotypes in the field.

In Mississippi (in the clover system) and in Minnesota (the lucerne system), preliminary results demonstrate a possible solution to the vexing problem (again pointed out by Prof. Cooper) of introducing new rhizobial strains into soils colonized by highly competitive indigenous populations. The solution is to breed a host plant for compatibility with a particularly competitive rhizobial strain that promotes effective N-fixation. Host manipulation may provide a way of maintaining the desirable *Rhizobium* in the root rhizosphere.

In closing I would like to reiterate and reinforce Prof. Cooper's thesis that it is now possible to specify some of the physiological and morphological traits of the plant that determine the efficiency of conversion of inputs. Teams of scientists combining the expertise of complementary specialites (exemplified by those in Prof. Cooper's Station) are the vehicle to further progress.

Questions and Responses

1. *Question:* Why is rate of dark respiration a useful selection criterion in perennial ryegrass whereas rate of net photosynthesis and rate of photorespiration are not? (Discussion Leader) *Response:* Measurements of net photosynthesis are usually made on a leaf area basis, and there is often a negative correlation between net photosynthesis/unit leaf area and leaf size. In a crop canopy, the rate of leaf area expansion may well be more important than the photosynthetic rate of the individual leaves, i.e., crop growth is sync-limited rather than source-limited. The extent of genetic variation of dark respiration rate of mature leaf tissues (maintenance respiration) is rather surprising, suggesting that there has been little natural selection for this characteristic in the past.

2. *Question:* Would selection for reduced losses due to senescence and decay be more successful than attempting to select for increased carbon inputs? (UK) *Response:* My colleagues have identified a mutant in tall fescue in which the leaves remained green rather than showing obvious senescence. More detailed work, however, revealed that this mutant only blocked the degradation of chlorophyll and that other aspects of senescence progressed normally.

There was, therefore, no improvement in carbon economy. A further search for quantitative genetic variation in senescence might well be of value.

3. *Question:* What is the role of crop physiology in assisting the plant breeder in improvement of tropical pasture and forage species? (Colombia) *Response:* The improvement of most tropical pasture and forage species is still in the stage of collecting and evaluating new species and ecotypes; and much progress has been and will continue to be made using this approach. There is evidence that the range of genetic variation for physiologically useful traits in tropical forages is similar to that in temperate species. As new breeding programs develop, the use of such characteristics should enable still further progress to be made.

4. *Question:* Do you think that more emphasis should be placed on developmental studies compared with research on photosynthesis in investigations of the genotype × environment interaction? (Australia) *Response:* In studies of crop improvement, it is difficult to draw boundaries between morphology, physiology, and genetics. The last 10 to 20 years have seen a great increase in research on crop photosynthesis, partly stimulated by the discovery of the C_4 photosynthetic system. The current realization that much of the variation in crop yield is based on the distribution and use of assimilates by the plant is already leading to increased interest in studies on plant development.

5. *Question:* What is your opinion on the use of the infrared reflectance technique for assessing forage quality? (Ireland) *Response:* This technique was initially developed for the evaluation of processed feeds, but it is now being used to assess digestibility of other nutritional factors in forages. A major problem in forage evaluation, however, is the assessment of intake characteristics both under grazing and in conserved feeds. It is not clear how far the IR technique can assist in this problem.

6. *Question:* Forage yield and quality are usually negatively correlated. How far can we go in attempting to improve both? (U.S.A.) *Response:* Maximum yields of dry matter are usually obtained from older stands such as those harvested at an advanced stage of reproductive development. In grasses, however, it has already proved possible to select for improved quality of the flowering stems.

7. *Question:* What are the major advances in forage legume improvement during the last few years in Western Europe? (Australia) *Response:* The use of forage legumes in Western Europe has been greatly influenced by the availability of cheap fertilizer nitrogen. Even so, advances have been made in lucerne, particularly in disease resistance, and in white clover varieties that are more compatible with grasses and can tolerate moderate inputs of nitrogen fertilizer. A major deterrent to the use of improved varieties of white clover is difficulty of seed production and hence high price of seed.

8. *Question:* Did selection for reduced stomatal frequency decrease net photosynthesis rates? (Netherlands) *Response:* Rather surprisingly, there was no appreciable decline in net photosynthesis associated with selection for

fewer stomata, possibly because in ryegrass as in other C_3 species, most of the diffusive resistance to CO_2 transport is in the mesophyll cells, with a comparatively small contribution from the stomata.

9. *Question:* You did not comment on the physiological limitations to forage seed yields. What is the role of physiology in the seed production of forages? (UK)

Response: The emphasis in the work discussed has been on improving the agronomic performance of existing varieties that already possess good seed yield, hopefully without any concurrent decline in seed yield. Many of the physiological and morphological features that determine flowering and seed production also show considerable genetic variation and are amenable to selection.

Resource Allocation for Animal-Grassland Systems

G.W. THOMAS

President, New Mexico State University,
Las Cruces, N.M., U.S.A.

Summary

The production of animal products from grasslands is dependent upon an adequate resource base—particularly the availability of land, water, and energy. In addition, range and pasture production systems must be designed to adjust to climatic variations and to minimize adverse environmental impacts.

Several new developments will have a significant impact on the future allocation of resources for animal agriculture: (1) increased competition for land and water, which is leading to transfers from agricultural to nonagricultural uses; (2) increasing costs and decreasing availability of "cultural energy" (particularly fossil fuels); (3) more pressure to divert grains and feed concentrates from animals to direct human food, thus increasing the dependence on roughages; (4) regional and worldwide changes in types and classes of livestock and systems of production; (5) greater environmental awareness and more recognition of the role of wildlife and other multiple-use aspects of grasslands; (6) impacts of fast-food chains, new grading standards, more concern for human health, and appropriate diets with less animal fat; and (7) the emerging role of the less-developed countries as they attempt to offset substantial population increases by producing more food, moving toward "appropriate technology," and utilizing new innovations for animal disease control and production.

KEY WORDS: natural resources, environment, water, land, energy, climate, vegetation, animal agriculture.

INTRODUCTION

During this XIV International Grassland Congress, several plenary papers and supporting discussions have centered on the contribution of forages to worldwide food production and the role of grasslands in animal agriculture. I do not intend to duplicate the material already presented but will try to focus on some issues in regard to resource allocation that have not been adequately covered.

The basic resources for animal production systems include land, climate, water, and energy. All of these resources interact to produce the vegetation that serves as food for the grazing animal.

THE RESOURCE BASE—LAND

The amount of land required to support animal agriculture varies with the class of livestock, system of production, climate, soil conditions, and the design of physical facilities. Land requirements vary from a fraction of a ha/animal under confinement to more than 50 ha/animal on some rangelands.

In the more intensive systems, most of the feeds originate on cultivated lands and are transported to the confined animals. Types of feed and associated resource requirements for these systems vary from location to location. It should be emphasized that there are many interrelationships among cropland, improved pasture, and rangelands for most animal-production systems. Therefore, a thorough analysis of the land requirements must relate the feed mix back to the original land base needed to produce the life-long ration.

On a worldwide basis, the largest amount of land is uncultivated rangeland—uncultivated because of severe restrictions in soils, slope, location, climate, or lack of water. In the U.S.A. the latest estimates show about 630 million ha of range. Data from a recent USDA study show that rangelands contributed 213 million animal-unit months (AUMs) of grazing each year, while pasture, cropland pasture, and aftermath contributed 763 million AUMs (USDA Forest Service 1980).

Land ownership and land-tenure policies are very important considerations in grassland agriculture. In the contiguous U.S. the federal government owns over 154 million ha, while 330 million ha remain in private ownership. Alaska is largely federally owned and controlled. Federal lands, for the most part, must be planned for multiple-use values, placing livestock in competition with uses such as for timber, wildlife, and recreation. The use of private land, on the other hand, is under the control of the landowner.

In many countries of the world, grazing is on communal or public lands, making it difficult to encourage proper grazing management. I have just completed a study of this problem in the Sahelian-Sudanian Zone of Africa, where land tenure and poor grazing management are contributing to the problem of desert encroachment (Thomas 1980). To get the most out of the land resource requires careful vegetation management, and too often land-tenure policies do not provide the necessary incentive for resource conservation.

THE RESOURCE BASE—CLIMATE

Climate is very important to animal-production systems because it affects the animal directly and influences the amount of forage and feed production. Climate is often defined as average weather. But this definition is too restrictive for resource planning. A more complete consideration of climate as a resource begins with these average values of weather events for a given time and place and includes the probability distributions of the weather occurrences.

Climatic variations have many time and place dimensions. Elements vary on a daily, seasonal, annual, or longer basis. In space, climatic elements vary in the immediate area surrounding the animal, over the farm or range, region, continent, and globe. The predicted climatic change caused by the continued rise in atmospheric CO_2 may produce changes in climate risks and, thus, induce changes in animal stress for any time or place. Therefore, the assessment of the possible range of climatic risks is essential to grasslands and future animal agriculture.

A major concern of climatic variation is the impact of drought. Droughts vary in their frequency, intensity, and length. Strong evidence indicates that droughts are cyclic in many continental areas and agricultural regions of the world. This is particularly true for the Great Plains of North America (Smith et al. 1980).

Drought indexes should be developed for each major agricultural and ecological system on each continent. Each year millions of kilograms of animal products are not achieved due to adverse effects of weather. Control over these unusual weather events must be developed in the coming decades. These methods of control must be less dependent on fossil fuels and designed to take greater advantage of solar radiation, winds, and other specific natural features.

THE RESOURCE BASE—WATER

The amount and quality of water are critical to all aspects of grassland agriculture. While the direct consumption of water to satisfy the thirst of the animal is relatively small, large quantities of water are used in getting animal products to the consumer.

The largest demand for water, however, is for production of forage and feeds. Some estimates show that about a metric ton of water is needed to produce one kg of grain. On part of the Western Range, over 100 metric tons of water are associated with the production of one kg of beef (Thomas 1977). Much of the water involved in this process on rangelands and pasture lands evaporates from the soil surface or is transpired through brush and undesirable weeds. Therefore, improved water management and better water conservation practices are essential to the improvement in quality vegetation. Herein lies an opportunity for research.

Irrigated cropland in the world has increased dramatically in the last 2 to 3 decades. McNamara (1980) stated recently that "in the last 10 years, roughly 40% of all increases in developing country food production has come from expanded irrigation." Much of the alfalfa production and part of the production of other feeds and forages are on irrigated land. Unfortunately, some of these irrigated areas are drawing water from geological aquifers that are not subject to recharge, and thus the forage and feed products will diminish in time. More research is needed on the interactions between irrigated lands and range lands as we look to the future of animal agriculture.

THE RESOURCE BASE—ENERGY

The role of energy in the world's grasslands can be analyzed by examining the two major energy-flow patterns: (1) the capture of solar energy by vegetation through the process of photosynthesis, the movement of this energy through ecosystems, and the ultimate utilization of a small fraction of this photosynthetic energy by man as a food product; and (2) the flow of cultural energy (energy subsidies) required to "run" the food and fiber ecosystems. This latter source of energy includes manpower, horsepower, hydroelectric power, large amounts of fossil fuels, and certain other energy subsidies.

Understanding the complicated interrelationships between these two energy-flow patterns is not important only to energy conservation and efficient utilization; indeed, it may be the key to man's survival as a viable organism on the planet Earth.

Most life is supported either directly or indirectly by the solar energy captured primarily by vegetation in the process of photosynthesis. This chemical reaction involving carbon dioxide, water, and sunlight energy to produce food and release oxygen and water (through transpiration) is the most important chemical reaction in the world. Even the fossil fuels (coal and petroleum) resulted from over 400 million years of photosynthetic activity. One estimate states that, on all of the land areas on earth, some 16 billion tons of carbon each year are fixed by photosynthesis. How can man influence this total capture capability? And, once the carbohydrates are formed, how can man influence the distribution process? These are questions that relate directly to both the "food" energy cycle and the "fuel" energy problem.

A complete analysis of energy-flow patterns on range and forest lands must take into consideration the capture of energy by photosynthesis, the dissipation pattern of energy flow, and the inputs of outside or "cultural" energy to harvest and utilize the vegetation. Fig. 1 presents a concept of energy flow on grasslands. All the photosynthetic energy captured by the producer organisms is eventually dissipated or consumed by animals (including man) and the organisms of decay.

The statement has been made that, on the average, about 1% of the sunlight energy falling on the earth is captured by vegetation. Recent studies show that 1% is much too high for arid and semiarid areas. For example, at the Pawnee grassland site in Colorado in 1972, only 0.3% of the usable radiation was captured by vegetation growth (Van Dyne 1974). For the desert biome in the southwestern U.S., the efficiency of utilization of solar energy ranged from 0.10% down to 0.03 percent over a 3-year period (Pieper et al. 1982). These photosynthetic capture rates should be similar to those of the Sahelian Zone of Africa. These measurements contrast with capture capabilities for mechanical collectors, which have tested efficiencies up to 50%–70%. Mechanical collectors capture heat only, whereas vegetation captures the energy in a complex chemical form.

Once the energy is captured by vegetation (producer organisms), the dissipation process begins. Respiration and growth take place, and primary consumers feed on the vegetation. On rangelands, the ruminant animal, both domesticated and wild, through its ability to convert roughage to edible meat, is the primary means of making productive use of these areas. But, here again, some interesting results are being developed by the International Biological Program (IBP) studies that change, rather substantially, some of the old assumptions. For example, termites in both the desert and grassland biomes appear to be far more important in vegetation harvest than one might assume. Even in the southwestern U.S., termites

consume 10 times more biomass than livestock. If this is true in the U.S., where termites are barely visible, the role of termites in sub-Saharan ecosystems and on other rangelands has probably been underestimated. In addition to termites, other forms of insects and microorganisms at all stages in the food chain divert a large amount of energy. For example, locusts or grasshoppers consume nearly all the available green vegetation under certain conditions in most range areas.

Van Dyne (1978), in summarizing the IBP studies, states that a characteristic of these range ecosystems "which seems surprising is the relatively small proportion of net primary production that enters the grazing food chain. Thus, when one summarizes the energy flow through the Shortgrass Prairie System (less than 800 mm of rainfall), of the total solar energy input in terms of the 5 month growing season, cattle capture only 0.0003%," and Van Dyne adds, "but this is a very important and tasty percentage." The important point here is that we need to know far more about energy flow to man through animals, and we need to plan how man can best tap these grassland ecosystems, on a sustained-yield basis, for food needs.

The fossil fuel input or "cultural energy" requirements for livestock production on range and pasture lands is relatively small because there is little energy required for supplies and production and a minimal use of mechanized equipment. Yet, it still "costs" lots of energy to grow that calf or lamb and get it to market. Certainly, the cultural energy requirements from fossil fuels to produce cattle and sheep may be low on grasslands, but the net energy analysis is incomplete without a consideration of the feeding, processing, and delivery systems.

In addition to the need for reductions in cultural energy—particularly fossil fuel—there are vast opportunities for improvement in biological efficiency. Biological efficiency can be measured in terms of the units of human digestible food produced/unit of plant energy consumed by the animal. A number of analyses have also evaluated animal energy efficiency by determining the energy value of the food produced relative to the cultural energy input. Such a determination, although useful in comparing the relative energy efficiency of various production practices for producing comparable food products, is of little use when comparing the values of differing foods. Meats, for instance, play an entirely different role in the dietary needs of people than do grains or fruits. Thus, the criteria for evaluation should encompass other parameters that measure total food value, including amino acids, proteins, and vitamins.

RESOURCE ALLOCATION FOR ANIMAL-GRASSLAND SYSTEMS: THE FUTURE

Several present trends and indications for the future cannot be ignored in planning for animal-grassland systems. These are: population pressure, increased energy costs, increased competition for land and water, changes in livestock production and marketing systems, and the emerging role of the less-developed countries (LDCs).

Fig. 1. A schematic diagram of solar energy capture and utilization on uncultivated lands. Fossil-fuel energy subsidies are relatively small compared with intensive crop production.

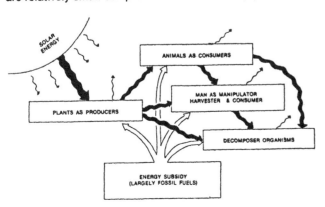

Population Pressure

Two factors in the population picture have an impact on resource requirements and the environment: (1) the growing numbers of people—United Nations projections still indicate between 5.8 and 6.5 billion by the year 2000, with an optimistic possibility of eventual leveling about the mid–21st century at about 12 billion—and (2) higher levels of affluence combined with the growing aspirations of the masses. We know that wealthy people place more pressure on the resource base than do poor people. We also know that improved communications and better education help to create unrest. More and more people realize that there is a better way of life "somewhere out there," and they are demanding a more nearly equal share of the earth's resources.

There are striking examples in both the developed and underdeveloped worlds of the direct and indirect impact of population on animal agriculture. Two of these may serve to illustrate my point.

1. In sub-Saharan Africa, population is doubling about every 20 years. Kenya's fertility rate is now about 8.1 births/woman, and the country's population, 16 million, will double in 17 years (Population Reference Bureau 1980). In Nigeria the population momentum will carry the nation from the present level of 90 million people to nearly one-half billion before it can plateau—assuming that replacement level will be reached by the year 2040 (van der Tak et al. 1979). Other sub-Saharan countries have similar problems. The impact of these increases is particularly critical to animal agriculture. Pastoralists, who often are nomadic, are forced into a smaller land base by the encroachment of cultivation from the south and man-accelerated desertification from the north (Thomas 1980). There is no economic incentive to control livestock numbers or to protect the resource base. The next drought south of the Sahara, which will start at an unpredictable future date, will hasten the advance of the desert and cause further serious livestock losses.

2. The second example is from the U.S.A. Both sheep and cattle numbers have been declining on the vast U.S. federal land base. This decline is not associated with human population numbers per se, but with increased affluence, increased environmental awareness, more interest in setting aside land for wilderness or single-use purposes, and the gradual elimination of the "tools" for the land manager to manipulate the vegetation complex. Indeed, grassland agriculture is coming under more restrictions as a land-use practice in both the developed and the underdeveloped worlds due to population numbers, changing attitudes of the public, and higher levels of affluence.

Increased Energy Costs

I find very few people, either in the scientific community or in the livestock industry, who are adequately facing up to the inevitable increased costs and limited supplies of traditional fossil fuels. Philip Abelson (1980), editor of *Science* magazine, put it this way:

We and the rest of the world have reached the end of a pleasant era of increasing standards of living and of cheap energy. . . . We now face a period of declining standards of living and the jarring and costly transitions to other energy forms. . . . economic adjustments will affect the purchasing power of consumers and, in turn, their preferences for food and especially for meat.

To meet the challenge presented by the energy crisis, more innovative approaches must be used in our research and development activities. One example of the possible future directions has been proposed by Dr. Earl Ray of New Mexico State University. Dr. Ray's preliminary evaluations indicate a fossil-fuel savings of over 30% by starting with a "Mor-Lean"[1] calf. This is a calf sterilized by the short-scrotum technique, which allows the continued release of hormones and produces the high feed-conversion and growth rates of the bull, thus saving feed and energy and eliminating the controversial use of DES (diethylstilbestrol). The Mor-Lean calf goes from the range to the feedlot for a short feeding period. At the time of slaughter, the hot carcass is subjected to electric shock treatment to create tenderness, eliminating aging and cold storage. The hot carcass is then deboned, saving labor and energy over cold deboning. While the carcass is still hot the hamburger is ground and vacuum-packaged; other cuts are also vacuum-packed. Cookers are located adjacent to the abattoir. The various cuts are dropped into hot water (preheated by solar or geothermal energy) and cooked to an internal temperature of 140°F. Cooking in hot water saves nearly 30% of the energy required for oven cooking. The precooked meat is then distributed directly to fast-food chains or supermarkets. Biological deterioration is minimized by precooking, and shelf life and handling problems are reduced. The final touch is seasoning by the chef or homemaker and heating to perfection in the microwave oven, which again saves energy over conventional oven cooking. The homemaker is supposed to be pleased with the "built-in maid service" and the reduced cost due to energy savings. Way out, you say? Not so, if you really look at the handwriting on the wall concerning our energy future.

Increased Competition for Land and Water

There is a continuing trend to transfer private land from agricultural uses (including animal agriculture) to industry, housing, urban development, and other uses. In the U.S. alone, this rate of transfer is now in excess of 1.2 million ha/yr, with about 400,000 ha classified as prime farmland. Other nations are also observing the growth of urban areas on land that once served for grazing or livestock forage production.

On a worldwide basis many millions of acres of grazing land are being converted to direct food crop production; other areas have grazing restrictions to protect wildlife or watersheds or are being reserved for parks, recreation, or wilderness values.

Water is being transferred out of forage and feed production to other uses or to urban development and industrialization.

The move toward alternative energy sources will mean an additional commitment of land and water for energy production. Perhaps 10% of our agricultural land base will be reserved for alcohol production, or "fuel farming," by the year 2000.

Changes in Livestock Production and Marketing Systems

The livestock industry will undergo substantial changes in the next two decades. Some of these changes include adjustments to a more limited resource base and an increase in the cost of available energy.

Worldwide inflation is also adversely affecting grassland agriculture, as is the rather phenomenal increase in land value — usually not associated with the animal productive capacity.

I am predicting changes in the percentage of livestock classes, with further reductions in sheep and goat numbers in favor of cattle. However, Ward (1980) states that the decline in U.S. sheep and dairy-cattle numbers appears to be over.

More emphasis on food grains and less diversion of grains and concentrates to livestock will place further pressure on roughage and grassland resources. We will see shorter feeding periods and more grass-fat livestock production. The need to divert more grains to direct human use will be partially offset by the increased aspirations of the people of most countries to add meat to their diets. The comprehensive study by Wheeler et al. (1981) on the food grain system predicts, "By 1985, worldwide use of grain for livestock feed will surpass that for human use by about 10 million tons." The U.S. and Canada will supply about 80% of total world grain exports by 1985.

More attention to wildlife and environmental protection will place further restrictions on animal agriculture and increase the cost of production.

New and improved techniques for livestock insect and disease control will allow livestock grazing in the higher-rainfall zones of the tropics. This is where much flexibility lies for taking care of some of the population pressure (such as in sub-Saharan Africa and South America). Many LDCs do not have access to mineral and petroleum wealth and, therefore, must build on their agricultural resource base. Properly managed animal-grassland systems have an important role to play in this expansion, since the ruminant animal offers the primary means of converting roughage to a usable human food.

The impact of fast-food chains and wider distribution of supermarkets will alter some of the traditional food patterns. Middle class people in the developed countries are eating more and more meals outside the home — with ground meat as the base. Changes in grading standards and preparation techniques also affect livestock production systems and resource requirements.

The Emerging Role of the Undeveloped World

To an increasing extent, the developed countries of the world are assisting the LDCs with food and nutrition problems. This assistance creates more attention to balanced nutrition and tends to emphasize the importance of animal products, protein, and vitamin deficiencies. At the same time, animal traction is being promoted as the next logical step in "appropriate technology" for many nations. All too often I find that the countries moving toward more animal agriculture have not assigned an appropriate land base to support the additional animals. From an ecological perspective, many of the so-called underdeveloped countries are already overdeveloped if one examines their resource base and vegetation management systems. Vegetation management is, after all, the key to both the productivity and the stability of most of these ecosystems.

LITERATURE CITED

Abelson, P.H. 1980. Animal agriculture and human needs in the 21st century. *In* W.G. Pond, R.A. Merkel, L.D. McGilliard, and V.J. Rhodes (eds.) Animal agriculture: research to meet human needs in the 21st century. Westview Press, Boulder, Colo.

McNamara, R.S. 1980. Food as a determinant of world development. Congressional Roundtable, Washington, D.C., 4 March 1980.

Pieper, R.D., J.C. Anway, M.A. Ellstrom, C.H. Herbel, R.L. Packard, S.L. Pimm, R.J. Raitt, E.E. Staffeldt, and J.G. Watts. 1982. Structure, function, and utilization of North American desert grassland ecosystems. New Mex. Agric. Expt. Sta. Res. Rpt. 39, Las Cruces, N.M.

Population Reference Bureau. 1980. Kenya's 4% population growth rate is world's highest. INTERCOM, October.

Smith, L.W., G.W. Thomas, A.B. Carr, D. Badger, J.S. Bartholic, R.D. Dunlop, J.P. Fontenot, R. Harris, H.K. Heady, W. Krejei, J.R. Miner, J.G. Morris, J.E. Newman, L. Rittenhouse, L.B. Safley, J.N. Walker, and R.K. White. 1980. Resources and environment. *In* W.G. Pond, R.A. Merkel, L.D. McGilliard, and V.J. Rhodes (eds.) Animal agriculture: research to meet human needs in the 21st century. Westview Press, Boulder, Colo.

Thomas, G.W. 1977. Environmental sensitivity and production potential of semi-arid rangelands. ICASALS. Tex. Tech Univ. Pub. 77-2:43.

Thomas, G.W. 1980. The Sahelian/Sudanian zones of Africa: profile of a fragile environment. Rpt. to Rockefeller Found. September.

USDA Forest Service. 1980. An assessment of the forest and rangeland situation in the United States. January.

van der Tak, J., C. Haub, and E. Murphy. 1979. Our population predicament: a new look. Pop. Ref. Bur. Vol. 34, No. 5. December.

Van Dyne, G.M. 1974. Unpublished data from the grasslands biome. Colo. State Univ., Ft. Collins, Colo.

Van Dyne, G.M. 1978. A grassroots view of the grassland biome. Colo. State Univ. Rpt.

Ward, G.M. 1980. Energy, land and feed constraints on beef production in the 80s. J. Anim. Sci. 51:1051–1064.

Wheeler, R.O., G.L. Cramer, K.B. Young, and E. Ospina. 1981. The world livestock product, feedstuff and food grain system. Winrock Int. Livest. and Train. Cntr. Rpt. to USDA.

NOTES

1. A technique patented by New Mexico State University to increase the efficiency of meat production.

DISCUSSION

Remarks of Discussion Leader, V.L. Lechtenberg

Dr. Thomas has identified several key resource areas that will be especially important in planning future forage-animal-production systems. I would like to add only a few thoughts to what Dr. Thomas has developed in his paper. My comments mainly emphasize several of the points that Dr. Thomas has discussed.

The land-resource base available for future forage-animal production will undoubtedly have a significant impact on livestock production systems. This resource is finite; in fact, the land available for animal production is likely to decrease in the future as competitive and alternative uses develop.

The interrelationship between grain-crop and forage-crop production is worthy of emphasis. The world demand for cereal grains in recent years has created short-term economic incentives that have resulted in an expansion of grain crop production onto marginal lands that are hilly and often highly erosive. These lands are best suited to the production of perennial forages. This shift in land use is especially prevalent in the humid areas where climatic conditions are favorable for grain production. This trend is likely to continue during the next several years, and the consequences of this shift in terms of soil erosion and loss in potential soil productivity merits the serious attention of grassland researchers as well as producers. The major problem appears to be one of short-term vs. long-term economic incentives.

Dr. Thomas has referred to several important considerations regarding the limitations imposed by climate and water. His comments have dealt primarily with the arid and semiarid regions of the world. These resources are also important production factors in temperate and tropical regions. However, the limitations imposed by climate and water are much less severe in these regions than in more arid regions. Previous Congress speakers have alluded to the fact that the potential exists to double or even triple production in humid areas and to do so within the constraints imposed by the water and climatic resources.

Dr. Thomas has discussed the importance of scarce fossil energy resources in relation to livestock systems. Several other speakers at this Congress have also pointed out that the scarcity of this resource will have an important impact on future production systems. I would like simply to emphasize the importance of two items that Dr. Thomas and others have mentioned. It is extremely important, as Dr. Thomas has pointed out, that the energy efficiency in forage, animal, or general crop-production systems not be calculated and compared solely on the basis of output:input ratios. Total energy production from the system or per-hectare is much more important.

In his discussion Dr. Thomas has pointed out that the move toward alternative energy sources could result in the production of energy crops on as much as 10% of our agricultural land in the future. Earlier in this Congress Dr. Hinsell and Dr. Edwards reminded us that crop residues, forages, and animal manure are good energy sources and that they provide much of the fuel in many developing countries. Thus, fuel from crops is not new; however, if research dealing with techniques for converting cellulosic crops into fuels leads to successful processes, the new demand for forage could have far-reaching impact on animal production systems.

The last comment I wish to make deals with the "feed" or "plant" resource. This includes the possibility of future changes in both the qualitative and quantitative aspects of livestock feed resources resulting from species shifts, genetic and breeding improvements, chemical and physical modifications of feeds, byproduct feeds and improved forage-animal-management systems. While the constraints in these areas are not as easy to define as those of land, water, and climate, they do represent areas in which great progress in animal production systems is possible. Progress in these areas is possible within the constraints imposed by land, climate, water, energy, and other natural resources.

Response to Questions

In response to questions, Dr. Thomas pointed out that the relative productive potential of the more humid areas is far greater than that of arid and semiarid regions. However, achieving increased productivity in humid areas is complicated by disease, insect, and parasite problems. These complications become more serious as one increases animal numbers and intensifies production to accommodate higher population densities. Many of the measures taken to help the inhabitants of the very arid areas, such as well-drilling, do not really lead to solutions. People simply stay in areas where productivity is insufficient to support them. Relocation of the people to more productive areas would be more likely to solve the starvation problems in these very arid regions. The productivity potential of the more humid regions is sufficient to accommodate the greater population density without serious environmental impact. Higher levels of affluence usually cause a greater negative impact on environment than do numbers of people.

Increasing productivity without further damaging the environment and jeopardizing future productivity is a great challenge worldwide. Approximately 40% of the increased food production in the last several years has been attributed to irrigation. Much of this irrigation is based on a nonrenewable water source. Thus, irrigation in these areas cannot be practiced indefinitely. There must be economic incentives to stop environmental deterioration. Governments must assume some of this responsibility. Difficulties and problems arise from political boundaries, nomadic life styles, and unstable political systems. Incentives are needed if people are to increase production and also preserve resources. Aid programs need to place more emphasis on institution building and on teaching people to help themselves. This is the only long-term solution.

Competitive, Adaptive, and Evolutionary Process in Grassland Ecosystems

B.D. PATIL

Director, Indian Grassland and Fodder Research Institute,
Jhansi, U.P., India

Summary

In the tropics, grasslands or grazing lands are found where the forest vegetation climax has been subjected to fire, grazing, felling, or other disturbances. Grazing lands in the tropics form various ecosystems under the major climatic types. Delineation of such ecosystems is possible through remnants of the climax vegetation of undisturbed or least-disturbed areas. Several subecosystems exist as seral stages under each ecosystem, the floristic composition of which is a reflection of the edaphic and soil-moisture-fertility conditions. Biological production in terms of energy for grazing animals varies among these ecosystems and subecosystems. Reconstruction of the adaptable competitive flora with woody browse components reveals significant upward trends in harvestable biomass production (energy) for grazing animals. In the evolution of new competitive and adaptive ecosystems in grasslands, examples of Indian work are cited. Natural grassland ecosystems in India provided much less energy than the reconstructed adaptive, floristic combinations with a browse component. The latter systems yielded more than double the energy of the others. Additionally, the complementary and associative effects of the woody browse component in the reconstructed ecosystems influence the micro-environmental changes beneficial to the herbaceous component of the system. Several reconstructed adaptable competitive floristic ecosystems, even for microsites of a major ecological region, are possible and are suggested to yield higher biomass. For ecosystems—namely, *Themeda-Arundinella, Phragmitis-Saccharum-Imperata, Dichanthium-Cenchrus-Lasiurus,* and *Sehima-Dichanthium*—provided energy to the extent of 1,700, 3,400, 2,215, and 2,550 kcal/m^2/yr (in the second case, of course, not all is consumable because of coarseness). The reconstructed adaptable and competitive ecosystem with woody browse provides energy as high as 3,697 kcal/m^2/yr. When compared with desert grasslands, low mountain, and mixed prairies, these reconstructed adaptive competitive ecosystems provided far more energy.

Tropical and subtropical grazing-land ecosystems could be usefully modified by either substitution or integration of several new adaptable ecosystems with browse as a component. Such studies can be extended to the waste areas of the tropics (over 944 million ha in Africa, Latin America, and the Indian subcontinent; several hundred million ha in Southeast Asia; and around 50% of the grazing areas of Australia). This paper emphasizes the utilization of a combined herbage-browse pastoral system in the evolution of competitive adaptive grazing-land ecosystems.

KEY WORDS: reconstructing, grassland ecosystems, tropics, competition, adaptation, evolution.

INTRODUCTION

Grasslands, appropriately termed grazing lands in the tropics, have been providing resources toward balancing of rural development programs. Optimal development of these programs, their maintenance and progressive production, has attracted the attention of many researchers. Basic to such optimal development is sound understanding of the formation of various ecosystems, their floristic composition, their functioning under a set of ecological and biotic influences, and the trends in their evolutionary process. Understanding of the fundamental role of climate, man, and man's beasts in the formation of such ecosystems is of equal importance.

Tropical grazing-land ecosystems occupy over 944 million ha in Africa, Latin America, India, and Pakistan. Also, several hundred million ha of such lands are in Southeast Asia, and in Australia 50% of the geographical area is under grazing. For many of these areas, major climatic vegetation types have been outlined by many

workers, along with descriptions and floristic composition of grazing-land vegetation. UNESCO (1979) brought into focus the present state of knowledge on this subject and identified some of the major gaps and priorities to be attended to if we are to get the best out of these grazing lands on a sustainable basis. Such gaps and priorities are identified under several headings. (1) Under ecosystem description and floristic composition are included the nomenclature of grazing-land ecosystems; methods of vegetation study, inventory, and mapping; vegetation dynamics; and geographical distribution of species. (2) Under climatic factors are the mechanisms determining climatic influences; the water balance, seasonal distribution, and intensity of water deficit; characterization and qualifying of the relationships among climate, weather, and particular farming systems; and development of conceptual models of biological systems. (3) Under water resources are surface runoff and erosion, inventory of surface-water reserves, and knowledge of underground water position. (4) Under soil studies are included soil survey data, soil erosion hazards and soil conservation practices on grazing lands, and detailed studies of soil water and nutrient status and requirements specific to each system. (5) In primary production and phenology of the ecosystem are the precise method of estimating above-ground production, the identification of components utilizable by livestock, clear identification of the role of trees and shrubs, long-term studies of physical and living systems, and scope to reconstitute with some certainty some parts within the overall functioning of ecosystems. (6) In secondary production, the role of secondary production agencies in the structure and evolution of ecosystems is to be understood with precision.

However, considerable research has been done on the evolution, development, and present distribution of the natural grasslands of temperate latitudes (Van Dyne et al. 1974). Their ecosystem structure, function, and utilization are known fairly well.

GRAZING-LAND ECOSYSTEMS OF INDIA

In evolutionary processes in tropical and subtropical India four major grazing-land or grassland ecosystems are recognized. These are *Themeda-Arundinella*, in the mountain region; *Phragmitis-Saccharum-Imperata*, in the sub-tropical high humid region; *Dichanthium-Cenchrus-Lasiurus*, in arid regions; and *Sehima-Dichanthium*, in central and southern semiarid regions.

These optimum-condition grazing-land ecosystems are by and large subclimax vegetation interspersed with woody shrubs, climatically, forest being the ecological climax vegetation in tropical and subtropical India as in all monsoon Asia. Throughout the grazing-land ecosystems are secondary formations created as a result of fire, overgrazing, tree-felling, etc. With progressive increases in intensity of these factors, the grassland formations (covers) break down to seral stages toward retrogression. On a broad basis, the retrogressive and progressive stages of these covers, along with the present and potential harvestable biomass and available energy, are given in Table 1.

Most of the grazing lands given these potential covers are in the third or fourth subecosystem (seral degraded stage). They are covered only to the extent of remnants of the climax ecosystem. However, the position with regard to *Phragmitis-Saccharum-Imperata* is different. This cover is predominant, owing to its being largely unacceptable to grazing animals except at very young growth. The harvestable production from each of the major ecosystems and subecosystems has been less than half of their poten-

Table 1. Retrogressive and progressive seral stages in subclimax tropical-subtropical grazing lands of India, along with levels of production and energy: A broad picture.[1]

Grass covers (ecosystems)			Harvestable biomass		Energy K Cal/m²/yr.
			Actual production g/m²	Potential production g/m²	
1. Themeda-Arundinella (Submountain regions)					
Themeda — Arundinella			220	400	1700
Protection	Grazing	Burning Dimeria			
Arundinella — Chrysopogon			220	300	1275
Protection	Grazing				
Heteropogon — Bothriochloa			200	350	1487
Protection	Grazing				
Cynodon			50	100	425
2. Phragmitis-Saccharum-Imperata (Subtropics — high humid)					
Phragmitis — Saccharum — Imperata			500	500	3400
Protection	Burning and Cutting	Grazing			
Saccharum — Imperata — Sclerostachya			450	750	3187
Protection	Burning and cutting				
Desmostachya — Imperata (Depauperate)			200	300	1275
Sporobolus Paspalum — Chrysopogon			100	150	637

Table 1, cont.

| Grass covers (ecosystems) | Harvestable biomass | | Energy K Cal/m²/yr. |
	Actual production g/m²	Potential production g/m²	
3. *Dichanthium-Cenchrus-Lasiurus* (Arid tracts)			
Dichanthium — *Cenchrus* — *Lasiurus*	330	500	2125
Protection Grazing			
Cenchrus — *Lasiurus*	250	500	2125
Protection Grazing			
Sporobolus — *Cynodon* — *Eleusine*	100	200	850
Protection Grazing			
Chloris — *Cenchrus biflorus*	100	150	637
Compact soil Loose soil			
4. *Sehima-Dichanthium* (Central and southern, semiarid; burning — eroding slopes — gravelly soil)			
Sehima — *Dichanthium*	350	600	2550
Increasing soil moisture			
Themeda — *Cymbopogon* —*Iseilema*			
Ischaemum indicum			
Grazing			
Chrysopogon — *Bothriochloa*	250	500	2125
Protection Grazing			
Heteropogon — *Eremopogon*	150	260	1105
Protection Grazing			
Aristida — *Eregrostis* — *Melanocenchrus*	20	40	170

[1]Grassland Survey of India. ICAR. 1954–1962.

tial production, primarily because of degraded soil fertility, erosion, and soil impoverishment resulting from scanty vegetation subjected to heavy grazing and tree-felling, among other factors.

Instead of the usual procedure of restoring degraded stages to their optimum through protection and judicious grazing practices, which is fairly time-consuming, we restructured the ecosystem-subecosystem of the floristic composition of adaptable, complementary, and non-competitive plant species by adding a woody plant component. This had shown promise in the competitive and adaptive process of modifying an ecosystem. An example of this, under Jhansi, India, conditions, on sites normally occupied by a *Heteropogon-Eremopogon* subecosystem, is presented in Table 2.

By adoption of a reconstructed adaptable browse-herbaceous ecosystem in the evolutionary process, it is possible to reconstruct such mixed adaptable floristic composition even for microsites encountered in a major ecosystem region. Examples of such microsites and the adaptable floristic composition of a browse-herbaceous system for each are given in Table 3.

From comparison of harvestable energy among the four major ecosystems described earlier, on the one hand, and the reconstructed adaptable mixed browse, grass, and legume forage vegetation ecosystem on the other, it may be observed that larger gains are possible, the reconstructed ecosystem yielding more energy than natural ecosystems. Further, such competitive and adaptive reconstructed ecosystems show superiority even over dryland cropping gains under similar edaphic and climatic conditions at Jhansi, India, in the context of animal production. Such comparisons are shown in Tables 4 and 5.

The superiority of the reconstructed mixed floristic composition in a reconstructed adaptable ecosystem could be due to varied operational factors, some of which are species adaptations while others are complementary and

Table 2. Production potential and energy of modified grazing-land ecosystems against a natural seral stage—*Heteropogon-Eremopogon* of *Sehima-Dichanthium* major cover at Jhansi, India.

Subecosystem	Maximum harvestable production g/m²	Energy K Cal/ m²/yr
Seral-stage (natural) *Heteropogon contortus—Eremopogon feoveolus*	260	1105
Silvipastoral combination reconstructed ecosystems.		
(a) *Cenchrus ciliaris* + *Leucaena* L. + Cowpea	870	3698
(b) *Cenchrus ciliaris* + *Chrysopogon fulyus* + *Albizzia lebbek* wild soybean	790	3358
(c) *Sehima nervosum* + *Cenchrus ciliaris* + *A. lebbek* + *Clitoris terneata*	735	3124
(d) *Setaria anceps* + *Brachiaria mutica* + *Sesbania grandiflora* + *Phaseolus calcaretus*	590	2508
(e) *Cenchrus ciliaris* + *Albizzia procera* + *Phaseolus*	760	3230
(f) *Cenchrus ciliaris* + *Prosopis juliflora* + *Dolichos lablab* Var. *Lignosus*	710	3018
(g) *Chrysopogon fulvus* + *Acacia tortalis* + *Dolichos lablab* Var. *Lignosus*	515	2189
(h) *Albizzia lebbek* + *Heteropogon contortus*	410	1742

(a) to (h) = Different restructural floral systems tested.

Table 3. Reconstructed floral structures on grazing lands in specific microsites with adaptable and evolutionary ecosystems of a major ecological *Sehima-Dichanthium* semiarid grass-cover region of India.

Edaphic and moisture features of microsites	Modified/reconstructed adaptable floral suggested ecosystem
Stony, gravelly, and dry habitats	*Acacia tortilis* (T)[1] *Cenchrus ciliaris* (G) *Sehima nervosum* (G) *Chrysopogon fulvus* (G) *Macroptilium atropurpureum* (L) *Atylosia scaraboides* (L) *Heylandia latabrosa* (L) *Indigofera linifolia* (L)
Gravelly dry habitats	*Albizzia amara* (T) *Acacia tortilis* (T) *C. setigerus* (G) *Chrysopogon fulvus* (G) *Stylosanthus hamata* (L)
Dry red soils	*Albizzia amara* (T) *Sehima nervosum* (G) *Heteropogon contortus* (G) *Chrysopogon fulvus* (G) *Stylosanthus hamata* (T)
Dry black soils	*Albizzia amara* (T) *Sesbania sesban* (T) *Dichanthium annulatum* (G) *Stylosanthus gracilis* (L)
Deep, good black soils with good drainage	*Leucaena leucocephala* (T) *Sesbania sesban* (T) *Cenchrus ciliaris* (G) *Stylosanthus gracilis* (L)

[1] T = tree; G = grass; L = legume

associative symbiotic growth-promoting effects for mutual benefit between the grass-legume and woody browse grown in association in a grazing-land ecosystem in tropical and subtropical regions.

CONCLUSIONS

With increased understanding of complementry and noncompetitive effects of the associative flora, the ecosystems could be progressively upgraded in the evolutionary process. Their energy production is not merely dependent on natural, time-consuming processes, but may be affected also by superimposing permissible and economical management inputs.

Insertion of adaptable woody plants or woody browse in an ecosystem has specific advantages, since the woody species are known to play a fundamental role in maintaining the equilibrium of natural ecosystems, in anchoring soils, in maintaining soil quality, in recycling minerals, and in maintaining favorable micro-climates and meso-climates (UNESCO 1979). Such reconstructed adaptable ecosystems or subecosystems with woody browse as a component have shown very relevant advantages in livestock rearing, apart from increasing the energy production of a system. This system, in comparison with a natural one, prolonged forage availability beyond the monsoonal period and added significantly in increased mineral availability and vitamin supply to rangeland animals (Patil and Pathak 1978). Under the present ecosystems rangeland animals continue to suffer from deficiency, as manifested by nutritional disorders and affected reproduction physiology.

Table 4. Comparative assessment of dry-matter and energy production in different ecosystems.

Grassland type/ ecosystems[1][2]			Above-ground production g/m² (harvestable)	Energy K Cal/m²	Comments
Desert grassland (1)[2]			—	1177	Within growing season
Low mountain (1)			—	2707	-do-
Mixed prairie (1)			—	3150	-do-
A.	(i)	*Themeda-Arundinella* (1 −)	400	1700	Total harvestable biomass leaving stubbles
	(ii)	*Phragmitis-Saccharum = Imperata* (2 −)	800	3400	-do-
	(iii)	*Dichanthium-Cenchrus = Lasiurus* (1 +)	500	2125	-do-
	(iv)	*Sehima-Dichanthium* (1 +)	600	2550	-do-
B.	Reconstructed adaptable mixed browse, grass, and legume forage strata ecosystems (at Jhansi, India) (1 +)		870	3698	-do-
C.	Dryland cropping ecosystem (at Jhansi, India) (1 −)		727	3198	67 days' growth rest period fallow

[1]Data for first three ecosystems taken from Sims and Singh; other data from present study.

[2]Animal feeding values (grading): (1 +) = excellent; (1) = very good; (1 −) = good; (2) = fair; (2 −) = poor except in very early growth stage.

Table 5. Yields of stover, grain, and energy from dryland cropping ecosystems.

Crop		Stover yield g/m²	Grain yield g/m²	Total g/m²	Energy K Cal/ m²/yr.	Growth duration and fallow/year days
Sorghum var.	SPV 289	585	0	585	2487	105
	MSH 33	437	121	558	2476	83
	SPV 245	537	62	599	2598	105
Pearl millet	MSH 110	600	127	727	3198	67
	BJ 104	562	100	662	2900	72

LITERATURE CITED

Patil, B.D., and P.S. Pathak. 1978. Canopy manipulation and species management for enhancing biomass production from wastelands in semi-arid and arid regions. Proc. Symp. Arid Zone Res. and Dev. CAZRI.

Sims, P.L., and J.S. Singh. 1971. Herbage dynamics and net primary production in certain ungrazed and grazed grasslands in North America. *In* N.R. French (ed.) Preliminary analysis of structure and function in grasslands. Range Sci. Dept. Sci. Ser. 10:59-124. Colo. State Univ., Fort Collins, Colo.

UNESCO. 1979. Tropical grazing land ecosystem. United Nations, Paris.

Van Dyne, G.M., F.M. Smith, R.L. Czaplewski, and R.G. Woodmansee. 1974. Analyses and syntheses of grassland ecosystem dynamics. First Int. Cong. of Ecol., The Hague, Netherlands.

DISCUSSION

Remarks of Discussion Leader, L.R. Humphreys

I congratulate Dr. Patil on his paper. I am pleased he has referred both to successional stages in Indian grasslands and to "reconstructed" pastures, which involve the planting of legume shrubs and of cultivated pasture plants.

I hope the discussion will distinguish the ecosystem that has minimum human interference in its operations and the managed farm or range environment in which farmer objectives may run quite counter to the normal adaptive processes and directions of grassland evolution. I invite you to consider the implications of this dichotomy for the objectives of plant improvement and grassland management.

Grassland scientists need to understand the adaptive mechanisms employed in plant evolution that lead to ecological success. One reason for this is that adaptive mechanisms may be inimical to management success. I am not sure if I am in the right counties of the U.S.A. to be popular in taking this rather non-Darwinian line and in exhibiting skepticism about the inevitable glories of evolution. Human creativity is needed to identify those plant processes leading to "farm success," as distinct from "ecological success," in environments undisturbed by man. We might reassess the criteria by which we judge plants.

Grassland scientists are committed to improving dry-matter yield, dominance of desired plants, leafiness, and resistance to environmental stresses and to maintaining a grazing "disclimax." Does our pursuit of these criteria of merit inhibit farm success? High growth rates are often associated with reduced nutritive value.

The seasonal distribution of growth may be more important in minimizing animal stress than the total dry-matter yield, but a long seasonal duration of growth may confer no ecological advantage.

When will we be prepared to escape from the tyranny of dry-matter yield and promote cultivars on the basis that they grow more slowly but have higher nutritive value?

Secondly, in seeking dominance we hope to have the most productive variety in control of the environmental growth factors.

In humid environments, height confers dominance. (We might note that the Chairman of the Program Committee, who has a peculiarly difficult role at this Congress, is Dr. Gordon Marten, a tall man). Height may not favor nutritive value or yield.

Mixed pastures provide legume compatibility; greater continuity of yield; insurance against shifts in weather conditions, unpredicted disasters, changes in management inputs, and emergence of new patterns of disease and pest attack; and greater variety of diet. Do we need to make the grassland world safer for diversity?

Third, both plant improvement and grassland management are often directed to increasing leafiness.

Does this commit us to perennials rather than annuals?

Many of the plants we are domesticating have unsatisfactory seed-production characteristics. Apparently perennial plants such as siratro and lotononis are quite short-lived. Seed production may in fact determine the persistence of yield. We are also recognizing the dietary value of many seeds.

Should we give more attention to the Goddess of Fertility?

Fourth, resistance to stress—climatic, edaphic, and biotic—is needed. Usually adaptive mechanisms providing plant persistence are sought by the farmer, except in connection with plant quiescence and plant avoidance by animals.

Finally, in this discussion we should consider the relative seral species with their proven animal productivity and the grasses belonging to high successional stages.

Questions and Responses

1. *Question:* Is it necessary to have more than one species of animal to utilize the restructured ecosystem, specifically the taller shrubs and trees? (Australia) *Response:* Taller trees and shrubs can be harvested manually. Sheep and goats can utilize the lower forage, and specific systems can be tailored for certain kinds of animals.

2. *Question:* Which legumes may be used in arid and semiarid areas? (U.S.A.) *Response:* There are many species that might be utilized, but the ones I could offer would be *Desmanthus varigatus*, *Stylosanthes* spp., and *Itylosia scaribdie.*

3. *Question:* In mixed cropping systems, does one wait until after seed production to begin grazing, and what animals should be used? (Australia) *Response:* Specifically, goats would be used for type one. If you mix goats and camels in type one, the camels would eat the grass and the goats would browse.

INVITED PAPERS

Grassland Development in Japan and Other Asian Countries

I. NIKKI

Professor of Grassland Science, Miyazaki University, Japan

Summary

Only since World War II has attention been directed to grassland in Japan. Forage crops were introduced on dairy farms for feeding cattle. Acreages have increased and now equal 15% of the total field area. Forage crops are second to rice as a major crop. The New Grassland Law, approved in 1950, has led to "grassland improvement" and development of established pastures. Japan, situated in the most northern part of the Asian Monsoon Area, is warm as a rule and has much rainfall. *Miscanthus sinensis* Steud. and *Zoysia japonica* Steud. are the dominant representative species of natural grassland. Most of the grassland soil, supposedly improved, is of volcanic ash. Red clover (*Trifolium pratense* L.), orchardgrass (*Dactylis glomerata* L.), etc., are being sown in pastures. Warm-season species such as bahiagrass (*Paspalum notatum* Flugge) are more suited to the warmer regions. Because of the Japanese climate, silage has been preferred to hay. Making silage of high quality that is high in calories is of utmost importance to grassland farmers. Research organizations (National Grassland Research Institute is one of these) have been established in the last 25 years. For higher education, departments of grassland science were established at Obihiro and Miyazaki Universities. Large areas of mountain land remain that could be utilized by grazing of cattle after some improvement. According to a government survey, 1.29 million ha of land is regarded as suitable for grassland. Japan imports huge amounts of feed grains, which constitutes a serious problem from the viewpoint of world food and also from the standpoint of national security. Although the climate of Japan, as a rule, is favorable for producing forage, various problems remain to be solved. Most of them cannot be resolved without systematic research. Of the many Asian countries, China, the Mongolian People's Republic, the Philippines, and Thailand were selected as examples for the present paper. In recent years there has been a remarkable increase in demand for milk and meat. Thus, grassland farming is attracting great interest from researchers and producers in those countries. Judging from past experience in these countries, grassland farming must be promising. Grassland farming in Asia has unique characteristics and many difficult problems to be solved. To those of us involved in developing grassland farming, nothing could be of greater help than the international exchange of knowledge, information, and research ideas.

KEY WORDS: grassland development, Asia, Japan, Mongolia, Philippines, China, Thailand.

THE BEGINNING OF GRASSLAND FARMING IN JAPAN

Only since World War II has milk production become intensified and attention directed to grassland in Japan. Conventional grassland of Japan, so-called natural grassland, had been used mainly as meadows to provide hay or bedding for farmers' draft cattle or horses or as pasture for grazing cattle during summer. As most of the native grass species included in the grassland were tough, high in fiber, poor in nutrition, and low-yielding, the natural grassland, as it existed, was considered unsuitable for grazing dairy cattle. Thus, because of the need for feeding dairy cattle it was necessary to improve productivity of native grassland by establishing improved pastures or by growing forage crops in the field.

The reasons for neglecting grassland farming until recently can be summed up as follows: (1) Japanese people, because of their customary diet, had not demanded much meat, milk, and milk products; and (2) from the viewpoint of nutrient yield/unit area, grassland farming was considered unsuitable because the yield of milk or meat was lower than that realized from growing cereals in such a country as Japan, which had limited land acreage. Also, in past years, grasses were assessed low in significance because of the use of inappropriate methods for evaluating many foreign forage crops.

Since World War II, however, conditions have changed. Demand for animal protein food, instead of fish and soybeans, has markedly increased, and during the same time marked changes have occurred in the animal industry.

First, the numbers of dairy cattle, pigs, and poultry have markedly increased, while those of draft cattle and horses have decreased with the popularization of small tractors for small land holdings.

Second, in order to meet the increased demand for concentrated feeds, huge tonnages of feed grains are being imported.

Third, along with changes in animal industries, forage crops were introduced on dairy farms for feeding cattle. The New Grassland Law, approved in 1950, has encouraged "grassland improvement" by the government's subsidizing establishment of improved pastures. The amount of land in forage crops has increased year by year, amounting to 875,000 ha, equivalent to 15% of the total field area. Now forage crops consisting of grasses, legumes, corn (*Zea mays* L.), sorghum (*Sorghum vulgare* Pers.), oats (*Avena sativa* L.), and others rank second only to rice as the major crops. According to government reports on production costs, milk-production costs are still rising because most dairy farmers still must purchase much of their concentrate feeds.

ENVIRONMENT OF GRASSLAND FARMING

Japan, being situated in the warmer area of the Temperate Zone and in the most northern part of the Asian Monsoon Area, is warm as a rule and has much rainfall. Since the territory ranges from lat 24° to 46°N, large differences in climate are experienced. In northern regions, such as Hokkaido Island and the northeastern region of the main island, the climate is cool and similar to that of Europe, while in the warmer regions the climate, on the average, is warm but with large yearly differences in temperature. There is a comparatively long period of cold days with the mean temperature below 5°C, as well as hot days with temperatures above 20°C, all within the same year.

The mean annual precipitation is between 1,000 mm and 3,000 mm, and extremely dry seasons are rare. June, July, and September constitute the rainy season. Because of variable seasonal rainfall, the pattern of forage production has a tendency to concentrate during June and July. Therefore storing of forage is especially important as a means of more evenly distributing supply. Because of high rainfall, silage is more effective than hay-making as a means of storage.

Most of the supposedly improved grassland soil is of volcanic ash, with heavy clay soils or peat soils in Hokkaido. Methods of amelioration and fertilization of volcanic ash soils have been developed step by step.

Natural grasslands, though not extensive in area, are classified according to the dominant species, as follows: (1) Miscanthus type (*Miscanthus sinensis* Steud.), (2) Zoysia type (*Zoysia Japonica* Steud.), (3) Sasa type (*Pleioblastus* spp.), (4) weed type, (5) brush type, (6) woodland type, and (7) barren type. Among those seven types, the first three are representative and most useful.

GRASSLAND DEVELOPMENT IN JAPAN

In cool areas such as Hokkaido, the northeastern regions of the main island, and high-altitude regions, cool-season species, including red clover (*Trifolium pratense* L.), white clover (*Trifolium repens* L.), orchardgrass (*Dactylis glomerata* L.), timothy (*Phleum pratense* L.), tall fescue (*Festuca arundinacea* Schreb.), and perennial ryegrass (*Lolium perenne* L.) are gaining in prevalence.

In warmer regions, introduced cool-season species suffer from summer killing, diseases, and pests. Warm-season species such as bahiagrass (*Paspalum notatum* Flugge) and dallisgrass (*Paspalum dilatatum* Poir.) have been introduced. Since they are subject to winter kill and growing periods are too short to enable them to grow

freely, they are not regarded as flawless species for these warmer regions. Therefore, in order to ensure grassland production under the prevailing climatic conditions of these regions, combinations of both warm- and cool-season species must be used. Combinations of perennial warm-season grasses and annual cool-season grasses have been developed. Also combinations of annual warm- and annual cool-season species are highly productive. Since such systems are practicable only under intensive management, they are not suitable for mountain grasslands. Because of the variable climatic conditions of the warmer regions, it is necessary to select proper mixtures. For the unique conditions of these regions, it is also essential to explore zealously both domestic and foreign flora, in addition to breeding for more suitable new varieties than present forage plants.

Since native grass species on natural grassland have been regarded as low in quality for dairy cattle, introducing new superior forage species is regarded as necessary to effectively utilize the grasslands for dairy cattle. The natural grasslands were not only very low in grazing capacity, but most of them were contaminated severely with such diseases as Piroplasmosis. Improvement is considered essential for use by grazing.

Mechanized procedures have been adopted for plowing, mixing of improving material with soil, harrowing, fertilizing, and seeding; and peculiar to Japan, it is necessary to apply abundant quantities of ground limestone and fused phosphate in order to improve the volcanic ash soils. In certain cases, recently, ploughing has been substituted for harrowing or for so-called hoof cultivation.

In established pastures, forage production is generally concentrated in the months of June and July, favored by rainy season and temperature. Consequently seasonal distribution of forage production is inevitably unbalanced, and of course unfavorable for grazing. This tendency may be worsened by warmer climatic conditions. Thus, balancing the forage distribution is one of the major concerns for efficient grassland management.

Even in cases of high productivity, yields of established grasslands are apt to decline year by year until renovation is necessary, because of increases in weeds, diseases, pests, and deteriorated soil conditions, etc. Though this tendency may not be peculiar to Japan, it seems much worse there than in other countries. Within Japan, the problems appear to be greater in the warmer regions.

Recently, grazing has become more popular in Japan. The reasons are as follows: (1) grassland on a larger scale and on steeper land has been increasing, and these grasslands are not fully utilized without grazing; and (2) more dairy farmers have enlarged their scale and so find it necessary to save labor. One of the most serious problems of grazing, however, is, as mentioned above, the seasonal imbalance of forage production. Animal disease is another problem. In comparing performance results of grazing among different localities, it can be said that performance in warmer regions is as a rule inferior to that in cooler regions. This situation is considered to be due to lower quality of forage in warmer areas and to the stress of severe summer heat on the animal. The period when stored forage is needed ranges from 3 to 7 months, according to different temperatures. This means that between 25% and 70% of the total consumed forage must be made into hay or silage. Since the rainy season, *baiu*, coincides with the most important harvesting period, it is very difficult to make high-quality hay. For this reason silage has been preferred to hay, and on most farms silage is the only method of storing. Therefore techniques for making high-quality silage are of major importance to dairy farmers. A year-round silage system has been popular among dairy farmers, as it can be more efficient, particularly for those farms with high-producing dairy herds. In any case, making silage high in quality and in calories must be of major importance.

Organizations for grassland and forage crop research have been established in the last 25 years. The specialized National Grassland Research Institute was established in 1970. Nationally there are six Regional Agricultural Experiment Stations, of which three have grassland divisions. The remaining three stations have grassland laboratories. In each prefectural experiment station of agriculture or animal production, experimental work on grassland and forage crops is being conducted. In addition to those stations, there are 16 breeding stations where research is in progress on 15 forage species, including white clover, orchardgrass, Italian ryegrass (*Lolium multiflorum* Lam.), and bahiagrass. For higher education, departments of grassland science were established at Obihiro University of Agriculture and Veterinary Medicine in Hokkaido and at Miyazaki University in Kyushu; and, at most institutions having faculties of agriculture, lectures on grassland farming and forage-crop production are being given.

A VIEW OF GRASSLAND FARMING IN JAPAN

The low ratio of grassland to total crop cover plus the markedly high ratio of forest may be peculiar to Japanese agriculture. Such unbalanced land utilization is due mainly to the mountainous topography of this country. In some countries, mountainous land is utilized to a greater extent for grazing cattle. In Japan there remain large areas of mountainous land that could be utilized for grazing cattle after some improvement. According to a recent government survey, 1.29 million ha of land is regarded as suitable for grassland.

Because of decreased consumption of rice, the government policy has been to convert up to 30% of the land used for rice production to production of other upland crops such as feed grains, soybeans, and forage crops. Rotation systems, including both rice and forage crops, may be inevitable for the future. Thus the introduction of forage crops and grassland farming into the Japanese farming system is playing a pivotal role in the reorganization of Japanese agriculture. Although most farmers are landowners, they own very small acreages. Most of them have turned to part-time farming, under which circumstances farm work is usually carried out on holidays or, very often, by housewives; "Sunday peasants" and

"mama farming" are popular everywhere in Japan. On the other hand, it can be said that those part-time farmers and the rural exodus have supported the high economic growth of Japan. Full-time farmers constitute only 12.9% of all farmers. Most dairy farmers and cattlemen who are interested in grassland, however, belong to the full-time farmers' group.

Japan, the world's largest food-importing country, has been importing huge amounts of feed grain, equivalent to 12.9% of the total world grain available for international trade. From the viewpoint of world food supply, it must be considered a major problem for a single country to purchase such large amounts of food grain for feed while some other countries may be suffering from want of food. From a national point of view, such dependency on imports must be a most serious threat to national security. For Japan, grassland farming is believed to be one of the most efficient countermeasures to that problem.

As mentioned above, the climate of Japan, as a rule, is favorable for producing forage. However, this potential has not been fully realized because of some negative factors to be resolved. In the near future, with research progress, an intensive method of grassland farming and forage production will surely be developed. However, various problems remain and most of them cannot be resolved without systematic research.

GRASSLAND DEVELOPMENT IN OTHER ASIAN COUNTRIES

Next, I would like to refer to grasslands of Asian countries other than Japan. They extend from the tropics to the arctic and vary widely because of their localities; but some significant similarities to Japan can be found. Of the many Asian countries I have selected four as examples.

1. The People's Republic of China has vast areas of forest, grassland, and desert. Because its herbaceous resource is rich, being favored with the conditions in which stock farming can be developed, this country intends to meet the increasing demand for meat and dairy products, to accelerate modernization of agriculture, and to raise the national living standard by utilizing its abundant herbaceous resources. Great efforts have been made to promote systematic research on grassland, and four Grassland Research Institutes have been established at the National Academy of Agriculture, Kirin Normal University, Sichuan, and Kansu. In the field of higher education, departments of grassland science have been established at four universities. At most other agricultural colleges, laboratories for grassland studies have been established. Research work is being carried out in the following areas: development of the grassland resource, breeding of forage species, establishment of sown grassland, storing of forage, and biological control of pests, diseases, rats, etc.

With the intent of stimulating research work and exchanging knowledge, the Chinese Society of Grassland Science, as the national organization of grassland scientists, was organized in January 1980. It is issuing a quarterly magazine, *Zhonguo Caoyuan,* in Chinese.

2. The Mongolian People's Republic, situated in dry Asia extending from lat 41° to 52°N. with a mean altitude of 1,580 m, has extensive grasslands. The climate is continental, with temperatures fluctuating severely not only within the year but also with the day. Annual rainfall ranges between 100 mm and 400 mm according to locality. Of the total territory, 89.4% is covered with different types of steppe. The forest steppe type, occupying 25.2% of the land, is the main farming area, with fertile soils and rivers. Dairy cattle and fine-wool sheep are concentrated in this zone. The steppe-type zone, occupying 26.1% of the land, is the main stock-raising zone and is favored with abundant herbaceous resources. Beef cattle and semifine-wool sheep are mainly raised here. The desert steppe zone, 27.7% of the land, is very poor in vegetation because of extremely low rainfall. Mainly camels and goats are reared there. Grassland is grazed generally throughout the year, in spite of severe climatic changes, by moving herds or flocks from one unit of divided grazing land to another with the changing of the season.

With the intention of increasing forage productivity, irrigation for forage crops has been popularized. In spite of the year-round grazing system, storing of forage has been regarded as important. The main types of storage are hay, silage, and briquet. For silage, sunflower (*Helianthus annus* L.), maize (*Zea mays* L.), and soybean (*Glycine soja* Sieb. et Zucc.) are utilized; and, for making hay, oats have been popular for the cultivated areas. Hay is made mainly of native herbages with mechanized power. Briquet making has been advanced as a system for storing forage. The briquets are made primarily from wheat straw, although some are made of high-quality forage mixed with grain and minerals, for feeding camels and horses.

Grassland research is being carried out at agricultural colleges and institutes. The authorized research areas for grassland and forage crops are as follows: (1) quality of forage, (2) productivity of grazing stock, (3) grazing, (4) improvement of grassland, (5) improvement of native grasses, (6) introduction of leguminous plants, and (7) introduction of winter wheat.

3. The Philippines lie between lat 5° and 20°N with a maritime climate, being under the primary control of monsoons. Annual precipitation ranges from 750 mm to 5,000 mm. The Philippines can be divided into three areas with respect to conditions relative to pasture production. The three types are (1) dry season of 6 months; (2) dry season of 3–4 months; and (3) no dry season. According to the FAO Production Yearbook, in 1978 there were 1.82 million head of cattle and 5.3 million head of buffaloes. The native pastures consist of native vegetation, mainly *Imperata*, *Themeda*, and *Saccharum*. The carrying capacity is ¼ to ½ animal unit/ha. The native pastures are usually set on fire so as to get rid of overgrown *Imperata*, and the regrowth is grazed as intensively as possible.

Dairy production has not yet met the huge domestic demand for milk and milk products. Open grassland and managed pastures amount to 3.48 million ha, roughly half under alienated land and half under public land. However, the area actually under grazing or managing is only 0.49 million ha. The most economical way to bring more areas under grazing is to burn off the mature stands to encourage regrowth. In fact, burning is about the only management practice applied on pastures. And the only practicable low-cost method of improving grasslands is by the introduction of adapted legumes into the native sward. Improved pastures constitute a very small percentage of the total in this country, the majority being pure stands of napiergrass (*Pennisetum purpureum* Schumach.), paragrass (*Brachiaria mutica*), and guineagrass (*Panicum maximum* Jacq.).

As research priorities, three pasture development projects have been identified, namely: (1) improvement of native grasslands, (2) development of moderately to intensively managed pastures under coconuts, and (3) development of intensively managed pastures for dairy and special-purpose beef operations.

4. In Thailand efforts have been made to increase beef and dairy production. The aim is to keep the production cost low enough that farmers can made a profit, with prices at a level the average consumer can afford. Since there are few improved pastures in Thailand as a whole, the general practice is to graze the animals on forages available in natural meadows, forests, and along the roadsides and canals. After rice and upland crops are harvested, the animals are permitted to graze the stubble, straw, and any weed grasses that may come up. Recently strong emphasis has been placed on forage crop production and pasture management research by all of the authorities concerned. The livestock department, the Land Development Department, Kasetsart University, and Chieng Mai University have initiated and are carrying out various forage trials. Many forage species have been introduced from tropical countries all over the world to be planted and tested. By and large, only a few grasses are being used, on both pasture and field. Practically, most modern livestock farmers depend on *Brachiaria mutica*, *Panicum maximum*, *Pennisetum purpureophoides*, and *Sorghum almum* Parodi.

Though there are strong indications that pasture development will be fast, moving along with Thailand's Development Scheme, many forage problems remain to be solved, particularly the breeding work, seed production, and proper management.

Last, it can be said that most of the Asian countries, as the pattern of their land utilization shows, have practiced very little grassland farming. A recent tendency, however, in most countries is that demand for milk and meat has been increasing markedly. Thus, grassland farming is attracting great interst.

Judging from the results of past experiences in these countries, growth rates of herbaceous plants are, as a rule, high; therefore, grassland farming must be promising. If grassland farming in Asia is to increase it must have unique characteristics applicable to the peculiar environment of Asia; and there are many difficult problems to be resolved through systematic research.

To those of us involved in establishing grassland farming systems, nothing can be of greater help than international exchange of knowledge and information.

Livestock Resources in the World Food Supply

J.A. PINO

The Rockefeller Foundation, New York, N.Y., U.S.A.

I would like to express my appreciation to the Governing Board of the XIV International Grassland Congress, to Dr. Robert Barnes, its President, and especially to Dr. John E. Baylor, Chairman of the Governing Board of the XIV International Grassland Congress, for the kind invitation to speak at this distinguished gathering. I am pleased also to be in the state of Kentucky, a recognized leader in pasture and forage research and practice, a recognition attested to by this Congress. I regret that my own schedule did not allow me to participate in the various scientific and technical sessions. I understand, however, that these have been well attended and of exceptional quality. I strongly encourage your organization to seek every means to intensify the research as well as the developmental aspects of grassland use.

According to the information I have received, the first Grassland Congress was held in Leipzig, Germany, in 1927. Following the third meeting, it was decided to internationalize the group beyond the European countries, and subsequent meetings have been held in the U.S., New Zealand, the UK, Brazil, Finland, Australia, the USSR, and the most recent one, the XIII International Grassland Congress, in the GDR in 1977, at which time a Constitution was adopted.

The international character of these gatherings becomes even more significant in today's settings when problems of population, health and hunger, resource use and abuse, economic development, personal and national security have all required dialogue and action between and among nations and peoples of the world. As one watches the attempts made through the political process to ameliorate the continuing problems of food scarcity, poor health and high infant morality, war, security, migration, and refugees, it is easy to become cynical about the political process as a means of resolving these pressing issues. One has only to examine, for example, the limited achievements resulting from the World Food Congress held in Rome in 1974 to become less than enchanted with the results, notwithstanding the tireless efforts of some excellent people such as John Hannah and Maurice Williams. I do not mean to disparage the political process. On the contrary, I believe that the resolution of some of the major problems facing all nations requires an even higher degree of political commitment and skill, and especially interaction between science and technology and policy and politics, than that which exists today.

These Congresses, perhaps more than any other national or international body, have as a matter of course emphasized the critical importance of resource use and ecological balance. Although the importance of natural resources in support of man and his institutions has long been recognized, people have only recently (in historical terms) come to recognize the finiteness, the fragility, the susceptibility of the environment to irreparable damage caused by man's activities and have taken steps to understand the extent of man's impact on the earth's resources. Indeed, in many traditional systems, for example, the inclusion of domestic animals into the otherwise normal environment would seem to cause very little disturbance in the natural habitat. But "traditional systems" must not be equated with being more desirable, acceptable, or more protective of the resource base. So-called traditional systems of dependence on land resources may be very destructive. Today one can observe both extremes of livestock husbandry with respect to its impact on the land resource, that is on the one hand "traditional or extensive" systems, causing severe resource damage, and on the other hand the highly productive systems of rangeland use that even tend to improve the resource base.

There is little doubt that due to increased human requirements for food, there are and will continue to be dramatic shifts in the relative proportions of land used for cultivation, grazing, and forests and that remaining for nonagricultural uses (recreational purposes, cities, roads, watersheds, protected coastal areas, and uninhabitable areas). While one may question the absolute accuracy of the data relative to land usage worldwide, the trends and implications are quite clear indeed. Expansion of cropland areas will take first priority and show the greatest increase proportionately, relative to current use. We know that increased food supply can come from increased crop yields, increased productivity/unit area, and/or expansion of area under cultivation and reduction of losses and waste once crops are produced. Generally speaking, area expansion would seem to be the most economic alternative for increasing food production. Nevertheless, the cost of land clearing, drainage, irrigation, desalinization, and other procedures for expanding the crop area is becoming a major factor in the shifting balance between investment costs and priorities and potential benefits. High crop productivity resulting from new technology has become a highly significant factor in the matter of where investment priorities will be made. Yet very frequently it is precisely the areas coming under expansion pressures for which the technology base is completely inadequate. This is true for crops as it is for livestock. As human population numbers grow and as demand for all food and feed commodities intensifies, the implications for the livestock sector are quite clear. Existing grazing areas must become more productive and new areas will be added. *Both will require additional technology and elaboration of production systems appropriate to the particular ecosystems.*

Contrary to some of the doomsday futurologists, I am optimistic that science and technology will continue to provide new opportunities for increased crop productivity with increased efficiency of energy and other inputs at greater economic advantage. Additionally, the 3,000 million ha of permanent pasture and meadowland is a vast nutrient-producing resource. According to the best estimates, this area is likely to be expanded by about 25%. Of one thing we can be certain: The pressures to produce more food for expanding populations will result in a need for more intensive land use and will, as well, cause a shift in land-use patterns, e.g., from forest and nongrazing land to grazing land and agriculture use and from grazing to crop production and mixed farming.

I would like now to look briefly at the potential grassland areas and then give my views as to some of the initiatives that must be undertaken if that potential is to be realized.

There is considerable variability in the basic plant-growing capacity of the 3,000 million ha presently classified as grazing land. Consequently, animal carrying capacity varies greatly. Not all areas can become highly productive. Even so, only a small fraction of the existing grassland areas have ever had even the minimum of improved management and technology applied to them. Even in the U.S.A., the status of rangelands leaves much to be desired. According to Rummell (1981), U.S. forest and rangelands annually provide 213 million animal-unit months (AUMs) of livestock grazing. Expectations are that grazing requirements will increase to 300–365 million AUMs by the year 2000. U.S. ranges and forests are estimated to have the capacity to produce 566 million AUMs of range grazing annually. The 48 contiguous states have 263 million ha of rangeland in the following conditions:

Good	38.8 million ha
Fair	82.6 million ha
Poor	98.8 million ha
Very Poor	42.9 million ha

Although only slightly less than half of the rangeland areas in the U.S. are in fair to good condition, most of the technology is available for improvement of the remainder. If these technologies are not economically feasible today, I suspect that in the years to come, as we tend to become more and more dependent on grasses and forage to produce meat, improvement practices will become more economically attractive. Unfortunately, the consequences of poor returns only tend to exacerbate the deterioration of agricultural and grazing lands. As farmers are unable to purchase inputs, further deterioration of land resource occurs, and as deterioration occurs, returns diminish. Rangeland and grassland development requires long-range investment. Returns are not immediate. Credit institutions are reluctant to provide the capital investment necessary to establish range-improvement practices. Farmers cannot afford to carry high-interest loans. The result is that technology is not applied. This vicious cycle

can be broken and must be broken if we are to preserve the productive capacity of the resource base.

Some of the major areas where the grassland and grazing potential will be achieved only if the research effort is intensified can be found in Africa, South America, Southeast Asia, and the China mainland. An excellent description of the Sahelian-Sudanian Zone of Africa was recently completed by Dr. Gerald W. Thomas (1980). This is a vast region of disturbed grassland and mixed brush and forest in which most of the present-day livestock husbandry is practiced. Thomas describes it thus; "The most alarming portrait one gets from a study of the vegetation complex is the evidence of continuous overgrazing. There are striking indications of retrogression: changes in plant species composition, marked reduction in total plant cover in the more arid zones, trends toward annual grasses in rainfall zones . . . and soil disturbance." Thadis Box (1981), in his section of the Winrock report prepared for this Congress, has nicely described the characteristics of these arid regions. He says, speaking of these arid systems, "Most rangeland ecosystems are rather delicately balanced and this balance is easily disturbed by abusive grazing."

One is impressed, too, with the apparent absence of the application of pasture-improvement practices and with the low level of husbandry and consequent low productivity. One is even more impressed with the enormous difficulty of introducing changes in the existing systems of that region. As Box also points out, "Restoration of damaged rangelands is at best difficult because of the rigorous climatic and biological constraints imposed upon the system." Curiously, as pointed out by Mott and Popenoe (1977), "In Central and South America, as well as Australia, most artificial pastures are being planted to grasses which have their origin on the African continent." It would appear that the livestock potential is even greater in the more humid sub-Sahelian region of Central Africa. Most of the area is presently excluded from domestic livestock grazing because of the prevalence of the tsetse fly, vector of the blood protozoan parasite trypanosoma, which causes sleeping sickness in man and nagana in cattle. There are other diseases; there are other problems. The fact is that little or no technology for livestock production is available or being generated or being applied in this area except for some disease-control measures. The International Laboratory for Research on Animal Diseases (ILRAD) is making progress toward the possible control of trypanosomiasis. Even if it succeeds, the technology for management of the grass and forage species and management of livestock systems for the humid and high-rainfall areas is woefully inadequate. ILCA (International Livestock Centre for Africa) has properly focused its attention on those areas presently used for livestock grazing. Reduction of the grazing pressures in the Sahelian-Sudanian area will ultimately depend on the north-south stratification of livestock production—utilizing the potentially more productive humid tropical regions. In spite of the efforts of the affected African nations alone or in association (i.e., CILSS —Comité Permanent Interétats de Lutte Contre la

Sécheresse dans le Sahel), and of the international donor community, one would have to conclude that the grassland situation today in Africa can only be described as disastrous.

The prospects on the South American continent as well as in Central America and Mexico for substantially expanding and intensifying livestock productivity are considerably brighter than those of the African subcontinent. There are vast areas in South America for which additional technology is required, including the cerrados of Brazil; the savannas of Brazil, Guyana, and Surinam; and the llanos of Colombia and Venezuela. Fortunately, much excellent work, especially with forage species, has been underway in Brazil, Colombia, Costa Rica, and Mexico. Nevertheless, with the advent of the Centro Internacional de Agricultura Tropical (CIAT), there is greater structure and continuity to these efforts. In the vast llanos of Colombia, CIAT has been researching sustainable high-production systems, and the forage research has been especially exciting. Although further corroboration is needed, data collected over a 3-year period have shown 2-fold or more weight gain/animal and as much as a 20-fold increase in offtake/unit area comparing improved pastures over native savanna under experimental conditions (CIAT 1980).

Further description of the vast rangeland potential of different ecosystems is given in the Winrock report that I referred to earlier. In addition to the grassland potential, the conversion of plant residues from crop production, other byproducts, and recycled nutrients represent still not fully utilized resources. Crop productivity around the world has increased dramatically. As a direct consequence, the amounts of crop residues and byproducts have also increased.

We probably have not given adequate attention to the nutritive value of so-called crop residues in the crop-livestock systems that are so prevalent throughout the world. I recall that in a project in Mexico called the Puebla project, I pointed out to the agronomists that in calculating the benefits to be derived from improved crop production techniques they were paying little attention to the value of the added stover. Not only had grain yield increased 2, 3, and 4 times, but fodder offtake and quality had more than doubled. Now, both the farmers and technicians are working on corn plants that retain an even higher stalk value after the corn is harvested.

Another important dimension in livestock production is the use of grain. The world is producing approximately 1.4 billion tons of grain annually. Currently, approximately 450 million tons of grain are fed to livestock of all classes (including pigs and poultry). If current trends continue, the total grain requirement for animal feed (633 million tons) will exceed that used directly for human consumption by 1985 (623 million tons) (Wheeler et al. 1981). Of this amount, poultry and pigs together consume approximately 60% and represent the greatest increase in demand for grain use. A number of reasons are quickly apparent for the dramatic expansion in that livestock sector. They include: (1) urbanization and concentration of population on non-food-producing land; (2) high demand for animal products at all economic levels; (3) ease of adoption of production technology with rapid life-cycle animals, including controlled life-cycle feeding and management and disease prevention systems (labor skills needed can be quickly taught); (4) ease of concentrating production units near urban centers, thereby reducing infrastructure needs (e.g., transport, refrigeration); and (5) credit, usually readily available because of the rapid turnover time, though investments are often made by the wealthier sector.

Nevertheless, it is important to point out that poultry and pigs contribute approximately 40% of the food energy derived by people from the consumption of animal products while consuming 60% of the grain resources fed to animals. Ruminant livestock convert large quantities of forages, crop residues, and other nutrient resources not usable by man. In total, all such nutrients represent more than twice the amount needed by ruminant animals, according to Fitzhugh et al. (1978).

Expansion in the monogastric sector would seem to be a contradiction of the direction in which the animal industry should expand, given the increasing demand for cereals by expanding populations. The fact of the matter is that increases in output of ruminant products are likely to be modest over the next decade. Thus, we are likely to witness continued growth in grain feeding to monogastric animals in an attempt to meet market demands for animal products, even though the total grain deficit in the developing countries is likely to be double the current 30 million tons by 1985. Most of the grain in the developing countries is used for human food. The competition for use of cereals in either feed or food will be one factor in increasing prices of grains. It is apparent that the interrelationship among grain production, grain use for food or feed, and the relevant costs of each will be a dynamic one for the foreseeable future and one that will have significant positive implications for the development of the grassland potential.

What, then, are the strategies that must be adopted if the tremendous grassland resource is going to make the needed contribution to meeting the world's foreseeable food demands?

As scientists, we tend to think in terms of scientific and technological solutions to problems. We tend to forget that people, societies, and political systems are the context within which the physical and biological systems must be made to work. Yet the social systems we invent have to be compatible with what the earth is capable of sustaining. As professionals we have failed to bridge the gap between the biological and the social scientist. We have also fractionated our research efforts so that we look for the solution to the problem in pieces. Look at the livestock development programs financed by the international agencies. Even among economists, I believe, it is generally recognized that the rate of return of development loans has been much lower than expected, as calculated ex-ante.

There is the widespread impression that increased livestock productivity and expanded development will take place by rather simplistic approaches. It is rather surprising—perhaps I should say alarming—that the live-stock development plans of the LDCs are based on single-factor approaches. I need mention only the enormous investments in boreholes in Africa or the purchase of "improved" breeds for the hot humid tropics and the tragic consequences of these approaches. The tragedy is all the more alarming since often the recommendations to invest in such approaches come from the "experts" themselves in the assistance agencies.

It is time that we look holistically at livestock research and development. While it is inconceivable that we would structure a crop research program around a single specialty, we have failed to address livestock programs in a similar fashion. In part, the success of the international crop research centers is due to the interdisciplinarity and multidisciplinarity of their approach; more recently they have included social scientists among the teams. It is convenient and perhaps necessary at times to think in terms of a division of responsibility in order to achieve the depth of understanding necessary for a particular function, be that a specific disease of animals, soil-nutrient dynamics, or what-not. However, if the changes in productivity levels we hope to achieve are going to take place, we must develop more cohesive and comprehensive approaches than we have in the past.

In the course of my own experience, one of the greatest frustrations and a challenge at the same time was to try to build comprehensive animal research programs, especially bridging the chasm between the veterinary professional, the animal husbandry specialist, and the agronomist. I would like to think that we are now able to put the people, the disciplines, and the institutional structures together in an effort to get at development problems more effectively.

In reviewing some of the fascinating work done by soil scientists and others in following the dynamics of nutrients, (i.e. minerals, etc.) through the soil, plants, man, or animal systems, it would seem to me that this fundamental process is at the heart of our food, agriculture, and resource system. The dynamic process of nutrient materials moving through the system can be altered substantially, either through natural processes, such as fire, flood, or drought, or more especially through man's activities. It is also easy to get excited about many other aspects of research. The excellent work going on in forage improvement holds high hopes of making grassland areas more productive. Knowledge of animal nutrition, physiology, reproduction, genetics, and breeding is all changing very rapidly.

But that knowledge has to be put together, tested on the ground, and translated into practices that are doable by the farmers. Ultimately, when the farmer or the rancher is expected to use new information, it has to be made available in a form that he understands and has confidence in. He may not need to know and probably does not want to know why a particular practice works. But he does want to know if it has been tried and proven successful and if it will reduce risk and make a better return.

As I view the current grassland, livestock, and food situation, I see four needs:

1. The need to develop comprehensive strategies to deal with the major grassland resource areas. Although technology is often location-specific, it would be quite possible to design a development strategy that has all of the essential elements that have to be considered for a particular area. We have to find better ways to deal with the destructive effects on the grassland resource resulting from the failure to adopt range-improvement practices.

2. Research strategies must be devised to integrate scientific efforts across a broad spectrum of disciplines whose efforts are systematically and simultaneously applied to grassland development and conservation practices. We have yet to structure our research efforts in such a way that account is taken of the total biological, climatological, and sociological interaction and that in turn serves as the basis of appropriate policy formulation to enable the system to work. It is not my intention to review here the kinds of specific research needs in the many disciplinary areas. These have been amply identified by others elsewhere.

3. Support for research. It is easy to say that we need more support for the research efforts. Generally speaking, that is true, for it has been shown that returns on investment in agricultural research are quite high. I am particularly concerned, however, with the utilization of research findings in the development of workable systems for different ecosystems. The adoption of improved grassland technology, or indeed improved husbandry practices generally, has been quite disappointing.

I believe that national and international support will have to be redirected from some of the current approaches to better-structured comprehensive research programs and delivery systems.

4. Policy formulation that is directed to the judicious utilization of grasslands is urgently needed. Security of land ownership, controlled land-use patterns, infrastructure development, and support of research all require that policymakers understand the complexities of grassland resource use.

Specific effort must be made to achieve greater interaction between the scientific community and the policymakers. Scientific meetings, such as this one, might consider holding special sessions dealing with policy aspects of this sector. Certainly such efforts should be a part of institutions such as the International Livestock Centre for Africa (ILCA) and the International Food Policy Research Insitute (IFPRI).

Bringing about change in the grassland sector is not going to be easy. Generally it has low political "sex appeal" as an area of investment. Time is a factor. There is a paucity of trained people to conduct the testing, formulate production packages, deliver information, and interact with the support structures. There is great dependence upon natural phenomena, which are maddeningly unpredictable. Research over large areas is expensive, and variation compounds the problems of applicability.

But then I do not recall anybody saying it would be easy.

LITERATURE CITED

Box, T. 1981. Potential of arid and semi-arid rangelands for ruminant animal production. Winrock Rpt., Winrock Int., Morrilton, Ark., 81-92.

CIAT (Centro Internacional de Agricultura Tropical). 1980. Ann. Rpt., Cali, Colombia.

Fitzhugh, H.A., H.J. Hodgson, O.J. Scoville, T.D. Nguyen, and T.C. Byerly. 1978. The role of ruminants in support of man. Winrock Rpt., Winrock Int., Morrilton, Ark.

Mott, G.O., and H.L. Popenoe. 1977. Ecophysiology of tropical crops (P. de T. Alvim and T.T. Kozlowski, eds.), Ch. 6, 157-186. Academic Press, New York.

Rummell, R.S. 1981. Grazing land resources for meeting food and fiber needs. J. Anim. Sci. 52(3):644-649.

Thomas, G.W. 1980. The Sahelian/Sudanian zones of Africa—Profile of a fragile environment. Rpt. to Rockefeller Found., New York.

Wheeler, R.O., G.L. Cramer, and K.B. Young. 1981. The world livestock and food grain system. Winrock Rpt. Winrock Int., Morrilton, Ark.

SECTION PAPERS

Seedling Traits as Possible Selection Tools for Improving Seedling Emergence of *Astragalus cicer* L.

C.E. TOWNSEND
USDA-SEA-AR Crops Research Laboratory and
Colorado State University, Fort Collins, Colo., U.S.A.

Summary

Cicer milkvetch (*Astragalus cicer* L.), a nonbloating legume, has many of the attributes of a desirable forage species. Relatively poor seedling vigor, however, has limited its use. Our experiment's objectives were to determine the variations among selected progenies of cicer milkvetch in seedling vigor traits in the laboratory and in seedling emergence in the field and to discover whether the seedling traits could be used to predict seedling emergence in the field.

A total of 112 polycross progenies from 3 selected populations were evaluated for seedling vigor traits at temperatures of 15°/25° C in the laboratory and for seedling emergence at 2 field locations. In all instances the experimental design was a randomized complete block with 4 replications.

Differences among populations were relatively small for all seedling traits. Mean radicle length, hypocotyl length, hypocotyl diameter, total seedling length, and seedling weight ranged from 36 to 38 mm, 25 to 26 mm, 1.19 to 1.27 mm, 61 to 64 mm, and 3.29 to 3.88 mg, respectively. Within populations the progenies differed significantly (0.05 level) for all seedling traits.

Progenies within populations also differed significantly for seedling emergence at both locations. Seedling emergence at one location was about twice that at the other location.

Of the 30 possible correlation coefficients between the five seedling vigor traits mentioned above and seedling emergence at the two locations for three populations, only 4 were significant, and they were of low predictive value. Therefore, we concluded that seedling traits as developed in a laboratory environment at temperatures of 15°/25° C are not suitable for predicting seedling emergence of cicer milkvetch under field conditions.

KEY WORDS: cicer milkvetch, *Astragalus cicer* L., seedling emergence, seedling traits, seedling vigor.

INTRODUCTION

Cicer milkvetch (*Astragalus cicer* L.), a nonbloating legume, has many of the attributes of a desirable forage species. Relatively poor seedling emergence and subsequent stand establishment, however, have limited its use. In an attempt to overcome these deficiencies, we have evaluated selected progenies for speed of germination in the laboratory and for seedling emergence in the greenhouse and in the field. In comparison with some forage species, cicer milkvetch is relatively slow to germinate (Townsend and McGinnies 1972), but variability exists within the species for both speed of germination and seedling emergence (Townsend 1974). Seed germination in

the laboratory has been significantly correlated with seedling emergence in the field (Townsend 1974), but laboratory germination is a better indicator of field emergence under favorable than under unfavorable soil-moisture conditions. Also, polycross progenies of cicer milkvetch differed significantly in seedling dry weight, indicating that variability exists in seedling vigor (Townsend 1976).

We have studied the potential advantages of using larger seeds because in many species of forage legumes large seeds tend to produce more vigorous seedlings than small seeds do during the early stages of growth (Cooper 1977). We found that large seeds of cicer milkvetch give better seedling emergence from deep depths of planting in the field than do small seeds, but at shallow depths of planting there are no differences among progenies of different seed-weight classes (Townsend 1979). In addition, various seedling traits, such as hypocotyl diameter, hypocotyl length, radicle length, total length, and dry weight,

C.E. Townsend is a Research Geneticist with USDA-SEA-AR. This is a contribution of the USDA-SEA-AR in cooperation with the Colorado State University Experiment Station, Scientific Series No. 2605.

Table 1.   Growth measurements of 7-day-old seedlings in a 15–25° C environment for 112 polycross progenies in three populations of cicer milkvetch.

Population	Radicle length		Hypocotyl length		Hypocotyl diameter		Total Seedling length		Seedling weight	
	Mean	Range	Mean	Range	Mean	Range	Mean	Range	Mean	Range
				mm					mg	
One	37	27-46	25	19-32	1.19	1.02-1.32	62	50-69	3.29	2.89-4.16
L.S.D. (0.05)		6		5		0.08		9		0.24
Two	36	28-43	25	17-34	1.19	1.12-1.24	61	46-75	3.51	2.99-4.05
L.S.D. (0.05)		6		6		0.06		10		0.31
Three	38	29-47	26	17-34	1.27	1.12-1.37	64	49-75	3.88	2.96-4.41
L.S.D. (0.05)		6		5		0.06		10		0.24

are significantly influenced by seed weight, and the high-seed-weight progenies tend to rank higher for these attributes than low-seed-weight progenies.

The objectives of this study were to evaluate selected progenies of cicer milkvetch for various seedling traits in the laboratory and to determine the relationship, if any, between these traits and seedling emergence in the field.

METHODS

Three populations of polycross progenies were evaluated. Population 1 was derived from clones whose polycross progenies ranked high for seedling emergence in diverse environments. Populations 2 and 3 were second-cycle recurrent selections for high seed weight (Townsend 1977). Mean seed weights of populations 1, 2, and 3 were 3.98, 4.25, and 4.62 g/1,000 seeds, respectively. There were 33, 39, and 40 polycross progenies in populations 1, 2, and 3, respectively. Because of the large number of progenies, the three populations were evaluated in three separate but identical experiments in both the laboratory and the field.

Seedling growth characteristics were evaluated by placing 40 scarified seeds on blotter paper in a plastic box (13 × 13 × 4 cm) containing creped cellulose and 100 ml of tap water. Each box was a replication for that progeny. The experimental design was a randomized complete block with 4 replications. The boxes were placed in a dark germinator at alternating temperatures of 15° and 25°C with 12 hours at each temperature. We retained 15 seeds whose radicle extension began on the same day. On the seventh day we selected the 10 most vigorous seedlings of the 15 and measured hypocotyl diameter, hypocotyl length, radicle length, total length, and dry weight.

Field studies were conducted at the Agronomy Research Center near Fort Collins and at the Central Plains Experimental Range near Nunn, Colorado. The soil at the Agronomy Research Center is a Nunn clay loam (mesic aridic Argiustoll) and the soil at the Central Plains Experimental Range is a Vona sandy loam (mesic ustollic Haplargid). The same 112 progenies that were used in the laboratory phase plus one large-seeded lot (4.53 g/1,000 seeds) that served as a reference progeny were seeded in the field study. Plantings were made with a single-row cone seeder equipped with a double-disk opener and

packer wheel. Scarified seeds were inoculated with *Rhizobium* sp. and planted at the rate of 55/m of row to a depth of 2 cm. A plot consisted of a single row 3.6 m long. Spacing between rows was 0.6 m. The experimental design was the same as that of the laboratory study, a randomized complete block with 4 replications. Plantings were made on 15 April 1975 at the Central Plains Experimental Range and on 13 April 1976 at the Agronomy Research Center.

Seedling emergence counts were taken from the center 3 m of each row when the earliest progenies were emerging well. Seedling emergence was expressed as a percentage of that of the reference progeny.

RESULTS AND DISCUSSION

Differences among populations were relatively small for all seedling traits (Table 1). Population 3, the large-seed population, however, ranked the highest for each trait, supporting the findings of a previous study (Townsend 1979). Mean radicle length, hypocotyl length, hypocotyl diameter, total seedling length, and seedling weight ranged from 36 to 38 mm, 25 to 26 mm, 1.19 to 1.27 mm, 61 to 64 mm, and 3.29 to 3.88 mg, respectively. There were, however, significant differences (0.05 level) among progenies within populations for all seedling traits.

Seedling emergence of the three populations varied in the two locations (Table 2). Population 1, which was

Table 2.   Seedling emergence* for 112 polycross progenies in three populations of cicer milkvetch at two locations.

Population	Agronomy Research Center		Central Plains Experimental Range	
	Mean	Range	Mean	Range
One	103	79–138	80	28–146
L.S.D. (0.05)		21		26
Two	90	71–110	149	100–211
L.S.D. (0.05)		26		47
Three	95	74–121	143	75–175
L.S.D. (0.05)		24		33

*Based on percentage of the large-seeded lot. Average number of seedlings/m of row for the reference progeny was 26 and 13 at the Agronomy Research Center and Central Plains Experimental Range, respectively.

Table 3. ,Simple correlation coefficients between seedling emergence at two field locations and five seedling traits for three populations of cicer milkvetch.

Population		Total seedling length		Radicle length		Hypocotyl length		Hypocotyl diameter		Seedling weight	
		ARC†	CPER	ARC	CPER	ARC	CPER	ARC	CPER	ARC	CPER
One	(31df)	+0.36*	−0.11	+0.25	+0.11	+0.14	−0.36*	+0.20	+0.07	−0.13	+0.55**
Two	(37df)	+0.10	+0.08	+0.21	+0.22	−0.01	−0.08	−0.06	+0.04	−0.02	+0.08
Three	(38df)	−0.06	+0.01	+0.02	+0.13	+0.02	−0.05	+0.30	+0.02	+0.02	+0.36*

** and * Significant at the 0.01 and 0.05 levels, respectively.
†ARC—Agronomy Research Center
CPER—Central Plains Experimental Range

selected for improved seedling emergence, gave the highest mean emergence at the Agronomy Research Center but the lowest emergence at the Central Plains Experimental Range. Low emergence at the range site was attributed to the extremely dry conditions that immediately followed seed germination of the earliest progenies. Such progenies were at an extremely critical stage of germination and could not withstand desiccation. Progenies within all populations differed significantly in seedling emergence, indicating that genetic diversity existed within these populations.

Correlation coefficients between seedling emergence and the five seedling traits were low (Table 3). Only 4 of 30 coefficients were significant, and they were of low predictive value. Seedling weight of populations 1 and 3 was significantly and positively correlated with emergence at the Central Plains Experimental Range. Total seedling length of population 1 also was positively and significantly correlated with emergence at the Agronomy Research Center. Hypocotyl length was negatively and significantly correlated with emergence at the Central Plains Experimental Range. Such seedling traits as radicle length or total seedling length may influence emergence and survival only under conditions in which the soil surface dries more rapidly than the radicle of the slower-growing seedling is able to extend. Therefore, these traits may have no effect on emergence or survival in certain environments. High correlation coefficients for these traits were not expected in this study.

Seedling traits as developed in a 15°/25°C laboratory environment proved not to be suitable for predicting seedling emergence of cicer milkvetch under field conditions. Seedling development at lower temperatures may be a better indicator of seedling emergence. Selecting for vigorous seedlings in high-seed-weight progenies might be worthwhile. Depth of planting may also play a role in the relationship between seedling traits and seedling emergence.

LITERATURE CITED

Cooper, C.S. 1977. Growth of the legume seedling. Adv. Agron. 29:119–139.

Townsend, C.E. 1974. Selection for seedling vigor in *Astragalus cicer* L. Agron. J. 66:241–245.

Townsend, C.E. 1976. Combining ability for seedling dry weight and forage yield in cicer milkvetch. Crop Sci. 16:480–482.

Townsend, C.E. 1977. Recurrent selection for high seed weight in cicer milkvetch. Crop Sci. 17:473–476.

Townsend, C.E. 1979. Associations among seed weight, seedling emergence, and planting depth in cicer milkvetch. Agron. J. 71:410–414.

Townsend, C.E., and W.J. McGinnies. 1972. Temperature requirements for seed germination of several forage legumes. Agron. J. 64:809–812.

Selection for Bluegreen Aphid Resistance in Subterranean Clover

D.J. GILLESPIE and J.D. SANDOW
Department of Agriculture, Western Australia

Summary

Bluegreen aphid (BGA) (*Acyrthosiphon kondoi* Shinji), recently arrived in Western Australia, poses a serious threat to 7 million ha of legume-based pastures, 93% of which is sown to subterranean clover (*Trifolium subterraneum*).

Two experiments were conducted in the summer of 1980 to investigate the resistance of commercial cultivars and breeding lines to BGA. In the first experiment, 18 commercial subterranean clovers, 3 commercial medics (*Medicago* spp.), and 1 serradella (*Ornithopus compressus*) were tested as 7-week-old plants. The second experiment tested a total of 67 selections, breeding lines, and commercial cultivars as seedlings.

Plants were grown in aphid-proof cages in controlled environment conditions with 18°/10°C day/night temperatures. Aphids were applied at the beginning of each experiment, and plant damage was monitored at weekly intervals. After 4 weeks, almost 80% of the varieties tested showed damage symptoms ranging from some dead leaves to all dead plants. Symptoms did not develop at the same rate on all varieties. One commercial cultivar and one introduction showed a high level of tolerance to BGA.

KEY WORDS: bluegreen aphid, *Acyrthosiphon kondoi* Shinji, resistance, subterranean clover, *Trifolium subterraneum.*

INTRODUCTION

Bluegreen aphid (BGA) (*Acyrthosiphon kondoi* Shinji) was first detected in Australia in May 1977 and subsequently in Western Australia in June 1979. In Australia, as in the U.S.A., the initial effects of BGA were first observed on lucerne (*Medicago sativa* L.), but other pasture legumes have also been seriously affected. The cool climatic conditions that favor multiplication and development of BGA (Marble 1978) are common in southern Australia in late autumn, winter, and spring, the growing season for annual temperate legumes. Aphid numbers in excess of 500/stem have been recorded in annual *Medicago* species (Sandow 1980), and in the 4 years that BGA has been present in New South Wales, damage to subterranean clovers (*Trifolium subterraneum*) has steadily increased (Wolfe, personal communication). Instances of serious damage resulting in the death of entire pastures and complete cessation of seed set in subterranean clover have already been recorded in Western Australia.

Subterranean clover is the most important single component of annual pastures in southern Australia and the key factor in the improvement of soil fertility in cereal farming and mixed farming (Francis 1981). Donald (1970) estimates that as much as 80% of the 20 million ha of established pastures in southern Australia has been sown to subterranean clover. In Western Australia, 93% of the 7 million ha of legume-based pastures has been sown to subterranean clover.

The arrival of BGA in Australia precipitated numerous projects aimed at effective control, including biological control and assessment of legume varieties for resistance to the aphid. Much of this work has been similar to that initiated in California in 1975. Until now resistance testing has largely involved lucerne and annual medics (Franzmann et al. 1979, Lloyd et al. 1980, Mathison 1980); however, identification of subterranean clovers carrying genes for resistance is a high priority in Australia.

This paper describes the techniques adopted for aphid-resistance testing and the results of initial trials.

MATERIALS AND METHODS

Two experiments were conducted during the summer of 1980. In the first, 18 commercial subterranean clovers, 3 commercial medics, and 1 serradella (*Ornithopus compressus*) were tested as 7-week-old plants. The second experiment tested a total of 67 selections, breeding lines, and commercial cultivars as seedlings. All except 2 of the varieties tested in experiment 1 were included in experiment 2.

Plants were grown in a sand-sawdust mixture and housed in controlled-environment rooms maintained at 18°C (day) and 10°C (night). These temperatures are known to be suited to rapid aphid multiplication and development (Marble 1978), and they simulate plant-growing temperatures in autumn and early spring in southern Australia. Thermoperiod and photoperiod were both set at 12 hours. Artificial light was provided by high-output fluorescent tubes and approximately 20% tungsten light. Light intensity at the top of the plant canopy ranged from 200–300 μ e/m^2/second. Relative humidity, not controlled, varied from 60% to 70% throughout the experiments.

In both experiments a tolerant and a susceptible control variety were used—Clare and Daliak, respectively, chosen because of previous field and greenhouse observations. One row or pot of each control was used in each screening box in both experiments.

Experiment 1 included 5 replications sown in 1-l plastic pots. Ten scarified seeds were sown/pot, and each pot was thinned to 5 after germination. After 7 weeks, pots were randomly placed in plastic boxes (420 mm × 320 mm × 140 mm) so that 4 test varieties and 2 controls were present in each. Insect-proof muslin cages were fitted over the boxes, and on day 1 BGA were added to each pot at the rate of 1.7–1.9 g/box (approximately 0.3 g/pot). A further application of 0.2 g/pot was added on day 7.

In the second experiment 67 varieties were sown in rows directly into plastic boxes and were then screened as seedlings. Five replications were used. Six test varieties and 2 controls were sown/box at the rate of 50 germinable seeds/row. After 17 days, all rows had fully emerged; they were thinned to the 30 oldest and healthiest seedlings/row. Boxes were enclosed in insect-proof cages as in experiment 1, and aphids were applied at the rate of 1.7–1.9 g/box (approximately 0.2 g/row).

The BGA used for these experiments were bred in controlled environments on Hunter River lucerne. Environ-

D. J. Gillespie is a Research Officer in Agronomy and J.D. Sandow is an Entomologist with the Department of Agriculture of Western Australia.

The authors wish to thank M. Grimm for invaluable advice in the initial planning of these experiments. They also gratefully acknowledge the technical assistance of L.G. Sharpe, R.N. Emery, and P.L. Wilson.

mental conditions were similar to those during screening. Aphids were collected from the lucerne by cutting the plants and sieving them through a 4-mm mesh stainless-steel screen. The sieved aphids were weighed and applied to each box by shaking them through a piece of stainless-steel screen. This technique had the effect of separating the aphids and distributing them evenly among the plants.

Assessments were made weekly from the time the aphids were applied until termination of each experiment. Each of the experiments was terminated when the susceptible control (Daliak) was completely dead. Termination occurred after 28 days in experiment 1 and after 32 days in experiment 2.

In experiment 1, assessments were made of plant damage, aphid density, and aphid distribution over the plant. In experiment 2, only plant damage was assessed because of the difficulty of recording aphid density and distribution on seedling plants. Plant damage was recorded on a subjective 0–5 scale with 0 representing no effect and 5 representing all dead or dying plants. Aphid density was recorded on a subjective 0–4 scale with 0 representing no aphids found and 4 representing a dense aphid population.

RESULTS

Plant damage was first visible as yellowing of the oldest trifoliate leaves in experiment 1. This yellowing occurred about 10–12 days after application of aphids. In experiment 2 yellowing of cotyledons occurred after 6–8 days; yellowing of trifoliate leaves occurred at 10–12 days.

The varieties developed damage symptoms at different rates, and in many cases early assessments did not reflect the final damage scores (Table 1). For example, the two controls, Clare and Daliak, had similar scores on day 14, but at termination Daliak was more severely damaged than Clare. On the other hand, varieties with similar final susceptibility (e.g., Seaton Park and Nangeela) often had quite different damage levels at the earlier assessments.

Damage at termination ranged from the yellowing of a few leaves for the most tolerant cultivar (an overseas introduction, 12693 A [*T. subterraneum* ssp. *subterraneum*], damage score 1.4) to death of all plants for the most susceptible cultivars (*T. subterraneum* ssp. *yanninicum*— Yarloop, Larisa, and 70088B, damage scores 5.0).

Weekly ratings of aphid density in experiment 1 indicated a steady buildup in aphid numbers on most varieties; maximum scores were common at termination of the experiment. The most notable exception to this trend was serradella var. Pitman, on which low aphid numbers were consistently recorded. Even when aphid numbers on nearby varieties were very high, serradella never exceeded a density score of 2. Highest aphid densities were generally recorded on the medics and on all members of *T. subterraneum* spp. *yanninicum*. Assessment of the distribution of aphids over the plants in experiment 1 indicated little differences between plant varieties, with aphids tending to concentrate around the growing plants.

DISCUSSION

A wide range of susceptibility to BGA was recorded among the 67 legume varieties. Almost 80% had a dam-

Table 1. Plant damage on some annual legumes caused by BGA.

Variety	Species[1]	Days from aphid application			
		14	21	28	32
Pitman	O. comp.	0[2] (0)	0.2 (0.6)	1.8 (1.8)	(2.4)
Dwalganup	T. sub. sub.	0.8 (0.2)	1.2 (1.0)	2.2 (1.4)	(2.0)
Nungarin	T. sub. sub.	1.4 (1.0)	2.2 (1.4)	2.6 (2.0)	(2.8)
Bacchus Marsh	T. sub. sub.	1.4 (0.8)	2.6 (2.2)	3.0 (2.8)	(2.8)
Mt Barker	T. sub. sub.	1.8 (1.0)	3.2 (2.2)	3.2 (2.6)	(2.8)
Clare	T. sub. br.	2.2 (1.2)	3.0 (2.0)	3.2 (2.5)	(2.8)
Northam	T. sub. sub.	1.4 (1.0)	3.0 (2.0)	3.2 (2.4)	(3.2)
Geraldton	T. sub. sub.	1.0 (1.0)	1.6 (2.2)	3.2 (2.8)	(3.4)
Tallarook	T. sub. sub.	2.4 (0.8)	3.2 (1.4)	3.4 (2.4)	(2.2)
Seaton Park	T. sub. sub.	1.2 (0.4)	2.6 (2.0)	3.4 (2.8)	(3.6)
Nangeela	T. sub. sub.	2.2 (1.2)	2.8 (1.6)	3.6 (2.0)	(3.0)
39327YB	T. sub. yan.	2.2 (1.4)	3.0 (2.2)	3.6 (3.2)	(3.8)
Trikkala	T. sub. yan.	1.0 (1.8)	2.6 (3.4)	3.8 (4.2)	(4.6)
Esperance	T. sub. sub.	2.6 (1.0)	3.2 (2.2)	3.8 (3.6)	(4.2)
Woogenellup	T. sub. sub.	2.8 (1.6)	3.2 (3.2)	4.0 (3.6)	(4.2)
Larisa	T. sub. yan.	2.2 (0.8)	3.2 (2.8)	4.0 (4.8)	(5.0)
Cyprus	M. trunc.	1.8 (2.0)	3.6 (3.0)	4.6 (4.2)	(4.8)
Tornafield	M. tornata	2.2 (2.0)	3.2 (4.0)	4.6 (4.8)	(4.8)
Daliak	T. sub. sub.	2.4 (1.2)	3.8 (3.0)	4.8 (4.4)	(4.7)
Yarloop	T. sub. yan.	1.6 (1.2)	3.2 (3.6)	4.8 (4.8)	(5.0)

[1]T. sub. sub.   = *Trifolium subterraneum* ssp *subterraneum*
T. sub. br.   = *Trifolium subterraneum* ssp *brachycalycinum*
T. sub. yan.   = *Trifolium subterraneum* ssp *yanninicum*
M. trunc.   = *Medicago truncatula*
[2]Ratings for vegetative plants with rating for seedlings in parentheses. See text for scale. L.S.D. (P ± .05) was 0.74 for vegetative plants at 28 days and 1.01 for seedlings at 32 days.

age rating of 3.0 or higher, indicating at least partial plant death due to BGA. Eleven of the 18 commercial subterranean clovers account for about 99% of the area sown to clover in Western Australia (Quinlivan, personal communication), and 9 of them (82%) had ratings of 3.0 or higher. BGAs therefore have the potential to damage a very large proportion of the subterranean clover pastures of Western Australia, and there is an urgent need for replacement clovers possessing some resistance to this aphid.

The assessment technique used in these experiments is based on both plant damage and plant survival and seems ideally suited to handling large numbers of test varieties. Other indicators of varietal susceptibility, such as dryweight yield, are laborious, requiring the use of untreated controls when varieties differing in growth rate are being assessed. Such methods considerably reduce the number of varieties that can be processed when space and labor are limited.

A strong correlation needs to be established between seedling and flowering-plant resistance before large-scale resistance screening is done, because seedlings can be considered a worthwhile indicator of field resistance. No investigations of susceptibility at different growth stages have been attempted with subterranean clover, but differences have been reported in lucerne (Wynn-Williams and Burnett 1977) and with spotted alfalfa aphid (*Therioaphis trifolii* Monell) in medics (Mathison et al. 1978).

The applicability of seedling screening trials to field infestations does not depend only on a comparable response from seedlings and vegetative or flowering plants. Field infestations of differing duration or severity could result in plant effects that differ from those in greenhouse trials where constant or increasing aphid pressure is artificially maintained. For example, our results indicate that damage develops at different rates; thus, BGA infestations of

short duration will affect only varieties that respond quickly to aphid attack. Field susceptibility under conditions of severe or prolonged BGA attack is likely to be similar to the results reported.

Our own observations and other evidence accumulated from southern Australia during 1980 indicated that varieties identified as being highly susceptible in these experiments were indeed the varieties suffering severe damage in field infestations, notably Yarloop, Trikkala, and various *Medicago* species.

LITERATURE CITED

Donald, C.M. 1970. Temperate pasture species. *In* R.M. Moore (ed.) Australian grasslands, 303–320. Austral. Natl. Univ. Press, Canberra.

Francis, C.M. 1981. Breeding better clovers for Australian farms. Proc. Austral. Fm. Mangt. Soc. Conf., Merredin.

Franzmann, B.A., W.J. Scattini, K.P. Rynne, and B. Johnson. 1979. Lucerne aphid effects on eighteen pasture legumes in southern Queensland: a glasshouse study. Austral. J. Expt. Agric. Anim. Husb. 19:59–63.

Lloyd, D.L., J.W. Turner, and T.B. Hilder. 1980. Effects of aphids on seedling growth of lucerne lines. I. Bluegreen aphid in field conditions (*Acyrthosiphon kondoi* Shinji). Austral. J. Expt. Agric. Anim. Husb. 20:72–76.

Marble, V.L. 1978. Lucerne aphids—a worldwide threat. World Fmg. (May) 20:10–11, 24–26.

Mathison, M.J. 1980. Aphid resistance screening of annual medics. Austral. Pl. Breed. Genet. Newsletter 30:78–82.

Mathison, M.J., E. Kobelt, and G. Baldwin. 1978. Medic and sub clover susceptibility to S.A.A. and B.G.A. Fact Sheet, 28/78. S. Austral. Dept. Agric. and Fisheries.

Sandow, J.D. 1980. Bluegreen aphid—a pest of pasture legumes. Farmnote No. 81/80. West. Austral. Dept. Agric.

Wynn-Williams, R.B., and P.A. Burnett. 1977. Effects of bluegreen aphid on seedling lucerne in a glasshouse. Proc. 30th N.Z. Weed and Pest Contr. Conf., 152–154.

A Study of the Breeding Potential of a Population of Alfalfa (*Medicago sativa* L.)

S.R. BOWLEY and B.R. CHRISTIE
University of Guelph, Guelph, Ontario, Canada

Summary

Improved cultivars are the goal of all forage-breeding programs. In alfalfa (*Medicago sativa* L.), great improvements have been made in pest resistance and stability of performance, but little has been accomplished in improving yield potential. This lack of improvement in yield potential may be due to selection for yield under noncompetitive conditions. The objective of this study was to evaluate the breeding potential of an alfalfa population when selection was based on performance under competition.

A heterozygous alfalfa population, consisting of F_1 families from crosses between synthetics BW1 and BW9, was used in the study. The F_1 families and polycross progenies of random plants from within each F_1 family were evaluated in a simulated sward under growth-room conditions. Dry-matter yield was measured over three harvests. The additive and digenic variances and the narrow-sense heritability for dry-matter yield of the F_1 population were estimated from the analyses of variance. A number of

6-clone synthetics were developed from the F_1 population to determine the response to selection.

The estimates of additive variance were significant for dry-matter yield at the second harvest and for total yield. None of the digenic variance estimates was significant. This lack of significance indicates that the genetic variance in this population was mainly additive, with a narrow-sense heritability for total dry-matter yield of 15%.

The genetic coefficient of variation was greatest within the original population of F_1 families. Significant position correlations between the F_1 population and polycross progeny yields and the presence of significant additive variance indicated a potential for improving the F_1 population through phenotypic mass selection.

Comparisons between the experimental synthetics and an unselected bulk of all F_1 families indicated that no response to selection was apparent. Although the mean yield of 4 synthetics at the Syn-3 generation was 7% above that of the bulk, this difference was not significant. Further cycles of selection or a change in the experimental design may be required to detect differences.

Judging by these results, selection gains for yield could be made in this population of alfalfa. Further cycles of mass selection, conducted under sward conditions, are warranted.

KEY WORDS: alfalfa, synthetics, forage, selection, yield.

INTRODUCTION

There has been little or no increase in yield of forage cultivars for a number of years. Persistence and yield stability of species such as alfalfa (*Medicago sativa* L.) have been improved, but yield itself has remained unchanged (Elliot et al. 1972, Busbice and Gurgis 1976). Although a potential for improvement exists, it has not been exploited fully. This lack of significant gain has prompted breeders to reevaluate existing practices and to develop new approaches to improving forage yield potential.

The lack of yield improvement could be due to a number of factors. One commonly given reason is that the autopolyploid nature of alfalfa reduces or prevents selection gains (Dessureaux 1977). Admittedly, tetrasomic inheritance may create difficulties in selection, but it should not prevent improvement of the population. Instead, our present methods of selection may be at fault. For instance, a breeder usually makes selections from among relatively few plants grown at wide spacings. Selection in this manner is quite effective for highly heritable traits, but it is not effective for yield (Christie and Keoghan 1979).

This study was undertaken to determine the breeding potential of an alfalfa population developed by Dr. L. Dessureaux, Agriculture Canada, Ottawa. The objectives were to estimate variance and heritability and to determine response to selection for dry-matter yield when the population was selected under competitive conditions.

MATERIALS AND METHODS

Ninety F_1 families, resulting from crosses between 2 broadly based alfalfa synthetics (BW1 and BW9) and 71 polycross (P ×) families were evaluated in this study. The experimental method and statistical analysis are described in detail elsewhere (Bowley and Christie 1981).

From an analysis of variance, the genetic variance of the F_1 and P × families was estimated and the covariance

between F_1 and P × performance calculated for dry-matter yield. Correlations between relatives were used to partition these variances into additive and digenic effects (Bowley 1980). Narrow-sense heritabilities and the expected response to selection were estimated for the F_1 population.

Sixty 6-clone synthetics were developed from the F_1 and P × trials, with total dry-matter yield as the selection criterion. Many of these synthetics were advanced to the Syn-2 generation and a few to the Syn-3 for testing. A bulk of all F_1 families was used as an unselected check. The synthetics were evaluated in the same manner as were the F_1 and P × families. Realized heritabilities, determined by dividing the response to selection by the selection differential, were compared to the estimated heritability.

RESULTS AND DISCUSSION

There were significant differences among families both in the F_1 and the P × trials at each harvest and in total dry-matter yield. From these analyses, the estimates of genetic variance and phenotypic variance were obtained and the genotypic coefficient of variation was calculated (35.5% and 24.8%, respectively). As expected, the genetic variation was greatest among the F_1 families.

Correlations between relatives were used to partition the genetic variance (among the F_1 families) into additive and digenic effects and to calculate the narrow-sense heritability ($H_n^2 = 15\%$). The additive variance was significant ($P < 0.05$) for the second harvest and for total dry-matter yield. Although the additive variance estimates were numerically lower than the digenic variance estimates, the latter were not significant. This difference indicated that the additive component was more important in the genetic variance of dry-matter yield in this population of alfalfa. Nonadditive variance may exist, but it was not detected in this experiment (Bowley and Christie 1981).

There were significant differences among synthetics for dry matter at each harvest and for total yield. Only 4 synthetics will be discussed, those from a phenotypic mass selection within the F_1 population. One synthetic was developed from the highest-yielding plant, one from the

S.R. Bowley is a Graduate Student at the University of Kentucky and former Research Associate at the University of Guelph. B.R. Christie is a Professor of Crop Science at the University of Guelph.

Table 1. Mean dry-matter yields, selection differentials and responses to selection (as a % of the checks), and realized heritabilities of the phenotypic mass selections for total dry-matter yield within the F_1 trial†.

	Syn-1	Syn-2	Syn-3
Yield (% of the checks)††			
Bulk of the population	124 a	105 a	103 a
Mean of 4 (6 parent) synthetics	123 a	120 a	110 a
Response to selection of the four synthetics			
Selection differential	22.9	19.9	10.0
Realized heritability	16%	14%	7%

†The mean dry-matter yield of the checks (BW1, BW9, Angus and Saranac = 100% (3.14g/plant).
††Means followed by the same letter are not significantly different from each other at the 5% level of probability as determined by Duncan's new multiple range test.

second-highest-yielding plant, and 2 from a random plant from within each of the 6 top-yielding F_1 families. These synthetics were compared to the bulk of the population (Table 1). Numerically, there were no yield differences at the Syn-1 generation. When the plants advanced, the bulk synthetic declined more than the selected synthetics, with the selected synthetics 7% greater at the Syn-3. However, these differences were not statistically significant.

The realized heritability for total dry-matter yield at the Syn-1 generation (Table 1) corresponded very well with the estimated heritability of 15%. At the Syn-2 and Syn-3 generations, the realized heritabilities were lower, indicating that the heritability estimate was biased upwards.

The results of selection for total dry-matter yield in this population were inconclusive. The mean response to selection within the F_1 population was similar to that predicted from the covariance of relatives. However, the yields of these synthetics were not significantly different from the yields of the bulk of the population. As the gain in yield through the cycles of selection would be relatively small, a number of cycles of selection would be required to obtain a measurable yield increase. Considering the sizable amount of additive variance that is present and the trends that were observed, selection for improved yield within this population of alfalfa appears possible.

LITERATURE CITED

Bowley, S.R. 1980. A study of heterozygous population of alfalfa (*Medicago sativa* L.). M.S. Thesis, Univ. of Guelph.

Bowley, S.R., and B.R. Christie. 1981. Inheritance of dry matter yield in a heterozygous population of alfalfa. Can. J. Pl. Sci. 61:313–318.

Busbice, T.H., and R.Y. Gurgis. 1976. Evaluating parents and predicting performance of synthetic alfalfa varieties. Agric. Res. Serv., USDA, ARS-S-130.

Christie, B.R., and J.M. Keoghan. 1979. The use of simulated swards in breeding alfalfa (*Medicago sativa* L.). Can. J. Pl. Sci. 59:701–706.

Dessureaux, L. 1977. Allelic selection in alfalfa. Forage Notes 22:2–9.

Elliot, F.C., I.J. Johnson, and M.H. Schonhorst. 1972. Breeding for forage yield and quality. In C.H. Hanson (ed.) Alfalfa science and technology, 319–333. Am. Soc. Agron., Madison, Wis.

Breeding for Disease Resistance in Red Clover

R.R. SMITH
USDA-SEA-AR and the University of Wisconsin,
Madison, Wis., U.S.A.

Summary

Diseases incited by fungi and viruses cause substantial annual losses in the production and quality of red clover (*Trifolium pratense* L.) forage. The most feasible means of controlling red clover diseases is incorporating genes for disease resistance into agronomically acceptable cultivars or developing new germ plasm with disease resistance. Phenotypic recurrent selection was applied to red clover populations to develop germ plasm resistant to northern anthracnose (causal agent *Kabatiella caulivora* [Kirch.] Karak.), powdery mildew (causal agent *Erysiphe polygoni* DC ex St. Amans), target spot (causal agent *Stemphylium sarciniforme* [Cav.] Wiltshire), leaf rust (causal agent *Uromyces trifolii* [Hedrow. f. ex DC] Lev. var. *fallens* Arth.), and clover rot caused by *Fusarium* spp. The specific steps in the procedure are outlined, and the methods of preparing and applying the respective inoculum are presented. After five cycles of selection for resistance, 100% of the plants were resistant to northern anthracnose and 85% were resistant to leaf rust. The population was improved for resistance to target spot after two cycles and to powdery mildew after one cycle. For those diseases for which resistance is controlled by dominant genetic factors, progeny testing within each cycle would result in more rapid progress/cycle. In addition, parental selections could be evaluated for agronomic performance while progeny testing was being conducted.

KEY WORDS: forage breeding, *Trifolium pratense* L., phenotypic recurrent selection, disease resistance.

INTRODUCTION

Red clover (*Trifolium pratense* L.) is one of the leading forage legumes in the U.S.A., Canada, and northern and eastern Europe. A forage of excellent quality, it is used for silage, hay, and pasture. However, each year diseases incited by fungi and viruses cause substantial losses in production and quality. The most practical means of controlling red clover diseases is to incorporate genes for disease resistance into agronomically acceptable cultivars or to develop new germ plasm with disease resistance.

Disease resistance in red clover generally is controlled by one or a few genes (Taylor and Smith 1979), with most sources of resistance occurring at low frequencies in existing germ plasm. Phenotypic recurrent selection has been effective in breeding for resistance to northern anthracnose (causal agent *Kabatiella caulivora* [Kirch.] Karak.) (Maxwell and Smith 1971, Smith and Maxwell 1973, Smith et al. 1973), powdery mildew (causal agent *Erysiphe polygoni* DC ex St. Amans) (Owen 1977, Smith and Maxwell 1980), target spot (causal agent *Stemphylium sarciniforme* [Cav.] Wiltshire) (Smith and Maxwell 1980), leaf rust (causal agent *Uromyces trifolii* [Hedrow. f. ex DC] Lev. var. *fallens* Arth.) (Smith and Maxwell 1980), and clover rot (Ludin and Jonsson 1974). Breeding procedures used to develop disease-resistant red clover germ plasm at the University of Wisconsin are described in this paper.

METHODS

Breeding Procedures

The most important fungal diseases of red clover in the north central region of the U.S. are northern anthracnose (NA), powdery mildew (PM), target spot (TS), leaf rust (R), and root rots, the latter being caused by *Fusarium* spp. and *Rhizoctonia solani*. The USDA–University of Wisconsin program has concentrated on all of these diseases except that caused by *R. solani*.

Phenotypic recurrent selection (Fig. 1) has been the primary breeding scheme used for the development of resistant germ plasm in this program. During earlier years the procedure was applied to a specific population for each disease separately. Once resistance was established in the population for that disease, a second disease was studied. Currently, selection is being practiced within populations for several diseases in a sequential manner. Inoculations and selections are conducted in the greenhouse during winter months. Intercrossing of selected plants is done under isolation in the field to produce the next generation.

R.R. Smith is a Research Geneticist with USDA-SEA-AR and Professor of Agronomy at the University of Wisconsin. This is a joint contribution of USDA-SEA-AR and the Department of Agronomy of the University of Wisconsin.

Fig. 1. Diagrammatic scheme of phenotypic recurrent selection procedure.

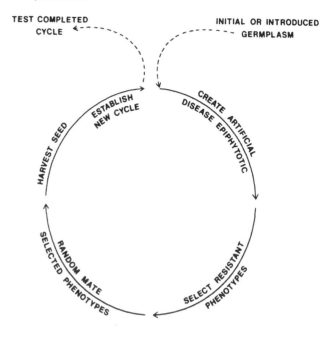

TEST COMPLETED CYCLE

INITIAL OR INTRODUCED GERMPLASM

CREATE ARTIFICIAL DISEASE EPIPHYTOTIC

ESTABLISH NEW CYCLE

HARVEST SEED

RANDOM MATE SELECTED PHENOTYPES

SELECT RESISTANT PHENOTYPES

Inoculation Procedures

Inoculum preparations, inoculation procedures, and symptom expression scales used are according to methods described by Maxwell and Smith (1971) for NA, Murray et al. (1976) for TS, Engelke et al. (1975) for R, Hanson (1966) for PM, and Leath and Kendall (1978) for *Fusarium*. For NA, PM, and TS, a composite of isolates is used from each pathogen separately. Single-spore isolates are used for R and root rot.

Progress Tests

To evaluate the progress of the various cycles of phenotypic recurrent selection for resistance to NA, TS, and R, a replicated test is established in the greenhouse. Germ plasm representing each cycle of selection is usually evaluated for dry-matter production and disease reactions in preliminary field plot tests. Specific details of all procedures can be obtained from the author.

RESULTS AND DISCUSSION

The percentage of resistant plants (Disease Severity Index [DSI] = 1) was substantially increased after five cycles of phenotypic recurrent selection for resistance to NA and R (Table 1). Initially, the Syn-0 had a moderate degree of resistance to NA (44% of plants DSI = 1; average DSI = 1.8); however, further improvement was realized after five cycles. The original population had essentially no resistance to R, but after five cycles more than 85% of the plants showed resistance. Selection for resistance to TS was not begun until cycle 3, but improvement

Table 1. Response of two or five cycles of phenotypic recurrent selection for resistance to northern anthracnose, target spot, and leaf rust in red clover.

Disease and cycle	Percent of total plants with DSI* of					Avg DSI
	1	2	3	4	5	
Northern anthracnose (NA)						
Syn O	44	37	13	6	0	1.8
Syn V	92	8	0	0	0	1.1
Arlington†	49	36	10	5	0	1.7
Leaf rust (R)						
Syn O	0	3	22	22	53	4.2
Syn V	70	15	12	3	0	1.5
Arlington	10	5	23	20	41	3.8
Target spot (TS)						
Syn O(Syn III)	0	0	5	27	68	4.6
Syn II(Syn V)	0	3	36	56	5	3.6
Arlington	0	0	10	44	46	4.4

*DSI = Disease severity index: 1-5: 1 = healthy, 5 = severely infected.
†Arlington is relatively resistant to NA but is susceptible to R and TS.

Table 2. Average disease severity index (DSI) of two or five cycles of phenotypic recurrent selection for resistance to northern anthracnose (NA), target spot (TS), leaf rust (R), and powdery mildew (PM) in red clover.

Cycle	Average DSI*			
	NA	R	TS	PM
Syn O	1.8	4.4	—	67
Syn I	1.4	2.5	—	38
Syn II	1.3	2.2	—	36
Syn III†	1.2	2.1	4.6	22
Syn IV	1.2	1.5	3.8	37
Syn V	1.1	1.5	3.6	32
LSD(5%)	0.3	0.4	0.4	3

*DSI scale is 1-5; 1 = healthy, 5 = severely infected for NA, TS, and R; As percent infected for PM.
†Began selection for TS in Syn III of NA and R germplasm.

was realized after two cycles of selection, as was shown by the shift in number of plants in the respective DSI classes.

The phenotypic recurrent selection scheme should be most effective for those characters controlled by one or a few genes and less effective for characters controlled by many genes. Resistance to NA is controlled by a few dominant genes (Sakuma et al. 1973, Smith and Maxwell 1973). In the moderately resistant initial population the most improvement occurred in cycle 1 of selection (Table 2). Although a significant improvement was realized between cycles 1 and 5, the progress was less rapid.

A similar response was expressed in selection for resistance to R. This character is controlled by a single dominant gene (Diachun and Henson 1971, 1974). The greatest improvement occurred after one selection cycle, but continued selection was effective in improving resistance in the population. The response to selection for resistance to PM was similar to that for resistance to NA and R, but the Syn-5 did not have as high a level of resistance as the plants selected for NA or R. Even though resistance to PM is controlled by a single dominant gene, specific races of the pathogen have been identified that create varying responses in different red clover clones (Hanson 1966, Stavely and Hanson 1967). As a result of the variability in the pathogen, the inoculum used from one cycle to the next may not have been uniform in representation of the various races.

Inheritance of resistance to TS is not yet understood. Slow progress in other populations that have been subjected to five cycles of selection for resistance to TS would suggest that resistance to this disease is controlled by many genes (Smith, unpublished data).

Selection for resistance to *Fusarium*-caused root rot was not effective after one cycle. Further efforts to obtain re-

sistance to this important disease will be attempted.

In most cases, "vertical" resistance in red clover appears to be conditioned by dominant genes. As a result, the susceptible alleles are carried by the heterozygotes in the population. Therefore, this procedure of phenotypic recurrent selection will not produce a strain with 100% resistant types. With random mating of the population after the final selection, the susceptible types will remain in the population at a low frequency if the final cycle contains the level of resistance expressed for NA and R in the current germ plasm. Under field conditions, this low level of susceptibility appears to present no problem and would provide some plasticity to the strains in the event that new pathogenic races occur.

Phenotypic recurrent selection is an effective, inexpensive procedure to apply for the development of disease-resistant germ plasm in red clover. Some form of progeny testing would undoubtedly result in more rapid progress, especially in selection for disease resistance controlled by dominant genes. However, progeny testing would be more expensive and would result in longer periods between cycles. If, on the other hand, attention were given to the characters that affect adaptation, yield, and persistence, it might be beneficial to evaluate the selected material for perenniality while conducting the progeny testing.

LITERATURE CITED

Diachun, S., and L. Henson. 1971. Resistance to rust in red clover. Phytopathology 61:889.

Diachun, S., and L. Henson. 1974. Dominant resistance to rust in red clover. Phytopathology 64:758-759.

Engelke, M.C., R.R. Smith, and D.P. Maxwell. 1975. Evaluating red clover germ plasm for resistance to leaf rust. Pl. Dis. Rpt. 59:959-963.

Hanson, E.W. 1966. Disease resistance in species of clover and alfalfa. Proc. X Int. Grassl. Cong., Sec. 3:734-737.

Leath, K.T., and W.A. Kendall. 1978. *Fusarium* root rot of forage species: pathogenicity and host range. Phytopathology 68:826-831.

Ludin, P., and H.A. Jonsson. 1974. Weibull's Britta—a new medium late diploid red clover variety with high resistance against clover rot. Agric. Hort. Genet. 32(1/4):44-54.

Maxwell, D.P., and R.R. Smith. 1971. Development of red clover germ plasm resistant to *Kabatiella caulivora*. Pl. Dis. Rpt. 55:920-922.

Murray, G.M., D.P. Maxwell, and R.R. Smith. 1976. Screening *Trifolium* species for resistance to *Stemphylium sarcinaeforme*. Pl. Dis. Rpt. 60:35-37.

Owen, C.R. 1977. Red clover breeding in Louisiana. La. Agric. Expt. Sta. Bul. 702.

Sakuma, T., T. Shimanuki, and K. Suginobu. 1973. Observations on the degree of susceptibility to *Kabatiella caulivora* of F_1 progenies derived from artificial crosses of red clover. Jap.

Soc. Grassl. Sci. 7:242-244.

Smith, R.R., and D.P. Maxwell. 1973. Northern anthracnose resistance in red clover. Crop Sci. 13:271-273.

Smith, R.R., and D.P. Maxwell. 1980. Registration of WI-1 and WI-2 red clover (Reg. No. GP 31 and GP 32). Crop Sci. 20:831.

Smith, R.R., D.P. Maxwell, E.W. Hanson, and W.K. Smith. 1973. Registration of Arlington red clover (Reg. No. 16). Crop Sci. 13:771.

Stavely, J.R., and E.W. Hanson. 1967. Genetics of resistance to *Erysiphe polygoni* in *Trifolium pratense*. Phytopathology 57:193-197.

Taylor, N.L., and R.R. Smith. 1979. Red clover breeding and genetics. Adv. Agron. 31:125-154.

Resistance to *Kabatiella caulivora* in *Trifolium subterraneum*

P.E. BEALE
South Australia Department of Agriculture

Summary

Clover scorch caused by the fungus *Kabatiella caulivora* is a serious problem in southern Australia on subterranean clover (*Trifolium subterraneum*). The disease is particularly troublesome in higher-rainfall areas (> 500 mm/yr), where resistant cultivars are urgently needed. The aim of this research program was to examine variability in resistance to clover scorch in subterranean clover, to investigate its inheritance, and to determine strategies for breeding resistant cultivars.

The pattern of disease development was studied in detail in the field, involving the calculation of 3 disease parameters: total disease development (TDD), rate, and time required to reach 50% disease development (Delay). Several genotypes exhibited levels of resistance adequate to ensure minimal losses in pasture production. Considerable variation in the pattern of disease development in different genotypes was observed. A single measure of disease resistance in the middle of the epidemic was highly correlated with TDD.

Some variation in pathogenicity of the fungus was observed, but no interaction between isolates of the fungus and host genotypes was detected.

The relative levels of resistance of most genotypes were fairly stable over a range of environments. An analytical technique, based on analysis of variance followed by a ranking process, was developed. This technique permits a comparison of rankings between environments and utilizes quantitative variability to determine significant rank order. The results of the study show that potential new cultivars need to be evaluated over a range of environments.

Diallel analysis of F_2 swards indicated that variation in disease resistance is heritable and that genes for resistance are dominant. A study of F_2 and backcross populations, measured as single plants in the field, indicated that there is a single major gene for resistance that exhibits dominance. Differences in resistance between moderately resistant and susceptible lines appeared to be determined by at least four genes of minor effect. Effective resistance may best be combined with superior agronomic characteristics by backcrossing.

KEY WORDS: clover scorch, *Kabatiella caulivora*, subterranean clover, *Trifolium subterraneum*, resistance, backcrossing.

INTRODUCTION

Clover scorch, caused by the fungus *Kabatiella caulivora*, is the first major disease to threaten the use of subterranean clover (*Trifolium subterraneum*) since the plant's agronomic value was recognized by Howard in 1889 (Hill 1936). Clover scorch has caused widespread damage in Australia during the last decade (Beale 1976), being most prevalent in the higher-rainfall regions (>500 mm/yr). No satisfactory cultural techniques have been found to control clover scorch (Beale 1976), and although spraying with fungicides can successfully control the disease, the cost of this technique is such that it is impractical on grazing land, suitable only for short-term protection of hay and seed crops. The solution to the problem apparently lies in breeding resistant cultivars. The aim of this research was (1) to examine variability in resistance to *Kabatiella caulivora*, mainly in subspecies *yanninicum*; (2) to investigate the inheritance of resistance; and (3) to determine strategies for breeding resistant cultivars.

VARIATION IN THE HOST

The development of a clover scorch epidemic was studied at Denmark, Western Australia, using 62 genotypes of subterranean clover (Beale and Thurling 1979). Total disease development (TDD), intrinsic rate of disease development (Rate), and time required to reach 50% disease development (Delay) were calculated. These 3 parameters, along with graphs of disease development over time, describe the pattern of disease development, whereas previous methods assessed scores at only one or two random dates during an epidemic. Considerable variation in the pattern of disease development was found (Table 1). Several genotypes exhibited levels of resistance that we considered adequate to ensure minimal losses in pasture production from clover scorch.

Single measures of disease damage obtained during the middle and at the end of the epidemic were significantly correlated with TDD and Delay, respectively. Screening for seedling resistance in the greenhouse is less precise and needs further refinement (Beale 1980).

VARIATION IN THE PATHOGEN

Variation in pathogenicity was demonstrated in 18 isolates of *Kabatiella caulivora* derived from infected plant material collected over a wide area in the southwestern coastal region of Western Australia (Beale and Thurling 1980a). No interaction between genotypes of subterranean clover and isolates of *K. caulivora* was observed. It should not be necessary, therefore, to take into account variation in the pathogen when breeding for resistance, and it seems unlikely that resistance will break down

The author is a Senior Research Officer in Agronomy in the South Australia Department of Agriculture.

Table 1. Disease parameters of 7 subterranean clover genotypes (selected from Beale and Thurling 1979).

Genotype	TDD (disease weeks)	Rate (units/day)	Delay (days after first assessment)
14742	139	–	–
Yarloop	118	0.109	20
39314YA	83	0.057	42
Larissa	47	0.038	74
39327YB	26	0.063	85
47308D	17	0.087	98
39357YB	16	0.011	242

rapidly as a result of the appearance of new strains of the fungus.

GENOTYPE x ENVIRONMENT INTERACTIONS

Genotype × environment interactions for resistance to clover scorch in subterranean clover were studied by growing 11 subterranean clover genotypes at three sites in Western Australia and one site in South Australia (Beale and Goodchild, 1980). Clover scorch damage was assessed by scoring at fortnightly intervals, and TDD was calculated from the results of this scoring. The analytical technique was based on an analysis of variance followed by a ranking process to determine the significant rank order (SRO) for each site. This technique permits a comparison of rankings among sites, thus eliminating problems associated with large differences in general disease levels among environments, and it utilizes quantitative variability measured at each site to determine the SRO. Interactions based on SRO were demonstrated, although the relative levels of resistance of most genotypes were fairly stable over the range of environments tested. The results clearly indicated that potential new cultivars need to be evaluated over the range of environments for which they are intended prior to commercial release. Interactions between genotypes and seasons were also observed, indicating that replication over years is also desirable.

GENETIC CONTROL OF RESISTANCE

Diallel Analysis

A 10 × 10 complete diallel cross of subterranean clover subspecies *yanninicum* genotypes was evaluated for resistance to clover scorch as F_2 swards (Beale and Thurling 1980b). Analysis of variance of the diallel-cross data indicated that variation in 10-October disease score, TDD, Rate, and Delay were heritable. Coefficients of regression of covariances (W_r) on variances (V_r) were significantly greater than 0 but not significantly different from a value of 1 for all 4 disease parameters (b = 0.91 ± 0.09 for TDD). This inconclusive evidence suggests that gene interactions are relatively unimportant. The four most resistant parents (39334YA, 14198, 19476Y, and 39327YB) were clustered close to the origin, suggesting that

that a high proportion of the relevant genes in these parents were dominant. Estimates of dominance-variance components were, with the exception of that for Delay, substantially greater than the estimate of additive genetic component. Values of F were positive for all parameters, and, except for Delay, values of $\sqrt{(H_1/D)}$ were greater than 1, indicating that dominant alleles are more numerous than recessive alleles among the parents and that dominance is complete at those loci exhibiting dominance. Heritabilities were substantially greater for TDD and 10-October disease score than for Rate and Delay.

Variation in Segregating Populations

Variation in resistance to *K. caulivora* in F_2 and backcross populations was examined in six crosses among genotypes representative of the known range of resistance (Beale and Thurling 1980c). The distribution of disease-score frequencies in six populations—parents, F_1, F_2, and the backcross to each parent—indicated that resistant genotypes were differentiated from the other genotypes by a single major gene, resistance being dominant to susceptibility. Populations derived from a cross between two moderately resistant lines and a cross between two susceptible lines exhibited marked transgressive segregation towards susceptibility in the former and higher resistance in the latter. At least four genes of relatively minor effect were presumed to be contributing to differences in resistance among the moderately resistant and the susceptible lines.

Analyses of population means for each of the six crosses showed that gene interactions were not significant in any of the crosses. Significant additive gene effects were detected only in crosses between the resistant line on the one hand and a moderately resistant and a susceptible line on the other.

The genetic analyses described here have provided a sound foundation upon which to base planning of a program for transferring clover scorch resistance to agronomically superior cultivars of subspecies *yanninicum*, such as Trikkala. Trikkala (Francis 1976) has a low estrogenic isoflavone content and characteristics that enable it to adapt well to the waterlogging-prone areas of southern Australia, but it has only a moderate level of resistance to clover scorch. Confirmation of the presence of a single dominant gene for resistance in 39327YB provides reasonable grounds for expecting that the superior characters of Trikkala could best be combined with effective clover scorch resistance by backcrossing.

CONCLUSIONS

The results of experiments conducted during this research program have provided information that may be utilized in devising an efficient strategy for breeding new cultivars of subterranean clovers with adequate levels of resistance to clover scorch. Production of cultivars resistant to clover scorch is currently being undertaken, utilizing the techniques devised in this program.

LITERATURE CITED

Beale, P.E. 1976. *Kabatiella* research—a progress report. J. Agric. So. Austral. 79:49–52.

Beale, P.E. 1980. Genetic variation in resistance to *Kabatiella caulivora* in *Trifolium subterraneum* subspecies *yanninicum*. Ph.D. Thesis, Univ. of West. Austral.

Beale, P.E., and N.A. Goodchild. 1980. Genotype × environment interactions for resistance to *Kabatiella caulivora* in *Trifolium subterraneum* subspecies *yanninicum*. Austral. J. Agric. Res. 31:1111–1117.

Beale, P.E., and N. Thurling. 1979. Genotypic variation in resistance to *Kabatiella caulivora* in *Trifolium subterraneum* subspecies *yanninicum*. Austral. J. Agric. Res. 30:651–658.

Beale, P.E., and N. Thurling. 1980a. Reaction of *Trifolium subterraneum* genotypes to different isolates of *Kabatiella caulivora*. Austral. J. Agric. Res. 31:89–94.

Beale, P.E., and N. Thurling. 1980b. Genetic control of resistance to *Kabatiella caulivora* in *Trifolium subterraneum* subspecies *yanninicum*. I. Diallel analysis of variation in disease development parameters. Austral. J. Agric. Res. 31:927–933.

Beale, P.E., and N. Thurling. 1980c. Genetic control of resistance to *Kabatiella caulivora* in *Trifolium subterraneum* subspecies *yanninicum*. II. Variation in segregating populations. Austral. J. Agric. Res. 31:935–942.

Francis, C.M. 1976. Trikkala—a new safe clover for wet areas. J. Agric. West. Austral. 17:2–4.

Hill, R. 1936. Subterranean clover—its history in South Australia. J. Agric. So. Austral. 40:322–330.

Subterranean Clover Improvement: An Australian Program

W.R. STERN, J.S. GLADSTONES, C.M. FRANCIS, W.J. COLLINS,
D.L. CHATEL, D.A. NICHOLAS, D.J. GILLESPIE, E.C. WOLFE,
O.R. SOUTHWOOD, P.E. BEALE, and B.C. CURNOW

Summary

At the XI International Grassland Congress, we reported on a subterranean clover improvement program being developed jointly by the University of Western Australia, the Western Australian Department of Agriculture, and the CSIRO Division of Plant Industry, Canberra. Since 1970, the program has grown to become the National Subterranean Clover Improvement Program, and collaboration now extends to departments of agriculture in the states of New South Wales, South Australia, and Victoria. The program is supported by wheat, wool, and meat producers, with contributions from federal and state government sources.

Since 1970, the objectives of the program have become more sharply defined as a result of agronomic and physiological studies and farmers' experience. Factors contributing to the adaptation of strains can now be better delineated.

In addition to the major selection criteria of environmental adaptation and low formononetin, further criteria have been incorporated, including tolerance to *Kabatiella caulivora*, root rots, and aphids. Research to support the breeding program is being undertaken on a wide range of topics.

A gene bank of over 4,000 lines has been developed. Observations are maintained in accordance with the practices of the International Board of Plant Genetic Resources.

Since 1975, four cultivars (Esperance, Nungarin, Trikkala, and Larisa) have been developed and released, and several lines are at advanced stages of testing.

KEY WORDS: subterranean clover, *Trifolium subterraneum*, spp. *subterraneum*, *brachycalycinum*, *yanninicum*, breeding, genetic resource, EXIR, formononetin, clover scorch, *Kabatiella caulivora*, aphids, root rot.

INTRODUCTION

Subterranean clover (*Trifolium subterraneum* L.) is an annual pasture legume that is widely used as a feed for ruminants and a source of nitrogen in the farming system of southern Australia (Donald 1970, Rossiter and Ozanne 1970). Subterranean clover improvement, described by Francis et al. (1970) and Underwood and Gladstones (1979), is now incorporated into the National Subterranean Clover Improvement Program, which extends across several states. The program is supported by the Wheat Industry Research Council and the Wool Research Trust Fund (which receive money from primary-producer levies and government contributions), state wheat industry research committees (which operate from grower levies), and state departments of agriculture, particularly in Western Australia. The present arrangements are informal rather than by negotiated agreement. Appropriate collaboration can meet a wide range of requirements without duplicating breeding programs, as long as the edaphic, climatic, and cultural requirements are clearly understood.

EARLIER OBJECTIVES AND NEW REQUIREMENTS OF GENETIC RESOURCES

Low estrogenic activity continues to be a primary selection criterion, all material being screened initially for low formononetin content. The early maturing cultivar Trikkala, suited for waterlogged conditions, was released in 1976. Less attention has been given to resistance to clover stunt virus, the importance of which in the field is still unclear.

Selection for improved adaptation in the general sense is a less tangible objective. Of the agronomic characters, commencement and duration of flowering, burr burial, and hardseededness remain important selection criteria. Material from the Mediterranean region, locally occurring strains, and crossbreds from the breeding program continue to provide sources of variability.

New requirements include resistance to the causal

W.R. Stern is with the Department of Agronomy at the University of Western Australia, Nedlands, W.A. 6009. J.S. Gladstones, C.M. Francis, W.J. Collins, D.L. Chatel, D.A. Nicholas, and D.J. Gillespie are with the Western Australian Department of Agriculture, Jarrah Road, South Perth, W.A. 6151. E.C. Wolfe is with the Agricultural Research Institute, New South Wales Department of Agriculture, Wagga Wagga, N.S.W. 2650. O.R. Southwood is with the New South Wales Department of Agriculture, P.O. Box K220, Haymarket, N.S.W. 2000. P.E. Beale is with the Turretfield Research Centre, South Australian Department of Agriculture, Rosedale, S.A. 5350. B.C. Curnow is with the Regional Office of the Victorian Department of Agriculture, P.O. Box 125, Bendigo, Vic. 3550.

organism of clover scorch, *Kabatiella caulivora* (Kirch.) Karak. Resistance has been found in the local ecotypes Daliak and Guildford D and in a few introductions from Italy (Chatel et al. 1973). Regular field testing of introductions and progenies from crosses is carried out at Denmark, Western Australia (lat 35°S long 117°E). A single major dominant gene has been shown to be responsible for resistance to clover scorch in the ssp. *yanninicum* (Beale and Thurling 1980).

Widespread outbreaks of bluegreen aphid (*Acyrthosiphon kondoi* Shinji) in 1977 resulted in a search for sources of resistance. Several lines with a fair degree of tolerance have been observed.

Root rots are recognized as causing degeneration in older pastures, especially where autumn rainfall is high. Several fungi appear to be involved, especially species of *Fusarium* and *Pythium* (MacNish et al. 1976). Screening has revealed tolerance in cv. Dinninup and Larisa and in some introductions from Italy and Turkey.

The present Australian collection contains over 4,000 lines, mostly of Mediterranean origin (Francis and Gladstones 1981). The accessions are mostly in the subspecies *subterraneum* but also include about 100 lines of the ssp. *yanninicum* and over 500 lines of the ssp. *brachycalycinum.* All the collected material has been grown in Perth, and the main diagnostic and agronomic characters have been recorded and entered into a computerized data retrieval system, EXIR, developed for the International Board of Plant Genetic Resources.

BREEDING STRATEGIES AND SUPPORTING RESEARCH

The early releases met immediate needs, but a growing complexity of identifiable requirements called for a continuing, broader-based central breeding program with supporting research. The improvement effort has been directed at established subterranean clover areas, but future attention might be given also to environments regarded as marginal for the species or to farming systems where the species might be advantageous.

Examples of specific requirements include adaptation to calcareous soils, using the ssp. *brachycalycinum,* and selection for a long growing season for Victoria, Tasmania, and parts of New South Wales. Where summer rainfall stimulates untimely germination and is followed by adverse weather conditions, the seed supply in the soil may be reduced to such an extent that the establishment of a satisfactory sward at the beginning of the growing season is jeopardized. In areas where the opening of the growing season is irregular, the capacity to maintain a sufficient seed reserve and the ability of seedlings to withstand a "false break" are important selection criteria.

Present breeding objectives may be summarized as follows:

1. Very early and early maturity group (flowering 70–85 and 85–100 days from emergence): low formononetin content, high seed production and hardseededness to ensure persistence through drought or cropping phases, and good burr burial and dormancy to minimize out-of-season germination.

2. Midseason maturity group (flowering 100–115, 115–130, and 130–145 days from emergence): resistance to clover scorch, root rots, aphids, and red-legged earth mites; resistance to intermittent waterlogging; winter vigor in districts with cold winters; and high physiological embryo dormancy and greater hardseededness for protection against unseasonal summer rains. In this maturity group a balance is required between disease and insect resistance and agronomic characters related to establishment and maturity; the specific climatic environment or the prevailing farming system becomes important.

3. Late and very late maturity group (flowering 145–160 and 160 + days from emergence): strong resistance to clover scorch and root rots, a moderate degree of hardseededness, and hardiness and vigor in winter.

The stages in the current program are summarized in Table 1. Material low in formononetin narrowly missing individual plant selection will be sown at a few selected sites in bulk mixtures for natural selection over a period of years.

An improvement program with such wide-ranging objectives needs supporting research. Identification of resistance or tolerance to clover scorch and aphids and studies of the genetics of resistance to clover scorch and of the inheritance of hardseededness, embryo dormancy,

Table 1. Simplified representation of stages in the National Subterranean Clover Improvement Program, Australia.

Generation	Year	Activity
F_1–F_2	1,2,3,4	Single plant selections. Screen for leaf markers, low formononetin and maturity characteristics.
F_3	4,5	Row or small plot testing at several locations. Selections for clover scorch and aphid resistance, for agronomic characters such as burr burial and hard-seededness.
F_4–F_6	6,7,8	Single plant selections of 100 genotypes with desired characters.
F_7	9,10,11	Uniform line comparisons in interstate trials. Retain and seed increase 40 superior lines.
	12	Rogue and seed increase 20 best lines. Early release of outstanding lines.
	13,14,15	District trials including comparative grazing test, tests for persistence, tests for competition with existing cultivars.
	16,17,18	Final seed increase, registration and release.

and leaf markers are in progress or contemplated. Seed dynamics in subterranean clover requires further study, and data are needed on seed losses and regeneration under various farming systems in different environments. Recent studies on seed production have shown marked variety × defoliation interactions (Collins 1978). Investigations of burr burial (Collins et al. 1976), prolongation of available moisture beyond the flowering period (Collins and Quinlivan 1980), and withholding water at various stages of flowering (Andrews et al. 1977) are helping to identify the relevance of specific characters as selection criteria. Analysis of competition between strains may indicate desirable characters for introducing new genotypes into existing stands (Burch and Andrews 1976). Information on variation in existing strains would help clarify evolutionary changes in subterranean clover and suggest how these changes might be used in a breeding program; studies have begun on the variability in populations of old and new stands of cv. Dwalganup and Seaton Park (W.J. Collins and R.C. Rossiter, personal communication). The identification of strains more effective in *Rhizobium* symbiosis, more efficient in nutrient uptake, or more tolerant of either lower levels of essential elements or higher levels of salinity need to be constantly explored.

TESTING, REGISTRATION, AND RELEASE

Testing involves small plot and field trials at multiple locations, preferably under grazing. Such testing is carried out by state departments of agriculture and usually proceeds for 2, 3, or more years, especially if superiority is difficult to establish. Registration of a prospective cultivar with the Australian Herbage Plant Registrar is necessary and is subject to the approval of the various state herbage plant liaison committees (Anon. 1970). When results differ between states, the state wishing to use the material arranges for its release.

NEW CULTIVARS

Since 1975, four cultivars have been released:

1. *Trikkala* (1976): a member of the ssp. *yanninicum,* derived from a cross between Neuchatel (a putative early flowering mutant of Yarloop) and the low-estrogen, late-maturing cv. Larisa. An early midseason variety with good tolerance to waterlogging and considerable resistance to clover scorch.

2. *Larisa* (1977): a direct introduction of ssp. *yanninicum* collected in Greece in 1965. It has some resistance to clover scorch and root rots. Late flowering and thus confined to high-rainfall areas.

3. *Nungarin* (1977): a selection from a cross between cv. Daglish and Northam A2; superior to the widely grown cv. Geraldton in earliness of maturity and hardseededness. Although bred for the drier margin of the wheatbelt of Western Australia, it has been sown extensively across southern Australia. It appears to persist and produce well in drier areas, especially under grazing.

4. *Esperance* (1978): selected from a cross between cv. Daliak and Bacchus Marsh; resembles Daliak in growth form and matures about 4 weeks later. A potential replacement for the susceptible cv. Woogenellup and Seaton Park in scorch-prone areas, and the first scorch-resistant cultivar released.

CONCLUSION

Low estrogenicity and the development of cultivars with appropriate maturity for specific regions remain major objectives of the program. However, characters associated with persistence in different environments and management systems and resistance to clover scorch and aphids are becoming increasingly important. Supporting research remains an integral part of the program.

LITERATURE CITED

Anon. 1970. The making of a new pasture variety. J. Agric. West. Austral. 11:124–127.

Andrews, P., W.J. Collins, and W.R. Stern. 1977. The effect of withholding water during flowering on seed production in *Trifolium subterraneum* L. Austral. J. Agric. Res. 28:301–307.

Beale, P.E., and N. Thurling. 1980. Genetic control of resistance to *Kabatiella caulivora* in *Trifolium subterraneum* subspecies *yanninicum.* I. Diallel analysis of variation in disease development parameters. II. Variation in segregating populations. Austral. J. Agric. Res. 31:927–933, 935–942.

Burch, G.J., and P. Andrews. 1976. Replacement of cv. Yarloop in swards by new genotypes of *Trifolium subterraneum* L. subspecies *yanninicum* and the effects of defoliation. Austral. J. Agric. Res. 27:217–226.

Chatel, D.L., C.M. Francis, and A.C. Devitt. 1973. Varietal variation in resistance to clover scorch (*Kabatiella caulivora* [Kirch.] Karak) in *Trifolium subterraneum.* Dept. Agric. West. Austral. Tech. Bul. 17.

Collins, W.J. 1978. The effect of defoliation on inflorescence production, seed yield, and hard-seededness in swards of subterranean clover. Austral. J. Agric. Res. 29:789–801.

Collins, W.J., C.M. Francis, and B.J. Quinlivan. 1976. The interrelation of burr burial, seed yield, and dormancy in strains of subterranean clover. Austral. J. Agric. Res. 27:787–797.

Collins, W.J., and B.J. Quinlivan. 1980. The effects of a continued water supply during and beyond seed development on seed production and losses in subterranean clover swards. Austral. J. Agric. Res. 31:287–295.

Donald, C.M. 1970. Temperate pasture species. *In* R.M. Moore (ed.) Australian grasslands, 303–320. Austral. Natl. Univ. Press, Canberra.

Francis, C.M., and J.S. Gladstones. 1981. Exploitation of genetic resources in the selection and breeding of new cultivars of *Trifolium subterraneum* L. *In* Genetic resources of forage plants. Symp. 6–11 May 1979, Townsville, Qld. CSIRO, Melbourne.

Francis, C.M., J.S. Gladstones, and W.R. Stern. 1970. Selection of new subterranean clover cultivars in southwestern Australia. Proc. XI Int. Grassl. Cong., 214–218.

MacNish, G.C., M.J. Barbetti, D. Gillespie, and K. Hawley. 1976. Root rot of subterranean clover in W.A. J. Agric. West. Austral. 17:16–19.

Rossiter, R.C., and P.G. Ozanne. 1970. Southwestern temperate forests, woodlands, and heaths. *In* R. Milton Moore (ed.) Australian grasslands, 199–218. Austral. Natl. Univ. Press, Canberra.

Underwood, E.J., and J.S. Gladstones. 1979. Subterranean clover and other legumes. *In* G.H. Burvill (ed.) Agriculture in Western Australia, 139–156. Univ. West. Austral. Press, Nedlands.

Second-Generation Progeny Tests for Forage Breeding

R.R. HILL, JR.

USDA-SEA-AR, Regional Pasture Research Laboratory,
University Park, Pa., U.S.A.

Summary

Second-generation progeny testing was studied as part of a continuing search for selection methods that (1) are at least as effective as polycross progeny test selection for all levels of heritability, (2) do not require production of large quantities of seed by hand pollination, and (3) permit storing much of the germ plasm as seed instead of as clones. Second-generation progeny tests are defined as the evaluation of polycross progenies of plants from a set of first-generation families. Selection was done in two steps—identification of first-generation families with superior average second-generation family performance and selection of superior second-generation parents within the superior first-generation families. Remnant seeds of superior first-generation families could be used to initiate improved populations and cultivars.

Theoretical changes in gene frequency with second-generation progeny tests were computed for diploid and autotetraploid populations in which the first-generation families were polycross progeny ($P \times P \times$), full-sibs ($FsP \times$), and selfed families ($S1P \times$). The methods used were an extension of those that appear in the quantitative genetic literature.

Theoretical response relative to that of the single-generation methods commonly used in forage breeding depended on the number of parents used/first-generation family. Theoretical response with $P \times P \times$ and $FsP \times$ selection was always greater than that with polycross progeny test or full-sib family selection, respectively, when the number of parents/first-generation family was greater than six. Differences in relative theoretical responses for $P \times P \times$ and $FsP \times$ selection were small. The required minimum number of parents/first-generation family could not be computed for $S1P \times$ selection.

KEY WORDS: progeny test, breeding methods, heritability, forage breeding.

INTRODUCTION

Theoretical responses for many different methods of selection have been evaluated for diploids (Empig et al. 1972) and for autotetraploids (Hill and Haag 1974, Hill and Byers 1979). The most effective of the methods studied by Hill and Haag (1974) were selfed-progeny test (S1Pt), full-sib family (FsF), and polycross-progeny test (P × Pt) selection. The study reported here describes theoretical responses for second-generation progeny test selection in forage breeding. Second-generation progeny test selection is conducted in three steps:

1. Produce seed of m families from clones to be included in the selection program. Families in stage 1 may be selfed, full-sib, or polycross. These methods are coded S1P ×, FsP ×, and P × P ×, respectively.

2. Produce polycross seed of n individuals from each of the m families formed in stage 1. The n individuals from each family are selected at random.

3. Evaluate the mn stage 2 families for traits to be considered in the selection program. The evaluation should be done in a nested design with an analysis of variance, as shown in Table 1. The stage 1 families with superior second-generation offspring performance are identified, and the superior stage 2 parents within the best stage 1 families are selected for the next cycle of selection.

Second-generation progeny test selection in forages is similar in many respects to modified ear-to-row selection used in *Zea mays* L. (Empig et al. 1972), except that the within-family selection is based on a progeny test. Davis (1955) proposed a selection program similar to S1P × selection for alfalfa (*Medicago sativa* L.), but a theoretical evaluation has not been conducted for any of the second-generation methods.

METHODS

The approach used was an extension of the procedures used by Comstock et al. (1949), Empig et al. (1972), and Hill and Haag (1974); details will not be presented here. Responses were calculated for diploid and autotetraploid populations. The initial population was assumed to be in random mating equilibrium and to have only two alleles

R.R. Hill, Jr., is a Research Agronomist with USDA-SEA-AR and Adjunct Professor of Plant Breeding at Pennsylvania State University. This is contribution no. 8011 of the U.S. Regional Pasture Research Laboratory, USDA-SEA-AR, University Park, Pa.

Table 1. Analysis of variance for family structures in second generation progeny test selection.

Source	Degrees of freedom[†]	Mean square expectations
Among first generation families	$m-1$	$\sigma_e^2 + r\sigma_{PX/F}^2 + rn\sigma_F^2$
Second generation families in first generation	$m(n-1)$	$\sigma_e^2 + r\sigma_{PX/F}^2$
Error	$(r-1)(mn-1)$	σ_e^2

Phenotypic variances:
Among first generation family means;

$$\sigma_{P(F)}^2 = (1/rn)\sigma_e^2 + (1/n)\sigma_{PX/F}^2 + \sigma_F^2$$

Second generation family means in first generation families;

$$\sigma_{P(PxF)}^2 = (1/r)\sigma_e^2 + \sigma_{PX/F}^2$$

[†]m = the number of stage-1 families, n = the number of stage-2 parents per stage-1 family, and r = the number of replications.

at the locus affecting the trait under selection. Forces other than selection that may change gene frequency were assumed to be absent.

Techniques similar to those presented by Falconer (1960) were used to express response with P x P x and FsP x selection in terms of that for P x Pt and FsF selection, respectively. The response ratios for P x P x /P x Pt and FsP x /FsF were then calculated. Comparable ratios for S1P x /S1F could not be calculated.

RESULTS

Except for S1P x selection in autotetraploids, response was proportional to a constant × pqα, where p is the frequency of the desirable allele, q is 1 – p, and α is the additive genetic parameter of the original population (Table

2). The α* for S1P x selection in autotetraploids could not be expressed in terms of parameters in a random-mating equilibrium population. Theoretical response for the first generation of S1P x selection was greater than that for P x P x or FsP x selection. Response to selection among second-generation parents within first-generation families was greater for P x P x than for FsP x or S1P x selection. Neither of the second-generation methods by itself had a response greater than those of conventional single-generation methods. The combined effects of two-stage selection and differences in phenotypic variances must be considered for a valid comparison, however.

The ratio of response with selection among first-generation families to that with single-generation selection was greater than 1.0 when the number of P x parents/first-generation family was greater than six for P x P x /P x Pt and FsP x /FsF comparisons (Fig. 1). The ratio tended to be larger for smaller values of heritability and slightly larger for the P x P x /P x Pt than for the FsP x /FsF comparison. A replication effect of the number of P x parents/first-generation family was responsible for the increased efficiency of second-generation progeny testing. Although the ratio could not be calculated for the S1P x /S1Pt comparison, a similar replication effect should be present. A ratio greater than 1.0 illustrates that second-generation progeny testing would be more effective than single-generation methods. This fact holds even if the response from selection of P x parents within first-generation families is ignored.

DISCUSSION

Several modifications could be made in the second-generation progeny test procedure. One modification would be to bulk second-generation seed according to first-generation family before evaluation. This modification would usually provide ample seeds for small-plot evaluations in a number of replications at a number of locations. The increased precision from additional replication could offset the loss from the elimination of the second stage of selection.

Table 2. Theoretical response equations for different selection methods.

Method[†]	Single generation[††]	Second generation tests	
		Among first generation	Second generation within first
PxPx	$\dfrac{k(1/2)pq\alpha^{\S}}{\sigma_P}$	$\dfrac{k(1/4)pq\alpha}{\sigma_{P(F)}}$	$\dfrac{k(3/8)pq\alpha}{\sigma_{P(PxF)}}$
FsPx	$\dfrac{k(1/2)pq\alpha}{\sigma_P}$	$\dfrac{k(1/4)pq\alpha}{\sigma_{P(F)}}$	$\dfrac{k(1/4)pq\alpha}{\sigma_{P(PxF)}}$
S1Px	$\dfrac{k(1)pq\alpha^{*\P}}{\sigma_P}$	$\dfrac{k(1/2)pq\alpha^{*\P}}{\sigma_{P(F)}}$	$\dfrac{k(1/4)pq\alpha^{*\P}}{\sigma_{P(PxF)}}$

[†]Type of first generation parent family. Second generation parents polycrossed in each method.
[††]From Hill and Haag (1974).
[§]k = a constant proportional to selection intensity, p = frequency of desirable alleles, q = 1 – p, and α = the additive effect of desirable alleles.
[¶]Value of α* for selfed progeny is an approximation. The approximation varies with progeny type for autotetraploids. α* = α for second generation tests with diploids.

Fig. 1. Ratio of response with selection among first-generation parents to that of single-generation selection for polycross and full-sib methods, with indicated heritability and family size values. Shape of response surface would be the same for diploids and autotetraploids.

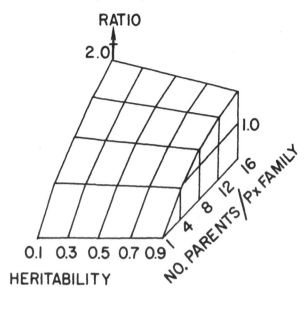

PxPx/PxPt

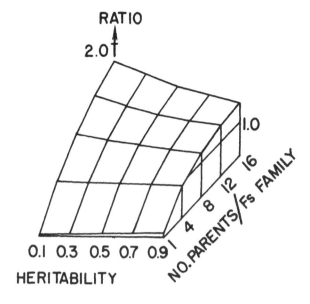

FsPx/FsF

Hill and Haag (1974) concluded that differences in response with P × Pt and FsF selection would be small. Differences in response with P × P × and FsP × should also be small, and the breeder could choose the method best suited to resources and objectives of the breeding program. For most forage species, the choice would probably be P × P × selection.

LITERATURE CITED

Comstock, R.E., H.F. Robinson, and P.H. Harvey. 1949. A breeding procedure designed to make maximum use of both general and specific combining ability. Agron. J. 41:360–367.

Davis, R.L. 1955. An evaluation of S₁ and polycross progeny testing in alfalfa. Agron. J. 46:572–576.

Empig, L.T., C.O. Gardner, and W.A. Compton. 1972. Theoretical gains for different population improvement procedures. Neb. Agric. Expt. Sta. Bul. MP26 (rev.).

Falconer, D.S. 1960. Introduction to quantitative genetics. Ronald Press, New York.

Hill, R.R., Jr., and R.A. Byers. 1979. Allocation of resources in selection for resistance to alfalfa blotch leafminer. Crop Sci. 19:253–257.

Hill, R.R., Jr., and W.L. Haag. 1974. Comparison of selection methods for autotetraploids. I. Theoretical. Crop Sci. 14:587–590.

Effects of Root Diameter, Nematodes, and Soil Compaction on Forage Yield of Two Tall Fescue Genotypes

C.B. WILLIAMS, C.B. ELKINS, R.L. HAALAND,
C.S. HOVELAND, and R. RODRIGUEZ-KABANA
USDA-SEA-AR and Auburn University, Auburn, Ala., U.S.A.

Summary

The effect of soil compaction and nematode infestation on large- and small-rooted fescue (*Festuca arundinacea* Schreb.) genotypes was evaluated. A factorial 2 × 2 × 2 randomized-block experiment was conducted in a soil with a tillage pan and a nematode infestation. The large- and small-rooted tall fescue genotypes were planted in two soil treatments in 4 combinations:

conventional tillage (plow [20-cm depth]–disk) without fumigation; conventional tillage with methyl bromide fumigation; deep tillage (subsoil [38-cm depth]–disk) without fumigation; and deep tillage with methyl bromide fumigation. The genotypes exhibited no differences in water loss (leaf diffusive resistance) or soil water extraction ability. On a tillage pan, the large-rooted genotype had superior root penetration, forage yield, persistence, and sward cover. Nematodes preferred the large-rooted to the small-rooted genotype. These data indicate that forage yields of tall fescue can be increased by breeding for specific root-system characteristics.

KEY WORDS: tall fescue, *Festuca arundinacea* Schreb., root size, forage yield, nematodes, tillage pan.

INTRODUCTION

Tall fescue (*Festuca arundinacea* Schreb.) is an important cool-season perennial forage grass in the U.S.A. (Buckner and Cowan 1978). In the southeastern U.S.A. it provides an economical pasture from October to June. Well-managed, disease-free tall fescue pastures result in impressive, economical beef yields and provide excellent erosion control (Hoveland et al. 1979). However, the yield potential of tall fescue is not often realized because of poor physical condition of soil, low soil fertility, and the presence of plant pathogens and nematodes (Hoveland et al. 1979, Elkins et al. 1977, Elkins et al. 1979, Jordan and Monk 1980).

Elkins et al. (1979), in a growth-room study, identified a small-rooted (median root diameter 1.0 mm) tall fescue genotype with water-use efficiency superior to that of a thicker-rooted (median root diameter 2.2 mm) genotype in soils either with or without high nematode populations. Both genotypes resulted from progeny selection of plant introductions of Mediterranean origin. In greenhouse studies, Elkins et al. (1977) found that the thicker-rooted genotype was better able to penetrate compacted soil layers. The results of their studies suggest that the detrimental effects of soil in poor physical condition and of nematodes on yield may be overcome by specific morphological and physiological root characteristics that can tolerate these adverse environmental conditions.

The objective of this research was to evaluate under field conditions the growth and performance of the large- and small-rooted tall fescue genotypes identified by Elkins et al. (1979). Knowledge of the characteristics of a superior shoot and root system should lead to more efficient plant selection in forage-breeding programs.

MATERIALS AND METHODS

A field experiment was conducted in central Alabama on a Cahaba loamy sand (Typic Hapludults, fine-loamy, siliceous, thermic) with a tillage pan at a depth of 25 to 35 cm and a high endemic nematode population. Nutrients were added according to the soil-test recommendations of the Auburn University Soil Testing Laboratory. Two soil

treatments were used in four combinations before planting: conventional tillage (plow [20-cm depth]–disk) without fumigation; conventional tillage with methyl bromide fumigation; deep tillage (subsoil [38-cm depth]–disk) without fumigation; and deep tillage with methyl bromide fumigation. During November 1978 clones of the large- and small-rooted tall fescue genotypes were planted immediately after tillage on 0.3 × 0.23–m spacing. The experimental design was a factorial 2 × 2 × 2 randomized block with 5 replications. Plot size was 1.5 × 5 m.

Forage was harvested six times, beginning in autumn 1979 and ending in late autumn 1980. Forage harvest intervals were variable, depending on environmental conditions and forage growth. Soil water content was determined to a depth of 1.5 m with a neutron probe at periodic intervals throughout the summers of 1979 and 1980. Soil and plant nematode populations were sampled during the early summers of 1979 and 1980 (Rodriguez-Kabana and King 1975). Leaf water loss (leaf diffusive resistance) was measured twice with an autoporometer during the summer of 1980. Forage or sward cover rating was subjectively evaluated throughout the growing seasons. Percentage of plant survival was determined by comparison of the number of dead or absent plants with the original number of plants. Root growth patterns and soil profiles were observed by digging a 1-m-deep pit into each plot of one replication during early summer 1980.

RESULTS

The large- and small-rooted tall fescue genotypes produced equivalent total forage (Table 1). Subsoiling and fumigation increased forage yields of initial harvests but not of later harvests. The late autumn 1980 forage yield of the large-rooted genotype was 1,160 kg/ha, whereas the small-rooted genotype produced no forage. The large-rooted genotype was more persistent than the small-rooted genotype, as is illustrated by stand survival and sward cover rating during autumn 1980 (Table 2). The genotype × tillage interaction for total yield indicates that the small-rooted fescue was more responsive to subsoiling than the large-rooted fescue genotype (Table 3).

Soil water content and leaf diffusive resistance data indicated no difference between genotypes or treatments. Total plant parasitic nematode population levels were low in 1979. Data from the 1980 growing season indicated a moderate level of nematodes in soil samples (25 plant parasitic nematodes/50 cm³ soil) and a low population of root-inhabiting parasitic nematodes (9 plant parasitic

C.B. Williams is an Agronomist with USDA-SEA-AR; C.B. Elkins is a Soil Scientist with USDA-SEA-AR; R.L. Haaland is a former Associate Professor of Cool Season Grasses at Auburn University; C.S. Hoveland is a Professor of Forage Crops Ecology at the university; and R. Rodriguez-Kabana is a Professor of Plant Nematology there as well.

Table 1.   Forage yields of two tall fescue genotypes grown in four soil treatments.

Treatment	Yield, kg/ha						
	1979 Early autumn	1979 Late autumn	1980 Early spring	1980 Late spring	1980 Early summer	1980 Late summer	Total yield
	*	*	*	*	*	*	
Large-rooted	590	310	1500	2360	1820	1160	7740
Small-rooted	2070	1060	1800	1400	1220	0	7750
	*	*	*	*			*
Subsoiled	1980	910	1850	2070	1550	520	8880
Conventional	680	460	1450	1680	1490	660	6420
	*	*	*				*
Fumigated	1930	960	1950	1980	1550	560	8930
Nonfumigated	730	410	1340	1770	1500	630	6380

*Means within a block are significantly different at the 5% level as determined by the F-test.

Table 2.   Percent plant survival and sward cover rating of a large- and small-rooted tall fescue, after 2 years of growth, autumn 1980.

Treatment	% plant survival	Sward cover rating†
Large-rooted	83*	4.3*
Small-rooted	51	1.3
Subsoiled	75	3.0*
Conventional	60	2.5
Fumigated	66	2.9
Nonfumigated	68	2.6

* Means within a block are significantly different at the 5% level.
† 0 = bare ground; 5 = excellent sward cover.

Table 3.   Genotype × tillage effect upon total forage yield between autumn 1979 and late autumn 1980.

Tillage	Genotype	
	Large-rooted	Small-rooted
	----------- kg/ha-----------	
Deep-tilled	8440	9320
Conventional	7030	5800

nematodes/10 g root). No difference was found between soil treatments. However, analysis of soil samples indicated a preference by the parasitic nematodes for the large-rooted genotype. Nematode counts on the root samples suggest that there were more stunt (*Tylenchorhychus* spp.) nematodes in the large-rooted genotype.

Rooting patterns of the 2 genotypes differed when plots were dug in early summer 1980. All treatments resulted in deep root systems for the large-rooted genotype. Roots of this genotype were observed to 1 m depth, but root development extended deeper. The small-rooted genotype produced very few roots deeper than the top of the tillage pan (approximately 25 cm) except with subsoiling. With subsoiling, the penetration of the small-rooted genotype into the subsoil was sparse.

DISCUSSION

This study was based on the response of two tall fescue genotypes with different root characteristics, grown under field conditions. The genetic implications of these genotype differences indicated great potential for improving stand persistence and forage yield of tall fescue.

The large-rooted tall fescue genotype produced less forage than the small-rooted genotype for the first two harvests; however, the large-rooted genotype was more productive at later harvests. We speculate that photosynthate of the large-rooted genotype is initially directed toward the root system and that, once the root system is established, shoot growth ensues.

The small-rooted genotype does not appear to have the genetic potential to produce a deep root system or to penetrate a tillage pan (Elkins et al. 1977).

The effect of nematodes on forage yield was not clearly determined, but increased numbers of soil nematodes in the large-rooted tall fescue indicated some phytological preference by the parasites, as reported by Elkins et al. (1979). The poor persistence of the small-rooted genotype is probably due to several environmental and genetic factors. However, we think that the inability of the small-rooted genotype to penetrate the plow pan, produce a deep-rooted system, and exploit a greater soil volume was the primary cause of death when this genotype was subjected to severe drought during the summer of 1980. The large-rooted genotype was able to penetrate the tillage pan, produce a deep-rooted root system, and extract water from deep in the soil, thus escaping severe drought injury. These large roots penetrating deeply into the soil may act as water conduits to the crown, preventing severe water stress during drought periods. Continuation of forage production is largely dependent upon the ability of the plant to tolerate a multitude of environmental extremes over time.

Leaf diffusive resistance often increases with decreasing soil-water content; therefore, differences would be expected between the large- and small-rooted tall fescue genotypes for leaf diffusive resistance and soil water content. We surmise that the root system of the large-rooted genotype penetrated to a soil depth beyond that measured for soil water loss. Observation of root growth deep in the soil profile of the large-rooted genotype is evidence of this theory. Other physiological factors (e.g. drought-induced dormancy) may also explain water use by the genotypes. Drought periods, compounded with nematode infestation and a tillage pan, create adverse growing conditions for tall fescue, reducing persistence and/or forage yield of some tall fescue genotypes.

This experiment emphasizes the need for plant breeders to consider the whole plant in improvement programs. The diversity of tall fescue offers a great spectrum of morphological and physiological plant characteristics. A synthesis of superior root and shoot characteristics should promote more economical forage yields.

LITERATURE CITED

Buckner, R.C., and J.R. Cowan. 1978. The fescues. *In* M.E.

Heath, D.S. Metcalfe, and R.F Barnes (eds.). Forages: the science of grassland agriculture, 3rd ed., 297–306. Iowa State Univ. Press, Ames, Iowa.

Elkins, C.B., R.L. Haaland, and C.S. Hoveland. 1977. Grass roots as a tool for penetrating soil hardpans and increasing crop yields. Proc. 34th South. Past. Forage Crop Impr. Conf., 21–25.

Elkins, C.B., R.L. Haaland, R. Rodriguez-Kabana, and C.S. Hoveland. 1979. Plant-parasitic nematode effects on water use and nutrient uptake of a small- and large-rooted tall fescue genotype. Agron. J. 71:497–500.

Hoveland, C.S., R.L. Haaland, C.C. King, W.B. Anthony, J.A. McGuire, L.A. Smith, H.W. Grimes, and J.L. Holliman. 1979. Steer performances on AP-3 phalaris and 'Kentucky 31' tall fescue pasture. Agron. J. 72:375–377.

Jordan, W.R., and R.L. Monk. 1980. Season and genetic variations in root patterns of sorghum. Agron. Abs., Ann. Mtg., ASA-CSSA-SSSA, Detroit, Mich.

Rodriguez-Kabana, R., and P.S. King. 1975. Efficiency of extraction of nematodes by flotation-sieving using molasses and sugar by elutriation. J. Nematol. 7:54–59.

Pooling the Genetic Resources of the Crested Wheatgrass Species-Complex

K.H. ASAY and D.R. DEWEY
USDA-SEA-AR and Utah State University,
Logan, Utah, U.S.A.

Summary

The crested wheatgrasses — *Agropyron cristatum* (L.) Gaertn., *A. desertorum* (Fisch. ex Link), and related taxa — are composed of a series of diploid (2n = 2× = 14), tetraploid (2n = 4× = 28), and hexaploid (2n = 6× = 42) forms. Dewey (1974) concluded that the same basic genome, modified mainly by structural rearrangements, occurs in the three ploidy levels. He proposed that breeders should expand the genetic base of their breeding populations by treating all species in the complex as a single gene pool.

Gene flow to the 2× level from tetraploid *A. desertorum* (N4×) has been restricted by the failure of N4×-2× triploids to produce functional reduced gametes. Dewey (1971) found that triploids from crosses between colchicine-induced tetraploids (C4×) and 2× *A. cristatum* produced a higher proportion of functional reduced gametes than their N4×-2× counterparts. Dewey (1977) successfully used these triploids in a backcrossing scheme to transfer a recessive marker gene from the 4× to the 2× level. The procedure lends itself to introgression of other agronomic traits to the highly uniform 2× level.

Several approaches apparently can be used to convey germ plasm from other ploidy levels to the 4× level. Dewey (1971) concluded that stable 4× populations could be derived from (N4×-2×)-N4× backcross progenies. The most promising interploidy breeding populations have stemmed from C4×-N4× crosses. Dewey and Pendse (1968) found that many C4×-N4× hybrids were substantially more vigorous than their parental clones. Also, the C4×-N4× hybrids were relatively fertile, and preliminary results indicated that selection for improved fertility would be effective. Dewey (1969) reported that stable 4× populations could be obtained from 6×-C4× and 6×-N4× pentaploids through selection and hybridization with 4× derivatives. Dewey (1973) and Asay and Dewey (1979) found that 6×-2× tetraploids could successfully be used in crosses with N4× tetraploids to combine germ plasm from three ploidy levels at the 4× level.

Fully fertile 6× progenies have been obtained in first-generation 6×-4× backcross progenies (Dewey 1969, 1974). Results from Dewey's (1974) studies also indicate that 5×-5× crosses will yield fertile 6× progenies. However, the 6× material is agronomically inferior and probably does not warrant a major breeding effort.

KEY WORDS: *Agropyron cristatum* L., *Agropyron desertorum* (Fisch. ex Link), grass breeding, introgression, polyploidy, autopolyploidy, sterility, interspecific hybridization, cytogenetics.

INTRODUCTION

Crested wheatgrasses—*Agropyron cristatum* (L.) Gaertn., *A. desertorum* (Fisch. ex Link) Schult., and related taxa—have proven to be a valuable source of early spring forage on the rangelands of western America. Tzvelev (1976) described 10 species and 11 subspecies in the crested wheatgrass complex, which consists of a series of diploid (2n = 2× = 14), tetraploid (2n = 4× = 28), and hexaploid (2n = 6× = 42) forms. Dewey and Pendse (1968) considered the crested wheatgrass species to be an autoploid series. Others (Taylor and McCoy 1973, Schulz-Schaeffer et al. 1963) have proposed that alloploidy or segmental alloploidy was involved in the development of the polypoid forms in the complex. Dewey (1974) has concluded that the same basic genome, modified primarily by structural rearrangements, occurs in all three ploidy levels, and he has proposed that grass breeders should expand the genetic base of their breeding populations by treating all species in the complex as a single gene pool.

Although crossing barriers exist, all combinations of crosses have been made among the 2×, 4×, and 6× ploidy levels (Knowles 1955, Dewey 1969, 1973). In addition, genetic transfer among ploidy levels has been greatly enhanced in crested wheatgrass by induced polyploidy (Dewey and Pendse 1968). The objectives of this review are to discuss the application of induced polyploidy and hybridization schemes to combine the genetic resources of the three ploidy levels into a common gene pool.

GENETIC TRANSFER TO THE DIPLOID LEVEL

Gene transfer from hexaploid (6×) *A. cristatum* or natural tetraploid (N4×) *A. desertorum* to (2×) *A. cristatum* through 6×-2× and N4×-2× crosses and subsequent backcrosses does not appear to be feasible. Triploids from (6×-2×)-2× and N4×-2× crosses are difficult to obtain and are highly sterile (Asay and Dewey 1979, Dewey 1971). Moreover, Dewey (1971) found that more than 75% of the functional eggs produced by triploids from N4×-2× crosses were unreduced. The preponderance of functional unreduced gametes contributed by the triploids in backcrosses restricted the gene flow from the tetraploid to the diploid level. A similar phenomenon was encountered in *Dactylis glomerata* L. by Zohary and Nur (1959) and Jones and Borrill (1962).

One-way gene flow is particularly unfortunate in crested wheatgrass. Although diploid *A. cristatum* is leafier and presumably of higher forage quality, it is somewhat less productive and usually has smaller seeds than tetraploid *A. desertorum*. Also, diploid *A. cristatum* is a much more uniform taxon than are the tetraploid forms. An added influx of genetic diversity to complement the meager amount available to grass breeders working at the diploid level would make a major contribution.

Dewey and Pendse (1967) and Dewey (1971) observed that triploids derived from crosses between colchicine-induced tetraploids (C4×) and 2× *A. cristatum* produced a higher proportion of functional reduced gametes than triploids of N4×-2× parentage. Accordingly, most progenies from these triploids reverted to the diploid level. On the basis of these results, Dewey (1977) reasoned that an increase in the proportion of *A. cristatum* chromosomes in the 4×-2× triploids would probably result in a concomitant increase in the production of functional reduced gametes. He tested this theory in a scheme to transfer a recessive gene responsible for the absence of anthocyanin pigments from the N4× to the 2× level. His procedure involved the production of triploids from (N4×-C4×)-2× crosses. In subsequent backcrosses to the 2× parent, the triploids produced a sufficient number of functional reduced gametes to complete the genetic transfer.

Pentaploids (5×) from 6×-N4× and 6×-C4× crosses appear to be of limited value in genetic transfer from the 6× to the 2× level. Dewey (1974) had difficulty making the 5×-2× cross, and the few progenies obtained were meiotically irregular and highly sterile. The feasibility of using triploids from crosses between (6×-2×)-C4× tetraploids and 2× *A. cristatum* as a bridge between the 6×-2× levels merits additional study. These tetraploids have been found to be relatively fertile (Asay and Dewey 1979), and triploids derived from them should produce an adequate number of functional reduced gametes to effect the transfer. If not, the (6×-2×)-C4× tetraploid could be backcrossed to the C4× parent to increase the frequency of *A. cristatum* chromosomes.

Triploids of *A. desertorum* are apparently of limited value as a genetic bridge. Dewey and Pendse (1967) and Dewey (1971) found that 3× *A. desertorum* plants, derived from polyhaploids, failed to produce functional reduced eggs.

GENETIC TRANSFER TO THE TETRAPLOID LEVEL

Dewey (1971) obtained progenies that were at or near the 4× level from backcrosses of N4×-2× triploids with the parental species. He concluded that with continued selection, stable tetraploid populations could be derived from these backcross progenies. Triploid *A. cristatum* also hybridized with 6× *A. cristatum* to produce reasonably fertile 5× progenies, many of which should stabilize at the 4× level. Dewey concluded that although triploids offered some promise as a bridge for gene flow from lower to higher ploidy levels, they were difficult to obtain and were meiotically irregular.

Tai and Dewey (1966) found that C4× derivatives were nearly as fertile as their 2× prototypes and that they crossed readily with N4× clones. Dewey and Pendse (1968) later reported that the fertility of C4×-N4× hybrids, although highly variable (0.02 to 2.30 seeds/spikelet), compared favorably with that of the parents. Furthermore, preliminary results from Dewey and Pendse's studies indicated that selection for improved fertility

K.H. Asay and D.R. Dewey are both Research Geneticists with USDA-SEA-AR, at Utah State University.

would be effective. Many of the C4× -N4× hybrids studied by Dewey and Pendse were substantially more vigorous than their parental lines, and populations subsequently derived from them have displayed exceptional agronomic merit.

Dewey (1969) had little difficulty obtaining 5× hybrids from 6× -C4× and 6× -N4× crosses. Many 5× progenies were more vigorous than either parent and appeared to have direct value in a breeding program. The fertility of the hybrids approached that of natural tetraploids, ranging from 0.54 to 4.62 seeds/spikelet. In a subsequent study, Dewey (1974) concluded that pentaploids could be effectively used to transfer chromosomes or entire genomes from the 6× to the 4× level. Backcrosses to the 4× parent were easily made, and some fully fertile 4× progenies were obtained.

Tetraploids from 6× -2× hybrids appear to have particular promise in hybridization schemes designed to combine germ plasm from all three ploidy levels at the tetraploid level. Dewey (1973) obtained a limited number of 4× hybrids from 6× -2× crosses. The 4× hybrids, which had little agronomic merit in their own right, produced an average of 16 seeds/spike in backcrosses to the 6× parents and more than 9 seeds/spike in crosses with the N4× and C4× tetraploids.

Asay and Dewey (1979) studied the fertility and meiotic relationships of the 5× and 3× backcross progenies from the 6× -2× hybrid and of the 4× progenies from (6× -2×)-N4× and (6× -2×)-C4× crosses. They concluded that (6× -2×)-N4× hybrids, although less fertile than their (6× -2×)-C4× counterparts, would be of immediate value to the breeder. Intense selection coupled with backcrosses to the N4× parent should effectively establish fully fertile breeding populations.

Asay and Dewey (1979) found pentaploids from (6× -2×)-6× crosses to be of less value in an interploidy breeding program than were the 6× -4× pentaploids studied by Dewey (1974). The 6× -C4× and 6× -N4× pentaploids were much easier to obtain and were more fertile in subsequent crosses that had been designed to produce stable 4× germ plasm pools. The (6× -2×)-2× triploids were the least fertile of the hybrids studied by Asay and Dewey. They produced an average of less than 1 seed/spike under open pollination; however, they may offer some potential in crosses with N4× tetraploids, such as those done by Dewey (1971).

GENETIC TRANSFER TO THE HEXAPLOID LEVEL

Stabilization of germ plasm pools at the hexaploid level is possible but not very practical. The sample of natural hexaploids and the hexaploid hybrids observed to date have a relatively coarse texture, and in general they appear to have less agronomic merit than the diploids or tetraploids. Furthermore, no hexaploid crested wheatgrass cultivars have been developed by plant breeders.

Pentaploid hybrids derived from 6× -4× crosses by Dewey (1969) apparently can be used to convey germ plasm from the 4× to the 6× level. Dewey (1974) obtained some fully fertile 6× progenies in the first generation of backcrosses to the 6× parent. Two of the 86 progenies studied were 2n = 42 with 80% stainable pollen and seed sets of 34 and 86 seeds/spike. One clone with 41 chromosomes produced 115 seeds/spike, a fertility level greater than those of many tetraploid cultivars of crested wheatgrass. Dewey (1974) also obtained one reasonably fertile hexaploid from 44 5× -5× hybrid plants. Hexaploids from 5× -5× progenies may be preferred in a breeding program because they would have a higher proportion of N4× germ plasm represented in the genetic complement than the (6× -4×)-6× derivatives.

DISCUSSION

Although the same basic genome, altered mainly by structural rearrangements, occurs in the three ploidy levels of crested wheatgrass, crossing barriers have restricted gene flow within the complex. Induced polyploidy coupled with hybridization schemes has substantially reduced these obstacles, and the taxa within the complex can realistically be treated as one breeding pool.

Germ plasm from the 2× , 4× , and 6× levels can be combined most effectively at the 4× level. Progenies from C4× -N4× hybrids have displayed exceptional promise in breeding programs (Knowles 1955, Dewey and Pendse 1968). Selection at the 2× and 4× levels should provide improved parentage for future hybrid populations. Tetraploid breeding populations also could be generated by crossing 6× -2× tetraploids and 6× -4× pentaploids with C4× , N4× , and N4× -C4× selections. Hybrids from these programs then could be combined in a breeding pool comprising germ plasm from the three ploidy levels.

Genetic transfer to the diploid level has been particularly difficult because of the failure of N4× -2× triploids to produce an adequate number of functional reduced eggs in backcrosses to the 2× parent. However, the procedure proposed by Dewey (1977) offers considerable promise to breeders endeavoring to infuse genetic variability into the highly uniform 2× populations. Dewey's triploids derived from (C4× -N4×)-2× crosses produced enough functional reduced gametes to reverse the previously encountered one-way gene flow.

Introgression of germ plasm from the 2× and 4× levels to the 6× and perhaps to higher ploidy levels is possible but probably is not as productive as other alternatives would be. The hexaploids would be more difficult to manipulate in a breeding program, and their adaptation to western range sites has yet to be demonstrated. Ploidy levels above 6× would present even more complex inheritance patterns, and the fact that such levels are not found in nature indicates that chromosome numbers above 6× may be above the optimum level for crested wheatgrass.

LITERATURE CITED

Asay, K.H., and D.R. Dewey. 1979. Bridging ploidy differences in crested wheatgrass with hexaploid × diploid hybrids. Crop Sci. 19:519–523.

Dewey, D.R. 1969. Hybrids between tetraploid and hexaploid crested wheatgrass. Crop Sci. 9:787–791.

Dewey, D.R. 1971. Reproduction in crested wheatgrass triploids. Crop Sci. 11:575–580.

Dewey, D.R. 1973. Hybrids between diploid and hexaploid crested wheatgrass. Crop Sci. 13:474–477.

Dewey, D.R. 1974. Reproduction in crested wheatgrass pentaploids. Crop Sci. 14:867–872.

Dewey, D.R. 1977. A method of transferring genes from tetraploid to diploid crested wheatgrass. Crop Sci. 17:803–805.

Dewey, D.R., and P.C. Pendse. 1967. Cytogenetics of crested wheatgrass triploids. Crop Sci. 7:345–349.

Dewey, D.R., and P.C. Pendse. 1968. Hybrids between *Agropyron desertorum* and induced-tetraploid *Agropyron cristatum*. Crop Sci. 8:607–611.

Jones, K., and M. Borrill. 1962. Chromosomal status, gene exchange, and evolution in *Dactylis*. Genetics 32:296–322.

Knowles, R.P. 1955. A study of variability in crested wheatgrass. Can. J. Bot. 33:534–546.

Schulz-Schaeffer, J.R., P.W. Allerdice, and G.C. Creel. 1963. Segmental alloploidy in tetraploid and hexaploid *Agropyron* species of the crested wheatgrass complex (Section *Agropyron*). Crop Sci. 3:525–530.

Tai, W., and D.R. Dewey. 1966. Morphology, cytology, and fertility of diploid and colchicine-induced tetraploid crested wheatgrass. Crop Sci. 6:223–226.

Taylor, R.J., and G.A. McCoy. 1973. Prposed origin of tetraploid species of crested wheatgrass based on chromatographic and karyotypic analyses. Am. J. Bot. 60:576–583.

Tzvelev, N.N. 1976. Poaceae URSS. Tribe 3. Triticeae Dum. USSR Acad. Sci. Press, Leningrad.

Zohary, D., and U. Nur. 1959. Natural triploids in the orchardgrass, *Dactylis glomerata* L., polyploid complex and their significance for gene flow from diploid to tetraploid levels. Evolution 13:311–317.

Breeding for Higher Magnesium Content in Orchardgrass (*Dactylis glomerata* L.)

C.G. CURRIER, R.L. HAALAND, C.S. HOVELAND, C.B. ELKINS, and J.W. ODOM

Auburn University, Auburn, Ala., U.S.A.

Summary

Grass tetany is a metabolic disorder of lactating ruminants caused by low Mg levels in the blood serum. During winter and spring, pasture soils in the southeastern U.S.A. often become wet or flooded. Flooded soil conditions affect the relative amounts of Mg, Ca, and K present in the soil solution and also reduce cation uptake by cool-season perennial forage grasses.

Orchardgrass, *Dactylis glomerata* L., an important cool-season perennial forage grass, often contains concentrations of cations that are suboptimal for animal requirements. An experiment was conducted (1) to compare Mg, Ca, and K uptake and production of plant dry weight and roots of four orchardgrass cultivars grown in a nonflooded and in a flooded soil, and (2) to determine if a selection program would be effective in isolating genotypes capable of high Mg uptake under these conditions.

Significant differences in Mg, Ca, and K uptake and in plant dry weight were found in the 4 cultivars. No significant differences for root number were found. Individual genotypes varied greatly in their response to flooded and nonflooded soil for the characters measured.

Broad-sense estimates of heritability were high. Heritability estimates of cation content indicated that progress in selection to increase the Mg and Ca content of orchardgrass could be made. The potential to change K content, plant dry weight, and number of roots also existed.

Breeding orchardgrass for improved Mg uptake in both nonflooded and flooded soil conditions should produce a cultivar that would help alleviate grass tetany problems in orchardgrass pastures.

KEY WORDS: orchardgrass, *Dactylis glomerata* L., grass tetany, magnesium, hypomagnesemia, flooded soil, broad-sense heritability, oxygen.

INTRODUCTION

Orchardgrass is an important cool-season perennial pasture grass in the southeastern U.S.A. During winter and spring in this area, high amounts of rainfall generally cause water-saturated soils. Cool water-saturated pasture soils have low soil oxygen levels, ranging from 0% to 10% oxygen (Elkins et al. 1978). Low soil oxygen is one factor identified as being responsible for reduction of cation uptake by cool-season pasture forages (Elkins et al. 1978, Haaland et al. 1978). Cattle grazing pastures under these conditions often have inadequate amounts of magnesium (Mg) (Elkins and Hoveland 1977). A lactating cow consuming plant tissue with low levels of Mg probably will have a low blood-serum Mg level, a condition that often causes hypomagnesemia (grass tetany), which may be fatal to the animal. A possible solution to this problem would be to develop an orchardgrass cultivar that is able to take up sufficient amounts of Mg in both nonflooded and flooded soil conditions. An experiment was conducted to determine if a selection program would be effective in altering Mg, calcium (Ca), and potassium (K) content in four orchardgrass cultivars in a nonflooded and in a flooded soil.

METHODS

Four orchardgrass cultivars of diverse origin—Boone, Frode, Hallmark, and KO-191—were represented by 18 to 23 genotypes to give a total of 80 genotypes. The crown of each genotype was divided into 10 vegetative propagules. Two large benches were prepared inside a greenhouse. One bench was lined with polyethylene plastic. Both benches were filled with a 1:3 mix of peat and Cahaba sandy loam. The vegetative propagules were planted in each bench in a randomized complete-block design with 5 replications of each genotype in each bench. The plastic-lined bench was kept flooded (soil oxygen content less than 1%). The other bench was watered when the soil surface became dry (soil oxygen content 20%).

After 6 weeks, top growth was clipped, dried, weighed, digested using a perchloric–nitric acid mix, put into solution, and assayed for cation content. Cation concentrations found in the forage were expressed as percentages on a plant dry-weight basis.

Plant roots were extracted from the soil with approximately 588 cm³ of soil surrounding the roots. Primary roots 5 cm below the crown were counted.

C.G. Currier is a Research Associate at Auburn University; R.L. Haaland is a former Associate Professor at the university; C.S. Hoveland is a Professor in the Department of Agronomy and Soils at the university; C.B. Elkins is a Soil Scientist with USDA-SEA-AR; and J.W. Odom is an Assistant Professor in the Department of Agronomy and Soils at Auburn University.

This is a contribution of the Department of Agronomy and Soils, Auburn University.

All data were subjected to an analysis for variance. Duncan's Multiple Range Test was used for separation of means. Broad-sense estimates of heritability were calculated, using the method of Burton and Devane (1953).

RESULTS

Cultivar Comparisons

In the nonflooded soil, Boone and KO-191 had the highest and Frode the lowest mean for Mg (Table 1). Frode also had the lowest mean for Ca, with no differences among the other cultivars. Differences among culivars were not significant for K, plant dry weight, and number of roots in the nonflooded soil.

In the flooded soil, Boone and Hallmark had the highest Mg and KO-191 the lowest Mg and Ca (Table 1). Hallmark had the highest and KO-191 the lowest K. Hallmark and KO-191 produced the highest and Boone the lowest plant dry weight in the flooded soil. There were no significant differences among cultivars for root number.

Broad-sense estimates of heritability were calculated using selection units based on phenotypic means for 5 plants and means for single plants (Table 2). Heritability estimates were generally high for all characters measured. In general, estimates for Mg and Ca were relatively higher than those obtained for K, plant dry weight, and root number.

DISCUSSION

On the whole, the cultivars responded differently to the nonflooded and flooded soils. The minimum range values for Mg, K, and plant dry weight were generally lower for the cultivars grown in flooded soil, indicating that cation uptake and plant growth of certain genotypes were reduced by low soil oxygen. However, several genotypes were apparently unaffected by flooding. These genotypes would be useful in a breeding program to develop an orchardgrass cultivar that would contain high levels of Mg under almost all pasture moisture situations.

The broad-sense estimates of heritability indicated that selection would be effective for changing the characters measured. The 5-plant and the single-plant mean heritability estimates (Table 2) are presented to show the increased precision in identifying superior genotypes that is gained by using replication. The heritability estimates indicate that Mg and Ca can be increased and K can be decreased through selection to obtain a suitable mineral content in the forage.

Increased Mg in orchardgrass should provide sufficient amounts of Mg to the pastured animal throughout the grazing season, especially during periods when soils are water-saturated and cation uptake by forage species is reduced.

Table 1. Mean, range, and standard deviations of Mg, Ca, K, plant dry weight, and root number of four orchardgrass cultivars grown in a nonflooded and flooded soil.

Cultivar	Plants/cultivar	Nonflooded soil				
		Mg	Ca	K	Plant dry weight	Root number
	no.	%	%	%	g	no.
Boone	90	0.29 a*	0.22 a	4.34 a	0.93 a	37 a
Range		0.22-0.37	0.16-0.30	2.78-5.26	0.32-1.77	15-90
Frode	105	0.26 c	0.21 b	4.28 a	0.96 a	40 a
Range		0.19-0.34	0.15-0.33	2.48-5.82	0.23-2.71	13-81
Hallmark	115	0.27 b	0.22 a	4.35 a	0.96 a	35 a
Range		0.11-0.37	0.11-0.29	1.28-5.38	0.36-2.39	15-73
KO-191	90	0.29 a	0.22 a	4.34 a	1.06 a	40 a
Range		0.17-0.39	0.15-0.32	3.38-5.46	0.35-2.14	11-98
Standard deviation		0.031	0.039	0.469	0.311	12.6

Cultivar	Plants/cultivar	Flooded soil				
		Mg	Ca	K	Plant dry weight	Root number
	no.	%	%	%	g	no.
Boone	90	0.25 a	0.41 a	3.09 ab	0.60 b	73 a
Range		0.16-0.39	0.25-0.65	2.07-4.97	0.10-1.25	18-158
Frode	105	0.24 ab	0.43 a	2.93 ab	0.67 ab	79 a
Range		0.09-0.39	0.15-0.63	0.90-4.00	0.30-1.53	13-171
Hallmark	115	0.25 a	0.41 a	3.15 a	0.73 a	75 a
Range		0.14-0.40	0.26-0.57	2.06-4.70	0.24-1.58	8-171
KO-191	90	0.23 b	0.37 b	2.84 b	0.70 a	78 a
Range		0.14-0.44	0.23-0.63	1.92-6.27	0.01-17.2	14-179
Standard deviation		0.046	0.074	0.629	0.224	28.4

* Means within a column followed by the same letter are not significantly different (1% level of Duncan's Multiple Range Test).

Table 2. Broadsense heritabilities based on phenotypic means for five plants and single plant means.

Plant characteristic	Nonflooded soil		Flooded soil	
	Five-plant mean	Single plant	Five-plant mean	Single plant
Mg	0.86	0.54	0.81	0.46
Ca	0.76	0.38	0.83	0.50
K	0.67	0.29	0.79	0.44
Plant dry weight	0.67	0.29	0.68	0.30
Root number	0.61	0.24	0.73	0.35

LITERATURE CITED

Burton, G.W., and E.H. Devane. 1953. Estimating heritability in tall fescue (*Festuca arundinacea*) from replicated clonal material. Agron. J. 45:478–481.

Elkins, C.B., R.L. Haaland, C.S. Hoveland, and W.A. Griffey. 1978. Grass tetany potential of tall fescue as affected by soil O_2. Agron. J. 70:309–311.

Elkins, C.B., and C.S. Hoveland. 1977. Soil oxygen and temperature effects on tetany potential of three annual forage species. Agron. J. 69:626–628.

Grunes, D.L., P.R. Stout, and J.R. Brownwell. 1970. Grass tetany of ruminants. Adv. Agron. 22:331–347.

Haaland, R.L., C.B. Elkins, and C.S. Hoveland. 1978. A method for detecting genetic variability for grass tetany potential in tall fescue. Crop Sci. 18:339–340.

Spaced Plants in Swards as a Testing Procedure in Grass Breeding

G.E. VAN DIJK

Foundation for Agricultural Plant Breeding (SVP),
Wageningen, The Netherlands

Summary

Grassbreeders like to observe plants separately, and therefore they make use of wide spacing. In this paper a procedure is described in which plants are spaced in a sward of another grass. In this way plants can be assessed separately throughout the test, although they are growing in a dense crop.

When plants are tested under wide spacing in the sward of another species, the differences that exist between plants and between progenies show up earlier and are more pronounced than in plants tested under noncompetitive spacing. Plants that are weak performers in the sward will disappear.

Ease of assessment depends on suppression by the companion crop and on that crop's height and color. The companion crop has a greater influence on the production level than on the yield ranking of the entries, the latter not being substantially changed as a rule.

Experimental varieties obtained by selecting perennial ryegrass plants in swards of timothy and of cocksfoot proved to be highly persistent. They showed good regrowth after cutting, with few plants disappearing. As a consequence the sward remained dense and productive. This kind of resistance to sward deterioration is very important at high production levels.

KEY WORDS: spacing, competition, grass breeding, perennial ryegrass, *Lolium perenne* L., timothy, *Phleum pratense* L.

INTRODUCTION

Reliable assessment of separate plants is important in breeding. Assessment requires ample space/plant, a situation that deviates from the dense stand culture required in farming. Perennial grasses grown in very dense swards and under high competitive pressure are considerably different from plants grown in wide spacing. Various suggestions for improving the reliability of testing spaced plants have been made, such as reducing the distance between plants (Lazenby and Rogers 1964) or searching for plant traits correlated with production of the plants when growing in dense swards (Rhodes and Mee 1980). This article describes wide-space planting in swards of other grass species. Data from earlier studies (van Dijk 1973, van Dijk and Winkelhorst 1978) are given and are supplemented by the results of an experiment in which perennial ryegrass (*Lolium perenne* L.) plants were assessed in swards of other grasses.

METHODS

Seedlings raised in a greenhouse, 6–8 weeks old, were set out at wide spacing, 50 × 50 cm, in a field sown to another grass species at a normal seeding rate. For tests

involving clones, rooted tillers were planted in a newly sown competitive grass. In most cases, sowing and planting were done on the same day. After both species were fully established, an initial assessment was made of individual plants in sward conditions, followed by further evaluations in subsequent years. Plants could have been weighed, but visual assessment was more practical (van Dijk 1973). When families were tested, both the number of surviving plants and the total family performance were evaluated. Data reported here are mainly from two experiments.

Experiment I. Eight perennial ryegrass clones were planted in the spring of 1974 in plots consisting of 5 ramets replicated twice. Treatments were monoculture and competing swards of browntop (*Agrostis tenuis* Sibth.) cv. Holfior, cocksfoot (*Dactylis glomerata* L.) cv. Holstenkamp, Chewings fescue (*Festuca rubra* L.) cv. Koket, small cat's tail (*Phleum bertolonii* DC.) cv. S50, timothy (*Phleum pratense* L.) cv. Pastremo, and smooth-stalked meadowgrass (*Poa pratensis* L.) cv. Fylking. Plant spacing was 50 × 50 cm in monoculture and 45 × 45 cm in the freshly sown grasses. Four cuts were taken in 1975 and two in 1976. Nitrogen (N) fertilizer was applied after each cutting, totaling 338 kg/ha in 1975 and 208 kg/ha in 1976.

Experiment II. In March 1972 56 late-flowering perennial ryegrass families and 8 cultivars were planted. The planting design was 20 spaced plants/plot, 2 replications, with 60 × 60-cm spacing in monoculture and 50 ×

G.E. van Dijk is a Senior Research Officer with the Foundation for Agricultural Plant Breeding.

50–cm spacing in swards of cocksfoot cv. Holstenkamp and timothy cv. Pastremo. In 1973 and 1974 plant performance was assessed visually. In September 1974 dry-matter (DM) yield of families and cultivars grown in timothy was recorded. N fertilization was 450 kg/ha each year.

In 1974 the following ryegrass selections were made, and seed was harvested in 1975.

SVP-1: 99 plants from the cocksfoot sward; the seed of 92 plants was mixed. Seven plants gave few seeds.

SVP-10: 15 to 18 plants from each of the four best families in the timothy sward; the seed of 66 plants was mixed.

SVP-P: the individuals of SVP-1 and SVP-10 belonged to 37 families. The 37 parent clones were planted in isolation, and the seed of 35 clones was mixed to reconstruct the original population.

Two cultivars favored in the Netherlands were added: Vigor, with high persistence, Splendor, with moderate persistence. Persistence was defined as the ability to survive and to contribute to yield.

Populations and cultivars formed part of an experiment with broadcast plots. The 1 × 3–m plots were seeded at 35 kg/ha in the spring of 1976 with 4 replications. Nitrogen was applied at a rate of 468 kg/ha/yr, and four dry-matter harvests were taken both in 1978 and in 1979. A sandy soil with over 6% organic matter and a pH-KCl of about 5 was used in all experiments. Basic applications of phosphate and potassium were made as needed.

RESULTS

Experiment I demonstrated that the competing sward determines the yield level of the perennial ryegrass clones. Relative DM yields of perennial ryegrass clones in monoculture and in swards in 1975 were: monoculture, 100% (882 g DM/clone); browntop, 33%; small cat's tail 22%; smooth meadowgrass, 20%; timothy, 16%; Chewings fescue, 10%; and cocksfoot (two cuts), 4%. In 1976 data were similar, but no clones survived in the sward of cocksfoot. Table 1 presents the Spearman rank correlations among treatments for clonal DM yields from two cuts. Most correlations were high and significant but were somewhat lower when monoculture was involved.

Experiment II was started in 1972 with 2,560 plants/ treatment. In the autumn of 1973 the percentage of surviving plants was 99.8% in monoculture, 94.5% in timothy, and 46% in cocksfoot. Differences among entries in survival and plant vigor were more apparent in the swards, where they became visible sooner. In September 1974 DM production of cultivars in timothy ranged from 54 to 214 g and that of all entries from 54 to 446 g (least significant range [LSR] 0.05 = 210 g).

The range of surviving plants in timothy averaged from 5 to 16.5 plants out of 20 for cultivars and from 6.5 to 20 plants for families (LSR 0.05 = 6.2). Families with the highest number of survivors generally scored highest for yield, but exceptions did occur.

SVP-1 and SVP-10 were obtained by selecting for productive surviving plants. DM production of the two populations was compared with that of two cultivars and that of SVP-P, which served as the parent population, with a difference of one cycle of selection. In 1977 the differences among entries were small, but in 1978 differences in the third cut were large (Table 2). SVP-1 yielded 164% and SVP-10 128% of SVP-P, the parent population. Higher production resulted from quicker regrowth and a denser sward. The heavier third cut of the selections did not affect production in the next cut. In 1979 differences were small, but in the third cut, SVP-1 and SVP-10 yielded, respectively, 3% and 9% more than SVP-P.

DISCUSSION

Grassbreeders often utilize widely spaced plants from which highly heritable traits, such as plant type, uniformity, and heading date, may be assessed. Unfortunately, competition among plants in wide spacing is not comparable to competition in the dense swards used by farmers. A procedure was developed whereby the test plants are space-planted in the sward of another grass species. These transplants grew in a dense stand but still could be observed separately. In comparison with widely spaced plants, those in the sward of another grass species developed less profusely due to more competition. Differences among individual plants in the year after planting were more apparent in competing swards than in monoculture.

Table 1. Rank correlation coefficients between dry-matter yields of two cuts of perennial ryegrass clones tested in monoculture and in competing swards.

	Mono-culture	A.tenuis	Phl.bert.	P.prat.	Phl.prat.
A.tenuis	0.64				
Phl.bert.	0.73*	0.76*			
Poa prat.	0.24	0.84**	0.51		
Phl.prat.	0.69	0.91**	0.76*	0.78*	
F.rubra	0.80*	0.84**	0.78*	0.67	0.84**

* P < 0.05.   ** P < 0.01.

Table 2. Dry-matter production in g/plot of 3 populations and 2 cultivars in broadcast plots (4 replications, 3m²) 1978.

	11-5	15-6	2-8	4-10	Total
SVP-1	629	1188	993	549	3359
SVP-10	630	1119	776	543	3068
SVP-P	578	1160	604	550	2892
Vigor	606	1166	591	544	2907
Splendor	515	1044	265	463	2287
F*)	8.45**	NS	13.00**	NS	12.14**
LSD5%*)	54.6		154		236
LSD1%*)	73.7		207		312

NS = Nonsignificant.
** P < 0.01.
*) Calculated for 10 entries.

With timothy in perennial ryegrass, root competition was more important than shoot competition (van Dijk 1973), as it probably would be in other combinations (Snaydon 1978). Testing in the sward would impose selection pressure for a good root system.

The ease of assessment depends on suppression by the companion crop and on that crop's height and color. Of the grasses mentioned in Table 1, browntop provides the easiest background for assessment of perennial ryegrass, whereas cocksfoot provides the most difficult. Strong competition from cocksfoot results in rapid identification of vigorous perennial ryegrass genotypes and is attractive from the viewpoint of shortening a breeding cycle.

The competing sward affects the level of production more than the yield ranking of genotypes.

Use of a competing sward results in much lower survival, even with persistent cultivars, than does wide spacing in monoculture. Disappearance of plants during establishment is a well-known problem, but it also occurs in established pastures, especially when high N fertilization is used to obtain a high production level with heavy cuts (Ennik et al. 1980). Snaydon (1978) has remarked that, surprisingly enough, little attention has been paid to surviving plants. In experiment II surviving plants in swards of cocksfoot and timothy formed very persistent and pro-ductive populations. Ennik et al. (1980), who confirmed the high persistence of SVP-1 and SVP-10, also found that the percentage of weeds in swards of the populations was lower than in swards of the two control cultivars.

LITERATURE CITED

Ennik, G.C., M. Gillet, and L. Sibma. 1980. Effect of high nitrogen supply on sward deterioration and rootmass. *In* W.H. Prins and G.H. Arnold (eds.) The role of nitrogen in intensive grassland production, 67–75. Pudoc, Wageningen.

Lazenby, A., and A.A. Rogers. 1964. Selection criteria in grass breeding. J. Agric. Sci. Camb. 62:285–298.

Rhodes, I., and S.S. Mee. 1980. Changes in dry matter yield associated with selection for canopy characters in ryegrass. Grass and Forage Sci. 35:35–39.

Snaydon, R.W. 1978. Genetic changes in pasture populations. *In* J.R. Wilson (ed.) Plant relations in pastures, 253–269. CSIRO. E. Melbourne, Austral.

van Dijk, G.E., 1973. Het beordelen van planten en families bij de veredeling van Phleum pratense (Assessment of plants and families in the breeding of *Phleum pratense*). Eng. summary Agric. Res. Rpt. 103.

van Dijk, G.E., and G.D. Winkelhorst. 1978. Testing perennial ryegrass (*Lolium perenne* L.) as spaced plants in swards. Euphytica 27:855–860.

Use of Temperature and Moisture Indices To Delineate the Global Adaptation of Two Tropical Pasture Species

D.A. IVORY

Department of Primary Industries, Queensland, Australia

Summary

The success of pasture-development programs depends to a large extent on the selection of species that are well adapted to the climatic and edaphic environment of the project area. The study reported here examined the use of temperature and moisture indices to define agroclimatic zones for rhodesgrass (*Chloris gayana* Kunth.) and green panic (*Panicum maximum* var. trichoglume Eyles) within Queensland, Australia, and in other countries as well. Areas within these agroclimatic zones that would be unfavorable because of edaphic and topographic features were not excluded.

Three moisture and two temperature indices were used to define the boundaries of agroclimatic zones for these grasses:

Arid Index: significant species instability occurred when the precipitation/potential evaporation (Thornthwaite estimate) ratio (P/E) in the previous year was less than 0.45 for green panic and 0.55 for rhodesgrass. Because of the variance in annual P/E, values of 0.6 and 0.7 were used to define the arid boundary of adaptation for these species.

Humid index: defined by P/E values of 1.0 and 1.2 for rhodesgrass and green panic, respectively.

Rainfall distribution index: defined by the ratio of P for the 6 warmer-to-colder months being greater than 1.0.

Thermal index: an annual mean thermal index greater than 0.3 was used to define a suitable thermal environment for the grasses.

Frost index: defined by the minimum temperature for the coldest month of 0°C and 3°C for rhodesgrass and green panic, respectively.

Global agroclimatic zones were mapped for these species. There was general agreement between this classification of areas of adaptation and an independent assessment of the pastoral potential of these species in Queensland. Significant areas of adaptation were indicated in a number of countries in Central and South America, Asia, and Africa. In the tropics the boundaries were delineated by the arid and humid indices, but at high latitudes and altitudes low temperature and frosts provided additional constraints. Green panic was more widely adapted than rhodesgrass.

The use of these indices to define agroclimatic zones could be extended to other species.

KEY WORDS: *Panicum maximum* var. trichoglume Eyles, *Chloris gayana* Kunth., temperature, moisture, global adaptation.

INTRODUCTION

Large areas of the tropics and subtropics are being sown to improved pastures. The ability to select adapted species on the basis of documented climatic and edaphic criteria, without recourse to the preliminary evaluation of a range of potential pasture species, would hasten implementation of the development program and provide economic advantages.

This paper examines the use of temperature and moisture indices generated from accessible meteorological measurements to define agroclimatic zones for rhodesgrass (*Chloris gayana* Kunth.) and green panic (*Panicum maximum* var. trichoglume Eyles) within Queensland and on a global scale. There is no attempt to exclude areas because of unsuitable edaphic or topographic features.

METHODS

Moisture Indices

Three indices were used to define a suitable moisture environment for the two species. These indices define the arid and humid boundaries of adaptation within a predominant summer rainfall regime.

Arid index. The effect of moisture stress on species survival and sward stability was determined from the performance of rhodesgrass and green panic sown in pure swards in a semiarid (Surat) and subhumid (Dalby) environment in the subtropics of Australia. The stability of a species, and hence its level of adaptation, was assessed from the annual change in the ratio of yield of the sown species to total yield of the sown species plus invading native grasses, termed the sown-species index (SI).

In order to identify conditions that caused instability in sown species, the annual percentage of reduction in SI was related to the annual precipitation/potential evaporation (Thornthwaite estimate) ratio (P/E) for the previous year (Fig. 1). Species instability occurred when the P/E ratio was less than 0.45 for green panic and 0.55 for rhodesgrass. These values represent the marginal boundaries of adaptation. The well-adapted arid boundaries were set one standard deviation higher, at 0.6 and 0.7, respectively.

Humid index. On the basis of general experience with these species (McCosker and Teitzel 1975, Winter 1976),

P/E values of 1.0 for rhodesgrass and 1.2 for green panic were used to define their humid boundaries of adaptation.

Rainfall distribution index. For introduced species to be productive and persistent, rainfall must coincide with favorable temperatures for growth. Thus, within the boundaries of the humid and arid indices, a suitable moisture environment for these species was arbitrarily defined as areas in which the ratio of P for the 6 warmer-to-colder months is greater than 1.0.

Temperature Indices

Two temperature criteria were used to define agroclimatic zones for green panic and rhodesgrass, with the aim of eliminating environments in which seasonal temperatures are inadequate for reasonable growth or too cold for plant survival.

Thermal index. The thermal index was expressed as the ratio of growth rate for any given day and night temperature to maximum growth rate at optimum temperatures. Mathematical functions for each species are given in Ivory and Whiteman (1978a).

Agroclimatic zones were defined as those in which the annual mean thermal index exceeded 0.3 for both species.

Frost index. A common global statistic that may be used to define an approximate frost boundary is the mean minimum air temperature for the coldest month. This is set at 0°C and 3°C for rhodesgrass and green panic, respectively, based on species survival in the field in southern Queensland (Jones 1969) and on differences in freezing resistance under controlled conditions (Ivory and Whiteman 1978b, 1978c).

RESULTS AND DISCUSSION

In Queensland the temperature and moisture indices defined areas of climatic adaptation of green panic and rhodesgrass. In the tropics boundaries were defined by the arid and humid indices, and in the subtropics frost and seasonal rainfall indices provide additional constraints.

There was general agreement between this classification of areas of adaptation in Queensland and an assessment of potential areas for use of these species, based on climate and soils, presented in a resources survey of each shire (Weston, Harbison, and Leslie, personal communication).

Global zones of adaptation for green panic and rhodesgrass were similarly derived (Fig. 2). Apart from Australia, significant areas of climatic adaptation for these species were found in Central America (Mexico and Caribbean countries), South America (Venezuela, Brazil,

———————
D.A. Ivory is Senior Agrostologist with the Department of Primary Industries.

The author wishes to thank H.A. Nix of the CSIRO Division of Land Use Research, Canberra, for providing global data for temperature, precipitation, and evaporation.

Fig. 1. Relationship between annual percentage of reduction in the sown species index (SI) of rhodesgrass and of green panic and the precipitation/potential evaporation ratio (P/E) in the previous year at 2 sites in the subtropics. SI is the ratio of the yield of the sown species to sown species plus invading native species.

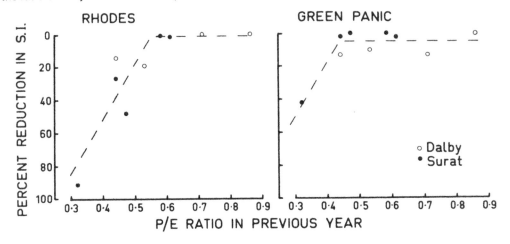

Fig. 2. Agroclimatic zones for rhodesgrass and green panic.

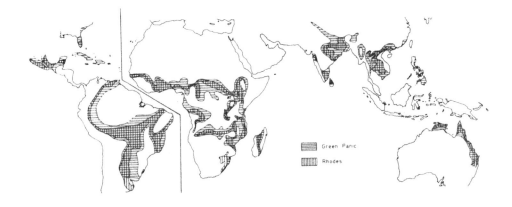

and the Gran Chaco areas of Argentina, Paraguay, and Bolivia), Asia (India and northern Pakistan, southern China, Thailand, Cambodia, and Laos, and parts of the Philippines, Burma, and Indonesia), and large areas of Africa (South Africa, Angola, Zimbabwe, Zambia, Mozambique, Madagascar, Tanzania, Uganda, Kenya, Ethiopia, Sudan, Central African Republic, Cameroon, Nigeria, Upper Volta, Mali, Ghana, Togo, Dahomey, Ivory Coast, and Senegal). Green panic showed wider climatic adaptation than rhodesgrass.

Agroclimatic zones only delineate potential areas of adaptation. In a country or region, actual areas of adaptation for pasture species can be delineated by using local soil maps to modify the climatic boundaries on the basis of the moisture characteristics of the soils and to eliminate areas in which the soils are known to be unfavorable, either physically or chemically, to a particular species.

The prediction of agroclimatic zones from simple mois-

ture and temperature indices corresponds to the grasses' areas of adaptation in Australia and the natural distribution of rhodesgrass (Bogdan 1969) and short *Panicum maximum* types (Strickland, personal communication) in Africa. This correlation provides encouragement to define the appropriate moisture and temperature indices for other tropical grasses and legumes on the basis of their known areas of adaptation in experimental or commercial situations. The definition of agroclimatic zones for a wide range of commercial pasture species for tropical regions of the world would enable the selection of species for areas in which preliminary evaluation has not occurred.

LITERATURE CITED

Bogdan, A.V. 1969. Rhodesgrass. Herb. Abs. 39:1–13.

Ivory, D.A., and P.C. Whiteman. 1978a. Effects of temperature on growth of five subtropical grasses: I. Effect of day and night temperatures on growth and morphological development. Austral. J. Pl. Physiol. 5:131-148.

Ivory, D.A., and P.C. Whiteman. 1978b. Effects of environmental and plant factors on foliar freezing resistance in tropical grasses: I. Precondition factors and conditions during freezing. Austral. J. Agric. Res. 29:243-259.

Ivory, D.A., and P.C. Whiteman. 1978c. Effects of environmental and plant factors on foliar freezing resistance in tropical grasses: II. Comparison of frost resistance between cultivars of *Cenchrus ciliaris*, *Chloris gayana* and *Setaria anceps*.

Austral. J. Agric. Res. 29:261-266.

Jones, R.M. 1969. Mortality of some tropical grasses and legumes following frosting in the first winter after sowing. Trop. Grassl. 3:57-63.

McCosker, T.H., and J.K. Teitzel. 1975. A review of Guinea grass (*Panicum maximum*) for the wet tropics of Australia. Trop. Grassl. 9:177-190.

Winter, W.H. 1976. Preliminary evaluation of twelve tropical grasses with legumes in northern Cape York Peninsula. Trop. Grassl. 10:15-20.

Procedure for Selecting Subterranean Clover Cultivars in South Western Australia

D.A. NICHOLAS and D.J. GILLESPIE
Department of Agriculture, Western Australia

Summary

Nineteen cultivars of subterranean clover (*Trifolium subterraneum* L.) have been registered for use over a wide range of environments in Western Australia. New cultivars are being sought to extend the range of environments in which clover can be grown and to improve on the already established cultivars.

The primary objective is the selection of low-estrogen clovers that are persistent under either a continuous grazing or a ley-farming system. While low estrogenicity is achieved by selecting clovers that are low in the isoflavone formononetin, persistency results from the combination of a number of characteristics, such as good seed yield, hardseededness, tolerance to flooding, and disease and insect resistance. Although the particular combination of characters required varies considerably, depending on the environment in which the cultivar may ultimately be grown, a five-stage procedure has been developed for general use.

Selection in the first three stages involves the assessment of a large number of introduced and crossbred clovers at two or three locations. Some testing for disease and insect resistance, either in the greenhouse or in the field, is included. For the final two stages, when the number of clovers may have been reduced to 10-20, selection is undertaken at a number of sites in the intended area of use. Persistency is finally assessed as the maintenance of a high reserve of clover seed, a high clover content in the pasture, and a dense regeneration each year of clover seedlings following cropping or grazing.

The characters being selected for in the early stages of the program can be reliably assessed from undefoliated rows or small swards, thus allowing large numbers of clovers to be screened. Final selection uses swards at a number of field sites. Little emphasis is placed on the direct measurement of herbage yield, and comparative animal productivity data are not required prior to a cultivar's being released for general use.

Five recently released cultivars are being grown successfully over a large area. The new cultivars constituted over one-third of the 1979-80 production of certified subterranean clover seed in Western Australia.

KEY WORDS: subterranean clover, *Trifolium subterraneum* L., selection, estrogens, persistence.

INTRODUCTION

Plant introduction, together with correct fertilizer use, has led to greatly increased carrying capacity and crop production from the temperate regions of southern Australia. The annual legume subterranean clover (*Trifolium subterraneum* L.) has been sown into some 80% of the 20 million ha of sown pasture (Donald 1970). In Western

Australia nineteen cultivars have been registered (Quinlivan and Francis 1977) for use over a wide range of environments—from short (4-month) to long (9-month) growing seasons, for soils prone to waterlogging, for ley farming, and for continuous grazing systems.

Selection from introduced and crossbred material is being employed in what has become a nationally oriented program to improve and extend the range of adaption of subterranean clover (Francis et al. 1970). The basic aim of the program has been the selection of low-estrogen cultivars capable of persisting under grazing. Low estrogenicity has been achieved by selecting for a low level of the

---

D.A. Nicholas and D.J. Gillespie are both Plant Research Officers with the Department of Agriculture in Western Australia.

isoflavone formononetin, the concentration of which in the leaves of some older cultivars exceeds 1% of the dry weight. Such concentrations can cause a severe decline in ewe fertility (Marshall et al. 1971). For persistence seven primary characters have been defined as important (Francis et al. 1976): suitable maturity, reliability of seed production, hard seed content, grazing tolerance, physiological dormancy, capacity for burr burial, and tolerance to pasture diseases and insects known to contribute to pasture deterioration, e.g., clover scorch (*Kabatiella caulivora*) (Bokor et al. 1978), south coast root rot (*Pythium* and *Fusarium* spp.) (MacNish et al. 1976) bluegreen aphid (*Acyrthosiphon kondoi*) (Gillespie and Sandow 1982).

In traditional forage-improvement programs, the direct measurement of herbage yield has received little emphasis, particularly in the early stages of selection. Rather, emphasis has been placed on clover persistence for the purpose of ensuring a reliable supply of nitrogen for growing either the associated nonleguminous species or subsequent crops (Watson and Lapins 1964).

SELECTION PROCEDURE

The particular combination of characters required in a cultivar varies considerably, depending on the environment in which the clover is to be grown. A five-stage selection procedure has been developed that facilitates the combining of different characters required in a cultivar.

Stage I

The initial aim is to measure important characters related to persistence, animal production, and identification. Clovers collected in Queensland and overseas, together with crossbred material, are hand sown in unreplicated, undefoliated rows 1-3 m long at one central location (Perth, lat 31° 57'S long 115° 51'E, annual rainfall 873 mm). Rows on land free of clover are kept weed free, well fertilized, and irrigated (if necessary). Large numbers of clovers are screened rapidly for easily identifiable characters, e.g., days to flowering, formononetin content (Francis and Millington 1965), hard seed content, dormancy (Anon. 1976), rate of breakdown of hardseededness (Quinlivan and Millington 1962), burr burial capacity, growth habit, and leaf markers. Testing may take 1-3 years, depending on the purity of the initial seed supply and the rate of buildup of seed.

Stage II

The aim of the second phase is to assess some of the characteristics measured in stage I in an environment closer to that in which the clovers may ultimately be used. Measures of genotype × environment interactions can be made at this stage. Two major centers are employed, Wongan Hills (146 km northeast of Perth, annual rainfall 350 mm) and Denmark (400 km south-southeast of Perth, annual rainfall 1,000 mm). Early flowering test clovers are hand sown at Wongan Hills in nonreplicated, undefoliated, dense rows 3-5 m long in land free of clover.

The characters assessed are days to flowering and seed maturity, burr burial, hardseededness, and vigor. Susceptibility to diseases and insects can also be noted.

Later-flowering clovers are hand sown at Denmark in nonreplicated 1-m² swards into virgin land (land recently cleared of native vegetation and not previously sown to pasture) to ensure pure swards and to allow early sowing. Swards are screened to assess the resistance of the clovers to clover scorch disease. Infection is obtained by spreading diseased clover material over the plots and trampling it into the growing clover (Chatel et al. 1973). Ratings of disease incidence and general growth are made in winter and spring. Reactions to other diseases and insects are also noted, and flowering date is checked. Over 1,000 clovers/yr have been tested by this technique.

Stage III

Clovers selected in stage II are screened for resistance to bluegreen aphid in the greenhouse (Gillespie and Sandow 1982) and for clover root rot. Little success has been achieved with greenhouse screening for root-rot resistance; however, a field test has been developed. Two field sites, in high-rainfall, root-rot-prone areas, are used. One hundred germinable seeds of up to 100 clovers are sown in 1-m rows and assessed after 6 weeks for survival and root damage.

Stage IV

The aim of this step is to test the selected clovers in the target region—a region having a particular range of soil types and usually differing from the test location by less than one month in length of growing season. Until recently, the selection pressure applied in stages II and III was sufficient to reduce the number of remaining clovers for a specific region to less than 20. However, with the identification of resistant parent clovers, crossbreeding is rapidly increasing the number of clovers available for subsequent testing.

During stage IV the clover is sown in 2-6-m² replicated swards at a seeding rate of 50 kg/ha, a rate designed to simulate normal sward densities. The swards are established at 3 to 10 sites within the defined region on land on which a pasture has been established for a number of years but on which the existing cultivars have failed. At such sites the selected clovers are exposed to competition from weeds, diseases, and insects. Swards are defoliated by grazing and/or mowing. In areas with short growing seasons, defoliation is necessary to control the growth of weedy species. Where the growing season is longer and there is greater herbage growth, additional defoliation is required to simulate normal grazing. Seed production (typically from 0.2-m² areas), hardseededness, burr burial, seedling regeneration, and clover content ('t Mannetje and Haydock 1963, Jones and Hargreaves 1979) are measured over 2 to 3 years. In grazing regions herbage production is also measured two to three times/yr either by a calibrated rating (Campbell and Arnold 1973) or a plate meter (Earle and McGowan 1979) or by mowing strips and drying and weighing the herbage collected.

Stage V

For the final stage less than 10 clovers are selected for a given region. Sites, like those in stage IV, are on established but deteriorated pasture. Commercial equipment is used to prepare and sow replicated (3 or 4) 100–200-m² plots. A commercial seeding rate of 10 to 15 kg/ha is used. The trials are cropped and/or grazed in common with the remainder of the field. Seed production and seedling regeneration are the principal factors measured; some estimates of herbage yield and botanical composition are made in regions having longer growing seasons. Reactions to soil type, diseases, and insects are again noted. The trials are continued for 2 to 4 years, with additional plantings in subsequent years.

RELEASE

Following the assessments made in stages I to V, a clover is selected, registered, and released for commercial use without the collection of animal productivity data. Objections to the use of grazing experiments to compare clovers similar in morphology and maturity have been detailed by Gladstones (1975). They include the limitations in time, number of sites, and number of clovers tested, as well as the high cost and the possibility of cumulative mishap.

DISCUSSION

The program has resulted in the selection of five new cultivars. Indeed, the obvious potential of some experimental clovers made testing at all stages unnecessary. Two cultivars, cv. Trikkala and cv. Esperance, were released without being tested at stage V because of their good early performance and some of their particular attributes, e. g., greatly improved tolerance to clover scorch disease and lowered formononetin content. Cultivars released after they were fully tested have included two early-maturing clovers, cv. Nungarin and cv. Northam, and one very late clover, cv. Larisa. Almost 1,300 tonnes of the seed of the five cultivars released since 1975 was produced in Western Australia from the 1979 seed crop, a figure that is 37% of the total production of certified subterranean clover seed (Paterson, personal communication).

The selection program outlined above has proved capable of testing clovers from overseas and from local collections, together with some crossbred material. Modifications of the program are now being made because of the rapidly increasing number of crossbred clovers and because of an increasingly national approach (Stern et al. 1982). In addition, although selection for such simply assessed characters as disease and insect resistance will continue, attempts are being made to replace some already established and reasonably well-adapted cultivars, e.g., cv. Seaton Park. Possible changes include a reselection cycle for crossbreds after stage III, testing throughout Australia in stages IV and V, and the identification of successful competitive clovers from sowings of complex mixtures.

LITERATURE CITED

Anon. 1976. International rules for seed testing. Seed Sci. and Technol. 4:1–180.

Bokor, A., D.L. Chatel, and D.A. Nicholas. 1978. Progress in clover scorch research. J. Agric. West. Austral. 12:58–62.

Campbell, N.A., and G.W. Arnold. 1973. The visual assessment of pasture yield. Austral. J. Expt. Agric. Husb. 13:263–267.

Chatel, D.L., C.M. Francis, and A.C. Devitt. 1973. Varietal variation in resistance to clover scorch (*Kabatiella caulivora*) in *Trifolium subterraneum* L. Dept. Agric. West. Austral. Tech. Bul. 17.

Donald, C.M. 1970. Australian grasslands, 303–320. Austral. Natl. Univ. Press, Canberra.

Earle, D.F., and A.A. McGowan. 1979. Evaluation and calibration of an automated rising plate meter for estimating dry matter yield of pasture. Austral. J. Expt. Agric. Anim. Husb. 19:337–343.

Francis, C.M., J.S. Gladstones, and W.R. Stern. 1970. Selection of new subterranean clover cultivars in south Western Australia. Proc. XI Int. Grassl. Cong., 214–218.

Francis, C.M., and A.J. Millington. 1965. Varietal variation in the isoflavone content of subterranean clover: its estimation by a microtechnique. Austral. J. Agric. Res. 16:557–564.

Francis, C.M., B.J. Quinlivan, N.J. Halse, and D.A. Nicholas. 1976. Subterranean clover in Western Australia. II. Characteristics required for success. J. Agric. West. Austral. 17:26–31.

Gillespie, D.J., and J.D. Sandow. 1982. Selection for bluegreen aphid resistance in subterranean clover. Proc. XIV Int. Grassl. Cong.

Gladstones, J.S. 1975. Legumes and Australian agriculture. Austral. Inst. Agric. Sci. J. 41:227–240.

Jones, R.M., and J.N.G. Hargreaves. 1979. Improvements to the dry-weight rank method for measuring botanical composition. Grass and Forage Sci. 34:181–189.

MacNish, G.C., M.J. Barbetti, D.J. Gillespie, and K. Hawley. 1976. Root rot of subterranean clover in Western Australia. J. Agric. West. Austral. 17:16–19.

Marshall, T., H.E. Fels, H.G. Neil, and R.C. Rossiter. 1971. Pasture legume varieties and ewe fertility. J. Agric. West. Austral. 12:110–112.

Quinlivan, B.J., and A.J. Millington. 1962. The effect of a Mediterranean summer environment on the permeability of hard seeds of subterranean clover. Austral. J. Agric. Res. 13:377–387.

Quinlivan, B.J., and C.M. Francis. 1977. Registered cultivars of subterranean clover in Western Australia—their origin, potential use and identification. West Austral. Dept. Agric. Bul. 4012, 1–19.

Stern, W.R., J.S. Gladstones, C.M. Francis, W.J. Collins, D.L. Chatel, D.A. Nicholas, D.J. Gillespie, E.C. Woolfe, C.R. Southwood, P.E. Beale, and B.C. Curnow. 1982. Subterranean clover improvement: an Australian program. Proc. XIV Int. Grassl. Cong.

't Mannetje, L., and R.P. Haydock. 1963. The dryweight rank method for the botanical analysis of pasture. J. Brit. Grassl. Soc. 18:268–275.

Watson, E.R., and P. Lapins. 1964. The influence of subterranean clover pastures on soil fertility. Austral. J. Agric. Res. 15:885–894.

Improving the Efficiency of Forage-Crop Breeding

G.W. BURTON

USDA-SEA-AR, Coastal Plain Station, University of Georgia,
Tifton, Ga., U.S.A.

Summary

It is the purpose of this paper to describe briefly ways of improving the efficiency of forage-crop breeding. Breeding superior forages requires good germ plasm, proper objectives, efficient breeding methods, adequate support, assistance from other disciplines, cooperation with animal science, time, and hard work. Methods that reduce the labor and cost/unit of advance are described. Among them are recurrent restricted phenotypic selection (RRPS), which quadruples the efficiency of mass selection in improving Pensacola bahiagrass (*Paspalum notatum*) yields; a modification of RRPS for pearl millet (*Pennisetum americanum*) improvement; backcrossing procedures to transfer major genes into pearl millet lines in 2 years; and four methods of putting the F_1 hybrid on the farm.

KEY WORDS: bermudagrass, *Cynodon dactylon*, Pensacola bahiagrass, *Paspalum notatum*, pearl millet, *Pennisetum americanum*, forage breeding.

INTRODUCTION

Breeding superior forages requires good germ plasm, proper objectives, efficient breeding methods, adequate support, assistance from other disciplines, cooperation with animal science, time, and hard work. It is costly. The purpose of this paper is to describe briefly methods and procedures that can increase the efficiency of forage-crop breeding by reducing the time and cost/unit of advance.

METHODS AND DISCUSSION

Germ plasm is the forage breeder's most valuable natural resource (Burton 1979). The breeder should begin with the best. To do so, he must collect and evaluate the best cultivars, land races, and wild relatives of the species he wishes to improve. Visual assessment of a grow-out of the germ plasm under *uniform* conditions usually indicates which material should become part of an active breeding program. This grow-out should produce the selfed and intercrossed seed increase necessary for further evaluation, use, and maintenance. Selfed seed is the most efficient storage form for future use.

Germ plasm collections may yield superior varieties without modification. Crested wheatgrass (*Agropyron desertorum* [Fisch. ex Link] Schult.) and pangolagrass (*Digitaria decumbens* Stent.) are examples of varieties that have made substantial contributions to man.

Every forage-breeding program should have objectives. The most important objectives pertain to dependability, quality, yield, efficiency, pest resistance, propagation, and weed potential.

Efficient genetic improvement of a species requires knowledge of its mode of reproduction. If such knowledge is not available, it can be discovered from a study of selfed and crossed progenies to detect self incompatabilities, apomixis or sexuality, self- or cross-pollination, and inbreeding depression as an index of probable heterosis. Chromosome counts and genome relationships are useful, particularly in programs using wide crosses.

Commercial propagation methods influence breeding procedures. Generally, annuals and bunchgrasses must be propagated by seeds. Perennial stoloniferous species such as bermudagrass (*Cynodon dactylon*) can be propagated vegetatively, greatly simplifying breeding procedures.

Time and labor are the most costly elements in forage breeding. Reducing time or labor/unit of advance increases the efficiency of any forage-breeding operation. Efficient forage breeding requires enough suitable land close to the office and the laboratory to save the time and money required to travel long distances for essential field operations. Efficient forage breeding also requires adequate greenhouse space to permit the breeder to grow plants every day of the year if his program requires him to do so.

Efficient forage breeding requires hands—more and less costly hands than the breeder has. Honest, dependable, industrious people who like the work and are taught the things they need to know by the breeder can supply the hands. (College training is not necessary; it only adds to the cost.) Well-qualified people, often selected from a temporary labor force, must be rewarded with words as well as pay. Such thinking technicians have helped

---

The author is a Research Geneticist with USDA-SEA-AR.

materially to improve the efficiency of forage-breeding programs.

Forage breeders hybridize to transfer and combine desirable traits, to increase variation, and to exploit hybrid vigor. A new book, *Hybridization of Crop Plants* (Fehr and Hadley 1980), contains many crossing procedures of value to the forage breeder. Pensacola bahiagrass (*Paspalum notatum*) culms with a bit of attached stolon that are ready to flower will bloom and mature seed when placed in water. This characteristic facilitates the production of hybrids and polycrosses (Burton 1948).

Because bermudagrass hybrids are propagated vegetatively, the forage breeder needs only to produce a superior plant (Burton 1947). Progeny testing is not necessary. To maximize hybrid vigor, the breeder is usually concerned with the production and evaluation of F_1 hybrids between good parents having desirable traits. He soon learns that some superior plants produce better hybrid offspring than others. Coastal bermudagrass and Tift 23 pearl millet (*Pennisetum americanum*) are examples of excellent parents.

Isolation of the better plants from large variable populations requires efficient screening methods. We discovered in the selection of Coastal bermudagrass from 5,000 F_1 hybrids that copious measurements taken on each spaced plant a number of times during the first season were of very little value (Burton 1947). Visual selection of the plants with the greatest accumulated growth at the end of the first year would have included Coastal. However, several other replicated plot tests were required to isolate Coastal from the other better bermudagrass hybrids.

Locating the space-planted nurseries on a uniform deep sand of low fertility helps to screen plants with deep root systems and improve drought tolerance. This technique was used in the original screening of Coastal and Coastcross-1 bermudagrass, both of which have excellent drought tolerance. Such a screening system may help to select plants with greater efficiency in fertilizer and water use—characteristics of Coastal and Coastcross-1 bermudagrass (Burton et al. 1954).

Mass selection based on phenotype response is the most efficient breeding method for population improvement of characters with high heritability. Forage yield usually gives low heritability values. However, recurrent restricted phenotypic selection (RRPS), which quadruples the progress possible from mass selection, has been effective in increasing the forage yields of individual plants as well as a population of Pensacola bahiagrass (Burton 1974).

Since 1974 we have improved RRPS efficiency by adding three additional restrictions:

1. Transplanting to 5-cm clay pots in the greenhouse in January the 7 largest plants in ± 100 seedlings (in a row in flats of soil) from polycrossed seed of each selection gives a population of 20,000 and permits the screening for seedling yield of one chance recombination between each (200) selection (the formula is n(n − 1) − 2 = 19,900).

2. Keeping the selection number with the 7-plant progeny in the greenhouse until set in the field in April permits elimination of progenies of plants that perform poorly after the polycross is made in July and allows one cycle/year.

3. Improved cultural practices, including fumigating the soil with methyl bromide to control weeds, permit making the RRPS polycross in July of the year planted and cut the time required/cycle from 2 years to one. The 1,000-plant space planting for any RRPS cycle isolated by mowing from other bahia can produce breeders' seed for a new variety.

The F_1 hybrid seed of Pensacola bahiagrass can be produced by harvesting all seed from a field planted vegetatively (with mechanical sprig diggers and planters) to alternate strips of two self-sterile, cross-fertile clones. Discovering the best clones requires making replicated plot tests of selected clones and their hybrids. A seedling or vegetative-increase plant of Pensacola bahiagrass planted in 5-cm clay pots in January and set in the field in mid-April will spread by short stolons to cover a circle 0.4 to 0.6 m in diameter by autumn with moderate fertilization and good weed control. Thus, 1.5 × 4.3-m plots can be established in one season by setting 12 potted (5-cm) plants 0.3 m apart in the center of each plot. If care is taken to maintain uniformity, meaningful yields of vegetation and seed can be obtained the first year. To establish the 5 replications of each two-clone hybrid, we need only 60 seedlings. The few hybrid seeds for such a test can be produced from one mutual hybridization of the clones, as described above. Yield trials with in-vitro dry-matter digestibility (IVDMD) analyses are usually conducted over a 3-year period. We have two two-clone experimental hybrids that yielded 30% more than the check; they are being evaluated with steers in replicated pastures.

Pearl millet plants produce several culms that complete anthesis in 6 to 10 days. We make polycrosses of selected plants by pollinating single heads on each selection with a mixture of pollen from the selected plants. The mixture is prepared in an air-conditioned laboratory as follows: pollen that has accumulated in glassine bags placed on heads starting to flower a day earlier is collected, screened through a 100-mesh soil screen, thoroughly mixed, and measured into 7.5 × 35-cm kraft bags. One of these bags is then used to pollinate a single head on each of the selected plants. A mixture of equal quantities of the hybrid seed from each plant makes the polycross seed for the next RRPS cycle.

The recessive d_2 gene in pearl millet reduces height 50% and dry-matter yield 15 to 20% but increases leaf percent, leaf yield/ha, animal daily gains, and live-weight gains/ha when d_2 pearl millet is grazed. We have been able to transfer d_2 and other important recessive genes to other millet lines in less than 2 years of backcrossing in the following manner: The recurrent parents are planted at 10-day intervals in the field or greenhouse around the calendar to insure a pollen source at all times. The greenhouse is kept above 28°C, a temperature that, with short days, hastens anthesis and seed maturity. The F_1 hybrid between the d_2 line and the recurrent parent is grown in the field in summer, or in the greenhouse in other seasons, and is backcrossed to the recurrent parent as soon as the first heads appear. Dormancy in the mature backcrossed seed is broken by soaking the seed in an aqueous solution with 1% dichloroethanol plus 0.5% sodium hypochlorite for 1 hour at 25°C. Seven F_1 plants are then selfed and backcrossed to the recurrent parent. (Seven gives a probability of 0.01 that at least one F_1 will carry the recessive gene.) Seven plants of each F_1 are planted for the next backcross and 7-plant selfed progenies of each F_1 are grown to detect which F_1s are carrying the recessive d_2 gene. Seven F_1s of these backcrosses must be selfed and backcrossed to the recurrent parent to produce the next backcross generation. In approximately 18 months, we have selfed seed of the fifth backcross from which d_2 plants carrying 98% of the recurrent-parent genes can be selected. Dominant major genes can be transferred in a similar period of time without making the selfed progeny test of the F_1s.

F_1 hybrids can yield more and tolerate more stress than varieties. How to put the F_1 hybrid on the farm economically is a major challenge facing the forage breeder. If vegetative propagation is practicable (as with bermudagrass), the breeder need only cross superior diverse parents, screen large F_1 populations, and select the best plants for vegetative propagation. Few seed heads, to improve quality, and sterility, to reduce weed potential, are desired traits.

The first-generation chance-hybrid method used in the development of Gahi-1 pearl millet could be used for other cross-pollinated species. In this method inbred lines are developed and tested in diallels until four that give high-yielding hybrids in all combinations are found. A mixture of equal quantities of seed of these four lines is used to plant seed-production fields. Bulked seed harvested from these fields contains 75% of the six possible hybrids of the four lines and 25% of their selfs and sibs. Millet grown from this seed will perform as well as a double cross involving these four inbreds because the more vigorous hybrids crowd out the inbreds in the seedling stage.

When cytoplasmic male sterility can be found in a species, the breeder need only transfer the plants to a good seed-producing line and discover a male parent that will produce a high-yielding hybrid with good forage qualities. Male parents that fail to restore fertility are preferred because they reduce the weed potential of the hybrid and improve forage quality, longevity, and yield under stress.

Inbred lines for the production of F_1 forage hybrids need not be as homozygous as inbreds used to produce hybrids of many other crops. Consequently, they need only be selfed long enough to fix desired characteristics. Heterozygosity left in the lines will add vigor and allow possible improvement through selection.

Apomictic species, such as *Panicum maximum* and tetraploid *Paspalum notatum,* have an ideal mechanism for fixing hybrid vigor and other desired traits in the F_1. Here the challenge is to find or develop sexual females and then produce desirable F_1 hybrids that are apomictic (Burton and Forbes 1960).

The improvements in the efficiency of forage breeding presented here are examples of the kinds of improvement every forage breeder can make. Every improvement described can itself be improved by continually searching for a better way.

LITERATURE CITED

Burton, G.W. 1947. Breeding bermudagrass for the southeastern United States. J. Am. Soc. Agron. 39:551–569.

Burton, G.W. 1948. A method of producing chance crosses and polycrosses of Pensacola bahiagrass, *Paspalum notatum.* J. Am. Soc. Agron. 40:470–472.

Burton, G.W. 1974. Recurrent restricted phenotypic selection increases forage yields of Pensacola bahiagrass. Crop Sci. 14:831–835.

Burton, G.W. 1979. Handling cross-pollinated germplasm efficiently. Crop. Sci. 19:685–690.

Burton, G.W., E.H. DeVene, and R.L. Carter. 1954. Root penetration, distribution, and activity in southern grasses measured by yields, drought symptoms, and P^{32} uptake. Agron. J. 46:229–233.

Burton, G.W., and I. Forbes, Jr. 1960. The genetics and manipulation of obligate apomixis in common bahiagrass (*Paspalum notatum* Flugge). Proc. VII Int. Grassl. Cong., 66–71.

Fehr, W.R., and H.H. Hadley. 1980. Hybridization of Crop Plants. Am. Soc. Agron., Madison, Wis.

Improvement of Pollen and Seed Fertility in
Lolium-Festuca Hybrids

Y. TERADA

Yamaguchi Agricultural Experiment Station,
Yamaguchi-City, Japan

Summary

In order to introduce desirable characteristics from other species, *Lolium multiflorum* and *L. perenne* were crossed with *Festuca arundinacea* and *F. pratensis,* and three kinds of F$_1$ hybrids were backcrossed or tricrossed with five species of *Lolium* and *Festuca* to the F$_4$ generation.

F$_1$ hybrids of *L. multiflorum* × *F. arundinacea* were mostly male sterile, with only 0.2% to 2.5% stainable pollen and 0.5% to 2.4% viable seeds from backcrosses or tricrosses with five species. F$_1$ hybrids of *L. perenne* × *F. arundinacea* and *L. perenne* × *F. pratensis* produced extremely sterile pollen with rudimental anthers and 0% to 0.6% viable seeds in backcrosses or tricrosses with four species.

Hybrid derivatives with *Festuca* species as recurrent parents had very low fertility, less than 0.3% fertile pollen and 0% to 1.3% viable seeds, even after the F$_4$ generation. On the contrary, *Lolium*-like hybrid derivatives produced by pollination with *Lolium* species increased in pollen and seed fertility as the generation advanced to F$_4$, and enough progeny plants were obtained in the F$_3$ and F$_4$ generations, even with interpollination within hybrid plants.

Pollen and seed fertility of the hybrid derivatives varied, depending on the combination of backcross and tricross or on hybrid generation. A correlation coefficient of r = 0.802 between pollen and seed fertility indicates that selection for higher pollen fertility should lead to improvement of seed fertility of hybrids.

From these results, I suggest that the breeding technique may be considered effective to improve *L. multiflorum* and *L. perenne* if (1) intergeneric hybrids are backcrossed or tricrossed with *Lolium* species, (2) highly fertile pollen plants are polycrossed within hybrid derivatives, and (3) progeny plants are selected for seed fertility and other desirable characters.

KEY WORDS: *Lolium multiflorum, Lolium perenne, Festuca arundinacea, Festuca pratensis,* intergeneric hybridization, crossability, pollen fertility, seed fertility.

INTRODUCTION

Scientists (Carnahan and Hill 1961) have made efforts to produce desirable intergeneric hybrids of *Lolium* and *Festuca* in order to combine the palatability and rapid growth of *Lolium* with the winter-hardiness, summer tolerance, and persistency of *Festuca*. These two genera, *Lolium* and *Festuca,* are each recognized as valuable grasses in Japan. Italian ryegrass (*Lolium multiflorum* Lam.) is used as a winter-annual grass in the uplands and drained paddy fields of southwestern Japan, while *Festuca* species, such as tall fescue (*F. arundinacea* Schreb.) and meadow fescue (*F. pratensis* Huds.), are grown as perennial grasses in the northern and southern parts of Japan.

The successful production of new types of plants that would combine the desirable characteristics of annual *Lolium* and perennial *Festuca* would contribute greatly to the development of livestock production in Japan. *Lolium-Festuca* hybridization should be one of the high priorities in grass-breeding programs. In Japan, however, only a few scientists have worked on the intergeneric hybridization of *Lolium* and *Festuca*. Nagamatsu et al. (1964) studied the crossability and cytological analysis of F$_1$ hybrids. Hayashi (1968) also researched their crossability.

Since 1971, I have carried out experiments with intergeneric hybrids of *Lolium* and *Festuca* that have emphasized pollen and seed fertility in the hybrid derivatives. This paper reports my approach to ryegrass breeding and some results that have been applied to the actual breeding of Italian ryegrass at our Agricultural Experiment Station.

MATERIALS AND METHODS

Intergeneric crosses, backcrosses, and tricrosses were carried out with *L. multiflorum* Lam. (2n = 14, 28), *L. perenne* L. (2n = 14, 18), *F. arundinacea* Schreb. (2n = 42), *F. pratensis* Huds. (2n = 14), and hybrid *Festulolium* strains produced in the Netherlands. Emasculation of female inflorescences for the intergeneric crosses was carried out by hand-emasculation and by hot-water pollen-killing. Other plants were not emasculated. Pollination was controlled for crossing by paraffin bags or by plant isolation in the greenhouse or the field.

Pollen fertility was determined by iodine staining with a potassium iodide solution, and seed fertility was evaluated by X-ray photography. Chromosomes were counted in root-tip cells by Feulgen's method, and fluorescence character was investigated by applying an ultraviolet beam to the roots.

RESULTS

Crossability Between *Lolium* and *Festuca*

L. multiflorum (2n = 14) was crossed with *F. arundinacea*, using the two-pollen emasculation treatment and nonemasculation of the female plants. Although as many as 4,175 florets were tried, the percentage of successful hybrids obtained was 2.4% from hand-emasculated plants, 0.2% from nonemasculated plants, and 0% from plants treated by hot-water pollen-killing. When nonemasculated *L. perenne* was crossed with *F. arundinacea*, true hybrids were found at a rate of 9.3% in the progeny plants from diploid *L. perenne*.

In order to improve the screening technique to detect whether or not the plant was hybridized, the fluorescence response of the cross between *L. perenne* and hybrid *Festulolium* strains was examined. In the progeny plants of female parents of unresponsive fluorescence, the fluorescence-responsive plants were 10.5% (diploid) and 15.3% (tetraploid). No fluorescence was found in the selfed progenies of female plants.

The results of these experiments show that *L. multiflorum* and *L. perenne* are able to cross with *Festuca* species in isolated conditions without emasculation and that some hybrids can be identified on the basis of fluorescence response, inflorescence type, and pollen fertility in the progeny plants.

Pollen and Seed Fertility of F_1 Hybrids

F_1 hybrids of *L. multiflorum* (diploid) and *F. arundinacea* have 28 chromosomes with irregular meiosis and have *Festuca*-like panicles. Pollen fertility in these hybrids was found to be extremely low, ranging from 0.2% to 2.5%, perhaps because of the lagging univalent chromosomes in the pollen mother cells. Seed fertility resulting from self-pollination between F_1 plants was 0%. However, 1.5% and 2.4% seed sets were obtained from the backcrosses of *L. multiflorum* and *F. arundinacea*, respectively, and 0.5% seed set was obtained in the tricrosses.

Pollen sterility in F_1 hybrids from *L. perenne* × *F. arundinacea* and *L. perenne* × *F. pratensis* was a result of rudimental anthers. Therefore, the seed fertility of these hybrids was also extremely low—for example, only 0.6% of the seeds from the backcross with *F. arundinacea* were fertile.

In spite of extremely low viable pollen, several F_2 plants were obtained by backcrossing and tricrossing under suitable pollinating conditions, such as continuous pollination in control rooms.

Improvement of Pollen and Seed Fertility In the F_2 to F_4 Generations

Most hybrid derivatives (F_2 to F_4) were produced by backcrosses and tricrosses of *L. multiflorum*, *L. perenne*, *F. arundinacea*, *F. pratensis*, and hybrid *Festulolium* strains, with F_1 hybrids of *L. multiflorum* × *F. arundinacea*. As Table 1 shows, *Festuca*-like plants that were pollinated with *F. arundinacea* or *F. pratensis* as the recurrent parents had extremely low pollen fertility, similar to that of the F_1 hybrids— 0% to 0.3% fertility in the F_2 to F_4 generations, with most plants having rudimental anthers. The seed fertility of these progenies was extremely low, less than 1.3%, and fertility was not restored until the F_4 generation.

In contrast, the progenies of F_1 hybrids with *Lolium* species as the recurrent pollen parents were considerably improved in pollen fertility with each advancing generation (Table 1). In the F_4 generation, fertile pollen percentages of *Lolium*-like hybrid derivatives averaged 59.4%, with more than 80% fertility in some individuals. In addition, improved seed fertility was evident, with rates of 12.0% in the F_2, 16.1% and 30.4% in the F_3, and 25.3% to 75.0% in the F_4 generations.

Table 1. Pollen and seed fertility of *Lolium—Festuca* hybrid derivatives.

	Genomic constitution[1]	Pollen fertility %	Seed fertility[2] %	pollen
F_1	LmFa	0.6	1.5	Lm
	LmFa	0.6	2.4	Fa
	LpFa	0.0	0.2	Lp
	LpFa	0.0	0.0	Fa
	LpFp	0.0	0.2	Lp
	LpFp	0.0	0.6	Fp
F_2	LmFa Fa	0.1	0.5	Fa
	LmFa Fp	0.0	0.6	Fp
	LmFa Lp	0.0	12.0	Lp
	LmFa Lm	13.0		
	LpFa Fa	0.2	0.0	Fa
	LpFa Lp	13.0		
F_3	LmFa Fa Fa	0.3	1.3	Fa
	LmFa Fa Fp		0.8	Fp
	LmFa Fp Ft	8.3		
	LmFa Lp Lp	24.3	30.4	Lp
	LmFa Lm Lm	51.7	16.1	Lp
F_4	LmFa Fa Fa Fa	0.3	0.2	Fa
	LmFa Lm Lm Ft	69.5	25.6	Open
	LmFa Lm Lm Lp	64.1	75.0	Lp
	LmFa Lm Lm[3]	72.5	25.3	Open
L.	multiflorum	85.5	62.7	Open
L.	perenne	89.6	74.4	Open
F.	arundinacea	72.0	61.6	Open
F.	pratensis	92.0	84.0	Open
Ft		62.0	35.5	Open

[1]Lm = *Lolium multiflorum*. Lp = *L. perenne*. Fa = *Festuca arundinacea*. Fp = *F. pratensis*. Ft = Hybrid Festulolium strains.
[2]Seed fertility was evaluated by X-ray photography and hybrid plants were pollinated with pollen parent.
[3]Polycross progeny plants from F_3 hybrids.

The relationship between pollen fertility and seed fertility—correlation coefficient of r = 0.802, regression equation of Y (seed) = 0.605 exp(0.056 X)—indicated that pollen fertility improvement was prerequisite to the production of hybrid plants with high seed fertility.

DISCUSSION

In order to improve seed fertility of *Lolium-Festuca* hybrids, several researchers have attempted to produce amphidiploids by the colchicine treatment (Buckner et al. 1965, Hertzsh 1966) or by meiotic doubling (Dijkstra and De Vos 1975). The amphidiploid-making method may be excellent because of the possibility of maintaining the good characteristics of F_1 hybrids. However, the backcross-and-tricross method should also be effective for extending genetic variance or introducing valuable genes from other species.

In this experiment, the backcrosses and tricrosses of the *Lolium* species considerably increased the pollen and seed fertility. These data confirmed Sulinowaski's (1966) report. In addition, the variation among individual plants was greater in the hybrids than in their parents, and this increased variation may be an advantage for selecting highly fertile plants.

From these experimental results, the breeding techniques to improve *L. multiflorum* and *L. perenne* may be described as follows: (1) intergeneric F_1 hybrids are backcrossed or tricrossed with *Lolium* species as the recurrent pollen parent, (2) highly fertile pollen hybrids are intercrossed among the hybrid derivatives in the F_2 and F_3 generations, and (3) intercross progeny plants are selected for seed fertility and other desirable characteristics in following generations.

LITERATURE CITED

Buckner, R.C., H.D. Hill, A.W. Howin, and P.B. Burrus. 1965. Fertility of annual ryegrass × tall fescue amphiploid and their derivatives. Crop Sci. 5:395–397.

Carnahan, H.L., and H.D. Hill. 1961. Interspecific and intergeneric hybridization. Bot. Rev. 27:1–131.

Dijkstra, J., and A.L. De Vos. 1975. Meiotic doubling of chromosome number in Festulolium. Euphytica 24:743–749.

Hayashi, K. 1968. Physiological studies in the crossability between three species of *Lolium* and *Festuca arundinacea*. Jap. J. Breed. 18. Add. Vol. 2:109–110.

Hertzsh, W. 1966. Intergeneric and interspecific hybrids between *Lolium* and *Festuca*. Proc. X Int. Grassl. Cong., 638–685.

Nagamatsu, T., T. Omura, and H. Mine. 1964. Studies on the interspecific and intergeneric hybridization between *Lolium* and *Festuca*. Sci. Bul. Faculty Agric. Kyushu Univ. 21:36–46.

Sulinowaski, S. 1966. Intergeneric hybrid *Lolium multiflorum* Lam.(2n = 14) × *Festuca arundinacea* Schreb.(2n = 42). Part 1. Genetica Polonica. 7:81–98.

Selection Effect for Digestibility of Summer Regrowth in Orchardgrass

S. SAIGA

Ministry of Agriculture, Forestry and Fisheries, Japan

Summary

This study was conducted to assess selection effect for the digestibility of second growth in a hay-and-aftermath harvest treatment. For estimation of digestibility, the one-step cellulase method was used, and in-vivo digestibility values were predicted by the regression equation Y = 0.83X + 27.68.

From 511 plants, ranging from 58.0% to 75.4% in digestibility, 4 plants with high digestibility (HD) and 5 plants with low digestibility (LD) were selected in 1975. Progenies of these two groups were space-planted with their parent plants and evaluated in 1978 and 1979. Differences in digestibility between the progenies of HD and LD plants were smaller than those between the parents. Realized heritabilities, estimated only for second-growth forage, were 0.48 in 1978 and 0.42 in 1979.

Two experimental strains derived from 4 HD plants and 5 LD plants were compared with 16 other cultivars under harvest treatments of 3, 4, and 6 annual cuts. Digestibility of the HD strain was greater than that of any other entry in the trial in each of the three harvest treatments.

Repeatability of digestibility estimates was measured on plants from 6 different cultivars. In the first growth, some negative correlations were obtained between two yearly estimates. The negative correlations indicate that selection in a certain direction may result in negative response. However, high positive correlations were obtained for all combinations with second-harvest estimates.

Genotypes selected for improved digestibility of second-harvest forage also had higher digestibility as first-harvest forage. Therefore, it is suggested that selection for improved digestibility in orchardgrass be conducted on second-harvest forage within gene pools of known maturity.

KEY WORDS: orchardgrass, *Dactylis glomerata* L., summer regrowth, digestibility, cellulase, heritability.

INTRODUCTION

Orchardgrass (*Dactylis glomerata* L.) is one of the most important forage grasses in Japan because of its wide adaptability, high yield, good response to fertilizer, and good regrowth. However, this grass has the disadvantages of low palatability and low digestibility. Relatively large estimates of heritability led Christie and Mowat (1968) and Cooper et al. (1962) to suggest that digestibility could be improved by breeding.

Dent and Aldrich (1966) found that digestibility differences among varieties were greater under hay management than under simulated grazing management. The reduction in digestibility with delay in date of harvest was 0.64%/day for first-harvest forage and 0.35%/day for second-growth forage (Saiga, unpublished). The large changes in digestibility of first-harvest forage are thought to be associated with changes in maturity. Further investigations (Saiga, unpublished) indicated that digestibility of orchardgrass forage was lower in August than in other months of the growing season. Thus, the need for improved digestibility and for stability of digestibility are greater in aftermath than in first-harvest forage. The study reported here was conducted to determine the effects of selection for digestibility in second-growth orchardgrass forage.

METHOD

Estimation of Digestibility

To estimate digestibility the one-step cellulase method (Saiga and Hojito 1977) was used. A 500-mg sample of dried and milled forage grass was incubated at 40°C for 6 hours in a shaking apparatus with c. 45 ml of a solution of 1% cellulase dissolved in an acetate buffer adjusted to pH 4.0. After incubation, the solution was filtered with No. 5A filter paper, and the residue was dried and weighed. The regression equation used to predict in-vivo dry-matter digestibility (Y) from the value obtained from the one-step cellulase method (X) was Y = 0.83X + 27.68 (r = 0.89, rsd = 2.09). All digestibility values appearing in this paper are the estimated in-vivo dry-matter digestibility obtained by this equation.

Selection and Evaluation

The experiment was conducted at Sapporo (lat 43°N, annual temperature 7.0°C). A population of 511 plants derived from ecotypes of Hokkaido (the northern island of Japan) was planted in the spring of 1974. The first and second growths were harvested on 20 August 1975; digestibility analysis was conducted only on second-growth forage. The population ranged from 58.0% to 75.4% in digestibility, and 4 clones with digestibility greater than 72.6% and 5 with digestibility less than 60.5% were selected. These plants were interpollinated within groups in isolated greenhouses in the spring of 1976.

Progenies were planted in the spring of 1977 in a spacing of 0.8 × 0.8 m with their parent plants propagated vegetatively, and the progenies were evaluated for digestibility in 1978 and 1979. Fertilizer applied in early spring and after the harvest of the first growth was 120 kg N, 90 kg P₂O₅, and 120 kg K₂O/ha.

High-digestibility (HD) and low-digestibility (LD) strains were synthesized by builking an equal weight of seed from each parent in each group. The two strains were included in an experiment with 16 other orchardgrass cultivars and evaluated in 1979 and 1980. The experimental strains were replicated twice and the other cultivars replicated 4 times in plots that were 5 m². Fertilizer was applied at the rate of 150 kg N, 113 kg P₂O₅, and 150 kg K₂O/ha. The experiment was divided into three and six harvest treatments in 1980. Because of the unbalanced data, statistical analyses were not conducted.

Correlation Coefficient Digestibility Between Years

Consistency of digestibility estimates was measured on plants from different cultivars: Akimidori, Kitamidori, Okamidori (Japan), Latar, Pennlate (U.S.A.), Chinook (Canada), Tammisto (Sweden), and Tenderbite (Netherlands). The plants were established in a spaced-plant nursery (0.5 × 0.5 m) in 1976 and evaluated from 1977 to 1980. First growth was sampled 10 days after individual ear emergence. After samplings were completed, all plants were harvested uniformly in late June. Second growth was sampled at one time in late August.

RESULTS

Evaluation of Selected Plants and Their Progenies

Although the parents differed in digestibility by more than 12% when selections were made, differences in digestibility were only 4.8% in 1978 and 1979 (Table 1). Differences between the progenies of HD and LD plants were smaller than those between the parents. Larger differences in digestibility were observed in the August than

The author is an Examiner in the Ministry of Agriculture, Forestry and Fisheries and a former Research Agronomist at Hokkaido National Agricultural Experiment Station.

Table 1. Digestibility of selected plants and their progenies under space-planted conditions.

Generation	Group	Number evaluated	Digestibility (%)		
			August 1978	June 1979	August 1979
Parents	HD	4	66.2	58.5	67.1
	LD	5	61.4	57.2	62.3
	HD-LD	—	4.8*	1.3	4.8**
Progenies	HD	40	63.9	58.5	62.3
	LD	38	61.6	57.0	59.0
	HD-LD	—	2.3**	1.5**	3.3**

* P<0.05, ** P<0.01.

in the June harvests. Selection for increased digestibility of second-harvest forage increased the digestibility of first-harvest forage.

Realized heritabilities were estimated as $h_{BN}^2 = (M_1' - M_2')/(M_1 - M_2)$, where M_1 and M_2 are means of digestibility of selected plants in each HD and LD group and M_1' and M_2' are means of progeny plants in each HD and LD group respectively. Estimates, made only for the second-growth forage, were 0.48 in 1978 and 0.42 in 1979.

Evaluation of Experimental Strains

Digestibility of the HD strain was greater than that of any other entry in the trial in each of three harvest treatments (Table 2). Some commercial cultivars had digestibility values that were less than those of the LD strain.

Repeatability of Digestibility Estimates

In the first growth, positive correlations of digestibility estimates were obtained between 2 years among 1978, 1979, and 1980 (Table 3). However, correlations of 1977 estimates with those of the other 3 years were all negative. This fact indicates that selection in a certain direction may result in negative response in some years. High positive correlations were obtained for all combinations with second-harvest estimates.

DISCUSSION

Genotypic differences in maturity affect estimates of digestibility of first-harvest forage. If all genotypes are harvested on the same day, late-maturing genotypes have higher digestibility than early-maturing ones (Carlson, 1974). However, late-maturing genotypes have lower maturity when all are harvested at ear emergence (Dent and Aldrich 1966). Associations between maturity and digestibility probably caused the negative correlations with measurements made in 1977 in my experiment. The late-maturing genotypes flowered earlier than usual in 1977 (Saiga 1979), and this early flowering significantly affected their normal digestibility values. Christie (1977) reported genetic differences for in-vitro digestibility of first-harvest bromegrass (*Bromus inermis* L.) and orchardgrass forage, but he did not obtain a significant response to selection. Environmental effects and differences caused by variation in maturity make selection for improved digestibility of first-harvest forage difficult.

Repeatable high estimates of digestibility on second-

Table 2. Comparison of digestibility of two experimental strains with 16 other cultivars.

Strain or variety	Digestibility (%)			
	4 annual cuts, 1979	3 annual cuts, 1980	6 annual cuts, 1980	Average of 3 treatments
HD	66.8	61.8	68.3	65.6
LD	63.2	59.4	66.0	62.9
Kitamidori	63.4	59.1	66.0	62.8
Range of the other 15 varieties†	62.5–66.2	57.5–60.7	65.1–67.4	62.2–64.7

†Okamidori, Frontier, Hayking, Hokuren-imp., Hokkaido (Japan), Masshardy, Crown, Dayton, Napier, MTC 4-5 (U.S.A.), Dorise (Netherlands), Phyllox (Denmark), Kay (Canada), Frode (Sweden) and Kamp (Germany).

Table 3. Correlation coefficients between years for digestibility of the first and the second growth.

Year	Correlation coefficient†							
	First growth				Second growth			
	1977	1978	1979	1980	1977	1978	1979	1980
1977		−0.68	−0.22	−0.46		0.78*	0.68	0.82*
1978	−0.26**		0.68	0.83*	0.44**		0.77*	0.88**
1979	−0.01	0.34**		0.76*	0.53**	0.43**		0.82*
1980	−0.06	0.25*	0.31**		0.56**	0.44**	0.50**	

† Upper-right half and lower-left half are correlations based on cultivars (n = 8) and individual plants (n = 96), respectively.
*P<0.05, **P<0.01.

harvest forage could be obtained, however. Genotypes selected for improved digestibility of second-harvest forage also had higher digestibility for first-harvest forage. Therefore, it is suggested that selection for improved digestibility in orchardgrass be conducted on second-harvest forage within gene pools of known maturity.

LITERATURE CITED

Carlson, I.T. 1974. Correlations involving *in vitro* dry matter digestibility of *Dactylis glomerata* L. and *Phalaris arundinacea* L. Proc. XII Int. Grassl. Cong. 3:732–738.

Christie, B.R. 1977. Effectiveness of one cycle of phenotypic selection for *in vitro* digestibility in bromegrass and orchardgrass. Can. J. Pl. Sci. 57:57–60.

Christie, B.R., and D.N. Mowat. 1968. Variability of *in vitro* digestibility among clones of bromegrass and orchardgrass. Can. J. Pl. Sci. 48:67–73.

Cooper, J.P., M.A. Tilley, W.F. Raymond, and R.A. Terry. 1962. Selection for digestibility in herbage grasses. Nature 196:1276–1277.

Dent, J.W., and D.T.A. Aldrich. 1966. The *in vitro* digestibility of herbage species and varieties and its relationship with cutting treatment, stage of growth and chemical composition. Proc. X Int. Grassl. Cong., 419–424.

Saiga, S. 1979. Application of estimated digestibility from some plant characters to individual selection of orchardgrass, *Dactylis glomerata* L. J. Japan Grassl. Sci. 125:97–102.

Saiga, S., and S. Hojito. 1977. *In vitro* digestibility of orchardgrass clones in relation to some plant characters. J. Japan. Grassl. Sci. 23:177–182.

A Simple Method of Evaluating Forage Digestibility and Its Application to Sorghum Breeding

I. TARUMOTO and Y. MASAOKA

National Grassland Research Institute, Tochigi, Japan

Summary

Breeding to improve digestibility of forage sorghum requires methods for rapid evaluation of the digestion rate. A simple cellulase digestion (SCD) method, in which in-vitro dry-matter digestibility (IVDMD) of the leaf samples, kept almost in their original form, is determined by a one-stage procedure, has been developed. In this report, the applicability of this one-stage SCD method to the evaluation of digestibility of various sorghums was examined.

IVDMD values of the top fully expanded leaves of each of 22 sorghum cultivars sampled simultaneously at the early heading stage of CK-60 were determined both by the weighing method of the SCD and by the two-stage sodium sulfite–cellulase method (experiment 1). The correlation coefficient between IVDMD values by the above methods was r = 0.726***. Sweet Sudan with nonglossy leaf and tan plant color, bmr-18 with brown midrib, bl-Sudan with bloomless trait, and Tan GR-3 with tan plant color were selected as the cultivars with high digestibility by the SCD method as well as by the two stage method. These results suggest that the SCD method may be used to evaluate the digestibility of cultivars in the selection of recombinant parents.

The top fully expanded leaves of each of 24 sorghum cultivars and one corn cultivar were taken simultaneously at each of four leaf stages. Their scores and IVDMD values were determined by the scoring and weighing methods of the SCD (experiment 2). The scores of the cultivars correlated well with each other and with the IVDMD values at each of four sampling times. F$_3$-5-8 (grain sorghum) with brown midrib and bloomless trait and F$_3$-2-38(sudangrass) with brown midrib and nonglossy leaf were confirmed as the lines with high digestibility by the above two methods. These results suggest that the SCD method could be used to evaluate the digestibility of breeding lines in the early generations.

The SCD method thus appears applicable to the breeding program of improving forage quality in sorghum. Application of the SCD method to evaluation of digestibility both of collected cultivars in the selection of recombinant parents and of the breeding lines in early generations is considered to be most effective.

KEY WORDS: forage sorghum, *Sorghum bicolor* (L.) Moench., digestibility, forage quality, breeding.

INTRODUCTION

Recent demands of dairy and beef-cattle farmers in Japan require that sorghum breeders improve both yield potential and forage qualities, such as digestibility. To breed forage sorghum with high digestibility, a method for rapid evaluation of the digestion rate or potential value of plant selection is necessary. For this purpose, a simple cellulase digestion (SCD) method, in which in-vitro dry-matter digestibility (IVDMD) of leaf samples kept almost in their original form is determined by a one-stage technique, has been developed (Tarumoto and Masaoka 1978, 1980; Masaoka and Tarumoto 1979, 1980). In this paper the applicability of this one-stage SCD method to the evaluation of digestibility of various sorghums is examined.

MATERIALS AND METHODS

Experiment 1 (weighing method, SCD): 22 sorghum cultivars were grown in the field during the summer of 1978 (Fig. 1). The top fully expanded leaves of each cultivar were taken simultaneously on 1 August (early heading stage of CK-60). From each cultivar, two sets of five

Fig. 1. Relationship between in-vitro dry-matter digestibility (IVDMD) of leaves determined by the simple cellulase digestion method and by the two-stage sodium sulfite-cellulase method.

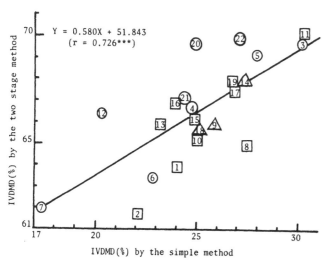

IVDMD(%) by the simple method

Notes: (1) Cultivars tested: 1. bloomless (*bl*) CK-60, 2. Bloom (*B*) CK-60, 3. *bl*-Sudan, 4. *B*-Sudan (GA337), 5. *bmr*-18, 6. N-18, 7. Tan GR-2 (Red), 8. Sudax-11, 9. Sendachi (605A × Sweet Sudan), 10. 605A, 11. Sweet Sudan, 12. Hiromidori (390A × Regs. Hegari), 13. 390A, 14. Regs. Hegari, 15. Chugoku ko-4 (378A × Zairai-Tokin), 16. 378A, 17. Zairai-Tokin, 18. Tozan ko-2, 19. Senkinshiro, 20. Brandes, 21. Sugar Drip, 22 Tan GR-3 (GA).
(2) Stages of the cultivars at 1 August sampling time: △ – flowering, □ – heading, ○ – before heading.
(3) Cultivars with sweet stalk.

The authors are both Research Scientists with the National Grassland Research Institute.

7-cm-long leaf sections of the center part of the leaf blades were prepared, dried, and weighed. They were digested without shaking in 1% cellulase in 0.1-M acetate buffer (pH 4.0) for 48 hours at 40°C. After digestion, each set of leaf sections was washed, dried, and weighed, and IVDMD values were calculated. The IVDMD values of dried and ground leaf samples from the same parts of leaves used in the SCD method were determined by the sodium sulfite–cellulase method developed by Abe et al. (1972).

Experiment 2 (scoring method, SCD): 24 sorghum cultivars and one corn cultivar were grown in the field during the summer of 1979 (Table 1). The top fully expanded leaves of each cultivar were taken simultaneously at each of four sampling times (Table 1). From each sample, two sets of five or ten fresh leaf sections were digested by the same digestion procedure (SCD) as described in experiment 1. After digestion, the leaf sections were stained in 1% solution of Safranin O. The stained sections were scored from 0 (nonstained or just a few flecks) to 5 (90% to 100% of leaf area stained red).

RESULTS

Experiment 1. The relationship between IVDMD values of 22 cultivars determined by the simple and the two-stage methods is shown is Fig. 1. Although IVDMD values of a few cultivars were overestimated or underestimated by the SCD method, the correlation coefficient between IVDMD values obtained from the two methods was positively significant ($r = 0.726$, $P < 0.001$). The range in IVDMD values determined by the SCD method was greater than that for IVDMD values determined by the two-stage method by about 5 percentage units of IVDMD. The larger magnitude of difference among digestibilities of cultivars permits better separation among entries by the SCD method, which also uses simpler procedures and requires smaller samples than the two-stage method. The cultivars ranked high in digestibility by the SCD method were also ranked high by the two-stage method. Sweet Sudan with nonglossy leaf(*Gl*) and tan plant color(*pp*), bmr-18 with brown midrib(*bmr*), *bl*-Sudan with bloomless trait(*bm*), and Tan GR-3 with tan plant color(*pp*) were selected as cultivars with high digestibility. The above results suggest that the SCD method would be useful for evaluating the digestibility of collected and/or introduced cultivars in the selection of recombinant parents.

Experiment 2. The correlation coefficients between scores and IVDMD values of 25 cultivars at each of four sampling times and the cultivars with high digestibility selected by the SCD method are shown in Table 1. The scores at each of the first three sampling times (eighth to flag leaf stages) correlated well with each other and with IVDMD values at each of the four sampling times. The cultivars ranked high in digestibility by the SCD score method at the first three sampling times had high IVDMD values at each of the four times. The F_3 grain sorghum line with brown midrib and bloomless trait

Table 1.   Correlation coefficients between scores and/or IVDMD values of 25 cultivars at each of four sampling times (I, II, III, IV), and the cultivars selected by scores or IVDMD values.

		Score[1]				IVDMD(%)[1]			
		I	*II*	*III*	*IV*	*I*	*II*	*III*	*IV*
Score	II	.85***[2]							
	III	.72***	.89***						
	IV	.24	.36	.49*					
IVDMD(%)	I	.88***	.66***	.53**	−.46*				
	II	.89***	.81***	.79***	.29	.82***			
	III	.75***	.64***	.72***	.47*	.68***	.83***		
	IV	.71***	.63***	.68***	.60**	.53**	.72***	.89***	
Cultivars selected at each treatment[3]		S>3.0	S>2.0	S>0.9	S>1.0	D>60%	D>40%	D>34%	D>34%
		1 5 8	1 5	1 3 5	1 5 9	5 8	5 11	5 16	1 5
		13 16	14 16	16 18	11 16	13 14	14 16	18 19	18 19
		18 19	18 19	19 20	18 19	16 17	18 19	20 21	20 21
		20 21	20 24	24	21 *23*	18 19	20 21	24	24
		24			25	20 21	24		
						24			

[1]Date taken samples and the leaf stages of CK-60: I:June 25, 8th; II:July 7, 11th; III:July 20, 14th; IV:July 31, 17th(Heading).

[2]*, **, and *** indicate statistical significance at the 5%, 1%, and 0.1% levels, respectively.

[3]Cultivars tested: 1.bloomless CK-60, 2.Bloom CK-60, 3.bmr-18, 4.N-18, 5.F₃-5-8(*bm,bmr*), 6.Tan GR-2, 7.605B, 8.390B, 9.3048B, 10.Regs.Hegari, 11.Senkinshiro, 12.Af.Millet, 13.Mn752, 14.Brandes, 15.Tracy, 16.*bl*-Sudan, 17.*B*-Sudan(GA337), 18.F₃-2-38(*bmr,G1*), 19.Sweet Sudan, 20.Sudan GR-1, 21.Sendachi, 22.Hiromidori, 23.Tozan ko-2, 24.NB7035, 25.Takanewase(Maize).

(F₃–5–8) and the F₃ sudangrass line with brown midrib and nonglossy leaf (F₃–2–38) were confirmed as the lines with high digestibility. The above results suggest that the SCD method would be useful for evaluating the digestibility of the breeding lines in the early generations.

DISCUSSION

Selection of parent plants by the tests of digestibility and phenotypic characters, following seed multiplication, is necessary for breeding of high-quality annual forage crops with few tillers, like sorghum, corn, and millets. Rapid and simple evaluation of digestibility of a number of small samples by using the SCD method proved to be possible. Sampling of the top fully expanded leaf/plant in the SCD method did not hinder the phenotypic expression of plants and their seed-setting potential. Therefore, the SCD method appears applicable to practical programs of breeding for improved forage quality. Application of the SCD method is considered to be most effective for evaluating digestibility both of collected cultivars for selecting recombinant parents and of breeding lines for selection in the early generations.

Among the major genes evaluated in experiment 1, *bmr,* the brown midrib gene, was reported to show the pleiotropic effect on digestibility by Porter et al. (1978) and Tarumoto and Masaoka (1978), and *bm,* the bloomless gene, to be related to forage quality and digestibility by Cummins and Sudweeks (1976), Hanna et al. (1974), and Tarumoto and Masaoka (1978, 1980). Tarumoto et al. (1981) reported that the nonglossy leaf gene (*Gl*)

showed higher digestibility than the glossy leaf (*gl*), the gene detected by Tarumoto (1980). Utilization of the genotypes with the above major genes seems to be most promising for improving quality of sorghum forage.

LITERATURE CITED

Abe, A., S. Horii, and K. Kameoka. 1972. Development and application of cellulase hydrolysis for predicting digestibility. III. Application of two step method for predicting dry matter digestibility. Jap. J. Zootech. Sci. 43:175–180.

Cummins, D.G., and E.M. Sudweeks. 1976. *In vivo* performance of bloom and bloomless sorghum forage. Agron. J. 68:735–737.

Hanna, W.W., W.G. Monson, and G.W. Burton. 1974. Leaf surface effects on *in vitro* digestion and transpiration in isogenic lines of sorghum and pearl millet. Crop Sci. 14:837–838.

Masaoka, Y., and I. Tarumoto. 1979. Influence of cellulase concentrations and incubation periods in *in vitro* digestibility of fresh sorghum leaf by simple cellulase digestion method. J. Jap. Grassl. Sci. 24:337–344.

Masaoka, Y., and I. Tarumoto. 1980. Comparison of several variations in the simple *in vitro* method using cellulase digestion of sorghum leaf section. (In Japanese). J. Jap. Grassl. Sci. 26:231–232.

Porter, K.S., J.D. Axtell, V.L. Lechtenberg, and V.F. Colenbrander. 1978. Phenotype, fiber composition, and in vitro dry matter disappearance of chemically induced brown midrib (*bmr*) mutants of sorghum. Crop Sci. 18:205–208.

Tarumoto, I. 1980. Inheritance of glossiness of leaf blades in sorghum, *Sorghum bicolor* (L.) Moench. Jap. J. Breed. 30:237–240.

Tarumoto, I., and Y. Masaoka. 1978. Digestibility of fresh sorghum leaf by simple cellulase digestion method. J. Jap. Grassl. Sci. 24:1–9.

Tarumoto, I., and Y. Masaoka. 1980. Influence of length of cut and quantity of leaf sample per test tube on digestibility determined by digesting dried leaves in cellulase solution. J. Jap. Grassl. Sci. 26:233–235.

Tarumoto, I., Y. Masaoka, and K. Isawa. 1981. Pleiotropic effects of glossiness in sorghum. 1. Effects on digestibility and leaf blight resistance. Jap. J. Breed. 31 (Suppl. 1):166–167.

Fermentation Potential of Vegetation from Various Herbaceous Plants Following Enzymatic Hydrolysis

D.H. SMITH, J.C. LINDEN, and R.H. VILLET
Colorado State University, Fort Collins, Colo., and
Solar Energy Research Institute, Golden, Colo., U.S.A.

Summary

Vegetation from herbaceous plants has been suggested as a possible feedstock for production of liquid fuels. Our objective was to determine the fermentation potential of vegetation from various herbaceous species following enzymatic hydrolysis.

Preliminary evidence with sorghums (*Sorghum bicolor* [L.] Moench) suggested that enzymatic cellulose hydrolysis could be improved either by using brown-midrib (*bmr*) mutants or by ensiling forage materials. Subsequent studies indicated that rates of cellulose conversion to glucose of *bmr* mutants were not always superior to those of their normal counterparts even though the lignin content of these mutants was usually lower. Averages from *bmr* and normal sorghum harvested on two different dates indicated that from 29.4% to 36.5% of the total dry matter could be enzymatically converted to fermentable hexoses. The contribution of nonstructural carbohydrates to the total ranged from 62% to 65%. The noncrop species were substantially higher in lignin content than the sorghums, corn (*Zea mays* L.) or pearl millet (*Pennisetum americanum* [L.] Leeke). Cellulose hydrolysis of these species ranged from 13.5% of the cellulose present for kochia (*Kochia scoparia* [L.] Schrad.) to 25.8% for Jerusalem artichoke (*Helianthus tuberosus* L.). The cellulose conversion rates for the grasses ranged from 16.7% for corn at physiological maturity to 30.4% for forage sorghum harvested during seed filling.

Variability associated with species, varieties, stage of maturity, and ensiling produced substantial variation in enzymatic cellulose hydrolysis of herbaceous plant vegetation. In addition, quantitative estimates of lignin content were not always associated with the variation observed in cellulose hydrolysis.

KEY WORDS: cellulose, enzymatic hydrolysis, fermentable carbohydrates, lignin, *Sorghum bicolor* (L.) Moench, brown-midrib.

INTRODUCTION

Liquid-fuel shortages in the U.S. have stimulated investigation of various agricultural crops as potential sources of fermentation feedstock for ethanol production. Sources of fermentable carbohydrate from vegetative components of herbaceous species include nonstructural carbohydrates as well as structural polysaccharides.

Our objective was to determine the fermentation potential of various herbaceous species, using enzyme pretreatment alone. A preliminary screening involved evaluation of a limited number of samples of sorghum and sudangrass (*Sorghum bicolor* [L.] Moench). Results of further systematic studies of brown-midrib (*bmr*) sorghums, sudangrasses, and other herbaceous species are also reported.

MATERIALS AND METHODS

Preliminary Screening

Methods used in the 1979 preliminary screening have been previously described (Linden et al. 1980).

Bmr Study

Seed of *bmr* sorghum lines (*bmr*-6, *bmr*-18, and their normal counterparts designated as N-6 and N-18, respectively) and two *bmr*-6 sudangrasses (Piper *bmr*-6 and Greenleaf *bmr*-6) were obtained from Purdue University.

D.H. Smith is an Assistant Professor of Agronomy and J.C. Linden is a Research Assistant Professor with the Department of Agricultural and Chemical Engineering at Colorado State University. R.H. Villet is Program Director for the Biological/Biochemical Conversion Group at the Solar Energy Research Institute.

This is a joint contribution of the Departments of Agronomy and Agricultural and Chemical Engineering, Colorado State University, and the Solar Energy Research Institute. Research was supported in part by Colorado State University Experiment Station, Solar Energy Research Institute, and Colorado Gasohol Commission. Scientific Series Manuscript No. 2626.

The authors are grateful to Drs. J.D. Axtell, R.P. Cantrell, and V.L. Lechtenberg for providing the seed of the bmr *sorghum mutants.*

Each of these entries along with Piper (Piper-N) and Greenleaf (Greenleaf-N) were established in 3 replicate plots using a randomized complete-block design at Fort Collins, Colorado, in May 1980.

Three uniform plants within each sudangrass plot were harvested at heading on 31 July and 15 September. Sorghum entries were harvested on 15 September (approximate date of anthesis for each entry) by collecting 3 plants within each plot.

Fresh plant samples were coarsely ground, subsampled, and frozen in liquid nitrogen. Frozen samples were taken to the laboratory, allowed to thaw for 2 hours at room temperature, dried at 100°C for 1 hour and 70°C for an additional 48 hours, and ground to pass a 20-mesh screen in a Wiley mill. Ground samples were subsequently subjected to analysis for fiber composition, nonstructural carbohydrate content, and cellulose hydrolysis.

Herbaceous Species Screening Study

Commercial hybrids of corn, forage sorghum, and sorghum-sudangrass; open-pollinated selections of Tift pearl millet (*Pennisetum americanum* [L.] Leeke), Jerusalem artichoke (*Helianthus tuberosus* L.), pigweed (*Ameranthus retroflexus* L.), and kochia (*Kochia scoparia* (L.) Schrad.) were established in May 1980 at the Central Great Plains Field Station, Akron, Colorado. Plot design was a randomized complete block with 4 replications.

Plant material from plots of each species was collected from a 5-m² area within each plot during the period from 28 August to 3 September. Grain-producing ears of corn, forage sorghum, sorghum-sudangrass, and pearl millet were removed prior to harvest and discarded. Subsamples were collected and processed as described for the *bmr* study and subjected to analysis for fiber composition and cellulose hydrolysis.

Laboratory Analysis

The cellulase/cellobiase enzyme preparations and procedures used to determine cellulose hydrolysis have been previously described (Linden et al. 1980). Net cellulose hydrolysis, expressed as percentage of anhydride glucose, was calculated by subtracting buffer-extractable glucose from glucose content of the enzyme hydrolysate and multiplying by 0.9. When glucose derived from cellulose hydrolysis was used to calculate total fermentable carbohydrate, glucose was expressed in the hydrated form. Cellulose conversion rates were calculated by dividing net hydrolysis values by apparent cellulose content and multiplying by 100.

Nonstructural carbohydrates were determined using amyloglucosidase (Sigma) enzyme digestion as described by Greub and Wedin (1969). Following mild acid hydrolysis, aliquots of the enzyme incubation filtrate were analyzed for reducing sugar content with 3,5-dinitrosalicylic acid (Miller 1959), using glucose as a standard. Apparent cellulose and either permanganate lignin or acid detergent lignin (ADL) estimates were obtained, using procedures described by Goering and Van Soest (1970).

RESULTS AND DISCUSSION

Preliminary Screening

A detailed report of the results from these studies has been presented previously (Linden et al. 1980). Data obtained from forage sorghum indicated that with advancing maturity enzymatic cellulose hydrolysis declined substantially and lignin content increased. In addition, enzymatic conversion of cellulose to glucose in ensiled forage sorghum was much higher than in unensiled samples. Since the lignin content of the silage was equal to that of the mature material, factors resulting from ensiling apparently improved accessibility of enzymes to the fibers.

The *bmr* mutants of sorghum exhibited the best potential for conversion of cellulose to glucose. Conversion rates for the 3 mutant sorghums ranged from 63.0% to 80.5% of the total cellulose present, higher rates than that obtained from the sudangrass (50%). These higher conversion rates for the *bmr* sorghums were not totally attributable to differences in lignification, since the *bmr*-6 mutant was higher in lignin than the sudangrass harvested at a similar stage of development.

Bmr Study

All samples from this study were collected when plants were more mature than the plants sampled in 1979. Maturity was reflected in both cellulose content (higher in 1980) and cellulose conversion rates (lower in 1980). Results obtained from the sudangrass entries harvested on 31 July showed no differences in cellulose hydrolysis between *bmr* and normal entries (Table 1). This was true even though the ADL values indicated lower lignin content in the Greenleaf *bmr*-6 line as compared with Greenleaf-N.

Analysis of samples harvested on 15 September revealed greater enzymatic hydrolysis of cellulose to glucose for *bmr* mutants in three of the four comparisons between mutant and normal lines (Table 1). However, increased cellulose conversion resulted in higher glucose yields in only one of the comparisons (*bmr*-18 to N-18) due to the lower cellulose content of the *bmr*-6 mutants of Greenleaf and Piper than those of their normal counterparts. In all comparisons showing increased cellulose hydrolysis for the *bmr* mutant as compared with the normal counterpart, ADL content was lower in the mutant line. These results are compatible with findings reported by Porter et al. (1978).

Averages from all entries indicated that 36.5% and 29.4% of the total dry matter from the 31 July and 15 September harvests, respectively, could be enzymatically converted to fermentable carbohydrate (Table 1). Where differences in total fermentation potential were noted between *bmr* mutants and their normal counterparts, the potentials of the *bmr* sorghums were higher. These differences, however, were largely attributable to higher quantities of carbohydrate from nonstructural sources rather than from cellulose hydrolysis. Fermentable carbohydrate

Table 1. Glucose derived from cellulose hydrolysis, lignocellulose composition, theoretical conversion of cellulose to glucose, non-structural carbohydrate (CH_2O) content, and total fermentable CH_2O for normal (N) and brown-midrib (*bmr*) sudangrass and sorghum lines. Data are from two harvest dates in 1980.

Harvest Date/Entry	Glucose Yield[1]	Apparent Cellulose	ADL[2]	Cellulose[3] Conversion	Non-Struct. CH_2O	Total Ferment. CH_2O[4]
	---------------- % of dry matter ----------------			%	----- % of dry matter -----	
July 31						
Greenleaf *bmr*-6	11.3	29.9	2.4	38.4	23.8	36.4
Greenleaf-N	10.4	32.2	4.2	32.6	19.8	31.4
Piper *bmr*-6	11.2	28.1	2.4	39.9	26.9	39.3
Piper-N	13.5	30.2	3.3	44.7	23.9	38.9
LSD (P = .05)	N.S.	2.7	1.0	N.S.	2.9	5.8
September 5						
Greenleaf *bmr*-6	11.6	29.6	2.3	39.1	16.0	28.9
Greenleaf-N	10.8	32.1	3.5	33.8	16.2	28.2
Piper *bmr*-6	9.5	31.1	2.9	30.5	19.2	29.8
Piper-N	9.7	37.9	4.1	25.7	12.6	23.4
bmr-6	9.8	27.6	3.1	35.3	23.8	34.7
N-6	8.4	25.4	3.8	33.3	20.2	29.5
bmr-18	11.2	28.7	2.5	39.2	19.3	31.7
N-18	9.5	29.8	3.3	31.7	18.4	29.0
LSD (P = .05)	1.4	2.2	0.8	4.3	4.2	4.1

[1]Calculated as total glucose after enzymatic hydrolysis minus buffer extractable glucose. Glucose yield expressed as anhydride form ($C_6H_{10}O_5$) to correct for hydrolysis.

[2]Acid detergent lignin

[3](Glucose Yield ÷ Apparent Cellulose) × 100

[4]Hydrated glucose from cellulose hydrolysis plus nonstructural CH_2O

of nonstructural origin was substantially higher than that derived from cellulose hydrolysis in all entries at both harvest dates. Nonstructural carbohydrates accounted for 65% and 62% of the total fermentable hexoses in materials harvested on 31 July and 15 September, respectively.

Herbaceous Species Screening Study

Substantial variation in lignin content and cellulose conversion was observed among the different species in this study (Table 2). Comparisons within the grass species

indicated that corn and pearl millet had higher ADL content and lower cellulose hydrolysis values than forage sorghum and sorghum-sudangrass. These findings were not surprising, since corn and pearl millet were harvested at more mature stages of development than the sorghums. Cellulose conversion rates for corn and pearl millet were of magnitudes similar to those for forage sorghum and sorghum-sudangrass.

Among the noncrop species, cellulosic conversion and resulting glucose yields upon hydrolysis were lower for kochia than for Jerusalem artichoke and pigweed. How-

Table 2. Glucose derived from cellulose hydrolysis, lignocellulose composition, and theoretical conversion of cellulose to glucose for vegetative components of various species grown at Akron, Colorado in 1980.

Species	Stage of Harvest	Glucose Yield	Apparent Cellulose	ADL	Cellulose Conversion
		---------------- % of dry matter ----------------			%
Corn	Physiologic maturity	5.9	35.4	3.6	16.7
Forage Sorghum	Seed filling	7.7	25.4	2.2	30.4
Sorghum-sudangrass	Seed filling	7.2	26.2	2.8	27.7
Pearl millet	Physiologic maturity	7.3	36.1	4.1	20.4
Jerusalem-artichoke	Tubers initiated	6.4	24.8	8.7	25.8
Pigweed	Seed filling	6.8	34.4	9.1	20.0
Kochia	Seed filling	4.3	32.8	7.0	13.5
LSD (P = .05)		1.0	3.5	1.3	4.5

See footnotes, Table 1 for explanation of variables.

ever, the ADL content of kochia was lowest of these three species.

Comparisons between the grasses and the noncrop species further illustrate the lack of relationship between lignin content and cellulose hydrolysis (Table 2). Noncrop species contained substantially higher quantities of lignin than the grasses did. However, cellulose conversion rates for the noncrop species were either comparable to or greater than those for the grasses in several instances. For example, cellulose conversion rates for the Jerusalem artichoke and sorghum-sudangrass were similar, yet the ADL content of Jerusalem artichoke was more than 3 times that of sorghum-sudangrass. Further comparison revealed that cellulose hydrolysis of the Jerusalem artichoke was more than 50% higher than that of corn, while lignin content of the former was more than twice as great as in the corn.

These results suggest that the effect of the lignin-hemicellulose matrix on enzymatic degradation of cellulose is not simply related to the amount of lignin present. Reasons for this are unclear but probably involve the lack of methodology for accurate chemical characterization of plant cell walls. Another factor could be variation in the degree of crystallinity of cellulose (Cowling and Kirk 1976).

LITERATURE CITED

Cowling, E.B., and T.K. Kirk. 1976. Properties of cellulose and lignocellulosic materials as substrate for enzymatic conversion processes. Biotechnol. and Bioeng. Symp. 6:95–123.

Goering, H.K., and P.J. Van Soest. 1970. Forage fiber analysis (apparatus, reagents, procedures, and some applications). U.S. Dept. Agric., Agric. Handb. 379.

Greub, L.J., and W.F. Wedin. 1969. Determination of total available carbohydrates in forage legume roots by extraction with takadiastase, amyloglucosidase, or sulfuric acid. Crop Sci. 9:595–598.

Linden, J.C., A.R. Moreira, D.H. Smith, W.S. Hedrick, and R.H. Villet. 1980. Enzymatic hydrolysis of the lignocellulosic component from vegetative forage crops. Biotechnol. and Bioeng. Symp. 10:199–212.

Miller, G.L. 1959. Use of dinitrosalicylic acid reagent for determination of reducing sugar. Anal. Chem. 31:426–428.

Porter, K.S., J.D. Axtell, V.L. Lechtenberg, and V.F. Colenbrander. 1978. Phenotype, fiber composition, and *in vitro* dry matter disappearance of chemically induced brown midrib (*bmr*) mutants of sorghum. Crop Sci. 18:205–208.

Cyanogenesis in Dhurrin-Containing Forage Grasses

H.J. GORZ, F.A. HASKINS, and K.P. VOGEL
USDA-SEA-AR and University of Nebraska, Lincoln, Nebr., U.S.A.

Summary

Cyanogenesis, the ability of organisms to produce hydrocyanic acid (HCN), has been reported in approximately 1,000 species of plants representing 250 genera and 80 families, but the identity of the cyanogenic compound(s) involved has been established for less than 100 of these species. The first cyanogenic compound identified in the grass family was dhurrin, which was isolated from sorghum plants in 1902. Many other grasses were subsequently reported to be cyanophoric, but until recently *Sorghum* remained the only grass genus in which the presence of dhurrin had been demonstrated unequivocally.

A precise, rapid, and highly specific spectrophotometric assay for dhurrin in seedling leaves was developed at Nebraska in 1977, primarily for use in breeding for low dhurrin content in sudangrass and sorghums. This procedure has been used successfully in selecting sudangrasses with reduced levels of hydrocyanic acid potential (HCN-p). Two cycles of divergent selection for high and low HCN-p in Greenleaf sudangrass have resulted in an average realized heritability of 0.26 and an average broad-sense heritability estimate of 0.84. The high- and low-HCN-p populations each differed from Greenleaf by about 20%. Dry-matter yields for the high- and low-HCN-p populations, obtained during one growing season after only one cycle of selection, were 7,440 and 6,680 kg/ha, respectively, compared with 6,910 kg/ha for unselected Greenleaf.

The spectrophotometric assay also has been used to screen seedlings of 72 grasses representing 39 species, 14 genera, and 2 tribes for the presence of dhurrin. Among genera other than *Sorghum*, dhurrin was found only in the genus *Sorghastrum*. Levels were in the same range as for *Sorghum* seedlings considered to be high in dhurrin content.

The toxicity of a cyanophoric forage grass is influenced by various factors, one of which is the ability of the grazing animal to detoxify the HCN it encounters. In ruminants, HCN is rapidly detoxified in the rumen and liver by reactions with sulfide or cystine, and the resulting thiocyanate is excreted in the urine. The low rates of gain that are sometimes observed when cattle or sheep graze pure stands of sorghum or sudangrass may, in some cases, be explained in terms of an induced sulfur deficiency in the animal due to the priority requirement of sulfur to detoxify HCN. Therefore, sulfur supplements may reduce the potential for HCN poisoning, replace sulfur lost in detoxification of HCN, and insure an adequate dietary level of sulfur for animal needs. Recent studies have shown that released cyanide also may protect animals from the toxic effects of selenium by forming a compound with selenium that is apparently readily excreted in the urine. However, this same reaction may cause selenium deficiency in livestock in areas where selenium levels are low.

KEY WORDS: cyanide, prussic acid, HCN, indiangrass, sorghum, sudangrass, selenium, sulfur.

INTRODUCTION

Cyanogenesis, the ability of organisms to produce hydrocyanic acid (HCN), has been reported in approximately 1,000 plant species representing 250 genera and 80 families, but the identity of the cyanogenic compound(s) involved has been established for less than 100 of these species (Conn 1978). The first cyanogenic compound isolated from a member of the Gramineae was dhurrin [(S)-*p*-hydroxymandelonitrile β-D-glucopyranoside] (Tjon Sie Fat 1977), which was isolated from sorghum (*Sorghum bicolor* [L.] Moench) plants in 1902 by Dunstan and Henry (1902). Many other grasses were subsequently reported to be cyanophoric (containing cyanogenic compounds), and it was generally assumed that these grasses contained dhurrin or an amygdalin-type compound, but until recently *Sorghum* was the only grass genus in which the presence of dhurrin had been shown unequivocally (Tjon Sie Fat 1977). Although the danger of cyanide poisoning of livestock grazing sudangrass or sorghums is greatly reduced by proper management practices, losses still may occur even when the recommended practices are followed (Hunt and Taylor 1976). Therefore, low hydrocyanic acid potential (HCN-p) is an important objective in breeding programs involving sudangrass and sorghum-sudangrass hybrids.

SPECTROPHOTOMETRIC ASSAY OF HCN-p

To facilitate breeding for low HCN-p in sorghums, a simple and rapid procedure was developed for the determination of HCN-p (Gorz et al. 1977). The procedure, which was based in part on earlier studies by Akazawa et al. (1960), consists of excising and weighing first leaves of sorghum seedlings, autoclaving the leaves in water, diluting the extracts in 0.1 N NaOH, and reading the absorbance of the resulting solutions at 330 nm. The procedure is effective because dhurrin is readily extracted and hydrolyzed when the tissue is autoclaved in water; HCN, glucose, and *p*-hydroxybenzaldehyde (*p*-HB) in equimolar amounts are the products of this hydrolysis (Akazawa et al. 1960); and *p*-HB in alkaline solution has a strong absorption maximum at 330 nm. When sudan-

H.J. Gorz is a Supervisory Research Geneticist with USDA-SEA-AR and Professor of Agronomy at the University of Nebraska, F.A. Haskins is George Holmes Professor of Agronomy at the university, and K.P. Vogel is a Research Agronomist with USDA-SEA-AR and Associate Professor of Agronomy at the university as well.

This is a contribution of USDA-SEA-AR and the Nebraska Agricultural Experiment Station. Journal Series Paper No. 6194.

grass and sorghum seedlings were grown as described (Gorz et al. 1977), high-HCN-p lines yielded extracts whose spectra closely resembled the spectrum of *p*-HB, and the 330 nm maximum characteristic of *p*-HB also was readily apparent in extracts of low-HCN-p lines (Haskins et al. 1979a). The ranking of genotypes for HCN-p as determined by this spectrophotometric procedure agreed well with rankings based on other assay methods (Gorz et al. 1977).

In a later study (Haskins et al. 1979a), the spectrophotometric procedure was used to determine the HCN-p of first leaves of individual chamber-grown sorghum seedlings, and subsequently the same procedure was used to assay young leaves from tillers of the plants grown from these seedlings in the field. Highly significant correlation coefficients were obtained in most comparisons of seedling and tiller HCN-p. However, absorption spectra indicated that interfering constituents were more abundant in tiller extracts than in seedling extracts. It was concluded that the spectrophotometric method was most reliable for the assay of young seedlings grown under controlled conditions.

EFFECT OF NUTRIENT LEVEL ON HCN-p

Sorghum seedlings were subjected to various combinations of nutrients in solution culture, after which leaves were assayed for HCN-p (Clark et al. 1979). Increases in HCN-p were most consistently associated with increased levels of ammonium salts. This effect was not surprising, inasmuch as ammonium nitrogen can be readily incorporated into the amino group of tyrosine (and other amino acids), and Conn (1979) has indicated that the nitrile nitrogen of dhurrin is derived from the amino nitrogen of tyrosine. In a recent report from Conn's group (Kojima et al. 1979), dhurrin was shown to be localized in the vacuoles of epidermal cells of young sorghum leaves, whereas the enzymes responsible for the release of cyanide from dhurrin were localized in mesophyll cells. This physical separation of dhurrin and the degradative enzymes doubtless accounts for the fact that in healthy, intact leaves, little if any free cyanide is present.

ALTERATION OF HCN-p IN *SORGHUM* BY BREEDING

The spectrophotometric assay was used to evaluate the effectiveness of recurrent selection for increasing or decreasing HCN-p in two cycles of individual plant selection in the sudangrass cultivar Greenleaf. In cycle 1, HCN-p means of the high and low populations of Greenleaf sudangrass were higher and lower, respectively, than for the unselected lot of Greenleaf, but only the low-HCN-p population was significantly different from Greenleaf. In cycle 2, mean HCN-p values for both populations differed significantly from Greenleaf. The average realized heritability for the two cycles was 0.26, while broad-sense heritability estimates averaged 0.84. This difference indicates that nonadditive genetic effects were probably of considerable importance in determining HCN-p in the sundangrass cultivar used in these experiments. After two cycles of selection, the low- and high-HCN-p populations each differed from Greenleaf by about 20%.

Dry-matter yields for the high- and low-HCN-p populations, obtained during one growing season after only one cycle of selection, were 7,440 and 6,680 kg/ha, respectively, compared to 6,910 kg/ha for unselected Greenleaf. These preliminary results suggest that selection for low HCN-p may cause a reduction in forage yield while selection for high HCN-p may increase forage yield. Seed increases of the second-cycle high- and low-HCN-p Greenleaf populations, grown in 1980, will be used for planting additional yield tests and for establishing replicated pasture trials with beef cattle in 1981.

DHURRIN IN *SORGHASTRUM* SEEDLINGS

The spectrophotometric procedure was used to screen seedlings of a number of grasses other than sorghums for the presence of dhurrin (Haskins et al. 1979b). Among 72 nonsorghum entries representing 39 species and 14 genera in the grass tribes Andropogoneae and Tripsaceae, only entries of the genus *Sorghastrum* were found to contain dhurrin. That the compound was indeed dhurrin was shown by ultraviolet absorption spectra of the purified material before and after hydrolysis in alkali, by paper chromatographic comparison with authentic dhurrin in eight solvent systems, and by determination of the melting point and nuclear magnetic resonance spectrum of the pentaacetate of the isolated material (Gorz et al. 1979). *Sorghastrum* is thus the second grass genus in which the presence of dhurrin has been conclusively shown.

Dhurrin was found in seedlings of each of the 10 entries of indiangrass (*Sorghastrum nutans* [L.] Nash) and the 3 entries of *S. pellitum* included in the study. In a separate study, the HCN-p of seedlings of 5 indiangrass cultivars ranged from about 900 ppm for Holt to about 1,200 ppm for Llano and Oto. Dhurrin was also detected in young tillers of field-grown plants of indiangrass, but tiller extracts contained more interfering substances than seedling extracts, in agreement with previous observations on sorghum seedlings and tillers (Haskins et al. 1979a).

Indiangrass is a warm-season, tall, perennial species widely distributed in the U.S., and it is a common constituent of the hay meadows, pastures, and rangelands of the eastern Great Plains. Although the danger of grazing livestock on *Sorghum* under some conditions is well recognized, no reports have been found of livestock loss resulting from the grazing of indiangrass, even though the HCN-p levels in indiangrass seedlings were in the same range as values for *Sorghum* seedlings that are considered to be high in HCN-p. The absence of livestock loss on indiangrass may result from this grass's nearly always being grazed in pastures containing a mixture of species, whereas *Sorghum* is usually grazed in pure stands. Also, warm-season pastures usually are not grazed until the plants are past the stage when higher HCN concentration would be expected.

FACTORS INFLUENCING TOXICITY

Although the potential of a plant to produce HCN is important in determining its toxicity, Conn (1978) has called attention to other factors that also must be considered, including size and kind of consuming animal, speed of ingestion, type of food ingested simultaneously with the toxic plant, possibility of the plant's degradative enzymes remaining active in the animal's digestive tract, and ability of the animal to detoxify the HCN it encounters. A single oral dose of about 2 mg HCN/kg of body weight may be lethal for cattle and sheep, resulting in death from the general anoxic state produced by the inhibition of cytochrome oxidase. With nonfatal doses, the inhibition of cellular respiration is reversed because of removal of HCN by respiratory exchange or by metabolic detoxification (Conn 1978).

Wheeler et al. (1975) reported that HCN is rapidly detoxified in the rumen and liver of ruminants by reactions with sulfide or cystine, and the resulting thiocyanate is excreted in the urine. They also stated that the sulfur content of sorghum forage is usually low, and since 1.2 g sulfur/g HCN is required for detoxification, a significant fraction of the sulfur ingested by animals may be rendered unavailable for productive purposes. These authors believe that the low rates of gain often observed when cattle or sheep graze pure stands of forage sorghums, sorghum-sundangrass hybrids, or sudangrasses may, in some cases, be explained in terms of an induced sulfur deficiency in the animal due to the priority requirement of sulfur to detoxify HCN. Archer and Wheeler (1978) cite studies in which sheep, beef cattle, and dairy cattle responded significantly to sulfur supplementation when grazing sorghum. However, sheep grazing cyanophoric white clover did not respond to sulfur supplementation, probably because productive white clover has a sulfur content 2 to 3 times as high as that of most sorghum forage (Wheeler et al. 1975). Sulfur supplements may reduce the potential of HCN poisoning, replace sulfur lost in detoxification of HCN, and insure an adequate dietary level of sulfur for animal needs.

Recent studies by Palmer (1980) have shown an interaction of cyanide and selenium similar to that described for cyanide and sulfur. As early as 1941, workers at South Dakota State University recognized that linseed oilmeal could protect laboratory animals from selenium toxicity. Not until 1980, however, was the protective agent found to be cyanide liberated from cyanogenic glycosides present in flax. The cyanide formed a compound with selenium that apparently was readily excreted in the urine. Thus, in areas where selenium occurs in relatively high concentrations, consumption of forages such as sudangrass or sorghum that contain subtoxic levels of HCN could conceivably protect animals from selenium toxicity. In low-selenium areas, however, livestock consuming these same forages may develop selenium deficiencies.

LITERATURE CITED

Akazawa, T., P. Miljanich, and E.E. Conn. 1960. Studies on cyanogenic glycoside of *Sorghum vulgare.* Pl. Physiol. 35:535–538.

Archer, K.A., and J.L. Wheeler. 1978. Response by cattle grazing sorghum to salt-sulphur supplements. Austral. J. Expt. Agric. Anim. Husb. 18:741–744.

Clark, R.B., H.J. Gorz, and F.A. Haskins. 1979. Effects of mineral elements on hydrocyanic acid potential in sorghum seedlings. Crop Sci. 19:757–761.

Conn, E.E. 1978. Cyanogenesis, the production of hydrogen cyanide by plants. *In* R.F. Keeler, K.R. Van Kampen, and L.F. James (eds.) Effects of poisonous plants on livestock, 301–310. Academic Press, New York.

Conn, E.E. 1979. Biosynthesis of cyanogenic glycosides. Naturwissenschaften 66:28–34.

Dunstan, W.R., and T.A. Henry. 1902. Cyanogenesis in plants. II. The great millet *Sorghum vulgare.* Roy. Soc. London, Phil. Trans., Ser. A 199:399–410.

Gorz, H.J., W.L. Haag, J.E. Specht, and F.A. Haskins. 1977. Assay of *p*-hydroxybenzaldehyde as a measure of hydrocyanic acid potential in sorghums. Crop Sci. 17:578–582.

Gorz, H.J., F.A. Haskins, R. Dam, and K.P. Vogel. 1979. Dhurrin in *Sorghastrum nutans.* Phytochemistry 18:2024.

Haskins, F.A., H.J. Gorz, and R.L. Nielsen. 1979a. Comparison of the hydrocyanic acid potential of *Sorghum* seedlings and tillers. Agron. J. 71:501–504.

Haskins, F.A., H.J. Gorz, and K.P. Vogel. 1979b. Cyanogenesis in indiangrass seedlings. Crop Sci. 19:761–765.

Hunt, B.J., and A.O. Taylor. 1976. Hydrogen cyanide production by field-grown sorghums. N.Z. Expt. Agric. 4:191–194.

Kojima, M., J.E. Poulton, S.S. Thayer, and E.E. Conn. 1979. Tissue distributions of dhurrin and of enzymes involved in its metabolism in leaves of *Sorghum bicolor.* Pl. Physiol. 63:1022–1028.

Palmer, I.S. 1980. Selenium: villain or hero? S. Dak. Fm. and Home Res. 31:7–10.

Tjon Sie Fat, L. 1977. Contribution to the knowledge of cyanogenesis in angiosperms. I. Communication. Cyanogenesis in some grasses. Proc. Koninkl. Nederl. Akad. Wetensch. 80(C):227–237.

Wheeler, J.L., D.A. Hedges, and A.R. Till. 1975. A possible effect of cyanogenic glucoside in sorghum on animal requirements for sulphur. J. Agric. Soc. Camb. 84:377–379.

Interspecific Hybrids Between Apomictic Forms of *Poa palustris* L. x *Poa pratensis* L.

W. NITZSCHE
Max-Planck-Institut für Züchtungsforschung,
Federal Republic of Germany

Summary

Kentucky bluegrass (*Poa pratensis* L.) has been hybridized with many other *Poa* species, but interspecific hybrids between fowl bluegrass (*P. palustris* L.) have not been reported so far. In contrast to most other *Poa* species, *P. palustris*, like *P. pratensis*, is a commercially used forage crop, and we expected better results in such hybrids than in those described earlier. The obligate apomict *P. palustris* was X-irridiated, and two-seed-descendant X_2 plants were grown in the nursery. A total of 47 pairs with different partners were selected and checked for sexuality. One plant with a segregating progeny was used for hybridization. Flowering time of the two *Poa* species differed by about one month. This difficulty was overcome by storage of the *P. pratensis* pollen at a temperature of –80°C. Hybridization was done without emasculation on single plants in isolation chambers. Among the progeny of 200 plants, we found up to 27 hybrids with *P. pratensis*, the remainder being pure *P. palustris*.

In 1972, the first recorded hybrid plants were pure apomicts and stable in multiplication, the interspecific heterosis being fixed in the evolved strains. One strain is a pasture-and-lawn type, another a meadow type. Both strains have the subterranean runners of *P. pratensis* and the leaf color of *P. palustris*. One strain has hairless seeds like the *P. palustris* parent. Disease resistance and yield quality of these strains are much better than those of *P. longifolia* × *P. pratensis* interspecific crosses. One of the new hybrids looks so promising that it is under registration proceedings in West Germany.

KEY WORDS: *Poa palustris* L., *Poa pratensis* L., apomixis, interspecific hybrids, mutation.

INTRODUCTION

Of the bluegrasses, *Poa pratensis, P. palustris, P. trivialis,* and *P. compressa* are used as forage crops. Known interspecific hybrids of these species are *P. compressa* × *P. pratensis* and *P. palustris* × *P. compressa.* Kappert and Rudorf (1959) listed nine different interspecific hybrids with bluegrasses. Considerable work has been done with the hybrid *P. longifolia* × *P. pratensis,* but none of these hybrids was used in practice at any time.

In Germany, Kentucky bluegrass (*Poa pratensis* L.) and fowl bluegrass (*Poa palustris* L.) are economically the most important bluegrasses. The two species have complementary agronomic characters. Kentucky bluegrass has runners and is well adapted in pastures. Fowl bluegrass has high yielding capacity and hairless seeds. Both species are adapted to the climate and epidemiological conditions of West Germany. We thus expected better results from hybrids between fowl bluegrass and Kentucky bluegrass than from those hybrids mentioned earlier.

MATERIALS AND METHODS

Registered types of bluegrass were used in the hybridization experiments. Fowl bluegrass cv. Hauna was chosen as the female parent because of its hairless seeds and Kentucky bluegrass cv. Ottos as the other parent on the basis of its good performance in German pastures. The direction of hybridization is conditioned by the later flowering time of fowl bluegrass. Hauna is an obligate apomict with no known aberrants, while Ottos is a facultative apomict with about 2% aberrants.

X-ray mutagenesis was used for production of sexual fowl bluegrass. In December 1967, fowl bluegrass was X-irradiated with 65 kr, the dose being determined for 20% survival according to the method given by Gaul (1963). The X_1 generation was cultured in 4-cm multipots, and the seeds were harvested individually.

Sexuality and apomixis can be recognized by the segregation of progenies. Therefore, instead of the single-seed-descendant method commonly used in mutation breeding, the X_2 generation was grown as double-seed descendants, i.e., two progeny seeds/X_1 plant. Differences between progeny plants can be caused either by sexuality of the mother plant or by a chimera. Therefore, a second screening in X_3 was necessary.

Flowering times of Kentucky bluegrass and fowl bluegrass differ by more than 4 weeks. Although flowering time of the male parent could be delayed by culturing it at 5°C, the procedure could not be used because of the simultaneous induction of male and female sterility. It was necessary, therefore, to use stored pollen. Storage was done according to a method described by Nitzsche (1970), but at a temperature of −80°C. Hybridization was accomplished without emasculation of the female parent. The first pollination was done in 1970 on X_2 plants in the field, and a second one was done in 1978 on

The author is a Professor of Plant Breeding at the Max-Planck-Institut für Züchtungsforschung.

Table 1.   Breeding schedule of fowl bluegrass hybrids.

Year	Operation	Generation	Material
1967	X-irradiation		seeds
1968	cultivation	X_1	plants
1969	2-seed descendants	X_2	pairs
1970	screened pairs		
	first pollination	X_2	clones
1972	hybrid screening	F_1	plants
1974	homogeneity test	F_1	bulks
1976	multiplication	F_1	bulks
1978	start of		
	registration procedure	F_1	cultivar
	2nd pollination	X_3	plants
1980	hybrid screening	F_1	plants

X_3 plants in isolation chambers.

Hybrid plants were identified by marker characters in the field. Plants were evaluated for growth type, runners, ligules, and hairiness. The breeding schedule is given in Table 1.

RESULTS

After X-irradiation of seeds, the X_1 generation was grown in very small pots that were nevertheless big enough for every plant to develop 2 to 5 panicles. For the next generation, only two seeds/plant were necessary. In the X_2 generation, 2,256 pairs of progeny plants were grown in the nursery. In most of the pairs the partners were equal, and only 47 could be selected as segregating. Only one plant of all the pairs could be recognized as sexual in progeny tests involving 20 plants/progeny. The progeny of this plant consisted of 11 fowl bluegrass types and 7 hybrids, 2 of which died before evaluation. Two strains were developed from the hybrids.

The hybrids are very similar to Kentucky bluegrass except that the ligule is longer and the plants, especially the glumes, are more intensively colored. One hybrid has hairless seeds and short runners, whereas the other has hairy seeds and long runners. Both hybrids are pure apomicts.

We started seed multiplication and lawn testing in 1976 with the best strain of the first selection. Seed yield of this hybrid is about 500 kg/ha, which is especially good under our climatic conditions and particularly with a low agricultural input. First results of forage yield were obtained in 1980. Both the best strains of the first selection were tested for dry-matter yield in three cuts. They yielded 114% and 107% of the two most-used German varieties, Ottos and Union.

Official registration procedures started in 1978. The name of the new cultivar will be Ikone. To verify the 1970 results, a second hybridization program was carried out on 16 X_3 plants in isolation chambers in 1978. The results of this program, obtained from progenies grown in the nursery in 1979–1980, are given in Table 2. The frequency of the hybrids in different strains differed significantly; the calculated Chi-square (χ^2) value (172 at 15 df.) is much higher than expected (25.0 at 5% and 37.7 at 0.1% probability levels) in the case of homogeneity.

Table 2.   Number of hybrids from 16 X_3 plants of *P. palustris* after pollination with *P. pratensis*.

Strain	Plants	Hybrids
1030	180	1
1032	200	4
1034	200	0
1036	200	0
1038	200	3
1040	200	29
1042	200	0
1044	200	9
1046	200	4
1048	200	2
1050	200	1
1052	200	2
1054	200	2
1056	200	4
1058	200	0
1060	200	1
Total	3180	62

DISCUSSION

The breeding procedure demonstrated here is a successful combination of mutation breeding, interspecific hybridization, and breeding apomicts. In apomictic Kentucky bluegrass, the induction of sexuality was done by Julén (1958) with X-rays. Interspecific apomictic hybrids have been produced mainly in *Paspalum* (Bashaw et al. 1970). The pollen-storage method was developed in *Lolium* (Nitzsche 1970). The combination of all these methods resulted in the above-described breeding program.

The two-seed-descendant method for selection of sexual plants after mutagenic treatment of the apomictic fowl bluegrass was very effective. From only 2,256 X_1 plants and 47 X_2 pairs, the wanted type could be selected. Additionally, the selection of only one sexual plant does not necessarily mean that only one mutation to sexuality has occurred. But in this breeding program, one sexual plant was enough. The differences in frequency of hybrids in X_3 strains are significant, but it is nevertheless not clear whether the differences depend on hereditary or environmental factors since we had no replications. The differences could depend not only on hereditary factors but also on differences in flowering stage or pollen quality. Nevertheless, the best strain of fowl bluegrass, 1040, has been given to the German gene bank at Braunschweig for general availability. This strain is a facultative apomict with about 15% sexuality.

Heterosis results from the combination of different genomes, and it decreases with subsequent inbreeding. In our interspecific F_1 hybrids, a high degree of heterogenesis is created, and large heterotic effects may be expected. The results since the onset of the field tests agree with the expectations. Recurrence of apomixis in F_1 prevents a decrease of heterosis during multiplication. Thus, the combination demonstrated here resulted in a quick and successful breeding method requiring relatively little work.

LITERATURE CITED

Bashaw, E.C., A.W. Hovin, and E.C. Holt. 1970. Apomixis, its evolutionary significance and utilization in plant breeding. Proc. XI Int. Grassl. Cong, 245–248.

Gaul, H. 1963. Mutationen in der Pflanzenzüchtung. Z. Pflanzenzüchtung 50:194–307.

Julén, G. 1958. Über Effekte der Röntgenbestrahlung bei Poa pratensis. Der Züchter. 28:37–40.

Kappert, H., und W. Rudorf. 1959. Handbuch der Pflanzenzüchtung. Vol. 4: Züchtung der Futterpflanzen. Paul Parey, Berlin.

Nitzsche, W. 1970. The storage of grass pollen. Proc. XI Int. Grassl. Cong., 259–260.

Improvement of Forage Quality of Tall Fescue Through *Lolium-Festuca* Hybridization

R.C. BUCKNER, L.P. BUSH, P.B. BURRUS II, J.A. BOLING,
R.A. CHAPMAN, R.W. HEMKEN, and J.A. JACKSON, JR.
University of Kentucky, Lexington, Ky., U.S.A.

Summary

Annual ryegrass (*Lolium multiflorum* Lam.) × tall fescue (*Festuca arundinacea* Schreb.) hybrid derivatives were used to develop G1-307, a strain with low perloline alkaloid content and improved succulence and digestibility. Animals consuming G1-307 under high ambient temperature stress exhibited the "summer syndrome" malady. *Epichloë typhina*, an endophytic fungus, was associated with the occurrence of N-acetyl and N-formyl loline (pyrrolizidine) alkaloids in G1-307. The pyrrolizidine alkaloids appear to be related to the summer syndrome condition. Our objectives were to investigate the variability and heritability of perloline, N-acetyl loline, and N-formyl loline and the association of *E. typhina* and the lolines.

Twenty-eight parental clones of G1-307, the product of two cycles of recurrent selection for low perloline content and improved succulence and digestibility, were grown in two separate isolated polycross blocks. Clones in one block were untreated, and those in a second block had been treated with benomyl, a systemic fungicide, to remove *E. typhina*.

Polycross progenies of untreated clones, the G1-307 strain, and commercial cultivars were evaluated for perloline content in replicated sod plots arranged in a randomized block design. Equal amounts of seed of each clone of treated and untreated blocks were blended to form the G1-307 strain. Seeds of treated G1-307 and untreated G1-307 were used to establish 2.2- and 3.3-ha swards, respectively.

Means of untreated parental clones and polycross progenies (pcp) differed significantly for perloline, N-acetyl loline, and N-formyl loline content. Broad-sense heritability estimates using parental variance of perloline, N-acetyl loline, and N-formyl loline were 0.88, 0.67, and 0.76, respectively, and the same estimates using pcp variance were 0.50, 0.33, and 0.79, respectively. Narrow-sense heritability estimates for perloline, N-acetyl loline, and N-formyl loline content were 0.45, 0.51, and 0.91, respectively. Infection by *E. typhina* and content of pyrrolizidine alkaloids were reduced 71.5% and 67.9%, respectively, when parental clones were treated with benomyl. Hay from swards established from seed of benomyl-treated and nontreated clones had pyrrolizidine alkaloid levels of 115 and 875 µg/g and infection estimates by *E. typhina* of 44% and 95% respectively.

KEY WORDS: alkaloids, perloline, N-acetyl loline, N-formyl loline, digestibility, succulence, heritability, summer syndrome, *Epichloë typhina*, animal performance.

INTRODUCTION

Plant variability within tall fescue (*Festuca arundinacea* Schreb.) is insufficient for major forage-quality improvement. *Lolium-Festuca* hybridization offers new sources of germ plasm with increased variability from which superior germ plasm may originate. Intergeneric hybridization of annual ryegrass (*Lolium multiflorum* Lam.) and tall fescue is used to transfer the forage quality of annual ryegrass to tall fescue while maintaining the excellent agronomic qualities of tall fescue (Buckner et al. 1979).

Erratic performance by cattle feeding on tall fescue during summer and exhibiting reduced rate of gain and/or milk production, rough hair coat, rapid breathing, increased body temperature, and a generally unthrifty condition is generally referred to as "summer syndrome" (Bush et al. 1979).

Tall fescue alkaloids have a seasonal accumulation maximum coinciding with summer syndrome, and observations from in-vitro and in-vivo experiments suggest that elevated alkaloid concentrations in the forage and poor animal performance are related (Bush et al. 1979). Perloline, a diazaphenanthrene alkaloid, was shown to reduce in-vitro cellulose digestibility and in-vivo protein and cellulose digestibility and to be under genetic control with sufficient genetic variability to facilitate selection within ryegrass–tall fescue hybrid derivatives (Bush et al. 1979). Ruminants consuming G1-307, an experimental strain with low perloline content, had lowered feed intake and exhibited the summer syndrome (Hemken et al. 1979).

Bacon et al. (1977) suggested that *Epichloë typhina*, a systemic phytopathogen, might be involved in the summer syndrome of cattle. Jones et al. (1980) reported that *E. typhina* was associated with pyrrolizidine alkaloid production in tall fescue. Jackson et al. (1980) recently showed a direct positive association between pyrrolizidine alkaloid content and the summer syndrome when young cattle under ambient temperature stress were fed tall fescue seeds containing different levels of the alkaloid.

In this paper we present the variability and heritability of perloline and of N-acetyl and N-formyl loline (pyrrolizidine) alkaloids in *Lolium-Festuca* hybrid derivatives and the association of *E. typhina* and pyrrolizidine alkaloids.

MATERIALS AND METHODS

Origin and selection of the 28 parent clone lines used in this study were described by Buckner et al. (1979). Establishment and management of the parent clones in an isolated polycross block, designated test 377, and of their polycross progenies (pcp) and G1-307 synthetic strain (formed by blending equal amounts of seed of each pcp) in a test, designated test 404, that also included Kentucky 31, Kenhy, and Fawn cultivars, were described by Buckner et al. (1981).

In the chemical analyses, perloline content of the samples was determined as described by Bush and Jeffreys (1975), and N-acetyl and N-formyl loline content was determined as described by Kennedy (1980).

In pathological studies, the clonal lines described in test 377 were treated during 1978 for a minimum of 10 days in a 1:1,000 benomyl:soil mixture (w/w) in a greenhouse in an effort to remove *E. typhina* from the clones before transplanting them to test 494, which was arranged identically to test 377. Sufficient seed was harvested from tests 494 and 377 to establish by broadcast seeding (14.4 kg/ha) 2.2 and 3.3 ha, respectively, of G1-307 for forage production. Plants of all tests were examined for presence of *E. typhina* by microscopic examination of pith scrapings from longitudinally slit culms after the culms had been stained with aqueous aniline blue (Bacon et al. 1977).

Broad- and narrow-sense heritability were estimated by the statistical methods described by Buckner et al. (1981).

R.C. Buckner is a Research Agronomist with USDA-SEA-AR; L.P. Bush is Professor of Agronomy at the University of Kentucky; P.B. Burrus II is a Research Agronomist with USDA-SEA-AR; J.A. Boling is Professor of Animal Sciences at the university; R.A. Chapman is Professor of Plant Pathology at the university; R.W. Hemken is Professor of Animal Sciences at the university; and J.A. Jackson, Jr., is a Research Fellow at the university.

The investigation reported in this paper (No. 80-3-5-11-231) is in connection with a project of the Kentucky Agricultural Experiment Station and is published with approval of the Director.

Table 1. Mean and range of perloline, and of N-acetyl and N-formyl loline alkaloids of 28 parents (Test 377) and polycross progenies (Test 404) during August 1974-1976.

Constituent	Test	Mean	Range	PR>F	C.V. (%)
			------------μg/g------------		
Perloline	377	254	45-660	.0001	55.8
	404	227	164-314	.0017	32.8
N-acetyl loline	377	1837	1177-3206	.0001	21.9
N-formyl loline	404	1619	1143-2271	.0001	17.6

RESULTS AND DISCUSSION

Mean alkaloid content of parental lines (test 377) and pcp (test 404) differed significantly (Table 1). The pcp means for all constituents were lower than those for parents. The highly significant differences among means for perloline suggest that considerable total genetic variability remains after two cycles of selection, and, therefore, further improvement may be possible by additional selection. However, means of parents (254) and pcp (227) were low, indicating that progress had been made by selection. Since test means for perloline were lower than 500 μg/g, further selection may not be needed because in-vitro studies (Bush et al. 1979) indicate that levels of 500 μg/g or above seem to be necessary for the significant inhibition of rumen microflora and the concomitant reduction of digestibility in the ruminant. The highly significant differences among means for N-acetyl and N-formyl loline content are perhaps a reflection of the degree of infection of the clones by *E. typhina* and its influence on pyrrolizidine alkaloid levels in the plant. Jones et al. (1980) reported a direct relationship between incidence of *E. typhina* infection and content of N-acetyl and N-formyl lolines in the plant.

Broad-sense heritability estimates using parental variance resulted in heritabilities ranging from 0.67 to 0.88. Broad-sense heritability estimates among parent clones were relatively high, indicating that a considerable portion of the total variability was genetic. The heritability estimates (0.33-0.79) were lower for pcp, except for N-formyl loline (0.79), than for parents (0.76). Narrow-sense heritability estimates obtained by regression of pcp on parents were relatively high for perloline (0.45) and N-acetyl loline (0.51) and especially high for N-formyl loline (0.91).

Although the G1-307 strain had higher forage quality than Fawn and Kentucky 31 cultivars, Hemken et al. (1979) showed that lactating dairy cows receiving the G1-307 strain exhibited summer syndrome symptoms. Poor performance of the cows fed G1-307 was not related to perloline levels (200 μg/g) or to digestibility of the forage. Thus, in the selection for low perloline content, apparently the pyrrolizidine alkaloids that were more toxic and detrimental to animal performance were accentuated in G1-307.

Treatment of 14 parental clones with benomyl, a systemic fungicide (test 377), reduced *E. typhina* infection and pyrrolizidine alkaloid content 71.5% and 67.9%, respectively (Table 2). Hay from swards established from seed of benomyl-treated and untreated clones of synthetic G1-307 had pyrrolizidine alkaloid levels of 115 and 875 μg/g and infection by *E. typhina* of 44% and 95%, respectively.

LITERATURE CITED

Bacon, C.W., J.K. Porter, J.D. Robbins, and E.S. Luttrell. 1977. *Epichloë typhina* from toxic tall fescue grasses. Appl. and Envir. Microbiol. 34(5):576-581.

Buckner, R.C., P.B. Burrus, II, P.L. Cornelius, L.P. Bush, and J.E. Leggett. 1981. Genetic variability and heritability of certain forage quality and mineral constituents in *Lolium-Festuca* hybrid derivatives. Crop Sci. 21:419-423.

Buckner, R.C., L.P. Bush, and P.B. Burrus, II. 1979. Succulence as a selection criterion for improved forage quality in *Lolium-Festuca* hybrids. Crop Sci. 19:93-96.

Bush, L.P., J.A. Boling, and S. Yates. 1979. Animal disorders. *In* R.C. Buckner and L.P. Bush (eds.) Tall fescue monograph, Ch. 13. Am. Soc. Agron., Madison, Wis.

Bush, L.P., and J.A.D. Jeffreys. 1975. Isolation and separation of tall fescue and ryegrass alkaloids. J. Chromatog. 111:165-170.

Hemken, R.W., L.S. Bull, J.A. Boling, E. Kane, L.P. Bush, and R.C. Buckner. 1979. Summer fescue toxicosis in lactating dairy cows and sheep fed experimental strains of ryegrass-tall fescue hybrids. J. Anim. Sci. 49(3):641-646.

Jackson, J.A., R.W. Hemken, J.A. Boling, L.S. Bull, L.P. Bush, and R.C. Buckner. 1980. Summer fescue toxicosis as a result of feeding tall fescue seed. J. Dairy Sci. 63. Suppl. 1:133.

Jones, T.A., R.C. Buckner, L.P. Bush, R.A. Chapman, P.B. Burrus, II, and D.R. Varney. 1980. Association of the endophytic fungus, *Epichloë typhina*, with loline alkaloid content of tall fescue. Agron. Abs. 1980:126.

Kennedy, C.W. 1980. Accumulation and metabolism of alkaloids in ryegrass × tall fescue hybrids. Ph.D. dissertation, Univ. of Ky.

Table 2. *Epichloë typhina* infestation and N-acetyl and N-formyl loline content of 14 benomyl-treated (Test 494) and untreated (Test 377) genotypes of *Lolium-Festuca* hybrid derivatives.

	Infection (%)		Alkaloid content (μg/g)	
	Treated	Untreated	Treated	Untreated
Mean	26.2	91.7	1044	3250
Range	0-45	0-100	19-1580	24-5475
Percent reduction	71.5		67.9	

Breeding of Apomictic *Eragrostis curvula*

P.W. VOIGT and B.L. BURSON

USDA-SEA-AR Grassland, Soil and Water Research Laboratory,
Temple, Tex., U.S.A.

Summary

Eragrostis curvula (Schrad.) Nees and related species are important pasture and range forage plants in the semi-arid and arid southwestern U.S.A. Types within these polymorphic, apomictic species have different adaptations and characteristics. Breeding cultivars with new combinations of characteristics should greatly increase their usefulness. Our objectives were to determine the relationship between data from cytological studies of hybrids and from progeny test classifications of mode of reproduction, to evaluate mode of reproduction in lovegrass hybrids, and to develop an efficient breeding scheme. In general, results of cytological analysis of megasporogenesis and embryo sac development and spaced-plant tests of progeny variability gave similar evaluations of mode of reproduction. Results indicated that 40% of hybrids were predominantly apomictic, 20% were intermediate, and 40% were predominantly sexual in tetraploid sexual × tetraploid apomict crosses. In crosses of tetraploid sexual × one 2n = 69 apomict, 90% of the hybrids were predominantly apomictic. We propose a breeding scheme using sexual × apomict hybridization to produce apomictic cultivars.

KEY WORDS: weeping lovegrass, *Eragrostis curvula* (Schrad.) Nees, Lehmann lovegrass, *Eragrostis lehmanniana* Nees, mode of reproduction, diplospory, hybrids, cytology, progeny, breeding scheme.

INTRODUCTION

Eragrostis curvula (Schrad.) Nees is a vigorous, polymorphic, diplosporous-apomictic species (Brown and Emery 1958, Leigh 1961, Streetman 1963). Types within this species, e.g., weeping and boer lovegrass, are adapted to different environments of the arid to semiarid southwestern U.S.A. (Crider 1945). However, weeping lovegrass can be low in forage quality (Duble et al. 1971) and can be killed by drought under improper management (McIlvain and Shoop 1970). Boer lovegrass is less winter-hardy than weeping lovegrass but relatively drought-resistant (Crider 1945). Lovegrass strains of the robusta type were reported to be higher in in-vitro dry-matter disappearance (IVDMD) than weeping cultivars such as Morpa and Ermelo (Voigt and Tischler 1979). Thus, development of procedures to manipulate apomixis in *E. curvula* and to allow gene exchange and selection of recombinants could be of great value in breeding of improved cultivars.

Eragrostis curvula was originally believed to be an obligate apomict (Streetman 1963). Voigt (1971) reported discovery of sexual reproduction in *E. curvula* var. conferta and successfully hybridized sexual and apomictic plants (Voigt and Bashaw 1972). Subsequent study of aberrant plants from a tetraploid apomictic strain (Brix 1974) and from lovegrass hybrids (Voigt and Bashaw

1976) showed that the species must be considered a facultative rather than an obligate apomict. Objectives of the present research were (1) to determine the relationship between cytological and progeny test evaluation of mode of reproduction, (2) to evaluate mode of reproduction in hybrids, and (3) to develop an efficient breeding scheme for *E. curvula*.

METHODS

All hybrids studied were obtained from mutual pollination in parchment bags using highly sexual, tetraploid (2n = 40) plants of *E. curvula* var. conferta as females and highly apomictic plants as males. The apomicts used were curvula types common (representative of strain A-67), Ermelo, and Morpa (all 2n = 40); robusta types Renner (2n = 60) and R-1 (2n = 69); and (CHL-1) (2n = 40), classified as *E. lehmanniana* (Nees). Classification of CHL-1 as *E. lehmanniana* was confirmed by E.E. Terrell, Botanist of the USDA-SEA-AR Plant Taxonomy Laboratory in Beltsville, Maryland. CHL-1 is more robust, upright, and winter-hardy than typical Lehmann lovegrass and has been referred to as cold-hardy Lehmann.

Hybrids were evaluated for mode of reproduction by cytological analysis of megasporogenesis and embryo sac development (Voigt and Bashaw 1976) and by unreplicated progeny tests of plants spaced about 1 m apart in field plantings. A hybrid was classified apomictic if the progeny contained about 30% aberrant plants or less, sexual if no uniform phenotype was seen, and intermediate if a few possibly uniform plants were observed. Most

---
The authors are both Research Geneticists with USDA-SEA-AR. This is a contribution of USDA-SEA-AR in cooperation with the Texas Agricultural Experiment Station.

Table 1. Comparison of two methods of classification, cytological and progeny, for determining mode of reproduction in 40-9-69FQ X CHL-1 lovegrass hybrids.

Cytological†	Progeny†† Apomictic	Intermed.	Sexual	Total
	------------------------- Plants, no.-------------------------			
Sexual	0	1	17	18
Intermed.	1	3	4	8
Apomictic	16	0	1	17
Total	17	4	22	43
	------------------Sterile embryo sacs, %§------------------			Mean
Sexual	–	49.0	14.0±3.0	15.9±3.4
Intermed.	16.0	31.0±13.5	32.5±12.9	29.9±7.7
Apomictic	15.4±3.1	–	25.0	15.9±2.9
Mean	15.4±2.9	35.5±10.6	17.9±3.5	18.5±2.4

†Megasporogenesis and embryo sac development classified as apomictic, 0-25% sexual; intermediate, 26-75% sexual; and sexual, 76-100% sexual.

††Progeny variability classified as apomictic, about 30% aberrant plants or less; intermediate, a few possibly uniform plants; and sexual, no uniform phenotype observed.

§Mean percentage of sterile ovules ± the standard error where n > 1.

progenies were rated from 10 to 15 individuals, but a few were rated from as few as 7 to 9 plants.

RESULTS AND DISCUSSION

Results of both cytological study and progeny evaluation of the cross 40-9-69FQ × CHL-1 showed a predominance of extreme types (apomictic and sexual) and a few intermediates (Table 1). Thus, most hybrids were either highly sexual or highly apomictic. The ratio of apomictic to sexual hybrids, ignoring intermediates, was 1.0:1.1 and 1.0:1.3 by cytology and progeny test, respectively. Misclassification in progeny tests resulted primarily from inclusion of too many intermediates in the sexual category. Additional observations from other crosses are needed to confirm the cytology-progeny test relationship reported here.

The presence of a small intermediate class might be explained in part by sterility (Table 1). The percentage of sterile ovules—the percentage of ovules classified sterile out of a total of those classified as sexual, apomictic, or sterile—was higher for both the intermediate progeny class and the intermediate cytological class than for the extreme classes. Thus, the masking effect of sterility could have prevented the expression of a more extreme reproductive performance (predominantly apomictic or sexual) in some hybrids. However, not all intermediates were high in sterility.

Mode of reproduction was determined by progeny test for 1,301 lovegrass hybrids (Table 2). Most hybrids using tetraploid apomicts as males (common, Morpa, Ermelo, and CHL-1) produced apomictic:sexual ratios ranging from 1.0:1.1 to 1.0:1.6. The cross 398-18-69 × CHL-1 was an exception that produced only 1.0 apomict for every 3.5 sexuals. The reason this cross was exceptional is not known. CHL-1 hybrids gave similar ratios to common and Morpa in crosses with 40-9-69FQ. Also, 398-18-69 is known to be highly sexual (Voigt and Bashaw 1976).

Crosses of tetraploids with the hexaploid Renner as the male parent produced results similar to crosses among tetraploids. However, crosses with the 2n = 69 R-1 produced an average of 12 apomicts for each sexual. The 40-9-69FQ × R-1 cross produced 8 apomictic hybrids for each sexual compared to a 1:1 ratio for the 40-9-69FQ × common cross. These results suggest that apomixis may be dominant to sexuality and indicate a possible dosage effect. Additional work is needed to clarify the inheritance of apomixis in *E. curvula*.

A simple apomictic breeding scheme can be developed from our results (Fig.1). Hybridization of sexual × apomictic tetraploids should produce about 40% apomictic, 10% intermediate, and 50% sexual hybrids by progeny test classification. However, if the observed relationship between progeny and cytological classification holds (Table 1), the actual proportions should average closer to 40% apomictic, 20% intermediate, and 40% sexual. Selected highly sexual hybrids can be crossed again with the same or different apomictic plants to continue the breeding process. Following each cycle of hybridization, the more uniform apomicts can be evaluated as potential new cultivars. On the average, we found that about 60% of the apomictic hybrids (25% of all hybrids) were highly uniform (less than 13% aberrant plants). Many of these should have sufficient uniformity for use as forage cultivars. Plants intermediate for mode of reproduction could be used as parents in further hybridization. However, when they are used as females, these plants would be less efficient in producing hybrids than in producing sexual plants, and when they are used as males, they might produce a lower degree of apomixis in their apomictic hybrids. We have had only limited success using the more

Table 2. Mode of reproduction of lovegrass hybrids classified by progeny performance.

| Cross | n | Number | | | Ratio |
		Apomictic	Intermed.	Sexual	Apo:Sex†
57-4-69 X common	150	57	11	82	1.0:1.4
60-7-69 X common	26	9	4	13	1.0:1.4
40-9-69FO X common	372	159	41	172	1.0:1.1
Total	548	225	56	267	1.0:1.2
40-9-69FQ X Morpa	449	160	36	253	1.0:1.6
109-7-70W X Ermelo	18	8	4	6	
109-18-70W X Ermelo	59	25	3	31	1.0:1.2
Total	77	33	7	37	1.0:1.1
58-2-69 X CHL-1	3	2	1	0	
398-18-69 X CHL-1	30	6	3	21	1.0:3.5
40-9-69FQ X CHL-1	55	23	4	28	1.0:1.2
267-10-72E X CHL-1	6	2	0	4	
Total	94	33	8	53	1.0:1.6
60-7-69 X Renner	19	7	2	10	
40-9-69FQ X Renner	14	5	3	6	
98-1-70E X Renner	15	7	0	6	
119-3-70W X Renner	9	2	1	6	
Misc. X Renner	22	8	1	13	
Total	77	29	7	41	1.0:1.4
40-9-69FQ X R-1	28	24	1	3	8.0:1.0
98-2-70E X R-1	2	1	0	1	
104-1-70E X R-1	17	16	1	0	
109-3-70W X R-1	8	7	0	1	
Total	55	48	2	5	9.6:1.0

†Ratios of apomicts:sexuals presented only where n = 25 or more.

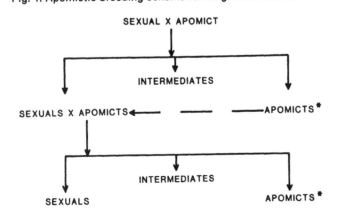

Fig. 1. Apomictic breeding scheme for *Eragrostis curvula*.

SEXUAL X APOMICT

INTERMEDIATES

SEXUALS X APOMICTS ← APOMICTS *

INTERMEDIATES

SEXUALS   APOMICTS *

* PLANTS TO BE EVALUATED AS POTENTIAL CULTIVARS

facultative plants as female parents.

If a higher polyploid is used as the apomictic male parent, the efficiency of producing highly apomictic hybrids can sometimes be greatly increased. Approximately 75% of the R-1 hybrids (tetraploid × 2n = 69) were considered uniform enough to have potential as apomictic cultivars. However, the reduced number of sexual plants and the possible adverse effects of higher ploidy on their fertility could cause problems if recurrent hybridization is necessary.

Although many questions concerning apomixis in *E. curvula* remain, it is clear that apomixis is no longer an important barrier to breeding in the *E. curvula* complex. It is an asset.

LITERATURE CITED

Brix, K. 1974. Sexual reproduction in *Eragrostis curvula* (Schrad.) Nees. Z. Pflanzenzüchtung 71:25–32.

Brown, W.V., and W.H.P. Emery. 1958. Apomixis in the Gramineae: Panicoideae. Am. J. Bot. 45:253–263.

Crider, F.J. 1945. Three introduced lovegrasses for soil conservation. U.S. Dept. Agric. Cir. 730.

Duble, R.L., J.A. Lancaster, and E.C. Holt. 1971. Forage characteristics limiting animal performance on warm-season perennial grasses. Agron. J. 63:795–798.

Leigh, J.H. 1961. Leaf anatomy in certain strains of *Eragrostis*

(Beauv.). J. So. Afr. Bot. 27:141–146.

McIlvain, E.H., and M.C. Shoop. 1970. Grazing weeping love-grass for profit—8 keys. *In* R.L. Dalrymple (ed.) Proc. 1st Weeping Lovegrass Symp., 110–117. Samuel Roberts Noble Foundation, Ardmore, Okla.

Streetman, L.J. 1963. Reproduction of the lovegrasses, the genus *Eragrostis*. I. *E. chloromelas* Steud., *E. curvula* (Schrad.) Nees, *E. lehmanniana* Nees and *E. superba* Peyr. Wrightia 3:41–51.

Voigt, P.W. 1971. Discovery of sexuality in *Eragrostis curvula* (Schrad.) Nees. Crop Sci. 11:424–425.

Voigt, P.W. and E.C. Bashaw. 1972. Apomixis and sexuality in *Eragrostis curvula*. Crop Sci. 12:843–847.

Voigt, P.W., and E.C. Bashaw. 1976. Facultative apomixis in *Eragrostis curvula*. Crop Sci. 16:803–806.

Voigt, P.W., and C.R. Tischler. 1979. Variation for digestibility and wax content in a less winterhardy lovegrass population. Proc. Grass Breed. Work Planning Conf. 25:7.

Use of Embryo Culture with Nurse Endosperm for Interspecific Hybridization in Pasture Legumes

W.M. WILLIAMS and E.G. WILLIAMS
Grasslands Division, DSIR, Palmerston North, N.Z.

Summary

Progress in the breeding of pasture species depends on the use of base populations with maximum variation, particularly in the breeding of species for extension into marginal areas. Interspecific hybridization offers one possibility for extending the range of variability beyond that offered by a single species.

A technique of in-vitro embryo culture has been developed to overcome endosperm failure after interspecific pollination. This technique involves sterile culture of a hybrid embryo after transplantation into a nurse endosperm dissected from a normal ovule.

Materials produced include *Trifolium ambiguum* × *T. repens*, *T. ambiguum* × *T. hybridum*, *T. repens* × *T. uniflorum*, *Ornithopus sativus* × *O. compressus*, *O. pinnatus* × *O. sativus*, and *Lotus pedunculatus* × *L. corniculatus*.

The *O. sativus* × *O. compressus* and *L. pedunculatus* × *L. corniculatus* hybrids were fully fertile in advanced generations and showed good agronomic potential. The *T. ambiguum* × *T. repens* and *T. repens* × *T. uniflorum* hybrids were particularly fertile and are likely to respond to selection for increased fertility. The *O. pinnatus* × *O. sativus* hybrids were fertile but required embryo culture to overcome defective endosperm development. The cross *T. ambiguum* × *T. hybridum* has so far given completely sterile progeny.

A feature of the embryo-culture technique using nurse endosperm is its high success rate for obtaining hybrids. Although hybrid numbers are small, the hybrids arose from only a few hundred pollinated flowers, and generally fewer than 100 putative hybrid embryos were cultured. Therefore, we conclude that prospects for large-scale application of this method are very promising.

KEY WORDS: interspecific hybridization, nurse endosperm, embryo culture, white clover, *Trifolium repens* L., *Trifolium ambiguum* M. Bieb., alsike clover, *Trifolium hybridum* L., *Ornithopus sativus* Brot., serradella, *Ornithopus compressus*, *Ornithopus pinnatus*, *Trifolium uniflorum* L.

INTRODUCTION

Interspecific hybridization in forage legumes frequently fails as a result of endosperm abortion and subsequent starvation of the hybrid embryo (Williams and White 1976). In such cases embryo culture can sometimes be used to achieve successful development of the hybrid embryo. An endosperm transplant technique has been developed (Williams and de Lautour 1980a) and used in the genera *Trifolium*, *Lotus*, and *Ornithopus*.

The authors would like to thank their colleagues Dr. Derek White, Isabelle Verry, and Michael Baker for helpful comments.

METHODS

Pollination procedures have been described by Williams (1978, 1980) and Williams and de Lautour (1980b). Embryo culture and partial ovule culture with transplanted nurse endosperm have been described in detail by Williams and de Lautour (1980a). The standard technique is based on insertion of a hybrid embryo into a cellular endosperm dissected from a normally developing seed and aseptic transfer of embryo and endosperm together to the surface of a sterile nutrient agar medium. The preferred source of nurse endosperm is the female parent species, although it has also been found possible to use the male parent or even a third species (de Lautour et al. 1978). Young plants are transferred from sterile culture to soil at the 3- to 5-leaf stage.

RESULTS AND DISCUSSION

Trifolium ambiguum M. Bieb. × T. repens L.

This cross was reported for the first time by Williams (1978). Tetraploid ($2n = 4 \times 32$). *T. ambiguum* was used as female parent with *T. repens* ($2n = 32$) as male.

The main aim of this cross was to transfer rhizomatous habit and virus resistance (Barnett and Gibson 1975) from *T. ambiguum* to *T. repens*. Large-scale crossing between these species has not been attempted, and so far only 4 hybrid plants have been raised to maturity. One is partially fertile, has produced small numbers of F_3 and backcrosses to *T. repens*, and has given promising results in grazed field trials. Most of the F_2, F_3, and backcross individuals have the stoloniferous habit of *T. repens*, but a few appear to be rhizomatous. Testing for virus resistance, drought tolerance, and rhizobial affinity has not yet been carried out.

Trifolium repens × T. uniflorum L.

Trifolium repens and *T. uniflorum* have been hybridized by Pandey (1957) and Gibson et al. (1971) without embryo culture, but embryo-culture techniques have since been used to obtain further F_1 hybrids. *T. repens* has a comparatively shallow, fibrous root system that may contribute to its susceptibility to drought and root-chewing insects. *Trifolium uniflorum*, on the other hand, has thick taproots but lacks the vigor and stoloniferous habit of *T. repens*. The main aim of this cross was to confer a more taprooted habit on white clover.

Several F_1 hybrids have been grown to maturity, and a number of F_2 plants and backcrosses to *T. repens* have been obtained. Fertility of the hybrids is variable and generally low. Initial selection is being carried out for improved fertility. The plants generally show the prostrate small-leaved type of *T. uniflorum* and have thick taproots.

Trifolium ambiguum × T. hybridum L.

Hybrids between these species have previously been obtained by Keim (1953) and Evans (1962) using embryo culture. So far our crosses have been successful only with tetraploid *T. ambiguum* ($2n = 4 \times = 32$) pollinated by diploid *T. hybridum* ($2n = 16$) (Williams 1980). Hybridization has been attempted in order to combine the tall habit and vigor of *T. hybridum* with the virus resistance of *T. ambiguum*. Of 10 putative hybrids produced by embryo culture, 2 have flowered. Both show very low pollen fertility (0.5% and 3.1%) and have produced no progeny. One plant is very weak, but the other is growing well in grazed-field trials. Both are triploid, with one *T. hybridum* genome and two genomes from *T. ambiguum*. Phenotypically the plants resemble *T. ambiguum* and lack the tall-growing stems of *T. hybridum*.

Lotus pedunculatus Cav. × L. corniculatus L.

Several hybrids were produced by embryo culture after pollination of tetraploid *L. pedunculatus* (derived by colchicine treatment) with pollen from several cultivars of *L. corniculatus*. A selected F_3 population has been evaluated as the breeders' line Grasslands 4712. Although no detailed study of pollen fertility was carried out in the early generations, there have been no problems with seed production in later generations.

Lotus corniculatus performs well under intensive farming. *L. pedunculatus*, on the other hand, is of value on soils of low fertility and pH (Levy 1970), is resistant to chewing insects (Farrell and Sweney 1974), and contains foliar tannins that help to prevent bloat in grazing ruminants (Jones and Lyttleton 1971). The cross was made to combine these desirable characteristics.

Extensive agronomic evaluations have indicated that the hybrid population, although variable in type, predominantly resembles *L. corniculatus* in plant habit but is inferior in yield and persistence to several populations of *L. corniculatus* evaluated recently (Charlton et al. 1978). The hybrid contains condensed tannins in the leaves (Ross and Jones 1974) but lacks resistance to chewing insects (Farrell and Sweney 1974).

Ornithopus sativus Brot. × O. compressus

This hybrid was first reported by Williams and de Lautour (1980a) and was obtained by partial ovule culture with transplanted endosperm from the legume *Ulex europaeus*. The aim of this cross was to combine the prostrate growth form and hard seed of *O. compressus* with the productivity and easy seed hullability of *O. sativus* (Williams et al. 1975). Only one hybrid was raised to maturity. This hybrid was morphologically abnormal, with less than 10% pollen fertility. However, from this single F_1, an F_6 experimental breeders' line has been obtained that shows high fertility and variation in flower color and in the characters under selection. Some families within this population combine all the desired characteristics of the two parent species.

Ornithopus pinnatus (Mill.) Druce × O. sativus

Williams and de Lautour (1980b) crossed these species by means of embryo culture with nurse endosperms from *O. compressus* and *O. pinnatus*. Four F_1 hybrids were produced, but all were sterile. One which was chromosome-doubled with colchicine attained partial fertility ($< 10\%$ pollen fertile), but no viable seeds were produced because of continuing endosperm failure. Embryo culture with nurse endosperm was repeatedly required to obtain F_2 to F_5 generations. The cross was made to combine the prostrate habit and foliar tannins of *O. pinnatus* with the productivity of *O. sativus*. Although the F_2 to F_5 populations were functionally sterile because of endosperm failure, most individuals combined the desired characteristics of the parent species.

CONCLUSION

Infertility is an apparent barrier to the success of only two of the six hybrid combinations discussed in this paper. *O. sativus* × *O. compressus* and *L. pedunculatus* × *L. corniculatus* appear to suffer from no fertility barriers to their potential success. *T. ambiguum* × *T. repens* and *T. repens* × *T. uniflorum* have reduced fertility but show some prospects for improvement in fertility by selection. *O. pinnatus* × *O. sativus* is functionally sterile, owing to endosperm failure. However, the production in later generations of this cross of normal embryos accompanied by occasional endosperm development suggests that this type of infertility might prove to be temporary. The triploid *T. ambiguum* × *T. hybridum* hybrids so far produced are highly sterile. Their sterility may be overcome by crossing parents of the same ploidy or by chromosome-doubling the triploid F_1 hybrids.

A feature of the hybrid program using embryo culture with transplanted nurse endosperm is its high success rate for obtaining hybrids with a relatively undefined technique. All the examples cited have been achieved with no more than a few hundred flowers pollinated and generally fewer than 100 putative hybrid embryos cultured on exploratory media. Prospects for refinement and increased use of this method are promising. Apart from the *O. sativus* × *O. compressus* hybrid population, which already appears to be worthy of agricultural acceptance, the other examples should all benefit from production of larger numbers of F_1 hybrids based on a wider range of selected parents.

LITERATURE CITED

Barnett, O.W., and P.B. Gibson. 1975. Identification and prevalence of white clover viruses and the resistance of *Trifolium* species to these viruses. Crop Sci. 15:32–37.

Charlton, J.F.L., E.R.L. Wilson, and M.D. Ross. 1978. Plant introduction trials. Performance of *Lotus corniculatus* introductions as spaced plants in Manawatu. N.Z. J. Expt. Agric. 6:201–206.

de Lautour, G., W.T. Jones, and M.D. Ross. 1978. Production of interspecific hybrids in *Lotus* aided by endosperm transplants. N.Z. J. Bot. 16:61–68.

Evans, A.M. 1962. Species hybridization in *Trifolium*. I. Methods of overcoming species incompatibility. Euphytica 11:164–176.

Farrell, J.A.K., and W.J. Sweney. 1974. Plant resistance to the grass grub *Costelytra zealandica* (Coleoptera: Scarabaeidae). III. Resistance in *Lotus* and *Lupinus*. N.Z. J. Agric. Res. 17:69–72.

Gibson, P.B., C.-C. Chen, J.T. Gillingham, and O.W. Barnett. 1971. Interspecific hybridisation of *Trifolium uniflorum* L. Crop Sci. 11:895–899.

Jones, W.T., and J.W. Lyttleton. 1971. Bloat in cattle. XXXIV. A survey of legume forages that do and do not produce bloat. N.Z. J. Agric. Res. 14:101–107.

Keim, W.F. 1953. Interspecific hybridisation in *Trifolium* utilizing embryo culture techniques. Agron. J. 45:601–606.

Levy, E.B. 1970. Grasslands of New Zealand. Government Printer, Wellington, N.Z.

Pandey, K.K. 1957. A self-compatible hybrid from a cross between two self-incompatible species in *Trifolium*. J. Hered. 48:278–281.

Ross, M.D., and W.T. Jones. 1974. Bloat in cattle. XL. Variation in flavanol content in *Lotus*. N.Z. J. Agric. Res. 17:191–195.

Williams, E. 1978. A hybrid between *Trifolium repens* and *T. ambiguum* obtained with the aid of embryo culture. N.Z. J. Bot. 16:499–506.

Williams, E.G. 1980. Hybrids between *Trifolium ambiguum* and *T. hybridum* obtained with the aid of embryo culture. N.Z. J. Bot. 18:215–220.

Williams, E.G., and G. de Lautour. 1980a. The use of embryo culture with transplanted nurse endosperm for the production of interspecific hybrids in pasture legumes. Bot. Gaz. 141:252–257.

Williams, E.G., and G. de Lautour. 1980b. Production of tetraploid hybrids between *Ornithopus pinnatus* and *O. sativus* using embryo culture. N.Z. J. Bot. 19:23–30.

Williams, E. and D.W.R. White. 1976. Early seed development after crossing of *Trifolium ambiguum* and *T. repens*. N.Z. J. Bot. 14:307–314.

Williams, W.M., G. de Lautour, and W. Stiefel. 1975. Potential of Serradella as a winter annual forage legume on sandy coastal soil. N.Z. J. Expt. Agric. 3:339–342.

Differential Interspecific Compatibilities Among Genotypes of *Trifolium sarosiense* and *T. pratense*

N.L. TAYLOR, G.B. COLLINS, P.L. CORNELIUS, and J. PITCOCK
University of Kentucky, Lexington, Ky., U.S.A.

Summary

Interspecific compatibilities were investigated among genotypes of *Trifolium sarosiense* Hazsl., used as females, and *T. pratense* L. (red clover), used as males. This research was undertaken to introduce the perenniality of *T. sarosiense* into *T. pratense* by hybridization and thus to increase the longevity of stands of red clover for hay, pasture, and soil improvement. Objectives of the study were to further the knowledge of interspecific hybridization barriers in *Trifolium* and to isolate clones of *T. sarosiense* and *T. pratense* that could aid in their eventual hybridization, either directly or by embryo rescue.

Twenty-three clones of *T. sarosiense* were used as females to cross with 8 clones of *T. pratense* as males. Plants were maintained in a growth chamber at 15°–18°C at a photoperiod of 16 hours, so that all plants flowered continuously for 3 years. Seeds were dissected from pods and visually classified into four categories of seed size. Class I seeds were largest and, based on grow-out tests, were considered to be selfs.

Parent genotypes were significantly different in seed set, but no combination produced hybrid seeds that would germinate in vivo. *T. sarosiense* × *T. pratense* interaction effects, while significant, were not as important as the female effects.

T. sarosiense and *T. pratense* clones that produced relatively large amounts of class II hybrid seeds (the largest among the hybrid seeds) also tended to produce the largest amounts of class III seeds (the next largest) but did not often produce large amounts of class IV seeds (the smallest). It was inferred that production of class II and class III shriveled seeds was of greater importance than producion of class IV seeds. Class II and III embryos, although not capable of germinating in vivo, were capable of being rescued via tissue culture because of their delayed abortion.

Recurrent selection for viable seeds probably would be effective for interspecific hybridization. In the present hybrid, however, recurrent selection is not necessary since embryos are being produced that are sufficiently developed for rescue by tissue culture.

KEY WORDS: red clover, perenniality, hybridization, embryo rescue, *Trifolium sarosiense, Trifolium pratense.*

INTRODUCTION

Hybridization of red clover, *Trifolium pratense* L., with perennial *Trifolium* species was attempted to incorporate greater perenniality into red clover. The only hybrids that had been made before this experiment involved annual species and thus did not contribute to greater longevity (Taylor et al. 1963).

Hybrids have been obtained among perennial species closely related to red clover. None of these hybrids would cross with *T. pratense*. However, shriveled seeds were produced from the cross of *T. sarosiense* × *T. pratense* (Quesenberry and Taylor 1976, 1977, 1978).

Investigations of interspecific barriers (Taylor et al. 1980) have suggested techniques for enhancing hybridization with annual species but not with perennial species. However, one barrier not examined to this point is the differential compatibility of genotypes. Since preliminary crosses of *T. sarosiense* × *T. pratense* produced shriveled seeds, it seemed possible that crossing a wide range of genotypes of these species might aid in eventual hybridization, either directly or with aid of embryo rescue techniques.

METHODS

The plant materials used in this study were previously studied by Quesenberry and Taylor (1978). For females, 23 genotypes of *T. sarosiense* (P.I. 292827, origin Sweden, 2n = 48) and for males, 8 genotypes of *T. pratense* (Kenstar, origin Kentucky, 2n = 14) were chosen. Of 184 possible parental hybrid combinations, 163 crosses were attempted.

N.L. Taylor is a Professor of Agronomy, G.B. Collins is a Professor of Agronomy, P.L. Cornelius is an Associate Professor of Agronomy and Statistics, and J. Pitcock is a Graduate Assistant at the University of Kentucky.
The investigation reported in this paper (No. 81-3-28) is in connection with a project of the Kentucky Agricultural Experiment Station and is published with the approval of the Director.

Each parental plant was increased vegetatively to several plants that flowered continually when grown at 15°C to 18°C with a photoperiod of 16 hours at approximately 13,000 lux. The number of *T. sarosiense* flowers pollinated/head ranged from 12 to 89 (averaging about 35), and the number of crosses/parental combination ranged from 1 to 13. Similar intraspecific *T. sarosiense* and *T. pratense* crosses were made for controls. Approximately one month after the crosses were made, heads were harvested and dried, and seed pods were individually dissected. Seeds produced were counted and visually classified: class I, normal, with seed weight of about 2.5 mg; class II, normal or dark brown, shriveled, with weight of about 0.4 mg; class III, shriveled, with weight of about 0.02 mg; and class IV, stimulated ovules with weight of about 0.01 mg.

Nonfertilized seeds (coats) were not counted but weighed 0.003 mg. These clones and the crossing techniques described above were also used by Collins et al. (1982) for embryo rescue investigations that are reported in these proceedings.

For each head used, the proportion (p) of shriveled seeds in classes II, III, and IV was calculated and transformed to arcsin ($\sqrt{p}$). A zero proportion was assigned a value of $\frac{1}{4}n$ before transformation, where n is the number of flowers minus the number of seeds in class I. Flowers that produced class I seeds were omitted because such seeds subsequently were found to be selfs, not hybrids. Least squares analyses, with each value of p weighted by its value of n, of both the transformed and the untransformed data were calculated to determine the significance of *T. sarosiense* and *T. pratense* parental effects and the genotype interactions.

RESULTS

Intraspecific crosses of *T. sarosiense* and *T. pratense* produced 73% and 91% viable seeds, respectively, of flowers crossed. Both species were self-incompatible and pro-

Fig. 1. Seeds dissected from pods of crosses of *Trifolium saro-siense* × *T. pratense*. Left, class I seeds (selfs); top right, left to right, class II and class III; bottom right, left to right, class IV and nonfertilized seed coats.

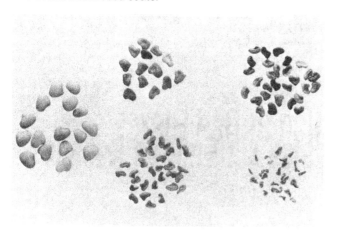

generally were significant, the range in seeds/flower pollinated was considerably less than that for females. Ranges in seeds/flower were 0.0033 to 0.0084, 0.0819 to 0.1440, and 0.3315 to 0.4567 for classes II, III, and IV respectively. Analysis of variance indicated that class II means were not significantly different. However, except for the 2 lowest-seed-setting clones, *T. pratense* effects were nonsignificant if analyzed as either transformed or untransformed data and differed significantly from 0.

With respect to class III seeds, only 3 clones differed significantly from the 2 lowest-ranking clones. As was the case for female effects, superior clones for classes II and III were not the same as those for class IV.

Male x Female Effects

Interactions of females × males were significant in some seed classes but were small relative to the magnitude of the *T. sarosiense* clone effects in all classes and the *T. pratense* clone effects in classes III and IV.

duced few seeds via selfing. Nonpollinated flowers of *T. sarosiense* female parents produced only empty seed coats (Fig. 1). Analysis of variance indicated significant (P < 0.05) female and male main effects, and female × male interactions for most classes of hybrid seed.

Female Effects

The number of heads crossed/female *T. sarosiense* plant varied from 2 to 107. Only shriveled hybrid seeds (classes II to IV) were produced, none of which would germinate in vivo. Ranges in seeds/flower were 0 to 0.0237, 0.0024 to 0.2673, and 0.1738 to 0.5309 for classes II, III, and IV, respectively. Significant differences based on pairwise t-tests indicated that high-ranking clones in class II were 16, 21, 12, 15, and 1, and these were the only clones that differed significantly from 0. In class III shriveled seeds, high-ranking clones were 21, 6, 4, 16, and 15, and the 4 lowest-ranking clones (23, 19, 22, and 13) did not differ significantly from 0. In class IV seeds, all clones differed significantly from 0. High-ranking clones were 14, 11, 13, and 22, and low-ranking clones were 23, 20, 19, 15, and 5.

Comparison of clonal ranking in classes II and III indicated clones 16, 21, 12, and 15 were superior. These clones do not rank high in class IV, a fact that shows that production of extremely small shriveled seeds may be under different genetic control than that of the larger seeds (classes II and III). On the other hand, with the exception of clone 15, high-ranking class II and III clones did not rank exceptionally low in class IV. Clones 23, 19, and 20 were low in all classes.

Male Effects

Only 8 *T. pratense* clones were used as males on the *T. sarosiense* clones, and although differences among clones

DISCUSSION

Genotypic differences in interspecific incompatibilities were not of sufficient magnitude to produce completely viable hybrid seeds. However, considerable differences in class II and III seeds for both female and male effects were shown, a finding that may have important implications for eventually overcoming the barrier to hybridization of *T. sarosiense* with *T. pratense*. High frequencies of class II shriveled seeds are thought to be important because with these seeds abortion occurs rather late in development. An increase in frequency of class IV seeds that abort prior to 10 to 12 days after pollination would not be beneficial.

Rescue was successful on embryos from crosses of some of the same clones used in this study (Collins et al. 1982), and it is quite likely that these rescued embryos were of class II size. Of 16 embryos rescued, the most frequently successful female and male clones were 1 and 21, respectively. *T. sarosiense* clone 1 ranked fifth among class II seed-producing clones, and *T. pratense* clone 21 ranked second in class II seed production. However, a complete comparison is not valid because more pollinations and rescue attempts were made from the combination of 1 × 21 than from any other combination. On the other hand, it probably would be safe to conclude that crosses with *T. sarosiense* clone 23 would have been unsuccessful even with embryo rescue.

LITERATURE CITED

Collins, G.B., G.C. Phillips, and N.L. Taylor. 1982. Hybridization of red clover with perennial *Trifolium* species using in-vitro embryo rescue. Proc. XIV Int. Grassl. Cong.

Quesenberry, K.H., and N.L. Taylor. 1976. Interspecific hybridization in *Trifolium* L. Sect. *Trifolium* Zoh. I. Diploid hybrids among *T. alpestre* L., *T. rubens* L., *T. heldreichianum* Hausskn., and *T. noricum* Wulf. Crop Sci. 16:382–386.

Quesenberry, K.H., and N.L. Taylor. 1977. Interspecific hybridization in *Trifolium* L. Sect. *Trifolium* Zoh. II. Fertile polyploïd hybrids between *T. medium* L. and *T. sarosiense* Hazsl. Crop Sci. 17:141–145.

Quesenberry, K.H., and N.L. Taylor. 1978. Interspecific hybridization in *Trifolium* L. Sect. *Trifolium* Zoh. III. Partially fertile hybrids of *T. sarosiense* Hazsl. × 4 × *T. alpestre* L.

Crop. Sci. 18:536–540.

Taylor, N.L., R.F. Quarles, and M.K. Anderson. 1980. Methods of overcoming interspecific barriers in *Trifolium*. Euphytica 29:441–450.

Taylor, N.L., W.H. Stroube, G.B. Collins, and W.A. Kendall. 1963. Interspecific hybridization of red clover (*Trifolium pratense* L.). Crop Sci. 3:549–552.

Successful Hybridization of Red Clover with Perennial *Trifolium* Species via Embryo Rescue

G.B. COLLINS, N.L. TAYLOR, and G.C. PHILLIPS

University of Kentucky, Lexington, Ky., U.S.A.

Summary

The objective of this research was to hybridize long-lived perennial *Trifolium* species with the agronomically desirable forage species red clover (*Trifolium pratense* L.). Numerous previous attempts to hybridize red clover with perennial *Trifolium* species using conventional crossing procedures have been unsuccessful. In the present research, in-vitro embryo culture methods were used to facilitate the recovery and growth of hybrid embryos prior to in-situ embryo abortion. The major effort was directed toward the cross of *T. sarosiense* Hazsl. with red clover. Preliminary efforts have been made to cross zigzag clover (*T. medium* L.) and *T. alpestre* L. with red clover.

Crosses were made with the perennial species serving as female parents. Embryos were dissected aseptically from pollinated florets at 12–19 days after the crosses were made. The immature embryos, principally at heart stage, were cultured on agar-solidified nutrient media modified from that developed previously for tissue and cell cultures of red clover. The initial culture period was 8–14 days on a medium containing a high concentration of sucrose, a moderate level of auxin, and a low cytokinin activity. The high level of sucrose prevented precocious germination of radicles from the embryos, and the levels of growth regulators used allowed for the maturation of the embryos. Embryos that survived this culture treatment were transferred to a medium with low auxin and moderate cytokinin levels, which encouraged shoot germination. Hybrid shoot numbers were increased on a low-auxin, high-cytokinin medium and were subsequently rooted before transfer to soil in the greenhouse.

Numerous hybrid plants have been obtained from the cross *T. sarosiense* × *T. pratense*. About 10% of the cultured embryos were rescued on the optimal culture sequence. Five F_1 lines of the hybrid were grown successfully to maturity. Rescued plants were verified to be hybrid on the basis of cytological and genetical evidence. Preliminary efforts with the crosses of *T. medium* and *T. alpestre* with red clover have produced a few plants using the same methods, but verifications have not been performed.

The embryo rescue procedures utilized in these studies have produced the first successful, verified cross of a perennial *Trifolium* species with red clover. These procedures should be useful for producing other interspecific hybrids of red clover. The transfer of genes for the perennial character to red clover from interspecific hybrids would be of great value in efficient production of red clover by improving field persistence.

KEY WORDS: red clover, persistence, *Trifolium*, hybridization, in vitro, embryo rescue.

INTRODUCTION

Red clover (*Trifolium pratense* L.) is a valuable forage legume around the world, but it is severely limited in production by its lack of field persistence (Taylor et al. 1963). Numerous attempts to hybridize red clover with perennial

species using conventional crossing procedures have not been successful (Taylor and Smith 1979, Taylor 1980). Some of these perennial species can be hybridized among themselves, but only with difficulty (Maizonnier 1972, Quesenberry and Taylor 1976, 1977, 1978). Hybrid failure between red clover and the perennial species seems to be enforced by postzygotic mechanisms of isolation (Quesenberry and Taylor 1976, 1978; Kazimierska 1978; Taylor and Smith 1979; Taylor et al. 1980). Red clover has been hybridized successfully only with two annual species, which have provided no new genetic variation for persistence (Taylor and Smith 1979, Taylor 1980).

G.B. Collins is a Professor of Agronomy, N.L. Taylor is a Professor of Agronomy, and G.C. Phillips is a Graduate Research Assistant at the University of Kentucky.

The investigation reported in this paper (No. 81-3-21) is in connection with a project of the Kentucky Agricultural Experiment Station and is published with the approval of the Director.

Postzygotic barriers to interspecific and intergeneric hybridization have been overcome in a number of cases by using the method of in-vitro culture of embryos prior to abortion (Raghavan 1977). Attempts at embryo culture of interspecific red clover hybrids have provided only limited success (Keim 1953, Evans 1962).

Our laboratory has recently developed a basal medium that was adjusted in both inorganic and organic composition for optimal red clover tissue growth in vitro (Phillips and Collins 1979a). We also have identified specific growth regulator source and level requirements for red clover manipulation in vitro, including plant regeneration from cells and callus (Phillips and Collins 1979a, 1979b, 1980). The auxin source of preference for shoot growth is picloram (4-amino-3,5,6-trichloropicolinic acid), described previously (Collins et al. 1978). Adenine has been used as a cytokinin with low activity, and N6-benzyladenine has been used as a cytokinin with high activity.

The objective of this research was to hybridize red clover with *T. sarosiense* Hazsl., using improved cultural methods for embryo rescue. The development of an efficient cultural method provides the basis for rescue of other interspecific hybrids of red clover. Preliminary efforts involving hybrids with zigzag clover (*T. medium* L.) and *T. alpestre* L. were undertaken in order to indicate the potential use of the *T. sarosiense* × *T. pratense* system as a model for the production of other red clover hybrids.

METHODS

Plants of Kenstar red clover (2n = 14) were used as a pollen source for hand pollination of *T. sarosiense* (2n = 48), zigzag clover (2n = 64–80), and *T. alpestre* (2n = 16) flowers in a greenhouse following the methods of Taylor (1980). Parent clones used in the *T. sarosiense* × *T. pratense* cross were the same as those described by Taylor et al. (1982). Pollinated florets were collected 12–19 days after pollination and surface sterilized. Immature embryos were dissected aseptically from the ovules and cultured on modifications of the L2 basal medium (Phillips and Collins 1979a) that varied in concentration of sucrose, level of auxin, and activity of cytokinin. After 8–14 days, viable embryos were transferred to modified media containing the standard level of 2.5% (w/v) sucrose and varying in levels of auxin and cytokinin. Germinated embryos were propagated in vitro as shoot cultures following the methods developed for red clover meristem-tips (Phillips and Collins 1979b). All cultures were maintained under continuous, diffuse light. Rooted plants were potted in soil and vermiculite mixtures, returned to the greenhouse, and grown to maturity.

Plants rescued from the cross of *T. sarosiense* with red clover were evaluated to determine their hybridity. Somatic chromosome numbers were determined from root-tip squashes following the methods of Sharma and Sharma (1972). Rhizomatous root habit is a dominant gene marker for the perennial parent (Quesenberry and Taylor 1976, 1977). The occurrence of a central,

V-shaped leaf-mark is a dominant gene marker for the male parent, red clover (Wexelsen 1932).

RESULTS

The cross of *T. sarosiense* × *T. pratense* yielded a 29% frequency of embryo formation. Only globular, heart, and early torpedo stages of embryos were obtained in the cross, and heart-stage embryos predominated. These immature embryos were cultured with a 20% survival rate on modified L2 medium containing 12.5% sucrose, 25 nM picloram (a moderate level), and low cytokinin activity in the form of 15 μM adenine. The high level of sucrose prevented precocious germination of radicles from the immature embryos; other levels of sucrose did not promote survival as efficiently. Higher levels of growth regulators promoted loss of organization. After 8–14 days in culture, embryos turned yellow or white and ceased growth. Young cotyledons were generally present.

Prior to degeneration on the initial culture medium, *T. sarosiense* × *T. pratense* embryos were transferred to modified L2 medium with the standard level of sucrose, containing 4 nM picloram (a very low level) and 0.66 μM 6-benzyladenine (a moderate level). Shoots germinated from the embryos within 2–4 weeks at a 45% frequency. Another 40% of the embryos survived this culture treatment but suffered a loss of organization. Germinated embryo shoots were cultured on modified L2 medium containing 12 nM picloram (a low level) and 2.2 μM 6-benzyladenine (a high level). This medium was appropriate for shoot number increase from all germinated embryos.

Spontaneously rooted plants were returned to the greenhouse and grown to maturity. Only half the F$_1$ lines rescued in vitro survived to maturity. Plants rescued from the cross of *T. sarosiense* with red clover possessed 31 somatic chromosomes, the expected F$_1$ number (24 from *T. sarosiense* and 7 from red clover). The hybrid exhibited the leaf-mark trait inherited from the male parent and the rhizomatous root habit inherited from the female parent.

Immature embryos obtained from crosses of red clover with zigzag clover and *T. alpestre*, produced in fewer numbers, also responded to the embryo rescue method described above. The same media were appropriate for immature embryo growth and germination, although some adjustments in auxin and/or cytokinin were required for shoot number increase. However, no verifications have been performed yet to prove their hybridity.

DISCUSSION

Embryo rescue methods were successful in this hybridization study where previous efforts have failed. One reason for this success was that the basal medium used had been optimized experimentally for optimal growth of red clover tissue in vitro. This basal medium allowed for maximal growth rates of immature embryos and minimal toxicity reactions. A second reason for this success was that specific growth-regulator source requirements were known, and levels were adjusted experimentally for optimal ontogenetic development of hybrid embryos. Auxin was required for the maturation of young embryos, which were highly sensitive to active cytokinins. Germination of shoots from these embryos required a reduction of auxin and an increase in cytokinin activity. Shoot numbers were increased on a low-auxin, high-cytokinin medium. And, third, the concentration of sucrose was adjusted experimentally, following classical methods, for the optimal physical effect on the growth and maturation of the heart-stage embryos.

Using the optimal cultural sequence, more than 10% of the developing *T. sarosiense* × *T. pratense* embryos were rescued in vitro. Preliminary results obtained with two other perennial species hybrids of red clover indicate the potential use of this improved cultural technique for obtaining other wide hybrids with red clover. The field persistence of red clover may be greatly improved by the transfer of genes for the perennial character from interspecific hybrids produced by the embryo rescue methods presented here.

LITERATURE CITED

Collins, G.B., W.E. Vian, and G.C. Phillips. 1978. Use of 4-amino-3,4,5-trichloropicolinic acid as an auxin source in plant tissue cultures. Crop Sci. 18:286–288.

Evans, A.M. 1962. Species hybridization in *Trifolium.* I. Methods of overcoming species incompatibility. Euphytica 11:164–176.

Kazimierska, E.M. 1978. Embryological studies of cross compatibility in the genus *Trifolium* L. I. Hybridizations of *T. pratense* L. with some species in the subgenus *Lagopus* Bernh. Genet. Polon. 19(1):1–14.

Keim, W.F. 1953. Interspecific hybridization in *Trifolium* utilizing embryo culture techniques. Agron. J. 45:601–606.

Maizonnier, D. 1972. Hybridization between four species of perennial clover. Ann. Amélior. Pl. 22:375–387.

Phillips, G.C., and G.B. Collins. 1979a. *In vitro* tissue culture of selected legumes and plant regeneration from callus cultures of red clover. Crop Sci. 19:59–64.

Phillips, G.C., and G.B. Collins. 1979b. Virus symptom-free plants of red clover using meristem culture. Crop Sci. 19:213–216.

Phillips, G.C., and G.B. Collins. 1980. Somatic embryogenesis from cell suspension cultures of red clover. Crop Sci. 20:323–326.

Quesenberry, K.H., and N.L. Taylor. 1976. Interspecific hybridization in *Trifolium* L., Sect. *Trifolium* Zoh. I. Diploid hybrids among *T. alpestre* L., *T. rubens* L., *T. heldreichianum* Hausskn., and *T. noricum* Wulf. Crop Sci. 16:382–386.

Quesenberry, K.H., and N.L. Taylor. 1977. Interspecific hybridization in *Trifolium* L., Sect. *Trifolium* Zoh. II. Fertile polyploid hybrids between *T. medium* L. and *T. sarosiense* Hazsl. Crop Sci. 17:141–145.

Quesenberry, K.H., and N.L. Taylor. 1978. Interspecific hybridization in *Trifolium* L., Sect. *Trifolium* Zoh. III. Partially fertile hybrids of *T. sarosiense* Hazsl. × 4 × *T. alpestre* L. Crop Sci. 18:536–540.

Raghavan, V. 1977. Applied aspects of embryo culture. *In* J. Reinert and Y.P.S. Bajaj (eds.) Plant cell, tissue, and organ culture, 375–393. Springer-Verlag, New York.

Sharma, A.K., and A. Sharma. 1972. Chromosome techniques: theory and practice, 2nd ed. Univ. Park Press, Baltimore.

Taylor, N.L. 1980. Clovers. *In* W.R. Fehr and H.H. Hadley (eds.) Hybridization of crop plants 261–272. Am. Soc. Agron. and Crop Sci. Soc. of Am., Madison, Wis.

Taylor, N.L., G.B. Collins, P.L. Cornelius, and J. Pitcock. 1982. Interspecific compatibilities among genotypes of *Trifolium sarosiense* and *T. pratense.* Proc. XIV Int. Grassl. Cong.

Taylor, N.L., R.F. Quarles, and M.K. Anderson. 1980. Methods of overcoming interspecific barriers in *Trifolium.* Euphytica 29:441–450.

Taylor, N.L., and R.R. Smith. 1979. Red clover breeding and genetics. Adv. Agron. 31:125–154.

Taylor, N.L., W.H. Stroube, G.B. Collins, and W.A. Kendall. 1963. Interspecific hybridization of red clover (*Trifolium pratense* L.). Crop Sci. 3:549–552.

Wexelsen, H. 1932. Segregations in red clover (*Trifolium pratense* L.). Hereditas 16:219–240.

Phylogenetic Investigations of *Paspalum dilatatum* and Related Species

B.L. BURSON
USDA-SEA-AR Grassland, Soil and Water Research Laboratory,
Temple, Tex., U.S.A.

Summary

Common dallisgrass (*Paspalum dilatatum* Poir) is an important forage grass throughout many of the warmer regions of the world, including the southern U.S.A. Apomixis has been an important factor in the origin and preservation of this species. It is a 50-chromosome natural hybrid with three genomes that pair as 20 bivalents and 10 univalents at meiosis. Because it is a pentaploid obligate apomict, plant breeders have been unsuccessful in improving the species. The apomictic barrier has not been altered by radiation or interspecific hybridization. In an effort to circumvent this apomictic barrier, a phylogenetic investigation was initiated to identify its progenitors, to eventually resynthesize the species, and to allow breeding progress. A sexual tetraploid, yellow-anthered biotype is closely related to common dallisgrass, and its chromosomes are homologous with 40 chromosomes of common dallisgrass. The yellow-anthered biotype was used as a cytological substitute for common in the phylogenetic program and was crossed with other *Paspalum* species. Two diploid species, *P. intermedium* Munro. ex. Morong and *P. jurgensii* Hackel, have genomes homologous with a yellow-anthered dallisgrass genome. The genome formulas II, JJ, and II JJ have been assigned to *P. intermedium*, *P. jurgensii,* and yellow-anthered dallisgrass, respectively. Findings to date indicate that *P. intermedium* and *P. jurgensii* or closely related species are the ancestors of yellow-anthered dallisgrass. Information gained from hybrids involving other species indicates that the I and J genomes are widespread in the genus. One or both of the genomes or evolved forms have been identified in species of the Dilatata, Paniculata, Quadrifaria, and Virgata groups but are absent in the Disticha, Notata, and Setacea groups. Common dallisgrass has the genome formula II JJ X; X is an unknown genome. It was proposed that the grass originated from a cross between a sexual tetraploid with the genome constitution II JJ and an apomictic hexaploid II JJ XX. Efforts are underway to identify the hexaploid species.

KEY WORDS: common dallisgrass, yellow-anthered dallisgrass, *Paspalum intermedium*, *Paspalum jurgensii*, chromosome pairing, interspecific hybrids, genomes, species relationships.

INTRODUCTION

Paspalum is a large, diverse genus with more than 400 species. With few exceptions, their chromosome numbers are multiples of 10. Apomixis is prevalent within the genus and has been an important factor in the evolution of many of the species. Apomixis, once considered an evolutionary dead end, is now recognized as an important factor in the origin and preservation of many grass species (Bashaw et al. 1970). Common dallisgrass (*Paspalum dilatatum* Poir), an important forage grass native to southern Brazil, Uruguay, and northeastern Argentina, has been preserved since its origin by apomixis. The grass is a natural hybrid with 50 chromosomes that pair at meiosis as 20 bivalents and 10 univalents (Bashaw and Forbes 1958). Because it is an obligate apomict (Bashaw and Holt 1958), sterility was circumvented and its genome constitution was preserved, although it is meiotically irregular.

Common dallisgrass is an important forage grass in the southern U.S.A. and the warmer regions of the world, but it has poor seed fertility and is susceptible to ergot (*Claviceps paspali*). Consequently, the species needs to be improved to increase its use as a forage grass. Although apomixis has been responsible for the preservation of this grass, apomixis is also a barrier to improvement. Improving an apomictic species requires a sexual or partially sexual cross-compatible plant to be used as a female parent in hybridization with the apomict. To date all attempts to discover naturally occurring sexual forms of common dallisgrass through plant exploration or to create sexuality by radiation (Bashaw and Hoff 1962) or interspecific hybridization (Bennett and Bashaw 1966) have been unsuccessful.

A different approach to circumventing this apomictic barrier was initiated in 1970. Because common dallisgrass is a natural hybrid with three distinct genomes (10II +

10II + 10I), we hoped sexual forms could be synthesized once the grass's progenitors were identified. Consequently, a phylogenetic investigation was begun to determine those progenitors.

RELATIONSHIP BETWEEN DALLISGRASS BIOTYPES

The most important plant in the phylogenetic program is a dallisgrass biotype that was introduced to the United States from Uruguay. This biotype is morphologically and cytologically distinct. It has yellow anthers, a more erect growth habit, and more pubescent spikelets, and it is more fertile than the common biotype. However, it is a poor forage grass. It reproduces sexually (Bashaw and Holt 1958) and has 40 chromosomes that pair as 20 bivalents during meiosis (Bashaw and Forbes 1958).

This sexual, yellow-anthered biotype was crossed with apomictic common dallisgrass. The F$_1$ hybrid was sexual and had 45 chromosomes that paired as 20 bivalents and 5 univalents at meiosis (Bennett et al. 1969). The bivalent pairing indicates a close homology between two genomes of both biotypes, suggesting that the yellow-anthered biotype might be used as a substitute for common dallisgrass in a phylogenetic program. To determine the validity of this assumption, yellow-anthered dallisgrass and the yellow-anthered × common dallisgrass F$_1$ hybrid were both crossed with two diploid species (2n = 20), *P. jurgensii* Hackel and *P. notatum* Flugge. Meiotic chromosome pairing in the hybrids was similar whether the yellow-anthered biotype or the F$_1$ hybrid with common dallisgrass was used as the female parent (Burson and Bennett 1972b, Burson et al. 1973). Homology between the genomes of the 2 biotypes and the chromosomes of the diploids substantiates the assumption that the yellow-anthered biotype can be substituted cytologically for common dallisgrass in a phylogenetic study.

RELATIONSHIP BETWEEN DALLISGRASS AND DIPLOID SPECIES

Yellow-anthered dallisgrass was hybridized with several diploid (2n = 20) species in an effort to identify in-

The author is a Research Geneticist with USDA-SEA-AR. This is a contribution of USDA-SEA-AR in cooperation with the Texas Agricultural Experiment Station.

Fig. 1. Phylogenetic relationships among different *Paspalum* species and their involvement in the origin of *P. dilatatum.*

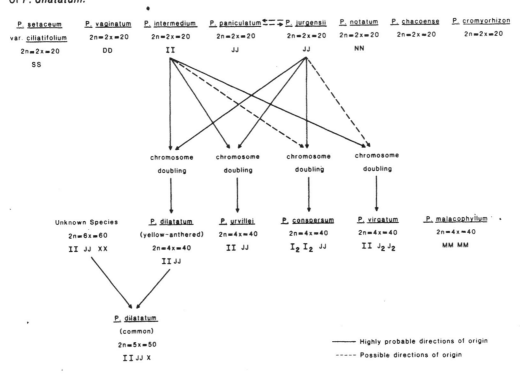

dividual genomes. Cytological analyses of the F$_1$ hybrids revealed that the yellow-anthered biotype has a genome homologous with that of *P. intermedium* Munro. ex Morong, *P. jurgensii*, and *P. paniculatum* L., but not with that of *P. chacoense* Parodi, *P. cromyorhizon* Trin. ex Doell., *P. notatum*, or *P. vaginatum* Swartz (Fig. 1, Burson et al. 1973, Burson 1979, unpublished data). In order to establish which of the yellow-anthered genomes were homologous with the different diploid genomes, crosses were made between *P. jurgensii* and *P. intermedium*. It was determined that these species have different genomes (Burson 1981); therefore, chromosomes of the two diploid species paired with different dallisgrass genomes. The chromosomes of the two yellow-anthered dallisgrass genomes differ slightly in size, and *P. intermedium* chromosomes appear to pair with the smaller members while those of *P. jurgensii* pair with the larger chromosomes. The *P. intermedium* genome was designated I and the *P. jurgensii* genome was assigned J (Fig. 1, Burson 1978). Therefore, yellow-anthered dallisgrass has the genomic formula of II JJ (Fig. 1, Burson 1978). From taxonomic and ecological data, *P. jurgensii* and *P. paniculatum* appear to be closely related with a common genome (Burson 1979).

Because *P. intermedium* and *P. jurgensii* each has a genome in common with dallisgrass, their involvement in the origin of dallisgrass is important. Contrary to cytological findings, morphological and ecological data indicate the lack of a close relationship among the three species. Phenotypically, *P. intermedium, P. jurgensii,* and their F$_1$ hybrids do not closely resemble yellow-anthered or common dallisgrass. *Paspalum intermedium* and *P. jurgensii* are not sympatric in their natural habitats, and neither

common nor yellow-anthered dallisgrass occurs in the same areas as *P. intermedium*. Therefore, considering all factors, closely related species or ancestral forms of *P. intermedium* and *P. jurgensii* may be the progenitors of yellow-anthered dallisgrass rather than of the two species as we know them today.

RELATIONSHIP BETWEEN DALLISGRASS AND TETRAPLOID SPECIES

Distribution of the I and J genomes within the genus appears relatively widespread. We crossed *P. intermedium* and *P. jurgensii* with tetraploid (2n = 40) species in an effort to determine their relationship to dallisgrass. This approach is necessary when the species of concern will not hybridize with yellow-anthered dallisgrass or when their F$_1$ hybrid does not survive until floral initiation. An excellent example is yellow-anthered dallisgrass and vaseygrass (*P. urvillei* Steud.), which are members of the Dilatata group and considered closely related taxonomically. The F$_1$ hybrids from controlled crosses between these two species die before flowering. However, in their native habitat in southern Brazil and Uruguay the species apparently hybridize readily, and the hybrids are extremely vigorous, a fact that supports their taxonomic classification. Vaseygrass (2n = 40) was crossed with *P. intermedium* and *P. jurgensii*. Chromosome pairing in the F$_1$ hybrids indicated that the I and J genomes were homologous with the two vaseygrass genomes (Burson and Bennett 1972a, Burson 1979). Thus, vaseygrass has the genomic constitution of II JJ (Fig. 1), the same as yellow-anthered dallisgrass. Their genomic constitution supports the taxonomic treatment of these two species (Burson 1979).

Paspalum intermedium and *P. jurgensii* were also used to study the relationship between dallisgrass and *P. virgatum* L. (2n = 40). This species, a member of the Virgata group, is morphologically different from dallisgrass. Attempts to cross *P. virgatum* and yellow-anthered dallisgrass were unsuccessful. *Paspalum intermedium* and *P. jurgensii* were crossed with *P. virgatum*. Chromosome pairing in the F_1 hybrids indicated that one *virgatum* genome was homologous to the *intermedium* genome and the other was partially homologous to the *jurgensii* genome. Consequently, *P. virgatum* has the genomic formula II J_2J_2 (Fig. 1).

Another member of the Virgata group, *P. conspersum* Schrad. ex Schult. (2n = 40), was crossed with yellow-anthered dallisgrass. The species had one homologous genome and one partially homologous genome (Burson and Bennett 1976). To identify the two genomes, *P. conspersum* was crossed with *P. intermedium* and *P. jurgensii*. Chromosome pairing in these hybrids showed that the *intermedium* genome was partially homologous with one genome and the *jurgensii* genome was homologous with the other *conspersum* genome. *Paspalum conspersum* was given the genomic formula I_2I_2 JJ (Fig. 1, Burson 1978). The partial homology is apparently the result of structural changes that have occurred within members of the genomes.

These findings indicate that the II and JJ genomes are fairly widespread within the genus. They appear to have played a major role in its evolution. To date, one or both of the I and J genomes or evolved forms have been identified in species of the Dilatata, Paniculata, Quadrifaria, and Virgata groups but are absent in the Disticha, Notata, and Setacea groups (Burson 1978, 1979, unpublished data). Additional research is needed to extend our knowledge of the distribution of the two genomes in other species within the genus.

ORIGIN OF COMMON DALLISGRASS

The presence of three distinct genomes in common dallisgrass indicates that it was derived directly or indirectly from three different parental sources. Of several proposed theories concerning its origin (Bennett et al. 1969), the most feasible one involves natural crosses between tetraploid and hexaploid species. The yellow-anthered biotype or a closely related sexual form with the genome constitution of II JJ is a probable candidate for the tetraploid parent, and the male parent was probably an apomictic hexaploid species with the genomic formula II JJ XX, with X being an unknown genome (Fig. 1). Apomictic common dallisgrass with the genome constitution of II JJ X probably resulted from this cross and has remained essentially the same since its origin.

Efforts are underway to identify the hexaploid species and the X genome. In its native habitat in South America, dallisgrass is sympatric to several species of the Quadrifaria group. In addition, some morphological similarities in the species suggest that the members of the Quadrifaria group are probable candidates as the hexaploid species. Apomixis and high ploidy levels, including hexaploidy, are also prevalent within species in this group.

When the unknown X genome is identified and the distribution of the I and J genomes is better understood, sufficient information should be available to attempt to resynthesize common dallisgrass. Through a dynamic interspecific hybridization program coupled with induced amphiploidy, new forms of dallisgrass should become available for use in improving this important forage grass.

LITERATURE CITED

Bashaw, E.C., and I Forbes, Jr. 1958. Chromosome numbers and microsporogenesis in dallisgrass *Paspalum dilatatum* Poir. Agron. J. 50:441–445.

Bashaw, E.C., and B.J. Hoff. 1962. Effects of irradiation on apomictic common dallisgrass. Crop Sci. 2:501–504.

Bashaw, E.C., and E.C. Holt. 1958. Megasporogenesis, embryo sac development and embryogenesis in dallisgrass, *Paspalum dilatatum* Poir. Agron. J. 50:753–756.

Bashaw, E.C., A.W. Hovin, and E.C. Holt. 1970. Apomixis, its evolutionary significance and utilization in plant breeding. Proc. XI Int. Grassl. Cong., 245–248.

Bennett, H.W., and E.C. Bashaw. 1966. Interspecific hybridization with *Paspalum* spp. Crop Sci. 6:52–54.

Bennett, H.W., B.L. Burson, and E.C. Bashaw. 1969. Intraspecific hybridization in dallisgrass, *Paspalum dilatatum* Poir. Crop. Sci. 9:807–809.

Burson, B.L. 1978. Genome relations between *Paspalum conspersum* and two diploid *Paspalum* species. Can. J. Genet. Cytol. 20:365–372.

Burson, B.L. 1979. Cytogenetics of *Paspalum urvillei* × *P. intermedium* and *P. dilatatum* × *P. paniculatum* hybrids. Crop Sci. 19:534–538.

Burson, B.L. 1981. Cytogenetic relationships between *Paspalum jurgensii* and *P. intermedium*, *P. vaginatum*, and *P. setaceum* var. *ciliatifolium*. Crop Sci. 21:515–519.

Burson, B.L., and H.W. Bennett. 1972a. Cytogenetics of *Paspalum urvillei* × *P. jurgensii* and *P. urvillei* × *P. vaginatum* hybrids. Crop Sci. 12:105–108.

Burson, B.L., and H.W. Bennett. 1972b. Genome relations between an intraspecific *Paspalum dilatatum* hybrid and two diploid *Paspalum* species. Can J. Genet. Cytol. 14:609–613.

Burson, B.L., and H.W. Bennett. 1976. Cytogenetics of *Paspalum conspersum* and its genomic relationship with yellow-anthered *P. dilatatum* and *P. malacophyllum*. Can. J. Genet. Cytol. 18:701–708.

Burson, B.L., H. Lee, and H.W. Bennett. 1973. Genome relations between tetraploid *Paspalum dilatatum* and four diploid *Paspalum* species. Crop Sci. 13:739–743.

Cytology and Breeding Behavior of Cicer Milkvetch

R.L. LATTERELL and C.E. TOWNSEND

USDA-SEA-AR and Colorado State University, Fort Collins, Colo., U.S.A.

Summary

Meiosis was investigated in the putative octoploid cicer milkvetch (*Astragalus cicer* L.) to determine whether irregularities of the reproductive process could affect the breeding behavior of that species. Microsporocytes from plants of several breeding populations showed a variable but generally low frequency of multivalents at metaphase I of meiosis. Meiotic irregularities that occurred in conjunction with multivalent associations, including univalents at metaphase I, lagging chromosomes, and dividing univalents at anaphase I and II, also were observed. The relatively low incidence of these events indicates that they are an improbable source of significant genotypic instability, infertility, or phenotypic variation in this species. The essential regularity of meiosis in cicer milkvetch, despite the tendency to form multivalents, indicates that this species is probably an autoallooctoploid that has been diploidized by natural selection.

KEY WORDS: cicer milkvetch, *Astragalus cicer* L., meiosis, multivalents, meiotic irregularities, autoallooctoploidy.

INTRODUCTION

Cicer milkvetch (*Astragalus cicer* L.) is a vigorous, rhizomatous, perennial legume native to, and widely dispersed over, central and southern Europe, including European Russia and the Caucasus. This species possesses a number of attributes that make it a promising forage crop for the semiarid regions of the central and northern Great Plains of the U.S.A.

Cicer milkvetch (2n = 64) is evidently an octoploid because European diploid species of *Astragalus* have 16 somatic chromosomes with a basic number of 8 (Ledingham and Rever 1963). The marked inbreeding depression and high fertility of cicer milkvetch (Townsend 1971, 1972) suggest that the species is an allooctoploid in which meiosis is characterized by bivalent associations, regular disjunction, and random segregation of homologues capable of conditioning a rapid increase in homozygosity. To our knowledge, information does not exist on the cytology and related phases of reproductive biology. Therefore, the objectives of this investigation were to study meiosis and to consider any special features of that process that might account for the breeding behavior of the species.

MATERIALS AND METHODS

Inflorescences were collected from cicer milkvetch plants grown in the greenhouse February to May 1980

and in the field May to July 1980. Collections were made between 1100 and 1200 hours and were considered to be random samples of open-pollinated plants of several breeding populations.

Chromosomes were small and difficult to stain with either Feulgen's method or the acetocarmine method, and they tended to clump excessively. The best and most consistent results were obtained from fixation in HgCl-saturated Carnoy's fluid (4 parts absolute methanol : 1 part glacial acetic acid : 2 parts chloroform) for 24 hours and staining in Snow's methanolic hydrochloric acid–carmine for 2 to 3 weeks. Anthers were squashed in 45% acetic acid and made semipermanent with Rattenbury's medium.

Microsporocytes were examined by both bright field and phase contrast microscopy. Later stages of meiosis I provided the most useful data for cytogenetic analysis.

RESULTS AND DISCUSSIONS

Eleven of 14 cicer milkvetch plants showed 2n = 64 chromosomes, as did plants of that species studied by Ledingham and Rever (1963). Two plants, one mature (no. 6) and one a seedling, were hypoploid with 2n = 63 chromosomes, whereas another mature plant (no. 1) was hyperploid with 2n = 65 chromosomes (Table 1). This relatively high incidence of aneuploidy within a small sample suggested that meiosis in this species might be less regular than previously reported chromosome counts and the species' breeding behavior suggested.

Analysis of microsporocytes revealed a number of irregularities in meiosis (Table 1). Although bivalent association was the norm in all plants studied, multivalents and univalents occurred with a variable but usually relatively low frequency. Predictably, univalents

R.L. Latterell is an Associate Professor at Shepherd College in Shepherdstown, W. Va., and C.E. Townsend is a Research Geneticist with USDA-SEA-AR, Colorado State University.

This paper (Scientific Series No. 2617) is a contribution of USDA-SEA-AR in cooperation with the Colorado State University Experiment Station.

Table 1. Types and frequencies of chromosome associations at metaphase I in microsporocytes of cicer milkvetch.

Plant no.	Chromosome no.	Total no. cells	Average no./cell					
			Univ.	Biv.	Triv.	Quad.	Pent.	Hex.
2	64	40	0.12	29.30	0.01	1.22	0.00	0.02
5	64	16	0.06	30.31	0.06	0.69	0.00	0.06
11	64	18	0.06	28.67	0.06	1.44	0.00	0.11
6	63	7	0.86	27.14	0.14	1.43	0.00	0.29
1	65	105	1.03	30.49	0.04	0.67	0.01	0.03

were far more prevalent in both of the aneuploid plants than among the euploid (2n = 64) group. However, multivalent frequencies were generally similar for euploids and aneuploids.

Quadrivalents were the predominant type of multivalent observed. The number of quadrivalents/cell at metaphase I ranged from 0 to 7. However, average frequencies of these configurations/cell for individual plants were characteristically low (Table 1). Alternate orientations of quadrivalents predominated slightly to moderately in the 4 plants where comparisons of frequencies of alternate versus adjacent orientations were possible. Quadrivalents of all types showed a strong tendency toward symmetrical orientation relative to the metaphase plate. Hence, irregular disjunction of quadrivalents as a consequence of their orientation at metaphase I does not seem to be a probable cause of any high incidence of chromosomal imbalance.

Chromosome distribution during anaphase I in microsporocytes of euploid plants provided further evidence of the essential regularity of meiosis in cicer milkvetch, including regular 2:2 disjunction of quadrivalents. In 40 of 55 cells from 8 euploid plants, the meiotic chromosomes separated into two groups of 32 chromosomes each. The remaining 15 cells exhibited various irregularities, including lagging univalents, dividing univalents, and bridges. However, 10 of these 15 aberrant cells were from one plant (no. 12) (Table 2). The frequency of meiotic irregularities was actually lower among microsporocytes produced by aneuploid plants where unpaired univalents were a nearly universal feature of metaphase I (Table 1). In the hypoploid plant (no. 6), chromosome disjunction was 31:32 in 12 of 13 cells. A lagging univalent was present in the single abnormal cell. In the hyperploid plant

(no. 1), chromosome separation was 32:33 in 19 of 20 cells, with a dividing univalent in the single abnormal cell. These relatively low frequencies of anaphase I irregularities were most probably products of unequal disjunctions of trivalents and higher multivalents that occurred with correspondingly low frequencies. Lagging univalents due to failure of synapsis and/or chiasma formation were evidently not a significant factor leading to aneuploid nuclei in cicer milkvetch.

Impressions regarding the essential regularity of the meiotic process were corroborated by more abundant, though less precise, data from late anaphase–early telophase of meiosis I. Frequencies of abnormalities at these stages (Table 2), which were mainly lagging or dividing univalents, were low compared with those observed at early anaphase I. There was considerable variation among plants in their frequencies of abnormalities. However, the overall frequency of late anaphase I–early telophase I irregularities among euploid plants was only 7%.

Meiosis II, particularly in the earlier stages, yielded little relevant information because of excessive clumping of chromosomes throughout. Examination of microsporocytes in late meiosis II revealed types (mainly lagging univalents) and frequencies of abnormalities comparable to those observed during later stages of meiosis I (Table 3). Anaphase II abnormalities were rare among microsporocytes of the hypoploid plant, no. 6. This rareness was consistent with the observation that unpaired univalents, characteristic of metaphase I in that plant, segregated regularly at anaphase I and showed little tendency either to lag or to divide.

Formation of multivalents at metaphase I of meiosis indicates that cicer milkvetch is at least partly autoploid.

Table 2. Meiotic irregularities during meiosis 1 in microsporocytes from 6 euploid plants of cicer milkvetch.

Plant no.	Stage	No. of cells		Abnormal cells (%)
		Normal	Abnormal	
7	Anaphase I	49	2	3.9
8	Anaphase I	61	6	9.0
10	Anaphase I	11	3	21.4
12	Anaphase I	26	7	21.2
13	Anaphase I	108	6	5.3
12	Telophase I	95	5	5.0
9	Telophase I	139	7	4.8

Table 3. Meiotic irregularities during Ana-Telophase II in microsporocytes of cicer milkvetch.

Plant no.	Chromosome no.	No. of cells		Abnormal cells (%)
		Normal	Abnormal[1]	
8	64	46	4	8.0
12	64	32	17	34.7
9	64	41	9	18.0
6	63	49	1	2.0

[1]Univalents or bridges.

The high ratio of bivalents to miltivalents at metaphase I of meiosis, however, indicates that the species has undergone extensive, though incomplete, diploidization. Further, whereas diploidization has tended to stabilize chromosomal segregation, the autoploid component of the genotype would tend to buffer that species against the potentially disruptive effects of meiotic accidents. The commonest types of aneuploids observed or expected among progeny of polyploids are those that involve loss or gain of single chromosomes. Consequently, the phenotypic effects of these types of aneuploidy would be minimized by the genetic redundancy present in the autoploid component of the genotype. Analogous buffering of genotypes has been reported among fertile aneuploid progeny of tetraploid and hexaploid derivatives of species of *Glycine* (Newell and Hymowitz 1980).

LITERATURE CITED

Ledingham, G.F., and B.M. Rever. 1963. Chromosome numbers of some southwest Asian species of *Astragalus* and *Oxytropis* (Leguminosae). Can. J. Genet. Cytol. 5:18–32.

Newell, C.A., and T. Hymowitz. 1980. Cytology of *Glycine tabacina*. J. Hered. 71:175–178.

Townsend, C.E. 1971. Self-compatibility studies with *Astragalus cicer* L. Crop. Sci. 11:769–770.

Townsend, C.E. 1972. Comparison of S_1 and open-pollination progenies of *Astragalus cicer* L. for certain agronomic characters. Crop Sci. 12:793–795.

Differences in Photosynthetic Types in the Laxa Group of the *Panicum* Genus: Cytogenetics and Reproduction

J.H. BOUTON and R.H. BROWN
University of Georgia, Athens, Ga., U.S.A.

Summary

Several grass species related to *Panicum milioides* Nees ex Trin., a species with intermediate photosynthetic characteristics, were collected in South America. *Panicum prionitis* Griseb. was found to possess C_4 photosynthesis. *Panicum schenckii* Hack. and *Panicum decipiens* Nees ex Trin. were found to have intermediate photosynthesis, like *P. milioides*. *Panicum laxum* Sw., *Panicum hylaeicum* Mez., and *Panicum rivulare* Trin. were found to have C_3 photosynthesis. In this study, we describe the cytogenetic and reproductive characteristics of these species for future studies of hybridization among distinct photosynthetic types.

A basic chromosome number of 10 was common for all species. Diploid (2n = 2× = 20), tetraploid (2n = 4× = 40), and hexaploid (2n = 6× = 60) accessions were found. *P. prionitis* possessed a diploid and tetraploid accession. *P. milioides* was diploid, *P. schenckii* was hexaploid, and *P. decipiens* exhibited one diploid and one hexaploid accession. *P. laxum*, *P. hylaeicum*, and *P. rivulare* were all tetraploid. There was stable and primarily bivalent pairing at metaphase I in all species, which resulted in a high percentage of stainable pollen. Embryo sac analyses showed normal sexual development in all plants except one accession of *P. prionitis*, which was found to possess multiple embryo sacs at anthesis. Selfpollinated seed set was high in all species.

An intraspecific hybrid resulted from a cross of two *P. laxum* accessions. Although this hybrid possessed more multivalent pairing than either of its parents did, it did produce large amounts of viable pollen and seed.

Similar and stable cytogenetic characteristics and similar taxonomy indicate possible success in producing hybrids among these species for a genetic study of photosynthetic characteristics in grasses.

KEY WORDS: embryo sac analysis, C_3 photosynthesis, C_4 photosynthesis, intermediate photosynthesis, hybridizations, metaphase I chromosome configurations, pollen stainability, chromosome number.

INTRODUCTION

Panicum milioides Nees ex Trin. is native to Central and South America and was taxonomically classified as a member of the informal group Laxa (Hitchcock and Chase 1910). This species was reported to possess intermediate photosynthetic characteristics when compared to C_3 and C_4 species (Brown and Brown 1975).

Panicum species were collected in South America based on their taxonomic proximity to *P. milioides*, and each species was examined for various photosynthetic characteristics (Morgan and Brown 1979, 1980). Three of the

J.H. Bouton is an Assistant Professor of Agronomy and R.H. Brown is a Professor of Agronomy at the University of Georgia.

This is a contribution of the Department of Agronomy, University of Georgia.

The research was supported by state and Hatch funds allocated to the Georgia Agricultural Experiment Stations and USDA-SEA AR under Grant No. 5901-0410-8-018-0 from the Competitive Research Grants Office.

species examined, *Panicum laxum* Sw., *Panicum hylaeicum* Mez., and *Panicum rivulare* Trin., exhibited characteristics typical of C₃ species—that is, about 30% inhibition of photosynthesis by 21% O_2, a CO_2 compensation concentration of about 50 μ 1/1, and non-Kranz leaf anatomy. *Panicum prionitis* Griseb. had characteristics typical of C₄ plants, its photosynthesis being insensitive to O_2 and its CO_2 compensation being less than 10 μ 1/1. Its leaves also were characterized by features of Kranz anatomy. The remaining species, *Panicum schenckii* Hack. and *Panicum decipiens* Nees. ex Trin., were like *P. milioides*, falling between C₃ and C₄ types in the photosynthetic characteristics mentioned above.

The purpose of this investigation was to describe the cytogenetic and reproduction characteristics of these *Panicum* species for future studies of hybridizations among distinct photosynthetic types.

METHODS

Chromosome counts were made on root tips of plants growing in greenhouse pots. These root tips were fixed in Carnoy's fluid (6:3:1), hydrolyzed in 5N HCl at room temperature for 15 minutes and stained by Feulgen's method. Meiosis was studied on pollen mother cells from anthers fixed in Carnoy's, crushed in acetocarmine, and examined with light microscopy. The percentage of stainable pollen was derived by staining dehisced pollen with aniline blue in lactophenol and counting the number of stained pollen grains in 20 fields of 100 grains. Embryo sac analyses were conducted by fixing ovaries in FAA (90 ml 95% EtOH:5 ml 37% formaldehyde:5 ml glacial acetic acid), embedding in paraffin, sectioning at 10–12 μm, staining with safranin-fast green, and observing with light microscopy. The amount of selfing was determined by comparing seed set in both self- (bagged) and open-pollinated conditions. Crosses were made by mutually bagging seedheads from two plants together and harvesting seed from the desired female parent.

RESULTS AND DISCUSSION

Root-tip chromosome counts (2n) and meiotic analyses for all accessions are presented in Table 1 (Bouton et al. 1981, unpublished data). Accession identification numbers indicate each collected plant that we have maintained as clonal material in greenhouse pots. The only exception was PI 310043, which was obtained from the USDA Southern Regional Plant Introduction Station, Experiment, Georgia.

Panicum prionitis accession 126 was diploid (2n = 2 × = 20), while *P. prionitis* accession 106 was tetraploid (2n = 4 × = 40). Accession 101 and PI 310043 of *P. milioides* both were diploid (2n = 2 × = 20). *P. schenckii* accession 109 was hexaploid (2n = 6 × = 60). *P. decipiens* exhibited two ploidy levels; accession 136 was diploid (2n = 2 × = 20), and accession 142 was hexaploid (2n = 6 × = 60). Accessions 104, 124, and 137 of *P. laxum*, accessions 121, 127, and 122 of *P. hylaeicum*, and accession 107 of *P. rivulare* were all tetraploid (2n = 4 × = 40). Bivalent pairing at metaphase I was common in all accessions execpt *P. prionitis*, accession 106, which averaged 7.35 tetravalents per cell. A stable meiosis was demonstrated in the high percentage of stainable pollen, which ranged from 70.2%

Table 1.   Root tip chromosome numbers (2n), average number of metaphase I chromosome configurations, and % stainable pollen of seven Panicum species.

Species and accession	2n	Metaphase I chromosome Configurations (average per cell)			Stainable pollen %
		I	II	IV	
		--------------------------------- *no.*---------------------------------			
P. prionitis					
Accession 106	40	0.48	5.00	7.35	90
Accession 126	20	–	9.64	–	91
P. decipiens					
Accession 136	20	–	9.67	–	96
Accession 142[1]	60	–	–	–	–
P. milioides					
Accession 101	20	0.01	9.95	–	97
PI 310043	20	–	9.93	–	96
P. schenckii					
Accession 109	60	1.03	29.22	–	90
P. laxum					
Accession 137	40	0.04	18.10	0.05	90
Accession 124	40	0.60	18.10	0.12	70
Accession 104	40	0.20	19.60	–	–
P. hylaeicum					
Accession 121	40	0.37	19.25	–	88
Accession 127	40	0.23	18.87	–	94
Accession 133	40	0.32	18.95	–	97
P. rivulare					
Accession 107	40	–	18.55	0.43	94

[1]Accession 142 has never flowered; therefore no PMC's were available for meiotic analysis.

to 97.0% over all accessions (Table 1).

Panicum is a large and diverse genera of grasses, but species taxonomically similar to the species in this collection have been assigned to the informal group Laxa (Hitchcock and Chase 1910). However, Brown (1977) reclassified the Laxa group and removed *milioides, decipiens,* and *schenckii* from *Panicum,* placing them in the genus *Steinchesma* Rof. Our chromosome counts showed these species to possess a basic chromosome number of 10. This is a unique number for *Panicum;* the majority of species have a basic number of 9 (Brown 1977). The similar chromosome number, as well as similar taxonomic characteristics (Rosengurtt et al. 1970), indicate that Brown's reclassification may have been premature. It may be best to leave all the species in this collection in the Laxa group (Hitchcock and Chase 1910) until further cytotaxonomic studies are completed.

Embryo sac development was normal and sexual, while seed set under bag was high in all accessions. The only exceptions to these observations were *P. prionitis* accession 106, which possessed multiple embryo sacs at anthesis, and *P. laxum* accession 124 and *P. prionitis* accession 126, which did not set seed under bagged conditions (Bouton et al. 1981).

This low seed set under bag in *P. laxum* accession 124 and *P. prionitis* accession 126 may represent self-incompatibility. Therefore, mutual crosses with these and several other accessions in the collection were attempted. In the case of *P. prionitis* accession 126, several plants were recovered, but all were exactly like accession 126 in both morphology and chromosome number. *Panicum laxum* accession 124 also produced several plants with similar morphology and chromosome number. However, one intraspecific hybrid was isolated from the cross *P. laxum* accession 124 × *P. laxum* accession 137. Accession 124 has an open panicle and pubescent glumes, while accession 137 has a closed panicle and smooth glumes. The hybrid was intermediate in panicle expansion and had pubescent glumes. It also was tetraploid ($2n = 4 \times = 40$), produced a great deal of stainable pollen (91.8%), and, like accession 137, set a large amount of viable selfed seed. A meiotic analysis showed the hybrid averaged 0.20 I, 15.33 II, 0.60 IV, and $0.93 \geq$ V configurations/cell during metaphase I.

It is possible that *P. laxum* accession 124 is self-incompatible and may be useful in producing hybrids. The intraspecific hybrid we found from this accession did not show more multivalent pairing than its parents. This pairing does not warrant separation of these two accessions, 124 and 137, into separate species because normal chromosome disjunction apparently still results, as is evidenced by its high percentage stainable pollen and viable seed set. In all, accession 124 appears to be a stable hybrid.

The higher photosynthetic capacity of C_4 plants may give these plants advantages in certain ecological situations, although the degree of advantage of the C_4 cycle in crop production is still not clear. Since C_4 plants have higher photosynthetic capacity, it is important to determine the genetic inheritance of some of the characters associated with C_4 metabolism. This knowledge would increase the possibility of incorporating higher photosynthetic potential into crops of agronomic importance. The existence of these photosynthetically distinct *Panicum* species provides an excellent opportunity to study the genetic basis of photosynthesis in grasses. Breeding hybrids among these species would help clarify their relationships and provide germ plasms to study the genetics of photosynthesis. Their similar and stable cytogenetic and taxonomic characteristics indicate interspecific hybridizations may be successful.

LITERATURE CITED

Bouton, J.H., R.H. Brown, J.K. Bolton, and R.P. Campagnoli. 1981. Photosynthesis of grass species differing in carbon dioxide fixation pathways. VII. Chromosome numbers, metaphase I chromosome behavior, and mode of reproduction of photosynthetically distinct *Panicum* species. Pl. Physiol. 67:433–437.

Brown, R.H., and W.V. Brown. 1975. Photosynthetic characteristics of *Panicum milioides*, a species with reduced photorespiration. Crop. Sci. 15:681–685.

Brown, W.V. 1977. The Kranz syndrome and its subtypes in grass systematics. Mem. Torr. Bot. Club. 23:197.

Hitchcock, A.A., and A. Chase. 1910. The North American species of *Panicum*. Contr. U.S. Natl. Herb. 15:1–396.

Morgan, J.A., and R.H. Brown. 1979. Photosynthesis in grass species differing in CO$_2$ fixation pathways. II. A search for species with intermediate gas exchange and anatomical characteristics. Pl. Physiol. 64:257–262.

Morgan, J.A., and R.H. Brown. 1980. Photosynthesis in grass species differing in carbon dioxide fixation pathways. III. Oxygen response and enzyme activities of species in the Laxa group of *Panicum*. Pl. Physiol. 65:56–159.

Rosengurtt, B., B.A. de Maffei, and P.I. de Artucio. 1970. Gramineaes Uruguayas. Univ. de la Republica, Montevideo, Uruguay.

Breeding Challenges in Apomictic Warm-Season Grasses

E.C. BASHAW, P.W. VOIGT, and B.L. BURSON

USDA-SEA-AR and Texas Agricultural Experiment Station,
College Station, Tex., U.S.A.

Summary

Efforts to breed apomictic grasses met with little success in the past, but results of more recent research show significant progress with some species and new prospects for improving apomicts. Early research uncovered the cytological basis for apospory and diplospory, found the common apomictic mechanisms in grasses, and showed that apomixis is transmitted from parent to offspring. Discovery of sexual or partially sexual (facultative apomicts) plants in several species formerly considered obligate apomicts allowed for hybridization and genetic studies. Results of subsequent research confirmed inheritance of apomixis, and data from the more conclusive studies indicate that method of reproduction is simply inherited.

Demonstration of control and manipulation of obligate apomixis through hybridization of sexual and apomictic plants and production of true-breeding hybrids established the potential for using apomixis in a breeding program. Soon after, obligate apomictic hybrid cultivars were developed for commercial use. Reports of progress in the breeding of some facultative apomicts are considered very significant. Scientists found that, contrary to some earlier contentions, highly apomictic hybrids can be recovered following hybridization of plants capable of both apomictic and sexual reproduction. Efforts to improve apomictic *Cenchrus ciliaris*, *Eragrostis curvula*, and *Paspalum dilatatum* are discussed.

KEY WORDS: buffelgrass, weeping lovegrass, dallisgrass, obligate apomixis, facultative apomixis, method of reproduction.

Apomixis is the predominant mode of reproduction of several important forage grasses, and it has been reported in more than a hundred species. Fortunately, many apomictic species are polymorphic, and selected natural apomictic ecotypes of several species have been used successfully for pasture and range production. However, improvement of these species through breeding has been extremely difficult because of the apomictic barrier to hybridization or inability to control variation in the progeny of facultative apomicts. In recent years, as scientists have developed a better understanding of the genetic basis of apomixis, progress has been made in the breeding of some apomictic grasses. Gustafsson (1947) and Carnahan and Hill (1961) have given excellent reviews of early research on apomixis. A recent review emphasizes opportunities for utilizing apomixis in breeding (Bashaw 1980). The objective of this paper is to review some important developments in the genetics and breeding of apomictic grasses and to report our experiences with apomictic species in three genera.

Burton (1962) has discussed the problems posed by obligate apomixis in conventional breeding of *Pasaplum dilatatum* Poir. He has noted that until a dependable method for breaking apomixis can be found, evaluation of large numbers of naturally occurring ecotypes may be the most fruitful method for improving the species. Much research has been devoted to the search for sexuality in apomicts and for ways to break apomixis and control method of reproduction in a breeding program. Radiation has created potentially useful mutants in some apomictic species but has not changed method of reproduction in obligate apomicts (Bashaw and Hoff 1962, Burton and Jackson 1962). However, some facultative apomictic plants have been recovered following radiation of a sexual species (Hanna and Powell 1973). Fertilization of an unreduced gamete can permit transfer of valuable traits to the apomict but apparently does not function to break apomixis. We have observed this phenomenon frequently in *Paspalum* and *Cenchrus* species, but the progeny were always obligate apomicts. Hybridization between sexual and apomictic plants seems to be the only effective method for breaking the apomictic barrier. Little progress was made in breeding apomicts until it was realized that apomixis is genetically controlled.

Gustafsson (1947) noted that most genetic studies with *Poa* clearly indicated a genetic basis for apomixis. Facultative apomixis prevented development of conclusive inheritance data, but the results showed that plants with a high frequency of apomixis could be recovered following hybridization. Funk and Han (1967) reported that about 14% of 280 *Poa pratensis* L. hybrid progenies were highly apomictic. They found this frequency high enough for practical development of useful

---

E.C. Bashaw is a Research Geneticist with USDA-SEA-AR, College Station, Tex., and P.W. Voigt and B.L. Burson are Research Geneticists with USDA-SEA-AR, Grassland, Soil and Water Research Laboratory, Temple, Tex.

This is a contribution of USDA-SEA-AR in cooperation with the Texas Agricultural Experiment Station.

hybrid turf cultivars. Harlan et al. (1964) reported genetic control of facultative apomixis in the *Bothriochloa-Dichanthium* agamic complex. They noted that apomixis is probably rather simply inherited and is controlled by no more than one gene/genome. Apomixis was dominant to, but independent of sexuality. They found that crosses of sexual × apomict and apomict × sexual resulted in a ratio of 3 apomicts:1 sexual. This proportion of apomictic hybrids would be adequate for efficient development of apomictic cultivars.

Identification of sexual plants in apomict *Panicum maximum* Jacq. provided germ plasm for new genetic studies and breeding of this species. Smith (1972) found 5 sexual plants among plant introductions and observed that their progenies segregated for method of reproduction. Hanna et al. (1973) isolated obligate sexual plants from plant introductions and crossed them with apomictic plants. Approximately equal numbers of sexual and apomictic hybrids were recovered. The data from selfed and hybrid progenies indicated that sexuality is dominant and is controlled by at least two loci. Savidan (1982) conducted hybridization studies with *P. maximum,* using as female parents sexual tetraploids derived from sexual diploids, and sexual hydrids from the breeding program. Apomictic and sexual hybrids were recovered at a ratio of approximately 1:1. Savidan stated that some hybrids were obligate or nearly obligate apomicts and were being evaluated as prospective new cultivars. Results from those studies show that facultative apomixis can be manipulated in a breeding program.

Inheritance of obligate apomixis and production of true-breeding F_1 hybrids have been demonstrated in *Paspalum notatum* Flugge and *Cenchrus ciliaris* L. Burton and Forbes (1960) succeeded in breaking apomixis in *P. notatum* by doubling the chromosome number of a sexual diploid and crossing the induced autotetraploids with an obligate apomictic tetraploid. Both obligate apomictic and completely sexual F_1 hydrids were recovered. Burton and Forbes's data from F_2 progenies of sexual hybrids indicated that obligate apomixis is recessive in this species and is controlled by a few recessive genes. There was no evidence of facultative apomixis in segregating progenies. Heterosis was maintained in progenies of apomictic hybrids but was rapidly lost in succeeding generations from sexual hybrids. Breaking apomixis in the naturally occurring tetraploid released many genes previously unavailable to the plant breeder and demonstrated the potential of obligate apomixis for use as a breeding tool.

Discovery of a sexual plant enabled Taliaferro and Bashaw (1966) to study inheritance of obligate apomixis in *Cenchrus ciliaris* and to develop a scheme for successful breeding of this species. The sexual plant proved to be heterozygous for method of reproduction, and it produced selfed progeny that were either completely sexual or obligate apomicts (13 sexual:3 apomicts). Hybridization of the sexual plant with obligate apomictic accessions of *C. ciliaris* and closely related *C. setigerus* Vahl. gave a ratio of 5 sexual:3 obligate apomictic F_1 hybrids. These data fit ratios expected if mode of reproduction is controlled by 2

genes and epistatis and the genotypes of the sexual and apomictic parents are AaBb and Aabb, respectively. We hypothesized a dual action of dominant gene (*A*) that caused normally senescent nuclear cells to become meristematic and form functional unreduced embryo sacs (apospory), with simultaneous suppression of the sexual mechanism. We assumed epistasis of the second gene (*B*), which restores sexuality when it is dominant. The expected ratios of sexual to apomictic F_1 hybrids have been confirmed in other studies, but the dual action of gene *A* remains uncertain.

The breeding scheme developed for *C. ciliaris* illustrates the efficiency of an apomictic breeding program with a species in which obligate apomixis and obligate sexuality are the only alternatives for reproduction and cross-compatible sexual plants are available for hybridization. In *C. ciliaris* a vast number of obligate apomictic ecotypes, typically highly heterozygous, are available for use as male parents. After hybridization we identify the apomictic hybrids as soon as possible through embryo sac studies of each F_1. Since the apomicts breed true, dominant, simply inherited traits can be transferred and fixed in the F_1. Promising apomictic hybrids are selected for further evaluation as potential cultivars, and often enough seed is obtained in the first season for replicated plot tests the second year.

Sexual hybrids can be selfed or crossed back to the male parent to recover traits that are not dominant. One apomict cultivar, Higgins buffelgrass, was selected from selfed progeny of the original sexual plant. We have noted considerable loss in vigor in succeeding selfed generations from sexual plants. Since genotype and heterosis are permanently fixed in apomictic F_1 hybrids, and the parents are highly heterozygous, we concentrate most efforts on developing and evaluating large numbers of hybrids using many different apomictic male parents. This breeding scheme has been very effective with *C. ciliaris,* and two excellent apomictic F_1 hybrid cultivars, Nueces and Llano buffelgrass, have been developed in the breeding program.

Unviable pollen of pseudogamous apomicts or cross-incompatibility of sexual and apomictic plants can prohibit breeding apomicts, even when sexual biotypes are available. For example, there has been no progress in the breeding of obligate apomict common dallisgrass, a natural pentaploid trihybrid with 50 chromosomes (20II + 10I) and three distinct genomes. The sexual biotype, a tetraploid (20II) with two different genomes, is almost completely incompatible with apomictic common; only a few hybrids of no value have been produced. In the hybrids, genomes representing the 20 bivalents of common were homologous with the two genomes of the sexual type, indicating a close relationship between the parents. Since common is a natural trihybrid, Burson and Bennett (1972) hypothesized that it might be possible to synthesize new useful forms of common if its progenitors could be identified. Burson (1982), who has crossed the sexual type with numerous *Paspalum* species, reports success in identifying the genomes of the sexual type in 2 sexual diploids.

He notes that the genome represented by the 10 univalents in common dallisgrass is not yet identified. This novel approach could provide a way to improve apomicts when all other efforts fail.

Critical genetic data are lacking for diplospory, but results of recent research with *Eragrostis curvula* (Schrad.) Nees indicate that this apomictic mechanism also is inherited. Voigt (1971) discovered sexual plants of *E. curvula* and crossed these with diplosporous apomicts. Both sexual and diplosporous apomictic F_1 hybrids were recovered. Although many of the hybrids were facultative apomicts, most were either highly sexual or highly apomictic; plants with intermediate levels of sexuality were rare. The ratio of highly sexual to highly apomictic plants was about 1:1 for tetraploid × tetraploid hybrids. For a tetraploid × octoploid cross, most hybrids were apomictic. Voigt and Burson (1982) proposed a scheme for breeding *E. curvula* and succeeded in developing numerous apomictic strains for evaluation as potential cultivars.

Breeders of apomictic grasses will continue to face problems in the future, but research with some species has shown that apomixis can provide an opportunity to develop a remarkably efficient breeding program. Success depends largely on availability of sexual or partially sexual cross-compatible plants for use in hybridization with the apomicts. The discovery of sexual plants in several grasses that were formerly believed to be obligate apomicts suggests that sexuality probably exists in all species even though obligate apomicts dominate in the natural habitat. Progress in the breeding of some facultative apomicts is particularly encouraging, for we once thought it was not possible to develop obligate or even highly apomictic hybrids in these species. Results of most inheritance studies indicate that apomixis is genetically controlled in both obligate and facultative apomicts and that we should be able to manipulate method of reproduction in most apomictic grasses.

LITERATURE CITED

Bashaw, E.C. 1980. Apomixis and its application in crop improvement. *In* W.R. Fehr and H.H. Hadley (eds.) Hybridization of crop plants, 765. ASA Press, Madison, Wis.

Bashaw, E.C., and B.J. Hoff. 1962. Effects of irradiation on apomictic common dallisgrass. Crop Sci. 2:501–504.

Burson, B.L. 1982. Phylogenetic investigations of *Paspalum dilatatum* and related species. Proc. XIV Int. Grassl. Cong.

Burson, B.L., and H.W. Bennett. 1972. Genome relations between an intraspecific *Paspalum* hybrid and two diploid species. Can. J. Genet. Cytol. 14:609–613.

Burton, G.W. 1962. Conventional breeding of dallisgrass, *Paspalum dilatatum* Poir. Crop Sci. 2:491–494.

Burton, G.W., and I. Forbes. 1960. The genetics and manipulation of obligate apomixis in common bahiagrass (*Paspalum notatum* Flugge). Prox. VIII Int. Grassl. Cong., 66–71.

Burton, G.W., and J.E. Jackson. 1962. Radiation breeding of apomictic prostrate dallisgrass, *Paspalum dilatum* var. *Pauciciliatum.* Crop Sci. 2:495–497.

Carnahan, H.L., and H.D. Hill. 1961. Cytology and genetics of forage grasses. Bot. Rev. 27:1–162.

Funk, C.R., and S.J. Han. 1967. Recurrent intraspecific hybridization: a proposed method of breeding Kentucky bluegrass, *Poa pratensis*. N. J. Agric. Expt. Sta. Bul. 818, 3–14.

Gustafsson, A. 1947. Apomixis in higher plants. Haken Ohlson's Boktryekeri, Lund, Sweden.

Hanna, W.W., and J.B. Powell. 1973. Stubby head, an induced facultative apomict in pearl millet. Crop Sci. 13:725–728.

Hanna, W.W., J.B. Powell, C. Millot, and G.W. Burton. 1973. Cytology of obligate sexual plants in *Panicum maximum* Jacq. and their use in controlled hybrids. Crop Sci. 13:695–697.

Harlan, J.R., M.H. Brooks, D.S. Borgaonkar, and J.M.J. de Wet. 1964. The nature and inheritance of apomixis in *Bothriochloa* and *Dichanthium*. Bot. Gaz. 125:41–46.

Savidan, Y. 1982. Genetics and utilization of apomixis for the improvement of guinea grass (*Panicum maximum* Jacq.). Proc. XIV Int. Grassl. Cong.

Smith, R.L. 1972. Sexual reproduction in *Panicum maximum* Jacq. Crop Sci. 12:624–627.

Taliaferro, C.M., and E.C. Bashaw. 1966. Inheritance and control of obligate apomixis in breeding buffelgrass, *Pennisetum ciliare*. Crop Sci. 6:473–476.

Voigt, P.W. 1971. Discovery of sexuality in *Eragrostis curvula* (Schrad.) Nees. Crop Sci. 11:424–425.

Voigt, P.W., and B.L. Burson. 1982. Breeding of apomict *Eragrostis curvula*. Proc. XIV Int. Grassl. Cong.

Genetics and Utilization of Apomixis for the Improvement of Guineagrass (*Panicum maximum* Jacq.)

Y.H. SAVIDAN

Office de la Recherche Scientifique et Technique,
Outre-Mer (ORSTOM), Abidjan, Ivory Coast

Summary

Natural polymorphism in guineagrass (*Panicum maximum* Jacq.) has been stabilized in tetraploid biotypes by apomixis. Sexual diploids were found in East African populations. Our objective was to determine whether apomixis could be related with a simple genetic background and manipulated in an improvement program.

Apomicts of guineagrass are characterized by a 4-nucleate embryo sac structure, and this trait was used in genetical analyses to differentiate apomictic from sexual hybrids. Progeny tests were also conducted as controls. Ten different types of progenies were analyzed: tetraploid sexual (from colchicine treatment of diploids) × apomictic F_1 hybrids; selfed and backcrossed progenies of sexual F_1 hybrids; three-way hybrids combining information from one sexual and two different apomicts; selfed and backcrossed progenies of sexual three-way; sib-mating progenies (crosses between sexual hybrids or sex × apo combinations); testcrossed progenies; and selfed progenies of facultative apomicts.

All the data fit with a single gene model for the inheritance of apomixis, in which the sexual diploids would have the genotype aa, and the apomictic tetraploids the genotype Aaaa. Apomixis is dominant.

Apomixis in *Panicum maximum* is often facultative, but with a low percentage of residual sexuality. Eighty natural apomicts and 80 apomictic hybrids were compared to show that this rate of sexuality remains low following hybridization.

According to these cytogenetical and embryological studies, along with other data from plant evolution studies, an improvement scheme was drawn up and successfully tested. From each cross combination one-half of the hybrids are apomicts that can be immediately propagated by means of seeds if they are found to be promising.

We have conclusively demonstrated that apomixis could be simply inherited in guineagrass and thus easily manipulated in an improvement program. Different hybrids adapted to the humid tropical lowlands of West Arica have been selected. The conservation of a large germ plasm of guineagrass and related species has also been realized.

KEY WORDS: guineagrass, *Panicum maximum*, apomixis, embryo sac analysis, crop improvement.

INTRODUCTION

In early 1969, 500 accessions of *Panicum maximum* (guineagrass) were established in the south of Ivory Coast, following the exploration, by Combes and Pernès (1970), of natural populations in East Africa (Kenya and Tanzania). Twenty-two sexual diploids (2n = 16) were found in this material. Nearly all of the other plants were tetraploids (2n = 32), reproducing by apomixis.

This discovery of sexuality has led us to the consideration of new breeding procedures for guineagrass. Many genes heretofore unavailable to the plant breeder could potentially be released and promising new types created by hybridization.

Preliminary studies showed that hybridization between sexual and apomictic plants gives rise to sexual as well as to apomictic hybrids (Combes 1972, Savidan 1978, 1980). Our major objectives were to study the inheritance of apomixis and sexuality in *Panicum* and thus to determine whether these modes of reproduction could be easily manipulated in an improvement program.

METHODS

Since genetical analyses require a large number of progenies, cytological methods were preferred, although small progeny tests were periodically used as controls. A microscopic technique, using phase contrast, appeared to be especially rapid and convenient. It was made possible by the fact that apomixis and sexuality are closely related, with two different embryo sac structures in *Panicum maximum* as well as in most grasses of the *Panicoideae* subfamily. Sexual biotypes contain only 8-nucleate embryo sacs. Apospory is characterized by a 4-nucleate structure. In our material, apospory appeared to be linked with parthenogenesis so that the observation of some 4-nucleate

The author's current address is G.P.D.P., Centre National de la Recherche Scientifique, 91190 Gif sur Yvette, France.

Table 1. Segregation for modes of reproduction in ten different cross combinations.

Cross combinations	No. plants observed			Theoretical ratio expected	Chi²
	total	sexual	apomictic		
1. F1 hybrids (S1xA1)	133	62	71	1:1	.48
2. F1 sexual selfed	126	126	0	1:0	0
3. F1 sexual back-crossed (S1xA1xA1)	26	12	14	1:1	—
4. Three-way hybrids (S1xA1xA2)	279	144	135	1:1	.23
5. Three-way sex selfed	57	57	0	1:0	0
6. Three-way sex backcrossed by A2	113	60	53	1:1	.32
7. Sib mating (SxS)	82	82	0	1:0	0
8. Sib mating (SxA)	60	34	26	1:1	1.06
9. Test cross S1 × apom. hybrid	23	10	13	1:1	—
10. Apo × apo	71	18	53	1:3	0

Chi² are indicated when N>30.

embryo sacs could be considered to be sufficient proof of an apomictic behavior.

Cytological and embryological techniques have already been described in previous papers (Herr 1971, Savidan 1975). All cross combinations were made between tetraploids, using the high self sterility of the sexual progenitors. The 970 plants reported in Table 1 are tetraploid. The 10 progenies studied were the F_1 progeny from a sexual × apomictic cross (1), the selfed progeny of sexual F_1 hybrids (2), the backcross progeny of the same sexual F_1 hybrids (3). Some of these sexual F_1 hybrids were crossed with a second apomict (A2) to give the so-called three-way hybrids (4). Sexual plants from this progeny were selfed (5), backcrossed with A2 (6), or crossed with other hybrids from the same progeny (7 and 8). Some apomictic hybrids were crossed with the original sexual progenitor S_1 (9). Finally, using a facultative apomict with 40% off-types in progeny tests, apomictic × apomictic crosses were made (10). The segregation reported for this cross combination was observed among the off-types.

RESULTS

Modes of reproduction appeared to segregate in our progenies (Table 1) according to the following rules: sexual × sexual hybrids (or sexual selfed) are always sexual; sexual × apomictic hybrids segregate in a 1:1 ratio; apomictic × apomictic hybrids segregate in a 1:3 ratio (apomictic hybrids are 3 times more numerous).

Apomixis in *Panicum maximum* is often facultative, and the possibility of using such a reproductive process in plant breeding is sometimes questioned. The sexual potential of 80 apomictic hybrids was compared to that of 80 natural apomicts (Table 2). Obligate apomixis was observed in 21 hybrids, and many others were nearly obligate.

DISCUSSION

All the segregations we observed fit perfectly into a single gene model for the inheritance of apomixis in *Panicum maximum*. In such a model, sexual diploids would

Table 2. Distributions of percentage sexuality in natural and hybrid apomicts of guineagrass (*Panicum maximum*).

	No. plants with percentage sexuality* of									
	0	0-10	10-20	20-30	30-40	40-50	50-60	60-70	70-80	80-90
Natural apomicts (N=80)	10	54	6	4	4	1	—	1	—	—
Hybrid apomicts (N=80)	21	49	5	1	1	—	1	—	1	1

*Sexuality: percentage ovules with a single 8-nucleate embryo sac (100 ovules observed/plant).

have the genotype aa and the tetraploids arising from colchicine treatment of these diploids the genotype Aaaa. The 1:1 ratios observed in all sexual × apomictic progenies could be explained by an Aaaa genotype for apomicts. Thus, apomixis of guineagrass is dominant over sexuality.

Crosses between apomicts with a high percentage of sexuality gave one-fourth of sexual plants among the offtypes, as the model predicted. The fact that true sexuality can arise from apomictic accessions could explain the discovery of sexual tetraploids from South African apomictic introductions, as reported by Smith (1972) and Hanna et al. (1973). Some of these South African introductions in our field nursery showed an especially high percentage of sexuality.

Apomicts collected in East African populations usually exhibit a low potential for sexuality. Our data show that highly apomictic hybrids can be recovered following hybridization between sexual and apomictic accessions, contrary to earlier statements.

Data from cytogenetical studies, along with other data from biological and plant evolution analyses (Combes 1972, Pernès 1972), led us to propose an improvement scheme for *Panicum maximum* that could be used in similar agamic complexes (Pernès et al. 1975). Several series of sexual × apomictic crosses were attempted, following a tetraploidization process using the $2\times \times 4\times$ crosses made in natural mixed populations of Tanzania. These crosses have led to a rapid extension of the sexual tetraploid pool and the regular release of apomictic hybrids. Some of the latter have already been tested in Ivory Coast. About 10 of them exhibited grass productions higher than 30 tons dry matter/ha/yr, with a seed production of 200 to 500 kg/ha. National societies for the development of cattle production in Ivory Coast are now growing several *Panicum maximum* varieties selected by

ORSTOM. The Institute is also engaged in the conservation of a large germ plasm pool including apomictic as well as sexual material.

LITERATURE CITED

Combes, D. 1972. Polymorphisme et modes de reproduction dans la section des *Maximae* du genre *Panicum* (Graminées) en Afrique. Thesis, Univ. of Paris (France): Mém. ORSTOM 77, Paris (1975).

Combes, D., and J. Pernès. 1970. Variations dans les nombres chromosomiques du *Panicum maximum* Jacq. en relation avec le mode de reproduction. C.R. Acad. Sci. Paris, sér. D. 270:782–785.

Hanna, W.W., J.B. Powell, J.C. Millot, and G.W. Burton. 1973. Cytology of obligate sexual plants in *Panicum maximum* Jacq. and their use in controlled hybrids. Crop Sci. 13:695–697.

Herr, J.M. 1971. A new clearing squash technique for the study of ovule development in angiosperms. Am. J. Bot. 58:785–790.

Pernès, J. 1972. Organisation évolutive d'un groupe agamique: la section des *Maximae* du genre *Panicum* (Graminées). Thesis, Univ. of Paris (France): Mém. ORSTOM 75, Paris (1975).

Pernès, J., R. Réné-Chaume, J. Réné, and Y. Savidan. 1975. Schéma d'amélioration génétique des complexes agamiques du type *Panicum*. Cah. ORSTOM, sér. Biol. 10:67–75.

Savidan, Y. 1975. Hérédité de l'apomixie. Contribution à l'étude de l'hérédité de l'apomixie sur *Panicum maximum* Jacq. (analyse des sacs embryonnaires). Cah. ORSTOM, sér. Biol. 10:91–95.

Savidan, Y. 1978. L'apomixie gamétophytique chez les Graminées et son utilisation en amélioration des plantes. Ann. Amélior. Pl. 28:1–9.

Savidan, Y. 1980. Chromosomal and embryological analyses in sexual × apomictic hybrids of *Panicum maximum* Jacq. Theor. Appl. Genet. 57:153–156.

Smith, R.L. 1972. Sexual reproduction in *Panicum maximum*. Crop Sci. 12:624–627.

Preliminary Evaluation of Legume Germ Plasm in the Cerrados of Brazil

D. THOMAS and R.P. DE ANDRADE

CIAT, Colombia, and EMBRAPA-CPAC, Brazil

Summary

The cerrados of the west central region of Brazil occupy 180 million ha and are classified as well-drained, tropical savannas. The Brazilian cattle population numbers 110 million, almost 60% of which is found in the cerrados. To improve the nutrition of the grazing animal, attention has been focused on cultivated pasture species, particularly tropical legumes. None of the commercial legume cultivars can be recommended for large areas of the cerrados. Species such as leucaena (*Leucaena leucocephala* [Lam.] de Wit) are not adapted to these acid, infertile soils with a high aluminum content, while *Stylosanthes* Sw. species are highly susceptible to the fungal disease anthracnose (*Colletotrichum gloeosporioides* Penz.).

In 1978 an evaluation program was initiated to select legumes that (1) grow and produce seed on acid soils under aluminum and water stress, (2) persist under grazing, and (3) tolerate pests and diseases. The work is being conducted at the Cerrados Agricultural Research Center, near Brasilia, at lat 15°S and altitude 1,000 to 1,100 m. Annual rainfall is 1,500 mm, mostly distributed in a 6-month wet season, and mean annual temperature is 21°C. Nine hundred legume accessions from thirteen genera have now been established as spaced plants on the 2 major soil types of the region. The pH of the soils is 4.6, with aluminum saturation in excess of 70%. Known nutrient deficiencies were corrected. Results are reported for 352 of the accessions over a 2-year period.

The most promising genus is *Stylosanthes*. Seventeen accessions of *S. guianensis* (Aubl.) Sw., *S. capitata* Vog., *S. scabra* Vog., and *S. viscosa* Sw. have combined good adaptation to acid soil conditions and good tolerance to pests and diseases. All *S. guianensis* accessions belong to a distinctive group of fine-stemmed, highly viscous ecotypes currently referred to as the "tardio" group. Their outstanding attribute is tolerance to anthracnose, the major limiting factor to the use of the genus in the region. Seed of the selected accessions is being multiplied for evaluation under grazing. The number of accessions of *Stylosanthes* has been increased further within the program.

None of the other genera have shown the potential of the genus *Sylosanthes*. New *Calopogonium* Desv. and *Galactia* P. Br. accessions were no more productive than commercial control cultivars. *Aeschynomene* L. accessions were highly susceptible to anthracnose, while those of *Pueraria* DC., *Teramnus* P. Br., *Vigna* Savi, and *Centrosema* (DC.) Benth. species grew relatively poorly. *Zornia* J.F. Gmel. accessions were very vigorous, but all plants were very susceptible to an insect/virus/fungus complex. Tolerance has been found in new ecotypes of *Z. brasiliensis* Vog. Most *Desmodium* Desv. species showed poor vigor and were affected by the disease little-leaf mycoplasma. Observations are continuing with these other genera, but introductions are being made on a more limited scale.

KEY WORDS: savannas, cerrados, Brazil, anthracnose, *Colletotrichum gloeosporioides* Penz., *Stylosanthes*, adaptation.

INTRODUCTION

The cerrados of the west central region of Brazil occupy 180 million ha and are classified as well-drained, tropical savannas. Almost 60% of the Brazilian cattle population is found in the cerrados, which makes the region one of the most important beef production areas in the tropics.

To improve the nutrition of grazing animals, attention has been focused on cultivated pasture species, particularly tropical legumes. However, none of the existing commercial cultivars can be confidently recommended for large areas of the cerrados. Most of these cultivars either are not adapted to the extremely acid soils or are susceptible to pests and diseases. Commercial legumes such as leucaena (*Leucaena leucocephala* [Lam.] de Wit) require substantial quantities of lime and other nutrients. Only relatively small areas of the cerrados can be improved in this way. Commercial cultivars of *Stylosanthes* are adapted to soils of low pH with aluminum stress—e.g., *Stylosanthes guianensis* (Aubl.) Sw. (formerly *S. gracilis* Kunth), IRI 1022 (Buller et al. 1970)—but are highly susceptible to the fungal disease anthracnose (*Colletotrichum gloeosporioides* Penz.). This disease causes loss of vigor, defoliation, and death of susceptible plants in subtropical and tropical areas.

In 1978 we began a pasture-evaluation program to select, under cerrados conditions, legumes that would (1) grow and produce seed on acid soils under aluminum and water stress, (2) persist under grazing, and (3) tolerate pests and diseases. The work is being conducted at the Cerrados Agricultural Research Center (CPAC), lat 15°S. Rainfall is 1,500 mm/yr with 90% falling from October to March inclusive. Mean annual temperature is 21°C, and altitude is 1,000 to 1,100 m.

This paper reports on the performance of 352 legume accessions established in 1978 for which observations over 2 years are available. Fifty percent of these accessions were *Stylosanthes* species; the remainder were accessions from twelve other genera.

METHODS

Accessions were established as spaced plants on the 2 most important soils of the cerrados. Where possible, commercial cultivars were included for comparison. The soils, both oxisols, were the dark-red latosol (DRL) and the red-yellow latosol (RYL) (Anon. 1976). The pH of the soils was 4.6, with more than 70% aluminum saturation. The site on RYL was 100 m higher than that on DRL. The original pH of the soils was maintained, but nutrients were added to correct known deficiencies. These were 35 kg/ha of phosphorus, 50 kg/ha of potassium, 48 kg/ha of sulfur, 77 kg/ha of calcium, 4 kg/ha of magnesium, 2 kg/ha of zinc, and 0.25 kg/ha of molybdenum. Single superphosphate was the source of phosphorus, sulfur, and calcium.

Observations were made on phenology, dry-matter (DM) yield, regrowth, nutritive value, seed production, and tolerance to pests and diseases. Anthracnose was assessed on a scale of 1 (no anthracnose) to 5 (plant death) during the wet and dry seasons. Accessions scoring less than 3 were selected for continued evaluation.

D. Thomas is a Senior Scientist in Pasture Agronomy with the Tropical Pastures Program of CIAT, and R.P. de Andrade is a Pasture Agronomist with EMBRAPA-CPAC.

RESULTS

Stylosanthes Sw. Species

Seventeen accessions of *Stylosanthes guianensis, S. capitata* Vog., *S. scabra* Vog., and *S. viscosa* Sw. have shown good adaptation to cerrados conditions (Table 1). The accessions of *S. guianensis* produced appreciably more dry matter on both soils than the control Australian commercial cultivar Endeavour. However, all accessions of the species were more productive on the DRL soil. In contrast to *S. guianensis*, the accessions of *S. capitata* grew best on the RYL soil. *S. scabra* accessions showed variation in their adaptation to soil type, while *S. viscosa* CIAT 1094 grew satisfactorily on both soils. Most selections combined good productivity with a higher nutritive value than the control cultivars. All accessions were readily consumed by cattle. Seed of the 17 accessions is being multiplied and observations will continue over the next 2 years.

A major consideration in this selection has been the incidence of anthracnose. None of the accessions of *S. humilis* HBK or *S. hamata* (L.) Taub. were selected because of their high susceptibility to the disease.

Other Genera

Accessions of *Calopogonium* Desv. and *Galactia* P. Br. species were well adapted to the acid soils but proved to be no more productive than the Brazilian commercial control cultivars. *Pueraria* DC., *Teramnus* P. Br., *Vigna* Savi, and *Centrosema* (DC.) Benth. species grew relatively poorly at both sites. *Zornia* J.F. Gmel. accessions were very vigorous and excellent in regrowth but suffered from leaf shed in the dry season. The control, *Zornia latifolia* DC. CIAT 728, remained the most productive accession. With the exception of the browse species *Desmodium gyroides* (Roxb. ex Link) DC. CIAT 3001, accessions of *Desmodium* Desv. were poor in vigor and were subject to severe leaf shed in the dry season.

A number of diseases were prevalent among species of some genera. *Aeschynomene* L. accessions were highly susceptible to anthracnose, which killed most of the plants. All *Zornia* accessions were very susceptible to an insect-virus-fungus complex, common on native species, and several plants were killed. Little-leaf mycoplasma was a problem in *Desmodium* species. *D. gyroides*, despite good adaptation to the acid soils, was severely affected by the disease in the second season. Nematode problems were recorded in *D. ovalifolium* Guill. et Perr. CIAT 350.

DISCUSSION

These preliminary observations demonstrate the potential of the genus *Stylosanthes* for the cerrados of Brazil. A number of species and accessions, collected in the savanna areas of Latin America, have proved to be well adapted to

Table 1.   Performance of selected *Stylosanthes* accessions at CPAC.

CIAT No.	Season of[†] flowering	Disease[††] score	DM Production (g/plant) DRL	RYL	Crude protein[o] %	DDM(in vitro)[o] %
S. guianensis						
Endeavour[x]	Late	4.0	40	16	11.1	40.6
1095	Late	1.5	136	86	10.6	47.1
2191	Mid	1.5	147	58	11.7	43.2
2203	Late	1.5	183	75	10.9	39.9
2243	Late	1.5	118	71	10.4	42.3
2244	Late	1.0	140	126	10.8	37.9
2245	Late	1.5	111	42	10.9	40.9
S. capitata						
1405[x]	Mid	4.0	62	83	10.3	41.3
1686	Mid	1.5	53	103	12.4	51.2
1728	Late	1.5	34	165	11.5	47.6
1943	Late	1.5	60	104	10.6	50.4
2246	Late	1.5	61	91	11.6	45.8
S. scabra						
4	Mid	1.5	135	189	12.5	47.3
2554	Mid	2.0	117	114	9.6	42.4
1047	Mid	1.0	83	104	13.3	48.4
2299	Mid	1.5	148	114	12.0	45.2
2304	Early	2.0	167	76	12.1	44.6
2308	Mid	1.0	84	132	12.1	48.4
S. viscosa						
1094	Mid	1.0	92	94	11.8	37.3

[†]Early (December, January); mid (February, March); late (April or later).

[††]Anthracnose rating (1.0 = no symptoms; 5.0 = plant death).

[o]Samples collected in May.

[x]Control species.

the DRL and RYL soil types.

The primary limiting factor to the genus is the incidence of anthracnose. Because this disease is widespread and endemic in the tropics of Latin America, plant tolerance is the only practical means of control. Fortunately, variation in anthracnose tolerance exists within the various species of *Stylosanthes*. All 17 selected accessions scored below the critical value of 3 in anthracnose rating. The *S. guianensis* accessions belong to a distinctive group widely distributed in the cerrados of Brazil and the savannas of Venezuela. This group is currently referred to as the "tardio" group to distinguish the ecotypes from the "common" types, such as the commercial cultivars Endeavour and IRI 1022. In contrast to the "common" types, these "tardio" accessions are predominantly late-flowering, fine-stemmed, and highly viscous. Their outstanding attribute is good tolerance of anthracnose, a trait deficient in the "common" types.

Accordingly, emphasis has been placed on increasing the *Stylosanthes* accessions at CPAC. The accessions under evaluation now number 900, of which 70% are *Stylosanthes*. These have included new accessions of *S. capitata*, *S. scabra*, *S. macrocephala* M.B. Ferr. et S. Costa, and the "tardio" ecotypes of *S. guianensis*. Accessions of these key *Stylosanthes* species will now be grown at other locations in the cerrados through a regional trial network.

None of the other genera have so far shown as much potential as *Stylosanthes*. Nevertheless, observations are being continued and new accessions brought in on a more limited scale. Brazilian four-leaved ecotypes of *Z. brasiliensis* Vog., recently introduced to CPAC, appear to be tolerant of the virus disease affecting other *Zornia* species as well as showing excellent vigor, good regrowth potential, and a better retention of leaves in the dry season. New accessions of *Centrosema macrocarpum* (DC.) Benth. also have potential for the cerrados region.

LITERATURE CITED

Anon. 1976. Centro de Pesquisa Agropecuária dos Cerrados (CPAC). Relatório técnico anual 1976, Brasilia, Brazil, 11-15.

Buller, R.E., S. Aronovich, L.R. Quinn, and W.V.A. Bisschoff. 1970. Performance of tropical legumes in the upland savannah of central Brazil. Proc. XI Int. Grassl. Cong., 143-146.

Adaptation and Utilization of Three Legumes in China

HUANG WENHUI

Ministry of Agriculture, People's Republic of China

Summary

China is a vast territory with extremely varied climate, soil, and topography, and it is very rich in pasture grass resources. The main legumes used for forage are alfalfa (*Medicago sativa* L.), yellow alfalfa (*Medicago falcata* L.), and shadawang (*Astragalus adsurgens* Pall.).

Alfalfa has been cultivated for some 2,000 years in China. Total area cultivated at present is about 0.4 million ha; alfalfa is the major species in the area of planted forages. Alfalfa is distributed in more than ten provinces located between lat 35° and 45° N. It is adapted to areas where the annual mean temperature is 5°-12°C, the annual accumulated temperature is 2,700°-5,000°C, and the annual rainfall is about 400-600 mm. It does survive the winter areas where the annual mean temperature is below 5°C.

Yellow alfalfa is tolerant of low temperature and can survive the winter safely in the northern part of the Neimenggu Autonomous Region (Inner Mongolia) and Heilongjiang Province (Manchuria). Yellow alfalfa is used as parent material for alfalfa breeding.

Shadawang is a perennial legume domesticated from wild ecotypes. Over 3,000 ha of shadawang has been planted in recent years in the areas between lat 33° and 43° N. It is tolerant of cold weather, drought, and poor soil and is resistant to wind erosion and desertification. Aerial seeding on the loess plateau in northwest China has been successful. It is adapted to those areas where the annual rainfall is about 300-500 mm, the annual accumulated temperature is below 3,600°C, and the frost-free period is less than 150 days.

KEY WORDS: legume, introduction, utilization, adaptability, distribution.

INTRODUCTION

China is a large country with a total area of 9.6 million km². The distance extends about 5,000 km from east to west and about 5,500 km from south to north. The country's climate, soil, and geographical conditions are so diversified that widely different species and various kinds of ecotypes of forage plants have formed. The pasture region, situated in the western and northern plateau, is characterized by an arid and cold alpine climate providing conditions favorable for development of the animal industry. The eastern and southern parts of China have warm, humid weather and soil that is relatively fertile and suitable for cultivation. Throughout the country there are large areas of rolling, hilly lands suitable for growing cash crops and keeping livestock.

Because of the complex climatic, soil, and geographical conditions, a large number of grasses and legumes have evolved, although no more than a dozen have been widely used. The main legumes in frequent use are alfalfa (*Medicago sativa* L.), yellow-flowered alfalfa (*M. falcata* L.), and shadawang (*Astragalus adsurgens* Pall.).

ALFALFA (*MEDICAGO SATIVA* L.)

Alfalfa is an ancient cultivated legume grown for more than 2,000 years since it was first introduced from Iran and Persia during the West Han Dynasty (Shi Maqian 1959). Originally planted in the Yellow River valley, it is now widely distributed over Xinjiang, Gansu, Ningxia, Neimenggu, Shaanxi, Shanxi, Shandong, Hebei, Henan, and the northern part of Jiangsu and Anhui, generally extending between lat 35° and 45°N. Up to 1979, the total area under alfalfa was 0.4 million ha, about half being in hay and half in pasture. Generally speaking, alfalfa is one of the most widespread forages grown in China. In farming zones, it is grown mainly in rotation with grain crops, with a view to developing soil fertility and enhancing yield of herbage. Stems and leaves of alfalfa are commonly used as forage for animals. In recent years, the areas under alfalfa have been expanding; the species is widely used as one of the most valuable supplements to pasture in cold seasons.

Alfalfa is distributed on chernozem, chestnut soil, and desert soil in northeast, northwest, and north China, respectively. It is adaptable for growing in areas where the annual mean temperature ranges from 5°C to 12°C, the annual average rainfall fluctuates between 400 and 600 mm, and the annual accumulated temperature is about 2,700°–5,000°C (Huang Wenhui 1980). Hay yield varies with time of cutting, being about 3,750 to 15,000 kg/ha. Generally, hay is mowed two to four times a year. Seed yield is estimated at about 150 to 225 kg/ha.

It is necessary to select winter-hardy cultivars in northern China, especially in Neimenggu Autonomous Region, Heilongjiang Province, and many alpine and subalpine areas where the annual mean temperature falls

below 5°C. Some of the cultivars are not likely to overwinter without being frozen. Annual mean temperature (Table 1) in Hailar in Neimenggu is – 2.1°C; in Xilinhot, 1.8°C; and in Qiqihar in Heilongjiang Province, 3.2°C. In Qilianshan, a subalpine area of Gansu Province, winter temperature drops to – 20°C. Cultivars such as Zhaodong and Variegated, as well as the introduced Canadian Rambler, have the ability to overwinter in those regions.

We have two excellent cultivars, Xinjiang Large-Leaved and Gongnong No. 1, with fast-growing habit and high yielding ability. Among those introduced, Hunter River from Australia has luxuriant growth, quick regrowth, and high yielding ability in comparison with our two inland cultivars.

Planting areas under alfalfa in Xinjiang are relatively large. Alfalfa grows and develops relatively well, and its yield of hay is estimated at about 7,500 kg/ha in Hetian County, where the annual mean temperature reaches as high as 12.1°C, annual accumulated temperature is about 4,612°C, and rainfall is as low as 32.1 mm (Table 1).

YELLOW-FLOWERED ALFALFA (*M. FALCATA* L.)

Yellow-flowered alfalfa has many wild cultivars with different ecotypes distributed over the valley of Ili, West Xinjiang Uighur Autonomous Region, the northern part of Neimenggu, and areas in the north of Heilongjiang and Hebei where the cultivars grow in natural plant communities, frequently in mixtures with smooth bromegrass (*Bromus inermis* Leyss.). Its adaptation area is generally the same as that of common alfalfa, though it is much more winter-hardy. Wherever common alfalfa cannot survive the winter, yellow-flowered alfalfa will take its place in wild habitat. It has been grown in high altitudes and in pastoral areas having severely cold winters. Because of its winter-hardiness and drought resistance, it has been used as parent material to cross with *M. sativa*. Legume breeders have developed and released a hybrid alfalfa with valuable traits of drought resistance, winter-hardiness, and high productivity. Its area of cultivation has expanded somewhat, and now this hybrid strain appears promising in Tongde Range of the Qinghai Plateau at an elevation of 3,200 m, where it is able to survive the winters.

SHADAWANG (*ASTRAGALUS ADSURGENS* PALL.)

"Shadawang" is the Chinese name for wind-resistant Loco (Huang Wenhui et al. 1976). The species is being released for forage use and is distributed widely over such provinces as Shandong, Hebei, Henan, Neimenggu, Shanxi, and Northern Jiangsu. Generally, the geographical position is between lat 33° and 43°N. It is well adapted to those areas where the annual mean temperature is 8°–15°C, the annual average rainfall ranges from 300 to 500 mm, and the accumulated temperature varies from 3,600°C to 5,000°C. It grows

---

The author is an Associate Professor in the General Bureau of Animal Husbandry.

Table 1. Air temperature and total rainfall in areas of China where alfalfa is grown.

| Location | Altitude (m) | Temperature (°C) | | | Annual accum. | Annual mean rainfall (mm) |
		Annual mean	Jan. mean	July mean		
Qiqihar	145.9	3.2	− 19.3	22.6	3125.0	433.2
Changchun	236.8	5.6	− 17.2	23.4	3408.9	571.6
Xilinhot	989.5	1.8	− 20.3	18.8	2758.9	269.3
Lanzhou	1517.2	8.9	− 7.3	22.0	3764.6	331.5
Siling	2261.2	5.6	− 8.6	17.2	2730.5	371.2
Beijing	31.2	11.6	− 4.7	26.1	4549.4	584.0
Taiyuan	777.9	9.4	− 6.5	23.4	3901.8	494.9
Hetian	1374.6	12.1	− 5.2	25.3	4612.2	32.1

well on drab soil and can persist in soils with pH values ranging from neutral to slightly alkaline. It is a fine-quality legume, used as a mainstay for soil and water conservation and for foraging and green-manuring.

Shadawang normally grows 5 to 8 years. It has a penetrating taproot, sometimes reaching a depth of 1 to 4 m during a 4-year growth period in the loess plateau. Usually, it attains an aerial height of 70 to 150 cm. Shadawang was domesticated from a wild habitat. By the end of the 1970s, its area under cultivation had reached 2,500 to 3,300 ha. Experiments have indicated that it is highly resistant to drought stress and possesses a stable and high yielding ability. Hay yields of 7,500 to 15,000 kg/ha are commonly recorded. Nutrient value is high, ranking equivalent to that of purple alfalfa and containing about 17% crude protein. Soil fertility may be increased by growing shadawang, which has dry-matter content of N, 1.1%–3.8%; P, 0.15%–0.36%; and K, 1.4%–1.6% in the succeeding crop. When shadawang is used as a green-manure crop, total N content/500 kg fresh weight is estimated to be equivalent to 9 to 28 kg of ammonium sulfate.

In recent years, this legume has been successfully planted by air-sowing for control of soil erosion on loess plateau and for improving deteriorated grassland. Its establishment is slow, however, the vegetation cover being better in the second year.

Shadawang has the primary disadvantages of low palatability and low seed yields. Because its green matter usually tastes bitter, animals prefer hay to fresh forage. Late maturity and a prolonged flowering stage result in late or no ripening of seeds in high latitudes. Seed production is so limited that yields as low as 150 to 225 kg/ha (unacceptable commercially) are common.

In areas with annual mean temperatures below 10°C, annual accumulated temperatures below 3,600°C, and frost-free periods of less than 150 days, shadawang will bear very few, if any, seeds (Huang Wenhui 1980). Because of its improved winter-hardiness over alfalfa, it has expanded further north to the highlands of Qinghai, Xinjiang, Heilongjiang, and Neimenggu (Table 2), where it rarely blooms and has very poor seed yield. We are selecting early-maturing cultivars and studying improved cultivation techniques in order to improve seed production.

LITERATURE CITED

Huang Wenhui. 1980. A search for the rules in forage introduction. Feed Res. 3:7–12.

Huang Wenhui, Hou Yougzhang, Chen Xiaoquan, and Wang Quoxian. 1976. Alfalfa, shadawang. *In* Good varieties of forage, 1:1–14, 20–36. Chinghai People's Publishing House, Cilin.

Shi Maqian. 1959. The biography of Dawan. History Record, 10:3173–3174. China Publishing House, Beijing.

Table 2. Air temperature and total rainfall in areas of China where *Astragalus adsurgens* is grown.

| Location | Altitude (m) | Temperature (°C) | | | Annual accum. | Annual mean rainfall (mm) |
		Annual mean	Jan. mean	July mean		
Fuxin	144.0	7.4	− 12.1	24.0	3691.6	570.4
Yulin	1057.5	7.9	− 9.9	21.3	3648.7	451.2
Changwei	51.6	14.3	− 1.4	27.6	4686.6	689.4
Zhengzhou	110.4	14.3	− 0.2	27.8	5238.6	697.6

Hedysarum mongolicum Turcz.:
An Important Protein Resource Legume on Dry Sandy Land

LI MIN and A.W. HOVIN

Institute of Grassland, Chinese Academy of Agricultural Sciences,
Hohehot, Neimenggu Autonomous Region, People's Republic of China

Summary

In searching for plants that can both increase the protein production and improve the sandy-soil conservation of the grassland in the arid, windy regions of North China, so far, from among many native plants, we have identified *Hedysarum mongolicum* Turcz. as the best species to fulfill our dual purposes. We investigated its distribution, planted its native seeds in the Maowusu desert, and studied it under its native and cultivated growing conditions from 1957 to 1964.

Chinese farmers call this perennial subbrush legume of the sweetvetch genus "Yang Cai," meaning "sheep like to eat." It can thrive in the stable or substable sand-dune areas between lat 38° and 48° N with annual rainfall less than 300 mm.

Under favorable growing conditions, its main green stems may be 100 to 150 cm tall and form a canopy 100 to 150 cm in diameter. Its taproot goes 2 m deep into sandy soil. Below its crown, 5 to 6 m long, strong rhizomes creep out in all directions from which new branches shoot up as high as 1 m, and new roots grow down 20 to 30 cm below sandy soils. We found a single native Yang Cai plant on a sand dune with creeping roots spreading over an area 30 m² and total root length of 108 m. The roots had also formed 81 vegetative clumps above the ground.

Because the seed coat of Yang Cai is impermeable to water, the seeds should be properly scarified before seeding. Early-spring seedings usually are more successful than late-summer seedings. In the year of establishment, very few flowers bloom, because most stems are vegetative. In the second year, branches begin to flower. Beautiful purple-red flowers and green bushes cover the yellow sand in July and August. The annual yield of hay can reach 4,500 kg/ha and that of seeds can reach 100 kg/ha under cultivation.

Yang Cai is a high-protein forage with a crude-protein content of more than 20% of total dry matter in the flowering phase. Its several amino acids compare favorably with those of alfalfa. Its high protein content and fine stems lead most animals to devour it green or dry. It has improved the nutritive value of the vegetation of China's sandy soil. Many seed-production regions have been established in Neimenggu Autonomous Region (Inner Mongolia) where many natives in the dry steppes have expanded the acreage of its plantation.

KEY WORDS: sweetvetch, *Hedysarum mongolicum*, legume, protein, resource, dry land, sandy land, amino-acid composition.

INTRODUCTION

It has been a long-cherished hope of Chinese scientists and herdsmen to identify plants that can both increase the protein production and stabilize the sandy soils of grassland in the arid, windy regions of North China. These regions of the steppes have a none-too-certain annual rainfall averaging less than 300 mm and contain immense areas of sandy soils and sand dunes.

Since establishment of the Chinese Steppe Research Station, Ordors, Neimenggu Autonomous Region (Inner Mongolia), we have searched for species, particularly leguminous ones, with highly developed root systems. Among plant collections from desert regions we identified a few species (Anon. 1955) of genus *Hedysarum* (sweetvetch) that appeared to serve our purpose, such as *H. scoparium*, *H. fruticosum*, and *H. mongolicum*. Among these, *H. mongolicum* has been the most promising.

Zhong and Li (1956) wrote that *H. mongolicum* had been cultivated in the Zungore region and that it was an excellent subbrush legume for both hay and grazing on dryland sites. Liu Ying-xin (1958) noted that sweetvetch was an important genus for stabilizing the shifting sand dunes and made good use of *H. scoparium* to fix the sand dunes in the Sha Po Tou section along the Pao-lan railway. Li (1964) reported that *H. mongolicum* was a well-adapted, permanent pasture legume that had been used successfully to improve 10 million ha of unproductive deserts in the Inner Mongolian steppe. Since then, this species has been planted by natives in many localities in Inner Mongolia. *Hedysarum* has been recommended as a good feed for sheep, cattle, and horses.

Li Min is Head of the Grassland Crops Department of the Institute of Grassland, Chinese Academy of Agricultural Sciences, and A. W. Hovin is Associate Director of the Montana Agricultural Experiment Station.

Table 1. Nutrient compostition of *H. mongolicum* at different growth phases compared with three grades of dehydrated alfalfa.

Constituent	*H. mongolicum**			*Dehydrated alfalfa***		
	I	II	III	I	II	III
	----------------- % dry matter-----------------					
Moisture	12.13	13.44	11.18	7.14	6.88	6.90
Protein	25.41	20.35	9.43	22.47	20.64	15.22
Fat	2.84	3.78	3.64	3.69	3.58	2.32
Crude fiber	21.57	25.45	–	18.54	20.15	26.41
N-free extract	31.82	29.79	–	37.85	38.42	40.77
Ash	5.93	7.19	7.28	10.32	10.34	8.40
Calcium	2.40	1.19	1.43	1.47	1.47	1.23
Phosphorus	0.58	0.52	0.13	0.28	0.27	0.22

*Growth phase of I = Veg. growth, II = Flowering, III = Seed, fruiting.

**Source: American Dehydrators Assoc. (1969). Grades I, II, and III correspond to crude protein concentration of 22%, 20%, and 15%, respectively.

METHODS

A large number of plants were collected from desert regions of Maowusu, Kunshandak, and Kubugi. The native seeds of *H. mongolicum* collected here were also used to establish vegetative material in the Maowusu desert from 1957 to 1964. At the Ordors Grassland Improvement Research Station, studies were made of the origin and regional distribution of various species, wild seed characteristics, and morphological and biological features of roots, stems, leaves, flowers, and seedling growth, and of cultivation techniques and production.

Data on nutrient composition were determined in the laboratory by the Inner Mongolian College of Agriculture in 1958 (Table 1), and the amino-acid contents of *H. mongolicum* were analyzed by the Animal Husbandry Institute of Beijing in 1980 (Table 2).

Table 2. Composition of amino acids for cornmeal, 15% crude protein grade of dehydrated alfalfa, and *H. mongolicum*.

Amino acid	Corn-meal*	Dehydrated alfalfa**	*H. mongolicum*	
			Leaves and branches	Seeds
	----------------------- % dry wt-----------------------			
Arginine	0.40	0.58	0.90	1.79
Histidine	0.23	0.30	0.34	0.57
Leucine	1.03	1.05	1.19	1.11
Lysine	0.32	0.60	0.84	0.84
Methionine	0.12	0.23	0.12	0.21
Phenylalanine	0.38	0.66	0.90	0.73
Threonine	0.34	0.60	0.72	0.74
Valine	0.41	0.84	0.99	0.86
Glycine		0.72	0.82	1.02
Alanine		0.79	0.80	0.77
Serine		0.62	0.72	1.23
Proline		0.63	1.17	0.56
Glutamic acid		1.40	1.62	2.81
Aspartic acid		1.63	1.93	2.57

*Source: Feed Science of Animals, China.

**Source: See Table 1, based on 15% crude protein grade.

RESULTS

Natural Distribution and Adaptation

Hedysarum mongolicum is known by its Chinese name, "Yang Cai," which means "sheep like to eat." The species is widely distributed from lat 38° to 48°N. It has been found also in the southern steppes of Mongolia (Yunatov 1954). It will grow well in fixed or subfixed sand-dune areas.

Usually, *H. mongolicum* grows in communities with *Caragana, Atraphaixis, Artemisia*, etc., but sometimes it covers a pasture by itself. Under favorable growing conditions the main stems may reach a height of 100 to 150 cm, and the canopy diameter generally reaches 100 to 150 cm.

In July and August, the purple-red flowers and green bushes of the Yang Cai communities make the desert exceedingly charming. The planted acreage of Yang Cai is expanding in some areas of Inner Mongolia, especially in Qinshuihe County, where it is grown on approximately 2,000 ha by farmers.

Morphological Characteristics

The leaves of this species are pinnately compound with six to ten pairs of leaflets plus a terminal one. The blades are very small, but their functions are taken care of by the green branches. The small blades help to reduce wind resistance and evaporation. The stems have three kinds of branches on the same bush; most are green, but some are yellow or brown. Their flowers, with two petals and a keel, are big and beautiful, like purple-red butterflies. The seedpods are like round beads; each seed is encased in a very thick net-wrinkled coat. It is therefore difficult for moisture to permeate them.

The plant has a long taproot, going 2 m deep into the soil. Below its crown several buds of creeping-rooted shoots send out 5- to 6-m-long, strong rhizomatous branches 20 to 30 cm below the surface of sandy soils. They also can produce new shoots from adventitious nodes.

We have studied its root systems in native sand dunes. The creeping roots of one plant expanded to some 30 m² with a total length of 108 m. The shoots above the ground had 81 clumps and formed a vast vegetative cover.

Growth and Development

The plants of Yang Cai can develop only slowly from seeds, but they are vigorous, hardy, and drought-resistant once they have become established in a pasture or hay field. They are less tolerant of alkaline or poorly drained soils. The species grows best on loose sandy and sandy loam soils. During its first year, few branches develop flowers; most remain vegetative. In the second year, the plants begin to flower in summer. It takes 36 days to form an inflorescence and to flower. One flower needs 12 hours to bloom; tripping the keel at noon promotes seed set.

Naturally, seed germination averages only 14%, but

the rate can be raised to 46% if the seed coats are hulled. The 1,000-seed weight is 11.4 g. The germination rate of seeds from cultivated plants is about 60%.

The establishment time of Yang Cai, from seeding to seedling, is 7 to 8 days. The length of the seedling above the ground is only 2 to 3 cm one month after seeding, while that of the taproot is about 15 cm. A pair of leaflets appears at the same time. The roots grow very rapidly and absorb water and nutrients for the stems and leaves. At the end of the first year, the taproot may be 60 cm deep. The crown below ground will produce several buds with creeping roots over the years.

Culture and Management

Yang Cai plants have greater tolerance than alfalfa to drought and infertile soil. The seed coats are impermeable to water, and the seed should be properly scarified before seeding. In the country of Qinshuihe, early-spring seedings are usually more successful than late-summer seedings. Scarified seeds sown at the rate of 20–30 kg/ha give satisfactory stands.

When seeds are drilled, care must be taken not to seed too deep, preferably 1 to 2 cm on sandy soils. Broadcast seedings are most successful in spring on a rainy day. Sometimes the seeders let flocks of sheep tread through the land just after broadcasting the seed.

Yang Cai can grow on infertile, dry, or poor sandy soils, but it responds well to fertilization and irrigation. The cultivated pasture must be protected in the first 2 years after planting, and it should not be grazed or mown during this period.

Productivity and Nutritive Value

Yang Cai seldom makes much growth before early summer, but it grows rapidly and vigorously during the midsummer rainy season. Therefore, the general time of harvest is in the fall during the mature stage. The crop can be cut for hay once annually. New shoots and stems will grow out from the crown by the following spring. The one harvest/yr will not influence production of forage or seeds the next year. The annual yield of hay reaches 4,500 kg/ha and that of seeds 100 kg/ha.

Yang Cai is a high-protein-producing plant; its crude-protein content is more than 20% of dry matter in the flowering phase (Table 1). Most animals like Yang Cai when it is fed as hay in winter, and they also like to graze it in pasture.

Concentrations of several amino acids, such as arginine, glutamic acid, lysine, and proline, are higher than the 15% crude-protein grade of dehydrated alfalfa (Table 2). Most amino acids present in corn meal are at a lower concentration there than in Yang Cai.

CONCLUSION

So far, our studies have shown that *Hedysarum mongolicum* is the species we have long hoped to find for the grassland of the arid, windy regions of North China, for the following reasons:

1. Its high protein content, tender texture, and palatable forage lead sheep, cattle, and horses to devour it green or dry. The feeding value of the grassland can be quickly improved by interplanting this legume.

2. Its deep taproot and long, strongly rhizomatous root system enables it to protect sandy soils from wind erosion and to stabilize the moving sand dunes.

3. It can thrive on sandy soils and sand dunes with an annual rainfall of less than 300 mm.

4. Many natives of such regions already have extended the cultivation of this legume because they have learned its value as a feed and as a soil protector.

Our further studies will seek to determine the crop's digestibility, the physiological and biochemical basis of its drought tolerance, and the process of nitrogen fixation.

LITERATURE CITED

Anon. 1955. The diagrams of Chinese legume plants. Bot. Inst. Acad. Sinica. Science Press, Beijing.

American Dehydrators Association. 1969. A detailed nutrient analysis of commercial samples of dehydrated alfalfa. *In* Dehydrated alfalfa assay report, 3d ed. Am. Dehydr. Assoc., Shawnee Mission, Kans.

Li Min. 1964. Culture and adaptation of four wild forage species. *In* Report of the Chinese Steppe Research Station, Ordors. Vol. 3.

Liu Ying-xin. 1958. A report of a systematic investigation in the desert regions. *In* Desert institute of Academy Sinica. Vol. 1. Science Press, Beijing.

Yunatov, A.A. 1954. Forage plants of pasture and meadow in the People's Republic of Mongolia (Chinese translation by Wu Zho-hua). Science Press, Beijing.

Zhong Yong-an and Li Min. 1956. A plant, *Hedysarum mongolicum,* of the sweetvetch genus.

Rhizoma Peanut: Perennial Warm-Season Forage Legume

G.M. PRINE

University of Florida, Gainesville, Fla., U.S.A.

Summary

Rhizoma peanut is the common name given species in section *Rhizomatosae* of the genus *Arachis*. Florigraze rhizoma peanut (*A. glabrata* Benth.) has been released as a cultivar by the Florida agricultural experiment stations and the USDA Soil Conservation Service. Florigraze is adapted to the well-drained soils of Florida and the warm humid tropics and subtropics around the world. Rhizoma peanuts are propagated by rhizomes, which in Florida are best planted during winter. The slowness of coverage during early establishment is the biggest shortcoming of rhizoma peanuts. If they are established first, rhizoma peanuts will grow well in mixture with bermudagrass (*Cynodon dactylon* [L.] Pers.), digitgrass (*Digitaria decumbens* Stent.), and bahiagrass (*Paspalum notatum* Flugge). Average seasonal yield of Florigraze grown in mixture with a sparse stand of Pensacola bahiagrass over five seasons was 10.5 metric tons (mt)/ha. Bahiagrass without peanut yielded only 2 to 3 mt/ha annually. Pure Florigraze peanut cut at hay stage normally has a protein content of 14% to 16% and in-vitro organic-matter digestibility (IVOMD) above 60%. Florigraze cut every 2 weeks had an annual dry-matter yield of 5 mt/ha, 22% protein, 74% IVOMD, and 93% leaf in forage. Over 2 seasons the dry-matter yield (10 mt/ha) was similar at 6-, 8-, 10-, and 12-week cutting intervals. Protein, IVOMD, and leaf percentages of Florigraze rhizoma peanut progressively decreased as length of cutting interval increased from 2 to 12 weeks. Rhizoma peanuts can be used for pasture, hay, greenchop, high-quality dehydrated hay and leaf meal, creep grazing, ornamental ground cover, and possibly as a living mulch in which other crops and trees can be planted. Weeds are the biggest pest problem with rhizoma peanuts, especially during establishment. So far, there have been no serious insect, disease, or nematode problems reported with rhizoma peanuts in Florida.

KEY WORDS: rhizoma peanut, *Arachis glabrata* Benth., perennial legume, perennial peanut, cutting frequency.

INTRODUCTION

The peanut genus, *Arachis*, in which the common peanut (*A. hypogaea* L.) is classified, has approximately 50 species (Gregory and Gregory 1979). The common name "rhizoma peanut" has been given to the perennial rhizomatous *Arachis* species in the section *Rhizomatosae* (Prine et al. 1981).

Arachis is native to South America, perhaps originating in the northern part of Argentina or in Peru, where some species have been cultivated since 1000 B.C. (Smith 1968, Martin et al. 1976). Higgins (1951) states that the wild species of *Arachis* are found from the Amazon river in Brazil through Bolivia, Paraguay, Uruguay, and the northern regions of Argentina. The wide distribution of the wild species of *Arachis* in South America is believed to have been caused by flooding rivers carrying seed pods and plant parts from southern branches of the Amazon river in Paraguay (Gregory et al. 1973).

The Florida agricultural experiment stations and the USDA Soil Conservation Service (SCS) have jointly released Florigraze rhizoma peanut (*A. glabrata* Benth.) (Prine et al. 1981). Foundation rhizome stocks have been distributed to growers in the winters from 1978 to the present, to plant approximately 50 ha for commercial rhizome production.

Three other rhizoma peanuts—Arb, PI 118457; Arblick, PI 262839; and Arbrook, PI 262817—have been released by the SCS Plant Materials Center at Brooksville, Florida, and small quantities have been grown for testing. Small acreages of these accessions still exist on the test farms, but little increase has taken place. Arbrook (*A. sp.*) and a stoloniferous perennial peanut, *Arachis benthamii* (PI 338282), are presently being evaluated and increased by the Florida agricultural experiment stations and the SCS for possible future joint release as cultivars.

Annual hay yields of 5 to 7 mt/ha under low-fertility conditions were obtained at the University of Florida with Arb perennial peanut (Prine 1964). Blickenderfer et al. (1964) reported yields of rhizoma peanut ranging from 5.5 to 7.3 mt/ha. Prine (1964) obtained crude-protein values for Arb ranging from 10.7% to 16.7%. In other experiments, Prine (1973) found the average crude-protein content of forage to be 12.8%, 13.5%, and 14% and of in-vitro organic-matter digestibility (IVOMD) to be 60.8%, 57%, and 62% for Arb, Arblick, and Florigraze cultivars, respectively.

The author is a Professor of Agronomy at the University of Florida.

PROPAGATION AND ADAPTATION

As rhizoma peanut usually produces few seeds, it must be propagated from rhizomes. When planted in winter months (December to March), rhizoma peanut begins to spread through rhizome growth about midsummer of the planting year. If planting is delayed until the summer rainy season (June through August) there is usually no spread of peanut rhizomes until midsummer of the next year.

Rhizomes are dug by cutting the rhizome mat into ribbons 22 to 38 cm wide with coulters and digging with a modified potato digger. They can also be dug with bermudagrass sprig harvesters. Sprig harvesters break the rhizomes into shorter pieces than are needed for best growth of new plants but allow mechanized planting of the rhizomes with bermudagrass sprig planters. The rhizomes should be inoculated and completely covered 6 cm deep in sandy loam and heavier textured soils and 8 cm deep in sands.

Dependable first-season coverage has required 12 to 14 hectoliters of rhizomes/ha. Peanut hills should be planted 90 cm or less apart in every direction. When rhizoma peanut is planted with bermudagrass sprig planters, rows of peanut should not be more than 50 cm apart.

Rhizoma peanut grows best under warm temperatures and humid conditions. It will grow on all types of well-drained soils. Rhizoma peanut is adapted where the warm growing season has well-distributed rainfall in quantities equal to or greater than evapo-transpiration and temperatures are above 16°C for 5 months or longer and never fall below −10°C. Established rhizoma peanut can tolerate long periods of drought, but growth is retarded, and under severe drought conditions the shoots will sometimes die back to the rhizomes.

In the continental U.S.A., rhizoma peanut is limited to Florida and warmer portions of the other southernmost states. Rhizoma peanut should be adapted to warm humid tropics and subtropics around the world.

MIXTURES

Florigraze, Arb, and Arblick rhizoma peanuts have been grown in mixture with common, Coastal and Coastcross-1 bermudagrasses (*Cynodon dactylon* [L.] Pers.), pangola and transvala digitgrasses (*Digitaria decumbens* Stent.), and Pensacola and Argentina bahiagrasses (*Paspalum notatum* Flugge). However, it is necessary to establish the peanut before the grass. In Florida, the peanut is established in winter, and grass seed is sparsely broadcast or the grass sprigged between rows during the first rainy season, June through August. No nitrogen fertilizer should be applied to peanut or grass. Nitrogen retards growth of peanut rhizomes (Adjei and Prine 1976) and increases growth and competition of the grass. The peanut should dominate a mixture and make up two-thirds or more of forage of a peanut-grass mixture. Productivity of mixtures depends upon the productivity and nitrogen fixation of the peanut. When peanut and grass grow intermingled, livestock graze both grass and legume at the same time.

FORAGE PRODUCTION

Dry-matter yields of Arb, Arblick, and Florigraze rhizoma peanuts in mixture with sparse Pensacola bahiagrass for a 5-year period at Gainesville, Florida, were 7.6, 8.5, and 10.5 mt/ha, respectively (Table 1). Experiment plots were broadcast fertilized in March with 560 kg/ha of 0–10–20 (N-P_2O_5-K_2O), which also contained 15 kg/ha of fritted trace elements (FTE 503). Forage was cut twice annually the first 3 years and three times annually the following 2 years.

A well-established stand of Florigraze grown in Arredondo fine sand at Gainesville, Florida, was harvested at six cutting intervals ranging from 2 to 12 weeks during 24-week periods in 1978 (Breman 1980) and 1979 (Beltranena 1980). Dry-matter yields, crude protein, IVOMD, nitrogen recovered in hay, and leaf percentage of hay are given in Table 2.

We recommend that Florigraze be cut for hay three times a year in peninsular Florida. This schedule corresponds to the 8-week cutting interval (Table 2) and seems a good compromise to obtain high yields of good-quality hay. Florigraze hay cut three times a year has had IVOMD and protein above 60% and 14%, respectively. Because Florigraze holds its quality and yield well, some farmers may elect to cut only twice a year for hay. Florigraze should probably not be grown in mixture with perennial grasses for hay since the grasses need to be cut four or more times a year for good-quality hay.

Table 1. Hay yields of three rhizoma peanut cultivars growing in mixture with Pensacola bahiagrass on Kanapha fine sand at Gainesville, Fla. during 5 growing seasons.

Peanut cultivars in mixture with bahiagrass	Dry-matter yield[a], mt/ha					
	1970	1972	1973	1974	1975	5-season average
Florigraze	10.5a	7.2a	12.3a	11.9a	11.2a	10.5a
Arb	7.6b	5.6c	7.2c	8.3b	9.4b	7.6b
Arblick	8.1b	6.3b	8.5b	8.3b	10.7ab	8.5b
Bahiagrass only	−	−	−	2.4c	2.3c	−

[a]Yield averages within a column followed by the same letter are not significantly different at .05 level.

Table 2. Seasonal average dry-matter yields, % protein, in vitro digestible organic matter, nitrogen contents, and leaf percentage of Florigraze rhizoma peanut over 1978 and 1979 seasons.[a]

Cutting interval weeks	Dry-matter yield mt/ha	Crude protein (%)	IVOMD (%)	Hay Nitrogen (kg/ha)	Hay Leaves (%)
2	5.7b	21.9a	74.3a	180	93
4	6.6b	20.1b	72.9b	210	90
6	10.0a	17.9c	70.4c	250	84
8	10.0a	16.0de	67.4d	240	79
10	10.0a	15.0e	64.0e	225	68
12	10.1a	14.7e	64.8e	230	72

[a]Values within each column followed by the same letter are not significantly different at .05 level.

OTHER USES

Rhizoma peanut offers primary uses as hay and pasture, but it could also be used for greenchop, silage, high-quality dehydrated hay, leaf-meal pellets, and ornamental ground cover. Rhizoma peanut can be grown as a living mulch between fruit and nut trees and grapevines. Row crops, vegetables, annual ryegrass, small grains, and cool-season annual forage legumes possibly can be planted by minimum tillage in establishing and established rhizoma peanut sods. Another promising use of rhizoma peanut is as creep pasture for calves. (Saldivar et al. 1981).

PESTS

Weeds are the biggest pest problem affecting rhizoma peanut, particularly during the first year of establishment. Herbicides recommended for common peanut have generally proved to be effective in controlling weeds in rhizoma peanut. Tall weeds should be mowed just above the peanut shoots. Once a thick coverage of peanut develops there is little problem with weeds, except for perennial grasses where 100% peanut is desired. Insects and nematodes have not been a problem on rhizoma peanut. This peanut is apparently immune or very resistant to the *Cercospora* leaf spots that plague common peanut. Two leaf-spot diseases caused by *Phyllosticta* and

Stemphylium fungi have been identified on rhizoma peanut, but they have not caused long-term or serious damage. So far, no virus diseases have been reported in rhizoma peanut in Florida. Florigraze and other rhizoma peanut accessions are single genotypes and therefore vulnerable to any deadly pest or pestilence that may attack them. Rhizoma peanut can be easily eradicated if some pest problem makes it necessary to do so.

LITERATURE CITED

Adjei, M.B., and G.M. Prine. 1976. Establishment of perennial peanuts (*Arachis glabrata* Benth.). Proc. Soil and Crop Sci. Soc. Fla. 35:50-53.

Beltranena, R. 1980. Yield, growth and quality of Florigraze rhizoma peanut (*Arachis glabrata* Benth.) as affected by cutting height and frequency. M.S. Thesis, Univ. of Fla.

Blickenderfer, C.B., H.J. Haynsworth, and R.D. Roush. 1964. Wild peanut is promising forage legume for Florida. Crops and Soils 17(20):19-20.

Breman, J.W. 1980. Forage growth and quality of Florigraze perennial peanut (*Arachis glabrata* Benth.) under six clipping regimes. M.S. Thesis, Univ. of Fla.

Gregory, M.P., and W.C. Gregory. 1979. Exotic germplasm of *Arachis* L. interspecific hybrids. J. Hered. 70:185-193.

Gregory, W.C., M.P. Gregory, A. Krapovickas, B.W. Smith, and J.A. Yarbrough. 1973. Structures and genetic resources of peanuts. *In* Peanut culture and uses, 47-133. Am. Peanut Res. and Ed. Assoc. Inc., Stillwater, Okla.

Higgins, B.B. 1951. Origin and early history of the peanut. *In* The peanut—the unpredictable legume, 18-27. Natl. Fert. Assoc., Washington, D.C.

Martin, J.H., W.H. Leonard, and D.L. Stamp. 1976. Principles of field crop production, 3rd ed., 739-786. Collier, Macmillan Canada Ltd., Don Mills, Ont.

Prine, G.M. 1964. Forage possibilities in the genus *Arachis*. Proc. Soil and Crop Sci. Soc. Fla. 24:187-196.

Prine, G.M. 1973. Perennial peanuts for forage. Proc. Soil and Crop Sci. Soc. Fla. 32:33-35.

Prine, G.M., L.S. Dunavin, J.E. Moore, and R.D. Roush. 1981. "Florigraze" rhizoma peanut, a perennial forage legume. Fla. Agric. Expt. Sta. Cir. S275.

Saldivar, A.J., W.R. Ocumpaugh, G.M. Prine, and J.F. Hentges. 1981. Creep grazing "Florigraze" rhizoma peanuts with beef calves. Abs. of Tech. Papers, South. Branch ASA with Assoc. South. Agr. Workers. 1981 Mtg., 13.

Smith, C.E., Jr. 1968. The new world centers of origin of cultivated plants and the archeological evidence. Econ. Bot. 22(3):253-256.

Performance and Variability of Agronomic Characters in Populations of *Stylosanthes guianensis* (Aubl.) Sw.

P.S. MARTINS and N.A. VELLO

University of Sao Paulo, Brazil

Summary

Stylosanthes guianensis (Aubl.) Sw. shows wide variations in morphological and physiological characteristics, and basic information on genetic variability is needed for future breeding programs. The objectives of this paper were to evaluate 25 native Brazilian populations of *S. guianensis* for forage characteristics and to estimate genetic and phenotypic parameters for these characters.

The experiment was conducted in a duplicated 5×5 balanced lattice design with plots that contained 4 individuals spaced on 2-m centers. The following characters were evaluated after the beginning of flowering period: days to flowering (DF), mean diameter (MD), basal area (BA), growth habit (GH), foliage (F), and green- and dry-matter yields (GMY and DMY).

Genetic differences among populations were detected for all characters. Coefficients of genotypic determination indicated that GH and DF should respond readily to selection but improvement of GMY and DMY would be more difficult.

The correlations between means and intrapopulational variances indicated the existence of populations with high means and variances for MD, BA, GMY, and DMY. For GH the correlation ($r = -0.530$) indicated that the erect populations were less variable than the prostrate populations.

Correlations between populations means for pairs of characters were positive and significant for the 6 correlations between GMY, DMY, MD, and BA. All 9 other correlations were negative. Only the correlations of F with MD and BA were negative and significant. Therefore, it would be possible to improve DMY indirectly by selecting for increased GMY, MD, and BA, among which MD would be the preferred trait as it is easy to evaluate and has the largest coefficient of genotypic determination.

KEY WORDS: *Stylosanthes guianensis* (Aubl.) Sw., phenotypic parameters, genetic parameters.

INTRODUCTION

Beef production in Brazil is based mostly upon native pastures, particularly in areas with cerrados vegetation. Nearly 60% of Brazilian cattle population is concentrated in these areas (Buller et al. 1970), which are characterized by low forage quality, marked seasonal fluctuation in forage production, low soil fertility, and a high concentration of aluminum in the soil. This situation causes a low stocking rate of about 1 head/5 ha (Santos et al. 1979).

Use of legumes in association with grasses promises to be a method of improving forage quality and extending the grazing season in the cerrados areas (Hymowitz 1971). Legumes in grass mixtures would supply the needed protein for animal diet and the needed nitrogen for increased growth of grasses.

Brazil contains an enormous germ plasm reservoir of indigenous legumes and is considered the center of diversification of several genera and species (Mehra and Magoon 1974). The native Brazilian legumes offer valuable material from which plant-breeding programs designed to improve forage production and quality can be initiated. *Stylosanthes guianensis* (Aubl.) Sw., widely used in the tropical and subtropical regions of the world, shows large variation in morphological and physiological characters (Burt et al. 1971). The purposes of this study were to evaluate several populations of *S. guianensis* for forage characteristics and to estimate genetic and phenotypic parameters for these characters.

MATERIALS AND METHODS

Twenty-five population samples of *S. guianensis* from the states of Sao Paulo, Minas Gerais, Rio de Janeiro, and Goias were used in this study. The experiment was conducted at the Anhembi Experimental Station, Piracicaba, Sao Paulo. A duplicated 5×5 balanced lattice with 4 individuals/plot spaced on 2-m centers was used. Days to flowering, mean plant diameter, basal area, growth habit (visually scored from 1 = prostrate to 5 = erect), foliage (visually scored from 1 = small to 3 = large), and green- and dry-matter yields were measured. Visual scores for various characters were transformed to $(x + 1)^{1/2}$ for statistical analysis. Therefore, the data were obtained at one location and were based on one harvest/season.

The authors are both Assistant Professors at the Escola Superior de Agricultura, "Luiz de Queiroz."

Support for this project was provided by PIG/CNPq and FAPESP, Brazil.

Table 1. Mean squares (MS) and estimates of parameters of indicated characters of *Stylosanthes guianensis*.

Source of variation	df	E(ms)	MS						
			Days to flowering	Mean diameter, cm	Basal area, dm²	Growth habit, scores	Foliage	Green-matter yield	Dry-matter yield
Replications	11	—	285.50[a]	2145.19[a]	5368.03[a]	0.0066	0.0884[b]	591,752[a]	74,347.2[a]
Populations	24	$\frac{1}{K}\sigma_d^2 + \sigma_e^2 + RV_p$	2587.98[a]	5336.09[a]	11556.73[a]	0.8344[a]	1.3147[a]	814,149[a]	88,914.8[a]
Residual	264	$\frac{1}{K}\sigma_d^2 + \sigma_e^2$	85.28	228.52	608.06	0.0046	0.0427	66,604	8,123.2
Within	(K − 1)300	σ_d^2	—	444.66	1241.10	0.0064	0.0546	104,897	12,653.8
K (harmonic means of number of plants/plot)	—		—	3.64	3.64	3.68	2.00	2.00	2.00
General mean			162.48	99.48	84.207	2.207	1.987	553.92	202.91
CV (%)			5.68	15.2	29.3	3.22	10.4	46.6	44.4
Efficiency of lattice (%)			100.0	107.9	106.3	100.1	100.3	107.4	104.4
Coefficient of genotypic determination (b)			0.71	0.44	0.38	0.88	0.60	0.34	0.32

[a] P ≤ 0.01

[b] P ≤ 0.05

The analysis of variance for these characters was done according to Table 1. Based on the expected values of mean squares, phenotypic variance (s_F^2) and the coefficient of genotypic determination ($\hat{b}$) were calculated as follows:

$$s_F^2 = \hat{V}_p + \hat{\sigma}_e^2 + \hat{\sigma}_d^2 \text{ and } \hat{b} = \frac{\hat{V}_p}{s_F^2} \qquad (1)$$

The coefficient of genotypic determination is similar to that of broad-sense heritability.

Spearman rank correlations were estimated between population means and between intrapopulational variances for pairs of characters X and Y, and between mean and intrapopulational variance for each character. In all cases, 25 pairs (25 populations) of values were used in estimates of correlation coefficients.

RESULTS AND DISCUSSION

The overall mean plant survivorship was 90.2% indicating satisfactory adaptation of the different populations to the local conditions of the experiment, especially low soil pH (4.7), high soil aluminum content (1.8 meq/100 ml), and intense drought during 1978. This adaptive behavior is confirmed by the good vegetative development of plants, with mean dry-matter yield of 202.9 g/plant (estimated at 211 days after sowing) and range from 68.3 to 444.8 g/plant.

Highly significant differences were observed among populations for all characters analyzed (Tables 1 and 2).

The potential improvement by interpopulational selection depends on the character considered. On basis of the b values (Table 1), growth habit and days to flowering are expected to be more easily alterable through selection,

Table 2. Spearman rank correlations for indicated characters.[1]

Characters	Mean diameter	Basal area	Growth habit	Foliage	Green-matter yield	Dry-matter yield
Mean diameter	<u>.525[a]</u>	.847[a]	.332	.600[a]	.674[a]	.526[a]
Basal area	1[a].	<u>.849[a]</u>	.485[a]	.653[a]	.825[a]	.686[a]
Growth habit	−.199	−.199	<u>−.530[a]</u>	.608[a]	.651[a]	.718[a]
Foliage	−.344[b]	−.344[b]	−.107	<u>−.288</u>	.658[a]	.517[a]
Green-matter yield	.891[a]	.891[a]	−.085	−.125	<u>.817[a]</u>	.735[a]
Dry-matter yield	.924[a]	.924[a]	−.124	−.234	.935[a]	<u>.773[a]</u>

[1]Underlined diagonal values indicate correlations among means and intrapopulational variances; values above diagonal are correlations among intrapopulational variances; values below diagonal are correlations among means of 85 populations of *Stylosanthes guianensis*.

a = P ≤ 0.01; b = P ≤ 0.05; unilateral test.

and green- and dry-matter yields appear to be the most difficult traits to improve through selection.

The correlations between means and intrapopulational variances were positive and significant for mean diameter, basal area, and green- and dry-matter yield (Table 2). For growth habit the negative correlation (r = −0.530) indicated that the erect populations are less variable than the prostrate populations. The correlations between interpopulation variances indicated that populations with large variance for any character tended to have large variances for the other characters.

Positive and significant values were obtained for green- and dry-matter yields, mean diameter and basal area, and green- and dry-matter yields with mean diameter and basal area (Table 2). All the other 9 correlations were negative. The negative correlations were significant for foliage with mean diameter and basal area.

In conclusion, it should be possible to improve dry-matter production indirectly by selection for increased green-matter yield, mean diameter, or basal area. Mean diameter would be the preferred trait because it is easy to evaluate and has the largest coefficient of genotypic determination. However, in selecting for larger diameter, care must be taken not to select less leafy plants because of possible negative effects on forage quality.

LITERATURE CITED

Buller, R.E., S. Aranovich, L.R. Quinn, and W.V.A. Bisschof. 1970. Performance of tropical legumes in the upland savannah of central Brazil. Proc. XI Int. Grassl. Cong. 143–146.

Burt, R.L., L.A. Edye, B. Grof, and R.J. Williams. 1971. Assessing the agronomic potential of the genus *Stylosanthes* in Australia. Proc. XI Int. Grassl. Cong., 219–228.

Hymowitz, T. 1971. Collection and evaluation of tropical and subtropical Brazilian forage legumes. Trop. Agric. 48:309–315.

Mehra, K.L., and M.L. Magoon. 1974. Gene centers of tropical and subtropical pasture legumes and their significance in plant introduction. Proc. XII Int. Grassl. Cong., 908–913.

Santos, C.A., S. Estermann, P. Estermann, and A. Estermann. 1979. Aproveitamento da pastagem nativa do cerrado. V Simposio sobre o Cerrado, 421–435.

White Clover: An Old Crop with a Promising Future

P.B. GIBSON, W.E. KNIGHT, O.W. BARNETT,
W.A. COPE, and J.D. MILLER
USDA-SEA-AR and Clemson, Mississippi State, and
North Carolina State Universities, U.S.A.

Summary

White clover (*Trifolium repens* L.) is a widely adapted, important forage legume for grazing in the U.S. The palatability and quality of the forage, the wide adaptation of the plants to growth in different climates and under various grazing conditions, and the role it plays in supplying nitrogen to pastures make white clover our finest pasture legume. The importance of white clover's contributions as a component of pastures has increased markedly in recent years, primarily because of the rising cost of energy. The higher cost of energy has resulted in an increase in the cost of nitrogen fertilizer and has thus increased the importance of symbiotic nitrogen fixation by clovers and other legumes.

The need for a reliable, inexpensive source of protein feed is causing everyone concerned with the livestock industry to consider growing more forage legumes. There is universal acceptance of the fact that clover forage is highly digestible. This digestibility may range from 60% to 80% digestible dry matter. Some of the less obvious, yet highly economic, benefits of clover in pastures are related to animal health, milk flow, calf-weaning weights, and conception rate. In addition, problems with fescue-foot, grass tetany, and similar maladies are minimized when grass-clover mixtures are properly utilized.

Surveys, observations, and results of research show that virus diseases of white clover are prevalent. They cause severe damage to white clover and reduce persistence. Accordingly, breeding for virus-resistant white clover has top priority in the USDA white clover program. The main effort is a recurrent selection program for resistance to peanut stunt (PSV), clover yellow vein, and alfalfa mosaic viruses, with priority of emphasis in the order listed. In one test for progress made in the breeding program, the incidence of infection with PSV resulting from mechanical inoculation was about 70% for Ladino compared to 30% for selected material. In a study of the nature of resistance, results from a 16-clone diallel cross indicate that susceptibility to PSV is conditioned by both specific and general combining ability. Interspecific hybridization is another viable source of virus resistance. The hybrids of *Trifolium repens* and *T. ambiguum* obtained by Williams in New Zealand provide an alternative and possibly a superior source of virus resistance. *Trifolium ambiguum* is highly resistant if not immune to the important viruses that infect white clover.

The use of adapted white clover cultivars should result in large economic gains to the livestock industry in the U.S.A. Beef and dairy farmers would profit from reduced need for nitrogen fertilizer, an extended grazing season, increased forage production and quality, better use of land resources, stimulation of milk flow, higher calf-weaning weights, and better calving percentages.

KEY WORDS: *Trifolium repens* L., virus resistance, longevity, breeding, recurrent selection.

INTRODUCTION AND REVIEW

White clover (*Trifolium repens* L.), cattle, and man have been associated since antiquity. Apparently, this association existed at the time migrating animals were used for food by early man. The grazing animals scattered white clover seed that remained viable as it passed through the animals' digestive systems. On fertile soil where tall plants were controlled by grazing, white clover grew with grass and supplied nutritious feed for the animals. White clover, by plan or accident, came to America with the early settlers. Once here, it spread rapidly and actually preceded the colonists into fertile valleys, such as that of the Ohio River. Over the centuries, selection pressures conditioned by the grazing of animals and by some human management have influenced the evolution of white clover. In these ways, the evolution and spread of white clover have been closely associated with the domestication of cattle and the actions of man.

The history of white clover provides suggestions for its successful management. It evolved in grazed areas, on fertile soil supplied with ample moisture; therefore, it is adapted to survival in areas where such conditions prevail. Its survival in a frequently clipped lawn is evidence of this adaptation. Controlling the height of grass in a clover-grass pasture to permit an adequate amount of sunlight to reach the clover is an essential part of management.

White clover makes a twofold contribution in pastures by providing quality forage and fixating nitrogen. Cattle grazing white clover consume only the leaves. The stems (stolons) remain on the ground and bear the buds that produce new leaves. Leaves are high-quality feed. They are low in fiber, high in minerals, highly digestible, and, on a dry-weight basis, a protein concentrate. This high quality promotes animal health, milk flow, and rapid growth. The high quality results in higher grades of grass-fed cattle, production of high calving percentages, and a reduction in the incidence of fescue toxicity, grass tetany, and milk fever. By means of its symbiotic relationship with bacteria (*Rhizobium trifolii* Dangeard), the clover supplies nitrogen both for its growth and for that of an associated grass. In this relationship, energy from the sun is used to fix nitrogen rather than energy from fossil fuel.

Although favorable characteristics of white clover are impressive, the use of the species in the U.S.A. has declined for approximately the last 3 decades. This decline was associated with (1) inexpensive nitrogen (N) fertilizer and its application to pastures, (2) increased use of clover in mixtures with highly competitive grasses (tall fescue and bermuda), (3) errors in utilization, especially failure to control the height of aggressive grasses, and (4) the spread of viruses, in particular peanut stunt virus (PSV). The reason given by most farmers for the decline in white clover use was that the stand of clover did not persist with the grass. By understanding the crop and its growth requirements, we can choose management practices that alleviate or correct factors 1, 2, and 3. Correct management often involves judgment decisions that are affected by edaphic, climatic, and biological conditions. Therefore, the manager should be well versed in the physiology of the plants, the conditions required for their growth, and ways to compensate for various stresses. The practical solution for factor 4 is the development of resistant cultivars.

The longevity of the primary axis of white clover varies with the favorability of the environment and genotype of the plant. Usually, the primary axis does not function through the second year (Westbrooks and Tesar 1955). The continuance of clover in a pasture depends on volunteer seedlings and natural asexual propagation, i.e., new plants from growth centers that are rooted stolons of the original plant (Hollowell 1966). At Clemson University, we have found that if a seedling and a rooted stolon start growth in the fall, the resulting plants are remarkably similar the following spring (Gibson and Trautner 1965). Beginning in the second year after establishment of a clover grass pasture, the population of clover consists of varying amounts of the two kinds of plants.

Continuance of stands by means of asexually propagated plants has the advantage of bypassing the small seedling stage, which is vulnerable to climatic stresses and pests. The asexual propagule is a full-size stolon supporting leaves with full-length petioles. The size of this propagule favors survival, new growth, and forage production. Thus, in the absence of systemic diseases, asexual propagation results in a more continuous production of forage. Cultivars vary in their tendency to persist by means of asexual propagation and volunteer seedlings.

In its evolution, white clover acquired a range of characteristics that enable it to continue to live under various environmental conditions. It produces seed the first year and therefore can be a reseeding annual. The longevity of the primary axis and the tendency to survival by natural asexual propagation vary. Production of hard

P.B. Gibson is a Research Agronomist with USDA-SEA-AR, Clemson University; W.E. Knight is a Research Agronomist with USDA-SEA-AR, Mississippi State University; O.W. Barnett is a Plant Pathologist at Clemson University; W.A. Cope is a Research Agronomist with USDA-SEA-AR, North Carolina State University; and J.D. Miller is a Research Agronomist with USDA-SEA-AR, Georgia Coastal Plains Agricultural Experiment Station.

seed enables white clover to reappear when conditions become favorable in areas where it has previously grown. Because of this versatility, white clover frequently grows and volunteers in grazed areas where soil fertility and moisture are adequate. Because this clover is not planted, it frequently is called wild or common white clover. This volunteer clover makes a contribution that usually is attributed to some other forage. Although the total contribution of volunteer clover is large, the contribution/ha is low in comparison with the yields of improved cultivars in research plots. Volunteer plants often are small to intermediate in size and blossom profusely. The flowering results in fewer leaves and lower forage production.

Ladino was the first cultivar to offer major improvements for white clover in pastures of the U.S.A. The larger plants produced more leaves and fewer flowers in the areas where Ladino was used in pastures. The primary axis persisted longer, and the tendency to natural asexual propagation in competition with aggressive grasses was high. These characteristics generated hopes of high production compatible with modern intensive farming and of improved survival with tall grasses. However, Ladino did not live up to these expectations. Furthermore, re-establishment of Ladino in a pasture often gave results that were disappointing in comparison with the original planting.

We interpret the results of our research with virus diseases of white clover to indicate that these diseases played a major role in Ladino's not performing as expected. We have found Ladino plants to be highly susceptible to viruses. For example, incidence of infection with one or more viruses of 100 spaced plants in the field at Clemson University was 100% by the end of the first year, and the average incidence of infection with PSV resulting from mechanical inoculation of 1,200 plants in the greenhouse was 72%. Surveys have shown that PSV and other viruses that infect white clover are widely distributed in the U.S. (Barnett and Gibson 1975). Results from research conducted in greenhouses, controlled-environment chambers, and in the field show that viruses reduce yields of forage and seed (Barnett and Gibson 1977, Kreitlow and Hunt 1958, Kreitlow et al. 1957). Impairment of asexual reproduction is especially important with Ladino because it produces few seeds for volunteering. Ladino's continuance in a pasture largely depends upon natural asexual propagation.

The incidence of infection and the damage caused by viruses are greater in the second and later years of the life of a stand of clover. A stand established on a clean seedbed starts virus free and may remain virus free for a few months. Normally, migrating aphids cause initial infections. Thereafter, crawling and flying aphids spread the virus; also, some spread may be caused by machinery.

METHODS

Information presented in the foregoing review caused us to give highest priority in our white clover research to solving the virus problem. Concluding that plant resistance would be the most logical approach, we embarked on a recurrent selection program to identify resistance and to concentrate it into desirable lines of white clover. Recurrent selection seemed to be the most logical approach because the nature and the inheritance of resistance were not known and classification of plants was not absolute. Our classification of plants includes mechanical inoculations, aphid inoculations, and field exposure combined with virus indexing (assays).

RESULTS

Recent advances in virus technology, including adaptation of the enzyme-linked immuno-sorbent assay (ELISA) for virus research on the forage legumes, have enabled us to handle relatively large numbers of plants. Using these techniques, we have screened white clover from various sources for resistance to alfalfa mosaic virus (AMV), clover yellow vein virus (CYVV), and peanut stunt virus (PSV), with priority given to PSV. We have made progress. The average incidence of infection with PSV resulting from mechanical inoculation of 4,200 plants representing progenies of clover selected for resistance was 27.5% compared with 72% for Ladino. This difference in susceptibility indicates that usable levels of resistance are available in *Trifolium repens*. Another source of resistance is available in *T. ambiguum* Bieb. This species is highly resistant to most of the viruses that infect white clover, and it is a potential source of multiple resistance. Williams (1978) has obtained the *T. repens* × *T. ambiguum* species hybrid that is the first step toward using this resistance.

We have gained some information on the nature of resistance to PSV. Results of a diallel cross involving 16 clones and the classification of more than 4,200 plants indicate that resistance is conditioned by both specific and general combining ability. This information supports our belief that potential parent clones of a cultivar should be evaluated in a diallel cross.

We are optimistic about controlling viruses by developing resistant cultivars, and we believe the cultivars will make a major contribution to pastures by increasing the longevity and production of clover. The importance of this contribution will increase with increases in cost of energy used for manufacturing N fertilizer. In our opinion, viruses are the top deterrent to increasing the contribution of white clover in pastures of the U.S.A. Development of virus-resistant cultivars should assure a bright future for the association of white clover, cattle, and man and for the part this association plays in grassland agriculture.

LITERATURE CITED

Barnett, O.W., and P.B. Gibson. 1975. Identification and prevalence of white clover viruses and the resistance of *Trifolium* species to these viruses. Crop Sci. 15:32–37.
Barnett, O.W., and P.B. Gibson. 1977. Effect of virus infection on flowering and seed production of the parental clones

of Tillman white clover (*Trifolium repens*). Pl. Dis. Rpt. 61:203-207.

Gibson, P.B., and J.L. Trautner. 1965. Growth of white clover without primary roots. Crop Sci. 5:477-479.

Hollowell, E.A. 1966. White clover, *Trifolium repens* L., annual or perennial? Proc. X Int. Grassl. Cong., 184-187.

Kreitlow, K.W., and O.J. Hunt. 1958. Effect of alfalfa mosaic and bean yellow mosaic viruses on flowering and seed production of ladino white clover. Phytopathology 48:320-321.

Kreitlow, K.W., O.J. Hunt, and H.L. Wilkins. 1957. The effect of virus infection on yield and chemical composition of ladino clover. Phytopathology 47:390-394.

Westbrooks, F.E., and M.B. Tesar. 1955. Tap root survival of ladino clover. Agron. J. 47:403-410.

Williams, E. 1978. A hybrid between *Trifolium repens* and *T. ambiguum* obtained with the aid of embryo culture. N.Z. J. Bot. 16:499-506.

Genetic Variability for Morphological and Agronomic Characters in *Desmodium uncinatum* (Jacq.) D.C. and *D. intortum* (Mill.) Urb.

E.M.P. OLIVEIRA, P.S. MARTINS, and A.B. CRUZ, FILHO

EMBRAPA, Minas Gerais, and
Instituto de Zootecnia, Nova Odessa, Brazil

Summary

Objectives of this research were to study the performance of eight populations of *D. uncinatum* and ten populations of *D. intortum*, with relation to morphological and agronomic characters; to estimate the genetic variation between populations and between plants within populations; and to estimate genetic and phenotypic parameters for these characters.

The experiment was conducted at the National Research Center for Dairy Cattle (EMBRAPA), Minas Gerais, Brazil. It was set up in a randomized complete-block design, with 10 replications. The plots, composed of the two species, were subdivided in 18 subplots. Each supblot contained 4 individual plants spaced 1.5 m apart. The following characters were evaluated at the beginning of the flowering period: length of main stem (LMS), number of lateral branches (NLB), dry-matter percentage (DMP), fiber content (FC), and crude-protein percentage (CPP).

Results obtained in analysis of variance showed significant differences between species for the characters LMS, NLB, DF, DMY, and CPP. In addition, significant differences were observed for all characters analyzed in *D. uncinatum*. In *D. intortum* there were populations that differed only in respect to LMS, DF, DMY, and DMP.

Analysis of variance within subplots indicated the largest variability among plants within populations for DF. On the other hand, as values of the coefficient of genotypic determination (b) were higher in *D. uncinatum*, it should be easier to obtain changes through selection in the populations of that species than in *D. intortum*. On the basis of the b values, DF, FC, and CPP are expected to be the characters more easily alterable by selection in *D. uncinatum*.

KEY WORDS: *Desmodium uncinatum* (Jacq.) D.C., *Desmodium intortum* (Mill.) Urb., phenotypic parameters, genetic parameters, forage quality, forage yield.

INTRODUCTION

The legumes *Desmodium uncinatum* (Jacq.) and *Desmodium intortum* (Mill.) Urb. have been utilized successfully as fodder or pasture plants in many tropical and subtropical regions of the world. Two cultivars, *D. uncinatum* cv. Silverleaf and *D. intortum* cv. Greenleaf, were selected in Australia (Bryan 1969).

Although Brazil is considered a center of diversity for both species (Mehra and Magoon 1974), very little research has been done to evaluate the genetic variability of morphological and agronomic characters of native populations.

Our research objectives were to study the performance of different populations of *D. uncinatum* and *D. intortum* in relation to morphological and agronomic characters, to estimate the genetic variation within and between populations, and to estimate genetic and phenotypic parameters for the characters analyzed.

E.M.P. Oliveira is a Research Agronomist with EMBRAPA, Instituto de Zootecnia; P.S. Martins is an Assistant Professor, ESALQ-USP; and A.B. Cruz, Filho, is a Research Agronomist with EMBRAPA-CNP-GL.

MATERIALS AND METHODS

The experiment was conducted at the National Research Center for Dairy Cattle (EMBRAPA), Minas Gerais, Brazil. Eight populations of *D. uncinatum* and ten populations of *D. intortum* were analyzed. The cultivars Silverleaf and Greenleaf were included for comparison with other populations that were natives of different regions of Minas Gerais.

A split-plot design was used, with 10 replications of the main plots, each containing the two species. The main plots were subdivided into 18 subplots. Each subplot had 4 individual plants spaced 1.5 m apart. In each subplot 2 plants that were not cut were used for evaluation of growth habit (GH), length of main stem (LMS), number of lateral branches (NLB), and days to flowering (DF). The other 2 plants, which were cut periodically, were used for evaluation of dry-matter yield (DMY), dry-matter percentage (DMP), fiber content (FC), and crude-protein percentage (CPP).

The characters DMY, DMP, and FC (Van Soest 1963) and CPP (AOAC 1965) refer to the means of 2 plants/replication.

The analysis of variance was done as is shown in Table 1. For the characters LMS, NLB, and DF, the analysis of variance was based on individual plant data, and the coefficient of genotypic determination (b) was obtained as follows:

$$\hat{b}_1 = \frac{\hat{V}_{p_1}}{(\hat{V}_{p_1} + \hat{\sigma}^2_{d_1} + \hat{\sigma}^2_{e})} \quad (1)$$

For the characters DMY, DMP, FC, and CPP, the analysis of variance was based on combined samples of 2 plants in each subplot, and the coefficient of genotypic determination (b) was obtained as follows:

$$\hat{b}_1 = \frac{\hat{V}_{p_1}}{(\hat{\sigma}^2 + \hat{V}_{p_1})} \quad (2)$$

RESULTS AND DISCUSSION

In regard to GH, all populations of both species were prostrate except for one erect population of *D. uncinatum*.

The combined analysis of variance showed highly significant differences beween species for the characters LMS, NLB, DF, DMY, and CPP. In addition, significant differences were observed for all characters analyzed in *D. uncinatum*. In *D. intortum*, populations differed only with respect to LMS, NDF, DMY, and DMP. The within-subplots variances indicated significant differences for NLB and NDF in *D. uncinatum* and for LMS and NDF in *D. intortum*. Therefore, the largest variability among plants within both populations was observed for DF, an indication of a good possibility of selection for early- and late-flowering cultivars.

The mean for the characters and respective Δ values (Tukey 5%) indicated the existence of populations of *D. uncinatum* with good forage characteristics: large plants, good branching, large dry-matter yield, and high crude-protein content (Table 2). In relation to yield and forage quality, the significant differences between populations can be attributed to population 10, the only erect population of *D. uncinatum*, which was lowest in DMY, DMP, and FC and highest in CPP. This population appeared to be daylength-insensitive, flowering from February to November, while the other populations showed a definite and short flowering period.

Among populations of *D. intortum* much less variation was observed for the characters analyzed than among those of *D. uncinatum* (Table 3). This difference is reflected, too, by the estimated b values for both species (Table 4).

Studies showing different patterns of variation for pairs of congeneric species have been mainly with introduced grasses in California (Jain 1969, Jain and Marshall 1967, Jain and Rai 1974). Those studies explain the observed differences in genetically based morphological diversity between populations in terms of the genetic base of the introduced material and genetic as opposed to plastic

Table 1.   Method used in analysis of variance.

Source of variation	df	E(MS)	MS	F
Blocks	R − 1	$\frac{1}{K}\sigma^2_d + \sigma^2_e + PE\sigma^2_b$	M_7	$M_7{:}M_2$
Species (S)	S − 1	$\frac{1}{K}\sigma^2_d + \sigma^2_e + P\sigma^2_{tb} + PRV_t$	M_6	$M_6{:}M_5$
Residual (a)	(R − 1)(S − 1)	$\frac{1}{K}\sigma^2_d + \sigma^2_e + P\sigma^2_{tb}$	M_5	$M_5{:}M_2$
Populations/S$_1$	P_1 − 1	$\frac{1}{K}\sigma^2_{d_1} + \sigma^2_e + RV_{p_1}$	M_4	$M_4{:}M_2$
Populations/S$_2$	P_2 − 1	$\frac{1}{K}\sigma^2_{d_2} + \sigma^2_e + RV_{p_2}$	M_3	$M_3{:}M_2$
Residual (b)	(R − 1)(P$_1$ + P$_2$ − 2)	$\frac{1}{K}\sigma^2_d + \sigma^2_e$	M_2	$M_2{:}M_1$
Within	(K − 1)(P$_1$ + P$_2$)R	σ^2_d	M_1	

P is the harmonic mean between P$_1$ and P$_2$

Table 2.   Means for the indicated characters in eight populations of *Desmodium uncinatum* (Pop 6: cv Silverleaf).

Pop.	LMS[1](cm)	NLB $(x+1)^{1/2}$	DF days	DMY g	DMP %	FC %	CP %
3	254.1a^{2}	8.0^a	145.6^a	310.9^a	22.7^{ab}	59.7^a	12.8^b
4	138.2^b	4.9^b	125.3^c	229.4^a	24.9^a	52.2^b	12.8^b
5	172.0^b	5.4^b	130.0^b	247.3^a	23.3^{ab}	58.4^{ab}	14.1^b
6	238.5^a	5.6^a	145.0^a	266.1^a	23.2^{ab}	56.8^{ab}	12.9^b
7	227.5^a	7.1^a	145.1^a	216.6^a	22.9^{ab}	55.5^{ab}	13.9^b
9	235.5^a	7.6^a	146.0^a	237.1^a	23.4^{ab}	54.5^b	13.4^b
10	62.8^c	5.0^b	78.0^d	5.2^b	20.8^b	36.8^c	25.3^a
11	232.4^a	4.6^b	134.6^b	189.3^a	29.9^b	55.8^{ab}	14.7^b
Mean	195.1	6.0	131.2	212.7	22.9	53.7	14.7

[1]The abbreviations in headings represent population, length of main stem, number of lateral branches, days to flowering, dry-matter yield, dry-matter percentage, fiber content, and crude-protein content.

[2]Means within a column not having same superscript differ significantly.

Table 3.   Means for the indicated characters in ten populations of *Desmodium intortum* (Pop. 6: cv Greenleaf).

Pop.	LMS[1] cm	NLB $x-1$	DF days	DMY g	DM %	F %	CP %
2	209.5b^{2}	4.3	169.1^b	302.5^b	22.9^a	53.8	10.8
3	239.0^{ab}	4.2	175.6^{ab}	347.1^{ab}	22.7^{ab}	54.8	11.2
4	246.5^{ab}	3.9	175.2^{ab}	431.4^{ab}	21.2^a	54.9	9.8
5	225.5^{ab}	3.3	175.0^{ab}	418.8^{ab}	19.2^b	54.6	11.0
6	228.6^{ab}	3.7	178.0^a	347.8^{ab}	23.2^a	53.7	10.6
8	263.0^a	4.4	174.2^a	438.8^{ab}	23.2^a	54.4	10.3
9	251.0^{ab}	4.5^a	173.6^{ab}	471.3^a	23.7^a	53.3	10.9
10	258.2^a	4.2	174.0^{ab}	384.6^{ab}	22.5^{ab}	55.6	10.4
11	250.5^{ab}	3.9	171.4^{ab}	463.6^a	23.5^a	54.2	9.4
12	237.8^{ab}	3.3	169.7^b	424.2^{ab}	24.5^a	54.2	10.5
Mean	241.0	4.0	173.6	403.0	23.0	54.4	10.5

[1]The abbreviations in headings represent population, length of main stem, number of lateral branches, days to flowering, dry-matter yield, dry-matter percentage, fiber content, and crude-protein content.

[2]Means within a column not having same superscript differ significantly.

Table 4.   Coefficients of genotypic determination for indicated characters in eight populations of *D. uncinatum* (b_1) and ten populations of *D. intortum* (b_2). CNPGL, Coronel Pacheco, M.G., Brazil, 1978.

Characters	b_1	b_2
Length of main stem	0.78	0.06
Number of lateral branches	0.26	0.01
Days to flowering	0.95	0.19
Dry-matter yield	0.40	0.15
Dry-matter percentage	0.18	0.23
Fiber content	0.85	0.00
Crude-protein percentage	0.83	0.02

adaptation to the habitats. The two native species of *Desmodium,* collected in the same region of Brazil, differed strikingly in genetic variability, suggesting the necessity for detailed studies of the genetic and plastic mechanisms of adaptation to the local environment.

The possibility of populations of *D. uncinatum*'s being changed through selection depends on the character considered. Based on the b values obtained for the characters evaluated (Table 4), DF, FC, and CPP should be most easily altered by selection, LMS should be intermediate in response, and NLB and DMY should be the least responsive to selection. For *D. intortum* the values of b were very low, indicating less responsiveness to selection due to the genetic uniformity of all characters studied.

LITERATURE CITED

Association of Official Analytical Chemists (AOAC). 1965. Official methods of analysis, 10th ed. Washington, D.C.

Bryan, W.W. 1969. *Desmodium intortum* and *Desmodium uncinatum*. Herb. Abs. 39:183–191.

Jain, S.K. 1969. Comparative ecogenetics of *Avena fatua* and *A. barbata* occurring in central California. Evol. Biol. 3:73–118.

Jain, S.K., and D.R. Marshall. 1967. Genetic changes in a barley population analyzed in terms of same life cycle com-

ponents of selection. Genetica 38:355–374.

Jain, S.K., and K.N. Rai. 1974. Population biology of *Avena*. IV. Polymorphism in small populations of *A. fatua*. Theoret. Appl. Genet. 44:7–11.

Mehra, K.L., and M.L. Magoon. 1974. Gene centres of tropical and subtropical legumes and their significance in plant introduction. Proc. XII Int. Grassl. Cong., 908–913.

Van Soest, P.J. 1963. Use of detergents in the analysis of fibrous feeds. I. Preparation of fiber residues of low nitrogen content. J. Assoc. Agric. Chem. 46:825–829.

Value of White Sweet Lupin in Production of Protein Fodder and Feeding of Animals

O. SARIĆ and I. RAMOŠEVAC

Faculty of Agriculture, Sarajevo, Yugoslavia

Summary

The potential of white sweet lupin (*Lupinus albus* L.) as a grain or forage feed for livestock was investigated in a series of experiments. Grain yield of lupin was greater than or equal to that of soybeans (*Glycine max* [L.] Merrill) in most evaluations. The quality of lupin seed compared favorably with that of soybean meal and fishmeal.

Lupin is capable of producing large amounts of forage in a short period, making it a good prospect for forage production in regions that have a short growing season.

KEY WORDS: white lupin, *Lupinus albus* L., production, quality, digestibility, live-weight gain, sheep.

INTRODUCTION

White lupin (*Lupinus albus* L.) chould be of great value for forage production under some conditions. It is free of alkaloids, tolerant to acid soils (Bacevskij 1966), and adapted to short growing seasons (Withers et al. 1976, Reeves et al. 1977), and both the seed and the forage of lupin have high protein content (Heuser et al. 1934, Rüther 1953, Mihajlec and Levickij 1978). The feed value of lupin seeds is similar to that of soybeans (Kirsch and Kasprzik 1935, Offutt and Davis 1973, Davis and Offutt 1975, Lebas 1980).

MATERIALS AND METHODS

The properties of Bosna white sweet lupin were investigated in four experiments:

Experiment I. The production of seed of lupin was compared with that of soybeans (*Glycine max* [L.] Merrill cv.

Manchu) in the lowlands and hilly regions of Bosnia. The lowland site had an elevation of 107 m, an average temperature of 11.4°C, an average precipitation of 878 mm, and a 261-day average growing season. The hilly site had an elevation of 518 m, an average temperature of 9.3°C, an average precipitation of 944 mm, and a 248-day average growing season. A randomized complete-block design with 5 replications of 12-m² plots was used. The plots were seeded during early April at the rate of 60 seeds/m² in rows 30 cm apart. Treflan and 27 kg/ha of N, 54 kg/ha of P_2O_5, and 54 kg/ha of K_2O were applied before seeding.

Experiment II. The effect of seeding time on grain and forage production was studied in the mountainous region. The site had an elevation of 872 m, an average temperature of 6.9°C, an average precipitation of 797 mm, and a 219-day growing season. A randomized complete-block experiment with 10-m² plots was used. Plots were seeded at the rate of 70 to 90 seeds/m² in rows 20 cm apart.

Experiment III. The digestibility of lupin silage was investigated in a feeding trial with wethers. Three treatments—pure lupin in the pod stage, lupin plus 10%

O. *Sarić is a Professor of Forage Crop Production and I. Ramoševac is an Assistant Professor on the Faculty of Agriculture.*

Table 1. Grain yields at 14% moisture (mc/ha).

	Lowland				Hilly land			
	1974	*1975*	*1976*	*Av.*	*1974*	*1975*	*1976*	*Av.*
Lupin cv. Bosna	18.3	23.3	22.0	21.2	15.5	11.5	7.0	11.3
Lupin cv. Blanca	17.2	21.5	17.7	18.8	20.2	13.2	12.0	15.1
Soybean cv. W. Manchu	13.7	8.9	7.6	10.1	11.9	12.9	11.9	12.2
LSD 0.05	4.0	7.7	3.2	2.7	2.8	2.8	4.1	1.8
0.01	5.3	10.4	4.3	3.7	3.7	3.0	6.0	2.4

barley grain, and lupin plus 20% barley grain—were compared.

Experiment IV. The value of lupin grain was investigated in the diets of three groups of twelve lambs each. The trial was conducted for 66 days. Three diets, in which the protein component was provided by soybean meal, lupin grain, or fishmeal, were evaluated.

RESULTS

Lupin grain yields were greater than soybean yields in the lowland region (Table 1). Soybean yields were intermediate between the yields of the two lupin cultivars in the hilly region. Bosna yielded more than Blanca in the lowland and less than Blanca in the hilly region.

Protein content of soybeans was less than that of the lupin cultivars on the lowland site (Table 2). Differences in protein content between soybeans and lupin were small on the hilly site. Oil content of soybean seeds was much greater than that of lupin seeds (20.2% compared with 10.2%).

Amino-acid contents were generally greater in lupin than in soybeans (Table 3). Exceptions to this generality were alanine, phenylalanine, tryptophan, and lysine. Concentrations of methionine and tryptophan were below standards set by FAO.

Forage yield was high in both years, and no significant differences among seeding dates were observed (Table 4). Seed yield decreased with later seeding dates. Seed yields were very low in 1979, when conditions for seed production were not favorable.

The coefficients of digestibility for dry matter, organic matter, fiber, and nitrogen-free extract (NFE) were in-

creased significantly by the addition of barley to lupin silage (Table 5). Differences between 0% and 10% added barley were greater than those between 10% and 20% added barley. Digestibility of crude protein was not increased by the addition of barley.

Protein-supplement source did not significantly affect fattening of lambs fed concentrate diets (Table 6). Food-conversion factors were slightly more favorable for lupin grain than for fishmeal and considerably more favorable than for soybean meal.

DISCUSSION

Lupin was found to have several characteristics that show that it has good potential as an animal feed. Its seed yield was greater than that of soybeans in several trials. Its ability to produce high yields of forage in a short period makes it a promising forage crop for regions that have a short growing season. Quality of lupin silage compares favorably with that of other forage legumes.

Table 2. Crude-protein content (%) in lupin grain and soybeans (%)

	Lowlands	Hilly land
Lupin cv. Bosna	38.08	38.58
Lupin cv. Blanca	39.52	40.85
Soybean cv. W. Manchu	33.85	41.33

Table 3. Amino-acid content in lupin and soybean grain (g/100 g protein).

Amino acids	Lupin	Soybean
Asparagine	4.61	3.90
Threonine	1.24	1.22
Serine	1.90	1.49
Proline	1.53	1.07
Glutamic acid	7.72	6.30
Glycine	1.47	1.37
Alanine	1.26	1.48
Valine	1.60	1.49
Cystine	0.71	0.53
Methionine	0.66	0.46
Isoleucine	1.44	1.26
Leucine	2.86	2.19
Tyrosine	0.96	0.45
Phenylalanine	1.24	1.72
Tryptophan	0.34	0.49
Lysine	1.70	2.23
Histidine	0.76	0.68
Arginine	3.57	2.28

Table 4. Influence of sowing dates on yields of green fodder and lupin grain (t/ha).

Sowing date		Green fodder		Dry matter		Grain	
1979	1980	1979	1980	1979	1980	1979	1980
20 Mar	28 Mar	32.2	32.8	6.12	5.19	.44	1.87
2 Apr	9 Apr	37.9	39.6	6.87	4.97	.17	1.39
12 Apr	19 Apr	35.1	33.9	6.22	6.00	.02	.34
21 Apr	29 Apr	25.0	33.5	5.23	5.59	–	.43
28 Apr	9 May	33.2	26.9	6.55	4.58	–	.09
Average		32.7	31.3	6.20	5.26		.82
LSD	P<0.05	7.01	5.04	1.08	.71		.71
	P<0.01	9.45	6.81	1.46	.95		.95

Table 5. Coefficients of digestibility for nutrients in Lupin silages.

% Barley in silage	Dry matter	Organic matter	Crude prot.	Crude fat	Crude fiber	Nitrogen Free extract
0	56.0	56.7	75.8	60.0	44.3	58.4
10	69.1	69.2	74.9	80.4	52.6	75.8
20	70.6	70.3	74.4	72.2	53.3	77.2

Table 6. Effect of soybean meal, lupin grain, and fishmeal as supplement to concentrate diets on live-weight gain (LWG) of lambs.

	Soybean	Lupin grain	Fishmeal
No. of animals	12	12	12
Initial weight (kg)	17.71	17.95	17.85
Final live-weight (kg)	39.33	40.84	41.15
Total live-weight gain (kg)	21.62	22.89	23.30
Average live-weight gain (g)	284.4	301.2	306.7
Conversion of food kg diets/1 kg LWG	4.43	3.86	3.96

The digestion efficiency of lupin seeds compares favorably with that of soybean meal, fishmeal, and other seed legumes. It would not be recommended for non-ruminants, however, because of low concentration of methionine and tryptophan (Mironenko 1975, Lenoble 1979).

LITERATURE CITED

Bačevskij, S. 1966. Reakcija pocvi i produktivnost ljupini. Zemljedelie. 12:47–49.

Davis, G.V., and M.S. Offutt. 1975. Nutritive value of sweet white lupine for ruminants. Ark. Agric. Exp. Sta. Bul. 792.

Heuser, W., K. Boekholt, and G. Ulich. 1934. Der Gehalt der Samen von Lupinus albus in Eiweiss, Fett und Alkaloiden im Vergleich zu anderen Lupinenarten und unter dem Einfluss äusserer Bedingungen. Pflanzenbau. II:129–138.

Kirsch, W., and B. Kasprzik. 1935. Der Wert der "Süsslupine." II. Erträge und Nährstoffgehalt der Süsslupine als Körnerfutter. Mitt. Landw. 50:25–26.

Lebas, F. 1980. Essai d'emploi du lupin chez le lapin en croissance. Laboratoire de recherches sur l'élévage du lapin, INRA-C.R. de Toulouse.

Lenoble, M. 1979. Objectifs et méthodes de sélection du lupin blanc (L. albus L.). Le Sélectionneur français. 27:35–38.

Mihajlec, V.I., and M.R. Levickij. 1978. Kormova ocinka riznih sortiv ljupinu. Visnik siljskogospod. Nauki. 3:53–56.

Mironenko, A.V. 1975. Biohimija ljupina, 17–35. Nauka i tehnika, Minsk.

Offutt, M.S., and G.V. Davis. 1973. Nutritive value of sweet white lupine forage. Ark. Fm. Res. 22(1):4.

Reeves, T.G., A.K. Broundy, and D.H. Brooke. 1977. Phenological developments studies with Lupinus angustifolius and L. albus in Victoria. Austral. J. Expt. Agric. and Anim. Husb. 17(87):637–644.

Rüther, H. 1953. Die Anbauwürdigkeit von Lupinus albus als Fett- und Eiweisspflanze. Z. für Acker- und Pflanzenbau. 97:253–260.

Withers, N.J., I.P.M. McQueen, and W.K. Clark. 1976. A preliminary study of lupins on pumice soils. Proc. Agric. Soc. N.Z. 6:71–73.

Selection for Specific Leaf Weight in Reed Canarygrass and Its Effects on the Plant

I.T. CARLSON, D.K. CHRISTENSEN, and R.B. PEARCE
Iowa State University, Ames, Iowa and Colorado State University,
Fort Collins, Colo., U.S.A.

Summary

Specific leaf weight (SLW, dry weight/unit leaf area) may serve as a selection criterion for forage yield through a positive relationship with photosynthesis. Two cycles of phenotypic selection for high and low SLW were conducted in reed canarygrass (*Phalaris arundinacea* L.) to investigate direct and correlated responses, particularly those related to yield and quality of forage.

In a space-planted test, average direct responses from two cycles of selection for high and low SLW were 23% and 19%, respectively. Significant correlated responses were obtained for ten traits. Compared with the low-SLW cycle-2 population, the high SLW cycle-2 population averaged 52% higher in SLW, 38% greater in leaf thickness, 13% greater in leaf width, 12% higher in CO_2-exchange rate, 24% higher in leaf elongation rate, 38% lower in tillers/plant, 83% higher in dry weight/tiller, 39% higher in dry-matter yield/plant, 15% greater in plant height, 55% higher in seed yield/plant, and 22% higher in 100-seed weight. These results suggested that SLW reflected some basic aspect of plant growth and development.

In solid stands, the high- and low-SLW cycle-1 populations did not differ significantly in forage yield when harvested either three or five times/yr. No consistent differences were found between high- and low-SLW populations in percentages of in-vitro dry-matter digestibility, neutral and acid detergent fiber, hemicellulose, cellulose, permanganate lignin, silica, and insoluble ash in nonflowering regrowth.

KEY WORDS: reed canarygrass, *Phalaris arundinacea* L., recurrent selection, specific leaf weight, correlated responses.

INTRODUCTION

Selection for high and low specific leaf weight (SLW, dry weight/unit leaf area) was initiated in reed canary-grass (*Phalaris arundinacea* L.) to investigate its usefulness in forage-yield improvement. Positive correlations between SLW and photosynthesis had been reported previously in forage crops (Pearce et al. 1969, Wilson and Cooper 1969, Carlson et al. 1970).

Gains of 11% and 10% were obtained by one cycle of phenotypic selection for high and low SLW, respectively, in the germ plasm source NCRC1 of reed canarygrass (Topark-Ngarm et al. 1977). Changes in SLW were accompanied by changes in other leaf characteristics and agronomic components of forage yield. Compared with the low-SLW population, the high-SLW population had thicker, darker green, and more upright leaves with a higher CO_2-exchange rate; fewer but heavier tillers/plant; and higher dry-matter yield/plant.

The objectives of this research were to study direct and correlated responses to a second cycle of phenotypic selection for high and low SLW and to evaluate the cycle-1 populations for yield and quality of forage in solid stands harvested three and five times annually.

METHODS

Methods used to develop and evaluate high- and low-SLW cycle-2 populations were essentially the same as those used by Topark-Ngarm et al. (1977) for development and evaluation of cycle-1 populations. We selected 52 plants for high SLW and 52 for low SLW from approximately 1,000 Syn-2 plants of respective cycle-1 populations. Selections within each group were intercrossed in the greenhouse, and cycle-2 populations were formed by randomly combining an equal number of polycross progeny from nearly all selections in each group.

The cycle-2 populations, the Syn-2 generation of cycle-1 populations, and the original germ plasm source, NCRC1, were evaluated in a space-planted test near Ames, Iowa, during 1978–1980. Plants were started in a greenhouse in April and transplanted to the field on 11 May 1978. The planting plan consisted of 10-plant plots of the 5 entries in a randomized complete-block design with 15 replications. Plants in the single-row plots were spaced 102 cm apart within and between plots.

Specific leaf weight, leaf thickness, leaf width, and CO_2-exchange rate were determined on leaf blade sec-

I.T. Carlson is a Professor of Agronomy at Iowa State University; D.K. Christensen is a Research Technician at Colorado State University; and R.B. Pearce is a Professor of Agronomy at Iowa State University.

This is Journal Paper No. J-10177 of the Iowa Agriculture and Home Economics Experiment Station, Ames. Project 1755.

208 Carlson et al.

tions with the methods described by Topark-Ngarm et al. (1977). These determinations were made on first growth of individual plants in July 1978 and on spring growth in May 1979. Tiller number, dry weight/tiller, and dry-matter yield/plant were determined 3 days after determination of leaf characteristics in July 1978, following Topark-Ngarm et al. (1977). Leaf elongation rate was determined in August 1978 by measuring length from the tip of a developing leaf to the collar of the leaf immediately below on two tillers on each of 5 plants/plot in 5 replications. The elongation rate was determined from the first and last measurements during a 3-day period. Leaf width was measured on the last day.

In 1979, date of anthesis, plant height, dry-matter and seed yield/plant, and 100-seed weight were determined. Anthesis was recorded as the date when three panicles on a plant had started shedding pollen. Plant height at anthesis was measured as height of the tallest culm to the tip of the panicle. Yield of dry matter was determined during seed ripening in 10 replications and after seed harvest in the remaining 5 replications.

The Syn-2 generation of the cycle-1 populations and NCRC1 were included in a forage-yield test conducted near Ames, Iowa, during 1977–1980. Entries were subplots in a split-plot design with two harvest frequencies (three and five harvests/yr) as whole plots. A subplot consisted of 6 rows spaced 23 cm apart. In 1978–1980, harvests were taken in early June, mid-July, and mid-September under the three-cut management and at about monthly intervals starting in late May under the five-cut management.

Forage-quality determinations were made on samples of nonflowering summer regrowth from both spaced plants and solid stands.

RESULTS

Significant direct responses to selection for SLW were obtained in each cycle in both directions (Table 1). The changes in average SLW over dates of determination from the first and second cycles of selection were, respectively, 0.45 and 0.46 mg/cm² for high SLW and 0.46 and 0.26 mg/cm², respectively, for low SLW. The smaller change in the second cycle of selection for low compared with high SLW was associated with a smaller phenotypic variance in the low cycle-1 population (Topark-Ngarm et al. 1977).

In the space-planted test, direct responses to selection for SLW were accompanied by significant positively correlated changes in leaf thickness, leaf width, CO_2-exchange rate (CER), leaf elongation rate, dry weight/tiller, plant height, dry-matter and seed yield/plant, and 100-seed weight (Table 1). Significant negatively correlated changes were found in tiller number. No significant changes were found in date of anthesis. The changes in CER from selection for high SLW usually were small and not significant. The substantial responses in leaf thickness, tiller number and weight, dry-matter yield, plant height, and 100-seed weight were closely associated with the responses to selection for SLW. The correlation coefficients on population means for 1978 and 1979 SLW with tiller number were, respectively, – 0.979 and – 0.991, and they ranged from 0.952 to 0.999 for SLW with the other 5 traits.

The correlated responses in leaf thickness, CER, tiller number and weight, and dry-matter yield/plant in the cycle-1 populations were in the same direction as those reported by Topark-Ngarm et al. (1977); however, the magnitude of the responses differed in some instances.

Table 1. Means for indicated traits determined in a space-planted test of five reed canarygrass populations, Ames, Iowa.

| | | Population | | | | | L.S.D., 0.05 |
| | | Low SLW | | NCRC1 | High SLW | | |
Trait	Date	Cycle-2	Cycle-1		Cycle-1	Cycle-2	
Specific leaf weight, mg/cm²	25 to 28 July 1978	3.63	3.91	4.57	5.15	5.77	0.16
	14 to 18 May 1979	2.70	2.92	3.20	3.54	3.81	0.13
Leaf thickness, mm	25 to 28 July 1978	0.162	0.176	0.193	0.219	0.234	0.007
	14 to 18 May 1979	0.222	0.243	0.264	0.287	0.294	0.008
Leaf width, mm	25 to 28 July 1978	14.4	15.7	16.4	17.0	16.6	0.6
	14 to 20 Aug. 1978	12.4	15.0	14.2	15.7	16.3	1.6
	14 to 18 May 1979	15.0	16.2	16.5	16.8	16.5	0.5
CO₂-exchange rate, mg CO₂/dm²/hour	25 to 28 July 1978	19.6	21.1	22.6	23.0	21.1	1.4
	14 to 18 May 1979	14.5	15.6	16.6	16.6	16.8	0.9
Leaf elongation rate, mm/day	14 to 20 Aug. 1978	34.3	39.0	37.7	38.2	42.7	4.1
No. of tillers/plant	31 July to 1 Aug. 1978	200	179	169	137	124	15
Dry weight/tiller, g	31 July to 1 Aug. 1978	0.64	0.77	0.89	1.06	1.17	0.09
Dry matter yield/plant, g	31 July to 1 Aug. 1978	128	139	149	154	159	13
	21 June & 6 July 1979	573	663	730	783	812	42
Anthesis date	June 1979	7.3	6.8	7.4	7.2	7.2	NS
Plant height, cm	June 1979	143	151	154	156	164	5
Seed yield/plant, g	1979	37	47	51	65	57	9
100-seed weight, mg	1979	80	85	86	89	97	5

For example, responses in dry-matter yield were greater in both populations, and the response in CER was less in the high-SLW cycle-1 population than in those reported previously. In this study, tiller weight was more important than tiller number in dry-matter yield of first growth of the plants, whereas tiller number was more important in the previous study.

In solid stands, the high- and low-SLW cycle-1 populations did not differ significantly in forage yield over 3 years when harvested either three or five times/yr. The interaction of strains with harvest frequencies was not significant in the analysis of variance of total yield over years. Under both harvest frequencies, the high-SLW population yielded slightly more than the low-SLW population. The greater tillering of spaced plants of the low-SLW population was not an advantage in yield and persistence in solid stands clipped frequently.

Differences between high- and low-SLW populations in forage quality characteristics of nonflowering regrowth usually were not significant. The high-SLW population tended to be slightly lower in acid detergent fiber, cellulose, and silica, and slightly higher in hemicellulose than the low SLW population.

DISCUSSION

The sizable correlated responses to selection for SLW suggest that SLW reflects basic aspects of growth and development of the grass plant. The results seem to fit ideas proposed by Grafius (1978) regarding physiological and developmental relations between characters. He indicated that the size of the stem, leaves, seed head, and floral parts and the maximum seed size in grasses might be proportional to the size of the shoot apex on tillers. In our study, high SLW, thick and wide leaves, high tiller weight, and high 100-seed weight might be developmentally associated with large shoot apices. In addition, plants in high-SLW populations were observed to have larger culms and panicles than plants in low-SLW populations. The larger panicles probably contributed to higher seed yield/plant.

As a corollary to Sinnott's law that the size of an organ is proportional to the size of the meristem from which it develops, Grafius (1978) proposed that number and size of organs tend to have an inverse relationship. In our

research, a strong negative correlation (r = -0.994) was found between tiller number and tiller weight of first growth of spaced plants of the five populations in 1978. Larger tillers more than compensated for fewer tillers in determination of dry-matter yield/plant at that time. A negative correlation between tiller number and weight also was found by Nelson et al. (1977) in tall fescue (*Festuca arundinacea* Schreb.); however, tiller number was more important than tiller weight in determining yield/plant under low competition. Nelson et al. indicated that tiller weight would be the more important yield component under sward conditions.

The higher tiller weight of the high-SLW populations compared with NCRC1 may be explained by more rapid leaf area development due to a larger supply of photosynthate in larger shoot apices. The larger supply of photosynthate probably was not due to a higher CER because the high-SLW populations were similar to NCRC1 in CER. More photosynthate probably was due to better light interception (more upright leaves) and better partitioning of photosynthate to leaf enlargement.

The significant yield advantage of the high-SLW cycle-1 population over the low-SLW population in the space-planted test was not expressed in solid stands. Thus, plant density had a strong influence on yield expression.

LITERATURE CITED

Carlson, G.E., R.H. Hart, C.H. Hanson, and R.B. Pearce. 1970. Overcoming barriers to higher forage yields through breeding for physiological and morphological characteristics. Proc. XI Int. Grassl. Cong. 248–251.

Grafius, J.E. 1978. Multiple characters and correlated response. Crop Sci. 18:931–934.

Nelson, C.J., K.H. Asay, and D.A. Sleper. 1977. Mechanisms of canopy development of tall fescue genotypes. Crop Sci. 17:449–452.

Pearce, R.B., G.E. Carlson, D.K. Barnes, R.H. Hart, and C.H. Hanson. 1969. Specific leaf weight and photosynthesis in alfalfa. Crop Sci. 9:423–426.

Topark-Ngarm, A., I.T. Carlson, and R.B. Pearce. 1977. Direct and correlated responses to selection for specific leaf weight in reed canarygrass. Crop Sci. 17:765–769.

Wilson, D., and J.P. Cooper. 1969. Apparent photosynthesis and leaf characters in relation to leaf position and age, among contrasting *Lolium* genotypes. New Phytol. 68:645–655.

Effective Selection for Tolerance to Grass-Killing Herbicides in Perennial Ryegrass (*Lolium perenne* L.)

C.E. WRIGHT and J.S. FAULKNER
Department of Agriculture for Northern Ireland
and The Queen's University, Belfast, Northern Ireland, UK

Summary

The aim of grassland reseeding is to establish a high-quality sward of improved herbage cultivars. To obtain the maximum value from the improved sward, weed invasion should be prevented. Herbicides are the main tool available. There are many herbicides for selectively controlling dicotyledonous, but not graminaceous, weeds. The control of graminaceous weeds would be possible in swards of grass cultivars selected for tolerance to grass-killing herbicides.

Two cycles of recurrent selection for glyphosate tolerance in *Lolium perenne* L. were carried out by spraying large seedling populations with critical dosages in the greenhouse and intercrossing the survivors. It was found that two populations selected in this way were about 50% and 67% more tolerant than two control cultivars.

To combine the paraquat tolerance of the *L. perenne* cultivar Causeway with the agronomic characters of cultivar Aberystwyth S 23, crosses were made between the two cultivars and selection for tolerance was practiced on seedlings of the F_2, F_3, and F_4 generations. The F_3 and especially the F_4 were found to be even more tolerant than Causeway.

These results illustrate the feasibility of breeding for herbicide tolerance in grasses. Herbicide-tolerant cultivars would facilitate the establishment and maintenance of weed-free pastures and seed crops, the maintenance of varietal purity, and the distinction of the cultivars from others in the species.

KEY WORDS: perennial ryegrass, *Lolium perenne* L., herbicide, glyphosate, paraquat, selection, tolerance.

INTRODUCTION

Deteriorating swards may be replaced by reseeding with improved herbage cultivars or may be renovated by various means. The use of hormonal-type herbicides during renovation can remove dicotyledonous weeds, but there is still no chemical that can eliminate most of the less desirable grasses that invade sown swards as a result of damage caused by stock, disease, pests, poor management, or extremes of weather.

Several chemicals have been proposed as capable of selectively reducing the proportion of various unsown grasses in grass swards. For example, ethofumesate—first developed for use on sugar beet crops—controls various grass weeds in pastures (Minter 1974), herbage seed crops (Oswald and Haggar 1974), and newly sown leys (Griffiths et al. 1978).

The alternative strategy of breeding grass cultivars for tolerance to broad-spectrum grass-killing herbicides, suggested by Wright (1966), has been shown to be feasible by Faulkner (1976, 1978a), who has develped and registered cultivars of *Lolium perenne* L. tolerant to paraquat and dalapon. Swards sown with such cultivars can be maintained free of both grass weeds and broadleaf weeds by spraying with the broad-spectrum herbicide to which they are tolerant. This strategy is economically preferable, the cost of development of a new herbicide (£10 million, Anon. 1979) being much greater than the cost (estimated at £0.5–1.0 million) of a breeding program to produce a herbicide-tolerant cultivar.

This paper reports progress in perennial ryegrass breeding programs (1) to select for tolerance to the broad-spectrum herbicide glyphosate and (2) to transfer the paraquat tolerance character from Causeway, the first cultivar bred for resistance to a grass-killing herbicide, to more high-yielding genetic backgrounds.

MATERIALS AND METHODS

A two-stage screening procedure is normally used in selecting for tolerance to foliar herbicides. When the main flush of seedlings, from seeds sown in trays of fine soil–based compost at 15 mm spacing in the greenhouse, reaches the two-leaf stage, the seedlings are sprayed by a laboratory sprayer with a critical dose of the herbicide calculated to kill all but a small proportion of seedlings.

Those remaining are either potted singly or transplanted into the field for the second-stage screening. The

---

C.E. Wright is Head of the Field Botany Research Division of the Department of Agriculture for Northern Ireland and a Professor of Agricultural Botany at The Queen's University of Belfast; J.S. Faulkner is a Forage Breeder at the Northern Ireland Plant Breeding Station, Department of Agriculture for Northern Ireland, Loughgall.

dosage used is one expected to have visible effects on most plants but to kill only a few, including any "escapees." Selection is based on degree of herbicidal damage, in contrast to the first stage, where it is essentially a matter of survival. The rationale underlying this procedure has been discussed by Faulkner (1982).

Selection for Glyphosate Tolerance

Five thousand seedlings of each of six diploid varieties of *L. perenne* L. were sprayed with 0.4 kg a.e./ha of glyphosate, the survivors mated in six polycross groups corresponding to the parent varieties, and 4,000 seedlings of each progeny group subjected to another cycle of selection using 0.5 kg a.e./ha of glyphosate. Of several hundred resprayed survivors, 30 plants were selected and mated in two polycross groups—one (GRP/Ea) derived from two early cultivars, and one (GRP/La) from four late cultivars.

The level of glyphosate tolerance in the progeny of these two polycrosses was compared in the greenhouse with that of two control cultivars—LY/L and Callan—that represented the tolerant and susceptible extremes, respectively, of normal diploid varieties. Sixteen trays were sown with 70 seeds of each cultivar and selected population, allowing 4 replicate trays for each of four glyphosate dosages (0.20, 0.283, 0.40, and 0.566 kg a.e./ha). The herbicide was applied at the two-leaf stage, and the dosages required to kill 50% of seedlings (ED_{50}) were calculated by regression.

Transfer of Paraquat Tolerance to a Superior Genetic Background

Ten plants of the paraquat-tolerant *L. perenne* L. cultivar Causeway were mated in pairs, without emasculation, with an equal number of plants of the widely used agricultural cultivar Aberystwyth S 23. Up to 42 seeds/cross of F_1 seed from the S 23 parent were sown in seed trays. A low dosage of paraquat, 0.04 kg a.i./ha, was applied at the two-leaf stage to remove any seedlings produced by self-fertilization. Six of the strongest surviving seedlings/cross were selected and interpollinated with seedlings from other crosses. F_2 seed was harvested in bulk. Intensive selection for paraquat tolerance was practiced on the F_2 generation. Twelve thousand seedlings were grown in seed trays and sprayed with 0.10 kg a.i./ha of paraquat. Four hundred survivors were space-planted in the field and reselected by spraying at 0.60 kg/ha after establishment. Using agronomic criteria, the F_2 generation was finally reduced to 17 plants, which were polycrossed to produce an F_3 generation. A similar selection procedure, using slightly higher dosages of praquat, was followed to produce the F_4 generation.

Assessments were made of the tolerances of the F_2 to F_4 generations as compared with cv. Causeway and cv. Barlenna (which has a level of tolerance similar to that of S 23). The method was analogous to that used for glyphosate tolerance. The tolerance of the F_2 generation was investigated separately from those of the F_3 and F_4.

RESULTS

The two selected populations showed a consistently higher survival rate over the range of glyphosate dosages used (Fig. 1A). On the basis of ED_{50}s, GRP/La was about 50% more tolerant and GRP/Ea about 67% more tolerant than the controls. The difference between the two controls was very small in relation to that between con-

Fig. 1. Survival of seedlings (A) of two selected populations (GRP/La and GRP/Ea) and two control cultivars (Callan and LY/L) under four glyphosate application rates (B) of an F_2 hybrid population, and (C) survival of the selected F_3 and F_4 generations compared to that of their tolerant parent (Causeway) and a susceptible cultivar (Barlenna) under three paraquat treatments.

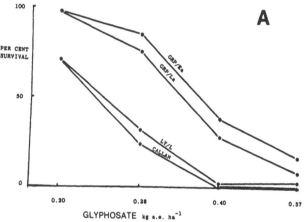

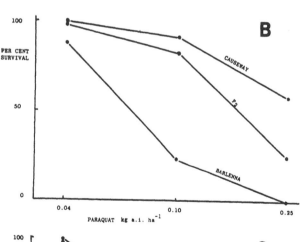

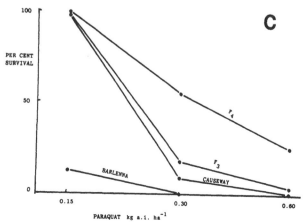

trols and selected populations.

There was, as expected, a wide gap between the tolerance levels of Causeway and Barlenna (Fig. 1B). In tolerance, the F_2 generation was somewhat closer to Causeway than to Barlenna. A large gain in tolerance occurred (Fig. 1C) between the F_2 and F_3 generations, such that the latter was slightly more tolerant than the tolerant parent. A further gain took place in the next generation to produce an F_4 population substantially more tolerant than Causeway. This transgressive segregation is a pointer to heterozygosity in the parents.

DISCUSSION

It is clear that sufficient variation exists in perennial ryegrass, the most widely sown grassland species in the UK, to permit selection of plants tolerant to the broad-spectrum herbicides paraquat and glyphosate, and that intermating of these tolerant plants permits concentration of the tolerance genes by selection among the progeny. Parallel work in amenity grasses has proved equally successful (Lee and Wright 1981).

Cultivars tolerant to broad-spectrum herbicides under field conditions have several advantages.

1. Areas sown with such cultivars can be sprayed with the appropriate herbicide shortly after seedling emergence to eliminate competing grass weed and dicotyledonous weed seedlings, thereby ensuring a weed-free environment to permit vigorous establishment of the sown cultivar (Faulkner 1978b).

2. When sward deterioration eventually takes place, spraying of the mature sward should eliminate undesirable grasses and dicotyledonous weeds. The frequency of such sprayings might be once each 2 to 3 years, but it would depend on the proportion of unwanted species that could be tolerated without affecting the output of the sward and the ability of the sown cultivar fully to recolonize the sward after spraying.

3. Stands of tolerant cultivars for seed multiplication could be sprayed to remove weed plants, ensuring virtually clean seed and making harvesting and seed cleaning easier and less costly.

4. The removal of susceptible volunteer plants or rogues of the same species by spraying would preserve the purity of the cultivar, thereby helping to ensure seed certification.

5. Because of its unique herbicide-tolerance character the cultivar would be clearly distinct, and disputes about the authenticity of seed could be easily resolved.

6. Experimentation to gain a better understanding of the value of pure stands as opposed to swards invaded by weed grasses should be easier.

Fears that swards of herbicide-tolerant grasses will be indestructible are unwarranted. Herbicide tolerance is specific to one herbicide or group of herbicides. Thus, swards tolerant to paraquat can be destroyed if desired by spraying with an alternative herbicide, e.g., glyphosate, as well as by cultivation.

Trials are now underway to determine the yield potential of the new material with a view to releasing cultivars for agricultural use.

LITERATURE CITED

Anon. 1979. Royal commission on environmental pollution. 7th Report, agriculture and pollution, Sept. 1979. H.M. Stationery Off., London.

Faulkner, J.S. 1976. A paraquat resistant variety of *Lolium perenne* under field conditions. Proc. 1976 Brit. Crop Protect. Conf.—Weeds, 485–490.

Faulkner, J.S. 1978a. Dalapon tolerant varities—a possible basis for pure swards of *Lolium perenne* L. Proc. 1978 Brit. Crop Protect. Conf.—Weeds, 341–348.

Faulkner, J.S. 1978b. The use of paraquat for controlling weeds in seedling swards of paraquat resistant *Lolium perenne* L. Proc. 1978 Brit. Crop Protect. Conf.—Weeds, 349–355.

Faulkner, J.S. 1982. Breeding herbicide tolerant cultivars by conventional means. *In* H.M. Le Baron and J. Gressel (eds.) Herbicide resistance in plants. John Wiley & Sons, New York.

Griffiths, W., C.H. Hammond, and C.J. Edwards. 1978. Weed control in new leys and established pastures with ethofumesate. Proc. 1978 Brit. Crop Protect. Conf.—Weeds, 309–316.

Lee, H., and C.E. Wright. 1981. Selection for tolerance to a grass-killing herbicide in temperate amenity grass species. Proc. 4th Turfgrass Res. Conf., 41–46.

Minter, L.K. 1974. Control of barley grass and other pasture weeds with ethofumesate (NC 8438). Proc. 27th N.Z. Weed and Pest Contr. Conf., 82–84.

Oswald, A.K., and R.J. Haggar. 1974. The tolerance of ten grass varieties to six herbicides with a potential for wild oat control in herbage seed crops. Proc. 12th Brit. Weed Contr. Conf., 715–721.

Wright, C.E. 1966. Some implications of genotype-herbicide interactions in the breeding of *Lolium perenne*. Euphytica 15:229–238.

Comparison of Italian Ryegrass and Meadow Fescue with F_2, F_3, and F_4 Hybrids of Both Grass Species

H. KALTOFEN and E. WOJAHN

Institute of Forage Production, GDR

Summary

In forage-grass breeding there is an increasing interest in hybrids of Italian ryegrass (*Lolium multiflorum* Lam.) and meadow fescue (*Festuca pratensis* Huds.). The objective of this study was to investigate some agronomic properties of such *Festulolium* hybrids with special reference to the question of whether these hybrids change from one generation to the next.

In a pot experiment with four levels of nitrogen fertilization, Italian ryegrass and meadow fescue as well as F_2, F_3, and F_4 hybrids of the two grass species were examined for 2 years. In each year four cuts were taken. All the F_2, F_3, and F_4 hybrids were descendants of a single F_1 plant. Therefore, some inbreeding effects must be taken into account in spite of open pollination of the hybrids. The fertility of the hybrids increased from one generation to the next.

In the first year, after the grasses were sown in spring, meadow fescue produced substantially lower yields than Italian ryegrass at the higher nitrogen doses and died off almost completely at the highest nitrogen level. The yields of the hybrids exceeded those of meadow fescue and, at lower nitrogen supply, even those of Italian ryegrass. Very high nitrogen doses were tolerated much better by the hybrids than by meadow fescue but not quite as well by the hybrids as by ryegrass.

In the second year, after wintering, all grasses formed culms, with Italian ryegrass showing the most abundant culm development. The hybrids outyielded meadow fescue and ryegrass in the second year nearly without exception. In general, the yields of the hybrids were distributed more evenly among the various cuts than were those of Italian ryegrass. In both years, the hybrids mostly gave noticeably higher yields than either of the parent species.

Within the different nitrogen levels, the crude-protein content was lower and the content of water-soluble carbohydrates usually was higher as the yield of dry matter increased. The crude-fiber content of the hybrids often proved to be somewhat higher than that of Italian ryegrass but not higher than that of meadow fescue.

In regard to yield, quality, and response to nitrogen, no well-defined changes occurred with the succession of hybrid generations. There was no evidence that the hybrids reverted to one of the parent species. These results may encourage plant breeders to direct greater attention to such hybrids.

KEY WORDS: *Festulolium*, Italian ryegrass, meadow fescue, yield, quality, nitrogen.

INTRODUCTION

In forage-grass breeding there is an increasing interest in hybrids of Italian ryegrass and meadow fescue. In the GDR the most important aim with the intergeneric hybridization of these grass species is to combine the high yields and the favorable forage quality of Italian ryegrass with the persistence and winter-hardiness of meadow fescue to obtain a high-yielding, high-quality forage grass suitable for mowing and ensiling that may be grown for 2 or 3 years.

Experiments were conducted to determine whether intergeneric hybrids of meadow fescue and Italian ryegrass indeed exhibit the desired agronomic properties, whether these hybrids change from one generation to the next, and whether the hybrids tend to revert to one of the 2 parental species.

METHODS

Italian ryegrass (*Lolium multiflorum* Lam. cv. Marino) and meadow fescue (*Festuca pratensis* Huds. cv. Bundy) as well as F_2, F_3, and F_4 hybrids of these grass species were examined for 2 years in a pot experiment involving four different levels of nitrogen fertilization (0.5, 1.0, 2.0, and 3.0 g N/pot). In each year four cuts were taken. The F_2, F_3, and F_4 hybrid plants were descendants of a single F_1 plant (*Festulolium braunii* K.A.). Therefore, despite open pollination of the hybrids, some inbreeding effects must be taken into account. All the grasses were sown in spring. Judging from the emergence of the plants, the fertility of the hybrids increased from one generation to the next.

Seven seedlings were planted in each pot; the pots measured 20 cm in diameter and were filled with 6.8 kg of quartz sand. The pots were fertilized yearly with appropriate amounts of phosphorus, potassium, calcium, magnesium, and micronutrients.

RESULTS AND DISCUSSION

In the first year, after sowing in spring, all the grasses remained in the vegetative stage. At first, meadow fescue grew rather slowly in comparison with Italian ryegrass or the hybrids.

The highest nitrogen rate impaired the growth of meadow fescue to such an extent that this grass species died almost completely during the first season (Table 1). The hybrids were more tolerant of high nitrogen rates than was meadow fescue but less tolerant than was Italian ryegrass. On the other hand, at low nitrogen levels, the tiller density of Italian ryegrass decreased considerably during the first season.

During the first year, the total yields of the hybrids proved to be substantially higher than those of meadow fescue. The superiority of the hybrids compared with meadow fescue increased with increasing levels of nitrogen fertilization. The unfavorable influence of very high nitrogen rates on meadow fescue is clearly shown by its reduced total yield. In the third and fourth cut of the first season, the yield of meadow fescue at the highest nitrogen level was near zero because most plants of this grass species had died.

At nitrogen levels of 0.5 and 1.0 g/pot the hybrids tended to produce somewhat higher yields than Italian ryegrass did. At nitrogen rates of 2.0 g/pot, on the other hand, Italian ryegrass and the hybrids exhibited approximately the same yields in the first year. At the highest nitrogen level, the hybrids were partially damaged; however, the yields of the hybrids were distributed more evenly among the various cuts than were those of Italian ryegrass.

In the second season, after wintering, all grasses of the trial formed culms and inflorescences, with culm development on Italian ryegrass being the most abundant. Tiller death of Italian ryegrass continued during the second year, at first mainly in the case of lower nitrogen supply but later at higher nitrogen rates, too. Thus, in the last two cuts of the second season, the yields of Italian ryegrass were very poor. Obviously, the persistence of Italian ryegrass was diminished by nitrogen deficiency.

If the data for the highest nitrogen level, which was detrimental to plants, are omitted, the hybrids produced higher yields than Italian ryegrass or meadow fescue. The superiority of the hybrids in this year was most pronounced at low and moderate nitrogen rates. At these nitrogen rates, the poor growth of Italian ryegrass was very striking. Contrasted to Italian ryegrass, yields of the hybrids were fairly high even in the last two cuts of the second season. At this time the tiller density of the hybrids was not lower than that of meadow fescue, indicating a comparatively high persistence for the hybrids.

Averaged over both years, the hybrids produced considerably higher yields than did meadow fescue at all

Table 1.   Total dry matter yields (g/pot).

g N/pot and cut	Italian ryegrass	Meadow fescue	Hybrids		
			F_2	F_3	F_4
First year					
0.5	63.6	62.9	95.5	83.4	91.4
1.0	132.8	101.2	150.5	137.6	142.2
2.0	148.3	75.6	131.0	136.6	152.4
3.0	134.4	38.2	136.0	71.8	94.9
Second year					
0.5	38.1	56.1	98.0	83.2	92.0
1.0	59.0	95.5	149.0	145.9	127.5
2.0	122.0	122.1	159.0	184.1	175.3
3.0	147.1	0.0	174.0	81.6	117.6
First and second years combined					
0.5	50.9	59.5	96.8	83.3	91.7
1.0	95.9	98.4	149.8	141.8	134.9
2.0	135.2	98.9	145.0	160.4	163.9
3.0	140.8	19.1	155.0	76.7	106.3

nitrogen levels (Table 1). At the lower nitrogen rates, the hybrids were clearly superior to Italian ryegrass as well. At nitrogen rates of 2 g/pot, the hybrids also tended to produce somewhat higher yields than Italian ryegrass. Only at the highest nitrogen rate was there no superiority of the hybrids in comparison with Italian ryegrass. The hybrid populations examined in this experiment were not the result of any purposive selection process. On the other hand, Italian ryegrass and meadow fescue were represented in this trial by known high-yielding varieties.

There were no significant differences in crude-ash content among the examined grasses; thus, ash data are not presented. The crude-fiber content of the hybrids averaged somewhat higher than that of Italian ryegrass but not higher than that of meadow fescue (Table 2). This higher crude-fiber content may be related to the fact that the hybrids, on the average, yielded more than did Italian ryegrass, since the portion of cell-wall constituents in high-yielding plants often is somewhat higher than in low-yielding ones. However, the genotypic correlation between dry-matter yield and crude-fiber content is not high, a fact that would suggest that selection for high-yielding genotypes with comparatively low crude-fiber content is a possibility.

The highest lignin contents were generally found in meadow fescue. On the average, Italian ryegrass and the hybrids were similar in lignin content.

At the same nitrogen-fertilization rates, crude-protein content was negatively correlated with dry-matter yield. Crude-protein content was highest in meadow fescue and lowest in the hybrids, with protein content being inversely related to yield. The hybrids were distinguished by high internal efficiency of nitrogen; i.e., they produced a high amount of dry matter/unit nitrogen taken up, resulting in a comparatively low crude-protein content.

Crude-protein and water-soluble carbohydrate con-

Table 2. Average composition of harvested forage.

Item	Number of comparisons	Italian ryegrass	Meadow fescue	Hybrids		
				F_2	F_3	F_4
Crude fiber	22	18.9	21.1	20.8	20.2	19.9
Lignin	13	4.8	6.1	4.7	4.9	4.8
Crude protein	27	18.2	22.4	16.6	17.4	16.4
Water-soluble carbohydrates	15	18.5	13.9	21.0	21.2	22.0

tents were negatively correlated. At the same nitrogen supply, positive correlations existed between dry-matter yield and content of water-soluble carbohydrates. Thus, the low-yielding meadow fescue on the average exhibited the lowest content of water-soluble carbohydrates. The high-yielding hybrids, on the other hand, had the highest content of water-soluble carbohydrates, even though they had only slightly higher crude-fiber content than meadow fescue. In the GDR, forage grasses with high levels of water-soluble carbohydrates are desirable as they are par-

ticularly suitable for making high-quality grass silage.

From the results of this experiment, it may be concluded that hybrids of meadow fescue and Italian ryegrass show rather favorable agronomic properties. In respect to yield, quality, persistence, and response to nitrogen, no well-defined, systematic changes were observed with the succession of hybrid generations. There was no evidence that the hybrids reverted to one of the two parental species. These results may encourage plant breeders to direct greater attention to such hybrids.

Breeding *Centrosema pubescens* Better Adapted to the Acid Infertile Soils of South America

E.M. HUTTON

Centro Internacional de Agricultura Tropical, Colombia

Summary

Continued use of *Centrosema pubescens* Benth. (centro) in tropical South America depends upon development of lines with increased tolerance to high acidity and aluminum (Al). This study classified levels of acid tolerance among *Centrosema* introductions and crosses of promising introductions.

Initially, greenhouse and field evaluations in Carimagua oxisol (pH 4.5, 90% Al saturation) were used. Introductions of *C. schottii* and *C. pascuorum* died, while most *C. virginianum* and *C. acutifolium* and a few *C. pubescens* had low acid tolerance. Most *C. pubescens* introductions had medium acid tolerance, whereas *C. macrocarpum* and *C. schiedeanum* were highly acid tolerant. The degree of acid tolerance in introductions was negatively correlated with Al uptake.

Several promising *C. pubescens* introductions, and one each of *C. macrocarpum* and *C. schiedeanum*, were selected for crosses aimed at increasing the acid tolerance of *C. pubescens*. Most crosses obtained were between *C. pubescens* introductions, there being none with *C. schiedeanum* and only one with *C. macrocarpum*.

Rapid greenhouse screening of F_2s for acid tolerance was achieved, first in sand culture and then in Carimagua soil. More efficient sand-culture screening followed reduction of Al from 8 to 4.5 ppm and maintenance of 10 ppm nitrogen (N). None of the best *C. pubescens* introductions, nor their F_2s, had more than medium acid tolerance.

The fertile *C. pubescens* × *C. macrocarpum* hybrid has provided the basis for selection of highly acid-tolerant lines, combining the best feature of both parents. The high acid tolerance of *C. macrocarpum* was inherited in about 20% of F_2 plants.

KEY WORDS: *Centrosema* spp., acid soils, adaptation, acid tolerance, aluminum, selection.

INTRODUCTION

Centrosema pubescens is one of the important legumes commonly used, with varying success, in South American pastures. Current cultivars are poorly adapted to the acid (pH 4.2–4.7), infertile oxisols and ultisols predominating in tropical South America and covering more than 500 million ha (T. Cochrane, personal communication). Increased use of *C. pubescens* in South America is dependent on selecting or breeding lines with greater tolerance to low fertility and high levels of acidity and Al.

This paper reports the selection in greenhouse and field experiments of some promising introductions of *C. pubescens* and of other *Centrosema* species showing high acid tolerance. Crosses between these species aimed at increasing the adaptation of *C. pubescens* to very acid soils. A greenhouse method involving sand culture and Carimagua soil in sequence was used for rapidly selecting the most acid-tolerant F_2 plants.

MATERIALS AND METHODS

Commencing early in 1978 at CIAT headquarters, near Palmire, Colombia, the acid tolerance of a range of *Centrosema* introductions was studied in two successive greenhouse pot experiments. The second utilized 135 *Centrosema* introductions (23% of the CIAT collection) and included the 24 in the first experiment. Each pot (15 cm in diameter) contained 1,730 g of unsterilized Carimagua red-brown oxisol with a pH of 4.5 and 90% Al saturation. There were 5 plants/pot and 2–3 pots (replicates)/ introduction in a randomized layout. Each pot received CIAT *Rhizobium* 590, double deionized water, and weekly amounts of solution containing all essential nutrients (as in sand culture), the total applied being the equivalent of 10 kg/ha in the first experiment and 18 kg/ha in the second. At harvest, after 3 months' growth in the first experiment and 4 months' in the second, acid tolerance of each introduction was rated as follows: 1 (low), plants small (4–6 cm high), yellow, and often necrotic; 2 (medium), plants of medium vigor (10–15 cm high), often with yellowing; 3 (high), plants large (20 cm or higher) and green.

In the second experiment, 4 introductions were selected randomly from each rating, the tops from each replicate being bulked, dried, and weighed. The dried tops from the replicates of each introduction were then bulked and analyzed for Al, N, phosphorus (P), calcium (Ca), and magnesium (Mg).

Field selection of acid-tolerant *Centrosema* introductions was done at Carimagua Research Station at lat 4.7° N in Colombia, south of the Meta River, in typical llanos. The predominant red-brown oxisol is deep and friable. The climate is lowland tropical with a rainfall of 2,000 mm/yr

The author is Visiting Scientist at CIAT.

The author gratefully acknowledges the help of Dr. K. Schultze-Kraft with introductions, Dr. A. León with chemistry, and José Isván Giraldo with technical aspects.

and a hot dry season from December to April. Thirty-five *Centrosema* introductions in the Carimagua plots were classified for acid tolerance on the basis of yield and persistence at the end of the dry seasons in 1979 and 1980, respectively.

Centrosema Crossing Program

A series of half-diallel crosses was attempted between *Centrosema* introductions selected from the greenhouse and field experiments. CIAT introductions in the program included the promising *C. pubescens* 5052, 5203, 5206, 5209, 5210 and the highly acid-tolerant *C. macrocarpum* 5062 and *C. schiedeanum* 5066. F_1 plants were grown in the field during 1979 at CIAT headquarters to produce seed for selection of acid-tolerant F_2 plants.

Selection of F_2s for Acid Tolerance in Sand Culture and Carimagua Soil

A sand-culture unit (Andrew 1974) had 4 troughs, each with 17 square pots (14 cm²) of pure sand, irrigated with nutrient solution each hour by a pump connected to a time clock. The solution contained all essential nutrients in the following ppm concentrations: P, 0.5; sulfur (S), 5.0; potassium (K), 10.0; Ca, 1.0; Mg, 1.0; iron (Fe), 0.5; manganese (Mn), 0.01; molybdenum (Mo), 0.01; copper (Cu), 0.01; zinc (Zn), 0.05; boron (B), 0.05. Al was maintained at 8 ppm in initial evaluations but was then reduced to 6 ppm and finally to 4.5 ppm.

A replicated randomized experiment in a unit (25 seedlings/pot) included up to 3 F_2s (>300 seedlings/ cross) and the parents (each 100 seedlings). F_2s were repeated in different experiments. Initially *Rhizobium* 590 and 4 ppm N were added to promote nodulation. Adjustment of solution levels and pH to 4.2 was made daily, and that for the Al level was made 3 times weekly. After 4–6 weeks' growth, plants were harvested, but only those at 6 ppm Al or less could be rated: 1 (low acid tolerance), small and yellow with poor, blackened roots; 2 (medium tolerance), height 5 cm, leaves yellow-green, and roots 8 cm long, branched and brown; 3 (high tolerance), height 6–8 cm, leaves green, and roots 10–11 cm long, well branched and light colored.

The highly acid-tolerant selections from sand culture were transplanted singly into pots (15 cm in diameter) of Carimagua soil after roots were covered with a peat suspension of *Rhizobium* 590. Controls included the parents and hybrid plants from ratings 1 and 2. After 6–8 weeks' growth, plants were rated for acid tolerance as in the earlier experiments with *Centrosema* introductions. Weekly nutrient applications were adjusted to give the equivalent of 10 kg/ha of total nutrients during the growth period. At Carimagua Research Station in early July 1980, rows of F_2s of 4 *C. pubescens* crosses and 5052 × 5062 together with parents were established from seed. Minimal amounts of essential nutrients were applied, including 10 kg P/ha. After 5 months' growth, plants were rated for yield.

Table 1. 135 *Centrosema* introductions classified for acid tolerance after growth in Carimagua oxisol (2d experiment) and an association between acid tolerance and Al content of tops.

Acid tolerance rating† and number of introductions	Species†† and number of introductions	Four random introductions* Mean dry wt g/pot (Mean Al ppm)
Dead (16)	C. schottii (10)	—
1 (30)	C. pascuorum (3)	
	C. virginianum (10)	1.04 (764)
	C. pubescens (5)	
	C. acutifolium (3)	
2 (79)	C. pubescens (49)	2.03 (198)
	C. plumieri (4)	
3 (10)	C. macrocarpum (5)	2.74 (53)
	C. schiedeanum (1)	

†1,2,3 = Low, medium, high acid tolerance, respectively.

††Only main species listed.

*Least significant difference 0.45 (P < 0.001) calculated from analysis of variance. Correlation coefficient (r) between Al and dry weight = − 0.96, d.f. 133.

RESULTS

Only results of the second experiment with *Centrosema* introductions are presented in Table 1, as the results from the first were similar. Most *C. schottii* and *C. pascuorum* introductions were highly intolerant of the acid soil and died. *C. virginianum* and *C. acutifolium* introductions had low tolerance, and nodulation was poor or absent. In *C. pubescens* some introductions had low acid tolerance, but most had medium tolerance, a trend present in the few *C. plumieri* introductions. A small group, including *C. macrocarpum*, *C. schiedeanum*, and unidentified species, was highly acid tolerant. Introductions with medium or high acid tolerance were well nodulated.

Introductions with low acid tolerance (Table 1) had the highest Al content (764 ppm), whereas those with medium and high acid tolerances had 26% and 7%, respectively, of this high Al level. Contents of other minerals were within the usual limits.

At Carimagua, by the end of their second dry season in March 1980, some *Centrosema* introductions had died. *C. macrocarpum* 5062 and 5065 had outstanding vigorous green growth, large coarse leaves, and poor seed production. *C. schiedeanum* 5066 had good growth and persistence. Fifteen *C. pubescens* introductions showed promise, including 5052, 5203, 5206, and 5210.

Centrosema Crosses

Most *Centrosema* crosses were achieved, except those with 5066 and 5209. Only one *C. pubescens* × *C. macrocarpum* cross resulted, 5052 × 5062. Its F_1 and F_2 were fertile, and in a CIAT field trial 20% of F_2 plants gave similar or better seed yields than the heavy-seeding parent 5052 (104 g seed/plant).

Table 2. Acid tolerance of F_2's and parents of *Centrosema* crosses in sand culture and Carimagua Soil.

Experiment and lines	Carimagua Soil No. plants in ratings†† 1	2	3	% Original population highly acid tolerant
Expt. B†				
5052 (C.p)	6	18	0	0
5062 (C.m)	0	8	16	67
F_2 5052 × 5062	333	63	102	20
$\chi^2(4) = 120.99$; P<0.01				

Experiment and lines	Sand Culture No. plants in ratings†† 1	2	3	% Original population highly acid tolerant
Expt. F†				
5052 (C.p)	0	97	0	0
5062 (C.m)	0	36	67	65
5210 (C.p)	0	98	0	0
F_2 5052 × 5062	250	265	132	20
F_2 5052 × 5210	410	201	0	0
$\chi^2(8) = 730.24$; P<0.01				

†Sand culture Al levels: Expt. B, 8ppm; Expt. F, 4.5 ppm.

††1,2,3 = Low, medium, high acid tolerance, respectively.

C.p *Centrosema pubescens;* C.m *Centrosema macrocarpum*

Distribution of Acid Tolerance in F_2 Populations

None of the 6 *C. pubescens* crosses obtained gave highly acid-tolerant F_2 segregates. In all greenhouse screenings, similar inheritance patterns for acid tolerance have been obtained in the F_2 of 5052 × 5062. As shown in Table 2, 20% of this F_2 had high acid tolerance in 2 of the screenings, B and F, even though sand-culture Al levels differed markedly.

In experiment B, the sand-culture plants at 8 ppm Al were small, yellow, and nonnodulated. Selection for acid tolerance was possible only when the plants were grown and nodulated in Carimagua soil. Similar results occurred in other experiments with sand-culture Al at 8 ppm.

In experiment F, with sand culture at 4.5 ppm Al, nodulation was still poor, so 10 ppm N was maintained. The lower Al level allowed satisfactory screening and comparison of the 2 F_2 populations. No highly acid-tolerant segregates occurred in the *C. pubescens* cross, 5052 × 5210. In experiment F, reactions of the selections and controls in sand culture and Carimagua soil were similar.

Expression of high acid tolerance in all 5062 plants in B and F was prevented by delayed germination and slow seedling growth in a portion of the population.

In the 1980 field experiment at Carimagua Research Station, after 5 months' growth the 4 *C. pubescens* F_2s showed medium acid tolerance. Compared with the vigorous 5062 parent, 25% of the 5052 × 5062 F_2 plants had a similar yield and high acid tolerance but finer leaves and stolons.

DISCUSSION

The evaluations of *Centrosema* introductions for acid tolerance in Carimagua soil in the greenhouse experiments and in Carimagua field trials agreed. They confirmed that a small group, including *C. macrocarpum* and *C. schiedeanum*, had high acid tolerance. There were no highly acid-tolerant introductions in *C. pubescens*. This is understandable, as Schultze-Kraft and Giacometti (1979) found that *C. pubescens* usually occurs naturally in less acid soils.

Table 1 shows that introductions like *C. virginianum* do not appear to restrict Al uptake, which inhibits growth and nodulation. Al uptake is restricted in the more tolerant introductions, especially in highly tolerant ones like *C. macrocarpum*. The degree of acid tolerance was negatively correlated with Al uptake.

Breeding highly acid-tolerant lines from *C. pubenscens* parents appears to be impossible, a conclusion already reached by R.J. Clements and colleagues (personal communication). In this study (Table 2), only 5052 × 5062 has provided a basis for breeding highly acid-tolerant lines with desirable agronomic characters. The fertility of this *C. pubescens* × *C. macrocarpum* cross shows genetic compatibility between these species. Clements and Williams (1980) found that *C. pubescens* grades taxonomically into several species, including *C. macrocarpum*.

In sand culture, pH 4.2 inhibited nodulation, and 8 ppm Al stunted growth. By reducing Al from 8 ppm to 4.5 ppm (Table 2) and maintaining 10 ppm N, sand culture gives better definition of plants likely to show high acid tolerance in Carimagua oxisol. Attention to germination in 5062 should reduce its variability in expression of acid tolerance.

Field testing could commence with highly acid-tolerant F_3 lines from rapid greenhouse screening of F_2 populations, since greenhouse and field results are similar. With 5052 × 5062, 20% of F_2s in greenhouse experiments B and F and 25% in the 1980 Carimagua field trial had the high acid tolerance of 5062.

LITERATURE CITED

Andrew, C.S. 1974. An automatic subirrigation sand culture technique for comparative studies in plant nutrition. Lab. Pract. 23:20–21.

Clements, R.J., and R.J. Williams. 1980. Genetic diversity in *Centrosema*. *In* Summerfield and Bunting (eds.) Advances in legume science, 559–567. Roy. Bot. Gardens, Kew.

Schultze-Kraft, R., and D.C. Giacometti. 1979. Genetic resources for forage legumes for the acid, infertile savannas of tropical America. *In* P.A. Sanchez and L.E. Tergas (eds.) Pasture production in acid soils of the tropics, 55–64. CIAT, Cali, Colombia.

Evaluation of Within-Half-Sib Family Selection from Leo Birdsfoot Trefoil

D.T. TOMES, O.U. ONOKPISE, and B.E. TWAMLEY
University of Guelph, Guelph, Ontario, Canada

Summary

This research was undertaken to evaluate the effectiveness of within-half-sib family recurrent selection for seedling vigor, seed yield, and forage yield in birdsfoot trefoil (*Lotus corniculatus* L. cv. Leo). The effectiveness of selection was measured by comparing progeny performance, genetic variance, and heritability in selected and unselected populations. In addition, experimental synthetics (Syn-2 generation) from the selected population were compared to Leo for forage yield and seed yield in several trials.

Seedling vigor and yield in 1979 and forage yield in 1980 of polycross progeny in the selected population were significantly higher than those of similar progeny in the unselected population. Forage yield in 1979 of polycross progeny was similar in the selected and unselected populations. Seedling vigor of selfed progeny in the selected progeny was at a level similar to that of the polycross progeny of the unselected population. Although selfed-progeny performance in the selected and unselected populations was similar in magnitude for seed yield and forage yield, relative decline was greater in the selected population.

Estimates of genetic variance indicated significant and similar levels of additive variation for forage yield in row plots and nonsignificant dominance variation in both populations. Heritabilities were similar in the populations tested.

Both the cycle-3 and cycle-4 populations advanced to the Syn-2 generation were significantly higher yielding in the year of seeding (a measure of seedling vigor), similar in the second year, and significantly higher in the third year when compared with

Leo. Seed yield was below that of Leo for the cycle-3 population and above that of Leo for the cycle-4 population. Selections from cycle-4 for seedling vigor and seed yield were based on progeny testing. The seedling-vigor selection had significantly higher forage yield in the year of establishment than Leo, while the seed-yield selections had about 40% higher seed yield than Leo.

Lack of observed significant levels of dominance variation and greater inbreeding depression in the selected population suggest that additional populations of diverse genotypic background would be necessary to obtain optimum performance when elite genotypes are used as parents for new cultivars.

Performance of experimental synthetics from the selected population, which was significantly above that of Leo for forage yield in the establishment year and much better than Leo's for seed yield, is evidence of the success of this selection method in birdsfoot trefoil.

KEY WORDS: birdsfoot trefoil, *Lotus corniculatus* L., recurrent selection, progeny testing, seedling vigor, forage yield, seed yield, forage legumes.

INTRODUCTION

Birdsfoot trefoil (*Lotus corniculatus* L.) is an important forage legume in the U.S. and eastern Canada, but it suffers from poor establishment due to weak seedlings, lack of competitiveness, and erratic seed yield (Seaney and Henson 1970). Separate recurrent selection for seed yield and seedling vigor was initiated in 1969 from the cultivar Leo. The two populations were combined in 1976, and an additional cycle of selection including forage yield was carried out. From 1977 to 1980, we evaluated progress from selection to determine how much genetic variability remained in the population and to compare the performance of experimental synthetics with Leo's for forage and seed-yield potential.

MATERIALS AND METHODS

Within-half-sib family selection was practiced in each generation by testing polycross progeny for the trait(s) under selection. Phenotypic selection for robust plant type was restricted to those progeny that had above-average progeny performance for the trait(s) under selection.

The progress of selection and relative levels of genetic variability in the selected population and in an unselected Leo population were determined by examining the parent, selfed-progeny, and polycross-progeny performance of randomly chosen plants in the cycle-4 progeny and in a certified lot of Leo (Onokpise 1980). Seedling vigor (measured by top growth after 6 weeks), forage, and seed yield were measured, and proportions of additive and dominance genetic variation and heritability were calculated, using an autotetraploid model according to Onokpise (1980). In addition, genetic variance and heritability for forage yield were calculated in a similar manner for four diverse genotypes originally selected for a nectar-yield study but not selected for other agronomic traits (Murrell 1980).

Experimental synthetics consisted of the entire cycle-3 population for seedling vigor and the cycle-4 population increased two generations by random mating. Superior genotypes of the fourth cycle of selection identified by progeny testing for seedling vigor and seed yield, respectively, were intercrossed to form the Syn-1, and open-pollinated seed was harvested from the Syn-1. Forage yield was measured for 3 successive years in a 4-replicate, randomized complete-block design for the cycle-3 and cycle-4 populations and for the year of establishment in the progeny-tested synthetics. Seed yield was taken in 1979 in a 4-replicate, randomized complete-block design for cycle 3 and cycle 4 and in a 2-replicate row-plot trial for the two progeny-tested synthetics from cycle 4.

RESULTS

Seedling vigor and seed yield in 1979 and forage yield in 1980 were significantly higher in the polycross progeny of the selected population than in the unselected population. Selfed-progeny performance for seedling vigor of the selected population was similar to polycross progeny performance of the unselected population. The two populations had similar selfed-progeny seed yield and forage yield in 1979 and 1980, although the relative decline in performance was greater in the selected population (Table 1). Additive genetic variance was significant in both selected and nonselected populations for forage yield, with similar estimates of broad-sense heritability. Dominance variance was significant only in the genotypically diverse population (Table 2).

The experimental synthetics had significantly higher performance than Leo for forage yield in the year of seeding for cycle 3 and cycle 4 and for the cycle-4 seedling-vigor selections. Forage yield in the second year was similar to that of Leo, but it was significantly higher in the third year. Seed yield was relatively lower among selections for seedling vigor and higher for the cycle-4 and seed-yield selections in cycle 4 (Table 3).

DISCUSSION

The superior performance of the cycle-4 population for seedling vigor and seed yield (Table 1) compares with similar superior performance in earlier cycles of selection for seed yield (Sandha and Twamley 1973) and seedling vigor (Twamley 1974). The similar estimates of additive

D. T. Tomes is an Associate Professor at the University of Guelph; O. U. Onokpise is a Research Scientist with the Rubber Research Institute of Nigeria; and B. E. Twamley is an Associate Professor, Retired, from the University of Guelph.

Table 1.   Mean seedling vigor (1979), seed yield (1979), and forage yield (1979 and 1980) of parents, self, and polycross progenies of selected and unselected populations of birdsfoot trefoil.

Population	Progeny type	Number genotypes	Seedling vigor†	Seed yield	Forage yield 1979	Forage yield 1980
					----------g/plot----------	
Unselected	Parents	6	–	1.0a††	86a	156a
	Cross	6	1.24a	17.8b	296b	248b
	Self	6	.81b	7.9c	199c	208b
Selected	Parents	21	.71b	12.5b	282b	381c
	Cross	21	1.51c	20.5d	303b	354c
	Self	21	1.12a	5.8c	175c	293b
Check	Leo	–	1.36	18.0	240	391

†Seedling vigor measured by top growth after 6 weeks postseeding.

††Means within the same column followed by different letters are significantly different, T test (0.05 level).

genetic variance and heritability between selected and unselected populations (Table 2) suggest that a relatively high level of genetic variability remains in the selected population, as in earlier studies (Sandha et al. 1977, Twamley 1974). The combination of relatively greater inbreeding depression in the selected population (Table 1) and observed significant dominance variation only in a genotypically diverse population (Table 2) indicates that additional performance due to specific combining ability will be achieved only if parallel selection in genotypically distinct populations is carried out. With such parallel selection, elite genotypes from two or more populations can be identified as potential parents for synthetic cultivars whose performance would excel for both additive and dominance genetic interactions. However, the superior performance of experimental synthetics (Table 3) from the within-half-sib family-selection scheme indicates that progress has been made that would be expressed after at least two generations of seed increase. Within-half-sib family selection appears to be a worthwhile selection method for forage legumes such as birdsfoot trefoil.

LITERATURE CITED

Murrell, D.C. 1980. Nectar production and floral attractiveness to honey bees of varieties of birdsfoot trefoil (*Lotus corniculatus* L.). M.S. Thesis, Univ. of Guelph.

Onokpise, O.U. 1980. Evaluation of self (S1) and polycross progeny performance in two populations of birdsfoot trefoil (*Lotus corniculatus* L.) cultivar Leo. M.S. Thesis, Univ. of Guelph.

Sandha, G.S., and B.E. Twamley. 1973. Recurrent selection for seed yield improvement in *Lotus corniculatus*, cultivar Leo. Can. J. Pl. Sci. 53:811–815.

Sandha, G.S., B.E. Twamley, and B.R. Christie. 1977. Analysis of quantitative variability for seed yield and related characters in *Lotus corniculatus* L. cv. Leo. Euphytica 26:113–122.

Seaney, R.R., and P.R. Henson. 1970. Birdsfoot trefoil. *In* N.C. Brady (ed.) Advances in agronomy, 119–157. Academic Press, New York.

Twamley, B.E. 1974. Recurrent selection for seedling vigour in birdsfoot trefoil. Crop Sci. 14:87–90.

Table 2.   Estimates of genetic variance and broad-sense heritability (H) for forage yield in selected and unselected populations in birdsfoot trefoil.

Population	Additive	Dominance	H
	----------variance----------		%
Cycle 4	*	–	21
Leo (unselected)	*	–	32
Diverse††	*	*	45

*Statistically significant when estimate exceeded twice standard error.

††One genotype from cultivars Carroll, Leo, Maitland and germplasm source Wallace.

Table 3.   Forage and seed yield of experimental synthetics (Syn 2) from third and fourth cycle of recurrent selection in birdsfoot trefoil.

Cycle of selection	Selection criterion	Year of seeding	Forage yield First	Forage yield Second	Forage yield Third	Seed yield
			----------%----------			
Cycle 3	Combined	1976	112†	105	120*	
	Combined	1978	142*	86	121*	70
Cycle 4	Combined	1978	130*	93	109*	116
	Seed vigor	1980	104*	–	–	120
	Seed yield	1980	99	–	–	143

†Data shown as percentage of Leo check in each trial.

*Significantly above Leo check in the respective tests (LSD at 0.05 level).

Evolution of Selection Techniques in Breeding for Bloat-Safe Alfalfa

B.P. GOPLEN, R.E. HOWARTH, G.L. LEES,
W. MAJAK, J.P. FAY, and K.-J. CHENG
Agriculture Canada Research Stations,
Saskatoon, Kamloops, and Lethbridge, Canada

Summary

This report summarizes the Canadian breeding program to develop a bloat-safe alfalfa (*Medicago sativa* L.). This program has experienced an evolution of selection methods in attempting to identify bloat-safe alfalfa genotypes. In chronological order, the methods involved selection for fraction I (18S) protein, soluble proteins, foam volume, tannins (flavolans), saponins, initial rate of digestion (IRD), sonication, purified enzymes, gas production, and leaching.

Although soluble herbage proteins are correlated with the occurrence of pasture bloat, we have experienced slow progress in a selection program to reduce soluble-protein concentration in alfalfa. We have found that foam volume is not correlated with either soluble protein or bloat incidence. After an extensive search of 33 annual and 25 perennial species of *Medicago*, we have found no tannins. Similarly, a mutation approach to induce flavolan production in alfalfa has failed to reveal tannin (bloat-safe) phenotypes. Seedcoats of alfalfa contain tannins. The saponin theory of bloat has been proven to be invalid.

A comparison of bloat-safe and bloat-causing forage legumes revealed slower initial rates of digestion and lower concentrations of intracellular leaf constituents in rumen fluid when bloat-safe legumes are ingested. Selection is under way to develop contrasting alfalfa synthetics, one with a low initial rate of digestion (bloat-safe synthetic) and another with a high initial rate of digestion. Preliminary data indicate genetic progress for change in initial rate of digestion with no change in overall digestibility as measured by in-vitro dry-matter digestibility.

Several new complementary techniques are under development and evaluation to select bloat-safe alfalfa plants. They are all devised to measure basic parameters involved in the etiology of bloat (e.g., cell-wall strength, enzymatic breakdown of cells, microbial degradation, gas production, leaching).

KEY WORDS: pasture bloat, alfalfa, *Medicago sativa* L., cattle, selection techniques, breeding.

INTRODUCTION

Alfalfa (*Medicago sativa* L.) is a remarkable forage crop with very wide adaptation, high productivity, and unsurpassed nutritional qualities. However, its propensity for causing pasture bloat often limits its use in pastures, especially under extensive grazing conditions where conventional methods of prevention or treatment of bloat are not practical. The development of a bloat-safe cultivar would be a major breakthrough leading to large annual economic savings from bloat losses and would also lead to major expansion in pasture acreage of this highly productive, nitrogen-fixing legume crop.

There has been no shortage of theories on the cause of pasture bloat. General agreement exists that this type of bloat is caused by the formation of intraruminal foam that inhibits the eructation of gases produced by microbial fermentation. Research on the foaming agents has implicated plant proteins, saponins, salivary mucoproteins, and lipids. More recent research has indicated cell-wall rupture and initial rate of digestion as important factors in the etiology of bloat (Howarth et al. 1982). Thus, alfalfa-breeding programs have experienced an evolution of selection techniques in efforts to identify bloat-safe alfalfa genotypes. A major effort has been devoted by Agriculture Canada to produce a bloat-safe alfalfa cultivar, and this report primarily summarizes that breeding program since its inception in 1970. The selection techniques involved are reviewed and discussed in chronological order.

SELECTION TECHNIQUES

Fraction I (18S) Protein

The work of McArthur and Miltimore (1966) implicated fraction I as the main soluble-protein fraction

B.P. Goplen is a Senior Research Scientist in Genetics, R.E. Howarth is a Senior Research Scientist in Biochemistry, and G.L. Lees is a Research Scientist in Physiology at the Saskatoon Station; W. Majak is a Research Scientist in Biochemistry at Kamloops; and J.P. Fay is a Postdoctoral Fellow in Rumen Microbiology and K.-J. Cheng is a Senior Research Scientist in Rumen Microbiology at Lethbridge.

The current address of J.P. Fay is Departamento de Produccion Animal, Estacion Experimental, INTA, 7620 Balcarce, PCIA Buenos Aires, Argentina.

responsible for pasture bloat. This role of fraction I was supported by Stifel et al. (1968). However, the exclusive fraction I theory was criticized by Jones and Lyttleton (1972), and reinvestigation of the foaming properties of alfalfa proteins implicated both fraction I and fraction II as factors responsible for bloat (Howarth et al. 1973a, 1977). Thus, the fraction I theory of bloat was discarded and further screening for fraction I was suspended in 1973 in favor of screening for the soluble proteins (fraction I plus fraction II).

Soluble Proteins

Howarth et al. (1973b) developed a rapid and reliable test for soluble proteins (SP) based on the colorimetric biuret method. This method was used from 1972 to 1977 in a routine screening program to develop contrasting alfalfa populations of low SP content (bloat-safe) and high SP content (to serve as a check on genetic progress). Progress was slow in developing low-SP and high-SP alfalfa populations, as was expected from relatively low (h^2 = 0.23–0.31) regression heritability estimates for soluble protein (Gutek et al. 1976). After two cycles of selection, the mean SP values (one-cut, dry-matter basis) of these two polycross-progeny populations were: low-SP progenies, 11.07%, 1,500 plants analyzed; high-SP progenies, 13.76%, 1,300 plants analyzed.

Concurrent investigations of the nitrogen and protein fractions and their relationship to ruminant bloat (Howarth et al. 1977) led to the conclusion that at least a 50% reduction in SP concentration would be required to develop a bloat-safe alfalfa cultivar. This breeding program of selection based on SP was suspended in 1980 when the approach based on cell-wall rupture and initial rate of digestion looked more promising.

Foam Volume

Working with a variety of forage species, Kendall (1966) observed that more foam was produced by bloat-causing forages than by bloat-safe forages. Cooper et al. (1966) reached the same conclusion after examining 27 herbaceous species. Rumbaugh (1969) simplified the foam-volume technique and, working with alfalfa, found a high heritability (h^2 = 0.73) for foam volume, indicating the feasibility of breeding a bloat-safe cultivar by selecting for low foam volume. However, because experiments in our laboratory indicated no correlation between foam volume and soluble protein (Goplen and Howarth 1977), and none between foam volume and bloat incidence (Howarth et al. 1976), the foam-volume theory of bloat was considered invalid.

Tannins

Kendall (1966) first suggested that tannins in nonbloating legume forages may be responsible for preventing pasture bloat. Experimental evidence indicates that condensed tannins prevent bloat by acting as protein precipitants (Goplen et al. 1980). Sarkar and Howarth (1976) introduced a control reagent to ensure

that the vanillin-HCl screening test is specific for condensed tannins. Screening of 33 annual species and 28 perennial species of *Medicago* was negative. In addition, extensive screening of common varieties, breeding populations, and 2n and 4n populations of mutagen-treated *M. falcata* and *M. sativa* revealed no plants with tannins. Tannins were absent in vegetative tissues, but Goplen et al. (1980) found condensed tannins in the seed-coat of all common alfalfa seeds examined, indicating that the genetic mechanism for tannin production may already exist in alfalfa.

Saponins

Although herbage proteins are generally regarded as the main foaming agents in pasture bloat, saponins have often been suggested (Cheeke 1976) as secondary actors, either as foaming agents or as the cause of toxic inhibition of reticulo-rumen activity. The role of saponins in pasture bloat was re-examined by feeding rumen-fistulated cattle fresh herbage of high-saponin and low-saponin near-isogenic strains of alfalfa. No difference in bloat incidence was found, and it was concluded that the saponin theory of bloat was invalid (Majak et al. 1980).

Initial Rate of Digestion

Recent experiments have provided strong evidence that disruption of leaf tissues and the consequent release of intracellular constituents (foaming agents) are of primary significance in the etiology of pasture bloat (Howarth et al. 1978, 1979). Thus a nylon bag (nb) digestion screening technique was developed (Howarth et al., unpublished) to select alfalfa plants with slower initial rates of digestion (IRD). This selection technique is the main screening method now being used to develop a bloat-safe cultivar at Saskatoon. It is regarded as the most reliable selection test to date since it measures relative digestion rates in vivo, combining all the factors involved in ruminant digestion (e.g., mechanical disruption, enzymatic breakdown, microbial degradation). Experiments with cattle indicate a moderate association between IRD and daily bloat incidence, supporting the validity of this nbIRD screening technique. It should be noted that assessment by the Saskatchewan Feed Testing Laboratory showed the low-IRD and high-IRD selections were equal in in-vitro dry-matter digestibility. It is estimated that a 20%–30% reduction in the IRD of alfalfa will be required to produce a bloat-safe cultivar.

Sonication

Studies based on a comparison of some bloat-safe and bloat-causing legumes have shown that the ease of mechanical cell disruption may be one of the factors contributing to increased bloat potential in a plant (Lees et al. 1981). A method of selecting alfalfa plants having a higher resistance to mechanical cell disruption has been developed and is currently being evaluated for use in our breeding program. In this procedure, whole leaves are sonicated, and cell disruption is determined by comparing chlorophyll released during sonication to chlorophyll originally present in the leaves.

Enzyme Digestion

An in-vitro laboratory procedure has been developed (Lees, unpublished) for preliminary evaluation of alfalfa seedlings for resistance to cell disruption by enzymes. For this assay, disks of leaf tissue are incubated with a combination of pectin and cellulose-degrading enzymes. Digestion rates are determined by measuring the relative amount of chlorophyll lost from the disks.

Gas Production

Using a gas-production technique, Fay et al. (1980) showed that the bloat-causing legume species produced more gas than the bloat-safe species. However, in order to be useful as a bioassay to differentiate between bloat-safe and bloat-causing selections in a plant-breeding program, the method needs further modifications to reduce its inherent variability.

Leaching

Leaching of intracellular constituents of plant cells appears to be an important factor influencing the rate of rumen bacteria colonizing the leaf surface (Fay et al. 1981) by way of a chemotactic response. Thus, the extent of dry-matter loss by leaching may influence the initial rate of digestion. A leaching bioassay has been developed (Fay et al., unpublished) that measures dry-matter loss under anaerobic conditions.

DISCUSSION AND CONCLUSIONS

Research in our laboratory from 1970 to 1980 has led to the overthrow of a number of theories on the causes of pasture bloat—the fraction I theory, the foam-volume theory and the saponin theory. New techniques on selection for bloat-safe alfalfa have evolved in concert with our most recent experimental findings on the cause(s) of pasture bloat. The nylon bag digestion screening technique is regarded as our most valid current method of selecting for bloat-safe genotypes. Some of the newer techniques now under development and evaluation include sonication, enzymatic breakdown, gas production, and cell leaching. It is anticipated that these new techniques may be useful for initial screening or as complementary tests but that the nylon bag digestion technique will continue to be the main screening method because it combines all the causal factors in bloat resistance in the natural rumen.

LITERATURE CITED

Cheeke, P.R. 1976. Nutritional and physiological properties of saponins. Nutr. Rep. Int. 13:315–324.

Cooper, C.S., R.F. Eslick, and P.W. McDonald. 1966. Foam formation from extracts of 27 legume species *in vitro*. Crop Sci. 6:215–216.

Fay, J.P., K.-J. Cheng, M.R. Hanna, R.E. Howarth, and J.W. Costerton. 1980. *In vitro* digestion of bloat-safe and bloat-causing legumes by rumen microorganisms: gas and foam production. J. Dairy Sci. 63:1273–1281.

Fay, P.J., K.-J. Cheng, M.R. Hanna, R.E. Howarth, and J.W. Costerton. 1981. A scanning electron microscopy study of the invasion of leaflets of a bloat-safe and a bloat-causing legume by rumen microorganisms. Can. J. Microbiol. 27:390–399.

Goplen, B.P., and R.E. Howarth. 1977. Breeding a bloat-safe alfalfa (*Medicago sativa* L.) cultivar. Proc. XIII Int. Grassl. Cong., 355–358.

Goplen, B.P., R.E. Howarth, S.K. Sarkar, and K. Lesins. 1980. A search for condensed tannins in annual and perennial species of *Medicago*, *Trigonella*, and *Onobrychis*. Crop Sci. 20:801–804.

Gutek, L.H., B.P. Goplen, and R.E. Howarth. 1976. Heritability of soluble proteins in alfalfa. Crop Sci. 16:199–201.

Howarth, R.E., K.-J. Cheng, J.P. Fay, W. Majak, G.L. Lees, B.P. Goplen, and J.W. Costerton. 1982. Initial rate of digestion and legume pasture bloat. Proc. XIV Int. Grassl. Cong.

Howarth, R.E., B.P. Goplen, J.P. Fay, and K.-J. Cheng. 1979. Digestion of bloat-causing and bloat-safe legumes. Ann. Rech. Vét. 10:332–334.

Howarth, R.E., B.P. Goplen, A.C. Fesser, and S.A. Brandt. 1978. A possible role for leaf cell rupture in legume pasture bloat. Crop Sci. 18:129–133.

Howarth, R.E., B.P. Goplen, W. Majak, and D.E. Waldern. 1976. Relationships between bloat and proteins in alfalfa herbage. Proc. Fed. Am. Soc. Exp. Biol. 35:580.

Howarth, R.E., J.M. McArthur, and B.P. Goplen. 1973a. Bloat investigations: determination of soluble protein concentration in alfalfa. Crop Sci. 13:677–680.

Howarth, R.E., J.M. McArthur, M. Hikichi, and S.K. Sarkar. 1973b. Bloat investigations: denaturation of alfalfa Fraction II proteins by foaming. Can. J. Anim. Sci. 53:439–443.

Howarth, R.E., W. Majak, D.E. Waldern, S.A. Brandt, A.C. Fesser, B.P. Goplen, and D.T. Spurr. 1977. Relationships between ruminant bloat and the chemical composition of alfalfa herbage. I. Nitrogen and protein fractions. Can. J. Anim. Sci. 57:345–357.

Jones, W.T., and J.W. Lyttleton. 1972. Bloat in cattle. XXXVI. Further studies on the foaming properties of soluble leaf proteins. N.Z. J. Agric. Res. 15:267–278.

Kendall, W.A. 1966. Factors affecting foams with forage legumes. Crop Sci. 6:487–489.

Lees, G.L., R.E. Howarth, B.P. Goplen, and A.C. Fesser. 1981. Mechanical disruption of leaf tissues and cells in some bloat-causing and bloat-safe forage legumes. Crop Sci. 21:441–448.

McArthur, J.M., and J.E. Miltimore. 1966. Pasture bloat and the role of 18S protein. Proc. X Int. Grassl. Cong., 518–521.

Majak, W., R.E. Howarth, A.C. Fesser, B.P. Goplen, and M.W. Pedersen. 1980. Relationships between ruminant bloat and the composition of alfalfa herbage. II. Saponins. Can. J. Anim. Sci. 60:699–708.

Rumbaugh, M.D. 1969. Inheritance of foaming properties of plant extracts of alfalfa. Crop Sci. 9:438–440.

Sarkar, S.K., and R.E. Howarth. 1976. Specificity of the vanillin test for flavanols. J. Agric. Food Chem. 24:317–320.

Stifel, F.B., R.L. Vetter, R.S. Allen, and H.T. Horner, Jr. 1968. Chemical and ultrastructural relationships between alfalfa leaf chloroplasts and bloat. Phytochemistry 7:335–364.

Development of Gama Medic (*Medicago rugosa* Desr.) as an Annual Leguminous Species for Dryland Farming Systems in Southern Australia

E.J. CRAWFORD

Department of Agriculture, South Australia

Summary

Annual medics (*Medicago* spp.) are the basic legume for nitrogen fixation in the alkaline soils of southern Australia. Not all individual species are well adapted to all soil types, one difficult soil being the highly alkaline clay loam where annual rainfall is less than 500 mm.

The first gama medic (*M. rugosa*) cultivar, Paragosa, was well adapted to the clay-loam soil type but proved to be susceptible to germination after summer thunderstorms, and seedlings died in the following drought. Many gama medic accessions were then grown in an attempt to select a genotype capable of maintaining seedcoat impermeability during the summer but having enough permeable seed by mid-April to allow dense regeneration. Other essential criteria included good winter herbage production, earliness of flowering, and high seed yield. These accessions were grown to maturity in the field nursery and sampled at regular intervals over the summer and autumn to measure changes in seedcoat permeability. Subsequent sward sowings in different environments allowed measurement of the other criteria.

Screening of 1,364 genotypes of 13 species revealed the variability of gama medic compared with other species. Although gama medic from the eastern Mediterranean region maintained a high level of seedcoat impermeability, accession types from Greece, Italy, and Portugal displayed a range of changes between maturity and mid-April. Evaluation of this and other criteria resulted in the development of the cultivar Paraponto.

Species other than those previously recognized may have application to specific environments. A more intensive search for, and evaluation of, genotypes of relatively rare species, such as gama medic, as well as other potentially useful species could result in more productive and better adapted cultivars in the future.

KEY WORDS: gama medic, *Medicago rugosa* Desr., alkaline clay loams, seedcoat permeability, herbage production, seed production, flowering time.

INTRODUCTION

Annual medics (*Medicago* spp.) play an important role as nitrogen fixers in the pasture phase of cereal rotations in the Mediterranean climate zone of southern Australia. They are adapted to the wide expanse of alkaline soils in the 250–500-mm mean annual rainfall regions. Species such as *M. minima* (L.) Bart., *M. polymorpha* L., and *M. truncatula* Gaertn. are naturalized and widespread but either are not well adapted to all environments or have some undesirable attribute, such as spiny seedpods, which become a vegetable fault in wool.

Gama medic (*M. rugosa* Desr.) is an uncommon species. In the world's largest collection of annual medics, now exceeding 10,000 genotypes (the South Australian collection), only 100 samples from 13 countries are gama medic. This rareness may be partly because gama medic has spineless pods that restrict its mode of dispersion

The author is Senior Plant Introduction Officer in the Department of Agriculture, South Australia.

(Heyn 1963). However, in its native habitat its distribution is limited to heavy-textured soils, which suggests a requirement for such soils (Heyn 1963, Anon. 1967).

Banyer (1966) has shown that gama medic is particularly well adapted to the grey and black self-mulching clay loams of cereal-growing areas in South Australia. The cultivar Paragosa (Barnard 1972) was selected from 2 genotypes available from Portugal at that time.

Paragosa seed proved to be too permeable in summer in the South Australian environment, high seedling populations being lost through premature germination following summer rains (Crawford 1977). The need for seedcoat impermeability in summer and autumn followed by a rapid change to permeable seed by early winter became evident.

From the screening of 1,364 genotypes of thirteen annual medic species between 1968 and 1971, Crawford (1976) reported that gama medic was the only species consistently to have shown between-genotype variability in seedcoat permeability; indeed, in one study, Crawford (1977) showed that 48% of the 79 *M. rugosa* genotypes tested had 30% permeable seed by mid-April of the year after seed production. Although only 50 genotypes of gama medic were available by 1971, these were compared agronomically with Paragosa in an attempt to develop a cultivar better able to adapt to the local environment. Such a cultivar could extend the use of gama medic to wider climatic and edaphic horizons in southern Australian farming systems.

METHOD

Seeds of 50 genotypes from four major regions of the Mediterranean Basin (Greece, Israel, Italy, and Portugal) were used in the program. Under the principles of evaluation outlined by Crawford (1979), two experiments were conducted on heavy-textured, clay-loam soils in the 400–500-mm mean annual rainfall regions of South Australia.

Experiment 1. Small replicated swards were sown at 22.4 kg/ha of scarified seed, the high sowing rate ensuring a good sward in the first year. All genotypes were allowed to grow unimpeded, and total spring herbage production and seed production were measured in the year of establishment. Plant populations were counted after germinating rains in late summer and autumn of the second year of the experiment and again after winter rains. Dead seedlings (resulting from early germination) were counted where appropriate. Flowering time was compared to that of the control cultivar, Paragosa. Seasonal herbage production during the second winter and spring gave a measure of capacity to recover after defoliation.

Experiment 2. As a result of intensive selection for a combination of high seedcoat impermeability during the summer and early autumn with low impermeability in late autumn and early winter, good early-winter vigor, early maturity, and the capacity to produce high seed yields, 3 genotypes were chosen for further evaluation. They were sown at 10 kg/ha of scarified seed and subjected to a pasture-fallow-cereal-pasture rotation over 3 consecutive years along with the control cultivar, Paragosa. This experiment was grazed in the preflowering and postmaturity stages, the sheep being used as defoliators rather than to measure animal production. Plant populations and herbage and seed production were determined throughout the life of the experiment.

RESULTS AND DISCUSSION

Natural Regeneration

In experiment 1, late-summer rain resulted in varying populations of up to 1,500 plants/m² in the year after sowing, many seedlings in the denser populations dying before followup rains were received. Midautumn followed by early-winter rains resulted in further populations of up to 3,700 plants/m², the low figures being genotypes with impermeable seed at both the late-summer and the early-winter rain period and the high figures being genotypes with low permeable-seed levels at the time of late-summer rains and high permeable-seed levels by early winter.

Fourteen, including all 8 Portugese genotypes had more permeable seed (> 500/m²) by late summer than by early summer and performed similarly agronomically to Paragosa, while 4 from Greece and Italy remained mainly impermeable and germinated densely in early winter (> 1,900 plants/m²). All 22 Israeli genotypes remained impermeable through both the late-summer and the early-winter rain period. A range of seedcoat permeabilities was evident in genotypes from the isles of the Greek archipelago.

In experiment 2, summer and autumn rain was insufficient to bring about a germination of the only summer-permeable-seeded genotype, Paragosa. However, following winter rainfall, the 2 autumn-permeable-seeded genotypes regenerated with 32% and 34% of their available seed, respectively, while the third, more impermeable-seeded genotype, had only a 4% regeneration rate. Paragosa's rate was 31%. After fallowing and cropping, regeneration of the surviving seeds resulted in populations of 210–240 plants/m² compared with 170 plants/m² for Paragosa. These plants represented from 4% to 16% of the initial seed reserves produced 3 years earlier, the remainder being lost during fallowing and cropping.

Resowings in the second year failed to survive 2 consecutive years of drought. The third year of sowing also proved to be a drought year, resulting in very low seed yields. However, populations of the 3 genotypes ranged from 220 to 310 plants/m², compared with 30 for Paragosa—a range of 53% to 58% of the available seed produced in the previous year, compared with Paragosa's 8%. This difference was because half the seed reserves of Paragosa were lost during postmaturity grazing compared with 27% to 31% for the other 3 genotypes. Regeneration following the fallow and cereal years resulted in very low populations of 13 to 22 plants/m² of all 4 genotypes; however, these plants grew vigorously through winter and set sufficient seed to maintain pasture stability.

Herbage Production

Herbage yield was measured with particular emphasis on early-winter production. Although there were significant differences in total production between genotypes in experiment 1 in the year of sowing, and in winter production in the subsequent year of regeneration, only one yielded better than Paragosa in the second year. Following regeneration in the next year, 2 genotypes from the Greek archipelago outyielded Paragosa, but these had produced poorly in the previous year due to low plant numbers, a result of low seed permeability.

In experiment 2, herbage yield was not measured in the year of sowing since only high-yielding genotypes had been selected for further testing.

The second year, the year of regeneration, proved to be a drought year, and no genotype significantly outyielded Paragosa. After consecutive fallow and crop years the 3 selected genotypes significantly outyielded Paragosa both in the first 9 weeks after germination (200–330 kg/ha compared with Paragosa's 20 kg/ha) and in the subsequent 6 weeks following grazing (280–400 kg/ha compared with Paragosa's 100 kg/ha). Where initial seed set was low due to drought, there were no significant differences between genotypes in early winter production.

Flowering Time

Flowering time within the 50 genotypes ranged from 28 days before to 32 days after Paragosa, which flowered 116 days after germination. Crawford (1975) reported that flowering within 125 days of germination was optimal for southern Australia, as is indicated by the mean range of 102–124 days for the existing commercial cultivars.

Seed Production

In experiment 1, significant differences in seed yields were obtained, but no genotype was higher yielding than Paragosa. Seed yields ranged from 575 to 1,337 kg/ha (least significant difference [LSD] = 225, $P < 0.05$).

In experiment 2, the 3 selected genotypes (831–1,006 kg/ha) significantly outyielded Paragosa (651 kg/ha) (LSD = 169, $P < 0.05$) in the first year; there were significant differences, but none performed better than Paragosa in the second year, while in the third year all 3 again significantly outyielded Paragosa, both the second and the third year being drought years.

CONCLUSIONS

Although South Australia enjoys a Mediterranean-type climate, the incidence of summer rain is greater than it is where gama medic is native. This variability can be seen when comparing ombrothermic diagrams from Adelaide (South Australia), Lisbon (Portugal), Athens (Greece), and Tel Aviv (Israel), important sites of genetic variability in gama medic (Anon. 1963). Seedcoat impermeability in summer and early autumn followed by a rapid breakdown by early winter was evident within 4 genotypes from Greece and Italy.

Although none of the genotypes significantly outyielded Paragosa in early-winter herbage production, 14 yielded

an equal amount in the year of regeneration in the first experiment. In the second experiment, the 3 selected genotypes were superior to Paragosa in early-winter herbage production and in recovery after defoliation.

The range of flowering time and its impact on seed production allowed ample scope for selection for a range of environments.

Two genotypes excelled in seed production, early-winter herbage production, and capacity to regenerate densely by early winter. One, however, had too high a level of permeable seed in late summer to satisfy the objective of impermeability in late summer and permeability by early winter. As a result of this program one genotype from Italy has been released as a new cultivar named Paraponto. Fortunately, in common with other gama medics, it is tolerant of the recently arrived legume aphids, resistance to which is now a major criterion in annual medic selection programs.

LITERATURE CITED

Anon. 1963. Bioclimatic map of the Mediterranean Zone. UNESCO-FAO, Paris and Rome.

Anon. 1967. Commonwealth Plant Introductions. Pl. Intro. Rev., CSIRO, Canberra, Austral. 4:2, 91a–92a.

Banyer, R.J. 1966. Paragosa gama medic—a superior medic for heavy black soils. So. Austral. J. Agric. 70:35–39.

Barnard, C. 1972. Register of Australian herbage plant cultivars. Div. Pl. Indus., CSIRO, Canberra, Austral.

Crawford, E.J. 1975. Time of flowering in the annual species of the genus *Medicago*. Australasian Pl. Breed. Genet. News 25:76–79.

Crawford, E.J. 1976. The search for permeable seed in species of the genus *Medicago* L. Austral. Seed Sci. News 2:21–24.

Crawford, E.J. 1977. Changes in seedcoat permeability in annual species of *Medicago* with special reference to the variability in *M. rugosa* Desr. Austral. Seed. Res. Conf., Canberra, Austral. 2(a):18–21.

Crawford, E.J. 1979. Evaluating annual *Medicago* species. Proc. Genet. Res. Forage Pl. Sym.

Heyn, C.C. 1963. The annual species of *Medicago*. Scripta Hierosolymitana XII, Magnes Press, The Hebrew Univ., Jerusalem.

Productivity of Forage Legumes on Rice-Paddy Walls in Northeast Thailand

R.C. GUTTERIDGE

Khon Kaen University, Khon Kaen, Thailand

Summary

Rice stubble left standing in paddy fields after rice harvest is the major source of forage for livestock in the dry season in northeast Thailand. Although the quantity of this stubble is generally not limiting, its quality as a forage is very low, and substantial live-weight losses in livestock usually occur by the end of the dry season.

A trial was conducted to evaluate a range of forage legumes that could be grown on rice-paddy walls to supplement rice stubble as a dry-season feed. Seven legumes were planted at two sites in villages close to Khon Kaen in northeast Thailand. The dry-matter yield and nitrogen content of each species were measured in 2 consecutive years just prior to the rice harvest. In both years, Shrubby stylo (*Stylosanthes scabra* cv. Seca), Verano stylo (*S. hamata* cv. Verano), and Viscosa stylo (*S. viscosa*) were the most productive species, with mean dry-matter yields of 4,800, 3,320, and 2,340 kg/ha/yr, respectively. A simple cultivation treatment improved legume establishment and growth, particularly in the first year. Nitrogen (N) contents ranged from 0.8% N in Shrubby stylo to 3% N in leucaena (*Leucaena leucocephala*). Although the yield of Shrubby stylo was the highest, a large proportion of its yield was inedible thick-stem material, so its effective yield was considerably lower. Verano stylo was considered the most suitable legume for growing on paddy walls. It established and regenerated easily, persisted well, was readily eaten by livestock, and substantially increased the quality and quantity of fodder available to livestock from the paddy walls.

KEY WORDS: rice stubble, forage legumes, *Stylosanthes* spp., Verano stylo, Shrubby stylo, paddy walls, northeast Thailand.

INTRODUCTION

The northeast region of Thailand has the highest population of ruminant livestock in the country (57% of the nation's 4.7 million cattle and 45% of its 5.4 million water buffalo). Most of these animals are raised in small numbers by small-holder village farmers, mainly for draft purposes. Because farmers are primarily crop growers, livestock raising is always of secondary importance, and inputs under the traditional system are minimal since ruminants can graze crop residues and unimproved natural pasture.

There are distinct seasonal management patterns that influence livestock feed supply. During the wet season (May-November) much of the rice-paddy land is cropped, and ruminants are grazed on communal land, along roadsides and forest margins, and on private land under fallow. During the dry season (December-April) they are herded onto the paddy fields where they graze rice stubble left standing after rice is harvested.

Although the quantity of rice stubble is generally not limiting, its quality as a forage is very low, and livestock usually lose weight by the end of the dry season (Rufener 1971). Because of this annual weight loss, the dry-season forage supply has become a focal area for improvement, and several simple techniques have been investigated to try to improve feed quality.

One such technique involves sowing forage legumes on the paddy walls (or bunds) at the time of rice transplanting and allowing the leguminous material to be grazed in conjunction with the rice stubble after harvest. This paper reports on the productivity of a range of forage legumes established in this manner.

METHODS

Trials were conducted at two sites in village paddy areas close to Khon Kaen, northeast Thailand (lat 16°20'N long 102°50'E, altitude 152 m), where the climate is tropical savanna with a mean annual rainfall of 1,150 mm, 85% of which falls mid-April to mid-October

The author is a Research Agronomist with the Khon Kaen Pasture Improvement Project.

through the influence of the southwest monsoon.

At both sites the soils were classified as low humic gleys of the Roi Et series, which are widely used for paddy rice production in northeast Thailand (Moorman et al. 1964).

At both sites the legume species Verano stylo (*Stylosanthes hamata* cv. Verano), Endeavour stylo (*S. guianensis* cv. Endeavour), Schofield stylo (*S. guianensis* cv. Schofield), Shrubby stylo (*S. scabra* cv. Seca), *S. viscosa* CPI 34904, leucaena (*Leucaena leucocephala*), and siratro (*Macroptilium atropurpureum* cv. Siratro) were sown into either cultivated or noncultivated plots 4 m in length. Cultivation was the removal of the native grasses and weeds from the paddy wall using a chip hoe with minimal soil disturbance. Verano, Endeavour, Schofield, and siratro were sown in mid-July 1978 directly onto the plots at a rate of 15 kg seed/ha. Because of limited seed supplies, Shrubby stylo, *S. viscosa*, and leucaena were first sown in small pots in a greenhouse and then transplanted to the sites at the time of sowing the other species. Seedling plant density of seed-sown species was recorded 4 weeks after sowing. As rice had already been transplanted into the adjacent paddies at the time of sowing, all plots were not grazed until after the rice harvest in late November 1978, allowing a 5-month "grazing-free" establishment period. Just prior to rice harvest the plots were sampled for dry-matter yield of legumes and other species. All plots were heavily grazed over the dry season, and on several occasions individual species were rated for grazing performance.

Plant density of all legumes was recorded in June 1979, and dry-matter yield of legumes and associated species was again measured prior to rice harvest in November 1979. At both harvest times the naturalized species outside the plots were also sampled for dry-matter yield. The dried legume material from both sites for both years was analyzed for nitrogen content, and in the second year rice stubble adjacent to the plots was also sampled and analyzed for nitrogen.

RESULTS

Establishment

All legumes established well except for siratro and leucaena. Verano stylo had the highest mean seedling density, 110 seedlings/m² at site 2, while siratro had the

Table 1. Dry matter yield (kg/ha) over two years of seven forage legumes and other species grown on cultivated and on uncultivated paddy walls.

Species	Site 1				Site 2				Forage species mean	
	Cultivated		Uncultivated		Cultivated		Uncultivated			
	1978	1979	1978	1979	1978	1979	1978	1979	1978	1979
Verano	2280	2570	1330	1480	2810	9140	820	6160	1810	4840
Endeavour	260	90	450	50	1380	750	450	960	640	460
Schofield	1640	590	940	670	2290	1770	870	690	1430	930
Shrubby style	4830	5000	1340	7150	3640	5640	5160	4670	3740	5620
Viscosa	2070	4000	1080	2200	2690	1610	2280	1780	2030	2400
Siratro	30	0	50	0	520	990	120	910	180	470
Leucaena	20	0	20	0	350	450	260	1040	160	370
Legume (mean)	1590	1750	1030	1650	1950	2910	1420	2320		
Other species (mean)	1280	1140	1530	1200	1210	4100	2350	3830		

lowest density, 8 seedlings/m² at site 1. At both sites there was a trend toward higher seedling density in the cultivated treatments.

In 1979, the three most productive legumes exhibited three characteristic patterns with respect to plant density. Verano stylo had a high density of seedlings (140 seedlings/m²) but only a low percentage of mature plants surviving from the previous year (12 plants/m²). With Shrubby stylo almost 100% of the mature plants survived into the second year, but there were very few seedlings. *S. viscosa* had a high seedling density (65 seedlings/m²) and high proportion of perennating plants, and the remaining species had low densities of both seedlings and surviving plants.

Dry-Matter Yields

The three species, Shrubby stylo, Verano, and *S. viscosa*, were the most productive legumes over the 2-year period (Table 1). In the first year, yields of all three species were high, and they increased in the second year, especially that of Verano at site 2. In general, the yields of siratro and leucaena were poor; both these species disappeared from site 1 in the second year. Cultivation improved legume yields, particularly of those species sown as seed, the effect being more pronounced in the first than in the second year.

Yields of native grasses and weeds outside the plots were only slightly higher than their yields inside the plots in both years, indicating a direct additive benefit to dry-matter yield from legume oversowing. Cultivation reduced the yield of native species at both sites in the first year, but they recovered by the second year.

Nitrogen Content

In both years the nitrogen (N) content of most legumes was relatively low. This low level may have been associated with the time of sampling, as most species had ceased growing and nitrogen content generally decreases

as plants reach maturity. The exception was leucaena, which was still growing and had the highest nitrogen content (3% N) at site 2 in the first year. Shrubby stylo had the lowest nitrogen content (0.8% N). Most of the other species averaged 1.2% to 1.5% N in the first year. In the second year Verano and *S. viscosa* had the lowest nitrogen content (1.1% N) at site 1, and leucaena was again the highest, with 2.7% N at site 2.

The mean nitrogen content of rice straw was similar at the two sites (0.33% N), indicating rice straw's poor quality as a sole livestock diet.

Grazing Preference

Although all plots at both sites were heavily grazed through the 1978–1979 dry season, some preferential grazing effects were noticeable, particularly early in the season. Leucaena was always most favored, followed, in order of preference, by Verano, Schofield, siratro, Endeavour, native grasses, and Shrubby stylo. *S. viscosa* was grazed only very late in the dry season when all other species had been virtually grazed out. Only the leafy material and finer stems of Shrubby stylo were grazed, and tall, thick stems were left standing in the plots.

DISCUSSION

Verano stylo was the most suitable legume for oversowing onto paddy walls. It was high-yielding, it regenerated easily and persisted well, and it was accepted well by livestock. The combined yield of Verano and native species from a plot was on average 2 to 3 times higher than the yield of native species outside the plot, and, assuming that the protein content of the leguminous fodder was at least 1.5 times that of the native species, both quantity and quality of forage available from paddy walls could be increased by oversowing with Verano. Verano was more reliant on seed reserves for regeneration and persistence than on survival of mature plants. This characteristic is perhaps an advantage in a situation where exceptionally heavy grazing or adverse environmental conditions during the dry season may seriously reduce plant survival, even of well-adapted species. Seed reserves would not be affected by adverse conditions, especially if flowering and seed set were well advanced before the onset of grazing. This is the case with Verano, as is evidenced by the large number of seedlings recorded in the plots early in the 1979 wet season.

In the assessment of grazing preference, Verano was the most favored species after leucaena. There were no acceptability problems, and virtually all the plant was consumed by livestock. In addition, Verano, in contrast to the tall-growing Shrubby stylo, did not restrict movement along the paddy wall, which is an important consideration to village farmers.

Shrubby stylo could be considered an alternative or supplemental species to Verano. It outyielded Verano in both years, but a large proportion of its yield (70% in 1979) was inedible thick-stem material. Thick stems enabled it to persist through the dry season, but its effective yield was substantially reduced. In addition, it was transplanted into the field as vigorous, well-established seedlings well able to compete with the natural vegetation. If it were planted directly as seed, its productivity, especially in the first year, might not be as high.

In terms of dry-matter yield and persistence, *S. viscosa* was also a suitable species. However, its acceptability to livestock was poor, and, because of its sticky, viscous foliage and shrubby habit, it could restrict movement along the paddy wall.

The two cultivars of *S. guianensis* (Endeavour and Schofield) were not well adapted to the environment or to the management conditions imposed, and it is recommended that they not be used in this context.

Siratro and leucaena performed quite poorly, particularly at site 1. In 1978 the soil in the paddy walls was waterlogged for most of the rice-growing season, and the moist conditions favored the spread of the fungus disease *Rhizoctonia solani*, affecting all siratro plots at both sites. In addition, it has been reported that siratro persistence is often poor in waterlogged or inundated areas (Rees et al. 1976). Although the native variety of leucaena was used, it was not adapted to the acid soils, and this, together with the waterlogged conditions in the first year, did not favor good growth.

Oversowing of paddy walls with an adapted forage legume such as Verano stylo is a simple technique that could be easily adopted by village farmers. Verano has potential to improve the quantity and quality of dry-season feed available to livestock, and oversowing with it is a low-input improvement that requires only legume seed and labor.

LITERATURE CITED

Moorman, F.R., S. Montrakun, and S. Panichapong. 1964. Soils of northeastern Thailand: a key to their identification and survey. Land Dev. Dept., Soil Surv. Div., Bangkok, Thailand.

Rees, M.C., R.M. Jones, and R. Roe. 1976. Evaluation of pasture grasses and legumes grown in mixtures in Southeast Queensland. Trop. Grassl. 10:65–78.

Rufener, W.H. 1971. Cattle and water buffalo production in villages of Northeast Thailand. Ph.D. Thesis, Univ. of Ill.

Alfalfa Breeding for Forage Yield and Low Saponin Content

P. ROTILI and L. ZANNONE
Istituto Sperimentale per le Colture Foraggere,
Lodi, Italy

Summary

The objective of our breeding program is to build alfalfa (*Medicago sativa* L.) varieties having good vigor and persistency, high protein level, and low saponin content. The breeding method utilizes a selfing phase combined with selection in competitive conditions. Leaf-saponin content was analyzed by the *Trichoderma viride* bioassay. Parent populations (cv. Leonicena and Cantoni), S_0, S_1, and S_2 families were evaluated under competitive and frequent cutting conditions. In both populations, some S_2 families yielded more than their respective S_1 family. Variation in yield and mortality increased with the level of selfing, and some S_2 families showed lower minimal values of mortality than the S_0 families. These persistent S_2 families were the earliest in flowering. A positive correlation existed between S_1 families and S_0 mother plants and between S_2 families and S_1 mother plants. Yield depression is a parameter that depends to some extent on the experimental conditions of plant density and cutting frequency.

KEY WORDS: alfalfa, autotetraploid, competition, inbreeding, *Medicago sativa* L., selection, saponin.

INTRODUCTION

Our breeding work is designed to develop alfalfa (*Medicago sativa* L.) cultivars for intensive management systems. Such cultivars must exhibit high yield, good persistency, high level of leaf protein, and low saponin content.

Our previous research (Rotili 1976, 1977) showed that selfing combined with selection in competitive conditions effectively improved forage yield, and research on the evolution of the alfalfa canopy (Rotili 1979) detected a positive correlation between the persistency of the alfalfa crop and the homogeneity of its constituents (the individual plants) for such characteristics as regrowth, velocity of growth, and flowering.

Considering the low variation in protein content of leaves (Rotili and Zannone 1976, Rotili et al. 1978, 1979), we adopted the strategy of improving this character by selecting for increased tolerance to frequent cutting. Cutting at the blue-bud stage provides, in fact, a product of better quality because of its higher protein level. Tolerance to frequent cutting could provide a higher amount of protein/ha and a forage yield/ha not lower than that obtained from normally managed cultivars.

METHODS

Two parent populations, cultivars Leonicena and Cantoni, were chosen for this research. Two generations of selfing were conducted within each population. In each generation, plants with dry-matter weight exceeding that of the cultivar mean by two standard deviations were selected. At each cycle of selection the plants were grown in the greenhouse in concrete containers 80 cm high, 25 cm wide, and 60 cm long.

Seventy mother plants were selected in each parent population and self-pollinated. Some S_0 plants were also autocrossed, and their progenies were compared with the respective S_1 and S_2 progenies for forage yield and mortality. Parent populations, S_0, S_1, and S_2 progenies were all tested together at a density of 66 plants/linear m. Six cuttings were made when the whole material was 20% blue-bud stage. The time between the cuttings was 22–24 days.

The material was also analyzed for leaf-saponin content by the bioassay that utilizes the fungus *Trichoderma viride* (Zimmer et al. 1967).

RESULTS

The inbreeding depression for dry-matter yield was very similar in the two populations, based on the mean value or the range of variation using average values for six cuttings (Table 1). In both populations some S_2 families yielded more than their respective S_1 families, indicating

P. Rotili is Director of the Institute and L. Zannone is Director of the Biology Section of the Institute.

Table 1. Inbreeding depression in competitive conditions and under frequent cuttings; average of 6 cuttings.

Cultivar		First cycle[1]	Second cycle
Leonicena			
	Mean	− 28	− 32
	Range	− 9 to − 38	15 to − 78
Cantoni			
	Mean	− 28	− 31
	Range	− 12 to − 43	10 to − 61

[1]Inbreeding depression computed as $(S_1 - S_0)$ 100/S_0 and $(S_2 - S_1)$ 100/S_1 for the first and second cycle, respectively.

that selection in S_1 was very efficient. The range of variation increased with the level of selfing, indicating that selection at the S_2 level may be more efficient than at the S_1 or the S_0 level.

The inbreeding depression increased from the first to the last cutting (Fig. 1). At the first and second cuttings, the trend of increase followed the theoretical values expected for autotetraploid species, while at the fifth and sixth cuttings, the trend was closer to the pattern expected for diploid species, as happens when inbreeding depression is studied on spaced plants. The increase in selfing effect through the cuttings was observed also in spacing, but in this case the cumulative effect was less pronounced than in a dense sward (Fig. 2).

Fig. 1. Effect of selfing in S_1 and S_2 families on dry-matter yield in competitive conditions and under frequent cuttings.

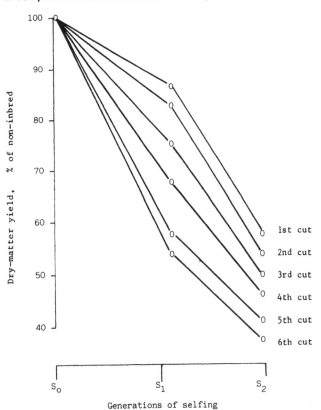

Fig. 2. Comparison of selfing effects in S_1 families on dry-matter yield of alfalfa grown in dense swards and in spacing.

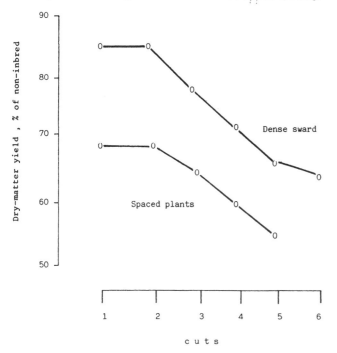

Table 2. Percent mortality at the 6th cutting in competitive conditions and under frequent cuttings.

	Cultivar					
	Leonicena			Cantoni		
	Generation of selfing			Generation of selfing		
Item	S_0	S_1	S_2	S_0	S_1	S_2
	%---					
Mean	17.9	19.7	41.7	14.0	26.7	42.4
Min.	13.6	10.2	5.7	6.8	2.2	5.7
Max.	37.5	62.5	89.7	35.2	71.6	92.0

about a loss in vigor (Fig. 1) and an increase in mortality (Table 2). The inbreeding depression appeared dependent to a large extent on such experimental conditions as density of growing (Fig. 2), number of cuttings considered, and frequency of cutting.

Saponin content of leaves decreased with an increasing level of selfing (Table 3), as was previously observed in a study with single and double crosses (Rotili et al. 1977) where the greater heterozygosity of the material was associated with a higher saponin content.

In the present study, material was not selected for saponin content of leaves. Bioassay results revealed a high heritability of this character. The absence of a correlation between saponin content and forage yield indicated the possibility of selecting for low saponin content without negative consequences on dry-matter yield (Table 4).

Mortality at the sixth cutting increased with the level of selfing. Some S_2 families had lower mortality than S_0 families (Table 2). These persistent S_2 families were the earliest in flowering and the most homogeneous for earliness.

The leaf-saponin content, expressed by the percentage of growth inhibition in *Trichoderma viride* colonies, was similar for the two parent populations. The same was true for the progenies at any level of selfing (Table 3). The saponin content decreased with increased level of selfing. The decrease was highly significant ($P < 0.01$) from the S_1 to the S_2 level.

A positive correlation for leaf-saponin content existed between S_1 and S_0 mother plants (r = 0.82 and 0.83 for Leonicena and Cantoni, respectively) as well as between S_2 families and S_1 mother plants (r = 0.81 and 0.84 for Leonicena and Cantoni, respectively), indicating the possibility of successful selection for this character.

No correlation was detected between leaf-saponin content and forage yield at any level of selfing (Table 4).

Table 3. Percent growth inhibition of *Trichoderma viride* colonies. Mean values and coefficients of variation (C.V.) among progenies.

	Cultivar					
	Leonicena			Cantoni		
	Generation of selfing			Generation of selfing		
Item	S_0	S_1	S_2	S_0	S_1	S_2
	%---					
Mean	42.5	38.5	32.4	40.8	40.1	32.7
C.V.	21.4	26.4	25.0	18.6	23.0	27.1

DISCUSSION

The data on inbreeding depression (Table 1) indicated that selfing enhances the efficacy of selection in competitive conditions. The frequency of cutting, every 22–24 days when all the material (S_0, S_1, and S_2) was about 20% blue-bud stage, has to be considered among the causes of the variation observed in the estimates of inbreeding depression. The depression increased from the first to the sixth cutting because the S_1 and S_2 families have a cumulative delay in root reserve restoration that brings

Table 4. Coefficients of correlation between leaf-saponin content and forage yield.

Cultivar	S_0 mother plants	S_1 families	S_2 families
Leonicena			
n − 2	68	53	30
r	0.09	−0.07	0.17
Cantoni			
n − 2	68	57	38
r	0.01	−0.01	0.03

LITERATURE CITED

Rotili, P. 1976. Performance of diallel crosses and second generation synthetics of alfalfa derived from partly inbred parents. I. Forage yield. Crop Sci. 16:247–251.

Rotili, P. 1977. Performance of diallel crosses and second generation synthetics of alfalfa derived from partly inbred parents. II. Earliness and mortality. Crop Sci. 17:245–248.

Rotili, P. 1979. Contribution à la mise au point d'une méthode de sélection de la luzerne prenant en compte les effets d'interférence entre les individus. I. Etude expérimentale de la structure de la luzernière. Ann. Amélior. Pl. 29 (4):351–379.

Rotili, P., and L. Zannone. 1976. Protein and saponin content in single and double crosses of lucerne (*Medicago sativa* L.). *In* Protein quality from leguminous crops, 377–386. EEC, Dijon.

Rotili, P., L. Zannone, and G. Gnocchi. 1979. Effect of inbreeding on vigor of alfalfa measured in competitive and frequent cutting conditions. Rpt. Fodder Crop Sect. Mtg. Eucarpia, Perugia, 111–119.

Rotili, P., L. Zannone, G. Gnocchi, S. Proietti, and A. Serena. 1978. Teneur en saponines et proteines de plantes mères et descendances S_0 et S_1. Rpt. of *Medicago sativa* Group Mtg. Eucarpia, Osijek, 59–69.

Zimmer, D.E., M.W. Pedersen, and C.F. McGuire. 1967. A bioassay of alfalfa saponin using the fungus *Trichoderma viride*. Crop Sci. 7:223–224.

Forage Quality of Barley Straw as Influenced by Genotype

D.W. MEYER, D.O. ERICKSON, and A.E. FOSTER

North Dakota State University, Fargo, N. Dak., U.S.A.

Summary

Cereal straw is an untapped carbohydrate source for ruminant animal feeds, but little information is available on genotypic differences in the feeding value of barleys (*Hordeum vulgare* L.) commonly grown in the northern Great Plains of the U.S.A. Our objective was to determine if barley genotypes differ in straw forage quality and if straw-quality parameters are associated with agronomic characters.

Straw from 37 and 31 barley genotypes was sampled in 1977 and 1978, respectively. Straw from 10 barley genotypes grown at four locations was sampled in 1979. Standard laboratory procedures were used to determine the quality parameters measured. Correlations among straw-quality parameters and agronomic characters were determined in 1977 and 1978.

Barley straw tested in 1977 and 1978 averaged 5.6% crude protein (CP), 42.0% in-vitro dry-matter disappearance (IVDMD), 50.7% acid detergent fiber (ADF), 71.8% neutral detergent fiber (NDF), 6.6% acid detergent lignin (ADL), 21.1% hemicellulose, 11.4% ash, 0.13% phosphorus (P), 0.58% calcium (Ca), and 0.20% magnesium (Mg). The average feeding value was different among environments with higher-quality straw produced in low-grain-yield environments.

Significant differences among barley genotypes were found for most straw-quality parameters measured. Six-rowed barley straw was generally lower in forage quality than 2-rowed barley straw. Straw from Karl barley, a low-grain protein cultivar, was higher in CP and digestibility and lower in ADF and NDF than most genotypes. Significant genotype × environment interactions were detected for most straw-quality parameters except CP. Heading date and plant height correlations with straw-quality components varied among years. Grain, straw, and protein yields; lodging; and grain-protein percentage were not consistently correlated with a straw-quality component.

We have demonstrated potentially useful differences in straw-forage quality among barley genotypes. The data suggest that genotypes high in straw quality can be selected without loss of desirable agronomic characters.

Key Words: *Hordeum vulgare* L., straw, forage quality, barley genotypes, straw digestibility.

INTRODUCTION

The potential feed value of spring barley (*Hordeum vulgare* L.) straw is incompletely documented, with little information available on quality differences among genotypes commonly grown in the U.S.A.'s northern Great Plains. Kernan et al. (1979) in western Canada and White et al. (1981) in Montana found significant differences among and within wheat (*Triticum aestivum* L.), oat (*Avena sativa* L.), and barley for crude protein (CP) and straw digestibility.

In-vitro dry-matter disappearance (IVDMD) of barley straw ranged from 40% to 49% among several genotypes of European and American spring barleys (Alderman 1976, White et al. 1981). Genotypic differences in straw protein percentages were found also. Published data on

D.W. Meyer is an Associate Professor of Agronomy, D.O. Erickson is a Professor of Animal Science, and A.E. Foster is a Professor of Agronomy at North Dakota State University.

This is a contribution of the Agronomy Department, Agricultural Experiment Station, North Dakota State University. Journal Article No. 1146.

other nutritive-value characters as influenced by the genotype are limited or lacking. Our objective was to determine genotypic differences in forage quality of barley straw and their relationship to agronomic characters.

MATERIALS AND METHODS

Barley straw was sampled from two uniform regional trials in North Dakota. One, planted at Langdon in 1977, included 37 cultivars and advanced breeding lines (27 6-rowed and 10 2-rowed barleys). The second trial, planted at Fargo in 1978, included 31 genotypes, of which 19 were 6-rowed (including 9 used in the 1977 trial) and 12 were 2-rowed barleys (including 6 used in the 1977 trial). Among these were 7 feed barley entries classified as unsuitable for malting purposes. All cultural practices were consistent with a 5,500 kg/ha yield goal. A randomized complete-block design (RCBD) with 3 replications was used. Straw samples consisting of all harvested plant parts except the grain and chaff were collected at physiological maturity or beyond.

Eight barley cultivars (Larker, Beacon, Morex, Glenn, Summit, Karl, Shabet, and Hector) and 2 advanced lines (M-32 and ND2199) were planted at four locations in 1979 to further evaluate the genotype × environment (G × E) interaction. A randomized complete-block design with 4 replications was used at each location. Straw from each experimental unit was sampled randomly at physiological maturity.

All straw samples were oven dried at 60°C or less and ground to pass through a 1-mm screen. Subsamples were analyzed in duplicate by standard procedures.

RESULTS AND DISCUSSION

Significant differences among barley genotypes were found for all forage-quality parameters measured on straw in 1977 (Table 1), 1978 (Table 1), 1979 (Table 2), and 1977–1978 except for hemicellulose in 1977, CP in 1978, and IVDMD, ADL, and P in 1977–1978. Several genotypes had adequate CP content to meet the requirements of a mature beef cow during the midgestation period. Six-rowed barleys generally were lower in forage quality than 2-rowed barleys in 1977 and 1978. Six-rowed

barleys were significantly lower in CP (1978 excepted), P, hemicellulose, and IVDMD (1977 excepted) and higher in ADF, ADL, ash, Ca, and Mg than 2-rowed barleys. Kernan et al. (1979) also found 2-rowed barleys to have a higher CP percentage than 6-rowed barleys. Few differences in forage quality were noted among straw of feed and malting barleys.

Straw from Karl barley averaged 8.7% or 1.8 percentage units higher than the next highest genotype in 1977 and 1979 (Table 2). Karl is a genotype with low grain protein and average grain yield, features that suggest that it is a poor translocator of plant protein. Karl also averaged 2.4, 2.2, and 0.6 percentage units less in ADF, NDF, and ash, respectively, and 2.7 units higher in estimated digestibility than the next closest desirable genotype in 1979. Belfort (a forage barley) appears to have straw-quality characteristics similar to those of Karl.

The straw-quality components differed according to environment. Barley straw from Fargo, North Dakota, in 1978 was higher in CP, ash, P, and IVDMD and lower in ADF, NDF, ADL, and hemicellulose than barley straw from Langdon in 1977. Grain yields were twice as high at Langdon, North Dakota (6,450 kg/ha), as at Fargo (3,252 kg/ha), possibly indicating that higher-forage-quality straws are produced in low-yield environments. Ash, ADF, and ADL were higher and estimated digestibility and hemicellulose lower in high-yield than in low-yield environments in 1979. CP and NDF were similar among yield environments.

Significant G × E interactions were detected for P, NDF, and hemicellulose quality characters for the 15 barleys tested in 1977–1978. These data agree with those of White et al. (1981) and Kernan et al. (1979), who found that genotypic differences in straw digestibility and CP were consistent across years. However, G × E interactions were detected for all straw-quality characters (except CP) measured in 1979. These data suggest that G × E interactions should be anticipated for most straw-quality characters when testing is done at several locations. However, potentially useful genotypic differences still can be detected.

Simple correlation coefficients between agronomic and straw-quality characters generally were nonsignificant or inconsistent among the years. Heading data and plant

Table 1. Mean and range of barley straw-quality characters in 1977 and 1978.

Quality character	1977		1978	
	Mean	Range	Mean	Range
	---------------- % dry weight----------------			
IVDMD	41.7±2.52	34.3 – 49.6	42.8±1.15	38.3 – 46.1
Crude protein	4.9±0.51	3.7 – 8.6	6.2±0.63	5.2 – 7.9
ADF	52.1±1.06	47.0 – 54.9	49.3±1.28	45.7 – 52.0
NDF	77.0±1.57	71.7 – 81.1	66.6±1.13	60.6 – 71.2
ADL	6.9±0.46	5.8 – 8.0	6.2±0.35	5.3 – 7.1
Ash	9.0±0.31	7.8 – 10.5	14.0±0.42	12.5 – 15.8
P	0.06±0.01	0.04 – 0.10	0.20±0.02	0.15 – 0.30
Hemicellulose	24.9±1.37	19.3 – 28.4	17.3±0.97	13.8 – 20.9
Ca			0.58±0.05	0.48 – 0.78
Mg			0.20 + 0.02	0.14 – 0.31

Table 2.   Forage quality of straws from 10 barley genotypes at four North Dakota locations in 1979.

Genotype	CP[†]	Ash	ADF	NDF	ADL	HEMI[†]	ED[††]
				— % dry weight —			
M-32	6.8b§	11.4a	45.0e	71.1b	5.2d	26.2bc	48.4b
ND2199	6.5bc	10.9b	47.0b	72.6ab	6.1b	25.6c	46.1de
Beacon	6.4bc	10.7bc	47.7ab	73.5a	6.4a	25.8c	45.2ef
Larker	6.7bc	10.4cd	46.5cd	72.7ab	6.4a	26.2bc	46.6ed
Glenn	6.2c	10.4cd	47.4ab	72.7ab	6.3ab	25.4c	45.6ef
Morex	6.6bc	10.7bc	48.0a	73.6a	6.5a	25.6c	44.9f
Summit	6.4bc	10.1d	46.1d	72.0ab	5.5c	26.0bc	47.1c
Karl	8.7a	10.3c	42.6f	68.9c	5.7c	26.3bc	51.1a
Shabet	6.5bc	9.5e	45.0e	73.3a	5.6c	27.3a	47.2c
Hector	6.9b	10.0d	47.2abc	74.0a	6.1b	26.9ab	45.9de
Mean	6.8	10.4	46.4	72.5	6.0	26.1	46.8

†CP = crude protein, HEMI = hemicellulose, ††ED = estimated digestibility.

§Means within a column followed by different letters are significantly different at the 5% probability level.

height were associated with several straw-quality characters in 1978 but generally not in 1977. IVDMD increased with late-maturing genotypes, probably due to a decreased ADL content. ADL increased as plant height increased. The percentage of dry matter at harvest was associated positively with ADF and NDF but generally not with other straw-quality characters. These data suggest that differences in forage quality among barley straws in 1978 might have been affected by maturity differences among genotypes. However, Kernan et al. (1979) found that CP, crude fiber, and straw digestibility changed little from the middough stage until 10 days after the crop could be swathed.

Grain, straw, and protein yields; lodging; and grain protein percentage were not consistently associated with any straw-quality character. These data, which agree with results of White et al. (1981), suggest that barley genotypes high in straw quality may be selected without sacrificing desirable agronomic characters.

LITERATURE CITED

Alderman, G. 1976. Feeding value of cereal straws. Grass: J. Brit. Assoc. Green Crop Driers 16:12.

Kernan, J.A., W.L. Crowle, D.T. Spurr, and E.C. Coxworth. 1979. Straw quality of cereal cultivars before and after treatment with anhydrous ammonia. Can. J. Anim. Sci. 59:511–517.

White, L.M., G.P. Hartman, and J.W. Bergman. 1981. Straw digestibility of winter wheat, spring wheat, barley, and oat cultivars in eastern Montana. Agron. J. 73:117–121.

Neutral Detergent Fiber and Protein Levels in Diploid and Tetraploid Ryegrass Forage

L.R. NELSON and F.M. ROUQUETTE, JR.
Texas A&M University Agricultural Research and Extension Center,
Overton, Tex., U.S.A.

Summary

Nutritive value of 18 ryegrass (*Lolium multiflorum* Lam.) cultivars was studied at Overton, Texas, during a 3-year period to determine the variation in protein content and neutral detergent fiber (NDF) of 9 diploid and 9 tetraploid ryegrass cultivars as influenced by season. Comparisons were also made within diploid and tetraploid cultivars.

Samples for analysis were collected from forage variety tests for the growing seasons of 1976 to 1979. Ryegrass cultivars were drilled into conventionally tilled seedbeds prior to mid-September and fertilized at a rate of 67 kg/ha each for N, P_2O_5, and K_2O. Additional N was applied in split applications in November and February for a total seasonal rate of 235 kg/ha. The forage was clipped to a uniform height of 5 cm with a flail-type harvester. Forage samples were oven dried and ground to pass through a

0.42-mm screen. Plots were harvested four times during each of the first 2 years and five times in the third year. In 1977–1978 and 1978–1979, NDF and protein analyses were run on samples from 2 replications. All forage samples were analyzed for protein via the micro-Kjeldahl technique. Percentages of NDF were obtained via Van Soest extraction procedures.

A seasonal increase in neutral detergent fiber (NDF) content and a decrease in protein percentage was observed. Differences in NDF were observed among cultivars and were most apparent during April and May. The differences probably were related to maturity levels of individual cultivars since earlier-maturing cultivars tended to have higher NDF levels. Tetraploids were lower in NDF, probably because they matured later than did the diploids. Protein levels decreased during the growing season from about 25 % in December to about 13 % in late May. Differences among cultivars and between ploidy levels were not significant for protein percentage.

KEY WORDS: *Lolium multiflorum* Lam., forage quality, total structural carbohydrate.

INTRODUCTION

The cow-calf industry of eastern Texas has been formed around the basic production attributes of warm-season perennial grass. These grasses, such as bermudagrass (*Cynodon dactylon* L. Pers.), provide adequate forage for grazing from late April to early November, but an alternate forage source and/or concentrates must be supplied during other months. Hay has been one of the most widely used forages for wintering cattle in the southeastern U.S. However, in order to maximize calf weaning weights, cows must calve in the fall-winter period, and in order to satisfy their nutritional requirements, a supplementary feed relatively high in nutritive value must be supplied. The use of cool-season annual forages in a bermudagrass sod is one wintering alternative. There is a real need, however, to overseed such pastures with a reliable, high-producing annual forage for maximum net returns. In order for this cool-season forage to be complementary to the sod and valuable for the livestock, it must possess some of the following traits: (1) good seedling vigor in the presence of sod, (2) consistently available grazing by mid-January, (3) continued grazing until 1 June, and (4) nutritional value comparable with those of currently available small grain and ryegrass (*Lolium multiflorum* Lam.) cultivars. The study described in this paper is presented as a method of evaluating new selections that may be superior in overall forage quality to existing diploid cultivars. It was conducted to determine the variation in percentage of protein and in NDF content of 9 diploid and 9 tetraploid ryegrass cultivars as influenced by season.

METHODS

Ryegrass cultivars were planted in conventionally tilled seedbeds prior to mid-September each year. A broadcast application of fertilizer at a rate of 67 kg/ha each of N, P_2O_5, and K_2O was applied in late August. Additional N was applied in split application in November and February to make the total N applied 235 kg/ha.

Ryegrass forage samples were collected for analysis from forage-cultivar tests in the growing seasons of

1976–1977, 1977–1978, and 1978–1979. All ryegrass plots were clipped to a uniform height of 5 cm with a flail-type harvester. Forage samples were oven dried and ground to pass through a 0.42-mm screen. Plots were harvested four times the first 2 years and five times during the third year of the study. In 1976–1977, analyses for neutral detergent fiber (NDF), by the method described by Van Soest et al. (1967), and protein, by micro-Kjeldahl analysis, were made on samples from 3 replications of each cultivar/harvest. In 1977–1978 and 1978–1979, analyses were run on samples from 2 replications. Duplicate analyses were run on any samples that appeared abnormally high or low in either NDF or protein.

RESULTS

Results from the NDF analyses (Table 1) indicate significant changes in the fibrous fraction of all cultivars with the normal change in seasonal conditions. Early in the growing season NDF was low. Later, however, the fibrous portion of the plant increased to more than 45 % in April and to more than 55 % in May. Neutral detergent fiber values for mature ryegrasses were lower than those often reported for most warm-season perennial grasses. Minson et al. (1964) reported that at a given cutting date, late-maturing grasses were more digestible than early-maturing cultivars.

Differences in NDF content among cultivars were not significant for December and March harvests (Table 1). Billion and Tetragulf had the lowest NDF content in the April harvest. Florida Rust Resistant had the highest NDF content, 48.7 %. The primary reason for the increase in NDF was that the early-maturing varieties such as Florida Rust Resistant were beginning to form seed heads, and therefore a higher proportion of the available forage was in the form of stems rather than in leaves. Although the NDF values may indicate an acceptable forage quality, the high stem : leaf ratio during maturation has a detrimental effect on the grazing pattern of livestock. The amount of selective grazing increases at this time and is negatively related to the efficiency of pasture utilization. In-vivo studies by Minson et al. (1960, 1964) with 2 cultivars of ryegrass and one of orchardgrass (*Dactylis glomerata* L.) and an in-vitro experiment by Terry and Tilley (1964) with the same forage

The authors are both Associate Professors in the Department of Soil and Crop Science at the Center.

236   *Nelson and Rouquette*

Table 1.   Percent neutral detergent fiber (NDF) contents of ryegrass (*Lolium multiflorum* Lam.) forage.

Cultivar	Years tested	Dec.[1]	March	April	Early May	Late May[2]
Gulf	3	35.2a[4]	38.4a	44.7abc	60.7ab	58.1ab
Magnolia	2	29.1a	39.3a	46.5ab	59.1a-c	54.4bc
Common	2	39.4a	37.2a	47.0ab	59.7a-c	54.8bc
Marshall	1	32.5a	31.3a	48.0a	57.6a-d	56.3bc
Tx-0-R-78-2	2	32.4a	36.2a	45.3abc	56.9a-d	63.7a
Tx-0-R-78-3	2	32.3a	33.7a	45.2abc	57.0a-d	60.8ab
Florida Rust R.	1	—[3]	42.3a	48.7a	64.3a	—[3]
Exp Col Sta 171-1	1	—	39.7a	47.3a	64.0ab	—
Exp Col Sta 29-27[5]	1	—	43.3a	48.3a	58.7a-c	—
Ninak[6]	2	30.6a	32.3a	41.1c	50.3de	49.9c
Tetragulf	1	31.9a	30.7a	39.3c	44.8e	—
Billion	1	32.5a	30.5a	39.3c	49.1de	—
Tetrablend 444	2	29.6a	49.9a	46.2ab	57.2a-d	60.0ab
Charleston	1	26.9a	32.0a	41.2bc	47.9de	—
Tetrone	2	30.9a	33.0a	41.7bc	49.0de	57.3ab
Furone	1	—	45.7a	40.7c	55.3b-c	—
Aubade	1	—	43.7a	43.0abc	59.3a-c	—
NAPB-28	1	—	38.3a	47.0ab	52.7c-e	—
Diploid mean		33.5a	37.9a	46.8a	59.8a	58.0a
Tetraploid mean		30.4a	36.2a	42.2b	51.7b	55.7a

[1]December data are from last 2 years of study only

[2]Late May data are from 1979 only.

[3]Cultivar not tested.

[4]Yields within each column followed by the same letter are not significantly different at the 5% level as judged by Duncan's multiple range test.

[5]Cultivars above dotted line are diploids.

[6]Cultivars below dotted line are tetraploids.

species with individual results for leaf lamina, leaf sheath, stem, inflorescense, and dead material indicated that the digestibility of leaf-lamina fractions decreased 0.13%/day with advancing maturity. Leaf sheaths and stem fractions, however, decreased more rapidly than lamina with increasing maturity, 0.4% and 0.7%/day, respectively. In the early- and late-May harvests, we continued to observe significant differences among cultivars associated with maturity. Later-maturing cultivars such as Ninak, Tetragulf, Billion, Charleston, Tetrone, and NAPB-28 generally had lower NDF values.

Diploid cultivars had significantly greater NDF values than did the autotetraploids in the April and early-May harvests (Table 1). This difference was 4.6% for April and 8.1% for early May. Since the tetraploid cultivars were usually later in maturing, their nutritive value remained high into early May. By late May no real differences existed since both the diploid and the tetraploid cultivars were no longer vegetative. These results probably explain why it has been reported that the tetraploids have produced higher animal gains.

Protein levels (Table 2) show important differences for time of harvest, but differences among cultivars were not significant for any harvest. Protein levels of nearly 25% were found for the December harvest. The March protein levels decreased to about 22% and the April levels to 17%. A low of 12% was noted for early May and a slight increase to about 14% for late May. This increase may have been caused by warm-season weed species, which were beginning to make good growth in some plots by late May. The most important trend noted in these analyses was the extreme decrease in protein percentage over time. Six of the cultivars tested had protein values of less than 9% in early May, a level that may restrict performance of certain classes of livestock. These cultivars should therefore be planted either with a companion legume or in a warm-season grass sod.

LITERATURE CITED

Minson, D.J., C.E. Harris, W.F. Raymond, and M. Milford. 1964. The digestibility and voluntary intake of S 22 and H 1 ryegrass, S 170 tall fescue, S 48 timothy, S 215 meadow fescue, and germinal cocksfoot. J. Brit. Grassl. Soc. 19:298–305.

Minson, D.J., W.F. Raymond, and C.E. Harris. 1960. Studies in the digestibility of herbage. VIII. The digestibility

Table 2. Percent protein of ryegrass (*Lolium multiflorum* Lam.) forage.

Cultivar	Years tested	Dec.[1]	March	April	Early May	Late May[2]
Gulf	3	24.7[4]	22.8	18.4	12.7	11.4
Magnolia	2	25.8	22.2	16.1	13.1	14.2
Common	2	28.5	22.0	16.6	12.1	13.7
Marshall	1	23.6	23.4	19.3	16.9	13.7
Tx-0-R-78-2	2	26.4	25.4	19.5	16.9	13.1
Tx-0-R-78-3	2	24.1	23.4	17.2	15.5	14.1
Fla Rust Res.	1	—[3]	21.4	14.0	8.3	—[3]
Exp Col Sta 171-1	1	—	21.4	17.3	8.8	—
Exp Col Sta 29-27[5]	1	—	17.6	15.0	9.8	—
Ninak[6]	2	24.6	25.2	21.7	17.7	14.3
Tetragulf	1	23.8	24.5	18.5	13.1	—
Billion	1	23.7	22.8	20.4	14.0	—
Tetrablend 444	2	26.1	22.8	14.7	11.9	14.0
Charleston	1	22.6	24.5	21.9	12.7	—
Tetrone	2	24.5	24.3	19.1	16.1	15.2
Furone	1	—	19.6	14.2	8.3	—
Aubade	1	—	18.6	14.1	8.5	—
NAPB-28	1	—	20.5	12.9	8.3	—
Diploid mean		25.5	22.2	17.0	12.7	13.4
Tetraploid mean		24.2	22.5	17.5	12.3	14.5

[1]December data are from last 2 years of study only.

[2]Late May data are from 1979 only.

[3]Cultivar not tested.

[4]There are no significant differences between cultivars within dates.

[5]Cultivars above dotted line are diploids.

[6]Cultivars below dotted line are tetraploids.

of S 37 cocksfoot, S 23 ryegrass, and S 24 ryegrass. J. Brit. Grassl. Soc. 15:174–180.

Terry, R.A., and J.M.A. Tilley. 1964. The digestibility of the leaves and stems of perennial ryegrass, cocksfoot, timothy, tall fescue, lucerne and Sainfoin as measured by an *in vitro* procedure. J. Brit. Grassl. Soc. 19:363–372.

Van Soest, P.J., R.E. Wine, and L.A. Moore. 1967. Use of detergent in the analysis of fibrous feeds. IV. Determination of plant cell-wall constituents. Assoc. Off. Agric. Chem. J. 50:50–55.

Collections of Western Wheatgrass and Blue Grama and Associated Nematode Genera in the Western Dakotas

R.E. BARKER, J.D. BERDAHL, J.M. KRUPINSKY, and E.T. JACOBSON
USDA-SEA-AR Northern Great Plains Research Laboratory, Mandan, N. Dak.,
and USDA-SCS Plant Materials Center, Bismarck, N. Dak., U.S.A.

Summary

Western wheatgrass (*Agropyron smithii* Rydb.) and blue grama (*Bouteloua gracilis* [Willd. ex H.B.K.] Lag. ex Griffiths) are widely distributed throughout the U.S.A. They are two of the predominant native grasses in rangelands in the northern Great Plains. Because suitable cultivars are not always available for reseeding disturbed areas, we collected plants of the two species and evaluated phenotypic variability to determine potential for breeding improvement. We also sampled the nematode population associated with the collections.

Sampling sites were systematically chosen by establishing a grid pattern 19.3 km apart in the western halves of North Dakota and South Dakota. Five collections, including a portion of the root system, were made of each species at each of about 1,040 sites in 1977 by SCS field technicians. Individual plants were extracted from samples and transplanted to the field for agronomic evaluations. Nematodes were extracted from the soil fraction of the collection samples.

Data collected on individual plants were averaged over all plants collected at one site (entries). Relative maturity of western wheatgrass entry means ranged from 178.2 to 195.0 Julian days and from Julian date 169 to 204 for individual plants. Average fertility for entry means was 35.6%, and mean seed yield from 5 heads was 0.59 g. Awn length ranged from none to >5 mm. Fewer characteristics were measured of blue grama than of western wheatgrass. The maturity range for blue grama was 13.5 days for entry means and 15 days for individual plants.

Plant parasitic nematodes were found in 59% of the 3,101 soil samples analyzed. The most prevalent genera found on both blue grama and western wheatgrass were *Helicotylenchus*, *Paratylenchus*, *Xiphinema*, and *Tylenchorhynchus*. *Paratylenchus* and *Helicotylenchus* were the most prevalent genera in North Dakota and in South Dakota, respectively.

We concluded that sufficient phenotypic variability existed in western wheatgrass to warrant initiating a breeding program. Blue grama varied in maturity, but as other characteristics were not identified, initiation of a breeding program will be delayed until variability is investigated further. Plant parasitic nematodes commonly found in range sites might be a limiting factor on range productivity.

KEY WORDS: selection, grass breeding, vegetative collection, germ plasm evaluation, western wheatgrass, *Agropyron smithii*, blue grama, *Bouteloua gracilis*, nematodes.

INTRODUCTION

Western wheatgrass (*Agropyron smithii* Rydb.) and blue grama (*Bouteloua gracilis* [Willd. ex H.B.K.] Lag. ex Griffiths) are widely distributed throughout the U.S.A. and are frequently dominant or codominant in the short- and mixed-grass prairies and rangelands of the northern Great Plains. Native grasses generally tend to be less productive than introduced or "naturalized" grasses when grown under intensive management (Gomm 1974), but native grasses have been shown to be as productive or more productive than the naturalized grasses in long-term evaluations (McWilliams and van Cleave 1960).

Native grasses have considerable genetic diversity, but only limited research has been directed toward characterization of this variation (Crowston and Goetz 1976). Like many other native grasses, western wheatgrass and blue grama have received only minor breeding attention. Johnston et al. (1975) collected 462 ecotypes of western wheatgrass from southern Alberta and Saskatchewan in Canada. Evalution of agronomic characteristics identified sufficient variability to assure potential progress in improvement through a breeding program.

Our objectives were to collect western wheatgrass and blue grama in a defined region, to evaluate phenotypic variability, and to determine potential for improvement through breeding. We also sampled the nematode population associated with western wheatgrass and blue grama in their native environment.

R.E. Barker and J.D. Berdahl are Research Geneticists and J.M. Krupinsky is Plant Pathologist with USDA-SEA-AR in Mandan, and E.T. Jacobson is Plant Materials Specialist with USDA-SCS in Bismarck. They would like to thank Pat Donald for nematode identification and Mary Kay Tokach and Russell Haas for technical assistance.
This is a contribution of the USDA-SEA-AR Northern Great Plains Research Laboratory and the USDA-SCS Plant Materials Center.

MATERIALS AND METHODS

Collections were made in the western halves of North Dakota and South Dakota in section 36 of alternate townships. Because all sections 36 were designated state school lands by the U.S. Public Lands Survey, a high proportion remains as native range. Thus, collection sites were in a grid pattern 19.3 km apart and on land not previously cultivated.

Five samples, collected at least 75 m from each other within a site, were made of western wheatgrass (1,042 sites) and blue grama (1,039 sites) in 1977 from late August to early October by Soil Conservation Service field technicians. Vegetative samples, with intact root systems, were removed with a spade. The root ball was trimmed to 10 × 10 × 15 cm deep and placed in a plastic bag and cardboard box for transporting. Individual plants were extracted from the samples, grown in a greenhouse, and transplanted to the field in 1978.

Plants were arranged in the field in the order in which they were received at Mandan, North Dakota. Western wheatgrass was planted on 2-m and blue grama on 1-m centers. Collection samples were replicated twice but were not intentionally randomized within replicates. Data collected the year following transplanting for western wheatgrass were: maturity, determined as the Julian date when at least 50% of the heads of a plant were in anthesis; vigor, scored from 1 to 3 (3 = best); plant or rhizome spread, scored from 1 to 5 (5 = rhizomes spread to all parts of a 1.2-m square); density of the spread, scored from 1 to 5 (5 = completely filling the 1.2-m square); plant color, scored from 1 to 3 (1 = slate, 3 = green); auricle color, scored from 1 to 5 (1 = no color, 5 = purple pigmentation throughout leaf collar); pubescence on upper leaf surface, scored from 1 to 3 (1 = none); and leaf angle, scored from 1 to 3 (3 = most erect). Seed characteristics of western wheatgrass were measured on two plants from each collection site. Seed yield was the weight of threshed seed from 5 heads, and fertility was the percentage of seed yield/weight of the unthreshed heads. Ergot infestation, caused by *Claviceps purpurea* (Fr.) Tul., was determined by the number of sclerotia from 5 heads, and awn length was visually classified into 5 groups (0 = no awns and 4 = awns 4 mm or longer). For blue grama, maturity of individual plants was determined as with western wheatgrass, plant vigor was scored from 1 to 9 (9 = best), and growth type was scored from 1 to 9 (1 = erect leaves and culms, 9 = decumbent habit).

After individual plants had been extracted from each collection sample, soil from the root ball was screened through a 6-mm wire mesh screen and prepared for nematode extraction. Before nematode extraction, 100-ml soil samples were stored in a refrigerator at 6°C ± 1°C for 12 to 72 hours. Nematodes were extracted from 3,101 samples selected randomly, with each county represented at least twice for each grass species. The Baermann funnel technique was used to extract the nematodes. Nematodes were killed by heating in a water bath at 58°–60°C for 1 minute and then were preserved in FAA (6 parts formalin : 1 part glacial acetic acid : 20 parts 95% alcohol : 40 parts distilled water). After being stirred with a vortex mixer, the contents of the test tube were stored in vials until nematodes were identified by light microscopy.

Phenotypic data obtained in 1979 on the 5 individual plants collected at each site were averaged to constitute an individual entry, and descriptive statistics were calculated on weighted entry means to account for unequal numbers of plants among entries. For the nematode study, the plant parasitic nematode genera were identified, but no attempt was made to identify the species or count their number in the sample.

RESULTS AND DISCUSSION

Maturity of western wheatgrass entry means ranged from Julian 178.2 to 195.0 (Table 1). The earliest-flowering plant was in anthesis on day 169 and the latest on day 204. There were 327 of the 7,855 surviving plants in the field that did not produce sufficient heads to be considered in the maturity analysis. Mean fertility for entries was 35.6% and seed yield/5 heads averaged 0.59 g. Approximately 3% of the entries that flowered produced no seed; the highest entry produced an average of 2.03 g. Entries had an average of 2.6 ergot sclerotia/5-head lot. Average awn length was about 2 mm (score = 1.9), but some plants had awns longer than 5 mm. Summarizations of other characteristics measured are in Table 1.

Phenotypic variability was adequate to indicate that a breeding program for western wheatgrass could be initiated in our population. Dewey (1975) reported that western wheatgrass is an allooctoploid, $2n = 56$, and that bivalent pairing is predominant. Even though chromosome behavior may be similar to that of a diploid, we would expect plant improvement to be slow because of the high chromosome numbers and complexity of the genomes.

Maturity of blue grama entries ranged from 190.3 to 203.8 Julian days (Table 1). Individual plants with extreme vigor and growth-type scores were found, but 77% of the plants fell in the 4–6 score range for vigor and 5–8 score range for growth type. We were unable to identify other meaningful characteristics in blue grama because most plants were strikingly similar. It is not unusual for blue grama to show a lack of phenotypic variability (Riegel 1940). Observations from this study indicate more intensive studies of blue grama ecotypes of more

Table 1. Descriptive statistics for entry means of western wheatgrass and blue grama.

Variable	Number of entries	Mean	Minimum value	Maximum value	Standard deviation
Western wheatgrass					
Maturity (Julian date)	1027	187.5	178.2	195.0	1.7
Fertility (%)	1031	35.6	0	69.6	10.4
Seed yield (g)	1031	0.59	0.02	2.03	0.23
Ergot (Number)	1031	2.6	0	18.3	2.7
Awn (Score)	1031	1.9	0	4.0	0.6
Vigor (Score)	1028	2.1	1	3	0.2
Spread (Score)	1028	3.1	1	4.5	0.4
Density (Score)	1028	2.7	1	4	0.4
Plant color (Score)	1028	1.9	1	3	0.3
Auricle color (Score)	1028	2.5	1	4.3	0.5
Pubescence (Score)	1028	1.4	1	3	0.3
Leaf angle (Score)	1028	1.9	1	2.4	0.2
Blue grama					
Maturity (Julian Date)	1029	194.0	190.3	203.8	1.7
Vigor (Score)	1029	5.1	2.7	7.3	0.8
Growth type (Score)	1029	6.1	3.2	8.8	0.8

Table 2. Plant parasitic nematode (PPN) genera identified in soil samples from vegetative collections of western wheatgrass (WWG) and blue grama (BG) made in North Dakota and South Dakota.

Genus	North Dakota		South Dakota		Total
	WWG	BG	WWG	BG	
	---------------------------- Number of Samples----------------------------				
Helicotylenchus	194	221	211	224	850
Paratylenchus	273	320	103	99	795
Xiphinema	132	125	99	67	423
Tylenchorhynchus	128	105	83	72	388
Hoplolaimus	11	14	3	2	30
Criconemoides	4	9	1	11	25
Pratylenchus	7	2	2	1	12
Heterodera	2	3	2	4	11
Total samples with PPN	529	570	358	364	1821
Total samples analyzed	855	942	638	666	3101

diverse origin are needed to identify variability adequately. Further studies will be done with our populations.

Plant parasitic nematodes were found in 59% of the 3,101 soil samples processed. Some individual samples or collections contained more than one genus of plant parasitic nematode (Table 2). *Helicotylenchus* and *Paratylenchus*, the most abundant genera, were extracted from 47% and 44%, respectively, of the collections containing plant parasitic nematodes. *Helicotylenchus* was the most prevalent genus (60% of the samples) isolated from collections in South Dakota, followed by *Paratylenchus* (28%). *Paratylenchus* was the most prevalent genus (54%) isolated from collections in North Dakota, followed by *Helicotylenchus* (38%). *Xiphinema* and *Tylenchorhynchus* were isolated from 23% and 21%, respectively, of the collections containing plant parasitic nematodes. There were no apparent differences between the blue grama and western wheatgrass collections or between North Dakota and South Dakota for *Xiphinema* or *Tylenchorhynchus*. Overall, the genera *Pratylenchus*, *Criconemoides*, *Hoplolaimus*, and *Heterodera* were found in less than 2% of the collections containing plant parasitic nematodes. Further research is needed to ascertain the effect of plant parasitic nematodes on these grasses.

LITERATURE CITED

Crowston, W.A., and H. Goetz. 1976. Fertilizer and clipping effects on ecosystems of western wheatgrass and blue grama. N. Dak. Agric. Expt. Sta. Bul. 502.

Dewey, D.R. 1975. The origin of *Agropyron smithii*. Am. J. Bot. 62:524–530.

Gomm, F.B. 1974. Forage species for the Northern Intermountain Region—a summary of seeding trials. U.S. Dept. Agric. Tech. Bul. 1479.

Johnston, A., S. Smoliak, and M.D. MacDonald. 1975. Agronomic variability in *Agropyron smithii*. Can. J. Pl. Sci. 55:101–106.

McWilliams, J.L., and P.E. van Cleave. 1960. A comparison of crested wheatgrass and native grass mixtures seeded on rangeland in eastern Montana. J. Range Mangt. 13:91–94.

Riegel, A. 1940. A study of the variations in the growth of blue grama grass from seed produced in various sections of the Great Plains region. Kans. Acad. Sci. Trans. 43:155–171.

Test Adaptation and Evaluation Trials of Forage Plants in Major Ecological Regions of Ethiopia and Kenya

K.M. IBRAHIM and A. ORODHO
National Agricultural Research Station,
Kitale, Kenya

Summary

Kenyan and Ethiopian natural grazing lands support more than 20% of the cattle population of the African continent. Many of the natural pastures of these countries are severely overgrazed, as unfavorable grazing conditions prevail for over half the year. Objectives of adaptation trials were to evaluate and identify introduced or indigenous forage and fodder species or ecotypes having genetic potential for high production of digestible dry matter for direct pasture and range improvement and for future use in breeding programs. Over 4,000 indigenous and 3,000 introduced ecotypes were evaluated for 24 vegetative characters in six ecological regions between 1971 and 1980.

Promising species in the cool wet highlands situated between 2,300 and 3,000 m with rainfall ranging between 1,000 and 1,250 mm were oats (*Avena sativa* L.), woolypod vetch (*Vicia villosa* Roth.), perennial ryegrass (*Lolium perenne* L.), Kentucky bluegrass (*Poa pratensis* L.), cocksfoot (*Dactylis glomerata* L.), and tall canarygrass (*Phalaris aquatica* L.). Promising species in the temperate wet highlands ranging between 1,850 and 2,000 m with rainfall between 1,000 and 1,270 mm were woolypod vetch, oats, rhodesgrass (*Chloris gayana* Kunth.), setaria (*Setaria anceps* Stapf ex Massey), tall guineagrass (*Panicum maximum* Jacq.), napiergrass (*Pennisetum purpureum* Schumach), and fine stylo (*Stylosanthes guianensis* [Aubl.] Sw.). In the warm wet medium altitudes of 1,500 to 1,700 m with rainfall between 1,250 and 2,000 mm the promising species in the wetter areas were napiergrass, rhodesgrass, tall guineagrass, and tall setaria. In drier areas, the most productive species were buffelgrass (*Cenchrus ciliaris* L.), colored guineagrass (*Panicum coloratum* L.), Wilman lovegrass (*Eragrostis superba* Peyr.), glycine (*Glycine wightii* [Grah. ex Wight & Arn] Verdc.), siratro (*Macroptilium atropurpureum* [D.C.] Urb.), silverleaf desmodium (*Desmodium uncinatum* [Jacq.] D.C.), and fine stylo. The hot dry lowland region in Kenya is between 150 and 900 m with average annual rainfall between 450 and 850 mm. Species adapted to this region included Wilman lovegrass, buffelgrass, scabra stylo (*S. scabra* Vog.), koa haole (*Leucaena leucocephala* [Lam.] de Wit), glycine, siratro, and centro (*Centrosema pubescens* Benth.). The hot humid coastal region in Kenya has an average annual rainfall of 1,015 to 1,270 mm, and species adapted to this region included napiergrass, rhodesgrass, tall guineagrass, tall setaria, buffelgrass, Wilman lovegrass, stargrass (*Cynodon dactylon* [L.] Pers.), cowpea (*Vigna unguiculata* [L.] Walp.), fine stylo, glycine, centro, and siratro.

The potential forage production of the different ecological regions could be substantially increased by sowing highly productive forage and fodder species and by managing these lands properly.

KEY WORDS: Ethiopia, Kenya, forage crops, adaptation, evaluation, East Africa.

INTRODUCTION

Eastern Africa is a region of great climatic and ecological extremes, with altitudes ranging from sea level to over 5,000 m and mean rainfall from less than 250 mm to over 2,000 mm (Table 1). About 80% of Kenya's 58 million ha and 55% of Ethiopia's 122 million ha are generally considered to be too dry for cultivation but suitable for livestock production. About 50% to 60% of the Kenyan and Ethiopian cattle population (about 35 million) and about 35% of the sheep population (15 million) are kept in these areas. About 92% of Kenya's grade cattle are found in cultivated areas with high production potential.

The Kenyan and Ethiopian natural grazing lands support more than 20% of the cattle population of the African continent. Grazing is the predominant feeding practice, and much of the natural pasture is severely overgrazed.

The East African Livestock Survey undertaken by a United Nations team (FAO 1967) indicated that in areas where tick-borne diseases are controlled, nutrition is the main factor limiting efficient animal production. The team suggested that priority be accorded to nutritional investigations and methods of improving pasture productivity.

Potential production of a large part of the land masses of Kenya and Ethiopia could be substantially increased by sowing highly productive forage and fodder species and

properly managing the improved forage-producing areas. Indigenous species have been used more commonly than introduced materials for improving pastures and rangelands in Kenya. Bogdan (1965) and Mwakha (1967, 1971) have listed promising cultivars resulting from local collection followed by subsequent selections between 1950 and 1970.

Objectives of our adaptation trials were to identify introduced and indigenous forage and fodder species or ecotypes with genetic potential for high production of digestible dry matter for direct pasture and range improvement and for future use in breeding programs. Over 3,000 ecotypes of grasses and legumes were introduced and tested at 10 sites in Ethiopia between 1971 and 1976, and over 4,000 indigenous and introduced ecotypes of grasses, legumes, and fodder shrubs were tested at 15 sites in Kenya between 1974 and 1980 (Ibrahim 1981a, 1981b, 1981c, Ibrahim and Chabeda 1978).

METHODS

Grasses and legumes were sown in replicated 3- or 4-row plots 5 m long. The distance between rows varied (30–50 cm) according to edaphic and other ecological conditions, growth habit of the species, and weed problems in the region. Unreplicated rows were used in testing large numbers of species and when seed quantities were small. Fodder shrubs were planted 3–5 m apart. Every entry was evaluated for 24 vegetative characters: emergence data and rate; seedling and plant vigor; growth habit; tillering; foliage canopy and inflorescence height; density, texture, and layering of foliage; stem diameter; flowering date and period; length of growing season; drought resistance; dor-

K.M. Ibrahim is FAO Project Manager for Forage Seed Multiplication and Pre-Extension, and A. Orodho is Senior Pasture Research Officer at the station.

Table 1. Ecological regions in Ethiopia and Kenya.

Regions	Altitude m	Annual rainfall mm	Av. annual Min. temp. °C	Av. annual Max. temp. °C	Months of Max. rainfall
Cool wet highlands	2150 – 3000	1000-1250	<6	<18	April & July
Temperate wet highlands	1850-2000	1000-1270	6-15	18-26	August
Warm wet medium altitude	1500-1700	1250-2000	10-14	26-30	March & May-August
Warm dry medium altitude	1100-1600	600-900	14-18	26-30	April & Oct.
Hot dry lowland	150-900	450-850	≥20	30-35	April & Oct.
Hot wet coastal land	0-150	1000-1250	≥22	26-30	April & Sept.

mancy; regrowth; leaf, stem, and seed diseases; seed production; and insect and bird damage. Persistence, regeneration of perennials, and self-reseeding of promising annuals were recorded over a 2-year period. Promising species were subjected to more vigorous testing of compatibility with other grasses and legumes, fertilizer responses, seeding rate and cultural trials, sowing data and harvesting stage trials, and tolerance to different grazing pressures. Plot sizes varied between 4 and 9 m², and the number of replications was from 3 to 5, depending on the statistical analyses used in these different studies.

RESULTS

Cool Wet Highland Region (Ethiopia)

Two zones of this region in Ethiopia can be distinguished: (1) central highlands over 2,400 m average altitude (under 2,760 m), and (2) central highlands between 2,760 and 3,000 m altitude. Of over 9,600 entries tested or screened in the first of these zones, the most promising annual legume was woolypod vetch (*Vicia villosa* Roth.), which reseeded itself and became naturalized in native pasture. The 6 most productive and disease-resistant oat accessions were Jasari, Lampton, CI-8235, CI-8237, CI-82351, and CI-8257. None of the perennial legumes survived after the first season. The promising grasses were perennial ryegrass (*Lolium perenne* L.), tall canarygrass (*Phalaris aquatica* L.), meadow fescue (*Festuca pratensis* Huds.), Kentucky bluegrass (*Poa pratensis* L.), and cocksfoot (*Dactylis glomerata* L.).

In the second zone, central highlands between 2,700-3,000 m altitude, over 500 entries were tested. Performance of the annual legumes and grasses, except for the oats, was rather poor. The most promising perennial grasses were those reported above for the slightly lower altitude zone (over 2,400 m).

Temperate Wet Highland Region (Kenya)

About 480 grasses and legumes, one fodder shrub, and the world collection of oats were tested. Woolypod vetch

and oats were the most promising annuals. The most productive and best-adapted perennials were rhodesgrass (*Chloris gayana* Kunth.), setaria (*Setaria anceps* Stapf ex Massey), tall guineagrass (*Panicum maximum* Jacq.), napiergrass (*Pennisetum purpureum* Schumach), fine stylo (*Stylosanthes guianensis* [Aubl.] Sw.), and silverleaf desmodium (*Desmodium uncinatum* [Jacq.] DC).

Warm Wet Medium-Altitude Region

In Kenya, 109 ecotypes of perennial legumes and grasses were tested. The most promising were glycine (*Glycine wightii* [Grah. ex Wight & Arn] Verdc.), fine stylo, centro (*Centrosema pubescens* Benth.), siratro (*Macroptilium atropurpureum* [DC.] Urb.), rhodesgrass, setaria, tall guineagrass, and napiergrass. In Ethiopia, 104 ecotypes were tested, including one fodder shrub. Silverleaf desmodium proved outstanding and was fire-resistant; fine stylo also showed excellent adaptation. Koa haole (*Leucana leucocephela* [Lam.] de Wit) was a promising fodder shrub, two-thirds of the whole plant being edible (Kernich 1980, FAO, Rome, personal communication). Promising grass species included napiergrass, outstanding for both grazing and cutting; colored guineagrass (*Panicum coloratum* L.); and rhodesgrass; *Brachiaria mutica* (Forsk.) Stapf showed outstanding promise in swamp areas.

Warm Dry Medium-Altitude Region

In Central Kenya, two subregions were recognized. In the drier subregion, 297 ecotypes and in the wetter subregion, 1,315 ecotypes, including 5 fodder shrubs, were tested. The most productive and best-adapted species in both subregions were rhodesgrass, tall guineagrass, buffelgrass (*Cenchrus ciliaris* L.), Wilman lovegrass (*Eragrostis superba* Peyr.), siratro, glycine, and *Macrotyloma axillare* (E. Mey.) Verdc. Colored guineagrass and scabra stylo (*S. scabra* Vog.) were adapted and productive in the semiarid areas, while napiergrass, tall setaria (*Setaria splendida* Stapf), fine stylo, and silverleaf desmodium were productive in the wetter subregion.

In the Rift Valley of Ethiopia, a total of 570 ecotypes,

including one fodder shrub, were tested. The most promising annual legume was pigeonpea (*Cajanus cajan* [L.] Hutt). Several very promising selections of sorghum also were identified. The most promising perennial forages were rhodesgrass, guineagrass, buffelgrass, siratro, and lablab (*Dolichos lablab* L.).

Hot Dry Lowland Region

A total of 860 ecotypes of perennial grass and legume species were tested. The most promising species in this region were buffelgrass and Wilman lovegrass. In areas receiving the upper limits of rainfall, tall guineagrass and rhodesgrass were well adapted and productive. In good years, napiergrass was well adapted and productive; in dry years, its production was very low. Glycine, siratro, centro, scabra stylo, *M. axillare,* and koa haole were the most productive legumes. All the selected species showed marked drought tolerance and higher productivity than the local native vegetation. The real contribution to this semiarid region is likely to occur from vast introductions of fodder shrubs.

Hot Humid Coastal Region

A total of 622 ecotypes of perennial grasses and legumes were tested. The outstanding species were tall guineagrass, buffelgrass, Wilman lovegrass, stargrass (*Cynodon dactylon* [L.] Pers.), napiergrass, *Digitaria milanjiana* (Rendle) Stapf, *Vigna vexillata* (L.) A. Rich., cowpea (*V. unguiculata* [L.] Walp.), fine stylo, *Clitoria ternatea* L., glycine, centro, and siratro.

DISCUSSION

The most successful and widely accepted forage species are those having a wide range of adaptation. Species with narrow ecological adaptation are unlikely to find their way into the infant seed industry of the developing countries even if they are highly productive. Other important characters include ease of establishment, potential long growing season, tolerance to heavy grazing, disease and drought resistance, high forage and seed production, and high nutritive value. A great number of these characters could be realized by selection at the ecotypic level within promising species. A still greater number of these

characters could be realized by proper mixtures of grasses, legumes, and fodder shrubs. The incompatibility of tropical grasses and legumes poses management problems. In early trials legumes of high palatability, such as *Desmodium* spp., could not be maintained under field grazing. Highly productive and palatable legumes could be maintained as a source of high-quality forage if planted and managed separately from grasses.

Recent trials demonstrated that grass and legume mixtures could be maintained for a longer period under field-grazing conditions if legumes such as *Stylosanthes* spp. that have relatively lower palatability than *Desmodium* spp. were used.

In adaptation trials on fodder shrubs in the dry areas the shrubs remained green and productive the year round. Fodder shrubs usually are lightly browsed during the growing season, but during the dry season or in dry years they provide excellent forage for livestock. Thus, fodder shrubs contribute to the stability of fodder production, both quantitatively and qualitatively, in semiarid areas, and they could also be utilized as forage reserves for livestock raised in adjacent dry areas.

LITERATURE CITED

Bogdan, A.V. 1965. Cultivated varieties of tropical herbage plants in Kenya. East Afr. Agric. For. J. 30:330–338.

Food and Agricultural Organization of the United Nations. 1967. East African livestock survey. UNDP/FAO 1:134–135. FAO/SF:21 REG. ROME. FAO, Rome.

Ibrahim, K.M. 1981a. Pasture and fodder collection. FAO project forage collection and evaluation for animal production. Natl. Agric. Res. Sta., Kitale, Kenya. Rpt. 1:79.

Ibrahim, K.M. 1981b. Forage plant introduction. FAO project forage collection and evaluation for animal production. Natl. Agric. Res. Sta., Kitale, Kenya. Rpt. 2:42.

Ibrahim, K.M. 1981c. Evaluation of indigenous and introduced species in Kenya. FAO project forage collection and evaluation for animal production. Natl. Agric. Res. Sta., Kitale, Kenya. Rpt. 3:16.

Ibrahim, K.M., and A. Chabeda. 1978. Forage exploration and evaluation in Kenya. *In* Proc. 1st Int. Rangeland Cong., 339–342. Soc. Range Mangt., Denver, Colo.

Mwakha, E. 1967. Progress in pasture plant introduction in Kenya. CSIRO Pl. Intro. Rev. 4:24–27.

Mwakha, E. 1971. Potential of pasture legumes in Kenya. Kenya Fmr. 181:11.

Growth and Development of Two *Echinochloa* Millet Species in a Warm Temperate Climate

D.K. MULDOON

Department of Agriculture, Agricultural Research Station,
Trangie, N.S.W., Australia

Summary

Most summer forage crops fall into either the sorghum or the millet species group. Included in the millet group are two species, *Echinochloa utilis* and *Echinochloa frumentacea*, commonly known as Japanese barnyard millet and Indian barnyard millet, respectively. These species have a negligible HCN potential and higher nutritive value than *Sorghum* species. However, they generally flower early, yield less than forage sorghum, and lack vigor in regrowth. In this study the pattern of growth and development of two *Echinochloa* millets was examined as part of a larger program of improving the production of these forage species.

E. utilis cv. Shirohie and *E. frumentacea* cv. White panicum were grown under surface irrigation at Trangie (lat 32°S long 148°E). Mean weekly maximum and minimum temperatures ranged from 37°/23° C to 28°/14° C during the growing period. Uninterrupted growth was measured by sampling every 2 weeks; regrowth from sampled areas was measured at the end of the growing season. Leaf area was determined, and the dried plant fractions were weighed and analyzed for nitrogen, sodium, and sulfur.

The maximum growth rate and dry-matter yield of *E. utilis* (30 g/m²/day and 11.7 tonnes/ha, respectively) were greater than those of *E. frumentacea* (11 g/m²/day and 5.9 tonnes/ha, respectively). Morphological features including a prostrate growth habit and smaller leaf size were associated with the low yields of *E. frumentacea*.

Late flowering in *E. frumentacea* (20 days later than *E. utilis*) indicated delayed elevation of the apical meristem; regrowth was better in this species. When the extent of tiller decapitation was equivalent in the two species, regrowth yields were similar. To obtain good regrowth, it is necessary to cut or graze a crop of *E. utilis* before 7 weeks. Alternatively, regrowth may be improved by otaining cultivars that flower later and/or tiller more extensively.

E. utilis was higher in sodium and sulfur than *E. frumentacea*; nitrogen contents were similar. Mineral concentrations changed with time and, especially with sodium, differed between plant parts. For both species, mineral levels exceeded animal requirements, although the N/S ratio for *E. frumentacea* was marginal.

Important agronomic differences between the two *Echinochloa* millets have been elucidated. The characterization of growth and development has demonstrated how high forage yields in these nutritionally desirable species may be obtained.

KEY WORDS: *Echinochloa utilis, Echinochloa frumentacea,* barnyard millet, growth, development, morphology.

INTRODUCTION

Summer annual forage crops are often used to supplement pasture production in animal grazing systems. *Echinochloa* millets and forage sorghums both are widely grown for this purpose. Forage sorghums, which have been improved by hybridization, tend to dominate forage-crop production. However, recent investigations (Stobbs 1975, Hedges et al. 1978) have revealed serious deficiencies in the quality of sorghum forage for animal production.

The main limitation of *Echinochloa* millets is their relatively low dry-matter production in both primary growth and regrowth. There is little agronomic or physiological information on which to base suggestions as to the cause of low production. The present study was aimed at gaining a greater understanding of the growth and development of these millets, a prerequisite for any improvement in this forage. The two species examined were *E. utilis* cv. Shirohie and *E. frumentacea* cv. White panicum, commonly known as Japanese barnyard millet and Indian barnyard millet, respectively. The taxonomic distinction of these species was made by Yabuno (1971). However, agronomically important differences have not previously been described.

MATERIALS AND METHODS

The millets *E. utilis* cv. Shirohie and *E. frumentacea* cv. White panicum were sown at the Agricultural Research Station, Trangie (lat 32°S long 148°E), at the rate of 7 kg seed/ha. The soil was a heavy, black, self-mulching soil

The author is a Research Agronomist at the Agricultural Research Station.

(pH 8.0); fertilizer phosphorus (30 kg P/ha) as single superphosphate and nitrogen (68 kg N/ha) as ammonium nitrate were incorporated before sowing. Spray irrigation was used initially to obtain an even germination, and then the plots were flood irrigated at 8–10-day intervals. During the growing season only 114 mm of rainfall was received. Mean weekly maximum and minimum temperatures ranged from 37°/23°C in December to 28°/14°C in March, and daylength decreased from 14.2 to 12.1 hours.

Plots (9 × 10 m) were replicated 3 times in a randomized-block design. Primary growth was measured by sampling 1 m² from each plot every 2 weeks, commencing 5 weeks after sowing. Plants were cut at 2–3 cm, and a subsample was separated into green leaf, stem, and inflorescence or head fractions. Height was measured to the top ligule and to the apical meristem of the tallest tiller. Leaf area was determined on an electronic area meter. The plant fractions were dried at 80°C and milled. Nitrogen, sodium, and sulfur contents were determined by the Kjeldahl, atomic absorption, and tubidimetric methods, respectively.

Regrowth from the sampled areas was measured 15 weeks after sowing, when the growing season was almost complete.

RESULTS

Dry-Matter Accumulation

The rate of dry-matter accumulation was much greater in *E. utilis* (Eu) than in *E. frumentacea* (Ef) (Fig. 1). Mean growth rates from 5 to 13 weeks were 19 and 8 g/m²/day, respectively, while the maximum growth rates over a 2-week period were 30 and 11 g/m²/day, respectively. Although Ef continued dry-matter accumulation for 2 weeks longer than Eu, its maximum dry-matter yield was only half that of Eu. Differences between these species were also apparent in the relative proportions of leaf, stem, and inflorescence in the total dry weight.

The regrowth yield of Eu fell sharply when it was cut 7 weeks from sowing or later (Fig. 1). Regrowth yields of Ef were superior to those of Eu, except at 5 weeks. The combined primary growth and regrowth yields of Ef were similar throughout the growth period, and between 7 and 11 weeks they were only 1–2 tonnes/ha lower than those of Eu. Regrowth characteristics of these two species were associated with morphological differences.

Morphology and Development

The number of plants/m² at 5 weeks was 82 ± 15 for Eu and 53 ± 6 for Ef. Eu tillered early, completing tiller production by the time sampling commenced at 5 weeks (Table 1). At this time Ef had fewer tillers/unit area (although a similar number of tillers/plant), but it continued tiller production until 9 weeks, when the total of over 600 tillers/m² was 25% greater than the total reached by Eu.

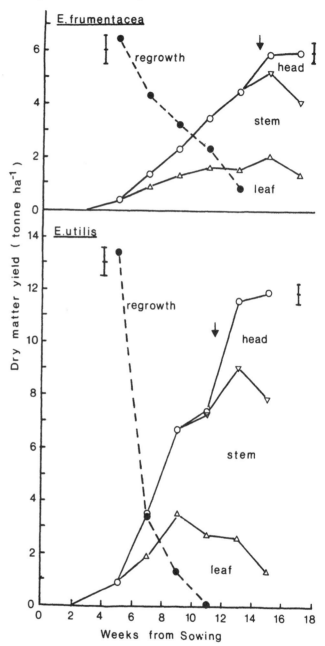

Fig. 1. The pattern of dry-matter accumulation in different plant fractions during primary growth, together with the regrowth yield of two *Echinochloa* species. The vertical bar represents twice the mean standard error. Time of head emergence is indicated by an arrow.

The two millets exhibited markedly different habits of growth. Eu rapidly increased in height once it was established and grew to a height of 131 cm. Ef remained prostrate until shortly before inflorescence emergence and then reached a height of only 82 cm. Underlying the differences in plant height were changes in the height of the apical meristem. In Eu the apical meristem had begun to rise by 7 weeks; cutting at 8 weeks resulted in half of the tillers' being decapitated (apical meristem removed). In Ef, elevation of apical meristem did not commence until

Table 1. Changes in tiller number, leaf number and leaf area with time in *E. utilis* (Eu) and *E. frumentacea* (Ef)

	Species	Weeks from sowing					
		5	7	9	11	13	15
Tillers	Eu	465	410	270	246	172	145
/m²	Ef	266	544	623	452	255	327
Leaves	Eu	17	18	15	16	12	7
/dm²	Ef	11	19	24	23	12	22
Leaf-area index	Eu	1.7	5.0	7.3	7.6	5.8	2.3
	Ef	0.7	2.0	2.8	3.9	3.1	4.5

Mean SE for tillers/m², 35; leaves/dm², 1.5; and leaf-area index, 0.4.

10 weeks, so 50% decapitation did not occur until 12 weeks. This feature was advantageous to the regrowth of Ef.

Increases in leaf area in the two species followed increases in dry weight except in the late stages of growth. After 11 weeks the leaf area of Eu decreased rapidly due to senescence of the basal leaves (Table 1). The poor early growth and lower dry-matter yield of Ef can be attributed to its poor leaf-area development. Leaf area increment between 5 and 11 weeks was 20.0 dm²/m²/day for Eu and 7.7 dm²/m²/day for Ef. This difference, approximately 60%, was the same as the difference in dry-matter increments. The maximum leaf-area index attained by Eu was 7.6 compared to 4.5 for Ef. Ef therefore was handicapped by restricted leaf-area development. This disadvantage resulted not from fewer leaves but from a smaller individual leaf size. Mean individual leaf sizes at 7 weeks, when leaf numbers were similar (Table 1), were 27 and 10 cm² for Eu and Ef, respectively. Maximum leaf sizes attained were 48 and 20 cm², respectively. Leaves of Ef were more noticeably keeled and shorter than those of Eu.

Mineral Content

The nitrogen content of Eu leaves and stems decreased linearly with time (Fig. 2). In Ef the decrease was similar for the first 9 weeks; thereafter, a much slower rate of decrease was apparent. Consequently, at the time of head emergence the nitrogen contents of the two species were similar. The sodium content of the two species differed greatly. For both species sodium levels peaked during the middle of the growing period. Eu was also higher in sulfur than Ef, with no significant ($p = 0.05$) difference between the leaf and stem fractions. The sulfur level initially decreased but increased at the end of flowering. This increase indicates continued uptake of sulfur beyond the time of head emergence.

DISCUSSION

E. utilis (Eu) accumulated a large bulk of dry matter. The yield of 5 tonnes/ha at 8 weeks is well above previ-

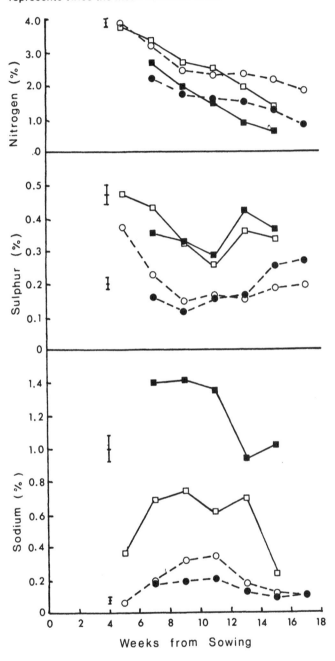

Fig. 2. Changes with time in the nitrogen, sulfur, and sodium concentrations in leaf and stem material of two *Echinochloa* species. *E. utilis* leaf is represented by □ and stem by ■; *E. frumentacea* leaf by ○ and stem by ●. The vertical bar represents twice the mean standard error.

ously recorded yields from nonirrigated sites (Hedges et al. 1978). The yield at the end of the growing season was 11.7 tonnes/ha, a level that would be close to the maximum attainable in this environment for this species. The crop was sown early enough to allow full growth and development, as was evidenced by the slight interruption of dry-matter accumulation just before head emergence. A delay in flowering would only lengthen this interruption. To obtain a greater bulk of dry matter, taller types would be required.

Cutting usually results in extensive removal of apical meristems and hence poor regrowth in Shirohie and other cultivars of Eu (Hedges et al. 1978). Decapitated tillers do not regrow and, in this species, produce few new basal tillers. An examination of other cultivars of this species, of which over 60 exist (Yabuno 1971), may reveal cultivars in which stem elongation occurs later in the growing season. Alternatively, cultivars that tiller more profusely may be found, and these would also improve the regrowth of Eu.

Dry-matter accumulation in Ef was slow, and the maximum yield obtained was only 6 tonnes/ha. Stobbs (1975) obtained a comparable yield, but the pattern of accumulation differed in his later-sown crop; he recorded a higher peak growth rate as the crop rapidly matured late in the growing season. In my study a temporary ceiling yield appears to have been reached prior to head emergence, as in Eu. The growth of Ef was restricted by poor leaf-area development, a consequence of small leaf area. A type of Ef having larger leaves and a more upright habit is to be preferred.

In Ef head emergence occurred 100 days after sowing, some 20 days later than in Eu. Later flowering was associated with later elevation of meristems in Ef and this habit favored regrowth in this species. However, at an equivalent severity of decapitation Ef was not superior in regrowth.

Eu should be cut or grazed before 7 weeks if substantial regrowth is to be expected. This conclusion is similar to those of other studies in which cutting at 8 or 9 weeks resulted in poor growth (Hedges et al. 1978). For Ef the critical time for cutting or grazing is 11 or 12 weeks. Obviously, environmental factors and height of cutting modify these recommendations.

The high nutritive value of *Echinochloa* millets is becoming increasingly recognized. At Trangie, sodium levels in Eu were much higher but sulfur contents were similar to those found in previous studies (Hedges et al. 1978). The nitrogen/sulfur ratio ranged from 5 to 10, as was found by Hedges et al. (1978), and was relatively lower in the stem fraction. In Ef the sulfur content quickly fell to a level as low as that found in *Sorghum* species. The N/S ratio exceeded the suggested limit of 10:1 (Wheeler 1980); however, this marginal sulfur deficiency would not be exacerbated, as it is in *Sorghum* species, through sulfur utilization in detoxifying HCN in rumen.

LITERATURE CITED

Hedges, D.A., J.L. Wheeler, C. Mulcahy, and M.S. Vincent. 1978. Composition and acceptability to sheep of twelve summer forage crops. Austral. J. Expt. Agric. Anim. Husb. 18:520–526.

Stobbs, T.H. 1975. A comparison of zulu sorghum, bulrush millet and white panicum in terms of yield, forage quality and milk production. Austral. J. Expt. Agric. Anim. Husb. 15:211–218.

Wheeler, J.L. 1980. Increasing animal production from sorghum forage. World Anim. Rev. 35:13–22.

Yabuno, T. 1971. A note on barnyard millet. SARRAO Newsletter 3:43–45.

Summer Dormancy in Italian Populations of *Dactylis glomerata* L.

F. LORENZETTI, A. PANELLA, and E. FALISTOCCO
University of Perugia, Perugia, Italy

Summary

Mediterranean types of perennial grasses are winter-growing and summer-dormant, while central and northern European types are winter-dormant and summer-growing. Dormancy severely limits the annual production and largely determines the seasonal distribution pattern of the perennial grasses. Summer dormancy was studied in 81 Italian populations of *Dactylis glomerata* L. collected from north (48°N) to south (38°N) on the Italian peninsula. Dormancy was measured at Perugia (43°05'N) on 60 plants for each population. The percentage of summer-dormant plants was negatively correlated with the mean score for summer growth (r = − 0.75**). The results distinguish three main groups of populations with increasing levels of summer dormancy, namely, northern, central, and southern. Winter growth and greenness were negatively correlated with summer growth (r = − 0.80**). Among Italian populations of *Dactylis glomerata* L., variability for dormancy, both in summer and winter, is so high that breeding of adapted varieties with the desired growth rhythm should not be difficult.

KEY WORDS: orchardgrass, *Dactylis glomerata* L., dormancy, Italian populations.

INTRODUCTION

Summer dormancy may be defined as the development of dormant buds following the death of all leaves; it occurs in several populations of perennial grasses grown in typical Mediterranean environments at the beginning of the summer season. This type of dormancy is not merely a restrained growth, because when dormant buds are formed, plants can survive adverse conditions for many weeks (Hoen 1968). Following seed maturity the leaves senesce, leaving only stems that die back slowly, often retaining green tissue in the basal internodes throughout the entire summer (McWilliam 1968).

In Mediterranean environments, vegetative tillers of *Lolium perenne* contribute little to summer survival (Silsbury 1964), but northern European types have a contrasting rhythm of dormancy and growth because they are summer-active and winter-dormant (Cooper 1963).

In perennial grasses, summer dormancy has received relatively little attention (Laude 1953, Vegis 1964), perhaps because many agricultural environments are not so extreme as to necessitate dormancy (Wilson 1979).

Although dormancy is a powerful means of persistence, it severely limits annual yield and largely determines seasonal yield distribution of orchardgrass (*Dactylis glomerata* L.) in Italy. In our breeding programs, therefore, special attention must be paid to the variability for dormancy among and within populations before they are used as basic material for new varieties.

METHODS

Italy, with its longitudinal extension and its variable climate, provided useful materials for a geographical survey of dormancy. In recent years, 81 populations of orchardgrass were collected at sites scattered from north (48°N) to south (38°N) in the Italian peninsula; 27 populations will be referred to as northern populations, 27 as central populations, and 27 as southern populations.

All populations were tested at Perugia (43°05′N), where summer dormancy was measured on 60 plants − 10 plants of each population distributed in 6 replicates. On 7 August 1979, we scored the plants as being dormant (1) or having varying amounts of green foliage (from 2 to 9) and classified the populations according to percentage of dormancy (ratio between plants scored 1 and total number of plants).

F. Lorenzetti is a Professor of Agricultural Genetics at the University of Perugia, A. Panella is a Professor of Plant Breeding at the university, and E. Falistocco is a Fellow of the Italian Research Council.

On 18 December 1979 the same plants were scored for winter dormancy and growth following the same criteria described for summer dormancy, and data were collected on heading time in the following spring.

RESULTS

All the northern populations had a summer dormancy percentage between 0% and 20%, while central and southern populations had percentages ranging from 10% to 40% and from 0% to 100%, respectively (Table 1). The southern populations as a group were much more variable than the other groups.

The score means of summer growth were 3.8 ± 1.68, 3.1 ± 1.61, and 1.8 ± 1.23 for groups from north, center, and south, respectively. These means agree with data in Table 1 as well as with the reported number of population with a given mean (Table 2).

The distribution of populations for winter growth complemented that for summer growth, because the sums of summer and winter scores for each population tend to be constant. This complementary relationship holds true also for variability, the level of which is maximum in the southern populations for summer growth and in the northern populations for winter growth.

Heading time showed an unexpected trend, with northern populations being the earliest. The overall means for this character were 25 April, 3 May, and 2 May for northern, central, and southern populations, respectively.

DISCUSSION

It is not difficult to relate growth and dormancy rhythms of the Italian populations of *Dactylis glomerata* L. with rainfall and temperature of their area of origin. The climatic diagrams for Cavaglià, Vercelli (45°20′N), in the north and for San Cataldo, Lecce (40°22′N), in the south, chosen as typical sites for north and south, respectively, give a clear picture of the situation (Fig. 1). In Italy, lack of rainfall in summer and low temperatures in winter seem to be the main factors influencing dormancy, because maximum daily temperatures are similar during summer and precipitation is never a limiting factor during winter. The correlation coefficient between summer growth at Perugia and May-June rainfall in the place of origin is 0.70**, and the correlation coefficient between winter growth and average minimum temperature of the coldest month (January) is 0.77**. As a consequence of the climatic situation of the collection sites, there is a negative correlation of − 0.80** between summer and

Table 1. Populations of *Dactylis glomerata* from three geographic areas of Italy classified according to percentage of summer dormancy at Perugia, 1979.

	Number of populations with percentage dormancy of:										
Origin	*100*	*90*	*80*	*70*	*60*	*50*	*40*	*30*	*20*	*10*	*0*
North	–	–	–	–	–	–	–	–	1	16	10
Center	–	–	–	–	–	–	1	7	10	9	–
South	4	3	4	1	–	1	2	2	4	5	1

Table 2. Populations of *Dactylis glomerata* classified on the basis of mean score for summer and winter growth at Perugia, 1979.

	Population mean score								
Origin	*1**	*1.1-2*	*2.1-3*	*3.1-4*	*4.1-5*	*5.1-6*	*6.1-7*	*7.1-8*	*8.1-9*
	--No--								
Summer growth									
North	–	–	3	13	11	–	–	–	–
Center	–	–	10	16	1	–	–	–	–
South	4	14	8	–	1	–	–	–	–
Winter growth									
North	2	2	5	6	6	4	2	–	–
Center	–	–	–	1	4	11	10	1	–
South	–	–	–	–	2	–	3	8	14

*Score 1 = dormant, 9 = very active.

Fig. 1. Climatic diagrams of representative sites of northern, central, and southern Italy and mean scores for summer and winter growth of local populations of *Dactylis glomerata* L.

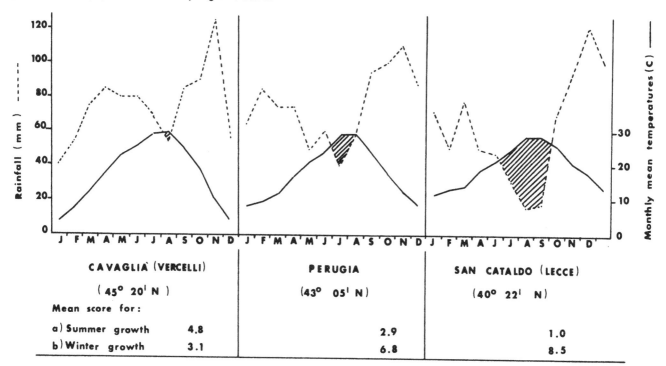

winter growth. The percentage of dormant plants in summer was correlated with summer growth scores of nondormant plants (r = − 0.75**). This correlation indicates that populations may be classified for dormancy when the climatic conditions at the evaluation site do not permit a sharp and arbitrary distinction between dormant and nondormant plants, but further research is needed on this point.

In Mediterranean environments, dormancy of natural populations begins at seed maturity and coincides with onset of the dry period. In these environments, natural selection has favored plant types that spread their seeds and develop dormant buds at the base of their reproductive stems just at the beginning of summer when the climatic conditions prevent seed germination and the development of basal tillers. This perfect adaptive mechanism may also explain the unexpected lateness of heading time observed in central and southern populations as compared with populations from northern Italy.

Natural selection has achieved too high a summer dormancy level in present orchardgrass pastures. Our results demonstrate that opportunities exist in plant breeding to utilize parent plants or populations selected from less severe environments in development of more productive varieties.

LITERATURE CITED

Cooper, J.P. 1963. Species and populations differences in climatic response. *In* L.T. Evans (ed.) Environmental control of plant growth, 381–404. Academic Press, New York.

Hoen, K. 1968. Summer dormancy in *Phalaris tuberosa* L. Austral. J. Agric. Res. 19:227–239.

Laude, H.M. 1953. The nature of summer dormancy in perennial grasses. Bot. Gaz. 114:284–292.

McWilliam, J.R. 1968. The nature of the perennial response in Mediterranean grasses. II. Senescence, summer dormancy and survival in *Phalaris*. Austral. J. Agric. Res. 19:397–409.

Silsbury, J.H. 1964. Tiller dynamics, growth and persistency of *Lolium perenne* L. and *Lolium rigidum* Gaud. Austral. J. Agric. Res. 15:9–20.

Vegis, A. 1964. Dormancy in higher plants. Ann. Rev. Pl. Physiol. 15:185–224.

Wilson, D. 1979. Physiological and morphological characteristics of herbage crops in relation to tolerance to climatic stress. *In* Breeding for stress conditions. Rpt. of Mtg. Eucarpia Fodder Crops Section, Perugia, Italy, 4–6 September 1979, 57–76.

Trials on Time of Nitrogen Application in the Spring to Various Grasses Grown for Seed Production

A. NORDESTGAARD

State Research Station, Roskilde, Denmark

Summary

With most of the grasses grown for seed in Denmark, a series of trials has been carried out on increasing nitrogen rates in the autumn combined with increasing rates in the spring. The results of these trials show that an autumn application of nitrogen is very important to most of the grasses and necessary if the optimum yield of seed is to be achieved.

In order to find the best time to apply nitrogen in the spring when producing grass seed, a number of trials have been carried out in recent years with various grasses. Up to and including 1980, the following trials have been carried out: 29 with *Festuca rubra*, 28 with *Dactylis glomerata*, 4 with *Poa pratensis*, 13 with *Lolium perenne*, 23 with *Festuca pratensis*, 22 with *Phleum pratense*, and 6 with *Phleum bertolinii*.

The first application was made as soon as winter was over and it was possible to drive in the field; the second was made when growth had started; the third application was made when the new tillers were about 15 cm long, with the average dates of application to *Festuca rubra* being 2 March, 1 April, 30 April (these are very near the averages for the other grasses).

The results showed that time of nitrogen application in the spring is of great importance for development of the grass. Delays in nitrogen application to most of the species resulted in negative effects of varying magnitudes on the number of fertile tillers/unit area but positive effects on seed weight and number of seeds/fertile tiller.

The result of these effects on the yield components was a decrease in seed yield of *Festuca rubra*, *Dactylis glomerata*, and *Poa pratensis* but an increase in yield of *Phleum pratense* and *Phleum bertolinii*. As for *Lolium perenne* and *Festuca pratensis*, the positive and negative effects on yield cancelled each other out, and the seed yield did not depend on time of application.

On the basis of these results it is recommended that nitrogen be applied in the spring to *Festuca rubra*, *Dactylis glomerata*, and *Poa pratensis* as soon as winter is over and to *Festuca pratensis* and *Lolium perenne* as soon as growth has started but to *Phleum pratense* and *P. bertolinii* (which ripen late) only when growth is well under way and the grass has tillers about 10 cm long.

KEY WORDS: seed production, nitrogen, time of application, grasses.

INTRODUCTION

In grass-seed production, as in any other kind of plant production, yield depends on a number of factors. Some of these factors, such as climate, air composition, solar energy, and soil, cannot be changed by the grower, but such factors as the amount of available nutrients in the soil can, to a very large extent, be modified by the grower. Nitrogen, generally, is the nutrient applied to Danish farmland that has the largest influence on development of seed grasses. Many trials on the use of nitrogen in grass-seed production have been carried out in recent years.

The author is an Agronomist and Head of Experiments on Seed Production of Grasses and Vegetables with the State Research Station.

INCREASING NITROGEN RATES

In the various seed grasses, trials have been carried out on increasing nitrogen (N) rates in the autumn combined with increasing rates in the spring. Because of these trials, we now have a good knowledge of the yearly nitrogen requirements of the various seed grasses and also of the best way of splitting this yearly nitrogen rate into an autumn application and a spring application. The trials have shown that the optimum N rate in the spring depends on the amount of N applied to the seed field the preceding autumn. In seed production of smooth-stalked meadow grass and red fescue, under Danish conditions, application of at least 50% of the yearly N rate in the autumn is recommended (Larsen and Nordestgaard 1969, Nordestgaard and Larsen 1971). In the case of cocksfoot, meadow fescue, and timothy, application of about 30% of the yearly N rate in the autumn and about 70% in the spring is recommended (Nordestgaard 1972, 1974). For perennial ryegrass, the whole N rate should be applied in the spring (Nordestgaard 1977). Trials have proved that in the case of various seed grasses under Danish conditions, the autumn application of N should be made preferably in the second half of September.

TIME OF SPRING APPLICATION

In order to find the most appropriate time for spring application of N to various seed grasses, trials have been carried out in 1972–1980 according to the following factorial plan with a total of 9 combinations: (a) as soon as winter is over, (b) when the grass begins vernal growth and has tillers about 5 cm long, and (c) when new tillers are 15 to 20 cm long. Level of spring nitrogen application: x = 1N, y = 2N, z = 3N. For perennial ryegrass, cocksfoot, meadow fescue, and timothy 1N = 45 kg N/ha, and for red fescue and smooth-stalked meadow grass 1N = 30 kg N/ha. In the autumn, the trials were fertilized with nitrogen in the second half of September with 60 kg N/ha for red fescue and smooth-stalked meadow grass, 50 kg N/ha for cocksfoot, and 40 kg N/ha for meadow fescue and timothy.

As the effect of increasing N rates in these trials was similar to the effect of increasing N rates in the trials formerly mentioned and the effect of changing the time of application was, in most trials, independent of the use of 1N, 2N, or 3N, the effect of changing time of application will be shown only as an average of the three nitrogen amounts.

Trials were carried out on first- and second-year seed crops and, in the case of red fescue, cocksfoot, and meadow fescue, also on third-year seed crops. Response to the time of nitrogen application was similar in first-, second-, and third-year crops; therefore, the average of all trials is presented.

A delay in nitrogen application in the spring negatively affected the number of fertile tillers/unit area in all seed-grass species, although in timothy the negative effect was small and not significant. The negative effect was partly compensated for by the fact that a delay in N application affected the number of seeds/fertile tiller and the seed weight positively (Table 1). In meadow fescue and perennial ryegrass, the negative and positive effects on the yield components balanced each other. In red fescue, cocksfoot, and smooth-stalked meadow grass, the increases in seed weight and in number of seeds/fertile tiller were not large enough to neutralize the negative effect on the number of fertile tillers, and thus a delay in nitrogen application caused a decrease in seed yield. In timothy, where a delay in N application had no significant effect on the number of fertile tillers, the delay brought about an increase in seed yield.

In other Danish trials, similar results have been obtained for cocksfoot and smooth-stalked meadow grass (Nordestgaard and Larsen 1974). In addition to this, other researchers have shown that a delay in N application in the spring in perennial ryegrass reduces the number of fertile tillers (Hebblethwaite et al. 1980, Hebblethwaite and Ivins 1978, Langer 1980, Ryle 1964, Schöberlein 1972).

In these trials, it was observed that in all grasses a delay in N application also led to an increasing number of secondary vegetative tillers. In a trial with red fescue, carried out in concrete frames in 1974, and in a similar trial with

Table 1.   The effect of time of nitrogen application on mean seed-yield components and seed yield in several species of grasses.

Characteristics	Time of application							
	a*	b*	c*	LSD₉₅	a*	b*	c*	LSD₉₅
	Red fescue (29 trials)				Meadow fesc. (23 trials)			
Date of application	2/3	1/4	30/4		1/3	1/4	30/4	
No. of fertile tillers/m²	3289	3071	2792	92	1705	1625	1577	58
No. of seeds/fertile tiller	28	30	31	2	41	42	43	2
Seed weight, mg/seed	1.19	1.21	1.24	0.02	1.81	1.81	1.90	0.03
Seed yield, hkg/ha	9.9	9.7	9.3	0.2	11.3	11.2	11.4	0.3
	Perenn. ryeg. (13 trials)				Cocksfoot (28 trials)			
Date of application	1/3	30/3	28/4		2/3	1/4	30/4	
No. of fertile tillers/m²	2425	2337	2152	101	865	779	693	23
No. of seeds/fertile tiller	36	37	40	2	127	139	145	8
Seed weight, mg/seed	1.69	1.71	1.73	0.05	1.06	1.07	1.10	0.02
Seed yield, hkg/ha	14.3	14.4	14.4	0.2	10.9	10.5	9.8	0.2
	Sm. st. mead. gr. (4 trials)				Timothy (22 trials)			
Date of applications	10/3	4/4	4/5		2/3	2/4	1/5	
No. of fertile tillers/m²	1983	1772	1578	155	820	800	806	33
No. of seeds/fertile tiller	126	133	127	13	171	187	188	14
Seed weight, mg/seed	0.32	0.32	0.34	0.02	0.36	0.36	0.37	0.01
Seed yield, hkg/ha	8.1	7.7	6.7	0.3	4.6	4.8	5.2	0.2

*a = as soon as winter is over; b = when grass begins vernal growth and has tillers about 5 cm long; c = when new tillers are 15–20 cm long.

Table 2. Effect of time of nitrogen application on number of fertile and vegetative tillers per m² in red fescue (1974) and cocksfoot (1977).

Date of N-application	Red fescue, 60 kg N/ha		Cocksfoot, 90 kg N/ha		
	22/2	26/4	7/3	25/3	3/5
No. of fertile tillers/m²	4513	3900	734	730	525
No. of vegetative tillers/m²	1662	3100	69	79	116
Ratio fertile: vegetative tillers	2.72	1.26	10.6	9.2	4.5

cocksfoot in 1977, both fertile and vegetative tillers were counted (Table 2). With a decrease in the number of fertile tillers caused by delaying the N application and an increase in the number of vegetative tillers, the fertile tiller:vegetative tiller ratio was markedly reduced. Similar results have been found in other Danish trials with cocksfoot and smooth-stalked meadow grass (Nordestgaard and Larsen 1974) and in English trials with perennial ryegrass (Hebblethwaite and Ivins 1978).

SPLIT APPLICATIONS OF NITROGEN

Trials were carried out with the same grasses on splitting the optimum nitrogen amount in the spring into an early application and another one just before earing. Splitting the nitrogen amount did not bring a higher seed yield than applying all the nitrogen at one time in the early spring. The late application increased secondary vegetative tillering. Hebblethwaite and Ivins (1978) have also demonstrated that there is no advantage in splitting optimum levels of nitrogen in perennial ryegrass and that nitrogen applied late substantially increases the amount of secondary vegetative tillering. These tillers are undesirable in seed crops, as they compete with fertile tillers for metabolites, water, and light until they may become self-supporting (Hebblethwaite 1977), and they also make harvesting more difficult.

CONCLUSION

Even if grass-seed crops are well supplied with nitrogen in the autumn, the time and level of spring nitrogen application is important in most species. A delay in spring nitrogen application decreases fertile tiller number but increases seed weight and number of seeds/fertile tiller in red fescue, smooth-stalked meadow grass, cocksfoot, meadow fescue, and perennial ryegrass. However, these responses in yield components significantly decrease seed yield in red fescue, cocksfoot, and smooth-stalked meadow grass. A delay in nitrogen application in timothy, however, does not lead to a decrease in the number of fertile tillers and increases seed yield. In most species, delay in nitrogen application results in heavier lodging and increased secondary vegetative tillering.

On the basis of these results, it is recommended that nitrogen fertilizer should be applied to red fescue, smooth-stalked meadow grass, and cocksfoot as soon as possible in the spring and to meadow fescue and perennial ryegrass as soon as growth begins. In timothy, a delay until growth has taken place for some time and tillers are about 10 cm long is recommended.

LITERATURE CITED

Hebblethwaite, P.D. 1977. Irrigation and nitrogen studies in S.23 ryegrass grown for seed. I. Growth, development, seed-yield components and seed yield. J. Agric. Sci., Camb. 88:605–614.

Hebblethwaite, P.D., and J.D. Ivins. 1978. Nitrogen studies in *Lolium perenne* grown for seed. II. Timing of nitrogen application. J. Brit. Grassl. Soc. 33:159–166.

Hebblethwaite, P.D., D. Wright, and A. Noble. 1980. Some physiological aspects of seed yield in *Lolium perenne* L. *In* P.D. Hebblethwaite (ed.), Seed production, 71–90. Butterworths, London.

Langer, R.H.M. 1980. Growth of the grass plant in relation to seed production. *In* Herbage seed production, Grassl. Res. and Pract., Ser. 1, N.Z. Grassl. Assoc., 6–11.

Larsen, A., and A. Nordestgaard. 1969. Experiments on the use of increasing amounts of nitrogenous fertilizer on smooth meadow grass (*Poa pratensis*) for seed, distributed in autumn and spring. Tidsskr. Planteavl 73:45–56.

Nordestgaard, A. 1972. Experiments with autumn and spring application of increasing amounts of nitrogenous fertilizer to cocksfoot (*Dactylis glomerata*) for seed growing. Tidsskr. Planteavl 76:625–645.

Nordestgaard, A. 1974. Experiments with autumn and spring application of increasing amounts of nitrogenous fertilizer to meadow fescue (*Festuca pratensis*) for seed growing. Tidsskr. Planteavl 78:395–407.

Nordestgaard, A. 1977. Experiments with autumn and spring application of increasing amounts of nitrogenous fertilizer to perennial ryegrass (*Lolium perenne* L.) for seed growing. Tidsskr. Planteavl 81:187–202.

Nordestgaard, A., and A. Larsen. 1971. Experiments with autumn and spring application of increasing amounts of nitrogenous fertilizer to red fescue (*Festuca rubra* L.). Tidsskr. Planteavl 75:27–46.

Nordestgaard, A., and A. Larsen. 1974. Seed growing experiments in frames with cocksfoot (*Dactylis glomerata*), meadow fescue (*Festuca pratensis*) and smooth meadow grass (*Poa pratensis*). Tidsskr. Planteavl 78:116–130.

Ryle, G.J.A. 1964. The influence of date of origin of the shoot and level of nitrogen on ear size in three perennial grasses. Ann. Appl. Biol. 53:311.

Schöberlein, W. 1972. Zur Frühjahrsdüngung mit Stickstoff im Grassamenbau. Saat-und Pflanzgut 13:26–27.

Postharvest Residue Burning as a Management Tool in Grass-Seed Production

D.O. CHILCOTE, H.W. YOUNGBERG, and W.C. YOUNG III

Oregon State University, Corvallis, Oreg., U.S.A.

Summary

Postharvest residue burning, initiated in Oregon in the early 1940s to control blind seed disease in perennial ryegrass (*Lolium perenne* L.), has been found to be a most effective field sanitation method that also enhances seed yield. It is now a general practice in Oregon, and its use is expanding to other regions.

Questions of how burning beneficially affects grass-seed crop production have been the focus of many studies conducted over several years in the Willamette Valley. These studies were given impetus by the concern for clean air in certain locations in Oregon. The objectives of the research were (1) to document the yield response of different species and varieties to postharvest residue burning over a several-year period, (2) to compare burning to alternative methods, such as mechanical removal, for effectiveness in maintaining seed yield, (3) to evaluate relationships of time of burning to subsequent crop yield, and (4) to determine effects of burning on growth characteristics related to seed yield.

Investigations carried out over several years examined response to burning and mechanical residue removal of different grass species and varieties in experimental plots at various locations in the Willamette Valley. Results indicated that most species and varieties respond positively in terms of seed yield; however, some show a greater yield response than others. Forage-type perennial ryegrass, orchardgrass (*Dactylis glomerata* L.), tall fescue (*Festuca arundinacea* Schreb.), and bentgrass (*Agrostis tenuis* Sibth.) are less responsive to residue management than is fine fescue (*Festuca rubra* L.). Varieties within species respond differently to burning, but in general more effective residue removal results in higher seed yield. Increases in fall tillering and earlier emergence of a larger number of panicles are the basis for seed-yield increases in response to burning. Weed control is improved by postharvest burning. Greater numbers of weeds survive where mechanical methods of residue removal are practiced.

Burning early, when straw moisture is low and before regrowth starts, gives superior yields. Burning late, when regrowth has developed and straw moisture is greater, can cause stand injury and is less effective in removing residue. Alternate-year burning is undesirable from the standpoint of seed yield and disease control. Smoke management by burning during meteorological conditions favorable for smoke dispersion is aiding the continuation of the practice of postharvest residue burning by reducing the impact of smoke on population centers.

The wide-spectrum nonresidual pest control and yield stimulation provided by postharvest burning make the practice an extremely important tool for grass-seed crops in Oregon and elsewhere where climate permits.

KEY WORDS: cool-season grasses, seed yield, tillering, panicle emergence, cultivars.

INTRODUCTION

Postharvest residue burning was initiated in Oregon in the late 1940s as a control for blind seed disease, which threatened to destroy the perennial ryegrass seed crop (Hardison 1964). The burning treatment proved extremely effective for control of this and other diseases in grass species grown for seed. Increased seed yield was usually observed after burning, and subsequently burning became an established management practice in grass-seed production in Oregon. It is also being used effectively in other states and other regions of the world.

Several benefits from the practice of burning residue after harvest for grass-seed production have been cited (Chilcote et al. 1974, Chilcote 1976, Hardison 1964). Burning is an inexpensive method of removing residue that, if left, would inhibit light interception and interfere with regrowth. Control of seed-crop pests is another established benefit. Also, in several species there is a significant stimulation of seed yield in the seed harvest after burning. Minimum-tillage establishment of annual ryegrass (*Lolium multiflorum* L.) is possible where burning removes the residue, making direct grassland drilling practical.

The relationship between postharvest burning and beneficial effects on grass-seed crop production has been the focus of many studies conducted over a period of several years in the Willamette Valley of Oregon. The effect of emissions from residue burning on air quality combined with a greater concern for cleaner air in the Wil-

D.O. Chilcote is a Professor of Crop Physiology, H.W. Youngberg is a Professor of Extension Agronomy, and W.C. Young III is a Research Assistant at Oregon State University.

lamette Valley has given impetus to these studies. The objectives of the research were (1) to document the yield response of different species and varieties to postharvest residue burning, (2) to compare burning with alternative methods, such as mechanical residue removal, for effectiveness in maintaining seed yield, (3) to evaluate relationships of time of burning to subsequent crop yield, and (4) to determine effects of burning on growth characteristics related to seed yield. The procedures involved investigating the response to burning or different mechanical residue removal methods over a period of several years on species and varieties of grass grown for seed in experimental plots at various locations in the Willamette Valley.

RESULTS AND DISCUSSION

Species Response

Nearly all of the cool-season grass-seed crops showed some yield stimulation in the year following postharvest residue burning, even in the absence of pest problems. The yield response of selected species after burning compared with yield after no straw removal or after flail-chop

removal is summarized in Table 1. The need for residue removal and the response to burning is greater in fine fescue. Seed production of forage-type perennial ryegrass, orchardgrass, tall fescue, and bentgrass is less responsive to postharvest residue management. Postharvest burning can cause damage to stands of some species, such as *Agrostis palustris* Huds.

Varietal Response

Interactions were found among varieties of orchardgrass, perennial ryegrass, bluegrass (*Poa pratensis* L.), and fine fescue in response to burning treatment. A knowledge of varietal response is needed before a residue removal practice can be applied.

Growth and Development Effects

The basis for seed-yield increase from burning residue appears to be related to changes in the microenvironment that enhance autumn tiller development and floral induction. Fig. 1 shows the increased number of tillers produced during the late autumn and winter by red fescue in burned plots. As a result of burning, more tillers are ex-

Table 1. Seed yield of different grass species following various post-harvest residue removal methods (1969).

Post-harvest treatment	Seed yield (kg/ha)				
	Chewings fescue	Red fescue	Perennial ryegrass	Colonial bentgrass	Orchard-grass
No straw removal	282	409	908	434	967
Flail-chop straw and stubble removal[a]	362	747	1035	402	1178
Burning (early)	1035	1243	1278	596	1263

[a]Removed to approximately 7 cm height.

Fig. 1. Seasonal variation in tillering of red fescue (*Festuca rubra* L.) in burned and unburned plots during 1970–1971. Each point represents the mean tiller populations/25 core samples (22.82 cm² each).

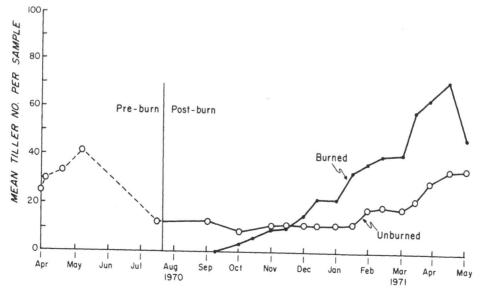

Fig. 2. Inflorescence emergence pattern of burned and unburned plants of red fescue (*Festuca rubra* L.). Average of nine similar-sized plants.

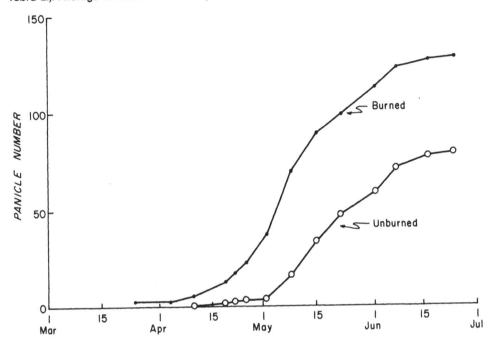

posed to floral inductive conditions for a longer period. Panicle emergence begins earlier, and greater numbers of panicles are produced in burned plots (Fig. 2). Increased seed yield from burning appears to be due to the larger number of reproductive tillers and seeds/fertile tiller (Table 2).

Weed Control

Weed control by burning is superior to control by mechanical removal of residue. Burning left fewer surviving annual ryegrass weeds in perennial ryegrass than mechanical methods of removal, probably because it left fewer weed seeds and enhanced effectiveness of pesticides because of less adsorption on residue.

Time of Burning

Burning fields soon after harvest in Oregon is preferable in terms of seed yield and residue removal. Normally there is little plant growth after harvest during the dry summer season. Burning late in the season (late September–October), when regrowth has occurred and straw moisture is higher, reduces the effectiveness of residue removal and can cause stand injury because newly developed tillers are destroyed. Of course, soil moisture, the amount of green regrowth, and straw moisture are dependent upon seasonal and daily weather patterns. Therefore, proper timing of a burn must depend on the particular conditions for that year and season.

Alternate-Year Burning

Smoke pollution problems from the large acreage burned in the Willamette Valley each year promoted an evaluation of postharvest burning of perennial grass-seed crops on an alternate-year basis. Preliminary results show that any reduction from annual burning will probably result in a reduced seed yield; however, recovery of seed yield by burning in the following year may be possible. Disease control would also be reduced if burning in alternate years were practiced.

Table 2. Effects of burning and mechanical removal of post-harvest residue on seed yield and yield components of red fescue (1972).

Treatment	Seed yield (kg/ha)	Fertile tillers/m²	Seeds/fertile tiller	500-seed weight (g)
Burned	552	582	424	0.5551
Mechanically removed	296*	392**	337*	0.5502 NS

*Significantly different at 5% level of probability from 'burned' treatment.

**Significantly different at 1% level of probability from 'burned' treatment.

NS: No significant difference from 'burned' treatment.

Degree of Residue Removal

A machine was designed to "close clip" stubble to crown height and vacuum residue from fields after harvest. This technique was compared to burning and to flail-chopping residue from the field with standard equipment. Seed yields were significantly higher for burning followed by "close clip," which was superior to flail-chop. The more nearly complete the residue removal is, the better the subsequent seed yield will be. Flail-chopping leaves a greater amount of stubble and residue on the soil surface. Burning is most effective, since it eliminates crown residue and removes old surviving tillers along with the stubble.

Smoke Management

Alternative methods of postharvest residue removal have not been as effective as burning from the standpoint of subsequent seed yield or pest control, and they are much more expensive. However, smoke emissions are an air-quality problem. Under certain meteorological conditions dispersion of smoke is more favorable. Field burning has been restricted to times of favorable meteorological conditions, reducing smoke accumulation in populated areas. This procedure has been quite successful. Weather conditions early in the season are often more favorable for smoke dispersion, and early burning is also beneficial from the agronomic standpoint. However, awaiting favorable weather conditions frequently delays burning

into the late-season category, increasing risk of crop injury. Continued refinement of weather predictions for timing burning will be beneficial in maintaining this important management tool for grass-seed production in the Willamette Valley.

Climate

Open-field burning has been particularly beneficial to growers in the Pacific Northwest because of the area's mild, dry summers, which contribute to the effectiveness of burning as a residue removal–sanitation treatment. In areas of the world where grass-seed crops are grown under summer rainfall conditions, the effectiveness of open-field burning may be reduced. However, postharvest residue burning for field sanitation and seed-yield stimulation remains one of the most important management tools for the grass-seed producer wherever climatic conditions are favorable.

LITERATURE CITED

Chilcote, D.O. 1976. Grass seed field burning: air quality and agronomic aspects. Ann. Mtg. Soil Conserv. Soc. Am., 124–126.

Chilcote, D.O., P.C. Stanwood, and S. Kim. 1974. The nature of the post-harvest residue burning effect on perennial grass seed yield. Proc. Oreg. Seed Grow. League, 15–19.

Hardison, J.R. 1964. Justification for burning grass fields. Proc. Oreg. Seed Grow. League, 93–96.

Effect of Autumn and Spring Defoliation and Defoliation Method on Seed Yield of *Lolium perenne*

P.D. HEBBLETHWAITE and T.G.A. CLEMENCE
University of Nottingham, Loughborough, Leicester, UK

Summary

Trials in 1979 and 1980 evaluated the effect of autumn and spring defoliations on fertile apex removal, tiller survival, seed-yield components, and seed and forage yield in two amenity- and one agricultural-type perennial ryegrasses (*Lolium perenne*), established in the autumn. Methods (reciprocating mower or sheep), height (3–5 cm or 6–8 cm) of defoliation, and the addition of supplementary nitrogen (N) after defoliation were evaluated in the variety Royal in 1980.

In both years and all varieties, autumn defoliation had no significant effect on seed yield. With the exception of Royal in 1980, seed yields were reduced by cutting, if it was carried out after spikelet initiation, due to an increase in apex removal, particularly in autumn-emerged tillers. In Royal in 1980, survival of spring-emerged tillers as mature ears at final harvest was increased by defoliation at floret initiation, and this increase compensated for tiller loss.

Seed-yield reductions were accompanied by decreases in all yield components except 1,000-seed weight in both years and seeds/spikelet in 1980.

Supplementary N failed to compensate for yield loss, but cutting high (6–8 cm) at floret initiation produced 1.2 tons(t)/ha of dry matter with reduced apical loss.

Defoliation in autumn-established crops is unnecessary. However, growers with forage requirements may safely defoliate up to spikelet initiation and obtain approximately 2 t/ha of dry matter from these varieties.

KEY WORDS: ryegrass, seed production, defoliation, tillering.

INTRODUCTION

Detailed information on the effect of time and method of defoliation on perennial ryegrass (*Lolium perenne* L.) seed yield is limited, particularly with regard to autumn-sown crops and amenity types grown for seed in the UK. In spring-sown forage types, autumn defoliation has been shown to benefit seed production through a decrease in "winter burn," a delay in lodging (Roberts 1958), and increased reproductive tillering after both autumn and early-spring defoliation (Roberts 1958, 1966; Hayward 1959; Hill and Watkin 1975). Roberts (1966) and Hill (1971) have shown that late-spring defoliations, which produce maximum forage yields, seriously decrease the crop's potential for seed production, but they have not provided detailed information on apex removal in relation to tiller origin. Such information is necessary to predict the effects of time and methods of defoliation in any year and on all types of ryegrasses. Further information is also required on possible compensatory management treatments at or after defoliation.

Our paper describes experiments carried out in 1979 and 1980 to assess the effects of time and method of defoliation on the forage and seed-yield potential of one forage and two amenity-type autumn-sown ryegrass seed crops.

EXPERIMENTAL METHODS

Trials were carried out in 1979 and 1980 on the amenity varieties Royal and Majestic and the forage variety Morenne, sown in late August–early September on the University of Nottingham Farm, Sutton Bonington, Loughborough, Leicester, UK. Details of soil type, establishment, management, and experimental methods are reported elsewhere (Hebblethwaite et al. 1978, Wright and Hebblethwaite 1979, Clemence 1981). Defoliation treatments were:

- No cut: no cut throughout
- Autm: autumn only
- DR: autumn and at 70% "double-ridge" stage
- SI: autumn and at 70% spikelet initiation

- FI: autumn and at 70% floret initiation
- EE: autumn and at first ear emergence
- SI (S): autumn and at 70% spikelet initiation (grazed by sheep)
- FI (S): autumn and at 70% floret initiation (grazed by sheep)
- FI (H): autumn and at 70% floret initiation (cut high at 6–8 cm)
- FI (N): autumn and at 70% floret initiation (with compensatory nitrogen, equivalent to the amount removed by defoliation)

RESULTS

Apical Differentiation

The stage of apex differentiation and the consequent date of defoliation in relation to treatment are presented in Table 1.

Seed Yield and Seed-Yield Components

Seed yields for all treatments up to SI were not significantly different in each year and for each variety (Table 2), although yields for these treatments in 1979 were greater than those in 1980 as a result of more seeds/spikelet (Table 3). Defoliations after spikelet initiation significantly decreased yields of Majestic and Morenne in both years and of Royal in 1979, particularly when carried out at ear emergence (EE), but time and method of defoliation had no effect on yield of Royal in 1980 (Table 2). Table 3, which for clarity presents only data for significantly differing treatments represented by the mean of all varieties, indicates that yield reductions were accompanied by decreases in potential and realized seed-yield components (Hebblethwaite et al. 1980), except for 1,000-seed weight in both years and seeds/spikelet in 1980. Supplementary N did not compensate for yield losses where these occured.

Loss of Apices at Defoliation

In all varieties, more apices were removed by defoliation the later it was carried out in the spring, except where cutting height was raised to 6–8 cm (Table 4). Percentage of loss of apices in 1979 was twice that in 1980, which indicates that the start of elongation was slightly later in 1980 than in 1979.

The ringed-tiller study carried out on Royal in 1980 showed that of the 11% of tillers removed by defoliation at FI, 96% had originated in the period 4 September–16 December 1979.

P.D. Hebblethwaite is a Senior Lecturer in agronomy and T.G.A. Clemence is a Research Student in agronomy at the University of Nottingham.

The authors gratefully acknowledge the technical assistance of Mrs. S. Manison and Mr. J. Travers, as well as the statistical advice of Mr. J. Craigon. They would also like to thank Mommersteeg International (UK) Ltd for financial support of this project and the Frank Horne Memorial Fund for providing funds to attend this Congress.

Table 1.   Dates of attainment of growth stage/defoliation treatments, autumn defoliation, and supplementary N application; and levels of supplementary N application.

Apical state of tiller pop.	Royal 1979	Royal 1980	Majestic 1979	Majestic 1980	Morenne 1979	Morenne 1980
70% double ridge	27 Mch	–	18 Apr	–	24 Apr	–
70% spikelet initiation	18 Apr	14 Apr	1 May	12 May	1 May	1 May
70% Floret initiation	1 May	28 Apr	16 May	19 May	16 May	12 May
Ear emergence	22 May	–	31 May	–	31 May	–
Autumn cuts	(13 November 1979; 10 December 1980)					
Supplementary N	–	29 Apr	–	23 May	–	15 May
N level (kg/ha)	–	45	–	70	–	85

Table 2.   Effect of defoliation on dry-seed yield (t/ha) in all varieties, 1979 and 1980.

Defoliation treatment	Royal 1979	Royal 1980	Majestic 1979	Majestic 1980	Morenne 1979	Morenne 1980
No cut	1.8	0.8	2.1	0.7	2.0	0.7
Autm	1.7	0.7	2.0	0.8	1.9	0.8
D.R.	1.7	–	2.0	–	2.0	–
S.I.	1.8	0.8	1.8	0.8	1.9	0.7
S.I. (S)	–	0.8	–	–	–	–
F.I.	1.0	0.8	1.0	0.5	0.8	0.5
F.I. (S)	–	0.8	–	–	–	–
F.I. (H)	–	0.8	–	–	–	–
F.I. (N)	–	0.7	–	0.6	–	0.5
E.E.	0.6	–	0.4	–	0.3	–

	d.f.	S.E.D.
1979	45	0.11
1980 Royal only	45	0.08
1980 All varieties	36	0.07

Tiller Survival

Most autumn-formed tillers survived to final seed harvest when the crop was not defoliated in the spring (autm.) (Table 5). Cutting at FI considerably reduced autumn-emerged tiller survival, but more spring-emerged tillers produced fertile ears, and there was a corresponding reduction in spring-emerged tillers remaining vegetative (Table 5). Secondary tillers arising in April from decapitated tillers produced a greater proportion of mature ears than secondaries arising simultaneously from intact tillers (Table 5), although they were fewer in number.

Forage Dry-Matter Yield

Spring-forage dry-matter yield increased from about 0.7 t/ha when cut at DR to 1.5 and 3.5 t/ha, respectively,

Table 3.   Effect of autumn defoliation and further cutting at F.I. and E.E. on the seed yield components in 1979 and 1980.

Defoliation treatment	Number of fertile tillers/ 0.1m² 1979	1980	Number of spikelets/ fertile tiller 1979	1980	Number of florets/ spikelet 1979	1980	Number of seeds/ spikelet 1979	1980	1000 seed weight (g) 1979	1980	% seed set 1979	1980
Autm	320	318	20.2	21.7	7.79	8.02	1.62	0.71	1.86	1.65	20.7	8.8
F.I.	257	243	17.4	17.8	6.09	6.82	1.27	0.95	1.80	1.59	26.1	13.9
E.E.	157	–	16.4	–	6.81	–	1.06	–	1.81	–	15.6	–
d.f.	45	32	45	32	45	32	45	36	45	36	–	–
S.E.D.	23	29	0.32	0.32	0.29	0.30	0.23	0.23	0.02	0.02	–	–

Table 4.   Loss of apices due to defoliation as % total tillers at defoliation in all varieties in 1979 and 1980.

Defoliation treatment	Royal 1979	Royal 1980	Majestic 1979	Majestic 1980	Morenne 1979	Morenne 1980
D.R.	0	—	0	—	0	—
S.I.	2	0	1	5	5	5
S.I. (S)	—	1	—	—	—	—
F.I./F.I. (N)	40	11	37	11	28	15
F.I. (S)	—	15	—	—	—	—
F.I. (H)	—	1	—	—	—	—
E.E.	57	—	45	—	49	—

	d.f.	S.E.D.
1979	6	5.0
1980 Royal only	15	1.5
1980 All varieties	6	2.3

Table 5.   Tillers surviving at final harvest (%) from samples ringed when newly-emerged in the periods specified, in Royal treatments Autm. and F.I.

Period of tiller emergence	Vegetative tillers		Mature reproductive tillers	
	Autm.	F.I.	Autm.	F.I.
4 Sept–16 Dec	0	3	70	22
17 Dec–28 Jan	11	6	27	38
29 Jan–3 Mch	24	13	14	41
4 Mch–31 Mch	35	15	7	32
1 Apr–9 May*	62	45	0	7
1 Apr–9 May†	—	13	—	10

*Tillers arising from intact parent tillers.

†Tillers arising from decapitated parent tillers.

when cut at SI and FI. Raising the cutting height from 3–5 cm (FI) to 6–8 cm (FI [H]) decreased yield to 1.3 t/ha.

DISCUSSION

Autumn or spring defoliation in crops sown in August or September was found to be unnecessary to obtain maximum seed yield in all varieties. This may be of benefit to growers without livestock or other methods of defoliation.

These experiments show that excess forage in all varieties can be successfully exploited by growers, either by cutting or grazing, without decreasing seed yield, provided that defoliation is neither too late (after SI) nor too close. Economic dry matter (1.3 t/ha) may be harvested by cutting high (6–8 cm) as late as FI, without undue apical and seed-yield loss.

LITERATURE CITED

Clemence, T.G.A. 1981. Seed production studies in amenity ryegrass *Lolium perenne*. Ph.D. Thesis, Univ. of Nottingham.

Hayward, P.R. 1959. A grower's experience on management. J. Natl. Inst. Agric. Bot. 8:518–524.

Hebblethwaite, P.D., A. Burbidge, and D. Wright. 1978. Lodging studies in *L. perenne* grown for seed. I. Seed yield and seed yield components. J. Agric. Sci., Camb. 90:261–267.

Hebblethwaite, P.D., D. Wright, and A. Noble. 1980. Some physiological aspects of seed yield in *Lolium perenne* L. (perennial ryegrass). *In* P.D. Hebblethwaite (ed.), Seed production, 71–90. Butterworths, London.

Hill, M.J. 1971. Closing ryegrass crops for seed production. N.Z. J. Agric. 123(2):43.

Hill, M.J., and B.R. Watkin. 1975. Seed production studies on perennial ryegrass, timothy and prairie grass. I. Effects of tiller age on tiller survival, ear emergence and seed head components. J. Brit. Grassl. Soc. 30:63–71.

Roberts, H.M. 1958. The effect of defoliation on the seed producing capacity of bred strains of grasses. I. Timothy and perennial ryegrass. J. Brit. Grassl. Soc. 13:255–261.

Roberts, H.M. 1966. The seed productivity of perennial ryegrass varieties. J. Agric. Sci., Camb. 66:225–232.

Wright, D., and P.D. Hebblethwaite. 1979. Lodging studies in *Lolium perenne* grown for seed. III. Chemical control of lodging. J. Agric. Sci., Camb. 93:669–679.

Seed Production for the Establishment of *Pennisetum americanum* x *P. purpureum* F₁ Hybrid Pastures

M.E. AKEN'OVA and H.R. CHHEDA
University of Ibadan, Ibadan, Nigeria

Summary

Several genotypes among the interspecific F_1 hybrids between Maiwa, a short-day, photoperiod-sensitive local cultivar of pearl millet (*Pennisetum americanum* [L.] Leeke), and elephant grass (*P. purpureum* Schumach) produced at the University of Ibadan, Nigeria, combine the high dry-matter yield potential and perenniality of elephant grass with greater acceptability to livestock and increased digestibility, resulting in higher live-weight gains than elephant grass. Though sterile, the F_1 *Pennisetum* hybrids are easily propagated vegetatively. Distribution over a wider area of adaptation can be facilitated by large-scale F_1 seed production utilizing cytoplasmic male-sterile pearl millet.

Cytoplasmic male-sterile Maiwa was developed by transferring male-sterility factors from Tift 23A pearl millet, using a backcrossing procedure.

Male-sterile Maiwa was grown between rows of 6 elephant grass ecotypes that combine well with Maiwa. To ensure coincidence of flowering with elephant grass, Maiwa was sown at the end of August.

Seed-head diseases, the most serious being ergot (*Claviceps* sp.), resulted in heavy losses; 3.17 kg of F_1 hybrid seed was recovered from a 358 m² plot. This is equivalent to a production of 88.5 kg/ha of hybrid seed, adequate to establish 8 ha of pasture when the seed is sown at the rate of 11 kg/ha in rows 90 cm apart. At Ibadan, under ideal conditions, a 1.60-ha multiplication nursery is required to establish the same acreage vegetatively.

The study has shown that large-scale production of the F_1 *Pennisetum* hybrid seed is practical. A much higher seed yield can be expected if diseases prevalent in the humid environment of Ibadan can be contained. Seed production in less humid regions where pearl millet is normally cultivated should minimize the problem.

KEY WORDS: *Pennisetum* hybrids, seed production, seed yields.

INTRODUCTION

In much of the tropics, lack of seed for improved forage genotypes places a serious constraint on increasing the acreages of productive sown pastures or improving the species composition of existing rangelands. Seed production of most perennial forage grasses is attended with such problems as low inflorescence density, poor seed setting, poor synchronization of seed-yield components evidenced by prolonged flowering and seed-head shattering, suboptimal management, and diseases.

The F_1 hybrid between the annual pearl millet (*Pennisetum americanum* [L.] Leeke) and the perennial elephant grass (*P. purpureum* Schumach) is an important perennial fodder in parts of the tropics and subtropics. At the University of Ibadan, research has shown that a short-day photoperiod-sensitive pearl millet type such as Maiwa, a local cultivar, is a suitable parent for producing well-adapted F_1 *Pennisetum* hybrids for the low-altitude humid tropics. As a result, hybrid genotypes that combine the high dry-matter yield potential of elephant grass with greater acceptability to livestock and increased digestibility, resulting in higher live-weight gains compared to elephant grass, were identified (Aken'Ova and Chheda 1973, 1976; Chheda et al. 1973).

Distribution of the hybrids over a wider area of adaptation can be facilitated by large-scale F_1 seed production using cytoplasmic male-sterile pearl millet (Powell and Burton 1966). To accomplish this, during 1970–1976, the male-sterility factors of cytoplasmic male-sterile Tift 23A pearl millet were incorporated into Maiwa using a backcrossing procedure. The newly developed male-sterile Maiwa was then used to ascertain the feasibility of large-scale F_1 *Pennisetum* hybrid seed production.

METHODS

In a 358-m² plot of land, crown splits of 6 local elephant grass ecotypes selected on the basis of good seed set in crosses with Maiwa and on the basis of the compatibility and superior performance of the resulting F_1 progenies in mixed stands were planted at 90 cm spacing in rows 3.90 m apart. Each row contained one ecotype. Ten weeks later, in the last week of August 1976, the plants were cut

M.E. Aken'Ova is Senior Lecturer and H.R. Chheda is Professor of Agronomy at the University of Ibadan. Their study was supported by grants from the Rockefeller Foundation and the University of Ibadan Senate.

15 cm above ground level, and male-sterile Maiwa seed was sown in hills 60 cm apart between the elephant grass rows so that 2 Maiwa rows, 90 cm apart, were established between any 2 elephant-grass rows, with each Maiwa row 1.50 m from the nearest elephant grass row. The August planting date ensured coincidence of flowering with elephant grass.

RESULTS

Seed set on male-sterile Maiwa heads varied widely, ranging from very poor (less than one seed/cm of head) to excellent or as good as open-pollinated male-fertile Maiwa. Very heavy infestation of Maiwa heads with ergot (*Claviceps fusiformis* Loveless), as well as the incidence of smut (*Tolyposporium penicillariae* Bref.), reduced seed yields drastically. Nearly 70% of the heads were severely affected by these diseases. Bird damage, which is serious in some years, was relatively mild.

The plot yielded 3.17 kg of clean seed with a germination rate of about 45%.

DISCUSSION

The hybrid-seed yield recorded in this study is equivalent to 88.5 kg/ha, adequate to establish 8 ha of hybrid pasture. Our experience has shown that satisfactory establishment of the *Pennisetum* hybrid pasture by drilling seed in rows 90 cm apart requires about 11 kg/ha of the hybrid seed. This moderately high seed requirement is necessary because of low seed germination and the occurrence of weak plants in the hybrid population. On the other hand, to establish a similar acreage of 8 ha vegetatively, under ideal conditions at Ibadan, a multiplication nursery 1.60 ha in size with plants spaced 90 × 60 cm is required, assuming each plant produces on an average enough material to establish 5 stands in the pasture. Thus, a considerable savings in land area required for producing propagating material is achieved if seed production is adopted. More savings may be expected with improved seed yields.

The study suggests that the level of seed yields can be raised considerably by:

1. locating the seed-production plots in the less humid pearl millet–growing regions of the country, a practice that might also minimize disease problems. On the basis of the area actually occupied by Maiwa in the seed-production plot (138 m²), the seed yield in this study is equivalent to 230 kg/ha. This yield is much lower than the yields of 450–700 kg/ha obtained in Nigeria's pearl millet–growing areas.
2. developing and utilizing male-sterile Maiwa lines that are resistant to the common seed-head diseases.
3. improving management on the seed-production plot, for example by manually shaking flowering inflorescences of elephant grass to aid pollen dissemination and consequently improve seed setting. Making more pollen available for more rapid pollination could also reduce the incidence of ergot through the pollen-based escape mechanism described by Thakur and Williams (1980).

The perennial nature of elephant grass requires the establishment of pearl millet alone in any particular year in which seed production is desired. Other factors of advantage associated with seed production for pasture establishment include ease of long-distance transportation and possibilities for long-term seed storage.

Finally, the fact that pearl millet, a commonly cultivated, highly evolved, and relatively large-seeded grass, is used as a seed parent also helps to minimize or avoid many problems, such as uneven flowering, seed ripening over an extended time period, and difficulties in seed cleaning and processing, that are commonly associated with seed production of most tropical forage grasses.

In conclusion, the study has demonstrated that it is practicable to produce quantities of the F_1 *Pennisetum* hybrid seed economically for widespread distribution.

LITERATURE CITED

Aken'Ova, M.E., and H.R. Chheda. 1973. Interspecific hybrids of *Pennisetum typhoides* S & H × *P. purpureum* Schum. for forage in the low-altitude humid tropics of West Africa. Seedling studies and preliminary evaluation. Niger. Agric. J. 10(1):82–90.

Aken'Ova, M.E., and H.R. Chheda. 1976. Beef production on rotationally grazed F_1 *Pennisetum* hybrid and elephant grass (*Pennisetum purpureum* Schum.) pastures. Niger. J. Anim. Prod. 3(2):20–24.

Chheda, H.R., M.E. Aken'Ova, and L.V. Crowder. 1973. *Pennisetum typhoides* S & H × *P. purpureum* Schum. hybrids for forage production in the low-altitude humid tropics. Crop Sci. 13:122–123.

Powell, J.B., and G.W. Burton. 1966. A suggested commercial method of producing an interspecific hybrid forage in *Pennisetum.* Crop Sci. 6:378–379.

Thakur, R.P., and R.J. Williams. 1980. Pollination effects on pearl millet ergot. Phytopathology 70:80–84.

Effect of Different Systems of Seed Treatment, Packing, and Storage on Vigor and Germination of Five Tropical Forage Legumes

R. CABRALES and J. BERNAL

Semillas La Pradera, Bogotá, Colombia

Summary

Production of tropical forage-legume seeds is expensive. Very often, the establishment of mixtures is difficult because of lack of uniformity in vigor and germination of the legume seed. Treatment with chemical compounds has been reported to be useful in breaking dormancy. The packing material and temperature of storage have also been reported to affect seed vigor and germination.

The following treatments were applied to recently harvested seeds of tropical kudzu (*Pueraria phaseoloides* [Roxb.] Benth.), clitoria (*Clitoria ternatea* L.), centrosema (*Centrosema pubescens* Benth.), siratro (*Macroptilium atropurpureum* DC. Urb.), and desmodium (*Desmodium intortum* [Mill.] Fawc. and Rendle): concentrated sulfuric acid (96%) for 5 minutes at the beginning of the experiment (prestorage), concentrated sulfuric acid for 5 minutes at planting time, and hot water at 80°C for 20 minutes at planting time, with untreated seeds being used as controls.

The seeds were packed in paper and plastic (polyethylene) bags and stored in the greenhouse (28°C average temperature) and in the coldroom (8°C average temperature). The seeds were germinated in a moisture- and temperature-regulated germinator (27°C and 95% moisture) every 20 days for a period of 6 months. Results were recorded 5 days after planting (vigor) and 20 days after planting (germination).

Results indicated that vigor and germination increased with scarification treatment in most of the studied species. The best overall results were gained by soaking the seeds in hot water for 20 minutes.

In general, seeds stored in plastic bags at high temperatures broke dormancy faster than seeds stored in paper bags or at low temperatures.

Treatment with sulfuric acid does not present any significant advantages over soaking in hot water for most species.

KEY WORDS: tropical kudzu, *Pueraria phaseoloides* [Roxb.] Benth., clitoria, *Clitoria ternatea* L., centrosema, *Centrosema pubescens* Benth., siratro, *Macroptilium atropurpureum* DC. Urb., desmodium, *Desmodium intortum* [Mill.] Fawc. and Rendle, seeds of tropical forage legumes, seed treatment, packing, storage, vigor, germination.

INTRODUCTION

Quality of tropical forages is relatively low but can be improved by introducing tropical legumes in the grasslands. One of the main problems in establishing good mixtures is the lack of uniformity in vigor and germination of the legume seed. Most of these seeds present dormancy from a hard cover that reduces germination (Purcell 1974). Thickness of cover and permeability to water and oxygen vary with the species (Mayer and Shain 1974). The cover may be weakened by prolonged storage or by means of chemical or physical scarification. Strong acids can be used for this purpose (Varner 1965). Changes in temperature have also proven very efficient in promoting modifications of the integuments (Strickland 1971).

Tropical kudzu (*Pueraria phaseoloides* [Roxb.] Benth.), centrosema (*Centrosema pubescens* Benth.), clitoria (*Clitoria ternatea* L.), desmodium (*Desmodium intortum* [Mill.] Fawc. and Rendle), and siratro (*Macroptilium atropurpureum* [DC.] Urb.) are among the most important tropical forage-legume species. Different treatments were applied to seeds of these species to break dormancy.

METHODS

Recently harvested seeds of kudzu, centro, clitoria, desmodium, and siratro produced near the Atlantic coast of Colombia were scarified to measure vigor and germination.

The scarification treatments applied were: concentrated sulfuric acid (H_2SO_4, 96%) for 5 minutes at the beginning of the experiment (prestorage), concentrated H_2SO_4 for 5 minutes at planting time, hot water at 80°C for 20 minutes at planting time, and untreated seeds used as controls.

R. Cabrales is an Agronomist and J. Bernal the Manager at Semillas La Pradera.

Table 1. Effect of scarification treatments on vigor of seeds of five tropical forage legumes.*

| | | Scarification treatments | | |
Species	Control	H_2SO_4 pre-storage	H_2SO_4 at planting time	Hot water at planting
Kudzu	4.05a	7.42a	13.84b	20.29c
Centro	10.72a	25.32b	32.03c	34.07d
Clitoria	28.44a	34.15b	29.78a	36.62c
Desmodium	93.03c	93.35c	91.51b	84.74a
Siratro	6.70a	11.11b	35.97d	22.19c

*Values are comparable horizontally. Comparable values with a letter in common do not differ at the P = 0.01 level.

The seeds, packed in paper or plastic (polyethylene) bags, were stored in the greenhouse (28°C average temperature) and in the coldroom (8°C average temperature). The seeds were germinated in a humidity- and temperature-regulated germinator (27°C and 95% relative humidity) every 20 days for a period of 6 months. Germination was conducted in flats using paper towels kept permanently moist. Results were recorded 5 days after planting (vigor) and 20 days after planting (germination). The experiment was arranged in randomized blocks with 4 replications. Every replication consisted of 100 seeds.

RESULTS

Results are summarized in Tables 1 to 4.

1. *Tropical kudzu.* There were significant differences for vigor and germination of kudzu resulting from the different scarification treatments. The best results were obtained with hot water, followed by H_2SO_4 at planting time. Both vigor and germination were higher when the seed was stored in the greenhouse than when it was stored in the coldroom. Plastic bags were better than paper bags for wrapping the seeds.

2. *Centro.* Centro showed the same tendency as kudzu in response to the different scarification treatments. No significant difference was found between greenhouse and coldroom storage. Packing in plastic bags resulted in higher vigor and germination than packing in paper bags.

3. *Clitoria.* Scarification with hot water was significantly superior to the other treatments for vigor and ger-

mination. Storage in the coldroom was superior to storage in the greenhouse, and packing in plastic bags increased vigor and germination as compared with packing in paper bags.

4. *Desmodium.* Vigor decreased when the seed was treated with H_2SO_4 or hot water at planting time. Germination decreased when the seed was scarified with hot water but was not affected by treatments with H_2SO_4. There was no response to differences in storage (greenhouse vs. coldroom) or packing (plastic vs. paper bags).

5. *Siratro.* The best response in vigor and germination was obtained when the seed was treated with H_2SO_4 at planting time, followed by treatment with hot water. When the seed was scarified with H_2SO_4 and then packed and stored, the increase in vigor and germination was remarkable, compared with the control. Storage had no effect on either vigor or germination, but packing in plastic bags resulted in higher vigor and germination than packing in paper bags.

DISCUSSION

Scarification with hot water was the best treatment for kudzu, centro, and clitoria, but treatment with H_2SO_4 at planting time was superior for siratro. Desmodium seems not to have dormancy, and it was in general negatively affected by scarification. These results indicate that the external cover is different among the studied species, as is the response to the different scarification treatments. From the practical and economical point of view, the best results are obtained by soaking the seed in hot water at

Table 2. Effect of scarification treatments on germination of seeds of five tropical forage legumes.*

| | | Scarification treatments | | |
Species	Control	H_2SO_4 pre-storage	H_2SO_4 at planting time	Hot water at planting
Kudzu	14.61a	19.68b	23.32b	55.59c
Centro	22.62a	37.14b	46.11c	59.76d
Clitoria	49.52a	49.64b	50.31a	55.55b
Desmodium	96.53b	96.40b	95.72b	92.54a
Siratro	12.78a	21.47b	54.08d	46.12c

*Values are comparable horizontally. Comparable values with a letter in common do not differ at the P = 0.01 level.

Table 3. Effect of storage and packing on vigor of seeds of five tropical forage legumes.*

Species	Storage		Packing	
	Greenhouse	Coldroom	Polyethylene	Paper
Kudzu	15.53b	7.28a	15.68b	7.13a
Centro	23.99a	26.42a	37.76b	13.80a
Clitoria	30.17a	34.25b	57.25b	7.41a
Desmodium	90.02a	89.97a	89.95a	90.51a
Siratro	18.35a	19.68a	23.81b	14.20a

*Values are comparable horizontally. Comparable values with a letter in common do not differ at the P = 0.01 level.

Table 4. Effect of storage and packing on germination of seeds of five tropical forage legumes.*

Species	Storage		Packing	
	Greenhouse	Coldroom	Polyethylene	Paper
Kudzu	35.85b	24.31a	34.68b	21.92a
Centro	40.26a	42.67a	54.21b	28.91a
Clitoria	49.51a	53.00b	80.62b	21.90a
Desmodium	92.79a	92.64a	92.90a	93.26a
Siratro	33.71a	33.55a	40.22b	27.04a

*Values are comparable horizontally. Comparable values with a letter in common do not differ at the P = 0.01 level.

planting time, a treatment that can be done at farm level. Desmodium seed should be planted untreated.

Storage in the greenhouse apparently helped to break dormancy in kudzu, but not in centro, desmodium, or siratro; clitoria seed tended to break dormancy when stored at a low temperature (coldroom).

Packing in plastic bags tended to break dormancy for all species except desmodium, which was not affected. There was a significant interaction between storage and packing; vigor and germination tended to be higher when the seed was stored in the greenhouse and packed in plastic bags, perhaps because the temperatures developed were higher in that case. After 6 months under those conditions, two of the species showed a decrease in germination (clitoria and siratro), indicating that dormancy is broken by storing seeds at high temperatures in plastic bags, but viability apparently decreases after a relatively short time. For long-term storage, the use of bags and low temperatures would seem to be appropriate.

LITERATURE CITED

Mayer, A.A., and Y. Shain. 1974. Control of seed germination. Ann. Rev. Pl. Physiol. 25:167–193.

Purcell, D.L. 1974. Producción, almacenamiento y tratamiento de semillas. *In* Caja de Crédito Agrario Industrial y Minero, Bogotá. La producción de ganado de carne mediante el uso de pastizales en el clima cálido de Colombia. Manual Técnico 4, 1–20.

Strickland, R.W. 1971. Seed production and testing problems in tropical and subtropical pasture species. Proc. Int. Seed Test. Assoc. 36:189–199.

Varner, J.E. 1965. Seed development and germination. *In* J.Y. Bonner and E.J. Varner (eds.), Plant biochemistry, 763–792. Academic Press, New York.

Inefficient Conversion of Floret Populations to Actual Seed Harvested in Grass-Seed Crops

K.R. BROWN

Grasslands Division, DSIR, Christchurch, N.Z.

Summary

Obtaining more than 34 machine-dressed seeds from every 100 florets in trials at Lincoln, New Zealand, was uncommon. In ryegrass (*Lolium* sp.), Grasslands Matua prairie grass (*Bromus catharticus* Vahl), and G17 cocksfoot (*Dactylis glomerata* L.) crops, the larger the floret population, the lower the number of seeds obtained from every 100 florets. Nitrogen (N) improved the return, and in prairie grass the return increased as the rate of N increased beyond 20 kg N/ha. In prairie grass, floret populations decreased as row spacings increased. With this seed, recovery was highest at wide row spacings, but in cocksfoot the reverse was true. For crops with similar floret populations, seed recovery was greater in crops with fewer but larger seed heads. Research aimed at inhibiting reduction of or increasing seed recovery offers a promising approach for increasing the efficiency of grass-seed production.

KEY WORDS: ryegrass, *Lolium perenne,* prairie grass, *Bromus catharticus,* cocksfoot, *Dactylis glomerata* L., seed yield, floret populations, seed recovery.

INTRODUCTION

In trials with Grasslands Nui perennial ryegrass (*Lolium* sp.), Grasslands Matua prairie grass (*Bromus catharticus* Vahl) and G17 cocksfoot (*Dactylis glomerata* L.), only rarely were more than 34 seeds for every 100 florets recovered. Conversion of florets to machine-harvested seed was inefficient.

Some factors influencing seed recovery are abortion, shattering, and harvest losses, but whatever the cause, small individual effects may accumulate into major changes in seed yields. For example, reducing seed recovery from 30 to 25 seeds/100 florets in a prairie-grass crop of 50×10^3 florets/m² and 1,000-seed weight of 13 g would reduce yield by 325 kg/ha.

The purpose of this paper is to draw attention to the significance of inefficient floret conversion.

METHODS

Following maximum seed-head emergence, heads/m², spikelets/head, and florets/spikelet were counted, and from these figures florets/m² were calculated. Following harvest, seeds/m² were calculated from seed yields and 1,000-seed weights.

RESULTS

Figures 1 to 5 are diagrammatic representations without extrapolation, based on the appropriate regression analysis.

Overall Recovery

Fifty percent of the samples had between 5 and 20 seeds/100 florets, 30% between 21 and 29, and only 5% more than 34 (Fig. 1).

Ryegrass

Floret density and recovery were negatively related ($r^2 = 0.87$), but there was a positive elongation N effect (Fig. 2). "Elongation N" is nitrogen applied at the start of stem elongation (Brown 1980).

Matua Prairie Grass

In second-year crops, seed recovery fell from 26 to 21 to 11 as row spacing narrowed from 60 to 45 to 15 cm, and floret density increased from 61 to 72 to 166×10^3/m² (Fig. 3).

Within each N rate, seed recovery decreased as floret density increased, but beyond 20 kg N/ha seed recovery increased with increasing rate of N (Fig. 4).

Cocksfoot (Orchardgrass)

As with the other cultivars, seed recovery declined as floret density increased (Fig. 3).

The author is a Research Agronomist in the Grasslands Division of DSIR. The author is indebted to Mr. W. Archie and Mr. J. Hogland for technical assistance and computer operation and to Mrs. M. Hanson and R. Lamberts for preparing the figures.

Fig. 1. Frequency histogram for seed recovery.

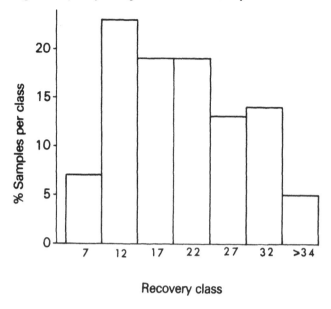

DISCUSSION

Three widely different cultivars reacted similarly to increases in floret density. However, the level at which similar responses occurred was significantly lower for the large-seeded prairie grass than for the smaller-seeded ryegrass and cocksfoot (Fig. 3). These results suggest that the level at which certain responses occur may be a function of seed (floret) size, with individual cultivars falling into broad but differing groups determined by gross differences in seed (floret) size. However, within any one group response is likely to be independent of constituent cultivars. Thus, effects obtained with one cultivar may be extended to others within the same seed-size group. Seed-recovery studies may be done on a few representative cultivars.

There can be no doubt that management practices aimed at increasing floret populations may have detrimental effects on crop efficiency, with yields failing to improve or differing by far less than the increased floret population warrants (Brown 1980). In most practical situations increased floret populations are associated with increases in seed-head numbers rather than increases in seed-head size. From our rate-of-N work with prairie

Fig. 2. Seed recovery in ryegrass.

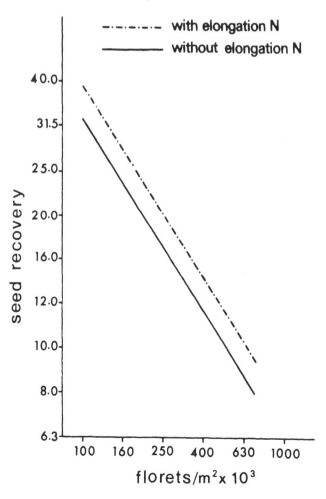

Fig. 3. Seed recovery in 3 different grasses.

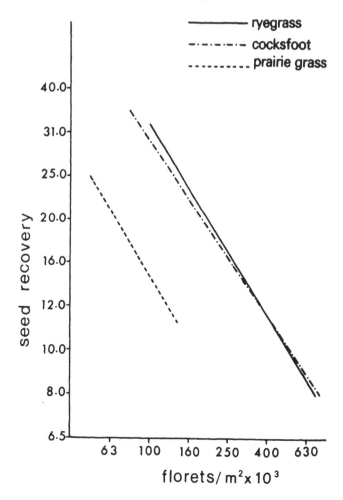

Fig. 4. Rate of elongation N and seed recovery in Matua prairie grass.

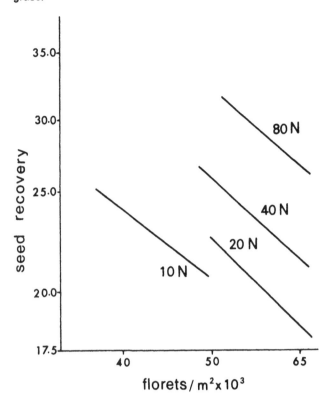

Fig. 5. Seed recovery and increases in seed-head size alone or seed-head numbers alone.

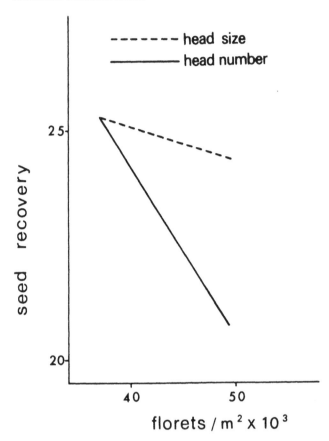

grass there is evidence to indicate that the effect of increased floret population was more severe if the increase was the result of more heads than if it was the result of increased seed-head size (Fig. 5).

Recovery can be improved by applications of N at elongation, when N-induced increases in floret densities are brought about by larger seed heads rather than by more seed heads. The data from the prairie-grass rate-of-N work indicates that if medium-to-high rates of N are used there will be sufficient N available not only to overcome the N-induced floret increase but to increase seed recovery overall—i.e., there will be an overcompensation effect.

Three generally applicable points may be made:

1. Crop efficiency and seed yields could be increased if management practices aimed at increasing floret densities were primarily designed to increase seed-head size, with increases in seed-head numbers secondary.
2. Rates of N should be high enough to obtain the overcompensation effect.
3. Rates of elongation N should be higher for crops with

larger seed-head populations of the same floret density.

Although the factors outlined in this paper account for apparent small differences in seed recovery, they have been shown to make important differences in seed yields and are the first steps toward understanding and overcoming the inefficient nature of floret conversion to seed harvested. Further progress could be made by studying and fully understanding the nature of the possibly larger effects on seed recovery of losses incurred at harvest and cleaning.

LITERATURE CITED

Brown, K.R. 1980. Seed production in New Zealand ryegrasses. II. Effects of N, P, and K fertilisers. N.Z. J. Expt. Agric. 8:33–39.

Selection for Seed Retention in *Phalaris aquatica* L.

J.R. MCWILLIAM and C.N. GIBBON
University of New England, Armidale, N.S.W., Australia

Summary

Seed shattering is a problem in most pasture grasses, especially in those that have a history of recent domestication. In many of these grasses, losses from shattering are aggravated by uneven ripening of inflorescences and a low percentage of viable seed.

Fragility of the rachilla appears to be the most common cause of seed shattering in grasses, and once this connection is broken, the seed is free to shed from the inflorescence. In some of the more domesticated species, the nature and configuration of the glumes and the manner in which they are packed on the inflorescence play an important part in retaining seed.

Seed shattering in phalaris (*Phalaris aquatica*) has been overcome through the discovery of a naturally occurring seed-retaining mutant possessing a strong nonbrittle rachilla that anchors the seed in the spikelet at maturity. This character has now been incorporated into a seed-retaining cultivar which can be harvested without suffering any significant loss of seed for a period extending at least three weeks after full stand maturity.

It seems possible that mutations of the type found in phalaris that prevent seed shattering may be present in many grass species, but because these mutants are at a selective disadvantage in the wild, the frequency of the genes controlling them in populations will be low. Despite this low frequency it should be possible to locate such natural mutants in other important grasses, thereby overcoming one of the major problems inhibiting the domestication of these species for use in agriculture.

KEY WORDS: phalaris, *Phalaris aquatica*, seed retention.

INTRODUCTION

Seed dispersal mechanisms, well developed in all wild grasses, represent important adaptive mechanisms in these species. The actual shedding of seed involves the disarticulation of the rachis either above or below the glumes, a process that is usually associated with the formation of an abscission layer at or about the time of seed maturity.

One of the most important consequences of the domestication of wild grasses, involving both conscious and unconscious selection, has been the development of cultivars, such as in the major cereals, that retain their seed in the inflorescence at maturity. Development has been achieved through strengthening of the rachis or rachilla and elimination of the abscission layer, which is usually controlled by one or a few dominant or recessive genes (McWilliam 1980).

In contrast to the cereals, the cultivated pasture grasses, with few exceptions, have received little attention from plant breeders, and seed shattering remains a serious problem.

J.R. McWilliam is Professor of Agronomy and C.N. Gibbon is Senior Technical Officer in the Department of Agronomy and Soil Science of the University of New England.

SEED RETENTION IN PASTURE GRASSES

Although little conscious effort has been made to improve seed retention in pasture grasses by selection, certain genera, including *Lolium*, *Festuca*, and *Dactylis*, have a long history of cultivation and do have acceptable levels of seed retention.

Grasses that have been introduced into agriculture in relatively recent times, such as the Mediterranean perennial *Phalaris aquatica* (syn. *P. tuberosa*) and tropical perennials such as *Panicum maximum*, *Setaria anceps*, and *Brachiaria mutica*, retain the seed-shattering characteristic of their wild ancestors. In many of the tropical species, this characteristic is often aggravated by uneven ripening and low seed viability (McWilliam 1963, Humphreys 1978). As a consequence there often are serious harvesting problems, resulting in low yields and often inferior-quality seed. These factors, in turn, are responsible for the relatively high cost of such seed and for some of the difficulties associated with seedling establishment (McWilliam and Schroeder 1974).

The genetic control of seed retention in cereals is well documented (McWilliam 1980), but less is known about pasture grasses. Also, few accounts have been published on mutations in pasture grasses that prevent shattering, such as have occurred in association with the domestication of cereals. Hertzsch (1961) has reported that he has used natural mutants in developing a nonshattering vari-

Fig. 1. Details of the basal attachment of the fertile lemma enclosing the caryopsis (seed) within the spikelet of phalaris. (a) Spikelet of seed-retaining cultivar of *P. aquatica* with glumes removed, showing the stout nonbrittle rachilla, which remains intact after maturity and anchors the fertile lemma to the pedicle at the end of the branched rachis (× 48). (b) Base of the mature fertile lemma of the Australian cultivar of *P. aquatica* (seed shedding) showing the two sterile lemmas (× 40). (c) Overlapping bases of the sterile lemmas in the Australian cultivar, showing the small gap through which the slender, fragile rachilla connecting the caryopsis to the pedicle is inserted (× 150). (d) Basal view of the mature fertile lemma of *Phalaris canariensis* with glumes removed, showing the prominent sterile lemmas and the rachilla firmly attaching the entire structure to the pedicle at the end of the branched rachis (× 40).

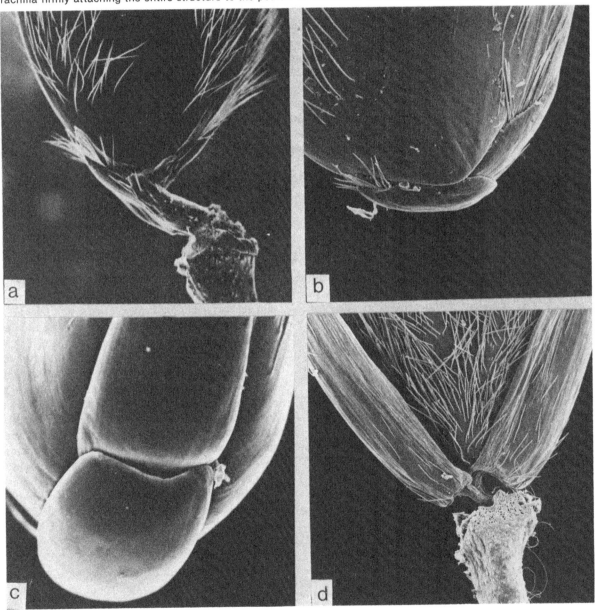

ety of *Arrhenatherum elatius* (oat grass) and also that he has improved the seed-retaining properties of *Phalaris arundinacea* by selecting among X-ray-induced mutations of this species (Hertzsch 1957). In neither case was the mechanism of seed retention explored.

One grass that may have evolved its seed-retaining characteristics as a consequence of mutation is *Phalaris canariensis* (annual canarygrass), which is widely grown for bird seed and is one of the few wild grasses that has achieved the status of a commercial grain crop, with seed yields up to 2,000 kg/ha (Anderson 1961).

SELECTION FOR SEED RETENTION IN PHALARIS (*P. AQUATICA*)

Seed shattering, a major problem in phalaris, is a reflection of the relatively recent domestication of this perennial pasture grass. Spikelets making up the condensed panicle contain a single fertile floret. At maturity the caryopsis (seed) disarticulates, owing to the fracturing of the fragile rachilla, and lies free within the glumes. As the inflorescence dries further the glumes gape open, and the seed is shed by any slight movement of the inflorescence (Anderson 1961, McWilliam 1963).

Strategies employed by seed producers to minimize the problem include cutting and windrowing or stooking prior to maturity, repeated "trough" harvesting of mature seed, and spraying inflorescences with an adhesive plastic lacquer prior to seed maturity to minimize shattering (McWilliam and Schroeder 1974). All these techniques can improve seed yield, but the additional costs involved have limited their adoption. The most effective solution to the problem is selection for nonshattering genotypes that will retain all seed at full maturity.

A selection from a cultivar introduced from Argentina, with stiff, inflexible inflorescences and densely packed spikelets, showed superior seed retention, and it was released as the high-seed-retaining cultivar Seedmaster (McWilliam and Schroeder 1965).

More recently, a single plant of phalaris was identified in a large experimental population of the Australian cultivar that retained all seed firmly in the inflorescence at full maturity. The remainder of the population, by comparison, lost an average of 50% of its seed within 2 weeks of the appearance of the first mature seed.

This natural seed-retaining mutant was self-incompatible and was allowed to cross at random with the other members of the population. Plants raised from the first-cross seed were intercrossed in isolation, and four seed-retaining genotypes were recovered from the half-sib progenies (segregation ratio approximately 256:1), suggesting the possibility of four recessive genes's controlling the expression of seed retention in this population.

These four nonshattering genotypes, plus the original mutant, showed no further segregation when intercrossed, giving progeny that retained all seed at maturity. Four further cycles of recurrent selection to improve inflorescence quality and plant type have confirmed the sta-

bility of this homogenous recessive genotype, the first full seed-retaining cultivar of *P. aquatica*.

THE NEW SEED-RETAINING CULTIVAR

The new cultivar has a well-developed nonbrittle rachilla that remains intact after maturity and anchors the fertile lemma containing the caryopsis (seed) firmly in the spikelet, irrespective of the shape and configuration of the glumes (Fig. 1a). This mechanism contrasts with that of the existing seed-shattering cultivars of phalaris, in which the fragile, ribbon-like rachilla fractures at maturity, leaving the seed lying free in the glumes (Fig. 1b, c).

The stout rachilla of the new cultivar, which varies somewhat in length depending on the genotype, and the well-developed sterile lemmas are also features of the seed of canarygrass (*Phalaris canariensis*), the annual phalaris species that appears to have been domesticated by man in modern times (Fig. 1d). The similarity of the seed-retention mechanism in these 2 species of *Phalaris* and in other temperate cereals suggests that this mutation for seed retention has played an important role in the domestication of these wild grasses. Incorporation of the seed-retaining character into the Australian cultivar of phalaris results in a significant increase in the yield and quality of harvested seed.

A trial to evaluate the effect of the seed-retaining character on yield of harvestable seed was conducted under commercial conditions in the field. Two harvests were taken from replicated plots of the Australian and seed-retaining cultivars, the first 30 days after midanthesis of each cultivar, when approximately 90% of the seed in the inflorescences had reached physiological maturity (i.e., was capable of maturing into viable seed), and the second 20 days later. Yields of viable seed obtained at the two harvests are given in Table 1.

At the time of first harvest the Australian cultivar had lost an average of 20% of the ripe seed in the upper part of the inflorescences. Earlier harvesting is not desirable because of the presence of immature seed in the lower positions of the inflorescences. This implies that the potential seed yield of the two cultivars was similar. Over the following 20 days the Australian cultivar lost a further 53% of its seed, making a total of 73% loss at day 50,

Table 1. Yields of seed from two phalaris cultivars (kg/ha) harvested 30 days and 50 days after midanthesis and percent losses due to seed shattering.

Cultivar	Days after midanthesis		% Loss
	30	50	
Australian	494(616)*	169	73
Seed Retaining (SR)	611	578	5
S.E. = 72			

*Actual and (potential) seed yield of the Australian cultivar at first harvest (30 days).

whereas the seed-retaining cultivar lost virtually no seed (average 5%) over the same period (Table 1). The seed-retaining cultivar flowered approximately 7 days later than the Australian cultivar, but in other respects there were no morphological or agronomic differences between the two.

These results indicate that the seed-retaining cultivar can be harvested at any time up to 3 weeks after full stand maturity without suffering any significant loss of seed. This character will increase the flexibility and ease of harvest and the yield and quality of seed obtained.

LITERATURE CITED

Anderson, D.E. 1961. Taxonomy and distribution of the genus *Phalaris*. Iowa State J. Sci. 36:1–96.

Hertzsch, W. 1957. Mutationsversuch mit Rohrglanzgras. Z. Pfl. Zücht 37:263–279.

Hertzsch, W. 1961. Die Bedeutung, Erfolge und Probleme der Züchtung auf festen Kornsitz bei *Phalaris arundinacea*. *In* Neue Ergebnisse Futterbaulicher Forschung, 133–141. DLG-Verlag, Frankfurt (Main).

Humphreys, L.R. 1978. Tropical pastures and fodder crops, 135. Longman, London.

McWilliam, J.R. 1963. Selection for seed retention in *Phalaris tuberosa* L. Austral. J. Agric. Res. 14:755–764.

McWilliam, J.R. 1980. The development and significance of seed retention in grasses. *In* P.D. Hebblethwaite (ed.), Seed production, 694. Butterworths, London.

McWilliam, J.R., and H.E. Schroeder. 1965. Seedmaster—a new cultivar of phalaris with high seed retention. Austral. Inst. Agric. Sci. J. 31:313–315.

McWilliam, J.R., and H.E. Schroeder. 1974. The yield and quality of phalaris seed harvested prior to maturity. Austral. J. Agric. Res. 25:259–264.

Improved Processing for High-Quality Seed of Big Bluestem, *Andropogon gerardii*, and Yellow Indiangrass, *Sorghastrum nutans*

R.R. BROWN, J. HENRY, and W. CROWDER

USDA SCS, Columbia, Mo., and USDA SCS
Plant Materials Center, Elsberry, Mo., U.S.A.

Summary

The recent reintroduction of native prairie warm-season grasses for livestock forage in the Corn Belt presents seed-handling problems with some species. Seeding big bluestem (*Andropogon gerardii* Vitman) and indiangrass (*Sorghastrum nutans* [L.] Nash) through commonly available seeding equipment is difficult with seed from commercial sources. Commercial seed contains significant amounts of leaf and stem particles as well as awns and other appendages still attached to the seed. These seed appendages and inert matter prevent an even flow of seed through the seeding equipment. Specially designed rangeland drills handle the seed rather well, but the limited use of this equipment makes buying it difficult to justify.

Our objective was to improve seed quality and thereby increase the acceptability and use of big bluestem and indiangrass in pasture forage systems in the Corn Belt.

Seed from the combine was placed in a fresh-air bin dryer, which reduced seed moisture to levels adequate for storage in 2 to 3 days. The dried bulk material was processed through a debearder and then through a 3-screen seed cleaner to separate clean seed from the other plant material. The mixing action of the debearder rubs or breaks off awns and other seed appendages and increases separation of seed from other plant parts in the seed cleaner.

Debearding and cleaning seed improved both purity and germination. Field planting of debearded seed was successful with grain drills, gravity and broadcast seeders, and rangeland grass drills. As debearded seed in storage maintained its viability up to 3 years, increased processing apparently caused no seed damage.

Development of a commercial supply of debearded big bluestem and indiangrass seed that can be planted with commonly used equipment should greatly enhance acceptability and use of these species in pasture forage systems in the Corn Belt.

KEY WORDS: indiangrass, *Sorghastrum nutans* (L.) Nash, big bluestem, *Andropogon gerardii* Vitman, processing, seed quality, debearder.

INTRODUCTION

The reintroduction of warm-season native grasses to the Corn Belt area of the U.S.A. may present seed-handling problems. The grass species being emphasized are switchgrass (*Panicum virgatum* L.), big bluestem (*Andropogon gerardii* Vitman), and indiangrass (*Sorghastrum nutans* [L.] Nash). These species are the most productive and widely adapted on the deep soils of the tallgrass prairie.

Switchgrass presents no problem with seed handling, as its seed is quite firm, smooth, and similar to that of other forage species in flow-quality characteristics. Commonly used grain drills, roller-packer seeders, or broadcast seeders readily distribute the seed, and adequate stands can be obtained with these planters.

Big bluestem and indiangrass seed have awns 1 to 2 cm, and each spikelet has basal pubescent appendages about 1 cm. These appendages increase the difficulty of separating seed from other plant parts during harvest and cleaning. They also cause the seed to become entangled and therefore not easily seeded through forage-seeding equipment commonly used in the Corn Belt. This handling problem reduces the acceptability of these species for pasture forage systems.

In the Great Plains area immediately west of the Corn Belt, large acreages of native prairie containing indiangrass and big bluestem have been maintained. Seedings of these species in that area have been enhanced with the development of special drills to handle seed containing considerable leaf, stem, and awned seed material. The seed is sold on a pure-live-seed (PLS) basis (germination × purity/100 = percentage of PLS). Commercial seed producers process the seed minimally because of the leaf, stem, and awned seed material.

Farmers in the Corn Belt accustomed to planting forage seeds with excellent flow quality find the chaffy seed of the native grasses unacceptable as it is now sold commercially. Commonly used equipment cannot handle these seeds. The specialized equipment needed to plant the chaffy seed has no alternative use to justify its purchase. Rental equipment is not desirable or available, and no adequate alternative seeding method has been developed thus far.

Identifying the need for warm-season grass forages in the Corn Belt has caused a renewed use of native grass species. This interest and the poor handling qualities of commercial seedlots stimulated an investigation to improve the processing (cleaning) to produce a high-quality seed for planting.

A literature search has revealed little experimentation with seed processing of the species in question. Schwendiman et al. (1940) discussed the use of a hammer mill to remove seed awns and other appendages. The Plant Materials centers of the USDA SCS have used this system

on many plant species to process seed to a high quality. Atkins and Smith (1967) emphasized the importance of controlling the speed of the hammer mill to prevent excessive seed damage. The hammer mill is designed to grind, and it can easily damage seed unless it is properly operated. Use of the hammer mill can produce seed of good quality but is practical only on a small scale, as it is time-consuming and requires great care by the operator to prevent seed damage. Atkins and Smith (1967) mention a debearder that can "remove awns, break up straw, and remove outer glumes on some species." They do not discuss details or hazards of the process.

METHODS

The manager of the SCS Plant Materials Center, Elsberry, Missouri, purchased a debearder in 1976 in an attempt to develop a seed-processing system to improve the flow characteristics of big bluestem and indiangrass seed.

The debearder is constructed of a steel drum with a motor-driven rotor. The rotor consists of several steel fingers operating offset and among several stationary fingers. The resulting stirring action, unlike the striking action of a hammer mill, rubs and stirs the seed, breaking off the bothersome appendages.

Processing procedures that improve seed purity were developed. The seed was harvested by combine from standing grass plants and immediately transported to a fresh-air bin for drying. After about 3 days in the bin dryer, the dry seed was processed in the debearder at full capacity to break off awns, remove other seed appendages, and break up leaf and stem parts for separation. Seed from the debearder was then processed through a 3-screen fanning mill to separate the clean seed from other plant parts.

Seed was tested for purity, germination, and bulk density by the Kansas State Seed Testing Laboratory with standard procedures.

RESULTS

Seed of indiangrass and big bluestem was very successfully processed with the debearder and was easily separated from foreign matter and other plant parts with the fanning mill. Seed quality was significantly increased by processing in this manner. Indiangrass seed tested at 98% purity and 83% germination. Four-year averages were 94.1% purity and 61.5% germination, with 57.9% PLS. For big bluestem seed processed with the debearder, 4-year averages were 92.5% purity and 70.3% germination, with 65.0% PLS. The standards of the Association of Official Seed Certifying Agencies for certifying big bluestem and indiangrass require only 25% PLS, and many commercial seedlots contain 25%–30% PLS.

Test results for seed purity and germination are presented in Table 1. The low purity value of the 1977 big bluestem was probably due to improper adjustment of cleaning equipment. The decrease in germination of the 1978 indiangrass seed retested in 1979 is not understood;

R.R. Brown is a Plant Materials Specialist at the USDA SCS, Columbia. J. Henry is the Manager and W. Crowder a Biological Technician at the USDA SCS Plant Materials Center in Elsberry.

Table 1. Purity and germination of seed of big bluestem and indiangrass with and without debearding treatment.

Seed/ year	Not debearded			Debearded		
	Purity (%)	Germ. (%)	Test date	Purity (%)	Germ. (%)	Test date
Big bluestem						
1979	80.7	71	1979	92.5	73	1979
1978				91.9	64	1979
1978				91.9	50	1978
1977 (Reclean)				95.1	74	1979
1977				52.0	78	1977
1976				90.5	64	1979
Indiangrass						
1979	80.8	44	1979	98.1	57	1979
1978				97.0	72	1979
1978				97.0	83	1978
1977				91.5	69	1979
1977				91.5	62	1977
1976				88.0	58	1977
1976				89.9	44	1976

seed of these species is generally expected to increase in germination while in storage. Indiangrass seed of 1976 and 1977 did so.

Bulk density of seed was increased 2.9 to 3.4 times with the debearder–fanning mill system. Table 2 presents the increase in bulk density of seed after cleaning.

DISCUSSION

Processing seed of indiangrass and big bluestem with a debearder–fanning mill system yielded clean seed with high bulk density and excellent handling qualities. The increased processing produced high PLS values with little or no seed damage. Storage generally increased germination percentage, as is usual with these species.

Debearded and cleaned seed can be planted with grain drills, packer seeders, broadcast equipment, or any equipment designed to seed grass. The special rangeland drills must be modified to plant the debearded seed at recommended rates. Gear ratios on the metering mechanism must be modified.

The availability of debearded big bluestem and indiangrass seed should enhance the use of these species for forage production in the Corn Belt. Some interest has been shown and questions have been raised concerning the feasibility of and need for processing seed to this degree. Most concern relates to added cost and possible loss of viable seed from the increased processing. No cost estimate has been prepared by commercial producers; however, studies are being conducted to determine whether viable seed is lost. When properly installed and operated, the debearder and fanning mill can be operated simultaneously by one operator.

Table 2. Bulk density of combined seed and of variously processed seed.

Seed	Bulk density (g/liter)		
	Combined	Fanning mill	Debearder/ fanning mill
Big bluestem	66.2	101	192
Indiangrass	69.6	126.4	236.6

LITERATURE CITED

Atkins, M.D., and J.E. Smith, Jr. 1967. Grass seed production and harvest in the Great Plains. U.S. Dept. Agric. Fmrs. Bul. 2226.

Schwendiman, J.L., R.F. Sackman, and A.L. Hafenrichter. 1940. Processing seed of grasses and other plants to remove awns and appendages. U.S. Dept. Agric. Cir. 558.

Seed-Production Potentials of Eight Tropical Pasture Species in Regions of Latin America

J.E. FERGUSON, D. THOMAS, R.P. DE ANDRADE, N. SOUZA COSTA, and S. JUTZI

Tropical Pastures Program CIAT, Cali, Colombia, and Brasilia, Brazil; EMBRAPA-CPAC, Brasilia, Brazil; EPAMIG, Sete Lagoas, Brazil; and COTESU, Santa Cruz, Bolivia

Summary

Early identification of geographic regions providing high and consistent seed yields for species of interest, especially legumes, would be very advantageous. Our objectives were (1) to identify potential seed yield and the principal determinants thereof for a range of species, (2) to determine progressively the relative seed-production potential of distinct geographic regions, and (3) to stimulate emphasis upon seed-production issues within pasture programs.

Eight grass and legume species, either promising accessions or local cultivars, were established in replicated pure stands at five locations. Compound fertilizers were applied, including nitrogen for the grasses. Seed was harvested once each time a seed crop matured. Weight of clean dry seed was recorded, and a purity estimation was made. Observations were made upon phenology and problematic weeds, diseases, and insects. Data are available for the first 2 years of the field experiment.

Maximum annual seed yield (kg/ha) and multiplication rate (ha/yr) were recorded, respectively, as follows: *Desmodium ovalifolium* CIAT 350, 220 and 73; *Pueraria phaseoloides* common kudzu, 135 and 34; *Stylosanthes capitata*, average of CIAT 1315 and 1405, 962 and 190; *Zornia latifolia* CIAT 728, 680 and 170; *Andropogon gayanus* CIAT 621, 143 and 70; *Brachiaria decumbens* cv. Basilisk, 366 and 146; *Brachiaria humidicola* common, 400 and 160; *Panicum maximum* cv. Petrie, 263 and 88.

Determinants of potential seed yield recorded during 1979 and 1980 were (1) temporal weed dominance, (2) lack of persistence, (3) failure to flower, (4) reduced vegetative and reproductive vigor, (5) mismanagement, and (6) reduced genetic purity of native plant populations of *Zornia* spp. at two locations. *S. capitata* CIAT 1405 was killed by anthracnose (*Colletotrichum* spp.) at Brasilia. *D. ovalifolium* and *D. phaseoloides* did not flower in 1979 at Brasilia and Felixlandia because of moisture stress before induction. Diseases such as anthracnose and sphaceloma scab and root-knot nematode affected *S. capitata*, *Z. latifolia*, and *D. ovalifolium*, respectively, while *Urocystis* and *Ustilago* spp. affected *D. decumbens* and *P. maximum*. Prolonged high soil moisture caused sparse flowering in *P. phaseoloides*, and moisture stress had a similar effect on *D. ovalifolium*. Some established stands of *A. gayanus*, when fertilized with nitrogen but without a preflowering cut, lodged after flowering.

The biological potential for seed production in each region appears favorable for particular species only, e.g., Brasilia, *Z. latifolia* and all four grasses; Chimore, *D. ovalifolium*; Felixlandia, *S. capitata*, *Z. latifolia*, and three grasses; Quilichao, *S. capitata*, *Z. latifolia*, and *A. gayanus*; Sete Lagoas, *S. capitata*, *Z. latifolia*, and three grasses. These initial results will assist researchers and producers to optimize localization and management of future production efforts with these relatively unknown species.

KEY WORDS: seed yield, seed-yield determinants, *Desmodium*, *Pueraria*, *Stylosanthes*, *Zornia*, *Andropogon*, *Brachiaria*, *Panicum*.

INTRODUCTION

Within tropical Latin American, pasture seeds are produced essentially as byproducts of pastures within cattle-production regions (Ferguson 1978). As demand for seed increases and as new species and cultivars become available, there is a basic need to identify the major determinants of high and consistent seed yield for each species. Only then can the species' broad geographic reactions be predicted and seed production be encouraged within the most favorable regions. Hopkinson and Reid (1978) described the influence of climate upon several legumes in tropical Australia and predicted geographic regions in other continents with high seed-production potentials.

In the developmental stages of a national seed-production effort, early identification of appropriate geographic regions for the species involved would provide the following advantages: higher seed yields, reduced risk to new growers, earlier attainment of a demand-supply balance at reasonable prices, an increased spectrum of species that can attain seed-yield levels to make commercial production viable, and increased efficiency of research effort.

This paper provides a progress report toward our objectives of (1) defining the biological seed-yield potentials and principal determinants thereof for eight grass and

The authors are all Research Agronomists. J.E. Ferguson is at CIAT, Cali, Colombia; D. Thomas is at CIAT, Brasilia, Brazil; R.P. de Andrade is at EMBRAPA-CPAC, Brasilia, Brazil; N. Souza Costa is at EPAMIG, Sete Lagoas, Brazil; and S. Jutzi is at COTESU, Santa Cruz, Bolivia.

legume species; (2) determining progressively the relative potentials of distinct geographic regions based upon climatic, edaphic, agronomic management, and economic factors; and (3) generally stimulating emphasis on seed production, assuming future releases of new cultivars and expansion within the sector generally.

MATERIALS AND METHODS

The species and accessions, nominated jointly, were of interest to either local or CIAT scientists.

A field experiment is in progress at five locations (Table 1). Planting took place between October 1978 and January 1979. All species were planted from seed in pure-stand plots of 10 × 8 m with 3 replicates/location. A randomized complete-block design was used for both legumes and grasses at each location. A compound, preplant, fertilizer dressing was applied consistent with local knowledge and practice. Nitrogen (50 kg/ha as urea) was applied to the grasses at the beginning of each regrowth period. Weeds were controlled manually at most sites. Grasses with multiple annual harvests were cut back to 15 cm immediately after each harvest. Records were kept of phenology and the incidence of problematic diseases, insects, and weeds. Seed was harvested once each time a seed crop matured, by direct manual collection, mechanical cutting and threshing, or a combination of both. Weight of clean dry seed was recorded, and an estimation of purity was made, consistent with ISTA-type procedure. Year of establishment and second-year data were recorded in 1979 and 1980.

RESULTS AND DISCUSSION

Because flowering data in 1979 were confounded with establishment effects, 1980 data are emphasized here. Tentative conclusions about flowering control mechanisms are confined to the data from the four higher-latitude (15°–19°S) locations.

Pure-seed yields of the legumes at the five locations are presented in Table 2. *S. capitata*, at lat 15°–19°S, commenced flowering in mid-April and matured in late May, whereas at Quilichao it flowered and seeded in both rainy seasons each year. A short-day flowering response is tentatively assumed. Seed yields in 1979 were high at four locations. In 1980 anthracnose reduced yields at both Brasilia and Chimore, but all plants of CIAT 1405 were killed at Brasilia. Seed yields in 1980, however, were high at Quilichao, Felixlandia, and Sete Lagoas, indicating that the species is a prolific and versatile seeder, depending upon the prevalent races of anthracnose. The maximum yield for the two accessions averaged 962 kg/ha (Felixlandia) in 1980. Assuming a planting density of 5 kg/ha, a maximum multiplication rate of 190 ha/yr is derived.

D. ovalifolium did not flower at Brasilia or Felixlandia in 1979, probably due to the onset of moisture stress before induction. In 1980, time of commencement of flowering was erratic among locations at lat 15°–19°S, ranging between early March and early May. Seed yields were highly variable, ranging from 0–123 and 2–220 kg/ha in 1979 and 1980, respectively. The species is highly sensitive in vegetative and reproductive vigor to moisture stress; high seed yield was recorded only in the presence of prolonged moisture availability. Maximum seed yield was 220 kg/ha (Chimore) in 1980. Assuming a planting density of 3 kg/ha, a maximum multiplication rate of 73 ha/yr is indicated. Root-knot nematodes were recorded at Brasilia and Sete Lagoas.

P. phaseoloides did not flower at Brasilia or Chimore in 1979. In 1980 flowering commenced in late April at Brasilia and late May at Chimore, indicating a weak short-day photoperiodic response reduced by high moisture availability. Seed yields were highly variable, ranging from 0–38 and 0–135 kg/ha in 1979 and 1980, respectively. Maximum seed yield was 135 kg/ha (Brasilia) in 1980. Assuming a planting density of 4 kg/ha, a maximum multiplication rate of 34 ha/yr is derived.

Z. latifolia commenced flowering in mid-February and matured in June at lat 15°–19°S, whereas at Quilichao it flowered and seeded in both rainy seasons. Seed yields in 1979 and 1980 ranged from 14–175 and 16–691 kg/ha, respectively. A maximum seed yield of 680 kg/ha was recorded in 1980 (average of Brasilia and Felixlandia). Assuming a planting density of 4 kg/ha, a maximum multiplication rate of 170 ha/yr is derived. Sphaceloma scab reduced yield in 1980 at Quilichao. Plants of native *Zornia* spp. prevalent at Brasilia and Felixlandia would

Table 1. Climatic summary of regional locations in field experiment.

| Geographic region | | Latitude | Altitude, m | Temperature, °C[1] | | Frost risk[2] | Precipitation, mm | |
Location	Country			High	Low		AAR[3]	DSC[4]
Brasilia	Brazil	15° 35′ S	998	23.1	19.5	0	1,580	639
Chimore	Bolivia	18° S	226	26.0	21.0	0	4,000	1,000
Felixlandia	Brazil	18° 45′ S	614	23.8	18.5	L	1,235	398
Quilichao	Colombia	3° 06′ N	990	23.5	22.3	0	1,844	–
Sete Lagoas	Brazil	19° 28′ S	732	22.6	17.4	M	1,402	360

[1]Average daily mean temperature. 'High' is for hottest and 'Low' for coolest month.

[2]Local estimates of frequency of damaging frost. 0 = zero; L = low, <20%; M = medium, <33%.

[3]AAR = Average annual rainfall.

[4]DSC = Dry-season component, or precipitation outside four wettest long-day months.

Table 2. Pure seed yield of five legumes at five locations in two consecutive years.

Species/accession	Year	Pure seed yield (kg/ha)				
		Brasilia	Chimore	Felixlandia	Quilichao	Sete Lagoas
Desmodium ovalifolium CIAT No. 350	1979	0	57	0	75	
					123	—[1]
	1980	12	220	2	170	
Pueraria phaseoloides common Kudzu	1979	0	0	—	38	—
	1980	135	0	—	70	—
Stylosanthes capitata CIAT No. 1315	1979	150	135	266	110	
					546	
	1980	26	100	867	355	314
					190	
Stylosanthes capitata CIAT No. 1405	1979	199	114	297	97	
					296	
	1980	0	60	1057	224	309
					140	
Zornia latifolia, CIAT No. 728	1979	175	79	14	63	
					42	
	1980	691	60	670	16	135

[1]Implies not present at that location.

complicate maintenance of varietal purity.

Pure-seed yields of the grasses at each location are presented in Table 3. *A. gayanus* flowered several times in 1979 and 1980 at Quilichao but only once each year at higher-latitude locations. At lat 15°–19°S in 1980, flowering commenced in mid-April with seed maturity from late May onwards. A short-day flowering response is indicated, as has been reported by Tompsett (1976). In 1979, seed yields were relatively high and consistent over all locations, from 26 to 159 kg/ha. In 1980, however, harvested yields declined markedly, with a range of 8 to 50 kg/ha. At the higher-latitude locations, this decline was due to inappropriate defoliation management: Nitrogen was applied at the start of the rainy season (late November), and plants, not having been cut, grew excessively tall and lodged postflowering. A precut or grazing in approximately midseason should avoid this problem in future years. Maximum seed yield was 143 kg/ha (average of Brasilia and Felixlandia in 1979). Assuming a planting density of 2 kg/ha pure seed, a maximum multiplication rate of 70 ha/yr is derived.

B. decumbens flowered sparsely at Quilichao in both years, whereas once it was established at lat 15°–19°S it commenced flowering in mid-December and produced

Table 3. Pure seed yield of four grasses at five locations in two consecutive years.

Species/Accession	Year	Pure seed yield (kg/ha)				
		Brasilia	Chimore	Felixlandia	Quilichao	Sete Lagoas
Andropogon gayanus, CIAT No. 621	1979	128	53	38	46	
					58	159
					26	
	1980	45	12	21	8	17
					50	
Brachiaria decumbens, cv Basilisk	1979	147				
		17	0	27	0	100
	1980	262	0	282	0	186
		178		70		121
Brachiaria humidicola, common	1979	18	—[1]	—	—	—
	1980	232	—	—	—	—
		168				
Panicum maximum[2]	1979	90			21	
		40	—	103	3	128
		101			5	
	1980	263	—	4	0	0
		17			15	

[1]Implies not present at that location.

[2]At Brasilia, cv Petrie (green panic); at Felixlandia and Sete Lagoas EPAMIG No. 483 (guinea); at Quilichao, CIAT No. 604 (guinea).

two seed harvests. This finding would indicate a day-neutral flowering response. In 1979 pure-seed yield ranged from 0 to 147 kg/ha. In 1980, although yields of 0 were recorded (Quilichao and Chimore), maximum yields (Brasilia and Felixlandia) averaged 270 kg/ha/harvest, and the average annual yield (Brasilia, Felixlandia, and Sete Lagoas) was 366 kg/ha. Assuming a planting density of 2.5 kg/ha pure seed, a maximum multiplication rate of 146 ha/yr is derived. A smut (*Urocystis* sp.) was recorded at several locations.

B. humidicola was present at Brasilia only. Once it was established, it commenced flowering in December and resulted in two seed harvests in 1980. A day-neutral flowering response is indicated. In 1980, maximum seed yield was 232 kg/ha/harvest and total annual yield was 400 kg/ha. Assuming a planting density of 2.5 kg/ha, a maximum multiplication rate of 160 ha/yr is derived.

P. maximum was represented by three cultivars, a fact that restricts interpretations. Multiple annual flowering peaks were recorded, consistent with a day-neutral flowering response. Seed yields at Quilichao were low and progressively declined because of soil-fertility problems. Maximum seed yield was 263 kg/ha (Brasilia). Assuming a planting density of 3 kg/ha pure seed, a maximum multiplication rate of 88 ha/yr is derived. Smut (*Ustilago* spp.) was common at several locations.

While maximum multiplication rates (ha/yr) ranged between 34 and 190 for legumes and 70 and 160 for grasses, low and no seed yields were recorded frequently. Important determinants of seed yield were foliar diseases, soil moisture effects on reproductive vigor, and weeds for legumes and defoliation management, soil-fertility limitations, and inflorescence diseases for grasses.

The biological seed production potential of each geographic region was species specific. Some favorable combinations appeared as follows: Brasilia, *Z. latifolia*, and all four grasses; Chimore, *D. ovalifolium;* Felixlandia, *S. capitata*, *Z. latifolia*, and three grasses; Quilichao, *S. capitata*, *Z. latifolia*, *A. gayanus;* Sete Lagoas, *S. capitata*, *Z. latifolia*, and three grasses.

LITERATURE CITED

Ferguson, J.E. 1978. Systems of pasture seed production in Latin America. *In* P.A. Sanchez and L.E. Tergas (eds.), Pasture production in acid soils of the tropics, 385–395. CIAT, Cali, Colombia.

Hopkinson, J.M., and R.R. Reid. 1978. Significance of climate in tropical pastures/legume seed production. *In* P.A. Sanchez and L.E. Tergas (eds.), Pasture production in acid soils of the tropics, 343–360. CIAT, Cali, Colombia.

Tompsett, P.B. 1976. Factors affecting the flowering of *Andropogon gayanus* Kunth. Response to photoperiod, temperature and growth regulators. Ann. of Bot. 40:695–705.

Phosphorus Utilization in Grassland Ecosystems

J. KARLOVSKY
Ruakura Soil and Plant Research Station, Hamilton, N.Z.

Summary

The process of phosphorus (P) fixation by soil was long thought to result in large losses of P, an idea that led to a common belief that plants can recover only about 10% to 25% of applied P. Because such small recoveries were considered unrealistic and contrary to common farming experience, the efficiency of P utilization in agricultural ecosystems has been investigated. This was done by using (1) a balance sheet approach in long-term P maintenance trials that had reached a steady state, and (2) published data fitted to a common nutrient cycle for nonsteady-state conditions. The efficiency of P utilization in steady-state ecosystems was defined as the amount of P taken up by the plants as a percentage of the total amount of P required to maintain the available soil P pool at equilibrium. The efficiency of P utilization in nonsteady-state ecosystems was defined as plant P uptake, plus gains or losses in the available soil pool, as a percentage of the total P inputs to the soil.

New Zealand evidence shows that P utilization in grassland ecosystems ranges from 70% to 90% of the added and recycled P, the remaining 10% to 30% being lost in and from the soil. These findings have been incorporated into fertilizer recommendation schemes that are currently undergoing field testing in New Zealand. The maintenance fertilizer requirements are first determined for 90% of the maximum production and then calculated for other levels of production by the following equation:

$$Pm = \left[\log_{10}\left(\frac{100}{100\text{-}RY} \right) \right] \times P_{90}$$

where Pm is the fertilizer required, kg P/ha/yr, for a given relative yield (RY) and P_{90} is fertilizer maintenance requirement for 90% of the maximum yield, kg P/ha/yr.

Balance-sheet studies based on nutrient cycles indicate that P is utilized efficiently in most ecosystems. This finding is in conflict with traditional and still-maintained theories of large-scale P fixation by soil. The concept of nutrient efficiency defined in this paper is of practical importance since it can be used to calculate nutrient inputs required to maintain steady production in any agro-ecosystem.

KEY WORDS: phosphorus utilization, phosphorus cycling, fertilizer requirements.

INTRODUCTION

There is a traditional belief in soil science that the major part of phosphorus (P) applied as fertilizer is irreversibly fixed in the soil. Although by the late 1950s (Sokolov 1959) and early 1960s (Russell 1961) there were suggestions that this traditional idea was not realistic, many scientists have continued to regard P fixation as a serious problem (Watkinson 1974, Wilson 1968). The traditional views probably originated early in the 20th century, when short-term experiments suggested that plants recovered considerably less applied P than they did applied

potassium (K) and nitrogen (N) (Kuhn 1931). Studies involving P adsorption in soil added support to fixation theories, as did measurements of efficiency in which utilization was defined either in terms of P in salable products relative to fertilizer input (Watkinson 1974) or in terms of "consumable output" relative to "farm input" (Frissel and Kolenbrander 1977).

METHODS

Steady-State Method

When P is applied to P-deficient grasslands at a fixed rate annually, pasture production increases until an

The author is a Research Scientist at the Ruakura Soil and Plant Research Station.

equilibrium is established (Karlovsky 1961, 1962). When equilibrium is achieved, the effective P utilization (U) can be expressed as:

$$U = \frac{Pup}{Pr + Pm} \times 100 \qquad (1)$$

where Pup is the total P absorbed by the plant (shoots and roots), Pr is recycled P in dung, litter, and roots (Pr is Pup less P removed in animal produce and by transfer in animal wastes), and Pm is maintenance fertilizer P, all in kg P/ha/yr.

Nonsteady-State Method

Equation 1 can be modified to account for ecosystems in which changes in the available soil P pool have been quantified:

$$U = \frac{(Pup \pm Pb)}{Pt} \times 100 \qquad (2)$$

where Pb is the gain or loss in the available soil P pool, kg P/ha/yr, and Pt is the total amount of P supplied from all sources, kg P/ha/yr. It is assumed that for gains or compensation of losses of x quantity of P in the soil available pool, the same quantity of P has to be added to the soil compartment as is needed to compensate for x quantity of P taken up by plants.

To apply this concept to practical farming, New Zealand soils have been divided into three groups based on maximum production potential, and each group has been subdivided into three categories based on P utilization (Cornforth and Sinclair 1979). The maintenance fertilizer requirements were estimated for all soil groups at 90% of the maximum production. The maintenance requirements for any other relative yield can then be calculated from:

$$Pm = \left[\log_{10}\left(\frac{100}{100 - RY} \right) \right] \times P_{90} \qquad (3)$$

where Pm is fertilizer required, kg P/ha/yr, for a given relative yield (RY) and P_{90} is fertilizer maintenance requirement for 90% of the maximum yield, kg P/ha/yr.

RESULTS

Animal P Losses

The amount of P lost via animals has been studied by several workers in New Zealand (During and O'Connor 1975, Karlovsky 1975). Studies have indicated that the losses under dairy farming and sheep or cattle farming are 18 and 14 kg P/ha/yr, respectively, on high-producing soils with 60 kg P/ha/yr uptake (shoots and roots).

Steady-State Method

The balance-sheet method shows that P utilization ranges from 70% to 90% (Karlovsky 1961, 1962, 1975). For example, in New Zealand, high pasture production with plant P uptake of 60 kg P/ha/yr (shoots and roots) can be maintained by applying 35 kg fertilizer P/ha/yr to soils in the medium–P utilization group. Average animal losses are 10 kg P/ha/yr in produce and 8 kg P/ha/yr transferred in wastes. Recycled P is the difference between uptake and animal losses (60 − 18 = 42 kg P/ha/yr). Using these values in Equation 1 gives:

$$U = \frac{60 \times 100}{42 + 35} = 78\% \qquad (4)$$

Nonsteady-State Method

In a study (Jacquard 1977) of a French farm on which leys are rotated with arable crops, used here as an example, P uptake of 60 kg/ha/yr (shoots and roots) was obtained by adding 80 kg fertilizer P/ha/yr. Animals removed 12 kg P/ha/yr, 48 kg/ha/yr was recycled, and the gain in available soil pool was 37 kg P/ha/yr. Using these values in Equation 2 gives:

$$U = \frac{60 + 37}{48 + 80} \times 100 = 76\% \qquad (5)$$

DISCUSSION

Recent research in many parts of the world has confirmed that a very high degree of P is utilized in agro-ecosystems (Sadler and Stewart 1974, Fotyma 1978, Middleton and Smith 1978). The most convincing evidence that the traditional thinking is incorrect was presented at the 1976 Amsterdam Symposium of the Royal Netherlands Land Development Society, published in Agro-Ecosystems Vol. 4 (1977/1978). Fig. 1 shows the effective P utilization of 39 ecosystems, presented at the above symposium, calculated by Equation 2. The efficiency of P utilization is between 70% and 100% in 29 ecosystems. Eight systems reported by Newbould and Floate (1977) and two systems reported by Thomas and Gilliam (1977) form a group in which the efficiency of P utilization is much lower. The reason for this is that most, or all, of the recycled P was considered to be immobilized in the soil organic pool. The apparent gains in this pool increased apparent losses from the available soil nutrient pool to such an extent that in three native ecosystems (Newbould and Floate 1977), P utilization became negative, ranging between −9% and −16%. Another reason for very low utilization in three intensive ecosystems (Newbould and Floate 1977) was an estimate that 71% of the fertilizer P was fixed in the soil mineral pool. This estimate was based on the results of a study by Widdowson and Penny (1973)

Fig. 1. Plant P uptake plus P balance in the available soil pool vs. total P input. Solid lines connect points with equal efficiencies (100 × plant uptake + balance/input). Dotted lines connect points with equal absolute losses (kg P/ha/yr). Systems with P uptake + balance below 0.2 kg P/ha/yr are grouped near abscissa and indicated with < 0.2.

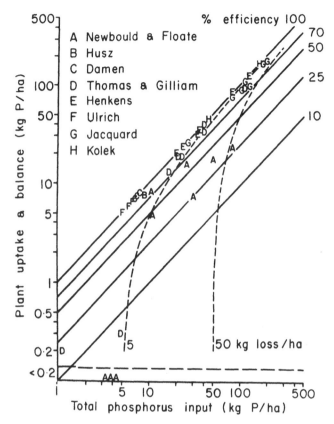

new experimental evidence has been presented, and the conclusion that "fixation of P has been overrated as a problem in most soils" (Engelstad and Parks 1976) and that "attitudes towards soil P and crop use of applied P need thorough re-examination" (Sadler and Stewart 1976) appear fully justified.

It is hoped that after 20 years of controversy in this important field, the correct interpretation of the P cycles presented at the Amsterdam Symposium will bring a world-wide realization that the traditional idea is a complete fallacy that has led to "totally erroneous conclusions" (Elliot 1962). This realization will have great significance in relation to fertilizer research, fertilizer education, and fertilizer use throughout the world.

LITERATURE CITED

Cornforth, I.S., and A.G. Sinclair. 1979. A new approach to maintenance fertiliser requirements. Proc. Ruakura Fmrs. Conf., 74–80.

During, C., and M. O'Connor. 1975. Fertilisers for New Zealand sheep and cattle farms. Proc. Ruakura Fmrs. Conf., 30–34.

Elliott, I.L. 1962. Presidential address. Proc. N.Z. Grassl. Assoc. 24:1–7.

Engelstad, O.P., and W.L. Parks. 1976. Build up of P and K in soils and effective use of these reserves. Proc. TVA Fert. Conf., Cincinnati, Ohio, 50–58.

Fotyma, M. 1978. The principles of phosphate fertilisation. Phos. Agric. 72:11–22.

Frissel, M.J., and G.J. Kolenbrander. 1977. The nutrient balances. Agro-Ecosystems 4:277–292.

Jacquard, P. 1977. Agro-ecosystems in France. Agro-Ecosystems 4:107–126.

Karlovsky, J. 1961. Phosphate utilisation and phosphate maintenance requirements. Proc. Ruakura Fmrs. Conf., 142–151.

Karlovsky, J. 1962. Method of assessing the utilisation of phosphorus on permanent pasture. Trans. Int. Soc. Soil Sci., Com. IV and V, 726–730.

Karlovsky, J. 1975. How much phosphate do we really need? N.Z. Agric. Sci. 9:146–161.

Kuhn, V. 1931. Racionelní Pouzití Hnojiv. (Rational use of fertilizers). A. Neubert, Praha.

Middleton, R.K., and G.S. Smith. 1978. The concept of climax in relation to fertiliser input of a pastoral ecosystem. Pl. and Soil 50:595–614.

Newbould, P., and M.J.S. Floate. 1977. Agro-ecosystems in the U.K. Agro-Ecosystems 4:22–70.

Russell, E.W. 1961. Sixty years of Saxmundham. II. Problems still unsolved. Agriculture, London 66:431–435.

Sadler, J.M., and J.W.B. Stewart. 1974. Residual fertiliser phosphorus in western Canadian soils. Sask. Inst. Pedology Pub. 136.

Sokolov, A.V. 1959. Supply of available soil phosphates and their accumulation when phosphorus fertilisers are applied. Sov. Soil Sci. 1:117–123.

Thomas, G.W., and J.W. Gilliam. 1977. Agro-ecosystems in the U.S.A. Agro-Ecosystems 4:182–243.

Tinker, P.B. 1977. Economy and chemistry of phosphorus. Nature 270:103–104.

in which 29% of the added P in fertilizer was recovered by crops in 15 years, but Newbould and Floate did not take into account Widdowson and Penny's conclusion that "the balance presumably remained as recoverable P residues."

Such overestimates of phosphorus fixation and immobilization lead to misleading conclusions and to such statements as "due to process of P fixation, inputs (as fertilizer) exceed outputs (as harvested crops) by a factor of between two and ten," and "The central problem in the world phosphate economy . . . consists largely of digging out phosphorus at one place and storing it at another" (Tinker 1977). Neither experimental evidence nor the experience and practices of many of the world's farmers support these views.

CONCLUSIONS

A historian in soil science may justifiably call the 20th century "*an era of P-fixation problems in agriculture.*" Perhaps no other field of soil science has received so much attention and such extensive study as inorganic P fixation. While it is difficult to discover where and when these ideas originated, it is even harder to understand why they persisted so long. Fortunately, during the last 20 years, much

Watkinson, J.H. 1974. Phosphate fixation. Proc. Ruakura Fmrs. Conf., 212-213.

Widdowson, J.P., and F.V. Penny. 1973. Yields and N, P, and K contents of the crops grown in Rothamsted Reference Ex-

periment, 1956-70. Rothamsted Expt. Sta. Rpt. 1972, Part 2, 111-130.

Wilson, A.T. 1968. The chemistry underlying the phosphate problem in agriculture. Austral. J. Sci. 31:55-61.

Availability of Soil Phosphate to Tropical Pasture Species

F.W. SMITH
Division of Tropical Crops and Pastures,
CSIRO, Australia

Summary

Quantitative measurements of the amount of soil phosphate available to plants were made by an isotopic-dilution technique (L-value). Following very thorough mixing and equilibration of ^{32}P with soil P, sequential harvests of perennial pasture species were used to monitor the available P pools in two alkaline clay soils. In initial experiments with *Cenchrus ciliaris* (buffelgrass), an apparent increase was noted in the size of the available P pool over a number of sequential harvests during a 3-month period. This finding suggested that P was being transferred from a previously unavailable pool to the available pool during the experimental period. Planting buffelgrass after various periods of incubation of ^{32}P with soils established that the apparent increase was not driven by P removal from the soil by growing plants since it also occurred in the absence of plants. The increase was linearly related to time. It may represent a natural cycling of P that is occurring continuously in the soils.

A further experiment was run to determine whether species that differ in their efficiency of using soil P may influence this rate of transfer. The tropical legumes *Stylosanthes hamata*, *Stylosanthes guianensis*, *Macroptilium atropurpureum*, and *Desmodium intortum* were used in addition to buffelgrass. Measurements of the initial sizes of the available P pools were similar with all species. Furthermore, the apparent rate of increase in size of the available P pools was the same for all species in spite of large differences in both the rates of P removal and the total amounts of P removed by the various species. Differences between these species in their efficiency in taking up and using soil P were therefore due to differences in their ability to exploit the same pool of available P. Differences were not due to a capacity of any of the species to expand this pool during their growing period by increasing the rate of transfer of P from previously unavailable sources.

KEY WORDS: phosphorus, isotopic dilution, tropical pastures, buffelgrass, *Stylosanthes*, *Desmodium*, siratro.

INTRODUCTION

Plants differ widely in their ability to utilize low levels of soil phosphorus (P). These differences have been attributed to a wide range of morphological, physiological, and metabolic factors (Gerloff 1976, Loneragan 1978). Among the more important factors that should be considered is the possibility that more efficient species may have access to soil P from sources that remain unavailable to less efficient species. These studies, encompassing a number of tropical pasture species known to vary in their efficiency to utilize soil P, were designed to examine this possibility.

METHODS

Availability of soil P was assessed by an isotopic-dilution technique based on the L-value method (Larsen 1952). Two P-deficient soils were used: an alluvial clay (pH 8.1, 212 ppm total P) and a clay-loam derived from tertiary sediments (pH 8.0, 138 ppm total P). Great care was taken to label the soils uniformly by slowly atomizing ^{32}P orthophosphate onto the dry soil (<2mm) as it was tossed for 30 minutes in a small inclined revolving drum. Approximately 9.6 mbq of ^{32}P contained in 10 ml of $10^{-4}M$ KH_2PO_4 was sprayed onto each 2-kg batch of soil. Batches of the same soil were then bulked and further mixed in a concrete mixer for 3 hours. During this period finely powdered $CaH_4(PO_4)_2$ was also mixed through the soils of those treatments with added soluble P fertilizer. Soils were then weighed into pots and subsamples taken for determination of total P and ^{32}P.

The author is Principal Research Scientist in the Division of Tropical Crops and Pastures, CSIRO.

The author wishes to thank P.J. Vanden Berg for technical assistance and Consolidated Fertilizers Ltd. for generous financial support.

All pots received a basal nutrient dressing containing sufficient N, K, S, Zn, Cu, B, and Mo to ensure adequate plant growth during the experiment. Additional N and K were applied after each harvest. Four plants were established in each pot, and soils were maintained at pF2 ($\log_{10}$ cm water tension) by daily watering. Plants were harvested 2 cm above the soil surface at each harvest, dried at 70°C, and weighed. They were then ashed for 16 hours at 480°C, and the ash was taken up in 6M HCl, dried, and redissolved in 0.1M HCl. Aliquots were taken for total P determination (Murphy and Riley 1962) and for measurement of radioactivity by liquid scintillation counting. L-values were calculated from these data, corrections being applied for the contribution of seed P (Equation 1) and for P removed in previous harvests (Equation 2) (Probert 1972).

$$L = \frac{(^{31}P \text{ plant} - {}^{31}P \text{ seed})}{^{32}P \text{ plant}} \times {}^{32}P \text{ added to soil} \quad (1)$$

$$L'_n = L_n \left(1 - \sum_{i=1}^{n-1} \frac{P_i}{L_i} \right) + \sum_{i=1}^{n-1} P_i \quad (2)$$

where $L_1 \ldots L_n$ is measured L value at each harvest from Equation 1 and $P_1 \ldots P_n$ is the P removed in the plant tops at each harvest.

Experiment 1, designed to obtain a quantitative measure of the available P pools in these two soils, was conducted in the greenhouse using *Cenchrus ciliaris* cv. Gayndah (buffelgrass) as the test species. Each pot contained 1,700 g of alluvial soil or 1,800 g of tertiary soil. Six replicates were established, and plants were harvested six times over a 147-day period.

Experiments 2 and 3 were conducted in a growth room (14-hour day, 27.5°C night, 75% relative humidity, 500 $\mu E/m^2$/second light intensity). Experiment 2 compared five pasture species: buffelgrass, *Macroptilium atropurpureum* cv. Siratro (siratro), *Desmodium intortum* cv. Greenleaf (desmodium), *Stylosanthes hamata* cv. Verano (Caribbean stylo), *Stylosanthes guianensis* cv. Cook (Cook stylo). Two levels of P fertilization were established, 0 or 50 mg P/kg soil. Each pot contained 1 kg of dry soil, and there were 4 replicates of each treatment. Plants were harvested at appropriate intervals over a 153-day period.

Experiment 3 examined the effects of P removal by plants. Buffelgrass was the test species, and the two levels of P fertilization used in experiment 2 were included. Pots contained 1 kg of dry soil, and there were 4 replications. Treatments were established by planting at three different times after all pots were initially watered: 2, 49, and 93 days. Unplanted pots were maintained at pF2 and treated exactly the same as planted pots. Plants were harvested at appropriate intervals.

RESULTS

L-values increased at each successive harvest in experiment 1. The increase was linearly related to time since wetting up (Fig. 1) and also linearly related to the cumula-

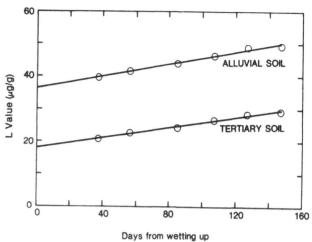

Fig. 1. Relationships between L-values and time from wetting up using buffelgrass as test species in experiment 1.

tive amount of P removed in the plant tops (alluvial soil, $L = 35.51 + 0.889 \Sigma P$ [$r = 0.96$]); tertiary soil, $L = 17.01 + 1.374 \Sigma P$ [$r = 0.96$]).

Similar increases in L-value with each successive harvest were noted with all species in experiment 2. These were linearly related to time since wetting up (Fig. 2) but not to the cumulative amount of P removed in the plant tops. Tests of homogeneity of the regression lines for the no-P treatment revealed that within each soil there were no significant differences between the slopes of the regression lines for different species. Although this test yielded statistically significant differences between the intercepts for various species within a soil, the differences were small (less than 1.8 μg P/g for the alluvial soil and less than 1.0 μg P/g for the tertiary soil) and not considered agronomically significant. The P fertilizer treatment, introduced in experiment 2 to check on the ability of the technique to recover added soluble P, indicated that excellent P recoveries were obtained with all species (mean over all species: alluvial soil, 101.8%; tertiary soil, 96.3%). Slopes of regression lines of L-value against time since wetting up were not affected by the P fertilizer treatment.

Results from experiment 3 showed that the linear increase in L-values with time from wetting up was not related to the amount of P removed and occurred even in the absence of plants.

DISCUSSION

Quantitative measures of the labile-P pool in the native soils were obtained from the extrapolated Y intercepts of plots of L-values against time. The robustness of the technique is indicated by the fact that these intercepts remained constant for each soil over all experiments and that excellent recoveries of added soluble P fertilizer were obtained. Although L-values should be independent of the growing period (Larsen 1967), results indicating increases with time (Figs. 1 and 2) have also been found in

Fig. 2. Relationships between L-values in no-P treatments in soils from experiment 2 and time from wetting up. Buffelgrass ×; siratro, solid circles; desmodium, triangles; Carribean stylo, squares; Cook stylo, open circles. Overall correlations, ignoring species: alluvial soil, L = 34.46 + 0.121T (r = 0.91), tertiary soil, L = 19.49 + 0.62T (r = 0.89).

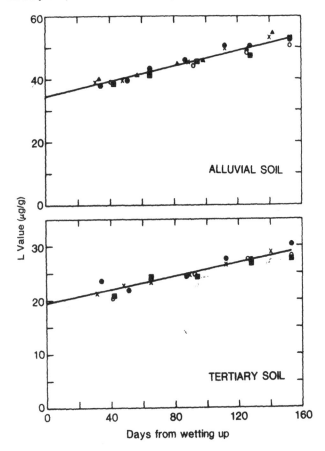

other experiments in which a series of harvests of perennial species has been used (Andersen et al. 1961, Probert 1972). Since L-values increased at the same rate in treatments receiving added soluble P fertilizer as they did in no-P treatments, the increase could not be due to labeled P being converted to inaccessible forms. It could be due to failure to attain isotopic equilibrium between labeled P and the soil labile P or to uptake of P from fractions that were initially unlabeled. However, it is unlikely that the increase could be solely due to failure to attain isotopic equilibrium because of the precautions taken to thoroughly mix ^{32}P through the soil and the excellent linear fit of L-value against time in all experiments. If failure to attain equilibrium were the major reason, a curvilinear relationship in which the rate of increase decreased with time could be expected in experiments of this duration (Larsen and Sutton 1963). The data suggest, therefore, that the increase was at least partially due to transfer of P from a previously unlabeled pool to the labeled pool over the experimental period.

Probert (1972) arrived at a similar conclusion from experiments in which the increase in L-value over the growing period was linearly related to P uptake (ΣP). He postulated that depletion of P from the rooting zone caused mobilization of P from previously nonlabile forms. A similar linear relationship between L-values and ΣP was found in experiment 1, but when P stress on the soil was increased by decreasing the pot size in experiments 2 and 3, the linear relationship broke down, and L-values tended to increase exponentially with ΣP. The relationship also varied between species due to varying rates of P removal by different species, and occurred even in the absence of P uptake by plants. In all experiments, time since the initial wetting up of the soil was the unifying factor (Figs. 1 and 2). This fact suggests that, in these soils, the increase may be due to a continuous cycling of P between labile and nonlabile forms. Organic P fractions could provide a P pool for such cycling.

Interpretation of results from the species comparison in Fig. 2 involves two points. First, if a more efficient species had had access to labile-P fractions that were unavailable to other species, the Y intercept for that species should have been higher than those for the others. Second, had any species been able to obtain additional P by expanding the soil labile-P fraction during growth through effects occurring in the rhizosphere (for example, root exudates or pH changes close to the root), this ability would have been reflected in an increase in the slope of the plot for that species in Fig. 2. Since neither of these occurred to a degree that would have biological significance, it is concluded that all the tropical pasture species studied drew upon the same labile-P pool. However, there was variation among species in the amounts of P they removed from this pool during the growing period. For no-P treatments this ranged from 15.9 μg P/g soil for buffelgrass to 21.1 μg P/g soil for siratro in the alluvial soil and from 7.6 μg P/g for buffelgrass to 11.0 μg P/g soil for siratro in the tertiary soil. Differences between these species in their efficiency in taking up soil P were therefore due to differences in their ability to exploit the same labile-P pool. Other studies (Nye and Foster 1958, Keay et al. 1970, Probert 1972) support this conclusion. However, differences in L-values between species reported by Russell et al. (1958) and Andersen and Thomsen (1978) suggest that this principle may not be applicable in all experimental situations or for all soils.

LITERATURE CITED

Andersen, A., B. Gregers-Hansen, and G. Nielsen. 1961. Determination of the phosphate condition of soils by means of radioactive phosphorus in pot experiments. Acta Agric. Scand. 11:270–290.

Andersen, A.J., and J.D. Thomsen. 1978. Efficiency in absorption and utilization of phosphorus of four plant species. *In* Isotopes and radiation in research on soil-plant relationships, 499–509. Int. Atomic Energy Agency, Vienna.

Gerloff, G.C. 1976. Plant efficiencies in the use of nitrogen, phosphorus, and potassium. *In* M.J. Wright (ed.), Plant adaptation to mineral stress in problem soils, 161–174. Cornell Univ., Ithaca, N.Y.

Keay, J., E.F. Biddiscome, and P.G. Ozanne. 1970. The comparative rates of phosphate absorption by eight annual pasture species. Austral. J. Agric. Res. 21:33–44.

Larsen, S. 1952. The use of ^{32}P in studies on the uptake of phosphorus by plants. Pl. and Soil 4:1–10.

Larsen, S. 1967. Soil phosphorus. Adv. Agron. 19:151–210.

Larsen, S., and C.D. Sutton. 1963. The influence of soil volume on the absorption of soil phosphorus by plants and on the determination of labile soil phosphorus. Pl. and Soil 18:77–84.

Loneragan, J.F. 1978. The physiology of plant tolerance to low phosphorus availability. *In*. G.A. Jung (ed.), Crop tolerance to suboptimal land conditions., 329–343. Am. Soc. Agron. Spec. Pub. 32.

Murphy, J., and J.P. Riley. 1962. A modified single solution method for the determination of phosphate in natural waters. Anal. Chem. Acta. 27:31–36.

Nye, P.H., and W.N.M. Foster. 1958. A study of the mechanism of soil-phosphate uptake in relation to plant species. Pl. and Soil 9:338–352.

Probert, M.E. 1972. The dependence of isotopically exchangeable phosphate (L-value) on phosphate uptake. Pl. and Soil 36:141–148.

Russell, R.S., E.W. Russell, and P.G. Marias. 1958. Factors affecting the ability of plants to absorb phosphate from soil. II. A comparison of the ability of different species to absorb labile soil phosphate. J. Soil Sci. 9:101–108.

Manurial Value of Liquid, Anaerobically Digested Sewage Sludge on Grassland in the West of Scotland

K.F. EDGAR, R.D. HARKESS, and J. FRAME
The West of Scotland Agricultural College,
Auchincruive, Ayr, Scotland, UK

Summary

The manurial value of liquid, anaerobically digested sewage sludge is under assessment in west Scotland to evaluate the effects of disposal on grassland. The concentrations of plant nutrients in sludges have been measured with sludge total N being considered as two fractions, volatile N and organic N. The herbage yield response to sludge N, relative to fertilizer N, has been measured in a number of field trials evaluating the effects of sludge composition and season of application.

The concentration of organic N, P, and K in sludge depends on the dry-matter (DM) concentration, which was quite variable (3–100 g/kg). Volatile N concentration averaged 40%–50% of the total N content but was largely independent of sludge DM concentration. Relative yield response to sludge N at the first harvest after application was, on average, proportional to the volatile N content but varied with season of application and DM concentration in the sludge. Late-winter and late-summer applications gave the best yield response to sludge N relative to fertilizer N. This is possibly explainable in terms of seasonal rainfall pattern, with the rainfall increasing the relative response to sludge N. Residual response to sludge N during the season of application was equivalent to about 20% of the sludge organic N.

This basic information on the manurial value of digested sludge and the results of the ongoing work will eventually be used to optimize the benefits of its disposal on agricultural land in Scotland.

KEY WORDS: manurial value, organic N, volatile N, sewage sludge, grassland.

INTRODUCTION

Environmental and economic pressures to dispose of sewage sludge on agricultural land have been increasing. Research in many countries is aimed at evaluating the benefits and dangers of this practice. About 90% of sewage sludge from west Scotland towns, including the most industrially polluted sludges, is dumped at sea; the remainder, in anaerobically digested liquid form, is applied mainly to grassland. To assess the manurial value of sludge nitrogen (N) under west Scotland conditions a series of trials began in 1979 comparing sludge N with inorganic fertilizer N when applied on perennial ryegrass (*Lolium perenne* L.) swards.

METHODS

Plots were established at three sites. Site 1 was a poorly drained clay soil (pH 5.9), site 2 a sandy loam (pH 6.3), and site 3 a clay loam (pH 5.6). A 3-m³ slurry tanker with

K.F. Edgar is a Research Agronomist, R.D. Harkess is a Senior Agronomist, and J. Frame is Head of the Agronomy Department at the West of Scotland Agricultural College.

The authors wish to thank the UK Water Research Centre, Stevenage, Hertfordshire, for financial support and the West of Scotland Agricultural College Agricultural Chemistry Division for chemical analyses.

the spreader blanked off allowed thorough mixing of sludge for sampling and spreading. Sludge samples were analyzed for dry matter (DM), total N, volatile N (N lost on drying at 100°C), phosphorus (P), and potassium (K). Sludge was spread with 10-l watering cans.

Trials were of the replicated, randomized-block type. Plot size was 5 × 2 m. Fertilizer N was applied as Nitrochalk (26% N). Phosphate fertilizer (superphosphate, 20% P_2O_5) for fertilizer N treatments was applied at half the rate of sludge phosphate. Potash fertilizer (muriate of potash, 60% K_2O) was given to all treatments so that a ratio of 0.5 kg K_2O/kg N was attained. Herbage production was assessed by cutting three silage crops/season. Analyses included dry matter (DM), crude protein (CP), and in-vitro digestibility. Results of selected trials are given.

RESULTS

Nutrient Composition of Sludges

Variability in DM concentration between tankloads of sludge from one sewage treatment plant was large (3–100 g/kg), but nutrient composition of the sludge solids was consistent (20 g/kg organic N; 10g/kg P; 5 g/kg K). Volatile N (0.55 g/l), being present in solution, was largely independent of variability in sludge DM within the range studied. Small amounts of P (0.04 g/l) and K (0.04 g/l) were also present in solution.

Growing-Season Application

In three trials sludge total N was compared with fertilizer N at rates of 0, 120, 240, 360, and 480 kg/ha/yr, applied in three equal applications during the growing season. Herbage DM yield response to sludge N relative to fertilizer N was calculated for each harvest (Table 1). Also shown are the proportions of volatile N in sludge total N and rainfall during each growth period.

In spring, yield response to sludge total N compared with fertilizer N was equivalent to the proportion of volatile N. In autumn, yield response was higher than the percentage of volatile N indicated, but in the summer of 1979 it was lower, and in the summer of 1980 it was higher. These responses may have been dependent on rainfall,

which was lowest in spring, highest in autumn, and very high in the summer of 1980. So far, there have been no differences between sludge and fertilizer treatments in herbage quality (CP and in-vitro digestibility).

Proportion of Volatile N

Two sludges that contrasted markedly in DM concentration, and therefore in proportion of volatile N to total N, were mixed in various proportions to produce six sludges containing a range of proportions of volatile N. All sludges were applied at 45 kg/ha sludge total N and compared with fertilizer N at 0 and 45 kg/ha. A single application was carried out on replicated 0.4 × 0.8–m plots at site 2 in 1979.

Over the range of sludges, yield response relative to fertilizer N increased as the proportion of volatile N in the sludge increased (Table 2), but yield response relative to the proportion of volatile N in sludge total N decreased as the proportion of volatile N decreased.

Winter Application

Two trials, at sites 1 and 3, began in the autumn of 1979. Sludge N was compared with fertilizer N on four application dates during the winter. The same sludge was used at both sites on all application dates, and its chemical analysis did not alter during storage; the DM was high, 60 g/kg, and the proportion of volatile N, 30%, was low. Sludge volatile N and fertilizer N application rates were 18, 36, and 54 kg/ha.

Yield responses (meaned over N rates) are shown relative to the yield response to the fertilizer N applied on 1 April (Table 3). In general the yield response to fertilizer N increased with successive application dates, but yield response to sludge N was highest for January (site 3) or February (site 1) application, and at site 3 yield response was relatively poor when sludge N was applied in April.

DISCUSSION

Manurial value of sludge N is known to depend on sludge composition, volatile N content, and time of ap-

Table 1.   Herbage yield response to sludge,[1] volatile-N in sludge,[2] and rainfall during growth periods,[3] 1979–1980.

	Harvest					
	1 (spring)		2 (summer)		3 (autumn)	
	1979	1980	1979	1980	1979	1980
Relative yield response†	18	62	56	52	51	66
Volatile-N (%)†	20	58	77	40	29	38
Rainfall (mm)	105	60	86	250	184	245

[1]Herbage DM yield response to sludge total-N relative to fertilizer-N = 100.

[2]Proportion of volatile-N in sludge total-N.

[3]Rainfall/growth period.

†Mean of sites 1 and 2 in 1979 and sites 1 and 3 in 1980.

Table 2. Herbage DM yield response to sludge total-N relative to fertilizer-N = 100 for six sludges with a range of DM concentrations and proportions of volatile-N in sludge total-N.

DM g/kg	60	30	17	10	6	3	
Volatile-N (%)	26	38	50	61	73	84	SE
Relative yield response	18	33	47	54	70	97	11

plication. The results indicate the importance of these factors in west Scotland.

Variability in Sludge Composition

The maximum recommended rate for single applications of slurry in Scotland is 55 m³/ha (Scottish Agricultural Colleges 1980). The average nutrients supplied by this amount of liquid digested sludge from different sewage treatment plants in west Scotland vary from 42 to 105 kg N, 21 to 84 kg P_2O_5, and 7 to 24 kg K_2O (I. Shaw, personal communication). Volatile N as a proportion of total N may also vary.

Sludge settling during storage can cause variability in DM content among tanker loads taken for spreading. On farm scale this will cause variability within and between fields in the initial response to volatile N (Table 2) and in the evenness of distribution of organic N, P, and K in sludge DM.

Proportion of Volatile N

On average the initial yield response to sludge total N relative to fertilizer N, when applied during the growing season, was 45% and was equal to the average proportion of volatile N in the sludge total N. The initial yield response, however, was equivalent to 60%–132% of the volatile N content, depending on time of application and sludge DM content. About 20% of the organic N became available during the year of application so that the availability of sludge total N in the year of application was 50%–60%.

These apparent availabilities of the sludge N fractions agree with the findings of other workers (Dowdy et al. 1976, Coker 1978, O'Riordan 1980). Initial yield response to sludge volatile N is better than the response to ammoniacal N reported for cattle slurry (Tunney 1980). Lower DM content and lower concentration of ammonia

in sludge may result in less volatilization of ammonia from sludge than from cattle slurry.

Time of Application

Yield response to sludge N relative to the proportion of volatile N in the sludge was lowest for spring applications and highest for late-summer applications (Table 1). This pattern of response to growing-season applications may depend on rainfall and temperature patterns. Mineralization will increase with successive additions of sludge organic N and also as soil moisture content and soil temperature increase (Bartholemew 1965). Mineralization will be greatest when high soil temperatures are combined with high rainfall, as for the summer harvest of 1980 and the autumn harvests of 1979 and 1980 (Table 1). Rainfall may also overcome certain disadvantages associated with sludge DM application. Washing the sludge DM off the herbage and into the soil could prevent volatilization of ammonia, smothering of herbage, and drying of the sludge to form a physical barrier to grass growth. High rainfall in autumn may also cause leaching of fertilizer N, thus increasing the relative response to sludge N. Rainfall in west Scotland is high (> 1000 mm/yr at the trial sites) and the manurial value of sludge may be higher there than in regions with lower rainfall.

In contrast to fertilizer N, yield response to January and February application of sludge N was as good as or better than the response to April application (Table 3). This result agrees with the findings of Coker (1978) but may have been caused partly by the sludge's having had a high DM concentration and partly by rainfall's being low in January 1980. Frost action may increase the amount of sludge organic N subsequently mineralized.

The initial results from these ongoing trials show that for farm spreading operations, the sludge nutrient composition must be known for each sewage treatment plant. Disposal authorities should measure volatile N concentration in digested sludges, as initial yield response to sludge N depends on this fraction. A more accurate estimate of N manurial value of the digested sludge could also be achieved by taking into account the time of application and mineralization of sludge organic N. The initial yield response to sludge N will, however, be somewhat unpredictable, since it depends on the weather and variability in sludge DM concentration. Variability in sludge DM concentration can also result in uneven distribution of nutrients present in the sludge DM, including organic N. Unless measures can be taken to minimize the variability in DM concentration, sludge is best suited to less intensive grassland production systems where predictable responses to fertilizer input are less critical.

Table 3. Yield response to winter application of sludge-N and fertilizer-N relative to fertilizer-N applied on 1 April = 100.

Application date	Nov 23	Jan 10	Feb 28	Apr 1	SE
Site 1					
Sludge	33	94	99	94	
Fertilizer	20	75	73	100	15
Site 3					
Sludge	101	115	101	83	
Fertilizer	67	97	74	100	11

LITERATURE CITED

Bartholemew, W.V. 1965. Mineralisation and immobilisation of nitrogen in the decomposition of plant and animal residues. *In* W.V. Bartholemew and F.E. Clark (eds.), Soil nitrogen, 285–306. Amer. Soc. Agron. Madison, Wis.

Coker, E.G. 1978. The utilisation of liquid digested sludge.

Proc. Conf. Util. of Sewage Sludge on Land, Oxford. Water Res. Cent. 1979, 117–129.

Dowdy, R.H., R.E. Larson, and E. Epstein. 1976. Sewage sludge and effluent use in agriculture. Land application of waste materials. Proc. Soil Conserv. Soc. Amer. Symp. Ankeny, 1976, 138–153.

O'Riordan, E. 1980. Use of sewage sludge on agricultural land. Fm. Food Res. 11:135–136.

Scottish Agricultural Colleges. 1980. Handling and utilisation of animal wastes. Pub. 16.

Tunney, J. 1980. Fertiliser value of animal manures. Fm. Food Res. 11:78–79.

Waste-Water Application, Dry-Matter Production, and Nitrogen Balance of Rhodesgrass Grown on Fine-Textured Soil or on Sand Dunes

T. KIPNIS, A. FEIGIN, I. VAISMAN, and J. SHALHEVET

Agricultural Research Organization, the Volcani Center, Bet Dagan, Israel

Summary

The shortage of irrigation water in the semiarid zone of Israel necessitates the use of partially treated municipal effluents for irrigation of forage grasses. Irrigation with sewage effluents is being considered very carefully with the aim of maximizing production and minimizing the pollution hazard.

A high nitrogen (N) input originating from intensive application of effluent water may well increase dry-matter production, but it also increases the pollution hazard. Frequent application of relatively small quantities of effluent water was suggested as an efficient method of improving N uptake by rhodesgrass (*Chloris gayana* Kunth.) swards. It was expected that nitrate pollution would be reduced and nitrogen fertilizer could be saved.

In field experiments rhodesgrass was sprinkle-irrigated at different frequencies with three sources of water in combination with four levels of N fertilization.

Irrigation according to 100% class A pan evaporation (approximately 1,000 mm) revealed the superiority of twice-a-week application of effluents on fine soil. This schedule resulted in high yields and high N recovery as well as no response to additional N fertilization. Differences between irrigation frequencies diminished as N input increased, and effluent application was reduced to 80% of class A pan evaporation. Total N input higher than 660 and 380 kg/ha did not increase dry-matter production on fine and sand dune soils, respectively, but did increase N losses. Dry-matter production and N-balance values are discussed in relation to quantity and frequency of water application and N loss due to denitrification or leaching.

KEY WORDS: rhodesgrass, *Chloris gayana* Kunth., waste water, nitrogen, yield, pollution hazard.

INTRODUCTION

Land treatment is a favored method for the control of potential environmental hazards inherent in the increasing volumes of municipal waste water (Day 1973, Bouwer and Chaney 1974, Harlin 1978). In Israel, the shortage of good-quality water is so severe as to impose a threat to the future of intensive agriculture. Therefore, efforts have to be made to ensure optimal use of waste water, agriculturally, technologically, and ecologically.

Special attention should be given to land application of waste water when tropical grasses such as rhodesgrass (*Chloris gayana* Kunth.) are irrigated and fertilized intensively.

Rhodesgrass is very efficient in taking up nitrogen (Kipnis et al. 1979) in spite of its fairly shallow active root system. However, large amounts of nitrate and other potential pollutants might leach out of this zone if irrigation were uncontrolled.

The objectives of this study were to determine the long-term effect the application of effluents has on the soil-plant system and to evaluate the significance of effluent nitrogen (N) as a nutritional element and a potential pollutant.

T. Kipnis is a Crop Physiologist at the Institute of Field and Garden Crops and A. Feigin, I. Vaisman, and J. Shalhevet are Soil Scientists at the Institute of Soils and Water of the Agricultural Research Organization.

METHODS

Field experiments of 5 and 2 years' duration were conducted with rhodesgrass grown on a fine-textured soil (grumusol) and on a sand dune soil, respectively. Partially treated (secondary) municipal effluents were used for irrigation in both experiments. The experiment on fine-textured soil was sprinkle-irrigated with effluents containing approximately 50 ppm of ammonium N once or twice a week, or irrigated with fresh water once a week. The grass growing on sand dune soil was sprinkle-irrigated three times a week with effluents containing approximately 20 ppm of NH_4-N. Averages of 1,000 and 820 mm of water were evaporated from class A pan throughout the growing season in the fine-textured and the sand dune soil experiment, respectively. The fine-textured soil was irrigated according to 100% or 80% of class A pan evaporation, whereas quantities of 60%, 80%, 100%, and 120% of class A pan evaporation were applied to the sand dune soil. Aliquots of N fertilizer, $(NH_4)_2SO_4$, were applied immediately after each cutting (every 21 days) amounting to 0, 160, 320, and 480 and 0, 250, 500, and 700 kg N/ha/season to the fine-textured soil and sand dune soil experiments, respectively.

RESULTS

Dry-Matter Production

The experiments were based on the hypothesis that by frequent irrigation an optimal moisture status would be obtained in the upper soil layers. This moisture status would increase N uptake and recovery and would decrease nitrate leaching.

Consistent dry-matter (DM) yields were obtained on the fine-textured soil irrigated with effluents over the 5-year period. In contrast, a gradual yield decline was recorded for the fresh-water treatment even when 480 kg N/ha/season was supplied (Table 1). During the first 2 years, water was applied according to 100% of class A pan evaporation. Application of effluents twice a week gave results superior to those with application once a week. Differences in yield between these treatments amounted to approximately 3 tons/ha when no N fertilizer was added and diminished as supplementary N levels increased.

During the last 3 years of the experiment the effluent was applied according to 80% pan A evaporation. This treatment resulted in a significant response to N fertilizer without yield differences between the irrigation treatments. The data for the 5-year period show a clear response to a total N input (effluent + fertilizer) up to 660 kg N/ha and a production efficiency of 26 kg DM/1 kg N.

Production efficiency/unit of N of established sward growing on the sand dune soil was similar to that obtained on the fine-textured soil. Total yield, however, was lower on the sand because of fewer cuts. In each of the N-fertilization treatments a moderate response to water quantity was observed. The response to nitrogen, however, was much more pronounced. Optimal yields were obtained when irrigation was applied at a rate of 80% of class A pan evaporation and total N input amounted to 380 kg/ha.

Nitrogen Balance and Pollution Hazard

Nitrogen balance calculated for the first 2 years of the fine-textured-soil experiment indicated the advantage of the twice-a-week irrigation frequency in terms of N uptake and N recovery, especially when no N fertilizer was

Table 1. Effects of water quality, irrigation frequency, and N fertilization on dry-matter production (ton/ha ± S.E.) of rhodesgrass grown on fine-textured soil.†

Water	Irrigations per week	Fertilizer-N kg/ha	1975	1976	1977	1978	1979	Average
					Experimental season			
Effluent	1	0	18.1	15.5	17.2	16.0	15.0	16.4±0.6
		160	20.3	18.6	18.8	21.5	20.2	19.9±0.5
		320	20.6	19.3	22.7	26.3	21.9	22.2±1.2
		480	22.6	19.3	24.6	26.1	22.5	23.0±1.2
		Average	20.4±0.9	18.2±0.9	20.8±1.7	22.5±2.4	19.9±1.7	20.4±1.7
Effluent	2	0	21.3	18.5	19.4	18.3	15.0	18.5±1.0
		160	21.4	18.9	22.2	21.7	18.4	20.5±0.8
		320	22.7	19.9	24.5	25.0	19.0	22.2±1.2
		480	21.7	20.7	24.6	25.4	18.4	22.2±1.3
		Average	21.8±0.3	19.5±0.5	22.7±1.2	22.6±1.7	17.7±0.9	20.9±1.3
Fresh	1	0		14.1	4.9	2.0	1.9	5.7±2.9
		160		16.3	9.8	5.8	5.2	9.3±2.6
		320		17.7	13.2	9.0	11.4	12.8±1.8
		480		19.4	17.5	15.0	10.9	15.7±1.8
		Average		16.9±1.1	11.3±2.7	8.0±2.8	7.3±2.3	10.9±2.3

†Eight cuts/season.

Table 2. Effects of effluent application and N fertilization on N balance of two soil-grass systems.

Soil	Treatments		Total N input kg/ha	N† uptake kg/ha	N† recovery %	N† balance kg/ha
	Irrigations per week	Fertilizer-N kg/ha				
Fine	1	0	492	330(375)	67.0(56.6)	162(218)
		160	652	451(455)	69.2(55.4)	201(292)
		320	812	517(495)	63.7(50.4)	295(391)
		480	972	563(578)	58.0(50.6)	409(442)
Fine	2	0	492	384(489)	78.0(73.9)	108(103)
		160	652	448(531)	68.7(64.5)	204(221)
		320	812	519(555)	63.9(56.5)	293(364)
		480	972	556(580)	57.2(50.9)	416(494)
Sand Dune††		0	130	132	98.5	2
		250	380	275	72.3	105
		500	630	272	43.2	358
		700	830	255	30.7	575

†Numbers in brackets are results of the first two years of the experiment; those without brackets represent results of the whole experimental period.

††Irrigation according to 80% of class A pan evaporation.

added (Table 2). With increasing levels of N fertilization, N uptake increased; however, percent of recovery dropped to lower levels in both irrigation treatments. Nitrogen balance for the whole experimental period shows similar values for the two irrigation treatments except for the case of no fertilization (Table 2). Between 201 and 416 kg N/ha were lost from the soil-plant system as total N input increased from 652 to 972 kg/ha. This means that at the highest rate of N fertilizer application, approximately 86% was not taken up by the grass. Application of 250 kg N/ha [(NH$_4$)$_2$SO$_4$] to sand dune soil approximately doubled the N uptake recorded for no fertilizer application. Higher rates of N fertilization did not increase uptake but resulted in a steep decline in N recovery, indicating N loss.

DISCUSSION

The high dry-matter yields (Table 1) and the higher N recovery (Table 2) obtained from the twice-a-week irrigation–no fertilization treatment on fine-textured soil indicate that effluent N is readily available for plant growth and can be utilized efficiently.

Adequate water application to a shallow-rooted grass with a high water requirement is difficult. Evidently, N loss was greatly influenced by the level of N fertilizer as well as by the frequency and quantity of application of the effluent. It is concluded that with increasing levels of N fertilizer or with effluent application lower than 100% pan evaporation, the advantage of frequent irrigation is lost.

Although high N losses were evident, not more than 5 ppm of NO$_3$-N were found down to 180 cm soil depth at the end of the growing season (Feigin et al. 1978). The NO$_3$-N not taken up by the grass growing on fine-tex-

tured soil was lost, apparently through denitrification. This loss was probably favored by the high organic-matter content originating from waste water as well as from root decomposition of frequently cut grass and also by the very low (4%) O$_2$ concentration of the soil atmosphere (Kipnis et al. 1979). It seems, therefore, that the hazard of nitrate pollution under these conditions is low. Denitrification takes place no matter which kind of water is used for irrigation. However, N loss under effluent application was compensated for by the frequent application of N through the irrigation system, which ensured sustained high yields.

Free-drainage lysimeters installed in the sand dune experiment revealed the high potential for nitrate pollution, as most of the leached N was collected. When the nitrogen input was increased to 830 kg/ha a very steep and dangerous rise in N loss was evident (Table 2).

Our results indicate that approximately 160 kg fertilizer N/ha to supplement effluent water containing 50 ppm N and 250 kg fertilizer N/ha to supplement water containing 20 ppm N are required for achieving optimal and comparable rhodesgrass yields on fine-textured and sand dune soil, respectively. These applications would account for a savings of approximately 45%–75% of the cost of N fertilizer and would provide a reasonable safety margin for avoiding the hazard of pollution with nitrates.

LITERATURE CITED

Bouwer, H., and R.L. Chaney. 1974. Land treatment of waste water. Adv. Agron. 26:133–176.

Day, A.D. 1973. Recycling urban effluents on land using annual crops. *In* Proc. natl. workshop land application of municipal sludge effluent, 155–160.

Feigin, A., H. Bielorai, Y. Dag, T. Kipnis, and M. Giskin.

1978. The nitrogen factor in the management of effluent-irrigated soils. Soil Sci. 125:248–254.

Harlin, C.C., Jr. 1978. Land treatment methods in perspective. Int. conference on developments in land methods of waste water treatment and utilisation. Paper 1. Int. Assoc. Water Pollut. Res., Melbourne, Austral.

Kipnis, T., A. Feigin, A. Dovrat, and D. Levanon. 1979. Ecological and agricultural aspects of nitrogen balance in perennial pasture irrigated with municipal effluents. Prog. Water Technol. 11:127–138.

Effect of Converting Chaparral to Grassland on Soil Fertility in a Mediterranean-Type Climate

M.B. JONES, R.L. KOENIGS, C.E. VAUGHN, and A.H. MURPHY
Hopland Field Station, University of California,
Hopland, Calif., U.S.A.

Summary

In the region of Mediterranean-type climate in California, brush is converted to grassland to reduce fire hazards, increase feed for herbivores, and increase water yields. There is little information on how this change of cover type affects soil fertility over a long period. Research was initiated to measure the long-term effect of converting brush to grass on the availability of nutrients in the soil.

In 1956 chaparral growing on Los Gatos soil (fine-loamy, mixed mesic family of typic argixerolls) was crushed, burned, and seeded to grasses and clovers, and half the area was fenced to exclude herbivores. Sprouting brush either was treated periodically by burning or with herbicides or was not treated at all. Twenty-three years after the initial burn, about 3% of the cover on reburned plots was brush, virtually none was brush where herbicides were used, and nearly 100% was brush where nothing was done to control regrowth. Herbaceous plants were abundant where brush sprouts had been controlled. Soil samples were taken from the six treatments for chemical analysis and a pot experiment. Soft chess (*Bromus mollis* L.) was grown in pots fertilized with all combinations of nitrogen (N), phosphorus (P), and sulfur (S). The plants were analyzed for N, P, S, potassium (K), calcium (Ca), and magnesium (Mg).

In the pot study, uptake of N, P, K, S, Ca, and Mg was greater from soils where brush regrowth was controlled than from soils where brush had regrown. These differences were much greater for soils under grazing. Extractable nutrients in the soils from the grazed area were also higher from grassland plots than from brush plots.

The grazed grasslands had 81% ground cover by living plants, of which 14% was annual legumes, 25% annual grasses, 30% *Erodium* spp., and 11% perennial grasses. The ungrazed plots had 25% ground cover and only 3% legumes, 7% annual grasses, 1% *Erodium* spp., and 9% perennial grasses.

Nutrient availability was greater after a 23-year period in brush-soil converted to grassland than where brush regrew, and this difference in soil fertility was enhanced by grazing. The difference in soil fertility on the grazed grassland plots may be due to (1) the shallower, more fibrous root systems of grassland species (as compared with brush) resulting in a retention of mineral nutrients in the surface soil; (2) increased rate of cycling, and thus nutrient availability, due to grazing animals; (3) the contribution of annual legumes to soil N; and (4) a reduction in surface-soil erosion, there being much bare soil and an erosion pavement under the brush.

KEY WORDS: annual grasslands, extractable soil inorganic N, P, K, S, Ca, and Mg.

INTRODUCTION

Chaparral, and evergreen-brush cover-type, grows on about 6 million ha of land in California. Before Europeans arrived, the native Americans burned the brush to drive out wildlife in hunting and to increase accessibility of the land (Sampson 1944). Now chaparral is often coverted to grassland to help control fire, increase feed for livestock and wildlife, increase water yield, and increase accessibility (Murphy et al. 1975). Reduced sheet erosion is often another benefit (Wischmeier and Smith 1978).

Burning is the usual method of removing the large volume of woody material from brushlands, although mechanical and chemical methods have also been used. Use of fire also increases the availability of nutrients in the soil immediately after the burn (DeBano et al. 1979, Dunn et al. 1979, Vlamis and Gowans 1961).

The objective of this study was to compare the long-term effects of brush conversion and brush regrowth, with and without grazing, on the availability of soil nutrients.

METHODS

In 1956, chaparral growing on Los Gatos soil (fine-loamy, mixed mesic family of typic argixerolls) was crushed and burned. The area was seeded to grasses (*Phalaris tuberosa* var. Stenoptera, *Dactylis glomerata* var. Palestine, *Bromus mollis*, and *Lolium multiflorum*) and clovers (*Trifolium subterraneum* and *T. hirtum*), and half the area was fenced to exclude herbivores. Resprouting brush was then treated periodically as follows, with 2 replications: (1) reburning, (2) herbicide treatment, and (3) no followup treatment. In reburned plots it was necessary to reburn in 1959 and 1963 because of brush-sprout growth. Additional burning treatments were needed in the area protected from grazing in 1967, 1970, 1974, and 1978. The herbicide treatment was a standard brush killer (low volatile esters of 2,4-D and 2,4,5-T) at 4.5 kg active ingredient/ha in 378 l of water with 1% diesel oil. The herbicide was first applied in June 1958. In 1959 and 1963 the surviving sprouts were treated again, which was sufficient to control all brush growth.

Soil samples for chemical analysis and a pot study were taken from the 2 replications of each treatment in spring of 1979. In the pot experiment, N, P, and S were applied separately and in all possible combinations. The N and P were each applied at the rate of 100 ppm and S at 50 ppm. The 1-kg pots were seeded with soft chess (*Bromus mollis* L.) at the rate of 0.2 g/pot.

Grass was harvested, ground, and analyzed for total N, P, K, S, Ca, and Mg. Field-collected soils were sieved (2-mm) and analyzed for Bray No. 1 available P; total and inorganic N; total S; exchangeable K, Ca, Mg, and Na; and organic matter.

M.B. Jones is an Agronomist, R.L. Koenigs is a Postgraduate Researcher, C.E. Vaughn is Staff Research Associate, and A.H. Murphy a Specialist at Hopland Field Station.

RESULTS

Table 1 shows some of the chemical properties of soils sampled from the field plots. The major influence of the followup brush treatments was on the inorganic N, followed by total S and Bray No. 1 P; there were only small differences in total N and no significant difference in organic matter. The largest differences were on the grazed plots between soils with regrown brush and soils with grass cover.

Exchangeable Ca varied from 5.0 to 7.0, Mg from 1.1 to 1.6, K from 0.6 to 1.1, and sodium (Na) from 0.1 to 0.3 meq/100 g soil. Soil variability was such that treatment differences were not significantly different.

Table 2 shows the pot yields and the uptake of nutrients by soft chess as influenced by added fertilizers and by brush-control treatments and grazing. The column labeled "Grass" is an average of fire and herbicide followup treatments, since there was little difference in grass yields or nutrient uptake between these two treatments. The largest increase in yield of grass occurred on grazed grass-covered soils. Yields were about 6 times greater than those on the brush soils when no N was applied and about 3 times greater than when no P or S was applied. When N, P, and S were applied together the increase in grass yield due to change in cover type was 1.3-fold (3.5/2.7 g/pot), indicating that most of the deficiencies were satisfied by N, P, and S. However, other nutrients may be involved, or different rates of N, P, and S might bring additional response. Differences among soils from the ungrazed plots were much smaller.

Uptake of N by unfertilized soft chess was about 6-fold greater from grazed-grass soil than from brush soil. The increase was 1.8-fold when N, P, and S were applied. Uptake of N from ungrazed grass-cover soils was about 1.5 times greater than from ungrazed-brush soil when no fertilizer was applied and 1.2 times greater than when N, P, and S were applied. On grazed plots the increases in P and S uptake due to brush conversion to grass were about 4- to 5-fold with no fertilizer applied and about 1.3-fold when N, P, and S were added. The uptake of Ca, Mg,

Table 1. Chemical properties of soils sampled in 1979 from plots converted from brush in 1956 as affected by follow-up treatment.

Follow-up treatment	Inorganic				Bray No. 1 P
	OM %	N %	N ppm	S ppm	P ppm
Ungrazed					
Ck (brush†)	5.6	.13	5.1	83	7.0
Fire (grass†)	5.2	.13	7.9	97	7.2
Herbicide (grass†)	4.7	.13	7.2	71	4.8
L.S.D. (.10)	N.S.	N.S.	2.1	8	N.S.
Grazed					
Ck (brush†)	6.7	.17	7.8	94	5.5
Fire (grass†)	6.1	.19	44.1	116	10.6
Herbicide (grass†)	7.0	.20	35.6	123	12.8
L.S.D. (.10)	N.S.	.03	23.6	22	6.8

†Cover type in 1979.

Table 2. Yield and nutrient uptake by soft chess in pots as affected by fertilization, cover type conversion, and grazing treatments.

Fertilizer	Ungrazed			Grazed		
	Brush	Grass†	L.S.D. .05	Brush	Grass†	L.S.D. .05
	Grass yields (g/pot)					
NPS	2.6	2.9		2.7	3.5	
PS	0.3	0.4		0.3	1.7	
NS	1.0	0.8	0.3*	0.6	1.9	0.5*
NP	0.9	1.6		0.9	2.5	
Check	0.2	0.3		0.2	1.3	
	N-uptake (mg/pot)					
NPS	42	51	6*	41	72	14*
Check	4	6		4	27	
	P-uptake (mg/pot)					
NPS	9.0	7.7	0.8*	8.2	10.4	1.3*
Check	0.6	0.7		0.5	2.5	
	S-uptake (mg/pot)					
NPS	5.6	6.8	0.7*	4.8	6.8	0.8*
Check	0.3	0.6		0.4	1.7	
	K-uptake (mg/pot)					
NPS	53	66	11*	53	81	13*
Check	5	8		5	34	

†Mean of herbicide and fire-control follow-up treatments of brush.

*L.S.D.'s within each yield and nutrient uptake section apply to the fertilizer and brush control treatment interactions.

Table 3. Botanical composition (live ground cover) in May 1979 of grass plots converted from brush in 1956 as affected by follow-up treatment (herbicide vs. fire) on ungrazed and grazed plots.

	Ungrazed		Grazed	
	Herbicide	Fire	Herbicide	Fire
	% live ground cover			
Annual grass[1]	11	3	22	27
Perennial grass[2]	12	6	10	11
Erodium spp.	2	0	30	29
Legumes[3]	5	1	18	11
Other forbs	4	4	1	1
Brush	0	2	0	3
Total	34	16	81	82

Dominant species:
[1]*Bromus rubens* in ungrazed, *B. mollis* in grazed.
[2]*Phalaris tuberosa* and *Dactylis glomerata* var. Palestine.
[3]Native annual *Lotus* spp. in ungrazed, native annual *Trifolium* spp. in grazed.

and K increased 4.5 to 6.8 times when no fertilizer was applied and 1.5 to 1.7 times when N, P, and S were added.

DISCUSSION

When soil cover was converted from brush to grassland and grazed by herbivores for 23 years, available soil nutrients were higher than in comparable soil where the cover type reverted back to the original brush. Nitrogen was the nutrient with the greatest increase, but P and S were also made more available, and Ca, Mg, and K were taken up in greater amounts from grazed grassland soils than from brush soils.

Several factors may have contributed to the increase in soil fertility. The fire treatments may have increased the availability of soil nutrients (Vlamis and Gowans 1961, Dunn et al. 1979). These nutrients were then held in the surface soils by the shallower fibrous-root systems of the grassland cover (Table 3). A more rapid cycling in the grazed plots may have increased the level of available nutrients. Dawson (1974) reported that soils sampled from grazed subterranean clover–grass pastures yielded much more ryegrass in a pot study than did soils from ungrazed pastures. Grazing in our study and in Dawson's resulted in large increases in legumes (Table 3) and thus

in additional N fixation, accounting for a substantial increase in available N. The direct effect of the animals was probably to cause a loss of N due to volatilization from urine patches (Ball and Keeney 1981) and the export of all nutrients from the plots to camping areas (Hilder 1964). Finally, brush soils have virtually no herbaceous ground cover beneath the brush, and erosion is often clearly visible. Such an environment is not conducive to nutrient retention or soil formation.

LITERATURE CITED

Ball, P.R., and D.R. Keeney. 1981. Nitrogen losses from urine-affected areas of a New Zealand pasture under contrasting seasonal conditions. Summaries of Papers, XIV Int. Grassl. Cong., 152.

Dawson, M.D. 1974. Recycling nitrogen and sulphur. Sulphur Inst. J. 10(2):2–5.

DeBano, L.F., G.E. Eberlein, and P.H. Dunn. 1979. Effects of burning on chaparral soils. I. Soil nitrogen. Soil Sci. Soc. Am. J. 43:504–509.

Dunn, P.H., L.F. DeBano, and G.E. Eberlein. 1979. Effects of burning on chaparral soils. II. Soil microbes and nitrogen mineralization. Soil Sci. Soc. Am. J. 43:509–514.

Hilder, E.J. 1964. The distribution of plant nutrients by sheep at pasture. Proc. Austral. Soc. Anim. Prod. 5:241–248.

Murphy, A.H., O.A. Leonard, and D.T. Torell. 1975. Chaparral shrub control as influenced by grazing, herbicides and fire. Down to Earth 31(3):1–8.

Sampson, A.W. 1944. Plant succession on burned chaparral lands in northern California. Univ. of Calif., Berkeley, Calif., Bul. 685.

Vlamis, J., and K.D. Gowans. 1961. Availability of nitrogen, phosphorus and sulfur after brush burning. J. Range Mangt. 14:38–40.

Wischmeier, W.H., and D.D. Smith. 1978. Predicting rainfall erosion losses—a guide to conservation planning. U.S. Dept. Agric., Handb. 537.

Effect of N-P-K Fertilization on Mineral Composition of *Poa pratensis* L. Grown on a Shallow Muck Soil

J.W. LIGHTNER, C.L. RHYKERD, G.E. VAN SCOYOC,
D.B. MENGEL, C.H. NOLLER, and B.O. BLAIR
Purdue University, West Lafayette, Ind., U.S.A.

Summary

More than 1 million ha of muck soil suitable for agriculture exists in the northern portion of the Corn Belt in the U.S.A. Corn (*Zea mays* L.) and soybeans (*Glycine max* [L.] Merrill) are the major grain crops grown on these soils.

Because of problems associated with producing grain crops (e.g., wet conditions, early frosts, excessive weed competition), many farmers are using these muck soils for pasture. Since very little information is available on the fertility requirements of the pastures, a 2-year study was initiated in the spring of 1979 on Edwards muck soil (Limnic Medisaprist) on the Pinney-Purdue Agricultural Center at Wanatah in northwestern Indiana. Eight combinations of nitrogen (N)–phosphorus (P)–potassium (K) fertilizer were applied with 3 replications to a permanent Kentucky bluegrass (*Poa pratensis* L.) pasture. Dry-matter yield, percentage of crude protein, and mineral composition were determined for each of four cuttings taken during the 1979 and 1980 growing seasons. The harvested forage was from a 1.5-m^2 cage randomly located in each replication.

Application of N fertilizer had a pronounced effect, but P and K had little, if any, effect on dry-matter yield and crude-protein production over the 2-year period. The application of 168 kg/ha of N increased dry-matter yield of Kentucky bluegrass from 6.3 to 8.2 metric tons (mt)/ha in 1979 and from 6.4 to 7.8 mt/ha in 1980. Crude-protein production was increased from 1,153 to 1,983 kg/ha in 1979 and from 1,206 to 1,974 kg/ha in 1980.

Nitrogen significantly increased N concentration in plants, especially during the earlier part of the growing season. A slight increase in P concentration resulted from the application of P fertilizer; adequate levels of P were present, however, in control forages. Potassium levels in the plants increased with application of K fertilizer, but, again, K levels were adequate in plant tissue from the unfertilized treatment.

Although P appeared to be adequate, the application of P with N increased the concentration of N over that of N fertilizer alone. The addition of N fertilizer resulted in a small, but highly significant, increase in K concentration.

KEY WORDS: N-P-K fertilization, *Poa pratensis* L., Kentucky bluegrass, mineral concentration of N-P-K, muck soil, permanent pasture.

INTRODUCTION

Kentucky bluegrass, (*Poa pratensis* L.), a perennial cool-season grass, is extremely well adapted to the muck soils found in the northern portion of the Corn Belt. Birdsfoot trefoil (*Lotus corniculatus* L.), a perennial legume, also grows well on muck soils of this region. However, Kentucky bluegrass usually crowds out birdsfoot trefoil after a few years.

A major reason that Kentucky bluegrass–birdsfoot trefoil pastures are difficult to maintain is that livestock producers often fail to remove the early growth of bluegrass, or they do not have an adequate number of animals to graze off a sufficient quantity of the bluegrass to enable the birdsfoot trefoil to compete successfully with the bluegrass. Consequently, most pastures on muck soils eventually become nearly pure bluegrass.

Very little is known about the fertilizer needs of Kentucky bluegrass grown on muck soils in the northern Corn Belt. Consequently, a study was undertaken to determine the effect of N-P-K fertilization on dry-matter and crude-protein yield and concentration of minerals in the forage of Kentucky bluegrass.

MATERIALS AND METHODS

In the spring of 1979, a 2-year experiment was begun at the Pinney-Purdue Agricultural Center at Wanatah, Indiana, to determine the response of Kentucky bluegrass to N-P-K fertilizer when grown on an Edwards muck soil (Limnic Medisaprist). Dry-matter yields have already been reported (Lightner et al. 1979, 1982). The effect of N, P, and K concentrations are presented in this paper.

J.W. Lightner is a Research Assistant, C.L. Rhykerd is a Professor of Agronomy, G.E. Van Scoyoc is an Associate Professor of Agronomy, D.B. Mengel is an Assistant Professor of Agronomy, C.H. Noller is a Professor of Animal Sciences, and B.O. Blair is a Professor of Agronomy at Purdue University. This paper is a contribution of the university's departments of Agronomy and Animal Sciences (Journal Paper No. 8588).

The experimental design was a randomized complete block with eight fertilizer combinations of N-P-K and 3 replications. The eight fertilizer combinations, expressed as kg/ha of N-P-K, were 0-0-0, 0-99-0, 0-0-372, 0-99-372, 168-0-0, 168-99-0, 168-0-372, and 168-99-372.

All fertilizers were applied annually in a single broadcast application in early spring of 1979 and 1980. The source of N fertilizer was ammonium nitrate in 1979 and urea in 1980. The P and K fertilizer sources were triple superphosphate and muriate of potash, respectively.

Soil-test results indicated that the organic matter was 46%, soil pH was 5.1, available P was 37 kg/ha, and available K was 209 kg/ha.

The plots were established on a permanent Kentucky bluegrass pasture. Each plot was approximately 3.3 × 33 m. Since the pasture was grazed by beef cattle, a 1.5-m² cage was placed randomly in each plot in order to obtain forage samples for yield estimate and mineral analysis.

The caged areas were harvested four times during the 1979 and 1980 growing seasons in May (at the early-heading stage), June, August, and September, with a lawn mower equipped with a collection bag. Nitrogen concentration of the forage samples was determined using the technique outlined by Nelson and Sommers (1973) and Bremmer and Edwards (1965). Concentrations of P and K were determined by IMC Agronomic Services Laboratory, Terre Haute, Indiana, employing an emission spectrograph.

RESULTS AND DISCUSSION

Application of N fertilizer had a pronounced effect on dry-matter yield and crude-protein production over the 2-year period while application of P and K had little, if any, effect. As reported previously (Lightner et al. 1979, 1982), the application of 168 kg/ha of N increased dry-matter yield from 6.3 to 8.2 mt/ha in 1979 and from 6.4 to 7.8 mt/ha in 1980. Crude-protein production was increased from 1,153 to 1,983 kg/ha in 1979 and from 1,206 to 1,974 kg/ha in 1980.

Nitrogen Concentration

The effect of N-P-K fertilization on N concentration is shown in Table 1. With no N fertilizer, the concentration of N ranged from 2.77% to 2.92% when averaged over the four cuttings for the 2 years. The addition of 168 kg/ha of N annually increased the respective N concentration to between 3.58% and 3.78% when averaged over four cuttings and 2 years.

As also shown in Table 1, K does not appear to affect the concentration of N in the forage either with or without the application of N fertilizer. However, the application of P along with N tended to increase the N level in the bluegrass forage.

The concentration of N in forage tended to decrease from the first through the fourth cutting when averaged

Table 1. Effect of NPK fertilization and cutting on N concentration (dry-matter basis) in Kentucky bluegrass forage (2 year average).

Treatment N-P-K	Cutting				
	1	2	3	4	x̄
	---------------%N[1]---------------				
0-0-0	2.91	2.88	2.94	2.93	2.92
0-0-372	2.70	2.83	2.87	2.79	2.80
0-99-0	2.83	2.63	2.86	2.77	2.77
0-99-372	2.88	2.77	2.94	2.67	2.82
168-0-0	4.05	3.59	3.40	3.36	3.60
168-0-372	3.87	3.67	3.34	3.43	3.58
168-99-0	4.27	3.84	3.47	3.55	3.78
168-99-372	4.14	3.92	3.40	3.46	3.73
x̄ (average)	3.46	3.27	3.15	3.12	

[1]Effect of N and N × P significant, $P < 0.01$.

over the eight treatments (Table 1). When the average concentration of N in the forage of the four cuttings of the four treatments receiving N fertilizer was compared with that of those not receiving N fertilizer, it was found that the average N levels were 4.08%, 3.76%, 3.40%, and 3.45% for cuttings 1 through 4, respectively, as compared with 2.83%, 2.78%, 2.90%, and 2.79%, respectively, for the four treatments not receiving N fertilizer. When no N fertilizer was applied the percentage of N in the forage was similar for all cuttings; but where N fertilizer was applied in early spring, the N concentration of the first cutting was considerably higher and that of the second cutting slightly higher than that of the third and the fourth cuttings, which were approximately the same.

It should be pointed out that the Kentucky bluegrass in this experiment was cut at the early heading stage. If time of cutting is delayed beyond this stage the percentage of N in the forage tends to drop rather rapidly. In an earlier experiment on this same farm under drought conditions and cutting at the fully headed stage, it was found that N concentration of the Kentucky bluegrass forage was only 1.5%.

It is not uncommon for Kentucky bluegrass forage grown in summer on muck soils in the northern portion of the Corn Belt to contain N concentrations of 2.5% to 3.0%. When no N fertilizer is applied, the first cutting of forage tends to be low in percentage of N, especially if it is allowed to mature past the early heading stage. Only the first cutting of Kentucky bluegrass produces a seed head. During the remainder of the season, Kentucky bluegrass produces only vegetative growth, which tends to be fairly high in protein. This is due to the fact that Kentucky bluegrass is a nonjointing grass and consequently only leaf blades are harvested when it is cut or grazed following the first cutting. Since the major portion of the N is in leaves rather than stems, the later cuttings, even when not fertilized with N or when grown in association with a legume, tend to be higher in percentage of N than later cuttings of jointing cool-season grasses.

Table 2. Effect of NPK fertilization and cutting on P concentration (dry-matter basis) in Kentucky bluegrass forage (2 year average).

Treatment	Cutting				
N–P–K	1	2	3	4	x̄
	--------------- % P[1] ---------------				
0-0-0	0.35	0.28	0.26	0.26	0.29
0-0-372	0.37	0.29	0.27	0.29	0.31
0-99-0	0.45	0.35	0.37	0.38	0.39
0-99-372	0.50	0.38	0.37	0.39	0.41
168-0-0	0.33	0.26	0.26	0.26	0.28
168-0-372	0.36	0.29	0.27	0.26	0.30
168-99-0	0.50	0.34	0.34	0.34	0.38
168-99-372	0.49	0.35	0.34	0.34	0.38
x̄ (average)	0.42	0.32	0.31	0.32	

[1]Effect of P significant, $P < 0.01$.

Table 3.   Effect of NPK fertilization and cutting on K concentration (dry-matter basis) in Kentucky bluegrass forage (2 year average).

Treatment	Cutting				
N–P–K	1	2	3	4	x̄
	--------------- % K[1] ---------------				
0-0-0	2.51	2.10	2.19	2.30	2.28
0-0-372	3.11	2.54	2.43	3.06	2.78
0-99-0	2.85	2.37	2.50	2.77	2.62
0-99-372	3.53	2.67	2.65	3.34	3.05
168-0-0	2.81	2.31	2.27	2.38	2.44
168-0-372	3.74	2.44	2.69	3.35	3.06
168-99-0	3.06	2.20	2.54	2.55	2.59
168-99-372	4.07	2.57	2.96	3.33	3.23
x̄ (average)	3.21	2.40	2.53	2.88	

[1]Effect of N, P, and K significant, $P < 0.01$.

Phosphorus Concentration

Application of P fertilizer significantly increased the P concentration in Kentucky bluegrass forage (Table 2). The P fertilizer was applied in early spring, and the first cutting tended to be somewhat higher in P than the three later cuttings, which contained essentially identical concentrations of P.

Table 2 illustrates that the mean annual P value for each cutting is not affected by application of N. This lack of effect may be a result of the N fertilizer's increasing dry-matter production and thereby diluting the P concentration.

Potassium Concentration

Although the sufficiency level of K in Kentucky bluegrass forage has not been precisely established, it is probably in the range of 1.8% to 2.0%. All the K concentrations reported in Table 3 are in the range of approximately 2% to 4%, depending upon the fertilizer treatment and cutting. The major factor affecting K concentration in the forage was K fertilizer (Table 3). As is generally the case, the first cutting tended to be higher in K than later cuttings. Also, the application of N and P fertilizers tended to increase K concentration slightly. These observations have been reported previously for Kentucky bluegrass (Hojjati et al. 1977, Walker and Pesek 1963), but those studies were conducted on mineral soil. Schoper

et al. (1979), working on an organic soil, reported that N and P applications did not affect K concentration in quackgrass (*Agropyron repens*).

LITERATURE CITED

Bremmer, J.M., and A.P. Edwards. 1965. Determination and isotope-ratio analysis of different forms of nitrogen in soils. I. Apparatus and procedure for distillation and determination of ammonium. Soil Sci. Am. Proc. 29:504–507.

Hojjati, S.M., W.C. Templeton, Jr., and T.H. Taylor. 1977. Changes in chemical composition of Kentucky bluegrass and tall fescue herbage following N fertilization. Agron. J. 69:264–269.

Lightner, J.W., C.L. Rhykerd, B.O. Blair, B.J. Hankins, V.L. Lechtenberg, and J.M. Hertel. 1979. Response of *Poa pratensis* L. to NPK on shallow muck soil. Proc. Ind. Acad. Sci. 89:400–403.

Lightner, J.W., C.L. Rhykerd, E.L. Hood, G.E. Van Scoyoc, and D.B. Mengel. 1982. Effect of NPK fertilization on dry matter yield and crude protein concentration of *Poa pratensis* L. grown on muck soil. Proc. Ind. Acad. Sci. 90:423–427.

Nelson, D.W., and L.E. Sommers. 1973. Determination of total nitrogen in plant material. Agron. J. 65:109.

Schoper, R.P., C. Simkins, G. Malzer, and R. Farnham. 1979. Fertilization of grasslands on organic soils. Soil Ser. Dept. Soil Sci. Univ. Minn. 95:189–206.

Walker, W.M., and J. Pesek. 1963. Chemical composition of Kentucky bluegrass (*Poa pratensis* L.) as a function of applied nitrogen, phosphorus, and potassium. Agron. J. 55:247–250.

Some Effects of Applied Nitrogen on Grass Growth in Field Swards at Different Times of Year

D. WILMAN and P.T. WRIGHT

Department of Agriculture, University College of
Wales, Aberystwyth, UK

Summary

Results from 21 field experiments are examined to discover the effects of level of applied nitrogen (N) on the regrowth of grass over 8 weeks following a cutting and the situation after about 6 weeks' regrowth, at different times of year.

Percentages of N and nitrate N in herbage generally increased in the early stages of regrowth and then declined. The better the growing conditions were, the faster was the decline. The positive effect of applied N on relative growth rate tended to be greatest and most short-lived when growing conditions were most favorable.

Response to applied N increased as growing conditions improved in the spring with respect to yield, number of tillers, and leaf size and declined as growing conditions deteriorated in the autumn, except in respect to number of tillers. Area/green leaf blade tended to be greatest in September, and response to N was large at this time. Response to the final increment of N was very low after September, as measured by area/blade and yield of blade.

KEY WORDS: nitrogen application, time of year, dry-matter yield, nitrogen yield, nitrate N, tillers, leaves, stem.

INTRODUCTION

We have combined the results from 21 field experiments carried out during the period 1961–1978 in order to seek a better understanding of the effects of applied nitrogen (N) on grass growth at different times of year.

METHODS

In all experiments the plots had been sown with grass alone, mostly perennial ryegrass (*Lolium perenne* L.) or Italian ryegrass (*Lolium multiflorum* Lam.) and kept free of legumes by hand weeding. Area/plot was at least 7 m² except in the case of Wilman (1980). Some of the data for the period April to September in Fig. 1 are from experiments carried out in Cambridge. The remaining data in Fig. 1 and all the data in Fig. 2 are from experiments at Aberystwyth.

RESULTS AND DISCUSSION

In Fig. 1 are summarized some of the changes that have occurred during 8-week periods of regrowth at different times of year. In each case the plot was cut to leave 4–5 cm stubble and raked clear, and fertilizer was applied at the beginning of the 8-week period. Results have been grouped to show the position at four times of year: (1) late winter–early spring, (2) during the main part of the growing season, (3) early autumn, and (4) late autumn. The results for (1) and (4) are from the experiment of Wilman (1980); those for (2) are from a 1961 experiment (Wilman, unpublished) and from the experiments of Wilman (1970, 1975) and Wilman and Wright (1978); and those for (3) are from experiments of P.R. Bannister (unpublished) in 1971 and 1972, Ojuederie (1974), and Wilman and Wright (1978). Nitrogen, in the form of Nitrochalk, was applied at the following rates: (1) and (4), 0, 84, and 126 kg N/ha; (2) and (3), mostly 28, 84, and 140 kg N/ha.

Percentage of N in herbage dry matter generally increased in the early stages of regrowth, particularly at the higher levels of applied N, and subsequently, except in late autumn, declined. The results for percentage of NO_3-N were somewhat similar to those for percentage of N, but percentage of NO_3-N generally began to decline earlier than percentage of N, and there was a relatively large positive response to the final increment of applied N in the case of percentage of NO_3-N compared with that of N. The timing and rate of decline in percentages of NO_3-N and N seemed to be related to the growing conditions, i.e., with the highest level of applied N, percentage of NO_3-N returned to the week-1 value after approximately 14 days during the main part of the growing season (average solar radiation during those 14 days being 14.9 MJ/m²/day) compared with 33 days (10.5 MJ/m²/day) in later winter/early spring, 37 days (7.0 MJ) in early autumn, and 54 days (3.3 MJ) in late autumn. The effect

D. Wilman is Senior Lecturer in Agriculture and P.T. Wright is a Research Associate at the University College of Wales.

Fig. 1. Effect of applied N on changes during 8-week periods of regrowth at different times of year in % N and % nitrate-N and in relative growth rate. In (a) and (b) the lower lines represent 0 [in the case of (1) and (4)] or 28 [in the case of (2) and (3)] kg N/ha, the intermediate lines 84 kg/ha, and the upper lines 126 [(1) and (4)] or 140 [(2) and (3)] kg N/ha. In (c) and (d) the unbroken line represents 0 or 28 and the dashed line 126 or 140 kg N/ha.

Fig. 2. Effect of applied N on changes during the year in yields, numbers of tillers, and area/leaf blade on swards repeatedly cut and fertilized. In each section of the figure the lower line represents no applied N, the intermediate line 53 kg N/ha/6 weeks, and the upper line 105 kg N/ha/6 weeks.

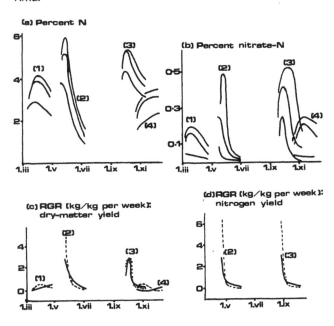

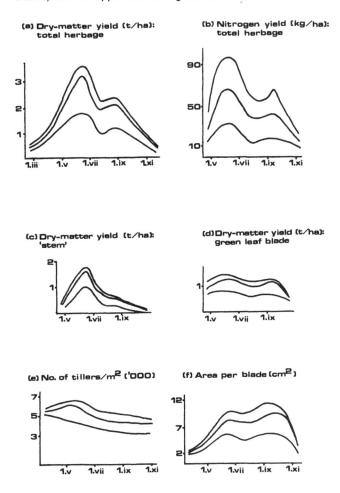

of applied N on relative growth rate (RGR) was positive in the early stages of regrowth and tended to be negative in the late stages. The positive effect of applied N on relative growth rate tended to be greatest and most short-lived when growing conditions were most favorable.

In Fig. 2 are summarized some changes during the year in swards that were repeatedly cut and fertilized. The experiments compared different intervals between cuts, with an average interval being about 6 weeks. Any point on any of the lines represents the approximate value that would have been obtained 6 weeks after cutting and applying fertilizer. Results from the experiment described by Wilman et al. (1976a, 1976b, 1976c) have been used in all six sections of Fig. 2. Results from the experiment of Wilman and Mohamed (1980b, 1981) have been used in sections (a), (b), (e), and (f). Results from the experiment of Wilman (1980) and Wilman and Mohamed (1980a) have been used in sections (b), (e), and (f). In Fig. 2 the rates of application of N (as Nitrochalk) are taken to be 0, 53, and 105 kg N/ha/6 weeks, with approximately 5 applications/yr.

Response to applied N increased as growing conditions improved in the spring in all aspects of the crop reported here: yields (measured as extra kg/kg N applied) of N and of total herbage, green leaf blade and stem dry matter, number of tillers, and area/green leaf blade. Response declined as growing conditions deteriorated in the autumn except in the case of number of tillers. Response

to applied N diminished with increasing rate of application, most markedly in the case of the dry-matter yields and least in the case of N yield. The graph for the number of green leaf blades harvested (not presented here) had a similar shape to that for number of tillers, except that response to the final increment of N was relatively low in the case of number of leaves. Area/green leaf blade tended to be greatest in September, and response to N was large at this time. Evidently in late summer, when the plants are not diverting assimilates to stem production, they have the ability, provided there is sufficient moisture available, to produce relatively large (and relatively heavy) leaves and to maintain the number of tillers, and these processes are aided by an ample supply of N. Response to the final increment of N was very low in the autumn after September in respect of both area/green leaf blade and yield of blade. Whereas yield of total herbage, both dry matter and N, and yield of stem had a marked peak in June, yield of leaf blade and number of tillers had no very major peak.

LITERATURE CITED

Ojuederie, B.M. 1974. Effects of nitrogenous fertilizer on grass growth. Ph.D. Thesis, Univ. Coll. of Wales, Aberystwyth. Welsh Plant Breeding Station. 1980. Rpt. 1979, 254.

Wilman, D. 1970. The effect of nitrogenous fertilizer on the rate of growth of Italian ryegrass. III. Growth up to 10 weeks: nitrogen content and yield. J. Brit. Grassl. Soc. 25:242–245.

Wilman, D. 1975. Nitrogen and Italian ryegrass. II. Growth up to 14 weeks: nitrogen, phosphorus and potassium content and yield. J. Brit. Grassl. Soc. 30:243–249.

Wilman, D. 1980. Early spring and late autumn response to applied nitrogen in four grasses. I. Yield, number of tillers and chemical composition. J. Agric. Sci., Camb. 94:425–442.

Wilman, D., D. Droushiotis, A. Koocheki, A.B. Lwoga, and J.S. Shim. 1976a. The effect of interval between harvests and nitrogen application on the proportion and yield of crop fractions in four ryegrass varieties in the first harvest year. J. Agric. Sci., Camb. 86:189–203.

Wilman, D., A. Koocheki, and A.B. Lwoga. 1976b. The effect of interval between harvests and nitrogen application on the proportion and yield of crop fractions and on the digestibility and digestible yield and nitrogen content and yield of two perennial ryegrass varieties in the second harvest year. J. Agric. Sci., Camb. 87:59–74.

Wilman, D., A. Koocheki, A.B. Lwoga, D. Droushiotis, and J.S. Shim. 1976c. The effect of interval between harvests and nitrogen application on the numbers and weights of tillers and leaves in four ryegrass varieties. J. Agric. Sci., Camb. 87:45–57.

Wilman, D., and A.A. Mohamed. 1980a. Early spring and late autumn response to applied nitrogen in four grasses. II. Leaf development. J. Agric. Sci., Camb. 94:443–453.

Wilman, D., and A.A. Mohamed. 1980b. Response to nitrogen application and interval between harvests in five grasses. I. Dry-matter yield, nitrogen content and yield, numbers and weights of tillers, and proportion of crop fractions. Fert. Res. 1:245–263.

Wilman, D., and A.A. Mohamed. 1981. Response to nitrogen application and interval between harvests in five grasses. II. Leaf development. Fert. Res. 2:3–20.

Wilman, D., and P.T. Wright. 1978. The proportions of cell content, nitrogen, nitrate-nitrogen and water soluble carbohydrate in three grasses in the early stages of regrowth after defoliation with and without applied nitrogen. J. Agric. Sci., Camb. 91:381–394.

Effect of Lime on Lucerne in Relation to Soil Acidity Factors

G.P. MAHONEY, H.R. JONES, and J.M. HUNTER

Department of Agriculture, Victoria, Australia

Summary

Most soils in northeastern Victoria are acid, and problems commonly occur with lucerne (*Medicago sativa*) establishment and production. Aluminum and manganese have been recognized as major components of soil acidity. Liming to reduce exchangeable aluminum (Al) has been suggested as a rational approach to determining the lime requirements of lucerne growing in acid soils.

This study was undertaken to clarify the relationship between soil acidity factors and lucerne growth and to provide a basis for liming recommendations in northeastern Victoria. The aim was: (1) to determine the effect of a range of rates of lime on establishment, production, and nutrient status of lucerne cv. Hunter River, and (2) to relate these parameters to changes of pH, Al, and available manganese (Mn) in the topsoil (0–15 cm) and subsoils.

Three field experiments were established on the acid granitic soils of the Strathbogie ranges, where average rainfall is approximately 900 mm, with the major incidence in winter-spring. Each experiment was a replicated split-plot design with three main plots treated with ground agricultural limestone (80% CaCO₃) drilled with the seed and six subplots with broadcast agricultural limestone applied and incorporated two months prior to sowing.

Average annual lucerne dry matter over 3 years ranged from 1.1 to 6.5 metric tons (t)/ha on the unlimed plots and from 5.1 to 9.8 t/ha on the plots limed at the heaviest rate.

In the topsoil, Al was virtually eliminated by the highest rate of lime application but was largely unaffected in the subsoils. At the site where Al in the subsoil was higher than 1.0 meq/100 g, lucerne production was considerably lower than at the other sites.

Levels of Mn were reduced in the topsoil by lime; however, lucerne seedlings at all sites showed symptoms of severe manganese toxicity. The severity of the symptoms decreased with increasing rates of application of lime. Under any treatment, levels of manganese in mature lucerne varied throughout the year but were reduced by lime additions.

In this series of experiments, the determination of Al alone in the topsoil was not sufficient to determine the lime requirements of lucerne. In these soils, manganese toxicity occurred in seedling and mature lucerne when Al had been eliminated in the topsoil. Reductions in lucerne yield also occurred as a result of high levels of Al in the subsoil, and these levels were not affected by surface applications of lime.

Lucerne yield was highly correlated with Al in the topsoil and subsoil and Mn in the topsoil. The pH of the topsoil and subsoil was rejected as a significant contributor to yield in the stepwise regression model used to analyze these data.

Key Words: lucerne, *Medicago sativa* cv. Hunter River, soil acidity, exchangeable aluminum, available manganese, pH.

INTRODUCTION

Increases in animal production can be achieved by the incorporation of lucerne (*Medicago sativa*) into a farming system based on annual pastures (Reeves and Sharkey 1980). Because most of the soils in northeastern Victoria are moderately to strongly acid, problems are frequently encountered in growing healthy, productive lucerne.

The adverse effects of soil acidity on lucerne growth have been ascribed to many factors, which have been reviewed by White (1967). Lime is generally used to overcome problems associated with soil acidity. However, the transportation costs of lime and the unreliability of its effect discourages the widespread use of lucerne in this region. The lime requirements of lucerne in northeastern Victoria are based on the pH of the surface soil. However, even when lime is applied, lucerne often fails to grow satisfactorily. Some instances have also been observed in the region of lucerne growing well on fairly acid soils without the use of lime.

Measurement of exchangeable aluminum (Al) has been suggested as a reliable indicator of the lime requirements of some acid soils (Kamprath 1970). Hoyt and Nyberg (1972), however, claimed that liming recommendations based solely on Al do not always account for manganese (Mn), which can occur in toxic concentrations in some acid soils.

This study was undertaken to clarify the relationship between factors associated with soil acidity and lucerne growth in order to provide a more reliable basis for liming recommendations in northeastern Victoria.

METHODS

Three field experiments were established on acid granitic soils in the Strathbogie ranges in northeastern Victoria. Average annual rainfall is approximately 900 mm, with the major incidence in winter-spring. Each experiment was a split-plot design with three main treatments (ground agricultural limestone, 80% $CaCO_3$, drilled with the seed at sowing at 0, 0.1, and 0.5 t/ha) and six subtreatments (ground agricultural limestone broadcast at 0, 0.6, 1.3, 2.5, 5.0, and 10 t/ha, and incorporated to 15 cm 2 months prior to sowing). Each plot was 2 x 5 m

and there were 4 replications. All plots were harvested when most were at 10% flowering and also at the end of autumn–winter for the 3 years of the experiments.

Plots were topdressed annually with superphosphate (9% P) and muriate of potash (50% K), each at 125 kg/ha. Molybdenum was applied at 120 g/ha in the first year of the experiments, and a dressing of boron was applied at 10 kg/ha in the second year.

Soil pH was measured with a glass electrode in 1 : 5 soil water suspension. The Al was determined according to Oien and Gjerdingen (1972). Manganese was extracted by shaking the soil in 0.01 M $CaCl_2$ (1 : 2) for one hour and determined by atomic absorption. Plant samples were wet-digested and analyzed for manganese using atomic absorption spectrophotometry.

The relationship between average annual lucerne production and Al, Mn, and pH in the topsoil, subsoil 1 (15–30 cm), and subsoil 2 (31–45 cm) were analyzed statistically, using stepwise regression algorithm operating at the 5% level. Interaction terms and site variables were included in the equation.

RESULTS

Effect of Lime on Soil Acidity and Related Factors

Each rate of broadcast lime increased pH and reduced Al and available manganese (Mn) in the topsoil (Table 1). The highest rate of broadcast lime increased the pH of the topsoil by approximately one unit and virtually eliminated Al and Mn in the topsoil. At all sites, subsoil 1 and subsoil 2 were moderately to strongly acid. Liming the topsoil resulted in small increases in pH of subsoil 1 and subsoil 2 but had very little effect on Al and Mn in these zones. Levels of Al increased with depth while those of Mn decreased.

Effect of Lime on Lucerne Establishment

Broadcast lime at the higher rates significantly increased the number of lucerne plants that established at Strathbogie and Shean's Creek. At Gooram, surface crusting after sowing resulted in patchy emergence and may have masked any treatment effects. Drilled lime at 0.5 t/ha significantly increased lucerne establishment at Strathbogie (results not presented).

G.P. Mahoney is a District Agrostologist, H.R. Jones an Agrostologist, and J.M. Hunter an Agricultural Officer in the Victoria Department of Agriculture.

Table 1. Effect of lime on pH, exchangeable Al meq./100g (AL), and available Mn ppm (MN) in top- and subsoil and on average annual lucerne production t/ha, (Y).

Lime t/ha	Strathbogie				Gooram				Shean's Creek			
	pH	AL	MN	Y	pH	AL	MN	Y	pH	AL	MN	Y
						Top soil, 0–15cm						
0.0	4.7	1.9	14	1.1a*	5.1	0.7	20	6.5a	5.8	0.1	7	6.2a
0.6	4.8			2.0b	5.2			6.8a	5.8			7.1b
1.3	4.9			3.0c	5.2			8.1b	6.0			7.5b
2.5	5.0	1.0	7	3.5c	5.4	0.2	9	9.1bc	6.3	0.0	2	7.4b
5.0	5.2	0.4	5	3.8c	5.7	0.1	4	9.3bc	6.7	0.0	0	7.8b
10.0	5.7	0.1	1	5.1d	6.0	0.0	1	9.8c	7.0	0.0	0	7.5b
						Subsoil 1, 16–30cm						
0	4.7	2.3	4		4.7	1.1	12		5.2	0.3	2	
2.5	4.9	2.3	5		4.9	1.0	9		5.4	0.0	0	
5.0	4.9	2.3	3		5.0	0.9	7		5.5	0.0	0	
10.0	5.0	2.2	4		5.3	0.8	7		5.6	0.0	0	
						Subsoil 2, 31–45cm						
0	4.8	2.8	3		4.8	1.0	10		5.3	0.1	4	
2.5	4.9	2.5	4		5.0	1.1	10		5.5	0.1	4	
5.0	5.0	2.6	3		5.1	1.0	8		5.7	0.1	5	
10.0	5.1	2.7	3		5.3	0.9	7		5.4	0.1	2	

*Yield values followed by the same letter are not significantly different at 5% level.

Effect of Lime on Lucerne Production

Average annual lucerne production increased with increasing rates of lime up to the maximum rate, except at Shean's Creek, where the highest yields occurred at 5 t/ha (Table 1). Average annual dry-matter production ranged from 1.1 to 6.5 t/ha on the unlimed plots and from 5.1 to 9.8 t/ha on plots limed at the maximum rate. A significant response to drilled lime at 0.5 t/ha was obtained at Strathbogie and Gooram, but not at Shean's Creek, which was the least acid site.

Effect of Lime on Manganese Content of Lucerne

Young lucerne plants, at all sites, showed symptoms of manganese toxicity. Lime at 10 t/ha reduced manganese levels in young lucerne plants from 1,319 to 449 ppm at Gooram and from 764 to 335 ppm at Strathbogie. Young lucerne herbage was not collected at Shean's Creek because leaf drop caused by manganese toxicity occurred prior to sampling. The severity of the symptoms and the level of manganese in the herbage at each site was reduced by each additional increment of lime. Manganese levels in mature lucerne herbage varied throughout the year but were substantially lower than the levels found in young lucerne and were reduced by each increment of lime.

The range of manganese concentrations detected in mature lucerne herbage from the 0 and 10 t/ha lucerne plots at one particular sampling at Strathbogie, Gooram, and Shean's Creek was 489–174 ppm, 316–106 ppm, and 214–87 ppm, respectively. These figures indicate that manganese levels were sufficiently high in some periods of the year to be toxic, and they would explain part of the response to lime in lucerne production.

Effect of Soil Acidity Factors on Lucerne Yield

When data were analyzed by stepwise regression to determine the relationship between average annual lucerne production and pH, Al, and Mn in the topsoil and subsoils, the following prediction equation was obtained:

$$Y = 8.04 \quad 0.61 \, (A_1 + A_3) - 0.17 \, M_1 \quad (1)$$
$$(0.27) \qquad\qquad (0.02)$$

$$(100R^2 = 89.8)$$

Y = average annual lucerne production (t/ha); A_1 = Al in topsoil (meq/100g); A_3 = Al in subsoil 2 (meq/100g); M_1 = Mn in topsoil (ppm). Figures below the line in parentheses are standard errors at 0.05%.

DISCUSSION

Kamprath (1970) proposed that realistic lime rates to alleviate acid soil conditions for sensitive crops could be based upon the level of Al in the soil. Using the formula that Kamprath suggested (meq Al/100g × 2 = meq $CaCO_3$/100g), the lime requirements for the Strathbogie, Gooram, and Shean's Creek sites would be 5.4, 1.5, and 0.3 t/ha, respectively. At each site, however, significant responses in lucerne production were obtained to rates of lime in excess of the calculated rates (Table 1).

These additional responses could be explained by the effects of manganese. Simon et al. (1974) showed that high levels of Mn in the soil and manganese toxicity in lucerne could be induced, despite the use of lime, in moderately acid soils under extreme climatic conditions of moisture and temperature similar to those commonly experienced in northeastern Victoria.

In the experiments reported here manganese toxicity was responsible for death of lucerne seedlings and for reduced productivity of mature plants. Liming to eliminate Al did not account in full for climatically induced Mn, although it did reduce its effect.

The concept of liming to reduce surface-soil Al also fails to account for the deleterious effect of subsoil acidity on lucerne production, which has been demonstrated by Pohlman (1946). In our series of experiments, Al increased with depth, and surface applications of lime had little effect on subsoil levels.

The differences in the maximum yields that occurred between Strathbogie and Gooram (Table 1) reflect in part the large differences in subsoil levels of Al. When Al was in excess of about 1.00 meq/100 g in the subsoil, lucerne yields were unsatisfactory, even though surface soil levels of Al and Mn had been eliminated by lime.

In this series of experiments variations in lucerne yield were well explained by levels of Al and Mn in the topsoil and Al in the subsoil. It would seem appropriate that liming recommendations for lucerne growth in northeastern Victoria should be based on a determination of Al and

Mn in the topsoil and that an assessment of the suitability of a site for lucerne production should not be made before the level of Al in the subsoil has also been determined.

LITERATURE CITED

Hoyt, P.B., and M. Nyberg. 1972. Use of dilute calcium chloride for the extraction of plant-available aluminium and manganese from acid soil. Can. J. Soil Sci. 52:163–167.

Kamprath, E.J. 1970. Exchangeable aluminium as a criterion for liming leached mineral soils. Soil Sci. Soc. Am. Proc. 34:252–254.

Oien, A., and K. Gjerdingen. 1972. Extended use of atomic absorption in soil analyses. Acta Agric. Scan. 2:173–177.

Pohlman, G.G. 1946. Effect of liming different soil layers on yield of alfalfa and on root development and nodulation. Soil Sci. 62:255–266.

Reeves, J.L., and M.J. Sharkey. 1980. Effect of stocking rate, time of lambing and inclusion of lucerne on prime lamb production in North-east Victoria. Austral. J. Expt. Agric. Anim. Husb. 20:637–653.

Simon, A., F.W. Cradock, and A.W. Hudson. 1974. The development of manganese toxicity in pasture legumes under extreme climatic conditions. Pl. and Soil 41:129–140.

White, J.G.H. 1967. Establishment of lucerne on acid soils. *In* R.H.M. Langer (ed.), The lucerne crop, 105–113.

Liming for Tropical Legume Establishment and Production

G.H. SNYDER and A.E. KRETSCHMER, JR.
University of Florida, Gainesville, Fla., U.S.A.

Summary

Liming has been declared unnecessary for tropical legume production because tropical legumes (1) are efficient in extracting Ca from soils and are tolerant of excess Al, Fe, and Mn, and (2) are readily nodulated by alkali-producing *Rhizobium* that can function in acid soils. Numerous research reports from Australia, Uganda, Brazil, and Colombia lend credence to this thinking, since no lime responses were observed in those studies. In other work, liming actually depressed tropical legume growth. However, positive responses have been reported in Australia, Hawaii, and Brazil. In Florida, we have observed considerable improvement in the production of Siratro (*Macroptilium atropurpureum* Urb.), *Desmodium heterocarpon* DC., *Stylosanthes guianensis* Swartz., *Centrosema pubescens* Benth., and Aeschynomene americana L. with increasing lime rates up to 2,000–3,000 kg/ha in pot studies, plot studies, and in commercial pastures.

Perhaps tropical legume production could be increased in other regions through a combination of liming, correcting nutrient deficiencies, and inoculating with *Rhizobium* adapted to moderate soil pH. At least there now is sufficient evidence to suggest that lime responses should be evaluated when tropical legumes are being introduced into new areas.

KEY WORDS: lime, tropical legumes, soil pH.

INTRODUCTION

Do tropical legumes differ from temperate legumes in ability to grow in acid soils and in requirements for lime? A number of reports indicate that fundamental differences between these legume groups do occur. Andrew and Norris (1961) observed that tropical legumes in a very calcium (Ca)-deficient, acid (pH 5.5) soil did not show any visible symptoms of Ca deficiency, whereas temperate legume growth was very poor and Ca deficiency symptoms were evident. Prior to this, Norris (1958) cited several reports that lime did not improve nodulation of certain tropical legumes. He concluded that those interested in the establishment of tropical legumes in tropical soils should "1) forget completely the catch-cry 'lime for legumes,' and the concept of Lime Requirement, and 2) inoculate with an appropriate effective strain of *Rhizobium*, but do not worry that nodulation will fail because of low pH unless acidity is extreme." In subsequent papers, Norris (1965, 1967, 1971) reinforced these ideas. He postulated that slow-growing, alkali-producing strains of *Rhizobium* ("cowpea type") are associated with tropical legumes, which have evolved on acid soils low in Ca. Faster-growing, acid-producing, mutant forms of *Rhizobium* are at a disadvantage in acid soils but not in higher pH soils where the temperate legumes are found. Andrew and Hutton (1974) have shown that many tropical legumes are more efficient than temperate legumes at extracting Ca from soil and that tropical legumes are more tolerant of excess aluminum (Al) and manganese (Mn). Thus, the conclusion was drawn that tropical legumes grow in acid soils because they can extract Ca efficiently, they are not adversely affected by Al and Mn, and they are associated with alkali-producing *Rhizobium*. Therefore, they do not require liming.

INTERNATIONAL RESEARCH FINDINGS

In agreement with this theory, there are many cases in which tropical legumes have shown little or no response to lime (Olsen and Moe 1971, de Franca et al. 1973, da Eira et al. 1972, Spain 1975, Truong et al. 1967, Russell 1966, Probert et al. 1979, Kerridge et al. 1972, de Carvalho et al. 1971). In other cases liming actually depressed tropical legume growth (Olvera 1976, Zaroug and Munns 1980, Thomas 1973, Munns and Fox 1976). However, there are some reports in the literature that tropical legume growth is improved by liming (von Stieglitz et al. 1963, Dobereiner 1978, Moomaw and Takahashi 1958, Wolf 1974, Munns and Fox 1976). In fact, inspection of a paper of Andrew and Norris (1961) reveals that the tropical legumes produced maximum yields at about the same lime rate as was optimum for the temperate legumes. However, since the tropical legumes made appreciable growth without liming but the temperate legumes did not, the liming response was less spectacular

G.H. Snyder is a Professor of Soil Science and A.E. Kretschmer, Jr., is Professor of Agronomy at the University of Florida.

for the tropical legumes. Nevertheless, liming was required for optimum production of both types of legumes.

LIME RESPONSES IN FLORIDA

Our own experience with liming tropical legumes in south Florida has caused us to question the notion that liming is fundamentally unnecessary for tropical legume production. For example, in a pot study using Oldsmar Fs, a 2,511 kg/ha application of lime increased *Stylosanthes humilis* HBK. top weight 3-fold over the unlimed check, and siratro (*Macroptilium atropurpureum* Urb.) top growth increased 10-fold (Kretschmer 1970). In another pot study, Urrutia (1972) showed that top growth of *Centrosema pubescens* Benth. grown in Leon Fs was increased 5-fold at 1,200 ppm lime over the unlimed check. In the field, we have shown positive lime responses up to at least 2,000 kg/ha for siratro, *Desmodium heterocarpon* DC., *S. guianensis* Swartz., and *C. pubescens* (Snyder et al. 1978). These responses were most evident in the presence of added 20% superphosphate. All treatments received potassium (K), copper (Cu), magnesium (Mg), zinc (Zn), boron (B), and molybdenum (Mo). In a separate study, maximum 2-year total yields of siratro, *D. heterocarpon*, and *Aeschynomene americana* L. were obtained at 3,000 kg/ha lime (Snyder et al. 1979). In Florida, lime is sufficiently inexpensive that liming at rates near the agronomic maximum are economically justifiable. The soils used in these studies were sandy and low in Ca, aluminum (Al), manganese (Mn), and pH. The legumes were inoculated with "cowpea" *Rhizobium* prior to planting. Responses to lime were observed at rates in excess of those that should satisfy Ca nutritional requirements or neutralize exchangeable Al. In another (unpublished) study conducted on native range (*Aristida stricta* Michx. and *Serenoa repens* [Bartr.] Small) in two south Florida counties, *D. heterocarpon* was seeded in the summer of 1979 following lime applications at 300 or 2,000 kg/ha. (This was a cooperative study with C.G. Chambliss, an assistant professor of agronomy at the University of Florida.) All plots received 45% superphosphate at 40 kg P/ha. There were 16 replications in Okeechobee County and 32 in De Soto County. In the fall of 1979 counts were made of legume plants that could readily be found within a circle 6 m in diameter in the center of each plot, and mature plants were marked to determine overwinter survival. Counts also were made in the fall of 1980. In most cases, significantly more plants and greater overwinter survival were associated with the higher lime rate (Table 1). These experiences led us to believe that lime usage cannot always be discounted for establishing and maintaining tropical legumes.

DISCUSSION

In addition to the hypothesis that tropical legumes inherently do not need liming, there are several explanations of why they may fail to respond to lime or why growth may be depressed by liming. Growth may be limited by nutritional, moisture, or other environmental

Table 1. Number of *Desmodium heterocarpon* plants/plot observed in native range pastures in two Florida counties, following liming and seeding.[1]

| Lime (kg/ha) | Okeechobee | | DeSoto | | Winter survival | |
	1979	1980	1979	1980	Okeechobee	DeSoto
300	3.7	1.9	3.4	1.3	0.9	0.7
2000	5.9	6.7	5.0	3.6	3.2	1.8
Significance	NS†	**	⨎	**	**	*

[1]Phosphorus was applied to all plots at 40kg P/ha.

† **, *, ⨎, and NS refer to significance by the t-test at 0.01, 0.05, 0.1 and no significance, respectively.

factors so that lime responses cannot be observed. In certain tropical soils, the liming itself may be responsible for nutrient deficiency or other unfavorable soil conditions (McLean 1971, Kamprath 1971). The question in these cases is whether some growth without liming is preferable to better growth with lime plus other inputs that might be used either in addition to liming or, in some cases, as a result of the lime. Also, even though certain tropical legumes can nodulate with acid-tolerant *Rhizobium*, would better growth and more certain nodulation be obtained if the soil were limed so that the faster-growing, acid-producing *Rhizobium* could be utilized? Even natively occurring "cowpea" group rhizobia have been shown to nodulate better following liming (Morales et al. 1973). Economics will determine the practical application of the answers to these questions, but the agronomic researcher should demonstrate production potential under optimum conditions, since economic factors fluctuate. There is considerable interest in breeding tropical legumes to withstand low-pH, low-Ca, high-Al, Fe, and Mn soils. But this will not necessarily lead to maximum production. For example, Kang and Fox (1980) recently demonstrated that whereas cowpea (*Vigna unguiculata* Walp.) cultivars can be found that are able to grow in a manganiferous oxisol in Hawaii (pH 4.8), high yield potential was usually associated with cultivars that responded to lime rather than with cultivars that were tolerant of acid environments. It now seems apparent that tropical legumes differ in their ability to nodulate and produce in acid soils, a fact that should lead to hesitancy about making sweeping conclusions about liming tropical legumes. Andrew (1976) recently stated that the need for added lime for tropical legumes should be determined by experimentation for each legume species in each locality, a conclusion with which we agree.

LITERATURE CITED

Andrew, C.S. 1976. Effect of calcium, pH and nitrogen on the growth and chemical composition of some tropical and temperate pasture legumes. I. Nodulations and growth. Austral. J. Agric. Res. 27:611–623.

Andrew, C.S., and E.M. Hutton. 1974. Effect of pH and calcium on the growth of tropical pasture legumes. Proc. XII Int. Grassl. Cong., 2:23–28.

Andrew, C.S., and D.O. Norris. 1961. Comparative responses to calcium of five tropical and four temperate pasture legume species. Austral. J. Agric. Res. 12:40–55.

da Eira, P.A., D.L. de Almeida, and W.C. de Silva. 1972. Nutritional factors limiting legume growth in a red-yellow podzolic soil. Pesq. Agropec. Braz. Ser. Agron. 7:185–192.

de Carvalho, M.M., G.E. de Franca, A.F.C. Bahia, Filho, and O.L. Mozzer. 1971. Exploratory experiment on fertilization of six tropical legumes on dark-red latosol from a "cerrado" area. Pesq. Agropec. Braz., Ser. Agron. 6:285–290.

de Franca, G.E., M.M. de Carvalho, J.G. Ferreira, and H.L. dos Santos. 1973. Effect of calcium and phosphorus levels on the establishment and production of *Centrosema pubescens* Benth. and *Hyparrhenia rufa* (Nees.) Stapf. PIPAEMG. Programa bovinos relatorio annual, 68–72. Minas Gerais, Brazil.

Dobereiner, J. 1978. Potential for nitrogen fixation in tropical legumes and grasses. *In* J. Dobereiner, R.H. Burns, and A. Hollaender (eds.), Limitations and potential for biological nitrogen fixation in the tropics, 13–14. Plenum Press, N.Y.

Kamprath, E.J. 1971. Potential detrimental effects from liming highly weathered soils to neutrality. Soil and Crop Sci. Soc. Fla. 31:200–203.

Kang, B.T., and R.L. Fox. 1980. A methodology for evaluating the manganese tolerance of cowpea (*Vigna unguiculata*) and some preliminary results of field trials. Field Crops Res. 3:199–210.

Kerridge, P.C., C.S. Andrew, and G.G. Murtha. 1972. Plant nutrient status of soils of the Atherton Tableland, North Queensland. Austral. J. Expt. Agric. Anim. Husb. 12:618–627.

Kretschmer, A.E., Jr. 1970. Production of annual and perennial tropical legumes in mixtures with pangolagrass and other grasses in Florida. Proc. XI Int. Grassl. Cong., 149–153.

McLean, E.O. 1971. Potentially beneficial effects from liming: Chemical and physical. Soil and Crop Sci. Soc. Fla. 31:189–196.

Moomaw, J.C., and M. Takahashi. 1958. Initial response of pangola grass–giant spanish clover pasture to lime and fertilizer. Hawaii Fm. Sci. 7(2):4–5.

Morales, V.M., P.H. Graham, and R. Cavallo. 1973. Influencia del método de inoculación y el encalamiento del suelo de carimagua (Llanos Orientales, Colombia) en la nodulación de leguminosas. Turrialba 23:52–55.

Munns, D.N., and R.L. Fox. 1976. Depression of legume growth by liming. Pl. and Soil 45:701–705.

Norris, D.O. 1958. Lime in relation to the nodulation of tropical legumes. *In* E.E. Hallsworth (ed.), Nutrition of the legumes, 164–182. Proc. 5th Easter School, Nottingham Univ., Butterworths Sci. Pubs., London.

Norris, D.O. 1965. Acid production by *Rhizobium*: a unifying concept. Pl. and Soil 22:143–166.

Norris, D.O. 1967. The intelligent use of inoculants and lime pelleting for tropical legumes. Trop. Grassl. 1:107–121.

Norris, D.O. 1971. Seed pelleting to improve nodulation of tropical and sub-tropical legumes. II. The variable response to lime and rock phosphate pelleting of eight legumes in the field. Austral. J. Expt. Agric. Anim. Husb. 11:281–289.

Olsen, F.J., and P.G. Moe. 1971. The effect of phosphate and lime on the establishment, productivity, nodulation and persistence of *Desmodium intortum, Medicago sativa* and *Stylosanthes gracilis*. East Afr. Agric. and Forest. J. 37:29–37.

Olvera, S.E. 1976. Response to Ca and P application in the establishment of three tropical forage legumes in the savanna of Huimanguillo, Tabasco. M.S. Thesis, Colegio Superior de Agricultura Tropical, Cardenas, Tabasco.

Probert, M.E., W.H. Winter, and R.K. Jones. 1979. Plant nutrition studies on some yellow and red earth soils in northern Cape York Peninsula. III. Effects of liming and placement on responses to applied phosphorus. Austral. J. Expt. Agric. Anim. Husb. 19:583–589.

Russell, J.S. 1966. Plant growth on a low calcium status solodic soil in a sub-tropical environment. I. Legume species, calcium carbonate, zinc, and other minor element interactions. Austral. J. Agric. Res. 17:673–686.

Snyder, G.H., A.E. Kretschmer, Jr., and J. Alvarez. 1979. The effect of lime and P on the production of three tropical legumes in south Florida. Agron. Abs. 1979:49.

Snyder, G.H., A.E. Kretschmer, Jr., and J.B. Sartain. 1978. Field response of four tropical legumes to lime and superphosphate. Agron. J. 70:269–273.

Spain, J.M. 1975. Forage potential of Allic soils of the humid lowland tropics of Latin America. *In* Tropical forages in live stock production systems, 1–8. Am. Soc. Agron. Spec. Pub. 24., Madison, Wis.

Thomas, D. 1973. Nitrogen from tropical pasture legumes on the African continent. Herb. Abs. 43:33–39.

Truong, N.V., C.S. Andrew, and P.J. Skerman. 1967. Responses by Siratro (*Phaseolus atropurpureus*) and white clover (*Trifolium repens*) to nutrients on solodic soils at Beaudesert, Queensland. Austral. J. Expt. Agric. Anim. Husb. 7:232–236.

Urrutia, R.V.M. 1972. Effects of lime, P and other nutrients applied to tropical soils and a Florida spodosol on growth and mineral composition of forage legumes. Ph.D. Dissertation, Univ. of Fla.

von Stieglitz, C.R., A. McDonald, and L.R. Wentholt. 1963. Experiments in utilization of Wallum country in southeastern Queensland. Qld Agric. Sci. 20:307–316.

Wolf., J.M. 1974. Agronomic-economic research on tropical soils. N.C. State Univ. Ann. Rpt., 99–100.

Zaroug, M.G., and D.N. Munns. 1980. Screening strains of *Rhizobium* for the tropical legumes *Clitoria ternatea* and *Vigna trilobata* in soils of different pH. Trop. Grassl. 14:28–33.

Residual Effects of Two Years of Very High Nitrogen Applications on Clay Soil under Grass in a Humid Temperate Climate

W.H. PRINS

Institute for Soil Fertility, Haren (Gr.), The Netherlands

Summary

In recent field trials on all-grass swards on clay soils, fertilizer rates of 480 kg/ha and higher showed distinct residual effects in the following year. This paper analyzes the considerable residual effects of one treatment on heavy clay soil over 2 experimental years, after very high rates of N had been applied during 3 pretreatment years (totaling 800 kg N or more/ha/yr). The residual effect was measured through the increase in yield over a reference treatment, at one rate of N application (40 kg N/ha/cut, totaling 200 and 240 kg N/ha in the first and second experimental years, respectively).

Over the 3 pretreatment years the amount of N applied exceeded the amount of N harvested by 941 kg N/ha. The excess N was reflected in accumulation of mineral N in the soil.

At the beginning of the first experimental year the soil contained 314 kg mineral N/ha against the reference treatment 21 kg N. The accumulated mineral N was nearly depleted by the end of the first season. A residual effect was evident at every cut, and over the whole growing season the increase in yield was 6 metric tons (t) dry matter (DM) and 199 kg N/ha.

In the second experimental year, there was no difference in soil mineral N status of the treatments. Still, a residual effect was observed through an increase in yield of 1.1 t DM and 21 kg N/ha.

The study showed that on heavy clay soils—despite N losses from the soil, especially through leaching in winter—part of the applied N may remain in the soil-plant ecosystem and will be available for grass growth in the following season(s).

KEY WORDS: grassland, nitrogen fertilization, residual nitrogen, soil mineral nitrogen, yield, nitrogen uptake.

INTRODUCTION

It has recently been shown in field trials on all-grass swards of mainly perennial ryegrass (*Lolium perenne* L.) that at rates of 480 kg N/ha and more in one season, residual effects occurred in the following season (Prins et al. 1981). These residual effects were very small or absent on sandy soil but considerable on clay soil. One treatment showed a particularly high residual effect following very high N applications during the pretreatment years 1975 to 1977. As visual observations indicated that residual effects may have lasted longer than one season, it was decided to extend the study over two successive seasons, referred to as experimental years 1978 and 1979.

METHODS

In 1975 a long-term fertilizer trial (IB 2244) with different rates of N application was laid out on a heavy clay soil (in the upper 20 cm 52% of the mineral fraction was smaller than 0.002 mm), in Finsterwolde, in the northeastern part of the Netherlands. The site had been sown to grassland in 1974 after arable cropping. Three treatments (A, B, and C) were selected to study the residual effect (Table 1). The residual effect of treatments B and C was determined through the increase in herbage DM and N yields over the reference treatment, A. For this purpose all treatments received one rate of N during the experimental years, namely 40 kg N/ha. Full experimental details have been reported by Prins et al. (1981).

RESULTS AND DISCUSSION

Pretreatment Years

The very high rates of N of treatments B (1977 only) and C (1975–1977) gave much higher DM and N yields than the N rate of reference treatment A (Table 1). It is notable that in 1976 yields were much lower, mainly due to drought (Table 3).

In treatment C, N applied exceeded N harvested (N harvested = N yield = herbage N uptake) in all 3 pretreatment years. This condition was reflected in accumulation of mineral N in the soil (Table 1). Although part of the accumulated N disappeared from the 0–100 cm layer between growing seasons, at the beginning of the first experimental year (1978) the soil contained over 300 kg mineral N/ha.

Treatment B received a very high rate of N in one season only, so the amount of accumulated N was less.

No accumulation occurred in reference treatment A.

Experimental Year 1978

At one rate of N application, treatments B and C showed considerably higher DM and N yields than treat-

Table 1. Annual N application, herbage DM yield and herbage N yield, and soil mineral N in the layer 0–100 cm in pretreatment years 1975–1977 and experimental years 1978 and 1979.

Treatment	N-appl. kg/ha	DM yield t/ha	N yield kg/ha	Soil mineral N, kg/ha Start	End
Pretreatment years					
1975					
A = B[1]	200(5)[2]	12.6	248	28[3]	14
C	884(7)	17.4	664	28[3]	136
1976					
A = B	160(4)	8.2	143	NR	11
C	800(5)	12.0	427	NR	388
1977					
A	200(5)	11.6	212	39	8
B	720(6)	15.3	539	44	52
C	960(6)	15.3	612	263	518
Experimental years					
1978					
A	200(5)	9.5	187	21	20
B	200(5)	12.4	268	77	30
C	200(5)	15.5	386	314	37
1979[4]					
A	240(6)	9.5	218	19	14
C	240(6)	10.6	239	23	11

[1]Treatments A and B were identical in 1975 and 1976.

[2]In parentheses, number of cuts.

[3]28 kg N/ha in layer 0–50 cm. No data of 50–100 cm available.

[4]In the second experimental year treatment B was no longer included.

ment A (Table 1). As has been reported elsewhere (Prins et al. 1981), part of the residual effect can be explained through the accumulated soil mineral N from preceding very high rates of N application. During the first experimental year soil mineral N was determined in spring and at the start of regrowth following each cut. Fig. 1a shows the relationship between the amounts of mineral N in the upper 1 m of soil and the DM yield of each of the five cuts. Throughout the season the effect was clear: the larger the amount of mineral N in the soil, the larger the DM yield. A similar relationship existed between soil mineral N and herbage N content, as can be deduced from the trends of DM yield and N yield in Fig. 1a and 1b, respectively.

Fig. 2 shows changes in the amount of mineral N in various layers of the soil in the course of the growing season. The largest change occurred during spring growth when more than half of the soil mineral N disappeared. Herbage N uptake was large (Fig. 1b), and possibly some leaching losses occurred because rainfall was higher than normal during March. Fig. 2 shows that the amount of soil mineral N decreased during the season, particularly with treatment C. The last cut was on 25 October, and on 7 November the soils of treatments B and C still contained slightly more mineral N than the soil of reference treatment A (Fig. 2).

The author is an Agronomist at the Institute for Soil Fertility, supported in part by the Agricultural Bureau of the Netherlands Fertilizer Industry (LBNM).

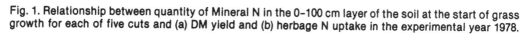

Fig. 1. Relationship between quantity of Mineral N in the 0–100 cm layer of the soil at the start of grass growth for each of five cuts and (a) DM yield and (b) herbage N uptake in the experimental year 1978.

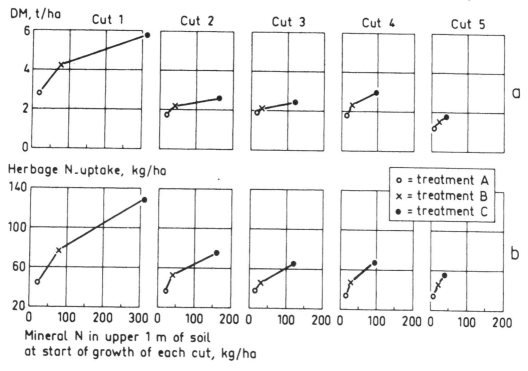

Fig. 2. Changes in the quantity of mineral N in the soil profile down to 100 cm in the course of the growing season in experimental year 1978.

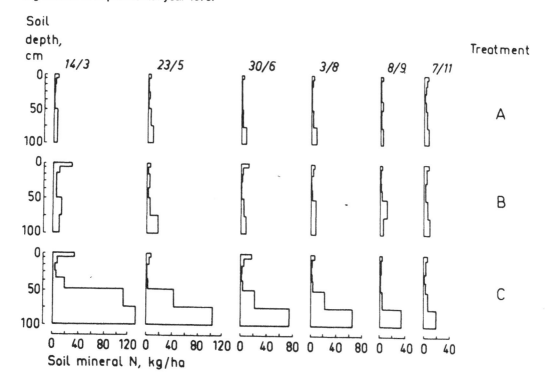

Table 2. Herbage DM yield and herbage N-content of each of six cuts of treatments A and C in the second experimental year.

Cut	DM yield, t/ha		N-content, %	
	Treatment A	Treatment C	Treatment A	Treatment C
1	2.2	2.4	1.8	1.9
2	1.3	1.8	2.6	2.4
3	2.0	2.3	2.1	2.0
4	1.8	2.0	2.3	2.2
5	1.4	1.4	2.5	2.5
6	0.8	0.7	3.3	3.4
	9.5	10.6		

Table 3. Precipitation (mm) at Finsterwolde during the periods 1 March-31 October (roughly coinciding with the growing season) and 1 November-28 February.

Period	Pretreatment years			Experimental years	
	1975	1976	1977	1978	1979
March-October*	453	278	459	506	531
November-February*		321	302	268	181

*Mean precipitation is about 500 and 250 mm, respectively, in these periods.

Compared with treatment A, the increase in N yield over the season was 81 and 199 kg/ha for treatment B and C, respectively (Table 1). The increase for treatment B was higher than the amount of accumulated mineral N in the soil in spring (56 kg N, Table 1). Other sources of residual N in the soil-plant ecosystem, like N in roots and stubble, have been discussed elsewhere (Prins et al. 1981).

Experimental Year 1979

At the beginning of the second experimental season the soil mineral N status was about the same for treatments A and C (Table 1). (Treatment B was no longer included in the study of residual N). Nevertheless, during the season treatment C showed a DM increase of 1.1 t/ha compared with reference treatment A. At every cut except the last one, DM yields of treatment C were higher. Only in the first cut, however, was there a clear indication of a specific residual N effect of treatment C, since herbage N content was also higher than for treatment A (Table 2). Although N uptake in the following cuts was higher, herbage N contents were lower. Together with the higher DM yields, these data suggest an improved sward condition and a better N utilization by treatment C. Because of the average lower N content in treatment C, the increase in total N yield was only 21 kg N/ha compared with treatment A (Table 1).

With a total excess of 941 kg N/ha applied over N harvested in treatment C during the pretreatment years 1975-1977, the total yield increase for the 2 following years was 7.1 t DM and 220 kg N/ha in comparison with reference treatment A (Table 1). This residual effect is remarkable considering the possibility of N being lost during these years, especially during the winter periods (Table 3), as has also been indicated by van der Paauw (1963) in his study of the residual effect with arable cropping in the Netherlands. For treatment C these losses of N are reflected in the differences in quantity of mineral N in the soil between the end of 1976 and the beginning of 1977 and also in those between the end of 1977 and the beginning of 1978 (Table 1).

CONCLUSION

The results clearly show that on heavy clay soil under grass not all N applied in excess of N taken up is lost under the climatic conditions of the Netherlands. Part of the N remains in the soil-plant ecosystem and is available for grass growth in the following season. Even in a second season there can be a residual effect through a combination of residual N and improved sward condition.

LITERATURE CITED

Prins, W.H., G.J.G. Rauw, and J. Postmus. 1981. Very high applications of nitrogen fertilizer on grassland and residual effects in the following season. Fert. Res. 2:309-327.

van der Paauw, F. 1963. Residual effect of nitrogen fertilizer on succeeding crops in a moderate maritime climate. Pl. and Soil 19:324-331.

Impact of Phosphorus and Potassium Fertilization on Maintaining Alfalfa-Orchardgrass Swards in Hokkaido, Japan

H. OOHARA, N. YOSHIDA, K. FUKUNAGA, W.G. COLBY, M. DRAKE, W.F. WEDIN,
G. KEMMLER, H.R. ÜXEKÜLL, and M. HASEGAWA

Summary

In spite of the importance of alfalfa for all kinds of livestock, particularly for dairy cattle, the cultivation of alfalfa has not yet expanded in Hokkaido, Japan. A comprehensive research project has been undertaken, based on observations of long-term effects of fertilization on longevity and productivity of alfalfa-orchardgrass mixtures. This report is concerned with (1) the dry-matter production of alfalfa and orchardgrass, cultivars of which were introduced from the U.S.A. and (2) the ratio of alfalfa to orchardgrass and uptake of nitrogen (N), phosphorus (P), potassium (K), calcium (Ca), and magnesium (Mg). The research began in 1960 and is continuing with close communication among coresearchers in Japan, the U.S.A., and the Federal Republic of Germany. Included in the project is the introduction of technical equipment, e.g., band seeder. Experimental results can be summarized as follows:

1. Phosphorus applied as a basic fertilizer was very important for the growth of forages, particularly in the seedling stage.
2. Alfalfa topdressed with K produced a higher dry-matter yield and maintained a higher percentage of alfalfa in the sward, as compared with alfalfa not fertilized with K.
3. Percentage of P in forages decreased gradually and came to indicate a critical level after 10 years. As the result of application of P, however, the dry-matter yield increased and percentage of P recovered to the normal level. The application of P and K was necessary for both the normal growth of forages and the production of a high dry-matter yield.
4. Application of P and K enhanced stand longevity of alfalfa during 20 years.
5. Application of P and K stimulated the activity of N-fixing bacteria and increased percentage of N in forage dry-matter.

KEY WORDS: alfalfa, orchardgrass, longevity, dry-matter yield, phosphorus application, potassium topdressing.

INTRODUCTION

Grassland farming in Hokkaido, the northernmost island of Japan, is marked by the overwhelming popularity of a legume-grass mixture on an acreage representing as much as 50% of the total farmland used to raise livestock on the island. The productivity, persistence, and longevity of the stand as well as the nutrient composition of the forages produced, however, are greatly influenced by a number of factors, including species, cultivars within species, management, climatic conditions, and soils. Alfalfa is one of the most important legumes in Hokkaido, but its production has not expanded substantially. To elucidate the effect of P and K topdressing on an alfalfa-orchardgrass mixture, a long-term study began in 1960 and is still in progress in Hokkaido. The sites are on sandy-loam soils originating from volcanic ash. Present and future production of alfalfa, singly or in mixture with grasses, on soils of this nature are of primary importance to dairy farming in Hokkaido. Data were recorded for dry-matter production, ratio of alfalfa to orchardgrass, and uptake of N, P, K, Ca, and Mg over four 5-year periods from 1960–1979.

EXPERIMENTAL PROCEDURE

Climatic conditions at the experimental field indicated marked yearly variations in temperature, hours of sunshine, and rainfall. The mean daily temperature for the growing season, lasting 153 days from 1 May to 30 September during 1960–1979, was 15.9° C, with the maximum 17.8° C in 1975 and the minimum 14.4° C in 1971, respectively. The average number of sunshine hours was 833, with a maximum of 1,017 and a minimum of 685. For accumulated rainfall, the average was 529

H. Oohara is at the Dairy Research Institute, Japan; N. Yoshida and K. Fukunaga are at Obihiro University of Agriculture and Veterinary Medicine, Japan; the late W. G. Colby was at the University of Massachusetts, U.S.A., and M. Drake is at the University of Massachusetts; W.F. Wedin is at Iowa State University, U.S.A.; G. Kemmler is at Büntehof Agricultural Research Station, Federal Republic of Germany; and H.R. Üxeküll and M. Hasegawa are at the Potash Research Association, Japan.

The authors wish to state that mention of proprietary products implies no endorsement of them.

Table 1. Fertilizer topdressing (K, P, dolomitic limestone) applied in addition to basic fertilizer at seeding.

Treatment no.	Fertilizer applications (kg/ha)
1	No topdressing (1960–1979)
2	250 K (1961–1979)
3	250 K (1961–1979), plus 44 P (1970–1979)
4	500 K (1961–1979)
5	500 K (1961–1979), plus 44 P (1970–1979)
6	250 K (1961–1979), plus dolomitic limestone at 1 mt/ha (1979)
7	250 K, 44 P, plus dolomitic limestone at 1 mt/ha (1979)

mm, the maximum 807 mm, and the minimum 326 mm. Soil at the experimental field is an Obihiro fine sandy-loam soil of pH 6, derived from volcanic ash and containing approximately 10% organic matter, but deficient in available N, P, K, and Mg.

A Planet Junior seeder from the U.S.A. was used for seeding at the following rates (kg/ha): alfalfa (DuPuits), 5; orchardgrass (Masshardy), 4.

Basic fertilizer broadcast, in kg/ha, was: dolomitic limestone, 5,000; K, 120; and B, 7.5. Two hundred kg/ha ammonium phosphate (11-21-0) was disked into the surface 15 cm. As the experiment continued, fertilization treatments were altered as indicated in Table 1.

Topdressing of K at either the 250 or the 500 kg/ha level was split to follow each of three harvests annually. To correct for P deficiency observed in period II, an early, annual application of P was added in period III. In 1979, dolomitic limestone was applied at the rate of 1 t/ha.

Plot size was 1.65 × 9.144 m, with areas of 0.914 × 9.144 m harvested, using a Jari mower. There were 3 replications. In 1960 (seeding year) plots were harvested twice, with the stage of maturity at first cut being first bloom to 1/3 bloom in alfalfa. In subsequent years, plots were harvested three times annually. Analyses for N, P, K, Ca, and Mg were made in all years, using standard laboratory analyses. Data on chemical analyses presented here are for 1979 only.

RESULTS AND DISCUSSION

Yearly changes in dry-matter yield of the components, alfalfa and orchardgrass, over 20 years (1960–1979) are shown in Fig. 1. In the early stage after seeding, the growth of alfalfa and orchardgrass was responsive to the first increment of precision-banded P, and the response helped to sustain high yields over the 5-year period (Oohara et al. 1963). In later years, dry-matter yield was increased by topdressing of K. After period II, symptoms of P deficiency appeared in some plots. The critical point for percentage P had been reached. Therefore, topdressing of P was begun from the eleventh crop year (period III), and it resulted in an increase in dry-matter yield. The topdressing of P and K brought an increase in ratio of alfalfa to orchardgrass as well. This result suggests that K fertilization stimulates the dry-matter production of alfalfa and orchardgrass.

The N, P, K, Ca, and Mg percentages of alfalfa and orchardgrass in 1979 are presented in Fig. 2. Percentages of N, P, K, Ca, and Mg varied with the harvest year, species, cutting, etc., but responded markedly to topdressing of P and K. The ratios of some mineral nutri-

Fig. 1. Dry-matter yield (t/ha) of alfalfa and orchardgrass.

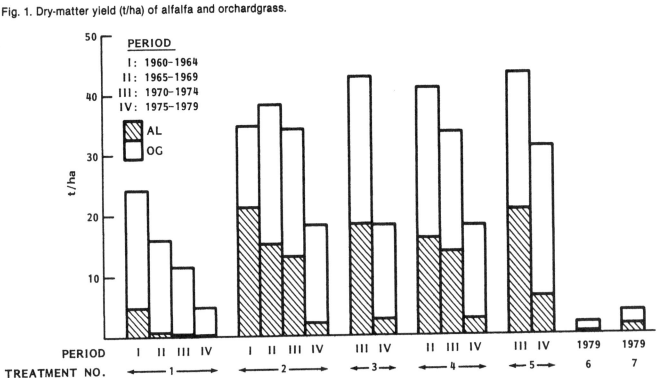

Table 2. Uptake in kg/ha of N, P, K, Ca, and Mg, by treatment.

Species	No topdressing (1960-1979) Trt. 1					250 K (1961-1979) Trt. 2				
	N	P	K	Ca	Mg	N	P	K	Ca	Mg
AL	1	1	1	1	1	21	1	14	6	1
OG	10	1	6	1	1	37	4	74	6	5
Total	11	2	7	2	2	58	5	88	12	6

Species	250 K (1961-1979) + 44P (1970-1979) Trt. 3					500 K (1961-1979) Trt. 4				
	N	P	K	Ca	Mg	N	P	K	Ca	Mg
AL	55	3	38	20	5	20	1	16	6	1
OG	53	9	80	8	6	33	3	59	3	3
Total	108	12	118	28	11	53	4	75	9	4

Fig. 2. Percentages of N, P, K, Ca, and Mg of swards (1979).

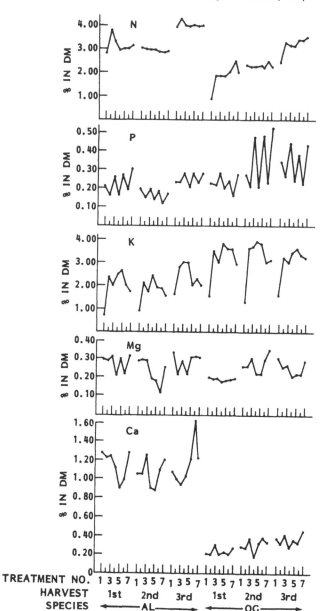

TREATMENT NO. 1 3 5 7 1 3 5 7 1 3 5 7 1 3 5 7 1 3 5 7 1 3 5 7
HARVEST 1st 2nd 3rd 1st 2nd 3rd
SPECIES ◄————AL———— ◄————OG————►

ents, of K: (Ca + Mg), and of Ca:P are of a great significance to all kinds of livestock, particularly to dairy cattle. Table 2 shows the amounts of major mineral nutrients—N, P, K, Ca, and Mg—taken up by alfalfa and orchardgrass plants based on 1979 yields and analyses. Effects of the topdressing of P and K on the uptake of N, P, K, Ca, and Mg were computed from Table 2 and are shown in Table 3.

As indicated in Table 3, topdressing of K was associated with greater uptake of the mineral nutrients. However, application of 500 kg/ha of K/yr seems less effective than application of 250 kg/ha of K. The topdressing of 250 kg/ha of K was associated with 47 kg/ha of N in the forage from either N fixation or soil release.

The rate of K uptake was 32.5% when 250 kg/ha of K was topdressed, and the rate increased to 44.3% when both P and K were topdressed. However, the topdressing of 500 kg/ha of K decreased the rate of K uptake to 13.6% of K applied. The rate of P uptake was 19.2% when 44 kg/ha of P was topdressed, and it was increased to 26.5% by the effective topdressing of P and K. These tables suggest that topdressing of P and K would have a great effect on dry-matter yield and uptake of mineral nutrients in alfalfa and orchardgrass grown in mixtures in Hokkaido.

LITERATURE CITED

Oohara, H., N. Yoshida, K. Fukunaga, and M. Drake. 1963. Effects of the basic application of phosphorus and the topdressing of potassium on the productivity, some chemical compositions, and the digestibility of forage crops. Obihiro Zootech. Res. Bul. Ser. 1, 4:82-108.

Table 3. Calculated uptake of nutrients (kg/ha).

Nutrients	250 kg/ha K	44kg/ha P	250 kg/ha K + 44kg/ha P	500 kg/ha K
N	47	50	97	53
P	4	7	11	4
K	81	30	111	75
Ca	10	15	26	9
Mg	5	5	10	4

Suitability of Early Pleistocene Sandy Lowlands for the Cultivation of Alfalfa (*Medicago sativa* L.)

E. WOJAHN, G. MUNDEL, and W. KREIL

Institute of Forage Production, Paulinenaue, GDR

Summary

The areas with sandy soils in the Pleistocene lowland of the GDR are being ameliorated on a large scale. In these lowlands, low bogs appear together with hydromorphic sandy soils. The latter are situated higher above the groundwater level. Because of the lowering of groundwater by drainage, these sandy soils are no longer suitable for intensive and high-yielding grass cultivation. Especially during summer the comparatively shallow-rooted grasses cannot gain their water requirement from ground water. Therefore, we investigated whether and under what conditions higher and more reliable forage yields could be obtained on such soils by cultivation of the deeper-rooted alfalfa (*Medicago sativa* L.).

In addition to forage-yield experiments, water consumption and removal of macronutrients were determined with ground-water lysimeters. Several soil types were identified that differed in substratum, content of calcium carbonate, carbon, nitrogen, pH value, degree of base saturation, and amplitude of ground water. The ground-water level varied from 40 to 250 cm, depending on soil type and weather conditions. The relationships between these factors and the value of alfalfa for cultivation were quantified. The experiments demonstrated that new spring seedlings were not able to use the ground water at 70 or 100 cm of depth until the beginning or the end of June, respectively. Although old alfalfa plants can utilize deeper ground water, new seedlings require water in the topsoil and subsoil. Medium ground-water levels of about 100 cm appear especially favorable for alfalfa production. In the vegetative growth period, higher ground-water levels are tolerated well. At the same water supply and the same weather conditions, evapo-transpiration by alfalfa is similar to that of cultivated grasses. At a ground-water level of 120 cm, alfalfa removed 20% to 70% of its total water consumption from the ground water, depending upon weather conditions. Five to 11 mm/day were removed from the ground water at this level in an obviously vast utilization of the soil-water store during dry periods. In practice, reasonable alfalfa yields were still obtained at ground-water levels of 150 to 220 cm. Depending upon the type of sandy soils, annual yields of 6.0 to 8.5 metric tons (t) of dry matter/ha were achieved on large areas.

KEY WORDS: hydromorphic sandy soils, soil characteristics, water consumption, alfalfa.

INTRODUCTION

The areas of the Pleistocene lowland of the GDR with sandy soils are being ameliorated on a large scale. In these lowlands, low bogs are interspersed with hydromorphic sandy soils (gleys, haplaquolls, calciaquolls). The latter are located on the higher parts of the soil relief and have, therefore, the deepest ground-water levels. Due to the easily drainable substratum, the large-scale amelioration causes a profound lowering of the ground water. Consequently, these sandy soils are no longer suitable for intensive, high-yielding grass cultivation. Especially during dry summer months, the comparatively shallow-rooted grasses can no longer gain their water requirement from ground water. Therefore, it was desirable to determine whether and under what conditions higher and more reliable yields of protein-rich forages could be obtained on such soils by cultivating the deeper-rooted alfalfa. Moreover, growth of alfalfa would result in a saving of nitrogen fertilizer.

METHODS

An area of the Havelländisches Luch was examined by soil mapping and by measuring alfalfa-forage yields in a combined investigation to determine which of the above-mentioned soils are particularly suitable for cultivating alfalfa. Additionally, water consumption and demand for macronutrients by alfalfa were estimated by means of ground-water lysimeters.

RESULTS

Soil Characteristics and Their Influence on Growth of Alfalfa

Table 1 shows the most important characteristics of the examined soils, considering only the soil types that cover the major part of the area. The approximate equivalents of the soil classification of the U.S.A. and the USSR are presented along with the soil types of the GDR (Table 2).

Table 1. Characteristics of the tested soils.

Character	Soil Type		
	Sand-Graugley	Kalksand-Humusgley	Sand-Anmoor
Approximative equivalent of soil classification in the USA	typic Aquipsamments	typic Calciaquolls	histic Humaquepts
Texture	medium and fine sand	stratified slightly loamy sand strongly sandy loam	medium and fine sand
$CaCO_3$ in subsoil, %	0	3–20	0
Carbon content, kg/m²	4–12	12–20	26–30
Soil nitrogen, kg/m²	0.3–0.5	0.8–1.2	1.1–2.0
pH value KCl			
Surface layer	4.7–6.5	7.0–8.0	5.6–6.1
Subsoil	6.9–7.6	8.0–8.5	7.0–7.4
Maximum ground-water amplitude in cm below surface	90–250	50–200	40–150
Suitability for alfalfa	unsuitable	very good	moderately good

The geological parent materials are typical glacio-fluviatile sands. They consist mainly of medium and fine sands. The silt and clay content of some of the soils correlates closely with the calcium carbonate content. In general, the texture is very similar in all soil types. There are, however, distinct differences with regard to thickness of topsoil, accumulated carbon or humus, and nitrogen content. In comparison with anhydromorphic soils, all types have a larger reserve of carbon and nitrogen. With regard to soil reaction and base saturation, there are also more or less distinct differences among the soil types. In general, the soil reaction changes from acidic in the topsoil to neutral or slightly alkaline in the subsoil. In addition, Table 1 presents the seasonally conditioned amplitude of ground water for the individual soil types. As mentioned above, ground-water levels have been lowered generally by amelioration, and consequently the extent of hydromorphy of these soils is diminished.

The suitability of the tested soils for alfalfa production is predicted by the soil factors together with the factor of ground water. Medium ground-water levels of 90 to 120 cm in the growing season appear to be especially favorable for growth of alfalfa. In some soil types, the highest ground-water levels in spring limit the persistence of alfalfa. Higher ground-water levels are better tolerated when ground-water-free topsoil is dry.

The differing accumulations of humus and nitrogen in the topsoil affects new seedlings. High supplies of humus and nitrogen insure better seedling growth and better initial development of alfalfa. The differing water-holding capacities of the topsoil and the delivery of nitrogen seem to be responsible for this phenomenon. Therefore, the grey gley soil type is not suitable for cultivation of alfalfa. Besides, with these soils, wind erosion frequently occurs in spring.

Neutral soil reactions and higher base saturation also affect the efficiency of alfalfa. The presence of calcium carbonate in some soil types influences soil acidity. The

Table 2. Drymatter yields of *Medicago sativa* on different soils.

Soil type	Approximate equivalents of soil classification		DM yield metric ton/ha
	In the USA	In the USSR	
Sand-Humusgley	Aqueptic Haplaquolls	Gleevaja počva	6.5–7.0
Kalksand-Humusgley	Typic Calciaquolls	Gleevaja počva	6.5–8.0
Kalksand-Anmoor	Histic Calciaquolls	Lugov-bolotnaja počva	7.0–9.5
Silt-Schwarzerde	Typic bzq. haplic Vermudolls	Cernozem	8.0–11.0
Silt-Fahlerde	Typic Glossoboralfs	Dernovo-podzolostaja počva	8.5–10.0
Tieflehm-Fahlerde	Glossoboralfs	Dernovo-podzolostaja počva	5.0–9.5
Kalk-Rendzina	Lithic bzw. typic Rendolls	Rendzina	4.0–8.0
Sand-Braunerde	Normipsamments	Buraja lesnaja počva	4.0–6.0

Fig. 1. Comparison of evapo-transpiration from the ground water (120 cm in depth) of *Medicago sativa* and *Lolium multiflorum* in the 1978 growing season (precipitation 358 mm).

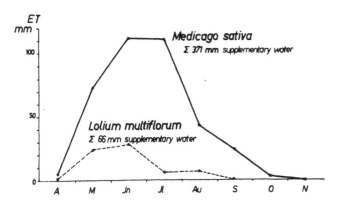

Fig. 2. Relationship of evapo-transpiration and dry-matter yield.

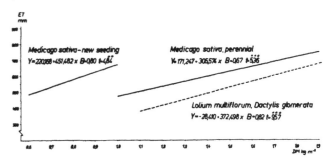

high demand of alfalfa plants for Ca is met in all these soils by ground water rich in Ca.

Water Requirement

Although the reserve of water in the upper layers of ground-water-containing soils is relatively unimportant for established alfalfa stands, the water capacity in the ground-water-free soil horizon of such sites is critically important for new seedlings. Our lysimeter experiments demonstrated that new seedlings (seeding date 12 April) reached the ground water to depths of 70 and 100 cm and were able to use it to a noteworthy extent by the beginning and the end of June, respectively.

Fig. 1 compares the monthly water consumption from ground water at a depth of 120 cm by a 2-year-old alfalfa stand with that of a stand of Italian ryegrass sown in the autumn of the preceding year. Although medium amounts of rain fell during the growing season, the alfalfa stand took more than half of the consumed water from the ground water. On the other hand, ground-water removal by Italian ryegrass was insignificant. During growing seasons with low precipitation, alfalfa stands take more than two-thirds of the utilized water from ground water and produce remarkable yields. Forage grasses are not as efficient in utilizing ground water, and they yield considerably less.

Because estimates from the literature on water requirements by alfalfa vary greatly, water consumption was examined using ground-water lysimeters during several growing seasons. Since, in lysimeter studies, transpiration and evaporation cannot be measured separately, it is not possible to determine water consumption in terms of transpiration coefficient. However, it was evident with all annual and perennial forage plants that the level of organic-matter production clearly influenced the rate of evapo-transpiration. With rising yields, evapo-transpiration also increased. In this case, evapo-transpiration was determined with reference to the area.

Fig. 2 presents the regression of mean evapo-transpiration on dry-matter yield of alfalfa and a mixture of two grasses. It is quite evident that the evapo-transpiration of new seedlings of alfalfa differs from that of established stands. With new seedlings, the evapo-transpiration is high despite the lower dry-matter yield, and it rises more rapidly with increasing yields than in old stands. The mean evapo-transpiration of grasses is lower than that of alfalfa by approximately 80 to 100 mm under identical meteorological conditions, ground-water levels, and yields. Experimental grasses were Italian ryegrass and cocksfoot. The cultivated grasses of more or less mesophytic type have similar rates of evapo-transpiration under otherwise equal conditions. The somewhat higher evapo-transpiration of alfalfa with the same yield seems to be due to luxury consumption of water. We are not certain that this difference between forage grasses and alfalfa occurs under all meteorological conditions.

Yields

Table 2 shows annual yields of perennial alfalfa in practice. The soils tested in our experiments were compared with anhydromorphic soils. The yields of alfalfa growing in early Pleistocene, medium-to-slightly hydromorphic sandy soils equal or even surpass the yields of that growing in most anhydromorphic soils. These yields are distinctly higher than those of alfalfa growing in anhydromorphic soils or most of the limy soils. Only in the traditional alfalfa soils (loess) are the yields greater.

Phytotoxic Response of Five Range-Grass Species to Six In-Situ Fossil-Fuel Retort Waters

Q.D. SKINNER, J.C. SEXTON, T.S. MOORE, and D.S. FARRIER
University of Wyoming and Laramie Energy Technology
Center, Laramie, Wyo., U.S.A.

Summary

Oil shale, tar sands, and underground coal gasification represent viable sources of fossil fuel to assist in meeting U.S. energy needs. Development of fossil-fuel-processing technologies has progressed to a point where commercialization may soon become a reality. Concurrent with this development must be the assessment and solution of environmental problems. A particular and significant problem resides in the handling, containment, treatment, disposal, use, and eventual fate of waters recovered with the oil, gas, and subsurface waters associated with synfuel-processing techniques. The magnitude of plant responses and potential phytotoxic effects resulting from exposure to these produced waters have not been evaluated, but they need to be quantified for purposes of risk assessment. To address this problem, research was conducted (1) to develop or test quantitative short-term, cost-effective, diagnostic screening procedures for testing vegetation against synfuel-produced waters; (2) using such procedures, to quantify the response of plants to these waters, and (3) to provide information for developing controlled technology. Growth responses (leaf area, dry weight, root:shoot ratio, and leaf:shoot ratio) as well as germination were parameters monitored to test produced waters against vegetation under greenhouse conditions using hydroponic techniques. Grass species respond differently to an individual water and in turn may react in changing orders of response to several waters originating from retorting various fossil fuels. Growth response varies with dilution and may be more valuable as an indicator of phytotoxicity than germination. Plant-species response to various waters can be separated and ranked to allow risk assessment, to facilitate selecting the correct species for reclamation, and, in the future, to assess the success of water-treatment procedures.

KEY WORDS: fossil fuel, water, germination, growth response, vegetation.

INTRODUCTION

Oil shale, tar sands, and underground coal gasification represent viable sources of fossil fuels to assist in meeting U.S. energy needs. Development of fossil-fuel-processing technologies has progressed to a point where commercialization may soon become a reality. Concurrent with this development must be the assessment and solution of environmental problems. A particular and significant problem resides in the handling, containment, treatment, disposal, use, and eventual fate of waters associated with synfuel-processing techniques. Process waters recovered with the oil, gas, and subsurface waters associated with production zones represent examples of aqueous byproducts affiliated with in-situ processing of fossil fuels. Possible mechanisms for bioenvironmental contact are accidental spills resulting from failure of a containment, transfer, or treatment process or contamination and subsurface migration of ground waters to a surface outlet. The magnitude of plant responses and phytotoxic effects resulting from such exposures has not been, but needs to be, quantified for purposes of risk assessment. This need becomes apparent when it is realized that during in-situ processing of oil shale at least an equal volume of process water is coproduced with each barrel of shale oil (Farrier et al. 1978). During in-situ processing the amount of water produced increases as a result of recovery of ground water intruding into the retort zone and consequently recovered with the oil (Fox et al. 1978, Farrier et al. 1977). In-situ coal gasification and tar sands processing will create similar water-related concerns.

The initial objectives of this study were (1) to develop screening procedures for testing vegetation against synfuel-produced waters; (2) using such procedures, to quantify the response of plants to process waters; and (3) to provide information for developing controlled technology. In the latter objective, quantitative data on plant responses would provide rationale for water treatment, would permit assessment of treatment efficacy for waters destined for surface disposal, and would provide scientific logic for species selection and design for reclamation of disturbed lands.

Q.D. Skinner is an Associate Professor of Range Management, J.C. Sexton is a Research Scientist, T.S. Moore is an Associate Professor of Botany at the University of Wyoming; and D.S. Farrier is a Scientist in the Department of Energy, Laramie Energy Technology Center.

The authors wish to state that mention of proprietary products does not imply endorsement of them.

METHODS

Growth response and seed germination assays were utilized for screening produced waters from fossil fuels and vegetation. A complete description of methods are described by Skinner et al. (1979).

Growth Response Assay

Growth response was measured, utilizing plants grown in a hydroponic system under greenhouse conditions. Individual plants were transplanted from sterile vermiculite after the appearance of the radicle, cotyledon, and first leaf into 2-l steam-sterilized pots containing thrice-rinsed horticulture-grade perlite. Pots were then attached to a self-feeding hydroponic system. Controls were fed Hoagland's salt solution (Hoagland and Aron 1950) as modified by Miller (1963) one-half hour/day. Produced waters were fed 1% and 5% by volume with modified Hoagland's nutrient solution to individual plants. Four repetitions were carried out for each control, 1% by volume and 5% by volume of produced water. Plants were harvested 10 weeks after transplant stabilization had occurred.

Leaf area, total dry weight, root weight, shoot weight, and root:shoot and shoot:leaf area ratios were the parameters measured. Dry weights were determined after oven drying of whole plants or plant parts for 48 hours at 80°C. Leaf area was determined either by use of a Lambda Instruments Corporation (Lincoln, Nebraska) leaf-area meter model L1–3000 or by length-width measurements and subsequent calculations of the area. Leaf-area measurements were taken every 2 weeks over a 10-week period.

Germination Assay

Germination response was tested by utilizing a sample of produced water appropriately diluted to known concentrations in distilled water. Two ml of the diluted sample were added to 10 seeds sown on filter paper in a covered 10-cm petri dish. Controls were treated with 2 ml of distilled water. Five replicates were used for each experiment. All experiments were carried out at 21°C and kept in the dark. Seeds were examined for germination, and germinated seeds were removed at regular time intervals. Germination was determined by the visible appearance of the radicle tip.

Produced Waters

Fossil-fuel process waters utilized included: (1) Omega 9, a water collected from an in-situ oil shale burn near Rock Springs, Wyoming; (2) Oxy-6-Condensate, a pilot-plant condensate produced from a vertical in-situ oil shale retort near DeBeque, Colorado; (3) Geokinetics-17, a water produced from fracturing a shallow oil shale formation followed by an in-situ burn near Vernal, Utah; (4) 150T-SI17, a shale-oil process water produced during run 17 of the 150-ton capacity simulated in-situ retort,

Laramie, Wyoming; (5) tar sand, a water produced during a true in-situ burn of tar sands near Vernal, Utah; and (6) Hanna IV, a water produced during an underground coal gasification burn, Hanna, Wyoming.

Vegetation

Five grass species were selected for testing effects of produced waters on vegetation. Species tested (common names taken from Beetle 1970) included (1) western wheatgrass (*Agropyron smithii*), (2) bluebunch wheatgrass (*Agropyron spicatum*), (3) basin wildrye (*Elymus cinereus*), (4) Nuttall alkaligrass (*Puccinellia airoides*), and (5) alkali sacaton (*Sporobolus airoides*).

RESULTS

Fig. 1 illustrates the response of five grass species to one water. Other graphs for species against individual waters

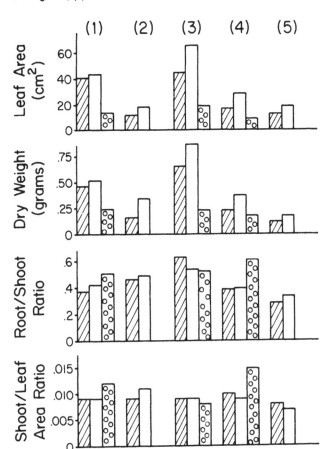

Fig. 1. Response of five grass species to in-situ produced waters from coal gasification. Crosshatched bars = controls; open bars = 1% Hanna IV by volume; bars with circles = 5% Hanna IV by volume. (1) = western wheatgrass; (2) = bluebunch wheatgrass; (3) = basin wildrye; (4) = Nuttall alkaligrass; (5) = alkali sacaton.

Fig. 2. *Sporobolus airoides* response to in-situ produced waters from retorting oil shale, tar sands, and coal. Cross-hatched bars = control; Water 1 = Omega 9; Water 2 = Oxy-6; Water 3 = Geokinetics-17; Water 4 = 150T-SI17; Water 5 = tar sand; Water 6 = Hanna IV.

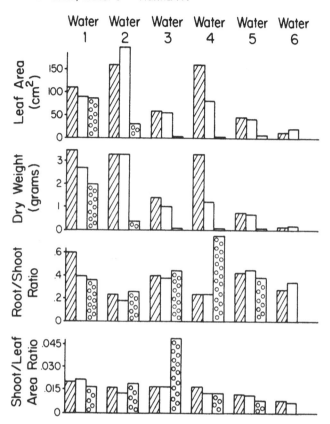

Table 1. Comparison between growth response and germination of five species of grass to Hanna IV produced water.

Response	Species
Growth	
Best	Basin wildrye
	Western wheatgrass
	Nuttall alkaligrass
	Bluebunch wheatgrass
Worst	Alkali sacaton
Germination	
Best	Alkali sacaton
	Bluebunch wheatgrass
	Nuttall alkaligrass
	Western wheatgrass
Worst	Basin wildrye

6. Germination success of seeds subjected to different produced waters may be misleading as an indicator for measuring toxic effects. Results indicate that a plant may have a successful germination percentage but will not respond well in terms of growth and survival.

7. Parameters measured separate responses of plant species to produced waters obtained from retorting fossil fuels. Plant species can be ranked for response. Species selection for maximizing potential reclamation success, using a measurement of phytotoxicity, is possible. In addition, a program now exists to screen produced waters against water-treatment procedures.

show similar trends. Species tolerance, however, may change from one water to another. Response of an individual grass species to six waters is illustrated in Fig. 2.

Comparisons between growth parameters and germination (Table 1) rank species according to response to one water. Again, similar trends exist when seeds and plants are subjected to other synfuel waters.

DISCUSSION

Results may be summarized as follows:

1. Plant species respond differently to any one water.
2. The order of plant response may change when plants are subjected to different waters produced from different technological processes for obtaining oil from shale.
3. The order of plant response may change when plants are subjected to different waters obtained from processing different fossil fuels.
4. Plant tolerance to in-situ produced water varies from species to species.
5. An increase in growth response may occur at the 1%-by-volume level but decreases toward 5%.

LITERATURE CITED

Beetle, A.A. 1970. Recommended plant names. Wyo. Agric. Expt. Sta., Res. J. 31.

Farrier, D.S., R.E. Paulson, Q.D. Skinner, J.C. Adams, and J.P. Bower. 1977. Acquisitions, processing and storage of environmental research of aqueous effluents derived from *in situ* oil shale processing. Proc. 2d Pacific Chem. Cong., Denver, Colo., 11:1031.

Farrier, D.S., J.E. Virgona, T.S. Phillips, and R.E. Paulson. 1978. Environmental research for *in situ* oil shale processing. 11th Oil Shale Symp., Colo. School of Mines Press, Golden, Colo., 81.

Fox, J.P., D.S. Farrier, and R.E. Paulson. 1978. Chemical characterization and analytical consideration for an *in situ* oil shale process water. Laramie Energy Technol. Cntr. Rpt., Invest. LETC/RI-78/7. Laramie, Wyo.

Hoagland, D.R., and D.I. Aron. 1950. The water culture method for growing plants without soil. Calif. Agric. Expt. Sta. Circ. 347, 32.

Miller. C.O. 1963. Kinetin and kinetin-like compounds. *In* K. Paech and M.V. Tracey (ed.), Moderne Methoden der Pflanzenanalyse, 6:194. Springer-Verlag, Berlin.

Skinner, Q.D., T.S. Moore, R.O. Asplund, J.C. Sexton, and D.S. Farrier. 1979. Plant responses to aqueous effluents derived from *in situ* fossil fuel processing. I. Development of screening methods. Laramie Energy Technol. Cntr. Rpt., Invest. LETC/RI-79/4. Laramie, Wyo.

Prediction of the Supply of Soil Nitrogen to Grass

D.C. WHITEHEAD
The Grassland Research Institute, Hurley,
Maidenhead, Berkshire, UK

Summary

Predictions of the supply of soil nitrogen (N) to grass over a growing season requires an assessment of nitrate N plus ammonium N (1) present in the soil at the beginning of the season and (2) likely to be released from the potentially mineralizable organic N during the season. Although a number of methods have been proposed for predicting soil N supply, and in particular the component arising from the potentially mineralizable N, many are too time consuming for routine use, and others have not been proven effective over a range of soils. However, in recent studies, the measurement of nitrate N plus ammonium N, extracted either (1) by boiling with 1M or 2M KCl or (2) by shaking with 0.05M Ba(OH)$_2$, was found to account for 80% of the variation in the yield of N in the herbage of perennial ryegrass (*Lolium perenne* L.) grown in pots on 21 soils under uniform environmental conditions. The prediction of soil N supply in the field, at over 18 sites in the UK, was substantially improved when the results of soil analysis were adjusted by a factor, calculated for each site, based on mean values of temperature and soil-water status.

KEY WORDS: nitrogen, soil, mineralization, perennial ryegrass, *Lolium perenne* L., analysis.

INTRODUCTION

The supply of soil nitrogen (N) available for uptake by plants over the course of a growing season can be regarded as arising from two sources: (1) nitrate (NO$_3$-N) plus ammonium (NH$_4$-N) initially present in the soil at the beginning of the season, and referred to hereafter as N$_i$, and (2) potentially mineralizable organic N, hereafter referred to as N$_m$, some of which will mineralize during the season with the release of NO$_3$-N and NH$_4$-N.

To be generally applicable, any method for predicting soil N supply must include the measurement of both N$_i$ and N$_m$ either separately or together. Under grass swards in humid temperate regions, the proportion of the total soil N supply that arises from N$_m$ is usually much greater than that from N$_i$, but the proportions will vary with soil type, weather, and sward management.

The analysis of a soil sample for N$_i$ is relatively simple. NH$_4$-N and NO$_3$-N can be extracted with a salt solution (e.g., 2M KCl), and both ions can be estimated by automated colorimetry. Analysis to estimate the supply of N from N$_m$ is more difficult, for two reasons. First, the organic N in soils consists of several fractions, including fresh plant residues, microbial biomass and residues, and humified organic matter. These fractions, although not entirely distinct from one another, differ in their susceptibility to mineralization, with humified organic matter being the least susceptible. N$_m$, which includes a proportion of each of these fractions, also cannot be precisely defined and will vary to some extent with the method of measurement. Second, in the field, only some of the N$_m$ will actually be mineralized, the proportion being influenced by several factors, especially temperature and water status.

It is generally considered that the most accurate methods for estimating N$_m$ involve incubating a soil sample under favorable and standard conditions for a period of weeks and measuring the NH$_4$-N and NO$_3$-N released, but such methods are too time-consuming for routine use. Although a number of chemical extraction methods have been proposed as alternatives to incubation, more evidence of their effectiveness in predicting the uptake of N by plants, particularly for grassland, is needed. Evidence is also needed on whether the effects of temperature and water status can be quantified in a way that enables the accuracy of the prediction of soil N supply to be improved.

EFFECTIVENESS OF CHEMICAL EXTRACTION METHODS

As indicated by Dahnke and Vasey (1973), very few chemical extraction methods have been adequately tested. Some have been compared only with the results of incubation methods, and others, even though tested against N uptake by plants, have been examined with only a small number of soils.

In a recent investigation, therefore, five chemical extraction methods (see Table 1) were examined and compared with the measurement of total soil N in terms of their ability to predict the supply of soil N to grass grown at 18 sites of a multicenter field trial in the UK (Whitehead et al. 1981). Predicted values of soil N supply were calculated from the results of each of the five extrac-

Table 1. Percentage of the variation in herbage N yield of ryegrass (from field plots at 18 sites in U.K.) accounted for by the results of six methods of soil analysis, with and without adjustment for soil temperature and water status.

Method of analysis	% variation accounted for	
	Without adjustment	With adjustment
1. NH_4-N extracted by autoclaving with 0.01M $CaCl_2$ (Stanford, 1977)	40	51
2. NH_4-N extracted by acid 0.25M $KMnO_4$ (Stanford and Smith, 1978)	24	26
3. 'Glucose' extracted by 0.05M Ba $(OH)_2$ (Jenkinson, 1968)	43	65
4. Non-NO_3-N extracted by 0.01M $NaHCO_3$ (MacLean, 1964)	20	36
5. UV absorbance at 260 nm of 0.01M $NaHCO_3$ extract (Fox and Piekielek, 1978)	16	33
6. Total soil N	23	34

Table 2. Percentage of the variation in herbage N yield, from perennial ryegrass grown in pots in 21 soils, accounted for by seven methods of soil analysis.

Method of analysis	% variation accounted for
1. (see Table 1)	63
3. (see Table 1)	79
6. Total soil N	52
7. $NO_3-N + NH_4-N$ extracted by boiling for 1 hour with 1M KCl	80
8. $NO_3-N + NH_4-N$ extracted by boiling for 1 hour with 2M KCl	80
9. $NO_3-N + NH_4-N$ extracted by shaking for 3 hours with 0.05M $Ba(OH)_2$	80
10. $NO_3-N + NH_4-N$ extracted by autoclaving for 8 hours with 0.01M $CaCl_2$	72

tion methods, taking into account the results of analysis for N_i as appropriate, and with and without adjustment for soil temperature and water status (see below). The effectiveness of each method was then assessed by calculating, using linear regression, the extent to which the soil analysis results accounted for variation among the sites in soil N supply. For each site, soil N supply was assumed to equal the amount of N, summated over six harvests, in the herbage of perennial ryegrass (*Lolium perenne* L.) from plots that had received no fertilizer N. This ranged over the sites from 14 to 136 kg N/ha/yr. Of the five extraction methods examined, three involved the measurement of a fraction of soil N, but two (nos. 3 and 5) depended on the extraction of a readily degradable fraction of the soil organic matter, measured either as glucose or by UV absorption. For either of these two methods to be effective, the amount of glucose extracted, or the UV absorption of the extract, must be closely related to N_m.

A summary of the results is shown in Table 1. The best prediction was given by the method based on glucose extracted by 0.05M Ba$(OH)_2$ (method 3) and the next best by the method based on NH_4-N extracted by autoclaving with 0.01M $CaCl_2$ (method 1). The other three extraction methods gave no better predictions than that based on total soil N. With the soils in this investigation, the effectiveness of prediction by methods 3 and 1 was not appreciably improved by the inclusion of values for the NO_3-N component of N_i (as extracted by 2M KCl), but the application of either method to soils that had received N from fertilizer or animal excreta would require the inclusion of this NO_3-N. Such inclusion would lengthen

both methods and reduce their suitability for routine use. Therefore, in a second phase of the investigation, attention was directed toward developing an extraction and analysis method that would enable N_i and N_m to be determined simultaneously (Whitehead 1981). With such a method, N_m must of course be estimated from a measurement of NO_3-N and/or NH_4-N, not, for example, of glucose. In addition to speed of operation, this method has the advantage that increases in N_i occurring between sampling and analysis are less likely to influence the value for predicted soil N supply. A number of possible extraction methods, including boiling with KCl (methods 7 and 8), extraction with Ba$(OH)_2$ (method 9), and autoclaving with $CaCl_2$ (method 10) were therefore examined for the estimation of N_i plus N_m in the forms of NH_4-H and NO_3-N. In this second phase of the investigation, perennial ryegrass was grown, in a controlled-environment cabinet, in pots of soil from 21 experimental sites in the UK, and values for predicted soil N supply based on the various methods were correlated with the amounts of N in the herbage summated over 6 harvests. A summary of the results is shown in Table 2. An important feature is that methods 7, 8, and 9, developed for speed of operation and suitability for routine use, were at least as effective as method 3, the best of the methods tested earlier.

ADJUSTMENT OF ANALYTICAL RESULTS FOR SOIL TEMPERATURE AND WATER STATUS

It has been known for many years that temperature and water status influence the rate of mineralization of the potentially mineralizable soil organic N. However, these factors had not been taken into account in the prediction of soil N supply until the development by Stanford's team in the U.S.A. of an adjustment factor that could be applied to the results of soil analysis (Stanford 1977, Stanford et al. 1977). In order to calculate the factor for a particular site, Stanford's team divided the growing season into periods of 4 weeks and calculated mean values for soil temperature and soil water status, expressed as a percentage of available water capacity. Then, on the basis of relationships derived from incubation studies, they calculated the percentage of N_m expected to be mineralized during each period. The adjustment was employed in studies aimed at predicting the uptake of N by sugar beets at a range of sites in Maryland, but no indication has been given of whether it improved the prediction (Stanford et al. 1977). The effect of such an adjustment was therefore examined in the investigation, briefly described above, in which five chemical extraction methods were evaluated in terms of their ability to predict the supply of soil N to grass (Whitehead et al. 1981). Soil temperature was measured, and information on rainfall, soil water-holding capacity, and potential evapo-transpiration enabled the water status for each site to be estimated, in a manner similar to that described by Stanford et al. (1977), for successive periods of 4 weeks. For all methods, adjustment of the results considerably improved the prediction (Table 1). In this calculation, the adjustment factors were obtained retrospectively, but, for advisory purposes, it would be necessary to use average meteorological data. For 16 of the sites, values of the average monthly rainfall over a 60-year period and the monthly air temperatures over a 30-year period were available, and the effect of calculating an adjustment factor (F) based on these average values was therefore examined:

$$F = \frac{t \times r \times d}{200}$$

where t = mean air temperature in °C (April-September inclusive), r = rainfall in mm (April-September inclusive), and d = rooting depth in m with a maximum of 1.0 m. Inevitably, F was somewhat less effective than the adjustment factor based on data for the particular year, but the difference was small. With the best method (3), 62% instead of 65% of the variation in herbage N yield was accounted for.

CONCLUSION

Although further testing and calibration against the uptake of N in the field are required, the results presented here indicate that the following methods will provide a good prediction of soil N supply: (1) NO_3-N plus NH_4-N extracted by boiling for 1 hour with 1M or 2M KCl; or (2) NO_3-N plus NH_4-N extracted by shaking for 3 hours with 0.05M $Ba(OH)_2$. With both methods, prediction over a range of sites with different weather and soil conditions is likely to be improved by adjustment of the results for temperature, rainfall, soil depth, and possibly texture.

LITERATURE CITED

Dahnke, W.C., and E.H. Vasey. 1973. Testing soils for nitrogen. *In* L.M. Walsh and J.D. Beaton (eds.), Soil testing and plant analysis, 2d rev. ed., 97–114. Soil Sci. Soc. Am., Madison, Wis.

Fox, R.H., and W.P. Piekielek. 1978. A rapid method for estimating the nitrogen-supplying capability of a soil. Soil Sci. Soc. Am. J. 42:751–753.

Jenkinson, D.S. 1968. Chemical tests of potentially available nitrogen in soil. J. Sci. Food Agric. 19:160–168.

MacLean, A.A. 1964. Measurement of nitrogen supplying power of soils by extraction with sodium bicarbonate. Nature 203:1307–1308.

Stanford, G. 1977. Evaluating the nitrogen-supplying capacities of soils. Proc. Int. Sem. on Soil Envir. and Fertility Mangt. in Intens. Agric., Tokyo, 412–418

Stanford, G., J.N. Carter, D.T. Westerman, and J.J. Meisinger. 1977. Residual nitrate and mineralizable soil nitrogen in relation to nitrogen uptake by irrigated sugarbeets. Agron. J. 69:303–308.

Stanford, G., and S.J. Smith. 1978. Oxidative release of potentially mineralizable soil nitrogen by acid permanganate extraction. Soil Sci. 126:210–218.

Whitehead, D.C. 1981. An improved chemical extraction method for predicting the supply of available soil nitrogen. J. Sci. Food Agric. 32:359–365.

Whitehead, D.C., R.J. Barnes, and J. Morrison. 1981. An investigation of analytical procedures for predicting soil nitrogen supply to grass in the field. J. Sci. Food Agric. 32:211–218.

Soil Nutrient Constraints for Legume-Based Pastures in the Brazilian Cerrados

W. COUTO and C. SANZONOWICS

CIAT and EMBRAPA, Centro de Pesquisa Agropecuaria dos Cerrados, Brazil

Summary

Soil-nutrient deficiencies in two cerrados soils have been investigated as a preliminary to small-plot research oriented to estimate fertilizer requirements for establishment of legumes in grass-legume associations. The studies were conducted in the greenhouse, in 2-kg pots, with surface-soil samples taken from two sites representative of the dominant soils in the region. These are the dark red latosol (DRL) and the red-yellow latosol (RYL), which qualify as Haplustox and Acrustox, respectively. Two kinds of experiments were carried out. One consisted of a 1/4 replicate of a 2^8 factorial including potassium (K), calcium (Ca), magnesium (Mg), sulfur (S), zinc (Zn), molybdenum (Mo), copper (Cu), manganese (Mn), and boron (B), in which Cu and Mn were applied together as a single factor. Levels of these nutrients were 40, 100, 10, 15, 1, 0.13, 1, 1, and 0.5 parts/million parts of soil (ppm), respectively. All pots received monocalcium phosphate $[Ca(H_2PO_4)_2 \cdot 2H_2O]$ at a rate of 50 P ppm. A second experiment consisted of four levels of P (25, 50, 100, and 200 ppm) and four levels of Ca (0, 20, 100, and 200 ppm) as $CaCO_3$, with basal application of all other nutrients and rates used in the previous experiment. Test plants used were calopo (*Calopogonium mucunoides* Desv.), a *Centrosema* (Benth.) species CIAT 438, stylo (*Stylosanthes guianensis* Aubl.) CIAT 2243, and gambagrass (*Andropogon gayanus* var. bisquamulatus [Hochst.] Hack) CIAT 621.

Results showed a clear response to P up to 100 P ppm in both soils, when gambagrass or stylo was used, and up to 200 P ppm when the *Centrosema* sp. or calopo was used. Responses to the highest levels were more evident in the RYL soil. In all species, Ca improved plant responses to low levels of P. Maximum yields were also increased except in the case of stylo.

In the second experiment, it was observed that S was deficient in both soils for all test plants, as measured by plant responses to applied S. Potassium responses were also observed in both soils, while Ca response was more common in the RYL soil. Magnesium response was observed only in the RYL soil when gambagrass was used as a test plant. Molybdenum and Zn deficiencies were observed in both soils with some legumes.

The results demonstrate the importance of S, Ca, Mg, K, Mo, and Zn deficiencies after P deficiency has been corrected. On the basis of these observations, field experiments have been established to define the required amounts of S and K and to confirm under field conditions the need for Ca, Mg, Mo, and Zn.

KEY WORDS: soils, sulfur, micronutrients, lime, phosphorus, legumes, *Andropogon gayanus* Kunth.

INTRODUCTION

The cerrados of Brazil, which occupy about 200 million ha, are characterized by acid soils of low fertility with high phosphorus (P)–fixing capacity. Intensive agriculture occurs in the region, but most of the area is covered by grasses and shrubs or trees in varying proportions, and extensive beef-cattle grazing is the most important activity in this part. The 6-month dry season severely limits cattle performance because of the poor feeding value of native pasture. Over many years, large areas have been improved by incorporating more productive grasses. Grass-legume associations, however, have not persisted under grazing in most cases. Several factors have been mentioned as being responsible for the lack of persistence or poor establishment of legumes, among them soil-nutrient deficiencies.

Previous findings have shown important soil-nutrient deficiencies in some cerrados soils. McClung et al. (1961) and Jones and Quagliato (1970) demonstrated sulfur (S) deficiencies, while Jones et al. (1970) found plant response to S, calcium (Ca), magnesium (Mg), and micronutrients. De Franca and de Carvalho (1970) also reported a marked response to micronutrients. However, most previous findings were obtained in soils that are not found within the area of interest of the pasture program, or they have limited application to low-fertility demands or high-aluminum-tolerant pastures because of one or more of the following experimental conditions: (1) soil pH had been previously corrected by lime application; (2) there was no separate use of single micronutrients; (3) test plants used were not tropical legumes adapted to acid soils; and (4) no nutrient interactions were studied.

The purpose of this paper is to investigate possible soil-nutrient deficiencies in two cerrados soils. Results are from greenhouse studies conducted to provide preliminary information for field work.

MATERIALS AND METHODS

Surface soil (0–20 cm) samples were taken from two sites representative of two soils in the Brazilian cerrados. They classify as dark-red latosol (DRL) and red-yellow latosol (RYL), which belong to the Haplustox and Acrustox great groups, respectively, according to U.S. Soil Taxonomy (Soil Survey Staff 1975).

Two kinds of experiments were conducted with both soils. An exploratory one consisted of a 1/4 replicate of a 2^8 factorial in two blocks in which potasssium (K), Ca, Mg, S, zinc (Zn), molybdenum (Mo), copper (Cu), manganese (Mn), and boron (B) were either present or absent. Where these elements were applied, they were applied at rates of 40, 100, 10, 15, 1, 0.13, 1, 1, and 0.5 parts/million parts of soil (ppm), respectively. Sources were potassium chloride (KCl), calcium carbonate $(CaCO_3)$, magnesium chloride $(MgCl_2 \cdot 6H_2O)$, sulfuric acid (H_2SO_4), zinc chloride $(ZnCl_2)$, ammonium molybdate $[(NH_4)_6Mo_7O_{24} \cdot 4H_2O]$, copper chloride $(CuCl_2 \cdot 2H_2O)$, manganese chloride $(MnCl_2 \cdot 4H_2O)$, and sodium borate $(Na_2B_4O_7 \cdot 10H_2O)$, all of analytical reagent grade. A basal application of P as $Ca(H_2PO_4)_2 \cdot 2H_2O$ was applied to all plots at a rate of 50 P ppm. A second experiment conducted under the same conditions consisted of four levels of P (25, 50, 100, and 200 P ppm) and four levels of Ca (0, 20, 100, and 200 ppm). Eleven levels of P and Ca were arranged factorially in a randomized complete-block design with 2 replicates. Sources of Ca and P were the same as in the previous experiment. All other nutrients were applied to all pots at

W. Couto is a Soil Scientist in the Tropical Pastures Program, CIAT, and C. Sanzonowics is a Research Agronomist at EMBRAPA-CPAC.

Table 1.   Means (n = 32) of dry-matter yield (g/pot) in the presence (+) or absence (−) of nine nutrients, on two soils.

Nutrients[1]		DRL Soil				RYL Soil			
		Calopo	Centro-sema	Stylo	Gamba grass	Calopo	Centro-sema	Stylo	Gamba grass
K	−	(B)[1]	N.S.[2]	(S)	7.0	3.1	(S)	N.S.	N.S.
	+				9.0	3.6		13.6	7.0
Ca	−	N.S.	N.S.	(S,Mo)	N.S.	2.3	N.S.	13.6	7.0
	+					4.3		15.2	8.1
Mg	−	(Zn)	N.S.	N.S.	N.S.	N.S.	(S)	N.S.	(S)
	+								
S	−	5.4	6.6	11.2	6.8	3.1	2.6	12.0	6.2
	+	8.2	9.0	18.4	9.3	3.6	5.1	17.0	8.9
Mo	−	(S)	N.S.	(S,Ca)	N.S.	N.S.	(S)	N.S.	N.S.
	+								
Zn	−	6.6	N.S.	N.S.	N.S.	N.S.	N.S.	13.7	N.S.
	+	7.0						15.1	
Mn and Cu	−	(−Mg)	N.S.	N.S.	N.S.	N.S.	(−K)	N.S.	N.S.
	+								
B	−	(K)	N.S.	15.1	N.S.	N.S.	N.S.	N.S.	N.S.
	+			14.6					

[1]Significant response only in the presence or the absence (−) of the nutrient indicated within parenthesis.

[2]N.S. = no significant difference between levels of nutrients, according to the F test ($P \leqslant 0.05$).

the same levels as in the previous experiment. Water was then applied to the weight required to reach a water content approximately equivalent to 0.3 bars of water tension. De-ionized distilled water was used.

Each experiment was conducted separately with four test plants: calopo (*Calopogonium mucunoides* Desv.); a species of *Centrosema* (Benth.), CIAT 438; stylo (*Stylosanthes guianensis* [Aubl.]) CIAT 2243, and gambagrass (*Andropogon gayanus* var. bisquamulatus [Hochst.] Hack.) CIAT 621. Gambagrass was sown at a rate of 10 to 15 seeds/pot, from which four healthy plants were selected. All other species were sown at a rate of 4 pregerminated seeds/pot, previously inoculated with *Rhizobium* strains of known efficiency. Plants were grown until approximately

10 g of dry matter/pot were obtained or until the first signs of flowering. Plants were cut at about 5 cm aboveground, oven-dried at 60°C for 72 hours, and weighed. All plants were cut once, except gambagrass, which was cut twice.

RESULTS AND DISCUSSION

Table 1 summarizes results obtained from fractional replication experiments on both soils. Responses to S were observed with all test plants in both soils. In some cases, dry-matter production was doubled with S application, showing the low S-supplying capacity of these soils. Plant responses to Ca were most marked in the RYL soil

Table 2.   Dry-matter yields (g/pot) in the presence (+) or absence (−) of K, Ca, Mg, and Mo, as affected by S level.

Soil and test plant	K		Ca		Mg		Mo	
	−	+	−	+	−	+	−	+
DRL soil								
Calopo								
− S	N.S.[1]		N.S.		N.S.		5.4	5.4
+ S							7.7	8.7
Stylo								
− S	11.2	11.3	11.5	11.0	N.S.		11.2	11.3
+ S	17.9	18.9	17.8	19.0			17.8	18.9
RYL soil								
Centrosema								
− S	2.4	2.7	N.S.		2.7	2.4	2.6	2.5
+ S	5.9	4.3			4.8	5.4	4.6	5.5
Gamba grass								
− S	N.S.		N.S.		6.3	6.1	N.S.	
+ S					7.7	10.1		

[1]N.S. = No significant interaction among factors, according to F test ($P \leqslant 0.05$).

L.S.D. = Least significant difference between means (n = 16; $P \leqslant 0.05$) equals 0.5, 0.5, 0.6, and 1.0 for Calopo, stylo, centrosema, and Gamba grass, respectively.

with several test plants, and K responses were observed in both soils with two of the test plants. Magnesium responses were observed only with gambagrass in the RYL soil. Interactions of S with K, Ca, Mg, and Mo were also observed in both soils. Table 2 shows mean yields of two levels of these factors for each level of sulfur.

In relation to micronutrients, plant responses to Mo and Zn were observed in both soils with some legume plants (Table 1). No main effects of Cu and Mn were observed, but there were interactions with Mg and K.

In the second experiment, calopo and *Centrosema* sp. responded to P over the whole range of P levels. Gambagrass and stylo, on the other hand, showed most of their response to P levels up to 100 P ppm. There was also a clear response to Ca applied at levels of 100 or 200 ppm Ca. Response to Ca was more apparent in the RYL soil with all species, showing the severe Ca deficiency of this soil. In most cases, Ca applied at levels of 100 or 200 ppm improved not only dry-matter production at lower levels of P but plant response/unit of P applied. Soil analysis performed after harvesting showed that pH increased

from 4.2 to 4.5 in the DRL soil and from 4.6 to 4.8 in the RYL soil with the highest rates of applied $CaCO_3$.

LITERATURE CITED

de Franca, G.E., and M.M. de Carvalho. 1970. Exploratory trial on fertilization with five tropical legumes on a cerrado soil. Pesq. Agropec. Braz. 5:147–153.

Jones, M.B., and J.L. Quagliato. 1970. Response to sulfur levels by four tropical legumes and alfalfa. Pesq. Agropec. Braz. 5:355–363.

Jones, M.B., J. Quagliato, and L.M.M. de Freitas. 1970. Response of alfalfa and tropical legumes to applied mineral nutrients in three campo cerrado soils. Pesq. Agropec. Braz. 5:209–214.

McClung, A.C., L.M.M. de Freitas, D.S. Mikkelsen, and W.L. Lott. 1961. Cotton fertilization on campo cerrado soils of State of Sao Paulo, Brazil. IRI Res. Inst., Inc. Bul. 27.

Soil Survey Staff. 1975. Soil taxonomy. A basic system of soil classification for making and interpreting soil surveys. U.S. Dept. Agric., SCS, Agric. Handb. 436.

Pasture Production and Changes in Soil Fertility on a Long-Term Irrigated Superphosphate Trial at Winchmore, New Zealand

B.F. QUIN and D.S. RICKARD
Ministry of Agriculture and Fisheries, N.Z.

Summary

The Winchmore trial was commenced in 1952 to investigate the superphosphate requirements of irrigated ryegrass–white clover pasture under grazing. Treatments receiving 0, 188, and 376 kg/ha annually of single superphosphate have produced average annual yields of 4.3, 10.0, and 11.2 metric tons (t)/ha dry matter, respectively. The control production is being maintained by continuing weathering of soil P and by the input of S in irrigation water and rainfall.

Topsoil organic C, N, and S reached plateaus on all treatments after 10–15 years. Organic P (P_o) showed a linear increase on all treatments until the trial was limed (4 t/ha) in 1972, which increased the pH from 5.8 to 6.6. Following liming, accumulation of P_o ceased. There was a subsequent positive change in P_i and Truog P but not in either Olsen P or in increased plant uptake. This finding can be explained partly by the short-term precipitation of phosphate on lime particles and partly by a decline in the quality of superphosphate used since the early 1970s.

The very large accumulation of P_o prior to liming above pH 6.0 has led to investigations to determine: (1) whether this accumulation is general to New Zealand's climate; (2) the mechanisms, and forms of P_o, responsible for the accumulation, (3) the extent to which the accumulation can be reduced by various means, including liming; and (4) the consequent savings in P fertilizer.

KEY WORDS: superphosphate requirements, irrigated pasture, organic phosphorus, phosphorus cycling.

INTRODUCTION

Winchmore is located on the Canterbury Plains on a shallow (300–450 mm), moderately weathered soil derived from accumulation of greywacke alluvium and loess.

Average rainfall is 750 mm, which in the long term is evenly distributed throughout the year. Mean monthly temperatures range from 5°C in July to 16°C in January.

EXPERIMENTAL DESIGN AND MANAGEMENT

The area was ploughed out of low-fertility browntop (*Agrostis tennuis*) in 1948, prepared for border-strip irrigation, and sown to permanent ryegrass–white clover pasture in 1950. Between 1948 and the commencement of the trial, a total of 4 t/ha lime and 420 kg/ha superphosphate was applied.

The trial had 4 replicates of five treatments. Each replicate was a separately fenced border strip of 0.07 ha. The superphosphate treatments were: (1) 0 control; (2) 188 kg/ha/yr (188 p.a.); (3) 376 kg/ha in 1952–1957, none since (376 residual); (4) 564 kg/ha in 1952–1957, none since (564 residual); and (5) 376 kg/ha/yr (376 p.a.). Since 1965 the total P content of the superphosphate has declined from 9.5% to 8.0% and the citric acid solubility from 90% of the total to 70%. The remaining 30% is largely unreacted rock phosphate. The superphosphate applications have not been adjusted for the decline in quality.

Lime was applied at a rate of 4 t/ha to the whole trial area in 1972. No fertilizers other than superphosphate have been applied.

Each treatment was grazed by a separate flock of dry ewes that rotated around the 4 replicates of that treatment only. The trial was irrigated as required—an average of five irrigations/yr. Pasture production was measured by an enclosure technique.

ANALYTICAL METHODS

Soil samples (0–75 mm) were analyzed for Truog P and Olsen P (which correlate closely with soil calcium P and aluminum P, respectively), total inorganic P (P_i), SO_4-S, and organic carbon (C), nitrogen (N), phosphorus (P_o), and sulfur (S_o). Plant samples were analyzed for total P and S. Details of all these methods may be obtained from the authors.

RESULTS

Pasture Production and Botanical Composition

Production of the control, averaging 4.3 t/ha dry matter, has been supported by the weathering of soil P (3–5

kg P/ha/yr) and the input of S in rainfall and irrigation water (3 and 6 kg S/ha/yr). Long-term relative yields have been used to obtain a Mitscherlich equation of yield vs. fertilizer input (Table 1).

Since cessation of topdressing in 1957, production of the 564 and 376 residual treatments has declined, but they are still producing about 50% more than the control. Plot trials superimposed on these treatments (unpublished) have shown that this decline was due initially to S deficiency and ultimately to P deficiency as well.

In 1952, the pasture was predominantly perennial ryegrass (*Lolium* spp.) and white clover (*Trifolium repens*)—60% and 24%, respectively. Minor components were red clover (*T. pratense*), cocksfoot (*Dactylis glomerata*), timothy (*Phleum pratense*), dogstail (*Cynosurus cristatus*), and browntop. Pastures of the treatments receiving superphosphate remained predominantly ryegrass and white clover. The percentage of clover varied between 20% and 40%, but there were no long-term trends. The percentage of weeds and weed grasses in the control increased to 60% of the sward by about 1966 and remained at that level. This change was accompanied by a decline in the proportion of both clover and higher-fertility grasses. Similar trends occured after superphosphate applications were stopped on the two residual treatments.

Changes in Soil Organic Matter

Organic C increased for about 10 years on all treatments before reaching a plateau in the early 1960s; N and S_o tended to increase for another 5 years until the late 1960s.

Unlike other organic constituents, P_o continued to increase linearly on all treatments (Fig. 1 and Table 2), even where P_i was declining (Table 2). Studies of N.Z. pastures of varying ages (Jackson 1964) had provided some evidence for continued accumulation of P_o after C, N, and S_o have reached plateaus.

The linear increase in P_o ceased after the trial was heavily limed (4 t/ha) in 1972, when the pH had declined to 5.8. The effect of liming on P_o accumulation has not previously been examined in New Zealand.

Changes in Inorganic Phosphorus Parameters

Control, 376 and 564 residual treatments. These treatments had shown a steady decline in P_i, Truog P, and Olsen P prior to liming (Table 2), reflecting the continuous conversion of soil P_i and residual fertilizer P to P_o. Following liming, P_i increased considerably. This increase was reflected by an increase in Truog P but not in Olsen P (Table 2).

Treatments 188 and 376 p.a. Prior to liming, the 188 p.a. treatment was just maintaining P_i, Truog P, and Olsen P (Table 2). The 376 p.a. treatment had shown a steady increase in all three parameters (Table 2). Since liming, P_i has increased at a faster rate, reflected in Truog P but not in Olsen P.

The authors are Scientists in Soil Fertility at the Winchmore Irrigation Research Station.

Table 1. Average pasture production, uptake and soil test data.

Superphosphate and(P)[a] applied kg/ha/yr	Dry-matter prodn, t/ha/yr 1952-80	Uptake kg/ha/yr 1968-75		Soil test µg/g		
		P	S	Olsen P	Truog P	SO$_4$-S
Control(0)	4.3(38)	9	8	7	8	2
188 p.a.(17)	10.0(88)	32	35	11	20	7
376 p.a.(34)	11.2(98)	48	44	15[b]	28[b]	10
564 res.(51)[c]	11.4(100)[c]	18	14	20[b]	32[b]	—

Mitscherlich relative yield: RY = 100 $\left[1 - 10^{-(4.8 + 0.037 \text{ applied P})} \right]$

[a]Not corrected for decline in quality of superphosphate.

[b]At time when maximum production was first attained on these treatments.

[c]Prior to cessation of topdressing (i.e., 1952-1958 only).

Uptake of Phosphorus and Sulfur by Herbage

The pasture uptake data (Table 1) showed large increases with increasing fertilizer application. The difference between treatments 188 and 376 p.a. was mainly a reflection of a difference in concentration of these nutrients. The 1972 liming had no consistent effects on P or S uptake by pasture.

Cycling and Losses of Phosphorus and Sulfur

Data for annual uptake of P and S by pasture (Table 1); inorganic and organic P and S in litter, roots and feces; and changes in soil P$_i$ and P$_o$ (to be fully reported elsewhere) have been used to construct preliminary estimates of annual P and S cycles with an input of 200 kg/ha of good-quality superphosphate, which is calculated to be sufficient to maintain labile soil P and Olsen P levels and to give 90% of maximum pasture production (Fig. 2).

The figures in the annual P cycle represent the situation before the 1972 liming. Periodic liming to control P$_o$ ac-

cumulation could reduce annual P requirements by up to 10 kg P/ha. The annual S cycle represents the situation after the attainment of plateaus for soil S$_o$. Excess SO$_4$-S is leached from the soil.

DISCUSSION

Pasture production reached typical levels for the respective treatments soon after the start of the trial (unpublished), despite the large changes taking place in soil

Table 2. Average annual changes in soil pH units and P parameters (µg/g soil, 0-75 mm depth).

Parameter	Value in 1952	Treatment	Preliming[1] (1952-1972)	Postliming (1972-1979)
pH	6.2	all	-0.01*	-0.09*[2]
Organic P	330	1[3]	7*	-1
		2	10*	-2
		3	10*	-8
		4	8*	-3
		5	9*	-3
Inorganic P	267	1	-2*	7*
		2	0.6*	10*
		3	6*	19*
		4	5*	7*
		5	8*	7*
Truog P	16	1	-0.4*	1.2*
		2	0	2.4*
		3	1.0*	3.8*
		4	-0.6	1.4
		5	-2.4	1.0
Olsen P	10	1	-0.1*	0
		2	0	0
		3	0.7*	0
		4	-0.4*	0
		5	-0.7*	0

[1]Preliming changes on original treatments are for 1958-1972.

[2]pH at start of postliming period was 6.6.

[3]Superphosphate treatments were (1) nil, control; (2) 188 kg/ha/year; (3) 376 kg/ha/year; (4) 376 kg/ha/year 1952-1957, none since; (5) 564 kg/ha/year 1952-1957, none since.

* = P < 0.05.

Fig. 1. Changes in soil organic C, N, P$_o$, and S$_o$ for the control and the average of the 188 and 376 p.a. treatments. The two residual treatments followed the 188 and 376 p.a. curves.

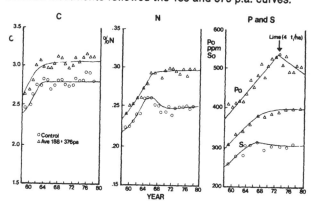

Fig. 2. Preliminary estimates of annual P and S cycles for an annual superphosphate application rate calculated to maintain soil P_i and Olsen P levels against losses and to give a pasture production level within 90% of the maximum (i.e., 200 kg/ha). The figures in the P cycle represent the situation in kg P/ha/yr before the 1972 liming. The annual S cycle, in kg S/ha/yr, is based on the S input in the equivalent superphosphate application, and represents the situation after the attainment of a plateau for soil S_o.

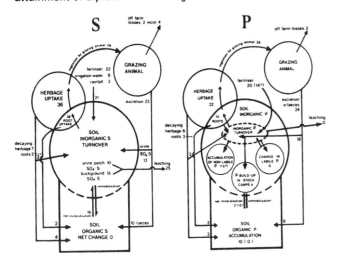

organic matter. This fact suggests that most of the recycling of N, S, and P_o necessary for plant growth occurs relatively soon after incorporation in the soil as animal excreta and plant residues. The more resistant organic compounds will continue to accumulate until the soil biological population adjusts to the new levels of plant production on the various fertilizer treatments. In the case of P_o, this adjustment occurred only after the trial was heavily limed, increasing the pH from 5.8 to 6.6. This liming had a long-term effect on P_o (Fig. 1), probably by increasing the solubility and therefore susceptibility to mineralization of the large inositol phosphate component of P_o.

The rapid deterioration in sward quality on the control and the 376 and 564 residual treatments was not reflected in a decline in soil organic components. In these treatments, nutrient cycling in the soil must therefore become a slower, more closed cycle, with less inorganic N, S, or P being available at any time for plant uptake. The recommencement of fertilizer applications on these treatments, as demonstrated in superimposed plot trials (unpublished), produced a rapid recovery in pasture growth by stimulating clover production.

The increased net mineralization of P_o following liming should have been reflected in higher concentrations of the plant-available component of P_i, such as indicated by Olsen P, particularly on the 188 and 376 p.a. treatments. Any increase appears to have been prevented by short-term precipitation of P_i on lime particles and, more importantly, by a decline in total P and citric acid soluble content of the superphosphate since the late 1960s. For example, the amount of citric acid soluble P in the 188 p.a. treatment declined from 17 kg/ha to 11 kg/ha between about 1965 and 1975. The increase in Truog P reflects the accumulation of unreacted rock P.

The very large accumulation of P_o on this trial prior to liming above pH 6.0 has led to further investigations to determine (1) whether this accumulation is general in New Zealand's temperate climate; (2) the mechanisms, and forms of P_o, responsible for the accumulation; (3) the extent to which this accumulation can be reduced by various means, including liming; and (4) the consequent savings in P fertilizer requirements.

LITERATURE CITED

Jackson, R.H. 1964. Accumulation of organic matter in some New Zealand soils under permanent pasture. I. Patterns of change in organic carbon, nitrogen, sulphur and phosphorus. N.Z. J. Agric. Res. 7:445–471.

Interactions Between Grasses and Rhizosphere Nitrogen-Fixing Bacteria

M.H. GASKINS, D.H. HUBBELL, and S.L. ALBRECHT

USDA-SEA-AR and University of Florida, Gainesville, Fla., U.S.A.

Summary

Sorghum and *Digitaria* cultivars grown in soil or nutrient solutions were inoculated with a strain of *Azospirillum* in short-term greenhouse experiments. Carbon compounds lost by roots in solution culture experiments were determined in order to estimate the quantity of energy substrates that might be obtained by rhizosphere organisms from growing plants. Other plants were treated with plant growth hormones, and responses were compared to those of plants treated with killed *Azospirillum* cells. The similarity of the responses suggests that growth-hormone production by the inoculant organism, rather than nitrogen (N) fixation, is responsible for part or all of the increase in plant growth rate.

The estimated quantity of reduced carbon compounds (lysate) lost by roots to the nutrient medium would support substantial N-fixation activity if all were used for that purpose, but that clearly is not the case. The available evidence indicates instead that relatively little may be so used. Thus, the recognized low N-fixation efficiency of free-living heterotrophs and the demonstrated response of plants to bacterial growth hormones suggest that benefits from inoculation, even though they may be highly important in some circumstances, may not always involve N fixation.

KEY WORDS: plant growth hormones, nitrogen fixation, soil bacteria, root lysate, rhizosphere, grasses.

INTRODUCTION

Accurate measurement of nitrogen (N)–fixation rates achieved by root-associated bacteria has proven difficult. Some reports suggest that rapid fixation, comparable to that in legume root nodules, may take place in rhizospheres of cultivated plants, but most investigators doubt that this comparison should be made. Some of the questions involved have been reviewed recently (van Berkum and Bohlool 1980).

Plant growth, nevertheless, increases dramatically in some circumstances following inoculation with nonsymbiotic bacteria. Particularly interesting results were reported recently from Israel (Cohen et al. 1980). Clearly, when substantial yield increases are achieved by a simple and inexpensive treatment, as reported recently by the Israeli investigators and earlier by others, continued investigation is appropriate to understand the biological phenomena involved. It seems quite reasonable that some

N fixation occurs, as is clearly indicated by experiments with [15]N (Ruschel et al. 1978, Rennie 1979). However, the assumption that favorable responses depend entirely on assimilation by the plant of N fixed by the bacteria seems tenuous, particularly in view of the rapid treatment response that often occurs.

Brown (1974) and Barea and Brown (1974) indicate that bacterial production of plant growth-promoting hormones explained the responses of plants inoculated with *Azotobacter* spp. We have presented data to indicate that this also was the case when a sorghum cultivar was inoculated with a strain of *Azospirillum* (Gaskins and Hubbell 1979). Increased plant growth and tuber yield following treatment of potato seedpieces with a *Pseudomonas* strain led to the conclusion that the favorable effect resulted from suppression of growth by the inoculant organism of other rhizosphere organisms harmful to the plants (Burr et al. 1978). Many other interactions not yet identified might be involved separately or in combinations when favorable plant reactions occur (Hubbell and Gaskins 1980).

This paper reports some further experiments in a series conducted to develop more information about interactions between plants and beneficial root-associated bacteria that result in improved plant growth.

M.H. Gaskins is a Plant Physiologist of USDA-SEA-AR and a Professor of Agronomy at the University of Florida; D.H. Hubbell is a Professor of Soil Science at the University of Florida; and S.L. Albrecht is a Plant Physiologist of USDA-SEA-AR and an Assistant Professor of Agronomy at the University of Florida.

METHODS

Plant Inoculation

Plants of *Digitaria decumbens* cv. Pangola and Transvala were grown on greenhouse benches in pots containing 3.0 l of sandy-loam soil. Bacterial suspensions for inoculation were grown, washed, and applied as described previously (Gaskins and Hubbell 1979). The *Digitaria* plants were clipped to 6 cm above the soil line, and 20 ml of the bacterial suspension (10^8 cells/ml) were added. Water to fully saturate the root system was then added. The *Sorghum bicolor* plants were inoculated by the same procedure 7 days after shoot emergence. The inoculation procedures were carried out three times at 7-day intervals. The *Digitaria* plants were clipped three times in a period of 65 days before the experiment was ended. The sorghum plants were grown for 40 days, and tops were then removed and weighed. Acetylene reduction rates were not determined because in earlier experiments they had not correlated with bacterial treatments.

Root Lysate Studies

Sorghum plants were grown gnotobiotically in 50 × 300–mm glass tubes fitted with stainless steel supports on which seeds were held above the nutrient solutions. The roots grew into 50 ml of ¼-strength Hoagland's mineral nutrient solution. After 3 weeks of growth, plants were harvested and weighed and solutions were lyophilized for later analysis. Carbohydrates were determined by the phenosulfuric acid method, amino acids by the ninhydrin method, and organic acids by titration with $0.1N$ NaOH. For the purpose of this discussion the resulting data were combined to provide an estimate of total reduced-carbon compounds that might be used as energy sources by rhizosphere bacteria. The term "root lysate" as proposed by Martin (1976) is used without distinguishing between plant metabolites excreted by functional cells and those lost to the solution by rupture of senescent root hairs or other cells.

Effects of Growth Hormones

In an additional experiment, sorghum seedlings were grown in small pots containing 0.5 kg of sand and were irrigated daily with ¼-strength Hoagland's solution. Plants were treated with auxin, kinetin, gibberellic acid, or combinations of these growth hormones and with killed *Azospirillum* cells. Effects of the treatments were determined by comparing weights of plants at harvest, 20 days after the seeds were started.

RESULTS AND DISCUSSION

Plant Inoculation

The *Digitaria* plants showed high sensitivity to inoculation treatments in this experiment. Dry-weight yields

Table 1. Growth of *Digitaria* and sorghum inoculated with an *Azospirillum* culture.

| | Dry weights of tops at harvest[1] | | |
| | Digitaria | | Sorghum, Funk's G522 (g) |
Treatment	Transvala (g)	Pangola (g)	
Inoculated	54.9*	41.6	17.7*
Control	28.8	38.0	22.5

[1]Yields are combined weights of 3 clippings from a 65-day period for the *Digitaria* and from a single harvest at 40 days for the sorghum.

*Means marked by asterisks differ significantly from the appropriate control (P = .05).

were substantially increased by the bacteria (Table 1). The reverse was true, however, with the sorghum seedlings. Their final weights were reduced more than 20% by the treatments with live bacteria. This finding contrasts with results from an earlier experiment (Gaskins and Hubbell 1979) in which the same cultivar (Funk's G522) responded positively to *Azospirillum* inoculation. The *Digitaria* plants, which must be propagated vegetatively, had been established 4 months prior to inoculation. The plants were root-bound, and their condition was similar to that of a sod in the field. The sorghum plants, on the other hand, newly germinated from seeds, were growing rapidly as juvenile plants under nutrient-limited but otherwise favorable conditions. The difference in conditions of growth, rather than the genetic difference between the experimental plants, is suggested as the reason for opposite responses to the bacterial treatments.

Root Lysate Studies

The combined weights of sugars, organic acids, and amino acids collected from root solutions of 22-day-old plants were between 3.2 and 3.6 mg/plant, and plant weights were 175–200 mg. These quantities are small but not insignificant, relative to the growth requirements of microorganisms. As pointed out by Barber and Martin (1976) and substantiated by us (Gaskins, unpublished), plant growth is slower under gnotobiotic conditions than otherwise. However, for a period of about 3 weeks plants under these conditions can sustain rates of dry-matter accumulation 70%–80% of those achieved by normally grown plants. To compensate for restricted growth in the closed containers, we have doubled the figures above to estimate performance of normally grown plants.

Total root weight later in the life of sorghum plants has been found to be 80- to 100-fold greater than the weight at 3 weeks (Gaskins, unpublished). If loss of carbon compounds to the rhizosphere is proportional to weight, an additional correction factor of 100 might be appropriate to estimate the potential of a typical root system to support bacterial growth. Assuming 80% of the carbon materials identified were produced in the final (third) week of seedling growth, and applying the adjustments described above, a weekly production rate by a well-nourished plant

Table 2. Effect of growth hormones and *Azospirillum* cells on weight gain of sorghum seedlings.[1]

| | Dry-weight yield after 20 days | | |
Treatment	Tops (mg)	Roots (mg)	Total (mg)
Gibberellic acid 10^{-3}M	437	217	654
Kinetin 10^{-4}M	515	279*	794*
Azospirillum	539	208*	747*
Control	432	136	568

[1]Plants were grown 20 days in washed sand, irrigated with 1/4-strength Hoagland's solution. Treatments were applied twice weekly, 5 times. Bacteria were prepared as described in the text, and killed by heating before application.

*Means marked by asterisks differ significantly from the appropriate control (P = .05).

might be estimated in the range of 3 g. This is sufficient to support a significant rate of microbiological growth, but in the highly competitive environment of the rhizosphere, at best only a modest part of the energy derived from this substrate would be used in the reduction of free N to ammonia. In view of the energy substrate requirements of this process, usually cited as $10-30$ μg N fixed/g of carbon source consumed where free-living heterotrophs are involved (Hill 1978), the process would fall far short of meeting the N requirements of rapidly growing plants. Further, it is reasonable to assume that N fixed under these conditions would be released slowly from the microbial biomass, since the organisms involved are not known to excrete ammonia. A contribution of N by the bacteria need not be doubted, but it is necessary to consider other responses in the search for a fully adequate explanation of plant growth responses.

Effects of Growth Hormones

Both the gibberellin and kinetin treatments increased growth (Table 2). The greatest response was to kinetin to 10^{-4} M, which increased dry weight 40%. In all cases where auxin was included (data not shown), yields were reduced. Addition of killed *Azospirillum* also produced a substantial dry-weight increase. These cells obviously could not fix N but have been shown to contain growth hormones (Tien et al. 1979), and it seems highly probable that their growth-hormone content brought about the dry-weight increase shown.

CONCLUSIONS

The results described here show that in some circumstances bacterial treatments can substantially accelerate plant growth rates, whether or not N fixation oc-

curs. The condition of the plants at the time of treatment is one of the variables that determine whether they respond. Further efforts to identify cause-and-effect relationship should not be confined to N-fixation studies, since other phenomena may be equally significant.

Although achieving maximum N fixation by root-associated bacteria may remain the most important long-term objective, the other potential benefits of synergistic plant-bacterial combinations should not be slighted in research. Moderately improved root growth resulting from the presence of bacteria that supply a hormonal stimulus or other growth factors, supply enzymatic activity to mobilize nutrients, or improve resistance to harmful organisms might, in many situations, significantly increase plant yields.

LITERATURE CITED

Barber, D.A., and J.K. Martin. 1976. The release of organic substances by cereal roots into soil. New Phytol. 76:69–80.

Barea, J.M., and M.E. Brown. 1974. Effects on plant growth produced by *Azotobacter paspali* related to synthesis of plant growth regulating substances. J. Appl. Bact. 37:171–174.

Brown, M.E. 1974. Seed and root bacterization. Ann. Rev. Phytopathology 12:181–198.

Burr, T.J., M.N. Schroth, and T. Suslow. 1978. Increased potato yields by treatment of seedpieces with specific strains of *Pseudomonas fluorescens* and *P. putida*. Phytopathology 68:1377–1383.

Cohen, E., Y. Okon, J. Kigel, I. Nur, and Y. Henis. 1980. Increase in dry weight and total nitrogen content in *Zea mays* and *Setaria italica* associated with nitrogen-fixing *Azospirillum* spp. Pl. Physiol. 66:746–749.

Gaskins, M.H., and D.H. Hubbell. 1979. Response of nonleguminous plants to root inoculation with free living diazotrophic bacteria. *In* J.L. Harley and R.S. Russell (eds.), The soil-root interface, 175–182. Academic Press, New York.

Hill, S. 1978. Factors influencing the efficiency of nitrogen fixation in free-living bacteria. Ecol. Bul. (Stockholm) 26:130–136.

Hubbell, D.H., and M.H. Gaskins. 1980. "Cryptic" plant root–microorganism interactions. What's New in Pl. Physiol. 11:17–20.

Martin, J.K. 1976. Factors influencing the loss of organic carbon from wheat roots. Soil Biol. Biochem. 9:1–7.

Rennie, R.J. 1979. ^{15}N-isotope dilution as a measure of dinitrogen fixation by *Azospirillum brasilense* associated with maize. Can. J. Bot. 58:21–24.

Ruschel, A.P., R.L. Victoria, E. Salati, and Y. Henis. 1978. Nitrogen fixation in sugarcane. Ecol. Bul. (Stockholm) 26:297–303.

Tien, T.M., M.H. Gaskins, and D.H. Hubbell. 1979. Plant growth substances produced by *Azospirillum brasilense* and their effect on the growth of pearl millet (*Pennisetum americanum* L.). Appl. Env. Microbiol. 37:1016–1024.

van Berkum, P., and B.B. Bohlool. 1980. Evaluation of nitrogen fixation by bacteria in association with roots of tropical grasses. Microbiol. Rev. 44:491–517.

Selection of *Rhizobium* Strains for Enhanced Dinitrogen Fixation in Forage Legume Production

C. HAGEDORN and W.E. KNIGHT
Mississippi State University, Mississippi State, Miss.,
and USDA-SEA-AR, U.S.A.

Summary

A set of procedures for the selection of *Rhizobium* strains for enhanced dinitrogen fixation in forage legumes is presented. Although forage legumes are among the most efficient members of the Leguminosae regarding biological nitrogen fixation (BNF), they are generally neither productive nor persistent without effective nodulation by appropriate rhizobia. The following methods define a selection and evaluation process to obtain appropriate *Rhizobium* strains. These include (1) regional survey of the production area, (2) strain collection and soil analyses, (3) strain characterization, (4) strain selection and manipulation, and (5) field evaluation of strains.

The initial procedure is to survey the region where the host legume is produced. This is necessary to obtain soil and plant samples, to become familiar with the major soil and climatic zones within the region, and to select potential field sites. Specific sampling sites chosen within each zone, based on the history of the legume host, are used to collect soil and plant samples for analysis and nodule isolations. At least 10 *Rhizobium* isolates should be recovered from each site, and the plant-infection assay should be used to determine the density of the indigenous *Rhizobium* populations.

The effectiveness of the *Rhizobium* isolates on those host plant cultivars that are grown in the region is the most important selection criterion. Other methods that may help in selecting strains for field trials might include growth in acidified media, salt tolerance, and survival under moisture/temperature extremes. In order to measure strain performance in the field it frequently becomes necessary to mark *Rhizobium* strains genetically so that they can be identified when recovered from nodules. Antibiotic resistance and/or serology are the techniques of choice, although there are problems with both. Eventually field plots will become cross-contaminated with different strains, and then identification of each strain is the only way that the evaluations can proceed.

Field plots should be designed with at least 4 replicates and appropriate uninoculated controls. Data should consist of plant dry weights, nitrogen content, nodulation score, and root and leaf stage from samplings every 6 or 8 weeks. Nodule-occupancy tests should focus on those strains that over time are found to be most productive in BNF and host-plant responses. The best strains from the field tests should be considered for inclusion in inoculants or used in some distribution system for wider testing at the producer level.

KEY WORDS: clover, *Rhizobium*, forage legumes, nitrogen fixation.

INTRODUCTION

Forage legumes are among the more efficient agronomic members of the Leguminosae regarding biological nitrogen fixation (BNF). Because several forage species possess the capacity to fix nitrogen in excess of their own needs, BNF rates can be relatively high (Gibson 1977). This situation is a result of the effective symbiosis between legume genotypes and specific *Rhizobium* strains, although forage legumes are neither productive nor persistent without effective nodulation by appropriate rhizobia.

Those forage cultivars of agronomic interest are the result of selections mainly for production, disease resistance, and hardiness. BNF has appeared as a breeding and selection concern only recently; it awaits exploitation. To obtain the necessary *Rhizobium* strains an evaluation and characterization process concurrent with host plant selections becomes mandatory.

No standardized procedures have been developed to accomplish this process, although there are sufficient examples of such efforts from several countries to illustrate the most useful approach. This approach can be divided among five categories; (1) regional survey of the production area, (2) strain collection and soil analysis, (3) strain characterization, (4) strain selection and manipulation, and (5) field evaluation of strains. This paper will briefly address the essential methods, guidelines, and approaches for each category, and the authors' experiences with annual clovers will be used to provide examples.

C. Hagedorn is an Associate Professor of Agronomy of Mississippi State University and W.E. Knight is a Research Agronomist at USDA-SEA-AR and a Professor of Agronomy at Mississippi State University. This paper is a contribution of the Department of Agronomy of Mississippi State University and the Forage Research Unit of USDA-SEA-AR; the research was supported in part by Grant No. 7800102 from the Competitive Grants Research Office of USDA-SEA-AR.

REGIONAL SURVEY OF THE PRODUCTION AREA

A survey, which must include the entire area where the host legume is grown, should be conducted in the planning stages. The objective is to insure that all important soil types that represent major areas are included. These may be grouped on the basis of soil parent materials, geographical zones (Roughley et al. 1976), or microclimate regions (Hagedorn 1978). Considerable travel may be necessary to obtain soil and plant samples, select potential field sites, and become familiar with the history of the host plant over the area.

STRAIN COLLECTION AND SOIL ANALYSIS

Specific sampling sites should be chosen within each major subregion and characterized on the basis of agronomic history, current use, soil types, drainage, and chemical composition. Sampling ought to include those sites that (1) presently contain the host species, (2) have previously been used for the host species, and (3) have never been planted with the host species. Soil analyses should be as complete as possible for both physical and chemical properties. Many of these are standardized (Black 1965). Emphasis should be placed on soil pH, N, P, K, and exchangeable bases. Soil analyses will serve as guidelines for fertilizer requirements later when the field sites are established.

The soil samples must be examined for indigenous rhizobia that can nodulate the host species. Plant-infection assays in either pouches or tubes will provide an adequate estimate of population density (Vincent 1970). *Rhizobium* isolates should be taken from nodules on plants in the field and from soil cores (for those sites where no plants occur) on which plants are grown. Some isolates should be recovered from the plant-infection assay tests since the dilution will separate only those strains that occur in very large numbers. At least 10 isolations/sampling site should be included, all from nodules on the healthiest plants that can be found. It is often valuable to analyze soil data against the *Rhizobium* and host plant measurements, as this analysis can yield information on the levels of fertility needed to maintain productive stands and the types of climate-related stress that will occur most frequently (Hagedorn 1978).

STRAIN CHARACTERIZATION

Testing the effectiveness of the isolated strains on those host-plant cultivars that are grown in the region is highly important. Other *Rhizobium* strains, from any source, should be included in the evaluations. These diverse rhizobia will usually produce wide variation in effectiveness tests when examined on different cultivars of the same host. Effectiveness ratings of isolates can be based on an arbitrary scale of dry-matter yields and should be examined on both a sampling-site (Hagedorn 1978) and a host-plant basis (Hagedorn and Caldwell 1981).

Many techniques may be used further to characterize the most effective isolates. The objective is to screen the strains with tests that will be useful in selecting strains for field trials. Procedures might include growth in acidified media (Hagedorn and Caldwell 1981), salt tolerance, survival at moisture and/or temperature extremes, and persistence in sterilized soil cores from the sampling sites (Keyser and Munns 1979). The soil analyses are frequently useful in determining which type of stress situation to use in order to select those strains that can persist under field conditions. A rating should be developed for each strain, based on effectiveness and stress survival. This rating should reduce the number of strains suitable for consideration in field trials.

STRAIN SELECTION AND MANIPULATION

In order to measure strain performance in the field adequately it is necessary to determine what proportion of the host plant's nodules are occupied by the inoculum strain. These strains must be characterized in some manner so they can be differentiated both from each other and from indigenous rhizobia. Serology (Vincent 1970) and antibiotic resistance (Davis 1962) have been most widely used, although with forage legumes the antibiotics have received more attention because of cross-reaction problems with serology. A wide range of antibiotics appear to be suitable for markers in the rhizobia (Brockwell et al. 1977, Hagedorn 1979a, Schwinghamer and Dudman 1973), although each strain must be examined for marker retention (after mutant selection) through plant passage and persistence with time and stress. Resistance to intrinsic levels of antibiotics is a promising technique since mutant selection is avoided (Josey et al. 1979). There are serological procedures, such as immunogel-diffusion, that can distinguish between closely related but nonidentical strains (Roberts et al. 1980) although such procedures are not readily adaptable to assays of large numbers of nodules. It may not be essential to mark the strains if the host plant species will be inoculated and planted annually. Regardless of the precautions taken to avoid the cross-contamination of field plots, if volunteer stands are used after the initial planting the strains will eventually become mixed, and identification of each strain is then the only way that evaluations can proceed.

FIELD EVALUATION OF STRAINS

For most purposes the strains should be grown in broth, added to sterilized, neutral-pH peat, and then inoculated onto the seed with an adhesive (gum arabic). Seed should then be lime-pelleted as a general rule, even if the soils at the field sites are not highly acidic. Small plots are preferable, with a minimum of 4 replicates/treatment. Randomized-block designs are frequently used, although it may be necessary to group each treatment in a nonrandom fashion in order to minimize cross-contamination between strains. This is feasible if the soil at the field site does not vary greatly across the plot area (Vincent 1970). Uninoculated control plots should generally be placed at least several meters away and ought to consist of (1) no fertilizer, (2) fertilizer, and (3) fertilizer plus added nitrogen (Hagedorn 1979b). A compatible non-legume should be planted along the borders between plots.

Plant samples should be collected every 6–8 weeks, and the data for each strain ought to consist of dry weight, plant N content, nodulation score, and root and leaf stage (Knight and Hollowell 1973, Vincent 1970). Nodule-occupancy tests as a function of time are needed to determine whether these factors affect any of the strains. All plots should be properly clipped and maintained in good condition for at least 3 years to evaluate strain persistence on reseeding stands. Frequently, large differences will be found between strains, and these should correlate well on both an individual plant and a plot-yield basis. In the clovers, those *Rhizobium* strains intended to provide maximal symbiotic effects should occupy at least 75%–80% of the host plant's nodules (Hagedorn 1979b).

The best strains from the field trials are then considered for inclusion in inoculants that may be regionally distributed to growers. It might be necessary to develop an inoculum with more than one strain, although this can result in problems. Commercial firms may participate in strain distribution, but a local production arrangement for a particular area may be quite satisfactory.

LITERATURE CITED

Black, C.A. (ed.). 1965. Methods of soil analysis, Parts 1 and 2. Agronomy 9 and 10. Am. Soc. Agron., Madison, Wis.

Brockwell, J.E., A. Schwinghamer, and R.R. Gault. 1977. Ecological studies of root-nodule bacteria introduced into field environments. V. A critical examination of the stability of antigenic and streptomycin-resistance markers for identification of strains of *Rhizobium trifolii*. Soil Biol. Biochem. 9:19–24.

Davis, R.J. 1962. Resistance of rhizobia to antimicrobial agents. J. Bact. 84:187–188.

Gibson, A.H. 1977. The influence of the environment and managerial practices on the legume-*Rhizobium* symbiosis. *In* R.W.F. Hardy and A.H. Gibson (eds.), A treatise on dinitrogen fixation, Section IV, 393–450. John Wiley and Sons. New York.

Hagedorn, C. 1978. Effectiveness of *Rhizobium trifolii* populations associated with *Trifolium subterraneum* L., in southwest Oregon soils. Soil Sci. Soc. Am. J. 42:447–451.

Hagedorn, C. 1979a. Relationship of antibiotic resistance to effectiveness in *Rhizobium trifolii* populations. Soil Sci. Soc. Am. J. 43:921–925.

Hagedorn, C. 1979b. Nodulation of *Trifolium subterraneum* L. by introduced rhizobia in southwest Oregon soils. Soil Sci. Soc. Am. J. 43:515–519.

Hagedorn, C., and B.A. Caldwell. 1981. Characterization of diverse *Rhizobium trifolii* isolates. Soil Sci. Soc. Am. J. 45:513–516.

Josey, D.P., J.L. Beynon, A.W.B. Johnston, and J.E. Beringer. 1979. Strain identification in *Rhizobium* using intrinsic antibiotic resistance. J. Appl. Bact. 46:343–350.

Keyser, H.H., and D.N. Munns. 1979. Tolerance of rhizobia to acidity, aluminum, and phosphate. Soil Sci. Soc. Am. J. 43:519–523.

Knight, W.E., and E.A. Hollowell. 1973. Crimson clover. Adv. in Agron. 25:47–76.

Roberts, G.P., W.T. Leps, L.E. Silver, and W.J. Brill. 1980. Use of two-dimensional polyacrylamide gel electrophoresis to identify and classify *Rhizobium* strains. Appl. Envir. Microbiol. 39:414–422.

Roughley, R.J., W.M. Blowes, and D.F. Herridge. 1976. Nodulation of *Trifolium subterraneum* by introduced rhizobia in competition with naturalized strains. Soil Boil. Biochem. 8:403–407.

Schwinghamer, E.A., and W.F. Dudman. 1973. Evaluation of spectinomycin resistance as a marker for ecological studies with *Rhizobium* spp. J. Appl. Bact. 36:263–272.

Vincent, J.M. 1970. A manual for the practical study of root-nodule bacteria. IBP Handbook No. 15. Black Sci. Pubs., Oxford.

Photosynthate Supply and Nitrogen Fixation in Forage Legumes

P.M. MURPHY

Agricultural Institute, Johnstown Castle Research Center,
Wexford, Ireland

Summary

Photosynthate supply is known to limit nitrogen fixation in legumes, and the resulting shortfall in energy supply necessary for maximum crop production may be aggravated by the energy-wasteful process of hydrogen (H_2) evolution by the legume N_2 fixing system. There is less information available on these aspects of the nitrogen-fixing process in forage legumes, such as white clover (*Trifolium repens*). The objectives of the present work were to determine how $N_2(C_2H_2)$ fixation and H_2 production are influenced by carbohydrate supply in forage legumes.[1]

Plants were grown under controlled environmental conditions, and nodule energy supply was varied by adjusting either photoperiod or atmospheric carbon-dioxide concentration. The effect of these treatments on $N_2(C_2H_2)$ fixation and H_2 evolution both by intact plants and by nodulated root segments was determined.

Light-induced diurnal variation in $N_2(C_2H_2)$ fixation by intact S100 white clover (at a constant temperature of 20°C) is related to the duration of the prior photoperiod. Short-photoperiod (6-hour) plants show a diurnal effect. Plants in a 14-hour photoperiod showed a constant $N_2(C_2H_2)$ fixation rate throughout the light and dark periods. Fixation rate of these plants was double that of the 6-hour plants during the photoperiod, and this difference was related to increased nodule mass. Specific activity (n moles $N_2(C_2H_2)$/mg nodule/hr) was similar for both treatments during the light cycle. Activity of nodulated root segments from both light treatments showed diurnal variation. Response to light was greater for the long-photoperiod plants.

Plants growing in the short photoperiod evolved more H_2 to $N_2(C_2H_2)$ fixed than did long-photoperiod plants.

Atmospheric enrichment with carbon dioxide did not affect the $N_2(C_2H_2)$ fixation of a number of forage legume species when measurements were made within days of commencement of treatment. After 3 weeks of enrichment, activity/plant was doubled. This increase was associated with an increased nodule mass. Specific $N_2(C_2H_2)$ fixation was not affected.

It is concluded that length of photoperiod and integrity of the plant system are important determinants of light-induced diurnal variation in $N_2(C_2H_2)$ fixation. Photosynthate supply limits nodulation rather than $N_2(C_2H_2)$ fixation per se. Short-photoperiod plants have reduced efficiencies of $N_2(C_2H_2)$ fixation in terms of H_2 evolved relative to $N_2(C_2H_2)$ fixed.

The results show that constant rates of $N_2(C_2H_2)$ fixation in forage legumes can be maintained during the light and dark periods by carbohydrate reserves in the plant and that photosynthate supply is not limiting for nodule activity. Conditions that deplete these reserves decrease rates of dark fixation. Energy loss through H_2 evolution is not reduced even in carbon-stressed plants.

KEY WORDS: nitrogen fixation, forage legume, photosynthate supply, hydrogen production.

INTRODUCTION

The importance of photosynthesis as an energy source for nitrogen (N_2) fixation by legumes has long been recognized. Demonstration by Hardy and Havelka (1976) that energy supply limits N_2 fixation in the soybean stimulated further research on the interrelationships of photosynthesis and N_2 fixation in legumes. The present studies were undertaken to obtain a greater understanding of the role of photosynthate supply in controlling N_2 fixation in forage legumes. Both variable photoperiod and increased atmospheric carbon dioxide (CO_2) concentration were used to alter the supply of photosynthate available to the nodule for N_2 fixation.

Another aspect of the energy requirement for N_2 fixation relates to H_2 evolution by the nodule. Information on the intimate linking of N_2 fixation and H_2 production by legumes (Hoch et al. 1957) was followed by the realization that H_2 evolution indicates a serious energy wastage by the N_2 fixing system. The work of Schubert and Evans (1976) showed that the amount of H_2 evolved by different root nodule associations varies greatly, depending on whether an uptake hydrogenase was present. Significant H_1 recycling has not been reported for clover nodules. This study investigated the relationship between carbon supply and H_2 evolution in this species.

METHODS

Plants were grown under controlled environmental conditions in pots containing perlite and supplied with a

The author is a Senior Research Officer at the Agricultural Institute. The author wishes to thank C.L. Masterson for helpful criticism during the preparation of this paper and M. Murphy for technical assistance.

defined nutrient solution. A 14-hour photoperiod was normally used, with 26,000-IX illumination and day/night temperatures of 20°/15°C. Plants were inoculated with an effective strain of *Rhizobium*. All assays were done at the vegetative growth stage when plants were 8–12 weeks old.

The effect of photoperiod on N_2 fixation was studied by dividing 6-week-old plants into two treatments, one continuing in the 14-hour light period and the other with the photoperiod reduced to 6 hours. After 2 weeks N_2 fixation was determined both on intact plants and on nodulated root segments by the acetylene reduction method [$N_2(C_2H_2)$]. H_2 evolution was determined with nodulated root segments only.

Enriched atmospheric level of CO_2 (1,200 ppm) was maintained during the 14-hour photoperiod with 6-week-old plants, and the effect of this treatment on $N_2(C_2H_2)$ activity was determined over a period of 3 weeks.

RESULTS

Effect of Photoperiod

$N_2(C_2H_2)$ fixation by intact plants. Plants growing in the 14-hour photoperiod showed no diurnal variation in $N_2(C_2H_2)$ fixation (Fig. 1). This activity in plants receiving 6-hour light decreased in the dark but increased sharply on recommencement of the light period. Because of their responses to light, plants from both treatments had similar specific $N_2(C_2H_2)$ fixation, although activity/plant was reduced by over 50% in the short photoperiod. This reduction was accompanied by a similar decrease in nodule mass.

$N_2(C_2H_2)$ fixation by root segments. The pattern of $N_2(C_2H_2)$ fixation by root segments differed from that ob-

Fig. 1. Effect of photoperiod on $N_2(C_2H_2)$ fixation by intact S100 white clover plants at constant temperature (20°C).

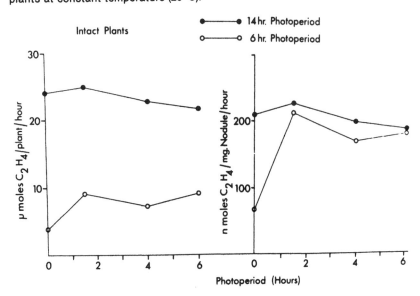

tained with whole plants (Fig. 2). A diurnal variation in specific activity was found in both treatments. Specific activity of the 14-hour root segments increased with photoperiod to reach a value similar to that obtained for whole plants. A much slower response to light was noted in root samples from the 6-hour photoperiod, and specific activity remained depressed throughout.

H_2 production by root segments. Results in Fig. 2 show H_2 production increasing as the photoperiod progressed. This increase coincided with increased rates of $N_2(C_2H_2)$ fixation. The ratio of H_2 produced to $N_2(C_2H_2)$ reduced was significantly increased in the short-photoperiod plants over that in the 14-hour plants during the light period.

Effect of CO_2 Enrichment

There was no short-term effect of increased levels of CO_2 on $N_2(C_2H_2)$ fixation. After 3 weeks of treatment, activity/plant was significantly increased; this increase corresponded to an increase in nodule mass (Table 1). Specific $N_2(C_2H_2)$ fixation was not significantly increased in any of the forage legumes, although this activity was increased in the case of the pea.

DISCUSSION

The results obtained with intact white clover plants revealed that no diurnal variation in $N_2(C_2H_2)$ fixation occurred in 14-hour-photoperiod plants. Under these conditions carbohydrate reserves present in the plant are sufficient to maintain a constant rate of $N_2(C_2H_2)$ fixation throughout the light/dark cycle. In a short photoperiod, reserves are depleted, photosynthate becomes limiting, and $N_2(C_2H_2)$ fixation is reduced during the dark period.

Fixation activity of nodulated root segments from plants growing in both photoperiods was depressed in the absence of photosynthesis, and in this instance separation of the shoot from the plant was an additional factor interrupting carbohydrate supply to the nodule. These results, particularly those obtained with the 14-hour plants, suggest that both shoot and root reserves function in maintaining $N_2(C_2H_2)$ fixation during the dark period. Thus, length of photoperiod and integrity of the plant system are important determinants of light-induced diurnal variation in $N_2(C_2H_2)$ fixation in white clover.

Atmospheric CO_2 enrichment increased root nodulation and $N_2(C_2H_2)$ fixation/plant. There was no significant effect of CO_2 on nodule $N_2(C_2H_2)$ fixation per se with the exception of the pea, where both nodule weight and specific activity were increased. These results, indicating a different response to CO_2 enrichment by forage and grain legumes, show that photosynthate supply in forage legumes limits $N_2(C_2H_2)$ fixation as a result of its effects on nodulation only. Increased photosynthate supply enables the plant to support a greater nodule mass without affecting the $N_2(C_2H_2)$-fixing activity of the individual nodules.

Short-photoperiod plants evolved more H_2 to $N_2(C_2H_2)$ fixed than did long-photoperiod plants. The short-photoperiod plants were carbon-stressed, as is evidenced by the observed diurnal variation in $N_2(C_2H_2)$ fixation. Under these conditions photosynthate supply was limiting, and energy loss through H_2 evolution continued at rates even greater than those of the 14-hour plants. No evidence was obtained, therefore, to suggest that energy conservation through reduced H_2 production occurred in photosynthate-limited white clover nodules with the particular plant-*Rhizobium* combination used.

Fig 2. Effect of photoperiod on $N_2(C_2H_2)$ fixation – and H_2 production – by nodulated S100 white clover roots at constant temperature (20°C).

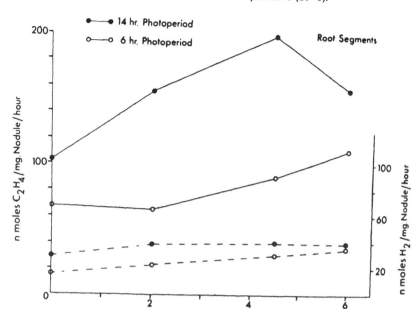

Table 1. Effect of long-term (3 weeks) CO_2 enrichment on $N_2(C_2H_2)$ fixation and root-nodule mass of legumes.

	Ambient – 350 ppm CO_2			Enriched – 1200 ppm CO_2		
	$N_2(C_2H_2)$		Nod. Wt.	$N_2(C_2H_2)$		Nod. Wt.
	(a)	(b)	(c)	(a)	(b)	
Legume						
Blanca	13	225	290	21***	232	449***
S100	12	277	264	19**	303	373**
Hungaropoly	16	242	330	24***	284	426***
Lucerne	10	356	145	16***	344	225***
Pea (cv Meteor)	18	192	191	37***	266*	278***

(a) = μ moles C_2H_4/plant/hour. (b) = n moles C_2H_4/mg nod./hour.
(c) = mg dry weight.

*P ≤ 0.05, **P ≤ 0.01, ***P ≤ 0.001

LITERATURE CITED

Hardy, R.W.F., and U.D. Havelka. 1976. Photosynthate as a major factor limiting nitrogen fixation by field-grown legumes with emphasis on soybeans. *In* P.S. Nutman (ed.), Symbiotic nitrogen fixation in plants. Int. Biol. Prog. 7:421-439. Cambridge Univ. Press.

Hoch, G.E., H.N. Little, R.H. Burris. 1957. Hydrogen evolution from soybean root nodules. Nature 179:430-431.

Schubert, K.R. and H.J. Evans. 1976. Hydrogen evolution: a major factor affecting the efficiency of nitrogen fixation in nodulated symbionts. Proc. Natl. Acad. Sci. 73:1207-1211.

NOTES

1. Abbreviations: $N_2(C_2H_2)$ denotes measurement of nitrogen (N_2) fixation by the acetylene (C_2H_2)-reduction method. C_2H_4 is ethylene, H_2 is hydrogen, and CO_2 is carbon dioxide.

Symbiotic Nitrogen Fixation of Alfalfa, Birdsfoot Trefoil, and Red Clover

G.H. HEICHEL, C.P. VANCE, and D.K. BARNES

USDA-SEA-AR and University of Minnesota, St. Paul, Minn., U.S.A.

Summary

Forage legumes such as alfalfa (*Medicago sativa* L.), birdsfoot trefoil (*Lotus corniculatus* L.), and red clover (*Trifolium pratense* L.), grown in crop rotations, can provide an inexpensive source of nitrogen (N) to succeeding crops. Few quantitative measurements of seasonal patterns of N fixation in managed communities of perennial forage legumes have been undertaken, chiefly because of the lack of satisfactory methods. Therefore, there is little knowledge of how N fixation varies with crop development, yielding ability, and weather patterns. Quantitative measurements of N fixation are important in choosing the appropriate legume for use in a crop rotation and for understanding the constraints to N fixation in normally managed crop communities. Our objectives were to compare the N-fixation patterns of alfalfa, birdsfoot trefoil, and red clover under field conditions in 2 successive years and to relate these seasonal changes in symbiotic activity to dry-matter yield, leaf area, and precipitation.

Nitrogen fixation was measured by the ^{15}N isotope-dilution technique, in which the isotope composition of the nitrogen of the legume is compared by mass spectrometry with that of a perennial grass used as the non-fixing control.

In the seeding year the proportion of plant N derived from symbiosis ranged from 27% to 50% among species at the first harvest and from 66% to 80% among species at the second harvest. Seasonal N fixation was greater for alfalfa (15.9 g N/m^2) than for either red clover (11.9 g N/m^2) or birdsfoot trefoil (10.6 g N/m^2). N fixation varied significantly among harvests and among the species at individual harvests.

In the second year the proportion of N from symbiosis ranged from 0% to 62% at the first harvest and 32% to 60% at second harvest. Alfalfa was the only species harvested a third time, and it contained 35% N from symbiosis. Seasonal N fixation of alfalfa (17.1 g N/m^2) was again greater than that of red clover (9.8 g N/m^2) or birdsfoot trefoil (7.6 g N/m^2). N fixation of alfalfa was more sensitive to heavy midsummer rainfall than was that of birdsfoot trefoil or red clover.

On a seasonal basis, alfalfa, the highest-yielding species, fixed more N/unit of land area than birdsfoot trefoil, the lowest-yielding species. Within an individual harvest, N fixation was not closely associated with yield because the high yields common early in the growing season contained less N from symbiosis than did lower yields obtained later in the season.

The results clearly show that N-fixation capability varies significantly with species of forage legume. N fixation also varies significantly with management and environmental factors that affect plant development and total N yield.

KEY WORDS: nitrogen fixation, isotope-dilution technique, perennial legumes, *Medicago sativa* L., *Lotus corniculatus* L., *Trifolium pratense* L.

INTRODUCTION

Forage legumes are important components of cropping systems in the north central U.S.A., where they provide a significant source of feed for beef and dairy animals. With the continual increase in the price of nitrogen (N) fertilizers, there is an increasing interest in the role of perennial forage legumes as inexpensive sources of N in rotations. Little quantitative information is available on the seasonal N-fixation capabilities of perennial legumes in managed communities. There is also little knowledge of how N fixation varies with crop development, crop yielding ability, and weather patterns. This information is essential for finding the most effective use of perennial legumes as N sources in cropping systems. The objectives of this research were to compare the N fixation of alfalfa (*Medicago sativa* L.), birdsfoot trefoil (*Lotus corniculatus* L.), and red clover (*Trifolium pratense* L.) in 2 successive years and to relate N fixation to seasonal changes in dry-matter yield and precipitation.

G.H. Heichel and C.P. Vance are Plant Physiologists and D.K. Barnes is a Research Scientist of USDA-SEA-AR, and all three authors are Professors of Agronomy at the University of Minnesota. This paper is a joint contribution of USDA-SEA-AR and the Minnesota Agricultural Experiment Station (Scientific Journal Series No. 11602).

The authors wish to thank Dr. C.M. Cho of the Department of Soil Sciences of the University of Manitoba, Canada, for the mass spectrometric analyses.

METHODS

Seeds of Saranac alfalfa, Carroll birdsfoot trefoil, and Arlington red clover were sown in rows 18 cm apart in 2 × 10-m plots arranged in 4 replicates of a randomized complete-block design. Reed canarygrass (*Phalaris arundinacea* L.) and tall fescue (*Festuca arundinacea* Schreb.) were sown as the nonfixing controls. Experimental plots were located on Port Byron silt loam at the Rosemount Experiment Station, University of Minnesota.

Ten days before sowing, a 1-m² subplot in the center of each plot was uniformly sprayed with 350 ml of water containing 0.42 g ^{15}N from $(^{15}NH_4)_2 SO_4$ and 0.6 g additional N as $(NH_4)_2 SO_4$. Immediately after application of the isotope, the upper 13 cm of soil was thoroughly mixed and lightly irrigated. In mid-April of the second year a similar amount of isotope in water was applied by spraying into trenches 2 cm wide × 15 cm deep between the established rows of plants in the subplots. The solution thoroughly moistened the sidewalls and bottom of the trenches, which were subsequently backfilled with soil to prevent volatilization of labeled ammonium. Nitrogen fixation was measured throughout the year by the ^{15}N isotope-dilution technique (McAuliffe et al. 1958, Legg and Sloger 1975, Heichel et al. 1981).

Nitrogen fixation of crop communities was expressed either as N fixed/unit of land area (N_f) or as the proportion of N in the plant derived from symbiosis (N_{sy}). N_f (in g N/m²) was calculated as (total N yield/m²) (N_{sy}), where the total N yield is the product of total dry-matter or herbage yield and the N concentration, and $N_{sy} = 100 - 100[(\%$ atom excess ^{15}N in legume)/(% atom excess ^{15}N in control species)]. In the seeding year, all species were harvested twice, based upon flowering criteria. In the second year, alfalfa was harvested three times and the other legumes twice, following locally recommended practices. Herbage from each harvest was analyzed for total N, and the condensed Kjeldahl distillate was subsequently analyzed for atom excess ^{15}N by mass spectrometry. Unlabeled plants growing adjacent to the subplots were sampled to determine total N partitioning between herbage and roots. This technique allowed adjustment of N-fixation values calculated from herbage analyses to a whole-plant basis.

RESULTS

Significant (P < 0.01) differences among species were observed in total herbage yield from the two harvests in the seeding year; alfalfa produced 655, red clover 595, and birdsfoot trefoil 496 g/m² of dry matter. There were also significant (P < 0.01) differences among species in herbage yield at individual harvests, and a significant (P < 0.01) species × harvest interaction.

Alfalfa was harvested three times in the second year and the other species twice. Significant (P < 0.01) differences among species were again observed in total herbage yield, with alfalfa producing 1,075, red clover 863, and birdsfoot trefoil 748 g/m² of dry matter. Significant (P < 0.01) differences in herbage yield among species occurred at each of the first two harvests, and the species × harvest interaction was significant (P < 0.01). In all species, the herbage yield declined with successive harvests.

The N_{sy} of alfalfa, birdsfoot trefoil, and red clover varied significantly with species (P < 0.05) and harvest (P

Fig. 1. The proportion of N from symbiosis and N fixation/m² ground area based upon isotope analyses of three forage legumes harvested at the end of two growth intervals during the seeding year. Harvests were based upon recommended flowering criteria. The duration of each growth interval is shown beneath the bar for each species in the histogram. The rainfall accumulated during each growth interval for each species is shown at the top of the histogram.

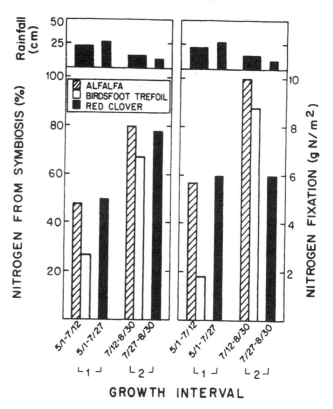

< 0.01) in the seeding year (Fig. 1). The species × harvest interaction was also significant (P < 0.10). Alfalfa and red clover derived a larger proportion of their N from symbiosis earlier in the year than did birdsfoot trefoil, but trefoil exhibited the largest relative increase in N_{sy} between the first and second harvest, so the N_{sy} of the three species was very similar at the second harvest. All three species derived more of their N from symbiosis in midsummer than in early spring.

When differences in yielding ability and N concentration in dry matter were combined with measurements of N_{sy}, N_f differed significantly (P < 0.01) among harvests for alfalfa and birdsfoot trefoil but not among those for red clover (Fig. 1). This difference occurred because the increase in N_{sy} of red clover from the first to the second harvest was offset by a yield decline between the two harvests. Alfalfa fixed 15.9, red clover 11.9, and birdsfoot trefoil 10.6 g N/m² over the two growth intervals.

In the second year, N_{sy} again varied significantly in both harvests and species (P < 0.01). Although the first growth interval in the second year terminated 5 to 6 weeks earlier than in the first year, N_{sy} of alfalfa was signifi-

Fig. 2. The proportion of N from symbiosis and N fixation/m² ground area based upon isotope analyses of three forage legumes harvested at the end of two (red clover and birdsfoot trefoil) or three (alfalfa) growth intervals during the second year of growth. Note that birdsfoot trefoil fixed no N during growth interval 1. The rainfall accumulated during each growth interval for each species is shown at the top of the histograms. See Figure 1 for explanation of dates.

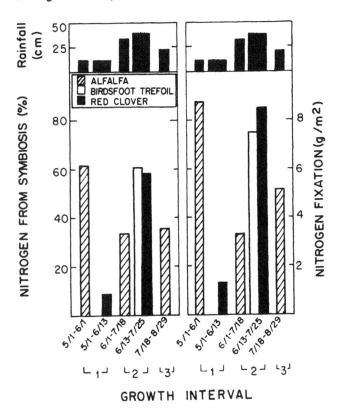

cantly greater than that of birdsfoot trefoil and red clover (Fig. 2). This fact suggests that the N fixation of alfalfa commenced earlier in the spring than that of the other two species. No N fixation was measured in the first growth interval for birdsfoot trefoil.

In the second year, N_{sy} of red clover and of birdsfoot trefoil increased in the second growth interval (Fig. 2), as in the seeding year. In the second year, the N_{sy} of alfalfa declined in June and July, while that of trefoil and clover increased (Fig. 2). This decline in alfalfa N_{sy} coincided with an abnormally high amount of rainfall during midsummer of the second year in comparison with the first year (Figs. 1 and 2), during which N_{sy} of alfalfa did not decline (Fig. 1).

Significant ($P < 0.01$) differences in N_f among species and harvests were also apparent in the second year, and their interaction was significant (Fig. 2). Birdsfoot trefoil fixed less N on a seasonal basis (7.6 g N/m²) than did either alfalfa (17.1 g N/m²) or red clover (9.8 g N/m²) because its abundant first-harvest yield contained no symbiotically fixed N. Despite the excessive rainfall during midsummer, alfalfa fixed the most N of the three species

during the second year because it is adapted to a three-harvest management that permitted greater annual productivity.

DISCUSSION

Significant differences in the seasonal N fixation among and within annual and perennial legumes have long been reported, although the values vary widely with location, method of measurement, and species. The application of the isotope-dilution technique in field experiments has permitted an increased understanding of the reasons for seasonal differences in N fixation and for the differences in performance among species.

Alfalfa and red clover averaged 64% and birdsfoot trefoil 46% N from symbiosis over the seeding year (Fig. 1). In the second year, alfalfa averaged 43%, red clover 33%, and birdsfoot trefoil 30% N from symbiosis. In comparison, soybean (*Glycine max* [L.] Merr.) obtained 39% to 66% of its seasonal N needs from symbiosis (Deibert et al. 1979, Ham 1978, Legg and Sloger 1975).

These results illustrate that the seasonal N fixation of forage legumes depends upon a number of independent factors. The N_{sy} of a species varies with stage of crop development and with prevailing patterns of precipitation. The symbiotic metabolism of alfalfa is apparently less tolerant of excessive midsummer rainfall than that of the more shallow-rooted birdsfoot trefoil and red clover.

On a land-area basis, N_f varied with N_{sy} and with any factor that affected total N yield. Total N yield depends upon dry-matter yield and N concentration, both of which depend upon species, management, and weather. Additionally, dry-matter yield and N concentration in dry matter often vary independently of N_{sy}, so that accretion of biomass or leaf area at individual harvests in perennial legumes may not directly reflect their N-fixation capability. Variation of symbiotic activity (N_{sy}) with factors like stage of growth that average the influences of a variable preceding field environment is often greater than variance due to genetic factors, i.e., plant species.

LITERATURE CITED

Deibert, E.J., M. Bijeriego, and R.A. Olson. 1979. Utilization of ¹⁵N fertilizer by nodulating and non-nodulating soybean isolines. Agron. J. 71:717–723.

Ham, G.E., 1978. Use of ¹⁵N in evaluating symbiotic N₂ fixation of field-grown soybeans. *In* Isotopes in biological dinitrogen fixation, 151–162. Int. Atomic Energy Agency, Vienna.

Heichel, G.H., D.K. Barnes, and C.P. Vance. 1981. Nitrogen fixation of alfalfa in seeding year. Crop Sci. 21:330–335.

Legg, J.O., and C. Sloger. 1975. A tracer method for determining nitrogen fixation in field studies. *In* E.R. Klein and P.D. Klein (eds.), Proc. 2nd Int. Conf. Stable Isotopes, Oak Brook, Ill., 661–666.

McAuliffe, C., D.S. Chamblee, H. Uribe-Arango, and W.W. Woodhouse. 1958. Influence of inorganic nitrogen on nitrogen fixation of legumes as revealed by ¹⁵N. Agron. J. 50:334–337.

Effects of Grazing Management on Seasonal Variation in Nitrogen Fixation

J.L. BROCK, J.H. HOGLUND, and R.H. FLETCHER

Grasslands Division, DSIR, Lincoln and Palmerston North, N. Z.

Summary

Using the acetylene reduction assay, field studies in recent years have investigated the variation in N fixation in grazed systems, and some of the principles involved in the interaction with available soil N have been described. This experiment extends this work to include the effect of grazing management on the factors influencing N fixation.

Two self-contained sheep-grazed units, one a 10-paddock rotationally grazed system and the other set stocked, were established on a high-fertility ryegrass–white clover pasture, with a basic stocking rate of 22 sheep/ha. Over the 1975–1978 period, both units were monitored for pasture growth, N fixation, and associated herbage and soil parameters.

Annual N fixation under set stocking was 10% higher than under rotational grazing. The difference, which occurred during the dry summer-autumn period, reflected a greater ability of the set-stocked pastures to respond to short-term alleviation of moisture stress. Legume growth rate was the principal factor controlling N fixation in both systems. Available soil N appeared to be of greater significance in controlling N fixation under rotational grazing. This finding was consistent with the generally lower herbage NO_3-N levels found under set stocking. These differences were attributed to differences in sward structure and physiological stress between the two grazing systems.

KEY WORDS: N fixation, available soil N, legume growth, grazing management, herbage nitrate.

INTRODUCTION

Atmospheric N fixation can be considered a function of legume N requirements (growth rate) minus soil mineral N uptake. In general, as soil mineral N supply increases, N fixation decreases (Hoglund 1973).

In pastures, legumes are generally grown with companion species that through direct competition reduce legume growth, but the concurrent reduction in soil mineral N tends to neutralize the effect on N fixation (Sears et al. 1965). Pasture management can manipulate these competitive effects by altering the grass/legume balance (Brougham 1960) on either a seasonal or a long-term basis, a change that could have marked effects on N fixation. This paper contrasts annual and seasonal N fixation in set stocked and rotationally grazed pasture systems.

METHODS AND MATERIALS

Site

The soil of the experimental site at DSIR, Palmerston North, was a Manawatu fine sandy-loam topsoil (Dystric Fluventic Eutrochrept), low in total N (0.183% N or 3,400 kg N/ha/0–150 mm), with a C:N ratio of 9.5 and a pH of 5.7, overlying deep sands and gravels.

Grazing Systems

The two self-contained grazing systems, each of 0.4 ha, one set stocked (SS) and the other of 10 rotationally grazed paddocks (RG), were established in 1973 on a hybrid ryegrass (*Lolium hybridum* Hausskn. cv. Grasslands Ariki)–white clover (*Trifolium repens* L. cv. Grasslands Huia) pasture sown in 1971. Basic stocking rate on both systems was 22 mature wethers/ha, and stocking rate and rotation length varied seasonally to simulate stocking patterns on sheep-breeding farms.

Measurement Techniques

All measurements were made at 1- to 2-week intervals using procedures for herbage growth, acetylene reduction assay of N fixation, soil mineral N (NH_4-N and NO_3-N), and herbage NO_3-N, similar to those detailed by Hoglund and Brock (1978).

Measurements continued for 3 years (1975–1978). Each year was divided into three periods on the basis of general environmental conditions: spring, mid-August to mid-December (125 days), warm, moist; summer-autumn, mid-December to mid-May (150 days), warm, dry; winter, mid-May to mid-August (90 days), cool, wet.

J.L. Brock is a Scientist at the Grasslands Division of DSIR, Palmerston North; J.H. Hoglund is a Scientist at the Grasslands Division of DSIR, Lincoln; and R.H. Fletcher is a Scientist at the Applied Mathematics Division of DSIR, Palmerston North, N.Z.

The experiment was terminated during the third year owing to an abnormal summer drought.

RESULTS AND DISCUSSION

On average, N fixation was 37% higher under the SS system than under the RG system from the late spring to the early winter, particularly in the drier second and third years (65% difference). For the remainder of the year, N fixation was similar. This difference developed consistently in late spring (November), when growth rates were high and similar for the two systems (Fig. 1). This was approximately the time when potential evapo-transpiration (PET) rates reached 3.5-4 mm/day consistently, 5-cm noon soil temperatures were about 20°C, and soil moisture, although still relatively high (50%-70% available soil water, Fig. 2), was declining. These data suggest that increasing environmental stress may be the initiating factor.

During the following dry summer-autumn period, the N-fixation raw data indicated that when soil moisture was low, the difference in N-fixation rates between the systems decreased or disappeared. Complementary increases in N fixation 1-3 weeks after rain had significantly raised soil moisture were greater under set stocking (Fig.

2). These differences disappeared when late-autumn rain returned soil moisture to near field capacity. Differences in plant structure resulting from the contrasting defoliation patterns were probably responsible. Infrequent severe defoliation of rotationally grazed plants imposes a higher degree of physiological stress on plants than the frequent though lenient defoliation of set stocking. Consequently, the rotationally grazed plants would be relatively less responsive to short-term alleviation of moisture stress.

Of the factors measured as indicators of soil N availability to the legume (soil NO_3-N and NH_4-N, herbage NO_3-N), only herbage NO_3-N levels demonstrated significant differences between the systems, being consistently higher in both grass and legume in all seasons (except winter) in the RG system.

Using the least squares multiple linear regression, a model of significant factors controlling N fixation was developed for each system, as follows:

N fixation (RG) = [- 0.3375 + 0.0388 legume growth rate (kg/ha/day)] + 0.0112 grass growth rate (kg/ha/day) + 0.00072 grass herbage NO_3-N (ppm) - 0.00132 legume herbage NO_3-N (ppm).

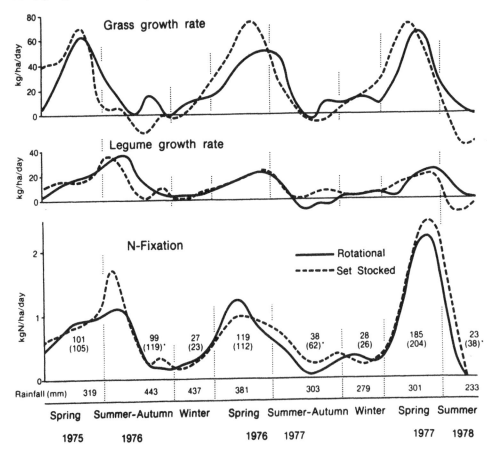

Fig. 1. Seasonal variation in herbage growth rates and N fixation (*significantly different N fixation [kg/ha]; set-stocked in parentheses).

N fixation (SS) = [−0.33 + 0.0307 legume growth rate (kg/ha/day)] + 0.1305 PET (mm/day) + 0.0137 soil moisture % (measured at the previous assay)

Fig. 2. The relationship between fluctuations in soil moisture and N fixation in rotationally grazed and set-stocked pastures.

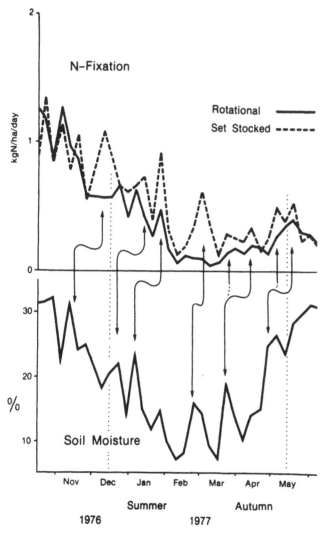

In both cases the primary factor was the host-legume growth rate, but thereafter the additional significant factors fell into separate groups for each system, again as a result of the different defoliation patterns. The high physiological stress imposed by rotational grazing causes underground release of N from senescing tissues in a quantity that could be as high as that being returned via the grazing animal (Dilz and Mulder 1962, Hoglund and Brock 1978). These high concentrations of available N occur when the plants are relatively inactive, a fact that explains the dominance of soil N-related factors in the N-fixation (RG) model. Hoglund and Brock (1978) produced similar models for seasonal and regrowth N-fixation patterns under rotational grazing. Under set stocking the less concentrated N return (above and below ground) is consistent with lower herbage NO_3-N levels and the importance of the environmental factors (PET and soil moisture).

The 10% higher annual N fixation under set stocking suggests that (1) one or both of the systems studied was not in a steady state; (2) set stocking is accumulating more N (as soil organic N), or (3) set stocking is losing more N than rotational grazing. From the N-fixation models developed in this experiment, option (3) is the least likely to be operating. This fact would imply that set stocking in some form may have advantages in the long-term maintenance of soil fertility at high production levels. Further work on this aspect is being conducted.

LITERATURE CITED

Brougham, R.W. 1960. The effect of frequent hard grazings at different times of the year on the productivity and species yields of a grass-clover pasture. N.Z. J. Agric. Res. 3:125–136.

Dilz, K., and E.G. Mulder. 1962. Effect of associated growth on yield and nitrogen content of legume and grass plants. Pl. and Soil 16:229–237.

Hoglund, J.H. 1973. Bimodal response by nodulated legumes to combined nitrogen. Pl. and Soil 39:533–545.

Hogland, J.H., and J.L. Brock. 1978. Regulation of nitrogen fixation in a grazed pasture. N.Z. J. Agric. Res. 21:73–82.

Sears, P.D., V.C. Goodall, R.H. Jackman, and G.S. Robinson. 1965. Pasture growth and soil fertility. VIII. The influence of grass, white clover, fertilisers, and the return of herbage clippings on pasture production of an impoverished soil. N.Z. J. Agric. Res. 8:270–283.

Nitrogen Losses from Urine-Affected Areas of a New Zealand Pasture, Under Contrasting Seasonal Conditions

P.R. BALL and D.R. KEENEY

Grasslands Division, DSIR, Palmerston North, N.Z.,
and University of Wisconsin, Madison, Wis., U.S.A.

Summary

Grazing ruminants aggregate large quantities of forage nitrogen (N) into urine patches. Pasture plots treated with urine N were studied to assess its fate under contrasting seasonal conditions: cool-moist (representing winter–spring); warm-moist (late spring–early summer), and warm-dry (late summer–autumn).

A permanent ryegrass (*Lolium perenne* L.)–white clover (*Trifolium repens* L.) sward on free-draining alluvial soil was studied. Treatments were replicated in randomized blocks. Urine N was applied at 30 or 60 g N/m². Relevant measurements were then continued until treatment differences disappeared. Apparent N balances for urine treatments were compiled.

Fixation by white clover was always greatly reduced in urine-treated swards. Urine N increased herbage N yield only when moist conditions favored plant growth. Overall, apparent recovery of urine N by pasture averaged 32% and ranged from 55% under cool-moist to 11% under warm-dry conditions. Soil total N was unaffected by treatments.

Some two-thirds of urine N was lost from the soil-plant system during these studies. Extremes in average loss were 45% and 80% under cool-moist and warm-dry conditions, respectively.

Nutrient recycling in the excreta of grazing animals is generally considered beneficial to pasture production and the major pathway for transfer of clover-fixed N to grass associates. The authors dispute those views with respect to N. Rather, they contend that grazing ruminants cause substantial N losses from developed pastoral ecosystems.

Sheep or cattle grazing well-managed grass-clover pastures in New Zealand excrete several hundred kg urine N/ha/yr. The more intensively utilized systems may be in negative N balance, largely because of N escape from urine patches.

KEY WORDS: N cycle, N balance, N losses, urinary N.

INTRODUCTION

Ruminants grazing intensively managed pastures ingest several hundred k N/ha/yr and excrete the bulk of it in urine (Ball 1979). The fate of urinary N is therefore very important to the N economy of pastures.

Urinary N is concentrated (c. 10 g N/l) and deposited in discrete patches. Depending on diet and animal, between 30 and 60 g N/m² is deposited in a urine patch (Whitehead 1970), where it undergoes complex physical, chemical, and biochemical reactions (Ball et al. 1979).

Research reported here was undertaken to determine the fate of urinary N in a ryegrass–white clover pasture. Experiments were conducted in the field and repeated three times under contrasting seasonal conditions.

P.R. Ball is a Scientist at the Grasslands Division of DSIR and D.R. Keeney is a Professor in the Department of Soil Science of the University of Wisconsin.

The authors wish to thank P.W. Theobald and P. Nes for technical assistance. D.R. Keeney was assisted by a National Research Advisory Council (N.Z.) senior research fellowship.

METHODS

Site

The studies were undertaken at Palmerston North. Annual climatic averages are: rainfall, 1,000 mm, fairly evenly distributed; potential evapo-transpiration, about 700 mm; and mean air temperature, 13.2°C. The soil is Karapoti silt loam (Dystric Eutrochrept), formed from recent alluvium and free draining. Soil total N content was 0.224% at 0–15 cm, and C : N ratio was about 10. Surface soil pH (distilled water) was 5.5. A more detailed description has been given previously (Ball et al. 1979).

An established sward of perennial ryegrass and white clover was studied. It had been grazed by sheep but was repeatedly mown starting 2 months before these measurements began.

Experimental Approach

Urine solution (urine N) was applied to 3-m² plots at rates simulating those caused by sheep and cattle. Relevant measurements were then continued until treatment differences disappeared. Comparison with control plots allowed apparent N balances to be compiled.

Nine plots were established for each experimental run, using a randomized block design. Treatments were: control (NO), 30 g N/m² (N30), and 60 g N/m² (N60).

This experiment was repeated three times throughout one year under contrasting seasonal conditions on separate sites located at random within the study area.

Materials and methods used are reported elsewhere (Ball and Keeney 1981), together with greater detail of meteorological conditions throughout the three experimental runs.

RESULTS

Meteorological Conditions

Cool-moist. Treatments were imposed in early winter, and measurements continued for 138 days. Relatively low temperatures characterized this period. Weighted averages for mean air and 10-cm soil temperatures were 10.0°C and 8.6°C, respectively. Soil moisture was near field capacity (FC) throughout the profile when urine N was applied, and leaching conditions prevailed.

Warm-moist. A 55-day experimental run was made under late spring–summer conditions (see Ball et al. 1979). Average temperatures were 16.3°C and 16.1°C, respectively. The soil profile was near FC at the outset. Drying conditions occurred during the initial 2 weeks, followed by a pronounced leaching phase.

Warm-dry. A 96-day run was made from late summer

through autumn. Temperatures were comparable to those during the warm-moist run until week 6 and coincided with a negative water balance. Urine N was applied to drier soil (about 0.5 FC at 0–2.5 cm and 0.7 FC at > 2.5 cm). Thereafter, temperatures fell, and a positive water balance was reinstated. Leaching conditions prevailed from week 9.

Apparent N Recovery in Herbage

Under all conditions urine N markedly reduced fixation by white clover (Table 1). Herbage N yields were increased over control plots under cool-moist and warm-moist but not under warm-dry conditions. Estimates for apparent N recovery (i.e., enhanced removal from soil as a proportion of urine N applied) averaged 55%, 30%, and 11% for the respective seasonal periods and were always lower at the higher N rate. Recovery was especially low under warm conditions, irrespective of N rate.

Soil Total Nitrogen

Addition of urine N did not significantly affect soil total N, measured prior to and at the finish of each experimental run.

DISCUSSION

Urine N Recovery by Swards

Swards were scorched by urine N treatment in warm weather. Damage was more pronounced at the heavier

Table 1. N fixation and herbage N yields, and apparent N recovery in urine-treated swards.

Treatment	N fixation	Herbage N yield	Nett soil N removal	Apparent N recovery
	---------- g N/m² ----------			(%)
(a) Cool-moist				
NO	14.5	15.6	1.1	—
N30	5.9	26.4	20.5	65
N60	4.0	32.5	28.4	46
S.E.	±1.0	±1.0	±0.8	±1
Result†	**	***	***	**
(b) Warm-moist				
NO	8.0	14.3	6.3	—
N30	1.5	18.8	17.3	37
N60	0.5	20.2	19.7	23
S.E.	±0.8	±1.0	±1.1	±4
Result	**	*	**	ns
(c) Warm-dry				
NO	3.8	7.4	3.6	—
N30	1.1	9.0	7.9	14
N60	0.9	9.0	8.1	8
S.E.	±0.4	±0.6	±0.5	±2
Result	*	ns	**	ns

†ns, *, **, *** = $P > 0.05$, < 0.05, < 0.01, < 0.001.

rate and remained obvious for a week or longer. Inspection of grazed swards confirmed a similar influence in urine patches.

This ryegrass-clover pasture did not recover urine N effectively: Recovery averaged only 32% over the experiments. Clearly, N application rates were greatly in excess of plant requirements, and surplus N was at risk of loss.

Losses of Urine N

Generally poor N recovery by plants, coupled with the fact that soil total N did not accumulate, establish that most of the urine N was lost from the soil-plant complex.

Patterns for transformation and loss of urine N are detailed elsewhere (Ball and Keeney 1981). Volatilization of NH_3-N was the major pathway of loss only under warm-dry conditions. When moist soil conditions prevailed, much urine N was lost through an undefined combination of leaching and denitrification. Movement of NO_3-N to subsoil under leaching conditions was confirmed.

Implications for Pastoral Agriculture

Support for the generality of these results arose from an independent study (Ball 1979). Nitrogen balances for intensively utilized, sheep-grazed pastures revealed unaccounted-for outgoings of several hundred kg N/ha annually. Two-thirds loss of urinary N would account for more than 80% of the unidentified losses from those grass-clover systems.

In general, formation of urinary N will increase as herbage utilization increases. More intensively farmed grass-clover systems in New Zealand may be in negative N balance — a view confirmed experimentally (Ball 1979). Similarly, Woodmansee (1978) concluded that semiarid grassland ecosystems in North America may be losing N if they are moderately stocked, largely through N escape from animal urine.

Nitrogen Cycling in Animal Urine Reconsidered

Benefits to pasture production from the recycling of nutrients by grazing animals are generally recognized. Sears (1956) and Levy (1970) refer to the "shower of fertility" in animal excreta, considered to be vitally important in the transfer of symbiotically fixed N from clovers to grasses in mixed pastures (Whitehead 1970).

The authors dispute those views with respect to N. Rather, we have developed an alternative hypothesis that grazing animals cause substantial N outgoings from productive, well-managed pastoral ecosystems. Ruminants ingest relatively proteinaceous forage. The bulk of ingested N is separated from the carbon with which it was associated in herbage and is excreted in urine. Biologically labile N is aggregated into urine patches at rates that are far too high to allow effective plant utilization. As resident soil organic matter is of the mull type (Ball 1979), limited availability of carbon substrates restricts microbial immobilization. Accordingly, substantial N losses from urine patches are inevitable.

LITERATURE CITED

Ball, P.R. 1979. Nitrogen relationships in grazed and cut grass-clover systems. Ph.D. Thesis, Massey Univ., N.Z.

Ball, P.R., and D.R. Keeney. 1981. Transformation, apparent recovery and losses of nitrogen in urine-affected areas of a ryegrass-white clover pasture. N.Z. J. Agric. Res.

Ball, P.R., D.R. Keeney, P.W. Theobald, and P. Nes. 1979. Nitrogen balance in urine-affected areas of a New Zealand pasture. Agron. J. 71:309–314.

Levy, E.B. 1970. Grasslands of New Zealand. Government Print. Wellington, N.Z.

Sears, P.D. 1956. The effect of grazing animals on pasture. Proc. VII Int. Grassl. Cong., 92–102.

Whitehead, D.C. 1970. The role of nitrogen in grassland productivity. Commonw. Agric. Bur. Bul. 48. Farnham Royal, Bucks., UK.

Woodmansee, R.G. 1978. Additions and losses of nitrogen in grassland ecosystems. BioScience 28:448–453.

Annual Losses of Ammonia from a Grazed Pasture Fertilized with Urea

V.R. CATCHPOOLE, L.A. HARPER, and R.J.K. MYERS

Division of Tropical Crops and Pastures, CSIRO, St. Lucia, Queensland, Australia

Summary

Nitrogen-balance studies have shown losses of applied nitrogen (N) from grazed pastures in southeastern Queensland. Losses as high as 80% have been observed from urea N broadcast at 376 kg N/ha/yr over 8 years on a *Setaria sphacelata* cv. Nandi pasture. Management techniques aimed at reducing this loss and thereby increasing the efficiency of use of N by pastures cannot be devised until the pathways of loss are known. The objective of this research was to estimate the annual ammonia (NH_3) loss by convective transport from a urea-fertilized pasture at Samford Pasture Research Station in southeastern Queensland.

Urea was broadcast onto a grazed pasture at 94 kg N/ha in February, May, August, and November 1978 and February 1979. Ammonia fluxes above the pasture were estimated by the momentum balance method, which uses micrometeorological measurements, and the NH_3 concentration profiles in the air above the pasture. Fluxes were measured for 24-hour periods on the day before urea application, on the day of application, and then on the second, fourth, seventh, tenth, fourteenth, and fiftieth days. Supporting soil data, including temperature, pH, water, and mineral N contents, were also obtained.

The annual net loss was estimated to be 90.7 kg N/ha, equivalent to 24% of the applied N. Of that loss, 82% occurred within 2 weeks of application. The greatest loss was from the May fertilizer application, when in 2 weeks losses were 42% of the applied N, while on the other occasions losses ranged from 9% to 13%.

Days when there was a net gain of NH_3–N by the pasture from the atmosphere were common. This fact demonstrates that the pasture is both a source and a sink for atmospheric NH_3.

The estimated loss of 24% of applied N does not fully explain losses observed in previous work. Other loss pathways, including denitrification and leaching, now need to be assessed.

KEY WORDS: Ammonia volatilization, *Setaria sphacelata*, urea, cattle, nitrogen loss.

INTRODUCTION

Large areas of grazed pasture in tropical Australia are of low nitrogen (N) status. Productivity may be increased either by encouraging N fixation by legumes or by the addition of nitrogenous fertilizer. However, at Samford in southeastern Queensland, Catchpoole and Henzell (1975) estimated losses as high as 80% from urea applied over 8 years to a grazed *Setaria sphacelata* pasture. Since action to improve fertilizer use efficiency depends initially on identification of the major pathways of loss, further knowledge of the fate of nitrogenous fertilizer applied to these tropical pastures was required. Surface applications of urea are known to result in substantial losses of N as NH_3 to the atmosphere (Terman 1979).

Losses of NH_3 from land surfaces can now be estimated using micrometeorological methods to determine NH_3 fluxes (Denmead et al. 1974). These methods offer several advantages over previously used chamber methods—the measurement is made over a relatively large area, minimizing the effects of field spatial variability; the site suffers only minimal disturbance; animals may continue to graze the site; and natural weather and micrometeorological conditions are maintained. The aim of this work was to use these techniques to measure losses of NH_3 from a grazed pasture receiving urea fertilizer every 3 months and to estimate the annual loss of N as NH_3.

MATERIALS AND METHODS

The research site was at Samford Research Station, near Brisbane, Queensland. Site management and soil characteristics are given in Vallis et al. (1982) and Harper et al. (1981).

Measurements to calculate fluxes of NH_3 above the pasture were made for 1 day immediately before each application of urea commencing at 1200 hours, continued for the first, second, fourth, seventh, tenth, and fourteenth days after application and again for 2 days approximately 50 days after application of urea. This sampling sequence was varied slightly in response to marked weather changes.

Fluxes of NH_3 were calculated, using the momentum balance equation (Lemon 1967) from measurements of profiles of wind speed, ambient temperature, and atmospheric concentrations of NH_3. All microclimate and NH_3 concentration measurements were made near the center of the pasture with fetch length in any direction. (For a description of measurements, see Harper et al. 1981.) Ammonia concentrations were measured by drawing air at the rate of 0.75 m^3/hr through H_2SO_4 traps. These were changed every 2 hours, and NH_3 was determined colorimetrically (Weier et al. 1980). Ammonia concentrations for the first series were measured at the same heights as the wind speeds and at the center of the pasture. The NH_3 concentrations demonstrated satisfactory profile development up to 80 cm over a range of conditions. Subsequent measurements were made at heights of 20 and 40 cm at three places near the center of the pasture, to avoid single-site problems due to uneven fertilizer spread or to urine patches near the sensors.

Net daily figures for NH_3-N loss or gain were the sum of the fluxes over 12 2-hour periods. They were net figures since the pasture commonly gained some NH_3-N during the night and lost some during the day. Ammonia fluxes were estimated for the periods between measurements by curvilinear extrapolation.

Soil-surface temperature and water content of the 0–0.5-cm layer were measured throughout each series of measurements.

RESULTS

The patterns of NH_3 influx or efflux over the year of measurement are given in Fig. 1. After each of the five urea applications there was a period of rapid NH_3 loss. This period peaked within 2 days, then declined.

V.R. Catchpoole is an Experimental Officer at the Division of Tropical Crops and Pastures of CSIRO; L.A. Harper is a Soil Scientist at the Southern Piedmont Conservation Research Center, USDA; and R.J.K. Myers is a Principal Research Scientist at the Division of Tropical Crops and Pastures of CSIRO.

Ultimately this decline resulted in efflux rates as low as 10 g/ha/day, or in several cases resulted in influxes of NH_3 into the soil-plant system.

The maximum rates and duration of the efflux peaks varied between different urea application dates. Rates of NH_3 loss were highest in May-June and lowest in August-September. High rates (>1,000 g N/ha/day) were maintained for long periods (~7 days) in May-June and August-September and for short periods (~2 days) in February-March, November, and February. Soil-surface

Figure 1. Daily ammonia fluxes, surface soil temperature, and soil water data from a grazed, fertilized pasture in southeastern Queensland. Triangular symbol indicates dates of application of urea. On temperature graph, upper line denotes maximum temperature and lower line denotes minimum temperature.

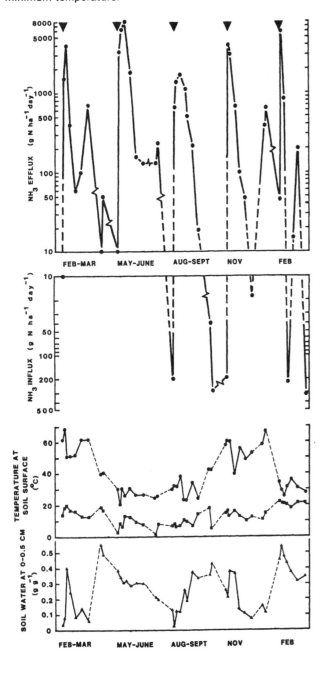

Table 1. Estimates of ammonia losses from a grazed pasture over several time periods.

Date of urea application		Measurement period (days)	Loss of ammonia	
			g N/ha	% of application
1978,	Feb 22	0-14	11300	12
		15-91	8360	9
	May 24	0-14	39400	42
		15-91	5900	6
	Aug 23	0-14	11800	13
		15-78	6740	-8
	Nov 8	0-14	8160	9
		15-91	12200	13
1979,	Feb 7	0-14	8160	9

temperatures and soil-moisture trends (Fig. 1) indicate that urea was broadcast on relatively dry soil in February-March, August-September, and November and on very wet soil in May-June and February. The soil-surface temperatures were very high in February-March and November.

The largest ammonia loss in the 14 days after application occurred in the cooler weather (Table 1). However, after day 16, daily averages were between a loss of 76 g N/ha and a gain of 100 g N/ha. The larger of these losses were in summer and the smaller in winter. The total loss for the year (excluding the second February application) was estimated to be 90.7 kg N/ha, equivalent to 24% of the N applied.

Days when there was a net gain of NH_3-N by the pasture from the atmosphere were common. All series except May-June were preceded by, were followed by, or contained such days.

DISCUSSION

We associate the large variation in NH_3 loss between different times of application with variation in soil-surface moisture and temperature. In particular, it appeared to be influenced by the water content at time of application. Thus, when urea was applied to dry soil (February-March and August-September), the ammonia lost in the first day was less than 1,500 g/ha, in comparison with more than 3,200 g/ha when urea was applied to wet soil. The subsequent pattern of loss was influenced by rainfall, which sharply reduced the rate of ammonia loss.

Periods when the net movement of NH_3 was from the atmosphere into the soil-plant system were common. Whether this sorption was by the soil or by the plant is not obvious from our data. However, Denmead et al. (1976) have demonstrated sorption of atmospheric NH_3 by the plant canopy.

At other times, net losses of NH_3 occurred. Immediately after urea application, this NH_3 would have come directly from the urea, but this source is unlikely for measurements made 50 and 90 days later. Volatilization of NH_3 from urine patches was possibly a major factor in the 20 kg/ha/yr loss estimated for these periods. Losses of a similar magnitude from urine patches in the same pasture for the same year were measured with semiclosed chambers placed over the urine patch (Vallis et al. 1982).

Our estimate of annual loss, although substantial, accounts for only one-third of previous estimates of total fertilizer N losses on this pasture system (Catchpoole and Henzell 1975, Henzell 1972). Obviously, other pathways of N loss are operating, and further work is required to trace losses of N due to leaching and denitrification and to more recently recognized pathways, including gaseous losses from senescing leaves and losses accompanying nitrification.

LITERATURE CITED

Catchpoole, V.R., and E.F. Henzell. 1975. Losses of nitrogen from pastures. CSIRO Trop. Agron. Div. Rpt. 1974-75, 83.

Denmead, O.T., J.R. Freney, and J.R. Simpson. 1976. A closed ammonia cycle within a plant canopy. Soil Biol. Biochem. 8:161-164.

Denmead, O.T., J.R. Simpson, and J.R. Freney. 1974. Ammonia flux into the atmosphere from a grazed pasture. Science 185:609-610.

Harper, L.A., V.R. Catchpoole, and I. Vallis. 1982. Gaseous ammonia transport in a cattle-pasture system. Proc. Int. Symp. Nutrient Cycling in Agric. Ecosystems, Athens, Ga., Sept. 21-26, 1980.

Henzell, E.F. 1972. Loss of nitrogen from a nitrogen-fertilized pasture. Austral. Inst. Agric. Sci. J. 38:309-310.

Lemon, E.R. 1967. Aerodynamic studies of CO_2 exchange between the atmosphere and the plant. In A. Dietro, F.A. Greer, and T.S. Army (eds.), Harvesting the sun: photosynthesis in plant life, 203-290. Academic Press, New York.

Terman, G.L. 1979. Volatilization losses of nitrogen as ammonia from surface-applied fertilizers, organic amendments, and crop residues. Adv. Agron. 31:189-223.

Vallis, I., L.A. Harper, V.R. Catchpoole, and K.L. Weier. 1982. Volatilization of ammonia from urine patches in a subtropical pasture. Agron. J. Austral. J. Agric. Res. 33:97-107.

Weier, K.L., V.R. Catchpoole, and L.A. Harper. 1980. An automated colorimetric method for the determination of small concentrations of ammonia in the air. Trop. Agron. Tech. Mem. 20.

Fate of Nitrogen Applied to Grassland in Animal Wastes

M. SHERWOOD

Agricultural Institute, Johnstown Castle Research
Centre, Wexford, Ireland

Summary

The experiment was designed to determine the effects of land spreading of animal manures on the nutrient content of surface runoff water and infiltrating soil water. This report deals only with nitrogen (N) and attempts to quantify the losses through runoff, leaching, and volatilization of ammonia (NH_3) as well as N uptake in the grass over a 3-year period.

Experimental plots sited on grassland, on two soil types (a moderately drained loam and an impermeable gley), were equipped to collect surface runoff water. Ceramic probes were installed in each plot at 15-, 30-, 60-, and 100-cm depths to extract infiltrating soil water. Pig slurry was applied at three different rates three times each year. Inorganic fertilizer and control plots were included. Volatilization of NH_3 and nitrification were determined by daily analyses of soils following slurry application. Grass was harvested three times each year.

The results indicate that 40%-80% of the slurry ammonium nitrogen ($NH_4^+ - N$) was lost through volatilization of NH_3 within ~ 7 days of application. In the moderately drained soil, nitrification of the remaining N occurred as shown both by soil analyses and by subsequent appearance of nitrate nitrogen ($NO_3^- - N$) in the soil water. The leaching losses increased as rate of slurry application increased and were estimated to account for ~ 5% of the total applied N at the lowest rate of slurry application and ~ 13% at the highest rate. Leaching losses from fertilizer were comparatively higher than from slurry.

At the impermeable site $NO_3^- - N$ was absent from the soil water of slurry treatments during the first 2 years of the experiment. It was not clear whether nitrification was inhibited in the waterlogged soil, but there was evidence to suggest that nitrification in the top few centimeters of soil may have been followed by rapid denitrification in the anaerobic subsurface soil.

Rainstorms causing runoff did not occur for several weeks following most of the treatment application dates, and therefore losses in runoff were small. The highest recorded loss was ~ 8% of the ($NH_4^+ - N$) applied when rainstorms occurred soon after application at the impermeable site in November 1978.

Nitrogen uptake in grass from slurry was very poor, ranging from 20% to 24% recovery from the low rate to 14% to 17% from the high rate. Uptake from fertilizer was 52%-55% in the same period.

The results show clearly that spreading animal slurry on grass is a very inefficient method of recycling its nitrogen content. Improved efficiency would depend on alternative methods of land spreading, such as injection, where losses through volatilization of NH_3 would be greatly reduced.

KEY WORDS: nitrogen, infiltration, runoff, ammonia volatilization, animal wastes, nitrification, denitrification.

INTRODUCTION

The quantities of both pig and cattle slurry to be disposed of by land spreading have increased in Ireland in recent years due to changes in housing patterns. The main objective of this research was to study the effects of such disposal on water quality—both of surface runoff water and of infiltrating soil water. During the course of the experiment it became very obvious that the recovery of slurry nitrogen (N) in grass was extremely poor, and consequently the fate of this N was also studied. This paper reports the findings on N transformations and losses following application of pig slurry to grassland.

METHODS

Experimental grass plots (2 × 30 m) were located on sloped sites (~10% slope) on two different soil types, a moderately drained loam (permeability 0.12 m/24 hours) known as Hoarstone Field and on impermeable gley (permeability of 0.03 m/24 hours) at Castlebridge. The plots were equipped to collect all the runoff water, and ceramic probes were installed at depths of 15, 30, 60, and 100 cm at two locations in each plot to extract the infiltrating soil water.

The plot treatments at both sites were as follows: plot 1, no treatment; plot 2, pig slurry, 22.5 m³/ha; plot 3, pig slurry, 45 m³/ha; plot 4, pig slurry, 90 m³/ha; and plot 5, fertilizer N, phosphorus (P), and potassium (K), 120, 42, 50 kg/ha. The N fertilizer was calcium ammonium nitrate in 1976 and ammonium sulphate on all other dates.

Each treatment was applied to the same plot three times each year for 3 years at dates that were chosen to give maximum information about runoff or infiltration rather than for optimum plant nutrient uptake (Table 1). Slurry was applied on a volumetric basis, and as the dry matter varied the quantities of nutrients applied on different dates also varied.

Runoff water was measured and sampled after each runoff event and analyzed for biochemical oxygen demand (BOD), ammonium N ($NH_4^+ - N$), nitrate N ($NO_3^- - N$), P, and K (Byrne 1979). Soil water was sampled every 2 weeks and analyzed for $NO_3^+ - N$.

Ammonia volatilization and N transformations were determined by taking soil samples on days 1, 2, 3, 7, 14, and 21 after treatment application and determining

$NH_4^+ - N$ and $NO_3^- - N$ by steam distillation (Bremner 1965).

Grass was harvested three times each year by cutting to 5 cm, and it was analyzed for dry matter, N, P, and K (Byrne 1979).

RESULTS

Volatilization of Ammonia

At the end of one week approximately 65% of the slurry inorganic N could not be found in the soil samples as shown in Table 1. Although some of this N may have been immobilized by mircoorganisms, it is more likely that most was lost through volatilization of ammonia. The losses were very similar at both sites. The mean $NH_4^+ - N$ concentration of all the pig slurry samples applied over the 3 years of the experiment was 56% of total N. Loss of 65% of the applied $NH_4^+ - N$ would therefore result in loss of 36% of the total applied slurry N.

Runoff

Rainfall was 1,135, 1,081, and 1,040 mm, exceeding evapo-transpiration by 710, 607, and 443 mm for 1976-1977, 1977-1978, and 1978-1979 respectively. Approximately 10% of the surplus water was collected as runoff at Hoarstone Field, whereas 40% of the surplus was collected at Castlebridge. The amount of inorganic N lost in runoff water is mainly dependent on the time interval between the treatment application and the rainstorm that causes the runoff (Sherwood and Fanning 1981), and as the rainstorms often did not occur for several weeks after application, the mean loss of N in runoff water was < 1% at Hoarstone Field and ~2% at Castlebridge. The highest recorded loss was 8% of the $NH_4^+ - N$ applied where storms occurred soon after application at Castlebridge in November 1978.

Infiltration of Nitrate

At Hoarstone Field the mean $NO_3^- - N$ concentration in the soil water in each slurry treatment plot was related to its rate of application. Concentrations of all readings for the two sites at the 100-cm depth, averaged over the moisture surplus period for each of the three seasons, are shown in Table 2. An approximation of the amount of N leached at Hoarstone Field was obtained by assuming that 10% of the surplus water was lost as runoff water and the remainder leached. It was also assumed that the

---

The author is a Research Scientist in the Department of Soil Biology of the Agricultural Institute.

Table 1. Changes in inorganic-N concentration of soil (μg/g @ 0 – 15 cm) with time, following application of 90 m³/ha pig slurry (Hoarstone Field).

Application date	N applied		Soil inorganic N concentration					
			Day 0*		Day 1**		Day 7**	
	Total N	$NH_4^+ - N$	$NH_4^+ - N$	$NO_3^- - N$	$NH_4^+ - N$	$NO_3^- - M$	$NH_4^+ - N$	$NO_3^- - N$
30 June 1976	401	224	112	1	–	–	–	–
3 Dec 1976	558	263	131	1	–	–	55	17
25 March 1977	736	355	178	2	118	5	49	2
18 August 1977	346	287	144	1	160	1	73	12
22 Nov 1977	349	239	120	1	–	–	18	7
29 March 1978	330	183	92	1	14	0	8	2
14 June 1978	503	339	170	2	33	1	15	13
17 Nov 1978	550	299	149	2	95	2	41	7
26 April 1979	359	166	83	3	39	1	12	2

*Calculated from the volume and inorganic-N content of the slurry applied.

**Measured by soil analysis.

$NO_3^- - N$ concentration of the leaching water was similar to that of the soil water at 100-cm depth. Based on these data, 128, 218, 196 kg N/ha were leached from the 90 m³/ha slurry treatment during the 3 years, or ∿13% of the total slurry N, whereas a smaller fraction, ∿5%, was leached from the 22 m³/ha treatment.

At Castlebridge, during the first two seasons there was virtually no $NO_3^- - N$ in the soil water at any depth under the slurry treatments. In the third season, slurry was applied in September when there was a 28-mm moisture deficit, and nitrification of slurry N was well advanced before subsequent rain leached it down through the soil where it was detected at all depths. With 40% of the surplus water collected as runoff in that season, the estimated maximum amount of N that could have leached was 84.5 kg N/ha from the 90 m³/ha pig slurry treatment, or ∿5% of the total N applied to that season.

N Recovery in Grass

Recovery of total slurry N in grass was very poor over the 3 years, ranging from 20% to 24% from the 22.5 m³/ha treatment and from 14% to 17% recovery from the 90 m³/ha treatment. The overall recovery in grass from fertilizer N applied on the same dates was 52% at Hoarstone Field and 55% at Castlebridge. The recovery data from some of the treatment plots at Hoarstone Field

are shown in Table 3. The trends at the other site were very similar.

DISCUSSION

Volatilization of ammonia was estimated at ∿36% of the total applied N, a figure that is consistent with the losses from dairy manure observed by Lauer et al. (1976). This loss was to be expected, since most of the $NH_4^+ - N$ present in slurry is associated with carbonate/bicarbonate anions derived from hydrolysis of urea, and both these compounds decompose on drying.

Nitrogen losses through surface runoff were small during the 3 years of the experiment, although more recent studies at the same site have shown that up to 20% of the $NH_4^+ - N$ can be lost through runoff where heavy rainstorms occur within 48 hours of spreading (Sherwood and Fanning 1981).

Infiltration accounted for up to 13% of the slurry N at the more permeable site. It was negligible during the first 2 years at the impermeable gleyed site. It was not possible to determine if nitrification was inhibited in this very wet soil or if, in fact, nitrification took place in the top few centimeters and was followed by rapid denitrification before it reached the 15-cm probe. Incubation studies in the laboratory showed that slurry was a very suitable carbon source for denitrification under waterlogged conditions at both 5° and 20°C (Sherwood 1980).

The balance sheet for recovery of slurry N over the 3 years at Hoarstone Field for the 90 m³/ha pig slurry treat-

Table 2. $NO_3^- - N$ concentration in soil water at 100-cm depth averaged over the moisture surplus period.

	Mean $NO_3^- - N$ conc. (mg/l) in soil water at 100-cm				
	Plot 1	Plot 2	Plot 3	Plot 4	Plot 5
Hoarstone Field					
1976–77	0.4	2.4	11	20	28
1977–78	0.6	5.8	13	40	19
1978–79	1.5	3.1	26	49	10
Castlebridge					
1976–77	0.6	1.0	1.0	1.1	3.1
1977–78	0.5	1.0	1.0	1.8	2.3
1978–79	1.9	4.0	10	32	5.9

Table 3. Recovery of nitrogen in grass at Hoarstone Field.

	N harvested in grass (kg/ha)			
	Plot 1	Plot 2	Plot 4	Plot 5
1976–77	123	143	221	268
1977–78	79	194	291	303
1978–79	79	197	352	275
Total	281	534	864	846
Total N applied (1976–79)	0	1259	4132	1080
Recovery (%)		20	14	52

ment was as follows: volatilization of ammonia, 36%; grass uptake, 14%; leaching loss, 13%; runoff, < 1%, with 36% unaccounted for through immobilization and possibly some denitrification. The corresponding figures for Castlebridge were 38, 17, 2, and 2, respectively, with 41% unaccounted for. It is very likely that denitrification was higher at this site.

LITERATURE CITED

Bremner, J.M. 1965. Inorganic forms of nitrogen. *In* C.A. Black (ed.), Methods of soil analysis, Part 2. Agron 9: 1179–1251. Am. Soc. Agron., Madison, Wis.

Byrne, E. 1979. Chemical analysis of agricultural materials. An Foras Taluntais, Dublin, Ireland.

Lauer, D.A., D.R. Bouldin, and S.D. Klausner. 1976. Ammonia volatilisation from dairy manure spread on the soil surface. J. Environ. Qual. 5:134–141.

Sherwood, M. 1980. The effects of land-spreading of animal manures on water quality. In J.K.R. Gasser (ed.), Effluents from livestock, 379–390. Applied Sci. Pub. Ltd., Essex, Eng.

Sherwood, M., and A. Fanning. 1981. Nutrient content of surface runoff water from land treated with animal wastes. *In* J.C. Brogan (ed.), Nitrogen losses and surface run-off from landspreading of animal manures. 5–17. Martinus Nijhoff/Dr. W. Junk Pub. London.

Progress in Development of Conservation Plant Cultivars by the USDA Soil Conservation Service

H.W. EVERETT

USDA SCS, Washington, D.C., U.S.A.

Summary

The USDA Soil Conservation Service (SCS) operates a coordinated network of 22 plant materials centers in the U.S.A. to assemble, evaluate, select, cooperatively release, and provide for the commercial increase of native and introduced plants for the conservation and improvement of soil, water, and related resources. Standardized procedures have been developed and are being used for this comparative plant-testing program. Evaluation data can be retrieved from an automatic data-processing system through standardized reports. Over 140 SCS-released cultivars of conservation plants are available commercially for use in range, pasture, and other grassland improvement to reduce sediment movement and improve wildlife habitat. These improved plants are also used to solve other erosion problems including the reclamation of surface-mined land, roadside development, and sand dune and shoreline stabilization. A number of new cultivars for use in grassland plantings are now available.

KEY WORDS: Soil Conservation Service, grasses, legumes, shrubs, forbs, cultivars.

INTRODUCTION

Since the 1930s, the USDA Soil Conservation Service (SCS) has operated a coordinated network of plant materials centers to evaluate, select, release, and introduce into commercial use new plant cultivars for use in the conservation of soil, water, and related resources. There are now 22 plant materials centers strategically located within the U.S., each serving two or more major land-resource areas (USDA 1981) of similar soils, climate, and conservation problems. The centers are operated by SCS or receive some management and guidance from SCS personnel.

A short review of the procedures in plant evaluation and release (Fig. 1) and comments on a number of new cultivars of interest to our grasslands-oriented friends are two objectives of this presentation.

PROCEDURES

Operation of any plant materials center begins with a documentation of the conservation problems, their scope, and their priority. SCS personnel at the state and county levels identify conservation problems and priorities by which the center will direct plant assembly, evaluation, selection, and release of superior cultivars. Representatives of other agencies and soil conservation districts assist in this process. Improved plants are needed for such problems as range and pasture improvement, grassed waterways, gully stabilization, roadside development, streambank protection, reclamation of mined land, dune and shoreline stabilization, and wildlife food and cover.

Plant materials centers assemble seed or plants from local native ecotype collections, from plant-introduction stations, from private seed or plant companies or researchers, and from foreign collection trips by other agencies. Collected plant materials are grown in rows or blocks, and performance data are collected and tabulated for one to 10 years. Comparisons are made using available commercial strains or cultivars. Data on vigor; production and quality of forage, seed, or biomass; insect and disease resistance; adaptation; and other plant characteristics are collected and analyzed. All plant materials work is conducted in cooperation with state agricultural experiment stations and other federal or state agencies.

Selection of superior accessions or ecotypes is made. These accessions are increased at the center and evaluated on controlled sites away from the center to determine adaptation and performance. These field-evaluation plantings are often done on state experiment station sites or on land under the control of other agencies. Again, any available commercial strains or cultivars are evaluated as a standard. Data are recorded for 1 to 5 years, and selec-

The author is a Staff Plant Materials Specialist at USDA SCS.

Fig. 1. SCS plant materials program plant evaluation steps.

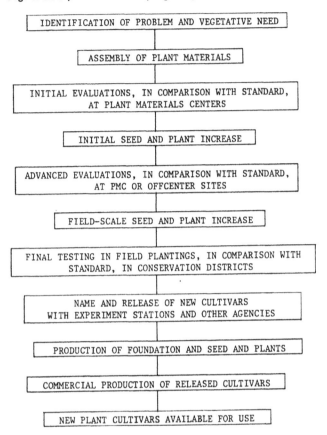

dation seed project associated with the state experiment station. SCS shares and exchanges seed or plants with state experiment stations and may distribute them through soil conservation districts to qualified local growers for production of certified seed or plants. SCS or state experiment stations do demonstration plantings and distribute information to encourage the purchase and use of new cultivars. SCS recommends the new cultivars for use in solving local conservation problems.

SCS has strengthened the evaluation process by using a more statistical evaluation of plant performance, by increasing the number of cooperative evaluations by agricultural experiment stations and other federal and state agencies, by developing a more accurate field evaluation, and by implementing a data-processing system (USDA 1977).

RESULTS: RECENT CULTIVAR RELEASES

Although most of the cultivars released since 1943 have been grasses, legumes, and forbs for use in grassland conservation plantings, several shrub and tree cultivars have been released for windbreak and wildlife food and cover plantings (USDA 1980b). Over 140 are available (USDA 1980a). A few recent examples of such cooperatively released cultivars are given below:

- Hachita blue grama, *Bouteloua gracilis* (Willd. ex H.B.K.) Lag. ex Griffiths, a warm-season native bunchgrass, was released for range and critical area seedings in the southwestern U.S. It will be commercially available in late 1981.
- Bigalta, Greenalta, and Redalta limpograsses, *Hemarthria altissima* (Poir.) Stapf and C.E. Hubb., were released in 1978 for use as forage plants. These limpograsses, introduced from South Africa in the early 1970s, proved to be well adapted to much of Florida, Puerto Rico, and Hawaii, as well as to a portion of Texas.
- Three cultivars—Arriba, Barton, and Rosana—of western wheatgrass, *Agropyron smithii* Rydb., a native cool-season sod-forming grass species, were released in the 1970s for use in range seedings and critical area plantings in the midwestern and western U.S. All are commercially available.
- Several cultivars of *Atriplex* species have been released for use in range seedings, critical area stabilization, and wildlife food and cover plantings. Marana and Wytana fourwing saltbush, *Atriplex canescens* (Pursh) Nutt. and *Atriplex canescens* var. aptera (A. Nels.) C.L. Hitchc., respectively; Corto Australian saltbush, *Atriplex semibaccata* R.Br.; and Casa quailbush, *Atriplex lentiformis* (Torr.) Wats., are now available for use in the various parts of the western U.S.
- Maximillian sunflower, *Helianthus maximilianii* Schrad., is a tall native perennial useful for range mixtures in much of the southern Great Plains. The cultivars Aztec and Prairie Gold are available commercially. Several other forbs have been released

tion of the best-performing accession is made. The selected accession is increased at the plant materials center for a final evaluation step—field plantings.

A unique and most effective step in the development and introduction of a new conservation plant is that of field plantings on farms or ranches of soil conservation district cooperators or on the land of other official cooperators, such as state departments of natural resources, fish and game agencies, or state highway departments. Field planting is the final check on the adaptation and performance of a potential conservation plant under actual use conditions. It is a direct comparison of the accession being tested and the plant currently recommended for a specific site, soil, and conservation problem. SCS conservationists, state experiment station workers and other agency personnel, farmers, and ranchers collect data to determine specific adaptation and performance. Based upon the successful performance of field plantings, a new plant may be recommended for release as a superior cultivar.

SCS, agricultural experiment stations, and other cooperating agencies assemble the plant-evaluation data, name the plant, prepare a release notice for it, and jointly recommend its release as a new cultivar. The agency that developed the new cultivar then assumes responsibility for the maintenance of breeders' seed or plants. Foundation seed or plants are distributed to commercial seed or plant growers through a crop improvement association or foun-

since 1970 for use in mixtures for rangeland reseeding and in critical area stabilization. These include Kaneb purple prairie clover, *Petalostemum purpureum* (Vent.) Rydb.; Sunglow grayhead prairie coneflower, *Ratibida pinnata* (Vent.) Barnh.; Nekan pitchersage, *Salvia azurea v. grandiflora* Benth.; Eureka thickspike gayfeather, *Liatris pycnostachya* Michx.; and Bandera Rocky Mountain penstemon, *Penstemon strictus* Benth.

SCS has recently been concentrating on native plant evaluation and release, but introduced species, such as the limpograsses, a number of lovegrasses, buffelgrasses, and wheatgrasses, often play an important role in helping solve soil, water, and related resource conservation problems because they are adapted to and productive on grasslands in the United States (USDA 1979).

CONCLUSION

Uniform procedures for plant collection, evaluation, selection, and release enable evaluation data to be tabulated, entered, and retrieved from a data-processing system. Procedural handbooks are available from SCS.

To date, the centers have released more than 200 cultivars of native and introduced plants for specific conservation purposes; more than 140 of them are available from commercial sources. The tabulated benefit:cost ratio of the center's work is 9:1, on the basis of the retail value of the commercially available cultivars, and there is potential for treatment of approximately 0.6 million ha/yr using commercially available, cooperatively released cultivars.

LITERATURE CITED

U.S. Department of Agriculture (USDA). 1977. National plant materials handbook. Soil Conserv. Serv.

U.S. Department of Agriculture (USDA). 1979. Plant materials for conservation. Prog. Aid 1219. Soil Conserv. Serv.

U.S. Department of Agriculture (USDA). 1980a. Commercial production of SCS released plant varieties – 1979 production, retail value, and potential acres to be treated. Soil Conserv. Serv.

U.S. Department of Agriculture (USDA). 1980b. Improved plant materials cooperatively released by SCS through 1980. Soil Conserv. Serv.

U.S. Department of Agriculture (USDA). 1981. Land resources regions and major land resources areas of the United States. U.S. Dept. of Agric. Handb., rev., Soil Conserv. Serv.

Effects of Topsoil Depths and Species Selection on Reclamation of Coal-Strip-Mine Spoils

W.J. MC GINNIES and P.J. NICHOLAS

USDA-SEA-AR Forest and Range Research and Colorado Agricultural Experiment Station, Fort Collins, Colo., U.S.A.

Summary

The research objectives were (1) to determine the effects of topsoil on establishment of seeded vegetation on strip-mine spoils, (2) to determine the effects of depth of topsoil placed over spoil on growth of herbage and roots, and (3) to evaluate 17 grass species grown on topsoil and mine spoil.

In a field study in northwestern Colorado, a seed mixture of range grasses, forbs, and shrubs was drilled on five (0, 10, 20, 30, and 46 cm) depths of topsoil placed over surface coal-mine spoil, and the effect of topsoil depth on species establishment was measured. In a greenhouse study, the effects of five depths of topsoil placed over spoils on herbage and root yields of intermediate wheatgrass (*Agropyron intermedium* [Host] Beauv.) and wheat (*Triticum aestivum* L.) were determined. In a second greenhouse study, 17 grasses were planted in 25 cm of topsoil over spoil and in spoil alone to evaluate species differences in herbage yield, root yield, and root distribution in topsoil and underlying spoil.

In the field planting, stand establishment averaged 1.9 times better on all topsoil depths than on spoil alone, and the density of the seeded stand increased linearly as topsoil depth increased. In the greenhouse, herbage yield increased linearly with the increase in topsoil depth. Root production by both species was greater in topsoil than in underlying spoil; however, intermediate wheatgrass produced two to four times more root weight in spoil than did wheat. In the second greenhouse study, the 17 species were divided into 5 groups based upon productivity, rooting patterns, and other characteristics.

Depth of topsoil and species selection should be considered for optimum revegetation success. The greater the topsoil depth is, at least up to 46 cm, the denser the seeded stand will be in the field, and the greater the herbage and root production will be in the greenhouse. Species can be selected that will provide high root production, good rooting into the underlying spoil, maximum herbage yield, or a combination of these qualities.

KEY WORDS: topsoil, reclamation, coal mine, strip mine, mine spoils, species selection.

INTRODUCTION

The importance of using topsoil for reclamation of coal strip mines has been demonstrated, but little is known about the effects of thickness of the replaced topsoil on establishment and subsequent growth of seeded species in the central Rocky Mountain region. Much is also known about above-ground growth of many of the adapted species when grown on rangeland, but, again, little has been reported concerning patterns of root growth when these species are grown in topsoil placed on mine spoil. Such information is necessary for understanding the ecology of seeded species.

The objectives of the studies reported here were (1) to determine the effects of various thicknesses of replaced topsoil on stand establishment and growth of seeded species and (2) to evaluate the responses of 17 grass species when grown in topsoil and in spoil.

METHODS

A field study was established on the Energy Fuels Mine, located 32 km southwest of Steamboat Springs, Colorado. The elevation was 2,135 m, average annual precipitation was 41 cm, soil was a fine montmorillonitic Typic Argiboroll, and native vegetation was dominated by sagebrush (*Artemisia tridentata* Nutt.), oak (*Quercus gambelii* Nutt.), serviceberry (*Amelanchier alnifolia* Nutt.), and associated grasses. A 32-species mixture of perennial range grasses, forbs, and shrubs was drilled in October 1976 on 0, 10, 20, 30, and 46 cm of topsoil placed on regraded coal-mine spoil. In August 1978 stand establishment was assessed, using a scale of 0% to 100% where the best possible stand was assigned a value of 100%.

In the first greenhouse study, intermediate wheatgrass (*Agropyron intermedium* [Host] Beauv.) and wheat (*Triticum aestivum* L.) were planted in cans 15 cm in diameter on 0, 10, 20, 30, and 46 cm of topsoil, which in turn were placed on mine spoil for a combined depth of 53 cm. Herbage was harvested after 3 months and 5 months; the two harvests were combined to obtain a total herbage yield. The cans were then split open, the roots washed from the soil, and the roots divided into those growing in topsoil and those growing in the underlying spoil.

W. J. McGinnies is a Range Scientist and P. J. Nicholas is a Range Technician in USDA-SEA-AR Forage and Range Research. This paper is a contribution of the USDA-SEA-AR and the Colorado Agricultural Experiment Station, Fort Collins, and is published with the approval of the Director of the Colorado Agricultural Experiment Station (Scientific Series Paper No. 2614).

The Environmental Protection Agency, the Bureau of Land Management, and Energy Fuels Corporation also provided funding or other support for the research.

The second greenhouse study was conducted in a similar manner. However, the topsoil was 25 cm thick over 28 cm of spoil, and the spoil-alone treatment was 53 cm of spoil material. Seventeen grass species were evaluated in this study.

The soil for both field and greenhouse studies was a Routt loam, which is a montmorillonitic Typic Argiboroll with a pH of 6.0. The spoils are from the Williams Fork Formation of the Upper Cretaceous Mesa Verde Group consisting of mixed beds of shale and sandstone. The pH of the spoils was 7.2, the sodium adsorption ratio was 0.75, electrical conductivity was 1.05 mmhos/cm, total N was 1,411 ppm (mostly unavailable), and $NaHCO_3^-$ extractable P was 4 ppm. The spoil contained no phytotoxic levels of any metals or minerals.

RESULTS AND DISCUSSION

Field Study

Stand establishment was poorer than is usually observed on the mine because only 56% of the average amount of snow fell during the winter and early spring of 1976–1977. In 1977, seedlings were more abundant on the topsoiled than on the nontopsoiled plots, but results were inconclusive with respect to effects of topsoil thickness on stand density. In 1977–1978, snowfall exceeded the average by 34%, and the seeded species in 1978 averaged 1.9 times higher stand ratings on topsoiled treatments than on spoil alone; the ratings increased linearly (r = 0.90) with topsoil thickness. (Stand ratings were 20%, 31%, 35%, 38%, and 49% for topsoil depths of 0, 10, 20, 30, and 46 cm, respectively.) On the average, meadow brome (*Bromus biebersteinii* Roem and Schult.), smooth brome (*B. inermis* Leyss.), desert wheatgrass (*Agropyron desertorum* [Fisch] Schult.), intermediate wheatgrass, streambank wheatgrass (*A. riparium* Scribn. and Smith), western wheatgrass (*A. smithii* Rydb.), orchardgrass (*Dactylis glomerata* L.), timothy (*Phleum pratense* L.), alfalfa (*Medicago sativa* L.), cicer milkvetch (*Astragalus cicer* L.), and crownvetch (*Coronilla varia* L.), as a group, made up 85% of the total seeded stand.

First Greenhouse Study

In the first greenhouse study, herbage production of wheat and intermediate wheatgrass increased linearly (r = 0.95) as topsoil thickness increased (Table 1). Root production of both species was much greater in topsoil than in spoil. Intermediate wheatgrass produced 78% more root weight in topsoil and over 400% more root

Table 1. Herbage and root production of intermediate wheatgrass and wheat in the greenhouse on thicknesses of topsoil from Energy Fuels Mine.

Species	Topsoil depth (cm)	Greenhouse study no. 1			
		Herbage (g)	Roots in topsoil (g)	Roots in spoil (g)	Total roots (g)
Intermediate wheatgrass	0	1.47	–	2.72	2.72
(*Agropyron inter-*	10	4.18	2.88	3.58	6.46
medium (Host) Beauv.)	20	7.51	6.34	3.93	10.27
	30	11.78	10.52	4.28	14.80
	46	16.56	13.24	4.12	17.36
Wheat (*Triticum*	0	2.50	–	1.11	1.11
aestivum L.)	10	5.72	2.22	0.93	3.15
	20	10.03	3.36	0.87	4.23
	30	14.56	4.01	0.93	4.94
	46	22.17	8.92	0.98	9.90

weight in spoil than did wheat, but wheat averaged 32% more herbage production than intermediate wheatgrass. The increases in yields related to topsoil thickness resulted from the higher quantity of nutrients (particularly nitrogen and phosphorus) available in the topsoil as compared to the spoil.

Second Greenhouse Study

In the second greenhouse study, average herbage yields were 7 times greater and average root yields 6 times greater when the 17 grasses were grown in topsoil than when they were grown in spoil alone. However, there were great differences in both herbage and root yields among species and even greater differences in proportion of roots growing into the spoil under the topsoil. On the basis of differences in root growth, productivity, and other characteristics, each of the 17 species was placed into one of five groups. Group I contained two fine-leaved fescues, hard fescue (*Festuca ovina duriuscula* [L.] Koch) and Arizona fescue (*F. arizonica* Vasey). Both species produced an extremely dense mat of fine-textured roots near the soil surface, but relatively few roots penetrated the spoil below the 25 cm of topsoil (Fig.1). Group II contained two high-yielding introduced pasture grasses, timothy and orchardgrass. Both species produced a moderately high number of roots in the topsoil but very few roots in the underlying spoil. Group III consists of two short-lived native species, slender wheatgrass (*Agropyron trachycaulum* [Link] Malte) and mountain brome (*Bromus marginatus* Nees). Although total root production by group III species was relatively low in the topsoil, root penetration into the underlying spoil was better than in groups I and II. Group IV consisted of tall fescue (*Festuca arundinacea* Schreb.), Russian wildrye (*Elymus junceus* Fisch.), intermediate wheatgrass, pubescent wheatgrass (*Agropyron trichophorum* [Link] Richt.), and western wheatgrass. These five species were characterized by moderate-to-high root production in topsoil with 35%–45% of their total root mass in the spoil underlying the topsoil. Of all

the grasses studied, tall fescue produced the second greatest quantity of herbage and total root growth on topsoil over spoil and in spoil alone. Western wheatgrass was the only native species in this group, and it did well on both topsoil and spoil. Group V contained thickspike wheatgrass (*Agropyron dasystachum* [Hook.] Scribn.), beardless wheatgrass (*A. inerme* Scribn. and Smith), streambank wheatgrass, desert wheatgrass, smooth brome, and basin wildrye (*Elymus cinereus* Scribn. and Merr.). These species do not otherwise fit into groups I through IV. The species in group V were average or below average in both herbage and root production compared with the other grasses studied. Desert wheatgrass and smooth brome are introduced species that are usually considered well adapted to western Colorado. Beardless wheatgrass had the lowest herbage production of all species on spoil alone.

Timothy, tall fescue, western wheatgrass, thickspike wheatgrass, and Arizona fescue had the highest herbage yield when grown in topsoil. Hard fescue, tall fescue, Arizona fescue, Russian wildrye, and smooth brome had the highest total root production. On spoil alone, western wheatgrass, tall fescue, orchardgrass, mountain brome,

Fig. 1. Root and herbage growth on topsoil over spoil and on spoil only in the second greenhouse study.

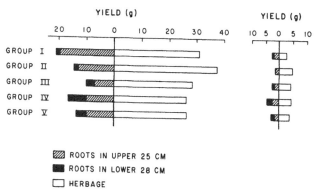

YIELD (g)

20 10 0 10 20 30 40

YIELD (g)

10 5 0 5 10

GROUP I
GROUP II
GROUP III
GROUP IV
GROUP V

▨ ROOTS IN UPPER 25 CM
■ ROOTS IN LOWER 28 CM
☐ HERBAGE

and desert wheatgrass had the highest total herbage production, and intermediate wheatgrass, tall fescue, Russian wildrye, hard fescue, and pubescent wheatgrass had the greatest root production.

CONCLUSIONS

These studies show the importance of topsoil and the need for using as thick a topsoil layer as possible. Judging by field and greenhouse studies, the thicker the topsoil layer is (up to the 46 cm used in these studies), the better the seeded stand and the higher the yield of herbage and roots will be.

High herbage production, a dense root system in the topsoil, and good root penetration into the underlying spoil are desirable growth characteristics of grasses that enhance soil stabilization and soil development. Because a single species seldom possesses all of these characteristics, a mixture of species will probably be best for revegetation of disturbed lands.

Savanna Ecosystems in Tropical South America: Findings from a Computerized Land-Resource Survey

T.T. COCHRANE

CIAT, La Paz, Bolivia

Summary

A land-resource survey is being made in tropical South America with the objective of creating a base for development and transfer of improved food-crop and pasture-plant germ plasm–based technology. Preference was given to study of the 200 million ha of native savannas, although the survey now stretches across the Amazon forests to southern Brazil.

Land information is reduced to a common base, with land systems defined as repetitive patterns of climate, landscape, and soils. They are delineated directly onto satellite and side-looking radar imagery following field work and climatic analyses. Within land systems the vegetation and soils are separately described and estimated according to their occurrence on landscape facets that follow the principal topographic, vegetation, and soil sequences within the land systems. Land-resource information is recorded in a computerized system for storage, retrieval, and production of analytical maps and data-printouts in order to facilitate speedy and comprehensive analyses.

Savanna regions are found from Venezuela into southern Brazil. They can be separated on the basis of soil drainage into poorly drained and well-drained savannas. In spite of their widespread geographical distribution, well-drained savannas occupy a well-defined habitat delimited by the climatic potential for growth of natural vegetation; they can be discriminated from other physiognomic vegetation classes on the sole basis of total wet-season potential evapo-transpiration. Wet-season mean monthly temperatures separate lowland from higher-land savannas.

Savanna landscape and soil features were compared. The well-drained savanna soils are mainly oxisols; their fertility status is generally very poor.

The findings of the survey are discussed in terms of their contribution to the development of improved germ plasm–based pasture technology to help increase cattle production in the savannas of tropical America.

KEY WORDS: savannas, potential evapo-transpiration, land systems, oxisols, land-resource survey, tropical South America.

INTRODUCTION

CIAT,[1] in conjunction with national agencies,[2] is evaluating the agricultural land resources of tropical South America in order to create a base for more effective development and transfer of food-crop and pasture-plant germ plasm–based agrotechnology. The work was commissioned in response to increasing recognition on the part of agricultural scientists in the tropics that a lack of understanding of environmental factors is largely responsible for the failure of the Green Revolution to fulfil expectations (Metz and Brady 1980). The work started in 1977 as a survey of the savanna regions (CIAT 1978), with the primary objective of formulating guidelines for development of improved pastures genetically suited to the various climate-soil environments. The study now encompasses a large part of the lowlands of tropical South America (Cochrane et al. 1981).

There are many opinions on the nature and properties

The author is a Land Resource Specialist at CIAT.

of the savannas; the very reason why savannas rather than forests should exist is an enigma that has evoked considerable controversy. Since the classic studies of Warming (1892), climatic, edaphic, geomorphic, biotic, and even anthropic explanations have been advanced. Confusion and contradictory opinions exist mainly because of the limited geographical extension of most studies.

METHODOLOGY

Land-resource information is put into a comparable geographical context by delineating land systems, following the approach of Christian and Stewart (1953). A land system was defined as "an area or group of areas throughout which there is a recurring pattern of climate, landscape and soils." Inherent to their delineation is the treatment of environmental parameters in the following categorical order:

1. Climate—(a) radiant energy received, (b) temperature, (c) potential evapo-transpiration, (d) water balance, (e) other climatic factors.
2. Landscape—(f) topography, (g) hydrology, (h) vegetation.
3. Soil—(i) soil physical characteristics, (j) soil chemical characteristics.

Long-term (covering more than 20 years) climatic data obtained from stations throughout tropical America were compiled by Hancock et al. (1979). Potential evapo-transpiration (POT ET) was calculated to determine the water balance and growing seasons and to assess the amount of energy available for plant growth. Estimates of monthly POT ET were made according to Hargreaves's method (1977), which is based on solar radiation and temperature.

The wet season for natural vegetation was judged to be that part of the year with a monthly moisture availability index (MAI) greater than 0.33 (Hargreaves 1975). MAI is a moisture adequacy index at the 75% probability level of precipitation occurrence.

To provide an estimate of the total amount of energy natural vegetation can use for growth on nonirrigated soils, the total wet-season potential evapo-transpiration (WSPE), was calculated by summing the POT ET of the wet months. This method assumes that little or no dry-season growth takes place. Other climatic parameters were also calculated on a seasonal basis.

The landscape was subdivided into land systems that were delineated onto 1:1,000,000 satellite and side-looking radar imagery (Cochrane 1980). Land-systems maps were collated and computerized to serve as the basis for thematic map production. A limited amount of field work was carried out to provide on-the-ground control and to study the variation of landscape features within land systems. These variations, although not mapped because of scale limitations, were described as land facets, and the proportion of each land facet within the land systems was estimated. In this way landscape and soil features were computed on the basis of the land-facet subdivision.

The most extensive soils in each land facet were classified as far as the Great Group category of the U.S. Soil Taxonomy system, then described in terms of their main physical and chemical properties. Vegetation classes were identified following the criteria of Eyre (1968) for tropical forests and Eiten (1972) for tropical savannas.

The methodology of the study has been detailed by Cochrane et al. (1979). That report includes a description of the computerized data-management system that was set up to facilitate storage, retrieval, and analysis of information, along with thematic-map production.

RESULTS

Savannas are distributed throughout lowland tropical South America, from Venezuela to southern Brazil. They can be subdivided into two broad classes: predominantly well-drained savannas, covering 153 million ha; and predominantly poorly drained savannas, covering 49 million ha. The poorly drained savannas invariably have an edaphic condition that causes them to be waterlogged for prolonged periods of each year. The wide geographical separation of both the well-drained and the poorly drained savannas would suggest that they are found in different climatic regimes; however, this theory was not true in the case of the well-drained savannas.

To investigate the dependency of vegetation of the well-drained soils on climate, Cochrane and Jones (1981) compared many combinations of different climatic parameters, using Hancock et al.'s (1979) meteorological data sets from the 251 stations located as evenly as possible throughout the region, by discriminant analyses. Fig. 1 summarizes the dependency of vegetation classes on total wet-season potential evapo-transpiration (WSPE) and wet-season mean temperature (WSMT). The clustering of vegetation classes can readily be seen; the posterior probability of correct assignment for the vegetation classes was high. In the case of the well-drained savannas, it may be seen that they can be discriminated on the basis of WSPE alone; they form a narrow band right across the center of the diagram. As WSPE provides an estimate of the amount of energy vegetation might use for growth during the annual growing cycle on well-drained soils as accorded by the moisture-balance pattern, it is clear that the savannas occupy a well-defined habitat delimited by the climatic potential for growth.

The possible variation of WSPE between well-drained savannas with different wet-season lengths was also examined. The meteorological data sets from the 61 stations located in the savannas were subdivided into three groups with 6, 7, and 8 months of wet season, and the POT ET values of the groups were compared. There were no significant differences among the WSPE values for savannas having a 6-, 7-, or 8-month wet season ($P < 0.2$). On the other hand, the monthly average wet-season POT ET values were significantly different ($P < 0.001$); these values decrease with increasing wet-season length.

Fig. 1. Cluster diagram of vegetation classes by total wet-season potential evapo-transpiration (WSPE) and wet-season mean monthly temperature (WSMT). A = deciduous forests, B = well-drained savannas, C = poorly drained savannas (not discriminated), D = tropical semievergreen seasonal forest, E = tropical rain forests, F = subtropical semievergreen forests, G = subtropical evergreen forests. Lines of equiprobability of assignments are shown solid. Dashed lines are 95% confidence elipsoids for the vegetation classes. Source: Cochrane and Jones, 1981.

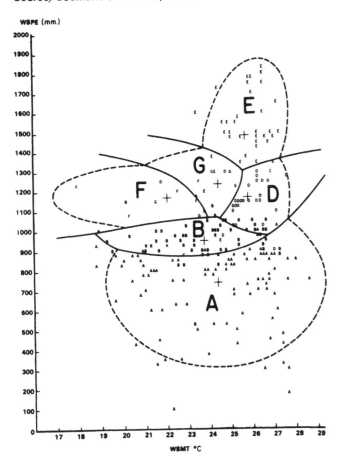

The well-drained savannas were subdivided into two major groups according to wet-season mean temperature (WSMT): equal to or greater than 23.5°C or less than 23.5°C (CIAT 1979, 1980). This division effectively separates the lowland savannas from the higher-land and higher-latitude savannas.

After savanna climates were compared, their landscape and soil features were examined. Of particular importance to ranch management is the variation in amounts of poorly drained lands in the predominatly well-drained savannas and vice versa.

The soils of the well-drained savannas were examined in some detail. Oxisols predominate in the lowland savannas; these are acidic, and their high-percentage aluminum (Al) saturation levels are a potential problem. Major nutrient deficiencies include phosphorus (P), potassium (K), and calcium (Ca). Magnesium levels are often low, and both sulfur deficiencies and some trace-ele-

ment problems are likely. The higher-land savannas have even poorer soils. Once again, oxisols predominate; Al toxicity and low P, K, and Ca levels are major constraints. Phosphorus fixation may prove to be a problem in some soils. Sulfur deficiencies appear to be common-place (CIAT 1980). The new formula for liming soils developed by Cochrane et al. (1980) would provide a low-cost solution to correct Al toxicity in the savannas.

DISCUSSION

The study has led to the finding that well-drained savannas may be separated from forested regions on the basis of WSPE, which effectively equates the total amount of energy that natural vegetation can use for growth, as accorded by annual water-balance pattern.

It should be emphasized that the results presented summarize only a few aspects of the work being carried out. As a result of the studies, CIAT has been able to locate a strategic number of trial sites for screening pasture germ plasm to take major climate and soil variability into account (CIAT 1980). In this way, limited resources are being focused in an effort to develop improved germ plasm-based pasture agrotechnology for increasing cattle production on the savannas of tropical South America.

LITERATURE CITED

Christian, C.S., and S.A. Stewart. 1953. Survey of Katherine-Darwin Region, 1946. Land Res. Ser. 1. Austral.

CIAT. 1978, 1979, 1980. Ann. Rpts. CIAT, Cali, Colombia.

Cochrane, T.T. 1980. The methodology of CIAT's land resource study of tropical America. *In* VI Internatl. symp. on soil information systems and remote sensing and soil surv., 227–231. Purdue Univ., Lafayette, Ind.

Cochrane, T.T., and P.G. Jones. 1981. Savannas, forest and wet-season potential evapotranspiration in tropical South America. Trop. Agric. (Trinidad) 58 (3):185–190.

Cochrane, T.T., J.A. Porras, L. Azevedo de G., P.G. Jones, and L.F. Sanchez. 1979. An explanatory manual for CIAT's computerized land resource study of tropical America (rev. ed.). CIAT, Cali, Colombia.

Cochrane, T.T., J.G. Salinas, and P.A. Sanchez. 1980. An equation for liming acid mineral soils to compensate crop Al tolerance. Trop. Agric. 57:133–140.

Cochrane, T.T., L.F. Sanchez, J.A. Porras, L. Azevedo de G., and P.G. Jones. 1981. Land in tropical America: an agronomist's guide to climates, landscapes, and soils. Vol. 1. The land-systems map of the Amazon, Andean Piedmont, Central Brazil, and Orinoco regions. CIAT, Cali, Colombia.

Eiten, G. 1972. The cerrado vegetation of Brazil. Bot. Rev. 38(2):201–341.

Eyre, S.R. 1968. Vegetation and soils. A world picture. 2nd ed., 195–258. Edward Arnold, Ltd., Gt. Brit.

Hancock, J.H., R.W. Hill, and G.H. Hargreaves. 1979. Potential evapotranspiration and precipitation deficits for tropical America. CIAT, Cali, Colombia.

Hargreaves, G.H. 1975. Water requirements manual for irrigated crops and rainfed agriculture. Utah State Univ., Logan.

Hargreaves, G.H. 1977. World water for agriculture. Utah State Univ., Logan.

Metz, J.F., and N.C. Brady. 1980. Priorities for alleviating soil-related constraints to food production in the tropics. Pt. V. IRRI, Los Baños, Philippines.

Warming, E. 1892. Lagoa Santa. Et. Bidrag til den Biologiske Plantegeografi. K. Danske Vidensk. Selsk. Skr. 6. Copenhagen.

NOTES

1. Centro Internacional de Agricultura Tropical, Apartado Aéreo 67-13, Cali, Colombia.

2. Ministries of agriculture or hydraulic resources of most Latin American countries and the Empresa Brasileira de Pesquisas Agropecuarias (Centro de Pesquisa Agropecuaria dos Cerrados) in Brazil.

Hail as an Ecological Factor in the Increase of Pricklypear Cactus

W.A. LAYCOCK
USDA-SEA-AR Forest and Range Research and
Colorado State University, Fort Collins, Colo., U.S.A.

Summary

Detached joints or pads of plains pricklypear cactus (*Opuntia polyacantha*) can root and form new plants. However, few data exist to quantify this phenomenon or to describe conditions under which rooting takes place. A severe hailstorm at the Central Plains Experimental Range in northeastern Colorado on 31 July 1978 broke segments from plants of plains pricklypear. This occurrence provided an opportunity to determine the amount of joint dispersal and rooting from a natural destructive event.

Cactus joints detached and scattered by the force of the hail were counted on permanent belt transects established in two pastures having different pricklypear populations. One year later, the transects were sampled again to determine the number and proportion of joints that had successfully rooted and become established.

In one pasture, where the average ground cover of pricklypear was approximately 1.2%, the hail detached 4,100 cactus segments/ha. In the other pasture, pricklypear totaled about 3.3% of ground cover and hail detached 19,100 segments/ha. In the pasture with less pricklypear, 34% of the segments had rooted by August 1979, while only 13% of the segments had rooted in the other pasture. However, the population of rooted segments was equivalent to 1,400 and 2,400 new plants/ha in the pastures with less and more pricklypear, respectively.

Precipitation from the 1978 hailstorm through the first 4 months of 1979 was less than 100 mm (67% of the long-term average). Some joints had short roots on the underside in late winter of 1978-1979, but no roots reaching the soil were found in April. In May and June, more than 200 mm of precipitation fell, almost twice the average for that period.

The detached pricklypear joints that eventually rooted survived 9 months of unfavorable moisture conditions and then were able to root successfully after moisture became available. This remarkable adaptability may be one of the reasons why pricklypear is so prevalent on the Great Plains.

KEY WORDS: *Opuntia*, pricklypear, hail, roots, control, re-establishment, drought, precipitation, ecology.

INTRODUCTION

Pricklypear cactus (*Opuntia* spp.) is widespread throughout the Great Plains of the U.S.A. and in other parts of the world. Proper management of rangeland is difficult where pricklypear occurs in dense stands, because forage in and near clumps of pricklypear is relatively unavailable to grazing animals.

The knowledge that detached pricklypear joints can root and form new plants is widely recognized, but the phenomenon has never been studied in detail. Rooting of joints has been used in propagation of pricklypear for forage. Griffiths (1908, 1912) described methods of propagating *Opuntia lindheimeri* and thornless pricklypear (*Opuntia ficus-indica*) from single joints.

Because of the extensive range of pricklypear, many different mechanical control methods, such as grubbing, scraping with a blade, or railing, have been tried. Pricklypear plants are easily broken or uprooted by these

The author is a Range Scientist of USDA-SEA-AR. This is a contribution of USDA-SEA-AR and the Colorado Agricultural Experiment Station, Fort Collins. It is published with the approval of the Director of the Colorado Agricultural Experiment Station (Scientific Series Paper No. 2636).

methods, and new plants can start from the scattered joints. The problem of reinfestation by such new plants has been observed in mechanical control trials in Colorado (Costello 1941), Texas (Hoffman and Darrow 1955, Hoffman 1967), New Mexico (Gay, undated), and Wyoming (Hyde et al. 1965). In Arizona, Martin and Tschirley (1969) reported large increases in numbers of plants of jumping cholla (*Opuntia fulgida*) and other cholla species the first year after cabling as a result of the establishment of the scattered joints and fruits as new plants.

A severe hailstorm at the Central Plains Experimental Range (CPER) in northeastern Colorado broke numerous segments from plants of plains pricklypear. This storm provided an opportunity to study the damage to pricklypear, the subsequent development of new plants, and the possible ecological and management implications of this naturally occurring destructive event.

METHODS

The hailstorm occurred in the evening on 31 July 1978. Areas in two adjacent pastures near the center of the storm path were chosen to sample the damage done to pricklypear by the hail. Both had a sandy-loam surface soil and were dominated by blue grama (*Bouteloua gracilis*). One pasture had a relatively small amount of pricklypear, and the other had a moderate stand. Three weeks after the storm, permanent belt transects 1 m wide × 30 m long were established in both pastures. Ten transects were established in an exclosure and ten in adjacent grazed areas in each pasture.

Segments of pricklypear plants detached and scattered by the force of the hail were counted within each transect. Cover of pricklypear was measured in each area by the line intercept method using the edges of each 30-m belt transect.

In August 1979 the permanent transects were sampled again by the same methods. Successfully rooted joints were easily identifiable because they were turgid and green and resisted considerable force exerted to dislodge them. Joints that were not rooted had partially or completely dried, making them easy to distinguish. Some of the successfully rooted plants had already produced one or more small new joints. These plants were becoming difficult to distinguish from other small pricklypear plants.

RESULTS AND DISCUSSION

The hailstorm traveled through CPER in a generally southeasterly direction. The rain gauge at CPER headquarters accumulated 18 mm of water, which undoubtedly represented only a portion of the actual precipitation because a large amount of the hail would have bounced out of the gauge. Most of the hail was 18–20 mm in diameter, and the hail was accompanied by strong winds that intensified the damage.

A report by the Convective Storms Division of the National Center for Atmospheric Research, headquartered in Boulder, Colorado, stated for 31 July 1978: "Late development occurred WNW of CYS [Cheyenne, Wyoming]. This storm produced golfball-size hail at CYS and heavy hail at Briggsdale [Colorado]. Visually this was a large storm, tops near 47,000 ft from radar summary" (Griffith M. Morgan, Jr., NCAR, Boulder, Colorado, personal communication). In Cheyenne, Wyoming, 40 km NNW of the study area, insurance adjusters reported heavy hail damage from this storm, with one company reporting 1,000 home-roof damage claims and almost 1,300 automobile damage claims (Ed Doebbling, State Farm Insurance, Greeley, Colorado, personal communication). Briggsdale, Colorado, is 40 km southeast of the study area.

Hail Damage

Pricklypear made up approximately 1.2% of the ground cover in the pasture with the sparse stand (Table 1). The pasture with the moderate stand was 3.3% pricklypear ground cover. Total above-ground standing crop of pricklypear was 350 kg/ha (dry weight) and 1,340 kg/ha in the sparse and moderate pricklypear pastures, respectively. Average cover of vegetation in both pastures, including blue grama crowns, was approximately 44%, litter was 14%, and bare soil was 42%.

In the sparse pricklypear pasture, hail detached 4,100 segments/ha consisting of one or more complete and connected joints (Table 1). Most detached pricklypear segments consisted of one or two complete joints. The moderate pricklypear pasture suffered almost five times as much hail damage, with an average detachment of 19,100 segments/ha consisting of one or more complete joints.

Table 1. Numbers of pricklypear cactus segments detached from plants by hail and subsequent rooting success, Central Plains Experimental Range, Colorado.

	Characteristics of pricklypear stand in 1979		Pricklypear detached by hail in 1978		Rooted in 1979	
	Above-ground standing crop (kg/ha)	% ground cover	No. of segments /ha consisting of at least one complete joint	Biomass detached (kg/ha)	Total no. /ha	% of detached joints rooted
Pasture						
Moderate pricklypear	1340	3.3	19100	125	2400	13
Sparse pricklypear	350	1.2	4100	25	1400	34

Estimates of total pricklypear oven-dry biomass removed by hail were 25 kg/ha in the sparse pricklypear pasture (approximately 7% of the standing crop) and 125 kg/ha in the moderate pricklypear pasture (approximately 9% of the standing crop).

Climatic Conditions

The 9 months following the hailstorm were dry. Excluding the 18 mm of rain measured from the hailstorm, the total precipitation from August 1978 through April 1979 was 98 mm, about two-thirds of the long-term average of 147 mm. During the winter the ground was free of snow most of the time, and the soil surface remained dry most of the period through April. In May and June 1979, 205 mm of precipitation fell, almost twice the average for this period.

Origin of Roots

Periodic observations of dislodged pricklypear joints were made. As early as February, some roots had formed on the underside of a few joints, but no roots reaching the soil surface were found as late as April. The stiff spines kept the joints from touching the soil surface, and the roots had to form and extend several mm from the bottom of a joint to the ground. The abundant moisture, high humidity, and favorable temperature conditions during May and June apparently were sufficient for roots to form, elongate, contact the soil surface, and become established. Rooted joints were found only on bare or partially litter-covered spaces between patches of blue grama or other vegetation. Live plant material beneath a detached joint apparently prevented successful rooting.

The only reference found concerning details of root formation was that of Griffiths (1908). He reported that "roots spring from the areoles or cushions of spines and spicules distributed regularly over the surface of the stems (joints). . . . The areoles in the center of the lower surface in contact with the ground . . . will form roots in a very short time." In the present study, contact of the joint surface with the ground did not appear to be necessary for root development.

Rooting Success

In August 1979, no segments consisting of less than one complete joint had rooted, and the numbers of these partial joints are not reported. Of the segments consisting of one or more joints, 34% had rooted in the sparse pasture by 1979, and 13% had rooted in the moderate pricklypear pasture. The population of rooted segments was equivalent to 1,400 and 2,400 new pricklypear plants/ha, respectively. Thus, more new plants were established in the pasture that already had the most pricklypear, even though a lower percentage of the detached joints were rooted. Further observations indicated that most rooted joints eventually became vigorous established plants. The number and percentage of joints that rooted were not greatly different in the exclosures and grazed areas.

The only comparable data found in the literature were presented by Martin and Tschirley (1969). One year after a mature stand of cholla was cabled, numbers of plants increased from 1,110 to 18,500/ha for jumping cholla and other cholla species. The increases were the result of rooting of scattered joints and fruits. However, 3 years after cabling, most of the new plants had died.

ECOLOGICAL AND MANAGEMENT IMPLICATIONS

This study illustrates the remarkable survival ability of plains pricklypear and indicates one mechanism by which pricklypear can be spread by natural means. Hail is a common phenomenon on the Great Plains, and establishment of new plants after severe hailstorms must be a periodically recurring event. The center of the area in the U.S. having the highest average number of days with hail (8/yr) is near Cheyenne, Wyoming (U.S. Department of Agriculture 1941). How frequently a storm as destructive as the one reported could be expected to occur in a specific area is not known, but such storms are not uncommon on the central Great Plains. It is possible that local heavy infestations of pricklypear that cannot be attributed to other causes might be the result of one or more severe hailstorms in the past.

This study also provides information pertinent to mechanical pricklypear control measures. Previous studies have implied that little reinfestation will take place if mechanical control is done during dry periods (Hoffman and Darrow 1955, Gay, undated). In this study, the detached joints survived 9 months of dry conditions, and then a large number rooted when moisture became favorable. This adaptability may be one of the reasons why pricklypear is so prevalent and persistent on the Great Plains. Even if all of the old pricklypear plants had been removed, densities of new plants established would represent the beginning of a significant stand of pricklypear.

LITERATURE CITED

Costello, D.F. 1941. Pricklypear control on the short-grass range in the central Great Plains. U.S. Dept. Agric. Leaflet 210.

Gay, C. Undated. Mechanical and chemical methods of controlling pricklypear cactus. New Mex. State Univ., Ext. Serv. Livest. Guide 4008-801.

Griffiths, D. 1908. The prickly pear as a farm crop. U.S. Dept. Agric. Bur. Pl. Indus. Bul. 124.

Griffiths, D. 1912. The thornless prickly pears. U.S. Dept. Agric. Fmrs. Bul. 483.

Hoffman, G.O. 1967. Controlling pricklypear in Texas. Down to Earth 23(1):9-12.

Hoffman, G.O., and R.A. Darrow. 1955. Pricklypear—good or bad? Tex. Agric. Expt. Sta. Bul. 806.

Hyde, R.M., A.D. Hulett, and H.P. Alley. 1965. Chemical and mechanical control of plains pricklypear in northeastern Wyoming. Univ. Wyo. Agric. Ext. Serv. Cir. 185.

Martin, S.C., and F.H. Tschirley. 1969. Changes in cactus numbers after cabling. Prog. Agric. in Ariz. 21(1):16-17.

U.S. Department of Agriculture. 1941. Climate and Man. USDA Yearbook. Govt. Print. Off., Washington, D.C.

Nutritional Quality of Leaf Proteins Prepared from Crops Containing Phenolic Compounds and Polyphenolase

T. HORIGOME and S. UCHIDA

Okayama University, Tsushima, Okayama, Japan

Summary

Many studies have indicated that leaf protein prepared by green crop fractionation affords good potential as a protein supplement. It has been noted, however, that there were some marked differences between crops in the nutritional quality of leaf protein in spite of a similarity in amino-acid composition between leaf proteins made from different crops.

We previously showed that brown casein, which was produced by the interaction of milk casein with phenolics undergoing enzymic oxidation, had lower nutritional value than original casein. In addition, it has been suggested that leaf proteins prepared from crops containing phenolics and polyphenolase were brown after extraction with acetone. It can therefore be presumed that the browning of leaf protein affects its nutritional quality.

Italian ryegrass, red clover, sorghum, and alfalfa were used for leaf-protein preparation. Fresh leaves were pulped in the presence or absence of a reducing agent (sodium ascorbate or $NaHSO_3$) and green juice was separated. Protein was coagulated by adjusting the pH of the juice to 4; then the juice was heated and washed with acetone. The biological evaluation of leaf proteins was carried out by the growth method with male rats weighing about 45 g.

Italian ryegrass, red clover, and sorghum had o-diphenolic contents and polyphenolase activity. The leaf proteins from Italian ryegrass, red clover, and sorghum were brown when leaves were pulped in the absence of a reducing agent. On the other hand, alfalfa had neither o-diphenolics nor polyphenolase, and hence the alfalfa leaf protein did not brown during pulping even in the absence of a reducing agent. The brown leaf protein from Italian ryegrass had lower digestibility than the leaf protein protected from browning, although there were no differences in growth-promoting effect and protein-efficiency ratio (PER) between the two leaf proteins. The feeding of brown leaf protein from red clover resulted in the lowering of weight gain, digestibility, and PER, and all the measurements including diet intake were lowered by feeding the brown leaf protein from sorghum. In the case of alfalfa leaf protein, there was no difference in nutritional quality between the two leaf proteins made with and without an attempt to prevent browning.

The results mentioned above indicate that the occurrence of phenolics and polyphenolase in a crop is responsible for the browning of leaf protein and that the browning of leaf protein causes its nutritional impairment.

KEY WORDS: leaf protein, diphenolic compound, polyphenolase, browning, Italian ryegrass, red clover, sorghum, alfalfa.

INTRODUCTION

Leaf protein prepared by green-crop fractionation has been studied by many researchers in many countries, and it has been indicated that leaf protein affords good potential as a protein supplement (Bickoff et al. 1975, Morris 1977, Pirie 1978). It has been noted, however, that there are some marked differences among crops in nutritional quality of leaf protein in spite of a similarity of amino-acid composition among leaf proteins from different crops.

In a previous paper, Horigome and Kandatsu (1968) showed that brown caseins that were produced by the interaction of milk casein with phenolic compounds undergoing enzymic oxidation had lower nutritional value than the original casein. It was also shown that many crops contained phenolic compounds and polyphenolase (Horigome and Kandatsu 1966). Accordingly, it can be assumed that the browning of leaf protein is attributable to phenolics and that such browning depresses its nutritional quality. The primary aim of the investigation reported here was to determine whether the browning of leaf protein affects its nutritional quality.

METHODS

Leaf-Protein Preparation

The forage crops used for preparation of leaf proteins were alfalfa (*Medicago sativa* L.), Italian ryegrass (*Lolium multiflorum* Lam.), red clover (*Trifolium pratense* L.), and sweet sorghum (*Sorghum vulgare*). These crops were harvested before flowering or before earing. Diphenolic (Arnow 1937) and tannin (Broadhurst and Jones 1978) contents and polyphenolase (o-diphenol : oxygen oxidoreductase) activity (Horigome and Kandatsu 1966) in the

T. Horigome is a Professor of Animal Nutrition and S. Uchida is an Associate Professor of Applied Nutrition at Okayama University.

leaves of these crops were determined. Laboratory protein preparation was performed as follows: Fresh leaves were pulped with a small amount of water and fractionated into green juice and pressed leaves. To prevent the browning of protein, either sodium ascorbate (20g/l) or $NaHSO_3$ (5g/l) was added as a reducing agent during pulping. The green juice was adjusted to pH 4 with 10% HCl and then heated to 80°C. This treatment gave a voluminus coagulam of protein. The coagulam was separated from brown juice and washed repeatedly with acetone. This acetone-washed leaf protein was subjected to further experiments. The amino-acid composition of leaf proteins was determined by hydrolysis in constantly boiling HCl under vacuum at 110°C for 24 hours, followed by ion-exchange chromatography of an automatic 2-column analyzer. Methionine and cysteine were analyzed as methionine sulphone and cysteic acid, respectively, after performic acid oxidation of leaf protein according to Moore (1963).

Animal Experiments

Growth experiments were performed with weanling male rats of the Wista strain weighing about 45 g each. The basal protein-free diet contained (g/100 g): cellulose powder, 1.0; cane sugar, 10.0; mineral mixture, 4.0; vitamin mixture, 1.0; methionine, 0.2; corn oil, 10.0; and cornstarch, 73.8. Each leaf protein was incorporated into the basal diet in replacement of an equal weight of cornstarch to provide about 8% protein (N × 6.25). Methionine was added to all diets, since it is the first limiting amino acid of leaf proteins.

Each diet group included five rats; they were allowed ad-libitum access to the diets for a 14-day period. Protein-efficiency ratio (PER) was calculated after 14 days as g body-weight gain/g protein eaten. The apparent digestibility of leaf proteins was determined by analyzing the nitrogen intake and fecal nitrogen excreted during the last 5 days of the period.

Table 1. Phenolic compound content and polyphenolase activity of leaves.

Crop	o-Diphenol[1]	Condensed tannin[2]	Polyphenolase[3]
Alfalfa	0	0	0
Italian ryegrass	0.36	trace	20.1
Red clover	0.95	trace	21.6
Sorghum	0.75	<0.06	6.0

[1]Expressed in terms of chlorogenic acid (% of dry matter).

[2]% of dry matter.

[3]mg purpurogallin produced/g leaf acetone-powder.

RESULTS AND DISCUSSION

Diphenolic and tannin contents and polyphenolase activity of forage crops are shown in Table 1. Alfalfa had neither diphenolics nor polyphenolase, and hence the leaf protein was not brown even when no reducing agent was added during pulping. However, Italian ryegrass and red clover had diphenolic content and polyphenolase activity, so browning occurred in the leaf proteins of both crops when the leaves were pulped in the absence of a reducing agent. In particular, red clover contained approximately 2.6 times the phenolic content of Italian ryegrass, and the red clover leaf protein was very brown. Sorghum had very low activity of polyphenolase, but the sorghum leaf protein also became brown in the absence of a reducing agent during pulping. When a reducing agent was added during pulping, the leaf protein obtained from the forage crops tested was not browned. In addition, it can be seen from Table 1 that every forage crop had a negligible or very small amount of tannin.

Table 2 shows the results of nutritional evaluation of leaf proteins. In the case of alfalfa leaf protein, the presence or absence of a reducing agent during pulping did not affect the nutritional quality of the leaf protein.

Table 2. Diet intake, weight gain, digestibility, and protein efficiency ration (PER) of male rats given a diet containing leaf protein.

Leaf protein source		14-Day diet intake g	14-Day weight gain g	Digestibility of protein %	P E R g gain/ g protein intake
Alfalfa	(A)	97.4±0.2	26.3±2.2	79.5±1.9	3.27±0.13
	(B)	105.6±4.3	29.6±2.8	81.4±1.8	3.37±0.30
Italian ryegrass	(A)	113.7±0.2	36.9±1.0	86.4±1.4	3.85±0.11
	(B)	113.7±0.1	36.2±2.0	80.9±0.6**	3.74±0.20
Red clover	(A)	114.9±0.1	42.3±2.1	86.3±0.4	4.65±0.23
	(B)	114.9±0.1	37.2±1.7*	79.4±1.4**	3.95±0.18**
Sorghum	(A)	113.3±0.4	32.4±1.8	77.6±0.8	3.30±0.17
	(B)	82.0±5.7**	14.9±4.4**	67.0±0.9**	2.10±0.48**

(A) Leaf protein with attempt to prevent browning.

(B) Leaf protein with no attempt to prevent browning.

*Significantly different from the corresponding value of (A) at the 0.01 level.

**Significantly different from the corresponding value of (A) at the 0.001 level.

On the other hand, Italian ryegrass leaf protein that was not protected from browning was significantly lower in digestibility than leaf protein protected from browning. In red clover, weight gain and PER of rats fed brown protein were depressed as compared to those of rats fed protein prevented from browning. Digestibility of brown protein was lower as well. These results indicate that the nutritional quality of leaf protein is lowered by the browning reaction and that a large amount of diphenolic and a high activity of polyphenolase in a forage crop depresses nutritional quality.

In spite of very low activity of polyphenolase in sorghum leaves, however, a large depression of nutritional quality occurred in sorghum leaf protein that was not protected from browning. It is particularly noteworthy that intake by rats fed on the sorghum leaf protein exposed to browning was low. The cause of this low intake could not be determined from the data of these experiments, but the results suggest that, in addition to the amount of phenolics present in leaves, the type (or kind) of phenolics may be an important factor in the nutritional impairment of leaf protein.

The results of amino-acid analyses showed that amino acids of leaf proteins were scarcely damaged by browning. Lysine, histidine, and cysteine were very slightly impaired in red clover leaf protein prepared with no attempt to prevent browning. Therefore, amino-acid composition, as determined by the methods presented in this paper, does not provide an explanation for the nutritional impairment of leaf protein by the browning reaction.

In conclusion, the results presented have demonstrated that the occurrence of phenolics and polyphenolase in a crop is responsible for browning of leaf protein and that browning causes nutritional impairment. However, identification of phenolic compounds present is important for elucidating the mechanism of nutritional impairment.

LITERATURE CITED

Arnow, L.E. 1937. Colorimetric determination of the components of dopa-tyrosine mixture. J. Biol. Chem. 118:531.

Bickoff, E.M., A.N. Booth, D. de Fremery, R.H. Edwards, B.E. Knuckles, R.E. Miller, R.M. Saunders, and G.O. Kohler. 1975. Nutritional evaluation of alfalfa leaf protein concentrate. *In* M. Friedman (ed.), Protein: nutritional quality of foods and feeds, Part 2, 319–340. Marcel Dekker, New York.

Broadhurst, R.B., and W.T. Jones. 1978. Analysis of condensed tannins using acidified vanillin. J. Sci. Food Agric. 29:788–794.

Horigome, T., and M. Kandatsu. 1966. Lowering effect of phenolic compounds and o-diphenoloxidase on the digestibility of protein in pasture plants. J. Agric. Chem. Soc. Jap. 40:449–455.

Horigome, T., and M. Kandatsu. 1968. Biological value of proteins allowed to react with phenolic compounds in the presence of o-diphenoloxidase. Agric. Biol. Chem. 32:1093–1102.

Moore, S. 1963. On the determination of cystine as cysteic acid. J. Biol. Chem. 238:235–237.

Morris, T.R. 1977. Leaf-protein concentrate for non-ruminant farm animals. *In* R.J. Wilkins (ed.), Green crop fractionation, 67–82. Grassl. Res. Inst., Harrogate, Yorks.

Pirie, N.W. 1978. Leaf protein and other aspects of fodder fractionation, 1–137. Camb. Univ. Press. Cambridge.

Evaluation of Biological Components in Decision-Making in Forage Allocation

J.W. SKILES and G.M. VAN DYNE
Colorado State University, Fort Collins, Colo., U.S.A.

Summary

Forage allocation to mixtures of large herbivores is accomplished by minimizing the difference between the available herbage and the required forage for animals on rangeland. Availability is determined in part by the plant's physiological tolerance to grazing or its allowable-use factor and the animal preference for that plant, measured in part by the proper-use factor. The mix of large herbivores on the rangeland must also be considered. The grazing requirements for animals are dependent on animal preference for plant species and on the total forage intake rates. This paper presents an analysis of these concepts and published information. The concept of proper-use factor is widespread, but published methodologies for its formulation are lacking. Allowable-use factors, determined primarily in clipping studies, are generally higher than proper-use factors, but the exact relationship between them is unclear at best. Preference for plants by herbivores, as defined in the literature, varies in meaning from percentage eaten by forage class (grasses, forbs, and shrubs) to rankings by percent of plant species in herbivore diets to actual calculation of a preference index. The method used to calculate a preference index can yield varying results and interpretations. Forage intake-rate studies that use animals under grazing rather than under pen conditions appear to be confined to domestic sheep and cattle. Some 52 intake-rate values for cattle averaged 2.2% of body weight/day; 53 intake-rate values for sheep averaged the same. We also report forage intake rates for pronghorn antelope, bison, bighorn sheep, mule deer, horses, and elk. Through understanding of these variables, effective management strategies may be derived whereby the rangeland resource can be used prudently without degradation.

KEY WORDS: Forage intake rate, preference, proper-use factors, forage allocation, optimization.

INTRODUCTION

Allocation of the herbage on grazing land to various large herbivores requires consideration of many factors. The problem results in a mathematical programing-problem objective function (Van Dyne and Kortopates 1981) to select X so as to minimize:

$$\left[\sum_i (H \times U_i (S_i + G_i)(1 - L_i) - D \sum_j X_j \times C_j \times P_{ij}) \right] \quad (1)$$

where H = ha, i = plant species or group, U = allowable-use level, S = standing crop at the beginning of the season or grazing period, G = gain in standing crop, L = proportion of standing crop lost to shattering and weathering, D = days, j = animal species, C = daily forage consumption, and P = relative preference the animal expresses for each plant species or group. These items are known and are provided in the analysis. The X_js, the numbers of individuals of different species of animals, are unknown. There are also various constraints on the selection of X_j. We have reviewed the scientific literature for western North American situations (Skiles et al. 1980) and world grazing lands (Van Dyne et al. 1980) on U, C, and P (above) and on botanical composition of diets. These reports include extensive synthesis and citation of the scientific literature and detailed tabulations of data.

The objective of the present paper is to discuss the logical and biological basis of U, P, and C.

USE FACTORS

The use factor U (equation 1) is an integral part of the model conceptualization, but scientific information on U is not abundant. A related but different concept, the proper-use factor (PUF), which has been used by resource-management agencies for 40 years, provides insights. The PUF is defined as the percentage of use that is made of a forage species when the range as a whole is used properly (Stoddard et al. 1975, Sampson 1952).

The proper-use factor is defined for a known range composed of certain plants in given abundances *and* grazed by a specific herbivore species *and* in a particular season. A PUF value can vary for a variety of reasons, e.g., plant community composition, species of herbivore

grazing the range, season, year-to-year climatic variation, previous grazing effects, differential plant grazing by a herbivore species, and degree of animal familiarity with the forage plants.

Many factors affecting a PUF are difficult to quantify, particularly because neither the herbage base nor the herbivore population is monospecific. Rangelands have many plant species and high variability and are utilized by several competing herbivores exhibiting preference for one forage species over another. In practice the PUF values must be established for each herbivore for each forage species. Seasonality adds another dimension. Consequently, various workers report highly different PUF values for the same plant species, leading to confusion from a management viewpoint. A manager determines PUF values by looking at the range. Most techniques of estimating proper use of a range, and by inference the PUF values for the plant species, are observational, qualitative, and not repeatable.

Several individuals have used the PUF concept in their work; e.g., Nelson (1977) used the concept of proper use in his examination of the work of Smith (1965), who addressed himself to a "utilization standard" of a single plant species when determining the tradeoffs in herbivore stocking rates. Neither worker described the method for PUF determination, though both authors present models in which PUF values play a major role. Earlier work by Cook (1954) and by Hopkin (1954), who reworked Cook's data, consider the economic benefits from using PUF values in stocking sheep and cattle on a Utah range. Both assume PUF values, and allude to their formulation but do not say explicitly how they were determined.

The allowable-use factor (AUF) is used to measure a plant's physiological response to grazing. It is the amount of above-ground plant parts that can be grazed with the plant still recovering. AUF formulation is presumably derived from clipping or grazing studies. The transition between the allowable level of use in kg/ha and the percentage or proportion of AUF is unclear; the literature lacks description of detailed methods to make the transformation. An AUF varies with season.

Several precepts concerning the AUF may be elucidated. Generally, the AUF for a given plant species is greater than the PUF. On a rangeland with one plant species the AUF equals the PUF. Frequently, for efficient use of the range, some highly palatable forage plants that are present in small amounts will have PUF values greater than AUF values. These plants are then sacrificed.

There is conceptual linkage between AUF values and PUF values, but the real-world link is not at all well estab-

J.W. Skiles is a Research Associate and G.M. Van Dyne, deceased in August of 1981, was a Professor of Biology at Colorado State University.

lished. And PUF and AUF formulation methodologies are lacking in the literature and range science textbooks. The presence of AUF and PUF lists in government agency files does indicate that at some time persons in the agencies gave thought to the problem, but generally there is no satisfactory discussion of the data or information base from which the lists were developed.

PREFERENCE

The mechanisms by which an animal selects diet items are not simple. Stoddard et al. (1975) define preference as the selection of plants by animals. They distinguish preference from palatability, which is the attractiveness of a plant to the animals. But "palatability" is often used when "preference" is meant.

Preference, electivity, and selectivity are related to metabolic body size (which dictates the energy requirements of the animal) and can be modified by an animal's physiological state (resting, running, etc.) and reproductive state (pregnant, lactating, etc.). Also, forage size, quality, and novelty may influence preference. These variables are important to the animal prior to ingestion of a food item.

After ingestion of a food, the food's chemical makeup may control continued ingestion or rejection. Sugars and soluble carbohydrates, proteins, acid detergent fiber or cell-wall structure, mineral concentrations, vitamin contents, and plant pigments all have been reported to affect preference. Caswell et al. (1973) have proposed that herbivores that evolved in temperate climates tend to avoid plants with the C_4 photosynthetic pathway; presumably this tendency is related to difference in chemical composition and leaf morphology.

Incorporating a preference measure (P_{ij}) into equation 1 requires the following assumptions: (1) P_{ij} is not dependent on the number or kind of other herbivores on the range; (2) forage intake rates for each herbivore remain constant within any particular grazing period or season; and (3) preferences do not alter as the grazing season or period progresses.

There are many measures of animal preference—percentage of time spent grazing a plant, percentage of grazed plants in the diet, animal density in a specific plant community, plant use, calculated use factors and availability factors, preference indices, forage weight in the community, site use, averages and statistical tests, and rankings (Skiles 1981). Unfortunately, most studies that report on preference use different measures. Unless the same index is used, comparisons between studies are not possible.

Calculation of six preference measures using a common set of diet data (Skiles 1981) shows that the methods used yield results giving differences in interpretations. A preference measure or index does not explain the mechanism by which a herbivore selects one plant over another, although it is a convenient predictive tool for range management. Its inclusion in a mathematical programing scheme for forage allocation (as in equation 1) is essential.

FORAGE INTAKE RATE

The rate at which an animal consumes forage (C in equation 1) influences the forage allocation decision. We have summarized and standardized consumption-rate data for large herbivores (Skiles et al. 1980). Data from grazing trials for cattle and sheep are common, but little information on wild herbivores exists. Forage intake rates reported for wild herbivores are generally derived from pen feeding trials. All studies found in the literature lacked consumption values for all four seasons, a factor of prime importance in making decisions on forage allocation.

Averaged over all available data, the mean live body weight in kg and the daily forage intake as percentage of body weight are: cattle, 343 and 2.2%; sheep, 56 and 2.2%; bison, 348 and 1.7%; bighorn sheep, 62 and 2.6%; pronghorn antelope, 41 and 2.1%; deer, 67 and 2.2%; elk, 181 and 2.4%; and horses, 378 and 1.7%.

CONCLUSIONS

The allocation of forage to large herbivores in grazing-land situations would seem to be a straightforward minimization of the difference between the herbage available and the forage required. The allocation problem, however, is complex when the needed information is considered in depth.

Through a review of the scientific literature, we have attempted to quantify some of the variables useful in models for forage allocation. Meanings of PUF and AUF values are unclear, since the methodology of their formulation is not specified. The many preference measures are rarely comparable, and their interpretation is obscure. Published forage intake rates for large herbivores are generally pasture-specific or region-specific, lack seasonality, and, except for domestic animals, are derived from feeding trials. We conclude that more work is needed by resource-management agencies and range-land researchers on factors affecting preference and intake and on use of this information in forage allocation models. A conceptual framework has been developed, and enough data exists and has been tabulated, however, to start the decision-making process.

LITERATURE CITED

Caswell, H.F., F. Reed, S.N. Stephenson, and P.A. Werner. 1973. Photosynthetic pathways and selective herbivory: hypothesis. Am. Nat. 107:465–480.

Cook, C.W. 1954. Common use of summer range by sheep and cattle. J. Range Mangt. 7:10–13.

Hopkin, J.A. 1954. Economic criteria for determining optimum use of summer range by sheep and cattle. J. Range Mangt. 7:170–175.

Nelson, J.R. 1977. Maximizing stocking rates with common-use and proper-use grazing. Col. Agric. Res. Center, Wash. State Univ. Bul. 856.

Sampson, A.W. 1952. Range management: principles and practices. John Wiley and Sons, New York.

Skiles, J.W. 1981. A review of animal preference. *In* National Research Council. Forage allocation. Natl. Acad. Sci., Washington, D.C.

Skiles, J.W., P.T. Kortopates, and G.M. Van Dyne. 1980. Optimization models for forage allocation to combinations of herbivores for grazingland situations: a critical review of forage intake; a critical review of preference and its calculation; a critical review of proper use factors; a critical review and evaluation of dietary botanical composition. Dept. Range Sci., Colo. State Univ. Tech. Rpt.

Smith, A.D. 1965. Determining common-use grazing capacities by application of the key species concept. J. Range Mangt.

18:196–201.

Stoddard, L.A., A.D. Smith, and T.W. Box. 1975. Range management. 3rd ed. McGraw-Hill Book Co., New York.

Van Dyne, G.M., N.R. Brockington, Z. Szocs, J. Duek, and C.A. Ribic. 1980. Large herbivore subsystems. *In* A.I. Breymeyer, and G.M. Van Dyne (eds.), Grasslands, systems analysis, and man, 269–537. Camb. Univ. Press, Cambridge.

Van Dyne, G.M., and P.T. Kortopates. 1981. Quantitative frameworks for forage allocation. *In* National Research Council, Forage Allocation. Natl. Acad. Sci., Washington, D.C.

Pasture Evaluation in the Desertic Zone of Venezuela for the Development of Goat Production

A. GALLARDO and A. LEONE

Estacion Experimental El Cují, Venezuela

Summary

The desertic zone of Venezuela is an area of approximately 4.1 million ha that represents 4.8% of the total national territory, on which there are two million goats. Goats represent the main, if not the only, income source for 60,000 rural families.

According to UNESCO (1973), the desertic zone of the country comprises two vegetation types, evergreen scrub and deciduous scrub. In each of these zones the plant species consumed by goats were collected and, for those plant parts consumed by the animals, chemically analyzed to determine nutritive value. Forage plants from other tropical countries were also introduced and evaluated.

Several species from the natural vegetation have a high crude protein content. Outstanding among these are: *Acacia flexuosa* (21.7%), *Prosopis juliflora* (27.8%), *Setaria macrostachya* (13.5%), *Sporobolus pyramidalis* (16.4%), *Lippia alba* (11.2%), *Lippia origanoides* (22.0%), *Malphigia glabra* (23.2%), and *Wedelia caracasana* (18.7%).

Previous work at El Cují Experimental Station, where we have been introducing and evaluating forage and browse species, permitted us to select the most promising ones, such as: *Panicum coloratum* (Bambatsii), *Cenchrus ciliaris* (Malopo, Gayndah, Biloela), *Cynodon dactylon* (SR-954, Gigante), *Pennisetum purpureum* (Taiwan A-146, Mineiro), *Glycine javanica*, and *Leucaena leucocephala*.

The results of this study provide a valuable guide for managing vegetation for desirable forage production and, at the same time, offer alternatives for repopulating these areas with species of high nutritive value that could contribute to a better use with goats.

KEY WORDS: arid grasses, palatability, goats, dry matter, protein content.

INTRODUCTION

The inhabitants of the desertic tropical zones use sheep and goats extensively. This statement is corroborated by the fact that 54% of the world's goat population exists on pasture and browse lands of the dry areas of tropical countries, where few other domestic species could survive. Sheep and goats are therefore valuable animals for thousands of rural families.

Venezuela's desertic zone occupies 41,023 km², which represents 4.8% of the national territory. Plant species include *Agave sisalana, Anana comosus, Aloe vera,* and pasture species. This range ecosystem maintains the herds of sheep and goats that often constitute the main or the only source of income for the human population. Productivity of the goat ranches is very low, because, for one reason, the Spanish goat breeds that have been in the country for hundreds of years have not been improved by selection; for another, there is complete absence of rational forage-management practices and lack of veterinary care. However, the Spanish goat is a hardy animal, a good selective

---

The authors are Research Agronomists at FONAIAP-MAC.

grazer of herbaceous plants and shrubs, and a good converter of low-quality plant material.

At El Cují, experimental studies have been directed at discovering the natural vegetation associations that compose the range ecosystems and at quantifying several characteristics of the vegetation. Studies have been made of plants offering high-quality biomass that goats will consume, whose number and production level is a direct function of dry-matter production, nutrient value, and distribution during the year. With regard to cultivated pastures, studies have been of different forage species brought from other countries with similar conditions.

METHODS

The behavior of free-grazing goats was observed. Plants preferred by goats were sampled and analyzed to determine the dry-matter content, crude protein, ether extract, and crude fiber. Native plants also were botanically identified.

Cultivated grass species were introduced and observed under field conditions. The most promising ones were tested to evaluate cutting frequency, desirable fertilization practices, annual dry-matter production, and crude-protein yield/ha.

RESULTS

From chemical analyses of the plant parts that were consumed by goats, we can conclude that the plant material ingested by these animals is of excellent nutrient value, as is shown by its high content of crude protein (Table 1).

Of the introduced forage species, bermudagrass hybrid SR-954, was superior to hybrid SR-947 and to Coastal bermudagrass, as measured by annual production of dry matter and crude protein (Table 2).

Panicum coloratum (Table 3) has an optimal cutting frequency of 56 days, due to its high dry-matter and crude-protein content just before flowering. Nitrogen fertilization has a positive effect on both dry-matter and protein yield (Table 4). Phosphorus or potassium fertilization had little if any effect on production of the forage tested.

A comparative study of nine cultivars of buffelgrass is summarized in Table 5. Molopo, Biloela-559, Lignée No. 2, and common were outstanding in terms of dry-matter and crude-protein yield.

In a comparison of seven cultivars of elephantgrass, yields of dry matter and crude protein tended to be higher for hybrids Taiwan A-121, Mineiro, Taiwan A-148, and Taiwan A-146 (Table 6).

DISCUSSION

The range ecosystem of the desertic areas of Venezuela supports about two million goats. The vegetation is of low quality; it must be upgraded or improved by the use of native or introduced forage or browse plants to permit effective use, either in intensive or in extensive management systems. The native vegetation is diverse; it includes several legumes and other species of high protein content. The native grasses in the swards are mainly annuals; perhaps the perennial species have been eliminated over centuries of goat pasturage. There are also several plants of the Compositae family that withstand grazing by goats.

Herbs are very rare in the area, as this portion of the

Table 1.   Chemical analysis of vegetable species most consumed by goats (dry basis).

Name	DM %	C.P. %	C.F. %	Ether extract (%)	Ashes %	Nitrogen-free extract (%)	Ca %	P %	Rel. Ca/P	Part analyzed
Acacia Flexuosa		21.7	31.2	5.9	10.2	31.0	1.48	0.16	9.3	YL
Acacia Flexuosa	49.3	15.4	23.7	4.9	5.9	50.1	2.83	0.16	18.0	AL
Alternanthera halimifolia	21.9	22.9	12.1	1.7	14.8	49.1	—	—	—	YS, L
Bulnesia arborea		22.2	25.4	3.1	5.8	43.5	—	—	—	F
Caesalpinia coriaria	51.8	17.0	16.6	3.9	4.3	58.3	0.84	0.15	5.7	L
Caesalpinia coriaria	53.2	4.6	6.5	0.4	2.2	86.2	—	—	—	F
Capparis odoratissima	53.9	19.0	29.2	5.6	10.7	35.6	1.69	0.15	11.4	L
Cassia fruticosa	—	17.9	25.3	3.2	7.5	46.1	—	—	—	
Cercidium praecox	—	30.9	14.1	1.4	5.5	48.1	—	—	—	Fl
Jacquinia aciculata	52.2	9.4	31.7	3.2	11.5	44.2	2.81	0.08	36.4	L
Lippia alba	90.1	11.2	10.9	1.9	5.8	70.2	0.9	0.2	—	L, Fl, S, F
Lippia origanoides	38.3	22.1	13.8	3.0	9.9	51.2	3.15	0.21	15.0	B, L
Malpighia glabra	36.1	21.5	22.6	4.4	11.2	40.3	0.3	0.3	—	L, Fl, F
Mimosa arenosa	—	27.8	12.3	5.0	4.6	50.3	0.86	0.15	5.7	L
Prosopis juliflora	34.2	27.8	21.9	3.6	8.8	37.8	2.13	0.36	5.8	L
Setaria macrostachya	20.4	13.5	28.4	1.6	15.2	41.3	0.2	0.3	—	L, S, Fl
Sporobolus pyramidatus	36.8	16.4	18.2	2.4	15.0	48.0	—	—	—	S, L, Fl
Tabebuia bilbergii	—	15.0	17.2	0.9	5.7	61.2	—	—	—	DL
Wedelia caracasana	34.7	25.7	13.6	6.2	14.4	40.1	2.58	0.25	10.4	L

L – Leaf   DL – Dead leaf   Fl – Flowers
YL – Young leaf   S – Stem   F – Fruit
AL – Adult leaf   YS – Young stem   B – Bud

vegetation has been readily consumed by goats. By introducing aggressive grasses and controlling goat numbers, ground cover could be re-established. Progress has been made toward the selection of forage species for introduction into this desertic countryside. Such species must be adaptable to existing climatic and soil conditions and future management plans. During extended dry periods, a minimum forage production must be maintained through use of irrigated plots of tall grasses that can be harvested.

LITERATURE CITED

UNESCO. 1973. Clasificación internacional y cartografía de la vegetación, 93, Paris.

Table 4. Annual response of *Panicum coloratum* to applications of Nitrogen (N), Phosphorus (P), and Potassium (K).[1]

Treatments	DM (kg/ha)	CP (%)	CP (kg/ha)
N	5997	9.22	553
P	4561	7.42	338
K	5023	8.18	410
NP	6021	8.50	511
NK	6233	8.68	539
PK	4622	7.75	358
NPK	5858	8.20	480
Control	5706	7.70	443

[1]Averages (for 4 cuts) at 50-day intervals.

Table 2. Annual dry-matter and protein production by bermudas (*Cynodon dactylon*) hybrids.[1]

Hybrid	DM (kg/ha)	CP (%)	CP (kg/ha)
SR-954	5956	9.0	536
SR-947	4427	9.0	398
Coastal	4357	12.0	522
Gigante	4145	12.0	497
SR-949	4129	12.0	495

[1]Cutting frequency, each 50 days.

Table 5. Annual dry-matter and protein yields of nine cultivars of *Cenchrus ciliaris*.[1]

Cultivar	DM (kg/ha)	CP (%)	CP (kg/ha)
Molopo	8716	7.8	682
Biloela-559	8500	9.3	790
Lignée-2	6359	10.2	652
Common	6193	8.6	531
Gayndah-Q-412	5747	9.1	524
Lignée-3	5736	10.2	588
Biloela Straind-D	4085	8.1	331
West Australian	3510	9.8	343

[1]Averages for 4 cuttings at cutting frequency of 50 days.

Table 3. Determination of cutting frequency in *Panicum coloratum*: annual dry-matter and crude-protein yields (averages for seven cuttings).

Cutting interval, days	DM (kg/ha)	CP (%)	CP (kg/ha)
42	3445	7.00	241
56	4691	8.03	376
70	5570	4.96	276

Table 6. Comparative annual yield of seven cultivars of *Pennisetum purpureum*.[1]

Cultivar	DM (kg/ha)	CP (%)	CP (kg/ha)
Taiwan A-121	14,590	7.10	1035
Mineiro	13,528	6.40	866
Taiwan A-148	13,448	6.40	857
Taiwan A-146	13,391	6.87	920
Taiwan A-144	10,549	6.22	656
Pinda Gigante	10,167	7.17	756
Pinda Cubano	9,981	6.50	642

[1]Average for 4 cuttings of 60-day cutting frequency.

Grassland Development in the Southern Highlands of Peru

E.J. STEVENS and P. VILLALTA

Joint New Zealand–Peru Project

Summary

This paper describes grassland development in Peru. Vegetation of the high alpine region of Peru, which occupies some 4 million ha, is dominated by native grasses normally producing less than 2,000 kg/ha dry matter annually. On land with adequate soil moisture, pastures based on perennial ryegrass (*Lolium perenne* L.) and white clover (*Trifolium repens* L.) were established successfully, but on drier sites alfalfa proved superior, producing over 12,000 kg/ha/yr in some trials. Alfalfa and clovers were also successfully introduced into improved native pastures by mechanical direct seeding and by hand sowing. With direct seeding, dry-matter production increased from about 600 to 6,000 kg/ha in 2 years and to more than 10,000 kg/ha in older swards. Inoculation of legumes proved essential. The effectiveness of the inoculants was enhanced by increasing the rate of inoculant, using lime-inoculated and -pelleted seed, or sowing 500 kg/ha lime with the seed. Excellent results were obtained from sowing turnips (*Brassica rapa* L.), 8,000 to 10,000 kg dry matter/ha being produced in 5 to 6 months.

Pasture establishment and development have resulted in production increases up to 10-fold and animal-carrying capacity and performance have been increased correspondingly.

KEY WORDS: grassland development, high alpine, Peru, alfalfa (*Medicago sativa* L.), white clover (*Trifolium repens* L.), inoculation, lime, pelleting, phosphate.

INTRODUCTION

Intensive development of unimproved pastures offers one means of greatly increasing agricultural production of approximately 4 million ha situated between 3,400 and 4,800 m above sea level in southern Peru. Yearly rainfall averages 700–800 mm, with over 70% falling in December-March and very dry conditions prevailing in May-September, leading to acute forage shortages. Cold temperatures are frequent in winter, and frosts are common.

The natural vegetation is dominated by unpalatable native grasses that normally produce less than 2,000 kg/ha/yr of dry matter. Arable crop production is restricted, usually limited to favorable climatic zones that are traditionally used for producing annual crops, mainly cereals for green fodder, potatoes, chenopodium, and beans.

Prior to 1974, less than 40 ha of improved pastures and alfalfa had been sown in the southern highlands, and the cultivation of forage root crops had not been introduced. Since then 4,000 ha of improved pastures and alfalfa have been sown, and there have been annual sowings of 800 ha of forage turnips.

Successful development of low-fertility native pasture

in New Zealand was achieved by introduction of legumes such as white clover, topdressing of major and trace elements, legume inoculation, and suitable grazing management. These practices were tested on both a plot and a farm scale throughout the southern Peruvian highlands.

METHODS

From 1974 to 1978 integrated research-demonstration areas were established within different agro-ecological zones over a 200-km² area in the Altiplano near Puno. At each location, plot trials were laid down, and proven practices were demonstrated in field sowings.

Major areas of research and demonstration included (1) introduction and screening of different pasture and forage species, varieties, and cultivars together with determination of major and trace element requirements and legume inoculation responses; and (2) grassland development by traditional pasture cultivation and establishment techniques and by oversowing, mechanical seeding, and hand sowing of alfalfa, N.Z. white clover, red clover (*Trifolium pratense* L.), and alsike clover (*T. hybridum* L.) directly into unimproved native pastures.

RESULTS AND DISCUSSION

Legume Inoculation

Successful establishment of and sustained production from improved pastures were shown to be dependent on effective legume inoculation. Establishment was readily achieved on neutral soils of medium phosphate status, but

E.J. Stevens is an Agronomist (current address, Hill Farming Development Project, Muzalfarabad, Azad Kashmir, Pakistan) and P. Villalta is an Agronomist in the Ministry of Agriculture, Peru.

The authors wish to acknowledge the help of the Peruvian and New Zealand governments and the staff involved in the New Zealand–Peru project and the assistance of Mr. N.A. Cullen, Director, Ruakura Soil and Plant Research Station, Hamilton, N.Z.

Table 1. Effect of lime on alfalfa establishment.

Application rate of lime at sowing	Dry matter yield three months after sowing kg/ha
0	575
2 tonnes/ha	1278
4 tonnes/ha	1296
8 tonnes/ha	2056
450 kg/ha (applied in rows)	1176
Lime pelleted + inoculated seed	1604

LSD 5% = 527 kg/ha; 1% = 826 kg/ha.

legumes did not establish readily on acid soils (pH 4.5–5.5).

The effectiveness of inoculants was enhanced by increasing the rate of inoculant, liming the soil before sowing, using inoculated and lime-pelleted seed, or drilling inoculated seed with 450 kg lime/ha. The effect of lime on alfalfa establishment is shown in Table 1. Establishment from sowings made on 13 January 1978 into shallow acidic soils (pH 5.4), which overlaid a neutral profile, was similar to that from applications of 2 and 4 tonnes of lime incorporated into the soil prior to sowing and use of inoculated and lime-pelleted seed, or from drilling of 450 kg of lime/ha with the seed at the time of sowing. Eight tonnes of lime/ha incorporated prior to sowing gave a significant increase in production during the first 3 months of establishment when compared with 0, 0.45 (applied in rows), 2, and 4 tonnes of lime/ha. There was no significant difference between yields obtained from 8 tonnes of lime/ha and lime-pelleted seed. Plant vigor was best where 10 times the recommended rate of inoculant was applied. Thirteen months after sowing, no significant differences were recorded in annual yields (7,000 kg/ha, four cuts) and vigor among the lime treatments, while control plots failed to establish. Alfalfa failed to establish under all treatments on soils where the entire profile was acidic.

Fertilizer Requirements

In their undeveloped state, the native pastures were severely deficient in nitrogen, and large responses were obtained to nitrogen fertilizer on pasture and forage crops. Sowing of legumes in the pasture proved a successful alternative to applying commercial nitrogen fertilizer.

Small amounts of lime-reverted superphosphate fertilizer drilled with 6 kg of alfalfa seed/ha at time of seeding directly into unimproved grassland enhanced legume establishment (Table 2). In this trial, sown at 4,100 m on a soil of pH 6.0, alfalfa provided over 25% ground cover 16 months after sowing with an application of 12 kg P/ha and over 50% cover after 16 months with 50 kg P/ha. In trials with ryegrass–white clover pasture, high initial rates of phosphate improved white clover production and ground cover in the first year. With good grazing management

and regular maintenance applications of phosphate, grass species began contributing significantly to total dry-matter production during the third and fourth years of development following a buildup of available soil nitrogen.

No major responses were obtained to applications of potash, sulfur, or trace elements.

Cultivation

Permanent high-producing pastures based on ryegrass and white clover were successfully established using traditional cultivation techniques on land with adequate moisture. Apart from relatively few wet-lying and seepage areas, the pastures required irrigation. Under dryland conditions, cultivated improved pasture associations of N.Z. Wairau alfalfa and cocksfoot (*Dactylis glomerata* L.) established well and gave sustained production.

Mechanical Direct Drilling and Hand Seeding

N.Z. Wairau alfalfa, Huia white clover, Montgomery red clover, and alsike clover gave good results when seeded directly into unimproved native grassland by mechanical direct drilling and by hand sowing into grooves made by using hand grubbers.

Fig. 1 shows the seasonal characteristics of dry-matter production on undeveloped and developed grassland over a range of altitudes, and it highlights how effectively the spread and peak of dry-matter production may be altered by grassland development. Total annual dry-matter production from grasslands developed through the direct introduction of improved pasture species that increased from less than 1,000 kg/ha to 6,000 kg/ha after 2 years and reached more than 10,000 kg/ha in older swards.

Annual forage crops provided one means of overcoming feed shortages in the dry season. Fig. 2 highlights the production potential of several annual root crops and greenfeed cereals. Best results were obtained with turnips, particularly from early spring sowings.

CONCLUSION

Pasture and crop development in the highlands of Peru shows excellent potential, offering a means of increasing agricultural production severalfold.

Table 2. Effect of phosphate on the establishment of alfalfa direct-seeded into unimproved grassland.

	kg P/ha			
	0	12	25	50
Number of plants/m²				
3 months after sowing (2 Jan 1978)	32	43	39	34
16 months after sowing	1	21	33	35
Vigor (score 0–30)				
3 months after sowing	3	12	19	25
16 months after sowing	3	16	19	24

Fig. 1. Mean seasonal distribution of dry-matter production, 1977–1979 pasture and alfalfa associations.

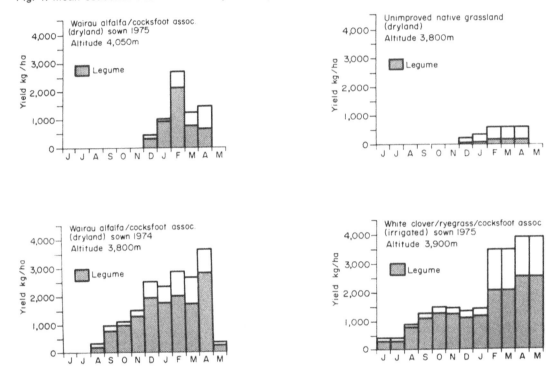

Fig. 2. Effect of time of sowing on dry-matter yield of turnips, swedes, and greenfeed cereals (mean, 3 years' data, 1977–1979).

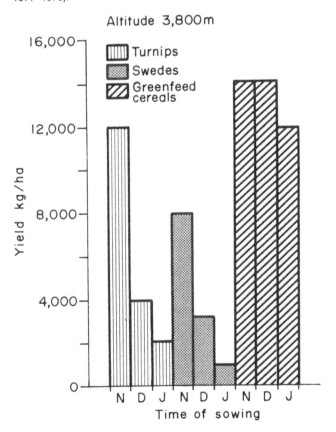

Interception of Rainfall by Creosotebush (*Larrea tridentata*)

J.M. TROMBLE

USDA-SEA-AR, Jornada Experimental Range, Las Cruces, N.M., U.S.A.

Summary

The objective of this study was to examine interception by creosotebush (*Larrea tridentata* [DC.] Cov.) of artificially applied rainfall for improved understanding of this phenomenon in hydrologic processes. This research was conducted near Las Cruces in southern New Mexico. Forty-four creosotebush shrubs were collected to obtain a representative sample of shrub size classes. Simulated rainfall was applied at the rate of 6 cm/hr. Measurements taken for each shrub were: (1) shrub height, (2) canopy area, (3) shrub volume, (4) number of stems, (5) leaf area, (6) green weight of stems, (7) oven-dry weight of stems, (8) green weight of leaves, (9) oven-dry weight of leaves, and (10) shrub green weight. Shrub height and the maximum and minimum diameter of the crown canopy were measured for determining canopy cover and shrub volume. The shrubs were severed at the base of the stem, weighed, and placed in a metal holding device. Shrub weight was again taken after 30 minutes of simulated rainfall, and the difference in weight was recorded as the amount of rainfall that was intercepted. Canopy cover of the creosotebush community was determined from 10 line transects 30.48 m (100 ft.) long. A stepwise regression analysis was performed on the data to determine from the collection of independent variables which have the best relationship to the dependent response variable. It was determined that leaf area was most highly correlated with rainfall interception, followed by number of stems, crown canopy area, and weight of oven-dry leaves. The average interception capacity of the 44 creosotebush shrubs was 1.2 g/cm², expressed in g of water held/unit area of crown canopy. Expressing the amount of water intercepted as a function of leaf area shows 0.54 g/cm². The annual rainfall in the southwestern U.S.A. is produced from storms of small amounts. Thus, interception by desert shrubs is of significant importance, since a high percentage of precipitation from these storms is "lost" from interception and subsequent evaporation back into the atmosphere. Twenty percent of the artificially applied rainfall was intercepted by creosotebush. For the native stands of creosotebush that had 30% crown cover, the loss of rainfall by interception would equal 22%. These data clearly demonstrate that light showers (< 5 mm) do little to replenish soil water.

KEY WORDS: interception, creosotebush, *Larrea tridentata* (D.C.) Cov.

INTRODUCTION

The hydrologic cycle has been the subject of a large number of experiments because of its extreme importance in dryland ecosystems, and it is probably the best known of the abiotic cycles. Interception, a process affecting the disposition of water in the hydrologic cycle, can be defined as the process of aerial redistribution of precipitation by vegetation (Collins 1970). Although some information is available (Zinke 1966, Helvey 1967, Helvey and Patric 1965) concerning interception by trees, there is a general paucity of information on interception by arid and semi-arid rangeland shrubs. Reasons for this lack of information may be the small, inconspicuous stature of shrubs compared to trees, and the limited total vegetation cover of shrubs, often less than 50%, giving the appearance of individual plants rather than of a solid block such as is presented by a dense stand of trees. Also, arid rangelands, characterized by low amounts of precipitation, have not been the focal point for hydrologic investigations that forested lands have been.

The few available studies dealing with interception indicated that saltbush (*Atriplex argentea* Nutt.) 46 cm high and in full bloom occurring in dense pure stands intercepted 50% of a 15-cm rain applied in 30 minutes, while burning bush (*Kochia scoparia* [L] Schad.) 76 cm high in a pure stand intercepted 44% (Collins 1970). Hull (1972) and Hull and Klomp (1974) studied big sagebrush (*Artemisia tridentata* Nutt.), using 10-cm-diameter gauges in dense stands at two locations in Idaho. Comparing gauges in heavy brush and brush-free areas, they indicated that the heavy brush intercepted about 30% of the rainfall between 1 April and 30 October. By spraying 10 individual plants with water they determined the potential interception/rainfall event to be 0.11 cm.

West and Gifford (1976) determined mean interception rates of individual plants of big sagebrush and shadescale (*Atriplex confertifolia* [Torr. and Frem.] Watts) to be 0.15 cm for both species, averaged over three sampling dates and two intensities. Utilizing this information and the average rainfall for 1 April to 30 November for north-

The author is a Research Hydrologist of USDA-SEA-AR.

ern Utah, but ignoring storm events less than 0.15 cm, they determined that an average of 0.59 cm of rainfall was intercepted by big sagebrush and shadescale communities. This amount was about 4% of the total precipitation that fell as rain.

The objective of this study was to examine interception by creosotebush (*Larrea tridentata* [DC.] Cov.) of artifically applied rainfall for elucidating and improving the understanding of this phenomenon in hydrologic processes.

METHODS

The interception studies were performed near Las Cruces in southern New Mexico. Forty-four individual plants representing various volume classes were subjected to simulated rainfall from a Purdue-type rainfall simulator. Rainfall intensity was 6 cm/hr. This high intensity was selected to insure that water loss by evaporation would be minimized, since we were interested in actual rainfall interception and storage on the canopy. Parameters determined for each shrub included (1) crown cover, (2) shrub height, (3) shrub green weight, (4) green weight of stems, (5) oven-dry weight of stems, (6) green weight of leaves, (7) oven-dry weight of leaves, (8) number of stems, (9) leaf area, and (10) shrub volume. As the shrub crowns were elliptical rather than circular in shape, both maximum and minimum diameters were measured for determining crown area. Each shrub was then severed at the soil surface, weighed, and subjected to simulated rainfall. At the end of 30 minutes the shrub was reweighed, and the difference in weight was recorded as intercepted rainfall. Leaves were stripped from the stems, and the leaf area was determined. Green weight of leaves and stems was measured, and the leaves and stems were oven-dried at 60°C for 24 and 48 hours, respectively, and reweighed. Crown cover for each shrub was calculated, using the formula for an ellipse. Shrub volume was calculated by multiplying crown area by shrub height.

The average crown cover of the creosotebush community was determined from 10 line transects 30.48 m long. Utilizing the interception storage data determined from individual shrubs and data from the line transects, rainfall interception was calculated for the creosotebush community.

RESULTS AND DISCUSSION

The average rainfall interception by the 44 individual shrubs was 0.12 cm, as determined from the crown cover.

Linear regression analysis (Table 1) was used to examine the effects of plant parameters on interception. Nine of the 10 parameters were highly significant with respect to the amount of water intercepted. Shrub height was the only parameter not correlated with the amount of water intercepted. Green weight of leaves (r = 0.52) and leaf area (r = 0.52) were the two most highly correlated parameters, closely followed by oven-dry weight of leaves (r = 0.51). The total shrub green weight had a correlation coefficient of 0.67, followed in descending order by oven-dry weight of stems (0.60), number of stems (0.50),

Table 1. Correlations of interception vs. plant parameters for creosotebush.

Plant parameter	r^2	Equation
Crown cover	0.24	y = 1.25(x) + 88.5
Stem-green wt.	0.20	y = 8.99(x) + 1796.0
Stem-oven-dry wt.	0.36	y = 13.38(x) + 1421.4
Leaf-green wt.	0.52	y = 39.41(x) + 331.5
Leaf-oven-dry wt.	0.51	y = 58.00(x) + 257.9
Number of stems	0.25	y = 293.30(x) + 2242.6
Leaf area	0.52	y = 1.11(x) + 333.7
Shrub volume	0.19	y = 0.01(x) + 1691.7
Shrub-green wt.	0.43	y = 0.50(x) + 110.4

Note: Shrub height was not correlated with interception using a linear regression analysis.

crown cover (0.49), green weight of stems (0.45), and the volume of the shrubs (0.44).

Stepwise regression analysis for maximum r^2 improvement was also used to analyze these data further. This method determines the best 1-variable model, the best 2-variable model, and so forth for describing the influences of the measured variables on the water intercepted. Utilizing this technique, the best 1-variable model was leaf area, which accounted for 46% of the variability of the intercepted rainfall. A further example is the 4-variable model, which would account for 61% of the variability. This model includes crown cover, shrub height, oven-dry weight of leaves, and shrub volume.

The most important parameter in the interception process is the canopy storage capacity (Aston 1979). Leonard (1965) stated that storage capacity is a function of leaf area, leaf-area index, storm intensity, and surface-tension forces resulting from leaf surface configuration, liquid viscosity, and mechanical activity. Canopy storage is usually expressed as mm of water/crown projection area on the soil surface or as depth of water/unit area of the representative plant community.

The change in water detained on the canopy, assuming no evaporation, has been described by Aston (1979) as:

$$\frac{C}{t} = (1 - p) R - \exp(a + bC) \qquad (1)$$

where a = empirically determined constant, b = empirically determined constant, C = quantity of water detained on the canopy, R = rainfall intensity, p = proportion of rainfall passing through the canopy, and t = time.

It was considered that the leaves were the major tissues intercepting water and that it would be the depth of water on the leaf surface that determines the rate of water loss. Interception storage capacity is a function of the amount and nature of the intercepting leaf surfaces, and the storage is linearly related to leaf area. Under field conditions and natural rainfall the amount of intercepted water would be influenced by wind, and this influence would need to be assessed. The impact of raindrops may influence water flow across the leaf surface and so may the leaf angle. These factors, plus others that may influence the balance of leaf surface-tension forces with gravitational forces, will all affect water storage on the leaves.

Total interception loss is far from an insignificant quantity of water (Helvey and Patric 1965). Losses are proportionally smaller in regions of lower rainfall than in regions of higher rainfall, but losses may be more important in arid regions simply because less water is available. The amount of precipitation received from individual rainfall events is characteristically small (< 0.5 cm) in arid regions. Interception would unquestionably subtract a relatively large proportion of the total amount of rainfall received from these events.

The crown cover of a native creosotebush community was calculated to be 30.5% from 10 line transects 30.48 m long in the area from which the 44 shrubs were selected for this study. Thus, in a storm of size and intensity sufficient to wet these shrubs completely, they would intercept 0.36 cm of rainfall. Extrapolating the calculated interception loss of artificially applied rainfall to the native stand of creosotebush would indicate that approximately 12% of a 0.3 cm rainfall event would be lost. This amount could be quite important since the annual average precipitation for the experimental site is 23 cm. Approximately 55%, or 12.6 cm, of this amount is received from rainfall events of varying amounts and intensities during the summer when the creosotebush is in full leaf and has maximum interception potential. Although it would be erroneous to assume that 12% of all rainfall would be abstracted as interception, certainly a generous portion of the summer precipitation would be intercepted and would not be available for soil water replenishment and subsequently would not be available for the growth of beneficial forage plants.

The intercepted precipitation would be expected to be held above the surface of the soil in proportion to the amount of canopy cover and would be subjected to evaporative losses at a rate exceeding that of the soil surface. Regarding this, approximately 14 events greater than 0.36 cm occurred each year from the period from 1 May to 31 October for 1956 to 1965, according to National Oceanic and Atmospheric Administration records. Disregarding rainfall events of less than 0.36 cm, an average of 5 cm of rainfall was intercepted by the creosotebush community. This amount is 22% of the total annual precipitation.

LITERATURE CITED

Aston, A.R. 1979. Rainfall interception by eight small trees. J. Hydrol. 42:383-396.

Collins, D. 1970. Climate-plant relations affecting semi-desert grassland hydrology. *In* Simulation and analysis of dynamics of a semi-desert grassland. Range Sci. Dept., Series 6, Colo. State Univ., Ft. Collins.

Helvey, J.D. 1967. Interception by eastern white pine. Water Resource Res. 3:723-729.

Helvey, J.D., and J.H. Patric. 1965. Canopy and litter interception of rainfall by hardwoods of eastern United States. Water Resource Res. 1:193-206.

Hull, A.C., Jr. 1972. Rainfall and snowfall interception of big sagebrush. Utah Acad. Sci. and Letters 49:64.

Hull, A.C., Jr. and G.J. Klomp. 1974. Yield of crested wheatgrass under four densities of big sagebrush in southern Idaho. U.S. Dept. Agric. Tech. Bul. 1483.

Leonard, R.E. 1965. Mathematical theory of interception. *In* International symposium on forest hydrology. Pergamon Press, New York.

West, N.E., and G.F. Gifford. 1976. Rainfall interception by cool-desert shrubs. J. Range Mangt. 29(2):171-172.

Zinke, P.J. 1966. Forest interception studies in the United States. *In* International symposium on forest hydrology. Pergamon Press, New York.

Utilization of Hydromorphic and Alkaline Hydromorphic Soils in SAP Vojvodina by Establishing Cultivated Grassland

B. MIŠKOVIĆ, P. ERIĆ, and LJ. STEFANOVIĆ
University of Novi Sad and AIC "Bečej," Yugoslavia

Summary

Hydromorphic and alkaline hydromorphic soils are heavy and unsuitable for the growing of row crops. Their poor structure and water and air properties make cultivation expensive. The establishment of cultivated grassland for production of fodder for 4 or 5 years requires less expenditure than that necessary for the production of row crops.

Four years of experiments, conducted on hydromorphic and alkaline hydromorphic soils, included four grass cultivars and three legume cultivars. Yields, chemical composition and physical properties of hay, and soil properties were examined.

The results indicated that cultivating grasses for the production of fodder is a good use of swampy areas. *Dactylis glomerata* L., *Arrhenatherum elatius* L., with *Medicago sativa* L. and *Lotus corniculatus* L., resulted in the best performance. *Festuca pratensis* Huds. and especially *L. corniculatus* L. responded best to the occurrence of surface water for periods of 4 to 6 weeks. Birdsfoot trefoil was the most resistant to drought. Average annual hay yields were 8,100–8,989 metric tons (t)/ha on hydromorphic soil and 9,720–10,420 t/ha on alkaline hydromorphic soil. The highest yields on the two soils were 12,620 and 13,270 t/ha, respectively. The overall average was 10,933 t/ha.

Cultivated grasses improve the physical and chemical properties of the examined soil types, especially their structure and water-air properties, assuring a more successful cultivation of row crops in rotation.

Cultivated grasses are a form of long-term land use. They regenerate well, as we confirmed by obtaining three or four cuttings of excellent fodder yearly with crude-protein content of 16% to 17% and good digestibility.

Cultivating grasses on hydromorphic and alkaline hydromorphic soils not only assures the production of fodder and improves physical-chemical soil properties but also is an important factor in improving human environment.

KEY WORDS: hydromorphic soils, alkaline hydromorphic soils, black soils, utilization, cultivation, grasslands, fodder.

INTRODUCTION

Hydromorphic and alkaline hydromorphic black soils are heavy and unsuitable for growing row crops. The period for crop production is short, and mechanized cultivation is expensive because of unfavorable soil structure and water-air properties (Živković et al. 1972). The growing of row crops on such soils requires large expenditures of energy and machinery (Mišković and Erić 1978).

It has been concluded that the establishment of cultivated grasslands for production of fodder is the most rational method for utilizing swamps (Mihalić et al. 1980, Mišković et al. 1980). Cultivated grassland may be used for 4 to 5 years with minimum expenditures of energy and machinery.

The objective of our investigation was to determine whether establishment of cultivated grasslands on hydromorphic black soil for production of fodder is a justifiable way of using the area.

MATERIALS AND METHODS

Experiments were conducted on hydromorphic and alkaline hydromorphic soils during 1977–1980. Basic cultivation and fertilization were performed in the fall of 1976; 400 kg/ha N-P-K with the ratio 10–30–20 was applied, and 60 kg/ha of nitrogen in the form of calcium ammonium nitrate was applied before planting. The experiment included four grass mixtures: (A) *Dactyilis glomerata* L., *Festuca pratensis* Huds., *F. rubra* L., and *Lotus corniculatus* L.; (B) *F. pratensis* Huds., *D. glomerata* L., *Arrhenatherum elatius* L., *Poa pratensis* L., *Medicago sativa* L., and *L. corniculatus* L.; (C) *F. pratensis* Huds., *D. glomerata* L., *F. rubra* L., *M. sativa* L., and *L. corniculatus* L.; (D) *F. pratensis* Huds., *A. elatius* L., *D. glomerata* L., *F. rubra* L., *Trifolium pratense* L., and *L. corniculatus* L. The ratio of grasses to legumes was 70:30. Experimental plots were 1,000 m² in 4 replications. Distance between rows was 12 cm and planting depth 2 cm. Plots were cultipacked after planting. Natural grassland was used as the control.

Two cuttings were taken in the first and the fourth year and three cuttings in the second and third years. Yields of green and dry herbage in t/ha were determined. Samples for analyses were taken, and green herbage was botanically analyzed. The hay was analyzed for crude protein, cellulose, crude fat, nitrogen-free extract, ash, phosphorus (P_2O_5), potassium (K_2O), and calcium (CaO).

METEOROLOGICAL CONDITIONS

The region in which the experiment was conducted has a modified continental climate with warm and dry summers and cold winters without snow. The mean annual temperature for the 14-year experimental period was 11.6°C. The mean annual accumulated temperature was 3,860°C, which varied from 3,616°C in 1980 to 4,134°C in 1977. The mean accumulated temperature for the growing season (April-September) was 3,059°C, which varied from 3,178°C in 1977 to 2,927°C in 1980.

The mean annual rainfall for the experimental period was 582.2 mm, with a range of 664.3 mm in 1978 to 521.1 mm in 1977. The mean rainfall during the growing season (April-September) was 340.2 mm, with a range of 286.6 mm in 1977 to 451.0 mm in 1978. The rainfall was insufficient and unfavorably distributed, with the best distribution in 1978 and the worst in 1980.

SOIL PROPERTIES

Hydromorphic black soil contains: 1.10% $CaCO_3$, 70.00% clay, 27.00% silt, small quantities of sand, 3.77% humus (Kotcman), 10–11 mg P_2O_5/100 g of soil, and 29 mg K_2O/100 g of soil, and the pH ranges from 5.0 to 6.9. Alkaline hydromorphic soil contains: 2.0% $CaCO_3$, 68.0% clay, 30.0% silt, small quantities of sand, 3.5% humus (Kotcman), 13 mg P_2O_5/100 g of soil, and 30 mg K_2O/100 g of soil, and the pH ranges from 7.8 to 8.4.

B. Mišković is a Professor and P. Erić is an Assistant on the Agricultural Faculty of the University of Novi Sad, and Lj. Stefanović is an Agronomist at AIC "Bečej," Bečej.

RESULTS

Hay Yield

Hay yields are shown in Tables 1 and 2. There were small variations in yield within years, but there were large variations from year to year and between soil types. Mixture B produced higher yields than did mixtures C, D, and A. Differences in yields among the mixtures were insignificant, but the yield of the mixture was significantly higher than that of the control. In other studies, yields for cultivated grasslands were 4 to 6 times higher than that of natural grassland (Erić 1980).

Insufficient rainfall and its unfavorable distribution (drought in May and June 1977, only 15.1 mm of rain) adversely affected some plant species (*Poa pratensis, Festuca rubra, Trifolium pratense*). Some species were more tolerant of the drought (*D. glomerata, A. elatius*), especially when grown on alkaline hydromorphic soil. Meadow fescue and birdsfoot trefoil were not harmed after 4 to 6 weeks under water in microdepressions (Mišković et al. 1980, Solyukov 1977). Birdsfoot trefoil is very resistant to drought (Strelkov 1977), with only slight yield decreases (Mišković et al. 1980). Hay yields were satisfactory but would have been increased considerably by fertilization and irrigation. The biological productivity of these mixtures justifies their use on these and similar soils (Zaharev et al. 1977).

The comparison between yields of the grasses and of the legumes in mixtures B, C, and D showed that the proportion of legumes was larger. It confirms their adaptation to such conditions. Grasses (*D. glomerata, A. elatius*) dominated in the fourth year. Yield of birdsfoot trefoil increased throughout the test period, but yield of red clover

Table 2. Total and average hay yields, experiment 2.

Mixture*	Yields, kg/ha/year				Average annual yield, kg/ha
	1977	1978	1979	1980	
A	5,105	13,470	12,680	8,587	9,961
B	4,534	12,907	13,008	10,488	10,202
C	6,168	11,302	13,270	8,063	9,701
D	5,293	11,315	12,568	9,700	9,719
Control	2,010	2,410	2,304	2,263	2,247
Average	4,622	10,281	10,766	7,820	8,372
LSD 0.01					6,663
0.05					4,818

*The mixtures and the control are defined as follows:
  A: *Dactylis glomerata* L., *Festuca pratensis* Huds., *F. rubra* L., *Lotus corniculatus* L.
  B: *F. pratensis* Huds., *D. glomerata* L., *Arrhenatherum elatius* L., *Poa pratensis* L., *Medicago sativa* L., and *L. corniculatus* L.
  C: *F. pratensis* Huds., *D. glomerata* L., *F. rubra* L., *M. sativa* L., *L. corniculatus* L.
  D: *F. pratensis* Huds., *A. elatius* L., *D. glomerata* L., *F. rubra* L., *Trifolium pratense* L., *L. corniculatus* L.
Control: natural grassland.

declined to the third year, and the species could be found only sporadically in the fourth. The yield and quality of the mixtures were superior to those of natural grasslands (SGV 1978). Orchardgrass was the best, although tall oat grass was exceptionally good on alkaline hydromorphic black soil. Meadow fescue reacted best to anaerobic conditions for 4 to 5 weeks. Grasses with deeper, stronger roots and high competitiveness are important for the examined soils, especially in dry periods (Zaharev et al. 1977, Mišković and Erić 1978). These grasses withstand soil cracking and reduce the evaporation of water from the soil surface. Water permeability is higher than when row crops are grown (Mišković et al. 1980). Grasslands stimulate the mineralization process of organic matter by improving water-air soil properties that in turn improve the microbiological activity and thus the fertility of soil (Mihalić et al. 1980).

Chemical Composition of Hay

Table 3 shows the level of important nutritive substances. Crude-protein content was higher in mixtures B and C, due to the higher portion of legumes. Cellulose content, averaging 28.65% was lowest in mixture B and highest in mixture A. Fat content averaged 2.39%. The average content of nitrogen-free extract was 35.37%, with a range from 33.98% to 36.46%. Ash content averaged 9.26%, the highest amount being in mixture C, which had a high proportion of legumes in the mixture. The average phosphorus content was 0.33%, which is rather low. The average calcium content was 1.94%. Calcium leaches down intensively in hydromorphic soils. The average potassium content was 2.38%.

Table 1. Total and average hay yields, experiment 1.

Mixture*	Yields, kg/ha/year				Average annual yield, kg/ha
	1977	1978	1979	1980	
A	4,058	9,277	10,996	9,175	8,377
B	4,875	10,768	11,312	7,787	8,686
C	4,263	9,411	12,623	8,363	8,665
D	4,360	9,979	11,304	8,413	8,514
Control	2,596	2,837	2,915	2,629	2,744
Average	4,030	8,454	9,830	7,373	7,397
LSD 0.01					5,832
0.05					4,217

*The mixtures and the control are defined as follows:
  A: *Dactylis glomerata* L., *Festuca pratensis* Huds., *F. rubra* L., *Lotus corniculatus* L.
  B: *F. pratensis* Huds., *D. glomerata* L., *Arrhenatherum elatius* L., *Poa pratensis* L., *Medicago sativa* L., and *L. corniculatus* L.
  C: *F. pratensis* Huds., *D. glomerata* L., *F. rubra* L., *M. sativa* L., *L. corniculatus* L.
  D: *F. pratensis* Huds., *A. elatius* L., *D. glomerata* L., *F. rubra* L., *Trifolium pratense* L., *L. corniculatus* L.
Control: natural grassland.

Table 3.   Content of nutritive substances in hay[1] (% dry matter).

Mixture[2]	N	CP	C	CF	NFE	Ash	P_2O_5	CaO	K_2O
					-%-				
A	2.47	15.4	29.7	2.3	36.0	8.8	0.35	1.81	2.08
B	2.76	17.2	27.7	2.4	36.5	9.1	0.34	2.08	2.58
C	2.90	18.1	28.4	2.4	34.0	9.8	0.35	2.18	2.45
D	2.63	16.4	28.8	2.5	35.0	9.2	0.26	1.92	2.35
Average	2.69	16.8	28.6	2.4	35.4	9.2	0.33	1.98	2.37

[1] N—total N; CP—crude proteins; C—cellulose; CF—crude fats; NFE—N-free extract.

[2] See footnotes for Table 1.

CONCLUSION

Fodder production on cultivated grasslands is a good method of utilizing hydromorphic and alkaline hydromorphic black soils. Such production is less expensive than the production of row crops on these soils.

Managing soils as grasslands improves physical, chemical, and biological properties of hydromorphic and alkaline hydromorphic black soils, making them more suitable for subsequent growing of row crops. The yields of cultivated grasslands are significantly higher than those of natural grasslands, especially on alkaline hydromorphic soil. The average hay yield was 10,933 kg/ha. Mixtures B and C were more suitable for hay production, while mixture A was more suitable for grazing.

Cultivated grasslands on hydromorphic soil are an important factor in the improvement of the environment, especially when they are located near human settlements.

LITERATURE CITED

Erić, P. 1980. Proizvodne i kvalitativne karakteristike prirodnih travnjaka u Socijalističkoj Autonomnoj Pokrajini Vojvodini, Magistarski rad, Novi Sad.

Mihalić, V., N. Miljković, V. Nejgebauer, and I. Mušac. 1980. Problemi intenziviranja proizvodnje na normalnim zemljištima, uključujući odvodnju i druge meliorativne zahvate. JDPZ, VI Kongres, Novi Sad.

Mišković, B., and P. Erić. 1978. Savetovanje o proizvodnji i korišćenju stočne hrane kao daljeg faktora razvoja stočarstva u SAP Vojvodini, Novi Sad.

Mišković, B., P. Erić, and Lj. Stefanović. 1980. Establishment of cultivated grassland on hydromorphic black soil in SAP Vojvodina aimed at its better utilization. Proc. VIII Europ. Grassl. Cong. Zagreb.

Salyukov, P.A. 1977. Persistence of forage plants in seeded grassland on liman meadows in Kazakhstan. Proc. XIII Int. Grassl. Cong., 681–683.

Statistički godišnjak SAP Vojvodini (SGV). 1978. Novi Sad.

Strelkov, V. 1977. Biological and ecological properties of *Lotus corniculatus*. Proc. XIII Int. Grassl. Cong., 801–803.

Zakharev, N.I., A.A. Atabaev, L.A. Golovina, L.V. Koverga, N.G. Kotysheva, I.P. Lazarev, N.S. Sotnikova, and N.N. Sycheva. 1977. The biological performance and efficient utilization of artificial and natural grassland areas in the valleys and mountainous regions of the western part of northern Tien Shan. Proc. XIII Int. Grassl. Cong., 555–558.

Živković, B., V. Nejgebauer, Dj. Tanasijević, M. Miljković, L. Stojković, and P. Drezgić. 1972. Zemljišta Vojvodine. Institut za poljoprivredna istraživanja, Novi Sad.

Simple Mixtures for Dry Mountain Regions: Relations Between Tall Fescue and Birdsfoot Trefoil During Five Years

G. HAUSSMANN, R. PAOLETTI, and C. LOCATELLI
Istituto Sperimentale per le Colture Foraggere, Lodi, Italy

Summary

Grass-legume mixtures are of great interest, especially in central and northern Appennine regions with a summer dry season. Birdsfoot trefoil (*Lotus corniculatus* L.) cv. Franco and tall fescue (*Festuca arundinacea* Schreb.) cv. Manade were grown in mixture and in pure stand at the experimental farm of the Institute. Our objectives were to determine the seeding rates in the mixtures and the most suitable method of sowing. The two species were seeded in mixture and alone to 20 seeding treatments: (a) 16 mixtures in alternate rows 12 cm apart, in factorial combinations of four seeding rates (5-10-12-15 kg/ha of birdsfoot trefoil and 5-7-10-15 kg/ha of tall fescue); (b) 1 blended mixture at the highest seeding rate (15 kg/ha) in rows 12 cm apart; and (c) 3 pure stands, 1 of birdsfoot trefoil (15 kg/ha) and 2 of tall fescue (15 and 30 kg/ha), in rows 12 cm apart. The experimental layout was a randomized block with 4 replicates giving 80 experimental 2.4-m² plots. The trial lasted 5 years (1975-1979) without irrigation. The harvest of each cut was dried (105°C) and weighed; yields of mixtures were separated into their components.

The analysis of collected data showed that (1) all the mixtures yielded significantly more than the pure stands; (2) blend was significantly superior to seeding in alternate rows; and (3) total percentage of birdsfoot trefoil on a dry-matter basis was 24% and 26% in the blends and alternate rows, respectively. The mixtures of birdsfoot trefoil and tall fescue were noteworthy because of their high yields and good legume-grass balance. The two species together yielded a bulk of good fodder and stability of production during the growing season.

KEY WORDS: mixture, pure stand, birdsfoot trefoil, *Lotus corniculatus* L., tall fescue, *Festuca arundinacea* Schreb.

INTRODUCTION

Grass-legume mixtures have been studied throughout the world. References on this subject are numerous in Europe (Stebler 1883, 1885; Caputa 1948; Haussmann 1949-1975; Talamucci 1970). The Institute of Fodder Crops, located at Lodi (Po Valley), has actively contributed to this subject, particularly in regard to mixtures of short- and long-lived species for fodder crops and meadows suited to the different environmental conditions of Italy.

In the past 30 years, a great part of the Appennine hills and mountains has been abandoned by thousands of farmers. The only profitable utilization of these deserted areas seems to be cattle raising, which needs little manpower. Even with good management of the meadows or pastures, the swards must sometimes be renewed to assure good production. For these reasons, our study was undertaken to compare different methods and rates of sowing, in simple mixtures, two species (*Lotus corniculatus* L. and *Festuca arundinacea* Schreb.) that can grow in dry conditions.

METHODS

A 5-year (1975-1979) field study on long-lived grass-legume mixtures was conducted at the Lodi experimental farm (altitude, 92 m; lat 45°20'N long 9°15'E; mean precipitation (1950-1979), 88.9 cm/yr and 21.7 cm/May-July). Birdsfoot trefoil (*Lotus corniculatus* L.) cv. Franco and tall fescue (*Festuca arundinacea* Schreb.) cv. Manade were seeded in pure stand and in mixture to 20 seeding treatments, which consisted of: (a) 16 mixtures in alternate rows 12 cm apart, in factorial combinations of four different seeding rates (5-10-12-15 kg/ha of birdsfoot trefoil and 5-7-10-15 kg/ha of tall fescue); (b) 1 blended mixture of the two species at the highest seeding rates (15 kg/ha each) in row spacing 12 cm apart; (c) 3 pure stands, 1 of birdsfoot trefoil (15 kg/ha) and 2 of tall fescue (15 and 30 kg/ha), in row spacing 12 cm apart. The

experimental design was a randomized complete block, replicated 4 times. Plots measured 2 × 1.2 m, consisting of 10 rows, 12 cm apart, with an effective harvest area of 1.5 × 0.72 m.

Stands were established in April 1975 in sandy-loam soil previously fertilized at the rates of 300 quintals (q)/ha of farmyard manure and 50 kg/ha of N-P$_2$O$_5$-K$_2$O and ploughed (30-cm depth), harrowed, and rolled. During the year of seeding, plots were gently irrigated as needed. In the subsequent 4 years of culture no irrigation was applied. At the end of each winter a compound fertilizer was applied, consisting of 20 kg/ha of N-P$_2$O$_5$-K$_2$O.

For each treatment, the number of plants of the seeded species/m² 1 month after establishment and the dry-matter yields during the 5 years were recorded. A 1.5 × 0.72-m cut was mown with hand scythe at a cutting height of approximately 5 cm when the birdsfoot trefoil was in early bloom. The grass and legume components of the mixtures were separated by hand. Herbage from the pure stands and the mixtures was dried for 48 hours at 105°C and weighed.

RESULTS

Birdsfoot trefoil and tall fescue counts, taken 1 month after seeding, showed a better establishment of the legume than of the grass. The counts were 73% and 64% of viable sown seeds, respectively. Percentages of seedlings were inversely proportional to the seeding-rate increase. Both species in pure stand or in mixtures gave a higher percentage of seedlings than in the alternate-row mixtures. Annual and total dry-matter yields from the 20 treatments and total percentage of birdsfoot trefoil are reported in Table 1. In this table, "L" designates birdsfoot trefoil, "F" designates tall fescue, and the subscript numerals indicate the seeding rate in kg/ha.

Over the 5 years, all mixtures had greater total yields than the pure stands of the two species. All the mixed swards but the $L_{12}F_7$ were significantly different from the grass pure stands, while all the 17 mixtures were significantly better than the legume pure stand. Of the mixtures, the 1:1 blend $L_{15}F_{15}$ differed significantly from all the 16 alternate-row mixtures, among which the $L_{10}F_{10}$ and $L_{15}F_5$ treatments were best. The mean percentage of

The late G. Haussmann was Director of the Istituto Sperimentale. R. Paoletti is a Professor of Agronomy and C. Locatelli is an Agricultural Researcher at the Istituto.

Table 1.   Annual and total dry-matter yields of 20 birdsfoot trefoil (*Lotus corniculatus* L.) and tall fescue (*Festuca arundinacea* Schreb.) stands, quintals/ha.

Treatment	1975	1976*	1977*	1978	1979	1975–1979*
L_5F_5†	81.96	120.06def	78.93bcde	90.58	64.37	435.90de(29)**
L_5F_7	80.25	124.90bcd	76.72cde	84.98	60.04	426.89e(25)
L_5F_{10}	79.95	131.97b	78.78bcde	91.47	56.38	438.55de(27)
L_5F_{15}	82.26	128.79bc	79.59bcd	92.82	65.57	449.03bcd(22)
$L_{10}F_5$	82.34	118.54ef	73.89de	85.80	63.91	424.48e(33)
$L_{10}F_7$	85.27	115.25f	77.41bcde	96.25	57.82	423.00e(29)
$L_{10}F_{10}$	89.07	123.47cd	83.58b	97.43	65.45	459.00b(29)
$L_{10}F_{15}$	83.06	118.18ef	72.23e	89.27	66.01	428.75e(28)
$L_{12}F_5$	74.41	126.28bcd	80.54bcd	91.71	61.41	434.35de(26)
$L_{12}F_7$	70.86	126.22bcd	79.95bcd	81.90	51.16	410.09f(22)
$L_{12}F_{10}$	78.76	125.20bcd	74.32de	86.83	66.65	431.76e(26)
$L_{12}F_{15}$	80.64	124.76bcd	73.91de	86.03	62.80	428.14e(25)
$L_{15}F_5$	76.48	128.23bc	82.93bc	97.19	71.05	455.88b(29)
$L_{15}F_7$	87.64	127.49bcd	82.35bc	93.70	63.92	455.10bc(24)
$L_{15}F_{10}$	79.50	126.61bcd	75.09de	85.98	57.55	424.73e(24)
$L_{15}F_{15}$	84.75	124.16cd	76.50de	89.11	65.76	440.28cde(23)
$L_{15}F_{15}$(blend)	78.36	142.95a	93.46a	100.05	64.11	478.93a(24)
$L_{15}F_0$	88.11	112.26f	57.57f	62.40	23.03	343.37g(100)
L_0F_{15}	56.20	117.92f	76.81cde	86.97	59.63	397.53f –
L_0F_{30}	61.54	115.45f	78.04bcde	84.86	70.84	410.73f –

*Values not followed by the same letter in a column differ significantly at the 5% level of probability.

**Total percentage of birdsfoot trefoil on a dry matter basis.

†L = birdsfoot trefoil, F = tall fescue with seeding rates in kg/ha.

birdsfoot trefoil was 26% in the mixed swards, with higher values in $L_{10}F_5$, $L_{10}F_{10}$, and $L_{15}F_5$ (33%, 29%, and 29%), respectively, than in the most productive $L_{15}F_{15}$ blend (24%).

DISCUSSION

Differences between simple grass-legume mixtures and pure stands were noteworthy. Birdsfoot trefoil was favored by the spring sowing, but tall fescue established slowly and reached complete development only in the second year.

Among the mixtures tested, the $L_{15}F_{15}$ blend emerged as superior. This method of sowing is best suited to the two species. In fact, percentage of seedlings was higher in blend than in alternate rows, at the same rate of sowing. Sowing rates in blend at 15 kg/ha and in alternate-row mixture at 10 kg/ha gave the best dry-matter yields and persistence of the legume. A better legume-grass balance was obtained with the $L_{10}F_{10}$ treatment, which, though significantly lower-yielding than $L_{15}F_{15}$, contained 29% birdsfoot trefoil.

The foregoing satisfactory results were obtained in absence of irrigation and with low N-P-K dressing. The information thus should be useful in the Appennine districts.

LITERATURE CITED

Caputa, J. 1948. Untersuchungen über die Entwicklung einiger Gräser und Kleearten in Reinsaat und Mischung. Diss. ETH, Zurich.

Haussmann, G. 1949–1975. Relazioni della stazione di praticoltura e dell'Istituto Sperimentale per le Colture Foraggere (18 vol.).

Talamucci, P. 1970. Effetti delle modalità di semina sulla produttività e sull'equilibrio fra i costituenti di un miscuglio di erba medica-erba mazzolina. Fir. Agric. 33:8–19.

Stebler, F.G. 1883, 1885. Die Grassamen-Mischungen zur Erzielung des grössten Futterertrages von bester Qualität, 3rd and 4th eds.

Strategies of Seedling Development in Warm-Season Grass

L.E. MOSER and C.J. NELSON
University of Missouri, Columbia, Mo., U.S.A.

Summary

Establishment of warm-season prairie grasses is often difficult because of slow seedling growth and weed competition. Seedling development of four perennial warm-season grasses was compared with that of a vigorous annual grass, giant foxtail (*Setaria faberi* Herrm.) Big bluestem (*Andropogon gerardii* Vitman), switchgrass (*Panicum virgatum* L.), indiangrass (*Sorghastrum nutans* [L.] Nash), caucasian bluestem (*Bothriochloa caucasica* [Prin.] C.E. Hubb.), and giant foxtail were grown in a greenhouse. Leaf elongation rate, collaring date, and leaf length were measured daily. Carbon-exchange rate (CER) was determined. Plants were harvested at the sixth and eighth leaf stages, and yield components, leaf area, and tiller numbers were recorded.

Giant foxtail had a very rapid leaf elongation rate, developed quickly, and had little tillering up to the eighth collared-leaf stage. Big bluestem and caucasian bluestem developed nearly as quicky as giant foxtail, but leaf elongation rate was much slower, and leaves were shorter. Caucasian bluestem tillered at an early stage and developed additional photosynthetic area, although not much. Big bluestem allocated more of its assimilates to the root system and did not tiller as much as caucasian bluestem. Switchgrass and indiangrass developed much more slowly. Once they reached the fourth leaf stage, their leaf elongation rate was intermediate between those of giant foxtail and caucasian or big bluestem. Switchgrass tillered more than big bluestem. Carbon-exchange rate was lower for giant foxtail and caucasian bluestem than for the other grasses.

In a breeding program, selecting for rapid leaf elongation rate in caucasian bluestem and big bluestem could result in greater seedling vigor since more photosynthetic area would be produced. Switchgrass and indiangrass selected for rapid development prior to the fourth leaf stage with the same or greater leaf elongation rate that occurs later in their development could result in greater seedling vigor. Screening for CER probably would not result in greater seedling vigor.

KEY WORDS: *Andropogon gerardii* Vitman, *Sorghastrum nutans* (L.) Nash, *Panicum virgatum* L., *Bothriochloa caucasica* (Prin.) C.E. Hubb., *Setaria faberi* Herrm., leaf growth, tillering, carbon-exchange rate.

INTRODUCTION

Warm-season prairie grasses are slower and more difficult to establish than cool-season grasses. Because they are spring seeded, competition with annual weeds is keen. Improving seedling vigor would improve their ability to become established (Wright 1971).

Many factors affect seedling vigor, including seedling height, which has been proven to be quite important in influencing first-year yields (Voight and Brown 1969), and seedling leaf area (Perry and Moser 1975). Yield/tiller

and tillers/plant are the two major yield components, and there is ample genetic variability present for selection in tall fescue (*Festuca arundinacea* Schreb.). Leaf elongation rate affects yield/tiller and is highly associated with regrowth yield of tall fescue. However, CO_2-exchange rate (CER) and yield are not correlated with yield/tiller (Nelson et al. 1977).

The purpose of this research was to compare seedling development of four warm-season prairie grasses with a vigorous warm-season annual grass, giant foxtail (*Setaria faberi* Herrm.).

METHODS

Big bluestem (*Andropogon gerardii* Vitman) cv. Pawnee, switchgrass (*Panicum virgatum* L.) cv. Pathfinder, indiangrass (*Sorghastrum nutans* [L.] Nash) cv. Oto, caucasian bluestem (*Bothriochloa caucasica* [Prin.] C.E. Hubb.),

L.E. Moser is a Professor at the University of Nebraska and C.J. Nelson is a Professor at the University of Missouri. The research reported here was accomplished when the senior author was on sabbatical leave at the University of Missouri, Columbia. This is a contribution of the Department of Agronomy of the University of Missouri (Journal Article No. 8761).

Mention of a proprietary product does not imply endorsement of it.

and giant foxtail were chilled (5°C) for 3 weeks and germinated on filter paper, and three seedlings were transplanted to deep pots containing equal parts of Mexico silt loam (Udollic Ochraqualf) and sand. Plants were placed in a greenhouse with a 14-hour photoperiod and temperatures ranging from 33°C (day) to 18°C (night). The experiment was a completely randomized design with 6 replications of paired plants with three plants in each pot. Daily measurements were made to document new tiller appearance, leaf collaring date, and final leaf length. The daily leaf elongation rates of leaves 6 and 7 were determined.

When 50% of the plants of a particular species had collared the sixth leaf, harvest 1 (one pot of each pair) was taken. The second half of the plants of that species was harvested (harvest 2) when 50% of the eighth leaves were collared. Leaf area was determined with a LI-Cor Model LI 3000 leaf-area meter. Plants were separated into roots, leaf blades, 2.5 cm of the lower stem base, and remaining herbage. Carbon-exchange rate (CER) was measured on 3 replications when the plants were in the fifth and eighth collared-leaf stages with an infrared CO_2 analyzer. Two leaves were used for each measurement, and measurements were made outside from 1000 to 1400 hours on a clear day.

RESULTS

Giant foxtail had the most rapid leaf elongation rate of any of the grasses (Table 1), and much new photosynthetic area was produced each day. Switchgrass and indiangrass had intermediate rates, and caucasian bluestem and big bluestem had the lowest leaf elongation rates. Average leaf-blade length at collaring was similar for giant foxtail, switchgrass, and indiangrass but less for caucasian bluestem and big bluestem. Giant foxtail had the shortest plastochron (time interval between collaring of successive leaves), and indiangrass had the longest.

Development of leaves 4 through 8 occurred in a linear manner during this experiment (Fig. 1). Giant foxtail reached the fourth collared-leaf stage first. Switchgrass

and indiangrass reached the fourth collared-leaf stage much later. Caucasian bluestem and big bluestem developed nearly as rapidly as giant foxtail, but because of their lower leaf elongation rates and, thus, shorter leaves, they had much less photosynthetic area at any stage. Across species, seed size was not associated with rapidity of development.

At any morphological stage giant foxtail had fewer tillers than the perennial grasses (Table 1). Caucasian bluestem tillered quite rapidly. It had more tillers and was harvested earlier than switchgrass or indiangrass.

Although caucasian bluestem and big bluestem developed morphologically nearly as rapidly as giant foxtail, they were much smaller plants when harvested, as is reflected by the leaf area present and weight of the plant components (Table 2). Switchgrass and indiangrass produced large leaf area and plant weight by the time they reached the sixth and eighth collared-leaf stages, but they were slow in reaching these stages as compared with giant foxtail. The leaf-area ratio indicated that giant foxtail and caucasian bluestem had the greatest amount of photosynthetic area/unit of total plant weight. The shoot:root ratio showed that big bluestem allocated more assimilates to the roots.

Giant foxtail and caucasian bluestem had a lower CER than did indiangrass, switchgrass, and big bluestem (Table 1). CER was associated with difference in specific leaf weight (SLW), but other factors were evidently involved. Stem-base total nonstructural carbohydrates were not associated with seedling growth.

DISCUSSION

The vigor and competitiveness of giant foxtail at the seedling stage was associated with rapid leaf development coupled with a high leaf elongation rate, so a large amount of photosynthetic area was produced by the main shoot after germination. The high leaf elongation rate appeared to be the major factor in the rapid development of leaf area.

Caucasian bluestem developed fairly rapidly mor-

Table 1. Leaf development parameters, harvest times, tiller numbers, carbon exchange rate (CER), and specific leaf weight (SLW) of warm-season grasses grown in greenhouse at Columbia, Missouri.

Species	Leaf 7 extension rate	Average final leaf blade length (leaves 4–8)	Average Plastochron	Harvest Time		Additional tiller numbers		CER	SLW
				6th collared leaf	8th collared leaf	6th collared leaf	8th collared leaf		
	mm/day	mm	days	days after	germination			mg CO_2 /dm²/hour	mg/cm²
Giant Foxtail	35a[1]	178a	3.5d	28	38	0.0b	0.9d	35.8b	3.9b
Switchgrass	30b	205a	5.3b	41	53	1.4a	3.4ab	62.5a	4.1b
Indiangrass	28b	181a	6.1a	46	56	0.9a	2.7bc	63.4a	4.8a
Caucasian Bluestem	15c	102b	4.1c	33	41	1.0a	4.3a	33.7b	3.1c
Big Bluestem	13c	98b	4.5c	33	41	0.9a	1.9cd	51.2a	3.9b

[1]Values within a column followed by the same letter are not significantly different using Duncan's New Multiple Range Test, p < .05.

Fig. 1. Leaf collaring time of giant foxtail (GF), caucasian bluestem (CB), big bluestem (BB), switchgrass (SG), and indiangrass (IG) grown under greenhouse conditions at Columbia, Missouri.

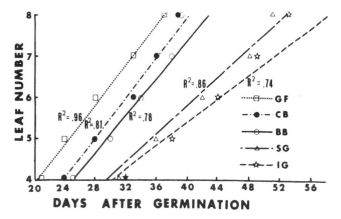

phologically, but with its slow leaf elongation rate and short leaves it had much less leaf area than giant foxtail. However, caucasian bluestem had more small tillers and hence more additional shoot apices than giant foxtail and most of the other grasses. It had a number of equal-sized tillers at harvest and did not allocate as much assimilate to the main tiller. Big bluestem was similar to caucasian bluestem, but did not tiller as early and evidently allocated more assimilates to the roots.

Switchgrass and indiangrass, even though high in CER, reached the fourth collared-leaf stage much more slowly, developed more slowly than the other grasses from the fourth to eighth collared-leaf stages, and had an intermediate leaf elongation rate. Tillering occurred at an earlier morphological stage than it did in big bluestem and giant foxtail, but chronologically tillering came later. CER could not be used to explain seedling vigor. Selection for a rapid rate of early leaf formation and high leaf elongation rate should make it possible to improve the seedling vigor of the warm-season prairie grasses by breeding. If big bluestem and caucasian bluestem had a more rapid leaf elongation rate, early seedling leaves would be longer, with more photosynthetic area. However, rapid leaf elongation rate and development of the main shoot may result in less or later tillering (Nelson et al. 1977). Root development should also be monitored to see how rapid early shoot development affects it.

In a similar study under cooler greenhouse conditions similar results were obtained, and there was evidence that barnyardgrass *Echinochloa crusgalli* [L.] Beauv. had more tolerance to cool temperatures than did the perennial warm-season prairie grasses. Selecting for cool-temperature tolerance of seedlings may be beneficial for improving establishment capability of warm-season grasses.

LITERATURE CITED

Nelson, C.J., K.H. Asay, and D.A. Sleper. 1977. Mechanisms of canopy development of tall fescue genotypes. Crop Sci. 17:449–452.

Perry, L.J., Jr. and L.E. Moser. 1975. Seedling growth of three switchgrass strains. J. Range Mangt. 28:391–393.

Voight, P.W., and H.W. Brown. 1969. Phenotypic recurrent selection for seedling vigor in sideoats grama, *Bouteloua curtipendula* (Michx.) Torr. Crop Sci. 9:664–666.

Wright, L.N. 1971. Drought influence on germination and seedling emergence. *In* K.L. Larson and J.D. Eastin (eds.), Drought injury and resistance in crops, 19–44. Crop Sci. Soc. Am. Spec. Pub. 2.

Table 2. Leaf area, dry weight, leaf area ratio, and shoot/root ratio of warm-season grasses grown in greenhouse at Columbia, Missouri, harvested when 50% of the 6th leaves had collared (Harvest 1) and when 50% of the 8th leaves had collared (Harvest 2).

Species	Total leaf area	Main shoot leaf area	Tiller leaf area	Total leaf blade weight	Total above ground weight	Total root weight	Total biomass weight	Leaf area ratio	Shoot/ root ratio
	----------------- cm²/plant-----------------			------------------------------------- mg/plant-------------------------------------				cm²/mg	mg/mg
Harvest 1									
Giant foxtail	21.1a[1]	19.6a	0.0b	49.3bc	74.3b	48.9b	123.2b	0.16a	1.84ab
Switchgrass	23.4a	18.2a	3.6a	92.2a	150.1a	81.1a	231.2a	0.10c	1.85ab
Indiangrass	13.3b	11.7b	0.7b	65.9b	115.5a	76.3a	191.8a	0.07d	1.51ab
Caucasian bluestem	5.3c	4.7c	0.4b	14.1d	23.8c	14.5c	38.3c	0.14b	2.19a
Big bluestem	6.1c	5.2c	0.4b	27.5cd	45.8bc	40.4b	86.2bc	0.07d	1.16b
Harvest 2									
Giant foxtail	63.7b	60.4a	3.1b	244.4b	387.2b	208.2bc	595.4b	0.11b	1.87a
Switchgrass	83.9a	44.6b	38.7a	426.6a	750.3a	422.1a	1172.5a	0.07d	1.79a
Indiangrass	37.9c	26.8c	10.8b	197.6b	362.9b	262.3b	625.2b	0.06e	1.35b
Caucasian bluestem	23.0c	16.0d	6.8b	58.6c	100.6c	67.2d	167.8c	0.14a	1.51b
Big bluestem	20.5c	16.2d	3.6b	75.8c	125.1c	106.9cd	232.0c	0.09c	1.17c

[1]Values within a column followed by the same letter are not significantly different using Duncan's New Multiple Range Test, p < .05.

Climatic Influences During Flowering on Seed Dormancy and Seed Formation of *Stylosanthes hamata* cv. Verano

P.J. ARGEL and L.R. HUMPHREYS

University of Queensland, St. Lucia, Australia

Summary

Effects of variation in temperature, soil moisture supply, and illuminance on seed dormancy and seed formation of *Stylosanthes hamata* cv. Verano were studied. Plants were grown in the open in successive seasons, in controlled-temperature cabinets, and in the greenhouse where shading and watering treatments were varied after the onset of flowering.

Hardseededness was strongly developed and was positively related to temperature during seed formation. After seed maturation, seed moisture content was the main factor governing hardseededness, and this quality was negatively associated with temperature during seed formation. High levels of hardseededness developed in all seeds when seed moisture content was below approximately 7%, but the equilibrium moisture content of seeds stored in atmospheres of varying relative humidity was modified by temperature during seed formation. This fact helps to explain variation in hardseededness related to differing provenances of seedlots.

Seeds formed under high temperature had higher content of lignin (which was concentrated in the counter-palisade cells) and of hemicellulose, lower content of cellulose, and shorter palisade cells; cutin content was independent of temperature. The seed-coat of hard seeds exhibited a more regular and organized structure and a more evenly recticulated surface than that of soft seeds. Seed color changed from dark to light as temperature of seed formation decreased from 27°C to 21°C, but hardseededness was found to be unrelated to seed color per se.

Embryo dormancy was transitory, and little developed. An inhibitor in the pod or testa restricted germination; this effect was independent of climatic conditions during seed formation and decreased after 120 days of seed storage. Dormancy was more strongly developed in the hooked upper articulation than in the hookless lower articulation.

Short durations of soil moisture stress that reduced leaf water potential to minimum values of about − 25 to − 28 bars, and shading treatments that reduced radiation to about 20% of full daylight reduced seed production, but hardseededness was not consistently related to these treatments.

Time of blooming, rate of floret appearance, total floret differentiation, duration of flowering, floret abortion, lower articulation formation, and time to pod maturity were sensitive to temperature conditions. Seed yield was maximal at 31°/24°C.

S. hamata cv. Verano is a short-lived plant, and persistence of sustained yields depends upon continual plant replacement from soil seed reserves. The levels of seed production and of hardseededness are increased by warm conditions during flowering, and this factor favors and may even restrict its adaptation to dry tropical environments.

KEY WORDS: *Stylosanthes hamata*, Verano, seed dormancy, climate, temperature.

INTRODUCTION

Stylosanthes hamata cv. Verano (Caribbean stylo) is a pasture legume that has been sown widely in monsoon climates in northern Australia, northeast Thailand, and elsewhere. Increased interest is being shown in the pathways of persistence of grazed tropical legumes and in environmental influences on survival mechanisms (Humphreys 1981). Verano is a short-lived perennial (Gardener 1978), and plant replacement through the sequence of flowering, seed formation, accretion to soil seed reserves, and seedling regeneration is essential to its per-sistence. Seed dormancy is a significant factor influencing the success of this cycle that determines the measures farmers need to take when sowing seed. The degree of hardseededness in Verano varies substantially in seedlots according to their spatial or temporal provenance (J.E. Butler, personal communication). Cameron (1967) observed in *S. humilis* differing levels of hardseededness in successive seasons and in lines with differing flowering times. For these reasons the effects of variation in temperature, soil moisture, and illuminance on seed dormancy and seed formation of Verano were studied.

METHODS

Verano plants were grown at St. Lucia, Queensland, in the field, in controlled-temperature cabinets (Sherer CEL 37–14), and in the greenhouse where shading and water-

---

P.J. Argel's current address is Centro Internacional de Agricultura Tropical, A.A. 6713, Cali, Colombia.

Mention of a proprietary product does not imply endorsement of it.

ing were varied after the onset of flowering. Individual inflorescences were tagged to record the sequence of blooming and seed formation. In Verano the inflorescence is a nondeterminate spike in which the loments each consist of an upper hooked articulation (pod) and a lower hookless articulation, each containing a single seed. Pods were germinated at 25°C in the dark; after 14 days about one-third of the seed was cut off in nongerminated pods. After a further 6 days the pod and tests of the remaining nongerminated podded seeds were carefully removed, and the naked seeds were incubated for a further 5 days.

RESULTS

Temperature and Hardseededness

Seeds developed from the first basal floret of inflorescences that bloomed between March and May on plants grown in the open exhibited differing degrees of hardseededness. In 1976, hardseededness was 80%–88% in early-formed seeds and 29%–32% in later-formed seeds. Autumn temperatures and hardseededness were higher in 1977, but differential hardseededness was again observed. Significant correlations for the combined data occurred between hardseededness and maximum, minimum, and mean temperatures during seed formation. Humidity did not show seasonal trends in the datum periods.

In subsequent controlled-environment experiments hardseededness was positively associated with temperature. For example, it was 97%, 97%, and 64% in 27°, 24°, and 21°C treatments, respectively, in the upper articulations after 30 days' storage. Few seeds were formed at 18°C.

Seed Characteristics and Hardseededness

Seed moisture. Hardseededness increased in seeds stored at 10°C and 25% relative humidity as seed moisture content decreased. The interaction between relative humidity of storage environment and temperature provenance of seeds was then studied, using saturated solutions of various salts to produce variation in relative humidity.

Table 1. Moisture content and hardseededness of Caribbean Stylo cv. Verano as influenced by temperature of provenance and relative humidity during storage for 120 days.

Temperature	Relative humidity, %			
	75	32	15	6
	Seed moisture content, %			
27°C	11.1	6.1	4.1	3.3
24°C	13.0	6.1	3.7	3.6
21°C	13.8	9.3	4.4	3.6
	Hardseededness, %			
27°C	76	99	99	99
24°C	25	100	100	100
21°C	0	78	94	95

Table 2. Seed composition of Caribbean Stylo cv. Verano as influenced by temperature of provenance.

Constituent %	Temperature, °C		
	27	24	21
Cellulose	9.7	13.4	16.2
Hemicellulose	19.4	9.2	11.5
Lignin	0.6	trace	trace
Cutin	4.1	3.9	4.3

Seeds formed at 27°, 24°, and 21°C, which exhibited 95%, 85%, and 12% hardseededness at harvest, respectively, were stored at 20°C in varying humidity conditions. Hardseededness (Table 1) developed to a high level in a dry atmosphere irrespective of temperature treatment. In this and other studies, hardseededness was well developed if moisture content was below about 7%. However, the equilibrium moisture content of seeds in any storage environment was negatively associated with their temperature provenance.

Seed anatomy. Hard and soft seeds from each temperature treatment were sectioned through the hilum, micropyle, and strophiole. Sections stained with phloroglucinol and hydrochloric acid revealed the counter-palisade cells in the hilum region as lignified. Cell measurements of the seed coat were made following maceration. Length of palisade cells was significantly varied by temperature treatment and averaged 38.9, 42.7, and 47.7 μ in the 27°, 24°, and 21°C treatments, respectively.

Studies of the surface structure of the seedcoat with the scanning electron microscope showed that hard seeds had a more regular and organized structure and a more evenly reticulated surface, while soft seeds showed deformations and structural irregularities.

Seed color was matched with colors in the Munsell Soil Colour Chart; a significantly higher proportion of very dusky and dark red seeds occurred at high temperature, while seeds formed at 21°C were predominantly light brown and yellow. However, within each temperature treatment there was no consistent association between color and hardseededness.

Seed composition. Analysis by conventional chemical techniques showed that some carbohydrate constituents of the seeds were influenced by temperature provenance. Lignin and hemicellulose contents (Table 2) were higher at 27°C, while cellulose content was lower at 27°C.

Seed Germination

Numerous observations showed that hardseededness exerted the main constraint on germination of Verano, and hardseededness was developed more strongly in the upper than in the lower articulation; for example, 95% and 85%, respectively, in one experiment. Immediately after harvest, seeds germinated readily as naked seeds. Embryo dormancy was usually of the order of 1% to 8% and rarely evident more than 30 days after harvest. On the other hand, inhibition of germination by the cut pod and testa often approached levels of about 70% in the first 40 days after harvest and decreased to negligible levels at 180 days, when hardseededness was still strongly evident. The level of inhibition was independent of temperature provenance.

Moisture Stress

After the commencement of flowering of Verano plants grown in the greenhouse, the soil was watered to pF2 at intervals of 1, 2, 3, and 4 days. This resulted in leaf water potentials as low as -25.1 and -27.7 bars developing in the 3- and 4-day treatments, respectively. Seed yield ranged 5.8 to 3.8 g/plant in the 1- and 4-day watering treatments, respectively. Infrequent watering reduced inflorescence density, floret number, and seed setting. Hardseededness averaged 90% and was not consistently related to treatment.

Irradiance

Verano plants in the greenhouse were shaded after the commencement of flowering to 0.3%, 0.5%, and 0.68%, treatments that reduced seed yield as compared with that of the control. Shading treatment inconsistently affected seed moisture content and hardseededness at harvest, but these differences disappeared during storage.

Seed Formation

Seed yield was maximal at a 31°/24°C day/night temperature regimen. Florets bloomed earlier in the day under warm conditions, and the interval between successive florets blooming on an individual inflorescence was greatly increased at low temperatures; for example, it averaged 2.2 days at 31°/24°C and 4.6 days at 20°/16°C. Duration of flowering on an inflorescence increased from 14.2 days at 35°/28°C to 40.0 days at 20°/16°C, and the interval from blooming to seed ripening was similarly affected. Seed setting and the formation of the lower articulation were especially sensitive to low temperature.

DISCUSSION

Hardseededness in Verano is well developed and persistent. Although relative humidity exerts the primary control over the development of hardseededness, low seed moisture content is promoted by warm temperature occurring during seed formation. This finding contrasts with that of Dotzenko et al. (1967) for *Medicago sativa*, but the apparent discrepancy may be related to unspecified differences in moisture content. It was expected from the data of Quinlivan (1965) with *Trifolium subterraneum* that moisture stress during seed development might impair the development of hardseededness, but impairment did not eventuate in Verano under a regimen of short-term intermittent stress. This question requires elucidation under the more slowly increasing stress conditions characteristic of field situations.

The low embryo dormancy reported in our study contrasts with data from Gardener (1975) regarding Verano, but there is an indication that an inhibitor was also present, based on the highly significant differences in germination between seeds with or without the pod removed. Hardseededness develops as the seed dries out; in this phase the seed is precluded from germinating by the activity of an inhibitor present in the pod or testa.

Verano has not been persistent in the subtropics or in high altitudes in the tropics. Seed production and hardseededness are favored by warm conditions, and this factor may well prove to be significant in favoring and perhaps restricting Verano's adaptation to dry tropical environments.

LITERATURE CITED

Cameron, D.F. 1967. Hardseededness and seed dormancy of townsville lucerne (*Stylosanthes humilis*) selections. Austral. J. Expt. Agric. Anim. Husb. 7:237–240.

Dotzenko, A.D., C.S. Cooper, A.K. Dobrenz, H.M. Laude, M.A. Massengale, and K.C. Feltner. 1967. Temperature stress on growth and seed characteristics of grasses and legumes. Colo. Agric. Expt. Sta. Tech. Bul. 97.

Gardener, C.J. 1975. Mechanisms regulating germination in seed of *Stylosanthes*. Austral. J. Agric. Res. 26:281–294.

Gardener, C.J. 1978. Seedling growth characteristics of *Stylosanthes*. Austral. J. Agric. Res. 29:803–813.

Humphreys, L.R. 1981. Environmental adaptation of tropical pasture plants. Macmillan, London.

Quinlivan, B.J. 1965. The influence of the growing season and the following dry season on the hardseededness of subterranean clover in different environments. Austral. J. Agric. Res. 16:277–291.

Seombadi (*Dystaenia takesimana* [Nakai] Kitagawa), a New Forage Plant: Its Germination and Early Growth Characteristics

D.A. KIM, S.H. HUR, Y.J. KIM, and C.M. PARK

Seoul National University, Jeonbug National University, Chonnam National University, and Rural Economics Institute, Korea

Summary

Seombadi (*Dystaenia takesimana* [Nakai] Kitagawa) is a member of the Apiaceae (Umbelliferae) family and is ecologically adapted to the island of Ulneung, situated 160 km off the east coast of the Korean peninsula. The plant is an important forage for livestock on the island; but because of lack of knowledge about its germination, and its slow establishment characteristics, its use on mainland Korea has been limited.

Objectives of our study were to examine the seed characteristics and factors affecting germination of seombadi and to evaluate depletion of seed food reserves and elongation of organs of seedlings during germination and initial seedling growth.

Lipids made up 28.7% of the total dry material in seombadi seeds. The imbibition rate of the seeds with wings intact, wings removed, and potassium hydroxide (KOH) applied for 5 hours were 29%, 40%, and 65%, respectively. Removal of wings significantly improved germination. Seed sections showed waxy layers between the wings and the seedcoat. Germination percentage improved with increasing seed weight. Seeds rapidly lost their viability 45 days after harvest, and after one year germination was 0%. Light neither inhibited nor improved germination. High temperature pretreatment increased percentage of germination from 44.3% to 91.3%. Soaking seeds with the wings removed in water for 6 days increased germination.

The higher imbibition and germination of seombadi seeds after wing removal, and with the seedcoat weakened by mechanical and chemical seed treatments, indicated that the high percentage of wax in the wings and seedcoat may inhibit seed germination.

Seedling growth of seombadi was slower than that of alfalfa (*Medicago sativa* L.) or orchardgrass (*Dactylis glomerata* L.). Seombadi seeds lost less weight during germination than did orchardgrass seeds.

These studies indicate that establishment of seombadi under Korean mainland environments could be improved by hastening germination with mechanical and chemical seed treatments.

KEY WORDS: seombadi, *Dystaenia takesimana*, forage plant, germination, growth characteristics.

INTRODUCTION

Since forage production has become important in Korea, much research has been expended on the selection and improvement of native herbage plants. Among these plants seombadi (*Dystaenia takesimana* [Nakai] Kitagawa) was chosen as a promising species because of its relatively high productivity and protein content (Lee 1973).

Seombadi is a member of the Apiaceae (Umbelliferae) family. Its wide distribution, earliness, and high production on Ulneung Island make it potentially valuable as stock fodder. Since seombadi is a cool-season plant, its spring growth is much earlier than that of alfalfa (*Medicago sativa* L.) and orchardgrass (*Dactylis glomerata* L.). However, the plant is difficult to establish, partly because of its low germination of seeds and slow early growth (Kim and Kim, unpublished). Therefore, to improve the chances for successful establishment, a series of experiments was carried out to examine the germination and seedling growth characteristics of the plant.

METHODS

Seeds were supplied by the Livestock Experiment Station, ORD, Suweon, Korea, and stored at room temperature. Seeds of seombadi, alfalfa, and orchardgrass were chemically analyzed to determine composition.

Ten g of seeds, wings intact, wings removed, and wings removed plus KOH treatment, were blotted with filter paper after soaking in water for 5 hours, and water imbibition was measured at 50-minute intervals. Ten g of fresh seeds were also soaked in water at 7°, 20°, and

D.A. Kim is a Professor of Pasture Management at Seoul National University, S.H. Hur is an Assistant Professor of Pasture Management at Jeonbug National University, Y.J. Kim is a Research Advisor at the Korea Rural Economics Institute, and C.M. Park is a Professor of Animal Science at Chonnam National University.

28°C. Water imbibition was measured at 3-hour intervals.

Two months postharvest, seeds were killed and fixed in a solution of formalin, acetic acid, and 95% ethyl alcohol, embedded in tissuemat, sectioned at 10 μ, and mounted on microscope slides as serial sections. Sections were stained in Safranin 0 and counterstained in fast green or sudan 111.

Interaction between seed weight and germination rate was investigated by grading seeds with $CaCl_2$ solution into four groups with specific gravities of 1.05, 1.10, 1.12, and 1.14. After specific gravity was determined, the seeds were immediately washed with distilled water and germinated at 20°C. Seeds stored at room temperature following harvesting were germinated at 15-day intervals to measure viability.

Three replications, 100 seeds each, with wings intact and wings removed, were germinated in petri dishes to determine the effect of mechanical removal of wings on germination. Seeds germinated were counted every 2 days.

Seeds were maintained at 40°C for 2 to 8 days and then tested for germination to determine pregermination temperature effects. Also, various periods of soaking were tested.

Seeds were germinated at 20°C in darkness using distilled water, and seedling length and weight were measured every 2 days. After hypocotyl, root, and shoot lengths were measured, seedling day weights were obtained.

RESULTS

Seed Characteristics and Factors Affecting Germination

Lipids made up 28.7% of the total dry material in seombadi seeds, while orchardgrass seeds contained only 7.7% lipids. Protein content was much lower in the seeds of seombadi than in the seeds of alfalfa, but other inorganic contents were generally high.

Imbibition of seombadi seeds with wings intact, with wings removed, and treated with KOH for 5 hours were 29%, 40%, and 65%, respectively (Table 1). Presumably the wings or seedcoat mechanically restrict water absorp-

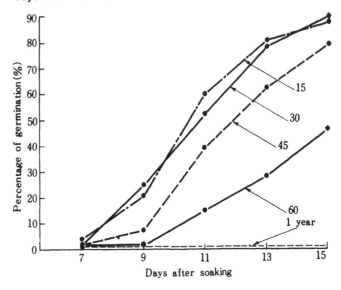

Fig. 1. Germination of seombadi seeds 15, 30, 45, 60, and 365 days after harvest.

tion. For 11 hours absorption appeared to be directly proportionate to temperature (7° to 28°C). After 11 hours imbibition was less at 28°C than at 20°C.

There was an obvious difference in structure between seombadi and alfalfa seedcoats. Waxy layers occur between the coat and wings of seombadi seed, and fats appear in the form of small globules dispersed in the endosperm. No waxy layers were found in the seedcoat of alfalfa.

The germination percentages of seombadi seeds with specific gravity of 1.05, 1.10, 1.12, and 1.14 were 61.6%, 67.1%, 84.5%, and 85.7%, respectively. Furthermore, weight increased with increased specific gravity. Seombadi seed viability decreased with time. After 2 months, germination was reduced to 50% and after one year to 0% (Fig. 1).

Seed Treatments

At 20°C, dewinging seombadi seeds increased rate of germination and germination percentage (Table 2). Presoaking for 0, 2, 4, and 6 days resulted in germination rates of 44.3%, 68.7%, 85.7%, and 88.1%, respectively. Germination of seeds kept in darkness and those exposed to continuous light were not significantly different. Pregermination temperature treatment, 40°C for 8 days, increased germination from 44.8% to 91.1% (Fig. 2).

Table 1. Imbibition rate of seombadi seeds with wings removed and KOH scarified.

Treatment	Time (minutes)					
	50	100	150	200	250	300
	----------g water/100 g seed----------					
Seeds with wings intact	–	–	0.6	18	28	29
Seeds with wings removed	11	20	23	32	36	40
Seeds with wings removed and KOH scarified	25	39	51	54	63	65

Table 2. Germination of seombadi seeds with wings removed.

Treatment	Days from start of test						
	7	9	11	13	15	17	19
	----------------------% germinated----------------------						
Seeds with wings intact	0	3.0	18.6	31.4	47.7	66.4	75.8
Seeds with wings removed	0	9.2	41.3	63.4	79.4	85.1	90.5

Fig. 2. Effects of various periods of high-temperature treatment of seombadi seeds on germination (numbers in figure indicate days).

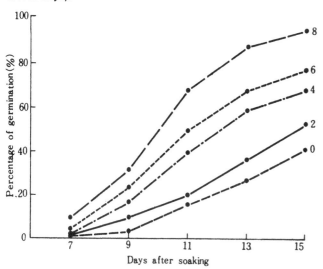

Fig. 3. Growth characteristics of seombadi, alfalfa, and orchardgrass seedlings.

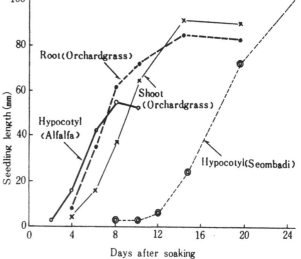

Growth Characteristics

Orchardgrass roots commenced growth and elongated quicker than did shoots, up to 12 days after germination (Fig. 3). The hypocotyl of seombadi elongated more slowly than that of alfalfa. Seombadi germinated the most slowly, and the seed lost the least weight of the three species examined (Fig. 4).

DISCUSSION

Seeds with wings removed or treated with KOH inbibed water faster than seeds with wings intact. Dungan (1924) reported that rapidity of water absorption was associated with more rapid germination.

In other experiments of these studies in which wings were removed, germination improved significantly. Springfield (1970) observed the same phenomenon in fourwing saltbush (*Atriplix canescens*), and Yang et al. (1977) reported similar results in seombadi. The stimulating effect of removal of the seedcoat and associated coverings on seed germination supports the view that the integuments of the seed are major germination inhibitors (Edwards 1969, Rost 1975).

Water uptake of seombadi seeds is influenced by the specific structures of wings and seedcoat, and thus the wings, seedcoat, and waxy layer between wings and seedcoat appear to be involved in controlling germination.

Soaking and high-temperature treatments increased final germination of seombadi seeds, apparently by weakening the seedcoat. Twitchel (1955) found that soaking fourwing saltbush seeds in water for several hours removed 90% of the chlorides and increased germination. Mott (1974) reported that *Aristida contorta* seeds required a high-temperature treatment for germination.

Hur et al. (1977) reported that early root growth of seombadi was very poor and that the growth curve was different from that of other pasture plants. Seombadi germinated slowly and had a slow rate of hypocotyl growth, while alfalfa was quicker in attaining its maximum length. Loss in weight during germination proceeded much slower in seombadi than in any other species. This is evidenced by the gradual curve, which leveled off to constant weight after 25 days. The reserves in seombadi seeds were apparently released slower than those in alfalfa or orchardgrass. This slow release may account for some of the difficulty in establishing stands of this species. Field establishment might be improved by selecting for hypocotyl growth.

LITERATURE CITED

Dungan, C.H. 1924. Some factors affecting the water absorption and germination of seed corn. Am. Soc. Agron. J. 16:473–481.

Edwards, M.M. 1969. Dormancy in seeds of charlock. II. The influence of the seed coat. J. Expt. Bot. 19:583–600.

Fig. 4. Weight changes in seedling organs of seombadi, alfalfa, and orchardgrass.

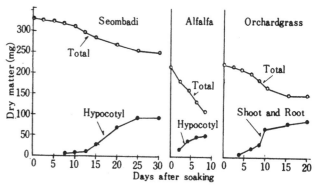

Hur, S.N., D.A. Kim, and H. Park. 1977. Studies on the growth characteristics of *Dystaenia takesimana*. Korean J. Anim. Sci. 19:172–179.

Lee, J.Y. 1973. Selection of new forage crop "*Dystaenia Takesimana*". Prompt Rpt. Agric. Res., O.R.D. 1:1.

Mott, J.J. 1974. Mechanisms controlling dormancy in the arid zone grass *Aristida contorta*. I. Physiology and mechanisms of dormancy. Austral. J. Bot. 22(4):635–645.

Rost, T.L. 1975. The morphology of germination in *Setaria lutescens* (Gramineae): The effects of covering structures and chemical inhibitors on dormant and non-dormant florets. Ann. Bot. 39:21–30.

Springfield, H.W. 1970. Germination characteristics of *Atriplex canescens* seeds. Proc. XI Int. Grassl. Cong., 586–589.

Twitchel, La F.T. 1955. Germination of fourwing saltbush seed as affected by soaking and chloride removal. J. Range Mangt. 8:218–220.

Yang, J.S., K.J. Park, H.J. Han, and J.Y. Lee. 1977. Studies to improve the germination of *Dystaenia takesimana* seeds. I. Influence of temperature and seed-wing removal on the germination of *Dystaenia takesimana* seeds. Res. Rpt. O.R.D. 19:99–103.

Responses of Alfalfa to a Simulated Midwinter Thaw

M. SUZUKI

Agriculture Canada, Charlottetown Research Station,
Charlottetown, Prince Edward Island, Canada

Summary

Effects of a freezing-waterlogging-freezing treatment, under controlled environmental conditions, on the viability and chemical components of alfalfa (*Medicago sativa* L.) were determined to assess physiological responses to midwinter thaw. Alfalfa plants grown in a field were more resistant to freezing but less resistant to waterlogging than those grown in a greenhouse. Both types of plants sustained severe damage from the second freezing treatment. An accumulation of ethanol and methanol, a solubilization of proteins, and a decrease in carbohydrates occurred during the waterlogging. Two important physiological characteristics for surviving a midwinter thaw appear to be an ability to maintain freezing resistance during the thawing period and an ability to remove ethanol quickly after waterlogging.

KEY WORDS: alfalfa, *Medicago sativa* L., winter-hardiness, winter thaw, waterlogging, freezing, ethanol, soluble protein.

INTRODUCTION

Midwinter thaws occur almost every winter in the Maritime Provinces of Canada. Although the midwinter thaw has been assumed to be one of the serious causes of winter injury to alfalfa (*Medicago sativa* L.), no experimental data are available to support this assumption. Two major stress factors associated with a midwinter thaw are (1) alternate freezing and thawing and (2) anaerobic environments due to waterlogging or ice encasing. The present study was conducted under controlled environments to determine physiological responses of alfalfa grown in a field and in a greenhouse to waterlogging between two cycles of freezing-thawing. The results are discussed in relation to field observations of midwinter thaw.

METHODS

A split-plot experiment was conducted with 4 replications, main plots being stress treatments and subplots being growing conditions (field vs. greenhouse). Sample plants of 2-year-old Iroquois alfalfa were taken from a greenhouse and a field at Charlottetown in early November. The greenhouse plants were hardened in a growth cabinet with a decreasing temperature from 5° to 0°C. Both the naturally hardened field plants and the artificially hardened greenhouse plants were washed free of soil and subjected to the first freezing treatment at $-6°C$ in a coldroom. Frozen plants were thawed and completely immersed in distilled water at 10°C. During waterlogging, the redox potential (Eh) of the water was recorded every day (Bohn 1971). The waterlogged plants were subjected to the second freezing at $-6°C$. Sample plants were taken at completion of each treatment stage, transplanted into soil, and grown in a greenhouse to assess viability. Fresh root-tissue samples were extracted with water, and the extract was analyzed for soluble protein (Lowry et al. 1951), alcohols (Supelco 1976), volatile fatty acids and lactic acid (Suzuki and Lund 1980), and cyanide (Gilchrist et al. 1967). Oven-dried root samples were analyzed for crude protein, phosphorus (P) and potassium (K) (Thomas et al. 1967) and total available carbohydrates (TAC) (Suzuki 1971). Toxicity, or LD_{50}

The author is a Plant Physiologist at Charlottetown Research Station. This is contribution no. 471 of the Charlottetown Research Station, Agriculture Canada.

(50% killing point) of internal concentration of ethanol and methanol, was determined by a method described by Andrews and Pomeroy (1979). A separate group of samples was subjected to a range of subfreezing temperatures, the lowest being $-16°C$, to determine 50% lethal temperature (LT_{50}) before and after waterlogging.

RESULTS

The first freezing treatment affected the greenhouse plants more than the field plants (Table 1, Fig. 1). The following waterlogging treatment caused a greater damage to the field plants, such that there was little or no difference in percentage of survival and viability index between the field plants and the greenhouse plants after the freezing-plus-waterlogging treatment. The second freezing after waterlogging imposed a very severe stress on both the field plants and the greenhouse plants. Effects of the combined stress treatment on percentage of survival and viability index were highly significant, but the differences in these criteria between the field plants and the greenhouse plants were not significant (Table 2). The LT_{50} values of the field plants before and after waterlogging were $-13°C$ and $-6°C$, respectively; thus, the decrease in freezing resistance was equivalent to 7°C.

The first freezing resulted in decreases of P and K in greenhouse plants, but the remaining treatments did not change the concentrations significantly (Table 1). Concentration of soluble protein on fresh-weight basis or crude-protein basis increased during waterlogging, while the TAC concentration tended to decrease (Table 2, Fig. 1). Redox potential of water decreased from $+518$ to -180 mV with time of waterlogging. The Eh value of greenhouse plants was slightly higher than that of field plants, the maximum difference being 100 mV, indicating that the former plants consumed less oxygen than the latter. Cyanide or other highly toxic compounds were not detected in any of the sample plants, but some fermenta-

tion products were found in the roots after waterlogging. The accumulation of fatty acids was not significant, as only 0.5 mg acetic acid/g fresh weight together with trace amounts of propionic, butyric, and lactic acids were detected. The accumulation of alcohols was much more evident. The concentration of ethanol increased during the first freezing and reached the maximum level, 1.8 mg/g fresh weight, after waterlogging (Table 2, Fig. 2). The effects of both stress treatments and growing condition on the concentration of methanol were significant, but the increase in methanol concentration during waterlogging was not as great as the increase in ethanol. Both alcohols showed a similar toxicity to alfalfa, as the LD_{50} of internal concentration was 12 mg/g fresh weight for either methanol or ethanol.

DISCUSSION

Frequency of occurrence of winter thaw in January and February and the duration of each thawing period varied from 0 to 22 times/season and 0 to 20 days, respectively, depending on the year and the location in Prince Edward Island, in the past 3 years. During midwinter thaws, overwintering alfalfa plants may be partially or competely immersed in water. Bolton and McKenzie (1946), McKenzie (1951), and Heinrichs (1970) reported that alfalfa (*M. media* Pers) survived early spring flooding for 10 to 15 days in western Canada. Our greenhouse experiment, however, showed that 19-month-old alfalfa plants grown in plastic pails where the soil environment was more anaerobic than that in the field survived up to 14 weeks of ice encasing or flooding with 3 cm of water above the soil surface (unpublished data). Thus, alfalfa plants may exhibit an adaptation to anaerobic environments under certain circumstances. In the present study, the greenhouse plants appeared to be more resistant to waterlogging than the field plants, but both field plants and greenhouse plants sustained severe damage from the second freezing.

Table 1. Effect of simulated midwinter thaw on the viability index, the concentration of P and K on dry-weight basis and soluble protein on crude-protein basis of alfalfa as influenced by growing condition.

Treatment stage	Growing condition†	Viability index††	P % dry wt.	K % dry wt.	Soluble protein % of crude protein
Hardening	F	10.0a§	0.28a	0.89a	19.27ab
	G	10.0a	0.47c	1.20b	17.52a
First-Freezing	F	7.9ab	0.27a	0.81a	18.95ab
	G	6.6bc	0.31ab	0.99a	18.04a
Waterlogging	F	4.6c	0.34ab	0.88a	31.52d
	G	5.4c	0.32ab	0.85a	26.53cd
Second-Freezing	F	1.0d	0.30ab	0.81a	28.77cd
	G	1.0d	0.37b	0.89a	24.16abc

†F = field; G = greenhouse.

††Viability index: Rated according to the speed of recovery growth (days required for 50% of plants to start recovery growth), using a scale of extremely low viability of 1.0 (10 days or more).

§Means followed by the same letter within a column are not significantly different at 5% of probability according to Duncan's Multiple Range Test.

Fig. 1. Effect of simulated midwinter thaw on percentage of survival, and concentration of total available carbohydrates (TAC), crude protein, and soluble protein of alfalfa grown in field and in greenhouse.

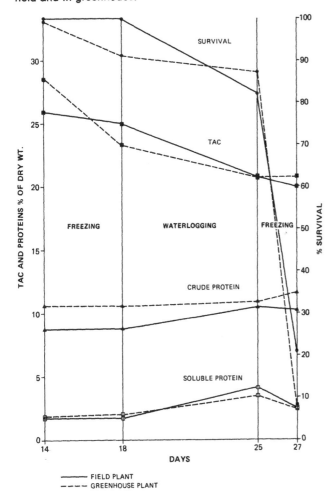

Table 2. Analysis of variance (F values).

Criteria	Stress treatments (A)	Growing conditions (B)	AXB
% Survival	96.44	5.89	0.87
TAC	19.63	1.29	1.73
Crude Protein	3.64	18.48	1.03
Soluble Protein	43.82	0.82	1.09
Ethanol	16.32	4.13	0.24
Methanol	8.40	72.31	0.41
P = 5%	3.16	10.13	3.16
P = 1%	5.09	34.12	5.09

resistance during thaw and effective ethanol removal. Cold-hardy alfalfa, *Medicago falcata* L., does not survive well in the region and does not seem to have the needed resistance to midwinter thaw.

LITERATURE CITED

Andrews, C.J., and M.K. Pomeroy. 1979. Toxicity of anaerobic metabolite accumulation in winter wheat seedlings during ice encasement. Pl. Physiol. 64:120–125.

Barta, A.L. 1980. Regrowth and alcohol dehydrogenase activity in waterlogged alfalfa and birdsfoot trefoil. Agron. J. 72:1017–1020.

Bohn, H.L. 1971. Redox potentials. Soil Sci. 112:39–45.

Bolton, J.L., and R.E. McKenzie. 1946. The effect of early spring flooding on certain forage crops. Sci. Agric. 26:99–105.

Gilchrist, D.G., W.E. Lueschen, and C.N. Hittle. 1967. Re-

These observations suggest that a considerable loss of freezing resistance, as much as 7°C, may occur during a midwinter thaw, and the loss is hardly minimized by altering growing conditions. One of the physiological requirements for surviving a midwinter thaw is an ability to maintain freezing resistance during the thawing period. Another requirement is an ability to remove ethanol. Under cool-temperature conditions waterlogged alfalfa roots accumulate ethanol more and eliminate it less than under warm conditions (Barta 1980). In winter the accumulated ethanol may remain in the tissue for a long period and may affect viability, although the concentration is at a sublethal level. Indeed, we detected 2.3 mg of ethanol/g fresh weight of alfalfa roots in a dry field 6 weeks after the occurrence of a midwinter thaw (unpublished data). To minimize the incidence of winter injury in the Atlantic region of Canada, it is essential to select alfalfa resistant to midwinter thaw. The present study indicated that this resistance is associated with two physiological characteristics: retention of freezing

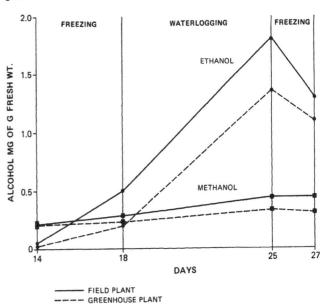

Fig. 2. Effect of simulated midwinter thaw on concentration of ethanol and methanol of alfalfa grown in field and in greenhouse.

vised method for the preparation of standards in the sodium picrate assay of HCN. Crop Sci. 7:267-268.

Heinrichs, D.H. 1970. Flooding tolerance of legumes. Can. J. Pl. Sci. 50:435-438.

Lowry, C.H., N.D. Rosebrough, A.L. Farr, and R.J. Randall. 1951. Protein measurement with the Folin phenol reagent. J. Biol. Chem. 193:265-275.

McKenzie, R.E. 1951. The ability of forage plants to survive early spring flooding. Sci. Agric. 31:358-367.

Supelco Inc. 1976. Carbopack. Bul. 738A.

Suzuki, M. 1971. Semi-automatic analysis of the total available carbohydrates in alfalfa roots. Can. J. Pl. Sci. 51:184-185.

Suzuki, M., and C.E. Lund. 1980. Improved gas-liquid chromatography for simultaneous determination of volatile fatty acids and lactic acid in silage. J. Agr. Food Chem. 28:1040-1041.

Thomas, R.L., R.W. Sheard, and J.R. Mayer. 1967. Comparison of conventional and automated procedures for nitrogen, phosphorus and potassium analysis of plant material using a single digestion. Agron. J. 59:240-243.

Comparative Flooding Tolerance of Tropical Pasture Legumes

P.C. WHITEMAN, M. SEITLHEKO, and M.E. SIREGAR
University of Queensland, Australia

Summary

Growth and persistence of many tropical pasture legumes can be limited by flooding or waterlogging. Selection of tolerant species is important for pasture development in the humid lowland tropics. The aim of this project was to compare flooding tolerance of seven tropical pasture legumes and to determine whether the pot technique we developed was reproducible and relevant to the field situation.

Plants were grown in pots and at the start of flowering were flooded to 15 cm above soil level. Water temperature and oxygen content were monitored regularly. Sets of four uniform pots were removed from flooding after 7, 14, or 21 days. After 7 days of recovery the dry weight of tops, roots, and nodules of the flooded and matched sets of nonflooded controls were compared. The ratio of dry weight flooded : dry weight control was an index of tolerance. The species were ranked by index values as follows: *Macroptilium lathyroides* (1.04 and 0.87), *Desmodium intortum* (0.85), *Pueraria phaseoloides* (0.76), *Stylosanthes guianensis* (0.62), *M. atropurpureum* (0.44), *Centrosema pubescens* (0.37), and *D. uncinatum* (0.34).

Root growth was more affected by flooding than shoot growth. Nodulation was drastically reduced in all species except *M. lathyroides*. The adaptation of *M. lathyroides* was related to a rapid production of adventitious roots from the submerged stem and rapid renodulation of these and other roots. *D. intortum* also had rapid adventitious root production but no renodulation. The screening technique appeared to yield reproducible results, and rankings were related to field observations.

KEY WORDS: flood tolerance, tropical pasture legumes, *Macroptilium, Desmodium, Stylosanthes, Centrosema, Pueraria.*

INTRODUCTION

Many areas in the humid lowland tropics with potential for sown pasture development are exposed to seasonal flooding. Although a number of tropical grasses are adapted to these situations (Anderson 1970), there appear to be few adapted tropical legumes. However, there are few data available on the comparative flooding tolerance of tropical legumes.

Assessment of comparative tolerance is valuable for the agronomic selection of species for flooding situations. Francis and Poole (1973) found a wide range in tolerance of flooding in annual *Medicago* species. In contrast, the perennial *Medicago sativa* is intolerant of flooding when actively growing (McKenzie 1951). In the tropical legumes McIvor (1976) compared 33 accessions of *Stylosanthes guianensis*. Waterlogged plants of the most tolerant accession produced 82% of the yield of control, while the least tolerant accessions produced only 5%. In this paper a range of commercially sown pasture legumes are compared for their tolerance of flooding.

P.C. Whiteman is a Reader, M. Seitlheko is a Research Scholar (current address, Ministry of Agriculture, Lesotho), and M.E. Siregar is a Research Scholar (current address, Lembaga Penelitan Peternakan, Jl. Raya Pajaran, Bogor, Indonesia) at the University of Queensland.

Mention of a proprietary product does not imply endorsement of it.

MATERIALS AND METHODS

Two experiments were conducted. In the first, the species *Macroptilium lathyroides, M. atropurpureum* cv. Siratro, *Centrosema pubescens* cv. common, and *Stylosanthes guianensis* cv. Schofield were compared. The second compared *M. lathyroides, Desmodium uncinatum* cv. silver leaf, *D. intortum* cv. greenleaf, and *Pueraria phaseoloides*.

Scarified, inoculated seed was planted into a loamy sand (pH 7.5) and thinned to three plants/pot 14 days after emergence. A basal fertilizer was applied as 200 kg superphosphate/ha, 50 kg K_2SO_4/ha, 7 kg $CuSO_4$/ha, 7 kg $ZnSO_4$/ha, and 0.5 kg $NaMoO_4$/ha.

Flooding treatments were imposed at the onset of flowering in each species. The pots were placed in 20-l plastic buckets and filled with water to 15 cm above the soil surface. Before flooding treatments were imposed, the pots of each species were divided into two uniform groups to become the flooded and control treatments. Control pots were watered to field capacity daily. Both groups were further subdivided into sets of four pots with a similar range of plant sizes in each set and allocated to a duration of flooding treatment. Four control pots were harvested on the day of flooding (day 0). Four pots were removed from the flooding treatments after 7, 14, and 21 days of immersion and allowed a 7-day recovery period before harvest. At each harvest four control pots also were harvested.

During flooding visible symptoms, plant mortality, and water temperature were recorded at 0900 hours daily, and the percentage of oxygen in the water was recorded weekly (Beckman Field-lab analyser). At harvest, oven-dry weight of shoots, roots, and nodules was measured.

RESULTS

Marked responses to immersion were evident for all species within 7 days. Thickening and suberization of the immersed stem was evident in all species, very markedly in *M. lathyroides* and *M. atropurpureum* but only to a minor extent in *D. uncinatum*. Adventitious root development was clearly evident in *M. lathyroides* and *D. intortum* by day 7, and evident to a smaller extent in *M. atropurpureum, C. pubexcens,* and *S. guianensis,* but was not evident in *D. uncinatum* until day 14 and in *P. phaseoloides* until day 21.

By day 21 the adventitious roots of *M. lathyroides* were well nodulated; *M. atropurpureum* also nodulated but to a lesser extent, and *D. uncinatum, D. intortum,* and *P. phaseoloides* did not nodulate.

Within 7 days all species showed some chlorosis of the older leaves. In *D. uncinatum* symptoms increased with time so that by day 21 a large proportion of leaves had died. In *M. atropurpureum* leaves died and abscissed. *M. lathyroides* recovered so that by day 28 leaf color was similar to that of the controls.

M. lathyroides maintained similar rates of growth in control and flooded treatments (Table 1). In marked contrast, *D. uncinatum* ceased growth at the onset of flooding and declined in weight, while *C. pubescens* ceased growth but did not decline. *M. atropurpureum* ceased growth after 7 days, and *S. guianensis* after 14 days. *D. intortum* and *P. phaseoloides* maintained some growth up to 21 days.

Root weight was usually more affected by flooding than shoot weight (Table 2). Nodule weights were drastically reduced by flooding in all species except *M. lathyroides* (Table 1). In experiment 2 nodule weights in the flooded treatment were higher than in the control as a result of

Table 1.   Effects of duration of flooding on total dry weight of tops + roots + nodules, and weight of nodules at day 21 in seven tropical pasture legumes.

Flood duration (days)	M. lathyroides		M. atropurpureum		C. pubescens		S. guianensis	
	Control	Flood	Control	Flood	Control	Flood	Control	Flood
Experiment 1								
	---------- Total dry wt. (g/pot) ----------							
0	7.4	6.4	16.8	17.2	9.8	10.2	24.8	24.6
7	17.4	17.1	38.8*	28.7	17.3	11.9	43.4*	21.5
14	22.7	18.4	47.8*	29.0	20.4*	13.8	62.2*	35.1
21	36.1	31.4	52.2**	22.9	30.8**	11.3	55.5*	34.6
Nodule wt.	1.4	0.7	1.55	0	2.2	0	–	0

Flood duration (days)	M. lathyroides		D. uncinatum		D. intortum		P. phaseoloides	
	Control	Flood	Control	Flood	Control	Flood	Control	Flood
Experiment 2								
	---------- Total dry wt (g/pot) ----------							
0	13.4	13.5	18.3	23.3	46.6	50.2	45.1	47.3
7	36.4	32.6	39.4**	21.6	70.7	59.9	68.5	59.9
14	46.4	40.1	49.3**	21.1	76.1	70.3	87.0**	63.6
21	49.2	51.2	57.6**	19.9	102.9*	87.3	90.8**	69.4
Nodule wt.	0.1	0.7	0.5	0	0.5	0	2.9	0.2

*, ** significant differences between flooded and control within species at p = 0.05 and p = 0.01, respectively.

Table 2. Ranking of tolerance of flooding based on ratio of total dry weight flooded to total dry weight control at day 21, and ratios for the dry weight of the components.

Rank	Species	Total, dry wt.	Shoot, dry wt.	Root, dry wt.	Nodule, dry wt.
1	*M. lathyroides* (2)	1.04	1.26	0.52	5.20
	M. lathyroides (1)	0.87	0.88	0.90	0.51
2	*D. intortum*	0.85	0.88	0.62	0
3	*P. phaseoloides*	0.76	0.86	0.59	0.05
4	*S. guianensis*	0.62	0.70	0.43	–
5	*M. atropurpureum*	0.44	0.51	0.27	0.06
6	*C. pubescens*	0.37	0.38	0.47	0.04
7	*D. uncinatum*	0.34	0.36	0.31	0

nodulation of the adventitious roots. In the other species nodules decayed and declined to negligible values.

As an index of relative tolerance to flooding, the ratios of dry weights flooded : dry weights control at day 21 are ranked in Table 2. *M. lathyroides* was clearly less affected by flooding and produced a higher shoot yield under flooding in experiment 2, although root growth was less than that of the control. *D. intortum* also performed well under flooding. *M. atropurpureum*, *C. pubescens*, and *D. uncinatum* were much less tolerant, while *P. phaseoloides* and *S. guianensis* formed an intermediate group. These rankings of relative yield under flooding were not closely related to absolute yields under flooding, where in experiment 1 *S. guianensis* gave the highest flooded yield (Table 1), and *D. intortum* gave the highest in experiment 2.

The water temperatures at 0900 hours were similar among species, in experiment 1, 30 ± 1.5°C and in experiment 2, 25 ± 1°C. However, oxygen content of the water varied quite widely among species. While *M. lathyroides* (17%–23% O_2) maintained relatively constant O_2 contents, the *Desmodium* species recorded low values on one occasion (16% O_2), and *M. atropurpureum* dropped to low values by day 7 and remained low (10%–12% O_2).

DISCUSSION

M. lathyroides was clearly the species best adapted to flooding, a finding that confirmed field evidence (Humphreys 1974), while *D. uncinatum* was least tolerant. Bryan (1969) suggested that *D. uncinatum*, although able to survive short-term waterlogging, preferred well-drained sandy soils. In contrast, *D. intortum* was quite tolerant of flooding, ranking second, followed by *P. phaseoloides*.

S. guianensis cv. Schofield was ranked fourth with an index value of 0.62. This value was very similar to the value of 0.60 recorded for this cultivar by McIvor (1976). This correspondence suggests that the technique of comparing dry weights of flooded and control plants after 21 days of flooding gives comparable results among experiments when temperature conditions are similar. The better growth and higher nodulation of *M. lathyroides* in experiment 2 than in experiment 1 might be related to the lower

water temperature (25 ± 1°C compared with 30 ± 1.5°C in experiment 1), which was also reflected in a higher mean oxygen content of the cooler water.

Varied physiological responses to the flooding treatments among species may be related to flooding tolerance. The most tolerant species, *M. lathyroides* and *D. intortum*, began adventitious root development from the flooded stem section rapidly, within the first 7 days. Development did not begin in *D. uncinatum* until day 14. However, rapid adventitious root development was not an absolute requirement, as *P. phaseoloides*, which was ranked third, did not begin this development until day 21. Gill (1975) noted that, while formation of adventitious roots on stems tends to be characteristic of woody species native to periodically flooded habitats, experimental evidence for their physiological role is lacking.

Root growth was more affected by flooding than shoot growth, apparently because of oxygen deficiency in the root environment. Harris and Van Bavel (1957) concluded that root respiration was the most sensitive to soil aeration of all plant activities. Impaired root respiration limits nutrient and water uptake as well as energy for root growth.

The fact that renodulation of adventitious roots after flooding was far greater in *M. lathyroides* suggests that this species was better able to maintain a favorable oxygen level within its roots than the other species. Bergersen (1971) considered that the factor most detrimental to the establishment and efficient operation of a legume-*Rhizobium* symbiosis is a lowering of oxygen availability to the nodules.

These results demonstrated a wide range of adaptation to flooding among species. A similar or wider range of responses has been shown among ecotypes within species (McIvor 1976). Thus, screening among and within species for flooding tolerance should yield legume cultivars better adapted to many lowland tropical environments.

LITERATURE CITED

Anderson, E.R. 1970. Effect of flooding on tropical grasses. Proc. XI Int. Grassl. Cong., 591-595.

Bergersen, F.J. 1971. Biochemistry of symbiotic nitrogen fixation in legumes. Ann. Rev. Pl. Physiol. 22:121-146.

Bryan, W.W. 1969. *Desmodium intortum* and *Desmodium uncinatum*. Herb. Abs. 39:183-191.

Francis, C.M., and M.L. Poole. 1973. Effect of waterlogging on the growth of annual *Medicago* species. Austral. J. Expt. Agric. Anim. Husb. 13:711-716.

Gill, C.J. 1975. The ecological significance of adventitious rooting in woody species with special reference to *Alnum glutinosa* (L.) Gaertn. Flora 164:85-96.

Harris, D.C., and C.H.M. Van Bavel. 1957. Root respiration of tobacco, corn and cotton plants. Agron. J. 49:182-185.

Humphreys, L.R. 1974. A guide to better pastures for the tropics and sub-tropics. Wright Stephenson and Co., Ltd., Brisbane, Australia.

McIvor, J.G. 1976. The effect of waterlogging on the growth of *Stylosanthes guianensis*. Trop. Grassl. 10:173-178.

McKenzie, R. 1951. The ability of forage plants to survive early spring flooding. Sci. Agric. 31:358-363.

Responses of Siratro (*Macroptilium atropurpureum*) to Water Stress

M.J. FISHER and M.M. LUDLOW

Division of Tropical Crops and Pasture, CSIRO, Australia

Summary

Experiments were conducted to seek understanding of the adaptive characters that contribute to the success of the twining tropical legume siratro (*Macroptilium atropurpureum*) as a component of sown pastures in the seasonally dry tropics and subtropics of northeastern Australia. Siratro was grown both in the field and in controlled environments and was subjected to sequences of water stress. Measurements were made of tissue-water relations, leaf gas exchange, leaf movement, and leaf energy balance.

Siratro avoids the effects of stress by maintaining its leaf-water potential greater than -20 bar, while grasses grown in association with it are about -60 to -80 bar. The tissues of siratro were not able to tolerate leaf-water potential less than about -24 bar. Siratro made only slight changes in its water relations and its stomatal response when exposed to a variety of water-stress treatments, either in the field or in controlled environments.

Siratro has three main mechanisms by which it prevents leaf-water deficits from developing and by which it maintains relatively high leaf-water status.

1. It has good stomatal control over water loss; the stomata close at relatively high water potential and in reponse to falling atmospheric humidity independent of leaf-water potential.

2. Paraheliotropic leaf movements which operate to minimize the radiation received by the leaves and can result in their being 8°–10°C cooler than leaves held horizontal during water stress. The lower temperature reduces the vapor pressure gradient between leaf and air, and hence reduces water loss.

3. Shedding leaves reduces the amount of transpiring tissue. As the level of water stress increases, the older leaves are shed; new leaves are smaller, thicker, dark green, and hairy. If stress continues, whole branches die until only the crown remains, most of which is under the soil surface.

While these mechanisms allow siratro to avoid severe leaf water deficits in many circumstances, its inability to withstand leaf water potentials less than -24 bar makes it vulnerable in extended drought periods. We believe that increasing the ability of siratro to tolerate water deficits could increase its survival and expand its utility.

KEY WORDS: siratro, *Macroptilium atropurpureum*, water stress, photosynthesis, water relations, leaf movement, adaptation.

INTRODUCTION

Siratro (*Macroptilium atropurpureum*) is a twining perennial legume used as a component of pastures in the tropics and subtropics of northeastern Australia. It is well adapted to a wide range of soils and to rainfall regimes from 1,800 mm to less than 700 mm in the subtropics (Shaw and Whiteman 1977), but in the tropics the lower limit is about 1,000 mm. Much of the drier portion of this range has seasonally dry weather in which rainless periods lasting many weeks are common. Therefore, siratro is necessarily well adapted to withstand periods of water stress.

Peake et al. (1975) found that during water stress the leaf water potential of siratro remained high compared with that of buffelgrass (*Cenchrus ciliaris*), one of the grasses with which it is commonly grown. They described siratro as a drought-avoiding species, using Levitt's (1972) terminology. Ludlow and Ibaraki (1979) hypothesized that the ability of siratro to maintain high leaf water potential during rainless periods is due to good stomatal control over water loss. We have undertaken a series of studies to test this hypothesis and to find out which other adaptive characters contribute to the success of siratro in seasonally dry environments.

METHODS

Data and conclusions presented here come from a number of experiments. Siratro was grown both in the field at Narayen Research Station (lat 25°14′ S long 150°52′ E) in southeast Queensland and in controlled environments (30°C and 25 mbar vapor pressure day and night, 14-hour light period, photon flux density [400–700

The authors are Principal Research Scientists in the CSIRO Division of Tropical Crops and Pasture.

nm] of 435 μE/m²/second). In each experiment, plants subjected to a sequence of water stresses were compared with well-watered controls.

Siratro was grown in the field both in pure swards and in mixtures with green panic (*Panicum maximum* var. *trichoglume*) on a duplex (acid podzolic) soil. Measurements were made during rainless periods of 6 and 7 weeks during summer and autumn 1974–1975 (experiment A), and also during spring 1977 when water was withheld for two cycles of 17 days each (experiment B).

In controlled-environment experiments, plants were grown in 15-cm pots containing 2 kg of soil in which the onset of water stress occurred rapidly (experiment C), or in 24-cm cylinders 90 cm deep containing 52 kg of soil in which stress occurred slowly, similar to the field (experiment D). In experiment C plants were subjected to seven cycles each of 5 days without water, followed by watering for 2 days. In experiment D plants were subjected to two cycles of 7 weeks without water, followed by watering for 2 weeks. At 2-week intervals during the 7-week stress cycles some cylinders were cut off 20 cm below the soil surface, and the plants, together with the top 20 cm of undisturbed soil, were placed in shallow cylinders and rewatered. They were then stressed rapidly to obtain relationships between leaf gas exchange and leaf water potential.

An open-system gas-exchange apparatus was used to measure the response of leaf gas exchange to changing vapor pressure (at 30°C, 300 μbar partial pressure CO_2, and photon flux density [400–700 nm] greater than 1500 μE/m²/second) in experiments B, C, and D, and to measure the response of stomatal conductance to leaf water potential (ψ_l) at 26 mbar vapor pressure in experiments C and D. The gas exchange of leaves allowed to track the sun and of leaves held horizontal (both on the same plant) was measured in experiment B at conditions maintained as close to ambient as possible.

Tissue-water relations were measured in experiment B (Wilson et al. 1980) and also the response of stomatal conductance (measured using an automatic porometer) to changes in ψ_l. In all experiments ψ_l was measured using either a pressure chamber or dew-point hygrometers.

RESULTS

Siratro consistently maintained its water potential above − 20 bar both in the field and in a controlled environment, although in the field ψ_l of associated green panic fell below − 70 bar (Fig. 1). When subjected to prolonged stress in small pots, plants died when leaf water potential reached about − 24 bar; we conclude, therefore, that tissues of siratro are not capable of withstanding ψ_l less than − 24 bar and that it is necessary for the plant to avoid low leaf water potentials in order to survive.

We next examined the physiological processes by which siratro avoids low leaf water potentials. In many plants the response of the stomata to ψ_l changes as stress is imposed ("stomatal adjustment") so that the stomata remain partially open at progressively lower ψ_l. We found only

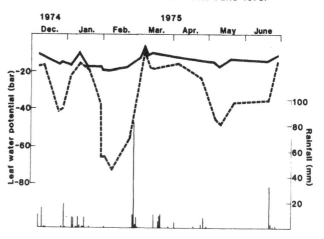

Fig. 1. Leaf water potential of siratro (solid line) and of green panic (broken line) in a mixed sward measured at 1430 hours, and rainfall between December 1974 and June 1975.

slight stomatal adjustment to ψ_l in siratro (less than 3 bar, Fig. 2). Moreover, the steep slopes of the curves in Fig. 2 show that the stomata of siratro are indeed very responsive to small changes in ψ_l below − 12 to − 13 bar. Stomatal adjustment is normally associated with osmotic adjustment (Ludlow 1980a). It is noteworthy that in experiment B there were less than 2 bars osmotic adjustment in siratro (Wilson et al. 1980), a finding that was consistent with the slight stomatal adjustment we recorded.

Fig. 2. Fitted curves to data of stomatal conductance (g_s) against leaf water potential (ψ_l) for well-watered controls after seven cycles of stress in 15-cm pots (---), two cycles of stress in 24-cm cylinders (—— = stressed; - - - = control), and two cycles of stress in the field (− · − · = stressed; ···· = control). The pot and cylinder data were fitted with the function $g_s = g_0/[1 \times (\psi_l/\psi_{1/2})^n]$, where g_0 is some estimate of maximum conductance and $\psi_{1/2}$ and n are parameters (Fisher et al. 1981). There were no estimates of maximum conductance in the field, so an exponential approximation ($g_s = a\psi^b$, where a and b are parameters) was fitted.

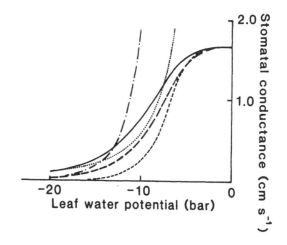

Fig. 3. Fitted relations between stomatal conductance (g_s, log scale) and the gradient of vapor pressure between leaf and air (Δvp) during two cycles of stress in the field and for well-watered controls. The slopes, which are a measure of the responsiveness of g_s to Δvp, do not differ significantly. ($\bullet$ = start of cycle 1; $\triangle$ = end of cycle 1, stressed; + = end of cycle 1, control; o = end of cycle 2, control; x = end of cycle 2, stressed.)

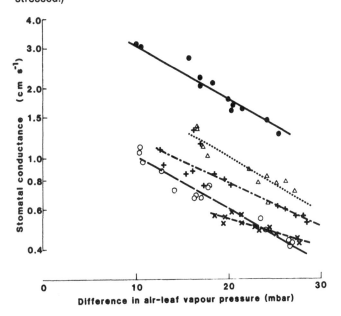

Stomata of siratro also close in response to decreasing atmospheric humidity (Sheriff and Kaye 1977). In experiment B, although maximum conductance of individual leaves varied widely, responsiveness of their stomata to changes in atmospheric humidity was independent of the stress the plants had experienced (Fig. 3).

We found that leaflets of well-watered siratro, like the leaves of some other plants (Ehleringer and Forseth 1980), track the sun such that their laminae are less than 18° from the normal to the elevation of the sun throughout the day (diaheliotropy). However, when the plants became stressed, the leaflets tracked the sun parallel to the incident radiation (paraheliotropy). Diaheliotropy increased the daily radiation integral on individual leaves by increasing the amount of radiation intercepted. Its effect on photosynthesis in experiment B was explained in terms of its effect on temperature and leaf irradiance. Paraheliotropy, on the other hand, occurred when the stomata were closed. It minimized irradiance on the leaves, kept leaf temperatures close to ambient, and thereby minimized transpiration. By comparison, similar leaves held horizontal were 8°–10°C above ambient. An increase of leaf temperature from 30°C to 40°C when the air vapor pressure was 25 mbar increased the gradient of vapor pressure between leaf and air from 17 mbar to 48 mbar and would increase water loss through both the cuticle and imperfectly closed stomata.

DISCUSSION

Because siratro is unable to tolerate leaf water potentials less than – 24 bar, it must avoid water deficits in order to survive. High leaf water potentials are maintained by stomata, which close in response both to falling ψ_l and to increasing vapor-pressure deficit, with little change in the response as stress increases. This is done by paraheliotropic leaf movements, by siratro's ability to reduce its leaf area, and possibly by deep rooting. The ability of siratro to maximize radiation integrals and hence photosynthesis by means of diaheliotropy explains why it is able to resume rapid growth when stress is relieved. However, during stress the plant continues to lose some water, and during extended drought it is obvious that the tissues will ultimately reach a critical level and die. For example, during a prolonged drought at Narayen Research Station only 33% of siratro plants survived (Peake et al. 1975).

The ability of tropical pasture plants to tolerate large water deficits seems to be associated with their capacity to exhibit osmotic and stomatal adjustment (Ludlow 1980a, 1980b). Indeed, siratro has low tolerance and exhibits little adjustment. If siratro had a higher tolerance of water deficits and the benefits of the associated adjustments, it might survive longer during dry periods, and its use might be extended into drier environments.

LITERATURE CITED

Ehleringer, J., and I. Forseth. 1980. Solar tracking by plants. Science 210:1094–1098.

Fisher, M.J., D.A. Charles-Edwards, and M.M. Ludlow. 1981. An analysis of the effects of short term soil water deficits on stomatal conductance to carbon dioxide and leaf photosynthesis by the legume *Macroptilium atropurpureum* (cv. Siratro). Austral. J. Pl. Physiol. 8:347–357.

Levitt, J. 1972. Responses of plants to environmental stresses. Academic Press, New York.

Ludlow, M.M. 1980a. Adaptive significance of stomatal responses to water stress. *In* N.C. Turner and P.J. Kramer (eds.), Adaptation of plants to water and high temperature stress, 123–138. Wiley Intersci., New York.

Ludlow, M.M. 1980b. Stress physiology of tropical pasture plants. Trop. Grassl. 14:136–145.

Ludlow M.M., and K. Ibaraki. 1979. Stomatal control of water loss in siratro (*Macroptilium atropurpureum* (DC) Urb.), a tropical pasture legume. Ann. Bot. 43:639–647.

Peake, D.C.I., G.B. Stirk, and E.F. Henzell. 1975. Leaf water potentials of pasture plants in a semi-arid subtropical environment. Austral. J. Expt. Agric. Anim. Husb. 15:645–654.

Shaw, N.H., and, P.C. Whiteman. 1977. Siratro—a success story in plant breeding. Trop. Grassl. 11:7–14.

Sheriff, D.W., and P.E. Kaye. 1977. Responses of diffusive conductance to humidity in a drought avoiding and a drought resistant (in terms of stomatal response) legume. Ann. Bot. 41:653–654.

Wilson, J.R., M.M. Ludlow, M.J. Fisher, and E.-D. Schultze. 1980. Adaptation to water stress of the leaf water relations of four tropical forage species. Austral. J. Pl. Physiol. 7:207–220.

Plant-Water Relationships of Crested, Pubescent, Slender, and Western Wheatgrasses

A.B. FRANK

USDA-SEA-AR Northern Great Plains Research Center,
Mandan, N. Dak., U.S.A.

Summary

Objectives were to characterize development of water stress in the field and in controlled environments; to determine stress effects on transpiration (E), stomatal conductance (C), leaf water potential (LWP), leaf osmotic potential (LOP), and photosynthesis (P); and to relate measured changes to drought tolerance in crested wheatgrass (*Agropyron desertorum* [Fisch.] Schult.), slender wheatgrass (*A. trachycaulum* [Link] Malte.), pubescent wheatgrass (*A. intermedium* var. *trichophorum* [Link] Haloc), and western wheatgrass (*A. smithii* Rydb.). Both LWP and LOP decreased for all species as soil water potential decreased, but not in a 1:1 ratio indicative of maximum osmotic adjustment. The LWP at zero turgor potential ranged from –28 and –27 bars for crested and slender, respectively, to –24 bars for western and –23 bars for pubescent wheatgrasses. Results from field plants suggest a greater degree of osmotic adjustment in crested, slender, and pubescent wheatgrasses than in western wheatgrass. For growth-chamber plants, based on rate of P, E, and C, western was more sensitive to decreasing soil water potential, followed by pubescent and crested wheatgrasses. Data on species response to water stress are important to development of growth and water-use relationships required for predictive models and in management and cultivar development.

KEY WORDS: transpiration, drought, water potential, osmotic adjustment.

INTRODUCTION

Accumulation of dry matter and leaf extension in plants grown under deficit soil water are dependent on maintenance of a high leaf turgor potential (Acevedo et al. 1979, Culter et al. 1980a). A plant's ability to maintain a high turgor potential with decreasing leaf water potential has been shown in plants differing in drought tolerance (Johnson 1978, Acevedo et al. 1979, Culter et al. 1980a). Maintenance of leaf turgor potential in response to soil water deficits requires a parallel reduction in leaf and osmotic water potential. Leaf extension rates of plants exhibiting osmotic adjustment have been reported to be less sensitive to soil water deficits than the rates of plants not capable of osmotic adjustment (Culter et al. 1980b). Plants capable of maintaining a high leaf turgor potential should have the potential for greater sustained productivity when grown under dryland conditions.

METHODS

Field Study

Plants from established plots of crested wheatgrass, pubescent wheatgrass, western wheatgrass, and slender wheatgrass were sampled on eight dates between time of collar formation on the second leaf from the spike to leaf physiological maturity. Leaf water potential (LWP) and leaf osmotic potential (LOP) were measured with thermocouple psychrometers on leaf segments collected at 1300 hours on cloud-free days. Four leaves from each of 2 replications/species were sampled. Soil water potential and soil temperature were measured using three screen-enclosed, double-junction-thermocouple psychrometers in each plot at 20- and 60-cm depths.

Controlled-Environment Study

Ramets of crested, pubescent, and western wheatgrasses were transplanted into pots 24 cm deep × 22 cm diameter containing 6.2 kg of four parts Parshall fine sandy loam to one part peat mixture. Gravimetric soil water content at 0.3 and 10 bars soil water potential were 31% and 16%, respectively. Fertilizer was added in the initial watering at 106 ppm nitrogen, 46 ppm phosphorus, and 57 ppm potassium. Pots were watered to maintain 0.2 bar or 37.5% water by weight. Plants were grown for about 60 days in the greenhouse at 15°C to 20°C with 14-hour photoperiods using 1,000-watt high-intensity-discharge (HID) sodium lamps to supplement daylight.,

The plants were then placed in a hermetically sealed controlled-environment chamber equipped with glove-port access for two dark periods and one light period before making measurements. Air temperature and water-vapor pressure were maintained at 19°C and 15°C ± 0.5°C and 12 and 10 ± 1.5 millibar during the light

The author is a Plant Physiologist of USDA-SEA-AR. This is a contribution of the USDA-SEA-AR Northern Great Plains Research Center.

and dark periods, respectively. The light system in the chamber consisted of equal numbers of alternately spaced 400-watt HID sodium and mercury lamps giving a quantum flux density of 1,150 E/second/m² (400–700 nm). The carbon dioxide level in the chamber was maintained at 320 ±7 μl/l. Photosynthesis (P), transpiration (E), and leaf temperature were measured on single leaves during 5 days of induced stress on four leaves from each of four pots/species. Stomatal conductance (C) was calculated using the methods of Gaastra (1959).

RESULTS

Field Study

Soil water potential averaged over the 20- and 60-cm depths declined slowly over the 45-day sampling period to −19 to −24 bars for all species (Fig. 1). Plant LWP showed an initial increase for 2 to 3 days followed by a general declining trend through the remainder of the sampling period (Fig. 1). Changes in LWP and LOP were more closely aligned in crested, slender, and western wheatgrasses than in pubescent wheatgrass. Similar changes in LWP and LOP are an indication of osmotic adjustment in leaves. Linear regression analysis of LWP on leaf turgor potential (LTP) produced intercepts (LWP at which LTP would be zero) of −28, −27, −24, and −23 bars for crested, slender, western, and pubescent wheatgrasses, respectively (Table 1). Slope comparisons indicated that western wheatgrass responded differently to development of plant water stress, resulting in a lower degree of osmotic adjustment.

Controlled-Environment Study

Crested and pubescent wheatgrass exhibited more tolerance (Curvilinear response) to stress development, as

Table 1. Regression characteristics with standard errors of leaf water potential (LWP) on leaf turgor potential (LTP).

Species	LWP at zero LTP	Slope	r^2
Crested wheatgrass	−27.8±2.00	0.50±0.12	.57**
Slender wheatgrass	−26.9±1.93	0.52±0.10	.69**
Western wheatgrass	−24.0±1.68	0.85±0.09	.75**
Pubescent wheatgrass	−22.7±1.00	0.64±0.10	.84**

**Significant at the 1% probability level.

was indicated by a slower rate of decrease in P, E, and C compared with a more linear decrease in all parameters for western wheatgrass (Fig. 2). Crested and, to a lesser extent, pubescent wheatgrass showed tolerance to decreasing P, E, and C until soil water content declined to about 26% (−7 bars). Below 26% soil water content, the reductions in P, E, and C were very rapid.

DISCUSSION

When subjected to water stress, leaves of some plants maintain a positive LTP by adjusting osmotically (Hsiao et al. 1976, Acevedo et al. 1979). Data presented in Fig. 1 indicate a greater trend toward osmotic adjustment for crested, slender, and possibly pubescent wheatgrasses than for western wheatgrass. Although both LOP and LWP decreased with decreasing soil water potential, the reduction in LOP was not in a 1 : 1 proportion to reduction in LWP for any of the four species. Therefore, maximum osmotic adjustment may not have occurred in these species. Although crested wheatgrass had the lowest average LWP, it also had the lowest average LOP, a situation that would favor maintaining a greater LTP. Culter et al. (1980a) suggested that in rice (*Oryza sativa* L.) the maximum extent of active osmotic adjustment was

Fig. 1. Leaf (Ψ), osmotic (π), and soil (ψ) water potential of four grass species grown under dryland field conditions.

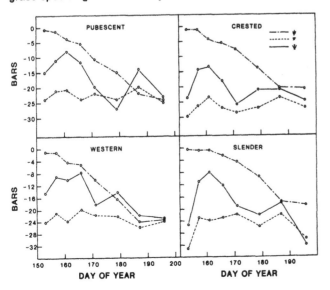

Fig. 2. Stomatal conductance (C), transpiration (E), and photosynthesis (P) of three grass species grown in controlled environment during a soil-drying cycle. Calculations are based on 37.5% soil water content. Vertical bars indicate standard errors.

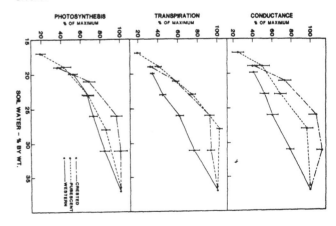

limited and that such adjustment was independent of the rate of development or duration of water stress. In this study, the slow decrease in soil water potential should have allowed maximum osmotic adjustment.

The LWP at zero LTP has been associated with observed drought resistance in the field by Johnson (1978). His plants were grown in 25°/15°C (day/night), and LWPs at zero LTP were 11.7, 11.7, and 9.5 bars higher for crested, slender, and pubescent wheatgrasses, respectively, than the presented values for field-grown plants of the same species as reported in Table 1 of this study. These differences could be related to differences in rate of development, duration, or intensity of primarily water stress resulting from growing plants in controlled environments vs. in the field. Species ranking for drought tolerance (defined as any plant parameter or process that enables a plant to endure periods of water stress) based on LWP at zero LTP would be, for most to least tolerant: crested, slender, western, and pubescent wheatgrass. General field observations would rank western wheatgrass as the most drought tolerant, followed by crested, pubescent, and slender wheatgrasses.

In the controlled-environment study each species responded somewhat differently to decreasing soil water content. Rate of decrease in P, E, and C was greater initially for western than for crested and pubescent wheatgrasses (Fig. 2). Crested maintained rates of P, E, and C near maximum until soil water decreased to about 26% (– 7 bars); after that the rate of decrease was greater than for either western or pubescent wheatgrass. Response of pubescent wheatgrass to decreasing soil water content was intermediate when compared with those of crested and western wheatgrass. The lack of sensitivity of

crested and, to a lesser extent, pubescent wheatgrass to decreasing soil water content indicates a potential for maintaining productivity during periods of moderate water stress.

Results of this study indicate that greater leaf water stresses develop for crested and slender than for western and pubescent wheatgrasses when grown at similar soil water potentials. Sensitivity to water stress was greater for western than for crested and pubescent wheatgrass. This response may indicate that western undergoes a more rapid conditioning to water stress than either crested or pubescent wheatgrass, a trait that may be related to drought tolerance.

LITERATURE CITED

Acevedo, E., E. Fereres, T. C. Hsiao, and D.W. Henderson. 1979. Diurnal growth trends, water potential, and osmotic adjustment of maize and sorghum leaves in the field. Pl. Physiol. 64:476–480.

Culter, J.M., K.W. Shahan, and P.L. Steponkus. 1980a. Dynamics of osmotic adjustment in rice. Crop Sci. 20:310–314.

Culter, J.M., K.W. Shahan, and P.L. Steponkus. 1980b. Influence of water deficits and osmotic adjustment on leaf elongation in rice. Crop Sci. 20:314–318.

Gaastra, P. 1959. Photosynthesis of crop plants as influenced by light, carbon dioxide, temperature, and stomatal diffusion resistance. Meded. Landbowhogesch, Wageningen 59:1–68.

Hsiao, T.C., E. Acevedo, E. Fereres, and D.W. Henderson. 1976. Stress metabolism: water stress, growth, and osmotic adjustment. Roy. Soc. London, Phil. Trans., Ser. B. 273:479–500.

Johnson, D.A. 1978. Environmental effects on turgor pressure response in range grasses. Crop Sci. 18:945–948.

Leaf Growth and Grain Yield of Wheat in North-South and East-West Rows

M.B. KIRKHAM

Kansas State University, Manhattan, Kans., U.S.A.

Summary

Few experiments report the effect of row direction on growth of wheat in windy areas. The objective of this research was to determine the optimal row direction for winter wheat (*Triticum aestivum* L. em. Thell.) in the southern Great Plains of the U.S.A. for minimizing stress from wind, optimizing leaf growth (forage yield), and maximizing grain yield.

During a 3-year period (1976–1979), Osage wheat was grown in north-south (NS) and east-west (EW) rows. Wind, soil water content, stomatal resistance, leaf water potential, leaf osmotic potential, elemental concentration of straw and grain, height, leaf area, and grain yield were measured.

Prevailing winds came mainly from the south in spring, summer, and autumn and from the north in winter. Soil water content was greater in EW than in NS rows. Stomatal resistance, leaf water potential, leaf osmotic potential, elemental concentration of straw and grain, and height of plants in NS rows were not significantly different from those of plants in EW rows. Leaf area of plants in NS rows ranged from 5% to 20% more than leaf area of plants in EW rows. During the 3-year study, plants in EW rows yielded an average of 11% more grain than plants in NS rows.

The results shows that, where prevailing winds come from the north and south, wheat should be oriented in NS rows for maximum leaf area (forage yield) and in EW rows for maximum grain yield.

KEY WORDS: winter wheat, *Triticum aestivum* L. em. Thell., leaf area, grain yield, soil water content, stomatal resistance, plant water potential, plant osmotic potential, height, elemental concentration.

INTRODUCTION

In the southern Great Plains of the U.S.A., winter wheat is grazed for forage during autumn and early spring. Animals are removed in mid-March to allow the wheat to produce again. The plants are shaken by strong winds. The effect of row direction on growth of wheat in the southern Great Plains has received little attention. This paper reports the results of a 3-year field experiment comparing wheat grown in north-south (NS) and east-west (EW) rows.

METHODS

The experiment was carried out between September 1976 and June 1979 in Stillwater, Oklahoma. Procedures used have been described previously (Erickson et al. 1979, Pearson 1979).

RESULTS

Averaged over the three growing seasons (October-June), the amount of wind from N, NE, E, SE, S, SW, and NW was 10,698; 2,384; 697; 5,695; 12,299; 3,850; 1,539; and 5,622 km, respectively. North-south was the prevailing direction.

Leaf area of plants in NS rows ranged from 5% more than leaf area of plants in EW rows (Fig. 1). The wheel design of the 1976–1977 and 1977–1978 studies (Erickson et al. 1979) showed that flag-leaf area of NS rows was about 13% higher than that of EW rows (p < 0.10).

Grain yield (p < 0.07) and soil water content (p < 0.15) were greater in EW than in NS rows (Fig. 2). Soil water content was not measured in 1976–1977, but grain yields in 1977 were 2,480 and 2,790 kg/ha for NS and EW rows, respectively. During the 3-year study, plants in EW rows yielded an average of 11% more grain than plants in NS rows.

Data obtained each year for stomatal resistance, leaf water potential, leaf osmotic potential, and height were averaged (data not given). Differences due to row direction were not significant (p = 0.05).

The author is an Associate Professor at the Evapotranspiration Laboratory of the Department of Agronomy of Kansas State University. This paper is contribution no. 81-273-A of the Evapotranspiration Laboratory.

Nitrogen in grain harvested in June 1977 was 1.5% for both NS and EW rows. In June 1978, nitrogen in NS and EW rows was 1.9 ± 0.2% and 1.4 ± 0.5%, respectively. In June 1979, even though differences in nutrient concentration in straw and grain were not significant, straw and grain from NS rows had lower and higher concentrations of nutrient elements, respectively, than did straw and grain from EW rows, except for manganese (Mn) and zinc (Zn) in straw (data not given).

DISCUSSION

Plants in the EW rows apparently were under greater mechanical stress since they faced into the wind. A plant

Fig. 1. Leaf area of winter wheat grown for 3 years in two row directions.

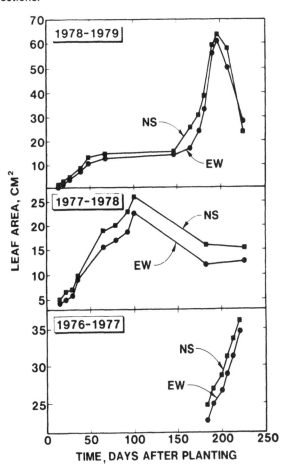

Fig. 2. Soil water content and grain yield of winter wheat for 2 years in two row directions.

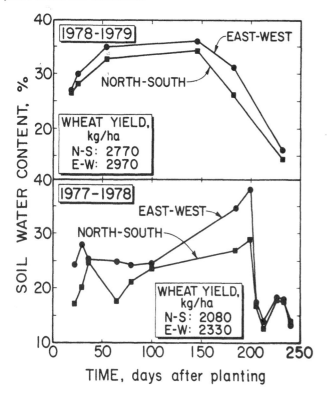

regularly shaken grows less than a plant in calm air (Parkhurst and Pearman 1972, Rees and Grace 1981). But they must have been under less water stress since EW rows had a higher soil moisture content than NS rows. Plants under stress flower earlier than plants that are not stressed (Newman 1965). Flowering was not observed in this study. However, it is speculated that, if plants in the EW rows flowered before plants in the NS rows, grain-filling time would have been longer, which would have permitted the high yields observed in EW rows.

LITERATURE CITED

Erickson, P.I., M.B. Kirkham, and J.F. Stone. 1979. Growth, water relations, and yield of wheat planted in four row directions. Soil Sci. Soc. Am. J. 43:570–574.

Newman, E.I. 1965. Factors affecting the seed production of *Teesdalia nudicaulis*. II. Soil moisture in spring. J. Ecol. 53:211–232.

Parkhurst, D.F., and G.I. Pearman. 1972. Tree seedling growth: effects of shaking. Science 175:918.

Pearson, C.H. 1979. Physiological measurements of winter wheat under stress. M.S. Thesis, Okla. State Univ.

Rees, D.J., and J. Grace. 1981. The effect of wind and shaking on the water relations of *Pinus contorta*. Physiol. Pl. 51:222–228.

Physiology of Growth of a Grazed Sward

E.L. LEAFE and A.J. PARSONS
The Grassland Research Institute, Berkshire, UK

Summary

To improve our understanding of grassland utilization under grazing, the physiology of a grazed sward was investigated. An enclosure method, infrared gas analysis, and radiocarbon were used to measure sward and single-leaf photosynthesis and assimilate utilization. Intake by sheep was also measured.

Expanding and recently expanded leaves had high rates of net photosynthesis but leaf sheaths, whose projected area in a "hard" grazed sward equaled that of the lamina, had low efficiency and contributed little to canopy photosynthesis. But because high leaf efficiency did not compensate for the small leaf area, canopy photosynthesis decreased markedly as grazing intensity increased.

Twice as much of the photosynthetic production was harvested as intake in hard grazing as in lenient grazing, but actual intake/ha was only one-third greater because the initial photosynthetic production was less. Conversely, much more carbon was lost to death and decay in the leniently grazed sward.

Maximum animal production requires the consumption of as much as possible of the young and most nutritious foliage, whereas maximum sward production requires a fully light-intercepting canopy of young, photosynthetically efficient leaves. The resolution of this conflict poses formidable problems in the continuously grazed sward, and rotational grazing appears to have provided only a partial solution.

KEY WORDS: perennial ryegrass, *Lolium perenne* L., physiology, photosynthesis, carbon balance, grazing, sheep, intake.

INTRODUCTION

Most studies of the physiology of grass crops have been made in swards harvested by cutting (e.g., Leafe 1978), but in practice, grazing is the more important harvest method. The present study was aimed at improving our understanding of sward growth under grazing and was based on an irrigated perennial ryegrass sward, fed with ample mineral nutrients and continuously grazed by sheep.

MATERIALS AND METHODS

The sward. A perennial ryegrass (*Lolium perenne* cv. S23) sward, sown in 1976, was used experimentally in 1978 and 1979. Fertilizer was applied at a rate equivalent to 55 kg N/ha/month during the growing season, and the sward was irrigated whenever the potential soil water deficit exceeded 25 mm.

Plots were grazed continuously with sufficient wether sheep to maintain a lamina area index (AI) of 1—hard grazed (H)—or 3—leniently grazed (L). In 1979 an overgrazed (O) treatment was added. Mean stock numbers/ha were 21.5 (L), 48 (H), and 125 (O). Measurements were confined to four 14-day periods in 1978 and three in 1979.

Canopy photosynthesis and respiration. An enclosure method was used (Stiles 1977) in which measurement areas, each 1 m², were defined before grazing began. The response of photosynthesis to irradiance was measured at the beginning and end of each period, using the natural range in radiation during a single day. Dark respiration was measured at the end of the day.

Leaf photosynthesis and assimilate distribution. Areas of the sward, 0.25 m², were allowed to assimilate $^{14}CO_2$ at 200 W/m² (400–700 nm), using the technique of Parsons (1981) and sampled after 5 minutes. The proportion of ^{14}C in leaves of different ages and in the sheath was determined, and their net CO_2 uptake was calculated by reference to the net uptake of CO_2 by the whole canopy. Distribution of assimilates was determined by assaying root and shoot fractions 48 hours later.

Biological and environmental measurements. Dry weight, expressed as hexose (CH_2O) equivalent after carbon analysis, and leaf area were determined by destructive sampling.

Air temperature at midsward height and incoming solar radiation were measured continuously, and air temperature, humidity, and irradiance inside the enclosure were measured during CO_2 flux measurements.

Intake. Herbage intake was measured by Cr_2O_3 marking and fecal collection.

RESULTS AND DISCUSSION

Dry weight, lamina, and sheath area. During the season, above-ground dry weight increased in the L sward but

The authors wish to thank Mr. P.D. Penning, Mr. W. Stiles, Mr. B. Collett, and Miss Jan Lewis, without whom the work could not have been done.

varied little in the H sward (Fig. 1a). Overgrazing severely reduced sward weight.

There were some departures from the intended lamina areas (Fig. 1c), but a marked difference was maintained between the H and L swards. In O swards, lamina AI declined to 0.3 or below. All swards maintained a sheath AI of about 1.0 throughout.

Leaf photosynthesis. Table 1 shows the contribution of leaves of different ages and the sheath to net photosynthesis (columns 1 and 4). Their contribution to the total leaf area is also shown (columns 2 and 5) and, by derivation, their mean rates of net photosynthesis (columns 3 and 6).

Over 75% of sward net photosynthesis was attributable to leaves 1 and 2, though they contributed less than half the leaf area. Shading was minimal, especially in the H sward—a factor known to enhance leaf photosynthesis (Woledge 1972). In contrast, sheaths made a negligible contribution—though in the H sward they accounted for half the light-intercepting area.

Sward photosynthesis. Gross photosynthesis irradiance relationship of the swards at the beginning and end of selected periods in 1979 was calculated as the sum of net photosynthesis and dark respiration (adjusted for temperature assuming a Q_{10} of 2.0). Differences between the swards were largely accounted for by differences in leaf area. Gross photosynthetic production over the measurement periods was calculated from response curves (Table 2).

Sward photosynthesis was markedly less at the greater intensities of grazing, and, clearly, the high photosynthetic efficiency of leaves in the H sward fell far short of compensating for the small leaf area index and incomplete light interception. Compared with the L sward, gross photosynthetic production was nearly one-third less in the H sward and almost halved in the O sward.

Assimilate utilization and respiration. Above-ground dark respiration is shown as a percentage of gross photosynthesis in Table 3 and the amount of ^{14}C found in the roots after 48 hours is shown as a percentage of ^{14}C in the whole plant. Assuming a similar efficiency of conversion of assimilate in root to that in shoot, respiration amounts to 5% of gross photosynthetic production in the roots and 2%–5% in the stubble. Total respiration loss was close to 45%—the figure that Robson (1973) obtained by direct measurement using simulated swards—and was the same in H and L swards.

Intake. Organic-matter (OM) intake, summarized in Table 4, is expressed as a percentage of gross photosynthetic production, assuming that OM corresponds closely to hexose equivalent. As expected, intake was greater in the H than in the L sward and much greater than both in the O sward. The apparent harvest efficiency of 63% in the O sward is unreal, since intake exceeded shoot production.

Death. Leaf death, estimated from the weight of the oldest leaf present (Hunt 1965), amounted to 28 and 51 kg CH_2O equivalent/ha/day in the H and L swards, respectively. When combined with tiller death, which in a

Fig. 1. (a) Dry matter (CH₂O equivalent), (b) projected sheath area index, and (c) lamina area index changes in leniently and hard-grazed swards in 1978 and 1979. The shaded bars indicate the periods of experimental measurements.

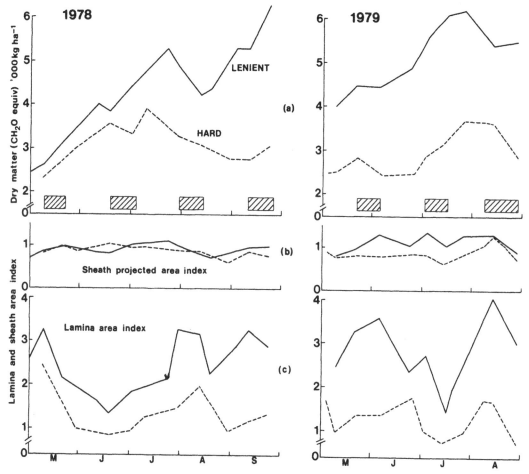

Table 1. The contribution of the sheath and leaves of different ages to sward net photosynthesis, measured in hard (H) and leniently (L) grazed swards in 1978.
Columns 1 and 4 — Percentage contribution of leaf age categories and sheath to net photosynthesis of the sward, H and L swards, respectively.
Columns 2 and 5 — Area of leaf age categories and sheath as a percentage of total lamina and sheath area.
Columns 3 and 6 — Mean rate of net photosynthesis of leaf age categories and sheath, g $CO_2/m^2/h$.

| | Grazing intensity | | | | | |
| | Hard | | | Lenient | | |
Category	1	2	3	4	5	6
Leaf 1 (expanding youngest leaf)	37.3	15.4	2.07	37.1	19.2	1.34
Leaf 2 (first expanded leaf)	39.1	25.2	1.37	42.7	27.7	1.00
Leaf 3 (second expanded + older leaves)	18.2	20.9	0.74	17.8	26.0	0.47
Leaf sheath (pseudostem)	4.6	38.5	0.11	2.0	27.1	0.05

perennial ryegrass sward approaches 1%/day (De Lucia Silva 1974), the resultant estimate of shoot death is very close to the estimate based on the carbon balance.

Carbon balance. Gross photosynthetic production in the L sward averaged 300 kg CH₂O equivalent/ha/day, 50% more than in the H sward but still considerably less than that of an undefoliated sward during reproductive growth in spring (Leafe 1978). Almost half the carbon assimilated was respired, but only about one-tenth was translocated to the roots.

Sheep consumed 38 kg OM/ha/day in the L sward against 53 kg OM/ha/day in the H sward — 13% vs. 25% of the gross photosynthetic production. Inevitably, in the L sward much more of the initially fixed carbon flowed into the final category — death (37% vs. 25%).

CONCLUSION

In grazing, there is a conflict between the factors needed to maximize animal production — namely, the consumption of the maximum amount of the young and most nutritious foliage — and those needed to maximize

Table 2. Gross photosynthetic production, kg CH$_2$O equiv./ha/day in leniently (L), hard (H), and overgrazed (O) swards during each measurement period and as a mean of each grazing intensity in 1979.

Grazing intensity	Measurement period			Mean
	1	2	3	
Lenient	294	357	248	300
Hard	199	238	189	209
Overgrazed	–	207	124	166

Table 3. (a) Above-ground respiration, measured by gas analysis, as a percentage of gross photosynthesis, and (b) the amount of ^{14}C in the roots as a percentage of ^{14}C in the whole plant after 48 hours. Leniently (L), hard (H), and overgrazed (O) swards during each measurement period and as a mean of each grazing intensity in 1979.

Grazing intensity	Measurement period			Mean
	1	2	3	
(a) Above-ground respiration as a % of 'gross' photosynthesis				
Lenient	38	26	28	31
Hard	38	28	35	34
Overgrazed	–	31	36	34
(b) % of ^{14}C in roots after 48 hours				
Lenient	11.4	8.6	9.3	9.8
Hard	7.6	7.2	14.65	9.8
Overgrazed	–	Not assessed		–

Table 4. Intake in kg OM/ha/day, and as a percentage of gross photosynthetic production (in parentheses), during each measurement period and as a mean of each grazing intensity in 1979.

Grazing intensity	Measurement period			Mean
	1	2	3	
Lenient	45 (15)	33 (9)	37 (15)	38 (13)
Hard	49 (24)	65 (27)	46 (24)	53 (25)
Overgrazed		103 (49)	95 (77)	99 (63)

sward production—the retention of a full canopy of photosynthetically efficient leaves. In practice, efficiency of harvesting is often sacrificed to achieve high individual animal performance, but the cost is reduced animal output/ha and massive losses of herbage to death.

Two questions arise. First, under continuous grazing, can the high efficiency of harvesting of the H sward be combined with the high productivity of the L sward, thereby increasing total animal production? In theory, this combination could be achieved if old and senescent leaves were grazed in preference to young ones, but this grazing pattern seems a remote prospect. Alternatively, grasses might be bred in which leaf photosynthesis does not decline as leaves age and in which the leaf sheath is more efficient photosynthetically. Second, can the same goal be achieved by separating sward growth and grazing in time—that is, by rotational grazing? It appears that sometimes there is greater animal production under rotational than under continuous grazing, but, at least in the all-grass sward, consistent improvements have not yet been achieved (Ernst et al. 1980).

LITERATURE CITED

De Lucia Silva, G.R. 1974. A study of variation in the defoliation and regrowth of individual tillers in swards of *Lolium perenne* L. grazed by sheep. Ph.D. Thesis, Univ. of Reading.

Ernst, P., Y.L.D. Le Du, and L. Carlier. 1980. Animal and sward production under rotational and continuous grazing management: a critical appraisal. Proc. Int. Symp. Eur. Grassl. Fed. on The Role of Nitrogen in Intensive Grassland Production 119–126.. Pudoc, Wageningen.

Hunt, L.A. 1965. Some implications of death and decay in pasture production. J. Brit. Grassl. Soc. 20:27–31.

Leafe, E.L. 1978. Physiological, environmental and management factors of importance to the maximum yield of the grass crop. Proc. Joint ADAS/ARC Symp. Maximising the yields of crops. Harrogate, HMSO London, 37–49.

Parsons, A.J. 1981. Carbon exchange and assimilate partitioning. *In* J. Hodgson, R.D. Baker, A. Davies, J.D. Lever, and A.S. Laidlaw (eds.), Sward measurement handbook. Courier Press, Haddington, Scotland.

Robson, M.J. 1973. The growth and development of simulated swards of perennial ryegrass. II. Carbon assimilation and respiration in a seedling sward. Ann. Bot. 37:501–518.

Stiles, W. 1977. Enclosure method for measuring photosynthesis, respiration and transpiration of crops in the field. Tech. Rpt. 18, Grassl. Res. Inst., Hurley.

Woledge, J. 1972. The effect of shading on the photosynthetic rate and longevity of grass leaves. Ann. Bot. 36:551–561.

Photosynthesis of Grass Swards Under Rotational and Continuous Grazing

B. DEINUM, M.L. 'T HART, and E. LANTINGA

Agricultural University, Wageningen, Netherlands

Summary

First-year results are presented on the photosynthesis of grass swards under intensive rotational or continuous grazing. In rotational grazing, crop photosynthesis was high in spring, declining during grazing but recovering rapidly during regrowth, possibly because of the high photosynthetic capacity of individual leaves and of the great area of active leaves left in the sward. In autumn, crop photosynthesis was rather high initially, declined rapidly during grazing, and recovered slowly during regrowth, possibly because photosynthetic capacity of the leaves was low, the decline with age was great, and the active leaf area in the stubble was small.

Crop photosynthesis under continuous grazing was rather high in spring despite the low leaf-area index, most likely because photosynthetic capacity of the many small leaves was high. In late summer, crop photosynthesis was low, possibly because the remaining leaves were old and less active.

Net photosynthesis of swards under continuous grazing, integrated over time at two radiant flux densities, was about the same in spring as for swards receiving rotational grazing. However, in autumn, photosynthesis was somewhat lower under continuous than under rotational grazing. This lower rate does not necessarily lead to a proportionally lower herbage production, as differences in respiration, translocation, and longevity may compensate for the differences. Research is being continued to compile an explanatory simulation model and to optimize grazing systems.

KEY WORDS: photosynthesis, rotational grazing, continuous grazing, *Lolium perenne* L., production, leaf-area index.

INTRODUCTION

Recent investigations by Ernst et al. (1980) have shown that animal production is similar in rotational and continuous grazing of well-fertilized pastures in the maritime climates of northwestern Europe. Much information on physiology and production of herbage is already available under conditions of uninterrupted growth (Alberda 1977, Leafe 1974). Some data may be extrapolated to rotational grazing but not necessarily to continuous grazing. We initiated research to improve our knowledge of physiology and production of herbage under grazing and to incorporate the information in simulation programs. We also hope to develop better grazing systems. Some initial results are presented here.

METHODS

Crop photosynthesis was used in this project as an estimate of herbage production together with measurement of sward characteristics, inasmuch as cutting techniques fail to characterize herbage production in continuous grazing.

The 1979 trial was on a 10-year-old pasture on river

clay dominated by *Lolium perenne*. It was divided into seven paddocks, one for continuous grazing and six for rotational grazing. N supply was about 400 kg/ha/yr, and supply of P, K, and water was adequate. Grazing management with yearling steers was as optimal as possible in both systems. Light-response curves (LRC) were estimated each work day for both systems inside water locks of 1 m². Exchange of CO_2 was measured with ordinary equipment.

Measurements of LRCs were made for rotational grazing before grazing, when about 50% and 80% to 90% of the herbage had been consumed, and in different stages of regrowth. Simultaneous measurements were made on the continuously grazed sward. Data were collected in May, June, August, September, and October. Daily measurements of sward structure were made outside the water locks on areas of the pasture that looked similar to the grass inside the water lock on which LRC was measured.

RESULTS

Rotational Grazing

Tiller density of the crop averaged about 12,000 tillers/m² during the growing season. Fig. 1a presents the typical results of LRCs during a grazing cycle in spring, when rate of production and of regrowth are high. Photosynthesis of the full crop with a leaf-area index

B. Deinum is a Research Agronomist, M.L. 't Hart is a Professor of Grassland Science, and E. Lantinga is a Student at Agricultural University.

408 *Deinum et al.*

Fig. 1. Some typical light response curves (photosynthesis expressed in kg CH$_2$O/ha/hr against radiant flux density (400–700 nm) of herbage under rotational grazing in the spring (a) and in autumn (b) of 1979. In spring (a) 1 = reading on 14 May; 2 = 16 May; 3 = 18 May; 4 = 22 May; 5 = 29 May. In Autumn (b) 1 = 28 Aug.; 2 = 30 Aug. (40% grazing); 3 = 30 Aug. (50% grazing); 4 = 5 Sept.; 5 = 11 Sept.; 6 = 26 Sept.; 7 = 8 Oct.

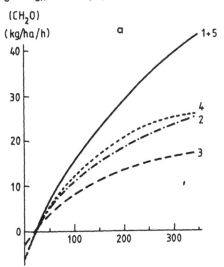
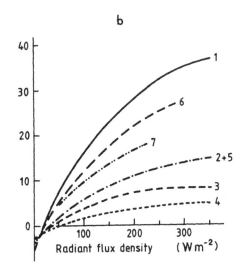

(LAI) of about 5 was high. After one day of grazing, LAI had dropped to about 3.5; photosynthesis declined considerably, presumably because many young, active leaves had been consumed and most of the less active leaf area was still present. The decrease in photosynthesis was even more pronounced after a second day of grazing (18 May, when only old leaves were present and little light was absorbed). Regrowth started so fast that after 4 days the curve was higher than the one after 50% grazing, even though LAI had only increased to 3. This rapid recovery was certainly due to leafy stubble and to formation of young, active leaves. After only 11 days of regrowth, the curve was back to the original level.

Fig. 1b presents results collected in late summer and autumn, when rates of production and regrowth are slower. The full crop still showed a rather high curve on 29 August. After one day of grazing the response had dropped considerably more than in spring, and after 6 days of light grazing, LRC was very low. Apparently only leaves with very low photosynthetic activity had been left. The rate of photosynthesis increased slowly during regrowth because of the slow appearance of new leaves. Final response never reached the original rate.

These differences between spring and autumn may be ascribed to differences in leaf activity. In spring, leaves form quickly and photosynthetic capacity starts high and declines slowly with time (Woledge 1979). In autumn, leaves form more slowly, and photosynthetic capacity declines faster with age.

Continuous Grazing

Several measurements of LRC were made under continuous grazing during the season at several sward heights (such as severely grazed spots, dung patches). Tiller density increased rapidly in this pasture up to about 20,000 tillers/m² and remained high during the season. The LAI was greater than under rotational grazing at the same sward height.

Fig. 2 presents typical results from spring and autumn measurements. Spring growth responses were somewhat lower than those of the full sward before a cycle of rotational grazing. However, responses were high in relation to LAI and sward height. At an LAI of 3, LRC of the continuously grazed sward was higher than LRC of the grazed crop under rotational grazing and similar to that of regrowth in rotational grazing. This may be explained by the large photosynthetic capacity and the slow decline in activity with age of the many small leaves that are continuously exposed to full irradiance (Woledge 1978) and that have a large assimilate demand (Deinum 1976).

In late summer and autumn, curves for continuously grazed swards were lower than in spring, possibly for the same reasons as for rotational grazing: greater decline of photosynthesis with age and slower leaf formation. At a given LAI, response was better with continuous grazing than with the grazed crop of rotational grazing and similar to that of regrowth (Figs. 1b and 2b).

Seasonal Photosynthesis Of Both Grazing Systems

An integration of seasonal production from the curves measured in the two grazing systems was accomplished by plotting photosynthesis at radiant flux densities of 85 and 250 W/m² (400–700 nm) against time and by subsequent planimetry of the area (Fig. 3).

In the conditions of 1979, net seasonal photosynthesis of the two grazing systems was similar in spring and early summer (Table 1). Photosynthesis was only slightly lower under continuous grazing in the short sward. In late summer and autumn, there were some advantages to rotational grazing. We are not yet sure whether these advantages increase or decrease if we take account of differences

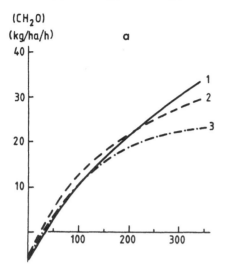
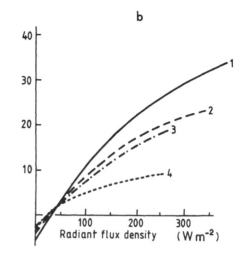

Fig. 2. Some typical light response curves of herbage under continuous grazing in spring (a) and in autumn (b) of 1979. In spring (a) 1 = reading on 15 May; 2 = 13 June; 3 = 25 June. In autumn (b) 1 = 3 Aug.; 2 = 10 Aug.; 3 = 10 Sept.; 4 = 28 Sept.

in daily radiation and possible differences in night respiration and assimilation distribution.

DISCUSSION

Our findings agree well with those of Leafe et al. (1979) for continuous grazing with sheep. They found that photosynthesis could be high even for very short-cropped grass if enough (40,000/m²) active green tillers were present. Youngest leaves were most active; older leaves had lower photosynthesis because of aging and possibly also because of shading. In general terms, grazed swards seem to behave similarly to plants in cutting trials and to those under undisturbed growing conditions.

Animal studies suggest that the same animal production can be obtained with continuous grazing even though photosynthesis is lower. Leafe et al. (1980) obtained similar results: gross photosynthesis was less in continuously grazed swards than in cut swards. However, after corrections for differences in respiration and distribution, they

calculate the same herbage mass production. Root production may be smaller under continuous grazing because of the continuous demand for replenishment of consumed leaves. The good performance of continuous grazing may be partly due to less death and decay of leaves than with rotational grazing. Under optimum conditions herbage production with continuous grazing may well be better than with rotational grazing.

From these early results and from those to be collected we plan to develop an explanatory simulation model to help us optimize grazing.

LITERATURE CITED

Alberda, T. 1977. Crop photosynthesis: methods and compilation of data obtained with a mobile field equipment. Agric. Res. Rpt. 865:4–11.

Deinum, B. 1976. Photosynthesis and sink size; an explanation for the low productivity of grass swards in autumn. Neth. J. Agric. Sci. 24:238–247.

Ernst, P, Y.L.P. Le Du, and L. Carlier. 1980. Animal and sward production under rotational and continuous grazing management; a critical appraisal. Proc. Int. Symp. Eur. Grassl. Fed. on The Role of Nitrogen in Intensive Grassland Production, 119–126. Pudoc, Wageningen.

Leafe, E.L. 1974. Physiological processes influencing the pattern of productivity on the intensively managed grass swards. Proc. XII Int. Grassl. Cong., 422–457.

Fig. 3. Estimated change in net photosynthesis at two radiant flux densities of swards under rotational (R.G.) and continuous (C.G.) grazing.

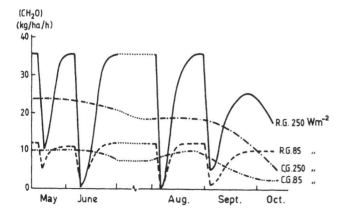

Table 1. Estimated net photosynthesis of swards from continuous grazing relative to rotational grazing (%) for different sward heights and radiant flux densities (400–700 nm).

Sward height (cm)	Relative net photosynthesis			
	May–June		Aug.–Oct.	
	85 W/m²	250 W/m²	85 W/m²	250 W/m²
6	90	85	75	65
8	99	90	78	70
10	100	100	80	80

Leafe, E.L., A.J. Parsons, W. Stiles, and B. Collett. 1980. The growth of the grass sward under grazing. Grassl. Res. Inst. Ann. Rpt. 1979, 29–32.

Leafe, E.L., W. Stiles, A.J. Parsons, and B. Collett. 1979. The growth of the grass sward under grazing. Grassl. Res. Inst. Ann. Rpt. 1978, 41–51.

Woledge, J. 1978. The effect of shading during vegetative and reproductive growth on the photosynthetic capacity of leaves in a grass sward. Ann. Bot. 42:1083–1089.

Woledge, J. 1979. Effect of flowering on the photosynthetic capacity of ryegrass leaves grown with and without natural shading. Ann. Bot. 44:197–207.

Photosynthesis and Dry-Matter Distribution in Altitudinal Ecotypes of White Clover

J. NÖSBERGER, F. MÄCHLER, and B.C. BOLLER

Swiss Federal Institute of Technology, Zurich, Switzerland

Summary

Productivity of white clover (*Trifolium repens* L.) is limited by plant characteristics that prevent efficient use of environmental inputs. In the temperate zone, low temperature often limits growth. Comparisons of ecotypes originating from sites with cool or warm climatic conditions (low or high altitudes) can show, when plants are grown in a common environment, the degree of adaptation of processes related to net photosynthesis and the distribution of dry matter. Our objectives were to determine the photosynthetic responses to temperature in relation to the thermal regime during growth and the effects of temperature and daylength on partitioning of dry matter and nonstructural carbohydrates.

Vegetatively propagated plants from 600 m (valley type) and 2,000 m above sea level (alpine type) were grown in the field and in growth chambers. Photosynthetic rate and assimilate distribution were determined. The alpine genotype and plants grown at low temperatures showed relatively high photosynthetic rates at 3°C but limited photosynthesis at 24°C, probably due to intracellular shortage of CO_2 and increased photorespiration. The valley genotype and plants grown at higher temperatures showed high photosynthetic rates at high temperatures but a relatively low rate at low temperatures, probably due to the decreased concentration of photosynthetic apparatus/unit leaf area. At low temperatures, pool sizes of metabolites of the carbon cycle suggest limitations by processes other than carboxylation. The high rate of photosynthesis of the alpine genotype was not associated with a high forage yield. The lower yield of the alpine genotype cannot be explained by leaf-area ratio but by leaf-area index and mean sward height. Growth temperature and particularly daylength modified the pattern of assimilate distribution more than did genotype. Plants grown at low temperatures showed increased content of nonstructural carbohydrates. This feature was associated with a greater decrease in rate of leaf appearance than in photosynthesis as temperatures were decreased. In addition, when daylength was reduced at low temperatures, 82% of all nonstructural carbohydrates were in the stolons and roots, compared with 55% to 65% in long days. Higher photosynthetic rate at low temperatures and a stronger shift to growth of storage organs of the alpine genotype are consistent with requirements of its ecological origin.

KEY WORDS: white clover, *Trifolium repens* L., photosynthesis, photorespiration, dry-matter distribution, nonstructural carbohydrates, temperature, adaptation.

INTRODUCTION

Plants of white clover (*Trifolium repens* L.) that orginate from permanent grasslands with dissimilar climatic conditions can show adaptations to the original environmental conditions and give information on the genetic resources. It is well known that photosynthetic capacity of single leaves in certain combinations of light, temperature, and CO_2 concentration does not necessarily reflect the productive performance of the plant stand, especially under natural conditions where environmental factors vary both temporally and spatially. To relate plant productivity to photosynthetic capacity one should know the dry-matter distribution and the responses of photosynthesis of different leaves in the canopy to both short-term and long-term changes in the environment. Major objectives of these studies were (1) to determine factors limiting leaf photosynthetic potential at different temperatures in relation to temperatures during earlier growth stages and (2) to determine the effect of temperature and daylength on partitioning of dry matter in white clover.

J. Nösberger is a Professor of Agronomy, F. Mächler is a Research Associate, and B.C. Boller was formerly a Graduate Research Assistant at the Swiss Federal Institute of Technology.

MATERIALS AND METHODS

Stolons from white clover were collected from permanent grassland at 600 (Chur, valley type = V) and 2,000 m (Arosa, alpine type = A) above sea level in the Swiss Alps and propagated vegetatively. The two sites represent contrasting mean and extreme temperatures and different lengths of growing season. Plants from A have shorter petioles and leaflet area about half those of V.

Stolon tips were cultured in growth chambers at day/night temperatures of 10°/7° and 20°/15°C or as indicated in the figures or tables. Photon flux density at canopy height during the 16-hour photoperiod was 350 $\mu E/m^2$/second. Gas exchange was measured by infrared gas analysis. Metabolite pools were estimated by treating leaves for 60 minutes with $^{14}CO_2$ of known specific activity and separating the metabolites by thin layer chromatography (Mächler and Nösberger 1978). Harvests were made when plants had 15 leaves on the main stolon. Dried samples were analyzed for total nonstructural carbohydrates by takadiastase digestion.

RESULTS AND DISCUSSION

The A genotype showed photosynthetic adaptation to the lower temperatures of higher altitudes (Fig. 1). Photosynthetic rate of A was high at low temperatures, and maximum photosynthesis was approached at a relatively low temperature, indicating rate limitation at high temperatures. The V genotype showed relatively low photosynthetic rates at low temperatures; however, photosynthesis increased rapidly as temperature increased throughout the range assayed. Photosynthesis was affected similarly by growth temperature. Plants grown at 11°/7°C showed rate limitation at high temperatures like A, whereas plants grown at 18°/13°C showed rate limitation at low temperatures like V.

Growth temperature response of photosynthesis was related to changes in specific leaf weight (Table 1). Specific leaf weights were similar for both genotypes. However, the photosynthetic apparatus was more concentrated in A than in V, as is indicated by high photosynthetic rate (Fig. 1) and greater pool sizes of several carbon-cycle intermediates (Table 1).

The increased photosynthetic rates of A and plants grown at low temperatures were associated with higher percentage of photorespiration. A higher percentage of ^{14}C occurred in glycine and serine after a 20-second exposure of leaves to $^{14}CO_2$ (Table 2). This fact suggests a close relationship between high photosynthetic rate and high percentages of photorespiration: As photosynthetic rate increases, CO_2 concentration in the chloroplasts is decreased. A decreasing ration of CO_2 to O_2 decreases carboxylase and increases oxygenase activity leading to higher photorespiration. The limitation of photosynthesis at high temperatures of the A genotype and of plants grown at low temperatures may be attributed partly to shortage of CO_2 and increased photorespiration.

Plants grown at 10°/7°C showed oxygen inhibition at

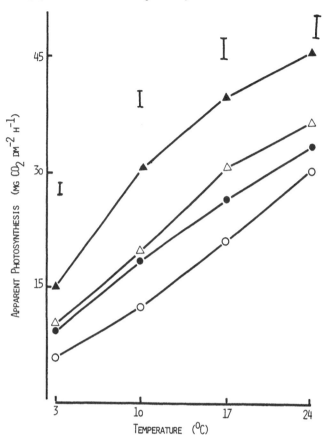

Fig. 1. Influence of temperature on apparent photosynthesis measured at 1,000 $\mu E/m^2$/second of the valley genotype (V, circles) and the alpine genotype (A, triangles). Plants were grown at 330 $\mu E/m^2$/second and at 18°/13°C (open symbols) and 11°/7°C (closed symbols) day/night temperatures, respectively (Mächler and Nösberger 1977).

Table 1. Influence of genotype and growth temperature on specific leaf weight (SLW) and pools of metabolites of the carbon cycle (P) (hexose monophosphates, 3-phosphoglyceric acid and ribose-5-phosphate) at 20°C (Mächler and Nösberger 1978).

| | Valley genotype (V) | | Alpine genotype (A) | |
	11°/7°C	18°/13°C	11°/7°C	18°/13°C
SLW, mg/cm^2	3.2	2.6	3.3	2.4
P, $\mu M^{14}C$/g D. W.	15.8	18.2	21.2	25.6
P, $\mu M^{14}C$/dm^2	5.1	4.7	6.9	6.0

Table 2. Influence of genotype and growth temperature on photorespiration estimated from percentage of total activity in glycine and serine after $^{14}CO_2$ treatment of leaves at 20°C for 20 seconds (Mächler and Nösberger 1978).

| | Valley genotype (V) | | Alpine genotype (A) | | |
	10°/7°C	20°/15°C	10°/7°C	20°/15°C	S.E.
	------------------- % -------------------				
Glycine	2.4	1.7	2.7	2.0	0.25
Serine	0.7	0.4	0.9	0.5	0.09

Table 3. Oxygen inhibition of photosynthesis at 10°C and pool sizes of 3-phosphoglyceric acid and ribose-5-phosphate in ecotypes grown and tested at 10°C or 20°C (Mächler and Nösberger 1978).

	Valley genotype (V)		Alpine genotype (A)	
	10°/7°C	20°/15°C	10°/7°C	20°/15°C
	---mg CO$_2$/dm^2/hour---			
O$_2$ inhibition	4.5	− 0.8	4.3	− 1.8
	---μM ^{14}C/g D.W.---			
In 3-PGA at 10°C	1.1	0.9	1.3	1.0
In 3-PGA at 20°C	1.2	1.8	1.9	2.5
In R-5-P at 10°C	3.0	4.5	4.3	5.7
In R-5-P at 20°C	2.3	3.3	2.7	4.5

10°C, whereas plants grown at 20°/15°C did not (Table 3). This difference was not due to a difference in photorespiration. Arrabaca et al. (1981) showed that under these conditions, the properties of carboxylase/oxygenase are masked by other factors. At 10°C, level of ribose-5-phosphate was higher, whereas level of 3-phosphoglyceric acid was lower than at 20°C (Table 3). As Arrabaca et al. (1981) suggested, the rate-limiting step in the carbon cycle seems to be located between ribose-5-phosphate and ribulose bisphosphate, possibly due to lack of adenosine triphosphate (ATP).

The high rate of photosynthesis of A was not associated with a high forage yield, a relationship also observed with other forages (Nelson et al. 1975). When grown in swards at low altitudes, A produced about one-third less harvestable biomass/yr than did V (Mächler 1976). This difference was not related to the leaf-area ratio, which was similar, or the rate of leaf appearance, which was higher in A than in V. However, V reached a higher leaf-area index and a greater canopy height, which may have allowed incoming light to be spread over a greater leaf area, hence increasing crop photosynthesis.

A allocated a greater percentage of dry matter to the stolons than did V when grown at 10°/7°C and 18°/13°C (Fig. 2). A more marked effect on the partitioning of dry matter was associated with daylength. Shorter days (12 vs. 16 hours) at low temperatures caused A and V to develop a higher percentage of dry matter in roots and stolons.

Concentrations of total nonstructural carbohydrates (TNC) in the various plant parts were highest in plants grown at cool temperatures. The decrease in day temperature from 18° to 10°C halved rate of leaf appearance in both genotypes (Boller 1980) but did not reduce photosynthesis by as much. Thus, less carbohydrate was used for expanding the photosynthetic tissue, allowing it to be accumulated. Shortening the daylength from 16 to 12 hours shortened the daily duration for photosynthesis but increased that for dark respiration. The TNC concentration of leaflets, petioles, and stolons at 12 hours was reduced to about one-third, one-half, and one-fourth respectively, of that at 16 hours, whereas TNC of the roots remained similar. About 80% of the TNC was in stolons and roots in short days, compared with 55% to

Fig. 2. Distribution of dry matter and total nonstructural carbohydrates of the valley genotype (V) and the alpine genotype (A) grown at different day/night temperatures and daylengths. L = leaves, P = petioles, S = stolons, R = roots, horizontal line = mean TNC content in the whole plant. Shaded areas signify percentage of total plant dry matter stored as TNC in the respective organs.

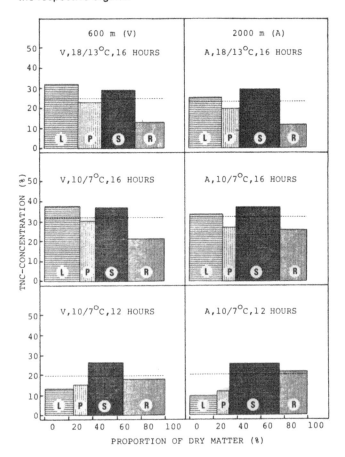

65% in long days. Obviously, sink hierarchy had shifted in favor of stolons and roots. This pattern of assimilate distribution is consistent with improving the prerequisite for persistence under long winter conditions.

CONCLUSION

Apparently, physiological and morphological constraints at each level of organization prevent the efficient use of various environmental inputs. Decreased carboxylation and increased photorespiration, both due to shortage of CO_2, tend to limit photosynthesis at high temperature, especially for the alpine genotype and for plants grown at low temperatures. Other reaction steps appear more important than carboxylation at low temperatures, particularly for plants grown at high temperature. The capability to store carbohydrate reserves is greater in alpine than in valley genotypes and is related inversely to leaf growth. Accumulation of reserves in stolons is increased when temperature is low and photoperiod is short.

LITERATURE CITED

Arrabaca, M.C., A.J. Keys, and C.P. Whittingham. 1981. Effect of temperature on photosynthetic and photorespiratory metabolism. Proc. V Int. Cong. Photosyn. Halkidiki, Greece.

Boller, B.C. 1980. Bestandesphotosynthese und Assimilatverteilung bei Oekotypen von Weissklee (*Trifolium repens* L.) unter verschiedenen Temperaturen und Tageslängen. Ph.D. Thesis, Swiss Fed. Inst. Technol. Zurich.

Mächler, F. 1976. Oekologische Aspekte der Photosynthese bei Oekotypen verschiedener Höhenlagen von *Trifolium repens* L. Ph.D. Thesis, Swiss Fed. Inst. Technol. Zurich

Mächler, F., and J. Nösberger. 1977. Effect of light intensity and temperature on apparent photosynthesis of altitudinal ecotypes of *Trifolium repens* L. Oecologia 31:73–78.

Mächler, F., and J. Nösberger. 1978. The adaptation to temperature of photorespiration and of the photosynthetic carbon metabolism of altitudinal ecotypes of *Trifolium repens* L. Oecologia 35:267–276.

Nelson, C.J., K.H. Asay, and G.L. Horst. 1975. Relationship of leaf photosynthesis to forage yield of tall fescue. Crop Sci. 15:476–478.

Using Leaf-Area Expansion Rate To Improve Yield of Tall Fescue

C.J. NELSON and D.A. SLEPER
University of Missouri, Columbia, Mo., U.S.A.

Summary

Several physiological characters have been proposed for use in breeding programs for higher yield potential. Our objectives were to evaluate recurrent restricted phenotypic selection for developing populations of tall fescue (*Festuca arundinacea* Schreb.) for low and high rates of leaf-area expansion (LAE). After four cycles, LAE was 181 and 74 mm²/day for the high (H) and low (L) populations, respectively. It appeared that LAE changes in early generations were due largely to increased leaf elongation rate (LER) of the H populations, whereas later progress was due largely to decreased leaf width (LW) of the L populations. Overall, LER had a larger effect on LAE than did LW. Number of tillers/plant was inversely related to LAE.

In field plots, yield of vegetative regrowth of the L_4 population was 25% lower than that of the control (C_0) population because the 2.1-fold increase in tiller density was more than offset by the 2.8-fold decrease in weight tiller. Tiller density was not reduced by selection for high LAE; in fact, the H_2 population yielded 39% more than the C_0 population because both tiller density and weight/tiller were increased about 20%. Later generations were lower yielding than the H_2 population, but higher yielding than the C_0 population. The H_2 population was superior, perhaps due to a complementary mixture of large and small plants. Alternatively, inbreeding depression may have limited further genetic advancement. Yields of Kenhy and Mo-96 were similar, while that of Ky-31 was lower than those of the C_0 population. Photosynthesis of single leaves was altered less than 10% by selection.

KEY WORDS: tall fescue, *Festuca arundinacea* Schreb., leaf-area expansion rate, leaf elongation rate, leaf width, tiller density, tiller weight, recurrent restricted phenotypic selection.

INTRODUCTION

For several years we have been investigating the photosynthesis-yield relationship of tall fescue (*Festuca arundinacea* Schreb.), especially during regrowth stages, when plants are in a vegetative condition and yield consists largely of leaf tissue. We found that rate of leaf elongation (Horst et al. 1978) was related positively but single-leaf photosynthesis (Nelson et al. 1975) was not related to genotypic differences in forage yield. Tiller density was found to be a more important determinant of yield than was weight/tiller in spaced plants (Nelson et al. 1977). Leaf elongation rate (LER) was associated positively with weight/tiller and was about 1.8 times more important than leaf width (LW) in determining rate of leaf-area expansion (LAE) (Nelson et al. 1977). Further, LER and LW were correlated positively among genotypes.

Studies on tall fescue showed that LAE had additive genetic variance with a narrow-sense heritability below 30% (Sleper et al. 1977). On the basis of the foregoing series of experiments and observations, a recurrent selection program was begun in 1975 to develop populations of tall fescue that had high and low LAE. Our objectives were to develop divergent populations and to evaluate a recurrent selection method for altering physiological characters of low heritabilty and high environmental influence. Further, we wanted to assess changes in yield and physiological factors that occurred during selection.

METHODS

Seeds from a broad-based population of tall fescue (C_0 generation) were germinated in petri dishes and then transplanted into soil in paper pots in April 1975. Seedlings were established in a greenhouse until late May, when plants and the soil mass were transplanted into the field at the Agronomy Research Center near Columbia, Missouri. The Mexico silt loam soil (Udollic Ochraqualf), fertilized with 114 kg/ha of nitrogen (N), tested high in phosphorus and potassium. Plants were randomly selected and spaced on 15-cm centers using a grid format to give semisward growth conditions yet allow retention of individual plant identity. We used recurrent restricted phenotypic selection to overcome limitations due to low heritability and large environmental influence (Gardner 1969, Burton 1974). For the C_0 population a total of 20 grids containing 40 plants each were evaluated. Subsequent generations were handled in a similar manner, except that 15 grids of 30 plants each were used for both the high (H) and the low (L) populations.

Each year seedlings were allowed to grow, receiving supplemental irrigation when needed, until about 15 August, when herbage was removed to leave a 5-cm stubble and discarded. Plots then received 45 kg N/ha. During

C.J. Nelson is a Professor and D.A. Sleper is an Associate Professor of Agronomy at the University of Missouri. This is a contribution of the Missouri Agricultural Experiment Station (Journal Series No. 8763).

September, LER was measured on two leaves/entry. Elongating blades were selected that were less than 30% of the length of the previous leaf blade. Distance between the tip of the elongating blade and the collar of the previous blade was measured 4 to 7 times over a 10- to 15-day period. The regression coefficient of blade length against time was reduced to estimate LER (Nelson et al. 1977). At time of the last length measurement the LW of the blade at the midlength was measured. The LAE was calculated as LER times LW. Data for the two leaves were averaged to evaluate each entry.

Plants were allowed to vernalize naturally in the field until early January, after which the upper two plants of the H and the lower two plants of the L populations were removed from each grid. Plants were potted and populations grown separately in the greenhouse where they were interpollinated, using forced air and hand agitation. Equal numbers of seeds from each maternal parent were used to begin the next generation in the greenhouse. In 1979, field plots were established of the C_0 to C_4 populations and cultivars Mo-96, Ky-31, and Kenhy. Preliminary data on vegetative regrowth between cuttings on 15 May and 6 November 1980 are presented.

RESULTS AND DISCUSSION

The recurrent restricted phenotypic selection method with high selection intensity was effective even though LAE has a low heritability and a high environmental influence. The large environmental effects on LAE and its components LER and LW are readily apparent from inspection of the data (Table 1). Reasons for the relatively more favorable growth responses in 1977 are not known, as temperatures and soil moisture were not markedly different that year from other years. However, in comparison with other years, the effect on LER was much greater than on LW, a finding that was consistent with other research (Robson 1974). After three generations LAE of the L population was only 41% of that of the H population. The C_4 populations did not appear to be more divergent than the C_3. However, the coefficient of variation for LAE of the C_4 populations was still about 75% of that of the C_0 population, suggesting that progress by selection should still be possible.

Assuming that population × environment interactions were minimal, we compared relationships between LAE, LER, and LW over population-years. There was a strong curvilinear association between LER and LW, showing a range in LW at low LER, but an increasing LW as environmental or genetic conditions led to a higher LER (Fig. 1). The relationship also suggests that LW reached a maximum asymptote of about 10 mm for this genetic population, while LER under low stress environments appeared to reach a minimum asymptote of about 10 mm/day. The 3-fold change in LAE was much more closely associated with the 3-fold range in LER than with the 1.5-fold range for LW (Fig. 2, A and B). The genetic control mechanism of LW has not been studied in great detail, but LW is related positively to culm diameter

Table 1. Leaf elongation rate (LER), leaf width (LW), and leaf-area expansion rates (LAE) of populations selected for high (H) and low (L) LAE.

Population (year)	LAE mm²/day			LER mm/day			LW mm		
	High	Low	L/H	High	Low	L/H	High	Low	L/H
C_0 (1975)	157	157	–	18.6	18.6	–	8.2	8.2	–
C_1 (1976)	106	91	0.86	13.0	11.4	0.88	7.9	7.6	0.96
C_2 (1977)	315	217	0.69	31.1	23.0	0.74	10.0	9.2	0.92
C_3 (1978)	179	74	0.41	17.8	10.4	0.58	9.9	6.8	0.69
C_4 (1979)	181	74	0.41	19.0	11.6	0.61	9.3	6.1	0.65

(Nelson and Sleper 1977). Genetic increases in LER are associated with both increased rate of cell division and increased cell size (Volenec and Nelson 1981).

The C_0 population had an annual forage yield of 9.4 metric tons (mt)/ha that was similar to the yield of Mo-96 and Kenhy. Yield of Ky-31 was lower than that of the C_0 population. Four generations of selection for low LAE led to a 25% lower yield during the vegetative regrowth period. The response was due to a 2.1-fold increase in tiller density that was more than offset by a 2.8-fold decrease in weight/tiller. In contrast with selection for low LAE, selection for high LAE gave a 39% increase in yield by the H_2 population over the C_0 population due to increases of about 20% for both tiller density and weight/tiller. Yields of H_3 and H_4 populations were intermediate between the C_0 and H_2 populations. This may be due partially to inbreeding depression, but we have carefully monitored parental lineages. The L_4 and H_4 populations are represented by descendants from 18 (9 each) maternals of the 80 plants selected originally from the C_0 population. Selection had little if any effect on photo-

synthesis/unit leaf area as the H_4, L_4, and C_0 populations had rates of 21.8, 23.7, and 24.2 mg CO_2/dm²/hr, respectively.

Yield performance of the H_2 population appeared to be better than those of other populations. Since the variance is still high in all populations, they consist of plants with a range of tiller densities and tiller weights. Other research (Zarrough and Nelson, unpublished) shows a complementarity among plants of several sizes that needs consideration in breeding and management decisions. The H_2 populaton may represent an ideal balance of large and small tillers for forage production under the given cultural and environmental conditons.

LITERATURE CITED

Burton, G.W. 1974. Recurrent restricted phenotypic selection increases forage yields of Pensacola Bahiagrass. Crop Sci. 14:831–835.

Gardner, C.O. 1969. The role of mass selection and mutagenic treatment in modern corn breeding. Proc. 24th Corn and Sorghum Res. Conf., 15–21.

Horst, G.L., C.J. Nelson, and K.H. Asay. 1978. Relationship of leaf elongation to forage yield of tall fescue genotypes. Crop Sci. 18:715–719.

Nelson, C.J., K.H. Asay, and G.L. Horst. 1975. Relationship of leaf photosynthesis to forage yield of tall fescue. Crop Sci. 15:476–478.

Nelson, C.J., K.H. Asay, and D.A. Sleper. 1977. Mechanisms of canopy development of tall fescue genotypes. Crop Sci. 17:449–452.

Nelson, C.J., and D.A. Sleper. 1977. Morphological characters associated with productivity of tall fescue. Proc. XIII Int.

Fig. 1. Association of leaf width and leaf elongation rate over four cycles of selection for low and high rate of leaf-area expansion. Data points are identified by direction and cycle of selection.

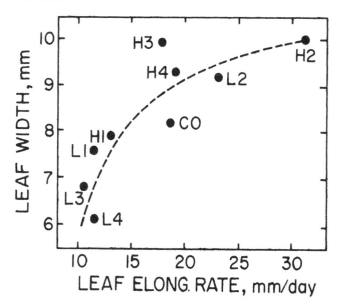

Fig. 2. Relationships between rate of leaf-area expansion and leaf width (A) and leaf elongation rate (B). Data points are identified by direction and cycle of selection.

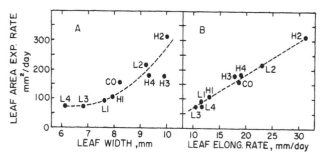

Grassl. Cong., Sect. 1, 13–19.

Robson, M.J. 1974. The effect of temperature on the growth of S 170 tall fescue (*Festuca arundinacea*, Schreb.) III. Leaf growth and tiller production as affected by transfer between contrasting regimes. J. Appl. Ecol. 11:265–279.

Sleper, D.A., C.J. Nelson, and K.H. Asay. 1977. Diallel and path coefficient analysis of tall fescue (*Festuca arundinacea*, Schreb.) regrowth under controlled conditions. Can. J. Genet. Cytol. 19:557–564.

Volenec, J.J., and C.J. Nelson. 1981. Cell dynamics in leaf meristems of contrasting tall fescue genotypes. Crop Sci. 21:381–385.

A Descriptive Scheme for Stages of Development in Perennial Forage Grasses

U. SIMON and B.H. PARK

Justus-Liebig-Universität Giessen, Federal Republic of Germany, and
Korean-German Grassland Research Project, O.R.D., Suweon, Republic of Korea

Summary

A descriptive scheme of morphological stages of development in forage grasses is urgently needed for scientific investigations as well as for practical purposes. It could be used for precise timing of management measures, such as application of fertilizers and herbicides, and in harvesting with respect to a certain level of forage yield and/or quality. Existing schemes, like the Feekes scale and its modifications, were devised for cereals. They do not adequately consider the peculiar stages of development of perennial grasses grown for forage.

We propose a scheme that consists of consecutive stages of the following phases of plant development: number of leaves, elongation of leaf sheaths, stem elongation, inflorescence emergence, anthesis, and seed ripening. In an attempt to standardize the schemes, we have adopted the 2-digit code and wherever appropriate have maintained the stages and terms of the scale of Zadoks et al. (1974). The complete descriptive scheme is given in the paper.

The association between a certain stage of growth and characteristics of nutritive value depends on the earliness of species and cultivars. At a fixed stage of plant development, nutritive value in terms of in-vitro digestibility, crude-fiber content, and crude-protein content is generally lower in late types than in early ones. Inversely, a certain level of nutritive value is reached in late types at an earlier stage of growth than in early ones.

KEY WORDS: descriptive scheme, development, forage, grass, growth stage, quality, yield.

INTRODUCTION

A descriptive scheme for classification of stages of development of forage grasses is urgently needed. It is a prerequisite for the precise timing of management practices, such as application of fertilizers and herbicides, or for determination of the exact time of grazing or cutting at a certain level of yield and/or nutritive quality. A system of sequential utilization of forage cultivars within their optimum utilization period, such as that proposed by Simon (1978), would also require an easily applicable growth scale. Feekes (1941) designed a scheme to record growth stages of wheat (*Triticum aestivum* L.). Zadoks et al. (1974) modified and adapted it to electronic data processing. A revised version of the Zadoks et al. scale was officially adopted for cereals in the Federal Republic of Germany

(Biologische Bundesanstalt 1979, Bundessortenamt 1979). In principle, the Zadoks et al. scale should be applicable to all Gramineae. Our experience, however, has been that this and other scales do not adequately consider the peculiar stages of development of perennial forage grasses.

Park (1980) studied the phenological changes of morphological characters during the ontogenesis of early and late cultivars of six important perennial grass species and proposed a scale for stages of primary growth. A generalized version on the basis of the Zadoks et al. coding system is presented here.

PROPOSAL OF A DESCRIPTIVE SCHEME FOR THE STAGES OF DEVELOPMENT IN PERENNIAL FORAGE GRASSES

Developing the Scheme

The variability at a certain stage of growth is much larger in cross-pollinating forage grasses than in cultivars

U. Simon is a Professor of Agronomy at Justus-Liebig-Universität and B.H. Park is a Research Agronomist at the Korean-German Grassland Research Project, O.R.D.

Table 1. Descriptive scheme for the stages of development in perennial forage grasses.

Code	Description	Remarks
	Leaf development	Applicable to regrowth of established
11	First leaf unfolded	(plants) and to primary growth of seed-
12	2 leaves unfolded	lings. Further subdivision by means of leaf
13	3 leaves unfolded	development index (see text).
•	• • • • • • • • •	
19	9 or more leaves unfolded	
	Sheath elongation	Denotes first phase of new spring growth
20	No elongated sheath	after overwintering. This character is
21	1 elongated sheath	used instead of tillering which is difficult
22	2 elongated sheaths	to record in established stands.
23	3 elongated sheaths	
•	• • • • • • • • •	
29	9 or more elongated sheaths	
	Tillering (alternative to sheath elongation)	Applicable to primary growth of seedlings
20	Main shoot only	or to single tiller transplants.
21	Main shoot and 1 tiller	
22	Main shoot and 2 tillers	
23	Main shoot and 3 tillers	
•	• • • • • • • • •	
29	Main shoot and 9 or more tillers	
	Stem elongation	
31	First node palpable	More precisely an accumulation of nodes
32	Second node palpable	Fertile and sterile tillers distinguishable.
33	Third node palpable	
34	Fourth node palpable	
35	Fifth node palpable	
37	Flag leaf just visible	
39	Flag leaf ligule/collar just visible	
	Booting	
45	Boot swollen	
	Inflorescence emergence	
50	Upper 1 to 2 cm of inflorescence visible	
52	1/4 of inflorescence emerged	
54	1/2 of inflorescence emerged	
56	3/4 of inflorescence emerged	
58	Base of inflorescence just visible	
	Anthesis	
60	Preanthesis	Inflorescence-bearing internode is visible. No anthers are visible.
62	Beginning of anthesis	First anthers appear.
64	Maximum anthesis	Maximum pollen shedding.
68	End of anthesis	No more pollen shedding.
	Seed ripening	
75	Endosperm milky	Inflorescence green.
85	Endosperm soft doughy	No seeds loosening when inflorescence is hit on palm.
87	Endosperm hard doughy	Inflorescence losing chlorophyll; a few seeds loosening when inflorescence hit on palm.
91	Endosperm hard	Inflorescence-bearing internode losing chlorophyll; seeds loosening in quantity when inflorescence hit on palm.
93	Endosperm hard and dry	Final stage of seed development; most seeds shed.

of self-pollinating cereals. The records of growth stages are, therefore, estimates of means derived from samples. The samples were prepared as follows: bunches of about 20 tillers each were cut close to the surface from four different sites and then combined; the mean of the 20 most advanced tillers provided the growth stage. The letter-number coding system of the original Park scale was transformed to the corresponding 2-digit code of the Zadoks et al. scheme. Wherever appropriate, the Zadoks et al. stages and descriptions were maintained (Table 1). The germination stage was omitted.

Stages of leaf development may be further subdivided by means of the aeropetal leaf-development index (LDI) (Haun 1973) as modified by Park (1980). For example, LDI 3.6.1 denotes 3 basal leaves unfolded, lamina of fourth leaf six-tenths emerged, lamina of fifth leaf one-tenth emerged. Stages of tillering are not exactly discernible in an ordinary grass sward. We use sheath elongation instead. Alternatively, in seedling stands tillering may be recorded. The first stages of stem elongation are characterized by palpable nodes. The first node is actually an accumulation of nodes, the number of which is equal to the number of nodes of the fully developed culm. One indication of late stages of seed ripening is seed shattering when the inflorescence is hit on the palm. Seed shattering would not apply to nonshattering species like Kentucky bluegrass (*Poa pratensis* L.). It should be stressed that not all consecutive stages considered in the scale must necessarily occur in any cultivar (Table 2).

Growth Stage and Nutritive Value

Analysis of the relationships between growth stages and constituents of nutritive value in ten cultivars of cocksfoot (*Dactylis glomerata* L.) and timothy (*Phleum pratense* L.) were

conducted by Simon and Daniel (1981). From their data, of which examples are presented in Table 2, it is concluded that there is no general association between nutritive values and certain stages of growth. At a fixed stage of plant development, the nutritive value as measured by in-vitro digestiblity of organic matter, crude fiber, and crude protein is generally lower in the relatively late timothy cultivar than in the relatively early cocksfoot cultivar. Inversely, a certain level of nutritive value is reached in a late type at an earlier stage of growth than in an early one.

LITERATURE CITED

Biologische Bundesanstalt für Land- and Forstwirtschaft. 1979. Ent-wicklungsstadien bei Getreide—ausser Mais. Merkblatt Nr. 27(1):1-18.

Bundessortenamt. 1979. Entwicklungsstadien bei Getreide—ausser Mais. Blatt für Sortenwesen 12, 79.

Feekes, W. 1941. De tarwe en haar milieu. Versl. XVII Techn. Tarwe Comm. 12:560-561.

Haun, J.R. 1973. Visual quantification of wheat development. Agron. J. 65:116-119.

Park, B.H. 1980. Untersuchungen zum Entwicklungsverlauf im Primäraufwuchs von perennierenden Futtergräsern. Dissertation, Univ. Giessen.

Simon, U. 1978. Sequential utilization of grass cultivars for optimum grassland production. Proc. 7th Gen. Mtg. Eur. Grassl. Fed., Gent. 1:233-240.

Simon, U., and P. Daniel. 1981. Knaulgras und Lieschgras. *In* P. Daniel, M.L. Rotermund, U. Simon, M. Wermke, and E. Zimmer (eds.), Die in vitro-Verdaulichkeit als Qualitätsmerkal bei der Sortenbeurteilung von Futtergräsern. Landbauforschung Völkenrode, 1981, 12-42.

Zadoks, J.C., T.T. Chang, and C.F. Konzak, 1974. A decimal code for the growth stages of cereals. Eucarpia Bul. 7, 42-52.

Table 2.   Relationships between stages of growth and crude fiber (CF), in-vitro digestibility of organic matter (IVD), and crude protein (CP).

Date, 1978	Code	Cocksfoot CF	Cocksfoot IVD	Cocksfoot CP	Code	Timothy CF	Timothy IVD	Timothy CP
		---- % ----				---- % ----		
25 Apr	22	17.4	82.9	27.2	22	16.6	83.9	27.9
28 Apr	22	18.5	81.2	25.7	22	17.5	82.7	26.8
2 May	23	19.8	78.9	23.8	22	18.7	81.1	25.3
5 May	31	20.9	77.2	22.3	23	19.7	79.9	24.2
9 May	32	22.2	74.9	20.4	23	20.9	78.3	22.7
12 May	32	23.3	73.2	19.0	24	21.8	77.1	21.6
16 May	45	24.7	70.9	17.0	31	23.0	75.5	20.1
19 May	52	25.7	69.2	15.6	32	23.9	74.3	19.0
23 May	53	27.1	67.0	13.7	32	25.2	72.7	17.5
26 May	55	28.1	65.3	12.2	32	26.1	71.6	16.4
30 May	58	29.5	63.0	10.3	33	27.3	70.0	15.0
2 June	60	30.5	61.3	8.8				
6 June					34	29.4	67.2	12.4
9 June					45	30.4	66.0	11.3
13 June					45	31.6	64.4	9.8
16 June					54	32.5	63.2	8.7

Effects of Maturity and Environment on Amount and Composition of Hemicellulose in Tall Fescue

S.L. FALES, D.A. HOLT, C.S.T. DAUGHTRY, and V.L. LECHTENBERG

Purdue University, West Lafayette, Ind., U.S.A.

Summary

Seasonal changes in amount, composition, and digestibility of fiber in vegetative tissues of forage grasses are usually associated with advancing tissue maturity. Our objective was to determine the relative effects of maturity and environment on amount and composition of hemicellulose in field-grown tall fescue (*Festuca arundinacea* Schreb.). Regrowth of identical chronological age was sampled weekly at different times during the growing season. Leaf samples were extracted with neutral detergent solution to isolate cell walls (NDF). The NDF was treated with 1.0 H_2SO_4 to remove and hydrolize hemicellulose. The hydrolyzates were analyzed for concentrations of xylose and arabinose, using high performance liquid chromatography.

Hemicellulose concentration decreased during regrowth in early summer, remained about the same during midsummer, and increased in late summer. Conversely, xylose concentration in hemicellulose increased during early summer and decreased during late summer, while arabinose concentration remained about the same throughout the season. These results indicate that the effect of time of season was greater than that of tissue age and imply that environment influenced the composition of the cell wall. Using growth data, weekly changes in hemicellulose composition of new growth were calculated. Differences due to season were greater than indicated by the whole-plant data, suggesting that composition of the newly synthesized cell wall changed considerably during the growing season, conforming more closely to trends in photoperiod than in temperature.

KEY WORDS: tall fescue, *Festuca arundinacea* Schreb., hemicellulose, maturity, environment, xylose, temperature, photoperiod.

INTRODUCTION

Forage grass fiber ranges from 20% to 80% digestible by ruminants. Digestibility decreases with increases in levels of cellulose, hemicellulose, and lignin (Waite 1970) and with increases in xylose:arabinose ratio in hemicellulose (Daughtry et al. 1978). Tissue age and temperature were positively correlated with these chemical changes in studies by Waite et al. (1964) and Wilson and Ford (1971). The purpose of this research was to determine the extent to which environmental variables interact with plant maturation in causing these changes.

MATERIALS AND METHODS

Two-year-old tall fescue (*Festuca arundinacea* Schreb.) located at the Purdue University Agronomy Farm, West Lafayette, Indiana, on a Russell silt loam (a fine-silty, mixed, mesic, Typic Hapludalf) was fertilized on 17 April 1979 with urea (45-0-0) at the rate of 112 kg N/ha. On 16 May, 14 June, and 19 July, strips 1.8 m wide × 6 m long were cut 8 cm from the soil surface with a flail-type forage harvester. Three replicates were cut on each date.

After each cut, the plots were divided into eight 0.75 ×

1.8-m subplots for serial sampling. At approximately 1-week intervals following each cut, height measurements were taken of regrowth within one subplot by measuring from the soil surface to the tips of extended leaves, and samples of regrowth above 8 cm from the soil surface were collected from three random locations within each subplot. Each subplot was sampled only once during the experiment. Immediately after removal from the field, all samples were frozen at −28°C and were later lyophilized and ground in a Wiley mill, using a 1-mm screen.

Neutral detergent fiber (NDF) (Goering and Van Soest 1970) from duplicate subsamples of dry, ground material was refluxed for 1 hour with 1.0 N H_2SO_4 to extract and hydrolyze hemicellulose. Xylose and arabinose in the barium carbonate–neutralized extracts were separated and measured using high-performance liquid chromatography (Ladisch and Tsao 1978).

The three growth periods, early, middle, and late season, were considered to be the whole plots of a split-plot experiment, and the weekly samplings were subplots arranged in a randomized complete-block design. All factors except blocks were assumed to be fixed. Trends and differences described in the next section were significant at $P \leqslant 0.05$.

S.L. Fales is a Postdoctoral Associate at the University of Georgia Experiment Stations; D.A. Holt is a Professor of Agronomy, C.S.T. Daughtry is a Research Agronomist (Laboratory for Applications of Remote Sensing), and V.L. Lechtenberg is a Professor of Agronomy at Purdue University. This paper is a con-

tribution of the Department of Agronomy of Purdue University (Journal Paper No. 8515) and of Purdue University Agricultural Research Station.

The authors wish to state that mention of a proprietary product does not imply endorsement of it.

RESULTS AND DISCUSSION

Hemicellulose concentration in the cell wall (HC/NDF) was generally highest in early-summer regrowth, and it decreased over time. It was lowest in late-summer regrowth but increased slightly over time. Hemicellulose percentage in midsummer regrowth remained about the same for most of the period of study but declined between the 6th and 7th week.

Xylose concentration in hemicellulose (XYL/HC), relatively low in early-summer regrowths, increased during 7 weeks, whereas xylose in late-summer regrowth decreased over time. Xylose concentration in midsummer regrowths was intermediate between the early- and late-summer concentrations. No significant changes in arabinose concentration in hemicellulose were observed. Because hemicellulose composition and trends were related more to time of season than to chronological age, it appears that environment was having an important effect, the mechanisms of which have been suggested by Wilson et al. (1971, 1976).

Our sampling procedure was such that each weekly sample contained all the tissue synthesized during the week prior to the sampling date plus all the tissue synthesized prior to that since the original cutting date. The composition of fiber produced early in the regrowth period was modified by meristematic activity, secondary cell-wall production, and lignification occurring under more recent environmental conditions.

We estimated composition of that portion of total fiber in each sample synthesized in each successive week of regrowth. Calculations were based on the assumption that once cellulose, hemicellulose, and lignin molecules were synthesized, that particular fiber did not change in chemical composition. Any subsequent changes in fiber composition observed were assumed to be due to the addition of more cellulose, hemicellulose, and lignin molecules during subsequent growth. The question was, if leaf-fiber composition at time 1 is measured, what must be the composition of fiber synthesized between time 1 and time 2 in order to result in the leaf-fiber composition observed at time 2? The final concentration can be expressed as:

$$C_2 = \frac{W_1C_1 + (W_2 - W_1)C_{2-1}}{W_2}$$

where C_1 and C_2 = concentrations of constituent at time 1 and time 2, respectively; C_{2-1} = concentration of constituent in fiber synthesized between time 1 and time 2; W_1 and W_2 = estimated weights of fiber in the sample at time 1 and time 2, respectively.

From this equation, C_{2-1}, the concentration of a constituent in the most recently synthesized fiber, can be calculated as:

$$C_{2-1} = \frac{W_2C_2 - W_1C_1}{W_2 - W_1}$$

We assumed that weight changes were proportional to measured height changes (Whitney 1974). Growth

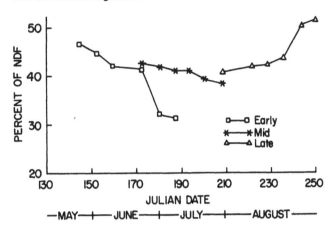

Fig. 1. Estimated trend in concentration of hemicellulose in newly synthesized cell wall (NDF) in early, mid, and late summer tall fescue regrowth.

(weight or height) was expressed as a fraction of total growth during regrowth to final sampling. Relative weight was then converted to units of NDF, hemicellulose, or xylose by multiplying by the appropriate measured concentrations. Relative units of fiber constituents were then plotted against time and smoothed within each regrowth period by linear regression. New data points were taken from the linear regressions and used to calculate the concentrations of constituents in fiber synthesized during each 1-week period, using the equations described above.

The trends (Figs. 1 and 2) are similar to those described for HC/NDF and XYL/HC, but the magnitude of differences is much greater. Under the assumptions described above, hemicellulose in newly synthesized cell walls clearly decreased during early-summer and midsummer regrowths and increased during late summer (Fig. 1). Xylose concentration in hemicellulose increased steadily through May and June in early-season growth

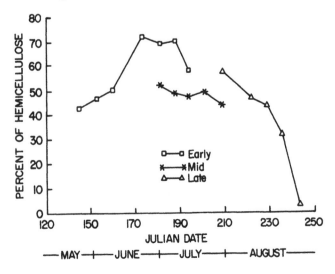

Fig. 2. Estimated trend in concentration of xylose in newly synthesized hemicellulose in early, mid, and late summer tall fescue regrowth.

and then declined in July, whereas xylose declined at a slow rate in midsummer growth but declined quite rapidly in late-summer growth (Fig. 2).

These results suggest that amount and composition of hemicellulose in grass tissue are related more to time of season than to maturity or tissue age. We could find no consistent relationship between mean air temperatures in the field and hemicellulose composition. However, the trends shown in Figs. 1 and 2 conform somewhat to the trend in photoperiod. This correlation might justify other studies to determine the effect of photoperiod on fiber composition of grasses.

LITERATURE CITED

Daughtry, C.S.T., D.A. Holt, and V.L. Lechtenberg. 1978. Concentration, composition and *in vitro* disappearance of hemicellulose in tall fescue and orchardgrass. Agron. J. 70:550–554.

Goering, J.K., and P.J. Van Soest. 1970. Forage fiber analysis (apparatus, reagents, procedures, and some applications).

U.S. Dept. Agric., Agric. Handb. 379.

Ladisch, M., and G. Tsao. 1978. Theory and practice of rapid liquid chromatography at moderate pressures using water as eluent. J. Chromatog. 166:85–100.

Waite, R. 1970. The structural carbohydrates and the *in vitro* digestibility of a ryegrass and a cocksfoot at two levels of nitrogenous fertilizer. J. Agric. Sci., Camb. 74:457–462.

Waite, R., M. Johnston, and D.G. Armstrong. 1964. The evaluation of artifically dried grass as a source of energy for sheep. I. The effect of stage of maturity on the apparent digestibility of ryegrass, cocksfoot, and timothy. J. Agric. Sci. 62:391–398.

Whitney, A.S. 1974. Measurement of foliage height and its relationship to yields of two tropical forage grasses. Agron. J. 66:334–336.

Wilson, J.R., and C.W. Ford. 1971. Temperature influences on the growth, digestibility, and carbohydrate composition of two tropical grasses, *Panicum maximum* var *Trichoglume* and *Seteria sphacelata,* and two cultivars of the temperate grass *Lolium perenne.* Austral. J. Agric. Res. 22:563–571.

Wilson, J.R., A.O. Taylor, and G.R. Dolby. 1976. Temperature and atmospheric humidity effects on cell wall content and dry matter digestibility of some tropical and temperate grasses. N.Z. J. Agric. Res. 19:41–46.

Effect of Active Roots of Forage Grass on Soil-Aggregate Formation

H. KOBAYASHI and I. YAMANE
Kitasato University and Tokyo University of
Agriculture and Technology, Japan

Summary

Half-rotted older roots and active younger ones both are present in abundance in forage grass, and both types assist in soil aggregate formation in their own rhizosphere. Italian ryegrass (*Lolium multiflorum* Lam.) was grown in quartz sand for 12 weeks in glass boxes placed in phytotron. In experiment 1, three boxes were opened after 4, 8, and 12 weeks of growth. Root weight was measured, and quantity of sand adhering to the roots was weighed. One hundred times more sand was found adhering to roots after 12 than after 4 weeks. When converted to root basis, weights at 8 weeks and 12 weeks both were around 2 g, or 10 times greater than at 4 weeks.

In experiment 2, the distribution of pectic substance on root surfaces and points in contact with the surrounding sand was ascertained by application of ruthenium-red solution. Adhesive strength of the pectic substance was measured using glass plates and was found to be 15–20g/single adventitious root. Total adhesive strength of the roots of each plant increased in proportion to root number. From the experiments, it was concluded that active forage grass roots excrete a pectic substance causing adhesion of soil particles in the surrounding area, acting as a first step in anchoring the plant. As the active root system expands, the quantity of pectic substance excreted increases rapidly, and the adhesive action affects a wider area, thus leading to aggregate formation.

KEY WORDS: Italian ryegrass, *Lolium multiflorum* Lam., root, soil aggregates, adhesive action, active roots.

INTRODUCTION

It is well known that roots are more abundant in permanent pasture than in other types of fields. It was reported in the case of Kuroboku soil of Japan that the dry weight of roots found between 0 and – 45 cm (i.e., 45 cm below soil surface) was 7.12, 6.31, 1.55, 1.40, and 0.44 metric tons/ha for orchardgrass (*Dactylis glomerata*), Italian ryegrass (*Lolium multiflorum* Lam.), Ladino clover (*Trifolium repens giganteum*), upland rice, wheat (*Triticum aestivum*), and soybeans (*Glycine soja*), respectively (Yamane 1963). Roots in permanent pasture consist of half-rotted older roots and active younger ones. In Japanese permanent pastures, the usual ratio of these at a depth of 0-10 cm is 2:1.

In the formation of soil aggregates, the half-rotted roots provide a continual supply of organic matter that serves as binding material of soil particles. Active roots make an aggregate more stable by, for example, partially drying it out through root networks surrounding the particles (Russell 1973, Kobayashi 1977).

Active roots also excrete organic compounds into the surrounding soil (Rovira 1962). Root cap cells and the epidermal cells (root hair) underneath them excrete an adhesive substance considered to be a type of polysaccharide (Kobayashi and Suzuki 1979). However, attention has not been paid to the role of the adhesive substance excreted from active roots in the formation of soil aggregates.

Two kinds of experiments were carried out to investigate soil aggregation by this adhesive substance.

H. Kobayashi is a Professor of Grassland Science at Kitasato University and I. Yamane is a Professor of Agronomy at Tokyo University of Agriculture and Technology.

Mention of a proprietary product does not imply endorsement of it.

METHODS AND MATERIALS

Washed particles of sand and quartz were sieved through a 1-mm mesh and then placed in glass boxes. Two sizes of boxes were used according to the cultivation period, one 30 cm deep × 30 cm long × 0.5 cm wide, and the other 40 × 30 × 2.5 cm. Week-old seedlings of Italian ryegrass were transferred from petri dishes to the boxes, one seedling/box, and the medium was soaked with liquid nutrient. Seedlings then grew for 12 weeks inside phytotron at 15°C, 55% relative humidity, and 7 hours of artificial lighting (30,000 lux) daily.

To measure adhesion of particles, three boxes were opened after 4, 8, and 12 weeks, respectively. Plants were grasped at the base and carefully uprooted. The quantity of sand adhering to the roots was weighed, and the adhesive strength of each plant was determined. Methylene blue liquid dye was used to distinguish roots from particles, and the manner of adhesion was observed by projector and stereoscopic microscope.

To measure adhesion of particles brought about by excreted substances, ruthenium-red solution was applied to the adventitious roots of growing plants to locate pectic substances. Distribution of excreted substance around the roots in the medium was observed by microscope. Next, the adhesive strength of the active roots was determined. The main shoot and tillers of each plant were then separated and left on glass plates (2 × 4 cm) kept moist to prevent the roots from drying out. After 2 days, the force needed to detach the roots from the plate was measured with a Jorry's spring balance.

RESULTS

Growth of plants was normal, with tiller numbers and general above-ground growth satisfactory (Table 1). A 10-fold increase occurred in number of roots/plant be-

Table 1.   Top and root of Italian ryegrass, and weight of catching sand by roots, at 4, 8, and 12 weeks after sowing. (mean ± S. D.)

	Weeks after sowing		
	4	*8*	*12*
Forage grass, top			
Tillers/plant (no.)	1.7±0.3	11.2±1.0	32.0±4.4
Maximum leaf length (cm)	17.1±2.0	33.0±3.1	42.8±4.6
Forage grass, roots			
Mean length (cm)	6.1±0.9	16.7±1.1	21.8±1.0
Maximum length (cm)	9.0±1.0	30.9±1.7	47.3±1.6
Total/plant (no.)	10.6±1.2	34.8±3.2	113.5±23.4
Dry weight (mg/plant)	6.8±2.2	234.9±46.1	1041.4±184.4
Catching sand			
Total weight (g/plant)	2.8±0.9	82.2±18.2	258.2±44.2
Weight/root (g)	0.2±0.6	2.2±0.5	2.3±0.4
Sample numbers	16	10	13

tween week 4 and week 12. The quantity of particles adhering to the roots of each plant increased in proportion to the increase in the number of roots, with the amount at 12 weeks 100 times greater than that at 4 weeks. However, the quantities adhering to each individual root at 8 weeks and 12 weeks were almost identical, both being around 10 times greater than at 4 weeks. This fact suggests that after the initial 4-week period, there is a limit to the adhesive strength of individual roots.

The distribution of pectic substance around roots was ascertained by dyeing the epidermis and root hairs of active roots and the contact points between soil and roots with ruthenium-red solution. The overall adhesive strength of roots of each plant increased exponentially with growth, while that of each adventitious root changed significantly beween week 4 and week 11 (Table 2). These data suggest that the increase in adhesive strength associated with growth of the plant is primarily attributable to the greater number of roots.

In fully grown forage grasses, root systems consist of large numbers of almost inactive aged roots as well as branch roots having highly active root hairs. Therefore, a relationship between the growth of active branched roots and adhesive strength was examined (Fig. 1). A close relationship existed between variation in the adhesive

Table 2. Total root numbers and root-adhesion strength of Italian ryegrass at various growth periods (mean±S.D.).

Weeks of growth	Total root numbers	Total adhesion strength, (g)	Adhesion strength/root (g)
4	12.0±0.7	195.5±10.6	15.9±1.6
6	19.3±0.9	265.8±14.9	13.7±4.4
8	39.7±0.3	760.8±38.0	19.1±3.0
11	139.3±27.0	3958.9±101.8	18.4±1.7

strength of each adventitious root and total length of branch roots. Therefore, it can be said that branch roots are associated with adhesive strength.

DISCUSSION

On the basis of the results of the preceding experiments active roots are considered to function as follows. The pectin-like substance excreted from growing juvenile roots causes adhesion of the soil particles in the surround-

Fig. 1. Distribution of branch roots and adhesive strength at 11 weeks' growth of Italian ryegrass.

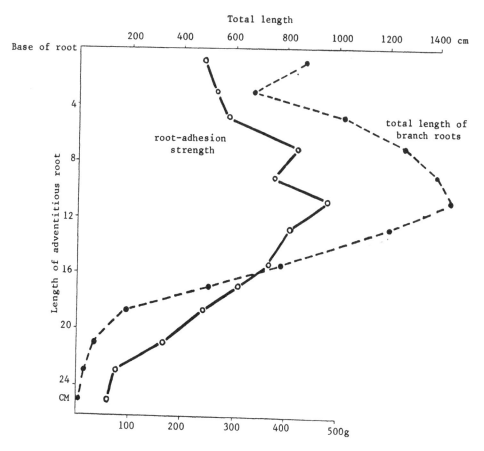

ing area, acting as a first step in anchoring the plant. As the active root system expands, the quantity of pectin-like substance excreted increases rapidly, and its adhesive action affects a wider area. At the same time water absorption, carried out chiefly by the root hairs of active roots, causes localized drying out of soils, and then the expanded root system compacts the surrounding soils. As a result soil aggregates become very stable against water.

However, the present work does not eliminate the possibility that half-rotted older roots and microorganisms may provide cementing matter for aggregate formation. Also, soil-inhabiting creatures and mycelium of microorganisms may make aggregates more stable against water.

LITERATURE CITED

Kobayashi, H. 1977. The influence of forage roots on soil water movement in grassland soil. (In Japanese.) J. Japan. Soc. Irrig. Drain. Reclam. Engin. 45:155–158.

Kobayashi, H., and N. Suzuki. 1979. Studies on polysaccharide-exudates with the advance of forage root growth. (In Japanese with English summary.) J. Japan. Grassl. Sci. 25:222–230.

Rovira, A.D. 1962. Plant-root exudates in relation to the rhizosphere microflora. Soils and Fert. 25:167–172.

Russell, E. W. 1973. Soil conditions and plant growth, 10th ed., 479–554. Longman, London.

Yamane, I. 1963. Soil and plant (6). (In Japanese.) J. Anim. Husb. 17:881–884.

Phenology and Culm-Weight Variation Between Two Wheatgrass Species: *Agropyron smithii* and *A. spicatum*

M.K. OWENS and H.G. FISSER
University of Wyoming, Laramie, Wyo., U.S.A.

Summary

Knowledge of plant phenology and live weight dynamics is necessary for intensive management of semiarid rangelands. Thirteen sites in Wyoming were sampled for plant phenological development and culm weights of western wheatgrass (*Agropyron smithii*) and bluebunch wheatgrass (*A. spicatum*). These species are sympatric and allopatric within these rangelands. Such a distribution allows site comparisons between these species. Environmental data were collected for use in differentiating sites.

Data on all variables were collected at 14-day intervals throughout the growing seasons of 1979 and 1980. Permanently marked plants were sampled for phenology, and plants of similar development were collected for culm-weight estimates. Cumulative precipitation, degree days, soil water potentials at three depths (0–15, 16–30, and 46–60 cm), soil temperature at 25 cm, and Julian dates were recorded.

A two-way analysis of variance performed on culm-weight and phenological data identified nine groups of sites. For culm weights, western wheatgrass formed four distinct groups, bluebunch wheatgrass formed two distinct groups, and both species together formed three groups. Phenological data revealed the same groupings of sites.

Discriminant analysis of micrometeorological variables showed that western wheatgrass produced the greatest culm weights at moderate amounts of degree-day accumulation and cumulative precipitation. Bluebunch wheatgrass culm weights were greatest at higher values of both variables. Phenological development was species related, with bluebunch wheatgrass developing more rapidly. Phenology on sites with low culm weights developed more rapidly than on sites with high culm weights.

Micrometeorological variables were utilized to distinguish sites with bluebunch wheatgrass and western wheatgrass according to culm weights and phenological development. Dominant factors affecting the species were Julian date, degree-day accumulation, and cumulative precipitation to the sampling site. These variables were able to account for 91% of between-site variability in culm weights and phenology.

KEY WORDS: western wheatgrass, bluebunch wheatgrass, phenology, culm weight.

INTRODUCTION

Two sympatric and allopatric perennial grasses on the semiarid rangelands of Wyoming are bluebunch wheatgrass (*Agropyron spicatum*) and western wheatgrass (*A. smithii*). To manage the forage resource represented by these species intensively, specific knowledge of their phenological and live-weight dynamics is necessary. Objectives of this study were to quantify the rate and degree of phenological development and live weight/culm. Environmental data were examined to account for differences in these responses.

METHODS

Thirteen study sites were located on a wide array of range sites in central and western Wyoming. Seven sites contained western wheatgrass, three sites contained bluebunch wheatgrass, and three sites contained both species. Phenology and culm-weight data were collected at 14-day intervals throughout the growing season. Twenty plants of each species were permanently located for phenological observation. The phenological scoring system, similar to that of West and Wein (1970), has been published elsewhere (Fisser et al. 1980a, 1980b). Separate plants representing the mean phenological stage were collected for culm-weight estimates. Samples were clipped at ground level, oven-dried at 65°C for 24 hours, and weighed to the closest 1 mg.

Environmental parameters were selected for sampling based upon their potential effects on plant growth. The micrometeorological characteristics were cumulative precipitation to the sampling date; average soil water potential at 0–15 cm, 16–30 cm, and 46–60 cm; soil temperature at 25 cm; degree-day accumulation; and Julian date.

Data analysis consisted of a two-way analysis of variance to determine significant differences in phenological development and culm weight. A multiple comparison was calculated via Duncan's New Multiple Range Test to identify significant differences between sites (Ott 1977). All statistical tests were calculated with $\alpha = 0.05$. Similar sites were arranged into groups for further analysis. A multiple discriminant analysis identified significant differences in environmental parameters between the groups of sites (Lindeman et al. 1980).

RESULTS

Because of an unequal number of sites for each species, three separate analyses were conducted on the peak standing-crop value for each species. The western wheatgrass analysis involved all ten sites on which this species was found. Four distinct groups of sites (A, B, C, D) were

Table 1. Group culm weights and phenology dates.

Species	Group	Mean culm weight (mg)	Mean Julian date of maximum vegetative phenology[1]
Western wheatgrass	A	68	203
	B	99	210
	C	125	217
	D	157	226
Bluebunch wheatgrass	F	88	195
	G	170	207
Both species	A	68	203
	E	151	198
	F	88	195

[1]The maximum vegetative phenology score is represented by the value before vegetative browning.

determined to be significantly different. The bluebunch wheatgrass analysis involved all six sites that contained this species. Two distinct groups of sites (F, G) were statistically separated. The final analysis involved the three sites that had both species present. Three distinct groups were recognized. The first corresponds to group A, the second to group E, and the third to group F.

Phenological data were analyzed by time and site. Groups of sites identified as similar were the same as the groups for the live weight/culm analysis (Table 1). The following analyses will, therefore, be based upon the culm-weight groups.

Analysis of micrometeorological variables revealed three significant discriminant functions, all of which were complex. The functions consist of factor-loading scores for each of the independent variables (Table 2). Although each function was a cumulative effect due to all of the variables, each was dominated by the variables with the largest absolute values.

Julian date, as expected, was a dominant factor in all functions. The first function was dominated by degree-day accumulation, the second by culm weight, and the third by cumulative precipitation. These variables were significant when their interactions with the remaining variables were considered. The combined functions accounted for 91% of between-site variation in culm weight and phenology.

DISCUSSION

In two (A, F) of the three groups that contained both species, culm weight was significantly greater for bluebunch wheatgrass than for western wheatgrass. In the remaining group (E), no difference was distinguished. These three groups (A, E, and F) reflected the entire range of culm weights for both species. Therefore, on similar sites within semiarid rangelands, bluebunch wheatgrass had a significantly higher culm weight than western wheatgrass.

M.K. Owens is a Research Assistant and H.G. Fisser is a Professor of Range Management at the University of Wyoming.

Table 2.   Standardized discriminant function coefficients for micrometeorological variables.

Variable	Function		
	1	2	3
Cumulative precipitation	.683	−.284	1.483
Soil-water potential at 0–15 cm	.044	−.118	−.021
Soil-water potential at 16–30 cm	.302	−.160	.039
Soil-water potential at 46–60 cm	.086	.108	.282
Soil temperature at 25 cm	−.555	−.173	−.077
Degree-day accumulation	3.857	−.008	−.165
Julian date	−4.027	−.366	−.665
Culm weight	.189	1.305	−.222
Percent of variability	51	25	15

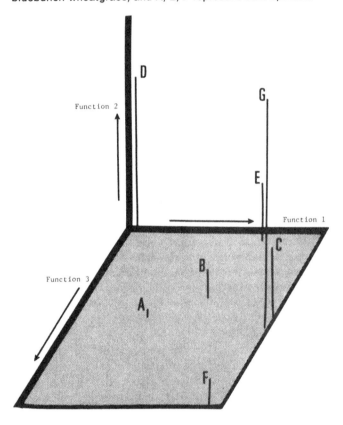

Fig. 1. Discriminant analysis demonstrating differences in groups based upon micrometeorological variables. Function 1: Degree days to Julian date. Function 2: Culm weight to Julian date. Function 3: Cumulative precipitation to Julian date. A through D represent western wheatgrass; F and G represent bluebunch wheatgrass; and A, E, F represent both species.

Bluebunch wheatgrass developed more rapidly phenologically than western wheatgrass in all of the groups. In addition, phenological development was accelerated in the low-culm-weight groups. Thus, culm weights were related to species and inversely related to rate of phenological development.

Differences in culm weight and phenology were examined by a discriminant analysis of environmental variables. Analysis of micrometeorological variables was performed on the groups designated significant by the multiple comparison. Results of the analysis are portrayed graphically in Fig. 1. Degree-day accumulation and cumulative precipitation were dominant factors in two disciminant functions; culm weight was dominant in the other. These variables are identified as function 1, 3, and 2, respectively. Culm weight was dominant because every measurement was partially dependent on the previous measurement.

Groups with greatest culm weights for each species (D, G) were considered the optimum groups for this analysis. Comparisons of other groups with these reference groups allowed differentiation of micrometeorological variables. Analysis also demonstrated that soil water potential at 46–60 cm was highest in these groups. Plant water stress should consequently have been lower, further indicating a favorable environment.

In western wheatgrass groups, the greatest culm weight was produced by group D at low levels of degree days and cumulative precipitation. This may be a reflection of the water-use capability of this species. Soil water potential was highest for this group. Group C had the highest

observations for degree days and cumulative precipitation but also had lower soil water potentials. The result was a lower culm weight than group D. Groups A and B had the same cumulative precipitation, but group A had a lower degree-day value. The culm weight was lower for group A, but soil water potential was higher. These data may reflect a degree-day threshold required for optimum growth.

Bluebunch wheatgrass groups also demonstrated the effect of varying degree-day and cumulative-precipitation value. Group E had a higher degree-day accumulation and lower cumulative precipitation than group G. These factors are expressed as lower soil water potential and lower culm weight. Group F had a lower degree-day value and a higher cumulative precipitation value than the optimum, which resulted in a lower culm weight. Culm-weight comparisons between groups E and F did not show a significant difference. This species tends to produce best on sites of moderate degree-day accumulation and cumulative precipitation.

Culm-weight differences of bluebunch and western

wheatgrass were separated on basis of climatic site characteristics. High production sites have degree-day and cumulative-precipitation combinations that result in higher soil water potentials. Regardless of species, the highest-culm-weight groups always had the highest water potential. Mean soil water potentials were higher at all depths for bluebunch wheatgrass sites than for western wheatgrass sites.

LITERATURE CITED

Fisser, H.G., M.K. Owens, J.W. Uhlich, W.J. Waugh, D.E. Rodgers, and N.E. Hargis. 1980b. Phenology and production studies on semi-arid shrub types. 1979 results. Vol. II: Precipitation, productivity and production section. Sect. III, IV, and V. Wyo. Agric. Expt. Sta. Sci. Rpt. 1106.

Fisser H.G., M.K. Owens, W.J. Waugh, N.E. Hargis, and J. Uhlich. 1980a. Phenology and production studies on semi-arid shrub types: 1979 results. Vol. I: Phenology study section. Univ. of Wyo. Coop. Res. Rept. to BLM. Wyo. Agric. Expt. Sta. Sci. Rpt. 1044.

Lindeman, R.H., P.F. Merenda, and R.Z. Gold. 1980. Introduction to bivariate and multivariate analysis. Scott, Foresman and Co., Dallas, Texas.

Ott, L. 1977. An introduction to statistical methods and data analysis. Duxbury Press, N. Scituate, Mass.

West, N.E., and R.W. Wein. 1970. A plant phenological index technique. Bioscience 21:116–117.

A Study of the Ecology of Yang-cao (*Leymus chinensis*) Grassland in Northern China

ZHU TING-CHENG, LI JIAN-DONG, and YANG DIAN-CHEN

Institute of Grassland Science, Northeastern Normal University,
Changchun, People's Republic of China

Summary

Leymus chinensis grasslands are distributed over the USSR, the People's Republic of Mongolia, and the People's Republic of China. This article presents three ecological characteristics of this grassland. The canopy is diagrammed, showing the structure of the grassland; the unassimilative:assimilative and underground:above-ground biomass ratios are given; and the life-form spectrum of the plants classified. The above-ground net primary productivity (ANPP) and its relationship with the quantity of rainfall has been calculated to indicate water-use efficiency of the *L. chinensis* grassland. The characteristics were found to be basically the same as those of the shortgrass prairie of North America. Seed production of *L. chinensis* grassland was also investigated, and the fructification ratio was found to be rather low. The distribution of buried seeds was examined, the results found being similar to those of the meadow steppe in the USSR. These results can form the theoretical basis for the management of this grassland.

KEY WORDS: Yang-cao, *Leymus chinensis*, canopy structure, net primary production, seed productivity, buried seeds.

INTRODUCTION

The Yang-cao (*Leymus chinensis* or *Aneurolepidium chinense*) grassland is found only in the eastern region of the Eurasian steppe zone, distributed mainly over three countries, including the outer Baikal area of the USSR, the northern and eastern parts of the People's Republic of Mongolia, and the northeast China plain, northern China plain, and Inner Mongolia Plateau of China (Zhu and Yang 1981). Yang-cao is highly tolerant of alkali and produces a high yield. It is suitable for grazing livestock and for hay land in lower and flat regions. Since there have been no ecological studies of Yang-cao, this paper describes its ecological characteristics. The study conforms to the International Biological Program.

Zhu Ting-cheng is Vice Chairman of the Grassland Science Society of China and a Professor of Grassland Ecology, Li Jian-dong is a Lecturer on Plant Sociology [Ecology], and Yang Dian-chen is an Engineer of Grassland Science at the Institute of Grassland Science of Northeastern Normal University.

ACHIEVEMENTS FOLLOWING A DECADE (1971–1980) OF OBSERVATIONS

Structure of *L. chinensis* Grassland

The canopy structure of *L. chinensis* grass for early August, studied on undisturbed plots by harvesting consecutive layers from the *L. chinensis* grassland, is shown in Fig. 1. The canopy structure is typical of the mature standing crop in early August. The layer 20–40 cm above ground surface is where the assimilative organs are amassed; light interception is above 40% in this layer. This figure shows that the *L. chinensis* grassland belongs to the upper lush type of medium height.

The average ratio of unassimilative (C) to assimilative (F) organs (C/F) is 1:4. The minimum value for *L. chinensis* appears at flowering, which is at the end of June. The ratio of biomass underground (U) to above-ground (A) (U/A) varies from 2:5 to 8:0, with a ratio of 5:0 most frequent. This ratio is required to maintain stability.

Fig. 1. Canopy diagram of *Leymus chinensis*.

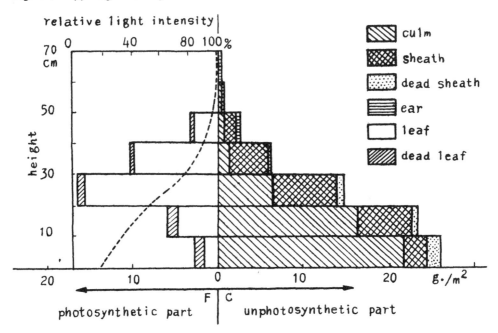

There is a dense network of rhizomes 5–20 cm below the ground surface. The rhizomes account for about 70% of the underground biomass and prevent other plants from invading. Consequently, the components of *L. chinensis* grassland are comparatively simple.

The life-form spectrum of the plants that make up the *L. chinensis* grassland, according to C. Raunkiaer's principle (Whittaker 1975), is 1.6% nanophanerophyte, 5.0% chamaephyte, 58.3% hemicryptophyte, 29.7% geophyte, and 5.4% therophyte. Hemicryptophytes and geophytes are the largest components (88%), and perennial grasses constitute the majority. The life-form spectrum is a reflection of the semiarid weather in this temperate zone and indicates that *L. chinensis* grassland is a typical prairie community.

In the middle of August, leaf-area index (LAI) is 0.8–1.4 m²/m² on chestnut soils and up to 2.0 m²/m² on the chernozems. The average transpiration efficiency of the community is 0.7 g/g/hr. The species belongs to the meso-xerophilous grassland.

Above-ground Net Primary Productivity

The above-ground net primary productivity (ANPP) as g of dry weight (DW)/m² of *L. chinensis* grassland is shown in Table 1. In *L. chinensis* pastures, the annual ANPP $\overline{X}$ = 170 g DW/m² (standard deviation [S] = 61.72 g and coefficient of variation [C.V.] = 22.7%). The most important factor affecting ANPP is the quantity of rainfall during the first days of the growing period. The mathematical relationship is y = 53.64 + 0.65 x, where x is the quantity of rainfall from January to July and y is the ANPP. The relationship is proportional in the same year (r = 0.91; p < 0.001). According to the regression formula it is possible to predict the level of net primary pro-

ductivity by simply using the quantity of rainfall. The ecosystem is sustained only if the quantity of rainfall from January to July is 82.7 mm. Above this quantity the yearly ANPP will be increased by 0.64–1.18 g DW/m² with each mm increase in rainfall. These rates coincide with rates of the shortgrass prairie (Webb et al. 1978).

The height of *L. chinensis* also influences ANPP. The average height (x) and yield (y) relationships of *L. chinensis* grass obtained from 60 1-m² samples from 15 communities are significant, expressed by logY = 0.929 + 0.036 x.

The N and P content of hay in summer is 69% and 42% higher, respectively, than in standing grass in the middle of the winter.

Seed Productivity and Seeds Buried

L. chinensis depends mainly on rhizomes to carry out asexual reproduction. The fructification of standing crops is low, and it is even lower when crops are grazed. The effect of grazing on fructification depends on the amount of stems grazed and the number of asexual branches pres-

Table 1. Above-ground standing crop of *Leymus chinensis* grassland (g DW/m²).

Item	Graze	Hay
a) All species	170.0 (100%)	226.4 (100%)
b) *L. chinensis*	128.7 (75.7%)	201.6 (84.7%)
c) Other species	41.3 (24.3%)	24.8 (15.3%)
d) Species number	8.5	5.1
e) Stem number *L. chinensis*	543.2	729.3
f) Sample size and number	1 m² × 28	1 m² × 32

Table 2. Seed productivity of *Leymus chinensis* grass.

		1 ear			1 m²	
	Spikelets	Seeds	Ratio of fructifi-cation*		Seeds	Weight (g/1000 grains)
Graze	16.5 ± 4.3	9.3 ± 6.1	36.3 ± 3.9		10,009 ± 3,217	2.5 ± 1.0
Hay	21.8 ± 6.5	15.3 ± 9.1	42.7 ± 4.3		12,215 ± 1,942	3.2 ± 1.6

*Fructification ratio is the percentage of fruit-bearing florets in all florets from 100 ears.

ent. Seed production/ear and /m² of Yang-cao grass is shown in Table 2.

L. chinensis seeds are very small; their rest period is very short; and 60%–80% will sprout within 2 to 4 months. The seeds on the ground surface can be carried away by wind, rain, birds, or animals. Seeds moved in these ways total 15%–25%. Few *L. chinensis* grass seeds become buried, and fewer than 5% become established plants. Since the number of seeds buried in the soil directly affects the regeneration of stands, the number of seeds of the species present in the soil was determined in 3-cm increments after the autumn "seed fall."

It can be seen clearly (Table 3) that the number of buried seeds under grazing is 5.5 times that under haying, and the ratio of monocotyl edonous seeds to dicotyledonous seeds is 1:2.2 under grazing and 1:6.4 from haying. The number of buried seeds under grazing *L. chinensis* is 10.2 times greater than in *Miscanthus*

grassland of IBP in Kawatabi, Japan (Hayashi and Numata 1975), but similar to that in the meadow steppe in the east part of Russia (18,823 to 19,623, Golbeva 1962). This fact suggests that the *L. chinensis* grassland is similar to the meadow steppe in respect to vegetation type.

Only three aspects of the ecological study of the *L. chinensis* grassland have been presented here. Through quantitative analysis of underground water and of soil, plants, and hair of horses, cattle, and sheep, an accumulation of titanium was found. Research on ecosystem cycling of *L. chinensis* grassland is in progress, and results will be used as the basis for studies on grasslands in temperate zones and will be the theoretical basis for grassland management.

LITERATURE CITED

Golbeva, L.V. 1962. Data on the reserves of living seed in the soil under steppe vegetation. Biul. Moskov. Obshch. Isp. P. Prirody otd. Biol. 67:76–89.

Hayashi, I., and M. Numata. 1975. Viable buried seed population in grasslands in Japan. JIBP Synth. Ecol. Japan. Grassl. Vol. 13:58–69.

Webb, W., S. Szarek, W. Lauenroth, R. Kinerson, and M. Smith. 1978. Primary productivity and water use in native forest, grassland, and desert ecosystems. Ecology 59(6):1239–1247.

Whittaker, H. 1975. Communities and ecosystems, 54–58. Macmillan. New York.

Zhu, Ting-cheng, and Dian-chen Yang. 1981. Some opinions on reconstruction of the meadow-steppe in Jilin Province. Acta Phytoecologia et Geobotanica Sinica 5(1):50–53.

Table 3. Seeds of *Leymus chinensis* buried in the grassland soil (grains/ 1 m × 1 m × 0.12 m).

Buried depth (cm)	Graze			Hay		
	M†	D†	Total	M	D	Total
0 - 3	4,900	9,617	14,517	250	2,666	2,916
3 - 6	333	1,566	1,899	100	466	566
6 - 9	150	783	933	33	183	216
9 - 12	83	500	583	150	100	250
Total	5,466	12,466	17,932	533	3,415	3,948

†M = monocotyl, †D = dicotyls.

Basic Types of Pasture Vegetation in the Songnen Plain

LI CHONGHAO, ZHENG XUANFENG, ZHAO KUIYI, and YE JUXIN
Changchun Institute of Geography, Academia Sinica, People's Republic of China

Summary

In this paper, the ecological conditions of pasture vegetation, the plant constituents of the Songnen plain, and their basic types and succession are discussed. An approach is also presented on ways to improve natural steppe artificially. In this region, there are 29 lignosa associations that are suitable for grazing and 450 plant species that are suitable forages. Among them are more than 30 species of perennial gramineous and leguminous plants.

KEY WORDS: ecological conditions, pasture vegetation, natural steppe, plant succession, pasture improvement, grazing range.

INTRODUCTION

The Songnen plain, situated in the central part of northeastern China, lies at lat 43°30′–48°40′N long 121°30′–127°00′E. It is located (Fig. 1) in the eastern end of the Eurasian steppe zone and covers an area of about 170,000 km². Many rivers crisscross this vast plain to provide abundant water, and the soil horizons are very thick. Since its grasses are superior both in quality and in quantity and poisonous herbages are scarce, it is one of the better steppes for the combination of animal husbandry, agriculture, and forestry. Since animal husbandry is of major importance, research on the pasture vegetation in the Songnen plain is highly significant in practice as well as in theory.

ENVIRONMENTAL CONDITIONS OF THE SONGNEN PLAIN

Climate. The Songnen plain is subject to the influences of continental monsoons of the temperate zone. Associated characteristics are periods of drought with frequent spring winds; warm, moist, and rainy summers; early autumn frosts; and little cold winter weather. The nonfrost period covers only 110 to 160 days. The region's main climatic characteristics are: mean annual temperature, 2°C to 4°C, with mean temperatures of −22°C to −28°C in January and 26°C to 28°C in July; mean annual precipitation 400 to 500 mm, with 65% of the total amount falling in July and August.

Landforms. In general, the topography is low and flat, with the altitude above sea level ranging from 150 to 200 m. The chief geomorphologic types are flood plain, low flat ground, smooth hillside fields, and hills and valleys.

Hydrology. There are many rivers and an ample water supply in the Songnen plain. In addition to the Nen River and the Songhua River flowing through the northern and central parts of the plain, respectively, there are many streams flowing into a network of large and small lakes. Groundwater also is plentiful.

Soil. The main soil types of the region are black soils, chernozems (meadow chernozem, sandy meadow chernozem, and sandy original chernozem), meadow soils, saline alkali soils, and swamp soils.

Fig. 1. Geographic location of the Songnen plain.

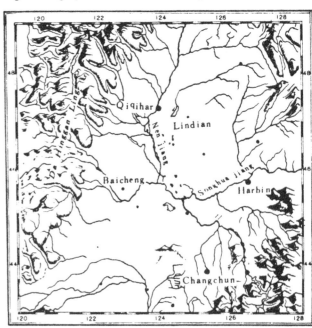

Scale 1 : 8000000

Li Chonghao is an Assistant Professor of Ecological Geobotany and the other authors are Lecturers in Ecological Geobotany at Changchun Institute of Geography of the Academia Sinica.

432

IMPORTANT PLANTS OF THE SONGNEN PASTURES

According to preliminary statistics, there are more than 600 species of seed plants in the region belonging to 82 families and 315 genera. The major families are the grass family (64 species), the composite family (71 species), and the pulse family (43 species). There are 450 species of fodder plants, which constitute about 75% of all plants. Among them are more than 30 species of fine perennial gramineous and leguminous herbages and only 11 poisonous species.

About 35 important species form different pasture associations. They are mainly of the grass (15 species) and sedge (7 species) families.

Of the important species, 57% belong to helophytes and mesophytes, which form meadow and swamp vegetations, and 42% belong to mesophytes-xerophytes and xerophytes, which make up the steppe vegetation. Thus, pastures in the Songnen are mainly mesophytes-xerophytes rhizomatous grass and tussock grass. The xerophyte tussock grasses are second in importance. Consequently, the Songnen steppe falls within the category of meadow steppe and differs from typical steppes by not consisting mainly of xerophyte, tussock grasses.

BASIC VEGETATION TYPES IN SONGNEN PASTURES

On the basis of the dominant species and their ecological and biological properties, and the environmental conditions, the vegetation in Songnen pastures may be classified into the following types.

I. Steppe.
   A. Meadow steppe with helophytic meadow.
      1. Stipa steppe (*Stipa baicalensis*) consists of three associations. Height of grass layer, 30 to 60 cm; degree of ground cover, about 40% to 50%. This steppe is good for grazing, with hay yields of about 926 kg/ha.
      2. Tansy steppe (*Tanacetum sibiricum*) consists of two associations. Height of grass layer, about 30 cm; degree of ground cover, about 30% to 40%. This steppe is good for grazing, with hay yields of about 652 kg/ha.
      3. Koeleria steppe (*Koeleria cristata*) consists of two associations. Height of layer, about 30 cm; degree of cover, about 30%. This steppe is good for grazing, with hay yield of about 692 kg/ha.
      4. *Aneurolepidium chinense* and Herbosa steppe consists of two associations. Height of grass layer, about 30 to 40 cm; degree of cover, 50% to 70%. This steppe is good for meadowland and grazing. The hay yield is about 1,649 kg/ha.
   B. Open-elm-forest steppe.
      1. The Siberian elm (*Ulmus pumila*) open-forest steppe generally has a density of about 0.3, with a few areas as high as 0.7, trees/m². Tree height is 3 to 5 m. These forests make good windbreaks. Height of grass layer under trees, about 30 cm; degree of cover, about 40%. The forests are good for grazing.
      2. Bigfruit elm (*U. macrocarpa* var. suberosa) bushwood steppe. Height of bigfruit elm is 1 to 3 m, and the density is about 0.3 to 0.4 trees/m². The components of the herb layer are similar to those of the elm steppe.
      3. Combination of open elm forest, sand vegetation, and farmland. This is the combination of the above two types after reclamation.
      4. Combination of sand vegetation and meadow. Height of grass layer, 30 to 40 cm; degree of cover, 40% to 50%. This combination is good for grazing and meadowland.
II. Meadow.
   A. Salt meadow.
      1. *Aneurolepidium chinense* meadow consists of eight associations. Height of grass layer, 50 to 60 cm; degree of cover, 50% to 80%. This is good meadowland with hay yields of 1,500 to 2,250 kg/ha.
      2. Seepweed (*Suaeda*) meadow has only one association. The grass cover is sparse and only suitable for seasonal grazing.
   B. True meadow.
      1. Grass meadow consists of three associations. Height of grass layer, 40 to 50 cm; degree of coverage, 40% to 50%. This meadow is suitable for grazing.
      2. Forb meadow has only one association. Height of grass layer, 30 to 40 cm; degree of cover, 40% to 60%. This meadow is suitable for grazing.
   C. Swamp meadow.
      1. Sagebrush meadow has only one association. Height of grass layer, 70 to 80 cm; degree of cover, 70% to 80%. This meadow is good for spring grazing.
      2. Flatsedge meadow consists of two associations. Height of grass layer, 30 to 60 cm; degree of cover, about 70%. This meadow is used as spring grazing range.
      3. Bushwood meadow has only one association—Bushwood *salix*. Height, 1.5 to 2 m. It is used as a temporary grazing range.
III. Swamp.
   A. Cattail (*Typha orientalis*) swamp has only one association. Height of grass layer, 80 to 120 cm; degree of cover, 60% to 70%. It is not good for grazing, but the leaves of cattail are useful for weaving.
   B. Reed (*Phragmites communis*) swamp has only one

association. Height of grass layer, 1.2 to 2 m; degree of cover, 70% to 90%. The reed is useful for papermaking. The meadows surrounding the swamps are used as spring grazing range.

The above summary shows that there are 32 plant associations in the Songnen pasture, of which 29 can be used for grazing. The area is about 6.05 million ha, making up 35.5% of the total area of the Songnen plain. Six plant associations, covering 1.46 million ha, are suitable as meadowland. Twenty-three of the plant associations are suitable for grazing only, and that area includes 4.59 million ha. The ratio of area of meadowland to area of grazing land is 1:3.14.

IMPROVEMENT OF THE SONGNEN PASTURE

In order to utilize and improve the vegetation, it is necessary to master the succession of meadow vegetation in the steppes.

Rapid development of animal husbandry and decrease of range area caused by reclamation have caused the pastures of the Songnen plain to be inadequate. For example, the range area of the Baicheng district in Kirin province has decreased 1 million ha since the late Fifties. Consequently, in some regions, the improvement of natural pasture has recently begun. The main improvement practices are discussed below.

1. Ameliorate the meadow of *Aneurolepidium chinense* by means of shallow ploughing and harrowing. This meadow is a major type of natural pasture on the Songnen plain. Overgrazing has caused its production to be rather low. After shallow ploughing and harrowing, the yield/unit area increases by 35% to 300%. This results from improved physiochemical properties of the soil increasing the water content by 30% to 90% and the porosity by 25%. Further, the temperature of soil increases by 1.5°C

to 2°C. These conditions are favorable for growth and propagation of rhizomes and stem of *Aneurolepidium chinense*, and they stimulate the increase of its density and height.

2. Establish semiartificial and artificial pasture. In order to increase forage yields, the area of sown pasture has been enlarged progressively on the Songnen plain. The area in some regions has reached over 50,000 ha. The main forages planted are *Medicago sativa, Astragalus adsurgens, Melilotus suaveolens, Aneurolepidium chinense,* and *Clenelymus dahuricus.* Seedings are made in pure stands. *Aneurolepidium chinense* has been successfully sown on both a large and a small scale. At the same time, overgrazed natural pastures, consisting of *Stipa* and *Tanacetum,* are of poor quality. Recently, a machine for shallow ploughing and for sowing seed has been tested.

3. Improve grazing lands by means of enclosures. Fencing has improved Songnen pasture productivity by 30% to 65%. Fencing is a valid means of establishing modern pastures so they can be managed rationally and scientifically. At present, fencing is being used on the vast Songnen plain as one of the most important methods to improve pasture and grazing lands. Over 60,000 ha have been fenced in the Kirin province.

4. Control and transform heavily alkalized *Aneurolepidium chinense* meadow. Overgrazing has progressively enlarged the alkaline patches of *Aneurolepidium chinense* meadow in Songnen pastures. Some meadows have shifted into alkaline land and become bare. Alkalized meadow can be improved by closing the field and leaving it fallow, by growing *Puccinella tenuiflora,* and by covering alkaline land with sand. The natural method of covering alkaline land by sand can be successful if banded ploughing is used that angulates with wind direction or if barriers are set up to accumulate sand on top of the alkaline land. As sand accumulates, *Aneurolepidium chinense* will invade, thereby reducing the alkaline patches until it alone occupies the land.

Grass-Leguminous Mixtures and Fertilization as Factors of the Ecosystem of Cultivated Grassland

M. MIJATOVIĆ, J. PAVEŠIĆ-POPOVIĆ, and S. KATIĆ
University of Belgrade, Zemun, Yugoslavia

Summary

On degraded soils in hilly-mountain regions, natural grasslands are impoverished, low-yielding, and of low quality. While such grasslands yield 1,870 to 2,430 kg/ha dry matter without fertilization, and 6,040 to 2,430 kg/ha with N-P-K fertilization, cultivated grasslands are more productive by far.

Experiments were conducted from 1971 to 1979 and from 1975 to 1980 in different ecological conditions of the hilly-mountain part of the country. Sixteen grass and grass-leguminous mixtures were used in the establishment of cultivated grasslands. Yields of dry matter/ha varied from 10,184 to 12,670 kg in hilly regions, 9,433 to 10,933 kg in transitional hilly-mountain regions, and 7,940 to 11,720 kg in mountain regions. Differences in yields among the examined mixtures were not always statistically significant ($P \leqslant 0.05$). It was found that cultivated grasslands with tall grasses and a limited number of components not including legumes were more vigorous and more productive.

Fertilization had a pronounced effect on the examined mixtures, especially on their yields, which ranged from 6,794 to 12,670 kg dry matter/ha, or 72% to 225% higher than the nonfertilized control. Differences were highly significant ($P \leqslant 0.01$). One kg of N-P-K nutrients produced from 10.8 to 20.0 kg of dry matter and in most cases ranged from 13.7 to 18.4 kg of dry matter/kg of N-P-K.

Fertilization also affected the composition of grass cover, the distribution of plant species and groups, and the quality of herbage.

KEY WORDS: cultivated grassland, mixture, fertilization, productivity, quality.

INTRODUCTION

Conventional cultural practices have a limited effect on highly degraded natural grasslands, especially in respect to yield increases. The floristic composition of natural grasslands (i.e., their quality) changes slowly. The eradication of such vegetation (by chemical or mechanical means) and subsequent establishment of cultivated grasses is important because they improve the production of forage and protect soil from further erosion.

The two factors important for successful establishment of cultivated grasslands are the selection of grass species or grass-legume mixtures and their fertilization. Klapp (1971), Minina (1972), Mijatović and Pavešić-Popović (1974), Mijatović et al. (1980), Baker (1978), and Troxler and Charles (1980) stated that the grass species selected and the composition of grass mixtures are of critical importance in establishing cultivated grasslands of high yields and high quality. On the other hand, Smelov (1966), Zürn (1968), Bauer (1977), Copeman (1978), Mijatović and Pavešić-Popović (1974), and Mijatović et al. (1980) stated that high-yielding cultivated grasslands cannot be successfully established and maintained without intensive fertilization, in spite of good grass mixtures used. In addition to yield, fertilization affects other important characters of grasslands.

MATERIALS AND METHODS

Experiments were conducted from 1971 to 1979 and from 1975 to 1980 in hilly and mountain regions, using 16 grass and grass-legume mixtures and fertilization as variables in the ecosystem of cultivated grassland. The composition of the examined mixtures is included with the research results. There were three fertilization variants in each experiment. A complete N-P-K fertilizer was applied in spring, and nitrogen was applied during the growing season.

Yield increases, growth and development of the

grasses, height and thickness of cover, and distribution of plant species and groups, especially weeds, were observed.

Forages were cut during the vegetative stage, and yields were calculated (kg/ha) as dry matter/test year. Dry-matter yields were statistically analyzed by routine analysis of variance. The LSD test was used to determine statistical differences.

CHARACTERISTICS OF THE SITES

Experiments were conducted in hilly, hilly-mountain, and mountain regions at three locations: (1) Hilly region—Kamenica: altitude 480 m; continental climate; mean annual air temperature 9.8°C; annual rainfall 870 mm (490 mm during growing season); pseudogley soil. (2) Transitional hilly-mountain region—Suvobor: altitude 800 m; continental climate with elements of mountain climate; mean annual air temperature 8.3°C; annual rainfall 820 mm (480 mm during growing season); rendzina soil. (3) Mountain region—Zlatibor: altitude 1080 m; moutain climate; mean annual air temperature 7.3°C; annual rainfall 910 mm (560 mm during growing season); brown mountain soil.

RESULTS

Hilly Region—Kamenica

The habitat of natural meadow *Agrostidetum vulgarae*, which yields, on the average, 2,430 kg/ha of dry matter without fertilization and 6,800 kg/ha with N-P-K fertilization. Average yields and composition of grass and grass-legume mixtures are shown in Table 1. The fertilizer levels used were: N,200; P, 140; and K, 140 kg/ha. Differences in yields were statistically significant among different rates of N-P-K fertilizers (Table 2).

Hilly-Mountain Region—Suvobor

The habitat of natural meadow *Danthonietum calycinae*, which yields 2,030 kg/dry matter/ha without fertilization and 6,430 kg/ha with N-P-K fertilization. The mixtures

M. Mijatović is a Professor, J. Pavešić-Popović is an Assistant Professor, and S. Katić is an Assistant on the Agricultural Faculty at the University of Belgrade.

Table 1. Composition and productivity of mixtures in the hilly region.

Species	Composition of mixtures (%)			
	A	B	C	D
Dactylis glomerata	40	30	30	25
Phleum pratense	30	20	30	20
Festuca pratensis	30	20	20	–
Arrhenatherum elatius	–	–	20	–
Lolium perenne	–	–	–	20
Lotus corniculatus	–	30	–	20
Trifolium repens	–	–	–	15
DM yield, kg/ha, avg.	12,670	11,080	12,000	10,184
Max.	13,070	12,325	13,800	11,040
Minimum	10,732	9,372	9,070	8,520
CP yield, kg/ha	2,097	1,830	1,910	1,743

Table 2. Effect of fertilization on the yield of cultivated grassland (mixture B).[1]

Fertilization rates	DM yield, kg/ha		kg DM/kg NPK
	Average	Range	
$N_{100}P_{60}K_{60}$	7,751	6,532 to 8,670	18.2
$N_{150}P_{100}K_{100}$	10,511	8,447 to 11,834	17.4
$N_{200}P_{140}K_{140}$	12,172	9,862 to 12,876	14.8
Control	4,476	3,921 to 4,860	–

[1]See table 1.

examined in the period 1976 to 1980 and the average yields of the mixtures during the test period are shown in Table 3. Fertilization levels were: N, 200; P, 150; and K, 150 kg/ha. The effect of fertilization on the yield of cultivated mixtures F and G during the period 1971 to 1979 are presented in Table 4. Differences in yields among fertilization variants were highly significant.

Mountain Region – Zlatibor

This is a region of natural and cultivated grasslands. The habitat of *Festucetum sulcatae* yields, on the average, 1,870 kg/dry matter/ha without fertilization and 6,040 kg/ha with N-P-K fertilization. The effects of fertilization of cultivated grass were examined in the period 1975 to 1980. The mixture evaluated included *Dactylis glomerata*, 25%; *Phleum pratense*, 20%; *Arrhenatherum elatius*, 20%; *Lotus corniculatus*, 20%; and *Trifolium repens*, 15%. The yields obtained are shown in Table 5. Dry-matter yields were significantly different for all fertilization variants.

Fertilization affected the composition of grass mixtures during the utilization period. The percentage of tall grasses increased, and those of short grasses and legumes decreased.

The chemical composition of herbage was affected as well. The crude-protein levels ranged from 16.4% to 21.2% within the same fertilization rate, from 16.7% to 21.7% among test years, and 16.2% to 18.1% among sites, for the same mixture. Chemical composition, especially crude-protein level, varied considerably from one fertilization rate to another. The average crude-protein level for the low, medium, and high dosages of N-P-K and for the control were 14.4%, 17.6%, 16.8%, and 13.7%, respectively. Total yield of protein ranged from 1,356 to 2,270 kg/ha, depending on the composition of mixture, experimental site, and fertilization rate.

LITERATURE CITED

Baker, R.J. 1978. The value of grass species and the need for changing sward composition. Occas. Symp. 10:235–237. Univ. of New York.

Bauer, U. 1977. Development of sward, yield dynamics, and productive life of seeded grassland on low-bog sites (Bestandesentwicklung, Ertrageverlauf und Leistungdauer des Saatgraslandes). Proc. XIII Int. Grassl. Cong. Sect. 6, 50–55.

Table 3. Composition and productivity of mixtures in the hilly-mountain region.

Species	Composition of mixtures (%)						
	A	B	C	D	E	F	G
Dactylis glomerata	60	65	30	40	–	40	–
Phleum pratense	40	–	20	30	–	10	–
Arrhenatherum elatius	–	–	–	–	–	15	–
Bromus inermis	–	–	20	30	–	–	–
Lolium perenne	–	–	–	–	30	–	20
Poa pratensis	–	–	–	–	30	–	20
Festuca rubra	–	–	–	–	–	–	25
Lotus corniculatus	–	35	30	–	20	20	20
Trifolium repens	–	–	–	–	20	5	15
Trifolium pratense	–	–	–	–	–	10	–
DM yield, kg/ha, avg.	10,256	10,933	10,426	10,080	9,433	10,460	9,360
Max.	12,665	12,250	12,275	11,343	10,343	12,870	12,517
Min.	8,957	9,252	9,337	8,806	8,095	8,046	7,452
CP, kg/ha	1,603	1,758	1,693	1,594	1,508	1,670	1,517

Table 4. Effect of fertilization on the yield of cultivated grassland (mixture F and G).

| Mixture | Fertilization | Yield, kg DM/ha | | kg DM/kg NPK |
		Average	Range	
F[1]	$N_{87}P_{60}K_{60}$	7,859	5,465 to 9,130	20.0
	$N_{174}P_{120}K_{120}$	10,308	7,228 to 12,403	15.9
	$N_{260}P_{180}K_{180}$	11,430	8,046 to 12,870	12.4
	Control	3,719	2,912 to 4,800	–
G[1]	$N_{87}P_{60}K_{60}$	6,794	5,367 to 7,860	13.7
	$N_{174}P_{120}K_{120}$	9,023	6,278 to 11,160	12.2
	$N_{260}P_{180}K_{180}$	10,680	7,452 to 12,517	10.8
	Control	3,957	2,952 to 4,112	–

[1]See Table 3 for composition of mixtures.

Table 5. Effect of fertilization on the yield of cultivated grassland, Mountain region.

| Fertilization | DM yield, kg/ha | | kg DM/kg NPK |
	Average	Range	
$N_{80}P_{50}K_{50}$	7,940	6,432 to 8,710	18.9
$N_{150}P_{100}K_{100}$	10,413	7,920 to 12,400	17.2
$N_{220}P_{150}K_{150}$	11,720	8,860 to 12,960	14.1
Control	4,470	3,070 to 5,343	–

Copeman, G.J.F. 1978. Causes of sward changes – climate. Occas. Symp. 10:151–155. Univ. of York.

Klapp, E. 1971. Meadows and pastures (Wiesen und Weiden). Verlag Paul-Parey, Berlin.

Mijatović, M., and J. Pavešić-Popović. 1974. Yielding capacity of planted meadows subject to sward composition and mineral fertilizers top dressing. Proc. XII Int. Grassl. Cong. 1:396–403.

Mijatović, M., J. Pavešić-Popović, and S. Katić. 1980. Grass mixtures as a yield factor of forage quality in the hilly mountainous regions of Serbia. Proc. VIII Eur. Grassl. Cong., Zagreb, 1:113–121.

Minina, N.P. 1972. Grass mixtures (Lugovie travosmesi). Kolos, Moscow.

Smelov, S.P. 1966. Theoretical principles of grassland production (Teoretičeskie osnovi lugovodstva). Kolos, Moscow.

Troxler, J., and J.P. Charles. 1980. Some aspects of grassland utilization of marginal land in the mountain area. Proc. VIII Eur. Grassl. Cong., Zagreb, 1:1–119.

Zürn, F. 1968. Modern grassland fertilization (Neuzeitliche Düngung des Grünlandes). DLG Verlags, Frankfurt.

Grass and Forb Production on Sprayed and Nonsprayed Mesquite (*Prosopis glandulosa* Torr.) Dunelands in South-Central New Mexico

R.P. GIBBENS
USDA-SEA-AR Jornada Experimental Range
Las Cruces, N.M., U.S.A.

Summary

Production of grasses and forbs was measured for 4 years, beginning in 1976, following aerial application of 2,4,5-T to 3,634 ha of mesquite (*Prosopis glandulosa* Torr.) dunelands on the Jornada Experimental Range in south central New Mexico. Plant production was also measured on a 3,318-ha control area. The mesquite dunelands are representative of large areas in the southwestern U.S.A. where mesquite has invaded former desert grasslands. Objectives of the study were to determine forage increases attributable to reduced mesquite competition and to evaluate the relationships between production and precipitation.

Both sprayed and control areas were grazed by cattle. Temporary exclosures were used to exclude cattle from three sampling sites on the sprayed and control areas each year. The sampling sites were all on sandy ranges with relatively deep soils (>50 cm to caliche layer). A dune-centered sampling system permitted the determination of differences in production on dune and interdune areas. The mesquite dunes, occupying 28% of the land surface, produced more annual forbs and fewer perennial grasses than the interdune areas. Stem kill of mesquite averaged 56% at sampling sites on the sprayed area. End-of-season harvests were used to compare treatments.

Perennial grass production was 7-, 8-, and 4-fold greater on the sprayed than on the control area in the first 3 years following treatment, respectively. Maximum perennial grass production of 642 kg/ha occurred in the first season following treatment. In the 4th year the control area received 49 mm more precipitation than did the sprayed area, and production of perennial grass was nearly equal on the two treatments. Mesa dropseed (*Sporobolus flexuosus* [Thurb.] Rydb.) contributed 66%–92% of the perennial grass production. Production of annual plants varied widely among years but was greatest in the 4th season when precipitation patterns favored annual plant growth, particularly on the control area where annual forb production was 543 kg/ha. Production of broom snakeweed (*Xanthocephalum sarothrae* [Pursh] Shinners) cycled from high to low and back to high on both treatments.

Mesquite control is an effective tool for improving forage production on arid rangelands, but the success of control efforts depends not only upon the degree of brush kill but also upon the posttreatment precipitation patterns.

KEY WORDS: mesquite, *Prosopis glandulosa* Torr., mesa dropseed, *Sporobolus flexuosus* (Thurb.) Rydb., 2,4,5-T, forage production.

INTRODUCTION

Mesquite (*Prosopis glandulosa* Torr.) has invaded and become dominant on large areas of once-productive rangeland in the southwestern U.S.A. Brush invasion has seriously reduced the carrying capacity of rangelands and has a serious impact upon the region's economy. It is imperative to control the invading brush and return the land to a productive condition if food and fiber demands are to be met.

The invasion of brush species in the Southwest during the past 100 years has been well documented (Brown 1950, Glendening 1952, Dick-Peddie 1966). Estimates indicate that mesquite is a serious problem on 37.6 million ha of rangeland (Platt 1959). Overgrazing in the late 1800s and early 1900s is often cited as a cause of brush invasion, but there is ample evidence that the invasion has continued even on well-managed, conservatively grazed ranges (Buffington and Herbel 1965). The invading brush species have been the object of control measures for many years (Fisher et al. 1959, Valentine and Norris 1960, Herbel and Gould 1970).

This study of primary production following mesquite control is part of a comprehensive program to determine the impact of mesquite control upon a grazing-land ecosystem. Included in the study were evaluations of the impact of mesquite control upon bird, insect, small mammal, and soil microorganism populations and livestock activities and production. It was conducted on the Jornada Experimental Range, located 37 km north of Las Cruces, New Mexico. The Experimental Range has an arid climate with 55% of the average annual precipitation of 231 mm occurring in July, August, and September.

METHODS

The mesquite control treatment consisted of two aerial applications of 0.56 kg 2,4,5-T [(2,4,5-trichlorophenoxy)acetic acid]/ha in a 1:7 diesel oil:water emulsion at a total volume of 9.4 l/ha. This treatment was applied to 3,634 ha of mesquite dunelands, and an adjacent area of 3,318 ha served as a control. The herbicide treatments, which were applied in early June, were begun in 1975 and were completed in 1978. Due to overlap of applications the sprayed sites monitored for grass and forb production

received herbicide applications in 1975, 1976, and 1977. However, average stem kill of mesquite at the sampling sites was only 56%, which was representative of mesquite kill on the entire sprayed area.

Grass and forb production was measured at three sites in both the sprayed and the control area. All study sites were located on a sandy range site with either Typic Haplargid (Onite series) or Calciorthid (Wink series), coarse-loamy mixed, thermic soils. These relatively deep soils are generally more than 50 cm to the caliche layer. Temporary exclosures 0.4 ha in area were built at each site to prevent livestock grazing during the sampling season. The exclosures were moved to a new location each year. Utilization of forage by livestock during 1975–1977 was light to moderate at each site. A recording, weighing-type rain gauge was installed at each exclosure. Within each exclosure the mesquite dunes were marked with numbered stakes. At each sampling date three dunes in each exclosure were drawn at random for sampling. Precipitation patterns determined sampling dates, which varied from three to four/season.

Since it was necessary to measure production on both dune and interdune areas, a dune-centered sampling system was used. The point of a wedge-shaped clipping frame was placed at the center of a dune, and the angle subtended by the sides of the frame (18.4°) included a proportional sample of the center and periphery of the circular-shaped dunes. Herbaceous and shrublike plants were harvested from the wedge-shaped area in each of the cardinal directions on each dune. At the base of the dune a 1-m-wide transect was clipped that extended across the interdune area halfway to the next dune, also in each of the cardinal directions. The dunes had an average height and diameter of 1 and 5.4 m, respectively.

All plants were harvested at ground level and segregated by species. The clipped samples were separated into current growth and standing dead components, and oven-dry weights were determined (60°C). Because fluffgrass (*Erioneuron pulchellum* [H.B.K.] Tateoka) is very difficult to separate into live and dead components, weights for this species are for total standing vegetation. The suffrutescent broom snakeweed (*Xanthocephalum sarothrae* [Pursh] Shinners) , and the shrublike soaptree yucca (*Yucca elata* Engelm.) were separated into live and dead components. Canopy of the shrublike species occurring within the plots was harvested; all other species were harvested only if rooted within the plots. Mesquite was not harvested. An average of 190 m² was

---

The author is a Range Scientist at USDA-SEA-AR Jornada Experimental Range.

sampled on each treatment/collection date.

Since plot size varied, yield/unit area was calculated for dune and interdune areas. On both treatments, production on the mesquite dunes differed from that on interdune areas. In general, there were more annual forbs and fewer perennial grasses on the dunes. Production for the treatments has been calculated on the basis of 28% dune area, which is the average value for dune area on the two treatments.

The study extended from 1976 through 1980. At this time production samples for the years 1976–1979 have been processed. The end-of-season harvest (late September or early October) is selected for presentation. Most of the annual plants on the study area function as summer annuals, and the end-of-season harvest represents the peak standing crop of all perennials and annuals with the exception of one annual forb.

RESULTS AND DISCUSSION

The mesquite dunelands support a very small flora. Only five perennial grasses, three annual grasses, twelve perennial forbs, twenty-two annual forbs, and seven shrubs or shrublike species were encountered in the 4 years of sampling. Not all annual species were present in any one year. The amount and timing of precipitation appeared to be the factor determining which of the annual species were present.

Annual precipitation in 1974 was 404 mm (long-term average, 231 mm), and many seedlings of mesa dropseed (*Sporobolus flexuosus* [Thurb.] Rydb.) became established. Defoliation of the mesquite by herbicide in 1975 permitted the seedlings to develop. Thus, in the first year following treatment (1976), perennial grass production was 7-fold greater on the sprayed area than on the control area (Table 1), and in 1977 it was 8-fold greater, although production from both treatments in 1977 was less than in 1976 because of low precipitation.

In 1978 the control area received 36 mm more precipitation than the sprayed area, resulting in increased production of mesa dropseed on the control area. However, mesa dropseed production was still 4-fold greater on the sprayed area than on the control area. Death of plants in

the mature population of broom snakeweed on both treatments caused a sharp drop in the production of shrublike species in 1978 (Table 1).

Normal precipitation occurred during the fall and winter of 1978–1979. This, plus favorable summer precipitation, led to the development of large populations of annual plants on both treatments. The control area received 49 mm more precipitation than the sprayed area in 1979. The additional soil water led to a tremendous stand of Russian thistle (*Salsola kali* L.) on the mesquite dunes in the control area, production of this species being as high as 396 g/m². Large numbers of broom snakeweed seedlings that became established in 1978 and 1979 grew rapidly, leading to a great increase in production of shrublike species on both treatments in 1979 (Table 1). Fluffgrass, the second most abundant perennial grass, increased on both treatments in 1979. Mesa dropseed production on the control area was nearly equal to that on the sprayed area in 1979. Mesa dropseed on the sprayed area began to die out in 1979; competition for soil water by the dense stand of annual plants was no doubt a factor contributing to the mesa dropseed mortality.

Overall, the herbicide treatment resulted in greatly increased production of mesa dropseed, which is an excellent forage species, for 3 years. The importance of amount and timing of precipitation in this arid region is evident from the fact that production of the control area in 1979 exceeded that of the sprayed area. Mesquite control is an effective tool for improving forage production on arid rangelands, but the success of such control depends not only upon the degree of brush kill but also upon post-treatment precipitation patterns.

LITERATURE CITED

Brown, A.L. 1950. Shrub invasion of southern Arizona desert grassland. J. Range Mangt. 3:172–177.

Buffington, L.C., and C.H. Herbel. 1965. Vegetational changes on a semi-desert grassland range from 1858 to 1963. Ecol. Monog. 35:139–164.

Dick-Peddie, W.A. 1966. Changing vegetation patterns in southern New Mexico. New Mex. Geol. Soc. 16th Field Conf., 234–235.

Fisher, C.E., C.H. Meadors, R. Behrans, E.D. Robinson,

Table 1. Production (kg/ha) of grasses, forbs, and shrub-like plants on sprayed and on control areas in the first 4 years following application of 2,4,5-T for control of mesquite. Production on both sprayed and control areas calculated on basis of 28% dune area and 72% interdune area. Crop-year precipitation (October–September) is shown.

	1976		1977		1978		1979	
	Sprayed	*Control*	*Sprayed*	*Control*	*Sprayed*	*Control*	*Sprayed*	*Control*
Mesa dropseed	589	80	289	25	442	109	329	326
Other perennial grasses	53	9	36	13	59	70	84	106
Annual grasses	1	5	1	5	T†	1	T†	T†
Perennial forbs	12	5	3	2	4	11	34	12
Annual forbs	30	9	97	11	13	26	263	543
Shrub-like plants	297	234	239	100	49	39	188	161
Total production	982	342	665	156	567	256	898	1,148
Precipitation (mm)††	284	240	180	187	211	247	333	382

†T = Trace, < 0.5 kg/ha

†† Long-term average for years 1915–1979 is 231 mm.

P.T. Marion, and H.L. Morton. 1959. Control of mesquite on grazing-lands. Tex. Agric. Expt. Sta. Bul. 935.

Glendening, G.E. 1952. Some quantitative data on the increase of mesquite and cactus on a desert grassland range in southern Arizona. Ecology 33:319–328.

Herbel, C.H., and W.L. Gould. 1970. Control of mesquite, creosotebush, and tarbush on arid rangelands of the south-western United States. Proc. XI Int. Grassl. Cong., 38–41.

Platt, K.B. 1959. Plant control—some possibilities and limitations. I. The challenge to management. J. Range Mangt. 12:64–68.

Valentine, K.A., and J.J. Norris. 1960. Mesquite control with 2,4,5-T by ground spray application. New Mex. Agric. Expt. Sta. Bul. 451.

Use of Fodder-Production Models for Mineral Soils on Grassland and Arable Land

J. BOCHNIARZ, B. KACZYŃSKA, M. KASPRZYKOWSKA, and B. ŁANIEWSKA

Institute of Soil Science and Plant Cultivation, Pulawy, Poland

Summary

In four regions of Poland experiments begun in 1973 and planned to last 12 years compared cutting and grazing yields of permanent grasslands over 20–30 years old to yields of seven field fodder-production models. The models had 3-, 4- and 6-year rotations. The experiments were located on mineral soils with 28%–50% silt and clay and 2.1%–3.5% humus. The water table depth varied from 44 to 170 cm in spring and from 115 to more than 175 cm in summer. Average precipitation was 375–515 mm during the growing season (April-October) and 523–658 mm annually. Average annual fertilization/ha was as follows: N, 360 kg; P_2O_5, 150 kg; and K_2O, 280 kg. Experiments used the randomized-block method with 4 replications of 36 m^2 plots (82 m^2 gross).

Results after 8 years showed that the average annual yields of green matter were 54–75 t/ha (t = 1000 kg); fodder beet roots, 116–135 t/ha; potatoes, 30–39 t/ha; and broad bean seeds, 3.1–3.5 t/ha. Calculated in comparable units, the yields of digestible protein were higher in pastures and meadows (1.6 t/ha) than in the other models (0.98–1.16 t/ha). Yields of starch from meadows were 7.2 t/ha; from pastures 6.0 t/ha; and from field models 5.8–7.6 t/ha. Starch and protein yields calculated in cereal units were 12.7 t/ha from meadows, 11.2 t/ha from pastures, 11.8 t/ha from the highest-yielding field model, and 9.6–10.8 t/ha from the other field models. These results showed that permanent meadows and pastures lying on mineral soils and under intensive fertilization and exploitation yielded as much as cultivated fields and still maintained a high level of productivity after 8 years of intensive exploitation.

KEY WORDS: meadow, pasture, grasses, kale, bean, beets, potatoes, maize.

INTRODUCTION

In most European countries with intensive agricultural production, the amount of natural grassland is decreasing in favor of arable land and leys. This change is due to higher and more valuable yields obtained from leys than from permanent grasslands (Lidtke 1975).

In Poland, 21.4% of all agricultural lands are grasslands (GUS 1980), 64% of which lie on mineral soils. To date, the progress in ploughing them up has been slow, but an increase in this rate can be expected. In this connection, it is necessary to examine the following questions: (1) Do natural grasslands truly have a worse overall production value than intensive field fodder-production models? (2) After how many years of intensive exploitation should natural grassland be renovated? (3) Which field fodder-production models on previously plowed grassland plots are more efficient?

METHODS

The experiments reported here have been carried out since 1973 in four experiment stations in different regions of Poland. They were located on permanent pastures (over 20–30 years old) on mineral soils with 28%–50%

The authors are Research Agronomists at the Institute of Soil Science and Plant Cultivation.

Table 1. Experiment objects and duration of rotations (years).

Year	A	B	C₆	D₄	E₃	F₃	G₄	H₆	I₆
1973	Permanent meadow	Permanent pasture	pt	pt	pt	pt	pt	pt	pt
1974	″	″	o	o	o	o	o	o	o
1975	″	″	+	+	+	×	×	×	×
1976	″	″	+	+	fb	fb	×	×	k
1977	″	″	+	fb	o	o	fb	×	b
1978	″	″	b	o	+	×	o	×	rm
1979	″	″	fb	+	pt	pt	×	fb	fb
1980	″	″	o	+	o	o	×	o	o
1981	″	″	+	pt	+	×	pt	×	×
1982	″	″	+	o	fb	fb	o	×	k
1983	″	″	+	+	o	o	×	×	b
1984	″	″	b	+	+	×	×	×	rm
1985	″	″	potatoes or fodder beets (2nd course)						

Key:

+ = Grasses, in objects C, D mixture: *Festuca pratensis* Huds. 7 kg + *Phleum pratense* L. 4 kg + *Bromus inermis* Leyss. 15 kg + *Lolium perenne* L. 6 kg + *Agrostis gigantea* Roth. 3 kg/ha. Object E: *Dactylis glomerata* L. 25 kg/ha.

× = Grasses-legumes mixture in Object F and I.: *Lolium multiflorum* Lam. 7 kg + *Trifolium pratense* L. 16 kg/ha; object G: *Festuca pratensis* 5 kg + *Phleum pratense* 5 kg + *Trifolium pratense* 12 kg + *Trifolium hybridum* L. 3 kg/ha., Object H; *Festuca pratensis* 4 kg + *Bromus inermis* 15 kg + *Trifolium pratense* 5 kg + *Trifolium hybridum* 5 kg. + *T. repens* L. 2.5 kg/ha.

b = broad bean, fb - fodder beets, k - marrow-stem kale, o - silage oats with underseeds, pt - potatoes, rm - rye as the winter catch crop and maize as the 2nd crop.

silt and clay and 2.1%–3.5% humus. Soil analyses were as follows:

	1973	1979
pH with KCl	4.8–7.2	5.3–7.1
P_2O_5 mg/100 g soil	0.9–8.7	7.8–23.0
K_2O mg/100 g soil	7.4–12.8	9.8–20.6
Mg mg/100 g soil	4.0–18.3	10.4–14.5

The water table depth was 44–170 cm in spring and from 115 to lower than 175 cm in summer. Average precipitation was 373–515 mm during the growing season (April-October) and 523–658 mm annually. Botanical composition of the sward (by weight) was similar in 1973 and in 1978: *Dactylis glomerata* L. + *Poa pratensis* L. + *Bromus inermis* Leyss. accounted for 54%–80%, and all grass species totalled 88%–99%.

In the experiments, yields of seven different field fodder-production models were compared with those of a meadow and a pasture. The full cycle is planned to last 12 years, from 1973 to 1984 (Table 1). The experiments were carried out using the randomized-block method with 36-m² plots (gross 82 m²) and 4 replications. Average annual fertilization with nitrogen (N) was 360 kg/ha (from 40 to 450 kg depending on the plant species); P_2O_5, 150 kg/ha; K_2O 280 kg/ha (from 150 to 300 kg). Managements used in the models were as follows: (A) two or three cuttings and one grazing; (B) three or four grazings and one cutting; (C-I) rye, oats, maize, and kale harvested for green fodder (silage); broad beans for seeds; grasses and mixed crops for hay. Because of the variety of crops, calculated yields were based on amounts of starch and digestible protein (according to Ziołecka 1974). Total cereal units were then calculated as 0.816 (total kg starch + 1.945 [total kg protein]).

RESULTS

The results of 8 years of the experiments are shown in Table 2. Average annual yields from four experiment stations and 8 years were: green matter, 54–75 t/ha; fodder beet roots, 116–135 t/ha; potatoes, 30–39 t/ha. Models A

Table 2. Mean yearly yields of the models.

	A	B	C	D	E	F	G	H	I
(a) Mean cereal unit yields tˣ/ha (average from 4 experiment stations)									
1973	11.63	9.35	7.25	7.25	7.25	7.25	7.25	7.25	7.25
1974	14.09	12.19	7.75	7.53	8.44	8.13	6.12	6.19	8.38
1975	13.41	12.41	14.73	15.14	18.71	10.92	14.46	13.65	11.42
1976	10.86	9.41	11.64	11.35	12.40	13.00	8.87	9.28	14.03
1977	13.02	12.30	12.78	9.16	8.69	8.90	10.04	11.48	5.99
1978	11.78	12.16	4.62	8.96	14.03	10.22	7.76	11.96	13.26
1979	11.57	8.89	13.76	10.54	12.54	12.72	9.03	13.05	12.51
1980	15.38	12.71	12.31	15.68	12.63	13.06	13.23	10.74	13.86
(b) Mean yearly yields in various units in t ˣ/ha (average from 4 experiment stations and 8 years)									
Starch	7.22	6.02	6.44	6.47	7.61	7.02	5.83	6.36	6.72
Digestible protein	1.62	1.60	1.14	1.16	1.06	0.98	1.03	1.12	1.09
Cereal units	12.72	11.18	10.61	10.70	11.84	10.53	9.60	10.46	10.84

tˣ = 1000 kg.

and B had very little yearly variation in production, whereas the other models displayed large variation. Although potatoes and oats with underseeds had high yields, they were less efficient than grasses, beets, kale, or rye and maize. Broad beans had a high yield of seeds (model I in 1977, 3.5 t/ha; and model C in 1978, 3.1 t/ha), but was the lowest-yielding crop in cereal units.

Yields averaged over four experiment stations and 8 years were very high (Table 2b). Starch yields were highest in model E and were 5.1% lower in model A. Models A and B produced the highest protein yields. The highest yields (calculated in cereal units) were obtained in models A, B, and E.

Results from 8 years of experiments on permanent grasslands with good botanical composition lying on mineral soils indicate that permanent grasslands intensively fertilized and utilized for cutting or grazing yield as much as field fodder-production models. The lack of change in botanical composition after 8 years of intensive exploitation indicates that there is no necessity to renovate.

LITERATURE CITED

Główny Urząd Statystyczny (GUS) 1980. Rocznik statystyczny. Warsaw.

Lidtke, W. 1975. Użytki zielone w uprawie polowej. Nowe Roln. 6:15–17.

Ziołecka, A. 1974. Wartość pokarmowa pasz dla przeżuwaczy. Normy żywienia zwierząt. PWRiL, Warsaw.

Productivity and Botanical Composition Response of a Stabilized Pasture to Different Utilization Patterns

E. RIVEROS V. and A. OLIVARES E.
University of Chile, Santiago, Chile

Summary

The effect of three different utilization patterns upon the productivity and botanical composition of typical natural pasture was studied during a year of growth. The trial consisted of cutting the swards of three plots of a natural pasture in a humid zone in the south of Chile when they reached an average height of 20 cm, leaving only a height of 5 cm. This process was carried out in three plots of 20 x 20 m using the following methods: plot 1, harvesting with bovines; plot 2, mechanized harvesting with a mower; plot 3, not harvested but sampled each time from the forage accumulated from the start of the trial. The botanical composition and other characteristics when starting the experiment were, in plots 1, 2, and 3, respectively: *Lolium* spp., 16.4%, 12.7%, 14.8%; *Dactylis glomerata*, 24.0%, 23.6%, 18.1%; *Anthoxanthum odoratum*, 0.7%, 4.3%, 6.2%; other grass spp., 2.1%, 2.3%, 1.9%; *Trifolium* spp., 4.5%, 2.7%, 7.4%; wide leaf spp., 12.9%, 7.7%, 8.9%; organic matter, 31.4%, 35.9%, 37.0%; and bare soil, 8.0%, 10.8%, 5.7%.

Unimportant variations in botanical composition were observed. Annual dry-matter yields were 10,442, 10,147, and 9,184 kg/ha in plots 1, 2, and 3, respectively.

The results suggest that pastures with several species and with a high homeostatis and stability grade can be utilized by different methods without affecting significantly the pasture performance in an annual growing period. This capacity is very important for farmers of these zones who must change management according to seasonal, environmental, and economic conditions.

KEY WORDS: naturalized pastures, botanical composition, management, ecology, humid ecosystems, productivity.

INTRODUCTION

A knowledge of ecosystems' functioning is essential for improving production. Many times grazing by animals is the only productive use that can be made of lands that are suboptimal for cultivation (Pendleton and Van Dyne 1980).

In multispecific pastures of dynamic ecosystems in which environmental factors such as temperature and rainfall determine productivity, pasture management through the year can affect the yield and its quality in the following seasons and consequently the subsequent production (Goić and Matzner 1977).

Gastó and Cañas (1975) determined that any ecosystem

The authors are Researchers and Professors of Pastures on the Faculty of Agronomy of the University of Chile.

has two kinds of principal attributes defining its state: architecture and functioning. Architecture is the element in which the functioning process is centered; therefore, a pastoral ecosystem can be improved by improving the architecture (Nava et al. 1976).

Perhaps the only way to know and solve the complicated problems of pasture systems is to consider the grazing-land area as a complete ecosystem. In describing factors influencing productivity and biological efficiency of the ecosystem, Teuber (1978) pointed out that the proportion of the forage supply removed by grazing animals is one of the main factors affecting productivity. The responses resulting from the interaction between plants and animals are to a great extent typical of predator-prey relationships (Abaturov 1979).

The main objective of this work was to study the influence of different pasture utilization patterns upon botanical composition and productivity of a natural pasture in humid southern Chile.

METHODS

The trial was carried out on the Estación Experimental Agronómica Campus Oromo of the Faculty of Agronomy of the University of Chile, located at lat 41° S long 73° W in a high-rainfall region (1,400 mm/yr) in the province of Osorno, Chile.

The experiment consisted of cutting the sward of three 20 × 20–m plots of a 12-year-old natural pasture of a perhumid zone on soils of volcanic origin when vegetation reached an average height of 20 cm, leaving a stubble height of 5 cm.

Harvesting was accomplished by three methods: (1) grazing by bovines; (2) mechanized harvesting with a mower; and (3) harvesting only once after a year of pasture growth. At each harvest plots were sampled from 1-m² quadrants to measure dry-matter production. The nonharvested plot was also sampled at each cutting time to measure accumulated forage yield throughout the total growing period. Botanical composition was determined each sampling time at 100 points on each plot.

RESULTS AND DISCUSSION

Dry-Matter (DM) Productivity

Dry-matter production did not differ significantly among treatments. There was a trend for increased productivity for all treatments in spring and summer. Similar

results were reported by Goić and Matzner (1977), who observed the largest growth rate during late spring. Some authors have reported that in the spring forages contain P, Mg, and Na in quantities that are often insufficient to meet the requirements of grazing dairy cows (Thompson and Warren 1979), a fact that should be kept in mind.

The yield of all treatments can be considered high for each yearly period of growth (Table 1), reaching over 10,000 kg DM/ha.

Similarly, Davies and Simons (1979) did not find significant differences for perennial ryegrass among spring defoliation patterns; however, the productivity was less for fall and winter harvests.

Although sward yields did not differ significantly among treatments, more leaves and tillers were present in the harvested swards than in the nonharvested pastures, a finding similar to the observations of Davies and Simons (1979). This fact suggests that forage nutritive value may be influenced by growth rate and seasonal environmental conditions, as was indicated by Jolliff et al. (1979) for bermudagrass.

Pike et al. (1979) also found pronounced differences between set-stocked and zero-grazed swards in their mean leaf : stem ratios, probably because of different morphological composition of stands resulting from grazing-animal selectivity. This finding was confirmed with cattle by Cowlishaw and Alder (1960). Pike et al. (1979) attributed the plant morphological changes to a marked delay in flowering time for set-stocked treatments, which was clearly evident, especially with grasses, in the present study.

Differences in dry-matter production between the plots mechanically harvested and those grazed by animals were not significant. Pike et al. (1979) and Dudzinski and Arnold (1973) found that animals harvested less forage because of trampling, deposition of dung, and the irregular height to which the sward is grazed. The results of the present trial suggest that mechanized cutting simulated animal grazing rather accurately.

Dry-matter production, under the conditions of the present work, leads to the conclusion that plasticity and high stability of this type of ecosystem are important attributes that permit wide variations in range management without the ouput's being affected in the short term.

Botanical Composition

According to Goić (1978), the predominant species of this zone are as follows: Grasses—*Lolium* spp., *Dactylis*

Table 1. Dry-matter production on each sampling (g/1-m² quadrant), and accumulative total (total g accumulated during each yearly period).

Harvesting method	Sampling dates, 1979–80						Sampling dates, 1980–81			
	14/6	31/8	18/10	26/11	3/1/80	10/3/80	19/6	16/8	20/10	16/1/81
Non-harvested	52.5	80.0	129.0	456.7	793.5	918.4	194.4	188.1	785.5	1,247.3
Mechanized harvested	28.8	34.0	135.0	204.5	296.0	316.4	170.7	219.2	504.4	416.0
Accumulated total		62.8	197.8	402.3	698.3	1,014.7		389.9	894.3	1,310.3
Harvested by bovines	42.9	54.0	76.6	228.1	366.0	276.6	98.7	66.3	584.0	533.3
Accumulated total		96.9	173.5	401.6	767.6	1,044.2		165.0	749.3	1,282.6

Table 2. Seasonal variations of botanical composition and other characteristics (%).

Species	Not harvested pasture 1979	1980	Mechanized harvested 1979	1980	Harvested by bovines 1979	1980	Not harvested pasture 1979	1980	Mechanized harvested 1979	1980	Harvested by bovines 1979	1980
	------------------------------- Fall (June) -------------------------------						------------------------------- Winter (August) -------------------------------					
Lolium spp.	14.8	12.1	12.7	18.4	16.4	10.4	17.9	11.8	16.8	10.7	14.8	14.9
Dactylis glom.	18.1	21.0	23.6	15.5	24.0	20.9	14.9	23.0	11.9	11.3	10.5	24.2
Anthox. odor.	6.2	11.6	4.3	19.3	0.7	9.7	12.7	22.3	14.3	41.1	12.5	14.7
Other grasses	1.9	9.7	2.3	12.8	2.1	9.9	6.5	2.5	7.8	2.4	5.4	1.4
Trifolium spp.	7.4	4.1	2.7	3.2	4.5	3.7	2.9	3.9	3.2	1.5	3.1	5.7
Wide-leaf spp.	8.9	13.5	7.7	6.5	12.9	17.0	12.3	18.0	13.3	10.2	14.5	16.5
Organic matter	37.0	13.3	35.9	12.6	31.4	15.4	16.9	11.4	17.9	14.2	23.1	15.9
Bare soil	5.7	14.7	10.8	11.7	8.0	13.1	15.9	7.0	14.8	8.7	16.1	6.7
	------------------------------- Spring (November) -------------------------------						------------------------------- Summer (March) -------------------------------					
Lolium spp.	20.7	6.8	9.2	7.1	9.3	16.3	10.4	7.4	11.1	10.4	15.7	16.7
Dactylis glom.	15.2	29.0	21.7	17.9	15.4	15.7	25.4	15.7	21.1	5.8	24.8	14.4
Anthox. odor.	12.3	25.3	10.5	41.7	9.7	26.8	13.2	24.0	5.6	13.5	2.4	8.4
Other grasses	2.9	4.9	11.9	3.1	10.1	3.7	7.5	8.3	13.1	18.3	0.1	9.4
Trifolium spp.	11.7	2.4	4.6	3.6	14.3	7.4	7.2	6.9	11.1	13.2	5.5	11.8
Wide-leaf spp.	14.1	17.7	16.9	13.0	23.4	20.6	17.9	24.1	20.9	21.8	21.7	28.0
Organic matter	10.6	7.6	10.3	8.7	9.3	5.1	14.3	7.5	12.2	11.4	16.3	6.8
Bare soil	12.5	6.4	14.9	4.9	8.5	4.4	4.1	6.1	4.9	5.6	13.5	4.6

glomerata L., *Holcus lanatus* L., *Agrostis tennuis, Anthoxanthum odoratum, Arrenatherum elatius* var. bulbosum (Willa) Spenner, *Bromus catharticus.* Legumes— *Trifolium repens* L., *Trifolium pratense* L., *Lotus uliginosus.* Others— *Plantago lanceolata, Plantago mayor, Taraxacum officinale, Hypochoeris radicata,* and *Erodium* spp.

Harvesting methods did not significantly affect botanical composition (Table 2). There was a clear grass dominance in all treatments, a composition that is very different from the recommended sward composition for highly productive grazing pastures. Boswell (1979) indicated that an optimum sward should contain 25% to 30% clover. The high P immobilizing power of the soil may have depressed clover growth.

The observations also suggest that soil temperature affects the growth of grasses and clovers during the cold seasons, resulting in differential growth rates among species. Similar observations were made by Morrow and Power (1979) for grass and by Boswell (1979) for clover.

In the nonharvested treatment, the wide-leaf species, such as *Plantago* spp., *Taraxacum officinale,* and *Hypochoeris radicata,* which are of low forage value, tended to increase in all seasons. This result can be explained by the fact that colonization of those species is mainly by seeds, which have the highest establishment success in uncut pasture.

The botanical composition did not vary significantly among treatments, suggesting that pasture plasticity is highly homeostatic in spite of differing external stimuli. The opposite is true of annual pastures that are affected by utilization patterns, the effects of which can last for a year or more following removal of grazing animals (Pitt and Heady 1979).

Even though botanical composition did not vary significantly, differences in nutritive quality of pasture, as measured by morphological plant composition, were observed throughout the seasons.

The results allow pastures to be classified as compound ecosystems with the ability to respond to external stimuli. This characteristic distinguishes stabilized pastures and allows important management variations without greatly affecting pasture conditions and animal production.

Management intensity can fluctuate over a relatively wide range while maintaining an ecological ratio between removed phytomass and pasture productivity. That ratio, according to Abaturov (1979), must not exceed 70%–75% of the above-ground phytomass, which is more than animals usually can remove, resulting in a typical predator-prey system.

LITERATURE CITED

Abaturov, B.D. 1979. Peculiarities of trophic interrelationships involving plant-animal interactions in pasture ecosystems. Agro-Ecosystems 5:317–327.

Boswell, C. 1979. Maintaining grass/clover balance. N.Z. J. Agric. 138(4):9–10.

Cowlishaw, S.J., and F.E. Alder. 1960. The grazing preferences of cattle and sheep. J. Agric. Sci., Camb. 54:257–265.

Davies, A., and R.G. Simons. 1979. Effect of autumn cutting regime on developmental morphology and spring growth of perennial ryegrass. J. Agric. Sci., Camb. 92:457–469.

Dudzinski, M.L., and G.W. Arnold. 1973. Comparisons of diets of sheep and cattle grazing together on sown pastures on the Southern Tablelands of New South Wales by principal component analysis. Austral. J. Agric. Res. 24:899–912.

Gastó C.J., and R. Cañas. 1975. Modelo simulado de funcionalmiento del ecosistema silvoagropecuario. Univ. Aut. Agr. Antonio Narro, Mexico. Monog. Téc. Cient. 1:1.

Goić, M.L. 1978. Mejoramiento de praderas naturales en la región de las lluvias. Inst. Inv. Agrop. Est. Exp. Remehue, Chile. Bol. Téc.17.

Goić, M.L., and M. Matzner K. 1977. Distribución de la producción de materia seca y característica de tres regiones de la zona de las lluvias. Av. Prod. An, Chile 2:23–31.

Jolliff, G.D., A. Garza, and J.M. Hertel. 1979. Seasonal forage nutritive value variation of coastal and coastcross-1 bermudagrass. Agric. J. 71:91–94.

Morrow, L.A., and J.F. Power. 1979. Effect of soil temperature on development of perennial forage grasses. Agric. J. 71:7–10.

Nava, C.R., J. Gastó C., and R. Armijo T. 1976. Arquitectura ecosistémica: fundamentos y génesis. Univ. Aut. Agr. Antonio Narro, Mexico. Monog. Téc. Cient. 2:738.

Pendleton, D.F., and G.M. Van Dyne. 1980. Livestock and world grazinglands: status and outlook. *In* Prediction of grazingland productivity under climatic variations. Working paper, U.S. Natl. Sci. Found. Proj.

Pike, A., T. McNeilly, and P.D. Putwain. 1979. Survivor populations of S 23 perennial ryegrass from zerograzed and set-stocked swards. Grass and Forage Sci. 34:89–94.

Pitt, M.D., and H.F. Heady. 1979. The effects of grazing intensity on annual vegetation. J. Range Mangt. 32:109–114.

Teuber, K.N. 1978. Evolución de la calidad de la pradera. Inst. Inv. Agrop. Est. Exp. Remehue, Chile.

Thompson, J.K., and R.W. Warren. 1979. Variations in composition of pasture herbage. Grass and Forage Sci. 34:83–88.

Mortality of Sorghum Plants after Cutting

Y. TAKASAKI, H. OIZUMI, and H. NOJIMA
Chiba University, Chiba-ken, Japan

Summary

Sorghum (*Sorghum* spp.) is usually harvested twice a year in most of Japan but three times a year in southern Japan. Some sorghum plants die after cutting, which leads to low aftermath yield. The purpose of this study was to clarify what cultural factors cause death and what plants die in the sorghum stands.

Three field experiments were carried out from 1977 to 1979 to investigate the effect of sowing date, planting density, nitrogen (N) level, and cultivar on mortality rate after cutting. In 1979 the effect of high temperature on mortality was investigated. Information about what plants die was obtained by examining all plants within 1 m² (2 replications) in the sorghum stands.

Plant mortality of sorghum was found to increase when the sorghum was cut at an early growth stage and to be higher as density and N level increased. Mortality among cultivars differed significantly; therefore, high temperature in the field should not cause high mortality. Regrowth was greater as temperature increased; Most dead plants belonged to the lower frequency distributions of plant length, plant yield, and Brix but were higher in N content. Positive correlations were found among plant length, plant yield, and Brix scale reading. N content was found to correlate negatively with these characters. Multiple regression, using yield of regrowth as the dependent variable and plant yield, Brix, and N content as the independent variables, accounted for 92% of the variation. Linear regression using only plant yield as the independent variable accounted for 90% of the variation and explained the relation as well as the multiple regression equation.

KEY WORDS: sorghum, *Sorghum* spp., mortality, Brix, regrowth, multiple regression analysis.

INTRODUCTION

Sorghum (*Sorghum* spp.) is usually harvested twice a year in most of Japan but three times in southern Japan. Some sorghum plants die after cutting, which leads to low aftermath yield (Plucknett et al. 1970, Chugoku Agricultural Experiment Station 1970). The height of cutting (Escalada and Plucknett 1977), nitrogen (N) level, and planting density are considered to be cultural factors affecting death after cutting. However, the cause of mortality following cutting is not entirely clear. The purpose of this study was to clarify what cultural factors cause death and what plants die in the sorghum stands.

MATERIALS AND METHODS

Throughout five experiments (1977–1979), sorghum plants were cut at 10 cm above ground level. Plant length, plant yield, Brix scale reading (Brix), and N content were recorded before, at, and after cutting.

Experiment 1. Effect of sowing date and fertilization rate. Sorgo (cv. Yukijirushi hybrid sorgo) was sown at 10-day intervals, on 2 May, 12 May, 22 May, 1 June, and 11 June 1977. The planting density was 44.4 plants/m². Two fertilization levels of 540 kg/ha and 270 kg/ha each of N, phosphorus (P₂O₅), and potassium (K₂O) were evaluated. The experiment was designed as a split plot with 3 replications. Each subplot was 4 m². The sorgo plants were cut on 10 August.

Experiment 2. Effect of planting density and N fertilization. Yukijirushi hybrid sorgo was sown on 8 May 1978 at two

Y. Takasaki is an Associate Professor of Crop Science, H. Oizumi is a Professor of Crop Science, and H. Nojima is a Lecturer on Crop Science at Chiba University.

densities and treated with three N levels. The N levels were 960 kg/ha, 240 kg/ha, and 0 kg/ha (in a split application). P_2O_5 and K_2O were applied equally to all plots at the rate of 40 kg/ha. The two densities evaluated were 100 and 11.1 plants/m². The experimental design was a split plot with 4 replications. Each subplot was 4 m², harvested on 27 July.

Experiment 3. Differences between cultivars. Four cultivars (Moso, Pioneer 988, Sweet sorgo, and Yukijirushi hybrid sorgo) were sown in a uniform field on 7 May 1977. The planting density was 25 plants/m². All plots received 250 kg/ha each of N, P_2O_5, and K_2O in a split application. The unreplicated area for each cultivar was 15 × 5 m. Harvesting was on 27 July.

Experiment 4. Effect of High temperature. A total of 180 Yukijirushi hybrid sorgo plants were grown individually in 1/5,000 Wagner pots. Each pot received 0.8 g each of N, P_2O_5, and K_2O in split applications. Ninety pots were subjected to temperature treatments following cutting on 20 July, the sixty-fourth day after planting. The remaining 90 pots were subjected to temperature differences on 9 August, the eighty-fourth day after planting. Before each cutting, half of the pots were shaded to 20% of full sunlight for 7 days. The temperature treatments were continued for 2 weeks, using three growth chambers set at 12-hour day/night temperatures of 35°/30°, 30°/25°, and 25°/20°C.

Experiment 5. Characteristics of dead plants. Yukijirushi hybrid sorgo plants were grown in a field at a density of 100 plants/m² (square planting) in two areas of 4 m². The sowing date was 21 May 1979. Plants received 150 kg/ha each of N, P_2O_5, and K_2O. The plants were cut on 6 August. The growth of individual plants within 1 m² of the center of each 4-m² plot was recorded before and at cutting. The regrowth to the fourteenth day after cutting was recorded.

RESULTS AND DISCUSSION

Experiment 1. There were no significant differences in mortality up to 14 days after cutting that could be attributed to the two fertilization levels. The pooled mortalities from the two fertilization levels were 5.6%, 1.9%, 4.2%, 7.6%, and 53.7% for planting dates of 2 May, 12 May, 22 May, 1 June, and 11 June, respectively. Mortality for the 11 June sowing was significantly higher than that of other sowing dates. Cutting sorghum plants too early in the growth stage caused high mortality.

Experiment 2. Mortality was severe in high-density plots with high and moderate N levels (Table 1). Almost no

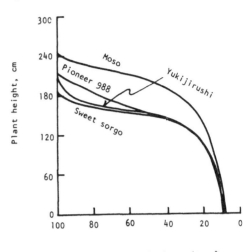
Fig. 1. Light extinction within the canopy of the four cultivars.

dead plants were found in low-density plots. Most of the dead plants were from the lower frequency distributions of plant length, plant yield, and Brix at cutting and were in the higher classes of N content.

Experiment 3. The mortality to 14 days after cutting was 0%, 0%, 32%, and 18%, for Sweet sorgo, Pioneer 988, Moso, and Yukijirushi, respectively. The growth stage at cutting was middle heading for Pioneer 988, early heading for Sweet sorgo, and before heading for Moso and Yukijirushi. The high mortality rate for Moso and Yukijirushi was attributed to differences in growth stage at time of cutting. The Brix of plants was low in Moso and Yukijirushi and associated with their high rate of mortality. Light-extinction curves for the cultivars (Fig. 1) also appeared to be associated with high mortality of Moso but not of Yukijirushi.

Experiment 4. During the two temperature treatments, only a few plants died after cutting. The yield from 14-day regrowths (Table 2) was increased by higher temperatures at each cutting and for each light intensity. The nonshaded plots gave higher yields than the shaded. Since temperatures as high as 35°/30°C are not experienced in natural field conditions, it is unlikely that high temperatures are a contributing factor to the death rate of sorghum plants.

Experiment 5. Plants that died generally belonged to the lower frequency distributions of plant length, plant yield, and Brix (Fig. 2) but to the higher frequency distribution

Table 1. Mortality rate (%) of sorghum plant to 14 days after cutting.

Density, plants/m²	N level, kg/ha		
	960	240	0
100	22.1	7.8	0.7
11.1	1.5	0.0	0.0

Table 2. Yield (g/plant) from 14-day regrowth.

Temperature, day/night	64 days after sowing		84 days after sowing	
	Nonshaded	Shaded	Nonshaded	Shaded
Brix at cutting	3.95	1.78	8.78	2.63
25/20 C	1.21a	0.34a	1.12a	0.61a
30/25 C	3.02b	1.04b	2.89b	1.35b
35/30 C	4.08c	1.19b	3.91c	2.41c

Numbers in the same column followed by the same letters are not significantly different at 0.05.

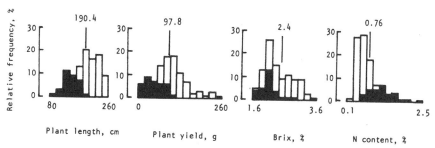

Fig. 2. Frequency distribution of plant characters at cutting. Darkened areas indicate plant mortality to 14 days after cutting. Vertical short lines indicate the mean values of each character.

of N content. However, not all plants that belonged to these critical classes died. Positive correlations were found between plant length, plant yield, and Brix. Nitrogen content correlated negatively with these characters. The yield from 14-day regrowth correlated with plant length, plant yield, and Brix at time of cutting. There was a negative correlation between the yield from regrowth and N content at cutting. Heavy plants with high Brix and low N content always made superior regrowth. The relationship in multiple regression form was

$$Y = -0.5315 + 0.0079x_1 + 0.1713x_2 - 0.0518x_3$$
$$R^2 = 0.924 \tag{1}$$

in which Y, x_1, x_2, and x_3 represented the dry yield from regrowth, the fresh yield, the Brix, and the N content respectively. Assuming x_1 and x_2 to be independent variables resulted in a multiple regression equation of the form

$$Y = -0.5963 + 0.0082x_1 + 0.1720x_2$$
$$R^2 = 0.923 \tag{2}$$

Selecting only x_1 as the independent variable gave a linear regression equation of

$$Y = -0.2593 + 0.0089x_1$$
$$R^2 = 0.906 \tag{3}$$

It was, therefore, considered that the linear regression equation including fresh yield alone explained the relation just as well as did the multiple regression equations.

LITERATURE CITED

Chugoku Agricultural Experiment Station. 1970. Study on the regional productivity of forage sorghum in Chugoku district. Coop. Res. in Chugoku Region 4:1-140.

Escalada, R.G., and D.L. Plucknett. 1977. Ratoon cropping of sorghum. III. Effect of nitrogen and cutting height on ratoon performance. Agron. J. 69:341-346.

Plucknett, D.L., J.P. Evanson, and W.G. Sanford. 1970. Ratoon cropping. Adv. in Agron. 22:286-326.

Seed and Seedling Dynamics of Annual Medic Pastures in South Australia

E.D. CARTER
University of Adelaide, Glen Osmond, South Australia

Summary

Annual species of *Medicago* (i.e., medics) are the key to successful cereal-pasture rotations in the integrated crop-livestock farming systems on the neutral to alkaline soils of southern Australia. However, many medic stands have deteriorated in recent years. This deterioration has prompted research on the effects of grazing sheep on the seed-seedling dynamics of annual medic-based pastures in the Mediterranean environment of South Australia, aiming to quantify the impact of medic seed reserves in or on the soil upon pasture establishment and productivity.

Animal-house experiments involving the feeding of purepod and pod-chaff diets to sheep and a field experiment have been used to measure the survival of medic seed following the ingestion of intact pods. In farm survey and sampling studies total seed reserves of annual medic species following cereal crops have been related to medic emergence and establishment, total pasture yield, and botanical composition.

In the field experiment on red-brown earth soil at the Waite Institute there was a highly significant linear decline in total pasture, medic pods, and medic herbage ($r = -0.998***$, $-0.992***$ and $-0.997***$, respectively) over a 56-day period of continuous grazing by merino wether sheep. Over the same period there was a highly significant linear decline in pod number/m^2 (from 4,167 to 619), pod weight (from 54.5 to 46.9 mg), seeds/pod (from 6.86 to 4.53), and seed weight (from 2.87 to 2.54 mg). Viable seeds/m^2 declined dramatically ($r = 0.987***$) from 23,715/m^2 on day 0 to 2,652/m^2 on day 56. Medic seed extraction from sheep feces showed a highly significant linear decline in seed throughput, from 6,155 to 3,251 viable seeds/sheep/day, and mean seed weight, from 2.91 to 2.60 mg, during the 56-day period.

The survey data from farms on solonized brown soils showed medic seed reserves in April varying from 16 to 282 kg/ha (mean 106.3). The mean emergence count at the end of April was 239/m^2, which population declined to 141/m^2 by the beginning of June. There was still some emergence in July, giving a mean establishment of 188/m^2 (range from 2 to 490/m^2), which was only 60% of the mean cumulative emergence of 313/m^2. The mean seed reserve in September had declined to 66.2 kg/ha with a negligible change in mean seed weight from 2.805 mg in April to 2.852 mg in September.

These studies have shown conclusively that grazing sheep can seriously deplete potential medic seed reserves and that this may result in grossly inadequate seedling density. Medic seed reserves in the top 5 cm of soil are a reliable indicator of potential density and productivity of medic pastures. For this reason core-sampling techniques are being developed to enable farmers to assess their needs for sowing additional medic seed before the autumn rains.

KEY WORDS: annual medics, *Medicago* spp., pods, sheep, seeds, emergence.

INTRODUCTION

In Australia some 40 million ha of crop and sown pasture land is dependent on legumes to provide soil nitrogen through fixation and to guarantee quantity and quality of livestock feed. Currently these legumes are worth at least U.S.$2.5 billion/yr to the Australian crop and livestock industries. The self-regenerating annual medic species (*Medicago* spp.) are the key to successful cereal-pasture rotations in the integrated crop-livestock farming systems on the neutral to alkaline soils of southern Australia. Yet many medic stands have deteriorated in recent years. This deterioration has prompted increased research at the Waite Agricultural Research Institute on the ecology and agronomy of annual medics (e.g., Carter 1976 et seq., Adem 1977, Dahmane 1978).

Recent animal-house experiments involving feeding, to sheep, pure pod and pod-chaff diets of barrel medic (*Medicago truncatula* cv. Jemalong), snail medic (*M. scutellata* cv. Commercial), strand medic (*M. littoralis* cv. Harbinger), and other medics showed large between-sheep differences in seed throughput and a seed survival rate of generally less than 2% (Carter 1980). Current research presented in this paper aims at identifying and quantifying the main effects of grazing sheep on the seed-seedling dynamics of annual medic pastures and assessing the impact of medic seed reserves on emergence, establishment, and productivity.

_____

The author is a Senior Lecturer in Agronomy at the University of Adelaide.

The author wishes to thank Mr. S. Challis and Mrs. S. Pohlenz for technical assistance. The research was supported by the Wool Research Trust Fund on the recommendation of the Australian Wool Corporation.

METHODS

Field Experiment – Waite Institute

Ten merino wethers, each fitted with fecal collection harness, were continuously grazed for 56 days from 1 March to 25 April 1979 on a 0.18-ha area of dry residues of annual medic on hard-setting red-brown earth soil. Initially the pods were 72.4% barrel medic and 27.6% burr medic (*Medicago polymorpha*) by number. Daily collections of feces were made at 0900 hr; these samples were dried in a forced-draft dehydrator at 45°C, and a subsample was threshed for estimating daily seed output, viability, etc. Weekly pasture samples (20 × 0.25 m^2 quadrats) were taken to assess changes in the availability of total pasture and its components. Data on pods and seeds in the field prior to ingestion by sheep were derived from the weekly herbage samplings. Pods and seeds were germinated in a humidified incubator at 19°C.

Farm Surveys – Mallala District

In 1980, sampling sites 25 × 25 m in area, each subdivided into 25 strata of 5 × 5 m, were located in 16 stubble paddocks on predominantly solonized brown soils of the Mallala cereal-sheep district some 60 km north of Adelaide. Sites were selected because they were intended to be put into the pasture phase of cereal-pasture rotation; however, some sites were lost through change of plans by farmers. Random soil cores were collected from each stratum in April and September to assess seed reserves of annual medics. Seed was extracted by the method of Carter et al. (1977). Medic emergence counts (one 0.1-m^2 quadrat/stratum) were made in April, June, and July,

Table 1. The components of mature medic pasture (kg DM/ha) in the Waite Institute field experiment in autumn 1979.

Component	Cumulative grazing period (days)								
	0	7	14	21	28	35	42	49	56
Total pasture	4,573	4,216	3,758	3,124	2,674	2,088	1,891	1,190	816
Medic pods	2,154	2,087	1,746	1,490	1,256	831	841	408	275
Medic herbage	2,293	2,100	1,997	1,607	1,407	1,241	1,034	771	538
Green weeds	126	29	15	27	11	16	16	11	3

and total pasture availability (kg dry matter [DM]/ha), botanical composition, and percentage of bare ground were measured three times from August to October.

RESULTS AND DISCUSSION

Field Experiment — Waite Institute

Table 1 summarizes herbage data. There was a highly significant linear decline with time in total pasture, medic pods, medic herbage, and the sum of pods and medic herbage ($r = -0.998^{***}$, -0.992^{***}, -0.997^{***}, and -0.997^{***}, respectively). At the start of grazing, medic pods were 47.1% of the total available pasture, but after 56 days pods were only 33.7% of the total pasture because of selective grazing of pods by the sheep.

Data in Table 2 show a highly significant linear decline during the experiment in the case of pod number/m^2, pod weight, seeds/pod, and seed weight ($r = -0.988^{***}$, -0.950^{***}, -0.964^{***}, and -0.883^{***}, respectively). Apparently, the smaller pods with fewer seeds escaped grazing longer than the larger pods. Though the data on viable seed percentage (based on germination of both the permeable and the impermeable component of the seed) appear somewhat erratic, the positive linear trend was significant ($r = 0.763^*$). Viable seeds/m^2 declined dramatically ($r = -0.987^{***}$) from 23,715/m^2 on day 0 to 2,652/m^2 on day 56.

Although Table 3 shows data for only 1 day in 7 these data are representative. There was a highly significant linear decline in seed throughput/sheep/day ($r = -0.742^{***}$) and mean seed weight ($r = -0.891^{***}$).

Farm Surveys — Mallala District

Variations in medic seed reserves in April were reflected in emergence (Table 4). Though mean emergence

of medics at the first count was 239/m^2, this population had declined to 141/m^2 by the second count. However, new plant numbers averaged 28/m^2, giving a total of 169/m^2. At the third count all medic plants could be regarded as established; establishment ranged from 2/m^2 on site 12 to 490/m^2 on site 10. Mean establishment was 188/m^2 in July, only 60% of the mean cumulative emergence of 313/m^2.

Drought conditions in the 1980 growing season and severe grazing pressure both caused abnormal reduction of medic seedlings and had far more influence on yield and botanical composition than did the initial medic populations. However, where grazing was less severe there were significant correlations between medic emergence-survival and April seed reserves, e.g., at site 14 for the first, second, and first + second emergence counts $r = 0.707^{***}$, 0.470^*, and 0.673^{***}, respectively.

The medic seed reserve in April consisted of both impermeable (hard) and permeable seeds averaging 2.805 mg/seed. In September the reserve was almost all hard seed, averaging 2.852 mg/seed. Thus, there was a negligible change in mean seed weight between April and September.

On 21 August, pasture availability on the 10 sites ranged from 5 to 919 kg DM/ha medic, from 51 to 1,788 kg DM/ha grass, and from 35 to 1,307 kg DM/ha other species. Total herbage availability ranged from 363 to 2,742 kg DM/ha at that time, while the amount of bare ground ranged from 10% to 74%. These low yields and high percentages of bare ground reflected both inadequate medic density and drought conditions.

The conclusion from this ongoing research is that grazing sheep may easily deplete potential medic seed reserves. Although medics are being selected and bred for improved performance (e.g., Crawford 1982), benefits from using new cultivars are negated by overgrazing of the pods. Regular replenishment of the seed reserve

Table 2. Data on medic pods and seeds before ingestion by grazing sheep in the Waite Institute field experiment in autumn 1979.

	Cumulative grazing period (days)								
	0	7	14	21	28	35	42	49	56
Pods (#/m^2)	4,167	3,866	3,111	3,040	2,601	1,812	1,878	917	619
Pod weight (mg)	54.5	54.7	54.1	50.4	49.8	47.6	47.1	46.2	46.9
Seeds/pod	6.86	6.82	6.35	6.43	6.30	5.74	5.55	5.13	4.53
Viable seed (%)	83.0	87.0	88.6	84.6	91.0	87.9	90.9	88.3	94.6
Seed weight (mg)	2.87	2.76	2.76	2.79	2.71	2.68	2.72	2.68	2.54

Table 3. Daily output of feces and medic seed/sheep and mean seed weight in the Waite Institute field experiment in autumn 1979.

| | *Cumulative grazing period (days)* | | | | | | | |
	7	14	21	28	35	42	49	56
Faeces/day (g DM)	664	681	811	826	762	828	651	706
Seeds/day (#)	6,310	4,763	3,596	2,854	2,602	2,727	3,357	3,298
Viable seeds/day	6,155	4,686	3,516	2,783	2,536	2,684	3,296	3,251
Seeds/day (g)	18.3	13.3	9.8	8.0	7.1	7.2	8.6	8.6
Seed weight (mg)	2.91	2.79	2.73	2.82	2.72	2.63	2.57	2.60

Table 4. Some medic seed and seedling data from pasture paddocks in the Mallala cereal-livestock farming district during the 1980 season.

| | *Seed reserve,*† *10–18 April* | | *Medic emergence*†† | | | | *Seed reserve,*† *23 September* | |
| Site | (kg/ha) | (#/m²) | *First* | *Second* | *Third* | *Total* (%) | (kg/ha) | (#/m²) |
			----------(plants/m²)----------					
1	19	878	79	2	66	16.7	13	547
5	282	10,820	477	6	80	5.2	186	6,925
6	204	6,354	419	18	0	6.9	142	4,524
7	29	1,508	209	41	0	16.6	14	742
8	43	1,644	89	20	56	10.0	24	974
10	240	8,498	479	26	200	8.3	121	4,547
11	16	595	35	16	4	9.2	10	374
12	53	619	9	9	0	2.9	54	550
13	27	1,106	283	7	52	30.9	9	395
14	150	5,870	312	136	0	7.6	89	3,633
Mean	106.3	3789.2	239.1	28.1	45.8	8.26	66.2	2321.1

†Soil cores for seed were c.14 cm diameter and 5 cm deep.

††The first, second, and third emergence counts were made 29–30 April, 11–12 June, and 15–16 July. The total of these three counts is expressed as a percentage of the April seed reserve.

through pod burial by shallow tillage for cereal crops is essential to ensure productive medic stands in the pasture phase of the rotation. Low medic density and consequent suboptimal nitrogen fixation is costing farmers at least U.S.$100 million annually in southern Australia. We intend to develop a simple soil-coring technique to allow farmers to assess their needs for sowing additional medic seed before the autumn rains.

LITERATURE CITED

Adem, L. 1977. Studies on the ecology and agronomy of annual *Medicago* species. M.S. Thesis, Univ. of Adelaide.

Carter, E.D. 1976. et seq. *In* Waite Agric. Res. Inst. Bien Rpts. 1974-75, 1976-77, and 1978-79.

Carter, E.D. 1980. The survival of medic seeds following ingestion of intact pods by sheep. Proc. Austral. Agron. Conf., Lawes, Qld., 178.

Carter, E.D., S. Challis, and I.G. Ridgway. 1977. The use of heavy solvents for separating seed from soil. Proc. XIII Int. Grassl. Cong., Sect. 5, 307-312.

Crawford, E.J. 1982. Development of gama medic (*Medicago rugosa* Desr.) as an annual leguminous species for dryland farming systems in southern Australia. Proc. XIV Int. Grassl. Cong.

Dahmane, A.B.K. 1978. The influence of soil, climatic and management factors on nitrogen accretion by annual *Medicago* species in a semi-arid environment of South Australia. Ph.D. thesis, Univ. of Adelaide.

Establishment of Grasses on Cracking Clay Soils: Seedling Morphology Characteristics

L.A. WATT

Soil Conservation Service Research Centre, Inverell,
N.S.W., Australia

Summary

Although native grasses establish readily on the cracking black clay soils of northwestern New South Wales, it has proven difficult to establish introduced summer grasses. Young grass seedlings were examined to see if morphological differences exist that may explain their differing capacities for establishment.

It has been shown that relative differences in the primary and secondary root growth of various species can have important effects on the agronomic characteristics of those species. Low soil moisture conditions near the soil surface will restrict greatly or possibly even prevent the growth of secondary roots of young seedlings. It is well known that the surface layers of this particular soil type dry very rapidly following rain.

Seedlings of four native and four introduced grasses were grown under favorable conditions, then destructively sampled 3, 4, and 5 weeks after planting. A number of morphological characteristics were examined, including weight of top growth, weight of root growth, length, and weight of primary and secondary roots.

There was considerable variation among the species in respect to both the amount and type of growth. Generalizing, native species tended to have smaller seedlings, lower ratios of weight of top growth to weight of root growth, and higher ratios of weight of primary root to weight of secondary root.

Given that secondary root growth is likely to be restricted by soil surface drying, it would seem the native grasses overcome this problem by having a better-developed primary root system and a less well-developed secondary root system. Also, they have a lower ratio of top growth to root growth, so that the roots should be better able to supply the plant during periods of moisture stress.

Results indicate that the selection of introduced species for use on this soil type should be made on the basis of well-developed primary root systems and a low ratio of top growth to root growth.

KEY WORDS: establishment of grasses, seedling morphology, primary roots, secondary roots, cracking clays.

INTRODUCTION

It has proven difficult to establish exotic summer grasses on the cracking black clay soils of northern New South Wales and southern Queensland (Leslie 1965). Despite this difficulty, it is evident that many of the native grasses, such as Queensland bluegrass (*Dicanthium sericeum* L), redgrass (*Bothriochloa macra* Steud, S.T.Blake), wallabygrass (*Danthonia linkii* Kunth), and native millet (*Panicum decompositum* R.Br.) do successfully establish on these soils.

Leslie (1965) considered that establishment problems were largely due to moisture stress caused by rapid drying of the surface layers, though other factors had an influence as well.

The primary root system (PR) of a grass seedling is often regarded as a temporary system that, under normal conditions, dies away after the secondary root system (SR) becomes established (Weaver and Zink 1945). However, several workers have shown that the PR of at least some cereals and perennial grasses remains alive and active for long periods, and in fact some plants can survive and set seed when grown on the PR alone if forced to do so (Locke and Clarke 1924, Pavlychenko 1937, Olmsted 1941, Sallans 1942, Weaver and Zink 1945, Boatwright and Ferguson 1967, Passioura 1972, Van Der Sluijs and Hyder 1974).

Under favorable conditions, it is usual for the secondary root system to start growth early in the life of the seedling. However, low moisture conditions at and near the soil surface will restrict greatly and possibly even prevent the growth of SR (Howard 1924, Pavlychenko 1937, Olmsted 1941, Reigel 1941, Boatwright and Ferguson 1967, Passioura 1972).

Different workers have argued that relative differences in the primary and secondary root growth of various species can have important effects on the agronomic characteristics of those species (Pavlychenko 1937, Salim

The author is the Officer-in-Charge of the Soil Conservation Service Research Centre.

et al. 1965, Passioura 1972, Hyder et al. 1971).

This study examined a range of grasses to determine if a consistent relationship could be found between seedling morphology, particularly root morphology, and establishment success.

METHODS

A selection of perennial grasses was made to include both native (N) and introduced (I) species (Table 1) and to cover a range from poor to good establishment capability. A complete objective assessment of the relative establishment capability of the species was not available, but the literature allowed a subjective assessment to be made. Species tested, listed in order of decreasing establishment capability, were Queensland blue grass, wallabygrass, redgrass, native millet, purple pigeongrass (*Setaria porphyrantha* Stapf), Bambatsi panic (*Panicum coloratum* L.), buffelgrass (*Cenchrus ciliaris* L.), and rhodesgrass (*Chloris gayana* Kunth).

Grasses were sown at a depth of 5 mm into divided sections of flower pots, four species/pot. Enough pots were sown to allow destructive sampling on three separate dates taking four replications/sampling date. The grasses were grown under favorable conditions of temperature and moisture, and all species made excellent growth. After 3,4, and 5 weeks of growth the pots were soaked in a soil dispersant for 2 hours, and then the plants were removed from the soil under a jet of water. Several plants were obtained from each pot. Two undamaged plants were selected from each replicate and measured for a number of parameters. After measurement these plants were dried by pressing and then were photographed on a 5-cm grid. Of the remaining plants several were divided into top growth, primary root growth, and secondary root growth, and these parts were oven-dried at 70°C and then weighed.

Mean measurements of the following parameters on a single-plant basis were arrived at: weight of top growth, length of primary root, weight of primary root including all branches, number of secondary roots, length of longest secondary root, and weight of secondary root system. This allowed calculation of weight of top growth:weight of total root and weight of primary root:weight of total root.

RESULTS

As space is limited, only the data from the first of the three harvests are presented here, as these are the most relevant to the overall objective of the study and the other samplings displayed similar data trends.

All seedlings produced a primary root (PR) and a mesocotyl, while none produced transitionary node roots (Hoshikawa 1969). In almost all seedlings the mesocotyl roots did not develop, and when they did they were very small. The SR (or secondary root) development varied from species to species.

There was considerable variation among the species, in respect to both the amount of growth and the type of growth (Table 1). For example, at 3 weeks the mean dry weight of the smallest plant was 2.9 mg, while that of the largest plant was 25.9 mg. Marked differences in the relative amounts of top growth, weight of primary root, and weight of secondary root were also evident. On the first sampling day Queensland bluegrass had the highest ratio of primary root to total root (1.0), while rhodesgrass had the lowest (0.42).

DISCUSSION

At the first sampling Queensland bluegrass, redgrass, and wallabygrass had the highest ratios of primary root to total root and also had the lowest ratios of top growth to total root growth (Table 1). These are native grasses known to establish readily on the black earth soils. At the second and third samplings the placement of the species was similar overall, although wallabygrass interchanged with purple pigeongrass to some extent.

Three introduced grasses (rhodesgrass, buffelgrass, and Bambatsi panic) that are difficult to establish on black earths had low ratios of weight of primary root to weight

Table 1.   Mean plant parameters expressed on per-plant basis after 3 weeks growth under favorable conditions.

SPECIES*	Qld blue-grass	red-grass	wallaby-grass	native millet	purple pigeon-grass	Bambatsi panic	buffel-grass	Rhodes-grass
Nat. (N) intr.(I)	N	N	N	N	I	I	I	I
Assessed Est. Capability	best							worst
Wt. plant (mg)	6.0	10.7	2.9	8.3	10.9	6.8	25.9	14.4
Wt. top growth (mg)	3.0	6.7	2.0	6.3	7.2	4.7	19.3	11.8
Wt. PR (mg)	3.0	3.9	0.8	1.5	3.0	1.7	4.4	1.1
Length PR (cm)	20.5	25.8	10.4	11.0	14.1	13.3	26.6	13.4
Wt. CNR (mg)	0	0.2	0.3	0.5	0.6	0.5	2.2	1.7
No. secondary roots	0.1	0.1	0.6	1.8	1.1	1.3	1.3	4.0
Longest CNR (cm)	0.5	0.4	3.1	9.5	6.5	6.5	7.5	11.1
Wt. top/total roots	1.13	2.07	2.05	3.39	2.39	2.5	3.07	4.85
Wt. PR/Total roots	1.00	0.95	0.86	0.76	0.79	0.76	0.74	0.42

*Species listed in order: Queensland bluegrass, redgrass, wallabygrass, native millet, purple pigeongrass, Bambatsi panic, buffelgrass, and rhodesgrass.

of total root and high ratios of weight of top growth to weight of total root growth.

Native grasses appear to put proportionately more of their early growth into the primary root and, indeed, into the total root system compared with the top growth than do the introduced species.

As these seedlings were grown under ideal well-watered conditions, plasticity of the grasses has not been well illustrated. The secondary root growth of the introduced grasses would have been enhanced over that expected under actual field conditions. However, the fact that the native grasses were so comparatively slow to respond to favorable conditions and to initiate secondary roots at an early stage may reflect a natural tendency to rely on the primary root system in the early stages of growth.

Because the surface layers of black earths dry out rapidly, grass seedlings may have difficulty producing secondary roots. Plants that have the best-developed primary root system may be best able to survive until conditions favorable to secondary root development occur. Second, the root system has to provide moisture and nutrients for the top growth. If dry conditions occur, one could expect that the grasses with the lowest ratio of top growth to root growth should be best able to survive.

The findings of this study support this general hypothesis and suggest that a breeding/selection program based on these seedling characteristics may provide a means of obtaining grasses that can survive drying seedbed conditions and reliably establish on black earths.

LITERATURE CITED

Boatwright, G.O., and H. Ferguson. 1967. Influence of primary and/or adventitious root systems on wheat production and nutrient uptake. Agron. J. 59:299-302.

Hoshikawa, K. 1969. Underground organs of the seedlings and the systematics of gramineae. Bot. Gaz. 130:192-203.

Howard, A. 1924. Crop production in India. Oxford Univ. Press.

Hyder, D.N., A.C. Everson, and R.E. Bement. 1971. Seedling morphology and seedling failures with blue grama. J. Range Mangt. 24:287-292.

Leslie, J.K. 1965. Factors responsible for failures in the establishment of summer grasses on the black earths of the Darling Downs, Qld. Qld. J. Agric. Anim. Sci. 22:17-38.

Locke, L.F., and J.A. Clarke. 1924. Normal development of wheat plants from seminal roots. Am. Soc. Agron. J. 16:261-268.

Olmsted, C.E. 1941. Growth and development in range grasses. I. Early development of *Bouteloua curtipendula* in relation to water supply. Bot. Gaz. 102:499-519.

Passioura, J.B. 1972. The effect of root geometry on the yield of wheat growing on stored water. Austral. J. Agric. Res. 23:745-752.

Pavlychenko, T.K. 1937. Quantitative study of the entire root systems of weed and crop plants under natural conditions. Ecology 18:62-79.

Reigel, A. 1941. Life history and habits of blue grama. Kan. Acad. Sci. Trans. 44:76-85.

Salim, M.H., G.W. Todd, and A.M. Schlehuber. 1965. Root development of wheat, oats and barley under conditions of soil moisture stress. Agron. J. 57:603-607.

Sallans, B.J. 1942. The importance of various roots to the wheat plant. Can. J. Agric. Sci. 23:17-26.

Van Der Sluijs, D.H., and D.N. Hyder. 1974. Growth & longevity of blue grama seedlings. J. Range Mangt. 27:117-120.

Weaver, J.E., and E. Zink. 1945. Extent & longevity of the seminal roots of certain grasses. Pl. Physiol. 20:359-379.

Legume Establishment in Grass-Dominant Swards: Concept and History in Kentucky

T.H. TAYLOR and W.C. TEMPLETON, JR.
University of Kentucky and U.S. Regional Pasture Research Laboratory, USDA-SEA

Summary

Grassland farming has been practiced for 2 centuries in central Kentucky and for shorter periods in other parts of the state. Savannas and tallgrass prairies were severely grazed by domestic animals and row-cropped by early settlers, destroying most of the indigenous plant species. By the early part of the 19th century, beef cattle were being produced on bluegrass (*Poa pratensis* L.) pastures and driven on foot to east-coast markets. During the first century of settlement, farmers learned to top-seed legumes into grass sods, and in the course of the last hundred years development of grassland technology for establishing legumes in grass-dominant swards has slowly and steadily improved. Highlights of research leading to current concepts of establishing and exploiting legumes in grass-dominant swards are given. Stand maintenance of birdsfoot trefoil (*Lotus corniculatus* L.) and bigflower vetch (*Vicia grandiflora* var. kitaibeliana W. Koch) through periodic natural reseeding is suggested.

KEY WORDS: grassland renovation, cool-season grasses, top seeding, minimum tillage, natural reseeding, overseeding.

INTRODUCTION

Savannas of central Kentucky and tallgrass prairies of the Pennyroyal region supported large numbers of wild ruminants prior to settlement by Europeans in 1774. Bluegrass (*Poa pratensis* L.) and white clover (*Trifolium repens* L.) were identified in the region (Gist 1750–51, Nourse 1775). Early white settlers brought cattle and horses to the area (Kincaid 1947). The native grasslands were grazed and row-cropped, destroying most of the indigenous species. By 1790 cattle raising was becoming established, and in the early 1800s fat cattle were being driven on foot to east-coast markets (Henlein 1959, Kincaid 1947).

Bluegrass came to be called Kentucky bluegrass between 1833 and 1859, to identify its origin when seeds were sold to other states (Fergus and Buckner 1973). Kentucky became known as the Bluegrass State during the 19th century because of the excellent bluegrass-clover pastures of the central part of the state.

ALFALFA PRODUCTION SCHEMES

Garman (1902) published cutting-management schemes for alfalfa (*Medicago sativa*), and, building on Garman's findings, Roberts et al. (1914) developed liming, fertilizing, inoculating, and cutting recommendations for alfalfa. They recognized that all nutrient and ecological needs of a forage crop should be provided, but, to this day, that concept is not appreciated by many grassland farmers. Mutchler (1914), who was in charge of the county agents, wrote, "We lay vigorous stress on the growth of legumes and emphasize the use of phosphates in acid or rock form. . . ." He reported that the Extension Service had 206 alfalfa demonstrations occupying 427 ha.

PLACE OF LEGUMES IN GRASSLANDS

Studies of the contributions of legumes to pasture and hay production continued at the Kentucky station during the 1920s and 1930s. Karraker (1925) showed that well-nodulated sweet clover (*Melilotus* sp.) added considerable amounts of biologically fixed N to the plant-soil system. Fergus (1935) and Roberts (1937) reported that the presence of legumes in grasslands increased production, improved vigor of the sod, reduced weed growth, and increased protein and mineral concentration of the forage. Fergus also demonstrated that clover could be established, increased, and maintained in grass sods by top-seeding under suitable grazing and/or clipping, liming, and fertilizing practices (Anon. 1939).

The authors are a Professor of Agronomy at the University of Kentucky and the Director of the U.S. Regional Pasture Research Laboratory, USDA-SEA, who was formerly a Professor of Agronomy at the University of Kentucky. The investigation reported in this paper (No. 81-3-27) is in connection with a project of the Kentucky Agricultural Research Station and is published with the approval of the Director.

KENTUCKY 31 TALL FESCUE

In the mid-1940s, Kentucky 31 tall fescue (*Festuca arundinacea* Schreb.) was released by the Kentucky station (Fergus 1952, 1972). Seed production increased rapidly, and by the early 1950s seed was available at reasonable cost to farmers throughout the U.S. Millions of ha in the southeastern U.S. were sown; however, farmers, county agents, and grassland researchers soon realized that pure stands of tall fescue did not provide high-quality feed as pasturage during summer.

RENOVATION OF TALL FESCUE SWARDS

Employing Fergus's (1935, 1952) concepts of grass-sod renovation, Jurado-Blanco (1952) demonstrated that legumes could be introduced into tall fescue swards without destruction of the grass. Other tall fescue renovation experiments were initiated in western Kentucky, where the grass had become the dominant forage species (Taylor et al. 1955). Extension publications were written on methods of introducing legumes into grass-dominant swards using common farm implements (Taylor et al. 1958, 1959). As in 1914, extension agents carried the research findings to farmers (Thompson et al. 1960).

Legume establishment proved successful on minimum-tilled tall fescue sods, provided that adequate rates of lime, P, and K were applied before seeding (Taylor et al. 1959). Minimum tillage consisted of drilling the seed into undisturbed sod with a grain drill or disking the sod lightly, then seeding with a grain drill. Development of volunteer legume stands from seed in the soil of check plots, managed in the same way as the legume-seeded plots, suggested natural reseeding as an alternative to mechanical renovation of swards. With these findings and observations on animal responses to grass-legume swards in the 1950s, experiments were conducted in the 1960s and 1970s to further develop procedures for grassland renovation through exploitation of legumes.

ANIMAL RESPONSE TO RENOVATED PASTURES

In a grazing trial conducted from 1962 to 1966, renovated bluegrass-alfalfa pastures gave 40% more grazing than did bluegrass–white clover (Templeton et al. 1970). In the later years of the trial, bluegrass-alfalfa was as productive as winter rye (*Secale cereale* L.), followed by sudangrass (*Sorghum bicolor* [L.] Moench) or sorghum-sudangrass hybrids with a total N application of 336 kg/ha/yr.

MINIMUM TILLAGE AND HERBICIDE STUDIES

Taylor et al. (1969) showed that minimum tillage and seed coverage enhanced stand establishment compared with top-seeding. In some instances, banding a grass herbicide over the seeded row improved legume stands. Seed coverage, 1.2 cm for alfalfa and 0.6 cm for white clover, was found to be the controllable factor that most consistently contributed to successful establishment.

DEVELOPMENT AND TESTING OF SEEDER

Additional overseeding studies were conducted (Taylor et al. 1972). Seed coverage in February, March, and April plantings enhanced stands and led to the conclusion that seeding rates could be reduced by 50% with appropriate seed coverage. Then a prototype field machine was designed and constructed to form a narrow furrow, drop seed in the furrow, firm soil over the seed, and apply herbicide in one pass over the land (Smith et al. 1973).

Employing the minimum-tillage seeder and herbicides, Taylor et al. (1977, 1978) seeded Vernal alfalfa and Kenstar red clover (*Trifolium pratense* L.) into established Kenblue Kentucky bluegrass and Kentucky 31 tall fescue for three successive years and observed each trial through the year after establishment. Four rates of N fertilizer were applied on grass plots not seeded to legumes. Use of herbicides did not improve red clover stands in either of the grasses; variable results were observed with alfalfa. Although tall fescue grown alone was higher yielding than bluegrass, the legumes were more easily introduced into fescue. Red clover-grass swards were more productive during the establishment year than alfalfa-grass mixtures, but the reverse was found the year after establishment. Alfalfa-grass swards were as productive in the post-establishment year as the grasses receiving 336 kg/ha or more of N, and red clover yields were only slightly lower.

NATURAL RESEEDING

Templeton et al. (1967) demonstrated that birdsfoot trefoil (*Lotus corniculatus* L.) stands may be maintained in Kentucky bluegrass under grazing through natural reseeding of the legume. Taylor et al. (1973) maintained excellent trefoil stands in bluegrass on plots permitted to set and shatter seed 2 out of 5 years. These results were in marked contrast to results in plots prevented from reseeding (102 compared with 22 plants/m², respectively). Herbage produced during the reseeding process was of acceptable quality and could be especially important as pasturage for certain classes of livestock in summer.

Templeton and Taylor (1975) found that bigflower vetch (*Vicia grandiflora* var. kitaibeliana W. Koch) could be successfully established in grass swards without seedbed preparation. Owing to ease of establishment and competitiveness with established grasses, they concluded that it could be used as a pioneer legume for grassland improvement where conventional seeding methods are impractical. In other experiments, Templeton et al. (1976) showed that bigflower vetch could be maintained with grasses through natural reseeding.

LITERATURE CITED

Anon. 1939. Effects of legumes on bluegrass. Ky. Agric. Expt. Sta. 52nd Ann. Rpt., 15.

Fergus, E.N. 1935. The place of legumes in pasture production. Am. Soc. Agron. J. 27:367-373.

Fergus, E.N. 1952. Kentucky 31 fescue—culture and use. Ky. Agric. Ext. Serv. Cir. 497.

Fergus, E.N. 1972. A short history of Kentucky 31 fescue. *In* Proc. 29th Southern Pasture Forage Crop Impr. Conf., 136-139. Clemson, S.C.

Fergus, E.N., and R.C. Buckner. 1973. The bluegrass and redtop. *In* M. E. Heath, D. S. Metcalfe, and R. F Barnes (eds.) Forages—the science of grassland agriculture, 3rd ed. 243-253. Iowa State Univ. Press, Ames, Iowa.

Garman, H. 1902. Kentucky forage plants—the clovers and their allies. Ky. Agric. Expt. Sta. Bul. 98.

Gist, C. 1750-51. Christopher Gist's journals, 55-62. J. R. Weldin, Pittsburgh, 1893.

Henlein, P.C. 1959. Cattle kingdom in the Ohio Valley, 1783-1860. Univ. of Ky. Press, Lexington, Ky.

Jurado-Blanco, B. 1952. Effects of clipping and tillage on the emergence and establishment of some legume species in Kentucky 31 fescue pastures. M.S. Thesis, Univ. of Kentucky.

Karraker, P.E. 1925. Note on the increased growth of bluegrass from associated growth of sweet clover. Am. Soc. Agron. J. 17:813-814.

Kincaid, R.L. 1947. The wilderness road. Bobbs-Merrill Co., Am. Trails Ser. Indianapolis, Ind.

Mutchler, F. 1914. The work of county agents. Ky. Agric. Exp. Sta. 27th Ann. Rpt., 16.

Nourse, J. 1775. Journey to Kentucky in 1775. *In* J. Am. Hist. 19(2):251-263. (1925).

Roberts, G. 1937. Legumes in cropping systems. Ky. Agric. Expt. Sta. Bul. 374.

Roberts, G., E.J. Kinney, and H.B. Hendrick. 1914. Alfalfa and sweet clover. Ky. Agric. Expt. Sta. Bul. 178.

Smith, E.M., T.H. Taylor, J.G. Casada, and W.C. Templeton, Jr. 1973. Experimental grassland renovator. Agron. J. 65:506-508.

Taylor, T.H., J.S. Foote, J.H. Snyder, E.M. Smith, and W.C. Templeton, Jr. 1972. Legume seedling stands resulting from winter and spring sowings. Agron. J. 65:535-538.

Taylor, T.H., E.M. Smith, and W.C. Templeton, Jr. 1969. Use of minimum tillage and herbicide for establishing legumes in Kentucky bluegrass (*Poa pratensis* L.) swards. Agron. J. 61:761-766.

Taylor, T.H., W.C. Templeton, Jr., P.L. Cornelius, E.M. Smith, and J.K. Evans. 1978. Yield, quality, and botanical composition of renovated cool-season grass sods. II. Year after establishment. Agron. Abs. 1978:105.

Taylor, T.H., W.C. Templeton, Jr., E.N. Fergus, and W.N. McMakin. 1958. Renovation of pastures. Ky. Agric. Ext. Serv. Leaflet 210.

Taylor, T.H., W.C. Templeton, Jr., and W.N. McMakin. 1959. Improve grass pastures by growing more legumes. Better Crops Pl. Food 43(1):32-38.

Taylor, T.H., W.C. Templeton, Jr., W.N. McMakin, and S.H. West. 1955. Establishment of legumes in an old tall fescue sod. Ky. Agric. Expt. Sta., 68th Ann. Rpt., 18-19.

Taylor, T.H., W.C. Templeton, Jr., E.M. Smith, and J.K. Evans. 1977. Yield, quality, and botanical composition of renovated cool-season grass sods. I. Establishment year. Agron. Abs. 1977:105.

Taylor, T.H., W.C. Templeton, Jr., and J.W. Wyles. 1973. Management effects on persistence and productivity of birdsfoot trefoil (*Lotus corniculatus* L.). Agron. J. 65:646-648.

Templeton, W.C., Jr., C.F. Buck, and N.W. Bradley. 1970.

Renovated Kentucky bluegrass and supplementary pastures for steers. Ky. Agric. Expt. Sta. Bul. 709.

Templeton, W.C., Jr., C.F. Buck, and D.W. Wattenbarger. 1967. Persistence of birdsfoot trefoil under pasture conditions. Agron. J. 59:385–386.

Templeton, W.C., Jr., and T.H. Taylor. 1975. Performance of bigflower vetch seeded into bermudagrass and tall fescue swards. Agron. J. 67:709–712.

Templeton, W.C., Jr., T.H. Taylor, J.W. Wyles, and P.G. Woolfolk. 1976. Seed production and stand regeneration of bigflower vetch, *Vicia grandiflora* var. kitaibeliana W. Koch, grown with cool-season forage grasses. Agron. J. 68:267–271.

Thompson, W.C., T.H. Taylor, and W.C. Templeton, Jr. 1960. Put legumes back in grass pastures. Ky. Agric. Ext. Serv. Leaflet 228.

Effects of Treading and the Return of Excreta on a Perennial Ryegrass-White Clover Sward Defoliated by Continuously Grazing Sheep

M.L. CURRL and R.J. WILKINS

Grassland Research Institute, Hurley, Berkshire, England, UK

Summary

Grazing sheep may affect the association between grass and clover in a mixed sward by defoliation, treading, and the circulation of nitrogen (N) via excreta. Our objective was to examine the effect of treading and excreta on a perennial ryegrass (*Lolium perenne*)-white clover (*Trifolium repens*) sward defoliated by sheep at high and low stocking rates.

Established swards of perennial ryegrass and white clover were continuously stocked from May to October with yearling wether sheep at 25 and 50 head/ha. Sward production and species composition were measured inside graze-through cages that allowed defoliation without treading and excreta return, and outside, where sheep grazed either fitted with harness to prevent excreta return or unharnessed to allow normal excreta return.

Herbage production from the defoliated sward was estimated at 5.42 and 8.99 metric tons (t) dry matter (DM)/ha at high and low stocking rates, respectively. Treading reduced these figures by 10% at high stocking rate (P < 0.05) and by 4% at low stocking rate (P < 0.05). Excreta return increased herbage production by 53% at high stocking rate and 26% at low stocking rate.

At low stocking rate, treading had little effect on botanical composition. Without the return of excreta, clover content increased from about 58% at the start of the experiment in spring to about 71% in autumn. Excreta return at this stocking rate reduced clover to 45% in autumn. The quantity of clover (kg/ha) present at the end of the experiment was higher than at the beginning in the treatments both with and without excreta return; the reduction in clover percentage where excreta were returned was primarily the result of an increase in the quantity of ryegrass present. Measurement of clover stolon changes showed that excreta return resulted in little change in stolon length, but without excreta return there was a substantial increase in stolon length.

At the high stocking rate, the percentage of clover present in autumn was 31% in the defoliated sward, 25% where this sward was trodden, and 18% where excreta were returned. Clover stolon length declined during the grazing season in all treatments, with the decline being greatest where excreta were returned.

In conclusion, at a stocking rate where treading by continuously grazing sheep had a negligible effect, N transfer via excreta significantly increased herbage yield and reduced the proportion of clover in a grass-clover sward. At a higher stocking rate, herbage yields were depressed, and the benefits of increased N transfer via excreta were outweighed by the detrimental effects of increased defoliation and treading.

KEY WORDS: perennial ryegrass, *Lolium perenne*, white clover, *Trifolium repens*, defoliation, treading, excreta return, sheep.

INTRODUCTION

Grazing sheep may affect the association between grass and clover in a mixed sward by defoliation, treading, and N circulation via excreta. The relative importance of each of these components of the grazing process may change with grazing management and stocking rate.

Several studies have considered the effect of feces and urine returned to grass-clover swards under a rotational grazing system of management involving short grazing periods, long rest periods, and an unspecified number of animals (e.g., Sears and Newbold 1942, Watkin 1954, 1957). Others (e.g., Edmond 1964) have simulated the effect of treading on grass-clover swards by use of a high-intensity, short-duration technique formulated by Edmond (1958).

As part of a study on the effects of set stocking by sheep on swards of perennial ryegrass (*Lolium perenne*) and white clover (*Trifolium repens*), the experiment described here examined the effect of treading on a defoliated sward and the effect of excreta on this defoliated and trodden sward.

METHODS

An established sward of perennial ryegrass (cv. Endura) and white clover (cv. Blanca) was continuously stocked from May to October with yearling wether sheep (55 ± 1.3 kg initial liveweight) at 25 and 50 head/ha. Sward growth and species composition were measured inside graze-through cages (Smith et al. 1971), which allowed defoliation without treading and excreta return, and outside where sheep grazed either harnessed to prevent the return of feces and urine or unharnessed to allow normal excreta return. Harnesses remained on the animals throughout the experiment, and the bags and tanks were emptied once or twice daily.

At 3-week intervals, the botanical composition of above-ground herbage in each sward was estimated from eight 78-cm² quadrats cut to ground level on sites selected using McIntyre's (1952) method of ranked sets. Movable cages were used to provide an estimate of pasture growth under grazing with the cages moved at 3-week intervals. Clover stolon length, leaf number, and mature leaflet width were measured in May and September from eight 78-cm² pasture cores selected according to McIntyre (1952).

Fecal and urine N output was measured over 8-day sampling periods at four equally spaced intervals during the experiment.

RESULTS

Observations, early in the grazing season, on the frequency of defoliation and its severity on marked ryegrass

M.L. Currl's *current address is Agricultural Research Institute, Wagga Wagga, N.S.W. 2650, Australia.*

Table 1. Effect of defoliation, treading, and excreta return at high and low stocking rates on herbage yield, May to October.

Treatment	Yield, t DM/ha		
	Total	Ryegrass	Clover
50 sheep/ha			
Defoliation	5.42	2.55	2.22
Defoliation + treading	4.87	2.46	1.82
Defoliation + treading + excreta	7.44	4.36	2.18
25 sheep/ha			
Defoliation	9.00	2.21	5.71
Defoliation + treading	8.64	2.20	5.31
Defoliation + treading + excreta	10.88	4.53	4.87
SE of difference	±0.24	±0.21	±0.19

tillers and clover stolons indicated that the graze-through cage did not interfere with the grazing behavior of the sheep.

Herbage yield for the grazing season was 66% greater (P < 0.001) in the defoliated sward at the low stocking rate than at the high stocking rate (Table 1). Treading reduced herbage yield by 10% (P < 0.05) at the high stocking rate and by 4% (not significant; P < 0.05) at the low stocking rate. However, the return of excreta increased herbage yield by 53% (P < 0.001) at high stocking rate and by 26% (P < 0.001) at the low stocking rate. Sheep grazing swards without feces and urine return excreted 225 and 191 kg N/ha during the grazing season at the high and low stocking rates, respectively.

Ryegrass and clover constituted at least 89% of all herbage present. Ryegrass percentage tended to change in a manner converse to that of the clover. At low stocking rate, treading had little effect on botanical composition. Without excreta return, clover content increased from about 58% at the start of the experiment in spring to 75% in the defoliated and trodden sward (Table 2). Excreta return at this stocking rate reduced clover content to 45% in autumn. The lower percentage of clover with excreta return was primarily the result of an increase in the quan-

Table 2. Effect of defoliation, treading, and excreta return at high and low stocking rates on botanical composition of green herbage on offer in autumn (27 August–8 October).[1]

Treatment	RG[2] %	Cl %	Oth %
50 sheep/ha			
Defoliation	59c	31d	10a
Defoliation + treading	64b	25e	11a
Defoliation + treading + excreta	74a	18e	8ab
25 sheep/ha			
Defoliation	23e	75a	2c
Defoliation + treading	26e	71b	3c
Defoliation + treading + excreta	51d	45c	4c

[1]For each component, means that do not have a common letter differ significantly (P < 0.05).

[2]RG = ryegrass; Cl = clover; Oth = other spp.

Table 3. Effect of defoliation, treading, and excreta return at high and low stocking rates on clover stolon length/unit area.[1]

Treatment	Stolon length, m/m²	
	17 May	21 Sept
50 sheep/ha		
Defoliation	85.0de	43.0g
Defoliation + treading	81.8cf	32.9h
Defoliation + treading + excreta	78.8f	25.4i
25 sheep/ha		
Defoliation	93.7c	173.4a
Defoliation + treading	93.4c	160.5b
Defoliation + treading + excreta	89.8cd	94.8c

[1]Values that do not have a common letter differ significantly (P < 0.05).

tity of ryegrass present; excreta increased the DM yield of ryegrass by more than twofold but reduced the DM yield of clover by only 8% (Table 1). With excreta return, clover stolon length showed little change (P < 0.05) during the grazing season, but there was a substantial increase without excreta return, the increase being marginally higher in the untrodden sward.

At the high stocking rate, the percentage of clover present in autumn was 31% in the defoliated sward, 25% when this sward was trodden, and 18% where excreta were returned (Table 2). Clover stolon length declined during the grazing season in all treatments (Table 3). At the end of the grazing season, stolon length was least with excreta return and greatest where the sward was defoliated only.

DISCUSSION

The difference in intensity of defoliation between stocking rates seemed to have a greater effect than excreta return or treading on sward performance.

The yield response and the encouragement given to ryegrass rather than to clover by feces and urine agree with the findings of Sears and Newbold (1942) for rotationally grazed swards. Watkin (1954 and 1957) also found that exreta encouraged ryegrass in the grass-clover sward, but they did not record a yield response. It appears that in the current experiment with set-stocked swards, stocking rates were high enough to provide sufficient quantity of reasonably uniformly distributed feces and urine to exert a significant effect on sward fertility; the cycling of readily available N via excreta appeared to more than match any possible increase in fixed N from a slightly higher content of clover without excreta return. If

the totals of gaseous and leaching losses of N from feces and urine are assumed to be 10% and 34%, respectively (Floate 1970, Whitehead 1970), the nonreturn swards were deprived of about 164 and 141 kg N/ha at the high and low stocking rate, respectively.

The effect of treading on sward yield and botanical composition was less than that recorded by others who simulated treading under grazing. At the low stocking rate there were presumably too few sheep to cause damage, or, alternatively, there was sufficient above-ground herbage to cushion the impact of the grazing animals' hooves. The greater sensitivity of clover to treading is in accord with the findings of Edmond (1966). However, the loss of clover at the high stocking rate can be attributed primarily to the intensity of defoliation, since the reduction over time in all treatments at this stocking rate was greater than any difference between trodden and untrodden swards or between swards with and without excreta return.

LITERATURE CITED

Edmond, D.B. 1958. The influence of treading on pasture: a preliminary study. N.Z. J. Agric. Res. 1:319-328.

Edmond, D.B. 1964. Some effects of sheep treading on the growth of 10 pasture species. N.Z. J. Agric. Res. 7:1-6.

Edmond, D.B. 1966. The influence of treading on pasture growth. Proc. X Int. Grassl. Cong., 453-458.

Floate, M.J.S. 1970. Decomposition of organic materials from hill soils and pastures. II. Comparative studies on the mineralization of carbon, nitrogen and phosphorus from plant materials and sheep faeces. Soil Biol. and Biochem. 2:173-185.

McIntyre, G.A. 1952. A method of unbiased selective sampling using ranked sets. Austral. J. Agric. Res. 3:385-390.

Sears, P.D., and R.P. Newbold. 1942. The effect of sheep droppings on yield, botanical composition and chemical composition of pasture. I. Establishment of trial, technique of measurement and results for the 1941-42 season. N.Z. J. Sci. and Technol. 24A:36-61.

Smith, A., R.A. Arnott, and J.M. Peacock. 1971. A comparison of the growth of a cut sward with that of a grazed sward, using a technique to eliminate fouling and treading. J. Brit. Grassl. Soc. 26:157-162.

Watkin, B.R. 1954. The animal factor and levels of nitrogen. J. Brit. Grassl. Soc. 9:35-46.

Watkin, B.R. 1957. The effect of dung and urine and interactions with applied N, P, K on the chemical composition of pasture. J. Brit. Grassl. Soc. 12:264-277.

Whitehead, D.C. 1970. The role of nitrogen in grassland productivity. Commonwealth Bur. Past. and Field Crops, Bul. 48. Farnham Royal.

Performance of a Subtropical Legume-Grass Pasture Under Different Grazing Management Systems

G.E. MARASCHIN, S.C. MELLA, G.S. IRULEGUI, and J. RIBOLDI

Departamentos de Fitotecnia and de Estatística, UFRGS, Brazil

Summary

Agronomic evaluation of the dynamics of tropical legume-based pastures is relevant in assessing the response of their components within systems of grazing management. The response of tropical legume-grass mixtures to periods of rest and grazing pressure and the combinations that result in meaningful legume content in the mixture need to be known.

A central composite design was used to study the performance of a *Desmodium intortum* (Mill.) Urb. and *Paspalum guenoarum* Arech. pasture mixture to 14, 28, 42, 56, and 70 days of rest combined with 2.0%, 3.5%, 5.0%, 6.5%, and 8.0% BW (i.e., kg dry matter (DM) on offer/day/100 kg body weight) as grazing pressure, in a rotational grazing system. Forage DM yield and botanical composition were assessed by a double-sampling procedure. Pasture parameters were analyzed by a second-degree polynomial, and canonical analysis facilitated the interpretation of pasture performance.

Daily pasture DM yield tended to increase under lenient grazing pressure, while higher percentages of contribution and survival of *P. guenoarum* and *D. intortum* were achieved with longer rest periods. The invasion of other grasses increased with short periods of rest and higher grazing pressure. Weeds were present at very low levels. The stocking rate increased with increasing grazing pressure, but forage consumption/animal tended to increase under lenient grazing pressure combined with long rest periods. *P. guenoarum* and *D. intortum* assure good-quality feed at long periods of rest.

Long rest periods are required to maintain *D. intortum* and *P. guenoarum* as active components of the pasture sward and can effectively contribute to forage feeding systems in the tropics.

KEY WORDS: *Paspalum guenoarum* Arech., *Desmodium intortum* (Mill.) Urb., grazing systems, days rest, grazing pressure, botanical composition, stocking rate, IVOMD.

INTRODUCTION

Legume-based pastures have played a minor role in livestock production throughout the tropics and subtropics because of poor persistence of tropical legumes under grazing. However, the ceiling for animal production from legume-grass pastures is determined by the legume content of the pastures, since the nitrogen supply to the soil-pasture-animal system is the main limiting factor. Furthermore, the legume component provides a diet of high intake of digestible energy and protein. The advantageous maintenance of the legume component in a legume-based pasture renders botanical stability an important aspect in pasture research.

One of the primary objectives in pasture management should be to define the plant-animal relationships affecting pasture and animal outputs. To study sward behavior it is necessary to offer the grazing animal a wide range of available forage in order to control as much as possible the amount of forage consumed/animal and to make a suffi-

cient number of meaningful measurements (Lucas 1962). In sown pastures the stability and contribution of the components in response to treatment (Bryan 1970) must yield results in terms of numbers of animals the area will support or in terms of animal production/unit area of land (Mott and Moore 1970).

This study investigated the response of a subtropical legume-grass mixture to combined levels of rest periods and grazing pressure, to determine which combination results in best legume content in the mixture.

MATERIALS AND METHODS

A grazing trial was conducted for two grazing seasons to evaluate the performance of a mixture of *Desmodium intortum* (Mill.) Urb. and *Paspalum guenoarum* Arech. sown in late summer of 1978 on a Plinthaquult soil at the Estacao Experimental Agronomica–UFRGS, Rio Grande do Sul, Brazil. The area was plowed, limed, and fertilized with 200 kg/ha of both P_2O_5 and K_2O, followed by 100 kg/ha of the same nutrients early in each spring. The agronomic evaluation was a combination of 14, 28, 42, 56, and 70 days of rest (X_1) with grazing pressures of 2.0, 3.5, 5.0, 6.5, and 8.0 kg dry matter (DM) on offer/day/100 kg body weight (X_2). Hereafter, grazing pressure is ex-

G.E. Maraschin is an Associate Professor in the Departmento de Fitotecnia, S.C. Mella is a Pasture Researcher at IAPAR-PR, G.S. Irulegui is a Graduate Student in the Departmento de Fitotecnia, and J. Riboldi is an Assistant Professor in the Departmento de Estatística of the UFRGS.

459

460 *Maraschin et al.*

pressed as percentage of body weight (% BW). The experimental design was a central composite with two factors.

Both a rotational grazing system and a continuous grazing treatment were used. Pasture sizes were adjusted for a minimum of 400 kg body weight for the rotation cycles with 7 days of grazing. Double-sampling procedure was used to estimate forage DM yield and botanical composition (Campbell 1966). Yearling heifers were utilized as grazing animals on a put-and-take basis. *D. intortum* and *P. guenoarum* were analyzed for in-vitro organic-matter digestibility (IVOMD) and nitrogen (N) content. Each parameter was analyzed using a second-degree polynomial to estimate the response surface by the model:

$$\hat{Y} = b_0 + b_1X_1 + b_2X_2 + b_{11}X_1^2 + b_{22}X_2^2 + b_{12}X_1X_2 \quad (1)$$

Two different expressions of grazing pressure were used in the analysis: (1) the actual values of X_2, and (2) the residual DM left after grazing. The quadratic surface was fitted, and the stationary point (SP) was determined. To study the response surface, the X_1 and X_2 variables were transformed into the canonical variables W_1 and W_2, respectively (Davies 1956). The direction of the response was determined by canonical analysis with W_1 and W_2 as the coordinate axes.

RESULTS AND DISCUSSION

The minimum daily pasture DM production ($\hat{Y}_s$) was 29.7 kg DM/ha/day for 25 days of rest and 6.2% BW. Production was greater under lenient grazing pressures. The percentage contribution of *P. guenoarum* was dependent on periods of rest. Its maximum contribution ($\hat{Y}_s$) was 30.4% at the SP of 61.0 days of rest and 4.8% BW. Deviating from this rest period reduced its contribution to pasture DM yield and well-being. Under continuous grazing, *P. guenoarum* was not maintained when grazed at 5% BW. Renewal of the aerial plant parts, at long intervals, is a sound management practice that will eliminate the grazing out of the plants (Humphreys and Jones 1975).

The contribution of *D. intortum* to total DM yield varied from 0.25% to 17.2%. This response was a minimax ($\hat{Y}_s$) of 5.15% (Fig. 1) at the SP of 36.0 days of rest and 2,780 kg of residual DM/ha. Longer periods of rest increase the percentage of *D. intortum*. The legume can also be retained by maintaining 36 days of rest and very lenient grazing pressure, but this treatment seems to be incompatible with good pasture utilization.

Despite the results obtained under cutting, in a grazing situation very few axillary buds are left for regrowth. Under the conditions of this experiment, *D. intortum* survived and contributed 17.2% of the DM yield when grazed ater 70 days of rest combined with 8.0% BW as grazing pressure. It is evident that a long rest period has the predominant effect on the response of *D. intortum* under grazing, as was observed by Maraschin (1976) and Serrão (1977). This fact suggests that rotational grazing is

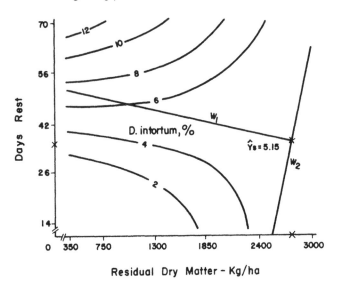

Fig. 1. Contours of *D. intortum* percentage as affected by days of rest and grazing pressure.

a suitable management for many tropical legumes, as was also verified for *Leucaena leucocephala* (Lam.) de Wit., with a simple two-paddock system of grazing (Jones 1973). Long rest periods might lessen the problems of tropical legumes under grazing. The missing information on their patterns of organic-food reserves and governing mechanisms for regrowth and survival warrants investigations to gain a better understanding of applied management practices (Blaser et al. 1962, Humphreys 1966) on such a highly palatable grass species (Ramirez 1954) and valuable forage legume.

The IVOMD of *P. guenoarum* showed a minimum response ($\hat{Y}_s$) of 55.7% at the SP for 58.5 days of rest when grazed at 3.4% BW (Fig. 2). Moving along the W_1 or W_2 axes tends to indicate increased response. At high grazing pressure, increases in IVOMD can be obtained

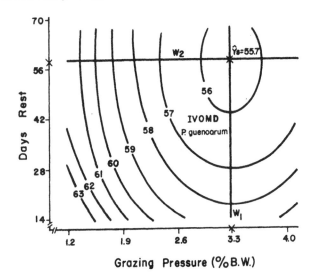

Fig. 2. Contours of *P. guenoarum* IVOMD as affected by days of rest and grazing pressure.

Fig. 3. Contours of stocking rate as affected by days of rest and grazing pressure.

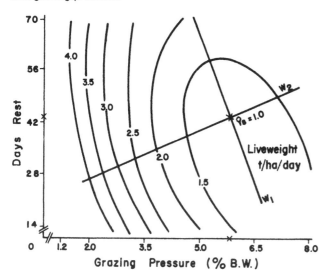

D. intortum and *P. guenoarum* can be maintained effectively under grazing if at least 56 days of rest are provided. A grazing-pressure level of 6.5% BW will benefit both species and will maintain native grasses and weeds at a very low level. This grazing-management system may not reduce to critical levels the forage nutritive value, but it is a compromise between daily DM yield of forage and live weight/ha/day and a greater opportunity for higher DM consumption/animal.

LITERATURE CITED

Blaser, R.E., J.R. Harlan, and R.M. Love. 1962. Grazing management. *In* Pasture and range research techniques, 11–17. Comstock Pub. Assoc., Ithaca, N.Y.

Bryan, W.W. 1970. Changes in botanical composition in some subtropical sown pastures. Proc. XI Int. Grassl. Cong., 636–639.

Campbell, A.G. 1966. Grazed pasture parameters. I. Pasture dry matter and availability in a stocking rate and grazing management experiment with dairy cows. J. Agric. Sci., Camb. 67:199–210.

Davies, O.L. 1956. Design and analysis of industrial experiments, 2nd ed. Hafner Pub. Co., New York.

Humphreys, L.R. 1966. Pasture defoliation practice: a review. Austral. Inst. Agric. Sci. J. 32:93–105.

Humphreys, L.R., and R.J. Jones. 1975. The value of ecological studies in establishment and management of sown tropical pastures. Trop. Grassl. 9:125–131.

Jones, R.J. 1973. Animal performance on *Leucaena leucocephala.* Austral. CSIRO Div. Trop. Agron. Ann. Rpt. (1972–73), 15.

Lucas, H.L. 1962. Determination of forage yield and quality from animal responses. *In* Range research methods: a symposium. U.S. Dept. Agric. Misc. Pub. 940.

Maraschin, G.E. 1976. Response of a complex tropical pasture mixture to different grazing management systems. Dissertation Abs. Int. B(1976) 36 (12, I) 5899, Univ. of Fla.

Minson, D.J. 1971. The nutritive value of tropical pastures. Austral. Inst. Agric. Sci. J. 37:255–263.

Mott, G.O., and J.E. Moore. 1970. Forage evaluation techniques in perspective. Proc. Natl. Conf. on Forage Eval. and Util., Lincoln, Neb.

Ramirez, J.R. 1954. El pasto rojas. Revista Argentina de Agronomia. 21(2):84–101.

Serrão, E.A.S. 1977. The use of response surface design in the agronomic evaluation of a grass-legume mixture under grazing. Dissertation Abs. Int. B(1977) 38 (2) 450–451, Univ. of Fla.

with shorter periods of rest. One important aspect that deserves attention is the relatively high value for IVOMD of *P. guenoarum* when compared with other tropical grasses (Minson 1971). The minimum N content for *P. guenoarum* (1.65%) and for *D. intortum* (2.78%) observed at 56 days of rest and the high levels of grazing pressure assure good-quality feed for the grazing animal.

Live weight/ha/day could be used to estimate the stocking rate necessary to attain the desired levels of grazing pressure and to describe the daily pasture DM yield close to the SP. The minimum response ($\hat{Y}_s$) was 1.0 metric ton (t) of live weight/ha/day (Fig. 3) at the SP of 42.6 days of rest combined with 5.9% BW, a fact that suggests that for higher levels of pasture utilization one has to move along the W_2 axis—i.e., increase the grazing pressure. The reestablishment of native-grass species (mainly *Axonopus affinis* Chase) in the heavily grazed treatments and the presence of litter in the leniently grazed ones might have influenced the response at the SP. The estimated forage DM that disappeared from the pasture, termed "forage DM consumption," /animal was reduced by increasing the grazing pressure. However, feasible levels of DM consumption/animal can be achieved with rest periods of 42 days or longer and a lenient grazing pressure.

Pattern of Defoliation by Cattle Grazing Crested Wheatgrass Pastures

B.E. NORTON and P.S. JOHNSON
Utah State University, Logan, Utah, U.S.A.

Summary

Utilization of a monotypic stand of perennial grass by cattle is generally assumed to be a random process that can be adequately measured by counting the number of plants grazed or by comparing biomass on caged and grazed plots. At a similar level of approximation, grazing is often simulated experimentally by clipping down to a prescribed height. The purpose of our study was to examine defoliation in more detail, following grazing severity and frequency on an individual plant basis with an interval of only 48 hours between observations.

The study was conducted on two 28-ha pastures of crested wheatgrass (*Agropyron cristatum* [L.] Gaertn.), originally seeded 25 years earlier, in a semiarid area (320 mm annual precipitation) located in west central Utah, 8 km southwest of Eureka. Nearly 300 permanent plots (0.5 m²) were involved in the project.

During 6 weeks of grazing by cattle in June and early July, 1979, the overall utilization was 46% by volume, with nearly 70% of all plants being grazed at least once. Only one-sixth of the grazed plants were defoliated on more than one occasion, and most of those were grazed just twice. Data analysis indicated a preference for moderate-sized plants, with less intensity of grazing on the smallest or largest size classes. Sixty percent of the smallest size-class plants were not grazed at all. For those plants that did experience defoliation, severity of grazing was inversely related to plant size.

Grazing on these crested wheatgrass pastures was neither random nor uniform. The cattle removed forage in a fashion that varied according to plant size. Very small plants were grazed in a manner similar to a clipping simulation. Plants greater than 200 cm² in basal area experienced a loss of less than half the plant volume, affecting less than half the plant cover, at an average defoliation event. The preponderance of biomass remaining in the large wolf plants at the end of the grazing period would suggest that estimating utilization by departure from amount of forage protected in ungrazed cages could underestimate actual utilization on the population of plants providing most of the consumed forage.

KEY WORDS: crested wheatgrass, *Agropyron cristatum* (L.) Gaertn., utilization, cattle grazing, grazing pattern, defoliation frequency.

INTRODUCTION

In the foothills of the intermountain region of the western U.S.A. range improvement over the last 3 decades has been effected by clearing sagebrush and juniper (*Artemisia* spp. and *Juniperus* spp.) and establishing fields of introduced grasses, chiefly crested wheatgrass (*Agropyron cristatum* [L.] Gaertn.), with higher forage production and greater tolerance to defoliation than native grasses (Cook 1966).

The present study on plant response to cattle grazing is part of an interdisciplinary research program exploring the potential for increasing the productivity and utilization of foothill ranges. These ranges provide critical forage for livestock and wildlife from the end of winter until melting snow permits access to high mountain pastures. In order to understand the impact of defoliation

on the forage resource it is necessary first of all to describe the defoliation activity on the basis of the plant—i.e., What is the likelihood that an individual plant will be grazed under a particular stocking rate for a particular duration? What is the probability of a plant's experiencing defoliation more than once? How much of the plant is likely to be harvested at each defoliation event?

Several workers have examined patterns of defoliation by sheep (Hodgson and Ollerenshaw 1969, Greenwood and Arnold 1968, Morris 1969, Hodgkinson 1980), but, apart from Gammon and Roberts' (1978) work in Zimbabwe, little attention has been given to the spatial and temporal impact of grazing by cattle on the plant or tiller level.

In the semiarid foothill environment it is possible to analyze patterns of defoliation on an individual-plant basis since the basal area is low (10% or less) and plants are easily distinguished, in contrast to those in a temperate grassland sward (Morris 1969). The experimental design consisted of permanent plots that were visited regularly during a normal early-summer grazing season for cattle.

B.E. Norton is an Associate Professor and P.S. Johnson is a Research Technician in the Department of Range Science of Utah State University. This paper is a contribution of the Department of Range Science and the Agricultural Experiment Station (project 771) of Utah State University.

METHODS

The study was conducted in early spring 1979 on crested wheatgrass pastures in foothill rangeland (1,830 m) managed by the Bureau of Land Management (U.S. Department of Interior). The area lies about 8 km southwest of Eureka, Utah, on the lower slopes of a broad alluvial plain characterized by mollisols and aridisols. The annual precipitation averages 320 mm, falling mostly in winter and spring (Cook 1966). In the 1950s much of the native vegetation, dominated by sagebrush and juniper, was cleared, and 24 improved pastures were established with better forage grasses, chiefly *Agropyron* spp. Two of the 28-ha pastures selected for the present study now consist of almost pure stands of crested wheatgrass.

Each pasture, 800 × 350 m, was divided into three equal zones approximately 270 m wide, and seven sampling grids were randomly located in each zone. A grid consisted of a 5 × 4-m matrix, with the corners of each square meter marked by 4-inch nails with painted flatheads driven almost flush with the soil surface. Sampling grids could be relocated by two pegs placed in line at each end of the western edge of the grid, but at least 5 m removed from it; the pegs also gave the grid an orientation. From the 20 squares in a grid, 7 were randomly chosen for study plots, giving a total of 147 plots/pasture. Each square plot of 0.5 m² overlapped the bottom left corner of its respective square m in the sampling grid, and the plot corners were similarly marked with flathead nails painted a different color. The use of inconspicuous nails rather than pegs was intended to avoid any interference with cattle grazing behavior.

At the end of May the basal outlines of all perennial grass plants on the study plots were mapped, and plant heights were recorded. Ungrazed plant heights were remeasured halfway through the study. Thirty-one heifers were introduced into each pasture on 31 May; they remained until 10 July. Defoliation was recorded by inspecting each plot every second day for 6 weeks, marking on the plot maps which part of a plant had been grazed, and noting the stubble height after grazing. These data allowed calculation of cover area grazed and plant volume removed. For purposes of analysis, any grazing activity observed at the end of the 2-day interval between inspections was designated as one grazing event, even though the same plant could conceivably have been grazed more than once in that period.

RESULTS

Plants were assigned to arbitrary size classes defined in terms of basal area (Table 1). The distribution of plants among the size classes was similar in the 2 replicates. The total number of plants did not differ greatly between pastures; about three-fourths of the plants were 50 cm² in area or smaller. The pasture with less available forage was grazed more intensively than the other, 76.4% of the plants being defoliated vs. 60.2%, but the distribution of grazed plants among the size classes matched quite

Table 1. Number of plants in both pastures, their distribution among size classes, and the proportion grazed in each class.

Size class (cm² basal area)	Number of plants in both pastures	Distrib. (%)	Percent grazed in size class
<0.5	445	9.6	40.0
0.5-10	1496	32.1	59.0
11-50	1579	33.9	76.3
51-100	491	10.6	83.1
101-200	357	7.7	84.6
201-300	140	3.0	87.9
301-400	61	1.3	85.2
401-500	38	0.8	73.7
>500	48	1.0	81.2
Total	4655		69.3

closely. For these reasons the data from both pastures were combined by size classes for purposes of analysis.

Overall, almost 70% of the plants received some degree of grazing pressure, the remainder being untouched over the 6-week period. A chi-square test ($\alpha \leqslant 0.05$) of grazing occurrence, ignoring magnitude of defoliation, indicated that there was a grazing preference for size classes with 11 through 400 cm² basal area. The probability of a plant's being grazed more than once was directly proportional to plant size. Of all the plants defoliated, only 16.6% were grazed more than once: 14.4% were grazed twice, 1.9% three times, 0.3% four times, and one plant was grazed five times.

Although larger plants were more likely to be grazed more than once, the severity of defoliation was inversely related to plant size. Whereas a single grazing event on small plants ($\leqslant 10$ cm² area) left an average stubble height of about 9 cm and affected more than 94% of the plant cover, the largest size class retained 21 cm stubble, and only 31% of the cover was grazed (Table 2). If defoliation is expressed in terms of harvested volume, grazed plants in the small size classes again lost the most volume and the largest class the least.

The last two columns in Table 2 compare the distribution of available forage (by volume) among the size classes with the distribution of harvested forage. The forage was harvested in proportion greater than its availability from plants up to 200 cm² in basal area; thereafter, the grazing pressure was proportionately less, with the largest plants being spared the most. The range of preference can therefore be reduced to a size distribution of 11 to 200 cm² (3.8 to 16 cm in diameter).

DISCUSSION

Of the total plant population (4,566 plants) examined in this study, only 11.2% was grazed in more than one 2-day period (16.6% of all grazed plants). These data support the use of the proportion of grazed plants as an index of utilization (Charlton 1968, Schmutz et al. 1963,

---

Table 2. Impact of grazing by size class in terms of plant cover, stubble height remaining, and plant volume (both pastures combined).

Size class (cm² basal area)	Average % cover impacted/plant at each grazing event*	Average plant height (cm)	Average stubble ht (cm) remaining on a grazed patch after each grazing event*	Percent of total plant volume harvested	Percent distribution of available forage volume	Percent distribution of harvested forage volume**
<0.5	99.2 (±0.9) a	12.7	9.2 (±0.6) a	56.5	0.05	0.06
0.6-10	94.4 (±1.1) b	15.6	9.0 (±0.3) a	60.5	1.48	1.95
11-50	83.2 (±1.4) c	17.9	10.3 (±0.3) b	55.9	11.75	14.25
51-100	71.3 (±2.6) d	20.8	11.5 (±0.5) c	54.6	9.16	10.85
101-200	61.5 (±2.9) e	22.6	13.5 (±0.7) d	49.4	22.05	23.63
201-300	45.9 (±3.9) fg	26.3	15.3 (±1.0) e	44.6	17.43	16.86
301-400	41.4 (±6.1) h	27.5	16.4 (±1.8) e	41.1	10.80	9.62
401-500	43.0 (±8.0) gh	26.9	16.2 (±1.9) e	43.0	7.32	6.83
>500	31.0 (±6.9) i	33.1	21.0 (±2.2) f	36.9	19.96	15.97

*Standard errors in parentheses. Numbers followed by the same letter in the column are not significantly different (p<0.05).

**Overall utilization is 46% of total available forage, by volume.

and others), at least for a relatively monospecific community.

At the end of the grazing treatment the general aspect of both pastures was dominated by the large crested wheatgrass plants, which made up only 1% of the plant population. This group appeared to have a significantly higher density of stem and foliage than the smaller plants, and therefore they probably carried a greater share of the available forage biomass and contributed a larger proportion of the harvested biomass than the percentages by volume in Table 2 would suggest. Nevertheless, these large plants were hardly affected by grazing. At an average defoliation event, 31% of the horizontal cover was grazed to a stubble height of 21 cm (reduced by a third), while multiple grazings late in the study were mostly on inflorescences, and altogether only 37% of the plant volume was harvested, leaving considerable shoot material and seed for regrowth and regeneration. In contrast, the average small plant up to 50 cm² in basal area (three-quarters of the total plant population) had its height reduced by 40% over 90% of its canopy area at each defoliation event. Altogether it lost 58% of the shoot volume. Total pasture utilization was 46% by volume.

Plants grazed in the smallest size class (≤ 0.5 cm²) had the top portion of their shoot system removed as if clipped but were not closely grazed; the average stubble height was 9.2 cm. This class also had the largest proportion of plants not grazed at all (60%). One can assume that, in company with the very large plants, the seedlings and juvenile plant populations are relatively protected from cattle grazing impact at the normal stocking rate imposed.

The pattern of defoliation encountered in this study complements the arguments of Spedding (1965) that challenge the use of clipping treatments to simulate grazing, especially when all shoot growth is clipped to pre-

scribed heights. The differential grazing pressure observed in relation to plant size classes leads to questions as to the value of procedures that rely on weight differences to measure utilization, as is done with cages, under the conservative normal grazing regime. As the grazing period proceeds, an increasing proportion of the biomass left outside the cage will be contributed by the large wolf plants, and true utilization of plants of the preferred size classes will be underestimated.

LITERATURE CITED

Charlton, M. 1968. Grazed plant utilization method. J. Range Mangt. 21:334–335.

Cook, C.W. 1966. Development and use of foothill ranges in Utah. Utah Agric. Expt. Sta. Bul. 461.

Gammon, D.M., and B.R. Roberts. 1978. Patterns of defoliation during continuous and rotational grazing of the Matopos sandveld of Rhodesia. Rhod. J. Agric. Res. 16:117–131.

Greenwood, E.A.N., and G.W. Arnold. 1968. The quantity and frequency of removal of herbage from an emerging annual grass sward by sheep in a set-stocked system of grazing. J. Brit. Grassl. Soc. 23:144–148.

Hodgkinson, K.C. 1980. Frequency and extent of defoliation in herbaceous plants by sheep in a foothill range community in northern Utah. J. Range Mangt. 33:164–169.

Hodgson, J., and J.H. Ollerenshaw. 1969. The frequency and severity of defoliation of individual tillers in set-stocked swards. J. Brit. Grassl. Soc. 24:226–234.

Morris, R.M. 1969. The pattern of grazing in 'continuously' grazed swards. J. Brit. Grassl. Soc. 24:65–70.

Schmutz, E.M., G.A. Holt, and C.C. Michaels. 1963. Grazed-class method of estimating forage utilization. J. Range Mangt. 16:54–60.

Spedding, C.R.W. 1965. The physiological basis of grazing management. J. Brit. Grassl. Soc. 20:7–14.

Consumption Rates of Desert Grassland Herbivores

R.D. PIEPER

New Mexico State University, Las Cruces, N.M., U.S.A.

Summary

Studies were conducted on a desert grassland range in southern New Mexico for 3 years to determine relative consumption rates of the main herbivores utilizing these range areas. Consumption of some groups (cattle, pronghorn antelope, and rabbits) was determined from direct estimates of intake/individual, and density was determined from censusing procedures or known stock rates. Consumption by other groups (birds, rodents, and invertebrates) was determined from census information relating to age structure of the population and metabolic rate. Consumption by all herbivores averaged less than 5% of the above-ground net primary productivity for the 3 years of the study. Highest consumption rates among the herbivore groups were for cattle, which consumed an average of 4.3 g/m² during the study. Consumption by rodents was fairly high during the first year of the study when rodent densities were high but declined during the second and third years in response to declines in rodent densities. In contrast, invertebrate consumption was highest during the third year of the study and actually exceeded that of rodents. Precipitation was favorable during the third year of the study and was reflected by the high above-ground net primary productivity that year. Apparently, invertebrate populations responded more quickly to favorable plant growth than rodent populations did. Consumption rates of pronghorn antelope, rabbits, and birds were relatively small. Rabbit consumption was nearly constant during the study period. Density of breeding bird populations was fairly low, although large buildups of migrant bird populations occurred late in the third year of the study. Energy flow to the herbivore populations was also small compared with above-ground net primary productivity. Total energy transfer to herbivores was only 24.3 kilocalories (kcal)/m² in contrast to 593 kcal/m² calculated for above-ground net primary productivity. Large herbivores, including pronghorn antelope and cattle, contributed most of the energy transferred during herbivory, followed by small mammals, invertebrates, birds, and rabbits. From the standpoint of range productivity, it appears that a relatively small portion of energy fixed by plants is harvested by man in the form of animal products. There appears to be a large, untapped source of energy on our rangelands that has the potential to be converted to food for a hungry world. How best to make use of this resource and still maintain stable ecosystems is a challenge facing range researchers and users.

KEY WORDS: consumption rates, desert grassland, herbivores.

INTRODUCTION

Rangelands are characterized by complex mixtures of plants and animals as well as other environmental variables. In the past, range workers have emphasized commercially valuable herbivores, including livestock or big game. Less attention was given to other herbivores, such as small mammals, rodents, birds, etc., unless their activities became destructive or they utilized forage needed by livestock. However, we are interested now in all components of range ecosystems, and especially herbivores that may have dietary overlap with domestic livestock. An understanding of range-ecosystem functioning should assist range managers in meeting overall management goals.

Consequently, the objective of this paper is to report on consumption rates of desert grassland consumers. This report is part of the International Biological Program on ecosystem analysis.

MATERIALS AND METHODS

The study was conducted on the Jornada Experimental Range, located about 37 km north of Las Cruces in southern New Mexico. All data reported were taken from a 10-ha exclosure located on a remnant of good-condition range. Black grama (*Bouteloua eriopoda*), soaptree yucca (*Yucca elata*), mesa dropseed (*Sporobolus flexuosus*), and broom snakeweed (*Xanthocephalum sarothrae*) were the main species contributing to the vegetational structure.

Although cattle did not graze in the exclosure during the study, intake rates were calculated on the basis of actual stocking rates on the rest of the pasture and daily intake rates of 8.6 kg/animal unit/day as reported by Hill

The author is a Professor of Range Science at New Mexico State University. This paper (Scientific Article 97) is a contribution of the Agricultural Experiment Station of New Mexico State University.

The author wishes to thank personnel at the Natural Resource Ecology Laboratory at Colorado State University, and especially Dr. Norman French, for analysis of small mammal intake rates. The research was supported by National Science Foundation grants GB 7824, GB 13096, GB 31826X, and GB 41233X to the Grassland Biome of the International Biological Program for "Analysis of Structure and Function, and Utilization of Grassland Ecosystems."

(1965). Pronghorn antelope intake rates were calculated from density estimates of Howard et al. (1973) and consumption estimates of 900 g/day during autumn and winter and 805 g/day during spring and summer, as reported by Wesley et al. (1973). Consumption rates of rabbits were calculated from density estimates made by Packard (1971, 1972) and consumption rates by Pieper et al. (1978). Rodent consumption was calculated by a program developed by French and Swift (1974) based on density, biomass distribution by age classes, and average metabolic rate adjusted for environmental conditions. Estimates of intake of invertebrates and birds were also calculated using density information and metabolic requirements for different phases of the life cycle (Pieper et al. 1978).

RESULTS

Consumption by all herbivores averaged less than 5% of the above-ground net primary productivity for the 3 years of the study (Table 1). Highest consumption rates among the herbivore groups were for cattle, which consumed an average of 4.3 g/m². Consumption by rodents was fairly high in 1970, when rodent densities were high, but declined in 1971 and 1972, when densities were low. In contrast, invertebrate consumption was highest in 1972 and actually exceeded that of rodents. Precipitation was favorable in 1972, as was shown by the high aboveground net primary productivity that year. Apparently, invertebrate populations responded more quickly to favorable plant growth than the rodent populations did (Ellstrom 1973).

Consumption rates of pronghorn antelope, rabbits, and birds were relatively small. Even though there were fluctuations in rabbit densities from 1970 through 1972, fluctuations were not reflected in consumption rates, which remained nearly constant during the study (Table 1). Density of breeding bird populations was fairly low, although large buildups of migrant bird populations occurred in late 1972 (Raitt and Pimm 1976).

Energy flow to the herbivore populations was also small as compared with above-ground net primary productivity (Fig. 1). Total energy transfer to herbivores was only 24.3 kcal/m² compared to 593 kcal/m² calculated for above-ground net primary productivity. Large herbivores, including pronghorn antelope and cattle, contributed most

Fig. 1. Annual energy flow in a desert grassland ecosystem averaged over 3 years.

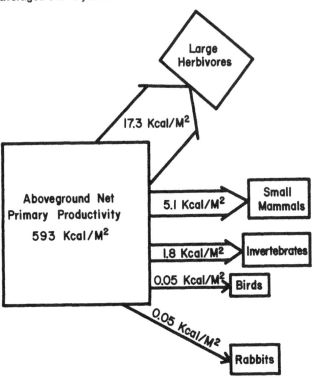

of the energy transferred during herbivory, followed by small mammals, invertebrates, and birds and rabbits.

DISCUSSION

Data from this study indicate that only a small portion of the total energy fixed during photosynthesis in desert grasslands is transferred to herbivores. Most of the plant biomass must be processed in the detritus food chain by decomposer organisms. However, we have not been able to account for all above-ground net primary productivity due to herbivory and decomposer activity (Pieper et al. 1978). Only about 300 g/m² were attributed to decomposer activity when decomposition was evaluated by CO_2 evolution techniques.

The relatively low consumption by herbivores in a desert grassland ecosystem leads one to speculate about possible empty food niches under such conditions. Other questions also arise concerning stability of desert grassland ecosystems with widely fluctuating driving variables of precipitation and temperatures. Perhaps these ecosystems can only support small populations of herbivores over time because of fluctuating availability of food sources.

However, data now appearing in the literature indicate that only a small portion of the net primary production is processed by herbivores in several grassland ecosystems. Results from other grassland types in the International Biological Program indicated that although cattle were the dominant herbivore, total herbivore consumption was less than 40% of above-ground net primary production for a

Table 1. Above-ground net primary production and consumption (g/M²) by various herbivore groups.

Item	1970	1971	1972	AVG
Annual primary production	134.0	125.0	186.0	148.0
Consumption:				
Pronghorn antelope	0.1	0.1	0.1	0.1
Cattle	5.0	4.4	3.4	4.3
Rodents	2.6	0.7	0.6	1.3
Rabbits	0.01	0.01	0.01	0.01
Birds	0.5	0.03	0.07	0.05
Total all herbivores	8.3	5.3	4.9	6.1

tallgrass ecosystem, 20% for a mixed prairie grassland, and 15% for a shortgrass ecosystem (Lewis 1971). Clary (1977) also reported that large plant biomass supported only a relatively small biomass of herbivores in ponderosa pine ranges in Arizona. Van Hook (1971) reported that arthropod populations in Tennessee grasslands consumed less than 10% of the net primary production.

From the standpoint of range productivity, it appears that a relatively small portion of energy fixed by plants on several rangelands is harvested by man in the form of animal products. On rangelands and pastures, when cultural inputs such as seeding, fertilization, etc., have been practiced, it is possible to transfer more of the net plant productivity to usable animal products (Williams 1966, MacFadyen 1964). There appears to be a large, untapped source of energy on our rangelands having the potential to be converted to food for a hungry world. How best to make use of this resource and still maintain stable ecosystems is a challenge facing range researchers and users. One approach that must be discarded is simply to increase stocking of game or livestock. Other approaches are necessary to allow us to make more efficient use of our range resources.

LITERATURE CITED

Clary, W.P. 1977. Producer-consumer biomass in Arizona Ponderosa pine. U.S. Forest Serv. Gen. Tech. Rpt. RM-56.

Ellstrom, M.A. 1973. Populations and trophic structures of a desert grassland invertebrate community. M.S. Thesis, N.Mex. State Univ.

French, N.R., and D.A. Swift. 1974. Program for calculating average production and respiration values for small mammals. U.S./IBP Grassl. Biome, Fort Collins, Colo.

Hill, K.R. 1965. Estimated intake and nutritive value of range forage grazed by Hereford and Santa Gertrudis cows. M.S. Thesis, N.Mex. State Univ.

Howard, V.W., Jr., C.T. Engelking, E.D. Glidewell, and J.E. Wood. 1973. Factors resticting pronghorn use on the Jornada Experimental Range. N.Mex. Agric. Expt. Sta. Res. Rpt. 245.

Lewis, J.K. 1971. The grassland biome: a synthesis of structure and function, 1970. *In* N.R. French (ed.) Preliminary analysis of structure and function in grasslands. Range Sci. Dept. Sci. Ser. 10. Colo. State Univ., Fort Collins.

MacFadyen, A. 1964. Energy flow in ecosystems and its exploitation by grazing. *In* D.J. Crisp (ed.) Grazing in terrestrial and marine environments. Brit. Ecol. Soc. Sym. 4. Blackwell Sci. Pub., Oxford.

Packard, R.L. 1971. Small mammal survey on the Jornada and Pantex sites. U.S./IBP Grassl. Biome Tech. Rpt. 114. Colo. State Univ., Fort Collins.

Packard, R.L. 1972. Small mammal studies on Jornada and Pantex sties, 1970-1971. U.S./IBP Grassl. Biome Tech. Rpt. 188. Colo. State Univ., Fort Collins.

Pieper, R.D., J.C. Anway, M.A. Ellstrom, C.H. Herbel, R.L. Packard, S.L. Pimm, R.J. Raitt, E.E. Staffeldt, and J.G. Watts. 1978. Structure, function and utilization of North American desert grassland ecosystems. U.S./IBP Grassl. Biome. Ms. Colo. State Univ., Fort Collins.

Raitt, R.J., and S.L. Pimm. 1976. Dynamics of bird communities in the Chihuahuan Desert, New Mexico. The Condor 78:427–442.

Van Hook, R.I., Jr. 1971. Energy and nutrient dynamics of spider and orthoptevan populations in a grassland ecosystem. Ecol. Monog. 41:1–26.

Wesley, D.E., K.L. Knox, and J.G. Nagy. 1973. Energy metabolism of pronghorn antelopes. J. Wildlife Mangt. 37:563–573.

Williams, W.A. 1966. Range improvement as related to net productivity, energy flow and foliage configuration. J. Range Mangt. 19:29–34.

Effects of Climatic Zone, Topography, Land Utilization, and Soil Condition on Nutrient Composition of Natural Grasses in Bali

I.M. NITIS and K. LANA
Udayana University, Denpasar, Bali, Indonesia

Summary

A survey based on climatic zone (zones B, C, D, E, and F), topography (coast, hill, and higher altitude), land utilization (rice field, dry land, and plantation) and soil condition (moist and dry surface) was carried out during the wet (January to February 1979) and dry (August to September 1979) seasons in 15 districts of 8 regencies in Bali. The objective was to study the quality of the natural grass commonly fed to livestock in the mixed-farming system.

During the wet season, the crude-protein content (on dry-matter basis) of the natural grass ranged from 7.4% (in climatic zone D on dry soil) to 13% (in climatic zone E at higher altitude); during the dry season, the protein content ranged from 3.64% (in climatic zone F on dry soil) to 13.6% (in climatic zone C at higher altitude). The crude-fiber content ranged from 22.14% (in climatic zone B at higher altitude) to 29.9% (in climatic zone F on moist soil) and from 23.3% (in climatic zone B on moist soil) to 30.5% (in climatic zone F on dry soil) during the wet and dry seasons, respectively. The gross energy content of the grass ranged from 3,482 kilocalories (kcal)/kg (in climatic zone B on moist soil) to 4,278 kcal/kg (in climatic zone F on dry soil) during the wet season, and during the dry season the value ranged from 3,637 kcal/kg (in climatic zone C on moist soil) to 3,992 kcal/kg (in climatic zone F on the coast).

The present data indicate that the four elements affect the nutrient composition of the natural grass in the following order from greatest to least: climatic zone, topography, soil condition, and land utilization. The crude-protein content was affected the most, and the gross energy content was affected the least, by the above-mentioned factors.

KEY WORDS: climatic zone, topography, land utilization, soil condition, nutrient, natural grass, seasons.

INTRODUCTION

In the traditional mixed-farming system of Bali, natural grass is still the major component of livestock feed during both the wet and the dry seasons (Nitis et al. 1980). It has also been shown that botanical composition, growth, and dry-matter production of the natural grass are affected to varying degrees by climatic zone, topography, land utilization, and soil condition.

The present paper describes the effect of climatic zone, topography, land utilization, and soil condition on the crude-protein, crude-fiber, and gross energy contents of the natural grass during the wet and dry seasons.

MATERIALS AND METHODS

Bali is an island of 5,621 km², located within lat 7°54′ to 8°30′ S and long 114°26′ to 115°43′ E. There are 101 rainy days annually with a total of 1,982 mm rain (10-year average), with 70% falling during the October-to-March rainy season and 30% falling during the April-to-September dry season.

A survey using stratified random sampling techniques, based on climatic zone, topography, land utilization, and soil condition, was conducted during the wet (January to February) and dry (August to September) seasons, 1979.

In each of the five climatic zones, 3 districts (*Kecamatan*) were selected at random, giving 15 districts in 8 regencies (*Kabupaten*). In each district, three types of topography, three types of land utilization, and two types of soil condition were selected at random for a total of 120 sites from which 600 samples were monitored in each season.

The climatic zones in Bali, based on the Schmidt and Ferguson classification (Manik et al. 1977), consisted of zones B ($14.3\% < Q^1 < 33.3\%$), C ($33.3\% < Q < 60\%$), D ($60\% < Q < 100\%$), E ($100\% < Q < 167\%$) and F ($167\% < Q < 300\%$). The topography is classified as coastal (0 to

500 m above sea level), higher altitude (500 to 1,400 m above sea level) and hilly (undulating land higher than the flat area). Land utilization is classified as rice field (under irrigation), dryland farming (dependent on rainfall), and plantation (plantation crop). The soil-surface condition is classified as moist (the surface wet most of the time) and dry (the surface dry most of the time) (Anon. 1980).

The grass was cut within a 0.5-m² quadrat. Five cuttings from each site were mixed thoroughly, and a 500-g subsample was dried in a forced-draft oven at 70° C to constant weight (dry weight). A required amount was ground to pass through a 1-mm screen and then stored in a sealed plastic bottle for chemical analysis.

Dry matter (DM) was calculated from the sample (dry weight) dried in the oven at 105° C to constant weight (AOAC 1970). Nitrogen (N) was determined by the semimicro Kjeldahl method described by Ivan et al. (1971), and crude protein (CP) was calculated as N × 6.25. Crude fiber (CF) was determined according to the method described by AOAC (1970) and gross energy (GE) by the automatic adiabatic bomb calorimeter.[2] All analyses were carried out in duplicate, and all results were expressed on a DM basis.

RESULTS

During the wet season, the average CP content of the natural grass at all locations was 9.7%, with highest value (13.1%) found at the higher altitude in climatic zone E and lowest value (7.4%) on dry soil in climatic zone D (Table 1). The CP content tended to decrease as the climatic zone and the soil surface became drier. The CP content at the higher altitude was higher than on the coast, whereas differences in land utilization exerted no effect on the CP content.

During the dry season the average CP content was the same as that in the wet season (7.4%), but the ranges were wider (13.6% at higher altitude in climatic zone C to 3.6% on the dry soil surface in climatic zone F). The effects of climatic zone, topography, and soil condition followed closely those of the wet-season trend, whereas the CP content of the grass in the rice field was higher than that in the plantation and dryland-farming areas.

The CF content of the natural grass during the wet

I.M. Nitis is a Professor of Animal Nutrition and Tropical Pasture Science and K. Lana is a Lecturer and Research Scientist at Udayana University. This is a contribution of the Department of Animal Nutrition and Tropical Pasture Science, FKHP, Udayana University. Part of the research was supported by grants from the Directorate General of Animal Husbandry, Jakarta, and IDRC Canada.

Mention of a proprietary product does not imply endorsement of it.

Table 1. Crude-protein and fiber content (% DM) and gross energy (kcal/kg DM) of the natural grasses in Bali.

Stratification	Crude protein		Crude fiber		Gross energy	
	Wet season	Dry season	Wet season	Dry season	Wet season	Dry season
Climatic zone						
B	10.0±0.9[1]	12.0±1.4	25.9±1.0	26.1±1.0	3874±213[1]	3835±78
C	9.7±1.5	11.1±1.8	26.5±2.2	25.3±1.3	4090±110	3809±112
D	9.5±1.0	9.9±1.8	28.2±0.8	28.9±1.3	3983±126	3871±79
E	9.9±1.5	9.4±1.3	27.1±2.4	26.1±1.9	3949±122	3839±77
F	9.6±1.2	6.7±3.1	27.8±1.0	27.9±2.3	3946±109	3863±128
Topography						
Coast	9.8±0.9	8.5±1.0	28.7±1.3	27.8±1.7	4031±185	3968±19
Hill	9.2±1.2	9.1±3.3	28.0±1.2	27.0±2.1	4060±65	3809±54
Higher altitude	11.2±1.3	10.3±3.6	26.1±2.7	27.0±1.8	3976±98	3825±103
Land use						
Rice field	9.5±0.3	11.0±2.0	26.8±2.1	26.6±2.2	3942±122	3861±108
Dryland farming	9.7±0.7	9.8±2.9	26.5±1.9	26.2±1.2	3988±96	3819±73
Plantation	9.9±1.1	9.6±3.6	26.8±1.1	26.8±1.5	3984±79	3825±114
Soil surface						
Moist	9.5±1.2	10.3±1.7	27.2±1.7	25.8±2.3	3770±178	3823±116
Dry	9.1±1.7	9.0±2.8	27.3±2.4	28.0±2.3	4027±200	3843±95
Average	9.7	9.7	27.1	26.9	3972	3844

[1]Mean and standard deviation

season ranged between 22.1% (higher altitude, climatic zone B) and 29.9% (wet soil, climatic zone F). As the climatic zone became drier, the CF content tended to increase. Grass at higher altitude contained lower CF than that on the coast. Land utilization and soil condition had no effect on the CF content of the grass.

During the dry season the average CF value was similar to that in the wet season, but it varied slightly more (23.3% on the moist soil in climatic zone B and 30.5% on the dry soil in climatic zone F). Unlike wet-season CF, dry-season CF content was not affected by topography.

The average GE content of the natural grass during the wet season was 3,972 kcal/kg; the highest value (4,278 kcal/kg) occurred on dry soil in climatic zone F, and the lowest value (3,482 kcal/kg) occurred on moist soil in climatic zone B. As the climatic zone and soil surface became drier the GE value tended to increase. Topography and land utilization exerted no effect on GE content.

During the dry season the average GE value was 3% lower, with the range varying between 3,636 kcal/kg (moist soil, climatic zone C) and 3,992 kcal/kg (on the coast, climatic zone F). During this season, climatic zone, topography, land utilization, and soil condition exerted no effect on the GE content of the natural grass.

DISCUSSION

The present results show that the average CP and CF contents of natural grass in Bali fall within the range of values reported in the Java, Madura, and Lombok islands of Indonesia (Anon. 1972). This correspondence was expected, since these islands are of similar latitude and utilize similar intensive traditional agricultural management systems.

Higher CP and lower CF contents of natural grasses in January and February than in June and July, as reported

by Weinman for Rhodesia (cited by Whyte et al. 1975), were not apparent in the present study. Difference was presumably due to differences in the management of natural grasses. In Bali, the grass is generally cut every 2 to 3 months for "cut and carry" livestock feeding. This cutting system and the mild dry season keep the grass young and actively growing throughout the year. Consequently, the CP, CF, and GE contents of the natural grass in the present study did not vary between the wet (January to February) and dry (August to September) seasons.

The differences in CP, CF, and GE content of the natural grass between climatic zones B and F, between the coast and higher altitude, between the rice field and plantation areas, and between the moist and dry soil may have been due to water stress. Growth stages may also have had some effect, since the grass in those land classifications ranged between the vegetative and reproduction stages. Botanical composition may also have caused these differences, since the dominant genus in each location varied among *Cynodon, Chrysopogon, Pennisetum, Paspalum, Polytrias, Schima, Eragrostis,* and *Cymbopogon*. The chemical composition of these genera does vary (Soerianegara 1970 and Bo Gohl 1975).

The results of the present study should be interpreted with much caution; however, there is indication the elements can be ranked from most to least effect on the nutrient composition of the natural grasses in the following order: climatic zone, topography, soil condition, and land utilization. The above parameters affected CP the most and GE the least.

LITERATURE CITED

Anon. 1972. Laboratory analysis of the roughage and concentrate in Java, Bali, and Lombok (translation), 15. Faculty of Anim. Husb., Gajah Mada Univ., Yogyakarta.

Anon. 1980. Monography of Bali island (translation), 2–111. Dinas Pertanian Propinsi Bali, Denpasar.

Association of Official Analytical Chemists (AOAC). 1970. Official methods of analysis, 11th ed. AOAC, Washington, D.C.

Bo Gohl. 1975. Tropical feeds. FAO, Rome, 7–147.

Ivan, M., D.J. Clack, and G.T. White. 1971. Improved nitrogen distillation apparatus. Laboratory Practise, 12. Univ. of New England, Austral.

Manik, G., I.G.N. Raka Haryana, and Ramli. 1977. Climatic zone of Bali based on the Schmidt and Ferguson classification (translation). Udayana Univ., FKHP. Bul. 081.

Nitis, I.M., K. Lana, I.B. Sudana, N. Sutji, and I.G.N. Sarka.

1980. Roughage supply and requirement of livestock in Bali (translation). Udayana Univ., FKHP. Denpasar.

Soerianegara, I. 1970. The status of range research and management in Indonesia. Rimba Indonesia 15:99–115.

Whyte, R.O., T.R.G. Moir, and J.P. Cooper. 1975. Grasses in agriculture, 5th print., 144–146. FAO, Rome.

NOTES

1. $Q = \dfrac{\text{total average precipitation dry month}}{\text{total average wet month}} \times 100$

2. Gallenkamp 1976 Autobomb, CB 100, Issue 9.

Effects of Water Stress on Herbage Quality

J.R. WILSON
CSIRO Division of Tropical Crops and Pastures,
Brisbane, Australia

Summary

Uncertainty exists about whether drought is detrimental to the nutritive quality of herbage. This paper tries to resolve this uncertainty by examining the influence of soil moisture on some herbage-quality characteristics. Unpublished data from my own work and review of published literature are presented.

Field plots of three tropical grasses, green panic (*Panicum maximum* var. trichoglume), buffelgrass (*Cenchrus ciliaris*), and speargrass (*Heteropogon contortus*), were subjected to periods of controlled drought, and herbage from these plots was compared with that from well-watered plots. Leaf and stem dry-matter digestibility of water-stressed herbage was either similar to or higher than that of unstressed herbage. Sward digestibility in plots subjected to drought was higher than that in the well-watered controls because of a decreased proportion of stem and a higher digestibility of both leaf and stem.

A review of the published literature indicates that low soil moisture rarely has an adverse effect on herbage quality. Available information leads to the conclusion that animal live-weight gain should be better than usual under conditions of low soil moisture unless low pasture yield seriously restricts intake. Information from a grazing experiment in Australia supports this contention.

KEY WORDS: tropical grasses, herbage quality, digestibility, water stress, animal performance, drought.

INTRODUCTION

Some observers (e.g., French 1961, Masuda 1977) have stated that drought lowers the nutritive quality of herbage. This statement is true for extreme droughts that kill all above-ground herbage. However, it is not clear how quality may change when water stress is moderate or when intermittent periods of low rainfall temporarily curtail growth without substantial herbage death. Various authors (Dent and Aldrich 1963, Van Soest et al. 1978) comment that herbage digestibility is increased under low-rainfall conditions.

Experiments were conducted in southeastern Queensland to compare the quality of herbage produced under controlled drought stress with that produced with adequate water. A review of the published literature on soil moisture effects on herbage digestibility is also presented.

METHODS

Plots of the tropical grasses green panic (*Panicum maximum* var. trichoglume), buffelgrass (*Cenchrus ciliaris*), and speargrass (*Heteropogon contortus*) were established at Narayen Research Station, Queensland (lat 25°42′ S).

The author is a Research Physiologist in the CSIRO Division of Tropical Crops and Pastures.

Experiments with controlled droughts were conducted over the summers of 1976–1979. Droughts of varying duration (2 to 6 weeks) were imposed by excluding rainfall with a rain shelter (dry treatment), and comparision was made with plots irrigated twice weekly to field capacity (wet treatment). At the end of stress periods samples were taken of leaves or stems of comparable age or type and also of the whole sward. In-vitro dry-matter digestibility (IVDMD) was determined as in Wilson and Ng (1975).

RESULTS

Compared with the net treatment, short periods of water stress had little effect on the IVDMD of the most recently produced leaf of grass tillers (Fig. 1a) but slowed down the decline in IVDMD as leaves aged (Fig. 1b). In another experiment, water stress markedly improved herbage quality through a restriction of stem development and a higher IVDMD of the various sward components (Table 1).

DISCUSSION

These results support most of the published literature summarized in Table 2. These papers generally indicate that low soil moisture has either little effect or a beneficial effect on herbage digestibility.

Other features of herbage quality may also be influenced by drought. Almost invariably the proportion of leaf is increased in stressed swards because stem development is restricted. Nitrogen content is either unaffected or higher in drought-stressed herbage, and the content of minerals and soluble carbohydrates is frequently increased. Animal intake of herbage is thought to be higher under drier conditions because of a lower water content of

Table 1. Some herbage quality characteristics of green panic harvested after 6 weeks drought (DRY) compared with plants irrigated twice weekly over the same period (WET). Minimum leaf water potentials were c.-55 bar. Standard error of mean is shown.

Fraction of sward	Proportion of dry-matter yield of (%)		Dry-matter digestibility (%)	
	WET	DRY	WET	DRY
Green leaf	37 ± 1	58 ± 1	64.0 ± 0.2	69.5 ± 0.7
Dead leaf	7 ± 1	15 ± 1	62.1 ± 0.8	68.0 ± 0.7
Stem	56 ± 1	27 ± 2	52.4 ± 1.0	65.4 ± 1.3

Table 2. Summary of published literature on effects of low soil moisture on herbage digestibility [increase (+), no effect (0) or decrease (−)]. Grass (G) and legume (L).

Dig. (%)	Plant material	Reference
+	G	Dent and Aldrich (1963)
0	G & L	Hiridiglou *et al.* (1966)
0/ +	L	Vough & Marten (1971)
+	G	Wurster *et al.* (1971)
+	L	Snaydon (1972)
−	G	Hutton (1974)
+	G stem	Wilson & Ng (1975)
0/ −	G leaf	
0/ +	L	Pesant & Dionne (1976)
0	G	Taylor *et al.* (1976)
0	G	Spurway et al. (1976)
0	G	Silcock (1978)
0/ +/ −	G	Garwood *et al.* (1979)

Table 3. Herbage available (HA) and animal daily gain (ADG) in relation to rainfall during the summer growing season (December-April) for three years at Narayen Research Station, S.E. Queensland. Buffelgrass/siratro (*Macroptilium atropurpureum*) pasture grazed at 1.1 head/ha. [Unpublished data courtesy of L't Mannetje].

	Wet(1973/4)	Dry(1974/5)	Very Dry(1979/80)
Wet days (rain, mm)	38 (507)	34 (225)	25 (282[2])
Mean HA[1] (kg/ha)	6984	3756	2767
Mean ADG (kg/head)	0.56	0.69	1.13

[1]At start of each month's grazing.

[2]142 mm in one week.

Fig. 1. Effect of a series of c. 14-day drying cycles (■) on leaf digestibility of (a) the uppermost fully expanded leaf on a tiller at the time of each sampling and (b) leaves that were identified as just fully expanded at day 0 and harvested progressively throughout the experiment as they aged. Data are averages for green panic, buffelgrass, and speargrass. Vertical bars indicate L.S.D., P = 0.05. Leaf water potentials (bar): wet (− 12 to − 14), dry cycles 1 (− 15 to − 20), 2 (− 20 to − 45), and 3 (− 20 to − 35).

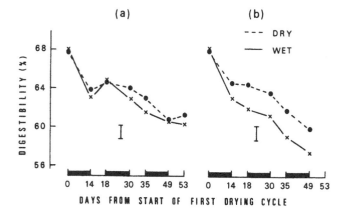

the tissues (e.g., Ketth and Ranawana 1971). Apart from tissue death, the only adverse quality factor consistently reported is a decrease in herbage phosphorus content under water stress (Pesant and Dionne 1976), and this decrease could influence animal performance in low-phosphorus areas.

It is concluded, based on the information presented, that, provided intake is not limited by low herbage yield, animal performance on forage produced under low soil moisture conditions should often be better than on forage produced under well-watered conditions because of the overall improvement in herbage quality. Comments are often made to this effect, but critical animal data are difficult to find. Table 3 presents information from a grazing experiment using steers at the Narayen Research Station. The higher rates of animal daily live-weight gain in the dry years support the conclusion that herbage grown under low soil moisture has improved nutritive quality.

LITERATURE CITED

Dent, J.W., and D.T.A. Aldrich. 1963. The inter-relationships between heading date, yield, chemical composition and digestibility in varieties of perennial ryegrass, timothy, cocksfoot and meadow fescue. Natl. Inst. Agric. Bot. J. 9:261-281.

French, M.H. 1961. Observations on the digestibility of pasture herbage. Turrialba 11:78-83.

Garwood, E.A., K.C. Tyson, and J. Sinclair. 1979. Use of water by six grass species. I. Dry matter yields and response to irrigation. J. Agric. Sci. 93:13-24.

Hiridiglou, M., P. Dermine, H.A. Hamilton, and J.E. Troelsen. 1966. Chemical composition and in vitro digestibility of forage as affected by season in northern Ontario. Can. J. Pl. Sci. 46:101-109.

Hutton, J.B. 1974. The effect of irrigation on forage yield, dairy cow production and intake under intensive grazing conditions in New Zealand. XIX Int. Dairy Cong., New Delhi, India. 1E:73-74.

Ketth, J.M., and S.S.E. Ranawana. 1971. Kikuyu grass: *Pennisetum clandestinum* Hochst. ex Chiou and its value in montane region of Ceylon. Trop. Agric. (Peradeniya) 127:93-103.

Masuda, Y. 1977. Comparison of the in vitro dry matter digestibility of forage oats grown under different temperatures and light intensities. J. Faculty Agric. Kyushu Univ. 21:17-24.

Pesant, A.R., and J.L. Dionne. 1976. Effets de la fertilisation et des régimes hydriques sur le rendement, l'utilisation d'eau et la composition chimique de la luzerne et du trèfle Ladino. Can. J. Pl. Sci. 56:293-302.

Silcock, R.G. 1978. Transpiration and water use efficiency as affected by leaf characteristics of *Festuca species*. Ph.D. Thesis, Univ. College of Wales.

Snaydon, R.W. 1972. The effect of total water supply and frequency of application upon lucerne. II. Chemical composition. Austral. J. Agric. Res. 23:253-256.

Spurway, R.A., D.A. Hedges, and J.L. Wheeler. 1976. The quality and quantity of forage oats sown at intervals during autumn: effects of nitrogen and supplementary irrigation. Austral. J. Expt. Agric. Anim. Husb. 16:555-563.

Taylor, A.O., R.M. Haslemore, and M.N. McLeod. 1976. Potential of new summer grasses in Northland. III. Laboratory assessments of forage quality. N.Z. J. Agric. Res. 19:483-488.

Van Soest, P.J., D.R. Mertens, and B. Deinum. 1978. Preharvest factors influencing quality of conserved forage. J. Anim. Sci. 47:712-720.

Vough, L.R., and G.C. Marten. 1971. Influence of soil moisture and ambient temperature on yield and quality of alfalfa forage. Agron. J. 63:40-42.

Wilson, J.R., and T.T. Ng. 1975. Influence of water stress on parameters associated with herbage quality of *Panicum maximum* var. *trichoglume*. Austral. J. Agric. Res. 26:127-136.

Wurster, M.J., J.G. Ross, L.D. Kamstra, and S.S. Bullis. 1971. Effect of droughty soil on digestibility criteria in three cool season forage grasses. Proc. S. Dak. Acad. Sci. 50:90-94.

Forage Quantity and Quality Contributions from a Grass-Legume-Shrub Planting on a Semiarid Rangeland

D.A. JOHNSON, M.D. RUMBAUGH, and G.A. VAN EPPS

USDA-SEA-AR and Utah State University, Logan, Utah, U.S.A.

Summary

Poor grazing management has led to deterioration of vast expanses of the world's semiarid rangelands. Although contemporary thought in range-management circles suggests that combinations of grasses, forbs, and shrubs are desirable in revegetating these areas, past revegetation mainly involved the establishment of monocultures of perennial grass. The objectives of this study were to determine the forage and protein yields of a grass-legume-shrub planting in a semiarid rangeland.

The study was conducted in central Utah on a rangeland receiving 32 cm of annual precipitation. Crested wheatgrass (*Agropyron cristatum*) was established alone and in combination with fourwing saltbush (*Atriplex canescens*), alfalfa (*Medicago sativa*), sicklepod milkvetch (*Astragalus falcatus*), and cicer milkvetch (*Astragalus cicer*). Belt transects were established, and all of the current year's above-ground production for each species was clipped, dried, weighed, and analyzed for total nitrogen.

Total forage and protein yields were greatly increased when crested wheatgrass was grown in association with either fourwing saltbush or legumes compared with when crested wheatgrass was grown alone. These enhanced yields in areas where crested wheatgrass was grown with other species were due to increases in (1) the high-quality forage provided by the associated species and (2) the forage and protein yields of the crested wheatgrass component. Revegetation of semiarid rangelands with combinations of crested wheatgrass, fourwing saltbush, and legumes (particularly alfalfa) should provide a more productive, higher-quality forage resource than crested wheatgrass alone.

KEY WORDS: crested wheatgrass, *Agropyron cristatum*, fourwing saltbush, *Atriplex canescens*, alfalfa, *Medicago sativa*, cicer milkvetch, *Astragalus cicer*, sicklepod milkvetch, *Astragalus falcatus*, forage yield, protein yield.

INTRODUCTION

Poor grazing management has led to deterioration of vast expanses of the world's semiarid rangelands. Many productive and high-quality forage species have been reduced or eliminated from many of these lands, resulting in a serious reduction in grazing capacity and range condition. In the past, revegetation of semiarid rangeland has largely involved establishing a monoculture of perennial grass. However, inclusion of legumes and shrubs with grasses in the revegetation of semiarid rangelands may have multiple benefits. Legumes produce high-quality forage and are capable of fixing nitrogen, often a limiting nutrient. Shrubs tap deep soil reserves not used by grasses and thus could provide forage further into the dry season than do herbaceous species alone.

The objective of the present study was to determine the forage quantity and quality dynamics of a grass-legume-shrub planting in a semiarid environment.

METHODS

The study was conducted at the Utah State University Nephi Field Station, 121 km south of Salt Lake City, Utah, at an elavation of 1,615 m. The soil at the station is a Nephi silt loam (fine-silty, mixed, mesic Calcic Argixerolls) with a nitrogren content of 0.1% in the surface foot. Long-term (1903–1979) annual precipitation is 32 cm with about one-third occurring in March, April, and May. Frost-free days average 110 days, with a daily maximum temperature in July in excess of 38°C. Free water surface evaporation for April to October averages 121 cm or about four times the total precipitation for the year. Precipitation in 1978 was 42 cm, or 10 cm greater than the long-term average. In 1979 it was slightly above average from January through May and somewhat below average for the remainder of the year, resulting in 28 cm, slightly lower than the long-term average. In both 1978 and 1979 precipitation during June and July was low, with 2-month totals of 0.03 cm and 0.74 cm, respectively.

Crested wheatgrass (*Agropyron cristatum* [L.] Gaertn.) cv. Fairway was seeded on the entire site on 24 November 1970 with rows spaced 53 cm apart. On 9 April 1971 single rows of bare-root stock of fourwing saltbush

D.A. Johnson is a Plant Physiologist of USDA-SEA-AR, M.D. Rumbaugh is a Research Geneticist of USDA-SEA-AR, and G.A. Van Epps is an Associate Professor of Range Science at Utah State University.

(*Atriplex canescens* [Pursh] Nutt.) were transplanted at approximately 3-m intervals continuously across the drill rolls of crested wheatgrass. On the same date, a 9-m-long double row of legumes was seeded parallel to the fourwing saltbush rows and 122 cm from them. The legume treatments included: (1) alfalfa (*Medicago sativa* L.) cv. Rambler, (2) cicer milkvetch (*Astragalus cicer* L.), (3) sicklepod milkvetch (*A. falcatus* Lam.), and (4) a mixture of the three legume species. The experimental design was a randomized complete block. After 7 years the all-legume mixture consisted almost exclusively of alfalfa. Because of the similarity of this mixture to the alfalfa treatment, the results will not be presented here.

A 107-cm-wide belt transect 6 m long was established at right angles to and centered on the rows of fourwing saltbush for each of the four legume treatments. Each transect was divided into 20 30-cm segments. Three replicate belt transects were newly established for each legume treatment on each sampling date so that each of the belt transects was harvested only once. For this paper three areas along the belt transect were examined. These included transect portions where crested wheatgrass was growing (A) alone (segments 1–4), (B) in combination with fourwing saltbush (segments 8–11), and (C) in association with a legume (segments 14–17). All of the current year's above-ground forage within a segment was harvested separately for each species and dried in a forced-draft oven at 38°C. Fourwing saltbush herbage harvested included both stem and leaf material produced during the current growing season. Forage yields were measured once in 1978 on 28–30 August and three times in 1979 on 21–25 May, 9–12 July, and 27–30 August. Protein yields were calculated by multiplying the oven-dry forage weights by the product of their nitrogen content multiplied by 6.25.

RESULTS

Alfalfa exhibited statistically greater forage yields than either sicklepod or cicer milkvetch on all sampling dates except August 1978. (Fig. 1). Forage yield of alfalfa was 1.5–8.0 times greater than that of the other legumes in any particular sampling date and was greatest during August 1979 with 171 g/m². Sicklepod milkvetch had lower yields than alfalfa but higher than cicer milkvetch. Both sicklepod and cicer milkvetch produced maximum forage yields during May 1979 with 96 and 46 g/m², respectively.

Fig. 1. Protein and forage yields from four sampling dates for alfalfa, sicklepod milkvetch, and cicer milkvetch forage grown in association with crested wheatgrass. Each bar represents the mean of 12 values; the same letter on top of the bars within a sampling date indicates no statistically significant differences at the 0.10 probability level.

Fig. 2. Protein and forage yields from four sampling dates for crested wheatgrass, fourwing saltbush, and alfalfa components. The capital letters A, B, and C correspond to the transect segments where crested wheatgrass was grown alone, with fourwing saltbush, and with alfalfa, respectively. Each bar represents the mean of 12 values; the same lowercase letters (a and b) at the bottom of the bars indicate no statistically significant differences within a sampling date in only the crested wheatgrass component at the 0.05 probability level. The same lowercase letters (x, y, and z) at the top of the bars indicate no statistically significant differences within a sampling date in the total forage at the 0.05 probability level.

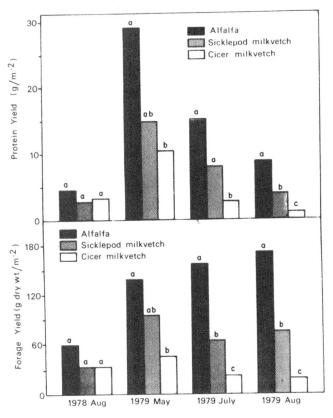

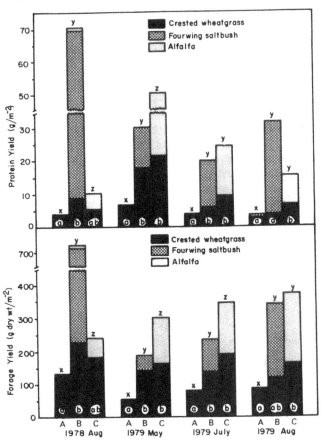

Alfalfa had the highest protein yields, followed by sicklepod milkvetch and cicer milkvetch (Fig. 1). Protein yields were highest for all three legumes during May 1979, with a decline in July and a further decrease in August. Maximum protein yields for alfalfa, sicklepod milkvetch, and cicer milkvetch were 29, 15, and 10 g/m², respectively. Because alfalfa yielded the most of all legumes tested, it was used in subsequent comparisons with crested wheatgrass alone and with crested wheatgrass grown with fourwing saltbush.

Total forage produced during August was more than 2-fold greater in 1978 than in 1979 in the transect segment where crested wheatgrass was growing in combination with fourwing saltbush, due mainly to the increased yield of fourwing saltbush (Fig. 2). Forage yields of crested wheatgrass grown alone were also higher in August 1978 than in August 1979; however, in the transect segment containing alfalfa, total forage yield was higher in August 1979 than in August 1978. Apparently the above-average precipitation received during January-April 1978 favored growth of fourwing saltbush more than the other species.

In 1979 total forage yields reached a maximum in August (Fig. 2). Highest total forage yields were produced by crested wheatgrass grown with alfalfa, followed

by crested wheatgrass grown with fourwing saltbush. Fourwing saltbush greatly outyielded the legume component in August 1978. The alfalfa component outyielded fourwing saltbush in May and July 1979, but no statistically significant differences were noted in August 1979. Crested wheatgrass alone yielded only 18%–23% of the other forage yields on all sampling dates. The increased yields where crested wheatgrass was grown with fourwing saltbush or alfalfa were due to the contributions of fourwing saltbush or of alfalfa plus increases in the crested wheatgrass component itself. For example, in May 1979 crested wheatgrass yielded only 55 g/m² where it was grown alone, but it yielded 142 and 163 g/m² where it was grown with fourwing saltbush or alfalfa, respectively.

Total protein yields during August for all three transect segments were higher in 1978 than in 1979 except for alfalfa (Fig. 2). Crested wheatgrass grown with fourwing saltbush produced 71 g/m² protein in August 1978, the highest of any sampling date. This high protein yield was mainly due to the 61-g/m² protein contribution of fourwing saltbush.

Total protein yields in 1979 were generally highest in the May sampling (Fig. 2). The combination of crested wheatgrass and alfalfa gave the greatest protein yield in May with 51 g/m². Crested wheatgrass grown with the shrub had the highest protein yield during August with 32 g/m². Even though crested wheatgrass in this segment showed a marked decline in protein yield in August, it was offset by an increase in protein yield of fourwing saltbush. On all sampling dates protein yield for crested wheatgrass grown alone lagged markedly behind those of the other two transect segments. Protein yields for crested wheatgrass were 6–17 times greater when grown with either fourwing saltbush or alfalfa than when grown alone.

DISCUSSION

Average forage and protein yields of crested wheatgrass were greatly increased when it was grown with either fourwing saltbush or alfalfa. These enhanced yields were due not only to the additional species but also to an increase in forage and protein yield in the crested wheatgrass component. Increased production in the crested wheatgrass–alfalfa mixture may be due to nitrogen fixation. However, factors other than nitrogen fixation are probably important in the fourwing saltbush associations.

Fourwing saltbush and alfalfa can influence the soil characteristics beneath their canopies and consequently affect the growth of the associated crested wheatgrass. Shrubs influence both the horizontal and vertical patterns of soil chemicals (e.g., Charley and West 1975), with minerals accumulating under them and concentrations declining in the surrounding area. Mineral accumulation can occur from an increase in canopy capture of wind-transported soils, animal activity, and nitrogen fixation by free-living microorganisms present in the increased litter mat under the shrub (Charley 1977). Fairchild and Brotherson (1980) showed that of six different shrubs, fourwing saltbush consistently exhibited the greatest concentrations of calcium, magnesium, potassium, and sodium beneath the shrub canopy.

Crested wheatgrass might use the upper soil profile for water and nutrients, while fourwing saltbush and alfalfa may utilize the greater soil depths more extensively. Shrubs and legumes may promote organic-matter development of soils under their canopy resulting in greater water-holding capacity. Also, the canopy of alfalfa and particularly of fourwing saltbush might lower soil temperatures through shading, thereby reducing evaporative soil water loss or providing a more nearly optimum temperature for growth of crested wheatgrass. The taller canopies could also cause greater snow accumulation for more favorable water relations.

Whatever mechanism or combination of mechanisms is involved, areas with combinations of crested wheatgrass, fourwing saltbush, and legumes (particularly alfalfa) provided a more productive, higher-quality forage resource than areas with monocultures of crested wheatgrass.

LITERATURE CITED

Charley, J.L. 1977. Mineral cycling in range ecosystems. *In* R.E. Sosebee (ed.) Rangeland plant physiology, 216–256. Soc. Range Mangt. Range Sci. Ser. 4.

Charley, J.L., and N.E. West. 1975. Plant-induced soil chemical patterns in some shrub-dominated semi-desert ecosystems in Utah. J. Ecol. 63:945–964.

Fairchild, J.A., and J.D. Brotherson. 1980. Microhabitat relationships of six major shrubs in Navajo National Monument, Arizona. J. Range Mangt. 33:150–156.

Age, Date of Cutting, and Temperature as Factors Affecting Chemical Composition of Berseem (*Trifolium alexandrinum* L.)

F. GUESSOUS
Hassan II Institute of Agronomy and Veterinary
Medicine, Rabat, Morocco

Summary

Chemical composition of berseem (*Trifolium alexandrinum* L.), produced in winter and spring under irrigation in Morocco, was investigated in a 4-year study. Berseem was planted at various dates, and after the first cut had been harvested, samples were taken from each regrowth at weekly intervals. Total plant samples and leaf and stem samples were analyzed for dry matter

(DM), ash, crude protein (CP), and crude fiber (CF). A total of 172 samples of above-ground regrowth were analyzed and used to calculate multiple regressions of chemical composition on age and date of cutting. Date of cutting was defined as number of days after August 1.

Regrowths remained vegetative from November to early May, after which time flowering occurred. Dry matter remained low (8% to 12%) during winter and spring and reached 20% only in June. Ash was always high (12% to 20%), while CF did not exceed 30%. Crude protein varied between 18% and 30%. At a given date of cutting, CF increased and ash decreased 0.19 and 0.06 percentage units/day of age, respectively. DM increased rapidly when plants grew old, while CP decreased while they were still young. At constant ages of cutting, regrowths showed a slight decrease in DM from November to the end of January and then increased until June. Ash decreased more rapidly than CP with the advancing season from November to June. Crude fiber increased dramatically during the same period. The effects of date of cutting appear to be a combination of effects of regrowth number, stage of maturity, and temperature. When samples of the same regrowth number and stage of maturity were compared, there was considerable variation from year to year of samples of similar ages for leaf percentage and chemical composition. For CF, variation exceeded 6 percentage units.

Multiple linear regressions between morphological or chemical composition, age, and average temperature showed that temperature partly explained these variations. Elevated temperature decreased leaf percentage for both the first and second regrowths, but not for the fourth. It also increased DM percentage rapidly in the last regrowth. Partial regression coefficients between CF and average temperature were 0.38, 1.75, and 0.92 for first, second, and fourth regrowths, respectively, while those with ash and CP were always negative. Leaves appeared to have nearly constant CF contents. Temperature appeared to increase CF of berseem by stimulating its maturity but also by a specific action on the fiber content of stems.

KEY WORDS: berseem, *Trifolium alexandrinum* L., chemical composition, age, season, temperature.

INTRODUCTION

Berseem clover (*Trifolium alexandrinum* L.) is an annual crop cultivated as a forage in most countries near the Mediterranean. Under irrigation in Morocco it can produce six or seven cuttings during winter and spring, and it is used mainly as a feed for cattle and sheep. However, little information is available on its chemical composition, particularly as influenced by season. Theriez (1965), Polidori and Galvano (1968), and Danasoury et al. (1971) have shown that as number of cuts increases, concentrations of dry matter (DM) and crude fiber (CF) increase, while those of ash and crude protein (CP) decrease. However, these data do not allow prediction of chemical composition of berseem, because it is normally sown over an extended period. Thus, the same cut number may be obtained in different seasons. The objective of the research reported here was to characterize effects of age and season on the chemical composition of berseem.

METHODS

A 4-year (1974–1978) study was conducted in which berseem was planted under irrigation at seven dates of sowing between August and November. After the first cut was harvested, samples were taken from each successive cut at weekly intervals. Total plant samples and leaf and stem samples were analyzed for DM, ash, CP, and CF according to methods of AOAC (1975). Data collected on 172 samples of entire above-ground plant parts were used

in a multiple regression model in which chemical composition was regressed on age, date of cutting, and regrowth number. Age was defined as the number of days of growth following the previous cutting. Date of cutting was defined as number of days after August 1.

RESULTS

Regrowths remained vegetative until early May. Regrowth number did not affect (P > 0.05) DM, ash, CP, or CF concentrations. Regression equations that show effects of age and date of cutting on chemical composition of berseem are presented in Table 1. As age increased, DM and CF increased while ash and CP decreased. With date of cutting held constant, ash decreased and CF increased at rates of 0.06 and 0.19 percentage units/day of age, respectively. Dry matter increased rapidly when berseem grew old, while CP decreased while it was still young.

At constant ages of cutting, regrowths showed a slight decrease in DM from November to the end of January; DM then increased rapidly until June. Ash decreased more quickly than CP with the advancing season from November to June. Crude fiber increased dramatically during the same period; for example, 42-day regrowths changed from 18% CF in November to 24% CF in June.

The influence of date of cutting on the chemical composition of berseem reflects effects of regrowth number, stage of maturity, and season. Regrowth number usually induced slight increases in CF and decreases in ash concentrations but had no effect on CP. Plants cut in May and June at a reproductive stage of maturity tended to have higher DM and CF concentrations than plants of the same age cut earlier at a vegetative stage.

When samples of the same regrowth number and stage of maturity were compared, there was considerable variation from year to year among samples of similar age cut at

The author is an Associate Professor of Animal Science at Hassan II Institute of Agronomy and Veterinary Medicine.

The author wishes to thank C. Demarquilly for help in the conception of this work, R.D. Goodrich for assistance with redaction, and P. Wittevrongel for help with chemical analyses.

Table 1. Effect of age (A) and date of cutting (D) on dry matter, ash, crude-protein, and crude-fiber concentrations in regrowths of berseem (dry-matter basis).[a]

| Dependent variable | Inter-cept | Regression coefficient with Independent variable | | | | | R | $S_{y \cdot x}$ |
		A	A²	$\frac{1}{A}$	D	D²		
DM, %	21.35		.0008**		−.148**	.00041***	.89	1.8
Ash, %	20.88	−.06**				−.000065***	.87	1.3
Crude protein, %	17.53			171.4***		−.000037**	.88	1.8
Crude fiber, %	9.00	.19***				.000084***	.91	2.0

A = Age = number of days of growth following the previous cutting.

D = Date of cutting = number of days after August 1.

[a]Multiple regression analyses based on 172 samples of entire aerial parts of berseem.

**P <.01.

***P <.001.

different seasons. For ash, CP, and CF, maximum differences were 4, 6, and 6 percentage units, respectively.

Regressing morphological and chemical compositions of berseem on age and average temperature (Table 2) showed that temperature and age partly explained this variation. High temperature had a negative effect on leaf percentages in the first and second regrowths but not in the fourth regrowth. Elevated temperature significantly increased DM only during the fourth regrowth; it also increased the CF percentages of berseem regrowth; partial regression coefficients were 0.38, 1.75, and 0.92 for the

first, second, and fourth regrowths, respectively. Such temperatures resulted in lower percentages of ash and CP in all three regrowths. However, change in temperature resulted in smaller percentage-unit changes in ash and CP than in CF.

Berseem stems showed influences of temperature on ash and CP similar to those in entire above-ground parts. As temperature increased, CF percentages increased to a greater degree for stems than for entire above-ground regrowths.

DISCUSSION

A striking characteristic of berseem in this experiment was its low DM content, which remained between 8% and 12% until April and reached 20% only in June. Ash was also high when compared with that of other forage crops (Demarquilly et al. 1978). Subsequent mineral analyses showed that berseem had high concentrations of potassium (2% to 5% of DM) and sodium (1% to 2% of DM). Crude fiber was always low even when plants flowered, while CP ranged between 18% and 30%. Similar characteristics have been reported from Tunisia (Theriez 1965), Italy (Polidori and Galvano 1968), and Egypt (Danasoury et al. 1971).

Age and stage of maturity had strong influences on the chemical composition of berseem. However, these factors tended to have more influence on chemical composition when plants were young than when they were old.

Chemical composition of berseem was also affected by season to reflect the influence of changing environmental conditions. Elevated temperatures stimulated maturation of plants and, hence, induced an increase in their fiber content. While some studies (Smith 1969, Marten 1970) did not indicate a significant effect of temperature on fiber content of alfalfa, others (Deinum et al. 1968, Faix 1974, Van Soest et al. 1978) made on both grasses and legumes produced results similar to those reported here. During the first and second regrowths, increasing temperatures reduced leaf percentage and increased CF. This effect was also found in the fourth regrowth, although leaf percentage was unaffected. High temperature appeared to have

Table 2. Effects of age (A) and temperature (T) on leaf percentage and chemical composition of regrowths of berseem (dry-matter basis).

| Regrowth number dependent variable | Inter-cept | Regression coefficient with independent variable -- | | R | $S_{y \cdot x}$ |
		A	T		
First regrowth, vegetative (n = 23)					
Leaves, %	93.1	−.44***	−1.65	.80	6.0
Dry matter, %	7.5	.05***		.62	1.2
Ash, %	22.8	−.09***	−.21	.78	1.3
Crude protein, %	29.6	−.13***	−.32	.84	1.5
Crude fiber, %	7.9	.19***	.38	.85	2.0
Second regrowth, vegetative (n = 30)					
Leaves, %	169.1	−.60***	−7.16***	.86	7.2
Dry matter, %	8.5	.06***		.54	1.6
Ash, %	34.8	−.07***	−1.35***	.85	1.1
Crude protein, %	41.7	−.17***	−1.07**	.90	1.5
Crude fiber, %	−8.8	.20***	1.75**	.83	2.5
Fourth regrowth, reproductive (n = 22)					
Leaves, %	68.2	−.55***		.80	6.7
Dry matter, %	−8.6	.13***	1.18***	.84	1.9
Ash, %	21.8	−.04***	−.48***	.87	.6
Crude protein, %	36.2	−.15***	−.65***	.85	1.5
Crude fiber, %	1.5	.21***	.92***	.92	1.5

A = Age = number of days of growth following the previous cutting.

T = Temperature = average temperature (C)

**P <.01

***P <.001

two effects—namely, to reduce leaf percentage and to stimulate fiber accumulation in stems. Leaves had nearly constant CF content regardless of age, season, or year.

Partial regression coefficients between CF and average temperature in the present study were higher (0.38 to 1.75) than the value of 0.26 obtained by Deinum et al. (1968) when light intensity, temperature, nitrogen fertilization, and stage of maturity were controlled. However, our results obtained under field conditions include the associative effects of daylength and light intensity, which varied with season.

These results suggest that the quality of berseem forage may depend more on season than on cut number per se. Thus, when making recommendations on the management of berseem, it is important to specifiy both of these conditions, plus stage of maturity and age.

LITERATURE CITED

Association of Official Agricultural Chemists (AOAC). 1975. Official methods of analysis, 12th ed. AOAC, Washington, D.C.

Danasoury, M.S., H.M. El Nouby, and A. Makky. 1971. Composition and feeding value of green berseem (Egyptian clover) and its hay cured by ground and tripod methods. Agric. Res. Rev., Cairo 49:131-139.

Deinum, B., A.J.H. van Es, and P.J. Van Soest. 1968. Climate, nitrogen and grass. II. The influence of light intensity, temperature and nitrogren on in vivo digestibility of grass and the prediction of these effects from some chemical procedures. Neth. J. Agric. Sci. 16:217-223.

Demarquilly, C., J. Andrieu, and D. Sauvant. 1978. Tableaux de la valeur nutritive des aliments. In R. Jarrige (ed.) Alimentation des ruminants 17:519-577. Inst. Nat. Rech. Agron., Paris.

Faix, J.J. 1974. The effect of temperature and day length on the quality of morphological components of three legumes. Ph.D. Thesis, Cornell Univ., Ithaca, N.Y.

Marten, G.C. 1970. Temperature as a determinant of quality of alfalfa harvested by bloom stage or age criteria. Proc. XI Int. Grassl. Cong., 506-509.

Polidori, F., and G. Galvano. 1968. Un triennio di osservazioni sulla produzione, composizione chimica e valore nutritivo del prato irriguo di trifoglio alessandrino (Trifolium alexandrinum L.) coltivato nella piana di Catania. Tec. Agric. 20:91-124.

Smith, D. 1969. Influence of temperature on the yield and chemical composition of "Vernal" alfalfa at first flower. Agron. J. 61:470-473.

Theriez, M. 1965. Valeur alimentaire des fourrages tunisiens. II. Composition chimique et digestibilité du bersim (Trifolium alexandrinum L.) Bul. de l'ENSA de Tunis 8-9:49-62.

Van Soest, P.J., D.R. Mertens, and B. Deinum. 1978. Preharvest factors influencing quality of conserved forage. J. Anim. Sci. 47:712-720.

A Systems Approach to Swedish Ley Production

J.F. ANGUS, A. KORNHER, and B.W.R. TORSSELL

Swedish University of Agricultural Sciences, Uppsala, Sweden

Summary

A simulation model for estimating yield of leys from standard meteorological observations is described. It is based on an S-shaped growth curve also accounting for phasic development and regrowth depression. Effects of temperature, radiation, and soil moisture are accounted for by a growth index. Soil moisture is simulated with a standard water budget.

The parameters in the model were estimated by comparing measured with calculated growth. The model then predicted final harvest with reasonable accuracy, but with underestimate of early season growth, presumably due to mobilization of carbohydrate from storage organs.

Observed ley yields from 49 Swedish parishes (1961-1978) were correlated with model calculations, showing a correlation coefficient of r = 0.9. North of lat 59°N the model underestimated the first hay cut, while south of this line, yields were overestimated. For the second cut the deviations were reversed. This geographical pattern suggests more assimilate storage and subsequent mobilization in northern regions with longer and more severe winters than in southern regions with milder winters.

KEY WORDS: forage production, growth prediction model, crop-weather interactions.

INTRODUCTION

Leys occupy over one-quarter of the arable land in Sweden, constituting the largest area of any crop. However, while the yields of other crops have risen rapidly during the past 30 years, ley yields have shown relatively little improvement (Åberg 1976). Ley yields also show great variability from year to year and from site to site, due partly to uncontrolled or environmental factors and partly to management factors controlled by the farmer. The controlled factors have been subjected to intensive investigation by agriculturalists, but environmental causes of yield variation are poorly understood and have rarely been quantified.

The purpose of this paper is to report some initial attempts to describe the effects of weather on yield variation of leys. A more extensive report has been made by Angus et al. (1980).

DEVELOPMENT AND TESTING OF THE MODEL

Our initial aim was to develop a model based on the growth index (GI) of Fitzpatrick and Nix (1970) to estimate the harvestable biomass of leys at the end of any growth interval, using standard meteorological observations and parameters describing the soil and crop.

The first version of this forage production model (FOPRO-1) is based on a classical growth curve modified to account for radiation, temperature, water supply, stage of phenological development, and regrowth depression.

Description of the Model

The model FOPRO-1 estimates the increase of harvestable biomass (W) from daily or weekly increments (t) by a difference equation that gives an S-shaped curve when environmental conditions are nonlimiting.

$$\frac{\Delta W}{\Delta t} = \alpha \times W \times AGE \times GI \qquad (1)$$

Where α is a relative growth rate (RGR) at time zero and AGE is a number scale between zero and one that when multiplied by α indicates the decline of RGR with developmental time. An empirical function is used for this relationship:

$$AGE = (1 - \exp - \gamma(1 - d)) + \beta/(1 + \beta) \qquad (2)$$

The pattern of decline in RGR is given by γ and β, and the development time, d, is estimated from temperature and photoperiod (Angus et al. 1980).

The growth index (GI) is the product of the multiplication of a radiation index (RI), a temperature index (TI), and a water index (WI), each of which is scaled between

J.F. Angus is a Visiting Agronomist (current address, CSIRO Division of Land Use Research, Canberra, 2061, Australia), A. Kornher is a Research Agronomist, and B.W.R. Torssell is a Professor of Agronomy at the Swedish University of Agricultural Sciences.

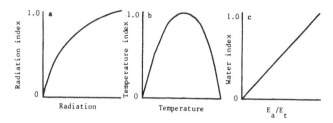

Fig. 1. Forms of the relationships linking (a) radiation index to radiation, (b) temperature index to mean temperature, and (c) water index to E_a/E_t, the ratio of actual to potential evapotranspiration.

zero and one, where zero represents growth totally limited by the factor and one represents no limitation by the factor:

$$GI = RI \times TI \times WI \qquad (3)$$

Each of the indices is calculated as a function of weather variables; the relationships are shown in Fig. 1. The radiation (global) and temperature (mean daily) indices are self-explanatory. The water index, based on the ratio of actual evapo-transpiration (E_a) to potential evapo-transpiration (E_t), is calculated as part of a conventional water budget that considers canopy cover, plant available soil water, rainfall, and potential evaporation (Johansson 1973).

Calibration of the Model

The biomass production of a pure stand of timothy (*Phleum pratense* L.) was measured by weekly serial harvests during two growth cycles during 1979 at the Ultuna bioclimatic station (near Uppsala). Soil water for the experimental plots was also determined weekly, using a neutron moisture meter.

The optimized model (Ross 1971) calculated final yield using daily growth steps of both cycles with reasonable accuracy, although it underestimated the rapid growth rate of the stand in mid-June. The model accurately simulated the growth of the second cycle, using a constant that expressed the maximum value of RGR as 85% of the rate during the first cycle.

Using a model calibration for growth curves based on a weekly time step for stands of timothy grown at the Ultuna bioclimatic station in 1967, 1968, and 1979 resulted in different growth patterns each of the 3 years. However, the model fitted to these data gave an adjusted r^2 of 0.92 (12 parameters, 39 data points), but there were systematic discrepancies between observed and calculated production. The observed rapid growth in early spring was consistently underestimated, and final yield was not accurately calculated in all cases. Overestimation ranged to 1.5 metric tons (t)/ha and underestimation to 0.8 t/ha.

The parameters obtained from optimizing the model over the 3 years at Ultuna were used in further testing the model on data collected throughout Sweden.

COMPARISON OF MODEL CALCULATIONS WITH YIELDS MEASURED ON FARMS AND FIELD EXPERIMENTS

For the first validation, the model was run with meteorological data from 49 selected weather stations in the Swedish meteorological network for the growing seasons 1961–1978. Selection insured representative samples of climatic conditions for crop production in Sweden. The model used weekly time steps after the weather data were aggregated into weekly means or totals. A maximum soil water storage of 150 mm was assumed.

Model calculations compared with farm data. Ley yields from 49 parishes throughout Sweden for the period 1961–1978 were obtained from surveys conducted by the National Central Bureau of Statistics (SCB). These surveys involved annual quadrat cuts in randomly selected farmers' fields. The selected parishes were located near the 49 weather stations.

Observed and calculated mean yields. Mean yields for the 1961–1978 period were calculated for each of the 49 parishes. The observed and calculated yields (first and second cut) are shown in Fig. 2. The correlation coefficient between observed and calcuated yield for the first cut was r = 0.67, for the second cut r = 0.37, and for the combined sample of first and second cut r = 0.90.

The deviations between observed and estimated yields formed a geographic pattern. North of lat 59°N the model generally overestimated the first hay cut, while south of this line, yields were generally underestimated. For the second cut the deviations were reversed, but the boundary line between northern overestimates and southern underestimates was approximately 60°N.

Observed and calculated annual yields. Comparisons of

observed and calculated yields for two parishes are shown in Fig. 3. The calculation for Lycksele in the north underestimates the first cut and overestimates the second. In contrast, the calculation for Lund in the south overestimates the first cut and underestimates the second cut.

Thus, the deviation between observed and calculated yields for individual years show a pattern similar to the deviations in Fig. 2.

Model calculations compared with experiment yields. The Regional Experimental Program is a source of data on ley productivity under well-managed conditions throughout Sweden. A series of similar experiments has been conducted for a number of years in many locations, with treatments such as fertilizer and cutting regimes. Since our interest at this stage was in environmental effects, we chose data from a well-fertilized treatment of an experimental series conducted in central Sweden with timothy. The model was run with weather data from the station nearest to the experimental sites.

The agreement between observed and calculated yields

Fig. 3. Comparisons of observed and calculated dry-matter yields for two parishes, 1961–1978.

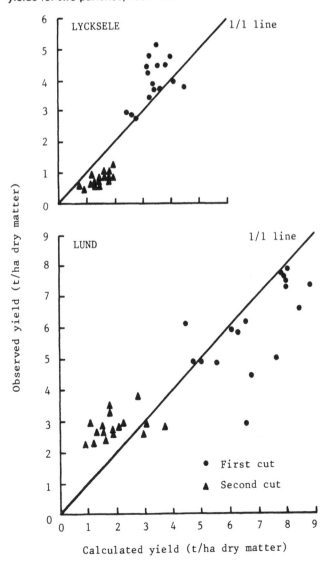

Fig. 2. Comparisons of observed and calculated dry-matter yields for 49 Swedish parishes.

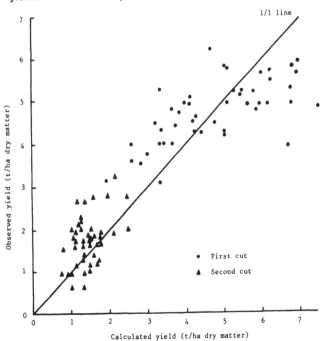

was similar to that found with the farm data, but there were no consistent under- or overestimates of yield for either first or second cut, presumably because the sites were in central Sweden.

DISCUSSION

The model shows reasonably good agreement with observed yields (1) under research conditions at Uppsala, (2) in regional experimental plots throughout central Sweden, and (3) in comparison with farmers' yields for 49 widely scattered parishes throughout Sweden. The last of these tests is particularly interesting as it suggests that there is no gap between yields on Swedish farms and yields under experimental conditions.

The deviations between calculated and observed yields found in all tests of the model suggest that overwintering reserves contribute to spring growth, particularly at more northerly latitudes. The accumulation of these reserves presumably occurs during autumn, again most markedly

in the north. The quantification of these processes is to be the next stage in this project.

LITERATURE CITED

Åberg, E. 1976. Framtida uppgifter för växtodlingen. Rapporter och avhandlingar 54. Institutionen för växtodling, Lantbrukshögskolan.

Angus, F.A., A. Kornher, and B.W.R. Torssell. 1980. A systems approach to estimation of Swedish ley production. Dept. of Plant Husb., Swedish Univ. of Agric. Sci. Prog. Rpt. 85, 1979-80.

Fitzpatrick, E.A., and H.A. Nix. 1970. The climatic factor in Australian grassland ecology. *In* R.M. Moore (ed.) Australian grasslands, 3-26. ANU Press, Canberra.

Johansson, W. 1973. Metod för beräkning av vatteninnehåll och vattenomsättning i odlad jord med ledning av meteorologiska data. Grundförbättring 26:57-153.

Ross, P.J. 1971. A computer program for fitting non-linear regression models to data by least squares. SCIRO Austral. Div. Soils. Tech. Paper 6.

Energy Flow and Conversion Efficiency in Grazing Grassland

T. AKIYAMA, S. TAKAHASHI, M. SHIYOMI, and T. OKUBO
National Grassland Research Institute and Nagoya University, Japan

Summary

Measurable energy storage on grassland ecosystems was examined for 5 years at the Fujinita site under the same rotation program with two grazing intensities. Daily global solar radiation and photosynthetic active radiation were measured as the energy sources in the ecosystem. Primary production by pasture plants and amount grazed by cattle were estimated by the biomass differences inside and outside of protection cages. For the estimation of energy produced or stored by plants or cattle, calorific values were examined for dominant plant species at the Fujinita site. Seasonal changes in digestibility were investigated to ascertain the energy of herbage digested by cattle. For the evaluation of stored energy in cattle, ARC (1965) feeding standards were adopted according to the increment of body weight. From energy accumulation, the energy conversion efficiency through the system was calculated. Out of 9.74×10^5 kilocalories (kcal)/m²/yr of global solar radiation, the rates of 6.50×10^{-3} and 5.78×10^{-3} were utilized in primary production for light (L) and heavy (H) grazing, respectively. Intensive grazing (H) resulted in greater utilization of photosynthate and a higher digestibility of pasture plants (0.600 and 0.695 for plot H and 0.523 and 0.657 for plot L). Overall efficiency beginning with solar radiation and terminating as stored energy in cattle was 7.43×10^{-5} and 9.61×10^{-5} for plot L and plot H, respectively.

KEY WORDS: grassland ecosystem, energy flow, cattle, global solar radiation, pasture plants, calorific values, digestibility.

INTRODUCTION

Dairy and animal production depend essentially upon herbage production, which is strongly affected by abiotic factors such as soil and radiation. However, detailed information concerning energy flow dynamics in a grassland ecosystem is limited.

The purpose of this paper is to present the results of quantitative ecological investigations to clarify the patterns of energy flow through the atmosphere, pasture plants, and cattle in a grassland ecosystem and also to present conversion efficiency through trophic levels that lead to practical considerations for the management of grassland and to the maximization of animal production in Japan.

T. Akiyama and S. Takahashi are Research Scientists and M. Shiyomi is the Chief Researcher at the National Grassland Research Institute; T. Okubo is an Associate Professor on the Faculty of Agriculture at Nagoya University.

METHODS

Description of Test Site and Rotation Program

Test site. The present study was conducted from 1974 to 1978 at the National Grassland Research Institute Experimental Pasture called the Fujinita site, in Tochigi Prefecture, which is situated at lat 36°55′ N long 136°58′ E, about 120 km north of Tokyo. The site covers an area of 6 ha and is divided into two sets of grazing fields, each consisting of four rotational paddocks.

Meteorological data. Climatic data, such as air temperature, radiation, and rainfall, were measured at the meterological station about 2 km west of the test site. The average global solar radiation (GSR) was 9.74 × 10⁵ kcal/m²/yr, and energy of 3.97 × 10⁵ kcal/m²/yr was estimated as the average photosynthetic active radiation. This amount is equivalent to 40.8% of GSR.

Rotation program. One set of pastures was assigned to light grazing (2 cattle/ha), and the other set was assigned to heavy grazing (4 cattle/ha). Pastures were rotationally grazed from April to October, with usually a 1-week stay in each paddock. We selected one paddock at each grazing level as the test paddock.

Sampling techniques. The standing-crop biomass, including the undergound parts, was measured just after completion of each grazing period on the test paddock. Movable protection cages (MP cages) were set in the test paddocks in order to measure the total production. Forage remaining outside the MP cages after grazing was subtracted from total production to determine the amount of forage consumed. Dry-matter yields were determined after oven-drying at 70°C. Samples were saved for energy and digestibility determinations.

Estimation of Energy Accumulation

Energy stored in plants. The calorific values of the plant samples from the tests paddocks were measured with a Shimadzu Oxygen Calorimeter. Energy storage in plant materials was calculated by multiplying the phytomass by the respective energy concentration.

Grazed energy. Grazed energy was determined by multiplying the respective calorific values by the postgrazing dry-weight differences between herbage inside and outside of MP cages.

Digestible energy. In order to estimate the amount of energy of digested grazed herbage, digestibility in vitro was examined by the neutral detergent and enzymatic method (Abe and Horii 1974). The amount of digestible energy was determined by the products of grazed energy and digestiblity of the respective types of herbage. Seasonal changes in digestibility occurred in the aboveground portion of pasture plants and in standing dead material sampled in the two levels of grazing intensity. Generally, the values are high in early spring and late fall. It was also observed that the plant materials collected in the heavily grazed plot had definitely higher digestiblity

than those in the lightly grazed plot from May to September.

Energy fixed by cattle. During the experiment, 12 to 16 head of fattening Holstein cattle (around 200 kg each) were used for weekly body-weight measurements when moved to the successive paddocks. Energy amounts fixed in cattle during the grazing period were determined by the ARC (Agricultural Research Council 1965) feeding standards, using a weekly increment of body weight and gain.

RESULTS AND DISCUSSION

As is shown in Fig. 1, energy conversion efficiency between global solar radiation and energy fixed by plants, ①→③, was 6.50 × 10⁻³ and 5.78 × 10⁻³ in the light (L) plot and in the heavy (H) plot, respectively. The rates between energy fixed by plants and grazed energy, ③→④, were higher in plot H (0.600) than in plot L (0.523). Higher digestiblity in plot H (0.695) resulted in higher digestible energy in plot H (2,348 kcal/m²/yr) than that in plot L (0.657 and 2,177 kcal/m²/yr). Energy stored by cattle was estimated as 72 and 94 kcal/m²/yr in plots L and H. Estimated total energy conversion rate through ①→⑨ was 7.43 × 10⁻⁵ and 9.61 × 10⁻⁵ in plots L and H, respectively.

Many problems are unsolved and unrealistic assump-

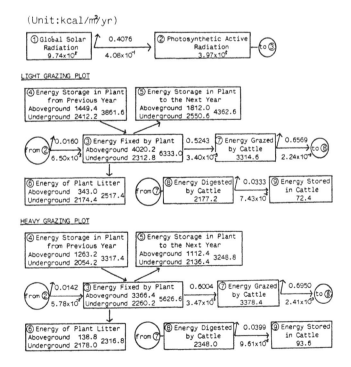

Fig. 1. Comparison of energy storage and heat conversion efficiencies from light and heavy grazing conditions observed at the Fujinita site over 5 years. The numbers inside each frame indicate energy storage in kcal/m²/yr. The numbers above the arrows indicate the conversion efficiencies from the previous to the next items, and the numbers below the arrows indicate overall efficiencies initiating from global solar radiation.

tions are made concerning estimation of energy storage in grassland-ecosystem investigations. The difficulties that must be overcome are (1) estimation of primary production and grazed amount of a whole rotation area from a small sample area; (2) estimation of the decomposition rate of underground parts; and (3) estimation of metabolized energy in grazing cattle.

LITERATURE CITED

Abe, A., and S. Horii. 1974. Cellulase hydrolysis of the cell wall and its application to estimate the nutritive value of the forage. J. Japan. Grassl. Sci. 20:16–21.

Agricultural Research Council. 1965. The nutrient requirements of farm livestock. II. Ruminants. A.R.C., London.

Simple Computer Simulation Models for Forage-Management Applications

G.W. FICK and D. ONSTAD
Cornell University, Ithaca, N.Y., U.S.A.

Summary

Dynamic computer simulation models of forage growth can be useful in interpreting physical and biological limitations on production and in designing management systems for testing in the field. The explanatory detail of such models makes them both more reliable and more expensive to use than simple empirical formulations. Simple simulation models reduce the expense. They can also serve as reference points for designing, evaluating, or selecting more complex versions and thus provide a more rational basis for using models in grassland research and management.

Our objectives were to develop and to test a simple dynamic simulation model of alfalfa (*Medicago sativa* L.) production using a combination of explanatory and empirical functions. Yield and forage quality in terms of digestibility and crude protein concentrations were predicted, and model testing was used to establish a reference point for evaluating more complex alfalfa models, several of which are available. The rationale of model development was to include the primary factors that cause variation in alfalfa yield. Genetic yield potential, moisture supply, temperature effects, and cutting management were selected as the most important variables. Cutting management was assumed to operate through its effect on root reserves in the control of regrowth potential and winter survival. Appropriate functional relationships were taken from other models or derived from data in the literature. Forage-quality prediction was based on an empirical relationship with time or growing degree days as the independent variable. The model was named ALSIM 1 (LEVEL 0).

The LEVEL 0 model simulated expected patterns of dry-matter accumulation and soil moisture supply for the northeastern U.S.A. It also indicated that temperature stress is an important element of yield determination in alfalfa and that the phenological development of at least some cultivars must depend on day length as well as on temperature. Critical tests of LEVEL 0 involved simulations of 60 observations from three harvest schedules over 2 years at three locations with differing supplies of soil moisture. The linear regression of observed yields on predicted yields rendered the equation $Y = 0.844 + 0.726X$ with $r^2 = 0.62$ and $S.E._y = 0.870$ metric tons/ha. There is considerable room for statistical improvement, but LEVEL 0 will be adequate for preliminary study of harvesting options. Corresponding analyses for digestibility and crude protein concentration rendered $r^2 = 0.25$ and 0.13, respectively, with $S.E._y$ on the order of 3 to 5 percentage units. The empirical, time-based quality model is not adequate.

KEY WORDS: alfalfa, lucerne, *Medicago sativa* L., soil water budget, temperature effects, genetic potential, forage quality prediction, model testing.

INTRODUCTION

Systems analyses of forage-based enterprises often use computer modeling for prediction and comparison of the probable results from management alternatives. Empirical statistically based models usually lack generality and the capacity to make reliable extrapolations to new situations. Simulation models based on cause-and-effect mechanisms are more suited to extrapolation, but they often contain too much detail for routine use in the con-

G.W. Fick is an Associate Professor of Agronomy and D. Onstad is a Graduate Research Assistant in the Department of Environmental Engineering of Cornell University. This is a contribution of the Department of Agronomy of the New York State College of Agriculture and Life Sciences of Cornell University (Agronomy Series Paper No. 1384).

This project was financed in part with federal funds from the Environmental Protection Agency under grant CR-806277-020. The contents do not necessarily reflect the views and policies of the Environmental Protection Agency.

Mention of trade names or commercial products does not constitute endorsement of them.

text of a farm-management service. Our first objective was to develop a simple alfalfa (*Medicago sativa* L.)-production simulation model suited for use in exploring alternative management programs for alfalfa. To do this, a combined empirical and explanatory approach was taken.

Our second objective was to test the model and establish a reference point for evaluating the merits of alternative models. Several simulation models of alfalfa production have been developed (Dougherty 1977, Fick 1975, Field and Hunt 1974, Gutierrez et al. 1976, Holt et al. 1975, Selirio and Brown 1979), and, in this paper, the predictive capacity of an additional simple model has been evaluated as the starting point for determining the achieved progress and suitable applications of the more detailed simulators.

METHODS

The simple alfala-production model, named ALSIM 1 (LEVEL 0), used a simulated time step of one day and required input data for genetic yield potential of the crop, water-holding capacity of the soil, and average monthly weather data for temperature, precipitation, and pan evaporation (which were converted to daily values). The model was designed to simulate management for hay production or close grazing of an established stand for any number of harvests at any location. Details and a listing of the program are available from the senior author.

Empirical or simple explanatory procedures were used to account for the primary factors that cause variation in alfalfa yield. The genetic yield potential of the crop was coded, using the sigmoid growth equation of Selirio and Brown (1979). Total seasonal yield, fraction of the total in each harvest, and average number of growing degree days to flowering in each harvest were needed to fit the equation. The simple soil-water budget model of Fitzpatrick and Nix (1969) was used to estimate soil moisture content of the rooting zone. The leaf-area factor for the model was derived for alfalfa using the empirical relationship of Bula and Hintz (1978). The effect of average daily temperature on alfalfa growth was taken directly from the model of Fick (1975). The effect of cutting management on alfalfa production was assumed to operate through the influence of root reserves on regrowth and winter survival. Root-reserve data were coded to generate growth-rate factors on a physiological time scale of growing degree days (for the first growth) or days since the last harvest (for regrowth). Autumn levels of root reserves controlled winter survival. The digestibility and crude-protein concentrations of the herbage were estimated, using the observed relationships between herbage age in growing degree days (first harvest) or days since the last cutting (later harvests) as measured by Liu and Fick (1975).

The model was evaluated by inputting specific harvest dates and simulating sequential years of management with two to ten harvests/year. Further testing was performed by simulating the alfalfa-quality and cutting-management trials of Sumberg (1977), conducted at

Ithaca, New York, in 1974 (5.5 cm below normal precipitation in July) and 1975 (normal or above-normal precipitation) with three harvest schedules at three locations with 14.5, 11.4, and 8.0 cm of available water at field capacity, respectively. For these trials yield potential was assumed to be 16 metric tons of dry matter/ha with a 7:5:3 distribution ratio in three harvests. The fit of the model was tested using the linear regression of observed or predicted data (Mood and Graybill, 1963).

RESULTS AND DISCUSSION

A simple model made it easy to realize the physiological consequences of the model theory. Determination of growth potential parameters for Ithaca, New York, indicated that 450, 600, and 570 growing degree days should be used for the first, second, and third harvests, respectively at that location. These data indicate that phenological development in at least some alfalfa cultivars must be controlled by photoperiod as well as thermoperiod. The importance of high temperature stress on alfalfa production could also be evaluated. The potential yield was reduced by 2.4%, 6.2%, and 5.1% in sequential harvests when the temperature function was added to LEVEL 0, confirming the findings of Field and Hunt (1974).

In the test format, LEVEL 0 generated expected patterns of alfalfa production and stand survival (Table 1), including better stands when harvests in the autumn (critical management period) are avoided. Comparison of predicted and observed production (Fig. 1) showed that LEVEL 0 had an r^2 value of about 0.62 and a standard error of the estimate (S.E.$_y$) of 870 kg/ha. Although further testing with data from other locations is planned, these results indicate that the genetic potential, soil moisture, temperature, and cutting-management functions incorporated into the simple model account for a large fraction of the variability in alfalfa yield data. There is con-

Table 1. Comparison of six simulated alfalfa cutting schedules applied over a 2-year period. Yield potential was set at 13.8 metric tons/ha, stand was initialized at 0.95, and average weather data for Ithaca, NY were used.

Harvest dates	Seasonal dry-matter yield		Initial third year stand[†]
	First year	Second year	
	—metric tons/ha—		
21/6, 30/8	12.1	12.1	0.95
10/6, 22/7, 10/9	12.3	12.3	0.95
31/5, 5/7, 16/8, 4/10	11.3	8.7	0.50
30/5, 29/6, 29/7, 28/8, 27/9	9.7	7.4	0.49
25/5, 19/6, 14/7, 29/7, 28/8, 17/10	7.4	6.8	0.65
25/5, 9/6, 24/6, 9/7, 24/7, 8/8, 23/8, 7/9, 22/9, 7/10	5.0	2.6	dead at 9/7 of year 2

†Stand is a model variable representing the combined attributes of alfalfa plant density, crown size, and vigor relative to that needed for maximum possible yield.

Fig. 1. The relationship between predicted and observed alfalfa dry-matter yields over three harvest schedules, three locations, and 2 years in the vicinity of Ithaca, New York.

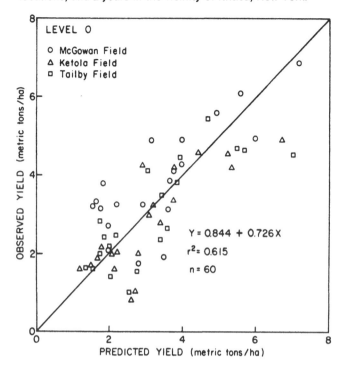

The corresponding equation for crude protein was

$$Y = 2.887 + 0.947X, n = 60, r^2 = 0.126** \qquad (2)$$

The $S.E._y$ values were 4.5% and 3.2%, respectively. This relatively poor performance demonstrates the inadequacy of an empirical time-based model for forage quality, even when that model includes physiological time. It will also make it possible to demonstrate progress quantitatively as better models of forage quality are developed for this important element of forage systems analysis.

LITERATURE CITED

Bula, R.J., and T.R. Hintz. 1978. Estimating alfalfa yield from stem height. Proc. Ann. Northeast. Invit. Alf. Insect Conf. 15:27–31.

Dougherty, C.T. 1977. Water in the crop model SIMED2, 103–110. DSIR Inf. Ser. 126, Wellington, N.Z.

Fick, G.W. 1975. ALSIM 1 (LEVEL 1) user's manual. Agron. Mimeo 75-20. Dept. Agron., Cornell Univ., Ithaca, N.Y.

Field, T.R.O., and L.A. Hunt. 1974. The use of simulation techniques in the analysis of seasonal changes in the productivity of alfalfa (*Medicago sativa* L.) stands. Proc. XII Int. Grassl. Cong. 1:357–365.

Fitzpatrick, E.A., and H.A. Nix. 1969. A model for simulating soil water regime in alternating fallow-crop systems. Agric. Meteorol. 6:303–319.

Gutierrez, A.P., J.B. Christensen, C.M. Merritt, W.B. Loew, C.G. Summers, and W.R. Cothran. 1976. Alfalfa and the Egyptian alfalfa weevil (Coleoptera: Curculionidae). Can. Entomol. 108:635–648.

Holt, D.A., R.J. Bula, G.W. Miles, M.M. Schreiber, and R.M. Peart. 1975. Environmental physiology, modeling, and simulation of alfalfa growth. I. Conceptual development of SIMED. Purdue Agric. Expt. Sta. Res. Bul. 907.

Liu, B.W.Y., and G.W. Fick. 1975. Yield and quality losses due to alfalfa weevil. Agron. J. 67:828–832.

Mood, A.M., and F.A. Graybill. 1963. Introduction to the theory of statistics, 2nd ed. McGraw-Hill, New York.

Selirio, I.S., and D.M. Brown. 1979. Soil moisture-based simulation of forage yield. Agric. Meteorol. 20:99–114.

Sumberg, J.E. 1977. Effects of managements and varieties on yield and quality of alfalfa (*Medicago sativa* L.) forage. M.S. Thesis, Cornell Univ.

siderable room statistically for improving LEVEL 0; but to constitute an improvement, an alternative should be simpler, have better general predictive power, or have special design features that justify application to a particular problem.

The model also contained equations to predict herbage digestibility and crude-protein concentration. When tested against corresponding data measured by Sumberg (1977), the regression equation of observed digestibility on predicted digestibility (%) was

$$Y = 20.098 + 0.744X, n = 60, r^2 = 0.253*** \qquad (1)$$

Spatial Simulation Modeling as a Tool in Multiple-Use Management of Renewable-Resource Systems

R.C. JUMP and G.M. VAN DYNE

Colorado State University, Fort Collins, Colo., U.S.A.

Summary

A conceptual model was developed using a modified-state variable approach, which involved the recognition and suitable description of specific objectives, processes, and situations considered relevant to the simulation of ecosystem dynamics. Six primary ecosystem subsystems were selected: terrestrial abiotic, floral and faunal, aquatic abiotic and biotic, and topographic land use. These subsystems are embodied by a geographic structure defined as an ecological response unit (ERU). Subsystem characteristics of an ERU are considered homogenous, but between ERUs the characteristics may differ greatly. A set of ERUs may be arranged into any mosaic pattern, a quality that affords a unique response behavior to both the individual ERUs and the mosaic pattern. Discrete management events, such as controlled burning or livestock stocking, and continuous events, such as oil-shale mining, may be scheduled for a complex of ERUs. As the model operates "mechanistically," the results are tractable, and they provide a framework for interpretation and discussion among resource managers.

A computer-based simulation model of the conceptual model, named An Ecosystem Simulation of Perturbations (AESOP), has been developed to provide resource managers with a methodology for analyzing the future consequences of multiple-use management scenarios. The computer model provides pseudoquantitative ecosystem responses as a consequence of various management strategies. AESOP allows a user to specify many types of scenarios involving both ecosystem characteristics and management practices, which may vary in location, timing, and magnitude.

AESOP was programed in the general-purpose computer language PASCAL because of the superior data-typing and program-structuring capabilities of the language, which allow model flexibility not possible with most models reported in the ecological literature. AESOP closely imitates the conceptual model through data-structuring techniques and derived mathematical functions representing ecosystem relationships. The block structure of the computer program provides ease in understanding and evaluation by the user and change or interchange of simulation procedures.

AESOP generates time-series data of the components and processes of the ERUs and indices of subsystem conditions. A separate analysis program summarizes the voluminous output into statistical tables and graphs.

The model is currently being tested by evaluating ecosystem response to multiple-use resource management of the sagebrush-grass areas in northwestern Colorado. This region is unique in its richness of energy-producing resources, its abundance of wild herbivores, and its livestock and agricultural practices.

Through analysis of the results from various management-policy scenario simulations, "hindsight" is available to the resource manager at the time of decision-making.

KEY WORDS: simulation, spatial, model, AESOP, data structure.

INTRODUCTION

Of the many possible spatial aspects in an ecosystem, several types are considered by the conceptual model, including elevation, slope, depth, height, distance, juxtaposition, and topographic boundaries. This paper focuses on the concepts and computer-programing strategy for addressing these types of spatiality in a simulation model. The computer program, An Ecosystem Simulation of Perturbations, AESOP (Jump et al. 1980), depends on data types and data-structuring techniques to imitate a conceptual four-dimensional continuum in a digital computer.

DATA STRUCTURE, RECORDS, AND POINTERS

A data structure is a hierarchical methodology used by a computer program to store and retrieve data values. It is analogous to a reprint filing system using several file cabinets with several drawers with several dividers with several folders containing both reprints and cross references to related reprints, folders, dividers, etc. This methodology allows a computer program to search and locate a data value without a priori "knowledge" of the location where the data value is stored.

Countless different data types can be constructed. Two of the basic types utilized in AESOP are (1) the record data type and (2) the pointer data type.

A record can be thought of as the folder from the reprint-filing-system analogy. The folder is the smallest unit containing the actual reprints; likewise, a record type may contain an assortment of different types of individual data elements. In addition, records can be aggregated such that a record may contain an assortment of other records.

The pointer data type provides much of the power of data structures. Pointers are used as cross references for indirect addressing. In direct addressing, a data value is stored at the location represented by a predefined variable. Indirect addressing is similar, except that the value of the predefined variable is the location where the data value is stored. By including pointers within a record, very elaborate structures can be created, such as hierarchical trees and networks.

A major advantage of these structures is that relationships among data elements can be implicitly defined by the structure methodology, rather than requiring additional information storage. Some relationships are more important than the actual data value—food webs, for example. Also, since related information is stored in discrete blocks, thus requiring less retrieval time, program execution times tend to increase logarithmically with linearly increasing simulation complexity.

AESOP DATA-STRUCTURE DEFINITION

The AESOP model includes several hundred state variables. Many processes may be calculated in updating the value of a single state variable, and several parameters

R.C. *Jump is a Research Associate and* G.M. *Van Dyne (deceased as of August 1981) was a Professor of Biology at Colorado State University.*

may be used in each calculation. Thus, the efficient storage and retrieval of information is important to the simulation.

Overall Organization

The AESOP data structure is elaborate, containing a variety of data types organized into a series of hierarchical trees and networks. The structure contains about 30 types of scalars along with 10 special subsets, 50 types of pointers, 86 types of records, and 19 input and output files. The highest-level entry point, named "GLOBAL," is a pointer to the entire data structure; thus, all the information in the simulation could be provided to an individual computer subroutine by passing in the variable named GLOBAL. However, there are many paths to the same information, and lower-level entry points provide much faster information retrieval.

GLOBAL points to a record of other pointers. This GLOBAL record is divided into fields representing the basic components of the conceptual model. These components are (1) the driving variables, such as climate, which are not affected by system responses; (2) the state variables, such as plant biomass, which indicate the condition of the system; (3) process variables, such as runoff volume, which represent the response of the system; (4) parameters, such as photosynthetic rate, which are used in calculating process variables; (5) intermediate variables, such as forage intake, which function as temporary state variables in calculating process variables; and (6) event descriptions, such as livestock stocking, which function as a special class of driving variable.

Each of these six fields points to other records that further divide the information into one of the six primary ecosystem subsystems selected for simulation, such as the soil moisture subsystem. The information is further aggregated (or disaggregated) according to many types of factors relevant to the conceptual model.

Spatial Organization of State Variables

The focus of the remainder of this paper is the spatial definition and organization of the state variables embodied by a record representing an ecological response unit, an ERU record. The ERU record is the basic geographic entity and simulation unit within the data structure.

The initial spatial organization is the definition of a region, which establishes a mapping correspondence between records in the data structure and real-world locations. The region is defined as a grid of landblock records, which might represent drainages, allotments, or districts. In our case, landblocks represented townships, such that an area of 96.6 × 96.6 km was within the grid. These landblocks may be either activated or inactivated by the user, so that a simulation may focus on a few areas of special interest or can expand its scope to the big picture.

Further, within each landblock record, lists of ter-restrial ecosystem response units (ERUs) are organized into elevation zones, and lists of aquatic ERUs are additionally arranged into either lotic or lentic types. Four elevation zones were used for describing the region; these were highland, upland, lowland, and bottomland.

Topographic characteristics and local climate conditions may be specified for each ERU's topography record, although the climate conditions may be shared by several ERUs. The topographic characteristics include (1) total acreage and, in this example, acreages of roadways, mining activities, and rehabilitated sites; (2) elevation; (3) average slope and aspect; (4) soil-horizon characteristics, such as thickness, soil class, and organic-matter contents; and (5) topographic relief, such as a cliff face.

Each ERU topography record also contains a list of its edges, so that a mosaic pattern of ERUs may be created within a given landblock. Each edge record contains a pointer to the adjacent ERU and a list of factors associated with that edge. These factors may include topographic relief or man-related obstacles, such as a fence line. If a certain edge, such as a canyon side, were considered very important in the conceptual model, that edge could be defined as an ERU and be described by its own appropriate characteristics.

Spatial Representation of Abiotic Components

Besides the geographic features, an ERU is characterized by the components of its subsystems. The dynamics of each of these subsystems are influenced by the spatial aspects already described and by the spatial aspects of the components. Thus, each ERU record contains its own individual set of subsystem and component records.

The dynamics of the terrestrial abiotic subsystems, both the soil moisture subsystem and the soil temperature subsystem, are closely related to the soil-horizon characteristics and intermediate variables, such as the calculated hydraulic conductivity for a soil layer. Each soil-layer record of an ERU has its own spatial aspects, namely, depth and soil class, that are used as an index to general soil properties relating to aggregate size, pore size, and swelling capacity. For example, infiltration rates are affected by soil class, thickness, and current temperature and moisture conditions of the surface soil layer. Infiltration rates are also indirectly affected by aboveground vegetative cover that intercepts a portion of the precipitation. The amount of interception is related to the intensity of the precipitation and to growth form of the vegetative cover. The growth form, such as grass, forb, or shrub, is used as an index to general shape and leaf area of the vegetation.

Runoff water, water that did not infiltrate, may move partly to adjacent, downhill ERUs as overland flow and partly into and through the aquatic network described by pointers within lotic ERUs. Erosion sediments may also be picked up and carried to lower, slower-moving lotic ERUs, where the sediments would be deposited.

Likewise, in addition to soil-horizon characteristics, the

soil temperature of the surface layer is directly affected by above-ground factors, such as canopy cover and snow depth. Most of the soil temperature subsystem dynamics are related to conditions of adjacent soil layers.

Spatial Representation of Autotrophic Components

Spatiality within the floral subsystems has both underground and above-ground dimensions. Each plant species has its own set of records for the various components. The under-ground aspects reflect temperature and moisture conditions of the soil layers. Thus, a plant's roots may extend down through the soil layers, so that there is a specific root biomass within each soil layer. Again, plant form is used as an index to the general type and to the maximum depth that the roots would reach under favorable conditions.

The above-ground aspects are also related to plant form. Plant height and cover are major factors in simulating the interaction (competition) between various plant species. Plant height also becomes important in determining an animal's habitat preference.

Another dimension of plant spatiality involves the allocation of photosynthate to structural and storage organs. For example, in addition to root biomass within each soil layer, biomass of leaves, green shoots, woody shoots, and storage organs are included within a shrub state-variable record. The translocation of biomass between these components is influenced by soil moisture and temperature conditions as well as by climatic conditions.

Finally, all plant species may germinate within all defined ERUs. Seed dispersal is assumed to be unrestricted. The resulting plant species distribution forms a subset mosaic of ERUs in which suitable conditions exist. This subset mosaic will change through simulation time as the conditions within each ERU change, perhaps as successional changes.

Spatial Representation of Heterotrophic Components

Dynamics of the faunal subsystems are closely related to several of the spatial aspects of the floral subsystem. Again, each animal species within an ERU has its own set of records for the various components.

One of the most basic faunal processes involves consumption of plant parts. Thus, the amounts of biomass stored within the leaves and green shoots is important in determining available herbage. The presence of desired plant species and surface water influences the suitability of an ERU for a particular animal species, and this suitability ultimately determines the population distribution of animal species within the ERU mosaic. Portions of animal populations within each ERU move to adjacent ERUs in accordance with relative suitabilities. Several spatial aspects of other subsystems influence the evaluation of suitability. For example, vegetation height and cover are important to animals attempting to escape predation or severe climatic conditions. After suitability is evaluated, the resulting animal distribution is affected by the edge barriers described by the ERU topography record. The effect of these barriers upon dispersal is evaluated differently for each species.

LITERATURE CITED

Jump, R.C., C.S. Loehle, G.M. Van Dyne, L.D. Watson, J. Davey, N.R. French, and J.W. Skiles. 1980. A.E.S.O.P., an ecosystem simulation of perturbations applied to shale oil development. Prog. Rpt. to U.S. Dept. of Energy. November 1980. Syst. Ecol. Group, Range Sci. Dept., Colo. State Univ., Fort Collins.

Effect of Experimental Methods on Results of Voluntary-Intake Experiments with Grass Cultivars

U. SIMON and P. DANIEL

Justus-Liebig-Universität, Giessen, Federal Republic of Germany

Summary

The amount of forage eaten by farm animals when forage is offered ad libitum is a major factor in the nutritive value of forage grasses. It is influenced by a number of internal and external factors. Experimental results of studies on the subject reported in the literature are often contradictory, partially because, perhaps, of the methods of evaluation employed. We studied whether and to what extent different experimental procedures influenced the results of intake experiments in grass cultivars.

Four cultivars of *Lolium perenne* differing in ploidy level and precocity were fed to sheep ad libitum, and voluntary intake was assessed using four different procedures: (1) fresh vs. dried forage; (2) grazing vs. barn feeding; (3) free choice vs. no choice between cultivars; and (4) feeding at different stages of growth vs. feeding at same stage of growth.

Highly significant differences between cultivars were found when cultivars were offered fresh for free choice on pasture and in the barn on the same dates. Late cultivars were preferred to early ones, and tetraploids were preferred to diploids. Essentially the same results were obtained when hay of the flowering stage was offered free choice. When hay of the cultivars was fed as single feeds, differences between cultivars varied with stages of growth. They were larger at earlier than at later stages, and sometimes interactions occurred. Hay intake decreased significantly with advancing stages of growth. Differences in stage of growth in all cultivars at harvesting of the feeds had a much larger effect than differences in cultivars.

With respect to voluntary intake, genetic differences among grass cultivars suggest the possibility of genetic improvement. The procedures for assessment of intake should be chosen according to the intended use of the forage.

KEY WORDS: intake, grass cultivars, *Lolium perenne* L., evaluation methods, sheep.

INTRODUCTION

The amount of forage eaten when forage is offered ad libitum to farm animals is one of the major determinants of the nutritive value of forage grasses. Differences among species and even among varieties have been shown (Lampeter 1974, Minson et al. 1964, Simon 1972, Thomson 1971, Walters 1974). Experimental results may vary, depending on the methods of evaluation used.

METHODS

Plots of four cultivars of perennial ryegrass (*Lolium perenne* L.) were established in 1972. The cultivars differ in earliness and ploidy level—namely, early diploid Printo, early tetraploid Reveille, late diploid N.F.G., and late tetraploid Havier. Average dates of heading are 4, 7, 19,

and 22 days, respectively, later than Gremie (Bundessortenamt 1979). The following experimental procedures were applied for assessment of forage intake from primary growth fed ad libitum to sheep: (1) forage grazed in situ at same dates with early and late cultivars at different stages of development, free choice; (2) same as (1) except barn feeding of fresh grass instead of grazing; (3) forage harvested at same stage of growth (end of flowering) offered as hay, free choice; (4) same as (3) but without choice among cultivars; (5) same as (4) but at beginning of flowering stage; (6) same as (4) but at heading stage; (7) same as (4) but at beginning of stem elongation stage.

The feeding experiments were conducted from 1972 to 1975. Wethers of the Deutsches schwarzköpfiges Fleischschaf, 18–24 months old, were used. In experiment 1, intake by flocks of six to eight animals was determined as the average difference between the dry-matter yields of the plots immediately before and after 2-day grazing periods, with 8 randomized replications each year. In experiments 2 and 3, three to four sheep had free access to four mangers that were daily filled at random with the

U. Simon is a Professor of Agronomy and P. Daniel is a Research Scientist at Justus-Liebig-Universität.

experimental feeds for a period of 3 weeks. In experiments 4 to 7, each forage was fed to three single sheep in individual pens for experimental feeding periods of 10 days. The same sheep were used in trials 4 to 7. In trials 3 to 7, intake was determined as the average daily difference between feed on offer and the remainder after 8 hours.

RESULTS

Experiment 1: grazing method. The 3-year average intake is shown in Fig. 1. Evident differences exist when cultivars are grazed in situ and animals can choose among them. More forage is eaten from tetraploid than from diploid grass, and at the diploid level the late cultivar is preferred to the early one.

Experiment 2: barn-feeding method. There were highly significant differences among all cultivars (Table 1). Most obvious is the strong preference at both ploidy levels for the late cultivars, of which two or more times as much forage was consumed as compared with the early ones. Also significant is the larger intake from tetraploid cultivars at both levels of earliness. The varietal difference was larger in 1973 than in 1974.

Experiment 3: hay-feeding method. When the forage is offered in form of hay of the end of the flowering stage, the ranking order among cultivars remains generally the same as in experiments 1 and 2, i.e., early diploid 95, early tetraploid 207, late diploid 412, late tetraploid 302 g dry matter (DM)/animal/day (least significant difference [LSD] 1% = 34).

Experiment 4: single hay-feeding method. When the forages used in experiment 3 were offered as single feeds without choice, differences among cultivars became negligible regardless of their earliness and ploidy level (Table 2, first column).

Experiments 5, 6, and 7: single hay of different stages of growth-feeding methods. At the beginning of stem elongation, the superiority of the tetraploids in both maturity groups was highly significant (Table 2). The intake from cultivars differing in earliness depends on their ploidy.

Table 1. Voluntary intake by sheep of primary growth of perennial ryegrass cultivars when cut grass is offered in the barn for free choice.

| Cultivar | | Intake | |
		1973	1974
		g DM/animal/day	
Early, diploid		70	78
Early, tetraploid		128	370
Late, diploid		635	577
Late, tetraploid		701	649
Average		383	419
LSD between cultivars	5 %	50	47
	1 %	66	73

Among diploids, more hay is ingested from the late type. The opposite is true among tetraploids. At the beginning of heading, intake of the early tetraploid exceeds that of the other cultivars, which do not differ significantly. At the beginning of flowering, significantly more hay of the early cultivars is consumed at both ploidy levels, and diploids are preferred to tetraploids at the same levels of earliness.

A notable result is also the decreasing average hay intake with advancing maturity stages of the forages, ranging from 1,713 to 932 g DM/animal/day from beginning of stem elongation to the end of flowering, respectively.

DISCUSSION

When fresh forage of cultivars differing in earliness is offered simultaneously for free choice during a period of growth ranging approximately from stem elongation to flowering stage, sheep generally eat more of the late types regardless of whether the grass is grazed or fed in the barn. The most obvious reason is that at any given date the late types are "younger" than the earlier ones, and the intake from "young" forage is higher than that from forage of advanced maturity. The superiority of tetraploids over diploids, which has been shown before (Simon 1972, Thomson 1971), is confirmed fully in the early-maturing

Fig. 1. Voluntary intake by sheep grazing the primary growth of perennial ryegrass cultivars offered on pasture for free choice.

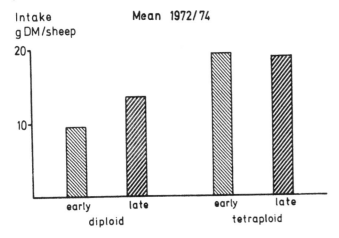

Table 2. Voluntary intake by sheep of primary growth of perennial ryegrass cultivars when hay is offered without choice, 1975.

| Cultivar | Stage of development | | | |
	End of flowering	Beginning of flowering	Beginning of heading	Beginning of stem elongation
	---------------- g DM/animal/day ----------------			
Early, diploid	927	1246	1550	1597
Early, tetraploid	950	1101	1714	1819
Late, diploid	952	1102	1506	1661
Late, tetraploid	900	1029	1532	1775
Average	932	1120	1576	1713
LSD 5%	ns	41	72	52
1%	ns	56	98	71

and partially in the late-maturing group. The same classification of cultivars is obtained when the animals can
choose among cultivars offered as hay of an advanced
stage of growth. If hays of the same growth stage are fed
as single feed without choice, the variability between
cultivars diminishes with advancing plant maturity, and
the ranking of the cultivars is partially reversed. It is concluded that the stage of maturity of the feeds and the lib-
erty of the animals to choose are factors that strongly affect the results of intake experiments with grass species
and cultivars.

LITERATURE CITED

Bundessortenamt. 1979. Beschreibende Sortenliste für Gräser
  und landwirtschaftliche Leguminosen, 1979. Alfred Strothe
  Verlag, Hannover.

Lampeter, W. 1974. Critical reflections on the determination of
  the field quality on the basis of ten-year investigations. Proc.
  XII Int. Grassl. Cong., 3(1):262–271.

Minson, D.J., C.E. Harris, W.F. Raymond, and R. Milford.
  1964. The digestibility and voluntary intake of S.22 and H.1
  ryegrass, S.170 tall fescue, S.48 timothy, S.215 meadow
  fescue and Germinal cocksfoot. J. Brit. Grassl. Soc.
  19:298–305.

Simon, U. 1972. Qualitätsprobleme bei Futterpflanzen.
  Ergebn. Landw. Forsch. Justus-Liebig-Univ., Giessen,
  12:117–129.

Thomson, D.I. 1971. The voluntary intake of diploid (S 22) and
  tetraploid (Tetila Tetrone) Italian ryegrass and white clover
  by sheep. J. Brit. Grassl. Soc. 26:149–155.

Walters, R.J.K. 1974. Variation between grass species and
  varieties in voluntary intake. Proc. 5th Gen. Mtg. Europ.
  Grassl. Fed., Uppsala, 1973, Växtodling, 28:184–192.

Yield and Forage Quality of *Panicum virgatum*

J.L. GRIFFIN and G.A. JUNG
USDA-SEA-AR Regional Pasture Research Laboratory
and Pennsylvania State University, University Park, Pa., U.S.A.

Summary

In the temperate areas of the U.S.A., high temperatures and low soil moisture in summer limit the production of cool-season
grasses and, consequently, the size of most beef cow-calf herds. Warm-season grasses, most productive in midsummer, may serve
as a supplement to cool-season grasses for grazing and hay. Switchgrass (*Panicum virgatum* L.), a warm-season species native to
the Great Plains of the U.S.A., was grown on soils low in available phosphorus at two locations in Pennsylvania to determine
changes in forage dry-matter yields, leaf (leaf blade):stem (sheath + stem) ratios, and quality of leaf and stem fractions
associated with increased maturation. Forage was harvested at 10-day intervals at both locations beginning at the three- or four-
leaf stage in late June and continuing until seed maturity in early August.

Forage yield at early head emergence averaged 4.72 metric tons/ha with no nitrogen fertilizer. Yield variability between locations and years was apparently due to previous stand management. Leaf : stem ratios declined as plants matured, and in early
August stems accounted for 63% of the total dry-matter yield. Leaf and stem forage-quality estimates at early head emergence,
respectively, were: in-vitro dry-matter disappearance (IVDMD), 58.8% and 47.2%; crude protein (CP), 8.5% and 3.8%;
neutral detergent fiber (NDF), 67.1% and 77.7%; and lignin, 4.7% and 8.3%. In concurrent feeding trials conducted using
sheep, IVDMD underestimated in-vivo digestibility by 17 percentage points.

Percentages of IVDMD and CP generally decreased with leaf and stem maturation. The decline was greatest at the early
harvests and less pronounced in leaves than in stems. In contrast, percentages of NDF and lignin generally increased with advancing growth stage, and changes were most rapid for stems at earlier harvests.

In conclusion, switchgrass can produce good yields of medium-quality forage at low soil fertility levels from late June to
August when production of temperate grasses is inadequate. Grazing switchgrass at more mature growth stages would provide
forage of adequate nutritional value for mature brood cows and stocker cattle and would also increase pasture carrying capacity.
In a hay program, harvesting at early head emergence would provide the best alternative when considering yield, quality, and
persistence factors.

KEY WORDS: switchgrass, *Panicum virgatum* L., warm-season grass, forage quality.

INTRODUCTION

In temperate areas of the U.S.A., cool-season grasses are used extensively for grazing and hay although summer production and quality are generally low. Livestock specialists agree that this summer shortcoming is a major factor limiting beef cow-calf herd size. Ideally, utilizing forages that differ in their period of maximum production should provide a more uniform supply of forage throughout the grazing period. Warm-season grasses, native to the Great Plains of the U.S.A., and generally most productive from June through August, may serve as a supplement to cool-season species in temperate areas.

In Missouri, the carrying capacity of switchgrass (*Panicum virgatum* L.) pastures from mid-June through September was reported by Rountree et al. (1974) to be 2 to 3 times higher than that of tall fescue (*Festuca arundinacea* Schreb.). Jung et al. (1978) reported that warm-season grasses in Pennsylvania produced 65% to 75% of their total yield during midsummer, in contrast to temperate species that have 60% to 66% of their production in May. Newell and Moline (1978), in evaluating forage quality of switchgrass, reported average crude protein (CP) and in-vitro dry-matter disappearance (IVDMD) in mid-July of 8.4% and 51.0%, respectively.

Few data are available on the potential productivity and forage quality of warm-season range grasses grown on soil sites with low pH and phosphorus (P) levels in the northeastern U.S.A. Our objective was to determine the influence of increased maturity of switchgrass on forage yield and on leaf and stem forage quality.

METHODS

Field studies were conducted at two locations in Pennsylvania in 1977 and 1978 to evaluate forage yield and quality of Blackwell switchgrass. A 9-year-old undisturbed stand was used at the first site, located at Allenwood. The Alvira silt loam soil (Aeric Fragiaquult; fine-loamy, mixed, mesic) had a pH of 5.8 and a Bray-1 extractable soil P level of 8 kg/ha. The second site was located at Delmont, where plantings were made in 1973 and grazed prior to initiation of the study. The Guernsey silt loam soil (Aquic Hapludalf; fine, mixed, mesic) had a pH of 5.4 and a Bray-1 extractable soil P level of 16 kg/ha.

A split-plot design in a factorial arrangement with grasses as whole plots and P levels as subplots was used. Phosphorus fertilization had little effect on any of the parameters measured; therefore, data are not presented. Nitrogen (N) fertilizer was not applied. Three replications were used at Allenwood and four at Delmont.

J.L. Griffin was formerly a Graduate Research Assistant at Pennsylvania State University and is now an Assistant Professor of Agronomy at Louisiana State University and G.A. Jung is a Research Agronomist of USDA-SEA-AR and Pennsylvania State University. This is contribution no. 8016 from the Regional Pasture Research Laboratory of USDA-SEA at University Park. It has been approved as Journal Series Paper No. 6165 of the Pennsylvania Agricultural Experiment Station.

Sampling was begun when plant height reached 46 cm (three- to four-leaf stage) and continued at 10-day intervals for five harvest dates at each location; regrowth was not sampled. Yield was determined by taking two subsamples from each plot, using a frame 61 × 61 cm^2 with a cutting height of 20 cm. Whole-plant subsamples were dried in a forced-draft oven at 60°C to measure moisture. Other plants were selected at random, botanically separated into leaf (leaf-blade only) and stem (sheath + stem) samples, frozen in liquid nitrogen, and freeze dried.

Total N content was determined by the standard Kjeldahl procedure (AOAC 1975). In-vitro dry-matter disappearance was determined by Barnes' modified Tilley-Terry method (Barnes 1967). Neutral detergent fiber (NDF) and lignin were determined according to the methods of Goering and Van Soest (1970).

RESULTS AND DISCUSSION

Forage Dry-Matter Yields

Switchgrass forage yields were generally higher at Allenwood than at Delmont in 1977 (P < 0.01 but were similar for those locations in 1978 (Table 1). In 1977 at Allenwood, the average yield was 6.3 metric tons/ha at early head emergence. In 1978, forage yields at Allenwood averaged 35% lower than those observed the previous year. In contrast, switchgrass forage yields at Delmont were similar both years of the study, averaging 4.3 metric tons/ha/yr at early head emergence. Previous stand management at Allenwood could account for the yield decrease that was observed in 1978. As stands had not been disturbed for a 9-year period prior to initial harvesting, the yield decrease may have been related to decreased availability of nutrients, since N and P removal were approximately 34% and 32% less, respectively, in 1978. In addition, harvesting stress may have adversely affected carbohydrate reserves, thereby decreasing both vigor and number of tillers the second year.

Nutritive Value

In-vitro digestibility of leaf tissue decreased an average of 0.19 points/day at Allenwood (P < 0.01); at early head emergence leaf IVDMD averaged 61.7% (Table 1). At Delmont, switchgrass leaf IVDMD did not change with maturation in 1977, but in 1978 it declined similarly to that observed at Allenwood. Stem IVDMD at Allenwood declined approximately 0.41 points/day in both years. At early head emergence, stems were approximately 50% digestible. At Delmont, similar trends were observed in 1978, but in 1977 stem IVDMD changed little until August. Digestibility of both stems and leaves was higher at Delmont in 1978 and was not associated with a decrease in NDF or lignin as would be expected. While the fiber estimates did not change between years, it is possible that digestibility of the fibrous fraction may have been greater in 1978, resulting in higher digestibility values.

Stem and leaf IVDMD averaged 51.3% and 61.1%,

Table 1. Forage dry-matter yields and leaf and stem in-vitro dry-matter disappearance (IVDMD), crude protein (CP), neutral detergent fiber (NDF), and lignin concentrations of Blackwell switchgrass harvested at two locations in Pennsylvania, 1977–1978.

Location, harvest date	DM yield mt/ha	IVDMD, % Leaf†	IVDMD, % Stem†	CP, % Leaf	CP, % Stem	NDF, % Leaf	NDF, % Stem	Lignin, % Leaf	Lignin, % Stem
1977									
Allenwood									
6/20	1.73	70.8	64.2	10.5	7.1	64.9	68.5	3.5	4.9
6/30	2.90	67.2	62.5	9.1	5.7	67.2	70.5	3.7	5.8
7/11	4.67	64.9	56.3	8.9	4.1	67.8	75.3	4.1	6.9
7/21#	6.31	61.8	49.6	8.1	3.4	67.4	77.1	5.2	8.7
8/1	6.89	62.7	44.1	8.3	3.1	66.4	78.9	4.3	8.5
Delmont									
6/28	2.26	54.4	49.5	9.0	5.4	66.6	71.6	4.2	6.1
7/8	3.11	51.1	44.5	8.3	4.8	67.1	73.6	3.9	6.2
7/18	3.68	49.5	44.0	8.5	4.2	68.0	75.9	4.4	7.4
7/28#	4.03	53.9	46.3	8.1	3.8	67.8	76.6	4.7	8.1
8/8	4.57	51.1	39.5	7.4	3.7	66.7	76.0	4.5	8.7
1978									
Allenwood									
6/26	1.06	70.9	60.5	10.8	7.4	64.5	69.6	3.6	5.4
7/6	1.51	66.7	60.2	9.5	6.3	65.2	70.0	3.5	5.3
7/17	3.08	64.8	55.7	8.4	4.4	66.7	75.1	3.5	6.1
7/27#	4.08	61.6	49.5	8.3	3.8	66.7	77.9	3.9	7.8
8/7	4.81	61.3	46.7	8.5	4.0	66.1	78.5	4.2	8.2
Delmont									
6/30	1.85	68.1	59.7	10.4	6.3	65.0	69.1	4.3	5.6
7/10	2.60	65.0	56.1	10.5	5.7	65.3	72.8	4.2	5.9
7/20	3.22	63.4	53.9	9.8	4.6	64.3	74.8	4.5	7.3
7/31#	4.49	58.0	43.2	9.5	4.0	66.3	79.1	4.9	8.5
8/10	4.83	55.6	40.4	9.1	4.0	66.1	78.9	4.9	8.6

†Leaf = leaf blade, stem = sheath + stem.

#Early head emergence.

respectively. While these values are low, it should be noted that feeding trials conducted with sheep have shown that switchgrass IVDMD underestimates in-vivo digestibility by 17 percentage points (Griffin et al. 1980). Stem digestibility declined more rapidly than leaf digestibility, suggesting that stem digestibility is the major determinant of whole-plant forage quality. On the average, stems accounted for 41%, 55%, and 63% of the total dry-matter yield at early vegetative stage, early head emergence stage, and seed-maturity stage, respectively.

At early head emergence, stems and leaves averaged 3.8% and 8.5% CP, respectively (Table 1). Rate of decline for stem protein was more than twice that for leaves. These data indicate that CP percentages of total plant tissue are associated with both leaf and stem maturation. These findings differ from those of Perry and Baltensperger (1979), who reported that leaf CP decline was primarily responsible for lower whole-plant protein levels.

Leaf NDF percentages were similar ($\bar{x}$ = 66.3%) for both locations both years of the study, and no significant change with advance in maturity was noted (Table 1). Fiber accumulated in stem tissue most rapidly at earlier harvests with little difference between locations. Averaged over all harvests, NDF levels of stems and leaves were 74.5% and 66.3%, respectively.

Lignin concentrations in leaves were lower at Allenwood than at Delmont (Table 1). However, at early head emergence, the difference between locations was small, and lignin averaged 4.7%. Lignin accumulation in stem tissue with increasing maturity was similar at the two locations, increasing most rapidly in early season. Lignin and digestibility at Allenwood were inversely related (r = −0.94), which could explain why IVDMD estimates for both leaf and stem tissue were higher there. At early head emergence, lignin levels in stems and leaves were 8.3% and 4.7%, respectively.

LITERATURE CITED

Association of Official Analytical Chemists (AOAC). 1975. Official methods of analysis, 12th ed. AOAC, Washington, D.C.

Barnes, R.F 1967. Brief outline of two-stage in vitro rumen fermentation technique. Purdue Univ. and USDA-ARS Mimeo.

Goering, H.K., and P.J. Van Soest. 1970. Forage fiber analyses. U.S. Dept. Agric., Agric. Handb. 379.

Griffin, J.L., P.J. Wangsness, and G.A. Jung. 1980. Forage quality evaluation of two warm-season range grasses using laboratory and animal measurements. Agron. J. 72:951–956.

Jung, G.A., C.F. Gross, R.E. Kocher, L.A. Burdette, and

W.C. Sharp. 1978. Warm-season range grasses extend beef cattle forage. Sci. Agric. 25(2):6.

Newell, L.C., and W.J. Moline. 1978. Forage quality evaluation of twelve grasses in relation to season of grazing. Nebr. Agric. Expt. Sta. Bul. 283.

Perry, L.J., Jr., and D.D. Baltensperger. 1979. Leaf and stem yields and forage quality of three N-fertilized warm-season grasses. Agron. J. 71:355–358.

Rountree, B.H., A.G. Matches, and F.A. Marz. 1974. Season too long for your grass pasture? Crops and Soils 26(7):7–10.

Use of Principal-Component Analysis in Summarizing Nutritive and Analytical Characteristics of Forages

O.T. STALLCUP, C.J. BROWN, Z.B. JOHNSON, D.L. KREIDER, and J.O. YORK

University of Arkansas, Fayetteville, Ark., U.S.A.

Summary

Principal component analysis and other multivariate analyses have not been used extensively to analyze forage data. Our objective was the synthesis and summary of 98 digestion trials on forages using a principal-component analysis.

The ten variables studied in all trials were dry matter (DM), crude protein (CP), digestible protein (DP), crude fiber (CF), ether extract (EE), nitrogen-free extract (NFE), ash (A), gross energy (GE), total digestible nutrients (TDN), and digestible energy (DE). Acid detergent fiber (ADF), acid detergent lignin (ADL), and cellulose (C) were summarized, in addition, for some forages.

When data on thirteen variables in 24 hays were summarized, the first two principal components accounted for 62.9% of the within-group variance. DE, TDN, CP, DP, EE, and A were positively correlated and ADF, CF, ADL, and C were negatively correlated with the first principal component. CP, DP, and ADL were positively correlated and NFE was negatively correlated with the second principal component. When data were plotted according to principal components 1 and 2, separation of grass hays from legume hays and hays made from cool-season and warm-season plants was observed. Two principal components accounted for 58%–77% of the within-group variation in five other groups of hays and silages.

Principal-component analysis permitted optimum summarization and clustering of desired information. It also permitted graphical presentation of variates according to the 2 principal components for easy diagnosis and evaluation. One may also identify latent clustering factors in the population.

KEY WORDS: principal component analysis, silage, hay, data bank, cattle.

INTRODUCTION

Efficient use of a forage analytical data bank depends in part on analysis and interpretation of stored information. As the number of characteristics studied increases, accurate knowledge and precise classification of all the relationships become unattainable. The correlation coefficient expresses the overall relationship, which may be the result of partial and even opposed relationships. The presence of unknown or unidentified distinct subpopulations can lead to erroneous interpretation of the correlation coefficient. The use of principal-component analysis may reduce the above-mentioned problems.

METHODS

The data used in this analysis consisted of chemical analyses of forages and digestion coefficients of these forages obtained in 98 digestion trials (Stallcup et al. 1964, 1975, 1976). Crude-protein, ether-extract, crude-fiber, and ash content of forage and feces were determined by AOAC methods (1960). Nitrogen-free extract was determined by subtraction. Moisture in silages was determined by a toluene distillation procedure (AOAC 1960), and energy of forage and feces by means of a Parr adiabatic bomb calorimeter. Acid detergent fiber and acid detergent lignin were determined by the method of Van Soest (1963). Cellulose was determined by the method of Crampton and Maynard (1936). The ten variables studied in all trials were: dry matter (DM), crude protein (CP), digestible crude protein (DP), crude fiber (CF), ether extract (EE), nitrogen-free extract (NFE), ash (A), gross energy (GE), total digestible nutrients (TDN), and

O.T. Stallcup is a Professor of Animal Sciences, C.J. Brown is a Professor of Animal Sciences, Z.B. Johnson is a Research Assistant in Animal Sciences, D.L. Kreider is a Research Assistant in Animal Sciences, and J.O. York is a Professor of Agronomy at the University of Arkansas.

Mention of a proprietary product does not imply endorsement of it.

digestible energy (DE). Acid detergent fiber (ADF), acid detergent lignin (ADL), and cellulose (C) were summarized, in addition, for some forages.

PRINCIPAL COMPONENTS

Principal-component analysis summarizes, with minimal loss, information included in a population of n observations on which p principals are measured. The technique basically attempts to discern those nonmeasurable factors that have generated the dependence structure among *p* measured responses. This technique consists of an orthogonal transformation of the coordinate axes of a multivariate system to new orientations through the natural shape of the scatter swarm of observation points. The response variates (X_i) are represented by a smaller number of "latent" factor variates (Y_j), defined as being mutually uncorrelated, and a linear combination of the response variates (X_i) (Morrison 1976). Analysis of the dependence structure centers around the estimation of the coefficients (a_{ij}) of the *j*th principal component (Y_j);

$$Y_j = \Sigma \ a_{ij} Z_i \qquad (1)$$

where $Z_i = X_i - \overline{X}_i / S_i$ and a_{ij} = coefficient for the *i*th variate in the *j*th component.

The first set of component coefficients should indicate the individual rotations required to provide a major axis for the system of *p* variates. The coefficients of additional components represent orthogonal sets that provide alternative axes in descending order of importance. This method has been used in analysis and interpretation of data in animal nutrition (Sauvant 1976) and in size and shape of animals (Wright 1932, Jolicoeur and Mosimann 1960, Carpenter et al. 1971, Brown et al. 1973).

RESULTS

Two principal components accounted for 59.8% of the variation among 34 sorghum silages (Table 1). CP, DP, TDN, and DE were positively correlated to principal component 1 (PC_1). CF was negatively correlated with PC_1. CP, DP, CF, and EE were positively correlated to principal component 2 (PC_2), while DM, NFE, and GE were negatively correlated.

Two principal components accounted for 77.3% of variance among 18 corn silages (*Zea mays*). CP, DP, EE, TDN, GE, and DE were positively correlated to PC_1. NFE and DM were positively correlated and CF, A, and EE were negatively correlated to PC_2 (Table 1).

The first two principal components accounted for 75.3% of within-group variation in 11 cereal silages. CP, CF, EE, A, GE, and DE were positively correlated to PC_1, unlike NFE. Gross energy (GE) value particularly depends on ash percentage, which appears to have some role of energy dilution, as observed by Bhatty et al. (1974) in various wheat and barley cultivars. This relationship between A and GE was not observed in corn or sorghum silages. There was a positive correlation between PC_2 and the variables CP, DP, TDN, and DE and a negative correlation to CF and DM.

Correlation coefficients between initial variables and the two principal components for pooled data on silages and hays are shown in Table 1. The data for 69 silages include data for 34 sorghum, 18 corn, and 11 cereal silages and 6 lots of sorghum-sudan (*Sorghum vulgare* Sudanese) silage. The first two principal components accounted for 58.3% of within-group variation. There was a clustering of variables having a positive relationship to PC_1. These included CP, DP, EE, TDN, and DE. NFE and DM were negatively correlated to PC_1. NFE, DM, and DE

Table 1. Coefficients of correlation between initial variables of (PC) silages and hays.

Variables	Sorghum silage PC_1	Sorghum silage PC_2	Corn silage PC_1	Corn silage PC_2	Cereal silage PC_1	Cereal silage PC_2	Silages without fiber analysis PC_1	Silages without fiber analysis PC_2	Silages with fiber analysis PC_1	Silages with fiber analysis PC_2	Hays with fiber analysis PC_1	Hays with fiber analysis PC_2	Hays without fiber analysis PC_1	Hays without fiber analysis PC_2
Dry matter	.56	−.64	−.47	.74	−.00	−.80	−.35	.48	−.27	−.76	−.46	.50	−.40	−.70
Crude protein	.69	.48	.92	−.12	.43	.72	.86	.07	.80	.24	.53	.77	.76	−.56
Dig. protein	.74	.52	.88	−.12	.22	.84	.88	.13	.82	.31	.66	.69	.85	−.42
Crude fiber	−.76	.49	−.18	−.81	.70	−.64	.13	−.88	−.36	.79	−.50	.54	−.07	−.52
Ether extract	.34	.67	.68	−.45	.87	.27	.50	.00	.42	.13	.68	.33	.75	−.15
NFE	.39	−.83	.06	.98	−.92	.24	−.62	.75	−.09	−.92	−.29	−.90	−.70	.63
Ash	.06	.26	−.67	−.54	.93	−.21	.31	−.48	−.22	.22	.56	.12	.42	.25
TDN	.85	.17	.88	.03	−.19	.74	.66	.62	.85	.17	.81	−.25	.70	.68
Gross E	−.07	−.33	.79	.32	.83	−.02	.28	−.03	.49	.03	.53	.24	.39	−.08
Digestible E	.75	−.10	.86	.10	.65	.44	.74	.43	.81	.17	.81	−.21	.74	.61
ADF	−	−	−	−	−	−	−	−	−.65	.69	−.83	.26	−	−
ADL	−	−	−	−	−	−	−	−	−.36	.23	−.43	.75	−	−
Cellulose	−	−	−	−	−	−	−	−	−.26	.87	−.52	.27	−	−
Variance accounted for by each PC, %	34.8	25.0	49.3	28.0	43.4	31.9	34.5	23.8	30.5	27.7	36.7	26.2	38.6	25.7
Variance accounted for by PC_1 + PC_2, %	59.8		77.3		75.3		58.3		58.2		62.9		64.3	
Number of feeds		34		18		11		69		31		24		29

were positively correlated and CF and A negatively correlated to PC$_2$.

Data including thirteen variables were collected on 31 silages. These included 14 sorghum, 11 corn, and 6 sorghum-sudan silages. Coefficients of correlation between these thirteen variables and PC$_1$ and PC$_2$ are shown in Table 1. The two principal components accounted for 58.2% of the within-group variance. There was a clustering of variables positively associated with PC$_1$. These included CP, DP, TDN, DE, and GE. ADF, CF, and ADL were negatively correlated to PC$_1$. C, CF, and ADF were positively correlated and DM and NFE were negatively correlated to PC$_2$.

Data for ten variables were available on 29 hays. These data are presented in Table 1. The two principal components accounted for 64.3% of within-group variance. CP, DP, EE, TDN, and DE were positively correlated and NFE and DM negatively correlated to PC$_1$. NFE and DE were positively correlated to PC$_2$. DM, CP, DP, and CF were negatively correlated to PC$_2$.

Data for thirteen variables associated with 24 hays are also in Table 1. Included were data on hays made from alfalfa (*Medicago sativa*), 2 lots; sorghum-sudan, 6 lots; soybean (*Glycine max.*), 6 lots; bermudagrass (*Cynodon dactylon*), 5 lots; and one lot each of bahiagrass (*Paspalum notatum* Flugge), smooth brome (*Bromus* spp.), fescue (*Festuca arundinacea*), johnsongrass (*Sorghum halapense*), and ryegrass (*Lolium multiflorum*). The first two principal components obtained from the thirteen variables accounted for 62.9% of the within-group variance. DE, TDN, EE, DP, CP, A, and GE were positively correlated and ADF, C, ADL, CF, and DM had a significant negative relationship to PC$_1$. CP, DP, ADL, CF, and DM were significantly correlated to PC$_2$, while NFE had a negative relationship to PC$_2$ (Fig. 1).

When hays were coded as legumes or nonlegumes, there was a definite clustering of legume hays, including those made from alfalfa and soybean, as contrasted with those made from grasses. This clustering was mainly due to higher NFE in the grasses (47.5%) than in the legumes (38.5%) and higher percentages of CP (15.9% vs. 11.0%), DP (10.5% vs. 6.3%), ADL (9.5% vs. 6.4%), and CF (34.4% vs. 31.5%) in the legumes than in the grasses, respectively.

When hays were coded as made from warm-season or cool-season crops, there was definite distinction between data from bermudagrasses, bahiagrass, and sorghum-sudan hybrids (classified as warm-season) and those from the remainder of the hays. This clustering (Fig. 2) was due mainly to higher NFE in the warm-season (48.9%) than in the cool-season crops (39.5%) and to a combination of comparatively higher values for CP, DP, ADL, and CF in hays made from cool-season crops (alfalfa, soybean, fescue, bromegrass, and ryegrass).

DISCUSSION

In interpretation of a data base, correlation coefficients express the whole relation between variables, which may, however, be the result of partial and/or even opposed relationships. The presence of unknown or unsummarized subpopulations can lead to erroneous interpretation of the correlation coefficient.

Principal-component analysis, by replacing the initial variates with a smaller number of latent variates (principal components), permits more concise summarization

Fig. 1. Correlations between analytical and nutritive characteristics of hays and principal components 1 and 2 (n = 24).

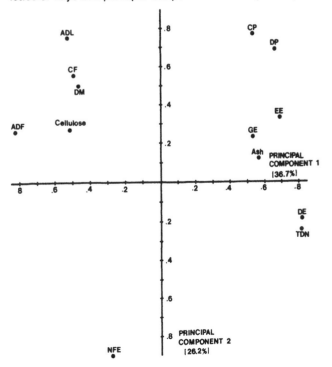

Fig. 2. Projection of the 24 hay samples on principal components 1 and 2 with code to designate whether made from cool-season or warm-season crop.

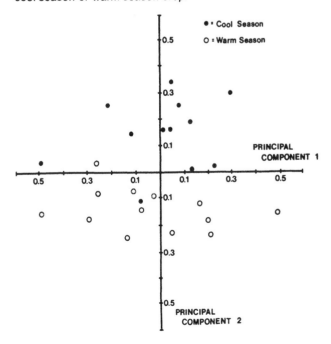

of data with minimal losses of information. Excellent examples of this quality are shown in Table 1, where subpopulations of silages made from different plants are differentiated by correlations of variables to principal components. Differences between silages and hays may also be observed.

LITERATURE CITED

Association of Official Agricultural Chemists (AOAC). 1965. Official methods of analysis, 10th ed. AOAC, Washington, D.C.

Bhatty, R.S., G.I. Christison, F.W. Sosulski, B.L. Harvey, G.R. Hughes, and J.D. Berdahl. 1974. Relationships of various physical and chemical characters to digestible energy in wheat and barley cultivars. Can. J. Anim. Sci. 54:419-427.

Brown, J.E., C.J. Brown, and W.T. Butts. 1973. Evaluating relationships among immature measures of size, shape and performance of beef bulls. I. Principal components as measures of size and shape in young Hereford and Angus bulls. J. Anim. Sci. 36:1010-1020.

Carpenter, J.A., Jr., H.A. Fitzhugh, Jr., T.C. Cartwright, A.A. Melton, and R.C. Thomas. 1971. Principal components for size of Hereford cows. J. Anim. Sci. 33:197 (Abs.).

Crampton, E.W., and L.A. Maynard. 1936. Relation of cel-
lulose and lignin content to the nutritive value of feeds. J. Nutr. 15:383-395.

Jolicoeur, P., and J.E. Mosimann. 1960. Size and shape variation in the painted turtle: a principal component analysis. Growth 24:339-354.

Morrison, D.F. 1976. Multivariate statistical methods. McGraw-Hill Book Co., New York.

Sauvant, D. 1976. Statistical methods for studying the variations of the feed analytical profile of data bank. First Int. Symp., Feed Comp., Anim. Nutr. Req. and Comput. of Diets, 39-44. Utah Agric. Expt. Sta., Utah State Univ., Logan.

Stallcup, O.T., G.V. Davis, and D.A. Ward. 1964. Factors influencing the nutritive value of forages utilized by cattle. Ark. Agric. Expt. Sta. Bul. 684.

Stallcup, O.T., P.V. Fonnesbeck, C.J. Brown, and Z. Johnson. 1976. Prediction equation for energy values in forages. First Int. Symp. of Feed Comp., Anim. Nutr. Req. and Comput. of Diets, 287-295. Utah Agric. Expt. Sta., Utah State Univ., Logan.

Stallcup, O.T., J.O. York, and C.J. Flynn. 1975. Nutritive value of corn and sorghum silage for milk production. Ark. Agric. Expt. Sta. Bul. 793.

Van Soest, P.J. 1963. The use of detergents in the analysis of fibrous feeds. II. A rapid method for the determination of fiber and lignin. Assoc. Off. Agri. Chem. J. 46:829-835.

Wright, S. 1932. General group and special size factors. Genetics 17:603-619.

Methodology for Determining Pasture and Animal Response Differences in Unbalanced Experiments with Intercorrelations Among Treatments

J.C. BURNS, R.W. HARVEY, and F.G. GIESBRECHT
USDA-SEA-AR and North Carolina State University, Raleigh, N.C., U.S.A.

Summary

Conventional statistical analyses require balanced data sets. Unbalanced data can be analyzed using regression procedures, but excessive adjustments may be needed for some treatment means. The objective of this paper is to evaluate an analysis of grazing-trial data using generalized least squares. The data was obtained from a 6-year study involving 12 treatments with not all treatments being made in all years. The analyses involved estimating variance components due to years, fields, and random error and then using these estimates to obtain adjusted treatment means. Variance-component estimates for the complete data set were similar to estimates from two balanced subsets of the data. For the most part, adjusted treatment means for cow-and-calf daily gain, gain/ha, and total digestible nutrients (TDN)/ha were similar to simple treatment means. Large differences that occurred had rational explanations. The statistical procedure used offers considerable promise for analyzing data from unbalanced experiments involving both fixed and random effects and large random disturbances. Further experience with this analysis using both real and simulated data is needed.

KEY WORDS: variance components, adjusted means, cow, calf, grazing.

INTRODUCTION

Studies involving small physical or biological experimental units such as pot, small-plot, or nonruminant studies can generally include a number of opportunity treatments, in addition to the major ones, all nicely balanced. However, in long-term forage-animal experiments, there is often a need to delete and replace unsatisfactory treatments. This paper presents results from a generalized least square analysis utilizing an estimated variance-covariance matrix to analyze responses from a 6-year, unbalanced cow-calf grazing trial involving 12 treatments. The variance-covariance matrix is estimated by minimum normal quadratic unbiased estimation (MINQUE) techniques. Variance components estimated for several variables from the 6-year unbalanced data set are compared with those from two internal 3-year balanced data sets, and adjusted and unadjusted means are compared. Statistical procedures and plant-animal responses will be the subject of a later article.

MATERIALS AND METHODS

The data are from a grazing experiment conducted at the Upper Mountain Research Station, Laurel Springs, North Carolina. The site is representative of the mid-Appalachian mountain region with predominant mountain-pasture species of Kentucky bluegrass (*Poa pratensis* L.) and white dutch clover (*Trifolium repens* L.) with some orchardgrass (*Dactylis glomerata* L.) present.

Agronomic and Animal Aspects

Treatments used with or imposed on improved (phosphate and lime applied) mountain pasture (IMP) included birdsfoot trefoil (*Lotus corniculatus* L.) and crownvetch (*Coronilla varia* L.) for midsummer grazing (Table 1). Variable stocking (put-and-take method), based on forage availability of the specific species (Burns et al. 1970), was used on pastures of 1.5 to 3.3 ha. Total digestible nutrients (TDN) were determined to measure pasture productivity (gain/ha and TDN/ha).

Hereford cow-and-calf test units were grouped according to cow age, sex of calf, and calf weight. Animal weights were obtained initially and at 14-day intervals. Early-weaned calves were provided a concentrate (ad libitum) while grazing IMP. Dry cows were stocked to either gain or maintain body weight. Cattle had free access to water, natural shade, and minerals.

Statistical Consideration

The experiment was set up as a randomized block with 3 replications to continue through 6 years. The replications represented three exposures (both to sun and topographic) of IMP. The appropriate model has the form:

$$Y_{hijk} = \mu + t_h + r_i + y_j + f_{ik} + e_{hijk} \qquad (1)$$

Both year (y_j) and field (f_{ik}) effects were assumed to be random, and treatment (t_h) and exposure (r_i) effects were fixed. All analyses discussed are based on variations of the basic model shown above.

Analyses of balanced data. Conventional analyses (Snedecor and Cochran 1967) using only the balanced subsets of the data allowed comparison of treatments 1, 3, 4, and 8 the first 3 years and treatments 1, 2, and 3 the last 3 years.

Analyses of unbalanced data. The total data set was unbalanced, ranging from treatments 5, 10, and 11, evaluated one year, to treatment 1, evaluated all 6 years. The analysis can be compactly outlined using the linear model

$$Y = X\beta + e \qquad (2)$$

where X is an (n × p) matrix of known constants with rank p, β is a vector of p parameters to be estimated, and e is a vector of n random variates with a mean of zero and a positive definite variance-covariance matrix V. Estimating the elements of V can be treated as a variance-component estimation problem since all nonzero elements are sums of variance components. Reformulating the basic equation (1) in matrix notation for convenience gives:

$$Y = X\beta + U_y y + U_f f + e \qquad (3)$$

where both treatment and exposure effect are incorporated in the product $X\beta$ with X having full column rank and U_y and U_f are matrices consisting of zeros and ones. The elements of y, f, and e are independent normal with zero means and variances σ_y^2, σ_f^2, and σ_e^2, respectively. The minimum norm quadratic unbiased estimation (MINQUE) techinque of Rao (1972) was used to estimate σ_y^2, σ_f^2, and σ_e^2. These were then used to obtain generalized least square estimates of β. A closely related method of analysis is given by Jennrich and Sampson (1976).

RESULTS AND DISCUSSION

Magnitude of Variance Components

The two balanced subsets and the unbalanced data set gave similar ratios of σ_y^2 and σ_f^2 to σ_e^2 for the four variables (Table 2). Generally σ_y^2 was larger than σ_f^2 for daily gains and smaller for pasture productivity (gain and TDN/ha) values. This relationship appears reasonably based on inherent field-to-field variation in forage productivity and proper use of variable stocking, i.e., controlled forage availability.

J.C. Burns is a Plant Physiologist of USDA-SEA-AR and a Professor of Crop Science at North Carolina State University; R.W. Harvey is a Professor of Animal Science and F.G. Giesbrecht is a Professor of Statistics at North Carolina State University. This is Journal Series Paper No. 6779 of the North Carolina Agricultural Experiment Station in cooperation with USDA-SEA-AR.

The authors acknowledge the vision and early encouragement of the late Dr. H.L. Lucas in applying generalized least square analyses to unbalanced data sets from grazing trials.

Table 1. Treatments imposed on improved mountain pasture (IMP) or seeded.[1]

	Cows with calves							Early weaned calves and dry cows IMP				
	IMP				Seeded legume			Nitrogen (kg/ha)				
	Nitrogen (kg/ha)				Rotation		Cont.	0	67	67	67	112
Item	0 (1)	67 (2)	112 (3)	224 (4)	BFT (5)	CV (6)	CV (7)	M (8)	M (9)	G (10)	S (11)	M (12)
Years treatments imposed:												
1	X	—	X	X	X	X	—	X	—	—	—	—
2	X	—	X	X	—	X	—	X	—	—	—	—
3	X	—	X	X	—	—	X	X	—	—	—	X
4	X	X	X	—	—	—	X	X	—	—	—	X
5	X	X	X	—	—	—	X	X	X	—	—	X
6	X	X	X	—	—	—	—	—	X	X	X	—

[1]BFT = birdsfoot trefoil; CV = crownvetch; M = maintenance; G = gain, and S = stockpiled.

Treatment Mean Adjustments in the Unbalanced Analyses

The estimates of treatment means computed via generalized least squares, using estimated variance components, can be interpreted as treatment means adjusted for replicate effects and intercorrelations introduced by year, field, and within-field variability. Results from such analyses are useful only if adjustments are biologically reasonable.

The magnitude of adjustment that occurred in the un-

Table 2. Comparison of the analyses of variance (ANOV) error mean square and year ($\hat{\sigma}_y^2$), and field ($\hat{\sigma}_f^2$) components for per-animal and per-ha responses relative to random error ($\hat{\sigma}_e^2$).

| Experimental error and variance components | Daily gain (kg) | | Production (kg/ha) | |
	Cow	Calf	Gain	TDN
A First 3-yr balanced data set				
Error (ANOV)	0.9	1.3	0.9	0.8
Components/$\hat{\sigma}_e^2$ ratio:				
$\hat{\sigma}_y^2 / \hat{\sigma}_e^2$	1.0	2.0	0.6	1.4
$\hat{\sigma}_f^2 / \hat{\sigma}_e^2$	0.1	0.9	0.7	2.6
$S_{\bar{x}}$	0.02	0.02	13.0	74.0
CV (%)	17.4	7.5	11.9	7.7
B Second 3-yr balanced data set				
Error (ANOV)	0.9	0.7	1.0	0.8
Components/$\hat{\sigma}_e^2$ ratio:				
$\hat{\sigma}_y^2 / \hat{\sigma}_e^2$	0.9	1.5	3.1	0.7
$\hat{\sigma}_f^2 / \hat{\sigma}_e^2$	0.2	1.0	3.1	1.6
$S_{\bar{x}}$	0.04	0.02	8.0	89.0
CV (%)	32.6	6.9	8.5	8.8
C Six-yr unbalanced data set†				
Error (ANOV)	1.2	1.1	2.0	3.1
Components/$\hat{\sigma}_e^2$ ratio:				
$\hat{\sigma}_y^2 / \hat{\sigma}_e^2$	0.4	0.7	0.7	0.9
$\hat{\sigma}_f^2 / \hat{\sigma}_e^2$	0.2	0.2	1.5	2.4
CV (%)	33.4	9.3	16.2	15.0

†Coefficient of variation (CV) obtained by ordinary least squares analyses ignoring years and fields.

balanced analyses (Table 3) compared with unadjusted mean in the balanced subsets for treatments 1, 2, 3, and 8 (data not shown) were small for all 4 variables. The largest adjustment in daily gain when considering all 12 treatments (Table 3) was 0.1 kg for calves in treatment 11 and 0.04 kg for cows in treatment 10. In both cases these treatments were evaluated only one year. Gain (kg/ha) from 11 of the 12 treatments (Table 3) was adjusted only 6.8% or less (average adjustment 8.9 kg/ha), but the mean of treatment 7 (Table 3) was adjusted 19.1% (reduced 59 kg/ha). This difference was attributed partly to the favorable rainfall in years 3 and 4, averaging 188 and 170 mm above normal, respectively, for June through August. The adjustment in TDN (kg/ha) means was 7.6% or less (mean adjustment 97 kg/ha) for 9 of the 12 treatments. The other 3 treatments (2, 6, and 7) were adjusted 9% to 14% (mean adjustment 336 kg/ha).

The mean $S_{\bar{x}}$s for the four variables calculated from the unbalanced analysis were 2- to 2.8-fold larger than those from the balanced (mean of both subsets) analyses (Tables 2 and 3). It is not clear if the difference is associated with underestimation of $S_{\bar{x}}$s in the balanced analyses (treatments exposed to similar year and animal variation) or with overestimation in the unbalanced analyses (treatments exposed to differing land, animal, and year variation). The magnitude of the adjustment of treatment means and $S_{\bar{x}}$s from the ordinary least squares analyses (years and fields fixed) compared with those from the MINQUE technique are of interest and will be the subject of a later manuscript.

LITERATURE CITED

Burns, J.C., R.W. Harvey, H.D. Gross, M.B. Wise, W.B. Gilbert, D.F. Tugman, and J.W. Bentley. 1970. Pastures and grazing systems for the mountains of North Carolina. N.C. Agric. Exp. Sta. Bul. 437.

Jennrich, R.I., and P.F. Sampson. 1976. Newton-Raphson and related algorithms for maximum likelihood variance component estimation. Technometrics 18:11–17.

Rao, C.R. 1972. Estimation of variance and covariance components in linear models. J. Am. Stat. Assoc. 67:112–115.

Snedecor, G.W., and W.G. Cochran. 1967. Statistical methods. Iowa State Univ. Press, Ames, Iowa.

Table 3.   Treatments applied and original and adjusted means of cow and calf daily gains and pasture productivity from the unbalanced six-year study.

| | Daily gains (kg) | | | | | | Production (kg/ha) | | | | | |
| | Cow | | | Calf | | | Gain | | | TDN | | |
Treatment†	Un-adj.	GLS‡	S$\overline{x}$	Un-adj.	GLS	S$\overline{x}$	Un-adj.	GLS	S$\overline{x}$	Un-adj.	GLS	S$\overline{x}$
1	0.42	0.43	0.05	0.83	0.84	0.03	251	240	25	2484	2423	211
2	0.34	0.34	0.06	0.80	0.78	0.04	340	317	29	3398	3090	252
3	0.35	0.36	0.05	0.77	0.77	0.03	335	330	25	3297	3270	226
4	0.36	0.37	0.04	0.74	0.76	0.04	323	334	29	3327	3581	249
5	0.59	0.63	0.08	0.83	0.87	0.05	234	230	33	2106	2133	272
6	0.48	0.50	0.06	0.80	0.83	0.04	225	237	26	2057	2303	217
7	0.59	0.60	0.06	0.90	0.86	0.04	308	249	28	3165	2712	240
8	0.27	0.25	0.05	0.94	0.93	0.04	363	364	25	3148	3176	209
9	0.30	0.28	0.07	0.80	0.82	0.05	334	332	27	3468	3335	226
10	0.42	0.46	0.09	0.67	0.75	0.06	311	303	33	3243	3177	261
11	0.41	0.44	0.09	0.81	0.91	0.06	208	209	32	2175	2081	260
12	0.18	0.15	0.06	0.90	0.88	0.04	413	438	25	3631	3817	215

†Treatments are described in Table 1.

‡GLS = General least-squares means adjusted for both random and fixed effects and associated S$\overline{x}$'s.

Rate and Extent of Fermentation and Potentially Digestible Fractions of Various Roughages

W.J. MAENG, H.S. YUN, and I.S. YUN

Chung-Ang and Kon-Kuk Universities, Seoul, Korea

Summary

To determine in-vitro digestion characteristics—such as rate of fermentation and amounts of potential digestibility of dry matter (PDDM), cell-wall constituents (PDCWC), acid detergent fiber (PDADF), and cellulose (PDC)—rice straw, barley straw, corn cobs, corn stover, orchardgrass, and alfalfa were digested in vitro for 6, 12, 24, 36, 48, 72, and 144 hours. PDDM, PDCWC, PDADF, and PDC were determined by differences from indigestible residues remaining after 144-hour in-vitro incubation. Fermentation rate (K) was determined as the slope of regression of the natural log of the residues on incubation time from 0 to 72 hours.

Potential digestibilities were 54%–80% for dry matter, 37%–73% for CWC, 42%–59% for ADF, and 52%–87% for cellulose. Fermentation rates of dry matter were 0.00661–0.01634/hour and of PDDM were 0.02928–0.04317/hour. Fermentation rates of CWC and PDCWC were 0.00698–0.01444/hour and 0.02505–0.01183/hour, respectively. Rates of ADF and PDADF were 0.00566–0.01183/hour and 0.01797–0.04547/hour. Fermentation rates of cellulose and PDC were 0.00955–0.02156/hour and 0.03529–0.05028/hour, respectively. Only fermentation rate of PDCWC was correlated significantly with percentage of CWC.

KEY WORDS: fermentation rate, potential digestibility.

INTRODUCTION

Considerable quantities of low-quality roughages are available throughout the world, but they have not been utilized well as a roughage source for ruminants because of low digestibility and low rate of digestion and because undigested residues remain in the rumen to limit further consumption of feed (Van Soest 1968). The extent of ruminal digestion is dependent on the combined influence of rate of fermentation and rate of ruminal turnover (Bull et al. 1979). Rate of fermentation is important in assessing forage quality, especially voluntary intake (Thorton and Minson 1972, Waldo 1969).

Little information is available in the literature on the dynamics of fermentation of low-quality roughages. Con-

W.J. Maeng is a Ruminant Nutritionist and Professor of Animal Science at Chung-Ang University, and H.S. Yun and I.S. Yun are Professors of Animal Science at Kon-Kuk University.

Mention of a proprietary product does not imply endorsement of it.

Table 1. Chemical composition (%) of dry matter of various roughages (mean of three replicates ± standard error).

Chemical composition	Rice straw	Barley straw	Corn cob	Corn stover	Orchardgrass	Alfalfa
Cell contents	23.4±0.62	18.1±0.14	17.4±0.40	31.6±0.10	54.6±0.65	45.1±0.60
Crude protein	5.5±0.24	4.6±0.17	4.6±0.10	6.4±0.20	20.5±0.23	14.7±0.08
Cell wall constituents	76.2±0.10	81.2±0.14	82.6±0.40	68.4±0.10	45.4±0.65	54.9±0.60
Acid detergent fiber	47.6±0.05	50.6±0.32	43.7±0.30	39.0±0.10	38.4±0.26	35.4±0.20
Hemicellulose	29.0±0.32	30.6±0.46	38.9±0.40	29.5±0.80	7.0±0.91	19.5±0.80
Cellulose	35.0±0.34	40.4±0.02	36.8±0.50	32.9±0.20	25.1±0.22	33.4±0.02
Lignin	8.8±0.78	10.5±1.21	6.5±0.62	10.9±1.12	4.8±0.93	10.1±0.78

sequently, the present study was conducted to measure the amount of potentially digestible and indigestible fractions and fermentation rate of various low-quality roughages, using an in-vitro method.

METHODS

Rice straw, barley straw, corn cobs, corn stover, orchardgrass, and alfalfa were ground through a 1-mm screen in a Wiley mill. Orchardgrass and alfalfa were harvested at late vegetative and early bloom stage, respectively.

In-vitro digestion was carried out as described by Tilley and Terry (1963) as modified by Maeng (1976). Rumen inoculum was prepared as described by Maeng et al. (1979). Four replicated samples were incubated for 6, 12, 24, 36, 48, 72, and 144 hours in a water bath maintained at 39°C with occasional shaking.

Potential digestibility of dry matter (PDDM), cell-wall constituents (PDCWC), acid detergent fiber (PDADF), and cellulose (PDC) were calculated by differences from indigestible residue remaining after 144-hour incubation. Fermentation rates (K) were determined as the slope of regression of the natural log of residue upon incubation times from 0 to 72 hours (Maeng and Baldwin 1976, Belyea et al. 1979). Crude protein was determined by the Kjeldahl method (AOAC 1975), and CWC, ADF, and lignin were determined according to Goering and Van Soest (1970). Cell contents were the difference between dry matter and CWC, and hemicellulose was the difference between CWC and ADF. Cellulose was determined according to Crampton and Maynard (1938).

RESULTS AND DISCUSSION

Chemical composition (Table 1), amounts of potential digestibility, and fermentation rates of roughages varied considerably among roughages. Potential digestibility of all fractions was higher for corn cobs and orchardgrass (Table 2), which had lower lignin content than other roughages (Table 1). Barley straw had higher potential digestibility than rice straw. In this study potential digestibility was calculated by differences from indigestible residues remaining after 144 hours of in-vitro incubation. Smith et al. (1971, 1972) estimated the indigestible fraction by 72 hours of in-vitro indigestibility, but Mertens and Van Soest (1972) suggested that using 72-hour in-

vitro residue leads to an overestimate of the indigestible fraction. We also observed that some digestion occurs beyond 72 hours of incubation. A wide variation in potential digestibility among various roughages is due to the morphological characteristics of plant tissue and the crystallinity of fibrous carbohydrates as well as to lignin and possibly silica contents that limit potential digestibility (Mertens 1977). Factors that affect fermentation rate can be divided into two categories: those inherent in the cell wall and those affecting the microbial population of their enzyme systems (Mertens 1977). It is generally believed that roughages that have low CWC ferment rapidly. However, corn cobs, which had the highest CWC content but less lignin, had the highest fermentation rate (Table 3), and alfalfa, which had low CWC, had a low rate of fermentation, as was observed by Mertens and Loften (1980), but fermentation rate of potentially digestible fractions of alfalfa was higher than that for other roughages and was similar to that for orchardgrass. Mertens and Loften (1980) also observed differences in digestion rate among the three grasses that had similar CWC, ADF, and lignin contents and suggested that other chemical or physical characteristics of the plant influence digestion rate. It was also observed in this experiment that fermentation rates of dry matter, PDDM, and CWC were not correlated with percentage of CWC in roughages but that only fermentation rate of PDCWC was significantly correlated with percentage of CWC.

It is possible that fermentation rate is controlled by physical or morphological characteristics of the roughages; if so, particle size reduction, surface area, and fermentation rate would significantly affect the dynamics of roughage fermentation.

Table 2. Potential digestibility (%) of dry matter, CWC, ADF, and cellulose of roughages (mean of four replications ± standard error).

Roughage	Dry matter	CWC	ADF	Cellulose
Rice straw	54.2±1.34	46.5±2.47	46.5±2.02	51.8±2.20
Barley straw	59.2±2.00	55.3±1.91	46.5±0.80	59.6±1.89
Corn cob	80.5±0.98	72.6±1.38	59.4±1.56	80.1±0.31
Corn stover	68.6±0.28	57.1±0.54	48.1±1.85	63.8±2.11
Orchardgrass	73.1±1.29	59.9±1.98	61.5±1.12	86.6±0.44
Alfalfa	57.3±0.75	37.4±0.43	41.6±1.36	63.5±0.27

Table 3. Fermentation rates (K) of constituents of roughages (mean of four replications).

Roughage	DM	PDDM	CWC	PDCWC	ADF	PDADF	Cellulose	PDC
				fermentation rate (K)/hr				
Rice straw	0.0075	0.029	0.0082	0.035	0.0080	0.034	0.0096	0.035
Barley straw	0.0091	0.031	0.0089	0.026	0.0072	0.030	0.0122	0.046
Corn cob	0.0163	0.030	0.0144	0.031	0.0118	0.045	0.0216	0.041
Corn stover	0.0099	0.029	0.0071	0.025	0.0057	0.018	0.0121	0.039
Orchardgrass	0.0121	0.032	0.0090	0.051	0.0100	0.034	0.0221	0.042
Alfalfa	0.0066	0.043	0.0070	0.044	0.0058	0.041	0.0117	0.051

LITERATURE CITED

Association of Official Agricultural Chemists (AOAC). 1975. Official methods of analysis, 12th ed. AOAC. Washington, D.C.

Belyea, R.I., B.K. Darcy, and K.S. Jones. 1979. Rate and extent of digestion and potentially digestible cell wall of waste papers. J. Anim. Sci. 49:887–892.

Bull, L.S., W.V. Rumpler, T.F. Sweeney, and R.A. Zinn. 1979. Influence of ruminal turnover on site and extent of digestion. Fed. Proc. 38:2713–2719.

Crampton, E.W., and L.A. Maynard. 1938. The relation of cellulose and lignin content to the nutritive value of animal feeds. J. Nutr. 15:383–395.

Goering, H.K., and P.J. Van Soest. 1970. Forage fiber analysis (apparatus, reagents, procedures, and some applications). U.S. Dept. Agric., Agric. Handb. 379.

Maeng, W.J. 1976. Improving feeding value of ligno-cellulosic materials. I. Effect of alkali treatment on the digestibility and chemical composition of barley straw. Korean J. Anim. Sci. 18:499–504.

Maeng, W.J., and R.L. Baldwin. 1976. Dynamics of fermentation of a purified diet and microbial growth in the rumen. J. Dairy Sci. 59:636–642.

Maeng, W.J., S.J. Oh, and P.I. Choi. 1979. Improving nutritive value of rice straw. I. Effect of alkali treatment on the chemical composition and in vitro digestibility of indica type rice straw. Korean J. Anim. Sci. 21:343–349.

Mertens, D.R. 1977. Dietary fiber components: relationship to the rate and extent of ruminal digestion. Fed. Proc. 36:187–192.

Mertens, D.R., and J.R. Loften. 1980. The effect of starch on forage fiber digestion kinetics in vitro. J. Dairy Sci. 63:1437–1446.

Mertens, D.R., and P.J. Van Soest. 1972. Estimation of the maximum extent of digestion. J. Anim. Sci. 35:286 (Abs.).

Smith, L.W., H.K. Goering, and C.H. Gordon. 1972. Relationships of forage composition with rates of cell wall components. J. Dairy Sci. 55:1140–1147.

Smith, L.W., H.K. Goering, D.R. Waldo, and C.H. Gordon. 1971. In vitro digestion rate of forage cell wall components. J. Dairy Sci. 54:71–76.

Thorton, B.F., and D.J. Minson. 1972. The relationship between voluntary intake and mean apparent relation time in the rumen. Austral. J. Agric. Res. 23:871–879.

Tilley, J.M.A., and R.A. Terry. 1963. A two-stage technique for the in vitro digestion of forage crops. J. Brit. Grassl. Soc. 18:104–111.

Van Soest, P.J. 1968. Chemical estimates of the nutritional value of feeds. Proc. Cornell Nutr. Conf., 38.

Waldo, D.R. 1969. Factors influencing the voluntary intake of forages. Proc. Natl. Conf. Forage Qual. Eval. Util., F1.

Bulk Volume: A Parameter for Measuring Forage Quality and Its Influence on Voluntary Intake

P.J. VAN DER AAR, H. BOER, B. DEINUM, and G. HOF

Agricultural University, Wageningen, The Netherlands

Summary

The experiment evaluated bulk volume (ml/g meal) as a method of estimating forage quality. Samples of *Lolium perenne* (Lp) and *Festuca arundinacea* and hay samples of Lp were collected at different stages of growth. Voluntary intakes of these forages by dairy cattle had been measured previously. Volumes of samples and volumes of cell-wall residues after 0, 6, 12, 24, and 48 hours in-vitro digestion were determined. Volumes were incorporated in a fill model for the rumen as described by Mertens (1973). This model was used to explain the observed variation in intake.

Original volumes of grasses gave low correlation coefficients with the different cell-wall constituents. For hay samples, higher-correlation coefficients were found, but only the one with NDF was significant (r = 0.90).

Volume of initial cell walls was highly correlated with NDF, hemicellulose, and lignin contents of grasses and hemicellulose content of hay samples. Volume of initial cell walls gave the best correlation with rate and extent of NDF digestion, indicating that cell-wall structure and its digestion are associated. During digestion, cell-wall residue volume/g residue increased, but absolute volume decreased.

The rumen fill model showed that volume of rumen fill explained less of the variation in intake than weight of rumen fill and that the cumulative fill, as a result of intake 24 hours prior to a meal, had more influence on intake than the fill caused by consumption 240 hours prior to a meal.

KEY WORDS: bulk volume, forage evaluation, voluntary intake, digestibility, physical characteristics.

INTRODUCTION

The main emphasis in forage evaluation, to date, has been on chemical composition and its relation to forage quality. However, several physical characteristics have also been used to evaluate forage quality. Quality has been correlated with physical characteristics like leaf tension strength (Wilson 1965), energy use during grinding (Chenost 1966), bulk density (Das and Arora 1976), and near-infrared reflectance (Norris et al. 1976). Baumgardt (1970) suggested that intake of a feedstuff with a low caloric density was limited by physical fill and that intake of feedstuffs with a high caloric density was limited by chemostatic regulation. Mertens (1973) developed a model to estimate rumen fill caused by digestion residues from previous consumption and showed that this model could be used to explain variation in intake.

The objectives of this study were (1) to evaluate the possibility of using forage volume or the volume of its cell-wall fraction as parameters for measuring forage quality; (2) to evaluate the fill model of Mertens (1973) for grasses grown in moderate climates; and (3) to incorporate volume of digestion residue data into the model so that both weight and volume of cumulative rumen fill could be estimated and thus be used to explain variation in intake.

MATERIALS AND METHODS

Thirteen samples of fresh-cut *Lolium perenne* (Lp) or *Festuca arundinacea* (Fa) and five samples of Lp hay were obtained from plots at three different experiment stations throughout the Netherlands. The ad-libitum intake of the forages had been measured previously. All the grasses had been harvested in the vegetative stage. Harvest yield varied from 1,900 to 3,500 kg/ha. The NDF content varied from 63% to 44% and the dry-matter digestibility from 68% to 82%. All samples were ground through a 1-mm screen. Samples were analyzed for neutral detergent fiber (NDF), acid detergent fiber (ADF), and acid detergent lignin (ADL) (72% sulfuric acid), using the methods described by Goering and Van Soest (1970).

Dry-matter (DM) and ash contents were also determined (AOAC 1975).

In-vitro true digestibility of the cell-wall fraction was measured after 0, 6, 12, 24, and 48 hours incubation. After digestion, the NDF residue from 3g of the original sample was isolated. Bulk volumes (ml/g) of whole samples and isolated cell-wall residues were measured by the method described by van der Aar and Van Soest (1982).

RESULTS

Internal Correlations

Correlation coefficients between bulk volume of the whole sample and of different cell-wall constituents (CWC) are given in Table 1. Positive correlations between bulk volume and several of the CWCs indicate that CWCs are associated with bulk volume. However, since these correlation coefficients are not significant within the groups of samples, the significance found for the total group of samples might be caused by the two different groups of samples.

Table 1 also presents correlation coefficients between CWC and bulk volume of cell walls of forages. Bulk volume of cell walls before digestion was negatively related to percentage of NDF (P < 0.01) in dry matter of grasses and in the total group of samples. However, both ADF and cellulose were not correlated with bulk volume

P.J. van der Aar is a graduate student and H. Boer is a Professor of Animal Nutrition at the University of Illinois; B. Deinum is a Professor of Grassland Science and G. Hof is a Professor of Animal Nutrition at Agriculture University, Wageningen, The Netherlands.

Table 1. Correlation coefficients between bulk volume of total forage or bulk volume of NDF prior to digestion and CWC in the dry matter.

Item	Bulk volume of Whole forage			Bulk volume of cell walls		
	Grass (n = 13)	Hay (n = 5)	Total (n = 18)	Grass (n = 13)	Hay (n = 5)	Total (n = 18)
NDF	0.28	0.90**	0.63**	−0.91**	0.68	−0.71**
ADF	0.18	0.56	0.89**	−0.53	0.28	−0.25
Hemi-cellulose	−0.14	0.75	0.26	−0.96**	−0.94**	−0.92**
Cellulose	−0.04	0.23	0.84**	0.50	0.50	0.28
Lignin	−0.29	0.39	0.56*	−0.99**	−0.59	−0.79**

*P<0.05
**P<0.01

Table 2. Correlation coefficients between lignin, bulk volume of whole forage, bulk volume of NDF, and the rate and extent of digestion and intake.

Item	Lignin content			Bulk volume (whole forage)			Bulk volume cell walls		
	Grass (n = 13)	Hay (n = 5)	Total (n = 18)	Grass (n = 13)	Hay (n = 5)	Total (n = 18)	Grass (n = 13)	Hay (n = 5)	Total (n = 18)
Rate of digestion (%/hr)	−0.38	−0.11	−0.42	0.42	−0.59	−0.31	0.67**	0.06	0.59**
Digestibility (%)	−0.80**	−0.60	−0.87**	−0.58*	−0.94*	−0.61**	0.82**	0.73	0.63*
Intake (kg/BW.75)	−0.61*	−0.41	−0.72**	−0.60*	−0.81	−0.39	0.56*	0.19	−0.10

*P < 0.05

**P < 0.01

of cell walls. The significant values found in the total group of samples appear to be true correlations.

Table 2 compares correlation coefficients between independent variables, lignin content, bulk volume of whole forage samples, and bulk volume of NDF with the dependent quality parameters, rate, and extent of digestion of the NDF fraction and intake/kg metabolic body weight. Rate of digestion was positively correlated with bulk volume of cell walls prior to digestion within grass samples (P < 0.01) and within the total group of samples (P < 0.05). Digestibility of grass cell walls was negatively correlated with both volume of whole forage sample (P < 0.05) and lignin content (P < 0.01), while a positive relation with bulk volume of NDF was found (P < 0.05). Within the group of grasses, bulk volume of cell walls explained the largest part of the variation in digestibility, while in the total group, lignin (percentage of DM) explained the most variation. Within the grass samples, lignin content, bulk volume of whole samples, and bulk volume of cell walls showed significant correlations with dry-matter intake/kg metabolic body weight (P < 0.05). For the total group of samples, only lignin content showed a significant negative relation (P < 0.01).

During digestion, total volume of cell-wall residues from 1 g of original material declined for the hay samples from 7.7 ml/g after 0 hours to 6.0 ml/g after 48 hours of digestion. For the grass samples these values were 6.1 ml/g after 0 hours and 3.7 m/g after 48 hours of digestion.

Fill Model

Mertens (1973) described a model to estimate the cumulative fill of the rumen caused by NDF digestion residues. It takes into account the number of meals/day (m), the rate of digestion (K_d), rate of passage (K_p), digestible cell walls (V_o), undigestible NDF (O_o), and lag time (t').

$$\text{Fill} = \frac{24}{m} \sum_{t=0}^{t=\infty} \left(V_o e^{-(K_d + K_p)(t - t')} + O_o e^{-K_p(t - t')} \right) \quad (1)$$

We modified this model by incorporation of the cell-wall volume and its change during digestion, so that a relative value for volume of rumen fill was obtained.

Both fill models, weight and volume, gave the highest

correlation coefficients with intake when only cumulative fill of digestion residues at 48 hours (t = ∞ changed to t = 48 hours) prior to feeding were taken into account. The model explained less variation when the cumulative fill caused by meals 240 hours prior to a specific meal was estimated than when fill caused by meals 48 hours prior was estimated.

Correlation coefficients of the fill model based on weight just before a meal and intake/kg metabolic body weight were −0.82 for the grass and −0.96 for the total group of samples, while these values for fill based on volume were −0.50 and −0.90, respectively. Correlation coefficients based on weight of fill after a meal, were −0.72 for grasses and −0.92 for the total group of samples, whereas, based on volume of fill after a meal, these values were −0.26 and −0.92, respectively. These values are based on the assumption that the animals consumed two meals/day. So, especially within the grasses, fill expressed as weight showed better agreement with intake than fill expressed as volume.

DISCUSSION

Selim et al. (1975) observed the digestion of a leaf of *F. arundinacea* through an electron microscope. They concluded that sparsely lignified large parenchymatous cells were digested faster than the thicker and more lignified cells in other tissues. The high correlation between NDF volume and rate and extent of digestion of NDF found in this experiment indicates that a high NDF volume might result from a high ratio of parenchyma to other tissues. A lower rate of digestion might be caused partially by thicker cell walls, a change in leaf : stem ratio, or the presence of more digestion-limiting lignin.

The high correlation within grasses between digestibility of the NDF fraction and bulk volume of NDF is probably due to the fact that both parameters are influenced by the same factor. Since, during maturation, lignin content increases and bulk volume of NDF decreases, a negative correlation is found, so bulk volume can be a parameter for estimating forage quality.

Total volume of cell-wall residues during digestion decreased. The decline in volume was more rapid for grasses low in lignin content (P < 0.05); thus, bulk volume reduction of the cell walls due to microbial digestion was influenced by the degree of lignification of the plant tissue.

Both models, weight and volume, showed the highest correlation coefficients with intake when the fill was calculated for time intervals of 0–32 and 0–48 hours prior to the meal, indicating that fill caused by meals eaten 0–36 hours before the meal had more influence on intake than fill caused by meals eaten 0–24 hours or 0–240 hours before a meal. These results confirm those of Weston (1966), who observed that animals partially fed intraruminally did not adjust their oral intake the first day they received intraruminal feed. After they had been fed intraruminally for 24 hours, they adjusted their intake, and fill of the rumen influenced the intake.

The model gave a better explanation for the variation in intake if the fill was calculated just before a meal than if it were calculated right after a meal, confirming the suggestion of Campling and Balch (1961) that critical fill of the rumen for intake is not at the end of a meal but rather when a constant amount of digesta is contained in the rumen just before the next meal.

The difference in the explanatory values of the volume model and the weight model seems to be in contradiction with the findings of Campling and Balch (1961), who suggested, on the basis of work with water-filled bladders placed in the rumen, that volume rather than weight of rumen fill is the controlling factor in physical intake regulation. However, this conclusion was based on dry-matter intakes, and it might be possible that the wet weight of rumen contents plays a role as well.

The correlation coefficient within the group of grasses for the weight model was similar to the one reported by Mertens (1973). Thus, the model has similar value for explaining voluntary intake for grasses grown in the U.S.A. as for grasses grown under moderate climatological conditions existing in the Netherlands.

LITERATURE CITED

Association of Official Agricultural Chemists (AOAC). 1975. Official Methods of Analysis, 12th ed. AOAC, Washington, D.C.

Baumgardt, B.R. 1970. Control of feed intake in regulation of energy balance. *In* A.T. Phillipson (ed.) Physiology of digestion and metabolism in the ruminant, 235–253. Oriel Press, Newcastle upon Tyne.

Campling, R.C., and C.C. Balch. 1961. Factors affecting the voluntary intake of foods by cows. I. Preliminary observations on the effect, on the voluntary intake of hay, of changes in the amount of reticulo-ruminal contents. Brit. J. Nutr. 15:523–540.

Chenost, M. 1966. Fibrousness of forages: its determination and its relation to feeding value. Proc. X Int. Grassl. Cong., 406–411.

Das, B., and S.K. Arora. 1976. Changes in cell wall carbohydrates, *in-vitro* dry matter digestibility, bulk density and hydration capacity of *Pennisetum pedicellatum* grass as affected by growth stage. Forage Res. 2:113–118.

Goering, H.K., and P.J. Van Soest. 1970. Forage fiber analysis. U.S. Dept. Agric., Agric. Handb. 379.

Mertens, D.R. 1973. Applications of theoretical mathematical models to cell wall digestion and forage intake in ruminants. Ph.D. Dissertation, Cornell Univ.

Norris, K.H., R.F. Barnes, J.E. Moore, and J.S. Shenk. 1976. Predicting forage quality by infrared reflectance spectroscopy. J. Anim. Sci. 43:889–897.

Selim, O.I., D. Wilson, and D.I.H. Jones. 1975. A histological technique, using cellulotic enzyme digestion, for assessing nutritive quality differences in grasses. J. Agric. Sci., Camb. 25:297–304.

van der Aar, P.J., and P.J. Van Soest. 1982. Measurement on bulk volume and hydration capacity of forages in relation to their stage of growth. J. Sci. Food Agric.

Weston, R.H. 1966. The effect of level of feeding on acetate tolerance in the sheep. Austral. J. Agric. Res. 17:933–938.

Wilson, D. 1965. Nutritive value and the genetic relationships of cellulose content and leaf tensile strength in Lolium. J. Agric. Sci. 65:285–292.

Using a Dynamic Model of Fiber Digestion and Passage To Evaluate Forage Quality

D.R. MERTENS and L.O. ELY

University of Georgia, Athens, Ga., U.S.A.

Summary

Assessing forage quality depends not only upon the characteristics of the plant and animal, but also upon the dynamic interaction of the forage, rumen microorganisms, and animal. Although static variables, such as composition or 48-hour in-vitro fermentation, are important for characterizing feeds and predicting in-vivo responses under standard conditions, they provide little information that can be used to assess forage quality in dynamic situations. Dynamic properties, such as rates of digestion and passage, can be used to predict digestion under varied circumstances but require an appropriate mathematical model. Our objectives were to describe a dynamic model of fiber digestion and passage in the ruminant and to illustrate its use in assessing forage quality.

In the model, neutral detergent fiber is divided into three fractions: fast-digesting (>0.02/hr), slow-digesting (<0.02/hr), and indigestible. While the indigestible fraction can escape the digestive tract by passage only, the digestible fractions can disappear by both digestion and passage. Forage particles enter the rumen into large-, medium-, or small-particle pools. As particle size is reduced, particles move from large to medium and from medium to small pools. Escape from the rumen can occur only from the small- and medium-particle pools. All rates of passage and digestion are assumed to be first order. The model, which requires 20 differential equations, has been implemented on the computer, using both CSMP and GASP IV simulation languages.

The model can be used not only to predict forage quality under varying situations of plant and animal characteristics but also to measure the relative importance of various factors in determining forage quality. For example, the dynamic model predicts that the indigestible fraction has the greatest influence on both digestibility and intake. Our simulations indicate that rate of passage is also an important variable influencing forage intake and digestibility. The dynamic model adds a new dimension for assessing forage quality because it provides flexibility in determining the effects of changes in both the plant and the animal on forage utilization.

KEY WORDS: digestion model, ruminant, forage quality, digestibility intake, passage, forage.

INTRODUCTION

In-vitro fermentation is a dynamic process, yet disappearance measured at 48 hours is a static variable that measures digestion at only one point in time and provides no information concerning the dynamic processes that occur when the forage is consumed, digested, and metabolized. Thus, 48-hour in-vitro disappearance can predict accurately in-vivo steady-state digestibility measured under standard conditions, but it is less accurate when the dynamic conditions of the animal are changed. The limits of static models are indicated by the observation that any condition of the animal that corresponds to a point other than 48 hours of fermentation will result in a different assessment of the relative value of forages (Fig. 1).

The dynamic property, rate of digestion, gives information about the forage and fermentation system that can be used to predict digestion and intake. However, use of kinetic properties for forage evaluation requires a dynamic model to evaluate the effects of change over time. Using rates of digestion as predictor variables in regression analysis does not take advantage of the dynamic information contained in rates; rather it treats rates as a static characteristic of the feed. Such regression analysis cannot predict changes with time or utilize information about the dynamic interactions between the plant and the animal to predict outcomes under different animal situations.

The objectives of this paper are to describe briefly a dynamic model of fiber digestion and to illustrate the use of the model in assessing forage quality. A complete description of the theoretical basis for the model and the sources of parameter estimates have been published by Mertens and Ely (1979).

D.R. Mertens is an Associate Professor of Animal and Dairy Science at the University of Georgia, and L.O. Ely is an Assistant Professor of Animal Science at Georgia Agricultural Experiment Station (currently at the Department of Animal Science of Colorado State University).

METHODS

Basically, the model is an expansion of the model of cellulose digestion proposed by Waldo et al. (1972), which assumes that cellulose can be divided into potentially digestible and indigestible fractions. Although the indigestible fraction can leave the digestive tract only by passage, the digestible fraction can disappear by both digestion and passage. Rates of passage and digestion are assumed to be first-order processes. The digestion aspects of the model of Waldo et al. (1972) were expanded by (1) substituting neutral detergent fiber (NDF) for cellulose, (2) dividing the digestible fraction into fast- and slow-digesting subfractions to describe the digestive process more accurately (Mertens 1977), and (3) including digestion in the lower digestive tract. The passage processes of the model of

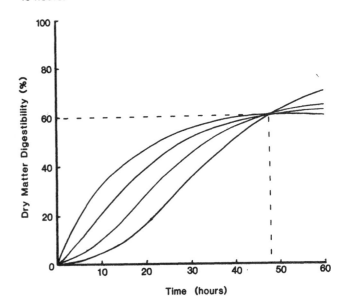

Fig. 1. Examples of digestion curves with different kinetic characteristics that obtain the same dry-matter digestibility at 48 hours.

Waldo et al. (1972) were also modified to include (1) rates of particle-size reduction (Matis 1972) and (2) rates of escape from the rumen and passage through the intestines (Blaxter et al. 1956).

The complete model uses 20 differential equations to describe the relationships between 20 pools and 13 fractional rate constants. The model has been implemented by using either IBM Continuous Systems Modeling Program (CSMP) (IBM Corporation 1972) or General Activity Simulation Program IV (GASP IV) (Pritsker 1974). Copies of the computer programs may be obtained from the authors of this paper. Input to the model consists of rates of passage and particle-size reduction obtained from the literature (Mertens and Ely 1979) and rates of digestion obtained from in-vitro determinations (Mertens 1973).

In addition to pool sizes in the various segments of the digestive tract, the model has been programmed to calculate (1) NDF digestibility in the rumen and total digestive tract, (2) dry-matter digestibility in the total digestive tract, (3) rumen lignin turnover time, and (4) maximum dry-matter intake, assuming that rumen capacity limits intake.

RESULTS AND DISCUSSION

Mertens (1977) suggested that the kinetics of fiber digestion of forages can be divided into three components: indigestible asymptote, rate of digestion, and digestion lag. Each component influences not only the extent of digestion in the dynamic in-vivo situation, but also the mass of material that must pass through the system, thereby limiting further intake. The model estimates the

relative effects of each kinetic characteristic upon digestion and intake and provides a basis for determining which factor is most critical for assessing forage quality.

Within a forage species, the fraction of the forage that is indigestible has greater influence upon both digestibility and intake than does the rate of digestion (Table 1). This greater influence suggests that forages should be selected for less indigestible fiber. Smith et al. (1972) have observed that the indigestible fraction is highly correlated with lignin content in forages, indicating that selection for lower lignin content would result in forage containing less indigestible matter. Information is needed concerning the variation in digestion rate and indigestible fraction within species to determine the relative selection intensity to be applied to each.

The relative importance of indigestibility and rate of digestion cannot be separated easily between species. Dry-matter digestibility (DMD) of Coastal bermudagrass, with 40.1% of the NDF being indigestible, is 52.5%, while DMD of alfalfa, with 44.1% of NDF being indigestible, is 62.7%. Similar trends are observed for dry-matter intake. These comparisons suggest that variables other than fiber indigestibility, such as NDF content or digestion rate, are important in affecting forage dry-matter digestibility and intake. Differences in digestion lag among forages cannot be tested using the model. The model was developed upon the assumption that the lag associated with digestion would be cancelled by the lag associated with passage under steady-state conditions. A revised model is being developed to evaluate digestion lag.

Rate of passage is one of the most variable and critical characteristics of the animal's digestive system that affects

Table 1. Effect of changing indigestibility of neutral detergent fiber (NDF), rate of NDF digestion, and rumen turnover time on the digestion and intake of alfalfa and Coastal bermudagrass.[a]

Independent variable[b]	Alfalfa			Coastal bermudagrass		
Dependent variable	Simulation 1	Simulation 2	Percent difference	Simulation 1	Simulation 2	Percent difference
Indigestible fraction (% NDF)	51.9	44.1	−15.0	40.1	26.1	−34.9
NDF digestibility (%)	42.7	50.0	+17.1	51.4	66.4	+29.2
Dry matter digestibility (%)	59.3	62.7	+ 5.7	52.5	63.0	+20.0
Maximum dry matter intake (% BW[c])	2.74	3.02	+10.2	2.01	2.58	+28.4
Max. dig. dry matter intake (% BW)	1.62	1.89	+16.7	1.06	1.63	+53.8
Rate of NDF digestion (per hour)	.0860	.1163	+35.2	.0782	.1056	+35.0
NDF digestibility (%)	45.2	47.2	+ 4.4	57.2	60.1	+ 5.1
Dry matter digestibility (%)	60.5	61.4	+ 1.5	56.5	58.6	+ 3.7
Maximum dry matter intake (% BW)	2.80	2.93	+ 4.6	2.17	2.34	+ 7.8
Max. dig. dry matter intake (% BW)	1.69	1.80	+ 6.5	1.23	1.37	+11.4
Rumen turnover time (hr)	29.4	45.8	+55.8	29.4	45.8	+55.8
NDF digestibility (%)	41.8	46.3	+10.8	52.8	58.9	+11.6
Dry matter digestibility (%)	58.9	61.0	+ 3.6	53.5	57.7	+ 8.5
Maximum dry matter intake (% BW)	4.28	2.87	−32.9	3.28	2.26	−31.1
Max. dig. dry matter intake (% BW)	2.52	1.75	−30.6	1.75	1.30	−25.7

[a]Adapted from Mertens and Ely (1979).

[b]Refers to the value of the independent variable changed between simulations. All other parameters within each forage were held constant in each simulation.

[c]Percentage of body weight consumed daily by a 500-kg steer.

fiber digestion and intake. The model indicates that increasing the rate of passage (decreasing rumen turnover time) reduces digestibility (Table 1). However, it has the opposite effect upon dry-matter intake, resulting in a net increase in digestible dry-matter intake. In this case the value of the forage is changed even though its composition and kinetic character remain constant, thus illustrating the value of dynamic assessment of forage quality. Use of the model may explain apparent discrepancies in forage evaluation by separating the effects due to the plant from those caused by the animal.

Since the rate parameters of the model can be varied independently, it is possible to compare quantitatively the digestion of forage under conditions that affect one or more of the rates. For example, the model can be used to predict in-vivo responses that would result from differences in in-vitro rates of digestion due to adding starch (Mertens and Loften 1980) or changing the pH of the system. Thus, the dynamic model of fiber digestion adds a new dimension to forage-quality evaluation. With accurate kinetic data for forages, the model can provide flexibility in determining the effects of many changes in both the plant and animal on forage utilization. Furthermore, it can provide a valuable tool for discovering the many facets of forage composition, morphology, and physical structure that limit forage utilization by ruminants.

LITERATURE CITED

Blaxter, K.L., N.McC. Graham, and R.W. Wainman. 1956. Some observations on the digestibility of food by sheep and on related problems. Brit. J. Nutr. 10:69.

IBM Corporation. 1972. System/360 Continuous System Modeling Program, 5th ed. IBM Technical Pub. Dept., White Plains, N.Y.

Matis, J.H. 1972. Gamma-time dependency in Baxter's compartmental model. Biometrics 29:597.

Mertens, D.R. 1973. Application of theoretical mathematical models to cell wall digestion and forage intake in ruminants. Ph.D. Dissertation, Cornell Univ.

Mertens, D.R. 1977. Dietary fiber components: relationship to the rate and extent of ruminal digestion. Fed. Proc. 36:187.

Mertens, D.R., and L.O. Ely. 1979. A dynamic model of fiber digestion and passage in the ruminant for evaluating forage quality. J. Anim. Sci. 49:1085.

Mertens, D.R., and J.R. Loften. 1980. Effect of starch on forage fiber kinetics in vitro. J. Dairy Sci. 63:1437.

Pritsker, A.A.B. 1974. The GASP IV simulation language. John Wiley and Sons, New York.

Smith, L.W., H.K. Goering, and C.H. Gordon. 1972. Relationship of forage compositions with rates of cell wall digestion and indigestibility of cell walls. J. Dairy Sci. 55:1140.

Waldo, D.R., L.W. Smith, and E.L. Cox. 1972. Model of cellulose disappearance from the rumen. J. Dairy Sci. 55:125.

Relationship of Forage-Evaluation Techniques to the Intake and Digestibility of Tropical Grasses

S.M. ABRAMS, H. HARTADI, C.M. CHAVES, J.E. MOORE, and W.R. OCUMPAUGH

USDA-SEA-AR and University of Florida, Gainesville, Fla., U.S.A.

Summary

Seventy-six grass hays, harvested at 2, 4, 6, and 8 weeks of regrowth, were fed ad libitum to wethers in a digestibility and intake trial. Range of organic-matter digestibility (OMD) was 54.1% to 72.3% and that of organic-matter intake (OMI) 17.3 to 27.4 g/kg. Lignin was the chemical component most highly related to OMD ($r = -0.72$). Neutral detergent fiber (NDF) was better than acid detergent fiber (ADF) as a predictor of OMD ($r = -0.66$ vs. -0.45). In-vitro NDF digestibility (IVNDFD) was superior to in-vitro organic-matter digestibility (IVOMD) and all other analyses as a predictor of OMD ($r = 0.84$). OMI was best predicted by ADF ($r = -0.54$). Lignin and p-coumaric acid were about equally well correlated with NDF digestibility, although p-coumaric acid represented an average of only 0.6% of NDF. Ferulic acid was not well related to any in-vivo measures of forage quality. In-vivo NDFD was more closely related to the amount of potentially digestible NDF than to NDFD predicted from estimated rates of passage and digestion ($r = 0.89$ vs. 0.75).

We conclude from these studies that (1) the relationship of fiber fractions to forage quality in these tropical grasses differs from the conventionally accepted relationship; (2) the influence of rate of passage on variation in digestibility of tropical grass hays, fed ad libitum, may be marginal; (3) in-vitro techniques remain the most accurate method for estimating digestibility; and (4) further investigation into the nature of plant phenols and their relationship to forage quality is merited.

KEY WORDS: sheep, ruminant, forages, intake, digestibility, fiber, phenols, rate of digestion.

INTRODUCTION

A major goal of animal nutritionists and agronomists is to develop laboratory techniques that will predict the quality of forages over a wide range of genotypes and maturities. A long-term project at the Florida Agricultural Experiment Station seeks to establish a representative set of tropical grass hays of known digestibility and intake for the purpose of evaluating traditional and new techniques of forage evaluation. The objectives of this study were (1) to evaluate the quality of ten cultivars of tropical grasses cut at four stages of regrowth; (2) to determine the relationship of forage chemical composition and in-vitro rates of digestion in in-vivo measurements of forage quality; and (3) to determine if use of a synthesized estimate of the rumen passage rate constant would improve predictions of digestibility.

METHODS

One cultivar of bermudagrass (*Cynodon dactylon*), six cultivars of digitgrass (*Digitaria decumbens* and *Digitaria* spp.), and three cultivars of bahiagrass (*Paspalum notatum*) were grown in upland, sandy soil in north central Florida during 1975, 2 field replicates/cultivar. Regrowths of 2, 4, 6, and 8 weeks, harvested from August to October, were artificially dried and chopped, for a total of 76 hays. The 4-week regrowth of some plots was heavily damaged by an infestation of striped grass loopers (*Mocis latipes*).

Organic-matter intake and digestibility (OMI and OMD, respectively) and ash-free neutral detergent fiber intake and digestibility (NDFI and NDFD, respectively) were determined by feeding each hay, ad libitum, to three mature wethers in individual pens. A 14-day preliminary period was followed by a 7-day collection period during which feed offered and refused and feces excreted were measured.

Samples were analyzed for dry matter (DM), organic matter (OM), crude fiber (CF), and crude protein (CP) by AOAC methods (1975). Ash-free neutral detergent fiber (NDF) and in-vitro NDF digestibility (IVNDFD) were determined by a modification of the procedures described by Goering and Van Soest (1970), and acid detergent fiber (ADF), permanganate lignin, and insoluble ash were determined by Goering and Van Soest's method. In-vitro OM digestibility was determined by the procedure of Moore and Mott (1974). Hemicellulose was calculated as the difference between NDF and ADF and cellulose as the difference between ADF and lignin. Thirty-eight hays (one field replicate of each hay) were analyzed for p-coumaric acid and ferulic acid using gas-liquid chromatography (Chaves et al. 1982).

Sixty of the 76 hays were incubated in vitro and NDF

S.M. Abrams is a Research Animal Scientist at USDA-SEA-AR in University Park, Pa.; H. Hartadi is an Animal Nutritionist at the Universitas Gadjah Mada in Yogyakarta, Indonesia; C.M. Chaves is an Animal Nutritionist at the Universidad de Costa Rica, San Jose; J.E. Moore is an Animal Nutritionist at the University of Florida; and W.R. Ocumpaugh is a Forage Agronomist at the University of Florida.

recovered after 24, 36, 48, 60, and 96 hours of fermentation. Residues at 96 hours were considered indigestible (U). Initial NDF minus U was considered to represent the potentially digestible fiber fraction (D). The digestion rate constant (k_d) and digestion lag were determined by a logarithmic transformation followed by regression analysis (Mertens 1973). The rate constant for passage (k_p) was estimated by solving the following model equation (Waldo et al. 1972) for k_p, using an estimated constant rumen NDF fill of 16 g/kg body weight: (NDFI/hour)/body weight $= 16/\{[D/(k_d + k_p)] + U/k_p\}$. Predicted ruminal NDFD (NDFDP) was then determined as: NDFDP $= Dk_d/(k_d + k_p)$.

RESULTS AND DISCUSSION

Forage quality declined with increasing forage maturity (Table 1), the most rapid decline occurring between the 2-week and 4-week regrowths. NDFD was highly correlated with OMD (r = 0.96), as was NDFI with OMI (r = 0.94). This suggests the overriding importance of cell-wall degradation in determining the quality of tropical grasses. Intake and digestibility were not closely related, the highest correlation being that between OMD and OMI (r = 0.47). Within-genus and within-cultivar correlations were also low. The average correlation between OMI and OMD within genera was 0.42, and within cultivars it was 0.48. Animal variation in OMI and NDFI was best removed by powers of body weight of 0.92 and 0.93, respectively.

Chemical Composition And In-Vitro Digestibility

In-vitro techniques were superior to chemical measurements of forage digestibility (Table 2). IVNDFD was superior to IVOMD for estimation of both OMD and NDFD. Among chemical analyses, lignin was the single best predictor of OMD (r = -0.72) and NDFD (r = -0.72). ADF was most closely related to OMI (r = -0.54), and NDF was superior to ADF in predicting both OMD and NDFD. This is the converse of the conventional relationship of these fiber fractions to forage quality (Rohweder et al. 1978). The use of quadratic equations did not alter this relationship.

Phenolic Acids

Content of p-coumaric and ferulic acid ranged from 0.24% to 0.78% and 0.25% to 0.57% of NDF, respectively. Content of p-coumaric acid increased with grass maturity, while that of ferulic acid tended to remain constant. Average p-coumaric acid content was twice as high as that reported for ryegrasses (Hartley 1972). Lignin and p-coumaric acid were equally well correlated with NDFD and OMD (Table 2). Ferulic acid was not closely related to any in-vivo parameters. The fact that p-coumaric acid, a compound representing only 0.45% of DM and 0.6% of NDF, was closely related to OMD and NDFD suggests that this phenol is involved in maintaining the integrity of

Table 1. Least squares means of intake and digestibility by genera and regrowth for 76 hays.[a]

Genera (cultivars)	Regrowth (wk)	OMD (%)	OMI (g/kg)	DOMI (g/kg.75)	NFD (%)	NDFI (g/kg)
Digitaria (6)	2	68.9	25.4	47.5	73.9	18.8
	4	62.4	22.9	37.6	66.3	17.8
	6	61.7	21.7	34.8	64.4	16.9
	8	59.7	22.5	34.4	61.4	17.0
Paspalum (3)	2	64.5	23.6	40.0	68.7	18.6
	4	55.8	22.2	32.6	58.6	18.4
	6	57.3	22.5	34.1	60.0	18.7
	8	57.1	21.4	31.7	60.5	17.9
Cynodon (1)	2	59.7	21.3	33.6	60.9	16.0
	4	58.1	17.3	26.4	61.1	14.2
	6	59.4	20.0	30.7	61.1	16.6
	8	55.2	21.5	31.2	55.4	17.8

[a]OMD = organic matter digestibility; OMI = organic matter intake; DOMI = digestible organic matter intake; NDFD = neutral detergent fiber digestibility; NDFI = neutral detergent fiber intake.

cell walls and may be intimately involved with the indigestible fraction of tropical grasses.

Model Estimates

NDFDP, D, and k_d declined with increasing maturity of the grasses (Table 3). Digestion lag showed no consistent maturity effects. With the exception of *Cynodon*, k_p increased between the 2- and 4-week regrowths, with little change thereafter. In-vivo NDFD should have been more highly correlated with NDFDP than with any of the rate parameters alone if the equation approximated in-vivo rumen NDF disappearance, since NDFDP takes account of variation in passage, digestion rate, and size of the pool amenable to digestion. However, NDFD was more

related to D than to NDFDP (r = 0.89 and 0.75, respectively).

Possible explanations for the inability to improve predictions of NDFD are (1) that the equation does not fully describe the complexity of rumen dynamics, (2) that NDF fill is not a constant percentage of body weight, or (3) that in-vitro parameters do not adequately reflect in-vivo conditions. An alternative explanation is that rate of passage of these grasses through the rumen and large intestine was sufficiently slow, and rate of digestion sufficiently fast, to allow for nearly complete digestion of D. Therefore, rate of passage, while important in determining intake variation, probably did not have a major effect on digestibility variation (W.C. Ellis, personal communication). This theory is supported by the fact that correlations of

Table 2. Correlation coefficients (r) of laboratory analyses with in-vivo measurements[a] of forage quality.

Item	OMD	OMI	DOMI	NDFD	NDFI
76 hays					
Crude protein	0.46	0.47	0.46	0.46	ns[b]
Crude fiber	−0.45	−0.41	−0.47	−0.33	ns
Neutral detergent fiber	−0.66	−0.38	−0.56	−0.57	ns
Acid detergent fiber	−0.45	−0.54	−0.57	−0.32	−0.27
Hemicellulose	−0.23	ns	ns	−0.24	−0.32
Cellulose	−0.40	−0.45	−0.49	−0.27	ns
Lignin	−0.72	−0.45	−0.64	−0.72	−0.24
Insoluble ash	0.32	ns	ns	0.38	ns
In-vitro neutral detergent fiber digestibility	0.84	0.43	0.68	0.84	ns
In-vitro organic matter digestibility	0.78	0.41	0.64	0.73	ns
38 hays					
Lignin	−0.74	−0.45	−0.66	−0.72	−[c]
p-Coumaric acid	−0.73	−0.39	−0.61	−0.75	−
Ferulic acid	−0.24	−0.38	−0.37	−0.27	−

[a]See Table 1.

[b]P>0.05.

[c]Correlation not determined.

Table 3. Means[a] of model estimates by genera and regrowth.

Genera (cultivars)	Regrowth (wk)	Lag (hr)	k_d (hr^{-1})	k_p (hr^{-1})	D (% NDF)	NDFDP (%)
Digitaria (6)	2	12.1	0.0663	0.0183	81.3	63.7
	4	6.7	0.0525	0.0241	72.2	49.5
	6	6.6	0.0470	0.0238	68.5	45.5
	8	10.9	0.0511	0.0246	66.3	44.8
Paspalum (3)	2	15.5	0.0558	0.0231	73.9	52.3
	4	15.0	0.0394	0.0304	63.6	35.9
	6	16.9	0.0439	0.0310	64.5	37.8
	8	14.5	0.0404	0.0293	64.7	37.5
Cynodon (1)	2	8.8	0.0693	0.0229	61.9	46.5
	4	10.8	0.0594	0.0205	62.0	46.1
	6	10.0	0.0515	0.0252	62.6	42.0
	8	6.8	0.0451	0.0213	54.0	36.7

[a]Lag = in-vitro digestion lag; k_d = in-vitro digestion rate constant; k_p = estimated rumen passage rate constant; D = potentially digestible neutral detergent fiber; NDFDP = predicted ruminal neutral detergent fiber digestibility.

IVNDFD with NDFD increased with increasing duration of fermentation: 24 hours < 36 hours < 48 hours < 60 hours < 96 hours (r = 0.45, 0.66, 0.73, 0.85, 0.89, respectively). Fiber digestion in the large intestine can compensate for variations in ruminal digestibility caused by altered ruminal retention times (Thomson et al. 1972).

LITERATURE CITED

Chaves, C.M., J.E. Moore, H.A. Moye, and W.R. Ocumpaugh. 1982. Separation, identification and quantification of lignin saponification products extracted from digitgrass and their relation to forage quality. J. Anim. Sci. 54:196.

Goering, H.K., and P.J. Van Soest. 1970. Forage fiber analyses (apparatus, reagents, procedures and some applications). U.S. Dept. Agric., Agric. Handb. 379.

Hartley, R.D. 1972. p-Coumaric and ferulic acid components of cell walls of ryegrass and their relationship with lignin and digestibility. J. Sci. Food Agric. 23:1347.

Mertens, D.R. 1973. Application of theoretical mathematical models to cell wall digestion and forage intake in ruminants. Ph.D. Thesis, Cornell Univ.

Moore, J.E., and G.O. Mott. 1974. Recovery of residual organic matter from the *in vitro* digestion of forages. J. Dairy Sci. 57:1258.

Rohweder, D.A., R.F Barnes, and N. Jorgensen. 1978. Proposed hay grading standards based on laboratory analyses for evaluating quality. J. Anim. Sci. 47:747.

Thomson, D.J., D.E. Beever, J.F. Coelho da Silva, and D.G. Armstrong. 1972. The effect of physical form on the sites of digestion of a dried lucerne diet. I. Sites of organic matter, energy and carbohydrate digestion. Brit. J. Nutr. 28: 31.

Waldo, D.R., L.W. Smith, and E.L. Cox. 1972. Model of cellulose disappearance from the rumen. J. Dairy Sci. 54: 1465.

Structural Characteristics Limiting Digestion of Forage Fiber

D.E. AKIN

USDA-SEA-AR and Richard B. Russell Agricultural
Research Center, Athens, Ga., U.S.A.

Summary

Cell walls of forage were studied for structural and chemical factors that limit digestion by rumen microorganisms. Techniques used to investigate digestion were histochemistry, electron microscopy, and anaerobic culturing of rumen populations on specific substrates.

Certain tissues, such as the xylem cells, totally resisted degradation by rumen microorganisms; these tissues stained positive for lignin with acid phloroglucinol (Weisner test). Other supportive tissues, such as the sclerenchyma in blades, stained positive for lignin with chlorine-sulfite (and also the Mäule test) and were also generally resistant to microbial digestion. Microscopic studies revealed that certain living tissues (i.e., those having cytoplasm and organelles indicative of metabolic activity) were only slowly or partially degraded in particular forage grasses, notably warm-season species.

Coastal bermudagrass and Boone orchardgrass, low- and high-digestibility grasses, respectively, were compared for the manner of attack by rumen microorganisms on digestible cell walls, using electronc microscopy. Tissues encompassing about 70% of the cross-sectional area and including the mesophyll, phloem, parenchyma bundle sheath, and inner part of the epidermis in orchardgrass leaf blades were degraded, often without the necessity of bacterial adherence, whereas only about 30% of the tissues (i.e., mesophyll and phloem) in bermudagrass blades were degraded in this manner. Although the percentages of the various types of adhering bacteria were similar in these forage species, more total bacteria associated with orchardgrass cell walls.

Anaerobic culturing on media containing xylan, pectin, or cellobiose of bacterial populations adapted to orchardgrass or bermudagrass fiber resulted in larger total bacterial counts for orchardgrass-adapted populations; numbers of xylan-using bacteria were significantly ($P \leqslant 0.05$) higher. Further, the hemicellulolytic protozoan *Epidinium ecaudatum* form *caudatum* degraded the mesophyll, parenchyma sheath, and epidermis of orchardgrass but did not attack tissues of bermudagrass.

The data derived from electron-microscopic and anaerobic-culturing studies indicated that carbohydrates in bermudagrass fiber were less available to rumen microbial digestion than those in orchardgrass fiber. Histochemical studies with the chlorine-sulfite test indicated that certain living tissues (i.e., parenchyma bundle sheath) that were only slowly or partially degraded in bermudagrass were lignified, which could explain the lack of availability of the carbohydrates within the cell walls to rumen microorganisms.

These data indicated that the carbohydrates, especially the hemicelluloses, in the digestible cell walls vary in their inherent availability to rumen microbial enzymes. Possibly, chlorine-sulfite lignin plays a significant role in limiting the rate of digestion of living tissue types in particular forages.

KEY WORDS: lignin, histochemistry, electron microscopy, digestibility, cell walls, rumen microorganisms.

INTRODUCTION

Information is needed on the biological and chemical structure of forages as it relates to quality. Indeed, research has been undertaken to place forage evaluation on a structural basis (Regal 1960). The cell walls that make up the tissue consist of cellulose, hemicellulose, pectins, proteins, and smaller quantities of other chemical compounds (Bailey 1973). However, the specific structure of the cell walls varies among forages (Metcalfe 1960). Further, lignin, a polymer of phenylpropanoid units, exists in particular cell walls and limits the availability of the structural polysaccharides for microbial fermentation (Van Soest 1973). Data on forage structure should be related to chemical, biochemical, and nutritional information to gain a full understanding of the factors involved in the complex process of forage digestibility.

TECHNIQUES TO EVALUATE STRUCTURE

The site and type of lignin in tissues can be determined with histochemical tests. The Weisner test (saturated phloroglucinol in 20% hydrochloric acid) has been reported to identify cinnamaldehyde groups (Vance et al. 1980). The chlorine-sulfite test (hypochlorite solution followed by sodium sulfite) and the Mäule test are both reported to indicate syringyl units in lignin (Vance et al. 1980).

Electron microscopy has been used to help establish a structural basis for forage digestibility. Transmission electron microscopy is useful in elucidating the specific interrelationships of plant cells and microorganisms and the manner of forage degradation. Scanning electron microscopy is used to establish relative rates and extents of digestion of specific tissue types.

Culturing of rumen bacteria on grasses that vary in digestibility has been useful in explaining differences. Rumen bacteria were evaluated for "total viable numbers," i.e., growth on the habitat-simulating medium 10 (Caldwell and Bryant 1966) or growth on media with specific substrates to enumerate fermentation types of bacteria.

RELATIONSHIP OF TISSUES TO RUMEN MICROBIAL ATTACK

Lignified tissues for the most part resist microbial digestion and make up a significant portion of the fecal residue (Drapala et al. 1947). However, differences exist in the degree of resistance of types of lignified tissues, suggesting that variations can occur in the complexing of lignin to structural carbohydrates. Tissues giving a positive acid-phloroglucinol reaction totally resist degradation and, in general, do not support a large number of adhering bacteria (Akin et al. 1974). These tissues include the xylem cells and the inner bundle sheath of vascular bundles in leaf blades. Research with other lignin-degrading bacteria indicates that only the intercellular layers are significantly attacked in these tissues, whereas the cell wall is separated into layers but not markedly degraded (Akin 1980b).

Sclerenchyma, a nonliving support tissue of several cell-wall layers, gives a positive lignin reaction with chlorine-sulfite. However, degradation at the periphery of the tissue shows that these cells are not totally resistant to rumen bacterial attack. Sclerenchyma cells can be extensively degraded by certain lignin-degrading bacteria (Akin 1980b).

Not only nonliving, lignified tissues, but some living tissues—i.e., those with cellular organelles and indicative of metabolic activity—resist digestion. Of particular significance are the epidermis and parenchyma bundle sheath, especially of the warm-season grasses. The sheaths are noteworthy in many warm-season grasses because of the amount of cross-sectional area (10%–25%) occupied by this tissue (Akin and Burdick 1975).

COMPARISON OF TISSUE DIGESTION IN GRASSES OF HIGH AND LOW DIGESTIBILITY

Differences in digestibility of leaf blades appear to be related to the relative ease of tissue digestion rather than to the amount of nondigestible, lignified tissue in Boone

---

The author is a Microbiologist of USDA-SEA-AR.

Fig. 1. Degradation of orchardgrass mesophyll cell wall (M), intercellular layer (arrow), and starch within chloroplasts after incubation with rumen fluid for 4 hours (8400 ×).

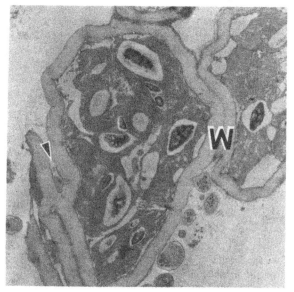

orchardgrass (*Dactylis glomerata* L.), a high-digestible cool-season grass, and Coastal bermudagrass (*Cynodon dactylon* [L.] Pers.), a low-digestible warm-season species. Tissues in orchardgrass are rapidly degraded, as is shown by the attack on mesophyll after 4 hours of in-vitro incubation with rumen fluid (Fig. 1). Further, other tissues—i.e., parenchyma bundle sheath and epidermis—are rapidly degraded, often by extracellular carbohydrases (Akin

Fig. 2. Bermudagrass mesophyll (M) and parenchyma bundle sheath (B) after incubation with rumen fluid for 4 hours showing only slight digestion in the mesophyll and the presence of few bacteria (arrow) (4200 ×).

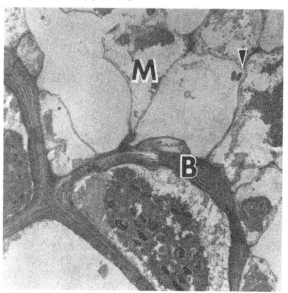

1980a). Tissues showing rapid degradation in this manner occupy about 65%–70% of the cross-sectional area of orchardgrass leaf blades (Akin and Burdick 1975). Bermudagrass tissues show a markedly slower digestion by rumen bacteria after 4 hours (Fig. 2). Further, although the mesophyll and phloem (about 30% of the cross-sectional area) were degraded by cell-free enzymes, the epidermis and parenchyma bundle sheath are degraded after direct adherence of fiber-digesting bacteria, but even then digestion is slower than that in similar tissues of orchardgrass (Akin 1980a).

Electron microscopy indicates that the relative proportions of the major fiber-digesting bacteria are not different for orchardgrass and bermudagrass but that more total bacteria adhere to orchardgrass fiber (Akin 1980a). Anaerobic culturing from a single inoculum source on medium 10 indicates that orchardgrass fiber supported 75% more "total" bacteria than bermudagrass and further establishes that the numbers of bacteria utilizing carbohydrates similar to those in plant cell walls are higher in populations grown with orchardgrass (Akin 1980a). Of particular importance in this study is the comparative number of xylanolytic bacteria, which is significantly (P ⩽ 0.05) higher from populations cultured with orchardgrass, although less xylan is present than in bermudagrass. The rumen entodiniomorph *Epidinium ecaudatum* form *caudatum* preferentially associated with cool-season grasses and degraded 11% of the dry matter in orchardgrass but only 3.7% in bermudagrass (Amos and Akin 1978). The protozoan attacked, degraded, and ingested orchardgrass cell walls (Akin and Amos 1979) and caused so much swelling that individual fibrils in the plant cell wall could be distinguished. The fact that *E. ecaudatum* had been shown to be xylanolytic and hemicellulolytic but not cellulolytic (Bailey et al. 1962) further indicates that carbohydrates, particularly hemicelluloses, are less rigidly bound and more available for microbial fermentation in orchardgrass than in bermudagrass cell walls.

LIGNIN IN "LIVING" CELL WALLS

When living tissues were stained for lignin with chlorine-sulfite and examined at 400 × magnification with the light microscope 1 to 2 minutes after staining, the parenchyma bundle sheath of several bermudagrass varieties, and some mesophyll, parenchyma bundle sheath, and epidermis of Kentucky-31 tall fescue (*Festuca arundinacea* Schreb.) that were partially or slowly degraded, gave a definite but temporary reaction for lignin with the chlorine-sulfite test. These tissues are slowly or partially degraded in bermudagrass and fescue, whereas similar tissues in orchardgrass and timothy (*Phleum pratense* L.) are rapidly degraded and give no reaction for lignin (Akin and Burdick 1981). These data suggest that syringyl-lignin tissues may influence the rate of digestion of certain living tissues, notably in warm-season grasses. Indeed, phenolic compounds that bind to carbohydrates have been reported to be prevalent in lignified and unlignified plant cell walls (Harris and Hartley 1976).

CONCLUSION

Plant tissues vary in their manner and rate of digestion by rumen microorganisms, and these factors could exert a significant influence on the utilization of forage by ruminants. By combining information on the structure of intact cell walls and digestion of these intact cell walls with data from other methods, researchers can expect to understand more fully the complex factors that limit forage utilization. Research is needed especially on the factors that bind the structural carbohydrates in warm-season grasses and limit their availability to rumen microorganisms.

LITERATURE CITED

Akin, D.E. 1980a. Evaluation by electron microscopy and anaerobic culture of types of rumen bacteria associated with digeston of forage cell walls. Appl. Environ. Microbiol. 39:242–252.

Akin, D.E. 1980b. Attack on lignified grass cell walls by a facultatively anaerobic bacterium. Appl. Environ. Microbiol. 40:809–820.

Akin, D.E., and H.E. Amos. 1979. Mode of attack on orchardgrass leaf blades by rumen protozoa. Appl. Environ. Microbiol. 37:332–338.

Akin, D.E., and D. Burdick. 1975. Percentage of tissue types in tropical and temperate grass leaf blades and degradation of tissue by rumen microorganisms. Crop Sci. 15:661–668.

Akin, D.E., and D. Burdick. 1981. Relationships of different histochemical types of lignified cell walls to forage digestibility. Crop Sci. 21:577–581.

Akin, D.E., D. Burdick, and G.E. Michaels. 1974. Rumen bacterial interrelationships with plant tissue during degradation revealed by transmission electron microscopy. Appl. Microbiol. 27:1149–1156.

Amos, H.E., and D.E. Akin. 1978. Rumen protozoal degradation of structurally intact forage tissue. Appl. Environ. Microbiol. 36:513–522.

Bailey, R.W. 1973. Structural carbohydrates. In G.W. Butler and R.W. Bailey (eds.) Chemistry and biochemistry of herbage, 1:157–211. Academic Press, New York.

Bailey, R.W., R.T.J. Clarke, and D.E. Wright. 1962. Carbohydrases of the rumen ciliate Epidinium ecaudatum (Crawley). Biochem. J. 83:517–523.

Caldwell, D.R., and M.P. Bryant. 1966. Medium without rumen fluid for nonselective enumeration and isolation of rumen bacteria. Appl. Microbiol. 14:794–801.

Drapala, W.J., L.C. Raymond, and E.W. Crampton. 1947. Pasture studies. XXVII. The effects of maturity of the plant and its lignification and subsequent digestibility by animals as indicated by methods of plant histology. Sci. Agric. 27:36–41.

Harris, P.J., and R.D. Hartley. 1976. Detection of the bound ferulic acid in cell walls of the Gramineae by ultraviolet fluorescence microscopy. Nature 259:508–510.

Metcalfe, C.R. 1960. Anatomy of the monocotyledons. I. Gramineae. Oxford Univ. Press, London.

Regal, V. 1960. The evaluation of the quality of pasture grasses by the microscopic method. Proc. VII Internat. Grassl. Cong., 522–524.

Vance, C.P., T.K. Kirk, and R.T. Sherwood. 1980. Lignification as a mechanism of disease resistance. Ann. Rev. Phytopathol. 18:259–288.

Van Soest, P.J. 1973. The uniformity and nutritive availability of cellulose. Fed. Proc. 32:1804–1808.

Grazing Techniques for Evaluating Quality of Forage Cultivars in Small Pastures

A.G. MATCHES, F.A. MARTZ, D.A. SLEPER, and R.L. BELYEA

USDA-SEA-AR and University of Missouri, Columbia, Mo., U.S.A.

Summary

Because plant beeders have few laboratory analyses of forage quality that are good predictors of animal performance when grazing pasture, we studied the feasibility of grazing small pastures with cattle in order to detect possible forage-quality differences among cultivars of tall fescue (*Festuca arundinacea* Schreb). Kentucky 31, Kenmont, Fawn, and Missouri 96 tall fescue cultivars and Kenhy, a *Lolium-Festuca* derivative, were grown in 0.47-ha pastures and grazed in separate periods ranging from 35 to 68 days during the spring, summer, and autumn of 1974–1976. Between grazing periods, cattle grazed a reserve pasture of the same cultivar. All test pastures were grazed at the same grazing pressure, and cattle were moved to a fresh feed each week.

Average daily gain (ADG) of heifers was 22% to 83% greater on Kenhy and Missouri 96 than on Kentucky 31. Differences in ADG occurred mainly in the spring and autumn. There were no measurable differences in herbage intake. Actual experimental errors for ADG and those estimated by the method of Petersen and Lucas (1960) were very similar in all but one period of grazing. This error-estimate method is a useful tool for designing grazing trials to evaluate forage cultivars in small pastures. Our results exhibit the importance of animal evaluation of new forage cultivars and the hazard of depending only on laboratory analyses as the basis of assessing forage quality.

KEY WORDS: tall fescue, *Festuca arundinacea* Schreb., cattle gain, experimental error.

INTRODUCTION

New forage cultivars are generally released with little or no data on how animals respond to feeding or grazing new selections as compared with their response to existing cultivars. Cultivars are usually evaluated and released on the basis of laboratory measurements of forage quality, which may include crude protein, neutral detergent fiber, acid detergent fiber, cellulose, lignin, mineral elements, and in-vitro dry-matter digestibility (IVDMD) (Barnes and Marten 1981, Minson 1981). However, few publications present data that show high correlations between chemical or IVDMD analyses and average daily gains (ADG) of cattle under grazing conditions.

In most forage-breeding programs, progress has been slow in breeding for improved animal performance. A grazing experiment was initiated in 1974 to determine the feasibility of grazing small pastures to detect possible forage-quality differences more effectively among new cultivars of tall fescue (*Festuca arundinacea* Schreb.).

METHODS

Kentucky 31, Kenmont, Fawn, and Missouri 96 cultivars of tall fescue and Kenhy, a *Lolium-Festuca* derivative, were seeded in the autumn of 1973 in 0.47-ha pastures at the University of Missouri Southwest Center, near Mount Vernon, Missouri. These cultivars represent a broad genetic base. Kentucky 31, widely grown in the southern Corn Belt, served as the control cultivar. The experimental design was a randomized complete block with 3 replications.

Pastures were irrigated only during the summers of 1974 and 1975. An average of 75 kg/ha of phosphorus and potassium was applied to all pastures each winter. During 1974 and 1975, 92 kg N/ha was broadcast in the winter, and 54 kg N/ha was broadcast in June and August. Because of no irrigation in 1976, only 84 kg N/ha was applied in late winter and 67 kg N/ha in August.

Yearling Hereford heifers that averaged 225, 204, and 254 kg live weight at the start of spring grazing in 1974, 1975, and 1976, respectively, were used to measure

A.G. Matches is a Research Agronomist of USDA-SEA-AR and a Professor of Agronomy at the University of Missouri; F.A. Martz is a Professor of Dairy Science, D.A. Sleper is an Associate Professor of Agronomy, and R.L. Belyea is an Assistant Professor of Dairy Science at the University of Missouri. This is a joint contribution of the USDA-SEA-AR and the Missouri Agricultural Experiment Station (Journal Series Paper No. 8789).

The authors would like to thank Stanley Bell and Marion Mitchell, Research Associates at the University of Missouri, for their expert assistance in carrying on this research.

Mention of a proprietary product does not imply endorsement of it.

animal performance. Grazing periods were from 35 to 68 days during separate periods in the spring, summer, and autumn (Table 1). Generally, three heifers grazed each pasture in the spring and autumn and two in the summer. Individual heifers grazed the same cultivar all season (both test and reserve pastures). Cattle were weighed biweekly following 16 hours of confinement without feed or water.

In 1974 and 1975 equal grazing pressure in all test pastures was maintained by adjusting the area grazed with forward and back electric fences. Cattle were allotted (disregarding sward growth) a daily amount of herbage (dry-matter basis) equivalent to approximately 2.5% of their live weight and were moved to a fresh strip each week. Pasture strips were sampled weekly (three 1-m² quadrats/strip cut to ground level) to estimate the amount of herbage available 1 day before grazing and the amount of residue remaining after grazing, with the difference approximating intake (Matches et al. 1981). In 1976, pastures were grazed in a three-paddock rotation, and equal grazing pressure was maintained with put-and-take animals. However, only one put-and-take animal was added to two of the three pastures of Fawn during the first 21 days of spring grazing.

RESULTS

Heifer Daily Gains

In preliminary small-plot tests (unpublished data), the forage quality of Kentucky 31, Missouri 96, and Kenhy

Table 1. Stocking rates, grazing periods, and means and variance of average daily gain of heifers on five cultivars of tall fescue.

Parameter	Spring 1974	Spring 1975	Spring 1976	Summer 1974	Summer 1975	Autumn 1974	Autumn 1975	Autumn 1976
Animals/ pasture	3	3	3	2	2	3	3	2
Days of grazing	42	45	68	48	49	49	35	42
Average daily gains, g								
Kenhy	717	763	636	300	359	458	622	590
Missouri 96	690	735	604	345	136	499	658	663
Kentucky 31	422	604	440	232	118	436	359	518
Fawn	409	545	336	481	104	404	418	804
Kenmont	386	617	522	281	241	363	377	649
Significance level	.01	.01	.05	.10	n.s.	n.s.	.20	n.s.
Coefficients of variation, %								
Estimated†	19	18	15	20	20	18	21	21
Actual	20	11	16	28	64	15	32	23

†By the method of Petersen and Lucas, 1960.

tall fescue was compared over two growing seasons. Average daily gains of heifers were significantly greater on Kenhy and Missouri 96 (means of 560 and 540 g, respectively) than on Kentucky 31 (390 g) pastures (Table 1) in five of the eight grazing periods by 22% to 83%. Average daily gain of heifers on Fawn and Kenmont (440 and 430 g, respectively) were not consistently higher or lower than gains on Kentucky 31.

Weekly sward measurements in 1974 and 1975 showed no differences in growth rates among cultivars. There were no differences in intake among cultivars, and the correlations between heifer ADG and intake were low (r = 0.11 to 0.57).

Experimental Error For Heifer Daily Gains

In the equation proposed by Petersen and Lucas (1960) for estimating the error of a treatment mean in grazing experiments, the error of a pasture mean for average daily production/animal (C) expressed as a coefficient of variation (CV) is:

$$C = \sqrt{\frac{(157.2)^2}{t} + \frac{(17.3)^2}{a} + \frac{(225.4)^2}{d}} \qquad (1)$$

where t = length of grazing period in days, a = number of different animals grazing the pasture, d = number of animal days the pasture is grazed, and C divided by the square root of the number of replications is the error of a treatment mean. In every grazing cycle except the summer of 1975, the actual and estimated CVs for ADG were similar (Table 1).

DISCUSSION

Under grazing, differences in animal gain among forage treatments are a function of the amount of herbage available, its nutritive value, and the amount consumed by the animals. To evaluate forage quality, herbage allowance/animal must be uniform among all treatments.

We do not know the optimum level of herbage allowance, but judging from recent literature (Matches et al. 1981), our herbage allowance of 2.5% of animal body weight may have been low. A dry-matter allowance of from 2 to 3 times expected consumption might have resulted in greater differences in daily gains of heifers among the five cultivars.

Small pastures are defined here as the minimum area needed for detecting differences in animal gains among treatments. By substituting hypothetical values in the equation variables of Petersen and Lucas (1960), error estimates for daily gain can be obtained. Estimates of variance can be compared and a decision reached as to an acceptable level of variance. Then pasture size will be dependent upon the area required to produce enough herbage to support the number of animals selected/pasture over the projected days of grazing.

Our results show that differences in daily gains of cattle in small pastures can be detected for some cultivars of tall fescue. We plan to continue using this system to screen potential cultivars of tall fescue until laboratory techniques are defined that can serve as good predictors of animal performance.

LITERATURE CITED

Barnes, R.F, and G.C. Marten. 1981. Recent developments in predicting forage quality. J. Anim. Sci. 48:1554–1560.

Matches, A.G., F.A. Martz, D.A. Sleper, and M.T. Krysowaty. 1981. Selecting levels of herbage allowance to compare forages for animal performance. *In* J.L. Wheeler and R.D. Mochrie (eds.) Forage evaluation: concepts and techniques, 331–339. Commonwealth Sci. and Indus. Res. Organ. and Am. Forage and Grassl. Council, Melbourne.

Minson, D.J. 1981. An Australian view of laboratory techniques for forage evaluation. *In* J.L. Wheeler and R.D. Mochrie (eds.) Forage evaluation: concepts and techniques, 57–71. Commonwealth Sci. and Indus. Res. Organ. and Am. Forage and Grassl. Council, Melbourne.

Petersen, R.G., and H.L. Lucas. 1960. Experimental errors in grazing trials. Proc. VIII Int. Grassl. Cong., 747–750.

Screening Perennial Forages by Mob-Grazing Technique

P. MISLEVY, G.O. MOTT, and F.G. MARTIN
University of Florida, Gainesville, Fla., U.S.A.

Summary

Mob-grazing allows the study of forage response to severe defoliation by grazing animals on a limited land area. This technique was used to evaluate two *Digitaria* spp., one *Paspalum* sp., and 13 *Cynodon* spp. at grazing frequencies of 2, 3, 4, 5, and 7 weeks.

Average dry-matter (DM) yield generally increased as grazing frequency decreased from 2 to 7 weeks. Decreasing the grazing frequency from 4 to 7 weeks resulted in a DM yield increase of 50% to 75% for most entries. Grazing at 4- and 5-week intervals resulted in the highest (42%) and lowest (22%) weed contamination, respectively, when compared with the other grazing intervals after 3 years. The 5-week grazing frequency resulted in high yields of quality forage with little weed contamination. These data suggest that several forage entries can be screened by using grazing cattle on a limited land area. Furthermore, the effects of management practices on various forage indices (yield, quality, weed invasion, etc.) can be measured under conditions of severe defoliation by grazing animals.

KEY WORDS: mob-grazing, forages, screening, *Cynodon*, *Digitaria*, *Paspalum*, technique.

INTRODUCTION

Where large numbers of forage genotypes are to be evaluated, clipping is often the easiest and most economical method. However, plant response of specific genotypes may differ between clipping and grazing (Quesenberry and Ocumpaugh 1979). Researchers in Florida believe that early evaluation of germ plasm by grazing animals will facilitate more rapid and efficient selection of germ plasm best adapted to grazing conditions (Quesenberry et al. 1977). Mott and Moore (1970) developed a scheme for testing and evaluating forages that emphasizes early animal testing of new experimentals. The scheme is divided into the following phases: (1) evaluation of plant introduction and breeders' lines by clipping, (2) regional adaptation in small-plot clipping trials, (3) forage responses to grazing animals, (4) animal response to forages, and (5) development of feeding systems.

When forage plants are evaluated under grazing conditions, the most palatable entries are more severely defoliated than those less palatable. One way to reduce this problem is by mob-grazing, which allows many cattle (150/ha) to consume rapidly all entries to a uniform height in a short grazing period. This type of defoliation is similar to intensive rotational grazing. It introduces the effects of trampling, pulling of plants, and deposition of feces and urine, reducing selectivity to a minimum. The purpose of this research was to use the mob-grazing technique to screen 16 perennial grass entries at grazing frequencies of 2, 3, 4, 5, and 7 weeks.

MATERIALS AND METHODS

All grasses were established on a sandy, siliceous, hyperthermic alfic Haplaquod (Eau Gallie fine sand) at the Agricultural Research Center, Ona, Florida. Bahiagrass was seeded at 45 kg/ha, and all other grasses were established from rooted stolons into 8-m² plots, 10 months before grazing. The field plot layout was a split-plot design with 4 replications. Grazing frequencies of 2, 3, 4, 5, and 7 weeks were main plots, with subtropical grass entries (Table 1) as subplots. When signs of vegetative

P. Mislevy is an Associate Professor of Agronomy at the University of Florida, Ona; G.O. Mott is a Professor of Agronomy and F.G. Martin is an Associate Professor of Statistics at the University of Florida, Gainesville. This is a contribution of the Agricultural Research Center of the University of Florida, Ona.

growth appeared, all grass entries except bahiagrass were sprayed with 1.1 kg/ha of Weedmaster® [0.28 kg/ha dicamba (3,6 dichloro-o-anisic acid) + 0.82 kg/ha 2,4-D amine (2,4-dichlorophenoxy acetic acid)] in 280 l water to control *Cyperaceae* and broadleaf weeds. Because all grass entries except bahiagrass were strongly stoloniferous, the herbicide paraquat (1,1'-dimethyl-4,4'-bipyridinium ion) was applied in 280 l water/ha to prevent spread between plots.

Nitrogen (N) was applied at 45 kg/ha to encourage plant establishment when developing tillers were 2 to 7 cm tall. In December of each year, when vegetative growth ceased, 0–25–93 kg/ha of N-P-K was applied. Two hundred twenty kg/ha N was applied during each summer season in four equal applications starting in March. Soil calcium (Ca) and magnesium (Mg) content was adequate, averaging 1,600 kg Ca/ha and 100 kg Mg/ha.

The grazing season was May to November each year. A strip of forage (0.5 × 3.1 m) was harvested from each plot with a rotary mower to a stubble height of 7.5 cm, except for bahiagrass, which was harvested at 4.0 cm, to determine total yield prior to grazing. Twenty-five yearling crossbred cattle consumed (mob-grazed) the forage to a stubble height of approximately 7.5 cm within 2 days or less. All grass entries were cut to a 7.5-cm stubble with a rotary mower following the removal of cattle. The percentage of ground cover occupied by weed species (mainly common bermudagrass, *Cynodon dactylon* [L] Pers. var. dactylon) was estimated visually in the spring and fall of each year.

RESULTS AND DISCUSSION

Dry-Matter Yield

Forage production May–November was generally highest for the *Cynodon* spp. regardless of grazing frequency (Table 1). Total seasonal dry-matter (DM) yield of the *Cynodon* spp. was an average of 64% and 36% higher than yields of digitgrass and bahiagrass, respectively. When DM yields were averaged over all grazing frequencies, 3-year average yields of *Cynodon* spp. ranged from a low of 10.1 metric tons/ha for Hybrid 6 bermudagrass to a high of 14.8 metric tons/ha for UF-5 stargrass. Dry-matter yield of all entries except Pensacola bahiagrass increased as grazing frequency decreased from 2 to 7 weeks. Because bahiagrass has a decumbent growth habit, its forage production did not increase as grazing frequency de-

Table 1. Dry-matter yields of subtropical grasses grazed at five frequencies, 1976–1978.

Entry number	Common name	Scientific name	Grazing frequency (weeks)				
			2	3	4	5	7
Cynodon spp.			---------------------- metric tons/ha/year ----------------------				
1.	Puerto Rico stargrass PR 2341†	*Cynodon nlemfuensis* Vanderyst var. *nlemfuensis*	7.6 de*	9.6 cd	10.8 ab	14.8 bc	20.4 a
2.	McCaleb stargrass (224152)‡	*C. aethiopicus* Clayton et Harlan	8.7 b-d	10.5 b-d	9.4 b-d	12.8 de	17.9 b-d
3.	Costa Rica stargrass	*C. nlemfuensis* var. *robustus* Clayton et Harlan	8.7 b-d	10.8 b-d	10.5 ab	13.4 c-e	19.0 a-c
4.	Cane Patch stargrass	*C. nlemfuensis* Vanderyst var. *nlemfuensis*	10.1 a-c	12.1 ab	11.4 ab	14.8 bc	19.9 a
5.	Sumner stargrass	*C. nlemfuensis* var. *robustus* Clayton et Harlan	9.9 a-c	12.5 ab	11.2 ab	14.1 b-d	19.3 ab
6.	Ona stargrass (224566)	*C. nlemfuensis* Vanderyst var. *nlemfuensis*	9.9 a-c	11.0 b-d	8.5 cd	15.0 bc	17.0 cd
7.	UF-5 stargrass (225957)	*C. aethiopicus* Clayton et Harlan	10.1 a-c	13.2 a	12.3 a	17.3 a	20.6 a
8.	Sarasota stargrass	*C. dactylon* var. *coursii* (Camus) Harlan et de Wet	10.3 ab	13.2 a	11.2 ab	16.1 ab	20.2 a
9.	Callie bermudagrass (290814)	*C. dactylon* var. *aridus* Harlan et de Wet	8.1 c-e	11.0 b-d	10.5 ab	14.3 b-d	17.5 b-d
10.	Alicia bermudagrass	*C. dactylon* var. *elegans* Rendle	11.7 a	12.3 ab	12.3 a	16.1 ab	19.3 ab
11.	Hybrid 6 bermudagrass [Coastal X Ethiopian (225957)]	*C. dactylon* var. *dactylon*-X-*C. aethiopicus*	6.5 ef	9.2 d	6.3 ef	11.7 ef	16.8 d
12.	Hybrid 38 bermuda-grass [Coastal X Kenya 61 (255450)]	*C. dactylon* var. *dactylon*-X-*C. nlemfuensis* Vanderyst var. *nlemfuensis*	9.4 b-d	11.4 a-c	12.1 a	13.4 c-e	17.7 b-d
13.	Hybrid 41 bermuda-grass [Kenya 61 (255450) X Coastal]	*C. nlemfuensis* Vanderyst var. *nlemfuensis*-X-*C. dactylon* var. *dactylon*	10.1 a-c	12.1 ab	9.6 b-d	16.1 ab	18.6 a-d
Digitaria spp.							
14.	Pangola digitgrass	*Digitaria decumbens* Stent.	4.9 f	4.9 f	6.1 ef	6.9 g	13.0 e
15.	Transvala digitgrass (299601)	*D. decumbens* Stent.	5.6 f	7.2 e	4.9 f	10.5 f	14.3 e
Paspalum sp.							
16.	Pensacola bahiagrass	*Paspalum notatum* Flügge	10.1 a-c	9.2 d	7.8 de	9.9 f	10.5 f

*Means within a column followed by the same letter are not significantly different at the 0.05 level of probability according to Duncan's Multiple Range Test.

†Designates Puerto Rico number.

‡Numbers in parenthesis = USDA (PI) numbers.

creased beyond 2 weeks. Puerto Rico stargrass increased in dry-matter yield 95% and 168% when grazing frequency decreased from 2 to 5 weeks and 2 to 7 weeks, respectively.

Species Persistence

The amount of weeds found growing in association with a desirable grass may be a reflection of species competitiveness. However, time required for establishment, adaptation, fertility, environmental factors, management, and allelopathy are all involved in the extent of weed infestation in perennial grasses.

After the study was mob-grazed for three summers, ground cover by weeds in *Digitaria* spp. was 92.5% (Table 2). The *Cynodon* and *Paspalum* spp. averaged 24% and 2% weed ground cover, respectively, after 3 years. It appeared that the *Digitaria* spp. could not successfully com-

pete under grazing when low quantities (1.2%) of common bermudagrass were present prior to grazing (Mislevy 1979). Cattle tend to pull and tear grass plants while grazing. Because Pangolagrass and Transvala have poorly developed root systems, many bare-ground areas remained after grazing, and these were invaded by common bermudagrass. However, two *Cynodon* entries, Cane patch stargrass and Puerto Rico stargrass, averaged only 3% and 6% ground cover by weeds, respectively, after 3 years of grazing. These small averages may be due to an allelopathic effect associated with these entries (Rice 1974) or to their superior competitiveness.

When grasses are screened by mob-grazing, a more severe management is imposed than that imposed by mowing or normal grazing conditions. Therefore, entries surviving this screening technique persist and compete more effectively under commercial grazing conditions. The major disadvantage of the mob-grazing technique is

Table 2. Ground cover occupied by weeds (%) in stands of subtropical grasses after 3 years of grazing management.

Entry	Initial† weed cover	Grazing frequency (weeks)				
		2	3	4	5	7
Cynodon spp.						
1	0.3	8.0 ab*	3.8 ab	16.3 a-c	0.8 a	2.8 a
2	12.0	32.0 c	14.3 a-c	46.0 ef	12.5 a	16.0 a-c
3	0.1	9.3 ab	10.5 ab	28.0 c-e	6.5 a	22.3 a-d
4	0.1	2.8 ab	1.0 a	7.0 ab	0.8 a	2.0 a
5	0.3	10.8 ab	25.0 bc	26.8 c-e	10.8 a	12.0 a-c
6	3.4	16.0 a-c	21.5 a-c	44.0 ef	8.3 a	26.0 b-d
7	0.1	23.5 bc	6.3 ab	39.8 d-f	2.3 a	1.0 a
8	6.2	12.8 a-c	6.5 ab	33.8 c-e	4.0 a	9.3 ab
9	1.3	4.3 ab	5.5 ab	20.5 b-d	2.8 a	16.8 a-c
10	1.0	22.5 a-c	33.8 cd	45.0 ef	21.3 a	31.3 cd
11	9.1	83.3 de	92.0 e	94.0 g	75.5 b	90.0 e
12	6.3	67.5 d	48.3 d	57.5 f	19.3 a	37.8 d
13	15.7	14.3 a-c	9.3 ab	43.8 ef	6.0 a	21.0 a-d
Digitaria spp.						
14	2.0	84.8 de	95.0 e	81.5 g	99.0 c	99.0 e
15	0.3	91.5 e	92.3 e	99.0 g	88.8 bc	93.0 e
Paspalum sp.						
16	0.1	1.5 a	0.3 a	0.0 a	0.5 a	6.0 ab

†After perennial grass plants were established, but before grazing commenced.

*Means within a column followed by the same letter are not significantly different at the 0.05 level of probability according to Duncan's Multiple Range Test.

that cattle may group on a specific grass entry, causing excessive trampling and deposition of feces. The possibility also remains that this severe screening technique could help eliminate a highly palatable entry from further evaluation. Therefore, the researcher must constantly monitor forage-quality data and species persistence to prevent the possible loss of a desirable entry.

LITERATURE CITED

Mislevy, P. 1979. Encroachment of common bermudagrass (*Cynodon dactylon* L.) in subtropical and tropical perennial grasses. *In* Proc. 36th South. Pasture and Forage Crop Impr. Conf., Beltsville, Md.

Mott, G.O., and J.E. Moore. 1970. Forage evaluation techniques in perspective. *In* R.F Barnes (ed., chm.) Proc. Nat. Conf. on Forage Quality Evaluation and Utilization, Nebr. Ctr. for Continuing Educ. Lincoln, Nebr., L1-L10.

Quesenberry, K.H., and W.R. Ocumpaugh. 1979. Persistence, yield and digestibility of limpograss genotypes under clipping and grazing. Agron. Abs. Amer. Soc. Agron., 1979:108.

Quesenberry, K.H., R.L. Smith, S.C. Schank, and W.R. Ocumpaugh. 1977. Tropical grass breeding and early generation testing with grazing animals. *In* Proc. 34th South. Pasture and Forage Crop Impr. Conf., Auburn Univ. Auburn, Al., 100-102.

Rice, E.L. 1974. Allelopathy. Academic Press, New York.

Regeneration of Windmill Grass Pastures Following Severe Grazing in Semiarid Southeastern Australia

D.L. MICHALK and G.E. ROBARDS
Department of Agriculture, New South Wales, Australia

Summary

Stability of windmill grass pastures in west central New South Wales was examined by measuring regeneration of two important grasses (*Enteropogon acicularis* and *Chloris truncata*) on two range sites grazed at three stocking rates (2.5, 3.7, and 4.9 sheep/ha) following 5 months of abusive grazing at 25 sheep/ha.

Severe grazing reduced density and size of mature *E. acicularis* plants on both sites, particularly when heavy grazing (4.9 sheep/ha) was reintroduced during the period of pasture recovery. However, during the study *E. acicularis* density remained relatively stable because seedling cohorts compensated for plant losses in all treatments. Moderately grazed plots were considered the most demographically stable pastures because significantly more large plants (>9 cm) were present.

In contrast, *C. truncata* populations were decimated by abusive grazing, and although they produced enormous cohorts subsequently, few plants survived one season. The combination of small plants and short life span renders *C. truncata* populations more susceptible to sudden catastrophe and therefore unstable.

From a range science point of view, these findings mean that *E. acicularis* should be the focus of management programs, since the maintenance of this species should result in greater stability of pastoral resources.

KEY WORDS: regeneration, perennial grass, severe grazing, semiarid, sheep.

INTRODUCTION

In semiarid rangelands (< 500 mm rainfall/yr) of west central New South Wales, windmill grasses (*Enteropogon acicularis* Lazar. and *Chloris truncata* R. Br.) are important disclimax perennials for sheep-grazing enterprises (Breakwell 1923, Biddiscombe 1953). A significant result of a long-term grazing experiment conducted on windmill grass pastures was the marked stability of *E. acicularis* (the dominant species) under continuous heavy grazing (4.9 sheep/ha) (Robards and Michalk 1978). Although some mature plants disappeared under heavy grazing, new seedlings replaced these losses, maintaining a stable population of *E. acicularis* (Michalk 1980).

A major concern of this observation, however, is that a series of favorable seasons may have produced only temporary demographic stability in a continuing long-term process of grazing-induced pasture degeneration. If *E. acicularis* density declined suddenly (as may occur during protracted drought), would more competitive, but less stable, annual species invade vacated perennial habitats, or would the observed replacement fidelity of *E. acicularis* continue unchanged, maintaining a steady-state pasture of the type described by Williams (1968) for other Australian rangelands?

One definition describes community stability as the "state whereby species densities when perturbed from equilibrium will quickly return to equilibrium when external perturbation is removed" (May 1979). If *E. acicularis* is as stable as the demographic results suggest, then according to May's definition, the species should quickly return to its former demographic state following severe grazing. Our experiment was designed to examine the influence of stocking intensity on the regeneration and stability of windmill grass pastures following abusive grazing with sheep.

MATERIALS AND METHODS

The experiment was conducted on two range sites at the Trangie Agricultural Research Station (lat 31°59′ S long 147°57′ E) in semiarid New South Wales (480 mm rainfall/yr). Site 1, a sandy loam, had about equal proportions of windmill grasses (*E. acicularis* and *C. truncata*), annual grass (*Hordeum leporinum* L. and *Vulpia myuros* [L.] Gmel.), naturalized medics, and forbs, whereas perennial grasses (mainly windmill grasses with some *Digitaria divaricatissima*]R. Br.] Hughes and *Paspalidium gracile* [R. Br.] Hughes) dominated the clay soil of site 2 when the study commenced.

To produce the severe grazing, six 0.8-ha plots on each range site that had previously been grazed for 7 years at 2.5, 3.7, and 4.9 sheep/ha (Robards and Michalk 1978) were abusively grazed at 25 sheep/ha for a summer-autumn growing season (January to May 1975). After a 3-month recovery period, plots were restocked at their

former stocking rates until March 1976, when the study terminated.

Number, diameter, and basal area of *E. acicularis* and *C. truncata* plants were carefully mapped in 12-m² quadrats/plot before severe grazing was initiated in January 1975. Maps were checked in April 1975 and March 1976, and the number of fatalities and additions were recorded for both grasses. Plant size distributions were established for each treatment by grouping plants into categories of <3, 4-6, 7-9, and >9 cm, based on diameter.

RESULTS AND DISCUSSION

E. acicularis

Throughout the study, site 2 had about 3 times more *E. acicularis* plants than site 1 at equivalent stocking rates (Table 1). On site 2, heavily grazed plots consistently had 20% fewer plants than moderately or lightly grazed plots, but on site 1 grazing intensity had little effect (Table 1). The preferred grazing by sheep of *H. leporinum* and naturalized medics that dominated site 1 between autumn and spring probably relieved grazing presure on *E. acicularis* at all stocking rates.

Some *E. acicularis* plants that appeared to be dead in 1975 following the severe grazing were found to be alive in 1976. This phenomenon seems to have been a combination of the severe grazing, which removed all above-ground plant material, and a summer-autumn drought in 1975 that triggered drought dormancy mechanisms (Whalley and Davidson 1969).

This apparent revival does not mean, however, that severe grazing and subsequent grazing intensity had no effect on *E. acicularis* density. On site 2, for example,

Table 1. Survival and establishment of *E. acicularis* and *C. truncata* plants mapped on sandy loam (site 1) and clay soil (site 2) between 1974 and 1976 (plants/m²).

Site and Year	*E. acicularis* 2.5	3.7	4.9[+]	*C. truncata* 2.5	3.7	4.9[+]
Site 1						
1974	3.5a	4.5a	4.9a	4.7b	10.5a	6.5b
1975	2.9a	4.2a	4.5a	0.6a	1.1a	2.2a
1976	1.9a	3.0a	3.0a	0.5a	0.4a	0.4a
	ns	**	ns	**	**	**
1976*	0.6b	1.0b	2.6a	11.2b	11.0b	19.0a
Site 2						
1974	15.0a	13.7ab	12.1b	11.3b	12.6a	13.1a
1975	13.7a	13.0a	10.6b	3.5b	5.9a	6.2a
1976	9.0a	9.2a	5.2b	0.3a	0.4a	1.5a
	**	**	**	**	**	**
1976*	4.5b	4.2b	5.6a	45.0b	40.0b	92.0a

[+] Stocking rate, sheep/ha.

*New plants recorded in 1976.

**and ns, P<0.01 and nonsignificant effects of time within stocking rates (1974 to 1976, original plants only)

a,b,c - Within grass species and sites, means followed by the same letter are not significantly different (P<0.05) when grazed at different stocking rates.

D.L. Michalk is a Research Agronomist and G.E. Robards is the Director of Animal Production Research in the New South Wales Department of Agriculture.

about 60% of plants mapped in 1974 did not survive until 1976 in heavily grazed plots, while about 40% of plants were lost from moderately and lightly grazed plots (Table 1).

Delayed effects of reduced carbohydrate levels may explain these losses. Under normal seasonal conditions, root reserves expended in spring and summer growth are probably replenished by *E. acicularis* in autumn when sheep shift their preference to annuals germinating in autumn and maturing in spring. In this study, however, carbohydrate reserves were below optimum when severe grazing concluded because drought limited germination of annuals, and sheep depended largely on perennial forage the following spring, preventing the reserve replenishment that would normally occur. This explanation is supported by changes in distributions of plant size measured between 1974 and 1976 for *E. acicularis*. While size distribution patterns changed little at lower stocking rates, significantly more small plants (<3 cm) were found in heavily grazed plots on both sites in 1976. Part of this shift was due to seedling recruitment (Table 1), but since regeneration was also observed at moderate and light stocking rates, it is apparent that grazing during recovery at 4.9 sheep/ha continued the decline in plant size (and, by inference, carbohydrate reserves; Biddiscombe 1953) initiated by abusive grazing. Since small plants are more vulnerable to damage by grazing and trampling, both sites became less stable and thereby more susceptible to retrogression at 4.9 sheep/ha than when they were moderately or lightly grazed.

Although *E. acicularis* density and size of original plants declined during the period after abusive grazing, new cohorts compensated for losses incurred in all treatments. A significant result was greater seedling establishment on heavily grazed plots on both sites (Table 1). On site 1, results reflect the lower proportion at 4.9 sheep/ha of *H. leporinum* and naturalized medics, species shown to be fierce competitors of *E. acicularis* (Michalk 1980). The smaller cohorts at lower grazing intensities on site 2, however, were possibly influenced by the density of mature plants, which exceeded the saturation density of about 10 plants/m² delineated for *E. acicularis* (Michalk 1980).

C. truncata

Like that of *E. acicularis*, *C. truncata* density was greater on site 2 than on site 1 (Table 1). However, in contrast to *E. acicularis*, more *C. truncata* plants were concentrated in heavily grazed pastures, a result confirming Biddiscombe's (1953) assertion that *C. truncata* is indicative of grazing-induced retrogression.

Unlike *E. acicularis*, populations of *C. truncata* plummeted and rebounded throughout the experiment (Table 1). Of the plants mapped in 1974, only 20% to 50% survived severe grazing, and few of these plants were present in 1976. Although vulnerable to abusive grazing, *C. truncata* responded to favorable seasonal conditions with large seedling cohorts being produced on each treatment in 1976.

The erratic behavior of *C. truncata* populations is consistent with the autecology of the species. Unlike *E. acicularis*, which can endure abusive grazing and drought by virtue of a robust root system, large protective crowns, and dormancy mechanisms (Whalley and Davidson 1969), *C. truncata* grows rapidly during favorable seasons and produces copious quantities of fertile seed as a means to perpetuate its kind (Breakwell 1923). In ecological terms, however, the combination of small plants and a short life span make *C. truncata* populations susceptible to sudden catastrophic change and therefore quite unstable in the range sense, because it leaves the ground surface unprotected during times of stress when erosion potential is greatest.

CONCLUSION

Based on Hurd et al.'s (1971) observation that "the magnitude of the response to perturbation is an inverse measure of stability," it is concluded from this study that *E. acicularis* is a more stable species than *C. truncata*. The responses suggest that *E. acicularis* acts as a primary grass or K-species, whereas *C. truncata* responds as a secondary or R-species.

A practical corollary to this conclusion is that the status of the *E. acicularis* population (that is, density, age, structure, and plant size) should be used to design management programs. For example, following periods of severe grazing, as may occur during protracted drought, moderate grazing (3.7 sheep/ha) should be imposed on regenerating windmill grass pastures at Trangie. At this stocking intensity, competition from annuals is minimized by selective grazing, but at the same time grazing pressure is not high enough to inflict undue stress on weakened and establishing *E. acicularis* plants.

LITERATURE CITED

Biddiscombe, E.F. 1953. A survey of natural pastures of the Trangie district, New South Wales, with particular reference to the grazing factor. Austral. J. Agric. Res. 4:1-8.

Breakwell, E. 1923. The Grass and fodder plants of New South Wales. Govt. Print. Sydney, Austral.

Hurd, L.E., M.V. Mellinger, L.L. Wolf, and S.J. McNaughton. 1971. Stability and diversity on three trophic levels in terrestrial successional ecosystems. Science 173:1134-1136.

May, R.H. 1979. The structure and dynamics of ecological communities. *In* R.M. Anderson and B.R. Turner (eds.) Popular dynamics. Oxford, Eng.

Michalk, D.L. 1980. Natural pastures of the Macquarie Region of New South Wales: their origin, composition and management. Ph.D. Dissertation, Utah State Univ.

Robards, G.E., and D.L. Michalk. 1978. Natural and sown dryland pastures for sheep production in central western New South Wales. N.S.W. Dept. Agric., Tech. Bul. 20.

Whalley, R.D.B., and A.A. Davidson. 1969. Drought dormancy in *Astrebla lappacea, Chloris acicularis* and *Stipa aristiglumis*. Austral. J. Agric. Res. 20:1035-1042.

Williams, O.B. 1968. The uneasy state between animal and plant in the manipulated situation. Proc. Ecol. Soc. Austral. 3:167-174.

A Mathematically and Conceptually Unified Approach to Grazing-Management Terminology

D.L. SCARNECCHIA and M.M. KOTHMANN

Texas A&M University, College Station, Tex., U.S.A.

Summary

Expressions such as "stocking density," "stocking rate," "grazing pressure," "herbage allowance," "grazing intensity," "stocking intensity," and "stocking pressure" have long been used to describe animal-pasture systems. In general, these expressions describe relationships among the basic variables of animal demand, forage quantity, pasture area, and grazing duration. Our objective was to develop a conceptually dynamic mathematical framework of expressions summarizing all of the meaningful relationships among these variables. Elementary differential and integral calculus were used to develop this dynamic approach.

The resulting expressions are dimensionally valid and are capable of accurately describing dynamic animal-pasture systems. Adoption of this system would standardize terms and eliminate dimensionally invalid expressions found in the literature.

KEY WORDS: grazing terminology, animal unit, animal-unit equivalent, stocking rate, grazing pressure, grazing modeling.

INTRODUCTION

Efforts to describe basic quantitative relationships between grazing animals and pasture have produced numerous expressions, including "stocking density," "stocking rate," "grazing pressure," "herbage allowance," "grazing intensity," "stocking intensity," "stocking pressure," and others (Booysen 1967, Society for Range Management 1974, Hodgson 1979).

In general, four variables are involved in describing animal-pasture relationships. These variables are the pasture area (A), usually expressed in ha, the forage dry matter (DM), usually expressed in kg, the duration of grazing (t), usually expressed in days or months, and the forage demand rate (D), here expressed in animal units (AU). The variable that is functionally important in describing the potential effect of the animal component on a pasture is neither animal live weight nor animal numbers alone, but rather the animals' rate of demand for forage, which might be termed the potential rate of forage intake. This potential rate of intake is a function of animal numbers, weight, physiological state, and other animal-related factors.

THE ANIMAL UNIT AND THE ANIMAL-UNIT EQUIVALENT

If we are to use the animals' demand rate for forage in describing animal-pasture systems, we must have an appropriate unit. The Society for Range Management

(1974) has defined one animal unit to be "one mature 454-kg cow or its equivalent based on an average forage consumption of 12 kg of dry matter per day." Here, we simply define an animal unit as a unit of animal demand rate and as representing an animal with a rate of demand for forage equal to 12 kg DM/day. Any animal may be represented as a certain fraction or multiple of the animal unit, based solely on its quantitative demand rate for forage. An animal that has a demand rate more or less than 12 kg DM/day will have an animal-unit equivalent (AUE) that is a proportionate fraction or multiple of one animal unit. The usefulness of the animal-unit concept is obvious when more than one class or species of animal is grazing a pasture, since all animals present can be expressed in common terms with respect to their potential function in the animal-pasture system.

In the past, the assigning of animal-unit equivalents to animals has usually been very approximate, and has often been based on live-weight (LW) or $LW^{0.75}$. If the approach discussed in this paper is to be of maximum use, accurate estimates of animal-unit equivalents are important. Animal-unit equivalents might be estimated using published dry-matter requirements for each particular kind, class, and weight of livestock like those published by the U.S. National Research Council (NRC 1976). Such published data might serve as a starting point from which further refinement, based on other specific animal characteristics, could be attempted. The use of controlled feeding experiments using appropriately chosen preferred forage would be useful in providing further details concerning potential dry-matter demand. Detailed tables and mathematical functions could be developed to obtain animal-unit equivalents quickly, based on reasonable correlates in animal characteristics. In any case, the

D.L. Scarnecchia is a Research Assistant in Range Science and M.M. Kothmann is a Professor of Range Science in the Department of Range Science of Texas A&M University.

researcher should describe in detail how animal-unit equivalents were obtained.

Within any infinitely small time interval, an animal at pasture represents a rate of forage demand on the pasture. The integration of this demand rate over a time interval t represents an amount of forage demanded:

$$\text{Amount of forage demand} = \int_{t_0}^{t} \text{Demand rate} \times dt \quad (1)$$

In practice, the summation form of equation 1 would be used:

$$\text{Amount of forage demand} = \sum_{i=1}^{n} (\text{Mean demand rate})_i \times t; \quad (2)$$

where each (mean demand rate)$_i$ is the demand rate assigned to subperiod t_i. If a mean demand rate is assumed over the period t,

$$\text{Amount of forage demand} = \text{Demand rate} \times t. \quad (3)$$

The most common units of this quantity (the amount of forage demand) are the animal-unit day (AUD), the animal-unit month (AUM), and the animal-unit year (AUY). Each of these units represents a specific amount of forage, i.e., 1 AUD = 12 kg DM, 1 AUM = 360 kg DM, and 1 AUY = 4,380 kg DM.

STOCKING DENSITY AND STOCKING RATE

The Society for Range Management (1974) defined stocking density as the relationship between the number of animal units and the land area at any instant in time. The typical unit is AU/ha.

Booysen (1967) and the Society for Range Management (1974) defined the stocking rate as the number of animal units/unit of land over a period of time. Mathematically, for a grazing period (t − t_0),

$$\text{Stocking rate} = \int_{t_0}^{t} \text{Stocking density} \times dt \quad (4)$$

In its discrete form, equation 4 becomes

$$\text{Stocking rate} = \sum_{i=1}^{n} (\text{Mean stocking density})_i \times t_i \quad (5)$$

where each (mean stocking density)$_i$ is the mean stocking density of each particular subperiod t_i making up the grazing period t. If a mean stocking density is assumed over the grazing period t = Σt_i,

$$\text{Stocking rate} = \text{Mean stocking density} \times t \quad (6)$$

The units of stocking rate must be in terms of an amount of forage demand/unit area, typically AUD/ha, AUM/ha, or AUY/ha. The annual stocking rate within a pasture is the sum of the stocking rates of individual grazing periods within a year, i.e.,

$$\text{Annual pasture stocking rate} = \sum_{j=1}^{m} (\text{Stocking rate})_j \quad (7)$$

where m is the number of grazing periods within a year. An annual pasture stocking rate can have units of AUD/ha, AUM/ha, or AUY/ha.

Since complex grazing systems are becoming increasingly common, it is also useful to express the stocking rate of a number of pastures managed under a grazing system. The stocking rate for an entire grazing system can be expressed as

$$\text{System stocking rate} = \frac{\int_{t_0}^{t} (\text{Animal demand rate}) \times dt}{\text{Total system area}} \quad (8)$$

or in summation form as

$$\text{System stocking rate} = \frac{\sum_{i=1}^{p} (\text{Mean animal demand rate})_i \times t_i}{\text{Total system area}} \quad (9)$$

where p is the number of pastures in the system and each (mean annual demand rate)$_i$ is the mean number of animal units within each calculation period i. The system stocking rate is usually of interest on an annual basis, and, for clarity, the term "annual system stocking rate" should be used when the annual rate is being described. When system stocking rates are calculated, for example, monthly, then

$$\text{Annual system stocking rate} = \sum_{i=1}^{n} \text{Monthly system stocking rates} \quad (10)$$

Typical units for both the system stocking rate and the annual system stocking rate are also AUD/ha, AUM/ha, and AUY/ha.

GRAZING PRESSURE AND GRAZING-PRESSURE INDEX

The Society for Range Management (1974) defines grazing pressure as the animal:forage ratio at a given instant. Grazing pressure, like stocking density, is defined at any instant, and thus its units have no time dimension. Typical units are AU/kg and AU/ton.

The integration over time of grazing pressure yields a meaningful, though previously undescribed, expression, the grazing-pressure index. Thus, mathematically,

$$\text{Grazing-pressure index} = \int_{t_0}^{t} \text{Grazing pressure} \times dt \quad (11)$$

In its summation form, equation 11 can be represented as

$$\text{Grazing-pressure index} = \sum_{i=1}^{n} (\text{Mean grazing pressure})_i \times t_i \quad (12)$$

If a mean grazing pressure is assumed over the grazing period $t = \Sigma t_i$,

Grazing-pressure index =
$$\text{Mean grazing pressure} \times t. \quad (13)$$

The fact that the expression we have termed the grazing-pressure index is in fact an index will be obvious if the reader recalls that the product of a demand rate and a time period is equivalent to an amount of forage. Values of the grazing-pressure index can range from zero to infinity, but most actual grazing situations will have grazing-pressure indices between zero and one. This index is the ratio of integrated animal demand : forage supply over a time period.

HERBAGE ALLOWANCE AND CUMULATIVE HERBAGE ALLOWANCE

While the grazing pressure expresses the ratio of animal demand : forage weight at any instant, the inverse of this ratio, the herbage allowance, expresses the ratio of weight of forage available : amount of animal demand. Thus, the single animal demand : forage relationship in a pasture at any time may be expressed either as a grazing pressure or as an herbage allowance. A typical unit of herbage allowance would be tons/AU.

Just as the integration over time of grazing pressure yields a grazing-pressure index, the integration of herbage allowance over time yields an analogous expression, the cumulative herbage allowance. Mathematically,

Cumulative herbage allowance =
$$\int_{t_o}^{t} \text{Herbage allowance} \times dt \quad (14)$$

The summation form of Equation 14 that would be used in practice is

Cumulative herbage allowance =
$$\sum_{i=1}^{n} (\text{Mean herbage allowance})_i \times t_i \quad (15)$$

If a mean herbage allowance is assumed over the time period t ($= \Sigma t_i$), equation 15 can be simplified to

Cumulative herbage allowance =
$$\text{Mean herbage allowance} \times t \quad (16)$$

The cumulative herbage allowance expresses the relationship between forage supply and animal demand over a time period and has typical units of kg/day/AU or ton/month/AU.

TWO OTHER USEFUL RELATIONSHIPS

If the stocking density is differentiated with respect to time, the result is an expression describing the rate of change in the stocking density. This relationship can be expressed as

Rate of change in stocking density (RCSD) =
$$\frac{d(\text{Stocking density})}{dt} \quad (17)$$

or in discrete form

$$\text{RCSD} = \frac{\Delta \text{ Stocking density}}{\Delta t} \quad (18)$$

The RCSD would have units of, for example, AU/ha/hour or AU/ha/day. This change in the number of animal units could be the result of (1) the addition or removal of livestock by the grazing manager or (2) a change in the animal-unit equivalents of the animals already present, corresponding to a change in the potential dry-matter intake/animal.

An expression analogous to the RCSD is obtained if grazing pressure is differentiated with respect to time. This differentiation is represented mathematically as

Rate of change in grazing pressure (RCGP) =
$$\frac{d(\text{Grazing pressure})}{dt} \quad (19)$$

The summation form of equation 19 is

$$\text{RCGP} = \frac{\Delta \text{ Grazing pressure}}{\Delta t} \quad (20)$$

and typical units for this expression would be AU/kg/day or AU/ton/day. This rate of change can result either from changes in the number of animal units in a pasture or from a change in the amount of forage available. Two analogous expressions exist describing the rate of change in herbage allowance (RCHA) with respect to time.

FORAGE-AREA RELATIONSHIPS

In order to describe the behavior of animal-pasture systems accurately, it is necessary to have expressions that describe the dynamic nature of the forage component. The useful expressions describing the forage-area relationships are shown in Table 1, which summarizes the meaningful relationships among the variables of animal demand, forage, area, and time.

SUMMARY OF THE EXPRESSIONS DESCRIBING ANIMAL-PASTURE SYSTEMS

The description of dynamic animal-pasture systems requires the use of dynamic variables expressions. The animal demand variable, with units of animal units, is suitable for describing these systems. The conceptually dynamic framework of expressions offered here adequately describes the relationships among the variables of animal forage demand, forage quantity, area, and duration of grazing, and should prove useful in the refinement of future grazing research and management.

Table 1. Summary of relationships among the variables of animal demand, forage quantity, pasture area and time.

ANIMAL/AREA	*ANIMAL/FORAGE*	*FORAGE/ANIMAL*	*FORAGE/AREA*
STOCKING RATE $= \int_{t_0}^{t}$ stocking density $\cdot$ dt Units: aud/ha aum/ha auy/ha	GRAZING PRESSURE INDEX $= \int_{t_0}^{t}$ graz. pressure $\cdot$ dt Units: Ratio of animal demand to forage over a period of time	CUMULATIVE HERBAGE ALLOWANCE $= \int_{t_0}^{t}$ herb. allowance $\cdot$ dt Units: kg $\cdot$ days/au tons $\cdot$ months/au	
STOCKING DENSITY = Animal demand per unit area at any instant Units: au/ha	GRAZING PRESSURE = animal demand per unit weight of forage at any instant Units: au/kg au/ton	HERBAGE ALLOWANCE = weight of forage per unit animal demand at any instant Units: kg/au tons/au	STANDING CROP = weight of forage standing per unit area at any instant Units: kg/ha g/m²
RATE OF CHANGE IN STOCKING DENSITY = $\dfrac{d(\text{stocking density})}{dt}$ Units: au/ha/day au/ha/hr	RATE OF CHANGE IN GRAZING PRESSURE = $\dfrac{d(\text{grazing pressure})}{dt}$ Units: au/kg/day au/kg/hr Etc.	RATE OF CHANGE IN HERBAGE ALLOWANCE = $\dfrac{d(\text{herbage allowance})}{dt}$ Units: kg/au/day kg/au/hr Etc.	NET FORAGE ACCUMULATION RATE = $\dfrac{d(\text{standing crop})}{dt}$ Units: kg/ha/day kg/ha/hr Etc.
			RATE OF CHANGE IN NET FORAGE ACCUMULATION RATE = $\dfrac{d(\text{net forage acc. rate})}{dt}$ Units: kg/ha/hr/hr kg/ha/day/day

1 au = 1 animal-unit = 12 kg forage/day in animal demand

LITERATURE CITED

Booysen, P. DeV. 1967. Grazing and grazing management terminology in South Africa. Proc. Grassl. Soc. So. Afr. 2:45–57.

Hodgson, J. 1979. Nomenclature and definitions in grazing studies. Grass and Forage Sci. 34:11–18.

U.S. National Research Council. National Academy of Sciences. Subcommittee on Beef Cattle Nutrition (NRC). 1976. Nutrient requirements of beef cattle, 5th ed. Natl. Acad. Sci., Washington, D.C.

Society for Range Management. 1974. A glossary of terms used in range management, 2nd ed. Soc. Rge. Mgmt., Denver, Colo.

Influence of Sward Structure upon Herbage Intake of Cattle Grazing a Perennial Ryegrass Sward

M.H. WADE and Y.L.P. LE DU

INTA, Balcarce, Argentina, and Grassland Research Institute, Hurley, Berkshire, England, UK

Summary

To examine the influence of structural distribution of herbage within a sward on intake by grazing animals, two swards of contrasting structure were established at two periods in the grazing season by either infrequent mowing (C) or hard sheep grazing (G) of a perennial ryegrass (*Lolium perenne*) pasture. The plots were then allowed to grow undisturbed for about 5 weeks before being grazed by young cattle for periods of 2 weeks at two levels of herbage allowance, 30 and 60 g dry matter/kg live weight/day, in July (1) and in September (2). Detailed measurements of sward structure were made by stratified cutting of the pasture. The herbage intake of the cattle was measured both by the chromic oxide dilution technique and by before-and-after clipping. In period 1, sward C was apparently more accessible than G (50% vs. 30% of herbage mass above 5 cm), but in period 2 this accessibility was reversed (G, 60% vs. C, 30%). In spite of this reversal, herbage intakes were 10% greater in the C than the G treatment in both periods, the difference being greater at the low than at the high allowance (16% vs. 5%). It was concluded that the presence of sheep excreta from the pretreatment had a greater effect on the herbage intake of the calves than did the spatial distribution of the swards.

KEY WORDS: cattle, grazing, ryegrass, intake, sward structure.

INTRODUCTION

The concept of herbage allowance may be deficient if the herbage offered is described only in terms of herbage mass/unit weight of animal. Although Baker (1978) defined an exponential relationship between herbage intake and allowance, both expressed relative to animal weight, there remained considerable variation to be explained. This variation could be due to a number of factors that may affect herbage intake as a sward is grazed down (Jamieson and Hodgson 1979). To understand the harvesting process, it is necessary to know what these factors are and what proportion of herbage is really available to the grazing animal. Chacon and Stobbs (1976) have shown the importance of spatial distribution and ease of prehension of tropical swards, but no effects are reported for cattle grazing temperate swards. The work described in this paper was undertaken to examine the response in herbage intake by calves to swards of differing spatial distribution offered at two herbage allowances.

METHODS

A well-established perennial ryegrass pasture (*Lolium perenne* cv. S23), fertilized with 1.5 kg/ha/day of N, was cut (C—interval of about 8 weeks) or grazed by sheep (G—continuous, variable stocking rate) from mid-March to early June and from June to August. The grass in each area was then allowed to grow undisturbed for about 5 weeks before being grazed during 2 weeks in July and 2 weeks in September by 6–8-month-old calves (mean live weight [LW] 138 and 170 kg, respectively). Twelve cattle grazed at herbage allowances of 30 (L) or 60 (H) g dry matter (DM)/kg animal LW/day, in a strip-grazing system. The following measurements were made before (IN) and after (OUT) each day's grazing: (1) herbage mass; two quadrats, 200 × 20 cm clipped to ground level with electrically powered sheep shears; (2) pasture spatial distribution; six strata > 20, 20–15, 15–10, 10–5, 5–3, and < 3 cm above ground level, clipped in a 25 × 25-cm variable-height quadrat using shears; samples from alternate pairs of days were either separated into live and dead fractions or analyzed for organic-matter disgestibility (OMD) and N content in each stratum; and (3) ryegrass tillers within a 16-cm-diameter frame were counted after herbage was clipped to ground level.

The herbage intake of the grazing cattle was calculated from the difference between the estimates of herbage mass made before and after grazing. It was also derived from estimates of fecal output obtained using the chromic oxide dilution technique and assessments of the in-vitro digestibility (Tilley and Terry 1963) of samples of herbage plucked by hand to simulate the material being grazed.

RESULTS

Sward Characteristics

Differences between the swards produced by the two pretreatments are shown in Fig. 1. Differences in the spatial distribution may be compared more simply by expressing them in terms of the proportion of herbage situated above 5 cm and the bulk density of that herbage. In 1C and 1G these were 60% vs. 30% and 127 vs. 64 kg/cm (height)/ha, respectively. In 2C and 2G they were 30% vs. 50% and 74 vs. 117 kg/cm/ha, respectively. Changes in N concentration appeared to follow changes in the proportion of live material more closely than did OMD. Within each period N concentration was greater in G than in C swards.

Herbage Intake

Significant effects of both sward treatment and herbage allowance on intake were demonstrated by the chromic oxide technique (Table 1). Similar though nonsignificant trends were detected by clipping, although clipping gave a 16% lower overall estimate of intake. Animals grazing sward G consumed 8% and 12% less herbage than those grazing C in July and September, respectively (means of clipping and chromic oxide techniques). These differences were greater at the low (17% and 15%) than at the high (0% and 10%) allowance. The intakes at the low allowance were 21% less than those at the high allowance.

DISCUSSION

Sward Characteristics

In period 1, two swards of similar weight but different vertical distribution of herbage were produced, with more herbage nearer the ground following grazing by sheep (Fig. 1). However, in period 2, poor growth on treatment C resulted in a reversal of herbage distribution. The

Fig. 1. Histograms showing the spatial distribution, herbage mass, and tiller number of ryegrass swards in July (1) and September (2), before (IN) and after (OUT) grazing by calves at high and low allowance (H 60 and L 30 g DM/kg LW/day, respectively). Live and dead herbage bulk density, N concentration, and OMD are given by vertical strata.

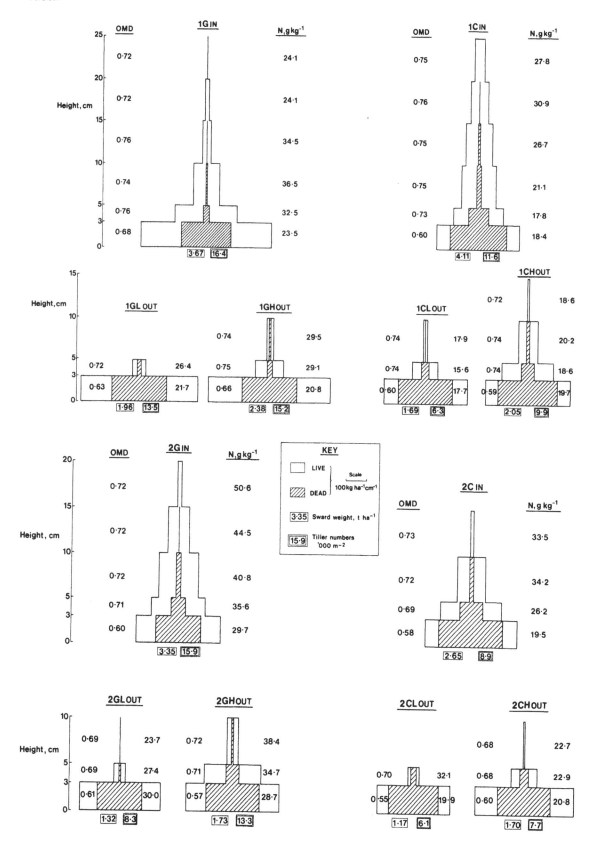

Table 1. Daily herbage organic-matter intake, estimated by clipping before and after grazing or by chromic oxide-digestibility techniques.

Period	Allowance	Clipped		Chromic oxide		Clipped, mean	Chromic oxide, mean
		Previously cut	Previously grazed	Previously cut	Previously grazed		
	g DM/kg LW	-- intake, g/kg LW --					
1 (July)	30	17.3	14.3	22.0	18.0	15.8 NS	20.0 ***
	60	23.0	21.9	23.7	24.9	22.5	24.3
	Mean	20.3 NS	18.1	22.8 *	21.4		
2 (September)	30	17.3	15.5	23.2	18.9	16.4 *	21.0 ***
	60	21.4	22.1	27.1	21.6	21.8	24.3
	Mean	19.3 NS	18.8	25.1 ***	20.2		

NS, *, **, or *** between two means indicates means are not significantly different or are significantly different at the 10, 5, or 1% level of probability, respectively.

higher N content of the G than of the C sward in both periods can probably be attributed to the recycling of N through the grazing sheep. Tiller number was also higher in the G swards due to the previous management.

Herbage Intake and Sward Characteristics

At the lower allowance, intakes of the cattle grazing swards previously cut were 16% higher than intakes of the cattle grazing previously grazed swards. In period 1 the results were consistent with the greater accessibility of herbage in the C swards. However, in period 2 the higher intake was found in the least accessible swards, suggesting that accessibility may not be important under these conditions. No other measured factor satisfactorily explains the observed differences in intakes; indeed the swards in the C treatments contained more dead material and were of a lower OMD below 5 cm than those of G (Fig. 1), characteristics that should tend to restrict intake. After grazing, slight visible signs of rejection were observed on the G swards close to sheep dung. It seems likely that the

presence of sheep excreta was a major factor restricting the herbage intake of cattle on the G treatment.

In future work the confounding effect of all animal excreta must be excluded or else be quantified and evaluated relative to other factors.

LITERATURE CITED

Baker, R.D. 1978. Beef cattle at grass: intake and production. *In* Grazing: sward production and livestock output, 2.1–2.7. Proc. Brit. Grassl. Soc. Winter Mtg., London, 1978.

Chacon, E., and T.H. Stobbs. 1976. Influence of progressive defoliation of a grass sward on the eating behaviour of cattle. Austral. J. Agric. Res. 27:709–772.

Jamieson, W.S., and J. Hodgson. 1979. The effect of daily herbage allowance and sward characteristics upon the ingestive behaviour and herbage intake of calves under strip-grazing management. Grass Forage Sci. 34:261–271.

Tilley, J.M., and R.A. Terry. 1963. A two-stage technique for the *in vitro* digestion of forage crops. J. Brit. Grassl. Soc. 18:104–110.

Forage Analysis with Near-Infrared Reflectance Spectroscopy: Status and Outline of National Research Project

W.C. TEMPLETON, JR., J.S. SHENK, K.H. NORRIS, G.W. FISSEL,
G.C. MARTEN, J.H. ELGIN, JR., and M.O. WESTERHAUS
USDA-SEA-AR Regional Pasture Research Laboratory and
Pennsylvania State University, University Park, Pa., U.S.A.

Summary

Objectives of this paper are to review the status of near-infrared reflectance spectroscopy (NIRS) in forage-quality evaluation and to indicate the approach being used in a national research effort to further test, validate, and develop the technology for use in forage breeding, production, and utilization programs. Use of NIRS assay of forage quality was first reported 5 years ago. Research has now been initiated at many locations.

A national research project, involving personnel at six widely dispersed locations in the U.S., was established in 1978. Computerized, high-precision near-infrared reflectance spectrophotometers are employed. A set of 30 forage samples provided experimental materials to ascertain magnitudes of errors involved in chemical, in-vitro digestibility, and NIRS analyses at the six laboratories.

The six instruments showed different average spectral curves, but normalization of the data resulted in curves essentially alike except for the bands affected by water, an effect caused by differences in moisture content of the samples at the different sites. Our studies indicate that the magnitude of the errors associated with NIRS analysis compares favorably with those found in routine chemical and in-vitro procedures.

Examples of research findings involving different plant species, harvests, growth environments, and drying procedures are presented. The results show that for maximum accuracy the calibration samples should be representative of the forages for which quality will be predicted by NIRS.

KEY WORDS: forage quality, chemical analysis, crude protein, neutral detergent fiber, acid detergent fiber, acid detergent lignin, in-vitro dry-matter disappearance.

BACKGROUND

The findings of Norris et al. (1976) generated wide interest in use of NIRS as a rapid, nonconsumptive technique for assessing forage quality. The technique involves application of radiant energy in the near-infrared portion of the spectrum (1,100 to 2,500 nm) to dry, ground samples of forage and the detection of energy reflected from the sample, relative to the instrument response. The amount of energy reflected from the sample depends on its chemical/physical properties. Finally, empirical relationships between the reflected energy (R), expressed as log (1/R), and quality parameters are determined statistically.

Some researchers have used instruments employing filters to obtain forage reflectance data at 3 to 19 specific wavelengths (Barton and Burdick 1978, 1979, 1980). Shenk and Barnes (1977) used a monochromator to simulate data acquisition by a filter instrument, employing the six wavelengths proposed for forages by Norris et al. (1976). It was inferred from the magnitude of the errors when the six specific wavelengths were used that filter instruments would be useful for some, but not all, applications.

NATIONAL RESEARCH PROJECT

By 1978 there was evidence that NIRS could provide a fast, relatively accurate method of assessing certain nutritive properties of forages (Shenk and Hoover 1976, Shenk and Barnes 1977, Shenk et al. 1978a, 1978b). In March 1978, research was initiated on a national basis, with cooperating laboratories at University Park, Pennsylvania;

Athens, Georgia; Beltsville, Maryland; St. Paul, Minnesota; El Reno, Oklahoma; and Logan, Utah.

Objectives of the project are:

- to develop and test computer programs that provide continuing advances in data processing and mathematical treatment of near-infrared data to maximize prediction accuracy;
- to further define and measure plant, environmental, and other factors affecting determination of chemical composition and animal response;
- to identify chemical and physical characteristics of forages that determine their spectral properties;
- to test the usefulness of NIRS in forage-breeding, forage-management, and animal-utilization research; and
- to establish a reference forage-sample library for use in NIRS calibration and other forage-evaluation studies.

Researchers at all locations are using a high-precision near-infrared reflectance spectrophotometer (Neotec Model 6100) for reflectance measurements in the 1,100 to 2,500 nm region and a Digital Equipment Corporation PDP-11 computer system. Such instruments offer the advantages of utilizing the full spectrum, providing more flexibility, and increasing computer capabilities.

A set of 30 forages, consisting of both cool-season and warm-season species, was used to compare chemical and spectral assays within and among locations. Each laboratory collected reflectance data for each of the samples on 3 consecutive days.

Average spectral curves from the six instruments were noticeably different in reflectance level and in amplitude of absorption bands, but a simple normalization of the data resulted in curves essentially alike except for the water regions (Fig. 1). Differences in the normalized

W.C. Templeton, Jr., is at the USDA-SEA-AR Regional Pasture Research Laboratory, University Park; J.S. Shenk is in the Department of Agronomy of Pennsylvania State University; K.H. Norris is at the USDA-SEA-AR Beltsville Agricultural Research Center, Beltsville, Md.; G.W. Fissel is at the USDA-SEA-AR Regional Pasture Research Laboratory, University Park; G.C. Marten is at the USDA-SEA-AR at the University of Minnesota, Saint Paul; J.H. Elgin, Jr., is at the USDA-SEA-AR Beltsville Agricultural Research Center, Beltsville, Md.; and M.O. Westerhaus is in the Department of Statistics of Pennsylvania State University. This paper is contribution no. 8017 of the USDA-SEA-AR Regional Pasture Research Laboratory, University Park.

The authors would like to express their appreciation to .M.J. Anderson, F.E. Barton II, S.W. Coleman, J.L. Halgerson, M.R. Hoover, F.P. Horn, R.J. Kerns, W.G. Lynch, J.B. Powell, and W.R. Windham for data and/or assistance in the national project.

Mention of a proprietary product implies no endorsement of it by the authors or by their agencies or institutions.

Fig. 1. Near-infrared reflectance spectral curves obtained with 30 forage samples, with normalization of data to insure indentical values at 1,300 and 2,100 nm.

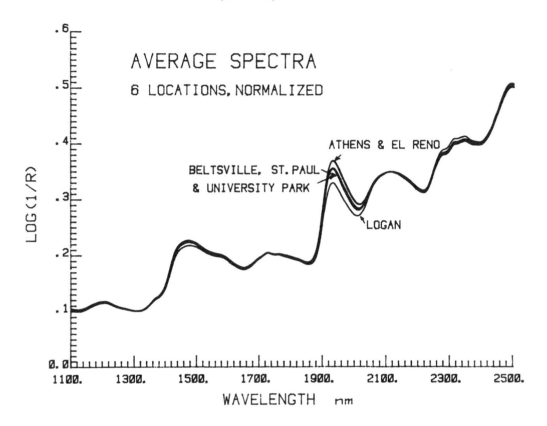

curves, especially noticeable in the 1,940 nm region, were caused primarily by differences in sample moisture at the six locations.

It appears that the errors associated with NIRS analysis compare favorably with those obtained by routine chemical procedures (Table 1). Moreover, the between-lab variation was no greater for NIRS than for chemical assay.

Scientists at University Park conducted a study to determine the effect of plant species, harvests, and drying methods on accuracy of prediction equations (Table 2). The bottom line of the table shows standard errors of prediction when values from half of the samples (n = 105)

were used to predict the remaining half. The other lines indicate magnitudes of the errors when a predictive equation, developed for a particular species, drying temperature, or harvest, was used to predict samples different from those used in the calibration set. The results show that when the calibration and prediction samples are similar, predictive accuracy improves.

At Saint Paul, 700 forage samples from two cultivars each of oats (*Avena sativa* L.), barley (*Hordeum vulgare* L.), spring wheat (*Triticum aestivum* L.), and triticale (X *Triticosecale,* Wittmack), harvested at six growth stages at two locations during 2 years, were evaluated. Prediction equations were developed from forages of one replicate

Table 1.   Standard errors (%) associated with analyses of five quality parameters of 30 forage samples at six laboratories.*

Methods of analysis and effects	Quality parameters				
	Crude protein	Neutral detergent fiber	Acid detergent fiber	Acid detergent lignin	In vitro dry-matter disappearance
Chemical					
Between labs	0.75	1.82	2.24	1.01	2.08
Error	0.43	1.70	1.40	0.90	3.67
NIRS					
Between labs	0.83	2.45	1.81	1.07	1.33
Error	0.56	1.12	0.60	0.63	0.83

*Statistical models are not the same for the two analytical methods; therefore, direct comparisons between them should be made with caution.

Table 2. Standard errors of prediction (%) associated with a single value of five forage-quality parameters predicted using NIRS.

Prediction equation derived using	Prediction equation applied to	Quality parameters				
		Crude protein	Neutral detergent fiber	Acid detergent fiber	Acid detergent lignin	In vitro dry-matter disappearance
Birdsfoot trefoil, *Lotus corniculatus*	Alfalfa, *Medicago sativa*	0.40	1.75	2.93	0.69	2.14
Orchardgrass, *Dactylis glomerata*	Red clover, *Trifolium pratense*	1.67	15.90	4.24	2.08	3.88
Orchardgrass, *D. glomerata*	Timothy, *Phleum pratense*	0.40	1.35	1.57	0.54	2.28
Oven drying, 65°C	Oven drying, 75°C	0.50	1.62	0.93	0.62	2.23
Oven drying, 65°C	Field drying, 3 days	0.99	4.46	3.19	1.34	5.69
Harvest 1	Harvest 2	0.94	2.52	1.52	1.42	3.63
Half of samples	Other half of samples	0.66	1.97	1.02	0.59	1.80

grown in one year at a single location and used to predict forage-quality values of samples from other replicates, locations, and years. Standard errors of prediction, corrected for bias, ranged as follows: crude protein, 0.68% to 1.1%; neutral detergent fiber, 1.8% to 2.3%; acid detergent fiber (ADF), 0.98% to 1.6%; acid detergent lignin (ADL), 0.31% to 0.45%; and in-vitro dry-matter disappearance, 3.1% to 4.1%. Studies, also at Saint Paul, of fibrous constituents of corn (*Zea mays* L.) stover showed standard errors of prediction from 1.0% to 2.0% for ADF and 0.45% to 0.75% for ADL.

LITERATURE CITED

Barton, F.E., II and D. Burdick. 1978. Analysis of bermudagrass and other forages by near infrared reflectance. Proc. 8th Res. and Indus. Conf., Coastal Bermudagrass Processors Assoc., Athens, Ga., 45–51.

Barton, F.E., II and D. Burdick. 1979. Preliminary study on the analysis of forages with a filter-type near infrared reflectance spectrometer. J. Agric. Food Chem. 27:1248–1252.

Barton, F.E., II and D. Burdick. 1980. Prediction of crude protein in dehydrated coastal bermudagrass by NIR reflectance. Proc. 10th Res. and Indus. Conf., Coastal Bermudagrass Proc. Assoc., Athens, Ga., 103–106.

Norris, K.H., R.F Barnes, J.E. Moore, and J.S. Shenk. 1976. Predicting forage quality by infrared reflectance spectroscopy. J. Anim. Sci. 43:889–897.

Shenk, J.S., and R.F Barnes. 1977. Current status of infrared reflectance. Proc. 34th South. Pasture and Forage Crop Impr. Conf., Auburn, Ala., 57–62.

Shenk, J.S., and M.R. Hoover. 1976. Infrared reflectance spectro-computer design and application. 7th Technicon Int. Cong., New York, 2:122–125.

Shenk, J.S., M.O. Westerhaus, and M.R. Hoover. 1978a. Infrared reflectance analysis of forages. Proc. Int. Grain and Forage Harv. Conf., 241–244, 252. Am. Soc. Agric. Eng., St. Joseph, Mich.

Shenk, J.S., M.O. Westerhaus, M.R. Hoover, K.M. Mayberry, and H.K. Goering. 1978b. Predicting forage quality by infrared reflectance spectroscopy. Proc. 2nd Int. Green Crop Drying Cong., Univ. of Saskatchewan, Saskatoon, 292–299.

Prediction of Forage Quality with NIR Reflectance Spectroscopy

F.E. BARTON II and D. BURDICK

USDA-SEA-AR Southwestern Livestock and Forage Research Station, El Reno, Okla.,
and Field Crops Research Unit, R.B. Russell Agricultural Research Center,
Athens, Ga., U.S.A.

Summary

Near-infrafred (NIR) reflectance spectroscopy can be used to predict the quality of forages and their chemical composition. Studies were conducted on warm-season grasses with a filter-type NIR reflectance spectrometer to evaluate the NIR technique and the tilting filter spectrometer. Hay and drum-dehydrated samples of bermudagrass were used. A monochromator NIR reflectance spectrometer was used to narrow the possibilities of wavelengths to be investigated. The following compositional and quality parameters were determined: crude fiber (CF), fat, acid detergent fiber (ADF), neutral detergent fiber (NDF), crude protein (CP), acid-insoluble lignin (AIL), and in-vivo digestibility of dry matter (DDM).

The results of bermudagrass hays (20 for calibration, 22 for prediction) were similar, with standard errors of calibration and prediction of CP 1.15 and 0.84; ADF 2.46 and 1.42; NDF 2.06 and 1.39; AIL 0.50 and 0.51; and DDM 1.78 and 2.54, respectively. When drum-dehydrated samples were examined, new wavelengths were chosen and the standard errors of calibration and prediction were CP 0.87 and 0.89; CF 0.54 and 1.54; and fat 0.73 and 0.37, respectively, with 20 samples in calibration set and 26 in the prediction set. All of these analyses were made using a second-derivative mathematical treatment. The use of a delta OD at one wavelength divided by delta OD at another wavelength ($\Delta OD_1/\Delta OD_2$) mathematical treatment improved the standard errors of calibration and prediction of CP for drum-dehydrated bermudagrass (CP 0.58 and 0.61, respectively).

KEY WORDS: near-infrared reflectance spectroscopy, NIR, forage-quality analysis, forage compositional analysis.

INTRODUCTION

Near-infrared (NIR) reflectance spectroscopy was first used to analyze forages by Norris et al. (1976), Norris and Barnes (1976), and subsequently by Shenk and Hoover (1976) and Shenk et al. (1976, 1977, 1978). These authors used scanning monochromators, with stepping motors, interfaced to digital computers. Recent work on forage analysis using a tilting filter NIR spectrometer has been published by Barton and Burdick (1978, 1979, 1980) and Burdick et al. (1981).

Constituents predicted by the NIR spectroscopy technique include dry matter (DM), crude protein (CP), acid detergent fiber (ADF), neutral detergent fiber (NDF), crude fiber (CF), permanganate lignin (PML), acid-insoluble lignin (AIL), and fat. In-vitro dry-matter digestibility (IVDMD) and in-vivo digestibility (DDM) have been predicted by most of the workers in this area; Shenk et al. (1976) used the NIR technique to predict animal intake and average daily gain.

In all of the studies to date, derivative spectroscopy techniques were used to enhance resolution and separate overlapping bands. These digital-derivative spectra were then regressed against the analytical data to establish a calibration equation for each constituent or animal-performance parameter desired. Finally, these equations, when applied to the derivative spectra of an unknown set of samples, will predict the analytical value of the constituent or animal-performance parameter desired.

This paper summarizes results obtained to date with the tilting filter instrument and discusses some aspects of mathematical treatments and derivative spectroscopy as they affect the spectra and the predicted values.

METHODS

The grass samples used in these studies were Coastal and Coastcross-1 cultivars of bermudagrass (*Cynodon dactylon* [L.] Pers.). Samples were at 4 to 8 weeks of regrowth when harvested (May through October). The dry (<6–8% moisture) samples were ground through a 1-mm screen in a UDY cyclone grinder and packed in standard Neotec sample cups.

Samples and chemical analyses of the bermudagrass hays were obtained from Franklinton, Louisiana, for CP, ADF, NDF, ADL, IVDMD, and in-vivo digestibility (DDM) (Burdick et al. 1981). The drum-dehydrated sam-

---

F.E. Barton II is a Research Chemist at the USDA-SEA-AR Southwestern Livestock and Forage Research Station, and D. Burdick is a Research Chemist at the Field Crops Research Unit of the D.B. Russell Agricultural Research Center.

The authors would like to thank B.D. Nelson and Oliver Jarrell for providing sample sets and analytical data.

Mention of a proprietary product does not imply endorsement of it.

ples were analyzed for CP, CF, and fat by a commercial analytical laboratory.

NIR measurements were obtained with the Neotec FQA-51 and spectrocomputer (Barton and Burdick 1979). The Neotec FQA-51 is capable of utilizing six different mathematical treatments, but only two were used — D^2OD (second derivative) and $\Delta OD1_1/\Delta OD_2$, a measure of one slope divided by another slope at different wavelengths.

RESULTS AND DISCUSSION

Successful prediction of forage nutritive value involves calibration, mathematical-data transformation (mathematic treatments), and choice of the appropriate wavelengths at which to make a measurement (Shenk et al. 1979).

The greatest source of error for calibration is in the laboratory data (Shenk et al. 1978, Barton and Burdick 1980). The calibration and therefore the prediction are no better than the accuracy of the laboratory data. Table 1 contains the standard errors of calibration and prediction for the bermudagrass hays. The standard errors of prediction were very low, especially for DDM (Table 1). Another consideration for calibration is the population to be examined. If the NIR calibration is to predict analyses across several species, drying regimes, and stages of maturity, a set of equations based upon a large file, frequently updated by adding more samples (several hundred), covering this diverse population is needed. In such a case, one must accept larger prediction errors and smaller R^2 values in the calibration. If a plant breeder wishes to predict the analytical and digestibility values of genotypes of a single species grown under controlled conditions, a more specific calibration with the best laboratory data possible is required. In this case the desired genetic differences may reside in population "outliers." Regardless of the application, it is best if the calibration sample set represents a wider range than the prediction set.

Mathematical transformations are used to transform data or an equation into a form that can be more readily interpreted and/or solved.

In NIR analysis of forages the second derivative of the logarithm of reciprocal reflectance with respect to wavelength $[d^2 (\log 1/R) d\lambda^2]$ has been the most widely used

Table 1. Standard errors of calibration (Cal) and prediction (Pred) for determining constituents in bermudagrass hays with the near-infrared filter spectrophotometer compared to laboratory (Lab) errors.

Constituent	Cal (N = 20)	Pred (N = 22)	Lab
Crude protein	1.15	0.84	0.5
Acid detergent fiber	2.46	1.42	0.8
Neutral detergent fiber	2.06	1.39	1.0
Acid detergent lignin	0.50	0.51	0.4
In vivo digestibility	1.78	2.54	3.1

Table 2. Standard errors of calibration (Cal) and prediction (Pred) for dehydrated bermudagrass.

Analysis	Math	R^2	Cal[†]	Pred[‡]
Crude protein	$(\Delta OD_1/\Delta OD_2)$[⁂]	0.95	0.58	0.61
Crude protein	(d^2OD)[a]	0.87	0.89	0.71
Crude fiber	(d^2OD)	0.54	1.54	2.84
Fat	(d^2OD)	0.72	0.37	0.25

[†]20 samples.

[‡]26 samples.

[⁂]Sample correlation coefficient for two mathematical methods 0.94.

mathematical treatment. The results shown in Tables 1 and 2 were obtained using this mathematical treatment. Once the degree of derivatization necessary — i.e., second (2), third (3), etc. — has been determined, the wavelength interval/data point for sampling and the wavelength range over which to take the derivative must be chosen. In general, as $\Delta\lambda$ increases, the ratio of signal to noise (S/N) increases (noise is independent of λ), but the ability to discriminate between overlapping peaks is decreased. For the second derivative a $\Delta\lambda$ equal to the width of the absorbance band at half its maximum height is considered optimal (Cahill 1980).

A second "mathematical treatment," referred to as the "Norris math" (Norris and Williams, unpublished), uses the algorithm "delta logarithm of the reciprocal of reflectance," $\Delta\log 1/R$, referred to as ΔOD in Neotec manuals. It is also possible to use the normalized function $\Delta(\log 1/R_1)$, $\Delta(\log 1/R_2)$, referred to as $\Delta OD_1/\Delta OD_2$ in Neotec manuals. The ΔOD works very well for protein on uniform sample sets, while second derivative works best on widely varying sample sets. The $\Delta OD_1/\Delta OD_2$ seems to work well for protein when moisture and particle size vary. The results shown in Table 2 are superior to those obtained with the second derivative, giving a prediction error of only 0.61 and a calibration error just slightly less (0.58) and a very high R^2 (0.95).

The choice or the method of choosing wavelengths is the single element of prediction that limits the spectrometer in its use in predicting forage nutritive parameters. For both monochromator and filter spectrometers the main criteria are the same. The coefficients for the wavelengths of the spectral feature placed into the equations must be physically related to the constituent being measured. In the case of IVDMD, DDM, and other bioassay measurements, a wavelength should be employed that reflects a constituent that is part of the chemical parameter of interest. For example, since protein and fiber are digested, protein and fiber wavelengths should be included, or should be expected to be selected, for digestibility equations. The number of wavelengths should be no more than necessary to obtain adequate predictions. The ΔOD treatments have only one term; they are the simplest. If two wavelengths will predict protein adequately, three are not needed. The more wavelengths used, the greater will be the chance that the calibration sample set will not be representative of the population.

Analytical chemists now recognize predictive procedures, known as chemometrics, as valid analytical tools. New instruments, along with cheaper and more powerful computers, will help create completely new ways to accomplish chemical analyses. Near-infrared reflectance spectroscopy and its application to forage analysis is at the frontier of chemical research.

LITERATURE CITED

Barton, F.E., II and D. Burdick. 1978. Analysis of bermudagrass and other forages by near-infrared reflectance. Proc. 8th Res. and Ind. Conf. Coastal Bermudagrass Processors Assoc., Athens, Ga., 45–51.

Barton, F.E., II and D. Burdick. 1979. Preliminary study on the analysis of forages with a filter-type near-infrared reflectance spectrometer. J. Agric. and Food Chem. 27:1248–1252.

Barton, F.E., II and D. Burdick. 1980. Prediction of crude protein in dehydrated coastal bermudagrass by NIR reflectance. Proc. 10th Res. and Ind. Conf. Coastal Bermudagrass Processors Assoc., Athens, Ga. 103–106.

Burdick, D., F.E. Barton II, and B.D. Nelson. 1981. Prediction of bermudagrass composition and digestibility with a near infrared multiple filter spectrophotometer. Agron. J. 73:399–403.

Cahill, J.E. 1980. Derivative spectroscopy: understanding its application. Am. Lab. 12(11):79–85.

Norris, K.H., and R.F Barnes. 1976. Infrared reflectance analysis of nutritive value of feedstuffs. *In* P.V. Fonnesbeck, L.E. Harris, and L.C. Kearl (eds.) First international symposium on feed composition, animal nutrient requirements and computerization of diets, 237–241. Utah State Univ., Logan.

Norris, K.H., R.F Barnes; J.E. Moore, and J.S. Shenk. 1976. Predicting forage quality by infrared reflectance spectroscopy. J. Anim. Sci. 43:889–897.

Norris, K.H., and P.C. Williams. Optimization of mathematical treatments for the determination of protein in HRS wheat by near-infrared reflectance spectroscopy.

Shenk, J.S., and M.R. Hoover. 1976. Infrared reflectance spectrocomputer design and application advances in automated analysis. Technicon Int. Cong., 2:122–125, Indus. Symp., New York.

Shenk, J.S., W.N. Mason, M.L. Risius, K.H. Norris, and R.F Barnes. 1976. Application of infrared reflectance analysis to feedstuff evaluation. *In* P.V. Fonnesbeck, L.E. Harris, and L.C. Kearl (eds.) First international symposium, feed composition, animal nutrient requirements, and computerization of diets. 242–248, Utah State Univ., Logan.

Shenk, J.S., K.H. Norris, R.F Barnes, and G.W. Fissel. 1977. Forage and feedstuff analysis with infrared reflectance spectrocomputer system. Proc. XIII Int. Grassl. Cong., 1439–1442.

Shenk, J.S., M.O. Westerhaus, and M.R. Hoover. 1978. Infrared reflectance analysis of forages. Proc. Int. Grain and Forage Harvest. Conf., Am. Soc. Agric. Engin., St. Joseph, Mo., 242–244, 252.

Shenk, J.S., M.O. Westerhaus, and M.R. Hoover. 1979. Analysis of forages by infrared reflectance. J. Dairy Sci. 62:807–812.

Proposed Hay-Grading Standards Based on Laboratory Analyses for Evaluating Quality

D.A. ROHWEDER, N. JORGENSEN, and R.F BARNES
University of Wisconsin, Madison, Wis., U.S.A.

Summary

The progress report of the Forage Analysis Subcommittee of the Hay Marketing Task Force organized by the American Forage and Grassland Council (Rohweder et al. 1978) proposed new hay standards that would express feed value more adequately and communicate this value more fully. The subcommittee studied the potential for expanding the standards to include legume-grass mixtures; however, individual legume and grass equations gave more consistent predictions of quality. Digestibility and dry-matter intake by cattle and sheep fed the same forages were compared. Digestibility was nearly identical in the two livestock species; however, cattle were somewhat more efficient than sheep in utilizing low-quality legume and legume-grass forages, but not grasses. Regressions combining legumes and grasses are proposed.

KEY WORDS: acid detergent fiber, neutral detergent fiber, forage quality, market hay grades, relative feed value.

INTRODUCTION

Hay is a major North American crop, with nearly 25 million ha harvested annually. Production for the 1977–1980 period averaged more than 125.6 million tonnes, 60% of which was alfalfa and alfalfa-grass mixtures. Twenty percent or 25.5 million tonnes are annually sold off the farm with a cash value of $6.5 to $7.0 billion. The value of all hay produced is conservatively estimated at $15 to $16 billion.

We are faced with a continuing need for increased production, improved quality, and more precise methods of evaluating the quality present in the forage produced. The need for hay standards that will more adequately express the feed value in forages and that will communicate this value has long been recognized worldwide. This need is becoming more acute.

HAY MARKETING TASK FORCE

In 1972, the American Forage and Grasslands Council (AFGC) formed a Hay Marketing Task Force (1) to identify long-range marketing problems, (2) to determine priorities and practical solutions, and (3) to develop recommendations for action. A Forage Analysis Subcommittee was charged with establishing a system for pricing hay based on realistic measurements of feed value. Numerous reports of the subcommittee's findings and recommendations have been prepared and presented (Barnes 1975; Barnes et al. 1977; Rohweder et al. 1976a, 1976b, 1978; Moore 1977). The original proposals have been used and evaluated in several states since their publication in 1975. As a result several refinements are now proposed.

The data base was expanded to include acid detergent fiber, neutral detergent fiber, crude protein, in-vitro digestible dry matter, and dry-matter intake (DMI) measurements of additional samples from 12 states. Species included were alfalfa (*Medicago sativa* L.), medium red clover (*Trifolium pratense* L.), birdsfoot trefoil (*Lotus corniculatus* L.), smooth bromegrass (*Bromis inermis* Leyss.), orchardgrass (*Dactylis glomerata* L.), timothy (*Phleum pratense* L.), reed canarygrass (*Phalaris arundinacea* L.), tall fescue (*Festuca arundinacea* Schreb.), pangola digitgrass (*Digitaria decumbens* Stent.), Pensacola bahiagrass (*Paspalum notatum* Flugge,) and Suwannee, Coastal, and Coastcross-1 cultivars of bermudagrass (*Cynodon dactylon* [L.] Pers.). Several legume-grass mixtures also are included.

Acid detergent fiber (ADF) is now widely used in evaluating forages. Voluntary intake has received considerable attention as a parameter of forage quality, perhaps exceeding digestibility in many practical situations. Studies involving 187 forages of diverse species showed highest correlations for ADF with digestibility (−0.75) and cell walls or NDF with intake (−0.76) (Van Soest and Mertens 1977).

RESULTS AND DISCUSSIONS

Because most of the hay produced and sold is fed to cattle rather than to sheep and goats, the Committee investigated digestibility and intake relationships in cattle and sheep. Cattle exhibited slightly higher digestible dry-matter (DDM) values for legumes and early-cut temperate grasses, while sheep exhibited slightly higher values for subtropical grasses. In general, DDM was the same for cattle and sheep fed the same hays, a finding that agrees with data presented by Buchman and Hemken (1964) but does not completely agree with data by Playne (1978). However, dry-matter intake (DMI) values for hays fed at maintenance levels was considerably different for cattle and sheep fed the same hays, the cattle averaging 50% to 71% higher DMI than sheep for the several grades of legumes and grasses.

Linear regression values (b) for the DMI/NDF and DDM/ADF relationships of legumes and selected cool-season grasses are shown in Table 1. Differences between animal species were significant only in intake of grasses. But the intake relationship was significant for forage species with both sheep and cattle. The digestibility relationship was significant for legumes with cattle and for grasses with sheep.

Some have questioned whether the relationships in improved DDM and DMI from higher-quality forages exhibited by sheep and cattle hold true for horses, a major consumer of cash hay. Horses, although not ruminants, are like cattle in fiber utilization but more like swine in

Table 1. Changes (b) in dry-matter intake (DMI) per unit change in neutral detergent fiber and in-vivo dry-matter digestibility (DDM) per unit change in acid detergent fiber by sheep and cattle on identical legume and grass species.

Species	Dry-matter Intake		In-vivo DDM	
	b*	t**	b*	t**
Legume[a]				
Sheep	− .96	20%	− .37	n.s.
Cattle	− 1.24	20%	− .35	20%
Animal species	−	n.s.	−	n.s.
Grasses[b]				
Sheep	− .92	20%	− .54	20%
Cattle	− 1.77	20%	− .37	n.s.
Animal species	−	10%	−	n.s.
Forage species				
Sheep	−	n.s.	−	n.s.
Cattle	−	20%	−	n.s.

[a]Alfalfa, birdsfoot trefoil, and trefoil-timothy mixes.

[b]Orchardgrass, smooth bromegrass, timothy, and reed canarygrass

b*Slope of regression line.

t**T test for significance.

D.A. Rohweder is a Professor of Agronomy and N. Jorgensen is a Professor of Dairy Science at the University of Wisconsin, and R.F Barnes is the Associate Regional Administrator for the Southern Region of USDA-SEA-AR.

Table 2.   Correlation (R) of acid detergent fiber (ADF) with digestible dry matter (DDM) and of neutral detergent fiber (NDF) with dry-matter intake (DMI) from several trials conducted in the U.S.

Species	Description	R^2	R	SD[a]
In vivo DDM				
Alfalfa	DDM = 34.1080 + 2.6429 ADF% − .0499 ADF%^{2}	57	.76	3.95
All grass	DDM = 6.1069 + 3.9963 ADF% − .0663 ADF%^{2}	56	.75	6.20
All species	DDM = 59.0 − 2.26 ADF% + 14.2 $\sqrt{\text{ADF\%}}$	53	.73	5.21
DMI				
Alfalfa	DMI = 146.9547 + 1.0137 NDF% − .0302 NDF%^{2}	35	.59	3.97
All grass	DMI = 9.7914 + 4.8171 NDF% − .0508 NDF%^{2}	59	.77	8.22
All species	DMI = 84.7 − 3.69 NDF% + 32.37 $\sqrt{\text{NDF\%}}$	62	.79	7.67

[a]Standard error of regression.

use of protein. Therefore, horses also prefer earlier-harvested, more digestible hay (J.P. Baker, Department of Animal Science, University of Kentucky, personal communication, 1979).

The Committee investigated DDM/ADF and DMI/NDF relationships, using alfalfa, all legumes, legume-grass mixtures, temperate grasses, subtropical grasses, and all grasses. The relationship of DDM/ADF in cattle was comparable to that in sheep; however, correlation (r) values for cattle were higher. The DMI/NDF relationship also was comparable except that correlation (r) values were higher for legumes with cattle than with sheep.

Regressions for legume-grass mixtures were tested on known precise mixtures of alfalfa and orchardgrass. For practical purposes, predictions from the nearly pure legume and pure grass equations were comparable to those from the mixture equations. Mixture regressions therefore were eliminated.

Regressions predicting DDM and DMI for alfalfa and grasses with resulting DDMI and relative feed values for each hay grade (based on six grades) are shown in Table 5. The data were based on 41 observations of alfalfa and 230 of grass. Correlation (r), r^2, and standard-error-of-regression values are shown in Table 2.

Table 3.   Proposed market hay grades for legumes and legume-mixtures (Hay Marketing Task Force).

Grade		Stage of maturity inter- national term	Definition	Physical description	Typical chemical composition—[a]			Relative feed value %
					CP%	ADF%	NDF%	
Prime	Legume hay	Pre bloom	Bud to first flower; stage at which stems are beginning to elongate to just before blooming.	40 to 50% leaves[b]; green; less than 5% foreign material; free of mold, musty odor, dust, etc.	>19	<31	<40	>140
1	Legume hay	Early bloom	Early to mid bloom; stage between initiation of bloom and stage in which 1/2 of the plants are in bloom.	35 to 45% leaves[b]; light green to green; less than 10% foreign material; free of mold, musty odor, dust, etc.	17–19	31–35	40–46	124–140
2	Legume hay	Mid bloom	Mid to full bloom; stage in which 1/2 or more of plants are in bloom.	25 to 40% leaves[b]; yellow green to green; less than 15% foreign material; free of musty odor, dust, etc.	13–16	36–41	47–51	101–123
3	Legume	Full	Full bloom and beyond.	Less than 30% leaves[b]; brown to green; less than 20% foreign material; free of musty odor, etc.	<13	>41	>51	≤100
6	Sample grade[c]		Hay which contains more than a trace of injurious foreign material (toxic or noxious weeds and hardware) or that definitely has objectionable odor or is under-cured, heat damaged, hot, wet, musty, moldy, caked, badly broken, badly weathered or stained, extremely overripe, dusty, which is distinctly low quality or contains more than 20% foreign material or more than 20% moisture.					

[a]Chemical analyses expressed on dry-matter basis. Chemical concentrations based on research data from NC and NE states and Florida. Dry-matter (moisture) concentration can affect market quality. Suggested moisture levels are: Grades Prime and 1<14%, Grade 2 <18%, and Grade 3 <20%.

[b]Proportion by weight.

[c]Slight evidence of any factor will lower a lot of hay by one grade.

CP = crude protein; ADF = acid detergent fiber; NDF = neutral detergent fiber; relative feed value is based on digestible dry-matter intake. See Table 5.

Crude-protein concentration was dropped from prediction equations because this value did not improve the ability to predict DDM or DMI over ADF and NDF, respectively. However, the Committee recommends retaining this component in descriptive information for each lot of hay because protein concentration is important in hay selection.

TESTING THE CONCEPT

The concept has been tested in hay-marketing programs in Wisconsin, Oregon, Pennsylvania (Shenk and Baylor, personal communication), and Florida. Use of the grades with forage testing has improved the quality of hay purchased in Wisconsin, but the AFGC equations overestimate digestibility and underestimate intake of tropical grasses, especially bermudagrass, because ADF and NDF were often not closely related to digestibility and intake, respectively. More complex models may be in order for

the tropical grasses (Moore 1977).

J.S. Shenk and J.E. Baylor (Pennsylvania State University, personal communication, 1979) evaluated the proposed equations on 450 hay samples collected at three hay-marketing centers in Pennsylvania. Their findings confirm the recommendation that individual legume and individual grass equations were more precise than equations for mixtures in predicting feed value. They also discussed the difficulties associated with accurately determining the proportion of legumes to grass in a sample prior to selecting the proper regression equations. The Committee proposes correcting this problem by shifting the relative location of grass hays to legume hays at comparable stages of maturity in the grading scheme (Tables 3 and 4) as well as by combining legumes and grasses into one regression equation for predicting digestibility and one for predicting intake (Table 5). Regression r and standard error values are shown in Table 3. This change combines DMI and DDMI values for grasses and le-

Table 4. Proposed market hay grades for grasses and grass-legume mixtures (Hay Marketing Task Force).

Grade	Stage of maturity international term		Definition	Physical description	Typical chemical composition [a]			Relative feed value %
					CP[b]%	ADF%	NDF[c]%	
2	Grass hay	Pre head	Late vegetative to early boot; stage at which stems are beginning to elongate to just before heading; 2 to 3 weeks' growth.[e]	50% or more leaves[d]; green; less than 5% foreign material; free of mold, musty odor, dust, etc.	>18	>33	>55	>124–140
3	Grass hay	Early head	Boot to early head; stage between late boot where inflorescence is just emerging until the stage in which 1/2 inflorescences are in anthesis; 4 to 6 weeks' growth.[e]	40% or more leaves[d]; light green to green; less than 10% foreign material; free of mold, musty odor, dust, etc.				
4	Grass hay	Head	Head to milk; stage in which 1/2 or more of inflorescences are in anthesis and the stage in which seed are well formed but soft and immature; 7 to 9 weeks' regrowth.[e]	30% or more leaves[d]; yellow green to green; less than 15% foreign material; free of mold, musty odor, dust, etc.	13–18	33–38	55–60	101–123
5	Grass hay	Post head	Dough to seed; stage in which seeds are of dough-like consistency until stage when plants are normally harvested for seed; more than 10 weeks' growth.[e]	20% or more leaves[d]; brown to green; less than 20% foreign material; slightly musty odor, dust, etc.	8–12	39–41	61–65	83–100
6	Sample grade[f]				<8	<41	<65	<83

Hay which contains more than a trace of injurious foreign material (toxic or noxious weeds and hardware) or that definitely has objectionable odor or is under-cured, heat damaged, hot, wet, musty, moldy, caked, badly broken, badly weathered or stained, extremely overripe, dusty, which is distinctly low quality or contains more than 20% foreign material or more than 20% moisture.

[a]Chemical analyses expressed on dry-matter basis. Chemical concentrations based on research data from NC and NE states and Florida. Dry-matter (moisture) concentration can affect market quality. Suggested moisture levels are: Grade 2<18%, Grades 3, 4, and 5<20%.

[b]Fertilization with nitrogen may increase CP concentration in each grade by up to 40 percent.

[c]Tropical grasses may have higher NDF concentrations than indicated in this table.

[d]Proportion by weight.

[e]For grasses that do not flower or for which flowering is indeterminant.

[f]Slight evidence of any factor will lower a lot of hay by one grade, except Grade 5.

CP = crude protein: ADF = acid detergent fiber; NDF = neutral detergent fiber; relative feed value is based on digestible dry-matter intake. See Table 5.

gumes in a more consistent manner than the schemes suggested previously and in Table 3.

Evaluation of these standards will continue. Modifications may be made at the conclusion of planned research.

Table 5. Typical digestible dry-matter (DDM), dry-matter intake (DMI), and digestible dry-matter intake (DDMI) values for proposed market hay grades described in Tables 4 and 5[a].

Grade	DDM in vivo %	DMI gm/Wkg$^{0.75}$	DDMI gm/Wkg$^{0.75}$	Relative[b] feed value, %
Prime	>68	>142	>96.6	>136
1	66–68	134–141	87–96	123–135
2	63–65	122–133	76–86	107–121
3	60–62	114–121	68–75 (69)	95–106
4	57–59	106–113	60–67	85–94
5	<57	<105	<59	<84

[a]Formulas used to calculate relative feed value:
$DDM = 59.0 - 2.26 ADF\% + 14.2 \sqrt{ADF\%}$
$DMI = 84.7 - 3.69 NDF\% + 32.37 \sqrt{NDF\%}$
$DDMI = DDM \times DMI/100$
Relative feed value = 1.45 DDMI where DDM = in-vivo digestible dry matter; DMI = dry-matter intake; DDMI = digestible dry-matter intake.

[b]Relative feed value is an estimate of over-all forage quality. It is calculated from intake and digestibility of dry matter when forages of known composition were fed to cattle. The values are relative; however, they are equally appropriate for all classes of livestock. Relative feed value estimates the intake of digestible energy when the forage is the only source of dietary energy and protein.

LITERATURE CITED

Barnes, R.F 1975. Predicting digestible energy values of hays. Proc. Lab. Meth. and Serv. Workshop, 20–22 May, 1975, Salem, Oreg. Cosponsored by Assoc. of Am. Feed Cont. Off. Inc., Assoc. Off. Anal. Chem. (AAFCA-AOAC).

Barnes, R.F, D.A. Rohweder, and N. Jorgensen. 1977. The proposed establishment of hay standards. Proc. 34th South. Past. and Forage Crop Impr. Conf., Auburn, Ala., 120–128.

Buchman, D.T., and R.W. Hemken. 1964. *Ad libitum* intake and digestibility of several alfalfa hays by cattle and sheep. J. Dairy Sci. 47:861–864.

Moore, J.E. 1977. Southern forages in the new hay standards. Proc. South. Past. and Forage Crop Impr. Conf., Auburn, Ala.

Playne, M.J. 1978. Estimation of the digestibility of low quality hays by cattle from measurements made with sheep. Anim. Feed Sci. and Technol. 3(1):51–55.

Rohweder, D.A., R.F Barnes, and N.A. Jorgensen. 1976a. The use of chemical analyses to establish hay market standards. 1st Int. Symp. on Feed Consump., Anim. Nutr. Req. and Comput. of Diets, Logan, Utah, 249–257.

Rohweder, D.A., R.F Barnes, and N.A. Jorgensen. 1976b. A standardized approach to establish market value for hay. Agron. Abs. 1976:112.

Rohweder, D.A., R.F Barnes, and N.A. Jorgensen. 1978. Proposed hay grading standards based on laboratory analyses for evaluating quality. J. Anim. Sci. 47(3):747–759.

Van Soest, P.J., and D.R. Mertens. 1977. Analytical parameters as guides to forage quality. Proc. Int. Mtg. on Anim. Prod. from Temp. Grassl., 50–52. Irish Grassl. and Anim. Prod. Assoc., Dublin, Ire.

The Potential of Forage Legumes and Their Role in Scotland

J. FRAME, R.D. HARKESS, and A.G. BOYD
The West of Scotland Agricultural College, Auchincruive,
Ayr, Scotland, UK

Summary

Intensive forage production requires fertilizer nitrogen (N), the manufacture of which depends upon fossil-fuel energy. Increasing energy costs and possible scarcity raise doubts about the future. Thus, there has been renewed interest in forage legume production.

In ongoing research at The West of Scotland Agricultural College, the effects of applied N levels and closeness of cutting on white clover varieties of differing morphological types are being assessed under a simulated grazing regimen. Also, the influence on production of white clover, diploid and tetraploid red clover, and lucerne swards, sown alone or with each of five companion grasses, is being evaluated under a simulated conservation regimen with no applied N; in a third experiment, the same four legumes, sown alone and in all possible combinations, are evaluated under similar management.

In the white clover experiment, mean total herbage dry matter (DM) responded from 7.62 metric tons (t)/ha at no N to 12.20 t/ha at 360 kg/ha N; conversely, clover contribution declined. Close defoliation increased both total herbage and clover DM. The absence of variety × applied N interactions suggested that varieties did not differ in tolerance to applied N. In the legume–companion grass experiment, red clover at 15.99 t/ha DM outyielded lucerne (10.38 t/ha) and white clover (9.55 t/ha). Mean annual DM yields were 13.36 t/ha for legume-grass mixtures and 12.10 t/ha for pure-sown legumes. In the legume-mixture experiment, red clover dominated other legumes, and annual DM yields ranged from 14.46 to 16.16 t/ha. Red clover sown alone also outyielded lucerne and white clover. White clover increased organic-matter digestibility where it made up a substantial content of the sward.

The future role of forage legumes is to substitute for or complement manufactured N fertilizers and provide high-quality forage. Legume exploitation will also depend upon how well legumes can be integrated into farming systems.

KEY WORDS: forage legumes, herbage production, quality, future role.

INTRODUCTION

Of the forage legume seed used in the UK (1430 t in 1980), 60% is white clover (*Trifolium repens*), 30% red clover (*T. pratense*), 6% alsike clover (*T. hybridum*), 3% lucerne (*Medicago sativa*), and 1% others. Use has declined by 75% since 1960, while fertilizer nitrogen (N) usage on grassland has increased. Present estimates for Scottish grassland (excluding rough grazings) suggest that 100–150 kg/ha N is applied annually with a range of 0 to 400 kg/ha on individual fields.

Intensive forage production using fertilizer N now depends largely on fossil-fuel energy. With increasing costs and possible shortages of energy creating potential instability in price/supply of fertilizer N, interest in forage legumes has been renewed. Their N-fixing capability provides nutritious and highly acceptable protein-rich forage, superior to grass herbage for animal production (Thomson 1977). Improved varieties and new conservation techniques are additional factors in stimulating the reappraisal. This paper presents selected results from ongoing agronomic investigations in west Scotland.

METHODS

The effects of applied N (0, 120, 240, and 360 kg/ha/yr) and closeness of cutting (4 or 8 cm) on four white clover varieties varying in leaf size—Aberystwyth S 184 (small), Grasslands Huia (medium small), Linda, and Olwen

J. Frame is the Head of the Agronomy Department, R.D. Harkess is a Senior Agronomy Specialist, and A.G. Boyd is a Senior Agronomy Specialist at The West of Scotland Agricultural College.

Table 1.   Mean annual yield and digestibility data, 1979–80 (white clover variety x cutting height x fertilizer N level).

	Total herbage DM (t/ha)		Clover DM (t/ha)		OMD
Year:	1	2	1	2	1
White clover variety					
S 184	9.69	9.71	2.15	1.66	0.79
Huia	9.89	9.35	2.29	1.36	0.79
Linda	9.95	9.81	2.72	2.24	0.79
Olwen	9.90	9.75	2.63	2.26	0.79
SED (±)	0.164	0.159	0.185	0.150	0.002
F	NS	*	*	***	NS
Closeness of cutting					
4 cm	10.53	10.23	2.56	2.20	0.80
8 cm	9.19	9.08	2.33	1.56	0.79
SED (±)	0.116	0.113	0.131	0.106	0.002
F	***	***	NS	***	***
Fertilizer N (kg/ha/annum)					
0	7.39	7.84	4.54	3.92	0.78
120	8.82	8.89	2.69	2.42	0.79
240	10.70	10.00	1.56	0.83	0.80
360	12.52	11.88	1.01	0.34	0.79
SED (±)	0.164	0.159	0.185	0.150	0.002
F	***	***	***	***	***

NS, *, *** = nonsignificant or statistically significant at $P<0.05$ or 0.001, respectively.

Table 2.   Mean annual yields and botanical composition data (legume x companion grass), 1980.

Variety	Seed rate (kg/ha)	Total herbage DM (t/ha)	OMD	Legume (%)†
Legume				
Blanca white clover	4	9.55	0.71	63
Violetta red clover	16	15.93	0.63	90
Hungaropoly red clover‡	16	16.05	0.63	92
Europe lucerne	16	10.38	0.63	72
SED (±)		0.242	0.003	2.2
F		***	***	***
Companion grass				
None (legume only)	—	12.10	0.64	96
Timo timothy	3	12.71	0.65	77
Bundy meadow fescue	6	13.43	0.65	76
Deborah sweet brome	8	12.68	0.65	82
Talbot perennial ryegrass	6	13.26	0.66	71
Barlatra perennial ryegrass	6	13.70	0.66	74
SED (±)		0.297	0.004	2.7
F		**	NS	***

†fresh matter basis

‡tetraploid variety

(both medium large) — are examined in an experiment established in 1978 (Table 1). The clovers were drilled at 3 kg/ha seed with 10 kg/ha Talbot medium-late perennial ryegrass. Six defoliations were made during harvest years 1979–1980 at 4- to 5-week intervals. Phosphate and potash were applied at 180 and 360 kg/ha/yr, respectively. Fertilizer N was applied in six equal dressings over the season.

A second experiment (Table 2) investigates the production of white clover (cv. Blanca), diploid red clover (cv. Violetta), tetraploid red clover (cv. Hungaropoly), and lucerne (cv. Europe) sown alone or with diploid or tetraploid perennial ryegrass (*Lolium perenne*), timothy (*Phleum pratense*), meadow fescue (*Festuca pratensis*), or sweet brome (*Bromus carinatus*). The same four legumes, sown singly or in various combinations, are under assessment in a third experiment (Table 3). In experiments 2 and 3, silage-stage crops are cut in June and August, with an aftermath grazing crop in October. Each of the silage-stage harvests received 90 kg/ha P_2O_5 and 180 kg/ha K_2O. No fertilizer nitrogen is applied.

RESULTS

In experiment 1 fertilizer N increased total herbage, but at the expense of white clover (Table 1). Close defoliation increased total herbage and clover yields and also organic-matter digestibility (OMD). The larger-leaved Olwen and Linda were the highest-yielding white clovers, but lack of variety × applied N interactions show that they did not exhibit increased tolerance to N.

White clover swards were the lowest yielding of the various legume-grass mixtures (Table 2) and of those mixed-legume swards where white clover was present to any degree (Table 3); OMD values were highest in those latter swards.

In experiment 2, red clover was superior to lucerne and

Table 3.   Mean annual yields and botanical composition data, 1980 (legume mixtures).

Legume†	Seed rate (kg/ha)	Total herbage DM (t/ha)	OMD	Dominant legume (%)‡
B	4	8.64	0.72	B 17–76
V)		15.41	0.63	V 91–100
H)	16	16.52	0.62	H 99–100
E)		10.72	0.62	E 95–96
B:V)		14.80	0.63	V 97–99
B:H)	3:12	15.32	0.63	H 95–97
B:E)		9.47	0.68	
H:V)		16.16	0.62	HV 97–100ʹ·
E:V)	8:8	15.55	0.61	V 78–96
E:H)		15.37	0.62	H 88–96
B:V:H)		14.85	0.62	HV 97–98ʹ·
B:V:E)	3:6:6	14.46	0.63	V 84–90
B:H:E)		14.46	0.63	H 81–99
V:H:E	6:6:6	15.13	0.62	HV 96–100ʹ·
B:V:H:E	3:4:4:4	15.30	0.63	HV 97–99ʹ·
SED (±)		0.541	0.005	
F		***	***	

†B = Blanca white clover, V = Violetta diploid red clover, H = Hungaropoly tetraploid red clover, E = Europe lucerne.

‡Fresh-matter basis, range for the three harvests.

ʹNot possible to separate V and H when in combination.

white clover both in total herbage yield and in legume content (Table 2). The addition of a grass companion significantly increased total herbage by 9% on average over all the legumes (5% to 13% range) and decreased legume contribution by 21%.

Red clover-dominant swards, yielding up to 16.52 t/ha DM in experiment 3, were again superior to white clover and lucerne (Table 3).

Lucerne and lucerne-grass mixtures slightly outyielded corresponding white clover swards but not red clover swards (Tables 2 and 3). In lucerne-red clover swards, red clover was dominant.

DISCUSSION

White Clover

Increasingly, reliance has been placed on fertilizer N to boost sward yields in Scotland, with clover presence regarded as a bonus. At the upper limits of N use in practice (300–400 kg/ha/yr), annual DM yields achieved are 12–16 metric tons (t)/ha compared with 6–10 t/ha from a grass–white clover sward given no applied N and with an output limited by a relatively short clover growth period of 4–5 months. Reliance on clover is, however, a common feature of hill-upland grazing swards where farming economics cannot afford intensive fertilizer N use. The effect of fertilizer N on the yield and clover contribution of mixed swards is well documented where older clover varieties were used (e.g., Frame 1973). More recently a new generation of higher-yielding white clovers has been introduced, and some of these varieties may have potential tolerance to applied N (Reid 1976). If tolerance is confirmed, it would allow flexibility in management, and a higher clover content would give sward material of higher feeding value. The present experiment (Table 1) is showing that larger-leaved varieties do not necessarily have increased tolerance to applied N.

A common finding in research and practice is within- and between-season variation of the white clover contribution to grass-clover swards. For example, in current varietal evaluation work, under standard management over a number of years, grass-clover annual yields ranged from 6.3 to 10.4 t/ha DM, with clover DM yields from 2.1 to 7.4 t/ha. Various interactive factors, including variety, compatibility of companion grass, weather, defoliation intensity, soil nutrients, diseases, and pests, have been identified in the variability syndrome. The unpredictability of white clover production and persistence, therefore, makes it difficult to implement planned animal-production systems. However, the benefits white clover offers undoubtedly make it worthwhile to continue research to overcome its inherent disadvantages and create a more positive role for it in Scottish grassland farming.

Red Clover

Current work confirms the potential of red clover in the absence of applied N. To achieve similar yield levels from a grass conservation sward, fertilizer N at 250–350 kg/ha/yr would be required. The mixed red clover–grass association offers scope for exploiting red clover potential in early years and use of fertilizer N when clover presence declines (Frame 1976).

Medium-late perennial ryegrass has been suggested by several sources (e.g., Frame 1976) as the best companion grass and is under investigation (Table 2). The dominance of red clover when sown in legume mixtures is illustrated in Table 3.

Red clover is suitable as a silage crop and in short-term leys. It fits less well into long-term grassland farming systems since it requires renewal after 2 to 4 years. Its exploitation will depend on how well research can resolve certain pest and disease problems and further improve its reliability and on the ability of livestock farmers to integrate it into their farming systems.

Other Legumes

The potential of lucerne in Scotland has been restricted by climate, and by wet acid soils in the west. It may be more useful in drier eastern areas. However, reappraisal is warranted.

The potential of the indigenous common birdsfoot trefoil (*Lotus corniculatus*) and marsh birdsfoot trefoil (*L. uliginosus*) for improving hill and upland natural pastures has been investigated in west Scotland. It was concluded that they might be a substitute for white clover on dry and wet acidic pastures, respectively, where grazing pressure is not severe.

Forage peas (*Pisum sativum*) for west-of-Scotland conditions have been investigated. Two potential uses are seen: first, high yields (up to 8 t DM/ha) of protein-rich material can be produced in a short growing season from a pure sowing, and, second, the digestibility and crude-protein content of arable (cereal) silage can be improved by the inclusion of forage peas in the mixture.

LITERATURE CITED

Frame, J. 1973. The yield response of a tall fescue/white clover sward to nitrogen rate and harvesting frequency. J. Brit. Grassl. Soc. 28:139–148.

Frame, J. 1976. The potential of tetraploid red clover and its role in the United Kingdom. J. Brit. Grassl. Soc. 31:139–152.

Reid, D. 1976. The role of white clover in grassland production. Hannah Res. Inst. Rpt. 1976:66–69.

Thomson, D.J. 1977. The role of legumes in improving the quality of forage diets. Proc. Int. Mtg. Anim. Prod. from Temp. Grassl., Dublin, Irel. 131–135.

Techniques of Overdrilling for the Introduction of Improved Pasture Species in Temperate Grasslands

C.J. BAKER

Massey University, Palmerston North, N.Z.

Summary

Experiments over several years are summarized. Collectively, these experiments sought to improve the reliability of overdrilling techniques during growth periods of temperate swards. The studies sought first to identify the important physical requirements of seeds and seedlings in untilled soils, especially in unpredictably dry situations, and then to develop drilling machinery and techniques to fulfill these requirements more nearly than had hitherto been achieved in New Zealand.

The studies first involved closely controlling and monitoring the drilling and seed-groove-covering procedures, the soil moisture status and mechanical properties, and postdrilling climates, using a tillage-bin technique. Cereal seeds were drilled into large undisturbed blocks of soil contained in steel bins, which were then placed in controlled-climate rooms for seedling emergence and groove microequipment studies. Subsequently, field trials with pasture and crop species sought to extrapolate these data to the variable oceanic climate and soil conditions of New Zealand.

In drying soils, contrasting designs of groove openers and covering devices had less effect on seed germination than on subsurface seedling survival and emergence. The more traditional V- or U-shaped seed grooves consistently promoted less seedling survival (even when covered with loose soil) than grooves where soil disturbance was largely confined to a subsurface layer, with a narrow surface slit and an overlay of dead vegetative mulch.

Two designs of drill openers, a bar-covering harrow, and several press-wheel designs were developed. The main effect that this improved equipment had was to improve the in-groove seedling habitat. An experimental chisel opener (with subsurface wings) and bar harrow combination significantly ($P < 0.01$) decreased the rate of loss to the atmosphere of in-groove humidity by 45% compared to a triple-disk opener and 16% compared to a hoe opener, both in association with the bar harrow. Press wheels operating on the seeds at the bases of the grooves before covering improved seedling responses for the triple-disk and hoe openers. A moderate correlation coefficient, $r = 0.75$, between the mean loss of in-groove humidity for 6 days and subsurface seedling survival was established across the three opener designs.

Physical removal of a strip of the competing resident sward overlying the drilled grooves clearly contrasted with the desirability of retaining a mulch as a major component of the groove cover for moisture retention. A simultaneous band-spraying technique was evolved and tested, using paraquat (1,1'-dimethyl-4,4' bipyridinium ion) and/or glyphosate [N-(phosphonomethyl) glycine] applied in strips either ahead of or following the openers. Vigor of introduced species was related to the width and effectiveness of the herbicide bands.

KEY WORDS: drilled groove, seed habitat, seedling habitat, drilling equipment, band spraying.

INTRODUCTION

The predominantly temperate climate of oceanic New Zealand (which lies geographically between lat, 35° and 48° S) undoubtedly favors year-round pasture production. However, the infrequency of extreme growth peaks and troughs can also impose problems for overdrilling. Competition from even low-quality and damaged swards can affect survival and/or vigor of drilled seedlings, and the unpredictable occurrence of dry conditions can rapidly (if only temporarily) transform an optimal soil moisture habitat into a suboptimal one.

In order to decrease the dependence of overdrilling on narrow tolerance limits of climatic, soil, and competitive conditions, a program aimed at improving the equipment and techniques available was initiated at Massey University in 1969.

HABITAT CONFLICT

Earlier workers were not in a position to consider the important conflict that was arising between modifying the seed habitat to promote emergence and modifying the seedling habitat to lessen the influence of competition.

Seedling Emergence Habitat

There is a considerable body of evidence (Baker 1976a, Choudhary and Baker 1980, 1981a, 1982) that points to

The author is a Reader in Agricultural Mechanization at Massey University.

the advantages of retaining, as much as possible, existing dead vegetation as the main covering medium over the drilled seeds in order to insulate the seed groove from drying out. This evidence is not necessarily in conflict with reports of allelopathic damage to seeds from rotting vegetation (Haggar 1977, Squires and Elliott 1975) because, to be effective as a moisture barrier, these organic residues usually remain on top of the ground and well removed from seed contact or anaerobic decomposition (Baker 1976a, Butterworth 1980, Butson, personal communication, 1978). This mulch appears to slow the interchange of soil water vapor with the ambient atmosphere. While "clean" loose soil cover has been shown to impede drying more than no cover, a vegetative mulch appears to present an even higher effective impedence.

In-groove humidity has been shown to exert a relatively small influence on seed germination (Baker 1976a, Choudhary and Baker 1980, 1981a, 1982), as liquid-phase water is mostly responsible for imbibition. However, vapor-phase water has been shown to have a major effect on the survival of the seedlings for the few days between germination and emergence. When the in-groove humidity decreases too rapidly, the seeds may germinate, but many seedlings die before emergence. The capacity of a drilled groove to retain humidity was referred to by Choudhary and Baker (1980, 1981b) as the "Moisture Vapour Potential Captivity" (MVPC) of the groove. It was numerically equal to the reciprocal of the mean daily loss of percentage relative humidity from the groove.

Groove Opener Performance

Various opener types have been evaluated at Massey University since 1969, both in the field and with a controlled testing procedure. This latter procedure utilized 500 kg blocks of undisturbed soil that were extracted from the field in steel bins and later drilled indoors using a special support bed and tool-testing apparatus (Baker 1969). They were then usually removed to controlled-climate rooms for postdrilling measurements (Choudhary and Baker 1980).

The opener designs most critically tested were a commercially available triple-disk opener with plain front disk (which usually formed a neat V-shaped groove), a commercially available hoe opener preceded by a plain front disk (which usually formed a torn or shattered U-shaped groove), and an experimental chisel opener preceded by a plain front disk (which formed an inverted T-shaped groove, torn or shattered beneath the surface). The latter device was developed at Massey University.

Biological tests included the ability of each opener to promote seed germination, subsurface seedling survival, and seedling emergence when followed by a covering bar harrow (Baker 1970) and/or various press-wheel combinations (Choudhary and Baker 1980, 1981a, 1981b). Mechanical tests of smearing, soil structure, and compaction, together with fuel-use, draft, and penetration-force recordings, were also carried out but are not summarized here. Test conditions have included moist and dry soil

and climatic conditions, but the latter have clearly produced the largest contrasts between opener designs, in terms of seedling performance.

In the controlled-climate experiments, small-grained cereals were sown because the seeds were easily identified in the soil and the growth habits of the seedlings were not unrelated to those of many grass species. Although the vigor of these relatively large seeds might have been expected to be to their advantage under stress, the physical difficulty of covering them appeared to increase with seed size and more than compensated for their vigor (Baker 1976b). Field experiments included small-seeded legumes.

In a summary of 19 experiments conducted during the period 1976 to 1979 in which the three openers were compared, Baker (1979) noted that in 11 out of 13 experiments in dry soils, the chisel opener design promoted significantly more seedling emergence (or at least subsurface seedling survival) than the triple-disk opener and in most cases was also superior to the hoe opener. The magnitude of these differences ranged from 20% ($P < 0.05$) to 14-fold ($P < 0.001$). The performance of the hoe opener was almost always intermediate between those of the other two openers. In a further 2 dry-soil experiments, and in 6 experiments in moist soils, there were no significant differences among the three opener designs.

Seedling emergence of seed sown with the hoe opener (and, to a lesser extent, of that sown with the triple-disk opener) was significantly improved by narrow press wheels operating directly on the seeds at the bases of the grooves before covering. No improvement resulted from the action of wider press wheels operating on the covered grooves. In only one experiment, however, was any press-wheel action with either of these two openers able to lift the seedling emergence count to that of the chisel opener without a press wheel (Choudhary and Baker 1980). This experiment was with the hoe opener.

Competition Habitat

Clearly, the desirability of retaining vegetation close to (and even over) the groove opening as a barrier to moisture loss contrasts with the desirability of physically removing it to lessen competition. A logical way of achieving both objectives seemed to be to kill or suppress the vegetation in situ in a band by spraying (Blackmore 1962).

Several mechanical options, and herbicides such as paraquat and glyphosate, have been compared for band spraying. The relatively favorable performance of paraquat in limited New Zealand comparisons (Baker et al. 1979c) may have been associated with its more rapid action, the large water-application rate (up to 500 l/sprayed ha), the high interception velocity of droplets (discharged from only a few centimeters above the foliage), and the probability that unlike glyphosate it was not translocated out of the band.

Where only temporary (4–8 weeks) suppression was required, or vigorous annual species were being introduced, the rapid "knockdown" effect of paraquat in relatively narrow bands appeared to be of advantage. Different band widths had effects on vigor and tillering in overdrilling less vigorous perennial species. On the other hand, where longer-term suppression is required, or difficult-to-kill species are present, glyphosate may have an important role to play. In this respect H.T. Kunelius (personal communication, 1980) reported promising results from the use of glyphosate for band spraying in both Canada and New Zealand.

AVAILABLE MACHINERY

There continue to be available in New Zealand several drill openers capable of creating a range of shapes of seed grooves while simultaneously chopping (Dunbar et al. 1979), bursting (Blackmore 1962), or scuffing aside (Robinson and Cross 1960) surface vegetation; but these options may be performed at the expense of MVPC. The triple-disk and chisel openers have little physical effect on the competing vegetation and therefore not only benefit from band spraying but, in dense swards, are reliant upon it.

The chisel opener is now being produced commercially in three forms by Aitchison Industries Ltd., P.O. Box 27, Wanganui, New Zealand. In an adaptation of its simplest form (Baker 1976a), it is attached to a simple, deliberately low-priced overdrilling machine. For operation in trash-free pastures this machine has no front disks and limited contour-following and trash-handling capabilities, and it does not sow dry fertilizer. A second version is attached to a more conventional combine drill that does feature these capabilities. With both machines, groove closure is improved by the use of a bar harrow behind the drill (Baker 1970). When the machines are equipped for optional band spraying, the nozzles are mounted ahead of the openers (and front disks, where applicable).

The more sophisticated and complex form of the chisel opener (Baker et al. 1979a, 1979b) is attached to a large and durable drill. In this form it has the additional capabilities of precise and individually adjustable depth control, simultaneous placement of fertilizer and insecticide (displaced 2 cm to the side of the seed), self-closure of the groove (using zero-pressure-tire press wheels), and excellent trash-handling properties. This machine has the flexibility of being suitable for a wide variety of activities, ranging from steep hill-country overdrilling to cash-crop direct drilling. Because the grooves are closed by press wheels, band-spraying nozzles, when fitted, are attached behind the openers where they are less liable to damage. Experiments are also proceeding with rolling-on of the herbicide, using the smooth rubber-tired press wheels in a similar manner to a lawn marker.

LITERATURE CITED

Baker, C.J. 1969. A tillage bin and tool testing apparatus for turf samples. J. Agric. Engin. Res. 14:357–360.

Baker, C.J. 1970. A simple covering harrow for direct drilling. N.Z. Fmg. 91:62–63.

Baker, C.J. 1976a. Experiments relating to techniques for direct drilling of seeds into untilled dead turf. J. Agric. Engin. Res. 21:133–144.

Baker, C.J. 1976b. Some effects of cover, seed size, and soil moisture status on establishment of seedlings by direct drilling. N.Z. J. Expt. Agric. 5:47–53.

Baker, C.J. 1979. Equipment impact and recent developments. Proc. "Roundup" Conserv. Till. Sem., Christchurch, N.Z.

Baker, C.J., E.M. Badger, J.H. McDonald, and C.S. Rix. 1979a. Developments with seed drill coulters for direct drilling. I. Trash handling properties of coulters. N.Z. J. Expt. Agric. 7:175–184.

Baker, C.J., J.H. McDonald, C.S. Rix, K. Seebeck, and P.M. Griffiths. 1979b. Developments with seed drill coulters for direct drilling: III. An improved chisel coulter with trash handling and fertilizer placement capabilities. N.Z. J. Expt. Agric. 7:189–196.

Baker, C.J., E.R. Thom, and W.L. McKain. 1979c. Developments with seed drill coulters for direct drilling: IV. Band spraying for suppression of competition during overdrilling. N.Z. J. Expt. Agric. 7:411–416.

Blackmore, L.W. 1962. Bandspraying: a new overdrilling technique. N.Z. J. Agric. 104:13–19.

Butterworth, W. 1980. Direct drilling: a technical roundup from overseas. Agr. Engin. 61(1):30–32.

Choudhary, M.A., and C.J. Baker. 1980. Physical effects of direct drilling equipment on undisturbed soils: I. Wheat seedling emergence from a dry soil under controlled climates. N.Z. J. Agric. Res. 23:489–496.

Choudhary, M.A., and C.J. Baker. 1981a. Physical effects of direct drilling equipment on undisturbed soils: II. Seed groove formation by a "triple disc" coulter and seedling performance. N.Z. J. Agric. Res. 24:183–187.

Choudhary, M.A., and C.J. Baker. 1981b. Physical effects of direct drilling equipment on undisturbed soils: III. Seedling performance and in-groove micro-environment in a dry soil. N.Z. J. Agric. Res. 24:189–195.

Choudhary, M.A., and C.J. Baker. 1982. Effects of direct drill coulter design and soil moisture status on emergence of wheat seedlings. Soil and Till. Res.

Dunbar, G.A., R. Horrell, and E.J. Costello. 1979. Sod seeding trials with a modified rotary hoe. Tussock Grassl. and Mount. Lands Inst. Rev. 38.

Haggar, R.J. 1977. Herbicides and low-cost grassland establishment, with special reference to clean seedbeds and one-pass seeding. Proc. Int. Conf. Energy Conserv. Crop Prod. Massey University, 31–38.

Robinson, G.S., and M.W. Cross 1960. Improvement of some New Zealand grasslands by oversowing and overdrilling. Proc. VIII Int. Grassl. Cong., 402–405.

Squires, N.R.W., and J.G. Elliott. 1975. The establishment of grass and fodder crops after sward destruction by herbicides. J. Brit. Grassl. Soc. 30:31–40.

Reduced-Tillage Pasture Renovation in the Semihumid Temperate Region of the U.S.A.

S.K. BARNHART and W.F. WEDIN
Iowa State University, Ames, Iowa, U.S.A.

Summary

Studies conducted in Iowa have shown that precision seed placement, good seed-to-soil contact, and reduced competition from other species are just as important when introducing legumes into an existing grass sward as when seeding into a conventional seedbed.

Tillage of strips 5.1 cm wide or wider was necessary for adequate sward suppression prior to and during legume establishment. Paraquat (1,1' dimethyl-4,4' bipyridinium ion) at 2.8 or 5.6 kg/ha could substitute for strip tillage in sward suppression. Paraquat banded over the seeded row provided satisfactory sward suppression, but legume establishment was improved when paraquat was applied broadcast or glyphosate [N-(phosphonomethyl) glycine] was band applied. Improved stand densities were obtained when precision placement of seed and press wheels were used, compared with surface seeding with shallow coverage and no soil firming over the seed.

Alfalfa (*Medicago sativa* L.), birdsfoot trefoil (*Lotus corniculatus* L.), and crownvetch (*Coronilla varia* L.) have all been successfully established in existing pasture swards via reduced-tillage renovation techniques.

Management of sward height postseeding significantly influences the rate of legume establishment in the renovated swards. The greater availability of light for a *Bromus*-dominated sward managed at a 7.5-cm height resulted in greater density of legume stand, subsequent dry-matter yields, and percentage of legume in the sward than in swards managed at greater heights.

In the year following renovation, dry-matter yields of renovated legume swards were similar to those of swards fertilized with 75 kg/ha of N, with a greater proportion of the total yield produced during the summer and early autumn months.

KEY WORDS: pasture renovation, interseeding, paraquat, glyphosate, *Lotus corniculatus* L., *Medicago sativa* L., *Coronilla varia* L.

INTRODUCTION

The semihumid temperate region of the U.S.A. is frequently recognized for its production of feed and food grains. Over 60% (14 million ha) of this area is presently in cropland. Much of the remainder is in grazed woodland and unimproved open permanent pasture. It is estimated that about 75% of this open unimproved pasture and woodland pasture might be profitably renovated.

An effective method of pasture renovation has been partial destruction of the existing sward by mechanical tillage plus liming, fertilization, and seeding as required to establish or re-establish desirable forage without an intervening crop (Sprague 1960). An alternative renovation method is often sought, particularly on hilly land, where soils are subject to excessive wind and water erosion and timing of tillage and seeding operations is difficult. There was a need for pasture-renovation techniques specifically designed for hilly lands and for a machine that would deliver seed at a precise depth and provide good seed-to-soil contact with a minimum of sod disturbance.

Several research studies conducted in Iowa have added to the basic knowledge and development of the components of effective reduced-tillage pasture-renovation techniques. Other common names given to this technique are interseeding, sod-seeding, and no-till pasture renovation.

SUMMARY OF IOWA RESEARCH

A study initiated in 1969 was designed to evaluate the potential of a strip-tillage technique for reduced-tillage pasture renovation (Gittins 1970). The study was conducted in an unimproved permanent pasture in south central Iowa on an Aquic Argiudoll soil. Kentucky bluegrass (*Poa pratensis* L.) was the dominant grass, with redtop (*Agrostis alba* L.) as a minor grass species.

In this study, tillage of strips 2.5, 5.1, and 7.6 cm wide was accomplished with modification of the blade design of a Howard Rotovator. Alfalfa (cv. Vernal and Travois), crownvetch (cv. Emerald), and birdsfoot trefoil (cv. Empire) were seeded in the tilled strips. Seedings were made by surface seeding with shallow seed coverage or with a shoe-opener garden seeder with a press wheel.

Results of this study are presented in Table 1. Very poor legume establishment was achieved in the 1.5-cm-

---

S.K. Barnhart is an Assistant Professor and W.F. Wedin is a Professor of Agronomy at Iowa State University.

Mention of a proprietary product does not imply endorsement of it.

Table 1. Average percent stand of seeded legumes as influenced by tilled-strip width and seeding method (Gittins 1970).

| Width of tillage | Seeding method | |
	Surface-seeded shallow coverage	Precision placement with press wheels
	----------------- % Stand-----------------	
2.5 cm	1.2	–
5.1 cm	8.1	–
7.6 cm	7.9	79.1

Table 3. Visual estimates of birdsfoot trefoil percentage in an interseeded *Bromus*-dominated sward, as influenced by sward-height management during the seeding year (Barnhart 1979).

| Seeding year canopy height | Year | | |
	1976	1977	1978
	------ % Legume ------		
Uncut	5	13	23
15.0 cm	6	20	33
7.6 cm	11	33	40

wide tilled strip. A 7-fold greater legume establishment was found in either the 5.1-cm- or the 7.6-cm-wide tilled strips. This improved stand establishment may have been related to a greater reduction in competition from the grass sward with the wider tillage. Precision seed placement and improved seed-to-soil contact resulted in a 10-fold increase in legume stand establishment in a 7.6-cm-wide tilled strip compared with surface seeding with shallow coverage.

In a second study at the same location, alfalfa (cv. Spredor), crownvetch (cv. Emerald), and birdsfoot trefoil (cv. Dawn) were seeded with a Howard Rotovator modified to till strips 2.5-cm wide (Clark 1974). Seedings in this study were made by surface placement on the tilled strip with a heavy press wheel for assuring seed-to-soil contact. Seeding rates of 6.7, 7.3, and 11.2 kg/ha of pure live seed for birdsfoot trefoil, alfalfa, and crownvetch, respectively, were used as the low seeding rate compared with rates of 1.5 and 2.0 times the low rate. Paraquat (1,1'-dimethyl-4,4' bipyridinium ion), a contact herbicide, was used in this study to test its suitability as a useful component in reduced-tillage pasture renovation.

The two higher seeding rates did not significantly (P < 0.05) improve legume stands. Data relating to the use of paraquat and stand establishment are presented in Table 2. The use of paraquat significantly (P < 0.01) improved the percentage of stand of the legumes, the greatest increase being for establishment of birdsfoot trefoil. Effectiveness of paraquat for sward suppression was similar (P < 0.05) at both the 0.28 and 0.56 kg/ha rates. Legume establishment as measured by stand percentage was also similar (P < 0.05) at both the 0.28 and the 0.56 kg/ha rate.

In a third study at the same location, a Midland "Zip Seeder" was used to establish alfalfa (cv. Spredor) and birdsfoot trefoil (cv. Carroll) in an untilled sod (Barnhart

1979). Paraquat was applied at a 0.56 kg/ha rate, both as a broadcast application and as a band application over the seeded row. A band application of 1.1 kg/ha of glyphosate [N-(phosphonomethyl) glycine] was also evaluated as a sward-suppression treatment. In the *Poa*-dominated sward, equally good legume establishment was obtained in the swards suppressed by banded glyphosate and broadcast paraquat (P < 0.05). Banded paraquat was less satisfactory for sward suppression, resulting in 20% to 50% less legume in the resulting swards.

In a study conducted on a Typic Hapludolf soil in southeastern Iowa, birdsfoot trefoil (cv. Carroll) was successfully seeded into a sward dominated by smooth bromegrass (*Bromus inermis* Leyss.) using a John Deere "Powr Till" seeder and band-applied paraquat for grass suppression (Barnhart 1979). Swards were maintained uncut or were managed by clipping at 7.5-cm or 15-cm heights during the seeding year. Light penetration of the sward was monitored during the seeding year. Estimates of integrated light penetration were made, using an ozalid paper technique (Francis 1970).

During periods of rapid sward growth, light penetration was reduced by over 50% within 20 days following clipping. Maintaining a low sward height during the seeding year resulted in a significantly (P < 0.05) higher percentage of legume in the sward (Table 3). Swards were managed at 4.0 cm and 7.6 cm during the 2 years following seeding. The conclusion of the study was that close defoliation of a rapidly growing sward at 2- to 3-week intervals may be necessary to maintain satisfactory light availability to the establishing interseeded species.

Dry-matter production from *Poa*- and *Bromus*-dominated swards interseeded with birdsfoot trefoil are presented in Table 4. Forage production from successfully interseeded swards was similar (P < 0.05) to that of swards fertilized with 75 kg/ha of nitrogen by the year following seeding. Of greater importance to livestock producers is the relatively greater production from legume-renovated swards during the summer and early autumn months as compared with unimproved or nitrogen-fertilized swards.

Table 2. Average percent stand of seeded legumes as influenced by paraquat rate (Clark 1974).

| Paraquat rate | Legume | | |
	Alfalfa	Birdsfoot trefoil	Mean
kg/ha	--------------- % Stand---------------		
0.00	52.2	18.0	35.1
0.28	71.1	46.6	58.9
0.56	61.7	49.4	55.6

CONCLUSIONS

This series of studies has identified several rules for the successful inclusion of legumes into grass swards via reduced-tillage methods. These rules are the same as those to be followed in conventional pasture renovation and forage seedings. Some of them are: (1) competition from the existing sward must be reduced during establishment; (2) precision placement of seed and good seed-to-soil contact aid in seed germination and seedling establishment; and (3) availability of light to establishing legumes increases the apparent vigor of the legume seedlings and the percentage of legume in the sward.

Table 4. Dry-matter yields of birdsfoot trefoil interseeded *Poa*- and *Bromus*-dominated swards interseeded with birdsfoot trefoil compared with unimproved and nitrogen-fertilized swards (Barnhart 1979).

Sward type	Poa-*dominated*			Bromus-*dominated*		
	1976	1977	1978	1976	1977	1978
	------------------------ kg/ha ------------------------					
Unimproved	840	1570	2320	1830	2570	3520
Sward fertilized with 75 kg/ha of nitrogen	1480	3440	3560	2670	4410	5400
Birdsfoot trefoil interseeded	490	3660	5360	1410	3840	6180

LITERATURE CITED

Barnhart, S.K. 1979. Interseeding and post-interseeding management of *Poa*- and *Bromus*-dominated swards. Ph.D. Dissertation, Iowa State Univ.

Clark, D. 1974. Using strip tillage to establish legumes in *Poa*-dominated swards. M.S. Thesis, Iowa State Univ.

Francis, C.A. 1970. Modification of the ozalid paper technique for measuring integrated light transmission values in the field. Crop Sci. 10:321–322.

Gittins, L.L. 1970. Interseeding legumes in *Poa*-dominant swards. M.S. Thesis, Iowa State Univ.

Sprague, M.A. 1960. Seedbed preparation and improvement of unplowable pastures using herbicides. Proc. VIII Int. Grassl. Cong., 264–266.

Influence of Pesticide, Fertilizers, Row Spacings, and Seeding Rates on No-Tillage Establishment of Alfalfa

L.R. VOUGH, A.M. DECKER, and R.F. DUDLEY
University of Maryland, College Park, Md., U.S.A.

Summary

Two experiments were conducted to investigate seeding techniques associated with no-tillage establishment of alfalfa (*Medicago sativa* L.) in barley (*Hordeum vulgare* L.) stubble. The objective of experiment 1 was to evaluate the influence of carbofuran (CF) insecticide/nematocide and of P and N fertilizers on seedling weight, plant population, and forage yield. All materials were applied in the row with the seed. The largest seedlings at 10 weeks after seeding occurred with CF + P + N. The smallest seedlings were associated with the check and P treatments. The highest plant populations at 6 and 30 weeks after seeding occurred where CF and CF + P were applied. The smallest plant populations occurred with the check and P treatments. Highest dry-matter yields at first harvest were with CF and CF + P. Yields of the check and P treatments were significantly lower than those of any of the other treatments. Dry-matter yields for the four treatments containing carbofuran were not significantly different at second harvest, but they were about double the yields of the check and P treatments.

The objective of experiment 2 was to examine the relationships of seeding rates and row spacings to seedling weight, plant population, weed competition, and forage yield. Seeding rates affected plant population and forage yield but had no significant effect on seedling size as measured by seedling weight 10 weeks after seeding. The 33 kg/ha seeding rate resulted in a significantly higher plant population at 30 weeks after seeding; however, there was no significant increase in forage dry-matter yield over either the 16.5 or 8.2 kg/ha seeding rates. Alfalfa seeded at the 5.5 kg/ha seeding rate yielded significantly less than that at the other seeding rates. At a seeding rate of 16.5 kg/ha, there were no significant differences in seedling weight, plant population, or forage dry-matter yield due to row spacings of 12, 24, and 36 cm.

KEY WORDS: alfalfa, *Medicago sativa* L., carbofuran, no tillage, seeding rates, row spacings, nitrogen, phosphorus, pesticides.

INTRODUCTION

No-tillage forage seedings can reduce soil erosion, conserve soil moisture, and lower fuel requirements, but seeding failures sometimes occur. There is a critical need to identify techniques that increase the frequency of successful no-till alfalfa (*Medicago sativa* L.) seedings. Failures are frequently blamed on some aspect of weather, but insects, nematodes, and molluscan pests may be involved.

Several research workers have indicated in personal communications that they have observed positive responses of no-till seeded forage legumes and grasses to carbofuran (2,3-dihydro-2,2-dimethyl-7-benzofuranol methylcarbamate) insecticide/nematocide. Reasons for the response have not been identified. Some researchers indicated that responses have been observed even when above-ground insects were not a problem. Investigations of effects of insecticides/nematocides on no-tillage establishment of forage crops are in the early stages, and published data are not available. However, with conventional alfalfa seedings (prepared seedbed), Byers et al. (1977) found that potato leafhopper (*Empoasca fabae* [Harris]) control in July with carbofuran applied preplant in April or May generally resulted in increased plant height and plant density; however, there was no indication of increased forage yields.

Band placement of seed and fertilizer was superior to broadcast seeding only on soils with low fertility and during seasons when environmental conditions were unfavorable (Brown 1959, Brown et al. 1960, and Carmer and Jacobs 1963). Generally, seeding rates greater than 11.2 kg/ha have not resulted in an increase in forage yield (Sund and Barrington 1976, Willard et al. 1934, Williams 1917, Zaleski 1959). Zaleski (1959) obtained significantly lower alfalfa yields from a 5.6 kg/ha seeding rate than from 11.2 kg/ha. Carmer and Jacobs (1963) found that total seasonal alfalfa yields were higher when seeded at 9.0 kg/ha than when seeded at 4.5 kg/ha. At each of the three seeding rates, row spacings of 10.2 and 20.3 cm had little or no influence on plant population, but the individual plants were somewhat larger at the 10.2-cm spacing. Yields were not affected by row width. Barney et al. (1974) found significantly lower yields at seeding rates below 9 kg/ha.

In an effort to identify some of the critical seeding techniques associated with successful no-till alfalfa seedings, we investigated the influences of insecticides, fertilizers, row spacings, and seeding rates.

METHODS

Experiment 1

Seeding was made 11 September 1979 in barley (*Hordeum vulgare* L.) stubble to evaluate the influence of

L.R. Vough is an Associate Professor and A.M. Decker is a Professor at the University of Maryland; and R.F. Dudley is an Agricultural Engineer, USDA, Plant Physiology Institute, Agricultural Equipment Laboratory, Beltsville, Md. Mention of a proprietary product does not imply endorsement of it.

granular carbofuran (CF), phosphorus (P), and nitrogen (N) on alfalfa seedling size, plant population, and forage yield. Soil test results for the plot area were: pH = 5.8, P = 39 kg/ha (medium range), and potassium (K) = 165 kg/ha (high range). Plots were seeded with a Tye Pasture Pleaser drill equipped with a grain box and two small seed boxes. Carbofuran was applied through one of the small seed boxes and fertilizer through the grain box. In this way, seed, CF, P and N were applied together in the row. Treatments were (1) check (seed only); (2) 12 kg P/ha; (3) 2.2 kg CF/ha; (4) 2.2 kg CF + 12 kg P/ha; (5) 2.2 kg CF + 27.5 kg N/ha; and (6) 2.2 kg CF + 12 kg P + 27.5 kg N/ha. Experimental design was a randomized complete block with 4 replications.

Experiment 2

Seeding was made 12 September 1979 in barley stubble to examine the relationships of seeding rates and row spacings to seedling size, plant population, amount of weeds, and forage yield. Soil test results for the plot area were: pH = 6.2, P = 51 kg/ha (medium range), and K = 231 kg/ha (high range). Plots were seeded with a Moore Uni-drill. Row spacings were 12, 24, and 36 cm. Seeding rates were 5.5, 8.2, 16.5, and 33.0 kg/ha. Experimental design was a randomized complete block with 4 replications.

Data collection was the same for both experiments. Plant counts were made 6 weeks after seeding in three 15.2 × 61–cm areas selected at random in each plot. Population counts were made again in these same areas the following spring to determine winter seedling mortality. Dry weights of the top growth of 25 seedlings selected at random in each plot were determined 10 weeks after seeding. Yields of the first and second cuttings were measured 5 June and 9 July 1980, respectively. Representative samples of first-cutting yields were separated into alfalfa and weed components. Weeds were minimal on all treatments at the second cutting.

RESULTS AND DISCUSSION

Experiment 1

Plant populations for the CF and CF + P treatments were more than double those of the check and P alone treatments at 30 weeks after seeding. The two treatments containing both CF and N resulted in significantly (P = 0.05) heavier seedlings than the check and P alone treatments (Table 1). However, N seemed to have a depressing effect on plant population. Addition of N to CF containing treatments significantly (P = 0.05) reduced plant population. Thus, the heavier seedling weights with addition of N may have been due to the lower plant population. There was no significant response to P either in seedling weight or plant population. Winter mortality was in the range of 50% to 75%, with CF and CF + P having the highest survival rates.

Yields from all treatments containing CF were significantly higher (P = 0.05) than the check and P treatments

Table 1.   Influence of carbofuran (CF), phosphorus (P), and nitrogen (N) on alfalfa seedling size, plant population, seedling survival, and dry-matter yields of alfalfa and weeds.

Treatments[1]	Dry wt. of 25 seedlings	Plant population 10/21/79	4/24/80	Survival	1st Cutting Yield Alfalfa	Weeds	2nd Cutting Yield
	g	Plants/0.09m²		%	ton/ha		
Check	1.47 b[2]	16.8 b	5.9 c	35	0.87 d	1.06 a	0.55 b
P	1.47 b	20.8 ab	4.9 c	24	0.71 d	1.08 a	0.66 b
CF	1.94 ab	30.2 a	13.8 a	46	2.00 ab	0.34 c	1.10 a
CF + P	2.19 ab	27.0 a	12.6 a	47	2.14 a	0.30 c	1.08 a
CF + N	2.34 a	25.0 ab	9.2 b	37	1.50 c	0.71 b	0.94 a
CF + P + N	2.77 a	23.0 ab	8.8 b	38	1.68 bc	0.64 b	1.10 a

[1]Seeded 9/11/79. Seedling weights measured 10/21/79.

[2]Means within a column followed by the same letter are not significantly different at the 5% level of probability using Duncan's Multiple Range Test.

(Table 1). Yields of both the first and second cuttings of the CF and CF + P treatments were about double those of the check and P alone treatments. Addition of P, with or without CF, did not increase forage yield. The check and P treatments were similar, CF and CF + P were similar, and CF + N and CF + P + N were similar, in terms both of yield and of plant population. Addition of N significantly (P = 0.05) reduced first-cutting alfalfa yields when compared with similar treatments without N. It also resulted in a significantly (P = 0.05) greater amount of weeds when compared to CF and CF + P treatments. However, there were no significant differences in second-cutting yields among the treatments containing CF.

Experiment 2

Neither row spacings nor seeding rates had an effect on seedling weight (Table 2). Doubling the seeding rate at the 12-cm row spacing resulted in significantly (P = 0.05) higher plant population 30 weeks after seeding but did not increase yields (Table 2). There were no significant effects of row spacing on plant population or yield when seeding rate was 16.5 kg/ha. Reducing seeding rate from 16.5 to 8.2 kg/ha at 24-cm row spacing significantly (P = 0.05) reduced plant population 30 weeks after seeding but did

not significantly reduce yield. There were no significant yield differences the first year for seeding rates of 8.2, 16.5, and 33.0 kg/ha. However, the plant population at 8.2 kg/ha was only 6.0 plants/0.09 m² in contrast to 9.0 to 11.7 plants/0.09 m² at 16.5 kg/ha. A plant population of 4.6 plants/0.09 m² at the 5.5 kg/ha rate resulted in significantly (P = 0.05) lower yields. Plant mortality over the winter ranged from 65% to 79%. Carbofuran was not applied at seeding in this experiment. Also note in Table 2 that as seeding rates were reduced below 16.5 kg/ha, the amount of weeds increased, particularly at 5.5 kg/ha. At 16.5 kg/ha, row spacing had no effect on the amount of weeds. This finding is contrary to popular opinion that wide row spacings result in more weeds.

Based on the preliminary results of this work, it would appear that carbofuran, or a similar insecticide/nematocide/molluscicide, is beneficial in no-tillage alfalfa establishment. Increasing the seeding rate above the 16.5 kg/ha recommended for conventional seedings is not beneficial.

LITERATURE CITED

Barney, G., M.A. Massengale, and A.K. Dobreny. 1974. Effect of seeding rate and harvest management on yield and stand

Table 2.   Influence of row spacings and seeding rates on alfalfa seedling size, plant population and dry-matter yields of alfalfa and weeds.

Treatments[1] Spacing	Rate	Dry wt. of 25 seedlings	Plant population 10/21/79	2/24/80	Survival	1st Cutting Yield Alfalfa	Weeds	2nd Cutting Yield
cm	kg/ha	g	Plants/0.09m²		%	ton/ha		
12	16.5	1.72 a[2]	26.9 ab	9.0 bc	34	2.35 ab	.14 c	1.52 a
12	33.0	1.39 a	57.8 a	16.3 a	28	2.90 a	.14 c	1.56 a
24	16.5	1.52 a	33.4 ab	11.7 b	35	2.37 ab	.09 c	1.43 a
24	8.2	1.69 a	17.0 b	6.0 cd	35	1.91 b	.28 b	1.08 a
36	16.5	1.26 a	40.6 ab	11.1 b	27	2.53 ab	.12 c	1.29 a
36	5.5	1.26 a	22.2 b	4.6 d	21	.92 c	1.22 a	.64 b

[1]Seeded 9/12/79. Seedling weights measured 10/21/79.

[2]Means within a column followed by the same letter are not significantly different at the 5% level of probability using Duncan's Multiple Range Test.

persistence in alfalfa. J. Ariz. Acad. Sci. 9(2):47–50.

Brown, B.A. 1959. Band vs. broadcast fertilization of alfalfa. Agron. J. 51:708–710.

Brown, B.A., A.M. Decker, M.A. Sprague, H.A. MacDonald, M.R. Teal, and J.B. Washko. 1960. Band and broadcast seeding of alfalfa-bromegrass in the Northeast. Md. Agr. Exp. Sta. Bul. A-108.

Byers, R.A., J.W. Neal, Jr., J.H. Elgin, Jr., K.R. Hill, J.E. McMurtrey III, and J. Feldmesser. 1977. Systemic insecticides with spring-seeded alfalfa for control of potato leafhopper. J. Econ. Entomol. 70(3):337–340.

Carmer, S.G., and J.A. Jackobs. 1963. Establishment and yield of late-summer alfalfa seedings as influenced by placement of seed and phosphate fertilizer, seeding rate, and row spacing.

Agron. J. 55:28–30.

Sund, J.M., and G.P. Barrington. 1976. Alfalfa seeding rates: their influence on dry matter yield, stand density and survival, root size and forage quality. Univ. Wis. Res. Bul. R2786.

Willard, C.J., L.E. Thatcher, and J.S. Cutler. 1934. Alfalfa in Ohio. Ohio Agr. Exp. Sta. Bul. 540.

Williams, C.G. 1917. Alfalfa culture, essentials in growing the legume the first year. Ohio Agr. Exp. Sta. Bul. 18:173–177.

Zaleski, A. 1959. Lucerne investigation. IV. Effect of germination and seed rates on establishment, mortality, and yield of dry matter and protein per acre. J. Agr. Sci. 53:260–267.

Production of Big Trefoil on Acid, Low-Fertility Soils in New Zealand

R.S. SCOTT and W.L. LOWTHER
Invermay Agricultural Research Center,
Ministry of Agriculture and Fisheries, Mosgiel, N.Z.

Summary

On undeveloped, acid ($pH < 5.0$) soils in New Zealand, growth of the conventional legumes white clover (*Trifolium repens* L.), red clover (*T. pratense* L.), and alsike clover (*T. hybridum* L.) is limited by acidity. The objective of this research was to identify an alternative legume species better adapted to both acidity and low inputs of phosphorus (P) and to examine grazing management. Big trefoil (*Lotus pedunculatus* Cav.) cv. Grasslands Maku and birdsfoot trefoil (*L. corniculatus* L.) cv. Maitland were examined. Seed was inoculated with appropriate strains of rhizobia and sown into undisturbed tussock grasslands on acid soils of low P status at altitudes of 550–880 m. Lime and phosphorus treatments were broadcast. Herbage yields were measured by mowing or by pregrazing and postgrazing ground-level cuts in animal-grazing studies.

Big trefoil gave 2–3 times the yield of white and red clover and birdsfoot trefoil. While broadcast lime increased yields of white clover (but not of big trefoil), the yields remained lower than those from big trefoil. Big trefoil gave markedly higher yields than white clover at light rates of P (7.5–10 kg/ha) and was responsive to the heaviest rate applied (60 kg/ha). White clover did not respond to rates of P greater than 15 kg/ha. Lack of P response by white clover is thought to be due to high levels of exchangeable aluminium present in the soils.

A grazing-management study on established swards showed yields similar under close defoliation (2.5 cm) with sheep and lax (7.5 cm) defoliation by cattle.

These studies demonstrate a large potential in New Zealand for the development of native grasslands on acid soils of low P status by oversowing with big trefoil.

KEY WORDS: *Lotus pedunculatus* Cav., *L. corniculatus* L., *Trifolium repens* L., *T. pratense* L., acid soils.

INTRODUCTION

There are large areas of undeveloped, nonarable tussock grassland soils on the South Island of New Zealand on which growth of the conventional pasture legumes white clover (*Trifolium repens* L.), red clover (*T.* *pratense* L.), and alsike clover (*T. hybridum* L.) is limited by soil acidity (Lowther 1980).

This paper presents results of investigations into alternative legumes for increased production on these soils without high inputs of lime or fertilizer. Of particular interest was an autotetraploid of Grasslands Maku big trefoil (*Lotus pedunculatus* Cav.) that has been shown to be better adapted than white clover to low phosphorus (P) and high exchangeable aluminium (Al) levels in eroded acid subsoils (Nordmeyer and Davis 1977).

The authors are Research Agronomists at Invermay Agricultural Research Center.

METHODS

Experiments were on undeveloped tussock grasslands subjected to extensive grazing (c. 1 ewe/ha) and producing about 500–1,000 kg dry matter/ha of low digestibility. The vegetative cover consisted of native snow tussock (*Chionochloa rigida* Raoul, Zotov), adventitious grasses browntop (*Agrostis tenuis* L.) and sweet vernal (*Anthoxanthum odoratum* L.), and a native plant understory. The sites were all on Dystrochrept soils with pH (1 : 2.5 soil : water) 4.5–4.9, and meq percentages for calcium 0.30%–0.62%, potassium 0.15%–0.30%, and magnesium 0.42%–0.88%. Phosphorus (Olsen) levels were 2–6 ppm and Al (0.01 M $CaCl_2$ extractable) 47–70 ppm. Sites ranged in altitude from 550 to 880 m and annual rainfall ranged from approximately 700 to 900 mm.

In all trials 200 g/ha of sodium molybdate and 20–40 kg P/ha as superphosphate were applied at sowing except on the rate-of-P experiment. Superphosphate was reapplied annually. Fertilizer and seed (3–6 kg/ha) were applied by hand to the undisturbed soil surface. Seed was inoculated with commercial peat inoculant of the appropriate rhizobia strain, pelleted with 1 : 1 rock-phosphate : dolomite, and sown within 24 hours. Yields of dry matter were taken from established swards (2–4 years) to avoid effects of differences in rate of establishment. In the mowing experiments an 8- to 12-week regrowth period was allowed to maximize production of the various legumes. Yields in the grazing experiment were measured by ground-level quadrat cuts pregrazing and postgrazing.

RESULTS AND DISCUSSION

Comparison of Legume Species

Swards of Grasslands Huia white clover and Grasslands Pawera red clover were sown alone or as a mixture of equal seed numbers and compared with big trefoil and Maitland birdsfoot trefoil (*L. corniculatus* L.) in the absence of lime on three sites (Table 1). Highest yield of dry matter (adventitious grasses and legume) was ob-

Table 1. Comparison of total (legume + grass) dry-matter yields (t/ha) for different legume species (% legume in parentheses).

	Berwick pH 4.6	Waipori pH 4.6	Rocklands pH 4.9
White clover	3.2 (51%)	1.9 (42%)	1.4 (27%)
Red clover	3.6 (75%)	–	–
White and red clover	3.2 (65%)	–	–
Big trefoil	5.9 (85%)	5.9 (87%)	2.8 (59%)
Birdsfoot trefoil	–	2.6 (85%)	1.7 (54%)
SEM	0.4	0.3	0.2

Table 2. Effect of broadcast lime on legume dry-matter yield (t/ha) [Al contents (ppm) in parentheses].

Lime	0	1250 kg/ha
White clover	0.7 (169)	1.4 (150)
Big trefoil	3.3 (83)	3.7 (77)
White clover and big trefoil	2.2 (84%)*	1.9 (43%)*
SEM	0.2	

*Percentage of big trefoil in the legume component.

tained from the big trefoil sward, which also contained a higher percentage of legume.

Birdsfoot trefoil plants were small and yellow, indicating an effect of soil acidity on nitrogen (N) fixation, a result in line with the pH requirements for birdsfoot trefoil (McKee 1961). Although white and red clovers did not exhibit N-deficiency symptoms, chemical analysis of white clover dry matter suggested that plant growth was limited by Al toxicity, as the level (170 ppm) was considerably higher than the level in big trefoil (83 ppm) and within the range associated with Al toxicity (Andrew et al. 1973).

As earlier studies (Lowther 1980) demonstrated that broadcast lime aided big trefoil nodulation on the more acid soils (c. pH 4.6), an experiment was conducted to study the effect of lime on plant growth. White clover and big trefoil sown alone or as a mixture of equal seed numbers were compared in the presence or absence of 1,250 kg lime/ha (Table 2). Yields for big trefoil were markedly higher than those from white clover or the white clover–big trefoil mixture in both presence and absence of broadcast lime, even though broadcast lime doubled the yield of white clover while having no effect on that of big trefoil. The yield from the white clover–big trefoil mixture was significantly higher than that from white clover alone in the absence of broadcast lime. Although broadcast lime had no effect on the legume yield of the mixture, it reduced the contribution of big trefoil. Sowing big trefoil with white clover or introducing it into an area containing white clover may increase yield over that obtainable from white clover, but production will be lower than from big trefoil alone. Furthermore, on higher pH soils competition from white clover may greatly reduce production from big trefoil.

In the present experiment a rate of 1,250 kg lime/ha was chosen since legumes have failed to respond to higher rates in similar environments (Adams and Lowther 1970), a result attributed to the restricted zone of influence of broadcast lime.

Results (Fig. 1) from an experiment where various rates of P were applied annually for 4 years (as monocalcium phosphate) to oversown white clover and big trefoil showed that without fertilizer P there was little growth of either species. Big trefoil outyielded white clover at all levels of applied P, with the yield advantage being greatest at the highest level of P. This contrasts with the results of Brock (1973), who found that big trefoil

Fig. 1. Effect of rates of applied phosphorus on white clover (---) and big trefoil (—) dry-matter yields.

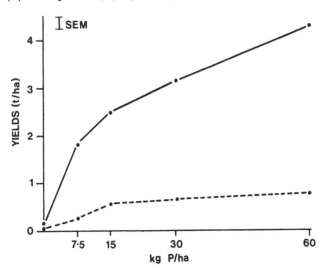

outyielded white clover only under conditions of low P. The poor growth from white clover in the present experiment can be attributed to the effect of soil acidity and associated high levels of Al limiting the response to applied P.

Although high rates of P are required for maximum production of big trefoil, its production is still considerably higher than that of white clover at low rates of P (7.5–15 kg/ha), which are the rates likely to be adopted by farmers applying fertilizer from aircraft to high-altitude sites.

Grazing Management Of Established Big Trefoil

Following a study that showed that rhizome development of big trefoil is increased by deferring grazing for 8 months following sowing (Scott and Mills 1981), a grazing experiment was conducted on established swards of big trefoil and white clover. It compared net herbage production from these swards when defoliated to either 2.5 cm with sheep or 7.5 cm with cattle. Common regrowth periods of about 9 weeks were given. Results (Table 3)

Table 3. Effects of sheep and cattle grazing on legume dry-matter yields (t/ha) two and three years after oversowing.

	Part Second Year		Third Year	
	Sheep	Cattle	Sheep	Cattle
Defoliation height:	2.5 cm	7.5 cm	2.5 cm	7.5 cm
Big trefoil	0.99	1.21	2.93	3.23
White clover	0.83	0.88	1.20	1.04
Significant effects:				
Species		*		*
SEM		0.07		0.09

*No significant difference.

demonstrated markedly higher yields from big trefoil but did not show any difference in herbage production of either species following grazing with either sheep or cattle. It therefore appears that big trefoil is suitable for grazing by either class of stock, a theory confirmed by the finding of Sheath (1980), who demonstrated similar net herbage production under lax and close defoliation.

Persistence is an important attribute of a species in permanent pasture. Brock and Charlton (1978) showed that big trefoil under high fertility and moderately high soil pH would not persist under grazing when sown in ryegrass mixtures, largely because of competition from volunteer white clover. Scott and Mills (1981) suggested that lack of persistence in these soils is due to low exchangeable Al and high P levels favoring white clover dominance. On acid soils big trefoil competed with white clover and formed the major component of the mixture, although the yields were lower than those from big trefoil alone (Table 2). However, small changes in soil acidity (e.g., by application of 1,250 kg lime/ha) influenced competitive ability. A competition study confirmed that big trefoil is the aggressor and white clover the subordinate on acid soils (Scott and Lowther 1980).

On these acid soils big trefoil persisted under practical farming. Following 10 years of grazing of variable intensity by both sheep and cattle, it totals over 60% of the legume component. Thus, present evidence suggests that big trefoil will be a persistent cultivar on appropriate soils.

LITERATURE CITED

Adams, A.F.R., and W.L. Lowther. 1970. Lime, inoculation, and seed coating in the establishment of oversown clovers. N.Z. J. Agric. Res. 13:242–251.

Andrew, C.S., A.D. Johnson, and R.L. Sandland. 1973. Effect of aluminium on the growth and chemical composition of some tropical and temperate pasture legumes. Austral. J. Agric. Res. 24:325–339.

Brock, J.L. 1973. Growth and nitrogen fixation of pure stands of three pasture legumes with high/low phosphate. N.Z. J. Agric. Res. 16:483–491.

Brock, J.L., and J.F.L. Charlton. 1978. *Lotus pedunculatus* establishment in intensive farming. Proc. N.Z. Grassl. Assn. 39:121–129.

Lowther, W.L. 1980. Establishment and growth of clovers and lotus on acid soils. N.Z. J. Expt. Agric. 8:131–138.

McKee, G.W. 1961. Some effects of liming, fertilisation, and soil moisture on seedling growth and nodulation in birdsfoot trefoil. Agron. J. 53:237–240.

Nordmeyer, A.H., and M.R. Davis. 1977. Legumes in high country development. Proc. N.Z. Grassl. Assoc. 38:119–125.

Scott, R.S., and W.L. Lowther. 1980. Competition between white clover 'Grasslands Huia' and *Lotus pedunculatus* 'Grasslands Maku'. I. Shoot and root competition. N.Z. J. Agric. Res. 23:501–507.

Scott, R.S., and E.G. Mills. 1981. Establishment and management of 'Grasslands Maku' *Lotus pedunculatus* Cav. in acid, low fertility tussock grasslands. Proc. N.Z. Grassl. Assoc. 42:131–141.

Sheath, G.W. 1980. Production and regrowth characteristics of *Lotus pedunculatus* Cav. cv. 'Grasslands Maku.' N.Z. J. Agric. Res. 23:201–209.

Effectiveness and Safety of Translocated Herbicides Applied to Pasture Weeds with a Rope-Wick Applicator

E.J. PETERS

USDA-SEA-AR and University of Missouri,
Columbia, Mo., U.S.A.

Summary

Applying herbicides with a rope-wick applicator is a new technique that permits treatment of tall weeds without treatment of low-growing forage species. The objective of this study was to determine the effectiveness of several herbicides and to evaluate their safety to forage species. The effectiveness of 2,4-D [(2,4-dichlorophenoxy)acetic acid], dicamba (3,6-dichloro-o-anisic acid), picloram (4-amino-3,5,6-trichloropicolinic acid), and glyphosate [N-(phosphonomethyl) glycine] in water-herbicide concentrations of 2:1, 4:1, and 8:1 were evaluated on gray goldenrod (*Solidago nemoralis* Ait) when applied with a rope-wick applicator in one and two directions on 15 June and 17 July 1979. The goldenrod was growing in a pasture containing primarily Kentucky bluegrass (*Poa pratensis* L.). Red clover (*Trifolium pratense* L.) was sown into the stand the winter after treatment to detect any residues of herbicides that might be present. Some injury to red clover occurred with picloram. Picloram, applied in two directions, killed more than 80% of the goldenrod, while the other herbicides killed 60% to 65%. Early treatments and treatment in two directions were most effective. Forage grasses were unaffected by the herbicides. This method of application effectively controlled goldenrod with smaller amounts of herbicides than with conventional broadcast applications.

KEY WORDS: 2,4-D, picloram, glyphosate, dicamba, *Solidago nemoralis*, rope wick.

INTRODUCTION

Broadleaf pasture weeds can be controlled by spraying with translocated herbicides, but injury to legumes frequently occurs. Also, some herbicides that are effective on weeds injure forage grasses. The chance of injury, the cost of application equipment, and the cost of herbicides frequently discourage farmers from spraying.

The rope-wick applicator was developed for use in row crops to apply glyphosate to the tops of weeds that extend above the crop. This application method shows promise for treating tall-growing pasture weeds while avoiding application to forage (Peters and Dale 1978). The applicators are economical to build, and they require less herbicide than broadcast spray applications because they treat only the weeds. Glyphosate was used initially, but the effectiveness of other translocated herbicides has not been fully evaluated. Also, the effectiveness of various ratios of water to herbicides has not been evaluated.

The objective of this research was to evaluate the effectiveness of the rope-wick applicator for applying four herbicides at three water-herbicide concentrations in one and two directions and on two dates during the summer.

METHODS

The rope-wick apparatus consisted of a polyvinyl chloride (PVC) pipe 7.5 cm in diameter × 2.4 m long with two rows of ropes on the leading side (Fig. 1). The ropes were cut into segments, and the ends were inserted into holes spaced 20 cm apart in the pipe. The ends of the pipe were fitted with handles to enable two men to carry the device at about 5 km/hour at the proper height for treating weeds.

The experiment was a complete factorial with plots 2.4 × 6 m arranged in a randomized complete-block design with 4 replications. Herbicides used were the butoxyethanol ester of 2,4-D [(2,4-dichlorophenoxy) acetic acid], dicamba (3,6-dichloro-o-anisic acid), the potassium salt of picloram (4-amino-3,5,6-trichloropicolinic acid) and glyphosate [N-(phosphonomethyl) glycine], each applied on concentrations of 1 part commercial formulation of herbicide to 2, 4, or 8 parts water. These solutions were applied on 15 June and 17 July 1979 to determine whether application dates or herbicide concentrations influenced control. Each herbicide treatment at each date was applied by wiping in one direction and compared with wiping in two directions.

The pasture used for the experiment, near Columbia, Missouri, contained primarily Kentucky bluegrass (*Poa pratensis* L.) with a dense stand of gray goldenrod (*Solidago nemoralis* Ait.) making up 50% to 60% of the canopy.

The author is a Research Leader of USDA-SEA-AR and a Professor of Agronomy at the University of Missouri. This a joint contribution of USDA-SEA-AR and the Missouri Agricultural Experiment Station.

Fig. 1. External view of rope-wick applicator showing the components assembled. Materials are as follows: (1) PVC pipe, 7.5 cm in diameter, (2) capped PVC fill spout, (3) end cap, (4) wicks of soft braided nylon marine rope 1.25 cm in diameter, and (5) rubber bushing or cement.

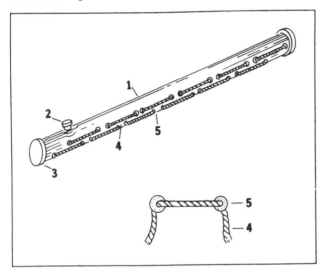

Fig. 2. Percentage of reduction of goldenrod stems in 1980 after treatments at two dates in 1979 with four herbicides at three concentrations and treated in one or two directions.

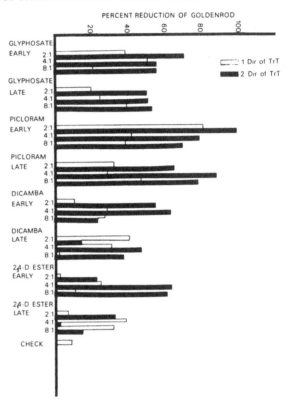

White clover (*Trifolium repens* L.) was present but died of summer drought. Live and dead stems of goldenrod were counted in 30-cm strips 6 m long at the time of the treatments and on 18 June 1980 to evaluate the effectiveness of treatments. In February 1980, red clover (*Trifolium pratense* L.) was broadcast on the surface of frozen soil at the rate of 10 kg/ha. The legume was sown to determine if any herbicide residues were present that might injure the clover after it emerged.

An estimate of the amount of herbicide applied with a rope wick was obtained on two large plots, each 0.4 ha in size, adjacent to the main experiment. The ropes of the applicator were saturated in a 1 : 2 glyphosate : water solution. The solution was then measured into the pipe and remeasured after treatment of each plot. The difference in volume was used to calculate the amount of herbicide applied. The herbicide was applied at a speed of about 5 km/hour with the rope wick mounted on a tractor.

RESULTS

Fig. 2 shows the percentage of goldenrod stems killed by the various treatments. Many of the stems that had been in contact with the rope containing glyphosate, dicamba, or 2,4-D died, but adjacent short stems that were not touched showed no visible injury symptoms. However, with picloram, many of the short goldenrod stems that did not come in contact with the rope died, as well as those that had received direct treatment. These results indicated that picloram was translocated to the base of the plants in sufficient quantity to kill most of the stems.

Statistically significant differences (0.05 probability) were found among the main effects of herbicides, direc-

tions of treatment, and dates of treatment. Picloram killed 15% to 20% more goldenrod stems than the other herbicides. Treatments in two directions killed significantly more goldenrod stems than treatment in one direction. Early treatments killed more stems than late treatments.

With glyphosate applied early in two directions, the 2 : 1 concentration killed about 18% more goldenrod stems than the other concentrations, but when glyphosate was applied late in two directions, there was little difference in effects among concentrations.

With picloram applied early, the 2 : 1 concentration was 20% to 30% more effective than the other concentrations; but when picloram was applied late, the 4 : 1 concentration was more effective than the 2 : 1 concentration.

Among the dicamba treatments, the 4 : 1 concentration was more effective than the other concentrations. The early-applied 2,4-D was more than twice as effective at the 4 : 1 and 8 : 1 concentrations than at the 2 : 1 concentration.

The amount of water : glyphosate solution (2 : 1) applied to the large 0.4-ha plots ranged from 130 to 160 ml/plot (average of 145 ml), or 48.3 ml of commercial glyphosate solution containing 480 g acid equivalent of glyphosate/l. This would amount to 58 g acid equivalent of glyphosate/ha.

Red clover became established in all plots, but plants in picloram-treated plots had abnormal leaves, indicating injury from the herbicide. No injury was apparent to red clover in plots treated with other herbicides. Whether or not the amount of picloram residue remaining was sufficient to kill red clover could not be determined because red clover in all plots died from drought in June 1980. The injury symptoms indicate that residues of picloram reached and persisted in the soil from 1979 to 1980. Picloram may have reached the soil by rain washing the herbicide from the treated stems or from decomposing goldenrod tissue, or it may have been excreted from the roots of goldenrod in a manner similar to that reported by Bauer and Bovey (1969), Hurtt and Foy (1965), and Merkle et al. (1967).

DISCUSSION

With a few exceptions, treatments in two directions were more effective than treatment in one direction because a second contact of the rope with the weeds was achieved, evidently resulting in more herbicide being applied.

Among the treatments in two directions, picloram was more effective than other herbicides for killing goldenrod. Early treatments with dicamba at 2:1 and 4:1 concentrations and early treatments with 2,4-D at 4:1 and 8:1 concentrations controlled goldenrod almost as well as glyphosate.

As the ratio of water to herbicide increases, it would be expected that the amount of herbicide applied would decrease and the degree of weed mortality would decrease; however, the degree of mortality due to concentration varied with herbicides and dates of treatment. The variability may be related to the degree of absorption of the herbicide by the plant as influenced by water-herbicide concentration and to the properties of each herbicide interacting with the environment at time of treatment.

The solution measured on the large plots contained glyphosate, and if it is assumed that with other herbicides the volumes of solutions used would not be greatly different, it can be concluded that very small quantities of herbicides were used. The degree of mortality of goldenrod with the rope-wick applicator in this study (applied in two directions), using approximately 116 g acid equivalent of herbicide, compared favorably with the mortality obtained in other studies with broadcast applications of 0.5 to 2 kg acid equivalent (Peters and Lowance 1969).

LITERATURE CITED

Bauer, J.R., and R.W. Bovey. 1969. Distribution of root-absorbed picloram. Weed Sci. 17:524–527.

Hurtt, W., and C.L. Foy. 1965. Some factors affecting the excretion of foliarly applied dicamba and picloram from roots of Black Valentine beans. Plant Physiol. (Sup.) 40:48.

Merkle, M.G., R.W. Bovey, and F.S. Davis. 1967. Factors affecting the persistence of picloram in soil. Agron. J. 59:413–415.

Peters, E.J., and J.E. Dale. 1978. Rope applicators and recirculating sprayers for pasture weed control. Proc. N. Cent. Weed Contr. Conf., 127.

Peters, E.J., and S.A. Lowance. 1969. Gains in timothy forage from goldenrod control with 2,4-D, 2,4-DB, and picloram. Weed Sci. 17:473–474.

Renovation of Tropical Legume–Grass Pastures in Northern Australia

H.G. BISHOP, B. WALKER, and M.T. RUTHERFORD
Department of Primary Industries, Queensland, Australia

Summary

Live-weight gain of cattle on tropical legume-grass pastures is highly correlated with legume content. Legume-yield decline is a major problem in 200,000 ha of siratro (*Macroptilium atropurpureum* [DC] Urb.)-based pastures in northern Australia. This decline is largely associated with overgrazing. Resting for at least one year has been required to improve siratro yield substantially. A more rapid method of restoring the yield of siratro was required.

The effect of various forms of mechanical soil disturbance on siratro population and yields was compared with resting pasture (control) in 4 replicated experiments and on 3 commercial pastures near Mackay in coastal northeastern Australia. The renovation treatments were cultivation with tines, deep ripper, disk harrow, disk plough, and rotary hoe, with a control treatment of no cultivation at all sites. Three of the experiments contained an oversowing treatment of 2 kg/ha siratro seed. Associated grass was Kazungula setaria (*Setaria sphacelata* var. sericea [Schum.] Stapf & C.E. Hubb) at 6 sites and Rodd's Bay plicatulum (*Paspalum plicatulum*) at 1 site.

The main effect over all sites was for cultivation to increase the number of siratro seedlings over the controls from 3 to 10/m² and siratro yield from 263 to 1,289 kg/ha. Disk ploughing resulted in more seedlings (17_2/m²) and higher siratro yields (1,439 kg/ha) than tine ripping (9/m² and 501 kg/ha, respectively), the latter disturbing approximately 50% of the pasture surface. One advantage of disk ploughing was to reduce substantially the asssociated grass yield. Oversowing with 2 kg/ha siratro seed failed to increase populations or yield. An adequate reserve of siratro seed in the soil was found to be essential, and management of pastures should aim to achieve these reserves.

This study showed that cultivation was a rapid method of restoring the legume content of siratro-based pastures.

KEY WORDS: renovation, tropical pastures, siratro, legume decline, soil seed reserves, oversowing.

INTRODUCTION

Tropical legume-grass pastures have contributed greatly to increasing cattle production from coastal northern Australia (Hutton 1974). The legume component is most important as a dual contributor of nitrogen for pasture growth and protein for the animal diet. Animal production has been shown to be highly correlated with the legume content of mixed pastures (Evans 1970).

Siratro is the most widely sown perennial tropical legume in northeastern Australia, and the current estimated area sown is around 200,000 ha. The maintenance of a high legume content in these siratro-based pastures is highly dependent on stocking rate; under heavy grazing pressures siratro fails to persist (Jones and Jones 1978). This decline after 3 to 5 years has been experienced in the Mackay region of central Queensland where approximately 50,000 ha have been planted (Bishop and Walker 1980).

Resting pasture is a slow means of re-establishing siratro. Jones and Jones (1978) suggest that rest for a full growing season is required, and Walker and Bishop (unpublished data) found that neither a full growing season (January to June inclusive) nor various periods of 8 weeks' rest within the season were successful in restoring the legume content.

Pasture renovation was therefore examined as a rapid method of restoring the legume content of old siratro-grass pastures. This paper reports various comparisons of five cultivation methods, and a control of no cultivation, on the siratro content of old siratro-grass pastures in the Mackay region. The results from four experiments and three commercial pastures are reviewed.

METHODS

Sites

Four replicated experiments (sites I, II, III, and IV) and three commercial pastures (sites V, VI, and VII) were all situated on grazing properties within 80 km of Mackay (lat 21°09′ S long 149°11′ E) with average annual rainfall 1,672 mm. Associated grass was Kazungula setaria (*Setaria sphacelata* var. sericea [Schum.] Stapf & C.E. Hubb or M.B. Moss) at six sites and Rodd's Bay

H.G. Bishop and B. Walker are Pasture Agronomists and M.T. Rutherford is an Experimentalist in the Queensland Department of Primary Industries.

plicatulum (*Paspalum plicatulum* Michx.) at one site. All the pastures were over 6 years old and were growing on either solodic or soloth soils. All sites had a previous history of adequate fertilizer application.

Treatments

The cultivation treatments were tine ripping, tandem-disk harrowing, disk ploughing, and rotary hoeing, each to 10 cm deep, and deep ripping, 30 cm deep. Cultivation was carried out following early storm rains within the period late November to early January. Not all cultivation treatments were used at each site (Table 1). All sites except VII contained a control treatment of no cultivation. Pastures were rested from grazing between renovation and final yield sampling. Sites I, II, and III contained an oversowing treatment of 2 kg/ha siratro seed immediately after renovation.

Measurements

Siratro seedlings were counted in 20 0.25-m² quadrats/plot for replicated experiments and 120 quadrats/paddock for commercial pastures. Counts were made 4 to 6 weeks after germinating rains. Dry-matter yields were estimated by cutting 4 quadrats/plot, 1.0 m² at site I and 0.5 m² at site IV. Siratro yields in the commercial pastures at sites V and VI were estimated by the technique of Jones and Hargreaves (1979). Yield was not recorded at the remaining sites. Siratro seeds in the surface 10 cm were determined, using the method of Jones and Bunch (1977).

RESULTS

Rainfall for the November-April period at all sites was adequate, ranging from 987 to 1,545 mm year. Oversowing with 2 kg/ha siratro seed did not significantly increase siratro seedlings or yields.

Siratro Seedling Numbers

All cultivation treatments increased the number of siratro seedlings over the control at each site (Table 1). At site VII, where no reserves of siratro seed were found, seedling numbers were low. Disk ploughing at site IV and rotary hoeing at site II significantly ($P < 0.05$) increased seedling populations compared with other cultivation

Table 1. Initial siratro seed reserves (kg/ha) and siratro seedling numbers/m² following various renovation treatments at 7 sites.

Treatments	Sites						
	I	*II*	*III*	*IV*	*V+*	*VI+*	*VII+*
Siratro seed reserves, kg/ha	32	59	n.a.*	5	12	2	0
Siratro seedlings/m²							
Control	6.0	3.2	0.3	4.3	4.0	0.5	
Tine ripping	31.0	7.2	4.6	8.9	10.0	4.0	1.0
Deep ripping		4.0					
Tandem discs		5.5					
Disc ploughing				14.1	20.0		
Rotary hoeing		11.8					
L.S.D. (P = 0.05)	11.92	3.14	2.78	2.02			

*N.a. = not available

+ Counts from the unreplicated commercial pastures were not statistically analysed.

methods at those sites. Deep ripping had no advantage over ripping to 10 cm.

Yield

Tine ripping significantly (P < 0.05) increased siratro yield at site I, where mean yield after 123 days was 49 kg/ha for the control and 675 kg/ha for ripping. The estimated siratro yields from ripping at sites V and VI were 1,095 and 1,445 kg/ha, respectively. These amounts were significantly (P < 0.05) greater than the control treatments, which yielded less than 100 kg/ha of siratro.

At site IV disk ploughing produced the highest siratro yields and was successful in reducing the growth of companion grass early in the season (Table 2). Grass re-established from seed on the disk-ploughed plots, while about 50% of old grass clumps survived tine ripping. Disk ploughing produced a large increase in broadleaf weeds, principally *Sida rhombifolia* L. and *Urena lobata*. L.

Table 2. Dry-matter yield (kg/ha) of pasture components at site IV after 49 and 131 days.

Pasture components	Treatments			
	Control	Tine ripping	Disc ploughing	L.S.D. (P = 0.05)
Siratro				
49 days	1	3	44	32
131 days	213	501	1,439	241
Kazungula setaria				
49 days	2,585	2,220	19	667
131 days	6,483	5,298	1,226	1,105
Broadleaf weeds				
49 days		not recorded		
131 days	52	266	3,270	316
Total+ yield				
49 days	3,721	3,382	625	1,258
131 days	8,789	8,086	7,139	1,455

+ Includes yield of miscellaneous weeds not shown separately.

DISCUSSION

Pasture renovation was found to be a successful means of restoring the siratro content of old pastures where adequate soil reserves of siratro seed were available. The maintenance of adequate soil reserves of seed is an important management consideration. Pastures should be allowed to seed, preferably in the establishment year. The lack of siratro seed reserves at site VII (Table I) was expected, as this pasture had a management history of early and heavy grazing that prevented siratro from seeding. The failure of 2 kg/ha of oversown siratro seed to increase seedling density in experiments I and II is understandable when this amount is compared with the soil seed reserves of 32 and 59 kg/ha, respectively, for these sites.

Jones and Jones (1978) report most of the reserves of siratro seed in old pastures are in the soil surface between 2.5 and 7.5 cm, germination having depleted the surface 2.5 cm of seed. Soil disturbance is required to bring deeper seed up to the surface. Cultivation also appears to be important for siratro growth. Walker (personal communication) has attributed part of the decline in siratro yield to increases in soil compaction at heavy stocking rates. Part of the improved growth by siratro after cultivation, in this study, is believed to be a response to the loosening and aeration of the soil, increasing infiltration of early storm rains and causing mineralization of nutrients and thus better growing conditions.

The 49-day harvest in experiment IV demonstrated the vigorous growth of Kazungula setaria when it was rested from grazing at the start of the growing season. Ripping had little effect in reducing grass growth, although it stimulated growth of broadleaf weeds (Table 1). Only disk ploughing effectively removed competition to growth of grass. Local experience has shown that broadleaf weeds are a problem only in the first year of a pasture.

Pasture renovation offers an opportunity for introducing better-adapted species into old pastures. Oversowing with Graham stylo (*Stylosanthes guianensis* var. guianensis [Aubl.] Sw.) has been successful in the Mackay region (Bishop, unpublished data), and in southeast Queensland siratro–Nandi setaria pastures have been replaced with kikuyu (*Pennisetum clandestinum* Hochst. ex. Chiro)–white clover (*Trifolium repens* L.) pastures by oversowing during the renovation phase (Hurwood 1979).

Pasture renovation has been shown to be a suitable technique for rapidly restoring the siratro content of old overgrazed siratro-based pastures. Cultivation treatments should aim to disturb the top 10 cm of the pasture surface, with best results expected from the greatest disturbance.

LITERATURE CITED

Bishop, H.G., and B. Walker. 1980. Pastures for the Mackay wet coast. Qld. Agric. J. 106:340–361.

Evans, T.R. 1970. Some factors affecting beef production from subtropical pastures in the coastal lowlands of south-east Queensland. Proc. XI Int. Grassl. Cong., 803–807.

Hurwood, R. 1979. Deterioration and renovation of pastures at "Bungawatta." Trop. Grassl. 13:181–182.

Hutton, E.M. 1974. Tropical pastures and beef production. World Anim. Rev. 12:1-7.

Jones, R.J., and R.M. Jones. 1978. The ecology of Siratro-based pastures. *In* J.R. Wilson (ed.) Plant relations in pastures, 353-367. CSIRO, Melbourne.

Jones, R.M., and G.A. Bunch. 1977. Sampling and measuring the legume seed content of pasture soils and cattle faeces. CSIRO, Div. Trop. Crops and Pastures, Trop. Agron. Tech. Memo. 7.

Jones, R.M., and J.N.G. Hargreaves. 1979. Improvements to the dry-weight-rank method for measuring botanical composition. Grass and Forage Sci. 34:181-190.

New Methods for Establishing Grass in the Southern Great Plains

V.L. HAUSER

USDA-SEA-AR Grassland, Soil, and Water Research Laboratory,
Temple, Tex., U.S.A.

Summary

Native and introduced grasses are difficult to establish in the southern Great Plains. Seedling establishment is often prevented by rapid drying of the surface 2 or 3 cm of soil, in which most grass seeds are planted. Three new or nontraditional methods of avoiding the problem in establishing grass were studied.

The first method was punch planting, in which the seeds were placed in the bottom of a deep open hole where the soil remained wet longer than surface soil. In the second method, the seeds were germinated before planting; thus, they emerged quickly to avoid the rapid drying of the surface soil layers. The third method was transplanting live plants that were grown and transported in plastic bandoleer strips.

When the soil surface was kept wet there was little difference in seedling emergence between punch planting and conventional planting; when conditions were dry, substantially more plants were established by punch planting. Seeds germinated before planting produced significantly more seedlings than conventional planting. Seedlings grown in the bandoleers and transplanted to the field produced very satisfactory grass stands.

Planting germinated seeds may be the easiest and most economical method to use in the field. With the development of appropriate machinery, the bandoleer transplant method may become economical for grasslands. Punch planting is limited in field use by rainfall filling the open holes with soil and by the lack of satisfactory machinery for field planting.

A significant improvement in the probability of success in grass establishment would make true range and pasture management more feasible in the southern Great Plains.

KEY WORDS: grass, seed, planting, establishment, range, pasture, southern Great Plains.

INTRODUCTION

Grasses are often difficult to establish on both pasture and rangelands of the southern Great Plains. A 3-year study of grass seeding in the Great Plains (Great Plains Council 1966) showed that seeding failures were common in the southern Great Plains.

Most of the desirable grass species have small seeds; thus, the recommended planting depth generally lies between 0.6 and 2.0 cm. The soil water content of the seed zone is one critical factor affecting establishment of seedlings. However, the top 2 cm of soil can dry from field capacity to wilting point in as little time as one day. The soil at a depth of 3 cm dries slower than the surface layers, and the rate of drying decreases with depth. It is impossible to irrigate most of the lands where grass is to be planted; therefore, new grass establishment methods are needed to avoid the dry conditions often found in the top 2 cm of soil.

The objective of the research was to find new, nontraditional ways to establish grass plants. Three methods of establishment that have not been used for grass were studied: punch planting, planting germinated seed, and bandoleer transplants.

In the punch-planting method, the seeds were placed in the bottom of an open hole that was four times the normal planting depth. The soil at the punch-planting depth remained wet much longer than that at the conventional

The author is a Research Leader of USDA-SEA-AR. This is a contribution of USDA-SEA-AR in cooperation with the Texas Agricultural Experiment Station.

planting depth, and the emerging seedlings encountered no soil crust. Cary (1967) demonstrated that lettuce and carrots were more easily established by punch planting than by conventional methods.

Vegetable seeds that were germinated before planting emerged faster and produced greater numbers of plants than did dry or untreated seeds (Gray 1974). I studied the effect of germination of grass seeds before planting on the emergence of seedlings.

The bandoleer transplant concept was proposed by Brewer (1978). In the bandoleer method, plants are grown in cells in a continuous plastic strip much like the pockets in ammunition bandoleers. The bandoleer construction, filling, seeding, and germination could all be handled by machines, thus reducing greenhouse growing costs. Germination can occur under ideal temperature and moisture conditions with good control of seedling diseases, before the seedlings are transplanted to the field as live plants. The bandoleer permits seedlings to be handled by machine, thus reducing cost.

METHODS

I studied punch planting in the greenhouse. The seeds were planted in a sandy-loam soil profile that was 91 cm deep, to insure adequate drainage of the seed zone. The holes for punch planting were made with a bullet-nosed steel rod 0.6 cm in diameter. The conventionally planted seeds were placed in furrows of appropriate depth for the species and quickly covered to simulate field planting methods. The grasses tested were (1) weeping lovegrass (*Eragrostis curvula*), (2) kleingrass (*Panicum coloratum* L.), and (3) TAM wintergreen hardinggrass (*Phalaris aquatica*). Lovegrass and kleingrass are warm-season grasses, and hardinggrass is a cool-season grass. The wet treatment was irrigated to keep the surface damp, the dry treatment was not. The seeding depth for conventional planting was the appropriate field recommendation; the punch-planting depth was 4 times the conventional depth. Additional details are available in Hauser (1982).

I studied the effect of germination of grass seed before planting in the same greenhouse test units mentioned above. The hardinggrass and kleingrass seeds were germinated before planting in aerated distilled water held at 25°C. The maximum length of radicle at planting was 2 mm. Two germination treatments were imposed before planting: (1) 1-mm radicle on all seeds planted, and (2) a seed lot germinated until radicles were visible on 10% of the seeds. Both the germinated and the untreated seeds were planted in furrows to simulate field planting. Additional details about the methods are reported in Hauser (1980).

The bandoleer transplant concept was compared with dry seed and seeds germinated before planting in a field test. Field planting was delayed by drought so that hardinggrass was planted (or transplanted) in the first week of November, more than a month after the best time for fall planting. The plastic cells in the bandoleers were 0.6 cm in diameter × 6 cm long (small cells) or 2.9 cm in diam-

eter × 11.5 cm long (large cells). The cells were filled with lightweight commercial potting material, seeds were planted and covered, and the plastic bandoleers were formed all in one continuous, hand-operated process. The plants were growing rapidly when transplanted. The dry seeds were planted in the field with a hand-pushed cone seeding machine. The germinated seeds were suspended in a gel and injected into the furrow from a large syringe (driven by the planter); they were carried to the furrow by a tube.

The results of the tests were expressed as percentage of the live seed planted that produced a living plant aboveground (percent emerged) or, for the transplants, percentage of plants surviving. The number of live seeds was determined by standard germination tests.

RESULTS

Table 1 presents the results from two punch-planting experiments. When the soil surface was kept wet, seedling emergence did not differ between punch and conventional planting. However, where the surface was allowed to dry, punch planting established substantially more plants than conventional planting. In other greenhouse tests, I found that the best hole diameter was the smallest tested, 0.6 cm. The best hole depth varied by seed size and vigor; it was 1.3–2.5 cm for lovegrass and 2.5–5 cm for kleingrass.

Hardinggrass seeds that were germinated before planting produced more and faster plant emergence than untreated seeds under both wet and dry conditions for the 1.3-cm planting depth (Table 2). Kleingrass seeds that were germinated before planting produced significantly more plants than untreated seeds (Table 2).

Table 3 shows the results of a field test that compared untreated or dry seeds, germinated seeds, and transplants from both small and large bandoleers. The emergence from the untreated seeds was greater than normally expected in field plantings. The germinated seeds produced

Table 1. Percent of live seeds planted that produced plants from conventional or punch planting treatments in 0.6-cm diameter holes, under either wet or dry conditions.

Treatment	Percent emerged	
	Wet[1]	Dry[2]
Lovegrass		
Conventional, 0.6 cm deep	52 c[3]	4 b
Punch 2.5 cm deep	63 bc	33 a
Kleingrass		
Conventional, 1.3 cm deep	48 c	0 b
Punch 5.1 cm deep	42 c	11 b
Hardinggrass		
Conventional, 1.3 cm deep	90 ab	—
Punch 5.1 cm deep	96 a	—

[1]Well watered for first 20 days.

[2]Emergence at 14 days after planting, no water added.

[3]Within each column, values followed by the same letter do not differ significantly according to the Duncan multiple range test.

Table 2.   Percent of live hardinggrass and kleingrass seeds that produced plants from untreated (dry) seeds, 1 mm radicles on all seeds (1 mm), or seeds germinated until 10 percent had radicles emerged (10%).

Treatment before planting	Percent emerged		
	Hardinggrass		Kleingrass
	Wet	Dry	Dry
1.3 cm planting depth	Day 17	Day 16	Day 15
1 mm radicle	87 a[1]	53 a	44 b
10% germinated	74 ab	60 a	65 a
Untreated (dry)	69 bc	38 b	17 c
2.5 cm planting depth			
1 mm radicle	54 cd	18 c	40 b
10% germinated	41 de	15 c	50 b
Untreated (dry)	28 e	17 c	18 c

[1]Within each column, means followed by the same letter do not differ significantly according to the Duncan multiple range test.

significantly more plants than the untreated seeds, but the best treatment was the large bandoleer cell. The winter following the planting of this test was relatively mild; however, record low temperatures occurred on 1 March (−7°C) and on 2 March (−9°C). The seedlings were growing well in all treatments until the low temperatures in March killed all plants in all treatments except the large bandoleer cell. This mortality rate was dramatic evidence that large transplants are able to survive under adverse conditions.

DISCUSSION

Each of the three new methods for establishing grass has potential to greatly increase the probability of success in revegetation. Planting germinated seeds may be the easiest method to use in the field and the most economical in the immediate future. The bandoleer transplant method will require a high degree of mechanization for use on grasslands. There are two major problems that

Table 3.   Percent of live hardinggrass seeds that produced plants or percent of live seedlings transplanted (bandoleer cells) which survived to 1 month or 8 months in a field test.

Treatment	Percent	
	1 month	8 months
Untreated (dry) seed	48 a[1]	0
Germinated seed	60 b	0
Small bandoleer cell	80 c	0
Large bandoleer cell	97 d	95

[1]Means followed by the same letter do not differ significantly according to the Duncan multiple range test.

may limit the use of punch planting in the field: (1) the possibility that rain will fill the open holes with soil, and (2) the lack of good machinery for planting.

The ultimate use of these ideas will depend upon the quality and cost of machines that are yet to be developed. There is great potential for developing an improved method for establishing grasses under adverse conditions.

LITERATURE CITED

Brewer, H.L. 1978. Automatic transplanter system for field crops. Am. Soc. of Agric. Engin. Tech. Paper 78-1011. St. Joseph, Mich.

Cary, J.W. 1967. Punch planting to establish lettuce and carrots under adverse conditions. Agron. J. 59:406-408.

Gray, D. 1974. Some developments in the establishment of drilled vegetable crops. Proc. XIX Int. Hort. Cong., 407-417.

Great Plains Council. 1966. A stand establishment survey of grass plantings in the Great Plains. Great Plains Counc. Rpt. 23.

Hauser, V.L. 1980. Seed pregermination for grass establishment. Am. Soc. Agric. Engin. Tech. Paper 80-1553. St. Joseph, Mich.

Hauser, V.L. 1982. Punch planting to establish grass. J. Range Mangt. 35(3):294-297.

Grass Establishment: An Evaluation of Seedling Characteristics Relative to Winter Injury, Survival, and Early Growth

R.S. WHITE and P.O. CURRIE

USDA-SEA-AR Livestock and Range Research Station, Miles City, Mont., and Montana State University, Bozeman, Mont., U.S.A.

Summary

Stand establishment in northern latitudes presents a challenging problem that must be conscientiously dealt with to obtain optimum response from introduced grasses. Many variables have been studied in regard to establishment, but relatively little research has been done on early seedling growth and development per se. Yet early in growth is when plants are most vulnerable to winter injury, dessication, competition, grazing, and other adverse impacts. Our research has sought to identify some of the critical morphological features in seedlings that can influence grass establishment. Specifically, objectives were (1) to define and quantify relationships between morphological development and winter injury, survival, and growth, and (2) to develop a management strategy for stand establishment that considers early seedling growth and development.

Three perennial forage species — crested wheatgrass (*Agropyron desertorum* [Fisch.] Schult.), pubescent wheatgrass (*Agropyron intermedium* subsp. *trichophorum* [Link] Reichb. ex Hegi), and Russian wildrye (*Elymus junceus* Fisch.) — were seeded in late summer and early fall to obtain seedlings with diverse morphological characteristics. Individual seedlings that emerged were marked after reaching a winter-dormant condition or upon emerging the following spring. The number of leaves was counted and correlated with winter damage, spring and fall growth, and summer mortality.

Winter injury was inversely related to plant size in all three species and could best be expressed by a quadratic curve. As plant size increased, winter damage decreased. Spring and fall growth were directly related to both seedling size and winter injury. Greater growth and higher summer survival were found in seedlings that had more leaves in the early spring, while substantially less growth and survival were observed in plants that emerged in the spring from late autumn seeding. Growth relationships were readily quantified by an exponential curve.

With respect to management applications, it is highly desirable to minimize the grazing deferment interval so that forage utilization can begin sooner. Our results show that minimization is best accomplished by late-summer planting to assure that seedlings have three or more leaves before winter. Such seedling development results in reduced winter injury, better summer survival, and greater growth during the following season.

KEY WORDS: grass establishment, winter injury, seedling survival, morphogenesis, planting, stand development, growth.

INTRODUCTION

Introduced cool-season grasses are commonly seeded in northern latitudes to provide high-quality forage before native species attain adequate growth for effective grazing (Kilcher and Lawrence 1979, Houston and Urick 1972, Lodge 1970). Establishment of such species in the northern Great Plains of North America, however, normally requires 2 or more years of deferment prior to grazing (Vallentine 1971). This interval is frequently associated with relatively slow early seedling growth. Winter injury, desiccation, and competition effectively contribute to delayed stand establishment.

Other variables, such as soil fertility, soil water, and competition, have been studied in regard to establishment, but little attention has been directed toward early seedling growth and development. Hyder (1974) discussed seedling morphogenesis in general terms, and Plummer (1943) presented data on early growth in range grasses. More recently, Hassanyar and Wilson (1978, 1979) and Wilson and Briske (1979) made substantive contributions toward understanding germination and initial growth in grasses. Our research was directed toward further identifying early-growth factors that influence grass establishment. Specifically, objectives were (1) to define and quantify relationships between morphological development and winter injury, survival, and growth, and (2) to develop a management strategy for stand establishment that considers early seedling growth and development.

R.S. White is a Research Plant Physiologist and P.O. Currie is a Range Scientist of USDA-SEA-AR. Research was conducted at the Livestock and Range Research Station, Miles City, Montana, in cooperation with Montana State University. Publication of this paper (Journal Series No. 1125) has been approved by the Director of the Montana Agricultural Experiment Station.

METHODS

Three perennial forage species — crested wheatgrass (*Agropyron desertorum* [Fisch.] Schult.), pubescent wheatgrass (*Agropyron intermedium* subsp. *trichophorum* [Link] Reichb. ex Hegi), and Russian wildrye (*Elymus junceus* Fisch.) — were selected to examine morphological relationships in stand development. These species are commonly used for range improvement in the northern Great Plains. Research was conducted using field plantings under dryland conditions.

Each species was planted in 23 × 90-m plots at 2-week intervals between 30 August and 26 October. This planting interval provided seedlings with diverse morphological characteristics prior to overwintering. Four replications of each species were established in a randomized complete-block arrangement. Fall seedling emergence was observed on plots planted earlier than 26 October. Seedlings from the 26 October planting emerged the following spring.

Thirty individual plants of each species were randomly selected from each planting date within each of the 4 replications. Plants were marked with wire loops, and the number of leaves were counted to provide a morphological index of seedling size. Plants that emerged in the fall were marked in November, while those that emerged the following spring were marked in April.

Marked plants were re-examined in early April to determine survival and winter damage. This period allowed initial recovery from winter damage, yet it preceded onset of rapid spring growth. A visual estimate was made of the percentage of dead tissue on each plant.

Growth and survival measurements were made in early July and again in late October of the first growing season. These sampling dates corresponded to a spring and early-summer growth period and a fall regrowth period after summer dormancy. Growth measurements taken on each plant included basal intercept, number of tillers and leaves, and dry weight of individual seedlings. One-half of the plants were sampled in July and the other half in October.

RESULTS AND DISCUSSION

Winter Injury and Seedling Survival

Winter tissue injury was inversely correlated with plant size at the time of fall dormancy. As the number of leaves increased from one to four or more, there was a corresponding reduction in tissue damage. This relationship has been discussed in detail for each individual species by White and Currie (1980). When mean values of all three species were regarded collectively, a meaningful mathematical relationship was obtained that described winter tissue damage. The relationship was:

$$Y = 63.5 - 21.4\,X + 2.0\,X^2 \qquad (1)$$

where: X = number of leaves (1 to 5) after fall dormancy and Y = mean postwinter tissue damage (%). Since the coefficient of determination (R^2) was relatively high (R^2 = 0.95) for this equation, this functional relationship appears to have immediate application in mathematical modeling. It could be used to estimate mean winter injury from a population of crested wheatgrass, pubescent wheatgrass, or Russian wildrye seedlings. Thus, high winter damage could be predicted in a plant population with predominantly small seedlings.

Winter mortality was relatively low, and consequently it played a minor role in ultimate stand establishment. In plants with one fall leaf, mortality ranged from 3% in Russian wildrye to 5% in crested and pubescent wheatgrass. Larger plants had better survival. The low mortality could be partially attributed to an above-average snowpack that served to reduce frost heaving and desiccation. Winters with less snow cover may have shown higher mortality similar to that reported by Hull (1966). Winter tissue damage was not related to subsequent seedling survival. Heavily damaged plants showed about the same survival as undamaged plants through the succeeding growing season. They did have reduced growth, however.

Seedling survival during the first full growing season was closely related to seedling size at the beginning of spring and varied according to individual species (Table 1). During the spring growing period, there was little mortality in any species. Soil moisture and growing conditions were favorable during this time. Plants that had one leaf when growth began tended to have lower survival than other size categories, but this difference was significant only in pubescent wheatgrass. After July, soil moisture became more limiting, and plants entered a summer-dormant period for about 2 months until September rains initiated fall regrowth.

When survival determinations were made in late Oc-

Table 1. Relationship between seedling survival and plant size during the first full growing season after planting. Plants with 0 spring leaves had been seeded in late October and were marked shortly after emergence in the spring.

Species	Number of leaves in early spring			
	0	1	2	3+
Survival after Spring (%)				
Crested wheatgrass	93 a[1]	88 a	94 a	100 b
Pubescent wheatgrass	99 a	89 b	99 a	97 a
Russian wild ryegrass	93 a	89 a	87 a	94 a
Survival after Summer (%)				
Crested wheatgrass	91 ab	84 a	91 ab	97 b
Pubescent wheatgrass	75 a	74 a	96 b	96 b
Russian wild ryegrass	43 a	66 b	78 b	96 c

[1]Differences between leaf categories within a species are significant ($P<0.05$) when followed by a different letter according to Chi-square analysis.

tober, Russian wildrye showed the greatest mortality, and crested wheatgrass displayed the least (Table 1). Seedling size was especially critical in determining survival in Russian wildrye. Seedlings that emerged in the spring from the late-October seeding had more than 50% mortality, and plants with one or two early-spring leaves had about 30%. Less than 5% of the plants with three or more early-spring leaves died during this same interval. Pubescent wheatgrass seedlings, in contrast, exhibited distinctly improved survival when early-spring leaves exceeded one. Seedlings with only one leaf in early spring and seedlings that emerged in the spring from the late-October seeding sustained mortalities of about 25%, while larger seedlings had less than 5%. Crested wheatgrass had little seedling mortality throughout the spring and summer of the first growing season. Although plants with three or more leaves in early spring exhibited higher survival, such differences among size categories would have only a slight effect on stand density.

Early Seedling Growth

Seedling growth of crested wheatgrass, pubescent wheatgrass, and Russian wildrye was exponentially related to number of leaves at the beginning of spring growth. The nature of this relationship is illustrated by regression curves in Fig. 1 with respect to basal intercept at the end of spring growth. Similar curves also provided a good fit for leaf and tiller number and seedling weight. Mathematically this family of curves is expressed as

$$Y = ae^{bx} \qquad (2)$$

where x = leaf number at the beginning of spring, Y = growth variable measured (basal intercept, weight, leaf or tiller number), e = natural log, and a and b = coefficient constants. Both spring and fall growth were reasonably well defined by the exponential relationship. Values of R^2 ranged from 0.80 to 0.99 with spring growth and from 0.54 to 0.95 in fall growth, depending upon the variable measured.

Fig. 1. Spring seedling growth in three grasses as influenced by number of leaves at the beginning of spring. Plants without spring leaves had been seeded in late October and marked shortly after emergence the next spring. Regression curves were derived from mean values.

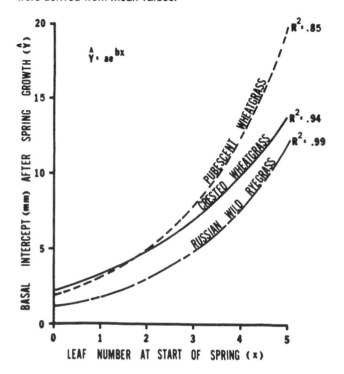

Seedlings that emerged in spring from the late-October seeding exhibited about the same growth as seedlings that had one early-spring leaf. After the number of early-spring leaves exceeded two, however, growth became more accelerated. Establishment of a mature stand was therefore facilitated by having seedlings in a three- to five-leaf growth stage at the beginning of the first full growing season.

The amount of winter injury also affected early seedling growth. As tissue damage increased, subsequent spring and fall growth decreased. The relative importance of winter injury, however, depended upon seedling size and the interaction between size and injury (unpublished data).

Management Applications

Stand establishment of crested wheatgrass, pubescent wheatgrass, and Russian wildrye is directly affected by early morphological development in seedlings. Winter injury and summer mortality are decreased as seedlings become larger, and growth is enhanced during the first full growing season. Stand development can therefore be expedited if seedlings reach a size of three or more leaves

before entering a fall-dormant condition. This can effectively shorten the deferment period by a full year between the time of initial planting and first grazing.

In the northern Great Plains, our results suggest a management practice of late-summer planting in a seedbed that has been summer fallowed to obtain good subsurface moisture. If intermittent late-summer rains occur, seeds will germinate, and roots can reach the fallowed moisture. Favorable temperatures would then allow seedlings to develop three or more leaves before fall dormancy. If late-summer rain does not occur, seeds will not germinate, and the planting would be analogous to a late-autumn seeding. Although this planting strategy is contrary to the more common practice of seeding in late autumn, it provides additional opportunity to shorten the grazing deferment period.

Our research also has important management applications for predictive purposes. When unfavorable temperatures or inadequate moisture result in small seedlings, an estimate can be made of winter injury, summer mortality, or subsequent growth. Such information is valuable for planning the amount of grazing deferment required.

LITERATURE CITED

Hassanyar, A.S., and A.M. Wilson. 1978. Drought tolerance of seminal lateral root apices in crested wheatgrass and Russian wildrye. J. Range Mangt. 31:254–258.

Hassanyar, A.S., and A.M. Wilson. 1979. Tolerance of desiccation in germinating seeds of crested wheatgrass and Russian wildrye. Agron. J. 71:783–786.

Houston, W.R., and J.J. Urick. 1972. Improved spring pastures, cow-calf production, and stocking rate carryover in the Northern Great Plains. U.S. Dept. Agric. Tech. Bul. 1451.

Hull, A.C., Jr. 1966. Emergence and survival of intermediate wheatgrass and smooth brome seeded on a mountain range. J. Range Mangt. 19:279–283.

Hyder, D.N. 1974. Morphogenesis and management of perennial grasses in the United States. *In* Proc. of Workshop of United States–Australia Rangelands Panel, Berkeley, Calif., 29 March–5 April 1971, 89–98. U.S. Dept. Agric. Misc. Pub. 1271.

Kilcher, M.R., and J. Lawrence. 1979. Spring and summer pastures for southwestern Saskatchewan. Can. J. Plant Sci. 59:339–342.

Lodge, R.W. 1970. Complementary grazing systems for the Northern Great Plains. J. Range Mangt. 23:268–271.

Plummer, A.P. 1943. The germination and early spring development of twelve range grasses. Am. Soc. Agron. J. 35:19–34.

Vallentine, J.F. 1971. Range development and improvements. Brigham Young Univ. Press, Provo, Utah.

White, R.S., and P.O. Currie. 1980. Morphological factors related to seedling winter injury in three perennial grasses. Can. J. Plant Sci. 60:1411–1418.

Wilson, A.M., and D.D. Briske. 1979. Seminal and adventitious root growth of blue grama seedlings on the Central Plains. J. Range Mangt. 32:209–213.

Influence of Grazing with Cattle on Establishment of Forage in Burned Aspen Brushland

R.D. FITZGERALD and A.W. BAILEY

University of Alberta, Edmonton, Alberta, Canada

Summary

The effect on establishment of introduced forages in aspen (*Populus tremuloides*) brushland of removal of suckers, by grazing with cattle early in the growing season or just before leaf fall, was investigated in central Alberta following burning of aspen forest in spring. In a year when moisture conditions were suitable, forage species were established successfully by broadcasting in spring shortly after the forest was burned. Suppression of competition from regenerating shrubs was necessary for forage seedling survival and improved establishment more when carried out early than when carried out late in the growing season. Cattle, when concentrated on the burned area, can suppress shrub competition and thereby enhance sown-forage establishment and subsequent productivity.

KEY WORDS: *Populus tremuloides*, alfalfa, brush control, grazing, forage establishment.

INTRODUCTION

The aspen parkland plant community of western Canada consists of an estimated 210,000 ha of aspen groves (*Populus tremuloides*) (Daviault 1977), interspersed with grassland, predominantly *Festuca hallii*. Since herbage yield under the forest canopy is as little as 10% of that in adjacent grassland (Bailey and Wroe 1974) the presence of forest rather than grassland depresses the carrying capacity of this rangeland for cattle.

Replacement of forest with grassland entails removal of the overstory, elimination or suppression of competition from tree and shrub suckers that sprout after overstory removal (Berry and Stiell 1978), and introduction of vigorous grasses.

Initial killing of the overstory can be accomplished by mechanical means, by herbicides, or by burning. Bailey (1972) found that, with mechanical clearing, successful germination and initial growth of seedlings could be achieved by sowing with a drill. However, if herbicides or fire are used to remove the tree overstory, the presence of logs and stumps prevents the placement of seed through a drill, and broadcasting would seem to be the only feasible method of sowing. Very much poorer establishment can be expected from broadcasting than from drilling (Nelson et al. 1970), but in some circumstances sufficient success has been achieved to make the procedure worthwhile (Campbell 1968, Cullen 1970). Following germination, competition from shrubs must be controlled to allow the

development of grass seedlings (Bryan and McMurphy 1968, Cullen 1970).

In this experiment we broadcast seed shortly after the burning of aspen forest and then imposed heavy grazing with cattle early or late in the growing season to suppress competition from regenerating brush species.

METHODS

The experiment was conducted at the University of Alberta Ranch, Kinsella, Alberta, on strongly undulating country. Average annual precipitation is 380 mm, of which about 70% falls in the growing season (April to September). Soils are dark brown to black chernozems of medium loam texture.

Mature aspen forest was burned in May 1972. The resulting stand of young aspen was burned on 15 May 1979. Three days later a cyclone seeder (in which seed is flung from a small hand-driven rotating disk) was used to broadcast a mixture of 2.8 kg/ha of alfalfa (*Medicago sativa*) and 7 kg/ha each of bromegrass (*Bromus inermis*), orchardgrass (*Dactylis glomerata*), and creeping red fescue (*Festuca rubra*).

The area was subdivided into six paddocks of approximately 0.4 ha each, three of which were stocked with 10 beasts/ha and heavily grazed from 5 to 17 July, soon after emergence of suckers of regenerating forest species. The remaining three paddocks were heavily grazed, using 30 beasts/ha, from 22 August to 1 September, i.e., towards the end of the growing season. All accessible edible material was consumed or trampled.

In 1980, early grazing took place from 31 May to 13 June and late grazing from 15 to 23 August.

R.D. FitzGerald is a Research Officer in the Department of Agriculture of New South Wales, Australia, and A.W. Bailey is a Professor of Range Management and Ecology at the University of Alberta.

Measurements

Ten 1-m² quadrat sites/paddock were established randomly and fixed in each paddock. Seedlings of sown species were counted at the conclusion of grazing in each treatment. Counts were made of alfalfa in 1979 and of all sown species in 1980.

In 1980 yield of sown species was estimated in May and August by harvesting, separating into components, and drying all plant material within 0.5-m² quadrats (10/paddock, 30/grazing treatment) situated 2 m from the fixed quadrats in a predetermined direction.

All counts and yield estimations were subjected to least squares analysis of variance using variation among the 30 samples/treatment to estimate error.

Ungrazed Comparison

Areas of forest adjacent to the experimental paddocks were burned and sown to forage species in a manner identical to that of the area within the experimental boundary. Estimates of grass and alfalfa yield are presented for comparison but were not included in the analysis.

RESULTS

Rainfall

Sowing preceded 15 days of dry weather followed by 22 days of almost continuously wet weather. Although rainfall for May (2.6 mm) and June (46.5 mm) was below average, the number of wet days for June (20 days) was almost double the average. Rainfall for July (120 mm) was 39% above average. Rainfall for the whole growing season was 20% below average in 1979 (259 mm) and 57% above average in 1980 (508 mm).

Plant Density and Yield

The density of each sown species was greater after early grazing than after late grazing (Table 1). Yields of sown species, and their contribution to total plant biomass in spring and late summer 1980, were consistently greater in

Table 1. Density of sown species (plants/m²) at completion of early or late grazing in 1979/80.

Grazing time	Alfalfa† 1979	1980	Orchard-grass 1980	Brome-grass 1980	Creeping fescue 1980
Early	16.8 **	11.4 **	25.0 *	23.7 **	10.5 **
Late	8.6	0.5	16.8	2.7	2.9
Standard error	1.9	1.9	2.6	2.4	1.3

*Significant (P<.05). **Significant (P<.01).

†Between-years difference significant (P<.01).

Table 2. Yield (kg/ha) and contribution to total biomass (%) of sown species in early or late-grazed paddocks in 1980.

Grazing time	Sampling time			
	May 31 '80		Aug 26 '80	
	Yield¹	%²	Yield	%
Early	381 **	20.4 **	1117 **	34.7 **
Late	175	13.3	661	18.0
Standard error	91	2.7	91	2.7

¹ ²Means within rows are significantly different (P<0.05).

**Treatment means significantly different (P<0.01).

the early grazed treatment (Table 2). The contribution to total biomass increased through the 1980 growing season in both treatments. Yield of sown species in ungrazed areas was 10 kg/ha or 0.3% of total biomass.

DISCUSSION

Establishment of forages by broadcasting following burning was highly successful in a year in which sowing preceded 3 weeks of almost continuously wet weather, after which generally good moisture conditions prevailed.

Surface-sown seeds are highly vulnerable to a period of drying weather after initial imbibition (Nelson et al. 1970). Hence, the success of any system of broadcasting will depend on the maintenance of moist soil in contact with the seed, either through the fortuitous continuation of wet or overcast weather or through the equally fortuitous falling of seed into cracks or depressions. The high seeding rates used here increased the chance of seeds falling into favorable niches.

Campbell and Swain (1973) achieved greater seedling germination and survival when seeds were dropped into mulch. Absence of mulch would appear to be a disadvantage for seedling establishment following burning.

Our work confirms the importance of removing competition early in the life of the seedling. Heavy grazing early in the growing season enhanced forage seedling establishment much more than grazing towards the end of the growing season. Establishment was negligible in the absence of grazing.

In the early grazing treatment in the first year, seedlings were too small to be grasped by animals, and thereby they escaped direct defoliation by grazing. Apparently, any trampling damage to the young seedlings at this stage was offset by the advantages of reduced competition from shrubs.

Cattle may be reluctant to graze shrubs under some conditions. In the early grazing of 1980, shrubs were not browsed until most of the gramineous material had been consumed. Under range conditions, some method of concentrating the cattle on regenerating shrubs, such as temporary electric fencing, may be necessary for early-season suppression of shrubs.

LITERATURE CITED

Bailey, A.W. 1972. Forage and woody sprout establishment on cleared, unbroken land in central Alberta. J. Range Mangt. 25:119-122.

Bailey, A.W., and R.A. Wroe. 1974. Aspen invasion in a portion of the Alberta parklands. J. Range Mangt. 27:263-266.

Berry, A.B., and W.M. Stiell. 1978. Effect of rotation length on productivity of aspen sucker stands. Forestry Chron. 54:265-267.

Bryan, G.G., and W.E. McMurphy. 1968. Competition and fertilization as influences on grass seedlings. J. Range Mangt. 21:98-101.

Campbell, M.H. 1968. Establishment, growth, and survival of six pasture species surface sown on unploughed land infested with serrated tussock *Nassella trichotoma.* Austral. J. Expt. Agric. Anim. Husb. 8:470-477.

Campbell, M.H., and F.G. Swain. 1973. Factors causing losses during the establishment of surface-sown pastures. J. Range Mangt. 26:355-359.

Cullen, N.A. 1970. The effect of grazing, time of sowing, fertilizer and paraquat on the germination and survival of over-sown grasses and clovers. Proc. XI Int. Grassl. Cong., 112-115.

Daviault, R. 1977. Selected agricultural statistics for Canada [1976]. Info. Div., Agric. Canada Pub. 77/10. June 1977.

Nelson, J.R., A.M. Wilson, and C.J. Goebel. 1970. Factors influencing broadcast seeding in bunchgrass range. J. Range Mangt. 23:163-170.

Silage Corn Cultivation on Degraded Pasture by Means of Minimum Tillage

T. GEMMA
Obihiro University of Agriculture and Veterinary Medicine, Obihiro, Hokkaido, Japan

Summary

The purpose of this study was to investigate the possibility of silage corn cultivation by means of minimum tillage for renovating degraded pasture in Hokkaido, Japan.

Corn (Wasehomare, a leading variety for silage in Hokkaido) was planted, in 4 replicates, on 23 May 1978 on the experimental farm of the Obihiro University of Agriculture and Veterinary Medicine under the conditions of no tillage, strip tillage, and conventional tillage. Plant density was 75 × 25 cm. Fertilizers applied were N, P_2O_5, and K_2O, at rates of 120, 150, and 100 kg/ha. Herbicides used were 5% EPTC, 10 kg/ha, and 47.5% atrazine, 2 kg/ha.

Dry-matter yield on the no-tillage plot was lower than on the strip-tillage or the conventional-tillage plot. There was no significant difference between the strip-tillage and conventional-tillage plots statistically. Percentage of vacant hills in the no-tillage plot was less than in the strip-tillage and conventional-tillage plots, though percentage of lodging in the no-tillage plot was greater. Dry weight of weeds decreased as tillage intensity increased. Intensity of light at ground level was lower in the no-tillage plot owing to the growth of weeds. Soil compaction was almost the same among these treatments.

Silage corn cultivation by means of strip tillage instead of by the plough-plant method was found to be applicable for renovating degraded pasture in Hokkaido with the use of effective herbicides. Strip tillage is advantageous in reducing tillage intensity and maintaining forage yield; also, it is useful for soil improvement and weed control.

KEY WORDS: silage corn, *Zea mays* L., degraded pasture, minimum tillage.

INTRODUCTION

It is obvious that the recent expansion of farm size and increase in livestock population in Hokkaido, Japan, will lead to a need for more feed. For this purpose, growing of annual fodder crops seems to be suitable and is thought to be effective not only for production of silage material but also for soil improvement and control of pests, diseases, and weeds.

Moldboard ploughing or similar deep-tillage operations have been considered the most advanced method of preparing the seedbed. However, reduction of tillage intensity has recently been emphasized, owing to increased ploughing costs.

The purpose of this study was to investigate the possi-

The author is a Professor of Forage Crops at Obihiro University of Agriculture and Veterinary Medicine.

The author wishes to thank the Japanese Ministry of Education for a Scientific Research Grant B that supported this study (Project No. 148066).

bility of silage-corn cultivation with minimum tillage prior to renovating degraded pasture in Hokkaido.

MATERIALS AND METHODS

Wasehomare, which is a leading variety for corn silage in Hokkaido, was planted in degraded pasture on the experimental farm of Obihiro University of Agriculture and Veterinary Medicine. The plots were prepared after mowing and applying granular EPTC (5% S-ethyl dipropylthiocarbamate) at 100 kg/ha. Three tillage treatments were used: no tillage (Z); strip tillage, which was tilled in rows 20 cm wide and 10–15 cm deep (M); and conventional ploughing (C).

Compound fertilizer—120 kg N, 150 kg P_2O_5, and 100 kg K_2O/ha—was applied on the soil surface as a basal dressing. Treatments were randomized in split plots with 4 replications. Seeds were planted on 23 May 1978 with in-row spacing of 25 cm and between-row spacing of 75 cm. After emergence, 2 kg/ha atrazine was applied for weed control.

RESULTS

Germination of seed. Germination of seed in the several plots was compared (Table 1). Percentages of germination in each plot were comparatively high, but there were significant differences between Z and M and between Z and C at the 1% level.

Plant height, number of leaves, and stem diameter as indices of corn growth. In early stages of growth, plant heights of Z and M were greater than those of C. However, 25 days after seed emergence, plant heights of C exceeded those of Z and M. There were no remarkable decreases in number of leaves or stem diameter from shading.

Percentage of vacant hills. Z had the least vacant hills resulting from differences in seedbed condition and seed placement (Table 1). There were significant differences between Z and M and between Z and C.

Percentage of lodging. M and C had lower percentages of lodging than Z (Table 1).

Relative irradiance in the community. There were great differences in relative irradiance at the bottom of the between-row areas. The differences between Z and M and between Z and C at 35 days after planting and at 70 days after planting were significant.

Weeds. There were significant differences among the treatments at the 1% level (Table 1). Species of weeds were orchardgrass (*Dactylis glomerata* L.), curled dock (*Lumex japonicus* Hout.), wild amaranth (*Amaranthus lividus* L.), Kentucky bluegrass (*Poa pratensis* L.), and common nightshade (*Solanum nigrum* L.).

Soil compaction. Soil compaction was measured in between-row areas on which corn was grown in the past year. There was no significant difference among treatments Z, M, and C.

Dry weight of plant parts at harvest. Dry weight of plant parts/10 plants at harvest is given in Table 1. There was no significant difference among the dry weight of leaves. However, there were significant differences in the weights of culm and grain between Z and M and between Z and C at the 1% level. Thus, there were significant differences in the total weight of plants between Z and M and between Z and C.

DISCUSSION AND CONCLUSION

Difference in percentage of seed germination could be attributed to difference in availability of water for absorption by seeds. Since soil capillary tubes surrounding seeds were readily formed in the soil of Z, which was not pulverized, treatment Z showed the highest percentage of germination.

Plant height in treatments Z and M at early stages was affected by shading by weeds and rapid pasturage recovery. Thus, it was considered to be necessary to apply herbicide as early as possible.

Although in M and C, as compared to Z, the percentage of vacant hills was higher, when these data are compared with the percentage of germination in all three treatments (Table 1), it is clear that there was no extensive damage after germination. But, in contrast, percentages of lodging in M and C were lower than that of Z. This difference could be attributed to the fact that in treatment Z, the grain was heavy enough to bend or break the lighter-weight culm (Table 1).

Relative irradiances in the community of M and C were significantly higher than that of Z.

Weight of weeds measured in plots on which corn was grown in the past year decreased with an increase in tillage intensity. There was no significant difference in soil compaction among these plots.

Dry weights of harvested plants/ha, in consideration of the percentages of vacant hills and lodged plants, were estimated as 9,780 kg in Z, 12,000 kg in M, and 12,130 kg in C, respectively. There were significant differences between Z and M and between Z and C, but there was no significant difference between M and C. Decrease in dry weight of plants in Z seemed to be derived from competi-

Table 1. Seed germination, vacant hills, lodging, weeds, and plant parts of silage corn grown on experimental farm in Hokkaido, Japan, 1978.

	Planting procedure		
	Z	M	C
Seed germination, %	96.9	81.3	82.3
Vacant hills, %	6.3	18.8	17.7
Lodging, %	15.5	6.4	6.3
Weeds, dry wt. (g/1.25m²)	642	443	333
Dry wt. of plant parts per 10 plants (g)			
Leaf blade	186	217	225
Leaf sheath	94	118	116
Culm	346	459	491
Tassel	31	31	32
Husk and cob	309	418	406
Grain	997	1153	1155
Total	1963	2396	2425
Grain/plant	.508	.481	.476

tion with weeds for light and nutrients. However, plants in M were not so subjected to competition. Thus, dry weight of harvested plants of M was as high as that of C.

Therefore, it was considered that silage corn could be successfully grown on degraded pasture by means of minimum tillage such as strip tillage. Minimum tillage results in saving of machinery and labor as compared with conventional tillage.

Increased Use of Southern Piedmont Land and Climatic Resources by Interseeding Small Grains in Dormant Coastal Bermudagrass

S.R. WILKINSON and J.A. STUEDEMANN

USDA-SEA-AR, Southeast Area, Southern Piedmont
Conservation Research Center, Watkinsville, Ga., U.S.A.

Summary

Experiments were conducted to assess interseeding small grains by no-till methods in dormant Coastal bermudagrass (*Cynodon dactylon* [L.] Pers.) for forage, silage, grain, or beef production. Wheat (*Triticum aestivum* L.) was interseeded in Coastal bermudagrass (CBG) sod. Grain yields of 1,695 kg/ha reduced CBG yields 14%, and silage yields of 6.7 tons/ha resulted in a 5% reduction in CBG yield. The nitrogen (N) requirements for tall fescue (*Festuca arundinacea* Schreb. L.) or rye (*Secale cereale* L.) interseeded in dormant CBG for 90% of maximum yield were similar at 166 and 175 kg N/ha, while that of rye aerially sown in soybeans at 10% leaf drop of soybeans was 99 kg N/ha. In each of 5 years, rye was no-till interseeded in each of two 1.7-ha dormant CBG fields in late October or early November and compared with tall fescue or rye after soybeans. Nitrogen was applied to each field in mid-November and in late January. Each 1.7-ha pasture was stocked with 7 steers for 140 days from early January through late May. Rye interseeded in dormant CBG sods produced 3,000 to 4,000 kg dry matter (DM)/ha/yr rye forage or 320 kg beef gain/ha/yr over a 140-day grazing period. This steer gain was 58% better than steer gains on tall fescue over the same 5-year period. Interseeding small grains for forage, grain, or beef production effectively increased use of land and climatic resources in the southern Piedmont while enhancing soil and water conservation.

KEY WORDS: Coastal bermudagrass, *Cynodon dactylon* L. Pers., *Secale cereale*, beef cattle, interseeding, tall fescue, *Festuca arundinacea* Schreb. L., *Triticum aestivum* L.

INTRODUCTION

Coastal bermudagrass (CBG) is a highly productive summer perennial that usually begins growth about 15 April and makes little growth after 1 October most years in the southern Piedmont. Its ability to grow during hot, dry summers makes it very useful for beef-production systems. Interseeding cool-season crops into CBG near the time of first frost (and dormancy of CBG) should permit fuller use of climatic resources and generally enhance economic returns from land.

A major factor affecting success of interseeding CBG sod with winter-annual grasses and legumes has been the limited amount of available soil water at seeding time. Robinson (1963) has indicated that the success of autumn sod-seeded crops is linked to rainfall and temperature patterns. By the time soil water is favorable, low temperatures limit growth of the interseeded crop. Welch et al. (1967) point out that the short effective autumn growth period is less serious in seeding for grain production than in seeding for winter grazing.

Wilkinson et al. (1968) and Hoveland et al. (1978b) have shown that tall fescue can be interseeded and maintained in CBG and in bahiagrass (*Paspalum notatum* Flugge) sods. Carreker et al. (1977) described the difficulties in overseeding Crimson clover (*Trifolium incarnatum* L.) and maintaining clover yields, particularly when CBG is fertilized with high levels of N. Hoveland et al. (1978a) have shown the positive benefits derived from including annual cool-season legumes on cow-calf production.

This report describes studies conducted from 1966 to 1976 to determine the N fertilizer requirements for rye and wheat interseeded into dormant CBG and harvested for forage, grain, or producing gain on stocker steers.

S.R. Wilkinson is a Supervisory Soil Scientist and J.A. Stuedemann is a Research Animal Scientist of USDA-SEA-AR. This is a contribution of USDA-SEA-AR Southern Piedmont Conservation Research Center.

METHODS

Experiment 1

Enclosures were made in pastures containing Kentucky 31 tall fescue, rye interseeded in dormant CBG sod, and rye seeded aerially in soybeans (*Glycine max.* L. Merr.) for the winter-spring seasons of 1971–72 and 1972–73. Rye was double-seeded at 94 kg/ha in a crossed pattern in 0.51-m wide rows in dormant CBG in late October or early November with a no-till planter. Aerial seedings into soybeans were done at approximately 10% leaf drop. Nitrogen levels of 0, 45, 90, 134, and 179 kg N/ha NH_4NO_3 were applied to plots in a randomized block design replicated 4 times. Forage was harvested about 10 January, 10 February, 15 March, and 18 April in each of the 2 years. Each pasture was grazed in the summer or cropped with soybeans, and a new plot area was established each year. These studies were conducted on Cecil soils (Typic Hapludult, clayey, kaolinitic, thermic) at Watkinsville, Georgia.

Experiment 2

For 4 years, wheat was seeded at 101 kg/ha with a grassland drill into dormant CBG, fertilized with 0, 45, 90, 134, and 179 kg N/ha, and harvested either as forage (early soft-dough stage of maturity) or grain. For purposes of comparison, an additional treatment of CBG not interseeded or fertilized during winter was included. The experimental design was a split-plot randomized complete block with 6 replications. Harvests as silage or grain were main plots, and N rates were subplots. Phosphorus (P) and potassium (K) were applied across all treatments to assure a medium soil test level of these elements. Coastal bermudagrass plots were fertilized with 358 kg total N/ha/yr in four equal applications. Harvests of CBG were made each year in June, July, September, and October. The last harvest was timed to remove most of the accumulated CBG topgrowth just before seeding.

Experiment 3

Pasture-grazing treatments included 140 days of tall fescue (F), 140 days on rye seeded in CBG (BR), and 105 days on rye aerially seeded into soybeans during 1972 and 1973, or rye seeded in a conventional seedbed following soybeans for years 1974–1976, followed by 35 days of tall fescue (SR) in all years. Nitrogen as NH_4NO_3 was applied to each pasture treatment in two applications of 45 kg/ha, with the first application in mid- to late November and late January for the years 1974–1976. In 1972 and 1973 the N rate was 100 and 134 kg/ha, respectively, for each of the three treatments.

Available forage was harvested with hand clippers at ground level from five random areas of 0.6 m² each/1.7 ha paddock. Rye was hand separated from other forages in the BR treatment. All estimates of available forage were made on an oven-dry basis (55°C).

Angus steers 9 to 12 months of age were randomly allotted to the three pasture treatments with 2 replications in each of 5 years (1972–1976). In early January of each year, 1.7-ha pastures were stocked with seven steers. Individual unshrunk weights, taken on two consecutive days at the beginning and end of the study, were averaged and used as the initial and final weights. Steers were weighed at 35-day intervals throughout each 140-day study. No supplemental feed was provided, but steers had access to a mineral mixture containing 1 part trace mineralized salt and 2 parts defluorinated rock phosphate.

RESULTS

Experiment 1

Total dry-matter yields of rye were similar at the 224 kg N/ha/yr rate (3,401 kg/ha) to those obtained from 224 kg N/ha applied to tall fescue (3,385 kg/ha) and only slightly less than the yields of the aerially sown rye (3,772 kg/ha). Ninety percent of maximum yield was achieved at similar N levels for rye interseeded in CBG and tall fescue (175 and 166 kg N/ha, respectively), while rye aerially sown into soybeans had a lower N requirement (99 kg N/ha) (Fig. 1).

Experiment 2

Fig. 2 gives the total annual yields of wheat interseeded into CBG and harvested as grain or silage and the yields of CBG affected by the interseeded treatment.

The top grain yields, 1,695 kg/ha (25 bushels/acre),

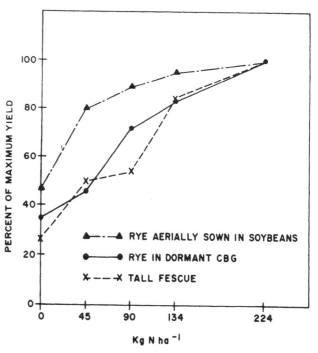

Fig. 1. Nitrogen requirement of rye interseeded in dormant Coastal bermudagrass (CBG) and tall fescue and rye aerially sown in soybeans.

Fig. 2. Dry-matter yields of Coastal bermudagrass (CBG) double-cropped with wheat for silage or with wheat for grain as affected by N fertilization rate to wheat (CO = non-overseeded CBG).

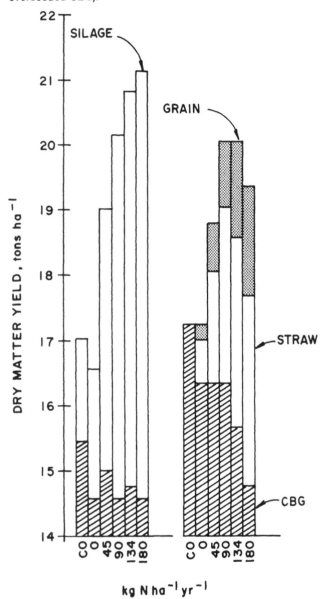

Table 1. Stocker-steer gains per/ha over 140 days as affected by winter pasture system and year (4.12 steers/ha).

Treatment Code[†]	Year					
	1972	1973	1974	1975	1976	Mean
	------- kg/ha -------					
F	234[b]	119[b]	254[b]	221[b]	182[c]	202[c]
BR	306[a]	107[b]	464[a]	312[a]	413[a]	320[a]
SR	278[a]	148[a]	368[b]	247[b]	243[b]	257[b]

[a, b, c]Means within the same column with different superscripts differ (P<.05).

[†]F = tall fescue, BR = rye interseeded in Coastal bermudagrass, SR = rye following soybeans plus 35 days of tall fescue, respectively.

cations of 90, 135, and 180 kg N/ha, respectively. Silage harvests resulted in minimal damage to CBG stands, but stand reductions under the grain-harvest system were substantial.

Experiment 3

Beef gains/yr are given in Table 1 for steers grazing rye interseeded in dormant CBG (BR), tall fescue (F), and rye for 105 days following soybeans plus a final 35 days of grazing on tall fescue (SR). The poor performance of steers on F was evident throughout the study and was reflected in a reduced rate of gain when steers were changed from rye to tall fescue for the final 35 days of grazing. The poorer performance on F is even more significant considering the higher levels of available forage on the tall fescue pastures (Table 2). The BR treatment produced the best gains 4 of 5 years when compared to F and 3 of 5 years when compared to R. A dry autumn and cooler winter temperatures in 1973 during the growing season

were obtained from 180 kg N/ha. The calculated N requirement for 90% of maximum wheat grain yield was 155 kg N/ha/yr. The N requirement for 90% of maximum yield of wheat seeded in CBG and harvested as silage was about 135 kg N/ha. Whether wheat was harvested for grain or silage, at least 90 kg N/ha was required to obtain satisfactory yields.

The most valid comparison for the effect of interseeded wheat on annual CBG yields is within the same harvest system. Reductions in CBG hay yields associated with interseeded wheat fertilized at 90, 134, and 179 kg N/ha were 0.9, 1.6, and 2.5 tons/ha, or 5%, 9%, and 14%, respectively. Comparable percentages of CBG hay-yield losses for silage harvests were 6%, 4%, and 6% for appli-

Table 2. Average available forage during the 140-day grazing period.

Treatment Code[†]	Year					
	1972	1973	1974	1975	1976	Mean
	------- Tons/ha -------					
F	1.6 (2.1)[¶]	1.5 (2.9)[¶]	1.7	1.5	1.5	1.6
BR[‡]	1.1 (3.8)	0.7 (1.7)	1.7	0.6	0.9	1.0
R	0.3	0.1	0.5	0.1	0.1	0.2
SR	0.5 (3.3)	0.4 (2.8)	0.4	0.4	0.2	0.3
F[§]	0.9	0.3	0.6	1.0	1.2	0.8

[†]F = tall fescue, BR = rye in Coastal bermudagrass, total yield, R = rye yield only, SR = rye following soybeans plus 35 days of tall fescue, respectively.

[‡]Includes dead CBG, rye, and volunteer weedy grasses and limited amounts of new CBG growth during the last growth period. Each value represents an average of 5 sampling dates.

[§]Tall fescue available forage for the last 35 days of the 140-day growth period.

[¶]Harvested yields from ungrazed, clipped field plots contained in an enclosure in 1972 and 1973, respectively.

apparently reduced growth of rye in dormant CBG more than tall fescue or rye following soybeans.

DISCUSSION

Utley et al. (1976) have reported similar animal average daily gains on ryegrass or oats either interseeded in sods or seeded in prepared seedbeds. However, the gain/ha on interseeded sods was less than half that obtained from prepared seedbeds because of reduced plant growth. Growth rates of interseeded annuals are usually lower in CBG sod than in prepared seedbeds because of fall drought, later dates of seeding, and CBG competition. Consequently, carrying capacities are greatly reduced. Prepared seedbeds may be affected by rainfall capture and erosion.

The major challenges facing fuller utilization of CBG land during the dormant season are (1) development of dependable reseeding legumes that produce forage in the autumn and winter, and/or (2) development of techniques that permit either earlier seeding of small grains or more rapid establishment so that more growth may accumulate before winter.

Interseeding permits increased flexibility of land-resource use through growing winter annuals for grazing, silage, hay, or grain. Because interseeding can be done without tillage in an established sod, these systems also provide good soil and water conservation practices for the highly erodible soils of the southern Piedmont.

LITERATURE CITED

Carreker, J.R., S.R. Wilkinson, A.P. Barnett, and J.E. Box. 1977. Soil and water management systems for sloping land. U.S. Dept. Agric. Spec. Pub. ARS-S-160.

Hoveland, C.S., W.B. Anthony, J.A. McGuire, and J.G. Starling. 1978a. Beef cow-calf performance on Coastal bermudagrass overseeded with winter annual clovers and grasses. Agron. J. 70:418-420.

Hoveland, C.S., R.F. McCormick, Jr., E.L. Carden, R. Rodriguez-Kabana, and J.T. Shelton. 1978b. Maintaining tall fescue stands in association with bahiagrass. Agron. J. 70:649-652.

Robinson, R.R. 1963. Rainfall distribution in relation to sod seeding for winter grazing. Agron. J. 55:307-308.

Utley, P.R., W.H. Marchant, and W.C. McCormick. 1976. Evaluation of annual grass forages in prepared seedbeds and overseeded into perennial sods. J. Anim. Sci. 42(1):16-20.

Welch, L.F., S.R. Wilkinson, and G.A. Hillsman. 1967. Rye seeded for grain in Coastal bermudagrass. Agron. J. 59:467-472.

Wilkinson, S.R., L.F. Welch, G.A. Hillsman, and W.A. Jackson. 1968. Compatibility of tall fescue and Coastal bermudagrass as affected by nitrogen fertilization and height of clip. Agron. J. 60:359-362.

Hydration-Dehydration as a Presowing Seed Treatment: Physiology and Application

W.M. LUSH, R.H. GROVES, and P.E. KAYE
Division of Plant Industry, CSIRO, Canberra City, Australia

Summary

Presowing hydration-dehydration treatments, in which seeds are partially hydrated and then dried back to their original weights before sowing, have been claimed to have wide-ranging and beneficial effects on seed germination and plant growth. We have assessed the physiological benefits and problems of hydration-dehydration treatments as they apply to pasture grasses, in particular to annual ryegrass.

Pretreatment increased the speed of germination because the embryos of pretreated seeds were in a more advanced developmental stage at sowing. Furthermore, the germinability of seeds was increased if hydration-dehydration coincided with exposure to dormancy-breaking conditions, but we found no evidence to support other claims made in favor of seed pretreatment.

Because the benefits of pretreatment are lost after exposure of seeds to high temperatures or to very dry conditions, the storage of pretreated seeds may be a problem. In addition, the unfavorable response of pretreated seeds to high temperatures may preclude their use in some places. Nevertheless, pretreatment could be a useful practice where species that germinate slowly are to be established in cool environments that dry rapidly after rain. This benefit was demonstrated in simulation studies in which establishment of seedlings was increased from 20% to 77% by using pretreated seeds.

KEY WORDS: ryegrass, *Lolium perenne*, *Lolium rigidum*, *Phalaris aquatica*, presowing seed treatment, seed germination, surface sowing, pasture establishment.

INTRODUCTION

The seeds of many species germinate more rapidly if they are hydrated and then dried back to their original weight before sowing (Kidd and West 1919). Presowing hydration-dehydration treatment, "seed pretreatment," may have other beneficial effects as well, such as more synchronous germination, the invigoration of seedlings, and the hardening of plants (Hegarty 1978). To date, most of the work on seed pretreatment has been done on agricultural and horticultural crop species, but some has been done on pasture plants (A-As-Saqui and Corleto 1978), and an increase has been demonstrated in the early dry-matter production of pastures as a result of using pretreated seeds (Bleak and Keller 1970). Many papers emphasize the beneficial effects of pretreatment on the quality of seeds or seedlings and urge the adoption of seed pretreatment. Nevertheless, it is not a common practice.

We suggest that the benefits of seed pretreatment are few and may be realized only within certain environmental constraints. We have demonstrated the value of pretreatment in conditions that favor its use.

PROCEDURE

Our standard pretreatment for annual ryegrass (*Lolium rigidum* Gaudin) was chosen after testing a number of pretreatment techniques. It resulted in the fastest germination without depressing the subsequent growth of seedlings (Lush et al. 1981). Seeds were weighed, and water equivalent to 55% of oven-dry weight (70°C) was added. After 2 days at 20°C, seeds were tipped onto trays and dried for a further 2 days at 20°C, after which seed weight had returned to within 1% of the starting weight. Following drying, seeds were sown in soil-filled pots for seedling

The authors wish to thank S. Craig for preparing the scanning electron micrograph.

W.M. Lush was supported initially by a CSIRO Postdoctoral Studentship and later by D.J. Cummings while on leave from the Commonwealth Department of Primary Industry.

studies or in petri dishes for germination tests. Perennial ryegrass (*Lolium perenne* L.) and phalaris (*Phalaris aquatica* L.) were pretreated by adding water equivalent to 45% of seed dry weight, incubating at 20°C for 2 or 4 days, and then drying at 20°C for 2 days before sowing.

RESULTS AND DISCUSSION

Benefits of Pretreatment

Pretreated seeds of annual ryegrass, perennial ryegrass, and phalaris germinated more rapidly than controls (Fig. 1). In annual ryegrass, this rapid germination was associated with elongation of the embryo during pretreatment such that the embryo emerged from the seed coat but not from the glumes (Fig. 2). At the time of sowing, the mean length of embryos in pretreated seeds was 1.2 mm, compared with 0.8 mm in controls.

The uniformity of germination, emergence, or establishment was not increased by pretreating seeds; i.e., pretreatment did not reduce variation in the timing of these processes between seeds. Seed germinability (the proportion of a seedlot that was capable of germinating in the conditions provided) could be increased if pretreatment hydration-dehydration were combined with exposure to dormancy-breaking conditions. For instance, the dark-dormancy of ryegrass was greatly reduced if seeds received two 8-hour periods of light during the hydration phase of pretreatment (Fig. 1c).

We found, as have others (e.g., Heydecker 1973/74), that the difference between the germination times of pretreated and control seeds is least at temperatures that promote rapid germination and increases at lower temperatures (Lush et al. 1981). But since lower temperatures increase the germination time of both seedlots by about the same factor—for instance, the germination time of seeds at 5°C was 4.5 times their germination time at 16°C—we suggest that the metabolic activity of pretreated seeds is not enhanced at low temperatures relative to that of controls.

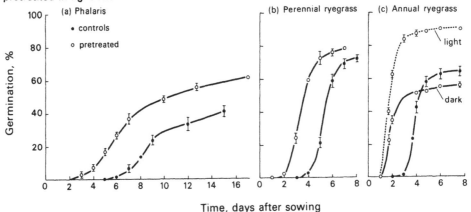

Fig. 1. Germination of pretreated and control seeds of (a) phalaris, (b) perennial ryegrass, and (c) annual ryegrass. All species were germinated in darkness; phalaris received 8 hours of light daily during pretreatment, perennial ryegrass was pretreated in dark, and annual ryegrass was pretreated in light and in dark.

Fig. 2. Scanning electron micrographs of control and pretreated seeds of annual ryegrass, seeds coated with gold in sputter coating unit. (A) control, (B) pretreated.

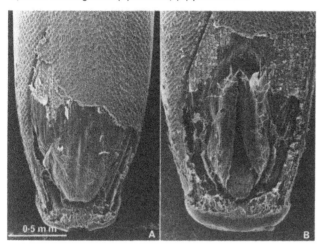

Table 1. Effects of storage conditions on germination time of control and of pretreated seeds of annual ryegrass.

Storage time, months	Storage conditions		
	20°–25°C ambient R.H.	20°–25°C 20% R.H.	4°–6° C[+]
Control			
0	3.9	—	—
1	3.7	3.8	3.8
3	3.7	4.4	3.9
Pretreated			
0	1.9	—	—
1	1.9	3.4	2.0
3	2.9	8.3	3.5

[+] Results the same in ambient and 20% R.H.

Problems Associated with Pretreatment

As the level and duration of seed hydration during pretreatment and the rapidity of drying back can be critical, pretreatment conditions may need to be defined not only for every species but also for every seedlot (Kidd and West 1919, Salim and Todd 1968, Heydecker 1973/4). In annual ryegrass, however, there was considerable latitude in the pretreatment conditions that resulted in an advance of germination (Lush et al. 1981). Furthermore, pretreatments effective in annual ryegrass were just as effective in perennial ryegrass.

Experiments on the storage of seeds after pretreatment have shown that the benefits of pretreatment decline with time, sometimes very rapidly (Heydecker 1973/74). We tested the effects on germination of annual ryegrass of storage at ambient temperature and humidity, at low temperatures, and at low humidity. The germination of controls was not greatly affected by storage conditions, but the effects of pretreatment diminished during storage, and germination was slower than that of controls if pretreated seeds were stored in low humidity at room temperature (Table 1). Seeds may also be "stored" in the seedbed if they are sown in anticipation of conditions favorable for germination. The viability of pretreated seeds of annual ryegrass was reduced much more than that of controls following exposure of air-dry or partially hydrated seeds to 50°C (Lush et al. 1981), and seedbed temperatures can be considerably higher than this (Lush and Groves 1981). These results indicate that problems may arise in bulk drying of seeds following pretreatment and in their subsequent storage.

Applications of Pretreatment

There is little to be gained by using pretreated seeds in favorable conditions, because differences in germination time are likely to be small and to become insignificant over a season or the lifetime of the crop. The advantages of pretreated seeds are said to be enhanced in nonoptimal conditions (Bleak and Keller 1970), but our results suggest that such statements require qualification. For instance, in environments where temperatures are below the optimum for germination, pretreated ryegrass seeds germinate and emerge well ahead of controls (Lush and Groves 1981), but in hot, dry environments, the establishment of pretreated seeds will be poor because of reduced viability. Increased germination speed was the main benefit of seed pretreatment in annual ryegrass. Seed pretreatment, therefore, is most likely to be worthwhile where rapid germination is vital. Such conditions may be found in the nonarable grazing land of the winter rainfall regions of Australia where temperatures are low during crop establishment and, since seeds must be surface sown, moisture levels in the seedbed are likely to change quickly. We tested this argument in controlled environments.

Annual ryegrass was surface sown in pots held at 12°C and watered only once, immediately after sowing. The seedbed was kept wet for 4 days by covering pots, after which time the covers were removed. Pretreated seeds germinated after about 2 days (compared with 4 days in controls) and 77 ± 5% of pretreated seeds succeeded in establishing (compared with only 20 ± 2% of controls). Dry-weight production/pot 18 days after sowing was greater from pretreated seeds than from controls (62 ± 2 mg compared with 13 ± 1 mg), mainly because of higher seedling numbers but also because pretreated seedlings were heavier than controls (Table 2), a difference accounted for, probably, by the earlier germination of pretreated seeds.

Table 2. Early growth of seedlings established from control and pretreated seeds.

Seedlot	Mean seedling dry wt. (mg)	
	18 days after sowing	16 days after germination
Control	1.3 ± 0.1	1.8 ± 0.1
Pretreated	1.6 ± 0.1	1.6 ± 0.1

CONCLUSIONS

The susceptibility of pretreated seeds of annual ryegrass to high temperatures or to very dry conditions indicates that their use may be restricted to cool, moist environments and that seeds must be stored carefully after pretreatment. Provided these constraints are recognized, the more rapid germination of pretreated seeds than of controls may be valuable where rapid germination is critical for seedling establishment.

LITERATURE CITED

A-As-Saqui, M., and A. Corleto. 1978. Effect of presowing hardening on seedling emergence of four forage species. Seed Sci. and Technol. 6:701–709.

Bleak, A.T., and W. Keller. 1970. Field emergence and growth of crested wheat grass from pretreated vs. nontreated seeds. Crop Sci. 10:85–87.

Hegarty, T.W. 1978. The physiology of seed hydration and dehydration, and the relation between water stress and the control of germination: a review. Pl., Cell and Environ. 1:101–119.

Heydecker, W. 1973/74. Germination of an idea: the priming of seeds. Univ. Nottingham School of Agric. Rpt., 50–67.

Kidd, F., and C. West. 1919. Physiological determination: the influence of the physiological condition of the seed upon the course of subsequent growth and upon yield. IV. Review of literature. Ch. 3. III. Ann. Appl. Biol. 5:220–249.

Lush, W.M., and R.H. Groves. 1981. Germination, emergence and surface-establishment of wheat and ryegrass in response to natural and artificial hydration/dehydration cycles. Austral. J. Agric. Res. 32:731–739.

Lush, W.M., R.H. Groves, and P.E. Kaye. 1981. Presowing hydration/dehydration treatments in relation to seed germination and early seedling growth of wheat and ryegrass. Austral. J. Pl. Physiol. 8:409–425.

Salim, M.H., and G.W. Todd 1968. Seed soaking as a presowing, drought-hardening treatment in wheat and barley seedlings. Agron. J. 60:179–182.

Maximum Production from Irrigated Rhodesgrass and Green Panic in a Mediterranean Environment

A. DOVRAT and T. KIPNIS
The Hebrew University, Rehovat, Israel,
and the Institute of Field Crops, Beit Dagan, Israel

Summary

Aspects of the production of rhodesgrass (*Chloris gayana* Kunth) and green panic (*Panicum maximum* var. trichoglume) as roughage for high-grade dairy cattle when land and water resources are limited are described. Results from some 10 years of research are highlighted.

The height of the canopy to which grasses should be allowed to regrow for maximum production is above the height that can be utilized efficiently when harvested in situ by cattle. Mechanical harvesting and indoor feeding provided 20%–25% more digestible dry matter/unit area of land (with a parallel increase in water-use efficiency) than strip grazing, almost without affecting the level of milk production. The efficiency of a single application of N fertilizer after cutting sharply decreased when the interval between irrigations was long, apparently due to leaching below the relatively shallow root system. Frequent irrigation and nitrogen application resulted in optimum uptake of nitrogen and high sustained yields during the season and over the years. Sewage effluent (containing about 50 ppm N) is an excellent source of water and of plant nutrients.

The rate of initial regrowth after cutting was inversely related to the length of the period before cutting, particularly during stem elongation and flowering, which is promoted by high temperature. Relatively long cutting intervals in spring are permissible and result in high dry-matter yields of relatively high digestibility. During summer, shorter cutting intervals are necessary in order to obtain comparable yields of digestible nutrients. Yields of 25 and 35 metric tons (t)/ha/season of rhodesgrass and green panic, respectively, were recorded from commercial fields. In-vitro digestibility of green panic was approximately 10% higher than that of rhodesgrass.

Key Words: *Chloris gayana* Kunth, *Panicum maximum* var. trichoglume, fertilization, cutting, digestibility, dairy cattle.

INTRODUCTION

In a warm climate under conditions of limited land and water resources, such as prevail in Israel, there is a great need for developing strategies that ensure maximum and sustained yields of forage with high metabolizable energy content to be fed to top-quality dairy cattle. Perennial C_4 grasses were chosen for this purpose because of their high potential for dry-matter yield/unit of area and unit of irrigation water, their versatility of utilization (barn fed, grazed, or conserved as hay) according to need, and their suitability as an efficient roughage supplement to a protein-rich diet for lactating cows (Dovrat and Arnon 1971, Tagari and Ben-Ghedalia 1977).

The climate is characterized by almost frost-free winters with varying amounts of rainfall according to region, followed by rainless summers with high radiation intensity (approximately 600 cal/cm²/day). Mean day summer temperature is 28°C, and daylength is between 12 and 14 hours. Soils are mostly of calcareous origin. Commercial fields are sprinkle irrigated, and the trickle system is being used experimentally for forage crops.

CUTTING, REGROWTH, AND FORAGE QUALITY

Cows with high milk-producing ability receiving large amounts of concentrated feeds supplemented with strip grazing on a rhodesgrass pasture, so that prolonged exposure to summer field conditions was avoided, produced milk yields comparable to those of cows on an exclusively soiling system not including rhodesgrass. In another experiment, Israeli Friesian cows strip grazed on irrigated rhodesgrass were compared with cows fed indoors with similar amounts of rhodesgrass. There was a distinct body-weight loss and a small but significant decrease in milk yield in the grazing treatment compared with the indoor feeding treatment, while feed intake was equal in the two treatments. With mechanical harvesting and indoor feeding between 20%–25% more dry matter/unit area of land was obtained (with a parallel increase in water-use efficiency) than with strip grazing (Amir et al. 1970).

Growth of the grasses is restricted to spring, summer, and autumn (April to November); during winter the grasses are more or less dormant.

Experiments were carried out to determine the potential production of rhodesgrass during a complete growing season (Dayan et al. 1981). Maximum growth rates of over 300 kg DW/ha/day (in spring and summer) and 200 kg DW/ha/day (in autumn) were maintained during 21–28 days after the canopy intercepted most of the incoming light. Data from these experiments were used for the evaluation of the crop growth simulation model BACROS (de Wit et al. 1978). In the field situation the maximum rate of dry-matter accumulation was predicted

with reasonable accuracy, as was the rate of water use. Measured and calculated rates of assimilation, respiration, and transpiration of an artificial sward also showed good agreement.

Flail-type mowers are used for cutting the grasses, the forage being chopped, hauled, and then fed in the barn. Cutting frequency varies according to season, regrowth potential after cutting, and digestibility of the dry matter.

High dry-matter yields of rhodesgrass during late spring and early summer induced slow initial regrowth after cutting and diminishing persistence of the grass in late summer and autumn (Dovrat and Cohen 1970). Pot experiments under controlled conditions showed that the optimum temperature for leaf growth was higher than for tiller initiation, and this high temperature brought about the suppression of tiller formation during the hot summer months. Tillering was also restricted when the cutting height above the ground was relatively high. The initial rate of regrowth following long cutting intervals was slower than the regrowth following short cutting intervals, since the longer growing period between cutting enhanced the number of tillers not capable of regrowth after defoliation. Common rhodesgrass starts flowering approximately 14 days after cutting throughout the season, whereas Katambora rhodesgrass produced culmed, nonflowering tillers during the greater part of the season (Dovrat et al. 1971, Dovrat et al. 1980). The absence of tiller regrowth after cutting was closely related to the developmental stage of the canopy.

To encourage tillering during summer, rhodesgrass should be cut frequently (every 16–20 days) to a height of approximately 6 cm above ground level. In spring, when temperature favors tillering and stem elongation occurs at a slow rate, cutting can be less frequent (approximately every 28 days).

The in-vitro digestibility (D) of Common rhodesgrass is greatly affected by age but only slightly affected by season. It ranged from approximately 60% to 50% in 14- and 35-day-old grass, respectively. In the later flowering cv. Katambora, this decline was less pronounced.

Green panic is relatively new to Israel and has been investigated less. Dry-matter production from stands managed in a way similar to that described for rhodesgrass outyielded the latter and amounted to approximately 35 t/ha/season. In summer, stem elongation started about 3 weeks after cutting, and we recommend cutting this grass at least every 35 days. In spring the growing period between cuttings can be extended to 50 days without an appreciable adverse effect on regrowth or digestibility.

The digestibility of green panic was higher than that of rhodesgrass but it was more affected by season. Digestibility values of 76%, 68%, and 64% were recorded in spring for 35-, 50-, and 65-day-old forage, respectively. Values of nearly 70% were recorded in summer provided the grass was not older than 35 days and was properly irrigated; otherwise, D values declined 5%–10%. In autumn the D of green panic was lower and resembled that of rhodesgrass.

A. Dovrat is a Professor of Pasture and Forage Crops on the Faculty of Agriculture of The Hebrew University, and T. Kipnis is a Crop Physiologist in the Agricultural Research Organization of the Institute of Field Crops.

WATER APPLICATION AND MINERAL NUTRITION

The quantities of irrigation water applied amount to 900–1,500 mm/year, depending on season and location. Up to the mid-1970s fields were irrigated about every 10 days between successive cuttings. Nitrogen (N) fertilizer, mainly $(NH_4)_2SO_4$, was applied in equal amounts immediately after each cutting. At this irrigation frequency, dry-matter yields did not increase when N was applied in excess of 450 kg/ha. Since approximately 80% of the root system was found in the 0–30-cm-deep soil layer, excessive amounts of irrigation and N fertilizer led to low water-use efficiency and to leaching of N into layers from which uptake is unlikely. However, by increasing the irrigation frequency to weekly intervals, the level of effective N fertilization rose to approximately 650 kg N/ha, giving efficiency values of approximately 25 kg DW to 1 kg applied N as well as higher water-use efficiency. These findings prompted the installation of permanent irrigation (instead of movable pipes) and fertilization systems facilitating frequent application of water and fertilizer.

Liquid NH_4NO_3 is usually added as fertilizer through the irrigation system. Urea-containing fertilizer, sprinkled on the canopy, resulted in a 30% reduction in yield compared with NH_4NO_3 solution or crystalline $(NH_4)_2SO_4$. However, when a solution of urea-containing fertilizer was applied to the soil in pot-grown rhodesgrass and direct contact with the foliage was prevented, no yield depression occurred.

In pot and field experiments using clay and sandy soils, it was found that NH_4^{++}-N and NO_3^--N were equally effective in dry-matter production of rhodesgrass, enabling flexibility in the use of different N sources on soils possessing different nitrification potential. An NO_3-N content of at least 0.15% (DW) in 21-day-old foliage was required for maintaining optimum growth and was closely related to an index based on nitrate reductase activity (Kipnis et al. 1975).

The use of increased amounts of N fertilizer was conducive to increased uptake of phosphorus (P) and potassium (K). The K content of the dry matter decreased with increasing N fertilization to approximately 0.5%. This amount was still found sufficient for maintaining maximum production of rhodesgrass. Phosphate fertilizer is regularly applied in order to maintain 0.3%–0.5% P in the dry matter of the foliage.

Rhodesgrass exhibits remarkable salt tolerance when grown on saline soils or when irrigated with saline water. This grass was able to take up 300 kg sodium Na/ha (originating from waste water) without showing any yield depression and produced 20 t/ha of hay when irrigated with 1 : 3 diluted sea water growing on sand dunes. In pot experiments it was found that a concentration of 50 mM NaCl was required for optimum growth, indicating the halophytic nature of this plant (Liphschitz et al. 1974). However, care must be taken in the use of N fertilizer when using saline water for irrigation, because of the interaction between N and salinity (Guggenheim and Waisel 1977, Shomer-Ilan et al. 1979).

The use of partially treated (secondary) municipal waste water for irrigation of rhodesgrass was studied intensively (Kipnis et al. 1982).

In the past rhodesgrass and green panic proved to possess considerable yield potential under conditions of proper management. Today research is directed toward exploration of the suitability of additional gene resources, such as pearl millet × napiergrass interspecific hybrids, which are expected to combine even higher yield potential with improved quality.

LITERATURE CITED

Amir, S., A. Dovrat, J. Leshem, and J. Kali. 1970. Grazing versus indoor feeding of Rhodes grass for dairy cows in a subtropical climate. Proc. 18th Int. Dairy Cong., Sydney, 1E:560.

Dayan, E., H. van Keulen, and A. Dovrat. 1981. Experimental evaluation of a growth simulation model. A case study with Rhodes grass. Agro-Ecosystems 7:113–126.

de Wit, C.T., et al. [sic]. 1978. Simulation of assimilation, respiration and transpiration of crops. Simulation Monog. Pudoc, Wageningen.

Dovrat, A., and I. Arnon. 1971. Pattern of intensive roughage production from irrigated forage and pasture crops in Israel. Bul. Rech. Agric., Genbloux, 601–605.

Dovrat, A., and Y. Cohen. 1970. Regrowth potential of Rhodes grass (*Chloris gayana* Kunth) as affected by nitrogen and defoliation. Proc. XI Int. Grassl. Cong., 552–554.

Dovrat, A., E. Dayan, and H. van Keulen. 1980. Regrowth potential of shoot and of roots of Rhodes grass (*Chloris gayana* Kunth) after defoliation. Neth. J. Agric. Sci. 28:185–199.

Dovrat, A., J.G.P. Dirven, and B. Deinum. 1971. The influence of defoliation and nitrogen on the regrowth of Rhodes grass (*Chloris gayana* Kunth). I. Dry matter production and tillering. Neth. J. Agric. Sci. 19:94–101.

Guggenheim, J., and Y. Waisel. 1977. Effects of salinity, temperature and nitrogen fertilization on growth and composition of Rhodes grass (*Chloris gayana* Kunth). Pl. and Soil 47:431–440.

Kipnis, T., A. Feigin, I. Vaisman, and J. Shalhevet. 1982. Waste-water application, dry-matter production, and nitrogen balance of rhodesgrass grown on fine-textured soil or on sand dunes. Proc. XIV Int. Grassl. Cong.

Kipnis, T., D. Levanon, and A. Dovrat. 1975. Assessment of the fertilizer nitrogen requirement of Rhodes grass for the control of soil nitrate accumulation. Proc. VI Conf. Israel Ecol. Soc., Tel Aviv, 158–166.

Liphschitz, N., A. Ilan, A. Eshel, and Y. Waisel. 1974. Salt glands on leaves of Rhodes grass (*Chloris gayana* Kunth). Ann. Bot. 38:459–462.

Shomer-Ilan, A., Y.B. Samish, T. Kipnis, D. Elmer, and Y. Waisel. 1979. Effects of salinity, N-nutrition and humidity on photosynthesis and protein metabolism of *Chloris gayana* Kunth. Pl. and Soil 53:477–486.

Tagari, H., and D. Ben-Ghedalia. 1977. The digestibility of Rhodes grass (*Chloris gayana* Kunth) in relation to season and proportion of the diet in sheep. J. Agric. Sci. 88:181–185.

Sprinkling Irrigation of Forage Plants Under Different Site Conditions in the GDR

W. BREUNIG, K. RICHTER, W. HENKEL, and G. SCHALITZ

Humboldt-Universität zu Berlin, Sektion Pflanzenproduktion, Berlin, GDR

Summary

Stability and continuity of forage production are primarily dependent upon the actual water supply and the level and distribution of nitrogen applications. Under the soil and climatic conditions in the GDR, in almost every year there are certain periods when the natural water supply does not satisfy the water needs of forage plants, leading to severe yield loss unless sprinkle irrigation is practiced. As sprinkle irrigation is very costly and water resources are limited, we must make full use of all potential for achieving high yields while keeping the water input at a minimum — i.e., maximizing overall efficiency of the supplemental water applied.

The effect of nitrogen fertilization and sprinkling on yield, continuity of production, and quality of multicut forage plants of different sites was examined. At all sites studied, sprinkle irrigation gave significant increases in forage yield, with yield increments decreasing with rising soil quality, from sand to loess. With regard to field grasses (*Lolium multiflorum*, *Dactylis glomerata*), the yield differences from irrigation between diluvial and loess sites were similar. Italian ryegrass (*L. multiflorum*) proved to be the most efficient field forage plant (17 t dry matter [DM]/ha) under sprinkling, and mixtures of red clover and grasses ranked second. Water utilization by forage plants improved as the need for sprinkling due to soil or climatic conditions increased. Yield increment was 30% to 40% (productive values > 20 kg DM/mm sprinkling water) on the diluvial sites (sandy and loamy soils) and 18% to 21% on loess. Energy, dry-matter, and crude-protein content, as parameters of forage quality, declined under sprinkling, and crude-fiber content increased slightly. The variation differed among forage plant species and among sites.

The significant results obtained were immediately made available to cooperative farms and State farms for crop production. The experimental results now serve as the basis for selecting forage plants that would give maximum and most reliable crop yields under sprinkle irrigation.

KEY WORDS: sprinkle irrigation, *Lolium multiflorum*, yield level, forage quality, forage quantity, perennial field forage plants.

INTRODUCTION

The growing demand for high-quality foodstuffs of animal origin calls for planned development of animal production. High and reliable forage production and top roughage quality are the main prerequisites to that end.

Stability and continuity of forage production are primarily dependent upon the natural water supply and upon the level and distribution of nitrogen applications. Under the soil and climatic conditions prevailing in the GDR, almost every year there are periods when the natural water supply does not satisfy the water needs of forage plants, leading to considerable yield loss unless supplemental water is applied.

At present about 40% of the supplemental water used for sprinkle irrigation in the GDR is applied to land under forage crops, with multicut field forage crops (Italian ryegrass, cocksfoot, clover-grass, alfalfa, and alfalfa-grass) being dominant.

As sprinkle irrigation is very costly and water resources are limited, we must make full use of all potential for achieving high yields while keeping water input at a minimum.

MATERIALS AND METHODS

The effect of nitrogen fertilization and sprinkle irrigation on yield, forage quality, and continuity of production of multicut forage plants at different sites was examined. Some of the results interpreted were from sprinkle irrigation experiments of several years' duration that had been conducted recently with forage plants in the GDR. The long-term experiments conducted by members of the Crop Production Section of Humboldt University of Berlin on the sprinkle irrigation test field of Berge (loamy sand), the experiments conducted by workers of the Müncheberg Research Center of Soil Fertility, Jena Branch (loess soils) of the Academy of Agricultural Sciences of the GDR, and those conducted by the National Variety Testing Center of Nossen provided the data for evaluation.

Table 1.   Effect of sprinkle irrigation on the dry-matter yield of *Lolium multiflorum*.

Item	D 1/D 2	D 3	D 4/D 5	Lö 1/Lö 2
			Sites	
Unsprinkled, t/ha	8.60	10.71	13.15	13.82
Sprinkled, t/ha	14.17	16.68	17.76	16.75
Yield increment, t/ha	5.57	5.97	4.63	2.93

Table 3.   Combining effect of nitrogen fertilization and sprinkle irrigation in *Lolium multiflorum*.

N/ha kg	DM (t/ha) Unsprinkled	Sprinkled	Yield increment %	kg DM/mm
240 (8)	13.2	15.7	19	17.7
360 (8)	14.1	17.2	22	21.1
480 (5)	14.4	19.0	32	25.6

( ) = number of years, differing according to experiment.

RESULTS AND DISCUSSION

At all sites reviewed, sprinkle irrigation significantly increased forage yields, with the yield increment decreasing markedly with rising soil quality, from sand to loess. With regard to field grasses (Italian ryegrass, cocksfoot), the soil-specific yield differences between medium-quality and better diluvial sites, on the one hand, and loess soils, on the other, were largely offset by sprinkle irrigation (Table 1).

Italian ryegrass (*Lolium multiflorum*) proved to be the most efficient field forage plant (about 17 tons DM/ha) under sprinkling; mixtures of red clover and grasses ranked second and cocksfoot (*Dactylis glomerata*) ranked third. Cocksfoot, in spite of its high dry-matter yields, yielded significantly less energy/ha on account of its inferior quality (energy concentration).

Response to supplemental water (kg DM/mm supplemental water) differed among forage plant species tested and also among sites. For all forage plants, however, water utilization improved as the need for sprinkling (due to soil or climatic conditions) increased. At the D4 site (diluvial, mainly loamy sand), production was highest for Italian ryegrass (27.7 kg DM/mm supplemental water), followed by silage maize, cocksfoot, and legume-grass mixture, with yield increments in response to sprinkling ranging as high as 30% to 40% (Table 2).

On the loess sites field grasses were also the crops that utilized supplemental water most effectively, but production levels (15 to 17 kg DM/mm) and yield increments (18% to 21%) were significantly less than those obtained on the diluvial soil. In the case of gramineous plants, the effectiveness of sprinkle irrigation is highly dependent on level of nitrogen fertility. The positive interaction of sprinkle irrigation and nitrogen fertilization is shown in Table 3. If supplemental water is applied, the optimum level of nitrogen fertilization is increased by 80 to 100 kg or to a total of about 360 kg N/ha.

Sprinkle irrigation lessens the variation in annual forage yield. Sprinkle irrigation gives higher yield increment in dry years and helps to stabilize forage production.

Criteria for assessing the yield-stabilizing effect of sprinkle irrigation include the minimum yield level and the deviations from the long-term average yield. The minimum yield level increases more on the lighter soils than on loess and thus has led to a pronounced reduction of yield variations on the diluvial sites (Table 4). This statement is corroborated by the coefficient of variation (s%), which declined by about 30% on the individual sites and by about 20% on loess.

Comparing forage-plant species, *Medicago sativa* and *Dactylis glomerata*—species more tolerant of drought and extreme temperatures—showed the highest yield stability. Sprinkle irrigation also had a positive influence on the continuity of forage production. Yield increases in the second, third, and fourth regrowths of multicut forage plants resulted in more even forage growth and in most years made it worthwhile to cut the last (fourth) regrowth.

Energy, dry-matter, and crude-protein content as parameters of forage quality declined as a result of sprinkle irrigation. While energy and crude-protein content declined only slightly, forage grass dry-matter content went down by 1.5% on the average, a decrease that required prolonged wilting prior to silage-making. Yield decline was greatest in cocksfoot and alfalfa. The excellent forage quality of Italian ryegrass and mixtures of red clover and grasses is maintained under sprinkle irrigation.

Table 2.   Relationships between supplemental water input, yield increment, and water utilization in forage plants.

Species	D 4 Yield t/ha	Increment %	kg/mm	Lö 1/Lö 2 Yield t/ha	Increment %	kg/mm
			Site			
Italian ryegrass	4.63	35	27.7	2.93	21	15.2
Alfalfa	1.92	16	13.5	0.15	4	3.2
Alfalfa-grass	3.75	30	22.1	0.90	9	7.7
Clover-grass	4.13	36	24.3	1.84	18	14.2
Silage maize	2.25	25	25.6	1.03	9	8.1

Table 4.   Effect of sprinkle irrigation on the minimum yield level of Italian ryegrass at different sites.

Site	Minimum yield level, t DM/ha Unsprinkled	Sprinkled	Rise in minimum yield t DM/ha	%
D 3	6.69	12.58	5.89	88
D 4/D 5	8.41	13.43	5.02	60
Lö 1/Lö 2	8.76	13.19	4.43	50

The results obtained were made available to our crop production enterprises (cooperative farms and State farms) to serve as the basis for selecting those forage plants that, with sprinkle irrigation, would give the highest and most reliable crop yields under the respective site conditions. Forage plants are selected to utilize the

supplemental water most efficiently in combination with other measures of intensification (nitrogen fertilization, time of cutting, cutting frequency).

The results provide a sound basis for maximizing the effective utilization of this very costly intensification factor, sprinkle irrigation.

Minimizing Disease Losses in Forage Crops Through Management

K.T. LEATH

USDA-SEA-AR Regional Pasture Research Laboratory,
University Park, Pa., U.S.A.

Summary

Diseases cause major losses in yield, quality, and stand life of forage crops. Strategies that are used to minimize such losses include the use of resistant varieties and crop-management practices that directly or indirectly affect diseases. Because of the perennial nature of both the crops and the diseases, there are many opportunities to implement management strategies to reduce disease. Implementation begins prior to the planting of the forage crop and continues through the actual planting and on through the production years. Insect control, weed control, and fertility maintenance are examples of crop management that also serve to minimize forage-crop losses to disease. Certain crop-management strategies, such as varietal selection, seedbed preparation, and seeding rate, are one-time opportunities, and if they are not applied at the appropriate time, disease losses may occur. Other management practices, such as fertilization and insect control, are recurrent. Failure to maintain sound management practices, even for a short period, may result in severe losses, because managing forage-crop diseases is basically therapeutic, not eradicative. Many diseases are chronic in forage crops, taking a small toll over several years. It is the rate of disease development that management affects. The future will bring an intensification of existing disease-management strategies because of increasing economic and environmental limitations. Our present system of managing forage-crop diseases is effective and economical and causes minimal disturbance to the environment, but it can be improved. Research into pathogen biology and disease development must be done before models for forage-crop diseases can be developed and integrated into master models of crop production.

KEY WORDS: disease management, crop management, disease-stress interaction.

INTRODUCTION

No estimate of the total amount of forage crops lost annually to disease is available; however, for one species, alfalfa (*Medicago sativa* L.), we believe the direct loss in yield of forage and seed to be around $400 million in the U.S.A. (Agricultural Research Service 1965). When losses in quality and the increased cost of premature reseeding are equated over the entire spectrum of forage crops, the cost must be enormous. Despite the size of these losses, they represent only a small fraction of what

disease losses would be without present efforts to reduce them. For example, losses prevented each year through use of resistant varieties are very significant, with resistance in alfalfa to bacterial wilt estimated as saving $100 million annually (Kehr et al. 1972).

Such costly disease losses are tolerated for several reasons. Forages are commonly sold in some form of bulk package, are of relatively low cash value/ha, and are used mainly on the farm where they are produced. Also, most producers do not realize the extent of the losses they incur. Diseases are often insidious, unseen, and chronic, taking a small toll in yield and quality over a long period of time. Losses are difficult to detect, because often the signs and symptoms of disease are below ground. Those producers who are aware of their disease losses must rely on host-plant resistance and management strategies to

The author is a Research Plant Pathologist and an Adjunct Associate Professor of Plant Pathology at the USDA-SEA-AR Regional Pasture Research Laboratory. This is contribution no. 8015 of the USDA-SEA-AR Regional Pasture Research Laboratory.

minimize those losses, because fungicides are not in general use on forage crops. They are not used because it is generally believed to be uneconomical to do so and because of concern for residues in animal products and possible adverse effects on nitrogen fixation.

CROP, DISEASE, AND STRESS RELATIONSHIPS

A forage stand consists of thousands of individual plants/ha growing en masse *in spite of* biological and physical stresses and *because of* a competent management system. Such a system manipulates stress factors to favor the crop species over the weed species. These forages are harvested several times during each growing season and then are expected to endure the winter and produce again the following year. Superimposed on the stress of harvest is that caused by diseases. If such stresses become too great, individual plants succumb, stands are thinned, and production decreases. Because forages are intensively managed over a period of several years, there are many opportunities to implement management strategies that reduce disease losses. Root rot begins shortly after planting and progresses gradually, increasing in severity with the age of the stand. It is this rate of disease increase that management can change. It is this close relationship between disease and management, this dependency of disease loss reduction on the correct and timely application of management strategies, that must not be overlooked.

Many fungi, bacteria, viruses, nematodes, and air pollutants contribute to the disease picture. Pathogens interact with each other and with other biotic and physical stresses to produce cumulative overall effect. Stresses need not occur at the same time to interact. Stresses separated temporally can and often do interact. For example, an autumn epidemic of foliar disease might limit the accumulation of carbohydrates in autumn and cause slower growth the following spring and thus make plants more prone to winter injury or devastation by some spring pathogen. Another example might be the effect of insect feeding in a young stand, limiting root growth and making the plants less able to resist or tolerate root rot or drought.

ROLE OF MANAGEMENT

The perennial nature of forage crops is of prime importance in an attempt to understand forage crop diseases, because diseases often are perennial, too. Root rots, viruses, and wilts are not transient. Once infection has taken place, the plant is diseased. The impact of that disease and the rate at which it develops are functions of climate and management. Since climate must be tolerated, management is the prime strategy. Those crop-management strategies acting most directly to reduce disease losses are listed in Fig. 1. The application of management strategies to reduce disease losses begins before the crop is planted. These early strategies can only be implemented once. If the opportunity to implement them is missed, the consequences must be suffered. Awareness and careful planning are needed to maximize the effects of

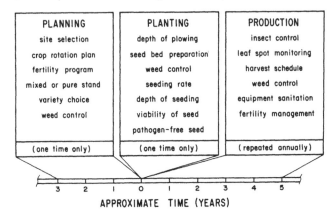

Fig. 1. Forage-crop management practices that directly or indirectly reduce disease losses and the temporal relationship of these practices to the stage of the crop.

these strategies. Often some of the accepted management practices are not appreciated from a disease-management viewpoint. Weed control, fertilization, and insect control all are disease-management strategies, because their application results in less loss from disease. Weeds not only compete with forage plants for growth requirements but also serve as alternative hosts for pathogens. The maintenance of potassium at levels that permit vigorous plant growth also results in reduced populations of potential root pathogens (O'Rourke and Millar 1966), and the control of leafhoppers reduces the impact of root disease (Fig. 2). Many other practices that are recognized mainly as production practices also minimize disease losses.

Because so much of our management strategy is preventive rather than therapeutic, of necessity it must be continuous. Any break in its application, however brief, may result in a disease situation that cannot be controlled, causing large losses. Against many pathogens there exists some inherent host-plant resistance, but this resistance does not constitute immunity, and the expression of this resistance can be modified by management practices (Leath and Ratcliffe 1974, Leath 1981).

Generally, a forage grower has two main objectives—production and stand persistence. The maximum

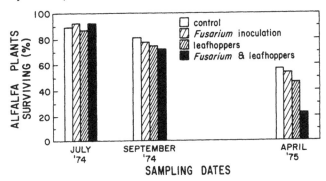

Fig. 2. Interaction of *Fusarium* root rot, potato leafhopper feeding, and winter stress on the survival of alfalfa (Leath and Byers 1977).

in each case is not necessarily attained by the same management system. Usually, the management system used represents a compromise that permits a satisfactory yield and a satisfactory stand life. Managing perennial forages is complex as compared with managing an annual crop, such as corn or soybeans, because only yield is of concern with annuals. What is acceptable longevity for a forage stand? Longevity goals must be realistic in terms of operational needs, land availability, and pest problems. Many potential diseases do not materialize simply because their respective pathogen populations increase slowly and the crop escapes. Nematodes are a good example of such a disease.

FUTURE NEEDS AND APPROACHES

The immediate future is not likely to bring major changes in approaches to reduce losses in forage crops. Our present system is effective and economical, and it causes minimal disturbance to the environment, but it can be improved. I believe that in the future present strategies will be intensified. New varieties will have resistance to more pathogens, and levels of resistance will be higher. Greater use of grass-legume mixtures, improved adaptation of crop to site, new formulations and methods of application of fertilizer, and increased efficiency of insect control all will contribute to greater reductions of forage losses from disease. Rapid evaluation of forage quality and the attendant emergence of forages as a market commodity will place protein content in its proper perspective and consequently increase interest in foliar disease control, perhaps employing the use of fungicides. Whether effects of new techniques, such as sod seeding and chisel plowing, will be beneficial or harmful from a disease standpoint remains to be determined.

With forage crops, as with many other crops, we do not have the knowledge needed to apply established principles of disease management in an optimum manner. We need reliable crop-loss data to determine the economic feasibility of applying control strategies. Much fundamental research in pathogen biology and disease development must be done before models for forage-crop diseases can be developed and incorporated into master models of crop production. Although we minimize disease losses in forage crops to a practical level through host-plant resistance and the judicious application of sensible crop-management strategies, we cannot afford to become complacent about our losses or to lose sight of the need to develop a more effective, consummate disease-management system.

LITERATURE CITED

Agricultural Research Service. 1965. Losses in agriculture. U.S. Dept. Agric. Handb. 291.

Kehr, W.R., F.I. Frosheiser, R.D. Wilcoxson, and D.K. Barnes. 1972. Breeding for disease resistance. *In* C.H. Hanson (ed.) Alfalfa science and technology. Agron. Monog. 15:335–354. Am. Soc. Agron., Madison, Wis.

Leath, K.T., 1981. Pest management—alfalfa diseases. *In* D. Pimentel (ed.) Handbook of pest management in agriculture. CRC Ser. in Agric. CRC Press, Boca Raton, Fla.

Leath, K.T., and R.A. Byers. 1977. Interaction of fusarium root rot with pea aphid and potato leafhopper feeding on forage legumes. Phytopathology 67:226–229.

Leath, K.T., and R.H. Ratcliffe. 1974. The effect of fertilization on disease and insect resistance. *In* D. Mays (ed.) Forage fertilization, 481–502. Am. Soc. Agron., Madison, Wis.

O'Rourke, C.J., and R.L. Millar. 1966. Root rot and root microflora of alfalfa as affected by potassium nutrition, frequency of cutting, and leaf infection. Phytopathology 56:1040–1046.

Improvement of Yield and Persistence of Italian Ryegrass Through Pest Control

R.O. CLEMENTS and I.F. HENDERSON
Grassland Research Institute, Hurley, Berkshire, England,
and Rothamsted Experimental Station, UK

Summary

Our objectives were to examine (1) whether increases in yield and persistence of Italian ryegrass varieties following frequent pesticide application are related to a reduction in frit-fly larval populations, and if so (2) whether the use of a nonpersistent pesticide would be a feasible method of reducing such damage.

In two field experiments, the yield and persistence of normally short-lived Italian ryegrasses (*Lolium multiflorum* L.) were greatly increased by application of large and frequent doses of the potent insecticide phorate. Much of the insect damage was attributed to stem-boring dipterans of the frit-fly complex (e.g., *Oscinella frit*).

In one experiment, small plots of 23 varieties of Italian ryegrass were either treated with phorate every 6 weeks or left untreated. Varietal differences in larval numbers and yield in response to pesticide use were compared. In a second experiment, chlorpyrifos was applied each year on three occasions to coincide approximately with the oviposition periods of frit-fly.

In the first experiment, larval numbers varied greatly among varieties, but there was a positive correlation between larval numbers and tiller density. Larval numbers were related to yield increase following pesticide application in the first year only, but there were large differences among varieties in their response to pesticide treatment in all 4 years. At the end of the third year of the experiment, many plants on untreated plots had died, but plants on treated plots in general remained vigorous.

In the second experiment, chlorpyrifos treatment, although not as effective as phorate, improved yields of the three varieties tested over a 3-year period by an average 25%, or 2.7 t/ha/yr.

It is concluded that there are large differences among varieties of Italian ryegrass in susceptibility to frit-fly that appear to be linked to yield response to pesticide treatment and sward longevity. This finding suggests that there are possibilities of breeding specifically for frit-fly resistance in order to increase yield and improve persistence. In the interim it may be feasible to use infrequent doses of nonpersistent pesticides to enhance yield and varietal persistence.

KEY WORDS: Italian ryegrass (*Lolium multiflorum* L.), yield persistence, pest control, frit-fly (*Oscinella frit*).

INTRODUCTION

Italian ryegrass (*Lolium multiflorum*), one of the important temporary grasses sown in the UK, is particularly prone to attack by frit-fly (Clements 1980). Treatment with the pesticide phorate led to marked gains in herbage yield and a dramatic increase in the persistence of this normally short-lived grass (Henderson and Clements 1979).

In the experiments described here the objectives were to determine whether increases in yield and persistence of several Italian ryegrass cultivars through frequent pesticide application are related to control of frit-fly larval populations (experiment 1) and whether a nonpersistent pesticide would be a feasible method of reducing such damage (experiment 2).

METHODS

Experiment 1

Small plots (2 × 6 m) of each of 23 Italian ryegrass cultivars and one perennial ryegrass cultivar (S24) were sown in May 1977 at Hurley, Berkshire, in each of six randomized blocks. The blocks were arranged in three pairs, and one block of each pair was selected at random and treated with pesticide (10% phorate granules applied at 3 kg a.i./ha every 6 weeks), commencing 6 weeks after the date of sowing. During the sowing year all plots received 250, 100, and 100 kg N, P, and K fertilizer/ha, respectively, and were cut on three occasions. In subsequent years, plots received 400, 75, and 75 kg of N, P, and K fertilizer/ha, respectively, and were cut on six occasions.

Frit-fly larval populations were assessed in January 1978, 1979, and 1980 by dissecting out the larvae from turf cores. An assessment of the proportion of ground covered by the grass species sown was made in February 1980.

Experiment 2

In a small-plot experiment at Wye, Kent, the effect on herbage yield of three applications of the organophosphorus insecticide chlorpyrifos was compared with the effect of frequent applications of phorate (3 kg a.i./ha, every 6 weeks) and a nil treatment. Chlorpyrifos (0.75 kg a.i./ha) was applied in early May, early July, and early September to coincide with the oviposition of frit-fly. All three treatments were applied to three cultivars of Italian ryegrass sown in spring 1977 in small plots (1 × 1 m) arranged in four randomized blocks. All plots received 420, 210, 210 kg N, P, and K fertilizer/ha/yr, respectively, and were cut on six occasions each year.

RESULTS

Experiment 1

There were significant differences in total annual herbage dry-matter yield among cultivars and a significant effect of pesticide (phorate) treatment in each of the first 3 years (1977–1979). By the end of the third year, many plants on untreated plots had died, and the area covered by them was small (Table 1).

Larval populations varied significantly among varieties (Table 1). In January 1979 the most susceptible variety, Dutch Landrace, had a population equivalent to 3,250 larvae/m² or 21.5% tillers infested, compared with Grasslands Manawa, which had a population equivalent to only 650 larvae/m² or 6.3% tillers infested. The number of tillers attacked/unit area was directly related to the number of tillers present/unit area in all 3 winters (r = 0.43* in 1978, r = 0.40* in 1979, and r = 0.47** in 1980).

Correlations among numbers of frit larvae, which peak in early winter (Clements and Henderson 1980), and the yield response to suppressing pest populations were sought. A significant correlation was found for the winter of 1977–78 (r = 0.47**). However, no such correlations

Table 1. Total DM yield (t/ha, 1977–80), and proportion of ground covered by sown species of insecticide treated and untreated plots of 23 cultivars of Italian ryegrass and proportion of tillers infected by frit in untreated plots.

Variety	Total DM yield (1977–80) t/ha		Ground cover Feb. 1980 %		Frit larvae Jan. 1978 (Untreated plots)	
	U	T	U	T	/m²	% tillers infested
S22	35.2	39.7	23	48	540	3.9
Combita	37.4	42.4	27	57	1460	10.0
Delecta	41.6	47.3	50	67	860	8.2
Dasas	36.3	43.2	28	53	520	5.2
Fat	37.5	42.0	35	43	980	7.4
Imperial	36.5	41.7	25	45	720	9.0
Stormont Ibex	32.4	40.3	17	37	640	5.7
Lema	39.4	43.2	33	67	820	5.1
Optima	36.6	41.8	35	67	500	5.1
RvP	43.1	50.0	52	83	560	5.3
Tiara	38.5	42.2	38	50	600	6.6
Vejrup	35.8	39.7	25	40	480	4.8
Trident	40.6	44.0	47	62	840	7.3
Dilana	32.5	37.0	10	25	1020	10.4
Bentra	37.8	42.2	32	73	520	5.8
Dalita	39.8	43.6	33	80	640	6.6
Sabalan	39.6	40.3	32	65	600	7.1
Tetila	38.1	41.2	27	60	380	5.4
Dutch Landrace +	27.1	31.9	5	7	520	10.2
Grasslands Tama +	27.8	34.0	5	5	740	10.0
Grasslands Ariki*	39.9	43.6	78	78	900	6.2
Grasslands Manawa*	38.1	42.6	47	67	920	6.1
Sabrina	40.3	44.2	55	78	520	7.1
S24†	41.2	42.9	68	70	580	4.1
SE of mean±	0.64	0.64	5.3	6.0	210	1.51

U = untreated; T = treated.

+ Westerwolds ryegrass.

* Italian × perennial ryegrass hybrid.

† Perennial ryegrass.

Table 2. Mean number of plants remaining/plot of each of three Italian ryegrass cultivars grown with and without phorate or chlorpyrifos.

Variety	Untreated	Phorate	Chlorpyrifos
RvP	16.5	21.5*	23.0*
S22	12.3	17.0	16.0
Delecta	10.3	18.8*	22.3*

*Significantly different from untreated (P≤0.05).

S.E. of mean ±1.79.

were found for the winters of 1978–79 and 1979–80, probably due to changes in sward composition observed on the untreated plots (Table 1).

Experiment 2

There were significant effects on herbage yield of both the chlorpyrifos and the phorate treatments (Table 2). Survival of the sown species was much better on the insecticide-treated plots than on the untreated plots (Table 3).

Table 3. Total annual dry-matter yields (t/ha) of three Italian ryegrass cultivars grown with and without phorate or chlorpyrifos.

Variety	1978			1979			1980		
	U	P	C	U	P	C	U	P	C
RvP	12.6	18.3	15.1	9.9	15.0	12.6	10.8	16.2	15.5
S22	12.5	15.5	13.1	9.9	12.2	11.2	8.1	12.1	11.2
Delecta	13.3	18.0	15.1	9.4	14.8	12.5	9.5	15.1	14.7
S.E. of mean ±		0.07			0.33			0.39	

U = untreated; P = phorate treatment; C = chlorpyrifos treatment.

On untreated plots more than half of the herbage harvested was made up of unsown species.

DISCUSSION AND CONCLUSIONS

In both experiments treatment with pesticides significantly increased yields. In the first experiment there was a link between frit numbers and yield increase, but there was considerable unexplained variation.

Besides depressing yield, pest damage is evidently a major cause of the rapid deterioration of Italian ryegrass swards. Observed differences in the persistence of cultivars were due, at least in part, to differences in their susceptibility to such damage.

The use of small amounts of some pesticides (e.g., chlorpyrifos) may be an effective method of increasing the annual yield and useful life of Italian ryegrass swards, but further larger-scale testing and development are required before such methods could be recommended for use in practice. It may be feasible to improve yields and longevity by the use of pesticides, but the cost and risks to the environment even of nonpersistent chemicals make this a less-than-ideal solution. In view of the evidence of genotypic differences in resistance to frit-fly, breeding for more resistant cultivars seems a more satisfactory objective for research.

LITERATURE CITED

Clements, R.O. 1980. Grassland pests—an unseen enemy. Outl. Agric. 10(5):219–223.

Clements, R.O., and I.F. Henderson. 1980. The importance of frit-fly in grassland. ADAS Q. Rev. 36:14–26.

Henderson, I.F., and R.O. Clements. 1979. Differential susceptibility to pest damage in agricultural grasses. J. Agric. Sci., Camb. 93:465–472.

Ladino Clover Persistence as Affected by Physical Management and Use of Pesticides

D.S. CHAMBLEE, L.T. LUCAS, and W.V. CAMPBELL
North Carolina State University, Raleigh, N.C., U.S.A.

Summary

Stand persistence is a major weakness of the widely used Ladino clover (*Trifolium repens* L.) in the U.S.A. The objective of these studies was to determine the influence of physical management practices and selected pesticides on longevity of Ladino clover stands. Ladino grown alone and in mixture with tall fescue (*Festuca arundinacea* Schreb.) was subjected to varying physical management regimes (clipping, grass density, and spacing) and insecticide and fungicide applications in two tests for 5 to 6 years. By the fall of the third year, the percentages of clover (average of two tests) in the Ladino clover–tall fescue untreated check and in the fungicide (F), insecticide (I), and F + I treatments were 12%, 34%, 50%, and 74%, respectively. These studies showed that physical management per se, although important in the short run, will not provide long-lived Ladino clover stands and that an entomological-pathological complex was largely responsible for the loss of stand and low productivity after the first 2 years.

KEY WORDS: Ladino clover, persistence, *Trifolium repens* L., pesticides.

INTRODUCTION

White clover is the most widely used perennial pasture legume in the humid regions of the U.S.A. Stand persistence is a major weakness of both intermediate and large types. When it is used for grazing, white clover is commonly seeded in mixture with a grass. Important principles for management of mixtures in some environments to enhance persistence of white clover have been summarized (Gibson and Hollowell 1966, Leffel and Gibson 1973, Smith 1975). The objective of these studies was to determine the effect of physical management (clipping, grass density, and spacing) and pesticides on longevity of stands of Ladino-type white clover grown alone and in mixture with tall fescue.

D.S. Chamblee is a Professor of Crop Science, L.T. Lucas is a Professor of Plant Pathology, and W.V. Campbell is a Professor of Entomology at North Carolina State University. This is Journal Series Paper No. 6766 of the North Carolina Agricultural Research Service, Raleigh.

METHODS

Two similar tests were established in successive years in different locations on a Cecil clay loam soil (a Typic Hapludult clayey, Kaolinitic, thermic) near Raleigh, North Carolina. The areas were fertilized according to soil test recommendations. Tillman Ladino clover was seeded on all plots in September at 5.5 kg/ha with and without tall fescue (cultivar Kentucky 31). Fescue was either broadcast (bdc) or seeded in rows 20 cm apart at rates of 6, 12, or 24 kg/ha. Most plots were cut from a height of 20 cm back to 5 cm (20–5) during the growing season. Other variables involved 20–2.5- and 10–5-cm cutting regimens and a treatment with a hay cut, 40 back to 5 cm (40–5), at the first harvest followed by harvests of 20–5 cm. A fall accumulation (fall rest) treatment was included (the tables show the exact treatments).

Pesticide treatments, applied alone and in combination, included an insecticide (I) carbofuran (2,3-dihydro-2,2-dimethyl-7-benzofuranyl methylcarbamate) and a fungicide (F), benomyl [methyl 1-(butylcarbomyl)-2-benzimidazole carbamate]. Carbofuran was applied monthly (May–October) at 1.1 kg a.i./ha/application of granular (10 G) plus a similar amount of 43.8% F carbofuran applied as a spray. Benomyl (F) was applied at 2-week intervals in test 1 and monthly in test 2 (1 April to 1 October) at 8.0 kg a.i./ha/application. Individual plots were 1.8 m wide × 6.1 m long with a 1.5-m alley between replications. Although the treatments for each test are presented in two separate tables, they were arranged in a randomized complete-block design in one experiment with 4 replications. A combined statistical analysis was made on all harvests for each year. Test 1 (seeded in 1971) and test 2 (seeded in 1972) were conducted for 5 and 6 years, respectively. All cutting management and pesticide treatments were imposed in the first year, but since excellent clover stands were maintained in all treatments the data are not reported. For the purpose of brevity, data from test 2, 1977, are not reported. Most treatments that did not include a pesticide showed sharply decreased clover stands by the end of the third year and were discontinued.

In test 1 visual dry-weight estimates (DW) of the individual species were made twice annually. In test 2 they were made for each harvest, and yield calculations were made for both the clover and tall fescue components (shown in the tables). Weed estimates were made at every harvest in both tests and are not included in yields. In addition, a modified line transect method was employed to determine ground cover of Ladino clover.

RESULTS AND DISCUSSION

Physical Management (Clipping and Fescue Spacing)

A spring-hay-crop management (first harvest 40–5 and others 20–5 cm) resulted in an increase in total forage pro-

Table 1. Effect of fescue spacing (row vs. broadcast) and clipping management on dry-matter yield of broadcast ladino clover grown with and without tall fescue.

Treatments†	Test 1		Test 2			
	1973	1974	1974		1975	
	M‡	M	M	C‡	M	C
	---------------------- metric tons/ha ----------------------					
Ladino-fescue						
Row (20–2.5)	–	–	5.2	2.1	3.4	0.8
Row (20–5)	7.8	3.2	3.9	1.4	2.8	0.4
Row (10–5)	6.5	2.2	3.4	1.1	2.4	0.4
Row 1st (40–5)						
Others (20–5)	8.7	.4.3	6.3	2.6	4.5	0.9
Row (20–5) with						
fall rest	8.0	2.6	3.0	1.0	2.4	0.2
Bdc (20–5)	7.7	2.9	2.7	0.8	2.3	0.2
Ladino						
(10–5)	5.0	1.1	–	2.5	–	0.5
(20–5)	6.5	1.4	–	3.3	–	0.8
L.S.D. (0.05)	1.2	0.7	0.8	0.6	0.7	0.4

†Fescue seeded in rows 20 cm apart or broadcast (bdc) as indicated under the mixture. Method of cut, for example, 1st (40–5), others (20–5), indicates first harvest cut back to 5 cm when growth reached 40 cm height, other harvests cut 20 back to 5 cm; (10–5) indicates all harvests cut back to 5 cm when 10 cm height reached.

‡Indicates total yield of mixture (clover and fescue); C indicates yield of clover component. First year's data not included.

duced by the mixture in 3 of the 4 years compared with the cut all season (20–5) (Table 1). However, there were no marked differences in percentages of Ladino clover between these two management treatments in either late summer or early fall in 3 of the 4 years (Table 2). Management for a spring hay crop appeared to improve percentage of Ladino in one year (test 2, 1974, Table 2). No advantages were realized in yield for a fall rest period (20–5 with fall rest). There were unexplainable fluctuations in percentage of clover as related to fall rest (Table 2). These fall seasons were relatively dry. Cutting the clover-fescue mixture more frequently to 5 cm (10–5) compared with 20–5 slightly reduced total yield in one year (Table 1), but had no marked effect on clover presence (Table 2). Reducing the stubble height to 2.5 cm (20–2.5) resulted in an increase in the yield of the mixture and the clover component in one of the 2 years and a marked increase in percent clover in both years (Table 2).

Cutting clover seeded alone more frequently (10–5 vs. 20–5) produced losses in yield in 2 of the 4 years. The greatest reduction was 1.5 metric tons/ha in test 1, 1973 (Table 1). Ladino seeded alone (20–5-cm treatment) had a stand of 80% and 75%, in the third fall in tests 1 and 2, respectively, yet the Ladino-tall fescue mixture (bdc, 20–5) had only 12% and 16% clover at the same date (Table 2). During the third year (1974, 1975) the yields of clover in the pure Ladino plots were only 1.4 and 0.8 metric tons/ha for tests 1 and 2, respectively (Table 1). Such low clover yields when fair to good stands occur indicate that the clover was unhealthy.

Table 2. Effect of fescue spacing (row vs broadcast) and clipping management on stands of ladino clover grown alone and with tall fescue.

	Test 1						Test 2	
	DW‡				L‡	LT‡		
	1973		1974		1975	1974	1974	1975
Treatments†	21 May	7 Aug.	8 May	11 Sept.	9 Apr.	11 Sept.	3 Sept.	7 Oct.
	---%---							
Ladino-fescue								
Row (20-2.5)	–	–	–	–	–	–	89	64
Row (20-5)	42	51	7	13	5	8	74	23
Row (10-5)	41	56	6	11	4	7	66	25
Row 1st (40-5) Others (20-5)	34	50	6	11	5	10	91	24
Row (20-5) with fall rest	39	52	11	20	5	22	74	9
Broadcast (20-5)	39	55	6	11	7	12	65	16
Ladino								
(10-5)	95	94	22	48	21	59	90	69
(20-5)	98	96	32	66	17	80	98	75
L.S.D. (0.05)	7	9	4	7	6	12	10	12

†See Table 1 for explanation of treatments.

‡DW = % dry weight estimate of clover.
 LT = % clover as determined by modified line transect.

Table 3. Effect of grass density and use of pesticides on dry-matter yield of a ladino clover-fescue mixture and the clover component.

| | Test 1 | | | | Test 2 | | | | | | | |
| | 1973 | 1974 | 1975 | 1976 | 1974 | | 1975 | | 1976 | | 1978 | |
Treatments†	M‡	M	M	M	M	C‡	M	C	M	C	M	C
	---metric tons/ha---											
Check	7.7	2.9	2.8	0.5	2.7	0.8	2.3	0.2	1.2	0.1	2.4	0.4
LG	8.2	3.8	–	–	4.1	1.5	2.9	0.4	–	–	–	–
HG	8.1	3.3	–	–	3.0	0.8	2.2	0.2	–	–	–	–
F	7.5	4.3	5.4	1.6	7.8	4.0	5.5	2.2	2.5	0.6	3.6	0.9
F+I	9.7	8.9	8.0	4.8	9.5	6.3	7.2	4.9	4.3	2.1	5.4	2.5
I	8.4	4.6	5.2	1.8	7.5	4.1	6.1	3.5	3.6	1.2	4.1	1.5
L.S.D. (0.05)	1.2	0.7	0.5	0.5	0.8	0.6	0.7	0.4	0.7	0.3	0.5	0.5

†All treatments above cut when 20 cm back to 5 cm all season. Fescue and ladino clover seeded broadcast (bdc). All treatments seeded at 13.4 kg/ha fescue except LG (low grass density) at 6.7 and HG (high grass density) at 26.8 kg/ha. Ladino seeded at 5.6 kg/ha. F and I indicate use of benomyl and carbofuran, respectively.

‡M indicates total weed-free yield of mixture (clover and fescue); C indicates yield of clover component. First year's data not included.

Table 4. Effect of grass density and use of pesticides on clover stands in a ladino-fescue mixture.

	Test 1								Test 2		
	DW‡						LT‡		LT‡		
	1973		1974		1975		1974	1975	1974	1975	1978
Treatments†	21 May	7 Aug.	8 May	11 Sept.	9 Apr.	7 Oct.	11 Sept.	7 Oct.	3 Sept.	7 Oct.	2 Oct.
	---%---										
Check	39	55	6	11	7	6	12	3	65	16	29
LG	42	59	16	19	10	–	21	–	75	16	–
HG	40	57	9	12	5	–	11	–	50	8	–
F	46	47	18	34	30	20	82	43	97	87	51
F+I	64	71	53	79	78	75	99	99	99	99	89
I	42	62	22	50	28	36	76	77	99	97	66
L.S.D. (0.05)	7	9	4	7	6	8	12	15	10	12	9

†LG and HG indicate low and high grass density. See Table 3 for other treatments.

‡DW = % dry weight estimate of clover. LT = % clover as determined by modified line transect.

Physical Management (Fescue Density) and Pesticides

The low seeding rate of tall fescue resulted in increased total yields (fescue + clover), clover yields, and percentages of clover in some years (Tables 3 and 4). Insects and diseases apparently drastically reduced the vigor and stands of Ladino clover even by the second or third year (Tables 3 and 4). In the third year, the yields of Ladino-fescue mixture treated with benomyl and carbofuran (F + I) were 8.9 and 7.2 metric tons (mt)/ha on tests 1 and 2, respectively, and only 2.9 and 2.3 mt/ha for the untreated checks. The clover percentage was above 50% on the treated stands (F + I) but less than 11% on the untreated checks. Excellent stands and production of Ladino clover in mixture with tall fescue were maintained on the benomyl plus carbofuran plots until the two experiments were discontinued (after 5 years for test 1 and 6 years for test 2, Tables 3 and 4). Plots treated only with carbofuran (I) usually produced better stands and yields of clover than those treated only with benomyl (F).

The role of the insects and disease was not determined. Narrow row spacings of tall fescue (row, 20–5) showed a yield advantage of the mixture over broadcast (bdc, 20–5) in one year (test 2, 1974, Table 1), but clover stands were not markedly affected (Table 2).

LITERATURE CITED

Gibson, P. B., and E.A. Hollowell. 1966. White clover. U.S. Dept. Agric. Handb. 314.

Leffel, R.C., and P.B. Gibson. 1973. White clover. *In* M.E. Heath, D.S. Metcalfe, and R.F Barnes (eds.) Forages, 167–176. Iowa State Univ. Press, Ames, Iowa.

Smith, D. 1975. Forage management in the north. Kendall/Hunt Pub. Co., Dubuque, Iowa.

Potential Grass Production in Ireland

A.J. BRERETON

Agricultural Institute, Johnstown Castle, Wexford, Ireland

Summary

Variation in weather geographically and between years can have a significant impact on intensive animal production systems dependent on grass as feed. This work was undertaken to quantify the effect of weather variations on grass production. The first objective was to relate grass growth quantitatively to radiation and temperature. These relationships are obscured by large changes in sward growth potential within the season, which are related to changes in the physiological status of the crop. Therefore, it was necessary to quantify this physiological effect.

Production data were drawn from production experiments, and weather data were drawn from the published records of the national meteorological network. A model is presented that relates grass accumulation to radiation received and to the efficiency of radiation use. The temperature effect is incorporated in the model through the measured effect of temperature on the efficiency of radiation use. A term is included in the model to represent the physiological factor.

It is suggested that the physiological factor is responsible for a 60% decline in production potential in June of each year and that this lower production potential persists until late December. Average annual production potential is estimated to be 12.5 tonnes/ha. Regional and year-to-year variation can cause this total to rise or fall by 13% and 15%, respectively.

KEY WORDS: grass, radiation, temperature, seasonal growth, potential production, regression.

INTRODUCTION

The primary objective of the work reported here was to quantify the effect of geographical and year-to-year variations in weather on annual grass production in Ireland. Attention is confined to radiation and temperature.

In spring and early summer, when the sward is in the reproductive state, growth potential is greater than it is later in the season. A secondary objective is to quantify the effect of physiological changes on sward growth potential.

METHODS

From Penman (1971),

$$QY = eR \tag{1}$$

where Y is yield (dry matter, kg/ha), R is radiation (joules [j]/ha), Q is heat formation of plant material (16.72×10^6 j/kg), and e is efficiency of radiation use. The effect of temperature on production was evaluated in terms of its effect on e in equation 1. Data from national grass production experiments (NGPE) carried out in 1967–1970 (Ryan 1974) were used to relate e to air temperature. The experiments were carried out at nineteen sites. Production at each site was measured by cuts in each year. Data from the first two harvests in the years 1967, 1968, and 1970 were used, the first in early April and the second in late May. In the first harvest it was assumed that growth began when the temperature exceeded 5°C. Initial yield was assumed to be zero. At each site production was compared at different fertilizer inputs. The data used was from plots receiving greater than 500 kg N/ha annually—the maximum-yield plots. These contained no clover.

Grass growth rates fall after May, when the crop changes from a reproductive to a vegetative state. Leaf extension rates measured in overwintering swards indicate that the reverse change in growth rate occurs at about the winter solstice (Parsons and Robson 1980). A term was inserted into equation 1:

$$QY = eRx \qquad (2)$$

where x represents the physiological factor, has the value 1.0 from January to May, and has a lower constant value at all other times.

To estimate the lower value of x, data were taken from the seasonal growth experiment (SGE) carried out at Johnstown Castle in 1963–1964 (Murphy 1963). The estimate was obtained from comparison of the regression of e on air temperature in the period before and after midwinter. In the SGE, a perennial ryegrass (*Lolium perenne* L.)–white clover (*Trifolium repens* L.) sward was grazed by sheep. At intervals starting in May 1963, individual plots were closed up from grazing, and cumulative regrowth over periods of up to 2 months was measured by successive harvests in 6 subplots of each plot. Plots were closed up from grazing at intervals of 1 to 2 weeks during summer and 4 weeks during winter. Data for the period late August 1963 to early May 1964 were used when the clover in the total dry matter was < 10%.

Usefulness of the method was evaluated by comparing e values measured in the rest periods of two grazing experiments with e values calculated from equation 2.

The two grazing experiments were carried out at Johnstown Castle in 1977 and 1978. In the 1977 experiment (Ryan 1977) two paddocks of a rotational grazing experiment were sampled for dry-matter yield immediately before and after each grazing, at about 20-day intervals. The crop was S24 perennial ryegrass grazed by steers at a high stocking rate.

In the 1978 experiment (O'Sullivan et al. 1978) Cropper perennial ryegrass was grazed by steers in a 21-day rotation at two stocking rates. In each 21-day grazing cycle about half the paddocks were sampled for dry-matter yield immediately before and after grazing. In both experiments grazing commenced about mid-April and ended in October and September in 1977 and 1978, respectively. Growth rates for the early spring period, before grazing commenced, were included in the analysis.

In both experiments N fertilizer was applied in February and at 3-week intervals after grazing commenced. Annual total was 250 kg N/ha.

The weather data used were taken from the records of the national network of meteorological stations.

RESULTS

A comparison of the relationship between leaf extension rate and temperature in perennial ryegrass (cv. Vigor) under field conditions during each of the 6 weeks before and after 1 January 1979 indicated that the midwinter change in sward physiology was abrupt. This finding agrees with recently published work (Parsons and Robson 1980). The change was accompanied by temporary acceleration of leaf senescence rate. Analysis of data from the SGE showed that efficiency of radiation use was greater after the event (Fig. 1). Efficiency of radiation use before midwinter was about 40% of the efficiency after midwinter. Therefore, the term x in equation 2 was given the value 0.4 for the period June–December. The effect of air temperature on efficiency of radiation use in spring with high N fertilizer is illustrated in Fig. 2 using data from the NGPE. Efficiency was higher at all temperatures compared with that in the SGE (postmidwinter). This higher efficiency reflects the higher N fertilizer input in NGPE. Because of the separation in temperature range for the first and second harvests, separate regressions were fitted. The results indicate that efficiency of radiation use responds sharply to temperature in the range 5–10°C and that the response declines at higher temperatures. The regressions were used to estimate e in equation 2; for temperatures in the range 5°–10°C, e = −0.8935 + 0.2037T, and at temperatures greater than 10°C, e = 0.3493 + 0.0702T. At temperatures less than 5°C, e = 0.

Fig. 1. Effect of air temperature on efficiency of radiation use in pasture, August 1963–January 1964 (•) and December 1963–May 1964 (o).

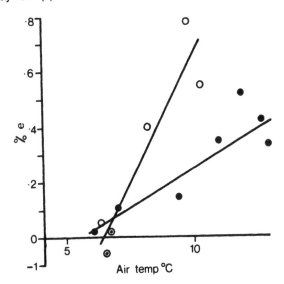

Fig. 2. Effect of air temperature on efficiency of radiation use in pasture, first harvest (•), and second harvest (o).

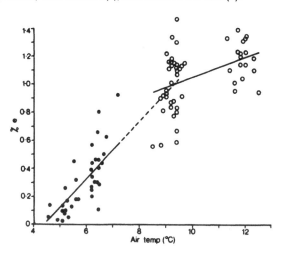

Comparison of e values observed in the two grazing experiments with e values calculated from equation 2 gave a correlation coefficient of r = 0.65. The regression of observed on calculated values [obs. = 0.0416 + 0.8337*** (± 0.1370) expt] indicated that observed values were generally 12% less than calculated values.

Radiation and temperature data recorded from 1971 to 1979 at 14 stations of the national network were used to calculate the expected average annual grass dry-matter production for Ireland. The results indicated that at the average site in the average year 12.5 t/ha would be expected. The range due to site location was ± 1.5 t/ha. The range due to year-to-year weather variation was ± 1.9 t/ha in 6 years out of 10 (i.e., when extreme years were omitted).

DISCUSSION

The decline in sward growth rates that occurs after May coincides with the elimination of reproductive tillers from swards by grazing or cutting. Insofar as the elimination of reproductive tillers occurs progressively over a period, the present procedure of changing the value of x abruptly at the end of May probably leads to an underestimation of growth rate in the following weeks. This error is less important in estimates of whole-season production. Available evidence indicates that the midwinter transition occurs more rapidly and at about the time of the winter solstice.

The effect of temperature on e shown in Fig. 1 agrees with general experience of temperature effects on grass growth (Alberda 1966) and on leaf photosynthesis (Monteith 1977). A comparison of the relationship between e and temperature over the range 5°–10°C in Figs. 1 and 2 shows that in the SGE (Fig. 1) e values were substantially lower. These lower values may be attributed to the fact that no nitrogen fertilizer was used in this experiment whereas the annual level of nitrogen use in the NGPE was in excess of 500 kg N/ha.

The predicated average annual dry-matter production for Ireland represents the expectation for heavily fertilized and irrigated grassland harvested by cutting. Irrigation of grassland is not commonly practiced in Ireland. The annual depression of yield due to summer drought has been estimated at about 1.5 t/ha for the country generally (Brereton 1978). Therefore, the average annual dry-matter production normally obtained may be estimated at 11.0 t/ha.

Total variation in production due to regional climatic differences is ± 13% of the national average. Variation among years is greater. The analysis showed that in 2 years out of 10, production will be more than 15% below average, and in a further 2 years out of 10, it will be more than 15% above average. The results indicate that some adjustment for regional location is necessary in intensive animal-production systems based on grass as feed. The scale of the year-to-year variation implies the need to develop a buffering capacity in the systems for year-to-year variation in grass production.

The overestimation of e values by equation 2 in the correlation analysis probably reflects the lower inputs of N fertilizer in the grazing experiments (observed) than in the NGPE on which the calculated values (expected) were based.

LITERATURE CITED

Alberda, T. 1966. Response of grasses to temperature and light. *In* F.L. Milthorpe and J.D. Ivins (eds.) The growth of cereals and grasses. Butterworths, London.

Brereton, A.J. 1978. Irrigation on dairy farms. Ire. Grassl. Anim. Prod. Assoc. J. 13:35–40.

Monteith, J.L. 1977. Climate and efficiency of crop production in Britain. Roy. Soc., Phil. Trans., Ser. B281, 277–294.

Murphy, W.E. 1963. Seasonal production. *In* Soils, 67. Ann. Rpt. Agric. Inst., Dublin.

Parsons, A.J., and M.J. Robson. 1980. Seasonal changes in the physiology of S24 perennial ryegrass. Ann. Bot. 46:435–444.

Penman, H.L. 1971. Water as a factor in productivity. *In* P.F. Wareing and J.P. Cooper (eds.) Potential crop production. Heinemann, London.

Ryan, M. 1974. Grassland productivity. Ire. J. Agric. Res. 13: 275–293.

Ryan, M. 1977. Cultivar trial. *In* Soils, 33. Ann. Rpt. Agric. Inst., Dublin.

O'Sullivan, M., B. McGrath, and M. Flynn. 1978. Herbage intake studies. *In* Soils, 46. Ann. Rpt. Agric. Inst., Dublin.

Stockpiling of Cool-Season Grasses in Autumn

C.T. DOUGHERTY

University of Kentucky, Lexington, Ky., U.S.A.

Summary

The practice of accumulating growth of cool-season grasses in the late summer and early autumn for subsequent grazing in late autumn and winter, known as stockpiling, has been widely accepted by beef cow-calf producers. The objectives of our study were to determine the growth characteristics of cool-season grasses in the autumn under nitrogen (N) fertilization and to investigate utilization of stockpiled forages.

Tall fescue (*Festuca arundinacea* Schreb.) has many attributes that make it highly suited to stockpiling. On Maury silt loams, Kentucky 31 tall fescue accumulated 6,500 kg/ha of dry matter from 16 August 1979 to 12 December 1979 with N at 100 kg/ha. Boone orchardgrass (*Dactylis glomerata* L.) accumulated 4,300 kg/ha, and Kenblue Kentucky bluegrass (*Poa pratensis* L.) accumulated 2,200 kg/ha. During the first month of stockpiling (mid-August to mid-September) orchardgrass outgrew tall fescue and Kentucky bluegrass with mean crop growth rates of 77, 65, and 43 kg/ha/day, respectively, but in the second month, with the onset of cooler weather, tall fescue accumulated dry matter at the rate of 84 kg/ha/day, and orchardgrass and bluegrass grew at 47 and 41 kg/ha/day, respectively. In the third month (mid-October to mid-November) tall fescue grew at twice the rate (33 kg/ha/day) of the other two grasses.

Water-soluble carbohydrates in stockpiled tall fescue increased with rate of N fertilizer and accounted for 19% of dry matter in mid-November. Silage made from tall fescue at this time (N = 100 kg/ha) had a pH of 4.4, and lactic acid accounted for 2.5% of silage dry matter. Conversion of tall fescue stockpiles into silage in late autumn seems to be a feasible method of using them in regions where severe weather impairs their utilization or accelerates their deterioration.

KEY WORDS: tall fescue, *Festuca arundinacea* Schreb., orchardgrass, *Dactylis glomerata* L., Kentucky bluegrass, *Poa pratensis* L., nitrogen, silage.

INTRODUCTION

Accumulation of growth of cool-season forages during the late summer and early autumn for subsequent grazing during the late autumn and winter is known as stockpiling and is widely practiced by beef cattle farmers in the grassland transition zone of the eastern U.S.A. As stockpiling is employed in the grasslands of Kentucky, pastures are closed to grazing about mid-August, and dry matter is accumulated until late autumn, when grazing with dry beef cows and heifers commences. Grazing continues until consumable forage is depleted or becomes inaccessible to livestock because of ice or snow cover. Growth of cool-season grasses in autumn stockpiles and the feasibility of using them as silage are discussed in this paper.

METHODS

A three-factor response surface in a central-composite second-order design in incomplete blocks (Cochran and Cox 1957) was used in these experiments on mature stands of tall fescue (cv. Kentucky 31), orchardgrass (cv. Boone), and Kentucky bluegrass (cv. Kenblue) located on Maury silt loams. Swards were clipped to 50 mm, and clippings were removed before fertilizers were broadcast on 16 August 1979. Nitrogen (N), phosphorus (P), and potassium (K) were supplied as ammonium nitrate (34% N), triple superphosphate (20% P), and potassium chloride (50% K) at five rates.

Forages were cut at 50 mm at approximately monthly intervals until mid-December to determine dry-matter yields and measure quality parameters. On 14 November 1979, tall fescue plots were mowed to 50 mm, wilted overnight, compacted, sealed in plastic, and ensiled at 20°C. Silage was assayed for organic acids by gas-liquid chromatography. All data presented were computed from statistically significant ($P < 0.05$) response surface equations.

RESULTS

Responses of autumn-stockpiled grasses to N show the advantages of tall fescue over the other major cool-season grasses in Kentucky, particularly at economic levels of N (< 100 kg/ha) (Table 1). Adequate rainfall ensured that soil moisture was nonlimiting during autumn of 1979 (Table 2). Applications of N up to 160 kg/ha resulted in

The author is an Associate Professor of Agronomy at the University of Kentucky. The investigation reported in this paper (No. 81-3-23) is in connection with a project of the Kentucky Agricultural Experiment Station and is published with the approval of the Director.

Table 1. Dry-matter yields (kg/ha) of tall fescue, orchardgrass, and Kentucky bluegrass, 12 December 1979.

	N (kg/ha)				
	0	40	100	160	200
Tall fescue	2150	4322	6486	7098	6685
Orchardgrass	2422	3256	4294	4987	5247
Kentucky bluegrass	1113	1612	2277	2791	3039

Table 3. Yield and quality parameters of stockpiled tall fescue at five rates of N harvested on 14 November 1979.

	N (kg/ha)				
	0	40	100	160	200
Yield (kg/ha)	1488	3300	5413	6606	6885
Tiller dry weight (g/tiller)	0.27	0.39	0.54	0.60	0.61
Nitrogen (%)	1.91	1.94	2.07	2.29	2.48
Water soluble carbohydrate (%)	17.2	18.0	19.1	20.2	20.8
Dead matter (%)	28.6	21.8	18.1	22.7	30.1

heavier tall fescue tillers by mid-November (Table 3). Tiller populations of stockpiles in mid-November were not significantly affected by N and averaged 1,311/m² in tall fescue and 1,028/m² in orchardgrass. Mean growth rates, calculated for harvest intervals (Table 2), show that tall fescue grew more slowly than orchardgrass during the warmer mid-August to mid-September period, but its growth rate accelerated to 84 kg/day with cooler weather. During the last period, tall fescue stockpiles increased in phytomass, while those of orchardgrass were static and those of bluegrass declined.

In tall fescue grown with N at 100 kg/ha, water-soluble carbohydrates rose from 11% of dry matter harvested in September to 19% in November before declining to 16% at the December harvest. Applications of N had a small, positive linear effect on levels of water-soluble carbohydrate (Table 3). Dead plant material accounted for 18% of the harvested tall fescue on 14 November, whereas in orchardgrass and Kentucky bluegrass it was 33% and 46%, respectively, when all were grown under N at 100 kg/ha. By mid-December dead material had increased to 49% for tall fescue, 68% for orchardgrass, and 75% for Kentucky bluegrass.

High-quality silage was made from stockpiled tall fescue harvested in mid-November when yield and quality were high (Tables 3 and 4). At intermediate levels of N (100 kg/ha), silage made from tall fescue had a pH of 4.4 and contained 2.5% lactic acid, 1.4% acetic acid, and 0.8% butyric acid. Neutral detergent fiber (NDF) and acid detergent fiber (ADF) averaged 55% and 33%, respectively. Increments of N raised silage yields, percentage of silage N, and content of lactic, acetic, and butyric acids. Silage pH declined with increasing rates of

N along with ADF and NDF. Only about 9% of silage N was in the form of volatile bases.

DISCUSSION

Cool-season grasses have limited productivity and mediocre quality during the high-temperature and moisture-stressed summers of Kentucky, but with the onset of cooler weather in the autumn, if there is adequate soil moisture and fertility, growth and quality improve dramatically. Of the perennial grasses adapted to this region, tall fescue seems to be the most suited for autumn stockpiling (Taylor and Templeton 1976, Matches 1979).

Yield of vegetative cool-season grasses is the product of tiller populations/unit area and the mean tiller dry weight. Growth of autumn stockpiles is characterized by an initial period of tillering followed by a self-thinning phase as temperatures, radiation, and photoperiod decline (Kays and Harper 1974). The cessation of tillering does not seem to be directly related to the amount of substrate because levels of water-soluble carbohydrate are very high at this time; however, as most of this water-soluble carbohydrate exists as fructosans (Smith 1968), it may be unavailable to the tiller bud. In the late autumn, low temperatures slow the rate of leaf and tiller appearance, and dry-matter accumulation results largely from leaf growth. During late autumn and early winter, self-thinning of tillers seems to be offset by accelerated leaf senescence, thus ensuring the overwintering of induced tillers.

Stockpiles of tall fescue may, depending on autumn weather, greatly exceed the winter feed requirements of cow-calf farmers. Inclement weather usually prevents the making of surpluses into high-quality hay. Stockpiles that are not utilized continue to deteriorate throughout the

Table 2. Climatic data and growth rates (kg/ha/day) of N-fertilized grasses (100 kg/ha) for four harvest intervals in 1979.

	Interval			
	15 Aug/ 17 Sept	17 Sept/ 15 Oct	15 Oct/ 14 Nov	14 Nov/ 12 Dec
High Temperature (°C)	26	19	16	11
Low Temperature (°C)	16	9	5	2
Precipitation (mm)	156	216	69	109
	---------------------- kg/ha/day ----------------------			
Tall fescue	65	84	33	37
Orchardgrass	77	47	16	−1
Bluegrass	43	41	16	−24

Table 4. Yield and quality parameters of tall fescue silage made on 15 November 1979 (dry-matter basis).

	N (kg/ha)				
	0	40	100	160	200
Silage (kg/ha)	1328	2907	4718	5777	6077
pH	4.9	4.7	4.4	4.4	4.4
Lactic acid (%)	0.2	1.2	2.5	3.5	3.8
Nitrogen (%)	1.9	1.9	2.1	2.3	2.5

winter, especially when they are covered by ice and snow. Their residues hinder growth in the spring, accelerate stand thinning, and hamper renovation operations (Taylor and Templeton 1976).

Stockpiles of N-fertilized tall fescue may be made into excellent silage if the grass is harvested before yields decline and quality deteriorates. High concentrations of water-soluble carbohydrate ensure sufficient substrate for lactic acid synthesis. Nitrogen fertilizer, which increases stockpile yield, also tends to elevate levels of water-soluble carbohydrate, presumably because low temperatures retard utilization more than assimilation (Wolf et al. 1979). Availability of substrates and the relatively low moisture content of autumn-stockpiled forages may eliminate the need for wilting and allow direct harvesting.

Silage made from stockpiled autumn pastures may be fed to livestock during severe winter weather or stored for feeding at times when grazing is unavailable or restricted. In tall fescue forage systems, autumn silage not consumed during winter feeding operations could be fed in late summer when productivity and quality of tall fescue are low (Van Keuren and Stuedemann 1979). In areas where snow is more frequent and abundant, farmers could opt to ensile some fraction of their area of stockpiled forages as insurance against extended periods of snow cover.

LITERATURE CITED

Cochran, W.G., and G.M. Cox. 1957. Experimental designs. John Wiley and Sons, New York.

Kays, W., and J.L. Harper. 1974. The regulation of plant and tiller density in a grass sward. J. Ecol. 62:97–105.

Matches, A.G. 1979. Management. *In* R.C. Buckner and L.P. Bush (eds.) Tall fescue. Agronomy 20:171–199.

Smith, D. 1968. Classification of several North American grasses as starch or fructosan accumulations in relation to taxonomy. J. Brit. Grassl. Soc. 23:306–309.

Taylor, T.H., and W.C. Templeton, Jr. 1976. Stockpiling Kentucky bluegrass and tall fescue for winter pasturage. Agron. J. 68:235–239.

Van Keuren, R.W., and J.A. Stuedemann. 1979. Tall fescue in forage-animal production systems for breeding and lactating animals. *In* R.C. Buckner and L.P. Bush (eds.) Tall fescue. Agronomy 20:201–232.

Wolf, D.D., R.H. Brown, and R.E. Blaser. 1979. Physiology of growth and development. *In* R.C. Buckner and L.P. Bush (eds.) Tall fescue. Agronomy 20:75–92.

Changes in Quality and Composition of Alfalfa During Autumn

S.C. FLEMING, M. COLLINS, and N.A. JORGENSEN

University of Wisconsin, Madison, Wis., U.S.A.

Summary

Autumn harvest of alfalfa (*Medicago sativa* L.) offers a potential for obtaining additional forage production during a particular season. Little is known about the quality or changes in the chemical composition of this forage. The objective of this research was to examine changes in chemical composition, dry-matter intake (DMI), and apparent digestibility of alfalfa during autumn in Wisconsin.

An established stand of Vernal alfalfa, early bud to midbud, was harvested using a flail harvester on 5 and 20 October and 5 and 18 November 1979. The forage was dried at 55°C for 72 to 96 hours in burlap bags. A total collection digestion trial was conducted utilizing four castrated Saanen goats. Apparent dry-matter digestibility and DMI were determined for each forage. Each sample was analyzed for cell-wall constituents (CWC), nitrogen (N), total nonstructural carbohydrates (TNC), phosphorus (P), potassium (K), calcium (Ca), magnesium (Mg), and sulfur (S).

The concentrations of neutral detergent fiber (NDF) and acid detergent fiber (ADF) in the forage declined by 3.4% and 5.8%, respectively, between 5 October and 5 November, and the cellulose concentration decreased 6.7% during the same period. The reduction in concentration of CWC probably represents a dilutive response to the 79.6% increase in total nonstructural carbohydrates (TNC) between 5 October and 5 November. Cellulose, ADF, and NDF concentrations increased significantly between 5 and 18 November.

A 4.8% increase in in-vitro dry-matter disappearance (IVDMD) accompanied the initial increase in TNC, but a 4.2% reduction in IVDMD occurred between 5 and 18 November. Total nitrogen concentration averaged 4.0%, 3.7%, 3.3% and 3.1% of dry matter at each successive harvest date, respectively.

Dry-matter intake (g/kg$^{0.75}$) increased 9.2% from 5 October to 5 November and then decreased 4.7% by 18 November. Digestible dry-matter intake (DDMI) increased from 51.7 g/kg$^{0.75}$ on 5 October to a high of 56.1 g/kg$^{0.75}$ on 5 November and then decreased slightly by the last harvest date. Decreases of 5.8% and 8.2% in the apparent digestibilities of NDF and ADF, respectively, occurred between 5 October and 5 November. This reduction in fiber digestibility may reflect an increase in rate of passage resulting from the increase in DMI. The apparent digestibility of cellulose also decreased 5.0% during this period. The apparent digestibility of cellulose, ADF, and NDF increased between 5 and 18 November. Apparent digestibility of nitrogen decreased slightly during autumn. These data indicate that the feeding value of alfalfa, as well as the ensiling potential, may increase initially during autumn due to increasing TNC concentrations during that period.

Key Words: alfalfa, *Medicago sativa* L., digestibility, dry-matter intake, forage quality, nutritive value.

INTRODUCTION

Autumn harvest of alfalfa (*Medicago sativa* L.) offers a potential for obtaining additional forage production during a particular season. Research in Kentucky by Collins and Taylor (1980) showed that at that location, alfalfa may be clipped on or after 1 November with little reduction in root carbohydrate levels.

Little information is available regarding changes that occur in the quality and chemical composition of alfalfa during autumn. Collins and Taylor (1980) reported decreases in leaf percentage, in-vitro dry-matter disappearance (IVDMD), and concentrations of nitrogen (N), phosphorus (P), potassium (K), calcium (Ca), and magnesium (Mg) in alfalfa as autumn and winter advanced.

The objective of this study was to examine changes in chemical composition, dry-matter intake (DMI), and apparent digestibility of alfalfa during autumn in Wisconsin.

METHODS

An established stand of Vernal alfalfa at the Arlington Experimental Farm (lat 43°N long 89°W) was used in this study. The alfalfa was at early bud to midbud by 5 October and did not mature further. Forage was harvested on 5 and 20 October and 5 and 18 November 1979, using a flail harvester. After chopping, the forage was dried at 55°C for 72 to 96 hours in burlap bags.

A total collection digestion trial was conducted using four castrated Saanen goats in a Latin square (4 × 4) design. An amount of feed equal to 110% of the average ad-libitum intake measured during days 5 to 10 was fed during days 11 to 15. Feed and fecal samples were dried at 55°C for 48 hours for dry-matter determinations, and urine samples were frozen for subsequent analysis. Feed, orts, and feces were ground to 1 mm for analysis.

Nitrogen concentration was determined using a colorimetric technique (Cataldo et al. 1974), and mineral analysis of the feed was determined by inductively coupled plasma–atomic emission spectrometry. In-vitro dry-

matter disappearance (IVDMD) was determined by the method of Barnes (1966), and total nonstructural carbohydrates (TNC) was determined by the method of Smith (1969). Neutral detergent fiber (NDF), acid detergent fiber (ADF), cellulose, hemicellulose, and lignin concentrations were determined by the method of Goering and Van Soest (1970). Standard analysis of variance tests were performed on the means of four observations and compared, using Duncan's Multiple Range Test.

RESULTS AND DISCUSSION

Forage Analysis

Total nonstructural carbohydrate concentrations in the alfalfa increased by 79.6% from 5 October to 5 November, with an additional 13.9% increase by 18 November (Table 1). This increase in TNC concentration might be expected to improve the fermentation characteristics of silage produced from alfalfa during autumn by increasing the supply of fermentable carbohydrate. A 4.8% increase in IVDMD accompanied the initial increase in TNC concentration. Nitrogen concentration decreased with each successive harvest between 5 October and 18 November. This reduction in N concentration (22.5%) may be attributed to the combined effects of leaf drop (Collins and Taylor 1980) and leaching. The lowest N concentration, measured on 18 November, was equivalent to 19.4% crude protein, indicating that alfalfa harvested even during mid-November would be an excellent source of protein for livestock feeding.

Concentrations of NDF and ADF in the forage declined by 3.4% and 5.8%, respectively, between 5 October and 5 November, and cellulose concentration decreased 6.7% during the same period. The reduction in concentrations of structural components appears to be related to the corresponding increase in TNC concentration. Cellulose, ADF, and NDF concentrations increased significantly between 5 and 18 November. A high negative correlation ($r = -0.99$, $P < 0.01$) was found between the ADF concentration and IVDMD.

Intake and Digestibility

Although apparent DM digestibility did not change significantly between harvest dates, the 9.2% increase in DMI from 5 October to 5 November resulted in an 8.5% increase in DDMI ($g/kg^{0.75}$) during the same period (Table 2). In spite of the relatively small change in NDF

S.C. Fleming is a Graduate Research Assistant, M. Collins is an Assistant Professor of Agronomy, and N.A. Jorgensen is a Professor of Dairy Science at the University of Wisconsin. This is a contribution of the University of Wisconsin Department of Agronomy. Research was supported by the College of Agricultural and Life Sciences.

The authors would like to thank Dr. R.M. Soberalske for his assistance in the laboratory analysis.

Table 1. Chemical analysis of autumn harvested alfalfa.

Date of harvest	TNC	IVDMD	N	NDF	ADF	Cellulose	Hemi-cellulose	Lignin
					% of dry matter			
5 October	4.4 c‡	68.7 b	4.0 a	35.2 ab	27.7 a	20.9 a	7.4 a	6.1 a
20 October	4.1 c	72.8 a	3.7 b	34.7 bc	26.2 b	20.5 ab	8.5 a	5.4 b
5 November	7.9 b	72.0 a	3.3 c	34.0 c	26.1 b	19.5 b	7.9 a	5.1 b
18 November	9.0 a	69.0 b	3.1 d	35.9 a	27.7 a	21.3 a	8.2 a	5.3 b

‡Means followed by the same letter are not different at the 5% probability level.

Table 2. Dry matter intake and apparent digestibility of autumn harvested alfalfa.

Date of harvest	Dry-matter intake g/kg$^{0.75}$	Apparent DM digestibility	Digestible dry-matter intake g/kg$^{0.75}$	Apparent digestibility					
				TNC	N	NDF	ADF	Cellulose	Hemi-cellulose
						% of dry matter			
5 October	77.2 a‡	66.9 a	51.7 a	95.0 a	76.2 a	58.6 a	57.1 a	71.6 a	64.1 a
20 October	81.0 a	67.7 a	55.0 a	94.8 a	75.8 a	58.6 a	55.4 ab	71.7 a	68.7 a
5 November	84.3 a	66.7 a	56.1 a	97.7 a	73.0 b	55.2 a	52.4 b	68.1 b	63.4 a
18 November	80.3 a	67.2 a	53.9 a	97.6 a	72.2 b	57.1 a	54.7 ab	70.6 ab	65.0 a

‡Means followed by the same letter are not different at the 5% probability level.

concentration, a high negative correlation (r = -0.99, P < 0.01) was found between NDF concentration and DMI. The increase in DMI may also be related to the considerable increase in TNC concentration during the same period.

The apparent digestibility of the TNC component of the forage ranged from 95.0% to 97.7% (Table 2). With each successive harvest the apparent digestibility of N decreased. The correlation (r = 0.99, P < 0.01) between N concentration and apparent digestibility of N agrees with the prediction of Holter and Reid (1959).

Decreases of 5.8% and 8.2% in apparent digestibility of NDF and ADF, respectively, occurred between 5 October and 5 November. This reduction in fiber digestibility may reflect an increase in rate of passage resulting from the increase in DMI. The apparent digestibility of cellulose also decreased 5.0% during this period, but apparent digestibility of cellulose, ADF, and NDF increased between 5 and 18 November.

Mineral Analysis

The concentrations of Ca, P, K, Mg, and sulfur (S) in the alfalfa declined during the harvest period (Table 3). Differences were apparent in the rate of decline in concentration of the elements measured. Phosphorus concentration decreased significantly with each of the first three harvests and then remained relatively unchanged through the 18 November harvest. Of the minerals studied, Ca concentration declined most slowly. This finding is not unexpected since Ca pectate, a major component of the middle lamella, is relatively insoluble in cold water. Furthermore, calcium oxalate, containing up to 33% of the calcium in alfalfa, is unavailable to cattle (Ward et al. 1979). K, found in plants in the K$^+$ form, declined rapidly during the harvest period, presumably a result of its high water solubility. Mg and S concentrations decreased at a similar rate during the harvest period. The overall decrease between 5 October and 18 November averaged 26% for Mg and 19% for S.

Table 3. Mineral concentrations in autumn-harvested alfalfa.

Date of harvest	Mineral				
	Ca	P	K	Mg	S
			% of dry matter		
5 October	1.57 a†	0.31 a	3.06 a	0.35 a	0.36 a
20 October	1.66 a	0.25 b	2.70 b	0.33 ab	0.34 ab
5 November	1.62 a	0.22 c	2.21 c	0.31 b	0.32 bc
18 November	1.40 b	0.20 c	1.95 d	0.26 c	0.29 c

†Means followed by the same letter are not different at the 5% probability level.

LITERATURE CITED

Barnes, R.F 1966. The development and application of *in vitro* rumen fermentation techniques. Proc. X Int. Grassl. Cong., 434-438.

Cataldo, D.A., L.E. Schrader, and V.L. Youngs. 1974. Analysis by digestion and colorimetric assay of total nitrogen in plant tissues high in nitrate. Crop Sci. 14:854-856.

Collins, M., and T.H. Taylor. 1980. Yield and quality of alfalfa harvested during autumn and winter and harvest effects on the spring crop. Agron. J. 72:839-844.

Goering, H.K., and P.J. Van Soest. 1970. Forage fiber analysis (apparatus, reagents, procedures, and some applications). U.S. Dept. Agric., Agric. Handb. 379.

Holter, J.A., and J.T. Reid. 1959. Relationship between the concentrations of crude protein and apparent digestible protein in forage. J. Anim. Sci. 18:1339–1349.

Smith, D. 1969. Removing and analyzing total nonstructural carbohydrates from plant tissue. Wis. Agric. Exp. Sta. Res. Rpt. 41.

Ward, G.W., L.H. Harbers, and J.J. Blaha. 1979. Calcium-containing crystals in alfalfa: their fate in cattle. J. Dairy Sci. 62:715–722.

Production Response of Russian Wildrye (*Elymus junceus* Fisch.) to Fertilizer and Clipping

L. HOFMANN and J.F. KARN

USDA-SEA-AR Northern Great Plains Research Center, Mandan, N.Dak., U.S.A.

Summary

Experiments were conducted on nitrogen (N) fertilization of long-term Russian wildrye (RWR) (*Elymus junceus* Fisch.) pasture and the effect of simulated spring grazing on quantity of forage available for fall use. A stand of RWR that had been fertilized with 45 kg N/ha and grazed annually for 10 years at Mandan, North Dakota, was divided into two experiments in 1978. The first evaluated effects from all combinations of 0, 45, 90, 135, 180, and 225 kg N/ha; 0, 45, and 90 kg phosphorus (P)/ha; and 0 and 135 kg potassium (K)/ha on dry-matter (DM) yields, whereas the second compared ten clipping treatments ranging from biweekly intervals to autumn (clipping after frost) harvests. Precipitation was above average in 1978, below average in 1979; in 1980 severe drought through June was followed by above-average precipitation during July and August. Yield increased with each increment of fertilizer N up to 180 kg/ha in 1978 and 1979. Yields ranged from 3,220 to 6,860 kg/ha in 1978 and 1,040 to 5,690 kg/ha in 1979. As yield on 17 July 1980 was only 110 kg/ha, treatment differences were of little practical significance. Yield of a harvest on 25 September 1980, after recovery from drought, ranged from 240 to 2,150 kg/ha for 0 through 225 kg N/ha, with each increment of N significantly increasing yield. Neither P nor K fertilization influenced yields. Biweekly clipping reduced yield each year and resulted in thin, weed-infested stands. Early July clipping led to maximum yields in 1978 and 1979 but not in 1980. Light spring grazing, simulated by a harvest at a 13-cm stubble height on 5 June 1978, led to 1,430 kg DM/ha at the autumn harvest compared with an autumn yield of 3,020 kg/ha where there was no spring grazing. A rate of 45 kg N/ha was insufficient to maintain optimum DM yield, especially during dry years. Limited spring utilization greatly reduced forage available for autumn grazing even though total production was increased. Biweekly harvests reduced overall yield and vigor.

KEY WORDS: Russian wildrye, *Elymus junceus* Fisch., fertilizer, clipping interval, fall harvest, precipitation.

INTRODUCTION

Russian wildrye (RWR) (*Elymus junceus* Fisch.) is often recommended for extending the grazing season in the northern Great Plains. Because of apparent nitrogen (N) deficiency on several 10-year-old, autumn-grazed RWR pastures, we began two studies in 1977 (1) to determine yield response of these pastures to N, phosphorus (P), and potassium (K) fertilization, and (2) to compare yield effects of 10 clipping treatments.

METHODS

A portion of a RWR pasture at Mandan, North Dakota, that had been seeded with 15-cm row spacing in 1968 on Temvik silt loam (Typic Haploborolls) was divided into a fertility study and a clipping study. Plots were 1.8 × 6.1 m in size. Pastures had been grazed each autumn and fertilized annually with 45 kg N/ha. Oven-dried dry-matter (DM) yields were determined from 0.9 × 3.6-m areas mowed at a height of 2.5 cm unless otherwise noted.

In the fertility experiment, all combinations of nitrogen as NH_4NO_3, P as triple superphosphate, and K as KCl were broadcast in all combinations at rates of 0, 45, 90,

L. Hofmann is a Research Agronomist and J.F. Karn is a Range Animal Nutritionist of USDA-SEA-AR.

135, 180, and 225 kg N/ha; 0, 45, and 90 kg P/ha; and 0 and 135 kg K/ha in a factoral arrangement, replicated 3 times, in October of 1977, 1978, and 1979.

Yields were taken 12 July 1978 and 1979 and 17 July and 25 August 1980. Height measurements and visual vigor estimates were made 11 May 1979.

Soil samples from 0 N and 45 kg N plots were taken at 0–15-cm and 15–30-cm depths on 25 April 1980 for determination of pH, P, and K. Gravimetric soil water content in all plots was measured to 180 cm in 30-cm increments on 8 May 1980.

In the clipping study, clipping treatments were randomized within 4 replicates. Harvest dates were as follows: (1) for biweekly clipping, 22 May, 5 June, 22 June, 5 July, and 7 August 1978; 31 May, 14 June, and 4 September 1979; and 27 October 1980; (2) for spring clipping, 6 June 1978, 14 June 1979, 14 July 1980; (3) for summer clipping, 5 July 1978, 29 June 1979, 14 July 1980; and (4) for autumn clipping, 20 October, 4 October, and 27 October for 1978, 1979, and 1980, respectively. Single spring, summer, and autumn harvests, and spring or summer combined with autumn harvests were taken. Harvests at 8-cm and 13-cm heights on 5 June 1978 were conducted to simulate moderate and light spring grazing; an autumn harvest at 2.5 cm height followed on 20 October 1978. All plots received 45 kg N/ha in October of 1977, 1978, and 1979.

Means were considered significant at the 0.05 probability level, using Duncan's Multiple Range Test.

RESULTS

For the period September 1977 through June 1978 precipitation was 35% above average (Table 1), the average being that for 1915–1979, and consequently 1978 RWR yield averaged 5,280 kg/ha, highest of the study. Precipitation for June 1978–June 1979 was 30% below average, and 1979 RWR yield was 3,570 kg/ha. For September 1979–June 1980, precipitation was 65% below average, and the 17 July 1980 harvest yielded only 100 kg/ha. Above-average July-September 1980 precipitation produced a 9 September 1980 harvest of 1,170 kg/ha.

Fertility Study

Extractable soil P averaged 6.7 and 5.6 kg P/ha and exchangeable K averaged 645 and 440 kg/ha for the 0-15- and 15-30-cm depths, respectively. Soil pH aver-

aged 6.7. Neither P nor K fertilization caused a visible or significant yield response.

As to height and vigor, the visible growth differences because of N application were evident in early spring each year. On 11 May 1979, both plant height and plant vigor were different (P < .05) among 0, 45, and 90 kg N/ha treatments (Table 2). Vigor and height differences apparent in April 1980 disappeared by May because of drought.

Annual yields differed because of precipitation, but within years N fertilization affected yields (Table 3). With 45, 90, 135, 180, and 225 kg N/ha, yields were 27%, 36%, 44%, 49%, and 53% greater, respectively, than with 0 N in 1978, and 49%, 69%, 75%, 80%, and 82% greater in 1979. Thus, N fertilization still benefitted RWR during a year with less-than-normal precipitation. Although yields were lower in 1979 than in 1978, the percentage of increase from each 45 kg N/ha increment was greater. Yields (Table 3) were dramatically reduced 17 July 1980 because of the September 1979 through June 1980 drought. The significant difference of 90 kg N/ha over 0 N meant little because yields of 110 and 30 kg/ha were so small. With the end of the drought, September 1980 yield increased with each increment of N fertilizer.

Soil water under the various treatments was measured 8 May 1980. Fertilization rate had no effect on soil water content nor on depth of water depletion.

Clipping Study

Originally, fourteen harvest treatments were planned, but insufficient growth, because of poor growing conditions, precluded harvesting according to schedule. A bi-weekly treatment, planned to simulate continuous grazing, produced an accumulated 1978 yield of 2,160 kg/ha, less than that of any other treatment except the spring-autumn combination. The bi-weekly plots became weedy and in 1979 and 1980 yielded 500 and 210 kg/ha, respectively. The highest single-harvest yield in 1978 was 4,150 kg/ha, obtained from the summer harvest, which did not differ greatly from the summer-autumn yield of 3,510 kg/ha. Other 1978 yields were lower, ranging from 2,950 to 3,340 kg/ha. In 1979 the single summer harvest again produced the highest yield, 1,700 kg/ha. Spring-autumn and summer-autumn combinations yielded less, 1,030 and 980 kg/ha, respectively. During 1980, all single-harvest plots were cut only on 14 July because of severe drought. Yield differences,

Table 1.   Precipitation (mm) at Mandan, N. Dak.; Jan. 1977– Sept. 1980.

Year	Jan-Mar	Apr	May	June	July	Aug	Sept	Oct-Dec	Total
1977	31	3	37	57	42	47	244	71	532
1978	12	51	107	72	44	13	53	45	397
1979	49	24	19	34	167	41	21	7	362
1980	26	12	21	37	68	153	31	–	–
Avg.[1]	37	40	55	86	60	41	38	45	402

[1] 1915–1979 average.

Table 2. Plant height and vigor (11 May 1979).

N applied (kg/ha)	Height (cm)	Vigor score[1]
0	6.8 a[2]	4.1 a
45	7.6 b	5.1 b
90	8.3 c	5.8 c
135	9.3 d	6.2 cd
180	9.6 d	6.6 d
225	9.5 d	6.8 d

[1]Rated from 1 to 9; 1 = yellow; 9 = dark green.

[2]Column means with same letter not different @ 5%.

Table 4. Fall dry matter yield (kg/ha) following scheduled pre-fall harvest.

Pre-fall harvest	Fall harvest yield		
	1978	1979	1980
None	3020 a[1]	[2]1040 ab	610 a
June @ 13 cm height	1430 b	1260 a	610 a
June @ 8 cm height	1140 b	1570 a	610 a
Early June	620 c	560 bc	230 b
Early July	150 d	150 c	250 b

[1]Column means with same letter not different @ 5%.

[2]Underlined yield indicates no pre-fall harvest that season.

averaging 230 kg/ha, were insignificant. Total yields of the spring-autumn and summer-autumn combinations were 260 and 270 kg/ha, respectively.

Table 4 summarizes the yield obtained from the autumn harvest of the two-harvest combinations. Even light removal of plant growth at a height of 13 cm in 1978 reduced autumn yield approximately 50%. In June 1979 and 1980, there was not enough RWR growth to harvest at 8 or 13 cm, and only autumn yields, underlined in Table 4, were obtained. The yield of the autumn portion of a two-harvest schedule was always less than that of a single autumn harvest.

DISCUSSION

Rogler and Lorenz (1970) reported more beef produced from 45 kg N/ha than from 0 N RWR pastures in the third and subsequent years of their study. This finding may suggest a slow depletion of soil N when N is not replaced with fertilization. In our fertility study, 90 kg N/ha produced 620 and 1,270 kg/ha more yield than 45 kg N/ha in 1978 and 1979, respectively. These increases on our 10-year-old pastures indicated rates higher than 35 to 45 kg N/ha (Rogler and Schaaf 1963) may be required to sustain high production of long-term RWR pasture. Our results confirmed those of Smika et al. (1960) that RWR responds to N rates higher than 45 kg N/ha.

Water use in our study was the same for all N rates

since precipitation and soil water depletion were similar for all plots; therefore, fertilizer N increased yield/unit of available water (water-use efficiency), even with lower-than-normal precipitation. Drought became so intense during the spring and summer of 1980 that N had little effect on yield. After the drought ended, RWR yield again was increased by N fertilization.

In the clipping study, weed-infested stands and low yields reflected stress of the biweekly clipping treatment. Thaine (1954) reported that first-crop RWR clipped at 3- or 6-week intervals produced higher forage yield than clipping once or twice/season; however, root yields and reserves diminished as clipping increased. Our biweekly treatment possibly depleted root reserves to the extent that RWR plants were unable to recover completely. Rogler and Lorenz (1970) grazed RWR with 1.51 steers/ha for 143 days/season for 7 years without stand loss. Our results showed that to simulate continuous grazing, a harvest interval of more than two weeks in length, a greater cutting height, or both must be used.

On the basis of our results, we do not recommend pre-autumn forage removal of RWR if maximum autumn yield is desired, because even light clipping in early June reduced autumn yield. Clipping results were obtained with 45 kg N/ha fertilization rates. In light of the fertilization-study results, perhaps higher rates of N fertilization would alter these conclusions.

LITERATURE CITED

Rogler, G.A., and R.J. Lorenz. 1970. Beef production from Russian wildrye (*Elymus junceus* Fisch.) in the United States. Proc. XI Int. Grassl. Cong., 835–838.

Rogler, G.A., and H.M. Schaaf. 1963. Growing Russian wildrye in the western states. U.S. Dept. Agric. Leaflet 524.

Smika, D.E., H.J. Hass, and G.A. Rogler. 1960. Yield, quality, and fertilizer recovery of crested wheatgrass, bromegrass, and Russian wildrye as influenced by fertilization. J. Range Mangt. 13:243–246.

Thaine, R. 1954. The effect of clipping frequency on the productivity and root development of Russian wildrye grass in the field. Can. J. Agric. Sci. 34:299–304.

Table 3. Dry matter yield (kg/ha) with six rates of N.

N applied (kg/ha)	Harvest Date			
	7/12/78	7/12/79	7/17/80	9/25/80
0	3220 a[1]	1040 a	30 a	240 a
45	4400 b	2050 b	70 ab	560 b
90	5020 c	3320 c	110 b	1080 c
135	5800 d	4100 d	100 b	1560 d
180	6340 e	5220 e	140 b	1920 e
225	6860 e	5690 e	140 b	2150 f

[1]Column means with same letter not different @ 5%.

China's Grassland Types: Their Locality and Improvement

CHIA SHEN-SIU

Beijing Agricultural University,
Beijing, People's Republic of China

Summary

China's natural grassland is mainly distributed over the temperate zone to the west and north of the isohyet of 400 mm annual rainfall and over the hilly subtropical areas in the southern part of the country, amounting to 37% of the country's total land area. Since 1949, comprehensive surveys have been conducted on several occasions in Tibet, Neimenggu Autonomous Region (Inner Mongolia), Xinjiang, and other regions. The resource of forage plants and grassland productivity has been studied. There are about 20 types and eight regions of grassland in China. On the natural steppe of the northern part of Mongolia-Xinjiang Plateau, the successive grassland types from east to west are as follows: meadow grassland, steppe, desert grassland, and desert. Where deserts extend into the mountains there are mountain grasslands and alpine meadows. In the Tibet Plateau there are alpine meadows, grassland, and alpine deserts. In the soggy and peaty regions all over the country, swampy meadows, wooded meadows, and tufty swamps have developed. In the tropical and subtropical areas there are hilly, scrubby, or wooded grasslands and savannas.

Salient features of the natural grassland include:

1. *Variance in productivity due to geographical location.* In general, in areas where annual rainfall is as much as 100 mm, grassland yield averages 750 kg of fresh grass/ha/yr. China's grassland can be classified into two types: (a) meadow, grassland, and steppe with yearly fresh-grass output > 1,500 kg/ha; and (b) desert grassland, desert, and alpine meadow with yearly fresh-grass output < 1,500 kg/ha. The ratio of total area of the former to the latter is 3:7.

2. *Longer winter-spring and shorter summer-autumn seasons and scarcity of pastures in the northern grassland region,* with low grazing capacity of the winter pasture.

3. *Interannual climatic changes that result in fluctuating grass production.* In the overused grasslands of Yanchi, a 2-year rehabilitation program resulted in a 181% increase of grass output. Disking and seeding also improved the *Aneurolepidium* grassland in the speed of permeability, hardiness, and water-holding capacity of the soil. The result was a 385% increase in grass output as compared with that of the undisked plots.

KEY WORDS: grassland, types, locality, improvement, China.

INTRODUCTION

China's natural grassland is mainly distributed over the temperate zone to the west and north of the isohyet of 400 mm annual rainfall and over the hilly subtropical region in the central and southern part of the country. The former consists of 290 million ha and the latter of 66 million ha, amounting to 37% of the country's total land area. Whereas the northern and western rangelands are composed mainly of natural steppe and desert, the southern grassy slopes are overgrown with secondary scrub-herb that cropped up after the original forest was destroyed years back.

Since 1949, comprehensive surveys and special-topic investigations have been conducted on several occasions in Tibet, Xinjiang, the northeast, and Inner Mongolia, and comprehensive studies and investigations have been carried out on the southern grassy slopes. Good forage plants have been studied, and their utility, nutritive

composition, and economic value have been analyzed and assessed. Grassland productivity and grazing capacity have been estimated, and seasonal nutrient composition of the forage plants and succession of the grassland ecosystem have been surveyed and determined. During the general survey, aerial photography and relief maps of medium scale were widely used. This effort resulted in the making of a map of the north grassland types of 1:500,000 scale and a map of the north grassland's seasonal pastures and their grazing capacity. On the basis of these, a national grassland map of 1:2,500,000 scale was prepared, and the proper ratios between different kinds and varieties of animals to be raised in certain types of grassland in the light of their nutritive composition were determined and put forward.

TYPES OF GRASSLAND

China is a country of typical monsoon climate — with a distinct winter and summer and a rainy and sweltering season — that deeply influences the growth of plants, the vegetation, and its laws of succession. There are 20 major

The author is a Professor of Grassland Science at Beijing Agricultural University.

types of grassland in China. On the natural steppe of the northern part of the Mongolia-Xinjiang Plateau, the successive grassland types from east to west are: meadow grassland, steppe, desert steppe, and desert, each showing a distinct vegetation zonation. There are extensive deserts in the inland area, taking up 64.4% of the total area of the northern grassland, where the steppe occupies only 31.2%. Where the desert has invaded the mountains, the distinct vertical zonation of vegetation from the base up ranges from mountain desert to mountain grassland, subalpine meadows, and alpine meadows. In the Qinghai-Tibet Plateau there are alpine meadows, grassland, and deserts in the alpine and frigid areas and dispersed swampy meadows, saline meadows, woodland meadows, and scrub-herb grounds in low-lying areas. In soggy and peaty regions all over the country, swampy meadows, grassy marshes, and tufty swamps have developed. In the vast tropical and subtropical regions where the hilly forests have long been destroyed and unrestored, scrubby grassland or sparsely wooded meadows, or occasionally wooded grassland, cover the slopes. Some of these are now used for grazing.

LOCALITIES OF THE GRASSLAND REGIONS

Eight grassland regions have been delineated in China, based on grassland composition and locality of each.

1. The northeast meadow grassland region. This includes nearly the whole of Liaoning, Jilin, and Heilongjiang provinces, and areas to the east of the Greater Khingan Range, where *Aneurolepidium chinense* predominates. In the low-lying grounds along the lower sections of the Heilong, Wusuli, and Songhua rivers, there are swamps and swampy meadows.

2. The Inner Mongolia steppe region. A major grassland of China, this region encompasses the Inner Mongolia Plateau and its extensions, with bunchgrass and needlegrass, such as *Stipa krylovii* and *S. grandis*, predominating.

3. The northwest desert region. Taking in the northern part of Xinjiang and Qinghai, the western part of Inner Mongolia, and part of Gansu and Ningxia, this region consists of huge stretches of the Gobi Desert. The predominating plants are xerophilous and saline shrubs and semishrubs such as *Resumeria soongarica* and *Salsala passerina*.

4. The Qinghai-Tibet alpine grassland region. Encompassing Tibet, Qinghai, and the northwestern part of Sichuan, the region abounds with stubby *Kobresia capillifolia*, *K. humilis*, and *Carex atrofusea*.

5. The north China scrub-herb region. Bounded by the Liaohe River to the north and the Huanghe-Huaihe Plain to the south, the region has *Bothriochloa ischaemum*, *Themeda triandran*, and other shrubby woodland growth as its major forage plants.

6. The south of Changjiang (Yangtze) shrubby grassland region. Encompassing the subtropical area south of the Qinling Mountains and the Huaihe River, the foot-hills of the Nanling Mountains, and the eastern part of the Sichuan Basin and Yunnan-Guizhou Plateau, this region of secondary shrubby undergrowth and woodland meadows, the aftermath of the depleted forests, has predominantly *Miscanthus sinensis*, *Arundinella hirta*, etc.

7. The southwestern canyon grassland region. Taking in the Yunnan-Guizhou Plateau and the southern fringe of the West Sichuan Basin, the canyon region gives a distinct vertical zonation of vegetation. *Polypogon contortus* is the representative grass of this shrubby woodland region.

8. The tropical wooded shrubby grassland region. Encompassing the southern part of Guangdong, Guangxi, the Yunnan-Guizhou Plateau, and Hainan Island, the region is covered with *Eulalia pallens*, *E. quadrinervis*, *Imperata cylidrica*, etc.

SALIENT FEATURES OF THE NATURAL GRASSLAND

1. *Variance in productivity due to geographical location.* In north China, differences in grassland output coincide, in general, with differences in isohyet with which the locale of each grassland is linked. As annual rainfall decreases from east to west in the northern part of the country, so does grassland productivity. In areas where annual rainfall is 100 mm, grassland yields average 750 kg of fresh grass/ha/yr. Starting from the Greater Khingan in the east, where annual rainfall is 400–500 mm, fresh-grass yields average 3,000–4,500 kg/ha yearly. In the west steppe, where annual rainfall barely averages 250–350 mm, yearly yield of fresh grass ranges from 1,500 to 3,000 kg/ha. In the west desert grassland, the output is even lower, a mere 750–1,500 kg/ha/yr.

China's northern grassland can therefore be classified into two types: (a) meadows, meadow grassland, and steppe with yearly fresh-grass output ≥ 1,500 kg/ha; and (b) desert grassland, desert, and alpine meadows with yearly fresh-grass output < 1,500 kg/ha. The ratio of total area of the former to the latter is 3:7.

Geographical location determines to a large extent the size and yield of the haymaking grassland of these parts. In Inner Mongolia, for instance, haymaking grassland takes up 10.3% of the total area of grassland in the eastern part, 4% in the central part, and a mere 0.7% in the western part. A small proportion of grassland for haymaking compared with vast grazing areas is another feature of China's grassland.

2. *Longer winter-spring and shorter summer-autumn seasons and scarcity of winter pastures in the northern grassland region.* If the autumn yield of grass of the steppe is considered as 100%, the summer yield of grass in this region will be 81% and the winter and spring outputs 69% and 44%, respectively. The contrast in grazing capacity during the two periods is even more striking. Grazing capacity of the Xinjiang grassland in summer-autumn is nearly double that in winter-spring. Grazing capacity of the Qinghai grassland in winter is less than one-fifth of that in summer, not to mention the decreased nutritive value of the winter grass, which contains only 13%–30% as much di-

gestible protein as the summer grass. Insufficient winter pastures hinder the growth of animal husbandry in China.

3. *Interannual climatic changes causing fluctuating grass output.* Generally speaking, if both sides of the Greater Khingan are favored by an annual rainfall of 400 mm, it is considered a good year; if the annual rainfall drops to 280–320 mm, the grass output is reduced one-half, and the year is a bad one. Generally, a year with annual rainfall of 300 mm is a bumper harvest year; one-fifth to one-third less than that amount, a normal year, and one-third to one-half less, a deficient year in the northern steppe. Variation of output of livestock farming with yearly rainfall is a detrimental factor in China's animal husbandry.

IMPROVEMENT OF THE GRASSLAND

For years, overgrazing has degenerated China's grassland. The 1976 data indicated that output of 23% of north China's grassland had declined 30% to 50% from its former level. Over the past few years effective measures have been adopted to improve the deteriorated grassland. These are:

1. Closing a degenerated area, deferring grazing, and setting up contrasting grass plots. As a result, the height, density, and proportion of reproductive individual forage plants in the sward increase. While quality improves, output swells also. In the overused grassland of Yanchi, Ningxia Hui Autonomous Region, for instance, grass output has increased by 181% at the end of 2-year rehabilitation, with the height and density of the grass increasing 30%. Closing the Kerkhin grassland area of Inner Mongolia to grazing for one year caused its grass output to increase by 57%, and closing it for 3 years caused an increase of 150%.

2. Ploughing of the topsoil and reseeding. This measure has proved effective in improving the rhizoming of the grasses in the Northeast, Inner Mongolia, Qinghai, and Gansu. The natural grass of the alpine northwest has formed a top layer of hard sod. Ploughing of the topsoil of the Qinghai grassland increased grass output 28.1% the first year after ploughing and 56.9% the second year. Disking and reseeding have also improved the *Aneurolepidium* grassland of the Northeast and Inner Mongolia. Before-and-after differences are shown in Table 1. Though the grass output drops in the year of disking and the next year, it begins to rise steadily the third year and peaks in the sixth year. At the same time, high-quality grass increases as poor-quality grass decreases, as is shown in Table 2. *Aneurolepidium* and other rhizoming grasses thrive after grassland is disked. Grass of such grassland, which can be used for quite a long period, is good for making hay.

To improve the grassland quickly, seeds of good forage and sand-fixing plants have been airsown to sand-eroded areas. For instance, 12,000 ha was sown to *Astragalus adsurgens* in 1979. Other plants sown are *Caragana* spp. The wasteland in dry-farming areas and badly eroded lands are also sown to indigenous wild forage plants. Not only is more fodder produced, but fallow land is also reclaimed and preserved from wind and rain erosion.

Table 1. Effects of disking on physical character of soil.

	Speed of permeability	Hardness of soil	Weight of soil (kg/cm³)	Water-holding capacity (%)
Before	0.20	2.60	1.50	100
After	0.83	1.60	1.60	130–200

Table 2. Effect of disking on quality of herbage, as measured by amount of *Aneurolepidium*.

	Before disking	2nd year	3rd year	5th year	9th year	13th year	14th year	22nd year
Percentage of *Aneurolepidium* in the grassland		54.2		86.5	73.0	67.7	31.6	4.9
Total grass output	100	70.0	358	357.0	168.0	132.0	116.0	101.0
Output of *Aneurolepidium*	100	413.0	910		935.0	577.0		314.0

Dynamics of Herbage Growth and Senescence in a Mixed-Species Temperate Sward Continuously Grazed by Sheep

J.S. BIRCHAM and J. HODGSON

Hill Farming Research Organization, Midlothian, Scotland, UK

Summary

Four perennial ryegrass (*Lolium perenne* L.)-*Poa annua* L.-white clover (*Trifolium repens* L.) swards with herbage mass ranging from 500 to 1,700 kg organic matter (OM)/ha were established and maintained in as near a steady state as possible by continuous but variable stocking with sheep in order to study the influence of herbage mass on rates of herbage growth (G), senescence (S), and net production (NP), and on species balance. Estimates of G, S, and NP were derived from measurements on individual grass tillers and clover stolon growth sites.

The mean contributions to total G were 54, 20, and 4 ($\pm$ 1.6) kg dry matter (DM)/ha/day for ryegrass, *Poa*, and clover, respectively. The low clover growth rates were due to low population densities, and *Poa* had lower population densities and lower individual-tiller growth rates than ryegrass. The ratio of senescence to growth was 0.47 : 1, 0.32 : 1, and 0.11 : 1 for *Poa*, ryegrass, and clover, respectively.

The combined species G increased with herbage mass in an asymptotic manner towards a maximum of 118 kg DM/ha/day, and S increased linearly from 17 kg/ha/day at 500 kg DM/ha to 41 kg at 1,700 kg OM/ha. NP increased rapidly to a predicted maximum of 70 kg DM/ha/day at 1,230 kg OM/ha and then declined slowly.

The results of this experiment suggest that there is limited scope, within practicable limits, for improving rates of net herbage production in temperate swards by manipulation of the herbage mass maintained under continuous stocking management.

KEY WORDS: herbage production, dynamics, perennial ryegrass, *Lolium perenne* L., *Poa annua* L., white clover, *Trifolium repens* L., continuous stocking, sheep.

INTRODUCTION

The control of frequency and severity of defoliation in grazing systems has long been advocated as a means by which both animal and pasture production can be increased, but annual net herbage accumulation in temperate grasslands appears to be relatively unresponsive to variations in grazing management or stocking rate (Hodgson and Wade 1978). Change in the amount of aerial plant tissue is a dynamic process involving the growth of new tissue (G) and the loss of old tissue by decay and decomposition (D), the balance between the two determining net gain or loss (net production, NP). All three can be measured as rates over time on a unit area or an individual plant unit (e.g., a grass tiller) basis. Since management of defoliation can influence both G and D on individual plant units, and also plant population density (Hodgson and Wade 1978), it is not surprising that there is no general agreement about the effects of grazing management on NP.

The objectives of this study were to investigate the influence of sward conditions under a continous stocking management on (1) plant growth and senescence and (2) the balance between the main constituents of a mixed-species sward. In grazed swards, net herbage accumulation (NHA) is the balance between NP and the consumption of tissue by grazing animals (C):

$$NHA = NP - C = G - (D + C) \qquad (1)$$

The universal measure of herbage production is net change in herbage mass in the absence of the grazing animal, where C = 0 and therefore NHA = NP. The procedure used here was to maintain swards in as near a steady state as possible by continuous but variable stocking with sheep so that NHA = 0 and NP = C. The measurement of D is not easy, and the alternative adopted was to measure senescence (S), the rate of transfer of tissue from the live to the dead state:

$$NP = G - S = C \qquad (2)$$

METHODS

The study was conducted in 1979 on a mixed sward, containing perennial ryegrass (*Lolium perenne* L.), *Poa an-*

J.S. Bircham is on study leave from Whatawhata Hill Country Research Station, Hamilton, N.Z.

nua L., and some white clover (*Trifolium repens* L.), which had previously been continuously stocked with sheep for 4 years. Swards of approximate herbage masses of 500, 700, 1,000, and 1,700 kg OM/ha, measured to ground level with electric clippers, were established and maintained in paddocks of 0.5 ha by continuous but variable stocking with ewes and lambs from May until mid-August. A compound fetilizer was applied to all paddocks in mid-April at a rate equivalent to 123 kg N, 37 kg P, and 37 kg K/ha.

Sward surface-height measurements were correlated with herbage mass and used as the basis for adjusting stock numbers. Population densities of grass tillers and clover stolon growing points were determined from 12 (10 × 5 cm) cores analyzed in the laboratory, a stolon growing point being defined as an aggregation of petioles associated with a site of active stolon formation.

Estimates of G, S, and NP corrected for tissue removed by defoliation were derived from twice-weekly linear measurements on grass lamina or clover petiole, grass pseudo-stem or clover stolon, and reproductive stem made over 2 weeks on 40 individually identified primary or primary + daughter aggregate units/species in each paddock. Measurements were made once in June and once in July. The primary unit for grass was an independent rooted tiller; a dependent attached tiller was the daughter unit. For clover, the primary unit was the terminal growing point of the main stolon, and any axillary growing points were daughter units. Constants for weight/unit area and unit length of lamina, pseudostem, petiole, and stolon, derived from measurements on randomly selected plant units in each period, were used to convert area and length measurements into estimates of mass. Estimates of G, S, and NP were obtained by multiplying unit values for each species by the population density of primary units.

RESULTS

As estimates of total sward G, S, and NP in June and July did not differ significantly, pooled results are shown in Table 1 and Fig. 1.

The levels of herbage mass achieved during the experiment were within ± 100 kg OM/ha of the desired mean levels and corresponded to the swave-height and leaf-area index values shown in Fig. 1. Species population densities are shown in Table 1. The mean proportion (percentage of dry weight) of ryegrass in the swards increased (33%–51%), whereas that of *Poa* decreased (54%–24%), with increasing herbage mass. The mean proportion of clover was 7%, inclusive of stolon tissue, and there was no consistent trend across treatments.

The contributions to estimates of total sward G (kg DM/ha/day) were in the order ryegrass > *Poa* > clover. Estimates of species NP followed the same order (Table 1). Ryegrass G increased with increasing herbage mass up to 1,000 kg OM/ha and then remained constant, whereas *Poa* G was highest in the 700 kg OM/ha sward. Highest NP for both ryegrass and *Poa* occurred at 1,000 kg OM/ha. Clover G and NP were relatively low, with no

Table 1. Population density (units/m²) and rates of growth, senescence and net production (kg DM/ha/day) of herbage in continuously stocked swards.

| | Treatment | | | | |
	500	700	1000	1700	S.E.
Population Density (units/m²)					
Ryegrass	14300 b†	28900 a	24700 a	20900 ab	2640
Poa	11900 b	21500 a	10800 b	8900 b	2690
Clover	1400 a	1000 a	700 a	1100 a	570
Growth (kg DM/ha/day)					3.1
Ryegrass	17 d	40 b	80 a	79 a	
Poa	7 e	30 c	24 cd	19 d	
Clover	4 e	3 e	3 e	5 e	
Senescence (kg DM/ha/day)					1.9
Ryegrass	6 c	14 b	17 b	32 a	
Poa	4 cd	18 b	7 c	9 c	
Clover	0.2 d	0.3 d	0.1 d	0.8 d	
Net Production (kg DM/ha/day)					3.4
Ryegrass	11 de	26 c	63 a	47 b	
Poa	4 ef	12 de	16 d	11 de	
Clover	4 ef	2 f	2 ef	4 ef	

†Values in each subtable without a common lower case letter are significantly different at the P < 0.05 level of probability.

Fig. 1. Combined species growth (G), senescence (S), and net production (NP) rates (kg DM/ha/day) vs. herbage mass (HM), sward surface height, and leaf-area index.

$$G = 118\,(\pm 5.4) - \frac{21749480\,(\pm 2313000)}{HM^2},$$

$$R^2 = 0.94^{***}, \, n = 8$$

$$S = 7\,(\pm 7.9) + 0.02\,(\pm 0.007)\,HM,$$

$$R^2 = 0.56^{*}, \, n = 8$$

$$NP = 113\,(\pm 29.1) - 0.02\,(\pm 0.02)$$

$$HM - \frac{21435530\,(\pm 6188800)}{HM^2},$$

$$R^2 = 0.83^{**}, \, n = 8$$

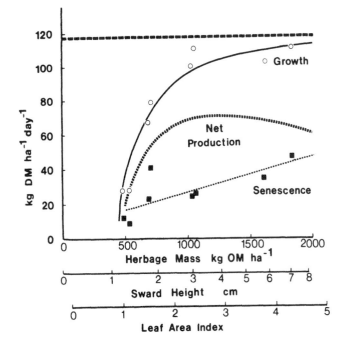

consistent trends across treatments.

Fitted functions for the regressions of the combined species estimates of G and S on herbage mass are shown in Figure 1. The increase in G was asymptotic, approaching a maximum of 118 kg DM/ha/day, and S increased linearly over the full range of herbage mass. The fitted NP curve, combining the functions for G and S, increased to a maximum of 70 kg DM/ha/day at a herbage mass of 1,230 kg OM/ha, and then fell slowly.

DISCUSSION

Stability was maintained in all except the 500 kg OM/ha sward, where uprooted grass tillers were a significant proportion of the animals' diet. This effect would exacerbate the decline in estimates of G and NP shown in Table 1 and Fig. 1 for that sward, which are derived from surviving tillers and stolons.

The differences in G, S, and NP among the three species can be explained in terms of species morphology, canopy structure, and population density. Inasmuch as the tips of the penultimate youngest *Poa* laminae were subtended below the tips of the corresponding ryegrass laminae on all swards, lower individual-tiller photosynthetic capacities (Woledge 1978) and growth rates would be expected for *Poa* than for ryegrass. As herbage mass increased beyond 700 kg OM/ha, ryegrass became the dominant species, making the major contribution to NP on all swards. The very low overall contribution of clover was due primarily to its low population density. The higher overall ratio of S to G for *Poa* (0.47 : 1) than for

ryegrass (0.32 : 1) reflected the relatively low proportion of *Poa* lamina consumed, which was directly attributable to the position of *Poa* leaf in the canopy. Most of the clover laminae were subtended into the grazed horizon and defoliated, leaving only petiole tissue to senesce, and consequently the overall ratio of S to G was only 0.11 : 1.

Both population density and NP/unit were severely depressed below 700 kg OM/ha. The rate of increase in individual tiller and stolon NP exceeded the rate of decline in population density between 700 and 1,000 kg OM/ha, but both NP/unit and population density tended to decline with further increase in herbage mass. The balance between the changes in population density and in the rates of G and S for the individual units of production meant that the predicted NP was equal to 90% or more of maximum over a range of herbage mass from 850 to 1,950 kg OM/ha. These results support the hypothesis that there is limited scope, within practicable limits, for influencing rates of net herbage production in temperate swards by manipulation of the herbage mass maintained under continuous stocking management.

LITERATURE CITED

Hodgson, J., and M.H. Wade. 1978. Grazing management and herbage production. Proc. Brit. Grassl. Soc. Winter Mtg. 1.1–1.12.

Woledge, J. 1978. The effects of shading during vegetative and reproductive growth on the photosynthetic capacity of leaves in a grass sward. Ann. Bot. 42:1085–1089.

A Comparative Evaluation of Six Ryegrass Cultivars Under Grazing and Conservation Managements

M. RYAN
Agricultural Institute, Johnstown Castle,
Wexford, Ireland

Summary

Meat and dairy products command overriding importance in the Irish agricultural economy. They are derived primarily from Irish grasslands, which occupy 89% of the arable land area. Use of improved grass cultivars is probably the best means of increasing grassland output once the basics, optimal fertilizer use, and high stocking rates have been met.

Objective identification of superior cultivars is the responsibility of state agencies. A recommended list, published annually in Ireland, presently contains 47 ryegrass (*Lolium perenne*) cultivars. There is some doubt, however, that results from small-plot cutting trials can be extrapolated to equivalent performance by grazing animals. Hence, it is desirable to evaluate some of the more promising grass cultivars using animal-production parameters. The objective of this research has been to evaluate six ryegrass cultivars in terms of their carrying capacity and persistence as grazed herbage and conserved winter feed. A simple linear model for the gain–stocking rate relationship was used wherein extra carrying capacity (i.e., the increased number of animals carried at a given level of performance) is regarded as a reflection of extra forage produced.

High-yielding cultivars were selected on the basis of heading date and ploidy. These were sown in monoculture and grazed by young steers stocked at moderate to high stocking rates over 3 years. Conserved herbage was fed to groups of heavier steers overwintering on slatted-floor accommodation. Animal live-weight or carcass gains were compared, and cultivar persistence was monitored annually via botanical analysis.

Results indicated superiority in carrying capacity and persistence of cv. Vigor over the control S24. A hybrid, Sabrina, showed inferior carrying capacity and persistence as grazed herbage but performed better when fed as silage. There was a striking contrast in seasonality of growth between S24 and Vigor as shown by the animal-performance data. Differences among cultivars were small in the silage-feeding experiment.

The performance of Vigor shows that this type of late-heading, persistent cultivar can be recommended to growers looking for long-term leys with good midseason production. Tetraploid hybrids (perennial–Italian) may produce well in the short term under conservation management, but they are not suitable for intensively stocked grazing-management systems.

KEY WORDS: ryegrass cultivars, carrying capacity, live-weight gain, persistence.

INTRODUCTION

Grassland occupies 89% of the 4.85 million ha of arable land in Ireland (Anon. 1980), and of this area 142,000 ha are sown annually to new pasture. Cultivar evaluation, using animal performance as the principal measuring parameter, has developed in the Agricultural Institute since 1972–1973 (do Valle Ribeiro 1974, Connolly et al. 1977). The experiment described here, covering 38 ha, is one of several large-scale trials presently underway designed to evaluate grass and legume cultivars.

METHODS

Six ryegrass cultivars—S24, Barpastra[1], Sabrina[1], Vigor, Cropper, and Oakpark—sown in monoculture were evaluated using Friesian steers over the 3 years, 1977–1979. The soil type varied from a loamy sand to an imperfectly drained gley. Plot size was 0.3 ha, and plots were randomly allocated to ten cultivar × stocking rate treatments in twelve blocks, six of which were always grazed, the other six contributing a silage cut before being grazed.

Soil fertility levels were maintained satisfactorily throughout. Nitrogen (N) was applied incrementally on grazed plots to a total of 259 kg/ha/yr. Soil pH was 6.8. Grazing of 150 steers (275 kg, 15/treatment) began in April each year and continued for approximately 185 days.

Statistical analysis of the animal-performance data (grazing) was based on a simple linear model for the gain–stocking rate relationship, in which extra carrying capacity of a treatment is regarded as a reflection of extra forage produced (Connolly 1976). Live-weight and carcass gains were compared in the grazing and silage evaluations, respectively.

Pasture measurements included species changes via tiller counts monitored annually on 10–15 turfs (10 × 10 cm)/plot. Pregrazing and postgrazing dry-matter (DM)

yields taking 3 to 6 quadrats (0.25 m²)/plot cut to ground level were estimated on a proportion of the blocks.

RESULTS

Carrying Capacity and Seasonality

Comparison of extra carrying capacity at a fixed level of animal performance is the best measure of treatment effect in this kind of experiment (Connolly 1976). Above the critical stocking rate the gain–stocking rate relationship for each cultivar is a straight line (Fig. 1). The ratios of the slopes of the regression lines describing these relationships represent the extra carrying capacity of each cultivar relative to S24 (the standard), and these are shown in Table 1, where S24 = 100. These data were based on average daily gain over the grazing season.

In 1977 the production from Barpastra, Sabrina, and Oakpark cultivars was significantly lower than that of S24 in terms of carrying capacity. In 1978 and 1979 only Sabrina and Vigor differed significantly from the control;

Fig. 1. Effect of stocking rate on average daily gain, grazing season 1978.

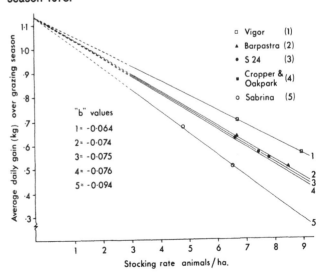

Table 1. Carrying capacity relative to S24 (= 100) (Total annual production).

Cultivar	1977	1978	1979
S24	100	100	100
Barpastra	88*	102	104
Sabrina	78**	80**	73***
Vigor	103	117*	109*
Cropper	94	99	104
Oakpark	89*	99	98

* ** ***Significantly different from S24 at 5%, 1% and 0.1% level of probability, respectively.

Table 2. Ryegrass content of swards (tiller numbers/100 cm²) and % ground cover (pregrazed herbage, cycle 3) 1978.

	Grazed, no. tillers	Conserved + grazed, no. tillers	Grazed, % ground cover
S24 ²LSR	39.1	49.6	78.3
S24 ²HSR	35.5	24.6	84.4
Barpastra LSR	24.6	24.0	85.6
Barpastra HSR	20.5	21.4	82.2
Sabrina LSR	10.5	9.6	70.6
Sabrina HSR	6.7	11.7	61.7
Vigor LSR	50.2	41.1	86.1
Vigor HSR	51.5	45.1	89.4
Cropper	50.2	47.4	86.7
Oakpark	36.0	36.1	85.0
F test	***	**	***
LSD (P = .05)	6.5	18.1	7.8

, * = significant difference among means at 1.0% and at 0.1% of probability, respectively.

Sabrina performance was poorer and Vigor performance was better.

An important difference among early- and late-heading cultivars is the distribution of production over the year. The contrast in seasonality of production for the early-and late-heading cultivars S24 and Vigor, respectively, was well illustrated at the high stocking rate (Fig. 2). The mean difference in stocking rate between S24 and Vigor was 9% at the high stocking rate.

Sward Measurements

Over the 3 years ryegrass content of the Sabrina plots was less than that of S24 (P = 0.05), while the Cropper and Vigor plots had similar or higher levels of ryegrass than the S24 plots (Table 2).

The percentage of ground cover in the Sabrina plots was consistently lower than that in the S24 plots. Analysis of 1978 data showed Sabrina lower and Vigor higher in ground cover than S24 in four cycles out of six (P = 0.05). Regression analysis of the same data revealed significant effects of ground cover on pregrazed DM yield for all treatments, cycles 2 to 5 and cycle 7. Data for one cycle in 1978 are shown in Table 2. Total DM yields reflected the carrying capacity (Table 1).

Gains by 450-kg steers fed cultivar silage and 2 kg meal daily differed significantly in only one 73-day trial, in

which Sabrina gave higher carcass gains than S24, Barpastra, or Vigor.

DISCUSSION

The most noteworthy feature of the results was the performance of Vigor and Sabrina relative to S24. Performance of the other cultivars, averaged over the 3 years, was similar to that of S24 (± 5%). Fig. 1 and Table 1 reveal an animal-production advantage of Vigor over S24 of 10%, averaged over 3 years. These results support the conclusion that Vigor is high yielding (Lee et al. 1977) and persistent (Camlin and Stewart 1978).

The high production obtained in June-July and the relatively low output achieved in April–early May (Fig. 2) suggest that Vigor is well suited to late districts and areas where moisture shortage in midsummer is not severe.

Sabrina performed poorly as grazed herbage in all years, achieving a mean of only 77% of the level of output of S24. This cultivar showed properties more characteristic of Italian ryegrass than of perennial, i.e., erect growth habit, low tillering capacity, and open sward (Table 2). Suprisingly, it did not flourish where herbage was primarily conserved and then grazed (second-year tiller count, Table 2). Decline in ryegrass content of the Sabrina plots was evident at the end of the first grazing season, when ryegrass content for S24, Sabrina, and Vigor plots was 84%, 38%, and 88%, respectively.

The results reported here are relevant to intensively stocked cattle-grazing systems receiving moderately high inputs of N. With low N inputs, greater dependence on clover, and lower stocking rates, the results could be quite different.

LITERATURE CITED

Anon. 1980. Irish Agriculture in Figures. An Foras Taluntais, Dublin.

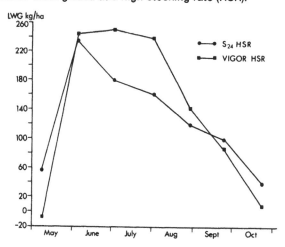

Fig. 2. Seasonality of production, S24 and Vigor, 1977–1979 inclusive when grazed at a high stocking rate (HSR).

Camlin, M.S., and R.H. Stewart. 1978. The assessment of persistence and its application to the evaluation of mid-season and late perennial ryegrass cultivars. J. Brit. Grassl. Soc. 33:275–282.

Connolly, J. 1976. The design of grazing experiments. I. A general linear model for the gain-stocking rate relationship. II. A simple linear model for the gain-stocking rate relationship. Ir. J. Agric. Res. 15:355–374.

Connolly, V., M. do Valle Ribeiro, and J.G. Crowley. 1977. Potential of grass and legume cultivars under Irish conditions. Proc. Int. Mtg. Anim. Prod. from Temp. Grassl., Dublin, 23–28.

do Valle Ribeiro, M.A.M. 1974. Ryegrasses and their use in Irish Agriculture. Ir. Grassl. and Anim. Prod. Assoc. J. 9:5–18.

Lee, G.R., L.H. Davies, E.R. Armitage, and A.E.M. Hood. 1977. The effects of rates of nitrogen application on seven perennial ryegrass varieties. J. Brit. Grassl. Soc. 32: 83–87.

NOTES

1. Tetraploid.

Effect of Sheep and Cattle Grazing on a Mixed Pasture of Perennial Ryegrass, Kikuyu, and White Clover

G.J. GOOLD
Ruakura Soil and Plant Research Station, Hamilton, N.Z.

Summary

Mixed pastures of perennial ryegrass (*Lolium perenne* L.), kikuyugrass (*Pennisetum clandestinum* Hochst.), and white clover (*Trifolium repens* L.) are commonly found within warm-zone areas of New Zealand but are frequently disliked because of problems associated with kikuyugrass dominance. A grazing experiment was conducted to provide basic information on the response of such swards to grazing by cattle and sheep. The pastures were rotationally grazed with young sheep or cattle at two stocking levels each year from 1970 to 1975, with stocking rates adjusted twice yearly according to seasonal pasture growth. The pastures grazed by cattle produced 18% more total annual dry matter (DM) than those grazed by sheep, the advantage being largely confined to summer and autumn when kikuyugrass was the active grass component of the swards. High stocking levels of both sheep and cattle reduced annual yield by 25%, although there was minimal effect in summer. The level-of-stocking effect was greater in winter in sheep pastures than in cattle pastures, but in other seasons there were no differential effects.

The botanical composition of the mixed pastures in spring was greatly altered by grazing treatment. Perennial ryegrass content was greatest in less heavily stocked sheep pastures, and kikuyugrass increased, at the expense of ryegrass, in the more intensely stocked cattle pasture. Annual Poa (*Poa annua* L.) content was higher in sheep than in cattle pastures; this component increased in all treatments over the experimental period. White clover content was variable between years but in general was lower in sheep pastures and occasionally lower, over all, in the more heavily stocked treatments.

The results of this experiment suggest that a balanced subtropical-temperate grass mixture cannot readily be maintained by moderately high stocking levels of either sheep or cattle and that the pastoral changes induced would require compensatory changes in farming practice if high production levels were to be sustained.

KEY WORDS: sheep, cattle, stocking rate, perennial ryegrass, *Lolium perenne* L., kikuyugrass, *Pennisetum clandestinum* Hochst., white clover, *Trifolium repens* L., mixed pastures.

INTRODUCTION

A combination of temperate and subtropical grasses can be used in some environments to exploit the seasonal advantages of each species. Within warm-zone areas of New Zealand, pasture mixtures of perennial ryegrass (*Lolium perenne* L.) and paspalum (*Paspalum dilatatum* Poir.), and of ryegrass-kikuyugrass (*Pennisetum clandestinum* Hochst.), based on white clover (*Trifolium repens* L.), are widespread (Lambert et al. 1977) although not always popular with farmers because of increased management requirements associated with these mixtures. Autumn dominance of kikuyugrass occurs readily with lightly stocked grazing and leads to low winter and spring production of pastures. Winter and spring are critical seasons in most New Zealand pastoral farm systems, and perennial ryegrass dominance at these times generally en-

The author is a Pasture Agronomist at Ruakura Soil and Plant Research Station.

sures high production levels. The control of species dominance is therefore essential in maximizing yield of a diverse pasture mixture such as ryegrass-kikuyugrass–white clover, although farmer experience suggests that this control is not readily attained by normal grazing management practices.

A grazing experiment was designed to examine the effects of grazing sheep and cattle on yield and production components of a mixed ryegrass-kikuyugrass–white clover pasture at a site near Dargaville in Northland, New Zealand. Since utilization of the pastures was thought to be important in species dominance, two sheep and cattle stocking levels, medium and high, were employed to ensure full pasture utilization and to maximize the grazing pressure on the pasture components.

METHODS

The pastures chosen for study were long-term mixtures, predominantly perennial ryegrass in winter and spring and kikuyugrass in summer and autumn, although they also contained significant amounts of annual Poa (*Poa annua* L.), paspalum, and white clover. The pastures had previously been dairy farmed, and the soil type was a high-fertility, winter-wet gley clay.

The pastures were divided into twenty-four paddocks allowing a six-paddock rotation of four treatment groups. Each group was maintained on an 18-day rotation throughout the year, and the animals were replaced each spring. No supplements were fed, and all animals were routinely treated for internal and external parasites.

One-year-old beef steers and wether sheep were stocked at two levels throughout each experimental year at mean annual stocking rates of 9 and 6 steers/ha and 45 and 30 sheep/ha. It was considered that one steer in the spring was equivalent to 5 sheep. Stocking rates were higher over the spring, summer, and autumn and lower in winter, according to wetness and pasture growth rates.

The animals were weighed monthly, sheep were shorn twice yearly and individual fleeceweights recorded. These data have been presented elsewhere (Goold 1981).

Dry-matter (DM) yields were estimated from two pasture frame cuts within each paddock according to the "rate-of-growth" technique, and botanical composition of the pastures was assessed each spring by point analysis along permanently sited line transects in each paddock.

RESULTS

Pasture Production

The effects of grazing by cattle and sheep at the two stocking levels on mean seasonal and annual total DM are shown in Table 1. Pastures grazed by cattle produced more than those grazed by sheep over summer and autumn and in annual total DM. The effects of sheep and cattle grazing on pasture yields over the winter and spring, when temperate species were dominant, were similar, although stocking level had a disproportionate effect on winter yields of sheep-grazed pastures. In the final

Table 1. Effect of stocking level and sheep or cattle grazing on mean seasonal and annual yield (kg DM/ha) of a mixed ryegrass/kikuyu/white clover pasture.

Stocking intensity	Season[+]				
	Spring	Summer	Autumn	Winter	Total
Cattle					
High (9/ha)	2910	3480	2000	1240	9630
Medium (6/ha)	3900	3510	2420	1630	11460
Sheep					
High (45/ha)	2780	2830	1080	990	7680
Medium (30/ha)	4080	2520	1570	1990	10160
SE of difference	220	310	240	150	560

[+]Winter = June, July, August.

year of the experiment the cattle-grazed pastures, while still producing close to average total annual yields, had become strongly oriented towards high summer and autumn and low winter and spring production.

Pastures stocked at the higher level produced 25% less annual DM than those stocked at the lower level, and the effect was consistent in all seasons except summer, when kikuyugrass growth was active (Table 1). The effect of stocking level on summer production was unusual in that there were strong suggestions (significant in the final summer) that DM yields were increased in pastures stocked at the higher level.

Pasture Composition

Major changes in botanical composition occurred in the pastures after the first year and continued thereafter. The ryegrass content of the spring pastures declined rapidly in the higher-stocked treatments and in the pastures stocked with cattle (Table 2). Changes in the kikuyugrass component were less rapid, but after the second year the less intensely stocked pastures contained significantly less kikuyugrass in each successive spring. After the second year, the sheep-grazed pastures contained less kikuyugrass in the spring than those grazed with cattle (Table 2). Coming of annual Poa into the pastures occurred after the third year; by the final spring Poa made up approximately 50% of the sward. In 3 of the 5 years there was significantly less Poa in the cattle pastures than in those grazed by sheep. Poa was little affected by either stocking level (Table 3). Effects of grazing on the white clover component of spring pastures was variable among years; generally there was less clover in the sheep pastures and in those stocked at the higher level (Table 3).

DISCUSSION

Pastures grazed by young beef cattle were 18% more productive in mean annual DM yield than those grazed by sheep. Previous trials conducted in New Zealand comparing the effects of sheep and cattle grazing on pasture productivity have shown contrasting results. Monteath et al. (1977) reported that in southern New Zealand

Table 2.   Effect of stocking level and cattle or sheep grazing on the perennial ryegrass (R) and kikuyu (K) content of mixed ryegrass/kikuyu/white clover pasture in spring.

Stocking intensity	Year											
	1970+		1971		1972		1973		1974		1975	
	R	K	R	K	R	K	R	K	R	K	R	K
Cattle												
High	22.5	12.8	4.8	28.2	4.7	29.4	3.3	27.2	1.6	20.0	4.4	24.1
Medium	16.8	24.6	8.2	26.5	8.4	14.8	6.8	16.3	8.7	12.2	12.5	10.0
Sheep												
High	20.0	27.8	3.5	25.8	4.2	18.8	11.2	19.7	6.3	14.3	6.8	12.0
Medium	17.0	22.5	33.8	23.3	20.1	9.6	7.0	23.5	12.3	22.3	4.5	
SE of difference	6.7	11.9	5.6	8.3	3.0	4.7	3.8	5.0	3.2	3.8	3.7	4.4

+ Pointings before experiment commenced.

ryegrass–white clover pastures under temperate conditions produced 28% more DM under sheep than under cattle grazing. Joyce (1970) reported higher yields from cattle-grazed pastures than in those grazed by sheep in the warm-zone Waikato region and noted that annual grasses replaced ryegrass in high-stocked sheep pastures.

The present experiment confirms that ryegrass and annual temperate-grass contents of pastures are favored by sheep grazing, while white clover content is likely to be reduced. The increase in temperate-grass content of sheep-grazed pasture results in lower summer and autumn production and a relatively larger proportion of growth in winter and spring. This pattern of pasture growth is acceptable for traditional forms of sheep production in New Zealand. Conversely, cattle grazing resulted in increased kikuyugrass and white clover in spring pastures and lower ryegrass content, leading to a sward with a pattern of growth dominated by high summer and autumn production. In a beef-breeding system, such a seasonal pattern of pasture growth would necessitate a later spring calving date than usual. Similar changes with sheep-breeding systems on kikuyugrass pastures have been suggested by Rumball and Boyd (1980).

Production and composition of the pastures were also modified by the stocking level imposed. At the higher stocking level annual DM yield was reduced by 25%,

with the greatest effects occurring over winter and spring. The high stocking level was well above average for the district, and the animal data confirmed that it was above optimum for per-head and per-ha animal performance (Goold 1981). At these stocking levels, the invasion of spring pastures by annual Poa was large and probably related to winter pugging damage.

Higher stocking levels did not depress pasture yield in summer; over the last two summers, yields were significantly increased under this treatment. The response was probably due to nitrogen (N) returned in animal excreta. Previous work suggests that N requirements of dominant kikuyugrass swards are large during active growth periods (Goold 1979). The use of N fertilizer on kikuyugrass swards over summer may be warranted in similar situations.

It is concluded that while mixed pastures of ryegrass-kikuyugrass–white clover are extremely sensitive to grazing management influences, the maintenance of a desirable species balance is not readily achieved, even at very high stocking levels of sheep and cattle grazing. The use of temperate-subtropical grass-clover mixtures in monospecific animal grazing enterprises is likely, therefore, to result in dominance of one of the grass components, which would necessitate appropriate management changes to sustain high levels of production.

Table 3.   Effect of stocking level and cattle or sheep grazing on the annual *Poa* (P) and white clover (C) content of mixed ryegrass/kikuyu/white clover pasture in spring.

Stocking intensity	Year											
	1970+		1971		1972		1973		1974		1975	
	P	C	P	C	P	C	P	C	P	C	P	C
Cattle												
High	14.9	13.5	3.9	7.7	31.6	15.5	28.6	24.2	28.9	16.0	42.0	13.4
Medium	6.4	10.4	7.7	11.3	37.1	19.4	29.5	24.0	28.2	21.8	45.0	19.2
Sheep												
High	10.7	6.9	12.7	3.7	45.7	5.6	40.3	12.0	36.6	8.4	57.3	12.3
Medium	11.0	4.7	6.5	14.7	50.2	8.2	29.3	15.5	34.3	7.9	47.6	13.4
SE of difference	5.0	2.5	5.0	3.8	4.7	3.9	5.0	4.5	4.5	4.0	5.4	4.0

+ Pointings before experiment began.

LITERATURE CITED

Goold, G.J. 1979. Effect of nitrogen and cutting interval on production of grass species swards in Northland, New Zealand. I. Kikuyu-dominant swards. N.Z. J. Expt. Agric. 7:353–359.

Goold, G.J. 1981. The effect of intensity of grazing on sheep and cattle performance in mixed pasture associations. Proc. N.Z. Soc. Anim. Prod. 41:95–100.

Joyce, J.P. 1970. Intensive beef and sheep production. Proc. N.Z. Grassl. Assoc. 32:168–179.

Lambert, J.P., P.J. Rumball, and A.R.J. Christie. 1977. Comparison of ryegrass and Kikuyu grass pasture under mowing. N.Z. J. Expt. Agric. 5:71–77.

Monteath, M.A., P.D. Johnstone, and C.C. Boswell. 1977. Effects of animals on pasture production. I. Pasture productivity from beef cattle and sheep farmlets. N.Z. J. Agric. Res. 20:23–30.

Rumball, P.J., and A.F. Boyd. 1980. Comparison of ryegrass-white clover pastures with and without paspalum and Kikuyu grass. II. Sheep production. N.Z. J. Expt. Agric. 8:21–26.

Level of Nitrogen Fertilizer in Annual Winter Pasture (Oats-Ryegrass) and Summer Pasture (Millet): Effects on Milk Production

C.J. OLIVO, I.L. BARRETO, and D.A. STILES

Brazil National Research Council and
Universidade Federale de Santa Maria, Santa Maria, Brazil

Summary

Two experiments were conducted with lactating Holstein cows to study the effects of different levels of nitrogen fertilizer (urea) on annual pasture: experiment 1, oats (*Avena byzantina* K. Koch) and ryegrass (*Lolium multiflorum* Lam.); experiment 2, millet (*Pennisetum americanum* [L.] Leche). Base fertilizer (8-28-18) was applied, and at the five-leaf stage three levels of nitrogen (N) (75, 150, and 225 kg/ha) were applied. The millet pasture followed the oat + ryegrass pasture, with the N treatments being maintained in the same areas. Two ha were planted for each treatment (total, 6 ha).

In experiment 1, twelve lactating cows were randomly assigned to one of the three oat + ryegrass N treatments in various sequences (treatments a, b, and c) according to Lucas' "incomplete block switch back designs" for an 84-day trial. Each cow was offered 1 kg ration (16% crude protein [CP])/3.5 kg of milk produced. Average milk production (kg), percentage of milk fat (MF), percentage of solids not fat (SNF), percentage of protein, and body weight (kg) were 11.3 kg, 3.7%, 8.1%, 3.3%, and 476 kg; 11.2 kg, 3.9%, 8.2%, 3.4%, and 481 kg; and 11.4 kg, 3.5%, 8.2%, 3.3%, and 490 kg for treatments a, b, and c, respectively. No significant (P = 0.05) differences were observed among treatments. Average pasture dry-matter availability/treatment during the trial was 2,900, 3,000, and 3,450 kg/ha/period for treatments a, b, and c, respectively.

In experiment 2, twelve lactating cows, randomly assigned to one of the three millet–nitrogen fertilizer sequences according to Lucas' "incomplete block switch back designs," served as tester cows. Pasture pressure was adjusted according to pasture availability with other lactating cows on a put-and-take basis. The trial was 63 days in duration. One kg of feed (16% CP) was offered/3.5 kg of milk produced. Average milk production, percentage of MF, percentage of SNF, percentage of protein, and weight were 11.1 kg, 3.6%, 7.72%, 2.93%, and 462 kg; 11.1 kg, 3.7%, 7.65%, 2.96%, and 457 kg; 11.2 kg, 3.2%, 7.8%, 2.5%, and 454 kg, respectively, for treatments a, b, and c. No significant differences were observed for these parameters, but in fat-corrected milk produced/cow, treatments a and b were superior to treatment c. Stocking rates/ha, milk production/ha, and pasture dry-matter availability/period were two cows, 1,350 kg, and 2,297 kg; three cows, 2,016 kg, and 3,734 kg; and five cows, 3,712 kg, and 4,836 kg for treatments a, b, and c, respectively. Treatment c was different from a and b at P < 0.01 for production/ha.

KEY WORDS: oats, *Avena byzantina*, ryegrass, *Lolium multiflorum* Lam., millet, *Pennisetum americanum* (L.) Leche, winter pasture, nitrogen fertilizer, dairy cows.

INTRODUCTION

Consumption of milk in Brazil is increasing faster than production. This difference is being offset by importing milk powders. Two factors that contribute to this problem are inexperienced milk producers and insufficient pasture. For the central part of the state of Rio Grande do Sul, the average milk production/ha does not exceed 400 l (Anon. 1978). The principal reason for low production is qualitative and quantitative nutritional deficiency. The same survey noted a 50% decrease in milk production during the winter months on a statewide basis. Also, during exceedingly dry summers, a similar drop in production occurs.

E.H. Moojen and M.J. Alvim (Santa Maria Department of Zootecnia, Universidade Federale do Santa Maria, personal communication, 1980) observed dry-matter production of ryegrass to be above 7,000 kg/ha at Santa Maria, Brazil, with the application of 150 kg of nitrogen (N)/ha. Seiffert and Barreto (1977) reported production of 13 to 15 metric tons of dry matter/ha with millet. Medeiros et al. (1978) obtained millet production of 7.8, 13.2, 17.7, and 18.2 metric tons/ha with 0, 100, 200, and 300 kg N/ha, respectively. This paper presents two experiments that evaluated the effects of fertilizer levels on pasture production and the resulting milk production of dairy cows grazing these pastures.

METHODS

Experiment 1

Twelve lactating Holstein cows were randomly assigned to one of three treatments in various sequences according to Lucas' "incomplete block switch back design." During the 84-day trial, each cow was offered 1 kg of 16% crude protein commercial dairy feed/3.5 kg of milk produced/day. The winter pasture was ryegrass broadcast and yellow oats planted in rows. Two planting dates (30 March and 20 April) were used to produce pasture of similar physiological age for rotational grazing.

Two hundred kg of 8-28-18 fertilizer was applied as starter fertilizer at the time of planting. N was applied at three levels (75, 150, and 225 kg/ha) at the fifth-leaf stage in a split application (treatments a, b, and c). Two ha were planted for each N rate. Pastures were further divided into plots with electric fences to manage the rotational grazing scheme. The area of oat-ryegrass pasture available offered a similar quantity of dry matter/cow/treatment. Shade and water were not available in the pastures. The cows had access to the pasture between 800 and 1200 and between 1700 and 2000 hours daily.

Water and mineral supplement were available when the cows were not pasturing. Each 28-day period was divided into a 14-day adaptation period and a 14-day data-collection period. Forage dry-matter availability and intake were estimated by an adaptation of the method outlined by Hodgson et al. (1942).

Composition of pasture samples (dry matter, crude protein, neutral detergent fiber, and acid detergent fiber) was determined. Milk production was recorded at each milking. Individual 24-hour composite milk samples were collected during the fourth week of each period. Samples were analyzed for fat (MF) and protein (MP), and for solids not fat (SNF) by the Golding bead method.

Experiment 2

Twelve lactating Holstein cows were randomly assigned to one of three millet–nitrogen treatment sequences according to Lucas' "incomplete block switch back design." Common millet was planted in rows 35 cm apart on 14 and 28 November. One hundred kg of 8-28-18 was applied as starter fertilizer. The three levels of N fertilizer (75, 150, and 225 kg/ha) were applied in a split application to the same areas as in experiment 1.

RESULTS

The results for both experiments are presented in Table 1. No significant differences (P = 0.05) were observed for daily milk production or for milk composition for cows on oat-ryegrass pasture in experiment 1. Also, no differences were detected in relation to body weight, although during the trial the cows were gaining more than 1 kg/day.

Pasture dry-matter availability increased with increased N levels, and crude-protein composition followed the same pattern, with 16%, 22%, and 27% for 75, 150, and 225 kg N/ha, respectively, during the three periods of the trial. Dry-matter consumption was estimated at 10 kg/cow.

Similar data for experiment 2 are found in Table 1. For daily milk production and milk composition, no significant differences (P = 0.05) were observed. However, treatment c, with 3.2% MF, was significantly lower in FCM than treatments a and b. No differences in body weights were observed during the 63-day trial. The cows gained an average of 24 kg. Crude-protein content of millet increased with the levels of N applied (10.1%, 12.5%, and 16.1%, respectively, for 75, 150, and 225 kg N). The increasing level of N resulted in greater dry-matter availability that supported two, three, and five cows/ha, respectively. More milk was produced during the 63-day trial by cows on treatment c than by those on treatment a or b. Forage dry-matter consumption was estimated at 9.6, 9.5, and 8.5 kg/cow/day for treatments a, b, and c.

DISCUSSION

Experiment 1

Application of N increased forage dry-matter production. The uniform level of dry-matter consumption of 10

C.J. Olivo is a Research Assistant for the National Research Council, I.L. Barreto is a Professor of Zootecnia and Researcher for the National Research Council, and D.A. Stiles is a Visiting Professor of Zootecnia at the Universidade Federale de Santa Maria.

This project was supported in part by a grant from the Foundation for Support of Research of the State of Rio Grande de Sul (FAPERGS).

Table 1. Average daily milk production, milk composition, body weight, dry-matter availability, production/ha/cow pasturing oat-ryegrass pasture (Exp. I) and millet (Exp. II) fertilized with three levels of nitrogen.

Treatment	Milk production (kg)	Fat (%)	SNF (%)	Protein (%)	Body weight (kg)	Forage dry matter/ ha/period (kg)	Milk Production per hectare (kg)
Exp. 1							
75 kg N/ha	11.3	3.7	8.1	3.3	476	2,900	–
150 kg N/ha	11.2	3.9	8.2	3.4	481	3,000	–
225 kg N/ha	11.4	3.5	8.2	3.3	490	3,450	–
Exp. 2							
75 kg N/ha	11.1	3.6	7.72	2.93	462	2,297	1,360
150 kg N/ha	11.1	3.7	7.65	2.96	457	3,734	2,016
225 kg N/ha	11.2	3.2	7.80	2.50	454	4,836	3,712

kg/cow/day may have been the result of inaccurate estimates of the growth rate of the ryegrass-oat mixture, but it could also have been the result of limited pasturing time (7 hours daily). The cows were probably not using 100% of the available time for grazing, although no observations were made. Fernando and Carter (1970) and Johnston-Wallace and Kennedy (1944), observed times of grazing of 5½ to 7½ hours, during 18–24 hour/day access to pasture. Another possible problem was the amount of available forage; the cows were restricted to areas such that about the same amount of forage dry matter was available/cow/day. Johnston-Wallace and Kennedy (1944) related the amount of forage dry matter available in the pasture to the amount of dry matter consumed. A considerable weight gain was experienced by the cows during the trial (more than 1 kg/day). This was due to an exceedingly difficult summer and fall and to the lack of conserved forage, causing a rather severe drop in weight prior to the experiment that was recovered during the trial. Fernando and Carter (1970) observed a weight loss in cows grazing oats with high levels of N fertilizer compared with lower levels. The reduced weight of the cows at the start of the trial probably overshadowed this possible response.

The failure to find differences in daily milk production was expected. Brumby (1959) noted a pasture relationship between grazing time and milk production. Since time as well as forage dry-matter availability was controlled, it would appear that the production level was also controlled. Bryant et al. (1961) reported that stocking rates, based on forage dry-matter availability, greatly influenced milk production/animal.

Experiment 2

As in experiment 1, increasing the amounts of N enhanced millet production. Consumption levels of 9.6, 9.5, and 8.6 kg/day were observed for treatments a, b, and c, respectively, which figures are in general agreement with the values of Mayton et al. (1965).

The discussion regarding time for pasturing (7 hours) in the oat-ryegrass pasture applies here also. The lack of

shade and water during the time period for grazing may have had a greater influence.

Crabgrass (*Digitaria sanguinalis* [L.] Scop and *Digitaria adscendens* [HBK.] Hevrard) were present in high quantities in the three pastures (72%, 61%, and 53% of the forage dry matter available). However, there was only a slight selection for millet during grazing (proportion of crabgrass in pasture remaining was 10% greater after grazing), so the intake of crabgrass was similar to that of millet (10%, 12.1%, and 15.7% crude protein in crabgrass in treatments a, b, and c, respectively). The planting of millet on two dates to maintain a similar physiological age, as well as the effect of rotational grazing, helped keep the neutral detergent fiber and acid detergent fiber relatively similar for the three treatments and the three periods. No differences (P = 0.05) were noted for average daily milk production or composition. Mayton et al. (1965) also did not observe a drop in milk fat composition.

With increasing levels of N the carrying capacity increased, with a maximum of five cows/ha for the 63-day trial. This rate is above the average obtained by Mayton et al. (1965) but still less than the high of eight cows encountered for a shorter, 1-month period. Bryant et al. (1961) found that a higher stocking rate increased output/ha. This increase was also noted in this study, with the highest production/ha during 63 days being for treatment c (3,712 kg/ha).

LITERATURE CITED

Anon. 1978. Levantamento do efetivo Bovino das Racas holandesas e Jersey existentes no Rio Grande do Sul, 34–62. Estado do Rio Grande do Sul, Porto Alegro.

Brumby, P.J. 1959. The grazing behaviour of dairy cattle in relation to milk production, live weight, and pasture intake. N.Z. J. Agric. Res. 2:797–807.

Bryant, H.T., R.E. Blaser, R.C. Hammes, Jr., and W.A. Hardison. 1961. Method for increased milk production with rotation grazing. J. Dairy Sci. 44:1733–1741.

Fernando, I.W.E., and O.G. Carter. 1970. The effect of level of nitrogen fertilizer applied to forage oats on the grazing behaviour of dairy cattle. Proc. XI Int. Grassl. Cong., 853.

Hodgson, R.E., J.C. Knott, V.L. Miller, and F.B. Wolberg. 1942. Measuring the yield of nutrients of experimental pastures. Wash. State Agric. Expt. Sta. Bul. 411.

Johnston-Wallace, D.B., and K. Kennedy. 1944. Grazing management practices and their relationship to the behaviour and grazing habits of cattle. J. Agric. Sci., Camb. 34:190–197.

Mayton, E.L., I.E. Hawkins, H.H. Blackstone, and J.A. Little. 1965. Forage systems compared for high producing cows. Auburn Univ. Agric. Expt. Sta. Bul. 363.

Medeiros, R.B., J.C. Saibro, and A.V.A. Jacques. 1978. Efeito do nitrogenio e da populacao de plantas no rendimento e qualidade do milheto (*Pennisetum americanum* Schum). Rev. Soc. Bras. Zoot. 7:276–285.

Seiffert, N.F., and I.L. Barreto. 1977. Forrageiras para ensilagem. I. Avaliacao de cultivares de milho (*Zea mays* L..), sorgo (*Sorghum* Sp) 3 milhetos (*Pennisetum americanum* Schum) na Regiao da Depressao Central do Rio Grande do Sul. Agron. Sulriograndense, Porto Alegra 13(1):205–214.

Animal Production with Controlled and Fixed Stocking and Managed Stocking Rates

R.E. BLASER, J.T. JOHNSON, F. MC CLAUGHERTY, J.P. FONTENOT,
R.C. HAMMES, JR., H.T. BRYANT, D.D. WOLF, and D.A. MAYS
Virginia Polytechnic Institute and State University, Blacksburg, Va., U.S.A.

Summary

Grazing experiments were conducted to implement and improve animal-grazing-management systems. The primary objectives were (experiment 1) to investigate animal production with fixed and controlled stocking and (experiment 2) to study animal production and canopy characteristics with three managed stocking rates.

Live-weight gains/head for the grazing season were similar for fixed and controlled stocking; during a summer period, gains/head were higher for controlled stocking. Gains/head were not associated with available herbage. The canopies with fixed stocking had 28% to 82% more forage than those with controlled stocking, the former being high in dead tissues and stems. Controlled stocking gave 61% more live weight/ha than fixed stocking because more of the flush high-quality forage was consumed.

Live-weight gains/head for the grazing season were higher for the low than for the two higher stocking rates. Gains/head during the summer-autumn season were similar for the three stocking rates even though available pasture varied, dry matter averaging 2,001, 880, and 369 kg/ha for low, medium, and high stocking. During spring, when canopies were similar in leafiness, live-weight gains/animal were allied with available pasture. Double stocking rates during the spring season caused high gains/ha. Values were lower for low than for medium and high stocking.

The findings imply that pastures and animals should be managed concurrently to supply the different nutritional needs for classes of ruminants by controlled grazing to maintain high yields of canopies of suitable structure and leafiness. Data suggest that farms should be stocked at high levels, but stocking rate within a farm should vary to control the quality, yield, and available growth for high productivity of responsive animals. Constant or fixed stocking ignores nutritional needs of ruminants and the dynamic characteristics of canopies during growth and grazing.

KEY WORDS: bluegrass, *Poa pratensis* L., white clover, *Trifolium repens* L., grazing pressure, stocking rates, pasture.

INTRODUCTION

Two grazing experiments were conducted to obtain findings that may improve the efficiency of animal production. Experiment 1 compared fixed with controlled stocking. Experiment 2 compared animal production and canopy characteristics of three managed-stocking rates.

METHODS

Experiment 1:
Controlled and Fixed Stocking

An area was blocked into 4 replications with two 0.81-ha subplots, each with similar areas of alfalfa

R.E. Blaser is a University Distinguished Professor Emeritus at Virginia Polytechnic Institute and State University (VPI & SU); J.T. Johnson is an Associate Agronomist at the University of Georgia; F. McClaugherty is an Assistant Profes-

sor, J.P. Fontenot is a Professor, R.C. Hammes, Jr., is an Assistant Professor, H.T. Bryant is an Associate Professor, and D.D. Wolf is an Associate Professor at VPI & SU; and D.A. Mays is a Project Coordinator, Tennessee Valley Authority. This is a contribution of Virginia Polytechnic Institute and State University.

(*Medicago sativa* L.) with orchardgrass (*Dactylis glomerata* L.), Ladino clover (*Trifolium repens* L.), white clover (*T. repens* L.), and Kentucky bluegrass (*Poa pratensis* L.). Grazing treatments were (a) fixed stocking at one rate to assure adequate forage during summer and (b) controlled stocking by adding or withdrawing steers (grazers) based on growth and in-situ forage. Two steers (testers) from groups of eight steers (295 kg/head) were randomized to each pasture. Grazers similar to testers served to calculate carrying capacity and live-weight gains/ha for controlled stocking. Only testers were used for gains/head. Grazers were added or withdrawn, if necessary, every 14 days, when all animals were weighed. Details have been reported elsewhere (Blaser et al. 1956).

Experiment 2: Managed Stocking Rates

Pasture growth is about twice as much during spring as during the rest of the year; hence, each of three stocking rates was doubled during spring. Spring stocking rates were four heifers (testers) at 227 to 350 kg/head for 1.22, 0.81, and 0.61 ha, or low, medium, and high stocking, respectively. The higher spring stocking based on in-situ forage lasted 90 to 120 days during the 3 years. The stocking rates were replicated in 3 randomized blocks. Pastures were sampled for forage dry matter by harvesting 15 areas 0.54 × 2 m/pasture at a 1-cm stubble monthly. Pastures in both experiments were grazed continuously and not mowed.

RESULTS

Experiment 1:
Controlled and Fixed Stocking

Daily gains/head (DG) of 0.78 and 0.71 kg for controlled (C) and fixed (F) stocking, respectively, were similar. (References to differences or no differences in value are based on statistical significance [P < 0.05].) During mid-summer, DGs were lower for F than C stocking (Fig. 1). DGs were substantially higher during spring than during summer and autumn. The more than 2-fold higher carrying capacity (CC) during spring with C than with F stocking was associated with flush growth (favorable moisture and temperature) in spring. For the season, CC was 29% higher for C than for F stocking. Live-weight gains/ha (LG) were 61% more for C than for F stocking. During 56 days of flush spring growth, LGs were much higher for C than for F stocking, 321 and 169 kg/ha, respectively.

When assuming similar growth rates of pasture and similar intake by animals, available pasture (AP) was 82% and 28% more for F than for C stocking during spring and summer-autumn, respectively. Canopies with C stocking were green and leafy; those with F stocking had closey grazed areas and ungrazed tall stemmy and dead herbage.

Experiment 2: Managed Stocking Rates

During 3 years, DGs were 0.55, 0.48, and 0.40 kg/head for low (L), medium (M), and high (H) stocking,

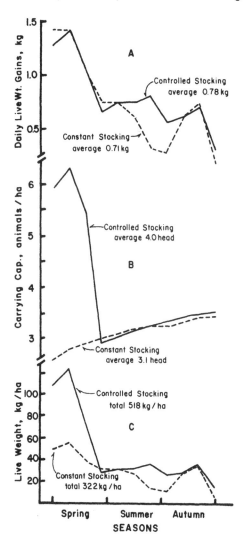

Fig. 1. (A) Daily live-weight gains, kg/head; (B) carrying capacity in days/ha calculated for steers weighing 318 kg; and (C) live-weight gains kg/ha during spring, summer, and autumn seasons at fixed (or constant) and controlled stocking.

the values for L being higher than for H stocking (Fig. 2, Table 1). During spring, DGs of 0.89 kg/head for L stocking were higher than 0.68 or 0.62 kg/head for M and H stocking, respectively. Summer-autumn DGs did not differ for stocking rates, but gains tended to be lowest with H stocking.

Average LGs were highest and similar for M (373 kg/ha) and H stocking (369 kg/ha); LGs were lowest for L stocking (297 kg/ha). The high CC and DG values caused high LGs during spring (Table 1).

Forage in situ, termed available pasture kg/ha (AP), averaged 640, 1,125, and 1,728 kg/ha for H, M, and L stocking (Fig. 2, Table 1). Stocking at double rates in spring maintained rather uniform and similar APs among each stocking rate. Although AP was similar during spring and summer-autumn seasons for respective stocking rates, DGs were much lower during the summer-autumn than the spring season. Also, available forage/head during the summer-autumn season was substantially higher

Fig. 2. Daily live-weight gains, kg/ha (above), and in-situ (available) pasture herbage (dry matter, kg/ha) at various seasons for three stocking rates (3-year average).

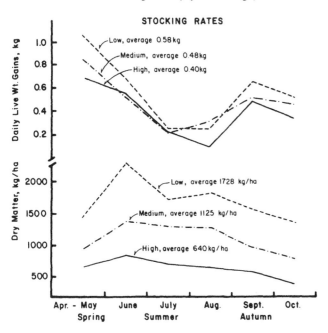

Table 1. Carrying capacity, liveweight gains per ha and per head, forage in situ per ha and per head, and botanical components in forage for three stocking rates and two seasons during three years.

Stocking rates	Grazing seasons[1]		Average or total/season
	Spring	Summer-autumn	
	Carrying capacity, 318 kg heifers, days/per ha		
Low	236 a	260 c	496 c
Medium	337 b	364 b	701 b
High	431 a	453 a	884 a
	Daily gains, kg/heifer		
Low	0.89 a	0.44 a	0.58 a
Medium	0.68 b	0.38 a	0.48 b
High	0.62 b	0.29 a	0.40 bc
	Liveweight gains, kg/ha		
Low	193 b	104 a	297 b
Medium	237 a	136 a	373 a
High	247 a	122 a	369 a
	Forage in situ, kg/ha		
Low	1896 a	1635 a	1728 a
Medium	1193 b	1095 b	1125 a
High	755 c	570 c	640 c
	Forage/318-kg heifer, kg		
Low	1325	2001	1633
Medium	570	880	724
High	330	369	349
	Weeds, % of in-situ forage[2]		
Low	12.3	15.0	13.5 a
Medium	7.0	8.8	8.0 b
High	6.0	6.5	6.5 b
	White clover, % of in-situ forage[2]		
Low	6.8	7.0	7.0 b
Medium	11.3	13.0	12.5 a
High	12.0	15.8	13.5 a

[1]The best quality and growth occurs during spring; hence, the data were divided with April-June (spring) and July-October (summer-autumn periods). Data in vertical columns with different letters differ significantly at $P<.05$.

[2]Data for last two years.

than during the spring, yet DGs were 49%, 56%, and 47% less during summer-autumn than during spring for L, M, and H stocking, respectively.

DISCUSSION

There is a direct relationship between output/head (OP) and digestible dry-matter intake (DDMI), which is usually allied with digestibility (D), leafiness, AP, retention in rumen, bite size, grazing time, species and mixtures, cool temperature, and favorable moisture; diseases, alkaloids, tetany, and bloat are adverse factors (Reid and Jung 1973, Laredo and Minson 1973, Blaser et al. 1975, Stobbs 1975a, Reed 1978). Conversely, Corbett et al. (1966) reported much higher net energy intake for spring than for summer pasture of similar D. Higher DGs during spring than in summer-autumn for experiments 1 and 2 were associated with D; declines in D with season (Reid et al. 1959, 1964) may be associated with temperature increases and water stress (Brown et al. 1963, Wilson and Ford 1972, Wilson and Ng 1975).

OP increases with added AP increments, finally reaching a plateau giving small animal responses with additional AP (Mott 1960). Increases in AP beyond OP responses have been reported (Willoughby 1959, Taylor 1966, Rodrieguez et al. 1974). Linear OP increases with AP were reported by Jones and Sandland (1974). In our two experiments, relationships between AP and OP were poor because of variations in leafiness, botanical components, and physiological effects among seasons and grazing managements. During spring when leafiness and clover contents were similar for stocking rates, DG declined with AP. During summer-autumn, DGs were

similar even though AP ranged from 570 to 1,635 kg/ha among stocking rates. In experiment 1, DGs were similar for C and F stocking, even though the latter had 82% and 28% more AP during the spring and summer seasons, respectively. For a brief period in summer, DG was less with F than with C stocking. With grazing practices that allowed stemmy and dead tissue accumulation after late spring in the two experiments, OP declined, not responding to AP. AP of a given species or mixture may or may not be associated with OP, depending on leafiness, canopy structure, and other factors (Stobbs 1975a, 1975b; Chacon and Stobbs 1976; Chacon et al. 1978). To be useful, AP should be controlled and characterized.

LGs were substantially higher for C than F stocking because stocking with C was increased during the spring to utilize most of the flush high-quality forage. C stocking also maintained leafy canopies with more legumes and better quality of AP than did F stocking. Doubling each of three stocking rates during spring to utilize most of the highly degestible flush growth caused substantial increases in LG. The M stocking rate with small declines in DG and large increases in LG would probably give the best economic returns.

Stobbs (1975a, 1975b), Chacon and Stobbs (1976), Chacon et al. (1978), McMeekan and Walshe (1962), and Blaser et al. (1974, 1975, 1980) imply that pastures and animals should be managed to provide nutritional needs for classes and cycles of production of ruminants. Managed pastures with high canopy yields of desirable species, structure, and leafiness are necessary for high OP. McMeekan and Walshe (1962) used constant stocking rates when evaluating rotational and continuous grazing; they found that variable stocking rates within each system along with conservation caused leafiness and augmented AP. Fixed stocking rates ignore nutrition of ruminants and dynamic canopies, changing in yield, quality, structure, and botanical components among seasons and years. Many forage species may be used flexibly, with combinations of grazing, conserving, and accumulating canopies for deferred grazing in animal-forage management systems. Judging and managing the pasture-animal complex for desirable canopy characteristics for high OP can be more objective than fixed stocking where canopy characteristics and potential OP are ignored.

LITERATURE CITED

Blaser, R.E., R.C. Hammes, Jr., H.T. Bryant, C.M. Kincaid, W.H. Skrda, T.H. Taylor, and W.L. Griffith. 1956. The value for forage species and mixtures for fattening steers. Agron. J. 48:508–516.

Blaser, R.E., R.C. Hammes, Jr., J.P. Fontenot, C.E. Polan, H.T. Bryant, and D.D. Wolf. 1980. Developing forage animal systems. Proc. 1980 Forage and Grassl. Conf., 217–245. Am. Forage Grassl. Counc., Lexington, Ky.

Blaser, R.E., E. Jahn, and R.C. Hammes, Jr. 1974. Evaluation of forage and animal research. *In* Systems analysis in forage crops production and utilization. Crop Sci. Soc. Am. Spec. Pub. 6.

Blaser, R.E., W.C. Stringer, E.B. Rayburn, J.P. Fontenot, R.C. Hammes, Jr., and H.T. Bryant. 1975. Increasing digestibility and intake through management of grazing systems. *In* Forage-fed beef: production and marketing alternatives in the South, 301–305. N. C. Agric. Expt. Sta. Bul. 220. South. Coop. Ser.

Brown, R.H., R.E. Blaser, and J.P. Fontenot. 1963. Digestibility of fall grown Ky 31 fescue. Agron. J. 55:321–324.

Chacon, E.A., and T.H. Stobbs. 1976. Influence of progressive defoliation of a grass sward on the eating behaviour of cattle. Austral. J. Agric. Res. 27:709–727.

Chacon, E.A., T.B. Stobbs, and M.B. Dale. 1978. Influence of sward characteristics on grazing behaviour and growth of hereford steers grazing tropical pastures. Austral. J. Agric.

Res. 29:89–102.

Corbett, J.L., J.P. Langlands, I. McDonald, and J.D. Pullar. 1966. Comparison by direct animal calorimetry of the net energy values of an early and late season growth of herbage. Anim. Prod. 8:13–27.

Jones, R.L., and R.L. Sandland. 1974. The relationship between animal gain and stocking rate. J. Agric. Sci. 83: 335–342.

Laredo, M.A., and D.J. Minson. 1973. The voluntary intake, digestibility and retention time by sheep of leaf and stem fractions of five grasses. Austral. J. Agric. Res. 24:875–888.

McMeekan, C.P., and S.H. Walshe. 1962. The interrelationship of grazing method and stocking rate in the efficiency of pasture utilization by dairy cattle. J. Agric. Sci. 61:147–166.

Mott, G.O. 1960. Grazing pressure and the measurement of pasture production. Proc. VIII Int. Grassl. Cong., 606–611.

Reed, K.F.M. 1978. The effect of season of growth on the feeding value of pasture. J. Brit. Grassl. Soc. 33:227–234.

Reid, J.T., W.K. Kennedy, K.L. Turk, S.T. Slack, G.W. Trimberger, and R.R. Murphy. 1959. Effect of growth stage chemical composition and physical properties upon the nutritive value of forages. J. Dairy Sci. 42:567–571.

Reid, R.L., and G.A. Jung. 1973. Forage-animal stresses. *In* Forages, ch. 58. Iowa State Univ. Press. Ames, Iowa.

Reid, R.L., G.A. Jung, and S. Murray. 1964. The measurement of nutritive value in a bluegrass pasture using *in vivo* and *in vitro* techniques. J. Anim. Sci. 23:700–710.

Rodriequez, Caprilles, J.M., and J. Hodgson. 1974. The influence of sward characteristics on the herbage intake of young growing cattle in tropical and temperate climates. Proc. XII Int. Grassl. Cong. 3:445–447.

Stobbs, T.H. 1975a. Factors limiting the nutritional value of grazed tropical pastures for beef and milk production. Trop. Grassl. 9:141–150.

Stobbs, T.H. 1975b. The effect of plant structure on the intake of tropical pastures. III. Influence of fertilizer nitrogen on the size of bite harvested by Jersey cows grazing *Setaria anceps* cv. Kazungula swards. Austral. J. Agric. Res. 26:997–1007.

Taylor, J.C. 1966. Relationships between the herbage consumption of carcass energy increment of grazing beef cattle and the quantity of herbage on offer. Proc. X Int. Grassl. Cong., 113–120.

Willoughby, W.W. 1959. Limitations to animal production imposed by seasonal fluctuations in pasture and management procedures. Austral. J. Agric. Res. 10:248–268.

Wilson, J.R., and C.W. Ford. 1972. Temperature influences on *in vitro* digestibilities and soluble carbohydrate accumulation of tropical and temperate grasses. Austral. J. Agric. Res. 24:187–198.

Wilson, J.R., and T.T. Ng. 1975. Influence of water stress on parameters associated with herbage quality of *Panicum maximum* var. trichoglume. Austral. J. Agric. Res. 26:127–136.

Factors Causing Greater In-Vitro than In-Vivo Digestibility of Sodium Hydroxide–Treated Roughages

L.L. BERGER, T.J. KLOPFENSTEIN, and R.A. BRITTON

University of Illinois, Urbana, Ill., and
University of Nebraska, Lincoln, Nebr., U.S.A.

Summary

Research on sodium hydroxide (NaOH) treatment of low-quality roughages generally indicates that in-vivo digestion is lower than in-vitro digestion. The purpose of this research was to identify factors that may cause less of the potentially digestible fiber to be digested in vivo than in vitro. Corn (*Zea mays*) cobs were raised to 60% moisture and treated to contain 0%, 2.5%, 5.0%, 7.5%, and 10.0% NaOH. Lambs were fed diets containing 80% cobs and 20% supplement, giving a complete mixed diet containing 0%, 2.0%, 4.0%, 6.0%, and 8.0% NaOH. Five abomasally cannulated lambs were assigned to a 5 × 5 Latin square to measure ruminal fiber digestion as affected by increasing level of NaOH. In-vitro digestibility of NDF reaching the abomasum increased linearly with increasing level of treatment. Rate of passage and rate of ruminal fiber digestion were measured in 15 ruminally fistulated lambs fed the same diet as in the previous trial to determine if these parameters could account for the difference in fiber digestion. Chromic oxide was used as an external marker, and rumen samples were collected at 3, 6, 12, 24, 48, and 96 hours postdosing to measure rate of passage. As level of NaOH increased, rate of passage increased linearly (P < 0.05). Mean ruminal retention time decreased from 46.8 hours for the control diet to 29.8 hours for the 8% NaOH diet. When rate of passage was regressed against NaOH level, the slope of the line was 0.142%/hour with r^2 = 0.733. Nylon bags containing 0.15 g cotton fiber were used to measure the rate of ruminal fiber digestion in lambs fed the five NaOH-treated diets. Bags were removed from the rumen after 12, 24, 36, and 48 hours, and the loss in weight was used to estimate ruminal digestion. As level of NaOH increased from 0% to 8.0%, rate of ruminal cotton fiber digestion decreased from 5.42% to 2.15%/hour. When rate of cotton digestion was regressed against NaOH level, the slope of the line was −0.488%/hour with r^2 = 0.934. These data suggest that the increased rate of passage and decreased rate of ruminal fiber digestion may explain many of the differences observed between in-vitro and in-vivo digestion of NaOH-treated roughages.

KEY WORDS: sodium hydroxide, fiber digestibility, in vivo, in vitro, rate of digestion, rate of passage.

INTRODUCTION

Sodium hydroxide (NaOH) treatment of low-quality roughages is a promising means of improving fiber digestibility, allowing ruminants to convert low-quality fibrous feedstuffs more efficiently to meat and milk. Improvement in fiber digestibility has often been related to level of NaOH treatment regardless of whether the digestibility was measured using the Tilley and Terry (1963) in-vitro procedure or an animal digestion trial. Greater in-vitro than in-vivo digestibility with increasing level of NaOH treatment has been reported by Krause et

al. (1968), Klopfenstein et al. (1972), Rexen et al. (1976), and Levy et al. (1977). An increased rate of passage from the rumen, decreased cellulolytic activity in the rumen, or a combination of the two are possible explanations for the observed difference. Apart from the work of McManus et al. (1976) little work has been done on the site and extent of digestion of NaOH-treated roughages in the ruminant digestive tract. The purpose of this research was to study changes in rate of passage and rate of ruminal fiber digestion with increasing level of NaOH treatment.

METHODS

Corn (*Zea mays*) cobs were ground in a hammer mill through a 0.64-cm screen, raised to 60% moisture, and treated to contain 0%, 2.5%, 5.0%, 7.5%, and 10.0% NaOH on a dry-matter (DM) basis. All diets were 80%

L.L. Berger is an Assistant Professor of Animal Science at the University of Illinois; T.J. Klopfenstein is a Professor of Animal Science and R.A. Britton is an Associate Professor of Animal Science at the University of Nebraska.

cobs, giving a complete mixed diet containing 0%, 2.0%, 4.0%, 6.0%, and 8.0% NaOH. The supplement was 85.0% brewers dried grains, International Reference Number (IRN) 5-02-141; 8.3% urea (IRN 5-05-070); 6.2% dicalcium phosphate (IRN 6-01-080); 0.3% limestone (IRN 6-01-069); 0.1% trace minerals; and 0.1% vitamin mix—all formulated to provide 12% crude protein in the complete diet.

In an abomasal trial, 5 abomasally cannulated wethers (41.8 kg/lamb) were randomly assigned to a 5 × 5 Latin square design. Each lamb was fed 25 g DM/hour in an attempt to maintain a steady-state flow. The same diet was used as described above. Approximately 200 ml of digesta was collected over 20 minutes on days 12, 14, and 16 following an 11-day adjustment period. Samples were centrifuged at 750 × gravity for 10 minutes, decanted, dried at 60°C, and ground prior to determination of neutral detergent fiber (NDF) and acid detergent fiber (ADF), as described by Goering and Van Soest (1970). In-vitro digestibility of abomasal samples was determined using the Moore modification of the Tilley and Terry in-vitro procedure as described by Harris (1970). The trial consisted of two periods with two tubes/sample/lamb/period.

For the rate-of-passage trial, 15 ruminally fistulated wether lambs (33.3 kg/lamb) were randomly assigned to the five sodium hydroxide–treated diets described previously. All lambs were fed 600 g DM once daily. Rate of passage was measured during two 19-day periods, each divided into a 15-day adjustment and a 4-day sampling period. Lambs were rerandomized to treatment for period 2. Thirteen g of a pellet containing 25% chromic oxide was given in a single dose via the rumen cannula just prior to feeding. Approximately 200 ml of digesta was taken at 3, 6, 12, 24, 48, and 96 hours postdosing. Chromium was determined by atomic absorption spectroscopy as described by William et al. (1962). Rates of passage were the regressions of the natural logarithms of the chromium concentrations on time, and mean retention times were 0.693 divided by rate of passage.

Rate of ruminal cotton-fiber digestion was measured using 15 ruminally fistulated wether lambs (44.9 kg/lamb). The trial consisted of two 12-day periods, each divided into a 10-day adjustment and a 2-day digestion period. The lambs consumed approximately 900 g DM/day of the diet used in the previous trial. Nylon bags (160 × 85mm) containing 0.15 g of cotton were suspended in the rumen of lambs fed the five cob diets. The nylon was 100 mesh (50 to 75 μm pore size), and the cotton was surgical cotton soaked 12 hours in 10% NaOH, rinsed, and dried at room temperature. The bags were removed after 12, 24, 36, and 48 hours of digestion. The loss in weight was used to calculate the amount of digestion occurring in the rumen. Rates of cotton digestion were regressions of the natural logarithm of the percentage of the potentially digestible cotton remaining on hours of digestion. Potentially digestible fiber was that digested in 96 hours. All data were statistically treated by orthogonal poynomial breakdown of the treatment degrees of freedom as described by Snedecor and Cochran (1967).

Table 1.   Effect of NaOH treatment on abomasal sample composition and in-vitro NDF digestibility.

Item	Diet % NaOH					SE
	0	2.0	4.0	6.0	8.0	
NDF	66.5	61.6	65.4	49.0	47.5	2.06
ADF	32.5	32.1	37.1	25.6	26.8	1.90
In-vitro digestion abomasal, NDF	23.7	25.6	42.0	63.3	70.0	2.26
In-vitro digestion diet, NDF	45.5	52.4	68.6	83.8	89.4	1.38

SE = Standard error.

RESULTS

In-vitro digestibility of NDF in the abomasal digesta increased linearly and cubically (P < 0.05) with increasing level of NaOH treatment, paralleling closely the increase in in-vitro dietary NDF digestibility (Table 1). The percentage of NDF and ADF in the abomasal digesta was fairly constant for the 0%, 2%, and 4% NaOH diets but lower for the 6% and 8% NaOH diets.

Rate of passage of chromic oxide from the rumen increased linearly (P < 0.05) with increasing level of NaOH treatment (Table 2). This resulted in the mean ruminal retention time of the marker decreasing from 32.4 hours for the control diet to 20.7 hours for the 8% NaOH diet. When rate of marker passage was regressed against NaOH level, the slope of the line was 0.142%/hour/unit of NaOH, with r^2 = 0.733.

Rate of ruminal cotton-fiber digestion decreased linearly (P < 0.05) with increasing levels of NaOH treatment (Table 2). The actual rates decreased from 5.42%/hour for the control diet to 2.15%/hour for the 8% NaOH diet. When the rate of cotton-fiber digestion was regressed against NaOH levels, the slope of the line was −0.488%/hour/unit NaOH, with r^2 = 0.934.

DISCUSSION

In-vitro digestiblity of the NDF reaching the abomasum increased sharply with greater than 2% NaOH in the diet. Some of this increase was due to the fact that NDF of the NaOH-treated diets was much more digestible initially. However, the increase does show that much of the potentially digestible fiber of the NaOH-

Table 2.   Effect of NaOH level on rate of passage and mean retention time and on rate of ruminal cotton digestion.[a]

Diet, % NaOH	Rate of passage, %/hr ± SE	Mean retention time, hr ± SE	Rate of digestion, %/hr ± SE
0	2.20 ± .18	32.4 ± 3.0	5.42 ± 0.68
2	2.38 ± .18	29.9 ± 2.5	4.74 ± 1.30
4	3.25 ± .20	21.3 ± 1.2	4.22 ± 0.36
6	2.82 ± .11	24.3 ± 0.8	2.32 ± 0.51
8	3.40 ± .24	20.7 ± 1.6	2.15 ± 0.37

[a]Six observations/treatment mean.

treated diets was escaping ruminal digestion. An increased rate of passage, decreased cellulolytic activity, or a combination of the two are possible explanations.

This explanation was partially confirmed in the rate-of-passage trial where lambs fed NaOH-treated diets had increased rates of passage of chromic oxide and decreased ruminal retention times compared with those fed the control diet. The increased chromic oxide flow is assumed to indicate reduced retention time of potentially digestible fiber. Mean ruminal retention times were considerably shorter than the 48 hours commonly used for in-vitro digestion trials and may partially explain the higher in-vitro and in-vivo digestibilities often observed with increasing level of NaOH treatment. Hogan and Weston (1971) fed lambs 8% NaOH-treated wheat straw at 400, 500, and 1,000 g/day. Mean ruminal retention time of chromium-EDTA (^{51}Cr-EDTA), a liquid marker, was 14.7, 14.0, and 10.0 hours for the low, medium, and high intakes, respectively. McManus et al. (1976) reported that the flow of whole digesta passing the abomasum was greater (P < 0.01) for pelleted rice hull–alfalfa diets containing 2.5% and 7.5% NaOH than for the untreated control.

As level of NaOH treatment increased, the rate of ruminal cotton digestion decreased. The nylon bag technique was used to measure the rate of ruminal fiber digestion on the basis of the work of Coombe and Tribe (1963), who showed that ruminal fiber digestion was correlated (P < 0.01) with the rate of cotton-thread digestion. A possible explanation for this decreased rate of fiber digestion may be that the increased water intake diluted the bacterial population, hindering substrate-enzyme contact and reducing fiber digestion (Baker and Harris 1947). Koes and Pfander (1975) reported that fiber digestibility decreased as water consumption increased in lambs. Another possible explanation may relate to the osmolarity of the rumen. Bergen (1972) reported that when osmolarity of the in-vitro media was increased above 400 mOsm/kg with sodium salts, in-vitro cellulose digestion was reduced by 80% or more. However, Cardon (1953) reported that feeding cows an alfalfa-hay diet containing 10.75% NaCl did not affect cellulose digestibility.

The increased rate of passage and decreased rate of ruminal fiber digestion may explain many of the differences between in-vitro and in-vivo digestibility that

have been observed with increasing levels of NaOH treatment.

LITERATURE CITED

Baker, F., and S.T. Harris. 1947. Microbial digestion in the rumen and caecum with special reference to the decomposition of structural cellulose. Nutr. Abs. Rev. 17:3–12.

Bergen, W.G. 1972. Rumen osmolarity as a factor in feed intake control of sheep. J. Anim. Sci. 34:1054–1060.

Cardon, B.B. 1953. Influence of a high salt intake on cellulose digestion. J. Anim. Sci. 12:536–540.

Coombe, J.B., and D.E. Tribe. 1963. The effect of urea supplements on the utilization of straw plus molasses diets by sheep. Austral. J. Agric. Res. 14:70–92.

Goering, H.K., and P.J. Van Soest. 1970. Forage fiber analyses (apparatus, reagents, procedures and some applications). U.S. Dept. Agric., Agric. Handb. 379.

Harris, L.E. 1970. Nutrition research techniques for domestic and wild animals. Vol. 1. L.E. Harris, Logan, Utah.

Hogan, J.I., and R.H. Weston. 1971. The utilization of alkali-treated straw by sheep. Austral. J. Agric. Res. 22:951–962.

Klopfenstein, T.J., V.E. Krause, M.J. Jones, and W. Woods. 1972. Chemical treatment of low quality roughages. J. Anim. Sci. 35:418–422.

Koes, R.M., and W.H. Pfander. 1975. Heat load and supplement effects on performance and nutrient utilization by lambs fed orchardgrass hay. J. Anim. Sci. 40:313–319.

Krause, V., T.J. Klopfenstein, and W. Woods. 1968. Sodium hydroxide treatment of corn silage. J. Anim. Sci. 27:1167. (Abs.).

Levy, D., Z. Holzer, H. Neumark, and Y. Folman. 1977. Chemical processing of wheat straw and cotton by-products for fattening cattle. I. Performance of animals receiving the wet material shortly after treatment. Anim. Prod. 25:27–37.

McManus, W.R., C.C. Choung, and V.N. Robinson. 1976. Studies on forage cell walls. IV. Flow and degradation of alkali-treated rice hull digesta in the ruminant digestive tract. J. Agric. Sci. 87:471–483.

Rexen, F., P. Stigsen, and F.V. Kristensen. 1976. The effect of a new alkali technique on the nutritive value of straw. Proc. IX Nutr. Conf. Feed Mfg., Univ. of Nottingham, 65–82.

Snedecor, G.W., and W.G. Cochran. 1967. Statistical methods, 6th ed. Iowa State Univ. Press, Ames, Iowa.

Tilley, J.M., and R.A. Terry. 1963. A two-stage technique for the *in vitro* digestion of forage crops. Brit. Grassl. Soc. J. 18:104–111.

William, C.H., D.J. Darid, and O. Ishmaa. 1962. The determination of chromic oxide in faeces samples by atomic absorption spectrophotometry. J. Agr. Sci., Camb. 59:381–385.

Effect of Solvent Treatments on Intake and Digestibility of Corn Stover

D.M. SCHAEFER, M.R. LADISCH, C.H. NOLLER, and V.L. LECHTENBERG

Purdue University, West Lafayette, Ind., U.S.A.

Summary

Treatment of crop residues with low levels of sodium hydroxide has been shown to improve organic-matter digestibility and animal performance. One objective of these experiments was to evaluate the effects of three treatments — water, sodium hydroxide, and chelating metal cellulose-swelling solution — on solubilization of cell-wall components in corn (*Zea mays*) stover. The two latter treatments were balanced for level of sodium hydroxide application, which was 3.2% of stover dry matter. The second objective was to determine the effects of these treatments on in-vitro and in-vivo digestibility as well as on consumption of corn stover. Twelve growing ram lambs were used in an intake and digestion trial composed of three periods.

Chelating metal cellulose-swelling solution was capable of dissolving purified cellulose. Sodium hydroxide solubilized 18.9% of the hemicellulose and chelating metal cellulose-swelling solution solubilized 11.6% of the hemicellulose and 12.8% of the cellulose in corn stover. Daily consumption of all three treated stovers was 23.7 g dry matter/kg body weight$^{0.75}$. Digestibility of neutral detergent fiber was improved in vitro and in vivo by treatment of stover with sodium hydroxide or chelating metal cellulose-swelling solution. The in-vivo digestibility of hemicellulose was improved 11.7% and 8.3%, and cellulose digestibility was improved in vivo by 9.4% and 5.6% by sodium hydroxide and chelating metal cellulose-swelling solution treatments, respectively. Nitrogen digestibility was not affected by the three treatments; however, nitrogen retention was depressed by the chelating metal cellulose-swelling solution. An explanation for this result is not available, but further evaluation of the results indicated that it was not due to differences in urinary nitrogen excretion. It is speculated that cellulose crystallinity is not the only major determinant of plant cell-wall digestibility.

KEY WORDS: corn, *Zea mays*, stover, sheep, intake, digestibility, fiber.

INTRODUCTION

Since feeding grain to ruminants is not always economical and since large quantities of crop residues are produced annually, chemical treatments have been evaluated as a means of improving the bioavailability of energy in crop residues (Klopfenstein 1978). Sodium hydroxide (NaOH) treatment has been reported to solubilize hemicellulose partially and to increase organic-matter digestibility in ruminants (Klopfenstein et al. 1979). Treatment of crop residues with 3% to 5% NaOH on a dry-matter basis has given the best improvements in animal performance (Klopfenstein 1978). The purpose of this experiment was to evaluate the ability of two chemical treatments, NaOH and chelating metal cellulose-swelling (CMCS) solution, to solubilize cell-wall components and to improve in-vitro and in-vivo fiber digestion. The

CMCS solution, which has been termed FeTNa by Browning (1967), is an aqueous alkaline complex of iron and sodium tartrate. Since a major component of CMCS is NaOH, these two treatments permit an evaluation of the effectiveness of iron and tartrate in CMCS.

MATERIALS AND METHODS

Corn (*Zea mays*) stover was collected into small, rectangular bales and sheltered from the weather prior to chopping with a forage chopper that was set to give a 4.8-mm length of cut. Batches of 2.27 kg of chopped stover were treated by spraying with either water, NaOH solution (99 g/l), or CMCS (Browning 1967). The application rate of CMCS was 0.4:1 (CMCS:corn stover dry matter, weight basis), while NaOH was added at a rate of 3.2% of corn stover dry matter to provide an amount of NaOH equivalent to that added in CMCS. Water was added as a control treatment at the same rate as for the CMCS treatment. The CMCS was prepared in several 8-l batches, and its ability to dissolve cellulose was confirmed by agitating a few grams of Avicel (50 μ, FMC Corp., Philadelphia, Pennsylvania) in 10 ml of CMCS. Dissolution of Avicel was complete in a few hours. All stover was treated at room temperature at least 3 days

D.M. Schaefer is an Assistant Professor of Animal Science, M.R. Ladisch is an Associate Professor of Agricultural Engineering and Chemical Engineering and a Group Leader in the Laboratory of Renewable Resources Engineering, C.H. Noller is a Professor of Animal Science, and V.L. Lechtenberg is a Professor of Agronomy at Purdue University. This is a contribution of the departments of Animal Science, Agricultural Engineering, and Agronomy of Purudue University (Journal Paper No. 8529).

Mention of a proprietary product does not imply endorsement of it.

prior to feeding and was stored in plastic bags at 4°C until fed. Samples of the treated corn stover that were used to determine the extent of solubilization of cell-wall components were obtained immediately after treatment and frozen until analyzed. Solubilization is defined as the reduction in concentration of the cell-wall component as a result of the treatment.

Digestibility of neutral detergent fiber (NDF) in the treated stover was determined in vitro. Rumen fluid was collected from two steers on alfalfa-hay diets, strained, and diluted 1:2 (v/v) with a buffered mineral solution (McDougall 1948). Thirty-ml aliquots of this mixture, which also contained final concentrations of 0.05% urea, 0.025% cysteine (HCl) H_2O, and 0.025% $Na_2S(9H_2O)$, were added under carbon dioxide to 50-ml centrifuge tubes containing 0.5-g samples of ground (1 mm) corn stover. Incubations were done in triplicate at 39°C for 48 hours and microbial activity was terminated by the addition of neutral detergent solution.

Twelve ram lambs averaging 26 kg were randomly allotted to six replicated treatment sequences in a switchback design for three treatments (Lucas 1956). Each of three periods lasted 23 days and consisted of a 15-day adaptation to the treated stover, a 1-day adaptation to metabolism stalls, and a 7-day total fecal and urine collection. Ingredient and chemical compositions of the diet are given in Table 1. A mixture of the supplemental protein, minerals, and vitamins was fed at 0800 daily just prior to the feeding of treated stover, which was available ad libitum until the next feeding. During the first 10 days of each period, the supplemental mix was fed in slight excess of that required to balance the ration at 12% crude protein, to prevent the supplemental nutrients from limiting intake and digestibility of corn stover during the adaptation phase. During the remainder of each period the supplement was fed according to the amount of stover consumed the previous day; this treatment served to balance the ration, as is indicated in Table 1. Orts, samples of fresh stover and supplement, urine, and feces were collected daily and frozen until analyzed.

The procedures of Goering and Van Soest (1970) were used for determination of NDF, acid detergent fiber (ADF), permanganate lignin (PL), and ash. Hemicellulose, lignin, and cellulose contents were calculated by difference. Nitrogen content was determined by the

Table 1. Ingredient and chemical composition of diet fed to ram lambs (dry-matter basis).

Ingredient:	
Corn stover, %	80.25
Soybean meal, %	19.15
Trace mineralized salt, %	0.41
Dicalcium phosphate, %	0.27
Vitamin supplement, %	0.01
Composition:	
Crude protein, %	12.0
Calcium, %	0.48
Phosphorous, %	0.25
Sulfur, %	0.20

Table 2. Solubilization of cell-wall components in corn stover.

Treatments	Percent hemicellulose	Percent cellulose	Percent lignin
Pretreatment composition	28.5	39.2	9.0
	---------(% units solubilized)---------		
H_2O	−1.2[a]	1.6[a]	2.5[a]
NaOH	5.4[b]	2.6[a]	0.1[b]
CMCS	3.3[b]	5.0[b]	1.2[ab]
SEM	0.7	0.4	0.4

[a,b]Unlike superscripts within columns indicate differences (P<.05).

Kjeldahl procedure (AOAC 1960). Corn stover pH was determined by suspending 10 g of ground corn stover in 100 ml of water. Treatment comparisons were evaluated by the Newman-Keuls test (Steel and Torrie 1960).

RESULTS AND DISCUSSION

Effects of the treatments on solubilization of cell-wall components in corn stover are given in Table 2. Treatment of stover with NaOH or CMCS resulted in the solubilization of more (P < 0.05) hemicellulose than did the water treatment. The CMCS treatment solubilized significantly more cellulose than did water or NaOH treatments. Solubilization of hemicellulose, but not of cellulose, by NaOH has been previously reported (Lesoing et al. 1981, Levy et al. 1977, Klopfenstein et al. 1979). More lignin was solubilized (P < 0.05) by water than by NaOH. No explanation for this difference is currently available.

Composition of the treated corn stovers that were fed to the lambs in the digestion trial is given in Table 3. The CMCS-treated stover contained less (P < 0.05) celluose than the other treated stovers. This finding supports the above results, which indicated that CMCS solubilized cellulose. The pHs of corn stovers treated with water, NaOH, and CMCS were 6.92, 9.61, and 9.32, respectively. There was no treatment effect on daily intake since consumption of all stovers was 23.7 g dry matter/kg body weight$^{0.75}$.

Digestibility of cell-wall components in corn stover was determined by in-vitro and in-vivo methods (Table 4). Treatment of stover with NaOH increased (P < 0.05) in-

Table 3. Composition of treated corn stover.

Treatment	Percent dry matter	Percent hemicellulose	Percent cellulose	Percent lignin
H_2O	77.0	27.5	36.2[a]	6.0
NaOH	75.4	23.1	37.1[a]	6.8
CMCS	77.8	24.2	33.2[b]	5.3
SEM	1.8	1.1	0.8	0.5

[a,b]Unlike superscripts within columns indicate differences (P<.05).

Table 4. Digestibility of cell-wall components in corn stover by in-vitro and in-vivo methods.

Treatment	In vitro NDF	In vivo[a] NDF	Hemicellulose	Cellulose	Lignin
			----(% digested)----		
H_2O	59.6[b]	60.2[e]	67.7[b]	68.2[b]	19.6
NaOH	63.7[c]	66.6[f]	75.6[c]	74.6[c]	27.6
CMCS	65.0[d]	65.0[f]	73.3[c]	72.0[c]	23.0
SEM	0.4	1.6	2.7	2.1	6.0

[a]Includes digestibility of cell-wall components in soybean meal supplement.

[b,c,d]Unlike superscripts within column indicate differences (P<.05).

[e,f]Unlike superscripts within columns indicate differences (P<.01).

vitro NDF digestibility, which was further improved upon (P < 0.05) by CMCS. Digestibility of NDF was also increased (P < 0.01) in vivo by NaOH and CMCS treatments, but these two treatments did not differ. Digestibility of NDF in vivo tended to be greater for the NaOH-treated than for the CMCS-treated stover, a result that was just opposite of results obtained in in-vitro incubations. The NaOH and CMCS treatments increased (P < 0.05) hemicellulose and cellulose digestibilities. Increases tended to be greater for NaOH, but NaOH and CMCS effects were not significantly different. The digestibility coefficient of the hemicellulose in the diet containing NaOH-treated stover agrees with that reported by Lesoing et al. (1981) for wheat straw treated with 4% NaOH. Digestibility values of cellulose in barley straw treated with 6% NaOH (Maeng et al. 1971) or 4.5% NaOH (Jayasuriya and Owen 1975) agree with these results (Table 4). Lignin digestibility was unaffected by the treatments (P > 0.05). The extent of digestion of corn stover lignin in this trial was similar to the results of Allinson and Osbourn (1970), Grant et al. (1974), Jayasuriya and Owen (1975), and Fahey et al. (1979), which were obtained with a variety of feedstuffs.

Utilization of nitrogen in the stover diets was also evaluated (Table 5). There was no significant treatment effect on nitrogen digestibility; however, nitrogen retention was reduced (P < 0.05) in those lambs consuming

CMCS-treated stover. These results and the fact that dry-matter intakes were equal for all treatments indicate that there were no differences in nitrogen absorption from the digestive tract and suggest that the excretion of urinary nitrogen was greater for the CMCS treatment. This suggestion is not supported by the data, since there was no treatment effect (P > 0.05) on urinary nitrogen excretion. An explanation of the observed reduction in nitrogren retention associated with the CMCS treatment is not available.

GENERAL DISCUSSION

Low levels of CMCS were applied to corn stover in this experiment because comparable levels of NaOH treatment have been reported to improve animal performance and because high levels of CMCS treatment might cause iron toxicity. The calculated iron content of CMCS-treated stover was 60 ppm, which is less than the dietary level of 500 ppm suggested to be the maximum tolerable level for sheep (National Research Council 1980).

Digestibility of corn stover cellulose was not different for NaOH and CMCS treatments at low treatment levels. This agrees with the results of Hamilton (1979), who investigated the conversion of Avicel, treated with NaOH or CMCS, to glucose by *Trichoderma reesei* cellulase. At levels of 5 ml of CMCS or 5 ml of 1.5M NaOH/g Avicel, he found no apparent improvement in conversion of cellulose to glucose by CMCS vs. NaOH treatment. However, at levels of 20.8 ml of CMCS or 20.8 ml of 1.5M NaOH/g Avicel, there was an 18% improvement in conversion of cellulose to glucose by CMCS over that by NaOH.

These results stimulate speculation on the role of cellulose crystallinity in plant cell-wall digestion. A purified form of cellulose was completely dissolved by CMCS but not by the NaOH solution used in this experiment. This dissolution by CMCS involved a considerable reduction in the crystalline nature of the cellulose. Even though low treatment levels were used here, the crystallinity of stover treated with CMCS may also have been reduced, although no direct evaluation of crystallinity reduction was made. Nevertheless, the lack of difference between NaOH and CMCS in digestibility of cell-wall components suggests that the degree of cellulose crystallinity may not be the only major limitation to cell-wall digestibility. A similar suggestion has been made by Bacon (1979), who noted that improved forage cell-wall digestibility could be realized by treatment with NaOH at concentrations below that needed to cause swelling and disruption of the molecular structure of cellulose.

Table 5. Nitrogen digestibility and balance associated with feeding diets containing treated corn stover.

Treatment	Nitrogen digestibility (%)	Nitrogen retention (g/d)	Urinary nitrogen (g/d)
H_2O	71.7	2.0[a]	56.8
NaOH	68.7	2.3[a]	62.6
CMCS	70.6	1.0[b]	60.1
SEM	2.0	0.4	3.0

[a,b]Unlike superscripts within columns indicate differences (P<.05).

LITERATURE CITED

Allinson, D.W., and D.F. Osbourn. 1970. The cellulose-lignin complex in forages and its relationship to forage nutritive value. J. Agric. Sci. 74:23–36.

Association of Official Agricultural Chemists (AOAC). 1960. Official methods of analysis, 9th ed. AOAC, Washington, D.C.

Bacon, J.S.D. 1979. Plant cell wall digestibility and chemical structure. Ann. Rpt. Rowett Res. Inst. 35:99–108.

Browning, B.L. 1967. Methods of wood chemistry. Interscience 2:542–543.

Fahey, G.C., G.A. McLaren, and J.E. Williams. 1979. Lignin digestibility by lambs fed both low quality and high quality roughages. J. Anim. Sci. 48:941–946.

Goering, H.K., and P.J. Van Soest. 1970. Forage fiber analyses (apparatus, reagents, procedures, and some applications). U.S. Dept. Agric., Agric. Handb. 379.

Grant, R.J., P.J. Van Soest, R.E. McDowell, and C.B. Perez. 1974. Intake digestibility and metabolic loss of napier grass by cattle and buffaloes when fed wilted, chopped and whole. J. Anim. Sci. 39:423–434.

Hamilton, T.J. 1979. The effect of FeTNa pretreatment on glucose yields from cellulosic materials. M.S. Thesis, Purdue Univ.

Jayasuriya, M.C.N., and E. Owen. 1975. Sodium hydroxide treatment of barley straw: effect of volume and concentration of solution on digestibility and intake by sheep. Anim. Prod. 21:313–322.

Klopfenstein, T. 1978. Chemical treatment of crop residues. J. Anim. Sci. 46:841–848.

Klopfenstein, T., L. Berger, and J. Paterson. 1979. Performance of animals fed crop residues. Fed. Proc. 38:1939–1943.

Lesoing, G., T. Klopfenstein, I. Rush, and J. Ward. 1981. Chemical treatment of wheat straw. J. Anim. Sci. 51:263–269.

Levy, D., Z. Holzer, H. Neumark, and Y. Folman. 1977. Chemical processing of wheat straw and cotton by-products for fattening cattle. Anim. Prod. 25:27–37.

Lucas, H.L. 1956. Switchback trials for more than two treatments. J. Dairy Sci. 39:146–154.

McDougall, E.I. 1948. Studies on ruminant saliva. I. The composition and output of sheep saliva. Biochem. J. 43:99–109.

Maeng, W.J., D.N. Mowat, and W.K. Bilanski. 1971. Digestibility of sodium hydroxide-treated straw fed alone or in combination with alfalfa silage. Can. J. Anim. Sci. 51:743–747.

National Research Council. 1980. Iron. *In* Mineral tolerance of domestic animals, 242–255. Natl. Acad. Sci., Washington, D.C.

Steel, R.G.D., and J.H. Torrie. 1960. Principles and procedures of statistics. McGraw-Hill, New York.

Evaluations of Low-Quality Roughages and Fibrous Cereal Grains Treated with Various Chemicals

G.C. FAHEY, JR., L.L. BERGER, M.I. MORA, and P.J. VAN DER AAR
University of Illinois, Urbana, Ill., U.S.A.

Summary

Large improvements in the feeding value of low-quality roughages have resulted from treatment of them with various chemicals. Chemical treatment of grains may also be a potentially useful practice. The research reported here was designed (1) to determine how Chelating Metal Caustic Swelling (CMCS), a nontoxic, alkaline, aqueous reagent, compares with sodium hydroxide (NaOH) in enhancing in-vitro cellulose digestion (IVCD) of cornstalks (CS) (*Zea mays*) and soybean residues (SBR) (*Glycine max.*), and (2) to evaluate the effects of NaOH and CMCS treatments of fibrous grains using in-situ techniques.

CMCS or 7.5% NaOH solutions were added to CS or SBR at chemical substrate ratios of 0:1, 0.25:1, 0.50:1, 0.75:1, 1:1, 3:1, and 5:1. The mixtures were allowed to react for 24 hours, after which time they were washed with water and used immediately for IVCD determinations. As concentration of chemical increased, CS IVCD also increased. At the higher chemical:substrate ratios, NaOH treatment resulted in marked increases in IVCD compared with CMCS. Treatment of SBR with CMCS resulted in either similar or lower IVCDs compared with controls. NaOH had no effect on SBR IVCD, regardless of concentration.

An in-situ digestion study was conducted to determine the effect of chemical treatment on ruminal digestion of whole oats (*Avena sativa*) and barley (*Hordeum vulgare*). In in-situ comparisons between 4% CMCS and 4% NaOH–treated whole oats and barley, CMCS was ineffective in enhancing digestion.

In conclusion, these experiments (1) point to the marked differences in reactivity that occur when low-quality roughages are treated with various chemicals and (2) show that NaOH treatment has potential as a grain-processing method.

KEY WORDS: cornstalks, soybean residues, oats, barley, sodium hydroxide, CMCS, digestibility, in vitro, in situ.

INTRODUCTION

Making more efficient use of available resources has become a prime motivating factor of R&D in all areas of industry, including agriculture. After quantities of fossil fuels, fertilizer, and solar energy have been expanded in producing a field of corn or soybeans, at least 40%–50% of the dry matter remains in the field (Krull and Inglett 1980). This biomass represents a large potential source of energy for ruminants and may become an increasingly important source of nutrients in the future. With possible increase in cost of grains due to mounting demands for grain as a source of human food and as a substrate for alcohol production, feeding grains as a source of energy for ruminants could become less economical. Grain processing has already become increasingly expensive because of the high energy requirement. Therefore, methodologies must be developed to improve the nutritive value of both agricultural biomass (crop residues) and cereal grains, especially those high in fiber, using low energy–requiring processes. This investigation was conducted to examine the feasibility of chemical treatment as a processing technique for crop residues and fibrous cereal grains.

METHODS

Experiment 1

Cornstalks (CS) and soybean residues (SBR) were ground through a 20-mesh screen and treated with various levels of sodium hydroxide (NaOH) or Chelating Metal Caustic Swelling (CMCS), a nontoxic aqueous reagent composed of 17% sodium tartrate, 6.6% ferric chloride, 7.8% sodium hydroxide, and 6.2% sodium sulfite. CMCS was prepared according to the procedure of Dale et al. (1978). The NaOH concentration used in this experiment was 7.5% to simulate the concentration of NaOH present in CMCS. NaOH or CMCS solutions were added to 75 g of CS or SBR at chemical solution:substrate dry-matter ratios of 0:1, 0.25:1, 0.50:1, 0.75:1, 1:1, 3:1 and 5:1. After the chemicals were added to the crop residues, a hand mixer was used at low speed to ensure that all of the residue was exposed to the chemical. The treated residue was then covered to avoid evaporative losses and allowed to react for 24 hours. All treated residues and an untreated sample were then washed extensively with water until the effluent from the last rinse had a light tea color. The rinsed residues were used immediately for in-vitro cellulose digestibility (IVCD) determinations.

For the in-vitro analyses (Tilley and Terry 1963), rumen fluid was collected from an alfalfa-fed fistulated steer, strained through 8 layers of cheesecloth, and diluted

1:1 (v/v) with a buffer-mineral medium (McDougall 1948). Incubations were conducted in triplicate, and each run was replicated 3 times. The amount of cellulose digested was determined according to Crampton and Maynard (1938).

Experiment 2

Dry whole oats and barley were treated with either a 50% solution of NaOH or a concentrated solution of CMCS to contain 4% chemical on a dry-matter basis. A 24-hour reaction time was allowed, after which approximately 3.5 g (DM) of the control and treated grains were placed in 6 × 15–cm nylon bags (150 μm–diameter pores) and suspended in the rumen of an alfalfa-fed fistulated cow for either 6, 12, 24, or 48 hours. Bags were removed, washed, and dried, and the loss in weight was used to calculate the amount of digestion. Three replicated runs were made using the same cow with three bags/grain × chemical treatment subclass/run.

Data from both experiments were analyzed using the FLSD test (Carmer and Swanson 1973).

RESULTS

Experiment 1

Washing did not significantly affect CS IVCD when compared with not washing (Table 1). As concentration of chemical increased, CS IVCD also increased. NaOH produced only slightly higher increases in CS IVCD at chemical:substrate ratios ranging from 0.25:1 up to 1:1. However, at the 3:1 and 5:1 levels, NaOH treatment resulted in marked increases in IVCD compared with

Table 1. IVCDs of CMCS and NaOH-treated cornstalks and soybean residues (experiment 1)[1]

Chemical:substrate ratio	Cornstalks IVCD, %		Soybean residues IVCD, %	
	CMCS	NaOH	CMCS	NaOH
Unwashed (0:1)	65.1[a,b]	65.1[a]	55.5[a,b]	55.5[a]
Washed (0:1)	64.5[a,b]	64.5[a]	59.3[b,c]	59.3[a]
0.25:1	61.9[a]	63.5[a]	57.0[a,b,c]	58.2[a]
0.50:1	62.9[a,b]	67.8[a,b]	57.8[a,b,c]	58.9[a]
0.75:1	67.5[b,c]	70.6[b,c]	57.3[a,b,c]	59.0[a]
1:1	70.9[c,d]	73.2[c]	61.7[c]	59.5[a]
3:1	71.2[c,d,A]	85.8[d,B]	56.8[a,b,c]	60.2[a]
5:1	74.7[d,A]	90.3[d,B]	53.1[a]	58.1[a]
SEM	± .7	± .6	± .7	± .7

[1]Both residues were thoroughly washed with water after 24 hours of reaction time and used immediately for IVCD analyses. Residues were not subjected to oven drying.

[a,b,c,d]Means in the same column followed by different superscripts are different (P <.05).

[A,B]Means in the same row followed by different superscripts are different (P <.05).

SEM = Standard error of the mean.

G.C. Fahey, Jr., is an Associate Professor of Animal Science, L.L. Berger is an Assistant Professor of Animal Science, M.I. Mora is a Research Assistant in Animal Science, and P.J. van der Aar is a Research Assistant in Animal Science at the University of Illinois.

Table 2. Effect of NaOH, CMCS and time on in-situ digestibility of whole oats and barley (experiment 2).

Grain	Hour	Control	Treatment 4% NaOH	Treatment 4% CMCS	SEM*
Oats	6	2.4[a]	8.6[b]	2.5[a]	±.4
	12	3.6[a]	12.9[b]	4.5[a]	±.4
	24	6.8[a]	24.0[b]	7.6[a]	±.4
	48	15.5[a]	52.6[b]	16.4[a]	±.4
Barley	6	− 0.3[a]	11.4[b]	1.6[a]	±.4
	12	0.5[a]	21.1[b]	2.3[a]	±.4
	24	2.1[a]	41.8[b]	4.8[c]	±.4
	48	6.6[a]	68.1[b]	9.9[c]	±.4

*SEM = Standard error of the mean.

[a,b,c]Means in the same row followed by different superscripts are different (P<.05).

CMCS. Washing of SBR resulted in greater IVCD compared with unwashed controls. Treatment with CMCS resulted in either similar or lower IVCDs of SBR compared with those of washed controls. NaOH had no effect on SBR IVCD, regardless of concentration. NaOH was generally somewhat superior to CMCS in improving IVCD of SBR.

Experiment 2

In-situ dry-matter digestibility coefficients of untreated whole oats ranged from 2.4% at 6 hours of digestion up to 15.5% at 48 hours of digestion (Table 2). Comparable values for barley were − 0.3% (6 hours) up to 6.6% (48 hours). Treatment of the whole grains with 4% CMCS resulted in much lower digestion coefficients than those obtained after treatment with 4% NaOH. Forty-eight-hour in-situ digestion coefficients for barley and oats treated with 4% NaOH were 52.5% and 68.1%, respectively.

DISCUSSION

Several structural features of cellulosic material determine its susceptibility to enzymatic degradation. These include degree of water swelling, crystallinity, molecular arrangement, and content of associated material such as hemicellulose and lignin (Fan et al. 1980). In our studies, we were particularly interested in two of the above-named properties, crystallinity of cell-wall carbohydrates and the association of cellulose with hemicellulose and lignin. In experiment 1 (Table 1), we noted significant increases in cellulose digestion of CS when this residue was treated with either NaOH or CMCS. CMCS was used in these experiments since it was reported (Ladisch et al. 1978) to be a nontoxic analog of Cadoxen, a chemical solvent that reacts with alpha-cellulose of agricultural residues to yield quantitative amounts of glucose. Cornstalks, bagasse, alfalfa, tall fescue, and orchardgrass pretreated with solvent gave over 90% glucose conversions on hydrolysis by

cellulose enzymes from *Trichoderma ressei*. Treatment with this reagent appears to have the same effect on residues as grinding or pulping (Dale et al. 1978). Since NaOH is a component of this solvent, we felt that the two chemical reagents should be evaluated for their ability to affect digestibilities of cellulose and dry matter of agriculture residues and fibrous cereal grains. It was noted in this experiment that at the higher chemical:substrate ratios (3:1 and 5:1), NaOH far surpassed CMCS in improving CS IVCD. SBR did not respond to treatment with either NaOH or CMCS. At certain concentrations, CMCS treatment resulted in lower IVCDs when compared with untreated controls. NaOH had no effect on IVCD of SBR. Morrison (1974) observed that a large proportion of the lignin-hemicellulose and cellulose linkages are alkali–labile ester bonds. Alkaline hydrolysis is believed to improve digestibility of cell-wall carbohydrates by increasing the solubility of hemicellulose and the availability of both hemicellulose and cellulose for fermentation. In this experiment, we were primarily interested in the effect of chemical treatment on cellulose digestion since most of the hemicellulose would be absent from the treated plant material due to washing. Data from this experiment support the work of Hartley and Jones (1976), who showed that few alkali–labile ester linkages could be identified in legumes such as clover and alfalfa. Therefore, little improvement in the fermentability of legume straws has been observed with alkali treatment. SBR, a collective term used to describe that part of the leguminous plant composed of leaves, stems, and pods, likewise did not respond to alkali treatment, probably for the same reason. It should be noted too that material used in this experiment was not dried before IVCD analyses but rather was used immediately after washing. This precaution was taken to prevent possible recrystallization of cellulose from oven drying.

In experiment 2 (Table 2), we investigated the ability of chemical reagents to disrupt the fibrous seed coat of whole, fibrous cereal grains, thus increasing grain digestibility. Oats and barley were the grains chosen since both contain relatively high concentrations of cell-wall constituents (22% and 31%, respectively) and relatively similar amounts of hemicellulose (14% and 15%, respectively), expressed on a dry-matter basis.

In enhancing in-situ dry-matter digestion of whole oats and barley, NaOH was markedly superior to CMCS (Table 2). Our results using NaOH are similar to those reported by Ørskov and Greenhalgh (1977), who treated whole cereal grains with 5% NaOH. Part of the improvement in digestion due to NaOH treatment undoubtedly results from solubilization of hemicellulose in the seed coat, rendering the starch portion of the kernel more available for microbial attack. Dry grains apparently respond to treatment differently than roughages; Jayasuriya and Owen (1975) found that low moisture levels decreased the response of roughages to NaOH treatment.

Our findings suggest that NaOH treatment has potential as a grain-processing method and that NaOH is probably the active ingredient of CMCS.

LITERATURE CITED

Carmer, S.G., and M.R. Swanson. 1973. An evaluation of ten pairwise multiple comparison procedures by Monte Carlo methods. J. Am. Stat. Assoc. 68:66-74.

Crampton, E.W., and L.A. Maynard. 1938. The relationship of cellulose and lignin content to the nutritive value of animal feeds. J. Nutr. 15:383-395.

Dale, B.E., M.R. Ladisch, T.J. Hamilton, and G.T. Tsao. 1978. High glucose yields from cellulosic material treated with a non-toxic agent. 176th ACS Natl. Mtg. 11-15 September 1978, Miami, Fla. (Abs.).

Fan, L.T., Yong-Hyun Lee, and D.H. Beardmore. 1980. Mechanism of the enzymatic hydrolysis of cellulose: effects of major structural features of cellulose on enzymatic hydrolysis. Biotech. and Bioengin. 22:177-199.

Hartley, R.D., and E.C. Jones. 1976. Diferulic acid as a component of cell walls of *Lolium multiflorum*. Phytochemistry 15:1157-1160.

Jayasuriya, M.C., and E. Owen. 1975. Sodium hydroxide treatments of barley straw: effect of volume and concentration of solution on digestibility and intake by sheep. Anim. Prod. 21:313-332.

Krull, L.H., and G.E. Inglett. 1980. Analysis of neutral carbohydrates in agricultural residues by gas-liquid chromatography. J. Agric. Food Chem. 28:917-919.

Ladisch, M.R., C.M. Ladisch, and G.T. Tsao. 1978. Cellulose to sugars: new pathway gives quantitative yield. Science 201:743-745.

McDougall, E.I. 1948. Studies on ruminant saliva. I. The composition and output of sheep saliva. Biochem. J. 43:99-109.

Morrison, I.M. 1974. Structural investigations on the lignin-carbohydrate complexes of *Lolium perenne*. Biochem. J. 139:197-204.

Ørskov, E.R., and J.F. Greenhalgh. 1977. Alkali treatment as a method of processing whole grain for cattle. J. Agric. Sci., Camb. 89:253-255.

Tilley, J.M.A., and R.A. Terry. 1963. A two-stage technique for the *in vitro* digestion of forage crops. J. Brit. Grassl. Soc. 18:104-111.

Improving Hay Quality by Ammoniation

K.J. MOORE, V.L. LECHTENBERG, K.S. HENDRIX, and J.M. HERTEL

Purdue University, West Lafayette, Ind., U.S.A.

Summary

Experiments were conducted to determine the effect of hay ammoniation on hay consumption and weight gain by steers when the hay was fed unsupplemented and supplemented with energy and protein. Mature orchardgrass (*Dactylis glomerata* L.) hay was baled in early June 1979 and treated with anhydrous ammonia at 30 g/kg of hay. Seven weeks after treatment the hay was uncovered and allowed to aerate for 3 days. It was then fed to weanling steer calves weighing approximately 225 kg each. Similar hay baled from the same field but not ammoniated was fed to separate groups of steers. Both the treated and the untreated hays were fed unsupplemented, supplemented with 1.80 kg corn/head daily, and supplemented with 1.35 kg corn and 0.45 kg soybean meal/head daily. The feeding trial lasted 90 days.

Ammoniation increased total nitrogen concentration of the hay from 1.12% to 1.96%. Apparent dry-matter digestibility was increased from 47.7% to 54.8%. Cellulose and hemicellulose digestibilities were increased from 49.5% to 62.1% and from 49.8% to 71.7%, respectively, by ammoniation. Digestibility of nitrogen added to the hay during treatment was slightly greater than digestibility of the native nitrogen.

Animals fed ammoniated hay consumed 17% more dry matter daily than animals fed untreated hay. Consumption averaged 66.2 and 77.1 $g/kg^{0.75}$, respectively, for the untreated and ammoniated hays. Animals consuming ammoniated hay gained 0.21 kg/day more than animals consuming untreated hay. Corn grain increased daily gain of animals consuming untreated hay from 0.16 to 0.45 kg. Similar supplementation for animals consuming ammoniated hay increased daily gain from 0.37 to 0.69 kg. Protein supplementation increased the weight gain of animals fed untreated hay but did not increase the weight gain of animals consuming ammoniated hay. Apparently the nitrogen added to the hay during ammoniation was utilized well enough to preclude the need for additional protein supplementation.

KEY WORDS: weight gain, digestibility, fiber, intake, hay supplementation.

INTRODUCTION

The structural carbohydrate concentration in mature grass forage is generally quite high, and the protein concentration is usually very low. Digestible energy percentage in these forages is often so low that ruminant animals consuming them are able to consume only enough digestible energy to meet their maintenance energy needs. In some cases energy consumption is so low that animals lose weight (Bula et al. 1977). A modest increase in the digestibility of fiber or dry matter of these low-quality forages usually results in a dramatic improvement in digestible-energy consumption and animal production.

A number of chemical treatments have been used experimentally to increase fiber digestiblity (Klopfenstein 1978). Treatments generally increased organic-matter digestibility by 10 to 12 percentage units and also increased feed consumption and animal weight gain. Arnason and Mo (1977) reported that the dry-matter digestibility of straw was increased from 47% to 64% by treatment with anhydrous ammonia. They also reported a 25% increase in animal weight gain. Horton and Steacy (1979) reported that ammoniation of barley, oats, and wheat straw increased dry-matter consumption by as much as 21% and increased organic-matter digestibility by 2.2, 3.7, and 6.3 percentage units, respectively. Buettner et al. (1982) observed that ammoniation of mature tall fescue (*Festuca arundinacea* Schreb.) hay increased dry-matter digestibility from 39% to 57% and increased hay consumption by cattle from 44 to 67 g/kg$^{0.75}$.

This study was conducted to determine the effect of treating mature orchardgrass (*Dactylis glomerata* L.) hay with anhydrous ammonia on hay consumption and weight gain of cattle. Dry-matter and fiber-component digestion coefficients were determined in a metabolism trial using sheep.

METHODS

Mature orchardgrass hay was baled in early June 1979 at the Southern Indiana Purdue Agricultural Center near Dubois, Indiana. A stack of approximately 1,000 rectangular bales was enclosed with polyethylene, and a quantity of liquid anhydrous ammonia equal to 30 g/kg of hay was injected into an open container placed in the center of the enclosed stack. The hay was left covered for 7 weeks. The stack was then uncovered and aerated for 3 days. A similar quantity of hay baled from the same field was stored in a barn and left untreated. This hay was used as the control hay.

The ammonia-treated and untreated hays were fed for 90 days to weanling steer calves weighing approximately 225 kg./head. Both the treated and untreated hays were fed in rations consisting of hay only, hay plus 1.80 kg corn

grain/head daily, and hay plus 1.35 kg corn grain and 0.45 kg soybean meal/head daily. Hay was fed ad libitum, and both the hay and supplement were fed once daily.

The treated and untreated hays were also fed ad libitum to ram lambs in a metabolism trial. The lambs, which weighed approximately 40 kg/head, were individually fed in metabolism crates. The untreated hay was fed unsupplemented and supplemented with a quantity of soybean meal equal to 10% of the total feed consumption. The ammoniated hay was fed unsupplemented and supplemented with a similar quantity of corn grain. Each of the four rations was fed to four lambs. Two weeks were allowed for the animals to become adapted to the rations, after which feces and urine samples were collected for 7 days, frozen, and stored for laboratory analysis. Hay and fecal samples were dried and ground, and the concentrations of neutral detergent fiber (NDF), acid detergent fiber (ADF), hemicellulose, cellulose, and lignin were measured (Goering and Van Soest 1970). Total nitrogen concentration in the hay, feces, and urine was determined using a micro-Kjeldahl procedure (Nelson and Sommers 1973).

RESULTS AND DISCUSSION

Hay Composition and Digestibility

Ammoniation increased the total nitrogen concentration of the hay from 1.12% to 1.96% (P < 0.01). The concentration of other constituents were not affected (P > 0.05) by ammoniation (Table 1). Supplementation of the untreated hay with soybean meal and supplementation of the treated hay with corn did not affect the digestibility of dry matter or of fiber constituents.

Ammoniation increased (P < 0.05) the digestibility of all constituents measured (Table 2). The greatest improvement in digestibility occurred in the hemicellulose fraction. These results are consistent with the findings of Buettner et al. (1982). Ammoniation increased apparent dry-matter digestibility by 7.1 percentage units but increased true dry-matter digestibility (Goering and Van Soest 1970) by 10.2 percentage units (Table 2). This finding suggests that either the rate of passage through the digestive tract was increased by ammoniation, resulting

Table 1. Effect of ammoniation on chemical composition of mature orchardgrass hay.

Constituent	Untreated	Ammoniated		Standard error
	---------- % of dry matter ---------------			
Total nitrogen	1.12	1.96	**	.078
NDF	71.2	73.2	ns[+]	.616
ADF	44.2	46.8	ns	.997
Hemicellulose	27.0	26.4	ns	1.29
Cellulose	34.6	36.0	ns	.552
Lignin	7.28	7.52	ns	.266

**Difference due to treatment significant (P < .01).

[+] ns = not significant.

K.J. Moore is a Research Assitant, V.L. Lechtenberg is a Professor of Agronomy, K.S. Hendrix is an Associate Professor of Animal Science, and J.M. Hertel is a Research Agronomist at Purdue University. This is a contribution of the Purdue University Agricultural Experiment Station (Journal Paper No. 8415).

Table 2. Effect of ammoniation on digestibility of dry matter and fiber components of mature orchardgrass hay.

Constituent	Untreated	Ammoniated	Standard error
	---------- % of dry matter ---------		
Dry matter (apparent)	47.7	54.8 **	0.97
Dry matter (true)	61.6	71.8 **	0.77
NDF	46.2	61.2 **	1.07
ADF	43.9	55.8 **	1.27
Hemicellulose	49.8	71.7 **	1.11
Cellulose	49.5	62.1 **	1.30
Lignin	27.7	34.2 **	2.03
Nitrogen+	53.7	58.5 **	1.50

**Difference due to treatment significant ($P < .01$).

+Unsupplemented rations only.

Table 3. Effect of ammonia treatment and supplementation on daily gain and hay consumption by steers.

	Ration components			Hay dry matter
Hay	Corn grain	Soybean meal	Gain (kg/day)	consumed (g/kg $wt^{0.75}$)
	------------------------------------- kg/head/day -----------------------------			
Untreated	0	0	0.16	66.2
Ammoniated	0	0	0.37	77.1
Untreated	1.80	0	0.45	58.1
Ammoniated	1.80	0	0.69	66.3
Untreated	1.35	.45	0.53	56.7
Ammoniated	1.35	.45	0.69	68.2
Standard error			.034	2.30

in the excretion of a greater quantity of digestible material, or metabolic excretion was greater with ammoniated hay. In-vitro dry-matter disappearance from feces of animals consuming ammoniated hay averaged 6 percentage units higher than for animals consuming untreated hay. This observation is consistent with the observed differences between apparent digestibility and true digestibility.

Ammoniation resulted in a moderate increase in the digestibility of total nitrogen (Table 2). This finding implies that the digestibility of the nitrogen added during ammonia treatment was greater than the digestibility of the nitrogen originally present in the hay. The nitrogen digestibility in the untreated hay was 53.7%.

If the digestibility of this native nitrogen was not affected by ammoniation, then the apparent digestibility of the added nitrogen was approximately 65%. Horton and Steacy (1979) reported similar results.

Dry-Matter Intake and Weight Gain

Consumption of ammoniated hay was 17% greater than consumption of untreated hay (Table 3). Supplementation with corn grain at 1.80 kg/head daily decreased consumption of untreated hay by 12.4% and of ammoniated hay by 11.9%. Supplementation with corn plus soybean meal did not further affect hay consumption. If the difference in apparent dry-matter digestibility observed in the sheep metabolism trial also existed in the cattle feeding trial, then ammoniation increased digestible

dry-matter consumption of hay by approximately 35%.

Ammoniation increased weight gain an average of 0.21 kg/head daily. Both energy and protein supplementation increased daily weight gain of animals on untreated hay; however, only energy supplementation increased gain of those on ammoniated hay. Apparently the nitrogen added to the hay during ammoniation was utilized by the animals well enough to preclude the need for additional protein supplementation.

LITERATURE CITED

Arnason, J., and M. Mo. 1977. Ammonia treatment of straw. Proc. 3rd Straw Util. Conf., Oxford, UK.

Buettner, M.R., V.L. Lechtenberg, K.S. Hendrix, and J.M. Hertel. 1982. Composition and digestion of ammoniated tall fescue (*Festuca arundaincea* Schreb.) hay. J. Anim. Sci. 54:173–178.

Bula, R.J., V.L. Lechtenberg, and D.A. Holt. 1977. Potential of temperate zone cultivated forages for ruminant animal production, 7–28. Winrock Rpt., Winrock Int. Livest. Res. and Train. Ctr., Morrilton, Ark.

Goering, J.K., and P.J. Van Soest. 1970. Forage fiber analysis (apparatus, reagents, procedures, and some applications). U.S. Dept. Agric., Agric. Handb. 379.

Horton, G.M.J., and G.M. Steacy. 1979. Effect of anhydrous ammonia treatment on the intake and digestibility of cereal straws by steers. J. Anim. Sci. 48:1239–1249.

Klopfenstein, T. 1978. Chemical treatment of crop residues. J. Anim. Sci. 46:841–848.

Nelson, D.W., and L.E. Sommers. 1973. Determination of total nitrogen in plant material. Agron. J. 65:109–112.

Effect of Delayed Sealing During Ensiling on Fermentation and Dry-Matter Loss

N. TAKANO, Y. MASAOKA, and T. MANDA

National Grassland Research Institute, Japan

Summary

Two experiments were carried out from 1976 to 1980 to ascertain the effect of delayed sealing of silo on fermentation, dry-matter (DM) loss, digestibility, and silage intake. The effect of delayed sealing (up to 72 hours) of grass and corn silage was tested in laboratory silos on fermentation quality with storage temperatures at 15°C and 30°C, moisture content from 49% to 85%, water-soluble carbohydrate (WSC) content ranging from 11.1% to 17.5%, and with and without an addition of 0.25% formic acid. In all cases, delayed sealing resulted in poor-quality silage with high pH values and low lactic acid content. Moreover, deterioration of silage quality was accelerated by increases in delay time, storage temperature, moisture content, and lowering of WSC content of the ensiled materials. Addition of formic acid before ensiling eliminated deterioration of silage quality caused by the delayed sealing.

High-moisture Italian ryegrass was harvested with a cylinder harvester at early heading stage and ensiled three times in each of two 4.5-m³ experimental tower silos. The green forage contained 14.1% DM, 15.1 ± 1.2% WSC, and 13.6 ± 0.8% crude protein on a DM basis. Silo 1 was sealed immediately after each ensiling by using the plastic-bag-liner method; silo 2 was sealed after 48 hours' delay at each ensiling (open for a total of 144 hours) using the same plastic-bag-liner method.

During ensiling in the silo where sealing had been delayed, maximum fermentation temperature in the middle layer was 34.9° ± 1.6°C and 50.8°C in the top layer. By comparison, in the silo that had been sealed immediately, maximum temperature was 22.0° ± 0.8°C. Delayed sealing produced silages of high pH values with high butyric acid and volatile-base nitrogen (VBN) and low lactic acid content. DM losses during ensiling were also high in these aerated silages.

The DM digestibility (in vivo) and intake of silage from the silo that had been sealed immediately was slightly higher, although not statistically significantly so, than the aerated silage.

From these experiments, it is concluded that delayed sealing of a silo not only has an immediate effect on the upper layers of the ensiled mass but also can cause deterioration of the lower layers even after sealing has been accomplished. Moreover, the effect of delayed sealing on fermentation and DM loss may be increased by attendant conditions of warm weather and low DM and WSC contents of the ensiled materials.

KEY WORDS: silage, delayed sealing, silage fermentation, dry-matter loss.

INTRODUCTION

In the process of making good-quality silage, the important factor is the presence and maintenance of air-free conditions. However, under actual farm conditions, it often takes 3 to 7 days to fill the silo. During such ensiling, the infiltration of air into the silage mass is a serious problem that will cause poor quality and loss of dry matter (DM). Experiments were carried out from 1976 to 1980 to ascertain the effect of delayed sealing on fermentation, DM loss, digestibility, and silage intake.

MATERIALS AND METHODS

Experiment 1

Trials were undertaken using 1.0-l laboratory silos with airlock rubber tubes. Italian ryegrass and corn were used,

The authors are Research Scientists at the National Grassland Research Institute.

chopped into 10-mm lengths. The main treatments were: storage period, 30 to 60 days; storage temperature, 15°C and 30°C in temperature-controlled chambers; moisture content of ensiled materials, high, medium, and low; water-soluble-carbohydrate (WSC) content of ensiled materials, high and low; and delay before sealing, 0, 24, 48, and 72 hours. Forage samples were analyzed for DM, crude protein, and WSC content. Silage samples were analyzed for pH, organic acids, volatile-base nitrogen (VBN), and DM.

Experiment 2

High-moisture Italian ryegrass was harvested by a cylinder-type harvester at early heading stage and ensiled in two 4.5-m³ silos three times at 48-hour intervals. Silo 1 was sealed immediately after each ensiling by closing the inside plastic-bag liner, while silo 2 remained open 48 hours after each ensiling (a total of 144 hours) and was then sealed. Thermocouples were inserted into the middle layer of each silo to ascertain fermentation temperatures.

Each silage was fed to four wether sheep in metabolic cages in order to measure digestibility and fed ad libitum to four dry cows in stalls to measure silage intake.

Analytical Methods

WSC was determined by the anthrone method. Organic acids were determined by gas-liquid chromatography. A toluene distillation method was used for estimation of silage DM. VBN was determined by distillation from potassium hydroxide.

RESULTS

Experiment 1

As is evident from Table 1, quality of silage in the silo where sealing was delayed 72 hours was very low with high DM loss. Quality of the silage stored at a high temperature (30°C) and low WSC (11.1%) content of ensiled materials was lower than that stored at a low temperature (15°C) and high WSC (17.5%) content, as shown by high pH, low lactic acid content, and low Flieg's score.

Unexpectedly, lower-quality silages were obtained from high-moisture materials than from materials of medium and low moisture. Silage-quality deterioration was accelerated by increases in the delay of sealing time and storage temperature.

In a test of effect of preservative, silage corn was harvested at dent stage, chopped into 10-mm length, and ensiled in laboratory silos with and without 0.25% formic acid on a fresh basis. Good-quality silages were obtained by the addition of formic acid as compared with the no-preservative silages under various delayed sealing times (Table 2).

Experiment 2

Fermentation quality of the silage indicated a better pattern in silo 1, which was sealed immediately, than in silo 2, in which sealing was delayed 144 hours, as is shown by pH value, lactic acid content, Flieg's score, and VBN

Table 1. Effect of 72 hours delayed sealing on high-moisture silage[1] stored at 15° and 30° C.

Storage temperature	Hours delay	pH	Lactic acid, % fresh	Flieg's score	DM loss, %
15°C	0	4.0	1.98	100	6.1
	72	4.7	1.19	45	10.2
30°C	0	4.0	2.47	100	8.9
	72	5.0	0.49	15	30.8

[1]Ensiled material contained 14.5% DM, 12.5% crude protein, and 16.5% WSC on DM basis. Stored for 60 days.

Table 2. Interaction of delayed sealing and formic-acid addition on quality of corn silage.[1]

Formic acid, % fresh	Hours delay	pH	Lactic acid, % fresh	VBN* of total nitrogen	Flieg's score
0	0	3.7	2.14	15	100
	24	3.9	1.24	18	45
	48	4.1	0.98	24	38
0.25	0	3.8	0.82	8	100
	24	3.8	1.05	10	98
	48	3.8	1.22	12	98

[1]Ensiled material contained 30% DM, 7.0% crude protein and 22.0% WSC on DM basis. Stored for 30 days at 30°C.

*Volatile-base nitrogen.

concentration (Table 3). Lower maximum fermentation temperature and higher DM recovery were obtained from silo 1 than from silo 2. DM digestibility and silage intake were slightly higher in the silage from the immediately sealed silo than in that from the 144-hour delayed-sealing silo, although the difference was not statistically significant.

DISCUSSION

In silage making, the first essential step is to achieve and maintain an air-free condition; the second step is to prevent clostridial growth by encouraging the rapid development of lactic acid bacteria. However, under actual farm conditions, it takes 3 to 7 days to fill a silo, depending on silo size and availability of labor and equipment. Under such conditions, the infiltration of air into the silage mass has a serious effect on silage fermentation. Henderson and McDonald (1975) reported that aeration from delayed sealing has an immediate effect only on the surface layers of herbage, but deterioration of the lower layers may occur even after sealing. The protein fraction is broken down, and sugars are oxidized, with a subsequent rise in temperature and changes in the bacterial population.

Reports to date on this type of research have apparently not dealt with factors in delayed sealing affecting silage quality. In all cases in the work reported here, delayed sealing resulted in the production of poor-quality silage

Table 3. Effect of delayed sealing on silage quality and DM recovery in high-moisture silage.

Hours delay	Maximum temperature	DM recovery, %	Lactic acid, % fresh	VBN* %	DMD** %
0	22.0°C	87.6	1.98	8.6	67.9
144	34.9°C	76.5	0.92	20.5	66.5

*Volatile-base nitrogen as percentage of total nitrogen.

**In-vivo digestible dry matter.

with high pH, low lactic acid content, and high DM loss. Moreover, silage-quality deterioration was accelerated by increases in delay time, storage temperature, and moisture content and by lowering of WSC content of the ensiled materials.

LITERATURE CITED

Henderson, A.R., and P. McDonald. 1975. The effect of delayed sealing on fermentation and losses during ensilage. J. Sci. Food Agric. 26:653–667.

Interrelationships Between Pattern of Fermentation During Ensilage and Initial Crop Composition

J.M. WILKINSON, P.F. CHAPMAN, R.J. WILKINS, and R.F. WILSON

Grassland Research Institute, Hurley, Berkshire, England, UK

Summary

Relationships between crop composition and silage composition were examined for 231 silages made at the Grassland Research Institute from 1968 to 1978 without additives.

A test was made of the model of Weissbach et al. (1974) in which silage characteristics were related to the percentage of dry matter in the crop (DM) and the ratio of sugar (Z) to buffering capacity (PK). Although these factors accounted for significant variance, the percentage accounted for was low at 36% for lactic acid (percentage of total acids), 26% for Flieg's score, 22% for ammonia N (percentage of total N), and 12% for butyric acid (percentage of total acids). Consideration of the separate effects of DM, Z, and PK accounted for broadly similar percentages of the variance. Variation in Z (percentage of fresh weight) appeared to be more important than variation in the other factors.

Cluster analysis was carried out according to silage characteristics, and the silages formed seven distinctive groups, three of which were considered well preserved and four poorly preserved. All the crop characteristics examined—DM, Z, PK, Z:PK, N (percentage of fresh weight), and pH—varied significantly with silage group, and mean values for Z were higher for the three well-preserved groups than for the other groups. The percentage of poorly preserved silages decreased with increase in Z; when Z contents were below 2% some 44% of the silages were poorly preserved, whereas when Z was above 2% only 5% of the silages were poorly preserved. It is concluded that Z alone can indicate in broad terms the probability of obtaining good preservation without the use of additives.

KEY WORDS: crop composition, fermentation pattern, ensilage, silage composition.

INTRODUCTION

In the absence of chemical additives, successful preservation of crops as silage requires fermentation of water-soluble carbohydrates to produce lactic acid and to reduce pH to a sufficiently low value to prevent secondary fermentation of that acid to butyric acid with concomitant degradation of plant proteins to ammonia.

Weissbach et al. (1974) proposed a simple model to describe the probable pattern of fermentation during ensilage. Using results from laboratory silos, they related fermentation quality to initial crop composition, considering dry-matter content (DM) and the ratio of water-soluble carbohydrates (Z) to buffering capacity (PK). By artifically altering DM and Z:PK of the crop, they showed that increases in both factors were associated with a reduction in the proportion of butyric acid and an increase in the proportion of lactic acid in the total acids of the silage. The model was not, however, tested with data from a range of crops ensiled at different contents of DM and Z:PK ratios.

The study described here sought to verify the extent to which, with crops ensiled without additive, the pattern of fermentation could be related to DM and Z:PK.

METHODS

Two types of statistical analysis of data for crop and silage composition derived from experiments at the Grassland Research Institute between 1968 and 1978 were carried out.

J.M. Wilkinson's current address is Chief Scientist's Group, Ministry of Agriculture, Fisheries and Food, Great Westminster House, Horseferry Road, London, SW1P 2AE, UK.

The first analysis, a test of Weissbach's model, consisted of stepwise multiple regression analysis for different silage characteristics and the crop components DM and Z:PK. Linear, quadratic, and cubic terms and interactions were considered, and the best-fit relationships used were those that accounted for the highest percentage of variance in the silage characteristic, and in which all regression coefficients were significantly different from zero.

The second part was a classification of the silages in terms of their similarity with respect to composition, using furthest neighbor cluster analysis (Gower 1967) and an examination of the type and composition of the crops prior to ensilage for each discrete group of silages.

A total of 231 silages made without additive were studied, of which 50% were from grasses (predominantly *Lolium perenne*), 30% were from legumes (predominantly *Medicago sativa*), and 20% were from annual forage crops (predominantly *Zea mays*). All crops were harvested with a precision-chop machine and ensiled for a minimum period of 60 days in laboratory silos or well-sealed experimental silos with capacity up to 20 metric tons (t) fresh weight.

RESULTS

Regression Analysis

Best-fit relationships between silage characteristics and initial crop composition are given for the Weissbach model in Table 1. Four indices of pattern of fermentation were chosen: lactic acid as a percentage of total acids (LAC), butyric acid as a percentage of total acids (BUT), ammonia nitrogen as a percentage of total N (NH_3-N), and Flieg's score.

The variance accounted for was low, ranging from 12.5% for BUT to 36.5% for LAC. Z:PK was included in all of the equations, but DM was included only in regressions with LAC and NH_3-N.

Regressions resulting from the simpler approach using the linear and quadratic terms for DM, Z (as percentage of fresh weight), and PK (m equiv/100 g fresh weight) as candidate independent variables are given in Table 2. The percentages of variance accounted for were broadly similar to those in equivalent regressions in Table 1. Z was a significant term in all equations.

Cluster Analysis

In this analysis the eight silage characteristics listed in Table 3 were taken equally into account in the determination of the degree of similarity between silages. Seven groups of silages were identified. Mean values for the silages according to group are given in Table 3, together with the composition of the initial crops according to silage group.

Group A included a small number of silages that had undergone restricted fermentation but were well preserved, as indicated by the absence of butyric acid and low NH_3-N. Groups B and C were also well preserved and were distinguished by a difference in the proportions of lactic and acetic acid. Groups D to G were poorly preserved, with high mean pH values in all four groups. Group D was distinguished by having a high proportion of acetic acid but a low proportion of butyric acid and only a moderate level of NH_3-N. Group E had high contents (above 20%) of the three acids (lactic, acetic, and butyric) whereas Groups F and G had very low proportions of lactic acid and were similar except for higher pH values, higher contents of acetic acid and ammonia, and lower total acid content in G than F.

All six crop characteristics differed significantly among the silage groups ($P < 0.001$), but silages could not be allocated clearly to individual groups according to measurements made before ensiling. Group A was distinct in having a high DM content, and the other groups of well-preserved silages, B and C, had higher Z contents than any of the groups D to G. There was no clear distinction in crop composition between groups B and C, but group B tended to have a higher Z content and consisted predominantly of perennial grasses. The poorly preserved groups all had lower Z contents than the well-preserved groups and also were lower in DM and Z:PK. There were distinctions between the poorly preserved groups,

---

Table 1.   Best-fit relationships between crop composition and silage characteristics, using Weissbach's model.

LAC	$= -31.6 + 345(Z/PK)^a - 319(Z/PK)^2 + 86.4(Z/PK)^3$ $(\pm 11.0)^b$ (± 53) (± 73) (± 31.4)	
	$-2.34DM - 6.98(Z/PK)DM + 4.05(Z/PK)^2DM$ (± 0.54) (± 2.10) (± 1.88)	VAC^c 36.5
BUT	$= 23.8 - 91.3(Z/PK) + 124(Z/PK)^2 - 51.3(Z/PK)^3$ (± 3.4) (± 22.7) (± 40) (± 18.8)	VAC 12.5
NH_3–N	$= 34.8 - 29.3(Z/PK) - 0.559DM + 0.684(Z/PK)^2DM$ (± 2.7) (± 5.6) (± 0.095) (± 0.175)	VAC 22.2
FLIEG	$= 9.25 + 288(Z/PK) - 379(Z/PK)^2 + 154(Z/PK)^3$ (± 7.54) (± 50) (± 88) (± 42)	VAC 25.7

[a] Z = water-soluble carbohydrate, PK = buffering capacity, and DM = dry matter.

[b] Figure in parenthesis represents the standard error of the regression coefficient immediately above.

[c] Percentage variance accounted for.

Table 2. Best-fit relationships between crop composition and silage characteristics considering independent effects of dry matter (DM), water-soluble carbohydrates (Z), and buffering capacity (PK).

$$LAC = 54.0 + 20.2Z - 1.86Z^2 + 8.75\ PK + 0.534\ PK^2 \qquad VAC^b\ 30.9$$
$$(\pm 10.4)^a \quad (\pm 2.3) \quad (\pm 0.27) \quad (\pm 2.80) \quad (\pm 0.183)$$

$$BUT = 16.2 - 5.34Z + 0.443Z^2 \qquad VAC\ 8.2$$
$$(\pm 2.3) \quad (\pm 1.38) \quad (\pm 0.160)$$

$$NH_3-N = 31.7 - 6.55Z + 0.605Z^2 - 0.293\ DM \qquad VAC\ 23.1$$
$$(\pm 2.4) \quad (\pm 1.15) \quad (\pm 0.129) \quad (\pm 0.104)$$

$$FLIEG = 80.9 + 20.5Z - 1.80Z^2 - 14.7\ PK + 0.908\ PK^2 \qquad VAC\ 24.6$$
$$(\pm 13.7) \quad (\pm 3.0) \quad (\pm 0.35) \quad (\pm 3.7) \quad (\pm 0.242)$$

[a] Standard error of regression coefficient immediately above.

[b] Percentage variance accounted for.

Table 3. Composition of seven groups of silages identified on the basis of similarity by cluster analysis and composition of crops according to silage group[1]

| | Fermentation quality | | | | | | |
| | Good | | | Poor | | | |
	Group A	Group B	Group C	Group D	Group E	Group F	Group G
Number of silages	7	128	46	19	9	20	2
Silage composition							
pH	4.79 ± 0.12	4.27 ± 0.04	4.23 ± 0.05	5.12 ± 0.05	5.40 ± 0.15	5.94 ± 0.08	6.83 ± 0.25
Z, % fresh wt.	5.12 ± 0.65	0.42 ± 0.04	0.45 ± 0.11	0.03 ± 0.02	0.11 ± 0.05	0.03 ± 0.01	0
Total acids, % fresh wt.	2.29 ± 0.43	2.48 ± 0.06	2.64 ± 0.05	1.92 ± 0.12	2.90 ± 0.35	2.31 ± 0.09	1.22 ± 0.52
Lactic acid[2]	62.3 ± 5.97	74.7 ± 0.93	52.4 ± 1.61	21.2 ± 2.32	39.0 ± 3.49	3.6 ± 0.87	0
Acetic acid[2]	37.5 ± 6.02	22.4 ± 0.86	46.1 ± 1.48	69.4 ± 1.63	24.4 ± 3.56	49.4 ± 3.20	60.7 ± 0.13
Propionic acid[2]	0	0.6 ± 0.10	1.1 ± 0.26	5.6 ± 0.69	5.0 ± 1.18	7.3 ± 0.46	14.1 ± 2.88
Butyric acid[2]	0.1 ± 0.12	2.3 ± 0.58	0.5 ± 0.40	3.8 ± 1.22	31.5 ± 3.40	39.7 ± 2.97	25.3 ± 3.01
NH₃-N, % total N	7.5 ± 1.24	10.1 ± 0.40	8.9 ± 0.61	16.8 ± 1.60	30.8 ± 2.92	41.9 ± 2.83	56.3 ± 0.05
Crop composition							
Dry matter, %	44.4 ± 4.49	21.4 ± 0.50	24.7 ± 1.33	16.1 ± 0.66	19.5 ± 1.97	16.3 ± 0.79	9.6 ± 1.20
pH	6.12 ± 0.05	5.94 ± 0.03	5.71 ± 0.03	5.77 ± 0.07	5.93 ± 0.13	5.75 ± 0.07	5.85 ± 0.05
Z, % fresh wt.	8.01 ± 0.69	3.03 ± 0.15	2.60 ± 0.25	1.31 ± 0.18	2.32 ± 0.49	0.87 ± 0.08	0.30 ± 0.04
PK, m equiv/100 g fresh wt.	11.8 ± 0.51	6.3 ± 0.22	5.9 ± 0.31	6.1 ± 0.49	6.6 ± 0.60	6.6 ± 0.43	4.0 ± 0.23
Z/PK	0.69 ± 0.08	0.54 ± 0.03	0.47 ± 0.04	0.27 ± 0.05	0.41 ± 0.10	0.14 ± 0.01	0.08 ± 0.01
N, % fresh wt	1.04 ± 0.12	0.53 ± 0.02	0.49 ± 0.03	0.48 ± 0.03	0.52 ± 0.06	0.48 ± 0.03	0.37 ± 0.03

[1] Means and standard errors.

[2] % total acids.

with Group E having the highest Z content (associated with the highest contents of lactic acid in the silage), and the two silages that made up Group G had particularly low DM.

DISCUSSION AND CONCLUSIONS

This study provided some support for the proposition by Weissbach et al. (1974) that pattern of fermentation is influenced by the Z:PK ratio and DM content of the crop prior to ensilage, but there was much variation not accounted for by these factors. In regression analyses, variation in Z appeared to be more important than variation in PK or DM, and the importance of Z was also highlighted in the cluster analysis.

The concept of a "critical" Z content for the production

of well-preserved silages has been adopted by the extension services in England and Wales (MAFF 1977), and the critical level currently suggested is 3% of the fresh-crop weight (MAFF 1978, Wilkins 1980). In Fig. 1 the well-preserved and poorly preserved silages are shown in relation to Z. The proportion of poorly preserved silages decreased with increase in Z, so that only 5% of silages made from crops with Z above 2% were poorly preserved compared with 44% of silages made from crops with Z below 2%.

Physical, chemical, and microbiological factors not assessed in this work could have contributed to the unexplained variation. More research is needed, but it appears that the analysis of Z alone can indicate in broad terms the probability of obtaining good preservation without the use of additive. Differences between practical and experimen-

Fig. 1. Number of well-preserved and poorly preserved silages in relation to content of water-soluble carbohydrates in the fresh crop (Z).

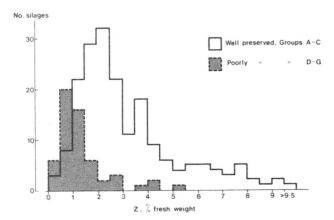

tal conditions probably justify the use of 3% as the Z content below which additives should be used to improve the probability of good preservation.

LITERATURE CITED

Gower, J.C. 1967. A comparison of some methods of cluster analysis. Biometrics 23:626–637.

Ministry of Agriculture, Fisheries and Food (MAFF). 1977. Silage. Liscombe Expt. Husb. Fm. Grass Bul. 2.

Ministry of Agriculture, Fisheries and Food (MAFF). 1978. Silage making. Profitable Fm. Enterprises, Booklet 9.

Weissbach, H.F., L. Schmidt, and E. Hein. 1974. Method of anticipation of the run of fermentation in silage making, based on the chemical composition of green fodder. Proc. XII Int. Grassl. Cong. 3:663–673.

Wilkins, R.J. 1980. Silage. J. Univ. Newc. Agric. Soc. 27: 6–13.

Addition of *Lactobacillus* sp. to Aid the Fermentation of Alfalfa, Corn, Sorghum, and Wheat Forages

N.J. MOON and L.O. ELY

University of Georgia, Experiment, Ga., and Colorado State University, Fort Collins, Colo., U.S.A.

Summary

As preservation of forage by ensiling has increased, so has the use of additives to enhance acid fermentation. Many of the fermentation aids currently on the market contain microflora to insure a lactic acid fermentation. The objectives of this study were to characterize the basic patterns of fermentation in a variety of crops and to evaluate the effects of adding selected microflora to forages to enhance the fermentation. Four crops were ensiled in each of three seasons. Alfalfa was harvested at 20%–40% bloom, second cut; wheat was cut in the early dough stage; corn was cut in the soft-dent stage; and sorghum was harvested at the early dough stage. Treatments used during the 3 years of trials included a commercially prepared dried preparation of *Lactobacillus acidophilus* and *Candida* sp., a frozen commercial preparation of *Lactobacillus plantarum*, or a *Lactobacillus plantarum* isolated from corn silage and prepared in the laboratory as a frozen concentrated culture. The inoculum was added at the time of ensiling at concentrations of viable populations of 10^5 to 10^7/g silage fresh weight. Fifty-five kg of freshly harvested forage was packed in 0.21-m³ steel drums. After 1, 2, 4, 8, 16, and 32 days of fermentation one drum each for the treatment and control silage was opened for analysis. A fresh sample was used to determine pH. A sample was oven-dried and ground for determination of dry matter, crude protein, crude fiber, crude fat, nitrogen-free extract, and soluble carbohydrate. Fresh samples were used for microbiological determinations including lactobacilli, facultative anaerobic bacteria, lactic acid cocci, and yeasts and molds. The silages differed in their rate and extent of fermentation, microflora, and recovery of nutrients. Corn and sorghum silages had high recovery of nutrients, while alfalfa and wheat fermentations were intermediate. Study of natural microflora of fresh forages indicated few (10^4/g) lactic acid bacteria are present in wheat, but higher numbers were recovered from alfalfa (10^5/g) and corn or sorghum (10^6/g) silages. Higher populations of lactic acid bacteria ensured faster acid production and pH decline in control silages. Addition of *L. acidophilus* and *Candida* sp. had no effect on the fermentation. Addition of *L. plantarum* caused increases in populations of lactobacilli during the fermentation and more rapid pH declines but only slight differences in recovery of nutrients.

KEY WORDS: silage, inoculation, lactobacillus, fermentation.

INTRODUCTION

The successful fermentation of green plants to produce silage preserves the starting material with minimal losses in nutritive value. Others (Bolson 1978) have recommended that bacterial inocula be added to silage to control the fermentation. Experiments were conducted during 3 years at the University of Georgia to determine the effect of inoculation of alfalfa, corn, sorghum, and wheat silages with *Lactobacillus acidophilus* or strains of *L. plantarum*.

MATERIALS AND METHODS

Details of procedures have been published elsewhere (Moon et al. 1980, 1981). The alfalfa was second cutting 20% to 40% bloom, wheat was in the early dough stage, corn was in the soft-dent stage, and sorghum was in the early dough stage. Fifty-five kg of forage was tightly packed in 6-mil polyethylene bags that were sealed and placed in 0.21-m³ steel drums. Seven drums for each control and for each inoculated silage were prepared, and the chemical and microbiological process of the fermentation was accessed by opening one drum for each treatment after 0, 1, 2, 4, 8, 16, and 32 days.

L. acidophilus and a yeast (Silagain) were applied in 1978 to alfalfa, wheat, and corn at 0.05% or 10^4 *L. acidophilus*/g silage. In 1980, a commercial frozen culture concentrate of *L. plantarum* was thawed, diluted, and sprayed on alfalfa, corn, sorghum, and wheat silages at levels of $10^{5.5}$/g (designated *L. plantarum* HL) and 10^7/g (designated *L. plantarum* HH). In 1979 and 1980 *L. plantarum* M isolated from corn silage was used to inoculate alfalfa, corn, sorghum, and wheat silages at 10^6/g.

Determinations of proximate composition, including fat, crude protein, ash, nitrogen free extract, acid detergent fiber–nitrogen (ADF-N), acid detergent fiber (ADF), crude fiber, protein, pH, and water-soluble carbohydrate (WSC), were made as described previously (Moon et al. 1981).

Microbiological analysis on 10-g samples was performed as described by Hausler (1972) and Moon et al. (1981).

The effect of the additive was evaluated statistically (Barr et al. 1976) using linear regression models as described previously (Moon et al. 1981). Tests were made for the significance at $P < 0.05$.

RESULTS AND DISCUSSION

The results are presented in Tables 1 and 2. Inoculation of the silages with *L. plantarum* significantly decreased the average pH over the 32-day fermentation period in alfalfa and wheat (Table 1). Inoculation with *L. acidophilus* had no effect on average pH in any of the silages and caused an undesirable high pH in wheat. In general, inoculation with *L. plantarum* was effective in controlling the overall

Table 1. Effect of addition of lactobacillus on average microbial populations during the 32-day fermentation period ($\log_{10}$ cell no./g) and average pH.

| Silage | Additive | Agar medium[a] | | | | pH |
		AZD	LBS	TSAN	YM	
Alfalfa	Control	8.26	8.07	9.00	4.79	4.92
	L. acidophilus	8.39	8.52[a]	9.09	4.54	4.81
	L. plantarum HH	8.74	8.86*[b]	9.19	4.07[b]	4.64*[b]
	L. plantarum HL	8.70	8.68*[b]	9.11*[b]	4.14[b]	4.64*[b]
	L. plantarum M	8.23	9.14*[b]	9.44[b]	4.42[b]	4.70*[b]
Corn	Control	7.99	8.05	8.56	5.48	3.82
	L. acidophilus	8.19	8.19	8.71	6.06	3.92
	L. plantarum HH	7.77	7.75[b]	8.00[b]	6.18	3.79
	L. plantarum HL	7.85	7.80[b]	8.15[b]	5.85	3.88
	L. plantarum M	7.84	8.57[b]	8.91[b]	5.47	3.79
Sorghum	Control	7.67	7.73	8.14	5.18	3.91
	L. acidophilus	—[c]	—	—	—	—
	L. plantarum HH	7.13	7.22	7.65[b]	5.42	3.91
	L. plantarum HL	7.00	7.22	7.55[b]	5.41	3.95
	L. plantarum M	7.66	8.28[b]	8.60[b]	5.07	3.84
Wheat	Control	7.78	7.36	8.74	4.93	5.07
	L. acidophilus	8.17	8.04[b]	9.04	5.35	5.45[b]
	L. plantarum HH	8.04	8.49*[b]	8.59	5.04	4.58[b]
	L. plantarum HL	7.77	7.81[b]	8.43[b]	4.61	4.69[b]
	L. plantarum M	7.68	8.99*[b]	9.21[b]	4.56	4.38*[b]

[a]AZD = streptococci, LBS = lactobacilli, TSAN = facultative anaerobes, YM = yeasts and molds.

*[b]Significantly different ($P < .05$) from control in (*) linear regression model or ([b]) Duncan's multiple range test of treatment means.

[c]Sorghum not inoculated with *L. acidophilus* in 1978.

pH rise (indicated by the mean pH) later in the fermentation. The inoculum was also effective in lowering the pH more rapidly initially (Table 2) in alfalfa and wheat. There was little difference between the two levels of inoculation of *L. plantarum* H (HH vs. HL), but inoculation with *L. plantarum* M produced slightly lower pH in wheat.

The effect of inoculation on microbial populations was observed mainly for lactobacilli and facultative anaerobic microflora (Table 1). Alfalfa and wheat silages had higher populations in general and lower populations of yeasts and molds (in alfalfa) when inoculated with *L. plantarum*. *L. plantarum* M had slightly higher populations in sorghum and corn silages than the other inocula. The greater effect of inoculation of wheat and alfalfa silages may be related to the population of lactobacilli in the fresh forage: 10^3–10^4/g for alfalfa and wheat vs. 10^6–10^7/g for corn and sorghum. Little difference was observed between the two levels of inocula of *L. plantarum*, HL vs. HH, but *L. plantarum* M silage had higher population.

The positive effect of the inoculum on nutrient recovery was also shown (Table 2), but in general differences were not as great as were observed for pH or microbial counts. This difference may be related to the extremely high recovery of nutrients in the controls.

N.J. Moon is an Assistant Professor in the Department of Food Science of the University of Georgia, and L.O. Ely is an Assistant Professor of Animal Science at Colorado State University.

LITERATURE CITED

Barr, A.J., J.H. Goodnight, J.D. Sall, and J.T. Helwig. 1976. A user's guide to SAS 76. SAS Inst., Raleigh, N.C.

Table 2.   Effect of inoculation[a] of silage on % recovery of nutrients at end of fermentation period in silages prepared in 1978–1980.

| Silage Component | Control | 1980 | | | 1979 | 1978 |
		LpHH[a]	LpHL[a]	LpM[a]	LpM	L. acid[a]
Alfalfa						
Dry matter	93.70	101.11	98.89	100.00	93.70	95.40
Protein	77.88	97.87	96.85	84.11	95.60	99.80
Soluble carbohydrate	98.88	108.32	115.42	111.29	60.80	—
Fat	194.12	189.98	182.31	216.64	100.20	—
Crude fiber	79.87	100.47	96.21	88.16	100.00	108.10
Nitrogen free extract	91.11	93.48	97.67	94.69	66.00	76.40
Corn						
Dry matter	100.00	101.32	101.32	98.68	99.50	88.90
Protein	108.87	101.26	99.71	105.63	100.00	91.80
Soluble carbohydrate	48.26	34.08	40.23	46.82	24.40	—
Fat	84.00	85.06	75.78	77.39	100.00	—
Crude fiber	69.86	80.06	75.81	66.62	100.00	91.20
Nitrogen free extract	138.30	135.01	136.36	137.52	94.00	86.10
Sorghum						
Dry matter	93.65	93.65	93.65	93.65	100.60	*[b]
Protein	88.23	91.13	90.81	87.33	100.50	*
Soluble carbohydrate	25.76	26.06	25.97	28.71	14.10	*
Fat	135.29	112.58	122.86	100.69	100.00	*
Crude fiber	88.24	91.13	87.42	84.59	100.00	*
Nitrogen free extract	91.27	93.36	93.03	93.89	99.00	*
Wheat						
Dry matter	100.00	104.06	101.65	101.32	100.00	89.70
Protein	93.55	100.05	94.74	96.29	100.00	86.40
Soluble carbohydrate	29.11	35.70	34.72	33.74	23.40	—
Fat	229.41	205.99	183.91	203.90	100.00	—
Crude fiber	107.79	88.67	83.02	101.12	100.00	97.50
Nitrogen free extract	72.92	98.73	91.41	81.05	97.00	83.10

[a]LpHH = *L. plantarum* HH (10⁷/g), LpHL = *L. plantarum* HL (10⁵/g), LpM = *L. plantarum* M, L. acid = *L. acidophilus*.

[b]Sorghum not inoculated with *L. acidophilus* in 1978.

Bolson, K.K. 1978. The use of aids to fermentation in silage production. *In* M.E. McCullough (ed.) Fermentation of silage—a review, 181–200. Natl. Feed Ingred. Assoc., W. Des Moines, Iowa.

Hausler, W.J. 1972. Standard methods for the examination of dairy products. Am. Pub. Health Assoc., Washington, D.C.

Moon, N.J., L.O. Ely, and E.M. Sudweeks. 1980. Aerobic deterioration of wheat, lucerne and maize silages prepared with *Lactobacillus acidophilus* and a *Candida* spp. J. Appl. Bacteriol. 49:75–87.

Moon, N.J., L.O. Ely, and E.M. Sudweeks. 1981. Fermentation of wheat, corn and alfalfa silages inoculated with *Lactobacillus acidophilus* and *Candida* spp. at ensiling. J. Dairy Sci. 64:807–813.

Effect of Formic Acid–Formaldehyde Treatments on the Nitrogenous Constituents in Annual Ryegrass Silage

M.L. FISHMAN, F. MC HAN, and D. BURDICK

USDA-SEA-AR, Eastern Regional Research Center, Wyndmoor, Pa., and
Richard Russell Agricultural Research Center, University of Georgia, Athens, Ga., U.S.A.

Summary

Prevention of proteolysis during ensiling is important in preserving the nutritive value of silage. Hence, the mechanism by which silage protein is protected against degradation was studied in wilted annual ryegrass (*Lolium multiflorium* Lam.) silage pretreated with formic acid, formaldehyde, and a formic acid–formaldehyde mixture.

Raw-wilted ryegrass, control ryegrass silage, and treated ryegrass silage were extracted and fractionated under conditions previously found to maximize yield of nitrogenous constituents and minimize proteolysis during the extraction of fresh grass. Extractions were by boric acid–borate buffer at pH 8.0 with and without detergent (1% sodium dodecyl sulfate). Sodium metabisulfite, an antioxidant, was included in all extractions. The fractions, obtained by differences in extractability, solubility, and molecular size, were analyzed for nitrogen content; in the case of the soluble fractions, first moment apparent molecular weights were determined by a specially designed automated gel filtration apparatus. For the forage extracted without detergent, the percentages of total nitrogen as nonextractable-insoluble residue were 72.5% for formic-formaldehyde-treated silage, 65.8% for the wilted control, 63.4% for formaldehyde-treated, 49.7% for formic acid–treated, and 41.0% for the untreated silage. The percentages of total nitrogen as extractable insoluble nitrogen were 17.1% for wilted control, 12.9% for formic-formaldehyde-treated silage, 7.6% for formaldehyde-treated silage, 5.8% for formic acid–treated silage, and 2.7% for the untreated control silage. The wilted control contained 21.7% soluble "protein," whereas the untreated control silage contained 47.8% soluble "protein." The formic acid–, formaldehyde–, and formic acid–formaldehyde-treated silages contained 38.1%, 28.1%, and 18.6% soluble nitrogen, respectively. Also, as was determined by gel chromatography, the treated silages had a higher relative concentration of high-molecular-weight polypeptides than the control silage. With the addition of detergent to the buffer 5%–14% more nitrogen was extracted and an additional 3%–5% nitrogen was solubilized.

Neither formic acid nor formaldehyde protects proteins normally soluble in the plant cell or eliminates anaerobic activity. Both treatments reduce the extractability and therefore the degradability of polypeptides associated with cell membranes and the plant matrix. At the levels employed, formaldehyde is about twice as effective as formic acid in protecting proteins, but because of the complementary nature of their mechanisms of protection, the effect on protection of combining both chemicals in one treatment is that of additivity.

KEY WORDS: annual ryegrass, *Lolium multiflorum* Lam., silage, formic acid–formaldehyde treatment, proteins, polypeptides.

INTRODUCTION

The objective of this study was to investigate the mechanism by which formic acid, formaldehyde, and a formic acid–formaldehyde mixture influence the ensiling process as determined by differences in extractability, solubility, and molecular-size distribution of the nitrogenous constituents in wilted annual ryegrass (*Lolium multiflorum* Lam.) ensiled with and without chemical treatment.

MATERIALS AND METHODS

Freshly harvested ryegrass was dried to 75% of its initial weight to simulate field wilting, and the "wilted" forage was chopped to 1-cm pieces. A subsample was immediately vacuum dried at ambient temperatures (raw-wilted). A second subsample was packed tightly into laboratory silos (McHan et al. 1979) without further treatment (control) and stored at a constant temperature, 33° ± 1°C. Other representative portions of the same forage were sprayed with 0.55% (fresh-weight basis) formic acid, 0.125% formaldehyde, and a mixture of 0.55%–0.125% formic-formaldehyde and then ensiled. After 30 days, the ensiled forage was removed from the silos and vacuum dried at ambient temperature. All dried samples were reduced to face-powder consistency by being ground in a Wiley mill to pass a 420-μm screen, then ball-milled for 72 hours. The powdered ryegrass samples (5 g) were extracted in duplicate with 100 ml of

buffer A at pH 8.0 and 0.2M boric acid–sodium borate containing 5mM sodium metabisulfite, an antioxidant. In addition, all but the raw-wilted samples were extracted in duplicate with buffers containing 1% w/v sodium dodecyl sulfate (SDS), a detergent. Extractions were for 1/2 hour in a beaker containing a magnetic stirrer and set in an ice bath. The extracted mixtures were fractionated as in the scheme in Fig. 1. Nitrogen contents of the unfractionated silages and their fractions were determined by micro-Kjeldahl analysis. Analytical gel chromatography was performed on soluble cuts 1–3 (Fishman and Burdick

Fig. 1. Flow diagram for the preparation of fractions from ryegrass samples.

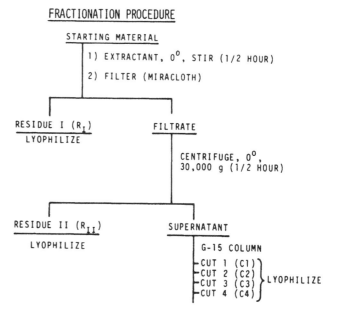

M.L. Fishman is a Research Chemist at the USDA-SEA-AR Eastern Regional Research Center, and F. McHan is a Microbiologist and D. Burdick is a Supervisory Research Chemist at the USDA-SEA-AR Richard Russell Agricultural Research Center.

Table 1.   Percent total nitrogen recovery (N_T)[1] of ryegrass treatment fractions based on micro-Kjeldahl analysis. (Extracted with and without sodium dodecyl sulfate, SDS.)

Fraction	% SDS	Control	Formic acid	Formaldehyde	Formic/formal.	Raw-wilted
Residue I (R_I)	0	41.0 ± 0.6	49.7 ± 2.0	63.4 ± 1.3	72.5 ± 4.9	65.8 ± 2.9
	1	38.5 ± 2.1	44.4 ± 2.7	58.9 ± 1.8	53.0 ± 0.6	N.D.[2]
Residue II (R_II)	0	2.7 ± 0.6	5.8 ± 1.2	7.6 ± 2.0	12.9 ± 0.5	17.2 ± 0.0
	1	6.1 ± 1.9	11.9 ± 1.8	11.6 ± 0.3	26.5 ± 0.1	N.D.
Cut 1 (C1)	0	1.7 ± 0.1	2.2 ± 0.1	3.2 ± 0.9	1.9 ± 0.2	10.1 ± 2.9
	1	4.8 ± 0.1	9.2 ± 1.8	5.7 ± 0.1	4.5 ± 1.0	N.D.
Cut 2 (C2)	0	20.3 ± 6.7	19.7 ± 1.3	12.8 ± 1.2	9.7 ± 0.9	7.4 ± 2.3
	1	25.7 ± 1.0	21.5 ± 1.5	18.0 ± 1.4	11.1 ± 1.6	N.D.
Cut 3 (C3)	0	25.7 ± 7.9	16.2 ± 0.5	12.1 ± 2.0	6.7 ± 0.3	3.2 ± 0.1
	1	17.5 ± 0.9	14.7 ± 0.1	6.9 ± 1.9	3.1 ± 1.1	N.D.
Cut 4 (C4)	0	0.00	0.00	0.00	0.00	0.9 ± 0.6
	1	1.3 ± 0.5	1.0 ± 0.0	0.00	0.4 ± 0.0	N.D.
Total	0	91.5 ± 1.4	93.6 ± 3.8	94.1 ± 0.7	103.9 ± 6.1	104.5 ± 3.0
	1	94.3 ± 4.4	102.8 ± 3.6	101.1 ± 2.1	98.6 ± 1.2	N.D.

[1]Duplicate analysis.

[2]N.D. - Not determined.

1977, Fishman 1980). First moment apparent molecular weights, M_{A_1} (Fishman 1976), were calculated from analytical gel chromatograms.

RESULTS

Soluble cuts eluted in the expected molecular weight order, with cut 1 averaging 4 to 2 × 10³ daltons, cut 2 averaging 1.4 to 0.8 × 10³ daltons, and cut 3 averaging 0.4 × 10³ daltons or less.

The percentages of starting nitrogen in each fraction N_T extracted with and without SDS are in Table 1. The percentage changes in N_T for the various fractions and ensiling procedures ($\Delta\%N_T$) over that for the comparable fractions of raw-wilted grass were calculated from the data in Table 1 and are in Table 2. A negative value of $\Delta\%N_T$ indicates a lower value of N_T in the silage fractions than in the comparable raw-wilted fractions, and a positive value of $\Delta\%N_T$ indicates the opposite. In the order control (C), formic acid (FA), formaldehyde (FE), formic acid–formaldehyde (FA/FE), the insoluble fractions R_I and R_{II} increased in $\Delta\%N_T$, whereas the soluble peptide fractions C2 and C3 decreased in $\Delta\%N_T$ (Table 2). The

Table 2.   Percentage change ($\Delta\%N_T$) in percentage total nitrogen (N_T) for various silage fractions over that for comparable raw-wilted fractions.

Fraction	Control	Formic acid	Formal-dehyde	Formic/Formal.
Residue I (R_I)	− 37.7	− 24.4	− 3.6	10.1
Residue II (R_II)	− 84.3	− 66.3	− 55.8	− 25.0
Cut 1 (C1)	− 83.1	− 78.2	− 67.6	− 81.1
Cut 2 (C2)	174.	166.	72.9	31.1
Cut 3 (C3)	703.	406.	116.	109.

soluble protein fraction, C1, was slightly higher with the formaldehyde treatment than with the others and was appreciably lower in N_T in all treated samples than in the raw-wilted grass (Tables 1 and 2). At the levels employed here, formaldehyde is about twice as effective as formic acid in preventing proteolysis, and the effects of the two in combination are approximately additive.

For more direct study of the effects of the various preensiling treatments on the hydrophobic proteins, the silages were extracted with buffer containing 1% SDS. Generally, SDS in the extractant increased the percentage of starting N in R_{II}, C1, and C2, and decreased it in R_I and C3 (Table 1).

Interestingly, the M_{A_1} value at 206 nm for buffer-extracted raw-wilted ryegrass (4 ± 2 × 10³ daltons) is only somewhat higher than the M_{A_1} values for buffer-extracted silage (2.0 ± 0.2 daltons). In contrast, buffer-extracted raw-fresh ryegrass has an M_{A_1} value of 28.7 × 10³ (Fishman et al. 1980); thus, it appears that wilting induces cytoplasmic proteolysis but not to the same extent as ensiling. Hence, the buffer-extracted raw-wilted ryegrass has a much higher value of N_T for cut 1 than any of the silages do. Furthermore, M_{A_1} for raw-wilted grass is somewhat higher at 206 nm than at 254 nm (i.e., 4.0 ± 2.0 against 2.0 ± 0.2 daltons), whereas the two values are virtually identical for the ensiled grasses (i.e., about 2.0 ± 0.2 daltons). It has been established that M_{A_1} is much larger at 206 nm than at 254 nm for raw-fresh Coastal bermudagrass (Fishman and Burdick 1977), whereas the values are virtually the same for ensiled Coastal (Fishman 1980). Raw-fresh ryegrass, also, has a much larger value of M_{A_1} at 206 nm than at 254 nm (Fishman et al. 1980). Differences between raw-wilted grass extracts at 206 nm and 254 nm and those between raw-wilted grass extracts and ensiled grass are shown in typical chromatograms of the entire apparent molecular weight distribution in Fig. 2.

Fig. 2. Typical gel chromatograms of cut 1 from silage.

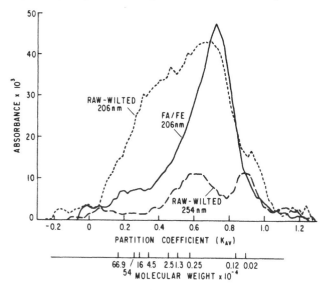

DISCUSSION

Our results are consistent with the hypotheses that formic acid prevents proteolysis by inhibiting microbial growth whereas formaldehyde protects by crosslinking and insolubilizing proteins. Thus, with detergent extraction, the formic acid–treated silage gives a considerably higher cut 1 value of N_T (Table 1) than either of the formaldehyde-treated silages. Moreover, total nitrogen solubilized by detergent is considerably less for formaldehyde-treated silage than for formic acid–treated silage.

New details of the mechanisms by which formaldehyde and formic acid protect proteins have been revealed. At the levels employed here neither formic acid nor formaldehyde appears to provide much protection for hydrophillic proteins normally soluble in the plant cell. Formaldehyde may not be present in sufficient quantity to kill significant numbers of anaerobes by crosslinking their constituent proteins, or it may kill them so slowly that the anaerobes still are able to degrade soluble protein. Furthermore, the quantity of formic acid may not be sufficient to inhibit the activity of the anaerobes until they have degraded most of the soluble protein. Still another possibililty is that extracellular enzymes produced by the anaerobes prior to their inhibition are degrading the soluble protein. Although not by the same mechanism, formaldehyde and formic acid both reduce extractability and thus the degradability of the hydrophobic polypeptides and proteins associated with cell membranes and the plant matrix.

Formic acid and formaldehyde, when used to treat grass prior to ensiling, appear to complement each other in their protective action. Formic acid, by preventing anaerobes from attacking hydrophobic proteins, increased protein availability for crosslinking by formaldehyde. Thus, the N in silage pretreated with formic acid–formaldehyde mixtures is no more soluble when extracted with detergent than the N in raw-wilted grass extracted with buffer containing no detergent.

LITERATURE CITED

Fishman, M.L. 1976. Semiautomated gel chromatography to characterize broad molecular weight distributions of cytoplasmic extracts from Coastal Bermuda grass. Anal. Biochem. 74:41–51.

Fishman, M.L. 1980. Extractability, solubility, and molecular size distribution of the nitrogenous constituents in Coastal Bermuda grass silage. J. Agric. Food Chem. 28:496–500.

Fishman, M.L., and D. Burdick. 1977. Extractability, solubility, and molecular size distribution of nitrogenous constituents in Coastal Bermuda grass. J. Agric. Food Chem. 25:1122–1127.

Fishman, M.L., E.L. Robinson, and D. Burdick. 1980. The nitrogenous constituents in annual ryegrass. Abs., 2d Chem. Cong. N. Am. Cont., Las Vegas, Nev. AGFD 29.

McHan, F., R. Spencer, J. Evans, and D. Burdick. 1979. Composition of high and low moisture Coastal Bermudagrass ensiled under laboratory conditions. J. Dairy Sci. 62:1606–1610.

Effects of Microwave Treatment on Drying and Respiration in Cut Alfalfa

S.A. SEIF, D.A. HOLT, V.L. LECHTENBERG, and R.J. VETTER
Purdue University, West Lafayette, Ind., U.S.A.

Summary

After forage crops are cut mechanically, they usually undergo some degree of natural drying before being stored or further processed. During early stages of drying, plant respiration continues. Treatments or management practices that hasten drying and reduce postharvest respiration are desirable, because they minimize quantity and quality losses. Our objective was to determine the effects of brief microwave treatment on subsequent natural drying rate and respiration of alfalfa (*Medicago sativa* L.).

Duplicate samples of field-grown Apollo alfalfa were subjected to focused microwave energy (2,450 MHz) generated at 0.25, 0.50, or 1.0 kilowatt of power for 3, 6, or 12 seconds. Control and microwave-treated samples were dried at 0% relative humidity in a drying chamber. Moisture percentage and respiration rates were measured immediately before and after microwave treatment and 18 hours after treatment.

A hypothetical example is given to show that small reductions in initial moisture percentage of forage are associated with relatively large water losses. Conversely, relatively small amounts of water must be removed to dry a sample that is already less than 50% moisture.

Increasing power and time of exposure to microwave energy increased the amount of water lost during the treatment and decreased percentage of moisture and respiration rates in samples measured immediately after treatment. The initial treatment removed 6% to 37% of the original leaf-sample water, depending on power and exposure time. Respiration was reduced slightly by weaker treatments but completely inactivated by longer, higher-power treatments.

All except the weakest microwave treatment resulted in samples that contained less than one-half as much moisture as the controls after 18 hours at 0% relative humidity. By 18 hours, posttreatment respiration rates of treated samples were one-half or less than one-half of the respiration rates of control samples, most of the treatments having completely or almost completely inactivated respiration by that time. Stem drying and respiration rates were more labile to microwave treatment than those of leaves.

We conclude that brief, focused microwave treatment removes substantial amounts of moisture from alfalfa leaves and stems, results in lower total drying time, and reduces or stops respiration. The treatments imposed had a range of effects, and the resulting data can serve as a base for future calculations of power requirements and economic thresholds.

KEY WORDS: microwave, alfalfa, *Medicago sativa* L., drying, respiration.

INTRODUCTION

Virtually all of the world's production of harvested forage undergoes some natural or artificial drying. Until the material is dry or ensiled, its metabolism continues, consuming some of the dry matter. The dry matter lost in this manner is the most valuable portion of the biomass, totally available to ruminant animals or fermentation microorganisms, and, if appropriately extracted, to nonruminant animals.

In the parts of the world where production of harvested forages is highest, conditions are generally too rainy and humid for ideal drying of this material. In this paper we report on the effects of microwave treatment on subsequent drying rate and postharvest respiration of alfalfa leaves and stems.

Hofman (1965), on the basis of preliminary experiments with a commercial microwave cooking range, concluded that microwave drying could be particularly useful where rapid killing and dehydration of herbage at a fairly low temperature (compared to oven drying) are desired, as would be the case when preparing samples for carbohydrate analysis. Jones and Griffith (1968), using a slightly modified commercial microwave cooking oven, were able to dry 400 g of fresh herbage in 15 to 20 minutes. Wolf and Carson (1973) reported that microwave heating of alfalfa herbage for 30 seconds inactivated respiration. Microwave treatment followed by low-temperature oven drying resulted in samples apparently as well preserved as those that had been freeze dried. Researchers agree, in general, that greater respiratory losses occur in cut material before it has dried below 40% moisture (Carter 1960, Wolf and Carson 1973). Russian workers, however, arrived at the opposite conclusion (Blagoveshensky at al. 1978).

The energy requirements of microwave generation would probably make it prohibitively expensive to dry forage completely with microwave energy in a commercial forage production. Microwave experiments to date have not been designed to investigate the effects of short exposures of forage to focused microwaves at various power levels nor to define the lower limits of power and time of exposure required to hasten subsequent natural drying and/or inactive respiration.

The objective of the study reported here was to study the effect of short exposures to microwaves generated at various power levels on drying and respiration rates in alfalfa.

MATERIALS AND METHODS

Apollo alfalfa (*Medicago sativa* L.) was grown on the Purdue Agronomy Farm on a Chalmers silty clay loam (Typic Haploquall, fine-silty, mixed, mesic); it had been fertilized with 200 kg/ha P_2O_5 and 600 kg/ha K_2O annually since its establishment in 1976. On 6 August 1980, 5 g alfalfa leaf and stem samples were placed in 8-cm styrofoam cups and exposed to microwave treatments. Stems had been cut in 3-cm lengths.

The microwave treatment device was described in detail by Medeiros et al. (1978). For purposes of this experiment, the microwave generator (Hl-1200, Holaday Industries, Hopkins, Minnesota) produced 0.25 or 1.0 kW of 2,450 MHz continuous-wave microwave radiation, which was transmitted to the sample chamber by a wave

S.A. Seif is a Graduate Student in Agronomy, D.A. Holt is a Professor of Agronomy, V.L. Lechtenberg is a Professor of Agronomy, and R.J. Vetter is a Professor of Bionucleonics at Purdue University. This is a contribution of the Department of Agronomy (Journal Paper No. 8516) of Purdue University and the Purdue University Agricultural Experiment Station.

Mention of a proprietary product does not imply endorsement of it.

guide (WR-284). Samples were placed on a rotating (5 rpm) platform in the treatment chamber. The treatments consisted of a factorial combination of leaves and stems, two power levels, and two exposure times, namely, 3 and 12 seconds.

Control and treated samples were placed in small 4 × 10-cm copper-screen containers. These containers, in turn, were placed in a drying chamber constructed by sealing the top of a large glass chromatography tank. The plexiglass lid fabricated for this purpose was fitted with a small fan to circulate air around the samples. A relative humidity (RH) of approximately 0% was maintained in the drying chamber by placing a dessicant ($CaCl_2$) in the bottom of the tank. The chamber temperature remained at about 25°C throughout the experiment.

Immediately before and after microwave treatment and 18 hours after treatment, the samples were weighed and their respiration rates were measured. Respiration rates were measured by monitoring the difference between CO_2 concentration in air entering and leaving small, sealed plexiglass chambers, each large enough to hold one alfalfa sample in its screen container. A Beckman infrared CO_2 analyzer was used for this purpose. After the experiment, samples were freeze dried to determine dry weight and original moisture content.

This experiment was analyzed as a split plot, with time of measurement as the whole plot. There were 2 replications of each treatment.

RESULTS AND DISCUSSION

It is useful to note, in interpreting the following results, that much more water must be lost to reduce the percentage of moisture of a moist sample 10 percentage units than to reduce a drier sample 10 units. For example, 43% of the total water in a sample must be removed to reduce the sample from 80% to 70% moisture, but only 5% to reduce it from 30% to 20% moisture. To the extent that the nature of the sample prevents the water from leaving the sample as freely as from a free water surface, this resistance becomes a significant limiting factor, and the pattern of moisture loss is altered. The results of our measurements on the drying process in microwave-treated alfalfa leaves and stems will be discussed in this context.

Increasing the power of microwave treatment of alfalfa leaf and stem samples from 0.25 to 1.0 kW caused more water to be lost during the microwave treatment and resulted in samples with lower moisture content 18 hours later (Tables 1 and 2). Increasing the time of exposure to microwave treatment from 3 to 12 seconds likewise increased initial water removal and decreased percentage of moisture of samples 18 hours posttreatment. In this experiment increasing exposure time caused greater differences in these parameters than increasing power.

All leaf samples started at approximately 80% moisture. After treatment, leaves subjected to shorter, lower-power treatments were still 76% to 78% moisture, while leaves exposed to the longest, highest-power treat-

Table 1. Percent moisture in control and microwave-treated alfalfa leaf and stem samples.†

Part	Treatment		% Moisture in Sample		
	KW	Seconds	Before	After	18 Hours
Leaves	0	0	80	78	63
	0.25	3	80	78	55
		12	80	76	34
	1.0	3	80	76	41
		12	79	70	28
Stems	0	0	85	83	63
	0.25	3	84	82	52
		12	84	82	29
	1.0	3	85	83	40
		12	85	75	22
	LSD	1		3	5

†Intermediate power and exposure time are not presented in Table 1 and 2 for sake of brevity.

ment were 70% moisture (Table 1). The former had lost only 6% to 10% of their moisture, but the latter had lost 37% of the original moisture. A similar pattern is evident in stem samples. This pattern illustrates that a large amount of water must be removed from a fresh sample to decrease substantially its moisture percentage. More water was removed from stem samples than from leaf samples by microwave treatment.

Even the weakest microwave treatment resulted in samples that were drier than the controls 18 hours later (Table 1). Samples with less than 44% moisture 18 hours posttreatment contained less than one-half as much water as the controls. All except the weakest microwave treatment caused at least this difference in moisture content 18 hours posttreatment.

The stronger microwave treatments denatured the respiratory mechanism totally or nearly so, immediately (Table 2). The weaker treatments had less immediate effect, but by 18 hours posttreatment they had reduced respiration rates to about one-half that of the controls. Increased power and longer exposure times seemed to be about equally effective in reducing respiration rates. The respiration rates recorded in this study were not high enough to represent a practically important dry-matter loss over the drying period, but in a field situation with

Table 2. Percent of total water lost during microwave treatment and respiration rates of microwave-treated alfalfa.

Part	Treatment		Water Lost	Relative Respiration		
	KW	Sec.	%	Before	After	18 Hr.
Leaves	0	0	2	100	85	59
	0.25	3	6	100	82	29
		12	10	100	53	6
	1.0	3	11	100	29	0
		12	37	100	6	0
Stems	0	0	5	100	88	14
	0.25	3	6	100	64	0
		12	16	100	0	7
	1.0	3	10	100	21	0
		12	42	100	0	0
	LSD		4		6	6

higher temperatures, these losses can be substantial (Dale et al. 1978).

Respiration rates in untreated stem samples averaged 82% of rates in untreated leaf samples. Stem respiration seemed to be somewhat more sensitive to microwave radiation than leaf respiration, possibly because the higher proportion of water in the stem couples more microwave energy. In only two of the treated stem samples was there any measurable respiration 18 hours posttreatment.

The reason the control material is not shown as having 100% original respiration rate is that some time elapsed between the original respiration measurements and the completion of the microwave treatments, at which time respiration rates were measured again. By that time, the respiration rate in control samples had decreased somewhat. All respiration data in Table 2 are expressed in units relative to the original respiration rates of material measured before the microwave treatments.

The results of the microwave experiment indicate that microwave treatment lasting only 3 or 12 seconds can remove substantial amounts of plant moisture and result in lower total drying time. Also, such treatments reduce or stop the respiratory process in alfalfa leaves and stems. The treatments we imposed had a range of effects, from relatively little reduction in drying time or respiration rate to halving the drying time and stopping respiration completely. These data should make it possible, therefore, for interested persons to estimate the power required to bring about a wide range of desired results in the alfalfa drying process.

LITERATURE CITED

Blagoveshensky, G.V., M.A. Maryna, and V.M. Sokolkov. 1978. Efficiency of methods of grass treatment before drying. Proc. 2d Int. Green Crop Drying Cong., Univ. of Saskatchewan, Saskatoon, Canada. Aug. 1978.

Carter, W.R.E. 1960. A review of nutrient losses and efficiency of conserving herbage as silage, barn dried hay, and field cured hay. J. Brit. Grassl. Soc. 15:220–230.

Dale, J.G., D.A. Holt, and R.M. Peart. 1978. A model of alfalfa harvest and loss. ASAE Tech. Paper 78-5030.

Hofman, M.A.J. 1965. Microwave heating as an energy source for the pre-drying of herbage samples. Pl. and Soil 23:145–147.

Jones, D.I., and G. Griffith. 1968. Microwave drying of herbage. J. Brit. Grassl. Soc. 23:202–205.

Medeiros, H., A.W. Kirleis, and R.J. Vetter. 1978. Microwave inactivation of thioglucosidase in intact crambe seeds. J. Am. Oil Chem. Soc. 55:679–682.

Wolf, D.D., and E.W. Carson. 1973. Respiration during drying of alfalfa herbage. Crop. Sci. 13:660–662.

Alternative Treatment Systems To Increase the Drying Rates of Green Forage Crops

W.E. KLINNER and O.D. HALE
National Institute of Agricultural Engineering,
Bedford, England, UK

Summary

Forms of conditioning treatment are being developed that aim to minimize the barrier effect of the crop cuticle without sacrificing too much of the structural strength of the herbage. Effective surface abrasion and favorable windrow structure can be achieved with steel-spoked rotors, which are now beginning to be offered commercially. More recently, plastic brushes have been developed because the higher population density and yielding ability of the crop-engaging elements, plus the absence of metal components, appeared to offer additional advantages.

Four different brush conditioner systems have been evaluated in the laboratory and in replicated field experiments. Tufted brushes using trilobar filaments lifted and conveyed the crops more effectively than full brushes and carried the conditioning effect deeper into the crop layer. With regard to improvement in rate of crop drying and digestible dry-matter yield, levels of effectiveness have been recorded that exceeded those of commercial conditioner systems.

Cost of the plastic elements is low. Durability, already near acceptance by farmers, has the potential of being further improved substantially.

KEY WORDS: mower conditioner, hay conditioning, plastic-brush conditioner, drying-rate improvement, green-crop treatment, forage.

INTRODUCTION

Severe conditioning to increase crop drying rates can bring disadvantages in terms of uneven drying and high dry-matter and nutrient losses. In view of this problem the concept of conditioning by surface abrasion was developed some years ago at the National Institute of Agricultural Engineering (NIAE) in England. It aims to minimize the barrier effect of the crop cuticle and thereby hasten the rate of moisture diffusion to atmosphere without destroying the crop structure and natural protection against wetting. Conditioning rotors using V-shaped steel elements devised for this purpose (Klinner 1976) are now finding favor also outside Europe (Sawyer et al. 1979). Ongoing R&D work has the objective of increasing treatment effectiveness further and improving the efficiency of energy use. One approach that is showing considerable promise is to scuff or stab the crop with stiff plastic brushes. Advantages are obtained from the much higher population density and yielding ability of the crop-engaging elements and the absence of metal components that could become detached and subsequently damage forage harvesters.

DESIGN AND FUNCTION OF BRUSH CONDITIONERS

Four different brush conditioner systems have been developed (Fig 1). All are suitable for combining with mowers or for use as separate pickup conditioners. In systems A to C the rotary brushes may be either of the full cylindrical type or of open construction made up of damage-protected tufts on transverse mounting bars. Much of the abrasive effect is achieved during acceleration of the crop and by the action of cooperating brushes so designed or driven that the fast-moving crop stream is retarded on one side relative to the other. Brush peripheral speed and the magnitude of the speed differential mainly determine the severity of treatment. Other variables include the clearances for crop to pass, the angular orientation of the static concave brushes (which are shown in Fig. 1 only in A but which may be added optionally to all other arrangements), and the acuteness of the bend in the front cover of C. System C is designed to carry the treatment through into the center of the crop layer by splitting it horizontally into two thin streams; therefore, system C should be particularly suited to heavy or windrowed crops. The intermeshing open brushes in system D must be synchronized but may be given different angular timing and peripheral overlap. When this arrangement is used without static brushes at the front, the lower primary brush accelerates the crop positively; in the overlap zone the material is subjected to spiking, crimping, scrubbing, and longitudinal distortion. Across the width of the brushes there are no "shadow" zones for

W.E. Klinner is a Principal Scientific Officer and Head of Department and O.D. Hale is a Higher Scientific Officer and Joint Project Leader at the National Institute of Agricultural Engineering.

Fig. 1. Operating principles of NIAE brush conditioner systems.

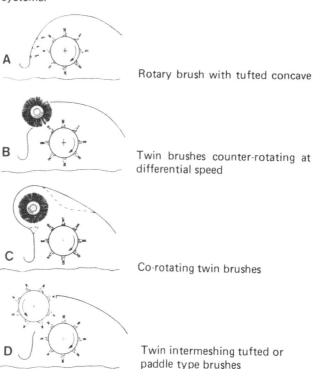

Rotary brush with tufted concave

Twin brushes counter-rotating at differential speed

Co-rotating twin brushes

Twin intermeshing tufted or paddle type brushes

appreciable quantities of crop to pass untreated, as with some rigid roll conditioners. In the three twin-brush systems the secondary brushes may be arranged in different angular relationships to the 11-o'clock position shown. The plastic filaments are made from inexpensive co-polymer polypropylene or nylon 66; in cross section they have a diameter of 6 mm and are trilobar.

LABORATORY EVALUATION

The results of initial evaluations of experimental brush conditioner systems have already been reported (Klinner and Hale 1980). More recently work has been carried out with brushes of improved design.

Apparatus and Procedure

To speed up development progress, extensive use is made of the NIAE conditioning rig shown diagrammatically in Fig. 2. It consists essentially of a 3.5-m-long trolley on which a simulated, reduced-width windrow can be placed. The trolley can be driven forward so that the crop is delivered into an equally reduced-width conditioning mechanism. This mechanism can take any form required; it treats the crop and replaces it onto the trolley so that samples can be taken from different points along the length of the windrow. For determining the effectivess of the treatment, samples are arranged in a thin layer on mesh trays for drying on a 12-unit laboratory drier. This system enables replicates to be processed simultaneously

Fig. 2. Laboratory conditioning rig: 1—trolley with simulated windrow, 2—linear motor, 3—conditioning brushes, 4—crop deflector.

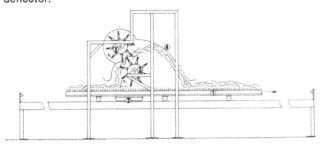

alongside untreated control material. The air supplied to the drying trays is evenly distributed and of constant temperature (24°C) and relative humidity (50%). By periodic weighing of the samples the drying curves for the materials are determined. Facilities are available for filming and measuring the power requirement of the conditioning unit on the rig.

Laboratory Results

Table 1 gives the greatest recorded reductions in crop-drying time above the equilibrium moisture content, together with the power requirements measured with all brush conditioning systems during 1980. The figures relate to crop through-put rates of 30 metric tons (t)/hour.

In the first instance the four systems were used normally, i.e., at speeds and clearances that gave visibly adequate conditioning and, in the case of tufted brushes, at a lateral tuft spacing of at least twice the tuft width. At those settings the peripheral speeds of the primary rotors were usually in the region of 22m/second.

All the brush systems are appreciably more effective than the steel-spoke rotor, the order of the brush systems being C, B, and D, and A (Table 1). However, during work with differing crops at different stages of maturity it was noticed that relatively little fragmentation occurred. Therefore the aggressiveness of the brushes was increased by reducing clearances to a minimum, increasing peripheral speeds by around 10%, and closing the lateral tuft spacing as much as possible. With these settings the results given under "Maximum effect" were obtained.

Table 1. Drying time decrease relative to untreated crop (%) and power requirement (kW) of conditioning systems.

System	Normal use		Maximum effect	
	%	kW	%	kW
Brush A	42.9	4.1	55.6	5.5
Brush B	50.7	4.6	53.5	5.8
Brush C	55.0	8.8	74.9	11.8
Brush D	54.5	5.0	61.5	7.7
Steel spokes (reference)	33.3	3.2	—	—

The least additional improvement was achieved with brush system B and the most with system C. Overall, the levels of maximum decrease in drying time recorded were high, and it is probably that further increases in peripheral speed and the use of even stiffer filaments will bring further improvement. The steel-spoke conditioning system used for reference purposes was not changed, to maintain the reference level of treatment. Brush system C had a power requirement that was disproportionately high relative to the other systems.

FIELD EVALUATION

Equipment and Procedure

Development trials and some detailed comparisons were carried out during the summer of 1980 with brush systems A and D each fitted to a 2-drum mower and with a trailed pickup conditioner capable of being fitted optionally with systems A, B, or C.

On two occasions systems A, B, and D were compared in plot experiments with the conventional system of tedding the crop periodically after mowing and with two commercial mower conditioners, namely, a disk mower with cleated rubber rolls and one with a steel-spoke rotor. The crops used for the experiments were Italian ryegrass (IR) (*Lolium multiflorum* Lam.) and perennial ryegrass (PR) (*Lolium perenne*). Weather conditions were changeable, with 20.4 mm of rain falling during the 5-day IR experiment and 0.45 mm during the 7-day PR experiment. Each of the treatments was replicated 4 times. The plots were sampled periodically for moisture and nutrient analysis, and dry-matter yields were determined at the beginning and end.

Results of Plot Experiments

To the 60% moisture level (wet basis), when crops are often harvested as wilted silage, the material treated by the simple brush system A dried the fastest in both crops. In one or the other crop the remaining brush systems ranked second and fourth and the two commecial mower conditioners third and fifth.

To the 35% moisture level (wet basis), when early baling can take place for barn drying or with the addition of preservatives, brush system D gave the greatest advantage in both crops. The ranking order of the other four treatment systems in IR and PR, respectively, was: brush system A, fifth and fourth; brush system B, third and second; steel-spoke conditioner, fourth and third; and cleated rubber rolls, second and fifth.

Total field losses of dry matter are given in Table 2. The crop of IR was the younger, and therefore loss levels were generally higher. In PR brush system A caused a significantly lower loss than any other form of treatment; apart from this, differences between systems were small.

Particularly when rain has fallen between mowing and baling, which was the case during both plot experiments, the differences in organic-matter digestibility of the

Table 2. Field losses incurred by machine treatment systems.

Treatment system	Loss, % dry matter	
	Italian ryegrass	Perennial ryegrass
Tedder (control)	23.6	19.5
Brush system A	23.2	9.7
Brush system B	23.3	16.5
Brush system D	21.3	18.8
Steel spoke rotor	22.1	16.2
Cleated rubber rolls	21.9	16.4

harvested hay are of interest. Analysis of digestible organic matter in the dry matter (DOMD), based on fiber analysis, gave no statistically significant differences between treatments.

Functional Performance

The structure of the windrows made with the brush conditioners, particularly system D, was loose and open and appeared to remain so noticeably longer than that of the windrow left by the crushing rolls.

Brief trials in alfalfa with brush system A have indicated that the amount of leaf detached from the plants during treatment is not very different from that when a steel-spoke conditioner is used. In a heavy crop of grass with approximately 50% red clover, brush system D appeared to cause very little fragmentation; this observation also applied to areas within the same field where the percentage of clover was considerably higher.

Of the two types of brush, the tufted version was more effective and more positive as a crop-conveying device than a full brush; it caused moderately severe conditioning and carried the treatment effect to a substantial depth into the crop layer. By contrast, full brushes caused very severe but shallow conditioning, probably because of the much greater population density of abrasive filament tips. For these reasons tufted brushes will be used in future exclusively as primary rotors in all four systems. The best application for a full brush is in system B where, as a secondary brush at low peripheral speeds, it acts as an effective crop-retarding device.

LITERATURE CITED

Klinner, W.E. 1976. A mowing and crop conditioning system for temperate climates. Amer. Soc. Agric. Engin. Trans. 19:237–241.

Klinner, W.E., and O.D. Hale. 1980. Engineering developments in the field treatment of green crops. Proc. Eur. Grassl. Fed. Conf., Brighton, 1979. Brit. Grassl. Soc. Occ. Symp. 11:224–228.

Sawyer, B.G., R.C. Gold, P. Illy, and R.D. Stephenson. 1979. John Deere 1320 mower-conditioner with rotary cutting and impeller conditioning. Am. Soc. Agric. Engin. Papers, Paper No. 79–1621.

Hastening Hay Drying

J.W. THOMAS, T.R. JOHNSON, M.A. WIEGHART, C.M. HANSEN, M.B. TESAR, and Z. HELSEL
Michigan State University, East Lansing, Mich., U.S.A.

Summary

The objectives of these studies were to increase the drying rate of cut alfalfa (*Medicago sativa*) by treating with chemical solutions and to examine factors that influence drying. Time to dry and drying rate of cut alfalfa were studied under laboratory and field conditions. Treatment consisted of spraying the cut alfalfa with methyl esters of long-chain fatty acids, an emulsifier alone, or a combination of an emulsifier and a potassium carbonate solution. Alfalfa sprayed with solutions of each chemical under specified conditions increased drying rate and decreased drying time to 60% or 75% dry matter (DM) or average percentage of dry matter during the trial. The combination of 0.2M potassium carbonate plus 2% to 4% methyl ester plus 0.25% emulsifier produced faster drying than either component of that concentrate alone. A successful application rate of that solution was 4% of the weight of the fresh-sprayed alfalfa.

Grass (*Bromis inermis*) had faster drying rates than did alfalfa, but the three-component spray solution did not enhance the drying of grass as it did that of alfalfa. Mature alfalfa dried faster than alfalfa in the one-fifth bloom stage. Field trials showed that spraying would hasten drying of alfalfa at both growth stages. Heavily crimped alfalfa dried more quickly than did lightly crimped alfalfa. When the crimped alfalfa was sprayed, drying was hastened, and drying rate for the two crimpings became about equal.

The drying measurements were made by weighing the treated alfalfa on a screen at intervals as it dried in the field or laboratory. Under field conditions significant correlation coefficients between water content (weight of water divided by weight of dry matter) and other variables at time "T" were (a) time since cutting (-0.82), (b) amount of sunshine (-0.24), (c) vapor-pressure deficit (-0.20), (d) initial moisture content ($+0.29$), and (e) relative humidity ($+0.19$). Multiple regression equations indicated that methyl esters and potassium carbonate hastened a decrease in moisture content and that the laboratory and field trials were from different sample populations probably because of differing environmental regimes during drying. In normal field-drying hay three independent variables (relative humidity, initial moisture content, and time since cutting) accounted for 75% of the variation in moisture content. Inclusion of eight to ten variables in the regression equation increased R^2 to 0.82. Spray treatment allowed improved quality of hay to be made. This type of approach should provide guidance in evaluating factors influencing the drying rate of forages.

KEY WORDS: alfalfa, drying, hay quality, methyl esters, potassium carbonate, drying forages.

INTRODUCTION

The quality of harvested forage dictates the success of dairy-cattle enterprises. High-quality forages will probably increase in importance for animal enterprises as the costs of energy used to produce crops increase. Harvested forage quality is influenced by length of time that the crop is at risk in the field between cutting and harvesting.

A promising technique of spraying with carbonates to hasten drying of alfalfa has been described (Tullberg and Angus 1978, Tullberg and Minson 1978, Wieghart et al. 1980). The objectives of these experiments were to determine the effect on forage drying rates of spraying alfalfa (*Medicago sativa*) with solutions of potassium carbonate (K_2CO_3) plus emulsified methyl esters of long-chain fatty acids (ME) under laboratory and field conditions.

METHODS

Greenhouse-grown alfalfa was cut; small amounts ($\pm$ 100 g) were placed on a plastic sheet and hand sprayed with the desired amount of the assigned solution. Unsprayed alfalfa served as control. The forage was then placed on a screen to dry and was weighed periodically until dry. A trial usually consisted of three or four treatments in quadruplicate. Application rates were expressed as amount of solution or chemical/unit of initial plant weight. Calculations of dry matter (DM) or water content (WC = g water/g DM) over time were calculated, and appropriate statistical evaluations made (Wieghart et al. 1980). Interval from cutting to attain a stated DM percentage has been used in evaluating treatment effects, along with average DM during the trial and calculated drying rates (Wieghart et al. 1980).

Field-grown alfalfa was cut with a commercial mower conditioner with moderate conditioning and left in a swath or windrow. Fresh-cut alfalfa (2 to 5 kg) was placed on screens (1.2 m²) in the same configuration as alfalfa occurred in the swath or windrow. Screens with alfalfa were weighed periodically for 2 to 3 days until field dry, and

then the entire amount of forage was oven-dried to obtain the total DM on that screen.

Treatments were imposed by spraying the alfalfa from nozzles on spray bars above and below the hay as it entered or left the conditioning rollers. A pump driven from the tractor hydraulic system was connected to a small supply tank and to the spray bars. Pressure was usually 20–30 psi. Application rate was varied by changing ground speed and was determined by weighing a portion of the swath cut and weighing the solution sprayed over the distance traveled. Treatment solutions consisted of K_2CO_3 (usually 0.2 M) plus various concentrations (2% to 10%) of ME with a commercial emulsifier (Wieghart et al. 1980).

RESULTS AND DISCUSSION

Grass Compared with Alfalfa

In two laboratory trials spraying caused very little increase in the rate at which the grass (*Bromis inermis*) dried but did increase the rate for alfalfa. Sprayed alfalfa had greater average DM and required less time to reach 60% DM than did unsprayed alfalfa. Grass dried faster than did alfalfa.

Carbonate Salts Compared

In five laboratory trials and one field trial, alfalfa treated with single-ingredient solutions of K_2CO_3 or sodium carbonate (Na_2CO_3) was compared with untreated alfalfa. In one trial drying was significantly faster for the K_2CO_3-sprayed alfalfa. In three trials there was a slight advantage for the K_2CO_3-sprayed alfalfa, and in one trial the Na_2CO_3-sprayed alfalfa dried somewhat faster. In all but one laboratory trial the carbonate-sprayed alfalfas dried significantly faster ($P < 0.01$ or < 0.05) than did the controls. In the one field trial the average DM for the K_2CO_3-sprayed alfalfa was more than that for the Na_2CO_3 ($P < 0.05$), and both were above that for control.

Various Carbonate Salts Combined with Methyl Esters

In four laboratory trials the Na or K carbonates were compared in a ME-plus-emulsifier solution. In three trials

J. W. Thomas is a Professor of Dairy Nutrition, T.R. Johnson is a Graduate Student in Dairy Nutrition, M.A. Wieghart is a Graduate Student in Dairy Nutrition, C.M. Hansen is a Professor of Agricultural Engineering, M.B. Tesar is a Professor of Crop and Soil Science, and Z. Helsel is an Assistant Professor of Crop and Soil Science at Michigan State University.

the drying rates were identical for the K or Na carbonate solutions. Three carbonates were compared (K, Na, Li), and all improved drying (P < 0.01) over control, but the K salt was superior. Large trials on the cost effectiveness of K or Na carbonates singly and combined appear desirable.

Methyl Esters or Emulsifier Used Alone

Alfalfa sprayed with ME plus emulsifier in one laboratory trial had DM values of 57% compared with the control's values of 30% after 24 hours. The application rate was high — 10% of a 3.7% ME solution giving about 4 g ME/kg fresh alfalfa. In four trials lower rates (2% ME solution applied at a rate of 5%) of 1.0 g/kg had a reduced, nonsignificant effect on average DM, and the treated alfalfa required 41 hours to attain 75% DM, while the control required 50 hours.

Carbonate and Methyl Ester Synergistic Activity

Time required to dry untreated alfalfa to 75% DM averaged 51.8 hours in three trials, and the alfalfa sprayed with 0.2M K_2CO_3 plus 2% ME plus emulsifier (1.1 g ME/kg forage) required only 26.1 hours. Average DM was also less (P < 0.05). Alfalfa sprayed similarly with only ME plus emulsifier required 40.9 hours to reach 75% DM. When the K_2CO_3 was added to a solution with over 10% ME the addition of K_2CO_3 had little or no synergistic effect. The combination had a longer-lasting effect; i.e., it increased the drying rate when WC was low compared with that of alfalfa sprayed with single-component solutions.

Influence of Amount of Methyl Esters

Several laboratory trials indicated faster drying rates when increased amounts of ME were added to the K_2CO_3-plus-emulsifier solution. Application rate in one field trial was 2% of a 5%, 10%, or 15% ME-K_2CO_3 solution to give 1, 2, or 3 g ME applied/kg forage. Interval required to attain 60% DM was 22 hours for control and only 5.7, 5.0, and 4.7 hours for the 1, 2, or 3 grams applied. Average DM contents were 40.8%, 54.8%, 59.9%, and 58.1%, respectively. Most trials showed significantly faster drying for the 3 g ME/kg rate than for the 1-g rate. The cost of ME is much greater than that of the inorganic salts, and several large trials may be required to obtain the most cost-effective amount of ME to be used in practice.

Application Rate

In laboratory and field trials drying was faster at a 4% application rate than at rates of 1% or 2%. General results indicate that a relatively large amount of liquid sprayed on the forage hastened drying more than did smaller amounts. The most effective application rate depends on yield plus spraying and environmental condi-tions. Less than 75 to 90 l/ha has been only marginally effective.

Growth Stage of Alfalfa and Drying Rate

In four field trials, spraying with the three-component solution increased drying rate of alfalfa at the one-fifth to one-half bloom and seed stages of maturity. In all trials the seed stage had a 2% to 4% greater initial DM and dried faster than the blooming alfalfa. In two trials control alfalfa at seed stage attained 75% DM in 33 hours, while sprayed alfalfa required only 26 hours. Respective values for the one-fifth bloom alfalfa were 34 and 27 hours.

Extent of Crimping and Spraying

In two trials heavily crimped alfalfa dried faster than did that lightly crimped (P < 0.01). Spraying increases the speed of drying so that drying rate for both extents of crimping were equal. Light crimping combined with spraying would achieve faster drying and reduce field losses compared to heavy crimping.

Quality of Stored Sprayed-Alfalfa Hay

In three trials the alfalfa was baled and stored when hay sprayed with the three-component mixture attained the appropriate DM percentage for baling. Measurements were made on quality of the final hay. In one trial both hays were baled 54 hours after cutting, just before rain oc-curred, when DM of sprayed hay was 76% and that of the untreated hay 65%. Maximum storage temperature was 26.2°C for sprayed and 46.3°C for untreated bales (P < 0.001). Bales from sprayed forage were normal when cured, but those from untreated forage were moldy. Respective DM loss during storage was 7.1% and 10.2%. Changes in analytical values during storage were (1) solu-ble carbohydrate had no change in sprayed but a 39% loss in untreated; (2) ash increased from 7.93% to 8.11% for sprayed but from 8.56% to 10.03% for untreated; (3) acid detergent fiber increased from 28.6% to 29.1% for sprayed but from 29.0% to 34.1% for untreated; (4) acid detergent insoluble nitrogen as percentage of total nitrogen increased from 7.1% to 7.8% for sprayed but from 8.0% to 11.9% for untreated; and (5) in-vitro DM disappearance values decreased 1.4 units for treated but 3.3 units for untreated. All these measures support the idea that spraying can improve hay quality by shortening the interval from cutting to harvesting.

Evaluating Variables Simultaneously

The variables in laboratory and field trials were used in multiple regression equations to predict moisture content (MC) at any given time after cutting. In field trials initial moisture content and relative humidity had small signifi-cant correlation coefficients with MC of 0.29 and 0.19. Significant negative correlations between MC and other variables were −0.82 for time; −0.24 for hours of sun-shine; −0.20 for vapor-pressure deficit; −0.14 for dry-bulb temperature; −0.12 for maturity and for g ME/kg applied; and − 0.09 for ME concentration and application rate.

Analysis of data indicated that the field and laboratory equations developed were different, representing two independent populations. The important parameters related to drying in the laboratory were different from those in the field. For control alfalfa in the field the three variables—initial moisture content, relative humidity, and time—accounted for 75% of the variability in MC. The best equations (those with eight to ten variables) explained 82% of the variability. The variables used explain 88% to 90% of the variability for day 1, but only 66% to 77% for day 2. The regression coefficients for spraying treatments were always negative, indicating that the treatments (K_2CO_3 and ME) consistently reduced MC,

with ME having a greater effect than K_2CO_3.

This approach will continue to define the important variables influencing hay drying and treatments that can increase drying rates.

LITERATURE CITED

Tullberg, J.W., and D.E. Angus. 1978. The effect of potassium carbonate solution on the drying of lucerne. I. Laboratory studies. J. Agr. Sci., Camb. 91:551

Tullberg, J.W., and D.J. Minson. 1978. The effect of potassium carbonate solution on the drying of lucerne. II. Field studies. J. Agri. Sci., Camb. 91:557.

Wieghart, M., J.W. Thomas, and M.B. Tesar. 1980. Hastening drying rate of cut alfalfa with chemical treatment. J. Anim. Sci. 51:1.

Wet Fractionation of Alfalfa:
A Direct Forage-Harvesting System

M. COLLINS, R.G. KOEGEL, N.A. JORGENSEN, and H.W. REAM
University of Wisconsin, Madison, Wis., U.S.A.

Summary

In this system, termed wet fractionation, fresh forage is macerated and pressed to remove approximately 50% of the moisture. The result is a forage product containing 32%–38% dry matter (DM) that can be ensiled directly. In addition, the protein contained in the plant juice can be coagulated and removed. The remaining deproteinized juice contains large quantities of mineral nutrients that can be returned to the soil.

This process has several potential advantages: (1) less weather-dependent harvesting, which minimizes harvest losses; (2) reduced harvesting time; (3) reduced fertilizer requirements; and (4) ability to harvest protein from land unsuited to row crops because of slope.

Maceration of the crop is accomplished by extrusion through small openings, resulting in rupture of plant cells. Juice expression is required to increase the DM percentage of the crop from an initial level of 18%–20% to 32%–38%, which is necessary for proper ensiling. A cone-type press is under development to accomplish juice expression. Steam has generally been utilized to raise the temperature of the juice to 80°C to coagulate the protein. New protein-coagulation methods being studied include fermentation and acidification by mineral acids.

Pressing reduces the crude-protein concentration of alfalfa forage and increases the neutral and acid detergent fiber concentration compared with field-wilted, ensiled alfalfa. The apparent digestibility of pressed-forage dry matter averaged 61.8%, compared with 62.8% for wilted silage. When formic acid (0.6% w/w) was added to the pressed forage at ensiling, the dry-matter intake was increased to a level similar to that of wilted alfalfa silage.

Deproteinized juice from alfalfa has been found to contain considerable quantities of nutrients that can be reapplied to alfalfa or other crops as a fertilizer. Deproteinized juice from the heat-coagulation process from first-crop alfalfa grown on high-fertility soil contained 0.021% phosphorus (P), 0.232% potassium (K), and 0.012% sulfur (S). Juice deproteinized by fermentation contained 0.083% P, 0.515% K, and 0.055% S but also contained 0.40% N compared with 0.09% N in heat-deproteinized juice. Research is underway to evaluate juice from other deproteinization methods.

KEY WORDS: alfalfa, *Medicago sativa*, forage harvesting, protein, fractionation, digestibility.

INTRODUCTION

Forages constitute the major portion of the diet of ruminant animals. In geographic regions with mild winter temperatures, grazing may provide most or all of the forage in the diet. In regions with climatic limitations to year-round grazing, stored forages must provide a significant percentage of the total forage consumed.

Preservation of alfalfa (*Medicago sativa* L.) forage has traditionally been primarily as field-cured hay. Losses both of dry matter (DM) and of nutritive value may be large when rain damage occurs during field curing (Collins 1982). Direct harvest and ensiling of alfalfa avoids these field losses; however, storage losses are high and quality is poor because of high concentrations of water and protein and low levels of fermentable carbohydrate. Dry-matter losses during harvest and storage are minimized by ensiling alfalfa in the 30%–50% DM range.

A wet-fractionation forage-harvesting system, designed to achieve the increase in forage DM percentage necessary to allow proper fermentation without field wilting and to produce a protein concentrate for animal feeding, is under development at the University of Wisconsin.

THE WET-FRACTIONATION SYSTEM

Wet forage fractionation is based on mechanical removal of water from the fresh crop, thus avoiding the need for field drying. This process has several potential advantages: (1) less weather-dependent harvesting, which minimizes harvest losses; (2) reduced harvesting time; (3) reduced fertilizer requirements; and (4) ability to harvest protein from land unsuited to row crops because of slope. The forage fractionation system comprises several individual components, including (1) tissue maceration, (2) juice expression, (3) protein coagulation and separation from the plant juice, (4) nutritional evaluation of the products of the process, and (5) utilization of the deproteinized juice as a fertilizer.

Maceration and Juice Expression

Effective juice extraction requires tissue maceration as a preparatory step in order to rupture a high percentage of the cells. Koegel et al. (1973) concluded that extrusion through orifices was the best method for maceration. The macerated forage is then dewatered by mechanical pressing. The pressed forage should contain 32%–38% DM compared with 18%–20% DM in the standing crop. To reach this percentage, more than 50% of the moisture in the fresh crop must be removed (Fig. 1).

The quantity of water, and thus of protein, extracted

Fig. 1. Wet plant fractionation.

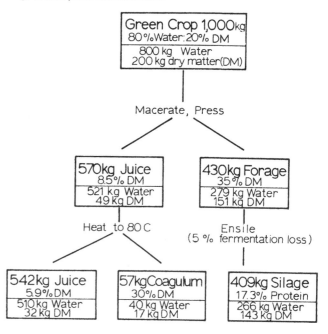

during pressing depends mainly on the pressure applied and the hold time for a given forage. Unpublished research (Collins 1980) indicates a high positive correlation (r = 0.84) between the initial DM percentage of alfalfa and the DM percentage of the resultant pressed forage when pressure and hold time remain constant.

Protein Separation from the Whole Juice

Coagulation of protein from the whole juice can be accomplished by several methods. Heating the juice to 80°C results in rapid coagulation and produces large particles that are easily separated from the deproteinized juice. Anaerobic fermentation of the whole juice reduces juice pH and results in protein precipitation. This method requires less energy than heat coagulation but would require storage capacity for large quantities of juice. The use of acids such as sulfuric and phosphoric to precipitate the protein appears to have some potential and is currently being evaluated.

Evalution of the Products of Wet Fractionation

Pressed Forage. Removal of protein during pressing results in a reduction in crude-protein (CP) concentration in the pressed forage compared with wilted silage not exposed to rain damage. Jorgensen (1977) reported CP, cell-wall constituents (CWC), and acid detergent fiber (ADF) concentrations of 18.5%, 47.2%, and 37.6%, respectively, in wilted alfalfa silage compared with 15.9% CP, 57.8% CWC, and 47.4% ADF in pressed alfalfa silage produced from the same crop. In-vivo apparent digestibility was 62.8% for the wilted silage and 61.8% for pressed forage. The similarity of digestibility in the

M. *Collins is an Assistant Professor of Agronomy, R.G. Koegel is a Professor of Agricultural Engineering, N.A. Jorgensen is a Professor of Dairy Science, and H.W. Ream is a Visiting Professor of Agronomy at the University of Wisconsin. This is a contribution of the departments of Agronomy, Agricultural Engineering, and Dairy Science of the University of Wisconsin. Research was supported by the University's College of Agricultural and Life Sciences.*

two forages was attributed to an increase in cellulose and hemicellulose digestibility in the pressed forage, presumably a result of tissue maceration.

Initial research indicated 7.6% lower milk production by cows fed pressed forage than by cows consuming wilted silage, due to lower intake of the pressed forage. Addition of formic acid (0.6% w/w) to the pressed alfalfa at ensiling resulted in similar forage intakes and similar milk-production levels for the wilted and the pressed silages (Lu et al. 1979).

Protein Coagulum. Protein coagulum from alfalfa harvested at four maturity stages was found to contain an average of 28% of the protein in the standing crop (Collins, unpublished). The value of alfalfa leaf protein concentrates is limited mainly by methionine and lysine, although the concentrations of these and other essential amino acids compare favorably with high-lysine corn protein. Protein coagulum from normal alfalfa varieties contains sufficient saponin to limit the growth of poultry and swine (Jorgensen 1977). Chicks consuming protein concentrate from normal alfalfa gained only 26% as much as chicks fed the same quantity of protein from a low-saponin alfalfa line.

The apparent digestibility of a high-saponin alfalfa protein concentrate (coagulation at 80°C, spray dried) by sheep ranged from 75.7% to 80.5% when the total diet contained 9% and 27% protein concentrate, respectively. While the high-protein concentrate did not affect apparent digestibility, bacterial N flow to the duodenum decreased with increasing levels of alfalfa protein concentrate (Lu and Jorgensen, unpublished, 1981).

Use of Deproteinized Juice as a Fertilizer. Potential uses for deproteinized alfalfa juice include (1) a medium for the growth of single-celled organisms, (2) a source of nutrients for monogastric animals, and (3) a fertilizer for crops. In on-farm operations reapplication of deproteinized juice to crops as a fertilizer appears to be most feasible. Deproteinized juice from the heat-coagulation process contains lower N concentrations than fermented or acid-coagulated juice. In addition, concentrations of P, K, Ca, S, and Mg are lower in heat-deproteinized juice than in other juice types.

A single harvest of 6,000 kg/ha of DM of alfalfa would return 17 kg N, 3.9 kg P, 43 kg K, and 2.2 kg S/ha for heat-deproteinized juice reapplied to an area equal to the harvested area. An equivalent application of fermented deproteinized juice would contain 74 kg N, 15.4 kg P, 95 kg K, and 10.7 kg S.

Application of high rates of heat-deproteinized juice, or more than 0.5-cm surface depth of fermented alfalfa juice, may result in plant death and yield reduction under field and greenhouse conditions (Collins, unpublished). Preliminary results indicate that removal of phenolic compounds from the juice can alleviate the toxicity and suggest that these compounds may be partially responsible for the observed toxicity.

LITERATURE CITED

Collins, M. 1982. The influence of rainfall during drying on the chemical composition of alfalfa, red clover, and birdsfoot trefoil. Proc. XIV Int. Grassl. Cong.

Jorgensen, N.A. 1977. Preservation and utilization of products resulting from green plant fractionation. *In* Nonconventional proteins and foods, 127–142. Natl. Sci. Found., Washington, D.C.

Koegel, R.G., V.I. Fomin, and H.D. Bruhn. 1973. Cell rupture properties of alfalfa. ASAE Trans. 16:712–716.

Lu, D.C., N.A. Jorgensen, and G.P. Barrington. 1979. Wet fractionation process: preservation and utilization of pressed alfalfa forage. J. Dairy Sci. 62:1399–1407.

Comparison of Conservation Methods Under Controlled Practical Conditions

H. HONIG, K. ROHR, and E. ZIMMER
Federal Research Center of Agriculture,
Braunschweig, Federal Republic of Germany

Summary

The most common forage conservation methods (direct-cut and low-moisture silage, barn-dried hay, and dehydration) have been compared under practical, but strictly controlled, conditions to determine their relative value for meeting the increasing demands of high-yielding ruminants. A grass mixture was used as forage, at different stages of maturity in different experiments. Parameters determined were changes of nutrient content; losses of DM and net energy; and, in three experiments, forage intake by dairy cows. From the results of the experiments, the following conclusions could be drawn: (1) roughage quality is influenced more by cuttng time than by preservation methods; (2) low-moisture silage and barn-drying of hay provide a forage equal in quality to dehydrated grass, and at equal losses; and (3) weather risk during prewilting should be reduced by new efforts to speed up this process.

KEY WORDS: silage, hay, dehydration, comparison, grass, nutrient content, losses, cutting date.

INTRODUCTION

Forage quality becomes more and more important in utilizing the increasing yield potential of ruminants, especially in dairy production. Therefore, comparisons of the most common methods of forage conservation have been carried out for several years to evaluate precisely their different effects on forage quality and system losses. Additionally, the cutting date—stage of maturity—was included as a further important factor influencing forage quality.

METHODS

The experiments were carried out on a practical scale. The first cut of a grass mixture—perennial ryegrass (*Lolium perenne*), meadow fescue (*Festuca pratense*), and timothy (*Phleum pratense*)—was the forage used, cut in one day for all treatments of one comparison. The conservation methods investigated were (1) direct-cut (DC) silage (tower silo); (2) low-moisture (LM) silage (40%–50% DM, gas-tight silo); (3) hay barn dried with heated air (5°C temperature rise); and (4) dehydrated grass (Mobile drier).

Eight comparisons covered all four methods, and two

covered only two methods at a time. Parameters determined in all experiments were nutrient content, digestibility, losses of DM, and net energy. In three experiments forage intake by dairy cows was measured.

Ten samples for crude-nutrient analysis and three samples for in-vitro digestibility determination of organic matter (OMD) were taken for each method before and after conservation. In parallel, a restricted number of in-vivo digestibility determinations with sheep were carried out, giving a set of digestibility data specific for these experiments, which were then used to calculate net energy content. Process losses were measured by input-output balance. Dry-matter (DM) losses in the field were calculated according to prewilting conditions (Honig 1980). The DM content was determined by oven drying at 105°C and corrected for volatiles.

RESULTS AND DISCUSSION

Nutrient Content at Mowing

Basis for all considerations was the nutrient content at mowing. The crude-protein content was rather low (Fig. 1); it decreased with increasing plant age from 18% to 9% DM. The content of crude fiber increased from 20% DM at the shooting stage to 24% DM at heading and to over 30% DM at full blossom.

Organic-matter digestibility and net-energy content (Fig. 1) decreased very rapidly after heading, by 10 points, from 80% to 70%, or by 1.4 mj/kg DM from 7.2

H. Honig is a Research Agronomist at the Institute of Grassland and Forage Research, K. Rohr is a Research Agronomist and Professor at the Institute of Animal Nutrition, and E. Zimmer is the Director and a Professor at the Institute of Grassland and Forage Research of the Federal Research Center of Agriculture.

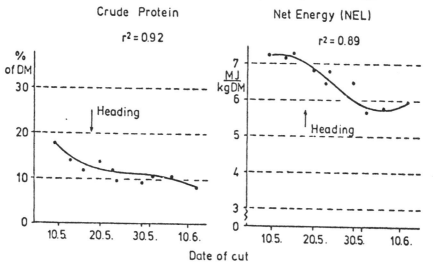

Fig. 1. Crude-protein and net energy content of forage grass at mowing as influenced by cutting date. Crude protein: $y = 29.153 - 1.787x + 0.059x^2 - 0.0007x^3$; net energy: $y = 5.852 + 0.259x - 0.014x^2 - 0.0002x^3$; x = days after 1 May.

to 5.8 mj/kg DM, respectively.

The experiments confirm the decisive importance of an early cut, which basically determines the quality of the forage.

Change of Nutrient Content During Preservation

During the experiments optimum conditions could not always be obtained. For LM silages in three of nine cases prewilting was disturbed by unfavorable weather, with up to 15 mm of rain. For hay unfavorable weather occurred in five cases. DC silages underwent suboptimum fermentation in two cases, as additives had not been used in the first experiments with the aim of interfering as little as possible with the process. The results have therefore been separately evaluated for optimum and for unfavorable conditions (Table 1). Cutting time did not significantly influence the changes in nutrient content during the conservation process and has therefore been omitted from the following discussion.

Crude-protein and Crude-fiber Content. The positive changes in crude-protein content were mainly a consequence of reduction of highly soluble nutrients in the carbohydrate fraction. When DC silage or hay was produced, the increase was not so high, as protein was lost in the effluent and by mechanical losses, as in fine leaves, during wilting. Under unfavorable conditions, the changes were not different during wilting but were clearly higher for the total process.

The changes of crude-fiber content (0.5% to 6% DM) were still higher, but with the same tendency as for crude protein, leading to crude-fiber contents of 25% to 30% DM at the optimum cutting stage.

Organic-matter Digestibility and Net Energy Content. Under favorable conditions prewilting did not affect OMD and net energy content. Silage fermentation reduced OMD and net energy more than barn hay drying. Effects of dehydration were negatively correlated with digestibility and net energy content, probably due to rather intensive drying. But on the whole, these reductions during conservation remained rather low, being much less than 10 days' delay in cutting time and leading for all methods to high final net energy content of 6.6 to 6.8 mj/kg DM at optimum cutting stage.

Greater differences may occur under unfavorable conditions. Suboptimum fermentation of DC silage caused the highest reduction of OMD and net energy measured in the experiments. Bad weather reduces OMD and net energy during wilting and also affects the conservation process adversely. Meeting the weather risk is thus the main task in making the prewilting methods safer.

Energy Losses

Total losses resulting from field and process losses are given in Table 2. Again, the influence of cutting time could be neglected.

Field losses remained small for methods using direct-cut material. Prewilting of silage was finished within one to 2 days, resulting in energy losses of 7%. Haymaking took one day longer, with losses increasing to 9.5%. Unfavorable weather conditions extended the field time to 3 days for silage but up to 7 days for hay, and thus increased field losses by 2-fold or even more.

The processing losses were highest in DC silage as a result of effluent formation and intensive fermentation, and they were lowest for hay and prewilted silage. They were rather high for dehydration, being partly due to the discussed high reduction of energy content. Unfavorable weather conditions do not affect process losses very much for LM silage and hay, but suboptimum fermentation does increase DC silage losses to a large extent.

Under optimum conditions equal total losses between 15% and 16% were reached for LM silage, hay, and dehydration. Only DC silage losses, at 25%, were clearly

Table 1. Changes in nutritive value caused by different conservation methods (material: grass).

	No. of Experiments	Crude prot., % DM	Crude fiber, % DM	OMD, %	NEL MJ, kg DM
		Optimum conditions			
Prewilting					
For silage	6	+0.7	+1.4	+0.1	0.0
For hay	4	+0.3	+0.9	+0.1	0.0
Total process					
DC silage	7	+0.5	+4.8	−2.7	−0.4
LM silage	6	+1.6	+3.3	−2.0	−0.2
Barn-dr. hay	4	+0.3	+1.6	−1.3	−0.1
Dehydr. grass	9	+0.1	+0.4	−1.1	−0.4
		Unfavorable conditions			
Prewilting					
For silage	3	+0.6	+1.5	−2.5	−0.3
For hay	5	+0.2	+2.5	−3.1	−0.4
Total process					
DC silage	2	+1.9	+6.5	−9.5	−1.2
LM silage	3	+1.9	+5.2	−4.3	−0.6
Barn-dr. hay	5	+0.7	+3.7	−3.7	−0.5
LSD 5% treatm./ condit.	−	0.7	1.4	1.9	0.2

Table 2. Net-energy losses for different conservation methods (material: grass).

Conditions and treatment	No. of Experiments	Losses %		
		Field	Process	Total
Optimum conditions				
DC silage	7	3.0	22.8	25.1
LM silage	5	7.0	8.6	15.6
Barn dr. hay	4	9.5	7.1	15.9
dehydr. grass	8	3.0	12.8	15.5
Unfavorable conditions				
DC silage	2	3.0	35.7	37.6
LM silage	3	14.0	10.8	23.3
Barn dr. hay	5	22.4	6.6	27.5
LSD 5% Treatm./condit.	−	2.1	3.0	3.2

Table 3. Influence of cutting date and ensiling method on forage utilization (material: grass).

		Cutting stage			
		Heading		Begin blossom	
		DC[a]	LM[b]	DC[a]	LM[b]
OM Digestibility	%	77		67	
Forage input	metr. t DM	7.5		7.5	
DM losses	%	23	12	18	12
Net-energy contact	MJ/kg DM	6.4	6.5	5.3	5.4
Daily silage intake	kg DM/cow	10.7	12.0	9.3	9.5
Milk from silage	kg FCM/day	11.1	13.9	5.0	5.5
Milk from input	kg FCM	5990	7650	3330	3840

[a]DC = direct-cut ensiling method.

[b]LM = low-moisture silage method.

higher. But with bad weather this order of magnitude can easily be reached by the prewilting methods.

The risk in DC silage production from very young grass must be reduced by the use of additives.

Forage Utilization

To allow a definitive conclusion on the relative preference of different methods, utilization by the animal has to be taken into account. In Table 3, the complete chain from field to milk yield was calculated for DC and LM silage at two cutting stages with a 10-day interval, using intake data determined in joint experiments. Silage intake was dependent on DM content but still more so on plant age. The lower net energy content in combination with a lower intake reduced the amount of energy available above maintenance and thus restricted milk production from silage considerably. This calculation on the basis of the described experiments emphasizes once more the enormous importance of an early cut and also the advantage of prewilting with respect to milk yield from forage.

CONCLUSIONS

Under optimum conditions equally high nutrient contents in conserved forage can be achieved by all the conservation methods, and these contents would be much higher than those found under practical farm conditions. This discrepancy may be explained first by cutting too late in farm practice but also by mismanagement of the conservation process, e.g., by poor prewilting conditions or poor ensiling technology. Losses do not differ much for the different methods except for DC silage, where they are markedly higher. Under unfavorable weather conditions losses for the prewilting methods, however, can easily reach those for DC silage; therefore, further efforts should be made to cut down this risk for the most effective conservation methods.

LITERATURE CITED

Honig, H. 1980. Mechanical and respiration losses during prewilting of grass. Occas. Symposium 11:201-204. Brit. Grassl. Soc., Hurley, Berks.

Harvesting and Processing Effects on Alfalfa Protein Utilization by Ruminants

R. BRITTON, D. ROCK, T. KLOPFENSTEIN, J. WARD, and J. MERRILL
University of Nebraska, Lincoln, Nebr., U.S.A.

Summary

Alfalfa (*Medicago sativa*) is an efficient and important means of producing supplemental protein for ruminants. This research was undertaken to characterize the value of alfalfa protein for ruminants and to show how the protein is affected by different harvesting or processing regimens. The effects of maturity (prebloom, one-tenth, one half, and full bloom) and processing method (freeze dried [FD], sun cured [SC], and oven heated [H]), on percentage of nitrogen (N), acid detergent insoluble N (ADIN), in-vitro rumen ammonia release (NH_3), and in-situ rumen protein degradation, using dacron bags (BAG), were studied on irrigated alfalfa harvested over 2 years. Plants were separated into leaves and stems after harvest or processing and analyses

were conducted on each. Nitrogen decreased (P < 0.05) and ADIN increased (P < 0.05) as plants became more mature. Leaves contained approximately 2.5 times as much N as stems. Processing (SC and H) reduced (P < 0.05) protein degradability measured by NH₃ and BAG. Heating produced greater decreases in protein degradability than SC. Leaves responded more to processing methods than stems. Two experiments were conducted to assess heat dehydration effects for steer calves (200 kg). Low-protein diets (60% corn cobs and 20% corn) were supplemented to 11.5% crude protein with urea or with combinations of soybean meal (SMB) and urea or dehydrated alfalfa (DEHY) and urea. Data for the two experiments were pooled. Daily gain (kg) and feed/gain were: 0.41, 15.23; 0.47, 12.60; and 0.54, 11.18 for urea, SBM, and DEHY, respectively. Protein efficiency values were calculated as the added gain of the test-protein animals above gains of the urea-control animals divided by the amount of test protein/day. Protein efficiency was 0.203 for SBM and 0.547 for DEHY. These values demonstrate that heat dehydration increases the value of alfalfa protein for ruminants by decreasing ruminal-protein degradability. The last set of experiments evaluated the effects of mechanical removal of solubles from alfalfa and subsequent heat dehydration on the value of alfalfa protein for ruminants. N-balance experiments with lambs showed that N digestibility of the press-cake alfalfa (PC) was reduced (P < 0.05) compared with that of regular DEHY, SBM, or urea but that N balance was not affected negatively compared with regular DEHY or SBM. The protein (P) present in the liquid from the mechanical squeezing of alfalfa can be harvested. The resultant liquid, called brown juice (BJ), is partially concentrated and can be dried on the PC. The last experiment tested the efficiency of protein utilization of PC, PC + BJ, PC + BJ + P, and regular DEHY in steer calves (10 animals/treatment). SBM and urea were included as positive and negative controls. Daily gain and feed/gain (in kg) were: urea, 0.63, 8.22; SMB, 0.66, 7.71; DEHY, 0.71, 7.81; PC, 0.77, 7.23; PC + BJ, 0.78, 7.12; and PC + BJ + P, 0.74, 7.29. Protein efficiency values were: SBM, 0.43; DEHY, 1.27; PC, 1.48; PC + BJ, 1.64; and PC + BJ + P, 1.37. These data demonstrate that removal of solubles from alfalfa does not decrease the value of the protein remaining for ruminants and confirm and emphasize the positive effect of heat dehydration of alfalfa protein for ruminants.

KEY WORDS: alfalfa, *Medicago sativa*, rumen protein degradability, heat dehydration, maturity.

INTRODUCTION

Ruminants absorb amino acids from the small intestine from three sources: microbial protein, endogenous protein, and undegraded dietary protein. In at least two physiological states (lactation and rapid growth) microbial protein is not sufficient to meet the animal's need for amino acids. In these instances, digestion of undegraded dietary protein in the small intestine is preferred to ruminal degradation of that protein (Chalupa 1975).

Recent research with alfalfa (*Medicago sativa*) has shown that heat treatment can increase the value of alfalfa protein in ruminant diets (Krause and Klopfenstein 1978, Klopfenstein et al. 1978). The purpose of this research was to determine the effects of maturity and processing methods on rumen degradability and efficiency of utilization of alfalfa protein by ruminants.

METHODS

Laboratory Experiments

Irrigated alfalfa was harvested at four maturities — prebloom (PRE), one-tenth bloom (1/10), one-half bloom (1/2), and full bloom (FB) — in four consecutive summer cuttings for 2 years (1978–1979). Processing methods were either freeze drying (F), sun curing (SC), or drying

in a forced-air oven at 95°C for 12 hours (H). Following processing, leaves and stems were hand separated and ground through a 1-mm screen. Results for the whole plant were computed from the leaf:stem ratio.

Analyses conducted on these samples were acid detergent insoluble N (ADIN) (Goering and Van Soest 1970), expressed as a percentage of the total N; in-vitro ammonia release (Britton et al. 1978), with 20 mg N from each sample/30 ml of rumen fluid incubated for 8 hours; and in-situ N degradation, estimated using 1-g samples in dacron bags (30–70 μm pore size) placed in the rumen of alfalfa-fed steers for 8 hours.

Animal Experiments

The first two animal experiments were growth trials using 200-kg Angus, Hereford, or Angus × Hereford steer calves individually fed 61% total digestible nutrients (TDN) and 11.5% crude-protein diets for 112 days. Nitrogen (N) sources evaluated in these experiments were urea, soybean meal (SBM), and dehydrated alfalfa (DEHY). SMB and DEHY diets were formulated so that either 20%, 30%, 40%, or 50% of the supplemental N came from each source. Ten animals were allotted to the urea treatment, and two steers were allotted/protein source and level for a total of ten steers/protein source. The rest of the trial was conducted as described by Waller et al. (1980). Protein efficiency is defined as the added daily gain observed above the urea control/unit of test protein source supplemented, and it was calculated according to the equation:

R. Britton is a Professor of Animal Science, D. Rock is a Graduate Assistant, T. Klopfenstein is a Professor of Animal Science, J. Ward is a Professor of Animal Science, and J. Merrill is a Graduate Assistant at the University of Nebraska. This is a contribution of the Nebraska Agricultural Experiment Station and is published with the approval of the Director (Journal Series Paper No. 6607).

$$\frac{\text{test protein daily gain} - \text{urea daily gain}}{\text{amount of supplemental test protein fed/day}} \quad (1)$$

The third and fourth animal experiments were designed to assess how mechanical treatment of alfalfa to remove a part of the solubles affects N utilization. Products of this processing are a fibrous press cake (PC), a high-protein precipitate (P), and a liquid residue called brown juice (BJ). PC used in these experiments was commercially dehydrated.

In the third animal trial, six wether lambs/treatment were fed diets containing either SBM, PC, direct-cut dehydrated alfalfa (DCD), or urea. Diets were formulated to 10.5% crude protein using corn cobs as the basal energy source. The experiment was conducted as described by Britton et al. (1978). The fourth experiment evaluated four alfalfa treatments: PC, PC + BJ, PC + BJ + P, and DCD. These alfalfa treatments were compared to SBM and urea in a growth experiment using sixty Angus, Hereford, and Angus × Hereford steer calves (209 kg). The experimental procedure and diet formulations were similar to those described for animal experiments 1 and 2 of this paper.

RESULTS AND DISCUSSION

Laboratory Experiments

As the plants matured, leaf content decreased from 55% at prebloom (PRE) and 52% at 1/10 to 47% at 1/2, and 40% at full bloom (FB). Leaves contained about 2.5 times as much N as stems. Decreases (P < 0.05) in leaf, stem, and computed whole-plant N occurred as maturity increased (Table 1). Leaves had lower ADIN than stems. Leaves, stems, and whole plants exhibited increases (P < 0.05) in ADIN from PRE to FB.

Nitrogen remaining in dacron bags after ruminal incubation in situ has been used as a method to predict ruminal degradation of feed proteins (Ørskov and McDonald 1979). Leaf dacron-bag N did not change significantly, but both stem and whole-plant dacron-bag N values increased with stage of maturity (Table 2). The changes in protein degradability with maturity in stems and whole plants are more apparent than real, as ADIN in these samples showed like increases. Leaf N was more degradable in rumen fluid than stem N. In-vitro ammonia release mirrored the changes observed in dacron-bag N.

Nitrogen components were affected more by processing (Table 2) than by maturity (Table 1). Heating yielded higher (P < 0.05) ADIN in both leaves and stems. Dacron-bag N and in-vitro ammonia release were lower (P < 0.05) for the H treatment. The SC alfalfa was intermediate for dacron-bag N and in-vitro ammonia release. Leaf protein responded more to heat effects than did stem protein. The values for whole plants closely follow the data obtained for leaves, emphasizing that the positive aspects of heat dehydration are mainly on leaf protein. Increased ADIN in the H treatment indicates some heat damage but does not account for all the changes in rumen degradability. Krause and Klopfenstein (1978) have shown that heating reduces rumen degradability of alfalfa protein.

Table 1. Maturity effects on alfalfa protein degradability.

Item	Maturity[a]			
	Pre-bloom	1/10 Bloom	1/2 Bloom	Full bloom
Leaves				
Nitrogen, % of dry matter	4.20[e]	3.94[e]	3.63[f]	3.38[g]
ADIN[b]	5.23[e]	6.40[ef]	6.84[fg]	7.70[g]
Dacron-bag nitrogen[c]	22.50	25.18	25.19	26.90
In-vitro NH₃ release[d]	14.08	13.19	13.02	12.60
Stems				
Nitrogen, % of dry matter	1.77[e]	1.61[f]	1.51[f]	1.50[f]
ADIN[b]	14.34[e]	16.29[ef]	17.67[f]	17.85[f]
Dacron-bag nitrogen[c]	43.35[e]	46.40[e]	47.78[ef]	48.40[f]
In-vitro NH₃ release[d]	6.57	5.93	5.30	5.26
Whole Plant				
Nitrogen, % of dry matter	3.11[e]	2.83[f]	2.50[g]	2.26[h]
ADIN[b]	9.34[e]	11.15[ef]	12.64[f]	13.80[g]
Dacron-bag nitrogen[c]	31.83[e]	35.20[ef]	37.01[ef]	39.92[g]
In-vitro NH₃ release[d]	10.73[e]	9.43[ef]	9.03[ef]	8.45[f]

[a]Four cuttings were harvested at each maturity in 1978 and 1979.

[b]Acid detergent insoluble nitrogen (N) is expressed as a percent of total N.

[c]N remaining in dacron bags after 8 hours ruminal incubation in situ expressed as a percent of total N in the alfalfa sample.

[d]Ammonia concentration (mg/100 ml) in rumen fluid after 8 hours in-vitro incubation.

[efgh]Means on the same line not bearing common superscripts are different (P<.05).

Table 2. Processing effects of alfalfa protein degradability.

Item	Processing[a]		
	F	SC	H
Leaves			
Nitrogen, % of dry matter	3.87[e]	3.67[f]	3.83[e]
ADIN[b]	5.29[e]	3.96[e]	10.38[f]
Dacron-bag nitrogen[c]	16.33[e]	21.42[f]	37.08[g]
In-vitro NH₃ release[d]	16.66[e]	13.81[f]	9.22[g]
Stems			
Nitrogen, % of dry matter	1.58	1.60	1.62
ADIN[b]	15.84[e]	11.63[f]	22.14[g]
Dacron-bag nitrogen[c]	38.93[e]	47.88[f]	52.88[f]
In-vitro NH₃ release[d]	9.47[e]	5.27[f]	2.55[g]
Whole Plant			
Nitrogen, % of dry matter	2.71	2.62	2.72
ADIN[b]	10.81[e]	7.89[f]	16.48[g]
Dacron-bag nitrogen[c]	27.95[e]	34.95[f]	45.07[g]
In-vitro NH₃ release[d]	12.97[e]	9.48[f]	6.53[g]

[a]Processing methods were freeze-dried (F), sun-cured (SC) and dried in a forced air oven (H).

[b]Acid detergent insoluble nitrogen (N) is expressed as a percent of total N.

[c]N remaining in dacron bags after 8 hours ruminal incubation in situ expressed as a percent of total N in the alfalfa sample.

[d]Ammonia concentration (mg/100 ml) in rumen fluid after 8 hours incubation in vitro.

[efg]Means on the same line not bearing the same superscript are different (P<.05).

Table 3. Performance of steers on diets supplemented with either urea, soybean meal, or dehydrated alfalfa (experiments 2 and 3 combined).

Item	Protein Source		
	Urea	SBM	Dehy
Daily gain, kg	.41[c]	.47[cd]	.54[d]
Daily feed, kg[a]	6.29	5.95	6.03
Feed/gain	15.23	12.60	11.18
Protein efficiency[b]		.203[e]	.547[f]

[a]Dry-matter basis.

[b]Daily gain of the test protein source minus the daily gain of the urea control divided by daily amount of test protein fed.

[cd]Means on the same line bearing different superscripts are different (P<.05).

[ef]Means on the same line bearing different superscripts are different (P<.10).

Animal Evaluation

In the first two animal experiments, DEHY-supplemented steers had increased (P < 0.05) gains as compared with the urea-fed animals; the SBM treatment was intermediate between the other two treatments (Table 3). These experiments were designed so that protein is limiting for these animals. If one protein source is degraded more slowly in the rumen more protein escapes to the small intestine, and we should measure differences in efficiency of use. The protein efficiency value for DEHY was higher (P < 0.05) than that for SBM, a finding that corroborates the laboratory data presented. Data gathered from these animal experiments show that heat dehydration produces changes in the alfalfa protein that enhance its utilization for meeting protein requirements that cannot be met by rumen microbial protein alone.

Nitrogen digestibility was reduced (P < 0.05) in the DCD treatment compared with that in the SBM and urea treatments (Table 4). The PC treatment reduced (P < 0.05) N digestibility more than did DCD. Heat dehydration or removal of soluble proteins could account for the reduced N digestibility cited. Nitrogen retention, however, was not affected negatively in either PC or DCD treatments and was approximately the same in the SBM treatment. All the other protein sources produced more (P < 0.05) retained N than did urea. Rumen ammonia concentrations in lambs fed DCD or PC were lower (P < 0.05) than in the lambs fed SBM or urea. This difference emphasizes that these proteins of low rumen degradability should be fed in combination with a highly degradable N source such as urea.

In the last steer growth trial, PC + BJ and PC + BJ + P were evaluated in addition to PC, DCD, SBM, and urea. Calves fed protein gained faster and were more efficient than calves fed urea (Table 4). Calves fed the alfalfa products had better gains and efficiency than those fed SBM. Protein efficiency figures from this experiment were higher for all alfalfa products than for SBM and were slightly higher for PC than for DCD. Although the absolute values for protein efficiency changed among the three steer growth trials, the relationship between SBM and DCD remained about the same. In all trials, the DCD was utilized better than SBM. Press cake, even after partial removal of soluble protein, was utilized well as compared with DCD and much better than SBM.

All experiments reported emphasize that alfalfa proteins respond to heating with decreased rumen degradability and increased efficiency of protein use in ruminants, if they are utilized properly in ruminant diets. Leaf protein responds more to heating effects, and so the greater the amount of leaves available, the greater the dehydration effect will be. In addition, newer processing methods that produce PC and partially harvest soluble protein do not diminish the positive performance response to heat dehydration.

LITERATURE CITED

Britton, R.A., D.P. Colling, and T.J. Klopfenstein. 1978. Effect of complexing sodium bentonite with soybean meal or urea on *in vitro* ruminal ammonia release and nitrogen utilization in ruminants. J. Anim. Sci. 46:1738.

Chalupa, W. 1975. Rumen bypass and protection of proteins

Table 4. Evaluation of various alfalfa products as protein sources for lambs and steers.

Item	Protein source[a]					
	Urea	SBM	DCD	PC	PC + BJ	PC + BJ + P
Dry-matter digestibility, %	67.1	67.4	67.1	64.2	–	–
N digestibility, %	76.4[b]	75.3[bc]	72.3[c]	68.4[d]	–	–
N retention % of N intake	24.2[b]	40.0[c]	43.8[c]	45.2[c]	–	–
Rumen NH₃ mg NH₃/100 ml	27.74[b]	8.97[c]	1.49[d]	.90[d]	–	–
Daily gain, kg	.63	.66	.71	.77	.78	.74
Feed/gain	8.10	7.66	7.57	7.19	7.05	7.20
Protein efficiency	–	.427	1.271	1.477	1.640	1.366

[a]There were six lambs/protein source in the digestion study and 10 steers/protein source in the growth experiment.

[bcd]Means bearing different superscripts are different (P <.05).

and amino acids. J. Dairy Sci. 58:1198.

Goering, H.K., and P.J. Van Soest. 1970. Forage fiber analysis (apparatus, reagents, procedures, and some applications) U.S. Dept. Agric., Agric. Handb. 379.

Klopfenstein, T., C. Dorn, R.L. Ogden, W.R. Kehr, and T.L. Hanson. 1978. Field wilted and direct cut dehydrated alfalfa as protein sources for growing beef cattle. J. Anim. Sci. 46:1780.

Krause, V., and T. Klopfenstein. 1978. *In vitro* studies on dried alfalfa and complementary effects of dehydrated alfalfa and urea in ruminant rations. J. Anim. Sci. 46:499.

Ørskov, E.R., and I. McDonald. 1979. The estimation of protein degradibility in the rumen incubation measurements weighted according to rate of passage. J. Agric. Sci., Camb. 92:499.

Waller, J., T. Klopfenstein, and M. Poos. 1980. Distillers feeds as protein sources for growing ruminants. J. Anim. Sci. 51:1154.

Round Baling from Field Practices Through Storage and Feeding

W.L. KJELGAARD, P.M. ANDERSON, L.D. HOFFMAN,
L.L. WILSON, and H.W. HARPSTER
Pennsylvania State University, University Park, Pa., U.S.A.

Summary

A model analysis compared labor and fuel requirements for round and rectangular bale systems. Round baling reduced harvesting, transport, and handling labor by one-half but increased fuel use by a factor of 1.6. When data from field studies of all dry matter (DM) losses were included, round baling required 80% as much labor and 200% as much fuel as did rectangular baling.

Baling losses for round balers and stackers averaged 10% compared with 3% for rectangular balers. Dry-matter losses from mowing through feeding of field-cured round bales, including 6 months of outdoor storage, totaled 41%. Round-baling losses decreased as baling rate increased by use of multiple windrows or higher field speed. However, greater raking losses while forming larger windrows caused overall field losses to be similar. Bale weight and density decreased as baling rate increased.

Preservative treatments using organic acids significantly reduced harvest and feeding losses compared with untreated field-cured hay. Treated hay and rectangular bales had higher acceptability in beef brood-cow feeding trials. It required approximately 10% more untreated than treated hay DM in similar packages to meet feeding requirements. It required about one-third more hay DM from untreated round bales and stacks than from treated rectangular bales.

KEY WORDS: hay harvesting, round bales, labor, energy, nutrition, preservatives, cattle feeding.

INTRODUCTION

This report on round baling of alfalfa hay will emphasize (1) labor and fuel requirements; (2) impact of windrow size and field speed upon field losses, bale density, and round-baling machine capacity; (3) nutritional quality changes during storage for chemically treated and untreated bales; and (4) feed acceptability and utilization by wintering beef cows. The research was conducted at the Agricultural Experiment Station at the Pennsylvania State University.

METHODS

Models of alternative hay-harvesting systems that integrated haymaking activities of mowing, baling, transport, and handling for estimates of labor and fuel requirements were assembled. Each system was assigned the same daily hay-harvesting quota. Typical capacity and work rates were used for the machines involved. The model estimated labor and energy need for those activities directly associated with field harvest, then added feeding components to give an overall comparison of various haymaking methods.

The influence of field practices on the properties and efficiency of round baling was investigated in replicated experiments using baler size, field speed, and windrow size as variables. Bale weight, land area harvested/bale, windrow yield, baling time, and bale dimensions provided

W.L. Kjelgaard is an Associate Professor of Agricultural Engineering, P.M. Anderson is an Associate Professor of Agricultural Engineering, L.D. Hoffman is a Senior Research Associate in Agronomy, L.L. Wilson is a Professor of Animal Science, and H.W. Harpster is an Assistant Professor of Animal Science at Pennsylvania State University.

Table 1.   Labor and energy requirements for making 18 ton/day of field-dried hay.

Method	Mow and rake		Baling		Wagons/movers		Avg. labor, Man-h/t	Avg. fuel** L/t
	Man-h	MJ*	Man-h	MJ	Man-h	MJ		
Round baler	2.6	68	2.0	214	3.2	134	.4	2.5
Baler with thrower	2.6	68	2.9	82	8.4	110	.8	1.6
Stack (1 t)	2.6	68	2.8	150	5.7	210	.6	2.6

*MJ = megajoules (Joules x 10⁶).

**Based upon 8.85 MJ available energy per liter of fuel.

data for determining baling losses, baling rates, density, and machine capacity.

Chemical-preservative studies compared treated high-moisture hay with untreated field-cured hay harvested as round bales, rectangular bales, and stacks. Treatments consisted of applying (while raking) an 80:20 propionic–acetic acid mixture or a 70:30 propionic-formaldehyde, methanol, water mixture at a rate of 9 kg/metric ton (t) of 25%-moisture hay. Hay was raked at 35%–40% moisture and packaged at 25%–30% moisture. Untreated hay was packaged at 15%–17% moisture. Round bales were divided between indoor and outdoor storage. Stacks were stored outdoors, and rectangular bales were stored indoors. Measurements of weight, moisture content, and nutritive quality were made before and after a 6-month storage period.

Hay was fed to wintering beef cows to evaluate the effects of package type, chemical preservatives, and covered vs. uncovered feeding structures on feed acceptability and utilization. Field, storage, and feeding DM losses as well as nutritive changes and feeding characteristics were evaluation criteria.

RESULTS

Labor and Fuel Requirements

Round baling required one-half the labor of the rectangular baler with thrower. Most of this labor difference occurred in the higher labor demand for transport and handling of rectangular bales. Handling 18 t/day used 8.4 manhours/day for rectangular bales compared to 3.2 manhours for round bales. Round baling, which required

power to turn the bale throughout the baling process, used more energy for the packaging function than either rectangular baling or stacking. Haystacks handled at 1 t/load required the highest transport energy. Overall, the rectangular-bale method had the lowest fuel requirement (Table 1).

The labor-saving benefits of large-package harvesting were diminished by DM and nutritive losses that occurred during the outdoor storage. An apparent 50% labor saving shrank to 20% by the time the round bales were consumed by livestock. In the same time, the fuel required/unit DM fed from round bales rose to twice that of rectangular bales.

Round Baling Field Practices

The influence of baler feed rate (kg/min) on harvest losses and bale density was examined. Baling rates were increased by doubling and tripling windrows and using field speeds of 5.6 and 8.1 km/h. Change in speed had much less effect than windrow size on feed rate, so data for field speeds were averaged together. The results are presented in Table 2.

As baler feed rate increased from 93 to 275 kg/min, bale density dropped 20%, and field losses decreased. However, the additional raking needed to produce double and triple windrows counteracted reduction in baling losses, resulting in only minimal gain. Bale densities were similar for large and small baling machines at a given feed rate. Baler capacity was not directly proportional to feed rates because the density reduction at higher feed rates caused more bales to be produced, which increased total tying time.

Table 2.   Influence of round baler diameter (1.4 m vs 1.7 m) and feed rate on capacity, bale density, and field losses.

Windrow	Baler feed rate (kg/min)	Baler capacity* (t/h)		Bale density (kg/m³)		Avg. baling losses (%DM)
		1.4	1.7	1.4	1.7	
Single	93	4.3	3.9	163	165	14
Double	193	7.2	8.0	135	134	12
Triple	275	10.8	10.6	134	130	5
Average	187	7.4	7.5	144	143	10

*Includes tying but neglects time losses in turning, etc.

Table 3. Effect of organic-acid treatment and hay package type on dry-matter losses.

Package type	Mow, rake, & curing loss	Packaging loss	Storage loss	Feeding loss	Total loss*
	%	%	%	%	%
Round bales					
Treated	7.1	9.1[c]	10.1[bc]	5.5[b]	28.3[b]
Untreated	13.2	10.2[cd]	11.7[c]	14.3[c]	41.1[c]
Rectangular bales					
Treated	7.1	2.6[a]	4.8[a]	0.4[a]	14.3[a]
Untreated	13.2	3.5[a]	7.8[ab]	5.2[b]	24.4[b]
Stacks					
Treated	7.1	6.8[b]	11.3[c]	5.5[b]	27.5[b]
Untreated	13.2	13.4[d]	16.6[d]	16.3[c]	47.6[d]

*Based upon standing DM yield.

[a,b,c,d]Within columns: indicate significant differences (P<.05).

Chemical Preservation

Chemical treatments containing 70%–80% propionic acid applied at the rate of 9 kg/t of 25%-moisture hay suppressed heating during storage and significantly reduced total DM losses (P < 0.05).

Unprotected outdoor storage of round bales resulted in 10%–15% DM loss. Most deterioration occurred in a 200-mm-deep surface layer. The interior portion of outdoor-stored bales was equal in quality to bales stored indoors.

Feeding

Feeding losses from round bales and stacks were greater than from chemically treated or untreated rectangular bales. The chemical treatment improved the acceptability of all hay packages over that of untreated hay (P ≤ 0.05). Feeding losses for treated round bales and stacks was about the same as for untreated rectangular bales. Thirty percent more DM/animal was required from untreated round bales and stacks than from rectangular bales because of high waste and refusal rates. Covered fenceline feeding structures improved the efficiency of forage utilization over similar uncovered structures for winter feeding of beef cows. Results from chemical preservation and feeding tests are summarized in Tables 3, 4, and 5.

DISCUSSION

Because of protection against weather, round bales can remain in the field for a few days following harvesting, thereby reducing harvest day labor. However, the energy required for spinning the entire hay mass during round bale formation increases fuel needs in comparison with conventional baling methods.

Increased feed rates for round balers tended to reduce machine losses and caused a reduction in bale density. Operator skill is required to maintain minimum baling losses yet produce firm, well-shaped bales that will with-

Table 4. Hay nutrients of treated and untreated hay (% DM) after 6-month storage, compared with windrow conditions just prior to packaging.

Package type	Digestible protein	ADF*	IVDMD*	Est. TDN*
Prepackaging	14.3	45.7[a]	57.8[a]	56.0[a]
Poststorage				
Round bales				
Treated	13.0	49.5[c]	53.1[b]	53.4[b]
Untreated	13.2	49.5[c]	50.9[c]	53.4[b]
Rectangular bales				
Treated	14.4	44.5[a]	56.2[a]	56.9[a]
Untreated	13.2	47.0[b]	57.0[a]	54.1[b]
Stacks				
Treated	13.6	48.2[c]	51.1[c]	54.3[b]
Untreated	13.4	47.2[c]	50.8[c]	53.6[b]

[a,b,c,d]Within columns: indicate significant differences (P<.05).

*ADF = acid detergent fiber; IVDMD = in-vitro dry matter digestibility; TDN = total digestible nutrients.

Table 5. Brood-cow acceptability of chemically treated high-moisture vs. untreated field-cured hay.

Package type	DM offered	DM refused	DM consumed	Feed index*
	---------------kg/head/day---------------			%
Rectangular bales				
Treated	10.6	0.1	10.5	100[a]
Untreated	11.2	0.6	10.6	106[b]
Round bales				
Treated	12.3	0.8	11.5	116[c]
Untreated	13.7	2.0	11.7	129[e]
Stacks				
Treated	13.0	0.9	12.1	123[d]
Untreated	14.1	2.2	11.9	133[e]

*DM needed to meet energy requirements, with treated rectangular bales as index base.

[a,b,c,d,e]Within columns: indicate significant differences (P<.05).

stand weathering and handling. When additional rakings were used for turning windrows together to obtain higher baling rates, increased raking losses tended to offset reduced baling losses for no net gain. Forming large windrows by using a V-hitch rake would generally help to minimize round bale harvest losses.

Chemical treatments containing 70%-80% propionic acid significantly improved DM recovery from harvest through feeding. The technical problems of uniform and efficient chemical application on round-baling machines have not been resolved. Where hay input shifts from side to side, as with round balers, chemical application is difficult.

Unprotected outdoor storage of round bales over winter resulted in weathering, rotting, and depressed quality of bale surface materials and caused high hay refusals at the feed bunk. In this case the cost of materials, labor, and fuel required for growing, harvesting, transporting and handling are allocated to a shrunken DM base that inflates the investment in terms of DM conserved and utilized. Also important, but unmeasured, were the labor and machine costs for subsequent removal of refused hay. For these reasons greater improvements are needed in handling, storage, and feeding systems for round bales if full labor and fuel efficiency and maximum utilization are to be realized.

Evaluation of Forages Ensiled in Concrete Stave and Oxygen-Limiting Silos

J. O'LEARY, R.W. HEMKEN, and L.S. BULL

University of Kentucky, Lexington, Ky., U.S.A.

Summary

This study evaluated corn silages stored in concrete stave and oxygen-limiting silos during 1978–1980. Factors considered included the effect of environmental conditions in each silo on silage stability and feeding value. Milk production was evaluated using Holstein cows, intake studies used Holstein heifers or steers. Weekly measurements of intake, milk production, silage temperature, pH, and yeast and mold content were made for each silage. Aerobic stability was determined monthly. Silo type had no effect on intake or milk production. Silage temperatures from the oxygen-limiting silo were higher, and the silage was less stable than that from concrete stave silos. No advantages of the oxygen-limiting structure over conventional concrete stave silos were observed for the parameters studied.

KEY WORDS: silage, corn, storage, stability, yeast, mold.

INTRODUCTION

Ensiling is a well-accepted method of preserving and storing forage crops. It also allows greater feed energy production/ha for small grain crops than other preservation systems. However, little information is available on the effect of silo type on feedout quality and aerobic stability of ensiled forages. This study compared two silo types as storage structures and examined the effect of environmental conditions on quality and stability of silage.

METHODS

Corn (*Zea mays*) was stored in two top-unloading concrete stave silos (4.3 × 15.2 m) or in a bottom-unloading oxygen-limiting silo (5.2 × 15.2 m) for feedout during the 1978–1980 seasons. Corn was harvested at 30%-45% dry matter (DM). In 1980 silos were filled with alternate loads from the same field. In 1978 and 1979, when one stave silo was filled at a different DM, the field was subdivided into plots, and three randomly selected plots were then used to fill that silo, while the other two were filled with alternate loads. The feedout period was from December to February.

Intake Studies

In 1978 corn silage was fed to eighteen Holstein cows in midlactation, six cows being fed from each silo. The silage

J. O'Leary is an Associate Professor of Animal Sciences, R.W. Hemken is a Professor of Animal Sciences, and L.S. Bull was formerly a Professor of Animal Sciences (now a Professor in the Department of Animal and Veterinary Sciences at the University of Maine) at the University of Kentucky. This is a contribution of the Kentucky Agricultural Experiment Station and is published with the approval of the Director (Journal Article No. 81-5-15).

Table 1. Mean yeast and mold counts, temperatures, pH, and stabilities of corn silage.

| | Concrete stave 1 | | | Concrete stave 2 | | Oxygen limiting | | |
	1978	1979	1980	1978	1979	1978	1979	1980
Yeast and mold (log$_{10}$ CFU/g)	4.17	6.92	6.05	3.79	5.84	4.51	6.70	6.19
Temperature (°C)	2.6	8.7	11.1	1.7	3.9	12.6	14.4	16.6
pH	3.9	4.3	3.9	3.7	3.9	3.9	4.1	3.9
Stability (days)	7	2.3	2	7	2	5	1.5	1

was supplemented with a corn-soy (24% protein) mineral concentrate to ensure adequacy of the diet. Rations were mixed for individual feeding in the ratio (DM basis) of 60:40 roughage:concentrate with corn silage as sole roughage. Cows were blocked by age and production and assigned at random within blocks to treatments. A 14-day preliminary period preceded the 70-day test period. Data collected included individual feed intake, daily milk production, and weekly fat percentages and body weights. In 1979 corn silage was fed for a 4-week period to yearling Holstein heifers (four per treatment) that had been blocked by weight and assigned randomly to a treatment. The heifers were then randomly assigned to another silage for an additional 4-week period. In 1980 corn silage was fed to Holstein steers (six/treatment) for an 8-week period. All animals were fed (twice daily) sufficient silage to allow a 5% to 10% weighback. Corn silage was supplemented with a soy mineral mix (35% protein) at 0.45 kg/animal/day. Data collected were amount of silage fed (twice daily), orts (daily), animal weight (weekly), silage temperature and pH (weekly), and chemical analysis of composite samples for each silage.

Microbiological Analysis

Yeast and mold counts and aerobic microbial counts of the silages were determined weekly on acidified potato dextrose (PDA) and APT agar, respectively, using procedures described by Marth (1978). All counts were recorded as colony-forming units (CFU)/g wet silage. Stability of each silage was determined monthly by exposing a 4-kg sample to air and following increases in yeast and mold counts, temperature, and pH. A silage was considered spoiled when a 5°–10°C rise in temperature occurred. These data were collected for each silo for 12–16 weeks.

RESULTS

The DM values are in Tables 2 and 3. The mean weekly values for temperature, pH, and yeast and mold counts are in Table 1. At the time of removal from the silo, the silage temperature was from 5.5°C to 11°C higher for material from the oxygen-limiting silo than from either stave silo and 10°C or more above ambient. The mean ambient temperature at the time of feedout was 1.8°C, 2.3°C, and 6.7°C for 1978, 1979, and 1980, respectively. The weekly variations in temperature for

1978 silage are in Fig. 1. The weekly ambient temperature was approximately the same as the temperature in stave silo 2. While there was considerable variation in temperature from week to week, the oxygen-limiting silage temperature was considerably higher than that of either of the stave silages, except for the last 3 weeks of the experiment when temperatures for the three silages were similar. Although silage temperatures were lowest in 1978, the overall weekly patterns were similar in 1979 and 1980. Higher silage temperatures coincided with the higher ambient temperatures in 1978 and 1979.

Silage yeast and mold counts (Table 1) ranged from a low of 0.2% of the aerobic population in 1978 for both stave silos to 38% for the oxygen-limiting silo in 1979. Mean yeast and mold counts were high for all silages, although the counts in 1978 were appreciably lower than in 1979 or 1980. There was considerable fluctuation in counts during the feedout period, with oxygen-limiting silage initially at a very low level in 1978 (Fig. 2) that rapidly increased in the next 2 weeks and then declined for most of the rest of the experimental period. Both stave silages had high initial yeast and mold counts that de-

Fig. 1. Weekly temperatures for corn silage in 1978 at removal from concrete stave or oxygen-limiting silos. Concrete no. 1, 40.8% DM; concrete no. 2, 33.1% DM; oxygen-limiting, 40.1% DM.

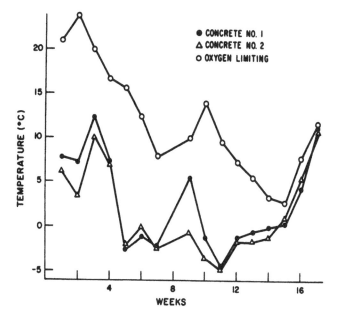

● CONCRETE NO. 1
△ CONCRETE NO. 2
○ OXYGEN LIMITING

Fig. 2. Weekly yeast and mold counts for corn silage in 1978 at removal from concrete stave or oxygen-limiting silos.

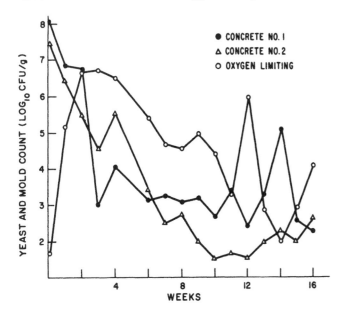

clined rapidly and then increased slightly in the latter part of the trial. Similar patterns were observed for yeast and mold counts in 1979 and 1980 although the counts were not as low.

Silage pH was low during all trials, with very little variation from week to week. The silage in 1978 was extremely stable (Table 1), due in part to the lower yeast and mold counts at the time the tests were conducted. The counts averaged 3.87 for stave 1, 3.18 for stave 2, and 4.46 for oxygen-limiting silage. In 1979 the counts were 7.25 for stave 1, 6.76 for stave 2, and 6.91 for oxygen-limiting silage, with a considerable decrease in stability. Results were similar for the 1980 silage.

Intake and production data are in Tables 2 and 3.

Table 2.   Performance data for Holstein cows fed corn silage in 1978.[a]

Silo type	CS 1	CS 2	OL
Silage dry matter (%)	40.8	33.1	40.1
Dry matter consumed (kg/day)	20.8	20.9	21.2
Milk production (kg/day)	24.7	27.0	25.8
Milk fat (%)	3.50	3.51	3.52

[a]There were no statistically significant (P<.05) differences among treatments.

Table 3.   Animal performance data for Holstein heifers or steers fed corn silage in 1979 and 1980.[a]

Silo	Year	Silage dry matter (%)	Intake (g DM/kg BW$^{.75}$)
Concrete stave 1	1979	45.9	90.0
	1980	34.6	117.3
Concrete stave 2	1979	34.7	92.9
Oxygen limiting	1979	45.4	91.9
	1980	35.7	122.4

[a]There were no statistically significant (P<.05) differences among treatments.

Neither moisture content nor silo structure influenced intake or milk production. Differences between treatments were not statistically significant. There was a trend in favor of the high moisture silage (stave 2) in 1978, but this trend was reversed in 1979.

DISCUSSION

Temperatures of silage from the oxygen-limiting silo were consistently higher than from either of the stave silos. This difference could be due partly to the insulating effect of the oxygen-limiting silo and partly to fungal metabolic activity in the silo. The drastic increase in fungal counts occurring shortly after opening of the oxygen-limiting silo and the lesser stability of its silage could also be indicative of fungal activity. All silages with fungal counts greater than 10^5 CFU/g were somewhat unstable, but the stability was subject to seasonal variations, with silage having greater stability for a given fungal count at a lower ambient temperature. Silage pH was stable before removal from the silo, but this fact does not in itself indicate no fungal activity. We have observed the folowing sequence in silage deterioration on exposure to air: rapid increase in fungal population followed by a rise in temperature. It may be as long as 2 days after the temperature rise before a rise in Ph is observed. Silo type did not affect intake, milk production, or milk fat. In our experiments the oxygen-limiting silo did not show any advantages that would recommend its use over conventional silos.

LITERATURE CITED

Marth, E.H. 1978. Standard methods for the examination of dairy products, 14th ed. Am. Pub. Health Assoc., Washington, D.C.

Relationship Between Silage Quality and 15N-Nitrate Reduction During Ensilage

K. ATAKU, M. HORIGUCHI, and T. MATSUMOTO

The College of Dairying, Rakuno-Gakuen, and Tohoku University, Japan

Summary

Nitrate in forage has been proven to occasionally poison ruminants. Our recent studies, however, have shown that a high concentration of nitrate inhibits butyric acid fermentation to the advantage of the silage quality. This work was undertaken to see the relationship between silage quality and the pattern of nitrate reduction during ensilage.

About 140 g of orchardgrass chopped 0.5 cm long was mixed with 0.05% (W/W) of $K^{15}NO_3$ (96 atom % excess) and put in a 250-ml glass bottle, which was immediately sealed to prepare "control silage" or sealed after 24 hours to prepare "air-exposed silage" with a rubber stopper fitted with a fermentation trap for gas absorption. "Glucose-added silage" was prepared in the same way as control silage except that 2% (W/W) glucose was added. The bottles were kept at 30°C and then opened after 1, 2, 4, 7, and 30 days. Nitrate, nitrite, ammonia, and nitric oxide were determined by the Devarda method and ^{15}N by the optical-emission spectrometric method.

The pH drop and the lactic acid production were significant only in the control and glucose-added silages after 2 days. Addition of glucose improved silage quality and exposure to air injured quality as evidenced by pH, organic acids, and ammonia. The nitrate-^{15}N decreased rapidly within 2 days to 0.3%, 0.5%, and 17.1% in the air-exposed, control, and glucose-added silages, respectively. Production of small amounts of nitric oxide-^{15}N was observed only in the control and glucose-added silages on the first day and the second day, respectively. Ammonia-^{15}N on the 30th day accounted for 97.7%, 71.5%, and 38.5% of total in the air-exposed, control, and glucose-added silages, respectively.

The pattern of nitrate reduction to nitrite, nitric oxide, and ammonia during ensiling is closely related to silage quality. The better the silage quality is, the smaller is the extent of nitrate reduction to ammonia and the larger is the production of nitric oxide.

KEY WORDS: orchardgrass, silage, nitrate, nitrite, nitric oxide, ammonia, nitrogen-15.

INTRODUCTION

The presence of nitrate in high concentration in forage has been proven to be detrimental in that it occasionally poisons ruminants. Although disappearance of nitrate during silage fermentation has been reported (Ohshima and McDonald 1978), the process of nitrate reduction has not been studied in detail. Our recent studies (Ataku et al. 1981, Ataku and Narasaki 1981) have shown that nitrate improves the quality of silage by inhibiting butyric acid fermentation during ensiling.

This work was undertaken to examine in detail the reduction of ^{15}N-labeled nitrate of orchardgrass during ensiling with reference to its effect on silage quality.

K. Ataku is an Associate Professor of Animal Nutrition at the College of Dairying; M. Horiguchi is an Associate Professor of Animal Nutrition and T. Matsumoto is a Professor of Animal Nutrition at Tohoku University.

Mention of a proprietary product does not imply endorsement of it.

METHODS

Materials

Fourth cutting of orchardgrass (*Dactylis glomerata*) containing 18% dry matter was harvested on 12 November 1979. The dry matter contained 17% crude protein, 7.4% water-soluble carbohydrate, and 0.12% nitrate nitrogen (NO^3-N).

About 140 g of the fresh grass was chopped 0.5 cm long, mixed with 0.05% (W/W) of $K^{15}NO_3$ (96 atom % excess), and put in 250-ml glass bottles. The bottles were either immediately sealed to give "control silage" or sealed after 24 hours to give "air-exposed silage." A third treatment was prepared in the same way as the control silage except that the grass was mixed with 2% (W/W) glucose prior to ensiling to give "glucose-added silage." Each bottle was sealed with a rubber stopper fitted with a fermentation trap containing 10 ml of 10M KOH to absorb nitric oxide. The bottles were kept in an incubator at 30°C and opened in triplicate for each treatment after 1, 2, 4, 7, and 30 days.

Measurements

Fifty g of the silage was made up to 200 ml with distilled water, left in a refrigerator (4°C) for 16 hours, and then filtered. The filtrate was subjected to analyses. The pH was measured with a glass electrode pH meter; lactic acid and volatile fatty acids (VFA) were determined by gas-liquid chromatography (GC 6A type gas chromatograph); nitrate, nitrite, ammonia, and nitric oxide were determined by the Devarda method (AOAC 1965); and ^{15}N was determined by emission spectroscopy (NIA-1 N-15 Analyser) on an aliquot of the ammonium sulfate solution obtained in the analysis of each nitrogen compound by the Devarda method.

RESULTS

Changes in pH and concentration of lactic acid during ensiling are shown in Fig. 1. The order of decreasing pH

Fig. 1. Changes in pH and concentration of lactic acid during ensiling.

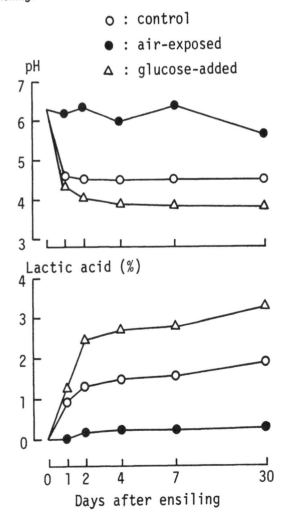

value of the silages was air-exposed, control, and glucose-added, as could be expected from the lactic acid concentration. The drop in pH and the production of lactic acid were observed only in the control and glucose-added silages after 2 days.

Table 1 shows characteristics of the silages at the thirtieth day. Although the control silage was excellent in quality, the glucose-added silage was better than the control, as was evidenced by a larger amount of lactic acid, a smaller amount of ammonia, and an absence of high VFA. The quality of the air-exposed silage was poor.

Disappearance over time of nitrate-^{15}N and production of nitric oxide-^{15}N and ammonia-^{15}N during ensiling are shown in Fig. 2. Nitrate-^{15}N decreased rapidly within 2 days to 0.3%, 0.5% and 17.1% in the air-exposed, control, and glucose-added silages, respectively. Production of nitric oxide-^{15}N in the control silage had a small peak on the first day, while 4% of total ^{15}N on the second day was nitric oxide-^{15}N with glucose addition. In both cases, an appreciable amount of nitric oxide-^{15}N was present after the fourth day. In contrast, production of nitric oxide-^{15}N in the air-exposed silage was negligible throughout. Ammonia-^{15}N increased rapidly within 2 days and then increased gradually to 97.7%, 71.5%, and 38.5% of the total ^{15}N at the thirtieth day in the air-exposed, control, and glucose-added silages, respectively. Nitrite could not be detected in any samples by the analytical method employed.

There were significant positive correlations between pH and disappearance rate of nitrate-^{15}N (r = 0.854) and production rate of ammonia-^{15}N (r = 0.893), whereas there was a significant negative correlation between pH and production rate of nitric oxide-^{15}N (r = – 0.800).

DISCUSSION

Silage fermentation is influenced by many factors, among which anaerobic conditions in the silo and high concentration of readily fermentable carbohydrates in ensiled materials have proven to be most effective. It has been observed by Ruxton and McDonald (1974) and Ohyama et al. (1975) that a prolonged exposure of ensiled materials to oxygen deteriorates the quality of silage. In agreement with these observations, the air-exposed silage prepared by delayed sealing was very poor in quality. Despite low concentration of sugars and high content of crude protein in the orchardgrass, the quality of the control silage was excellent. This may be explained in part by the high content of nitrate in the starting materials. Favorable effects of nitrate on silage fermentation have been observed by Ataku and Narasaki (1981). In accordance with our expectations, the quality of the glucose-added silage was best.

In the present study, nitrate-^{15}N was converted rapidly into ammonia-^{15}N during the first 2 days when the change in pH and the production of lactic acid were also rapid, a finding that agrees well with results of our previous studies. Ataku et al. (1976) found that a large

Table 1. Characteristics of the silages at the 30th day.

Silage	pH	Organic acids[a] (% of fresh silage)							Mark[b]	$NH_3 - N$[c]
		Lact.	Acet.	Prop.	But.	Val.	Cap.	Total		
Control	4.45	1.92	0.51	tr.[d]	tr.	tr.	0.02	2.45	97	11.0
Air-exposed	5.63	0.24	0.96	0.25	1.12	0.22	0.29	3.08	−6	46.4
Glucose-added	3.85	3.34	0.12	0.01	0	0	0	3.47	100	4.2

(a) Abbreviation for lactic acid, acetic acid, propionic acid, butyric acid, valeric acid and caproic acid, respectively. (b) According to Flieg's evaluation. (c) % of total nitrogen. (d) trace.

part of nitrate in grass was decreased by aerobic bacteria at the early stage of the ensiling process and that decrease of nitrate was prevented by growth of *Lactobacillus plantarum*, which was accompanied by the drop in pH.

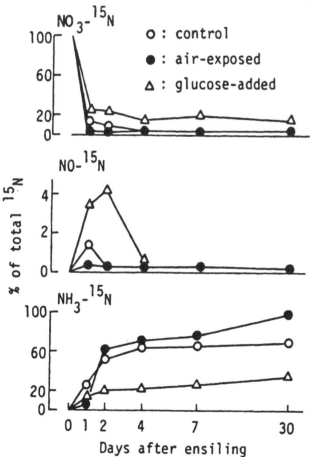

Fig. 2. Changes in the amount of nitrate-^{15}N (NO_3-^{15}N), nitric oxide-^{15}N (NO-^{15}N), and ammonia-^{15}N (NH_3-^{15}N) during ensiling.

Lamanna and Mallette (1965) reported that clostridia are also involved in nitrate reduction in butyrate silage.

Analysis of the results revealed positive correlations of the pH value with the disappearance rate of nitrate-^{15}N and the production rate of ammonia-^{15}N and a negative correlation between the pH value and the production rate of nitric oxide-^{15}N at the peak. These correlations strongly suggest that, in better-quality silages, the amount of nitrate reduced to nitrite is small, and the amount of the nitrate lost from the silo as nitric oxide gas is large; hence, the amount of ammonia production from the nitrate is smaller in good-quality silages than in poor-quality silages.

LITERATURE CITED

Association of Official Agricultural Chemists. (AOAC). 1965. Official methods of analysis, 10th ed., 17. AOAC, Washington, D.C.

Ataku, K., N. Narasaki, M. Kikuchi, and Y. Matsui. 1976. The influence of nitrate on silage fermentation. J. Jap. Grassl. Sci. 22 (Suppl.):95.

Ataku, K., N. Narasaki, and E. No. 1981. The effect of nitrogen fertilization on the quality of silage. J. Jap. Grassl. Sci. 27:100–105.

Ataku, K., and N. Narasaki. 1981. The influence of nitrate on the silage quality. J. Jap. Grassl. Sci. 27:308–317.

Lamanna, C., and M.F. Mallette. 1965. Basic bacteriology, 3rd ed., 844. Wiliams and Wilkins Co., Baltimore.

Ohshima, M., and P. McDonald. 1978. A review of the change in nitrogenous compounds of herbage during ensilage. J. Sci. Food Agric. 29:497–505.

Ohyama, Y., T. Morichi, and S. Masaki. 1975. The effect of inoculation with *Lactobacillus plantarum* and addition of glucose at ensiling on the quality of aerated silages. J. Sci. Food Agric. 26:1001–1008.

Ruxton, I.B., and P. McDonald. 1974. The influence of oxygen on ensilage. J. Sci. Food Agric. 25:107–115.

Effect of Herbage Mass and Allowance upon Herbage Intake by Grazing Dairy Cows

J.A.C. MEIJS

Institute for Livestock Feeding and Nutrition Research,
Lelystad, The Netherlands

Summary

An adequate supply of nutrients for grazing animals depends on the intake and nutritive value of herbage. Little information is available on factors affecting herbage intake of grazing cows, particularly for grazing systems that do not involve daily change of animals to a new pasture. Our objectives were to determine the effects of variation in herbage mass (due to variation in stage of maturity) and of herbage allowance on herbage intake by grazing dairy cows and on efficiency of consumption (herbage consumed as a proportion of herbage accumulated).

From 1976 to 1979 the herbage intake of lactating cows was determined at Lelystad during 23 grazing periods and during 28 stall-feeding periods. The precut swards were predominantly perennial ryegrass (*Lolium perenne*). A sward-sampling technique was used for estimating herbage intake by cows grazing swards for 3 or 4 days, with corrections for herbage accumulation during grazing. If herbage samples were cut both with a motor scythe and a lawnmower at the start and finish of grazing, accurate intake figures could be obtained.

The effect of herbage mass (M, kg/ha) was studied in three and four trials in 1976 and 1977, respectively. Within each trial, different levels of M were established by allowing parts of a sward to grow for periods of time of variable length. The daily herbage allowance (kg/day) was kept equal for all treatments. The treatments were compared in time within the same group of twelve cows. The experiment was performed both at pasture and indoors.

With advancing maturity, M increased significantly. There were no significant effects of higher levels of M on daily intake of organic matter (OM) from herbage, neither by grazing nor by stall-fed cows. In early summer, the herbage consumed by both groups showed a decline in digestibility at increasing maturity. As a consequence, daily intake of nutrients from herbage declined significantly at higher levels of M and also affected milk production negatively. However, in late summer, these effects were not significant. The efficiency of consumption was not affected significantly by M, so the mass of residual herbage was proportional to M.

In 1978 and 1979 three levels of daily herbage allowance (A) were achieved by varying the area grazed at equal levels of M for the treatments. Four groups of six cows each were used in an 8-week changeover design experiment performed twice each year. Higher levels of A had significant positive effects on daily intake of OM and of nutrients from herbage and on daily milk production/grazing animal. The efficiency of consumption declined at higher A levels. High amounts of residual herbage, however, had a strong positive effect on net regrowth of herbage, especially in early summer. In alternating grazing-cutting systems, a high allowance may be profitable, both for intake/animal and for the accumulation and consumption of herbage/unit area.

At a mean allowance level of 23 kg/day above 4.5 cm, our grazing cows consumed 13.6–14.8 kg/day of OM from herbage if no concentrates were fed. This amount was sufficient, at the quality of the herbage in our trials, for a daily production of 4% fat-corrected milk of 22–23 kg.

KEY WORDS: herbage consumption, efficiency of grazing, herbage accumulation, regrowth, herbage allowance, stage of maturity, herbage mass, perennial ryegrass, dairy cows.

INTRODUCTION

The choices of the moments of start and finish of grazing are two of the most important decisions to be made by the farmer in grazing management. The effect of the choice of both moments (in view of the available mass of herbage) on herbage intake/animal and on efficiency of consumption was examined in the trials reported. The mass of herbage at start of grazing was varied by lengthening the period of grass growth; the mass of residual herbage was varied indirectly by changing the daily herbage allowance.

MATERIALS AND METHODS

Materials

Spring-calving Dutch Friesian dairy cows with a mean annual production of 6,000 kg milk were used in the experiments. The twenty-four cows were classified in blocks of comparable age, calving date, and milk production; afterwards, they were allotted at random to a stall-fed and a grazing group (1976 and 1977) or to four grazing groups of six animals (1978 and 1979).

The main component of the sward on all pastures was perennial ryegrass (*Lolium perenne*) (78%-93%). The primary growth was pregrazed by sheep. All other swards were precut and were used only once during the grazing season. All swards received about 80 kg nitrogen/ha immediately after the previous harvest.

Methods

In 1976 and 1977 the treatments were different levels of herbage mass (M) at start of grazing, established by allowing parts of a sward to grow for variable times (ranging from 18 to 42 days). The higher the M was, the higher its maturity was also. Two comparable swards were chosen in each trial; one was used for grazing, and the other one was partially cut daily, and its herbage was fed indoors. In all, seven trials were performed. The treatments were compared within the same group of animals. Each cow was offered only 1 kg of concentrates/day. The stall-fed animals were fed fresh herbage six times daily in such amounts that at least 15% of the herbage was left uneaten. Individual daily intake could be determined indoors. The grazing period was 3 or 4 days. The daily herbage allowance of the grazing cows was kept constant by diminishing the area grazed at increasing M.

In 1978 and 1979, the treatments were different levels of daily herbage allowance (A), established by varying the area grazed for comparable groups of cows at the same number of animal-days and the same herbage mass. Three levels of A were compared. The treatments were compared between groups of animals during the same period. An 8-week changeover design was applied both in early and in late summer. The grazing period was 3 or 4 days; each week consisted of a 4-day preliminary period where the treatments were already applied, followed by a 3-day experimental period during which the intake was determined. In early summer 2 (1978) to 2.4 (1979) kg of concentrates/animal/day were supplied. In late summer, each animal was offered 1 kg of concentrates/day.

A sward-cutting technique was used to estimate the herbage intake of the grazing dairy cows. Strips of herbage were first cut with a motor scythe. After removing the cut herbage, a lawn mower of a smaller mowing width was used to cut the same strips again. It was shown that the herbage intake would have been overestimated by 10% if only a motor scythe were used for cutting. With the two-step cutting technique, however, reliable intake figures could be obtained under most conditions. On average, the herbage consumption could be estimated in our experiments with a coefficient of variation of 5.5%. Details of methodology are described extensively by Meijs (1981) and Meijs et al. (1981).

Herbage allowance and herbage intake were corrected for the accumulation of herbage taking place during grazing by using the equation of Linehan et al. (1952). The allowance and intake figures, presented in the next section, include herbage accumulation during grazing. All values of M and A are based on the cutting height of the motor scythe (> 4.5 cm), while changes in M in time (herbage consumed, herbage accumulated) have been measured by using motor scythe and lawn mower.

RESULTS

The four experiments have been extensively described in the doctoral thesis of Meijs (1981). In this paper, only some mean results are given without statistical analysis.

In 1976 and 1977, a total of 28 experimental periods were set up both indoors and with grazing cows. With advancing maturity, M increased significantly. In early summer, M of organic matter (OM) ranged from 1,400 to 4,800 kg/ha (> 4.5 cm), in late summer from 1,000 to 2,700 kg/ha. Because the range of M differed among the seven trials as a result of variable growing conditions, it is not possible to give mean figures for the treatments. Table 1 shows the main results of the grazing experiments by taking two distinct levels of herbage mass. At a standardized A, the daily intake of OM from herbage by the grazing cows was not affected significantly by M. Thus, the efficiency of consumption (herbage consumed divided by herbage accumulated) was not affected significantly by stage of maturity. As a consequence of the constant efficiency of consumption at increasing M, both the areic consumption (consumption/unit area) and the residual herbage mass were proportional to M (Table 1).

In early summer the herbage supplied to and consumed by the grazing cows declined significantly in digestibility at increasing maturity. As a consequence, the daily intake of nutrients declined significantly at higher levels of M and also affected 4% fat-corrected milk (FCM) production negatively. In later summer, however, the changes of digestibility were small and not significant at a smaller range of M, and no significant effects of M on intake of nutrients could be shown. All these effects obtained with the grazing cows were also established with the stall-fed cows.

In 1978 and 1979, 95 reliable intake periods were tried at pasture. The mean realized values of the three allowance treatments are reported in Table 2. A positive and linear effect of A on daily intake of OM from herbage was established. However, the extra supplied herbage at higher A was only partly consumed, and so the residual herbage mass increased at higher A. As the areic consumption of herbage declined at increasing A, the effi-

Table 1. Effect of herbage mass (due to maturity) on herbage consumption by grazing cows (all data expressed in organic matter).

		Early summer		Late summer	
Herbage mass at start of grazing	(kg/ha)	1500	3500	1500	2500
Residual herbage mass	(kg/ha)	500	1170	500	830
Areic herbage consumption	(kg/ha)	1200	2800	1200	2000
Daily herbage allowance	(kg/day)	22	22	22	22
Daily herbage consumption/ animal	(kg/day)	14.5	14.5	14.5	14.5
Efficiency of consumption	(%)	66	66	66	66
In vitro digestibility of organic matter of ingested herbage	(%)	82	77	78	78
Daily consumption of digestible herbage/animal	(kg/day)	11.9	11.2	11.3	11.3

ciency of consumption, at equal M, also declined. The digestibility of the herbage ingested was not affected by variation in A. Therefore, the effects of A on consumption and mass of nutrients were comparable to the effects of A on consumption and mass of organic matter. After a regrowth period of 19 days, M was again determined during 50 periods. The M after regrowth increased at higher A, an effect that could be attributed both to more residual herbage remaining after the prior grazing and to positive effects of the residual herbage on the herbage accumulation during the regrowth period. No effect of A on the in-vitro digestibility of M after regrowth could be shown.

The mean intake of OM from herbage averaged over all treatments was 13.5 kg/day, or 117 g/day/kg$^{0.75}$. It could be expected that the daily herbage intake would have been somewhat higher if no concentrates were consumed by our animals due to the lack of herbage substitution by concentrates. We therefore concluded that cows grazing for 3 to 4 days on precut pastures similar to those studied will consume 120 to 130 g/day/kg$^{0.75}$ of OM if no concentrates are fed at an A of 23 kg/day above 4.5 cm. This herbage consumption would be sufficient, at the quality of herbage used in our experiments, for a daily production of 22–23 kg of FCM.

The intake of nutrients from the total ration agreed well with the theoretical nutrient requirements for milk production, maintenance, and live-weight change.

DISCUSSION

Even at very high levels of M, the digestibility of the herbage ingested by both groups of cows in the herbage-mass experiments was always higher than 70%. This digestibility level probably explained the lack of any effect of M on daily herbage intake. In early summer the declining digestibility of herbage at increasing maturity, although it did not affect intake of organic matter negatively, affected the daily intake of nutrients from herbage negatively. Therefore, low levels of M are recommended in early summer if a high individual performance has to be achieved. In late summer a higher areic consumption of herbage can be achieved by allowing longer rest periods between grazing without affecting individual animal performance. Unfortunately, it was not possible to determine the effects of the higher levels of residual herbage resulting from higher levels of M on regrowth of herbage or on botanical composition of the sward on the long run in these short-term trials.

A strong positive effect of A on daily herbage intake was established. The decreasing intake at low A could not be explained by differences in digestibility among treatments. Herbage intake was restricted at low A by behavioral limitations; i.e., the major factor influencing herbage intake probably was the size of bites ingested.

On the single grazed swards, a negative effect of A on

Table 2. Effect of daily herbage allowance on herbage consumption by grazing cows and on herbage accumulation during regrowth (all data expressed in organic matter).

		Daily herbage allowance (kg/day)					
		Early summer			Late summer		
		14.7	20.9	30.3	15.6	22.2	32.5
Herbage mass at start of grazing	(kg/ha)	2520	2720	2570	1890	1980	1990
Residual herbage mass	(kg/ha)	490	1030	1500	320	570	970
Areic herbage consumption	(kg/ha)	2200	1860	1330	1590	1410	1160
Daily herbage consumption/animal	(kg/day)	11.5	12.6	13.7	11.7	13.9	16.4
Efficiency of consumption	(%)	78	60	45	75	63	50
Herbage mass after 19-day regrowth	(kg/ha)	1990	3010	3830	930	1440	1950
Herbage accumulation during 19-day regrowth	(kg/ha)	1490	1940	2360	620	890	1020

efficiency of consumption was compensated for by a strong positive effect of A on net regrowth. However, consumption of the herbage after regrowth by grazing could not be measured. Total areic consumption could be calculated only by combining grazing and cutting. Because of equal digestibility of M after regrowth among treatments, the assumption was made that daily intake of the cut herbage was not different among the treatments. It was shown then that both daily herbage consumption/animal and areic herbage consumption were positively affected by higher levels of A when the results of the grazing and regrowth periods were taken together. Therefore, alternating grazing and cutting combined with a high level of A seems to be a way to improve areic consumption of herbage; however, more research is needed on the ex-

pected equal daily consumption of the cut herbage.

Because of restricted space, our results are not compared here with data from the literature; a more extensive discussion is given by Meijs (1981).

LITERATURE CITED

Linehan, P.A., J. Lowe, and R.H. Stewart. 1952. The output of pasture and its measurements. Pt. III. J. Brit. Grassl. Soc. 7:73.

Meijs, J.A.C. 1981. Herbage intake by grazing dairy cows. Agric. Res. Rpt. (Versl. Landbk. Onderz.) 909. Pudoc, Wageningen.

Meijs. J.A.C., R.J.K. Walters, and A. Keen. 1981. Sward Methods. *In* J.D. Leaver (ed.) Herbage intake measurement handbook, 12. Br. Grassl. Soc.

Forage Intake of Cattle as Affected by Grazing Pressure

C.D. ALLISON, M.M. KOTHMANN, and L.R. RITTENHOUSE
New Mexico State University, Las Cruces, N.M., U.S.A.

Summary

A series of trials was conducted in 1977 to determine the effect of level of grazing pressure on organic-matter intake and the relationship between intake and forage disappearance. Levels of grazing pressure studied were 10, 20, 40, and 50 kg forage allowed/animal unit/day (kg/AU/day) for 14-day trials in April, July, and September.

A standing crop of forage was measured before, in the middle of, and immediately after each trial. Organic-matter intake was estimated by the fecal output:indigestibility ratio technique.

Organic-matter intake was highest in April; cattle also consumed more forage in September than in July. The combination of forage quality and animal physiological status explained the changes in intake by trial. Cattle consumed approximately 12 g/kg body weight$^{0.75}$ less organic matter by the end of a trial than at the beginning. A trend was exhibited for cattle under the 10 and 20 kg/AU/day treatments to consume less forage by the end of trials than at the beginning of a trial.

Total standing crop declined steadily during the grazing trials, with forage availability being significantly less at the end than at the beginning or middle of the trials. Averaged over the three trials, forage losses/AU/day were 8.5, 12.0, 12.7, and 16.3 kg for the 10, 20, 40, and 50 kg/AU/day grazing pressures, respectively. However, daily intake averaged across all treatments, periods, and trials was approximately 9 kg/AU/day. At the grazing-pressure level of 10 kg/AU/day, forage disappearance approximated the average daily intake, whereas under grazing pressures of 20, 40, and 50 kg/AU/day, forage disappearance exceeded intake by 28%, 48%, and 90%, respectively. These data indicate a possibility for a 2-fold increase in the efficiency of forage harvest by grazing cattle as grazing pressure is increased.

KEY WORDS: intake, forage disappearance, grazing pressure.

INTRODUCTION

Greenhalgh et al. (1966) stated that the relationship between herbage consumption and herbage allowance is probably curvilinear. When less herbage is offered than

the animal voluntarily consumes, incremental increases in herbage allowance are likely to produce increments of almost equal magnitude in herbage consumed. As the allowance increases further, the response is likely to become progressively smaller, and a point will be reached beyond which further increases have no effect on intake. An important point emphasized by Greenhalgh et al. (1966) is that an increase in the allowance may affect the quality as well as the quantity of herbage consumed, since opportunities for selective grazing are increased.

Gordon et al. (1966) stated that as grazing pressure was

C.D. Allison is an Assistant Professor of Range Science at New Mexico State University; M.M. Kothmann is a Professor of Range Science at Texas A&M University; and L.R. Rittenhouse is an Associate Professor of Range Science at Colorado State University. This is Scientific Paper No. 142 of the New Mexico State University Agricultural Experiment Station.

increased, cattle consumed more of the available forage. Their data were not on actual consumption but rather on forage disappearance as measured before and after a grazing period. Average grazing pressures of 11.1, 16.2, and 21.9 kg forage/cow/day resulted in dry-matter disappearance (DMD) values of 10.6, 13.4, and 15.0 kg forage/cow/day, respectively. It can be concluded from this research that forage disappearance for the most intensive grazing-pressure level closely approximates expected DM intake.

The objective of this study was to evaluate and compare forage disappearance and forage intake under four levels of grazing pressure on a *Stipa-Bouteloua* mixed-grass prairie in north central Texas.

METHODS

Field research for this study was conducted on the Texas Experimental Ranch near Throckmorton, Texas, during 1977. Average frost-free period is 220 days: average annual precipitation is 656 mm. Measured precipitation at the study area during 1977 was 538 mm. Three species, sideoats grama (*Bouteloua curtipendula* [Michx. Torr.]), Texas wintergrass (*Stipa leucotricha* Trin. and Rupr.), and buffalograss (*Buchloe dactyloides* [Nutt] Engelm) generally account for over 70% of the ranch vegetation (Kothmann et al. 1970).

Intake and standing-crop changes during a 14-day grazing trial were monitored under four levels of grazing pressure: 10, 20, 40, and 50 kg forage allowed/animal unit (AU)/day (kg/AU/day). Levels of grazing pressure were created by varying pasture size. Trials were conducted during April, July, and September 1977. Forage availability was measured before, in the middle of, and immediately after each trial. Organic-matter intake by cattle was estimated by the fecal output:indigestibility ratio technique, with fecal production measured by total fecal collection and digestible organic matter estimated by in-vitro techniques (IVDOM). Total fecal collections were obtained for days 2, 3, 4, 12, 13, and 14 during each 14-day trial. Esophageal extrusa and fecal samples were frozen and stored at − 20°C and subsequently lyophilized.

Grazing-pressure treatments were replicated twice. Eight cows with esophageal fistulae and eight intact Hereford cows were used during each trial. Animals were randomly assigned to a treatment within each replication at the initiation of each trial. Because of the limited number of animals/treatment, a preliminary intake trial was conducted on larger pastures 1 week prior to initiation of each of the three grazing-pressure trials. Intake estimates for individual cows during this preliminary trial were used to adjust their intake data from grazing pressure trials by analysis of covariance.

RESULTS AND DISCUSSION

Forage Disappearance

Forage disappearance, averaged over the three trials, was 236, 334, 355, and 457 kg/pasture for the 10, 20, 40,

Table 1. Average forage disappearance and organic-matter intake for four levels of grazing pressure (3 trials; 2 periods within each trial).

Grazing pressure (kg/au/day)	Intake[1] (kg/au/day)	Forage disappearance[2] (kg/au/day)	Efficiency[3] (%)
10	8.4	8.5	99
20	9.4	12.0	78
40	8.6	12.7	68
50	8.6	16.3	53

[1]Intake values are expressed as oven dry organic matter.

[2]Forage disappearance values are expressed as air-dry dry matter.

[3]Efficiency of harvest values are calculated as (intake ÷ forage disappearance) × 100.

and 50 kg/AU/day treatments, respectively. Expressed as disappearance/AU/day, these values were 8.5, 12.0, 12.7, and 16.3 kg/AU/day for the 10, 20, 40, and 50 kg/AU/day treatments, respectively (Table 1). Data such as these have been interpreted by some researchers to indicate that a greater amount of forage is consumed under lighter grazing pressures. However, average daily intake was approximately equal for all treatments. Estimated intake accounted for 99% of forage disappearance at the 10 kg/AU/day grazing pressure and 78%, 68%, and 53%, respectively, at grazing pressures of 20, 40, and 50 kg/AU day. These data support those of Gordon et al. (1966) and indicate the possibility for greater efficiency of forage harvesting as grazing pressure becomes more intense. With more intensive grazing pressure produced by greater stocking density, more of the forage disappearance fraction can be attributed to consumption by grazing animals, thus reducing the "invisible utilization" reported by Laycock et al. (1972).

Intake

Organic-matter intake varied with season; cattle had greater intake in April than in July or September. Cattle also consumed slightly more forage in September than in July (Table 2). High intake values observed in April were explained by the fact that cows were lactating and, therefore, had high nutrient requirements. Also, the high percentages of in-vitro digestible organic matter and crude protein (CP) in the forage during April was conducive to greater intake due to increased extent of digestion of the forage (Allison and Kothmann 1979). The greater intake values for cattle during September, when compared with July, were attributed to the fact that cows were in late gestation.

The period of a trial significantly affected intake. By the end of a trial, cattle were consuming approximately 12 g/kg body weight$^{0.75}$ less organic matter than at the beginning of a trial (Table 2). A period × treatment interaction was observed during all trials. Although the decline was not significant, intake by cattle on the 10 and 20 kg/ AU/day treatments declined by the end of the

Table 2.   Adjusted mean daily organic-matter intake for three trials, two periods within trials, and treatment x period interaction, expressed as grams organic matter intake per kg metabolic body weight (g/kg BW$^{.75}$).[1]

g/kg BW$^{.75}$			Trial					
			April 118.2[a]		July 88.8[b]		Sept. 94.7[b]	
					Period[2]			
g/kg BW$^{.75}$					1 106.3[a]		2 94.3[b]	
Grazing pressure[3]		10		20		40		50
Period	1	2	1	2	1	2	1	2
g/kg BW$^{.75}$	110.1[4]	85.2	117.0	100.3	94.5	94.5	99.8	97.4

[1]Means in the same row with a common superscript are not significantly different (P < .05).

[2]Periods 1 and 2 correspond to the average of days 2, 3, 4 and days 12, 13, and 14 of a 14-day trial, respectively.

[3]Grazing pressures are 10, 20, 40, and 50 kg forage allowed per animal unit per day.

[4]Treatment x period (within trial) interaction was non-significant by analysis of covariance.

trial, whereas intake by animals on the 40 and 50 kg/AU/day treatments remained relatively constant throughout a trial.

LITERATURE CITED

Allison, C.D., and M.M. Kothmann. 1979. Effect of level of stocking pressure on forage intake and diet quality of range cattle. Proc. West. Sec., Am. Soc. Anim. Sci. 30:174–178.

Gordon, C.H., J.C. Derbyshire, C.W. Alexander, and D.E. McCloud. 1966. Effects of grazing pressure on the performance of dairy cattle and pastures. Proc. X Int. Grassl. Cong., 470–475.

Greenhalgh, J.F.D., G.W. Reid, J.N. Aitken, and E. Florence. 1966. The effects of grazing intensity on herbage consumption and animal production. I. Short-term effects in strip grazed dairy cows. J. Agric. Sci. 67:13–23.

Kothmann, M.M., G.W. Mathis, P.T. Marion, and W.J. Waldrip. 1970. Livestock production and economic returns from grazing treatments on the Texas Experimental Ranch. Tex. Agric. Expt. Sta. Bul. 1100.

Laycock, W.A., H. Buchanan, and W.C. Krueger. 1972. Three methods of determining diet, utilization, and trampling damage on sheep ranges. J. Range Mangt. 25:352–356.

Combination Range-Pasture Management in South Florida for a Low-Energy Grassland Program

L.L. YARLETT
University of Florida, Gainesville, Fla., U.S.A.

Summary

The semitropical environment of south Florida is favorable for coordinated management of a wide variety of native grass species and improved pasture forages. The decade of the 1980s will emphasize change from forage programs of high energy utilization to those with a low requirement of fossil fuels. For 4 decades, grassland management in south Florida has been primarily pasture establishment created from the rangeland resource. The establishment and management required substantial inputs of fossil fuel. Use and management of rangeland was largely neglected. Ecological concepts as a basis for managing semitropical ranges were not accepted. In recent years practical programs have been developed combining rancher experience with range management, animal science, and agronomy. These programs coordinate the use of both rangeland resources and pasture forages. The physiological requirements of commercial cow-calf herds are likewise programmed into the yearly operation.

Implementation of a coordinated forage- and herd-management program that will result in a low input of fossil fuel requires planning to include (1) inventory of all available grassland resources, (2) growth processes of the major grass species, (3) cow and calf nutritional requirements and physiology and herd management as related to seasonal forage production, (4) range-management and pasture-conservation practices, and (5) grazing plans.

Forage inventories that include all forages available on the ranching unit are made. These forages usually include pasture species of pangolagrass (*Digitaria decumbens* Stent.), bahiagrasses (*Paspalum* spp.), and Coastal bermudagrass (*Cynodon dactylon* var.). Major range sites include flatwoods with species of bluestem (*Schizachyrium*) and *Andropogon, Paspalum,* and *Panicum.* Marsh range species include maidencane (*Panicum hemitomon*) and cutgrass (*Leersia* spp.).

Grazing systems are developed around the growth cycles and nutritional values of forage species and the physiological requirements of the cow-calf herd. Controlled breeding, calving periods, supplementation, and maintenance of cow weight are major animal husbandry considerations as an integral part of a combination range-pasture program. Major range-management practices include brush (namely saw palmetto [*Serenoa repens*] control, a planned deferred or rotation deferred grazing, and proper utilitization of the key species.

Commercial cow-calf operations have successfully managed to achieve over 90% calf crops, 60–90-day breeding season, and maintenance of 205–215 kg of calf/cow unit. Significant reduction in amount of fossil fuel materially contributes to a low-energy grassland program.

KEY WORDS: range, pasture, herd management, energy, Florida.

INTRODUCTION

Rangeland ecosystems of peninsular Florida are characterized by the availability of water, a long growing season, and a diversity of plants. The semitropical environment is ideally suited for management systems that utilize both native forage resources and improved pastures.

Changes in management of these resources began to develop in the late 1960s and early 1970s. Beef production from pastures that required large outlays of energy (fertilizer, labor, and equipment) was gradually being replaced by management practices requiring less of the costly production factors. Additionally, a buildup of environmental problems developed with a high use of fertilizer (Boyd 1970a, 1970b). These problems became widely recognized as a nonpoint pollutant in the major water resources of the south Florida region (Florida Department of Administration 1976).

Florida cattle have always relied on native forage produced by many semitropical vegetative associations. Conservation and productive management of these resources was largely neglected for 35 years following World War II. Ecological concepts as a basis for management were not generally accepted by those who would ultimately benefit — the producers. Rather, large acreages of native range were destroyed by being converted to improved pastures with introduced species.

In the mid-1960s and significantly since the initial boycott of oil imports into the U.S. in 1973, more practical and productive programs have emerged. The golden era of highly fertilized and often irrigated pastures was on the brink of decline. The energy crises brought in a new era of increased forage utilization with the proper use of the range resource as a factor in beef production. The resulting management system has combined rancher experience with the technologies of range management, animal sciences, and agronomy. Individual ranch operations have developed programs suited to their individual needs, coordinated the range resource, and improved pasture- and herd-management techniques.

DISCUSSION

Implementation of a combination range-pasture program initially requires an inventory of all available forage resources. In south Florida, the rangeland inventory includes the native forage sites, their present and potential production, season of use, and safe stocking rates. Major sites are flatwoods, having species of *Andropogon, Schizachyrium, Paspalum,* and *Panicum,* and marsh sites, with maidencane (*Panicum hemitomon*) and cutgrass (*Leersia* spp.). Animal-unit months and grazing seasons are indicated for the improved pastures. Introduced and adopted species include bahiagrass (*Paspalum notatum*), digitgrasses (*Digitaria* spp.), stargrasses (*Cynodon* spp.), and limpograss (*Hemarthria altissima*).

Grass growth and development and nutritional variation are key factors in a combination range-pasture-management program. In a semitropical environment, seasonal development of species follows much the same patterns as that of species adapted to the temperate regions. These characteristics are readily apparent among the native and improved pasture species. Dormancy, initial root and shoot growth, culm development, seed production, and fall maturity are apparent despite longer frost-free periods (Yarlett 1965, Yarlett and Roush 1970).

Proven animal husbandry practices are a major contribution towards the proper utilization of the forage species. Any single herd of brood cows must have a genetic potential for reproductive efficiency and growth and must be adapted to the semitropics. Adequate nutrition and a strict culling and breeding program are producing herds that breed in 60 to 90 days with a 95% calf crop. In south Florida, this production comes from herds that are predominantly British × Brahman crosses (Koger 1980).

Specific features of the combination range-pasture program are illustrated in Fig. 1. Variations are expected within individual operations. Cattle are essentially forage

L.L. Yarlett is a Range Conservationist in the Range Ecosystem Management School of Forest Resources and Conservation of the University of Florida.

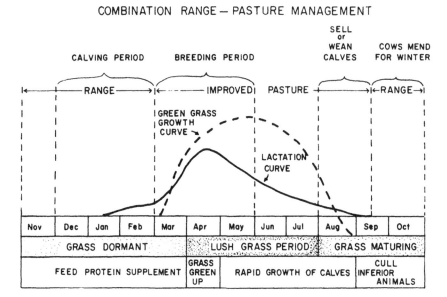

Fig. 1. Diagram of management coordinating the utilization of the range resource and improved pastures with herd management for a low-energy grassland program (adopted from USDA Soil Conservation Service).

consumers (Scoville et al. 1978), and for 3 to 6 months they fill that role on native forage during the cool season. Normally, 0.68 kg–0.90 kg of a basic 32% protein supplement/animal unit is supplied, thus stimulating the use of essential forage. Reserves of body fat and vitamin A are readily stored by healthy cows that have utilized improved pastures the previous summer. Mineral mixes are fed continually.

It is normal for brood cows on these programs to lose approximately 10–15% of their body weight while on native forages. The weight loss is recognized as a "controlled" loss, specifically for a reduction of fat surrounding the reproductive organs. The cow is thus conditioned for a short, successful, and natural breeding period in the spring.

There are many advantages to a controlled seasonal gain and weight loss, one of which is the longevity of the brood cow (Pope 1967).

The breeding season is regulated so that calves are born while cows are on native forages 6 to 8 weeks before improved pastures produce high-quality green forage. The native grasses are low in quality and not sufficient to produce a large quantity of milk. Newborn calves consume all the milk produced on that level of nutrition. When improved pastures are ready to graze in the spring, cows and calves are moved onto them for the summer. Milk flow increases substantially and is readily taken by the growing calf.

On returning to high-quality grass, the cows regain their weight losses rapidly. Reproductive organs are "flushed," and the cow comes into heat and breeds quickly. The result is a short breeding season, high-percentage calf crops, and uniform, marketable calves.

Calves are usually weaned in September. This date coincides with the period of declining quality of the improved pastures resulting from shorter days and cooler nights. At this time, milk production also declines. Cows are then returned to native grass to regain body stores. It is there that the next calf crop is born.

Conservation measures and the proper use of soil, water, and forage resources are readily implemented at the same time that a combination range-pasture program is formulated and executed. Deferred grazing or rest from grazing on range sites has been most effective when the most desirable native grasses are not grazed from spring until fall. Owners and managers report 5- and 6-fold increases in forage due to periods of deferred grazing. Control of brush species is normally most effective during the cool season, one pasture at a time, followed by rest for a full growing season. Rotational grazing systems are effective for both range and improved pastures. A system of rotational use of range units during the cool season and occasional light use during summer has been successful.

Combination range-pasture-management systems can be developed within 3 to 5 years. The time for implementation depends on a number of factors: size of operation, number of livestock, physical features, and managerial ability. With no fertilizer, well-managed ranges are equal in yield to improved pastures and are valued as much (Felton 1980).

Large inputs of energy have been profitable in recent years as improved pastures have been developed under low energy costs with several periods of high prices for livestock products. The present world energy crisis precludes continued neglect of the range resources and dictates that low-energy production methods be used.

Combination range-pasture-management systems presently being used in the south Florida region are practical low-energy systems.

LITERATURE CITED

Boyd, C.E. 1970a. Vascular aquatic plants for nutrient removal from polluted waters. Econ. Bot. 24:95–103.

Boyd, C.E. 1970b. Chemical analysis of some vascular aquatic plants. Arch. Hydrobiol. 67:78–85. Archabold Biol. Sta. Lake Placid, Fla.

Felton, E.R. 1980. Integrating native range and pasture. Symp. Range and Past. Mangt. Soil and Crop Sci. Soc. of Fla. Proc. 39:8–9.

Florida Department of Administration. 1976. Summary report on the special project to prevent eutrophication of Lake Okeechobee. Div. of State Planning, Tallahassee, Fla.

Koger, M. 1980. Reproduction and nutrition. Low Energy Beef-Forage Prog. Bartow, Fla. (unpublished).

Pope, L.S. 1967. Winter feeding and reproduction in cows. Beef Cattle Short Course, Univ. of Fla., Gainesville, Fla.

Scoville, O.J., H.J. Hodges, and H.A. Fitzhugh. 1978. The role of ruminants in support of man. Winrock Int. Livestock Res. and Training Cntr., Morrilton, Ark.

Yarlett, L.L. 1965. Important native grasses for range conservation in Florida. U.S. Dept. of Agric. Soil Conserv. Serv., Gainesville, Fla.

Yarlett, L.L., and R.D. Roush. 1970. Creeping bluestem (*Andropogon stolonifer* [Nash] Hitchc.) J. Range Mangt. 23(2): 117–122.

Cattle Preference and Plant Selection Within a Hybrid Grass

P.O. CURRIE, R.S. WHITE, and K.H. ASAY
USDA-SEA-AR Livestock and Range Research Station, Miles City, Mont.,
and Utah State University, Logan, Utah, U.S.A.

Summary

Recent development of hybrid forage grasses shows promise for advancing grassland agriculture. Previous selections within these hybrids have been based on morphological, reproductive, and adaptive attributes with only limited testing of animal acceptability. Testing of animal preference and selectivity was the primary purpose of the present study. Yearling cattle were tamed and observed for their selectivity in grazing individual plants and genetic lines of F_6 crosses between quackgrass (*Agropyron repens*) × bluebunch wheatgrass (*Agropyron spicatum*). Animals showed definite preference for individual plants within genetic lines and for different genetic lines. Different animals frequently preferred the same plant or lines with identical genetic backgrounds. Preferences were exhibited even though animals were provided access at different times or different locations to a large array of hybrid plants or given a choice between hybrids and other palatable forage species. Selectivity was not necessarily associated with observed agronomic or morphological traits. Animals preferred plants rated as caespitose to those ranked rhizomatous. They also selected plants that had a profusion of soft leaves as well as plants with stiff, erect leaves. The most significant feature was the consistency in selection for certain plants and genetic lines. Hybrid forages can be developed solely on the basis of plant-breeding criteria and human judgment, but this kind of development may result in plants with reduced utility for forage except under forced utilization. Conversely, developing hybrids entirely on the basis of animal preference may also limit utility. Plant breeding and forage use by animals need to be concurrently evaluated to assure the predetermined objective of plant improvement.

KEY WORDS: cattle preference, hybrid grass, *Agropyron repens* × *A. spicatum*, grazing preference, livestock forage, selection.

INTRODUCTION

Over 250 interspecific hybrids involving Triticeae grass species have been produced at Logan, Utah, in the last two decades (Dewey 1964, 1967, 1976). Among the more promising hybrids for supplying livestock forage is a fertile cross between quackgrass (*Agropyron repens* [L.] Beauv.) and bluebunch wheatgrass (*A. spicatum* [Pursh] Scribn. and Sm.) (Asay et al. 1978). These species, although differing in morphology, ecological adaptation,

P.O. Currie and R.S. White are Range Scientists at USDA-SEA-AR Livestock and Range Research Station, and K.H. Asay is a Research Geneticist at USDA-SEA-AR Crops Research Laboratory at Utah State University. This is a contribution of the Montana Agricultural Experiment Station and has been published with the approval of the Director (Journal Series Paper No. 1123). Research was conducted at the Livestock and Range Research Station, Miles City, in cooperation with Montana State University.

and origin, produced partially fertile F₁ hybrids (Dewey 1976). Thus, it was possible after several generations of selection to combine the vigor of *A. repens* with the drought tolerance, palatability, and rangeland adaptation exhibited by *A. spicatum* and to produce a new forage grass superior to either parent.

Perez-Trejo et al. (1979) tested forage yields, some quality characteristics, and phenological development of the *A. repens* × *A. spicatum* hybrid against the parent species. They found it was highly productive, was readily accepted by sheep, and had in-vitro dry-matter digestibility (DMD) comparable to that of the parental species. It tended to be intermediate between its parents for most features. Except for this research, most developmental work has been done with minimal consideration of animal use. Our objective was to evaluate the new hybrid for its grazing potential on rangeland. Initial testing was to determine adaptability of the hybrid to specific sites in the northern Great Plains. Next, we sought to determine whether animals would demonstrate a preference for the hybrid both among genetic lines (plants with same maternal genome) and also between the hybrid and other plant species.

MATERIALS AND METHODS

A study was conducted at the Livestock and Range Research Station, Miles City, Montana. Test areas were on light alluvial soils with good drainage and medium to high fertility along flood plains of the Tongue and Yellowstone rivers. One area, the Nursery, is on unirrigated dryland. Average long-term precipitation is approximately 34 cm annually. About 60% of this amount is received during May through August, with June being the wettest month. The other area, Field 8, is irrigated by a center-pivot sprinkler. Irrigation water is applied as required during the growing season.

Seeds of the genomic lines were limited to a few from an F₆ generation source. Seedlings were started in the greenhouse and transplanted to the field when plants were 6 to 8 weeks old. Plantings were made between 17 and 20 April 1978 at the Nursery and in early May at Field 8. At the time of planting, individual seedlings had several tillers between 10 and 20 cm tall.

Seedlings from 75 genetic lines with 50 seedlings/accession were planted on 90-cm centers in a 0.3-ha plot in the Nursery. Except for a very small border strip of crested wheatgrass (*Agropyron desertorum* [Fisch.] Schult.), no other species was planted at this location. The remaining seedlings were planted in Field 8, where 74 genetic lines were planted. Seedling numbers there ranged from as few as 14 up to 72 plants/accession. Most lines had about 50 seedlings. The remainder of the 1.5-ha study area was planted to blocks of Altai wildrye grass (*Elymus angustus* Trin.) or Sherman big bluegrass (*Poa ampla* Merr.) or left in a uniform block of pasture containing Regar meadow bromegrass (*Bromus biebersteinii* Roem. and Schult.), Eski sainfoin (*Onobrychis viciaefolia* Scop.), and Lutana cicer milkvetch (*Astragalus cicer* L.). Patches of volunteer yellow blossom sweet clover (*Melilotus officinalis* [L.] Lam.) and crested wheatgrass were also present.

Plants were allowed a full year to become established before grazing. Weeds were controlled by hand cultivation to allow full expression of lines that were rhizomatous. Grazing preference and degree of use were evaluated by the number of bites taken by each animal (Sanders et al. 1980). Prior to grazing in 1979, individual plants at the Nursery were rated for characteristics of growth, morphology, and appearance. Degree of vegetative spread, comparative vigor, and plant leafiness were rated on a scale of 1 to 9. For spread, a rating of 1 equalled a very caespitose appearance and 9 a very rhizomatous plant. Lowest vigor was rated 1 and highest vigor 9. A profusion of soft, drooping leaves was rated high on the scale, and plants with stiff, erect leaves were rated low. Average maximum leaf height was measured on each plant in 10-cm increments and combined with ratings used to reflect overall plant stature and to define factors contributing to yield. Using these rating criteria, 72 plants of the more than 3,500 growing in the Nursery were accorded recognition as being particularly desirable from a plant-breeding standpoint. These "ideal" plants had moderate spreading and rhizome activity, good vigor and plant height, and an abundance of leaves that appeared soft and somewhat drooping rather than stiff and erect.

Ratings and grazing preferences for individual hybrid plants were not made at Field 8. Instead, amount of time the animals spent grazing each species and the total number of bites taken from a genetic line were recorded when animals foraged on hybrids.

Grazing preference trials were conducted during June, July, August, and October 1979. Eight yearling steers were observed for 4 days during each trial. For each trial, two steers were released in either the fenced Nursery or Field 8 plots. Steers were released at different locations within the enclosures to start each trial.

Steers were kept off feed 4 to 8 hours before trials. They were then permitted to graze from one to several hours. Individual animals were numbered and rotated so that the same steers were not always paired. Also, they were rotated between morning and afternoon trials and by location. Thus, each steer had opportunity to forage in the morning, in the afternoon, at each location, and with a different grazing companion. All steers were used two or more times during a complete trial.

RESULTS AND DISCUSSION

Plant Adaptation and Characteristics

The hybrid species *Agropyron repens* × *A spicatum* (ARS) was found to be well adapted to the northern Great Plains. Under both irrigated and dryland conditions, all genetic lines survived the harsh winter and wet growing season of 1978 and the dry growing season of 1979 when precipitation was 65% of normal. Even under dryland conditions all lines attained good size and were still green when snow covered the plants in November 1978. No winter damage was observable, and when plants completed spring growth in 1979, most had attained a maximum height of 68 cm.

Average plant height ranged from 52 cm to nearly 80 cm. Rhizome development of the plants ranged from a 3 (or quite caespitose) to a very rhizomatous 9 rating. Yield and leaf characters varied more within lines than did the degree of vegetative spread. It was common for some plants within a line to rank low in yield and others in the same line to rank high. Leafiness and the erectness of leaves on individual plants within several lines ranked from 1 (stiff, erect) to 9 (soft, drooping) between individual plants in a genetic line. Many of the genetic lines had rankings from 3 to 8 for leaf characteristics.

Animal Preferences

At the Nursery location, individual plants most readily grazed and preferred by the cattle did not show a definable consistency in characteristics used for measuring plant attributes. For example, of 72 plants marked before grazing began in June and considered ideal, one plant was grazed moderately, 2 lightly, and 13 incidentally. None of the plants were among those highly preferred by the animals.

In a ranking of the 25 most preferred plants grazed during each season, between 20 and 99 bites were taken in the June trials. For other seasons, rankings ranged from 14 to 45 bites in July, 22 to 99 bites in August, and 41 to 94 bites in October. Many ranked plants were preferred in more than one season. Of those preferred 7 were grazed in four seasons, 13 in three seasons, and 27 in two seasons.

Height of preferred plants grazed ranged from 50 to 100 cm, and yield ratings ranged from 4 to 8 for the plants selected. Animals did not show preference for or against rhizome development, plant size, or leaf characteristics. Four of the plants preferred in more than one season had the stiff, erect leaf form and two had the drooping form.

It was common for preferred plants to be grazed by more than one animal; e.g., in June, only one plant obtained a high preference rating from being eaten exclusively by one animal. All other individual plants were grazed by at least five of the test animals. Even though the steers started from different points on different days, they would graze discriminately among the hybrids and return to preferred plants one or more times. Likewise, selectivity was such that different animals discriminately grazed the same plants.

Grazing selections between genetic lines indicated there were animal preferences associated with the lines tested. Forty of the genetic lines at the Nursery location had a cumulative total of less than 20 bites eaten from all plants within that line (Fig. 1). For the two lines contributing more than 100 bites, 12 plants were grazed in one line and 17 plants in the other. However, one plant contributed 45 bites in one line and 3 plants 51 bites in the other line, again indicating the high preference for individual plants within genetic lines.

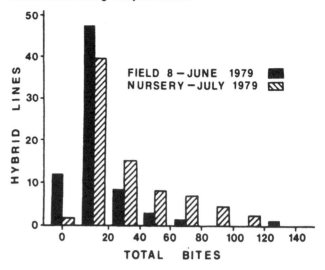

Fig. 1. Ranked preference of hybrid genetic lines: preferences based upon total number of bites taken by steers for all plants within a line during 4 days of trials.

Preference Among Species

During the June and July trials, preference for hybrid accessions in Field 8 followed a trend comparable to that observed in the Nursery. When hybrids were most preferred in June, 48 of the 74 genetic lines had fewer than 20 total bites taken by the steers, and 12 of them were not grazed at all (Fig. 1). The 14 remaining lines were most preferred. Since time individual plants were grazed was not recorded at this location, we do not know if select individual plants received the bulk of the grazing pressure as they did at the Nursery. However, three of the top-ranked lines at Field 8 were also among the preferred lines at the Nursery. Also, some of the lines most highly preferred in June in Field 8 were also among the preferred lines in July, a fact that indicates a consistency between time periods.

In a comparison with other species, the hybrid was grazed more than other species during the first two early summer trials but less later in the summer (Table 1). Animals spent 22% more time grazing the hybrid than they spent grazing crested wheatgrass, Sherman big bluegrass, and Altai wildrye, respectively. The only area on which they spent a greater amount of time was the pasture area composed of Regar meadow bromegrass, sainfoin, and cicer milkvetch; they spent 70% of their time on this area in August. By October, the animals spent a greater percentage of their time grazing Altai wildrye. This preference was a reflection of the palatability of this species late in the grazing season, as reported by Lawrence (1977) and Kilcher (1975).

CONCLUSIONS AND RECOMMENDATIONS

Dewey (1976) and Asay et al. (1978) suggested that hybrids could be developed for a variety of purposes and uses. Our present research supports this conclusion but emphasizes animal evaluations for defining the final product. Developing a hybrid based solely on animal

Table 1. Percent of time spent grazing hybrid plants and other species by yearling steers. (Average time for 4 days of trials during four time periods.)

Grazing period	ARS hybrids	Altai wild ryegrass	Sherman big bluegrass	Pasture mix	Crested wheatgrass
			%		
June	39	4	14	40	3
July	25	15	21	18	21
August	7	6	6	70	11
October	16	36	18	20	10
Average	22	15	15	37	11

preference may not be a suitable objective. Such an approach could produce a species too palatable for any use except in an absolute monoculture. If this were the sole intended use of the developed species, palatability alone might be an acceptable goal. However, for most applications, species with intermediate palatability probably would be more useful.

To ignore the animal factor is equally serious. Species can be developed based solely on plant-breeding criteria and human judgment. The species produced may do well in the environment selected but may be accepted by animals only under forced utilization. If an unpalatable species is grown in other than a strict monoculture, it may be avoided by the animals. This condition was demonstrated in the present study. A number of the genetic lines were left completely untouched or were incidentally grazed even during periods of the grazing season when hybrids were the preferred species. Thus, after desirable genetic traits are fixed in a population to obtain desired characteristics from a plant-breeding standpoint, additional selection using animal performance and testing data should be included in development of any new hybrid forage species.

LITERATURE CITED

Asay, K.H., D.A. Johnson, and D.R. Dewey. 1978. Breeding grasses for western range. Utah Sci., March, 3–5.

Dewey, D.R. 1964. Synthetic hybrids of New World and Old World Agropyrons. I. Tetraploid *Agropyron spicatum* × diploid *Agropyron cristatum*. Am. J. Bot. 51:763–764.

Dewey, D.R. 1967. Synthetic hybrids of New World and Old World Agropyrons: III. *Agropyron repens* × tetraploid *Agropyron spicatum*. Am. J. Bot. 54:93–98.

Dewey, D.R. 1976. Derivation of a new forage grass from *Agropyron repens* × *Agropyron spicatum* hybrids. Crop Sci. 16: 175–180.

Kilcher, M.R. 1975. Yield and quality of Russian and Altai wild ryegrass. Can. J. Pl. Sci. 55:1029–1032.

Lawrence, T. 1977. Altai wild ryegrass. Can. Dept. Agric. Pub. 1602.

Perez-Trejo, F., D.D. Dwyer, and K.H. Asay. 1979. Forage yield, phenological development, and forage quality of a *Agropyron repens* × *A. spicatum* hybrid. J. Range Mangt. 32:387–390.

Sanders, K.D., B.E. Dahl, and G. Scott. 1980. Bite count *vs* fecal analysis for range animal diets. J. Range Mangt. 33:146–149.

Relationship Between Botanical Composition of Herbage on Offer and Herbage Eaten by Grazing Beef Cattle

R.D.H. COHEN and D.L. GARDEN
University of British Columbia, Canada, and
Agricultural Research Station, Grafton, N.S.W., Australia

Summary

The relationship between botanical composition of the herbage on offer and herbage eaten is a key link in the modeling and simulation of grazing systems, yet there appear to be no data describing this relationship for multispecific or nontemperate swards. This paper describes one such relationship for cattle grazing multispecific swards in the subtropics.

Native or improved pastures were grazed by Hereford heifers fistulated at the esophagus. The native pasture, which had never received fertilizer, contained mixed grasses and no legumes. One of the improved pastures consisted of native pasture fertilized with superphosphate, and the other improved pastures were fertilized and oversown with white clover (*Trifolium repens*) alone or in conjuction with lotononis (*Lotononis bainesii*).

Samples of pasture and esophageal extrusa were collected on five occasions from June through February from 2 replications of each pasture. Extrusa were separated into green grass, green legume, and dead herbage components. Relationships were established between the total amount of green herbage on offer (GH, metric tons dry matter [DM]/ha) and the proportion of green herbage ingested (PG, g DM/g total extrusa DM). These relationships were similar for the four pastures and were described by a single pooled Mitscherlich equation

$$PG = 0.924 - 7.901e^{-3.7006\,GH} \qquad\qquad (r = 0.93)$$

with an approximate residual standard deviation (RSD) of $\pm$ 0.119.

Relationships were also fitted between the amount of green legume on offer (GL, metric tons DM/ha) in both of the grass-legume pastures and the proportion of green legume in the diet (PL, g DM/g extrusa DM). There were no significant differences between the relationships for the two pastures. The equation from the pooled data was linear, and the intercept did not differ significantly from zero.

$$PL = 0.432\,GL \qquad\qquad (SE_b \pm 0.070;\ RSD \pm 0.085;\ P < 0.001).$$

The proportion of green grass ingested (PGG, g DM/g extrusa DM) can be calculated by subtracting PL from PG.

KEY WORDS: botanical composition, herbage on offer, herbage eaten, grazing cattle.

INTRODUCTION

Quality of the diet of sheep and cattle is related to botanical components of the diet, particularly the green components (Dudzinski and Arnold 1973, Langlands and Sanson 1976). Hamilton et al. (1973) showed that, for several monospecific temperate grass swards, the proportion of green material in the extrusa from esophageal fistulae of grazing sheep was related to the amount of green herbage on offer. They fitted Mitscherlich equations to the relationships. Langlands and Sanson (1976) found that, for sheep grazing a pure sward of *Phalaris aquatica*, the relationship between dietary and available green herbage could be described with equal precision by a quadratic or a Mitscherlich equation but, for cattle grazing the same pasture, the precision of the estimate was lower, and a Mitscherlich equation was only slightly more precise than a linear equation.

The relationship between the botanical components of the herbage on offer and the herbage eaten is a key link in the modeling and simulation of grazing systems. There appear to be no data describing this relationship for multispecific or nontemperate swards. When multispecific swards are grazed by cattle (and they frequently are), the relationships for tropical grass species may differ from those for temperate species because of differences in canopy structure and ease of harvesting the available green material (Stobbs 1975). This paper describes one

such relationship for cattle grazing multispecific swards in the subtropics.

METHODS

The experiment was conducted at Fineflower, 65 km northwest from Grafton, New South Wales, lat 29°30'S. The environment is subtropical, with temperature extremes ranging from a winter minimum of 0°C to a summer maximum of 40°C, and 1,000 mm annual rainfall, most of which falls between December and May.

Pastures

Samples of the herbage on offer to grazing cattle were taken from four pasture types on five occasions at 45-day intervals between June and February. The pastures were unfertilized native grasses (U), native grasses fertilized with superphosphate (F), fertilized native grasses oversown with either the naturalized Clarence ecotype of white clover (*Trifolium repens*) (W) or with both Clarence white clover and lotononis (*Lotononis bainesii*) (WL). The main grass genera were *Hyparrhenia*, *Themeda*, *Imperata*, *Dichanthium*, *Bothriochloa*, *Capillipedium*, *Aristida*, *Cymbopogon*, *Axonopus*, and *Paspalum*. The pastures were continuously stocked with Hereford cattle at lenient stocking rates of 0.41 and 0.55 heifers/ha for pastures without and with legumes, respectively, to ensure adequate pasture bulk from which the cattle could select their diets. There were 2 replications of each pasture.

For pasture sampling, the paddocks were marked out into 4 strata, each 1.21 ha for grass pastures, and 3 strata of the same size for grass-legume pastures. There was no physical barrier between strata. At each sampling, four quadrats within each stratum were cut to ground level using a motor-driven, sheep-shearing handpiece. The

R.D.H. Cohen is an Assistant Professor of Animal Science at the University of British Columbia, and D.L. Garden is a Research Agronomist at the Agricultural Research Station. The research was done when R.D.H. Cohen was a Livestock Research Officer at the Agricultural Research Station.

The authors are grateful to Mr. G. Tarrant and the late Mr. W. McClymont for skilled technical assistance and to the Australian Meat Research Committee for financial assistance through project DAN 1.

quadrats were each 0.5 × 0.5 m and were located at random. Cut herbage from the four quadrats was bulked, sieved to remove soil, and hand separated into green grass, green legume, and dead herbage. These three components were dried in a forced-air oven at 80°C overnight and weighed. Available herbage dry matter (kg/ha) for each component was determined for each stratum and averaged to provide estimates for each paddock.

Experimental Animals

Eight Hereford heifers were fistulated at the esophagus as described by Cohen (1979) and used to provide samples of the grazed herbage (extrusa) in June, September, November, December, and February when the pasture quadrats were cut. Three extrusa samples were collected from both replicates of each pasture type during each sampling period. Collections were made on Monday, Wednesday, and Friday during the week in which pasture cuts were taken, using one fistulated heifer in each paddock. For each sampling period, the heifers were allocated at random to the pastures and grazed for one week prior to collection of samples. Extrusa were collected at 0800 hours following an overnight fast (Cohen 1979). Collections lasted for 20 minutes, during which time between 1 and 3 kg of herbage was extruded. The collection bags were made from vinyl-backed canvas, lined at the base with vinyl gauze to facilitate drainage of saliva. The extrusa were thoroughly mixed, and a subsample of 250 g was separated by hand with the aid of an illuminated magnifying lamp into green grass, green legume, and

dead herbage. The three fractions were dried at 105°C and weighed, and the proportions were expressed as a percentage of the total dry weight.

RESULTS

Linear, quadratic, and exponential relationships were established between the total amount of green herbage on offer (GH, metric tons DM/ha) and the proportion of green herbage ingested (PG, g/g total extrusa DM) in each pasture. These relationships were similar for the four pastures and were best described by a Mitscherlich equation that was fitted using a Gauss-Newton iterative procedure. A single pooled equation was fitted

$$PG = 0.924 - 7.901e^{-3.7006\,GH} \qquad (r = 0.93) \qquad (1)$$

This relationship, which is shown in Fig. 1, accounted for 86% of the variation and had an approximate residual standard deviation (RSD) of ± 0.119. There is no exact estimate of RSD because the equation was fitted by an iterative procedure. An approximation is given by the SD of the difference between observed and predicted values of PG ($SD_{y-\hat{y}}$).

Relationships were also fitted between the amount of green legume on offer (GL, metric tons DM/ha) in the grass-legume pastures and the proportion of green legume in the diet (PL, g/g extrusa DM). There were no significant differences between the relationships for the two grass-legume pasture types. The equation from the pooled data was linear, and the intercept did not differ

Fig. 1. The relationship between green herbage on offer (metric tons DM/ha) and the proportion of total green herbage (—) and green legume (···) in the diet of Hereford heifers grazing four pastures.

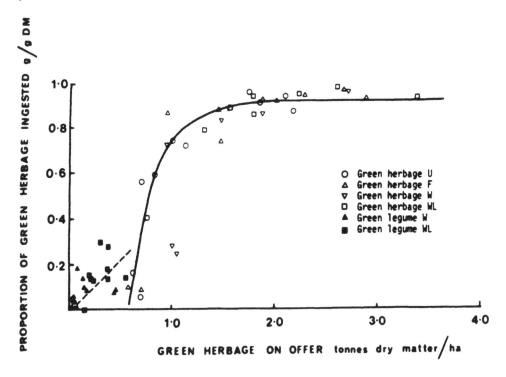

significantly from zero:

$$PL = 0.432\,GL$$
$$(SE_b \pm 0.070;\ RSD \pm 0.085;\ P < 0.001) \qquad (2)$$

This relationship also is shown in Fig. 1.

DISCUSSION

Cattle are less selective than sheep, and their diets are more strongly influenced by the amount of green herbage on offer (Langlands and Sanson 1976). The relationship between PG and GH for cattle was described by Langlands and Sanson (1976) as linear for a pure sward of *Phalaris aquatica* at levels of herbage availability between 0.5 and 2.1 metric tons/ha. They found little increase in precision when a Mitscherlich equation was fitted. A greater range of herbage availability was encountered in the present study, and the Mitscherlich function was more appropriate than a linear, quadratic, or other exponential equation, such as $Y = Ae^{-B/X}$.

In a general relationship for multispecific swards, such as we have presented, it is not logical to propose a single equation for the prediction of the proportion of green grass ingested (PGG) from the amount of green grass on offer (GG) because PGG will decline as the proportion of ingested green legume increases, and the decline in PGG will be independent of GG. Our results indicate that, at levels of green legume on offer below 0.6 metric tons/ha, the relationship between PL and GL is linear. Thus, any attempt to predict PGG directly will require a knowledge of the relationships between PGG and GG for all combinations of GG and GL. This requirement is not conducive to modeling exercises. However, PGG can be calculated as the difference between computed values of PL and PG in equations 1 and 2.

The relationship that we have presented in equation 2 would most probably be quadratic or exponential at higher levels of GL. The precision of the prediction of PL would probably also increase because errors in pasture sampling increase as availability decreases.

Hamilton et al. (1973) found no significant differences in the relationships between PG and GH for sheep grazing four temperate grass swards. Their data intercepted the Y axis closer to the origin than ours did. This difference is consistent with the findings of many workers, for example Langlands and Sanson (1976), that sheep select green herbage more effectively than cattle at low levels of GDM. We are not aware of any published relationships for multispecific swards grazed by cattle. In view of the results of Hamilton et al. (1973) and the lesser ability of cattle to select green herbage (Langlands and Sanson 1976), we believe that our results may have wide application for the modeling and simulation of cattle grazing systems.

LITERATURE CITED

Cohen, R.D.H. 1979. Factors influencing the estimation of the nutritive value of the diet selected by cattle fistulated at the oesophagus. J. Agric. Sci., Camb. 93:607–618.

Dudzinski, M.L., and G.W. Arnold. 1973. Comparisons of diets of sheep and cattle grazing together on sown pastures on the Southern Tablelands of New South Wales by principal components analysis. Austral. J. Agric. Res. 24:899–912.

Hamilton, B.A., K.J. Hutchinson, P.C. Annis, and J.B. Donnelly. 1973. Relationships between the diet selected by grazing sheep and the herbage on offer. Austral. J. Agric. Res. 24:271–277.

Langlands, J.P. and J. Sanson. 1976. Factors affecting the nutritive value of the diet and the composition of rumen fluid of grazing sheep and cattle. Austral. J. Agric. Res. 27:691–707.

Stobbs, T.H. 1975. Factors limiting the nutritional value of grazed tropical pastures for beef and milk production. Trop. Grassl. 9:141–150.

Diet Selection by Cattle on Tropical Pastures in Northern Australia

B. WALKER, M.T. RUTHERFORD, and P.C. WHITEMAN
Department of Primary Industries, Queensland,
and University of Queensland, Australia

Summary

Diet selection by steers in northern Australia on a continuously grazed pasture of *Setaria sphacelata* var. sericea (Kazungula), *Macroptilium atropurpureum* (siratro), and *Stylosanthes guianensis* var. guianensis (Schofield) was studied. The pasture treatment that supported the highest annual weight gains (254 kg/steer) in a stocking-rate experiment had 1.7 steers/ha. Mean annual rainfall was 1,490 mm, with 70% falling from December to March. Monthly average temperatures are highest in December (30.1°C) and lowest in July (11.9°C).

Pastures were sampled in 1973 (seven 3-day periods) and 1975-1976 (six 3-day periods) with three esophageal fistulated steers. Extrusa samples were analyzed for nitrogen (N), in-vitro digestibility (IVD), and botanical composition. Pasture yield and composition were determined and herbage samples were analyzed for N in 1973 and N and IVD in 1975-1976.

In 1973 the proportion of legume (mostly siratro) in the cattle's diets was lower than the proportion of legume in the pasture (8.2%), except in March, when it was 12.3%. The N content of the diet was always higher than the derived value of diet components. The IVD of extrusa samples exceeded 70% in five periods and was lowest (66.5%) in August 1973.

Except for December 1975, the cattle selected a higher proportion of legume (again mostly siratro) in their diets than the proportion of legume in the pasture, particularly in June (38.0% compared with 7.1%). Except for October 1975, the N content of the extrusa samples was greater than the derived values. The IVD of the diet was highest (67.7%) in April and lowest (52.3%) in October 1975.

Seasonal variations in pasture quality were apparent. Nitrogen and IVD values illustrated the relatively low quality of tropical pastures and the need for a high degree of diet selection to satisfy nutritional requirements. This degree of selection can be achieved only under lenient stocking pressures. On pastures continuously grazed at low stocking rates, preference for setaria over legumes during the early growing season is of practical importance in maintaining legumes. Higher legume yields can be obtained by resting pastures at critical times. In 1973, preference for setaria was associated with higher cool-season temperatures giving continued pasture growth.

KEY WORDS: diet selection, tropical pastures, cattle, legume, siratro, digestibility, nitrogen.

INTRODUCTION

Little is known of dietary selection by cattle on tropical mixed legume-grass perennial pastures in northern Australia. This information is needed to define more clearly the role of the legume and to develop management strategies to maintain or increase legume yields. The diet selection by cattle on lightly grazed tropical legume-grass pastures was studied for 2 years near Mackay in northern Australia to obtain such information.

METHODS

The pasture treatment that yielded the highest annual live-weight gains (154 kg/ha over 5 years) in a replicated stocking-rate experiment had 1.7 steers/ha. Three steers continuously grazed each plot and were replaced every November. Cultivars sown were *Setaria sphacelata* var. sericea cv. Kazungula and the legumes *Macroptilium atropurpureum* cv. Siratro and *Stylosanthes guianensis* cv. Schofield. Mean rainfall was 1,490 mm, with 70% falling from December to March. Monthly temperatures were highest in December (30.1°C) and lowest (11.9°C) in July. Data on soils and fertilizer applications are given by Walker and Potere (1974).

Three esophageal fistulated Brahman x Shorthorn steers sampled the pasture in 1973 for seven 3-day periods and in 1975-1976 for six 3-day periods. One extrusa sample/day/steer was collected in 1973, whereas in 1975-1976 up to six samples/steer were collected each day. Each bolus sample was thoroughly mixed and subdivided for analysis of botanical composition, in-vitro digestibility, and chemical content. Diet composition was

determined by microscopic technique. Categories identified were dead material, setaria (1973), setaria leaf and stem (1975-1976), siratro, and stylo. Three counts were made on each sample.

Pastures were sampled by quadrats (10 × 1 m in 1973 and 20 × 0.25 m in 1975-1976), cut 5 cm above ground level and sorted into dead material, setaria, siratro, and stylo. In 1975-1976, setaria was also separated into leaf and stem. The sorted components were dried at 90°C. Nitrogen (N) was determined following Kjeldahl digestion and IVD by the technique of Tilley and Terry (1963).

RESULTS

February to November 1973

Rainfall for the 9 months prior to January 1973 (127 mm) was well below recorded means, but in 1973 rainfall was above average, particularly in the September-November period. Cool-season temperatures were mild, with minimum temperatures going below 15°C on only 45 days.

Total pasture yields increased from February to December, apart from a slight fall in October (Fig. 1). The cattle continuously grazing these pastures gained weight at a constant rate except during August and September.

In diet selection in 1973 (Fig. 1), only in March was the proportion of legume in the cattle's diets (12.3%) higher than the proportion of legume in the pasture as a whole. The legume the cattle selected was c. 80% siratro. The highest dietary legume contents were in March and May 1973, while the proportion of grass in the diet was highest in February and November 1973, and dead material content was greatest in October 1973.

The percentage of N of the extrusa samples was higher

B. Walker is a Pasture Agronomist and M. T. Rutherford is an Experimentalist in the Department of Primary Industries, Queensland. P.C. Whiteman is a Reader in Pasture Agronomy at the University of Queensland.

Fig. 1. Live-weight gains, total dry-matter (DM) yields, pasture composition, percentage of nitrogen (N), percentage of in-vitro digestibility (IVD), and botanical composition of extrusa samples for seven sampling periods in 1973.

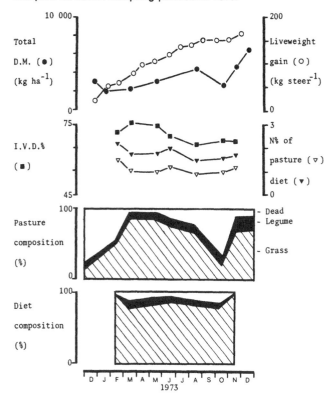

Fig. 2. Live-weight gains, total dry-matter (DM) yields, pasture composition, percentage of nitrogen (N), percentage of in-vitro digestibility (IVD), and botanical composition of extrusa samples for six sampling periods in 1975–1976.

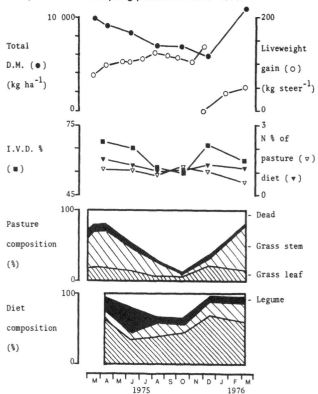

than the mean derived values of the available pasture at each sampling. Dietary N ranged from 1.86% to 2.24% until June 1973, declined in August and October, and increased in November. The IVD of extrusa samples was highest in March (76.3%) and May (74.6%) and lowest (66.5%) in August.

April 1975 to March 1976

Rainfall was above average, but cool temperatures began in May, two months earlier than in 1973, and there were 55 days with minimum temperatures lower than 15°C. Total pasture yields were much greater than in 1973, but live-weight gains were lower, with little weight gain between April and November (Fig. 2).

There was a greater seasonal variation in the botanical composition of the diet in 1975–1976, with higher proportions of legume and more dead material. At all times cattle selected more green matter, total setaria, and setaria leaf than was proportionally on offer (Fig. 2). Except for December 1975, a greater proportion of legume (again mostly siratro) was selected than that of the pasture as a whole, particularly in June (38.0% compared with 7.1%). The N content of the extrusa sample was greater than the derived values, except for October 1975. The IVD of extrusa samples was highest (67.6%) in April and lowest (52.3%) in October 1975. The IVD and percentage of N of the extrusa samples were lower than in 1973,

and differences between the percentage of N of the diet and derived values were narrower.

DISCUSSION

Results from the stocking-rate experiment agreed with data from the diet-selection studies. In 1973, when cattle did not preferentially select legume, annual live-weight gain was poorly correlated with legume yield (r = 0.44 nonsignificant, n = 20) and legume percentage of dry-matter yield (r = 0.50, P < 0.05). In 1975, however, when legume was preferentially grazed in all periods but one, weight gain was highly correlated with legume yield (r = 0.84, P < 0.001) and legume percentage (r = 0.78, P < 0.001).

The pattern in 1975–1976 of a high degree of legume selection in autumn (April and June 1975) agrees with results of Stobbs (1977) on siratro-setaria pastures and Gardener (1980) with a *Stylosanthes hamata* pasture. Palatability of legumes varies throughout the year and Stobbs (1977) showed that the acceptability and intake of siratro is much higher for autumn- than for summer-grown material.

The diet-selection pattern in 1973 differed from other published data on tropical pastures. This difference was probably due to differences in palatability associated with climatic effects. A late start of the wet season resulted in a shortened growing season and lower pasture yields with higher N contents than in 1975–1976. The mild autumn and winter and wet spring promoted young grass growth and some legume growth. Under such conditions cattle prefer grass to legume (Stobbs 1977).

The percentages of N and IVD in 1973 showed that a higher-quality diet was available than in 1975–1976, and this diet resulted in higher live-weight gains. The high IVD (over 70% for five of the seven periods) suggests that grass leaf would have contributed most to the diet and that it was selected in preference to stem (Fig. 2).

As cattle do not select legume preferentially during spring and summer, tropical legume-grass pastures can be continuously grazed at lenient stocking rates at these times. This lack of preference allows the legume to increase and helps its persistence. Higher legume yields can be obtained by resting pastures during the autumn period, when cattle preferentially select legume.

LITERATURE CITED

Gardener, C.J. 1980. Diet selection and liveweight performance of steers on *Stylosanthes hamata*—native grass pastures. Austral. J. Agric. Res. 31:379–392.

Stobbs, T.H. 1977. Seasonal changes in the preference by cattle for *Macroptilium atropurpureum* cv. Siratro. Trop. Grassl. 11:87–91.

Tilley, J.M.A., and R.A. Terry. 1963. A two stage technique for the *in vitro* digestion of forage crops. J. Brit. Grassl. Soc. 18:104–111.

Walker, B., and J.K. Potere. 1974. Effects of stocking rate on plant populations in tropical legume grass pastures. Proc. XII Int. Grassl. Cong., 388–394.

Seasonal Changes in Botanical and Chemical Composition and Digestibility of Diets of Large Herbivores on Shortgrass Prairie

G.M. VAN DYNE, J.D. HANSON, and R.C. JUMP

USDA-SEA-AR and Colorado State University, Fort Collins, Colo., U.S.A.

Summary

Studies were conducted from 1969 through 1974 on northeastern Colorado shortgrass prairie during different grazing seasons and under different grazing intensities to evaluate bison, cattle, sheep, and pronghorn antelope diets. The purpose was to examine seasonal changes in chemical and botanical composition and digestibility of diets of large herbivores.

Bison, cattle, and sheep diets were obtained from esophageal fistula samples. Fistula samples were analyzed microscopically to determine botanical composition. Pronghorn antelope diets were obtained by bite-count methodology. Data, presented for important plant species and species groups, were averaged for 28-day intervals to show typical seasonal trends in diets.

During all seasons, warm-season grass components were higher in bison diets than in cattle, sheep, and pronghorn diets. Shrubs contributed a greater portion of sheep diets than of bison diets in all seasons and of cattle and pronghorn diets except in autumn. Generally, sheep diets resembled cattle diets more closely than they did antelope diets. The greatest dietary difference was usually between pronghorn and bison.

Animals tended to select diets having high crude-protein content. Averaged across all species, dry-matter digestibility of diets did not vary greatly during the year but were somewhat higher in late spring and early summer. As rangelands are grazed more intensively in the future, there will be more emphasis on using several herbivores to maximize vegetation use. Present data are useful in providing information to make decisions for optimally allocating forage to different herbivore species.

KEY WORDS: diet evaluation, dietary botanical composition, dietary chemical composition, digestibility, large herbivores.

INTRODUCTION

Semiarid shortgrass and mixed-grass prairie originally occupied about 100 million ha from Mexico to Canada east of the Rocky Mountains. Historically, these grasslands were grazed by bison and pronghorn antelope. Recently, these two herbivore species have been largely replaced by cattle and sheep. However, increasing energy costs for supplemental feeds may provide opportunities for combinative grazing of all four herbivore species. Therefore, this paper is presented to examine the intra-annual dynamics of botanical and chemical composition and digestibility of four large herbivore diets on shortgrass prairie.

METHODS AND MATERIALS

Shortgrass-prairie data, collected from 1969 through 1974 in northeastern Colorado (Kautz and Van Dyne 1978), were summarized to provide information concerning diets of four herbivores. The study area receives about 310 mm annual precipitation; about 75% occurs as rain from April to September. Herbage availability varies from about 300 to 1,000 kg/ha for different seasons, years, and treatments. Grass and grasslike plants, forbs, shrubs and half-shrubs, and succulents make up about 23%, 14%, 19%, and 24% of the above-ground phytomass.

Diets of bison, cattle, and sheep were studied, using esophageal fistulas; diets of pronghorn antelope were determined by the bite-count method. Percentage weights of individual plant species for fistula samples were estimated microscopically. Unidentifiable dietary material was assigned to plant groups in proportion to identifiable group composition of the diet. Subsamples were analyzed for dry-matter digestibility, ash-free acid detergent fiber, crude protein, lignin, and cell-wall constituents. All chemical components were reported as percentages of organic matter, except digestibility, which was presented on a dry-matter basis. Samples were digested for 48 hours. Pronghorn antelope rumen liquor was used for digesting pronghorn diet samples, and cattle rumen liquor was used for digesting samples from all other herbivores.

Botanical composition was analyzed from 442 individual diet samples (for bison, cattle, and sheep) and herd diet estimates (for pronghorn) representing up to 52 sampling dates. Chemical components were analyzed from up to 35 composite samples for individual animal species, and 98 in-vitro digestibility determinations were made.

RESULTS

Dietary Botanical Composition

Data for March through November, the period when most of the samples were obtained, were averaged for 28-day intervals (Table 1). Several seasonal dietary trends were evident, the most important being the large cool-season grass component during the spring in the diets of bison, cattle, and sheep. Warm-season grasses became increasingly important in the diets of those herbivores by early summer to midsummer. Warm-season grass composition ranged from 27% to 76%, 11% to 55%, 5% to 43%, and 6% to 20% in the diets of bison, cattle, sheep, and pronghorn antelope (B, C, S, and P), respectively. In most of the herbivore diets, warm-season grasses decreased from midsummer to early fall and subsequently increased. Maximum values for bison, cattle, and sheep were reached in the July-August interval. There were no definite patterns of warm-season grasses in antelope diets. The primary warm-season grass in the diet was *Bouteloua gracilis*, which contributed over 50% of the bison diets in all periods except mid- to late spring. As warm-season grasses decreased in the diet, cool-season grasses and grasslike plants increased. Ranges of this dietary component were 18% to 67%, 15% to 56%, 6% to 62%, and 5% to 48% for B, C, S, and P, respectively. The principal species were *Agropyron smithii* and *Carex eleocharis* for bison, cattle, and sheep. Pronghorn were more varied in their dietary selection of cool-season plant species.

Forbs were highly variable in the diets of these large herbivores. Total cool-season and warm-season forbs in the diet ranged from a trace to 11%, 2% to 33%, 1% to 38%, and 12% to 83% for B, C, S, and P, respectively. Forb consumption peaked during midsummer for bison, early summer to midsummer for cattle, late summer for sheep, and midsummer to late summer for pronghorn. There were no clear shifts between warm- and cool-season forbs in the diets. *Sphaeralcea coccinea* was the most important forb, contributing up to 34% of the pronghorn diet in midsummer; concentrations of this species generally occurred in the summer diets of bison, cattle, and sheep.

Mean shrub concentrations in the diets ranged from 0% to 12%, 5% to 51%, 6% to 70%, and 7% to 30% for B, C, S, and P, respectively. Major shrubs and half-shrubs in the diets were *Artemisia frigida* and *Erigonum effusum* for bison and cattle diets, but at certain seasons sheep and pronghorn consumed considerable amounts of other shrubby species.

Chemical Composition and Digestibility

Mean dietary crude protein in herbivore diets ranged from 7% to 15%, 7% to 19%, 9% to 20%, and 10% to 23% for B, C, S, and P (Table 1). Highest values in the diets were found during the April-to-May interval. Cell-wall constituents ranged from 72% to 84%, 65% to 74%, 61% to 73%, and 47% to 73% in diets of B, C, S, and P, respectively. Seasonal patterns were not clear. Generally, high values for bison, cattle, and sheep were found from July to October; for cattle and sheep high values also occurred in April. Pronghorn diets had minimum cell-wall

G.M. Van Dyne (deceased August 1981) was a Professor of Biology at Colorado State University; J.D. Hanson is a Research Scientist of USDA-SEA-AR in Cheyenne, Wyo.; *and R.C. Jump is a Research Associate at Colorado State University.*

Table 1.   Seasonal averages of percentage dietary botanical and chemical composition and digestibility for bison, cattle, sheep, and pronghorn antelope from 1969 through 1974 on short-grass prairie.

	Midpoint dates of 28-day intervals*									
	12 Mar	9 Apr	7 May	4 Jun	2 Jul	30 Jul	27 Aug	24 Sep	22 Oct	19 Nov
Bison diets										
Plant groups										
Warm-season grasses	63	27	32	75	56	76	63	74	73	71
Cool-season grasses and grasslikes	24	64	67	18	22	18	27	16	22	18
Warm-season forbs	T	1	0	1	2	1	1	2	0	T
Cool-season forbs	T	4	1	7	12	3	9	3	3	1
Shrubs and half-shrubs	12	4	0	T	8	T	0	6	2	10
Important species										
Bouteloua gracilis	60	25	29	66	46	72	50	71	71	61
Agropyron smithii	20	37	25	7	13	10	3	8	2	9
Carex eleocharis	2	14	41	7	9	7	24	8	19	6
Sphaeralcea coccinea	T	3	1	5	11	2	9	3	3	1
Artemisia frigida	12	3	0	0	7	T	0	6	1	2
Eriogonum effusum	0	0	0	T	1	T	0	T	1	3
Chemical constituents										
Crude protein	8	13	15	10	12	11	12	9	10	7
Cell wall constituents	74	75	72	76	78	80	83	79	84	75
Ash-free acid detergent fiber	—	35	39	34	32	39	36	41	36	41
Lignin	—	10	11	7	7	10	11	11	13	11
In-vitro dry matter digestibility	—	40	40	53	46	43	24	33	46	—
Cattle diets										
Plant groups										
Warm-season grasses	32	12	11	27	34	45	45	30	27	42
Cool-season grasses and grasslikes	30	56	74	29	12	17	19	16	17	25
Warm-season forbs	T	0	T	8	14	9	14	3	2	8
Cool-season forbs	2	4	5	27	23	15	12	7	5	4
Shrubs and half-shrubs	35	29	9	7	16	14	9	45	49	17
Important species										
Bouteloua gracilis	18	10	6	20	26	39	39	29	24	29
Agropyron smithii	26	41	45	12	5	7	8	8	7	19
Carex eleocharis	9	13	23	8	6	8	10	6	5	4
Sphaeralcea coccinea	1	4	1	19	18	12	10	6	5	1
Artemisia frigida	22	28	6	3	15	8	2	33	49	13
Eriogonum effusum	T	0	1	3	1	3	2	10	0	1
Chemical constituents										
Crude protein	8	13	19	14	13	12	14	11	12	7
Cell wall constituents	69	74	65	68	68	73	74	70	70	72
Ash-free acid detergent fiber	50	39	31	38	37	36	38	46	42	44
Lignin	18	10	8	13	14	11	21	19	14	15
In-vitro dry-matter digestibility	38	46	55	51	46	39	35	31	38	42
Sheep diets										
Plant groups										
Warm-season grasses	22	5	10	29	29	39	24	24	20	36
Cool-season grasses and grasslikes	6	52	62	15	16	11	31	25	7	26
Warm-season forbs	1	T	T	6	6	9	2	6	1	1
Cool-season forbs	T	4	4	25	20	24	31	18	9	1
Shrubs and half-shrubs	70	39	23	25	28	18	12	28	63	33
Important species										
Bouteloua gracilis	2	5	7	27	25	36	20	23	19	34
Agropyron smithii	6	10	12	4	12	2	6	9	3	22
Carex eleocharis	T	41	39	10	4	8	25	15	4	4
Sphaeralcea coccinea	T	3	2	15	16	23	30	18	9	1
Artemisia frigida	50	38	16	0	10	2	1	21	62	30
Eriogonum effusum	T	0	3	24	18	15	11	6	T	1
Chemical constituents										
Crude protein	9	18	20	18	15	19	17	15	13	—
Cell wall constituents	65	73	63	65	71	72	66	64	61	65
Ash-free acid detergent fiber	46	41	28	36	34	36	31	39	40	40
Lignin	16	19	12	14	13	16	17	19	13	15
In-vitro dry-matter digestibility	33	35	57	48	50	45	39	36	45	45
Pronghorn antelope diets										
Plant groups										
Warm-season grasses	14	—	1	10	9	0	13	—	32	—
Cool-season grasses and grasslikes	52	—	22	3	3	0	4	—	38	—
Warm-season forbs	11	—	15	16	21	29	22	—	2	—

Table 1 (cont.)

	Midpoint dates of 28-day intervals*									
	12 Mar	9 Apr	7 May	4 Jun	2 Jul	30 Jul	27 Aug	24 Sep	22 Oct	19 Nov
Cool-season forbs	16	–	55	65	63	67	54	–	16	–
Shrubs and half-shrubs	8	–	7	6	4	4	7	–	12	–
Important species										
Bouteloua gracilis	12	–	1	9	8	T	6	–	24	–
Agropyron smithii	11	–	6	T	1	0	1	–	4	–
Carex eleocharis	0	–	1	0	0	0	0	–	0	–
Sphaeralcea coccinea	3	–	27	17	48	5	39	–	12	–
Artemisia frigida	18	–	1	2	1	0	2	–	16	–
Eriogonum effusum	2	–	6	4	4	3	7	–	9	–
Chemical constituents										
Crude protein	12	23	–	–	13	–	10	–	10	–
Cell wall constituents	65	47	–	–	47	–	47	–	66	–
Ash-free acid detergent fiber	31	17	–	–	31	–	22	–	31	–
Lignin	6	3	–	–	6	–	5	–	7	–
In-vitro dry-matter digestibility	44	66	–	–	58	–	57	–	46	–

*Data are means of measured values in 28-day intervals; a dash means no samples were taken; and T means < 1%.

constituents during April to September. Acid detergent fiber concentrations in the diets ranged from 32% to 41%, 31% to 50%, 28% to 46%, and 17% to 37% for B, C, S, and P. There was no clear seasonal trend in bison diets for this constituent, but for cattle and sheep diets values were highest in early spring. Pronghorn had their lowest fiber contents in early summer. Lignin values ranged from 7% to 13%, 8% to 21%, 12% to 19%, and 3% to 9% in the diets of B, C, S, and P. For cattle, sheep, and antelope the lowest values were in April to May, and lowest values for bison were in June and July. Highest values varied from late August for cattle to winter for pronghorn. Highest digestibility values were found in herbivore diets from early spring (pronghorn) to summer (bison). Lowest values were from late summer (bison) to midwinter (antelope). Means ranged from 24% to 53%, 31% to 55%, 33% to 57%, and 37% to 66% for B, C, S, and P.

General Considerations

There were observable intra-annual variations in botanical and chemical composition and digestibility of diets of the four large herbivores. It therefore seems conceivable to determine optimal grazing combinations that maximize forage consumption, digestible nutrient intake, meat production, and economic return. These data are being utilized in simulation and optimization analyses (Van Dyne et al. 1980, Van Dyne 1981, Van Dyne and Kortopates 1982) and should provide useful information concerning long-term planning and management of shortgrass prairies. We still need to develop mathematical functions depicting seasonal trends of dietary components, account for grazing intensity differences and to develop quantitative forage preference indices that relate dietary components to herbage availability and climatic conditions.

LITERATURE CITED

Kautz, J.E., and G.M. Van Dyne, 1978. Comparative analyses of diets of bison, cattle, sheep, and pronghorn antelope on shortgrass prairie in northeastern Colorado, U.S.A., 438–443. Proc. 1st Int. Rangeland Cong., Soc. Range Mangt., Denver, Colo.

Van Dyne, G.M. 1981. Optimal combination of four large herbivores for shortgrass prairie. Proc. XIII Int. Grassl. Cong. 1:629–631.

Van Dyne, G.M., N.R. Brockinton, Z. Szocs, J. Duek, and C.A. Ribic. 1980. Large herbivore subsystem. *In* A.I. Breymeyer and G.M. Van Dyne (eds.) Grasslands, systems analysis, and man, 269–537. Cambridge Univ. Press, Cambridge.

Van Dyne, G.M., and P.T. Kortopates. 1982. Quantitative frameworks for forage allocation. *In* Forage allocation. Natl. Res. Counc., Natl. Acad. Sci., Washington, D.C.

Nutritive Value of Cattle Diets on Fertilized and Unfertilized Blue Grama Rangeland

J.D. WALLACE, F.J. CORDOVA, G.B. DONART, and R.D. PIEPER

New Mexico State University, Las Cruces, N.M., U.S.A.

Summary

Low levels of nitrogen (N) fertilization can substantially increase forage production on native rangeland if adequate precipitation is available. The objective of this study was to compare nutritive value of diets selected by Hereford steers grazing unfertilized and N-fertilized (45 kg N/ha) rangeland in south central New Mexico.

The study area consisted of two adjacent pastures of approximately 60 ha each, containing comparable areas of bottomland, upland, and hillside sites. Stocking rates were 4.8 and 2.4 ha/yearling steer for control and N-fertilized pastures, respectively.

Dietary samples were collected from esophageally fistulated steers (average weight 308 kg) during early, middle, and late periods of two growing seasons and one dormant season. Sampling periods consisted of 4 consecutive days using four steers/pasture. Grazed samples were analyzed for organic matter (OM), crude protein, fiber, and in-vitro OM digestibility (IVOMD).

Nutritive value of grazed forage from both pastures followed seasonal trends — i.e., protein and IVOMD decreased while fiber increased with advancing plant maturity. During the growing seasons protein levels were higher ($P < 0.05$) in diets from the fertilized pasture than in those from the control pasture, with differences ranging from about 4 to 12 percentage units. Fiber values were usually about 4 percentage units higher in diets from the control pasture than in those from the fertilized pasture. IVOMD values favored the fertilized pasture during the latter part of the first growing season by 7 to 12 percentage units ($P < 0.05$), but during mid- to late dormancy, digestibility was higher ($P < 0.05$) on unfertilized range.

Increased nutritive value of rangeland forage selected by grazing cattle during the growing season, which was attributed to N fertilization, did not continue through dormancy. In order to be of benefit in terms of increased beef production on N-fertilized native ranges, such areas could be utilized during the growing season, thus relieving grazing pressure on unfertilized range until dormancy. Seasonal cattle gains recorded over a period of years on these pastures would substantiate such a management practice.

KEY WORDS: cattle, cattle diets, nutritive value, fertilized rangeland.

INTRODUCTION

Long-term studies have indicated that nitrogen (N) fertilization of native rangeland in south central New Mexico can substantially increase both total herbage and cattle production, although responses vary from year to year, depending on precipitation. Beef production (kg/ha) has generally been more than twice as much on fertilized as on unfertilized rangeland with variable stocking rates, but herbage production has usually been less than twice as much (Donart et al. 1978).

The present study was conducted to assess the nutritive value of diets selected by cattle grazing N-fertilized and unfertilized rangeland during different seasons of the year.

J.D. Wallace is a Professor of Animal Nutrition, F.J. Cordova was formerly a Graduate Assistant (current address, Mexican Embassy, 1 Beagle St., Canberra, A.C.T. 2603, Australia), and G.B. Donart and R.D. Pieper are Professors of Range Science at New Mexico State University. This is Scientific Paper No. 141 of the New Mexico State University Agricultural Experiment Station.

METHODS

Field data were collected from the Fort Stanton Experimental Ranch, located in a mountain foothill region of south central New Mexico. The study area consisted of two adjacent, similar pastures. One pasture (55 ha) had been fertilized annually (single application) with 45 actual N/ha just before the growing season for six years prior to this experiment. The other pasture (63 ha) received no fertilizer. Stocking rates were 2.4 and 4.8 ha/yearling steer on the fertilized and unfertilized pastures, respectively.

Elevation in the study area averaged 2,000 m above sea level. Annual total precipitation is 383 mm, and mean temperature is 11°C, with a mean maximum of 18.6°C and a mean minimum of 2.2°C. Approximately 63% of the total precipitation falls during the growing season (June through September). The average frost-free period is 161 days, with average first frost date of 10 October and average last frost date of 2 May. The major grass species in both pastures was blue grama (*Bouteloua gracilis*). Other vegetative and soil descriptions are given by Pieper et al. (1971).

Eight esophageally fistulated Hereford steers (average weight, 308 kg) were used throughout the study to provide diet samples for nutritive value analyses. The steers were randomly divided into two groups, with one group assigned to the fertilized pasture and the other to the unfertilized pasture. Steers became accustomed to their respective pastures for 2 weeks prior to the first sampling period and remained there during all subsequent interim periods.

Diet sampling periods consisted of 4 consecutive days with samples collected once daily starting about 0700 for 30 minutes. Samples were collected during the early, middle, and late growing seasons for 2 years and during the early, middle, and late dormant season for one year.

Nutritive value of diet samples was based on organic matter (OM), fiber, crude protein (CP), and in-vitro organic-matter digestibility (IVOMD). OM and N analyses were according to AOAC (1970) procedures. CP contents of cattle diets were calculated as N × 6.25. Fiber was determined by the acid detergent (ADF) method (Van Soest 1963); IVOMD analyses were the two-stage technique of Tilley and Terry (1963). Chemical composition and digestibility data are presented on an ash-free basis. Statistical analyses of all data were by analysis of variance (Steel and Torrie 1960).

RESULTS AND DISCUSSION

No differences (P < 0.05) were detected in the mean dietary OM levels between pastures, but differences (P < 0.05) were noted among sampling periods and for the interaction of pastures × sampling period (Table 1). Dietary OM was highest (87%) during the later part of the second growing season and lowest (69%) during the middle of the first growing season. Dietary OM values were lower than those reported from other studies in the western U.S.A. (e.g., Conner et al. 1963). Herbage growth was minimal until late in the first growing season (when lowest OM values were found) due to limited rainfall. Possibly, grazed forage was contaminated with soil, of which there was more and which was more variable, on

the fertilized pasture. Supporting evidence for this theory was reported by Allison et al. (1977).

Dietary fiber exhibited almost parallel seasonal trends for the two pastures (Table 1). Fiber was generally lower in diets from the fertilized pasture (x̄ = 38.7%) than in those from the control pasture (x̄ = 42.3%), but differences between pastures were not significant during all periods. The highest dietary fiber value observed throughout the study was about 48%, which occurred during mid-dormancy on the control pasture.

The diets of steers grazing fertilized forage was always higher (P < 0.05) in CP during both growing seasons than diets of those grazing the control pasture (Table 2). However, differences in dietary CP between pastures during mid- to late dormancy were negligible. Dietary protein was exceedingly high on the fertilized pasture during the first growing season. These high dietary protein levels were probably due to the high proportion of forbs in the diet (Allison et al. 1977).

During both growing seasons fertilized-pasture diets were more (P < 0.05) digestible than control diets, but during mid- and late dormancy this trend reversed (Table 2). This reverse trend in digestibility values across sampling periods caused a significant (P < 0.01) pasture × period interaction. Digestibility values for both pastures during the second growing season are slightly higher than those frequently reported for cattle diets on western range areas of the U.S. (Kartchner and Campbell 1979). The early "green up" effect of N-fertilized forage cited by Wight (1976) was apparent in the present study. That effect, as well as the longer period of consumption of more digestible diet, was undoubtedly caused by a more varied diet selection, as observed by Allison et al. (1977).

The lower (P < 0.05) digestibility of grazed forage from the fertilized pasture beginning in mid-dormancy (Table 2) may be partially explained by the findings of Kelsey et al. (1972), who reported that with blue grama, a large part of the increased production due to N fertilization occurred in the stem fraction, which is the least digestible plant part. Moreover, a companion study of forage intake by grazing cattle has shown that, during dormancy, intake on N-fertilized pastures was lower than that on control pastures (Wallace and Cordova 1980). A summary of

Table 1. Mean organic-matter and fiber contents of steer diets for each sampling period on N-fertilized and unfertilized rangeland.[1]

Season of sampling	Time of sampling	Organic matter (%)		Fiber (%)	
		Unfert.	N-Fert.	Unfert.	N-Fert.
First growing season	Early	81.2abc	78.7c	42.7ab	43.3a
	Mid	78.0ac	69.3d	41.1abc	36.1c
	Late	82.8abc	80.1bc	40.3bc	36.2c
Dormant season	Early	82.4ab	83.4abc	44.5a	40.6ab
	Mid	85.1ab	86.5a	47.4a	42.5a
	Late	82.5ab	84.8ab	44.0a	42.3a
Second growing season	Early	83.2ab	85.0ab	39.6bc	35.0c
	Mid	84.5ab	83.8abc	42.0abc	37.3bc
	Late	86.7a	87.0a	38.6c	35.2c
Overall mean		82.9	82.1	42.3	38.7

[1]Mean values within a constituent underscored by the same line and those for each column followed by the same letter are not significantly different (P<.05).

Table 2. Mean crude protein and in-vitro organic-matter digestibility (IVOMD) of steer diets for each sampling period on N-fertilized and unfertilized rangeland.[1]

Season of sampling	Time of sampling	Crude protein (%)		IVOMD (%)	
		Unfert.	N-fert.	Unfert.	N-fert.
First growing season	Early	15.3a	19.5bc	63bc	61b
	Mid	15.3a	27.3a	64bc	65b
	Late	15.4a	21.7b	60c	67b
Dormant season	Early	10.9b	16.7c	50d	62b
	Mid	8.4b	11.4d	48d	41c
	Late	10.7b	11.4d	48d	44c
Second growing season	Early	15.0a	18.8bc	68ab	74a
	Mid	12.4ab	19.3bc	74a	74a
	Late	12.2ab	17.0c	74a	76a
Overall mean		12.8	18.1	61	63

[1]Mean values within a constituent underscored by the same line and those for each column followed by the same letter are not significantly different (P<.05).

seasonal cattle gains on the pastures used in the present experiment clearly indicates that N fertilization improved gain during the growing season but may not be advantageous in this respect during dormancy. The combination of the above findings suggests that N-fertilized range could be utilized to the greatest benefit during the growing season. Such a practice would relieve the pressure on unfertilized range until dormancy.

LITERATURE CITED

Allison, C.D., R.D. Pieper, G.B. Donart, and J.D. Wallace. 1977. Fertilization influences cattle diets on blue grama range during drought. J. Range Mangt. 30:177-180.

Association of Official Agricultural Chemists (AOAC). 1970. Official methods of analysis, 11th ed. AOAC, Washington, D.C.

Conner, J.M., V.R. Bohman, A.L. Lesperance, and F.E. Kinsinger. 1963. Nutritive evaluation of summer range forage with cattle. J. Anim. Sci. 22:961-969.

Donart, G.B., E.E. Parker, R.D. Pieper, and J.D. Wallace. 1978. Nitrogen fertilization and livestock grazing on blue grama rangeland. Proc. 1st Int. Rangeland Cong. 1:614-615.

Kartchner, R.J., and C.M. Campbell. 1979. Intake and digestibility of range forages consumed by livestock. Mont. Agric. Expt. Sta. Bul. 718.

Kelsey, R.J., A.B. Nelson, G.S. Smith, R.D. Pieper, and J.D. Wallace. 1972. Effect of N-fertilization on blue grama parts. Proc. West. Sec. Am. Soc. Anim. Sci. 23:237-240.

Pieper, R.D., J.R. Montoya, and V.L. Groce. 1971. Site characteristics on pinyon juniper and blue grama ranges in south-central New Mexico. N.M. Agric. Expt. Sta. Bul. 573.

Steel, R.G.D., and J.H. Torrie, 1960. Principles and procedures of statistics. McGraw-Hill Book Co., New York.

Tilley, J.M.A., and R.A. Terry. 1963. A two-stage technique for *in vitro* digestion of forage crops. J. Brit. Grassl. Soc. 18:104-111.

Van Soest, P.J. 1963. Use of detergents in the analysis of fibrous feeds. II. A rapid method for the determination of fiber and lignin. Assoc. Off. Agric. Chem. J. 46:829-835.

Wallace, J.D., and F.J. Cordova. 1980. Forage intake by cattle grazing blue grama rangeland. Abs. 72nd. Ann. Mtg. Am. Soc. Anim. Sci., Cornell Univ., 27-30 July.

Wight, J.R. 1976. Range fertilization in the northern Great Plains. J. Range Mangt. 29:180-185.

Top-Grazing High-Protein Forages with Lactating Cows

H.R. CONRAD, R.W. VAN KEUREN, and B.A. DEHORITY

Ohio Agricultural Research and Development Center,
University of Ohio, Wooster, Ohio, U.S.A.

Summary

Two grazing systems, top-grazing of alfalfa (*Medicago sativa* L.) and high-intensity grazing of orchardgrass (*Dactylis glomerata* L.), were investigated with three and four levels of concentrate, respectively. The concentrates consisted of corn, oats, and minerals. High dry-matter (DM) alfalfa silage fed in the barn with 40% concentrate by weight was used for a positive control ration. Three- or 4-day grazing periods/paddock were used with 16- and 24-day rotation cycles for orchardgrass and alfalfa, respectively. Using alfalfa without concentrate resulted in mean daily yields of 20.2 kg of milk and 0.65 kg of protein. As amounts of concentrates were increased, daily milk and milk protein increased by 50%; but the proportion of protein edible for humans eaten by the cows increased from 0% to 43% of the milk protein produced. Consequently, the amount of essential amino acids returned to the human food supply in milk completely offset the amount consumed by the cows. With increasing concentrates the amount of fossil fuel used for feed production in cows grazing alfalfa rose from 1.6 to 3.2 megacalories (mcal)/day of herd life. All cows grazing alfalfa were fed between 10 and 20 g of poloxalene daily, which prevented pasture bloat. Milk production was less with orchardgrass, and the fossil fuel requirements for production were doubled. The average amount of protein consumed/ha of alfalfa/season was 1.8 metric tons in 8.1 tons of DM.

KEY WORDS: alfalfa, *Medicago sativa* L., cows, grazing, milk production, protein, human food, efficiency, bloat, concentrates.

INTRODUCTION

Alfalfa harvested in the prebud vegetative stage has been known to be of superior nutritive value since the turn of the century. Not only is it capable of meeting a much greater proportion of the cow's caloric and protein needs than are mature forages, but its relative value in terms of replacement feeds is high.

In an early study (Huffman 1939), alfalfa cut prebud was found to contain 25.3% crude protein and to have a coefficient of digestibility for organic matter of 76%. There are now many research reports documenting the greater nutritive value of young-cut forages than of mature forages (Barnes and Gordon 1972).

Long-range needs for conserving energy and the growing worldwide competition of humans for cereal grains dictate the use of minimal amounts of concentrate feeds commensurate with efficient utilization of available forages for lactating cows. The amount of forage a cow can eat in a 24-hour period is limited by the indigestible residue in the digestive tract. When digestibility is between 63% and 67% other factors than indigestible residue limit voluntary feed intake. In cows that possess the genetic capability for high milk production, feed intake is a major limiting factor on production (Conrad et al. 1964). The factor that influences forage quality to the greatest extent is the stage of maturity of the plant when it is harvested as pasture, hay, or silage (Conrad et al. 1961).

This paper considers the subject from the viewpoint of the amount of protein moved into the human food chain through grazing dairy cows.

MATERIALS AND METHODS

High-producing Holstein cows were used to compare several forage programs as the major component of the diet. One-half, limited, or no energy supplementation was provided. Alfalfa, *Medicago sativa* L., and orchardgrass, *Dactylis glomerata* L., were each grazed separately and also together (on alternate days). Two control systems, similar to current commercial dairy programs, were used: legume-grass pasture and legume-grass silage, each

H.R. Conrad is a Professor of Dairy Science, R.W. Van Keuren is a Professor of Agronomy, and B.A. Dehority is a Professor of Animal Science at Ohio Agricultural Research and Development Center of the University of Ohio.

fed with a high rate of a standard dairy concentrate mixture (11.6% total protein) providing 40% of the daily nutrient requirements.

After soil tests the fields were fertilized to maintain high productivity, and pesticides were applied as needed to control insects and weeds. Periodic subjective stand estimates of the pastures were made to determine botanical composition of the pastures and to follow any changes in the swards.

A short-term paddock system was used to provide a continuing supply of high-protein, immature forage for cows to graze throughout the summer. It combined the rotation and strip-grazing systems. Short grazing periods of 2 to 4 days were used, but pastures were divided into fixed areas by electric fencing.

Rotational pastures (0.6 ha) were used. Stocking rates ranged between 3.5 and 6.5 cows/ha. The put-and-take system of adding cows was used to graze excessive growth.

For alfalfa, the fields were divided into eight equal-sized paddocks. Each plot was grazed 3 days. This schedule allowed 21 days for regrowth of forages in each plot. Forage was 20 to 25 cm high when cows were removed, essentially a top-grazing procedure that had previously been shown to provide high-quality forage. Clipping was not practiced. If growth was excessive and could not be managed with extra cattle, an entire plot was ensiled, and rotation rate was accelerated by one period.

With orchardgrass the grazing was more intensive. Four plots were used; grazing was for 4 days/lot so that only 12 days elapsed in one rotation. Thus, the height of the orchardgrass was kept below 30 cm.

RESULTS AND DISCUSSION

The results are presented in the Tables. The alfalfa-Ladino grazing program was coupled with intensively grazed orchardgrass in which cows were rotated to new paddocks at 4-day intervals, completing a cycle of four paddocks in 16 days. The cows were in the alfalfa by day and the orchardgrass at night. The comparisons also included cows fed alfalfa silage, concentrates, and orchardgrass without alfalfa grazing. Milk production was generally higher in cows grazing alfalfa even though less concentrate was fed (Table 1). Even without concentrates, daily milk production averaged 20.2 kg/day at 145 days in lactation. Feeding concentrates resulted in in-

Table 1. Feed intake, total milk, and milk protein yield with single-crop grazing of alfalfa (A) or orchardgrass (OG).

Type of pasture or forage	Concentrate	Pasture	Total milk yield[1]	Milk protein
	------------------- (kg/day)-------------------			
A	0	18.6	20.2[a]	0.68
A	4.8	15.8	24.2[b]	0.81
A	9.6	9.9	31.1[c]	0.96
OG	0	17.2	13.2[d]	0.44
OG	6.0	15.2	17.7[e]	0.57
OG	9.6	10.6	21.4[a]	0.68
OG	10.6	9.1	29.4[cf]	0.94
AS[2]	8.0	10.1	28.4[f]	0.90

[1]Means having different superscripts were statistically different, P<.05.

[2]AS = alfalfa silage.

Table 3. Ratios of protein and fossil-fuel usage with single-crop grazing of alfalfa (A) or orchardgrass (OG).

Type of pasture or forage	Stocking rate[1]	Land for lactation	Fossil fuel[2] per day herd life	Ratio fossil fuel energy to milk energy
	(No./ha)	(ha/400 d)	(mcal/d)	
A	3.4	.86	1.56	1:7.4
A	3.9	.92	2.49	1:6.2
A	6.0	.97	3.18	1:6.2
OG	2.3	.92	2.47	1:4.0
OG	2.8	1.03	2.97	1:3.8
OG	4.0	.99	3.36	1:4.0
OG	4.6	.97	3.47	1:5.4
AS[3]	NA	1.01	3.41	1:4.8

[1]Mean grazing density on total area used during grazing season.

[2]Fossil-fuel energy values obtained from Keener and Roller (1977).

[3]AS = alfalfa silage.

creased (P < 0.05) milk production and milk protein. The increased milk protein obtained with increased concentrate feeding was almost equal to the digestible protein that would be available to people from the corn and oats. This tradeoff means that no real net loss of protein occurred inasmuch as low-protein concentrates, corn and oats, were all that was required for meeting total protein needs (Table 2).

Mean lactation day for the milk yields was 145 days. The experiments were initiated between 20 April and 8 May and terminated 10 September. Since declines in milk yield (persistence), which were essentially linear, ranged between 0.05 and 0.07 kg/day, it may be concluded that an average of 20 to 25 kg of 4% fat-corrected milk/day can be sustained in cows on pastures. Slightly more than 1,800 kg of protein and 8.1 metric tons of dry matter were consumed on each ha grazed.

In these studies, poloxalene (a bloat-preventing surfactant) was added to the concentrate ration so that each cow received between 10 and 20 g daily. Bloat was not observed in any cows. A significant problem was maintenance of electric fencing. In wet weather, electrical shorts to ground were numerous.

Grazing reduced gasoline requirements. Fossil fuel/day of herd life was less than that required for ensiled alfalfa and concentrate feeding in the barn (Table 3). The alfalfa grazing program resulted in less fossil fuel consumption/unit of milk output. With grazed alfalfa as the major source of protein, we obtained 6.2 to 7.4 megacalories (mcal) of energy output/mcal of fossil fuel used in growing. The results for stocking rates and land usage are summarized in Table 3 also. Only the pasture land was included in the computation for stocking rates, but the land required for producing corn and oats was included in computation of land requirements/lactation.

It was notable that the intensive grazing systems provided the total feed and high-protein forages on slightly less land than was required for cows fed concentrates. Thus, these systems are feasible alternatives without grain when a choice is required against high-cost fuel or for conservation of land. Such a decision would require acceptance of 29% less milk/ha (Tables 1 and 3).

Table 2. Grain intake, milk, and essential amino acids (EAA) produced with single-crop grazing of alfalfa (A) or orchardgrass (OG) and compared with alfalfa silage (AS).

Pasture	Grain	Milk yield	Feed EAA[1]	Milk EAA	Gain in milk EAA	Ratio of EAA in feed to milk
	------------------------------- (kg/day)-------------------------------					
A	0	20.2	0	0.33	0.33	0:1
A	4.8	24.2	0.08	0.39	0.06	1:3.8
A	9.6	31.1	0.14	0.48	0.15	1:2.4
OG	0	13.2	0	0.22	0.22	0:1
OG	6.0	17.7	0.10	0.29	0.07	1:1.9
OG	9.6	21.4	0.15	0.35	0.13	1:1.3
OG	10.6	29.4	0.24	0.48	0.26	1:1.4
AS	8.0	28.4	0.15	0.46	0.13	1:2.0

[1]Essential amino acids that would be available to humans if the cereal grains were consumed directly.

LITERATURE CITED

Barnes, R.F, and C.H. Gordon. 1972. Feeding value and on-farm feeding. In C.H. Hansen (ed.) Alfalfa science and technology. Am. Soc. Agron., Madison, Wis.

Conrad, H.R., A.D. Pratt, and J.W. Hibbs. 1961. Cutting date determines forage quality. Ohio Fm. and Home Res. 46:39.

Conrad, H.R., A.D. Pratt, and J.W. Hibbs. 1964. Regulation of feed intake in dairy cows. I. Change in importance of physical and physiological factors with increasing digestibility. J. Dairy Sci. 47:54

Huffman, C.F. 1939. Roughage quality and quantity in the dairy ration. J. Dairy Sci. 22:889.

Keener, H.M., and W.L. Roller. 1977. Maximizing energy efficiency of animal system. Proc. Ohio Dairy Day, Ohio Agr. Res. and Dev. Cntr., Wooster, 12 Aug.

Long-Term Feeding of Corn Silage as the Sole Forage in Diets of Lactating Dairy Cows

C.H. NOLLER, N.J. MOELLER, and C.L. RHYKERD

Purdue University, West Lafayette, Ind., U.S.A.

Summary

The long-term feeding of corn silage as the sole forage to lactating dairy cows has had only limited success due to problems with production, reproduction, and health after one lactation. Consequently, a long-term study was begun in August 1975 with two groups of 19 cows, 15 2-year-old heifers, and 4 3-year-old cows assigned at random to each group as they calved. One group was fed a diet with corn silage as the sole forage and the other group a diet with a 50:50 ratio of forage as corn silage (CS) and low-moisture alfalfa-orchardgrass (AO) silage. Both rations contained a 60:40 ratio of forage to concentrate on a dry basis, 15% crude protein, 1.25% calcium, 0.50% phosphorus, 0.52% trace mineralized salt, 0.22% sulfur, 0.25% magnesium, and 4,400 units vitamin A, 2,200 units vitamin D, and 44 mg vitamin E/kg. The diets were blended in a mixer wagon with electronic load cells and fed once daily as a complete mix in feed bunks. All cows remained on their respective diets the entire experiment, including the dry period when they were fed controlled amounts of the lactating-cow diet.

When the experiment was terminated in May 1980, there were 5 cows in the CS group and 3 in the other group. No cows were removed the first lactation. In succeeding lactations, cows were removed from the two groups, respectively, for the following: hardware, 2 and 0; aborted, 0 and 1; pneumonia, 1 and 0; udder problems, 3 and 3, leg problems, 3 and 2; low production, 2 and 3; failure to breed, 1 and 5; other, 2 and 2. There was little difficulty with digestive upsets and no problem with the fat cow syndrome. Milk production, fat tests, and fat production for 305-day lactations for the CS and the CS and AO group, respectively, were as follows: lactation: I, 6,686 kg, 3.62%, 238.5 kg (CS) and 6,535 kg, 3.67%, 237.7 kg (CS and AO); II, 7,462 kg, 3.64%, 271.9 kg (CS) and 6,800 kg, 3.63%, and 246.8 kg (CS and AO); III, 8,008 kg, 3.60%, and 288.3 kg (CS) and 6,989 kg, 3.68%, and 257.4 kg (CS and AO); IV, 8,338 kg, 3.51%, and 293.1 kg (CS) and 7,594 kg, 3.69%, and 280.4 kg (CS and AO). The differences were not significant (P > 0.05), although cows on the CS diet tended to produce more milk and fat.

Higher than normal levels of limestone were used in these diets to reduce loss of starch in feces and to improve gastrointestinal and fecal pH. Cows in the CS and the CS and AO groups had average fecal pH of 6.58 and 6.61, respectively. Calving weights of cows were slightly higher for the CS group. Calf weights were similar for the two groups. These data indicate that corn silage can be used successfully as the sole forage in complete diets of lactating cows when diets are properly balanced.

Key Words: corn silage, Zea mays L., dairy cows, milk production, limestone, buffering.

INTRODUCTION

In the North Central Region of the U.S.A., corn (Zea mays L.) silage is high yielding and easily grown, harvested, and stored, and it fits easily into mechanized feeding programs. Consequently, dairy farmers are interested in increasing the amount of corn silage fed to dairy cattle and in the possibility of using corn silage as the sole forage. Previous long-term experiments with corn silage as the sole forage for lactating cows have been at lower levels of milk production (Converse and Wiseman 1952) than are expected today. Also, there have been problems with unusually heavy losses of cows for unexplained reasons (Trimberger et al. 1972). Problems associated with the feeding of corn silage as the sole forage usually have not been noted until after a year or more on the program.

Many of the difficulties associated with the feeding of corn silage as the sole forage are not observed until the second lactation. Consequently, a study was initiated with lactating dairy cows fed corn silage as the sole forage for more than one lactation. Corn silage was used as the sole forage over the entire lactation and through the dry period to eliminate confounding effects from other feeds and to permit a more rigorous evaluation of the capability of cows to utilize corn silage as the sole forage.

MATERIALS AND METHODS

The experiment was begun at Purdue University in August 1975 with two groups of animals: one fed a diet with corn silage as the sole forage, and the other a diet with a 50:50 forage mixture of corn silage and low-

C.H. Noller is a Professor of Animal Sciences, N.J. Moeller is an Extension Dairy Specialist, and C.L. Rhykerd is a Professor of Agronomy at Purdue University. This is a contribution of the Purdue University Department of Animal Sciences and Agronomy (Journal Paper No. 8402).

Table 1. Ingredient composition of complete mixed diets for lactating dairy cows (dry basis).

Ingredients	Group	
	Corn silage	Corn silage and alfalfa-orchardgrass silage
Corn silage, %	60.00	30.00
Alfalfa-orchardgrass silage, %	–	30.00
Corn grain, %	19.96	29.29
Soybean meal (49), %	15.85	7.35
Dicalcium phosphate, %	0.99	0.93
Limestone, %	2.34	1.65
Trace mineralized salt, %	0.52	0.52
Magnesium oxide, %	0.10	0.05
Elemental sulfur, %	0.07	0.04
Vitamin A-D-E supplement, %	0.17	0.17
Total	100.00	100.00

moisture alfalfa-orchardgrass (*Medicago sativa* L.-*Dactylis glomerata* L.) silage (approximately equal parts of alfalfa and orchardgrass) (Table 1). Each group was composed of 15 2-year-old heifers and 4 3-year-old cows assigned at random to treatment as they calved. No attempt was made to balance groups for expected production; rather, all assignments were random depending on calving date. The cows were placed in their respective groups within two days after calving and remained on their respective treatments until they were removed from the experiment or the experiment was terminated. The experiment was terminated in May 1980 after four lactations.

All feed was fed as a complete mixed diet. No free-choice minerals or other feeds were available except for water. The grain portion of the diets was mixed, stored, and blended into the complete mixed diet in a mixer wagon with electronic load cells and was fed once daily in a feed bunk. The amount fed was adjusted to ensure essentially complete consumption each day.

Both diets contained a 60:40 ratio of forage to concentrate on a dry basis and 15% crude protein, 1.25% calcium, 0.50% phosphorus, 0.22% sulfur, 0.25% magnesium, 0.52% trace mineralized salt, and 4,400 units vitamin A, 2,200 units of vitamin D, and 44 mg vitamin E/kg feed. The ingredient composition of the diets varied because of changes in chemical composition of ingedients used, but the nutrient specifications remained the same throughout the experiment.

Feed intakes were ad libitum during the lactation phase. When a lactation was completed, the cows were placed in a separate area and fed approximately 8.5 kg dry matter daily of the same diet fed to the lactating cows. This level of feed intake was slightly above maintenance and was designed to reduce energy intakes and prevent overconditioning during the dry period. Immediately after parturition, the cows were returned to the milking groups. No adjustment period was used. No feed other than the experimental diets was available to the cows.

The cows were housed in free stalls with concrete lots

and outdoor feed bunks. They remained on concrete the entire time on experiment. They were milked twice daily in a milking parlor with no feed in the parlor. Body weights were taken biweekly and feed intakes and milk weights daily.

RESULTS AND DISCUSSION

Data obtained in this experiment indicate little difference between cows fed corn silage as the sole forage and those fed a forage mixture of corn silage and low-moisture alfalfa-orchardgrass silage. Milk-production data for the four 305-day lactations are presented in Table 2. Statistical analysis of the data showed no significant difference ($P > 0.05$) between the two groups in milk production, fat test, and fat production. However, cows fed the corn silage diet tended to produce more milk and fat each lactation. Although the diets were balanced to the same specifications, the energy content of the corn silage ration was slightly higher.

Milk-fat tests and fat production were not significantly different (Table 2). Low fat tests were a possibility because a fairly fine chop was used in making the corn silage. Crude-fiber values for the two respective diets were 12.8% and 15.4%, lower than recommended (NRC 1978). The average fat tests for animals in this study were similar to average fat tests for the milking herd from which the experimental animals were selected.

All experimental animals completed at least one lactation. As the experiment progressed cows were removed from the experiment for a variety of reasons. When the experiment was terminated in May 1980, there were 5 cows in the corn silage group and 3 in the corn silage and alfalfa-orchardgrass silage group. Cows were removed from the two groups, respectively, for the following reasons: hardware, 2 and 0; aborted, 0 and 1; pneumonia, 1

Table 2. Milk and fat production of dairy cows for four 305-day lactations with standard errors.

Lactation	Item	Group	
		Corn silage	Corn silage and alfalfa-orchardgrass silage
I	Cows, no.	19	19
	Milk, kg	6686±254	6535±254
	Fat, kg	238.5±6.9	237.7±6.9
	Fat, %	3.62±.10	3.67±.10
II	Cows, no.	16	15
	Milk, kg	7462±398	6800±410
	Fat, kg	271.9±8.5	246.8±8.7
	Fat, %	3.64±.10	3.63±.11
III	Cows, no.	8	13
	Milk, kg	8008±498	6989±365
	Fat, kg	288.3±15.3	257.4±11.2
	Fat, %	3.60±.11	3.68±.09
IV	Cows, no.	4	5
	Milk, kg	8338±491	7594±439
	Fat, kg	293.1±14.4	280.4±12.9
	Fat, %	3.51±.16	3.69±.14

Table 3. Cow weights and calf weights with standard errors.

	Group	
	Corn silage	Corn silage and alfalfa-orchardgrass silage
Age at calving, mo.		
First lactation animals (15)	24.6	24.5
Second lactation animals (4)	35.5	37.3
Calving weights, kg		
First calving	544±17	531±17
Second calving	635±15	613±15
Third calving	675±38	660±36
Fourth calving	690±23	685±30
Calf weights, kg		
First calving	42.2±1.6	40.9±1.6
Second calving	45.9±1.2	44.2±1.2
Third calving	42.5±1.9	43.8±2.0
Fourth calving	43.3±1.1	45.0±0.5

and 0; udder problems, 3 and 3; leg problems, 3 and 2; low production, 2 and 3; failure to breed, 1 and 5; other, 2 and 2. None of the cows was removed for reasons directly attributable to use of corn silage as the sole forage. Previously, Trimberger et al. (1972) reported extensive loss of cows from an all-corn silage group for unexplained causes.

Calf weights were similar for the two treatments (Table 3). Calving weights of cows increased similarly for both groups with each calving. Feed intakes of cows were restricted during the dry period to near maintenance to reduce energy intake and reduce weight gains. This approach to feeding counteracted the tendency of dry cows fed high levels of corn silage to have problems with excessive fattening associated with the postpartum fat cow syndrome.

In the present experiment, digestive upsets and displaced abomasums were a minor problem. There were two cases of displaced abomasums that required surgical intervention. The more serious health problems involved mastitis and leg problems associated with animals continuously on concrete. Retained placentas were a problem

at times, but the incidence was similar to that in the regular milking herd.

Higher levels of limestone were added to the diets to make them more like alfalfa forage in calcium content. The higher level of limestone was used to maintain a more favorable environment in the gastrointestinal tract (Noller 1978) and to reduce the incidence of problems related to digestive upsets. Previous research had shown the merits of limestone in reducing loss of starch in feces and in improving gastrointestinal tract and fecal pH (Wheeler and Noller 1976, 1977). The fecal pH of the two groups of lactating cows was 6.58 and 6.61, respectively, which is considered to be in the normal range.

Based on the results of this investigation, corn silage is a very satisfactory ingredient as the sole forage in diets of dairy cows provided the ration is supplemented with adequate protein, minerals, and vitamins and fed as a complete mixed ration to ensure consumption of a proper balance of nutrients. The lactating-cow ration can be used successfully in the dry period provided intake is restricted to near maintenance to prevent excessive fattening.

LITERATURE CITED

Converse, H.T., and H.G. Wiseman. 1952. Corn silage as the sole roughage for dairy cattle. U.S. Dept. Agric. Tech. Bul. 1057.

National Research Council (NRC). 1978. Nutrient requirements of domestic animals, No. 3. Nutrient requirements of dairy cattle, 5th rev. ed. Nat. Acad. Sci.-Natl. Res. Coun., Washington, D.C.

Noller, C.H. 1978. New developments in ruminant nutrition: the lower gastrointestinal tract. Proc. 25th Md. Nutr. Conf., 67-72.

Trimberger, C.W., F. Tyrrell, D.A. Morrow, J.T. Reid, M.J. Wright, W.F. Shupe, W.G. Merrill, J.K. Loosli, C.E. Coppock, L.A. Moore, and C.H. Gordon. 1972. Effects of liberal concentrate feeding on health, reproductive efficiency, economy of milk production, and other related responses of the dairy cows. N.Y. Food and Life Sci. Bul. 3.

Wheeler, W.E., and C.H. Noller. 1976. Limestone buffers in complete mixed rations for dairy cattle. J. Dairy Sci. 59:1788-1793.

Wheeler, W.E., and C.H. Noller. 1977. Gastrointestinal tract pH and starch in feces of ruminants. J. Anim. Sci. 44:131-135.

Energy Balance Studies with Lactating Cows Fed Fresh or Frozen Grass from Pasture Suitable for Grazing

Y. VAN DER HONING and A.J.H. VAN ES

Institute for Livestock Feeding and Research (IVVO), Lelystad,
and Agricultural University, Wageningen, The Netherlands

Summary

Information on feeding value and energy utilization of grass consumed by grazing cattle, especially lactating cows, is limited. It is nearly impossible to perform calorimetric studies with grazing dairy cows. Similar studies indoors often show low intakes of fresh grass and require much labor. Still, in much of the world, cattle graze at least half the year; therefore, knowledge of the feeding value of grass is important for practice.

In the Netherlands, each year (1973-1974) each of four permanent pastures in the grazing stage of maturity was cut twice daily and either fed fresh or stored at $-20°C$ to be fed later to four lactating cows in respiration chambers. Animals were adapted for 7-10 days before a 7-day collection period on fresh grass ad libitum plus 1 kg of concentrates. In a subsequent week, the same grass in the frozen form was used. Digestibility, N, and energy balances were measured to determine energy utilization from grass.

In 1977-1979, a total of seven similar calorimetric trials were conducted with four lactating cows fed ad libitum with grass after storage at $-20°C$. In some trials, half of the grass dry matter was substituted with concentrates. The grass was fed to dairy cows at production level as well as to wethers near the maintenance level. Because many feed-evaluation systems use sheep digestibility coefficients as their basis, differences in digestibility between sheep and cattle were determined. No differences in intake, digestibility, or energy utilization between fresh and frozen grass could be detected.

In the total of 83 energy balances, the digestibility of grass by cattle was high (organic matter 75%-83%, energy 71%-78%) but varied between pastures and months. Intake depended mainly on the pasture, the composition of the grass (percentage of DM), and/or the season. Intake varied from 10.2 to 16.0 kg organic matter, but was lowest in the later trials.

Despite higher feeding levels of the cattle, cow and wether digestibilities hardly differed, except for trials with extra concentrates in the diet. Utilization for maintenance and milk production of the metabolizable energy from grass agreed fairly well with efficiencies of diets of mixed conserved forage and concentrates, although some results tended to be lower.

It was concluded that grass from pastures suited for grazing is highly digestible and efficiently utilized for maintenance and milk production. Increased level of feeding caused ony a minor depression of digestibility in lactating cattle.

KEY WORDS: energy utilization, dairy cows, wethers, permanent pasture, prediction of net energy, fresh or frozen grass.

INTRODUCTION

Although lactating cattle graze standing pasture during much of the year, there is little information from calorimetric studies on utilization of fresh herbage, although there is some on grass in a preserved form, such as hay, silage, or artificially dried or freeze-dried grass (Jentsch et al. 1972, Corbett et al. 1966, Heany et al. 1966).

This paper presents results of calorimetric trials with lactating cows to determine: (1) energy utilization of fresh or frozen grass from Dutch permanent pastures in the grazing stage of maturity; (2) effect of feeding level on digestibility and metabolizability of grass compared with such effects in winter rations; and (3) the net energy of grass and of concentrates. Finally, predicted metabolizable and net energy, as described by van Es (1978), were compared with part of the experimental results.

METHODS

Digestibility, and N and energy balance of four lactating cows were measured over 7-10 days on a diet of grass ad libitum plus 1 kg of magnesium (Mg)-enriched concentrates to prevent hypomagnesia. Beginning in 1977, similar trials were carried out with four wethers, fed 1 kg of dry matter (DM) from the same grass as fed to the cows, at their approximate maintenance level of feeding. In 1978, trials with dairy cows were also performed on a diet where 50% of the grass was replaced by 7 kg of concentrates. Digestibility of these concentrates was mea-

Y. van der Honing is a Research Worker at IVVO; and A.J.H. van Es is the Deputy Director of IVVO, and a Professor of Energy Metabolism at Agricultural University.

sured with sheep near maintenance, using grass as a basal feed.

Grass was offered to cattle six to eight times a day and concentrates twice; feeds were offered to wethers, twice daily, also. Samples of the grass were freeze dried as well as dried at 65°C and were analyzed by the usual methods. In 1973 and 1974, a comparison of fresh and frozen grass was made eight times on a sward of mixed grass species harvested during 7 days (van Es and van der Honing 1976).

From 1977, at Lelystad, frozen grass (mainly perennial ryegrass) harvested on 205 consecutive days from reclaimed pastures was used to compare results from trials with dairy cows and sheep. The daily diet offered consisted of samples of the grass in proportion to the amount harvested on different days.

When digestibility coefficients of grass from sheep trials were not available, the metabolizable energy (ME) was predicted (van Es 1978) by using equation 1: $ME = 14.22 D_O + 5.86 D_{CP}$, where ME is kJ/g DM and CP, O, D_{CP}, and D_O are crude protein, organic matter, digestible CP, and digestible O, respectively, in g/kg DM. D_O was estimated from digestibility in vitro (modified Tilley and Terry method; van der Meer 1980), and D_{CP} was estimated from equation 2: $D_{CP} = 0.959 CP + 0.04$ ash − 40. Utilization of ME for maintenance and milk production (k_l) was predicted (van Es 1978) by equation 3: $k_l = 0.6 + 0.0024 (q - 57)$, in which q = ME as a percentage of gross energy (GE).

RESULTS

Dry-matter content (DM), crude protein, and gross energy in dry matter of the grass are presented in Table 1. DM as well as CP content varied considerably with season of harvest and fertilizer treatment (Table 1). Rainy

weather resulted in some lower DM values. The GE was fairly constant except for grass of 1977–1978 with a high ash content. The DM intake varied due to season, stage of lactation, and individuality of cows and experiments. No significant difference in intake between fresh and frozen grass was observed ($- 0.4 \pm 0.8$ kg DM). We found rather low values in the experiments since 1977 at Lelystad compared with intake figures obtained by Meijs (1981) from grazing cattle. No explanation could be found because in the earlier experiments a higher DM intake was observed in several trials.

The in-vitro D_O of the grass is presented in Table 1; D_O measured with sheep at about maintenance is in Table 2. The concentrates in 1978, measured in a difference trial with sheep at maintenance, had a D_O of 85.5%. These digestibility coefficients were used to predict digestibilities of dairy diets at maintenance, which were compared with D_O values from cows (Table 1, column 8); depression of digestibility due to increased level of feeding was then estimated (Table 2).

Digestibility of energy (D_E) was highly correlated with D_O; both were high for the grass. No significant differences in D_O, D_{CP}, and D_E were measured between fresh and frozen grass.

Metabolizability (q = 100 ME/GE) of dairy diets (Table 1) and the effect on q of feeding level, estimated from results with sheep at maintenance, are presented (Table 2). Difference between fresh and frozen grass in q was not significant (0.4 ± 1.8). Feeding level was calculated from ME intake expressed as a multiple of maintenance ($= 490$ kJ/kg$^{0.75}$ live weight). In per-unit increase in feeding level, the relative depression of D_O or q ($\% \Delta D_O$ or $\% \Delta q$) was greater for D_O than for q (Table 2). The effect on q was generally lower than the average of $- 1.8\%$ found with winter rations (van Es 1978).

Utilization of ME is shown in Fig. 1, where productive

Table 1. Time of harvest and composition of grass and mean intake (DMI), digestibility of organic matter (D_O), metabolizability (q = 100 ME/GE), and utilization of ME (k_l = net energy/ME × 100) of the diet consumed by lactating dairy cows in the balance trials.

	Grass				Dairy diet[1]			
Harvest (yr.-mo.)	DM[2] (%)	CP/DM (%)	GE/DM MJ/kg	D_O in vitro (%)	DMI (kg)	d_0 (%)	q (%)	k_l (%)
1973-5	15.2	22.7	18.9		14.7	81.5	66.3	59.7
1973-6	20.7	19.4	18.7		17.4	77.8	63.6	63.1
1973-8	20.3	24.4	18.9		13.9	77.3	60.6	58.1
1973-10	16.0	27.8	19.0		11.9	78.6	60.1	58.9
1974-5	21.1	17.9	18.8		14.6	81.8	66.6	63.1
1974-6	16.6	20.4	18.6		14.2	78.0	62.2	59.0
1974-8	18.2	24.4	19.0		14.4	77.6	60.6	59.7
1974-10	13.8	29.2	18.9		14.9	80.5	65.8	63.8
1977-9	18.2	25.2	18.2		14.1	80.0	62.2	
1978-5a	17.1	19.8	18.4	81.9	11.6	80.1	62.7	55.9
1978-5b		as in 5a			12.9	80.1	65.9	61.2
1978-8a	12.9	26.5	17.5	77.2	12.0	76.1	60.1	55.2
1978-8b		as in 8a			11.9	77.6	63.3	
1979-5	16.1	24.7	18.9	85.5	12.7	83.1	67.2	58.5
1979-8	14.2	29.5	18.4	82.0	12.3	80.4	63.6	54.7

[1]Ad lib grass + 1 kg of concentrates, except 1978-5b and 1978-8b with 7 kg of concentrates.

[2]DM = dry matter, CP = crude protein, GE = gross energy, ME = metabolizable energy.

Table 2.   Effect of level of feeding (FL) on digestibility of organic matter (ΔD_O) and metabolizability (Δq). Predicted D_O and q values at maintenance for dairy cows were estimated from D_O of sheep. Predicted values were compared with measurement of D_O and q at higher FL in dairy cows.

Experiment (yr.-mo.)	D_O of grass (sheep)	Predicted D_O at maintenance	ΔD_O (production maintenance)	q (production maintenance)	FL	\Delta per unit increase of FL as %	
						%ΔD_O	%Δq
1977–9	81.5	81.8	− 1.8	− 0.7	3.03	− 1.1	− 0.5
1978–5a	83.5	83.7	− 3.6	− 1.6	2.45	− 3.0	− 1.7
1978–5b		84.6	− 4.5	− 1.4	2.82	− 2.9	− 1.1
1978–8a	78.3	78.9	− 2.8	0.3	2.37	− 2.6	0.4
1978–8b		82.3	− 4.7	− 2.1	2.48	− 3.9	− 2.2
1979–5	85.3	85.3	− 2.2	0.2	3.03	− 1.3	0.1
1979–8	80.3	80.7	− 0.3	1.1	2.69	− 0.2	+ 1.0

net energy in milk plus body reserve tissue was plotted vs. ME (both/unit [$W^{0.75}$]). This was done assuming a constant net energy for maintenance of 293 kJ/kg$^{0.75}$ (van Es 1978). The results of T.E. Trigg (personal communication, 1981) were also plotted (Fig. 1). The figure indicates good agreement in utilization of ME from grass between N.Z. and Dutch experiments. Efficiency of utilization of ME (k_l) of dairy diets was calculated from production plus maintenance net energy/ME. These k_l values were approximately 4% lower than the values predicted from equation 3, varying from − 2.4% to + 11.2% in different experiments. Although in some trials k_l tended to be lower, in most trials ME of grass was utilized as well as ME from winter diets.

DISCUSSION

Depression of metabolizability by increased level of feeding seems to be lower in these grass diets with substantial ground concentrates. Probably the more gradual intake of grass fed six to eight times instead of twice daily as in use for winter diets has resulted in a lower initial rate of fermentation and of passage in the forestomachs. Although rather high concentrations of soluble carbohydrates may be present in grass, the intact cell wall at the moment of ingestion and the substantial proportion of structural carbohydrates are supposed to prevent a rapid increase of volatile fatty acids and too-low pH in the rumen fluid in most cases. Suboptimal rumen fermenta-

Fig. 1. Net energy in milk plus body reserve tissue (NE_p) vs. metabolizable energy (ME), both expressed/kg metabolic body weight ($W^{0.75}$) in kJ/kg$^{0.75}$. Data from N.Z. experiments from T.E. Trigg (personal communication).

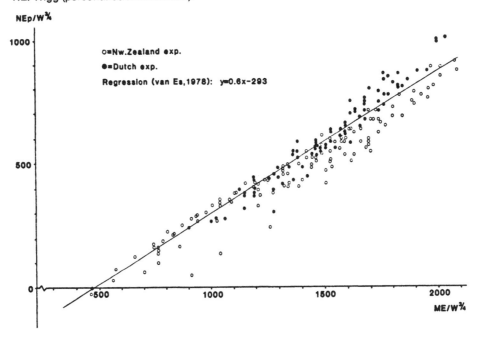

tion and increased rate of passage are believed to play a major role in the depression of digestibility and metabolizability found with ground feedstuffs.

The cows in our experiments were confined, and the lack of exercise and work of grazing decreased their requirement compared with the normal grazing situation. Therefore, the net energy produced in milk and body tissue may be slightly higher in such trials compared to practice.

LITERATURE CITED

Corbett, J.L., J.P. Langlands, I. McDonald, and J.D. Pullar. 1966. Comparison by direct animal calorimetry of the net energy values of an early and a late season growth of herbage. Anim. Prod. 8:13–27.

Heany, D.P., W.J. Pigden, and G.I. Pritchard. 1966. The effect of freezing or drying pasture herbage on digestibility and voluntary intake assays with sheep. Proc. X Int. Grassl. Cong., 379–384.

Jentsch, W., R. Schiemann, L. Hoffman, and H. Wittenburg. 1972. Die energetische Verwertung der Grünfutterstoffe im frischen and getrockneten Zustand durch Wiederkäuer (The energy utilization of fresh and dried green forage by ruminants). Arch. Tierernährung 22:17–40.

Meijs, J.A.C. 1981. Herbage intake by grazing dairy cows. Agric. Res. Rpt. 909. Pudoc, Wageningen.

van der Meer, J.M. 1980. Determination of the "in vitro" digestibility for the prediction of the "in vivo organic matter digestibility coefficient" of feeds for ruminants. Doc. Rpt. 67. Inst. Livest. Feeding Nutr. Res., Lelystad, The Netherlands.

van Es, A.J.H. 1978. Feed evaluation for ruminants. I. The system in use from 1977 onwards in the Netherlands. Livest. Prod. Sci. 5:331–345.

van Es, A.J.H., and Y. van der Honing. 1976. Energy and nitrogen balances of lactating cows fed fresh or frozen grass. In M. Vermorel (ed.) Energy metabolism of farm animals, E.A.A.P. Pub. 19:237–240.

Utilization of Energy of Fresh Pasture by Lactating Dairy Cows

T.E. TRIGG, C.R. PARR, and N.R. COX
Ruakura Animal Research Station, Victoria, Australia

Summary

The New Zealand dairy industry is almost totally reliant on grazed herbage for the production of milk solids, but few data exist on the efficiency with which lactating cows utilize energy of fresh herbage offered as the sole diet.

One hundred thirty-two energy balances have been conducted by indirect calorimetry using cows at two levels of feeding and at two stages of lactation over 3 years. Mixed pasture suitable for grazing and cut twice daily was the sole diet. Data for the utilization and partitioning of energy have been subjected to regression analysis on a within-cow basis.

Efficiency of utilization of metabolized energy (ME) for energy balance (milk + tissue) was higher in early lactation (spring) than in midlactation (autumn) (56% and 50%, respectively), and at both stages of lactation efficiency was lower than the 60% proposed by van Es (1975).

Reasons for reduced efficiency of utilization of ME were associated with higher heat increment of feed/unit intake in midlactation. The factors responsible included changing feed composition, reduction in ME content of the herbage, and increased energy cost of excretion of excess nitrogen in midlactation.

Key Words: fresh pasture, energy utilization, lactating dairy cows.

INTRODUCTION

Improved pastures provide the major source of food for dairy cows in New Zealand. These pastures follow a seasonal growth pattern, and although data exist on how factors related to nutritive value vary with season, most of these measures provide at best only an approximation of nutritive value (Ulyatt 1973). Few data exist on the metabolic efficiency with which nutrients, derived from pasture offered as a sole diet, are utilized by the dairy cow or on the effect of seasonality on this efficiency.

In this paper data from 132 energy balances with lactating cows have been used to compare energy utilization at two stages of lactation. Data were derived from an experiment that covered three consecutive lactations (Trigg et al. 1980) in which fresh pasture was the sole diet.

T.E. Trigg is a Scientist, C.R. Parr is a Senior Technical Officer, N.R. Cox is a Scientist (Biometrics) at the Ruakura Animal Research Station.

The guidance of A.M. Bryant in preparation of this paper is greatly appreciated. The invaluable assistance of many staff members at the Nutrition Centre, particularly M.J. Taylor, is also acknowledged.

METHODS

In 3 consecutive years a total of 15 sets of identical twin dairy cows of mixed breeds were used in an experiment designed to investigate the effects of underfeeding in early lactation on subsequent production. The trial design has been described previously by Trigg et al. (1980). Animals were housed for periods of 23 days on two occasions during weeks 8–18 of lactation (early) and during weeks 22–32 of lactation (midlactation). During these periods energy balance determinations were made on each cow at either ad libitum (A) or restricted (R) (approximately 75% of National Research Council requirements) levels of feeding.

Mixed pasture, consisting predominantly of perennial ryegrass, white clover, and paspalum and suitable for grazing, was cut twice daily to provide the sole diet offered during balance periods. At other times, cows were grazed so that intake was near maximum. Estimates of nutrient intake and apparent digestibility of some major nutrients were made during balance periods. Cows were milked twice daily, and lactation length averaged approximately 250 days.

Sample collection and analysis have previously been described in Bryant et al. (1977) and Trigg et al. (1980).

STATISTICAL ANALYSES

Partitioning of energy was subjected to regression analysis, made on a within-cow basis as described in Trigg et al. (1980). The marginal changes were determined in digested energy (DE) as gross energy (GE) changed; in urine, methane, and metabolized energy (UE, CH_4E and ME, respectively) as DE changed; and in heat, milk, and tissue energy (HE, LE and TE, respectively) as ME changed. These values, represented by the slope coefficient (Table 1), are referred to as incremental changes in partitioning. Slopes of regressions generated in early lactation and midlactation were tested for difference.

RESULTS

Mean calving date was 21 August. Dry-matter (DM) intakes during A feeding levels were 3.58 ± 0.36 and 3.36 ± 0.28 kg/100 kg LW for early lactation and midlactation, respectively; equivalent data during R feeding levels were 2.67 ± 0.31 and 2.30 ± 0.23 kg/100 kg LW. Fat-corrected milk (FCM) production for these periods was 18.9 ± 3.6 and 12.2 ± 2.5 kg/day during A in early lactation and midlactation, respectively; and equivalent data during R were 12.2 ± 2.5 and 8.8 ± 2.0, respectively.

Results for partitioning of energy for each stage of lactation are shown in Table 1. Within stage of lactation, incremental change in energy partitioning was not affected by the previous grazing treatments described by Trigg et al. (1980), and regression coefficients were pooled within stages of lactation. As intercepts differed significantly between treatments for all partitions in early lactation (Trigg et al. 1980), intercepts could not be pooled.

Incremental change in partitioning of GE to DE was greater in early lactation than in midlactation, although these differences were not statistically significant. DE partitioning to ME was significantly higher in early lactation than in midlactation due to an increase (P < 0.1) in increment of UE/unit intake in midlactation. Also, energy lost as methane at a common intake was higher in midlactation than in early lactation.

Incremental change of partitioning of ME for energy

Table 1. Regression analysis of energy partitioning at two stages of lactation, Y = bX + c.

		Early[†]				Mid[†]				Slope	
Y	X	b	c_H§	c_L	SE(d)¶	b	c_H	c_L	SE(d)¶	SE(d)	P
GE	DE	0.704 ±0.036	0.072	0.075	±0.004	0.647 ±0.024	0.079	0.075	±0.004	±0.046	NS
DE	UE	0.059 ±0.009	0.039	0.042	±0.001	0.070 ±0.006	0.027	0.039	±0.002	±0.009	*
	CH_4E	0.055 ±0.008	0.056	0.061	±0.002	0.067 ±0.005	0.037	0.039	±0.001	±0.010	NS
	ME	0.886 ±0.013	−0.095	−0.104	±0.004	0.855 ±0.006	−0.064	−0.068	±0.002	±0.013	*
ME	HE	0.440 ±0.039	0.322	0.303	±0.004	0.504 ±0.019	0.230	0.221	±0.006	±0.030	*
	RE††	0.560 ±0.039	−0.321	−0.303	±0.004	0.496 ±0.019	−0.230	−0.221	±0.006	±0.030	*
	LE	0.210 ±0.050	0.353	0.266	±0.015	0.146 ±0.024	0.207	0.196	±0.014	±0.079	NS
	TE	0.350 ±0.062	−0.675	−0.569	±0.018	0.350 ±0.036	−0.437	−0.417	±0.011	±0.040	NS

†Pooled slope (b) = intercept (c) of within-cow regression analysis, all divided by $LW^{.75}$ and in MJ.

††RE = milk + tissue E.

§H & L = Treatments imposed prior to energy balance measurements. For description see Trigg *et al.* (1980)

¶SE(d) of $c_H - c_L$.

Table 2. Fate of digestible nitrogen in 132 balance trials with lactating cows in early and midlactation.

	Early	Diff. (E-M)	SE(d)	Significance
Dig. N.I. (g)	219.4	27.7	7.7	P<0.01
Dig. N/MEI (g/MJ)	1.73	-0.28	0.05	P<0.001
Urine N/MEI (g/MJ)	1.19	-0.42	0.05	P<0.001
Bal N/MEI (g/MJ)	0.54	0.15	0.04	P<0.01

balance (RE) (milk + tissue) differed (P < 0.1) between stages of lactation. This difference was associated with an increase in heat increment/unit ME intake in midlactation, although differences in partition to milk and to tissue were not significant between stages of lactation. ME was utilized more efficiently in early lactation because RE (P < 0.1) and methane were lower at this time. At the same intake, therefore, more ME was available for milk production in early lactation. The effect of correcting energy balance data for tissue loss (van Es 1975) was to reduce the magnitude of the regression coefficients to 0.520 ± 0.035 and 0.468 ± 0.018 for early lactation and midlactation, respectively.

N as a proportion of ME intake (Table 2) available for milk + tissue synthesis (Bal N/ME intake) was less (P < 0.01) in midlactation than in early lactation despite the higher (P < 0.001) digestible N/ME intake. Larger urinary N losses in midlactation (P < 0.001) accounted for this lower N rate.

ME content of DM changed from September to February such that average (± SE) ME/GE (0.612 ± 0.007) and ME/DE (0.831 ± 0.002) in early lactation was lowered by 10% (P < 0.01) and 3% (P < 0.001), respectively in midlactation.

DISCUSSION

A feature of experimentation with animals offered fresh forages is that forage quality may vary with time. Separation of effects associated with seasonal changes in forage quality from those resulting from stage of lactation is impossible. Despite these disadvantages, several important points are highlighted by the data.

A small but significant difference existed in the ME/unit in DE intake from herbage offered at the different stages of lactation. This difference was due to an increase in the UE in midlactation and larger losses of CH_4E at that time. Increased methane losses in midlactation could arise from a number of factors, including variation in chemical composition between the feeds (Czerkawski 1969), increased retention time of feed in the rumen (Czerkawski 1969), and ruminal digestion of protein (Blaxter and Martin 1962). The former two are consistent with an observed reduced apparent digestibility of herbage offered in midlactation; the latter is supported by data in Table 2. Higher UE loss in midlactation was related to more urinary N excretion (Table 2).

The significantly higher heat increment reported for cows in midlactation has not previously been reported. Factors affecting heat increment associated with changing feed composition, including fermentation heat, heat production in the gut, and amounts of volatile fatty acids produced in the rumen, may all have contributed to this result. The high urinary N excretion/unit ME intake during midlactation, despite a larger intake of digestible N, may also be implicated, as excretion of surplus N requires extra energy (Tyrrell et al. 1970).

A small increase in HE due to the growth of the fetus and the observed reduction in q (ME/GE) (van Es 1975) of the herbage may also be involved. It is not known what effect the changing of partitioning of energy due to the reduction of milk yield would have on heat production, though this effect would be minimal if utilization of ME for milk or tissue synthesis occurs with equal efficiency.

A result of the reduced efficiency of ME utilization for ER and the higher heat increment in midlactation was that less ME/unit intake was partitioned to milk at this time; this result is at variance with findings of van Es (1975) and ARC (1965). Even though efficiency of ME above maintenance in early lactation and midlactation was within the range of most published values, both are lower than the value suggested by van Es (1975). Correcting the energy-balance data for tissue loss increases this discrepancy, suggesting that efficiency of utilization of ME from fresh long forage may be less than for other rations. The use of a common expression to derive ME requirements for energy balance with cows offered such rations at different stages of lactation may be invalid.

LITERATURE CITED

Agricultural Research Council (ARC). 1965. Nutrient requirements of farm livestock. II. Ruminants. H.M. Stationery Office, London.

Blaxter, K.L., and A.K. Martin. 1962. The utilisation of protein as a source of energy in fattening sheep. Brit. J. Nutr. 16:397–409.

Bryant, A.M., J.W. Hughes, J.B. Hutton, R.P. Newth, C.R. Parr, T.E. Trigg, and J. Young. 1977. Calorimetric facilities for dairy cattle at Ruakura Animal Research Station. N.Z. Soc. Anim. Prod. Proc. 37:158–162.

Czerkawski, J.W. 1969. Methane production in ruminants and its significance. World Rev. Nutr. Diet. 11:240–282.

Trigg, T.E., K.E. Jury, A.M. Bryant, and C.R. Parr. 1980. The energy metabolism of dairy cows under-fed in early lactation. Proc. VIII Symp. on Energy, Camb. EAAP Pub. 26:345–349.

Tyrrell, H.F., P.W. Moe, and W.P. Flatt. 1970. Influence of excess protein intake on the energy metabolism of the dairy cows. Proc. V Symp. on Energy Metab. EAAP Pub. 13:69–72.

Ulyatt, M.J. 1973. The feeding value of herbage. *In* G.W. Butler and R.W. Bailey (eds.) Chemistry and biochemistry of herbage, 131–178. Academic Press, London.

van Es, A.J.H. 1975. Feed evaluation for dairy cows. Livest. Prod. Sci. 2:95–107.

Intake of Grass Silage by Dairy Cows and Its Potential To Meet Requirements of Maintenance and Milk Production

J.P. DULPHY, C. DEMARQUILLY, and J.P. ANDRIEU

Laboratoire des Aliments, Institut National de la Recherche Agronomique,
Beaumont, France

Summary

The main objective of our research was to measure the intake and digestibility of grass silages in order to estimate their potential for meeting requirements for maintenance and milk production. Twelve silages were prepared by the following three techniques, which considerably improve the intake compared with those used 10–20 years ago: (1) fine chopping (1–3 cm), (2) adding formic acid as an efficient preservative (3.5 l/ton), and (3) covering the roughages with plastic to prevent oxygen from entering the silo. Three other silages were made simultaneously with 3 of the same forages, but without formic acid. All forages were directly cut and ensiled into 100-m³ bunker silos. All trials were carried out in winter, with silages fed ad libitum to groups of 8–12 cows from the second to fifth month after calving. The amount of concentrates fed was predetermined to meet the total requirements by taking into account silage intake, silage digestibility, milk production at the beginning of the experiment, and predicted milk production. Elsewhere 2 of 12 formic acid silages were also fed with 1 kg concentrate more and with 1 kg concentrate less. Experimental periods lasted 10–13 weeks, allowing variations in live weight to be measured.

The 12 silages with formic acid were well preserved and had a high organic-matter digestibility (72.4%). Ingested DM of silages and concentrates was 2.09 and 0.59 kg/100 kg live weight, respectively, for cows producing 19.5 kg fat-corrected milk with a daily live-weight gain of 126 g. When no formic acid was used, silage intake decreased considerably (– 12.6%), partly due to a greater distribution of concentrates. Indeed, for the two silages fed with different level of concentrates, DM intake dropped 0.5 kg when a 1-kg DM increase in concentrates was made.

Net energy provided by fine-chopped silages preserved with formic acid was enough to cover maintenance needs and a mean production of 12 kg of milk/cow (from 5 kg for late cut to 17 kg for early cut). For 5 silages of perennial ryegrass at first growth this milk-production potential was 15 kg for an organic-matter digestibility of 74.7%. Milk potential decreased by 2 kg when silages without formic acid were given and had to be compensated for by concentrates. Therefore, if grass silages are perfectly prepared at an early stage of growth, their feeding value can be very high. Moreover, these grass silages generally can sustain the same level of milk production with nitrogen as with net energy.

KEY WORDS: grass silage, feeding value, dairy cows.

INTRODUCTION

Development of grass-silage feeding systems has been slow, mainly because of problems with preservation and low animal performance due to low intake. Wilting had long been considered the best way to increase intake (Merril and Slack 1965). Now, direct-cut silages are readily ingested by bovines if they are prepared with an efficient additive (Waldo 1977) and if they are finely chopped (Dulphy and Demarquilly 1975a, Castle et al. 1979).

In many countries climate makes wilting techniques very difficult. Since 1971 we have researched direct-cut grass silages made with more efficient ways to preserve the feeding value of harvested forages.

METHODS

From 1971 to 1979, we prepared (Dulphy and Demarquilly 1973, 1975a, 1975b, 1975c, 1976; Dulphy and Andrieu 1976; Dulphy et al. 1980; plus three unpublished trials) 12 grass silages in several different trials as follows:

(1) First growth — 3 timothy, 5 perennial ryegrass, 1 natural grassland, and 1 timothy–red clover mixture.

(2) Second growth — 2 perennial ryegrass.

These forages were finely chopped (1 to 3 cm) and ensiled with 3.5 l/ton of commercial formic acid (85%). They were conserved in 3.5-m-high bunker silos (100 m³); filling was within 1 to 2 days. The forages were packed, then covered with a plastic sheet and straw. Three of these forages were also ensiled with no additive.

The following winter the experimental silages were fed ad libitum (5% to 10% refusals) to groups of 8–12 cows, from their second to their fifth month of lactation. Calving took place in November-December, and nonexperimental silage was always fed before calving. We used 222 cows of Friesian, Montbeliard, and Holstein breeding. Among them there were 24% primiparae with a mean weight of 550 kg and 76% multiparae with a mean weight of 600 kg.

Concentrates were fed with silages to meet requirements with a minimum of net energy (NE) and nitrogen (N) excess. Concentrate levels were predetermined after taking into account the following factors for each group of cows at the start of each experimental period (6–7 weeks after calving): (1) silage intake, (2) NE and N value of silage, and (3) amount of predicted milk production for each cow.

Predicted milk production was calculated by considering the maximum milk production and a weekly estimated decrease in milk production of 2% for adult cows and 1.5% for primiparae, when cows were adapted to their diets. Each experimental period lasted 10–13 weeks to allow an accurate measurement of live-weight variations.

Two of the formic acid silages were fed with 1 kg more and 1 kg less concentrates than the required calculated level in order to measure the effects of variation in concentrate levels (Dulphy and Demarquilly 1976).

The 3 silages without additive were fed to three groups of cows similar to those receiving the same silages with formic acid. Among these three comparisons, the first was conducted with two groups of cows receiving the same level of concentrates. In the second and the third the groups of cows fed the formic acid silages received a lower level of concentrates than the groups fed the silages without formic acid. The objective was to have the same performances as with the other silage treatments.

RESULTS

The 12 silages preserved with formic acid had pH values of 3.89 ± 0.15; $NH_3 - N$ as percentage of total N, 6.2 ± 1.6; and acetic acid concentration, 19.1 ± 5.2 g/kg DM. Dry-matter content of silages was 22.1% ± 0.8%, and crude-protein and crude-fiber concentrations (DM basis) were 15.1% ± 2.1% and 28.5% ± 2.9%, respectively. Mean organic-matter digestibility (measured in sheep) was 72.4% ± 3.2%. The following changes were found in unpreserved silages: + 0.12 for pH, + 3.3 for $NH_3 - N$ as percentage of total N, and + 21.6 g/kg DM for acetic acid.

The silage intake of primiparae was only 85% of that for adult cows (87% if expressed in kg DM as percentage of live weight). Silage intake increased when digestibility of the silage increased, but the differences were not statistically significant.

The feeding value of silages was evaluated by the milk production sustained by their net energy supply. To estimate this value we first calculated the total energy needs of the animals (maintenance + milk production + live-weight variation) and then reduced the net energy supplied by concentrates. Animal needs and concentrate values were taken from French tables (INRA 1978).

For the 12 silages preserved with formic acid, the fat-corrected milk sustained by silage intake varied from 5.0 to 17.6 kg. The sustained milk was 1.9 kg higher for adult cows than for cows at their first calving. The sustained milk (SM) varied most with organic-matter digestibility (OMD) and, therefore, with net energy value (expressed in feed unit × 100 for milk production = FU). For adult cows, the relationships were expressed as follows:

SM = 0.817 OMD – 46.9
(standard deviation 2.96, r = 0.685)

SM = 0.586 FU – 37.9
(standard deviation 2.30, r = 0.820)

The effects of formic acid were clear because the sustained milk was increased by 2.1 kg when the additive was used (Table 1). This substantial increase was also partly caused by the reduced level of concentrates. Indeed, intake of formic acid–preserved silages was 14.4% higher than that of unpreserved silages. This figure is higher than the 11% difference reported by Waldo (1977).

The feeding value of the 5 perennial ryegrasses at first growth and when preserved with formic acid was very high; the intake was 2.25 kg DM/100 kg of live weight, OMD was 74.7%, and sustained milk production was 15.4 kg/day for adult cows.

Finally, for silages fed with two levels of concentrate, an increase of 1 kg DM in form of concentrate, within a total supply of 1–4 kg/day, resulted in a 0.5 kg DM decrease in silage intake and a 0.35 kg increase in fat-corrected milk.

CONCLUSION

When grass silages are correctly prepared at an early stage of growth, their feeding value is high. In our trials, grass silages, with an OMD over 70%, were able to sustain milk production of from 10 to 17 kg. In comparison, Castle et al. (1977) found a 12–14 kg sustained milk production with an excellent grass silage.

Coarse laceration with a flail harvester does not ensure a high intake for a direct-cut silage (Dulphy and Demarquilly 1975a). If no preservative is added to the silage, approximately 1 kg of concentrates would be needed for compensation of lower intake, in order to obtain performances (milk production and daily live-weight gain) similar to those of cows fed the same silage with an additive.

LITERATURE CITED

Castle, M.E., C. Retter, and J.N. Watson. 1977. Silage and milk production: a comparison between additives for silage of

Table 1.   Intake and performance of cows (24% primiparae and 76% multiparae).

	Mean ± S.D.[a] 12 silages	Without additive, 3 silages	With formic acid, 3 silages
Intake (kg DM/100 kg LW)			
Silage	2.09±0.22	1.81	2.07
Concentrates	0.59±0.21	0.86	0.71
Live weight (kg)	587	594	598
Live-weight gain (g/day)	126±89	165	125
Milk production (kg/day)			
Total	20.3±2.6	22.2	22.0
Fat corrected	19.5±2.6	20.3	20.7
Milk composition (g/kg milk)			
Fat	37.3±1.7	34.2	36.1
Crude protein	31.7±1.5	30.4	31.3
Milk "sustained" by			
silage energy (kg/day)	11.9±3.9	9.0	11.1

[a]S.D. = standard deviation intergroup.

high digestibility. J. Brit. Grasl. Soc. 32:157-164.

Castle, M.E., C. Retter, and J.N. Watson. 1979. Silage and milk production: comparisons between grass silage of three different chop lengths. J. Brit. Grassl. Soc. 34:293-301.

Dulphy, J.P., and J.P. Andrieu. 1976. Utilisation par les vaches laitières des ensilages d'herbe enrichis en pulpes de betteraves déshydratées. Bul. Tech., CRZV Theix, INRA 25:25-31.

Dulphy, J.P., J.P. Andrieu, and C. Demarquilly. 1980. Étude de la valeur azotée d'ensilages d'herbe additionnés ou non d'acide formique pour les vaches laitières. Bul. Tech., CRZV Theix, INRA 40:27-33.

Dulphy, J.P., and C. Demarquilly. 1973. Utilisation des ensilages d'herbe par des génisses et des vaches laitières. II. Bul. Tech., CRZV Theix, INRA 10:5-15.

Dulphy, J.P., and C. Demarquilly. 1975a. Influence de la machine de récolte sur les quantités d'ensilage ingérées et les performances des vaches laitières. Ann. Zootech. 24 (3):363-371.

Dulphy, J.P., and C. Demarquilly. 1975b. Complémentation énergétique des rations à base d'ensilage d'herbe en cours de lactation. Bul. Tech., CRZV Theix, INRA 21:31-40.

Dulphy, J.P., and C. Demarquilly. 1975c. Incorporation de pulpes sèches dans l'ensilage d'herbe: utilisation par les vaches laitières. Bul. Tech., CRZV Theix, INRA 22: 45-52.

Dulphy, J.P., and C. Demarquilly. 1976. Complémentation énergétique d'une ration d'ensilage d'herbe en début de lactation pour les vaches laitières. Bul. Tech., CRZV Theix, INRA 26:29-34.

INRA. 1978. Alimentation des ruminants. INRA Pub., Versailles.

Merril. W.G., and S.T. Slack. 1965. Feeding value of perennial forages for dairy cows. Anim. Sci. Mimeo. Ser. 3. Cornell University.

Waldo, D.R. 1977. Potential of chemical preservation and improvement of forages. J. Dairy Sci. 60(2):306-326.

Effect of Supplements on the Live-Weight Gain of Beef Cattle Given Forage During the Dry Season

C. THOMAS and P.N. MBUGUA
Grassland Research Institute, Hurley, Berkshire, England, UK,
and Ministry of Agriculture, Kenya

Summary

The low quality of standing pasture grasses in the dry seasons presents a major problem in the intensification of beef production in the tropics. The objective of this research was to examine the role of maize and grass silage or standing napiergrass (*Pennisetum purpureum*) as sources of forage when given alone or with supplements to beef cattle.

In the first of three trials conducted at the National Agricultural Research Station, Kitale, Kenya, 36 Friesian castrates, initially 249 kg live weight (LW), were given silage of whole-crop maize either alone or supplemented with corn-and-cob meal. Each of these two basal diets was given ad libitum either without a nitrogen (N) supplement or with urea or cottonseed cake. In the second trial 17 Friesian steers, intially 158 kg LW, and 17 Hereford steers, initially 177 kg LW, grazed a napiergrass (*P. pur-*

pureum cv. French Cameroons) pasture throughout the dry season. The steers either received no supplement or were given concentrate at either 5 or 10 g/kg LW/day. In the third trial 25 Friesian steers, initially 161 kg LW, were given silage of rhodesgrass (*Chloris gayana* cv. Mbarara) ad libitum either alone or with concentrate at 0%, 20%, 40%, 60%, or 80% of total dry-matter (DM) intake. In the final two trials the subsequent performance of steers during the wet season was examined.

In the first trial the level of performance of steers given maize silage alone was not significantly different from maintenance. Supplementation with both corn-and-cob meal and N significantly ($P < 0.001$) increased LW gain to a mean of 1.25 kg/day. In the second trial the LW gain of steers grazing napiergrass was 0.35 kg/day, and this gain was increased to 0.70 kg/day ($P < 0.001$) when the animals were given concentrate at 10 g/kg LW. In the third trial steers given grass silage alone grew at only 0.10 kg/day. Supplementation of the silage increased LW gain up to a maximum of 1.40 kg/day when the concentrate was 80% of total DM intake. Examination of subsequent performance at grass in trials 2 and 3 showed that cattle given forage alone during the dry season grew significantly faster than those that had received concentrate. Further, the results of trial 3 showed that the marked differences in LW apparent at turnout were considerably narrowed by the end of the grazing season and also that there was no significant effect of previous dry-season treatment on LW at slaughter or carcass weight.

These results show that supplementation of forage with energy and/or N is required for high levels of LW gain during the dry season. However, in pasture-based systems in the tropics the need for such high levels of performance is questionable when compensatory growth can be exploited in a subsequent grazing season.

KEY WORDS: maize, *Zea mays*, *Pennisetum purpureum*, *Chloris gayana*, silage, beef cattle, intake, live-weight gain, carcass weight.

INTRODUCTION

The low quality of grasses in the dry season presents a major problem in the intensification of beef production in the tropics (e.g., Pfander 1971). Low growth rates by beef cattle during the dry season have been observed even in the high-potential areas of Kenya (Kidner 1966). Harker (1961) showed no benefit in saving tropical herbage produced during the wet season as standing hay, in terms of either dry-matter (DM) accumulation or quality. However, of the forages available, napiergrass (*Pennisetum purpureum*) has been shown to retain its quality as a standing crop over the dry season to a much greater extent than either natural or other planted pasture grasses (D.M. Thairu, unpublished data). Sheldrick and Thairu (1975) have shown that sown grasses, such as rhodesgrass (*Chloris gayana* cv. Mbarara), exhibit a high rate of DM accumulation after the onset of the rains, and at this time it may be feasible to conserve excess herbage as silage. Also, the potential of forage maize for producing high yields of DM for silage has been demonstrated in East Africa by Sheldrick (1975) and van Arkel (1977).

In the three trials reported here, rhodesgrass and maize, which were grown in the wet season and conserved as silage, and standing napiergrass were examined as sources of forage when they were given alone or with supplements to beef cattle during the dry season.

METHODS

The trials were conducted at the National Agricultural Research Station, Kitale, Kenya (altitude 1,800 m; average rainfall 1,191 mm; dry season, December-March).

Trial 1 – Maize Silage

Thirty-six Friesian castrates, initially 17 months of age and 249 kg live weight (LW), were offered maize silage ad libitum, either alone or supplemented with corn-and-cob meal at 33% of the total DM for a period of 70 days. Each of these two basal diets was given either without nitrogen (N) supplementation or with urea or cottonseed cake in quantities calculated to supply similar levels of N (1.2% and 8.4% of total DM for urea and cottonseed cake, respectively). The chemical composition of the feeds is shown in Table 1.

Trial 2 – Napiergrass

Seventeen Friesian steers (F), initially 8 months of age and 158 kg LW, and 17 Hereford steers (H), initially 8 months of age and 177 kg LW, grazed a sward of napiergrass (cv. French Cameroons) for a period of 90 days during the dry season. The steers received no supplement or concentrate (see Table 1, trial 3 for composition) at either 5 or 10 g DM/kg LW/day. Subsequently the steers were turned out onto a pasture of rhodesgrass (cv. Mbarara) at the commencement of the wet season and stocked at the rate of 3.4 animals/ha for a period of 190 days.

Trial 3 – Rhodesgrass Silage

The primary growth of rhodesgrass was cut on 5–7 June, wilted for 6 hours, and ensiled. Twenty-five Friesian steers, initially 7 months of age and 161 kg LW, were given grass silage ad libitum alone or with a concentrate supplement at 20%, 40%, 60%, or 80% of total DM intake. The chemical composition of the feeds is shown in Table 1. The diets were given for a period of 70 days during the dry season. Subsequently the animals were turned out onto a pasture of rhodesgrass and stocked at the rate of 2.5 animals/ha. Grazing was ended after 245 days when the animals were transferred to a feedlot for the final fattening phase lasting 120 days, after which they were slaughtered, and the weight of cold-dressed carcass was recorded.

Table 1.   Chemical composition of feeds used in trials 1 and 3.

		Trial 1			Trial 3	
		Maize silage	Corn-and-cob meal	Cotton-seed cake	Grass silage	Supplement†
DM content (g/kg fresh)		246	763	870	371	926
Total N (g/kg DM)		8.30	14.7	62.9	8.24	32.2
Ash (g/kg DM)		51.7	17.3	66.5	93.7	69.7
Silage pH		3.8	–	–	4.5	–
In-vitro D-value (%)		59.8	72.7	62.9	47.6	78.0

†Also used in Trial 2.

RESULTS

Trial 1 – Maize Silage

The supplementation of maize silage with either corn-and-cob meal or the N supplements led to a marked increase in total DM intake (Table 2). Further, the response in intake to the inclusion of corn-and-cob meal tended to be greater with the diet containing urea than with that containing cottonseed cake. Supplementation of the diet with N significantly increased LW gain (P < 0.001), and in this respect urea elicited LW gains similar to those achieved with cottonseed cake only when the diet contained corn-and-cob meal (Interaction P < 0.05) (Table 2).

Trial 2 – Napiergrass

LW gain (P < 0.001) increased progressively as a result of supplementation of napiergrass with concentrate (Table 3). Hereford steers grew, on average, at a significantly slower rate than Friesian steers (P < 0.001), and there was no interaction between breed and concentrate level. Steers given no concentrate during the previous dry season grew faster (P < 0.001) during the wet season than those that had received the supplement (0.66 and 0.51 kg/head/day for treatments without and with dry-season supplement, respectively).

Trial 3 – Rhodesgrass Silage

Total DM intake and LW gain were markedly increased by the inclusion of the concentrate supplement at 20% of total DM (Table 4). The response to further successive increments of concentrate diminished. During the wet season, the relationship between LW and time was curvilinear. Comparisons among treatments were effected, using orthogonal polynomials. Examination of the average LW gain (Table 4) showed that steers previously given silage alone grew significantly faster during the wet season than those that had received the concentrate supplement (P < 0.001). During the feedlot period the differences in LW gain were not significant and at slaughter there was no significant effect of previous-dry-season treatment on carcass weight (Table 4).

DISCUSSION

The maize and rhodesgrass silages were well preserved, judging by their pH values. However, the daily LW changes attained by the steers given the silages as the sole feed were low (– 0.05 and + 0.10 kg/head for maize and rhodesgrass silages, respectively) when compared with that of steers given napiergrass (0.35 kg/head).

Supplementation of maize silage with a source of nitrogen markedly improved intake and LW gain by steers. In

Table 2.   Effect of corn-and-cob meal inclusion and source of nitrogen supplement on voluntary intake and live-weight (LW) gain by steers given maize silage, trial 1.

	Maize silage			Maize silage + corn-and-cob meal		
N supplement	None	Urea	Cottonseed cake	None	Urea	Cottonseed cake
DM intake (g/kg LW/day)	14.7	24.4	25.0	19.0	27.2	25.8
LW gain (kg/head/day)	– 0.05	0.88	1.18	0.49	1.28	1.22
SE of mean†			±0.079/* **			

†In this and subsequent tables NS = not significant; *P<0.05, **P<0.01, ***P<0.001.

Table 3. Live-weight gain of steers given Napier grass alone or with a concentrate supplement, Trial 2.

Breed	Concentrate level (g DM/kg LW/day)			Mean[1]
	0	*5*	*10*	
Friesian	0.37	0.69	0.80	0.64
Hereford	0.33	0.48	0.59	0.48
Mean[2]	0.35	0.58	0.70	

[1]Standard error of mean = ±0.028 (P < 0.001).

[2]Standard error of mean = ±0.033 (P < 0.001).

this respect urea elicited LW gains similar to those achieved with cottonseed cake when it was given with the supplement of corn-and-cob meal. Responses to the supplementation of maize silage with nitrogen (e.g., Thomas et al. 1973) have not been so marked as those recorded here. The low N content of the maize silage (8.3 g/kg DM) may have severely limited intake and LW gain. However, the changes in the supply of N were unavoidably confounded with differences in energy supply.

The rhodesgrass silage for trial 3 was made from herbage cut at a relatively mature stage (c. 8 weeks' primary growth), and its low D value and N content reflected this maturity. The inclusion of the concentrate supplement at 20% of total DM intake led to only a small reduction in silage intake and a marked increase in total DM intake and LW gain. It is possible that this low marginal substitution rate was a reflection of low N content of the silage; the response in intake and LW gain diminished with further successive increments of the concentrate. A similar pattern was observed with grazed napiergrass in trial 2.

In this latter experiment the LWs of the steers were recorded for part of the subsequent wet season. During this period the steers that had received napiergrass alone during the dry season grew faster than those given concentrate. This aspect was examined more fully in trial 3 where grass silage was given in the dry season. The results showed that LW gain during the wet season (y) was inversely related to the gain achieved during the dry season (x): y = 0.547 - 0.167x, r = -0.76, P < 0.001. Thus, the maximum difference between treatment groups in

mean LW at turnout of 80 kg was reduced to 15 kg at yarding, and at slaughter the differences in carcass weight between the treatment groups were not significant. Compensatory growth effects (Wilson and Osbourn 1960) have been widely recorded in extensive (range) conditions in the tropics (Joubert 1954, French and Ledger 1957) and in temperate regions (Lawrence and Pearce 1964). However, the regression coefficient relating wet-season performance to dry-season LW gain (- 0.167) was low in relation to the value (- 0.776) determined by Lawrence and Pearce (1964), and this lower value may indicate that the poor quality of tropical pastures can limit the extent of compensatory growth. Further work is needed to examine the influence of stocking rate on the extent of compensation during the wet season.

The results of the trials show that high rates of LW gain can be achieved by beef cattle given grass and maize silages or napiergrass together with concentrate supplements during the dry season. However, the value of these supplements is questionable when compensatory growth can be exploited in a subsequent grazing season.

LITERATURE CITED

French, M.H., and H.P. Ledger. 1957. Liveweight changes of cattle in East Africa. Emp. J. Expt. Agric. 25:10-18.

Harker, K.W. 1961. A comparison of standing hay production from grasses at Entebbe. East Afr. Agric. For. J. 27:49-51.

Joubert, D.M. 1954. The influence of winter nutritional depressions on the growth, reproduction and production of cattle. J. Agric. Sci., Camb. 44:51-66.

Kidner, E.M. 1966. Beef production. I. Nutritional requirements for normal and rapid growth with fattening in relation to pasture composition. East Afr. Agric. For. J. 32:34-36.

Lawrence, T.J.L., and J. Pearce. 1964. Some effects of wintering yearling beef cattle on different planes of nutrition. J. Agric. Sci., Camb. 63:5-21.

Pfander, W.H. 1971. Animal nutrition in the tropics — problems and solutions. J. Anim. Sci. 33:843-849.

Sheldrick, R.D. 1975. Optimum cutting period for silage maize in Western Kenya. East Afr. Agric. For. J. 40:394-399.

Sheldrick, R.D., and D.M. Thairu. 1975. Seasonal patterns of grass growth at Kitale. Ministry of Agriculture, Kenya. Past. Res. Proj. Tech. Rpt. 3.

Thomas, C., J.M. Wilkinson, and J.C. Tayler. 1973. The utilization of maize silage for beef production. I. The effect of level and source of supplementary nitrogen on the utilisation

Table 4. Voluntary intake of total DM by steers given grass silage in the dry season and live-weight gains and final carcass weight, Trial 3.

Item	Level of supplement in dry season (% total DM intake)					SE of mean
	0	*20*	*40*	*60*	*80*	
Total DM intake in dry season (kg/head/day)	3.97	4.98	5.79	6.72	6.56	—
LW gain (kg/head/day)						
dry season	0.10	0.76	1.09	1.28	1.40	±0.061***
wet season	0.56	0.39	0.35	0.36	0.32	±0.027***
feedlot	0.85	0.72	0.85	0.69	0.78	±0.081[NS]
Carcass weight (kg)	201	206	216	213	214	±8.4[NS]

of maize silage by cattle of different ages. J. Agric. Sci., Camb. 84:353–364.

van Arkel, H. 1977. New forage crop introductions for the semi-arid highland areas of Kenya as a means to increase beef pro-

duction. Neth. J. Agric. Sci. 25:135–150.

Wilson, P.N., and D.F. Osbourn. 1960. Compensatory growth after under-nutrition in mammals and birds. Biol. Rev. 35: 324–363.

Effects of Feeding Level and Forage:Concentrate Ratio on the Digestibility of Pelleted Diets by Sheep and Cattle

G.M.J. HORTON, M.J. FARMER, B.E. MELTON, and S. SKLANE

University of Saskatchewan, Saskatoon, Canada, and
University of Florida, Gainesville, Fla., U.S.A.

Summary

Sheep are frequently used to determine the nutritive value of cattle feeds. However, there is some evidence that digestion coefficients for cattle and sheep may differ and that digestibility may be influenced by feeding level. The purpose of this study was to investigate the effects of feeding level and animal species on the digestibility of pelleted diets containing different ratios of alfalfa hay to barley grain. Five 40-kg yearling crossbred lambs and four 350-kg yearling steers were used to evaluate each diet. The six diets were isonitrogenous, containing the following proportions of barley grain and alfalfa hay (percentage as fed basis): (1) 0% and 98.0%, (2) 18.0% and 80.1%, (3) 36.0% and 62.0%, (4) 54.0% and 44.1%, (5) 72.0% and 25.6%, (6) 90.0% and 6.7%. The barley and alfalfa were ground through a 0.64-cm screen, and the complete feed was pelleted (1.59 cm in diameter). Feed intakes were 50, 75, and 100 $g/W^{0.75}$. Sheep offered 100 $g/W^{0.75}$ of the high barley diets were at approximately ad-libitum levels of intake. Average digestion coefficients for organic matter (OMD), crude protein (CPD), and acid detergent fiber (ADF) in the six diets were lower ($P < 0.05$) with sheep than with cattle: 70.4% vs. 71.8%, 71.2% vs. 73.4% and 31.9% vs. 35.3%, respectively. Increasing feed intake from 50 to 100 $g/W^{0.75}$ reduced ($P < 0.05$) OMD and CPD but did not affect the digestibility of ADF. OMD and CPD increased by 0.25 and 0.11 percentage units, respectively, for each percentage of increase of barley in the feed for OMD values ranging from 59.9% to 81.2% with sheep and from 61.0% to 83.3% with cattle.

The results of this investigation show that while digestion coefficients of the pelleted diets were significantly higher in cattle than in sheep, differences were not large. For routine feed evaluation the digestibility by cattle of pelleted feeds consisting of grain and hay may be estimated from values obtained with sheep, provided that relative feeding levels are similar.

Key Words: barley:alfalfa ratio, intake level, digestibility, sheep, cattle.

INTRODUCTION

Coefficients of digestibility reflect the availability of nutrients to the animal and are used to describe the nutritive value of feeds. However, digestibility of a given feedstuff may be affected by both animal species (Blaxter et al. 1966) and feeding level (Robertson and Van Soest 1975). Furthermore, intake/unit of metabolic size is often higher with cattle than with sheep (Blaxter et al. 1966, Horton 1978). Consequently, digestion coefficients obtained with sheep may not be suitable for evaluating feeds for cattle. The purpose of this study was to investigate the effects of feeding level and forage:concentrate ratio in pelleted feeds on apparent digestibility by cattle and sheep.

EXPERIMENTAL PROCEDURE

Thirty 40-kg Dorset crossbred wether lambs and twenty-four, 350-kg Hereford yearling steers were used. The animals were divided into six groups, each containing five lambs and four steers. Digestion coefficients for each of the six diets were obtained at three levels of intake, using the same groups of lambs and steers at each intake level.

Feed was offered in two equal amounts at 0830 and

G.M.J. Horton is an Associate Professor of Animal Science, B.E. Melton is an Assistant Professor in Food and Resource Economics, and S. Sklane is a Graduate Research Assistant in Food and Resource Economics at the University of Florida; M.J. Farmer is a Chief Technician at the University of Saskatchewan.

This study was financed by grants from Agriculture Canada and the Saskatchewan Agricultural Research Foundation. The skilled technical assistance of Judy Bourque, Ken Bassendowski, Graham Steacy, and Richard Harland is gratefully acknowledged.

1630 hours. The three feeding levels were 50, 75, and 100 g/$W^{0.75}$ (W = body weight, kg), calculated using the animal unfasted weights 6 days before the start of each digestibility measurement. Intake levels were expressed as g/$W^{0.75}$ to eliminate the effects of weight differences on relative intake. The 50 g/$W^{0.75}$ level of intake provided energy at approximately maintenance levels for wethers and steers.

The six diets contained the following proportions of barley grain and alfalfa hay (percentage as fed basis): (1) 0% and 98.0%, (2) 18.0% and 80.1%, (3) 36.0% and 62.0%, (4) 54.0% and 44.1%, (5) 72.0% and 25.6%, (6) 90.0% and 6.7%. Barley and alfalfa were ground through a 0.64-cm screen. The complete diet was pelleted with a 1.59-cm die to prevent selection by the animals. Urea was added to maintain crude-protein levels at approximately 13%. Following an 11-day adaptation and a 6-day intake measurement, the animals were offered feed at 100, 75, and 50 g/$W^{0.75}$ for 12-day consecutive periods, respectively. Total fecal collections were made during the last 5 days of each period. Feed and feces were analyzed by standard procedures (AOAC 1976) and acid detergent fiber (ADF) as described by Van Soest (1963).

RESULTS AND DISCUSSION

Voluntary feed consumption by sheep ranged from 1.385 to 2.224 kg DM/day for lambs fed diets containing 28% and 82% alfalfa, respectively. Sheep offered 100 g/$W^{0.75}$ of the high-barley diets were at approximately ad-libitum levels of intake. Consequently, 100 g/$W^{0.75}$ was established as the upper feeding level to facilitate inter-dietary and interspecies comparisons. Intakes for sheep and cattle fed 100 g/$W^{0.75}$ averaged 1.415 and 7.109 kg DM/day, respectively.

Organic matter digestion coefficients (OMD) for the six diets by sheep and cattle and at the three levels of intake are given in Table 1. Increasing (P < 0.05) OMD with decreasing levels of alfalfa and increasing barley content by both cattle and sheep are consistent with earlier findings (Horton 1978) and reflect the higher digestibility of barley grain than of alfalfa hay. Changes in OMD due to changes in alfalfa:barley ratio were not affected by feeding level, and they averaged 0.25 percentage units for each level of replacement of alfalfa by barley. Consistently lower OMD of high-alfalfa diets at high than at low feeding levels is primarily due to the increased rate of

Table 1. Effect of animal species, feeding level and alfalfa: barley ratio on the percent digestibility of organic matter of six diets.

Alfalfa: barley ratio	Species[x]		Feeding level, g/$W^{.75}$[y]		
	Sheep	Cattle	50	75	100
100:0	59.9[a]	61.0[a]	60.9[af]	61.9[a]	58.5[f]
82:18	66.0[b]	64.0[c]	65.6[b]	65.2[b]	64.2[b]
64:36	68.7[c]	71.6[de]	71.6[c]	70.9[c]	67.8[d]
46:54	70.8[d]	72.9[e]	72.2[c]	71.7[c]	71.7[c]
28:72	76.1[e]	78.0[f]	77.5[dg]	76.9[dg]	76.7[g]
10:90	81.2[f]	83.3[g]	82.9[e]	82.2[e]	81.7[e]

[a,b,c,d,e,f,g]Means with different superscripts in the same row or column for each subheading are significantly different (P<.05).

[x]Standard error of mean: sheep = .59, cattle = .66.

[y]Standard error of mean = .77.

passage at the higher level of intake (Robertson and Van Soest 1975). Organic-matter digestibilities were about 2 percentage units higher (P < 0.05) with cattle than with sheep, in accordance with Blaxter et al.'s (1966) findings.

Pooled data for the six diets revealed higher (P < 0.05) OMD values for cattle than for sheep at the low level of intake but not at the two higher levels (Table 2). Digestion coefficients for CP and ADF at 50 and 100 g/$W^{0.75}$ were also higher (P < 0.05) in cattle. The digestibility of OM by both cattle and sheep decreased (P < 0.05) with increasing feeding level, probably due to decreased retention time (Wheeler et al. 1975). Fiber digestibility was not affected by feeding level with steers, but with sheep ADF was highest (P < 0.05) at 75 g/$W^{0.75}$. Low digestibility of ADF may have resulted from the fine particle size due to grinding and pelleting and the consequent accelerated rate of passage through the reticulo-rumen.

Pooled data for the six diets and three intake levels show that digestibilities of OM, CP, and ADF were higher (P < 0.05) with steers than with lambs (Table 3). These differences may have been the result of a longer feed retention time in the alimentary tract of cattle (Rees and Little 1980). Lower digestibilities of both OM and CP at the higher level of feeding are in agreement with Galyean et al. (1979) and probably reflect depressions in both ruminal and postruminal digestion due to increased rate of passage (Wheeler et al. 1975).

Pooled digestibilities of OM, CP, and ADF in the six diets are presented in Table 4. Respective increases of

Table 2. Effect of animal species and feeding level on percent digestibilities.

Feeding level g/$W^{.75}$	Organic matter		Crude protein		Acid detergent fiber	
	Sheep	Cattle	Sheep	Cattle	Sheep	Cattle
50	70.7[ac]	72.9[b]	72.5[a]	75.8[b]	29.0[a]	35.3[b]
75	71.0[c]	71.9[bc]	71.2[ab]	72.6[ac]	35.7[b]	36.0[b]
100	69.6[ad]	70.6[d]	69.9[b]	71.8[c]	31.1[a]	34.8[b]
SEM	.42	.47	.55	.61	1.04	1.16

[a,b,c,d]Means with different superscripts in the same row or column for each subheading are significantly different (P<.05).

Table 3. Effect of animal species and feeding level on percent digestibility.

Item	Species		Feeding level, g/W·75		
	Sheep	Cattle	50	75	100
Organic matter	70.4[a]±.24	71.8[b]±.27	71.8[a]±.31	71.5[a]±.31	70.1[b]±.31
Crude protein	71.2[a]±.33	73.4[b]±.36	74.1[a]±.42	71.9[b]±.42	70.9[b]±.42
Acid detergent fiber	31.9[a]±.70	35.3[b]±.79	32.1[a]±.91	35.8[b]±.91	32.9[a]±.91

[a,b]Means with different superscripts in the same row are significantly different (P<.05).

Table 4. Effect of alfalfa:barley ratio on percent digestibilities.

Item	Alfalfa:barley ratio						
	100:0	82:18	64:36	46:54	28:72	10:90	SEM
Organic matter	60.4[a]	65.0[b]	70.1[c]	71.9[d]	77.0[e]	82.3[f]	.44
Crude protein	68.3[a]	69.9[b]	71.4[b]	71.5[b]	74.7[c]	78.0[d]	.58
Acid detergent fiber	35.5[a]	36.2[a]	36.0[a]	34.6[ac]	31.5[bc]	27.9[b]	1.10

[a,b,c,d,e,f]Means with different superscripts in the same row are significantly different (P<.05).

0.25 and 0.11 percentage units in OMD and CPD for each percentage of increase of barley in the feed are similar to earlier reports (Horton 1978). Digestibility of ADF was not affected by replacement of alfalfa by barley until the diet contained 28% or less alfalfa; then ADF declined (P < 0.05). High levels of grain are associated with low ruminal pH, and this low pH may depress the activity of the cellulolytic bacteria (Terry et al. 1969) and the digestion of cell-wall components in the rumen.

The linear regression equation of best fit was:

$$Y = 91.89 - 24.17A - 0.046L - 0.062X \quad (1)$$

where Y = OMD of diet given to cattle, A = proportion of alfalfa in the diet, L = feeding level (gDM/W$^{0.75}$), and X = OMD of diet by sheep (r = 0.97, SE = 5.04).

These data show that even though digestibilities of pelleted diets with different proportions of barley and alfalfa were significantly higher in cattle than in sheep, differences were not large, averaging 1.4 percentage units.

LITERATURE CITED

Association of Official Analytical Chemists (AOAC). 1976. Official methods of analysis, 12 ed. Assoc. Off. Anal. Chem., AOAC, Washington, D.C.

Blaxter, K.L., F.W. Wainman, and J.L. Davidson. 1966. The voluntary intake of food by sheep and cattle in relation to their energy requirements for maintenance. Anim. Prod. 8:75.

Galyean, M.L., D.G. Wagner, and F.N. Owens. 1979. Level of feed intake and site and extent of digestion of high concentrate diets by steers. J. Anim. Sci. 49:199.

Horton, G.M.J. 1978. Dehydrated alfalfa and barley in finishing diets for lambs and steers. Proc. 2nd Int. Green Crop Drying Cong., Saskatoon, Canada, 334.

Rees, M.C., and D.A. Little, 1980. Differences between sheep and cattle in digestibility, voluntary intake and retention time in the rumen of three tropical grasses. J. Agric. Sci., Camb. 94:483.

Robertson, J.B., and P.J. Van Soest. 1975. A note on digestibility in sheep as influenced by level of intake. Anim. Prod. 21:89.

Terry, R.A., J.M.A. Tilley, and G.E. Outen. 1969. The effect of pH on cellulose digestion under in vitro conditions. J. Sci. Food Agric. 20:317.

Van Soest, P.J. 1963. Use of detergents in the analysis of fibrous feeds. II. A rapid method for the determination of fiber and lignin. Assoc. Off. Agr. Chem. J. 46:825.

Wheeler, W.E., C.H. Noller, and C.E. Coppock. 1975. Effect of forage-to-concentrate ratio in complete feeds and feed intake on digestion of starch by dairy cows. J. Dairy Sci. 58:1902.

Simulation of Complementary Forage Cropping by Feeding Oat Grain to Ewes

J.L. WHEELER, D.A. HEDGES, and C. MULCAHY

CSIRO Pastoral Research Laboratory, Armidale, N.S.W., Australia

Summary

In many parts of the world where year-round grazing is feasible, complementary forages such as oats or sorghum are sown to impove the quantity and quality of the forage available at specific periods. Experiments in which these crops complement well-fertilized grassland have generally not shown an increase in animal production compared with all-grassland systems. The present experiment aimed to determine the effects of a simulated complementary forage crop on pasture yield and on the production of breeding ewes.

Eight groups of 10 sheep grazed 0.6-ha plots of *Phalaris aquatica* L.–*Trifolium repens* L., and eight groups of 10 sheep grazed 0.8-ha plots. To break the nexus between pasture and crop growth rate, the provision of a complementary forage crop was simulated by giving the ewes a ration of 250 g/ewe/day oat grain of c. 70% organic-matter digestibility, equivalent to almost half their basic maintenance requiremnt. It was supplied for 7-week periods at mating, during lambing, on both occasions, or not at all. The four treatments were replicated twice at each stocking rate. To simulate the effects of withdrawing part of a farm from grazing to grow a complemetary forage, stocking rates were increased by 10% for some weeks before each feeding period. Fine-wool merino ewes were used for 4 years, then crossbred ewes mated with Dorset rams were used for 2 years.

The treatments did not significantly affect forage availability, wool production, reproductive performance of the ewes, or weaning weight of their lambs. There were no interactions with stocking rate even in an abnormally dry year or with the high twinning rates of the last 2 years, although production/animal was generally lower at the higher stocking rate.

The experiment is considered in relation to other published comparisons. The importance of ensuring that the quantity and quality of the complement is consistently and materially superior to the pasture it replaces is briefly discussed.

KEY WORDS: complementary forage system, oats, supplement, ewes, lambs, wool.

INTRODUCTION

Complementary grazing systems, such as the combination of areas of oat forage with permanent pastures, are used in many parts of the world. A common system in southeastern Australia is to grow oats (*Avena* spp.) on a portion of the farm and to allow ewes to graze this area before and after lambing. There are few objective comparisons of the production from complementary systems with that from grassland alone. Generally, comparisons have been between fertilized grassland and complementary small-grain/grassland systems; these do not show consistent or appreciable differences in animal production (Wheeler 1980).

The published studies may be criticized on the ground that (1) researchers used 20%–50% of the model farm area for complementary forage, whereas commercial farmers use 5% or less; (2) research utilization schedules were rigid and predetermined, not flexible as in farming practice; and (3) researchers placed all model-farm stock on complementary forage, whereas farmers usually reserve this high-quality material for a small proportion of their stock. Another problem is that climatic conditions that favor growth of the complementary crop also often favor growth of the base pasture and hence reduce the need for the complement. Similarly, when the effects of moisture stress and low temperature on the grassland increases the need for a complement, crop growth may also be limited.

This paper describes a 6-year grazing experiment designed to meet the first of these criticisms and to overcome the difficulty of the positive relationship between the growth of the complementary forage and the base pasture by providing an assured supply of feed regardless of weather conditions. One possible means of improving the production of ewes that lamb in early spring is to provide supplementary feed in autumn, thereby possibly improving lambing rate and encouraging the winter growth of grass. The experiment, in effect, simulated the provision of a complementary forage to breeding ewes at mating, at lambing, on both occasions, or not at all.

METHODS

The experiment was conducted in a temperate upland environment at Armidale, New South Wales (lat 30° S, 1,030 m elevation). The six annual rainfalls were 911, 938, 768, 972, 846, and 883 mm. Sixteen groups of 10 sheep grazed plots of *Phalaris aquatica* L.-*Trifolium repens* L. fertilized with 23 kg phosphorus (P), 32 kg potassium (K), and 30 kg sulfur (S)/ha/yr. To break the nexus between growth rates of a complementary forage and the base grassland, we simulated the provision of a complementary forage by giving the sheep a ration of oat grain (c. 70% organic-matter digestibility: 250 g or 2.5 mJ/day) equivalent to 40%–50% of their basic maintenance requirement (ARC 1980). To simulate the effect of withdrawing 10% of a farm from grazing in order to grow the crop, one extra sheep was placed in each flock for several weeks before the supplementation period. As winter forages grow more slowly than summer crops, the increased stocking rate was imposed for longer in winter.

Four treatments were compared at two stocking rates (12.5 and 16.7/ha) with two blocks: (C), control, given no supplement and with no variation in stocking rate (SR); (M), oats for 33 days before mating began in April (autumn) and 16 days thereafter (7 weeks total); SR increased by 10% for the 2 months before feeding; (L), oats for 4 weeks before the expected median date of lambing (September) and for 3 weeks thereafter; SR increased by 10% for 4 months before feeding; and (ML), oats at mating and at lambing, a full combination of treatments M and L.

Each flock consisted of 9 ewes and 1 vasectomized ram and was increased by 1 castrated male for the periods specified above. Fine-wool merino ewes were used for 4 years, then the potentially more responsive Border Leicester × Merino ewes (crossbreds) mated with Dorset rams were used for 2 years. All flocks were replaced annually one month after weaning to reduce the errors in estimating reproductive performance. During the mating period a fertile ram replaced the vasectomized ram in each plot. Lambs were removed at a median age of 13 weeks. Internal parasites were controlled by regular use of anthelmintics. Observations made included pasture available (8 occasions/yr), live-weight change (monthly), wool growth (annually and by dyebanding during 6 periods/yr, and reproductive performance.

RESULTS

The mean annual quantity of pasture available varied considerably (1.6 to 4.6 t DM/ha) and was reduced by 40% by the higher SR (P < 0.01). There was generally more pasture available at the start of the summer period (4.0 t DM/ha) than at other periods (2.8, 2.4, 2.2), but there were no consistent effects of treatments.

A major criterion in assessing a grazing system for lamb production is the average weight of lambs when weaned. Neither breed group responded to the treatments (Table 1). Provision of the complementary forage at mating was associated with a reduction in weight of lamb weaned/ha (P < 0.05) in the Merinos, but this reduction appears to have arisen as a consequence of an inexplicably higher perinatal mortality (Table 1.)

There was an interaction between treatment means and periods in live-weight change (P < 0.05), but as differences were only c. 0.5 kg/sheep, they are not presented. Annual production of clean fleece wool, adjusted by covariance for wool production in a common pre-experimental period, was not affected by the treatments. An increase in SR depressed production slightly but not significantly (Merino, 2.87 vs. 2.73; crossbred, 3.55 vs. 3.39 kg). Although there was no direct effect of treatments considered over all 6 years, there was evidence that the response to the complement was related to the general level of pasture growth (Fig. 1).

DISCUSSION

Over 6 years the provision of the complement to grazing ewes at mating and/or at lambing failed to increase animal production. There was no response by either breed group at either 12.5 or 16.7 sheep/ha, despite evidence that the latter rate was sufficiently high to depress production/animal. This result is similar to that obtained in all five tests of complementary grazing systems published in Australia (see Wheeler 1980); however, the present simulation differed in two important respects: (a)

Table 1. Effect of treatments (T) and stocking rate (SR) on wool and lamb production.

| | Breed | Treatments | | | | | Significance††† | |
		C	M	L	ML	Mean	T	SR
Lamb deaths (%)†	Merino	17.3	25.7	15.3	25.0	20.8	*	ns
	Xbred	19.4	13.9	29.2	30.5	23.3	ns	ns
Lambs weaned (%)†	Merino	96	79	97	83	89	*	ns
	Xbred	144	154	147	139	146	ns	ns
Avg. weaning weight (kg)††	Merino	21.3	22.0	22.1	21.9	21.8	ns	*
	Xbred	24.9	24.5	24.9	25.7	25.0	ns	ns
Weight weaned lamb (kg/ha)	Merino	267	230	279	240	254	*	***
	Xbred	459	482	467	461	467	ns	*
Wool growth (kg clean/ewe)	Merino	2.81	2.77	2.80	2.82	2.80	ns	*
	Xbred	3.52	3.48	3.46	3.42	3.47	ns	*

†As % of ewes/plot.

††Adjusted for age at weaning.

†††*Denotes P < .05; ***, P < .001.

Fig. 1. Effect of mean annual pasture availability (control treatments) on response in annual wool production to treatment ML. Response (%) = 4.2%-0.055x green pasture available (kg DM/sheep/yr) (P < 0.05).

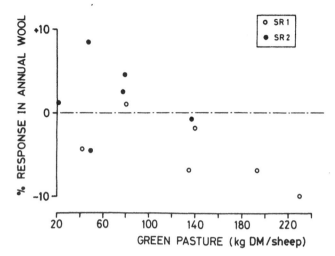

animals on the farm. The term "complement" is a useful one because of its connotation of interaction and mutual dependence (McIlvain and Shoop 1973), but it must be appreciated that in normal farm practice, where the area involved is 5% or less of the farm, the term may be a misnomer; all such an area can do is to provide a special-purpose feed for a small proportion of the total farm stock, not to complement the farm feed supply. As this feed is used opportunistically by farmers—at least in Australia—for one or more of several purposes, adequate experimental evaluation of the practice may never be possible.

There was evidence (Fig. 1) that the response in annual wool growth was related to the average amount of pasture available. The mean herbage availability at which positive responses ceased was equivalent to about 1,500 kg DM/ha. Where growth rates of the base pasture and the crop are likely to be depressed to a similar extent by low temperature or moisture stress, it should be ensured in practice that the proposed complement is not affected by the environment to the same extent as the base pasture. For example, in dry environments the complement may be grown more certainly if moisture can be accumulated by prior fallowing; in cold environments genotypes should be used that will grow at lower temperatures than the base pasture and will withstand frost without nutritional deterioration. Special fertilizer may be required. It should be noted that the most consistently successful complements used in world farming practice exploit the physiological differences between C_3 and C_4 plants rather than seeking to make use of the more limited differences available within these physiological types. Thus, systems using the C_3 winter wheat (*Triticum aestivum*) on farms dominated by the C_4 Coastal bermudagrass (*Cynodon dactylon*) have much more potential than those that use, say, oats (*Avena sativa*) on farms with basically *Phalaris aquatica* or *Festuca arundinacea* pastures.

stocking rate was raised by only 10% in the crop growth periods, and (b) quantity of supplement was constant regardless of the particular environmental conditions.

Previous experiments on complementary grazing systems have been designed to ensure an uninterrupted supply of high-quality forage for all sheep or cattle on the farm. Given the anticipated growth rates of small grains, researchers have calculated that 20%-50% of the farm should be sown with complementary crops, which necessarily increases stocking rate elsewhere on the property by 20%-50%. Unless this increase occurs at a time when there is a substantial surplus of feed available, its effect is normally to depress animal production, a penalty that is often difficult to recoup (Wheeler 1980).

The present experiment, conducted over 6 years of varied seasonal conditions, has shown that where only a small proportion of the farm (10%) is used for complementary forage, there is no penalty of any consequence. However, where the farm stocking rate is high, there are, no doubt, some circumstances in which the penalty may be biologically significant even with small areas of complementary crop.

The disadvantage of using only a small proportion of complement is that the area is unlikely to supply sufficient additional feed to affect production when it is used for all

LITERATURE CITED

Agricultural Research Council (ARC). 1980. The nutrient requirements of ruminant livestock. Commonwealth Agric. Bur., Farnham, UK.

McIlvain, E.H., and Shoop, M.C. 1973. Use of farmed forage and tame pasture to complement native range. Great Plains Agric. Counc. Pub. 63, L 1-19, Univ. Nebr., Lincoln.

Wheeler, J.L. 1980. Complementing grassland with forage crops. *In* F.H.W. Morley (ed.) Grazing animals, 239-260. Elsevier, Amsterdam.

Increasing Productivity and Extending Grazing Season of *Cynodon* Pastures for Steers with N and Overseedings of *Trifolium* and *Festuca* in the Southeastern U.S.A.

H.A. FRIBOURG, J.B. MC LAREN, and R.J. CARLISLE

University of Tennessee, Knoxville, Tenn., U.S.A.

Summary

A series of grazing experiments has been conducted 80 km east of Memphis, Tennessee, since 1969 on fine-silty, mixed, thermic, Typic Hapludalfs soils to evaluate productivity and quality of *Cynodon* pastures grown with N fertilizer or in combination with *Festuca arundinacea* and/or Ladino, *Trifolium repens*, and sometimes with *T. pratense*, for growing beef steers. Pastures were grazed by the put-and-take system, and forage growth and consumption were determined by the cage-and-strip method.

These experiments have shown that bermudagrass pastures can support many young beef animals for 5 months/yr if they are fertilized with N, can produce rapid gains in spring if legumes are present, and can provide grazing for 9 to 10 months/yr if they are overseeded with fescue and/or legumes. Combinations of bermudagrass, fescue, and clovers can be maintained for several years and can result in satisfactory beef production of over 450 kg/ha/yr.

KEY WORDS: tall fescue, orchardgrass, midland bermudagrass, Ladino clover, forage intake, beef production, humid mesothermal.

INTRODUCTION

In the midsoutheastern U.S.A., cool-season grasses and legumes have a lush growth period in spring followed by low production in summer and erratic regrowth in fall after temperatures have moderated and rainfall has occurred. Pastures of orchardgrass (*Dactylis glomerata* L.) mixed with Ladino clover (*Trifolium repens* L.) can support 2.5 steers/ha in April to August and produce about 500 kg beef/ha (Fribourg et al. 1979). However, a midsummer drought often reduces productivity of these cool-season species.

Pastures of tall fescue (*Festuca arundinacea* L.) produce well in spring, are relatively dormant in summer, and have moderate amounts of nutritious forage in fall and in winter when temperatures are favorable. Some warm-season grasses, such as bermudagrass (*Cynodon dactylon* [L.] Pers.) produce large amounts of forage and grow well in spring and summer. The high potential productivity of Midland has been demonstrated for alleviating the summer forage deficiency in beef-production systems. Improved bermudagrass cultivars have resulted in 370 to 525 kg beef/ha.

The objectives of this study in 1971–1973 were to determine the effects of various levels of nitrogen (N) fertilization of Midland on beef production/animal and beef production/ha and to further compare Midland and common bermudagrass (*C. dactylon* var. *dactylon*) to orchardgrass plus Ladino (OG-Lad) as forage for grazing beef steers. In 1975–1977, Midland with N was compared to Midland with fescue and N, to Midland with legumes without N, to fescue with legumes, and to OG-Lad. In the third phase, started in 1979, greater emphasis has been placed on establishing and maintaining mixtures of Midland, fescue, and legumes while sustaining acceptable levels of production for 9 to 10 months/yr without N fertilizer.

MATERIALS AND METHODS

Pastures of 1.2 ha were established in 1969 on Memphis silt loam (fine-silty, mixed, thermic, Typic Hapludalfs), 2% to 5% slopes. A randomized complete-block design with 2 replications was used. Treatments initiated in 1971 on Midland (Table 1) were: no N fertilizer (Midland-0), and 112, 224, and 448 kg N/ha/yr (Midland-112, Midland-224, Midland-448). Common bermudagrass was seeded and fertilized with 112 kg N/ha/yr (Common-112). Sixteen kg/ha of Boone orchardgrass plus 2.25 kg/ha of Regal Ladino seed resulted in the OG-Lad treatment. In April each year, there was an estimated 40% to 60% clover stand uniformly distributed in the OG-Lad pastures. As the season progressed, the clover decreased to 25% to 30%, occasionally to 15% to 20% in August. Nitrogen was applied as NH_4NO_3, and sufficient phosphorus (P) and potassium (K) were applied once each year to maintain soil test levels at a medium level or higher.

H.A. Fribourg is a Professor of Plant and Soil Science, J.B. McLaren is a Professor of Animal Science, and R.J. Carlisle is a Graduate Research Assistant in Plant and Soil Science at the University of Tennessee. This is a contribution of the Tennessee Agricultural Experiment Station, Knoxville.

Table 1. Animal performance and pasture productivity during the grazing season, April to September.
(CB = Common bermuda; MB = Midland bermuda; OL = Orchard ladino; F = fescue; L = legumes)

| Kg N applied to bermudagrass:† | Phase 1 - 1971 through 1973 | | | | | | Phase 2 - 1975 through 1977 | | | | | | |
| | Midland bermudagrass | | | | | | | MB+ L | F+ L | MB+ F (rows) | | | MB+ F†† |
	0	112	224	448	CB 112	OL 0	MB 224	0	0	224	CB 224	OL 0	224
Stocking rate, steers/ha§	3.7¶	5.9	8.9	12.4	4.9	4.0	6.3	4.4	4.3	7.6	5.2	4.1	4.5
Final weight, kg/steer	333	347	338	352	355	347	310	317	290	301	320	327	301
Gain, kg/steer	47	59	54	61	70	123	70	88	66	73	63	92	78
Daily gain, g/steer	354	418	395	445	499	849	518	576	527	478	511	796	515
Length of period, days	143	143	143	142	141	147	134	149	133	149	124	120	149
Total animal grazing days/ha	546	833	1290	1759	677	591	839	646	596	1127	645	505	672
Forage dry matter (DM), ton/ha	5.1	9.2	8.7	11.4	7.2	5.0	5.5	5.5	3.4	7.8	5.2	4.3	4.0
Forage DM intake, kg/steer/day	8.2	8.9	6.0	5.7	8.1	8.7	5.4	7.2	5.4	6.5	6.2	7.0	5.6
Forage DM intake, kg/kg gain	23.2	21.4	15.1	12.8	16.2	10.3	10.4	12.5	10.2	13.5	12.1	8.7	10.8
Beef production, kg/ha	161	282	354	604	308	561	391	389	310	520	310	377	341

†N applied to bermudagrasses in 3 equal installments on 15 March–1 April; 20 May–1 June; and 1–15 July; N applied to fescue in September (67 kg/ha) or September and 1–10 March (67 + 56 kg/ha).

††0.4 ha Midland + 0.8 ha fescue.

§Tester steers remained on their respective pastures continuously and additional steers were added or removed as needed to utilize forage production of individual pastures.

¶Mean of 2 replications.

Pasture treatments were changed during 1974, and the second phase was initiated in 1975 (Table 1). In one case, the 1.2-ha pastures were divided into one 0.4-ha pasture of Midland-224 and one 0.8-ha pasture of Kentucky 31 fescue fertilized with N in September and in early March (Fes-67-56). Another treatment consisted of a Midland-224 sod overseeded in the fall with fescue delivered by a drill into 25-cm rows at 17 kg/ha (Midland-F25). The Common-112 treatment was changed to Common-224 by doubling the N rate. A Midland sod was overseeded in late February (Midland-Leg) with 2.25 kg/ha of Ladino plus 4.5 kg/ha of Kenland red clover (*T. pratense* L.). Finally, a treatment (Fes-Leg) typical of the seeding mixture used on many pastures in Tennessee — fescue (17 kg/ha) plus Ladino (2.25 kg/ha) and red (4.5 kg/ha) clovers plus Kobe lespedeza (*L. striata* [Thunb.] H. and A.) (9 kg/ha) — was seeded in the fall.

Four phase-2 treatments were modified for the phase started in 1979. Midland and Common treatments were overseeded with Ladino and red clovers plus Kobe or with fescue in 50-cm rows plus the three legumes. The 0.4-ha supplement was changed to a forage *Sorghum bicolor* (L.) Moench each summer and *Secale cereale* L. plus *Lolium multiflorum* Lam. in the cool season.

Throughout, a put-and-take management system was followed. Yearling Angus steers, wintered uniformly, weighing 200–250 kg in the spring, were used. Individual weights of tester animals (three to eight depending on treatment) were taken at about 21-day intervals during the grazing season. Beef production for each period was calculated by multiplying total number of animal grazing days/weighing period by average daily gain (ADG)/

weighing period of the tester steers. At no time did the steers receive supplemental feed while on pasture.

Grazing of pastures with bermudagrass or fescue started each year when average forage height was 7.5 cm. Extra steers were added to a pasture whenever mean forage height had reached 10 cm and were removed when mean grass height was 3.5 to 4 cm. In OG-Lad pastures, height was allowed to reach 12.5 to 15 cm before grazing pressure was increased, and extra steers were removed when forage was 8 cm high. A lesser grazing pressure was used for OG-Lad because orchardgrass in Tennessee can be destroyed by heavy grazing pressures.

Forage yield and consumption were estimated by the cage-and-strip method. One cage (0.51 × 1.22 m) and one strip (0.51 × 6.10 m) were harvested at random from each one-sixth of each pasture at about 21-day intervals with a rotary mower set at a height equal to the lowest allowable grazing height for that pasture. Each sample was oven-dried at 65°C for 72 hours and weighed for dry-matter calculations.

RESULTS AND DISCUSSION

Length of the Grazing Season

Grazing of bermudagrass pastures was started between 15 and 30 April. When fescue or legumes were present it was possible for grazing to start 2 to 3 weeks earlier. Grazing of OG-Lad pastures also started about 15 days earlier than that of bermudagrass and ended in mid- to late August. Bermudagrass and fescue pastures usually provided grazing for 10 to 15 days more in fall than OG-Lad.

Fertilization of bermudagrass pastures with N did not affect length of grazing season, but presence of fescue lengthened it 2 to 3 weeks.

Animal Gains

During the entire grazing season in all phases, steers grazing OG-Lad pastures gained faster than those grazing the other pastures. There was a trend toward increased rate of gain in steers grazing Midland as N fertilization increased. Midland and Common pastures produced similar steer gains at the same N rate. Large seasonal differences were observed in ADG of steers grazing bermudagrass. Early-season ADG was similar for all pastures and higher than summer gains. In summer, ADG of steers grazing OG-Lad pastures was 3 to 6 times greater than that of steers grazing bermudagrass.

When legumes were present, bermudagrass or fescue pastures maintained gains in summer at a higher level than when they were absent, although ADG was less than with OG-Lad. Thus, satisfactory weight increases can be obtained in steers grazing bermudagrass pastures from spring until early July; then, either legumes or other feeds should be provided, or the steers should be marketed.

Pasture Productivity

Animal grazing days/ha increased significantly with each increase in N fertilizer. Midland with N supported more animal grazing days than OG-Lad, partly because the grazing season for OG-Lad was shorter. Midland-112 had greater productivity than Common-112, and Midland-224 produced more than Common-224, reflecting the greater potential productivity of the improved strain.

Midland-F25 resulted in more animal grazing days than Midland-224, partly because of the longer grazing season available from the combination than from each grass separately. The presence of legumes in the absence of N reduced carrying capacity about one-fifth.

Dry-Matter Intake

Dry-matter requirements/unit of gain were lower for steers grazing OG-Lad than for steers grazing other pastures. As N fertilization increased, dry-matter efficiency tended to increase; including legumes in Midland or fescue pastures resulted in intermediate dry-matter efficiencies.

Beef Production/Ha

Beef production/ha reflected both forage quality (expressed by ADG) and forage yield (reflected in total animal grazing days). Even though stocking rates were varied by adding or removing extra animals in order to utilize the forage production of each pasture, stocking rates remained relatively constant during the grazing season after they had been established in May. Stocking rates of OG-Lad and Fes-Leg were reduced about July 1 and remained constant until the grazing season ended.

Beef production from OG-Lad and from Midland-448 were similar in the first phase. These pastures produced more beef than the other pastures. In phase 2, Midland-Leg, Midland-224, and OG-Lad resulted in 375 to 390 kg/ha. Common-224 and Fes-Leg produced 65 to 80 kg/ha less, but Midland overseeded with fescue and with both grasses fertilized with N produced much more. Thus, bermudagrass pastures with legumes and no N produced as much beef/ha as N-fertilized grass pastures. The rapid early July decline in ADG of steers grazing N-fertilized bermudagrass pastures accounts for this similarity.

Animal-Forage Management

Animal-forage management had to be more intensive, and therefore more difficult, as N fertilization of bermudagrass pastures was increased. Increased N fertilization resulted in increased forage growth response to available moisture; concurrently, problems in establishing a stocking rate of individual pastures to maintain bermudagrass in a vegetative and actively growing stage also increased. Utilization of all forage production without periods of occasional overgrazing was difficult. Midland-448 required a high stocking rate in summer, resulting in extreme fouling of the pastures and uneven grazing. Animal-forage management of the other N-fertilized pastures was less intensive and posed fewer managerial problems.

It had been anticipated that difficulties might arise in the establishment and maintenance of fescue and/or legumes in bermudagrass sods. In fact, every overseeding into Midland sods was a success, and the mixed stands have persisted to date in the proportions desired. The introduction of fescue and legumes into common bermudagrass, which has a tighter sod than Midland, required two successive seasons to be successful. The maintenance of the desired combined stands is probably related directly to the management imposed on the pastures, where soil fertility, plant heights, and grazing pressures are monitored and adjusted as needed at relatively frequent intervals.

LITERATURE CITED

Fribourg, H.A., J.B. McLaren, K.M. Barth, J.M. Bryan, and J.T. Connell. 1979. Productivity and quality of bermudagrass and orchardgrass–ladino clover pastures for beef steers. Agron. J. 71:315–320.

Feeding Value of the Legume Component in Pasture and Silage

K.F.M. REED and P.T. KENNY

Victorian Department of Agriculture, Victoria, Australia

Summary

Grazing experiments were conducted to examine the feeding value of various pastures for grazing sheep and, in some cases, as silage. Recently, the use of legumes for improving the growth of weaned lambs during the dry summer and early autumn was investigated at Hamilton, Victoria.

Studies in Australia showed that the fleeceweights of weaner sheep grazed on eight pasture mixtures of varying grass species increased as the amount of clover in the pasture increased (r = 0.88**). In a separate study, young sheep grew faster on lucerne-grass pasture than on grass–subterranean clover pasture. Most recently, live-weight gains of 4-month-old lambs grazing legumes during 3 dry months in summer-autumn were 9.5, 9.4, 8.9, and 8.7 kg/head on red clover, giant shaftal clover (*T. resupinatum*), lucerne, and white clover pasture, respectively, greater than those obtained on perennial ryegrass–subterranean clover pasture (0.4 kg/head) or cocksfoot–subterranean clover pasture (2.2 kg/head).

In studies made in Ireland, store lambs offered unchopped wilted silage for 8 weeks had higher intakes of grass–red clover silage than of grass silage; corresponding carcass gains were 0.35 and 0.16 kg/week. Carcass gain/unit of digestible dry matter (DM) consumed was higher for grass–red clover silage than for grass silage. Mean retention time of DM in the reticulo-rumen was 19% shorter with the grass-clover silage.

Superior animal performance on legume-dominant pasture occurred, sometimes despite a higher DM production from grass-dominant pasture. Producers should aim to provide at least 30% legume in sheep pastures in southeastern Australia, especially in the dry summer when the feeding value of grass falls rapidly. Plant improvement should aim at providing perennial legumes suitable for all main soil and climate environments.

KEY WORDS: nitrogen fertilizer, lucerne, red clover, white clover, *Trifolium resupinatum*, *T. subterraneum*.

INTRODUCTION

During the last 2 decades the feeding value of pasture has been the subject of much research. In Victoria, early trials were concerned with identifying the most suitable sheep pasture for year-round production. Particular attention was given to alternative perennial grasses (Reed 1970, Cayley et al. 1974). Reviews indicated that a wide range of legume species have greater feeding value than grasses for both sheep and cattle (Reed 1972a, Ulyatt 1973, Demarquilly and Jarrige 1973). The effect of nitrogen fertilizer on the feeding value of grass-clover pasture was reviewed (Reed 1981). Recently, seasonal limitations of pasture quality have been given closer attention with a view to alleviating them with special-purpose legume pastures.

K.F.M. Reed is a Senior Research Officer and P.T. Kenny is an Agricultural Scientific Officer in the Victorian Department of Agriculture.

The work at Glenmoriston was supported by the Australian Dairy Research Trust. All other studies were supported by the Australian Sheep and Wool Research Trust. The studies at Canberra and in Ireland were made possible through the cooperation of Dr. F.H.W. Morley, Commonwealth Scientific and Industrial Research Organization, and Dr. A.V. Flynn, the Agricultural Institute, Dunsany, County Meath, respectively.

METHODS

Single Pastures Year-Round

At Glenormiston in southwestern Victoria, 10-week-old lambs were weaned onto one of eight pastures, which were then continuously grazed for 12 months at a stocking rate of 20 lambs/ha. The experiment was a 2^3 factorial design where the variable factor was the presence or absence of three perennial grasses — ryegrass (*Lolium perenne*), phalaris (*P. aquatica*), or cocksfoot (*Dactylis glomerata*). Each pasture contained white clover (*Trifolium repens*), strawberry clover (*T. fragiferum*), and subterranean clover (*T. subterraneum*).

At the CSIRO station at Canberra, four pastures were established: lucerne (*Medicago sativa*) and ryegrasses (*L. rigidum*; *L. perenne* × *L. multiflorum*), lucerne and phalaris, subterranean clover and ryegrasses, and subterranean clover and phalaris. Pastures were sown on separate areas within each of two adjacent blocks (a Latin square design). Each plot was split to facilitate two rates of stocking. Over 2 years, within each block, groups of young sheep rotationally grazed each set of four plots with the same species combination and stocking rate.

Special-Purpose Pastures

At Hamilton, western Victoria, weaned lambs (26 kg) were grazed on two grass-legume and four legume pastures during the summer-autumn in 1980 (January-April 1980 inclusive). Sheep were stocked at 14/ha, and they rotationally grazed 2 0.5-ha replicates of each pasture. For lucerne, plots were subdivided for a four-paddock rotation. Similar sheep were used on two additional treatments, i.e., a strawberry clover pasture growing on a reclaimed peat swamp, and a feedlot ration (clover hay ad libitum + lupins, 0.3 kg/day). Rainfall was 59, 4, 10, and 45 mm in January, February, March, and April, respectively; mean minimum and maximum temperatures were approximately 8.5°C and 23°C, respectively.

Grass-Clover Silage

At Dunsany in Ireland, store lambs were fed wilted silage for 8 weeks. Grass and grass–red clover crops were ensiled and compared in terms of digestibility, voluntary intake, and carcass-weight change. The mean retention time of dry matter in the reticulo-rumen (RTDM) was calculated from measurements of dry matter (DM) in the reticulo-rumen at slaughter and its relationship to recent intake. A similar experiment was conducted using beef cattle.

Botanical analyses reported below were made by hand sorting samples cut to ground level. Results were expressed on a DM basis.

RESULTS

Single Pastures Year-Round

Where perennial grasses were sown at Glenormiston, white clover contributed less than 2% to the pasture present on the plots (mean over three dates); elsewhere white clover contributed 31%. The total amount of all sown clovers ranged from 0.1 to 1.0 t/ha (c. 3%–30%). In most plots the main clover species was subterranean clover. Fleeceweight and live-weight gain (LWG) increased linearly with amount of clover; clover accounted for 77% and 66% of the between-plot variation in fleeceweight and LWG, respectively. Neither annual DM production, which was relatively high for some grass-dominant pastures, nor the amounts of other botanical fractions was significantly correlated with animal production. Sheep on white clover–dominant plots gained 12.8 kg/head during the summer-autumn months (December-April), but those on plots sown to perennial grass lost 0.3 kg/head ($P < 0.01$) (Reed 1972b).

At Canberra, LWG and wool production were 45% and 10% greater, respectively, on lucerne pasture (87% lucerne) than on grass-dominant subterranean clover pasture (8% subterranean clover); differences were greatest in summer. Mortality and supplementary feed requirement during drought were twice as great on grass-dominant as on lucerne pastures (Reed et al. 1972).

Special-Purpose Pastures

At Hamilton, LWG of weaned lambs indicated that, over the dry months, legume-dominant pasture was markedly superior to perennial grass pasture irrespective of whether the legume was a perennial or the dried-off annual *Trifolium resupinatum* (Table 1).

Grass-Clover Silage

Carcass gains for store lambs offered unchopped wilted silage were 0.35 and 0.16 kg/week with grass–red clover (*T. pratense*) silage (34% DM; 53% clover) and grass silage (38% DM), respectively. Grass-clover and grass

Table 1. Pasture descriptions and associated cumulative LWG of weaner sheep at Hamilton, 7 January–14 April 1980.

	Sown species +	Pasture DM 15 December 1979			LWG‡ (kg/hd)
		Total‡ (t DM/ha)	Green (% total)	Sown species (% green)	
1.	Perennial ryegrass, cv. Victorian + sub. clover cv. Mt Barker	3.08[bc]	67.5	21.0	0.40[e]
2.	Cocksfoot, cv. Porto + sub. clover, cv. Mt Barker	3.46[bc]	50.0	16.5	2.23[d]
3.	White clover, cv. Haifa	3.66[b]	86.5	31.5	8.65[b]
4.	Red clover, cv. Grasslands Hamua	3.27[bc]	85.8	47.8	9.48[b]
5.	Lucerne, cv. WL 318	1.83[c]	97.0	71.0	8.86[b]
6.	Giant Shaftal clover (*T. resupinatum*)	2.99[bc]	79.0	68.5	9.36[b]
7.	Strawberry clover growing on a reclaimed peat swamp	6.20[a]	95.0	89.0	11.68[a]
8.	Feedlot treatment: Strawberry clover hay *ad lib.* + lupins (0.3 kg/day)	—	—	—	6.70[c]

+ All treatments sown on silty loam except nos. 7 and 8.

‡ Figures followed by the same letter superscript are not different (P>0.05) by Duncan's multiple range test.

silages contained 71% and 76% digestible DM, respectively. Intake was 21% higher (P < 0.01) and RTDM was 19% lower (P < 0.01) for grass-clover silage relative to grass silage. Unwilted silages fed to sheep resulted in unsatisfactory levels of animal production (Reed 1979). A beneficial effect of clover on the feeding value of wilted silage for beef cattle was also noted (Reed and Flynn 1975).

DISCUSSION

The legume content of pasture was a major determinant of animal production in these experiments. The greater feeding value of legumes relative to grass was maintained through the ensiling process, at least when the crop was wilted. Additionally, the use of grass–red clover mixtures can increase the yield of silage relative to all grass crops receiving more N (Flynn and Reed 1974). The reviews cited indicate that the high feeding value of legumes is associated with high intake and more efficient utilization of nutrients of legume than of grass diets.

Current research into raising the legume content of pasture in Victoria includes the evaluation of both Mediterranean genotypes of white clover and cultivars of subterranean clover with good early-season growth. Cultivars of lucerne or other deep-rooted perennial legumes, capable of withstanding wet soils in winter, are being sought. Special-purpose legume pastures show promise for improving feeding value in summer, particularly where the quality of grass-dominant pasture declines rapidly (Reed et al. 1980). To realize the potential of legumes, efforts are needed in the areas of plant introduction and improvement. We need to improve establishment procedures. Research into pasture management and herbicide use is needed to help maintain legume-dominant pastures.

LITERATURE CITED

Cayley, J.W.D., A.H. Bishop, and H.A. Birrell. 1974. The role of perennial grass species in the Western District of Victoria. Proc. XII Int. Grassl. Cong., 3(1):97–104.

Demarquilly, C., and R. Jarrige. 1973. The comparative feeding value of grasses and legumes. Vaxtodling 28:33–41.

Flynn, A.V., and K.F.M. Reed. 1974. Comparison of sward type for silage production. Anim. Prod. Res. Rpt., 1974. An Foras Taluntais, Dublin, 28.

Reed, K.F.M. 1970. Variation in liveweight gain with grass species in grass-clover pastures. Proc. XI Int. Grassl. Cong., 877–880.

Reed, K.F.M. 1972a. The performance of sheep grazing different pasture types. *In* J. Leigh and J. Noble (eds.) Plants for sheep in Australia, 193–204. Angus and Robertson, Sydney.

Reed, K.F.M. 1972b. Some effects of botanical composition of pasture on the growth and wool production of weaner sheep. Austral. J. Expt. Agric. Anim. Husb. 12:355–360.

Reed, K.F.M. 1979. The effect of clover content on the feeding value of silage for store lambs. Anim. Prod. 28:271–274.

Reed, K.F.M. 1981. A review of legume-based vs. nitrogen-fertilized pasture systems for sheep and beef cattle: some limitations of the experimental techniques. *In* J.L. Wheeler and R.D. Mochrie (eds.) Forage evaluation: concepts and techniques, 401–417. Pub. CSIRO and Am. For. and Grassl. Coun., Melbourne.

Reed, K.F.M., and A.V. Flynn. 1975. Silage for beef cattle. Anim. Prod. Res. Rpt., 1975. An Foras Taluntais, Dublin, 21.

Reed, K.F.M., P.T. Kenny, and P.C. Flinn. 1980. The potential of pasture legumes for improving the quality of summer-autumn feed. Proc. Austral. Soc. Anim. Prod. 13:39–41.

Reed, K.F.M., R.W. Snaydon, and A. Axelsen. 1972. The performance of young sheep grazing pasture sown to combinations of lucerne or subterranean clover with ryegrass or phalaris. Austral. J. Expt. Agric. Anim. Husb. 12:240–248.

Ulyatt, M.J. 1973. The feeding value of herbage. *In* G.W. Butler and R.W. Bailey (eds.) Chemistry and biochemistry of herbage. 3:131–178. Academic Press, London.

Initial Rate of Digestion and Legume Pasture Bloat

R.E. HOWARTH, K.-J. CHENG, J.P. FAY, W. MAJAK,
G.L. LEES, B.P. GOPLEN, and J.W. COSTERTON
Research Stations at Saskatoon, Lethbridge, and Kamloops
and University of Calgary, Canada

Summary

Studies were conducted to provide the basic knowledge for breeding a bloat-safe alfalfa cultivar. Two working hypotheses of pasture bloat were tested. The cell-rupture theory proposed that leaf mesophyll cell rupture is an important event in the occurrence of pasture bloat. The initial-rate-of-digestion (IRD) theory proposed that susceptibility to rapid initial digestion by ruminal bacteria is a characteristic of bloat-causing forages.

The studies involved comparisons of bloat-causing and bloat-safe legumes: alfalfa (*Medicago sativa* L.), red clover (*Trifolium pratense* L.), arrowleaf clover (*Trifolium vesiculosum* Savi), birdsfoot trefoil (*Lotus corniculatus* L.), cicer milkvetch (*Astragalus* cicer L.), and sainfoin (*Onobrychis viciifolia* Scop.). IRD was measured in vitro and, with the nylon-bag technique, in fistulated sheep or cattle. Daily bloat incidence and rumen-fluid composition were recorded in fistulated cattle fed fresh alfalfa.

Bloat-causing legumes were digested more rapidly than bloat-safe legumes. Concentrations of plant cell constituents were higher in rumen fluid of sheep fed alfalfa than in sheep fed three bloat-safe legumes. Daily measurements of IRD by the nylon-bag technique were moderately correlated with daily bloat incidence, and the concentration of chlorophyll in rumen fluid was higher in bloated cattle than in nonbloated cattle.

Important events in IRD are (1) mechanical disruption prior to microbial digestion, (2) bacterial colonization and invasion of plant tissues, (3) maceration of plant tissues by pectinolytic enzymes, and (4) disruption of plant cell walls. Differences between bloat-causing and bloat-safe legumes have been identified in each of these events, but at this level of comparison the several bloat-safe legumes have different characteristics. These studies have provided the basis for breeding a bloat-safe alfalfa cultivar by selecting for low IRD.

KEY WORDS: digestion, pasture, bloat, alfalfa, birdsfoot trefoil, sainfoin, cicer milkvetch, arrowleaf clover, red clover, white clover.

INTRODUCTION

The soluble leaf proteins are generally accepted as the principal foaming substances responsible for legume pasture bloat, and tannins have been identified as antibloat forage constituents because they occur in many bloat-safe legumes and bind with soluble proteins to produce an insoluble protein-tannin complex (Jones and Lyttleton 1971). However, cicer milkvetch (*Astragalus cicer* L.), a bloat-safe legume, does not contain tannins (Sarkar et al. 1976), and its soluble-protein content is similar to that of alfalfa (*Medicago sativa* L.). Further, the correlation between the soluble-protein content of alfalfa and daily bloat incidence in cattle ($r^2 = 0.07$–0.17) was relatively low (Howarth et al. 1977). These discrepancies led to the formulation and testing of two new theories of legume pasture bloat: the cell-rupture theory and the initial-rate-of-digestion (IRD) theory.

WORKING HYPOTHESES

Cell-Rupture Theory

The soluble proteins of forage legumes are located inside mesophyll cells. The plasma membrane, which normally retains the soluble protein, should be completely inactivated in the anaerobic environment of the rumen, but an intact cell wall should retard passage of soluble proteins from the cell to the rumen fluid. Thus, cell-wall rupture may be an important event in the onset of pasture bloat, and cell walls of bloat-safe forages may be more resistant to rupture than cell walls of bloat-causing forages. Intracellular constituents may be released more slowly, and they may not accumulate in the rumen to a critical (threshold) bloat-causing concentration (Howarth et al. 1978).

Initial-Rate-of-Digestion (IRD) Theory

Cell-wall rupture can occur in chewing or by microbial digestion in the rumen. The basic premise of this theory is that bloat occurs with rapid initial digestion and release of cell contents.

EXPERIMENTAL STUDIES

IRD of Forage Legume Species

The proposal that bloat-causing legumes are digested faster than bloat-safe legumes has been verified in two laboratories using nylon-bag and in-vitro digestion techniques. The bloat-causing legumes were alfalfa, red clover (*Trifolium pratense* L.), and white clover (*T. repens* L.). The bloat-safe legumes were cicer milkvetch, birdsfoot trefoil (*Lotus corniculatus* L.), sainfoin (*Onobrychis viciifolia* Scop.), and arrowleaf clover (*T. vesiculosum* Savi). Dry-matter disappearance provided the best distinction between the bloat-safe and bloat-causing groups (Table 1). Gas production showed a similar pattern, but there was incomplete separation of the individual species, e.g., white clover (bloat-causing) and cicer milkvetch (bloat-safe).

To verify that plant cell constituents accumulate in the rumen fluid, fistulated sheep were fed fresh alfalfa, birdsfoot trefoil, cicer milkvetch, and sainfoin (Howarth et al. 1979). Both soluble protein nitrogen and chlorophyll were substantially lower in rumen fluid of sheep fed the bloat-safe legumes (Table 1). The low values for birdsfoot trefoil were due, in part, to low feed intake.

Daily Bloat-Incidence Studies

To determine whether day-to-day changes in the bloat-causing potency of alfalfa are correlated with IRD of the forage, rumen-fistulated cattle were fed fresh, chopped alfalfa. Correlation ranged from 0.11 to 0.60 (Majak, unpublished). Thus, IRD showed a low to moderate relationship with bloat occurrence. Average chlorophyll levels in rumen fluid were greater in bloated than in nonbloated cattle (Majak and Howarth 1980).

In summary, daily bloat incidence studies provided

R.E. Howarth is a Senior Research Scientist in Biochemistry at the Saskatoon Research Station; K.-J. Cheng is a Senior Research Scientist in Rumen Microbiology and J.P. Fay is a Postdoctoral Fellow in Rumen Microbiology at the Lethbridge Research Station; W. Majak is a Research Scientist in Biochemistry at the Kamloops Research Station; G.L. Lees is a Research Scientist in Physiology and B.P. Goplen is a Senior Research Scientist in Genetics at the Saskatoon Research Station; and J.W. Costerton is a Professor of Biology at the University of Calgary.

Table 1. Summary of data demonstrating differences between bloat-causing and bloat-safe legumes. (Results are expressed as a percentage of results with alfalfa.)

| | Bloat | Dry-matter disappearance | | Gas production in vitro§ | Sheep rumen fluid | |
		Nylon bag†	In vitro††		Soluble protein nitrogen¶	Chlorophyll¶
Alfalfa	†	100	100	100	100	100
Red clover	†	101	128	83	–	–
White clover	†	103	130	88	–	–
Birdsfoot trefoil	–	79	80	62	12	16
Cicer milkvetch	–	64	80	83	53	51
Sainfoin	–	26	47	58	23	28
Arrowleaf clover	Very little	74#	93#	71#	–	–

†Chopped leaves, 8 hours digestion in fistulated sheep, Howarth *et al.* (1979).

††Whole leaves, 22 hours digestion, Fay *et al.* (1981).

§Whole leaves, 10 hours digestion, Fay *et al.* (1980).

¶Two hours after feeding, Howarth *et al.* (1979).

#Fay and Howarth, unpublished results.

some evidence in support of the cell-rupture and IRD theories, but they were less convincing than comparisons of bloat-safe and bloat-causing legumes. However, other variables (e.g., changes in animal susceptibility) affect daily bloat incidence.

Factors Affecting Bacterial Digestion

Mechanical disruption of the epidermis provides sites for microbial invasion of forage tissues. Thus, IRD was increased when leaves were chopped (Howarth et al. 1979). Differences in IRD between bloat-safe and bloat-causing legumes were seen when chewed herbage was digested in vitro, but extensive disruption of leaves eliminated these differences (Fay et al. 1980), further corroborating the IRD theory.

Cheng et al. (1980) identified four key events during the initial digestion of fresh legume leaves: (1) bacterial colonization on the leaf surface, (2) bacterial penetration of the epidermal layer, (3) maceration of leaf tissue, and (4) bacterial penetration of mesophyll cell walls.

Bacterial colonization on the surface of whole leaves occurred more rapidly with alfalfa than with sainfoin, and bacterial proliferation around stomata clearly indicated a response to substances leaking from the stomatal openings (Fay et al. 1981). They suggested that the number of bacteria colonizing the leaf tissue may be determined by nutrients leached from the plant cells. Species differences in anatomy or cell-wall chemistry could affect the rate of dry-matter loss by leaching and thus could affect the IRD. The tannin vesicles in birdsfoot trefoil and sainfoin are located under the epidermal layer, frequently adjacent to the stomata (Lees, unpublished). Thus, tannins leaching from these vesicles might inhibit microbial digestion of these forages.

"Maceration" is used here in reference to softening and disruption of fresh herbage during bacterial penetration through the middle lamellae. Plant pathologists have demonstrated that maceration results from the action of pectinolytic enzymes and that it is a key event in the susceptibility of plants to pathogenic microorganisms. Maceration occurs earlier in bloat-causing legumes than in bloat-safe legumes (Fay et al. 1981). During this stage the epidermal layer separates from the mesophyll tissues, thus exposing new areas of mesophyll to microbial digestion. Separation of the epidermal layer is probably aided by gas bubbles that form inside the leaves. *Lachnospira multiparus*, a pectinolytic ruminal bacteria, macerated fresh legume leaves (Cheng et al. 1979). A similar result, including separation of the epidermal layer, has been obtained with a pectinolytic enzyme (Macerase) during in-vitro digestion of whole leaves (Lees, unpublished). With the exception of arrowleaf clover, bloat-causing legumes are more susceptible than bloat-safe legumes. The anatomy of the leaf veins in cicer milkvetch is particularly effective in resisting maceration and separation of the epidermal layer (Lees, unpublished).

Cell-Wall Rupture

Bacterial penetration of mesophyll cell walls in legume leaves occurred by a general disorganization of the cell wall (Cheng et al. 1980). Cell rupture may also occur during chewing. Mesophyll cells in bloat-causing legumes were more susceptible to rupture than those in bloat-safe legumes (Howarth et al. 1978). This is due to the strength of the cell walls per se and to adhesion between cells of bloat-causing legumes, which cause cell rupture in situ (Lees et al. 1981). There is no clear relationship between bloat potency and the content of the major cell-wall constituents (Howarth, unpublished).

In summary, resistance to cell rupture by crushing is a characteristic of bloat-safe legumes. We are uncertain whether this occurs directly during chewing or indirectly by resistance to microbial penetration of the cell wall.

DISCUSSION

In earlier bloat research much effort was given to the identification of plant foaming agents. Since this information was inadequate for fully understanding the occurrence of bloat, we took a new direction in bloat research. The cell-rupture theory provided the framework for a systematic comparison of bloat-safe and bloat-causing legumes. Studies of mechanical disruption, bacterial adhesion, maceration, and cell-wall rupture revealed a complex array of factors that apparently affect the IRD. At this level of comparison, each of the bloat-safe legumes has unique characteristics.

The strongest evidence in support of the cell-rupture and IRD theories comes from comparisons of the bloat-safe and bloat-causing legume species. To date we have examined three bloat-causing and four bloat-safe species. As time permits, we intend to examine more species, not only for further verification of the IRD theory but also to identify new bloat-safe characteristics that might be useful for breeding a bloat-safe alfalfa (Goplen et al. 1982).

LITERATURE CITED

Cheng, K.-J., D. Dinsdale, and C.S. Stewart. 1979. The maceration of clover and grass leaves by *Lachnospira multiparus*. Appl. Environ. Microbiol. 38:723–729.

Cheng, K.-J., J.P. Fay, R.E. Howarth, and J.W. Costerton. 1980. Sequence of events in the digestion of fresh legume leaves by rumen bacteria. Appl. Environ. Microbiol. 40:613–625.

Fay, J.P., K.-J. Cheng, M.R. Hanna, R.E. Howarth, and J.W. Costerton. 1980. *In vitro* digestion of bloat-safe and bloat-causing legumes by microorganisms: gas and foam production. J. Dairy Sci. 63:1273–1281.

Fay, J.P., K.-J. Cheng, M.R. Hanna, R.E. Howarth, and J.W. Costerton. 1981. A scanning electron microscopy study of the invasion of leaflets of a bloat-safe and bloat-causing legume by rumen microorganisms. Can. J. Microbiol. 27:390–399.

Goplen, B.P., R.E. Howarth, G.L. Lees, W. Majak, J.P. Fay, and K.-J. Cheng. 1982. The evolution of selection techniques in breeding for a bloat-safe alfalfa. Proc. XIV Int. Grassl. Cong.

Howarth, R.E., B.P. Goplen, J.P. Fay, and K.-J. Cheng. 1979. Digestion of bloat-causing and bloat-safe legumes. Ann. Rech. Vét. 10:332–334.

Howarth, R.E., B.P. Goplen, A.C. Fesser, and S.A. Brandt. 1978. A possible role for leaf cell rupture in legume pasture bloat. Crop Sci. 18:129–133.

Howarth, R.E., W. Majak, D.E. Waldern, S.A. Brandt, A.C. Fesser, B.P. Goplen, and D.T. Spurr. 1977. Relationships between ruminant bloat and the chemical composition of alfalfa herbage. I. Nitrogen and protein fractions. Can. J. Anim. Sci. 57:345–357.

Jones, W.T., and J.W. Lyttleton. 1971. Bloat in cattle. XXXIV. A Survey of forage legumes that do and do not produce bloat. N.Z. J. Agric. Res. 14:101–107.

Lees, G.L., R.E. Howarth, B.P. Goplen, and A.C. Fesser. 1981. Mechanical disruption of leaf tissues and cells in bloat-causing and bloat-safe forage legumes. Crop Sci. 21:444–448.

Majak, W., and R.E. Howarth. 1980. Further studies on the etiology of alfalfa pasture bloat. Can. J. Anim. Sci. 60:1039(abs.)

Sarkar, S.K., R.E. Howarth, and B.P. Goplen. 1976. Condensed tannins in herbaceous legumes. Crop. Sci. 16:543–546.

Role of Alkaloids and Toxic Compound(s) in the Utilization of Tall Fescue by Ruminants

J.A. BOLING, R.W. HEMKEN, L.P. BUSH, R.C. BUCKNER,
J.A. JACKSON, JR., and S.G. YATES
University of Kentucky, Lexington, Ky., U.S.A.

Summary

Tall fescue (*Festuca arundinacea* Schreb.) contains several alkaloids. Studies in our laboratory showed that perloline is a principal alkaloid in tall fescue during the summer and that concentration peaks at a time when cattle grazing tall fescue pastures frequently have severely reduced performance (summer syndrome or summer fescue toxicosis). The objectives of the following studies were (1) to determine the influence of perloline on in-vitro fermentation and in-vivo ruminant metabolism and (2) to determine the effects of two tall fescue strains selected for perloline concentration on performance of growing cattle and lactating dairy cows and on interrelationships with environmental temperature. In-vitro fermentation studies showed a concentration-dependent inhibition of cellulose digestion, volatile fatty-acid production, and growth of ruminal cellulolytic bacteria by perloline. Isolated perloline fed to lambs resulted in decreased cellulose digestion and nitrogen retention and increased body temperature. As a result of these studies, two strains of tall fescue were developed, one of which is high (G1-306) and the other low (G1-307) in perloline. A summer grazing study showed that gains of cattle on G1-307 were lower than those on Kentucky 31, Kenhy, or G1-306. In a 2-year study with lactating dairy cows fed Kentucky 31, G1-306, or G1-307 soilage, cows consuming

G1-307 had the lowest intake and milk production and highest respiration rate. These studies indicated that in the selection for low perloline, compound(s) more toxic and detrimental to animal performance were increased in G1-307. Further studies were conducted with dairy calves in environmental rooms to determine the influence of temperature and time of harvest of G1-307 on potential toxicity. Calves were fed July-cut G1-306 and G1-307 at 10°-13°C, 21°-23°C, and 34°-35°C. Intake and weight change in calves decreased, while body temperature and respiration rate increased on G1-307 at 34°-35°C, resulting in temperature × forage interactions. In a subsequent trial in which silage was fed in environmental rooms at 31°-35°C from May, July, and October harvests, gain and intake were decreased, and body temperature and respiration increased. Selection for low perloline in G1-307 resulted in concomitant increases in N-acetyl and N-formyl loline. Also, G1-307 N-acetyl and N-formyl loline concentrations were related to infestation with *Epichloë typhina*. Fields were established from seed of untreated and benomyl-treated G1-307 plants. Hay produced from benomyl-treated and untreated G1-307 had N-acetyl plus N-formyl loline concentrations of 115 and 875 μg/g and *E. typhina* infection estimates of 44% and 95%, respectively. Calves fed hay in 31°-32°C rooms from benomyl-treated G1-307 had higher intakes and lower temperatures and respiration rates than those fed hay produced from untreated G1-307. These studies show that the toxic factor(s) in G1-307 are present from May through October. N-acetyl and N-formyl loline concentrations are related to *E. typhina* infestation, and toxic effects to the animal are potentiated by elevated environmental temperature.

KEY WORDS: tall fescue, ruminants, performance, alkaloids, perloline, N-acetyl loline, N-formyl loline, *Epichloë typhina*.

INTRODUCTION

Tall fescue (*Festuca arundinacea* Schreb.) contains several different alkaloids. One of the first major alkaloids identified in fescue was perloline, which was initially isolated from perennial ryegrass (*Lolium perenne* L.). Perloline concentration in tall fescue usually peaks in late July or early August in Kentucky. Interest was stimulated in the possible role of this alkaloid in animal performance, since its peak concentration occurred at the time gains of cattle grazing fescue pastures are frequently depressed. This period of poor performance has been referred to as the summer syndrome, summer slump, or summer fescue toxicosis.

Bush et al. (1972) showed that perloline inhibited cellulose digestion and volatile fatty acid production at concentrations greater than 1.2×10^{-4} molar during 24-hour in-vitro incubations. We also found a perloline concentration-dependent inhibition of the rumen anaerobic cellulolytic bacteria *Bacteroides succinogenes*, *Butyrivibrio fibrisolvens*, *Ruminococcus albus*, and *R. flavefaciens*. Based on these data, a series of animal studies was conducted to evaluate the role of tall fescue alkaloids on ruminant animal performance.

METHODS AND RESULTS

Perloline, as the monohydrochloride salt, was isolated from tall fescue for an in-vivo study. Lambs were fed a control diet or the same diet with 0.5% added perloline in a digestion trial (Boling et al. 1975). In the perloline group, apparent crude-protein (P < 0.05) and cellulose digestibility (P < 0.08) coefficients were lower than those

of control lambs. Apparent digestibility of nitrogen-free extract, crude fiber, ether extract, and ash tended to be lowest in the perloline-fed lambs. Urine volume and urinary nitrogen tended to be higher and nitrogen retention lower in these lambs. The percentages of ruminal acetic, valeric, and iso-valeric acids were lower and propionic acid higher in the perloline-fed lambs. Also, body temperature was significantly elevated in the perloline group of lambs.

Since our data had shown an inhibitory effect of perloline on ruminal cellulose digestion and microbial growth, selection for low perloline concentration in our tall fescue breeding program was begun. Two experimental selections, G1-306, and G1-307, were developed for evaluation with cattle. G1-306 and G1-307 are ryegrass × tall fescue (*Lolium perenne* L. and *L. multiflorum* Lam.–*Festuca arundinacea* Schreb.) hybrid derivatives high and low in perloline concentrations, respectively. They were compared with Kentucky 31 and Kenhy tall fescue during a spring-summer grazing trial from 24 March–3 September (Steen et al. 1979). Pastures were seeded to pure stands of each variety. Daily gains during the grazing period averaged 0.42, 0.38, 0.29, and 0.40 kg/day for steers on Kentucky 31, G1-306, G1-307, and Kenhy, respectively. Gains of steers grazing the G1-307 low-perloline variety were lower (P < 0.05) than those of steers in the remaining treatment groups. Perloline concentrations for the four respective varieties averaged 172.9, 432.0, 66.5, and 164.2 μg/g during the study. Respiration rates were highest (P < 0.01) in steers grazing G1-307, while values for steers grazing the three remaining varieties were similar. Cattle grazing G1-307 spent more time under pasture shades and seemed extremely sensitive to heat stress.

G1-306 and G1-307 were also evaluated using lactating dairy cows in a 2-year study (Hemken et al. 1979). Pure stands of Kentucky 31, G1-306, and G1-307 were established in three replicated fields in the summer prior to the first-year trial. Holstein cows were fed 7.3 kg of a concentrate mix daily plus either Kentucky 31, G1-306, or

J.A. Boling and R.W. Hemken are Professors of Animal Sciences and L.P. Bush is a Professor of Agronomy at the University of Kentucky; R.C. Buckner is a Research Agronomist at USDA-SEA-AR; J.A. Jackson, Jr., is a Research Fellow at the University of Kentucky; and S.G. Yates is a Research Scientist of USDA-NRRC-SEA. This paper (No. 81-5-3-6) is published with the approval of the Director of the Kentucky Agricultural Experiment Station.

Table 1. Performance of calves fed tall fescue forages in temperature controlled rooms.[a]

Tall fescue	Room temp.	Weight change (21-day)	Dry matter intake[b]	Rectal temp.[b]	Respiration rate[b]
	(°C)	kg	g/kg B.W.$^{.75}$	(°C)	(counts/min.)
G1-306	10–13	+22.0	78.6	39.1	20.7
G1-307	10–13	+22.3	80.0	38.7	20.3
G1-306	21–23	+20.2	79.5	39.1	27.8
G1-307	21–23	+15.0	78.2	39.1	29.7
G1-306	34–35	+11.7	70.0	40.2	91.9
G1-307	34–35	−9.3	33.2	41.1	110.0

[a]Effects of temperature, forage, and forage × temperature were significant (P<.05) for all variables measured except for effect of forage on dry-matter intake.

[b]Average for second and third weeks of trial.

G1-307 harvested with a flail-type chopper and fed fresh ad libitum in tie stalls during July and August of each year. The cows consuming G1-307 had reduced intake and milk production, loss in body weight, and increased respiration rate when compared with those consuming Kentucky 31 and G1-306. Body temperature was significantly elevated in the second year in G1-307 cows. These data, consistent with the steer grazing data, prompted further investigation into the cause of increased toxicity of G1-307.

Since the grazing and lactating-cow trials were conducted during the summer when temperatures were high, a trial was conducted with calves in environmental rooms to determine whether the observed toxic syndrome was related to environmental temperature (Hemken et al. 1981). Calves were fed fresh harvested forage daily during July and were housed in rooms maintained at 10°–13°C, 21°–23°C, and 34°–35°C. Calves consuming G1-307 forage at the highest temperature had reduced forage intake, loss in body weight, increased body temperature, and increased respiration rate when compared with calves consuming G1-306 (Table 1). These same measurements were not different between treatments at 10°–13°C or 21°–23°C, resulting in a temperature × forage interaction (P < 0.01). A subsequent trial in which the forages were harvested in May, July, and October demonstrated that toxic substance(s) in fescue were present throughout the grazing season, and the deleterious effects were related to high environmental temperatures (31°–35°C).

In attempts to clarify the reason(s) for depressed performance in cattle consuming G1-307, Buckner et al. (1982) showed that G1-307 was high in N-acetyl loline and N-formyl loline, which had increased with concomitant selection for low perloline. It was also shown that the concentration of N-acetyl and N-formyl loline was related to infestation of the plants with the endophytic fungus *Epichloë typhina*. G1-307 parental plants were treated with the systemic fungicide benomyl for seed production and for establishment of fields of treated and untreated forage. Jackson and colleagues (J.A. Jackson, R.W. Hemken, J.A. Boling, R.C. Buckner, and L.P. Bush, unpublished data) fed hay from orchardgrass and the forages established from seed of benomyl-treated and untreated

G1-307 plants to calves in an environmentally controlled high-temperature room (31°–32°C). The hay from fields established from seed of benomyl-treated and untreated plants had N-acetyl plus N-formyl loline concentrations of 115 and 875 μg/g and infection estimates of *E. typhina* of 44% and 95%, respectively. Calves consuming forage from the benomyl-treated G1-307 had higher intake, lower body temperature and respiration rate, and less weight loss than calves consuming forage from untreated G1-307 (Table 2). Overall responses of calves in the environmental rooms at 31°–32°C were similar for the benomyl-treated G1-307 and orchardgrass forages.

The data presented here indicate that performance of cattle grazing tall fescue can be affected by two known classes of alkaloids. The influence of perloline seems to be related principally to inhibition of ruminal metabolism but is also involved in tissue metabolism. Selection for low perloline resulted in a simultaneous increase in plant N-acetyl and N-formyl loline concentrations, which seem to be potent alkaloids detrimental to animal performance. N-acetyl and N-formyl loline concentrations were related

Table 2. Responses of calves fed orchardgrass and G1-307 tall fescue hay from stands established to seed of benomyl treated and untreated plants.[a]

	G1-307 Untreated	G1-307 Treated	Orchard-grass	SE of mean
Feed intake (kg/day)	2.5	3.4	3.8	0.3[d]
Feed intake (gm/kg$^{.75}$)	51.1	73.1	77.8	31.0[c]
Rectal temperature (°C)	41.0	40.2	39.7	0.1[c]
Respiration rate (CPM)	103.9	75.2	72.6	28.0[b]
Body weight change, kg	−18.2	−2.3	1.5	11.0[c]

[a]Data are means of performance during last six days of a 15-day trial.

[b,c,d]Superscripts refer to probability level of differences among treatment means within rows: b = P < .01; c = P < .05; and d = P < .10.

to infestation of the plant by the endophyte *E. typhina*. Symptoms of the toxic syndrome observed with G1-307 were reduced intake and performance (milk production in cows and body weight in cows and calves) and increased body temperature and respiration rate.

LITERATURE CITED

Boling, J.A., L.P. Bush, R.C. Buckner, L.C. Pendlum, P.B. Burrus, S.G. Yates, S.P. Rogovin, and H.L. Tookey. 1975. Nutrient digestibility and metabolism in lambs fed added perloline. J. Anim. Sci. 40:972–976.

Buckner, R.C., L.P. Bush, P.B. Burrus, J.A. Boling, R.W. Hemken, J.A. Jackson, Jr., and R.A. Chapman. 1982. Improvement of forage quality of tall fescue through *Lolium-Festuca* hybridization. Proc. XIV Int. Grassl. Cong.

Bush, L.P., J.A. Boling, G. Allen, and R.C. Buckner. 1972. Inhibitory effects of perloline to rumen fermentation *in vitro*. Crop Sci. 12:277–279.

Hemken, R.W., J.A. Boling, L.S. Bull, R.H. Hatton, R.C. Buckner, and L.P. Bush. 1981. Interaction of environmental temperature and anti-quality factors on the severity of summer fescue toxicosis. J. Anim. Sci. 52:710–714.

Hemken, R.W., L.S. Bull, J.A. Boling, E. Kane, L.P. Bush, and R.C. Buckner. 1979. Summer fescue toxicosis in lactating dairy cows and sheep fed experimental strains of ryegrass-tall fescue hybrids. J. Anim. Sci. 49:641–646.

Steen, W.W., N. Gay, J.A. Boling, R.C. Buckner, L.P. Bush, and G. Lacefield. 1979. Evaluation of Kentucky 31, G1-306, G1-307 and Kenhy tall fescue as pasture for yearling steers. II. Growth, physiological response and plasma constituents of yearling steers. J. Anim. Sci. 48:618–623.

Beef Production from Tall Fescue Treated with Mefluidide, A Chemical Plant-Growth Regulator

T.W. ROBB, D.G. ELY, C.E. RIECK, S. GLENN,
L. KITCHEN, B.P. GLENN, and R.J. THOMAS
University of Kentucky, Lexington, Ky., U.S.A.

Summary

Mefluidide treatment of Kentucky 31 tall fescue (*Festuca arundinacea* Schreb.) in the spring will inhibit seed-head formation, maintain the forage in an immature and vegetative stage, and increase forage N and soluble-sugar content and in-vivo nutrient utilization in the ovine. Productivity of animals grazing mefluidide-treated tall fescue, however, has not been determined. The objective of this study was to evaluate agronomic characteristics of forage and productivity of yearling beef steers and heifers grazing mefluidide-treated (T) and nontreated control (C) pastures of Kentucky 31 tall fescue.

Experimental pastures of Kentucky 31 tall fescue (4 replications at location I and one replication at location II) were used to determine the effect of a spring application of 273.8 g mefluidide/ha on forage N, forage soluble sugars, and animal gain at 28-day intervals in two 140-day grazing seasons.

Mefluidide-treated forage had higher (P < 0.05) N content than the control on days 56, 84, and 112 of the grazing season in both years. Forage soluble-sugar content of treated fescue was greater (P < 0.05) than that of the control on days 28, 56, and 112 of the first year's grazing season and greater (P < 0.05) on day 28 of the second year. Animal gain from grazing treated forage was increased 9%, 22%, 380%, and 21% over control for 0 to 28-, 28 to 56-, 56 to 84-, and 84 to 112-day periods, respectively. Combining gain data and analyses for forage treatment, year, location, and 2- and 3-way interactions produced a positive (P < 0.01) effect for mefluidide treatment.

Mefluidide treatment of Kentucky 31 tall fescue inhibited seed-head formation, increased forage N and forage soluble-sugar content, and produced a forage that was more efficiently utilized by grazing animals than nontreated forage. This treatment resulted in a 27% increase in animal productivity/ha.

KEY WORDS: beef productivity, tall fescue, mefluidide, plant-growth regulator.

INTRODUCTION

There are presently more than 14 million ha of fescue grown in the United States (Buckner and Bush 1979). One of the major detriments associated with this cool-season forage, as well as other cool-season forages, is suboptimal animal digestibility and utilization (Tucker 1975) and thus reduced performance in summer (Hinds et al. 1974). A spring application of the chemical plant-growth regulator mefluidide (N-[2,4-dimethyl-5-[[(tri-fluoromethyl)sulfonyl]amino]phenyl]acetamide), has been shown to increase in-vivo dry matter, N, acid detergent fiber, and neutral detergent fiber digestion of Kentucky 31 tall fescue in the ovine (Robb 1980). Seed-head formation of treated tall fescue is suppressed (Glenn et al. 1980, Robb 1980) as N (Glenn et al. 1980, Robb 1980) and soluble-sugar (Glenn et al. 1980) levels are increased. Total yearly forage dry-matter production, however, is unaffected (Glenn et al. 1980). Therefore, grazing trials were initiated to evaluate performance of yearling beef steers and heifers grazing mefluidide-treated pastures of Kentucky 31 tall fescue.

METHODS

Grazing trials were conducted at two locations 80 km apart in 1978 and 1979. At location I, 273.8 g mefluidide/ha was applied in a spray to 4 1.6-ha pastures on 3 May 1978 and 2 May 1979. Four pastures received no mefluidide either year. Ammonium nitrate (36.1 kg ac-

T.W. Robb is an Assistant Professor of Animal Sciences at North Carolina State University; D.G. Ely is a Professor of Animal Sciences at the University of Kentucky; C.E. Rieck is an Agronomist at Du Pont Chemicals in Fletcher, Oklahoma; S. Glenn is an Assistant Professor of Agronomy at the University of Maryland; L. Kitchen is an Assistant Professor of Agronomy at Louisiana State University; B.P. Glenn is a Faculty Research Associate at USDA-ARS in Beltsville, Maryland; and R.J. Thomas is a Research Assistant at the University of Kentucky. This paper (No. 81-5-14) is published with the approval of the Director of the Kentucky Agricultural Experiment Station.

tual N/ha) was applied each year in split applications. Four 1.2-ha plots were used at location II. One was treated with the growth regulator, and one was untreated. Ammonium nitrate (36.1 kg actual N/ha) and mefluidide (273.8 g/ha) were applied on 3 May 1978 and 2 May 1979.

Thirty-two yearling crossbred beef steers were used each year at location I (4/pasture). Ten yearling heifers of similar breeding (5/pasture) were used each year at location II. Steers (294 kg) were allotted to experimental pastures on 10 May 1978 and on 7 May 1979 (301 kg). Heifers were allotted on the same dates. Average initial weights of heifers were 265 kg in 1978 and 249 kg in 1979.

Animals were individually weighed at the beginning and at 28-day intervals during each 140-day grazing season. Four forage samples/pasture were taken on each weigh date from within 1.2-m² wire cages randomly placed in each pasture. Cages were moved to a new location within each pasture on the first day of each weigh period. These samples were dried and ground through a 2-mm screen in a Wiley mill. N (AOAC 1975) and soluble sugars (Ough and Lloyd 1972) were determined on each sample.

RESULTS

Gains of steers and heifers grazing control and mefluidide-treated tall fescue are shown in Table 1. Steers grazing control pastures gained faster during the first 28 days of each grazing season than those grazing treated forage. Thereafter, except for the 84 to 112–day period in 1978, treated tall fescue produced numerically greater gains than control forage. Total heifer gain/ha was increased 56% in 1978 and 22% in 1979 by mefluidide treatment. Improvement in heifer gain from grazing treated tall fescue occurred during the 0 to 28– and 84 to 112–day periods in 1978 and during the 28 to 56–, 56 to 84–, and 84 to 112–day periods in 1979.

Animal performance (steer and heifer) from both loca-

Table 1.   Steer and heifer gains (kg) from grazing Kentucky 31 tall fescue treated with mefluidide in 1978 and 1979.

| | Steers, location I | | | | Heifers, location II | | | |
| | 1978 | | 1979 | | 1978 | | 1979 | |
Grazing season, day[a]	C[b]	T	C	T	C	T	C	T
0 to 28	24.4	21.6	24.1	21.5	.3	10.6	8.8	8.9
28 to 56	16.1	24.0	17.1	24.2	19.9	18.1	9.9	10.8
56 to 84[c]	8.3	15.3	7.1	11.7	–	–	– 13.6	– 4.1
84 to 112	8.3	4.5	3.8	6.5	2.5	5.8	13.6	17.6
112 to 140	12.6	14.7	12.2	14.8	16.9	16.3	10.1	1.8
Total gain/head (140 days)	69.7	80.1	64.3	78.7	39.6	50.8	28.8	35.0
Total gain/ha (140 days)	175.6	197.1	158.8	195.2	141.7	221.8	119.4	145.8
Gain/ha (% of C)		112.2		122.9		156.2		122.1

[a]Grazing season: 1978 = 10 May to 27 September; 1979 = 7 May to 24 September.

[b]C = control; T = mefluidide treated.

[c]Heifers not weighed on day 84 of 1978 grazing season.

Fig. 1. Weigh period gains of animals grazing Kentucky 31 tall fescue treated with mefluidide.

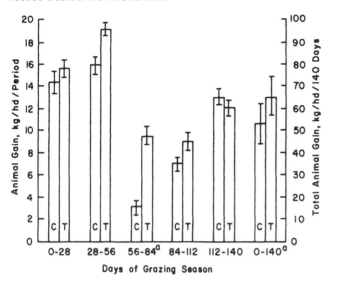

ᵃC Different (P<.01) from T.

tions and both years is summarized in Fig. 1. Animals grazing treated fescue gained 22%, 380%, and 21% more than those grazing control forage for the 28 to 56-, 56 to 84-, and 84 to 112-day periods, respectively. Animals grazing control forage, however, gained 8% more than those consuming treated tall fescue during the 112 to 140-day period. The only time interval in which animal gain was significantly (P < 0.01) different was the 56 to 84-day interval. Total gain/head from treated tall fescue was greater (P < 0.01) than from control. Total gain/ha increased from 149 kg for animals grazing control pastures to 190 kg for those grazing treated tall fescue, a 27% increase in productivity.

Nitrogen and soluble-sugar content of forages from both locations in both years are shown in Table 2. Treated tall fescue from location I had more (P < 0.05) total N than control on days 28, 84, and 112 in both years and on day 56 in 1978. Treated forage at location II contained numerically equal or more N than control at all measurement intervals except day 56 in 1978.

At location I, treated forage contained more (P < 0.05) soluble sugars than control on days 28, 84, 112, and 140 in 1978 and on day 28 in 1979. Although the amounts were not statistically significant, treated tall fescue at location II contained greater quantities of soluble sugars than control on days 56, 84, and 112 in 1978 and on days 56 and 112 in 1979.

DISCUSSION

Robb (1980) found that dry matter, N, and acid detergent fiber and neutral detergent fiber digestibility of untreated Kentucky 31 tall fescue were greater than those of mefluidide-treated forage one month after treatment. However, digestibility of treated forage was greater than that of control for the next 90 days. Similar trends in animal gain were found in this study, indicating that mefluidide treatment improves forage quality above control, resulting in greater animal production. This difference in animal gain between 149 kg and 190 kg/ha for control and treated tall fescue, respectively, may have resulted from the greater forage digestibility in midsummer and late summer associated with mefluidide-treated forage (Robb 1980). The use of mefluidide did suppress seed-head formation in both years at both locations, keeping the forage in an immature and vegetative state. Whether the altered physical form of the plant had any effect on palatability and/or consumption, and thus on utilization, was not determined.

When data were averaged across both locations and

Table 2. Nitrogen and water-soluble sugar content (% of air-dry) of Kentucky 31 tall fescue treated with mefluidide and grazed by beef animals in 1978 and 1979.

| | Location I | | | | Location II | | | |
| | 1978 | | 1979 | | 1978 | | 1979 | |
Grazing season, dayᵃ	Cᵇ	T	C	T	C	T	C	T
					Nitrogen			
28	1.5ᶜ	1.6ᵈ	1.6ᶜ	1.7ᵈ	1.1	1.4	2.1	2.3
56	1.3ᶜ	1.9ᵈ	1.2	1.3	1.4ᶜ	1.2ᵈ	2.2	2.3
84	1.9ᶜ	2.2ᵈ	1.9ᶜ	2.1ᵈ	1.8	1.9	1.5	1.7
112	2.0ᶜ	2.2ᵈ	1.6ᶜ	1.7ᵈ	1.9	2.2	2.0ᶜ	2.4ᵈ
140	2.0	2.0	2.2ᶜ	2.0ᵈ	2.1	2.5	2.1	2.1
					Sugar			
28	6.1ᶜ	7.8ᵈ	9.0ᶜ	10.1ᵈ	3.0	2.8	8.3	7.3
56	3.4	4.0	6.9	6.5	3.2	3.9	5.6	6.1
84	3.2ᶜ	3.8ᵈ	4.4	4.3	3.1	3.4	4.6	4.1
112	3.9ᶜ	4.6ᵈ	5.3	5.7	3.7	4.0	5.3	5.7
140	5.2ᶜ	6.4ᵈ	8.4	8.0	7.7	7.1	11.5	11.2

ᵃGrazing season: 1978 = 10 May to 27 September; 1979 = 7 May to 24 September.

ᵇC = control; T = mefluidide treated.

ᶜ,ᵈMeans with different superscripts within year are different (P<.05).

both years, treated forage contained 11%, 10%, 11%, and 13% more total N and 6%, 6%, 3%, and 10% more soluble sugars than did control tall fescue on days 28, 56, 84, and 112, respectively. Glenn et al. (1979) found that mefluidide may stimulate auxin production or act as an auxin, which in turn stimulates protein synthesis, while Robb (1980) found that mefluidide-treated Kentucky 31 tall fescue in midsummer and late summer contained greater amounts of total N, smaller amounts of nonprotein N, and greater amounts of hydrolyzable essential and nonessential amino acids than control forage. Glenn et al. (1980) also found that mefluidide treatment of tall fescue increases the soluble sugar content over untreated forage in midsummer and late summer. The combined increase of both these components dramatically improved forage quality, manifested by an improvement (P < 0.01) in total animal gain. (Fig. 1).

LITERATURE CITED

Association of Official Analytical Chemists (AOAC). 1975. Official methods of analysis, 12th ed. AOAC, Washington, D.C.

Buckner, R.C., and L.P. Bush. 1979. Preface. In R.C. Buckner and L.P. Bush (eds.) Tall fescue, xiii. Monog. No. 20, Agron. Ser., Am. Soc. Agron., Crop Sci. Am., Soil Sci. Am., Madison, Wis.

Glenn, S., C.E. Rieck, and N.W. Bugg. 1979. Evaluation of the mechanism of action of mefluidide. Proc. So. Weed Sci. Soc. 32:322.

Glenn, S., C.E. Rieck, D.G. Ely, and L.P. Bush. 1980. Quality of tall fescue forage as affected by mefluidide. J. Agric. Food Chem. 28:391–393.

Hinds, F.C., G.G. Cmarik, and G.E. McKibben. 1974. Fescue for the cow herd in Southern Illinois. Ill. Res. 16.(1):6–7.

Ough, L.D., and N.E. Lloyd. 1972. Automated determination of dextrose equivalent of cornstarch hydrolysates. Cereal Chem. 42:1. Quoted From technicon auto analyzer industrial method 142-71A/preliminary, May 1972.

Robb, T.W. 1980. Influence of plant growth regulation on Kentucky 31 tall fescue utilization by ruminants. Ph.D. Dissertation, Univ. of Ky.

Tucker, P.L. 1975. Ovine utilization of boone orchardgrass, Kentucky 31 tall fescue, Kenhy tall fescue and bigflower vetch. M.S. Thesis, Univ. of Ky.

Effect of Level of Nitrogen Fertilization of Kentucky 31 Tall Fescue (*Festuca arundinacea* Schreb.) on Brood-Cow Health

J.A. STUEDEMANN, S.R. WILKINSON, H. CIORDIA, and A.B. CAUDLE

USDA-SEA-AR and University of Georgia, Athens, Ga., U.S.A.

Summary

Studies were conducted to assess the long-term effects of nitrogen (N) fertilization of tall fescue on cow-calf performance and health problems, including grass tetany, fat necrosis, gastrointestinal parasitism, and nitrate poisoning. Kentucky 31 tall fescue pastures fertilized at three levels of N were grazed with Angus cows at 0.4 ha/cow on a year-round basis. Fertilization levels included high N (667 and 0 kg N/ha/yr from broiler litter in 1968-1974 and 1975-1977, respectively); moderate N (134 and 224 kg N/ha/yr from ammonium nitrate [NH₄NO₃] in 1969 and 1970-1977, respectively); and low N (84 and 74 kg N/ha/yr from NH₄NO₃ in 1972 and 1973-1977, respectively). Within years, cattle and pastures were managed similarly.

The overall mean available forage was 689, 1,287, and 2,167 kg/ha from 1973 to 1976 on the low-, moderate-, and high-N pastures, respectively. Mineral analyses of forage indicated a potential tetany hazard on all three pastures, yet from 1970 to 1977, 14, 9, and 0 cases of tetany occurred on the high-, moderate-, and low-N pastures, respectively.

Mean cow weights differed among N fertilization levels and averaged 362, 410, and 443 kg for the low-, moderate-, and high-N pastures, respectively. Respective actual weaning weights were 162, 182, and 188 kg. There was a tendency for lower (P = 0.11) conception rates (75%) on the moderate-N pasture than on the low- (90%) and high-N (86%) pastures. Supplementing cows with magnesium (Mg) during the grass tetany season increased adjusted 205-day calf weights across all pasture treatments and increased conception rate in cows grazing the moderate-N pasture from 46% to 79%.

At the stocking rate of 0.4 ha/cow, fertilization with 224 kg N/ha/yr or greater increased grass tetany and fat necrosis problems. Nematode parasitism was associated with the quantity of forage available for consumption and consequently decreased as the level of N fertilization increased.

KEY WORDS: tall fescue, *Festuca arundinacea* Schreb., nitrogen fertilization, beef cow, reproduction, grass tetany, fat necrosis, parasitism.

INTRODUCTION

Tall fescue (*Festuca arundinacea* Schreb.) has many outstanding agronomic attributes, including wide adaptability, ease of establishment, good persistence, and excellent seed- and forage-production characteristics. Unfortunately, these attributes mean very little in terms of animal production unless they result in acceptable animal performance. Beef producers associate tall fescue with a number of animal health problems, including fescue foot, summer syndrome, fat necrosis, and generally poor performance. Though they are not specifically identified with tall fescue, a number of other disorders have been associated with tall fescue or its management. These include grass tetany, nitrate poisoning, and poor reproductive performance.

The studies presented in this paper assessed the long-term effects of nitrogen (N) fertilization level of tall fescue on beef-cattle health problems including fat necrosis, grass tetany, reproductive performance, gastrointestinal parasitism, and nitrate poisoning.

METHODS

Pasture Treatments

Three Kentucky 31 tall fescue pastures fertilized at three levels of N were grazed by winter- or early-spring-calving Angus cows stocked at 0.4 ha/cow on a year-round basis. Pasture N fertilization levels included high N (667 and 0 kg N/ha/yr from broiler litter in 1968–1974 and 1975–1977, respectively); moderate N (134 and 224 kg N/ha/yr from ammonium nitrate [NH_4NO_3] in 1969 and 1970–1977, respectively); and low N (84 and 74 kg N/ha/yr from NH_4NO_3 in 1972 and 1973–1977, respectively). The low- and moderate-N pastures were fertilized with phosphorus (P) and potassium (K) whenever soil tests for these nutrients were less than "medium." Lime was applied only when soil pH levels dropped below 5.5.

Animal Management

Cattle on all pastures were managed similarly. Unshrunk cow and calf weights were obtained at approximately 6-week intervals. Adjusted 205-day weights were calculated (adjusted for age of calf, sex of calf-to-steer equivalent, and age of dam) from a weight obtained when the average calf age was near 205 days. Except for cows grazing the high-N pastures through 1971, all open cows, as judged by rectal palpation, were routinely replaced.

From 1970 to 1974 a portion of the cows in each pasture was given some form of magnesium (Mg) supplementa-

tion in order to study grass tetany prevention practices (Wilkinson and Stuedemann 1979). From 1975 to 1977 none of the cows received Mg supplementation. Nematode parasitism was monitored in cattle on all three pastures.

Available Forage, Forage Botanical and Chemical Composition

Available forage was estimated approximately monthly on all pastures by harvesting all forage above ground level. Botanical composition was estimated from the frequency with which 20 points 10.16 cm apart on a 203.2-cm bar came in contact with bermudagrass, fescue, clover, weeds, bare ground, feces, etc. Botanical composition was estimated approximately 25 times between April 1974 and December 1976, with a minimum of 36 estimates/pasture. Herbage samples were selected at random with a minimum of two different sampling dates each month from various parts of each pasture and were used to determine N, nitrate-N, P, K, calcium (Ca), and Mg.

RESULTS AND DISCUSSION

Available Forage, Forage Botanical and Chemical Composition

As the level of fertilization increased, the quantity of forage (Table 1) above ground level increased (P < 0.01). From early 1974 through early 1977, the average percentage of tall fescue was 65%, 74%, and 82% for the low-, moderate-, and high-N tall fescue pastures, respectively. These differences (P < 0.05) were comparatively consistent across seasons except on the low-N pasture, where the amount of tall fescue from July through September averaged 51%. As the amount of tall fescue decreased, the bermudagrass tended to increase. The high-N pasture contained less (P < 0.05) bermudagrass than either of the other pastures.

The N level in forage was higher (P < 0.05) for the high-N tall fescue (3.74%) than for either the moderate- (3.59%) or the low-N pasture (3.53%). All three pastures differed (P < 0.05) in levels of P, K, and Ca. The average forage Mg level on the moderate-N pasture was 0.31%, which exceeded (P < 0.05) both the low- and high-N pastures, each at 0.28%. The K:(Ca + Mg) ratio was highest on the high-N tall fescue (2.34), followed by 2.20 on the low-N tall fescue, which exceeded (P < 0.05) the ratio on the moderate-N pasture (1.76).

Cow-Calf Performance

Overall average cow weights (Table 2) differed (P < 0.05) among pastures (1973–1976, using weights in a given year only if a cow had been grazing a given pasture since before the preceding April 15). Actual weaning weights were higher (P < 0.05) on the moderate- and high-N than on the low-N pasture. The actual level of significance among the adjusted 205-day weights was

J.A. Stuedemann is a Research Animal Scientist, S.R. Wilkinson is a Supervisory Soil Scientist, and H. Ciordia is a Veterinary Parasitologist of USDA-SEA-AR; and A.B. Caudle is an Assistant Professor of Veterinary Medicine at the University of Georgia. This is a contribution of the USDA-SEA-AR Southern Piedmont Conservation Research Center in Watkinsville, Ga.; the USDA-SEA-AR Cattle Parasites Research Laboratory in Experiment, Ga.; and the University of Georgia College of Veterinary Medicine in Athens.

Table 1. Summary of the effect of level of nitrogen fertilization of Kentucky-31 tall fescue on available forage, botanical composition, and chemical composition.

| Item | Level of N fertilization | | |
	Low N	Moderate N	High N
Available forage, kg DM/ha			
1973–1976	689[c]*	1,287[b]	2,167[a]
Botanical composition,			
Feb. 1974–Jan. 1977			
Fescue, %	65[c]	74[b]	82[a]
Bermudagrass, %	14[a]	10[a]	4[b]
Clover, %	1[c]	5[a]	2[b]
Other, %	13[a]	6[b]	5[b]
Bare ground	8[b]	5[a]	6[a,b]
Forage mineral composition,			
Mar. 1973–Feb. 1977, across			
all seasons			
N, %	3.53[b]	3.59[b]	3.73[a]
P, %	.39[b]	.33[c]	.45[a]
K, %	3.10[c]	2.78[b]	3.40[a]
Ca, %	.29[c]	.33[a]	.32[b]
Mg, %	.28[b]	.31[a]	.28[b]
K:Ca + Mg	2.20[b]	1.76[c]	2.34[a]

*Means within the same row with different superscripts differ (P<.05).

0.075, with the same general trend as the actual weaning weight.

Cow Conception and Calving Date

Though it was not significant, a tendency (P = 0.11) existed for the conception rate of cows grazing the low-N (90%) and high-N (86%) pastures to be higher than that for the moderate-N (75%) pasture. Direct comparison of the three pastures is difficult because the trend was not progressive with increased N fertilization level; however, the high-N pasture was fertilized with broiler litter, so type of N and level of N fertilization were confounded. The overall trend (P = 0.32) in calving date was not in accord with the trend in conception rate; cows grazing the low-N tall fescue tended to calve later (day 52) than cows grazing the moderate-N (day 46) or high-N (day 48) tall fescue.

Grass Tetany

From 1970 to 1974, some of the cows grazing each level of N fertilization were given some form of Mg supplementation. From 1975 to 1977 no cows were supplemented with Mg. From 1970 to 1977, 0, 9, and 14 cases of tetany occurred on the low-, moderate-, and high-N tall fescue, respectively. Not only was the incidence of grass tetany greater on the high-N tall fescue, but the incidence of hypomagnesemia on it was greater on most blood-sampling dates during the tetany season, and hypomagne-

Table 2. Summary of the effect of level of nitrogen fertilization of Kentucky-31 tall fescue on cow-calf performance and health.

| Item | Level of N fertilization | | |
	Low N	Moderate N	High N
No. cows	16	24	21
Cow weight†, kg	362[c]*	410[b]	443[a]
Calf performance†			
Actual weaning wt., kg	162[b]	182[a]	188[a]
Adj. 205-day wt., kg	151[a]	170[a]	167[a]
Cow conception rate†, %	90[a]	75[a]	86[a]
Cow calving date†, day of year	52[a]	46[a]	48[a]
Parasitism, internal			
Cow egg count, 1973–75, eggs/g	5[a]	2[b]	3[a,b]
Calf egg count, 1973–75, eggs/g	426[a]	261[b]	137[c]
Total actual worms, 1973–75	52,331[a]	17,976[b]	7,374[c]

*Means within the same row with different superscripts differ (P<.05).

†Includes only data from cows and calves that had been on a given pasture treatment since prior to the preceding breeding season (e.g., would have included a cow's weight in January of a given year only if she had been grazing a given pasture since prior to the preceding April 15) from 1973–1976.

Table 3. Effect of magnesium (Mg) supplementation of the cow during a portion of the winter-spring grass tetany season on calf performance, cow conception, and calving date (1973 and 1974).

| Item | Level of N fertilization | | | | | | | |
| | Low N | | Moderate N | | High N | | Overall | |
	0 Mg	+ Mg	0 Mg	+ Mg	0 Mg	+ Mg	0 Mg	+ Mg
Adj. 205-day calf wt., kg	145	156	159	168	153	167	154	165
		(.28)*		(.17)		(.05)		(.01)
Cow conception, %	89	91	46	79	82	74	70	81
		(.40)		(.03)		(.50)		(.21)
Calving date, day of year	39	54	73	41	58	43	55	46
		(.03)		(.18)		(.12)		(.20)

*Represents the actual level of significance between means.

semia extended over a longer period than for either of the other two pastures. Mineral analyses of forage taken from each of the three pastures revealed that all could have been conducive to tetany, yet grass tetany occurred on only the moderate- and high-N tall fescue. Some of these and associated results have been presented in greater depth by Stuedemann et al. (1975) and Wilkinson and Stuedemann (1979).

Effect of Mg Treatment on Calf Performance, Conception, and Calving Date

Since cows were maintained on their respective pastures following completion of the grass tetany prevention studies, it was possible to monitor the effect of Mg treatments on other response criteria. As is shown in Table 3, in 1973 and 1974, cows receiving Mg supplementation during the grass tetany season in the form of a foliar-applied magnesium oxide (MgO)–bentonite-water slurry produced calves with higher (P = 0.01) adjusted 205-day weights (154 vs. 165 kg). The same general trend existed across all pastures.

In cows grazing the moderate-N tall fescue there was a dramatic improvement (P = 0.03) in conception rate among cows receiving Mg supplementation compared with those not receiving Mg (Table 3). Cows that received Mg supplementation on this pasture had an average conception rate of 84% compared with only 63% for those not receiving Mg, based on a 5-year period (1970–1974). There was a slight trend for this same effect in cows grazing the low-N tall fescue but little response in those on the high-N tall fescue. In terms of conception rate and calving date, the effect of Mg supplementation was not consistent across N fertilization level.

Fat Necrosis

The incidence of fat necrosis (hard fat), as determined by rectal palpation, increased as N fertilization level increased. At a similar fall palpation date from 1969 through 1977 (1972–1977 for the low-N pasture) the weighted percentage of cows with palpable lesions of hard fat was 3, 10, and 54 for the low-, moderate-, and high-N pastures, respectively. Five cows died or had to be removed from the high-N tall fescue as a result of intestinal

obstruction by necrotic fat. Other details regarding incidence, size of lesions, and chemical composition are presented elsewhere (Rumsey et al. 1979, Stuedemann et al. 1975).

Internal Parasitism

Although there were pasture differences (P < 0.05) among cows in terms of nematode egg counts, the mean levels were low, ranging from 2 to 5 eggs/g of feces for the moderate- and low-N pastures, respectively. The calves were more parasitized, with mean levels of 426, 261, and 137 eggs/g of feces for calves grazing the low-, moderate-, and high-N pastures, respectively. Average total actual worm counts obtained from necropsy of two calves at weaning from each pasture each year from 1973 through 1975 were 52,331, 17,976, and 7,374 for the low-, moderate-, and high-N pastures. As with egg counts, differences among pastures were significant (P < 0.05). Pasture differences in internal parasitism were very consistent across years and apparently were inversely related to the quantity of available forage, which in turn was directly related to the level of N fertilization (Ciordia et al. 1977, 1980).

Nitrate Toxicity

Plant levels of NO_3-N reached concentrations of 3,300 and 2,000 µg/g on the high- and moderate-N tall fescue, respectively, in August 1969. Nitrate-N levels were lower in 1970 and 1971 (Stuedemann et al. 1975). Although plant NO_3-N levels were not monitored during the entire study, there was no direct evidence of NO_3-N toxicity in the animals.

LITERATURE CITED

Ciordia, H., J.V. Ernst, J.A. Stuedemann, S.R. Wilkinson, and H.C. McCampbell. 1977. Gastrointestinal parasitism of cattle on fescue pastures fertilized with broiler litter. Am. J. Vet. Res. 83:1135–1139.

Ciordia, H., J.A. Stuedemann, J.V. Ernst, H.C. McCampbell, and S.R. Wilkinson. 1980. Effects of level of nitrogen fertilization of fescue pasture on gastrointestinal parasitism of beef cattle. Am. J. Vet. Res. 41:893–898.

Rumsey, T.S., J.A. Stuedemann, S.R. Wilkinson, and D.J.

Williams. 1979. Chemical composition of necrotic fat lesions in beef cows grazing fertilized Kentucky-31 tall fescue. J. Anim. Sci. 48:673–682.

Stuedemann, J.A., S.R. Wilkinson, D.J. Williams, H. Ciordia, J.V. Ernst, W.A. Jackson, and J.B. Jones, Jr. 1975. Long-term broiler litter fertilization of tall fescue pastures and health and performance of beef cows. *In* Managing livestock wastes, 264–268. Proc. III Int. Symp. Livest. Wastes. ASAE Pub. Proc., 275. ASAE, St. Joseph, Mich.

Wilkinson, S.R., and J.A. Stuedemann. 1979. Tetany hazard of grass as affected by fertilization with nitrogen, potassium, or poultry litter and methods of grass tetany prevention. 93–121. *In* V.V. Rendig and D.L. Grunes (eds.) Grass tetany, Am. Soc. Agron. Spec. Pub. 35.

Hydrocyanic Acid Potential in Stargrass (*Cynodon* spp.)

P. MISLEVY, E.M. HODGES, and F.G. MARTIN
University of Florida, Gainesville, Fla., U.S.A.

Summary

The acreage of stargrass and bermudagrass (*Cynodon* spp.) throughout tropical and subtropical regions has increased greatly. However, many *Cynodons* have a hydrocyanic acid potential (HCN-p) that may be toxic to livestock. The purpose of this study was to determine the HCN-p in fresh and hay samples of meristematic stem tips from nine *Cynodon* entries at different physiological stages of growth during the summer (19°–33°C) and autumn (11°–32°C). Two selected entries (Ona and UF-5 stargrass) were also separated into meristematic stem tips, leaves and sheaths, and nodes and internodes at different stages of maturity to monitor HCN-p, dry matter (DM), and percentage of whole plant contributed by plant components. Correlations were determined between HCN-p concentration and total solar radiation. All plant entries were staged at 7.5 cm and received 280, 25, and 93 kg/ha of nitrogen (N), phosphorus (P), and potassium (K), respectively, one week prior to commencement of summer and autumn studies.

Significant differences in HCN-p concentration in meristematic stem tips were obtained among the nine *Cynodon* entries over a 2-year period. The HCN-p concentration basically followed either high or low levels. High-level HCN-p grasses averaged 407 and 529 ppm on a fresh-weight (FW) basis during summer and autumn, respectively; low-HCN-p grasses averaged 121 and 131 ppm FW. Average HCN-p concentrations were lowest for Callie bermudagrass (75 ppm) and highest for Puerto Rico stargrass (506 ppm). Entries decreased with maturity during summer, but during autumn HCN-p remained high until temperatures fell below freezing. After a freeze HCN-p decreased from 92 to 43 ppm and 78 to 25 ppm in high- and low-level grasses, respectively. Meristematic tips dried to 85% DM decreased in HCN-p by 88% and 78% for high- and low-level grasses, when compared with fresh samples. Significant correlations ($r = 0.56$) were found between HCN-p and total solar radiation. Separating Ona and UF-5 grasses into plant parts for DM and HCN-p revealed similar DM levels in both grasses. However, HCN-p in UF-5 was consistently lower ($P < 0.05$) than HCN-p in Ona stargrass. The meristematic tips generally contained the highest ($P < 0.05$) HCN-p regardless of species. These data indicated differences ($P < 0.05$) in HCN-p concentrations between *Cynodon* entries that can be beneficial when managing tropical grasses.

KEY WORDS: hydrocyanic acid potential, HCN-p, stargrass, *Cynodons*, tropical grasses.

INTRODUCTION

Even though *Cynodons* have excellent forage potential, several contain cyanogenic glucosides that have the potential to liberate hydrocyanic acid (HCN) (DeLeon et al. 1977). Kingsbury, as cited by Eck (1976), reported that 200 ppm hydrocyanic acid potential (HCN-p) is the critical level for death to ruminant animals. Van der Walt and Steyn (1941), upon evaluating stargrass strains, found that the hydrocyanic acid (HCN) content exceeded that of johnsongrass (*Sorghum halepense* L. Pers.) and concluded that giant stargrass would constitute a danger to livestock. In Puerto Rico, Velez Santiago et al. (1979) indicated that the amount of HCN found in 10 *Cynodon* cultivars depended on physiological stage at harvest, ranging from 0 ppm to 333 ppm.

Various *Cynodon* spp. have been tested with grazing animals at the Agricultural Research Center (ARC), Ona, for more than 15 years without known animal health problems (Hodges et al. 1979). However, the

P. Mislevy is an Associate Professor of Agronomy and E.M. Hodges is a Professor of Agronomy at the University of Florida, Ona; and F.G. Martin is an Associate Professor of Statistics at the University of Florida, Gainesville. This is a contribution of the Agricultural Research Center of the University of Florida, Ona.

Mention of a proprietary product does not imply endorsement of it.

deaths of several cattle grazing stargrass of uncertain origin in Florida and Puerto Rico were diagnosed as HCN toxicity (Iverson 1975, DeLeon et al. 1977).

The purpose of this experiment was to monitor HCN-p concentration in nine *Cynodon* entries, to determine the correlation with solar radiation at various physiological stages, and to determine the HCN-p in plant parts at several maturity stages, after a freeze, and when preserved as hay.

MATERIALS AND METHODS

This investigation was conducted over a 2-year period on a coarse-loamy, mixed hyperthermic, Typic Ochraqualfs (Bradenton fine sand) soil at the University of Florida ARC, Ona. Field plot layout was a randomized complete block with nine *Cynodon* selections replicated 3 times (Table 1). Fertilization practices on these 2-year-old plots were 280–25–93 kg/ha N-P-K applied both in late May and in early September of 1978 and 1979 when plants were cut to a 7.5-cm stubble height. Soil calcium (Ca) and magnesium (Mg) content was adequate. Plant samples collected for HCN-p analysis were the upper three to four expanded leaves (apical meristem) of a tiller. Each field sample consisted of 40 to 50 randomly selected tips collected from three (1.2 × 6.1-m) subplots to form each replication. All samples were collected between 0800 and 1000, prepared for analysis (Schroder 1977), and expressed on a fresh-weight (FW) basis. Samples were analyzed biweekly for all nine grasses during summer (June to August) and autumn (September to November). Sampling started one week after fertilization and terminated after 10 weeks during summer and one week after the first freeze. Tips were also harvested at the 6-week sampling date and allowed to field cure until moisture dropped to 15%, simulating hay.

Beginning 2 weeks after fertilization in both summer

and autumn of 1979, samples of Ona and UF-5 were collected and separated into apical meristem tips, leaves and sheaths, and nodes and internodes and were analyzed, along with whole-plant samples, for HCN-p. Total solar radiation was recorded during summer and autumn of 1979, using a Lamda LI-550 printing integrator with a pyranometer sensor. Daily radiation for 1, 2, and 3 days before sampling and average daily radiation for 3 days and 7 days before sampling were used in statistical analysis. Correlations were obtained between radiation and HCN-p in apical meristem tips for the nine grasses.

Analysis for HCN-p was by the alkaline picrate colorimetric procedure (Schroder 1977) with one modification: standard HCN was prepared by dissolving 0.241 g HCN in a total volume of 1,000 ml distilled water.

RESULTS AND DISCUSSION

Plant Species

Concentrations of HCN-p in tips of nine *Cynodon* followed two distinct patterns during summer and autumn (Figs. 1 and 2). Since six *Cynodons* contained high concentrations of HCN-p at the same stages of growth, their values were combined at each sampling date to form a single high-concentration (HC) curve. The three remaining *Cynodons* (UF-5, Callie, and Sarasota) were low in HCN-p, consequently forming a low-concentration (LC) curve. Significant differences were found between the HC and LC curves at most sampling dates during summer.

The HCN-p concentrations of apical meristem tips dropped 73% and 83% for the HC and LC, respectively, over the 9-week summer. This decline in HCN-p may have been due mainly to a decrease in available N over time. Concentration differences between high and low curves appear to follow a similar pattern within years for

Table 1. Perennial grasses monitored for HCN-p.

Common name	Scientific name	PI†
Ona stargrass	*Cynodon nlemfuensis* Vanderyst var. *nlemfuensis* de Wet	224566
UF-5 stargrass	*C. aethiopicus* Clayton et Harlan	225957
Cane patch stargrass	*C. nlemfuensis* Vanderyst var. *nlemfuensis* de Wet	–§
Puerto Rico stargrass	*C. nlemfuensis* Vanderyst var. *nlemfuensis* de Wet	2341††
Costa Rica stargrass	*C. nlemfuensis* var. *robustus* Clayton et Harlan	–
Callie bermudagrass	*C. dactylon* var. *aridus* Harlan et de Wet	290814
McCaleb stargrass	*C. aethiopicus* Clayton et Harlan	224152
Sumner stargrass	*C. nlemfuensis* var. *robustus* Clayton et Harlan	–
Sarasota stargrass	*C. dactylon* var. *coursii* (Camus) Harlan et de Wet	–

†USDA plant introduction no.

§USDA PI unknown.

††University of Puerto Rico Agric. Expt. Sta. plant introduction number.

Fig. 1. Patterns of HCN-p concentration in *Cynodon* apical meristem tips during the summer as plants mature after staging at a 7.5-cm stubble in late May.

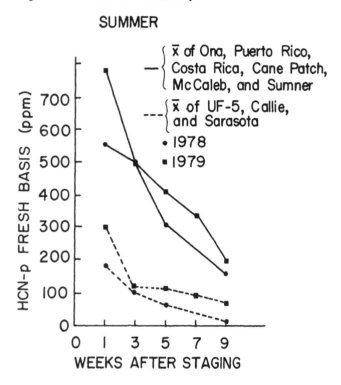

Fig. 2. Patterns of HCN-p concentration in *Cynodon* apical meristem tips during the autumn as plants mature after staging at a 7.5-cm stubble in early September.

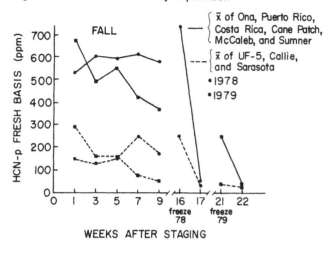

each growth period, indicating that temperature and light may influence HCN-p values. Differences ($P < 0.05$) between HC and LC curves were obtained during autumn for all sampling dates except for 7 days after the first freeze (9 Jan. 1979 and 19 Feb. 1980). Values from both curves then dropped to their lowest points. Similar results were also found with sorghum spp. (Harrington 1966, Loyd et al. 1971). Unlike summer values, autumn HCN-p values remained high as plants matured. This difference may have been due to cooler temperatures in autumn (11°–32°C) than in summer (19°–33°C). Tips harvested

6 weeks after staging and allowed to dry for hay followed a pattern similar to plants after a freeze. The HC and LC group averaged 387 and 65 ppm HCN-p fresh and decreased 88% and 78% ($P < 0.05$) in hay samples containing 85% DM.

Plant Parts

Whole plant, meristematic tips, and leaves and sheaths for 2- and 6-week growth periods during summer (Table 2) and 2-, 6-, and 10-week periods during autumn (Table 3) were different ($P < 0.05$) in HCN-p for Ona and UF-5 stargrass. Regardless of variety or growth period, HCN-p in nodes and internodes was low, ranging from 0 to 18 ppm. Allowing plants to mature 2 to 10 weeks after staging resulted in whole-plant, meristematic tips, and leaf and sheath HCN-p decreases of 91%, 80%, and 91%, respectively, in Ona, and 50%, 0%, and 63%, respectively, in UF-5 during the summer. Respective values

Table 2.   Grass entry and stage of growth on HCN-p concentrations, in whole and plant parts, during the summer (June to August).

Grass entry	Whole plant	Plant part		
		Meristematic tips	Leaves sheaths	Nodes— internodes
		---HCN-p fresh basis (ppm)---		
2-week growth				
Ona	336	488	379	0
UF-5	26	32	27	4
	*	*	*	NS
6-week growth				
Ona	197	310	257	11
UF-5	30	162	27	6
	*	*	*	NS
10-week growth				
Ona	30	97	34	3
UF-5	13	57	10	10
	NS	NS	NS	NS

*Significantly different at the 0.05 level of probability; NS = not significant.

Table 3. Grass entry and stage of growth on HCN-p concentrations in whole and plant parts, during the fall (September to November).

Grass entry	Whole plant	Plant parts		
		Meristematic tips	Leaves sheaths	Nodes— internodes
		HCN-p fresh basis (ppm)		
2-week growth				
Ona	321	587	455	25
UF-5	105	218	161	4
	*	*	*	NS
6-week growth				
Ona	265	424	312	13
UF-5	52	164	59	3
	*	*	*	NS
10-week growth				
Ona	144	222	96	10
UF-5	6	41	5	7
	*	*	*	NS

*Significantly different at the 0.05 level of probability; NS = not significant.

during the autumn decreased 55%, 62%, and 79% in Ona and 94%, 81%, and 97% in UF-5. Ona, which contained high HCN-p, decreased 80%–90% in concentration during summer. However, during autumn, concentrations decreased only 55%–79%. The UF-5 stargrass followed a reverse pattern, decreasing 0%–63% in summer and 81%–97% during autumn. Since grazing cattle tend to consume meristematic tips, that plant part may be the best indicator of HCN-p consumption. Further observations indicated meristematic tips (2 weeks after staging at 7.5 cm) made up 38%–42% and 28%–35% of the whole-plant weight in summer and autumn, respectively. These values decreased for both grasses after 10 weeks of growth to 5% and 14%, and 11% and 15% of whole-plant weight during summer and autumn, respectively. Leaves and sheaths were second highest, with plants ranging from 83% (UF-5) to 88% (Ona) of whole-plant weight in summer and 70% to 76% in autumn, respectively, after 2 weeks of growth. Even after a growth period of 10 weeks, leaf-sheath percentages were above 50% for both entries.

The nodes and internodes followed an inverse relationship to leaves and sheaths, increasing from 12% (Ona) to 17% (UF-5) and 24% (Ona) to 30% (UF-5) after a 2-week growth period during the summer and autumn, respectively. As plant parts matured, HCN-p concentration remained relatively constant (10% summer and 20% autumn) in nodes and internodes; however, HCN-p percentage of the whole plant by weight increased to nearly 50% after a 10-week growth period.

Percentage whole-plant dry matter of both Ona and UF-5 stargrass was similar, averaging 19% and 21% and 19% and 18% after a 2-week growth period during the summer and autumn, respectively. As whole plants matured, percentage of dry matter in both entries increased in a similar manner, until the 10-week growth period when Ona contained 10 percentage units more dry matter than UF-5.

HCN-p and Radiation

Correlations between solar radiation (watts/m²) and HCN-p concentration in meristematic tips were significant (P < 0.05) for the LC *Cynodons*. When solar radiation increased one day before sampling or when the 7-day average solar radiation increased before sampling, HCN-p increased. Correlations between radiation and HCN-p for the HC *Cynodons* were not significant except for Sumner, which was significant one day and an average of 7 days before sampling.

LITERATURE CITED

DeLeon, D., B. Salas, and J. Figarella. 1977. Hydrocyanic acid poisoning in dairy cows—a case report. J. Agric. Univ. Puerto Rico 61(1):106–107.

Eck, H.V. 1976. Hydrocyanic acid potential in leaf blade tissue of eleven grain sorghum hybrids. Agron. J. 68:349–351.

Harrington, J.D. 1966. Hydrocyanic acid content of Piper, Trudan I and six sorghum-sudangrass hybrids. Pa. Agric. Expt. Sta. Bul. 735.

Hodges, E.M., L.S. Dunavin, P. Mislevy, O.C. Ruelke, and R.L. Stanley, Jr. 1979. "Ona," a new stargrass variety. Fla. Agric. Expt. Sta. Circ. S-268.

Iverson, W. 1975. Diagnostic Lab. No. 51186. Fla. Dept. Agric. and Consumer Serv., Kissimmee, Fla.

Loyd, R.C., E. Gray, and E. Shipe. 1971. Effect of freezing on hydrocyanic acid release from sorghum plants. Agron. J. 63:139–140.

Schroder, V.N. 1977. Hydrogen cyanide from forage plants. Soil and Crop Sci. Soc. of Fla. Proc. 36:195–197.

Van der Walt, S.J., and Douw G. Steyn. 1941. The hydrocyanic acid content of *Cynodon plectostachyum* Pilger (Giant star grass) and its suitability as a pasture grass. Oenderstepoort J. Vet. Sci. and Anim. Indus. 17(1 and 2):191–199.

Velez Santiago, J., A. Sotomayor Rios, and S. Torres Rivera. 1979. Effect of three harvest intervals and two fertilizer rates on the yield and HCN content of ten *Cynodon* cultivars. J. Agric. Univ. Puerto Rico 63(1):35–44.

Review of Mechanisms of Toxicity of 3-Nitropropanoic Acid in Nonruminant Animals

D.L. BUSTINE and B.G. MOYER

USDA-SEA-AR Regional Pasture Research Laboratory,
University Park, Pa., U.S.A.

Summary

Crownvetch (*Coronilla varia* L.) now has limited use in the U.S.A. as a forage crop for ruminant animals. Even though crownvetch forage can contain high amounts of nutrients, intake by ruminants may be reduced. Toxicity problems upon feeding crownvetch to nonruminants have been traced to 3-nitropropanoic acid (NPA), which occurs esterified to glucose in the plant. Understanding mechanisms of toxicity in nonruminants will be helpful in avoiding livstock losses and in treating poisoned animals.

Symptoms of NPA toxicity are similar in rats, mice, meadow voles, chickens, and rabbits and probably reflect central nervous system disorders and oxygen deprivation. The other major symptom of NPA poisoning in animals is elevated serum levels of methemoglobin, a nonfunctioning form of hemoglobin, but this condition by itself does not cause death. NPA metabolism in nonruminants evidently produces nitrite ions, which combine with hemoglobin to form methemoglobin. Recent data indicate that NPA might be oxidized by a liver microsomal monooxygenase, thus producing nitrite. However, some additional mechanism is needed to explain NPA toxicity.

Investigators have found that approximately 1% of NPA in the circulation of intoxicated animals is converted to a dianionic carbanion and that the NPA carbanion concentration is sufficiently high to inhibit succinate dehydrogenase irreversibly. Under these conditions, NPA carbanion also competitively inhibits fumarase. These two mitochondrial enzymes catalyze sequential steps in the Krebs cycle, both essential to normal respiration. Two chemical mechanisms for inhibition of succinate dehydrogenase by NPA carbanion have been proposed; the mechanism least acceptable on chemical grounds accounts for nitrite appearance in animals (as methemoglobin), but the second, for which the chemical evidence is most convincing, does not account for nitrite release from NPA.

NPA apparently acts in two ways in exerting its toxic effects: (1) inhibition of two mitochondrial enzymes essential to respiration, succinate dehydrogenase (irreversibly) and fumarase (competitively); and (2) methemoglobin formation as a result of metabolic breakdown of NPA to nitrite ion.

KEY WORDS: crownvetch, *Coronilla varia*, *Astragalus* spp., 3-nitropropanoic acid, NPA, toxicity, methemoglobinemia, succinate dehydrogenase, fumarase.

INTRODUCTION

The carboxylic acid 3-nitropropanoic acid (NPA) is the basic building block for aliphatic nitro compounds identified in crownvetch (*Coronilla varia*) and several *Astragalus* spp. Although toxicity from ingestion of NPA in plant material is seldom seen in ruminant animals (Gustine 1979), toxicity often occurs in nonruminants. Understanding mechanisms of toxicity in nonruminants will be helpful in avoiding livestock losses and in treating poisoned animals. Gustine (1979) recently reviewed the toxicology and natural occurrence of NPA and of ruminal metabolism of NPA.

The symptoms produced by toxic doses of NPA are similar for chicks, mice, and meadow voles (Gustine 1979). They include depression of food intake, paralysis of the lower limbs, a tendency to fall down, difficulty in breathing, and head extension. These symptoms suggest depression of the central nervous system (CNS) and oxygen deprivation.

Symptoms similar to those described for NPA intoxication were described for sodium nitrite poisoning in mice, a condition accompanied by methemoglobinemia, the primary cause of death (Gustine 1979). Although methemoglobin levels in NPA-poisoned animals are elevated, toxicity was not due to methemoglobinemia alone (Gustine 1979).

MECHANISMS OF TOXICITY

Matsumoto et al. (1961) found that within 90–120 minutes after injection, about 2% of administered NPA in rats was excreted in the urine and that methemoglobin

D.L. Bustine is a Research Biochemist and B.G. Moyer is a Pathologist at USDA-SEA-AR Regional Pasture Research Laboratory. This is contribution no. 8013 of the USDA-SEA-AR Regional Pasture Research Laboratory.

Fig. 1. Mechanisms proposed for the inhibition of succinate dehydrogenase. Part A, Alston et al. 1977; part B, Coles et al. 1979.

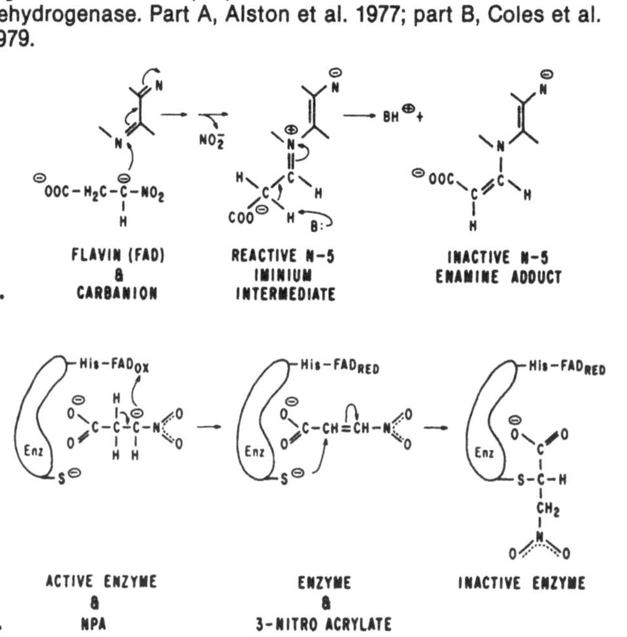

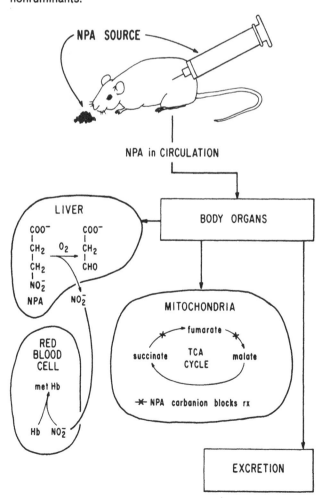

troacrylate, irreversibly inactivated the enzyme and, in contrast to NPA, did so rapidly. Since the data of Coles et al. (1979) did not support alkylation of the flavin N-5 moiety (Fig. 1A), they suggested the mechanism in Fig. 1B. This mechanism requires reaction of 3-nitroacrylate with a sulfhydryl group essential to SD activity. However, they were not able to provide direct evidence for the sulfur-linked NPA adduct shown in Fig. 1B. In their system, the NPA carbanion had a K_m of 1×10^{-5} M.

Porter and Bright (1980) demonstrated that a second mitochondrial enzyme, fumarase, was competitively inhibited by NPA carbanion ($K_i = 6.4 \times 10^{-8}$ M at pH 7.0). Although this inhibition is reversible, the concentration needed to produce fumarase inhibition is 150 times lower than for SD. Thus, fumarase inhibition probably represents a second biochemical lesion involved in NPA toxicity.

A toxic dose of NPA should attain a physiological concentration in all tissues of 0.02 mM NPA carbanion (based on data from Porter and Bright 1980). At a given instant, this concentration would only slightly inhibit SD, but since the effect is irreversible and NPA carbanion would be continuously formed, all the NPA would gradually react with SD and cause lethal inhibition of respira-

levels were 25% to 30%, suggesting extensive metabolism of NPA to nitrite. They also reported significant levels of NPA in the brain. They demonstrated that incubation of NPA in blood samples did not result in measurable loss of the toxin, indicating that metabolism must have occurred in other tissues. They concluded that nonmetabolized NPA was toxic, probably acting in the brain, and that some NPA was metabolized, thus accounting for both toxicity and methemoglobinemia.

NPA and succinate are chemically similar, and succinate is an essential metabolite for mitochondria respiration; therefore, inhibition of succinate dehydrogenase (SD) by NPA might represent a major biochemical lesion. Alston et al. (1977) studied the inhibition of SD by NPA in purified rat liver mitochondria. In their experiments, 1 mM NPA carbanion, prepared by treating NPA with NaOH at pH 11,

$$^{\ominus}OOC\text{-}CH_2\text{-}CH_2\text{-}NO_2 \xrightarrow{pH\,11} {^{\ominus}OOC\text{-}CH_2\text{-}}^{\ominus}CH\text{-}NO_2 \quad (1)$$

produced irreversible inhibition of mitochondrial SD activity but did not affect other mitochondrial functions. They proposed the mechanism shown in Fig. 1A. Since FAD is a covalently linked cofactor of SD, the final unreactive form of FAD irreversibly inhibits the enzyme activity. This mechanism, which is consistent with the results of Matsumoto et al. (1961), explains both the appearance of nitrite (and subsequently, methemoglobinemia) and NPA toxicity (inhibition of respiration).

Coles et al. (1979) found that soluble SD obtained from beef heart nonphosphorylating submitochondrial particles slowly developed complete inactivation in the presence of stoichiometric amounts of NPA carbanion. Also, NPA carbanion was slowly oxidized at a rate 0.1% that of succinate. The expected product of dehydrogenation, 3-ni-

Fig. 2. Mechanisms of NPA metabolism and toxicity in nonruminants.

tion. Furthermore, a carbanion concentration of 0.02 mM inside mitochondria would be sufficiently high to inhibit fumarase.

The inhibition mechanisms proposed by Coles et al. (1979) for SD, and Porter and Bright (1980) for fumarase, do not account for release of nitrite from NPA. Ullrich et al. (1978) reported the release of nitrite from 2-nitropropane by microsomal monooxygenases when the nitro compound was incubated with rat liver microsomes in the presence of NADP and O_2. Other aliphatic nitro compounds were also oxidized with the concomitant release of nitrite (Sakurai et al. 1980) in the same system. Although NPA was not tested in their system, presumably incubation of NPA would result in formation of nitrite and malonic semialdehyde.

We have summarized the present state of knowledge concerning NPA toxicity in nonruminant animals, and we present in Fig. 2 our current conception of the mechanisms involved. Further studies are needed to demonstrate whether mitochondrial SD is irreversibly inhibited in vivo and whether nitrite is released by NPA through microsomal oxidation.

LITERATURE CITED

Alston, T.A., L. Mela, and H.J. Bright. 1977. 3-Nitropropionate, the toxic substance of *Indigofera*, is a suicide inactivator of succinate dehydrogenase. Natl. Acad. Sci. Proc. 74: 3767–3771.

Coles, C.J., D.E. Edmondson, and T.P. Singer. 1979. Inactivation of succinate dehydrogenase by 3-nitropropionate. J. Biol. Chem. 254:5161–5167.

Gustine, D.L. 1979. Aliphatic nitro compounds in crownvetch: a review. Crop. Sci. 19:197–203.

Matsumoto, H., J.W. Hylin, and A. Miyahara. 1961. Methemoglobinemia in rats injected with 3-nitropropanoic acid, sodium nitrite, and nitroethane. Toxicol. Appl. Pharmacol. 3:493–499.

Porter, D.J., and H.J. Bright. 1980. 3-Carbanionic substrate analogues bind very tightly to fumarase and aspartase. J. Biol. Chem. 255:4772–4780.

Sakurai, H., G. Hermann, H.H. Ruf, and V. Ullrich. 1980. The interaction of aliphatic nitro compounds with the liver microsomal system. Biochem. Pharmacol. 29:341–346.

Ullrich, V., G. Hermann, and P. Weber. 1978. Nitrite formation from 2-nitropropane by microsomal monooxygenases. Biochem. Pharmacol. 27:2301–2304.

Relation of Animal Production to Stocking Rate on Cultivated Pastures in Cerrados Areas of Brazil

C.J. ESCUDER

Empresa de Pesquisa Agropecuária de Minas Gerais (EPAMIG), Brazil

Summary

The relationships among stocking rate, gain/animal, and gain/ha were compared on six grazed pastures. The experiment was established on an oxisol acrustox in Brazil. The pasture consisted of (1) jaraguagrass (J); (2) jaraguagrass plus legumes (J + L); (3) molassesgrass (M); (4) molassesgrass plus legumes (M + L); (5) pangolagrass (P); (6) Pangolagrass plus legumes (P + L); (7) mixture of grasses (jaragua, molasses, and pangola) plus legumes (JMP + L). Pasture of pangolagrass alone was discarded because of poor establishment. Four stocking rates were applied on each pasture, from 1.0 to 3.4 animals/ha. For all pastures, gains/animal decreased linearly (P < 0.05) with increasing stocking rate. No significant differences were found among the slopes of all the six linear regressions for each pasture, or among years. On the other hand, the adjusted means of the pastures of J, J + L, P + L, and JMP + L were significantly higher than those of the pastures of M and M + L. Live-weight gains were significantly related to pasture availability during the wet periods of 1975–1976 and 1976–1977 and during the dry period of 1976–1977. The highest stocking rates at which pastures of J, J + L, P, and JMP + L persisted for more than 5 years were 2.0, 1.4, 2.0, and 2.0 animals/ha, respectively. The experiment showed that some cultivated set-stocked pastures grazed at 2.0, animals/ha or less produced 5 times as much as the average of the region.

KEY WORDS: grazing, stocking rate, *hyparrhenia*, *melinis*, *digitaria*, tropical legumes.

INTRODUCTION

In Brazil, more than 150 million ha are covered by a characteristic vegetation called cerrado. Very poor soils with low pH, low phosphorus content, and high aluminum content, but with good physical properties, are characteristic of the region. Jaraguagrass (*Hyparrhenia rufa*) and molassesgrass (*Melinis minutiflora*) are spontaneous; some farmers are also clearing land to establish sown pastures with these grasses. However, very limited data are available on the animal productivity of monospecific pastures or of mixtures with legumes. To carry out such an evaluation, a range of stocking rates is necessary ('t Mannetje et al. 1976). Most authors have found that a linear relationship between animal gain and stocking rate fits satisfactorily over a wide range of stocking rates, both for temperature and for tropical pastures (Riewe 1961, Jones and Sandland 1974). The relationship between liveweight gain and stocking rate was used to analyze an unreplicated grazing experiment at several stocking rates.

EXPERIMENTAL PROCEDURES

The experiment was established in Felixlandia, Minas Gerais (Central Brazil), on an oxisol acrustox soil. The climate is tropical with mean annual rainfall of 1,300 mm, mainly in the summer. One-and-one-half tons of lime and 100 kg of P_2O_5/ha were applied to the area before planting (50% as superphosphate and 50% as rock phosphate of low solubility.) Micronutrients (Cu, Zn, B, and Mo) also were added. The plan was to compare seven pasture treatments: (1) jaraguagrass (*Hyparrhenia rufa*); (2) jaraguagrass plus legumes (*Macroptilium atropurpureum* cv. Siratro, *Stylosanthes guianensis,* and *Centrosema pubescens*); (3) molassesgrass (*Melinis minutiflora*); (4) molassesgrass plus legumes; (5) pangolagrass (*Digitaria decumbens*); (6) pangolagrass plus legumes; and (7) mixture of grasses (jaragua, molasses, and pangola) plus legumes.

Pastures were established in December 1972 and January 1973, and after intermittent grazing each of the pastures was grazed at four stocking rates, beginning in January 1975, with the exception of the pure pangolagrass pasture that was discarded because of poor establishment. As the area was already subdivided in 24 paddocks of 3.0, 3.5, and 4.0 ha, a different number of animals had to be used to obtain four similar, but not equal, stocking rates among pastures, ranging from 1.0 to 3.4 animals/ha. No replications were used. Each year new groups of castrated Zebu steers were introduced after the pastures were rested one or 2 months. Animals were sprayed periodically for external parasites and drenched for helminth control. Mineral salts were offered permanently. Cattle were weighed every 2 weeks at about 0800. Weight of forage over 5 cm was determined monthly by cutting 30 open quadrats of 0.5 m² on each paddock in 1975. In 1976 a double sampling visual technique was adopted.

RESULTS

Live-Weight Gains

After 252 days' grazing, because of the condition of pastures and animals, it was necessary to take out the animals from paddocks on pastures of M and M + L, indicating that even the lowest stocking rate used was too high for this type of pasture in this region. Similar results were found in the second year, and again it was necessary to suspend grazing on pastures of M and M + L. Also, pastures of J, J + L, P + L, and JMP + L at the highest

Fig. 1. Animal gains on pastures stocked at different rates.

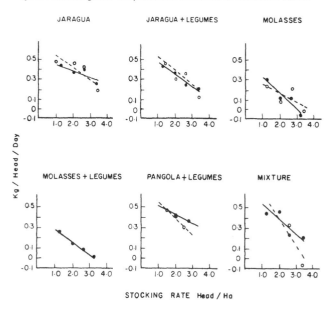

stocking rate became overgrazed and required stock reduction after the second year of grazing. Accordingly, linear regressions of gain/animal on stocking rate were calculated with the results of the 2 first years and compared by covariance analysis.

For all pastures, gain/animal decreased linearly with increasing stocking rate in both years (Fig. 1). No significant differences were found among the slopes of all six linear regressions for each pasture or among years (Table 1). On the other hand, the adjusted means of pastures of J, J + L, P + L, and JMP + L were significantly higher than those of pastures of M and M + L. The regression coefficients, optimum live-weight gains/ha, and optimum stocking rate calculated by the half-intercept method of

Table 1. Linear regression coefficients for daily live-weight gains (kg), weight gains (kg), and stocking rate (animals/ha), determination coefficients, calculated optimum stocking rates, and live-weight gains/ha.

Year and species	a	b	r^2	Optimum stocking rate,* anim./ha	Optimum calculated gain, kg/ha/yr.
1975–76					
J	0.553	−0.080	0.74	3.46	292
J + L	0.598	−0.122	0.92	2.45	224
M	0.504	−0.171	0.84	1.47	113
M + L	0.405	−0.126	0.98	1.61	100
P + L	0.593	−0.083	0.92	3.57	324
JMP + L	0.680	−0.148	0.82	2.30	239
1976–77					
J	0.665	−0.123	0.74	2.70	265
J + L	0.680	−0.153	0.84	2.22	223
M	0.343	−0.092	0.52	1.86	94
P + L	0.705	−0.156	0.97	2.26	235
JMP + L	0.913	−0.265	0.85	1.72	232

*Stocking rate with highest production.

Table 2. Mean live-weight gain/head and /ha of steers at stocking rates that persisted more than two years (1977-78/1979-80).

Pasture	Stocking rate head/ha	Gain/head kg/day	Gain/ha kg
J	1.25	0.457	173.3
	2.00	0.440	264.7
J + L	1.42	0.523	224.7
P + L	1.42	0.507	217.0
	2.00	0.507	307.3
JMP + L	1.25	0.483	182.0
	2.00	0.450	270.7

Jones and Sandland (1974) for each year and for the six pastures are shown in Table 1. On pastures that persisted after 2 years of grazing, apart from pasture J + L, a stocking rate of 2 head/ha yielded the highest gains on a per-ha basis (Table 2).

Pasture Availability and Animal Gain

Live-weight gains were related to the green forage available. The data were based on live-weight gains for each pasture at each stocking rate and were divided into two sections, one covering the dry season and the other the rainy season. No significant relationship was found during the dry period in 1975, but highly significant linear regressions were fitted for summer data in 1975-1976 and 1976-1977 and for the dry period in 1976. The equations were:

Wet summer 1975-1976:
$$y = 0.659 + 0.000235X \qquad r = 0.75** \qquad (1)$$

Wet summer 1976-1977:
$$y = 0.260 + 0.000350X \qquad r = 0.70** \qquad (2)$$

Dry winter 1976:
$$y = 0.244 + 0.000359X \qquad r = 0.70** \qquad (3)$$

where y is the live-weight gain (kg/day) and X is the amount of green forage available (kg/ha). No better fit was obtained with a quadratic relationship ($P < 0.05$). Relationships for each different stocking rate or pasture were not significantly different.

DISCUSSION

During the first 2 years of the experiment, animal production from pastures of M and M + L was lower than from the others ($P < 0.05$). The estimated carrying capacity also was generally lower (Table 2). These results agree with the lower dry-matter yields observed with molassesgrass in cutting experiments by Pedreira (1973).

Quinn et al. (1965), under similar conditions but with the put-and-take technique, found live-weight gains of 235.0, 240.0, and 91.0 kg/ha on pastures of pure jaragua-

grass, pangolagrass, and molassesgrass, respectively, which were comparable to results of this experiment. Gains/animal in the work of Quinn et al. (1965) were also similar to daily gains observed at Felixlandia. A live-weight gain of 250 kg/ha is approximately 5 times the production of the natural pastures found in a survey of this region (Campal and Carneiro 1977).

The percentage of legumes in the different pastures was generally low, and this condition may explain the lack of differences between the productivity of pastures with and without legumes. It was also observed that even in pastures with more than 40% legumes, the legume content in the extrusa during the dry period was less than 10% (Silva 1977). In addition, although live-weight gains during the dry period tended to be higher on pastures with a good percentage of legume, in the wet season the opposite occurred. These results suggest that benefits from legumes could be impaired by many factors and emphasize the need to select a germ plasm that is better adapted for grazing in this region.

Under other conditions, no significant relationship between cattle growth and pasture availability during the summer period has been reported (Yates et al. 1964). In contrast, highly significant relationships were found in this experiment, suggesting that selection by animals occurred even in periods of abundant rapid growth.

LITERATURE CITED

Campal, F.C., and J.M. Carneiro. 1977. Subsídios para a programacao do desenvolvimento da pecuaria bovina mineira. Comissao Estadual de Planejamento Agrícola de Minas Gerais, Belo Horizonte, Brazil.

Jones, R.J., and R.L. Sandland. 1974. The relation between animal gain and stocking rate. Derivation of the relation from the results of grazing trials. J. Agric. Sci. 83:335-342.

Silva, J. 1977. Dietas seleccionadas por novilhos azebuados em pastagens cultivadas no cerrado sob várias lotacoes. M.S. Thesis, Escola de Veterinária, Univ. Fed. de Minas Gerais.

't Mannetje, L.T., R.J. Jones, and T.H. Stobbs. 1976. Pasture evaluation by grazing experiments. In N.H. Shaw and W.W. Bryan (eds.), Tropical pasture research: principles and methods, 194-232. Commonwealth Agric. Bur. Bul. 51. Hurley, Berks., England.

Pedreira, J.V.S. 1973. Crescimento estacional dos capins coloniao (*Panicum maximum*, Jacq.), gordura (*Melinis minutiflora*, Pal. de Beauv.), Jaraguá (*Hyparrhenia rufa* (Nees) Stapf) e Pangola de Taiwan A-24 (*Digitaria pentzii*, Stent). Bol. Indús. Anim., Sao Pablo 30(1):59-145.

Quinn, L.R.C., G.O. Mott, W.V.A. Bisschoff, M.B. Jones, and G.L. da Rocha. 1965. Beef production of six tropical grasses in central Brazil. Proc. IX Int. Grassl. Cong., 1015-1020.

Riewe, M.E. 1961. Use of the relationship of stocking rate to gain of cattle in an experimental design for grazing trials. Agron. J. 53:309-313.

Yates, J.J., L.A. Edye, J.G. Davies, and K.P. Haydock. 1964. Animal production from a Sorghum almum pasture in South East Queensland. Austral. J. Expt. Agric. Anim. Husb. 4:326-335.

Intensive Beef Production in the Humid Tropics: An Evaluation of Technical and Economic Feasibility

R.O. BURTON, JR., and W.B. BRYAN

West Virginia University, Morgantown, W.VA., U.S.A.

Summary

Some parts of the humid tropics show great potential for intensive livestock production from forages. However, in many cases, economic analyses have not been performed.

In this study technical and economic feasibility of three strategies for developing intensive pasture for beef production are evaluated. These strategies are fast development (no capital limitations) using commercial nitrogen (N), slow development using commercial N, and slow development using legumes as a N source.

The strategies are based on experimental data and experience from the Orinoco Delta, Venezuela. Pastures of pangolagrass (*Digitaria decumbens*) and Swazi (*D. swazilandensis*) were rotationally grazed with unselected Criollo/Zebu cattle. Live-weight gains of 550 g/head/day and 1,000 kg/ha/yr were obtained.

Using 1974 prices and a 30-year period, net annual cash flow (NACF), debt payback period (DPP), net present value (NPV), and annual profit after debt is paid are used to evaluate economic feasibility. Sensitivity analyses are performed by varying interest and discount rates, investment and operating costs, and livestock receipts.

Under the alternative of fast pasture development with no capital constraints, NPV is highest and DPP is shortest. With slow pasture development using commercial N, NPV is negative and DPP is longest. NPV is positive with slow pasture development using legumes. NACF budgets indicate that receipts are greater than expenses earliest but accumulated debt is greatest with fast pasture development. All three strategies result in profitable farms once the developmental process is completed and the accumulated debt is paid. The farm in which legumes are substituted for commercial N is most profitable. However, the sensitivity analyses and other economic decision criteria indicate that this profitability must be interpreted cautiously.

Development of intensive beef production in the Orinoco Delta is technically feasible. Most economic criteria used indicate that a developmental strategy with fast pasture development is most desirable. Slow pasture development using commercial N is least desirable and may not be economically feasible. The economic feasibility of all three strategies is very sensitive to changes in interest rate, discount rate, operating costs, and livestock receipts. Development of intensive beef production could be encouraged by government subsidization of capital or by government policies that result in favorable levels of operating costs or livestock receipts.

KEY WORDS: beef, humid tropics, economic evaluation, pasture, legumes, investment.

INTRODUCTION

Some parts of the humid tropics show great potential for intensive livestock production from forages. However, few economic analyses of the limited production data have been performed. In this paper, experimental data from the Orinoco Delta of Venezuela are used as a basis to ex-amine three developmental strategies for an intensive beef-production grazing unit. Economic feasibility is evaluated using farm investment decision criteria and sensitivity analyses.

TECHNICAL FEASIBILITY

The Orinoco Delta lies between lat 8° and 10°N. Soils vary from silty loams to clays with internal peat layers. Rainfall (1,270 mm annual average) is well distributed with a 3- to 4-month dry season. Average monthly minimum and maximum temperatures ranged from

R.O. Burton, Jr., is an Assistant Professor of Agricultural Economics and W.B. Bryan is an Associate Professor of Agronomy at West Virginia University. This is Scientific Article No. 1677 of the West Virginia Agricultural and Forestry Experiment Station.

20°–24°C and 29°–33°C, respectively (Lárez et al. 1975).

Swazi (*Digitaria swazilandensis*) and pangolagrass (*D. decumbens*) produce well on clay loam soils. In a 4-year trial, pangolagrass yielded an average of 17.5 metric tons of dry matter (DM)/ha/yr using 800 kg/ha/yr of nitrogen (N) and 200 kg/ha/yr of phosphorus (P).[1] Swazi yielded 18 metric tons/ha/yr over approximately the same period. Over a 9-month period, the legumes *Stylosanthes guianensis* or *Centrosema plumieri* in association with pangolagrass produced an average of 10 metric tons of DM/ha with a pasture crude-protein content estimated at 17% (Velásquez and Bryan 1975).[2] For clay soils, annual yields of 27.5 metric tons of DM for Canarana (*Echinochloa pyramidalis*) and 19.5 metric tons for Para (*Brachiaria mutica*) are indicated (Lárez et al. 1975, Velásquez et al. 1975).

In a 2-year grazing experiment, 28 paddocks each of pangolagrass and Swazi were grazed, year-round, on a daily rotational basis. Average daily live-weight gains of animals weighing 200–400 kg were almost 550 g/head/day. Total live-weight production/ha averaged over 1,000 kg/yr.[3] The cattle were unselected; received only pasture, a salt-mineral mixture, and water; and were carried to slaughter on the experiment (Cunha et al. 1975).

Experimental data for the area are limited. Details of technical feasibility may change as more results become available.

PROPOSED INTENSIVE BEEF-FARM DEVELOPMENT

The proposed farm will consist of 100 ha. Seven ha will be undisturbed and used as weed-control barriers, quarantine pastures, living and working areas, and alleyways. The remaining 93 ha will be divided into 24 ha of Swazi, 21 ha of pangolagrass, 24 ha of Para, and 24 ha of Canarana. Pastures will be divided into 3–3.5-ha paddocks with electric fences. The farm will support an average of 6.13 animals/ha with an average daily gain of 0.55 kg/animal/day. A total of 570 head of cattle (steers and heifers) averaging 200 kg/head will be purchased/yr. Allowing a 3% loss, 553 slaughter animals will be sold.

The farmer wishes to improve his financial position, is managerially capable, lives on his farm, and starts with the land but no capital or animals. He may purchase animals as required. The entire farm is developed from natural vegetation free of Cambute (*Paspalum conjugatum*) and roaming cattle. Public roads and major drains are adequate. Climate does not vary appreciably from normal.

During the first year, enough land will be cleared, prepared, and planted to provide vegetative "seed" for the following year's pasture development. The "year" will begin in October. Pastures will be planted in June at the start of the major rainy season. One-half the number of animals will be carried during the first pasture-production year. A pasture area will be in full production 24 months after initial land cleanup.

STRATEGIES FOR FARM DEVELOPMENT

Three alternative development strategies for an intensive beef farm are explored. For the first strategy, which uses commercial N, it is assumed that there are no capital constraints. Enough land will be prepared and planted the first year to plant the entire pasture area the second year. One-half the total complement of cattle will be purchased the third year, and by the fourth year the farm will be in full production.

For the second and third strategies it is assumed that the farm is to be developed over a 16-year period. Each year 1.5 ha of each grass are planted (6.0 ha total). In the third strategy legumes will be sod-seeded into pastures during the year when commercial N applications would start for the second strategy. Ten percent fewer cattle are projected for the strategy using legumes.

ECONOMIC FEASIBILITY

Analytical Procedures

Annual cash flow and debt payment budgets for 30 years were prepared for each of the three development strategies. These budgets are based on the data discussed above and 1974 price levels. Slaughter animals weighing 400 kg are sold for $300/head. Expenses are categorized as either initial investment or operating expenses.[4]

Net annual cash flow (NACF), debt payback period (DPP), net present value (NPV), and annual profit after debt is paid (AP) are used to evaluate economic feasibility. NACF is defined as livestock receipts minus total expenses for a given year. DPP is defined as the number of years required to reduce the accumulated debt to zero.[5] After all pastures are developed and the accumulated debt is paid, NACF becomes AP. The concept of present value is used to place different levels of costs and returns, which occur over an extended time period, on an equal basis. Present value of investment costs (PVI), receipts minus operating costs plus farm resale (PVR), and NPV (PVR minus PVI) are calculated.[6]

Initial Analysis

An initial economic analysis used the budget data discussed above, an interest rate of 10%, and a discount rate of 12%. The farm operates for 30 years and then is sold for $50,000. Results indicate that NPV is highest when there are no capital constraints, positive with slow pasture development and legume-grass combinations, but negative with slow pasture development using commercial N. A similar ranking is obtained using the NACF and DPP criteria. However, total accumulated debt is greatest with fast pasture development. All three development strategies result in profitable farms once pasture development is completed and the accumulated debt is paid. The farm that uses legumes is most profitable.[7]

DISCUSSION

These results should be interpreted with some caution. The production data consist of estimates based on limited research data. Moreover, choice of interest rate, discount rate, farm resale value, and time period are somewhat arbitrary. To aid in interpretation and possible application of results, sensitivity analyses were performed. The overall ranking of the three development strategies was not affected. With interest rates of 9% or less, NPV values are all positive. However, it was revealed that if the opportunity cost of capital (discount rate) is 17% or greater, or if the interest rate is 14% or higher, none of the development strategies will result in wealth increases, as NPV values are negative. Impacts associated with changing operating costs or livestock receipts are much more severe than impacts associated with changing investment costs. If operating costs are decreased by 2% or livestock receipts increased by 2%, all NPV values are positive. If operating costs are increased by 9% or livestock receipts are decreased by 7%, all NPV values are negative, and the idea of beef-farm development should be abandoned. Development of intensive beef production could be encouraged by government subsidization of capital or policies that result in favorable levels of operating costs or livestock receipts.

Investment decision criteria indicate that fast farm development is much better than slow development. Even though initial investment costs and PVI are higher with fast development, the additional NACF generated by livestock receipts of the earlier years allows for a shorter DPP. Consequently, interest charges are reduced and the farm becomes profitable earlier. The initial-analysis NPV criteria indicate that an investor can increase his wealth using a fast-development strategy. However, a slow-development strategy using commercial N should not be attempted, as NPV is negative.

Investment decision criteria also indicate that using legumes as a N source is to be preferred over purchasing commercial N. The PVI for the two slow-development strategies is essentially the same. However, even with fewer cattle on the farm using legumes, the fact that there is no cash outflow for N results in a higher and in many cases an earlier positive NACF and a shorter DPP.

Numerous factors that may affect the feasibility of beef farm development are not considered here. Among these are bad weather, inflation, capital availability, terms of debt payment, management capability, cattle cycle, and taxes. The risk of favorable or unfavorable circumstances such as these should be considered along with the economic analysis before commitments to beef-farm development are made.

LITERATURE CITED

Cunha, E.R., F. Alvarez, O.R. Lárez, and W.B. Bryan. 1975. Pasture and livestock investigations in the humid tropics, Orinoco Delta—Venezuela. IV. Beef cattle and water buffalo grazing trials with native and introduced grasses. IRI Res. Inst., Inc. New York, Bul. 45.

Lárez, O.R., E.R. Velásquez, O. Parra, and W.B. Bryan. 1975. Pasture and livestock investigations in the humid tropics, Orinoco Delta—Venezuela. I. Observations on forage grasses and legumes. IRI Res. Inst., Inc. New York, Bul. 42.

Lee, W.F., M.D. Boehlje, A.G. Nelson, and W.G. Murray. 1980. Agricultural finance, 7th ed. Iowa State Univ. Press, Ames, Iowa.

Penson, J.B., Jr., and D.A. Lins. 1980. Agricultural finance: an introduction to micro and macro concepts. Prentice-Hall, Inc., Englewood Cliffs, N.J.

Velásquez, E.R., and W.B. Bryan. 1975. Pasture and livestock investigations in the humid tropics, Orinoco Delta—Venezuela. III. Grass-legume associations. IRI Res. Inst., Inc. New York, Bul. 44.

Velásquez, E.R., O.R. Lárez, and W.B. Bryan. 1975. Pasture and livestock investigations in the humid tropics, Orinoco Delta—Venezuela. II. Fertilizer trials with introduced forage grasses. IRI Res. Inst., Inc. New York, Bul. 43.

NOTES

1. Less nitrogen would be required for grazing. For the economic analysis a nitrogen level of 400 kg/ha/yr was used.

2. Data on long-term persistence of legumes in Orinoco Delta pastures are currently not available.

3. Stocking rates varied between 3.2 and 5.7 animal units/ha (1 animal unit = 400 kg live weight).

4. Budgets used are available upon request from the authors.

5. If NACF is negative, NACF is added to accumulated debt. The owner labor charge is assumed to be adequate to cover owner living expenses. Consequently, if NACF is positive, NACF is subtracted from the accumulated debt until the accumulated debt is totally paid off.

6. Further discussions of present value concepts and formulae used for calculations can be found in agricultural finance texts such as Penson and Lins (1980) or Lee et al. (1980).

7. A tabulated summary of numerical results is available upon request.

Pasture Research Project in the Brazilian Amazon

H.M. SAITO, H.W. KOSTER, E.J.A. KHAN, D.F. WINSLOW,
A.M. HOLLICK, P.M. PAOLLICHI, A.C. COSTA, and F.A. ROLIM
Instituto de Pesquisas IRI, Sao Paulo, SP, Brazil

Summary

The procedure for pasture establishment in the Brazilian Amazon has been to cut down and burn forest and then to seed guineagrass (*Panicum maximum* L.). Under normal management, animal production has dropped significantly as pastures have deteriorated and have been invaded by weeds. A dearth of knowledge of climate, soils, and plant species; the manner in which pastures have been planted and managed; and a complete absence of fertilization have led to overgrazing and complete degradation of the pastures.

This study was undertaken to identify soil problems, adapted forage species, pasture-management practices, and ways of regenerating pastures. Two regions were selected: northeast Mato Grosso and Paragominas/State of Pará, where cattle projects are concentrated.

The procedure was to install introduction garden plots, with over 100 forage species, and fertilizer and lime experiments. Plot studies were conducted to determine the principal nutrient problems. Under field conditions, experiments were installed with the best species and fertilized, and their performance was evaluated. Specific experiments aimed at the recuperation of pastures through fertilization. The investigation lasted 5 years.

Phosphorus was identified as the limiting factor in the development and normal persistence of guineagrass. Positive results were obtained with low level of phosphorus (50–75 kg/ha) alone or together with limestone for the recuperation of deteriorated guineagrass infested with weeds. Approximately 17 of the forage species were selected as the most promising. Fertilization of pastures in the Brazilian Amazon with phosphorus is essential if production and persistence are to be maintained. For renovation of guineagrass pastures, application of phosphorus and dolomitic lime is recommended. Approximatey 17 new vegetative species were identified as adaptable.

KEY WORDS: Amazon pasture, forage, tropical soils, guineagrass, *Panicum maximum* L.

INTRODUCTION

Occupation of the Amazon was based on the development of cattle raising. Initially, pastures were established utilizing the highly productive guineagrass (*Panicum maximum* L.). The first step, using the traditional slash-and-burn method, is deforestation followed by burning of the fallen timber. At the start of the rains the seed is broadcast, and cattle trample it into the soil. No fertilizer is applied.

After approximately 5 years, the pastures become degraded and infested with weeds. Faced with these conditions, SUDAM (Superintendencia do Desenvolvimento da Amazonia) contacted IRI to execute a 5-year research project, starting in 1975.

METHODS

Two ranches were selected: Fazenda Suiá Missu situated in the State of Mato Grosso and Fazenda Melhoramentos da Ligação in the State of Pará. The

H.M. Saito is General Manager and Agronomist and the other authors are Research Agronomists at the Instituto de Pesquisas IRI.

studies were on soil fertility, assessment and selection of forage grasses and legumes, and pasture management and recuperation. For soil fertility, experiments were conducted in the field and greenhouse and were accompanied by laboratory analyses. For selection of forage plants, introduction plots were established with potentially adaptable species. enriched with native material. The best introductions were submitted to fertilizer and grazing experiments.

RESULTS

Selection of Forage

From over 100 introductions, the following were selected for their potential based on adaptation to climate and soil, growth habit, apparent susceptibility to insects, and grazing preference. (1) Grasses: *A. gayanus; B. brizantha* IRI 594, 822; *B. decumbens* IRI 562; *B. humidicola* IRI 409; *C. ciliaris* cv. Beltsville; *C. gayana; C. dactylon* cv. Coastcross-1 IRI 650; *Cynodon nhenfluensis × nhenfluensis; D. decumbens* cv. Transval IRI 540; *D. pentzii; M. minutiflora; P. aquaticum* IRI 879; *P. maximum* cv. Trichoglume; *P. purpureum.* (2) Legumes: *C. mucunoides;*

Table 1. Nutrient levels of 3- to 15-year-old pasture soils, in Para and Mato Grosso. (150 samples)

Depth (cm)	Phosphorus, (ppm)				
	Low			Medium	High
	< 1	1	2-10	11-30	≥ 31
0 - 5	16	45	36	3	0
5 - 10	52	42	6	0	0
	Potassium, (ppm)				
	Low			Medium	High
	≤10	11-20	21-45	46-90	≥ 91
0 - 5	1	10	19	41	29
5 - 10	10	36	33	13	7
	Calcium & magnesium, (meq/100g)				
	Low		Medium		High
	≤ 1	1.1-2.0	2.1-5.0	5.1-10.0	≥ 10.1
0 - 5	16	30	51	3	0
5 - 10	55	30	14	0	0
	Aluminum, (meq/100g)				
	0.0-0.1		0.2-0.5	0.6-0.9	≥ 1.0
0 - 5	49		46	3	1
5 - 10	19		39	33	9
	pH				
	≤ 4.7	4.8-5.2	5.3-5.7	5.8-6.2	≥ 6.3
0 - 5	16	28	30	20	6
5 - 10	45	23	19	9	4

[1]Analyzed according to methods of Mehlich.

Table 2. Growth responses to 13 elements and lime in 8-year-old guineagrass-kudzu pastures with and without weeds, in Para and Mato Grosso.

Treatment	Forage production*	
	Melhoramentos-PA	Suia Missu
1. Check	15 [++]	88
2. Complete (P + K + Ca + Mg + S + Cu + Zn + B + Mo + Co)	100	100
3. Complete + N	106	84
4. Complete + lime	(96)[+]	118
5. Complete + Fe	(100)	86
6. Complete + Mn	60	82
7. Complete minus P	19 [++]	79
8. Complete minus K	(69)	78
9. Complete minus Mg	(100)	96
10. Complete minus S	55 [++]	85
11. Complete minus Cu	75	87
12. Complete minus Zn	69	99
13. Complete minus B	(70)	79
14. Complete minus Mo	96	90
15. Complete minus Co	118	118

*The forage production is expressed as a percentage in relation to the complete treatment.

[+] The values in brackets are visual estimates.

[++] The values are significantly different (P<.01) from the complete treatment. Wilcoxon test (Verdooren, 1963.)

C. pubescens; *D. adscendes*; *D. barbatum*; *G. wightii*; *I. hirsuta*; *L. leucocephala*; *P. phaseoloides*; *S. hamata* cv. Verano.

Soil Fertility

Analyses were made of soil from pastures established for more than 3 years, at depths of 0 to 5 and 5 to 10 cm (Table 1). The topsoil (0 to 5 cm) had higher levels of phosphorus, potassium, calcium, and magnesium, and also more favorable pH and aluminum levels. However, the levels of P were extremely low. In 61% of the samples of topsoil and in 94% of the samples from the 5-10 cm depth, the level of P was below 1 ppm. Levels of potassium, calcium, and magnesium in most of the topsoils were satisfactory.

Acidity and aluminum were tolerable in the topsoil. The acidity increased as well as the aluminum in proportion to the depth (5-10 cm).

Pasture Fertilization

Phosphorus. A representative experiment in Pará and Mato Grosso, in 8-year-old guineagrass pastures, studied the omission and addition of nutrients. The nutrients were: P, 26 kg/ha P_2O_5 as triple phosphate; K, 36 kg/ha K_2O as potash chloride; S, 30 kg/ha S as calcium sulphate (gypsum); Mg, 10 kg/ha Mg as magnesium carbonate; B,

2.3 kg/ha B as borax; Cu, 2.5 kg/ha Cu as copper sulphate; Zn, 2.3 kg/ha Zn as zinc sulphate; Mo, 400 g/ha Mo as sodium molybdate; Co, 200 g/ha Co as cobalt sulphate; Ca, 99 kg/ha Ca as triple phosphate/gypsum. Other nutrients added were: N, 68 kg/ha N as urea; Fe, 2 kg/ha Fe as iron sulphate; and Ca and Mg, 1,000 kg/ha of dolomitic limestone. The results are presented in Table 2.

At Fazenda Melhoramentos, where pastures were in worse condition than those of Suiá Missu, there was a significant response to P and S. Also, to a lesser degree, there was a slight, nonsignificant reduction in production in the absence of Zn and Cu. The addition of Mn lowered production, probably because of the toxicity of this nutrient.

In the treatments, minus P and check, no guineagrass foliage was visible. With P fertilization, kudzu (*Pueraria phaseoloides*) persisted, but there was a marked recuperation and growth of guineagrass. At Fazenda Suiá Missu, there were no differences stemming from treatments.

Phosphorus was the most important nutrient for guineagrass under the conditions studied; 60 kg/ha of P_2O_5 provided significant response and increased dry-matter production 5 to 10 times. S at a rate of 30 kg/ha was next in importance.

Among the micronutrients, Zn and Cu were the most noticeable elements, at rates of 2.3 kg/ha and 2.5 kg/ha, respectively.

Phosphorus and Lime. Under the same conditions as at

Table 3. Forage production in response to dolomitic limestone and phosphorus in 8-year-old weedy pasture of guineagrass-kudzu, at Fazenda Melhoramentos-PA.

Lime Kg/ha	Phosphorus* Kg/ha P_2O_5	Yield** %	Guineagrass %
0	0	2 a+	7
500	0	5 b	9
1000	0	7 b	6
0	75	56 c	73
500	75	46 c	66
1000	75	38 c	57
0	150	40 c	58
500	150	72 c	64
1000	150	100 d	68

*Phosphorus was applied in the proportion of 1 part of triple superphosphate to 5 parts of rock phosphate, by weight.

**The forage production is expressed as a percentage of the production from the treatment of greatest yield.

+Values followed by different letters are significantly (P<.01) different. Wilcoxon test (Verdooren, 1963.)

Fazenda Melhoramentos, also on a weedy 8-year-old guineagrass-kudzu pasture, an experiment was established to evaluate the effect of P and lime on dry-matter production. The treatment consisted of 0, 500, and 1,000 kg/ha of dolomitic limestone in all combinations with 0, 75, and 150 kg/ha of P_2O_5. The results are presented in Table 3.

When P was applied alone, there were yield increases of 20- to 28-fold, and guineagrass percentages increased from 7% to 73%. Lime topdressing alone had no effect on percentage of guineagrass in the stand. But at 150 kg/ha of P_2O_5, 1,000 kg/ha of dolomitic limestone produced the highest yield. Higher rates than 100 kg/ha of P_2O_5 required the utilization of at least 1,000 kg/ha of dolomitic lime. These results confirmed the need of P and lime for adequate guineagrass production on the Fazenda Melhoramentos.

Additional tests are needed in this and other Amazon areas to determine the extent of deficient P and lime soils and to better quantify the optimum rates for maximum forage response.

Pasture Research Results in the Brazilian Amazon

E.A.S. SERRÃO
EMBRAPA, Centro de Pesquisa Agropecuária
de Trópico Úmido, Brazil

Summary

Natural upland savanna grasslands are represented mainly by well-drained savannas, with the cerrado type predominant, and by poorly drained savannas with the campo alto–type being the most common. The main limitations are low forage-production potential and, especially, low forage quality.

About 3 million ha of rain forest has been replaced by improved pastures of guineagrass (*Panicum maximum*, 80%), jaraguagrass (*Hyparrhenia rufa*, 10%), and *Brachiaria* spp. and other grasses (10%). During the first few years after establishment, pastures are productive. However, with time, a gradual decline of productivity occurs, especially in *P. maximum* pastures. About 0.5 million ha is already in advanced stages of degradation. Limiting factors include climate and plant and soil factors, besides man's influence.

Research was conducted on 14 private ranches representing the most important improved- and native-pasture ecosystems of the Amazon region, with the objective of developing technology for (1) reclaiming sown pastures at varying degrees of degradation, (2) increasing the longevity of still-productive sown pastures in forest areas, and (3) increasing productivity of low-producing native pastures. Similar trials at all sites include the following: (1) introduction and evaluation of commercial forage species; (2) evaluation of grass-legume mixtures; (3) forage fertilization; (4) pasture reclamation, improvement, and management (grazing trials); and (5) adaptation of new forage germ plasm.

Results indicate that (1) maintenance of pasture productivity requires careful management of the soil-animal-plant system; (2) even though guineagrass has been planted on 2.5 million ha, other grasses can be more successful; (3) longevity of still-productive guineagrass pastures can be increased considerably by using appropriate grazing-management systems in combination with strategic use of phosphorus fertilization and legume introduction; (4) reclamation of guineagrass pastures in advanced stages of degradation can be achieved successfully by phosphorus fertilization and by introduction of low-demand grasses such as *Brachiaria humidicola* in combination with legumes, such as *Pueraria phaseoloides*.

KEY WORDS: Amazon region, tropical savanna, tropical forest, guineagrass, Quicuio-da-Amazonia grass, *Brachiaria humidicola*, *Pueraria phaseoloides*, oxisols, ultisols.

INTRODUCTION

The Amazon region of Brazil has great potential for cattle production in view of its natural grasslands, water systems, solar radiation, temperature, and rainfall. The Amazon area is in the humid climate group A of the Köppen classification. Except for the northern latitudes, the rainy season begins in November-December and ends in May-June. Sunshine hours range from 1,500 to 7,000 annually (35%–65% of potential radiant energy [Bastos 1972]). At present, the region's cattle population is estimated at 7 million head.

In the last 20 years, about 3 million ha of rain forest (which covers about 85% of the region) has been replaced by improved pastures: 80% guineagrass (*Panicum maximum*), 10% jaraguagrass (*Hyparrhenia rufa*), and 10% *Brachiaria* spp. and other exotic grasses. Pasture establishment includes felling of forest, burning of the vegetative biomass, and seeding of forage grasses.

For a few years after establishment, as a consequence of increased soil fertility obtained by incorporating the ash from the burned biomass into the soil (Falesi 1976, Baena 1977, Serrão et al. 1979, Falesi et al. 1980, Toledo and Serrão 1982), pastures are fairly productive. However, with time, the pastures, especially guineagrass, gradually decline. This decline is positively correlated with weed infestations, and, in many cases, an irreverisble degradation results (Hecht 1979). Currently, some 1 million ha of pasture is still productive, 1.5 million ha is in a medium stage of degradation, and 0.5 million is in advanced stages. The main causes of the decline of productivity include climatic, edaphic, and plant factors and management (Serrão and Falesi 1977).

The natural upland grasslands of the Amazon region are represented mainly by well-drained (WD) savanna ecosystems, generically called cerrados. Their main limitations are low forage production and, especially, low forage quality, due to low soil fertility and intrinsically low production potential of the herbaceous species.

PROCEDURES

PROPASTO/Amazonia, an aggressive pasture-research program, has been carried out by Brazil's agricultural research bureau, EMBRAPA (Empresa Brasileira de Pesquisa Agropecuária), since 1976 on selected private ranches—"experimental fields" (EF)—representative either of improved grassland in forest areas or of native upland grasslands. At each of 14 locations, a problem pasture area ranging from 50 to 150 ha was selected for the EF (Table 1 and Fig. 1). Nine EF are on improved pasture in forest areas, four on WD cerrado-type savanna grassland, and one on a poorly drained (PD) campo

The author is a Research Agronomist at EMBRAPA/CPATU.

The author wishes to thank his research colleagues at CPATU/Belém (Pará), UEPAE/Manaus (Amazonas), UEPAT/ Porto Velho (Rondônia), UEPAE/Rio Branco (Acre), and EMGOPA/Araguaína (Goiás) for the technical collaboration rendered, directly or indirectly, in furthering this study.

alto–type savanna (Island of Marajò, State of Pará). The existing pastures, ranging in age from 6 to 15 years, had been established in the traditional manner (clearing, burning, and grass seeding); they were in various stages of decline in productivity. Grasses were mainly in the genera *Andropogon, Trachypogon, Eragrostis, Paspalum,* and *Axonopus purpusii*. Table 1 shows the type and original vegetation of each EF and some climatic characteristics at the research sites.

Most of the improved pastures in forest areas have been established on oxisols, especially yellow latosol (YL) and red-yellow latosol (RYL) (Table 1). Large areas of improved pasture are also found on ultisols, especially distrophic red-yellow podzolics (RYP). On a rather minor scale, improved pastures also occur on concretionary lateritic soils (LAT), which are petric phases of oxisols and ultisols, and on deep acid sands (AS). In their natural state under primary forest, except for their clay content, the chemical properties of the above-mentioned soils do not differ to any large extent.

The predominant soil group in the WD cerrado-type savanna is also oxisol (especially YL and RYL). Groundwater lateritic (GWL) is the predominant oxisol in the PD campo alto–type savanna ecosystem. Both WD and PD upland savannas have very low soil fertility (organic matter, 1.0-2.5%; N, 0.05-0.1%; pH (H_2O), 4.5-5.0; Ca + Mg, 0.2-0.5 mE%; Al, 1.0-2.0 mE%; K, 10-30 ppm; P, 1-2 ppm).

RESEARCH AREAS

Similar trials were initiated at all sites along the following lines:

1. Introduction and evaluation of forages with and without phosphorus fertilization. About 20 grasses (including those most commonly used in each area) and 15 legumes were compared over a 4-year period for forage production, nutritive value, and disease, insect, and drought resistance. In general, phosphate fertilization was 50 kg P_2O_5/ha (50% each of readily soluble and less soluble phosphate).

2. Evaluation of grass-legume mixtures. Seven grasses and seven legumes were preselected on the basis of their probable success for each general area represented. Grasses and legumes were planted in mixtures of one grass to one legume with a base fertilization of 50 kg of P_2O_5/ha. The performance of each mixture was evaluated for about 4 years for compatibility between grass and legume and for forage yield and persistence. The mixtures were periodically grazed to introduce animal effects into the system.

3. Forage fertilization. In improved pastures in forest areas, fertilizer treatments consisting mainly of missing element or P level were superimposed on selected pastures representing two or three stages of productivity. Fertilizers were broadcast on native pastures or applied at planting time on cultivated pastures.

4. Pasture reclamation, improvement, and management (grazing trials). The experiment included three

Table 1. Climate, soil, and vegetation characteristics at different research sites.

Site	Köppen climatic type	Temperature Avg. (°C)	Temperature Range (°C)	Average relative humidity (%)	Average annual rainfall (mm)	Original vegetation	Soil type	Predominant pasture grass
Paragominas	Am	26	21–32	82	1950	Dense forest	$YL_{ma}{}^1$	Guineagrass
Marabá	Aw	26	22–31	87	1530	Open forest	LAT	Jaraguagrass
S.J. Araguaia	Aw	26	20–31	78	1900	Dense forest	$RYP_m{}^2$	Guineagrass
Conc. Araguaia	Aw	25	18–32	80	1650	Open forest	$GWL_a{}^3$	Guineagrass
Ponta de Pedras	Am	27	23–32	85	2200	PD savanna	$GWL_b{}^4$	Native grasses
Macapá	Am	26	23–31	84	2400	WD savanna	$YL_m{}^2$	Native grasses
Amapá	Am	27	23–31	84	3100	WD savanna	$LAT_m{}^2$	Native grasses
Itacoatiara	Am	27	23–32	82	2200	Dense forest	$YL_{ma}{}^1$	Guineagrass
Boa Vista	Aw	28	23–32	73	1700	WD savanna	$RYL_m{}^2$	Native grasses
Caracaraí	Aw	27	22–31	75	1900	Wd savanna	$YL_m{}^2$	Native grasses
Ji-Paraná	Aw	27	21–32	82	1870	Open forest	$RYP_m{}^2$	Jaraguagrass
Porto Velho	Am	25	21–32	86	2340	Dense forest	$YL_{ma}{}^1$	*B. decumbens*
Rio Branco	Am	24	20–31	86	1920	Dense forest	$RYP_m{}^2$	Guineagrass
Xambioá	Am	26	20–31	85	2030	Open forest	AS	Guineagrass

^{1}ma = very clayey texture. ^{2}m = loamy texture. ^{3}a = high phase. ^{4}b = low phase.

Aw = humid tropical with definite dry season.

Am = humid tropical with short dry season.

Fig. 1. Natural and improved grasslands of the Brazilian Amazon and pasture research locations (PROPASTO "experimental fields").

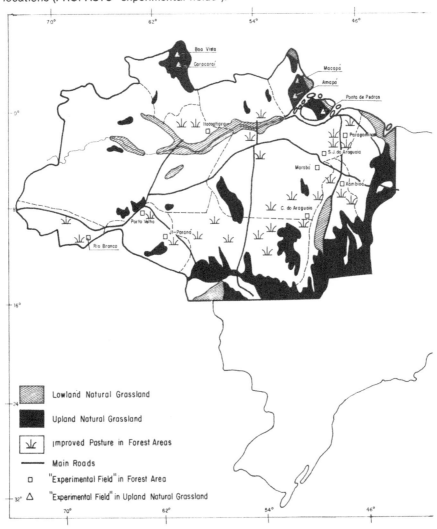

agronomic treatments, namely (1) control, improved or native local problem pasture (LPP); (2) LPP + P fertilization + legume introduction (L); and (3) LPP + P + L + introduction of Quicuio-da-Amazonia grass (*Brachiaria humidicola*). In forest areas, legumes and Quicuio-da-Amazonia grass were introduced in open spaces of the pasture where original grass was no longer present. In native pastures, the introduced legumes and grass either partially or completely replaced the native herbaceous vegetation. Each agronomic method was tested under continuous and/or rotational grazing, with stocking rates varying from low to medium to high, in a seasonal put-and-take grazing system

RESULTS

Improved Pastures in Forest Areas

In all EF in forest areas, the results were similar. The most productive and persistent grasses at most sites were Quicuio-da-Amazonia grass, sempre-verdegrass (*P. maximum* cv. Gongyloides), jaraguagrass, buffalograss (*P. maximum*), and setariagrass (*Setaria anceps* cv. Nandi). Guineagrass, the most commonly utilized grass, was not as productive and persistent as these grasses. The most productive and persistent legumes were *Pueraria phaseoloides, Centrosema pubescens, Leucaena leucocephala,* and *Stylosanthes guianensis.*

The results indicate that it is biologically feasible to maintain persistent high-quality guineagrass, sempre-verdegrass, setariagrass, jaraguagrass, and Quicuio-da-Amazonia grass pastures, each in association with either *P. phaseoloides, L. leucocephala, C. pubescens,* or *Stylosanthes* spp., in decreasing order of biological importance. Because of its aggressiveness, *P. phaseoloides* should be planted in strips in order to reduce its competitive effect on the companion grass.

Phosphorus was found to be the most limiting soil nutrient, particularly for guineagrass. It was also found that P limitation increases with increasing clay content of soil (above 40%). In general, the grasses of the genus *Brachiaria* were less demanding for phosphate than those of the genus *Panicum*. Responses to other nutrients were observed, but these were not as decisive and generalized as with phosphorus.

Grazing trials showed that productivity of guineagrass pastures at moderate stages of decline could be significantly improved and maintained by periodic (every 3 to 4 years) application of 50 to 75 kg of P_2O_5/ha, strategic weeding, and introduction of legumes, especially *P. phaseoloides* and/or *C. pubescens*. *P. phaseoloides* proved beneficial for controlling weeds. For pastures in more advanced stages of degradation, the introduction of Quicuio-da-Amazonia grass and legumes in bare spots proved to be a very efficient method of increasing productivity without interrupting pasture utilization for more than 4 to 5 months. Simple rotational grazing systems (three-pasture system, for example) enhanced the longevity of bunch-type grasses (such as guineagrass and jaraguagrass), especially under heavy stocking rates. Grazing

trials also demonstrated the importance of forage legumes in animal performance (especially during the dry season).

Preliminary results at all sites indicate that *Andropogon gayanus* CIAT 621 is a promising alternative grass for forest areas. Economic analyses indicate that establishment of new pasture involving the clearing of forest requires a higher investment than improvement or reclamation of existing pastures. Considering a pasture life expectancy of 15 years under traditional conditions (from establishment to advanced degradation), it would be necessary to clear forest and establish more pasture to offset the declining carrying capacity. Ecologically, further pasture establishment would be highly undesirable.

Improved Native Pastures

Results indicate that in native upland grassland, management, phosphorus fertilization, and grass-legume mixtures suitable for acid and low-fertility soils are very important in increasing productivity. The carrying capacity of these grasslands can be doubled or tripled with adequate management, such as grazing pressure and fire control, grazing deferment, and mineral and protein supplementation of grazing animals. Productivity can be increased 5 to 10 times by partial or total replacement of native and guineagrass pastures with less demanding grasses, such as Quicuio-da-Amazonia grass, *B. decumbens, A. gayanus,* or even jaraguagrass, and introducing legumes such as *Stylosanthes* spp., *C. pubescens,* or even *P. phaseoloides.*

In the WD cerrado-type savannas, productive pastures of Quicuio-da-Amazonia grass can be established with 50 kg of P_2O_5/ha. Results are even better when 50–75 kg N/ha are added during establishment. In PD campo alto–type savannas, excellent pastures of Quicuio-da-Amazonia grass have been established with little or no phosphorus fertilization.

At present, the most limiting factors for adopting these new technologies are high fertilizer prices and limited credit for cattle-raising activities. However, the new technologies will probably be adopted on a large scale over the next few years, to take advantage of the residual effect of fertilization of rice and other crops, as is already occurring in the savannas of Roraima.

LITERATURE CITED

Baena, A.R.C. 1977. The effect of pasture (*Panicum maximum*) on the chemical composition of the soil after clearing and burning a typical tropical highland rain forest. M.S. Thesis, Iowa State Univ.

Bastos, T.X. 1972. O estadual atual dos conhecimentos das condições climáticas da Amazônia Brasileira. *In* Bol. Tec. Inst. Pesq. Agropec. do Norte 54. Belém.

Falesi, I.C. 1976. Ecossistema de pastagens cultivadas da Amazônia Brasileira. Centro Pesq. Agropec. Trópico Úmido. Bol. Tec. 1. Belém.

Falesi, I.C., A.R.C. Baena, and S. Dutra. 1980. Consequência da experimentação. agropecuária sobra as condições físicas e químicas dos solos das microreggiões do nordeste paraense. Centro Pesq. Agropec. Trópico Úmido. Bol. de Pesq. 14. Belém.

Hecht, S. 1979. Spontaneous legumes of developed pastures of the Amazon and their forage potential. *In* P.A. Sanchez and L.E. Tergas (eds.) Pasture production in acid soils of the tropics, 65–78. Centro Int. Agric. Trop. (CIAT), Cali, Colombia.

Serrão, E.A.S., and I.C. Falesi. 1977. Pastagens do trópico úmido brasileiro. *In* IV Simpósio sobre manejo de pastagens. Esc. Sup. Luiz de Queiroz. Sao Paulo.

Serrão, E.A.S., I.C. Falesi, J.B. da Veiga, and J.F. Teixeira

Neto. 1979. Productivity of cultivated pastures on low fertility soils of the Amazon of Brazil. *In* P.A. Sanchez and L.E. Tergas (eds.) Pasture production in acid soils of the tropics, 195–225. Centro Int. Agric. Trop. (CIAT), Cali, Colombia.

Toledo, J.N., and E.A.S. Serrão. 1982. Pasture and animal production in Amazonia. *In* CIAT. Amazonia Agriculture and Land Use Research, 281–309. University of Missouri, Columbia, Missouri.

Advances in Pasture Research and Development in Santa Cruz, Bolivia

R.T. PATERSON and C.R. HORRELL
British Tropical Agricultural Mission (BTAM),
Santa Cruz, Bolivia

Summary

Cattle production is important in Santa Cruz but is low in productivity, due in part to low grazing value of natural pastures, particularly in the dry season. Other constraints to cattle production include low land values, poor management, high cost of inputs, and relatively low product prices. This program is aimed at providing higher-quality, low-cost pastures for complementary use within traditional grazing systems. Improved introduced species were identified for five agro-ecological zones, and commercial seed production of the following was encouraged: Petri guineagrass (*Panicum maximum*), *Brachiaria decumbens*, buffelgrass (*Cenchrus ciliaris*), glycine (*Neonotonia wightii*, formerly *Glycine wightii*), Archer axillaris (*Macrotyloma axillare*), lablab bean (L. *purpureus*), and stylo (*Stylosanthes guianensis*).

The value of legume-based pastures was measured and demonstrated in grazing trials, where dry-season access to legumes as mixed swards or as a protein reserve significantly improved animal production. Cattle were raised to 400 kg live weight in 30 months without supplementary feeding. These results have been used in model projections to view pasture and animal production in the whole-farm context and to illustrate the economic feasibility of pasture improvement.

Training and extension activities also were undertaken.

KEY WORDS: legumes, grasses, cattle, dry season, ranch production, seed production.

INTRODUCTION

Santa Cruz Department lies between lat 13° and 20° S and long 57° and 64° W and covers 370,000 km², one-third of all Bolivia. Population density averages only 2/km², and much land is undeveloped.

The cattle resource of 1.02 million head (CORDECRUZ 1982) derives from *Bos taurus* Criollo stock. In recent years, 70% of the stock has become crossbred with the *Bos indicus* Nellore and Gir breeds. Females 3 years old or more make up 44% of the population and are increasingly used for crossing in commercial dairy herds. If kept pure, they suffer environmental stress.

Isolated areas of natural grasslands have imperfect drainage, and other such areas have been formed by fire and grazing. The former contain grasses that are valuable for grazing when accessible, including *Paspalum* spp., *Panicum* spp., and *Leersia hexandra*. The fire-formed grasslands occur with scrub forest on sandy and shallow soils and consist of grasses of poor grazing value, such as *Sporobolus poiretti*, *Aristida* spf., *Bothriochloa saccharoides*, and *Andropogon bicornis*. Nutritional value is poor except in the young regrowth after burning (Vaca 1980), and carrying capacity is 3 to 5 ha/animal unit (AU).

Both authors are Technical Cooperation Officers of the Overseas Development Administration, London, UK. R.T. Paterson is a Pasture Agronomist and C.R. Horrell is an Agricultural Team Leader and Pasture Agronomist at BTAM with CIAT.

Many organizations have enabled this work to reach its present stage; foremost among them are CIAT, COTESU, the Universidad Boliviana "Gabriel René Moreno," and the Mennonite Central Committee. Permission to publish this paper was granted by CIAT and ODA.

Patches of induced *Paspalum notatum* pasture occur around homesteads.

The need for cultivated species has long been recognized. Jaraguagrass (*Hyparrhenia rufa*) and guineagrass (*Panicum maximum*) have been planted for more than 50 years on forest soils; they provide a carrying capacity of 1 AU/1–2 ha. Elephantgrass (*Pennisetum purpureum*) and pangola digitgrass (*Digitaria decumbens*) were planted after 1960, and buffelgrass (*Cenchrus ciliaris*) was used in the lower-rainfall areas. Later, in 1965, Petrie and Makueni guineagrasses, *Neonotonia wightii* (formerly *Glycine wightii*), and lablab bean (*Lablab*) were introduced, and by 1976 these and Coastal bermudagrass (*Cynodon dactylon*) were being planted, but only on a small scale because of lack of commercial seed supplies.

CORDECRUZ (1982) indicated that 6 million ha are used for cattle production by 20,300 farmers; 9,300 producers own less than 10 head; 7,800 own 11 to 50 head; only 313 possess more than 500 head. Cattle-farming systems are low in cost and generally low in productivity.

CIAT began pasture R&D with British technical assistance in 1976. The region has a variety of climate and soils. Only modest resources were directly available for research. The principal problem common to all areas was that existing pastures do not provide for reasonable levels of animal nutrition in the dry season. Solutions were required that lay within the context of low land values, unsophisticated management, low prices for animal products, and high costs of fencing, machinery, and fertilizers.

METHODS

Five agro-ecological zones were identified, for which different solutions to the principal pasture problem would be needed. They are:

1. Central Santa Cruz. Relatively fertile, light- to medium-textured soils of pH 6, imperfectly drained in parts and with a subzone of sandy Pampa. Altitude 460 m, rainfall 1,100 mm, formerly vegetated with tall forest. Sugar cane, mechanized annual crops, and dairying.
2. Yapacani. Less fertile alluvial soils of pH 5, rainfall 1,400 to 1,800 mm, tall rain forest. Small-scale upland rice, maize, and bananas.
3. Precambrian Shield. Soils red or dark over old metamorphic rock, rolling topography, pH 5 to 6. Unlike similar areas of Brazil, aluminum content not generally high. Altitude 500 to 1,000 m; rainfall 1,000 mm; semideciduous forest; areas of natural pastures; extensive ranching.
4. Mesothermic valleys. Fertile soils of neutral pH, altitude 1,000 to 2,000 m; rainfall 600 to 800 mm and erratic; scrub vegetation on hillsides. Small-scale maize, vegetables, pigs, cattle, and donkeys supported on the residues and hillsides.
5. Chaco-Cordillera fringe. Fertile alluvial soils, pH 7 to 8, altitude 300 to 400 m; rainfall 600 to 800 mm, irregular; deciduous and xerophylous woodland. Cattle mainly on browse species with carrying capacity 10 to 15 ha/AU.

For zones 1, 2, and 3 the aim was to find perennial forage legumes to supplement the existing natural and cultivated pastures because of the outstanding contribution they could make to dry-season production. Additional needs in zone 1 were a leguminous winter cover crop for rotation with summer-grown cotton and soybean and an annual leguminous forage for undersowing in maize in order to provide dry-season complementary grazing after harvest.

In zones 4 and 5, the objective was to introduce drought-resistant grasses to supplement crop residues, with hillside grazing in the former zone and native browse in the latter, to increase carrying capacity.

Forage conservation and supplementary grain feeding were considered to be economically inappropriate for general practice and were not investigated.

Species screening and formal trials were carried out. Simple grazing trials to measure effects on animal production, mainly in the dry season, were begun at an early stage. In zone 1, 3 trials measured live-weight gains from 10-year-old pastures of Petrie guineagrass with and without glycine. Two others compared the production of animals on rundown jaraguagrass pasture with and without access to portions of the paddock that had been ploughed, planted to leucaena (*Leucaena leucocephala*), and reserved for dry-season complementary grazing.

In zone 3, trials compared growth rates (once) and milk yields (twice) for animals in rundown jaraguagrass-*Paspalum* pastures with and without access in the dry season to reserved areas sown with an Archer-glycine mixture. In a further trial in zone 3, narrow parallel strips were ploughed in old pasture, cultivated, and sown with legumes. They were protected from grazing while the legumes became established, as were similar unimproved paddocks used for comparison. In all the zone 3 trials, the proportion of the grazing area sown to legumes was 25%.

In all trials the technique used was uniform fixed stocking and continuous grazing with 6 or more animals/treatment. The individual animals served as replicates. Results were projected in model calculations for different farm circumstances.

Training, demonstration, and extension work were carried on simultaneously, and efforts were made to ensure adequate seed supplies of recommended species.

RESULTS

Trials and observations (Paterson 1979) led to conclusions that the best species at present for cultivation, in the absence of fertilizers, are:

1. Central Santa Cruz; *Brachiaria decumbens*; *Cynodon* spp.; Petrie, Gatton, and Makueni guineagrasses; glycine cv. Tinaroo; lablab bean; leucaena; and Archer axillaris (*Macrotyloma axillare* cv. Archer).

2. Yapacani: *B. decumbens,* guineagrasses (as above), Archer axillaris, and tropical kudzu (*Pueraria phaseoloides*).
3. Precambrian Shield: (a) on red soils: *B. decumbens,* jaraguagrass (*Hyparrhenia rufa*), Archer axillaris (*M. axillare*), and stylo (*Stylosanthes guianensis*); (b) on dark soils: guineagrasses (as above), Tinaroo, and glycine.
4. Mesothermic valleys: *B. decumbens; Cynodon* sp. (African star); *Cenchrus ciliaris* cv. Cooper, Nunbank and Tarewinnabar; buffelgrasses (*Desmodium intortum*); Greenleaf desmodium; and glycine.
5. Chaco-Cordillera fringe: *C. ciliaris* (as above), buffelgrasses (as above), and *Panicum antidotale.*

In addition, *Brachiaria humidicola* and molassesgrass (*Melinis minutiflora*) were recommended for low-fertility soils and the former also for badly drained soils in zones 1, 2, and 3. Elephantgrass is useful for intensive forage production in zones 1 and 2 when required, and also for deep sandy soils with a high water table, provided that grazing is carefully managed.

Other species are still required for imperfect drainage conditions. *Hemarthria altissima* and *Digitaria* sp. are promising.

All grasses except *Brachiaria* spp. were compatible in mixture with legumes. Inoculation of legumes has not, so far, increased production. Siratro (*Macroptilium atropurpureum*), centro (*Centrosema pubescens*), and stylo were susceptible to various diseases, though stylo cv. CIAT 136 usually recovered from anthracnose attacks (Paterson and Horrell 1981).

In view of the background described, little was done with fertilizers. Rea (1978) applied nitrogen (N),

phosphorus (P), and potassium (K) to bermudagrass on a sandy Pampa site and found large dry-matter (DM) responses to N, especially in the presence of P. On the Precambrian Shield, Burgess (1979) found a large yield response to the application of sulfur (S) in glycine.

For winter legume fallow in annual cropping areas, 42 lines of 12 Mediterranean annuals were screened. Vetches (*Vicia villosa* and *V. sativa*) yielded up to 3 metric tons of dry matter with 20% crude protein (CP) when sown in late April and harvested in early October. It was found that they can be grazed up to 4 times during this period (Paterson 1979).

Lablab, to produce a protein reserve for winter grazing, was sown with maize-grain in the summer (Samur 1978). Maize-grain yield was not decreased, but manual labor for grain harvest was increased by 30%. The stover for dry-season grazing was increased from 1,637 to 4,886 (± 263) kg/ha and CP of the stover from 43 to 304 (± 29) kg/ha for maize alone and maize plus lablab, respectively. Lablab + maize stover provided 300 grazing days/ha in the driest period, July to October. Grazing trials are summarized in Table 1.

Prior to 1976 production of jaraguagrass and buffelgrass seed and planting material of guineagrass, bermudagrass, and elephantgrass were sufficient for only 600 ha/yr (Paterson 1979). Since then, in conjunction with Swiss Technical Cooperation (COTESU), seed production has increased steadily, and sufficient seed supplies are expected in 1980 to sow 2,500 ha of *B. decumbens,* 100 ha of buffelgrass, 2,000 ha of Petrie-glycine mixture, and 250 ha of Archer axillaris in association with jaraguagrass.

Commercial seed yields (kg/ha/yr), after cleaning to

Table 1. Animal performance from grass (G) versus grass-legume (GL) in dry-season grazing trials.

No.[+]	Period	Breed[‡]	Initial wt. (kg)	G	GL	Se*	Type of pasture and zone
Beef production (by day)							
1	May–Oct.	Br	188	0.16	0.40	0.031	Petrie versus Petrie +
2	June–Oct.	Br	178	0.12	0.24	0.044	glycine, zone 1
3	May–July	Br	173	0.19	0.24	0.034	
4	July	SG	240	0.23	0.70	0.052	Yaragua versus yara-
5	June–Oct.	SG	199	0.13	0.20	0.022	gua, with access to re-
	June–Oct.	ZC	175	0.27	0.35	0.033	serve leucaena, zone 1.
6	July–Aug.	ZC	145	0.19	0.32	0.034	G versus G + legume strips of stylo, Archer, and glycine, zone 3
7	July–Oct.	ZC	217	0.25	0.35	0.029	G v. G with access to Reserved legumes, Archer, and glycine, zone 3.
Milk production (by day)							
8	May	ZC	–	1.83	2.09	0.068	As above
9	Aug.	ZC	–	2.77	3.23	0.090	

[+] Trial numbers: 1, Paterson et al. 1979; 2, Paterson 1979; 3, unpublished; 4 and 5, Samur et al. 1982; 6 and 7, Paterson et al. 1982; 8 and 9, Paterson et al. 1981.

[‡] Br = Brangus, SG = Santa Gertrudis, ZC = Zebú/Criollo crossbreds.

* = Standard error.

Australian standards, are of the following order: glycine, Archer axillaris, and *B. decumbens*, 250–300; buffelgrass, 400; Petrie guineagrass, 150 kg/ha/yr.

Training inputs in pasture agronomy have included one overseas postgraduate scholarship, assistance to six undergraduate thesis students who worked in the program, and short courses for extension workers and development-project personnel.

DISCUSSION

This program has demonstrated that legume-based pastures can be grown in Bolivia without fertilizers and remain highly productive for at least 10 years.

Slaughter weights of cattle of 400 kg at 30 months of age have been achieved without conserved forage or concentrate feed when cattle have grazed these legume-grass pastures for two successive dry-season periods (Paterson 1979). Annual production of 1,800 l of milk/dairy cow/yr would be expected from similar pastures, while higher yields would require concentrate feed.

These production parameters are much above the norm in Santa Cruz, where steers commonly reach slaughter weight at 4–5 years, and dairy cows on pasture alone yield 1,000–1,200 l/yr (Burgess 1979, Wilkins et al. 1979).

Using the information obtained, projections have been made of the effect of legume-based pastures on whole-farm productivity in different farm situations.

In a ranching enterprise, for example, with 100 cows (322 animals and 227 AU), with fertility of 60%, calf mortality 5%, and selling steers at 4 years of age, the annual offtake is 36 young adults and 12 cull cows. If 18 ha of legume dry-season reserve is provided specifically for the weaned and second-year steers, these can be slaughtered at 30 months of age. This schedule permits an increase in the breeding herd to 119 cows (325 animals and 226 AU) without changing stocking rate. Annual sales will then be 46 young adults and 14 cull cows, an increase of 25%.

The use of similar illustrations in extension work is creating much interest and is reflected in the growing demand for seed.

LITERATURE CITED

(Publications of BTAM, Santa Cruz, are available at the Commonwealth Agricultural Bureau of Pastures and Field Crops, Hurley, Maidenhead, Berks., England, and at the library of CATIE, Turrialba, Costa Rica.)

Burgess, J.C. 1979. Dairy production in San Javier, Bolivia: investigation and development. (English and Spanish) BTAM, Santa Cruz.

CORDECRUZ. 1982. Cattle census, 1978. (Spanish).

Paterson, R.T. 1979. Tropical pastures in Santa Cruz: research and development. (English and Spanish) BTAM, Santa Cruz.

Paterson, R.T., J.C. Burgess, and J. Serrate. 1982. The low cost improvement of existing pastures in San Javier. (Spanish) Proc. III Natl. Sem. Agric. and Forest. Invest., Santa Cruz.

Paterson, R.T. and C.R. Horrell. 1981. Forage legumes in Santa Cruz, Bolivia. Trop. Anim. Prod. 6:44–53.

Paterson, R.T., C. Samur, and O. Bress. 1981. The effect of complementary grazing of a forage legume reserve on dry season milk production. Trop. Anim. Prod. 6(2).

Paterson, R.T., G. Sauma, and C. Samur. 1979. The growth of young bulls on grass and grass/legume pastures in subtropical Bolivia. Trop. Anim. Prod. 4:154–161.

Rea, W. 1978. Effect of fertilizers and frequency of cutting on bermuda grass. (Spanish) Ing. Agron. Thesis, Univ. Boliviana "Gabriel René Moreno."

Samur, C. 1978. Association of lablab with maize. (Spanish) Ing. Agron. Thesis, Univ. Boliviana "Gabriel René Moreno."

Samur, C., A. Vaca, and R.T. Paterson. 1982. Leucaena to improve the nutrition of animals on low quality pastures. (Spanish) Proc. III. Natl. Sem. Agric. and Forest. Invest., Santa Cruz.

Vaca, A. 1980. Potential of the low fertility soils of the Pampa Blanca for animal production. (Spanish) Ing. Agron. Thesis, Univ. Boliviana "Gabriel René Moreno."

Wilkins, J.V., G. Pereyra, A. Alí, and S. Ayala. 1979. Milk production in the tropical lowlands of Bolivia. Wild. Anim. Rev. (FAO) 32:25–32.

Development of a Pasture Research Program for the Tropical Savanna Region of Brazil

D. THOMAS, C.P. MOORE, W. COUTO, R.P. DE ANDRADE,
C.M.C. DA ROCHA, and D.T. GOMES.
Tropical Pastures Program, CIAT, Colombia, and EMBRAPA-CPAC, Brazil

Summary

The tropical savannas (cerrados) of west central Brazil contain almost 40% of the nation's cattle population. However, the low productivity and nutritive value of the native pastures is a serious limitation to cattle production. To improve the nutrition of the grazing animal, a research program has been developed with emphasis on the integrated use of cultivated pastures. The program aims (1) to select species adapted to the cerrados, (2) to determine their nutrient requirements, and (3) to develop methods of establishment and utilization. The program is being conducted at the Cerrados Agricultural Research Center (CPAC) at lat 15°S. Rainfall is 1,500 mm/yr distributed in a 5- to 6-month wet season from October to March.

A systematic three-stage pasture-evaluation scheme has been implemented. In stage I, there are 900 legume and 126 grass accessions under test on the two major soils of the region. The principal genera are *Stylosanthes* Sw., *Panicum* L., *Andropogon* L., *Brachiaria* Griseb., and *Melinis* Beauv. Observations are made on phenology, productivity, nutritive value, seed production, animal acceptance, and tolerance to pests and diseases. Promising accessions pass to stage II, in which simple grass-legume mixtures are evaluated in small, individually grazed plots for compatibility, persistence, and productivity. Currently, 19 accessions are included in stage II. In stage III, grazing trials are conducted in which year-round animal performance is monitored on the most promising mixtures at three stocking rates.

Soil fertility and plant nutrition studies are being undertaken in two stages. First, exploratory greenhouse experiments are set up with treatments arranged factorially to identify the most important soil nutrient deficiencies. Second, field experiments are designed to verify the greenhouse studies and to determine the amount and kind of fertilizer required for pasture establishment and maintenance.

A comprehensive program has been initiated to develop methods of pasture establishment. These methods include investigation of conventional planting techniques, undersowing of arable crops, and legume introduction into native and degraded cultivated pastures by oversowing and sod-seeding.

In the cerrados, 95% of beef producers are involved in cow-calf operations. Therefore, pasture utilization studies are designed to investigate the strategic use of new cultivated pastures for improving reproductive performance of cows, reducing age at first calving of heifers, and rearing early-weaned calves.

Preliminary observations are presented for the various studies in progress.

KEY WORDS: savannas, cerrados, Brazil, pastures, evaluation, fertility, establishment, utilization.

INTRODUCTION

The cerrados of west central Brazil occupy 21% of the country. They are classified as well-drained, tropical savannas. Almost 40% of the Brazilian cattle population is found in the area.

LIMITATIONS TO CATTLE PRODUCTION

The productivity and carrying capacity of the native pasture is low (Kornelius et al. 1979). During the long dry season the nutritive value of the native grasses also falls to very low levels. These limitations in feed supply have adverse effects on animal production (Table 1).

To improve the nutrition of the grazing animal, attention has turned to cultivated pastures. However, none of the existing commercial legume cultivars can be confidently recommended for the cerrados. Most of the cultivars either are not adapted to the acid, infertile soils or are susceptible to pests and diseases. The current practice is to establish pure stands of tropical grasses without the application of nitrogen. Carrying capacity can be increased initially, but after 2 or 3 years pasture productivity declines markedly because of nitrogen deficiency. Furthermore, pure grass pastures may result in weight losses in cattle in the dry season. Widely sown grasses such as signalgrass (*Brachiaria decumbens* Stapf.) are also highly susceptible to insect pests.

The Brazilian Agricultural Research Organization

D. Thomas is a Pasture Agronomist, C.P. Moore is an Animal Scientist, and W. Couto is a Soil Scientist at CIAT; R.P. de Andrade is a Pasture Agronomist, C.M.C da Rocha is an Animal Scientist, and D.T. Gomes is a Pasture Agronomist at EMBRAPA-CPAC.

(EMBRAPA) and the International Center of Tropical Agriculture (CIAT), based in Colombia, recently developed a joint pasture-research program. The work is being conducted at the Cerrados Agricultural Research Center (CPAC) near Brasilia at lat 15°S and an altitude of 1,000 m. Rainfall is 1,500 mm/yr with 90% falling from October through March.

PASTURE EVALUATION

The principal aim is to select species that will (1) grow and produce seed on acid soils under aluminum and water stress, (2) persist under grazing, and (3) tolerate pests and diseases. For this purpose, a three-stage evaluation program was implemented.

Stage I

Legume and grass accessions from germ plasm banks are evaluated against control commercial cultivars on the

Table 1. Beef cattle production parameters for the Cerrados of Brazil.

Calving rate (%)	40 to 50
Mortality to weaning (%)	10
Calf age at weaning (months)	8 to 10
Age at first calving (years)	3.5 to 4.5
Calving interval (months)	24
Daily liveweight gain from birth to slaughter (kg/head)	0.26
Age at slaughter (years)	4 to 5
Carcase weight (kg)	293
Annual turn-off rate (%)	10 to 12
Annual beef production (kg/ha)	20

SOURCE: Projeto Nacional de Bovinos, EMBRAPA (1974).

two major soil types. Known nutrient deficiencies are corrected in all stages of the evaluation program. Soil pH remains at 4.6, and aluminum saturation is 70%.

Currently, there are 900 legume accessions in stage I from thirteen genera, of which 70% are species of *Stylosanthes* Sw. Grass accessions number 126, mainly from the genera *Andropogon* L., *Brachiaria* Griseb., and *Panicum* L. Observations are made on phenology, dry-matter yield, regrowth, nutritive value, seed production, and tolerance to pests and diseases. Seed of promising germ plasm is multiplied, and these accessions pass into stage II for further evaluation.

Observations are showing that the most promising legume genus is *Stylosanthes*. Accessions of a number of species have shown good adaptation to both soils and good tolerance to the fungal disease anthracnose (*Colletotrichum gloeosporioides* Penz.), the major limiting factor to the genus in the cerrados.

Among the grass accessions, new ecotypes of gambagrass (*Andropogon gayanus* var. bisquamalatus [Hochst.] Stapf.) and guineagrass (*Panicum maximum* Jacq.) are showing promise.

Stage II

Accessions are evaluated in small, individually grazed plots. Each legume is sown separately with two grasses of contrasting growth habit—namely, gambagrass and signalgrass cv. Basilisk. Grasses are sown with stylo (*Stylosanthes guianensis* [Aubl.] Sw.). Observations are made of species compatibility, persistence, and productivity.

Currently, 14 legume accessions and 5 grass accessions are being evaluated. Species showing particular promise are accessions of *Stylosanthes capitata* Vog. and gambagrass CIAT 621.

Stage III

The best combinations from stage II are finally evaluated in grazing trials at three stocking rates (0.70, 1.40, and 2.10 animal units/ha). Paddocks are continuously grazed, but in the first season grazing is light to ensure good establishment and seeding of the species. Yearling heifers are used, and every year a new, homogeneous group of animals with a similar management history enters the trial.

Samples are taken periodically during the year for estimates of pasture availability and botanical composition. Animal live weight is measured every 28 days. Gambagrass CIAT 621 and *Stylosanthes scabra* Vog. cv. Seca are currently under evaluation.

SOIL FERTILITY AND PLANT NUTRITION STUDIES

These studies are conducted in two stages. First, in preliminary greenhouse experiments the most important soil nutrient deficiencies are identified. Treatments are arranged factorially in fractionally replicated designs and as complete blocks. Second, field experiments are conducted to verify the greenhouse studies and to determine the amounts and form of fertilizers required for pasture establishment and maintenance. This work is being conducted on the two major soil types, the dark-red latosol (DRL) and the red-yellow latosol (RYL) (Anon. 1976). Test plants now being used are those coming from the evaluation program.

The greenhouse experiments have shown that, after phosphorus, the most important deficiency is that of sulfur. Responses to calcium, magnesium, and potassium were recorded on the RYL. Molybdenum and zinc were also found to be limiting for some species. Field experiments on the RYL soil confirmed the greenhouse findings. On the DRL soil, preliminary field data have confirmed deficiencies of phosphorus and calcium.

PASTURE ESTABLISHMENT STUDIES

A comprehensive program has been undertaken to develop methods of pasture establishment. Included are investigations of (1) conventional sowing techniques, (2) undersowing of arable crops, and (3) legume introduction into native pasture and degraded signalgrass swards by oversowing and sod-seeding.

Preliminary data are available only for (3). Experiments have shown that legume establishment, following introduction into the two types of pasture, is poor when the legumes are oversown without soil disturbance. Oversowing following a light disk harrowing gave good results, particularly for *Stylosanthes capitata* in native pasture. Sod-seeding was a successful method for introducing legumes into a degraded 6-year-old signalgrass pasture.

PASTURE UTILIZATION STUDIES

In the cerrados 95% of beef producers are involved in cow-calf operations. Therefore, experiments in pasture utilization are designed to evaluate the strategic use of cultivated pastures for improving reproductive performance of Zebu (*Bos indicus* L.) cattle.

The most critical phase of the cycle for the cow is early lactation. The stress of lactation and the physiological effects of nursing the calf can significantly delay reconception in cows. Early weaning of the calf and/or provision of an adequate plane of nutrition for the cow in the breeding season can overcome this problem. However, suitable methods of calf rearing in the postweaning period have to be developed. Accordingly, the value of high-quality cultivated pastures for cows in the breeding season and for early-weaned calves is being investigated in long-term experiments. In preliminary work, strategically utilizing a cultivated pasture during a 90-day breeding season resulted in a 2-fold increase in daily live-weight gain of cows over a system of year-round grazing of native pasture. In addition, there was a 10-fold increase (from 0.2 to 2.0 animal units/ha) in carrying capacity. Results from another trial show that calves weaned at 3 months of age can be successfully reared on a gambagrass-stylo pasture without energy or protein supplementation.

Reproductive performance can be further improved by reducing age of heifers at first calving. In practice this age can be as high as 4.5 years. Experiments are being conducted with the aim of reducing this age to 3 years by increasing growth rates through use of cultivated pastures in the first year.

CONCLUSIONS

In the integrated pasture-research program for the cerrados, the pasture-evaluation studies aim to generate persistent grass-legume combinations showing good adaptation to local conditions and to provide recommendations on carrying capacity. The soil fertility studies are identifying limiting soil nutrients and will produce fertilizer recommendations for establishment and maintenance of newly selected species. Concurrently, methods are being tested for establishment of cultivated pastures and for legume introduction into native pasture and degraded pure grass stands. In the final phase, pasture utilization studies will determine the best way of integrating the new pastures into existing beef systems in the region.

LITERATURE CITED

Anon. 1976. Centro de Pesquisa Agropecuária dos Cerrados (CPAC). Relatório Técnico Anual 1976, Brasilia, Brazil, 11–15.

Kornelius, E., M.G. Saueressig, and W.J. Goedert. 1979. Pasture establishment and management in the Cerrado of Brazil. *In* P.A. Sánchez and L.E. Tergas (eds.) Pasture production in acid soils of the tropics, 147–166. Proc. CIAT Seminar, Cali, Colombia.

Diet-Quality Considerations in the Design and Management of Pastures in Seasonally Dry Tropics of Australia

R.L. MC COWN and C.J. GARDENER
CSIRO Division of Tropical Crops and Pastures,
Davies Laboratory, Townsville, Australia

Summary

Much of Australian tropical-pasture research in the last 20 years has been aimed at filling the "protein gap" of the dry season with a well-adapted legume. Several such legumes are now available, but their contribution to an improved nutritional regimen has been variable. This paper puts forward generalizations as to the contribution of pasture components to cattle nutrition and attempts to apply this information in devising regional strategies for pasture design and management.

The contribution of a forage component is a function of (1) its nutritive value and (2) its relative acceptability. Three components emerge as both highly nutritious and acceptable—i.e, young green grass, mature green legume, and legume litter (mainly leaf and seed).

Young grass dominates the diet from first storms until late in the rainy season. Drastic reduction in grass quality corresponds to peak yields of mature green legume, and diet composition changes correspondingly. Green legume remains the major diet component until soil water depletion triggers senescence and leaf shed.

During the dry-feed period, legume litter is capable of supporting live-weight gain unless rainfall causes spoilage. Sufficient rain to promote substantial grass growth can also have a net benefit. The most deleterious situation is one of small rainfalls, sufficient to cause spoilage but insufficient to support growth. Dry-season rainfall in the beef-producing regions of tropical Australia is highly variable, both geographically and temporally. Figures are presented that show (a) the variation in risk of rainfall damage to dry legume and (b) the probability of green grass availability. These provide a basis for design of pastures and management for various regions. Strategies for two contrasting environments are discussed.

Although the goal of a well-adapted, high-yielding pasture legume for the subhumid to semiarid tropics has been achieved, there remains a major task in optimizing utilization of such a plant in production systems over a wide range of environments.

KEY WORDS: tropical pastures, pasture legumes, *Stylosanthes,* beef cattle, diet selection, agro-climatology.

INTRODUCTION

The nutritional status of cattle on grasslands in the seasonally dry tropics cycles in phase with the seasonal cycle of rainfall. The nutritive value of tropical-grass leaf tissue declines drastically as the leaf ages (Wilson and 't Mannetje 1978), and live weight increases only during periods of grass growth (McCown 1981). In contrast, the quality of tropical-legume leaf declines little with age (Norman 1964), and even dry senesced leaf can support gain in cattle weight (Norman 1970). For this reason, much of the Australian tropical pasture research of the last 20 years has been aimed at filling the "protein gap" in the dry season with a well-adapted legume (McIvor et al. 1982). Several such legumes are now available, but their contribution to an improved cattle nutritional regimen has varied greatly. This paper draws on recent findings in generalizing the relative contribution to improved diet quality of the major herbage components in pastures of perennial grass and legumes of the genus *Stylosanthes*. It attempts further to apply this model in devising regional strategies for pasture design and management within the seasonally dry tropical zone.

PASTURE COMPONENTS OF DIETARY IMPORTANCE

From the onset of the growing season, cattle select a high-quality diet composed mainly of young grass, which is preferred to young legume tissue (Gardener 1980). Late in the growing season when grass quality declines during flowering, grazing preference shifts to the now-mature legume, whose nutritive value is enhanced by developing seed. Rates of live-weight gain, which fall at this time on native pastures, remain high as long as amounts of green legume leaf and seed are ample (Gardener 1980). As the soil dries out, growth ceases, and existing leaves senesce and are shed (McCown et al. 1981b). The rate of green leaf decline varies greatly among legume species (McCown and Wall, unpublished). In the absence of rain, dry leaf is readily licked up by cattle along with seed, providing a diet capable of supporting weight gains (Norman 1964, 1970). However, rainfall sufficient to mold the dry leaf litter results in a change of diet to mainly low-quality dry perennial grass, with consequent weight loss (Gardener 1980). If rainfall is sufficient, new green grass again becomes important in the diet, and cattle gain weight for a time; however, at this time legumes make a negligible contribution because of slow growth and low acceptability of new tissue (Gardener and McCown, unpublished).

It can be concluded, therefore, that in this environment three pasture components are both of high nutritive value and readily acceptable to cattle: young green grass, mature green legume, and nonmoldy legume litter (mainly leaf and seed). It follows that pastures should be

The authors are Senior Research Scientists at CSIRO.

designed and managed to optimize supply of these components.

REGIONAL VARIATION IN DRY-SEASON CLIMATE AFFECTING CATTLE NUTRITION

Dry-Season Length

A dry season is defined using a weekly pasture growth index; it corresponds to that portion of the year outside the main period of live-weight gain of cattle on native grass pastures (McCown 1981). Median values for the main beef-producing area of tropical Australia range from 10 to 28 weeks (Fig. 1).

Proportion of Dry Season Capable of Preserving Dry Legume

The deferred grazing of legume pasture until the dry season is an obvious strategy for reducing the impact of this dormant season. The duration of the contribution depends to a large degree on the time of occurrence of weather conditions that cause molding sufficient to reduce markedly acceptability of dry legume to cattle. The potential of various regions for preserving dry-legume quality can be compared (Fig. 2). This figure is based on simulated dry-season changes using models of McCown et al. (1981b) and McCown and Wall (1981) in which rainfall and potential evaporation drive the processes of leaf shed and molding. The climates at Darwin and Katherine are most favorable for this strategy, with no serious molding for the entire dry season in 6 to 7 years out of 10. In only 15% of years can the "good" period be expected to be less than half of the entire dry season. Weipa, Normanton, Wrotham Park, and Derby are progressively less favorable. At Woodstock the probability of dry legume's making a substantial contribution is less than 1 year in 10. In 5 years in 10, dry legume is valueless from the outset, having molded during leaf shed.

Pulses of Growing Conditions in Dry Season

Any beneficial effect of the dry-season rain depends on the duration of subsequent favorable growing conditions. Variation among sites is shown in Fig. 3. The variable is

Fig. 1. Map of median dry-season length (weeks).

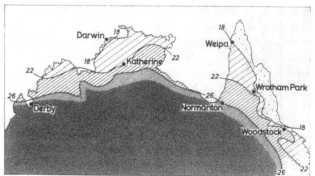

Fig. 2. Cumulative frequency distribution of the proportion of the dry season prior to serious molding of dry legume for stations identified and located in Fig. 1.

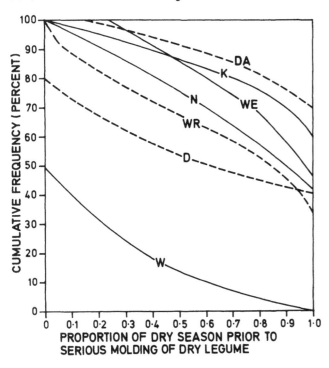

PROPORTION OF DRY SEASON PRIOR TO SERIOUS MOLDING OF DRY LEGUME

Fig. 3. Cumulative frequency distribution of the number of weeks of useful growth in the dry season for stations identified and located in Fig. 1.

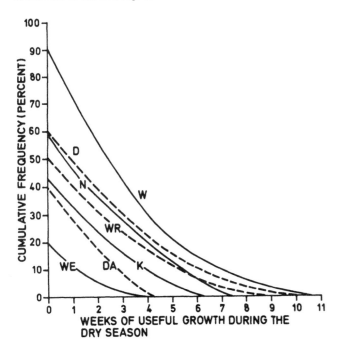

WEEKS OF USEFUL GROWTH DURING THE DRY SEASON

the number of weeks when the growth index exceeds 0.1—i.e., "green weeks" (McCown et al. 1981a). At the lower latitudes there is little expectation of rain and useful growth in the dry season. In the southeastern region, represented by Woodstock, pulses of pasture growth conditions are sufficiently common to be an important consideration in pasture design and management.

PASTURE DESIGN AND MANAGEMENT FOR DIFFERENT REGIONS

In the seasonally dry tropics of Australia, generally wherever a legume has been successfully established in native grassland and some minimal superphosphate applied, carrying capacity has increased, and the period of live-weight gain has been extended (McIvor et al. 1982). Recent research has, however, demonstrated that optimum design and management of pastures differ in major ways within this area.

Where the dry season is reliably dry (represented by Darwin, Katherine, and Weipa), senescence of deciduous legume at the end of the rainy season is rapid (McCown et al. 1981b). Because of a low molding risk, dry-season animal production is proportional to the quantity of dry legume (Norman 1970). Here grazing of pure legume pasture can be deferred until the dry season, allowing utilization of abundant native grass pastures when they are green. Although this strategy failed a decade ago because of the inability of Townsville stylo (*Stylosanthes humilis*) to compete with annual grasses, Caribbean stylo (*S. hamata*) has since proved to be much more competitive. An evergreen legume, i.e., *S. scabra* (Shrubby stylo), has not shown any nutritional superiority over the deciduous Caribbean stylo, and its substitution for Carribean stylo has some ecological drawbacks (W. Winter and J. Mott, personal communication).

In regions with more dry-season moisture (e.g., Woodstock), senescence is generally slower because of lower potential evaporation rates (McCown et al. 1981b). Legume litter is rarely consumed because it is generally too moldy or is mixed with other moldy or otherwise unacceptable residues (Fig. 2). In this environment, deferred grazing increases risk of spoilage, so legume should be consumed while it is green. A substantial perennial grass component is needed to provide a readily acceptable, although low-quality, bulk of feed in the dry season and to provide the prospect of some high-quality green growth in the event of dry-season rains. An intimate physical association of grass and legume can lead to greatly enhanced grass growth resulting from nitrogen from the legume (Gardener 1980). Here, use of evergreen legume species extends the duration of live-weight gains in the dry season beyond that of Caribbean stylo (Gardener, unpublished).

CONCLUSION

Until recently, the lack of well-adapted legumes for the seasonally dry tropics was the major barrier to legume-based agriculture in these regions. Although the search for new legumes for this climate continues, there is now an urgent need for increased understanding of how best to utilize the legumes at hand. Sound technology transfer is jeopardized by failure to give due consideration to variation in dry-season climate within this global climatic zone.

LITERATURE CITED

Gardener, C.J. 1980. Diet selection and liveweight performance of steers on *Stylosanthes hamata*-native grass pastures. Austral. J. Agric. Res. 31:379–392.

McCown, R.L. 1981. The climatic potential for beef-cattle production in tropical Australia. I. Simulating the annual cycle of liveweight change. Agric. Systems 6:303–317.

McCown, R.L., P. Gillard, L. Winks, and W.T. Williams. 1981a. The climatic potential for beef cattle production in tropical Australia. II. Liveweight change in relation to agroclimatic variables. Agric. Systems. 7:1–10.

McCown, R.L., and B.H. Wall. 1981. The influence of weather on the quality of tropical legume pasture during the dry season in northern Australia. II. Moulding of standing hay in relation to rain and dew. Austral. J. Agric. Res. 32: 589–598.

McCown, R.L., B.H. Wall, and P.G. Harrison. 1981b. The influence of weather on the quality of tropical legume pasture during the dry season in northern Australia. I. Trends in sward structure and moulding of standing hay at three locations. Austral. J. Agric. Res. 32:575–587.

McIvor, J.G., R.J. Jones, C.J. Gardener, and W.H. Winter. 1982. Development of legume-based pastures for beef production in dry tropical areas of northern Australia. Proc. XIV Int. Grassl. Cong.

Norman, M.J.T. 1964. The value of standing dry forage for winter cattle grazing at Katherine, Northern Territory, Australia. Ann. Arid Zone 3:38–43.

Norman, M.J.T. 1970. Relationships between liveweight gain of grazing beef steers and availability of Townsville lucerne. Proc. XI Int. Grassl. Cong. 829–832.

Wilson, J.R., and L. 't Mannetje. 1978. Senescence, digestibility and carbohydrate content of buffel grass and green panic leaves in swards. Austral. J. Agric. Res. 29:503–516.

Development of Legume-Based Pastures for Beef Production in Dry Tropical Areas of Northern Australia

J.G. MC IVOR, R.J. JONES, C.J. GARDENER, and W.H. WINTER

CSIRO Division of Tropical Crops and Pastures, Queensland, Australia

Summary

A program designed to select pasture legume species suitable for beef cattle grazing in the 500–1,500 mm annual rainfall zone of northern Australia is described. A detailed study of members of the genus *Stylosanthes*, involving spaced-plant studies and small-scale cutting trials at a number of sites, identified a number of persistent, high-yielding accessions.

Animal-production studies have shown that the introduction of a legume into native pasture generally results in an increase in carrying capacity, although production/head may not increase unless superphosphate is applied. The *Stylosanthes* accessions can produce moderate to high yields without superphosphate, and only small applications of superphosphate may be necessary to give large increases in animal production. The major contributions of the legumes are to prolong the period of live-weight gain and to reduce or eliminate dry-season weight losses.

This program has led to the release of cultivars in three species—*S. hamata, S. scabra,* and *S. guianensis*—and has shown that productive legume stands can be grown in dry tropical areas receiving more than 600 mm of rain/yr. Productive stands have also been grown in other countries.

KEY WORDS: legume evaluation, *Stylosanthes* spp., animal production.

INTRODUCTION

Extensive beef-cattle grazing is the major form of land use in the 500–1,500 mm annual rainfall zone of northern Australia. The native vegetation is usually a woodland with a *Eucalyptus* or *Melaleuca* tree story and an understory dominated by tall, perennial tussock grasses belonging mainly to the genera *Themeda, Heteropogon, Bothriochloa, Dichanthium, Chrysopogon, Aristida, Sehima,* and *Sorghum.* Native legumes are only minor components of the pastures. Most soils in the area are deficient in nitrogen and phosphorus and sometimes in other elements. The rainfall has a strong seasonal distribution with 75%–95% of the

The authors are Research Scientists at CSIRO Division of Tropical Crops and Pastures.

total occurring during the summer months, October to April.

This rainfall distribution and the low soil fertility levels result in patterns of pasture yield and quality that severely limit animal productivity. Pasture growth rates, which are high during the hot wet season, generally result in the production of large amounts of low-quality herbage. Grazing animals gain weight during the wet season but lose weight during the remainder of the year since the dry herbage is deficient in nitrogen and other elements and is of low digestibility. As stocking rates are low (1 animal/5–50 ha), the problem in most years is one of quality, not quantity, of feed. Numerous programs during the last 40 years have been aimed at introducing legumes into these systems to alleviate the problem.

EXPERIMENTAL STUDIES

Townsville Stylo Programs

It was shown more than 20 years ago (Norman and Arndt 1959, Shaw 1961) that the inclusion of the naturalized annual legume Townsville stylo (*Stylosanthes humilis*) could increase the animal productivity of these pastures. However, Townsville stylo is restricted to lighter soils in areas with annual rainfall above 800 mm, often fails to persist with grasses, and is severely affected by anthracnose (*Colletotrichum gloeosporioides*), and the nutritive value of the dead herbage can deteriorate over the dry season (Gillard and Fisher 1978, Edye and Grof 1981). Despite these deficiencies, the potential of Townsville stylo to increase animal production suggested that perennial members of the genus might provide useful pasture plants.

Perennial *Stylosanthes* Programs

Edye and Grof (1981) reviewed the early phases of these programs, which identified a number of persistent, high-yielding, perennial accessions adapted to areas with annual rainfall above 500 mm. Subsequently, Caribbean stylo (*S. hamata* cv. Verano), Shrubby stylo (*S. scabra* cv. Seca and Fitzroy), and stylo (*S. guianensis* cv. Graham) were released commercially.

These *Stylosanthes* accessions can produce moderate to high herbage yields on infertile soils in the absence of superphosphate application (Table 1). However, phosphorus levels in the unfertilized herbage were low, particularly in the stems.

Animal-Production Studies

Some perennial accessions have been compared with native pastures and Townsville stylo under grazing at a number of sites receiving more than 600 mm annual rainfall. In all experiments the animals grazed the plots continuously at fixed stocking rates and were weighed at intervals throughout the year. Comparisons at different stocking rates and different superphosphate levels have

Table 1. The effect of superphosphate application (400 kg/ha/an) on the yield and chemical composition of Caribbean stylo at Lansdown. Soil available P = 5 ppm.

Measurement	Unfertilized	Fertilized
Dry matter yield (kg/ha)	2710	4142
P concentration (%)		
Green leaf	0.15	0.25
Green stem	0.04	0.13
N concentration (%)		
Green leaf	3.04	3.55
Green stem	0.95	1.30

been included, and some annual live-weight gain results also are shown in Table 2. Annual live-weight gains on native pastures are generally less than 100 kg/head, and those on unfertilized legume–native pasture are similar, although the legume-based pastures support higher stocking rates. Application of superphosphate results in higher gains/head and a further increase in carrying capacity. Small superphosphate applications can result in large increases in animal production, e.g., at Katherine, 100 kg/ha at sowing and 25 kg/ha annually on a soil with available phosphorus level of 4 ppm gave a 250% increase in production/ha. In that study McLean et al. (1981) found that the addition of superphosphate markedly increased the selection of Verano in the diet, although the chemical composition of the legume was not altered (Winter, unpublished).

The major contributions of the legumes are to prolong the period of live-weight gain and to reduce or eliminate dry-season weight losses until rain reduces the quality of the dry herbage. Fig. 1 shows the changes in live weight throughout the year of steers grazing native pasture and Caribbean stylo–native pasture at Lansdown, both stocked at 0.6 animals/ha and not fertilized. Rates of live-weight gain were similar during the wet season, but animals grazing the legume-based pasture continued to gain weight for a longer period and maintained weight when those grazing native pasture had begun to lose weight.

Associated Studies

Evaluation of other legume genera has shown that *Macroptilium atropurpureum*, *Centrosema* spp., and *Alysicarpus* spp. can all give high annual production in some circumstances but generally require higher levels of soil fertility than the *Stylosanthes* species. *Leucaena leucocephala* grows well on the better soils and can provide valuable dry-season feed (Jones 1979).

For maximum animal performance it is essential that adequate levels of both legume and grass be retained in the pasture (Gardener 1980). However, legume-based pastures can support high stocking rates that sometimes lead to the elimination of the associated perennial native grasses. A number of introduced grasses have been grown at a range of sites to select species to grow with the

Table 2. Animal production from various pasture types in northern Australia.

Site	Pasture Type	Stocking rate beasts/ ha	Super phosphate Applica- tion kg/ha	Animal Production per head kg/an	per hectare kg/an	Reference
Swan's Lagoon (AAR = 800mm)	Native pasture	0.25	Nil	60	15	Winks *et al.* (1974)
	Townsville stylo-native	0.41	Nil	88	36	Winks (1973)
	Townsville stylo-native	0.41	125*/125+	143	59	
Kangaroo Hills (AAR = 650mm)	Townsville stylo-sabi grass	0.62	250/125	137	85	Gillard *et al.* (1980)
	Caribbean stylo-sabi grass	0.62	250/125	165	102	
Lansdown (AAR = 850mm)	Native pasture	0.6	Nil	98	59	Gardener (1980)
	Caribbean stylo-native	0.6	Nil	122	73	
	Caribbean stylo-native	1.2	Nil	93	112	
	Caribbean stylo-native	0.6	300/300	134	80	
	Caribbean stylo-native	1.2	300/300	112	134	
	Caribbean stylo-sabi grass	1.0	100/100	131	131	McIvor (unpublished data)
Katherine (AAR = 900mm)	Native pasture	0.06	Nil	75	5	Winter (unpublished data
	Caribbean stylo-native	0.43	Nil	55	24	
	Caribbean stylo-native	0.83	100/25	100	83	
	Caribbean stylo- Alysicarpus-sabi grass	2.0	200/50	130	260	

(*) Initial and (+) annual applications.

Fig. 1. Monthly rainfall and seasonal live-weight changes in cattle grazing native pasture and Caribbean stylo-native pasture (both unfertilized) at Lansdown near Townsville.

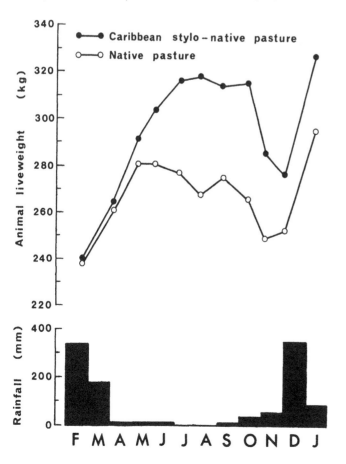

legumes (McIvor 1981), and research is continuing to enable stable productive grazing systems to be developed (McCown and Gardener 1982).

DISCUSSION

The research programs outlined above have shown that productive legume stands can be grown in dry tropical areas receiving more than 600 mm of rain/yr. In the lower-rainfall areas the perennial species are more productive than Townsville stylo. This greater productivity is reflected in superior animal performance and increased carrying capacity on fertilized pastures. Caribbean stylo and Shrubby stylo both have the capacity to produce high yields without superphosphate application, but nutrient levels in the herbage may be below those necessary for high levels of animal production. Research is currently in progress to determine whether animal productivity can be increased by mineral supplementation of the herbage.

Caribbean stylo in particular has been grown experimentally at many sites in northern Australia, and commercial pastures have also been established. In addition, it has been grown in a number of other countries (e.g., India, Thailand, Nigeria, Botswana) where its performance offers a potential for increased animal production.

LITERATURE CITED

Edye, L.A., and B. Grof. 1981. Selecting cultivars from naturally occurring genotypes: evaluating *Stylosanthes* species. *In* J.G. McIvor and R.A. Bray (eds.) Genetic resources of forage plants. CSIRO, Melbourne.

Gardener, C.J. 1980. Diet selection and liveweight performance of steers on *Stylosanthes hamata*-native grass pasture. Austral. J. Agric. Res. 31:379–392.

Gillard, P., and M.J. Fisher. 1978. The ecology of Townsville stylo-based pastures in northern Australia. *In* J.R. Wilson (ed.) Plant relations in pastures. CSIRO, Melbourne.

Gillard, P., L.A. Edye, and R.L. Hall. 1980. Comparison of *Stylosanthes humilis* and *S. hamata* and *S. subsericea* in the Queensland dry tropics: effects on pasture composition and cattle liveweight gain. Austral. J. Agric. Res. 31:205–220.

Jones, R.J. 1979. The value of *Leucaena leucocephala* as a feed for ruminants in the tropics. World Anim. Res. 31:13–23.

McCown, R.L., and C.J. Gardener. 1982. Diet-quality considerations in the design and management of semiarid tropical pastures. Proc. XIV Int. Grassl. Cong.

McIvor, J.G. 1981. Evaluation and ecology of sown grasses in grazed pastures. CSIRO, Div. Trop. Crops Past. Ann. Rpt. 1979/80.

McLean, R.W., W.H. Winter, J.J. Mott, and D.A. Little. 1981. The influence of superphosphate on the legume content of the diet selected by cattle grazing *Stylosanthes*-native grass pastures. J. Agric. Sci., Camb. 96:247–249.

Norman, M.J.T., and W. Arndt. 1959. Performance of beef cattle on native and sown pastures at Katherine, N.T. CSIRO, Div. Land Res. Reg. Survey. Tech. Paper. 4.

Shaw, N.H. 1961. Increased beef production from Townsville lucerne (*Stylosanthes sundaica* Taub.) in the spear grass pastures on central coastal Queensland. Austral. J. Exp. Agric. Anim. Husb. 1:73–80.

Winks, L. 1973. Townsville stylo research at Swan's Lagoon. Trop. Grassl. 7:201–208.

Winks, L., F.C. Lamberth, K.W. Moir, and Patricia M. Pepper. 1974. Effect of stocking rate and fertilizer on the performance of steers grazing Townsville stylo-based pasture in north Queensland. Austral. J. Exp. Agric. Anim. Husb. 14:146–154.

Growing Beef Cattle on Native Pasture Oversown with the Legume Fine-Stem Stylo in Subtropical Australia

K.G. RICKERT, G.M. MC KEON, and J.H. PRINSEN
Queensland Department of Primary Industries, Australia

Summary

Fine-stem stylo (FSS), *Stylosanthes guianensis* var. Intermedia, is a perennial legume suitable for sowing into native *Heteropogon contortus* pastures on well-drained soils in southeastern Queensland. A grazing system is described that aims to maintain a productive FSS pasture and grow steers to 480 kg live weight at 27 months of age. This goal requires gains of 300 kg/steer in 20 months after weaning. Results from a short-term field study of the system were analyzed for long-term repeatability using a simple model that predicted animal production from stocking rate and weather records.

The system was studied near Gayndah, Queensland, on 11.3 ha of FSS and *H. contortus* pasture invaded by *Rhynchelytrum repens*. Each June after weaning, seven steers (five-eights Hereford, three-eights Sahiwal) commenced a 20-month grazing cycle. Thus, two animal classes grazed together from June to January. From February to May, after the older animals were slaughtered, the stocking rate was halved to allow FSS to grow before winter. The mean annual stocking rate was 1.0 steer/ha.

Since 1976, three grazing cycles have been completed with total live-weight changes of 276, 326, and 315 kg/steer to give carcasses suited to local trade. FSS contributed 10% of pasture yield, but weeds progressively increased.

The animal production model predicted live-weight change (LWC) in winter, spring and summer, and autumn from pasture-growth index/head, the mean-growth index (GIX) for a season divided by stocking rate (SR). GIX was the product of separate indices (range 0 to 1) for temperature, solar radiation, and soil water supply. The regressions predicting LWC from GIX/SR were derived from results of the field study and two published studies; they accounted for 85% to 89% of variation in LWC.

Complete weather records (1958–1980) were processed to estimate GIX from seasonal rainfall and GIX in the previous season. This enabled LWC to be simulated from monthly rainfall from 1870 to 1980. Three mean annual stocking rates (0.74, 1.0, and 1.3 steers/ha) were considered over 109 grazing cycles. Frequency distributions of total live-weight gain showed that SR affected repeatability.

The model estimated long-term repeatability from a restricted data base but ignored possible influences of extreme feed shortages and changes in soil fertility and botanical composition of LWC. It supported the field study by indicating that a reduction in SR would improve repeatability of production and reduce weeds; it provided information that would otherwise have to come from a long-term grazing trial; it suggested that pasture stability in drought should be studied; and it indicated that the grazing system is suitable for commercial application.

KEY WORDS: fine-stem stylo, *Stylosanthes guianensis*, beef cattle, native pasture, simulation, growth index.

INTRODUCTION

Fine-stem stylo (FSS), *Stylosanthes guianensis* var. Intermedia, is a perennial, semiprostrate legume from Paraguay that can be sown into native speargrass pasture (*Heteropogon contortus*) on well-drained soils in southeastern Queensland (Stonard and Bisset 1970). It increases animal production from native pasture and tolerates frosts, fire, and heavy grazing (Bowen and Rickert 1979).

This paper describes a practical grazing system that aims to maintain a productive FSS pasture and grow steers to 480 kg live weight at 27 months of age. Steers grazing unimproved native pasture at 0.3 steers/ha reach this weight when 4 years old (Rickert and Winter 1980). Results from a field study are supported by an analysis of weather records to indicate long-term repeatability of animal production.

METHODS

Field Study

The grazing system was demonstrated near Gayndah (lat 25°38′ S long 151°41′ E, annual rainfall 710 mm) with 11.3 ha of FSS-speargrass pasture invaded by natal grass (*Rhynchelytrum repens*) on deep granitic sand. Superphosphate (9% P, 20% Ca, 10% S) was applied annually at 125 kg/ha from 1971 to 1974 inclusive (Bowen and Rickert 1979).

Each June from 1976, seven weaner steers (five-eighths Hereford, three-eighths Sahiwal) commenced a 20-month grazing cycle. Hence, two classes of cattle grazed together from June to January at 1.24 steers/ha. From February to May, after the older steers were slaughtered, the stocking rate was halved; thus, the mean annual stocking rate was 1.0 steer/ha. Cattle were weighed three times weekly, carcasses described at slaughter, and pasture yield and composition measured in May. Three grazing cycles have been completed (Table 1).

Computer Model

Seasonal live-weight change (LWC) was estimated from growth index/head (GIX/SR), a predictor that integrates the effects of climate and stocking rate (SR) on pasture growth (McKeon et al. 1980). The seasonal growth index (GIX) is the average product of separate daily indices for temperature, solar radiation, and soil water supply, all ranging from 0 to 1.0 (Fitzpatrick and Nix 1970). These indices for FSS were derived as follows: temperature index (TIX) from a growth study by 't Mannetje and Pritchard (1974) (TIX = 0.054 TM − 0.6, where TM is mean daily temperature); radiation index from Fitzpatrick and Nix (1970); and water supply index (the ratio of predicted transpiration to potential transpiration) from a water-balance model (Rickert and McKeon, unpublished).

The authors are Agrostologists in the Queensland Department of Primary Industries.

Table 1. Observed (O) and predicted (P) seasonal live-weight change (kg/steer) in three grazing cycles on fine-stem stylo pasture.

Commenced	1976 O	1976 P	1977 O	1977 P	1978 O	1978 P
Live-weight at start	188	180	172	180	178	180
1st winter, June–Aug.	7	6	6	−1	28	27
1st spring + summer, Sept.–Feb.	146	138	125	113	157	138
1st autumn, Mar.–May	58	58	70	57	57	58
2nd winter, June–Aug.	5	−16	6	12	−19	−8
2nd spring + summer, Sept–Jan.	60	60	119	84	92	77
Final live-weight	464	426	498	445	493	472
Total gain	276	246	326	265	315	292

Relationships were established using LWC from crossbred steers (Table 1) and from Hereford steers in an earlier study (1971 to 1976) at variable stocking rates (Bowen and Rickert 1979). Performance from Herefords was adjusted to a crossbred equivalent after Robbins and Esdale (unpublished): LWC(XB) = 0.864 LWC(H) + 168, g/head/day. Regressions for LWC (g/head/day) during the first year of grazing were established by least square methods:

Autumn: $\text{LWC} = 643(\pm 33) -$
$$1202(\pm 301)e^{-20.33(\pm 6.81)\text{GIX/SR}}$$
$$R^2 = 0.89 \text{ and } n = 14 \qquad (1)$$

Winter: $\text{LWC} = 534(\pm 226) -$
$$598(\pm 173)e^{-12.00(\pm 9.96)\text{GIX/SR}}$$
$$R^2 = 0.85 \text{ and } n = 10 \qquad (2)$$

Spring and summer: $\text{LWC} = 887(\pm 39) -$
$$1053(\pm 221)e^{-11.57(\pm 2.80)\text{GIX/SR}}$$
$$R^2 = 0.86 \text{ and } n = 15 \qquad (3)$$

where numbers in parentheses are the standard errors for the regression coefficients, e is the natural logarithm, n is the number of observations, and R^2 is the total variation accounted for. Spring and summer were combined since analysis showed that LWC in summer alone was not independent of LWC in spring. Prediction from equations 1 to 3 were adjusted for grazing in the second year because 2-year-old steers (LWC2) gained less than 1-year-old steers (LWC1):

$$\text{LWC2} = 0.981 \text{ LWC1} - 160 \text{ g/head/day} \qquad (4)$$

Repeatability of the grazing system was analyzed by predicting live-weight gain from long-term weather records. Complete daily weather records were available from 1958, but only monthly rainfall from 1870. To use this long-term data, multiple regressions were established on the daily weather data to estimate GIX from the seasonal rainfall and growth index of the preceding season. These regressions estimated GIX from 1870 to 1980 for use in equations 1 to 3. Predictions for 109 cycles of 20 months, adjusted for performance in the second year with equation 4, were used to establish the frequency distribution of live-weight gain in a cycle. Three mean annual stocking rates were considered: 1.0, 0.74, and 1.3 steers/ha, 1.0 being used in the field demonstration.

RESULTS

Field Study

Steers exceeded the target live-weight of 480 kg in cycles 2 and 3 after grazing FSS pasture for 20 months (Table 1). These animals dressed out at 52% on the basis of cold carcass weight, with a carcass length of 110 cm, eye muscle area of 65.4 cm², and 9.7 mm of fat over the fifth rib. However, LWC varied between and within seasons.

Yields of standing dry matter in May ranged from 1,500 to 2,100 kg/ha. These low values were due to low rainfall during the growing season and to the high stocking rate used. FSS contributed 10% of the standing dry matter. The population of a perennial weed *Sida cordifolia* increased during the three grazing cycles.

Model Predictions

Predicted LWC reflected observed values, but there was a tendency to underestimate LWC (Table 1). Over all seasons the average difference between observed and predicted LWC was 8 kg (SE = ± 3.1, n = 15); final live weights were underestimated as a result. Because the model is the best fit through all (1971 to 1980) animal-performance data from the pasture, the underestimation of LWC evident in Table 1 is unlikely to be a real feature of the model. If it is, the long-term predictions are conservative.

Long-term predictions are summarized in Fig. 1 as frequency distributions of live-weight gain in a 20-months grazing cycle. A gain of 300 kg/steer is needed to reach the target weight after weaning at 180 kg. The system is sensitive to stocking rate, as the probability of a 300-kg gain is 84%, 35%, and 7% for mean annual stocking rates of 0.74, 1.0, and 1.3 steers/ha, respectively.

DISCUSSION

The grazing system described evolved from results given by Bowen and Rickert (1979): relatively heavy grazing in spring to maintain a high population of FSS and relatively light grazing in autumn to allow FSS to set seed and to maximize LWC in autumn and winter. When demonstrated in the field, the system produced beef carcasses of age and condition acceptable to local trade. The legume component of the pasture was stable during the demonstration, but weeds increased. The system would be attractive commercially provided the results are reliable, a condition that implies long-term repeatability and botanical stability.

Testing the repeatability of the system highlighted several problems. Detailed information was not available on carryover feed from one season to the next or on temporal trends in soil fertility and botanical composition. All these factors could influence animal performance in the long term. Also, weather data were restricted to monthly rainfall. Because of these constraints our analysis expressed only the effects of variation in seasonal rainfall and stocking rate on LWC. More observations of the grazing system in summer drought would provide a better test of pasture stability and predictions from the model.

In spite of these limitations the model was useful because it estimated repeatability of a grazing system when rainfall in 109 grazing cycles ranged from 633 to 2,195 mm/cycle, compared with 1,056 to 1,249 mm in the three observed cycles. It supported the field study by indicating that a reduction in mean annual stocking rate to 0.74 steers/ha should give gains of 300 kg/steer in 84% of cycles. Production/ha would then be about 4 times greater than on unimproved native pasture (Rickert and Winter 1980), and the reduction in grazing pressure is likely to reduce the spread of weeds in the pasture. Thus, the simple analysis indicated a management change in the grazing system and increased confidence in recommending the system for commercial application.

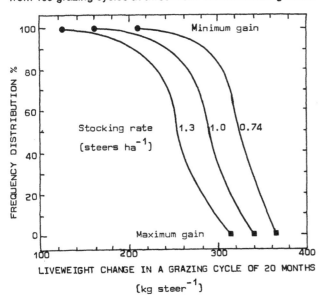

Fig. 1. Frequency distribution of predicted live-weight change from 109 grazing cycles at three mean annual stocking rates.

LITERATURE CITED

Bowen, E.J., and K.G. Rickert. 1979. Beef production from native pastures sown to fine-stem stylo in the Burnett region of south eastern Queensland. Austral. J. Expt. Agric. Anim. Husb. 19:140–149.

Fitzpatrick, E.A., and N.A. Nix. 1970. The climatic factor in Australian grassland ecology. *In* R.M. Moore (ed.)

Australian grasslands, 3-26. A.N.U. Press, Canberra.

McKeon, G.M., K.G. Rickert, G.B. Robbins, W.J. Scattini, and D.A. Ivory. 1980. Prediction of animal performance from simple environmental variables. Proc. 4th Bien. Conf. Simul. Soc. Austral., Brisbane, 9-16.

Rickert, K.G., and W.H. Winter. 1980. Integration of feed sources in property management: extensive systems. Trop. Grassl. 14:239-245.

Stonard, P., and W.J. Bisset. 1970. Fine-stem stylo: a perennial legume for the improvement of subtropical pasture in Queensland. Proc. XI Int. Grassl. Cong., 153-158.

't Mannetje, L., and A.J. Pritchard. 1974. The effect of day length and temperature on introduced legumes and grasses for the tropics and subtropics of coastal Australia. I. Dry matter production, tillering and leaf area. Austral. J. Expt. Agric. Anim. Husb. 14:173-181.

Value of Lucerne as a Grazing Supplement for Sheep Grazing Natural Pastures in a Semiarid Area of Australia

D.L. MICHALK, G.E. ROBARDS, E.A. ARMSTRONG, and J. LACY
Department of Agriculture, N.S.W. Australia

Summary

In the central areas of New South Wales, the natural pastures are dominated by annual species that grow during the cooler months, producing good-quality grazing that is particularly abundant during spring. However, these species mature rapidly in late spring and provide poor-quality summer grazing. The 4-year study reported here was designed to examine the value of sown lucerne, equivalent to one-third of the total area, as a supplement to be grazed mainly during late spring and summer. The availability of lucerne for spring grazing by ewes and lambs was of benefit only in one year when maturing grass seeds were abundant, and it adversely affected the growth rate of lambs on the natural pastures. As supplementary grazing did not affect production/ewe, the carrying capacity and productivity of an area in this semiarid environment can be calculated on the basis of the proportion of lucerne available if the carrying capacity and productivity are known for the two pasture types when grazed separately. For production/head to be increased it was concluded that a more intensive method of utilization of the supplementary grazing area is required but that such a system may not have practical relevance.

Key Words: lucerne, supplement, sheep, grazing, natural pastures, semiarid.

INTRODUCTION

In the marginal cereal-cropping area of New South Wales, domestic livestock are continuously grazed, predominantly on natural pastures. At Trangie in the central west, rainfall averages 480 mm/yr with a standard deviation of 180 mm, fairly evenly distributed throughout the year, with a peak of 49.0 mm (± 60.8) in February and a trough in September of 30.9 mm (± 25.1). However, evaporation is high during summer (302 mm in January), and only in the 6 months from May to October is rainfall likely to be sufficient for continuous plant growth. The natural pastures are dominated by annual species that germinate in autumn and provide livestock with high-quality feed through winter and early spring. However, following a peak of growth in early spring (September), the pastures mature rapidly, and their residues become the major component of summer grazing.

The study reported here was designed to test further the value of sown lucerne (Campbell et al. 1973), in this case as a grazing supplement for breeding ewes. It was assumed that the supplement would sustain high growth rates of lambs in spring because of the high quality of available forage and also because the lucerne areas would be relatively free of barleygrass, the maturing seeds of which can adversely affect lamb growth rates during spring (Campbell et al. 1972).

MATERIALS AND METHODS

Treatments

The 4-year (1971-1974) study consisted of six duplicated grazing treatments. Two treatments were sown lucerne stocked at 7.4 and 9.9 merino ewes/ha. The other four treatments consisted of natural annual pasture (4.86 ha) and sown lucerne (2.43 ha) combined into a pasture-management system as described by Campbell et al.

D.L. Michalk is a Research Agronomist, G.E. Robards is the Director of Animal Production Research, E.A. Armstrong is a Research Agronomist, and J. Lacy is a District Agronomist in the New South Wales Department of Agriculture.

(1973) and stocked at 4.1, 4.9, 5.8, and 6.6 ewes/ha. These stocking rates were derived as combinations of one-third of the rates (7.4 and 9.9) imposed on lucerne grazed alone and two-thirds of the rates (3.7 and 4.9) concurrently imposed on natural pasture grazed alone (Robards et al. 1978). These rates made it possible to compare the results obtained with the management treatments with values estimated from the treatments in which lucerne and natural pasture were grazed separately.

Pastures

The natural pasture areas were dominated by annual species, particularly barleygrass (*Hordeum leporinum*), medics (*Medicago* spp.) crowfoots (*Erodium* spp.), and silvergrass (*Vulpia myuros*), which germinate in autumn and mature in spring (Campbell and Beale 1973, Robards et al. 1978).

In December 1979 the sown pasture consisted of lucerne (mean density 32 plants/m²) with some broadleaf species and naturalized medics. The lucerne (var. Hunter River) had been sown in August 1969, at 3.4 kg/ha under a cover crop of oats sown at 22 kg/ha with an application of superphosphate at 56 kg/ha.

Sheep

The 396 medium-wool Merino ewes required for the experiment were randomly allocated to subgroups of 6 by stratified randomization on live weights recorded in November 1970. Experimental flocks were formed by random combination of necessary number of 6-member subgroups. Very dry conditions in autumn and winter 1971 made it necessary to remove all the ewes to other paddocks for the 5 weeks of lambing. In 1972 dry conditions again prevailed in the cooler months, and although the ewes remained in the experimental paddocks they were fed twice weekly for the 4 weeks prior to lambing at the rate of 400 g (5 MJ metabolizable energy) oats grain/ewe/day.

Management

Within the management-treatment groups a small number of sheep remained on the natural pasture or lucerne area throughout the year, while the majority of each experimental flock moved between the two pasture areas according to dry-matter availability and expected animal requirements.

During 1971, ewe live weights in all treatments were approximately 20% lower than in any of the other 3 years. The largest seasonal decline occurred in the winters of 1971 and 1972, when ewe live weights declined below 35 kg, and alternate grazing and supplementary grain feeding, respectively, were necessary to prevent pregnancy toxemia. For lucerne grazed alone, the lowest live weights were recorded in winter, with an additional reduction at the highest stocking rate. The integration of lucerne and natural pasture had a beneficial effect on live weight, the summer, autumn, and spring live weights ex-

ceeding those of ewes on natural pasture alone, although the magnitude of this effect was reduced with increasing stocking rate.

The number of lambs born/ewe mated and lamb birth weights were unaffected by the imposed treatments. However, there were more (P < 0.05) lambs born/ewe mated in 1974 than in 1971 (1.05 vs. 0.66), with 1972 and 1973 being intermediate. Lamb birth weights followed a similar pattern, with the lowest mean weight (3.6 kg/head) occurring in 1971. There were no significant differences in lamb birth weight among treatments.

Lamb growth rate was not affected by treatment when it was averaged over the 4 years; however, there were large differences within years.

During winter most of the sheep grazed the natural pasture portion of each treatment, whereas from the end of lambing (September) until late summer (March) most of the sheep grazed the lucerne areas.

Measurements

All ewes were weighed, and one-third of each experimental group were dyebanded at five weekly intervals to determine seasonal wool production. Total greasy fleeceweight was recorded at yearly intervals, and midside samples of the fleece were cleaned to determine yield and allow calculation of clean-wool production. Lambing records were maintained and lamb growth rates monitored by weighing within a day of birth and at regular intervals until weaning at 18 to 20 weeks of age.

RESULTS

Annual clean-wool production/ewe varied little among treatments (Table 1), and wool production/ha was in almost direct proportion to the stocking rates imposed. Spring proved to be the most reliable period for wool growth, followed by summer, winter, and autumn (Table 2).

Ewe live-weight differences associated with pasture conditions were confounded by pregnancy and lactation; however, some large differences among years were recognizable.

DISCUSSION

In a semiarid environment in eastern Australia, the provision of supplementary lucerne for late-spring–early-summer grazing did not substantially improve the per-head production of Merino ewes or their lambs. Lamb growth rates were generally only 5% higher on lucerne than on unimproved natural pastures during the spring. However, in one of the 4 years of the study, the natural pastures were dominated by barleygrass, the mature seed of which can adversely affect lambs (Campbell et al. 1972, Warr and Thompson 1976), and in this year the lambs grazing lucerne grew 20% faster than those grazing natural pastures.

No differences were noted in between-pasture treatments. Wool production/area was greater on lucerne

Table 1. Ewe clean-fleece weight (kg), lamb birth weight (kg), lambs born/ewe mated, and lamb growth rate (g/day) on natural pasture (2 units) plus lucerne (1 unit), on lucerne grazed alone, and on natural pastures grazed alone, at several stocking rates.*

Character		Natural pasture + lucerne at stocking rate of—				Lucerne alone, at stocking rate of—		Natural pasture alone†, at stocking rate of—		
		4.1	4.9	5.8	6.6	7.4	9.9	2.5	3.7	4.9
Ewe fleece weight	M	3.35	3.29	3.13	3.14	3.32	3.19	3.38	3.13	3.23
	E	3.36	3.31	3.27	3.22					
Lamb birth weight	M	3.87	3.96	3.92	3.79	3.86	3.97	3.84	3.78	3.67
	E	3.85	3.81	3.84	3.77					
Lambs born/ewe mated	M	0.82	0.75	1.03	0.92	0.80	0.90	0.92	0.91	0.89
	E	0.88	0.87	0.91	0.89					
Lamb growth rate	M	183	180	167	173	200	167	170	169	170
	E	180	179	168	169					

*Ewes/ha.

†Data from a concurrent experiment (Robards et al. 1978).

M = Measured values as recorded during the experiment.

E = Estimated values as derived by combining data from lucerne grazed alone and from natural pasture grazed alone, in the ratio 1:2 and dividing by 3.

(stocked at 7.4 and 9.9/ha) than on supplemented natural pasture (4.1 to 6.8/ha) or natural pasture alone (2.5 to 4.9/ha). Wool growth rate followed a similar pattern for ewes grazing each pasture type, with a maximum rate of production in spring due to abundant high-quality feed on both lucerne and natural pasture areas. The occurrence of minimum wool growth rates in autumn reflects a combination of the poor quality and low availability of pasture at the end of summer and the late germination of cool-season annuals that often occurs because of the high variability of rainfall.

Live weights were generally higher in spring and summer when the ewes were grazed on lucerne because of its response to a rainfall. There were few components of the natural pastures that could readily provide green feed during the warmer months. Alternatively, ewe live weights during the colder months were higher on natural pasture than on lucerne because the dominant natural annual species were growing, while lucerne growth was inhibited at low temperatures.

In the present study using a four-paddock rotation there was a marked deterioration in lucerne density at the highest stocking rate. This deterioration was reflected in animal production, with lamb growth considerably lower in the third and fourth years at 9.9/ha compared with 7.4/ha. Thus, although lucerne may be maintained in higher-rainfall areas with few paddocks in the rotation (McKinney 1974) to support up to 10 sheep/ha, in a semi-

Table 2. Ewe wool growth rate (WGR-g clean wool/day) and live weight (ELW-kg) for each season (mean of 4 years) on natural pasture (2 units) plus lucerne (1 unit), on lucerne grazed alone, and on natural pasture grazed alone, at several stocking rates.*

Season character	Natural pasture plus lucerne, at stocking rate of—				Lucerne alone, at stocking rate of—		LSD** 5%	Natural pasture alone†, at stocking rate of—		
	4.1	4.9	5.8	6.6	7.4	9.9		2.5	3.7	4.9
Summer										
WGR	10.4	10.1	10.2	9.9	10.3	10.3	1.2	10.2	10.1	10.5
ELW	40.9	40.2	39.8	38.9	41.1	39.9	4.2	39.7	39.6	37.3
Autumn										
WGR	8.3	8.5	7.7	7.7	7.8	7.1	1.2	8.9	8.4	8.4
ELW	42.0	41.6	40.4	39.5	41.1	39.6	3.2	41.3	40.9	39.1
Winter										
WGR	9.0	9.2	8.4	8.3	8.7	8.2	3.4	9.2	9.8	8.6
ELW	41.2	40.8	39.3	38.6	39.2	36.7	5.9	41.6	40.7	39.7
Spring										
WGR	12.0	11.5	11.5	12.0	12.9	11.4	2.7	11.5	10.8	10.8
ELW	40.9	39.4	38.4	38.0	40.3	38.9	3.2	39.3	37.8	36.9

*Ewes/ha.

**LSD only valid for comparison between lucerne only and lucerne plus natural pasture treatments.

†Data from a concurrent experiment (Robards et al. 1978).

arid environment, eight (Peart 1968) to ten (Robards 1967) paddock rotations are required if lucerne density is to be relatively stable.

CONCLUSIONS

For the semiarid environment of New South Wales it appears that the potential production from a given ratio of natural pasture and lucerne can be predicted and the value of developing supplementary areas of lucerne determined.

For supplementation to increase production/head it is likely that the movement of stock from natural pasture to supplementary lucerne area would need to occur at intervals of about four days (Robards and Flynn, unpublished). Such frequent stock movement represents an intensity of management that is not practical in the relatively extensive grazing conditions of west central New South Wales.

LITERATURE CITED

Campbell, R.J., and J.A. Beale. 1973. Evaluation of natural annual pastures at Trangie in central western New South Wales. II. Botanical composition changes with particular reference to *Hordeum leporinum*. Austral. J. Expt. Agric. Anim. Husb. 13:662.

Campbell, R.J., G.E. Robards, and D.G. Saville. 1972. The effect of grass seed on sheep production. Proc. Austral. Soc. Anim. Prod. 9:225.

Campbell, R.J., D.G. Saville, and G.E. Robards. 1973. Evaluation of natural annual pastures at Trangie in central western New South Wales. I. Sheep production. Austral. J. Expt. Agric. Anim. Husb. 13:238.

McKinney, G.T. 1974. Management of lucerne for sheep grazing on the southern tablelands of New South Wales. Austral. J. Expt. Agric. Anim. Husb. 14:926.

Peart, G.R. 1968. A comparison of rotational grazing and set stocking of dryland lucerne. Proc. Austral. Soc. Anim. Prod. 7:110.

Robards, G.E. 1967. Increased sheep numbers in a 17 inch rainfall. Wool Tech. Sheep Breed. 14:101.

Robards, G.E., D.L. Michalk, and R.J. Pither. 1978. Evaluation of natural pastures at Trangie in central western New South Wales. III. The effects of stocking rate on annual dominated and perennial dominated natural pastures. Austral. J. Expt. Agric. Anim. Husb. 18:361-369.

Warr, G.J., and J.R. Thompson. 1976. Live weight change and intake of lambs as affected by seed infestation. Proc. Austral. Soc. Anim. Prod. 11:173.

Verano Stylo in Semiarid Northeastern Thailand

A. TOPARK-NGARM

Khon Kaen University, Khon Kaen, Thailand

Summary

Stylosanthes hamata cv. Verano was first introduced to Thailand in 1975. In the northeastern part of the country where the climate is marginal and the soils are low in fertility, Verano appears to be one of the most suitable forage legumes used in pasture-improvement programs.

Verano can be grown satisfactorily on most soils in the northeast. Oversowing the seed at 3.1 kg/ha on unploughed areas resulted in a density of 5-30 plants/m², while in cultivated land the same amount of seed gave 20-45 plant/m². In a 3-year experiment under different seeding rates and cutting intervals Verano outyielded three commercial cultivars of *S. humilis* and *S. guianensis* more than 2-fold. Its dry-matter yield averaged over 5 tons(t)/ha/yr, and its competitive ability with natural weeds was high, permitting only 6-12% (by weight) weeds in most plots after the first year of establishment.

The highest seed yield recorded for Verano grown in experimental plots was 1,200 kg/ha of podded seeds, while under normal practice the average Verano commercial seed production in the northeast was 900 kg/ha. Heavy grazing could be imposed on Verano seed-production plots with no reduction in seed yield provided grazing were terminated before soil moisture became limiting.

Verano was affected by anthracnose (*Colletotrichum gloeosporioides*), but the damage was not serious. However, new accessions of this species are being evaluated for higher yield and better agronomic characters.

KEY WORDS: *Stylosanthes hamata* cv. Verano, forage legumes, establishment, seed production, anthracnose, accessions.

INTRODUCTION

The northeast region of Thailand lies between lat 14°–19° N and long 101°–108° E. The area is a slightly elevated plateau of 17 million ha approximately 100 to 300 m above sea level. Its tropical savanna-type climate (Koppen "AW") is influenced mainly by the southwest monsoon, giving an average annual rainfall ranging from 1,180–2,150 mm, of which about 85% falls during a 6-month wet period (May-October). Vegetation is mainly an open dipterocarp forest in which bamboo grass (*Arundinaria* spp.) is predominant. In heavily grazed areas around villages, the bamboo grass has been replaced by low-growing grazing-tolerant species. These include the grasses *Dactyloclenium aegyptum, Eragrostis viscosa, Digitaria ascendens, Perotis indica, Brachiaria miliformis, Eleusine indica, Chrysopogon aciculatus,* and *Heteropogon contortus,* and the legumes *Desmodium triflorum* and *Alysicarpus vaginalis* (Robertson and Humphreys 1976).

Most cattle and buffalo in the northeast are raised in small numbers by small-holder village farmers. The farmer inputs under the village production system are minimal, since bovines graze crop residues and unimproved natural pastures such as communal areas along roadsides, forest margins, private land under fallow, and upper paddy fields that are not planted to rice because of insufficient water. The seasonal stocking rates for a given number of cattle in a village are dependent on the proportion of lowland planted to rice and of upland planted to cash crops each wet season. Bovine production is thus intimately integrated with village cropping patterns. Improving natural pastures and increasing crop yields are likely to improve the feed supply for livestock through increased supplies of forage and crop residues, respectively. The selection of a suitable forage crop to be used in the communal grazing areas is a major concern for pasture improvement in the northeast.

METHODS AND RESULTS

Plant Introduction

Since 1962 several forage grasses and legumes have been introduced and evaluated at different locations in Thailand. Verano stylo, a commercial variety of Caribbean stylo (*Stylosanthes hamata*), was first introduced to the country in 1975 because of its good performance recorded in northern Australia (Edye et al. 1975). A number of trials have been conducted to test adaptation and forage potential of this legume in northeast Thailand, and some selected results are presented here.

Establishment

The establishment of a pasture species into communal grazing areas in the northeast requires a minimum of in-

put. Most farmers are reluctant to cultivate the area or use fertilizer in establishing pastures since they consider that forage crops, unlike other field crops they grow annually, will not give immediate cash returns.

It has been demonstrated that Verano can be established into upland grazing areas with simple land preparation (Wilaipon 1978). Seed was surface sown at a rate of 3.1 kg/ha on adjacent cultivated or uncultivated plots. The area was a farmer's private grazing field where the stocking rate, although the exact rate was unknown, was extremely heavy. Half of the experiment was fenced just after seed sowing to prevent grazing during early establishment, and the other half was subjected to free grazing from the start. Treatments also included application of P, S, and K fertilizers and an unfertilized control treatment. Plant density of the legume 175 days from sowing is presented in Table 1.

The plots that were neither fenced nor ploughed and that received no fertilizer had 16 plants/m². The plant density increased about 2-fold when the plots received either fencing or ploughing (30 and 29 plants/m², respectively) and about 3-fold when the two management treatments were combined (45 plants/m²). Fertilizer application had no marked effect on plant density of Verano stylo in this experiment. However, in another trial with Verano stylo (Gutteridge 1976), responses to applied P and S were obtained in a number of major soil types in the northeast.

Productivity

Since its introduction to Thailand in 1975, Verano has been tested for adaptation and yielding ability in several experiments under different methods or levels of management. In comparison with other herbaceous species of the genus *Stylosanthes, S. hamata* has adapted and yielded better in the low-rainfall and low–soil fertility conditions predominating in the region. In a 3-year experiment conducted on red latosol soil using different seeding rates and cutting intervals but no weeding, Verano outyielded three commercial cultivars of *S. humilis* and *S. guianensis* more than 2-fold (Table 2). Its annual dry-matter yield increased markedly from about 1.0 ton /ha in the first year of planting to 7.5 tons/ha in the second year and decreased to 6.6 tons/ha in the third year, giving on average about 5 tons/ha/yr. Verano's ability to compete with natural weeds was also high; only 6%–12% (by weight) of weed

Table 1. Plant density of Verano plots established in communal grazing area under different managements, 175 days after seed sowing. (Wilaipon 1978.)

Fertilizer	Land Preparation	
	Ploughed	Not ploughed
Fenced	Plants/m²	
Added	38	5
None	45	30
Unfenced		
Added	20	6
None	29	16

The author is an Assistant Professor in the Department of Plant Science on the Faculty of Agriculture of Khon Kaen University.

Table 2.   Dry-matter yield and weed percentage in plots of four *Stylosanthes* species grown at Khon Kaen University, 1976–1978. (Topark-Ngarm and Akkasaeng 1978.)

Stylosanthes spp.	1976		1977		1978	
	Yield*	% Weed**	Yield	% Weed	Yield	% Weed
	--tons/ha --					
S. humilis cv. Lawson	0.48	65	1.82	59	3.54	32
S. humilis cv. Paterson	0.65	52	1.46	64	3.07	40
S. hamata cv. Verano	1.04	46	7.53	12	6.63	6
S. guianensis cv. Schofield	0.07	92	2.98	35	4.49	38
C.V. (%)	33.6		15.1		22.4	
LSD (.05)	0.53		0.43		0.82	

*Average of 3 seeding rates and 2 cutting intervals.

**By weight.

was measured in most plots after the first year of establishment. In a separate experiment it was shown that the persistence and productivity of Verano stylo under heavy grazing was also satisfactory, at least in the first few years of establishment (Gutteridge 1978).

Seed Production

Verano's photoperiod nonsensitivity coupled with its flower prolificacy make it one of the heaviest seed-yielding tropical pasture crops. The highest seed yield recorded for Verano grown on an experimental plot at Khon Kaen University was 1,200 kg/ha (Wickham 1979), although the average commercial seed production under normal practice by farmers in the northeast was 900 kg/ha.

A reasonable amount of seed can also be obtained from grazed pure stands of Verano, providing the paddocks are grazed early enough in the wet season. Wilaipon (1977), using 26 native cattle (average weight 125 kg) to graze Verano paddocks at 3 different periods during a normal growing season, found no reduction in seed yield in the early grazing treatment as compared with the control (Table 3). It is suggested that because of the vigorous rapid early growth and long flowering period of Verano, farmers can utilize their seed-production areas more effectively by grazing them at least for short periods in the early wet season.

Varietal Improvement

Although Verano has several advantages over other legumes in the northeast environment, it is not entirely

Table 3.   The effect of different grazing times on the seed production of Verano stylo swards. (Wilaipon 1977)

Treatment	Seed Yield (kg/ha)
No grazing	374
Early grazing (17–30, August)	379
Mid grazing (7–20, October)	165
Late grazing (8–21, November)	85

CV (%)   = 114 kg/ha

LSD (.05) = 37

resistant to anthracnose. Vinijsanond and Topark-Ngarm (1979) found that although Verano grown under field conditions could have up to 100% infection, the symptoms apparent on the leaves or stems of infected plants were not severe. However, some accessions of *Stylosanthes hamata* introduced and tested for more desirable plant characters have performed better than Verano in yielding capacity and disease resistance in a preliminary cutting trial, as is shown in Table 4 (Topark-Ngarm 1979). These accessions will be further evaluated and, depending on their potential, may be used either in substitution for Verano, in combination with Verano, or as a source of genetic material to breed with Verano. Interspecific crosses between Verano and other species within the genus *Stylosanthes* may also be a possibility for varietal improvement (Cameron 1977).

LITERATURE CITED

Cameron, D.F. 1977. Selection and breeding programs with *Stylosanthes* species in northern Australia. Seminar, Past.

Table 4.   Performances in the 2nd year of growth of 14 *Stylosanthes hamata* accessions at Khon Kaen University, northeast Thailand. (Topark-Ngarm 1979)

S. hamata accession[1]	Total dry matter yield (ton/ha)	Yield relative to Verano (%)	Average weeds in the plots (%)	Anthracnose disease rating (1–5)[2]
CPI 55812	9.47	116	3	4
CPI 55820	7.14	87	11	4
CPI 55821	8.79	108	6	3
CPI 55822	8.73	107	4	3
CPI 55823	9.96	122	2	3
CPI 55824	8.31	102	6	4
CPI 55825	6.99	86	8	4
CPI 55826	5.54	68	13	5
CPI 55827	6.25	77	14	5
CPI 55828	8.31	102	3	5
CPI 55830	12.51	153	3	4
CPI 55831	8.06	99	8	5
CPI 61669	4.33	53	17	5
CV. Verano	8.16	100	8	3

[1]CPI = Commonwealth Plant Introduction number.

[2]1 = poorest; 5 = best.

Impr. for Livest. Prod. in Northeast. November 1977, Khon Kaen, Thailand (unpublished).

Edye, L.A., J.B. Field, and D.F. Cameron. 1975. Comparison of some *Stylosanthes* species in the dry tropics of Queensland. Austral. J. Expt. Agric. Anim. Husb. 15:655–662.

Gutteridge, R.C. 1976. Effect of phosphorus and sulphur fertilizers on the growth of *Stylosanthes* species on five soil types in northeast Thailand. Khon Kaen Univ. Past. Impr. Proj. 1976 Ann. Rpt., 84–90.

Gutteridge, R.C. 1978. The effect of heavy stocking rates on oversown legume persistence and productivity. Khon Kaen Univ. Past. Impr. Proj. 1978 Ann. Rpt., 67–70.

Robertson, A.D., and L.R. Humphreys. 1976. Effects of frequency of heavy grazing and of phosphorus supply on an *Arundinaria ciliata* association oversown with *Stylosanthes humilis*. Thai J. Agric. Sci. 9:181–188.

Topark-Ngarm, A. 1979. Preliminary yield trial of new introductions of *Stylosanthes hamata*. Khon Kaen Univ. Past. Impr.

Proj. 1979 Ann. Rpt., 18–21.

Topark-Ngarm, A., and R. Akkasaeng. 1978. Effect of seeding rate and cutting frequency on forage yield of four *Stylosanthes* spp. Khon Kaen Univ. Past. Impr. Proj. 1978 Ann. Rpt., 14–16.

Vinijsanond, T., and A. Topark-Ngarm. 1979. The incidence of anthracnose disease on *Stylosanthes* spp. grown in northeast Thailand. Khon Kaen Univ. Past. Impr. Proj. 1979 Ann. Rpt., 29–31.

Wickham, B. 1979. Effect of phosphorus and sulphur fertilizers on Verano stylo seed production. Khon Kaen Univ. Past. Impr. Proj. 1979 Ann. Rpt., 148–150.

Wilaipon, B. 1977. The grazing management of Verano stylo swards grown for seed production. Khon Kaen Univ. Past. Impr. Proj. 1977 Ann. Rpt., 83–85.

Wilaipon, B. 1978. The establishment of Verano stylo on private and communal grazing areas. Khon Kaen Univ. Past. Impr. Proj. 1978 Ann. Rpt., 42–45.

Animal Production from Brazilian Tropical Pastures

G.L. ROCHA, V.B.G. ALCÂNTARA, and P.B. ALCÂNTARA
University of Sao Paulo and Instituto de Zootecnia,
Nova Odessa (SP), Brazil

Summary

This review deals with the potential for beef production from nitrogen- and legume-based pastures in tropical Brazil. Possibilities for improving the beef industry supported by grass-legume pastures are very promising. According to the literature examined, legume-based pastures produced an annual average of 329 kg gain/ha, and nitrogen (N) fertilization resulted in 456 kg gain/ha.

Phosphorus (P) was the main fertilizer utilized in addition to N, followed by potassium (K), lime, and micronutrients. Annual rainfall averaged 1430 mm. Grasses used were of the genera *Melinis*, *Hyparrhenia*, *Panicum*, *Cenchrus*, *Digitaria*, *Cynodon*, *Brachiaria*, and *Pennisetum*, and the legumes were of the genera *Macroptilium*, *Centrosema*, *Glycine*, *Stylosanthes*, *Leucaena*, *Galactia*, *Calopogonium*, and *Cajanus*.

Animals used were Zebu, predominantly of Nelore breeding, and Zebu × European. Results are also presented for she-goats and ewes in semiarid northeastern Brazil.

High rainfall and the use of phosphate, other mineral nutrients, and locally adapted grasses and legumes contribute to greater live-weight gain.

KEY WORDS: tropical grasses, tropical legumes, live-weight gains, goats, sheep, cattle.

INTRODUCTION

There are approximately 572 million ha of oxisols and ultisols (CIAT 1977) under wet and monsoonal climates in tropical Brazil. Under the classification of cerrado (savanna-like vegetation), there are approximately 200 million ha of poor soils (oxisols), well drained with good physical properties, receiving no less than 1200 mm rainfall/yr (Ferri 1961). Tropical Brazil experiences a wide range of climatic conditions. Monsoonal rains predominate in the cerrado soil region, and winter temperatures below 15°C are common in the south central part, where frosts are rather frequent in the higher latitudes. This review presents results of 24 grazing experiments located in seven tropical Brazilian states, most of them on the cerrado soil. Two grazing trials with goats and sheep were located in the northeastern semiarid area. It is of para-

G.L. Rocha is a Professor (Credited), ESALQ, at the University of Sao Paulo, and V.B.G. Alcântara and P.B. Alcântara are Research Agronomists at the Instituto de Zootecnia. G.L. Rocha and P.B. Alcântara are working under CNP scholarships.

mount importance to find economic sources of nitrogen to supply to pasture ecosystem under such conditions.

METHODS

Twenty-four papers were selected from the literature; 70% of them involved 2- and 3-year grazing periods.

RESULTS AND DISCUSSION

Increase of beef production in Brazil has to be based on fertilizers, mainly phosphate, to secure the establishment of legume-based pastures. Table 1 summarizes the live-weight gain (LWG) responses for N fertilization (average of 456.4 kg gain/ha/yr) and for legume-grass pastures (average of 328.9 kg gain/ha/yr). Most of the commercially available tropical grass and legume species were successfully utilized.

Tropical Brazil has very favorable conditions for beef production on legume-based pastures, as is shown in Table 1, with a maximum annual LWG/ha of 580 kg and a minimum of 96 kg (Vilela et al. 1978, 1980b). Production from annual application of 100 kg N/ha ranged from 592 kg/ha (Vilela et al. 1980e) to about 350 kg/ha (Sartini 1975). As is shown in Table 1, the number of experiments summarized that used basic fertilizers was 18 for P, 10 for K, 7 for lime, and 4 for micronutrients. P was the only element used alone (six times). Frequency of K utilization seems to be a bit high and of lime too low, according to general soil analysis of the cerrado soils. At least 4 experiments were run in declivitous areas (10% to 30%) (Vilela et al. 1979, 1980a, 1980c), which could be considered wasteland. Performance of cattle (Zebu and Zebu × European) was satisfactory according to LWG/unit area. The sowing of *C. ciliaris* without fertilizer on native pasture in the semiarid region increased

Table 1. Cattle, goat, and sheep live-weight gains from fertilized and legume-based pastures.

Reference	State rainfall (mm)	Species	Fertilization (kg N/ha)	L.W.G. kg/ha	Comments (see footnotes)
QUINN et al., 1961	S. Paulo 1,200	*P. maximum* cv. Colonião	P,S,100N(1) 200N(2)	301(C) 494(1) 703(2)	2yr pt z
LIMA et al., 1965, 1968 and 1969	S. Paulo 1,330	Colonião, Pangola *P. purpureum*, Bermuda	44N(1)	346(C) 398(1)	3yr z pt z
SARTINI, 1975	S. Paulo 1,463	Pangola, *D. pentzii* cv. Taiwan, Bermuda	P,100N(1)	226(C) 350(1)	3yr pt z av. for the 3 grasses
ROLON et al., 1979	S. Paulo 1,415	Colonião, Centro, *C. wightii*, Siratro	P,Zn,Mo	125(C) 340(L)	2yr z
SARTINI et al., 1975	S. Paulo 1,463	*M. minutiflora* cv. Molasses, Centro	P,K, lime	86(L) 152(L) 207(L)	1.58yr rot. ze
LOURENCO et al., 1978	S. Paulo 1,423	Elephant, Centro, Siratro, *G. striata*	P,K, 50N(1) 100N(2) 150N(3)	370(L) 380(1) 508(2) 626(3)	3yr z
VILELA et al., 1980a, 1980b	M. Gerais 1,410	Colonião, Siratro, *G. wightii*, cv. Tinaroo	P	233(L) 44(C)	3yr pt ze decl 25–30%
VILELA et al., 1980c	MG 1,570	*B. decumbens*, *L. leucocephala* cv. Peru, Siratro, Centro	P,K,100N(1) lime	530(L) 592(1)	3yr pt-rot. decl 19%
VILELA et al., 1980d	MG 1,673	*P. maximum* cv. Guiné (g), *B. decumbens* (b) Leucaena, Centro	P,K, lime	353g(L) 363b(L)	2yr pt-rot. z
VILELA et al., 1980e	MG 2,017	Jaraguá(j) Guiné(g) Centro, Galactia	P,K	540g(L) 580j(L)	2yr pt-rot. z
VILELA et al., 1978	MG 1,200	Molasses Endeavour	P	24(C) 96(L)	1yr pt-cont. z

Table 1 (cont.).

Reference	State rainfall (mm)	Species	Fertilization (kg N/ha)	L.W.G. kg/ha	Comments (see footnotes)
ANDRADE & CAMPOS, 1978	MG	Colonião, Centro Glycine, *S. humilis,* Siratro	P, micronutrients	516(L) 117(C)	2yr set stock
FONSECA et al., 1980	MG	*C. ciliaris,* legumes	P	260[1](L) 263[2](L)	(1)2yr-av.2s. rates (2)3yr-av.4s. rates
ARONOVICH et al., 1970	Rio de Janeiro 1,300	Pangola, Centro	P,K, lime 100N(1)	349(C) 410(L) 531(1)	4yr ze
ROLON & MELO 1978a	Goiás 1,415	Jaraguá(j) *B. decumbens*(b) Stylo	P,K,lime	367[b](L) 117[j](C)	1yr z
ROLON & MELO 1978b	Goiás 1,365	Jaraguá(j) Green Panic(gp), Endeavour, Centro	P,K	332[gp](L) 168[j](C)	3yr cont. ze
COSTA & ROLON 1979	Goiás 1,600	Guiné(g) Glycine	P,K,Mo, lime	279[g](L) 89[g](C)	3yr cont. z
VILELA et al., 1979	E. Santo 1,090	Colonião, Tinaroo	P	494(L)	2yr cont. ze decl 10%
ROLON et al., 1978c	M. Grosso 1,454	(1)Green Panic, Glycine, Siratro, Galactia. (2)*B. decumbens,* Centro, *C. cajan.* (3) *B. decumbens.*	P,K,Mo, lime	234[1](L) 291[2](L) 158[3](C)	1yr z
CATUNDA et al., 1978a	Ceará	(a) native (b) native improved; (c) native improved + Buffel		15[a](C) 32[b] 48[c]	1yr she-goats
CATUNDA et al., 1978b	Ceará	(a) native (b) native improved; (c) native improved + Buffel		10[a](C) 25[b] 35[c]	1yr ewes

Footnotes

yr - year
pt - put and take
rot. - rotational

L - legume
decl - declivity
C - control
LWG - live-weight gain

cont. - continuous
z - zebu
ze - Zebu X european

LWG/ha/yr 2.2 times for she-goats and 2.5 times for ewes (Catunda et al. 1978a, 1978b), but still gains were relatively poor. Average annual rainfall for 17 experimental areas was 1430 mm (maximum 2110 mm, minimum 80 mm), adequate for a reliable beef enterprise based on legume-grass pastures. Different grazing systems were utilized, including rotational or continuous grazing, and these were combined with, or used without, put-and-take techniques. Stocking rates differed, resulting in noticeable differences among treatments (Fonseca et al. 1980, Vilela et al. 1979). A major need facing the beef industry in Brazil is for the economic availability of phosphate as a first priority, and lime, potassium, zinc, and molybdenum as secondary concerns.

LITERATURE CITED

Andrade, R.R.N., and J. Campos. 1978. Utilization of mixed pastures for the production of beef cattle. An. XV Reun. An. S.B.Z., 54–55.

Aronovich, S., A. Serpa, and H. Ribeiro. 1970. Effect of nitrogen fertilizer and legume upon beef production of pangola grass pasture. Proc. XI Int. Grassl. Cong., 796–800.

Catunda, A.G., F.H.F. Machado, D.F. Maciel, R.G. Serafim, S.M.S. Torres, J.A. Araujo Filho, and F.A.B. Menezes. 1978a. Continuous grazing trial with goats. An. XV Reun. An. S.B.Z., 107.

Catunda, A.G., R.G. Serafim, S.M.S. Torres, D.F. Maciel, F.A.R. Macedo, J.A. Araujo Filho, and F.A.B. Menezes. 1978b. Continuous grazing trial with sheep. An. XV Reun. An. S.B.Z., 106.

CIAT. 1977. Ann. Rpt., A1–114. Cali, Colombia.

Costa, N.A. and J.D. Rolón. 1979. Productivity of mixed guiné grass pasture in cerrado area, 3rd year. An. XVI Reun. An. S.B.Z., 378.

Ferri, M.G. 1961. Problemas de economia d'água na vegetação da caatinga e cerrado brasileiro. *In* Instituto de Zootecnia. Fundamentos de manejo de pastagens, 189–225. Sao Paulo.

Fonseca, D.M., C.J. Escuder, C.B. Carvalho, and J.C. Carmanin. 1980. Effect of stocking rates on Buffel grass (*Cenchrus ciliaris*) on the North of M.G. An. XVII Reun. An. S.B.Z., 439–440.

Lima, F.P., D. Martinelli, E.B. Kalil, H.J. Sartini, G.L. Rocha, and J.V.S. Pedreira. 1965. Beef production on grazing trials of four tropical grasses. B. Indús. Anim. 23: 83–90.

Lima, F.P., D. Martinelli, and J.C. Werner. 1968. Beef production on four tropical grasses. B. Indús. Anim. 25:129–137.

Lima, F.P., H.J. Sartini, D. Martinelli, P. Biondi, and M.F. Pares, Jr. 1969. Utilization of four tropical grasses in beef production on a typical red latosol soil. B. Indús. Anim. 26:199–214.

Lourenço, A.J., H.J. Sartini, M. Santamaria, and G.L. Rocha. 1978. A comparison of three nitrogen levels and grass (*Pennisetum purpureum* Schum.) in determination of stocking rates. B. Indús. Anim. 35(1):69–80.

Quinn, L.R., G.O. Mott, and W.V.A. Bischoff. 1961. Fertilization of colonial guinea grass pastures and beef production with zebu steers. 24 IBEC Res. Inst.

Rolón, D.J., F.C. Dias, and S.P. Canto. 1979. Effect of the association legume grass on the improvement of pastures established on a highly weathered soil. An. XVI Reun. An. S.B.Z., 2:335–336.

Rolón, D.J., and E.S. Mello. 1978a. Yields and productivity of a cultivated mixed green Panic pasture, in the 3rd year under continuous grazing. An. XV Reun. An. S.B.Z., 290–291.

Rolón, D.J., and E.S. Mello. 1978b. Productivity of *B. decumbens* on cerrado area. An. XV Reun. An. S.B.Z., 304–305.

Rolón, D.J., A.R. Oliveira, and S.P. Canto. 1978c. Productivity of mixed pastures in low fertility cerrado area. An. XV Reun. An. S.B.Z., 285–286.

Sartini, H.J. 1975. A comparison of four tropical grass species, with and without nitrogen fertilization, for beef production. B. Indús. Anim. 32(1):57–110.

Sartini, H.J., M. Santamaria, A.J. Lourenço, E.L. Caielli, and G.L. Rocha. 1975. Grazing trial utilizing 3 stocking rates in a mixture *Melinis minutiflora* and *Centrosema pubescens*. Zootecnia 13:219–228.

Vilela, H., S. Oliveira, A.B. Garcia, and E. Vilela. 1978. Liveweight gains and stocking rates with zebu in range and improved pasture in distrophic latosol. R. Soc. Bras. Zoot. 7(2):208–219.

Vilela, H., S. Oliveira, and J.A.A. Pires. 1979. Performance of crossbred holstein-zebu steers on range and improved pastures. Arq. Esc. Vet. U.F.M.G. 30(3):317–324.

Vilela, H., C.L. Paula, J.R.A. Silvestre, and J.A.A. Pires. 1980a. Pasture management in hilly country. An. XVII Reun. An. S.B.Z., 453.

Vilela, H., G. Ramalho, M. Almeida, and J.R.A. Silvestre. 1980b. Pasture management in cerrado area: guiné grass × *H. rufa*, plus legume. An. XVII Reun. An. S.B.Z., 452.

Vilela, H., A.S. Rodrigues, A.B. Garcia, and S.S. Rodrigues. 1980c. Pasture management in hilly country. An. XVII Reun. An. S.B.Z., 449.

Vilela, H., M.R. Santos, M. Almeida, and A.B. Garcia. 1980d. Pasture management in cerrado area: guiné grass × *B. decumbens*, plus legume. An. XVII Reun. An. S.B.Z., 455.

Vilela, H., E.J. Santos, J.O. Valente, and J.A.A. Pires. 1980e. Pasture management in cerrado area: grass plus nitrogen × legume-based pasture. An. XVII Reun. An. S.B.Z., 457–458.

Sown Grassland as an Alternative to Shifting Cultivation in Lowland Humid Tropics

H.R. CHHEDA, O. BABALOLA, M.A.M. SALEEM, and M.E. AKEN'OVA
University of Ibadan, Ibadan, Nigeria

Summary

Agriculture based on existing forms of shifting cultivation will not be able to support and feed adequately the rapidly increasing human population of the lowland humid tropics of Africa. The forest and natural fallows have shortened and even disappeared in some areas because of rising food demands. Crop yields are low and continue to decline. What, then, are the alternatives?

Using *Cynodon* as an example, an attempt is made to show that mixed farming offers an alternative whereby the nutrient status and physical properties of the soil can be maintained, erosion can be checked, weeds can be kept under control, and the buildup of pests and parasites minimized or avoided.

A synthesis of several years of intensive breeding and agronomic research at the University of Ibadan indicates:

1. Availability of improved robust, nonrhizomatous *Cynodon* genotypes with an annual dry-matter (DM) yield potential of around 15 tons (t)/ha, adequate for production of 650 kg/ha of live-weight gains by the local Zebu cattle when grazed from April to December.

2. Annual nutrient mobilization of over 122, 25, 175, 37, and 29 kg/ha of N, P, K, Ca, and Mg, respectively, by *Cynodon* pastures. Around 20% of the nutrients mobilized are "pumped up" from soil layers below 30 cm.
3. Possession by several *Cynodon* genotypes of high microbial population in the rhizosphere, capable of fixing varying quantities of N. IBX-7, an improved hybrid, is able to fix up to 2 kg/ha of N/week.
4. Increase in soil organic matter under temporary *Cynodon* pastures, replenishing a high proportion of the substantial loss of 63% in organic matter (OM) under 10 years of continuous cultivation.
5. Improvement of soil physical properties under *Cynodon* pastures as measured by soil aggregate size and aggregate stability, bulk density, and total porosity.
6. Protection against erosion with established *Cynodon*. *Cynodon*, when fully established, gives almost as much protection against erosion as does forest with closed canopy. On a soil with 3% slope, water runoff from *Cynodon* plots was 5.3% of the total erosive rain of 155 cm. Soil losses under pasture during the growing season were less than 10% of those observed in maize and cowpea plots.
7. Ability of root exudates collected from several *Cynodon* improved genotypes, when placed with seeds of common tropical weed species, to inhibit germination or reduce seedling growth. Aqueous extracts of *Cynodon* roots showed similar inhibitory effects. The inhibition factor disappeared after 27 to 45 hours of storage.
8. Resistance of *Cynodon* genotypes to the root-knot nematodes (*Meloidogyne* spp.). Growing of *Cynodon* in highly infested soils decreased the root-knot nematode population appreciably in 6 months and eliminated the nematodes completely within 18 months. Consequently, tomato yields increased by about 300%.

KEY WORDS: *Cynodon*, shifting cultivation, mixed agriculture, soil physical and chemical properties, allelopathy, nitrogen fixation in rhizosphere, pest control.

INTRODUCTION

For centuries peoples in the tropics have utilized the seemingly inexhaustible areas of forest and moist savanna under a system of so-called primitive agriculture known in West Africa as "shifting cultivation." It involved the use of digging sticks, cutlass, hoe, and axe for cutting the vegetation to be burned on small areas; planting of a variety of food crops for 1 to 4 years; abandoning the area when yields declined and weeds invaded for a 10- to 20-year fallow; and repeating the process in an adjacent locality.

When the human population density was less than 20 people/mile², shifting cultivation was a viable form of agriculture. Now the population density has increased. Many areas are facing acute food shortages and have no additional acreages to bring under cultivation, yet 40% of the potentially arable and grazing land in the tropics still remains under shifting cultivation.

Nye and Greenland (1960) discovered why shifting cultivation had been adopted in the tropics. Repeated cropping following the clearing of forests and plowing under of grasslands completely disrupts the ecological processes and results in reduction of nutrient status of soil, deterioration of physical condition of soil, decline in soil organic matter, erosion of topsoil, increase of weeds, multiplication of pests and diseases, and changes in number and composition of soil flora and fauna. These modifications lead to decline in crop yields and necessitate abandonment of the land to a long-term fallow, during which the natural regeneration of vegetation improves the physical and chemical properties of the soil.

To meet the rising demands for food, the interval of fallow has been decreasing steadily, causing declines in crop yields unless crops were fertilized. Forests are being transformed into woodland savannas, and in the middle belt of West Africa the fallow has completely disappeared.

What are the alternatives to shifting cultivation? Growing crops for food in the tropical forest zones? Destroying the existing natural and ecological balances in the humid tropical zones by using continuous-cropping technology? Suggesting an agricultural economy for the tropical forest based on intensive production of timber and plantation crops?

An alternative exists in much of the tropical forest zone and in the moist savanna areas for use of forage and fodder crops in the development of mixed-farming systems similar to the "regulated ley farming" of temperate countries. This paper summarizes the results of research on *Cynodon* at the University of Ibadan that has given some answers about the potentials of sown grasslands in tropical farming systems.

PASTURE POTENTIAL OF *CYNODON*

After several years of research, *Cynodon* genotypes (e.g., IB.8, IBX-7) have been identified that have high adaptability to the varying soil and climatic conditions of southern Nigeria. These were used to formulate management systems geared toward higher production of quality herbage required for efficient animal performance. Zebu steers, rotationally grazing 2- to 4-year-old *Cynodon* IB.8 pastures at Ibadan fertilized annually with 75 kg N/ha, gained up to 650 kg live weight/ha from April to December. Daily gains/steer ranged from 0.3 to 0.7 kg (Chheda 1974).

Tsetse fly, the vector of animal trypanosomiasis, is considered a primary deterrent to raising cattle and conse-

H.R. Chheda is a Professor of Agronomy, O. Babalola is a Reader, and M.A.M. Saleem and M.E. Aken'Ova are Senior Lecturers in the Department of Agronomy of the University of Ibadan.

quently to mixed farming in the humid tropics of West Africa. However, use of resistant or highly tolerant breeds of cattle such as N'Dama and Keteku, a high plane of nutrition, removal of vegetation bordering streams, and strategic use of chemotherapy and prophylactic drugs offer a distinct possibility of development of a viable livestock industry in the humid tropics.

IMPROVEMENT OF SOIL CHEMICAL PROPERTIES

A *Cynodon* IB.8 or IBX-7 pasture grown on continuously cropped soils will produce around 10 to 12 t/ha/yr of above-ground dry matter (DM). The herbage contains an average of 1.22% nitrogen (N), 0.25% phosphorus (P), 1.74% potassium (K), 0.37% calcium (Ca), and 0.34% magnesium (Mg), thus mobilizing annually over 122, 25, 175, 37, and 29 kg/ha of N, P, K, Ca, and Mg, respectively (Saleem 1972, Chheda 1974). To these amounts must be added an unknown but substantial quantity of these elements stored in the roots and stolons. With fertile soils or adequate N topdressing, DM yields of over 20 t/ha can be obtained, resulting in mobilization of much larger quantities of these elements. Burning the above-ground vegetation returns the elements to the soil, except carbon, nitrogen, and sulfur. Plowing under avoids losses and increases soil organic-matter reserves. Animal grazing recycles the plant nutrients to the soil with the exception of those utilized for meat and milk production.

Soil cleared on the University Farm in 1959 had an initial pH of 6.2. By 1972 pH declined to 4.2 with continued annual application of 100 to 200 kg/ha of ammonium sulfate. When *Cynodon* was established on this soil there was a steady rise in pH of the topsoil, reaching a value of 5.3 at the end of one growing season. A corresponding decline in pH occurred in the soil below the top 40-cm layer. Total Ca content in the topsoil increased. Low pH did not affect the DM production of *Cynodon*. Continuous death and decay of large amounts of fibrous roots, basal leaves, and stem stubble contributed to the increase of Ca and other basic ions in the topsoil layer. *Cynodon* was able to mobilize and translocate nutrients from the deeper soil horizons (Saleem et al. 1975).

Mackenzie and Chheda (1970) observed 14% of *Cynodon* IB.8 roots in soil layers below 30 cm, with some reaching depths of 1.75 m or more. Using ^{32}P, Nye and Foster (1960) demonstrated that a grass fallow derived about 30% of its P from layers below 25 cm, containing 19% of the roots. Thus, a conservative estimate of 20% nutrient uptake by *Cynodon* from soil layers below 25 cm gives an annual "pumping up" of over 24 kg N, 5 kg P, 35 kg K, 7 kg Ca, and 6 kg Mg/ha. This upward translocation of nutrients can be a very important factor in counterbalancing the nutrient losses from topsoil due to leaching.

Using *Cynodon* and several other tropical-grass genotypes, Solomon (1980) studied bacterial association and N fixation in their rhizosphere. Some *Cynodon* genotypes, particularly IBX-7 as well as *Brachiaria mutica*, stimulated

a large number of bacteria in the rhizosphere. These included *Bacillus* sp., *Enterobacter cloaceae*, and *E. aerogenes* capable of fixing atmospheric N. Quantitative estimates by the isotope dilution technique using ^{15}N-labeled organic matter (OM) indicated IBX-7 capable of fixing an equivalent of over 2 kg/N/ha/wk and thus contributing effectively to soil N gains.

IMPROVEMENT IN SOIL ORGANIC-MATTER CONTENT

The top 10 cm of fertile soils under tropical forests contains 2% to 5% OM, and the next 20 cm contains 0.5% to 2%. Under continuous cropping, the soil OM declines but is gradually built up again as a result of leaf and litter deposits and root residues of bush fallow.

In studies conducted at the University Farm, continuous cultivation under various conditions resulted in about 63% loss of OM in the topsoil, but growing *Cynodon* pastures for 4 years substantially increased the soil OM, from 0.85% to 2.3%.

IMPROVEMENT OF PHYSICAL PROPERTIES OF SOIL

Tillage operations tend to destroy soil structure; increase density; decrease total porosity, causing a decline in rainfall acceptance, infiltration, and percolation rate; and increase soil compaction. During bush fallow period soil structure and other physical properties are restored. Babalola and Chheda (1972) observed that the mean weight diameter (MWD) of soil aggregates in the upper 30 cm of soil under uncut forest averaged 1.003 mm as compared with 0.796 mm in soil continuously cropped for 20 years. A 4-year-old pasture of *Cynodon* IB.8 grown after 15 years of continuous cropping improved soil structure, the aggregate size being 1.124 mm. Stability of the soil aggregates taken from forest soil and permanent *Cynodon* pastures was superior to those from the short-term *Cynodon*, requiring about 160 to 140 drops of water falling from 10 cm to disperse them, as compared with 60 to 20 drops for soil from 4-year-old *Cynodon* pastures and 20-year continuous cultivation, respectively. Soil under *Cynodon* pasture when heavily grazed did not register improvement in bulk density and total porosity. In another study, however, the authors observed that 3 years of ungrazed *Cynodon* plots reduced bulk density significantly when compared with 3 years of continuous maize cultivation.

CONTROL OF EROSION

Cynodon is a sod-forming grass that, when fully established, gives almost as much protection against erosion as does forest with a closed canopy. In general, tropical soils having 6% to 8% slope and receiving 100 to 125 cm of rainfall lose 12 to 60 t/ha of soil annually if they are cropped continuously. In a study conducted at the University Farm (Babalola and Chheda 1975) on a soil with 3% slope, to determine soil and water losses from bare soil and from land cropped with mulched and un-mulched maize or cowpeas or covered with *Cynodon*, water runoff over a period of 18 months amounted to 18.3%, 13.7%, 9.6%, 10.4%, and 5.3% of the total erosive rain (155 cm), respectively. Soil losses in the 1970 growing season were in the ratio of 8:5:2:1 for bare soil, maize, cowpeas, and *Cynodon*, respectively. In 1971, after full establishment of *Cynodon*, the soil losses were in the ratio of 15:7:9:1 for unmulched and mulched maize, cow-peas, and *Cynodon*, respectively.

CONTROL OF WEEDS

One major reason for abandonment of land after a few years of cropping in shifting cultivation is the increase in weed infestation and resulting labor costs. Furthermore, there will be substantial increases in crop yields on newly cleared land.

Observations of the *Cynodon* nursery plots indicated that several genotypes were capable of suppressing weed growth to a considerable extent. Detailed records of *Cynodon* IB.8 pastures and experimental plots confirmed the existence of allelopathy. In a greenhouse study, four *Cynodon* genotypes were grown in sterile sand and sup-plied with nutrient solution at regular intervals. Root ex-udates collected inhibited seed germination or reduced seedling growth of *Crotolaria juncea*, *Euphorbia heterophylla*, *Cassia rotundifolia*, and *Sida corymbosa*. Aqueous extracts of *Cynodon* roots showed similar inhibitory effects. The in-hibition factor disappeared after 27 to 45 hours of storage. Thin-layer chromatographic separations (ethyl acetate ex-traction) from the *Cynodon* IB.8 root exudates revealed three (out of four) compounds that completely inhibited seed germination of the four species. *Cynodon* was useful for controlling weeds in highly infested areas following continuous cropping. The short-lived effectiveness of root exudates to suppress weed growth suggests that there will be no residual effects left to endanger the germination and growth of subsequent crops.

PEST AND DISEASE CONTROL

Continuous cropping tends to intensify pest and disease problems. On parts of the University Farm, growing of *Cynodon* IB.8 markedly reduced the nematode population. Tomato yields following *Cynodon* exceeded by 3-fold those from plots on which tomatoes were continuously grown, and the tomatoes showed little or no root galling (Adeniji

and Chheda 1971). Replicated greenhouse experiments indicated that all six varieties of *Cynodon* included in the trial were resistant to the root-knot nematode, lowered the nematode population in 6 months, and eliminated it within 18 months.

CONCLUSION

Agriculture based on the existing forms of shifting cultivation will not be able to support and feed adequately the ever-growing human population of the tropics. The use of *Cynodon* shows that mixed farming offers an alter-native whereby the nutrient status and physical properties of the soil can be maintained, erosion can be checked, and weeds, pests, and parasites can be controlled.

LITERATURE CITED

Adeniji, M.O., and H.R. Chheda. 1971. Influence of six vari-eties of *Cynodon* on four *Meloidogyne* spp. J. Nematol. 3:251–254.

Babalola, O., and H.R. Chheda. 1972. Effects of crops and management system on soil structure in a Western Nigeria soil. Niger. J. Sci. 6:29–35.

Babalola, O., and H.R. Chheda. 1975. Influence of crops and cultural practice on soil and water loss from a Western Nigerian soil. Ghana J. Sci. 15:93–99.

Chheda, H.R. 1974. Forage crops research at Ibadan. I. *Cynodon* spp. *In* J.K. Loosli, V.A. Oyenuga, and G.M. Babatunde (eds.) Animal production in the tropics, 79–94. Heinemann Ed. Books (Nigeria), Ibadan.

Mackenzie, J.A., and H.R. Chheda. 1970. Comparative root growth studies of *Cynodon* IB.8, an improved variety of *Cynodon* forage grass suitable for Southern Nigeria and two other *Cynodon* varieties. Niger. Agric. J. 7:91–97.

Nye, P.H., and W.N.M. Foster. 1960. The relative uptake of phosphorus by crops and natural fallow from different parts of their root zone. J. Agric. Sci. 56:299–306.

Nye, P.H., and D.J. Greenland. 1960. The soil under shifting cultivation. Commonw. Bur. Soils Tech. Com. 51.

Saleem, M.A.M. 1972. Productivity and chemical composition of *Cynodon* IB.8 as influenced by level of fertilization, soil pH and height of cutting. Ph.D. Thesis, Univ. of Ibadan, Nigeria.

Saleem, M.A.M., H.R. Chheda, and L.V. Crowder. 1975. Ef-fects of lime on herbage production and chemical composi-tion of *Cynodon* IB.8 and on some chemical properties of the soil. East Afr. Agric. Forage J. 40:217–226.

Solomon, M.G. 1980. Microbial associations and nitrogen fixa-tion in the rhizosphere of some tropical grasses. Ph.D. Thesis, Univ. of Ibadan.

Insect Pests of Tropical Forage Plants in South America

MARIO CALDERÓN C.

Tropical Pastures Program, CIAT, Cali, Colombia

Summary

The objectives of this continuing research are to make a general survey of the insect populations occurring on tropical forage plants in South America and to identify the most important pests.

To realize this investigation, pure stand plots 5 × 5 m were sampled every 14 days during 2 years using a D-Vac. Vacuum Sampler.

The following information was obtained:

- Most important families and/or genera of insects in tropical forage legumes and grasses.
- Insect group frequency for each season.
- Insect group frequency for each plant ecotype under study.
- Preliminary data on insect preference.
- Preliminary data on relationships between insects on forage plants and viral, fungal, and bacterial diseases.
- Most important pests of tropical forage plants in South America.

The important pests of tropical forage legumes include stemborer (*Caloptilia* sp.), leafhoppers (*Cicadellidae*, several genera), and leafeating beetles (*Crisomelidae*, several genera). Chinchbug (*Pentatomidae, Lygaeidae*) and sucking planthoppers (*Membracidae*) are under investigation as disease vectors; seed-eating beetles (*Curculionidae*) and budworms (*Stegasta bosqueella*) reduce seed production. In grasses, spittlebug (*Zulia, Aeneolamia*, and Deois spp.) causes serious damage, and seed-feeding leafhoppers (*Cicadellidae*) reduce seed production.

This ongoing research provides essential information about the most important insect pests, pests of secondary importance, potential pests, and beneficial insects of tropical forage plants in five countries in South America. This is the first complete study of the insect fauna of tropical forage plants, so it is an important one.

KEY WORDS: insect pests, tropical forage plants, South America.

INTRODUCTION

In the past, few reports have been made on insect pests of tropical forage plants. During the last 3 years, the Entomology Section of the Tropical Pastures Program of CIAT has conducted a continuing research project on insect populations of tropical forages in South America. Information has been collected on insect numbers, identities (order, family, genus, and species) and frequency and preference within different grasses and legumes. The most important pests of tropical forages have been identified and an understanding of population dynamics within the target area is developing. Methods are being developed to control these insect pests in the pasture situation.

MATERIALS AND METHODS

A D-Vac. Vacuum Sampler with a suction power of 9 m³/minute was used to collect samples every 14 days in 14 forages, 3 grasses, and 11 legumes. Each forage plot was replicated 3 times with a total of 42 plots, each being 25 m². A complete randomized-block design was used at the CIAT Research Station, Carimagua, Colombia.

Specimens collected in each sample were sorted taxonomically and quantitatively and codified as Co, Coleoptera; Di, Diptera; He, Hemiptera; Ho, Homoptera; Hy, Hymenoptera; L, Lepidoptera; N, Neuroptera, and O, Orthoptera.

RESULTS AND DISCUSSION

During 16 months of sampling, a total of 41,818 insects were collected and separated into order, genus, and species according to forage. Forty-seven percent of the population occurred on grasses and 53% on legumes. Although there were no significant differences in total insect population between the two groups of plants, there were differences with respect to orders of insects (Table 1). Homoptera, Diptera, Hemiptera, and Hymenoptera were most commonly associated with legumes, while Diptera and Homoptera were more common in grasses.

The author is an Entomologist in the CIAT Tropical Pastures Program.
Mention of a proprietary product does not imply endorsement of it.

Table 1. Groups of insects more frequently found in some forage legumes and grasses in South America.

Order	Frequency (%)	
	Legumes	Grasses
Homoptera	40.0	27.0
Diptera	17.0	51.0
Hymenoptera	12.0	6.0
Lepidoptera	8.0	7.0
Coleoptera	4.0	5.0
Hemiptera	18.0	4.0
Orthoptera	9.0	0.2

The frequency of Homoptera was higher in legumes than in grasses while the frequency of Diptera was higher in grasses than in legumes. Diptera families included Tachinidae, Syrphidae, and Sarcophagidae, which contain some excellent biological control genera. They were commonly recorded on *Andropogon gayanus*, a promising grass for South America (Anon. 1979).

Insect Pests of Legumes

The most important pests of legumes are stemborer (*Caloptilia* sp.) and budworm (*Stegasta bosqueella* Chambers). Many families of Homoptera and Coleoptera sucking and feeding on legumes are presently considered as secondary or potential pests. Since several genera are well known as viral, bacterial, and fungal transmitters, their importance may increase. The most important pest of grasses is spittlebug (*Aeneolamia* spp.), while yellow aphid (*Sipha flava* Forbes) and false army worm (*Mocis latipes* Guenné) are regarded as secondary pests.

Stemborer. The stemborer, *Caloptilia* sp., is the most important pest of *Stylosanthes* spp. in South America. Emphasis has been given to field screening, oviposition, and feeding preference tests to evaluate forage germ plasm for resistance rapidly.

Field screening has identified accessions of both *S. capitata* and *S. guianensis* with resistance to stemborer (Anon. 1977). Susceptible accessions possessed glandular trichomes, whereas resistant accessions did not (Anon. 1977). The oviposition preference test showed a reduction of 90.5% oviposition in accessions rated resistant in field screening in comparison with susceptible accessions (Anon. 1978). The use of artificial diets in feeding preference tests showed that pupae reared on a resistant legume-based diet were smaller than those reared on a susceptible legume-based diet (Anon. 1978). These results possibly indicate that the mechanism of resistance could be an antibiotic effect. Further studies are in progress.

Anatomical studies of stems of *Stylosanthes* revealed that susceptible plants possess thinner layers of sclerenchyma cells while resistant plants have thicker layers of cells (Anon. 1980). At the same time, studies on stem hardness have clearly shown that plants with thicker layers of sclerenchyma cells are harder than those with thinner layers of cells (Anon. 1980).

It is now clear that well-defined differences exist in percentage of infestation, degree of damage, and oviposition preference among ecotypes of *Stylosanthes* spp. Studies on physical and biochemical characteristics have shown that plants with soft and juicy stems are more susceptible to stemborer than those with hard stems. Physical characteristics such as sclerenchyma thickness, stem hardness, and possibly stem nutritional content may cause an additive or cumulative effect on insect development (antibiosis and antixenosis). In addition, the presence of glandular trichomes may attract females to oviposit. From these studies on host-plant resistance it will soon be possible to describe and identify the characteristics of resistant and susceptible plants.

Fig. 1. Infestation level of spittlebug nymphs (*Aeneolamia reducta*) in some grasses in Carimagua during 1979.

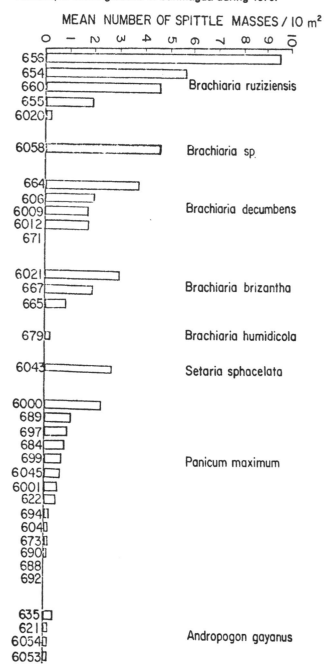

Budworm. The budworm, *Stegasta bosqueella* Chambers, is an important pest of *Stylosanthes* spp. and *Zornia* spp., damaging inflorescences and reducing seed production. In determining the critical level of damage caused by budworm of *Stylosanthes* spp., seed production capacity of the species is an important consideration. For example, the same population of larvae could cause more damage in smaller inflorescences of *S. guianensis* than in larger ones of *S. capitata.*

An experiment to determine the critical level of damage of budworm in *S. capitata* set three different levels of infestation, 10%, 20%, and 30% (Anon. 1980). The infestation level of 30% did not significantly reduce seed production in *S. capitata*. Studies are continuing on the critical levels of damage in *S. guianensis* and *Zornia* spp.

Insect Pests of Grasses

Spittlebug. The spittlebugs, *Aeneolamia, Deois,* and *Zulia* spp., are the most important pest of grasses in South America. In field screening at various sites to identify grasses with resistance, studies (Anon. 1979) to date have shown that *Brachiaria ruziziensis* and *B. decumbens* are most susceptible, while *B. humidicola* and *Andropogon gayanus* are resistant (Fig. 1).

After two years of evaluation, an infestation level of only 0.006 nymphs/m² has been found in *A. gayanus* (Anon. 1980). Although the basis of resistance is not yet known, it is possible that the growth habit of this grass, in which stems form dense compact clumps, offers a strong natural barrier to young nymphs attempting to reach soft growing points (Anon. 1979). In addition, characteristics such as stem hardness and hairiness may make *A. gayanus* an unsuitable host for spittlebug. Further studies are in progress.

Yellow Aphid. The yellow aphid, *Sipha flava* Forbes, has been observed on leaves of *A. gayanus*. Analyses show that forages affected by aphids are significantly lower in nitrogen and sulfur than unaffected forages (Anon. 1980).

Since aphids may cause a significant reduction in forage quality, population dynamics were studied. Three different behavioral phases of the aphid population were recognized. The survival phase occurs during the dry season, a small percentage of *A. gayanus* clumps being infested with a few aphids. During the multiplication phase, a rapid and progressive increase in the aphid population occurs in the few clumps that remained infested during the survival phase. This phase commences at the beginning of the wet season. In the dispersion phase, during the highest rainfall period, the population of aphids increases progressively and a high percentage of clumps are affected.

False Army Worm. Since 1975 the false army worm *Mocis latipes* Guenné has been reported as a sporadic pest of grasses. During 1980, this pest attacked *A. gayanus* at Carimagua and preliminary observations were initiated. Six years of records (Anon. 1980) reveal that sudden heavy outbreaks causing severe defoliation of grasses occur after a severe dry season when the wet season is initiated gradually with light showers. Little is known about this pest.

LITERATURE CITED

Anon. 1977. Ann. Rpt. CIAT, Cali, Colombia.
Anon. 1978. Ann. Rpt. CIAT, Cali, Colombia.
Anon. 1979. Ann. Rpt. Trop. Past. Prog., CIAT, Cali, Colombia.
Anon. 1980. Ann. Rpt. Trop. Past. Prog., CIAT, Cali, Colombia.

Some Initial Results of Field Trials Conducted in Establishing a Pasture-Seed Industry in Cojedes State, Venezuela

E.J.A. KHAN and W.H. MARK
IRI Research Institute, Inc., Venezuela

Summary

In starting a pasture-seed industry on a farm located at lat 9°50′ N that gets about 1,600 mm of rain yearly followed by a severe dry season from mid-December to April, four widely separated sites were selected to plant three grasses for commercial seed production in 1978. The objectives were to assess the fertility status of the soils, then to determine the fertilizer requirements of the grasses at each site and to establish an introduction garden with 70 cultivars of grasses and legumes to study their suitability for expanding seed production.

Based on the results of soil analyses, a modified missing-element technique was used to test the effects of 7 plant nutrients in a field trial with 10 fertilizer treatments in 4 randomized complete blocks at each of the four sites (two with *Panicum maximum* cv.

Colonial and one each with *Brachiaria humidicola* and *B. decumbens*). A replicated factorial experiment testing nitrogen (N) and phosphorus (P) at three levels was also established with Colonial at one site. In the second year, 12 grasses were planted in 4 randomized complete blocks to obtain forage yields and observe their acceptance by cattle.

Although the soil tests indicated that several plant nutrients might limit seed production, the results of the fertilizer trials showed a highly significant response only to P treatment at all four sites, and a significant response to N treatment at only one site in the first year. In the second year, spittlebug (*Aeneolamia reducta*) first appeared in *B. decumbens* only, and in 1980 this pest attacked all the introduced grasses except Colonial; fortunately, commercial seed yields were not seriously affected. *B. humidicola*, though attacked by spittlebug, recovered quickly and is regarded as the most outstanding grass introduced so far, with the potential to revolutionize Venezuelan pasture development. Cultivars of *Andropogon gayanus*, *Brachiaria brizantha*, and *Setaria anceps* appeared promising for future grass-seed production, as did certain cultivars of legume species, notably *Leucaena leucocephala*.

It has been conclusively shown that P fertilization is essential for pasture establishment and productivity; other nutrients are also low in availability and are being studied. Control of spittlebug is now necessary to prevent its increase and possible damage to grasses before seeding. A great potential exists for forage-seed production that can immensely benefit pasture improvement in Venezuela.

KEY WORDS: seed production, field trials, phosphorus, spittlebug, Venezuela.

INTRODUCTION

Venezuela does not produce any high-quality pasture seed, and importation of certified seed is increasing. Lárez et al. (1975) tested the adaptability of a large number of introduced grasses and legumes in the Orinoco Delta and savanna areas outside the delta. The present work was started in 1978 to establish a pasture-seed industry located at lat 9°50′ N in Cojedes State. The ranch receives about 1,600 mm of rainfall annually in a single rainy season from April to mid-December, followed by a severe dry season of at least 3 months, indicating an ustic soil moisture regime (Sanchez 1976).

METHODS

Soil Fertility Evaluation

Four widely separated sites were initially selected for commercial seed production of 3 grasses: guineagrass (*Panicum maximum*) cv. Colonial (sites 1 and 2), *Brachiaria humidicola* (site 3), and *Brachiaria decumbens* (site 4). Before mechanical clearing and land preparation for planting, soil samples were taken at 0 to 15 cm from each site and chemically analyzed. Based on the results, a fertilizer field experiment was designed to test the effects of 10 treatments (listed in Table 2) in 4 replicates of randomized complete blocks at each site. Rates of application (kg/ha) were: 100 of nitrogen (N) as urea, 43.8 of phosphorus (P) and 50 of sulfur (S) (single superphosphate, 8.8% P and 10% S), 83 of potassium (K) as potassium chloride, 2.3 of boron (B) as borax, and 4.6 of zinc (Zn) as zinc sulfate; calcium (Ca) (calcitic limestone) was applied at a rate of 1 ton/ha. In treatment 9, triple superphosphate (TSP) was used as the source of P. Where limestone was applied, N application was delayed for 2 weeks. Otherwise, fertilizer was applied by incorporation into the soil just before sow-ing grass seed at a rate of 5 kg/ha in 5 × 4-m plots. A factorial experiment testing N (urea) at 0, 30, and 60 kg/ha and P (triple superphosphate) at 0, 13.1, and 26.2 kg/ha in 4 replicates was also conducted with Colonial in site 1. Areas were established in June and July. Forage yields were obtained from plots at all sites (15-cm cutting height), and Colonial seed was harvested from site 1.

Introduction Garden

Seventy cultivars of 13 species each of grasses and legumes were established in 5 × 4-m plots.

Mob-Grazing Trial

In August 1979, 12 grasses were planted in 10 × 10-m plots in 4 replicates of randomized complete blocks to study their forage production and demonstrate to farmers their acceptance by cattle. The grasses were *Brachiaria brizantha*; *B. decumbens*; *B. humidicola*; *Cenchrus ciliaris*; *Cynodon plectostachyus*; *Dichanthium aristatum*; *Digitaria swazilandensis*; *Hyparrhenia rufa*; Colonial, green panic and Gatton panic guineagrasses; and *setaria* cv. Nandi. At planting, 46 kg/ha of N (urea), 40.3 kg/ha of P (TSP), and 24.9 kg/ha of K (KCl) were applied. Plots were grazed in December, and cattle preferences were noted. In April 1980 the grasses were cut at 15 cm and treated with 1 ton/ha of 15-6.6-12.5 (N-P-K) fertilizer. Dry-matter yields were obtained at 60-day intervals during the rainy season.

RESULTS

Chemical Characteristics of the Soils

Except at site 2 with high exchangeable Ca, the soils at the sites were strongly acid (Table 1)[1]. P was low at all four sites, and K was marginal at site 4. S was marginal at site 1 and low at site 4. Zn was low at sites 1, 3, and 4.

E.J.A. Khan is a Pasture Seed Specialist and W.H. Mark is the Director at the IRI Research Institute.

Table 1. Results of chemical analyses of soil samples taken at 0 to 15 cm from 4 sites.[1]

	Site number			
	1	2	3	4
E.A.[+][*]	1.4	0	1.7	1.7
Ca[*]	1.5	10.2	1.2	1.3
Mg[*]	1.0	3.7	1.7	2.8
K[*]	0.18	0.19	0.33	0.12
N[o]	20	10	20	10
P[o]	4	4	6	2
S[o]	12	14	20	6
B[o]	3.9	2.0	0.7	0.8
Cu[o]	4.3	10.1	3.0	2.6
Fe[o]	324	271	262	78
Mn[o]	32	31	26	47
Zn[o]	2.6	3.5	0.4	1.9
A.S. %[x]	34	0	35	29
pH	4.9	5.6	4.8	5.1
Organic matter %	1.5	1.7	1.3	1.4

[1]Analyzed by method of Mehlich (undated).

[+] E.A. = extractable acidity.

[*]meq/100 ml.

[o]μg/ml.

[x]Acid saturation.

Field Fertilizer Trials

Mean fresh forage yields of the 3 grasses and mean seed yields of Colonial only at site 1 are given in Table 2. Colonial was cut for forage and seed yields at site 1 a month earlier than at site 2, a difference that explains the larger yields from site 2. Seed was harvested at site 1 when the grass was 115 days old. No seed of *B. humidicola* and *B. decumbens* was harvested; these species were planted at the time of seed production. Results show that P was essential for high yields of forage and seed (Table 2), limiting grass productivity at all sites, and that N also limited productivity of *B. humidicola* (site 3) and *B. decumbens* (site 4).

In the factorial experiment with Colonial at site 1,

observations made one month after germination showed that plants receiving P averaged 50 cm high and had 8 tillers/plant compared with 10 and 0, respectively, for minus-P plants. Four months after planting, fresh forage yields were highly significantly larger for the 26.2 kg/ha P plots, and seed yields responded to the 13.1 and 26.2 P rates.

Introduction Garden

During the rainy season of 1978, apart from fungal infections in *C. ciliaris*, siratro, and stylo, all grasses and legumes were healthy and free of pests. In 1979, spittlebug (Homoptera: *Aeneolamia reducta*) occurred in *B. decumbens*, and in 1980 this pest attacked all the introduced grasses except Colonial. *C. ciliaris* was severely damaged, but several grasses showed good resistance to the pest. Forage species that showed good promise for future seed production were *Andropogon gayanus*, *B. brizantha*, Gatton panic, Kazungula setaria, and several cultivars of *Leucaena leucocephala*.

Mob-Grazing Trial

First choices by cattle were the *Digitaria* and *Brachiaria* species, then guineagrasses, especially Gatton panic, and last choices were *H. rufa* and *C. ciliaris*. Similar observations have been reported by Rolim et al. (1979) in Brazil. During the rainy season, *Brachiaria* produced the highest yields of dry matter in three cuttings at 60-day intervals; the highest yield was 37 tons/ha with *B. humidicola*. Other grasses produced between 20 and 27 tons/ha of dry matter, except *C. ciliaris* (15 tons/ha), and *D. aristatum* and Nandi setaria failed to survive the preceding dry season.

DISCUSSION

Vincente-Chandler (1975) reported that Colonial and other grasses responded well to N fertilization in Puerto Rico. It is surprising that a significant response to N was

Table 2. Mean yields of fresh forage from four sites and Colonial guineagrass seed from one site.

	Site 1, Colonial		Site 2,	Site 3,	Site 4,
Treatment[+]	Seed kg/ha	Forage	Colonial, forage	B. humidicola, forage	B. decumbens, forage
			---------------- metric tons/ha ----------------		
1. Control	19c[*]	2d	13b	0d	2h
2. All: N,P,K,S,B	226a	28ab	39a	10a	7bcde
3. All plus Ca	185ab	20bc	46a	7bc	6cdef
4. All plus Ca plus Zn	158ab	19c	47a	7bc	5defg
5. All plus Zn	176ab	27abc	40a	7bc	10a
6. All minus N	131b	22abc	41a	6c	3gh
7. All minus P	17c	2d	21b	0d	1h
8. All minus K	181ab	24abc	42a	8abc	8abc
9. All minus S	133b	24abc	50a	9ab	7bcd
10. All minus B	171ab	30a	46a	8abc	9ab

[+] Four replications.

[*]Means followed by a common letter are not different at the 5% level.

obtained only at site 4 with *B. decumbens* (Table 2), while in the trials Colonial did not respond. There was no explanation for this lack of yield response to N in the other trials. The highly significant response to P fertilization was very similar to that obtained by Koster et al. (1977) in Pará, Brazil, where degraded Colonial pastures were restored and their production increased 5 to 10 times with an application of only 26 kg/ha of P.

The high yields of forage harvested in the mob-grazing trial compare very favorably with those reported by Bogdan (1977).

Grazing cannot be used as a method to control the nymphs of spittlebug when grass is being grown for seed. Thus, for commercial seed production, as suggested by Treviño and Conrad (1975), control of spittlebug is essential for susceptible grasses.

B. humidicola is the outstanding grass introduced here because it suppresses weeds, recovers quickly from burning, resists drought and spittlebug, has high forage productivity suitable for making hay, and can be planted by stolons as well as seed. Large acreages are being planted in soils of low fertility in the humid tropics of Brazil (Rolim et al. 1979).

LITERATURE CITED

Bogdan, A.V. 1977. Tropical pasture and fodder plants. Longman, London.

Koster, H.W., E.J.A. Khan, and R.P. Bosshart. 1977. Programma e resultados preliminares dos estudios de pastagens na regiao de Paragominas, Pará, e nordeste de Mato Grosso. Min. do Interior, SUDAM, Belém, Brazil.

Lárez, O.R., E.R. Velásquez, O. Parra, and W.B. Bryan. 1975. Pasture and livestock investigations in the humid tropics Orinoco Delta—Venezuela. I. Observations on forage grasses and legumes. IRI Res. Inst. Bul. 42.

Mehlich, A. undated. Mimeograph. N.C. State Univ. Raleigh.

Rolim, L.R., H.W. Koster, E.J.A. Khan, and H.M. Saito. 1979. Alguns resultados de pesquisas agrostologicas na regiao de Paragominas, Pará, et nordeste de Mato Grosso. Min. do Interior, SUDAM, Belém, Brazil.

Sanchez, P.A. 1976. Properties and management of soils in the tropics. John Wiley & Sons, New York.

Treviño, R.G., and J.H. Conrad. 1975. Work group on pastures and forages/nutrition. Proc. Sem. on potential to increase beef production in tropical America, 283-284. CIAT, Cali, Colombia.

Vincente-Chandler, J. 1975. Intensive management of pastures and forages in Puerto Rico. *In* E. Bornemisza and A. Alvarado (eds.) Soil management in tropical America, 409-431. N.C. State Univ., Raleigh.

NOTES

1. Soil analysis was done by ± methods of Agro-services International, Inc. (Dr. W. Fitts), Orange City, Florida.

Potential of *Aeschynomene* sp. for Pastures in the Tropics

A.E. KRETSCHMER, JR., and G.H. SNYDER
University of Florida, Gainesville, Fla., U.S.A.

Summary

The objectives of this study were to assess the forage potential of readily available world germ plasm of the genus *Aeschynomene*, which comprises 160 species; to compare some attributes of the commercial American jointvetch (*Aeschynomene americana* L.) with those of other accessions of the same and of other species; and to compare the seasonal growth pattern and response to phosphorus (P) of commercial jointvetch with two perennial tropical legumes.

Evaluations were made of 342 accessions, including 25 identified species. Data collected from observational plots in 1978 and 1980 included initial flowering dates, plant vigor, and growth habit; from replicated plots, plant population and vigor; and from selected accessions, yield, nitrogen content, and digestibility. Comparison of yields at 20 and 60 kg/ha of P (2,000 kg/ha of lime) was made between jointvetch and siratro (*Macroptilium atropurpureum* Urb.) and Florida carpon desmodium (*Desmodium heterocarpon* L.).

A majority of *A. americana* accessions flowered after 1 October, and height distribution was from less than 51 cm (18 accessions) to greater than 152 cm. Thus, it should be possible to select a later-flowering and more leafy plant than the commercial jointvetch. *A. villosa*, similar to but with shorter stems than *A. americana*, appeared to have potential for grazing. *A. denticulata*, *A. falcata*, *A. elegans*, *A. evenia*, *A. fluitans*, and *A. paniculata* warrant further study, although their establishment and persistence when grown with *Digitaria decumbens* Stent. were not always good.

In June crude-protein content of commercial jointvetch (18.8%) was highest and digestibility among the best (52.1%) when compared with other accessions. These values ranged from 10.6% to 18.8% and 36.8% to 63.8%, respectively, among the accessions analyzed.

Establishment-year production of commercial jointvetch was better than that of siratro and of carpon desmodium but was much less in June of the subsequent year. Yield response of jointvetch to P at either 20 or 60 kg/ha was greater than that of carpon desmodium at either of the two rates.

KEY WORDS: *Aeschynomene*, jointvetch, evaluation, yield, growth habit, flowering, dates, quality.

INTRODUCTION

Kretschmer and Bullock (1980), in a review of the botanical and agronomic literature, determined that there were about 160 species in the nontoxic, primarily tropical legume genus *Aeschynomene*. Quesenberry and Ocumpaugh (1981) in north central Florida evaluated the forage potential, nitrogen (N), and percentage of digestibilities of 190 accessions. Diversity of species ranges from tree-like to small herbaceous forms, from annuals to perennials, from hydrophytes to xerophytes, and from widely distributed to isolated species that extend from sea level to about 2,800 m. *A. falcata*, a low-growing perennial, is being used commercially for grazing in New South Wales (Hennessy and Wilson 1974, Wilson 1980). American jointvetch, *A. americana*, a highly productive annual, is being successfully used as a grazing legume in south Florida (Hodges 1979), where about 5,000 ha have been planted.

The primary objective of this study was to assess the potential of the *Aeschynomene* genus for use as forage in the tropics. This assessment included the available world germ plasm of 115 introductions of *A. americana*, 152 accessions of 24 other species, and 75 unidentified accessions.

Observational and replicated data were obtained at the Agricultural Research Center, Fort Pierce, Florida (ARC-FP). Snyder et al. (1978) and Kretschmer and Snyder (1979) provide greater details of south Florida soils, forages, and environment.

MATERIALS AND METHODS

Observations for initial flowering date, general plant vigor, and growth habits were made from single row (1.5 to 2 m long) plants growing on research plots in the field at the ARC-FP (1978–1980). Observations for flowering were made weekly and were recorded when the first open flower was seen.

Preliminary data on plant populations, plant vigor, yield, and quality were obtained from 36 *Aeschynomene* accessions growing in combination with pangola digitgrass (*Digitaria decumbens* Stent.) on replicated research plots. Plots in a randomized complete-block experiment with 4 replications were seeded on 6 August 1979 and were

harvested on 26 June and 29 September 1980, leaving a 10-cm stubble. Plant population and vigor ratings were taken on 30 November 1979; on 17 March, 13 June, and 18 September 1980; and on 8 January 1981.

A randomized complete-block field experiment with 4 replications was initiated in 1977 to compare growth, at 20 and 60 kg/ha of P (2,000 kg/ha of lime), of commercial jointvetch and two perennial tropical legumes, siratro (*Macroptilium atropurpureum* Urb.) and Florida carpon desmodium (*Desmodium heterocarpon* L.) (Kretschmer et al. 1979). Yields were obtained by clipping plots in a conventional manner during 1977 and 1978. In addition to P and lime, K, Cu, Mn, Zn, B, and Mo at rates of 46, 2.2, 2.2, 2.2, 0.3 and 0.1 kg/ha, respectively, were applied (prior to seeding) and roto-tilled to a depth of about 15 cm.

In-vitro organic-matter digestibility (IVOMD) was determined by the method of Moore et al. (1972).

RESULTS

Observational Plots

Of *A. americana* accessions in 1978, 3 flowered prior to 1 September, 13 between 1 September and 1 October, 55 between 1 October and 1 November, and 27 after 1 November. Respective values for 1980 were 6, 31, 29, and 12. The commercial Florida jointvetch flowered on 11 September 1978 and on 21 September 1980.

Plant height of the 1978 planting at the ARC was reported by Kretschmer and Bullock (1980). On 3 October 1980, the commercial jointvetch plants were about 56 cm high. Height distribution of the other *A. americana* accessions were: less than 51 cm, 18 accessions; 52 to 76 cm, 18; 77 to 102 cm, 11; 103 to 127, 8; 128 to 152, 5; and greater than 152, 1 accession. These results confirm those of 1978; i.e., there is a good opportunity to select within *A. americana* for shorter but less stemmy types and to lengthen the fall grazing season by selecting for later-flowering types. In addition, some accessions appeared to be short-lived perennials.

Height of *A. villosa* accessions (17) followed the *A. americana* trend but were shorter, ranging from 5 to 56 cm with most in the 15 to 25–cm range on 3 October 1980. Vigor measured as leafiness and branching was rated fair to excellent. The diversity of *A. villosa* was wide and the potential as a grazing forage legume appeared good.

Accessions of the annuals *A. rudis*, *A. ciliata*, and *A. indica* did not appear to have agronomic attributes as desirable as those of the better *A. americana* accessions. Persistence of *A. falcata* was excellent but productivity and diversity low. *A. elegans* Schl. and Cham., *A. evenia* C.

A.E. Kretschmer, Jr., is a Professor of Agronomy at the University of Florida Institute of Food and Agricultural Sciences, Fort Pierce, and G.H. Snyder is a Professor of Soil Science at the University of Florida, Belle Glade. This is a contribution of the University of Florida, Gainesville (Journal Series No. 2915).

This research was supported in part by a grant from the USDA, TAD-406 7002-20108-0024.

Wright, *A. paniculata* Willd. ex. Vog., *A. denticulata* Rudd, and *A. fluitans* Peter warrant further study.

Replicated Plots

Commercial jointvetch contributed 29% of the dry matter of the mixture with pangola in June and 54% in September. Respective yields of 600 and 1,060 kg/ha were surpassed by 1 *A. americana* accession in June and 3 in September. Compared with commercial jointvetch, one accession each of *A. denticulata* (33% — 720 kg/ha), *A. rudis* (40% — 1,360), *A. histrix* (37% — 830), and *A. scabra* (40% — 1,460), and two *A. villosa* accessions (68% — 3,260 and 64% — 2,250), were better on 26 June. Only 2 *A. villosa* accessions were equal to jointvetch yields in September. Although *A. histrix, A. rudis, A. scabra,* and several *A. villosa* introductions were very vigorous in early spring, their productivity and plant population during the summer were lower than those of commercial jointvetch.

Crude-protein content of commercial jointvetch in June (18.8%) was highest (average of accessions was 15.1% and range was 10.6% to 18.8%), while IVOMD was slightly above (52.1%) the average (48.8%). Lowest crude protein (10.6%) occurred in *A. denticulata*, which already had seeds maturing and very low leaf : stem ratios. Low IVOMD was found in *A. denticulata* (39.9%) and *A. evenia* (36.8%), and an exceptionally high value of 63.8% was found in *A. histrix*.

Ratings of plant population and plant vigor were used to estimate the 36 accessions shortly after establishment and prior to each harvest. Establishment-year (November 1979) plant populations of *A. americana* accessions were considered adequate for seed production to supply better stands during the second year. By March 1980 most plants had died, although natural regeneration from seed was occurring by the June rating and continued during the summer.

On 30 November, 17 March, 13 June, 18 September, and 8 January, 2, 5, 2, 9, and 0 accessions, respectively, had plant populations rated as significantly ($P < 0.05$) better than commercial jointvetch.

Vigor ratings for *A. americana* accessions followed trends similar to those of populations, but even more pronounced differences occurred between the March and the June and September ratings. Rapid summer growth occurred after the June harvest. Vigor ratings in March, June, September, and January, respectively, were 3, 0, 0, and 0. These data indicate the excellent vigor of the commercial jointvetch, but they also show that natural regeneration of several accessions was greater.

One *A. villosa*; 2 *A. evenia* and 3 *A. villosa*; 2 *A. villosa*; 1 *A. villosa*; and zero accessions, respectively, had significantly higher populations than commercial jointvetch in November, March, June, September, and January. More vigorous accessions in March, June, September, and January included 1 *A. denticulata*, 1 *A. evenia*, and 3 *A. villosa*; 1 *A. evenia*; none; and none, respectively, compared with commercial jointvetch. There appears to be an opportunity to select other species that would compete with pangola and be more vigorous than commercial jointvetch, especially during the cool season.

Establishment, Yield, and Response to Phosphorus

On plots receiving either 20 or 60 kg/ha of P and 2,000 kg/ha of lime the 19 September 1977 dry-matter yield of commercial jointvetch (3,960 kg/ha) was much higher than those of carpon desmodium (190) and siratro (1,300). Jointvetch natural regeneration the second year, as measured by the 7 June 1978 yield (360), however, was delayed because of insufficient April-May rainfall. At this harvest the perennials produced 3,420 (desmodium) and 650 (siratro) kg/ha.

Rapid vegetative growth of jointvetch occurred primarily from late June (depending on seed germination) until early September. Productivity and leaf : stem ratios decreased drastically in autumn, during seed formation.

Where 20 and 60 kg/ha of P were applied, yields of jointvetch were 3,050 and 4,000 in 1977 and 2,900 and 7,000 kg/ha in 1978, respectively. In 1978, siratro responded slightly less and carpon desmodium much less to P than jointvetch. Growth of the legumes resulting from 3 kg/ha of P was insignificant. Maximum response of jointvetch to lime at the 60 kg/ha P rate was 3,000 kg/ha.

DISCUSSION

The morphological and agronomical diversity of *A. americana* was wide. Other species evaluated, except for *A. villosa*, seemed to be less diverse. Relatively few accessions were available, however, for comparison, which may account for the lack of diversity within the samples for these species. Botanical study of herbarium specimens, because no seed was available, indicated a wide diversity among the other 135 species not tested. There appear to be *Aeschynomene* species adapted to almost any environment from continuous flood to desert. It is urgent that plant collectors become cognizant of the potential forage value of this genus and that they collect seed whenever possible. Scientists in tropical America and Africa are urged to obtain information (from herbariums) on location sites for the various species within their area of responsibility so that they can collect seed during their normal travels.

The authors would be pleased to supply small quantities of seed of various species to researchers interested in initial evaluation tests.

LITERATURE CITED

Hennessy, D.W., and G.P.M. Wilson. 1974. Nutritive value of Bargoo jointvetch (*Aeschynomene falcata*) and companion grasses when fed to sheep. Austral. Inst. Agric. Sci. J. 40:82–83.

Hodges, E.M. 1979. Production and utilization of American jointvetch (*Aeschynomene americana* L.). Univ. Fla. 13th Ann. Conf. Livest. and Poultry in Lat. Am.

Kretschmer, A.E., Jr., J.B. Brolmann, G.H. Snyder, and S.W. Coleman. 1979. 'Florida' carpon desmodium, a perennial tropical forage legume for use in south Florida. Fla. Agric. Expt. Sta. Circ. S-260.

Kretschmer, A.E., Jr., and R.C. Bullock. 1980. *Aeschynomene* spp.: distribution and potential use. Soil Crop Sci. Soc. Fla. Proc. 39:145-152.

Kretschmer, A.E., Jr., and G.H. Snyder. 1979. Forage production on acid infertile soils of subtropical Florida. *In* P.A. Sanchez and L.E. Tergas (eds.) Pasture production in acid soils of the tropics, 227-258. CIAT, Cali, Colombia.

Moore, J.E., G.O. Mott, D.G. Dunham, and R.W. Omer. 1972. Large capacity in vitro organic matter digestibility procedure. J. Anim. Sci. 35:232.

Quesenberry, K.H., and W.R. Ocumpaugh. 1981. Forage evaluation of *Aeschynomene* species in north central Florida. Soil Crop Sci. Soc. Fla. Proc. 40:159-162.

Snyder, G.H., A.E. Kretschmer, Jr., and J.B. Sertain. 1978. Field response of four tropical legumes to lime and superphosphate. Agron. J. 70:269-273.

Wilson, G.P.M. 1980. Bargoo jointvetch: tough legume for tough country. Agric. Gaz. (N.S.W., Austral.) 91:51-53.

Yield and Quality of Tropical Legumes During the Dry Season: *Galactia striata* (Jacq.) Urb.

R.R. VERA, E.A. PIZARRO, M. MARTINS, and J.A.C. VIANA

Universidade Federal de Minas Gerais, Belo Horizonte, M.G., Brazil

Summary

A small-plot experiment was conducted during 2 consecutive years in the cerrados of Brazil to evaluate the effect of four dates of deferment (21 February, 7 and 21 March, and 4 April, respectively) on rate of growth, chemical composition, and in-vitro dry-matter digestibility (IVDMD) of a pure stand of *Galactia striata*, a naturalized legume. The chosen dates of deferment embraced the last 2 months of the rainy season. Yields were estimated at intervals of 28 days, over an experimental period of 237 days, ending with the beginning of the following rainy season. Asymptotic growth was observed throughout the experimental period. Yields during the critical August-September period ranged between 1,000 and 5,000 kg dry matter (DM)/ha, depending on treatment and year. Differences in yield between treatments and years were accounted for largely by differences in accumulated rainfall and maximum temperature, as was indicated by regression analysis. The contribution of leaves to dry-matter yield decreased from an initial 75% to about 35% at 168 days of age. This decrease was treatment-independent and strongly age-dependent. During most of the dry season, crude protein (CP) averaged 15% in the standing herbage; no treatment effect was noted when treatments were compared at the same age. Crude protein (CP) in leaves decreased from 32% at 28 days of age to about 20% at 112 days and remained constant thereafter. In the stems, CP was 15% initially, and it stabilized at 10% at 84 days. The IVDMD of the available forage fell very slowly during the dry season; in the most critical period it varied between 55% and 60%. In leaves, the IVDMD fell initally but soon stabilized around 65% at 56 days. Seed production was observed in the earliest dates of deferment, and it amounted to up to 280 kg/ha.

It was concluded that galactia is capable of providing high-quality forage during the dry season and that its yield depends mainly on the precipitation and temperature accumulated after the date of deferment.

KEY WORDS: galactia, *Galactia striata* (Jacq.) Urb., yield, deferment, dry season, nutritive value.

INTRODUCTION

In the cerrados of Brazil, the dry season extends for about 5 months, during which there is a pronounced deficit of quality forage. Milk producers and to a lesser extent beef producers resort to supplementation with maize silage, hand-cut elephantgrass, and sugar cane in addition to concentrates. Both maize silage and the standing crops are low in digestibility and nitrogen content (Pizarro and Vera 1979, Paiva et al. 1978). Hay made during the dry season with perennial tropical legumes is promising (Pizarro and Carvalho 1982), and a logical alternative would be the deferment of grazing of legume pastures at the end of the rainy season to allow them to grow and to accumulate forage into the dry period.

R.R. Vera is a Professor of Animal Science at the Universidade Federal de Minas Gerais (UFMG); E.A. Pizarro is a Research Agronomist at EPAMIG; M. Martins is a Graduate Student at UFMG; and J.A.C. Viana is a Professor of Animal Science at UFMG.

METHODS

A small-plot experiment was carried out in the Metallurgic region of the State of Minas Gerais (latitude 19°28′00″S, long 44°15′00″W) on a 3-year-old stand of galactia (*Galactia striata* [Jacq.] Urb.), a promising native legume, established on a red-yellow latosol. A split-plot design with 3 replicates, including four main treatments and ten subtreatments, was used. Main treatments were beginning dates of the deferment (A = 21 February, B = 7 March, C = 21 March, and D = 4 April), and subtreatments were dates of utilization, spaced at intervals of 28 days from the beginning of the deferment. At the time of deferment, the area was cut and fertilized with 500 kg/ha single superphosphate, 150 kg/ha K_2SO_4, 10 kg/ha each of $ZnSO_4$, $CuSO_4$, and borax, and 1 kg/ha calcium molybdate, on the basis of current recommendations.

The experiment was terminated on 16 October 1979 and 28 October 1980, after the occurrence of significant rains.

At sampling time, dry-matter yield (DMY) was measured, and samples were withdrawn for separation of leaves, stems, and pods; their contribution to yield was expressed on a DMY basis. Composite samples of the 3 field replicates were analyzed for crude protein (CP) and in-vitro dry-matter digestibility (IVDMD). Leaves and stems of the first deferment (A) were analyzed for soluble carbohydrates (CHO), which were extracted twice with boiling water for 5 minutes and determined by anthrone.

RESULTS

Yields in the two years were significantly different and are presented separately (Fig. 1). Rainfall during the first half of the 1979 dry season was higher than average, resulting in a mean DMY of 3,214 ± 739 kg/ha. Differences between treatments and dates of utilization, as well as the interaction of the two variables, were significant (P < 0.05). Yields inceased with time at a decreasing rate (Fig. 1) during the dry period, and regrowth began immediately with the beginning of the new rainy season. The second year was drier than average, with a mean yield of 1,282 ± 497 kg DM/ha. Only two treatments were sampled because of damage by unidentified nematodes.

In both years (all treatments), leaf DMY peaked at about 100–110 days of age with 1,959 ± 292 kg/ha produced in 1979, and 972 ± 176 kg/ha in 1980. Leaf DMY decreased linearly throughout the dry season, at the rate of 0.243 ± 0.0067%/day in 1979 and 0.182 ± 0.025%/day in 1980; differences between dates of deferment were not significant (P > 0.05).

Seed peaking occurred in treatments A and B in 1979 at 167 days of age with pod yields of 228 ± 84 kg/ha and 123 ± 21 kg/ha, respectively. Peak pod yield in 1980 was 280 ± 140 kg/ha and 185 ± 70 kg/ha for treatments B and C, respectively, at 127 days of age.

Crude protein in total DMY, leaves, and stems (DM basis) decreased curvilinearly with age. The fall was more

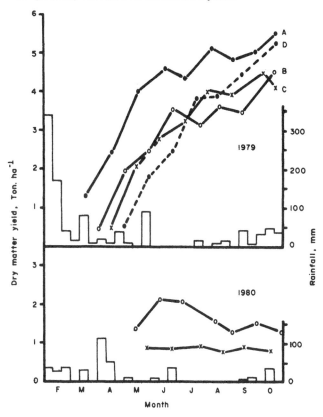

Fig. 1. Accumulated dry-matter yield of galactia, and rainfall, during the dry season of 2 consecutive years.

pronounced in leaves than in stems, but neither years nor treatments affected CP content (P > 0.05) in the fractions examined, when they were compared at the same age.

Except at very young ages, CHO content of leaves and stems was almost identical in the several treatments and increased slowly until the beginning of a new regrowth induced by rainfall.

IVDMD averaged 57.11% ± 4.1% and decreased linearly at the rate of 0.04% ± 0.02%/day in all treatments. The IVDMD of leaves was high and uniform throughout the dry period, with an average value of 64.13% ± 3.1%; IVDMD of stems averaged 45.41% ± 4.3%.

DISCUSSION

Galactia proved to have considerable potential for deferred grazing in the dry season. Yields compared favorably with those reported for year-round growth by Carvalho and Nascimento (1976) and Mattos and Werner (1975). The yield observed during the second year is similar to that reported by Pizarro and Carvalho (1982) for a 6-month-old stand of galactia cut for conservation when the regrowth was 90 days old. Nevertheless, the data of Mattos and Werner (1975) show a positive influence of pasture age on yield. In our test, differences between years and treatments were largely due to differences in soil water availability since stepwise regression analysis (using rainfall and air temperature as independent variables) removed yield differences between dates of deferment and years.

Since leaves are likely to be preferentially harvested by grazing animals, the fact that peak yield was obtained in all treatments at the same age suggests that by appropriately escalating dates of deferment in different paddocks, it should be possible to obtain a sustained yield over a period of 2 months, mid-July to mid-September, during the most critical part of the dry season. Werner et al. (1975) also have reported a larger rate of growth in galactia during the dry season than in other legumes, which resulted in more uniform seasonal distribution of yield (Mozzer et al. 1978).

Although the true value of galactia can be determined only from its effect on animal production, results indicate that the deferred herbage was of higher quality than that of any nonlegume conserved forage currently available (Pizarro et al. 1980, Miller and Rains, 1963). After an initial rapid decline, the quality of leaves and stems tend to remain constant regardless of date of deferment. The similarity of Pizarro and Carvalho's (1982) results to our own, obtained at a similar age but at the beginning of the dry season, shows the consistency of chemical composition. Rate of change of digestibility was somewhat lower than those reported by Milford and Minson (1966) for other tropical legumes.

LITERATURE CITED

Carvalho, L.J.C.M., and D. Nascimento, Jr. 1976. Competição entre leguminosas espontáneas do Brasil. Anais. Soc. Bras. Zoot. 13:378-379 (abs.).

Mattos, H.B., and J.C. Werner. 1975. Competição entre cinco leguminosas de clima tropical. Bol. Ind. Anim. 32:293-305.

Milford, R., and D.J. Minson. 1966. The feeding value of tropical pastures. In W. Davies and C.L. Skidmore (eds.) Tropical pastures, 106-114. Faber and Faber, London.

Miller, T.B., and A.B. Rains. 1963. The nutritive value and agronomic aspects of some fodders in Northern Nigeria. I. Fresh herbage. J. Brit. Grassl. Soc. 18:158-167.

Mozzer, O.L., M.J. Alvim, and R.M. Souza. 1978. Comparação entre cultivares de leguminosas forrageiras. Anais Soc. Bras. Zoot. 15:129 (abs.).

Paiva, J.A.J., E.A. Pizarro, N.M. Rodriguez, and J.A.C. Viana. 1978. Qualidade da silagem da região metalurgica de Minas Gerais. Arq. Esc. Vet. UFMG 30:81-88.

Pizarro, E.A., and L.J. Carvalho. 1982. Avaliação do feno de leguminosas espontaneas. Arq. Esc. Vet. UFMG.

Pizarro, E.A., J.O. Valente, and J.R.A. Silvestre. 1980. A produção de feno no Estado de Minas Gerais. Inf. Agrop., Belo Horizonte 6(64):3-5.

Pizarro, E.A., and R.R. Vera. 1979. Efficiency of fodder conservation systems: maize silage. In C. Thomas (ed.) Forage conservation in the 80's. Brit. Grassl. Soc. Occ. Symp. 11:436-441.

Werner, J.C., M.P. Moura, H.B. Mattos, E.L. Caielli, and L. Melotti. 1975. Velocidade de estabelecimento e produção de feno de dez leguminosas forrageiras e do capim gordura. Bol. Ind. Anim. 32:331-345.

Hemarthria altissima: A Pasture Grass for the Tropics

K.H. QUESENBERRY, W.R. OCUMPAUGH, and O.C. RUELKE
University of Florida, Gainesville, Fla., U.S.A.

Summary

Limpograss (*Hemarthria altissima* [Poir] Stapf. et C.E. Hubb), introduced into the U.S. in 1964, has been evaluated for use as a pasture grass. Under Florida conditions it has shown yield potential equal to or greater than other tropical pasture grasses. The cultivar Bigalta has higher in-vitro organic-matter digestion (IVOMD) than other tropical grasses at comparable growth stages. Early evaluations of limpograss involved only three clones, but additional genotypes were collected and introduced during the late 1960s and early 1970s, and an evaluation scheme for this germ plasm was undertaken.

Our objectives were to evaluate rapidly the available germ plasm of limpograss for genetic diversity in traits of importance as a component of a forage-livestock system, to compare clipping and grazing as evaluation techniques, and to select superior genotypes for animal performance experiments.

Following preliminary testing of 53 clones in greenhouse and small-plot clipping trials, 22 clones were selected for evaluation by clipping and 27 by grazing. Greater winter-stand loss was observed with grazing than with clipping, and by autumn of the second year some genotypes were eliminated. Others yielded from 880 to 15,800 kg/ha, and IVOMD ranged from 59% to 73% for 5-week regrowth forage.

Eight genotypes selected from these experiments were then evaluated at four frequencies of grazing (3, 5, 7, and 9 weeks) in a third experiment. Persistence was poorest at the 3-week grazing frequency. Total seasonal yields were similar at the 5- and 7-week grazing frequencies. The tetraploid genotypes Bigalta and P.I. 364888 had markedly higher yields at the 5- and 9-week grazing frequency. P.I. 364888 was generally the highest-yielding entry at all grazing frequencies and was not significantly lower in IVOMD than Bigalta. Bigalta was significantly lower in persistence at the 5- and 7-week rest interval than P.I. 364888.

From this research, it is evident that limpograss is a forage with good potential for utilization in the humid tropics on soils that may be intermittently flooded. It is often superior to other tropical grasses in subtropical regions because it initiates regrowth earlier in spring and grows later in autumn. Newly collected germ plasm shows genetic diversity for traits not present in the released cultivars. During the last 6 years over 50 introductions of limpograss have been evaluated, and one genotype (P.I. 364888) has been identified that has superior yield and persistence under grazing and is moderately high in digestion. The utilization of grazing animals in the early phase of evaluation has shortened the time necessary to advance plant germ plasm from plant introductions to cultivar release.

KEY WORDS: limpograss, *Hemarthria altissima* (Poir) Stapf. et C.E. Hubb., forage evaluation, forage quality.

INTRODUCTION

Limpograss (*Hemarthria altissima* [Poir] Stapf. et C.E. Hubb), a stoloniferous perennial tropical grass, has been collected in several countries around the Mediterranean, in southern Africa, and in Central and South America. Of four original introductions (Oakes 1964), three (P.I. 299993, 299994, and 299995) have been most thoroughly evaluated for agronomic use. These introductions were jointly released by the University of Florida and the USDA SCS as the cultivars Redalta, Greenalta, and Bigalta, respectively (Quesenberry et al. 1978). Limpograss seed production is low, and the grass is vegetatively propagated for commercial pasture establishment (Ruelke et al. 1978). Investigations of additional introductions revealed that broad genetic diversity exists among the more recent introductions in the U.S. collection (Oakes 1978, 1980; Oakes and Foy 1980; Quesenberry et al. 1982; Ruelke et al. 1976; Schank et al. 1973).

In many of the early clipping trials with limpograss, highest production occurred in early spring and late autumn (Kretschmer and Snyder 1979, Killinger 1971). Bigalta limpograss was generally the highest of all grasses in in-vitro organic-matter digestion (IVOMD), and Redalta was the lowest. Limpograss usually had a slightly lower protein content than other grasses. In preliminary grazing trials at Belle Glade, Florida (Quesenberry et al. 1978), cattle gain was usually higher on Bigalta limpograss than on Roselawn St. Augustinegrass (*Stenotaphrum secundataum* [Walt.] Kuntz). This experiment was terminated after one year due to poor summer persistence of Bigalta under grazing defoliation. In another trial at Ona, Florida, total animal gain was 585 kg/ha for a 350-day grazing period. Results from this study suggested that Bigalta could be grazed over much of the year in south central Florida.

Limpograss has been evaluated in many areas of the tropics and subtropics, as well as in more northerly locations in the U.S. Introductions are currently being evaluated in Brazil, Bolivia, Colombia, Ecuador, Hawaii, Malawi, Malaysia, Mexico, N.Z., Venezuela, the Virgin Islands, and other countries. In N.Z. limpograss did not persist well through a long summer dry season (Taylor et al. 1976).

The objectives of this research were to evaluate systematically and rapidly the available germ plasm of limpograss for use as a component of a forage-livestock system; to compare the persistence, yield, and quality of the grass under clipping and grazing as methods of evaluation; and to select superior genotypes for animal performance experiments.

MATERIALS AND METHODS

Two experiments were established at the Beef Research Unit near Gainesville, Florida, in 1976, and a third was established in 1978. All experiments were located on Pomona sand, a sandy siliceous hyperthermic Ultic Haploquod. Experiment 1 consisted of 22 limpograss genotypes plus Coastcross-1 bermudagrass planted in a randomized complete-block design of 5 replications of 2 × 4-m plots. Experiment 2 consisted of 27 limpograss genotypes planted in a randomized complete-block design with 2 replications of 2 × 7-m plots. The limpograss genotypes evaluated in these experiments were selected from previous small-plot clipping trials of 53 limpograss genotypes (Ruelke et al. 1976). Experiment 1 was harvested on five dates at 5-week intervals in 1977 and 1978 with a flail-type forage harvester to a height of 10 cm. Experiment 2 was grazed on five dates at 5-week intervals in 1977 and 1978 using the mob-grazing technique. Forage yield was estimated in experiment 2 using the simple disk-meter technique as described by Santillan et al. (1979). Second-year yields from experiments 1 and 2 were adjusted by percentage of pure limpograss. Samples for IVOMD were taken in 1977 from both experiments.

Experiment 3 consisted of eight limpograss genotypes selected from experiments 1 and 2 on the basis of at least one good agronomic characteristic, e.g., high yield, persistence, or IVOMD. These eight genotypes were established in 1978 in a randomized complete-block design with 2 replications of 4 × 7-m plots within a 0.1-ha pasture. Each 0.1-ha pasture was randomly assigned to a grazing frequency of 3, 5, 7, or 9 weeks. Forage yields before grazing in these plots were estimated using the simple disk-meter technique. Residual forage after grazing was also estimated with the simple disk, and the difference between the before and after values is reported as estimated forage yield. Persistence was judged at the end

K.H. Quesenberry is an Associate Professor, W.R. Ocumpaugh is an Assistant Professor, and O.C. Ruelke is a Professor of Agronomy at the Institute of Food and Agricultural Sciences of the University of Florida. This is a contribution of the Department of Agronomy Institute of Food and Agricultural Sciences of the University of Florida.

Table 1.   Performance of selected limpograss genotypes after two years of defoliation either by clipping or grazing.

		Clipping defoliation (exp. 1)		Grazing defoliation (exp. 2)	
	IVOMD[1]	Yield, MT/ha	Persistence[2] %	Yield, cm[3]	Persistence,[2] %
Redalta	59	10.9	87	68	95
Bigalta	68	4.5	28	00	00
347238	72	0.8	4	46	00
349748	64	11.1	71	80	50
349753	73	2.5	16	79	5
364869	66	7.6	84	20	58
364874	61	1.0	8	43	28
364884	59	—	—	78	88
364887	62	6.3	36	64	58
364888	67	15.8	81	72	83
365509	66	5.3	26	46	40
Coastcross I	65	9.2	67	—	—
LSD.05[4]	5.7	4.2	22.7	20.7	37.2

[1]IVOMD values determined from first year samples.

[2]Visual estimates of percent ground cover in the autumn.

[3]Relative yields expressed in total cm height using the simple disk meter technique. These values can be converted to approximate dry matter yields using the euqation yield (kg/ha) = disk height (cm) X 125. R^2 = 0.75 for total seasonal growth.

[4]LSD values calculated using data from all entries in each experiment, but only selected genotypes are presented for comparison.

of the first grazing year (1979) by visually estimating percentage of ground cover. Samples for IVOMD analysis were collected prior to the last grazing in the autumn of 1979.

RESULTS AND DISCUSSION

Yield and persistence of selected limpograss genotypes in experiments 1 and 2 are presented in Table 1. Persistence after one year of defoliation was higher in experiment 2 than in experiment 1. However, by spring 1978 several of the genotypes that showed good persistence in the autumn had suffered appreciable stand loss during the winter. Experiment 2 limpograss-percentage ratings taken in spring 1978 were generally equal to or lower than ratings of the same genotypes in experiment 1. Experiment 1 yields in 1978 were generally higher than in 1977, ranging from 800 to 15,800 kg/ha. P.I. 364888 yielded significantly more and persisted at least as well as other limpograsses in experiment 1. By October 1978 some genotypes (P.I. 347238 and P.I. 364874) had been essentially eliminated, whereas other genotypes (P.I. 364888 and P.I. 299993) were nearly full stands. Persistence in experiment 2 by October 1978 was generally less than or equal to persistence in experiment 1; thus, 1978 forage yields from experiment 2 were generally less than those from experiment 1. Responses of most genotypes were similar under the two methods of defoliation; however,

Table 2.   Yield estimates, persistence, and IVOMD of eight limpograss genotypes at four levels of rest between defoliation (experiment 3).

Cultivar or P.I. number	Chromosome number	Estimated Yield				Persistence[†]				IVOMD			
		3[‡]	5	7	9	3	5	7	9	3	5	7	9
		------------ cm[¶] ------------				-------------- % --------------				-------------- % --------------			
Redalta	18	24	32	32	36	45	70	80	85	66	68	52	58
Bigalta	36	26	51	38	73	45	65	50	70	80	78	72	74
349753	36	24	33	35	45	40	80	83	85	76	73	59	66
364874	18	18	32	34	40	15	50	65	38	75	70	66	67
364884	18	20	28	36	28	50	35	70	78	68	64	58	60
364887	18	22	31	32	39	28	60	48	88	76	70	62	72
364888	36	27	46	40	77	40	73	80	90	76	74	64	70
365509	36	19	37	40	44	20	53	55	65	76	70	68	70
Mean		22	36	36	48	35	61	66	75	74	71	63	67
LSD .05		nsd	11	nsd	14	nsd	32	nsd	26	4	7	10	6

[†]Visual estimate of percent ground cover in the autumn.

[‡]Grazing frequency in weeks.

[¶]Relative yields expressed in total cm height using the simple disk meter technique.

certain genotypes, such as Bigalta, P.I. 349748, and P.I. 364869, were less persistent under the grazing regime than under clipping.

A range of 59% to 73% in IVOMD of 5-week regrowth forage harvested in June 1977 (Table 1) indicates excellent opportunity for improvement through breeding and selection.

Table 2 summarizes the relative yields, persistence, and IVOMD of the eight selected genotypes after one year of experiment 3. Total seasonal estimated yields in experiment 3 were similar at the 5- and 7-week grazing frequency and at the 9-week grazing frequency were more than twice those at the 3-week frequency (Table 2). The tetraploid genotypes, Bigalta and P.I. 364888, had markedly higher yields at the 9-week grazing frequency. Most genotypes were lower-yielding at the 3-week grazing frequency. There was a marked decrease in percentage of stand at the 3-week grazing frequency. All genotypes except P.I. 364874 had acceptable persistence at the 9-week grazing frequency. P.I. 364888 was generally the highest-yielding entry at all grazing frequencies and was equal to or not significantly different than other lines in persistence at all grazing frequencies. Bigalta had the highest IVOMD at all grazing frequencies, but P.I. 364888 was not significantly lower than Bigalta.

LITERATURE CITED

Killinger, G.B. 1971. Limpograss (*Hemarthria altissima* [Poir] Stapf et C.E. Hubb), a promising forage and beverage grass for the south. Am. Soc. Agron. Abs., 56.

Kretschmer, A.E., Jr., and G.H. Snyder. 1979. Production and quality of limpograss in the subtropics. Agron. J. 71:37–41.

Oakes, A.J. 1964. Plant exploration in South Africa. New Crops Res. Branch, U.S. Dept. Agric., SEA-AR, CRD Mimeo., 158.

Oakes, A.J. 1978. Resistance in *Hemarthria* species to the yellow sugar-cane aphid, *Spiha flava* Forbes. Trop. Agric. (Trinidad) 55:377–381.

Oakes, A.J. 1980. Winter hardiness in limpograss, *Hemarthria altissima*. Soil and Crop Sci. Soc. Fla. Proc. 39:86–88.

Oakes, A.J., and C.D. Foy. 1980. A winter hardy, aluminum tolerant perennial pasture grass for mine spoil reclamation. Am. Soc. Agron. Abs., 103.

Quesenberry, K.H., L.S. Dunavin, Jr., E.M. Hodges, G.B. Killinger, A.E. Kretschmer, Jr., W.R. Ocumpaugh, R.D. Roush, O.C. Ruelke, S.C. Schank, D.C. Smith, G.H. Snyder, and R.L. Stanley. 1978. Redalta, Greenalta and Bigalta limpograss, *Hemarthria altissima*, promising forages for Florida. Fla. Agric. Expt. Sta. Bul. 802.

Quesenberry, K.H., A.J. Oakes, and D.S. Jessop. 1982. Cytological and geographical characterizations of *Hemarthria*. Euphytica 31(2). (Accepted for publication.)

Ruelke, O.C., K.H. Quesenberry, and W.R. Ocumpaugh. 1978. Planting technique effects of establishment, ground cover, production, and digestion of *Hemarthria altissima* (Poir) Stapf et C.E. Hubb. Soil and Crop Sci. Soc. Fla. Proc. 38:40–42.

Ruelke, O.C., K.H. Quesenberry, and D.A. Sleper. 1976. Comparison of greenhouse vs. field plot techniques for evaluating new germplasm of limpograss, *Hemarthria altissima* (Poir) Stapf et C.E. Hubb. Am. Soc. Agron. Abs., 112.

Santillan, R.A., W.R. Ocumpaugh, and G.O. Mott. 1979. Estimating forage yield with a disk meter. Agron. J. 71:71–74.

Schank, S.C., M.A. Klock, and J.E. Moore. 1973. Laboratory evaluation of quality in subtropical grasses. II. Genetic variation among *Hemarthrias* in *in vitro* digestion and stem morphology. Agron. J. 65:256–258.

Taylor, A.O., R.M. Haslemore, and M.N. Mcleod. 1976. Potential of new summer grasses in New Zealand. II. A further range of grasses. N.Z. J. Agric. Res. 19:447–481.

Optimizing the Use of Grazed Dryland Pasture in Western Oregon

T.E. BEDELL
Oregon State University, Corvallis, Oreg., U.S.A.

Summary

When they are properly fertilized, western Oregon dryland pastures of subterranean clover with grasses such as perennial ryegrass, tall fescue, and resident annuals produce total annual dry-matter yields averaging 6,500 kg/ha. Some sites are capable of over 9,000 kg/ha. Optimum use of this forage is complicated by its highly seasonal availability. Over 50% of the total growth occurs from late April to mid-June. Only about 10% takes place during autumn–early winter, and virtually none occurs during summer. Research shows livestock weight gains from 300 to over 700 kg/ha can be expected depending upon pasture mixture and kind of grazing use. Yearling cattle make the most efficient use of rapid spring growth. As grazing season lengthens, total animal production declines both because nutritive value declines and because old forage must be present for use during the summer dry and autumn-winter wet periods.

Four major livestock production programs are used: (1) sheep (winter lambing), (2) beef cows calved in autumn, (3) cows calved in spring, and (4) purchased calves or yearlings. Hay cut from excess spring growth is usually fed during winter except in some coastal locations where appreciable winter plant growth occurs.

By utilizing known characteristics of forage production and animal requirements, forage demand curves were developed. Yearlings stocked at 6–7.5/ha from mid-March to late June make highest weight gains/unit area as compared with cows-calves and ewes-lambs. Animal performance of about 100 kg/head or 730 kg/ha is possible on subclover–tall fescue in the Willamette Valley. Shorter duration of adequate forage quality with clover-ryegrass yields about 560 kg/ha. Pastures are capable of sustaining 10 ewes/ha, but few operators stock that heavily. Production of lamb/ha on clover-ryegrass pastures can be similar to that by yearling steers; but on clover–tall fescue or clover and other coarse grass, it will be less. Sheep do not consume coarse grasses such as tall fescue as well as cattle.

Calf production from cows calved in spring is the most common form of beef operation. But autumn-born calves and their dams make more use of spring forage than spring cow-calf pairs. From 27 to 45 kg more weaning weight can be expected. Wintering costs for this autumn cow-calf pair are an estimated $10–$35/head more. From a comprehensive management standpoint, ewes and lambs with either some purchased yearling cattle or autumn-calved cows appear to yield the most product from dryland and clover-grass pastures.

KEY WORDS: grazing, dryland pasture, Oregon, sheep, cow-calf, yearling cattle, clover-ryegrass.

INTRODUCTION

In the seasonally wet–dry environment of western Oregon (Mosher 1976) thousands of ha of brush and non-commercial forestland converted to improved pastures produce 5,000–8,000 kg/ha of nutritious clover-grass forage. Over 50% is produced from mid-April to June, but only 10% during the autumn-winter period. Sub-clover (*Trifolium subterraneum*) is commonly planted with perennial ryegrass (*Lolium perenne*) or tall fescue (*Festuca arundinacea*). Under improper management resident annuals often replace the perennial grasses, reducing total production (Mosher 1976). Wet soils limit winter use by cattle, and steep topography constrains haymaking. Matching stocking density to seasonally available forage is a formidable management challenge.

The author is an Extension Rangeland Resources Specialist at Oregon State University. This is a contribution of the Rangeland Resources Program of the Department of Animal Science of Oregon State University. It is Oregon Agricultural Experiment Station Technical Paper No. 5743.

Research Base

In the foothills of the Willamette Valley, high seasonal subclover-grass availability was efficiently utilized by yearling steers and ewes with twin lambs in a study to evaluate single and dual use (Bedell 1971, 1973). Pastures were set-stocked at rates designed to remove most of the available forage from late March to early August. Sheep preferred clover to grass except for immature perennial ryegrass. Cattle did not prefer clover to grass. As sheep preferentially removed clover, their performance decreased.

Average weight gains over 3 years as high as 542 kg of lamb and 732 kg of yearling/ha were achieved, depending on pasture mixture and kind, class, and ratio of livestock use. Gains were approximately 120 kg/ha more from clover-fescue than from clover-ryegrass. Cattle alone appeared to be better suited than sheep to utilize clover–tall fescue forage efficiently. Per-ha animal production tended to decrease as the sheep:cattle ratio increased on these pastures. But clover-ryegrass could be efficiently grazed by either or both kinds of stock. Clover did decline when high numbers of sheep were used, which gave some advantage to dual use on clover–perennial ryegrass pastures. During summer (July-August) steer gains were 0.34 kg/day on clover-ryegrass, compared to 0.83 kg/day (P < 0.05) on clover-fescue (Bedell 1971) indicating that for acceptable gains, stock should be removed before forage quality is reduced too greatly.

In a 1500-mm precipitation area in Coos County, Cannon (1976, 1980 personal communication) showed that sheep on a year-round basis can be effective and efficient utilizers of hill-land subclover–perennial ryegrass forage. Over an 8-year period, pastures were grazed at approximately 5 and 10 ewes/ha. The higher rate was the result of more complete utilization of undesirable plants and a buildup in soil nutrients. Lamb production averaged 164 and 270 kg/ha annually. With wool included, total gross value was $198 and $338/ha, respectively. One year's recent experience with ewes rotating through pastures rather than being set-stocked increased lamb production to 400 kg/ha.

On subclover–perennial ryegrass pastures in the Willamette Valley, ewes with their February-born lambs were stocked at 7.4, 10, and 12.4/ha from the time at least 900 kg/ha of green forage occurred in spring until they were taken to a lambing barn in late December (Sharrow et al. 1981). Production characteristics of both sheep and pasture were significantly affected by stocking rate. The heavy rate resulted in replacement of perennial ryegrass by annual grasses, lower total forage production, and a 3-5-week delay in spring turn-on. Lambing percentage decreased by approximately 10% for each stocking-level increase. Subclover declined in the 7.4 ewe/ha treatment because forage utilization was not sufficiently heavy for it to reestablish itself each fall. Pastures with 10 ewes/ha produced 10% more forage than either of the other treatments. Approximately 650 kg/ha of forage dry matter should be allocated/ewe/9-month grazing season. Either

underutilization or overutilization reduces clover-grass productivity.

PROPOSED SOLUTIONS

Using known forage-production patterns and recognizing that both animal performance and forage intake may be reduced when less than 785 kg/ha of dry matter is present (Bedell 1971), forage demand curves were generated for a number of uses (Fig. 1). A stocking density of 7.5 yearlings/ha could be accommodated in southwestern Oregon until mid-June, but in practice operators will not stock that heavily. In the Willamette Valley, with greater forage production for a longer time than shown in Fig. 1, 7.5 yearlings/ha can be reached, but turn-on must be delayed until April 1 or later.

Fall-calved cow-calf pairs will consume 30%–40% more forage than spring-calved pairs on pasture from mid-March until June. At the time of spring turn-on in mid-March to early April, fall calves are 5 to 6 months old and about 135 kg or more. They will easily gain one or more kg/day until forage dries in June or early July. Their dam's milk production increases when the dams are turned onto rapidly growing pasture. Conversely, the small spring-born calves have not yet developed the rumen microflora necessary to utilize green forage efficiently.

A relative benefit for sheep contrasted to breeding cattle

Fig. 1. Accumulative yearling steer and ewe-lamb forage intake at several stocking rates in relation to accumulative available forage on hill pastures (southwestern Oregon). Intake requirements computed for 250-kg steers on March 15 gaining 0.9 kg/day and 70-kg ewes with 130% lamb crop weighing 40 kg when weaned June 15.

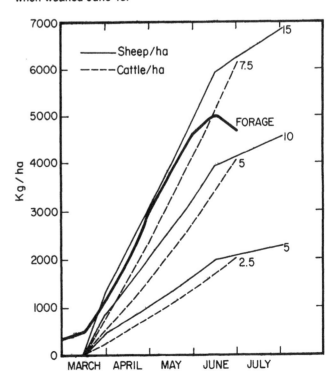

Table 1. Daily forage produced and projected forage intake with varying classes of livestock.

Forage growth, dry matter/ha/day		Forage demand, dry matter/animal/day			
	Steers	Fall calf + cow	Spring calf + cow	Ewe/ lamb[1]
			kg		
Mar 1–15	25	6.6	12.4	9.5	2.7
Mar 16–31	37	6.8	13.2	9.5	3.2
Apr 1–15	75	7.0	13.9	9.8	3.6
Apr 16–30	53	7.4	14.3	10.2	3.9
May 1–15	45	7.8	14.5	10.8	4.0
May 16–31	53	8.1	14.5	11.5	4.3
June 1–15	22	8.4	14.5	11.9	4.5
June 16–30	0	8.8	14.5	12.0	1.4[2]
July 1–31	0	[3]	7.7[4]	12.0	1.4

[1] 130% lamb crop, 70-kg ewes, lambs 40 kg on June 15.

[2] Lambs weaned June 15.

[3] Assumed to be sold or off pasture.

[4] Calf weaned July 1.

is that of greater forage demand concentrated in the spring with higher rates of multiple births. A number of sheep producers consistently raise 140%–180% lamb crops. The additional forage demand increases significantly during May and June; the 10 ewe/ha rate would probably fully utilize the forage produced (Fig. 1).

Data in Table 1 provide information to permit combination of kinds and classes of livestock. With intensive management, 75%–80% forage utilization can be achieved. Use of both cattle and sheep has some beneficial management advantages on pasture species composition (Bedell 1973). No animal production disadvantages were observed on subclover-ryegrass pasture. But on clover-tall fescue pastures, more animal production and fewer unfavorable botanical composition changes oc-

curred with all cattle or high cattle: sheep ratios. Since hill pastures in western Oregon contain some resident annual grasses, increasing stocking pressure in spring will reduce competition of weedy species, permitting subclover to benefit. Where possible, timely haymaking also reduces weeds.

In practice, economic and other considerations, such as predator management, constrain the most efficient use of forage. Sheep numbers are slowly increasing. Grazing yearling cattle is becoming more popular, but more economic risk is encountered with them than in retaining a base cow herd. Fall-born calves weaned in early July are significantly heavier than their spring-born mates weaned in October. Yet, most cows are bred to spring calve because lower winter feed and facility costs are incurred.

A vigorous extension education and demonstration program is showing that grazing either yearling cattle alone or in a combination of yearlings and ewes-lambs most efficiently uses available forage. However, until economic conditions become more favorable, the highest forage-utilization efficiency will not be realized.

LITERATURE CITED

Bedell, T.E. 1971. Performance by cattle and sheep on subclover-grass pastures. Oreg. Agric. Expt. Sta. Tech. Bul. 115.

Bedell, T.E. 1973. Botanical composition of subclover-grass pastures as affected by single and dual grazing by cattle and sheep. Agron. J. 65:502–504.

Cannon, L. 1976. Grazing and pasture management trial of Coos County, Oregon, U.S.A. *In* Proc. Hill Land Symp., 757–759. Morgantown, W.Va.

Mosher, W. 1976. Livestock production and forestry on western hill country. *In* Proc. Hill Land Symp., 661–665. Morgantown, W.Va.

Sharrow, S.H., W.C. Krueger, and F.O. Thetford. 1981. Effects of stocking rate on sheep and hill pasture performance. J. Anim. Sci. 52:210–217.

Pasture Demonstrations for Beef: Agronomic Considerations

N.P. MARTIN, P.R. HASBARGEN, R.L. ARTHAUD, and G.J. SULLIVAN
University of Minnesota, Saint Paul, Minn., U.S.A.

Summary

Beef cattle supply a major portion of farm income of northern Minnesota, but the major resource supporting this enterprise, forage, is vastly underutilized. A 5-year on-the-farm demonstration was conducted by the Agricultural Extension Service of the University of Minnesota and the Upper Great Lakes Regional Commission to improve income in this area by improving management of the beef enterprise.

We compared one or two improved pasture systems with an unimproved permanent grass pasture composed predominantly of Kentucky bluegrass (*Poa pratensis* L.), system A, on each of six ranches. Improved systems included use of a cool-season grass fertilized with 138 kg N/ha for spring and autumn grazing and either birdsfoot trefoil (*Lotus corniculatus* L.) or a legume-grass mixture (alfalfa [*Medicago sativa* L.], red clover [*Trifolium pratense* L.], smooth bromegrass [*Bromus inermis* Leyss.], and orchardgrass [*Dactylis glomerata* L.]) for summer grazing. All pastures were rotationally grazed, with excess forage harvested for hay. System B involved use of 6 ha of birdsfoot trefoil and 8 ha of fertilized grass. System C contained 9.6 ha of legume-grass and 6.0 ha of fertilized grass. A fourth system, AA, had Kentucky bluegrass fertilized once a year with 55 kg N/ha. All pasture systems carried 25 to 30 animal units (AU), and we used cows and calves, except in one case, in which we used lightweight yearlings, to measure productivity.

The introduced cool-season grass, smooth bromegrass, yielded more AUM (months)/ha than did the native Kentucky bluegrass when both received the same amount of N fertilizer. Legumes provided pasture during 70 to 90 days of the grazing seasons (100 to 132 days long). Harvesting of excess forage improved its utilization, but unless the excess was harvested early, fall pasturage was limited.

Animal live-weight gains/ha increased from the use of improved pasture-management systems on all six ranches. Gain increases ranged from 182% to 480% from the best system compared with system A. When extra harvested forage was converted to extra calf gain, at least two ranches produced more calf gain/ha (93 to 126 kg) than was shown at an agricultural experiment station.

KEY WORDS: Kentucky bluegrass, *Poa pratensis* L., smooth bromegrass, *Bromus inermis* Leyss., orchardgrass, *Dactylis glomerata* L., alfalfa, *Medicago sativa* L., red clover, *Trifolium pratense* L., birdsfoot trefoil, *Lotus corniculatus* L., cow-calf, yearlings, pasture systems.

INTRODUCTION

Beef cattle are a major source of farm income in Minnesota. More than half of the beef cows in Minnesota are in the Upper Great Lakes Region (northern 38 counties), where forage crops make up over one-half of the farmland. However, production of beef is far below potential.

In 1974, the University of Minnesota Agricultural Extension Service received a grant from the Upper Great Lakes Regional Development Commission to show, through on-farm demonstrations, more profitable ways to operate the beef cow-calf enterprise, ultimately improving northern Minnesota's economy.

The objectives of the pasture demonstrations were to demonstrate (1) the use of improved management practices, such as seeding improved cool-season grass and legume species, rotational grazing, fertilizing, controlling weeds, and harvesting excess forage for hay; and (2) that improved pasture-management practices can increase animal gain/unit of land area.

METHODS

Project Organization

The project was organized as an interdisciplinary effort during 1974–1979. An area extension agent with summer technical personnel and appropriate office staff was assigned the responsibility for directing the demonstrations on each farm. Specialists from agronomy, animal science, and agriculture and applied economics developed

the project plan, advised as to its execution, and assisted in evaluating the farm results.

Pasture Systems Comparison

All cooperators compared the improved system or systems to an unimproved native permanent-grass pasture, partially wooded and predominantly Kentucky bluegrass (*Poa pratensis* L.), designated system A. No fertilization, weed control, or rotational grazing was practiced in this system, except that animals on the Kenner ranch were rotationally grazed.

The first improved pasture-management system, system AA, involved rotational grazing of a native permanent-grass pasture fertilized with a spring application of 55 kg nitrogen (N)/ha.

Four cooperators selected an additional improved system for comparison. This system, system B, included 8 ha of fertilized native grass and 6 ha of birdsfoot trefoil (*Lotus corniculatus* L.). Birdsfoot trefoil was rotationally grazed, three 2-ha paddocks, 23 June–31 August. The fertilized grass pasture was rotationally grazed during spring, 10 May to 23 June, and in autumn.

Six cooperators selected the last improved system, system C, which was 6 ha of fertilized cool-season grass, either seeded smooth bromegrass (*Bromus inermis* Leyss.) or native permanent grass, and 9.6 ha of a legume-grass mixture—alfalfa (*Medicago sativa* L.), red clover (*Trifolium pratense* L.), orchardgrass (*Dactylis glomerata* L.), and smooth bromegrass. The fertilized grass pasture was grazed during early spring (20 to 31 May), followed by rotational grazing of legume-grass (split into three 3.2-ha paddocks with temporary electric fence) June through August. Then the fertilized grass pasture was divided into thirds for rotational grazing in September and part of October. Excess forage was harvested in June from the fertilized grass pasture and from one or two paddocks of legume-grass.

N.P. Martin is an Extension Agronomist and Associate Professor of Agronomy, P.R. Hasbargen is an Extension Economist and Professor of Agricultural Economics, R.L. Arthaud is an Extension Animal Scientist and Professor of Animal Science, and G.J. Sullivan is a retired Area Extension Agent at the University of Minnesota.

The grass pastures in systems B and C were fertilized with two applications of N, 82.5 and 55 kg N/ha, to provide spring and fall grazing. All pastures, except system A, received annual applications of phosphorus (P) and potassium (K) according to soil test recommendations.

Pasture systems were designed to carry 25 to 30 animal units (AU). All farms except one used cow-calf units to measure pasture performance. Each system used separate bulls. Cattle were weighed periodically during the pasture season to assess animal performance. Cow and yearling weights were adjusted to a standard 455-kg weight for determination of AU months (AUM). Extra forage harvested from each pasture was converted to cow-days by assuming 10 kg of hay were consumed/cow-day (or 8.6 kg for yearlings).

All improved species were seeded by cooperators using oats (*Avena sativa* L.) as a companion crop. Seedbeds were prepared by plowing or disking; lime and fertilizer were incorporated under current recommendations. Oats were removed as grain or hay.

RESULTS AND DISCUSSION

All cooperators obtained more live-weight gain/ha from the improved pasture systems than from their unimproved pasture, system A (Table 1). The increase in live-weight gain/ha, for the highest-yielding system on each ranch above system A, ranged from 186 to 289 kg/ha.

System C yielded more gain/ha than system B on all ranches except Kenner's. However, birdsfoot trefoil stands were excellent only at Kenner's ranch. The other three ranches lost significant portions of birdsfoot trefoil after the first grazing season, due to poor establishment under drought. System AA resulted in an increase of 69

kg/ha of gain over the unimproved pasture at Disterhaupt's (Table 1). This system included only fertilization, weed control, and rotational grazing practices. Systems C and B resulted in higher gain/ha due to the addition of pasture-management practices, such as introducing improved legume-grass species and harvesting excess forage as hay.

We predicted that system B would provide 132 grazing days, 62 from grass and 70 from trefoil, and that system C would provide 140 days (48 and 92 from grass and legume-grass, respectively). Note that the legume predictions were fairly accurate but the grass predictions were not, especially for smooth bromegrass (Table 2). The cool-season grasses did not produce in autumn as expected. Removal of excess forage and application of N fertilizer on bromegrass by Kenner and Savich were too late in summer for adequate fall accumulation of growth. Savich and Kenner harvested 84% and 61% of their AUM of forage as hay from their bromegrass pastures. Schmid et al. (1979) found that tall fescue (*Festuca arundinacea* Schreb.) or orchardgrass was a more productive cool-season grass species in the fall than was Kentucky bluegrass or smooth bromegrass.

Smooth bromegrass provided more AUM/ha grazing from equal N fertilizer than the native bluegrass (Table 2). The most productive legume-grass pasture was at Disterhaupt's because rainfall and soil productivity were best. Birdsfoot trefoil yielded 1.3 more AUM/ha than did the legume-grass at Kenner's because the alfalfa stand was severely depleted in 1979 while the trefoil stand was maintained as established.

Note that the carrying capacities (AUM/ha, Table 2) were similar for system C at the Savich and Disterhaupt ranches, but that live weight/ha (Table 1) was signifi-

Table 1. Animal performance on pasture systems demonstrated on six ranches in northern Minnesota 1976–1980.

Ranch	System[1]	Pasture ha	days	Average daily gain, kg cow	calf	Liveweight gain, kg/ha cow	calf	total
Disterhaupt	A*	45.6	126	.50	.86	44	64	108
	C*	15.2	132	.50	.95	180	217	397
	AA**	32.0	125	.54	.91	72	106	177
Kenner†	A*	27.2	126	.64		–	–	102
	B*	14.0	127	.64		–	–	288
	C*	15.2	121	.73		–	–	279
Lorensen	A**	32.0	88	.45	.91	36	64	100
	B**	14.0	113	.36	.68	76	141	217
	C**	15.6	120	.45	.82	134	158	293
Preisler	A**	43.2	127	.27	.77	26	56	82
	B**	14.0	120	.27	.82	85	163	248
	C**	15.2	128	.41	.77	125	140	265
Savich	A*	37.6	102	.54	.91	23	29	52
	C*	15.2	104	.59	.82	210	92	302
Seifert	A**	24.8	113	.23	.68	21	52	73
	B**	14.0	103	.54	.82	106	136	242
	C**	16.0	119	.54	.82	155	167	322

[1]A, unimproved pasture; B, fertilized grass and birdsfoot trefoil; C, fertilized grass and legume-grass; AA, native grass, fertilized.

*Means of four seasons, 1976–79.

**Means of three seasons, 1977–79 for Lorensen and Preisler, and 1978–80 for Seifert.

†Grazing was done with 204-kg yearling steers and heifers.

Table 2.   Seasonal use and carrying capacities for species components of three systems on three ranches, average of four years.

Ranch (season rainfall, cm)	Pasture system*	Species	Days grazed	AUM/ha		
				grazed	cut	total
Savich** (48.0)	C	S. bromegrass†	18	2.5	13.3	15.8
		legume-grass	83	8.3	3.3	11.6
			101			13.2‡
Disterhaupt (58.3)	C	K. bluegrass	62	11.6	—	11.6
		legume-grass	74	10.2	5.6	15.8
			136			14.0‡
Kenner (46.8)	B	K. bluegrass	61	7.7	3.0	10.7
		B. trefoil	71	8.7	—	8.7
			132			9.8‡
	C	S. bromegrass	36	5.0	7.8	12.8
		legume-grass	90	6.8	0.6	7.4
			126			9.3‡

*System A, unimproved pasture, yielded 1.4, 3.6, and 3.3 AUM/ha for each ranch listed top to bottom, respectively.

**Mean of three grazing seasons.

†N applied to grass species listed from top to bottom; 128, 138, 119, and 119 kg/ha, respectively.

‡Weighted mean for the system.

cantly higher at Disterhaupt's. Not harvesting the fertilized grass at Disterhaupt's allowed more calf gain/unit of area. Also, the Disterhaupt herd had calves with higher average daily gains (Table 1) for two reasons: (1) the herd was crossbred, as opposed to Savich's commercial breed, and (2) Disterhaupt's calving date was 30 days earlier than Savich's. Thus, crossbreeding improved calf gain, and the larger calves converted more forage to gain.

Rabas et al. (1973) demonstrated that legume-grass pastures could produce 330 kg/ha of calf gain for a 104-day grazing season in northern Minnesota. The most calf gain produced/ha in the project was 217 for system C at Disterhaupt's (Table 1). However, Rabas et al. (1973) used a put-and-take grazing system to provide extra animals to harvest excess forage in early spring. If we converted the extra forage harvested at the Disterhaupt's

(Table 2) into extra calf gain (4.5 kg hay/calf-day) instead of cow gain (method used in Table 1), calf gains would have been 423 kg/ha. Three of five ranchers equaled or exceeded 330 kg/ha calf gain; hence, ranchers were able to meet or exceed expectations indicated from grazing research.

LITERATURE CITED

Rabas, D.L., J.W. Rust, and R.H. Anderson. 1973. Evaluation of three legume-grass mixtures for pasture in northern Minnesota: four-year summary. Minn. Beef Cow-calf Rpt. C-4:17–23. Dept. Anim. Sci., Univ. of Minn.

Schmid, A.R., L.J. Elling, A.W. Hovin, W.E. Lueschen, W.W. Nelson, D.L. Rabas, and D.D. Warnes. 1979. Pasture species clipping trials in Minnesota. Minn. Agric. Expt. Sta. Tech. Bul. 317.

Grassland Societies in Manitoba

G. BONNEFOY
Manitoba Department of Agriculture, Canada

Summary

Farmers with forage and livestock problems were organized into grassland societies and cooperated with extension staff supplied by Manitoba Agriculture. The purpose of the grassland societies is to improve by demonstration the traditional management practices used by livestock producers for pastures and forage production for beef and dairy herds. Results over 5 years were clearly positive. Significant improvements were noted in both field and animal husbandry practices. The experience confirmed the principles of the diffusion-adoption process and its application to adult extension education.

KEY WORDS: information diffusion and adoption, beef producers, dairy producers.

INTRODUCTION

The Manitoba Department of Agriculture, recognizing the need for a total management package for forage and livestock management, organized a demonstration with producers in the Teulon area. The objectives included an improved understanding of known management practices related to good pasture and forage production along with good herd-management information for beef and dairy animals.

As a result of an early success with the Teulon experiment in 1972, five grassland societies were developed in 1973 jointly by the Manitoba Department of Agriculture and local livestock commodity groups. The projects were administered by a steering committee of producers and Department of Agriculture regional staff. Since the initial experiment in 1972, 15 grassland forage projects have been completed, involving some 2,000 producers in many Manitoba districts.

The structure of the grassland societies was similar throughout the Province. The St. Claude Grassland Society of dairy producers, for instance, represented people from (1) Dairy Herd Improvement Association (D.H.I.A.), (2) owner sampler program, (3) Record of Performance Association (R.O.P.), and (4) producers at large.

These representatives then elected officers to a Grassland Steering Committee, whose purpose was to organize a working project. The Grassland Steering Committee included 7 farmer producers and four Manitoba Department of Agriculture staff.

The Grassland Society established ground rules, including their own bank account, annual fees for memberships, and annual fees for grazing. They also purchased or rented some equipment.

The functions of the societies are:

- To outline goals and objectives.
- To provide interaction on a group basis as well as individually.
- To plan tours, field days, and meetings.
- To circulate technical production information to producers.
- To provide feedback about the dairy industry.
- To evaluate the program.
- To supervise society-owned equipment.
- To organize and attend short courses, seminars, etc.
- To disseminate technical and production information demonstrated on the project.
- To supply heifers to the project.
- To supply information about forage and livestock through project activities.

DIFFUSION AND ADOPTION

From the standpoint of adult extension education, the general experience of the grassland societies serves as a good example of the diffusion of new ideas among various producer groups.

In particular, the St. Claude experience lends itself to an in-depth study of the diffusion-adoption process. The project was designed to run for a 5-year term, and, fortunately, a formal evaluation was undertaken at its completion.

As was already mentioned, the St. Claude dairy farmers formed their steering committee and set out their aims and objectives in the spring of 1973.

A quarter section of land was leased for a period of 5 years. The land was seeded to tame grasses and legume mixtures. The members of the Grassland Society agreed to supply heifers for a grazing demonstration.

The demonstrations covered new grass and legume species, fertilizers, soil testing, legume inoculation, sod-seeding, chemical weed control, rotational grazing, mechanical grazing, corn silage, suspension fences, soil cement, portable and permanent corrals and chutes, open-face buildings, hay shelters, scale shelter, bunker silo, pocket-gopher control, insect and warble-fly control, AI breeding, pregnancy checking, chin ball markers, heat-mount detectors, synchronized cycling and breeding using prostaglandins, blood sampling, dehorning, pinkeye treatment, vitamin A, D, and E injections, complete ration formulation using corn silage and alfalfa hay, feed test analysis, micronutrient study of copper and zinc, forage-handling systems, round bale silages, and grinding and blending of forages and rations. The demonstrations included total management and integration of forages and livestock.

EVALUATION

In 1979 an impact study was conducted by random sampling of 61 members who had participated. The results were as follows:

In 1978, 54 producers were using established tame pastures, compared with 38 in 1972. Thirty-two are now inoculating legumes, compared with 22 initially. Seventeen are now testing soil prior to application of fertilizer, compared with 4 in the beginning. Eighteen are now using chemical weed control on their forages, compared with 3 before. Fifty of 61 are now using a rotational grazing system, compared with 17 before.

Forty-five are using fertilizer on the hay crop as compared with 15 before. Twenty-eight are now using silage, compared with 5 in 1973. By 1978, 19 of the farmer cooperators had established a corn silage program. Thirty are now getting their feed analyzed regularly, compared with 2 in 1973. Twenty-five are now using winter feed storage hay sheds, compared with none in 1973. Thirteen are using forage silos, compared with none in 1973. Grazing ration decreased on the average from 2.0 ha/head to 1.2 ha/head.

AI breeding increased from 25 to 54; pregnancy checking increased from 7 to 38, and reproductive-system examinations increased from 8 to 40. Forty-eight are now keeping first-breeding age records, compared with 5 in

The author is a Grassland Specialist in the Manitoba Department of Agriculture. Mention of a proprietary product does not imply endorsement of it.

1973. The average age at breeding is now 17.8 months; before the society project, the average breeding age was 27 months. Fifty-five are now using Vitamin ADE compound, compared with 2 before. Sixty-one are now using mineral supplements, compared with 45. Calving interval improved from 14.5 to 12.8 months.

Some 60 farmers are now utilizing insect control, compared with 43 before. Blackleg vaccination has increased from 25 to 46.

Some 40 of 55 farmers' herds had an average herd increase of 953 kg milk/cow in the last 5 years. The increase in herd size within the same period averaged 18 cows. Value of increased production was 90 members × 51 cows × 953 kg milk at $0.26/kg = $1,157,230.00/year.

Production of Grassland heifers averaged 10% higher than the St. Claude D.H.I.A. average.

THE PROJECT AS A TEACHING DEMONSTRATION

1. The project provided an opportunity for members to participate in a group or as individuals. By supplying his own cattle for the demonstration, a member became financially involved.

2. The members observed performance of their own cattle, comparing the performance with that of other members of the Society and with performance of the heifers on the home farm under the old regimen.

3. The project costs for technical-production improvements was not borne by the cooperators. Risk was minimal.

4. The project provided technical staff with complete control so that critical operations could be carried out in a timely manner.

5. The project provided an opportunity for nonmembers to become aware of the demonstration.

6. The project provided for frequent contact and interaction with other members and staff in a number of ways:

    a. The livestock specialist, project technician, and local vet visited the calf at the home farm.

    b. Delivery and pickup of the heifers became special occasions, as were the field days.

    c. The regular monthly newsletter with weights of the heifers on test and the short courses were valuable.

    d. Field trips and tours to Ontario, the U.S.A., and British Columbia were useful, as were annual dairy shows and a Christmas party.

    e. On request, Departmental staff members visited some of the member farms.

SOCIETY INVOLVEMENT

The Society accomplished the following objectives.

1. It defined the problems facing St. Claude producers: (a) Almasippi Sands of low organic matter and fertility; (b) lack of productive pastures (75–90 grazing days/sequence, 1.2–2.4 ha/cow); (c) lack of hay for winter feed supply; (d) poor average daily gains on cattle.

2. It stated the members' goals clearly: (a) pasture sustaining 1 head/acre; (b) 0.8 kg/day average daily gain; (c) 140 + grazing days; (d) cows 2 years of age at first calving.

3. It organized members to help build corrals and a bunker silo at the project.

4. It involved many young producers in the Society.

5. It cooperated with and involved the community with school poster competitions, dairy grassland "Queen" candidates, and prizes and a trophy at a local fair.

6. It sponsored well-organized tours to other parts of Canada and to the U.S.A.

Now that the project is completed, the Society maintains an active membership mailing list, and it plans to hold an annual grassland tour and a Christmas party.

REASONS FOR THE GRASSLAND SOCIETY'S SUCCESS

The Grassland Society concept incorporated most of the principles and techniques necessary for a successful extension program. The St. Claude experience closely approximated stages described in Rogers (1962) in the diffusion of new ideas and their eventual adoption by a number of producers. There were clearly identifiable stages of awareness, interest, trial, and evaluation before adoptions, as Rogers has shown.

The Society was an organizational structure involving people, forages, and livestock, while the project was the teaching demonstration used to change attitudes and skills.

LITERATURE CITED

Rogers, E.M. 1962. Diffusion of innovations. Free Press, New York; Collier-MacMillan, London.

Adoption of Leucaena for Cattle Grazing in Australia

J.H. WILDIN

Queensland Department of Primary Industries, Australia

Summary

The tree legume *Leucaena leucocephala* is a high-quality cattle forage in many tropical countries. For the past 30 years, cattlemen have not planted leucaena despite conclusive evidence from small grazing experiments that it improved beef production and was suited to a large area of northern Australia. Seed of four cultivars was available in the 1960s. In 1968 cv. Peru, an excellent forage type, was planted in a 24-ha block specifically for seed production. But little of this seed was used in Australia. There were several probable reasons for the nonadoption of leucaena. Adverse publicity on the extreme effects of mimosine toxicity, limited ecological adaptation, and problems with establishment created barriers to adoption. Most important, rotational grazing in short cycles advocated for leucaena hedgerows by researchers was in direct conflict with traditional continuous grazing management of native and improved pastures. Extension officers lacked confidence in leucaena and refrained from recommending it.

A management system in which the leucaena canopy was maintained permanently out of cattle reach allowed continuous grazing of accessible forage without detrimental effects on the leucaena plants and on cattle feeding on them. Since 1974 this management system has been effectively used by the producer with the 24-ha Peru cultivar. In November 1979 a promotional program on leucaena for cattle forage commenced in central Queensland. There was an immediate response from cattle producers. In 18 months 450 ha were planted, and the area is expected to increase rapidly in the next few years.

There may be several explanations for the recent commercial acceptance of leucaena. It was concluded that the main reasons were that the positive promotional program clearly defined the role and adaptation of leucaena and emphasized that its management could be simple and not in conflict with traditional practices; that establishment was inexpensive and uncomplicated; and that the inaccessible canopy ensured longevity and reliable seed production and was a valuable drought reserve. Fears of undesirable features were dispelled. The 24-ha commercial block instilled further confidence in leucaena.

The reasons for adoption of leucaena in Australia may be of benefit to other countries experiencing adoption problems. The transfer of grassland research findings to ranch practice is nearly always beset with difficulties. These difficulties are more pronounced on extensive ranches with traditional practices entrenched. Research and subsequent promotion of a ranch innovation should recognize features that lead to rapid adoption, important ones being its benefits, simplicity, compatibility with existing management, and its being observable on a commercial scale. For new crops, seed must be readily available when promoted.

KEY WORDS: leucaena, *Leucaena leucocephala*, grazing, cattle, extension, adoption.

INTRODUCTION

The transfer rate of grassland research findings to farm and ranch practice has often frustrated research workers, farm advisers, and organizations employing them. For high rates of adoption to occur, the innovation must have certain perceived attributes that producers consider important. Research findings must be seen by producers to be relevant to their situation, easy to implement, and practical. A commercial undertaking with which they can identify can often convince them to adopt a new innovation.

In northern Australia many exotic pasture species like buffelgrass (*Cenchrus ciliaris*), rhodesgrass (*Chloris gayana*), Townsville stylo (*Stylosanthes humilis*), and siratro (*Macroptilium atropurpureum*) were adopted rapidly. Leucaena, on the other hand, has been a promising cattle forage in Australia for about 30 years, but until recently it was not adopted commercially. This paper discusses leucaena in Australia, the likely reasons for nonadoption in the past, leucaena promotion and adoption, and some implications.

LEUCAENA IN AUSTRALIA

In many parts of tropical Australia the Hawaii type of leucaena is naturalized. The area of coastal and subcoastal Australia suited to leucaena exceeds 78 million ha (Hutton and Gray 1959, Wildin 1980). Formal research commenced when 33 accessions were examined near Brisbane by Hutton and Gray (1959). On the basis of peak flowering, vegetative vigor, and forage yield, three distinct types—Hawaii, Salvador, and Peru—were recognized. The new cultivar Cunningham, a Peru × Guatemala hybrid, is also a Peru type with desirable

The author is an Agrostologist in the Queensland Department of Primary Industries.

features for grazing purposes. Seed of cv. Peru has been commercially produced in Queensland since 1969, but nearly all this seed has been exported overseas. No commercial grazing areas of leucaena other than the 1968 Peru seed plot of 24 ha had been established before November 1979. Obviously, there were reasons why Australian cattleowners rejected leucaena.

BARRIERS TO LEUCAENA ADOPTION

Leucaena is a high-quality forage capable of very high production, but there were barriers to its adoption in Australia. Six distinct barriers were perceived by cattleowners and the absence of a large commercial grazing area reinforced the lack of confidence. Each of these barriers contributed to rejection of leucaena, but some of them had greater influence than others.

Mimosine Toxicity

Probably the best-known feature of forage leucaena is its mimosine toxicity, the adverse effects of which have been discussed by Jones (1979). The fear of toxicity is real, although toxicity has been observed only in a few grazing experiments where cattle consumed mainly leucaena for long periods. No animal deaths have been caused by mimosine toxicity in leucaena, unlike the effects of cyanogens, selenium, and saponigens (bloating agents) in temperate legumes like alfalfa and clover. Bloat and prussic acid poisoning cause acute effects in contrast to mimosine toxicity, yet the temperate legumes and sorghums are widely accepted. Leucaena has not caused carcinogenic or mutagenic effects (Brewbaker and Hutton 1979).

Climate and Soil Adaptation

The literature on leucaena gives the impression that climate and soil are very critical and that deep fertile tropical soils, neutral to alkaline in reaction and particularly calcareous ones are preferred. Otherwise lime-pelleted seed and correction of soil nutrient deficiencies are necessary (Jones and Jones 1979). Most of the soils in tropical Australia are acid and of relatively low fertility. In regions receiving 600–1,000 mm rainfall the neutral to alkaline soils are clays generally suitable for grain or annual forage cropping. Leucaena was therefore perceived as inferior to the latter alternatives and only suited to better soils. Experience has shown that leucaena will grow and persist on a wide range of well-drained soils over a wide climatic range.

Woody Weed Possibilities

In Queensland some cattleowners have had difficulty in controlling woody weeds that hinder cattle mustering and grass production. To a few people, leucaena seemed to pose a woody weed possibility. The high palatability of leucaena, however, discounts this possibility.

Seed Supply and Cost

Pasture adoption is impossible without seed. Leucaena seed has been available for many years, but in the 1979–1980 summer when central Queensland cattleowners began growing leucaena, the seed price moved from $6/kg (Peru) and $8/kg (Cunningham) to $20/kg for Cunningham in mid-1980. The high seed price became a slight barrier to adoption. One ton of Cunningham seed at $15/kg was sold in summer 1980, but some cattleowners intend to wait until seed price stabilizes at $6–$10/kg before establishing large blocks of leucaena.

Establishment and Early Growth

In experimental plantings field-establishment rate of leucaena has been relatively low, and several workers, including Shaw (1965), Cooksley (1974), and Jones and Jones (1979), have stressed weed control. Unfavorable publicity on poor field establishment created another barrier, although with seed scarification and improved agronomic practices, reliable leucaena establishment was assured (Wildin 1979).

Grazing Management

Nearly all Australian researchers used rotational grazing management. The reasons were that rationing of the available leucaena forage over a grazing period was best achieved by rotational grazing and that vigor and regrowth were adversely affected when all the edible forage was accessible to grazing (Jones 1973, Cooksley 1974, Jones and Jones 1979). Intensive rotational grazing management is foreign to Australian cattleowners and is suspected to be a major barrier to leucaena adoption because it clashes with traditional practice.

PROMOTION OF LEUCAENA AS FORAGE

Relevant Research and Development

Improvement in beef cattle from leucaena grazing was well known, but practical cattle and crop management was needed for Australian conditions. In central Queensland Wildin and Graham (unpublished) showed that continuous grazing of all the edible material during the active growing period adversely affected leucaena growth and that the forage on offer diminished each year. However, continuous grazing in winter had little effect on the vigor of leucaena, and a light, inaccessible canopy was produced. In early summer, leucaena offered little benefit over the grass pastures. One conclusion was that leucaena as a supplementary feed should be encouraged to grow into a tree with a canopy inaccessible to grazing and should be destocked for two to three months in the December–February period to allow maximum accumulation of edible material as lateral shoots and seedlings for grazing from March. This system was a commercial success at Mona Vale, north of Rockhampton (Wildin 1980).

Leucaena Extension

The extension program commenced in November 1979 with weekly awareness segments in the press for one month. A short television coverage increased awareness, and advisory leaflets on growing and grazing leucaena were printed to meet inquiries. A field day on a commercial ranch and an article with photographs in an extension journal aroused more interest. Five features important for rapid adoption of an innovation (Rogers and Shoemaker 1971) were highlighted. These features were on:

1. Relative benefits. Leucaena as a grazing supplement to grass pastures improves body weight of cattle from March to July and possibly to December. The tree crop is permanent. The inaccessible canopy provides a drought reserve and allows seed production options as well.

2. Compatibility with existing practice. The traditional continuous grazing has no detrimental effects on leucaena trees because a permanent canopy is inaccessible to foraging stock. However, destocking in the wettest 3 months of summer, when there is minimal nutritional stress, allows accumulation of high-quality feed for grazing from autumn, when feed quality generally declines. There is little change in management when leucaena is grazed as a feed supplement.

3. Ease of adoption. The simple program for leucaena presented to cattleowners included its wide adaptation (all well-drained soils where a seedbed can be prepared), ease of establishment (using agronomy guidelines), simple fertilizer management (25 kg/ha superphosphate drilled beside the planted row at sowing), and continuous grazing management.

4. Observable commercial stand. Eighty cattleowners attended a field day in December 1979 to inspect a mature leucaena stand growing on an acid soil and grazed as a commercial enterprise. In the 1979–1980 summer, 100 ha were successfully established on numerous infertile soil types on a number of ranches, and a further 350 ha were established in the 1980–1981 summer. These commercial-scale plantings should increase producer confidence in leucaena.

5. Trialability. Leucaena seed is readily available, and trial plantings on small blocks have been undertaken by many cattleowners.

LEUCAENA ADOPTION

Following the positive promotion of leucaena in November-December 1979, adoption in that summer amounted to 20 plantings, ranging from small trial plots to 30-ha blocks totalling 100 ha. There was an increase in the number of new plantings in the 1980–1981 summer, and the cumulative total area planted reached 450 ha. The awareness phase is continuing, and it is expected that, aided by the diffusion effect, the rate of adoption of leucaena in Australia will increase.

CONCLUSIONS

A major conclusion was that the adoption of leucaena on 450 ha by the end of summer 1980–1981 resulted from cattleowners perceiving the new innovation as being beneficial to their enterprise and as being simple, practical, and permanent.

The reasons for leucaena adoption in Australia may be valuable to other countries experiencing adoption problems. The transfer of grassland research findings to ranch practice is nearly always beset with problems. These problems are more pronounced on extensive ranches with traditional practices well entrenched. Research and subsequent extension of a new innovation should recognize features that lead to rapid adoption, the important ones being its relative benefits, simplicity, compatibility with existing management, and observability, particularly on a commercial scale. Finally, for a new forage, seed must be readily available commercially.

LITERATURE CITED

Brewbaker, J.L., and E.M. Hutton. 1979. Leucaena—Versatile tropical tree legume. *In* G.A. Ritchie (ed.) New agricultural crops, 207–259. Select. Symp. 38, AAAS. Westview Press, Boulder, Colo.

Cooksley, D.G. 1974. Growing and grazing leucaena. Qld. Agric. J. 100:258–261.

Hutton, E.M., and S.G. Gray. 1959. Problems in adapting *Leucaena glauca* as a forage for the Australian Tropics. Empire J. Expt. Agric. 27:187–196.

Jones, R.J. 1973. Animal performance on *Leucaena leucocephala*. CSIRO Div. Trop. Agron. Ann. Rpt. 1972-73:15.

Jones, R.J. 1979. The value of *Leucaena leucocephala* as a feed for ruminants in the tropics. World Anim. Rev. 31:12–23.

Jones, R.J., and R.M. Jones. 1979. Agronomy of *Leucaena leucocephala*. CSIRO Div. Trop. Crops and Past. Inform. Serv. Sheet 14-1:1–3.

Rogers, E.M., and F.F. Shoemaker. 1971. Communication of innovations: a cross-cultural approach. Free Press, New York.

Shaw, N.H. 1965. Weed control in *Leucaena leucocephala*. CSIRO Div. Trop. Past. Ann. Rpt. 1965:42.

Wildin, J.H. 1979. Growing and grazing leucaena. Q.D.P.I. Cent. Qld. Adv. Notes 1–4.

Wildin, J.H. 1980. A management system for leucaena. Qld. Agric. J. 106:194–197.

Alternative Forage-Production Patterns for Dairy Cattle Under Conditions of Limited Water Supply in Northern Mexico

R.A. MARTINEZ, M. QUIROGA, N. THOMAS, and K.F. BYERLY

Centro de Investigaciones Agricolas del Norte (CIAN),
Instituto Nacional de Investigaciones Agricolas (INIA), Torreon, Coahuila, Mexico,
and International Development Research Center, Canada

Summary

In the desert climate of northern Mexico ground-water resources are being exploited in forage production for the dairy industry. As rest-water levels are currently falling by 1.75 m annually, there is urgent need to examine the water-management efficiency index (WEI) of present forage cropping patterns and to replace the less efficiently managed species without decreasing the overall regional mean milk production of 14 l/cow/day. The present study describes the approach taken to achieve this goal and the implications of the approach for increased water-management efficiency in this region. Data are drawn from (1) on-farm surveys of 68 forage and 26 dairy-production units, (2) on-station feeding trials, and (3) small-plot studies. The principal analysis is the efficiency of transformation of water into forage DM and milk. Annual forages (corn, sudangrass, Italian ryegrass, oats) were compared individually with alfalfa in sole-forage diets for milk production. Although milk yields were lower with the alternative forages, ratios of WEI to milk and feed costs/kg milk produced were improved in every case. Milk production/cow did not decrease below the regional mean.

The current regional cropping pattern (1978) produces 1.6 million tons of green forage with an approximate WEI of 0.83 kg DM/m³ water. By replacing alfalfa (67% cropped area) with larger proportions of annual forage, WEI could be increased by 12%, with reduction of water input by only 19%. However, great uniformity of water requirement during the year makes a pattern based on annual forages attractive in a region where mean well extraction rates are 35 l/second.

Regional basin-irrigation technology is considered to be the greatest constraint to water-use efficiency. A 25% decrease in water depth/application (officially estimated at 0.16–0.20 ha-m) would increase WEI by 49% to 1.24 and decrease ground-water exploitation by 39%. As measurements determined the realistic value to be 0.22 m, the potential saving is even greater. It is concluded that savings of this magnitude would have a significant impact on the future prosperity of the region.

KEY WORDS: water, forage patterns, milk production, alfalfa, Italian ryegrass, annual forages, irrigation technology.

INTRODUCTION

The Comarca Lagunera, an area of northern Mexico (lat 26° N long 103° W, altitude 1,100 m, and mean annual rainfall 220 mm), has achieved importance over the last decade as the principal supplier of milk to Mexico City, about 1,000 km to the south. Because of the desert climate and the use of renewable water resources in the production of staple food and fiber crops, the dairy industry has been founded on the exploitation of ground water. Alfalfa (*Medicago sativa*) has been the principal forage grown under a basin-irrigation system.

Since the late 1960s, when the total area seeded to forage (approximately 7,000 ha) was only slowly increasing and rest-water levels (RWL) were practically constant, the increase in forage area has caused the net annual decline in RWL to reach 1.75 m. Alfalfa, the principal forage crop, has tripled in sown area since 1970. Farmers value alfalfa for its high yield and high-quality dry matter (DM). Though considerable differences are found among farmers, nine harvests can be made annually, total yields often being well above the regional mean of 18 t DM/ha reported officially (SARH 1978). Official estimates put the mean annual water input to the crop during a 3-year production cycle at 2.40 ha-m, resulting in an approximate water-management efficiency index (WEI) of 0.79 kg DM/m³ water. There is no government policy to encourage farmers to adopt more efficient irrigation systems.

The overall objective of the present study was to evaluate the difficult task of producing milk in a region with decreasing ground water. A specific objective was subsequently stated to be a search for increased efficiency

R.A. Martinez is an Animal Scientist at CIAN-INIA (current address, Technical Sub-Director, Centro de Investigaciones Agricolas del Norte Central, Calera, Zacatecas, Mexico), M. Quiroga is an Agronomist at CIAN-INIA, N. Thomas is a Technical Advisor at the International Development Research Center (current address, Small Ruminant CRSP, P.O. Box 210, Bogor, Indonesia), and K.F. Byerly is a Technical Sub-Director at CIAN-INIA.

of water management through modified forage-cropping patterns. This paper describes the work undertaken to achieve that objective.

METHODS

A combined approach was employed, of (1) on-farm sampling, (2) on-station feeding trials, and (3) small-plot studies. In component (1) a management survey of 68 forage-production and 26 dairy units was carried out, 40 and 9 respectively, being retained for detailed analysis. Data from (2) and (3) were obtained, using the forage and animal-management methods common to the region. The principal analysis was to determine the efficiency of transformation of irrigation water into forage and milk.

RESULTS AND DISCUSSION

The annual grasses, corn (*Zea mays*), sorghum (*Sorghum vulgare*), sudan (*Sorghum sudanensis*), and Italian ryegrass (*Lolium multiflorum*) achieve a higher WEI than alfalfa with similar irrigation management. The approach adopted was to determine whether alfalfa could be replaced to a large extent as the major component of dairy feeds without causing a decline in milk production. The average amount of forage on offer on the survey farms was determined to be 54 kg fresh matter and a further 7.4 kg of concentrate being fed/animal. The latter makes up 33% of the total ration, assuming 25% DM in the fresh forage. For a regional mean milk production of 14 l/cow/day it is considered that the animals are generally overfed. Alfalfa was the major DM component of the diets examined.

Italian ryegrass, oats (*avena sativa*), and sudangrass provide feasible greenchop alternatives to alfalfa, and corn silage can be substituted year-round. Each of these forages was compared with alfalfa as a basal diet for milk production (Table 1), with promising results. In each case the alternatives increased WEI measured in milk terms and reduced feeding costs. While all alternatives showed lower intakes and lower milk production than the compared alfalfa diet, especially corn silage, these lower intakes are considered to be largely due to the difference in form of forage offered. The use of green-chop alfalfa in-

stead of hay during summer would probably have reduced intake on these treatments. The low intake of corn silage can be attributed to its low crude-protein content (8.9%), partial substitution being made with concentrate. The productivity of animals on the oats diet is considered high; it probably reflects superior feeding prior to the study. This disparity would also be reflected regionally.

It is not envisaged that these species will be sole forage components in the cattle diets of the region. Because of regional farming patterns, alfalfa will continue to be a major DM source (it is important as a break crop for cotton). However, these results illustrate clearly that alfalfa can be substituted without jeopardizing milk production. Combined alfalfa and substitute diets could be expected to show intermediate productivity.

The mean water yield of ground-water wells in the region is 35 l/second, though most forage production units have more than one well. Based on the observed mean input of 0.22 ha-m of water/irrigation, Table 2 shows the water requirement of a standard 100 ha of the traditional forage pattern and an alternative based mainly on annual grasses. The traditional pattern requires more than 100 l/second during the critical time of the year when summer crops compete for this resource, WEI being 0.71. In the alternative WEI rises to 0.88, an increase of 22%. The greater uniformity of water input means this is a pattern suitable for wells of low output. As summer production always requires more water than winter, it is doubtful that the summer-winter difference could be reduced further.

The present regional cropping pattern produces 1.6 million t of green forage annually, at a cost of almost 500 million m³ of water, a WEI of 0.83, assuming 25% DM (Table 3). Alfalfa occupies 67% of the forage area but uses 81% of the applied water. The relationship between the increase in area of alfalfa (x = 1000 ha) and the decline in RWL (y) since their respective 1967-1970 values can be represented by y = 0.54 + 1.75 x (r² = 0.90). It does not seem possible that the fall in rest-water levels could be arrested, as the forage area is 3 times that of 1967-1970, but it should be possible to slow the fall by reducing the area of alfalfa below that of this period and replacing lost production with water-efficient annual crops. Adoption of the alternative farm pattern on a regional scale would increase WEI to 0.93, a 12% in-

Table 1. Milk production using alfalfa and annual forages.

| Trial
Basal forage | Winter | | | | Summer | | | |
| | I | | II | | III | | IV | |
	Alfalfa (hay)	Ryegrass (green)	Alfalfa (hay)	Oats (green)	Alfalfa (hay)	Sudan (green)	Alfalfa (hay)	Corn (silage)
DM production t/ha	18	13		7		16		14
Irrigation yearly, ha/m	2.40	1.50		1.00		1.20		0.80
WEI, kg DM/m³ H₂O	0.75	0.87		0.70		1.33		1.75
Intake, kg DM/cow/day	15.9	13.8	15.5	13.9	16.0	12.6	18.9	7.6
Concentrate, kg/cow/day	3.1	3.3	3.8	3.8	3.8	3.9	3.3	4.0
Milk production, kg/cow/day	21.8	18.8	22.7	21.0	16.7	14.7	21.3	13.1
WEI, kg milk/m³ H₂O	1.0	1.2	1.1	1.1	0.78	1.55	0.84	3.0
Feeding cost, pesos/kg milk	2.82	2.12	2.70	2.30	2.70	2.30	2.95	2.76

Table 2.   DM productivity and water requirement (l/second) of standard 100 ha of traditional and alternative forage-crop pattern.

	Crop	Area (ha)	DM (t)	
Traditional	Alfalfa	50	900	
	Corn	30	420	WEI = 0.71
	Oats	20	120	
Alternative	Alfalfa	10	180	
	Corn	20	280	
	Sudan	30	600	WEI = 0.88
	IR	30	330	
	Oats	10	60	

Month	Traditional (l/second)		Alternative (l/second)	
	Max	Mean	Max	Mean
J	50	42	41	37
F	45	45	45	43
M	72	64	103	85
A	97	95	103	80
M	97	97	72	72
J	127	94	78	63
J	60	60	48	44
A	69	55	36	25
S	69	68	72	46
O	63	58	54	51
N	50	50	43	42
D	50	50	41	41

crease but a reduction of only 19% in water input.

In nearly every case, the constraints imposed by basin-irrigation technology as practiced in the region result in irrigation beyond crop requirements. Calculations in this paper have used the measured 0.22 ha-m as the regional mean single application (other than Tables 1 and 3, which are based on official estimates), though significantly greater applications (by 25%–50%) have been measured. An improvement in technology, reducing the officially estimated application of 0.16–0.20 ha-m by 25%, would transform WEI to 1.24, a 49% improvement over the actual pattern, and would save 189 million m³, a decrease in ground-water extraction of 39% (Table 3). Given the higher observed value of 0.22 ha-m, potential savings are even greater.

CONCLUSIONS

The results of this study suggest that a significant increase in water-management efficiency in milk production could be gained through modified forage-production patterns, replacing alfalfa by annual forage species, without decreasing milk production/cow below the regional mean level. However, it is felt that significant savings in ground water use will occur only through adoption of improved irrigation technology.

LITERATURE CITED

SARH. 1978. Informe del patronato de sanidad vegetal. Secretaria de Agricultura y Recursos Hidraulicos, Comarca Lagunera Ciclo 1977–78.

Table 3.   Official forage-crop production data, Comarca Lagunera 1977-78, and estimated improvement in water management efficiency.

	1978 area	Annual irrigation (ha-m) (million m³)		Total DM production (1000 t)	Alternative area	Irrigation* (million m³)	Estimated DM prod. (1000 t)
Alfalfa	16368	2.40	392.8	309	3000	54	57
Corn	2212	0.80	17.7	24	6000	36	64
Sorghum	373	1.20	4.5	3.5	500	4.5	4.7
Sudan	565	1.20	6.8	5.8	4000	81	92
Ryegrass	3484	1.50	52.3	51	9000	101.7	132
Oats	1260	0.80	10.1	7.0	3000	18	17
	24262		484.2	400.3[1]	25500	295.2	366.7[2]

[1]WEI = 0.83 kg DM/m³ H_2O

[2]WEI = 1.24 kg DM/m³ H_2O.

*25% Less water/Irrigation.

Source: SARH (1978).

Legume Renovation of Grass-Dominant Swards: A Data-Based Educational Program That Led to Adoption of Practices by Kentucky Farmers

J.K. EVANS, G.D. LACEFIELD, and W.C. THOMPSON
University of Kentucky, Lexington, Ky., U.S.A.

Summary

Kentucky has about 2.83 million ha of forage crops, most of which are on land with moderate to severe erosion hazards. By the early 1950s, one cultivar (Kentucky 31) of tall fescue (*Festuca arundinacea* Schreb.) had been seeded on a relatively large area. Legumes grown with grasses had been shown to improve quality of forage markedly.

Legumes were lost from many Kentucky grass fields during the relatively dry summers of 1953 and 1954. Research on legume establishment was expanded at the University of Kentucky in 1955. By 1959 sufficient data had been accumulated to formulate recommendations and begin educational programs for farmers.

Counties were selected to receive intensive emphasis on educational programs in 1959 and in each year thereafter through 1965. County agents, county leaders, and extension forage specialists, meeting in each selected county, concluded that pasture production and quality were the major livestock problems needing attention.

Numerous demonstration seedings were made at easily accessible and highly visible locations in the participating counties. Information campaigns were conducted in the selected counties through radio, television, newspapers, farm magazines, and direct mail to create an awareness of (1) the need for legumes in grass fields, (2) the ability of legumes to be established without plowing, (3) locations of renovation demonstrations, and (4) locations where additional information could be obtained. Educational activities included county and neighborhood classes, field days, and tours.

The estimated area renovated statewide increased from about 7,000 ha in 1960 to 243,000 ha in 1967. The total number of cattle and calves in one of the early high-educational-emphasis (HE) counties had been increasing during the 1949–1969 period at about the same rate as in the district in which it is located. From 1969 to 1975, the HE country increased at a rate much faster than the district. In a comparison of 8 selected HE with 8 low-emphasis (LE) counties, the HE counties increased cattle numbers more slowly than the LE counties during the preprogram period, 1949–1959, but faster than LE counties after 1959. Although no cause-and-effect relationships between renovation and increased cattle numbers can be proven, it is probable that there was an effect due to increased feed production.

Prior to 1960, "renovation" was not a term used in connection with grass fields. However, in a 1980 survey conducted in 30 randomly selected Kentucky counties, 72% of the livestock farmers responding to the survey said that renovation is a common practice on their farms, and 73% said extension had had either "much" or "very much" influence on their decision to renovate. Thus, it would appear that extension educational programs have been influential in the renovation of grass fields becoming a routine practice on about three-fourths of Kentucky's livestock farms.

KEY WORDS: renovation, legumes, extension education, practice adoption.

INTRODUCTION

Kentucky has about 2.83 million ha of forage crops, most of which are on land with moderate to severe erosion hazards. By the early 1950s, one cultivar (Kentucky 31) of tall fescue (*Festuca arundinacea* Schreb.) had been seeded on a relatively large area; it is currently the base grass on about 2.43 million ha in Kentucky (Fergus 1971). Other cool-season grasses such as orchardgrass (*Dactylis glomerata* L.), Kentucky bluegrass (*Poa pratensis* L.), and timothy (*Phleum pratense* L.) provide the basic vegetation on the re-

J.K. Evans is an Extension Specialist in Forages and G.D. Lacefield is an Associate Extension Professor of Agronomy at the University of Kentucky; W.C. Thompson is a Manager of Forage Marketing and Planning of North American

Plant Breeders and was formerly an Extension Specialist in Forages at the University of Kentucky.

maining 0.4 million ha. Legumes grown with grasses usually improve both the quantity and the quality of forage (Roberts 1910, Seath et al. 1954, Hogg 1965, Spooner and Ray 1969).

Most legumes were lost from Kentucky grasslands during the dry summers of 1953 and 1954 (Templeton 1974). Subsequently, the Kentucky Legislature appropriated $100,000 for use by the University of Kentucky Agricultural Experiment Station to conduct forage research and extension programs dealing with low cattle performance on tall fescue. Research on legume establishment was expanded in 1955 and has continued to date.

By 1959 enough data and experience had been accumulated to formulate the following recommendations for renovating grass fields: (1) graze grass intensively; (2) test soil; (3) apply needed lime and fertilizer (not nitrogen); (4) disk or otherwise till to disturb vegetation, 40%–60% if seeding clovers, 80%–100% if seeding alfalfa; (5) smooth and firm soil; (6) inoculate legume seed with proper rhizobia; (7) seed and cover lightly (approximately ½ to 1 cm); (8) graze grass plants as needed to prevent shading of legume seedlings; and (9) when animals begin biting tops of seedlings, remove animals and allow time for legumes to become established before grazing or cutting for hay. These recommendations have been revised several times as new data have become available (Evans et al. 1978).

PLANNING AND CONDUCTING EDUCATIONAL PROGRAMS

The importance of county agents, community and neighborhood leaders, and friends in dissemination of information and adoption of practices has been long recognized (Lionberger and Coughenour 1957, Bohlen et al. 1961). With this factor in mind, the state extension forage specialist and his associates selected three counties in which to begin a renovation-education program in 1959-1960. County agents were asked to invite leaders from among the farmers, agribusinesses, and agencies in their counties to meet with the agent and state extension forage specialist to identify forage problems and to plan programs to help solve the problems. Each group concluded that pasture production and quality were the major livestock problems in its county.

Method-and-result demonstrations were conducted with several leading farmers in each county. Locations that needed renovation and that were easily accessible to view by the target audience were chosen. Signs were posted at each site identifying what had been done and by whom. Field days and tours were conducted so people from within and outside the county could see results of renovation.

County leaders promoted attendance at educational meetings and presented the renovation concept to farmers with whom they worked. They also requested that the federal government share renovation costs through the Agricultural Conservation Program (ACP).

Radio, television, newspapers, farm magazines, and direct mail were used to create an awareness of (1) the need for legumes in grass fields; (2) University of Kentucky research showing that legumes could be established without plowing; (3) locations where demonstrations were to be or had been conducted; and (4) locations where additional information could be obtained.

Essentially the same procedures were followed each year after 1959. Renovation continues to receive a strong emphasis in the statewide extension forage program. Since 1974 a major portion of time has been devoted to developing, testing, and extensively demonstrating renovation machines and chemicals.

RESULTS AND DISCUSSION

Objectives were to motivate farmers and to teach them how to renovate grass fields with legumes. Adding animals to utilize the increased forage production was emphasized throughout. It was felt, therefore, that two good measures of success would be ha renovated and cattle numbers.

Ha Renovated

Estimates of ha renovated were made annually during the period 1960-1967. ACP cost-sharing figures were available from county records. County agents were asked to estimate how much of the total seedings was cost shared. From these two values, total renovated ha was calculated.

Estimated ha renovated during the period 1959-1967 increased to a maximum of 243,000 ha in 1967. The method used in arriving at estimates was highly subjective; therefore, one may legitimately question the exact numbers obtained. Kentuckians agree, however, that there was a dramatic increase in ha of legumes seeded by the renovation method.

Cattle Numbers

Although cattle numbers show short-term cyclic response to price fluctuations, the general trends were upward in most Kentucky counties, the state, and the U.S. over the period 1949-1975. Numbers took a sharp turn downward in 1975 in response to the drastic price decline that occurred in 1973.

Percentage changes in total cattle numbers in 8 counties that received high emphasis (HE) and 8 that received lower emphasis (LE) are shown in Fig. 1. The 8 LE counties show percentage increases greater than those of HE counties during the period prior to the educational program 1949-1959, but in the 1959-1975 period the HE counties appear to have increased cattle numbers at a faster rate.

Livingston County was one of the first counties selected for program emphasis. Fig. 2 shows that cattle numbers increased at essentially the same rate in both this county and the district of which it is part during the period 1949-1969. In 1966 an additional extension agent was placed in the county to increase demonstration efforts

Fig. 1. Percentage change in numbers of total cattle and calves in 8 Kentucky counties that received high renovation-education emphasis and 8 counties that received low emphasis.

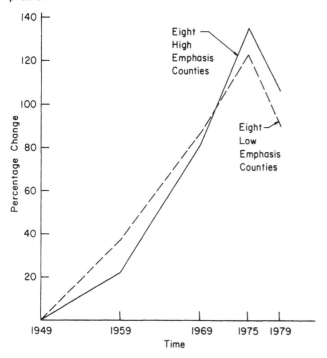

Fig. 2. Percentage change in numbers of total cattle and calves in Livingston County, Ky., and in Ky. crop reporting district 1.

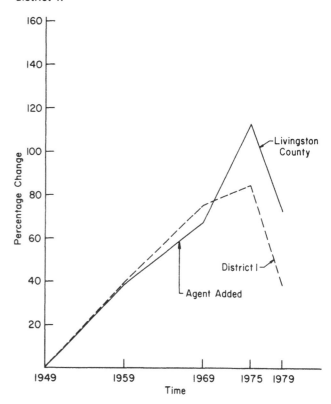

and provide more individual assistance. County cattle numbers increased more rapidly than district numbers from 1969 to 1975.

Percentage decreases during 1975–1979 were similar for all comparisons presented; however, numbers stabilized at high levels in all areas of HE.

Even though educational programs were started in only a few counties at the beginning, they were not confined to those counties. Intensive programs were started later in all districts. Also, farmers from other counties attended meetings, observed demonstrations, and received motivational mass-media messages aimed at farmers in the HE counties. It also has been shown that dissemination of ideas and adoption of practices may spread from leaders to followers, irrespective of geographic or political boundaries (Bohlen et al. 1961).

SURVEY OF KENTUCKY FARMERS, 1980

Prior to 1960, "renovation" was not a term used in connection with a grass field. Also, farmers did not at first widely use the renovation procedures outlined by our researchers and recommended in our educational programs. Consequently, a survey was conducted in December 1980 to determine (1) if renovation of grass fields is a common practice on farms and in neighborhoods of respondents, and (2) the degree of influence respondents believe extension had in getting farmers to renovate.

A random sample (30 of the 120 Kentucky counties) was selected for the survey. For a 95% confidence level (Summerhill and Taylor 1980) it was determined that 376

of 21,000 livestock farms in selected counties should be sampled. Questionnaires were mailed to farmers, using standard procedures for such surveys, and 78.5% of the questionnaires were returned.

Over 70% of those who returned the questionnaire said that renovation was a common practice on their farms and that extension had "much" or "very much" influence on their decision to renovate. Respondents were less certain about what others were doing. Thirty-four percent said renovation was not a common practice in their neighborhood, 41% said it was, and 25% were uncertain. Survey responses have reinforced the belief of the authors that renovation is now a commonly used practice in Kentucky.

LITERATURE CITED

Bohlen, J.M., C.M. Coughenour, H.F. Lionberger, E.O. Moe, and E.M. Rogers. 1961. Adopters of new farm ideas. N. Cent. Reg. Pub. 13.

Evans, J.K., G.D. Lacefield, T.H. Taylor, W.C. Templeton, Jr., and E.M. Smith. 1978. Renovating grass fields. Ky. Agric. Ext. Serv. AGR-26.

Fergus, E.N. 1971. Estimates of tall fescue acreages in various states. Univ. of Ky. Mimeo.

Hogg, P.G. 1965. Effect of legumes and nitrogen fertilizer on yield of cultivated cynodon (coastal bermudagrass) pastures in southern U.S.A. Proc. IX Int. Grassl. Cong., 1099–1101.

Lionberger, H.F., and C.M. Coughenour. 1957. Social structure and diffusion of farm information. Mo. Agric. Expt. Sta. Res. Bul. 631.

Roberts, I.P. 1910. The Roberts pasture. *In* Pasture in New York. Cornell Agric. Expt. Sta. Bul. 280.

Seath, D.M., C.A. Lassiter, G.M. Bastin, and R.F. Elliott. 1954. Effect of kind of pasture on the yield of TDN and on persistency of milk production of milk cows. Ky. Agric. Expt. Sta. Bul. 609.

Spooner, A.E., and M.L. Ray. 1969. Pasture fertilization and grazing management studies in southwest Arkansas. Ark. Agric. Expt. Sta. Bul. 741.

Summerhill, W.R., and C.L. Taylor. 1980. 4-H Accountability Workshop, Palm Beach, Fla. IFAS Fla. Agric. Ext. Serv.

Templeton, W.C., Jr. 1974. Pasture renovation in the upper south. So. Sect. Am. Soc. Anim. Sci. ASAW. Memphis, Tenn.

West Virginia University Allegheny Highlands Project: A Ten-Year Experiment in Technology Transfer

B.S. BAKER, P.E. LEWIS, E.K. INSKEEP, and R.H. MAXWELL
West Virginia University, Morgantown, W.Va., U.S.A.

Summary

West Virginia's livestock industry is characterized by relatively small units located predominantly on rough terrain not suited to highly mechanized or intensive agriculture. As a result of changes in West Virginia agriculture during the last 40 years, there has been a reduction in support from agribusiness, marketing alternatives, and profitability. Some agriculturalists contend that if present technology were adopted, beef and sheep farms in mountainous regions could be profitable and could contribute to the local economy.

The objective of the Allegheny Highlands Project (AHP) was to test the effectiveness of a method for delivering technological information to farmers to promote rural development. A multidisciplinary team of specialists in agronomy, animal science, farm management, and veterinary medicine worked together to provide producers with a package of recommendations to increase production and/or profitability.

Comprehensive production, inventory, and financial data were collected on 65 farms from 1971 through 1979. Linear regression analyses were used to assess the importance of years of project participation, year of entry into the project, farm size, farm enterprise, method of technology transfer, and time devoted to farm operation on changes occurring in farm production, inventory, and income. Hay yields/ha, 205-day weights of calves, market weights of calves, value of calf/cow, lambing percentage, marketable lamb-crop percentage, weight of lambs when sold, kg of lamb marketed/ewe, and value of marketable lambs/ewe all increased with years of participation in the project. Large farms without nonfarm income had higher hay yields/ha than other farms. The 205-day weights and market weights of calves were less on farms managed by part-time farmers who placed minor emphasis on farming. Large farms had more liabilities, receipts, expenses, and net income than smaller farms. Most production goals established on a farm basis were achieved, whereas production goals for the project area were not achieved.

The package approach used in the AHP was successful. Farmers benefitted from increased production and increased income. Farmer and community attitudes became more positive, and cooperation increased. The University benefitted from increased contacts with clientele, which stimulated research and teaching. Elements found necessary for a successful project were: goals and procedures, competent staff, appropriate disciplines to meet farmer needs, mutual confidence between farmer and field staff, community involvement, adequate funds, administrative support, continuity of staff, and adequate time. The authors concluded that much of the present technology is applicable but needs modification to meet individual needs, that one-to-one contacts are the most effective means of getting technology adopted, that people farm for different reasons, and, in general, that changes occur slowly.

KEY WORDS: package approach, small farms, technology transfer, Allegheny Highlands, rural development.

INTRODUCTION

West Virginia's livestock industry is characterized by small units located predominantly on rough terrain suited to neither highly mechanized nor intensive agriculture. Many of these units are part-time operations in which farming is a secondary family occupation. As a result of changes in agriculture during the last 40 years, there has been a reduction in support from agribusiness, in marketing alternatives, and in profitability.

In the late 1960s some agricultural leaders contended that technology existed that could make hill farming profitable and that agriculture could help meet the economic needs of rural West Virginia. Many essentials for expanding the beef and sheep industries were available, including raw materials, climate, and potential markets. Various faculty and administrators of West Virginia University's College of Agriculture and Forestry (WVU-CAF) recommended a package approach for dissemination of technology necessary to meet the needs of individual farmers, strengthen teaching and research programs, and revitalize the state's beef and sheep industries. With this objective the Allegheny Highlands Project (AHP) began in 1970. Cooperators were enrolled in subsequent years until the AHP encompassed a 9-county area with more than 80 cooperators.

The overall goal of the AHP was to integrate existing forage-livestock technologies into total management systems for individual cooperators and to determine the influences of the newly introduced technology on farm production, income, and community life over a 10-year period.

METHODS

Organization of the Project

A field team consisting of director, agronomist, animal scientist, farm economist, veterinarian, and support staff provided information to cooperators in a package approach designed to meet each farmer's needs and goals. A resident panel located on the campus of WVU served as liaison between campus faculty and field team. The State Department of Agriculture, Cooperative Extension Service, Soil Conservation Service, Farmers Home Administration, and Agricultural Stabilization and Conservation Service provided assistance to the field staff.

Cooperators were enrolled who met selection criteria. A memorandum of understanding outlining objectives, services, and commitments of each party was used as a formal agreement between cooperators and WVU-CAF.

B.S. Baker is a Professor of Agronomy, P.E. Lewis is an Associate Professor of Animal Science, E.K. Inskeep is a Professor of Animal Science, and R.H. Maxwell is an Associate Dean at the College of Agriculture and Forestry of West Virginia University. This is a contribution of the West Virginia University Allegheny Highlands Program and is published with the approval of the Director of the West Virginia Agricultural and Forestry Experiment Station as Scientific Paper No. 1679.

Cooperators, with assistance and supervision from field staff or cooperative extension service agents, kept production, income, and inventory records on their farms. Technical information was provided to cooperators through farm visits, newsletters, field days, educational meetings, workshops, tours, demonstrations, office visits, and telephone conversations. The only benefits received by cooperators were technology and the services necessary to demonstrate its use. Persuasion was the only means of getting cooperators to implement technology.

Description of Area and Cooperators

The AHP area was characterized by hilly terrain with narrow valleys adjacent to streams. Elevation varied from 300 to 1,200 m. Rainfall and temperature were favorable for forage and livestock production. The majority of the farms in the region were small, and employment in off-farm jobs by farmer or spouse was typical of farm families. The most common farm enterprises were beef and sheep.

Cooperators who enrolled in the AHP were typical of farmers throughout the area. Beef, the most common enterprise, was on 95% of the farms. Approximately half of all cooperators raised sheep. The mean farm size was 78 ha with 72 animal units (AU).

Analyses of Data

Data were analyzed using a multiple linear regression model in which changes over time were expressed as a function of year enrolled, method used to disseminate information, number of years of participation, farm enterprise, farm size as determined by animal units (AU), and amount of time and emphasis devoted to farming. Cooperators were enrolled in 5 different years, and 65 of the 81 cooperators enrolled provided data for 2 or more years that were used in analyses. Farmers who received information directly from field staff and enrolled between 1970 and 1973 were designated phase I cooperators. Farmers who enrolled in 1975 and 1976 and received information via county agents with less frequent field staff contact than phase I cooperators were designated phase II cooperators. There were 35 phase I and 30 phase II cooperators. Three enterprise groups were established: farms with sheep only, beef only, or a combination of beef and sheep. Four animal-number size groups were used: less than 30 AU, 30 to 54 AU, 55 to 99 AU, and 100 or more AU. Five time groups were utilized: full-time farmers with no outside income, full-time farmers with outside income, part-time farmers with major emphasis on farming, part-time farmers with minor emphasis on farming, and farmers who had retired from some occupation and were receiving benefits.

Subjective data were gathered during the AHP but were not submitted to conventional statistical analyses. Such data are presented as impressions and conclusions of the authors.

Table 1. Means and change in various production and income variables/year of project participation on 65 farms cooperating in the Allegheny Highlands Project.

Variable	Mean[†]	Change/yr[‡] of participation
Hay yield/ha (metric tons)	5.9	0.2
Adj. 205-day wt/calf (kg)	206	3.3
Market wt/calf (kg)	211	3.3
Kg calf marketed/cow	186	3.2
Kg calf marketed/day of age	0.9	0.02
Value calf/cow	$190	$24
Lambing percentage	143	2.8
Marketable lamb crop percentage	120	2.6
Value of marketable lamb/ewe	$53	$7
Kg lamb marketed/ewe	52	1.4
Total receipts[§]	$17,490	$2051
Total expenses[n]	$16,408	$1764
Operating income/animal unit	$195	$23

[†]Means presented are for the midpoint of the project.

[‡]Values for change/year of participation are regression equation estimates and are significant (P<.01) using t-test.

[§]Receipts for wool, lamb, and beef increased (P<.01).

[n]Numerous expenses, including fertilizer, lime, fuel, repairs, veterinary and drug supplies, and utilities, increased (P<.01).

RESULTS AND DISCUSSION

Areas in which major changes occurred during the AHP are presented in Table 1. Hay yields increased from 4,900 kg/ha to 6,900 kg/ha, or about 40%. At the beginning of the project hay yields were one major factor limiting livestock numbers, and the increase in production enabled several herds and flocks to expand. There was an increase in forage quality, indicated by forage tests. First-cutting hay averaged 9.2% crude protein and 54.6% total digestible nutrients during the first 3 years and 10.9% and 56.9%, respectively, during the last 3 years.

Adjusted 205-day weights of calves averaged 206 kg/calf at the midpoint of the project. There was an increase of 3.3 kg/yr of participation. Weights increased due to an increase in cross-breeding, use of improved sires, and selection of better replacement females (Baker et al. 1978). Phase I farmers had herds that produced heavier calves than phase II farmers even when adjustments were made for number of years and project participation. Calves on farms managed by part-time farmers who placed minor emphasis on farming had lower 205-day weights than calves on farms managed by other types of farmers. Market weights of calves, kg of calf marketed/cow and day of age followed the same pattern as adjusted 205-day weight. The value of calf/cow varied from year to year but exhibited an upward trend during the project.

Lambing percentage among sheep flocks increased 2.8%/year of participation in the AHP, and lamb weight marketed/ewe increased 1.4 kg/yr. This increase in lambing percentage was attributed to better management of ewe flocks, especially relative to ewe condition, flushing, and selection of replacements. The increase in lamb weight marketed/ewe was due to increased lambing per-

centage and sale of slightly heavier lambs, while death losses remained unchanged.

Expected increases in productivity were not achieved in all areas. Neither grain or silage yields nor beef cow numbers increased significantly. Death losses in calves (7%) and lambs (16%) were not reduced. Losses due to predators, principally dogs and bears, increased.

Operating expenses and receipts increased yearly. Net income/yr averaged $1,083/farm and did not change significantly, although there was an increase of $253/farm/yr. Farms with 100 or more AU had higher net income than smaller farms.

An evaluation (Balliet 1977) indicated a beneficial impact of the project on cooperators and the local community. More than 80% of the cooperators believed they had made improvements in their farm operations as a result of the AHP. Balliet projected that the benefits of the AHP to producers in the state had equalled its direct costs.

The AHP supplemented educational and training programs for students of WVU-CAF. WVU benefitted from the close association with producers and improved research programs in the area of animal agriculture.

The authors concluded from objective and subjective data that (1) the package approach was a successful way of disseminating information, and those working with the package approach need to view farms as multifaceted operations with many disciplines; (2) one-to-one contacts were the most effective way to get information adopted; (3) much of the information available to producers meets their needs if modified for individual situations; (4) because people farm for different reasons, the technology offered will be adopted only when it is in keeping with producer goals; and (5) because many farmers are reluctant to take financial risks or change traditional methods, changes occur slowly.

The following elements were found necessary for successful operation of the AHP: (1) the establishment of goals and procedures; (2) the assembling of a staff with expertise in agriculture and an ability to work together and with producers; (3) the ability to offer assistance and advice in critical areas of expertise; (4) the building of mutual confidence between producers and staff to permit free information flow in both directions; (5) the involvement of the total rural community in activities and of staff in nonproject activities; (6) adequate funds with sufficient flexibility to operate in a remote community; (7) institutional support that rewarded off-campus faculty for nontraditional contributions; (8) continuity of staff to give a feeling of permanence to participants; (9) suitable support staff; and (10) adequate time to implement programs.

LITERATURE CITED

Baker, B.S., M.R. Fausett, P.E. Lewis, and E.K. Inskeep. 1978. A progress report on the Allegheny Highlands Project. Agriculture. Coll. of Agric. and Forest. W.Va. Univ.

Balliet, L. 1977. Stimulating livestock production in the Allegheny Highlands: an economic, social and attitudinal evaluation of a university pilot program. Off. of Res. and Devel., Div. of Soc. and Econ. Devel., W.Va. Univ.

Effective Training for International Students

G.W. THOMAS

New Mexico State University, Las Cruces, N.M., U.S.A.

Summary

The observations made in this paper are based upon contacts with international students, both in the U.S. and in many other countries.

Two of the most serious problems in increasing world food production in the developing world are (1) lack of trained people, and (2) technology transfer. A third constraint, inadequate research, is also critical but not of the magnitude of the other two, since there is a great amount of research information available that is not being adequately utilized.

Training needs fall into two general categories in which deficiencies are obvious: (1) graduate-level programs, particularly in the applied fields relating to range ecology and grassland management, and (2) "technician"-level training, which may be below the B.S.-equivalent level.

Most international students, after returning to their home countries, expressed enthusiasm about their training in the U.S. They were less enthusiastic about training in other countries because it was "more basic" and "not as practical" or useful for their jobs at home. Also, there appeared to be more effective training at institutions within similar ecologic zones, whereas training—particularly in the agricultural sciences—in dissimilar zones was not as effective for their purposes. Careful selection of universities abroad is very important.

There was an almost unanimous feeling among foreign-trained students, as well as among agencies involved in development assistance, that the research required for a Ph.D. should be conducted in the country of the student's origin. This requirement may not be as critical if the research is more applied than basic and if the research is carried out in similar ecological environments. Even though "in-country research" poses some managerial and financial problems to foreign universities, its importance cannot be overemphasized.

While most of the less-developed countries (LDCs) require some service (usually 5 years) in the home country after the degree is conferred abroad, there is evidence of substantial migration of trained personnel from many of these LDCs.

The working environment in many LDCs is depressing for the returning professional. Some of these people cannot get foreign exchange to maintain their professional ties to such organizations as the Animal Science Society, the Agronomy Society, or the Society for Range Management. Some find it virtually impossible to communicate with other professionals in their field: there is no library and no intellectual stimulation. It is indeed sad to observe the trained mind in an environment that will not permit growth and development.

KEY WORDS: international, professionals, training, technology transfer.

INTRODUCTION

The observations made in this paper are based upon international student contacts, both in the U.S. and in many other countries. I have supplemented observations made over a 30-year period in higher education with interviews and contacts during a 3-month intensive study period in 1980 of the Sahelian/Sudanian zones of Africa.

THE CHALLENGE FOR EDUCATION IN THE DEVELOPING WORLD

There are more illiterate people, in actual numbers, in the world today than there were 5 or 10 years ago. World population is growing faster than our ability to educate the people. For example, the percentage of people who can read or write their own language in the countries of sub-Saharan Africa, where I conducted a study for the Rockefeller Foundation, ranges from about 40% in Kenya to only about 5% in Mali. "Few of the Sahelian countries have yet achieved more than an estimated 10 percent adult literacy rate. The estimated rate for all Africa is 17 percent." (USAID, 1980).

Education, in some respects, is a dangerous thing—that is, education and improved communication create unrest and some dissatisfaction among the masses. Education raises the aspirations of poor people for a better life and can cause problems, particularly for repressive governments. At the same time, adequate education and improved understanding are the only ultimate roads to world peace.

Education, in its broadest sense, is the first and most important step for progress in food production, for im-

The author is the President of New Mexico State University.

provement in the quality of life and for economic and political stability in the developing world.

Two of the most serious problems in increasing world food production in the developing world are (1) lack of trained people and (2) technology transfer. A third constraint, inadequate research, is also critical but not of the magnitude of the other two, since there is a great amount of research information available that is not being adequately utilized.

SOME OBSERVATIONS ON INTERNATIONAL EDUCATION

Training needs fall into two general categories in which deficiencies are obvious: (1) graduate-level programs, particularly in the applied fields relating to range ecology and grassland management, and (2) "technician"-level training, which may be below the B.S.-equivalent level. It is difficult to convince the developing country leadership of the value of vocational-technical training because such training does not carry the prestige of the traditional degree programs.

There is a strong feeling among foreign-trained students, as well as among agencies involved in development assistance, that *the research required for a Ph.D. should be conducted in the country of the student's origin.* This requirement may not be critical *if* the research is more applied than basic and *if* the research is carried out in similar ecological environments. Even though "in-country research" poses some managerial and financial problems to foreign universities, its importance cannot be overemphasized.

There is special need, particularly in Africa and South America, to blend the traditional European university systems with the American systems of higher education. Many students become confused by the differences and lose valuable time in preparing for graduate programs.

Higher priority should be given to the building of educational institutions within the countries themselves. The burden of the future leadership must eventually rest on people trained in their own environment, by their own people, and with locally designed curricula.

Some quotations from field interviews and literature from an African study by Thomas (1980) follow:

"There is a greater need in Africa for manpower development than in any other major world area. This shortage is seriously hampering economic and social development" (ACE 1980).

"The young nationals are poorly trained. . . . some are really concerned about the problems but the majority are interested only in their personal or professional status. . . . training is often too theoretical."

"Training is a special problem here because the French system . . . does not prepare them for entering graduate school."

"U.S. universities are very fussy about accepting students . . . probably lacking an understanding of the French system . . . and an unwillingness to adjust to the needs of the African student."

" . . . Many of the graduate students were trained at a

level which is too sophisticated for return to their country."

"The highest priority should be to build the national [education and research] structures."

"The farmer has very poor managerial capabilities to adapt to newer technology. . . ."

"The long-term social impact of education is not being adequately recognized by any of the development agencies."

" . . . The people of West Africa can be relied on to build a better future for themselves. . . . Lack of education divides the people more than anything else; . . . tribal affiliations are important and have to be taken into account" (Asante 1976).

EXTENSION AND TECHNOLOGY TRANSFER

While new information must be continually developed for the area through research, the problem of transferring this information to the field and adapting it to the problems of the farmer and pastoralist must receive a higher priority. The adaptation of new knowledge to improve on the traditional systems depends upon the availability of trained extension specialists who understand native cultures, speak native dialects, have something to say, and can gain the respect of the food producer and homemaker.

The extension effort should not only be aimed at the adult population but should serve as the core for agricultural and home management youth programs, such as the 4-H or FFA-type programs of the U.S. and other countries. The meager public school programs are not designed to do this practical youth training.

Extension skills are unique, and too few people have mastered this art. Also, financial support for extension activities is difficult to obtain in a marginal economy, and the rewards to the extension specialist are usually not as great as the rewards for other professional efforts.

A few comments on extension training from field interviews in the African study (Thomas 1980) follow:

"Progress depends a lot on the ability of the extension service. To build numbers of extension workers is not important, but to keep them working with farmers and livestock producers is the key."

"The extension services are poor, but the recent move toward specialization may be an improvement."

"Trained people will not work in the field . . . until you saturate the desk jobs, they won't go into the field."

" . . . The problem of transferring technology to the farmers is a major drawback."

"Extension here is horribly bureaucratic and there are not enough trained people."

THE WORKING ENVIRONMENT AT HOME

Most international students, after return to their home countries, expressed enthusiasm about their training in the U.S. They were less enthusiastic about training in European countries because it was "more basic" and "not as practical" or useful for their jobs at home. Also, there appeared to be more effective training programs at institutions within similar ecologic zones, whereas training, particularly in the agricultural sciences, in dissimilar zones was not as effective for their purposes. Careful selection of universities abroad is very important.

Many qualified professionals on their return home are placed in government jobs outside their profession and given management responsibilities for which they have received no formal training.

Students maintained their loyalties to their universities at home and abroad, usually through contacts with the alumni associations. They felt there was no discrimination in their home countries among professionals trained in Europe, the U.S., Australia, or other countries, although certain universities in certain areas seemed to be the "in" institutions.

While most of the less-developed countries (LDCs) require some service (usually 5 years) in the home country after the degree is conferred abroad, there is evidence of substantial emigration of trained personnel from many of these LDCs. As one professional in an LDC stated, "We are training our people for jobs in Saudi Arabia where the salaries are much higher than we can afford."

The working environment in many LDCs is depressing for the returning professional. Some cannot get foreign exchange to maintain their professional ties to such organizations as the Animal Science Society, the Agronomy Society, and the Society for Range Management. Some find it virtually impossible to communicate with other professionals in their field: there is no library and no intellectual stimulation. It is indeed sad to observe the trained mind locked into an environment that will not permit growth and development—especially when that expertise is so desperately needed to help solve the problem of world hunger.

LITERATURE CITED

American Council on Education (ACE). 1980. The African Manpower Development Project. Overseas Liaison Committee.

Asante, K.B. 1976. The future of West Africa. Proc. West Afr. Conf., Univ. of Ariz., Tucson.

Thomas, G.W. 1980. The Sahelian/Sudanian zones of Africa: profile of a fragile environment. Spec. Rpt. to the Rockefeller Found.

USAID. 1980. Sahel development program. Ann. Rpt. to Congress.

Innovations in Academic Curricula for International Education

D.D. DWYER
Utah State University, Logan, Utah, U.S.A.

Summary

A major contribution of U.S. AID to developing countries has been to support the education of host-country nationals. This form of assistance has not been without its problems, as both AID and the universities themselves have been less than totally effective in meeting the challenge of providing an education that best serves the student and the home country.

In most developing countries, the B.S. degree serves as an adequate scientific foundation for staffing agricultural programs. This is not to argue against higher levels of education but to state that faster progress could be made by spreading the educational efforts among more individuals at a training level better suited to the needs of the country. There is evidence that U.S. institutions have been guilty of prolonging the educational experiences of third-world students beyond their need.

When students from less-developed countries (LDCs) pursue graduate degrees, the U.S. institutions providing the education and the organization sponsoring the student should find a way for the required thesis or dissertation research to be done in the home country.

At the B.S. level, foreign nationals being educated in U.S. universities should not be subjected solely to curricula developed for U.S. students and U.S. conditions, as is now predominantly the case. It is necessary to relate the U.S. technology, principles, and concepts of agriculture and natural-resource management to the cultural and political realities of the LDC. Special courses must be developed to do this, and U.S. faculty must have the experience and willingness to develop and teach those courses properly.

This paper proposes some possibilities for undergraduate and graduate curricula in the U.S.A. to furnish more meaningful education for international students.

KEY WORDS: international education, academic curricula, education.

INTRODUCTION

Most U.S. universities and certainly all land-grant institutions are playing an increasingly important role in education of students from less-developed countries (LDCs). Despite the increased demand for education and training for these students over the past decade, and particularly since passage by the U.S. Congress of Title XII legislation, the educational process for international students has not improved as it should have. The amount and quality of "technology being transferred" are well below the potential of U.S. universities and will remain so until educators accept the fact that educational needs of international students vary greatly from those of their U.S. counterparts.

INTERNATIONAL STUDENTS ARE DIFFERENT

To say international students are different is both trite and obvious. If U.S. educators accept that statement, however, and still do not develop more creative and innovative means to deal with those differences, they are failing in a most fundamental and important component of their U.S. international mission.

Some of the more obvious ways in which most international students are different are: (1) they are older, (2) they are more experienced in the profession in which they seek education, (3) they are more highly motivated, (4) they are under more pressure to succeed, and (5) they are screened and selected from a large pool of qualified candidates. These factors are all tremendously advantageous to the educational process, but for the most part they are simply ignored as benefits. An international student's success in a U.S. institution is hampered by language inadequacies, cultural and religious differences, dietary restrictions, extended family separations, and nonfamiliarity with the U.S. university system.

Perhaps one of the more important problems standing in the way of successful completion of a U.S. university undergraduate or graduate degree by international students is the lack of relevance to their own country of the educational curriculum being pursued. For the most part foreign students are placed in curricula designed for U.S. students and U.S. agricultural problems with "like it or leave it" abandon. Failure of foreign students under these conditions can be easily attributed by a university advisor to the inability of the international student to handle the "advanced technical nature" of the material or to inadequate background education of the student in his home country. More frequently, the problems can be traced to a combination of language problems, inadequate or too-rapid acculturation, and the student's inability to rationalize the courses he is taking within the context of his home-country situation.

The author is a Professor and Head of the Range Science Department.

IMPROVING INTERNATIONAL STUDENT EDUCATION

Bettering the transfer of technology from one culture to another will require that U.S. universities give more serious attention to the specific needs of the international students and that U.S. educators try harder to understand their problems more fully. I am suggesting four possible avenues to do this.

Overcoming Language Problems

To assume that non-English-speaking students are ready for university classes when they achieve the minimum proficiency required of intensive English institutes or the TOEFL examination is wrong. The students cannot easily cope with the U.S. lecture-style format that typically, and especially in agricultural courses, is accompanied by slang, idioms, jargon, and an air of casualness that is well accepted by U.S. students. But the international students not only cannot readily incorporate nonformal English into their thought patterns, they cannot distinguish the unimportant from the important points. Not until the second term, and in many cases the third, have adjustments and adaptations been made by the foreign student to accommodate the system better. However, by then the student is often on probation or at best saddled with a record well below his potential, and much of the scholarship time has been expended before getting into the professional courses.

Three ways to deal with the language adjustment problems are as follows:

1. Provide experience in lectures and examinations during language training so that students get some idea of what will be expected in the U.S. classroom. Many U.S. university examination formats, such as multiple choice and true-false, manipulate the English language and seriously handicap the foreign students.

2. Develop one seminar-type course the first term for foreign students and teach it in a manner that recognizes the tremendous changes foreign students face. The course would permit a time for confidence to build and for students from various countries to share experiences. An oral report given by each student about his home country is necessary to help him gain experience expressing himself in English. A recording of the presentation allows each student to listen to himself. This format also serves to educate U.S. instructors about the various LDCs from which the students come. This experience on the part of the instructor will make him far more appreciative of the problems being encountered by the first-term international student and increase the teacher's awareness of the frailties of the communication process in technology transfer.

3. Make cassette tapes of all classes available, by which the foreign students can listen again to the lectures.

4. Ask each foreign student to visit the teacher early in the term of his professional courses to help the student feel

freer to seek help if it is needed. Foreign students are generally reluctant to ask questions in class or even to talk to the teacher during their first term—and that is when communication is most needed.

Relevant Courses for LDC Students

To improve the value of U.S. curricula for foreign students, certain specialized courses must be developed that have clear applicability to LDC situations. For example, in the field of range science two important courses in the curriculum are "Rangeland Improvement" and "Range Livestock Production and Management." It should be obvious that these two courses are very important to the international student studying range management, since most LDC rangelands need improvement and increased production of livestock from those lands is critical to the long-term livelihood of the people. What is less apparent is that an enormous amount of the information given in the classes is not applicable to the central rangelands of Somalia or western Sudan, for example. Agricultural courses are famous for their practical, how-to approach, which has worked well for the U.S. situation. This approach is less useful when it is applied to LDCs. Ways to improve rangelands in Sudan must be presented within the context of communal grazing in a nomadic subsistence-level pastoral system rather than by a system guided by the profit motive.

To overcome the handicap foreign students have in adapting U.S. technology to the LDC environment, a second course should be developed for those subject-matter areas that are application oriented. "Range Livestock Production and Management" (U.S. style) should have during the same term a course "Range Livestock Production and Management in Less-Developed Countries," which places the principles and concepts in the perspective of LDCs. This course would put a strain on the how-to approach, especially if the instructor had not had any experience in LDCs. One of the main benefits would be to provide LDC students the opportunity for better mental assimilation of the subject matter.

In addition, the "add-on" LDC-style course will be popular with U.S. students and cannot help but ultimately lead to a special kind of knowledge transfer from the foreign student to the U.S. student. Future international experts from the U.S. will be benefited by such exposure when they are students.

Title XII strengthening of grant funding should be used in part to provide experience in LDCs for university faculty who teach foreign students.

Graduate Degree Research in the Home Country

To suggest that research conducted for the graduate-level U.S. degree be conducted in the candidate's home country is not new. Smith and Dwyer (1981) have recently compared undergraduate and graduate degrees and their relative value to LDCs. They conclude that in most developing countries the Ph.D. degree is emphasized to the detriment of supplying adequate numbers of B.S.-level technicians. Where Ph.D. degrees are required or justified the authors make clear the merits of doing the required research in the home country to the degree-seeking individual, the country, and the interested U.S. institutions.

The innovation and the creativity of this method do not lie in the suggestions; they lie in the accomplishment. Though the conduct of research in the home country required for the graduate degree has been tried, the record of success is meager. The LDC institutional framework to support such a venture adequately rarely exists. Thus, the cost of not meeting requirements for the degree will normally not encourage a second try with a second candidate.

The U.S. Agency for International Development (AID) could help overcome the presently known disadvantages for such an obviously useful approach to the M.S. or Ph.D. for LDC candidates. AID should work with universities to provide the necessary funding and logistical support to accomplish the program without relying heavily on the LDC government, as is now most often the case. This could be accomplished by attaching a thesis research project directly to the development project so that advice and financial support are available. While they are most costly for U.S. AID, graduate programs that include doing research in the candidate's home country, travel expenses for the U.S. advisor, and logistical and financial support for the research seem to have benefits that outweigh the costs. The work itself should contribute to the overall mission of U.S. AID in the country, the data and recommendations coming from the research should help in solving LDC problems, and, perhaps most important, the U.S. university advisor gains experience in the LDC that would surely benefit other foreign students, broaden the perspective of the courses taught, and make more useful the contribution that the professor might make to his or her future efforts in other LDCs.

Centers for International Education

For the most part there are certain fields in agriculture in which degrees, short-term educational training, and consultancy assistance are consistently sought. Some examples are livestock production and health, crop production, agriculture and irrigation engineering, soil fertility and management, and rangeland management. For those subject areas that are frequently called upon to provide education for students from LDCs, universities known to have outstanding programs and faculties with experience in LDCs should be encouraged to develop centers for international education. By so doing, identity and visibility are gained for the program nationally and internationally, and perhaps simultaneously an obligation to provide more specialized and pertinent education for international students is incurred.

Financial support, on a pro-rata basis/student, for such

centers will need to come from U.S. AID, FAO, and other organizations that send students for training. As anyone who has supervised foreign students knows, the agency, country, or organization sponsoring the student is paying less than half the cost of the education for that student. That fact helps explain why the education LDC students now receive is not as specialized as it needs to be. In addition to this truth, it must be acknowledged that few faculty members are paid to teach foreign students or are promoted or tenured for developing programs uniquely applicable for them.

CONCLUSIONS

It is probably not too surprising that an educator would conclude that the solution to the world food crisis is better education of LDC students. It does seem plausible that knowledge transfer of the vast technological achievements of the U.S. in agricultural production will require that U.S. universities become resolute about making that knowledge transfer more efficient. As Roberts (1980) notes, university faculties ought to spend more time wrestling with worldwide problems in the educational process. This obligation implies that curricula taught to students from LDCs should include for them the relevance of U.S. technology to the LDC environment, cul-ture, tradition, and social institutions. Few U.S. agriculture courses now do that.

Special courses in various agricultural subjects designed to relate principles, concepts, and practices to LDCs would be helpful in making U.S. education more meaningful to LDC students. Graduate-student research programs for degree candidates from LDCs should be careful not to train a student in research techniques that have little application to the needs of the country or the position to which he will return. U.S. AID should support LDC graduate-student research in the home country, including support for the major professor to visit the country to work with the graduate student in setting up the thesis or dissertation research project. This work would require at least an annual trip to review progress. That experience would aid university instructors who teach foreign students to understand better the culture and level of technology from which students from LDCs come.

LITERATURE CITED

Roberts, N.K. 1980. The population-food squeeze: education for survival. 62d Honor Lecture, Utah State Univ.

Smith, A.D., and D.D. Dwyer. 1981. Range management training in developing African nations. J. Range Mangt. 34:11–13.

Role and Effectiveness of U.S. Universities in Training Professionals for Work in International Agriculture

D.F. FIENUP

Michigan State University, East Lansing, Mich., U.S.A.

Summary

Problems of hunger, malnutrition, and world food security became a major world concern in the 1970s. Two U.S. presidential reports have emphasized the seriousness of the situation and the need to expand food production and distribution, particularly in the less-developed countries (LDCs) where hunger is most severe.

The LDCs need to stengthen agricultural teaching, research, and extension in order to develop and disseminate the appropriate technology farmers need to increase food production. The most limiting factor is often the lack of trained scientists to provide local leadership and expertise. Here U.S. and other developed-country universities are needed to train LDC professionals as well as to provide technical assistance overseas.

U.S. universities have been encouraged to expand their international activities through Title XII legislation, passed in 1975. Although the Act has not been fully funded or implemented, it has been a major step toward the long-term support U.S. universities need to take on seriously the job of international agriculture. Demand for U.S. training from LDC students and the need for U.S. expertise overseas have continued to increase during the 1970s, while support for U.S. university–based work has substantially declined. Younger U.S. professionals in particular seem to be losing their interest in development.

These circumstances led the American Agricultural Economics Association (AAEA) to sponsor a major study on the training of foreign students in U.S. universities. Principal objectives were to obtain an evaluation from LDC alumni on the strengths and weaknesses of their U.S. training, to determine what the U.S. can do to assist in strengthening indigenous training and research capabilities in the LDCs, and to assess the future capacity of U.S. universities to meet LDC research and training needs.

Results indicate that LDC agricultural economists value their U.S. training highly; it has given them the confidence, analytical tools, and problem-solving skills they need to work in their own countries. They would like greater application of economic tools to their own situations. More U.S. professors need real knowledge and experience in developing countries, and students should be able to do their thesis research on a home-country problem. It is also important to maintain collaboration with their U.S. professors and departments after returning home. Cooperative research, short courses and seminars, and post-doctoral programs were suggested as means for continuing support.

Over 40% of young U.S. agricultural economists have deserted their interests in development. U.S. international capabilities in this discipline appear to be declining. Probably the most critical factor is the instability and low level of funding for university participation in international development activities. Increasingly stringent university budgets and little public support for assistance to LDCs have made it extremely difficult for U.S universities to recruit and maintain faculty in international development.

KEY WORDS: international training, U.S. universities, world food supplies, agricultural development, technology, hunger, malnutrition.

INTRODUCTION

The problems of hunger, malnutrition, and world food security became a major world concern in the 1970s. Shortly after the World Food Conference in Rome in 1974, U.S. President Gerald Ford asked the National Academy of Sciences "to make an assesssment of the problem and develop specific recommendations on how our research and development capabilities can best be applied to meeting this major challenge." The Academy's report, known as the World Food and Nutrition Study (WFNS), was delivered to President Jimmy Carter in June 1977. In September 1978 the President created a special Commission on World Hunger (CWH) "to identify the causes of domestic and international hunger and malnutrition, assess past and present national programs and policies that affect hunger and malnutrition, review existing studies and research on hunger, and recommend to the President and Congress specific actions to create a coherent national food and hunger policy." This report was delivered to President Carter in March 1980.

A strong recommendation was made by the WFNS to expand the agricultural research base in both the U.S. and developing countries as being "essential to all activities needed to increase the food supply, reduce poverty, and moderate the instability of supplies and prices." It was concluded that the developing countries must double their own food production by the end of the century in order to eliminate world hunger and malnutrition. In contrast to the emphasis on new technology by the WFNS, the World Hunger Commission indicted poverty as the primary cause of world hunger and emphasized social, political, and economic solutions to the problem. At the same time the Commission recognized that developing countries also need to expand their own food supplies greatly as an important part of the solution. The U.S. can play a vital role through experience in research and extension and in training research workers from developing countries to build their own research capabilities.

Concurrently, U.S. agricultural colleges and universities have been encouraged to expand their participation in international agricultural development activities through passage of the Title XII Famine Prevention and Freedom from Hunger Amendment to the Foreign Assistance Act of 1975. This legislation was based on the belief that much of the success of U.S. agriculture has resulted from the land-grant philosophy and practice of using an integrated approach in teaching, research, and extension to serve agricultural needs — and that a similar approach should be expanded in the LDCs. Particular emphasis has been placed on strengthening U.S. universities' capabilities and involvement in training foreign professionals in the U.S. and overseas; direct technical assistance to LDCs in research, extension, and institution building activities; and the establishment of international networks among agricultural scientists. The most significant aspect of Title XII is provision that U.S. universities through the Board for International Food and Agricultural Development (BIFAD) should have direct participation in USAID policy and program decisions. The intent is to provide long-term support for U.S. universities to take on the job of international agriculture as an integral part of their ongoing teaching, research, and extension activities.

Many U.S. universities built substantial teaching and research capabilities in international agricultural development during the 1960s. Faculty members had many opportunities for overseas assignments in technical assistance projects, and they were strongly encouraged to participate. It was also a period when increasing numbers of foreign students enrolled in U.S. M.S. and Ph.D. training programs. During the 1970s LDC demand for U.S. training has continued to increase, yet the support for U.S. university-based international work has substantially declined. Young U.S. agriculture professionals receive little encouragement to work in development. This decline has left the academic community with some sense of frustration as to how to maintain competencies in international agriculture needed for research and for training LDC students, who constitute up to one-third of graduate enrollments in many universities.

The author is a Professor in the Department of Agricultural Economics and the Institute for International Agriculture of Michigan State University.

RESEARCH OBJECTIVES

It was this set of circumstances that led the American Agricultural Economics Association (AAEA) to sponsor a major study on the training of foreign students in U.S. universities. Principal objectives were (1) to determine the extent to which LDC agricultural economists trained in the U.S. continue to work in their own countries; (2) to obtain an evaluation from LDC alumni on the strengths and weaknesses of their U.S. training, how it could be improved and made more relevant for developing country conditions; (3) to determine what the U.S. agricultural economics profession can do to assist in strengthening indigenous training and research capabilities in the LDCs; and (4) to assess the future capacity of U.S. universities to meet LDC research and training needs by studying the extent of involvement and participation by young U.S. professionals in the economics of international agricultural development. The problems and potential solutions in agricultural economics are considered to have relevance for other agricultural disciplines. Several other professional associations have already indicated interest in making studies similar to the one conducted by the AAEA.

The basic source of primary data for the AAEA study came from mail survey responses of 653 LDC agricultural economists who studied in U.S. universities over the last 15 years. This sample represented 79 countries and over 40 departments of economics and agricultural economics. In-depth studies were also conducted in nine developing countries in Asia, Africa, and Latin America with key professionals and major employers of agricultural economists. Mail survey responses were obtained from 108 young U.S. professionals who received their Ph.Ds in the last 10 years and had a major interest in the economics of development when they completed their degrees.

EMPLOYMENT AND RESIDENCE OF LDC ALUMNI

LDC agricultural economists educated in U.S. universities are generally working in jobs for which they were trained. Over 40% held university positions; another 40% work in government, including ministries of agriculture, national planning, and other state agencies; about 10% work for private businesses and as professional consultants and advisors; and 10% work for international agencies and foundations. Approximately 30% of those in LDC governments and universities have administrative positions. Eighty percent of LDC alumni of U.S. graduate schools are still living and working in their countries of origin. On a regional basis Asia has lost 31% of its U.S.-trained professionals, compared with a maximum of 15% in any other region.

EVALUATION OF U.S. TRAINING

Basically, LDC agriculturalists value their U.S. training highly. LDC agricultural economists felt they had gained important analytical skills in economic theory and quantitative methods that were useful in problem solving. Training in the scientific method of inquiry gave them confidence to tackle many development problems in their own countries. The breadth and depth of courses available as well as flexibility in the content of their graduate programs prepared them to teach and work in a wide range of agricultural economics subject areas. They were also impressed by the excellent faculty-student relationships and the favorable infrastructure for learning and research in U.S. universities.

At the same time, there was concern about the insufficient attention given to political, social, and institutional factors in development. LDC agricultural economists would like more application of theory and methods to LDC problems, and they felt that too little attention had been given to the problems of income distribution and other equity issues. They would also like more emphasis on primary data collection and analysis, project evaluation, agricultural planning, policy analysis, and public administration. Many of those trained in the U.S. move into administrative positions in their universities and research agencies when they return home; yet they receive little preparation in management and administration in U.S. graduate study.

Some suggestions received on ways to improve U.S. training of LDC students included having more professors in U.S. universities with real knowledge and experience in developing countries. This qualification is especially important for advising LDC students and helping direct their thesis research. It is important for LDC students to work on a thesis problem in their own countries whenever possible, or at least on a topic relevant to their own situations. It was also recommended that students be trained as broadly as possible within their disciplines. Most are not able to work within a narrow specialization; they should be prepared to work in a total farming-systems context with production scientists.

WAYS THE U.S. CAN HELP STRENGTHEN LDC PROFESSIONALS

The demand for agricultural economists in the developing countries considerably exceeds the capabilities of the countries to train M.S. and Ph.D. professionals. For the last 2 decades a large part of this training has been provided by universities of the U.S. and other developed countries. The need for U.S. training and collaboration in LDC institution building will continue at least through the 1980s. LDC enrollment in U.S. M.S. programs has continued to be high but should decline somewhat in the next decade, as more people will be trained at this level in the developing countries. Demand for Ph.D. training will remain strong, as will the need to form better linkages between the U.S. and developing-country professions.

When LDC agricultural scientists trained in the U.S. return home they often feel isolated and lack contact with their professional disciplines. Many are unable to keep up with current scientific developments while they are trying to build their own institutions and solve critical problems in their own countries. Most LDC professionals would like some support in this process through continuing collaborative relationships with U.S. scientists. Joint degree arrangements, shared thesis advising, and cooperative research projects developed between U.S. and LDC institutions are some of the ways that were suggested to establish these linkages. There is also a continuing need to increase LDC faculty competence through postdoctoral programs, short courses, and seminar activities.

DEVELOPING AND MAINTAINING U.S. PROFESSIONAL COMPETENCE

Responses of U.S. agricultural economists, who received their Ph.D.s between 1968 and 1977 and who had major career interests in international development, revealed that 40% no longer work in development. Most are employed by universities. The major reason given for leaving international agricultural development work was the lack of professional opportunities in either tenure-system university positions or longer-term career jobs with international agencies. Lack of support for research and development problems, difficulties in obtaining recognition and rewards for development work, and the need to simultaneously carry on domestically oriented activities were additional reasons cited for decreased involvement.

The pool of young U.S. professionals with skills in international development is small in relation to the need. Even so, the AAEA study suggests that existing talent is not fully utilized. Probably the most critical factor is the instability and low level of funding for university participation in international development activities. This factor, coupled with increasingly stringent university budgets and limited public and Congressional support for assistance to LDCs, has made it extremely difficult for universities to recruit and maintain faculty who specialize in international development.

CONCLUSIONS

U.S. universities have an important role to play in helping alleviate the critical problems of world hunger and malnutrition. This problem can only be solved by increased food production and improved distribution in the developing countries. The LDCs need well-trained agricultural scientists to staff and develop strong local institutions for agricultural teaching, research, and extension. The services of U.S. universities and universities of other developed countries are essential to help train LDC professionals and to assist directly in the institutional building process. Evidence exists that there is a desire for mutual collaboration in this process, yet U.S. university involvement in international research and related technical assistance has been declining.

A longer-term financial support base is essential if universities are to continue to provide high-quality training for substantial numbers of LDC and U.S. graduate students and to collaborate in research and institution building within the LDCs. The Title XII legislation is certainly a step in the right direction, but it must be adequately funded by Congress and implemented by AID and the Title XII universities. Much more effort is needed to educate and create greater public awareness of the problems of world hunger and U.S. interests in these problems. Land-grant universities also need to seek greater support at the state level for international activities that have a direct impact on their various clientele groups.

Development Style and Graduate Education in the U.S.A. for Students from the Developing Countries: Prospects for the Eighties

E.F. FUENZALIDA
Stanford University, Palo Alto, Calif., U.S.A.

Summary

The purpose of this paper is to examine the past and present relevance of graduate education in the U.S.A. for students from developing countries and to suggest a number of changes that would make that education better adapted to developing countries' needs.

The first section of the paper argues that in recent decades U.S. graduate education has been relevant for students from developing countries because most of those countries have adopted a particular style of development based on the modern transnational corporation. The requirements of this *transnational style of development* have led in turn to the need for an ever-increasing stream of graduates from developed-country institutions in order to service the transnational sector that is at the core of the development process.

The second part of the paper notes a number of important recent changes that have called the relevance of this kind of graduate education into question. On the basis of those changes, it is concluded that if U.S. institutions are to continue to attract large numbers of students from developing countries, they are going to have to make some major alterations in their mode of operation. A number of arguments are then offered as to why this course of action would be desirable.

The paper's final section sets forth some general guidelines for change. In particular, it is argued that complementary curricula should be drawn up that meet the needs of students from developing countries more effectively. Once developed, such curricula may serve as a stimulus to overall curriculum changes that would make graduate education more meaningful for students from all over the world.

KEY WORDS: transnational style of development, development process, guidelines for change, U.S. graduate education, developing countries.

HISTORICAL RELEVANCE OF U.S. GRADUATE STUDIES FOR STUDENTS FROM DEVELOPING COUNTRIES

It is relatively safe to say that graduate education in the U.S. has consistently tried to cater to the needs of the main agents of economic dynamism in this country, the big corporations and the federal government, and that by and large it has succeeded in this attempt. It has never considered part of its duties to be catering to the needs of other national economies, developed or developing.

In spite of the fact that such education has been exclusively geared to the needs of one type of advanced industrialized economy, over the last 2 decades, thousands of students from the developing countries have come to the U.S. to pursue graduate studies, and the numbers are on the increase (Spaulding and Flack 1976). One wonders why these students keep coming to institutions that obviously are concerned with problems (and solutions) that are miles apart from those that are endemic to developing countries.

In my opinion, the answer to this question lies in the characteristic style of economic and social development that emerged in the nonsocialist world in the years following World War II.[1] This style is based on the assumption that owing to the force of tradition, domestic social forces in the nonindustrialized world will never be able to bring development to their societies. Therefore, an external agent has to be introduced into the social organism to stimulate it to grow and diversify.

This external agent is the artful combination of capital, labor, technology, and organization represented by the contemporary transnational corporation. It is argued that if this remarkable social invention is allowed to operate upon the factors of production available in the developing countries, in due time it will accomplish the great transformation that once occurred in today's developed societies, and development will take place.

This style of development, based on the large transnational corporation, has historically been endorsed by the international financial organizations that deal with development in the nonsocialist world, as well as by most of the governments of the developed countries. On these grounds, and for lack of a better name, I proposed to call it the transnational style of development.

The transnational style of development makes several specific demands on developing countries. First, they must modernize their infrastructures, their administration, their schools, their laws (particularly tariff and tax laws), and their law enforcement agencies so as to cater to the needs of transnational corporations. In addition, they must prepare development plans and projects that satisfy the standards of international financial institutions. All these changes require a huge number of specialists in various disciplines. As there is no way to educate so many people in so many disciplines at home, governments have to look abroad for opportunities to educate their people.

Such governments soon discovered that the U.S. had the greatest experience with this kind of development and possessed high-quality educational facilities at the graduate level closely geared to the needs of the same big corporations that were a key component of the development package. Therefore, they decided to send their people to U.S. graduate institutions to study.

Nevertheless, 2 decades of development have already gone by, and the number of students from the developing countries continues to increase. If the original idea was that development would eventually come to be self-sustaining, how can this persistence be explained? The reason is that the kind of development induced by the transnational style has created a modern sector in the overall economy of developing countries that is far from being self-sufficient—on the contrary, it depends heavily on a constant flow of inputs from the developed countries, especially of technology. This dependence makes it necessary to keep sending students to U.S. graduate institutions in order to train the qualified personnel needed to undertake negotiations with the foreign suppliers of key inputs and to man administrative agencies and local corporations.

The previous reasoning points to the following conclusions: graduate education in a U.S. institution has been historically relevant for students from the developing countries precisely because those countries embarked on a style of development that made such education essential, and it will remain relevant as long as these countries continue to frame their development efforts in the transnational style.

PROBABLE IRRELEVANCE OF
U.S. GRADUATE STUDIES IN THE 1980s

Is it reasonable to expect that in the 1980s the situation will be, by and large, similar to that of the 2 previous decades as far as the relationship between development style and graduate studies in the U.S. is concerned? On the basis of the development experience of the last decade, I think not. There are a number of reasons for this conclusion.

1. Over the last 10 years it has become increasingly evident that the transnational style of development cannot fulfill its promise to bring about self-sustained development in the nonsocialist developing countries. Faith in the transnational style of development, so characteristic of the early 1960s, has therefore given way to feelings of intense frustration and rage and to a passionate search for domestic ways to develop (Sunkel and Fuenzalida 1979). This crisis of the transnational style of development has definite implications for the flow of students from the developing countries to the U.S. Let me illustrate these points with an example.

For at least 20 years, Iran was the most extreme example of total confidence in the transnational style of development. In spite of scores of "development projects," the living conditions of the great majority of Iranians did not improve. Instead, a segment of the Iranian population emerged that enjoyed all the advantages of modernity against a background of widespread misery. The polarization of the society reached such a level that hundreds of thousands of Iranians turned to a religious-political movement that is fundamentally opposed to the transnational style of development.

Nevertheless, the desire of the "fundamentalists" to revise Iran's prior development pattern will have to take account of some hard facts. One is that the productive structure of the country remains heavily dependent upon a steady flow of inputs and technology from the West, especially from the U.S. Another is that the vast majority of the technocrats who run both the productive structure and the specialized agencies of the State have been educated in the U.S. and look to this country for new ways of doing things.

Does this mean that Iranian students will continue to be sent to the U.S.? Not necessarily. Because of the obvious political implications of this issue, the government will need to show its supporters inside Iran that young Iranians sent to the U.S. will not be converted into Americans, and that their programs of study will deal with Iran's peculiar conditions. Therefore, if U.S. graduate schools want the Iranians back, they will have to reorient their programs to respond more effectively to Iranian needs and sensitivities.

Another major factor to be considered is the changing development style of one major socialist country, the People's Republic of China. After several decades of experience with a radically different strategy, the new Chinese leadership is committed to an ambitious program of modernization that includes several components of the transnational style of development, in particular the training of Chinese in scientific and technological disciplines. What the Chinese case shows is that, in spite of its overall crisis, *some* aspects of the transnational style of development still have considerable appeal for the developing world, an appeal that *may* be translated into a large stream of graduate students to the U.S. However, the Chinese are far from accepting the transnational style of development uncritically, and if U.S. institutions wish to attract the Chinese they will need to develop a way of imparting a portion of the "transnational package" in isolation from the whole.

The third major factor that has changed the relationship between development style and graduate education in the U.S. is the establishment in developing countries, with foreign help, of graduate programs that offer an education comparable to that attainable in the U.S.

Let me illustrate this point with an example in the field of agronomy in one developing country, Chile. Up to 1960, the teaching of agronomy and veterinary medicine at the University of Chile was characterized by the coexistence in the curriculum of theoretical and practical subjects, with very little connection between the two. After a student strike protesting this state of affairs, the authorities of the University initiated a revision of the curriculum that gave much greater importance to the disciplines dealing with manmade ecosystems.

Then in 1965, the University of Chile signed an agreement (*Convenio*) with the University of California that gave it an opportunity to continue and intensify the process of modernization already initiated in these fields. *Convenio*-supported activities included the training of young Chilean faculty at the University of Chile to do research with senior Chilean faculty. By the end of the *Convenio*, in 1978, 42 Chileans had obtained degrees from the University of California in agriculture and veterinary medicine (*Convenio* 1979).

Modernization of the disciplines of agriculture and veterinary medicine was further supported by a grant from the Inter-American Development Bank for construction of a new campus and the installation of teaching and research laboratories, as well as by grants from other sources. As a consequence of this support, by 1967–1968 the University of Chile was able to offer graduate-level programs in genetics and plant improvement and in food science and technology. These programs now compete with U.S. programs for Chilean graduate students.

Educators in the U.S. have not previously had to worry about the appropriateness of their graduate programs for students in developing countries, because students' sponsors had an unshaken faith in the transnational style of development and wanted foreign students to get exactly what was offered to U.S. students. In the 1980s, however, owing to a much more sophisticated understanding of the difficulties of development, as well as to a much more complex and ambiguous political situation, sponsors will be scrutinizing offerings for their relevance more carefully. Therefore, if graduate institutions in the U.S. want to maintain, or increase, the flow of students from developing countries, they will have to revise their offerings with an eye to their suitability to those countries' developmental needs.

But could U.S. graduate education simply forget about students from the developing countries and concentrate its attention on U.S. students and on foreign students from the developed countries? For several reasons, I think that U.S. graduate education institutions cannot afford to let this particular group of students go. The first reason is financial: in a world of rising costs and shrinking numbers of U.S. students, income contributed by foreign students from developing countries may make the difference between a healthy existence and mere survival.

A second reason is political. It is in the best interests of the U.S. to maintain as many links as possible with development experiences that may loom large on the future international scene and that in any case represent the majority of the population of the globe.

Another reason is intellectual. A new frontier for science and technology will emerge as countries like Iran and China struggle to reach higher levels of development through a combination of elements taken from the transnational style of development and from their own historical experience. Having students from these countries in U.S. institutions is the best way to be involved in the intellectual process of creating a new development style.

The final reason, related to the last one, is by far the most important. The crisis of transnational development is by no means restricted to the developing countries. The developed countries as well, and especially the U.S., are rapidly realizing that it is not possible to continue their development in the same way as in the past. In the ongoing search for alternative ways to further development, the experience of the presently developing countries should therefore be enlightening. And, as mentioned before, one of the best ways to keep track of their successes and failures is by accepting their students in U.S. graduate schools.

WHAT SHOULD BE DONE TO MAKE U.S. GRADUATE SCHOOLS RELEVANT TO A CHANGING WORLD?

Even if one maintains that graduate education institutions of the U.S. should think ahead and adjust their offerings to the new demands that are emerging from the developing countries, what *exactly* is to be done? Since by definition there is no clarity about what will replace the transnational style of development, this is an exceedingly difficult question to answer.

Nevertheless, I would suggest the following general orientations.

1. Curricula for developed-country students should *not* initially be changed. They have been tested over the years and are known to produce a well-qualified professional. In spite of the increasing importance of the students from the developing countries in their programs, U.S. graduate institutions have a primary responsibility toward the future employers of their graduates in the U.S.

2. A *complementary* curriculum should be made compulsory for students from developing countries, focused on two broad areas. In the first area, students would be asked to learn about the historical roots of underdevelopment, the structural characteristics of developing countries, and their role in the international division of labor. In the second, they would be exposed to the experiences of their own countries in attempting to develop in accord with the requirements of the transnational style, focusing on the contribution of the discipline or disciplines they are studying to the final outcomes of such experiences.

Such complementary curricula should be developed for each of the major development-related disciplines and for each major region of the developing world. In this effort, optimal use should be made of the resources accumulated over the years in terms of alumni and experienced faculty. Specialists in curriculum development, as well as social scientists who have studied the process of development and the contributions of their respective disciplines to it, should also be involved.

3. On the basis of the complementary curricula so developed, careful revision of the overall curriculum offerings should be initiated. The actual reformulation of the curriculum at each institution should take account of the needs of its particular clientele.

Since they must cover the wide diversity of disciplines and regions of the developing world, these recommendations are necessarily quite general. They are intended as a stimulus to the creativity of the faculty and administrators of U.S. graduate education in the belief that these educators will not fail to come up with innovative responses.

LITERATURE CITED

Spaulding, S., and M.J. Flack. 1976. The world's students in the United States, 306. Praeger, New York.

Sunkel, O., and E.F. Fuenzalida. 1979. Transnational capitalism and its national consequences. In J.J. Villamil (ed.) Transnational capitalism and national development, 67–93. Humanities Press, Atlantic City, N.J.

Universidad de Chile-University of California convenio (*Convenio*). 1979. Comprehensive report 1965–1978. Mimeo.

NOTES

1. The expression "development style" is widely used by Latin American social scientists working for the United Nations Economic Commission for Latin America (ECLA), following the lead of Oscar Varsavsky. See the articles by A. Pinto, J. Graciarena, and M. Wolfe in *CEPAL Review*, First Semester, 1976. My own conceptualization does not necessarily agree with theirs.

Income, Costs, and Profits Associated with Feeding Selected Forages to Dairy Cows

D. DORNFELD, D.A. ROHWEDER, and W.T. HOWARD
University of Wisconsin, Madison, Wis., U.S.A.

Summary

Over the last 10 years, dairy-farm enterprises have grown in size, with an apparent increase in use of improved scientific feeding and management techniques. There is growing emphasis on automated feeding and milking and on management strategies in an attempt to reduce labor needs and allow more cows/worker. The average herd size in the sample population was 76.8 cows. There was difficulty finding herds over 100 and under 40 head. It appears that the trend toward increasing growth will continue, inasmuch as sufficient capital, management ability, and debt service capacity exist to permit the growth to occur.

Forage management by Wisconsin dairy farmers also has changed over the last 10 years, with more emphasis on alfalfa (*Medicago sativa* L.) as a means of reducing purchased protein feeds and controlling feed costs. This change has led to increased soil testing, forage testing, and forage fertilization. Low-moisture silage (LMS) as a means of reducing labor and preserving forage quality has increased and has resulted in growing dairy-farm investments. Producers relate LMS forage systems to high-quality feed, lower costs of purchased feed, automated feeding, and increased profits.

KEY WORDS: dairy trends, herd size, forage systems, management methods, economic relations.

INTRODUCTION

Dairy farming is in a period of significant adjustment. Dairy farmers are faced with strong economic pressures to increase the dollar volume of their businesses and to improve the efficiency of their operations.

Limited information is available on the management requirements and the economic return from capital investment in the various types of forage production, storage facilities, and feeding systems. There is even less information on changes occurring in the dairy enterprise as a result of management shifts.

Objectives of this study were (1) to gather and analyze relevant production and economic data, (2) to determine the interrelationships between production and economics, (3) to identify the effects of unidentified changes on production and economics, (4) to determine how forage systems and herd size are related, (5) to determine if there are economies of scale in size of operation, and (6) to identify what dairy farmers are currently experiencing.

METHODS

A survey was conducted among 110 Wisconsin dairy farmers nominated by county agricultural agents and extension specialists. Each farmer selected had made significant changes in his forage-production program by 1977. Data furnished by the participants was analyzed after stratifying the farms according to herd size: 0–49, 50–74, and 75 + cows (Table 1). All dairy farmers were using low-moisture silage (LMS) in varying degrees.

The interview instrument was designed to gather specifically information that would address changes in (1)

Table 1. Number of cows, alfalfa seeding rates in kg/ha, percent of farmers cutting alfalfa in bud stage, and percent of alfalfa harvested as low-moisture silage (LMS) for three herd-size groups in 110 Wisconsin dairy herds, 1978.

Group	Number cows Range	Number cows Avg.	Seeding rate kg/ha	LMS %	Cutting in bud-%
3	75 plus	118	17.7	56	62
2	50–74	57	15.0	53	65
1	0–49	38	14.3	41	25
Sample mean	–	76.8	15.7	–	–

D. Dornfeld is a Graduate Assistant in Agronomy, D.A. Rohweder is a Professor of Agronomy, and W.T. Howard is a Professor of Dairy Science at the University of Wisconsin.

forage management, harvest, storage, and feeding; (2) dairy management, feeding, and performance; (3) forage investment and returns to operations; and (4) perception of dairy farmers of the effects of the forage system being used on progress and change in their operations.

Data were analyzed to determine how alfalfa acreages had changed, how cultural management of alfalfa had changed over time, the types of machinery and storage facilities currently being used, and the investment related specifically to the forage system. Investments were calculated using current replacement costs and, to facilitate comparison, were applied similarly to all farms. Annual machinery costs were calculated by multiplying machinery investment by 19%, and total forage handling and storage costs by multiplying investment by 11.5%, to represent a projected investment life of 7 and 20 years, respectively. Dairy income was calculated for all farms by applying a uniform base price to milk sold. These data were then related to investment and annual costs.

Correlation and regression analyses were applied to determine whether the data could serve as predictor of change and profit in Wisconsin dairy herds.

RESULTS AND DISCUSSION

Forage Changes

Alfalfa continued to increase in importance in dairy feeding systems in Wisconsin, and nearly 60% of the dairy producers surveyed reported that changing to the LMS system permitted them to manage their alfalfa more effectively and to increase the size of their dairy enterprise.

On the farms surveyed, an average of 62.3% of the total forage land was devoted to alfalfa, with 50% of that area harvested as LMS. This contrasts with only about 30% as an average for the state during the same period. Smaller herds (group 1) utilized the least LMS, with 41% harvested as haylage, followed by farms in groups 2 and 3, respectively (Table 1). This trend shows that farmers with larger herds placed greater emphasis on LMS systems. Similarly, Taylor and Barr (1968) found that greatest profit was obtained with the use of all LMS in herds 70 cows and above. While all of the farmers in group 1 were using LMS to some degree, they made the greatest use of dry hay.

Overall alfalfa management and harvest techniques also have intensified over time. More than 58% of the producers are using more alfalfa in their seed mixtures than 5 years earlier; 63% now use pure alfalfa seedings, 34% use mostly alfalfa with some grass, and only 2% use less alfalfa. The rate of alfalfa seeding/ha now averages 15.7 kg compared with 13.5 kg/ha 5 years ago. Larger farms in group 3 used the greatest amount of alfalfa in their seeding rates (Table 1). Not only did all respondents topdress alfalfa, but 76% increased the fertilization rates after going to LMS. The greatest emphasis on topdressing was noted among group 3 farmers, of whom 88% increased rates of topdressing, followed by group 2, with

75% increasing rates. Over 66% of all producers tested forage for protein and dry matter.

Forage Quality

Evidence of producer emphasis on forage quality was found in harvest management and equipment. More farmers in groups 2 and 3 than in group 1 were cutting alfalfa in the bud stage (Table 1) and were in agreement with the work of Cloud et al. (1968), who found that late cutting resulted in substantial net income declines.

Improved forage quality was either the first or the second reason given for shifting to the LMS system by all dairy farmers surveyed. Over 90% stated that quality had improved and that protein concentration had increased. The total LMS system permits the forage producer to take more effective advantage of high-legume swards, early and more frequent cutting, higher yield and quality through increased fertilization, and more rapid harvesting after cutting to reduce weather risks and conserve leaves. Taylor and Barr (1968), Cloud et al. (1968), Smith (1964), and Smith et al. (1974) recommended the practices respondents were using and concluded that quality improvement, as the data suggest, was to be expected. Fifty-eight percent purchased less protein, while only 10% reported an increase following shifting to the LMS system.

Storage Method

LMS must be stored in airtight structures to preserve the increased quality harvested in this forage system. The type of structure used was not statistically related to forage quality and milk production. Two-thirds of the silos were conventional tower silos, and one-third were oxygen limiting. No bunker silos were used in the study. Convenience also may have been a factor in selecting silo structure.

Dairy Production

Continuing increases in herd size paralleled much of the progress in shifting to LMS systems; 84% of the herds grew in size. The average number of cows on the study farms was 76.8 (Table 1).

Over 87% of the farms reported increased milk and fat production/cow. Most of the dairy farmers related the increase to their LMS systems. Production/cow in herds using LMS had increased at a more rapid rate compared with all herds. Corley et al. (1964) found that 25% of increases in milk and fat production was due to genetics, while 75% was due to environment. Of that amount, about one-third was attributed to improved forage quality, better feeding methods, and other management factors. Dairy herds in this study produced an average of 7,246 kg milk/lactation, compared with an average of 5,345 kg/lactation for all Wisconsin herds (1977–1979). In addition, 93% of respondents also reported increased butterfat production/cow after employing the LMS system. This is to be expected, as fat production is usually increased with increasing milk production unless problems occur in the feeding program. There was no significant difference between production levels for the three herd-size groups, and all were significantly higher than the state average.

Dairy Economics

The economics involved in operating the dairy enterprise was probably the key factor leading to herd increases. Hoglund (1973) found that labor was the second largest expense in the dairy industry. This is still true today. To avoid excessive labor costs, dairymen have moved to LMS systems where automation is possible in addition to achieving higher forage quality and improved production.

While there was no apparent production advantage/cow when comparing groups 1, 2, and 3, there was definitely an economic advantage in favor of the larger herds. Milk marketed averaged 272,776 kg, 417,962 kg, and 871,092 kg, respectively, resulting in gross income from milk marketings of $63,011, $96,549, and $201,222, respectively, for the three groups. Production and income generated/cow were nearly identical in the three groups, but smaller herds were at a decided disadvantage, as overhead costs/animal became excessive and limiting (Table 2).

Investment in the LMS harvest, storage, and feeding system averaged $111,589, $129,137, and $210,330 for groups 1, 2, 3, respectively. Annual gross income for farmers having herds of 75+ cows nearly equaled their LMS system investment, while producers with less than 50 cows grossed only 56% of their investment (Table 2).

Farmers in group 1 experienced lower total annual ownership costs but higher costs/cow than farmers in groups 2 and 3 (Table 2); hence, there was a decided advantage in larger herds.

The use of LMS for small herds may be questioned, based on this analysis, assuming that comparable quality

Table 2. Average number of cows, annual ownership costs, ownership costs as percent of gross income (OC/GI), and annual gross income as percent of low-moisture silage system investment (GI/LMS) for three herd-size groups in 110 Wisconsin dairy herds, 1978.

Group	Cows/herd, avg.	Annual ownership cost Total	Annual ownership cost Per cow	OC/GI %	GI/LMS % invest.
3	118	$32,542	$276	16.1	95.6
2	57	$21,765	$382	22.5	74.7
1	38	$15,686	$413	24.0	56.0

and yield can be obtained from hay. Taylor and Barr (1968) concluded that high-quality dry hay was a more economical approach. However, data indicate that achieving high-quality dry hay in the northern Midwest is difficult.

LITERATURE CITED

Cloud, C.C., G.E. Frick, and R.A. Andrew. 1968. An economic analysis of hay harvesting and utilization using a simulation model. N.H. Agric. Expt. Sta. Bul. 495.

Corley, E.L., G.R. Barr, D.A. Wieckert, E.E. Heizer, and C.E. Kraemer. 1964. Environmental influences on production in 46 dairy herds. Wisc. Agric. Expt. Sta. Res. Bul. 253.

Hoglund, C.R. 1973. Management, investments, and labor economics. J. Dairy Sci. 56:488–495.

Smith, D. 1964. Chemical composition of herbage with advance in maturity of alfalfa, medium red clover, ladino clover, and birdsfoot trefoil. Wisc. Agric. Expt. Sta. Res. Rpt. 16.

Smith, D., D.A. Rohweder, and N.A. Jorgensen. 1974. Chemical composition of three legumes and four grass herbages harvested at early flower during three years. Agron. J. 66:817–819.

Taylor, H.H., and W.L. Barr. 1968. An economic comparison of forage harvesting, storing, and feeding systems on Pennsylvania dairy farms. Pa. Agric. Expt. Sta. Bul. 751.

Estimation of Digestibility of Legume Forages by Means of Morphological Parameters

W. KÜHBAUCH

University of Kassel, Federal Republic of Germany

Summary

Forage quality is normally evaluated by time-consuming and costly laboratory analyses. It appears promising to use morphological parameters to evaluate forage quality in alfalfa and red clover. During progressive growth of white clover, red clover, and alfalfa, the leaf and stem portion was separated to determine the fiber fraction according to the Van Soest method and to estimate digestibility by the procedure of Tilley and Terry. Stem : leaf ratio and stand height as a measure of stem length were used as morphological parameters.

Leaves of the three legumes vary only slightly in concentration of cell-wall material and in digestibility within cuts. The leaf-supporting stems of white clover, in spite of being high in cell-wall concentration, are highly digestible. As a result, no practical relationship exists between morphology and food quality in this species. In the stems of tall-growing stands of alfalfa and red clover, however, the cell-wall concentration increases along with a decrease in digestibility. Significant inverse relationships were found between stem length or stand height and digestibility of both alfalfa and red clover, with r values of 0.95 and 0.80, respectively. These correlations were similar to those obtained between digestibility and fiber fractions.

In conclusion, it seems possible to estimate the quality of alfalfa and red clover by simple, nonchemical parameters. Since the leaf portion varies little in digestibility, it is possible to estimate total food quality by measuring stand height as a measure of stem length and leaf:stem ratio. Such a system might be of practical use for farmers.

KEY WORDS: white clover, *Trifolium repens* L., red clover, *Trifolium pratense* L., alfalfa, *Medicago* × *varia Martyn*, growth morphology, digestibility.

INTRODUCTION

Evaluation of forage quality usually is performed either by means of chemical procedures to separate the more and less available components of food before quantitative analyses, or with the entire material through biological systems, such as the Tilley and Terry digestibility procedure. Procedures that enable the farmer himself to estimate forage quality would be highly desirable. With alfalfa and red clover it seems promising to estimate food quality by simple nonchemical techniques. Previous research has shown that with progressive growth of alfalfa the decreasing digestibility is almost exclusively related to the stem, whereas the leaves remain highly digestible (Marten 1970, Kühbauch and Voigtländer 1979). Similar observations were made with red clover (Kühbauch, unpublished). With increasing stem length, the physical stresses increase stem material, and stem tissue is reinforced by less-digestible cell-wall material. In contrast, no comparable stress affects the legume leaves. As a consequence the stem length of tall-growing legumes could be a useful parameter to estimate food quality.

This paper reports on experiments with white clover, red clover, and alfalfa, during first and second growth, to determine the statistical correlations between morphology and quality. In addition, correlations between the digestibility of stem tissue and fiber fractions are presented.

METHODS

Plant Material

White clover (*Trifolium repens* L.), red clover (*Trifolium pratense* L.), and alfalfa (*Medicago* × *varia Martyn*) were harvested in 1977 after measuring the stand height of the respective plots. When stems and leaves were separated, the leaf-supporting stems of white clover (i.e., the harvested stem) and red clover were included in the stem portion, whereas, with alfalfa, the very small leaf-supporting stems remained with the leaf portion. Plant material was frozen with liquid nitrogen and stored at − 20°C.

The author is a Professor of Agronomy at the University of Kassel.

Quality Analyses and Biometric Calculations

Freeze-dried plant material was ground to particle sizes of less than 1.0 mm. In-vitro digestible dry matter (IVDDM) was measured with the two-stage procedure of Tilley and Terry (1963). Chemical constituents of the plant were analyzed by fiber fractionation (Goering and Van Soest 1970) into total cell wall (NDF), lignocellulose (ADF), and lignin (ADL). Values are expressed as percentage of dry matter. Partial correlations (r_p) between stem length and quality parameters were calculated using the BMDP2R program (Dixon and Brown 1979). The resulting regression equations will be published elsewhere (Kühbauch and Pletl 1981).

RESULTS AND DISCUSSION

In Table 1, stand height and dry-matter percentages of total cell wall (NDF) and digestibility (IVDDM) are shown for leaves and stem material of white clover, red clover, and alfalfa during first growth. Second-growth data are presented in Table 2. The leaves of the three legumes changed little in digestibility during plant growth. The high IVDDM values for leaves are associated with low percentages of total cell wall. The highest values for leaf NDF were found in white clover at about 23%. The digestibility of harvested stems of white clover varied little within first and second growth periods. Especially during first growth, digestibility values for white clover stems definitely exceeded those of the leaves, even though concentration of cell-wall constituents increased to around 30%. This finding suggests qualitative differences within the cell wall as a whole and/or within particular fiber fractions, such as have been shown for grasses (Waite et al. 1964). The stem of white clover, even with double the amount of cell-wall components, was no less digestible than the leaves; and an 8% increase (from 31% to 39%) in cell wall, during stem elongation, was associated with only 1.4% depression in digestibility (Kühbauch and Pletl 1981). However, the relatively short stand of white clover with a maximum height of 30 cm hardly results in physical stresses on plant tissue sufficient

Table 1. Stand height (H), cell-wall constituents (NDF), and in-vitro digestible dry matter (IVDDM) in leaves and stems of white clover, red clover, and alfalfa for the first growth in 1977.

| Date of harvest | Leaves | | Stem | | |
	NDF (%)	IVDDM (%)	H (cm)	NDF (%)	IVDDM (%)
White clover					
5/13	13.1	78.8	20	26.4	84.0
5/20	14.7	75.7	21	27.0	83.2
5/27	14.4	78.5	24	23.5	83.9
6/3	15.8	78.1	26	–	82.4
6/10	15.9	77.6	28	30.1	81.2
6/16	14.9	76.3	30	30.6	76.6
Red clover					
5/13	19.4	70.7	37	24.5	81.5
5/20	19.7	70.2	45	27.6	81.9
5/27	17.3	70.9	55	31.2	81.6
6/3	16.5	74.5	60	36.8	74.8
6/10	17.7	74.0	75	42.6	71.7
6/16	18.7	67.2	80	43.1	67.0
Alfalfa					
5/13	15.8	78.3	40	33.9	72.5
5/20	17.4	76.8	51	39.7	73.1
5/27	16.3	77.4	64	46.7	65.8
6/3	17.1	76.9	74	50.6	61.8
6/10	18.3	76.1	80	56.0	55.9
6/16	19.5	72.9	90	55.3	56.4

to initiate a distinct deterioration in food quality.

Red clover and alfalfa continue to grow up to a stand height of 80 to 90 cm (Tables 1 and 2, 1977, and Fig. 1, 1978). This growth is accompanied by marked increases in cell-wall percentages. During first growth, the cell wall (percentage of DM) increased to 43% for red clover and 55% for alfalfa, and digestibility (IVDDM) decreased from about 82% to 67% and 75% to 56% for red clover and alfalfa, respectively.

Table 2. Stand height (H), cell-wall constituents (NDF), and in-vitro digestible dry matter (IVDDM) in leaves and stems of white clover, red clover, and alfalfa for the second growth in 1977.

| Date of harvest | Leaves | | Stem | | |
	NDF (%)	IVDDM (%)	H (cm)	NDF (%)	IVDDM (%)
White clover					
6/29	16.6	77.5	21	–	79.4
7/6	20.6	77.5	21	28.2	74.2
7/13	20.3	76.0	26	33.3	73.8
7/20	18.0	75.7	27	30.4	72.3
Red clover					
6/29	–	73.9	20	–	77.2
7/6	16.8	68.7	30	27.2	77.6
7/13	22.0	71.1	50	32.6	78.5
7/20	18.1	64.6	60	34.9	74.2
Alfalfa					
6/29	15.3	73.9	26	–	76.0
7/6	16.5	74.9	45	35.9	71.2
7/13	16.6	77.5	66	49.8	62.8
7/20	16.6	73.9	77	47.8	62.4

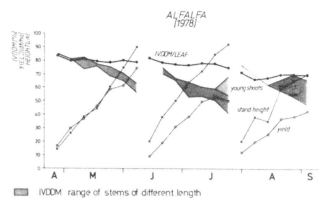

Fig. 1. Change of stand height, yield, and digestibility of alfalfa forage during growth, 1978.

IVDDM range of stems of different length

The small decrease in IVDDM values for red clover stems during second growth in 1977 may have been due to the fact that leaf-supporting stem material was not analyzed separately and may have contributed a high proportion of the total stem fraction.

The partial correlations (r_p) between stand height (H) and digestibility (IVDDM) for stem material of the legumes are given in Table 3, along with correlations between IVDDM and percentages of total cell wall (NDF), lignin (ADL), or lignocellulose (ADF). As expected from white clover morphology, r_p values for white clover stem material show no significant relationships with any of the factors tested. However, with red clover and alfalfa, highly significant negative correlations between stand height and digestibility r_p values were observed, –0.804 and –0.951, respectively (Table 3). In addition, a close relation between stand height and cell-wall synthesis is indicated by the high positive r_p values for both red clover and alfalfa. Cell-wall parameters and stand height, especially of alfalfa, are about equally precise for estimating IVDDM. The relationship between stem length and quality can vary because of the wide range in IVDDM within the stem portion, which occurs because shoots of different order and growth make up the total stem portion (Fig. 1).

Table 3. Partial coefficients of correlation (r_p)+ between stand height (H) and in-vitro digestible dry matter (IVDDM), and between IVDDM and percentage of cell-wall constituents (NDF), lignin (ADL), and lignocellulose (ADF), respectively, with stem tissue of white clover, red clover, and alfalfa, first and second growth 1977.

		IVDDM	NDF		n
White clover	H	–0.035	–0.083		12
Red clover	H	–0.804**	0.957***		9
Alfalfa	H	–0.951***	0.959***		9

		NDF	ADL	ADF	n
White clover	IVDDM	–0.66	–	–	12
Red clover	IVDDM	–0.899***	–0.709*	–0.833***	9
Alfalfa	IVDDM	–0.964***	–0.983***	–0.839***	9

+Significant at levels of 0.1%, ***; 1%, **; 5%, *.

From this investigation it appears that stand height provides a good estimate of IVDDM for stems. As for the leaf portion of the legume, it should be sufficient to take into account the stem:leaf ratio and assume a constant leaf digestibility. For example, with a given stem:leaf ratio of alfalfa of 60:40 at the end of first growth in 1977, we calculate (60 × estimated IVDDM of the stem + 40 × 77 as IVDDM of the leaf)/100 = estimated total IVDDM. This calculation, using stem:leaf ratio and stand height as the only parameters, may be useful as a quick estimate of forage quality.

LITERATURE CITED

Dixon, W.J., and M.B. Brown. 1979. BMDP-79. Biochemical computer programs P−series manual. Univ. Calif. Press, Berkeley.

Goering, H.K., and P.J. Van Soest. 1970. Forage fiber analyses (apparatus, reagents, procedures, and some applications).

USDA Dept. Agric., Agric. Handb. 379B.

Kühbauch, W., and L. Pletl. 1981. Berechnung der Futterqualität von Weißklee, Rotklee und Luzerne mit morphologischen Kriterien und/oder aus Pflanzeninhaltsstoffen. Z. Acker- und Pflanzenbau 30:271–290.

Kühbauch, W., and G. Voigtländer. 1979. Veränderungen des Zellinhaltes, der Zellwandzusammensetzung und der Verdaulichkeit von Knaulgras (*Dactylis glomerata* L.) und Luzerne (*Medicago* × *varia Martyn*) während des Wachstums. Z. Acker- und Pflanzenbau 148:455–466.

Marten, G.C. 1970. Temperature as a determinant of quality of alfalfa harvested by bloom stage or age criteria. Proc. XI. Int. Grassl. Cong., 506–509.

Tilley, J.M.A., and R.A. Terry. 1963. A two-stage technique for the in vitro digestion of forage crops. J. Brit. Grassl. Soc. 18:104–111.

Waite, M., J. Johnston, and D.G. Armstrong. 1964. The evaluation of artificially dried grass as a source of energy for sheep. I. The effect of stage of maturity on the apparent digestibility of ryegrass, cocksfoot and timothy. J. Agric. Sci. 62:391–398.

Pennsylvania's Alfalfa Growers Program: Four Years of Success

J.E. BAYLOR, L.E. LANYON, and W.K. WATERS
Pennsylvania State University, University Park, Pa., U.S.A.

Summary

Alfalfa (*Medicago sativa* L.) is the most important perennial forage legume grown in Pennsylvania. However, yields currently reported by farmers are considerably below what might be expected when all production inputs are utilized. Therefore, a program was developed (1) to measure how much alfalfa hay, protein, and energy can be produced on 1 ha of Pennsylvania land under nonirrigated conditions, (2) to determine the mineral content and uptake by high-yielding alfalfa crops, and (3) to obtain more realistic on-farm estimates of alfalfa production costs. To achieve the program goals, hay production for each harvest from 2.02-ha farm fields was estimated, using a method developed specifically for this program. Forage samples were taken at the same times and analyzed for quality and mineral content. Cost-of-production estimates were compiled from enterprise budgets prepared by each grower.

The average estimated yield for each of the four years of the program ranged from 12.3 to 13.0 metric tons (t)/ha (10% moisture) with the highest yield approximately 2½ times the lowest yield each year. The highest estimated production during the 4 years was: hay, 20.4 t/ha; crude protein, 3,898 kg/ha; and total digestible nutrients 11,648 kg/ha.

Mineral analysis of the forage samples indicated significant farm-to-farm and year-to-year variations in mineral content. Nevertheless, for the most part the values were in the normally expected range, and virtually no relationship was found between mineral content and yield. Annual mineral uptake also showed considerable variation. For example, P and K uptake ranged from 22 to 72 and from 162 to 624 kg/ha, respectively. The average annual mineral uptake for P, K, Ca, Mg, and S was 41, 355, 147, 25, and 27 kg/ha, respectively.

Average estimated annual production costs compiled for the 4 years of the program increased from $429 to $656/ha while the average net return decreased from $401 to $302/ha. The break-even yield during the period jumped from 6.3 to 8.3 t/ha, approximately 14%/yr, emphasizing the need for optimum yields to assure satisfactory returns. The major factors in the annual costs of production were: machinery, 39%; fertilizer and lime, 20%; labor, 17%; and land, 17%. Other production costs, including seed, spray materials, etc., totaled only 7%.

The Pennsylvania Alfalfa Growers Program has clearly shown the potential for higher production and greater profitability for alfalfa growers in the state through application of proven production practices and control of unnecessary production costs.

KEY WORDS: Alfalfa Growers Program, measuring yields, quality, mineral uptake, annual production costs, profitability.

INTRODUCTION

Dr. Robert Wagner, President of the Potash and Phosphate Institute, recently stated (1979), "Moving up to maximum economic yields is the best way—and about the only way—farmers can control their profit margins today. Research has given us good components or parts—high yielding varieties and hybrids, improved fertilizer practices, new pest control methods, better tillage and residue management practices, and many others. We can expect research to turn out still better individual components. But greater strides toward high yields will likely come through fitting those parts into combinations that produce positive interactions."

Over a 4-year period (1977–1980) more than 200 Pennsylvania alfalfa growers cooperated in a program to show how they put those parts together to attain maximum yields. The program had three objectives: (1) to measure how much alfalfa hay, protein, and energy could be produced on 1 ha of Pennsylvania land under nonirrigated conditions, (2) to determine, under farm conditions, mineral content and uptake by high-yielding alfalfa crops, and (3) to obtain more realistic on-farm estimates of alfalfa production costs.

METHODS

Yields were measured by taking six 1/2,470-ha random samples/harvest from the swath or windrow of a selected 2.02-ha farm field using a special measuring device designed for this purpose. Whole samples were weighed in the field, and subsamples were taken for moisture, forage quality, and mineral analysis. All harvesting, weighing, and recording of field data was under the direct supervision of local county agricultural advisors.

Yields were calculated for each harvest on the basis of metric tons/ha of hay at 10% moisture. Quality data by harvests included crude protein (CP), acid detergent fiber (ADF), total digestible nutrients (TDN), and mineral analysis for twelve elements. Total production of CP and TDN was calculated, as was total mineral uptake for all elements.

Cost-of-production data were calculated from budgets prepared by cooperating producers.

RESULTS

Yields

Average estimated hay equivalent yields and the range in yields for the 4 years for all participating growers are given in Table 1. The wide range in yields within years is a reflection of the wide range in soil and environmental conditions, along with other factors, encountered by growers throughout the State.

J.E. Baylor is a Professor of Agronomy Extension, L.E. Lanyon is an Assistant Professor of Agronomy, and W.K. Waters is an Associate Professor of Farm Management Extension at Pennsylvania State University.

Table 1. Average estimated hay equivalent yields and ranges in yields for participants in Pennsylvania's Alfalfa Growers Program.

Year	Average yield t/ha	Range t/ha
1977	12.3	7.4–18.1
1978	12.3	6.7–19.5
1979	13.0	8.5–20.4
1980	12.8	7.2–19.3

As Table 1 shows, maximum yields of hay equivalent recorded during the 4 years ranged from 18.1 to 20.4 t/ha. Estimated yields of CP for these top growers were: 3,507, 3,658, 3,898, and 3,817 kg/ha, respectively, for 1977, 1978, 1979, and 1980, while corresponding estimated yields of TDN were 10,389, 10,895, 11,648, and 10,935 kg/ha. In each of the 4 years the highest yield was approximately 2½ times the average yield for all producers participating in that year's program.

Mineral Relationships

The average mineral content and mineral uptake and the range in mineral content and uptake for the twelve minerals measured are given in Table 2. There was a wide range in the mineral content of alfalfa for each of the minerals measured. As expected, of the macronutrients and secondary nutrients N and K contents were highest, Ca intermediate, and P, Mg, and S lowest, whereas the micronutrients were present in much lower concentrations. Similar relationships also apply to the total mineral uptake by the crop. Wide differences in mineral content and uptake were observed from farm to farm and from year to year. Nevertheless, the average mineral content of

Table 2. Hay yield, mineral content, and uptake (1977–1980).

	Hay Yield			
	Average t/ha		Range t/ha	
	12.7		6.70–20.5	

	Mineral Content		Mineral Uptake	
	Average	Range	Average	Range
	%		kg/ha	
Nitrogen	3.25	2.45–3.79	372	174–626
Phosphorus	0.35	0.20–0.47	40.0	20.7–72.0
Potassium	3.05	1.42–4.05	347	162–624
Calcium	1.26	0.68–1.76	142	59.4–249
Magnesium	0.23	0.12–0.38	26.4	10.4–51.0
Sulfur	0.238	0.072–0.333	27.1	6.16–45.6
	ppm			
Boron	28	10–45	0.31	0.10–0.60
Copper	7.2	4.2–13.2	0.082	0.032–0.149
Zinc	25	17–42	0.28	0.13–0.55
Manganese	56	26–154	0.64	0.24–2.4
Iron	140	40–609	1.58	0.49–8.2
Aluminum	78	20–516	0.87	0.19–6.9

Table 3. Production and cost factors for those participants in the Pennsylvania Alfalfa Growers Program who had cost budgets.

	Per ha			
	1977	1978	1979	1980
Yield (t)	12.3	12.4	13.3	12.1
Crop Value[1]	$830	$847	$919	$958
Production Cost	$429	$487	$560	$656
Net Return	$401	$360	$359	$302
Break-Even Yield[2]	6.3	7.0	8.1	8.3

[1]For 1977–1979, crop value calculated @ $77/t hay equivalent times estimated net yield (measured yield less estimated harvest loss). For 1980, crop value calculated @ $80/t.

[2]Standing yield to cover all costs considering harvest losses.

P, K, and the micronutrients was generally in the normally accepted range for the entire alfalfa plant when plants were sampled at comparable stages of maturity. The average content of N, Ca, Mg, and S approached the lower limit of the published values. These marginal values may indicate potential deficiencies of these minerals with high crop removal, but they also suggest the overall lower mineral content of whole-plant samples taken under farm conditions.

Production Costs and Returns

Since participants in the program were above-average alfalfa growers, their cost data do not necessarily represent the costs and returns of producers in general. However, growers in the study did show a very favorable profit over the 4 years. Net margins/ha over costs, including land and labor, were $401, $360, $359, and $302 for 1977, 1978, 1979, and 1980, respectively (Table 3). An analysis of net returns at various yield levels indicates that profits are definitely better at higher yield levels. As an average for the 4 years, profit/ha ranged from $45 at the 8-t yield to $885 when the average yield was 19 t. The break-even yield during the 4-year period increased from 6.3 to 8.3 t/ha. Major factors in the annual costs of production were: machinery, 39%; fertilizer and lime, 20%; labor, 17%; and land, 17%.

DISCUSSION

One of the best indicators of practices leading to high yields and profits under farm conditions can be found in the production practices of top growers. Below is a brief summary of practices followed by the top 10 producers during each of the 3 years 1978-1980, the only years for which these data are available.

1. *Yield*. The overall average estimated hay equivalent yield for the top 10 growers in each of the 3 years was 16.7 t/ha, more than double the state average yield for all alfalfa. Average estimated production of CP and TDN for these same growers was 2,838 kg/ha and 8,439 kg/ha, respectively.

2. *Cultivars*. Over the 3-year period, nine high-yielding cultivars were used by the 30 growers, indicating the excellent germ plasm currently available. These cultivars were also consistently among the highest yielding in replicated research trials at various locations in the State.

3. *Soils and Fertility*. All top growers over the 3-year period planted alfalfa on well-drained soils, mainly of limestone origin, known to be well suited to the crop. Lime and fertility programs for establishment and maintenance were based on soil test and grower experience. Most top producers applied dairy-cow manure in the rotation before alfalfa. The lack of consistent relationship between postestablishment soil test results, fertilizer applications, and/or mineral content of the alfalfa and hay yield indicate that the fertility needs of the crop were adequately met, thus suggesting the importance these growers attach to soil fertility as a basic component of high-yielding crops.

4. *Establishment techniques*. Seventy-five percent of the top yields during the 3 years were made during the first full harvest year after the year of seeding and were from stands seeded in the spring of the year without a companion crop. On most of the fields, a herbicide was used to control weeds at establishment.

5. *Frequency of harvests*. Over the 3-year period, all of the top 30 growers made at least four cuttings/year, and 3 harvested their crop five times. In an area where three cuttings/yr are common, this move to more frequent harvest by top producers indicates the potential of more intensive management for high yields of high quality forage.

6. *Storage*. In 1978 3 of the top 10 growers stored all of their cuttings as silage, with 8 storing one or more cuts in this form. For the other 2 years all 20 top producers stored at least one cutting as silage, with 11 storing all cuts in the silo. Removing each crop as soon as possible after cutting is important in getting regrowth off to a fast start.

7. *Cutting interval*. The average cutting interval used by the top 10 growers in 1978 would be considered nearly ideal for Pennsylvania growing conditions—36, 39, and 45 days between cuts 1 and 2, 2 and 3, and 3 and 4, respectively. The cutting intervals in 1979 and 1980 were similar—approximately 38, 36, and 43 days, respectively, between cuts 1 and 2, 2 and 3, and 3 and 4, suggesting the less favorable harvesting conditions for alfalfa during those years as compared to 1978.

8. *Insect control*. In Pennsylvania controlling insects that attack alfalfa, especially potato leafhopper (*Emposa fabae* [Harris]), is essential in producing high yields of high-quality forage. In 1978 and 1980 all top growers sprayed their alfalfa crop at least once with an approved insecticide to control leafhoppers and other insects. Seven growers in 1978 and 6 growers in 1980 sprayed two or more times. In 1979, the frequency of spraying to control insects was less, with 3 growers reporting almost no insect damage for the entire growing season. Producer experiences reflect year-to-year changes in insect populations and the benefits of an insect management program geared to help producers better predict the buildup of troublesome insects in individual fields.

Information provided in this study verifies the findings

of research and certainly indicates that, in the words of Wagner (1979), ". . . greater strides toward high yields will likely come through fitting those parts into combinations that produce positive interactions."

LITERATURE CITED

Wagner, R.E. 1979. Where the action is. Better Crops 18:33. Potash and Phosphate Inst., Atlanta, Ga.

Pasture Demonstration Program for Beef: Economic Considerations

P.R. HASBARGEN, R.H. CRAVEN, and N.P. MARTIN
University of Minnesota, Saint Paul, Minn., U.S.A.

Summary

This study made an economic evaluation of several alternative pasture improvement programs on beef farms in northern Minnesota. Three questions were addressed relative to several pasture improvement systems. These were (1) Will it pay? (2) Is there a cheaper way? and (3) Is there a more profitable system for a particular farm resource situation? Partial budgeting and computerized linear programming techniques were used to analyze pasture cost and return data obtained from several ranches in northern Minnesota.

Study results show that rented pastures, when available, provided the cheapest source of additional summer feed, and that legume pastures provide cheaper sources of summer feed than fertilized bluegrass in northern Minnesota. However, if cropland is suitable for growing wheat that will average over 2,688 kg/ha or sunflowers that will yield over 1,334 kg/ha, bluegrass pasture should be fertilized so as to release acreage for cash crop production. Also, if hay is not available for purchase, or if hay prices jump significantly over the prices that prevailed in recent years (as they did after droughts of 1976 and 1980), grass fertilization becomes profitable. At the relatively low hay prices usually prevailing in northwestern Minnesota, it is more profitable to buy hay and feed it as supplement during short pasture periods than to fertilize bluegrass pasture.

If birdsfoot trefoil can be successfully established it can compete as a pasture crop with alfalfa in northern Minnesota. Using average feeder-cattle price relationships, the "best" beef program was consistently a cow-yearling program with heifer feeders sold in the spring and yearling steers sold in the fall. However, when the feeder price break comes, there usually are a couple of years when the cow-calf program is superior.

It appears from an economic standpoint that pasture renovation is superior to nitrogen fertilization of bluegrass pasture in northern Minnesota. However, pasture land that can be used for cropland might best be converted to other crops, and the remaining ha might then be fertilized in order to carry more beef cattle — at least during the stage of the cattle cycle when returns to the cow herd are likely to be above average.

KEY WORDS: Kentucky bluegrass, *Poa pratensis* L., smooth bromegrass, *Bromus inermis* Leyss., nitrogen, alfalfa, *Medicago sativa* L., birdsfoot trefoil, *Lotus corniculatus* L., beef pasture systems.

INTRODUCTION

Economic tradeoffs between alternative beef pasture-management systems are complex. More beef/ha does not always mean more profits/ha. Even if it did, a limited labor supply might preclude intensive management systems. Therefore, the "best" beef-forage systems will vary with the resource situation.

Three economic considerations for any management practice are (1) Will it pay? (2) Is there a cheaper way? and (3) Is there a more profitable system for this particular resource situation? To answer requires comparing (1) the added cost with the added return from the practice, (2) the cost of one practice with another that gets similar results, and (3) the net returns from several alternative ways of using a specific bundle of resources. The third comparison requires a total farm business analysis, as opposed to the partial budget procedure that can be used to answer the first two questions.

P.R. Hasbargen is a Professor and Extension Economist, R.H. Craven is a Graduate Research Assistant, and N.P. Martin is an Associate Professor and Extension Forage Specialist at the University of Minnesota.

SHOULD NITROGEN BE USED ON BLUEGRASS PASTURES?

To answer the question, Will it pay?, we need to compare the added costs of nitrogen fertilizer with the estimated value of the additional animal-unit pasture months that are gained. With nitrogen at $0.44/kg, the cost of obtaining an additional animal-unit month (AUM) from fertilizing bluegrass is about equal to the net value of beef that can be produced (using long-term price relationships)/AUM — making the answer to this question "Just barely."

To answer, "Is there a cheaper way?", alternatives such as pasture rental and pasture renovation can be considered. Pasture rental rates are only $4 to $8/AUM in northern Minnesota. Therefore, even if the renter incurs extra fence maintenance and cattle movement costs, the additional pasture often costs less than one-half as much as fertilizing grasses at recent nitrogen price levels.

OTHER IMPROVED SYSTEMS VS. BLUEGRASS PASTURES

The added costs of providing extra AUMs with "improved systems" on two farms in northern Minnesota are shown in Table 1. (For a description of this project, see our companion paper in Section XIII of these *Proceedings*: Martin et. al., "Pasture Demonstrations for Beef: Agronomic Considerations.")

Using feeder-cattle price relationships of the last 30 years and gains equal to those obtained by grazing yearling cattle on the Kenner ranch, the added returns/AUM would be about $18. Thus, the answer to "Will it pay?" is again "Just barely" — except for system C in northeastern Minnesota. The higher rainfall and better soil of this area made the legume-grass plus the nonuse of rotational grazing the best of the three. However, in northwestern Minnesota pasture renovation with birdsfoot trefoil is the best of the three at providing added pasture as ha of bluegrass previously rented are pulled into crop production.

SELECTION OF THE "BEST" BEEF FORAGE SYSTEM

Is there a better way? To find out, we set up an example farm under northwestern Minnesota conditions — with the resources and the crop and livestock program options shown in Table 2. We then used a profit-maximizing linear programming computer model to select "best" crop and livestock alternatives under various resource situations and price relationships. Table 3 shows the difference in net farm profits and in farm organization for three levels of pasture management when no cash crop or pasture rent is allowed (a specialized beef-forage farm). Note that pasture fertilization alone (medium management level) increased annual farm earnings about $800 and decreased hay purchases by 165 tonnes. When high-level management is used on alfalfa (three cuttings/yr), the

Table 1. Average and marginal costs of providing pasture in northwestern and northeastern Minnesota under several grazing systems 1976–1979.

	Pasture systems on Kenner ranch, Wannaska, Minn.		
	A 27.2 ha unimpr. bluegrass	B 8.0 ha fert. bluegrass, 6.0 ha birdsfoot trefoil	C 5.6 ha fert. bromegrass, 9.6 ha legume-grass
AU month/ha	3.32	9.95	9.17
Annual cost/ha	$50.00	$155.95	$158.05
Average cost/AUM	15.04	15.67	17.23
Added cost/ha	–	105.95	108.05
Added AUM/ha	–	6.63	5.85
Added cost/AUM	–	15.99	18.47
	Pasture systems on Disterhaupt ranch, Moose Lake, Minnesota		
	A 45.6 ha unimpr. bluegrass	AA 32.0 ha fert. rotated bluegrass	C 6.4 ha fert. grass 8.8 ha legume-grass
AUM/ha	3.63	6.00	14.00
Annual cost/ha	$62.50	$105.15	$185.70
Average cost/AUM	17.24	17.53	13.26
Added cost/ha	–	42.65	123.20
Added AUM/ha	–	2.38	10.37
Added cost/AUM	–	17.96	11.87

Table 2. Resources and crop and livestock alternatives available in the Roseau area linear-programming example farm.

Land
 Tillable - 110 ha   Total - 170 ha
 Maximum cash crop varies from 0 to 60–110 ha. Permanent pasture, 40 ha which can be fertilized and 20 ha which cannot.
Labor
 11 hours/day family (reduced for "part time" situation); additional labor at $3.50/hour (not permitted in some cases)
Livestock
 Maximum: 240 cows (90% calving rate, 15% cow replacement with raised stock). Options to sell calves ($1.76/kg, steers; $1.54/kg, heifers) or sell short yearling (spring-$1.72/kg, steers; $1.52/kg, heifers) or sell steers as yearlings in the fall for $1.61/kg.
Crop and Forage Alternatives
 Buy/sell alfalfa hay, $44/$35/tonne. Buy grass hay, $31/tonne. Rent permanent pasture @ $2, $4, and $6/ha; 48-ha limit.
 Permanent pasture: no nitrogen (N), 55/55 kg, and 112/112 kg split applications of N.
 Alfalfa pasture, after 1 or 2 cuttings of hay.
 Birdsfoot trefoil pasture, established every 10 years.
 Orchardgrass pasture: no N, 23/23 kg, and 45/45 kg split applications of N.
 Alfalfa and orchardgrass established every 5 years, timothy every 6 years.
 Alfalfa: 2 and 3 cut hay.
 Timothy: 2 cut hay - no N, 23 kg of N, 45 kg of N.
 Permanent pasture: 2 cut hay - no N, 23 kg of N, 45 kg of N.
 Sunflowers: 1,704 kg/ha yield.   Oats: 2,545 kg/ha yield.
 Wheat: 2,727 kg/ha yield.

beef herd is expanded, no bluegrass is fertilized, and earnings increase significantly. The alfalfa-grass pasture reduces supplemental hay feeding in June, July, and August when bluegrass pastures are poor.

To determine the effect different land : labor ratios, different beef price levels, and hay purchase limitations would have on the "best" forage-beef systems we varied resource situations and feeder prices as described in Table 4.

As labor becomes more limiting, first (column 2) cow numbers are cut in half, bluegrass pasture fertilization is discontinued, and more of the alfalfa is harvested as hay than as pasture. Then (column 3) cow numbers drop to 45 head, the bluegrass pasture is left unused, and pasture needs are met with 24 ha of orchardgrass and 6 ha of birdsfoot trefoil.

If feeder prices make a cyclical dip in the mid-1980s to the $1.32/kg level (in 1979 dollars), this farm should cut back to 17 cows if alfalfa hay can be sold for $35/tonne. This explains why cattle numbers dropped sharply during the late 1970s and why alfalfa hay stays relatively low priced in this area.

At 1979 price levels (steer calves at $2.20/kg) farm earnings were almost double what they are expected to average over a complete cattle cycle with calves at $1.76/kg and the alternatives selected remaining the same. If no hay is purchased, income drops sharply as cow numbers drop to 84, the permanent pasture is fertilized at the high rate, and alfalfa (three-cut pasture) is used.

Table 3. Farm earnings, cow numbers, and crop ha selected as the "best" beef-forage systems under three different levels of forage management intensity on a specialized beef-cow farm that cannot rent pasture.

	Level of forage management intensity		
	Low	Medium	High
Item	Will not fert. bluegrass; cut alfalfa hay two times/ year	Will fert. bluegrass; cut alfalfa hay two times/year	Will fert. bluegrass; cut alfalfa hay three times/year
Farm earnings, $	$17,324	$18,121	$22,435
Cows, number of head	200	200	240
Labor hired, hours	1,264	1,219	1,163
Pasture rented, ha	–	–	–
Grass hay bought, tonnes	627	462	921
Alfalfa hay bought, tonnes	–	–	–
Low fertility permanent pasture, ha	60	20	60
High fertility permanent pasture, ha	–	40	–
Medium fertility alfalfa hay, 2 cut, ha	92	92	–
Alfalfa pasture, ha	–	–	45
Alfalfa 1 cut pasture, ha	–	–	10
Alfalfa 2 cut pasture, ha	–	–	30
Alfalfa 3 cut pasture, ha	–	–	6
Alfalfa establishment, ha	18	18	18
Hay fed in June, tonnes	107	69	–
Hay fed in July, tonnes	92	58	–
Hay fed in August, tonnes	126	105	36
Hay fed in September, tonnes	116	75	142

Table 4. Farm earnings, cow numbers, and crop ha selected as the "best" systems for different resource and feeder-cattle price situations when no land is rented, northwestern Minnesota.

Item	"Best" plan 240 cows	No additional labor hired	No hired labor; 40 hrs/wk less family labor	Steer calf price $1.32/kg	No hay purchase
Farm earnings, dollars	$27,151	$21,679	$13,362	$11,551	$18,533
Cows, head	240	115	45	17	84
Heifers sold in spring, head	72	34	14	5	25
Steers sold as yearlings, head	108	52	20	8	38
Labor hired, hours	811	–	–	–	42
Buy grass hay, at $31/MT, tonnes	1,117	395	45	–	–
Buy alfalfa hay, at $44/MT, tonnes	105	–	–	–	–
Low fertility perm. pasture, ha	26	60	–	60	20
45 kg/45kg N. perm. pasture, ha	34	–	–	–	40
Alfalfa pasture, ha	34	18	5	1	–
Alfalfa, 1 cut, pasture, ha	8	–	–	–	–
Alfalfa, 3 cut, pasture, ha	–	23	12	40	42
Alfalfa establishment, ha	8	8	3	8	8
BFT - TRE pasture, ha	–	–	6	–	–
Raise wheat, ha	60	60	60	60	60
Orchardgrass, pasture, 23 kg/23 kg N, ha	–	–	24	–	–
Sell alfalfa hay at $35/tonne	–	–	–	278	–
June feed hay, tonnes	–	5	–	–	2
July feed hay, tonnes	–	–	–	–	–
August feed hay, tonnes	86	41	–	4	25
September feed hay, tonnes	130	62	–	–	1

Economics of Renovating Mountain Hay Meadows

J.J. JACOBS

University of Wyoming, Laramie, Wyo., U.S.A.

Summary

There are approximately 0.5 million hectares (ha) of irrigated native grass hay in the mountain and intermountain regions of California, Oregon, Idaho, Utah, Nevada, Montana, Wyoming, and Colorado. Often these lands are not producing at their full physical or economic potential. The objectives of this study were to identify alternative improvement procedures to increase hay production and to determine the economic feasibility of those alternatives.

The principal means of identifying alternative mountain hay-meadow improvement practices and their costs was personal interviews with ranchers. The ranchers interviewed were owners and/or operators of mountain ranches in Wyoming who were using or had used some meadow improvement practices. Another source of information was previous research on mountain hay-meadow improvements and hay production.

The interviews revealed different intensities of hay-meadow improvements. These intensities were: (1) farming, (2) farming and structures, (3) farming, structures, and leveling, and (4) farming and sprinklers. The ranch interviews revealed that farming and structures and farming, structures, and leveling were the most common intensities of meadow improvement.

To determine the economic feasibilty of these alternative meadow-improvement programs, the returns and costs associated with each level of meadow improvement were computed. By deducting the annual production and harvest costs from gross returns the net return/ha was obtained for each of the intensities of meadow improvement. The annual net returns/ha ranged from a low of $ - 140.05 to a high of $167.13.

Using the net return/ha and the initial capital investment, the economic feasibility and relative merit of each level of meadow improvement was determined by discounting the stream of net returns to a common base year. The largest net present value was for the farming option, which was $580/ha using a 10% discount rate and 20-year time horizon. The lowest net return was for the farming and sprinkler option with $ - 747/ha using the same discount rate and time period.

KEY WORDS: mountain hay meadows, meadow improvements, economics, western U.S.A.

INTRODUCTION

Mountain hay meadows play an essential role in ranching and livestock production in the western U.S.A. (Hart et al. 1980). Traditionally, mountain hay meadows round out the year-round forage supply by providing hay for livestock during winter. Mountain hay meadows can be characterized by the following general conditions: (a) short growing season because of the high elevation (altitudes above 1,829 m), (b) continuous flood irrigation during peak runoffs in spring and early summer but dry in late summer, and (c) native vegetation that is predominantly sedges, rushes, and grasses. The yield on these lands is about 2.7 metric tons (mt)/ha. These yields could undoubtedly be increased on much of this acreage through various improvement practices.

The questions investigated in this study were How might a rancher improve his mountain hay meadows? and Will improvements in mountain hay meadows pay for themselves? The answers to the above questions were sought through the study objectives of (1) identifying

alternative methods for improving meadow hay production and (2) determining the economic feasibility of the different methods of mountain hay-meadow improvements (Hutchins 1980).

PROCEDURE

During the summer of 1979, 45 ranchers in the mountain valley areas of Wyoming were interviewed. They were asked to describe procedures used in improving mountain hay meadows. These interviews indicated that there were at least four levels of hay-meadow improvement. The improvement of mountain hay meadows included farming the old meadows and planting new grass species, farming plus irrigation-control structures, farming plus structures plus leveling to grade, and farming plus installation of sprinklers. In each of the above meadow-improvement programs, the typical practice was to farm the meadow for 2 years and then reseed it.

The physical data obtained from these interviews were used to analyze the economic feasibility of alternative intensities of meadow improvement. Improvement and harvest operations were reconstructed from interviews to compute the costs of a rancher's improvement program.

The author is an Associate Professor of Agricultural Economics at the University of Wyoming.

These costs were computed for each of the four alternative improvement programs. The average cost was subtracted from the expected return to give a net return/ha for each of the meadow improvement alternatives. The net returns/ha are based upon a 5-year development period and a 20-year expected life.

MEADOW IMPROVEMENT METHODS

The four levels of improvement were designated as (1) farming; (2) farming and structures; (3) farming, structures, and leveling; and (4) farming and sprinklers.

In the economic analysis of mountain hay-meadow improvements, three groups were formed. The first two meadow improvement levels were combined since the operations performed and machinery used were very similar, the only difference between them being the installation of irrigation structures. This group consisted of 16 operators who improved a total of 174 ha.

Group 2 consisted of those operators who leveled their meadows with heavy earth-moving equipment. Eight ranchers were involved, and 130 hectares were improved. In group 3, 7 operators installed sprinkler irrigation systems on a total of 142 hectares.

COSTS AND RETURNS

Improvement costs were in two categories. The first included those expenses incurred from tillage operations performed in establishing improved grass species. The second consisted of capital improvements the rancher chose to make. Included in capital improvements were irrigation structures, leveling to grade, and sprinkler systems. In addition to improvement costs, there were the annual costs of producing the hay.

The per-ha cost for each tillage, grow, and harvest operation was computed and summed to determine the annual cost. To compare meadow improvement programs that were performed in different years, cost was calculated as though all improvements began in 1979.

A summary of the costs and returns over the 5-year period required for improved meadows to reach full production for two of the four alternative improvement programs are presented in Table 1. The fifth year's budget was used to represent the remaining 15 years of the 20-year period.

ECONOMIC RESULTS OF MEADOW IMPROVEMENT

Having computed the annual costs and returns for the period covered by the improvement cycle, the rancher or decision-maker will want two questions answered. The first question might be stated: "Based on the available information, will the meadow improvement program I am considering pay for itself?" A second question a rancher might ask is: "Which of the alternative meadow improvement options I am considering will increase my income the most?" A simple comparison of production costs and

Table 1. Per-ha costs of improving mountain hay meadows for improvement options: farming and structures and farming, structures, and leveling.

| Item | Year in improvement process | | | | |
	1	2	3	4	5
Farming and structures					
Variable cost	172.75	218.86	108.33	118.76	123.99
Fixed cost	96.03	98.23	43.52	43.52	43.52
Misc. costs[1]	35.06	39.90	21.50	29.59	32.62
Returns foregone on unimproved meadow	18.48	18.48	18.48	18.48	18.48
Total meadow improvement costs	322.32	375.47	191.83	210.35	218.61
Expected yield (MT/ha)	3.92	3.92	2.24	4.48	5.27
Revenue[2]	246.29	246.29	162.10	302.84	352.48
Net return from farming option	(76.03)	(129.18)	(29.73)	92.49	133.87
Annual cost of structures	2.74	2.74	2.74	2.74	2.74
Net return from farming and structures option	(78.77)	(131.92)	(32.47)	89.75	131.13
Farming, Structures, and Leveling					
Variable cost	147.77	202.53	108.33	118.76	123.99
Fixed cost	72.53	73.32	43.52	43.52	43.52
Misc. cost[1]	31.27	36.92	21.50	29.59	33.66
Annual cost of structures & leveling	7.79	7.79	7.79	7.79	7.79
Returns foregone on unimproved meadow	18.48	18.48	18.48	18.48	18.48
Total meadow improvement costs	277.84	339.04	199.62	218.14	227.44
Expected yield (MT/ha)	3.92	3.92	2.24	4.48	5.94
Revenue[2]	246.29	246.29	162.10	302.84	394.57
Net return	(31.55)	(92.75)	(37.52)	84.70	167.13

[1]Miscellaneous costs are 5% of variable and fixed costs, plus interest at 10% for 6 months on variable and miscellaneous costs, and management at 5% of expected gross revenue.

[2]Revenue is expected yield X $62.83/MT in year 1 and 2, plus $.34 MT/ha (hay equivalent of forage for grazing) in all other years.

returns for individual years over the life of the improvement is insufficient to answer the above questions. A proper evaluation requires the use of discounting.

Discounting is a procedure whereby future dollars are reduced, using a specific discount rate, to arrive at the present value (today's value) of these dollars over a certain time period. Present value is the sum of the discounted yearly net returns over a given time period at a specified discount rate. This procedure is needed to compare meadow improvement options since the streams of costs and returns occur over a 20-year period.

A summary of the discounted net returns for the four meadow improvement options is given in Table 2. The present value of net returns varies from a high of approximately $580/ha for farming to a low of – $747/ha for farming and sprinklers. The economic results in Table 2 suggest that each of the hay-meadow improvement options is economically feasible, except for farming and sprinklers. This does not mean that there are no situations for which farming and sprinklers may not be economically feasible. The results also suggest that the farming option would provide the greatest net return.

LITERATURE CITED

Hart, R.H., H.R. Haise, D.D. Walker, and R.D. Lewis. 1980. Mountain meadow management: 12 years of variety, fertilization, irrigation, and renovation research. U.S. Dept. Agric., SEA-AR, Oakland, Calif.

Hutchins, W. 1980. Economics of improving mountain hay meadows. M.S. Thesis, Univ. of Wyoming.

Table 2. Summary of discounted annual net returns for alternative mountain hay meadow improvement options ($/ha)

Year	Farming	Farming & structures	Farming, structures, & leveling	Farming & sprinklers
0[1]		(58.14)	(594.30)	(1076.86)
1	(69.12)	(71.61)	(28.68)	(85.13)
2	(106.76)	(109.02)	(76.57)	(115.74)
3	(22.34)	(24.40)	(28.19)	(58.22)
4	63.25	61.38	(57.93)	(7.92)
5	83.12	81.42	103.77	69.38
Cumulated				
6–20	632.24	619.30	789.32	527.72
1–20	580.39	498.93	223.34	(746.77)

[1]The values in year zero are the capital investments made/ha.

An Evaluation of the Black Walnut–Tall Fescue Pasture-Management System

H.E. GARRETT and W.B. KURTZ
University of Missouri, Columbia, Mo., U.S.A.

Summary

The economic and biological feasibility of a black walnut–livestock-grazing multicropping-management regime has been evaluated for comparison with a similar livestock-management regime conducted in the open. Pure tall fescue plots were established under open and black walnut–shade conditions. Forage yields on a dry-weight basis were estimated for a 3-year period. A concurrent grazing trial was conducted on similar sites using 1.2-ha plots. Black walnut seedlings were planted within the plots on a 3 × 12-m spacing and protected from grazing animals by electric fencing. The economic analyses of the alternate grazing systems were based on their respective present net worth and internal rates of return over a 60-year rotation.

The black walnut–livestock-grazing system yielded over 6-fold greater present net worth than the comparable system in the open. Tall fescue production under the canopy provided by black walnut trees was approximately one-third greater than in the open and was accompanied by an 11% increase in digestibility of spring-grown grass.

KEY WORDS: black walnut, *Juglans nigra* L., tall fescue, *Festuca arundinacea* Schreb., multicropping.

INTRODUCTION

The potential of a tall-fescue (*Festuca arundinacea* Schreb.)–black-walnut (*Juglans nigra* L.) multicropping regime was evaluated by studying its biology and economics. The goal of such a management system is maximum economic return to the land through the production of hay and forage for livestock and black walnut wood and nut products.

METHODS

Study Area

The general study area is located within a 40-km radius of Stockton, Missouri (lat 37°N long 94°W). Average annual precipitation is 110 cm, with about 60% falling during the spring and summer. Mean annual temperature is 14°C, and mean daily July temperature is 26°C.

Yield Trial

Individual 1.8 × 6.1-m plots of Kentucky 31 tall fescue were replicated 3 times in the open and in the shade of

The authors are Associate Professors in the University of Missouri School of Forestry, Fisheries, and Wildlife.

35-year-old black walnut trees thinned to a 9 × 9-m spacing. Forage yields on a dry-weight basis were estimated by mowing a 0.9 × 5.5-m strip down the middle of each plot (Matches 1966). Production yields were measured during the third week of July 1977, 1978, and 1979 and the first week of November 1977. Forage digestibility was compared, using in-vitro dry-matter digestibility (Tilley and Terry 1963).

Grazing Trial

Concurrent with the yield trials, a grazing trial was established on a similar site to test the feasibility of grazing young walnut stands. Black walnut seedlings were planted on a 3 × 12-m spacing in well-established tall fescue. Test plots and control plots (without walnut seedlings) measuring 1.2 ha and replicated 3 times were individually enclosed with barbed wire. Seedlings within each plot were protected by placing a charged electric fence 1 to 2 m on either side of a seedling row. Grazing pressure was regulated by maintaining two tester steers on each plot with additional animals available for put-and-take as dictated by forage growth.

Economic Analysis

Economic returns from a typical Missouri cow-calf enterprise and a black walnut-grazing regime were compared. Garrett and Kurtz (1980) and Callahan and Smith (1974) have shown multicropping to be profitable.

Present net worth (PNW) and internal rate of return (IRR) were determined for each management option for two qualities of land. Land of site indexes (tree height at 50 years) 20 and 24 m were included. Length of the financial planning period were 60 years, the period required to produce a veneer-grade black walnut log.

Grazing Model

The grazing regime was adapted from existing beef cow-calf enterprise budgets (University of Missouri 1979). An 87% calf crop is assumed with 15% replacement rate and 1% cow loss. On an annual basis, 1.8 ha of the lower-quality and 1.6 of the higher-quality land are required/cow.

Livestock receipts are projected to increase at an annual rate of 1% greater than costs. This figure is supported in part by Grimes (1980) in an examination of the U.S. cattle industry.

Multicropping Grazing Model

To allow for gains through selection, 270 trees/ha are planted initially on a 3 × 12-m spacing. Precommercial and commercial thinnings reduce the number of trees to 68/ha at rotation age. The livestock component is essentially the same as in the previously described grazing regime with early per-ha production reduced by 10% to account for the land allocated to trees.

An average annual increase in real price of 1.5% has been projected for stumpage returns (Hoover 1978). Further, a 1% annual price increase over costs was assumed for black walnut nuts. Neither land costs nor annual tax payments were included within the respective financial analysis models.

RESULTS AND DISCUSSION

Grazing of recently established black walnut plantations is feasible. It has provided structure for financial analysis.

Overall, first-year results revealed that total plant dry-matter production was significantly greater ($P \leqslant 0.05$) on open than on walnut shade plots (Table 1). However, much of the yield in the open consisted of weed species of low forage value. The purity of shade-grown fescue was approximately 95%; that of open-grown fescue was 50%. Spring and early-summer yields, which were greatly influenced by the first-year establishment process, were best under open conditions. Late-summer–fall yields reflected sufficiently better growth under the protective shade of black walnut. However, overall yields for the growing season were best under shade. Neel (1939) and Smith (1942) similarly demonstrated that pasture yields could be improved by the presence of widely spaced black walnut trees.

While the 1978 and 1979 fall harvests were lost to cattle, the spring–early-summer yields strongly supported the 1977 findings (Table 1). Tall fescue yields were significantly greater ($P \leqslant 0.05$) under shade conditions. A sharp decline in open yields between 1977 and 1978 was attributed to a 1977 summer drought. Moreover, substantial increases in pure fescue dry matter were observed from 1977 through 1979 on the shade plots but not on open plots. These increases are especially significant because at the end of the third year the yield under shade was approximately 100% pure fescue and that of the open plots only 60%. In addition to greater yields beneath black walnut, digestibility of the shade-grown fescue was superior to that of fescue grown in the open, although seasonal differences were apparent (Table 2).

Black walnut multicropping–grazing produced significantly greater returns than livestock grazing in the open; increases in PNW ranged from $1,000 to $3,000 more/ha (Table 3). The early returns from livestock grazing, the sizable intermediate returns from black walnut nuts, and the stable long-run growth of the investment in walnut stumpage makes the multicropping/grazing

Table 1. Pure tall fescue dry-matter yields, by light conditions and years.

| Light conditions | 1977 | | 1978 | 1979 |
	Spring	Fall	Spring	Spring
	------------------Kg dry weight/plot------------------			
Walnut shade	0.30[a]	0.62[a]	0.51[a]	1.04[a]
Open	0.52[b]	0.25[b]	0.31[b]	0.51[b]

[a,b]Within season values with different letters are significantly different at the 0.05 level.

Table 2. Percentage digestible dry matter of tall fescue grown in the open and beneath black walnut, during spring and fall.

	Sampling season	
Light condition	*Spring-early summer*[1]	*Late summer-fall*[2]
	---------------- % digestible dry matter ----------------	
Walnut shade	52[a]	57[a]
Open	41[b]	58[a]

[a,b]Within season values with different letters are significantly different at the 0.05 level.

[1]Harvested in mid-July.

[2]Harvested first week in November.

Table 3. Present net worth (PNW) and internal rate of return (IRR) for grazing and black walnut multicropping/grazing regimes on two different sites.

Site index	*PNW*[a]	*IRR*
	Dollars/ha	%
Grazing		
20	205.50	11.5
24	313.75	13.8
Multicropping/Grazing		
20	1264.00	10.9
24	3331.50	13.9

[a]Computed at a discount rate of 7.5 percent.

regime the most profitable management alternative. It must be pointed out, however, that such a management scheme, if adopted, will require significantly greater initial capital investment, to be exposed over a longer period of time than the strict grazing regime.

CONCLUSIONS AND SIGNIFICANCE

Our findings clearly indicate many positive benefits of the multicropping-grazing regime, including increased forage yields and digestibility. Increased fescue yields combined with improved digestibility translates to greater utilization by grazing animals and increases in beef gain and animal carrying capacities.

The black walnut multicropping-grazing concept is an economically viable land-use alternative. Through management intensification and diversification, increased economic productivity/unit area of land is possible. Multicropping represents a long-term investment where the effects of short-run fluctuations in livestock prices are dampened by the stability of the economic growth of the timber resource.

LITERATURE CITED

Callahan, J.C., and R.P. Smith. 1974. An economic analysis of black walnut plantation enterprises. Purdue Univ. Agric. Expt. Sta. Res. Bul. 912.

Garrett, H.E., and W.B. Kurtz. 1980. Walnut multicropping management. Walnut Coun. Bul. 7:15–19.

Grimes, G. 1980. U.S. cattle outlook. Dept. Agric. Econ. Paper 1980–21, Univ. of Mo., Columbia, Mo.

Hoover, W.L. 1978. Walnut log prices. Walnut Coun. Memb. Newsletter, March 1978 5(1):3–4.

Matches, A.G. 1966. Sample size for mower–strip sampling of pastures. Agron. J. 58:213–215.

Neel, L.R. 1939. The effect of shade on pasture. Tenn. Agric. Expt. Sta. Cir. 65.

Smith, R.M. 1942. Some effects of black locusts and black walnuts on southeastern Ohio pastures. Soil Sci. 53:14.

Tilley, J.M.A., and R.A. Terry. 1963. A two-stage technique for the in vitro digestion of forage crops. J. Brit. Grassl. Soc. 18:104–111.

University of Missouri. 1979. Missouri farm planting handbook. Manual 75, Table L-6: beef cow budget, fall calving. Coll. Agric. Ext. Div., Univ. of Mo., Columbia, Mo.

Potential Contribution of White Clover to the Economics and Support-Energy Inputs of Grass-Based Beef-Production Systems

T.A. STEWART, R.E. HAYCOCK, and A.S. LAIDLAW
Greenmount Agricultural and Horticultural College, Antrim;
Loughry College of Agriculture and Food Technology, Cookstown;
and Field Botany Research Division, Department of Agriculture,
Belfast, Northern Ireland, UK

Summary

Biological output, gross margins, and efficiency of support-energy usage in an 18-month beef-production system on grass–white clover swards receiving either 50 kg N/ha stocked at 3.3 head/ha (low N) or 300 kg N/ha stocked 4.5 head/ha (high N) were compared. To date, two production cycles (1977–1979 and 1978–1980) at two sites have been completed. Mean live-weight and carcass gains/ha were, respectively, 17.5% and 14% higher from the high-N than from the low-N system. Gross margin/head was higher from the high-N system in the first production cycle but not in the second. In the low-N system as compared with the high-N system support-energy output was 35.36 and 63.24 GJ/ha, and efficiency of usage was 22.5% and 15.4%, resulting in the low-N system's being 46% more efficient in energy utilization than the high-N system. Although a low-N grass-clover beef system will carry fewer animals and produce less beef/ha than will a high-N system, in the present economic situation it can be as profitable as similar systems that require high inputs of inorganic N and support energy.

KEY WORDS: *Trifolium repens* L., nitrogen, beef systems, economics, support energy.

INTRODUCTION

Research on the effects of environmental and management factors on grass-clover associations carried out in Northern Ireland for the last 30 years has not included assessment of the contribution of white clover under commercial systems of management and production. Consequently, using the information accumulated, a grass-based beef-production system relying principally on white clover (*Trifolium repens* L.) as its source of nitrogen (N) was synthesised, and output was compared with that from a similar system receiving high inputs of fertilizer N. An 18-month beef system using autumn-born calves was chosen because it was the best-defined grass-based complete beef-production system practiced in the UK. As well as biological, physical, and economic considerations, the efficiency of support-energy usage was also estimated, inasmuch as calculations have shown such efficiency to be markedly higher in systems that rely mainly on forage legumes than in systems that rely on inorganic N (Laidlaw and Wright 1980).

Data from two complete production cycles (autumn 1977 to spring 1979 and autumn 1978 to spring 1980) are available; they form the basis of the present paper.

METHODS

The trial, still in progress, consists of two 15-ha sites of contrasting soil type sown in August 1976 to a perennial ryegrass (*Lolium perenne* cv S.24) and white clover (cv. Blanca) mixture. Site A is located on a heavy loam soil at Greenmount Agricultural and Horticultural College, Antrim (lat 54°41′ N long 06°11′ W). Site B is located about 50 km west of Greenmount on a sandy soil at Loughry College of Agriculture and Food Technology, Cookstown.

The site at each center is divided into two 7.5-ha areas on which similar beef-production systems have been established. The low-N areas receive 50 kg N/ha annually as a single spring dressing and are stocked at 3.3 animals/ha; the high-N areas receive 300 kg N/ha annually, applied at intervals over the growing season, and are stocked at 4.5 animals/ha. Each area is rotationally grazed, with

T.A. Stewart is the Head of Research and Development at Greenmount Agricultural and Horticultural College; R.E. Haycock is the Head of Food Production Division at Loughry College of Agriculture and Food Technology; and A.S. Laidlaw is a Grassland Agronomist in the Department of Agriculture of Northern Ireland.

surplus herbage conserved as silage.

Autumn-born calves commence grazing in April, weighing approximately 195 kg. They are housed at the end of October and are sent for slaughter 4 to 5 months later when all the silage from each system has been fed. Small amounts of rolled barley are fed for a short time after turnout, before housing, and during the winter finishing period. For further details see Haycock and Stewart (1979).

Economic assessment, previously outlined by Stewart and Haycock (1979), is based on the average costs incurred (£ sterling) in operating the systems at the two sites from autumn 1977 to spring 1980. Support-energy values were mainly calculated from coefficients by the method of Leach (1976). The energy required in silage making was derived from figures by White (1980). Energy requirements of sward establishment were prorated over a 6-year period, and output energy was taken to be the energy content of the carcass at slaughter minus energy in the carcass of a 50-kg calf.

RESULTS

Seasonal pattern of clover growth was similar on the low-N areas at the two sites, starting with a clover content of 11% in April, rising to a peak of 50% in August, and declining to about 30% in October. Clover made little contribution to herbage output in the high-N area at site A in either year but contributed up to 18% of herbage DM in the August/September period at site B.

Estimates taken at site A in 1978 indicated that DM intake was similar throughout the season on the two areas, averaging 5.31 kg daily, as was utilization of grazed grass (low N, 78%; high N, 81%). A total of 8.5 metric tons (t) DM/ha (grazed + cut grass) was produced from the low-N area compared to 10.8 t DM/ha from the high-N area. More grass was conserved/animal from the low-N areas at both sites in both years (mean, 1.28 t DM/head from low-N area and 0.86 t DM/head from high-N area) resulting in a 2-4 week longer fattening period for the low-N groups of animals.

During grazing, animals on the low-N, clover-dominant swards gained more/day in both years at both sites than those on high-N swards (0.91 kg compared with 0.81 kg). Performance during the winter fattening stage was similar in the two groups. Consequently, mean carcass weight on the low-N system was 24 kg heavier at slaughter than on the high-N system.

The extra weight gain/animal on the low-N system did

Table 1. Animal production and economic comparison of low and high N systems for the first two production cycles (mean of 2 centers).

	Production cycle 1		Production cycle 2	
	Low N	High N	Low N	High N
Animal performance (kg/ha)				
Liveweight gain	811	969	869	1000
Carcass gain	480	572	520	572
Animal output (£/head)				
Carcass sold	375	358	366	323
− Calf cost	85	85	120	120
	290	273	246	203
Variable costs (£/head)				
Rearing calf to grass	60	60	70	70
Fertilizer	11	23	11	23
Cereal and other costs	30	31	36	33
	101	114	117	126
Gross margin (£)				
Per head	189	159	129	77
Per hectare	630	721	430	349

Table 2. Support-energy input and energy output in the low and high N systems.

	Support energy (GJ/ha)	
	Low N	High N
Inputs		
Calf rearing	14.92	20.46
Establishment of swards	2.04	2.04
Fertiliser and application	5.23	24.89
Machinery, fuel and buildings involved in silage making	5.79	6.55
Slurry application	0.34	0.46
Additive	2.92	3.94
Barley	3.68	4.30
Buildings	0.44	0.60
	35.36	63.24
Outputs		
Output of energy in carcass (10.17 MJ/kg dressed carcass)	7.96	9.71
Efficiency (output/input)	0.225	0.154

not entirely compensate for the greater number of animals that could be supported on the high-N system, with the result that live-weight gain/ha was, on average, 147 kg and carcass gain 70 kg/ha more from the high-N system (Table 1).

The gross margins shown in Table 1 were similar for the two systems when averaged over the two production cycles (low N, £530/ha; high N, £535/ha), but this similarity conceals the strong influence of market prices on profitability of the systems. The high-N system had a better gross margin than the low-N system in the first production cycle and a worse gross margin in the second. This change between cycles resulted mainly from an increase in calf prices from £85 to £120 and a drop in sale price from 155p to 146p/kg dead weight.

Although the energy output of the low-N system was only 82% that of the high-N system, the low-N system required only 56% of the support energy used by the high-N system (Table 2). Calf rearing accounted for 42% and fertilizer for 15% of the total support energy in the low-N system compared with 32% for calf rearing and 40% for fertilizer in the high-N system.

DISCUSSION

Tightly controlled comparisons are extremely difficult to undertake for farm-scale systems because of the number of variables involved. However, measurements taken of grass consumed at site A in 1978 would suggest that efforts to regulate grazing pressure to ensure equal grass intakes on the two systems were reasonably successful. Grass conserved/animal, however, was greater for the low-N area at both sites in both years, indicating that a slight adjustment of stocking rate would have increased the accuracy of the comparison.

The better individual animal performance on the low-

N system arose partly from better performance at grass and partly from a longer winter fattening period. It is not possible to conclude with certainty that the better growth rate at grass was caused by the presence of clover per se, which made up a sizable part of the diet during the latter half of the grazing season, or from greater intakes on the clover swards. Other research workers have found improvements in live-weight gain of Friesian calves similar to those recorded here when dry-matter intakes on grass and grass-clover swards were the same (Collins and O'Donovan 1970).

The overall live-weight gain/ha to date from the high-N system is similar to that set by various management standards for this type of beef system and to that recorded previously in a 4-year experiment at site A (McCullough and Stewart 1975). No data from a comparable low-N system are available, but work by Browne (1966) in the Republic of Ireland using a put-and-take method of assessing animal output from grazed grass showed that a grass–white clover sward receiving 52 kg N/ha in spring produced 820 kg live-weight gain/ha annually over a 5-year period. In the last 3 years of Browne's trial where white clover composed a significant proportion of the herbage, live-weight gains ranged from 799 to 1,015 kg/ha and increased by only 139 kg/ha from an additional annual N application of 360 kg/ha. In the present study, an increase of 147 kg live-weight gain/ha was obtained by the addition of an extra 250 kg N/ha.

The gross margin of the low-N system averaged over the two production cycles is similar to that quoted by Kilkenny (1980) for 87 commercial units in the UK (£479/ha) covering the autumn 1978–spring 1980 period where, on average, 159 kg N/ha was applied to the grassland. In the present study N fertilizer only accounted for 10% of the variable costs in the low-N and 20% in the high-N system and had only a small influence on profitability when compared with the effect of calf costs and final selling price.

Clearly, the efficiency of energy usage in the low-N system is much greater than in the high-N system. Even when herbage output and stocking rate are adjusted to that of the low-N system by reducing inorganic N to 160 kg/ha, the efficiency of energy usage of the high-N system is improved by only 4% (recalculated support-energy input 44.38 GJ/ha and output 7.12 GJ/ha).

Results from the present study suggest that the potential saving in support energy by exploiting white clover in a complete production system is not nearly as large as that calculated by Laidlaw and Wright (1980). Their study concerned the production of metabolizable energy and did not take into account loss of energy through conversion to animal product or the support energy associated with the concentrates necessary to make complete production systems viable.

Whereas in the economic assessment N fertilizer was not a major factor in determining cost differences between the two systems, it featured as a principal input in the high-N energy budget. At present, therefore, white clover as an alternative N source would appear to be more important as a means of saving energy than as a method of greatly improving gross margins in beef production systems. However, when calf prices are high, keeping fewer animals on a low-N, clover-dominant sward could be at least as profitable as maintaining higher stocking rates through increased inputs of fertilizer N.

LITERATURE CITED

Browne, D. 1966. Nitrogen use on grassland. I. Effect of applied nitrogen on animal production from a ley. Irish J. Agric. Res. 5:89–101.

Collins, D.P., and P.B. O'Donovan. 1970. An evaluation of white clover in swards for calf feeding. White clover research. Occ. Symp. 6, Brit. Grassl. Soc., 285–291.

Haycock, R.E., and T.A. Stewart. 1979. Beef production from grass/clover swards. Agric. North. Ire. 54:30–33.

Kilkenny, J.B. 1980. Results for grass/cereal systems—cattle slaughtered spring/summer 1980. Beef Impr. Serv. Data Sheet 80/8. Meat and Livest. Commission, London.

Laidlaw, A.S., and C.E. Wright. 1980. The advantages in energy terms of legumes in grassland systems. Grass and Forage Sci. 35:70–71.

Leach, G. 1976. Energy and food production. IPC Sci. and Technol. Press, Guildford.

McCullough, I.I., and T.A. Stewart. 1975. Beef from grass—a field grazing system for 18 month beef. Agric. North. Ire. 50:43–48.

Stewart, T.A., and R.E. Haycock. 1979. The economics of beef from grass clover swards. Agric. North. Ire. 54:186–191.

White, D.J. 1980. Calculating energy use in forage conservation. Span 23:120–123.

Efficiency of Dairying Under Contrasting Feeding and Management Systems in North America, Israel, Europe, and New Zealand

J.D.J. SCOTT
Ruakura Agricultural Research Centre, Hamilton, N.Z.

Summary

Four different systems of feeding and managing dairy cattle are seen in major milk-producing countries. These are (1) feeding purchased roughages and concentrates to zero-grazed, intensively penned cows; (2) home growing of roughages and most concentrates for zero-grazed animals; (3) feeding concentrates with conserved pasture and other roughages in winter and reduced concentrates to grazed cattle in spring-summer; and (4) feeding cattle on grazed pasture and conserved pasture throughout the year. The following data on the four systems were obtained from examples in southern California, Wisconsin, England/Wales, and New Zealand. Average fat-corrected milk (FCM)/cow is approximately 7,140, 5,300, 4,670, and 3,475 kg; output/ha 4,700–10,000, 4,400–5,300, 5,200–5,800, and 7,000 kg FCM; and hours/cow 60, 70, 48, and 30 in the California, Wisconsin, UK, and New Zealand examples, respectively. In energy output/input, the grazed pasture system is probably 3 to 4 times as efficient as any of the others.

Milk-production costs in New Zealand are only about 40% of those in the U.S.A. and UK, and it was calculated that dairy products from New Zealand could be landed in the U.S.A. at about two-thirds the cost of locally manufactured products. Much higher costs are incurred by countries such as Israel and Japan, which have to import feed and fuel. The protection of local dairy industries from competitive imports is leading to the need for subsidies, quotas on production, and a reduction in milk consumption and to a decline in the production of dairy products from temperate grasslands.

KEY WORDS: efficiency, feeds, dairy, North America, Israel, Europe, New Zealand.

INTRODUCTION

Four different systems of feeding, housing, and managing dairy cows are practiced in major milk-producing countries. Each system, and variations on it, is used in more than one country, but for convenience and greater simplicity, just one example of each is described. The examples used permit some comparison to be made of the efficiency of dairy production in the U.S.A., Israel, parts of Europe, and New Zealand, although it is realized that in a comparison of this kind, generalizations and simplifications are inevitable.

The systems are (1) feeding purchased roughages and concentrates to zero-grazed, intensively penned cows; (2) home growing roughages and most concentrates for zero-grazed animals; (3) feeding concentrates with conserved pasture and other roughages in winter and reduced concentrates to grazed cattle in spring-summer; and (4) feeding cattle on grazed pasture and conserved pasture products alone and without any housing. The data used were obtained from the systems as practiced in southern California, Wisconsin, England/Wales, and New Zealand.

Milk:feed price ratios and climate play major roles in determining which system is used. In recent years, U.S.A., the UK, and New Zealand milk:feed price ratios (kg of concentrates that can be bought with 1 kg milk) have been of the order of 1.5:1.0, 1.0:1.0, and 0.5:1, respectively, and with normal production responses it has been profitable to use concentrates in the UK and more profitable in the U.S.A. but not profitable in New Zealand.

Climate influences the amount, seasonal distribution, and quality of the fodder that can be grown. The pasture system can operate most efficiently only in a temperate climate. With more extreme climates there is a greater need for conserved products, machinery, storage, housing, and irrigation, and costs are consequently higher.

PERFORMANCE AND EFFICIENCY MEASURES

In the following comparisons of performance, attempts have been made to use average results from official surveys, or where such data were not available to present results from typical situations. Data in Tables 1, 2, and 4 were obtained from California Livestock Statistics (1979), Wisconsin Dairy Facts (1980), Israel Cattle Breeders' Association (1979), Milk Production in England and Wales (1979), New Zealand Dairy Board (1978, 1979), Voelker (1979), Bath et al. (1978), Williams et al. (1975), Holtman et al. (1977), Loew (1977), Castle and Watkins (1979), Watkins (1976), NRC (1978), Scott et al. (1980), and from personal communications from Bath, University of California; Wiekert, University of Wisconsin; Amies, Milk Marketing Board; Havron, Israel Cattle Breeders' Association; and Bryant, Ruakura Agricultural Research Centre.

The author is a Scientific Officer at the Ruakura Agricultural Research Centre.

Table 1. Production/cow, live weight, and consumption.

	Prod./cow, 4% FCM (kg)	Pre-calving live weight (kg)	Feed consumed/cow/year (kg DM) Rough-ages	Feed consumed/cow/year (kg DM) Concen-trates	DM intake/kg FCM (kg)
California	7,140	700-800	3,750	3,700	1.0
Israel	6,969	650-750	2,550	4,150	1.0
Wisconsin	5,305	650	4,000	1,980	1.1
England/Wales	4,673	600	3,600*	1,450	1.1
New Zealand	3,475	400	3,800**	–	1.1

*Includes an estimated 1500 kg/cow from pasture.

**All pasture and pasture products.

Production and Consumption

In Table 1, per-cow production figures are for tested herds in California and Israel and all herds in the other locations.

Data for feed consumption, and hence for efficiency of feed conversion, are difficult to obtain on a whole-herd, year-round basis. While in some studies, e.g., Trimberger et al. (1972), only 0.8 kg DM was required /kg FCM produced, the estimates shown in Table 1 were more common. They suggest that high- and low-producing herds differ by only 10%–15% in feed conversion efficiency, in contrast to the 20%–30% difference estimated by applying NRC (1978) tables or Broster and Alderman's (1977) formula. It should be noted, though, that where feed is home grown and grazed or stored, greater total nonusage or wastage is incurred than where feed is purchased.

Labor

The limited data available is summarized in Table 2. The estimates for Wisconsin and New Zealand agree well with those of Buxton and Frick (1976).

Much less labor is required for grazed than for housed cattle and for those milked under streamlined procedures, as in New Zealand. However, in terms of milk output/labor unit, very much higher cow yields can offset such advantages (see California and New Zealand results).

Nonsolar energy

Very few detailed measurements are known, and most results have been calculated from national statistics. Data

Table 2. Labor input and output.

	Total Labor hours/cow	Hours/1000 kg FCM
California/Israel*	60	8.5
Wisconsin*	70	13.2
England/Wales*†	48	10.3
New Zealand†	30	8.6

*Includes an allowance for growing purchased feed.

†Farm survey data.

Table 3. Energy use.

	Input (GJ/ha/yr)	Output (GJ/ha/yr)	Ratio, output:input
U.S.A.			0.5
England/Wales	45	17	0.4
New Zealand	8.5	17†	2.0

†Results for surveyed farms which had a low output of milk/ha.

are summarized in Table 3.

Comparison of data from different sources is very difficult because different energy values and inputs can be used. However, output:input ratios of about 2.0 up to 4.0:1.0 (McChesney, Lincoln College, personal communication; Dornom and Tribe 1976) for grassland farming are much higher than those for cropping systems of dairy production (Steinhart and Steinhart 1974, Leach 1976). Smith (1978) concluded similarly that the net efficiency of milk production was 8 to 14 times greater for a high-producing pasture than for any cropping system. On the other hand, products from a country like New Zealand are usually transported greater distances than those from the northern hemisphere countries, and the energy expended in shipping dairy products from New Zealand is approximately equivalent to 14.1 MJ/kg milk fat, amounting to about 40% of the energy inputs/ha (Pearson and Corbet 1976).

Costs and Returns

Resources needed and hence capital involved in the four systems vary greatly. At constant prices/commodity, capital involvement would be least with the feed-purchased system where cows are held in corrals (California) and greatest under the zero-grazed, home-grown (Wisconsin) situation.

In Table 4, direct running costs and returns from milk are compared. The American data are the estimates for 1978 from USDA (1978). The UK estimate for 1978/79 is from United Kingdom Dairy Facts and Figures (1979); and the New Zealand data are survey results for 1978–1979 (New Zealand Dairy Board 1979). Results are in dollars; at that time the U.S.A. and New Zealand dollars were approximately equal in value; and $1 = 50p UK.

While not all the figures are strictly comparable, the cost of producing milk, and the returns to farmers in New Zealand, are only about 40% of those in the U.S.A. and

UK. The major reason for low costs in New Zealand is that feed is produced much more cheaply.

TRADING

With such low costs of production, New Zealand produce can be marketed at competitive rates despite the disadvantage of distances. Buxton and Frick (1976) calculated that New Zealand butter, nonfat dry milk, and cheddar cheese could be landed at the U.S. East Coast at 56%–65% of the price of U.S. products at their manufacturing plants.

One might expect, then, good trading opportunities for New Zealand dairy products, especially with countries that have to import much of the animal feedstuffs and fuel they need, whose costs of production can be about 4 times those of New Zealand. However, as dairy products are one of the most protected commodities in international trade, the benefits of cheaper food, and provision of a measure of efficiency for the local industry, that the freer importation of relatively cheap produce would confer are unlikely to be realized, at least in the short term. These restrictions are unfortunate because where industries are allowed to develop without competition, costly and/or archaic methods of production are more likely to persist, with undesirable end effects of quotas being placed on production, subsidies being needed, and ultimately a decline in consumption.

CONCLUSION

Milk can be produced much more cheaply on grazed pastures in temperate climates than on crops and crop products. Consequently, where expansion of an industry is sought, careful consideration should be given to using temperate zones; and where imports are required, trading with countries that can produce manufactured products at genuinely lower prices has long-term advantages. Conversely, it is grossly undesirable that efficient, established dairy industries should decline as a consequence of restrictive importing policies.

LITERATURE CITED

Bath, D.L., F.N. Dickinson, H.A. Tucker, and R.D. Appleman. 1978. Dairy cattle: principles, practices, problems, profits. Lea and Febiger, Philadelphia, Pa.

Broster, W.H., and G. Alderman. 1977. Nutrient requirements of the high yielding cow. Livest. Prod. Sci. 4:263–275.

Buxton, B.M., and G.E. Frick. 1976. Can the United States compete with dairy exporting nations. J. Dairy Sci. 59:1184–1192.

California livestock statistics. 1979. U.S. Dept. Agric. Sacramento, Calif.

Castle, M.E., and P. Watkins. 1979. Modern milk production: its principles and applications for students and farmers. Faber and Faber, London.

Dornom, H., and D.E. Tribe. 1976. Energetics of dairying in Gippsland. Search 7(10):418–423.

Holtman, J.B., L.J. Connor, R.E. Lucas, and F.J. Wolak.

Table 4. Running costs and returns: $/100 kg milk.

	California	Wisconsin	U.K.	N.Z.
Costs				
Feed	10.9	9.1	11.4	2.5
Labor	4.4	6.4	3.1	3.1
Other direct costs			4.0	1.2
Total	15.3	15.5	18.5	6.8
Returns-Milk	22.6	21.6	21.3	8.7

1977. Material-energy requirements in alternative dairy farming systems. Mich. State Univ. Agric. Expt. Sta. Res. Rpt. 332.

Israel Cattle Breeders' Association. 1979. Summary of 1978 milk and butter fat recording. Tel Aviv, Israel.

Leach, G. 1976. Energy and food production. OPC Sci. and Tech. Press, Guildford.

Loew, B. 1977. Dairy farming in Israel. Ministry Agric.

Milk production in England and Wales. 1979. A handbook of physical inputs and outputs in the dairy sector. Milk Market. Bd., Thames Ditton, Surrey.

New Zealand Dairy Board. 1978. 54th farm production report.

New Zealand Dairy Board. 1979. An economic survey of factor supply dairy farms in New Zealand, 1978–79.

NRC. 1978. Nutrient requirements of dairy cattle. 5th rev. ed., Natl. Acad. Sci., Washington, D.C.

Pearson, R.G., and P.S. Corbet. 1976. Energy in New Zealand agriculture. Search 7(10):418–423.

Scott, J.D.J., N. Lamont, D.C. Smeaton, and S.J. Hudson. 1980. Sheep and cattle nutrition. Min. Agric. and Fisheries, N.Z.

Smith, N.E. 1978. The efficiency of energy utilization for milk production from intensive and extensive systems. 3rd World Cong. Anim. Feeding, Madrid 7:123–129.

Steinhart, J., and C.E. Steinhart. 1974. Energy use in the U.S. food system. Science 184:307–316.

Trimberger, G.W., H.F. Tyrrell, D.A. Morrow, J.T. Reid, M.J. Wright, W.F. Shipe, W.G. Merrill, J.K. Loosli, C.E. Coppock, L.A. Moore, and C.H. Gordon. 1972. Effects of liberal concentrate feeding on health, reproductive efficiency, economy of milk production, and other related responses of the dairy cow. N.Y. Food and Life Sci. Bul. 8, Cornell Univ., Ithaca, N.Y.

United Kingdom dairy facts and figures. 1979. Fed. UK Milk Market. Bds.

U.S. Department of Agriculture (USDA). 1978. Costs of producing milk in the United States. U.S. Dept. Agric., Statis. and Coop. Serv.

Voelker, D.E. 1979. Costs and returns. Iowa State Univ. Dy S-2184.

Watkins, P. 1976. Large scale dairying in California and Israel. Big Farm Mangt. December 1975–November 1976.

Williams, D.W., T.R. McCorby, W.W. Gunkel, D.R. Price, and W.J. Jewell. 1975. Energy utilization on beef feedlots and dairy farms. Proc. 7th Cornell Waste Mngt. Conf., 29–47. Ithaca, N.Y.

Wisconsin dairy facts. 1980. Div. Admin., Wis. Dept. Agric.

Impact on Energy Budgets in Grassland Production and Conservation of Possible Production of Energy from Biomass Crops

R.J. WILKINS

Grassland Research Institute, Hurley, Berkshire, England, UK

Summary

There is commonly a conflict between output/unit of area of land and output/unit of support-energy (SE) input. The possibility of using a proportion of the total land area for biomass energy production suggests an approach for calculating grassland output/ha unconfounded with differences in SE input. Output can be calculated for an area of land that is devoted to grassland and biomass energy crops in proportions that result in a net demand for SE of zero.

Outputs of utilized metabolizable energy (UME) are related to inputs of SE for (1) grazed grass with high application of nitrogen (N) fertilizer and (2) the conservation of grass by different methods. SE inputs in fuel, fertilizers, technical chemicals, and manufacture and maintenance of machinery and buildings are calculated. The biomass crop considered is a short-rotation forest (SRF) giving a yield of 12.5 t (1,000 kg) dry matter (DM)/ha and a net SE output of 195.5 GJ/ha.

For grass with N and grass-clover, the annual SE inputs are calculated as 33 and 2 GJ/ha respectively, giving UME outputs of 102 and 78 GJ/ha and UME output:SE input ratios of 3:1 and 37:1, respectively. The areas of SRF required would be 0.17 and 0.01 ha/ha grassland, respectively. The UME outputs/ha total land area (grassland + SRF) then become 87 and 77 for grass with high N and grass-clover, respectively.

A single cut of grass of 5 t DM/ha is considered to be conserved as silage, field-cured hay, or barn-dried hay. The energy inputs for production and conservation are 17, 12, and 25 GJ/ha, respectively, to give UME outputs of 40, 30, and 39 GJ/ha and UME output:SE input ratios of 2:3, 2:6, and 1:6, respectively. However, when adjusted for the required area of SRF, the UME outputs become 37, 28, and 35 GJ/ha for silage, field-cured hay, and barn-dried hay, respectively.

These calculations suggest that the adoption of grassland systems with a high ratio for UME output:SE input may not improve land use when the requirements both for animal feed and for SE are considered. If levels of UME output/ha are low, the area of land required for animal-feed production will be high. More intensive production systems using a smaller land area coupled with biomass energy production on the remaining area may be the most appropriate strategy.

KEY WORDS: grassland production, grass, legume, grazing, silage, hay, support energy, biomass, short-rotation forestry.

INTRODUCTION

There has been a large increase in agricultural research with the objective of increasing the efficiency of use of support enegy (SE). There is, however, commonly a conflict between output/unit of SE and output/ha. Output/unit of SE is often high in extensive agricultural systems with low output/ha (Wilson and Brigstocke 1980).

This paper describes an approach that allows consideration of the efficiency of use of land and SE on a single basis. The input of SE/ha is computed for different grassland systems, and then a calculation is made of the area of land that would need to be devoted to a biomass energy crop to provide an output of SE sufficient to make the net SE requirement for the total land area of grass plus biomass become zero. Figures for output/total ha of land are thus not confounded with differences in net SE requirements.

Comparisons using this approach are reported for (1) grazed perennial ryegrass (*Lolium perenne*) receiving high applications of nitrogen (N) fertilizer (Grass N) compared with grazed perennial ryegrass–white clover (*Trifolium repens*) without N fertilizer (grass-clover) and (2) different grass-conservation systems. The unit of output is utilized metabolizable energy (UME), and the biomass crop considered is a short-rotation coppiced forest (SRF) using poplar (*Populus*) or willow (*Salix*).

THE SYSTEMS STUDIED

Grass N vs. Grass-Clover

Yield estimates are based on unirrigated swards cut at 4-week intervals and receiving adequate phosphorus (P) and potassium (K). Grass N is assumed to receive N fertilizer to a level at which the incremental response to N is 10 kg DM/kg N (Morrison et al. 1980), whereas grass-clover is assumed to receive no N fertilizer. The yields were calculated to represent those that can be attained in average climatic conditions for lowland grassland in the UK (Wilkins et al. 1981). The output of UME is taken to be 80% of the output of metabolizable energy (ME), to allow for losses in grazing.

Grass Conservation

A single cut of grass at 5 t (1,000 kg) DM/ha is considered to be conserved as silage, field-cured hay, or barn-dried hay. The silage is assumed to be wilted to 25% DM and then harvested with a precision-chop harvester using formic acid additive at 2 l/t fresh-crop weight. The haymaking systems are described by White (1980). Figures for losses of DM and changes in ME are based on data appropriate for the UK given by Wilkins (1976) and Wilkinson (1981).

Short-Rotation Forestry

There is considerable variation in estimates of the output from SRF that can be sustained in northwestern Europe. Annual yields up to 20 t DM/ha have been reported (Callaghan et al. 1980), and Palz and Chartier (1980) give a figure of 17.5 t DM/ha, but the more conservative estimate of 12.5 t DM/ha, given by Neenan (1980), has been used here. Assuming a gross energy content of 16.9 MJ/kg DM, this estimate gives a gross yield of 211 GJ/ha. The annual SE inputs to cover establishment, fertilizers, field work, harvesting, and transport are estimated by Neenan at 15.5 GJ/ha, to give a net output of SE of 195.5 GJ/ha.

ENERGY BUDGETS

The SE inputs considered for the grassland systems are to cover fuel, fertilizers, technical chemicals, and the manufacture and maintenance of machinery and buildings. The data for grazing systems are based on Wilkins and Bather (1981), and those for conservation systems are derived largely from Wilkins (1976) and White (1980).

Grass N vs. Grass-Clover

Energy budgets are given in Table 1. The ratios for UME output : SE input are 3:1 and 37:1 for Grass N and grass-clover, respectively, suggesting much higher efficiency of SE use with the latter system. The area of SRF required to provide the quantity of SE used by one ha of grassland is 0.17 ha for Grass N and only 0.01 ha for grass-clover. Allowing for this extra area of land, the output of UME/ha of total land area (grassland + SRF) becomes 87 GJ for Grass N and 77 GJ for grass-clover. Thus, in the situation where the net use of SE is zero, higher output is calculated to be obtained with the Grass N system.

Table 1. Energy budgets for grazed grass with high-N fertilizer use and for grass-white clover with no fertilizer N.

	Grass N	Grass-clover
Input, GJ/ha		
Establishment[1]	1.0	1.0
Fertilizers[2]		
N	29.7	0
P_2O_5	0.4	0.4
K_2O	0.3	0.3
Fertilizer application[3]	1.3	0.4
	32.7	2.1
Output		
DM, t/ha	10.9	8.2
ME, GJ/ha	128	98
UME, GJ/ha	102	78
Output UME/Input SE, GJ/GJ	3.1	37.1
SRF required, ha/ha grassland	0.17	0.01
UME output/ha grassland + SRF, GJ/ha	87	77

[1]Sward life of 8 years for both systems.

[2]Applications (kg/ha) of N, P_2O_5 and K_2O of 371, 30, 30 for Grass N and O, 30, 30 for grass-clover. Energy equivalents in fertilizer taken as 80, 14, and 9 MJ/kg for N, P_2O_5, and K_2O, respectively.

[3]Number of applications per year taken as 6 and 2 for Grass N and grass-clover, respectively.

Table 2. Energy budgets for grass conservation as silage, field-cured hay, and barn-dried hay.[1]

		Hay	
	Silage	Field-cured	Barn-dried
Input, GJ/ha			
Fertilizers[2]	8.8	8.8	8.8
Fuel during conservation operations[3]	2.1	1.1	1.1
Energy equivalent of tractors and field machinery[3]	2.2	1.4	1.4
Storage[3]	1.5	0.3	0.3
Additives	2.6	0	0
Electricity	0	0	13.2
	17.2	11.6	24.8
Output[4]			
DM, tonnes/ha	4.25	3.50	4.25
UME, GJ/ha	40	30	39
Output UME/Input SE, GJ/GJ	2.3	2.6	1.6
SRF area required, ha/ha grass	0.09	0.06	0.13
UME output/ha grass + SRF, GJ/ha	37	28	35

[1]Herbage yield at cutting at 5 t DM/ha with ME of 9.75 MJ/kg DM.

[2]Application (kg/ha) of N, P_2O_5, and K_2O of 100, 25, and 50, respectively.

[3]From White (1980).

[4]Losses of DM for silage, field-cured hay, and barn-dried hay of 15, 30, and 15% respectively; ME values of conserved products 9.45, 8.55, and 9.15 MJ/kg DM, respectively.

Grass Conservation

Energy budgets are given in Table 2. The lowest SE input is with field-cured hay, and the UME output:SE input ratio is highest for field-cured hay followed by silage and barn-dried hay. However, when allowance is made for the area required for SRF, the output of UME/ha of total land area is ranked in the order silage, barn-dried hay, and field-cured hay. The figures are not directly comparable with those in Table 1, because they are based on the utilization of only a single cut. However, when different methods of using the remainder of the year's production were considered, the ranking of the systems in terms of output of UME/ha of total land area stayed the same.

DISCUSSION

The calculations made here indicate that the adoption of grassland systems with a high ratio for UME output:SE input may not result in improved land use when requirements for both animal feed and SE are considered. This apparent anomaly arises from (1) the lower output of UME/ha of grassland in the less energy-demanding systems and (2) the high net yield of SE that can be obtained from biomass crops, such as SRF. Thus, a relatively small area of land used for SRF could satisfy the requirements of even the most energy-intensive system considered here.

The ranking of the systems for UME output/ha of total land is apparently not highly sensitive to the net SE yield assumed for SRF; the Grass N system would be preferred to grass-clover until the net SE yield for SRF fell from 195.5 GJ/ha to below 110 GJ/ha.

It is not suggested that an individual farm should endeavor to become self-sufficient for SE but that the approach could be applied to plans for satisfying regional or national requirements for animal feed and SE, with different areas being devoted to grassland and to biomass crops.

Only a limited number of factors have been considered in the calculations made here. In particular, no cognizance has been taken of the relative values of different forms of SE or feed, and the other resources required for feed and energy production have been ignored. Also, socio-economic factors would affect the feasibility of devoting substantial areas of land to SRF.

The SE in the SRF biomass could readily contribute to a national energy supply. The harvested material would contain at least 50% DM, making possible its use for direct combustion in power stations, gasification, or pyrolysis. There are suggestions, however, that the ME in legume feeds is used more efficiently than that in grass feeds (Rattray and Joyce 1974), and allowance for this factor would increase the relative attraction of the grass-clover system.

The requirements for capital will vary between the systems considered here. For instance, the Grass N system implies capital investment in plant for industrial N fixation and also for the conversion of SRF biomass into useful energy. The investment associated with the grass-clover system would be much less.

The approach described here could contribute to regional land-use planning and could be used to consider other agricultural systems and potential biomass crops. The results suggest that the possibility of using land for either animal-feed production or for SE production could militate against grassland systems that have favorable ratios for UME output:SE input but low levels of UME output/ha, because such systems require a large area of land for animal-feed production.

LITERATURE CITED

Callaghan, T.V., A. Millar, D. Powell, and G.J. Lawson. 1980. A conceptual approach to plants as a renewable source of energy with particular reference to Great Britain. Nat. Envir. Res. Coun., Inst. of Terres. Ecol. Annual Rpt. 1979, 23–34.

Morrison, J., M.V. Jackson, and P.E. Sparrow. 1980. The response of perennial ryegrass to fertilizer N in relation to climate and soil. Grassl. Res. Inst. Tech. Rpt. 27.

Neenan, M. 1980. The production of energy by photobiological methods. In J. Monet (ed.) Energy: solar energy programme of the Commission of the European Communities, 141–167. Report prepared by the Directorate Gen. for Sci. Res. and Educ. EUR 6959 EN.

Palz, W., and P. Chartier. (eds.). 1980. Energy from biomass in Europe. Appl. Sci. Pub. Ltd., London.

Rattray, P.V., and J.P. Joyce. 1974. Nutritive value of white clover and perennial ryegrass. 4. Utilization of dietary energy. N.Z. J. Agric. Res. 17:401–408.

White, D.J. 1980. Support energy use in forage conservation. Occ. Symp. 11, Brit. Grassl. Soc., 33–45.

Wilkins, R.J. 1976. The potential for increased use of ensiled crops for animal nutrition with particular reference to support energy inputs. *In* J.C. Tayler and J.M. Wilkinson (eds.) Improving the nutritional efficiency of beef production. Com. Eur. Commun., Luxembourg, 224–240.

Wilkins, R.J., and M. Bather. 1981. Potential for changes in support energy use in animal output from grassland. *In* J.L.

Jollans (ed.) Grassland in the British economy, 511–520. CAS Paper 10, Cntr. for Agric. Strat., Reading.

Wilkins, R.J., J. Morrison, and P.F. Chapman. 1981. Potential production from grasses and legumes. *In* J.L. Jollans (ed.) Grassland in the British economy, 390–413. CAS Paper 10, Cntr. for Agric. Strat., Reading.

Wilkinson, J.M. 1981. Losses in the conservation and utilization of grass and forage crops. Ann. Appl. Biol. 98:365–375.

Wilson, P.N., and T.D.A. Brigstocke. 1980. Energy usage in British agriculture as a review of future prospects. Agric. Syst. 5:51–70.

Ecomomic Analysis of the Impact of Management Decisions on Production and Use of Silage for a Dairy Enterprise

A.J. CORRALL, H.D. ST. C. NEAL, and J.M. WILKINSON

Grassland Research Institute, Hurley, Berkshire, England, UK

Summary

The aim of this work was to examine the influence of variations in grass management and methods of ensiling on the economics of milk production for autumn-calving cows.

A simulation model was developed to combine knowledge of plant and animal responses and their interactions in order to calculate the effects of management options on economic balances. Input data were compiled for herbage production from cutting strategies that gave herbage at four digestibilities, at two levels of applied nitrogen, and in three climates. Experimental evidence was also collated to construct matrices of loss of dry matter and changes in digestibility and intake when herbage of different qualities is ensiled using any combination of the following options at harvest: flail or precision-chop harvester, wilting in the field or not wilting, and using or not using additive. The model combines the herbage-production data with the matrices for ensiling method to give values for the yield and quality of silage produced by each management option (192 combinations in all). A feeding program establishes for each silage the total requirements for silage and supplements of barley and soybean meal for cows achieving potential yield, maximizing silage use, and minimizing the cost of supplements. Margins are calculated for the value of milk minus the cost of supplements and silage in each option. Costs cover any items that vary between options.

When herbage yields appropriate to conditions in the UK were used in the model with costs and values current in the 1979–1980 season, greater margins resulted from 350 kg of N/ha than from 150 kg. In comparing cutting strategies, the highest margins/cow were often associated with the use of herbage cut at 66.5 D value (digestible organic matter as a percentage of dry matter), while on the other hand the highest margin/ha of grass came from 61.5 D value. When these margins were expressed/ha of grass plus land for the barley used as supplement, the optimum value for digestibility again rose to 66.5 or 64 D. The effect of digestibility at cutting was less at the high level of applied nitrogen than at the low level.

The model enabled factors with critical financial impact to be identified for a particular economic environment. It is of value in considering national strategies, in influencing the direction of future research, and in analyzing problems on a particular farm.

KEY WORDS: simulation model, economic balances, herbage production, ensilage, digestibility, manuring, dairy cows, perennial ryegrass, *Lolium perenne*.

INTRODUCTION

Many research projects have considered separately the influence of management factors, such as cutting frequency and nitrogen manuring, on herbage production and have expressed the results in terms of herbage yield. Some of them have extended the description of production to include such characteristics as digestibility and protein content. The next step is to consider animal responses to feeds of differing quality and to express results in terms of animal products, such as meat, milk, or wool. However, such studies still take no account of the differing demands on resources made by the various management options. These resources could be energy, land, labor, or finance. The work described here considers the influence of management options on economic balances. We have brought together experimental evidence about the parts of a production system that are normally considered in separate experiments. In this instance, the system considered is the use of grass silage as winter feed for an autumn-calving dairy herd.

METHODS

A model was constructed to combine data on yield and quality from a grass sward, given different cutting sequences and levels of manuring, with knowledge of the changes in yield and quality that would result when such herbage was made into silage by different methods. The model also incorporates requirements of dairy cows for nutrients at each stage of lactation and their response, in terms of intake, when presented with silage of particular characteristics. It represents the use of perennial ryegrass (*Lolium perenne* cv. S.24) in a system in which separate swards are used for cutting and grazing. The options considered are:

- Nitrogen applied at 150 or 350 kg N/ha/yr.
- Harvesting sequences chosen to give herbage at 70, 66.5, 64, or 61.5 D value (digestible organic-matter content as a percentage of dry matter).
- Use of a precision-chop harvester or flail harvester.
- Wilting or not wilting in the field.
- Use of silage additive at harvest or not.

These were considered in three climates varying in availability of water for summer growth, "Wet," "Dry," and "Very Dry." Most dairy farming in the UK is carried on in conditions between "Wet" (where water does not limit grass growth) and "Dry" (represented by average conditions at Hurley).

A computer program was composed to make calculations on input data. The silage yield and quality were calculated for each of the 192 combinations of the above factors. The program formulated rations that met the

J.M. Wilkinson's current address is Ministry of Agriculture, Fisheries and Food, Chief Scientist's Group, Great Westminster House, Horseferry Road, London SW1P 2AE, UK.

daily requirements for energy and protein of 100 cows calving at the beginning of the silage-feeding period and giving 5,620 kg milk/yr. The use of silage was maximized within the limitations of voluntary intake, and the cost of supplementary barley and soybean meal was minimized. The annual requirements for silage and supplements were generated for each silage option. Economic balances were calculated for the margin of milk value over those costs that vary from option to option, using prices and values prevailing in 1979. The costs included purchase of fertilizer, concentrate, and additive and cost of harvesting (including machinery, labor, and fuel). The margins can be expressed as £ sterling/cow or /ha. When considering the margins/cow a further cost for land was included because the options result in differing amounts of land being required for the herd. The land charge was approximately the rental value and was applied to the whole silage area plus an area for grazing. For the same reason, the margins calculated/ha included an allowance for the fixed costs associated with the herd, e.g., capital, housing, labor, etc. For the purpose of strategic planning, or for farmers who grow their own feed barley, it is appropriate that the area used for each option should include the area of land required to grow the barley given as supplement to silage. Therefore, the program also calculated the margin/ha on this basis.

A detailed description of the model, the prices used, and the method of calculating costs is presented by Corrall et al. (1982). That publication also gives the derivation of input data on crop yields and the biological relationships incorporated in the model.

RESULTS

Differences between silage-making methods varied with D value, although they were reasonably consistent over climate and nitrogen levels. However, this paper deals only with the effects of nitrogen level and digestibility at cutting in the three climates.

Nitrogen

When either margin/cow or margin/ha was considered, the advantage of 350 kg N was clear and consistent, although it was most marked with grass cut at 70 D (Fig. 1).

Digestibility

The effect of D value at cutting on the amount and quality of silage produced is illustrated in Table 1. This table shows how cost/tonne of silage falls with D value because of less-frequent cutting and higher yields. The table also indicates the effect of herbage quality on the amount of silage eaten and the concentrates required.

When averaged over all silage-making methods, the highest margins/cow for Wet and Dry were at 66.5 D, but for Very Dry the highest margin was at 64 D or 61.5 D (Fig. 1). Comparing margins/ha of grass for silage plus grazing, the highest margins came consistently from 61.5

Fig. 1. Margin/cow. (Milk, fertilizer, additive, harvesting, concentrates, and land at £45/ha.)

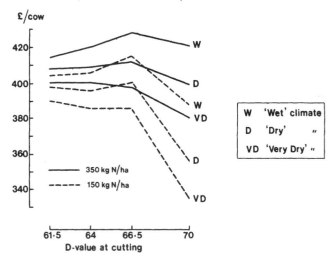

Fig. 2. Margin/ha: (a) grass area and (b) grass plus barley area. (Milk, fertilizer, additive, concentrates, harvesting, and herd at £170/cow.)

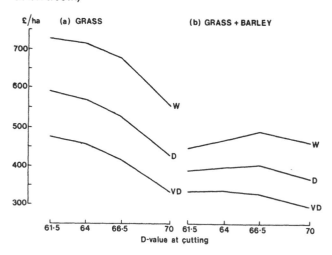

D (Fig. 2). However, with margin/ha of grass plus barley the optimum D value returned to 66.5 or 64 D depending on climate.

When the very small crops that occur at the end of the season were discarded, particularly at low N and in Very Dry climate, the position of 70 D material was improved but the overall ranking of D values was not altered.

Sensitivity

During compilation of the model it was clear that information required for the component relationships varied in its strength. However, in general, it was found that the ranking of silages in terms of economic margins would be affected very little by substantial changes in many of the biological relationships contained in the model. Price changes could have a major effect, but only if they were differential and of a considerable size.

DISCUSSION

As the model stands, using prices and values appropriate to 1979, certain broad conclusions can be

drawn. First, and most striking, is the advantage in margin/cow and /ha from 350 kg N/ha compared with 150 kg. This difference is large enough to survive in the margin/ha if the price of fertilizer were to be double the 1979 value. The optimum D value for harvesting depends on the margin considered, which in turn depends on the circumstances in which a decision is sought.

In practical terms, the margin/cow would be appropriate for the farmer with a fixed herd size but with the opportunity of varying the area of land used for silage by transfer of arable area or rental, etc. Such a farmer could profitably move from 61.5 to 64 or 66.5 D as appropriate. However, on a farm where the area available for silage is limited and stock numbers can be changed, margin/ha of grass is important, and the optimum D value would be 61.5 or below. The margin/ha of grass plus barley is appropriate in considering regional or national land use.

Differences between management options in the margins obtained have often been small, but it must be remembered that these differences would form a greater proportion of true profit than of the margins presented here.

USE OF THE MODEL

This model is limited by the number of factors considered and by the accuracy of the information on which it is built. Also, any use of prediction must take account of the dangers of extrapolating relationships into the future. However, such a model could play a part in several situations. It could be used to predict the probable result of a farmer's changing his management in given economic conditions. The opportunity given by the model to predict the probable impact of changes in price structure or management practice could be used in strategic planning of a large area or for government policy. A study of the sensitivity of the model to the values in particular biological relationships can help to determine priorities for

Table 1. The effect of digestibility at cutting on the annual yield, quality, and cost of silage and its contribution to the winter feeding of dairy cows – 183 days. (Dry climate, 350 kg N, chopped, wilted, plus additive).

	D value at cutting			
	70	66.5	64	61.5
Silage				
Yield of dry matter, tonne/ha	6.0	7.5	8.7	9.2
Digestibility, D value	67.5	64	61.5	59
Protein content, %	17	14	12	11.5
Cost (dry matter) £/tonne	65	52	47	45
Consumption by 100 cows, tonnes dry matter				
Silage	213	198	189	180
Barley	23	45	58	71
Soya bean meal	5	9	12	12
Area needed for silage, ha	36	26	22	20

research effort in the future.

An appraisal of this type could be used in other grass-land situations where the range of management options is large but the number of treatments that can be examined experimentally in terms of animal production and/or biological or economic efficiency is limited by the resources that would be needed.

LITERATURE CITED

Corrall, A.J., H.D. St.C. Neal, and J.M. Wilkinson. 1982. Silage in milk production: A simulation model to study the economic impact of management decisions in the production and use of silage in a dairy enterprise. Grassl. Res. Inst. Tech. Rpt. 29, Hurley, UK.

CONGRESS MEMBERS

ARGENTINA

CHIFFLET DE VERDE, Sonia, Calle 32 No. 938, Balcarce, Pcia, Buenos Aires 7620

IGLESIAS, Luis A., Isalema S. A., San Martin 232, Buenos Aires 1004

VERDE, Prof. Luis, INTA, EERA Balcarce, Casilla de Correo 276, Balcarce, Pcia, Buenos Aires 7620

WADE, Michael, EERA-INTA, Casilla de Correo 276, Balcarce, Pcia, Buenos Aires 7620

AUSTRALIA

BEALE, Dr. Philip, Turretfield Research Center, Department of Agriculture, Rosedale, South Australia 5350

BISHOP, Harry, "Hwaiyin" Glenthompson, Victoria 3293

CARTER, Edward, Department of Agronomy, Waite Agricultural Research Institute, Glen Osmond, South Australia 5350

CROFTS, Prof. Frank C., University of Sidney Farms, Werombi Rf., Camden, New South Wales 2570

CROFTS, Mrs. Frank C. (Lucinda)

FISHER, Myles, CSIRO, Mill Rd., St. Lucia, Queensland 4067

FISHER, Mrs. Myles (Ann)

FLETCHER, Elizabeth F., 6 Gray St., Hamilton, Victoria 3300

GRAMSHAW, Dr. David D., Department of Primary Industries, Box 201, Biloela, Queensland 4715

GUTTERIDGE, Ross C., Department of Agriculture, University of Queensland, St. Lucia, Brisbane 4067

HARTRIDGE, Frank, Department of Agriculture, Box 99, Armidale, New South Wales 2350

HENZELL, Dr. Edward F., CSIRO, Mill and Carmody Rds., St. Lucia, Queensland 4067

HILL, Michael J., Agronomy and Horticultural Science, University of Sydney, New South Wales 2006

HUMPHREYS, Dr. Leonard R., Department of Agriculture, University of Queensland, St. Lucia, Brisbane, Queensland 4067

HUMPHREYS, Mrs. Leonard R. (Lyle)

IVORY, Dr. David A., Queensland Wheat Research Institute, 13 Holberton St., Toowoomba, Queensland 4350

LUSH, Dr. Mary, Turf Research Institute, Ballarto Rd., Box 381, Frankston, Victoria 3199

MCCOWN, Robert L., CSIRO, University Rd., Townsville, Queensland 4815

MCIVOR, Dr. John G., CSIRO, Private Bag, PO, Aitkenvale Townsville, Queensland 4814

MCIVOR, Mrs. John G. (Margaret)

MCKEON, Dr. Gregory, Agriculture Branch, Department of Primary Industries, Box 46, Brisbane, Queensland 4001

MCWILLIAM, Prof. James, University of New England, Armidale, New South Wales 2350

MAHONEY, Gerard, Department of Agriculture, Bridge St., Benalla, Victoria 3672

MAHONEY, Mrs. Gerard (Ann)

MARLER, James W., CSIRO, 338 Blaxland Rd., Ryde, New South Wales 2112

MINSON, Dr. Dennis, CSIRO, Cunningham Laboratory, St. Lucia, Queensland 4067

MULDOON, Dr. Dennis, Agricultural Research Station, Trangie, New South Wales 2823

MYERS, Dr. Robert J. K., Division of Tropical Crops and Pastures, CSIRO, Cunningham Laboratory, St. Lucia, Queensland 4067

MYERS, Mrs. Robert J. K. (Margaret)

NICHOLAS, Donald A., Plant Research Division, Jarrah Road, South Perth, Western Australia 6151

NICHOLAS, Mrs. Donald A. (Christine)

PEARSON, Dr. Craig J., Agronomy and Horticultural Science, University of Sydney, New South Wales 2006

REED, Dr. Kevin F. M., Agriculture, Pastoral Research Institute, Box 180, Hamilton, Victoria 3300

REED, Mrs. Kevin F. M. (Rhonda)

RICHARDSON, Donald, Grazier, Willaroo, Coleraine, Victoria 3315

RICHARDSON, Mrs. Donald (Carol)

RICKERT, Dr. Kenwyn G., Brian Pastures, Primary Industries Department, Gayndah, Queensland 4625

ROBARDS, Dr. Geoff, Department of Agriculture, McKell Building, Rawson Pl., Sydney, New South Wales 2000

RUSSELL, C. L., 26 Panorama Cr., Freemans Reach, New South Wales

SAVAGE, Graham, F. J., Department of Agriculture, 51 McCartin St., Leongatha, Victoria 3953

SMITH, Dr. Frank, Division of Tropical Crops and Pastures, CSIRO, Cunningham Laboratory, Carmody Rd., St. Lucia, Queensland 4069

SMITH, Mrs. Frank (Janey)

STERN, Prof. Walter, Agronomy Department, Institute of Agriculture, University of Western Australia, Nedlands, Western Australia 6009

STERN, Mrs. Walter (Maida)

STEWART, Michael, Kinvonvie, Hamilton, Victoria 3300

STEWART, Mrs. Michael (Virginia)

TAYLOR, Graham B., Land Resource Management, CSIRO, Private Mailbag, PO, Wembley, Western Australia 6014

TIDEMAN, Arthur A. F., Agriculture, Grenfell Center, Box 1671, GPO Adelaide, South Australia 5001

TIDEMAN, Mrs. Arthur A. F. (Lesley)

WALKER, Barry, Agriculture, Department of Primary Industries, Box 689, Rockhampton, Queensland 4700

WATT, Lance, Research Center, Soil Conservation Service, Box 433, Inverell, New South Wales 2360

WHEELER, Dr. John L., CSIRO, Armidale, New South Wales 2350

WHEELER, Mrs. John L. (Dorothy)

WHITEMAN, Dr. Peter, Agriculture, University of Queensland, St. Lucia, Queensland 4067

WILDIN, John H., Primary Industries, Box 689, Rockhampton, Queensland 4700

WINTER, Dr. W. H., CSIRO, PMB 44, Winnellie, Northern Territory 5789

WINTER, Mrs. W. H. (Robin)

BELGIUM

BEHAEGHE, Prof. Tillo, Plant Husbandry, University Faculty of Agronomy, Coupure Links 533, Ghent B-9000

BISTON, Dr. Robert, Sta. de Haute Belgique, Ministere de l'Agriculture, 48, Rue de Serpont, Libramont 6600

DESWYSEN, Dr. Armand, Universite Catholique Louvain, Place Croix Du Sud, No. 3, Sc. 15 D 2, Louvain-La-Neuve

BOLIVIA

HORRELL, C. R., Trop. Pasture Agron., Casilla 359, Santa Cruz

PATERSON, Robert T., Mision Britanica en Agric. Trop., Casilla 359, Santa Cruz

BRAZIL

ABRAMIDES, Luis Guardia Pedro, Division de Nutri Cao Animal and Past., Institute de Zootecnia, Rua Heitor Penteado 56, Nova Odessa, Sao Paulo 13.460

ABRAMIDES, Mrs. Luis Guardia Pedro (Solamge)

ALCANTARA, Bardauil, Inst. de Zootecnia, Rua Heitor Penteado 56, Nova Odessa, Sao Paulo 13.460

ALCANTARA, De-Bem Gomes Velquiria, Inst. de Zootecnia, Rua Heitor Penteado 56, Nova Odessa, Sao Paulo 13.460

ANDRADE, Ronaldo R., EMBRAPA/CPAC, Caixa 70.0023, Planaltina, D.F. 73.300

BARRETO, Dr. Ismar, Department Zootecnia, UFSM, 97100 Santa Maria, R.S.

BUTTERFIELD, Alan, Private Rancher, CP 198, Guararapes, Sao Paulo 16700

CAMERON, Dr. Donald F., EMBRAPA/CPAC, Caixa Postal 70.0023, Planaltina, D.F. 73.300

COUTO, Dr. Walter, EMBRAPA/CIAT, Caixa Postal 70.0023, Planaltina, D.F. 73.300

Da ROCHA, Prof. Geraldo Leme, Inst. de Zootecnia, Rua Heitor Penteado 56, Nova Odessa, Sao Paulo

Da ROCHA, Mrs. Geraldo Leme (E. Marcarida Lima)

ESCUDER, Dr. Cesar Jorge, Zootecnia, Escola de Vet., (UFMG), Belo Horizonte, Minas Gerais 30.000

GOMES, Dr. Darci T., EMBRAPA/CPAC, Km 18 BRO20, CX Postal 70.0023, Planaltina, D.F. 73600

LEMOS, Joaquim, An. Sc., Centro Energia Nuclear Agric., Av. Centenario S/N, Caixa Postal 96, Piracicaba, Sao Paulo 13.400

MARASCHIN, Dr. Gerzy E., Fitotecnia-UFRGS, Univ. Fed do Rio Grande do Sul Ave., Bento Goncalves, 7712, Porto Alegre, R.S. 90.000

MELLA, Silvio Carlos, Centro de Producao e Exp.—Pvai, Jardim Ipe, Caixa Postal 564, Paranavai 87.700

MOORE, Dr. Patrick C., CIAT, EMBRAPA/CPAC, Cx Postal 70.0023, Planaltina, D.F. 73.300

PRIMO, Armando, Sain-Parque Rural, EMBRAPA, Brasilia 70.000

ROCHA, Carlos Magho C., EMBRAPA/CPAC, KM 18-BR020 cx Postal 70.0023, Planaltina, D.F. 70.300

SAITO, Mario H., IRI Res. Inst., Av. Vieira de Carvalho, 40-110 Andar, Sao Paulo 01210

SANTOS, Laerte Filho, Formazan Seeds Co., 580—conj. 21, Sao Paulo 3

SERRAO, Dr. Emanuel A., Propasto, CPATU/EMBRAPA, Cx Postal 48, Belem, Para 66000

SODERA MARTINS, Prof. Paulo, De Genetica, Escola Superior De Agric., "Luiz De Queiroz," C.P. 83, Piracicaba, Sao Paulo 13400

THOMAS, Dr. Derrick, CIAT, Shisul, QL 14, Conj. 10, Casa 18, Brasilia, D.F.

VERA, Dr. Paul, FAO Project BRA/79/010, EMBRAPA, Coronel Paceco, M.G. 36155

CANADA

BAILEY, Dr. L. D., Agriculture, Canada Research Station, Box 610, Brandon, Manitoba R7A 5Z7

BITTMAN, Shabtai, Agriculture Canada, Melford Research Station, Box 1260, Melfort, Saskatchewan S0E 1A0

BJORGE, Myron L., Alberta Agriculture, 5718 56th Ave., Bag Service 47, Lacombe, Alberta T0C 1S0

BONNEFOY, George, 25 Tupper St. N., Portage La Prairie, Manitoba R1N 3K1

CAMPBELL, Dave C., Soils and Crops Branch, Manitoba Department of Agriculture, 908 Norquay Building, Winnipeg, Manitoba R3N 1A2

CHILDERS, Dr. Walter, 232 Remic Ave., Ottawa, Ontario K1Z 5W5

CHILDERS, Mrs. Walter (Lynn)

CHOO, T. M., Box 1210, Agriculture Canada, Charlottetown, Prince Edward Island C1A 7M8

CLARK, Dr. Kenneth, Plant Science, University of Manitoba, Winnipeg, Manitoba R3T 2N2

CLARK, Mrs. Kenneth (Lea)

COHEN, Dr. Roger, Animal Science, University of British Columbia, Suite 248, 2357 Main Mall, Vancouver, British Columbia V6T 2A2

COULMAN, Dr. Bruce E., MacDonald Campus, McGill University, Plant Science, 2111 Lakeshore Rd., St. Anne de Bellevue, Quebec H9X 1C0

DEVERS, Clive, 2095 Roche Ct., Unit 114, Mississauga, Ontario LK5 2C8

FARIS, Dr. Mohamed A., Forage Section, Ottawa Research Station, Ottawa, Ontario K2E 5R3

FITZGERALD, R. Desmond, Plant Science, University of Alberta, Edmonton, Alberta T6G 2H1

FULLEN, Len, Production Economics, Alberta Agriculture, 9718 107th St., Edmonton, Alberta T5J 3K4

FULLEN, Mrs. Len (Jennifer)

GOPLEN, Dr. Bernard P., Research Station, Agriculture Canada, 107 Sc. Crescent, Saskatoon, Saskatchewan S7N 0X2

HOWARTH, Dr. Ronald E., Research Station, 107 Sc. Crescent, Saskatoon, Saskatchewan S7N 0X2

KILCHER, Mark R., Research Station, Box 1030, Swift Current, Saskatchewan S9H 3X2

LAWRENCE, Dr. Thomas, Research Station, Agriculture Canada, Box 1030, Swift Current, Saskatchewan S9H 3X2

LAWSON, Dr. Norman, MacDonald Campus, McGill University, 2111 Lakeshore Rd., St. Anne de Bellevue, Quebec H9X IC0

LESSARD, Dr. Raymond J., Agriculture Canada, Animal Research Center, Ottawa, Ontario K1A 0C6

MASON, Wesley N., Agriculture Canada, Box 90, Lennoxville, Quebec J1M 1Z3

PROULX, Dr. Julien G., Agriculture Canada, Research Branch, Experiment Farm, Kapuskasing, Ontario P5N 2X9

ROBERTSON, Dr. Alden J., Agriculture Canada, Research Station, Box 1240, Melfort, Saskatchewan S0E 1A0

ST-PIERRE, Dr. Jean-Claude, Agriculture Canada, Central Experiment Farm, Sir John Carling Building, Ottawa, Ontario K1A 0C5

STEWART, Fraser, Manitoba Agriculture, 250 1st Street, Box 50, Beausejour, Manitoba R0E 0C0

STORGAARD, Dr. Anna K., Plant Science, University of Manitoba, Winnipeg, Manitoba R3T 2N2

SUZUKI, Dr. Michio, Research Station, Agriculture Canada, Box 1210, Charlottetown, Prince Edward Island C1A 7M8

TOMES, Dr. Dwight T., Crop Science, University of Guelph, Guelph, Ontario N1G 2W1

UPFOLD, Richard A., Ministry of Agriculture and Food, Box 1330, Walkerton, Ontario N0G 2V0

WADDINGTON, Dr. John, Agriculture Canada, Research Station, Box 1240, Melfort, Saskatchewan S0E 1A0

WALTON, Dr. Peter D., Plant Science, University of Alberta, 14204 57th Ave., Edmonton, Alberta T6H 1B3

WILSON, Dr. Donald B., Agriculture Canada, Research Center, Lethbridge, Alberta T1J 4B1

WINCH, Prof. J. E., Crop Science, University of Guelph, Guelph, Ontario N1G 2W1

CHILE

PICHARD, Dr. Gaston, Department of Animal Science, School of Agriculture, Box 114-D, Santiago

SOTO, Patricio, Est. Exp. Quilama Pu, Inst. de Investigaciones Agro Pecuarias, INIA, Vicente Mendez d/n, Chillan, Casilla 426

COLOMBIA

ARGEL, Pedro, CIAT, Apartado Aereo 6713, Cali

BERNAL, Dr. Javier, Semillas La Pradera, Apartado Aereo, Bogota 100685

CALDERON, Dr. Mario, CIAT, Tropical Pastures, Apartado Aereo 67-13, Cali

COCHRANE, Dr. Thomas T., CIAT, Apartado Aereo 67-13, Cali

COCHRANE, Mrs. Thomas T. (Rose Elena)

FERGUSON, Dr. John E., CIAT, Apartado Aereo 67-13, Cali

GROF, Dr. B., CIAT, Apartado Aereo 67-13, Cali

HERRERA, Carlos, Cra 7N° 88-96, Bogota

HUTTON, Dr. E. Mark, CIAT, Apartado Aereo, 67-13, Cali

HUTTON, Mrs. E. Mark (Gwendolyn)

LENNE, Dr. Jillian M., Tropical Pastures Program, CIAT, Apartado Aereo 67-13, Cali

NORES, Gustavo, CIAT, Apartado Aereo 67-13, Cali

TOLEDO, Dr. Jose M., CIAT, Apartado Aereo 67-13, Cali

VALASQUEZ, Alvaro, Avenda 13, #13, 132-90 n 309, Bogota

WHITNEY, Dr. Sheldon, Tropical Pastures Program, CIAT, Apartado Aereo 67-13, Cali

COSTA RICA

COWARD-LORD, Prof. Juan A., Univ. de Costa Rica, San Pedro, San Jose

CZECHOSLOVAKIA

TOMKA, Dr. Ondrej T., Research Institute for Grassland, Mladeznicka 36, Banska, Bystrica 97421

DENMARK

HALDRUP, Jens, Bredgade 129, Logstor

NORDESTGAARD, Anton, Staten Forsogsstationn, Ledreborg Alle' 100, Roskilde 4000

PEDERSEN, Dr. Christian, Plant Breeding, L. Daehnfeldt Ltd., Box 185, 5100 Odense C.

TJAERBY, Erik, Vermundsgade 7 B, Viby J.

FEDERAL REPUBLIC OF GERMANY

BOEKER, Dr., Prof. Peter, Institute for Agronomy, University of Bonn, Katzenburweg 5, 5300 Bonn 1

BOEKER, Mrs. Peter (Marlies)

CAMPINO, Dr. Ignacio, Grunlandwirtschaft, Justus-Liegib-Universitat, Ludwigstrasse, 6300 Giessen

HONIG, Dr. Hans, Institute for Grassland and Forage Research, FAL, Bundesallee 50, Braunschweig D 3300

KUEBAUCH, Dr., Prof. Walter, University of Kassel, Nordbahnhofor la, 3430 Witzenhansen D-3430

SIMON, Dr. Uwe, Grunlandwirtschaft, Justus-Liegib-Universitat, Ludwigstrasse, 6300 Giessen

FINLAND

HAKKOLA, Heikki, Maatalouden tutkimuskeskus, Pohjois-Pohjanmaan, koeasema, Rukki 92400

HAKKOLA, Mrs. Heikki (Seija)

PULLI, Dr. Seppo, Department of Crop Science, Agriculture Research Center, Jokioinen 31600

RAURAMAA, Mrs. Aino, Valio Laboratory, Valio Finnish Cooperative Dairies Association, Kalevankatu 56 B, POB 176, Helsinki SF 00181

FRANCE

ANDRIEU, Jacques M., INRA THEIX, 63110 St. Genes Champanelle, CPTE 01 28060 000

CHEVALIER, Henri, Societe Commerciale Des Potasses de L'Azote, 2 Place Du General De Gaulle, Mulhouse 68100

CHEVALIER, Mrs. Henri (Francoise)

DELORT-LAVAL, Jean, INRA, Nantes 44072

GILLET, Dr. Michel, Amelioration des Plantes, Inst. National de la Recherche Agron., Sta. d'Amelioration des Plantes, Fourrageres, Lusignan 86600

HENTGEN, Dr. Andre, INRA, Route De Saint-Cyr, Versailles F78000

LOISEAU, Roger, Louseau Semences, Les Goderies ruaudin, 72230 Arnage

LOISEAU, Mrs. Roger (Louise)

PARNEIX, Pierre, ENSA, 65 Route De Saint-Brieuc, Rennes 35000

PFITZENMEYER, Claude, France Selection, Les Goderies Ruaudin, 72230 Arnage

SAVIDAN, Dr. Yves, C.P.D.P., C.N.R.S., Gif Sur Yvette 91190

GERMAN DEMOCRATIC REPUBLIC

BREUNIG, Dr., Prof. Willy, Humbolt University, Invaliden, Berlin

NITZSCHE, Werner, Max Planck Inst. fur Zuchtungsforschung, D 5, Koln 30

VOIGTLANDER, Dr., Prof. Gerhard, Inst. fur Grunlandlehre, Tech. Univ. Munchen, Freising-Weihenstephan, D-8050 Freising, Bayern, BR D-8050

VOIGTLANDER, Mrs. Gerhard (Ilse)

WOJAHN, Prof. E., Akademie der Landwirtschaftswissenschaften Der, DDR 1086, Berlin

INDIA

GHOSH, Dr. Abhijit N., Ministry of Agriculture, New Delhi

PATIL, Dr. B. D., Indian Grassland Fodder Research Institute, Jhansi U.P. 28400 3

SINGH, Nirmal T., Punjab Agricultural University, Ludhiana, Punjab 141001

INDONESIA

NITIS, Dr. I. M., F.K.H.P., Udayana University Jl., Jendral Sudirman, Denpasar, Bali

IRELAND

(See "Republic of Ireland" and "United Kingdom")

ISRAEL

DOVRAT, Dr. Amos A., Faculty of Agriculture, Box 12, Rehovot 76-100

KIPNIS, Dr. Tal, Institute of Field Crops, Agriculture Research Organization, POBG, Bet Dagan 50200

LESHEM, Dr. Yoel, Agriculture Research Organization, Bet Dagah

ITALY

ARANGINA, Renato, Via Carducci 18, Cagliari 09100

CENCI, Prof. Alberto, Inst. Allevamento Veg., Univ. degli Studi, Borgo XX Guigno, Perugia 06100

CORLETO, Dr. Antonio, Inst. di Agron. e Coltivazioni Erbacee, Universita degli Studi, Bari 70126

LORENZETTI, Prof. Franco, 1st Allevamento Veg., Univ. degli Studi, Borgo XX Guingno, Perugia 06100

PANELLA, Prof. Adelmo, 1st Allevamento Veg., Univ. degli Studi, Borgo XX Guingno, Perugia 06100

PAOLETTI, Renatto R., Agron., Inst. Sperimental Colture Foraggere, Viale Piacenze 25, Lodi (MI) 20075

PAOLETTI, Mrs. Renatto R. (Carla)

PIANO, Dr. Efisio, Inst. Sperimentale Colture Foraggere, Via Mameli, 118 Cagliari, Sardegna 09100

RIVEROS, Dr. Fernando, FAO, via delle Terme di Caracalla, Rome

ROTILI, Dr., Prof. Pietro P., Breeding Sect., Inst. Sperimentale Colture Foraggere, Viale Piacenze 25, Lodi (MI) 20075

ZANNONE, Miss, Dr. Luisa L., Bio, Sect., Inst. Sperimentale Per Le Colture Foraggere, Viale Piacenze 25, Lodi (MI) 20075

JAPAN

AKIYAMA, Dr. Tsuyoshi, Ecologist, National Grassland Research, 768 Nishinasuno, Tochigi 329-27

ATAKU, Kazuo, Dairy Science, College of Dairying, Rakuno Gakuen, Nishi Nopporo, Ebetsu, Hokkaido 069-01

GEMMA, Prof. Takuma T., Obihiro University of Agriculture and Veterinary Medicine, Inada-cho, Obihiro, Hokkaido 080

HARADA, Dr., Prof. Isamu I. H., Dairy Science, University of Rakuno Gakuen, 582 Nishi-Nopporo, Ebetsu, Hokkaido 069-01

HARADA, Mrs. Isamu I. H. (Ayako)

HAYASKI, Dr., Prof. Kenroku, Grasslands Research Laboratory, Tohoku University, Kawatasi, Narugo, Miyagi 989 67

HORIGOME, Dr., Prof. Takao H., Faculty of Agriculture, Okayama University, Tsushima, Okayama 700

KAWABATA, Dr. Syutaro K., National Grassland Research Institute, 768 Nishinasuno, Tochigi 329-27

KAWASAKI, Tsutomu, Grassland and Forage Sciences, Shintoku Animal Husbandry Experiment Station, Nishi 4-40, Shintoku, Hokkaido 081

KAWASHIMA, Sakae K., Experimental Farm, Tokyo University of Agriculture, 1737 Funako, Atsugi, Kanagawa Ken 243

KAYAMA, Dr., Prof. Ryosei, University Farm, Faculty of Agriculture, Nagoya University, 94 Hatasiri Morowa, Togocho, Aichi-ken 470-01

KIRITA, Dr. Hiromitsu, Ecologist, National Grassland Institute, 768 Nishinasuno, Tochigi 329-27

KOBAYASHI, Prof. Hiroshi, Grassland Research Laboratory, Kitasato University, Sanbongi, Towadi, Aomori 034

MAKI, Dr. Yoshisuke, Prof. of Agronomy, Akita Prefectural College of Agriculture, Minami 2-2 Ogata-Mura, Akita-Ken 010-04

MASUKO, Takayoshi M., Livestock Sciences, Tokyo University of Agriculture, 1-1 Sakuragaoka 1 chome, Setagayaku, Tokyo 156

MIYAZAWA, Yoshiharu, First Grassland Department, Hokkaido National Agricultural Experiment Station, Hitsujigaoka Toyohira-ku, Sapporo, Hokkaido 061-01

NIKKI, Dr. Iwao, Grassland Sciences, Miyazaki University, Funazuka 3-210, Miyazaki-shi 880

NISHIMURA, Dr. Shuichi, Kashi 1-12-12, Higashi-ku, Fukuoka 813

OGATA, Dr. Shoitsu, Hiroshima University, 2-16 Midori-cho, Fukuyama, Hiroshima Pref. 720

OKAMOTO, Meiji, Grassland Science Research Associate, Obihiro University of Agriculture and Veterinary Medicine, Inadamachi, Obihiro, Hokkaido 080

OOHARA, Dr. Hisatomo, Dairy Research Institute, Nishi 26 Chome, Kita 1 Jyo, Chuoku, Sapporo, Hokkaido 064

OOHARA, Mrs. Hisatomo (Teruko)

OOHARA, Dr. Yoichi, Obihiro University of Agriculture and Veterinary Medicine, Koen-higashimachi 3-11-2, Obihiro City, Hokkaido 080

OTANI, Tadashi O., Experiment Farm, Tokyo University of Agriculture, 1737 Funako, Atsugi, Kanagawa-Ken 243

SAIGA, Dr. Suguru, Seed Division, Ministry of Agriculture, Kasumigaseki 1-2-1, Chiyoda-ku, Tokyo 100

SHIMADA, Dr. Emisaku, Azabu University 1-17-71, School of Veterinary Medicine, Fuchinobe, Sagamihara, Kanagawa 229

SHIYOMI, Dr. Masae, Ecologist, National Grassland Research Institute, 768 Nishinasuno, Tochigi 329-27

SUGAWARA, Dr. Kazuo, Grassland Research Laboratory, Tohoku University, Kawatabi, Narugo, Miyagi 989-67

SUGINOBU, Dr. Ken-Ichi, Forage Division, National Grassland Research Institute, 768 Nishinasuno, Tochigi 329-27

SUZUKI, Shigeru, Division of Genetics, National Institute of Agricultural Science, Tsukuba 305

TAJI, Dr., Prof. Kuniyasu K. T., Faculty of Agriculture, Ehime University, Tarumi Cho., Matsuyama 790

TAKAHASHI, Shigeo, National Grassland Research Institute, 768 Nishinasuno, Tochigi 329-27

TAKAHATA, Shigeru, Forest Institute, Hokkaido Branch, Hitsujigaoka 1, Toyohiraku, Sapporo 061-01

TAKANO, Dr. N., Grass Plant, National Grassland Research Institute, 768 Nishinasuno, Tochigi 329-27

TAKASAKI, Dr. Yasuo, Faculty of Horticulture, Chiba University, Matsudo 648, Matsudo-shi, Chiba-ken 271

TAKEDA, Yoshihiko, Grassland and Forage Sciences, Shintoku Animal Husbandry Experiment Station, Nishi 4-36, Shintoku, Hokkaido

TAKIZAWA, Dr. Hirosada, Animal Breeding Section, Takikawa Animal Experiment Station, 735 Higashi, Takikawa, Takikawa-shi, Hokkaido 078

TARUMOTO, Dr. Isao T., Pasture Plant Division, National Grassland Research Institute, 768 Nishinasuno, Tochigi 329-27

TATSUO, Kanedo, 634 Naganumahara-cho, Chiba City

TERADA, Yasumichi, Yamaguchi Agricultural Experiment Station, Ouchi Mihari, Yamaguchi City 747-13

TSURUMI, Yoshiro, Kyushu Agricultural Experiment Station, Nishigoshi, Kumamoto 861-11

UCHIDA, Dr., Prof. Senji U., College of Agriculture, Okayama University, 1-1-1 Tsushima Naka, Okayama 700

UEHARA, Akio, Forage Crop Breeding, Snow Brand Seed Co., Ltd., Hokkaido Research Station, 1066 Horonai, Naganuma-cho, Yubari-gun 069-14

UENO, Dr. Masahiko, Grassland Science, Miyazaki University, Funazuka 3-210, Miyazaki-shi 880

WAKIMOTO, Dr. Takashi, Hokkaido Central Agricultural Experiment Station, Naganuma, Hokkaido 069-13

KENYA

BURZLAFF, D. F., Kiboko National Range Research Station, Box 12, Makindu

IBRAHIM, Dr. Kamal M., FAO United Nations, Box 1950, Kitale

WERE, Abel O., Kenya Seed Co., Ltd., Box 553, Kitale

MADAGASCAR

RASAMBAINARIVO, Dr. John Henri, Dept. de Recherches, Zootecnia et Veterinaires, Antananarivo

MALAWI

HODGES, Elver, Chitezde Agricultural Research Station, Box 158, Lilongwe

MEXICO

BORDIER, Jehu, Banco De Mexico S.A. — Fira., Yenustiano Carranza # 707, San Luis Potosi

CANUDAS-LARA, Eduardo, Forages, C.E.P., La Posto, Cristobal De Olid # 388, Veracruz

FLORES, Fernando, Forages/Fertilization, Ferticizandes Mexicanos, S.A., Roberto Gayor 87

GARZA, Dr. Ricardo, Fertimex, Morena 804, 12, D.F.

GONZALEZ, Dr. Martin H., Bufete Agro Pecuario S. A., Fresnos 157 Station, Engracio, Monterrey, N.L.

MARTINEZ, Ramon A., INIA, Apartado Postal 18, Calera, Zacatecas

MORALES, Jose L., Range Science, Escuela Sup. de Zootecnia, D. De Mendoza 1307, Chihuahua, Chih.

NERI, Oscar, Banco De Mexico, S.A. — Fira., Insurgentes sur 2375 40 Piso 20 Mexico City

QUIROGA, Hector M., CIAN, Apartado Postal 247, Torreon, Coahuila

TORRES, Bernado, Banco De Mexico, S.A. — Fira., Apartado 1087, Cuernavaca, Mor.

MOROCCO

ANEZIANE, Tayeb, Institute of Agronomy, BP 704, Agdal Rabat

GUESSOUS, Fouad, Hassan II Institute of Agronomy and Veterinary Medicine, BP 04, Rabat

NARJISSE, Hamid, E.N.A., BP S 40, Meknes

NETHERLANDS

ALBERDA, Dr. Thee, Kamperfoelielaan 2, Wageningen 6706 CW

ALBERDA, Mrs. Thee (Barensten Jozina)

BOONMAN, Dr. Joseph G., Zelder Breeding Station, Ottersum 6595 NW

BOONMAN, Mrs. Joseph G. (Elly)

BOSMA, Ate, Institute of Agricultural Engineering, Mansholtlaan 10-12, Wageningen

HINTZEN, J. J., Momersteeg Int. B. V., Postbus 1, 5250 AA Vlijmen

IPEMA, Albertus, Institute of Agricultural Engineering, Mansholtlaan 10-12, Wageningen

KEMP, Aart, C.A.B.O., Bornsesteeg 65, Wageningen 6708 PD

MEIJS, Dr. Jac, Institute for Livestock, Runderweg 2, Postbus 160, Lelystad 8200 AD

MINDERHOUD, Dr. Jan Willem, Department of Field Crops/Grassland, Agricultural University, Haarweg 33, Wageningen 6709 PH

NEUTEBOOM, Jan J. H., Department of Field Crops, Agricultural University, Wageningen-Netherlands, Haarweg 33, Wageningen 6709 PH

PRINS, Willem, Institute Soil Fertility, Box 30003, 9750 RA Haren (gr)

PRINS, Mrs. Willem (Josje)

't HART, Prof. Maarten M. L., Agricultural University, Haarweg 33, Wageningen

't HART, Mrs. Maarten (Tineke)

te VELDE, Harm Arend, RIVRO, Box 32, Wageningen, 6700 AA

THOMAS, Bertus, Rundveehonaery, Runderweg 6, 8219 PK

VAN BURG, Dr. Pieter, P.F.J., Landbouwkundig Bureau Van De Nederlands, 15 Verdieping, Thorbeckelaan 360, 2564 BZ Den Haag

VAN BURG, Mrs. Pieter (Dolly)

VAN CRUTCHTEN, Caspar C.J.M., Zwaan and de Wiljes B.V., Stat. 124, Box 2, Scheemda, Groningen 9679 ZG

VAN DER HONING, Dr. Ynze, Institute for Livestock, Runderweg 2, Postbus 160, Lelystad 8200 AD

VAN DER MEER, Hugo G., CABO, Box 14, Wageningen 6700 AA

VAN DIJK, Dr. Geert, Foundation for Agriculture Plant Breeding (SVP), Box 117, Wageningen 6700 AC

NEW ZEALAND

BAKER, Dr. C. John, Department of Agronomy, Massey University, Palmerston North

BAKER, Mrs. C. John (Marlene)

BALL, Dr. Roger, Grasslands Division, DSIR, Private Bag, Palmerston North

BARR, James M., RD 2, Whakatane

BARR, Mrs. James M. (Mae)

BROCK, John, Grasslands Division, DSIR, Private Bag, Palmerston North

BROUGH, Allan, Dalgety Travel, Box 1498, Wellington

BROUGHAM, Dr. Raymond W., Grasslands Division, DSIR, Private Bag, Palmerston North

BROWN, Kenneth, c/o Dalgety Travel, Box 1498, Wellington

CAMERON, Nicholas, Research Division, Yates Corporation, Box 16147, Hornby, Christchurch

CRESSWELL, Jim, Wairau Valley (N.Z. Grasslands Association), RD 1, Blenheim 22895

CULLEN, N. A., Ruakura Soil and Plant Research Station, Private Bag, Hamilton

DAVIS, Murray R., Forest Research Institute, Box 31011, Christchurch

DILLON, Julian L., N.Z. Grasslands Association, Wairu Valley Farm, Marlborough, Wairau Valley, RD 1, Blenheim

GOOLD, Gary, Ruakura Agricultural Research Center, Private Bag, Hamilton

GORDON, Alex, R. A., Rosyth Rd., RD 2, Waipu

GORDON, Mrs. Alex (Julie)

HARRIS, A. J., Grasslands Division, DSIR, Private Bag, Gore, Southland

HARRIS, Mrs. A. J. (Eleanor)

HOLDOM, Thomas, No. 29, RD, Kapongo, 5th Taranahi

HOLDOM, Mrs. Thomas (Elizabeth)

INGLIS, J., c/o Dalgety Travel, Box 1498, Wellington

JOHNSTON, George, South Otago Farm Improvement, Rongahere, No. 4, RD, Balclutha

JOHNSTON, Mrs. George (Gladys)

KARLOVSKY, Josef, Department of Agriculture, Ruakura Agricultural Research Center, Private Bag, Hamilton

KORTE, Chris, Department of Agronomy, Massey University, Palmerston North

LINTON, John, New Zealand Grasslands Association, RD 8, Tepuke

MCCAUSLAND, Doug, RD, Darfield, North Canterbury

MCCAUSLAND, Mrs. Doug (Joyce)

MORAHAN, Daniel, South Otago Farm Improvement Club, Roseneath Farm, Ltd., Wangaloa, RD 2, Kaitangata, South Otago

MORAHAN, Mrs. Daniel (Eleanor)

PANTALL, Alan, South Otago Farm Consultancy, Box 159, Balclutha

PETERSON, Donald, Oturoa Road, RD 1, Ngongotaha

PETERSON, Mrs. Donald (Janice)

QUIN, Dr. Bertram, University of Agriculture, Winchmore Irrigation Research Station, Private Bag, Ashburton

SAX, Raymond, RD 2, Whakatane

SAX, Mrs. Raymond (Betty)

SCOTT, John, Ruakura Agricultural Research Center, Hamilton

SCOTT, Myrtle, Dunedin

SCOTT, Dr. Robin S., Invermay Agricultural Research Center, Privage Bag, Mosgiel

SEARLE, Francis J., Poerua Rd., Box 45, Hari Hari, W. Coast

SEARLE, Mrs. Francis J. (Alma Margaret)

THOMSON, Blair, South Otago Farm Improvement Club, Inchclutha, RD, Kaitangata South Otago

WATKIN, B. R., 463 Ruahine St., Palmerston

WILLIAMS, Dr. Warren, DSIR, Private Bag, Palmerston North

WINTER, John, South Otago Farm Improvement Club, Greenfield No. 4 RD, Balclutha

NIGERIA

AFOLAYAN, Dr. Adeniyi, Forestry, University of Ibadan, Ibadan

AKEN'OVA, Dr. Michael E., Agronomy, University of Ibadan, Ibadan

CHHEDA, Hamchand, Agronomy, University of Ibadan, Ibadan

LAZIER, Dr. John R., ILCA % IITA, PM 13, 5320 Ibadan

LAZIER, Mrs. John R. (Elizabeth)

OKOLI, Dr. Bosa, School of Biological Science, University of Port Harcourt, Port Harcourt, Rivers State PMB 5323

OYENUGA, Prof. Victor, Animal Science, University of Ibadan, Ibadan

NORWAY

MARUM, Dr. Petter, S.F. Loken, 2942 Volbu

MO, Dr. Magne, Institute of Animal Nutrition, Agricultural University of Norway, Box 25, N-1432 As-NLH

OPSAHL, Dr., Prof. Birger, Farm Crops, Agriculture, University of Norway, Box 41, 1432 As-NLH

OPSAHL, Mrs. Birger (Kristie)

SIMONSEN, Dr. Oystein R., Agricultural Research Council of Norway, Box 3010, R.S. 8001, Vagones, Bodo

SIMONSEN, Mrs. Oystein R. (Olga Louise)

PANAMA

ORTEGA VEGA, Carlos M., IDIAP, Gualaca, Chiriqui

PEOPLE'S REPUBLIC OF CHINA

CHIA Shen-siu, Prof., Chairman, Grassland Society of China, Beijing Agricultural University

LI, Bo, University of Inner Mongolia, Huhehot

QING Cheng Chen, Biology Department, University of Lanzhou, Lanzhou

ZHU Ting-Cheng, Prof., Institute of Grassland Science, Northeastern Teachers' University, Stalin Ave., Changchun, Jilin Province

PORTUGAL

CRESPO, David G., National Plant Improvement Station, Elvas

SALGUEIRO, Teodosio, Rua Cidale Do Lobito, Lote 268-2oEsq., 1800 Lisbon

REPUBLIC OF IRELAND

BRERETON, Dr. Anthony, Johnstown Castle, The Agricultural Institute, Wexford

COLLINS, Dr. Dermot, Animal Management, Agricultural Institute, Grange, Dunsany Co., Meath

CONWAY, Dr. Aidan, Johnstown Castle, The Agricultural Institute, Wexford

CONWAY, Mrs. Aidan (Eileen)

FLANAGAN, Dr. Sean S., The Agricultural Institute, Belclare, Tuam, County Galway

GLEESON, Patrick A., The Agricultural Institute, Moorepark, Fermoy, Cork

MEANEY, Richard, ACOT, Frascati Road, Blackrock, County Dublin

MURPHY, William E., Johnstown Castle, The Agricultural Institute, Wexford

O'SULLIVAN, Dr. Michael, Johnstown Castle, The Agricultural Institute, Wexford

RYAN, Dr. Michael, Johnstown Castle, The Agricultural Institute, Wexford

RYAN, Mrs. Michael (Sheila)

SHERWOOD, Marie, Johnstown Castle, The Agricultural Institute, Wexford

VALLE RIBEIRO, Dr. Manuel, Plant Breeding, The Agricultural Institute, Oak Park, Carlow

VALLE RIBEIRO, Mrs. Manuel (Emilia)

SAUDI ARABIA

ABOHASSAN, Dr. Atalla A., College of Agriculture, Riyad

SOUTH AFRICA

DICKINSON, Edward, Gunson South Africa, Box 9861, Johannesburg 2000, 18 Pritchard, Johannesburg, Transvaal 2001

DICKINSON, Mrs. Edward (Peggy)

EDWARDS, Dr. Peter J., Pasture Science, PTE Bag X9059, Pietermaritzburg, Natal, Rep. 3200

WASSERMAN, Dr. Victor D., Department of Agriculture, Private Bag X1116, Pretoria 0001

SOUTH KOREA

KANG, Dr. Myun Hee, Animal Science, College of Agriculture, Korea University, Anamdong, Seoul 32

KANG, Seung, Dairy Extension Department, Seoul Dairy Coop., Seoul

KIM, Dr. Dong Am, Animal Science, College of Agriculture, Seoul National University, Suweon 170

KIM, Dr. Young J., Research, Korean Rural Economics Institute and President, Korean Society of Grassland Science, Seoul

LEE, Kee E., Chief, Dairyland and Forage Division, Ministry of Agriculture and Fisheries, Seoul

MAENG, Dr. Won Jai, Department of Animal Science, Chung-Ang University, Seoul 151

PARK, Dr. John Mahn, Animal Science, College of Agriculture, Chonman National University, Kwang-ju 500

YUN, Dr. Hi-Sup, Animal Science, College of Animal Husbandry, Kon-Kuk University, 93-1 Mojin-dong, Seongdong-gu, Seoul 133

YUN, Dr. Ik Suk, College of Animal Husbandry, Kon-Kuk University, Mojin-dong, Seongdong-gu, Seoul 133-1

SPAIN

MENDENDEZ DE LUARCA, Santiago, Inst. Tecnico Gestion Vacuno, Yanguas y Miranda 23-90, Pamplona, APDO 1118

MENDENDEZ DE LUARCA, Mrs. Santiago (Isabel)

SWEDEN

JONSSON, Dr. Nils, Department of Plant Husbandry, Swedish University of Agricultural Science, Uppsala S-750-07

KORNHER, Dr. Alois, Plant Husbandry, Swedish University of Agricultural Science, Uppsala S-750-07

KORNHER, Mrs. Alois (Eila)

LUNDAHL, Miss Gunilla, Forage Crop Svalof Ab., 5-26800 Svalov

RUFELT, Snorre, Swedish University of Agricultural Science, Department of Plant and Forest Protection, Box 7044, Uppsala S-750-07

SJODIN, Dr. Jan, Forage Crop Svalof Ab, Svalov 5-26800

SJODIN, Mrs. Jan (Elizabeth)

SWITZERLAND

GUYER-WYSS, Dr. Hans, Swiss Fed. Research Station for Agronomy, Box 412, Zurich

GUYER-WYSS, Mrs. Hans (Marianne)

NOESBERGER, Prof. Josef J., Fur Pflanzenbau, Eidg, Tech., Hochschule, Univ. 2, 8092 Zurich

SYRIA

CECCARELLI, Dr. Salvatore, Forage Programs, ICARDA, Box 5466, Aleppo

TAIWAN

CHIA Huang, CAPO, 37 Nanhai Road, Taipei ROC 107

THAILAND

CHANTKAM, Dr. Sanan, Agronomy, Kasetsart University, Paholyothin, Bangkok 9

GIBSON, Trevor, Faculty of Agriculture, Khon Kaen University, Khon Kaen

GIBSON, Mrs. Trevor (Nongnuch)

GWILDIS, Dr. Johannes H., Thai-Germany Settlement Promotion Project, Box 25, Chiang-Mai

TOPARK-NGARM, Dr. Anake, Plant Science, Faculty of Agriculture, Khon Kaen University, Khon Kaen

TRINIDAD

PERSAD, Nand K., Ministry of Agriculture, Crop Research Institute, Centeno

PERSAD, Mrs. Nand K. (Sita)

TUNISIA

MAJDOUB, Dr. A., Animal Science, INAR, 43 Avenue Charles Nicolle, Tunis

UNITED KINGDOM

ADAMSON, Alan, Nutritional Chemist, ADAS (MAFF), Government Building, Burghill Rd., Westbury-on-Tryn, Bristol, England BSID6NJ

BOYLE, Peter, Commonwealth Bureau of Pastures, Hurley, Maidenhead, England

CASTLE, Dr. Malcolm, Hannah Research Institute, Ayr, Scotland KA6 5HL

CASTLE, Mrs. Malcolm (Elizabeth)

CLEMENTS, Dr. Robert O., Grassland Research Institute, Hurley, Maidenhead, Berkshire, England RG9 IRE

COOPER, Prof. John, Welsh Plant Breeding Station, Plas Gogerddan, Aberystwyth, Dyfed, Wales SY23 3EB

CORRALL, A. James, Grassland Research Institute, Maidenhead, Berkshire, England SL6 5LR

DIBB, Colin, Rm. 123, ADAS, Great Westminister House, Horseferry Rd., London, England SWIP2AE

DIBB, Mrs. Colin (Hazel)

FRAME, Dr. John, The West Scotland Agricultural College, Auchincruive Estate, Strathclyde, Scotland KA6 5HW

GRANT, Miss Sheila A., Hill Farming Research Organization, Bush Estate, Penicuik, Midlothian, Scotland EH26 OPY

HAGGAR, Dr. Roger, Weed Research Organization, Oxford, England OX5 IPF

HAYES, William J., Agriculture and Food Science, Queens University, Newforge Lane, Belfast, Northern Ireland BT9 5PX

HEBBLETHWAITE, Dr. Paul, Agriculture and Horticulture, School of Agriculture, University of Nottingham, Sutton Bonington, Nr. Loughborough, Leicester, England LE12 5NL

HIDES, David, Seed Production Department, Welsh Plant Breeding Station, Plas Gogerddan, Aberystwyth, Dyfed, Wales SY23 3EB

HODGSON, Dr. John, Hill Farming Research Organization, Bush Estate, Penicuik, Midlothian, Scotland EH26 OPY

INGRAM, Jan, National Institute of Agriculture and Botany, Huntington Rd., Cambridge, England

JONES, Dr. Lloyd H.P., Grassland Research Institute, Hurley, Maidenhead, Berkshire, England SL6 5LR

KLINNER, Wilfred E., National Institute of Agricultural Engineering, Wrest Park, Silsoe, Bedford, England MK45 4HS

LAIDLAW, Dr. A. Scott, Field Botany Research Division, Department of Agriculture for Northern Ireland, Newforge Lane, Belfast, Northern Ireland BT9 5PX

LAZENBY, Prof. Alec, Grassland Research Institute, Hurley, Maidenhead, Berkshire, England SL6 5LR

LEAFE, Dr. Edward L., Grassland Research Institute, Hurley, Maidenhead, Berkshire, England SL6 5LR

LE DU, Dr. Yann, Grassland Research Institute, Hurley, Maidenhead, Berkshire, England SL6 5LR

LE DU, Mrs. Yann (Gillian)

LEWIS, D. A., Imperial Chemical Industries, Ltd., Box 1, Billingham, Cleveland, England TS23 1LB

LOWE, John J., 101 Ballynahinch Rd., Carryduff, Belfast, Co. Down, Northern Ireland B78 8DP

PEARSE, Joy, Department of Agriculture, University College of Wales, Aberystwyth, Dyfed, Wales SY23 3EB

PEEL, Stephen, Grassland Research Institute, Hurley, Maidenhead, Berkshire, England SL6 5LR

REID, Dr. David, Applied Studies, Hannah Research Institute, Ayr, Scotland KA6 5HL

REID, Mrs. David (Marian)

SMYTH, Walter, Richardsons Fertilizer, Ltd., Herdmann Channel Rd., Belfast, Northern Ireland BT3 9AP

SMYTH, Mrs. Walter (Evelyn)

THOMAS, Dr. Cledwyn, Grassland Research Institute, Hurley, Maidenhead, Berkshire, England SL6 5LR

WELLER, Richard, 25 Seymour Ave., Shinfield, Reading, Berkshire, England

WHITEHEAD, Dr. David C., Grassland Research Institute, Hurley, Maidenhead, Berkshire, England SL6 5LR

WILKINS, Dr. Roger, Grassland Research Institute, Hurley, Maidenhead, Berkshire, England SL6 5LR

WILMAN, Dr. David, Department of Agriculture, University College of Wales, Aberystwyth, Dyfed, Wales SY23 3DD

WRIGHT, Prof. Charles E., Field Botany Research Division, Department of Agriculture for Northern Ireland, Newforge Lane, Belfast, Northern Ireland BT9 5PX

WRIGHT, Dr. Peter, Department of Agriculture, University College of Wales, Aberystwyth, Dyfed, Wales SY23 3DD

YOUNG, Nigel, Grassland Research Institute, Hurley, Maidenhead, Berkshire, England SL6 5LR

UNION OF SOVIET SOCIALIST REPUBLICS

AFANASIEV, Dr. Rafail A., Institute of Fertilizer, Pzjanishnikova, Moscow

BESSARABOV, Aduard, Ministry of Agriculture, Orlekov 1/11, Moscow

BLAGOVESCHENSKY, Dr. German, Onz Vashnel, Sovetskaja, Leningrad

IGLOVIKOV, Dr. Vladilen C., All-Union Williams Fodder Research Institute, Moscow Region

IKDERM, Dr. Gubdrejm, Estonian Research Institute of Agriculture, Sarcu, Tallium

KADZIULIS, Dr., Prof. Leonas, Lithuanian SSR, Dotnuva-Akademija

KIRDYAVIN, Dr. Anatoliy, Institute of Grain, Celenogzad Shortondic

KOJENKOVA, Mrs. Raiss, Ministry USSR, Cadovo Tziuwfaljnaja 10, Moscow

KOZLOVA, Mrs. Klavdija, Ministry of Agriculture Seed Production, Sadovo-Tzcumfacjaaja 10, Moscow

LAUR, Dr. Voldemar H., Estonian Research Institute of Agriculture, Estonian SSR 203 300

MAKARENKO, Dr. Peter, Ukrainian Fozzyg Institute, Yogolo St. 30, Vinitso 287 100

MARCHENKO, Oleg, All-Union Research Institute of Agriculture Mechanics, VIM, Moscow

OLDER, Hindrek, Estonian Research Institute of Agriculture, Sarcu, Tallium, Estonian SSR 203400

OSTAPOV, Dr. Wladimir, Ukrainian Wate Institute, Nadnipjnke, Chersen 325 908

OUELTCHENKO, Vladimir S., Ukrainian Institute, Askonig-Wova 326 332

SAU, Dr., Asst. Prof. Arnold V., Estonian Agriculture Academy, Riia St. 112, Jartu Estonian SSR 202 400

SYTCHEV, Sergei, Grassland Dept., Ministry of Agriculture, Moscow

ZHIDONYTE, Dr. Yamtiva A., Ministry USSR Lithusnar Research Institute of Agriculture, Ruokio Y-12, Dotnuva-Akademija, Lithuanian SSR

UNITED STATES OF AMERICA*

Alabama

BALL, Dr. Donald M., Extension Agronomist, Auburn University, Extension Hall, Auburn, AL 36830

CURRIER, Cliff G., Department of Agronomy and Soils, Auburn University, Auburn, AL 36849

DANHO, Lucas A., Dept. of Agronomy and Soils, Auburn University, Auburn, AL 36849

HOVELAND, Dr. Carl S., Dept. of Agronomy and Soils, Auburn University, Auburn, AL 36849

KING, Dr. Cooper, Jr., Dept. of Agronomy and Soils, Auburn University, Auburn, AL 36849

MARTIN, Bill, Dept. of Agronomy and Soils, Auburn University, Auburn, AL 36849

PEDERSEN, Dr. Jeffrey F., Dept. of Agronomy and Soils, Auburn University, Auburn, AL 36849

RICKERL, Ms. Diane, Dept. of Agronomy and Soils, Auburn University, Auburn, AL 36849

SULLIVAN, Dr. Gregory, Auburn University, Auburn, AL 36849

WILLIAMS, Charles B., Dept. of Agronomy and Soils, Auburn University, Auburn, AL 36830

WILLIAMS, E. Darrell, Dept. of Agronomy and Soils, Auburn University, Auburn, AL 36849

WILLIAMS, Mimi J., Dept. of Agronomy and Soils, Auburn University, Auburn, AL 36849

Arizona

GUREVITCH, Jessica, Ecology and Evolutionary Biology, Biology Science East, University of Arizona, Tucson, AZ 85721

METCALF, Dr. Darrel, College of Agriculture, University of Arizona, Tucson, AZ 85721

RICE, Richard, Dept. of Animal Sciences, University of Arizona, Tucson, AZ 85721

SMITH, Dr. Dale, Plant Science, University of Arizona, Tucson, AZ 85721

SMITH, Mrs. Dale (Marian)

Arkansas

CHILD, Dr. Dennis, Winrock International, Route #3, Morrilton, AR 72110

HANKINS, Dr. Bishop J., Dept. of Agronomy, University of Arkansas, Little Rock, AR 72099

MOLINE, W. J., Dept. of Agronomy, University of Arkansas, Fayetteville, AR 72701

PEISCHEL, Dr. An, SRA, Box 34D, Homer, AR 99603

PETERSON, Dr. Roald A., 726 Rock Cliff Road, Fayetteville, AR 72701

PETERSON, Mrs. Roald A. (Carmen)

RAUN, Dr. Ned S., Winrock International (V.P.), Route #3, Morrilton, AR 72110

SPOONER, Dr. Arthur E., Dept. of Agronomy, Building 115, University of Arkansas, Fayetteville, AR 72701

STALLCUP, Dr. Odie T., Dept of Animal Sciences, University of Arkansas, Fayetteville, AR 72701

STALLCUP, Mrs. Larue, Fayetteville, AR 72701

California

BROOKS, William H., Cooperative Extension, County Agriculture Center, Court House, Ukiah, CA 95482

BROOKS, Mrs. William H. (Betty)

JONES, Dr. Milton B., Agronomy and Range Science Dept., Hopland Field Station, University of California, Hopland, CA 95449

KELLOGG, Ron, Ramsey Seed, Inc., Box 352, Manteca, CA 95336

KELLOGG, Mrs. Ron (Ruth)

LOVE, Prof. R. Merton, Graduate Group in Ecology, University of California, Davis, CA 95616

LOVE, Mrs. R. Merton (Eunice)

MURPHY, Alfred, Hopland Field Station, University of California, Hopland, CA 95449

MURPHY, Mrs. Alfred (Kathleen)

RHEA, Mark, Humboldt State University, Arcata, CA 95521

SMITH, Dr. Donald, Cal/West Seeds, Box 1428, Woodland, CA 95695

Colorado

CUANY, Dr. Robin, Dept. of Agronomy, Colorado State University, Ft. Collins, CO 80523

ELY, Dr. Lane O., Dept. of Animal Sciences, Colorado State University, Ft. Collins, CO 80523

JOYCE, Ms. Linda A., Dept of Range Science, Colorado State University, Ft. Collins, CO 80523

JUMP, Craig, Dept. of Range Science, Colorado State University, Ft. Collins, CO 80523

KINSINGER, Dr. Floyd, Society for Range Management, 2760 W. 5th St., Denver, CO 80204

LAYCOCK, Dr. William, USDA-SEA-AR,** Crops Research Laboratory, Colorado State University, Ft. Collins, CO 80523

MCGINNIS, Dr. William J., USDA-SEA-AR, Crops Research Laboratory, Colorado State University, Ft. Collins, CO 80523

RAFSNIDER, Dr. Giles T., Economics, Colorado State University, C307 Andrew G. Clark Building, Ft. Collins, CO 80523

SIEMER, Dr. Eugene, Mountain Meadow Research, Colorado State University, Box 598, Gunnison, CO 81230

SKILES, J., Dept. of Range Science, Colorado State University, Ft. Collins, CO 80523

SMEDLEY, Harold D., Agrigenetics Corporation, 1726 Cole Boulevard, Golden, CO 80401

SMITH, Dr. Dan, Dept. of Agronomy, Colorado State University, Ft. Collins, CO 80523

TOWNSEND, Dr. Charles E., USDA-SEA-AR, Crops Research Laboratory, Colorado State University, Ft. Collins, CO 80523

VAN DYNE, Prof. George, Dept. of Range Science, Colorado State University, Ft. Collins, CO 80523 (now deceased)

---

*Where it was known, the native country of graduate students is identified in parentheses after the name.

**The agency designation within the U.S. Department of Agriculture was changed from Science and Education, Agricultural Research (SEA-AR) to Agricultural Research Service (ARS) after the XIV Congress convened.

Delaware

BELL, Dr. Kenneth, Agriculture and Natural Resources, Delaware State College, Dover, DE 19901

CAULK, Lyndon D., Jr., Route #1, Box 118, Wyoming, DE 19934

CAULK, Mrs. Lyndon D. (Elizabeth)

JONES, Dr. Edward, Delaware State College, Dover, DE 19901

SWAIN, Harris R., Agriculture and Natural Resources, Delaware State College, Dover, DE 19901

Florida

ALLEN, Dr. Robert J., Jr., University of Florida, Drawer A, Belle Glade, FL 33430

ALLEN, Mrs. Robert J. (Evelyn)

ALLEN, Robert J., III, Belle Glade, FL 33430

BALTENSPERGER, Dr. David D., Dept. of Agronomy, University of Florida, Gainesville, FL 32601

BJORNDAL, Dr. Karen A., Zoology Dept., University of Florida, Gainesville, FL 32611

BLUE, Dr. William G., Soil Science Dept., University of Florida, Gainesville, FL 32611

BLUE, Mrs. William G. (Bernice)

BROLMANN, John, Dept. of Agronomy, University of Florida, Gainesville, FL 32611

CHAMBLISS, Dr. Carroll G., Dept. of Agronomy, University of Florida, 5007 60th St. East, Bradenton, FL 33508

CHRISTIANSEN, Scott, Dept. of Agronomy, University of Florida, Gainesville, FL 32611

CONRAD, Dr. Joseph H., Animal Sciences Dept., University of Florida, Gainesville, FL 32611

DA VEIGA, Jonas Bastos (Brazil), Dept. of Agronomy, University of Florida, Gainesville, FL 32611

DAVILLA, Ciro (Venezuela), Dept. of Agronomy, University of Florida, Gainesville, FL 32615

EUCLIDES, Valeria and Kepler (Brazil), Dept. of Agronomy, University of Florida, Gainesville, FL 32608

FRANCA-DANTAS, Mario (Brazil), Dept. of Agronomy, University of Florida, Gainesville, FL 32608

GILDERSLEEVE, Miss Rhonda R., Dept. of Agronomy, University of Florida, Gainesville, FL 32611

HODGES, Prof. Elver M. (Malawi), 1510 Highway 64 West, Wauchula, FL 33873

HORTON, Dr. Glyn, University of Florida, Agricultural Research Center, Ona, FL 33865

HORTON, Mrs. Glyn (Gerty)

JANK, Miss Liana L. (Brazil), Dept. of Agronomy, University of Florida, Gainesville, FL 32611

JONES, David W., Dept. of Agronomy, University of Florida, Gainesville, FL 32611

KALMBACHER, Dr. Robert, University of Florida, Ona Agricultural Research Center, Box 62, Ona, FL 33865

KORNELIUS, Euclides, Dept. of Agronomy, University of Florida, Gainesville, FL 32608

KRETSCHMER, Prof. Albert E., Jr., University of Florida, Agricultural Research Center, Box 248, Ft. Pierce, FL 33450

MISLEVY, Dr. Paul, University of Florida, Agricultural Research Center, Ona, FL 33865

MOORE, Dr. John E., Nutrition Laboratory, Dept. of Animal Sciences, University of Florida, Gainesville, FL 32611

MOTT, Dr. Gerald, Dept. of Agronomy, University of Florida, Gainesville, FL 32611

OCUMPAUGH, Dr. William R., IFAS, Dept. of Agronomy, University of Florida, Gainesville, FL 32611

PAZ, Espran A. (Venezuela), Dept. of Agronomy, University of Florida, Gainesville, FL 32608

PRINE, Dr. Gordon, Dept. of Agronomy, University of Florida, Gainesville, FL 32611

QUESENBERRY, Dr. Kenneth, Dept. of Agronomy, University of Florida, Gainesville, FL 32611

QUESENBERRY, Mrs. Kenneth (Joyce)

RODRIGUES, Luis R. (Brazil), Dept. of Agronomy, University of Florida, Gainesville, FL 32603

ROSERO, Oswaldo R., Animal Science Nutrition Laboratory, University of Florida, Gainesville, FL 32601

RUELKE, Dr. O. Charles, Dept. of Agronomy, University of Florida, Gainesville, FL 32611

RUSSO, Dr. Sandra, University of Florida, Gainesville, FL 32611

SALDIVAR-F, Abelardo J. (Mexico), Dept. of Agronomy, University of Florida, Gainesville, FL 32611

SANTILLAN, Paul, 305-23 Diamond Village, Gainesville, FL 32603

SCHANK, Prof. Stanley, Dept. of Agronomy, University of Florida, Gainesville, FL 32611

SCHANK, Mrs. Stanley (Sandra)

SNYDER, Dr. George H., AREC, Drawer A, University of Florida, Belle Glade, FL 33430

SOTO, Mr. Marco (Mexico), Dept. of Agronomy, University of Florida, Gainesville, FL 32611

WATSON, Dr. Earl C., Department of Research, U.S. Sugar Corporation, Box 1207, Clewiston, FL 33440

WEST, Dr. Sherlie, Dept. of Agronomy, University of Florida, Gainesville, FL 32608

WING, Dr. James M., Dept. of Dairy Science, University of Florida, Gainesville, FL 32611

YARLETT, Lewis L., 808 N.W. 39th Drive, Gainesville, FL 32605

Georgia

AKIN, Dr. Danny E., USDA-SEA-AR, Russell Research Center, Athens, GA 30613

BELESKY, David, USDA-SEA-AR, Southern Piedmont Conservation Research Center, Box 555, Watkinsville, GA 30677

BOUTON, Joseph H., Agronomy Dept., University of Georgia, Athens, GA 30602

BROOKS, Frank, Box 1221, Canton, GA 30114

BROWN, Dr. Harold, Agronomy Dept., University of Georgia, Athens, GA 30603

BURTON, Dr. Glenn W., USDA-SEA-AR, Agronomy Dept., Coastal Plain Experiment Station, Tifton, GA 31793

D'UVA, Paolo (Italy), Agronomy Dept., University of Georgia, Athens, GA 30602

FALES, Steven, 3 Charlotte Circle, Griffin, GA 30223

HANNA, Dr. Wayne W., USDA-SEA-AR, Agronomy Dept., Coastal Plain Experiment Station, Tifton, GA 31793

HARPER, Dr. Lowry A., USDA-SEA-AR, Southern Piedmont Conservation Research Center, Box 555, Watkinsville, GA 30677

KINCHELOE, Dr. Samuel, International Minerals and Chemical Corporation, 2201 Perimeter Center East, N.E., Atlanta, GA 30346

KINCHELOE, Mrs. Samuel (Carol)

MARTIN, Dr. Paul B., Entomology Dept., CPES, University of Georgia, Tifton, GA 37193

MERTINS, Dr. David R., Dept. of Animal and Dairy

Sciences, University of Georgia, Athens, GA 30602

MILLER, Dr. John D., USDA-SEA-AR, Agronomy Dept., Georgia Coastal Experiment Station, Tifton, GA 31793

MONSON, Dr. Warren G., USDA-SEA-AR, Agronomy Dept., Georgia Coastal Experiment Station, Tifton, GA 31793

MOON, Dr. Nancy J., Food Science Dept., University of Georgia, Experiment, GA 30212

McCORMICK, Dr. Michael E., Animal Science Dept., University of Georgia, Georgia Experiment Station, Experiment, GA 30212

McCULLOUGH, Prof. Marshall E., Animal Science Dept., Georgia Experiment Station, Experiment, GA 30212

McHAN, Dr. Frank, USDA-SEA-AR, Field Crops, Russell Research Center, Athens, GA 30613

ROBERTSON, Dr. James A., USDA-SEA-AR, Russell Research Center, Athens, GA 30613

STUEDEMANN, Dr. John, USDA-SEA-AR, Southern Piedmont Conservation Research Center, Box 555, Watkinsville, GA 30677

WILKINSON, Dr. Stanley, USDA-SEA-AR, Southern Piedmont Conservation Research Center, Box 555, Watkinsville, GA 20677

WILSON, Dr. John R., Agronomy Dept., University of Georgia, Athens, GA 30602

Hawaii

BURTON, Dr. Joe, University of Hawaii, Kamuela, HI 96743

SMITH, Dr. Burt, Cooperative Extension Service, University of Hawaii, Box 237, Kamuela, HI 96743

SPENCE, Earl W., Agronomy Dept., Parker Ranch, Box 458, Kamuela, HI 96743

TOMPKINS, John C., K M Seed Company, Box 117, Kapaau, HI 96755

Idaho

HAIGHT, John C., North American Plant Breeders, Box 1130, Nampa, ID 83651

SOURS, John M., Jacklin Seed Company, W. 5300 Jacklin Avenue, Post Falls, ID 83854

Illinois

BERGER, Dr. Larry L., Dept. of Animal Sciences, University of Illinois, Urbana, IL 61801

DIXON, Dr. Lee, Southern Illinois University, Carbondale, IL 62901

FAHEY, Dr. George C., Dept. of Animal Sciences, University of Illinois, Urbana, IL 61801

GRAFFIS, Dr. Don, Dept. of Agronomy, University of Illinois, Urbana, IL 61801

HARLAN, Jack R., Dept. of Agronomy, University of Illinois, Urbana, IL 61801

JONES, Dr. Joe H., Plant and Soil Sciences Dept., Southern Illinois University, Carbondale, IL 62901

KAISER, Dr. Clarence J., Dixon Springs Agriculture Center, University of Illinois, Simpson, IL 62955

OLSEN, Dr. Farrell J., Plant and Soil Sciences Dept., Southern Illinois University, Carbondale, IL 62901

SHI, Prof. Bo Hong, Prod. Training Dept., John Deere Intercontinental, Ltd., 909 3rd Avenue, Moline, IL 61265

STAUFFER, Dr. Mark D., Advanced Harvesting Systems, International Harvester Company, 4415 W. Harrison Street,

Suite 530, Hillside, IL 60162

VAN DER AAR, Petrus J., Dept. of Animal Sciences, University of Illinois, Urbana, IL 61801

VAN DER AAR, Mrs. Petrus J. (Ingetje)

VETTER, Dr. Richard, A. O. Smith Harvestore, Arlington Heights, IL 60004

VETTER, Mrs. Richard (Donna)

WALGENBACH, Dr. Richard P., Dept. of Agronomy, University of Illinois, Urbana, IL 61801

WILLIAMSON, Robert E., Prod. Training Dept., John Deere Intercontinental, Ltd., 909 3rd Avenue, Moline, IL 61265

WISIOL, Miss Karen, Dept. of Biological Sciences, University of Illinois, Chicago Circle, Box 4348, Chicago, IL 60680

Indiana

BUKER, Dr. Robert J., FFR Cooperative, 4112 E. St. Rd. 225, W. Lafayette, IN 47906

COLLINS, Steve, Pioneer Hi-Bred, 1000 W. Jefferson St., Tipton, IN 46072

EUBANK, Harold F., 1210 Hilltop Dr., Lowell, IN 46356

EUBANK, Mrs. Harold F. (Jeannine)

HEATH, Prof. Maurice E., Agronomy Dept., Purdue University, West Lafayette, IN 47907

HEATH, Mrs. Maurice E. (Florence)

HERTEL, Judith, Agronomy Dept., Purdue University, West Lafayette, IN 47907

HOLT, Dr. Donald A., Agronomy Department, Purdue University, West Lafayette, IN 47907

LECHTENBERG, Dr. Victor, Agronomy Dept., Purdue University, West Lafayette, IN 47907

LIGHTNER, John, Agronomy Dept., Purdue University, West Lafayette, IN 47907

MOORE, Kenneth, Agronomy Dept., Purdue University, West Lafayette, IN 47907

MOORE, Mrs. Kenneth (Marlyn)

NOLLER, Prof. Carl, Animal Sciences Dept., Purdue University, West Lafayette, IN 47907

SCHAEFER, Dr. Daniel, Animal Sciences Dept., Purdue University, West Lafayette, IN 47907

SEIF, Seif A., Agronomy Dept., Purdue University, West Lafayette, IN 47907

YORK, Vance, Pioneer Hi-Bred, 1000 W. Jefferson St., Tipton, IN 46072

Iowa

ALBRECHT, Kenneth, Agronomy Dept., Iowa State University, Ames, IA 50011

BARNHART, Dr. Stephen K., Agronomy Dept., Iowa State University, Ames, IA 50011

CARLSON, Dr. Irving T., Agronomy Dept., Iowa State University, Ames, IA 50011

HARTMAN, Dr. Wayne, 433 Hilltop Road, Ames, IA 50010

HURST, Steve M., North American Plant Breeders, R.R. #3, Ames, IA 50010

JENSEN, Dr. David O., Kemin Industries, 2100 Maury St., Des Moines, IA 50317

KALTON, Dr. Robert R., Land O'Lakes Research Farm, R.R. #2, Webster City, IA 50595

KALTON, Mrs. Robert R. (Vivian)

MILLER, Dr. Jonas W., Plant Breeding Division, Pioneer Hi-Bred International, Inc., Box 85, Johnston, IA 50131

MOUTRAY, Dr. Jim, North American Plant Breeders, R.R. #3, Ames, IA 50010

MUSIL, Miss Martha, Agronomy Dept., Iowa State University, Ames, IA 50011

ROBINSON, David, Harvesting B-2, Massey Ferguson, Inc., 1901 Bell Avenue, Des Moines, IA 50315

WEDIN, Dr. Walter F., Agronomy Dept., Iowa State University, Ames, IA 50011

WEST, Charles P., Agronomy Dept., Iowa State University, Ames, IA 50011

WOODWARD, Dr. Tim, North American Plant Breeders, R.R. #3, Ames, IA 50010

Kansas

BOLSEN, Dr. Keith K., Animal Sciences Dept., Weber Hall, Kansas State University, Manhattan, KS 66501

NUWANYAKPA, Mopoi, Y-21 Jardine Terrace, Manhattan, KS 66502

POSLER, Dr. Gerry L., Agronomy Dept., Kansas State University, Manhattan, KS 66506

Kentucky

ABSHER, Dr. Curtis W., Animal Sciences Dept., University of Kentucky, Lexington, KY 40546

APPEL, Mr. Paul, College of Agriculture, University of Kentucky, Lexington, KY 40546

BAKER, Dr. John, Animal Sciences Dept., University of Kentucky, Lexington, KY 40546

BARNHART, Dr. Charles E., Dean, College of Agriculture, University of Kentucky, Lexington, KY 40546

BARNHART, Mrs. Charles E. (Jean)

BARNHISEL, Dr. Richard I., Agronomy Dept., University of Kentucky, Lexington, KY 40546

BASTIN, Garland M., 613 Pasadena Dr., Lexington, KY 40503

BLEVINS, Dr. Robert L., Agronomy Dept., University of Kentucky, Lexington, KY 40546

BOLING, Dr. James A., Animal Sciences Dept., University of Kentucky, Lexington, KY 40546

BOSWORTH, S., Agronomy Dept., University of Kentucky, Lexington, KY 40546

BRADFORD, Dr. Garnett, Agricultural Economics Dept., University of Kentucky, Lexington, KY 40546

BRADLEY, Dr. Neil W., Animal Sciences Dept., University of Kentucky, Lexington, KY 40546

BUCKNER, Dr. Robert C., USDA-SEA-AR, 1849 Darien Drive, Lexington, KY 40504

BURRIS, Dr. Roy, Animal Sciences Dept., University of Kentucky, Research and Education Center, PO Box 469, Princeton, KY 42445

BURRUS, Mr. Paul, Jr., USDA-SEA-AR, 575 Halifax Drive, Lexington, KY 40503

BUSH, Dr. Lowell P., Agronomy Dept., University of Kentucky, Lexington, KY 40546

CALVERT, Dr. J. R., Agronomy Dept., University of Kentucky, Lexington, KY 40546

CHAPPELL, Dr. G.L.M., Animal Sciences Dept., University of Kentucky, Lexington, KY 40546

CHRISTENSEN, Dr. Christian, Entomology Dept., University of Kentucky, Lexington, KY 40546

COLLINS, Dr. Glenn B., Agronomy Dept., University of Kentucky, Lexington, KY 40546

COUGHENOUR, Dr. Charles M., Sociology Dept., University of Kentucky, Lexington, KY 40546

CRIST, Dr. William, Animal Sciences Dept., University of Kentucky, Lexington, KY 40546

CRIST, Mrs. William (Karen)

DAHLMAN, Dr. D. L., Entomology Dept., University of Kentucky, Lexington, KY 40546

DAVIS, Dr. Joe, Agricultural Economics Dept., University of Kentucky, Lexington, KY 40546

DIACHUN, Dr. Steve, Plant Pathology Dept., University of Kentucky, Lexington, KY 40546

DOUGHERTY, Dr. Charles T., Agronomy Dept., University of Kentucky, Lexington, KY 40546

EIZENGA, Dr. Georgia, Agronomy Dept., University of Kentucky, Lexington, KY 40546

ELY, Dr. Donald G., Animal Sciences Dept., University of Kentucky, Lexington, KY 40546

EVANS, Mr. J. Kenneth, Agronomy Dept., University of Kentucky, Lexington, KY 40546

FERGUS, Dr. Ernest N., 138 Jesselin Drive, Lexington, KY 40503

FIELD, Candace L., 1981 Favell Court, Lexington, KY 40503

FINKNER, Dr. V. C., Agronomy Dept., University of Kentucky, Lexington, KY 40546

FINKNER, Mrs. V. C. (Ruby)

FRYE, Dr. Wilbur, Agronomy Dept., University of Kentucky, Lexington, KY 40546

FURBEE, R., Agriculture Public Information, 131 Experiment Station Bldg., University of Kentucky, Lexington, KY 40546

GAY, Dr. Nelson G., Animal Sciences Dept., University of Kentucky, Lexington, KY 40546

HAYS, Dr. Virgil W., Animal Sciences Dept., University of Kentucky, Lexington, KY 40546

HAYS, Mrs. Virgil W. (Jackie)

HEERSCHE, Dr. George, Animal Sciences Dept., University of Kentucky, Lexington, KY 40546

HEMKEN, Dr. Roger W., Animal Sciences Dept., University of Kentucky, Lexington, KY 40546

HEMKEN, Mrs. Roger W. (Muzzie)

HIATT, Dr. A. J., Agronomy Dept., University of Kentucky, Lexington, KY 40546

HIATT, Mrs. A. J. (Elaine)

JACQUES, Dr. R. M., Agronomy Dept., University of Kentucky, Lexington, KY 40546

JENKINS, Jeff H., Western Kentucky University, Bowling Green, KY 42101

JOHNS, Dr. John T., Animal Sciences Dept., University of Kentucky, Lexington, KY 40546

JOHNSON, M., Plant Pathology Dept., University of Kentucky, Lexington, KY 40546

JUSTUS, Dr. Fred, Agricultural Economics Dept., University of Kentucky, Lexington, KY 40546

KEMP, Dr. James D., Animal Sciences Dept., University of Kentucky, Lexington, KY 40546

KEMP, Mrs. James D. (Helen)

KIRKLAND, Dr. Daniel L., Agriculture, Morehead State University, UPO 702, Morehead, KY 40351

KNAPP, Dr. Fred, Entomology Dept., University of Kentucky, Lexington, KY 40546

LACEFIELD, Dr. Garry D., Agronomy Dept., University of Kentucky, Research and Education Center, PO Box 469, Princeton, KY 42445

LANE, Dr. Gary T., Animal Sciences Dept., University of Kentucky, Lexington, KY 40546

LEGGETT, Dr. Everett, Agronomy Dept., University of Kentucky, Lexington, KY 40546

LITTLE, Dr. C. Oran, Associate Dean for Research, College

of Agriculture, University of Kentucky, Lexington, KY 40546

LITTLE, Mrs. C. Oran (Myrtle)

LOEWER, Dr. Otto, Agricultural Engineering Dept., University of Kentucky, Lexington, KY 40546

MASSEY, Dr. Herb, International Programs, College of Agriculture, University of Kentucky, Lexington, KY 40546

MASSIE, Ira, Agronomy Dept., University of Kentucky, Lexington, KY 40546

MIKSCH, Dr. Duane, Veterinary Sciences Dept., University of Kentucky, Research and Education Center, PO Box 469, Princeton, KY 42445

MIRANDA, Prof. Emilio (Nicaragua), Box 424, Western Kentucky University, Bowling Green, KY 42101

MITCHELL, Dr. George E., Jr., Animal Sciences Dept., University of Kentucky, Lexington, KY 40546

MITCHELL, Mrs. George E., Jr., (Carolyn)

MOODY, Dr. William, Animal Sciences Dept., University of Kentucky, Lexington, KY 40546

MURDOCK, Dr. Lloyd W., Agronomy Dept., University of Kentucky, Research and Education Center, PO Box 469, Princeton, KY 42445

NESMITH, Dr. William C., Plant Pathology Dept., University of Kentucky, Lexington, KY 40546

OLDS, Dr. Durward, Animal Sciences Dept., University of Kentucky, Lexington, KY 40546

O'LEARY, Dr. Joseph, Animal Sciences Dept., University of Kentucky, Lexington, KY 40546

OLSON, Dr. Ken, Animal Sciences, Dept., University of Kentucky, Lexington, KY 40546

PARKER, Dr. Blaine, Agricultural Engineering Dept., University of Kentucky, Lexington, KY 40546

PASS, Dr. Bobby C., Entomology Dept., University of Kentucky, Lexington, KY 40546

PASS, Mrs. Bobby C. (Ann)

PEASLEE, Dr. Doyle E., Agronomy Dept., University of Kentucky, Lexington, KY 40546

PENDERGRASS, George, Management Operations, College of Agriculture, University of Kentucky, Lexington, KY 40546

PIGG, Dr. Kenneth E., Sociology Dept., University of Kentucky, Lexington, KY 40546

PORTER, Annette S., Peregrine Farm, Salvisa, KY 40372

POWELL, Dr. Andrew J., Agronomy Dept., University of Kentucky, Lexington, KY 40546

POWELL, Antoinette P., Agricultural Library, University of Kentucky, Lexington, KY 40546

QUINN, Garvin W., Agricultural Public Information, University of Kentucky, Lexington, KY 40546

RAGLAND, Dr. John L., College of Agriculture, University of Kentucky, Lexington, KY 40546

RAGLAND, Mrs. John L. (Irene)

RASNAKE, Dr. Monroe, Agronomy Dept., University of Kentucky, Research and Education Center, PO Box 469, Princeton, KY 42445

REDMAN, Dr. John C., Agricultural Economics Dept., University of Kentucky, Lexington, KY 40546

ROBB, Dr. Thomas W., Animal Sciences Dept., University of Kentucky, Lexington, KY 40546

ROBERTSON, Dr. John, College of Agriculture, University of Kentucky, Lexington, KY 40546

RODRIQUES, Dr. L. D., Entomology Dept., University of Kentucky, Lexington, KY 40546

SCHAFER, James, 535 Mahaffey, C-7, Eastern Kentucky University, Richmond, KY 40475

SIGAFUS, Dr. Roy E., Agronomy Dept., University of Kentucky, Lexington, KY 40546

SMITH, Dr. Allan, 214 Shady Lane, Lexington, KY 40503

SMITH, Dr. Ed M., Agricultural Engineering Dept., University of Kentucky, Lexington, KY 40546

SPRINGER, Dr. Don, Agricultural Public Information, University of Kentucky, Lexington, KY 40546

TAYLOR, Dr. Norman L., Agronomy Dept., University of Kentucky, Lexington, KY 40546

TAYLOR, Dr. Timothy H., Agronomy Dept., University of Kentucky, Lexington, KY 40546

TEKRONY, Dr. D. M., Agronomy Dept., University of Kentucky, Lexington, KY 40546

THOMAS, Dr. Grant, Agronomy Dept., University of Kentucky, Lexington, KY 40546

THOMPSON, Elizabeth, 121 Dantzler Court, Lexington, KY 40503

THOMPSON, Warren C., 121 Dantzler Court, Lexington, KY 40503

TICHENOR, Dr. Doris, Home Economics Extension, University of Kentucky, Lexington, KY 40546

TUCKER, Dr. Ray E., Animal Sciences Dept., University of Kentucky, Lexington, KY 40546

VAUGHT, Dr. Harold C., Agronomy Dept., University of Kentucky, Lexington, KY 40546

WALKER, Dr. John, Agricultural Engineering Dept., University of Kentucky, Lexington, KY 40546

WELLS, Dr. Kenneth L., Agronomy Dept., University of Kentucky, Lexington, KY 40546

WITHAM, Debbie, Agricultural Public Information, University of Kentucky, Lexington, KY 40546

WOLF, Dean C., Agricultural Public Information, University of Kentucky, Lexington, KY 40546

WOOLFOLK, Dr. Patch G., Animal Sciences Dept., University of Kentucky, Lexington, KY 40546

WOOLFOLK, Mrs. Patch G. (Miriam)

YEARGAN, Dr. Kenneth V., Entomology Dept., University of Kentucky, Lexington, KY 40546

YUNGBLUTH, Prof. T. Alan, Dept. of Biology, Western Kentucky University, Bowling Green, KY 42101

Louisiana

BARNES, Dr. Robert F, USDA-SEA-AR, Southern Region, Box 53326, New Orleans, LA 70002

BARNES, Mrs. Robert F (Betty)

BOURDETTE, Vernon R., USDA-SEA-AR, Box 53326, New Orleans, LA 70002

CHERNEY, Dr. Jerry H., Baton Rouge, LA 70803

DE REMUS, Dr. Alan, West Louisiana Experiment Station, Louisiana State University, Box 26, Rosepine, LA 70659

DE REMUS, Mrs. Alan (Elaine)

FAW, Dr. Wade F., Extension Agronomist, Louisiana State University, Baton Rouge, LA 70803

GRIFFIN, Dr. James L., Louisiana State University, Experiment Rice Station, Box 1429, Crowley, LA 70526

GRIFFIN, Mrs. James L. (Carol)

HARVILLE, Dr. Bobby, 208 Agriculture Center, Louisiana State University, Crowley, LA 70526

MASON, Dr. Lee F., Southeast Louisiana Experiment Station, Louisiana State University, Drawer 567, Franklinton, LA 70438

MONDART, Dr. Clifford, Louisiana State University, Baton Rouge, LA 70821

NELSON, Dr. Billy, Louisiana State University, Baton Rouge, LA 70821

PEARSON, Dr. Henry, USDA Forest Service, 2500 Shreveport Highway, Pineville, LA 71360

PEARSON, Mrs. Henry (Dorothy)

ROBINSON, Dr. Donald L., Agronomy Dept., Louisiana State University, Baton Rouge, LA 70821

VERMA, Dr. Lalit R., Dept. of Agricultural Engineering, Louisiana State University, Baton Rouge, LA 70803

Maryland

BEARD, Dr. David F., W-L Research, 7625 Brown Bridge Road, Highland, MD 20777

CARLSON, Dr. Gerald E., USDA-SEA-AR, Room 314, Beltsville Agricultural Research Center-W, Beltsville, MD 20705

DECKER, Prof. Morris, Agronomy Dept., University of Maryland, College Park, MD 20742

ELGIN, Dr. James H., Jr., USDA-SEA-AR-BARC, Field Crops Laboratory, Beltsville, MD 20705

GLENN, Dr. Barbara P., USDA-SEA-AR, Ruminant Nutrition Laboratory, Room 217, Building 200, BARC-E, Beltsville, MD 20705

HALL, Thomas J., 9136 Edmonston Court, No. 301, Greenbelt, MD 20770

KATSIGIANIS, Dr. Ted, Jule Hall, University of Maryland, College Park, MD 20742

MARTINEZ, Mrs. Wilda H., USDA-SEA-AR, National Program Staff/PHST, Room 139 Building 005, BARC-West, Beltsville, MD 20705

VOUGH, Dr. Lester R., Dept. of Agronomy, University of Maryland, College Park, MD 20742

Massachusetts

HERBERT, Dr. Stephen, Dept. of Plant and Soil Sciences, University of Massachusetts, Amherst, MA 01002

Michigan

BREAUX, Mrs. Janice K., Dairy Science Dept., Michigan State University, East Lansing, MI 48824

FIENUP, Darrell, Agricultural Economics Dept., Michigan State University, East Lansing, MI 48824

ROTZ, Dr. Alan, Agricultural Engineering Dept., Michigan State University, East Lansing, MI 48824

SAVOIE, Philippe H., Agricultural Engineering Dept., Michigan State University, East Lansing, MI 48823

SAVOIE, Mrs. Philippe (Christiane)

TESAR, M. B., Crop and Soil Science Dept., Michigan State University, East Lansing, MI 48824

TESAR, Mrs. M. B. (Marian)

THOMAS, Dr. J. W., Animal Science Dept., Michigan State University, East Lansing, MI 48824

Minnesota

BRINK, Geoffrey, Agronomy Dept., University of Minnesota, St. Paul, MN 55108

EHLE, Dr. Fred R., Animal Sciences Dept., University of Minnesota, St. Paul, MN 55108

FRIDINGER, Dr. Tom, 3M Company, St. Paul, MN 55108

HEICHEL, Dr. Gary H., USDA-SEA-AR, University of Minnesota, St. Paul, MN 55108

HOVIN, Dr. Arne W., Dept. of Agronomy and Plant Genetics, University of Minnesota, St. Paul, MN 55108

MARTEN, Dr. Gordon, USDA-SEA-AR, Dept. of Agronomy and Plant Genetics, University of Minnesota, St. Paul, MN 55108

MARTIN, Dr. Neal P., Dept. of Agronomy and Plant Genetics, University of Minnesota, St. Paul, MN 55108

PAGE, Glenn A., Northrup King Company, RR #1, Stanton, MN 55081

PARKER, Deane, Agricultural Production, 3M Company, 223-6SE-04 3M Center, St. Paul, MN 55144

RIHANI, Nacif, Dept. of Animal Sciences, University of Minnesota, St. Paul, MN 55108

SHEAFFER, Dr. Craig C., Dept. of Agronomy and Plant Genetics, University of Minnesota, St. Paul, MN 55108

SULLIVAN, Dr. Timothy P., 3M Company, Building 223-6SE, St. Paul, MN 55144

Mississippi

HAGEDORN, Dr. Charles, Agronomy Dept., Mississippi State University, Box 5248, Mississippi State, MS 39762

KARAU, Paul K. (Kenya), Dairy Science Dept., Mississippi State University, Drawer NZ, Mississippi State, MS 39762

KNIGHT, Dr. William E., USDA-SEA-AR, Box 272, Mississippi State, MS 39762

KNIGHT, Mrs. William (Margaret)

LUSK, Dr. John W., Dairy Science Dept., Mississippi State University, Drawer DD, Mississippi State, MS 39762

SMITH, Gerald, Mississippi State University, Box 5248, Mississippi State, MS 39762

SMITH, Mrs. Gerald (Patricia)

Missouri

BEUSELINCK, Dr. Paul R., USDA-SEA-AR, Agronomy Dept., University of Missouri, Columbia, MO 65211

BUGHRARA, Suleiman S. (Libya), Agronomy Dept., University of Missouri, Columbia, MO 65211

COUTTS, John H., Agronomy Dept., University of Missouri, Columbia, MO 65211

CRANE, Dr. Charles, Agronomy Dept., University of Missouri, Columbia, MO 65211

DE SOUSA, Beni F., Agronomy Dept., University of Missouri, Columbia, MO 65211

GARRETT, Dr. Harold, School of Forestry, Fisheries and Wildlife, University of Missouri, Columbia, MO 65211

GERRISH, James R., RR #1, Linneus, MO 64653

HSU, FU-HSING, (People's Republic of China), Agronomy Dept., University of Missouri, Columbia, MO 65211

HUNT, Kenneth L., Agronomy Dept., University of Missouri, Columbia, MO 65211

JOHNSTON, Earl, MFA Milling Company, Box 757, Jewell Station, Springfield, MO 65801

KROTH, Dr. Earl, Agronomy Dept., University of Missouri, Columbia, MO 65201

KROTH, Mrs. Earl (Ruth)

LENTZ, Edwin, University of Missouri, Columbia, MO 65211

MATCHES, Dr. Arthur G., USDA-SEA-AR, Agronomy Dept., University of Missouri, Columbia, MO 65211

NELSON, Dr. Curtis J., Agronomy Dept., University of Missouri, Columbia, MO 65211

NGUYEN, Hunt T., Agronomy Dept., University of Missouri, Columbia, MO 65211

PAUL, Dr. Kamalendu B., Agriculture Dept., Lincoln University, Jefferson City, MO 65101

PETERS, Dr. Elroy J., USDA-SEA-AR, Agronomy Dept., University of Missouri, Columbia, MO 65211

REEDER, Louis R., Agronomy Dept., University of Missouri, Columbia, MO 65211

VOLENEC, Jeffrey, Agronomy Dept., University of Missouri, Columbia, MO 65211

WORSTELL, James Vard, Jr., Curtis Hall, University of Missouri, Columbia, MO 65211

ZARROUGH, Khames M. (Libya), Agronomy Dept., University of Missouri, Columbia, MO 65211

Montana

CURRIE, Dr. Pat O., USDA-SEA-AR, Rt. 1, Box 2021, Livestock and Range Research Station, Miles City, MT 59301

CURRIE, Mrs. Pat O. (Betty)

MILES, Arthur D., Lazy A-M Ranch, Box 985, Livingston, MT 59047

MILES, Mrs. Arthur D. (Lucille)

WHITE, Dr. Larry, USDA-SEA-AR, Northern Plains Research Center, Box 1109, Sidney, MT 59270

WHITE, Richard, USDA-SEA-AR, Rt. 1, Box 2021, Miles City, MT 59301

Nebraska

AINES, Glen, Dept. of Animal Sciences, University of Nebraska, Lincoln, NE 68583

ANDERSON, Dr. Bruce, Dept. of Agronomy, University of Nebraska, Lincoln, NE 68583

BRITTON, Dr. R. A., Dept. of Animal Sciences, University of Nebraska, Lincoln, NE 68583

FAULKNER, Dan, Dept. of Animal Sciences, University of Nebraska, Lincoln, NE 68583

GATES, Roger N., Dept. of Agronomy, University of Nebraska, Lincoln, NE 68583

GORZ, Dr. Herman J., USDA-SEA-AR, Dept. of Agronomy, University of Nebraska, Lincoln, NE 68583

KEHR, Dr. William, USDA-SEA-AR, Dept. of Agronomy, University of Nebraska, Lincoln, NE 68583

KLOPFENSTEIN, Dr. Terry, Dept. of Animal Sciences, University of Nebraska, Lincoln, NE 68583

MCDONNELL, Michael, University of Nebraska, Lincoln, NE 68583

MOSER, Dr. Lowell E., Dept. of Agronomy, University of Nebraska, Lincoln, NE 68583

VON BARGEN, Dr. Kenneth, 6110 Elkcrest Circle, Lincoln, NE 68516

WARD, Dr. John, Dept. of Animal Sciences, University of Nebraska, Lincoln, NE 68583

Nevada

EDWARDS, Mrs. Jeanne, Elko, NV 89801

New Hampshire

KOCH, Dr. David, Plant Sciences Dept., University of New Hampshire, Durham, NH 03824

New Jersey

EVANS, Dr. Joe L., Animal Sciences Dept., Rutgers University, New Brunswick, NJ 08903

SPRAGUE, Dr. Milton A., Soils and Crops Dept., Cook College, Rutgers University, New Brunswick, NJ 08903

New Mexico

ALLISON, Dr. Christopher, Cooperative Extension Serivce, New Mexico State University, Las Cruces, NM 88003

DONART, Dr. Gary, Dept. of Animal and Range Sciences, New Mexico State University, Las Cruces, NM 88003

GIBBENS, Dr. Robert P., USDA-SEA-AR, Box 697, Las Cruces, NM 88001

HERBEL, Dr. Carlton H., USDA-SEA-AR, Box 698, Las Cruces, NM 88003

LUGG, Dr. David G., Dept. of Agronomy, New Mexico State University, Las Cruces, NM 88003

PIEPER, Dr. Rex, Dept. of Animal and Range Sciences, New Mexico State University, Las Cruces, NM 88003

THOMAS, Dr. Gerald W., President, New Mexico State University, Las Cruces, NM 88003

TROMBLE, Dr. John M., USDA-SEA-AR, Box 698, Las Cruces, NM 88001

WALLACE, Dr. Joe, Dept. of Animal and Range Sciences, New Mexico State University, Las Cruces, NM 88003

New York

ALCONERO, Dr. Rodrigo, N.E. Regional Plant Introduction Station, Geneva, NY 14456

BART, Dr. Fiori, New York State Experiment Station, Geneva, NY 14456

BOUCK, Dan, Seed Division, Agway Inc., Box 4933, Syracuse, NY 13221

CONKLIN, Nancy, Animal Sciences Dept., Cornell University, Ithaca, NY 14853

FICK, Dr. Gary W., Agronomy Dept., Cornell University, Ithaca, NY 14853

GRUNES, Dr. David, USDA-SEA-AR, U.S. Plant, Soil and Nutrition Laboratory, Tower Road, Ithaca, NY 14853

MATTHEWS, David L., Agway, Inc., Box 4933, Syracuse, NY 13221

WALLACE, Dr. Linda, Biology Dept., Syracuse University, Syracuse, NY 13210

WOODWARD, Andrea, Cornell University, Ithaca, NY 14853

North Carolina

BURNS, Dr. George R., USDA-SEA-AR, Box 5847, Raleigh, NC 27650

BURNS, Dr. Joseph C., USDA-SEA-AR, Crop Science Dept., North Carolina State University, Raleigh, NC 27650

CHAMBLEE, Dr. Douglas S., Crop Science Dept., North Carolina State University, Raleigh, NC 27650

GREEN, Dr. James T., Jr., Crop Science Dept., North Carolina State University, Raleigh, NC 27650

HARYANTO, Budi (Indonesia), 123 Chamberlain Street, Raleigh, NC 27611

MOCHRIE, Dr. Dick, Animal Sciences Dept., North Carolina State University, Raleigh, NC 27650

MUELLER, Dr. J. Paul, 1207 Williams Hall, Box 5155, North Carolina State University, Raleigh, NC 27650

TIMOTHY, David H., Crop Science Dept., North Carolina State University, Raleigh, NC 27650

North Dakota

BARKER, Judy, North Dakota

BARKER, Dr. Reed E., USDA-SEA-AR, Northern Great Plains Research Laboratory, Box 459, Mandan, ND 58554

BERDAHL, Dr. John D., USDA-SEA-AR, Northern Great Plains Research Laboratory, Box 459, Mandan, ND 58554

DODDS, Prof. Duaine L., Animal Sciences Dept., North Dakota State University, Fargo, ND 58102

DODDS, Mrs. Duaine L. (Joan)

FRANK, Dr. Albert, USDA-SEA-AR, Box 459, Mandan, ND 58554

HOFMANN, Dr. Lenat, USDA-SEA-AR, Box 459, Mandan, ND 58554

MEYER, Dr. Dwain W., Agronomy Dept., North Dakota State University, Fargo, ND 58105

NYREN, Paul, Botany Dept., North Dakota State University, Dickinson Experiment Station, Dickinson, ND 58601

Ohio

CONRAD, Dr. Harry R., Dairy Science Dept., O.A.R.D.C., Wooster, OH 44691

HEINRICHS, Jud, O.A.R.D.C., Wooster, OH 44691

HENDERLONG, Dr. Paul R., Agronomy Dept., Ohio State University, Columbus, OH 43210

SHOCKEY, Bill, Dairy Science Dept., O.A.R.D.C., Wooster, OH 44691

Oklahoma

BOKHARI, Dr. Unab, USDA-SEA-AR, S.W. Livestock and Forage Research Station, Route 3, El Reno, OK 73036

BARTON, Dr. Franklin E., II, USDA-SEA-AR, S.W. Livestock and Forage Research Station, Route 3, El Reno, OK 73036

COLEMAN, Dr. Samuel W., USDA-SEA-AR, S.W. Livestock and Forage Research Station, Route 3, El Reno, OK 73036

DALRYMPLE, R. L., Agriculture Division, Noble Foundation, Route 1, Ardmore, OK 73401

DEWALD, Chester L., USDA-SEA-AR, Southern Plains Research Station, 2000 18th Street, Woodward, OK 73801

HART, Dr. Steven P., Southwestern Experiment Station, Route 3, El Reno, OK 73036

HART, Mrs. Steven P. (Sue)

HORN, Dr. Gerald, Animal Science Dept., Oklahoma State University, Stillwater, OK 74078

MCMURPHY, Dr. Wilfred E., Agronomy Dept., Oklahoma State University, Stillwater, OK 74078

ROMMANN, Dr. Loren M., Agronomy Dept., Oklahoma State University, Stillwater, OK 74078

Oregon

ANDERSON, Dr. Arthur R., Pioneer Hi-Bred International, 3930 S.W. Macadam, Portland, OR 97201

BEDELL, Dr. Thomas E., Rangeland Research, Oregon State University, Corvallis, OR 97331

CHILCOTE, Dr. David O., Crop Science Dept., Oregon State University, Corvallis, OR 97331

FENN, George S., Fenn Farms, Box 8, Henderer Road, Elkton, OR 97436

HANNAWAY, Dr. David, Crop Science Dept., Oregon State University, Corvallis, OR 97331

MCVEIGH, Dr. Kevin J., International Seeds, Inc., Halsey, OR 97348

MOSHER, Wayne D., 186 Whistler's Park Road, Roseburg, OR 97470

NIPPER, Dr. Allen, Animal Sciences Dept., Oregon State University, Corvallis, OR 97331

Pennsylvania

ABRAMS, Dr. Stephen, USDA-SEA-AR Regional Pasture Research Laboratory, Pennsylvania State University, University Park, PA 16802

ALGOEDT, Hugo, 910 Sperry New Holland, 500 Diller Avenue, New Holland, PA 17557

ALVA, Ashok K., Pennsylvania State University, University Park, PA 16802

BAYLOR, Dr. John E., Agronomy Dept., Pennsylvania State University, 106 Agriculture Administration Building, University Park, PA 16802

BAYLOR, Mrs. John E. (Henrietta)

BERG, Dr. Clyde C., USDA-SEA-AR, U.S. Regional Pasture Research Laboratory, Pennsylvania State University, University Park, PA 16802

BEST, Albert, Sperry New Holland, New Holland, PA 17557

BUSTO, Mario (Mexico), Dairy and Animal Science Dept., Pennsylvania State University, University Park, PA 16802

CARDINA, John, Pennsylvania State University, 119 Tyson, University Park, PA 16802

CLEVELAND, Dr. Richard, Agronomy Dept., Pennsylvania State University, University Park, PA 16802

GRIGGS, Thomas C., Agronomy Dept., Pennsylvania State University, University Park, PA 16802

HILL, Dr. Richard R., USDA-SEA-AR, U.S. Regional Pasture Research Laboratory, Pennsylvania State University, University Park, PA 16802

HILL, Mrs. Richard R. (Jean)

HINISH, Wayne, Pennsylvania State University, University Park, PA 16802

JUNG, Dr. Gerald A., USDA-SEA-AR, U.S. Regional Pasture Research Laboratory, University Park, PA 16802

KENDALL, Dr. William, USDA-SEA-AR, U.S. Regional Pasture Research Laboratory, University Park, PA 16802

KJELGAARD, Prof. William, Agricultural Engineering Dept., Pennsylvania State University, University Park, PA 16802

LANYON, Dr. Les E., Agronomy Dept., Pennsylvania State University, University Park, PA 16802

LEATH, Dr. Kenneth T., USDA-SEA-AR, U.S. Regional Pasture Research Laboratory, University Park, PA 16802

LEATH, Mrs. Kenneth T. (Marie)

LEWIS, Ms. Rebecca D., Kutztown, PA 19530

LUCAS, Mark, Pennsylvania State University, University Park, PA 16802

MASON, Douglas (Guyana), Pennsylvania State University, University Park, PA 16802

MOYER, Barton G., USDA-SEA-AR, U.S. Regional Pasture Research Laboratory, University Park, PA 16802

OBERHEIM, Robert L., Agronomy Dept., Pennsylvania State University, University Park, PA 16802

PANCIERA, Michael T., Agronomy Dept., Pennsylvania State University, University Park, PA 16802

SALEM, Miss Claire, 305 Fifth Avenue, New Kensington, PA 15608

SHERWOOD, Dr. Robert, USDA-SEA-AR, U.S. Regional Pasture Research Laboratory, University Park, PA 16802

SHERWOOD, Mrs. Robert (Ellen)

SHENK, Dr. John S., 1442 Westerly Parkway, State College, PA 16801

SHENK, Mrs. John S. (Gloria)

STONER, Roger R., Sperry New Holland, New Holland, PA 17557

STRINGER, Dr. William, Agronomy Dept., Pennsylvania State University, University Park, PA 16802

TEMPLETON, Dr. William C., Jr., USDA-SEA-AR, U.S. Regional Pasture Research Laboratory, University Park, PA 16802

TEMPLETON, Mrs. William C., Jr. (Martha)

WILKINS, Marion, 910 Sperry New Holland, 500 Diller Avenue, New Holland, PA 17557

ZEIDERS, Kenneth E., USDA-SEA-AR, U.S. Regional Pasture Research Laboratory, University Park, PA 16802

South Carolina

CROSS, Dr. Dee L., Animal Sciences Dept., Clemson University, Clemson, SC 29631

HU, Peter, Agronomy and Soils, Clemson University, Clemson, SC 29631

KRYSOWATY, Ms. Mary T., Animal Sciences Dept., Clemson University, Clemson, SC 29631

MCCLAIN, Dr. Eugene F., Agronomy and Soils Dept., Clemson University, Clemson, SC 29631

MORTON, Benjamin, Agronomy and Soils Dept., Clemson University, Clemson, SC 29631

RICE, Dr. James, Agronomy and Soils Dept., Clemson University, Clemson, SC 29631

South Dakota

BJUGSTAD, Dr. Ardell J., Forestry Research Laboratory, USDA Forest Service, South Dakota School of Mines and Technology, Rapid City, SD 57701

MOORE, Ray, South Dakota State University, Brookings, SD 57007

MOORE, Mrs. Ray (Marlys I.)

VIGIL, Dr. Rudolph, Plant Sci. Dept., South Dakota State University, Brookings, SD 57007

Tennessee

BURNS, Dr. Joe, Plant and Soil Science Dept., University of Tennessee, Knoxville, TN 37901

CARLISLE, Ricky, University of Tennessee, Knoxville, TN 37901

CONGER, Dr. Bob, University of Tennessee, Knoxville, TN 37901

FRIBOURG, Dr. Henry, Plant and Soil Science Dept., University of Tennessee, Knoxville, TN 37901

KLOSTERMEYER, Dr. Lyle E., Entomology and Plant Pathology Dept., University of Tennessee, Knoxville, TN 37916

RAY, Charles R., 10th Floor, American Trust Building, Nashville, TN 37202

REYNOLDS, John, University of Tennessee, Knoxville, TN 37916

Texas

ALLISON, Dr. Lonnie, Route 2, Box 430, Beeville, TX 78102

ALLISON, Mrs. Lonnie (Janice)

BADE, David H., College Station, TX 77840

BASHAW, Dr. E. C., USDA-SEA-AR, Soil and Crop Science Dept., Texas A&M University, College Station, TX 77843

BOYD, William, College Station, TX 77840

BURSON, Dr. Byron, USDA-SEA-AR, Box 7481, Temple, TX 76501

CONRAD, Dr. Bill, Texas A&M University, College Station, TX 77843

EVERS, Dr. Gerald W., Texas Agricultural Experiment Station, Angleton, TX 77515

EVERS, Mrs. Gerald W. (Eunice)

HOLT, Dr. Ethan, Texas A&M University, College Station, TX 77843

HUSSEY, Mark A., Soil and Crop Science Dept., Texas A&M University, College Station, TX 77843

MERRILL, John L., Ranch Management Program, Texas Christian University, Fort Worth, TX 76036

MICHAUD, Michael W., Soil and Crop Science Dept., Texas A&M University, College Station, TX 77843

NELSON, Dr. Lloyd R., Drawer E, Texas A&M University, Overton, TX 75684

NELSON, Mrs. Lloyd R. (Nancy)

PITMAN, Dr. William D., Soil and Crop Science Dept., Texas A&M University, College Station, TX 77843

ROUQUETTE, Monte, Texas A&M University, Drawer E Overton, TX 75684

SCARNECCHIA, David, Range Science Department, Texas A&M University, College Station, TX 77843

SCHUSTER, Dr. Joe, Texas A&M University, College Station, TX 77843

SMITH, Dr. Kenneth, Velsicol Chemical Corporation, 716 Motley Drive, Overton, TX 75684

SMITH, Mrs. Kenneth (Sue)

SWAKON, Dr. Doreen H. D., College of Agriculture, Texas A&I University, Kingsville, TX 78363

VOIGHT, Dr. Paul, USDA-SEA-AR, Box 7481, Temple, TX 76501

WILSON, Dr. Nick, Ranch Management Program, Texas Christian University, Fort Worth, TX 76129

Utah

ALLRED, Dr. Keith, Utah State University, Logan, UT 84321

ALLRED, Mrs. Keith (Maurine)

ARTZ, Neal E., Range Science Dept., Utah State University, Logan, UT 84321

ASAY, Dr. K. H., USDA-SEA-AR, Crops Research Laboratory, Utah State University, Logan, UT 84322

DEFFENDOL, Scotty F., Box 130, Midway, UT 84049

JOHNSON, Dr. Douglas A., USDA-SEA-AR, Crops Research Laboratory, Utah State University, Logan, UT 84322

NORTON, Dr. Brien E., Range Science Dept., Utah State University, Logan, UT 84322

O'ROURKE, Dr. James T., Range Science Dept., Utah State University, Logan, UT 84322

PETERSEN, Mark, Soil Conservation, 4012 Federal Building, 125 S. State Street, Salt Lake City, UT 84138

PETERSEN, Mrs. Mark (Katheline)

Vermont

GERMAIN, Leonard L., R.D. 2, Box 163, Springfield, VT 05156

SMITH, Albert, Animal Science Dept., University of Vermont, Burlington, VT 05401

Virginia

ALLEN, Dr. Vivien G., Agronomy Dept., Virginia Polytechnic Institute and State University, Blacksburg, VA 24061

BLASER, Dr. Roy E., 704 York Drive, N.E., Blacksburg, VA 24060

BRYANT, Dr. Harry, Virginia Polytechnic Institute and State University, Blacksburg, VA 24061

DYE, Dr. A., USDA-OICD, Springfield, VA 22150

FONTENOT, Prof. Joseph P., Animal Science Dept., Virginia Polytechnic Institute and State University, Blacksburg, VA 24061

FREDERICK, Mrs. Shirley, 6101 Eatsall Rd., No. 1508, Alexandria, VA 22313

GRIFFITH, Dr. William, Potash & Phosphate Institute, 865 Seneca Road, Great Falls, VA 22066

KIRBY, Bill W., Chevron Chemical Company, 2308 Middle Road, Winchester, VA 27601

LUEBBEN, Carl, Box 245, Harrisburg, VA 24061

PANDITHARATNE, Miss Sujatha (Sri Lanka), Animal Science Dept., Virginia Polytechnic Institute and State University, Blacksburg, VA 24061

TORRES, Rodolpho (Brazil), Animal Science Dept., Virginia Polytechnic Institute and State University, Blacksburg, VA 24060

WHITE, Dr. Harlan E., Virginia Polytechnic Institute and State University, Blacksburg, VA 24060

Virgin Islands

HEGAB, Dr. Ahmed, College of Virgin Islands, Box 920, Kingshill, St. Croix, U.S.V.I. 00850

Washington

KOES, Dr. Richard, Animal Science Dept., Washington State University, Pullman, WA 99163

MORRISON, Dr. Kenneth, Washington State University, Pullman, WA 99163

RINCKER, Clarence M., USDA-SEA-AR, Irrigated Agricultural Research Center, Box 30, Prosser, WA 99350

Washington, D.C.

BREWER, Frank L., SEA-E, Washingon, D.C.

BULLER, Roderic E., 4901 Rodman Street, N.W., Washington, D.C. 20016

BULLER, Mrs. Roderic E. (Maria)

CELIN, Daniel J., 907 6th Street, S.W., No. 601, Washington, D.C. 20024

CELIN, Mrs. Daniel J. (Lydia)

CHAPLINE, Dr. William R., Forest Service, 4225 43rd Street, N.W., Washington, D.C. 20016

DICKEY, James R., American Embassy-Bamako, Department of State, Washington, D.C. 20520

EVERETT, Herbert W., USDA, Soil Conservation Service, Box 2890, Washington, D.C. 20013

FREDERICK, Dr. Lloyd, Agency for International Development, Washington, D.C.

MACLAUCHLAN, Robert, 14th E. Independence, Washington, D.C.

SCHULZ, Dr. Karl-Heinz, Embassy of the German Democratic Republic, 1717 Massachusetts Avenue, N.W., Washington, D.C. 20036

SUTHERLAND, Dr. Donald, Agriculture and Rural Development Dept., The World Bank, 1818 H. Street, N.W., Washington, D.C. 20433

SUTHERLAND, Elizabeth, Washington, D.C.

SUTHERLAND, Frieda, Washington, D.C.

West Virginia

BAKER, Dr. Barton, West Virginia University, Morgantown, WV 26505

BAKER, Judy, West Virginia University, Morgantown, WV 26505

BALASKO, Dr. John, West Virginia University, Morgantown, WV 26505

BRYAN, Mrs. Carole, 647 Southview Drive, Morgantown, WV 26505

BRYAN, Dr. William B., Plant and Soil Science Dept., West Virginia University, Morgantown, WV 26505

BURTON, Robert, Division of Resource Management, West Virginia University, Morgantown, WV 26506

LATTERELL, Dr. Richard L., Biology Dept., Shepherd College, Shepherdstown, WV 25443

LEGG, Dr. Paul D., USDA-SEA-AR, Box 867, Beckley, WV 25801

REID, Dr. R. Leslie, Animal and Veterinary Science Dept., West Virginia University, Morgantown, WV 26506

STALLCUP, James, 6 E. Lubock Hills, Rt. 1, Washington, WV 26181

VAN EYS, Jan, Animal and Veterinary Science Dept., College of Agriculture and Forestry, West Virginia University, Morgantown, WV 26506

VONA, Miss Linda C., West Virginia University, Morgantown, WV 26506

Wisconsin

BRODA, Zbigniew L., Agronomy Dept., University of Wisconsin, Madison, WI 53706

BULA, Dr. Raymond J., USDA-SEA-AR, U.S. Dairy Forage Research Center, 1925 Linden Drive, University of Wisconsin, Madison, WI 53706

CASLER, Dr. Michael D., Agronomy Dept., University of Wisconsin, Madison, WI 53706

COLLINS, Dr. Michael, Agronomy Dept., University of Wisconsin, Madison, WI 53706

DJUNED, Dr. Harun, Agronomy Dept., University of Wisconsin, Madison, WI 53706

DORNFIELD, Dennis D., Agronomy Dept., University of Wisconsin, Madison, WI 53706

DOYLE, Miss Kristine M., Agronomy Dept., University of Wisconsin, Madison, WI 53706

FLEMING, Samuel C., Agronomy Dept., University of Wisconsin, Madison, WI 53706

GOMES, Mrs. Anabela M., Agronomy Dept., University of Wisconsin, Madison, WI 53706

HIGGS, Dr. Roger, College of Agriculture, University of Wisconsin, Platteville, WI 53818

HIGGS, Mrs. Roger (Francine)

KEENEY, Prof. Dennis R., Soil Science Dept., University of Wisconsin, Madison, WI 53706

LANG, David J., Agronomy Dept., University of Wisconsin, Madison, WI 53706

MC CASLIN, Mark, Cal/West Seeds, RR 1, Box 70, West Salem, WI 54669

ROHWEDER, Dr. Dwayne, Agronomy Dept., University of

Wisconsin, Madison, WI 53706
ROHWEDER, Mrs. Dwayne (Wilma)
SMITH, Dr. Richard R., USDA-SEA-AR, Agronomy Dept., University of Wisconsin, Madison, WI 53706
SMITH, Mrs. Richard R. (Shirley)

Wyoming

HANSON, Dr. Jon, 8408 Hildreth Road, Cheyenne, WY 82001
HART, Dr. Richard H., USDA-SEA-AR, High Plains Grassland Research Station, 8408 Hildreth Road, Cheyenne, WY 82001
JACOBS, Prof. James, Agricultural Economics Dept., University of Wyoming, Laramie, WY 82071
OWENS, Keith, University of Wyoming, Laramie, WY 82071
RODGERS, David, University of Wyoming, Laramie, WY 82071
SKINNER, Dr. Quentin, Range Management Dept., University of Wyoming, Laramie, WY 82071
SPECK, John E., 609 Russel, Laramie, WY 82070

VENEZUELA

ARIAS, Pio, Calle Vargas Sur 28, Maracay
GALLARDO, Prof. Augusto R., Apartado 592, FONAIAP— Estacion Exp. El Cuji, Colinas St. Rosa, carrera 9, No 1a 46, Barquisimeto, Lara 3001
GALLARDO, Mrs. Augusto R. (Raquel)

KAHN, Edward, Avenida Silva No 7, Tinaco, Cojedes
LEONE, Antonio, Apartado 102, FONAIAP-CIARCO, Aragua
OROPEZA, Hernan, Maracay, Aragua 2101
RODRIGUEZ, Prof. Santiago C., Ap. 4653, CENIAP, Maracay—Edo, Aragua

YUGOSLAVIA

BOSNJAK, Dr., Prof. Dragoljub, Institute of University of Osijek
CIZEK, Prof. Jan, Odjel za Krmno bilje, Osijek Box 1009, Agculture Faculty, Zagreb
KATIC, Slobodan, Poljoprivredni fakultet, Nemanjina 6, Zemun, SFR 11080
LUKAC, Svetislav, Plava Laguna, Porec Yugoslavia, Rade Knouar, Porec
MIJATOVIC, Dr. Milan, Inst. z ratarstvo, Poljoprivredni fakultet, Nemanjina 6, Zemun, SFR 11080
MISKOVIC, Borivoje, Institute, Field and Vegetable Crops, Faculty of Agriculture, University of Novi Sad, Veljka Vlahovica-2, 2100 Novi Sad
SARIC, Dr. Osman, Poljoprivredni fakultet, Zagrebaca 18, Sarajevo 71.000

ZAMBIA

KULICH, Dr. Jaroslav, Agricultural Research, Box 90, Choma
KULICH, Mrs. Jaroslav (Sonja)

ACKNOWLEDGMENTS

The organizing committee expresses sincere appreciation to the following people who volunteered their assistance to host tours, organize and provide transportation, assist with registration, project slides, prepare and serve the special meals, arrange and coordinate the special events, and assist with the many tasks associated with making the Congress a success:

Ova Alexander
Sommai Amonsilpa
Joey Armstrong
Ed Aylward
William Balden
Ellen Ballard
Paul Barney
Carol Baruc
Paul Beauchamp
Russell Beauchamp
Margaret Benson
Paul Bertsch
Tom Bishop
Barbara Black
Steven Bowley
Reuben Boyd
Patrick Boyle
Paul Boyle
David Brauer
David Bremel
Barry Bridges
Linda Brown
Jane Brown
Wayne Bugg
Teresa Burch
Myra Burchett
Barry Burton
Joe Cain
Frank Callahan
Jimmy Calvert
Larry Campbell
Roy Catlett
Robert Clay
Terry Clay
Ken Coffey
Jim Coulter
Rachel Cremers
Lori Crist
William H. Dalden
Lee Dame
Steve Davis
Paul Deaton
Randy Deaton
Larry Deetz

Bob DeGregorio
Joseph DeVerna
Elizabeth A. Douglass
Billy Downs
Steve Ebelhar
Melinda Edgerton
Andrew Elser
Maner Ferguson
Bonnie Flannery
Dennis Fox
R. Furbee
Bill Gallrein
Warren Gill
Mahendra Giri
Hudson Glimp
Larry Godfrey
William Graves
J. D. Green
Dan Grigson
Jude Grosser
Karen Grubbs
Richard Guffey
David Harrison
Richard Hartung
Eric Henton
Shelia Hibbs
James Hietholt
Douglas Hines
Hampton Hinton
John Hoehn
Charles Honeycutt
Robert Huffman
Tom Huntington
Glenn Ito
James Jackson
Steve Jackson
Mrs. Steve Jackson (Debbie)
Andrew Jacobs
Darrell Johnson
Glenn Johnson
Steve Johnson
Lynn Jones
Thomas S. Jones
John Kavanaugh

Carolyn Keene
Tom Keene
Harold Kitchen
Bernard Kitur
Ann Lane
Susan Lane
Leonard M. Lauriault
Bruce Leavitt
Mark Leopold
John Lew
Patricia Lewis
M. G. Long
Harold Love
James Luzar
Anne Maczulak
Maher Madhat
Alice Mann
Richard Mann
Fawn Marks
Rodney McCormick
Betsy McGee
Loel Meckel
Donald Miles
Joe Miles
John Miles
Charles Moore
Lonnie Mullins
Rosie Newsome
Michael Noe
Luyaku Nsimpasi
Diana Nunley
James Ostrem
Laura Parsons
Mark Pate
Tom Phelps
Paulette Pierson
Jim Powell
David Radcliffe
James H. Ragland
Charles Rice
Mike Richey
Bob Ritter
Carol Roberson
Louis Rodriquez

Pauline Rogers
Sid Roseberry
Malone Rosemond
Dan Rosenberg
John Rosene
Robert Ross
Steve Rosser
Tina Sandy
Bernice Schneider
Charlie Schnitzler
Kurt Schwinghammer
Dwight Seman
Scott Shepard
Shirley Shepherdson
Alvin Simmons
Janet Siria
Bill Sliski
Lanas Smith
Steven Smith
Rod Smothers
Prasomphol Somjintana
David Sparrow
Diana Spillman
Mac Stone
Charlotte Sutton
Pat Tackett
Linda Tapp
Bill Taylor
Kathy Taylor
William Taylor
Claire Teague
Kent Thompson
Laurel Tucker
John Venable
Mark Waggoner
Doug Waterman
Kenneth Watson
Larry G. Wells
Coleman White
Dennis Wickham
John Willard
John Williams
Calvin Willis
Lipyng Yan

AUTHOR INDEX

Abrams, S. M., 508
Aken'Ova, M. E., 261, 774
Akin, D. E., 511
Akiyama, T., 481
Albrecht, S. L., 327
Alcântara, P. B., 771
Alcântara, V.B.G., 771
Allison, C. D., 670
Anderson, P. M., 657
Andrieu, J. P., 702
Angus, J. F., 478
Argel, P. J., 384
Armstrong, E. A., 765
Arthaud, R. L., 795
Asay, K. H., 124, 675
Ataku, K., 663

Babalola, O., 774
Bailey, A. W., 564
Baker, B. S., 810
Baker, C. J., 542
Ball, P. R., 342
Barker, R. E., 237
Barnes, D. K., 336
Barnes, R. F, 534
Barnett, O. W., 198
Barnhart, S. K., 545
Barreto, I. L., 609
Barton, F. E., II, 532
Bashaw, E. C., 179
Baylor, J. E., 830
Beale, P. E., 113, 116
Bedell, T. E., 793
Belyea, R. L., 514
Berdahl, J. D., 237
Berger, L. L., 617, 623
Bernal, J., 263
Bircham, J. S., 601
Bishop, H. G., 555
Blair, B. O., 294
Blaser, R. E., 612
Bochniarz, J., 440
Boer, H., 502
Boling, J. A., 157, 722
Boller, B. C., 410
Bonnefoy, G., 798
Bouton, J. H., 176
Bowley, S. R., 108
Boyd, A. G., 539
Brereton, A. J., 587
Bruenig, W., 577
Britton, R. A., 617, 653
Brock, J. L., 339
Brougham, R. W., 48
Brown, C. J., 494

Brown, K. R., 266
Brown, R. H., 176
Brown, R. R., 272
Bryan, W. B., 741
Bryant, H. T., 612
Buckner, R. C., 157, 722
Bull, L. S., 660
Burdick, D., 532, 636
Burns, J. C., 497
Burrus, P. B., II, 157
Burson, B. L., 160, 170, 179
Burton, G. W., 138
Burton, R. O., Jr., 741
Bush, L. P., 157, 722
Bustine, D. L., 736
Byerly, K. F., 804

Cabrales, R., 263
Calderón C., Mario, 778
Campbell, W. V., 584
Carlisle, R. J., 714
Carlson, I. T., 207
Carter, E. D., 447
Catchpoole, V. R., 344
Caudle, A. B., 728
Chamblee, D. S., 584
Chapman, P. F., 631
Chapman, R. A., 157
Chatel, D. L., 116
Chaves, C. M.,508
Cheng, K.-J., 221, 719
Chheda, H. R., 261, 774
Chia Shen-siu, 598
Chilcote, D. O., 254
Christensen, D. K., 207
Christie, B. R., 108
Ciorda, H., 728
Clemence, T.G.A., 257
Clements, R. O., 581
Cochrane, T. T., 356
Cohen, R.D.H., 678
Colby, W. G., 309
Collins, G. B., 165, 168
Collins, M., 592, 648
Collins, W. J., 116
Conrad, H. R., 690
Cooper, J. P., 69
Cope, W. A., 198
Cordova, F. J., 688
Cornelius, P. L., 165
Corrall, A. J., 849
Costa, A. C., 744
Costerton, J. W., 719
Couto, W., 320, 753
Cox, N. R., 699

Craven, R. H., 833
Crawford, E. J., 224
Crowder, W., 272
Cruiz, Filho, A. B., 201
Curnow, B. C., 116
Currie, P. O., 560, 675
Currier, C. G., 127
Currl, M. L., 456

Daniel, P., 489
da Rocha, C.M.C., 753
Daughtry, C.S.T., 419
de Andrade, R. P., 184, 275, 753
Decker, A. M., 547
Dehority, B. A., 690
Deinum, B., 407, 502
Demarquilly, C., 702
Dewey, D. R., 124
Donart, G. B., 688
Dornfeld, D., 825
Dougherty, C. T., 590
Dovrat, A., 574
Drake, M., 309
Dudley, R. F., 547
Dulphy, J. P., 702
Dwyer, D. D., 815

Edgar, K. F., 285
Edwards, P. J., 64
Elgin, J. H., Jr., 528
Elkins, C. B., 121, 127
Ely, D. G., 725
Ely, L. O., 505, 634
Erić, P., 375
Erickson, D. O., 232
Escuder, C. J., 738
Evans, J. K., 807
Everett, H. W., 351

Fahey, G. C., 623
Fales, S. L., 419
Falistocco, E., 247
Farmer, M. J., 708
Farrier, D. S., 315
Faulkner, J. S., 210
Fay, J. P., 221, 719
Feigin, A., 288
Ferguson, J. E., 275
Fick, G. W., 483
Fienup, D. F., 818
Fisher, M. J., 396
Fishman, M. L., 636
Fissel, G. W., 528
Fisser, H. G., 424
Fitzgerald, R. D., 564

Other Titles of Interest from Westview Press

Future Dimensions of World Food and Population, edited by Richard G. Woods

Animal Agriculture: Research to Meet Human Needs in the 21st Century, edited by Wilson G. Pond, Robert A. Merkel, Lon D. McGilliard, and V. James Rhodes

Animals, Feed, Food and People: An Analysis of the Role of Animals in Food Production, edited by R.L. Baldwin

The Role of Centrosema, Desmodium, *and* Stylosanthes *in Improving Tropical Pastures*, edited by Robert Burt, Peter Rotar, and James Walker

Wheat in the Third World, Haldore Hanson, Norman E. Borlaug, and R. Glenn Anderson

Crop Reactions to Water and Temperature Stresses in Humid, Temperate Climates, edited by C. David Raper, Jr., and Paul J. Kramer

Climate's Impact on Food Supplies: Strategies and Technologies for Climate-Defensive Food Production, edited by Lloyd E. Slater and Susan K. Levin

Azolla as a Green Manure: Use and Management in Crop Production, Thomas A. Lumpkin and Donald L. Plucknett

Successful Seed Programs: A Planning and Management Guide, compiled and edited by Johnson E. Douglas

The Biology of Social Insects: Proceedings of the 9th Congress of the International Union for the Study of Social Insects, edited by Michael D. Breed, Charles D. Michener, and Howard E. Evans

World Soybean Research Conference II: Proceedings, edited by Frederick T. Corbin

† *The Myth of the Family Farm: Agribusiness Dominance of U.S. Agriculture*, Ingolf Vogeler

† Available in hardcover and paperback.

About the Book and Editors

Proceedings of the
XIV INTERNATIONAL GRASSLAND CONGRESS
Edited by J. Allan Smith and Virgil W. Hays

Approximately 1500 scientists from around the globe participated in the International Grassland Congress at the University of Kentucky in 1981, sharing existing knowledge of grasslands and exploring methods for increasing the productivity of livestock/forage systems so as to better feed mankind while maintaining or improving environmental quality. Of the nearly 500 papers presented on previously unpublished original research or experimental research and development projects, 273 were selected for inclusion in this book. They cover the current basic and applied research on production and utilization of forages from grasslands the world over.

J. Allan Smith, now retired, was editor of publications and head of the Department of Public Information in the College of Agriculture at the University of Kentucky. **Virgil W. Hays** is professor and chairman of the Department of Animal Sciences at the University of Kentucky.